VOLUME ONE

MEDICAL PHYSIOLOGY

VOLUME ONE

MEDICAL PHYSIOLOGY

Edited by

VERNON B. MOUNTCASTLE, M.D.

Professor and Director, Department of Physiology
The Johns Hopkins University
Baltimore, Maryland

FOURTEENTH EDITION

with 1668 illustrations

The C. V. Mosby Company

ST. LOUIS • TORONTO • LONDON 1980

FOURTEENTH EDITION

Printed in the United States of America

The C. V. Mosby Company
11830 Westline Industrial Drive, St. Louis, Missouri 63141

Library of Congress Cataloging in Publication Data

Mountcastle, Vernon B
Medical physiology.

Bibliography: p.
Includes index.
1. Human physiology. I. Title.
QP34.5.M76 1980 612 79-25943
ISBN 0-8016-3560-8

GW/VH/VH 9 8 7 6 5 4 3 2 1 01/A/053

Contributors

G. D. AURBACH

National Institutes of Health
Bethesda, Maryland

LLOYD M. BEIDLER

Florida State University
Tallahassee, Florida

JOHN D. BIGGERS

Harvard University
Boston, Massachusetts

F. J. BRINLEY, Jr.

National Institutes of Health
Bethesda, Maryland

JOHN R. BROBECK

University of Pennsylvania
Philadelphia, Pennsylvania

CHANDLER M. BROOKS

State University of New York
Brooklyn, New York

KENNETH T. BROWN

University of California
San Francisco, California

C. LOCKARD CONLEY

The Johns Hopkins University
Baltimore, Maryland

ROBERT D. DeVOE

The Johns Hopkins University
Baltimore, Maryland

ROBERT M. DOWBEN

The University of Texas
Dallas, Texas

ARTHUR B. DuBOIS

Yale University
New Haven, Connecticut

NORMAN GESCHWIND

Harvard University
Boston, Massachusetts

MOÏSE H. GOLDSTEIN, Jr.

The Johns Hopkins University
Baltimore, Maryland

H. MAURICE GOODMAN

University of Massachusetts
Worcester, Massachusetts

CARL W. GOTTSCHALK

University of North Carolina at Chapel Hill
Chapel Hill, North Carolina

JAMES D. HARDY

Yale University
New Haven, Connecticut

DALE A. HARRIS

Harvard University
Boston, Massachusetts

THOMAS R. HENDRIX

The Johns Hopkins University
Baltimore, Maryland

ELWOOD HENNEMAN

Harvard University
Boston, Massachusetts

JAMES C. HOUK

Northwestern University
Chicago, Illinois

KIYOMI KOIZUMI

State University of New York
Brooklyn, New York

CHRISTIAN J. LAMBERTSEN

University of Pennsylvania
Philadelphia, Pennsylvania

WILLIAM E. LASSITER

University of North Carolina at Chapel Hill
Chapel Hill, North Carolina

PETER C. MALONEY

The Johns Hopkins University
Baltimore, Maryland

JANICE W. MARAN

McNeil Laboratories
Fort Washington, Pennsylvania

THOMAS H. MAREN

University of Florida
Gainesville, Florida

DONALD J. MARSH

University of Southern California
Los Angeles, California

JEAN M. MARSHALL

Brown University
Providence, Rhode Island

LORNE M. MENDELL

Duke University
Durham, North Carolina

WILLIAM R. MILNOR

The Johns Hopkins University
Baltimore, Maryland

VERNON B. MOUNTCASTLE

The Johns Hopkins University
Baltimore, Maryland

WILLIAM L. NASTUK

Columbia University
New York, New York

BARRY W. PETERSON

Rockefeller University
New York, New York

JAMES M. PHANG

National Institutes of Health
Bethesda, Maryland

GIAN F. POGGIO

The Johns Hopkins University
Baltimore, Maryland

SID ROBINSON

Indiana University
Bloomington, Indiana

ANTONIO SASTRE

The Johns Hopkins University
Baltimore, Maryland

W. T. THACH, Jr.

Washington University
St. Louis, Missouri

PAOLA S. TIMIRAS

University of California at Berkeley
Berkeley, California

LESTER VAN MIDDLESWORTH

The University of Tennessee
Memphis, Tennessee

GERHARD WERNER

University of Pittsburgh
Pittsburgh, Pennsylvania

GERALD WESTHEIMER

University of California at Berkeley
Berkeley, California

VICTOR J. WILSON

Rockefeller University
New York, New York

F. EUGENE YATES

University of Southern California
Los Angeles, California

Preface

TO FOURTEENTH EDITION

The general principles by which this textbook is organized remain those described in the preface to its twelfth edition, namely, to present mammalian physiology as an independent biologic discipline as well as a basic medical science. Two new sections appear in this edition, the first on the principles of system theory as applied to physiology and the second on the physiology of development and aging. Sixty-five chapters of the present edition either are wholly new (ten) or have been extensively revised (fifty-five); twelve chapters remain essentially as they appeared in the thirteenth edition.

This edition has been written by forty-five authors, of whom twelve have joined this effort for the first time. Forty-one of these writers are continually engaged in research and teaching in physiology. Each has taken time from that dedicated life to summarize here the present state of knowledge in a particular field of interest. Whatever value this book possesses is wholly due to the contributors' depth of understanding, skill of exposition, and devotion to the task. For this I am indebted to each.

For them and for myself I wish to thank those authors and publishers who have allowed us to reproduce illustrations previously published elsewhere.

Vernon B. Mountcastle

Preface

TO TWELFTH EDITION

The twelfth edition of *Medical Physiology* presents a cross section of knowledge of the physiologic sciences, as viewed by a group of thirty-one individuals, twenty-three of whom are actively engaged in physiologic research and teaching. Each section of the book provides statements of the central core of information in a particular field of physiology, reflecting, by virtue of the daily occupations of its authors, the questioning and explorative attitude of the investigator and indeed some of the excitement of the search. These statements vary along a continuum from those with a high probability for continuing certainty to those that are speculative but, it is hoped, of heuristic value. An attempt has been made to maintain a balanced point of view. I hope this book will convey to the student who reads it the fact that physiology is a living and changing science, continuously perfecting its basic propositions and laws in the light of new discoveries that permit new conceptual advances. The student should retain for himself a questioning attitude toward all, for commonly the most important advances are made when young investigators doubt those statements others have come to regard as absolutely true. This is not a book that sets forth in stately order a series of facts which, if learned, will be considered adequate for success in a course in physiology. Many such "facts" are likely to be obsolete before the student of physiology reaches the research laboratory, or the student of medicine the bedside. Nor is it a book that provides ready-made correlations and integrations of the various fields of physiology necessary for a comprehensive understanding of bodily function. Those integrations are an essential part of scholarly endeavor not readily gained from books alone. It is my hope, however, that study of this book, combined with laboratory experience and scholarly reflection, will provide the student with a method and an attitude that will serve him long after the concepts presented here are replaced by new and more cogent ones.

The title *Medical Physiology* has been retained, for one of the purposes of this edition, in common with earlier ones, is "to present that part of physiology which is of special concern to the medical student, the practitioner of medicine, and the medical scientist in terms of the experimental inquiries that have led to our present state of knowledge." The scope of the book was and is still broader, however, and attempts to present mammalian physiology as an independent biologic discipline as well as a basic medical science. Mammalian physiology has its base in cellular physiology and biophysics, and it is from this point of view that many of the subjects treated here are approached. Above all, mammalian physiology must deal with problems of the interactions between large populations of cells, organs, and organ systems and, finally, the integrated function of an entire animal. Physiology thus must bridge the distance from cellular biology on the one hand to systems analysis and control theory on the other: each is important and any one is incomplete without the others. This approach to the problems of internal homeostasis, of reaction to the environment, and of action upon the environment is evidenced in several sections of this book.

Of the eighty chapters composing this book, twenty-nine are wholly new in this edition; forty-five from the last edition have been extensively revised either by their original authors or by new ones. Six have been allowed to stand substantially as previously written, for these seemed to comprise as balanced and modern a survey as any presently possible. The names and affiliations of my colleagues in this effort have been listed. They have taken time from busy lives to survey their fields of interest; for this I am greatly indebted to each. If this book possesses any worth it is in large part due to their continuing devotion to the task of its preparation.

For them and for myself I wish to thank those authors and publishers who have allowed us to reproduce illustrations previously published elsewhere.

Vernon B. Mountcastle

Contents

VOLUME ONE

MEDICAL PHYSIOLOGY

I

CELLULAR PHYSIOLOGY

ROBERT D. DeVOE and PETER C. MALONEY

1 Principles of cell homeostasis

Cells are semiautonomous units of tissue; isolated cells can survive for long periods of time in tissue culture media that mimic their normal environments. The intracellular environments are very different from the fluids around cells, however, and there are constant exchanges of metabolites, waste products, and other substances between a cell and its environment. For this molecular traffic to be possible, the cell cannot wall itself off altogether. On the other hand, it must have some barrier between its different internal and external environments simply in order to survive. The problem all cells face, therefore, is how to surround themselves with barriers (cell membranes) that allow desired substances to pass in and out while maintaining their own internal constancies. The maintenance of this constancy is what is meant by cell homeostasis.

Cells of epithelial tissues face additional problems. Parts of their surfaces border the body's relatively constant internal environment and parts the body's much more variable external environment. Across epithelial cells pass food, water, and oxygen for the body as well as wastes from the body. In the face of this molecular traffic, these cells must maintain their internal homeostasis, too. They do this by the same means as do nonepithelial cells. Indeed, the transepithelial traffic results from specializations of mechanisms used by all cells for homeostasis. Therefore cell homeostasis will be discussed primarily for the general case, with some of the epithelial specializations presented at the end of this chapter.

The beginning point here is this boundary between a cell and its environment. A "typical" cell (in fact, an epithelial cell) is depicted in Fig. 1-1, *A*, but for initial purposes it may be simplified to the hollow shell depicted in Fig. 1-1, *B*. This shell consists of a uniform cell membrane that surrounds a fluid of one composition and is itself surrounded by a fluid of a different composition. By way of illustration, the ionic compositions of intra- and extracellular fluids are given for a number of cells in Table 1-1 (with some other quantities that will be explained later). Proceeding from values given in this table and by reference to Fig. 1-1, *B*, the basic principles of cell homeostasis that will be developed in this chapter are sixfold.

First, water is in general in osmotic equilibrium across cell membranes and easily passes back and forth across them. Second, large internal organic molecules, both charged and uncharged, are retained within the cell. At nearly neutral intracellular pH values, their net charges are negative. They are designated collectively in Fig. 1-1, *B*, as P^-. Third, the presence of osmotically active organic molecules held within the cell by the membrane must be balanced by the external presence of some substance(s) impeded by the membrane from entering the cell. If this were not the case, the cell could not be in osmotic equilibrium. By and large, this external substance is sodium, shown in Fig. 1-1, *B*, in large concentration outside and in low concentration inside. Fourth, since the net charge on the internal organic ions is negative, some cation must be present to give electroneutrality within the cell. For most cells, the predominant intracellular cation tends to be potassium. This is shown in Fig. 1-1, *B*, as a large internal potassium concentration and a low external concentration. Fifth, there is a negative potential difference between the inside and the outside of the cell. This membrane potential, as it is called, is primarily due to the tendency of potassium ion to equalize its internal and external concentrations by diffusing out of the cell, thus upsetting electroneutrality across the membrane. The effect of the inside negative membrane potential, combined with internal indiffusible anions, is the greater or lesser exclusion of mobile negative ions, particularly chloride, from the cell interior. Conversely, further outward movements of potassium ion are

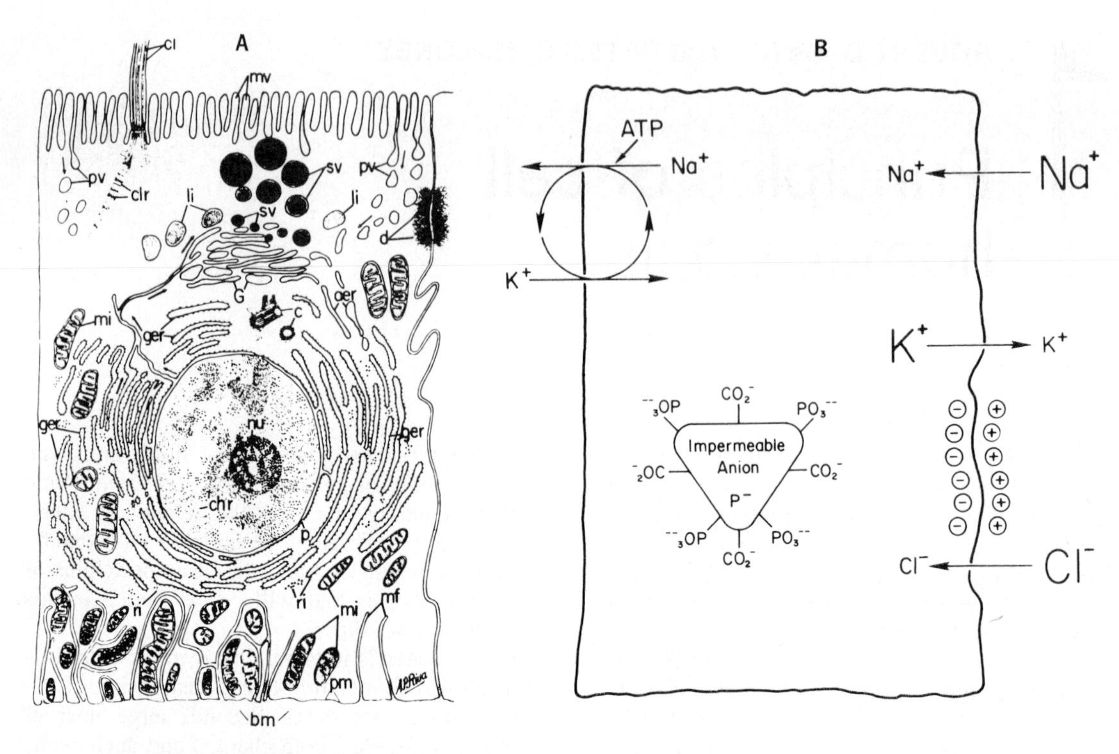

Fig. 1-1. A, Diagram of ultrastructure of ideal animal cell. (See De Robertis et al.[4] for key to details.) **B,** "Hollow-shell" depiction of plasma membrane around cell, ionic movements and electrical potential across membrane, and relative concentrations of substances on either side (shown by relative sizes of lettering). (**A** From De Robertis et al.[4])

Table 1-1. Some representative values for intracellular and extracellular ionic concentrations (in millimoles per liter cell water or extracellular volume) and equilibrium and resting potentials (in millivolts)

	Squid giant axon	Frog sartorius muscle	Human red blood cell
Intracellular concentrations (mM/L)			
$[Na^+]_i$	78.0	13.0	19.0
$[K^+]_i$	396.0	138.0	136.0
$[Mg^{++}]_i$	11.0	16.0	6.0
$[Ca^{++}]_i$	0.4	3.0	0.0
$[Cl^-]_i$	104.0	2.0	78.0
Extracellular concentrations (mM/L)			
$[Na^+]_o$	462.0	108.0	155.0
$[K^+]_o$	22.0	2.5	5.0
$[Mg^{++}]_o$	56.0	1.0	1.0
$[Ca^{++}]_o$	11.0	2.0	2.5
$[Cl^-]_o$	586.0	76.0	112.0
Equilibrium potentials (mV)			
E_{Na}	+45	+53	+55
E_K	−73	−101	−86
E_{Cl}	−44	−92	−9
Resting potentials—V_m (mV)	−73	−92	−6 to −10

retarded, whereas the slow tendency of sodium ions to enter the cell down its concentration gradient is accelerated. Sixth, and finally, in order to prevent even a slow net inward movement of sodium, which would upset the osmotic equilibrium, cells utilize metabolic energy to transport, or pump, the excess sodium out of the cell. In most instances, there is a linked inward movement of potassium ions to maintain the internal potassium concentration.

It can thus be seen that there are three points at which cells can control the states they will achieve in homeostasis by means of metabolic, synthetic, or other activity. These are (1) the permeabilities of their membranes to water, ions, and nonelectrolytes; (2) the osmolarities and amounts of charge of internal organic molecules; and (3) the rate of ion transport. Even in a given cell, one or more of these may be a variable, depending on physiologic activity. Thus nerve cells, which like other cells have low membrane permeabilities to sodium, transiently increase this permeability during the generation of action potentials. Cells such as those in kidney collecting tubules have water permeabilities that are under endocrine control. Oxygenation and deoxygenation of hemoglobin in erythrocytes, which occur as these cells transport oxygen and carbon dioxide to and from tissues, respectively, involve changes in internal anionic charge and hence in internal ionic distributions. Rates of ion transport may depend on internal ion concentrations or, in some epithelial cells, on hormones. It should thus be understood that there may be many different combinations of membrane permeabilities, internal anions, and ion transport mechanisms that cells use both to maintain themselves and to carry out their physiologic functions.

Whatever the particular combination that a given cell has evolved, it will result in homeostasis, a steady-state condition in which there are no *net* molecular movements of consequence into and out of the cell. There will be molecular traffic across the membrane, but the movement of osmotically active particles into the cell must be matched by an equal outward movement if the cell is to stay in osmotic equilibrium. In general, since cell membranes show selectivity toward each chemical species, this means that the inward movement of a given substance must be matched by an equal outward movement, unless it is consumed. Again, since the major osmotically active substances in body fluids are the three principal ions sodium, chloride, and potassium, the matched movements in question will be of these ions. Movement of any substance across a membrane per unit of time is called its *flux,* and the flux per 1 cm^2 unit area has the dimensions of moles per second per square centimeter. For a steady state, then, the outflux of each ion (and water) must be of the same size but opposite in sign (i.e., direction) to the influx; the net flux, their sum, must be zero. (This is as true for epithelial cells as for other cells, but in epithelial cells there may be a net transcellular flux. In this case, what enters at one cell border may leave at a different border.)

To complete this initial picture of cell homeostasis, recall that the *causes* of inward and outward movements of substances can be both the tendencies of these substances to distribute themselves according to concentration and electrical differences across the cell membrane and the use of metabolic energy by the cell to transport substances. The term "passive fluxes" is used to describe movements of substances due to kinetic forces, i.e., concentration gradients, electrical potential gradients (in the case of ions), and other gradients such as pressure and temperature. When influxes and outfluxes of a substance are both solely passive and equal to each other, that substance is in equilibrium across the membrane. As stated earlier, water is thought to be in equilibrium across cell membranes,[111] and in resting muscle in situ, chloride may likewise be in equilibrium.[11] However, there is, in addition, an *active transport* of ions such as sodium and potassium across cell membranes that results from the expenditure of metabolic energy by the cell. Intuitively it can be seen that if, for example, sodium ions are transported out of a cell interior, the negative organic anions are left behind. Therefore there will be movements of the other ions to reestablish electroneutrality. Thus active transport of only one ion out of a cell, with given amounts of internal indiffusible anion and with given membrane permeabilities to the various ions, can result in extensive redistributions of other ions as well. In general the process of this redistribution cannot be observed; what is usually seen is the final result. Nonetheless, cell homeostasis may be approached as the study of what active fluxes there are and hence what the requisite concentration and electrical gradients there must be in order to set up, across the membrane in question, the passive fluxes that will just balance the active fluxes. Once this is done, the distributions of ions (and water) can be determined from the necessity for electrical neutrality and for osmotic equilibrium, that is, for the distribution that will bring these other substances into equilibrium.

To proceed further, it is necessary to consider first the structure of membranes and how, to the extent known, they exert their selective effects on the movements of water, ions, and nonelectrolytes into and out of cells. Second, considerable emphasis will be given to the manner in which ions and water move under concentration and electrical gradients, which themselves are established by ion pumps. Initially, emphasis will be placed on the homeostasis of individual cells—their ionic and water distributions and how these affect membrane potentials and ionic equilibriums. Finally, the properties of the active transport mechanisms will be discussed, since these are the key to understanding the ultimate ionic and water distributions seen in cells.

COMPOSITION AND STRUCTURE OF CELL MEMBRANES

Cell membranes undeniably differ from one cell to another. A common structural basis now appears to be the fluid mosaic model,[15,115] diagrammed in Fig. 1-2. In this, about 70% of the membrane surface is a lipid bilayer[46] with integral, amphipathic proteins (having both hydro-

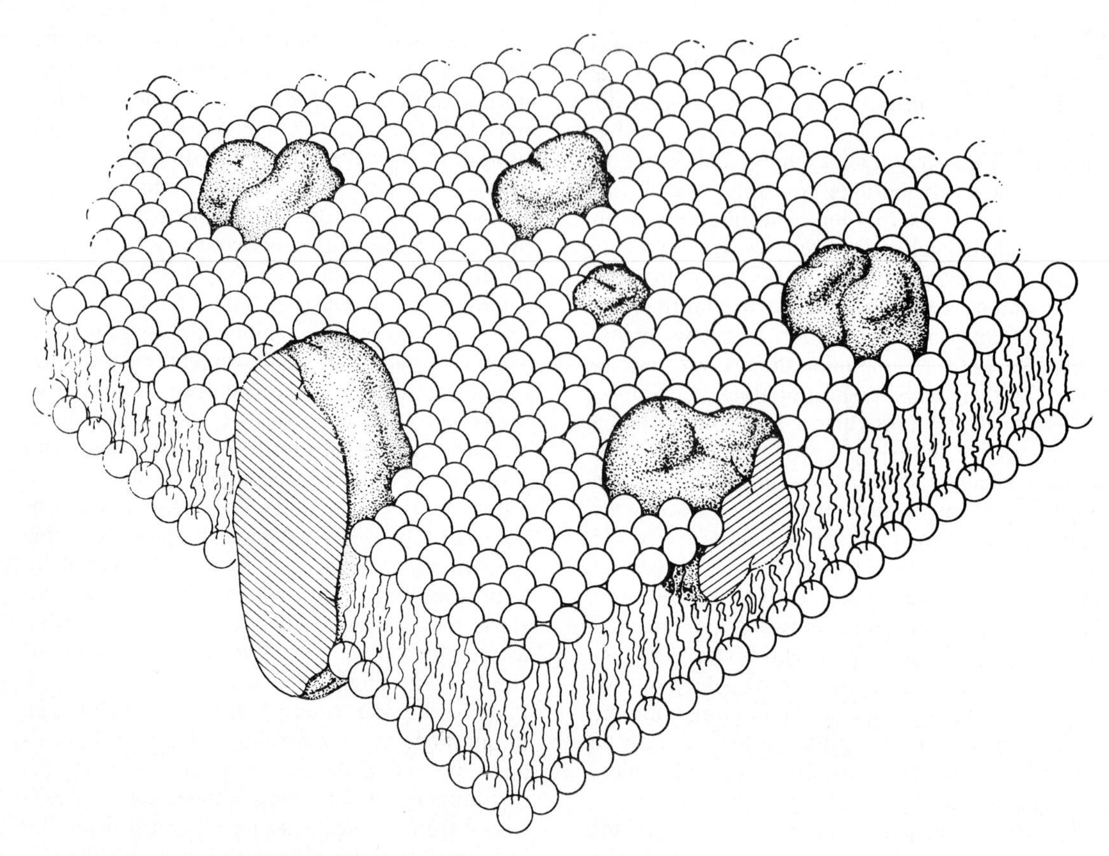

Fig. 1-2. Schematic three-dimensional and cross-sectional view of fluid mosaic membrane. Small spheres represent polar heads of lipids, whose hydrocarbon tails are indicated as thin wavy lines in membrane interior. Solid bodies with stippled surfaces represent globular integral proteins. (From Singer and Nicolson[15]; copyright 1972 by the American Association for the Advancement of Science.)

philic and lipophilic portions) randomly distributed and free to diffuse in the plane of the membrane. Short-range organization (aggregation), long-range organization (as at synapses, desmosomes, etc.), and cell-specific organization are superimposed on this random order. Of great physiologic interest is the fact that the properties of the lipid bilayers can be studied in vitro with artificial bilayers, alone or with protein inclusions. From such studies, it is often possible to learn what properties of biologic membranes may be due to the lipids and what to the proteins.

The present concept of membranes developed from a variety of physiologic, osmotic, electrical, and cytosurgical experiments[2,10,119] long before electron microscopy allowed membranes to be seen or before modern biochemistry allowed membrane components and their interactions to be characterized. First, electrical measurements of conductivities of packed suspensions of erythrocytes in the 1920s and of giant squid axons and frog sartorius muscles in the 1930s indicated that cell membranes might be very thin. It was found that these cells had highly conducting interiors but were bounded by poorly conducting layers. These insulating layers had capacitances on the order of 1 $\mu F/cm^2$ (Table 1-2).[2] The capacitance C of any insulating layer is related to its thickness a as follows:

$$C = \frac{\kappa \epsilon_0 A}{a} \tag{1}$$

where A is the area of the layer, κ is its dielectric constant, and ϵ_0 is 8.86×10^{-12} F/m. The problem in using such a relation to determine the thickness of the insulating membrane around cells is identifying the dielectric constant of this layer. By the beginning of this century, it was well known that lipid-soluble substances penetrated easily into cells.[10] In the 1920s it was argued that this was because the membrane itself was made up of lipids[59] (see later discussion).

Table 1-2. Comparison of some properties of artificial lipid bilayer membranes with biologic membranes*

Property	Biologic membranes	Artificial membranes
Electron microscope image	Trilaminar	Trilaminar
Thickness (Å)	60-100	40-90
Capacitance (μF/cm^2)	0.5-1.3	0.38-1.0
Resistance (ohm-cm^2)	10^2-10^5	10^6-10^9
Dielectric breakdown (mV)	100-200	150-200
Surface tension (dynes/cm)	0.03-1.0	0.5-2.0
Water permeability (μm/sec)	0.37-400	31.7

*Adapted from Henn and Thompson.[63]

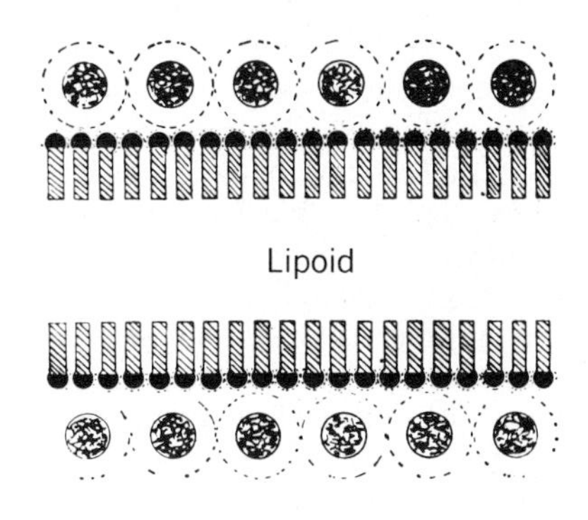

Fig. 1-3. Original paucimolar membrane model of Danielli and Davson. Proteins (circles) are shown associated with polar heads of lipids. (From Danielli and Davson.[39])

Thus if the membrane were taken to have a dielectric constant typical of a lipid, such as 3, the membrane thickness would be estimated at approximately 33 Å. Although such estimates of membrane thicknesses were off by a factor of two to three, they indicated both that the membrane thickness was of the order of the length of a single lipid molecule and that such high capacitances, far more efficient than man-made capacitors per unit area, were indicative of a high degree of molecular organization.

Further evidence that membranes had thicknesses on the order of the lengths of lipid molecules came from the work of Gorter and Grendel in 1925.[59] These workers extracted the lipids from erythrocytes and spread them out on an air-water interface (a Langmuir trough) to determine the area they would occupy as a monomolecular film when compressed. They used erythrocytes from dogs, sheep, goats, rabbits, guinea pigs, and man and in all cells calculated that the surface areas of extracted lipids were twice as great as the surface areas of the cells. They concluded from this that erythrocytes "are covered by a layer of fatty substances that is two molecules thick."[26,59] Later and quite independently, Danielli and Davson[39] proposed their now classic "paucimolecular" model of the cell membrane from entirely different considerations. They supposed, first, that since lipid-soluble materials passed through membranes easily, the membrane was most likely lipid itself; second, on the basis of chemical analysis, that these lipids had both polar and nonpolar regions[79]; third, on the basis of membrane capacitances, that the lipids might be arranged in as few as one to as many as three layers; and finally, because cell surface tensions were as low as 0.03 to 1.0 dyne/cm, compared to surface tensions as great as 9.0 dyne/cm for oil droplets, that the lipid layers were covered with surface-active proteins. Since the proteins, to have been surface active, would have had to be polar, Danielli and Davson suggested that the polar proteins were situated next to the polar ends of the lipid molecules at the outsides of the membrane, whereas the nonpolar lipid regions were associated together in the interior of what was thus at least a bimolecular lipid leaflet. This model is shown in Fig. 1-3.

In 1962 Mueller et al.[98] reported the fabrication of artificial lipid bilayers from mixtures of polar lipids and organic solvents, and it became possible to test the properties of such lipid bilayers for comparison with natural cell membranes. These artificial lipid bilayers have permitted tests of the lipid contributions to the physiologic properties of biologic membranes.[63,78,96,123] In Table 1-2 some of the comparisons between natural and such artificial bilayer membranes are listed. It is possible to show by optical means that these artificial membranes are about 72 Å or only two molecular layers thick. From the low surface tensions of these purely lipid membranes, it must be concluded that for energetic stability the polar groups (rather than the nonpolar groups) are in contact with water at the membranes' outer faces.[63] Thus these artificial membranes have the lipid arrangement postulated to underlie the paucimolecular membrane. However, their surface tensions are already as low as those of cell membranes, so that the original assumption of Danielli and Davson[39] that the low surface tensions of cell membranes are due to outer protein layers is unnecessary. Moreover, some of the other properties of naked lipid bilayers, such as their capacitances, dielectric breakdowns, and water permeabilities (to be discussed in the next section), are also very much like those of cell membranes. On the other hand, resistances of artificial membranes are much higher, that is, their permeabilities to ions as well as to other polar molecules are much lower than those of biologic membranes. Artificial membranes may be adulterated with various substances that, like the ionophore valinomycin, selectively increase the permeability of these membranes to potassium ions or that, like cholesterol, appear to diminish their permeability to water.[48] The role of "adulterants" in biologic membranes is played by proteins.

The original Danielli and Davson concept[39] that layers of protein are spread across both membrane faces is not only unnecessary to account for the low surface tensions of membranes but also may be impossible. For one thing, there may not be enough protein to cover both faces of the membrane[79] as "peripheral" proteins.[15] More importantly, there are amphipathic proteins "integral" to the membrane whose nonpolar, lipophilic portions can only be dislodged from the membrane by the use of detergents.[15,115] Energetically, the likelihood is thus very small that such hydrophobic portions of amphipathic proteins would be found outside the lipid matrix in sole association with polar groups of the lipids and with water. From such considerations came the model shown in Fig. 1-2, where the integral proteins are taken instead to be inserted *into* the lipid matrix.

Support for this lipid-matrix model comes from electron microscopy. Cross sections of biologic membranes have long been known to appear trilamellar.[109,110] Artificial lipid bilayers lacking protein likewise appear trilamellar, as shown in Fig. 1-4.[64] Thus the cross-sectional appearance of biologic membranes is well accounted for in the lipid-matrix portion of the fluid-membrane model.

In freeze-fractured membranes, where the plane of fracture can be in the middle of the lipid-matrix, particles are often seen in one or the other half of the membrane. An example is given in Fig. 1-5 for erythrocyte ghosts, some of which can be turned inside out to reveal different particles on the cytoplasmic face.[121] Electron-dense

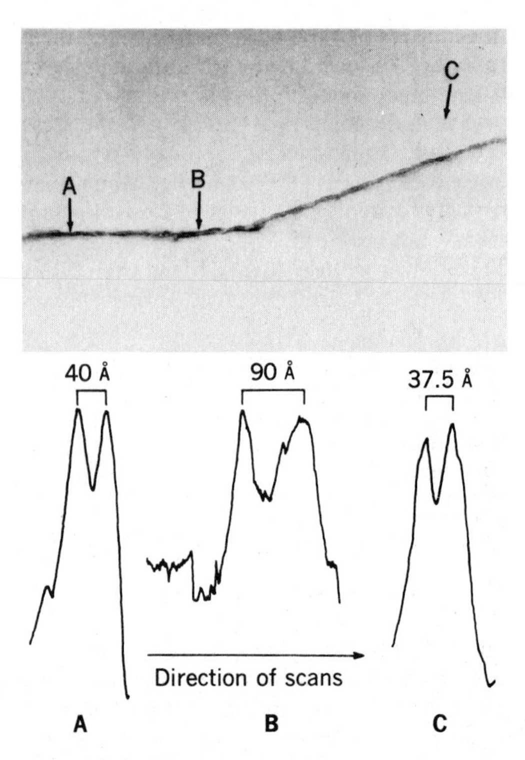

Fig. 1-4. Top: Electron micrograph of transverse section of artificial lipid bilayer membrane. Microdensitometer tracings shown at bottom were taken at points *A*, *B*, and *C*. Peak-to-peak distances between dark lines are given above each tracing. (From Henn et al.[64])

antibodies that had reacted with membranes prior to freeze fracture can be found attached to the membrane particles. This indicates that the particles are indeed proteins, whereas the distributions of the antibodies indicate that the polar (reactive) ends of the proteins protrude on the cytoplasm, on the extracellular side of the membrane, or on both.[115]

Integral proteins each appear to be tightly bound in the lipid matrix to 80 to 120 lipid molecules via the proteins' lipophilic portions.[78] This undoubtedly imposes some short-range ordering of the lipids surrounding the protein. Nonetheless, *some* integral proteins appear able to spin and diffuse laterally in the plane of the membrane, as though the lipid matrix had a relatively fluid viscosity of 1 to 10 poise.[33,43,44,85,105] (End-to-end tumbling is precluded energetically, however, since this would require polar *ends* of the proteins to pass through the lipid matrix.) The fluidity of membranes seems to increase the longer and the less saturated are the lipid tails, whereas fluidity decreases by addition of choles-

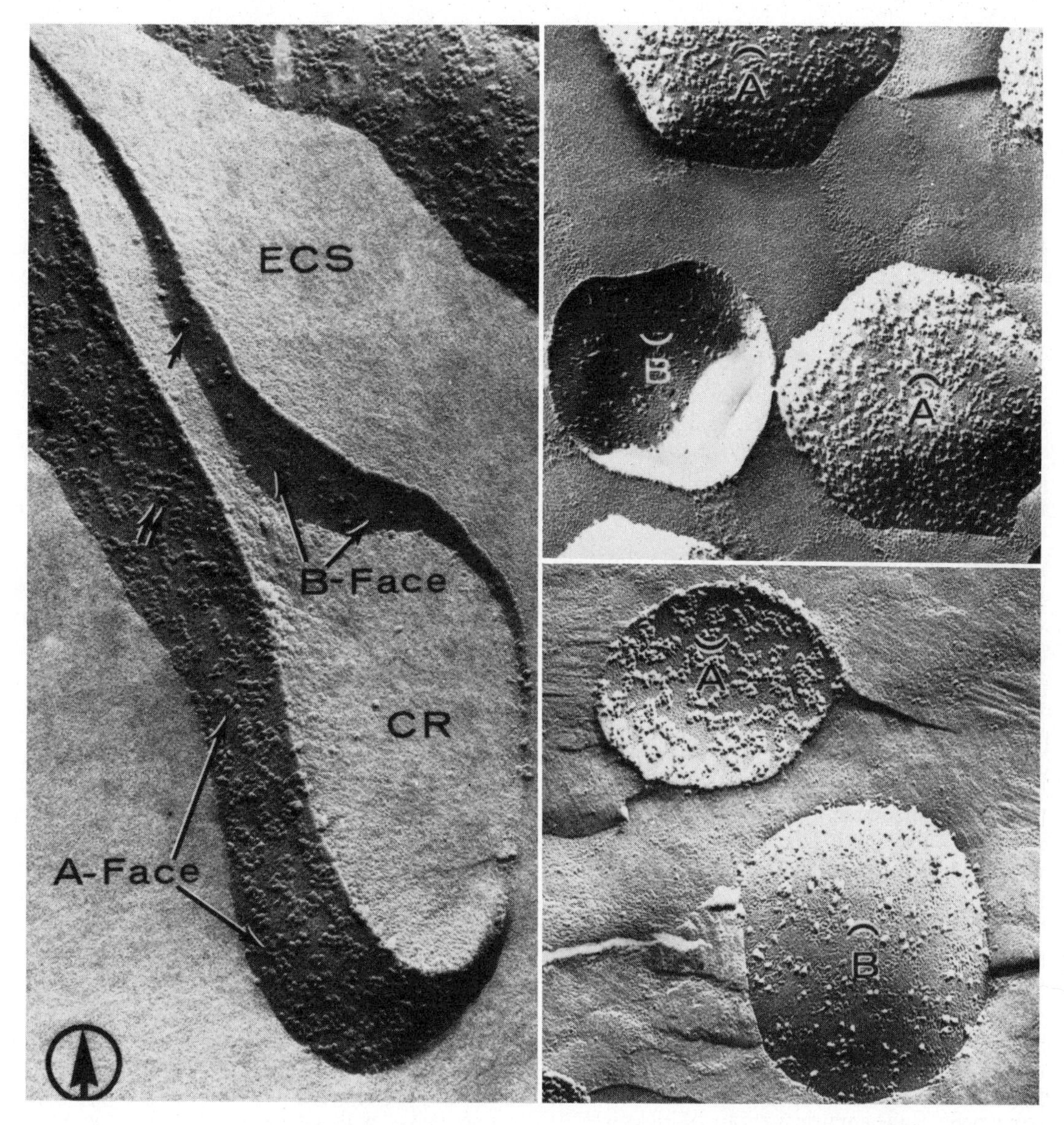

Fig. 1-5. Electron micrographs of freeze-etched red blood cell membranes. A-face is extracellular and contains many 100 Å particles (double arrows). B-face is intracellular and contains fewer particles. At left is a ghost; at upper right is a normal side–out vesicle; at lower right is an inside-out vesicle. Note that particle asymmetries are maintained in these vesicles. (From Steck et al.[121]; copyright 1970 by the American Association for the Advancement of Science.)

terol, for example.[103] Possibly, stiffness via cholesterol inclusion explains why other integral proteins do not spin.[58] Peripheral proteins that attach to and bridge integral proteins may also reduce lateral diffusion, as may glutaraldehyde (but not formaldehyde) fixation.[58,115] The overall view of cell membranes that emerges is, however, a dynamic one.

In what follows, physiologic properties of membranes are presented. In a few cases, it is possible to explain physiologic membrane properties using the properties of specific membrane components. Thus water and small nonpolar molecules may pass directly through the lipid bilayer.[47] Active transport of ions proceeds via specific ATPases that span the membrane and that can be tested in vitro in reconstructed lipid bilayer vesicles (as will be discussed later in this chapter). More often, it has been shown that specific membrane structures are inferred from the physiologic evidence. This was alluded to earlier, in the very origins of the membrane concept. It must be said, however, that the same evidence has and does compel some to reject the idea of a physiologic membrane altogether and to try to explain the distributions of water and ions by purported properties of cytoplasm instead.[77,86,88]

DIFFUSION AND PERMEATION ACROSS CELL MEMBRANES

Among the most important physiologic attributes of membranes are the ways in which substances can permeate, or pass through, them. The permeabilities of membranes describe their selectivities toward substances on the insides and on the outsides of cells and thus reflect the degree to which cell integrity depends on keeping these substances in or out or letting them pass back and forth. Metabolites must enter, waste products must leave, bur large osmotic shifts of water that could lead to lysis must be prevented.

The starting point for a study of cell permeabilities is the diffusion of substances in free solution. Permeation can be looked on as diffusion up to a membrane in a water phase, diffusion through the membrane, and diffusion away from the membrane on the other side. Permeation thus involves more than the mere passage through the membrane, and sometimes this passage is not the limiting factor, as, for example, when there are convoluted, restricted extracellular spaces around cells through which substances must diffuse to reach the membrane. In the simplest cases, however, substances will cross cell membranes, in the absence of metabolic activity, under the driving force of concentration gradients, or, more strictly, under the driving force of gradients in their chemical potentials (Appendix A, p. 36). The physical basis for such movements is the empiric first law of Fick, which relates the number of moles (dS) of a substance that will diffuse through a given area (A) in a given time (dt) under a concentration gradient (dc/dx):

$$\text{Flux} = \frac{dS}{dt} = -DA\,\frac{dc}{dx} \qquad (2)$$

Fick's law has the same form as Ohm's law, for in both a flow (flux, current) is related to a driving force (concentration gradient, potential difference) through a proportionality factor characteristic of the system: the diffusion constant (D) for diffusion, 1/resistance for Ohm's law. The dimensions of the diffusion constant can be found from the following:

$$\text{Moles/sec} = \frac{\text{cm}^2 \cdot \text{moles/cm}^3 \cdot D}{\text{cm}}$$

or

$$D = \text{Moles/sec} \cdot \frac{\text{cm}^3\ \text{cm}}{\text{cm}^2\ \text{moles}} = \text{cm}^2/\text{sec}$$

When it comes to cell membranes, however, neither the thickness (a) nor the concentration gradient (dc/dx) is generally known, although the thickness may be approximated as 75 to 100 Å. On the other hand, the concentrations in the solutions at either side of the membrane can be measured. Since the membrane, of uncertain thickness (a), is the boundary between the two solutions, the entire concentration difference must be developed across this thickness:

$$\frac{dS}{dt} = -DA\frac{(c_i - c_o)}{a} = -\frac{D}{a}\,A(c_i - c_o) \qquad (3)$$

where c_o is the external concentration and c_i is the intracellular concentration.

The ratio of the diffusion constant (D) to the unknown thickness (a) is called the *permeability constant* (P) and has the dimensions of a velocity (centimeters per second). It defines quantitatively the ease with which a given substance can penetrate a given cell membrane, and it is a property of the membrane.

The valid application of equation 3 to permeation across cell membranes requires that the actual concentrations at the cell membrane be known and that the permeating substances cross the membrane independently under simple passive kinetics. Some metabolites, such as sugars and amino acids, can exhibit saturation kinetics of permeation or be linked to sodium ion movements. Such movements will be discussed later in this chapter. They do not follow the rules of simple permeation discussed here. Similarly, some of the older measurements of permeabilities[3] were made by studying bulk entry of permeants into cells, with accompanying osmotic movements of water. When water enters simultaneously with permeant, it may drag with it more permeant than would move solely due to a concentration gradient (solvent drag).[17] This is not true when isotopic tracers are used for inextensible plant cells, where no water enters with permeant. However, even with tracers, it is essential to take account of tortuous paths through connective tissue or stagnant layers around membranes, as will be illustrated subsequently.

Valid permeability measurements indicate that the smaller the permeating molecule and the more lipid soluble it is, the more likely it is to cross membranes. Small molecules such as water, methanol, and dissolved gases cross membranes with great ease (the molecular sieve effect), as do lipid-soluble molecules such as ether.

Larger size need not result in lower permeability if the larger molecule has more nonpolar groups. In the plant cell *Chara,* ethylene glycol has a lower permeability than the larger propylene glycol, which is less polar because of its extra $—CH_2$ group.[3] Stein[17] has proposed a "lattice" model for permeation in an attempt to take into account a permeant's ability to leave the hydrogen-bonded, latticelike structure of water and enter the lipid matrix of the membrane.

Movement of water across membranes

By assuming a given thickness for membranes (say, 100 Å), it is possible to convert permeabilities to diffusion constants and thus compare how well a substance diffuses in a membrane as compared to free water solution. On this basis, even such highly permeable substances as water and methanol are found to diffuse three to six orders of magnitude less rapidly than in free solution. (The self-diffusion constant of water in water is about 2.4×10^{-5} cm^2/sec; the "diffusion constant" in biologic membranes is 4 to $4{,}000 \times 10^{-11}$ cm^2/sec.) A further complication, however, is that water permeabilities, in particular, can be more than twice as great under osmotic gradients as permeabilities measured using isotopic water.[6] The explanation is that during diffusive water flows, *unstirred layers* near the membrane effectively impede movements of the isotopes, whereas during osmotic flows the bulk movements of water flush out such unstirred layers. When great care is taken to stir the water next to membranes, these differences in osmotic and diffusive water permeabilities disappear.[38,48] They likewise disappear when diffusional permeabilities of the unstirred layers are measured and corrected for inside perfused cells.[60] All this implies that, in membranes, water molecules move independently of each other (as in diffusion) and not through pores. Indeed, it is possible to describe movements of water (and some other nonelectrolytes) in lipid bilayers as proportional to the solubility in the lipids of the membrane, the rate of diffusion through the lipid, and the mole fractions at each interface (i.e., to the concentration differences).[47,48] The water permeabilities of artificial membranes appear sufficient to explain the water permeabilities of biologic membranes, since they are both of the same order of magnitude (Table 1-2).

State of intracellular substances

To develop further the hollow-shell model of the cell shown in Fig. 1-1, *B*, it is important to consider if intracellular water and ions behave as they would in free solution or if some proportion is bound to cytoplasmic constituents. The state of intracellular water has been investigated in osmotic and nuclear resonance studies, whereas states of intracellular ions have been explored via measurements of intracellular activities and diffusivities.

Experiments involving cell shrinking and swelling in hypertonic and hypotonic solutions, respectively, indicate that cells behave as very good osmometers, provided little solute is gained or lost, that is, their contents obey the laws of simple solutions. To understand this, consider that the normal amount of osmotically active intracellular solute (n_i) equals its osmolarity (c_i) times the isotonic volume (v_i) of water in which it is dissolved, as follows:

$$n_i = (c_i v_i)_{\text{isotonic}}$$

If the cell shrinks or swells but retains its normal amount of solute, then:

$$n_i = (c_i v_i)_{\text{hypotonic}} = (c_i v_i)_{\text{hypertonic}} \tag{4}$$

Because water is in equilibrium across cell membranes, the internal osmolarity (c_i) will rapidly equal the external osmolarity (c_o), but from equation 4 the volume should change reciprocally:

$$v_i = \frac{n_i}{c_o} \tag{5}$$

This is not strictly true. Rather, it appears that the entire intracellular volume is not osmotically active, and equation 5 must be modified to include this osmotically inactive volume (b):

$$(v_i - b) = \frac{n_i}{c_o}$$

$$v_i = \frac{n_i}{c_o} + b \tag{6}$$

Hence a plot of cell volume vs the reciprocal of the external osmolarity (or of the tonicity) should be a straight line with an intercept of b for $\frac{1}{c_o} = 0$ (i.e., at infinite external osmolarity, all the cell water is withdrawn). An example of such a plot from frog muscle is illustrated in Fig. 1-6. Similar plots have been found for other cells.[6,89] Cells do indeed behave as osmometers over a range of external osmolarities.

It might be thought that the osmotically inactive space was simply that occupied by solids. However, in some cells the osmotically active space (v_i − b) is smaller than the volume of intracellular water. This has lead to suggestions that some of the water is bound. For example, in frog muscle the inactive volume is about 33%, but solids would account for only 13% to 15% of

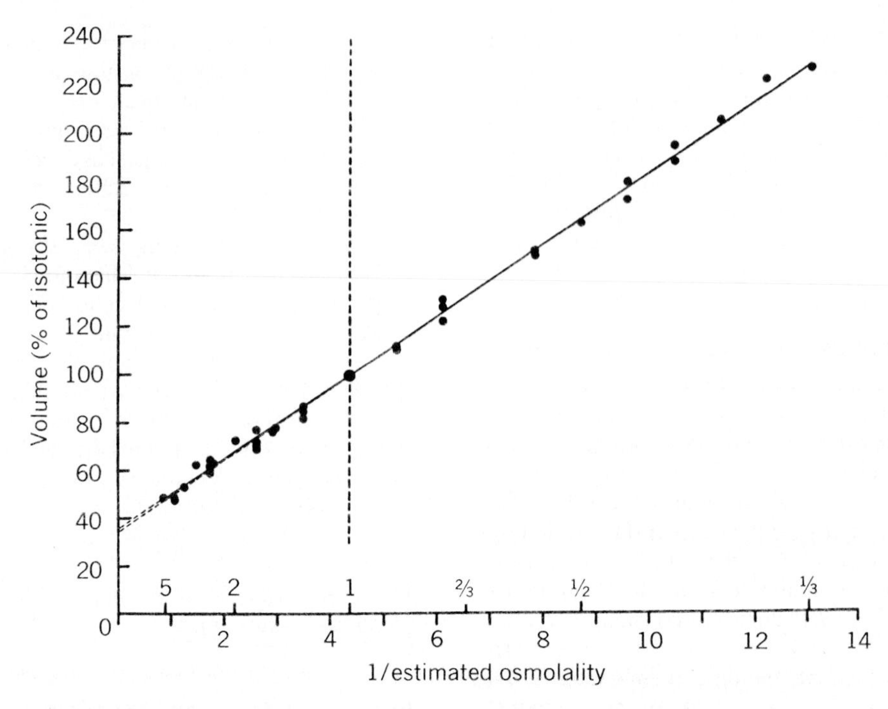

Fig. 1-6. Change in volume of single frog anterior tibial muscles with changes in external osmolarity. Reciprocal of osmolarity is plotted below abscissa, relative tonicity is plotted above abscissa. Extrapolation to infinite external osmolarity yields a value for osmotically inactive volume of 33% between two dashed lines. (From Blinks.[28])

this.[28,30] However, in hypertonic solutions (where the intercept b is measured) the volume of the sarcoplasmic reticulum increases as though it were part of extracellular space, and its volume nicely accounts for the differences between osmotically active space and the volume of intracellular water.[27] In the red blood cell, it can be calculated that the difference between the osmotically active space of about 62% ($b = 0.38$)[84] and the volume of intracellular water is due to changes in the osmotic coefficient of hemoglobin.[6] In both cells, it thus seems that all intracellular water is solvent water.

This conclusion is strengthened by nuclear magnetic resonance studies of the water component of the osmotically inactive volume. This water exists in two states. Relatively few water molecules are "bound" to, and move with, macromolecules. The rest of the water seems to form a loose hydration shell around ionic regions of macromolecules but to be able to exchange rapidly with bulk water and be available as a solvent.[35]

The state of intracellular ions in intracellular solvent water has been investigated by comparing concentrations measured chemically with activities measured with ion-specific intracellular electrodes. Even in free solution of physiologic strengths the activity is less than the concentration, with the ratio being the activity coefficient. The question then is whether activity coefficients are the same inside and outside the cell. If so, the ion may be said to be unbound. For potassium and chloride ions, this appears to be true, although some of the potassium in axons may be "sequestered."[126] For sodium, on the other hand, from 25% to 90% has been claimed to be bound, or compartmentalized, or otherwise not visible to the ion-selective electrode.[126] The larger percentages are found in muscle. From the kinetic rates of sodium exchange, it is argued that 96% to 98% of sodium in frog muscle is compartmentalized in the sarcoplasmic reticulum, in accord with the osmotic measurements presented previously.[112] This would be consistent with conclusions from nuclear magnetic resonance that all cellular sodium appears unbound,[45] that is, regardless of what compartment it was in, sodium could be in free solution even if not visible to an intracellular ion-selective electrode.

The final indication that intracellular substances are in a state of free solution is that they appear to be able to diffuse nearly as readily within cells as in free solution. Potassium in squid axons[70] and water, urea, and glycerol in barnacle muscles[31] have diffusion constants that are the same as in free solution, whereas diffusion of potassium, sodium, sulfate, and ATP ions as well as of sorbital and sucrose in frog muscles is reduced by about one half.[81] These reduced diffusivities are presumably due to physical (tortuosity) rather than chemical (binding) restraints, since the ions and nonelectrolytes were similarly affected.

In sum, therefore, it seems that intracellular substances may not always exist in the concentrations that can be measured by gross chemical means. However, those substances that are dissolved in the intracellular water appear to move as though they were in free solution under, at most, physical restraints. From this, it follows that the chemistry of solutions may indeed be applied to movements of substances within cells and across their membranes.

Diffusion potentials

Electrolytes diffuse in free solution and across cell membranes, just as do nonelectrolytes. However, because they are charged, it is energetically improbable that they diffuse across the lipid bilayers. More probably, they pass through polar pores in integral proteins, whose configurations determine the actual permeabilities.[66] They also move in solution under the influence of electrical potentials, and indeed, their very diffusion may set up electrical potentials. Such diffusion potentials can arise spontaneously (i.e., due to kinetic forces) whenever one ion species, for example, diffuses at a different rate than its counterion(s). An example of such diffusion would be at the junction between a salt bridge filled with a concentrated monovalent salt solution and a dilute bath of the same salt. The concentrated salt will tend to diffuse down its concentration gradient, according to Fick's law, into the more dilute salt across the liquid junction separating them. In general, one of the two ions (cation or anion) will diffuse faster, leaving its counterion behind. Such a charge separation immediately leads to an electrical potential. If the faster ion is the cation, the solution into which it is diffusing becomes positive, whereas the solution it is leaving becomes negative. This potential decelerates the cation and accelerates the anion; both finally move at the same rate, and equal numbers of both ions cross the liquid junction per unit time. The ultimate liquid junction potential is whatever is required to make the two ions move at equal rates, that is, there must be a potential if they are to move at the same rate, and if two ions are moving at the same rate, this does not mean that there can be no potential. The size of the potential is proportional to the difference between the rates at which each ion alone moves in free solution under a given force (i.e., proportional to the differences in ionic mobilities). The source of energy for the potential is, of course, the concentration gradient. If diffusion proceeds long enough to abolish the concentration gradient, the diffusion potential will likewise be abolished.

Membrane potentials

Concentration gradients of ions likewise exist across cell membranes. In general, none of the major extracellular ions (chloride, sodium, potassium) is at the same concentration on the two sides of cell membranes. As might be expected, therefore, there are diffusion potentials set up across cell membranes. The ultimate source of the concentration gradients, and hence of the membrane diffusion potentials, is the metabolic energy expended in active transport of ions. If this transport is abolished by metabolic poisons or by cold, the cell begins to "run down" as ion concentrations tend to equalize across its membrane. However, if the permeabilities of the membrane to ions are low and especially if the cell is large and contains much intracellular electrolyte, it may take many hours for a cell to run down. During this period of time, cells continue to exhibit membrane potentials, so that active transport per se is not necessary for the existence of a membrane potential (as opposed to its maintenance). For the moment, therefore, the emphasis will be on those membrane potentials that are created by the diffusion of ions down the concentration gradients that active transport processes have set up. Since these diffusion potentials depend only on kinetic movements of ions, they are called *passive* potentials.

Equilibrium potentials

Unlike a liquid junction a membrane may act as a physical restraint on the diffusion of ions to the extent that an ion permeating a membrane may not be followed by its counterion. In such a case a diffusion potential will still result, but the net movement of the permeating ion will rapidly stop as the developing potential difference across the membrane opposes further ions crossing the membrane. The situation is analogous to that at the liquid junction, where the diffusion potential arises from the differences in ionic mobilities. The mobilities of various ions in the membrane may, and generally do, differ, and so lead to diffusion potentials, too. (More strictly, it is the permeabilities of the ions that differ. However, the permeability constant is taken to be the diffusion constant of a substance in a membrane of unknown thickness, and the diffusion constant is proportional to the mobility [Appendix B, p. 36]). One important example is that large internal organic anions in cells cannot diffuse out through the membrane when their counterions, the mobile internal cations, do so, and a membrane potential opposing further net cation movement therefore results. Similarly, sodium permeates most resting cell membranes poorly (there are exceptions, however), so that if chloride in extracellular fluids (which contain much sodium chloride) permeates cells, it leaves sodium behind. The resulting membrane potential opposes further chloride entry.

For any given ion the membrane potential that just stops net diffusion of this ion across the membrane is called its equilibrium potential. Equilibrium potentials are denoted by E_{ion}. The three main equilibrium potentials of interest for cells are those of sodium (E_{Na}), potassium (E_K), and chloride (E_{Cl}). The equilibrium potential for an ion is found by equating the diffusion force on this ion (proportional to its concentration gradient) to the electrical force on this ion (proportional to the electrical field). The result (Appendix B, p. 36) is the well-known Nernst potential. As applied to membranes, this potential is taken to be the intracellular potential minus the extracellular potential. With this convention, and at 20° C:

$$E_{Na} = -58 \log \frac{[Na^+]_i}{[Na^+]_o} \text{ mV} \quad (7)$$

$$E_K = -58 \log \frac{[K^+]_i}{[K^+]_o} \text{ mV} \quad (8)$$

$$E_{Cl} = -58 \log \frac{[Cl^-]_o}{[Cl^-]_i} \text{ mV} \quad (9)$$

where the subscripts i and o represent the inside and the outside of the membrane, respectively.

If the membrane potential of a cell equals the equilibrium potential of a given ion, there will be no net movement of that ion into or out of the cell. (This does not mean that passive influxes and outfluxes of the ion across the membrane cease. The use of radioisotopic tracers has long since shown that there is a steady traffic of ions both ways across membranes. However, for an ion that is at equilibrium, the influx equals the outflux.) If an ion is not in equilibrium with the resting potential, there will be net passive movements of this ion (its permeability permitting) in the direction that would bring it into equilibrium. If the membrane potential changes during electrical activity, for example, so that it reaches the equilibrium potential of an ion not formerly in equilibrium, net passive movements of this ion cease. This is an important point, for the species of ions that move during electrical activity can often be identified by the membrane potentials at which their movements (currents) cease. These membrane potentials will be their equilibrium potentials, which can be compared with the Nernst potentials for the various ions whose concentrations are known.

In general, all ions do not have the same equilibrium potentials and therefore they cannot all be in equilibrium at any given cell membrane potential. However, the membrane potential of a "resting" cell (the resting potential) may be very close to the equilibrium potential of one or more ions. This can be seen from examples of equilibrium potentials and resting potentials of a number of cells in Table 1-1. In none of these cells is sodium at or near equilibrium, so there is a steady tendency for sodium to enter down its electrical and concentration gradients, a tendency that must be counteracted by active transport.

Ionic basis of membrane potentials

If an ion's equilibrium potential is nearly the same as the cell's membrane potential, the passive diffusion of this ion, until stopped by the potential it thereby sets up, could determine the membrane potential. On the other hand, an ion may simply distribute its concentration across the membrane until it comes into equilibrium ($E_{ion} \rightarrow V_m$) with a membrane potential set up by another ion. The test is to see if experimental changes in the concentration of any one ion change the membrane potential in a predictable manner. For example, it has long been known that increases in the concentration of extracellular potassium ions can depolarize nerve and muscle, and this has led to the belief that the diffusion of potassium ions results in the membrane potential in the first place. The advent of intracellular recording techniques made it possible to measure membrane potentials directly and test their dependence on potassium and other ions.

Resting bioelectrical potentials were measured earlier as the difference in potential between a "normal" part of a nerve or a muscle and a region that had been crushed, cut, or narcotized so as to make contact with the interior. These were called "demarcation potentials" or "injury potentials" and had their origins in the resting potentials. The injury potential rarely represents the full magnitude of the resting potential, however, since extracellular fluids may provide a short-circuiting path between the cut and normal tissue unless special precautions are taken. One such method for measuring the resting potential of single nerve fibers of frogs, using the demarcation potential, was devised by Huxley and Stämpfli.[72]

Resting potentials of large invertebrate nerve axons are measurable by threading an axial electrode into the axon from the cut end until it rests under normal membrane unaffected by the injury at the cut end.[68] However, it was the development by Ling and Gerard[87] of the glass micropipet electrode, with tips smaller than 0.5 μm, that made possible the measurement of resting and other bioelectrical potentials from a wide variety of cells. Such electrodes may be inserted through cell membranes with very little injury to the cell and little or no alteration of the membrane potential; membranes appear to "seal" around the electrode tip. Therefore the amount of short circuiting between the inside and the outside of the cell can be quite small, allowing accurate and reproducible measurements of resting potentials. Many of the measurements cited in this chapter were made using such micropipet electrodes.

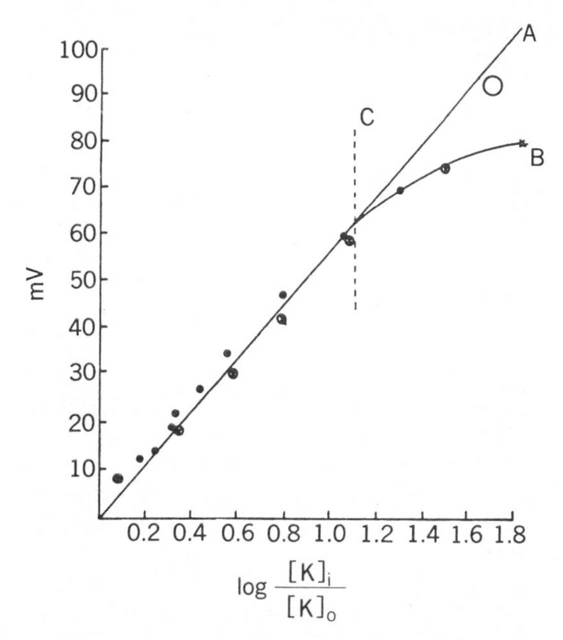

Fig. 1-7. Mean values of resting potentials of fibers in frog sartorius muscles at various ratios of $[K]_i/[K]_o$. Ordinate gives absolute values of measured membrane potentials. Filled circles and circles with crosses represent experimental points, whereas open circle represents initial membrane potential in 2.5 mM $[K]_o$ immediately after excision of muscle. Line *A* is theoretical line form from equation 8; vertical line *C* denotes 10 mM $[K]_o$. (From Conway.[34])

If normal resting potentials indeed arise as the diffusion potentials of potassium (and other ions), alteration of the potassium concentration ratios should produce parallel changes in the resting potential and the potassium equilibrium potential. Numerous experiments have been made on a variety of cells to test this hypothesis; several examples follow. Fig. 1-7 shows resting potentials of frog sartorius muscle fibers with varying concentration ratios of potassium across the muscle membrane.[34] Concentration ratios were altered from physiologic 2.5 mM $[K^+]_o$ by adding solid KCl to Ringer's fluid and allowing the muscle to reach a balanced state in the cold overnight. The solid line, A, represents the magnitude of the potassium equilibrium potential, and it can be seen that the membrane potential is the same as the potassium equilibrium potential only for external concentrations of potassium above 10 mM (vertical dashed line C). This is a common finding, for at near-physiologic external potassium concentrations, membrane potentials of excised tissues may deviate from the theoretical potential. However, the resting potential is closer to the potassium equilibrium potential immediately after excision (large circle in Fig. 1-7). Similar results have been found in muscles carefully dissected out in plasma. In another example of potassium determination of membrane potential, Kuffler et al.[80] altered potassium bathing the glial cells of the optic nerve of the mudpuppy *Necturus*. These cells were found to have a normal resting potential of 90 mV. Alterations of extracellular potassium due to activity of the optic nerve occur physiologically[102] (Chapter 2), and the membrane of the glial cells depolarizes as predicted from alterations of the potassium equilibrium potential (Fig. 1-8).

In these glial cells, as in molluscan neurons, discussed subsequently, chloride concentration changes have no effect on the membrane potential. This would occur if chloride permeability were negligible (so it could not diffuse across the membrane). In frog muscle cells, on the other hand, changes in external chloride concentrations cause *transient* but no steady-state alterations in membrane potential. This is not because chloride is impermeable (on the contrary, it is about twice as permeable as potassium[67]). Rather, $[Cl^-]_i$ is not metabolically controlled in these cells, whereas $[K^+]_i$ is (see later in this chapter). Thus the membrane potential at steady state is more nearly that of E_K, which is constant, and chloride redistributes itself passively to come into equilibrium. Numerous other examples of the dependence of membrane potential on external (and internal) potassium exist (squid,[25,37] frog muscle,[21,87,101] plant cells,[52,117] and many other cells).

Establishment of the resting potential

In order to get a better physical picture of what ions do at cell membranes to set up membrane potentials, consider what would happen if intra- and extracellular fluids were suddenly placed on either side of a membrane. At the initial instant, there would be no membrane potential, and there would be electroneutrality on each side of the membrane. Assume that only one ion species (such as potassium) is permeable, so that it alone can diffuse across the membrane. Thus, since it leaves its counterion behind, the diffusion potential that it generates very rapidly increases up to the equilibrium potential of the permeable ion, and its net diffusion comes to a stop. Then the amount of ion that moved across the membrane, an insulating dielectric, is balanced by an equal amount of counterion on the other side of the membrane. The membrane has literally charged up to the equilibrium potential and acts like a capacitor. Since most biologic membranes have capacitances of about 1 F/cm², the amount of charge on 1 cm² of membrane is:

$$Q = CV_m$$

where Q is the charge in coulombs.
For a typical 100 mV resting potential of a muscle, the charge is:

$$Q = 10^{-6}\ F/cm^2 \cdot 10^{-1}\ V = 10^{-7}\ coulombs/cm^2$$

Since there are about 10^5 coulombs per mole, the number of moles is:

$$Moles/cm^2 = 10^{-7}\ coulombs/cm^2 \cdot 10^{-5}\ mole/coulombs = 10^{-12}\ moles/cm^2$$

This quantity of ions, 10^{-12} moles, reappears often in calculations of amounts of ions involved in bioelectrical potentials and is termed a picomole (abbreviated pM). Such an amount of cations on one side of the membrane and an equal number of anions on the other side is that amount by which electroneutrality on each side fails: there is 1 pM/cm² more cations on one side than anions and vice versa on the other side. However, this may be an extremely small proportion of all ions in a cell. For example, in a muscle cell of 20 μm radius, length 1, internal potassium concentration of 140 mM, and with 85% of its volume occupied by fiber water (the rest being inactive volume), the proportion of potassium ions held in the surface ion clouds is as follows:

$$\frac{\text{Amount of ion on surface}}{\text{Amount of ion in cell}} = \frac{(10^{-12}\ mole/cm^2)\ (2\pi)\ (20 \cdot 10^{-4}\ cm)\ (l)}{(140 \cdot 10^{-3}\ mole/cm^3)\ (0.85)\ (\pi)\ (20 \cdot 10^{-4}\ cm)^2\ (l)} = \text{about } 10^{-5}$$

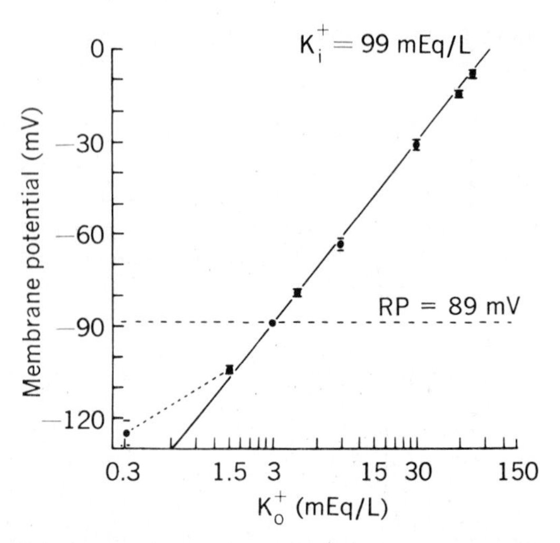

Fig. 1-8. Amphibian neuroglia. Dependence of membrane potentials of mudpuppy glial cells on external potassium concentrations. Solid line is theoretical line form from equation 8. (From Kuffler et al.[80])

Hence only one one hundred thousandth of this cell's potassium ion needs to diffuse across the cell membrane to set up a membrane potential of 100 mV. It is for this reason that the membrane may be "discharged" again and again, as during an action potential (described in the following chapter), and the membrane recharged again and again by the outward diffusion of potassium, without an appreciable change in the internal amounts of potassium.

The Donnan equilibrium

It has been shown that the membrane potential of frog muscle parallels the potassium equilibrium potential (at least at $[K^+]_o > 10$ mM), whereas chloride redistributes itself to stay in equilibrium. Since the equilibrium potential of each is equal to the membrane potential, they must be equal to each other. This leads to the following result in the case of frog muscle, for example:

$$E_K = -58 \log [K^+]_i/[K^+]_o = -58 \log [Cl^-]_o/[Cl^-]_i = E_{Cl}$$

$$[K^+]_i/[K^+]_o = [Cl^-]_o/[Cl^-]_i \qquad (10)$$

or

$$[K^+]_i[Cl^-]_i = [K^+]_o[Cl^-]_o \qquad (11)$$

General relations of this kind were derived by Donnan[41] in 1924 for equilibriums in inanimate systems in which solutions of nonpermeating organic electrolytes (such as phenol red) were initially separated by colloidin membranes from permeating solutions of salts. By way of illustration, consider such a system where the permeating salt is univalent (such as KCl) and the organic impermeant (P) has a valence of −1. The conditions can be depicted as follows for KCl:

Initial condition		**Equilibrium condition**	
c_1 K^+	K^+ c_2	$c_1 + x$ K^+	K^+ $c_2 - x$
c_1 P^-	Cl^- c_2	x Cl^-	Cl^- $c_2 - x$
		c_1 P^-	
SIDE 1	SIDE 2	SIDE 1	SIDE 2

The c's represent the concentrations. After an amount (x) of chloride has diffused into side 1, accompanied by a like amount of potassium, the resulting membrane potential stops further ion movements and equilibrium results (neglecting osmotic movements of water for the moment). At such an equilibrium, Donnan[41] showed that equations 10 and 11 hold. Inspection of the equilibrium condition shows that an asymmetry of ions across the membranes has been set up solely by the presence of the impermeant anion. Nonetheless, all ions but this impermeant (and water, as explained later) are in equilibrium, so that the maintenance of the ionic asymmetries requires no expenditure of energy.

It will now be seen to what extent equilibriums of the Donnan type exist across cell membranes. Inside, charged impermeable anions certainly exist; these include amino acids and proteins whose net charge is negative, phosphorylated intermediates of metabolism, etc. If two or more ions are near electrochemical equilibrium across such cell membranes, as Table 1-1 indicates some may be, these Donnan relations (equations 10 and 11) should apply to their intra- and extracellular concentrations. In 1941 Boyle and Conway[29] used chemical analyses of frog muscle to show that potassium and chloride could move across the muscle membrane and distribute themselves in accordance with the Donnan relations.[104] Table 1-3 gives results from muscles soaked overnight in Ringer's solution, to which solid KCl has been added. Over a 25-fold variation in extracellular potassium concentration, there was a near equality of the product of intracellular and extracellular concentrations. Below a $[K^+]_o$ of 10 mM, however, the predictions from the Donnan relations did not fit the experimental results as well.

Further evidence that the Donnan relations might apply to frog muscle was obtained by Hodgkin and Horowicz,[67] who altered the membrane potentials of these cells with changes in

Table 1-3. Donnan relations in frog muscle*

$[K]_o$	$[K]_o [Cl]_o \times 10^{-3}$	$[K]_i [Cl]_i \times 10^{-3}$	Ratio
12	1.05	1.00	1.05
18	1.69	1.72	0.98
30	3.18	2.99	1.06
60	8.16	8.61	0.94
90	14.90	15.80	0.94
120	23.50	24.20	0.97
150	33.90	34.40	0.99
210	60.00	52.80	1.14
300	112.80	118.70	1.05
			(average 1.01)

*Adapted from Boyle and Conway.[29]

external potassium ion charges while keeping the product $(K)_o$ $(Cl)_o$ constant. In this situation, both potassium and chloride ions should always be in equilibrium, and the membrane potential should be given by both equations 8 and 9. Fig. 1-9 indicates that this is nearly the case. The measured resting potentials agree well with those predicted, except at or below $[K^+]_o = 10$ mM. Thus there appear to be departures from simple equilibrium conditions at these low physiologic concentrations. These departures will be considered in the next section.

Two very important conclusions follow from all this. First, as mentioned earlier, concentration ratios of the permeant ions (potassium and chloride in muscle, anions in erythrocytes) can be maintained with little or no expenditure of energy, since the distribution of these ions is set by the concentration of the internal impermeant anion. However, there is another less permeant ion for most cells, and this is extracellular sodium. Boyle and Conway[29] felt that it was "practically perfectly excluded" in their experiments (i.e., it was also an impermeant). The only effect of impermeant sodium ion on the Donnan equilibrium would be to ensure that there would be more extracellular chloride than potassium ions to balance the sodium ions. However, the use of radioisotopes has shown that membranes are not impermeable to sodium ions, merely less permeable than to potassium and chloride in muscle, for example. Rather, it is the intervention of the sodium pump to eject the sodium that passively enters the cell that makes sodium seem *effectively* impermeable, at least in the resting state of most cells.

The second important conclusion to follow from the applicability of the Donnan systems to

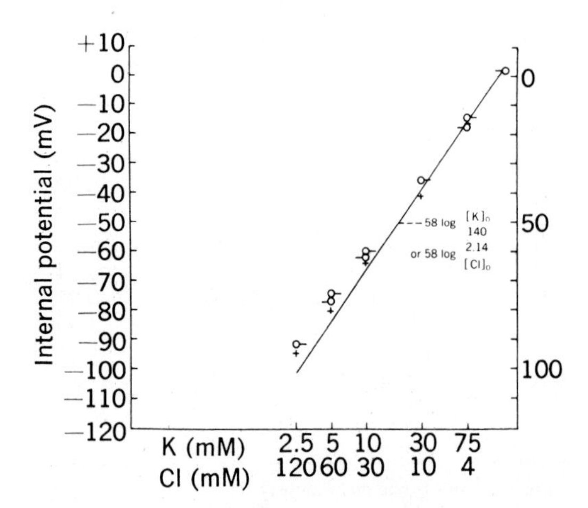

Fig. 1-9. Relation between membrane potential and log $[K]_o$ or $-\log\ [Cl]_o$ when using solutions with $[K]_o[Cl]_o = 300\ mM^2$. (From Hodgkin and Horowicz.[67])

cells is that no animal cell can be in osmotic equilibrium without extracellular sodium being effectively impermeant. This is because the osmolarity of the intracellular impermeant(s) must be balanced by some extracellular impermeant in order for an osmotic equilibrium to be present. No simple Donnan system of the type just depicted can reach osmotic equilibrium without the application of hydrostatic pressure to prevent osmotic entry of water.[41] Plant cells, with inextensible cell walls, could exert such hydrostatic pressures, but animal cells could not and they would swell. This has already been demonstrated in the behavior of animal cells as osmometers. What can now be added is that c_0 in equation 6 is almost entirely represented by extracellular sodium concentrations; in frog muscle, for example, the cell volume is inversely proportional to sodium alone.[29] However, the complete dependence of cell volume on extracellular sodium depends on the constancy of the intracellular impermeant; yet there are times when this undergoes normal physiologic changes. For example, during the oxygenation and deoxygenation of hemoglobin in erythrocytes (Chapter 62), the charge on the hemoglobin alters, as do the amounts of associated ions. With changes in the amount of associated ions, there are osmotic changes in cell water and hence in cell volume.

POLYIONIC MEMBRANE POTENTIALS

Taken all together, it is remarkable how closely ion distributions and resting membrane potentials seem to be explained on the basis of ions equilibrating across membranes. However, as previously emphasized, many ions (and especially sodium) are far from equilibrium across the resting membrane. In general, changes in sodium concentrations (or permeabilities), for example, may have only small effects on membrane potentials.[50,80,101,117] These effects are most significant precisely at the physiologic values of $[K^+]_o$, where membrane potentials deviate from the equilibrium potentials of potassium. Thus it might be possible to describe better these membrane potentials (and their deviations from equilibrium predictions) by including movements of ions far from equilibrium, such as sodium. Two ways in which this has been done will be illustrated here. The first of these results in the constant field equation for passive movements of ions. The second includes active fluxes due to electrogenic pumps.

Constant field equation[56,69]

The purpose of the constant field equation given subsequently is to assess how the distribution of a number of permeating ions, none at equilibrium, would determine the membrane potential. The approach is to consider the passive diffusion of a number of different ions and to predict the membrane potential that would cause no net membrane current. (If there were a 1:1 exchange of sodium and potassium ions, or if equal numbers of sodium and chloride ions, or combinations of these, moved across the membrane, there would be no net current. There would be net fluxes of each of the permeating ions, however.) The basic assumption used here in solving the diffusion equations is that the membrane potential, measured by determining the difference between the inside and the outside of the membrane, changes uniformly within the membrane (hence the term "constant field"). Other assumptions about the electrical field across the membrane can lead to similar equations for predicting ionic currents and membrane potentials.[18] The constant field derivation is valid for both anions and cations, however, whereas some of the others are not. For this reason, this derivation is the one given in Appendix C. For historical reasons the resulting equation will be called the constant field equation, although, as indicated previously, the results do not necessarily depend on the constant field assumption, which may have to be replaced by a more detailed view of the membrane's electrical field.

Originally,[69] it was supposed that there might be charges (dipoles) within the membrane that would ori-

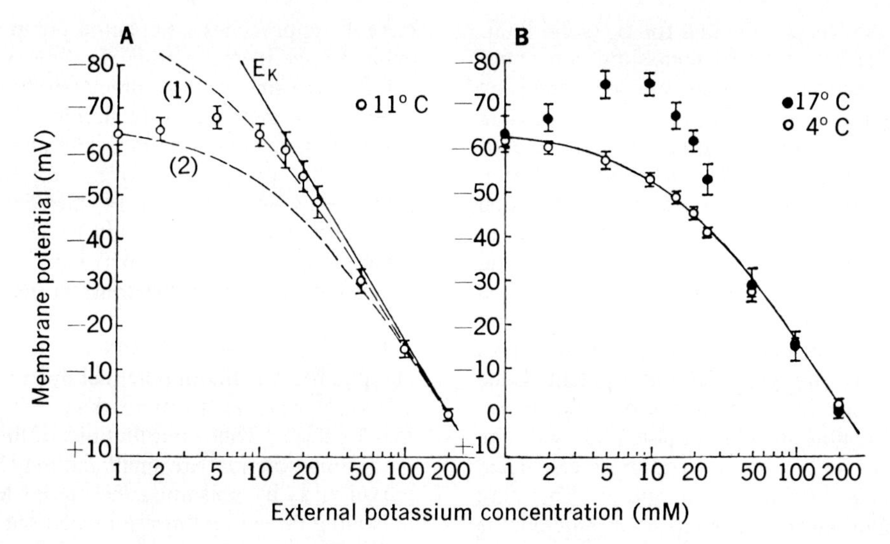

Fig. 1-10. Dependence of resting potential of molluscan neurons on $[K]_o$. **A,** Measured resting potentials at 11° C are not fitted by either potassium equilibrium potential, E_K, except at high $[K]_o$, or by constant field equation at two values of P_{Na}/P_K. **B,** At 4° C, points (open circles) *are* fitted by constant field equation—$P_{Na}/P_K = 0.033$; $[K]_i = 235$ mM for solid curve—but not at 17° C. (From Marmor and Gorman[91]; copyright 1970 by the American Association for the Advancement of Science.)

ent in the electrical field across the membrane. This would result in a uniform charge of the potential from one side to the other. The present view of ionic permeation is that ions move independently through charged pores in the membrane integral proteins.[66] However, these charges are likely to be relatively fixed carboxyl or amino groups on the proteins and probably would result in a uniform electrical field across the membrane.

Given the assumption of a constant electrical field across the membrane and the case of zero net ionic current, the membrane potential based on the passive movements of sodium, chloride, and potassium ions (the three most prevalent ions in biologic fluids) is predicted to be, at 20° C (Appendix C):

$$V_m = -58 \log \frac{P_K[K^+]_i + P_{Na}[Na^+]_i + P_{Cl}[Cl^-]_o}{P_K[K^+]_o + P_{Na}[Na^+]_o + P_{Cl}[Cl^-]_i} \quad (12)$$

The constant field equation has been applied to membrane potentials of squid axons,[69] frog muscles,[67] and molluscan neurons.[91,97] Several simplifications of the equation are possible. The membrane potentials of molluscan neurons appear to be nearly independent of external chloride ion concentrations, so chloride terms drop out ($P_{Cl} \approx 0$). Where an ion, such as chloride in frog muscle, is distributed passively at equilibrium, the terms for this ion also drop out (Appendix C).

Similarly, expressing all permeabilities with reference to the permeability to potassium, the equation may be rewritten as:

$$V_m = -58 \log \frac{[K^+]_i + P_{Na}/P_K[Na^+]_i}{[K^+]_o + P_{Na}/P_K[Na^+]_o} \quad (13)$$

Since in general the resting permeabilities of sodium ions are less than those of potassium, and the internal concentrations of sodium ions are low (especially compared with intracellular concentrations of potassium), the term $P_{Na}/P_K[Na^+]_i$ may be ignored. This means that if intracellular and extracellular ion concentrations are known, only the ratio of permeabilities (which may not be known) need be estimated. (If the intracellular concentrations are not known but the equation can be fitted to the data, then these concentrations, too, may be estimated.)[97] Finally, it can be seen that at high extracellular potassium concentrations the potassium concentration term in the denominator dominates and the equation simplifies to the Nernst potential.

Fig. 1-10 from Marmor and Gorman illustrates the application of the equation to resting potentials of neurons of a mollusc as well as the effects of electrogenic sodium pumps, which will be discussed subsequently. In Fig. 1-10, *A*, with high external potassium concentrations, the resting potentials of these neurons are indeed predicted by the Nernst potential for E_K (solid line). However, at low external potassium concentrations the resting potentials are described by neither the Nernst potential nor the constant field equation (dashed lines, two sets of parameters). On cooling (Fig. 1-10, *B*), the resting potential is well described by the constant field equation (solid line). The difference is seen to be due to an electrogenic sodium pump active at 17° and 11° C but not at 4° C. The pump is also blocked by ouabain (not shown). The membrane potential thus has two components in these experiments: a polyionic diffusion potential and a potential due to an electrogenic pump. It can be seen, moreover, that resting membrane potentials with the pump active lie *between* E_K and those calculated from the constant field assumption. Therefore the contribution of this pump to the membrane potential must be evaluated.

Contribution of electrogenic pumps to resting potentials

A full exposition of active transport (including "ion pumps") will be given in the next section. Here it suffices to emphasize that the sodium pump in such excitable tissues as nerve and muscle is a potassium pump, too, and the ion movements are linked. Were there to be a 1:1 exchange of sodium pumped outward for potassium pumped inward, the pump would be *electroneutral* and would make no contribution to the membrane potential. As was seen in Fig. 1-10, however, the pump does make a contribution. Such pumps are *electrogenic* and are in the majority. They pump more sodium outward than potassium inward, generally in the ratio of 3:2.[18]

The contribution of this asymmetry in pumping to the resting membrane potential can be estimated from an equation similar in form to the constant field equation.[18] This, recall, was only for *passive* ion movements. Moreover, these passive ion movements were for a cell running down. There were net fluxes of sodium inward and potassium outward. Adding for sodium and potassium the steady-state condition that active transport of ions in one direction exactly balances the passive movements in the other, all net fluxes are zero. Then for a ratio of r sodium ions pumped outward for every potassium ion pumped inward, the membrane potential is predicted to be (Appendix D, p. 39):

$$V_m = -58 \log \frac{r[K^+]_i + \frac{P_{Na}}{P_K}[Na^+]_i}{r[K^+]_o + \frac{P_{Na}}{P_K}[Na^+]_o} \qquad (14)$$

Here if r approaches 1 (a neutral pump or for the pump stopped), the predicted membrane potential is the same as that predicted from purely passive ion movements (equation 13). On the other hand, if only sodium is pumped, which is to say that potassium moves only passively, then $r \rightarrow \infty$ and equation 14 correctly predicts that resting membrane potentials with an electrogenic pump active will be between E_K and those calculated solely from the constant field assumption.

For a 3:2 ratio of sodium to potassium pumped the maximum potential due to the pump *at steady state* is no more than −10 mV and may be less.[18] Thus contributions of the pump to the resting potential are small and may be hard to measure, as by poisoning the pump. Moreover, poisoning the pump throws the cell out of steady state, as may injury during the very excision of a cell needed to make the measurement of a resting potential in the first place. Thus it may be coincidental that the average measured resting potential of 92 mV for excised frog muscle is closely predicted by equation 14 for a 3:2 ratio. In cells deliberately put out of steady state (as by injecting sodium intracellularly or following accumulations of sodium and loss of potassium in the cold), subsequent transient hyperpolarizations due to electrogenic pumps can greatly exceed the membrane potential values predicted by equation 14.[22] Natural activity of excitable cells also may put them out of steady state, with large afterpotentials or recovery potentials due to electrogenic pumps.[18,76,100] Considerations of such afterpotentials will be given in later chapters (e.g., Chapter 2).

ACTIVE TRANSPORT

This section focuses on the movements of relatively small molecules (sugars, ions, amino acids, etc.) across the cell membrane. One may distinguish two pathways by which such materials move through membranes. During simple passive diffusion, these substances move rather slowly through the lipid phase of the membrane. Such passive diffusion is often readily understood from considerations of the physical (e.g., molecular size and weight) and chemical (e.g., lipid solubility) properties of the diffusing substance.[17] Alternatively, and of more interest to the present discussion, molecules may also travel by pathways that reflect the biochemical properties of specific proteins embedded in the membrane. In turn, such "transport systems" catalyze one of two sorts of reactions: "facilitated diffusion" or "active transport." The distinction between these two classes of transport reactions is illustrated by a comparison between the transport of glucose by the mammalian erythrocyte and by the intestinal absorptive cell. In the erythrocyte, D-glucose enters much faster than one would expect on the basis of its molecular weight (and size) and lipid solubility, for a specific membrane protein(s) "facilitates" the movement of sugar. But the eventual internal concentration of glucose never exceeds that outside the cell. For this cell the end result of glucose transport is the same as if glucose had slowly entered by passive diffusion; thus the transport of sugar is termed facilitated diffusion. In the gut, something very different occurs. Here free sugar is accumulated within the absorptive cell to concentrations higher than those found in the external medium. Clearly in this case, transport involves something more than the simple catalysis of sugar movement across the membrane, for sugar is forced *out* of equilibrium and is accumulated within the cell. One then uses the term "active transport" to imply that during the transport reaction the cell must expend energy to perform the work of such accumulation. Active transport requires the expenditure of energy by the cell and results in movement of substrate against its electrochemical gradient. Passive or facilitated diffusion does not require an input of metabolic energy, and substrate equilibrates with its electrochemical gradient across the membrane.

Cells use active transport not only to accumulate needed substances, but also to extrude unwanted ones. In the "typical" mammalian cell (Fig. 1-1), internal sodium ion is maintained at a lower concentration than extracellular sodium ion. In addition, there is an electrical potential, negative inside, across the cell membrane. Thus there are two driving forces demanding inward movement of sodium ions—one, the chemical gradient, and the other, the electrical gradient. Active transport of sodium ions (in this case, the extrusion of sodium ions) must expend energy to overcome both these forces so that sodium is maintained as much *out* of electrochemical equilibrium as is normally found. In the same "typical" cell the maintenance of low levels of free intracellular calcium ion (about 10^{-6} M) in the face of both higher external calcium ion (about 10^{-3} M) and a membrane potential, negative inside, is also an example of active transport.

How does such active transport occur? Although one cannot yet answer this question at the most detailed levels of analysis, there are a few facts that do seem certain. One is that active transport reflects solely the operation of elements residing in the cell membrane. A good example of the experimental support for this statement comes from work done on the mammalian erythrocyte. After this cell has been osmotically lysed and freed of its internal constituents, it can be resealed so that the resulting vesicle retains either the original polarity or is "inside out." In either case, one can still demonstrate active transport of both sodium and potassium ions under the appropriate conditions (transport is opposite the usual direction for the everted vesicle, of course). A later section will present a more rigorous test of the idea that active transport reflects the activity of only membrane-bound proteins.

One also knows that there must be some link between active transport and cellular metabolism. For example, at low temperature or in the presence of metabolic inhibitors the normal sodium and potassium gradients tend to collapse. Raising the temperature or removing the inhibitors reverses this process, as metabolism once again fuels ion transport.[17,32] An early clue as to the nature of the energy source for sodium and potassium transport came from experiments using squid nerve, where suspected energy sources could be injected directly into the cell. Only injection of adenosine triphosphate (ATP), or compounds that could readily give rise to ATP, supported extrusion of sodium and accumulation of potassium.[32] Another important step in the analysis of this problem came with the identification of a membrane-bound enzyme that could hydrolyze ATP, but only in the presence of sodium and potassium.[16,116] It was already known that sodium and potassium ion fluxes were coupled in some way, for rates of exchange of one ion were markedly inhibited in the absence of the other ion. The discovery of a specific ATP-splitting enzyme (an ATPase) clearly suggested that ATP hydrolysis provided the driving force for a coupled exchange of sodium and potassium ions, one in which there was outward pumping of sodium and inward movement of potassium. Finally, the link between this identifiable chemical reaction and the transport of sodium and potassium ions came from a study in which the transport reaction was run backward by experimentally exaggerating the normal sodium and potassium gradients. When run in reverse, there was, in fact, synthesis of ATP.[53]

It has turned out that ATP is a fuel for the active transport of several ions, and later sections will center on two ATPases—one involved in hydrogen ion transport, the other one handling both sodium and potassium ions. These ATPases represent a direct link between cellular metabolism and active ion transport. But ATP is not the fuel for many other active transport processes (for certain other ions, sugars, or amino acids). Instead, these latter reactions are only indirectly linked to metabolism, for there is no chemical transformation of any of the participants in the reaction catalyzed by the transport protein. The nature of this secondary link to metabolism forms the subject of another section, later in this discussion, but it is useful now to anticipate the nature of such indirect coupling. In doing so, one should reconsider the implications of the fact that when the normal sodium and potassium gradients are exaggerated, the coupled exchange of sodium and potassium ions can be run in reverse to yield the synthesis of ATP. This means that, of themselves, ion gradients can serve as a reservoir of potential energy that can be used to do work (here the work done is represented by the synthesis of ATP). In turn, this implies that the ion gradients established normally by active sodium and potassium transport also represent a source of power. As noted later, this reasoning turns out to be especially fruitful in considering the role of sodium ion in mammalian cells or hydrogen ion in microorganisms. Thus at the expense of ATP hydrolysis, ions are extruded and maintained *out* of electrochemical equilibrium. For these ions, both the steady-state chemical gradient (low inside, high outside) and the electrical potential (negative inside) represent powerful driving forces for their reentry. As such, pushing these ions out of the cell, "uphill," stores potential energy now represented by their tendency to move inward, "downhill." It is when these ions subsequently reenter the cell through specific pathways that this stored potential energy can be used to drive energy-requiring reactions such as the active transport of metabolites. Thus one sees that in this circuit of ion flow, one element (an ATPase) provides the direct link to metabolism, for active ion extrusion is driven by hydrolysis of ATP. By forcing a specific ion out of equilibrium in this single reaction, the cell has also stored the power needed to drive a multitude of other reactions that may be linked to the reentry of that ion down its electrochemical gradient. It is with this general view in mind that one should approach the following discussions, for their purpose is to describe briefly those elements that participate in a circulation of ions and to illustrate the role of such an ion circuit in the physiology of the cell.

Active transport of hydrogen ion: the proton-translocating ATPase

It is appropriate to begin by describing transport of hydrogen ion by a membrane-bound ATPase, for such H^+-ATPases are found in virtually all living cells. In prokaryotes the H^+-ATPase is found on the plasma membrane of the cell; in eukaryotes it is found (usually) only in the membranes of organelles such as mitochondria and chloroplasts. The example treated here describes the physiologic role of this ATPase in certain bacteria (or fungi), where the ATPase serves to extrude hydrogen ion at the expense of ATP hydrolysis. However, in other cases (many bacteria, all mitochondria)[9] the usual function of this ATPase is to catalyze the reverse reaction, the synthesis of ATP coupled to the inward flow of hydrogen ion. But for the moment, consideration of the ATP*ase* aspect of this membrane protein is of primary interest because it will allow discussion of some general "rules" governing ion transport and because it will foster some instructive parallels to the Na^+, K^+-ATPase, the major ion transport device of animal cell plasma membranes.

The specific example given is that of hydrogen ion extrusion by bacteria of the genus *Streptococcus*. The diagram in Fig. 1-11 shows that in this "simple" cell, hydrogen ion is moved out-

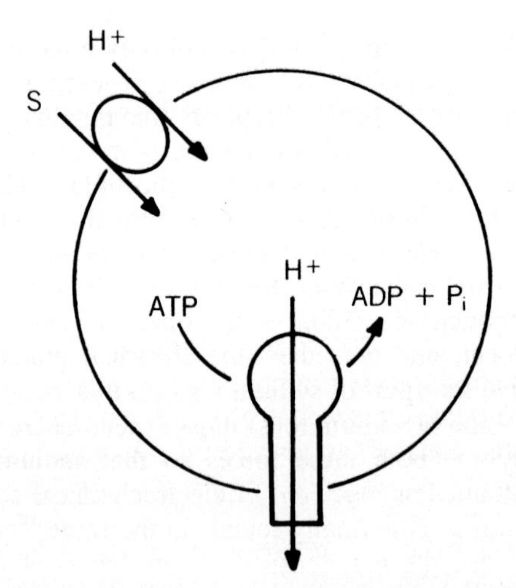

Fig. 1-11. Hydrogen ion extrusion coupled at ATP hydrolysis. In certain bacteria (e.g., *Streptococcus*), hydrogen ion is actively extruded by membrane-bound H^+-ATPase. In these and other cells, reentry of hydrogen ion may be coupled to inward flow of some substrate (S) by means of an independent membrane-bound transport system that catalyzes hydrogen ion/substrate "cotransport."

ward as ATP is split. If n protons are extruded for every ATP hydrolyzed, the chemical reaction is written as follows:

$$nH^+_i + ATP_i \rightleftarrows nH^+_o + ADP_i + Pi_i \qquad (15)$$

where the subscripts i and o indicate the inside and outside of the cell.

Since hydrogen ion is a positively charged particle, its transport from inside to outside displaces positive charge to the exterior of the cell. Thus the activity of this ATPase can support the development of a membrane potential (symbolized as V_m), negative within the cell. Transfer of hydrogen ion also means that the concentration of hydrogen ion (the pH) on either side of the membrane will be altered by activity of the enzyme. Clearly, then, the ATPase could also support the formation of a pH gradient (written as ΔpH), alkaline inside. The net result of this process is that energy "spent" in pumping hydrogen ion outward is not totally wasted. Instead, in large part it may be *conserved* by formation of an electrochemical potential gradient for hydrogen ion. (This is written as $\Delta\tilde{\mu}_{H+}/F$ and indicates the tendency for hydrogen ion to reenter the cell in this example.) In effect, potential energy available in chemical form (ATP) has been transformed into potential energy in the form of a gradient. One can write an expression indicating this "transduction," as follows (valid when the ion transport reaction has reached equilibrium; see Appendix F for the derivation):

$$n\,\frac{(-\Delta\tilde{\mu}_{H+})}{F} = Z\log K'eq + Z\log\frac{[ATP]}{[ADP][Pi]} \qquad (16)$$

In this equation, n represents the number of protons translocated per molecule of ATP split, Z gives the value of 2.303 RT/F (about 58 mV at 20° C), K′eq gives the equilibrium constant for the reaction of equation 15 when $\Delta\tilde{\mu}_{H+}/F = 0$ (K′eq is about 10^5), and the terms in brackets give the internal concentrations of ATP, ADP, and inorganic phosphate (Pi). The term to the left of the equal sign represents potential energy available as the disequilibrium of hydrogen ion. The term at the right gives the equivalent expression for energy available from ATP hydrolysis. (This term is also known as the "phosphate potential.") For convenience the common unit is given in electrical terms as the millivolt (mV).

Because $\Delta\tilde{\mu}_{H+}/F$ indicates the extent to which hydrogen ion is out of equilibrium, this term must take into account all driving forces tending to move hydrogen ion passively across the membrane. There are just two of these: one due to the chemical gradient (the pH gradient, ΔpH), the other due to the electrical gradient (the membrane potential, V_m). As shown in Appendix E, these two driving forces contribute to the total driving force on hydrogen ion as in the following expression:

$$\frac{\Delta\tilde{\mu}_{H+}}{F} = V_m - Z\Delta pH \qquad (17)$$

where ΔpH is equal to the inside pH minus the outside pH and, as before, the coefficient Z has a value of about 58 mV.

There are two questions that should be asked with regard to this active transport of hydrogen ion. First, one asks about the factors that determine the value of $\Delta\tilde{\mu}_{H+}/F$—how large a reservoir of potential energy can be formed as a result of the ATP-driven extrusion of hydrogen ion? Second, concerning the distribution of that potential energy between the electrical and chemical gradients—what determines the relative proportions of V_m and $Z\Delta pH$? An explicit mathematical treatment of these questions is found elsewhere.[94] Here these questions will be answered in a qualitative way simply to illustrate some important general principles.

There are two factors that influence the value of the electrochemical potential difference for hydrogen ion across the membrane. One has to do with the nature of the biochemical reaction itself: the number of protons translocated per molecule of ATP hydrolyzed (the value of n in equations 15 and 16). Thus as the ratio of H^+:ATP increases, the value of $\Delta\tilde{\mu}_{H+}/F$ will fall. This is clearly seen from equation 16, since the product $n(-\Delta\tilde{\mu}_{H+}/F)$ is constant for given levels of ATP, ADP, and inorganic phosphate within the cell. Another way of looking at this point is to consider what is meant by the term "equilibrium." At equilibrium the rate of the forward reaction (ATP hydrolysis and hydrogen ion export in this example) is equal to the rate of the reverse reaction (here, hydrogen ion entry coupled to the synthesis of ATP). If the rates of the two reactions are equal, the driving forces behind the two reactions are equivalent. For hydrogen ion extrusion the driving force is energy spent during ATP hydrolysis. For the reversal of the reaction, ATP synthesis, the driving force is n protons falling down their electrochemical potential gradient or $n(-\Delta\tilde{\mu}_{H+}/F)$. Thus as n rises, the value of $\Delta\tilde{\mu}_{H+}/F$ at equilibrium will decrease. It is important to understand this relationship between stoichiometry and the capacity of the transport reaction to drive an ion out of equilibrium. Intuitively, one's first thought may be that as more and more ions are pumped per molecule of ATP hydrolyzed, that ion must be forced further and further out of equilibrium. In fact, just the reverse must be true. This principle has few biologic examples to point to by way of illustration. However, current research suggests that such illustrations might come from comparisons of the H^+-ATPases found in mitochondria (H^+:ATP is 2[122]) and chloroplasts (H^+:ATP is apparently 3[106]).

Permeability of the membrane to protons themselves (or hydroxyl ions) is the second determinant of the extent to which hydrogen ions may be driven out of equilibrium. Clearly, if the membrane were exceedingly permeable to hydrogen ion, as soon as hydrogen ion was pumped outward it would reenter the cell at a different site. There would be a futile cycle of outward pumping and inward leaking, $\Delta\tilde{\mu}_{H^+}/F$ would not rise to any appreciable value, and energy dissipated by ATP hydrolysis would reappear as heat. In practice, although biologic membranes show considerable permeability to water, they display very low permeability to hydrogen (and hydroxide) ion.[12] This means that for membranes containing this ATPase, the inward leak of hydrogen ion is sufficiently small so that hydrogen ions may be pumped out of equilibrium nearly to the extent expected (the value given by equation 16) if there were no nonspecific pathways available for the reentry of these ions. For example, studies of the bacterium *Streptococcus faecalis*[61,62] indicate that V_m may be -150 mV and that $-Z\Delta pH$ can be -60 mV (a pH gradient of about 1 pH unit, alkaline inside). For such cells, $\Delta\tilde{\mu}_{H^+}/F$ may approach -200 mV or more. This is about the value expected if the stoichiometry of the bacterial H^+-ATPase is $2H^+$:ATP, for estimates of internal ATP, ADP, and inorganic phosphate indicate that the maximal value of $n(\Delta\tilde{\mu}_{H^+}/F)$ should be -400 to -500 mV.

One now asks about the factors that determine, not how far out of equilibrium hydrogen may be, but how this difference in electrochemical potential is partitioned between its electrical and chemical parts (equation 17). Here, some limiting cases will illustrate the rules. There are two factors determining the balance between the electrical and chemical gradients: (1) the buffering power of the internal and external phases and (2) the permeability of the membrane to the surrounding ions that are *not* protons. To visualize this, take as the first limiting case an instance where the membrane is impermeable to ions in general and buffering power is high on either side of the membrane. Here, as outward pumping of hydrogen ion proceeds and equilibrium is approached, the membrane potential will become large, whereas the pH gradient will remain small. This is because the external and internal buffering powers are high enough so that transfer of hydrogen ion into and out of these compartments makes little change in pH, as membrane capacitance is charged by hydrogen pumping. In the second limiting case, one sets a high value to membrane permeability for an ion (say, potassium) present at high concentration on either side of the membrane; in addition, one assumes that buffering power is not very substantial. In this case, outward movement of positively charged hydrogen ion may be compensated by inward movement of potassium ion. Therefore initial outward flow of hydrogen ion does not charge membrane capacitance appreciably, and the membrane potential remains near zero. Extrusion of hydrogen ion will then continue, and equilibrium will not be reached until the continued pumping of hydrogen ion has developed a sufficiently large pH gradient. At this equilibrium, of course, there will be a chemical gradient for both hydrogen and potassium ions ($[H^+]_i < [H^+]_o$ and $[K^+]_i > [K^+]_o$). But the assumption was that potassium ion was at high concentration on both sides of the membrane, so that the final gradient for potassium is not much changed from its original value of 1. Consequently, the membrane potential (reflected in the distribution of potassium ion) is small. Thus in this second case, $\Delta\tilde{\mu}_{H^+}/F$ is dominated by the $Z\Delta pH$ term. Except under special experimental circumstances, neither of the two limiting cases presented actually holds in detail, and the relative proportion of V_m and $Z\Delta pH$ is somewhere between the extremes given.[95] In any individual case the final result is a finely tuned compromise between buffering power and permeability to surrounding ions.

To illustrate some general principles about ion-translocating reactions, hydrogen ion extrusion coupled to ATP hydrolysis has been discussed. It is only fair to point out that, under most conditions, the physiologic role of the H^+-ATPase is to catalyze ATP synthesis coupled to proton entry. This is diagrammed in Fig. 1-12. The prerequisite for this mode of operation of the H^+-ATPase is only that the membrane have some other hydrogen ion "pump" so that protons can be moved sufficiently far out of equilibrium to drive the ATPase reaction in the reverse direction. In Fig. 1-12 this second hydrogen ion pump is shown as the series of reactions that couple proton extrusion to the oxidation of respiratory substrates in bacteria or mitochondria. This idea, first proposed by Mitchell in the early 1960s,[92] offers the best current explanation for

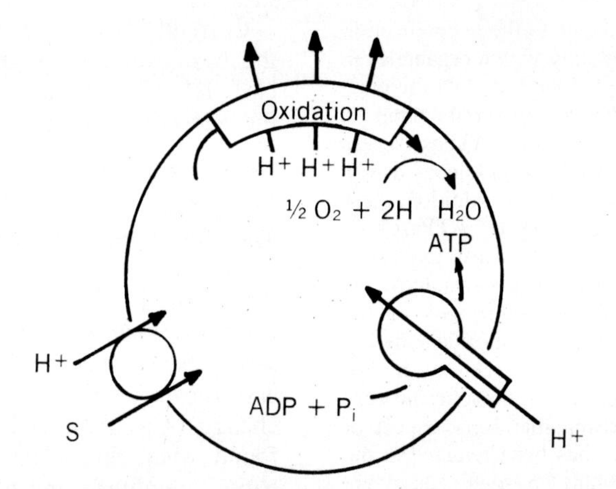

Fig. 1-12. Hydrogen ion extrusion coupled to oxidation. In mitochondria or in bacteria that carry out oxidative phosphorylation (e.g., *Escherichia coli*) there is an obligatory outward movement of hydrogen ion during oxidation of respiratory substrate. Two pathways for reentry of hydrogen ion are shown here. When hydrogen ion reenters by way of membrane-bound H^+-ATPase, net synthesis of ATP results (oxidative phosphorylation). Hydrogen ion may also reenter during co-transport with other substances (S).

the complex process of oxidative phosphorylation. Thus a difference in electrochemical potential for hydrogen ion is first established by export of protons during the oxidation of, for example, reduced nicotinamide adenine dinucleotide (NADH). Subsequently, protons reenter, moving down the electrochemical gradient, by way of the H^+-ATPase, giving rise to the net synthesis of ATP. A similar cycle takes place during photophosphorylation in chloroplasts, except that the polarity of proton translocations is reversed. A great deal of evidence supports this general view,[1,8,12] and for the present discussion a few major points should be summarized: (1) These various proton-translocating devices are all located in topologically closed structures (cells or organelles) and in membranes that show low permeability to hydrogen ion (and hydroxide ion). (2) Across each of these membranes, protons are maintained significantly *out* of electrochemical equilibrium by respiratory or light-driven hydrogen ion movements. (3) There is an H^+-ATPase of remarkably similar molecular structure in each of these membranes. (4) In each case the normal circulation of hydrogen ion (Fig. 1-12) can be bypassed, and one can show that the H^+-ATPase catalyzes net synthesis of ATP when sufficiently high gradients of hydrogen ion are artificially imposed. For example, in bacteria,[90] mitochondria,[108] or chloroplasts,[73] ATP synthesis occurs when a pH gradient of about 3.5 to 4.0 pH units is given.

In summary, in bacteria the proton-translocating ATPase is found on the plasma membrane, where its mode of operation is twofold. In cells lacking a respiratory chain (e.g., genus *Streptococcus)* the physiologic role of the ATPase is to extrude hydrogen ion at the expense of ATP derived from glycolysis. This drives protons out of electrochemical equilibrium so that other transport reactions, such as the accumulation of nutrients (discussed in a later section), can be driven by the reentry of hydrogen ion. When hydrogen ion is moved outward, the extent to which it is forced out of equilibrium is determined by two factors — the nature of the chemical reaction (the stoichiometry of H^+: ATP) and the permeability of the membrane to hydrogen ion itself. Other factors (buffering power and permeabilities to other surrounding ions) then come into play to determine whether this electrochemical potential gradient for hydrogen ion takes the form of a membrane potential, a pH gradient, or some mixture of the two. In bacteria with a respiratory chain (e.g., *Escherichia coli*) the H^+-ATPase may also operate as it does in mitochondria or chloroplasts, catalyzing synthesis of ATP when protons are first pumped out of equilibrium by other reactions.[8] In prokaryotes, then, the circulation of hydrogen ion across the plasma membrane is of primary importance. In eukaryotes, however, it appears that this circulation of hydrogen ion is largely restricted to organelles; during evolution the major hydrogen pumps have been segregated within the membranes of mitochondria and chloroplasts. But eukaryotic cells also carry out active transport of ions across the plasma membrane. In the following section, it will become clear that the general principles evident in the simpler prokaryotes have been preserved in the more complex eukaryotic cell.

There are certainly exceptions to the generalization that the H^+-ATPase is found only within organelles in eukaryotes. An H^+-ATPase is found on *both* the plasma and mitochondrial membranes in certain fungi.[118] It has also been suggested[55] that an H^+-ATPase serves to maintain the pH and electrical gradients across certain plant cell plasma membranes. In addition, there are good reasons for thinking that an H^+-ATPase participates in the secretion of acid by gastric mucosa.[83] (It appears that this latter ATPase may catalyze the exchange of hydrogen and potassium ions or the coupled extrusion of hydrogen and chloride ions.) If one assumes that ion transport in present-day microorganisms reflects ion transport as it arose in the very first cells,[20] it is not surprising that some aspect of ATP-driven proton pumping has been retained at the level of the plasma membrane in some eukaryotic cells.

Animal cells and the Na^+, K^+-ATPase

In animal cells the major ion pump of the plasma membrane is an ATPase that catalyzes the exchange of sodium and potassium ions, coupled to the hydrolysis of ATP. This Na^+, K^+-ATPase, just as the H^+-ATPase, mediates an electrogenic process, since more sodium ion is moved outward than potassium ion inward. However, unlike the H^+-ATPase, which transports hydrogen ion outward or inward according to whether ATP hydrolysis or synthesis is required, the Na^+, K^+-ATPase works only in one capacity under physiologic conditions: It operates solely to extrude sodium ion and accumulate potassium ion at the expense of ATP. The several important roles played by this ATPase have been indicated earlier. Because of the impermeant macromolecules within the cell, there is a constant tendency for the inward "leak" of small extracellular molecules (predominantly sodium and chloride ions). This brings water inward as well, so that the cell is faced with the problem of a constant tendency for swelling and lysis. By catalyzing net extrusion of sodium ion, effectively making extracellular sodium impermeant, the Na^+, K^+-ATPase assumes responsibility for the long-range osmotic stability of the cell. This ion pump also maintains the high internal potassium ion stores and low internal sodium ion levels required for many cellular activities. Finally, as outlined at a later point, this ATPase establishes the difference in electrochemical potential for sodium ion that allows active transport of solutes to be driven by the reentry of the sodium. One way to assess the importance of these duties is to note that some 20% to 30% of the ATP generated by metabolism is consumed during sodium and potassium ion transport by this enzyme.[42]

Currently (with a few exceptions, noted later) the best estimates of the stoichiometry of the Na^+, K^+-ATPase reaction allow one to represent the transport process as follows:

$$3Na^+_i + 2K^+_o + ATP_i \leftrightarrow 3Na^+_o + 2K^+_i + ADP + Pi_i \quad (18)$$

Thus for every ATP hydrolyzed, there are three sodium ions pumped outward and two potassium ions inward. These estimates of stoichiometry come from studies in many different cell types, as diverse as snail neurons and the mammalian erythrocyte.[18,54,65] The same asymmetric exchange of sodium and potassium ions is also found when the ATPase is extracted from its native membrane and placed in artificial membrane vesicles,[65,71] and the results of such "reconstitution" experiments have important implications for the way in which one should think about cell physiology. It has been known since the 1940s, when isotope methods were introduced into experimental biology, that animal cells (as all cells) continually exchange their internal ions with those in the external medium. Nevertheless, living cells tend to accumulate potassium and exclude sodium. The general assumption was that this preference for potassium over sodium reflected the active transport of these ions, and the discovery of an ATPase requiring both sodium and potassium for activity appeared to give strong experimental support to this view.[16] But an alternative to active transport was conceivable.[86] Suppose that the internal proteins of cells had a preference for the *binding* of potassium rather than of sodium. This might explain the ability of cells to maintain high internal potassium and low internal sodium levels despite the eventual exchange of internal and external materials. One can now distinguish between these alternatives because the Na^+, K^+-ATPase has been isolated, purified, and placed in artificial membrane vesicles that contain only salt solutions. Even in these elementary systems, one can demonstrate the ATP-dependent active transport of sodium and potassium ions. An example of this kind of experiment is given in Fig. 1-13, which shows results from work with purified Na^+, K^+-ATPase of the dogfish shark.[65] (No such Na^+, K^+-ATPase is found in the bacterial world. There the preference for internal potassium ion over sodium ion is apparently satisfied in an indirect manner, depending ultimately on the active extrusion of hydrogen ion.[23])

From all sources examined, the Na^+, K^+-ATPase of animal cells contains two different polypeptides,

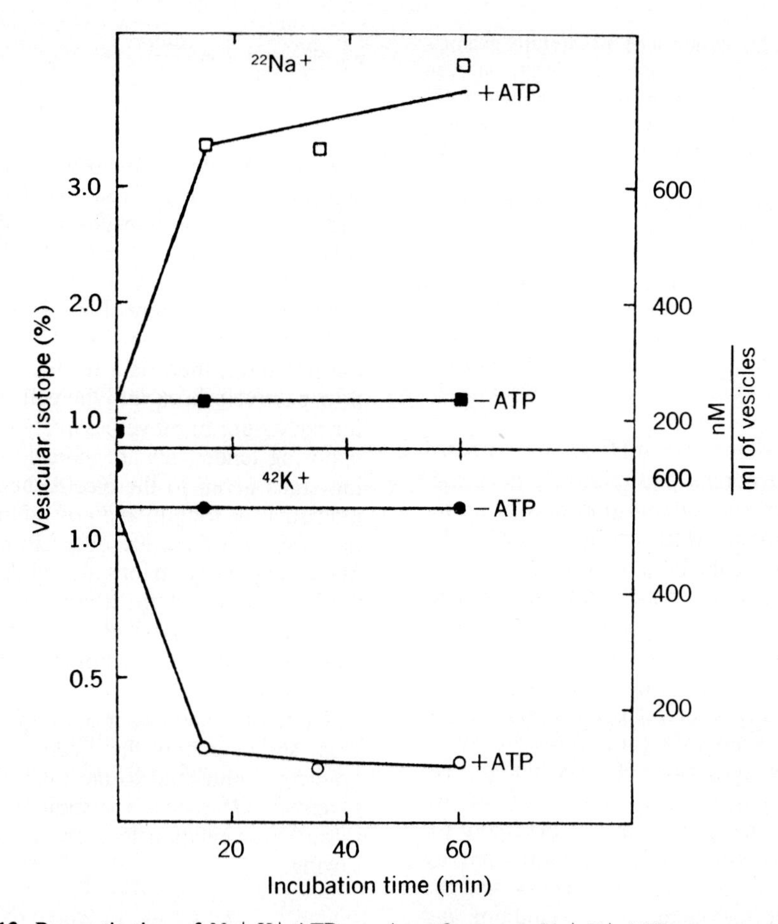

Fig. 1-13. Reconstitution of Na^+,K^+-ATPase. As a first step, Na^+,K^+-ATPase was purified so that 95% of protein was represented by α and β subunits of the enzyme. In the presence of cholate (a detergent) the enzyme was mixed with phospholipids, and dialyzed against detergent-free buffer containing 20 mM NaCl and 50 mM KCl. As detergent left the dialysis bag, lipid and aqueous phases separated, allowing formation of small membrane vesicles that had incorporated Na^+,K^+-ATPase into the lipid bilayer. Vesicles were then equilibrated in solutions containing Na^+ (20 mM NaCl) and K^+ (50 mM KCl) along with either $^{22}Na^+$ (top) or $^{42}K^+$ (bottom). After ATP was added (zero time), measurements of radioactivity in vesicles showed that Na^+ had entered and that K^+ had left. In a number of experiments of this kind, ratio of Na^+ moved inward to K^+ moved outward was 3:2. Because ATP did not penetrate vesicles during the experiment, ion transport was catalyzed only by ATPase molecules having an orientation opposite to that in the intact cell. (Data from Hilden and Hokin.[65])

α and β (of molecular weights about 90,000 and 40,000, respectively), and it seems likely that the active form of the enzyme has the structure $\alpha_2\beta_2$.[7,71] The series of experiments summarized by Kyte[82] has revealed that this enzyme spans the thickness of the membrane in such a way that different portions of the α subunit are exposed on the outer and inner surfaces of the membrane at the same time. Thus the fundamental asymmetry of the reaction catalyzed by this enzyme is reflected by (and presumably a reflection of) a polarity to the physical structure of the enzyme within the cell membrane. It is important that many of the characteristics of the Na^+, K^+-ATPase found in intact cells are also present in the reconstituted enzyme.[71] Thus (1) sodium ion is moved away from and potassium ion toward the side of the membrane at which ATP is split; (2) more sodium is moved than potassium, so the pump is electrogenic; (3) specific inhibitors (e.g., cardiac glycosides such as ouabain) block transport and ATP hydrolysis only when present on one side of the membrane—the outside of the cell or the inside of the vesicles in an experiment such as that of Fig. 1-13; and (4) various homologous exchange reactions occur (not active transport), for example, the exchange of internal and external sodium or internal and external potassium. These striking similarities between the in vivo and in vitro behavior of the enzyme suggest that one can learn things of physiologic significance by carefully examining the purified system. At the very least, one can now be certain that a specific membrane-bound enzyme is responsible for the active transport of sodium and potassium ion observed in the intact cell. Clearly the properties of elements residing in the cell membrane are sufficient to account for this phenomenon.

Successful reconstitution experiments have also been performed with the H^+-ATPase from both mitochondria[74] and bacteria.[120] Here, too, it is possible to show that active transport of an ion reflects the properties of the membrane-embedded enzyme rather than constituents present on either side of the original membrane.

In the discussion of H^+-ATPase, it was noted that because membranes are relatively impermeable to hydrogen ion, ion transport may be a conservative reaction in which energy dissipated by ATP hydrolysis could be stored as the electrochemical potential difference of hydrogen ion. This is also true of ion transport by Na^+, K^+-ATPase, and one may also ask questions of this latter process with regard to such energy transductions. Thus (1) what determines how far sodium and potassium ions can be forced out of equilibrium by operation of this pump? By analogy with what has gone before, one knows that the electrochemical potential gradients for sodium and potassium ions are influenced by both the nature of the chemical reaction and the permeability of the membrane to sodium and potassium. In addition, (2) for each of these electrochemical potential gradients, what determines the relative contribution of the electrical and chemical components? Again, by analogy with discussions of H^+-ATPase, one knows the answer to be (a) the permeability of the membrane to other ions and (b) the buffering power for sodium or potassium ions on either side of the membrane. (It may be assumed that buffering power for sodium and potassium ions is negligible. However, if one considers an ATPase that translocates calcium, as in muscle sarcoplasmic reticulum, this term could not be ignored. Biologic materials bind calcium rather well.) The reasoning behind the answers to these questions was outlined earlier with regard to H^+-ATPase; the comments made here should make it clear that these same principles hold for an ATPase that handles two ions rather than just one.

Appendix F gives the derivation of the following relationship for Na^+, K^+-ATPase, where n represents the number of sodium ions moved per ATP hydrolyzed, and m gives the stoichiometry of potassium ions:

$$n\left(\frac{-\Delta\tilde{\mu}_{Na^+}}{F}\right) + m\left(\frac{\Delta\tilde{\mu}_{K^+}}{F}\right) = Z\log K'_{eq} + Z\log\frac{[ATP]}{[ADP][Pi]} \quad (19)$$

As earlier, the terms to the right of the equal sign indicate energy made available for sodium and potassium ion transport due to hydrolysis of ATP. The terms to the left show that this available energy may be conserved as both an electrochemical potential gradient for sodium ion and one for potassium ion. Note that following ion pumping, there is a tendency for sodium to flow passively back into the cell, but a tendency for potassium to move passively outward. These opposing tendencies are indicated by the differing signs given to the electrochemical potential gradients for sodium and potassium ion. In addition, because there are different numbers of sodium and potassium ions moved during transport, the extent to which the energy dissipated by ATP hydrolysis is conserved in these ion gradients must be weighted according to the stoichiometries n and m.

As before, one can make a quantitative statement showing how the electrical and chemical gradients contribute to the total electrochemical potential differences for sodium and potassium ions. These relationships are shown by the following:

$$\frac{\Delta\tilde{\mu}_{Na^+}}{F} = V_m + Z\log\frac{[Na^+]_i}{[Na^+]_o} \quad (20)$$

$$\frac{\Delta\tilde{\mu}_{K^+}}{F} = V_m + Z\log\frac{[K^+]_i}{[K^+]_o} \quad (21)$$

With reference to equation 19, it is easy to see the effect of stoichiometry (n or m) on the size of the electrochemical potential gradients for sodium and potassium ions. If the number of sodium ions moved were to increase, the equilibrium value of $\Delta\tilde{\mu}_{Na^+}/F$ would fall—similarly for potassium ions and $\Delta\tilde{\mu}_{K^+}/F$. The effect of permeability to sodium or potassium is also apparent. For example, if permeability to potassium is very large (and potassium is present), $\Delta\tilde{\mu}_{K^+}/F$ would be diminished. Yet the transport reaction would still conserve energy dissipated by ATP hydrolysis if permeability to sodium remains low. The cycling of ion transport becomes fultile only when permeability to both transported ions is high, and this does not usually occur under physiologic conditions. In most animal cells, although permeability to potassium is high, permeability to sodium is low. Thus energy is conserved primarily by moving sodium ions out of equilibrium. This is true especially in muscle, where the ratio of potassium to sodium permeabilities may be as high as 100:1.[67] As pointed out earlier, the partition of this stored energy between electrical and chemical gradients is set by the permeability of the membrane to the variety of ions that surround it. Note that when considering $\Delta\tilde{\mu}_{Na^+}/F$ (equation 20), the relative contributions of the electrical and chemical components are determined by permeabilities (and concentrations) of ions that are not sodium—similarly for $\Delta\tilde{\mu}_{K^+}/F$ (equation 21). For example, of the total sodium ion in the cell, only a small fraction need move outward during ion transport to generate the electrical gradient required to bring the reaction to equilibrium. (A calculation of this kind was made earlier [p. 16] with regard to the passive movements of potassium ions.) Consequently, one might view the "natural" tendency of ATP-driven ion transport as one in which energy is conserved as a membrane potential alone. Only when other factors intervene (such as increased permeability to other ions) does continued ion pumping become necessary in an approach to equilibrium. In such cases, there may be a substantial change in the internal concentration of the transported ion, thus distributing energy dissipated by ATP hydrolysis to both the electrical and chemical gradients.

A good many of these points are illustrated by the experiment shown in Table 1-4. This table gives the value of the sodium electrochemical potential gradient in muscle after two kinds of experimental manipulations.[49] To obtain the data at the top of the table, external sodium level was varied without changing external potassium level. In this case, $\Delta\tilde{\mu}_{Na^+}/F$ did not change. Manipulation of external sodium ion level eventually altered internal sodium ion level so that the steady-state ratio of internal to external sodium was about the same in the different samples. As shown at the bottom of the table, when external potassium ion level was varied (but not sodium level), both the membrane potential and the ratio of internal to external sodium were changed. But, again, their new steady-state values were such that the electrochemical potential gradient for sodium ion remained constant; it did not change in the face of wide variation of its electrical and chemical substituents. Finally, one should note that changing external sodium ion concentration, with potassium ion concentration kept constant, did *not* alter the relative contributions made by the electrical and chemical components of $\Delta\tilde{\mu}_{Na^+}/F$. These proportions were varied only when other ions (potassium) were manipulated. These results are readily understood in terms of the preceding discussion, since for muscle the electrochemical potential difference for potassium ion is close to zero. This kind of experiment shows clearly that the cell seeks to regulate the total inward driving force on sodium ions rather than either the membrane potential or the chemical gradient for sodium ions alone. In the following section, one reason for this is made apparent, for it is the total inward driving force on sodium that determines the capacity of the cell to carry out other kinds of transport reactions. (Admittedly, the muscle cell used to illustrate this point about energetics is not a healthy one. Its ability to respond to a nerve impulse clearly depends on maintenance of normal resting potential. Nevertheless, its behavior under these extreme experimental conditions serves as a useful model for other cells less amenable to direct study.)

Table 1-4. Conservation of the electrochemical gradient for sodium in muscle*

Na^+_o (mM)	K^+_o (mM)	Na^+_i (mM)	V_m (mV)	$\Delta\tilde{\mu}_{Na^+}/F$ (mV)
140	4.7	30	−74	−106
70	4.7	17	−72	−110
25	4.7	7.5	−72	−107
10	4.7	3.2	−72	−106
115	4.7	23	−70	−114
115	9.7	22	−64	−109
115	20	15	−45	−100
115	35	9.6	−36	−102

*Data from Fozzard and Kipnis.[49] Rat diaphragm muscle was bathed in solutions containing varying concentrations of sodium and potassium ions. After the steady state had been attained, measurements were made of internal sodium ions and of the membrane potential (V_m), so that the difference in electrochemical potential for sodium ion could be calculated.

It has been assumed so far that there is an invariant stoichiometry for the coupled, asymmetric exchange of sodium and potassium ions. This may be an oversimplification. It seems clear that this ATPase almost always functions as an electrogenic pump, where more sodium is moved than potassium, but there are indications that the ratio of sodium to potassium ion transport can vary. When this question was investigated using the squid axon,[99] it appeared that if internal sodium ion concentration was maintained lower than normal (by perfusion), the relative stoichiometry of sodium to potassium ions was less than its usual value of 3:2. But when internal sodium ion concentration was set higher than normal, relative stoichiometry was elevated to as high as 4:1. This may reflect an important control mechanism that enables this cell to rapidly control levels of internal sodium ion.

Active transport coupled to ion gradients

In preceding sections, some general principles governing the energetics of ion-transporting reactions were outlined. The purpose of these discussions was to show that if the cell membrane is relatively impermeable to one or more ions, the active transport of those ions need not be entirely wasteful. Instead, such reactions are really energy transductions, allowing potential energy available in chemical form (e.g., ATP) to be transformed into potential energy in the form of gradients of electrochemical potential for specific ions. The discussions in the following sections show how this primary energy transduction is utilized by the cell in performing useful work as ions, which have been pumped out, reenter by specific pathways.

As discussed earlier in this chapter, cell membranes are effective barriers to the movements of most molecules. To increase the rate at which substances enter (and exit) the cell, the membrane contains a variety of proteins that catalyze the translocation of material. As for other enzyme-catalyzed reactions, these transport reactions may be characterized by stereospecificity, substrate specificity, "saturation" kinetics, and a sensitivity to various inhibitors of protein function.[17] The term "carrier" is often used in referring to this class of proteins, although it is unlikely that these catalysts physically "carry" substrates by diffusion across the thickness of the membrane. It is more reasonable to assume that subtle changes in protein structure alternately expose substrate-binding sites to one or the other side of the membrane.

Transport of glucose by the human erythrocyte is an example of facilitated diffusion. In this cell, and in most mammalian tissue, the movements of glucose are "facilitated," but the eventual internal concentration of sugar does not exceed its external level. Thus, as in passive diffusion, the driving force for glucose transport is solely the concentration gradient for this (uncharged) molecule. However, in some cell types (e.g., epithelial cells of the kidney or gut) the steady-state internal concentration of sugar may be many times higher than the external concentration, indicating the active transport of glucose. In studies of such transport in animal cells a clue to the mechanism was the finding that accumulation of sugar depended on the presence of sodium ion outside the cell. This led to the suggestion that the transport of sugar was coupled to the inward movement of sodium ion.[36] As one can now appreciate, in such a coupled reaction the driving force for glucose accumulation would be the electrochemical potential difference for sodium. In bacterial systems an analagous clue was that accumulation of sugar (in this case, lactose or its analogs) was blocked when the inward driving force on hydrogen ion was reduced to zero. This prompted the suggestion that there was cotransport of both hydrogen ion and lactose,[93] so that potential energy available as the electrochemical potential gradient for hydrogen ion could be used to power accumulation of substrate. These early suggestions have proved correct. For each of these very different transport events, it can be shown that (1) inward movement of sugar is accompanied by inward movement of positive charge[14,127,129] and (2) inward movement of sugar is paralleled by entry of the "coupling" ion, hydrogen or sodium.[57,114,128] The coupling between ion and substrate fluxes is verified in a striking way by experiments showing that when an electrochemical gradient for the ion is artificially imposed, sugar accumulation occurs. For animal cells an experiment illustrating this last point is given in Fig. 1-14. In that experiment,[24] membrane vesicles prepared from kidney epithelium were suddenly exposed to a chemical gradient for sodium ion ($[Na^+]_o > [Na^+]_i$). In response to this, there was a rapid accumulation of glucose within the vesicles. In the absence of sodium ion (or when sodium ion was at equal concentration on both sides of the membrane), glucose merely equilibrated across the vesicle membrane without an initial accumulation. In this experiment, there was only a transient accumulation of glucose in vesicles exposed to the chemical gradient for sodium ion. This is expected, for the empty vesicles cannot pump out the sodium ion that enters during cotransport with glucose or by nonspecific paths.

One can make a quantitative statement about the relationship between the capacity of cells to accumulate substrate (the ratio of internal to external substrate, $[S]_i/[S]_o$) and the size of the electrochemical potential gradient for the coupling ion ($\Delta\tilde{\mu}_{ION}/F$). If substrate (S) enters (and leaves) the cell only by way of the transport reaction:

$$0 = n\left(\frac{\Delta\tilde{\mu}_{ION}}{F}\right) + z_s V_m + \frac{RT}{F} \ln \frac{[S]_i}{[S]_o} \qquad (22)$$

where n is the stoichiometry of the reaction (n IONS per molecule S), z_S is the valence of substrate (S), and the other symbols are as defined earlier (see Appendix G for the derivation of this relationship). This expression simply states that at equilibrium, all driving forces on the move-

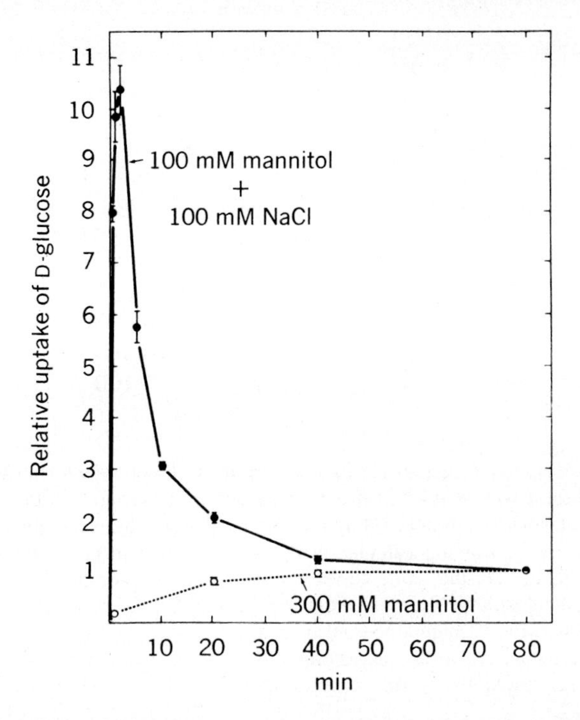

Fig. 1-14. Active transport of glucose driven by chemical gradient for sodium ion. Membrane vesicles were prepared from brush border of rabbit kidney epithelium. Such vesicles contain a sodium ion/glucose cotransport system. Vesicles were initially prepared in 0.3M mannitol and then diluted into media containing 0.1M mannitol and 0.1M NaCl (closed circles) or 0.3M mannitol (open circles), along with radioactively labeled glucose. Measurement of radioactivity within vesicles allowed calculation of the ratio of internal to external glucose as a function of time. (Data from Aronson and Sacktor.[24])

ments of substrate will sum to zero. To visualize this, imagine that a substrate has accumulated within the cell to its final steady-state level, high above medium concentration. This accumulation reflects a balance between inward and outward driving forces on the movements of substrate. In this case, there are three such driving forces to be considered: (1) Since substrate and ion both enter the cell during the same chemical (transport) reaction, an inward driving force is contributed by the electrochemical potential difference for the coupling ion (the term "$\Delta\mu_{ION}/F$"); this driving force is then "weighted" according to the stoichiometry of the reaction, the number of ions (n) falling down that electrochemical gradient. (2) There is an outward driving force, since substrate is at higher chemical concentration within the cell (the term "$RT/F \ln[S]_i/[S]_o$"). (3) Finally, if substrate is a charged molecule (valence z_S), the normal electrical potential, negative within the cell, will also influence the distribution of S; S will tend to move outward if it is an anion, but inward if a cation. This tendency is indicated by the term "$z_S V_m$." Each of these three different driving forces on substrate movements is given by one of the three terms in equation 22. (Note that still another verbal statement of equation 22 is that the electrochemical potential gradient for the coupling ion [term 1] will be equal and opposite to the electrochemical potential gradient for substrate [terms 2 and 3].) One may use this relationship in thinking of many kinds of transport reactions. For example, in facilitated diffusion, where no coupling to ion gradients occurs, n equals zero. For *co*transport, when the coupling ion and substrate move in the same direction, n has a positive value. For *counter*transport, since ion and substrate move in opposite directions, n may be given a negative sign (see Appendix G). In practice, it is not easy to determine if this relationship holds in any given situation. But this has been successfully done for the case of galactoside transport in bacteria.[75] One such example is given in Fig. 1-15 for the cotransport of hydrogen ion and thiomethylgalactoside (TMG).

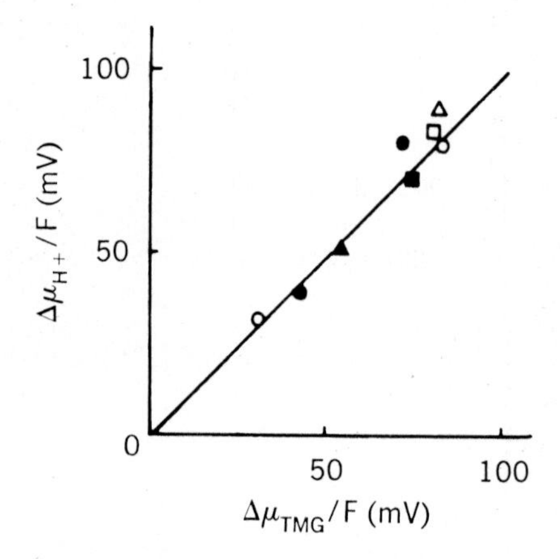

Fig. 1-15. Active transport of galactoside in a bacterium. Cells of *Streptococcus lactis* were given metabolizable substrates so that ATP from fermentations could support hydrogen ion extrusion by H^+-ATPase of the plasma membrane. Using indirect techniques, both membrane potential and pH gradient were measured, allowing calculation of the difference in electrochemical potential for hydrogen ion (ordinate). At the same time, accumulation of (nonmetabolizable) radioactively labeled thiomethylgalactoside (TMG) was monitored. From this measurement one can calculate the extent to which galactoside is maintained out of equilibrium across the membrane (abscissa). Different symbols give data from different experiments. Note that the slope of the derived relationship is 1, indicating stoichiometry (n) of cotransport as 1 H^+/TMG. (Data from Kashket and Wilson.[75])

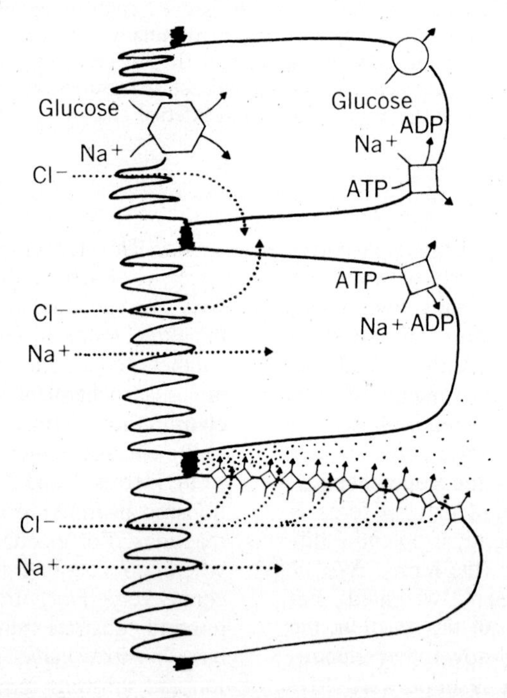

Fig. 1-16. Short segment of an epithelium engaged in active transport. Three adjacent cells of transporting epithelium are sketched. Net flow of material proceeds from outside (mucosa) to inside (serosa). Solid lines indicate movements of material catalyzed by specific membrane-bound proteins, as in sodium ion/glucose cotransport (hexagon), facilitated diffusion of glucose (open circles), or net extrusion of sodium ion from cell by electrogenic Na^+,K^+-ATPase (diamond shapes). Dotted lines indicate passive "leakage" of sodium and chloride ions. Density of dots between two lower cells reflects "standing osmotic gradient." Note that passive movement of ions between cells (through tight junctions) is not represented in drawing. In some epithelia, significant ionic traffic takes place at these junctions. See text for further details.

Here the balance sheet given by equation 22 is simplified because substrate is an uncharged molecule ($z_S = 0$).

The kinds of experiments shown in Figs. 1-14 and 1-15 make it clear that there is an important and instructive parallel between active transport of sugars in bacteria and animal cells. In each case the primary active extrusion of an ion (hydrogen or sodium) establishes an electrochemical potential gradient that is subsequently drawn on to power solute accumulation. This design is such that there need be no direct physical interaction between the various ion-translocating elements within the membrane. Instead, they communicate with each other only by means of a circulation of ions. In this way the physical and chemical properties of the membrane (its barrier function) allow electrochemical potential ion gradients to serve as true intermediates between the expenditure of chemical energy (e.g., ATP hydrolysis) and the eventual performance of osmotic work.

Earlier it was noted that the stoichiometry of the Na^+,K^+-ATPase seemed to be variable under certain conditions. The same has now been found for cotransport reactions in which hydrogen ion serves as the coupling ion.[107] Clearly, although one knows the general design of the circuitry of ion flow across cell membranes, there is still much to learn about the details of the individual reactions.

Definitive proof of ion/substrate cotransport is available only for a few well-studied systems in bacterial and animal cells. But because much active transport of solute (sugars, amino acids, and certain ions) in animal cells shows some sort of a sodium dependence, it is likely that sodium ion/substrate cotransport forms the basis for active transport at the level of the plasma membrane in these cells.[13]

Ion gradients and transepithelial movements

Much of the following discussion is based on the diagram given in Fig. 1-16, which illustrates three adjacent cells of an epithelium engaged in the active transport of material. This diagram emphasizes some of the important characteristic features of such cells in epithelia as diverse as the frog skin and the mammalian intestine. These cells display an anatomic polarity, for only their apical surface (the surface facing the lumen of the gut or the freshwater environment of the frog) has elaborate infoldings of the plasma membrane. These "microvilli" considerably increase the surface area of the membrane presented with material to be absorbed. In turn, this anatomic polarity is paralleled by a biochemical asymmetry, for only the basal and lateral membranes contain the electrogenic Na^+,K^+-ATPase. In addition, adjacent cells are cemented together by "tight junctions" that prevent the bulk movement of material into the tissue. In effect, these junctions impose a restriction (to varying degrees in different epithelia) on the free recirculation of sodium ion from the site at which it is extruded, the basal and lateral surfaces, to the site at which it may enter, the apical surface. Such restriction is of importance in directing the net flow of material across the thickness of the epithelium.

In Fig. 1-16 the upper cell is drawn to indicate the elements involved in the absorption of sugar. Here, to the asymmetries just noted, one may also add that the sodium ion/substrate cotransport systems for amino acids and sugars are located at the apical surface of the cell, whereas transport systems catalyzing the facilitated diffusion of sugar (and presumably others for amino acids) are found on the basal and lateral membranes. When glucose is present in the gut, it is actively transported into the cell during sodium ion/glucose cotransport. It can then leave the cell and enter blood plasma (down its concentration gradient) by facilitated diffusion across the basal and lateral membranes. The sodium ion that enters along with glucose also leaves the cell, as it must, being actively extruded by the Na^+,K^+-ATPase. The participation of sodium ion in this process is catalytic only to the extent that sodium ion travels back into the lumen of the gut and becomes available once again for cotransport. But the junctions between the cells are sufficiently "tight" so that there may be net movement of both sodium ion and glucose *across* the epithelium. In vivo the lumen of the gut contains chloride, and because the membranes of animal cells have a higher permeability to the chloride anion than to the sodium ion, the anion can then move passively across the epithelium in parallel with sodium. The net result, then, is the transepithelial movement of glucose and salt (NaCl). Since these are in solution rather than bound to plasma constituents, water must flow inward as well. Thus the anatomic and biochemical asymmetry of the tissue and the restriction placed on sodium ion recirculation allow chemical energy (as ATP) to drive the flow of nutrient, salt, and water into the body.

The middle and lower cells of Fig. 1-16 outline salt and water absorption in the absence of sodium ion/nutrient cotransport. Clearly, even in the absence of nutrient cotransport, sodium in the lumen of the gut will "leak" into the cell. Again, since it is pumped out at a site different from its entry, a restriction on its free recirculation will result in net transepithelial movements, so that both salt and water enter into blood plasma. (In mammalian intestine there is also a transport system that catalyzes the coupled entry of sodium and chloride ions at the luminal surface.[113] In effect, this can accelerate the inward "leak" of salt.) But there is an additional anatomic feature to these epithelia that makes this process more efficient. The analysis given so far requires only that the junctions between cells restrict free recirculation of sodium ion but does not specify the location of these junctions. It is significant that cells are adjoined at their apical ends rather than at their basal surfaces. This anatomic fact means that the spaces between cells (the lateral clefts) are open toward blood plasma. Because these clefts are of small dimension, active pumping of sodium into them can increase the actual concentration of sodium (and chloride) in this local compartment. This builds an osmotic gradient across the lateral surface membranes. Consequently, water flows from within the cell (and to a certain extent, through the junctions) into the clefts. Continual outpouring of water then dilutes the contents of the clefts so that toward the plasma opening, fluid tends to be more dilute (approaching isotonicity) than at the luminal (closed) end. In turn, this establishes a "standing osmotic gradient" extending from the luminal end (hypertonic) to the plasma end (near isotonic) of the lateral clefts. It is because of this local osmotic gradient that the tissue can function to give net water (and salt) absorption even in the absence of a gross osmotic gradient across the entire epithelium. A dramatic result of this is that during such water "transport" the distance between cells becomes progressively greater as distance along cell length increases away from the lumen. This is nicely shown by the electron micrographs given in Fig. 1-17, which illustrate the anatomy of an epithelium (gallbladder) in an inhibited state and during net water transport.[124] A complete analysis of this phenomenon is given elsewhere,[40] where it is made clear that this reasoning accounts for the observation that certain epithelia "actively" transport water against an osmotic gradient.

From the treatment given previously, one might predict that the primary inward movement of sodium ion would cause an electrical potential to develop across such transporting epithelia

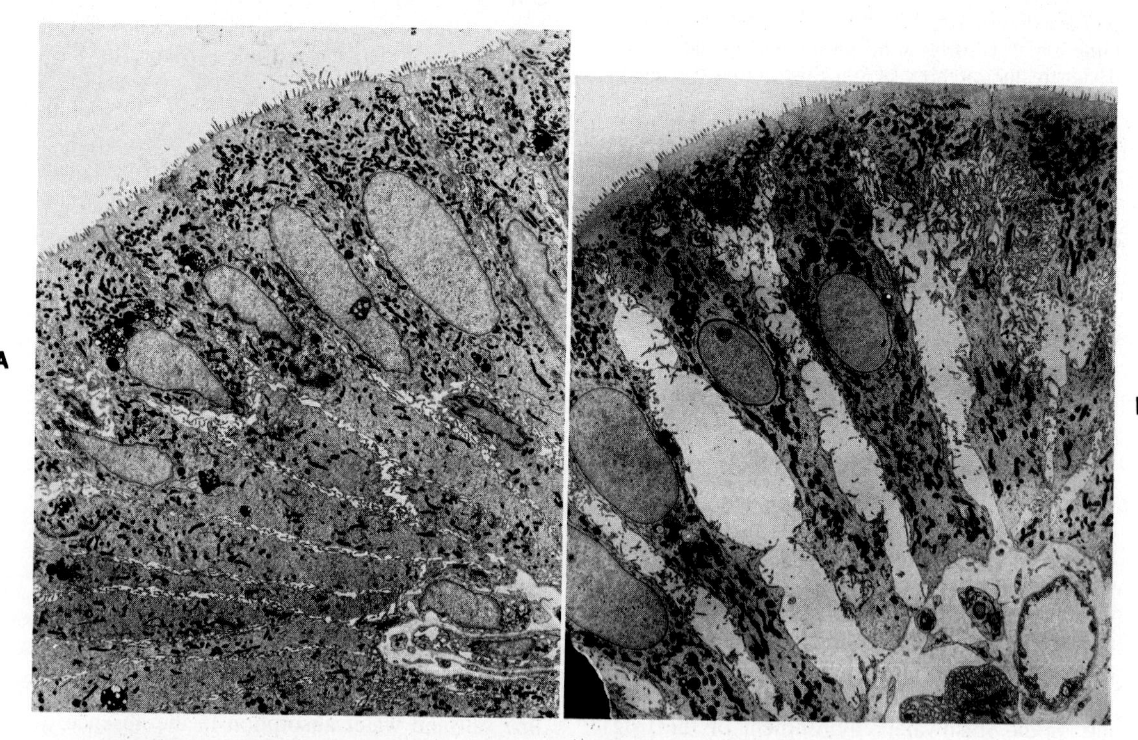

Fig. 1-17. Anatomy of gallbladder at rest and during water transport. **A,** Electron micrograph of gallbladder in which fluid transport has been blocked by treatment with ouabain, an inhibitor of Na^+, K^+-ATPase. **B,** Similar electron micrograph of gallbladder in which fluid transport was occurring at optimal rates before fixation. Most striking difference between two conditions is considerable dilatation of lateral clefts, between cells, in transporting epithelium. Despite this separation of adjacent cells, cells remain joined together at their apical surfaces. (From Tormey and Diamond.[123])

(e.g., negative inside the lumen of the gut). Although this is true, the absolute value of this potential difference is quite different for different epithelia. In some tissues (e.g., small intestine or gallbladder) the transepithelial potential difference is rather small, a few millivolts or so. In other cases, however, electrical potentials as high as 100 mV may be measured (frog skin, urinary bladder). This variability correlates well with the properties of the "tight junctions" that adjoin cells at their apical surfaces. When tight junctions are, in fact, rather leaky, a substantial electrical potential does not develop because cations (moving outward) and anions (moving inward) short-circuit the epithelium by their tendency to move between the cells.[51] Conversely, if the epithelium as a whole is not leaky to ions, as in frog skin, an electrical potential can develop. Diamond[5] has speculated that the variable tightness of tight junctions may reflect an important physiologic variability in the function of such epithelia. For example, in cases where junctions are really tight, epithelial function appears to be one of maintaining steep concentration gradients for sodium (and chloride) ions, as in absorption of salt from the freshwater environment by the frog skin or resorption of salt from the urinary bladder. On the other hand, leaky tight junctions are found when a major epithelial function centers on salt and water transport. In these latter instances (as in small intestine), one might reason that if the junctions allow significant ion traffic, they may also allow significant water flow as well. Thus as water flows inward through the junctions (as a result of the standing osmotic gradient in the lateral clefts), additional sodium and chloride ion will enter because of solvent drag. This may amplify net inward movement of salt and water.

In summary, in this discussion of active transport, initial emphasis was placed on the movements of material out of and into rather small compartments, bacteria or organelles (e.g., the mitochondrion). Later discussion focused on larger eukaryotic cells, where the emphasis was on events at the plasma membrane. Finally, in this section the point of view has been enlarged still more by considerations of active transport across epithelia. In fact, this pedagogic progression is the reverse of the historical development, for the very first active transport process to be adequately explored dealt with the movement of sodium ion across frog skin. Frog skin serves its host by actively transporting sodium from the freshwater environment into the body. Even when the excised skin is mounted between two chambers containing physiologic salt solutions (primarily NaCl), active transport of sodium ion is demonstrable. As discussed, a negative electrical potential (some 100 mV) is also observed at the outer surface of the epithelium. In principle, considering only the predominant ions in the system, these observations could arise from either (or both) the active inward transport of sodium ion or the active outward transport of chloride ion. Ussing and Zerahan[125] devised an ingenious technique to analyze this system, a technique that allowed a decision as to which ion was actively transported and which ionic flux formed the basis for the generation of the electrical potential across the epithelium. The frog skin was mounted between chambers containing identical concentrations of sodium and chloride ions. In addition, the electrical potential normally present was reduced to zero by applying an external current (electrons flowing from the outer to the inner surface of the skin). Measurement of this "short-circuit current" provided an exact determination of the net movement of charge across the epithelium. (This procedure is formally equivalent to the voltage clamp technique discussed in Chapter 2.) These manipulations eliminated the transepithelial differences in electrochemical potential for both sodium and chloride ion, and the decision as to which ion was actively transported could be made by using radioisotopes to follow the inward and outward fluxes of these ions. Such measurements showed that inward and outward flow of chloride proceeded at equal rates, so that there was no net transepithelial movement of the anion. In contrast, the tracer studies of sodium ion fluxes showed that there was a substantial imbalance between inflow and outflow, the net flux being inward. Thus because the electrochemical potential difference for sodium ion was zero under these short-circuit conditions, the net inward flux of sodium ion proved that sodium was actively transported inward. Under these same conditions the zero net flux of chloride ion showed that the anion was not actively transported. Moreover, the balance between measured sodium ion influx and efflux was exactly equal to the net inward current (represented as positive charge) measured by the short-circuit current. Thus the flow of sodium ion alone could account for the electrical properties of the epithelium.[125] From a variety of studies, Ussing and Zerahan postulated that the ultimate source of energy for this active transport was provided by a device, located at the inner surface of the epithelium, that pumped sodium outward in exchange for potassium. One now realizes that this "pump" is the electrogenic Na^+,K^+-ATPase.[19]

or

$$\frac{RT}{c}\frac{dc}{dx} = -zF\frac{dV}{dx} \qquad \text{(B-9)}$$

Such an equation states that the forces due to the gradient of the chemical potential and to the gradient of the electrical potential are equal and opposite. That is, there is no net force on the ions of this species.

For a given concentration ratio $\left(\frac{c_i}{c_o}\right)$, there may be calculated the equilibrium potential (E_{ION}) at which there will be no net force on these ions (the subscripts i and o represent inside and outside the membrane, respectively). Rearranging equation B-9:

$$-\frac{RT}{zF}\frac{dc}{c} = dV \qquad \text{(B-10)}$$

Integrating equation B-10 between the two sides of the membrane:

$$-\frac{RT}{zF}\int_o^i \frac{dc}{c} = \int_o^i dV$$

There results the well-known Nernst equation, in which the membrane potential ($V_m = V_i - V_o$) is, for this particular ion, the equilibrium potential (E_{ION}):

$$E_{ION} = V_i - V_o = -\frac{RT}{zF}\ln\frac{c_i}{c_o} \qquad \text{(B-11)}$$

Several simplifications of this equation are possible. First, the predominant ions in tissue are monovalent (K^+, Na^+, and Cl^-), and so z is ± 1. Second, it is more convenient to use common logarithms (log) than natural logarithms (ln), so equation B-11 becomes:

$$E_{ION} = \mp 2.303\frac{RT}{F}\log\frac{c_i}{c_o} \qquad \text{(B-12)}$$

Finally, at 20° C, $2.303\frac{RT}{F}$ is about 58 mV, whereas at 37° C, $2.303\frac{RT}{F}$ is about 61 mV.

APPENDIX C

Constant field equation[69]

The approach here follows that of Hodgkin and Katz[69] and begins with equation B-8. The primary assumption is that the electrical field $\left(\frac{dV}{dx}\right)$ across the membrane is constant and is given by:

$$\frac{dV}{dx} = \frac{V_m}{a} \qquad \text{(C-1)}$$

where:

a = Thickness of the membrane dielectric

Then:

$$I = -zu\left(RT\frac{dc}{dx} + zcF\frac{V_m}{a}\right) \qquad \text{(C-2)}$$

Rearranging:

$$I + \frac{uz^2cFV_m}{a} = -zuRT\frac{dc}{dx}$$

$$dx = -zuRT\left(\frac{dc}{I + \frac{ucFV_m}{a}}\right) \qquad \text{(C-3)}$$

where:

$z^2 = 1$ for univalent ions

Equation C-3 is to be integrated across the membrane to obtain the dependence of I (and V_m) on the concentrations:

$$\int_0^a dx = -zuRT\int_{c_o}^{c_i}\frac{dc}{\left(I + \frac{ucFV_m}{a}\right)} \qquad \text{(C-4)}$$

This integration is aided by the following substitution:

$$y = I + \frac{ucFV_m}{a}$$

$$\frac{dy}{\left(\frac{uFV_m}{a}\right)} = dc$$

Then:

$$\int_0^a dx = -\frac{uzRT}{\left(\frac{uFV_m}{a}\right)}\int_{I + \frac{uFV_m}{a}c_o}^{I + \frac{uFV_m}{a}c_i}\frac{dy}{y} \qquad \text{(C-5)}$$

The result of integrating and simplifying is:

$$I = -\frac{zRT}{FV_m}\ln\frac{I + \left(\frac{uFV_m}{a}\right)c_i}{I + \left(\frac{uFV_m}{a}\right)c_o} \qquad \text{(C-6)}$$

On rearranging and taking exponentials of both sides:

$$\exp\left(-\frac{FV_m}{zRT}\right) = \frac{I + \left(\frac{uFV_m}{a}\right)c_i}{I + \left(\frac{uFV_m}{a}\right)c_o} \qquad \text{(C-7)}$$

Equation C-7 may be solved for I:

$$I = \frac{uFV_m}{a}\left(\frac{c_o \exp\left(-\frac{FV_m}{zRT}\right) - c_i}{1 - \exp\left(-\frac{FV_m}{zRT}\right)}\right) \quad \text{(C-8)}$$

Since the diffusion constant (D) $= \frac{uRT}{F}$ for an ion and the permeability constant (P) $= \frac{D}{a} = \frac{uRT}{aF}$, equation C-8 becomes:

$$I = \left(\frac{F^2V_m}{RT}\right)P\left(\frac{c_o \exp\left(-\frac{FV_m}{zRT}\right) - c_i}{1 - \exp\left(-\frac{FV_m}{zRT}\right)}\right) \quad \text{(C-9)}$$

This is the basic constant field equation for passive ionic current under chemical and voltage gradients. For cations, the equation becomes:

$$I_{cat} = \left(\frac{F^2V_m}{RT}\right)P_{cat}\left(\frac{[cat]_o \exp\left(-\frac{FV_m}{RT}\right) - [cat]_i}{1 - \exp\left(-\frac{FV_m}{RT}\right)}\right) \quad \text{(C-10)}$$

For an anion:

$$I_{an} = \left(\frac{F^2V_m}{RT}\right)P_{an}\left(\frac{[an]_i \exp\left(-\frac{FV_m}{RT}\right) - [an]_o}{1 - \exp\left(-\frac{FV_m}{RT}\right)}\right) \quad \text{(C-11)}$$

Of initial interest is the membrane potential that will result in the steady state when there are net fluxes of cations and anions moving passively across a membrane (the cell is running down), but their charges neutralize each other. That is, there is no *net* current. The individual fluxes will be assumed to be made up of sodium, potassium, and chloride currents, since these are the predominant ions determining membrane potentials. Then:

$$I_{net} = I_{Na} + I_K + I_{Cl} = 0 \quad \text{(C-12)}$$

Call:

$$m = P_K[K^+]_i + P_{Na}[Na^+]_i + P_{Cl}[Cl^-]_o \quad \text{(C-13)}$$

$$n = P_K[K^+]_o + P_{Na}[Na^+]_o + P_{Cl}[Cl^-]_i \quad \text{(C-14)}$$

The net current is thus:

$$I_{net} = \left(\frac{F^2V_m}{RT}\right)\left(\frac{1}{1 - \exp\left(-\frac{FV_m}{RT}\right)}\right) \quad \text{(C-15)}$$

$$\Big[P_K[K^+]_o \exp\left(-\frac{FV_m}{RT}\right) - P_K[K^+]_i + P_{Na}[Na^+]_o \exp\left(-\frac{FV_m}{RT}\right) - P_{Na}[Na^+]_i + P_{Cl}[Cl]_i \exp\left(-\frac{FV_m}{RT}\right) - P_{Cl}[Cl^-]_o\Big] =$$

$$\left(\frac{F^2V_m}{RT}\right)\left(\frac{1}{1 - \exp\left(-\frac{FV_m}{RT}\right)}\right)\left[n \exp\left(-\frac{FV_m}{RT}\right) - m\right] = 0$$

Thus:

$$\exp\left(-\frac{FV_m}{RT}\right) = \frac{m}{n}$$

On taking logarithms of both sides:

$$-\frac{FV_m}{RT} = \ln\frac{m}{n} \quad \text{(C-16)}$$

Finally, on rearranging and writing out m and n, there results what has come to be known as the Goldman-Hodgkin-Katz (GHK) equation:

$$V_m = -\frac{RT}{F}\ln\frac{P_K[K^+]_i + P_{Na}[Na^+]_i + P_{Cl}[Cl^-]_o}{P_K[K^+]_o + P_{Na}[Na^+]_o + P_{Cl}[Cl^-]_i} \quad \text{(C-17)}$$

Note that if the concentrations of an ion adjust themselves so it is always at equilibrium, it will contribute nothing to the membrane potential (at steady state). Consider, for example:

$$E_{Cl} = V_m$$

Then:

$$-\frac{RT}{F}\ln\frac{[Cl^-]_o}{[Cl^-]_i} = -\frac{RT}{F}\ln\frac{P_K[K^+]_i + P_{Na}[Na^+]_i + P_{Cl}[Cl^-]_o}{P_K[K^+]_o + P_{Na}[Na^+]_o + P_{Cl}[Cl^-]_i} \quad \text{(C-18)}$$

and:

$$\frac{[Cl^-]_o}{[Cl^-]_i} = \frac{P_K[K^+]_i + P_{Na}[Na^+]_i + P_{Cl}[Cl^-]_o}{P_K[K^+]_o + P_{Na}[Na^+]_o + P_{Cl}[Cl^-]_i} \quad \text{(C-19)}$$

Call:

$$P_K[K^+]_i + P_{Na}[Na^+]_i = \{\ \}$$

$$P_K[K^+]_o + P_{Na}[Na^+]_o = |\ |$$

Then:

$$[Cl^-]_o|\ | + P_{Cl}[Cl^-]_i[Cl^-]_o = [Cl^-]_i\{\ \} + P_{Cl}[Cl^-]_i[Cl^-]_o \quad \text{(C-20)}$$

After cancelling, rearranging, and rewriting equation C-20 in the form of equation C-18, there results:

$$-\frac{RT}{F}\ln\frac{[Cl^-]_o}{[Cl^-]_i} = -\frac{RT}{F}\ln\frac{P_K[K^+]_i + P_{Na}[Na^+]_i}{P_K[K^+]_o + P_{Na}[Na^+]_o} \quad \text{(C-21)}$$

In other words, if chloride (or another ion) is

always at equilibrium, it makes no steady-state contribution to the membrane potential predicted by the GHK equation.

APPENDIX D

Contribution of an electrogenic pump to membrane potential[18]

In cell homeostasis, there are neither net currents nor net fluxes across a membrane. This is because passive net fluxes of ions out of electrochemical equilibrium are counterbalanced by active transport of these ions. The contributions to the membrane potential of pumps that are electrogenic can be calculated as follows: First assume that chloride fluxes are inappreciable, which is the case if chloride permeabilities are relatively very low or if chloride concentrations adjust themselves so chloride is always in equilibrium (equation C-21). Then if sodium and potassium ions are not at equilibrium, there will be passive currents (I) and these will be balanced by active transport currents (J); outward currents are positive:

$$I_{Na} + J_{Na} = 0 \qquad \text{(D-1)}$$

$$I_K + J_K = 0 \qquad \text{(D-2)}$$

Evidence presented elsewhere in this chapter indicates a ratio (r) of sodium ions pumped out for every potassium ion pumped in:

$$rJ_K + J_{Na} = 0 \qquad \text{(D-3)}$$

Eliminating the pumped currents (J_K, J_{Na}) from equations D-1, D-2, and D-3 yields:

$$rI_K + I_{Na} = 0 \qquad \text{(D-4)}$$

From equation C-10:

$$rP_K\left([K^+]_o \exp\left(-\frac{FV_m}{RT}\right) - [K^+]_i\right)f(V) + P_{Na}\left([Na^+]_o \exp\left(-\frac{FV_m}{RT}\right) - [Na^+]_i\right)f(V) = 0 \qquad \text{(D-5)}$$

where

$$f(V) = \left(\frac{F^2V_m}{RT}\right)\left(\frac{1}{1 - \exp\left(-\frac{FV_m}{RT}\right)}\right)$$

On rearranging and taking logarithms, there results:

$$V_m = -\frac{RT}{F} \ln \frac{rP_K[K^+]_i + P_{Na}[Na^+]_i}{rP_K[K^+]_o + P_{Na}[Na^+]_o} \qquad \text{(D-6)}$$

If the coupling ratio (r) approaches 1, the results of equation D-6 approach those of equation C-17, the GHK equation. If r approaches infinity (only sodium pumped), the result of equation D-6 approaches E_K, the Nernst potential, since this is the condition for zero net flux of potassium.

APPENDIX E

Difference in electrochemical potential

At constant pressure the general expression for the difference in electrochemical potential for J ($\Delta\tilde{\mu}_j$) between internal and external phases is given by:

$$\Delta\tilde{\mu}_j = RT\ln\{j\}_i - RT\ln\{j\}_o \qquad \text{(E-1)}$$

where use of { } denotes electrochemical activity. One may also write equation E-1 as:

$$\Delta\tilde{\mu}_j = z_jFV_m + RT\ln[j]_i - RT\ln[j]_o \qquad \text{(E-2)}$$

where use of [ ] indicates chemical activity ($\simeq$ concentration).

This equivalent expression shows that if j is a charged particle, $\Delta\tilde{\mu}_j$ is influenced by both the chemical gradient for j across the membrane (terms containing [j]) and the electrical field across the membrane (term containing V_m).

By collecting the terms in logarithms, one also sees that:

$$\Delta\tilde{\mu}_j = z_jFV_m + RT\ln\frac{[j]_i}{[j]_o} \qquad \text{(E-3)}$$

(Note that when $\Delta\tilde{\mu}_j = 0$, equation E-3 reduces to equation B-11, the Nernst equation.)

For hydrogen ion, since $z_{H^+} = +1$ and $pH = -\log_{10}[H^+]$, equation E-3 reduces to:

$$\Delta\tilde{\mu}_{H^+} = FV_m - 2.303\,RT\,\Delta pH \qquad \text{(E-4)}$$

where ΔpH refers to the pH difference between the internal and external spaces ($pH_i - pH_o$).

The units of equation E-1 through E-4 are kilocalories per mole. Since in this and subsequent chapters one deals largely with concentration gradients and membrane potentials, it will be useful to express the difference in electrochemical potential in electrical units, millivolts (mV). To do this, divide both sides of equations E-3 and E-4 by the Faraday (F) and substitute the symbol Z for 2.303 RT/F. Thus for sodium or potassium ion:

$$\frac{\Delta\tilde{\mu}_{ION}}{F} = V_m + Z\log\frac{[ION]_i}{[ION]_o} \qquad \text{(E-5)}$$

and for the proton:

$$\frac{\Delta\tilde{\mu}_{H^+}}{F} = V_m + Z\log\frac{[H^+]_i}{[H^+]_o} = V_m - Z\Delta pH \quad \text{(E-6)}$$

Under physiologic conditions, sodium or hydrogen ion is often significantly out of equilibrium across biologic membranes, since ion concentration is lower inside than outside and the normal membrane potential is negative inside. In these cases the difference in electrochemical potential has a negative value, indicating a tendency for these ions to flow passively into the cell.

APPENDIX F

Two cation translocating ATPases, and the relationship between ion gradients and the "poise" of the ATPase reaction

First consider the proton-translocating ATPase, following the derivation given by Mitchell.[94] For this ATPase the chemical reaction is as follows:

$$nH_i^+ + ATP_i + H_2O \leftrightarrows nH_o^+ + ADP_i + Pi_i \quad \text{(F-1)}$$

where n is the stoichiometry of the reaction (nH^+ translocated per molecule of ATP hydrolyzed or synthesized); ATP, ADP, and Pi have their usual meanings (adenosine triphosphate, adenosine diphosphate, and inorganic phosphate, respectively); and the subscripts i and o refer to the inside or outside, respectively, of the cell or organelle. The equilibrium constant (Keq) for the reaction is written as follows:

$$Keq = \frac{\{ADP\}_i\{Pi\}_i\{H^+\}_o^n}{\{ATP\}_i\{H_2O\}\ \{H^+\}_i^n} \quad \text{(F-2)}$$

Taking the activity of water as 1, rearranging terms, and using logarithms, then:

$$n\ln\frac{\{H^+\}_o}{\{H^+\}_i} = \ln K'eq + \ln\frac{\{ATP\}_i}{\{ADP\}_i\{Pi\}_i} \quad \text{(F-3)}$$

where K′ eq is the equilibrium constant for water activity of 1.

Since ATP, ADP, and Pi do not cross the membrane during the chemical reaction, one may replace the braces (indicating electrochemical activities) with brackets, showing chemical activities (taken as concentrations). But this may not be done for H^+. Since H^+ crosses the membrane, its tendency to participate in the chemical reaction is properly represented by its *electrochemical* activity. Thus:

$$n\ln\frac{\{H^+\}_o}{\{H^+\}_i} = \ln K'eq + \ln\frac{[ATP]}{[ADP]\,[P_i]} \quad \text{(F-4)}$$

From equation E-1:

$$\Delta\tilde{\mu}_{H^+} = RT\ln\{H^+\}_i - RT\ln\{H^+\}_o = RT\ln\frac{\{H^+\}_i}{\{H^+\}_o} \quad \text{(F-5)}$$

Combining equations F-4 and F-5 gives:

$$n(-\Delta\tilde{\mu}_{H^+}) = RT\ln K'eq + RT\ln\frac{[ATP]}{[ADP]\,[Pi]} \quad \text{(F-6)}$$

In the absence of ion transport ($n = 0$ or $\Delta\tilde{\mu}_{H^+} = 0$) the equilibrium position of the reaction described by equation F-1 is heavily in favor of ATP hydrolysis, so that from given initial conditions, the ratio [ATP]/[ADP] [Pi] approaches about 10^{-5} (K′eq is about 10^{+5}). The formulation given by equation F-6 shows that when ATP hydrolysis (or synthesis) is coupled to hydrogen ion transport, the equilibrium position of the coupled reaction now also depends on the values of $\Delta\tilde{\mu}_{H^+}$ and n. To express this in electrical units, divide by the Faraday (F) and express logarithms to base 10, so that:

$$n\frac{(-\Delta\tilde{\mu}_{H^+})}{F} = Z\log K'eq + Z\log\frac{[ATP]}{[ADP]\,[Pi]} \quad \text{(F-7)}$$

Finally, from equation E-6:

$$n\frac{(-\Delta\tilde{\mu}_{H^+})}{F} = -n(V_m - Z\Delta pH) = Z\log K'eq + Z\log\frac{[ATP]}{[ADP]\,[Pi]} \quad \text{(F-8)}$$

From this relationship, it is possible to understand how such an ATPase reaction, normally in favor of ATP hydrolysis, can be "poised" toward a reversal by the coupling to ion transport. This expression also explains the experimental observations[73,90,108] that indicate that there may be net synthesis of ATP when a sufficiently large pH gradient or membrane potential is imposed artificially.

Consider next the analogous derivation dealing with the Na^+, K^+-ATPase. Because this ATPase catalyzes the coupled exchange of sodium and potassium ions, one may write the chemical reaction as follows:

$$nNa_i^+ + mK_o^+ + ATP_i + H_2O \rightleftarrows nNa_o^+ + mK_i^+ + ADP_i + Pi_i \quad \text{(F-9)}$$

where n and m are the stoichiometries of sodium

and potassium ions translocated per molecule of ATP hydrolyzed. As before, the equilibrium constant for the reaction is:

$$Keq = \frac{\{ADP\}_i\{Pi\}_i\{Na^+\}_o^n\{K^+\}_i^m}{\{ATP\}_i\{H_2O\}\{Na^+\}_i^n\{K^+\}_o^m} \quad (F\text{-}10)$$

As for equation F-3:

$$n\ln\frac{\{Na^+\}_o}{\{Na^+\}_i} + m\ln\frac{\{K^+\}_i}{\{K^+\}_o} = \ln K'eq + \ln\frac{\{ATP\}_i}{\{ADP\}_i\{Pi\}_i} \quad (F\text{-}11)$$

Since ATP, ADP, and Pi participate in the reaction from the same phase (intracellular), only their chemical activities (≃ concentrations) need be considered. But, as earlier, electrochemical activities must be specified for all reactants or products that cross the membrane. From equation E-1:

$$\Delta\tilde{\mu}_{Na^+} = RT\ln\{Na^+\}_i - RT\ln\{Na^+\}_o = RT\ln\frac{\{Na^+\}_i}{\{Na^+\}_o} \quad (F\text{-}12)$$

and

$$\Delta\tilde{\mu}_{K^+} = RT\ln\{K^+\}_i - RT\ln\{K^+\}_o = RT\ln\frac{\{K^+\}_i}{\{K^+\}_o} \quad (F\text{-}13)$$

Thus combining equations F-11, F-12, and F-13 yields:

$$n(-\Delta\tilde{\mu}_{Na^+}) + m(\Delta\tilde{\mu}_{K^+}) = RT\ln K'eq + RT\ln\frac{[ATP]}{[ADP][Pi]} \quad (F\text{-}14)$$

Equation F-14 shows that at equilibrium, the differences in electrochemical potentials for both sodium and potassium ions are influenced by the ratio [ATP]/[ADP] [Pi], and vice versa. This relationship may also be expressed in electrical units as follows:

$$n\frac{(-\Delta\tilde{\mu}_{Na^+})}{F} + m\frac{(\Delta\tilde{\mu}_{K^+})}{F} = Z\log K'eq + Z\log\frac{[ATP]}{[ADP][Pi]} \quad (F\text{-}15)$$

Finally, in certain mammalian tissue, such as muscle, potassium ion is not much out of electrochemical equilibrium due to high permeability to potassium ion itself. In these cases, equation F-15 simplifies to:

$$n\frac{(-\Delta\tilde{\mu}_{Na^+})}{F} = Z\log K'eq + Z\log\frac{[ATP]}{[ADP][Pi]} \quad (F\text{-}16)$$

so that, using equation E-5:

$$n\frac{(-\Delta\tilde{\mu}_{Na^+})}{F} = -n\left(V_m + Z\log\frac{[Na^+]_i}{[Na^+]_o}\right) = Z\log K'eq + Z\log\frac{[ATP]}{[ADP][Pi]} \quad (F\text{-}17)$$

This simplification is useful in understanding experiments such as that presented in Table 1-4 of the text. Given that the ratio [ATP]/[ADP] [Pi] is constant (or nearly so) during such experimental manipulations, it is not surprising that the electrochemical difference for sodium ion remains constant despite wide fluctuations in absolute values of the membrane potential and the ratio of internal to external sodium.

APPENDIX G

Coupling of solute transport to ion gradients

As discussed in the section on active transport, it is now known that many systems in bacteria and animal cells couple the translocation of substrate molecules (S) to the movements of either hydrogen or sodium ions. Thus the potential energy available to drive active transport of S is represented by $\Delta\tilde{\mu}_{H^+}$ or $\Delta\tilde{\mu}_{Na^+}$. The formulations given here illustrate how, at equilibrium, the distribution of substrate depends on the electrochemical potential difference for the coupling ion.

One begins, as before, by writing the chemical reaction as follows:

$$nION_i + S_i^z \leftrightarrows nION_o + S_o^z \quad (G\text{-}1)$$

where ION stands for the coupling ion (hydrogen or sodium, for example), and z (or z_S in later equations) gives the valence of substrate (S). The reaction as written represents the *co*transport (also known as "symport")[8,12] of S and the coupling ion(s), since both cross the membrane in the same direction using transport. Formulations for the cases of countertransport ("antiport")[8,12] or facilitated diffusion ("uniport")[8,12] are easily derived after discussion of cotransport.

The equilibrium constant for the reaction described by equation G-1 is as follows:

$$Keq = \frac{\{ION\}_o^n\{S^z\}_o}{\{ION\}_i^n\{S^z\}_i} \quad (G\text{-}2)$$

Rearranging terms and taking logarithms gives:

$$n\ln\frac{\{ION\}_o}{\{ION\}_i} = \ln Keq + \ln\frac{\{S^z\}_i}{\{S^z\}_o} \quad (G\text{-}3)$$

From equation E-1 (as in the derivation of equations F-6 and F-14):

$$n(-\Delta\tilde{\mu}_{ION}) = RT\ln Keq + RT\frac{\{S^z\}_i}{\{S^z\}_o} \quad \text{(G-4)}$$

Because substrate S crosses the membrane during the reaction, equation G-4 continues to specify that the electrochemical activity of S determines its tendency to participate in the transport reaction. Only if S is an uncharged molecule can one replace {S} by [S].

The next step is to evaluate the constant Keq. This may be done by considering the case where $\Delta\tilde{\mu}_{ION} = 0$, so that:

$$-RT\ln Keq = RT\ln\frac{\{S^z\}_i}{\{S^z\}_o} \quad \text{(G-5)}$$

If $\Delta\tilde{\mu}_{ION} = 0$, it follows that at equilibrium $\Delta\tilde{\mu}_S = 0$. Thus from equations E-1 and G-5:

$$0 = \Delta\tilde{\mu}_S = RT\ln\frac{\{S^z\}_i}{\{S^z\}_o} = -RT\ln Keq = 0 \quad \text{(G-6)}$$

This means that Keq = 1, a general constraint for such transport reactions, since there is no chemical transformation of either ION or S during cotransport. Combining the results of equation G-6 with equation G-4:

$$n(-\Delta\tilde{\mu}_{ION}) = RT\ln\frac{\{S^z\}_i}{\{S^z\}_o} \quad \text{(G-7)}$$

Since the term to the right in equation G-7 gives the difference in electrochemical potential for substrate S, use of equation E-3 gives:

$$n(-\Delta\tilde{\mu}_{ION}) = z_S FV_m + RT\ln\frac{[S^z]_i}{[S^z]_o} \quad \text{(G-8)}$$

Finally, by rearranging terms and employing electrical units:

$$0 = n\frac{(\Delta\tilde{\mu}_{ION})}{F} + z_S V_m + Z\log\frac{[S^z]_i}{[S^z]_o} \quad \text{(G-9)}$$

Equation G-9 merely states that at equilibrium, all driving forces on the movement of S sum up to zero. This equation refers to the reactions of cotransport. In countertransport, where ION and S move in opposite directions, an equivalent expression arises, except that n has a negative sign. This is easily shown by changing the subscripts i and o for the ION terms in equation G-1 and carrying out the derivation as shown. In cases of facilitated diffusion of charged or uncharged substrates, n = 0, for there is no coupling to ION movements.

REFERENCES

General reviews

1. Boyer, P. D., et al.: Oxidative phosphorylation and photophosphorylation, Annu. Rev. Biochem. **46:**955, 1977.
2. Cole, K. S.: Membranes, ions and impulses, Berkeley, Calif., 1968, University of California Press.
3. Davson, H., and Danielli, J. F.: The permeability of natural membranes, ed. 2, London, 1952, Cambridge University Press.
4. DeRobertis, E. D. P., Nowinski, W. W., and Saez, F. A.: Cell biology, Philadelphia, 1965, W. B. Saunders Co.
5. Diamond, J. M.: The epithelial junction: bridge, gate and fence, Physiologist **20:**10, 1977.
6. Dick, D. A. T.: Cell water, Woburn, Mass., 1966, Butterworth (Publishers), Inc.
7. Guidotti, G.: The structure of membrane transport systems, Trends Biochem. Sci. **1:**10, 1976.
8. Harold, F. M.: Membranes and energy transduction in bacteria. In Sanadi, R. D., editor: Current topics in bioenergetics, New York, 1977, Academic Press, Inc., vol. 6.
9. Hinkle, P. C., and McCarty, R. E.: How cells make ATP, Sci. Am. **238:**104, 1978.
10. Jacobs, M. H.: Early osmotic history of the plasma membrane, Circulation **26:**1013, 1962.
11. Kernan, R. P.: Cell K, Woburn, Mass., 1965, Butterworth (Publishers), Inc.
12. Mitchell, P.: Vectorial chemistry and the molecular mechanics of chemiosmotic coupling: power transmission by proticity, Biochem. Soc. Trans. **4:**399, 1976.
13. Schultz, S. G., and Curran, P. F.: Coupled transport of sodium and organic solutes, Physiol. Rev. **50:**637, 1970.
14. Schultz, S. G., and Curran, P. F.: Sodium and chloride transport across isolated rabbit ileum. In Bronner, F., and Kleinzeller, A., editors: Current topics in membranes and transport, New York, 1974, Academic Press, Inc., vol. 5.
15. Singer, S. J., and Nicolson, G. L.: The fluid mosaic model of the structure of cell membranes, Science **175:** 720, 1972.
16. Skou, J. C.: Enzymatic basis for active transport of Na^+ and K^+ across cell membranes, Physiol. Rev. **45:**596, 1965.
17. Stein, W. D.: The movement of molecules across cell membranes, New York, 1967, Academic Press, Inc.
18. Thomas, R. C.: Electrogenic sodium pump in nerve and muscle cells, Physiol. Rev. **52:**563, 1972.
19. Ussing, H. H., Erlij, D., and Lassen, U.: Transport pathways in biological membranes, Annu. Rev. Physiol. **36:**17, 1974.
20. Wilson, T. H., and Maloney, P. C.: Speculations on the evolution of ion transport mechanisms, Fed. Proc. **35:**2174, 1976.

Original papers

21. Adrian, R. H.: The effect of internal and external potassium concentration on the membrane potential of frog muscle, J. Physiol. **133:**631, 1956.
22. Adrian, R. H., and Slayman, C. L.: Membrane potential and conductance during transport of sodium, potassium, and rubidium in frog muscle, J. Physiol. **184:** 970, 1966.
23. Altendorf, K. H., and Harold, F. M.: Cation transport in bacteria: K^+, Na^+ and H^+. In Bronner, F., and Klein-

zeller, A., editors: Current topic in membranes and transport, New York, 1974, Academic Press, Inc., vol. 5.
24. Aronson, P. S., and Sacktor, B.: The Na^+ gradient-dependent transport of D-glucose in renal brush border membranes, J. Biol. Chem. **250:**6032, 1975.
25. Baker, P. F., Hodgkin, A. L., and Shaw, T. I.: The effects of changes in internal ionic concentrations on the electrical properties of perfused giant axons, J. Physiol. **164:**355, 1962.
26. Bar, R. S., Deamer, D. W., and Cornwell, D. G.: Surface area of human erythrocyte lipids: reinvestigation of experiments on plasma membrane, Science **153:**1010, 1966.
27. Birks, R. I., and Davey, D. F.: Osmotic responses demonstrating the extracellular character of the sarcoplasmic reticulum, J. Physiol. **202:**171, 1969.
28. Blinks, J. R.: Influence of osmotic strength on cross-section and volume of isolated single muscle fibers, J. Physiol. **177:**42, 1965.
29. Boyle, P. J., and Conway, E. J.: Potassium accumulation in muscle and associated changes, J. Physiol. **100:**1, 1941.
30. Bozler, E.: Osmotic properties of amphibian muscles, J. Gen. Physiol. **49:**37, 1965.
31. Bunch, W. H., and Kallsen, G.: Rate of intracellular diffusion as measured in barnacle muscles, Science **164:**1178, 1969.
32. Caldwell, P. C., Hodgkin, A. L., Keynes, R. D., and Shaw, T. I.: Effects of injecting energy-rich phosphate compounds on the active transport of ions in the giant axons of *Loligo,* J. Physiol. **152:**561, 1960.
33. Cone, R. A.: Rotational diffusion of rhodopsin in the visual receptor membrane, Nature N. Biol. **236:**39, 1972.
34. Conway, E. J.: Nature and significance of concentration relations of potassium and sodium ions in skeletal muscle, Physiol. Rev. **37:**84, 1957.
35. Cooke, R., and Kuntz, I. D.: The properties of water in biological systems, Annu. Rev. Biophys. Bioeng. **3:** 95, 1974.
36. Crane, R. K.: Hypothesis of mechanism of intestinal active transport of sugars, Fed. Proc. **21:**891, 1962.
37. Curtis, H. J., and Cole, K. S.: Membrane resting and action potentials from the squid giant axon, J. Cell. Comp. Physiol. **19:**135, 1942.
38. Dainty, J., and House, C. R.: An examination of the evidence for membrane pores in frog skin, J. Physiol. **185:**172, 1966.
39. Danielli, J. F., and Davson, H.: A contribution to the theory of permeability of thin films, J. Cell. Comp. Physiol. **5:**495, 1935.
40. Diamond, J., and Bossert, W. H.: Standing-gradient osmotic flow. A mechanism for coupling of water and solute transport in epithelia, J. Gen. Physiol. **50:**2061, 1967.
41. Donnan, F. G.: The theory of membrane equilibria, Chem. Rev. **1:**73, 1924.
42. Edelman, I. S.: Transition from the poikilotherm to the homeotherm: possible role of sodium transport and thyroid hormone, Fed. Proc. **35:**2180, 1976.
43. Edidin, M.: Rotational and translational diffusion in membranes, Annu. Rev. Biophys. Bioeng. **3:**179, 1974.
44. Edidin, M., Zagyansky, Y., and Lardner, T. J.: Measurement of membrane protein lateral diffusion in cells, Science **191:**466, 1976.
45. Edzes, H. T., and Berendsen, H. J. C.: The physical state of diffusible ions in cells, Annu. Rev. Biophys. Bioeng. **4:**265, 1975.
46. Finean, J. B., Bramley, T. A., and Coleman, R.: Lipid layer in cell membranes, Nature **229:**114, 1971.
47. Finklestein, A.: Water and nonelectrolyte permeability of lipid bilayer membranes, J. Gen. Physiol. **68:**127, 1976.
48. Finklestein, A., and Cass, A.: Permeability and electrical properties of thin lipid membranes, J. Gen. Physiol. **52:**1455, 1968.
49. Fozzard, H. A., and Kipnis, D. M.: Regulation of intracellular sodium concentrations in rat diaphragm muscle, Science **156:**1257, 1967.
50. Freeman, A. R.: Electrophysiological activity of tetrodotoxin on the resting membrane of the squid axon, Comp. Biochem. Physiol. **40A:**71, 1971.
51. Frompter, E., and Diamond, J.: Route of ion permeation in epithelia, Nature N. Biol. **235:**9, 1972.
52. Gaffey, C. T., and Mullins, L. J.: Ion fluxes during the action potential in Chara, J. Physiol. **144:**505, 1958.
53. Garrahan, P. J., and Glynn, I. M.: Driving the sodium pump backwards to form adenosine triphosphate, Nature **211:**1414, 1966.
54. Garrahan, P. J., and Glynn, I. M.: Stoichiometry of the sodium pump, J. Physiol. **192:**217, 1967.
55. Giaquinta, R.: Possible role of pH gradient and membrane ATPase in the loading of sucrose into the sieve tubes, Nature **267:**369, 1977.
56. Goldman, D. E.: Potential, impedance, and rectification in membranes, J. Gen. Physiol. **27:**37, 1943.
57. Goldner, A. M., Schultz, S. G., and Curran, P. F.: Sodium and sugar fluxes across the mucosal border of rabbit ileum, J. Gen. Physiol. **53:**362, 1969.
58. Goldsmith, T. H., and Wehner, R.: Restrictions on rotational and translocational diffusion of pigment in the membranes of a rhabdomeric photoreceptor, J. Gen. Physiol. **70:**453, 1977.
59. Gorter, E., and Grendel, F.: On bimolecular layers of lipoids on the chromatocytes of the blood, J. Exp. Med. **41:**439, 1925.
60. Gutknecht, J.: Membranes of Valonia ventricosa: apparent absence of water-filled pores, Science **158:**787, 1967.
61. Harold, F. M., and Papineau, D.: Cation transport and electrogenesis by streptococcus faecalis. I. The membrane potential, J. Membr. Biol. **8:**27, 1972.
62. Harold, F. M., Pavlasova, E., and Baarda, J. R.: A transmembrane pH gradient in streptococcus faecalis: origin and dissipation by proton conductors and N,N′-dicyclohexylcarbodiimide, Biochim. Biophys. Acta **196:**235, 1970.
63. Henn, F. A., and Thompson, T. E.: Synthetic lipid bilayer membranes, Annu. Rev. Biochem. **38:**241, 1969.
64. Henn, F. A., Decker, G. L., Greenawalt, J. W., and Thompson, T. E.: Properties of lipid bilayer membranes separating two aqueous phases: electron microscope studies, J. Mol. Biol. **24:**51, 1967.
65. Hilden, S., and Hokin, L. E.: Active potassium transport coupled to active sodium transport in vesicles reconstituted from purified sodium and potassium ion-activated adenosine triphosphatase from the rectal gland of Squalus acanthias, J. Biol. Chem. **250:**6296, 1975.
66. Hille, B.: Ionic selectivity, saturation, and block in sodium channels, J. Gen. Physiol. **66:**535, 1975.
67. Hodgkin, A. L., and Horowicz, P.: The effect of sudden changes in ionic concentrations on the membrane

potential of single muscle fibers, J. Physiol. **148:**127, 1959.
68. Hodgkin, A. L., and Huxley, A. F.: Resting and action potentials in single nerve fibers, J. Physiol. **104:**176, 1945.
69. Hodgkin, A. L., and Katz, B.: The effect of sodium ions on the electrical activity of the giant axon of the squid, J. Physiol. **108:**37, 1949.
70. Hodgkin, A. L., and Keynes, R. D.: The mobility and diffusion coefficient of potassium in giant axons from Sepia, J. Physiol. **119:**513, 1953.
71. Hokin, L.: The molecular machine for driving the coupled transports of Na^+ and K^+ is an (Na^+ + K^+)-activated ATPase, Trends Biochem. Sci. **1:**233, 1976.
72. Huxley, A. F., and Stampfli, R.: Direct determination of membrane resting potential and action potential in single myelinated nerve fibers, J. Physiol. **112:**476, 1951.
73. Jagendorf, A. T., and Uribe, E.: ATP formation caused by acid-base transition of spinach chloroplasts, Proc. Natl. Acad. Sci. USA **55:**170, 1966.
74. Kagawa, Y., Kandrach, A., and Racker, E.: Partial resolution of the enzymes catalyzing oxidative phosphorylation. XXVI. Phospholipid specificity of the vesicles capable of energy transformation, J. Biol. Chem. **248:**676, 1973.
75. Kashket, E. R., and Wilson, T. H.: Protonmotive force in fermenting Streptococcus lactis 7962 in relation to sugar accumulation, Biochem. Biophys. Res. Commun. **59:**879, 1974.
76. Koike, H., Brown, H. M., and Hagiwara, S.: Hyperpolarization of a barnacle photoreceptor membrane following illumination, J. Gen. Physiol. **57:**723, 1971.
77. Kolata, G. B.: Water structure and ion binding: a role in cell physiology? Science **192:**1220, 1976. (Science **193:**528, 1976.)
78. Korenbrot, J. I.: Ion transport in membranes: incorporation of biological ion-translocating proteins in model membrane systems, Annu. Rev. Physiol. **39:**19, 1977.
79. Korn, E. D.: Cell membranes: structure and synthesis, Annu. Rev. Biochem. **38:**263, 1969.
80. Kuffler, S. W., Nicholls, J. G., and Orkand, R. K.: Physiological properties of glial cells in the central nervous system of amphibia, J. Neurophysiol. **29:**768, 1966.
81. Kushmerick, M. J., and Podolsky, R. J.: Ionic mobility in muscle cells, Science **166:**1297, 1969.
82. Kyte, J.: Structural studies of sodium and potassium ion-activated adenosine triphosphatase. The relationship between molecular structure and the mechanism of active transport, J. Biol. Chem. **250:**7443, 1975.
83. Lee, J., Simpson, G., and Scholes, P.: An ATPase from dog gastric mucosa: changes of outer pH in suspensions of membrane vesicles accompanying ATP hydrolysis, Biochem. Biophys. Res. Commun. **60:** 825, 1974.
84. LeFevre, P. G.: The osmotically functional water content of the human erythrocyte, J. Gen. Physiol. **47:** 585, 1964.
85. Liebman, P. A., and Entine, G.: Lateral diffusion of visual pigment in photoreceptor disk membranes, Science **185:**457, 1974.
86. Ling, G.: The role of phosphorus in the maintenance of the resting potential and selective ionic accumulation in frog muscle cells. In McElroy, W. D., and Glass, H. B., editors: Phosphorus metabolism, Baltimore, 1952, The Johns Hopkins University Press, vol. 2.
87. Ling, G. N., and Gerard, R. W.: The normal membrane potential of frog sartorius fibers, J. Cell. Comp. Physiol. **34:**383, 1949.
88. Ling, G. N., and Walton, C. L.: What retains water in living cells? Science **191:**293, 1976.
89. Lucke, B., and McCutcheon, M.: The living cell as an osmotic system and its permeability to water, Physiol. Rev. **12:**68, 1932.
90. Maloney, P. C., and Wilson, T. H.: ATP synthesis driven by a protonmotive force in Streptococcus lactis, J. Membr. Biol. **25:**285, 1975.
91. Marmor, M. F., and Gorman, A. L. F.: Membrane potential as the sum of ionic and metabolic components, Science **167:**65, 1970.
92. Mitchell, P.: Coupling of phosphorylation to electron and hydrogen transfer by a chemiosmotic type of mechanism, Nature **191:**144, 1961.
93. Mitchell, P.: Molecule, group and electron translocation through natural membranes, Biochem. Soc. Symp. **22:**142, 1963.
94. Mitchell, P.: Chemiosmotic coupling and energy transduction, Bodmin, England, 1968, Glynn Research, Ltd.
95. Mitchell, P., and Moyle, J.: Estimation of membrane potential and pH difference across the cristae membrane of rat liver mitochondria, Eur. J. Biochem. **7:**471, 1969.
96. Montal, M.: Experimental membranes and mechanisms of bioenergy transductions, Annu. Rev. Biophys. Bioeng. **5:**119, 1976.
97. Moreton, R. B.: An application of the constant-field theory to the behavior of giant neurones of the snail, Helix aspersa, J. Exp. Biol. **48:**611, 1968.
98. Mueller, P., Rudin, D. O., Tien, H. T., and Wescott, W. C.: Reconstruction of excitable cell membrane structure in vitro, Circulation **26:**1167, 1962.
99. Mullins, L. J., and Brinley, F. J., Jr.: Potassium fluxes in dialyzed squid axons, J. Gen. Physiol. **53:**704, 1969.
100. Nakajima, S., and Onodera, K.: Post-tetanic hyperpolarization and electrogenic Na pump in stretch receptor neurone of crayfish, J. Physiol. **187:**105, 1966.
101. Nastuk, W. L., and Hodgkin, A. L.: The electrical activity of single muscle fibers, J. Cell. Comp. Physiol. **35:**39, 1950.
102. Orkand, R. K., Nichols, J. G., and Kuffler, S. W.: Effect of nerve impulses on the membrane potential of glial cells in the central nervous system of amphibians, J. Neurophysiol. **29:**788, 1966.
103. Oseroff, A. R., Robbins, P. W., and Burger, M. M.: The cell surface membrane: biochemical aspects and biophysical probes, Annu. Rev. Biochem. **42:**647, 1973.
104. Palmer, L. G., and Gulati, J.: Potassium accumulation in muscle: a test of the binding hypothesis, Science **194:**521, 1976. (Science **198:**1281, 1977.)
105. Poo, M., and Cone, R. A.: Lateral diffusion of rhodopsin in the photoreceptor membrane, Nature **247:**438, 1974.
106. Portis, A. R., Jr., and McCarty, R. E.: Quantitative relationships between phosphorylation, electron flow, and internal hydrogen ion concentrations in spinach chloroplasts, J. Biol. Chem. **251:**1610, 1976.
107. Ramos, S., and Kaback, H. R.: pH-dependent changes in proton: substrate stoichiometries during active transport in Escherichia coli membrane vesicles, Biochemistry **16:**4271, 1977.
108. Reid, R. A., Moyle, J., and Mitchell, P.: Synthesis of

adenosine triphosphate by a protonmotive force in rat liver mitochondria, Nature **212:**257, 1966.
109. Robertson, J. D.: The ultrastructure of cell membranes and their derivatives. In Crook, E. M., editor: The structure and function of subcellular components, Biochem. Soc. Symp. **16:**3, 1959.
110. Robertson, J. D.: Unit membranes: a review with recent new studies of experimental alterations and a new subunit structure in synaptic membranes. In Locke, M., editor: Cellular membranes in development, New York, 1964, Academic Press, Inc.
111. Robinson, J. R.: Metabolism of intracellular water, Physiol. Rev. **40:**112, 1960.
112. Rogus, E., and Zierler, K. L.: Sodium and water contents of sarcoplasmic reticulum in rat skeletal muscle: effects of anisotonic media, ouabain and external sodium, J. Physiol. **233:**227, 1973.
113. Schultz, S. G., and Curran, P. F.: Sodium and chloride transport across isolated rabbit ileum, Curr. Top. Membr. Transport **5:**226, 1974.
114. Schultz, S. G., and Zalusky, R.: Ion transport in isolated rabbit ileum. II. The interaction between active sodium and active sugar transport, J. Gen. Physiol. **47:**1043, 1964.
115. Singer, S. J.: The molecular organization of membranes, Annu. Rev. Biochem. **43:**805, 1974.
116. Skou, J. C.: The influence of some cations on an adenosine triphosphatase from peripheral nerves, Biochim. Biophys. Acta **23:**394, 1957.
117. Slayman, C. L., and Slayman, C. W.: Measurement of membrane potentials in Neurospora, Science **136:** 875, 1962.
118. Slayman, C. L., Long, W. S., and Lu, C. Y. H.: The relationship between ATP and an electrogenic pump in the plasma membrane of Neurospora crassa, J. Membr. Biol. **14:**305, 1974.
119. Smith, H. W.: The plasma membrane with notes on the history of botany, Circulation **26:**987, 1962.
120. Sone, N., et al.: Electrochemical potential of protons in vesicles reconstituted from purified proton-translocating adenosine triphosphatase, J. Membr. Biol. **30:**121, 1976.
121. Steck, T. L., Weinstein, R. S., Straus, J. H., and Wallach, D. F. H.: Inside-out red cell membrane vesicles: preparation and purification, Science **168:** 255, 1970.
122. Thayer, W. S., and Hinkle, P. C.: Stoichiometry of adenosine triphosphate–driven proton translocation in bovine heart mitochondrial particles, J. Biol. Chem. **248:**5395, 1973.
123. Tien, H. T., and Diana, A. L.: Bimolecular lipid membranes: a review and a summary of some recent studies, Chem. Phys. Lipids **2:**55, 1968.
124. Tormey, J. M., and Diamond, J.: The ultrastructural route of fluid transport in rabbit gall bladder, J. Gen. Physiol. **50:**2031, 1967.
125. Ussing, H. H., and Zerahan, K.: Active transport of sodium as the source of the electric current in the short-circuited isolated frog skin, Acta Physiol. Scand. **23:**110, 1951.
126. Walker, J. L., and Brown, H. M.: Intracellular ionic activity measurements in nerve and muscle, Physiol. Rev. **57:**729, 1977.
127. West, I. C., and Mitchell, P.: Proton-coupled β-galactoside translocation in non-metabolizing Escherichia coli, J. Bioenergetics **3:**445, 1972.
128. West, I. C., and Mitchell, P.: Stoichiometry of lactose-proton symport across the plasma membrane of Escherichia coli, Biochem. J. **132:**587, 1973.
129. White, J. F., and Armstrong, W. M.: Effect of transported solutes on membrane potentials in bullfrog small intestine, Am. J. Physiol. **221:**194, 1971.

2

F. J. BRINLEY, Jr.

Excitation and conduction in nerve fibers

FUNCTION OF THE NERVE IMPULSE IN THE PERIPHERAL NERVOUS SYSTEM

The peripheral nervous system functions as a communications network, permitting transmission of information from one part of the organism to another. The message unit in this information transfer is the propagated or conducted nerve impulse.

Several physical changes occur in the nerve during the passage of an impulse, but the alteration most directly related to propagation occurs in the electrical properties of the membrane. The passage of an impulse along a nerve produces an abrupt, brief decrease in the steady resting potential across the axolemma and an associated flow of transmembrane electrical current (in the form of ions). The potential change is commonly referred to either as the "action potential" because it indicates activity in the fiber or as the "spike" because electrical recording of the action potential produces a brief spikelike deflection on an oscilloscope trace. The flow of current is referred to as the "action current." Typical action potentials recorded from nerve fibers and muscle cells are shown in Fig. 2-1. Although the shape of the action potential varies considerably in different cells, all are characterized by a very rapid depolarization (upward deflection) and somewhat slower repolarization to the resting or steady potential. Another characteristic feature is the transient reversal of polarity of the electrical potential at the peak of the spike, the inside of the cell becoming positive with respect to the extracellular fluid, whereas at rest the inside is negative.

Several important phenomena observed in excitable cells can be explained on the basis of certain selective ion permeability changes in the cell membrane and ion fluxes across the membrane. A comprehensive discussion of these phenomena is best deferred until after consideration of the classic experiments that have provided the basis for our modern understanding of the properties of excitable membrane. However, in order to provide the working vocabulary needed for discussion, it is necessary to define and illustrate certain terms commonly used in discussing electrical activity observed in peripheral nerve.

Excitability

The ability to produce an action potential is termed "excitability," the process of generating an action potential is referred to as "excitation," and the class of cells exhibiting such behavior is called "excitable." Although excitation is frequently regarded as a property solely of surface membranes, the process actually depends not only on events within the membrane but also on transmembrane ionic gradients. Broadly viewed, excitability is a property of the entire cell, including its immediate external environment. In mammals the only normally excitable cells are nerve cells and muscle fibers. In lower organisms, tissues other than nerve or muscle can become excitable. Under certain experimental conditions, for example, frog epithelium and algae can produce regenerative action potentials.

Stimulus

The event or process that elicits an action potential in excitable cells is called a stimulus. One of the most common experimental stimuli is electrical. By means of an electrical shock, current is passed across a membrane to produce a transient depolarization of the resting potential, which, if it is of sufficient duration and magnitude, can initiate the train of events that produces an action potential. Although electrical currents are a convenient laboratory means of initiating excitation, they are, strictly speaking, not physiologic stimuli. Examples of physiologic stimuli are as follows: hormonal (acetylcholine acting on the postjunctional membrane of the

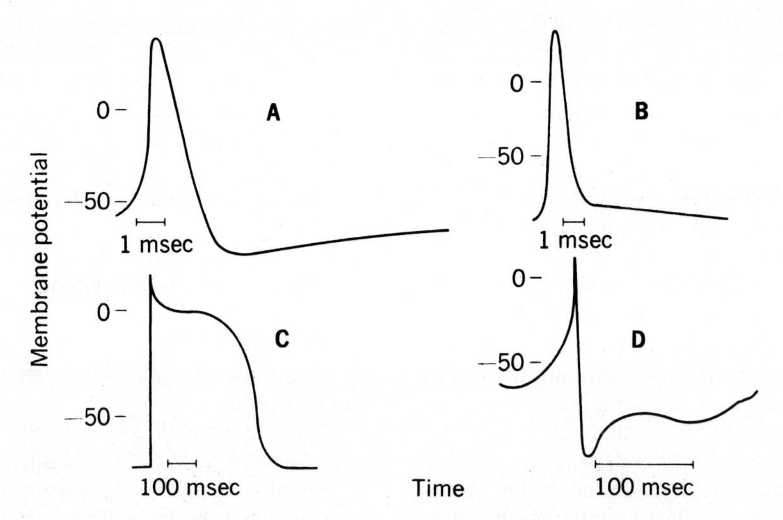

Fig. 2-1. Representative action potentials recorded from excitable cells with intracellular microelectrodes. **A,** Isolated squid axon. **B,** Frog sartorius muscle fiber. **C,** Dog ventricular muscle fiber. **D,** Guinea pig vas deferens. Zero level of potential is indicated on vertical axis. Beginning of trace indicates membrane potential (inside of cell negative) before excitation occurs. Note variation in time scale.

neuromuscular junction), thermal (skin temperature receptors), mechanical (displacement of outer lamellae of paccinian corpuscle, auditory hair cells), electromagnetic radiation (retinal rods), and chemical (protons and salts acting on gustatory receptors of the tongue).

Threshold

Not all physiologic or experimental stimuli will produce conducted action potentials. Only those stimuli with greater than minimum magnitude (intensity) and duration suffice. The minimum necessary intensity of a stimulus can be referred to as the threshold *stimulus*. (It is equally correct to define a threshold *duration* in a comparable way, but this is not generally done.) A stimulus of less than threshold intensity is referred to as *subthreshold,* or subliminal, one of greater than threshold intensity as *superthreshold,* or supraliminal.

The term "threshold" is commonly used to refer either to the absolute magnitude of the cell membrane potential at which an action potential is initiated or to the magnitude of the depolarization from resting potential required to initiate an action potential. The context usually makes the intended usage clear.

The threshold potential for excitation should not be considered a fixed parameter for all cells. The thresholds of different types of cells may vary considerably. Furthermore, the threshold of a single cell can change, either rapidly, as after a train of impulses, or more slowly, in response to metabolic or hormonal influences. The significance of the threshold phenomenon lies in the fact that it allows an excitable cell to function as a signal discriminator. Only those stimuli equal to or greater than the threshold intensity will produce information transfer in the peripheral nervous system.

Local, graded, or subthreshold (subliminal) response

Although a subthreshold stimulus fails to initiate an action potential, it still induces physiologically significant alterations in the membrane potential of a nerve fiber. The time course and magnitude of the stimulus determines, although in a mathematically complicated manner, the response of the membrane potential. These responses are not propagated or conducted along an excitable cell and are generally manifest only a very short distance from the stimulus point, hence the term "local response." The terms "local," "graded," or "subthreshold" as applied to these responses are essentially synonymous. The exact expression used serves to emphasize a particular aspect of the response. "Local" emphasizes that the response is not propagated, "graded" emphasizes that the configuration of the response is continuously variable or graded with the stimulus, and "subthreshold" indicates that the stimulus cannot initiate an action potential. Strictly speaking, local responses can be observed in any cell, whether excitable or not, since they depend only on the resistivity of the cell membrane and of the internal and external fluids and on the membrane capacitance (see discussion of cable theory, p. 52).

All-or-none response

The expression "all or none" describes the ability of a nerve fiber, once a suprathreshold response has been applied to its surface, to initiate an action potential whose configuration is determined solely by the properties of the cell independently of the precise configuration of the exciting stimulus, and to propagate such an action potential a very long distance along the nerve fiber *without variation of waveform* at essentially constant velocity. Although the condition of the fiber may change with time, the action potential configuration and conduction velocity are invariant for a single nerve fiber for at least a short period of time.

The expression "all or none" does not adequately describe the process of initiation of an action potential very close to the stimulus point where variations of configuration or velocity with stimulus may occur, but it does serve to emphasize a fundamental property of the peripheral nervous system: *the arrival of an action potential in the central nervous system (CNS) signals only that a suprathreshold stimulus (with respect to both magnitude and duration) has occurred in the periphery*. An action potential cannot signal the occurrence of a subthreshold stimulus, nor can a single impulse indicate either the magnitude or the duration of a suprathreshold stimulus. This latter type of information is coded in the interval between the action potentials. Information transfer within the nervous system, peripheral as well as central, is therefore *frequency* rather than amplitude modulated. Special cells called receptors, which exist in the periphery of the nervous system, function essentially as amplitude to frequency transducers. (Receptor physiology is discussed in Chapter 10.)

Summation

Under proper circumstances, two or more stimuli, each of which is subthreshold when occurring individually, may combine to cause excitation. This phenomenon is called summation. *Temporal* summation occurs when two subthreshold stimuli are applied in close succession. The local depolarizing response resulting from the second stimulus adds to the residual depolarizing response from the first stimulus. The net resulting depolarization of the membrane exceeds threshold, and excitation results. A second type of summation of great significance in the integrative function of the CNS is called *spatial* summation. Two subthreshold stimuli occur simultaneously but at different loci on a neuron. A local response is greatest at the point of stimulus application but does produce depolarization in adjacent regions. Therefore subthreshold responses arising from two (or more) loci can sum to produce threshold depolarization at another locus and cause excitation.

Refractory period

During the period in which an excitable membrane is producing an action potential in response to a suprathreshold stimulus the ability of the membrane to respond to a second stimulus of any sort is altered markedly. During the initial portion of the spike the membrane cannot respond to any stimulus, no matter how intense; this interval is called the *absolute* refractory period. Following the absolute refractory period an action potential can be produced first by very intense stimuli and then gradually by stimuli of progressively lesser magnitude. This interval is called the *relative* refractory period or sometimes the *subnormal* period. This refractory behavior of the excitable membrane is also frequently described as a change in threshold. Initially the threshold is infinite (absolute refractory period); then it declines (relative refractory period) to normal. In some cases following an action potential, there are intervals in which small, long-lasting changes in threshold occur; these are designated as subnormal (increased threshold) and supranormal (decreased threshold) periods.

Accommodation

Accommodation refers to the fact that the rate of membrane potential change during the application of a stimulus can affect the threshold voltage at which excitation finally occurs. The effect is illustrated in Fig. 2-2 for electrical stimuli with various rise times. The more slowly the stimulus depolarizes the membrane, the greater the depolarization required to initiate an action potential (i.e., the lower the absolute potential at which excitation occurs and the greater the total current required to stimulate). The membrane behaves as if it were becoming less excitable during the period of application of a stimulus and accommodates itself to the presence of the stimulus—hence the term "accommodation."

Electrotonic conduction

The potential changes associated with an action potential in excitable tissues propagate along the cell length with a velocity ranging from

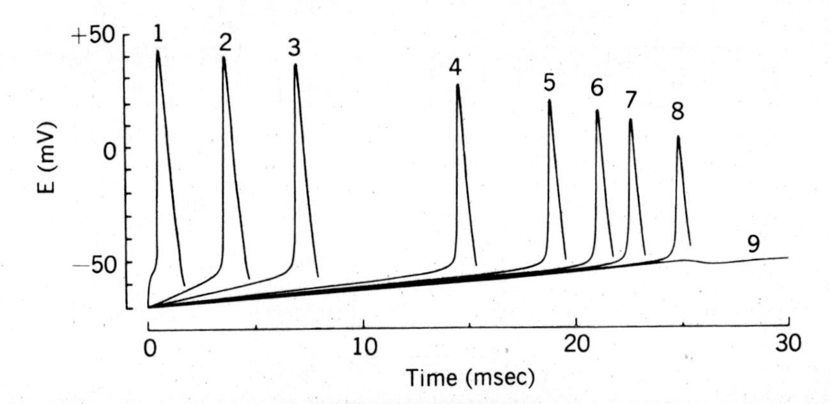

Fig. 2-2. Theoretical action potentials calculated to show effect of rate of rise of stimulus on final threshold for excitation (frog nerve). Traces show action potential produced by linearly varying electrical stimuli with various rates of rise. Threshold depolarization was 21 mV for step pulse (trace *1*) and 28 mV for slowest rising stimulus capable of producing an action potential (trace *8*). In trace *9*, rate of rise was too slow to initiate an action potential, and only a subthreshold response resulted. (From Frankenhaeuser and Vallbo.[8])

centimeters per second to tens of meters per second for different cells with no essential loss of amplitude. In sharp contrast, the alterations in potential induced by subthreshold depolarizations attenuate very rapidly with distance from the stimulus point, but the disturbance is manifested (i.e., propagated) extremely rapidly, the delay being due to the distributed membrane capacitance. The terms "electrotonic" conduction and "decremental" conduction are used to describe this disturbance produced in adjacent membrane by a localized subthreshold stimulus.

Stimulus artifact

The technical term "stimulus artifact" refers to any deflection of the recording trace that is produced by the stimulus itself and is not due to any response of the tissue being studied. Various arrangements of stimulating and recording devices are used to control the size of the stimulus artifact and to prevent distortion of the physiologic response. In this chapter we are primarily concerned with the electrical activity of nerve fibers recorded on an oscilliscope trace in response to an electrical stimulus. Under these conditions the stimulus artifact, or "shock artifact" as it is commonly called by electrophysiologists, appears as a large abrupt deflection of the baseline trace.

ELECTRICAL MANIFESTATIONS OF NERVOUS ACTIVITY

Study of the electrical signs of activity in excitable cells is important to an understanding of nervous phenomena as well as clinical medicine. This section will consider some aspects of recording and interpreting electrical phenomena in tissues.

Electrical recording of nervous activity

There are several techniques used for recording electrical activity in excitable cells. Since the action potential is manifest as a change in membrane potential, all these methods require, directly or indirectly, a measurement of the transmembrane potential. Several methods are illustrated schematically in Fig. 2-3. The easiest to understand conceptually is shown in Fig. 2-3, *A*. The transmembrane potential is measured directly as the potential difference between two electrodes, one inside the cell and the other outside. In most cases the membrane resistance is so large compared to the internal and external resistance that the potential recorded is essentially independent of the position of the recording electrode inside the cell. This method measures not only the action potential produced as the nerve impulse passes along the surface membrane but also the steady resting potential present in the absence of excitation. In fact, the importance of knowing the absolute magnitude of the resting potential was the impetus that led to the development of intracellular recording techniques. The first successful intracellular potential measurements were made as shown in Fig. 2-3, *A* by inserting a salt-filled glass capillary longitudinally along the axis of a nerve fiber. Given the minimum practical diameter of such long glass capillaries, on the order of 50 μ, this approach is feasible only with a few very large cells from marine invertebrates (e.g., the giant squid axon, the giant lobster axon, or the giant barnacle

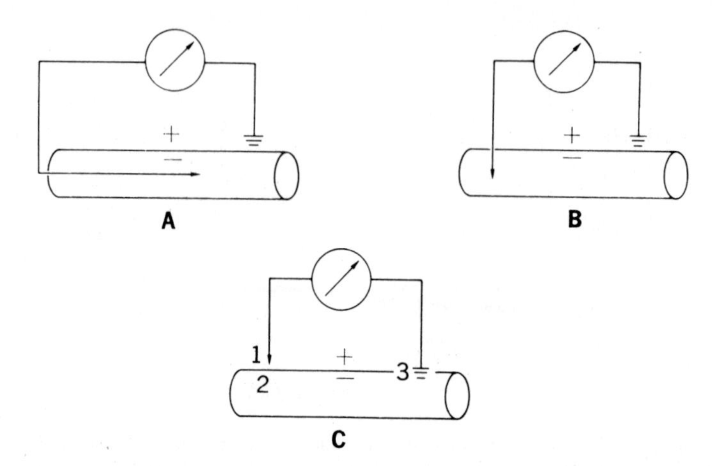

Fig. 2-3. Schematic illustration of three methods of measuring absolute size of resting potential. Potential is measured between tip of recording electrode (arrow) and an indifferent electrode (ground symbol). Membrane polarity inside is negative with respect to outside.

muscle fiber). A more common approach (Fig. 2-3, *B*) is to impale the cell transversely with a saline-filled glass microcapillary (called a microelectrode), which terminates in a very short, fine tip (approximately 0.1 μ in diameter). In this way the transverse membrane potential of a variety of small structures such as vertebrate muscle fibers, neuromuscular junctions, neurons, and glia has been determined. However, the microelectrode method does considerable damage to the membrane at the impalement site in very small cells and has not generally been satisfactory for use in vertebrate nerve fibers.

Fig. 2-3, *C*, illustrates an approach used to measure action potentials from these fibers. The method involves depolarizing the fiber at one recording site (position 3, Fig. 2-3, *C*), either by injury or by high concentrations of extracellular potassium, and placing the second recording electrode sufficiently distant so that the electronic depolarization from the cut end is negligible. If the external resistance is made very large, the transverse membrane potential between points 1 and 2 in Fig. 2-3, *C*, can be recorded, with negligible error, between the external electrodes at points 1 and 3. This method is generally described as the "air" or "sucrose" gap method, depending on the nature of the insulating material used to raise the external resistance.

This description of recording techniques designed to measure the absolute membrane potential belies the extreme technical difficulty of making such measurements. It is far easier and much more common to place a pair of external electrodes outside the cell and record external potential drop between active and inactive regions of the cell surface (Fig. 2-4). The amplitude of these externally recorded action potentials is considerably less than the intracellular action potential. Since the magnitude is a function of the current density at the two recording sites, it depends on the placement of the electrodes with respect to the membrane surface and also on the electrical resistance and geometry of the external medium. Because of the relative ease of this type of recording, it is frequently used when monitoring occurrence of action potentials rather than study of their waveform is required. It is also used in clinical situations, for example, electromyography, when more invasive techniques are not practical.

Several arrangements of external recording electrodes are shown in Fig. 2-4. The direction of impulse conduction is from left to right, and the region of axon undergoing depolarization is shown as stippled. *Monophasic* recording is illustrated in Fig. 2-4, *A*. One electrode is placed on healthy tissue and the other on a cut or damaged region. Since the resting potential is intact at position a and zero at position b, a steady "injury" current flows between the two points, giving rise to a steady "injury" potential measured in the external medium (stage 1). The magnitude of this injury potential depends on the external resistance. It is proportional, but not generally equal, to the resting potential. The direction of current flow in the external medium is indicated by the arrow in the ammeter. As the action potential approaches point a, the membrane becomes depolarized and the injury current flow in

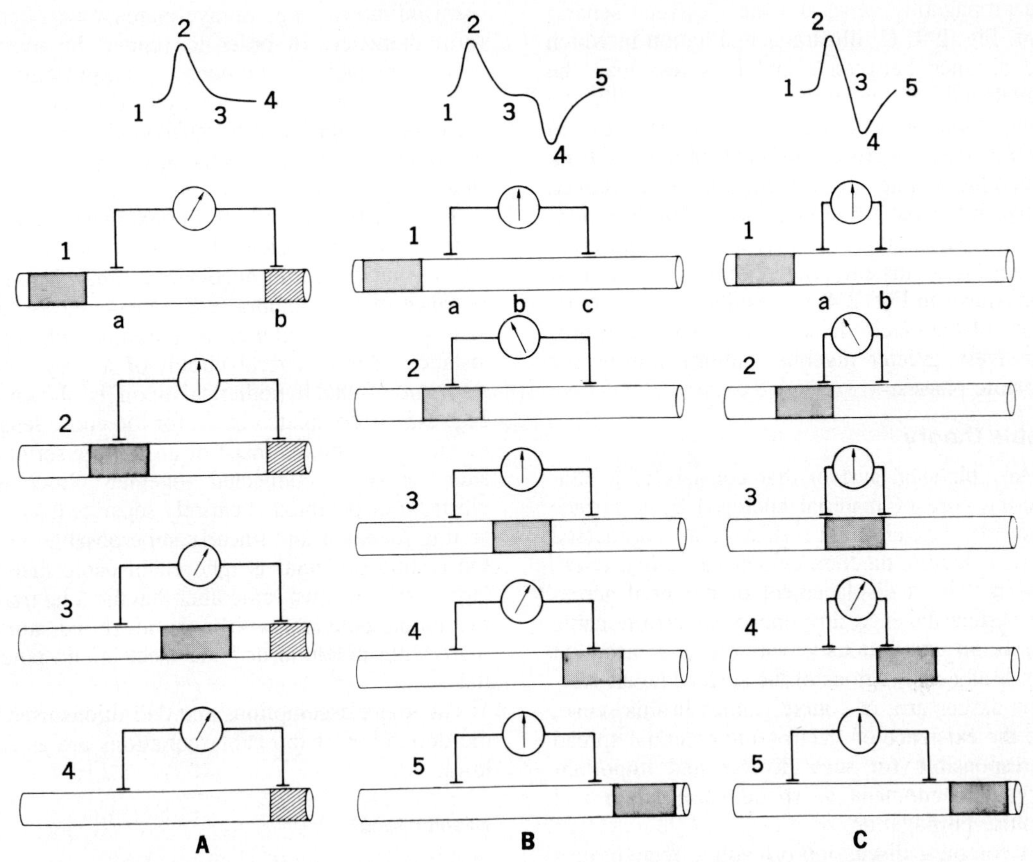

Fig. 2-4. Monophasic and diphasic recording of extracellular action potentials in nerve fibers. Curves at top of figure are recorded action potentials. Numbers on curves indicate instantaneous membrane potential when nerve impulse occupies various positions relative to recording electrodes, as shown in diagrams below. Stippled areas represent location of nerve impulse; coarsely hatched regions indicate permanently depolarized nerve. Direction of current flow is indicated by positions of arrows in ammeters.

the external circuit is greatly reduced. Consequently, the potential difference between points a and b largely disappears, as indicated by the position of the arrow (stage 2). As the wave of excitation passes point a, the membrane is repolarized and the injury current is reestablished, with its former magnitude and direction (stages 3 and 4). Since the current flow has not changed direction but only varied from positive to zero and then back to positive, the potential change is *monophasic,* hence the term for this method of recording.

A second type of external recording that does not involve nerve damage is illustrated in Fig. 2-4, *B*. Since both electrodes are placed on intact membrane, the steady current flow between them is zero, as indicated by the arrow (stage 1). As the action potential passes under point a, the deflection of the arrow is to the left (stage 2). When the active region is between the electrodes, at point b, the current flow from a to b and c to b will be opposite in direction and nearly equal in magnitude. Consequently, the net current flow between a and c and therefore the potential difference is nearly zero (stage 3). When the impulse reaches point c, the current flow is in the direction opposite that when the excitation passed under point a. The resulting potential deflection is therefore in the opposite direction, as indicated by the arrow (stage 4). As the action potential recedes from point c, the potential difference between the electrodes returns to zero (stage 5). Thus the complete sequence of potential change consists of two more or less symmetric phases of opposite polarity. This recording situation is therefore called *biphasic*.

The exact configuration observed depends, among other variables, on the relative length of the propagating wave and the electrode separation. Fig. 2-4, *C,* illustrates a situation in which the distance between a and b is less than the wavelength of the nerve impulse (actually the more common experimental circumstance). In such a case, there is no time interval during which the propagating wave is entirely contained between the recording electrodes. The isopotential interval of stage 3 in Fig. 2-4, *B,* is therefore reduced to an instant, and the biphasic configuration shown in Fig. 2-4, *C,* results. Further shortening of the electrode separation leads to a progressively greater algebraic summation of the separate phases.

Cable theory

Any physical system that consists of a conducting core of material enclosed by a surface layer of relatively high resistance and immersed in a conducting medium evidences *cable properties,* that is, a displacement of potential across the surface layer at any one point on the cable results in continuously variable displacement across adjacent regions of the surface layer. Nervous tissues are, of course, cables in this sense, and the existence of electrotonic potential spread is responsible for such diverse and important nervous phenomena as spatial summation and impulse propagation.

A complete discussion of cable effects in nervous tissue involves a consideration of complicated special cases beyond the scope of this text. Fortunately, an adequate insight into the physiologic significance of cable properties can be gained by examining the distribution of electrotonic potential following a square-wave voltage pulse applied to the axolemma of a hypothetical axon infinitely long, unmyelinated, and of uniform diameter. In order to render the mathematics tractable, a number of simplifying assumptions are made. Although arbitrary, these assumptions appear to be sufficiently realistic so as to introduce no serious error in interpretation.

The resistivities of both the axoplasm and external fluid are considered isotropic and time invariant, although not necessarily equal. The impedance of the membrane is considered as consisting of a capacitance in parallel with a resistance. The electrical circuit of a very small segment of the hypothetical axon is shown in Fig. 2-5. A complete circuit for the entire length of the axon would consist of an infinite series of such networks connected together. Since the electrotonic potential is entirely separate from the resting potential and linearly superposable on it, the resting potential is ignored in cable theory. However, one must remember that the total transmembrane potential will be the algebraic sum of the resting potential and the electrotonic potential.

The major assumptions and definitions used in the derivation of the cable equations are as follows:

Assumptions

1. Axon represented as infinite circular cylinder
2. Inside of cylinder (cytoplasm) of relatively low electrical resistance
3. Shell of cylinder (membrane) of relatively high electrical resistance
4. Region external to cylinder (extracellular space) of relatively low electrical resistance
5. Surface potential radially symmetric

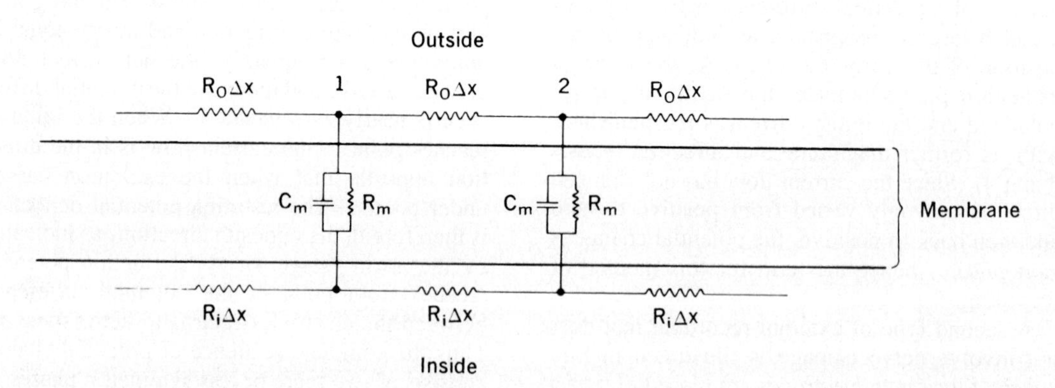

Fig. 2-5. Equivalent electrical circuit of cell. Membrane resistance and capacitance are represented as occurring in discrete packets for convenience; actually they are uniformly distributed along cell surface. Similarly, internal and external resistances are represented as discrete elements, although resistance is distributed throughout intra- and extracellular spaces.

6. Internal and external potentials independent of radial distance
7. Resistances ohmic (i.e., voltage and time independent)
8. Electrotonic potential linearly superposable on resting potential

Definitions

V_i = Inside electrotonic potential (V)
V_o = Outside electrotonic potential (V)
R_i = Internal longitudinal resistance (ohm/cm)
R_o = External longitudinal resistance (ohm/cm)
I_m = Radial membrane current (amp/cm)
I_i = Internal longitudinal current (amp)
I_o = External longitudinal current (amp)
$V_m = V_i - V_o$ = Electrotonic potential across membrane (V)
C_m = Membrane capacitance per unit length of cylinder (F/cm)
R_m = Membrane resistance per unit length of cylinder (ohm-cm)

In the external fluid (Fig. 2-5), current flow from point 1 to point 2 is I_o. The potential difference is as follows:

$$V_o^{(1)} - V_o^{(2)} = \Delta V_o = I_o R_o \Delta X \qquad (1)$$

$$(\text{amp} \cdot \frac{\text{ohms}}{\text{cm}} \cdot \text{cm})$$

If this is true for a finite increment (ΔX), it must be true for an infinitesimal increment (dx). Hence:

$$\frac{\partial V_o}{\partial x} = I_o R_o \qquad (2a)$$

and

$$\frac{\partial V_i}{\partial x} = I_i R_i \qquad (2b)$$

By subtraction, we get a relation between the electrotonic potential and longitudinal current flow inside and outside the cylinder:

$$\frac{\partial (V_o - V_i)}{\partial x} = \frac{\partial V_m}{\partial x} = I_o R_o - I_i R_i \qquad (3)$$

Next we derive an expression relating change in external longitudinal current (I_o) to the transmembrane current (I_m):

$$\Delta I_o = -\Delta I_i = I_m \Delta X \qquad (4)$$

Hence:

$$\frac{\partial I_o}{\partial x} = \frac{\partial I_i}{\partial x} = I_m \qquad (5)$$

By differentiating equation 3 and substituting into equation 5, we get a fundamental relationship:

$$\frac{\partial^2 V_m}{\partial x^2} = \frac{\partial (I_o R_o - I_i R_i)}{\partial x} = I_m (R_o + R_i) \qquad (6)$$

Now examine the components of I_m, the transmembrane current. The membrane behaves electrically as though it were a leaky condensor, that is, as though it were a pure capacitance connected in parallel with a pure resistance. The total transmembrane current (I_m) will be the sum of the current through the resistor plus the displacement current across the capacitor (Fig. 2-6). Thus:

$$I_m = I_R + I_C = V_m/R_m + C_m \frac{\partial V_m}{\partial t} \qquad (7)$$

Combining equations 6 and 7 gives a partial differential equation relating distance along the cylinder, time after initiation of signal, and known membrane parameters:

$$\frac{\partial^2 V_m}{\partial x^2} = \left(\frac{V_m}{R_m} + C_m \frac{\partial V_m}{\partial t}\right) \cdot (R_o + R_i) \qquad (8)$$

It is convenient to rearrange equation 8 and to define two new variables:

$$-\lambda^2 \frac{\partial^2 V_m}{\partial x^2} + \tau \frac{\partial V_m}{\partial t} + V_m = 0 \qquad (9)$$

where:

$$\tau = R_m C_m$$

$$\lambda^2 = \frac{R_m}{R_o + R_i}$$

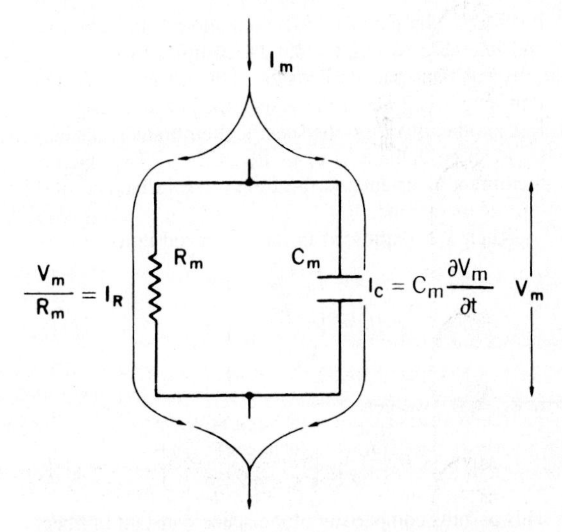

Fig. 2-6. Schematic diagrams showing passage of membrane current (I_m) across small patch of membrane. Part of membrane current passes across resistive element; this is ionic current (I_R). The other component (I_C) results from change in membrane potential across membrane capacitance. This charge, moving in response to potential change, does *not* physically cross capacitor but leaves (or enters) plates.

Note that the membrane time constant (τ) depends only on the membrane parameters, whereas the length constant (λ) involves not only the membrane resistance but also the internal and external resistance.

The forms of solutions of equation 9 depend on boundary conditions. One simple form is the time-independent displacement of potential along an infinite nerve fiber, which results when an initial displacement of potential (V_o) is made at some point on the nerve surface (x = 0).

$$V_m = V_o \exp(-x/\lambda) \qquad (10)$$

The total potential across the membrane equals the electrotonic potential with a spatial variation plus the resting potential, which is everywhere constant.

A note on units

R_m, R_i, and R_o have been defined in terms appropriate to a particular fiber. In order to compare data from different fibers, it is convenient to have formulas that contain dimensions and cell parameters explicitly. Consider a fiber of radius a and length L, and recall the definition of resistance, $R = \frac{L\rho}{A}$, where ρ is the resistivity (ohm-cm) of the interior medium, and A is the cross-sectional area. Therefore the interior resistance per unit length is $\frac{\rho}{A} = \frac{\rho}{\pi a^2}$. In most cases, fibers are immersed in large volumes of salt-containing solutions, so that the external resistance per unit length is negligible compared to R_i, even though the external and internal resistivities may be comparable.

If the fiber has a radius of a, its surface area per unit length is $2\pi a$. If the resistance per unit length is R_m, another quantity, the specific membrane resistance (r_m) can be defined as $r_m = R_m \times 2\pi a$. The specific membrane resistance is the transverse resistance of 1 cm^2 of membrane.

When λ is expressed in these derived units:

$$\lambda = \left[\frac{\frac{r_m}{2\pi a}}{R_o + \frac{\rho}{\pi a^2}}\right]^{1/2} \qquad (11a)$$

If $R_o \ll R_i$, we get:

$$\lambda = \left[\frac{r_m a}{2\rho}\right]^{1/2} \left[\frac{(\text{ohm-cm}^2)\cdot \text{cm}}{\text{ohm-cm}}\right]^{1/2} \qquad (11b)$$

This permits comparison of the space constant in fibers with different dimensions, membrane resistance, and internal resistance. In particular, note that the space constant increases as the square root of specific membrane resistance and fiber diameter.

As an example of the use of the cable theory, one can calculate the length constant for attenuation of the signal along a C fiber using equation 11b and the following typical values for the various parameters:

Specific membrane resistance (r_m)	1,000 ohm-cm^2
Internal resistivity (ρ)	110 ohm-cm
Fiber diameter	1 μ

The length constant is calculated to be about 330 μ, which means, from equation 10, that an electrotonic signal is attenuated ϵ-fold every 330 μ along the fiber. Clearly, electrotonic conduction is an ineffective way of propagating information along a nerve fiber, since the signal would be undetectable more than a few hundred microns away from the source.

ACTION POTENTIAL

Between 1950 and 1952, Hodgkin and Huxley[12,13] conducted experiments, the results of which permitted description of the action potential on the basis of specific sequential changes in the sodium and potassium permeability of the axolemma. They used a technique commonly referred to as a "voltage clamp" because the nerve membrane is held at a fixed arbitrary potential for a period sufficient to permit measurement of the ionic current that flows in response to the imposed potential. The initial experiments were performed on unmyelinated axons isolated from squid because the large diameter of these fibers (300 to 1,000 μ) facilitated the placement of electrodes inside the fiber. Subsequently, voltage clamp analyses of action potentials from other excitable tissues using different methods have led to the conclusion that regenerative (i.e., self-perpetuating) potential changes in excitable cells are generally explicable in terms of sequential changes in specific ion permeabilities, although the charge carriers need not be solely sodium and potassium.

In addition to providing a quantitative description of the nerve impulse, voltage clamp analysis of ion conductances in nerve membrane has also provided a satisfactory explanation for other rapidly occurring, physiologically important phenomena in excitable tissue (e.g., accommodation, refractory period, and pacemaker activity). The analysis does not provide, nor was it intended to, any explanation for other, slower phenomena such as adaptation or afterpotentials, which seem to have entirely different explanations.

Before presenting the somewhat complicated analysis necessary to derive the changes in ionic permeabilities from voltage clamp data, we will describe three observations antedating the voltage clamp experiments, which, when considered together, provide an intuitive basis for focusing on the important variables (sodium and potassium permeability) out of the many possible parameters that could have been investigated as an explanation of the action potential.

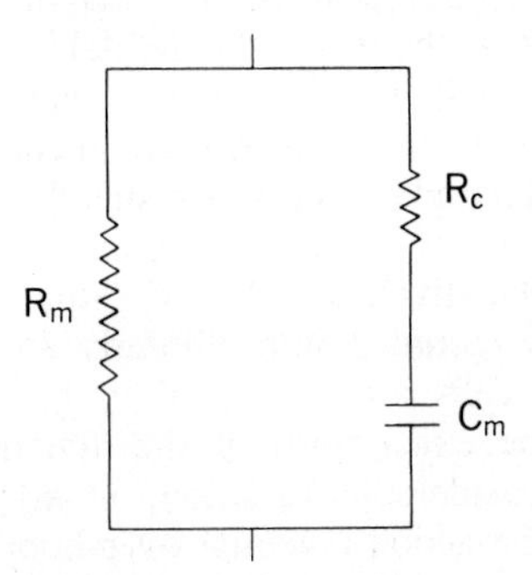

Fig. 2-7. Equivalent circuit of excitable membrane showing, in addition to main membrane resistance (R_m), a numerically smaller resistance (R_c) associated with capacitative elements (C_m).

Membrane impedance changes during action potential

It has already been shown (Chapter 1) that the membrane behaves electrically as if it were a resistance in parallel with a capacitance. A slight refinement, necessary for the present purpose, is the recognition that any capacitance element has a small resistive element in series with it. The circuit for the membrane will therefore be taken as shown in Fig. 2-7.

One standard approach to the analysis of such a reactive circuit is to determine the frequency dependence of the equivalent resistive and reactive elements and then to plot the impedance as a frequency-dependent locus in the R-X plane. For the circuit diagrammed in Fig. 2-7 the impedance locus is a segment of a circle, such as that illustrated in Fig. 2-8, *A*.

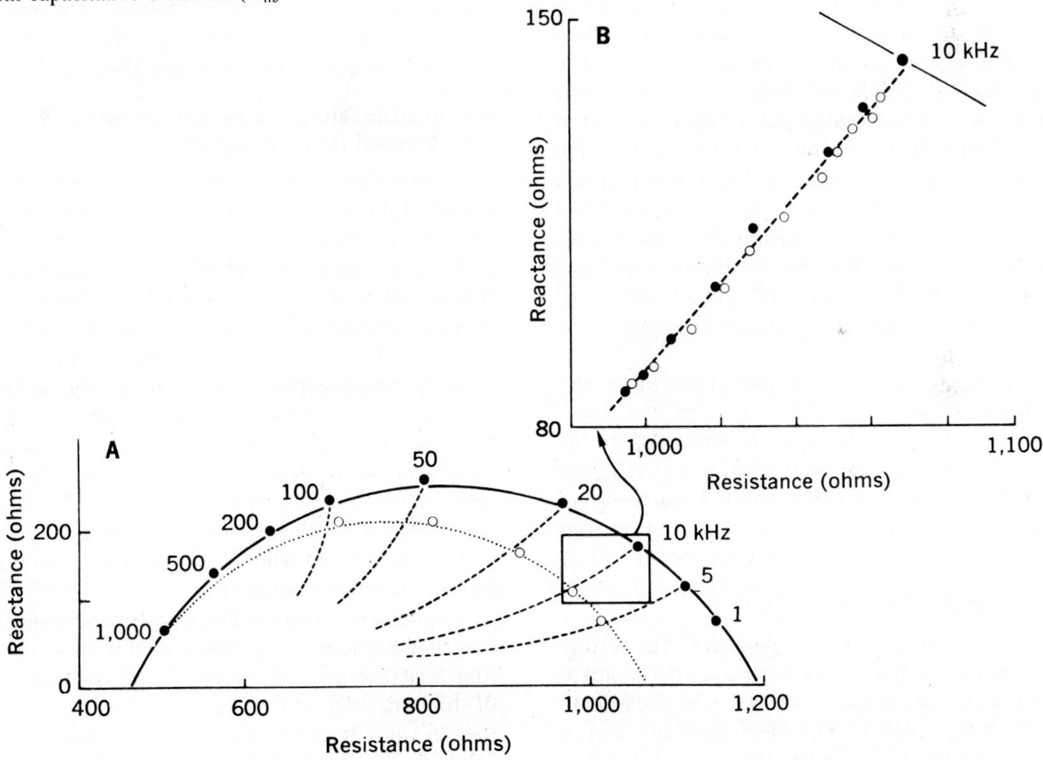

Fig. 2-8. Impedance loci for squid axolemma. **A,** Impedance locus as a function of frequency. Solid circles represent locus of resting membrane impedance; open circles represent locus at peak of action potential. Solid and dotted lines are theoretical loci calculated on the assumption that equivalent circuit of membrane is that shown in Fig. 2-7. Dashed lines are theoretical loci at a fixed frequency drawn on the assumption that only R_m changes during action potential. **B,** Detailed impedance locus at 10 kHz showing correspondence of theoretical curve and experimental points during rise (closed circles) and fall (open circles) of action potential. (From Cole[1]; originally published by the University of California Press; reprinted by permission of The Regents of the University of California.)

The experimental impedance loci for many cells closely approximate the circle loci shown in Fig. 2-8, *A*, which supports the circuit shown in Fig. 2-7 as one possible, although not unique, representation of the electrical behavior of the membrane.

Of greater importance to the present discussion is the fact that when the impedance locus was measured during the action potential, another circle diagram, with a different radius and center, was obtained. One result is shown in Fig. 2-8, *A* (open circles), as the impedance locus (determined at various signal frequencies) for the peak of the action potential. Other circle diagrams were obtained for the membrane impedance locus determined at other points on the action potential.

This result indicates that during the action potential there has been a change in one or more of the membrane parameters: R_m, R_C, or C_m. If the impedance at a fixed frequency (e.g., 10 kHz) is determined during the entire course of the action potential, the impedance locus is seen to follow the path shown as the insert in Fig. 2-8. The experimental points on both the rising phase (solid circles) and falling phase (open circles) of the action potential lie on the dotted line, which is the calculated impedance locus for a circuit in which *only* the membrane resistance (R_m) changes. Although not obvious from the figure, calculation shows that the membrane resistance during activity falls profoundly by a factor of 10 to 40, whereas the capacitance changes by a few percent at most.

The inference to be drawn from these impedance studies, therefore, is that the basic structure of the membrane, as indicated by the capacitance, remains intact during activity, but the ion traffic or concentration of charge carriers crossing the membrane, as indicated by the decrease in resistance, is greatly increased.

Overshoot of action potential

In 1902 Bernstein proposed that the resting membrane was permeable to potassium but not to other ions. The action potential was thought to result from a collapse of the membrane with a general increase in electrolyte permeability. At the time the hypothesis was proposed there was no way of directly measuring the intracellular potential. Since the theory provided a qualitatively correct description of the polarity of the resting potential and the monophasic action potential, it was generally accepted until the development of intracellular recording techniques about the time of World War II. The first published records of action potentials from isolated squid axons showed conclusively that the transmembrane potential did not approach zero during the peak of the action potential. Instead the membrane polarity was reversed, and the inside became transiently positive. The magnitude of this *overshoot*, as the reversed potential is called, was much too large to be ascribed either to a liquid junction potential between the recording electrode and axoplasm or to a residual diffusion of ions across a collapsed membrane.

Representative examples of resting and action potentials recorded intracellularly from several excitable cells are illustrated in Fig. 2-1. Although the exact form of the action potential shows considerable variation, in all cases the peak of the action potential overshoots the resting potential. These relatively straightforward, although technically difficult, measurements proved conclusively that during the action potential it was Bernstein's hypothesis that collapsed rather than the excitable membrane. The membrane does *not* become a nonspecific sieve during the passage of the nerve impulse.

Reversible block of action potentials by removal of external sodium

A third piece of evidence bearing on the mechanism of the action potential came to light at the end of the 1940s when it was shown that the action potential depended on the presence of external sodium in the bathing medium. Excitation was completely blocked if less than 10% of the normal sodium was present in the outside solution. At intermediate concentrations the amplitude of the overshoot appeared to correlate with the sodium equilibrium potential as calculated from the Nernst relation, implying that at the peak of the spike the membrane was selectively permeable to sodium, in sharp contrast to the resting situation in which the membrane was permeable mainly to potassium.

Although external sodium is a general requirement for excitability in mammalian tissues, there may be exceptions. Some work suggests that part of the ionic current flowing during action potentials in smooth muscle cells may be composed of calcium ions. In many invertebrate preparations, calcium, chloride, or other charge carriers are responsible for the initial displacement of the membrane potential from the resting levels.

Summary

Considered together, the existence of an action potential overshoot and the requirement for external sodium suggest that, at least during the initial phase of the action potential, there is an inrush of sodium that depolarizes the membrane and in fact transiently reverses the polarity. The studies on membrane impedance changes during the spike indicate that the ion flows occur in such restricted regions as to cause no great disorganization of membrane structure. Moreover, the fact that the conductance changes continuously during the entire spike suggests that intermediate potential during both the rise and the fall of the action potential might also be associated with increased conductance and ion currents through the membrane. Although the ion substitution technique was not suitable for accurate measurements of changes in relative ion permeabilities during the spike, such data were obtained by the voltage clamp technique described next.

VOLTAGE CLAMP

In addition to the experimental difficulty of dealing with an electrical potential change of very short duration, study of the propagated action potential also poses another problem. Because of the cable properties of nerve membrane and the finite conduction velocity of the nerve impulse, the action potential causes a continuous spatial variation in membrane potential along the nerve. Measurements of membrane conductance or current obtained from a length of membrane would therefore represent values averaged from regions of widely varying potential.

The problem of spatial variation can be solved either by limiting measurements to very short lengths of nerve or by shunting the core resistance with an internal longitudinal wire inside the axon. In the presence of a reasonably low external resistance the latter procedure effectively makes the membrane space constant infinite (i.e., the denominator of equation 11b becomes very small), thus achieving spatial uniformity of membrane potential. Since the technique of inserting an internal electrode is feasible only for very large axons, the first voltage clamp records were obtained from the giant axon of the squid, whose relatively large diameter permits such a procedure.

Description. The problem of temporal variation of potential was solved for squid axon through the use of the apparatus diagrammed schematically in Fig. 2-9. In addition to the shunt wire, A, used to achieve spatial uniformity of potential, a second electrode, B, is also inserted longitudinally into the axoplasm. The transmembrane potential recorded between this electrode and the reference electrode, C, can be held constant at any desired level for several milliseconds (i.e., achieving temporal clamping of the membrane potential) by passing currents of appropriate magnitude and direction between the shunt electrode, A, and the external current collecting electrode, D. The control and measurement of these currents is accomplished by complicated electronic circuitry involving multiple amplifiers but shown only as a single device in the figure.

The basic voltage clamp experiment consists of obtaining a set of membrane currents at constant potential for a series of voltage steps both above (hyperpolarized) and below (depolarized) the resting potential. Representative curves are shown in Fig. 2-10, *A*.

The numbers at left represent the absolute magnitude of the membrane potential. Inward currents are shown as downward on the diagram. The salient features of these curves are as follows:

1. At moderate depolarizations the ionic current is biphasic, initially inward followed by a sustained outward current.
2. At large depolarizations (i.e., positive potentials), there is no inward current. The current is outward at all times.
3. With hyperpolarizing voltage steps, there is little current flow in either direction.

When the experiment is repeated in the ab-

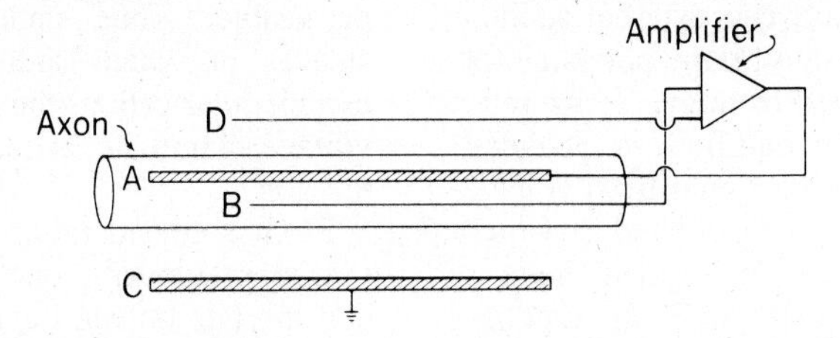

Fig. 2-9. Schematic diagram of voltage clamp technique as applied to squid axons. Membrane potential recorded between electrodes *B* and *C* can be held at any arbitrary level by passing appropriate currents between electrodes *A* and *D*.

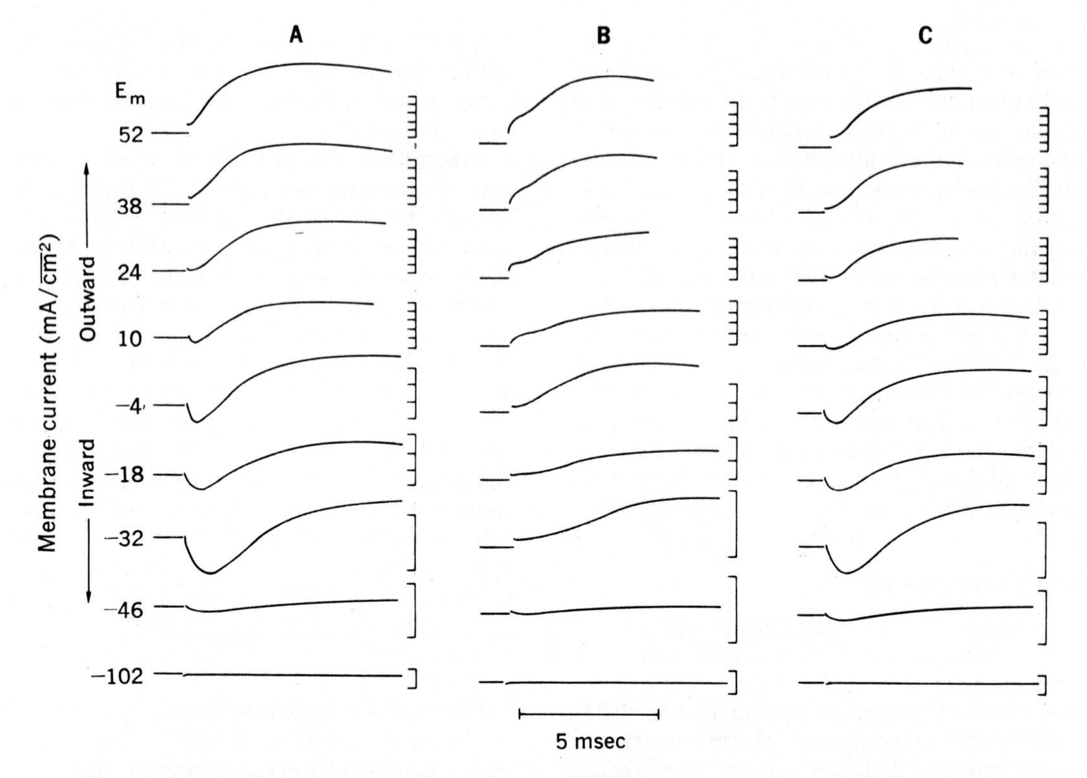

Fig. 2-10. Voltage clamp currents recorded from squid axon. Voltage-clamped membrane potentials is indicated opposite each record. Inward current is plotted downward. Each division on current scale to right of each column is 0.5 mA/cm² (note scale changes). Temperature of axon is 8.5° C. Records were taken as follows: **A,** Axon in seawater. **B,** Axon in sodium-free seawater (choline replacing sodium). **C,** Axon in seawater as control to show reversibility. (From Hodgkin and Huxley.[11])

sence of external sodium, an additional effect is obvious (Fig. 2-10, *B*). The inward phase of the current disappears, whereas the outward currents are essentially unchanged.

Identification of charge carriers comprising the ionic current. The dependence of the initial inward current on external sodium suggests that the charge carrier for this phase of the current is the sodium ion flowing inward across the membrane from a region of high electrochemical potential on the outside to a region of lower potential on the inside. This interpretation is supported by the relation between external sodium concentration and the equilibrium potential for the early current calculated from the Nernst relation. Experimentally the equilibrium potential for the initial current can be found from voltage clamp records by noting the potential at which the current flow initially is zero, that is, there is no flow of current in either direction. Equilibrium potentials determined in this manner are approximately linear with the logarithm of the external sodium concentration, as required by the theoretical Nernst relation, and also agree reasonably well with the equilibrium potentials calculated from analytic determination of the internal sodium concentration. The equilibrium potential for sodium in squid nerve is about +45 mV and probably somewhat lower for mammalian nerve. The selectivity of the process responsible for the early current is not absolute; other monovalent ions can pass through the early current channels. However, with the exception of lithium, which passes about as readily as sodium, the other monovalent cations are considerably less permeable. Thus under physiologic circumstances in which sodium is the predominant extracellular cation, the early inward current in a voltage clamp is carried almost exclusively by this ion.

From a consideration of the experiments discussed previously, one might have anticipated that the late outward current in voltage-clamped axons would be carried by potassium ions. Unfortunately the identification of the ionic species carrying the late currents is not entirely satis-

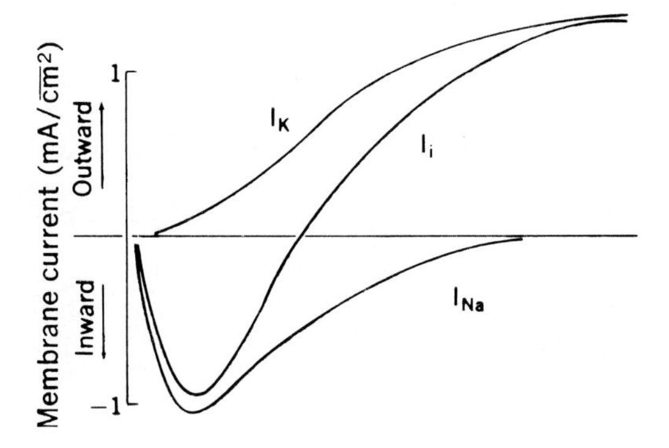

Fig. 2-11. Schematic drawing showing how total ionic current (I_i) is composed of two separate ionic currents, one due to potassium (I_K) and one to sodium (I_{Na}). Time course of third component, leak current, is not shown. In magnitude it corresponds to residual difference between I_i and I_K at end of long-duration voltage clamp of several milliseconds. (Redrawn from Woodbury.[5])

factory. Present opinion is that the process responsible for the late current involves primarily potassium ions. However, the relative selectivity of this process for potassium is not nearly as great as the relative selectivity of the early current process for sodium. It is possible that under physiologic conditions a significant portion of the late current is carried by monovalent ions other than potassium; however, for convenience, the late current will be considered the "potassium current."

In addition to the two major ion currents just described, there is a third current, called the *leakage* or *leak* current. This current does not appear to have the marked time dependence or ion specificity characteristic of early and late currents. As its name implies, it may be regarded as a nonspecific leak of ionic charge across the membrane. Although not of major theoretical interest, the leak current must be considered in calculations involving voltage clamp parameters.

The relative magnitudes and time courses of the two major components of the voltage clamp current are shown in Fig. 2-11 for a voltage step to approximately zero absolute membrane potential. Since the leak current does not change with time during the voltage clamp and is only a few percent of the total current, it is not distinguishable from the baseline. Fig. 2-11 shows clearly the different responses of the sodium and potassium currents to a voltage clamp. The sodium current begins to flow promptly in response to a voltage change but the response is transient, whereas the potassium current begins more slowly but persists for the duration of the pulse. For depolarizing steps to potentials less positive than the sodium equilibrium potential, the sodium current is directed inward across the membrane, whereas the potassium current is directed outward.

As has been discussed in Chapter 1, ion flow across a membrane depends on both the electrical and the chemical gradients existing across the membrane. Ion flow during a voltage clamp is no exception; the magnitude of the driving force on the sodium and potassium ions is the difference between the electrical and the chemical potentials calculated from the Nernst relation. The relation between specific ion flow and electrochemical potential gradient for that ion is conveniently expressed as the specific ion conductance, written as follows:

$$g_{Na} = \frac{I_{Na}}{V - E_{Na}} \tag{12a}$$

$$g_K = \frac{I_K}{V - E_K} \tag{12b}$$

$$g_L = \frac{I_L}{V - E_L} \tag{13}$$

The leak conductance is commonly formulated in this way simply to give it the same form as the specific ion conductances. E_L is an empirical quantity without theoretical significance.

The ionic conductances calculated according to equations 12a to 13 for squid axons are shown in Fig. 2-12; those for other excitable tissues are similar. The sodium conductance rises rapidly following a step depolarization, but falls to near zero within 1 to 2 msec. The dependence of the peak conductance on membrane potential is especially steep for very small depolarization, increasing nearly ϵ-fold for a depolarization of 4 mV (Fig. 2-12, *A*). In contrast to the behavior of the sodium conductance, the rise in potassium conductance (Fig. 2-12, *B*) following a voltage depolarization is much slower, rising to a plateau only after several milliseconds and remaining high for the duration of the pulse. The delay in rise of potassium conductance is especially noticeable at low depolarizations, although the peak potassium conductance has only a slightly less steep voltage dependence than does the sodium conductance.

Since these conductance data provide a reasonably complete description of nearly all short-term electrical phenomena in nervous tissue and in addition are related to permeability changes within the membrane, it is not surprising that a number of theoretical models have been devised in an attempt to explain these permeability

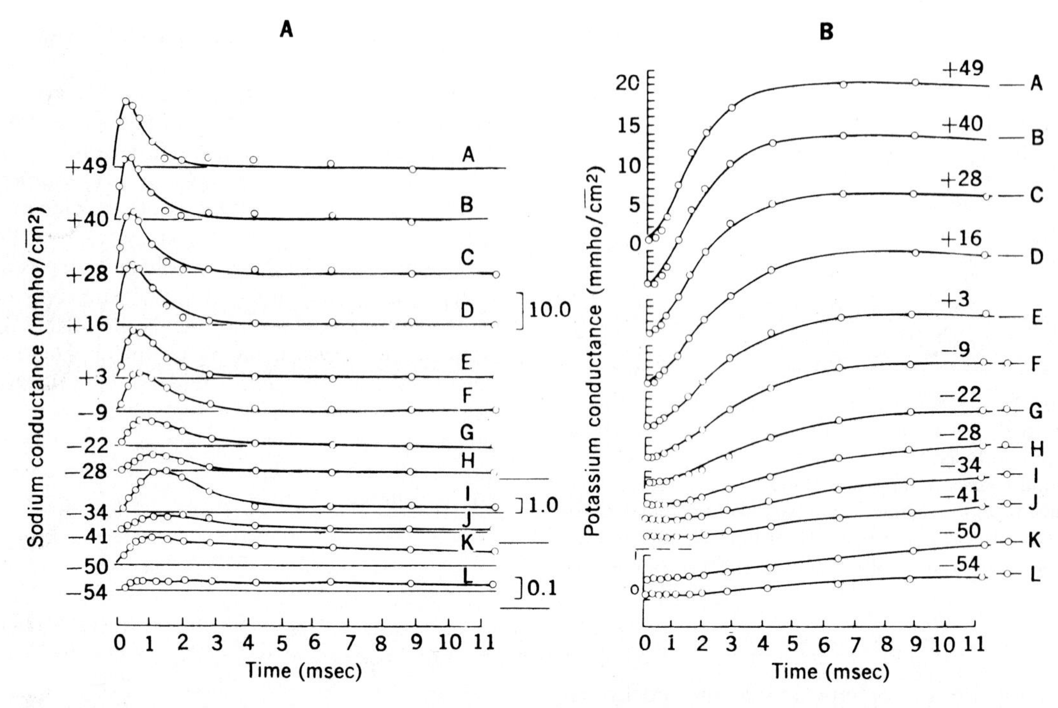

Fig. 2-12. Sodium and potassium conductance curves in voltage-clamped squid axons. **A,** Sodium conductance. **B,** Potassium conductance. Absolute value of clamped membrane potential is given opposite each curve. Note scale changes in conductance curves. Circles represent experimental points; solid curves are drawn according to theoretical equations developed in text. Temperature of axon is 6° to 7° C. (From Hodgkin and Huxley.[14])

changes on a molecular basis. Given the complexity of the phenomena to be explained and the paucity of information concerning membrane structure, it is equally predictable that such attempts have been largely unsuccessful. The model presented here is that originally devised by Hodgkin and Huxley to explain their voltage clamp data on squid axons. It seems adaptable with minor modification to other excitable tissues and is as physically reasonable as any other.

Before presenting the details of the mathematical analysis that follows from the model, the major underlying assumptions will be considered, since they are similar to those of other models.

1. The voltage-dependent ion flows are controlled by independent permeability mechanisms. This assumption seems intuitively reasonable in view of the marked difference in behavior of the two currents. It is also supported by the fact that certain drugs alter voltage clamp currents in such a way as to suggest that two distinct currents exist.
2. The charge is carried by sodium, potassium, and "leakage" ions. The experimental reasons for identifying the charge carriers as sodium, potassium and "leakage" have already been discussed.
3. The particles controlling the permeability changes reorient themselves with finite velocity in response to changes in the transmembrane electrical field. This assumption is little more than a restatement of the experimental results that the conductances are strong functions of the membrane potential but do not change instantaneously. No unique mathematical formalism follows from this assumption, which simply requires that the variables introduced have both a time and a potential dependence.

Formulation of the voltage clamp equations

The text follows the analysis developed by Hodgkin and Huxley. The equations giving the sodium and potassium conductances are as follows:

$$g_K = \bar{g}_K n^4 \qquad (14)$$

$$\frac{dn}{dt} = \alpha_n(1 - n) - \beta_n n \qquad (15)$$

$$g_{Na} = \bar{g}_{Na} m^3 h \qquad (16)$$

$$\frac{dm}{dt} = \alpha_m(1 - m) - \beta_m m \qquad (17)$$

$$\frac{dh}{dt} = \alpha_n(1 - h) - \beta_h h \qquad (18)$$

In these equations the $\bar{g}_K$ and $\bar{g}_{Na}$ represent the maximum possible values of the potassium and sodium conductances and are essentially scaling factors to allow the variables m, n, and h to range between 0 and 1.

The equations may be given a plausible, although not unique, physical basis by the assumption that passage of sodium or potassium ions across the membrane requires the cooperative interaction of several identical particles. The potassium conductance process is considered first because the mathematical form is simpler. One may assume that potassium ions can cross the membrane only when 4 N particles occupy some critical region of the membrane. The fraction of particles in this location is n, and the fraction elsewhere is 1 − n. The potassium conductance is therefore proportional to n^4. The α_n and β_n are voltage-dependent parameters that may be thought of as representing position coordinates of polar molecules whose orientation is determined by the electrical field across the membrane and that control the movement of the N particles into or out of the potassium-activating region. Although these polar particles reorient themselves nearly instantaneously in response to a step change in the electrical field (i.e., a voltage clamp), the N particles will still require a finite time to move into the proper position for carrying potassium ions. Therefore the conductance will not change instantaneously but will increase gradually with time, even though the rate constants α_n and β_n are not time dependent.

One must realize that the foregoing "explanation" of the potassium conductance process is little more than a verbal description of one set of equations that describe the time and voltage dependence of the potassium conductance. There is no a priori reason for assuming that 4 N particles are required. This number derives from the fact that the original voltage clamp conductance data were reasonably well fitted by an exponent of 4 in equation 14. Actually more recent data obtained with improved techniques are better matched by higher powers of the n parameter, but the improvement is minor.

The sodium conductance formula can be given a heuristic basis similar to that provided for potassium. Since the sodium conductance response to a depolarizing voltage step is transient, it is necessary to suppose that two classes of regulating particles exist—those whose association promotes sodium passage (i.e., the M particles), and those whose presence blocks passage (the H particles). In equation 17, m represents the fraction of activating particles present at the activating site for sodium conductance. In equation 18, h refers to the fraction of inactivating particles *not* present at the inactivating site. The defining relation for m is intuitive and is of the same form as for n, the potassium conductance–activating parameter. That for h was chosen so as to preserve the symmetry of the differential equations defining the dimensionless parameters m, n, and h.

Since the m or activation process is more rapid than the h or inactivation process, the initial response to a voltage depolarization is a rise in the sodium conductance. The h or inactivation process develops more slowly but eventually blocks all sodium movement, and the sodium conductance falls to very low levels.

As previously indicated, the equations for the conductances are essentially empiric. The sodium conductance, for example, could probably be fitted by a single variable obeying a second-order, differential equation describing a highly damped oscillator.

Electrical events preceding ionic conductance changes: gating currents

The pronounced voltage dependence of the sodium and potassium ionic conductances leads one to expect that these conductance changes might be the result of opening of channels in the nerve membrane resulting from the movement, under the influence of a changing electrical field, of macromolecules in the membrane having a large dipole moment. Movement of such macromolecules would change the orientation of the associated dipole moment and lead to changes in the charge distribution across the membrane. Since such charge redistribution would be confined to the nerve membrane, it would appear as changes in the capacitative or displacement current produced during a voltage clamp, rather than flow of ionic current *through* the membrane.

Several components of the capacitative current are shown in Fig. 2-13. Component a represents the conventional displacement current that flows when the voltage across any linear capacitor is changed. This component will be linear with potential change produced across the capacitor, and in particular the direction of this displacement current will change if the polarity of the applied

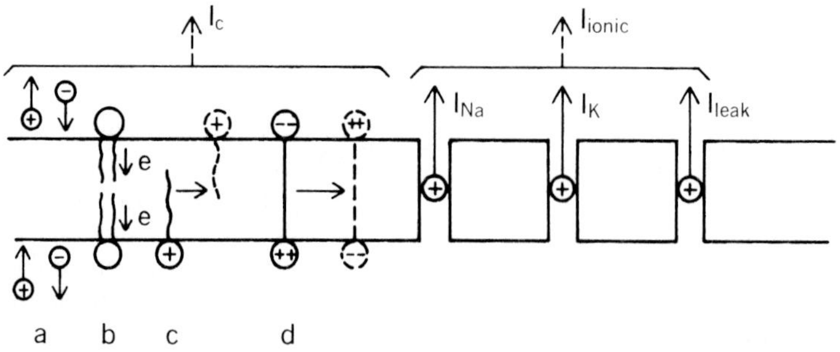

Fig. 2-13. Schematic diagram of various types of charged particle movement within or across membrane. I_C refers to capacitative charge movement, which moves only within confines of membrane. I_{ionic} refers to ionic current (shown as I_K, I_{Na} and I_{leak}, but which could be any mobile ion in either cytoplasm or bathing medium), which physically crosses membrane from bulk phase on one side to bulk phase on the other. (From Armstrong and Bezanilla.[6])

potential is changed. Components b, c, and d of the capacitative current represent current flow due to reorientation of particles of a kind not seen in parallel plate capacitors used in electrical circuits. These components represent possible contributions to the capacitative current due to the presence of unbound electrons on membrane lipid molecules (b) and the existence of charged molecules (c) or molecules with a dipole moment (d). If the electrical field across the membrane is changed (i.e., as in a voltage clamp or during an action potential), the charge distribution associated with molecules of types b, c, and d will be changed. This will appear electrically as a component of a capacitative current. Because of the complex nature of such molecules, it is presumed that the redistribution of charge due to their reorientation would be nonlinear (i.e., that the magnitude of the charge movement would not be directly proportional to the electrical field, and in particular the charge movement would not be symmetric when the polarity of the electrical field is reversed).

It is suspected that some fraction of this nonlinear capacitative current is a result of the motion of intramembrane macromolecules involved in opening and closing the ionic channels through which sodium and potassium flow during the course of an action potential. Since this charge movement reflects the movement of macromolecules necessary to open or close the ionic channels, such currents are commonly called "gating currents." They are also called "asymmetry currents" because the time course of the current is not symmetric with reversal of the potential field. The experimental trace of a "gating current" is shown in Fig. 2-14, *A*, where it may be compared with the sodium current shown for the same axon in Fig. 2-14, *B*. The following two features of this nonlinear capacitative displacement current are important for understanding their presumed role as gating currents:

1. *Direction of current flow.* The direction of the sodium current in *B* is inward in response to a depolarizing pulse. This is, of course, the direction the sodium current should flow in response to an increase of sodium conductance, since the electrochemical gradient for sodium is directed inwardly. However, the response of the nonlinear displacement current is in the opposite direction, directed outward in response to a depolarizing pulse. This is also the expected direction of positive current flow if the charges are confined to the nerve membrane. As the membrane is made less negative inside, any positive charge should be driven relatively more strongly to the outer surface, producing an outwardly directed displacement current.
2. *Rapidity of charge movement.* The displacement of the capacitative charge is virtually complete at the end of 300 to 500 μsec; the gating current has reached its maximum and returned virtually to the zero current baseline. In contrast, the inward sodium current is just beginning to rise substantially by the end of this period. These time relations between the displacement and ionic current are exactly what one would predict qualitatively if the displacement current in fact represents rearrangements of macromolecules to form or to open a channel in the membrane, because

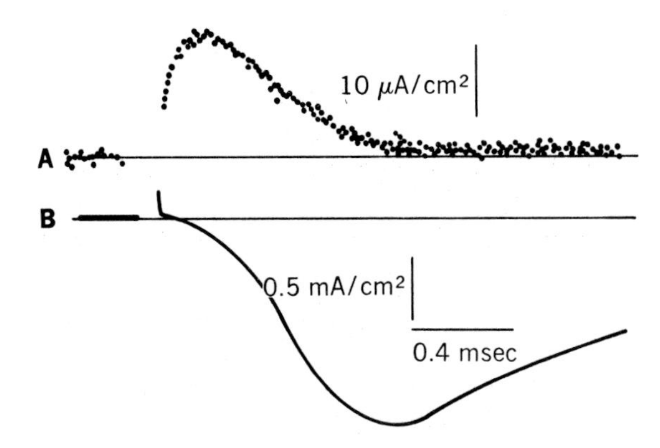

Fig. 2-14. Experimental voltage clamp records comparing magnitude and time course of sodium gating current, **A,** and sodium ionic current, **B.** Gating current rises much more rapidly than ionic current. Gating current is only about 1% of magnitude of ionic current and develops much more rapidly than ionic current. (From Armstrong and Bezanilla.[6])

the molecular rearrangements would have to be nearly complete (i.e., return of nonlinear displacement current to the baseline) before significant ionic flow could occur.

Although the full significance of gating currents is not yet clear, one hopes that analysis of these early currents, because they are presumably closely related to the physical movement of certain macromolecules critical for determining the sodium conductance, may permit a physical interpretation of the empirical Hodgkin-Huxley equations. Study of these currents has already indicated that some modification or reinterpretation of the original equations may be necessary, for example, coupling of activation and inactivation processes or controlling sodium conductance. The displacement currents may well provide information, such as the dipole moment and rotational speed of the permeability-controlling macromolecules, to lead ultimately to a truly molecular description of the permeability changes.

Channel density and conductance of single channels

Study of the electrical conductance of a single sodium or potassium channel is of interest because it permits estimates of the total number of ionic pathways involved during excitation. Estimates of the conductance of a single potassium channel can be obtained by an analysis of the random (i.e., noise) fluctuations in either membrane resistance or potential. Since the resting nerve membrane is much more permeable to potassium than to sodium ions, most of the resting conductance is due to potassium, and fluctuations in this resting conductance (or electrical membrane potential) reflect random variations in the number of potassium channels that are open at any instant. Analysis of noise data indicates that the conductance of a single channel is of the order of 10^{-11} mhos for a single channel in the frog node of Ranvier. This value is rather close to that obtained by direct measurement of the transmembrane pores induced in lipid bilayers by certain antibiotics (e.g., gramacidin), which are thought to have dimensions close to those existing in natural membranes. The channel density obtained, by dividing the maximum potassium conductance obtained in a voltage clamp by the single channel conductance, yields a value of 50 to 100 channels/μ^2 of surface membrane.

Sodium channel density has been measured by binding of the specific sodium channel blocker tetrodotoxin to nerve membranes. The results indicate a density of 2 to 50 channels/μ^2 in rabbit vagus nerve, with about 10-fold higher density for squid axons. The single channel conductance obtained by dividing the density into the maximum sodium conductance is about 1 or 2 × 10^{-11} mhos.

Despite the obvious uncertainties in the measurements and calculations, the data for both sodium and potassium channels indicate that the conductance of a single channel is approximately what one would expect from an aqueous hole of the appropriate dimensions in a lipid bilayer and that the total membrane area occupied by the maximum number of open sodium and potassium channels is an extremely small fraction of the total area. There are no direct measurements of the size of a single pore, but indirect measurements based on the size of particles that block the conductances suggest the diameter to be 0.3 to 0.4 nm.

Quantitative reconstruction of action potential

The conductance parameters can be used to calculate theoretical action potentials that are remarkably close to those observed experimentally. The method will be illustrated first by discussing a somewhat artificial situation in which the nerve is stimulated simultaneously everywhere along its length, producing what is called a *membrane action potential*. The considerations involved in calculating a more physiologic situation, the propagated action potential, are similar but the equations are more complicated. The total current across the membrane (I_m) is as follows:

$$I_m = I_C + I_i \tag{19}$$

that is, the sum of the capacitance current and the ionic current. The capacitative current is simply:

$$I_C = C_m \frac{\partial V}{\partial t} \tag{20}$$

whereas the ionic current is the sum of the sodium, potassium, and leakage currents. Thus:

$$I_i = I_K + I_{Na} + I_L \tag{21}$$

If we recall the definition of conductance used in the voltage clamp experiments and use the empirical conductance formula, we have for the membrane current:

$$I_m = C_m \frac{\partial V}{\partial t} + \bar{g}_K n^4 (V - E_K) + \bar{g}_{Na} m^3 h (V - E_{Na}) + \bar{g}_L (V - E_L) \tag{22}$$

For a membrane action potential the net current across the membrane (I_m) must always be zero. With $I_m = 0$, equation 22 becomes an ordinary differential equation capable of numerical solution by standard methods, giving V as V(t).

Calculated membrane action potentials for both unmyelinated (squid) and myelinated (toad) axons are shown in the upper part of Fig. 2-15 and may be compared with experimental curves obtained for similar depolarizations shown in the lower part of Fig. 2-15. The theoretical and experimental curves agree closely for suprathreshold as well as for subthreshold stimuli.

The time course during an action potential of the membrane conductance and the various specific ion conductance components as well as the m, n, and h parameters are shown in Fig. 2-16. During an action potential the transmembrane voltage is not maintained constant, as it is during the voltage clamp; therefore the conductance parameters vary not only with time but also in response to the continuing variation of m, n, and

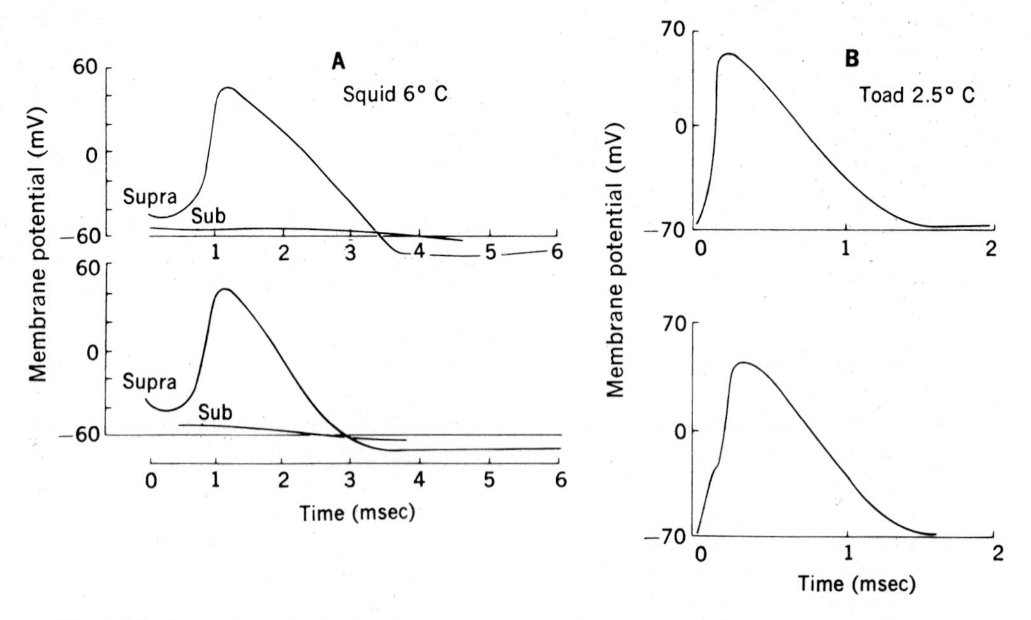

Fig. 2-15. Comparison of calculated (upper curves) and experimental (lower curves) action potentials. **A,** Unmyelinated (squid) nerve fibers. **B,** Myelinated (toad) nerve fibers. Considering complex shape of action potential, correspondence between calculated and experimental curves is remarkable. **A** includes both supra- and subthreshold responses for comparison. (**A** From Hodgkin and Huxley[14]; **B** from Frankenhaeuser and Huxley.[7])

h, as determined by the instantaneous voltage. For this reason the time course of the conductances shown in Fig. 2-16 relating to the action potential are quite different from those relating to voltage clamps, as illustrated in Fig. 2-12.

Fig. 2-16, *B*, shows that during the rising phase of the action potential the major part of the conductance increase is due to sodium. The potassium conductance does not increase appreciably until near the peak of the spike. Thereafter the sodium conductance decreases rapidly and a proportionately greater fraction of the total conductance is potassium. In squid axon the potassium conductance remains elevated for a short time after the end of the spike, thus producing the small postspike hyperpolarization of the membrane potential seen in Fig. 2-16, *A*.

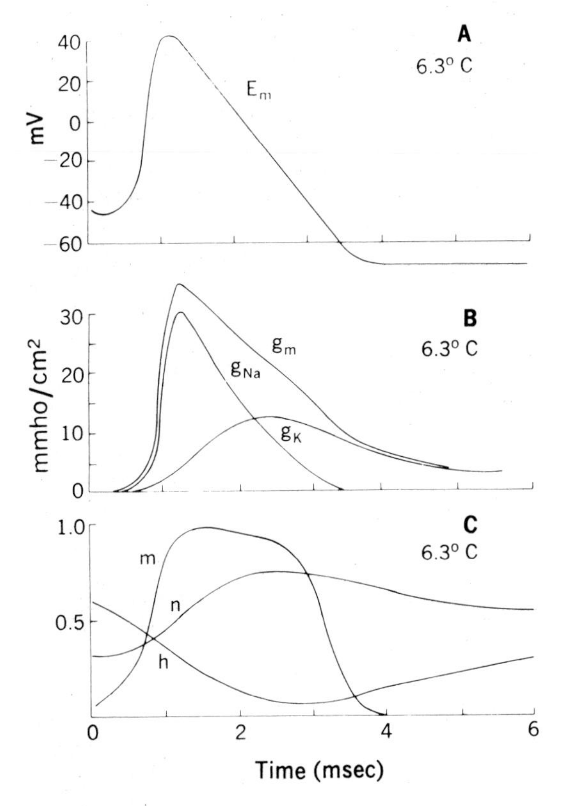

Fig. 2-16. Calculated parameters for squid nerve action potential. **A,** Membrane potential. **B,** Membrane conductance (g_m), sodium conductance (g_{Na}), and potassium conductance (g_K). Note that during rising phase of spike, increase in membrane conductance is due largely to sodium, whereas late on falling phase and during postspike hyperpolarization, increase in membrane conductance is due largely to potassium. **C,** Variation of m, n, and h parameters during spike. Note that whereas m remains high until well on falling phase of spike, h begins to fall immediately after stimulation. Net result of variations is reduction of sodium conductance to low level during falling phase of spike. The n parameter rises slowly but remains elevated until end of postspike hyperpolarization. (**A** and **C** from Cole[1]; originally published by the University of California Press; reprinted by permission of The Regents of the University of California. **B** from Michalov et al.[18])

The reason for the fall of the sodium conductance (Fig. 2-16, *C*) is a decrease in the h or sodium-inactivation parameter. Actually the m or sodium-activation process that tends to increase sodium conductance remains near its maximum value until well past the peak of the spike. The rather rapid development and persistence of sodium inactivation is important in determining the refractory behavior of nerve and will be considered further in that regard.

Propagated action potentials

The preceding discussion of the events during a membrane action potential referred to the situation in which the membrane potential of the fiber changed simultaneously along its entire length. In this situation the membrane current flow is entirely radial and there is no longitudinal component. The situation of physiologic interest, however, is the propagated action potential, during which the nerve impulse passes as a wave of depolarization at nearly constant velocity along the nerve. In consequence of the longitudinal variation in potential, current must flow longitudinally both inside and outside the nerve cylinder as well as transversely across the nerve surface. The existence of longitudinal current flow is most elegantly demonstrated in the case of myelinated nerve fibers, and evidence demonstrating the reality of such current flow will be considered in the section on saltatory conduction.

Since the wavelength of an action potential, on the order of a few centimeters, is rather short compared to the total length of the nerve, the longitudinal variation of transverse membrane potential and hence the longitudinal current itself is restricted to a correspondingly short segment of nerve. For this reason the currents are commonly referred to as "local currents."

The transverse membrane current flowing in the local circuit can be related to longitudinal potential variation by use of the cable theory (p. 52):

$$I_m = \frac{1}{R_o + R_i} \cdot \frac{\partial^2 V}{\partial x^2} \tag{23}$$

In the case of a propagated action potential, unlike the membrane action potential, the membrane current is not zero, but in fact provides the current for the local circuit.

This membrane current flow consists of a capacitative component and a resistive component in parallel:

$$I_m = C_m \frac{\partial V}{\partial t} + \frac{V}{R_m} = C_m \frac{\partial V}{\partial t} + I_i \qquad (24)$$

The resistive component is the sum of the ionic currents that flow across the membrane as the physical consequence of a change in membrane permeability. Their time course is given by the Hodgkin-Huxley equations.

The capacitative current, in contrast, does not result from a change in the physical properties of the membrane but only from the change in potential across the capacitative element. No ions cross the membrane as a result of capacitative current flow; rather, charge is removed from or added to either side of the membrane in response to the po-

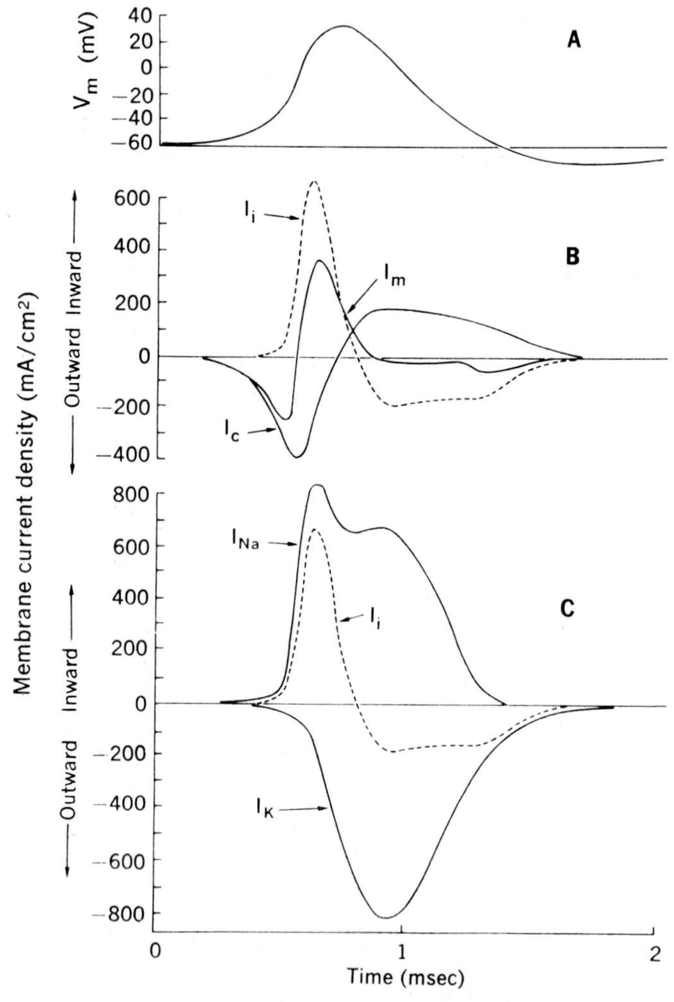

Fig. 2-17. Calculated time course of components of membrane current during propagated action potential in squid axon. Inward current is plotted upward. I_m is total membrane current, consisting of capacitative component (I_c) and ionic component (I_i). Ionic component is further subdivided into sodium current (I_{Na}) and potassium current (I_K). Note that initial flow of membrane current is *outward* (flowing to active membrane behind wave front) and is due entirely to displacement current leaking from membrane capacitance. This displacement of charge is responsible for initial membrane depolarization of nearly 10 mV, which reduces potential to point where sodium conductance increases (shown as rapid rise of sodium current, I_{Na}). Repolarization is accomplished by outward flow of potassium. During this period, charge gradually reaccumulates on membrane capacitance to return potential to resting level. (From Hodgkin and Huxley.[14])

tential change. The actual ionic species carrying the charge away from either membrane surface will be determined by the transference numbers and concentrations of ion present in the internal and external media.

The contribution of each of the components of the membrane current at various times during the course of the propagated action potential is shown in Fig. 2-17. In particular, note that the initial flow of membrane current at the beginning of the propagated spike is due entirely to the capacitative current and is actually outward, whereas the ionic current, which is initially inward, does not begin until the threshold has been reached (about 10 mV of depolarization). A second significant point is that during the falling phase of the spike the net membrane current is rather small, since the ionic current (mainly potassium) and the capacitative current are nearly equal and flow in opposite directions.

USE OF VOLTAGE CLAMP PARAMETERS TO EXPLAIN OTHER PHENOMENA IN EXCITABLE TISSUES

Subthreshold response

Subthreshold responses that closely resemble the experimental response (Fig. 2-15) can be calculated from the empirical equations by the same methods used for action potentials. Such responses can be understood by noting that the initial depolarization (i.e., stimulus), if it is to produce an action potential, must increase the sodium conductance sufficiently to cause a net inward current adequate to carry the membrane rapidly to still greater depolarizations. Otherwise the combined effects of sodium inactivation and potassium activation will act to reverse the net membrane current from inward to outward, and the membrane potential will gradually return to normal values. A calculated subthreshold response and the time course of the variables m, n, and h are shown in Fig. 2-18, together with corresponding parameters for a calculated action potential. The time courses of the Hodgkin-Huxley parameters, which determine the shape of the potential response, are similar for both subthreshold and suprathreshold stimuli. Thus both action potentials and subthreshold responses are produced by the same specific ionic permeability changes showing the same general time variation, although there are important quantitative differences in magnitude. In particular, noting that the m parameter controlling the turning on of sodium conductance (and hence promoting depolarization) is much more sensitive to a small increment in initial stimulus than is the n parameter controlling the turning on of potassium conductance (hence opposing depolarization).

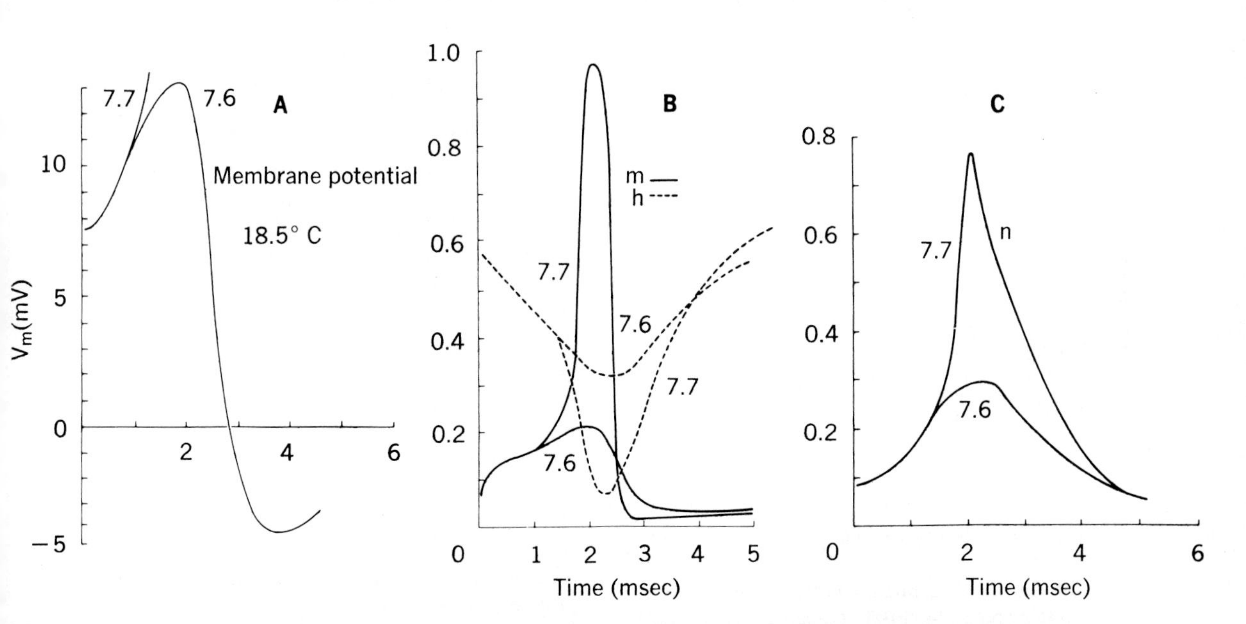

Fig. 2-18. Calculated responses of squid axon membrane to near-threshold electrical depolarization (subthreshold, 7.6 mV; suprathreshold, 7.7 mV). **A,** Membrane potential. Only foot of suprathreshold (7.7 mV) response is shown because complete spike is off scale, which was chosen to show subthreshold response. **B,** Time course of m and h parameters governing sodium conductance. **C,** Time course of n parameter governing potassium conductance. (From Michalov et al.[16])

Threshold

The Hodgkin-Huxley equations provide no physical insight into the mechanism of threshold, that is, how the sodium and potassium conductances are adjusted so as to permit the regenerative response to occur only in response to stimuli exceeding a sharp cutoff value. Calculations from these equations, however, serve better than any experiment to emphasize how sharp the threshold really is. Calculated responses to stimuli of various intensities are shown in Fig. 2-19. The transition from a subthreshold to a threshold response occurs over a stimulus range of less than 1 part in 5,000. Since a suprathreshold stimulus puts the membrane into a state in which the sodium conductance is strongly regenerative, the stronger the suprathreshold stimulus, the sooner the action potential begins, but the configuration of the response is essentially independent of the intensity of the stimulus. Detailed investigation of the Hodgkin-Huxley equations indicates that the membrane's response to a stimulus can be continuously graded from typically subthreshold to typically regenerative if the range of stimulus intensities varies by no more than 1 part in 10^{14}. The explanation for the marked qualitative differences between subthreshold and suprathreshold responses lies in the quantitative relations between the Hodgkin-Huxley parameters rather than in the existence of separate membrane mechanisms for the two types of responses. However, from a practical point of view, propagated graded action potentials do not exist because the minute voltage increments required to produce them are a great deal smaller than the gradations of stimulus intensity that occur physiologically.

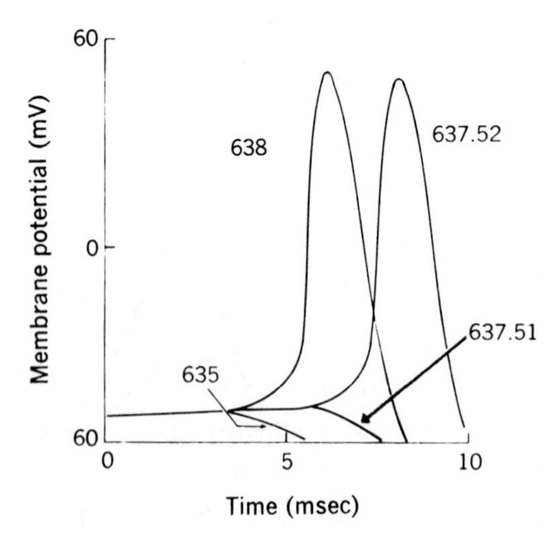

Fig. 2-19. Calculated action potentials for squid axons produced in response to closely graded electrical stimuli. Numbers opposite each curve are proportional to intensity of stimulation. (Redrawn from Cole[1]; originally published by the University of California Press; reprinted by permission of The Regents of the University of California.)

Refractory period

The refractory period following a single stimulation of a peripheral nerve can be adequately explained by the changes in potassium conductance and the h parameter component of sodium conductance. The time courses of the two variables in relation to the spike are shown in Fig. 2-16. Following a single stimulus, the h parameter falls rapidly toward zero, whereas g_K begins to rise, although more slowly. Both effects reach their maximum on the falling phase of the spike and tend to reduce the ability of the membrane to respond regeneratively to a second threshold depolarization. The low value of h reduces the level to which the sodium conductance can be raised by a second depolarization, thus reducing the inward current that can flow. The increased potassium conductance allows a larger than usual outward current to flow, thus tending to negate the effect of a transient surge of inward sodium current.

During the earlier part of the falling phase of the spike, these effects combine to prevent a regenerative response to any stimulus regardless of intensity, meaning that the threshold for excitation is infinite. During this interval of time the fiber is said to be *absolutely* refractory. After a short interval the fiber becomes able to respond to stimuli of greater than normal intensity, that is, the threshold for excitation has been increased but is not infinite. During this interval of time the fiber is said to be *relatively* refractory. During the relative refractory period the threshold declines toward the normal level, with a time course determined by many factors (e.g., temperature, nature of the fiber, and extent of previous excitation).

The time course of recovery of excitability for a single node of Ranvier of a frog nerve fiber is shown in Fig. 2-20. The threshold for a response relative to the control threshold is plotted against time after the first threshold stimulus. In this preparation the absolute refractory period *(R)* lasts about 1.5 msec. During the absolute refractory period the nerve is completely unresponsive, no matter how intense the second shock. By the end of the absolute refractory period the node has become slightly responsive. As the nodal membrane passes through the relative refractory period, the threshold for a second response declines, and after a few milliseconds the fiber has regained its normal excitability.

Nerve refractoriness is also expressed in terms of a parameter, the *excitability,* defined as the reciprocal of threshold. The complete time course for alteration of excitability for mammalian A fibers following a single test stimulus is shown in Fig. 2-21. On the time scale used in the figure the absolute and relative refractory periods are not shown in detail but can be seen to last a few milliseconds. Following the refractory periods the fiber is actually hyperexcitable for 10 to 20 msec (the supranormal period) and then becomes hypoexcitable for a longer period of about 50 msec (the subnormal period). Fig. 2-21 shows clearly that the refractory periods produce

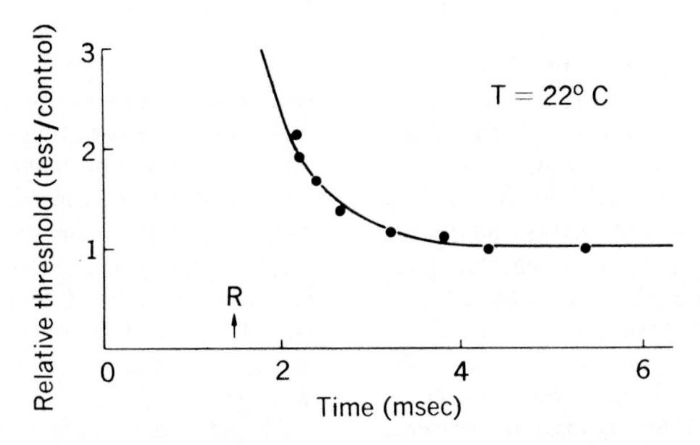

Fig. 2-20. Relative refractory period of frog node of Ranvier. Intensity of threshold test shock is plotted relative to intensity of preceding threshold control shock. End of absolute refractory period is indicated by arrow, *R*. During this period, node is unresponsive regardless of intensity of test shock. As node passes through relative refractory period, threshold for test shock rapidly decreases until, after 3 to 4 msec, fiber has regained normal excitability. (From Tasaki and Takeuchi.[17])

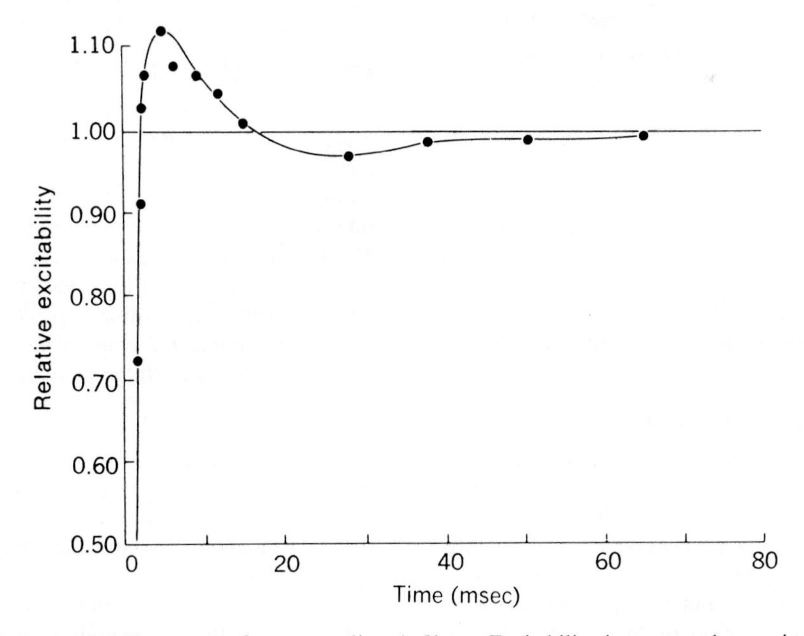

Fig. 2-21. Excitability curves for mammalian A fibers. Excitability is measured as reciprocal of relative threshold and is plotted relative to excitability of fiber at first shock. Absolute refractory period is not shown. Relative refractory period is represented only by first two points. It is followed by intervals of slightly altered excitability related to afterpotentials but not to refractory periods. (Redrawn from Erlanger and Gasser.[2])

the most profound alterations in nerve excitability but that they are transient phenomena succeeded by smaller, prolonged excitability changes. The explanations for these more slowly developing and longer lasting supranormal and subnormal periods are not related directly to the voltage-dependent conductance but are better considered in relation to the phenomena of afterpotentials discussed on p. 74.

Accommodation

Although accommodation can be studied and defined in several ways (see definition of terms, p. 48), the basic observation is that any subthreshold depolarization of the membrane of arbitrary duration and waveform decreases the membrane response to a second stimulus applied during the duration of the first, that is, the threshold to subsequent stimuli is increased. The phenomenon can be explained in the same general terms as refractoriness. The transient increase in inward sodium current induced by any subthreshold depolarization is ultimately overcompensated by the outward potassium current, which rises more slowly but is of greater magnitude. The net steady-state ionic current flowing across the membrane is therefore outward for all sustained depolarizations, tending to counteract the effect of an induced depolarization and raising the threshold for a regenerative response.

One convenient way of demonstrating accommodation is to stimulate a nerve with a linearly increasing current, that is, a ramp-shaped pulse (Fig. 2-2). As the slope of the ramp is decreased (i.e., a more slowly rising current), the threshold at which excitation occurs increases. Fig. 2-2 shows this effect for frog fibers and also shows that there is a minimum rate of rise below which no excitation will occur. The extent to which a fiber can accommodate and still produce an action potential is therefore somewhat limited. In Fig. 2-2 the threshold increased about 30%.

SALTATORY CONDUCTION

In unmyelinated fibers, nerve impulse propagation proceeds by a continuous progression of local circuit flow along the length of the fiber. In myelinated fibers the process by which excitation arises is nearly the same quantitatively as in unmyelinated nerves, but a major difference occurs during propagation of the impulse in that transmembrane ionic current does not flow across the myelin sheath but is constrained to flow across the axolemma only at the nodes of Ranvier. Myelinization has several consequences of tremendous significance for the development of the vertebrate nervous system.

Size. Myelinization permits many fibers of high conduction velocity to be contained in a relatively small volume of nerve trunk. Since conduction velocity in unmyelinated fibers is proportional to the square root of the diameter, an unmyelinated nerve (i.e., a C fiber) would need a diameter on the order of 4 mm if it were to conduct with the same velocity as the fastest myelinated A fibers, that is, around 120 m/sec.

Energy conservation. Since the ionic membrane currents only flow at the nodes, the quantity of sodium gain or potassium loss per centimeter of nerve per impulse is much less for myelinated fibers. For example, the conduction velocities of squid axons and frog fibers are comparable; however, the sodium gain in frog fibers (approximately 1.3×10^{-16} moles/cm/impulse) is about 5,000 times less than that for squid fibers. Consequently, the work done by ion pumps in restoring the concentration gradients will be correspondingly less.

Rapid repetitive firing. The large internodal regions act as a reservoir for diffusion of ions into and out of the node, thus effectively increasing the nodal volume somewhat and reducing the effect, on internal concentration, of the large surges of ionic current that occur during the spike and that might otherwise render the fiber inexcitable after relatively few impulses.

Evidence for local circuit flow

If local electrical currents flow during nerve impulse propagation, an external pathway for current must exist. In the absence of an external pathway the local circuit will be broken, no current can flow, and propagation cannot occur. The validity of this reasoning is demonstrated by a simple experiment illustrated in Fig. 2-22. A length of single nerve fiber is placed in two pools of Ringer's solution that are separated by a short air gap. In such a circumstance, conduction is blocked because local currents cannot flow. However, conduction can be restored reversibly simply by providing a conducting bridge of solution across the air gap. The fact that conduction is restored immediately when contact is made precludes diffusion of some activator substance in the external pathway as an explanation for the restoration of conduction. The design of the experiment with an internode in the air gap also demonstrates the absence of significant local current *within* the myelin sheath, for if there were such flow, removal of the saline bridge could not block propagation.

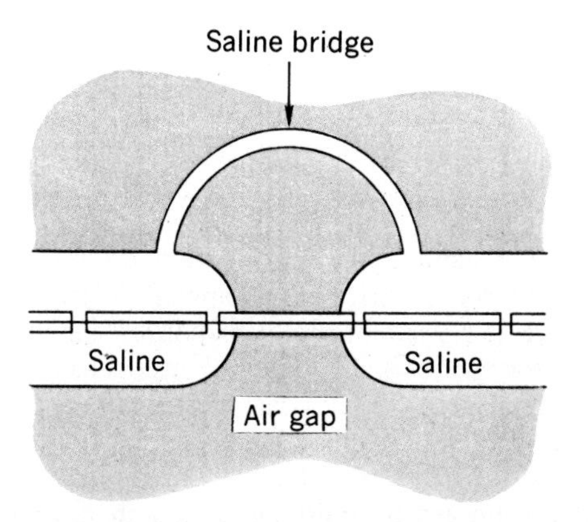

Fig. 2-22. Diagram of experiment demonstrating local circuit flow in single myelinated fibers. Short air gap separates two nodes bathed in saline solution. Conduction occurs only when electrical continuity between two saline pools is completed with external saline bridge.

Evidence for localization of membrane current flow to nodes

It will be recalled from the discussion on nerve impulse propagation that when a regenerative potential change occurs in a patch of nerve membrane, causing it to become active, the net membrane current becomes inward, whereas when the active region is elsewhere, the net membrane current through the patch is outward (Fig. 2-17). Applying these concepts to myelinated fibers allows identification of the active regions along a length of myelinated nerve simply by locating those areas in which inward membrane current flows occur during propagation. The results of such a survey are shown for a frog node (Fig. 2-23, *A*) and internode (Fig. 2-23, *B*). Only in the nodal regions does the membrane current have a large inward component, and it occurs during the rising phase of the action potential. The membrane current in the internodal region is

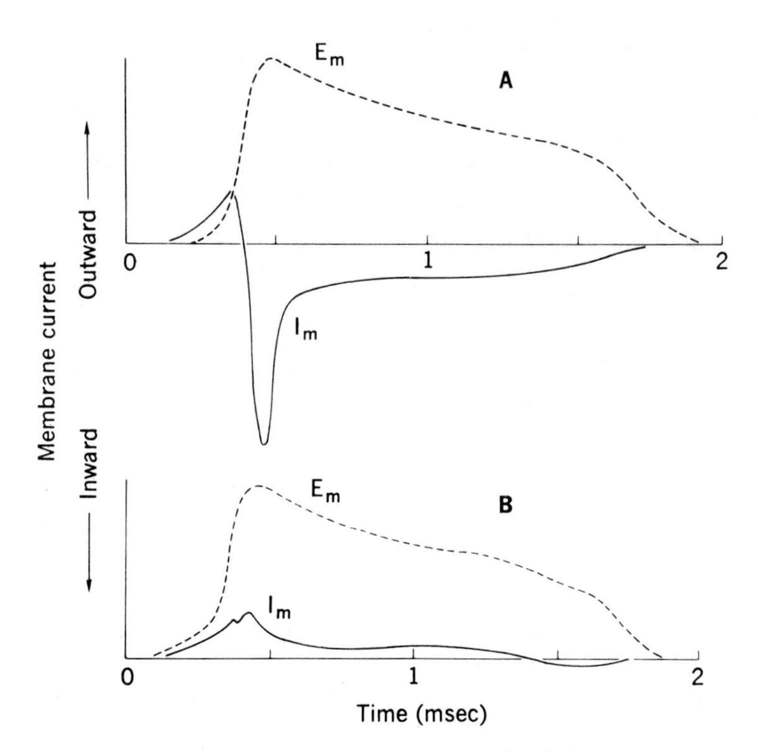

Fig. 2-23. Action potentials and membrane currents of single frog sciatic nerve fiber. **A,** Recorded at node. **B,** Recorded at internode. Note that large inward component of current occurs near peak of spike at node, implying ionic component to membrane current. In internodal region, net current is outward except during end of falling phase of spike, implying that internodal region supplies membrane current in form of displacement charge leak flowing from membrane capacitance, but does not pass inward ionic current. Compare with I_c and I_i in Fig. 2-17. (Redrawn from Huxley and Stämpfli.[15])

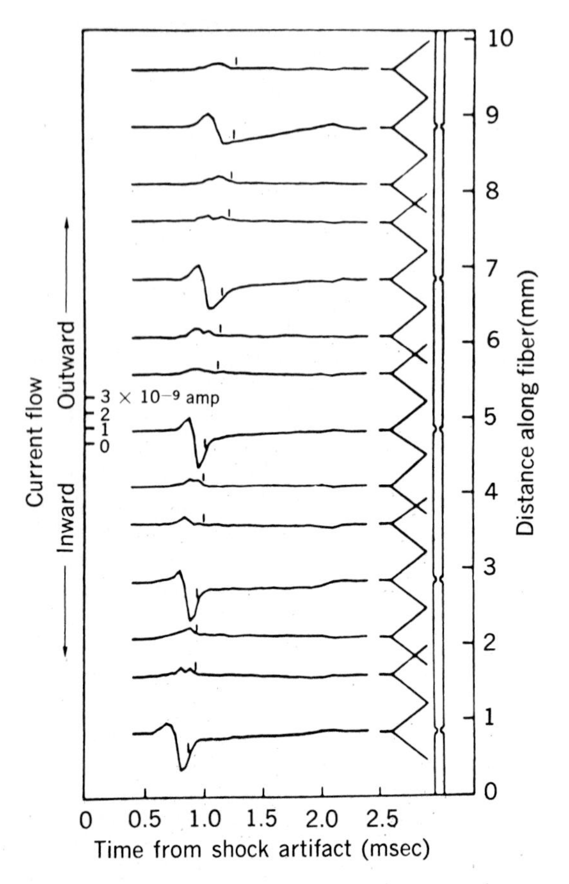

Fig. 2-24. Transmembrane current recorded from nodes and internodes along single myelinated fiber isolated from sciatic nerve of frog. Large inward current occurs opposite node but not internode (compare with Fig. 2-23). Vertical bars on current curves mark occurrence of peak of action potential. (From Huxley and Stämpfli.[15])

not only smaller in absolute magnitude but also flows essentially only in an outward direction.

The progress of the action current over a distance of several nodes is shown in Fig. 2-24, in which the inward component of membrane current, plotted downward, is seen only when records are taken opposite the nodes. One concludes that excitation must leap in a spatially discontinuous manner from node to node, thereby justifying the designation of the conduction process as saltatory.

Voltage clamp analysis of the action potential in myelinated fibers

Application of the voltage clamp technique to a single vertebrate node is less direct and technically much more difficult than its application to squid axons because the extremely small size of the node precludes direct use of internal electrodes or even direct micropuncture. Nevertheless, a complete voltage clamp analysis of toad and frog nodes using less direct techniques has been successfully accomplished by Frankenhauser and his colleagues. The computed action potentials agree closely with those found experimentally; as in the case of squid axons, the voltage clamp parameters serve to describe other short-term electrical phenomena in myelinated nerve (i.e., subthreshold responses and the refractory period). Although the results have largely confirmed the squid data, the experiments have been extremely important in demonstrating that the occurrence of selective sequential sodium-potassium permeability changes in the axolemma can be a general explanation of nervous activity in vertebrates and is not a unique phenomenon in an unusually large invertebrate nerve fiber.

The numerical value of the Hodgkin-Huxley parameters m, n, and h for frog nodes are remarkably close to those obtained for squid axons, especially considering that the nerve membranes in the two cases are bathed by solutions of markedly different ionic strengths. The major difference is that the node shows considerably greater sodium inactivation (i.e., lower value of h) at a given potential than does the squid axon, thus tending to reduce somewhat the sodium currents during an action potential.

Determinants of conduction velocity in myelinated fibers

Although the presence of a myelin sheath increases the velocity of conduction of a nerve impulse by confining the local circuits to the node, it does not follow that the node spacing is the sole or even the most important determinant of conduction velocity. There are several other factors to be considered.

Analysis of the problem indicates that there is a relation between the internode spacing, the fiber diameter, and the myelin thickness so that the fastest conduction velocity for a given external diameter occurs when the ratio of the axis cylinder diameter to the outside diameter is about 0.7, and that conduction velocity and internode spacing should be proportional to diameter.

These theoretical conclusions are generally substantiated by experiments. Fig. 2-25 shows that the observed ratio of axon diameter to fiber diameter is close to the theoretical optimum (0.7), at least for large myelinated fibers, although there appears to be a relative thinning of the myelin sheath in smaller fibers. Furthermore, in accordance with theory, the conduction velocity for myelinated fibers is roughly linear with diameter. Fig. 2-26 shows this relationship for both frog and cat nerves.

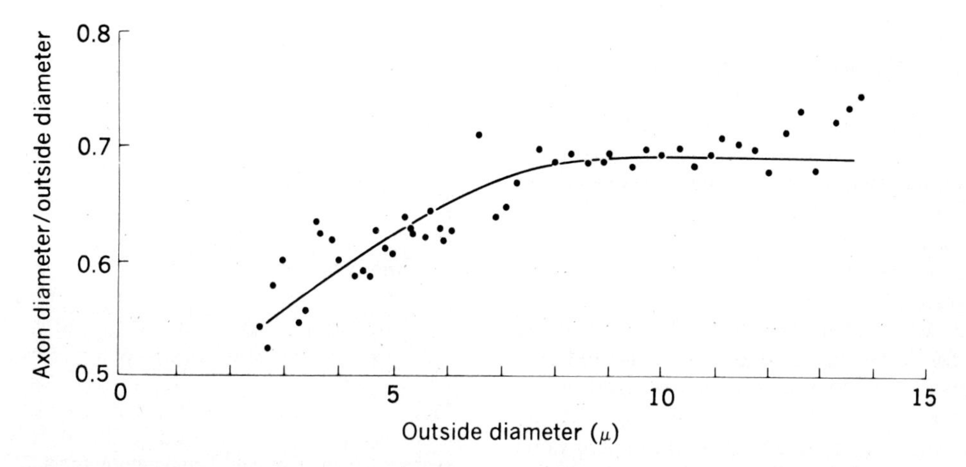

Fig. 2-25. Ratio of axon diameter to outside diameter for myelinated nerve fibers in saphenous nerve of cat. (From Gasser and Grundfest.[9])

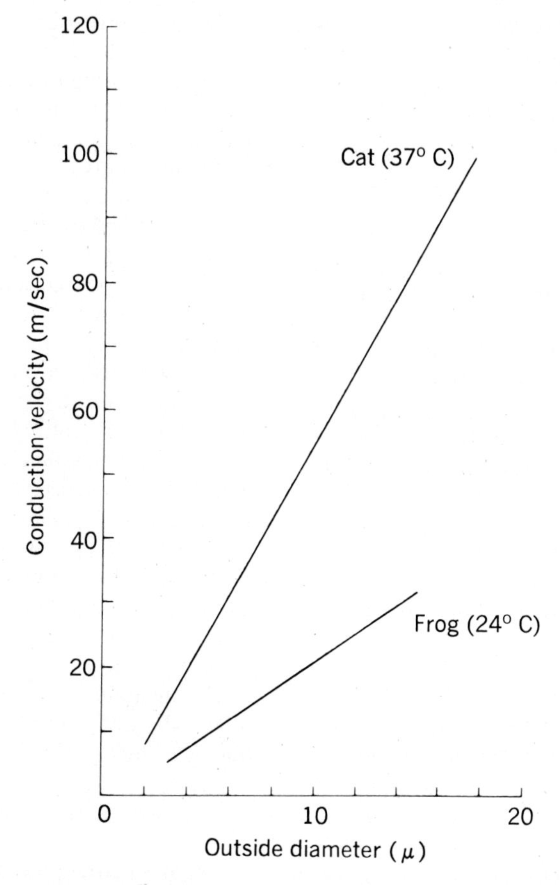

Fig. 2-26. Relation between outside diameter and conduction velocity at physiologic temperatures for myelinated nerve fibers of cat and frog.

Excitability of myelinated fibers in relation to diameter

Since the absolute membrane potential depolarization required for excitation is essentially the same for all myelinated fibers of a given species, excitability per se is not a function of fiber size. However, the usual method of exciting such fibers is not by direct transmembrane depolarization but rather by passing currents in the extracellular media between electrodes separated by distances much greater than the node spacing. Under these circumstances the magnitude of the stimulating current must vary inversely as the node separation if the voltage drop between the resting node and the excited node is to be the same for fibers of varying size. (The external resistance between nodes is proportional to the internode spacing. If the voltage drop between nodes necessary to induce excitation is the same for all fibers, then, in accordance with Ohm's law, the current necessary to induce this voltage drop will be inversely proportional to the resistance between nodes and hence the internode spacing.) Since the node spacing is roughly linear with diameter, for external stimulating arrangements the apparent excitability, as measured by the current required to excite, varies inversely with the diameter. In mammalian nerve fibers the ratio of conduction velocity to fiber diameter is 6 to 9 (m/sec)/μ.

AFTERPOTENTIALS

The term "afterpotential" refers to one of several small-amplitude, long-duration potential changes occurring subsequent to the action potential spike. The afterpotentials of a large mammalian myelinated fiber, together with the spike potential drawn to scale, are illustrated in Fig. 2-27.

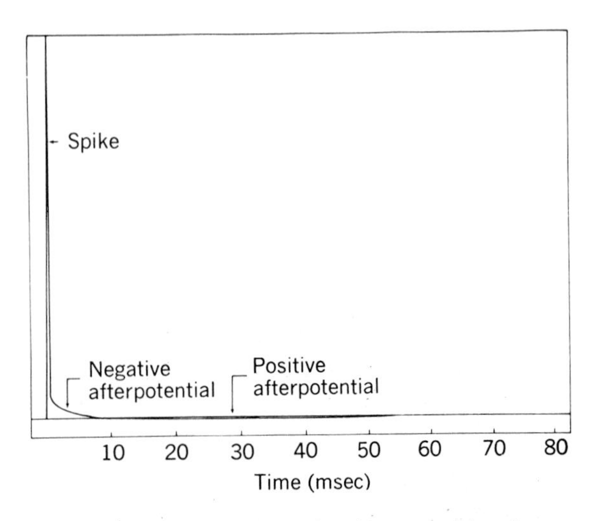

Fig. 2-27. Scale drawing of action potential and related afterpotentials recorded extracellularly from saphenous nerve of cat. Temperature about 25° C.

The first afterpotential is a residual depolarization of the membrane persisting after the spike. It is called the negative afterpotential, and it is followed by a hyperpolarization called the positive afterpotential. The relative magnitude and duration of the afterpotentials shown in Fig. 2-27 are those commonly observed. The negative afterpotential is generally of somewhat larger amplitude and shorter duration than the positive afterpotential. However, the shape of either one varies considerably, not only from one fiber type to another, but especially as a consequence of previous activity of the fiber and changes in metabolic condition.

There is in some fibers a third type of afterpotential not illustrated in Fig. 2-27. This deflection, occurring immediately after the falling phase of the spike, is a very transient hyperpolarization variously referred to as the positive phase, the positive underswing, postspike positivity, etc. (Fig. 2-15).

Certain properties of afterpotentials are listed in Table 2-1. Durations and magnitudes are given only to indicate orders of magnitude; they vary widely among different nerve fibers. Although each afterpotential has been assigned a single mechanism, it will become clear from the discussion of each that the mechanisms invoked as explanations for the afterpotentials act not sequentially but continuously and simultaneously following an action potential. The net effect on membrane potential of the processes acting in concert, and hence the type of afterpotential produced, will depend on their relative intensities.

Postspike positivity. The postspike positivity occurs after the falling phase of the spike when the relative potassium conductance of the membrane briefly exceeds the resting value. In consequence, the membrane becomes relatively more potassium permeable than it was before the spike, and the membrane potential approaches the potassium equilibrium potential more closely than it does at rest, producing a transient hyperpolarization relative to the resting level.

The magnitude of the postspike positivity will reflect the difference between the resting potential and the potassium equilibrium potential. Postspike positivity is therefore most obvious in cells in which V_m is relatively low compared to E_K.

Negative afterpotential. The negative afterpotential has been ascribed to a transient accu-

Table 2-1. Some properties of afterpotentials

Afterpotential	Typical duration	Typical maximum amplitude	Probable mechanism
Postspike positivity	5 msec	5 mV	Transient increase in potassium permeability
Negative afterpotential	30 msec	5 mV	Potassium accumulation outside axolemma
Positive afterpotential	200 msec	2 mV	Stimulation of sodium pump

mulation of the potassium ions released during an action potential in a region immediately external to the axolemma. If these ions are confined to a sufficiently thin layer around the axon, the immediately external potassium concentration will be increased some 20% to 30% above resting levels following a single impulse, an increment sufficient to depolarize the axon by a few millivolts. The initial magnitude of the negative afterpotential would therefore be determined by the amount of potassium rapidly dumped outside the axolemma during an action potential and by the dimensions of the extra-axonal space. The subsequent time course is determined by the rate at which potassium is removed from the space by either inward active transport into some cell or diffusion away from the axolemma.

In the case of the squid axon there is evidence that the space between the Schwann cell layer and the axolemma has the requisite dimensions and permeability properties to retain a small excess potassium concentration for the length of time required to explain the negative afterpotential. In other tissues (e.g., mammalian C fibers) the Schwann cell cleft appears to be too large to allow the postulated concentration increase, and it is necessary to postulate some sort of a diffusion barrier around the axolemma.

Although potassium accumulation in the juxtamembrane cleft is an adequate explanation of the negative afterpotential in nerve cells, it does not suffice for other tissues. In some muscle cells, for example, a slow negative afterpotential occurs although there is no histologically identifiable restraining layer. In frog skeletal muscle the explanation appears to be related to the existence of the tubular system, which may provide space for potassium accumulation or be the site of long-lasting conductance changes. Potassium may also accumulate in the glycocalyx that closely invests some muscle fibers.

Negative afterpotentials in nerve and muscle can also be produced by external application of the veratrum alkaloids. Although these substances do promote potassium leakage from cells, the explanation for the afterpotential appears to be related more to long-lasting conductance changes in the membrane produced by the drug than to potassium accumulation outside it.

Positive afterpotential. The positive afterpotential following a single nerve impulse is too small to study conveniently; consequently, most of the work relating to the positive afterpotential has actually dealt with the pronounced "posttetanic hyperpolarization" that develops following the last spike in a train of impulses, thought to be due to the same mechanism producing the positive afterpotential.

Since the positive afterpotential and the posttetanic hyperpolarization are increased by procedures that stimulate sodium transport and are reduced by treatments that inhibit the sodium pump, it is natural to suppose that the operation of a sodium pump is involved in the production of these postspike hyperpolarizations. In some cases the membrane may actually be hyperpolarized beyond E_K, suggesting that the effect may be due to an electrogenic sodium pump.

Sequence of afterpotentials. The mechanisms proposed for each of the separate afterpotentials also provide some explanation for their sequence and relative magnitudes. The postspike positivity, depending on transient conductance changes, might be expected to persist only a brief time after the spike. When the conductances return to approximately the resting level, the membrane potential will be determined by the ionic concentrations inside and outside the axon and the relative resting permeabilities. An accumulation of external potassium would therefore lead to a transient depolarization observable after the positive phase, the negative afterpotential. However, the sodium pump would be turned on immediately after the spike in response to the sodium load imposed by the action potential, and, if electrogenic, this pump would contribute a hyperpolarizing component to the membrane potential. The net effect on the postspike membrane potential at any moment will of course depend on the relative magnitudes of the depolarizing effect of potassium accumulation versus the hyperpolarizing effect of an electrogenic pump. One might expect the hyperpolarizing effect of the electrogenic pump to outlast the depolarizing

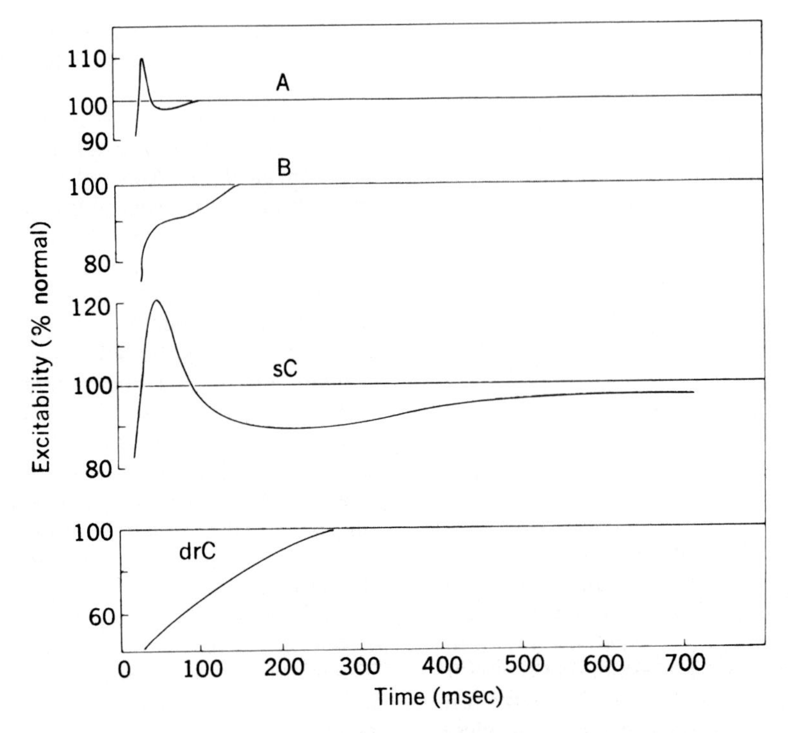

Fig. 2-28. Excitability curves of various fiber classes in myelinated nerve trunks of cat (temperature 37° C). Excitability, as measured by strength of extracellular test shocks, is plotted as percent of ratio (control shock/test shock). (Graphs for A, B, and sC fibers from Erlanger and Gasser[2]; graph for drC fibers from Grundfest and Gasser.[10])

Table 2-2. Some properties of mammalian nerve fibers

Property	Fiber type			
	A	B	sC	drC
Fiber diameter (μ)	1-22	<3	0.3-1.3	0.4-1.2
Conduction speed (m/sec)	5-120	3-15	0.7-2.3	0.6-2.0
Spike duration (msec)	0.4-0.5	1.2	2.0	2.0
Absolute refractory period (msec)	0.4-1.0	1.2	2.0	2.0
Negative afterpotential (%/spike amplitude)	3-5	None	3-5	None
Duration (msec)	12-20	—	50-80	—
Positive afterpotential (%/spike amplitude)	0.2	1.5-4.0	1.5	10-30*
Duration (msec)	40-60	100-300	300-1,000	75-100*
Order of susceptibility of asphyxia	2	1	3	3

*Refers to postspike positivity.

effect of potassium accumulation and thus lead to a late hyperpolarization because diffusion through the Schwann cell channels is faster than active transport across the axolemma. In accordance with this notion is the observation that most large invertebrate axons that do not have strong electrogenic pumps do not exhibit positive afterpotentials, although they do have negative afterpotentials. Since afterpotentials depend not only on membrane permeabilities but also on factors both internal and external to the axolemma, it is not surprising that they should be much more variable than the spike, which depends largely on potential dependent membrane permeability changes.

Significance of afterpotentials. Since the afterpotentials represent deviations of the resting potential either toward or away from threshold, one might expect them to be associated with postspike excitability changes. The excitability curves of mammalian A, B, and C fibers are illustrated in Fig. 2-28. There are several points of

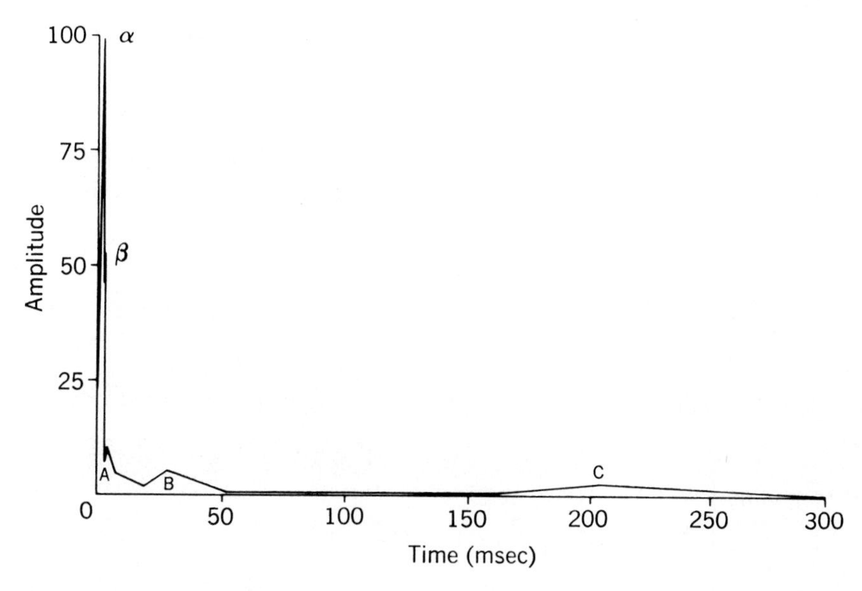

Fig. 2-29. Complete compound action potential recorded from sciatic nerve of bullfrog. Several components of A group of fibers are not clearly resolved because time scale of record was chosen to demonstrate B and C peaks. Amplitude of action potential is plotted relative to A-alpha peak. Room temperature. (From Erlanger and Gasser.[2])

correspondence between the excitability curves of these fiber types and the characteristics of their afterpotentials (Table 2-2). For example, in A and sC fibers the excitability sequence supranormal-subnormal correlates roughly with the duration of the negative and positive afterpotentials. In contrast, B fibers, which have no negative afterpotential and an intermediate-duration positive afterpotential, exhibit only an intermediate-duration period of subnormality. However, a precise correspondence between excitability and afterpotential cannot be expected. Because of the varied factors governing action potentials and afterpotentials, there need not be a direct correlation between the contribution of a given fiber to the spike (i.e., determining excitability) and to the afterpotentials.

Afterpotentials may have significance when they occur in those portions of the neuron that receive inputs resulting in subthreshold local responses (e.g., the synaptic regions of the neuron soma or dendritic tree). Here an afterpotential, depending on polarity, could sum either positively or negatively with an incoming postsynaptic potential and thus affect neuron excitability.

PROPERTIES OF MULTIFIBER NERVE TRUNKS

The compound action potential

Because of the practical difficulty of isolating single mammalian fibers, most of the electrical data on such fibers derive from studies of the extracellularly recorded compound action potential resulting from simultaneous activation of all or some of the population of fibers comprising the nerve tract. Since the compound action potential is merely the algebraic sum of individual fiber action potentials, most of the properties of the trunk can be inferred from a knowledge of single-fiber electrophysiology and the distribution of fiber sizes in the trunk. Close inspection of compound action potentials recorded from peripheral nerves indicates a number of components. A typical compound action potential, recorded from frog sciatic nerve trunks, is shown in Fig. 2-29. The record shows three major deflections: A (which can be further subdivided into A-alpha and A-beta peaks), B, and C. These elevations form the basis of one classification of nerve fibers. The exact relations between positions, size, and threshold of the various components depend on both the recording conditions and the fiber composition of the particular nerve studied. Generally the A elevation is the most prominent and the C fiber elevation is the smallest. Frequently the A and B elevations cannot be clearly distinguished. Because of the relations between conduction velocity, diameter, and threshold, the A fibers are the largest and also have the lowest thresholds. The fiber properties are further discussed in the section on fiber classification (Chapter 28).

Since the conduction velocity of single fibers depends on fiber diameter, the compound action

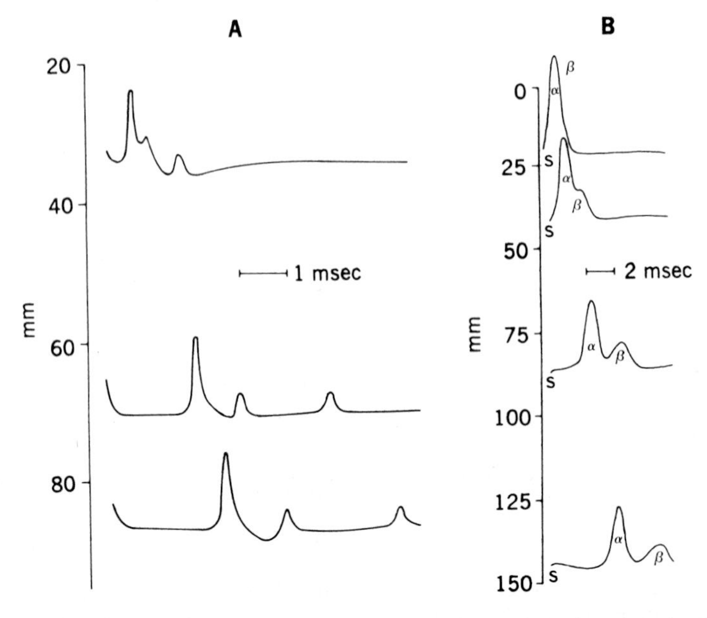

Fig. 2-30. Temporal dispersion of action potentials during propagation along length of frog sciatic nerve. Vertical scale indicates distance of recording site from point of stimulation. Note difference in time scale between **A** and **B. A,** Partially dissected nerve bundle containing only three excitable fibers. Conduction velocity of each fiber is constant. Because conduction velocities are different, separation between spikes increases at recording sites more distant from stimulating electrodes. Initial deflection at start of each trace is shock artifact. **B,** Compound action potential produced by A fibers in intact frog sciatic nerve with many hundreds of fibers. Action potentials cannot be resolved but add together to produce smooth envelope. In top trace, close to stimulating site, only single envelope can be seen. However, at progressively more distant recording sites, separation of compound action potential into two peaks (A-alpha and A-beta) occurs because of bimodal distribution of fiber sizes. (**A** redrawn from Tasaki[1]; **B** from Erlanger and Gasser.[2])

potential broadens as it propagates away from the point of stimulation. Fig. 2-30, *A,* shows individual action potentials recorded from a dissected frog sciatic nerve in which only three fibers were left intact. Since each fiber conducts impulses at a constant velocity proportional to diameter, the temporal dispersion of the spikes increases at progressively more distant recording sites. Fig. 2-30, *B,* shows the situation in an intact frog sciatic nerve with several hundred fibers. In this case, one cannot resolve action potentials from individual fibers because the limited length of the trunk does not permit adequate dispersion of the spikes, but one can clearly observe the separation of the compound action potential into two peaks, indicating a bimodal distribution of fiber sizes. The total area under the action potential remains constant, reflecting the fact that all fibers contribute to the compound action potential regardless of the location of the recording site relative to the stimulating site.

A compound action potential is probably never recorded from a peripheral nerve under physiologic conditions because there is normally both centripedal and centrifugal conduction of impulses from individual fibers in most nerve trunks. Moreover, physiologic stimuli never result in such nearly complete synchronization of activation as that produced by strong electrical stimulation. Physiologic stimulation is usually sufficiently asynchronous that individual spikes can be resolved on the oscilloscope trace. An example of discharge recorded from a peripheral nerve in response to physiologic stimulation during respiration is shown in Fig. 61-13.

Distribution of thresholds in myelinated fibers

Although the individual fibers in a nerve trunk each obey the all-or-none law with an extremely sharp threshold, the compound action potential usually appears to be a continuously graded response because the individual fiber diameters and thresholds vary by small increments and therefore form an essentially continuous distribution. However, when a clear-cut multimodal distribu-

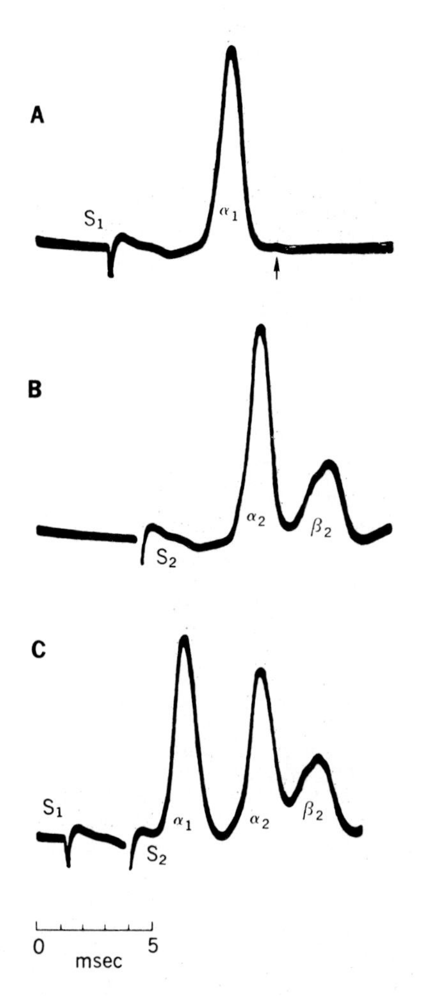

Fig. 2-31. Records showing that A-alpha and A-beta components of compound action potential represent conduction in different groups of fibers with different thresholds. Frog sciatic nerve. S_1 and S_2 represent shock artifacts. Arrow in **A** indicates small elevation in trace produced by stimulation of a few A-beta fibers. (Redrawn from Erlanger and Gasser.[2])

tion of fiber sizes occurs, it is possible by careful adjustment of stimulus intensity to activate the lower threshold (larger and faster) distribution of fibers independently of the higher threshold (smaller and slower) fibers. The intensity of stimulus S_1 in Fig. 2-31, *A*, has been adjusted to activate all the A-alpha fibers but essentially none of the A-beta fibers. (Close inspection of Fig. 2-31, *A*, shows a minimal deflection at the arrow, corresponding to the activation of a few of the largest beta fibers. This barely discernible deflection corresponds to the foot of the larger beta deflection in Fig. 2-31, *B*.) Stimulus S_2 (Fig. 2-31, *B*) is of greater intensity, sufficient to activate both alpha and beta groups of fibers. The alpha and beta elevations of the compound action potential represent conduction in entirely independent groups of fibers with different thresholds. This is demonstrated by close sequential application of both S_1 and S_2 (Fig. 2-31, *C*). The second alpha elevation (α_2) is somewhat reduced, as expected, since some of the alpha fibers will still be absolutely refractory after S_1 and unable to respond to S_2. In contrast, the beta fibers were not activated by S_1 and hence are not in a refractory state. These fibers are able to respond normally to S_2 even though S_1 has preceded, leading to the conclusion that the A-alpha and A-beta fibers have essentially nonoverlapping distributions of thresholds.

Classification of fiber types

Several systems are used in the classification of fiber types. One of the earliest was based on electrophysiologic properties, mainly the conduction velocity as revealed by peaks in the compound action potential and differences in afterpotential as revealed by excitability curves. This electrophysiologic classification divides fibers into three groups: A, B, and C. Some properties of these fibers are listed in Table 2-2.

The A and B fibers are all myelinated, whereas the C fibers are unmyelinated. A fibers can be further subdivided on the basis of mean conduction velocity, and hence fiber size, into several subgroups: alpha, beta, gamma, and delta. The B fibers cannot be distinguished from small A fibers either histologically or in terms of conduction velocity because both groups have similar diameters. However, a distinction can be made on the basis of differences in afterpotentials. A fibers have a short, pronounced negative afterpotential and minimal positive afterpotential. In contrast, B fibers show no negative afterpotential but have a large positive afterpotential. As might be expected, these differences in afterpotentials result in quite different excitability curves (Fig. 2-28). The main difference is that A fibers show a postspike hyperexcitability, whereas B fibers show only a relatively prolonged period of hypoexcitability.

C fibers can be readily distinguished from either A or B fibers because they are all unmyelinated and of small diameter and hence of very slow conduction velocity. C fibers have been further subdivided on the basis of postspike excitability changes into two classes: sC fibers (postganglionic efferent *sympathetic* C fibers) and drC fibers (afferent *dorsal root* C fibers). Sympathetic C fibers evidence a postspike sequence of hyper- and hypoexcitability, whereas

Table 2-3. Classification of afferent fibers

Fiber type	Diameter (μ)	Electro-physio-logic grouping	Origin
I-A	12-22	A-α	Annulospiral Golgi tendon organ
II (muscle, skin)	5-12	A-β	Flower spray, touch, pressure, vibratory receptors
III	2-5	A-δ	Free nerve endings, pain-temperature
IV	0.1-1.3	C	Pain, temperature, mechanoreceptors

the dorsal root C fibers show only a postspike hypoexcitability. As can be seen from Fig. 2-28, the postspike alternations of excitability are much more prolonged for C fibers than for either A or B fibers, in keeping with the more protracted afterpotentials.

Unfortunately it has been difficult to apply the A fiber classification to all types of nerves. In part the problem is technical; the original gamma elevation in the compound action potential appears to have been a recording artifact. More importantly, however, since the first elevation in the A part of the compound action potential was called A-alpha and represented the fastest conducting fibers, this designation could refer to groups of fibers of different conduction velocities in different nerves, depending on the distribution of fiber sizes. Thus the fastest identifiable fibers in afferents from the soleus muscle of the cat have diameters of about 17 μ, whereas in the saphenous nerve the largest fibers are only about 14 μ. Although these two groups of fibers are clearly distinguishable on the basis of conduction velocity, in the electrophysiologic classification, both would be designated as A-alpha. Despite the unsatisfactory nature of the A fiber group terminology, some vestiges remain and have to be learned. The small-diameter motor fibers to muscle spindles are called gamma fibers. The large motor fibers to extrafusal phasic muscle fibers are called alpha fibers. The fastest fibers in cutaneous nerves are sometimes called alpha cutaneous fibers.

A second classification, applied by sensory physiologists to afferent fibers, is based on a division of fibers into four groups, principally on the basis of fiber size, but also on the basis of fiber origin. Table 2-3 lists the four groups and the histologic structure at the afferent terminal of the fiber of each subgroup as well as the approximate corresponding A fiber subgroup. The functional significance of the various terminations is discussed in Chapter 11.

A similar size-function classification is possible for efferent fibers. Generally the largest motor fibers (12 to 20 μ) innervate the extrafusal muscle fibers, whereas the smaller efferent fibers (2 to 8 μ) innervate intrafusal fibers within the fusiform spindles.

REFERENCES

General reviews

1. Cole, K. S.: Membranes, ions and impulses, Berkeley, 1968, University of California Press.
2. Erlanger, J., and Gasser, H. S.: Electrical signs and nervous activity, Philadelphia, 1938, University of Pennsylvania Press.
3. Hodgkin, A. L.: The conduction of the nervous impulse, Springfield, Ill., 1964, Charles C Thomas, Publisher.
4. Tasaki, I.: Nervous transmission, Springfield, Ill., 1953, Charles C Thomas, Publisher.
5. Woodbury, J. W.: Action potential. Properties of excitable membranes. In Ruch, T. C., and Patton, H. D., editors: Physiology and biophysics, Philadelphia, 1966, W. B. Saunders Co.

Original papers

6. Armstrong, C. M., and Bezanilla, F.: Charge movement associated with the opening and closing of the activation gates of the Na channel, J. Gen. Physiol. **63:** 533, 1974.
7. Frankenhaeuser, B., and Huxley, A. F.: The action potential in the myelinated nerve fiber of Xenopus laevis as computed on the basis of voltage clamp data, J. Physiol. **171:**302, 1964.
8. Frankenhaeuser, B., and Vallbo, A. B.: Accommodation in myelinated nerve fibers of Xenopus laevis as computed on the basis of voltage clamp data, Acta Physiol. Scand. **63:**1, 1965.
9. Gasser, H. S., and Grundfest, H.: Axon diameters in relation to spike dimensions and conduction velocity in mammalian fibers, Am. J. Physiol. **127:**393, 1939.
10. Grundfest, H., and Gasser, H. S.: Properties of mammalian nerve fibers of slowest conduction, Am. J. Physiol. **123:**307, 1938.
11. Hodgkin, A. L., and Huxley, A. F.: Currents carried by sodium and potassium ions through the membrane of the giant axon of Loligo, J. Physiol. **116:**449, 1952.
12. Hodgkin, A. L., and Huxley, A. F.: The components of membrane conductance in the giant axon of Loligo, J. Physiol. **116:**473, 1952.
13. Hodgkin, A. L., and Huxley, A. F.: The dual effect of membrane potential on sodium conductance in the giant axon of Loligo, J. Physiol. **116:**497, 1952.
14. Hodgkin, A. L., and Huxley, A. F.: A quantitative description of membrane current and its application to con-

duction and excitation in nerve, J. Physiol. **117:**500, 1952.
15. Huxley, A. F., and Stämpfli, R.: Saltatory transmission of the nervous impulse, Arch. Sci. Physiol. **3:**435, 1949.
16. Michalov, J., Zachar, J., and Kostolanský, E.: Automatic computation of changes in membrane potential and underlying processes in the Hodgkin-Huxley model, Physiol. Bohemoslov. **15:**307, 1966.
17. Tasaki, I., and Takeuchi, T.: Weitere Studien über der markhaltigen Nervenfaser und über die elektrosaltatorische Übertragung des Nervenimpulses, Pfluegers Arch. **245:**764, 1942.

3 Contractility

ROBERT M. DOWBEN

WITH SPECIAL REFERENCE TO SKELETAL MUSCLE

Movement is an ubiquitous feature of living cells. Many kinds of prokaryotic cells are propelled about by flagella, but the flagellar proteins of different bacterial classes are genetically unrelated and probably represent separate instances of the evolution of contractile systems. On crossing the great evolutionary watershed from prokaryotes to eukaryotes, many similarities are found between widely differing contractile systems in various eukaryotic species.[157] Thus slime mold, algae, and protozoa contain contractile proteins responsible for cytoplasmic streaming whose amino acid composition and many other properties resemble those of contractile proteins extracted from mammalian heart and skeletal muscle.

Contractile systems are biologic machines that utilize chemical energy from the metabolism of food in the form of adenosine triphosphate (ATP) or guanosine triphosphate (GTP) hydrolysis to produce useful work. An understanding of contractility and muscle function requires an appreciation not only of the mechanical properties, but also of the energetics, that is, the processes by which chemical energy is transduced to useful work. Furthermore, to function effectively as integral units in an organism as complex as man, muscle function is exquisitely controlled, in part by intrinsic control, but more importantly by the central nervous system (CNS).

In higher organisms the function of movement is frequently carried on by specialized, highly differentiated cells of mesenchymal origin, the muscle cells, of which there are three kinds: skeletal muscle, primarily used for voluntary motion; smooth muscle, found in the walls of hollow viscera, blood vessels, and various ducts; and cardiac muscle. We know more about the most organized and highly differentiated contractile system, namely, skeletal muscle, than we do about less differentiated or even primitive contractile systems. Therefore the discussion of contractility in this chapter will emphasize skeletal muscle. The special properties of smooth muscle and cardiac muscle will be described elsewhere in this book, but the basic mechanisms are very similar to those of skeletal muscle.

The more undifferentiated contractile systems are also highly important in human physiology. Ciliated epithelia, such as the cilia in the epithelium of the respiratory tract, have important functions. Platelets contain contractile proteins that play an important role in blood clotting. Axonal flow, a form of cytoplasmic streaming, is necessary to transport vital substances from the bodies of the nerve cells where they are synthesized down the axons to the periphery where they are utilized.

STRUCTURE OF SKELETAL MUSCLE

Gross features of skeletal muscle

Skeletal muscle, which accounts for more than 40% of the body weight in man, consists of bundles of elongated, cylindric cells called *muscle fibers,* 50 to 200 μ in diameter and often many centimeters long. Bundles of muscle fibers, each called a *fasciculus,* are surrounded by a connective tissue covering, the *endomysium.* A muscle consists of a number of fasciculi encased in a thick outer layer of connective tissue, the *perimysium.* The connective tissue is much more resistant to stretching than the muscle fibers themselves, and, as a result, the amount of movement in a muscle is limited by its connective tissue. At both ends of a muscle the connective tissue melds into a tendon by which the muscle is attached to the bony skeleton.

There are several kinds of arrangements of muscle fibers with regard to the tendons (Fig. 3-1). In some muscles *(fusiform),* the muscle fibers run the whole length of the muscle between the tendons, which form at the opposite ends. In most muscles *(pennate),* one of the tendons penetrates through the center of the muscle; muscle

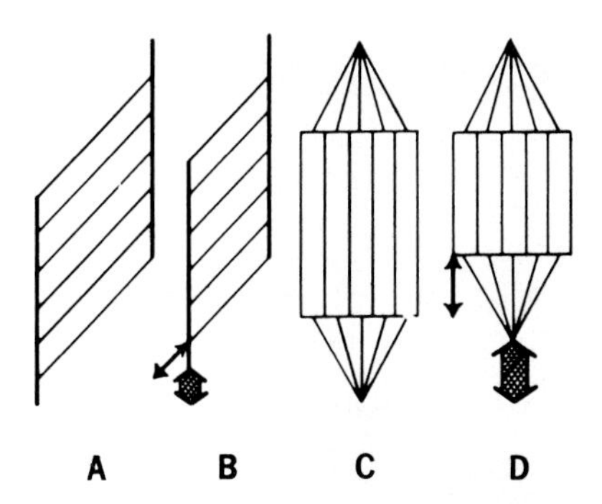

Fig. 3-1. Two arrangements of muscle fibers, both at rest and at contracted length. **A,** Pennate fiber arrangement at rest. Tendons are lines extending from two sides of parallelogram. **B,** Same muscle at maximum shortening. Fibers have shortened by one third rest length along axis indicated by narrow double arrow, whereas muscle as whole shortens only by amount indicated by length of and in direction of broad double arrow. **C,** Parallel fibers at rest. Lines radiating from rectangles to point at each end represent tendons. **D,** Same muscle at maximum shortening. Fibers have shortened by one third rest length (direction and magnitude indicated by narrow double arrow) and muscle shortens by same amount (indicated by broad double arrow). Thus although fibers in **A** and **C** are same length, shortening in **A** is less than shortening in **C.**

fibers run at an angle to the axis of the whole muscle from the central tendon to the perimysium. If the muscle fibers slant in toward the center to converge on a single tendon, they are called *unipennate*. If the fibers converge on both tendons, the arrangement is called *bipennate*. Compound muscles with several bellies of converging fibers are said to be *multipennate*. Because of the angular direction of the muscle fibers, pennate muscles have a mechanical advantage over fusiform muscles, generating more tension per unit of whole muscle weight, although the maximum tension per unit of muscle fiber cross section is the same. However, the fraction of rest length that pennate muscles can shorten is less than for fusiform muscles. Thus degree of shortening is traded for increased maximum tension in pennate muscles.

By and large, the number of muscle fibers does not increase after the neonatal period. The increase in size of a muscle during growth and physical training is due to enlargement of the individual fibers. Muscle does not have much capacity for regeneration. Gross injury to a muscle results in replacement by connective tissue and fat.

Light microscopy

Light microscopists have known for almost a century and a half that skeletal and cardiac muscle has *cross striations,* arising from alternate light and dark transverse bands.[14] The band that usually stains more darkly, or appears darker, when the muscle is viewed with polarized light is birefringent or *anisotropic* to light and is called the *A band.* The term ''birefringent'' means that the A band contains material that has a higher refractive index to light along the long axis of the muscle fiber than perpendicular to the long axis. The birefringence of the A bands accentuates their darkness when viewed by an ordinary light microscope slightly out of focus and makes them clearly visible in a phase-contrast microscope. The intervening lighter bands are *isotropic* to light and are called *I bands.*

Like other cells, muscle cells are surrounded by a cell membrane, the *sarcolemma. Myofibrils,* the *contractile elements,* are numerous parallel, lengthwise threads 1 to 3 μ in diameter that fill most of the muscle fiber. The cross striations are located in the myofibrils. A muscle fiber contains perhaps 100 or more myofibrils. Squeezed between the myofibrils and the sarcolemma is a small amount of cytoplasm, the *sarcoplasm,* in which are suspended multiple nuclei, numerous mitochondria (called *sarcosomes*), lysosomes, lipid droplets, glycogen granules, and other intracellular inclusions. The sarcoplasm contains glycogen, glycolytic enzymes, nucleotides, creatine phosphate, amino acids, and peptides.

In addition to muscle cells and fibroblasts in the connective tissue, a whole muscle contains fat cells and histiocytes.

Ultrastructure of skeletal muscle

Electron microscope studies in the 1950s[74,100] revealed a rich and complex submicroscopic structure in skeletal muscle (Fig. 3-2). As noted previously, each muscle fiber contains numerous myofibrils, 1 to 3 μ in diameter, that show cross striations owing to alternating A and I bands. The A and I bands in adjacent myofibrils are in register. The I band is bisected by a thin, very dark line, the *Z line.* The contractile unit, or *sarcomere,* is defined as extending from one Z line to the next Z line, usually a distance of 1.5 to 3.5 μ (Fig. 3-3).

Inside the myofibril, there are a great many longitudinal filaments of two kinds: *thick filaments* and *thin filaments.* The thick filaments are about 120 Å in diameter and about 1.8 μ long. They are located in the center of the sarcomere

Fig. 3-2. Electron micrograph of skeletal muscle cut longitudinally showing dense A bands and lighter I bands, the latter bisected by dense Z lines. A and I bands, which are in register, are formed by two types of myofilaments: thick and thin.

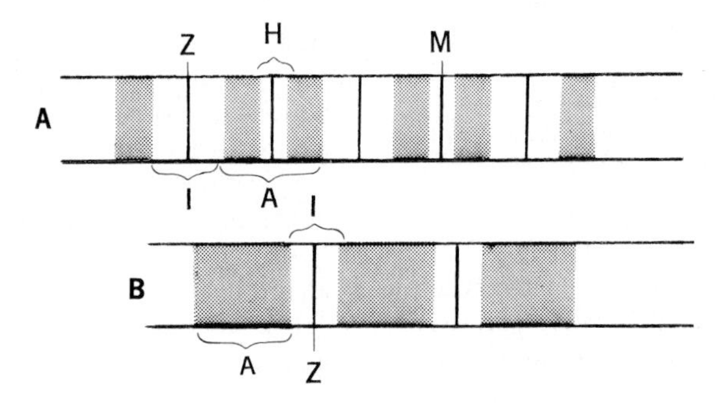

Fig. 3-3. Diagram of microscopic appearance of a myofibril in stretched state, **A,** and in the contracted state, **B.** Sarcomere extends from one Z line to the next. Sarcomeres are divided into dark A bands and light I bands. Z line transects I band, and H zone is lighter zone in midst of A band. H zone is bisected by a darker M line. On contraction, I band and H zone become narrower.

arranged in a hexagonal array about 450 Å apart. The thick filaments are birefringent and therefore responsible for the appearance of the A bands. The thin filaments, which are about 80 Å in diameter and about 1.0 μ long, extend from the Z line toward the center of the sarcomere through the I band and partway into the A band. The thin filaments are anchored into the transverse filaments forming the Z line.[68,112] At rest length the central portion of the A band, called the *H zone,* is devoid of thin filaments. The H zone is bisected by a dark line, the *M line*. The thick and thin filaments are made up of the proteins involved in the contractile process.

Cross-sectional electron micrographs show various patterns, depending on the location of the section (Fig. 3-4). Sections through the H zone show only thick filaments arranged in a hexagonal array. Sections through the I band show only thin filaments. Sections through the dense portion of the A band contain both thick and thin filaments. The thin filaments are interspersed between the thick filaments with remarkable geometric regularity. In mammalian muscle there are twice as many thin as thick filaments. Each thick filament is surrounded by six thin filaments. In some invertebrate muscles there are three or four times as many thin filaments as thick, yet the geometry of one thick filament surrounded by six thin filaments is preserved.

Each thick filament has two projections from opposite sides at about 143 Å intervals along its

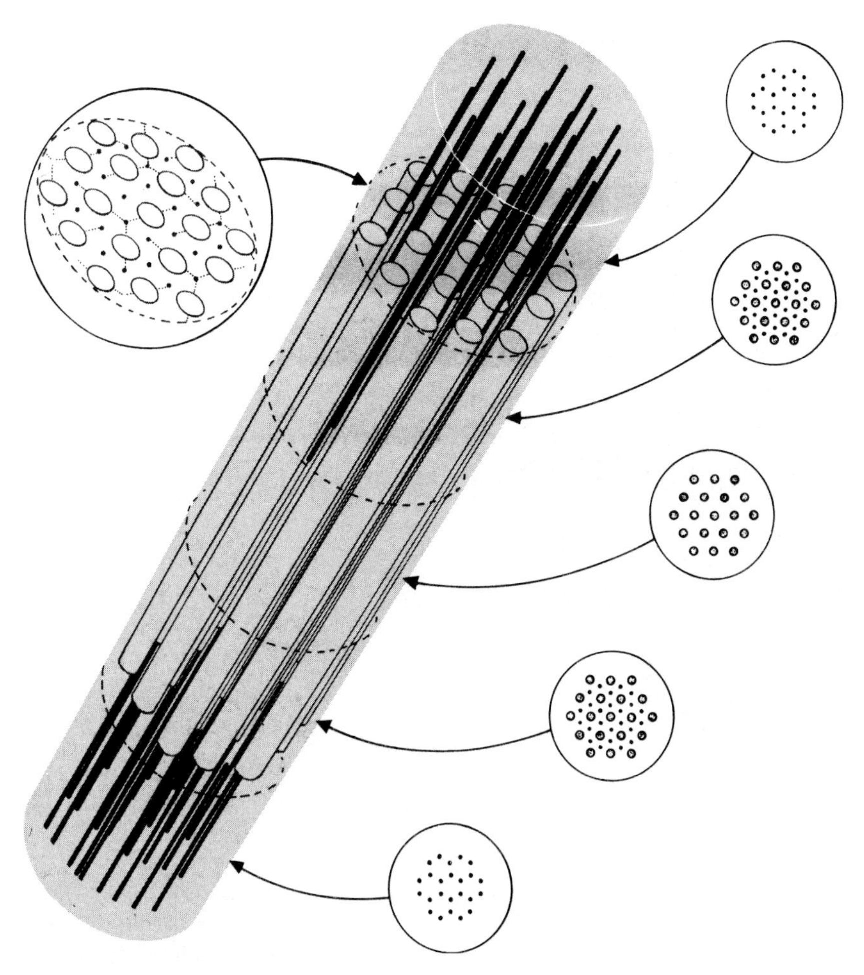

Fig. 3-4. Schematic representation of filament arrangement in skeletal muscle. Thin filaments extend from Z lines through I band and into A band as far as H zone. Thick filaments extend from one edge of A band to the other. Arrays in cross section at various levels of sarcomere are shown in insets. (From Dowben.[7])

length. These *cross bridges* extend from the thick filaments to touch one of the adjacent thin filaments (Fig. 3-5). In the array of thick filaments the levels of cross bridges are in register. The direction in which successive pairs of cross bridges project along the length of thick filaments undergoes a rotation of 120 degrees around the filament axis. Thus each thin filament receives cross bridges from the three adjacent thick filaments. At any given cross-sectional level the direction of the cross bridges is not uniform, but rather cross bridges from the next nearest thick filaments are rotated by ±120 degrees (Fig. 3-5). Thus cross bridges reaching one half of the thin filaments are coplanar and spaced at 430 Å intervals, whereas on the other half of thin filaments the cross bridges touch in sequence at 143 Å intervals in a helical fashion. The central 0.2 μ of the thick filaments is devoid of cross bridges.

During shortening the two sets of interdigitating filaments slide with respect to each other.[98,105] Thus during shortening the A band remains a constant width, but as the thin filaments move in toward the center of the sarcomere, the I bands and H zone become narrower (Fig. 3-3). To reiterate, the width of the A band remains constant whether a muscle fiber is relaxed, contracted, or stretched. On the other hand, the I band and H zone become narrower during shortening and wider during stretching. At *rest length,* the normal position of the muscle in the body, the thin filaments extend into the A band as far as there are cross bridges, but not into the central 0.2 μ region devoid of cross bridges. When a

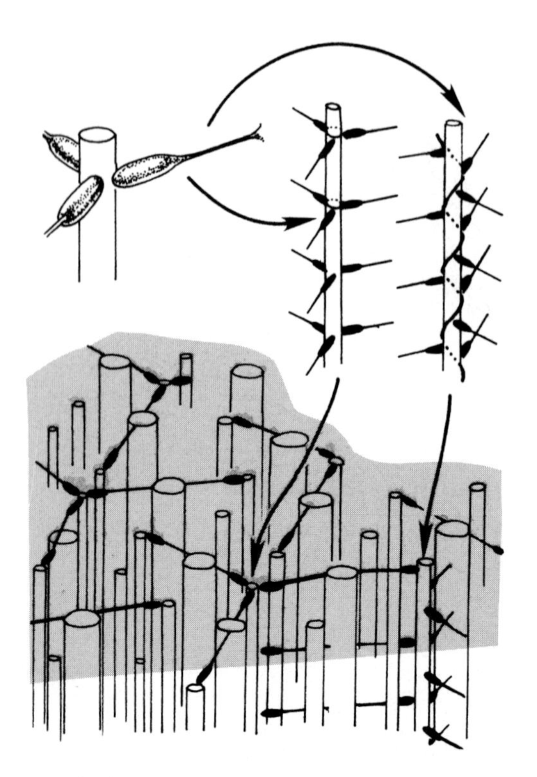

Fig. 3-5. Diagram of striated muscle showing arrangement of cross bridges. Each thin filament receives cross bridges from three adjacent thick filaments. On half of thin filaments, cross bridges are coplanar; on other half, cross bridges project serially and form a helix. (From Yu et al.[164])

muscle is made to shorten very markedly, the thin filaments arising from opposite Z lines may actually cross and overlap each other. With enormous shortening the A bands reach the Z lines, causing deformation of the ends of the thick filaments, giving rise to *contraction bands* at the Z lines. This process of muscle shortening involving progressive interdigitation of the two sets of filaments is frequently referred to as the *sliding filament mechanism of muscle contraction*.

It has been known since the mid-17th century that muscle maintains an almost exactly constant volume during shortening. In order to maintain a constant volume the lattice of filaments expands laterally during shortening, the filaments moving away from one another but maintaining geometric regularity.[34,59] An intact muscle fiber is required for the constant volume property.[123] The cross bridges readily accommodate to the expanding lattice, continuing to maintain contact with the thin filaments.

As to the other aspects of the ultrastructure, skeletal muscle has a well-developed endoplasmic reticulum, which in muscle is called the *sarcoplasmic reticulum*.[66] The sarcoplasmic reticulum forms an extensive hollow membranous system within the cytoplasm surrounding the myofibrils (Fig. 3-6). Periodically, there are branching invaginations of the sarcolemma called *T tubules* or transverse tubules.[69] The connection of the cavities of the T tubules to the extracellular fluid was first shown by soaking muscle fibers in a suspension of ferritin, a high molecular weight electron-opaque protein. Electron micrographs showed that the ferritin moved into the T tubule cavities. The sarcoplasmic reticulum bulges out on either side of the T tubules to form large *lateral cisternae*. The T tubule and two sets of lateral cisternae constitute a *triad*[68,133]; they are found at levels corresponding to the junction of the A and I bands in mammalian skeletal muscle; thus there are two triads per sarcomere. In heart muscle there is only one triad per sarcomere located at the Z lines. The triads, as will be explained subsequently, play an important role in the coupling of the depolarization of the sarcolemma with the development of the active, contractile state in the myofibrils.

Red and white muscle

Two types of muscle fibers are found in vertebrate skeletal muscle: red and white muscle fibers. The two types of muscle fibers are histochemically and functionally distinctive. Their names derive from their gross appearance, corresponding to dark and light meat in chicken and turkey. Many muscles are mixed, containing both types of fibers, which can be distinguished by various histochemical stains (Fig. 3-7).

Red muscle fibers respond to a stimulus with a

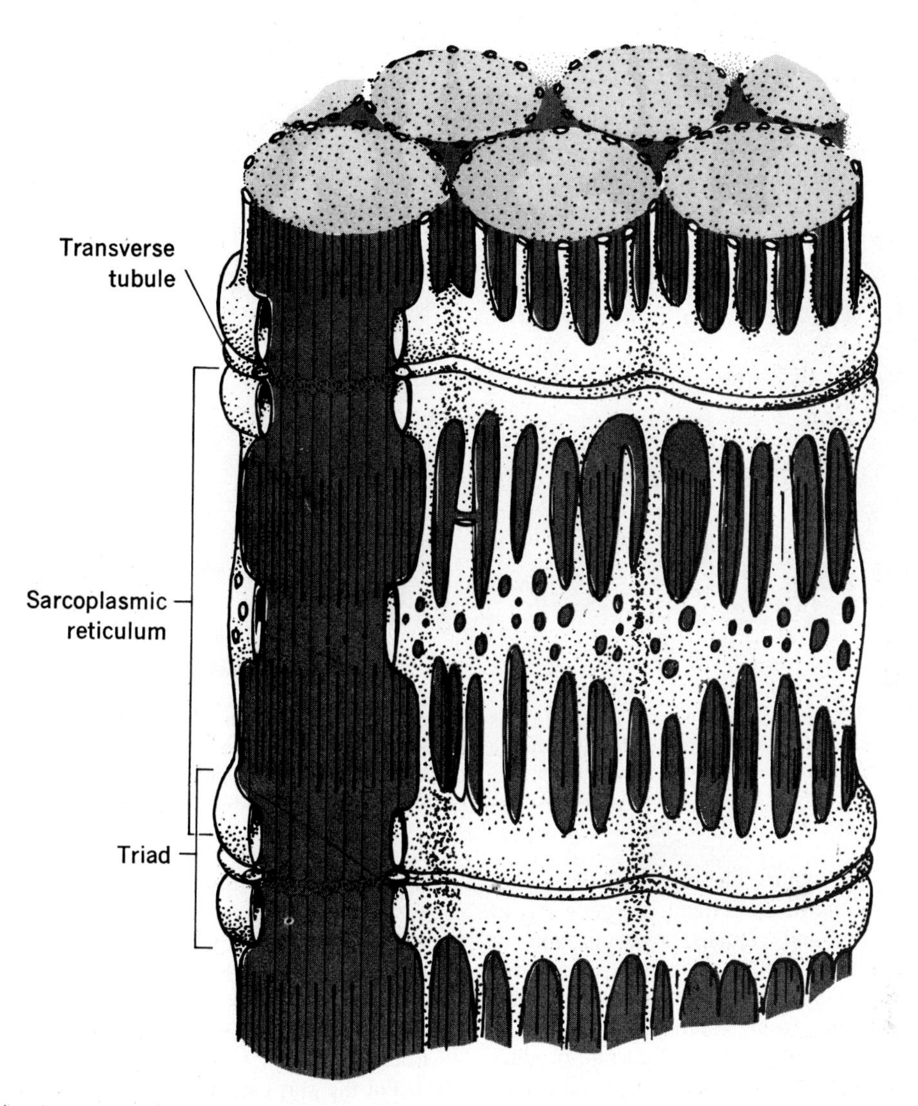

Fig. 3-6. Diagrammatic reconstruction of sarcoplasmic reticulum associated with several myofibrils taken from frog sartorius muscle. Note that transverse tubules are continuous and at level of Z lines. In mammalian muscle, triads are located at junction of A and I bands. (From Peachey.[133])

relatively slow twitch (maximum shortening velocity about 17 mm/sec) and therefore are sometimes called slow fibers, whereas white muscle fibers react to a stimulus with a rapid twitch (maximum shortening velocity about 42 mm/sec). Red muscle has a more extensive blood supply than white muscle. Red muscle fibers are able to sustain activity for long periods of time whereas white muscle fibers characteristically produce short bursts of great tension followed by the rapid onset of fatigue.

Whole red and white muscles differ in ATPase activity, and, indeed, the purified contractile protein myosin extracted from red and white muscle differs in ATPase activity, a finding associated with different myosin light chains (see subsequent discussion). White muscle and white muscle actomyosin show the greater ATPase activity.[38] The innervation of red and white muscle differs,[25] and, indeed, whether a given muscle is red or white results from trophic influences of the motor nerve (see subsequent discussion).

Slow (type I or red) muscle fibers are generally thinner and possess many sarcosomes (mitochondria) containing large amounts of respiratory enzymes, as well as copious quantities of the O_2-carrying protein myoglobin in the sarcoplasm and many lipid droplets. The numerous sarcosomes and high level of myoglobin give type I

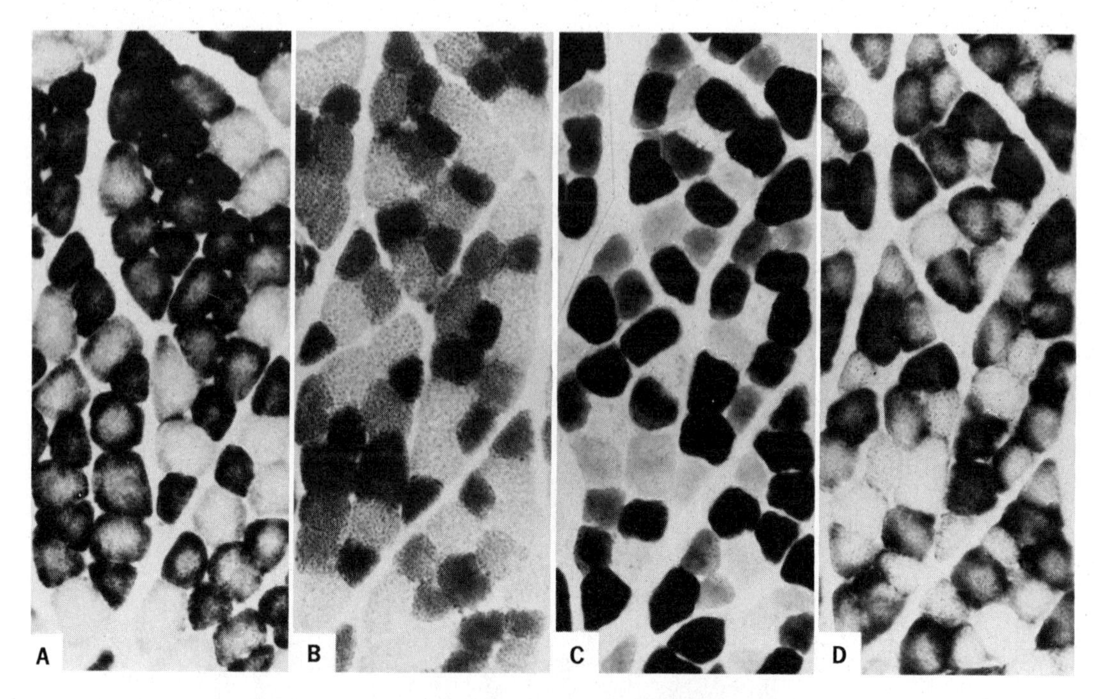

Fig. 3-7. Histochemical stains of serial cross sections of muscle fasciculus from rat plantaris muscle that contains both red and white muscle fibers. **A,** Stained for succinic dehydrogenase. **B,** Stained for esterase. Both of these stain red (type I) muscle fibers. **C,** Stained for phosphorylase. **D,** Stained for α-glycerophosphate dehydrogenase. Both of these stain white (type II) muscle fibers. (Courtesy F. C. A. Romanul.)

fibers their red color. Fast (type II or white) muscle fibers, on the other hand, are generally of larger diameter and contain large amounts of phosphorylase and glycolytic enzymes and large deposits of glycogen. Slow muscles derive energy predominantly from respiration, whereas in fast muscle fibers, glycolysis and lactate production are more prominent. The further subclassification of fiber types is beyond the scope of this chapter.

THE MOTOR UNIT

The Roman physician Galen appreciated that contraction of muscle requires an intact nerve supply. Anatomically the nerve and blood supply enters a muscle in a well-defined region known as the *neurovascular hilum*. The nerve contains both efferent and afferent fibers. The efferent fibers are myelinated axons of anterior horn spinal cord neurons. Every axon divides into 10 to 100 or more branches, each of which terminates on a single muscle fiber in a structure known as the *motor end-plate*,[25,51] which, together with neuromuscular transmission, will be discussed in detail in Chapter 5. A single anterior horn cell, its axon, the axonal branches, and the group of muscle fibers innervated constitute a *motor unit*. Motor units in small muscles, particularly those involved in fine movements, such as the extrinsic muscles of the eye or the laryngeal muscles, have few muscle fibers in a motor unit, whereas large muscles that perform gross movements, such as the thigh muscles, will have motor units containing as many as 500 muscle fibers.[65,96]

For the purposes of this chapter, it is sufficient to describe the motor end-plate as a complex structure composed partly of a proliferation of a motor nerve terminal and partly of correspondingly complex foldings of the muscle fiber sarcolemma, providing a large area of contact between the nerve and the sarcolemma.[51] The final ramifications of the axon terminals lie in grooves or troughs of the sarcolemma. Acetylcholine, the chemical transmitter of the nerve impulses, is liberated by the axon terminals and depolarizes the sarcolemma in the end-plate region. The end-plate is rich in the enzyme cholinesterase, which hydrolyzes acetylcholine.

The sequence of events leading to contraction in a motor unit is as follows. The motor nerve membrane depolarizes in response to a stimulus. The depolarization is propagated along a nerve fiber at a rate of about 75 m/sec. Like nerve, the muscle cell has a transmembrane potential of about −80 mV owing to selective ionic perme-

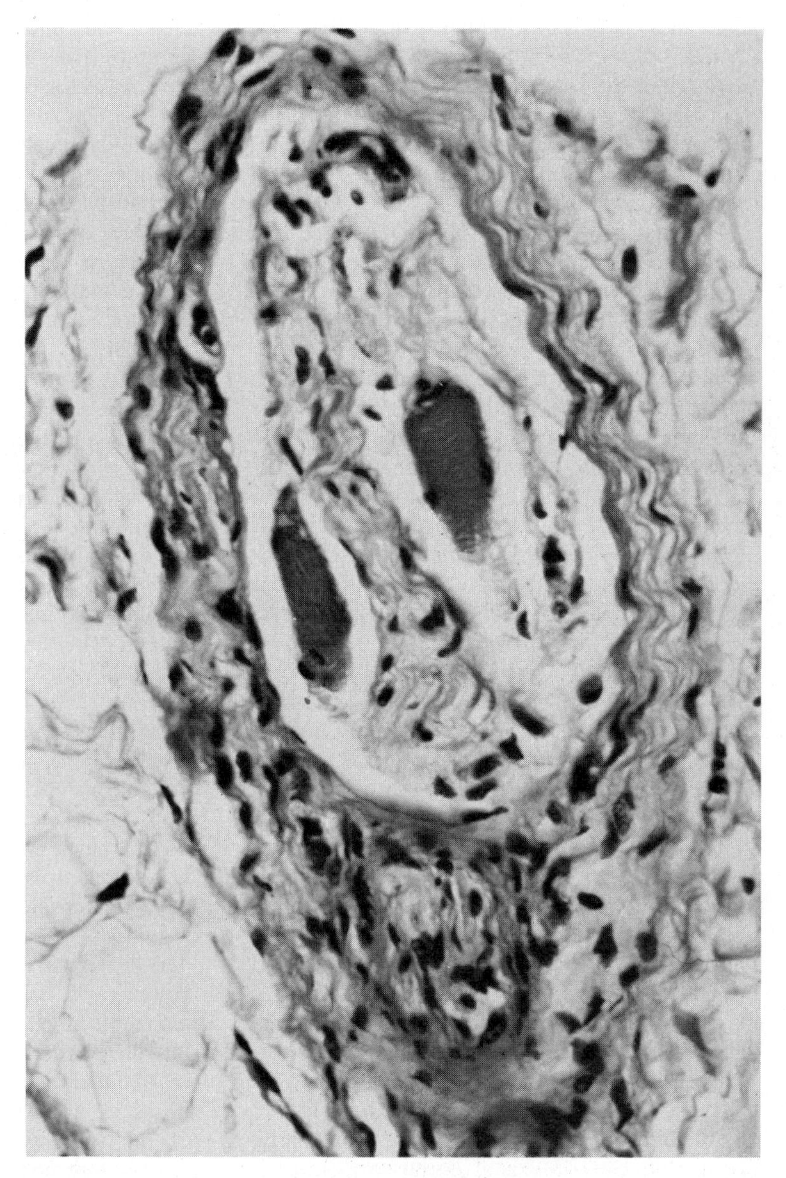

Fig. 3-8. Oblique section through human muscle spindle showing intrafusal muscle fibers surrounded by capsule that has nerve trunk embedded in lower portion. (Courtesy D. Vendrell.)

ability,[58,89] with the sarcoplasm negative to the exterior. The muscle transmembrane potential, like that of nerve, corresponds approximately to the potassium equilibrium potential arising from the high intracellular and low extracellular potassium concentrations (Chapter 1). Muscle membrane depolarizes in the same manner as nerve membrane. The chemical transmission of the impulse at the motor end-plate occupies about 0.3 msec. This is followed by depolarization of the motor end-plate and sarcolemma, which is then propagated toward the ends of the muscle fiber[43,64] at a velocity of about 3 m/sec. Thus several milliseconds elapse between stimulation of the motor nerve and depolarization of the sarcolemma. In amphibians (but not in mammals) there are some muscle fibers in which depolarization of the end-plate leads to a localized but not a propagated contraction.[115]

Muscle spindles

All muscles contain *muscle spindles,* long thin structures composed of several intrafusal muscle fibers with their own nerves and blood vessels surrounded by a connective tissue capsule (Fig. 3-8), through which they are connected to the tendons at either end of the muscle. Muscle spindles are primarily sensory structures that respond to muscle tension. Slow muscles contain more muscle spindles than fast muscles.

Two distinct types of fibers are found in muscle spindles[41]: nuclear bag fibers, one to three long muscle fibers of larger diameter containing a number of nuclei clustered in the equatorial region, and nuclear chain fibers, three to seven smaller, shorter muscle fibers in which the nuclei are distributed throughout the length of the fiber. Both types of intrafusal muscle fiber contained in the muscle spindle are cytologically and functionally distinct from the ordinary muscle fibers described previously.

The muscle fibers in the spindle have a double innervation. Although most muscle fibers are innervated by motor nerves of large diameter, about 12 μ, the intrafusal muscle fibers of the spindle are under independent motor control and are innervated mainly by thin γ fibers, about 3 μ diameter, with some slightly larger β axons. The spindle motor neurons not only stimulate spindle fibers dynamically, that is, when the whole muscle is contracting, but also statically, when the remainder of the muscle is at rest and relaxed.

The primary sensory receptor in a muscle spindle is a spiral structure surrounding both nuclear bag fibers and nuclear chain fibers and gives rise to large-diameter, fast-conducting α afferent (sensory) nerves that terminate in the spinal cord. The secondary sensory receptors in the spindle are also spirals, but these surround only nuclear chain fibers and give rise to smaller β afferent fibers. Both primary and secondary sensory receptors are sensitive to stretch, either by passive stretching of the intrafusal muscle fibers in the spindle by external forces or by active contraction of the intrafusal muscle fibers. The afferent impulses are important in the spinal stretch reflexes, as well as in the more complex CNS mechanisms for controlling movement and for maintaining posture and muscle tone, which are discussed in Chapters 20 through 25.

In addition to afferent nerve fibers arising from the muscle spindles, the nerve to a muscle will also contain afferent fibers arising from Golgi organs and Golgi-Mazzoni corpuscles, specialized tension receptors located mainly in the muscle tendons. Muscle also contains some afferent nerve fibers that terminate as free, bare nerve endings.

Trophic influence of the motor nerve

The motor nerve is not only responsible for stimulating the muscle it innervates by releasing acetylcholine from the efferent fibers at the motor end-plate, but it is also responsible for trophic stimuli, which determine many characteristics of the muscles. During embryonic development the differentiation of myoblasts, the development of myotubes, and the formation of myofibrils all occur without the influence of nerve fibers. Prior to innervation at 18 to 30 weeks of gestation, muscles in the human embryo are sensitive to acetylcholine over their entire surface. Embryonic muscles prior to innervation frequently depolarize spontaneously and contract. After innervation and development of the motor-end plate, acetylcholine sensitivity is restricted to the region of the end-plate, whereas the rest of the sarcolemma becomes quite insensitive to the local application of acetylcholine.[42] Thus innervation has the effect of markedly reducing the sensitivity of the sarcolemma to acetylcholine except at the motor end-plate.

Cutting the motor nerve in a mature animal sets off a sequence of changes in the muscle. The sensitivity of the sarcolemma to acetylcholine returns gradually over a period of a few weeks so that the application of acetylcholine anywhere on the muscle surface, not only at the motor end-plate, will cause depolarization of the sarcolemma. However, the end-plate is more sensitive to acetylcholine than other regions of the sarcolemma. The threshold required to trigger a depolarization gradually decreases.

About 5 to 15 days after denervation, spontaneous depolarizations of the muscle fiber called *fibrillations* appear. Fibrillations begin at the motor end-plate and are propagated in both directions to the ends of the muscle fibers.[106] Unlike normal nerve stimuli, which cause the muscle fibers to contract synchronously, fibrillations in adjacent muscle fibers are not synchronized or regularly spaced but occur randomly over a period of time at intervals of 2 to 10 seconds. The velocity of conduction of fibrillary potentials along the muscle fiber is significantly less than the rate of conduction of a normal depolarization in an innervated muscle. If the muscle is reinnervated, fibrillation and the increased sensitivity of the sarcolemma to acetylcholine disappear even before neuromuscular transmission is reestablished.[156] If the muscle is not reinnervated, the sensitivity of the sarcolemma gradually disappears, and, after a few months, fibrillation ceases.

After denervation, there is also a gradual loss of muscle bulk, a process called *atrophy*. Although the DNA content of muscle remains constant, the amount of protein and RNA progressively falls. Morphologic changes accompany the loss of bulk, first an apparent increase in the number of nuclei that move to the center of the now thinner muscle fibers and appear in clumps, followed by loss of cross striations and disintegration of the complex ultrastructural architecture. Gradually, muscle fibers are replaced by connective tissue and fat. Atrophy can be prevented, or at least minimized, by the direct electrical stimulation of a denervated muscle. Such direct electrical stimulation, however, does not affect fibrillation or the changes in sensitivity to acetylcholine.

Disuse of a muscle or prolonged immobilization, such as putting an injured limb in a cast, causes muscle atrophy but not the changes of irritability and not fibrillations.

When a muscle is made to contract greatly in excess of normal use, as in athletic training, it *hypertrophies;* it increases in size and weight and is capable of increased performance. In the mature person, work hypertrophy of muscle results mainly from an increased thickness of the muscle fibers rather than an increase in number. It is of interest that the muscle hypertrophy in certain diseases such as Duchenne muscular dystrophy demonstrates a trophic influence of the motor nerve because hypertrophy is found in weak muscles and disappears rapidly after denervation but not after immobilization. Muscle mass tends to be increased by androgens and related steroid hormones, accounting for the commonly known fact that males are more muscled than females.

Red and white muscles have different types of innervation.[42] A motor unit is composed of all red muscle fibers or all white but never a mixture, even in a mixed muscle. Red muscles are innervated by smaller nerve fibers with somewhat slower conduction velocities that end in grapelike clusters on the muscle fiber. White muscle fibers, on the other hand, are innervated by large nerve fibers with faster conduction velocities that terminate in plaquelike motor end-plates.[129] Whether a muscle fiber is red or white is determined by the motor nerve, but the mechanism of this trophic effect is not understood at the present time. If the two motor nerves to a red and white muscle are cut and reversed, so that on regeneration of the nerves the fast motor nerve innervates the red muscle and the slow motor nerve the white muscle, after a few months the muscles reverse their characteristics.[143] The former red muscle, now innervated by a fast motor nerve, becomes a white muscle, losing myoglobin and sarcosomes, increasing glycogen and glycolytic enzymes, and changing the ATPase activity of the myosin. The opposite changes take place in the former white cross-innervated muscle.

MECHANICS OF CONTRACTION

Time course of contraction

A consideration of the mechanics of muscle contraction will be clearer if we first note the time course of the sequence of events comprising a contraction. A skeletal muscle fiber responds to a sufficiently large direct stimulus or a sufficiently large stimulus to the motor nerve by a *twitch,* a brief period of contraction followed by relaxation. As noted previously, stimulation of the motor nerve is followed by a latent period of about 1 to 2 msec before depolarization of the sarcolemma begins. The depolarization of the sarcolemma reaches a peak about 2 msec after nerve stimulation in the end-plate region of the muscle fiber and a few milliseconds later farther away from the end-plates. The sarcolemma gradually repolarizes and has returned to the resting state after 5 to 10 msec, long before muscle tension reaches its peak.

Mechanical tension does not begin until well after the peak of sarcolemmal depolarization has passed. The time before maximum tension is reached (i.e., the latent period) depends on whether the muscle is slow, fast, or mixed (Fig. 3-9) and on other factors, such as the weight or load. Just before tension appears the muscle relaxes slightly; this transient decrease in tension is called the *latency relaxation*. Latency relaxation is thought to represent a molecular rearrangement prior to the onset of active contraction. In rat extensor digitorum longus, a fast white muscle, maximum tension is reached in about 14 msec at 35° C. On the other hand, in rat soleus, a slow red muscle, maximum tension is not reached until after about 40 msec at 35° C. The development of peak tension in mixed muscles is intermediate. In an actual contraction of muscle the time course depends, of course, on the combined compliance and inertia of the muscle itself and on the load.

The rates of tension development and relaxation are very temperature dependent. In frog muscle, which functions perfectly when cooled to 0° C, the rates of tension development and relaxation are slowed by a factor of almost 3 for every 10° C drop in temperature. Thus the time course of a twitch in frog muscle is 10 times longer at 0° than at 35° C.

The size of a single twitch is the same for a given set of conditions. However, if conditions change, the muscle fiber may respond with a twitch of a different size. Fatigue, for example, produces smaller twitches, whereas certain drugs will result in larger twitches. Stimulation of a muscle fiber shortly after a previous twitch has run its course will produce a second twitch that is larger than the first, a process called *facilitation*.

If a muscle fiber is stimulated a second time before the tension of the first twitch has completely decayed (this is possible because the sarcolemma has repolarized long before maximum tension develops), the second twitch will be grafted onto the first, resulting in significantly greater peak tension, and so on with a third or fourth stimulus. If a muscle fiber is stimulated repeatedly at a fast frequency, there results a smooth sustained contraction called a *tetanus* (Fig. 3-9). Individual twitches cannot be detected during a tetanus. The tension of a tetanus is much greater than the maximal tension of a single twitch, and it is maintained at a constant level as long as stimulation is continued or until the muscle fiber becomes fatigued. Twitches in slow red muscles fuse into a tetanus when the muscle is stimulated 20 times per second, whereas fast white muscles require 60 to 100 stimuli per second for fusion.

One of the important characteristics of cardiac muscle is that, except under the most unusual circumstances, it cannot be tetanized.

Elastic elements

Resting, relaxed muscle is elastic; it resists stretching beyond rest length. Rest length is the length of the muscle in the body at rest position. Resting muscle displays rubberlike thermoelastic properties, shortening when warmed and elongating when cooled. The elasticity of whole muscle that gives rise to the resistance to stretching is due to the meshwork of connective tissue surrounding the muscle fibers and to the sarcolemma and sarcoplasmic reticulum membranes. Thus when a muscle contracts, it must first stretch the connective tissue elastic element and generate a tension in it before tension appears in the tendons at the ends of the muscle. The connective tissue surrounding the muscle fibers acts as an elastic element in parallel with the contractile element.

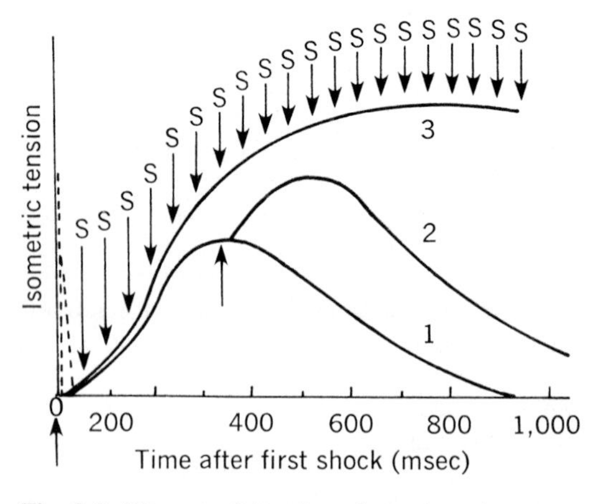

Fig. 3-9. Diagram of duration of muscle action potential (dashed line) and time course of tension development during single twitch *(1)*, during two twitches in which tension is summated *(2)*, and during tetanus *(3)*. *S* signifies stimulus.

The development of a force or tension by the *contractile elements* in the myofibrils is a characteristic property of a stimulated muscle fiber. The tension develops regardless of whether shortening takes place or whether the muscle is unable or prevented from shortening. As noted previously, the tension developed by the individual contractile elements is summed and transmitted by the connective tissue and tendons to produce the effective tension developed by the whole muscle. The contractile elements do not relax actively. To the extent that relaxation occurs spontaneously, it is due to the elasticity inherent in the connective tissue surrounding the muscle fibers. Muscles with more connective tissue are not only more resistant to stretching, but also show a greater tendency to return to rest length after contraction has terminated.

When a muscle fiber is stimulated, it stiffens markedly before any tension is developed, and the stiffness persists during the active or contracting state. The stiffness of muscle fibers during the active state is largely intrinsic because it is greater if the muscle is permitted to shorten and less when it is stretched. Thus the muscle behaves as though there were an elastic element in series with the contractile elements. The thermoelastic properties of muscle actively contracting are springlike, elongating when warmed and shortening when cooled.

The intrinsic stiffness is thought to arise mainly from the cross bridge attachments to the thin filaments during contraction. It can be quantified by forcibly stretching muscle fibers during the active state. The lengthening of a muscle during the active contractile state when confronted with a load greater than the maximum tension it can develop corresponds to many real life situations. For loads between the maximal tension and twice maximal tension the muscle yields slowly. For loads above twice maximal tension the muscle yields more rapidly and some internal structural damage occurs; when it is stretched to about twice rest length, the fibers actually begin to rupture.

The mechanical properties of muscle can therefore be represented (Fig. 3-10) as consisting of a *contractile element* (CE), a *parallel elastic element* (PEE), and a *series elastic element* (SEE). The elastic elements do not obey Hooke's law, that is, the resistance to stretching is not proportional to the amount of stretching.[107,108] A look at the curve of tension required to passively stretch a muscle (Fig. 3-11) shows that it becomes increasingly steeper the more a muscle is stretched. Similarly, when a muscle is stretched during contraction, it yields slowly at first and then more rapidly, but the curve is not linear.

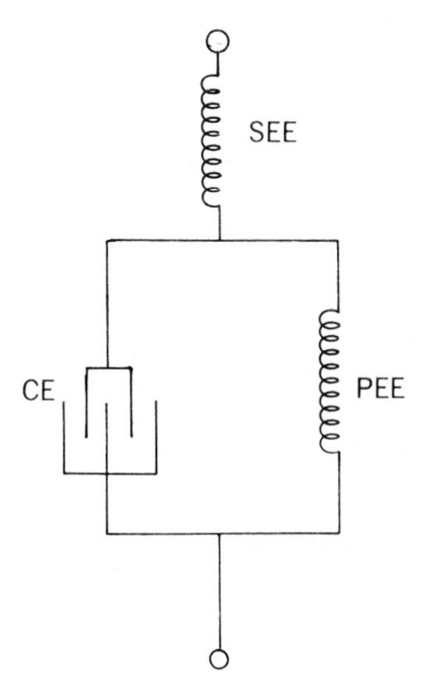

Fig. 3-10. Schematic diagram of mechanical components of muscle to explain gross viscoelastic properties. *CE,* Contractile element; *PEE,* parallel elastic element; *SEE,* series elastic element.

Although this formulation is overly simplistic and does not account for all observations of muscle contraction (e.g., when the load is varied during the midst of shortening[47,99,100,137]), it provides a very useful framework for understanding muscle mechanics.

Isometric contraction: length-tension curve

In order to study the mechanics of muscle contraction, an experimental arrangement such as that depicted in Fig. 3-12 is used. The number of variables is thereby reduced, and analysis is made easier. Muscle contraction in the whole animal is much more complex, of course, but the information obtained under these greatly simplified experimental conditions gives us very valuable insights into muscle function. The motor nerve is ignored and the muscle is stimulated directly by a series of electrodes along its entire length so that all CEs will be activated simultaneously. The muscle is anchored at one end, and, by means of a wire attached to the free tendon, it is connected to a lever system that measures muscle length and tension.

For one type of experiment the lever is immobilized in various positions, thereby fixing the length of the muscle. A contraction under circumstances such that muscle length is held constant is called an *isometric* contraction. In another important type of experiment the muscle is allowed to shorten while lifting a weight at-

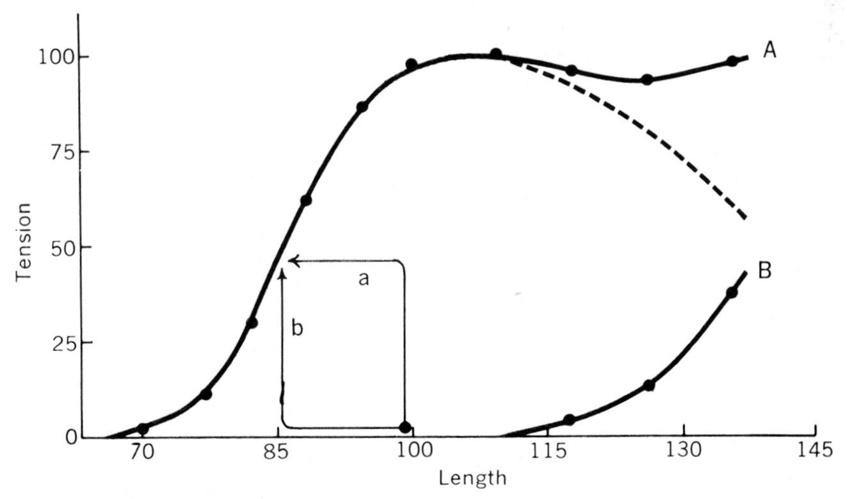

Fig. 3-11. Relation between length and tension. Point 100 is rest length and maximum tension. Monotonic increasing curve at lower right represents effect of passive stretch on tension. Upper solid curve is obtained from system as a whole. Broken line represents behavior of contractile elements when passive stretch curve is subtracted from upper curve. Set of thin-lined arrows beginning at zero tension and rest length indicates that, apart from effects of plasticity and whether muscle contracts isotonically, *a,* or isometrically, *b,* final relation between tension and length should be the same. (Courtesy K. L. Zierler.)

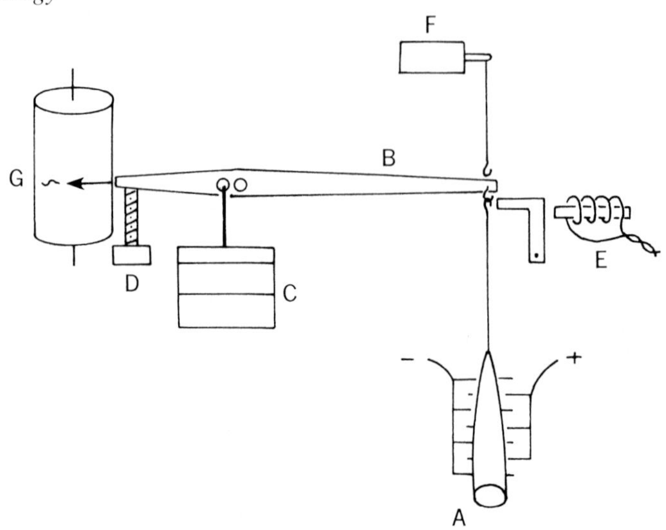

Fig. 3-12. Diagram of apparatus used for studying mechanics of muscle contraction. Muscle is stimulated by series of electrodes, *A*. Tendon is attached to a lever system, *B*, on which may be placed various weights, *C*. Lever system may be immobilized by adjustable stop, *D*, or quickly released by means of a solenoid-activated stop, *E*. Tensions are measured by means of a strain gauge, *F*, and movement is recorded by means of a pen attached to lever, *G*.

tached to the lever arm. This type of contraction is called *isotonic* because the tension developed by the muscle is determined by the weight and remains constant during shortening.

Referring to the length-tension curve depicted in Fig. 3-11, curve *B* shows the tension required to overcome the resistance of the PEE and to stretch the muscle beyond rest length. In an isometric contraction the variable measured is the maximum tension produced, which depends on whether the muscle is held at normal rest length, is permitted to shorten before the lever hits the mechanical stop, or is first stretched and fixed at a length greater than rest length. When the muscle is stretched, the total tension developed is the sum of the active tension produced by the CEs plus the passive tension of the stretched elastic elements. The net active tension is obtained by subtracting curve B from curve A, resulting in the dashed curve. Note that maximum tension is developed when the muscle is fixed at approximately rest length.

Connective tissue limits the extent to which a whole muscle can shorten or be stretched to 15% or sometimes as much as 30% of its rest length. This limitation can be overcome by studying tension developed by single muscle fibers from which all connective tissue has been dissected. Optical diffraction studies using a laser light source have shown that not all sarcomeres contract uniformly; those in the center of the muscle tend to shorten more than those at the ends.[39] An elegant electromechanical feedback system is used for experiments with single muscle fibers to maintain the sarcomeres in the central portion of the muscle fiber at constant length during the contraction. The tension developed can be compared to the position of the thick and thin filaments in the sarcomeres by fixing the muscle in situ, taking great care to avoid artifacts resulting from shrinkage and then obtaining and examining electron micrographs. Another method of measuring sarcomere length involves the analysis of optical diffraction patterns using a laser light source.

The results of such a series of isometric contractions in *single* muscle fibers[47,72,77] are shown in the length-tension curve of Fig. 3-13 and can be summarized as follows. Little or no tension is developed if the muscle fiber is stretched so much that there is no overlap of the thin and thick filaments (point A). For lesser degrees of stretching the maximum tension developed at various constant lengths is proportional to the amount of overlap until all cross bridges extending from the thick filaments are covered by thin filaments (point B). It can be shown that the activation of single muscle fibers by Ca^{2+} (see subsequent discussion) depends on the degree of overlap.[58] For somewhat shorter sarcomere lengths, as the thin filaments extend into the central area of the thick filaments devoid of cross bridges, the maximum tension remains constant (to point C). For still shorter lengths the tension

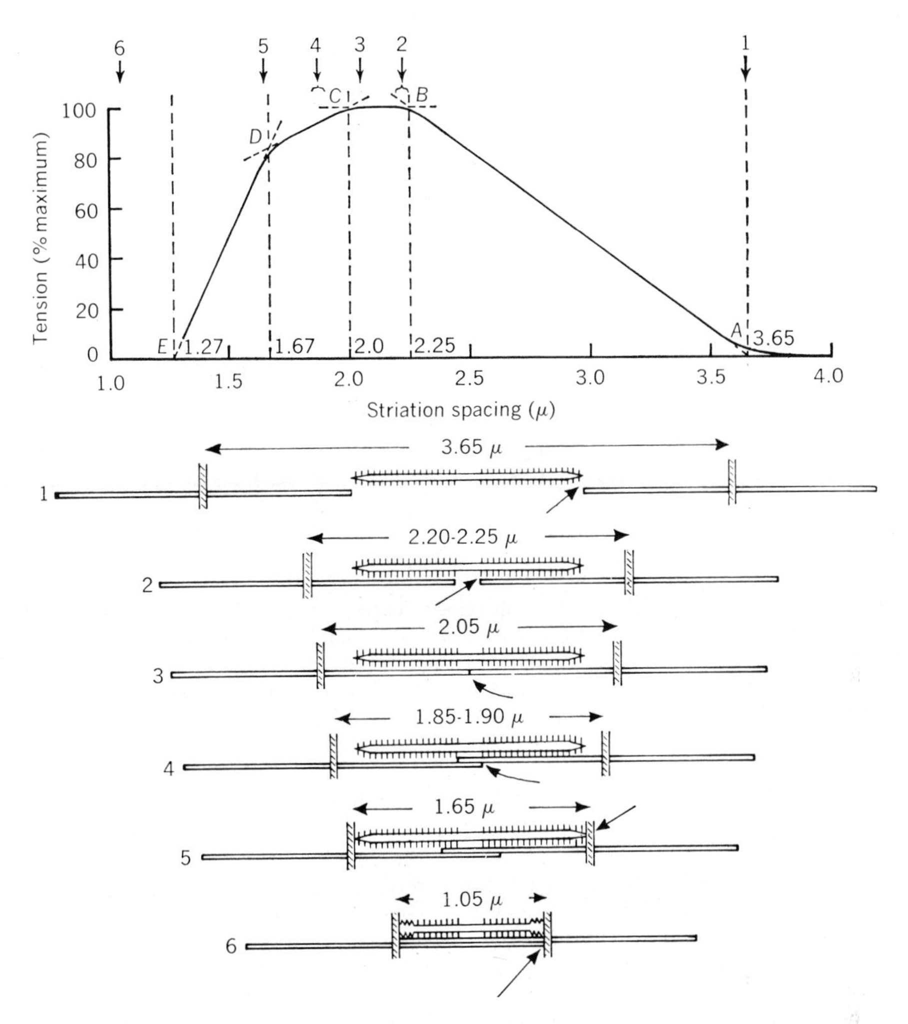

Fig. 3-13. Tension developed by single muscle fiber as function of length (upper part) compared to degree of overlap of filaments in sarcomere (lower part). From *A*, where thin and thick filaments just overlap to *B*, where overlap is sufficient to cover all cross bridges, tension developed is proportional to degree of overlap. The plateau corresponds to thin filaments overlapping the central zone of thick filaments devoid of cross bridges. Shortening greater than overlap of opposing thin filaments *C* is accompanied by a fall in tension. (From Gordon et al.[74])

falls as thin filaments from opposite Z lines overlap (to point D), probably because of interference with cross-bridge attachment to thin filaments. For very great shortening (beyond point D), tension falls off very rapidly, possibly because the rigidity of the thick filaments mechanically limits the development of tension and possibly because of failure of the activation mechanism.

Isotonic contraction: force-velocity curve

If a muscle at rest length is attached to a load and stimulated, it will shorten and lift the load. Shortening will continue until the load equals the maximum tension the muscle can generate, as found on the length-tension diagram.[28] The time course of such isotonic contractions will depend on the magnitude of the load (Fig. 3-14). After stimulation there is an interval of some milliseconds before shortening begins. The first part of this latent period is the absolute latent period before tension is generated in the CEs. Afterward, there is a relative latent period while tension develops and the elastic elements are stretched to a tension equal to that of the load before shortening can begin. The heavier the load, the longer the latent period. When shortening begins, the velocity is less for heavier loads.

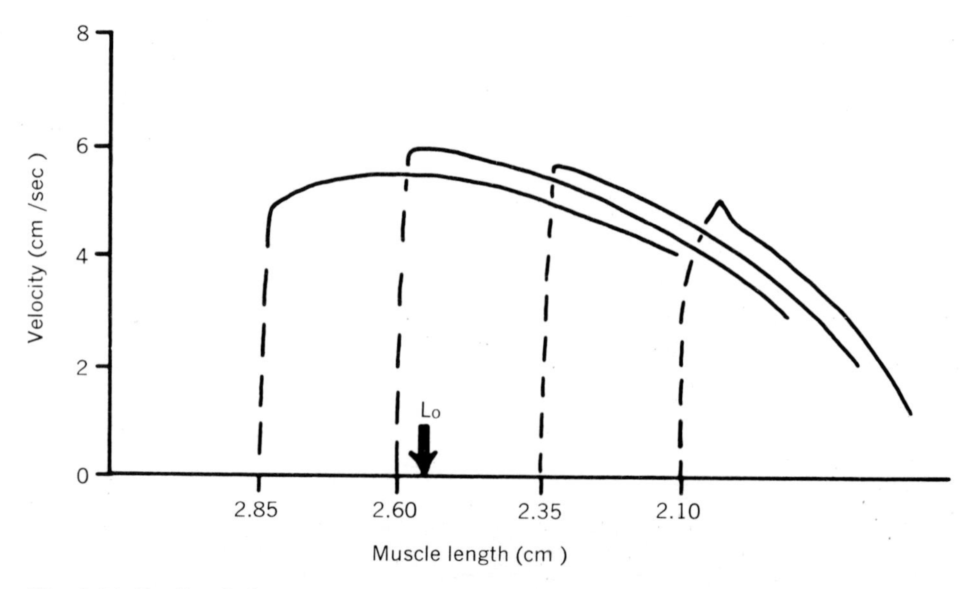

Fig. 3-14. Family of phase-plane or length-velocity curves obtained from gracilis muscle of rat showing effect of different initial lengths at same load. Initial length indicated on ordinate scale. L_0, shown by arrow, was 2.55 cm. Initial lengths corresponded to approximately 1.12, 1.02, 0.92, and 0.82 × L_0. Load was 6.94 gm, or 328 gm/cm² of muscle cross section. (Traced from unpublished photographic records obtained by A. S. Bahler.)

Also, the total amount of shortening that takes place is less for heavier loads.

Let us now look at the course of velocity of shortening (shortening speed) as shortening takes place (Fig. 3-14). Initially, as shortening begins, the velocity is approximately constant but falls off as shortening proceeds. The decrease in velocity is greater the heavier the load. If the muscle is first stretched so that shortening begins from an initial length greater than rest length, somewhat smaller velocities are achieved for comparable loads. Lower velocities are also observed for initial starting lengths shorter than rest length.[67] Using whole muscle again limits the experiments to 15% to 30% of rest length because of the connective tissue.

Isotonic contractions have been studied using single muscle fibers. Of course, markedly stretched fibers cannot lift very heavy loads, that is, they cannot lift loads heavier than the tension developed in an isometric contraction. For lightly loaded single muscle fibers starting from markedly stretched initial lengths the velocity of shortening is remarkably constant for most of the extent of shortening. For heavier loads the velocity of shortening is less and the velocity falls off more rapidly and sooner as shortening progresses.

If we plot the initial velocity of shortening developed for various loads in a series of isotonic contractions, we obtain a hyperbolic curve (Fig. 3-15). It should be emphasized that each point on the curve represents a single experiment. This *force-velocity curve* fits a relation empirically derived by A. V. Hill:

$$(P + a)v = (P_0 + P)b \qquad (1)$$

where P is the tension developed in an isotonic contraction, P_0 the maximum tension the muscle can develop in an isometric contraction, v the velocity of shortening, and a and b constants. The constant b depends on the intrinsic speed of the muscle and is numerically equal to about one fourth the maximum speed of shortening of an unloaded muscle. There are distinctive values of b for slow, fast, cardiac, and smooth muscle. The constant a has the dimensions of a force and is related to the coefficient of the heat of shortening (α), discussed subsequently.

The force-velocity curve depicted in Fig. 3-15 was obtained from a series of isotonic contractions starting at rest length. A family of force-velocity curves can be obtained by starting several series of isotonic contractions at lengths shorter or longer than rest length.[36,124] In these curves the velocity for a given load will be less the farther removed the initial length is from rest length, either shorter or longer. Such a family of force-velocity curves can be plotted as a three-dimensional surface (Fig. 3-16).

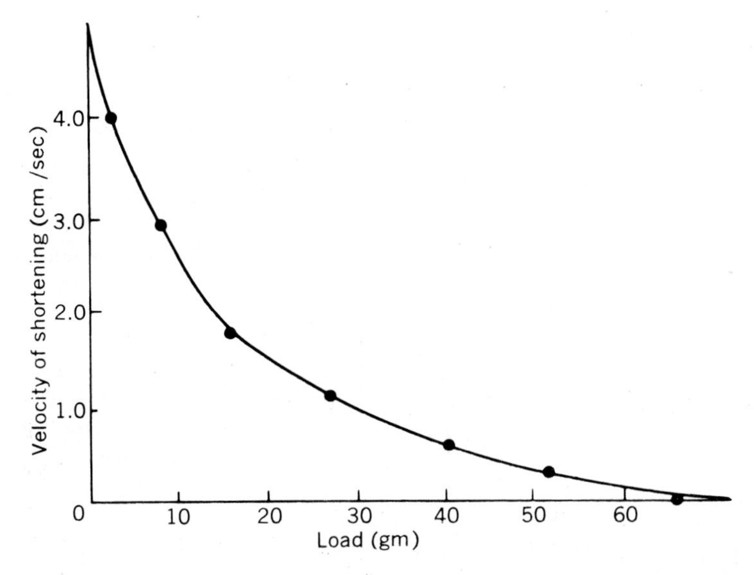

Fig. 3-15. Maximum velocity of shortening as function of load or force. (Redrawn from Hill.[80])

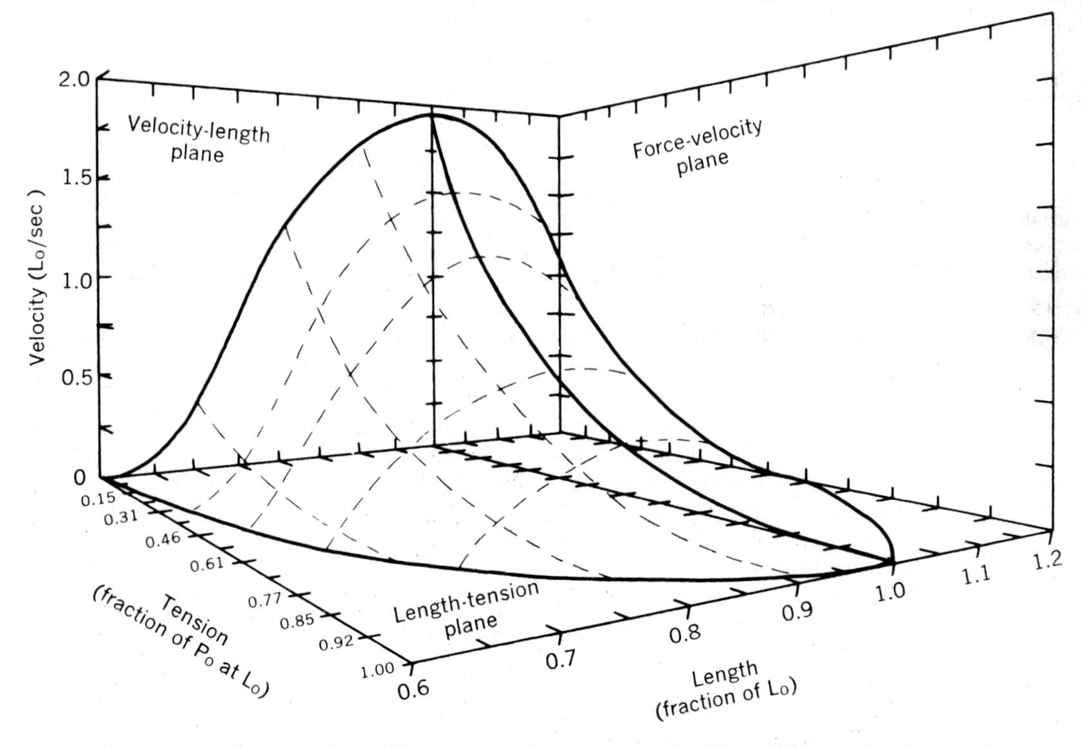

Fig. 3-16. Semischematic three-dimensional representation of interrelationship between length, tension, and velocity. Based on data from gracilis muscle of rat at approximately 16° C. Isometric length-tension diagram forms base of figure in length-tension plane. Force-velocity curve at initial length L_0 is given as heavy continuous line with its own velocity-length axes. Foot on velocity-length curve at short lengths and small loads is extrapolated beyond experimental data in order to meet isometric length-tension diagram. Other force-velocity and velocity-length relations are shown by broken lines to suggest curvature of surface. There is a family of such three-dimensional figures, depending on available chemical energy for contraction. (Courtesy K. L. Zierler.)

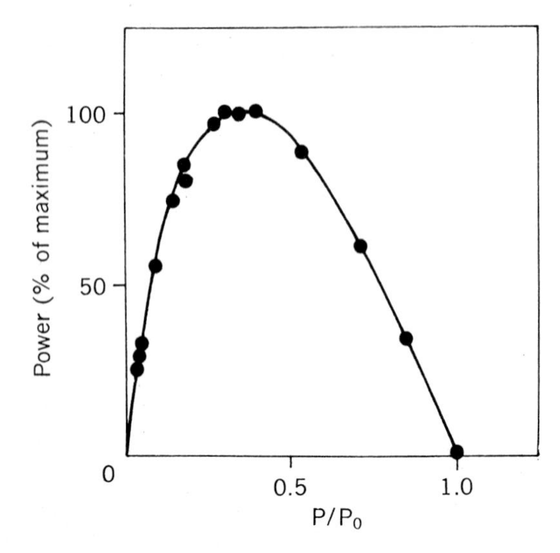

Fig. 3-17. Power output of frog sartorius muscle at 0° C as percent of maximum vs load expressed as fraction of maximum for isotonic contractions.

Power of shortening

The power output is equal to the product of force times velocity and is an index of mechanical efficiency.[43,116] For each point on the force-velocity curve representing a differently loaded isotonic contraction, the power of shortening can be calculated and plotted to give a curve like that of Fig. 3-17. For most muscles, power output reaches a maximum at about one third maximum tension. Thus a muscle operates most efficiently when it lifts a load corresponding to about one third maximum. For very heavy loads the work a muscle can perform and the efficiency with which it is performed are further limited by several additional factors: (1) the onset of shortening is delayed owing to lengthening of the latent period and (2) the distance through which shortening can take place is curtailed.

CONTRACTILE PROTEINS

The contractile or myofibrillar proteins, which make up the myofibrillar filaments, are responsible for contraction. The contractile proteins taken together are the second most abundant proteins in the body after collagen; they constitute about 60% of muscle proteins. Under physiologic conditions the contractile proteins in the thick and thin filaments are insoluble. They can be dissolved by strong salt solutions, such as 0.6 M KCl; their dissolution is accompanied by the disruption and disappearance of the filaments as seen on electron micrographs. The major contractile proteins and selected properties are listed in Table 3-1.

Thick filament proteins

The thick filaments are composed mainly of *myosin,* a threadlike molecule about 20Å in diameter and 1,600 Å long, with a bifurcated globular head at one end 50 Å in diameter and 210 Å long (Fig. 3-18). The myosin molecule has two "hinge" regions (regions of relatively great flexibility), one about 800 Å from the tail and one immediately behind the head.

Myosin from mammalian muscle has divalent cation-stimulated ATPase activity.[61] Myosin-like proteins isolated from other contractile systems show hydrolytic enzyme activity corresponding to the energy-rich phosphate (usually ATP or guanosine triphosphate [GTP]) used as the immediate energy source for contraction. The energy made available from ATP hydrolysis is used to perform mechanical work. Pure mammalian myosin in solution shows ATPase activity in the presence of 6 to 10 mM Ca^{2+}. These are nonphysiologic conditions because the concentration of Ca^{2+} in the sarcoplasm is ordinarily less than 10^{-6} M. In living muscle, myosin activity requires the presence of actin and about 3 mM Mg^{2+} (see subsequent discussion). The ATPase activity of red muscle myosin is about one half the ATPase activity of white muscle. Myosin ATPase activity is very temperature dependent, decreasing by a factor of about three for every 10° C drop in temperature. The ATPase activity of myosin from a poikilothermic animal like frog whose muscles can contract at 0° C is at that temperature about one tenth the activity at room temperature, 22° C.

Pure myosin is soluble in 0.3M phosphate buffer at neutral pH. It has a tendency to aggregate into long threads as the salt concentration or the pH of the solution is lowered. At first, two molecules come together tail to tail with the heads pointing in opposite directions. Gradually more molecules attach at the head ends and the thread grows in both directions.[103] It is believed that the thick filaments are formed in this manner. There are about 200 myosin molecules per thick filament. The body of the thick filament contains the tail portion of myosin molecules up to the first hinge region, whereas the remainder of the tail and the myosin head project outward to form the cross bridges (Fig. 3-19). The thick filaments also contain some minor proteins —one is the C protein and another the group of M line proteins—that are now the subject of active research; efforts are being made to purify them and to discern their function.

The myosin molecule has a molecular weight of about 460,000 daltons. It is composed of two

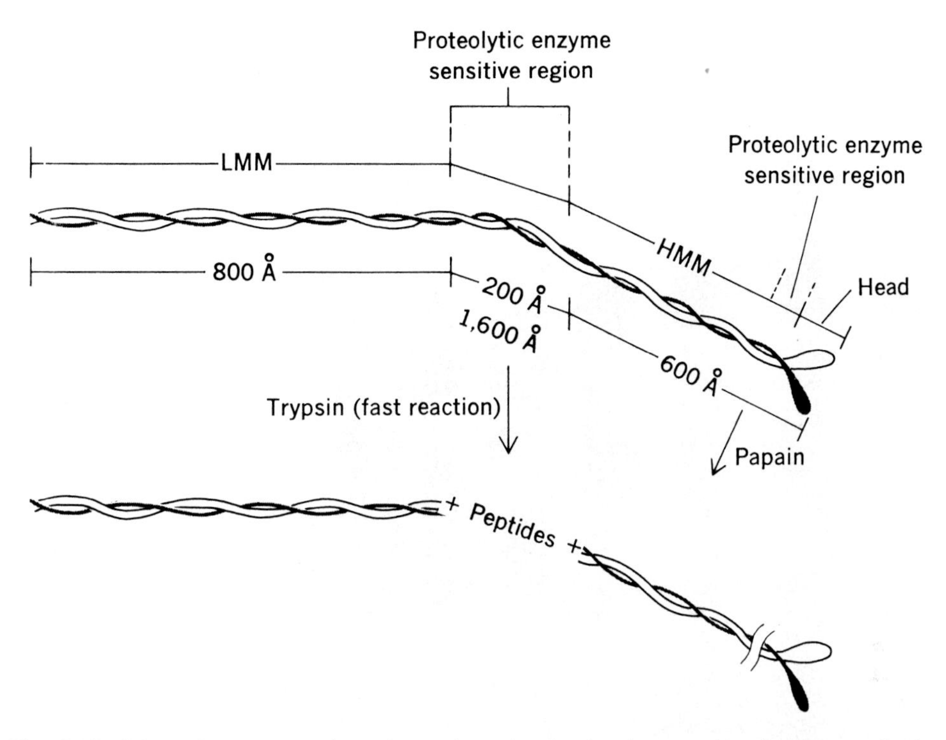

Fig. 3-18. Schematic representation of myosin molecule showing two identical large subunits wound in a superhelix. Small subunits are located in globular head. Partial proteolysis by trypsin cleaves molecule at hinge region into heavy meromyosin (HMM) portion and a light meromyosin (LMM) portion, each containing parts of original large subunits.

Table 3-1. Physical properties of the major contractile proteins

Protein	Molecular weight	Dimensions	Solubility
Myosin	460,000	20 Å diameter × 1600 Å 50 Å diameter × 210 Å (head)	Weak salt
Light meromyosin	150,000	20 Å diameter × 900 Å	Weak salt
Heavy meromyosin	300,000		Water
Subfragment 1	115,000	30 × 45 × 150 Å	
A-1 light chain	21,000		
A-2 light chain	16,000		
DTNB light chain	18,000		
G actin ATP	42,290	55 × 35 × 50 Å	Water
F actin		80 Å diameter × 1.2 μ	Insoluble
Actomyosin			Strong salt
Tropomyosin	65,516	20 Å diameter × 400 Å	Strong salt
Troponin complex	69,047		Weak salt
Troponin C	17,850	17 Å radius*	
Troponin T	30,503	21 Å radius*	
Troponin I	20,694	18 Å radius*	

*Radius calculated for a perfect sphere, assuming $\overline{V}$ = 0.73 ml/gm.

identical *large subunits*[8] of about 200,000 daltons each, which run the entire length of the molecule and are intertwined in a right-handed superhelix in the tail portion, and four *small subunits,* located in the heads only. Two of the small subunits are usually identical. The two identical small subunits have molecular weights of approximately 18,000 daltons, and the two others weigh about 25,000 daltons. The two 25,000-dalton light subunits are required for ATPase activity.

Different small subunits are found in myosin purified from embryonic, slow, fast, cardiac,

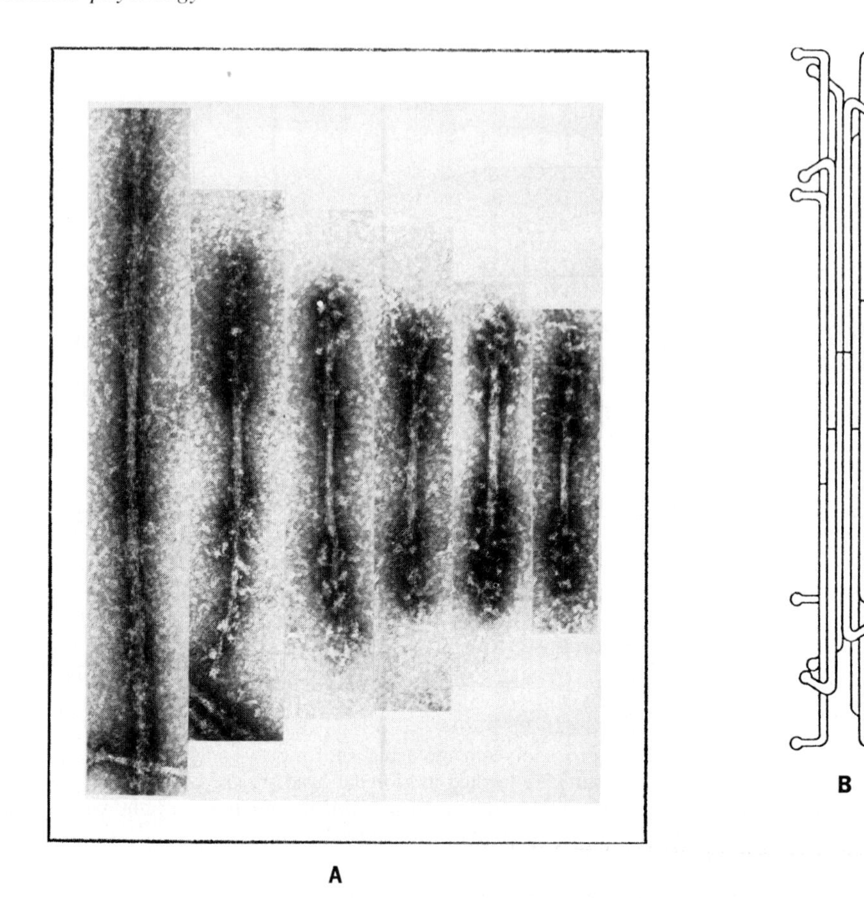

Fig. 3-19. Arrangement of myosin molecules in thick filaments. **A,** Electron micrograph of aggregation of myosin molecules into threads with projections near ends and thinner region in middle. **B,** Diagram showing arrangement of molecules with tails oriented toward center of thread and heads projecting toward ends. (**A** courtesy H. E. Huxley; negative stain with phosphotungstic acid.)

or smooth muscle; the different small subunits are believed to determine the different ATPase properties of these various types of muscle (see previous discussion). Although the correlation is not perfect, fast muscles have very high myosin ATPase activity, and slower muscles have lower myosin ATPase activity.

Myosin can be cleaved at the two hinge regions to form fragments. These fragments of myosin have been very useful in laboratory studies of the properties of myosin and the mechanism of contraction. Myosin can be cleaved at the first hinge region by limited proteolysis (i.e., brief exposure and digestion under suboptimal conditions for enzyme activity) using the proteolytic enzymes trypsin or chymotrypsin. The myosin molecule is broken into two large fragments called *heavy meromyosin* (HMM) and *light meromyosin* (LMM). LMM is the tail portion of the myosin molecule; it bestows the solubility properties on myosin. HMM is the cross bridge and head portion of the myosin molecule. HMM can be further cleaved at the second hinge region just behind the myosin head by limited proteolysis, using the proteolytic enzyme papain, into the two parts of the head portion called *subfragment 1* (S-1), and *subfragment 2* (S-2), which is the cross bridge portion of the myosin molecule, and possesses internal flexibility.[125] Each S-1 (two from each myosin molecule) has one ATPase catalytic site and one actin-binding site. Hydrolysis of ATP at these sites provides the energy for mechanical work. Whereas LMM (the tail portion) forms filamentous aggregates, the S-2 (cross-bridge) portion does not aggregate.

Thin filament proteins

Actin, tropomyosin, and *troponin* are the major contractile proteins that make up the thin filaments. Actin is a 42,000-dalton protein containing one nucleotide (ADP or ATP) and one divalent cation (usually Mg^{2+}) per protein mole-

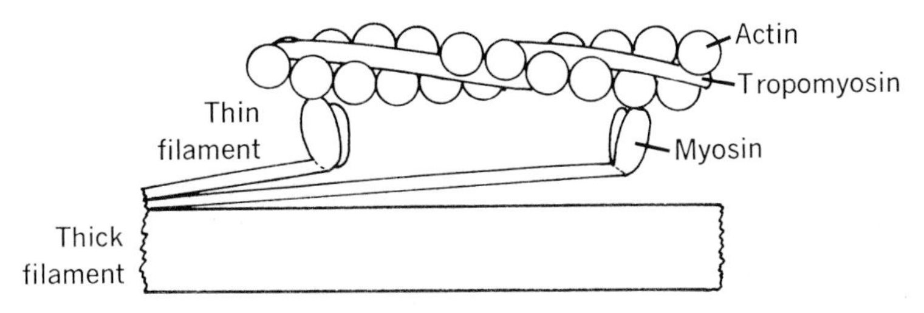

Fig. 3-20. Diagram showing organization of thin filament and relation of adjacent thick filament. Actin monomers are arranged to form two-stranded right-handed superhelix. Tropomyosin molecules lie in large grooves of superhelix. There is one tropomyosin molecule for approximately seven actin monomers.

cule. Actin and the other contractile proteins contain a few unusual methylated amino acid residues, methyllysine and methylhistidine. Pure globular actin (G actin) is soluble only in very weakly buffered distilled water containing a little ATP. On the addition of a very small amount of salt to G actin solutions, the actin salts out as fibrous actin (F actin). Two strands are intertwined in a right-handed superhelix with a pitch of about 700 Å containing about 13 G actin units per strand per turn. The double-stranded F actin filaments form the core of the thin filaments[77] (Fig. 3-20). During the aggregation of G actin to F actin the ATP attached to each actin monomer is hydrolyzed to ADP and inorganic phosphate:

$$n(\text{G actin ATP}) \underset{\text{ATP}}{\overset{\text{Salt}}{\rightleftharpoons}} (\text{F actin ADP}) + n(P_i) \quad (2)$$

The exact nature of the link between the monomers in F actin is not known. There is no evidence that the monomers in F actin are linked by a covalent chemical bond. Rather, they appear to be held together by multiple noncovalent (hydrophobic) interactions. The nucleotide does not appear to be involved in the linkage between actin monomers either, because G actin from which the nucleotide has been removed can polymerize perfectly well. From the results of fluorescent probe and other experiments, ordinarily the nucleotide seems to be buried in the interior of the actin molecule, away from the surface and not near the myosin-binding site. The ADP in the thin filaments is also functionally sequestered and does not exchange readily with the nucleotide pool in the sarcoplasm. Actin has the properties of interacting or associating with myosin and of markedly stimulating the Mg^{2+}-activated ATPase activity of myosin.

HMM binds to F actin filaments that in negatively stained electron micrographs appear as strands "decorated" with arrowheads at regular intervals. Such HMM-decorated strands have been used to discern many details of F actin structure and to localize F actin.

Tropomyosin, another thin-filament protein, is a slender molecule about 400 Å long of 68,000 daltons composed of two subunits. It is soluble as a monomer only in strong salt solutions. In the thin filaments there are two strands of tropomyosin molecules lying end to end in the major grooves of the actin superhelix. The dimensions of the molecules are such that there is one tropomyosin molecule for about every seven actin molecules. In the presence of troponin, tropomyosin inhibits the actin stimulation of Mg^{2+}-activated myosin ATPase activity. Thus in the presence of both troponin and tropomyosin the actin-Mg^{2+}–activated myosin ATPase activity is low.

Troponin is a 69,000-dalton spherical protein attached near one end of a tropomyosin molecule in the thin filament. Troponin binds Ca^{2+} avidly at four binding sites per troponin molecule. Troponin has a strong affinity for tropomyosin. Troponin by itself, in the absence of Ca^{2+}, does not affect the tropomyosin inhibition of the actin stimulation of myosin ATPase. In the presence of more than 10^{-7}M Ca^{2+}, troponin lifts the tropomyosin inhibition of the actin stimulation of Mg^{2+}-activated myosin ATPase activity.[9,10] Thus whole thin filaments that contain actin + tropomyosin + troponin stimulate Mg^{2+}-activated myosin ATPase activity when the Ca^{2+} concentration is greater than 10^{-7}M but not when the Ca^{2+} concentration is 10^{-8}M.

Troponin is composed of three subunits: troponin C (TN-C), the Ca^{2+}-binding subunit (molecular weight 17,850 daltons); troponin T (TN-T), the tropomyosin-binding subunit (molecular weight 30,500 daltons); and troponin I (TN-I), the inhibitory subunit (molecular weight 20,700 daltons). TN-C has four Ca^{2+}-binding sites, two high affinity sites that also bind

Mg^{2+} and two low affinity sites that do not bind Mg^{2+}. TN-I alone inhibits the actin stimulation of myosin ATPase in the absence of tropomyosin and more strongly in the presence of tropomyosin. Recent experiments indicate that TN-I can be phosphorylated; phosphorylation shifts the Ca^{2+} interaction curve of troponin to higher Ca^{2+} concentration. Phosphorylation may act as a mechanism for regulating the Ca^{2+} activation of contraction (see subsequent discussion).

Ca^{2+} binding by the troponin molecules must involve a series of changes in the structure and organization of the thin filament proteins so that the information is transmitted through the tropomyosin and actin to the myosin ATPase site.[138,147,155] Furthermore, because there are seven actin monomers for each tropomyosin and troponin molecule, the process must be cooperative for a change to be induced in each actin monomer.

The evidence for conformational and organizational changes in the thin filament proteins on Ca^{2+} binding by troponin comes from a variety of experiments, particularly x-ray diffraction experiments and experiments using electron spin or fluorescent probes. These experiments indicate that the tropomyosin molecules do not lie in the bottom of the troughs of the F actin superhelix, but rather that each tropomyosin strand is associated much more closely with one F actin filament than with the other (Fig. 3-21). In the relaxed state the tropomyosin molecules may be regarded as lying in a position that sterically hinders binding of myosin to actin. On binding of Ca^{2+} by troponin, the organization of the thin filament proteins changes so that the myosin-binding sites on the F actin filaments are uncovered.

Smooth muscle and many kinds of invertebrate muscle do not seem to contain troponin. Ca^{2+} activates smooth muscle also, but at higher concentrations. The myosin of smooth muscle contains special small subunits that bind Ca^{2+} and regulate activation. Thus Ca^{2+} also initiates contraction in smooth muscle, not by interaction with a troponin system, but by direct interaction with smooth muscle myosin. Some investigators think that Ca^{2+} may interact with myosin even in skele-

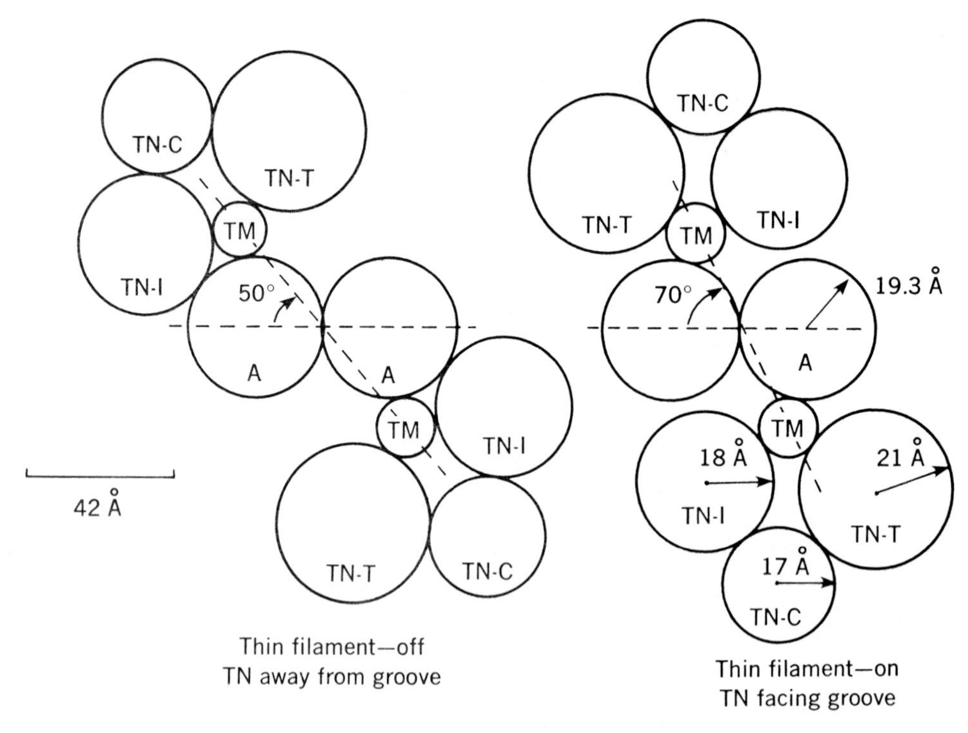

Fig. 3-21. Schematic diagram showing arrangement of protein molecules in cross section of thin filament at level of troponin molecules in "off" position (no calcium bound to troponin) and "on" position (calcium bound to troponin and myosin ATPase activated). *A*, Actin; *TM*, tropomyosin; *TN-T*, tropomyosin-binding subunit of troponin; *TN-C*, calcium-binding subunit of troponin; *TN-I*, inhibitory subunit of troponin; all are drawn to scale. Note movement of tropomyosin molecules in groove between actin molecules in shift from "off" to "on."

tal muscle and that this interaction may be physiologically important.

The thin filaments also contain some minor proteins—α-actinin and β-actinin are examples—whose properties and functions are the subject of active research.

Actomyosin

Pure myosin and pure actin free of traces of tropomyosin in <0.3M KCl interact in a 4:1 weight ratio to form a very viscous complex, *actomyosin*.[142] The viscosity of actomyosin is much greater than the sum of the viscosities of the pure myosin and pure actin measured separately. On mixing myosin and actin solutions, there is also a marked increase in light scattering. The interaction of myosin and actin can be studied in the laboratory by measuring viscosity and light scattering. In whole skeletal muscle, myosin is localized in the thick filaments and actin in the thin filaments, and their interactions are restricted. However, the many experiments performed on actomyosin have provided important clues about contraction, particularly in undifferentiated contractile systems where the contractile proteins are not separated into a highly structured submicroscopic filament system.

The addition of high concentrations of ATP (about 20 mM) to actomyosin in 0.6M KCl results in dissociation of the myosin and actin, a process that can be followed experimentally by the decrease in viscosity and light scattering. The ATP, of course, is slowly hydrolyzed by the myosin ATPase. When the ATP is almost completely hydrolyzed, the actomyosin complex re-forms. The process can be repeated by subsequent additions of ATP. Inorganic pyrophosphate will cause the same changes; therefore energy-yielding ATP hydrolysis is not required for dissociation of the actomyosin complex.

In more dilute salt solutions (<0.3M KCl), actomyosin forms a thick gel. The addition of small quantities of ATP, final concentration 2 mM or less, in the presence of 2 mM Mg^{2+} and a trace of Ca^{2+} results in superprecipitation, a marked contraction of the actomyosin gel with the extrusion of water (Fig. 3-22). Hydrolysis of ATP is essential for superprecipitation. Superprecipitation occurs to a much lesser extent with other energy-rich nucleotide triphosphates, but not at all with inorganic pyrophosphate. After the ATP is hydrolyzed the superprecipitated gel relaxes and slowly expands to its former volume. The process can be repeated by the addition of more ATP.

Threads of actomyosin can be formed by

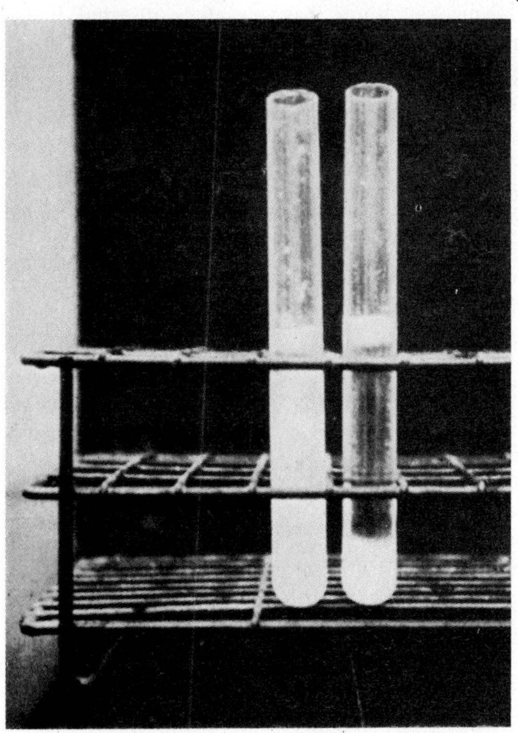

Fig. 3-22. Actomyosin gel in 0.15M KCl before and 20 min after addition of 2 mM final concentration ATP. Note supercontraction on right.

squirting an actomyosin solution in 0.6M KCl through a small orifice into a medium of low ionic strength containing 2 mM Mg^{2+} and a trace of Ca^{2+}. On the addition of ATP the actomyosin thread will shorten to less than one third its former length. Such supercontracted actomyosin threads will generate tension but not nearly as much as a muscle fiber of the same cross-sectional area.

EXCITATION-CONTRACTION COUPLING

It has already been indicated that contractile elements are activated when the Ca^{2+} concentration in the sarcoplasm rises above 10^{-7}M. The Ca^{2+} is bound by the troponin on the thin filaments and lifts the tropomyosin inhibition of actin-stimulated Mg^{2+}-activated myosin ATPase.

Relaxation of the contractile elements occurs as Ca^{2+} is subsequently removed from the sarcoplasm, and the concentration falls to about 10^{-8}M when troponin releases the bound Ca^{2+}. Ca^{2+} in the sarcoplasm is removed by an active transport system in the membranes of the sarcoplasmic reticulum, which is coupled to ATP hydrolysis in the presence of Mg^{2+}. Historically the discovery that relaxation followed Ca^{2+} removal from the sarcoplasm by an energy-requiring process in the sarcoplasmic reticulum was made long before it was realized that Ca^{2+} release into the sarcoplasm initiated the active contractile state. The relative amount of sarcoplasmic reticulum varies widely from muscle to muscle. As a general rule, the sarcoplasmic reticulum is much more developed and prominent in fast muscles than in slow muscles. Ca^{2+} does not accumulate in the cisternae of the sarcoplasmic reticulum ad infinitum; the sequestered Ca^{2+} is gradually transported through the sarcolemma to the extracellular fluid. Smooth muscles contain little or no sarcoplasmic reticulum and the Ca^{2+} is pumped out of the cell entirely by the sarcolemma.

Excitation-contraction coupling (i.e., the sequence of events that follows depolarization of the sarcolemma and culminates in activation of the contractile elements) has been a topic of intense research activity for the past two decades. The T tubular system is required for excitation-contraction coupling. Howell[95] showed that the connections of the T tubular system to the sarcolemma were disrupted by soaking a muscle in a hypertonic glycerol solution. When such muscle fibers are returned to Ringer's solution, the sarcolemma shows normal resting membrane potentials, depolarization, and conduction, but a normal propagated action potential does not lead to a twitch.[71]

The function of the T tubules is to depolarize and propagate the action potential into the core of muscle fiber. This function was first proposed purely on anatomic considerations. Then Huxley and Taylor[101] showed that subthreshold depolarizations of the sarcolemma produced by a microelectrode on the surface caused localized twitches if and only if the surface microelectrode was at the level of a T tubular system. Large depolarizations, of course, produce a propagated action potential along the muscle fiber.

It is not clear whether the depolarization of the T tubular membrane is (1) a self-regenerating propagated action potential characteristic of excitable membranes or (2) caused by electrotonic inward spread of currents produced by the depolarization of the sarcolemma.[32] The velocity of inward spread of activation in the T tubules, the sensitivity of T tubule action potential to temperature (changing by a factor of almost 3 for a change of 10° C), and analysis of the electrical properties of the T tubules all favor the first alternative, namely, that a depolarization is propagated inwardly from the sarcolemma by the T tubules to the center of the muscle fiber.[50] Recently the depolarization of T tubule membranes has been demonstrated directly using fluorescent dyes inserted in the membrane; these show a spectral shift and change in fluorescence intensity with a membrane potential change. It is of interest that the T tubule membrane is much less permeable to K^+ ions and more permeable to Cl^- ions than is the sarcolemma.[31,97] Contraction of a muscle fiber tends to begin at the periphery and spread centrally.[33,73]

Although the evidence is not conclusive, it is believed that the Na^+ and K^+ concentrations in the fluid in the cisternae of the sarcoplasmic reticulum resemble those found in extracellular fluid more closely than the concentrations found in sarcoplasm. The Na^+ space and sucrose space of muscle are about 12% larger than the extracellular space measured with inulin. The 12% difference corresponds very closely to the estimated volume of the sarcoplasmic reticulum. If this is the case, there should be an electrochemical gradient for cations across the sarcoplasmic reticulum membrane. Whether or not the sarcoplasmic reticulum is excitable and depolarizes is not known at present.

The depolarization of the T tubules causes the release of Ca^{2+} from the triads into the sarcoplasm, where it activates the contractile elements.[109,135,137,157,160] A quantitative analysis in terms of the electrical characteristics of sarcoplasmic reticulum and T tubule membranes indicates that it would be difficult for local currents arising from depolarizations of these membranes to account for all the Ca^{2+} released. To account for all the Ca^{2+} released into the sarcoplasm, it has been suggested that these membranes contain a Ca-K carrier that mediates by a process of facilitated diffusion the coupled influx of Ca^{2+} ions and efflux of K^+ from the sarcoplasm during the active depolarization of the triad membranes.

ENERGETICS OF CONTRACTION

Immediate energy source for contraction

The immediate energy source for contraction in mammals is ATP. Other nucleotide phosphates can also supply energy for contraction in the order ATP $\gg$ CTP $>$ UTP $>$ ITP $>$ GTP. Muscle contains about 2 μmole ATP/gram wet weight. The myosin head is the only site of major ATP hydrolysis in active muscle. The nucleotide in actin is sequestered and does not participate in any significant amount of turnover. At the concentrations of ATP, ADP, and inorganic phosphate present in the sarcoplasm, ATP hydrolysis yields about 11.5 kcal/mole. Approximately 0.3 μmole ATP/gm muscle is hydrolyzed by a single twitch of an unloaded frog sartorius muscle at 0° C. The amount of ATP hydrolyzed increases as more work is performed by the contracting muscle.[72,75,94,127]

Muscles also contain about 20 μmole creatine phosphate/gm wet weight. Creatine phosphate can phosphorylate ADP to form ATP in a reversible reaction catalyzed by the enzyme creatine kinase:

$$\text{CP} + \text{ADP} \underset{\text{Kinase}}{\overset{\text{Creatine}}{\rightleftharpoons}} \text{Creatine} + \text{ATP} \qquad (3)$$

Muscle contains large amounts of creatine kinase; it amounts to more than 25% of the soluble cytoplasmic protein. As soon as ATP is hydrolyzed, the ADP formed is very rapidly rephosphorylated by creatine phosphate and the ATP is regenerated. Thus creatine phosphate forms a reservoir of energy-rich phosphate bonds to quickly replenish the sarcoplasmic ATP. As a matter of fact, the regeneration of ATP by creatine phosphate is so rapid that in muscle poisoned with iodoacetate to prevent the synthesis of ATP from glycolysis, a series of twitches in frog sartorius muscle at 0° C produces a linear decrement in creatine phosphate concentration amounting to about 0.29 μmole/gm muscle per twitch.[44-46,158] Only after it was found that 2,4-dinitrofluorobenzene inhibits creatine kinase[53] was it possible to demonstrate unequivocally that ATP is the immediate energy source for contraction.

Intermediary metabolism

Ultimately, of course, ATP is produced by glycolysis and respiration. In glycolysis (Embden-Meyerhoff pathway), glucose is degraded to pyruvate, or to lactic acid in the absence of O_2, yielding 2 moles ATP/mole glucose metabolized. Muscle can be made to contract under anaerobic conditions in the laboratory, and this has often been done, but in the whole body, lactic acid is the main product of metabolism only at the beginning of contraction and after sustained work when fatigue is imminent. However, even in resting muscle in the whole animal, as much as 45% of the glucose utilized may be metabolized only as far as lactate. The lactate is transported to the liver and kidneys, where it is metabolized further.

Ordinarily the pyruvate formed by glycolysis is converted to acetyl-CoA, which is then oxidized in the mitochondria to carbon dioxide and water by the Krebs' tricarboxylic acid cycle. One molecule of glucose produces two molecules of pyruvate. The oxidation of two molecules of pyruvate reduces eight molecules of NAD to NADH, reduces two molecules of FAD to FADH, and phosphorylates two molecules of GDP to GTP. The reduced cofactors nicotinamide adenine dinucleotide (NADH) and flavin adenine dinucleotide (FADH) are oxidized in the mitochondria by a series of reactions coupled to the formation of 32 ATP molecules. These metabolic reactions are summarized in Fig. 3-23.

Intracellular glycogen granules provide a very readily available source of glucose. Muscles normally contain 9 to 16 gm/kg glycogen or, for a well-fed man of average height and weight, the total glycogen stores in muscle amount to 300 to 500 gm, with another 55 to 90 gm in the liver. Glycogen breakdown in muscle begins immediately on stimulation, and the amount of muscle glycogen depleted is proportional to the work done.

Glycogen is hydrolyzed by the enzyme phosphorylase a to glucose-1-phosphate, which then enters the glycolytic pathway. At rest, muscle contains mainly phosphorylase b, the inactive form of the enzyme. Phosphorylase is activated, that is, converted from phosphorylase b to phosphorylase a, by a second enzyme, phosphorylase kinase. Phosphorylase kinase, which also exists in resting muscle largely in an inactive form, can be activated in several ways; two appear to be most important physiologically: activation by Ca^{2+} and activation by 3′,5′-cyclic AMP in the

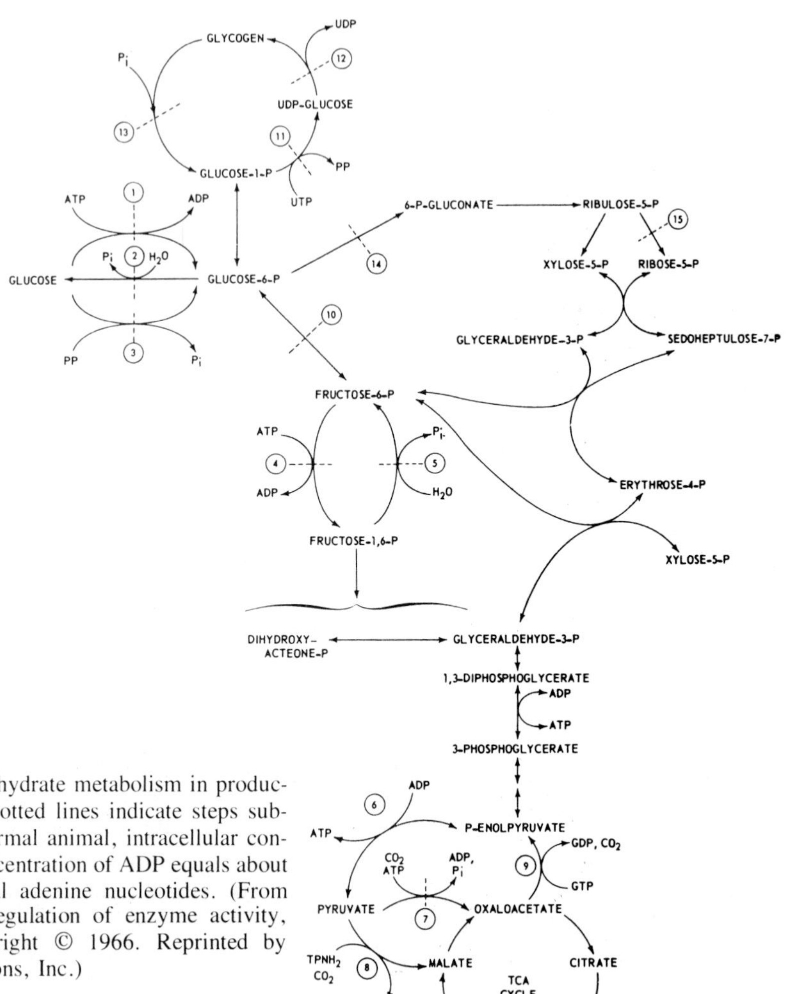

Fig. 3-23. Regulation of carbohydrate metabolism in production and utilization of ATP. Dotted lines indicate steps subject to allosteric control. In normal animal, intracellular concentration of ATP plus half concentration of ADP equals about 0.85 the concentration of total adenine nucleotides. (From Stadtman, E. R.: Allosteric regulation of enzyme activity, Adv. Enzymol. **28:**41, Copyright © 1966. Reprinted by permission of John Wiley & Sons, Inc.)

presence of a required protein called kinase-activating factor. Activation of phosphorylase kinase occurs at sarcoplasmic Ca^{2+} concentrations found in contracting muscle and is sometimes called the "contractile factor" of glycogen mobilization. The other important mechanism is extrinsic and involves (1) interaction of circulating adrenaline with β-receptors on the outer surface of the sarcolemma, (2) activation of the enzyme adenylcyclase on the inner surface of the sarcolemma, (3) conversion of ATP to 3′,5′-cyclic AMP by adenylcyclase, (4) activation of phosphorylase kinase, (5) conversion of phosphorylase b to phosphorylase a, and (6) mobilization of the glycogen stores.

Control of glucose metabolism occurs at several key points, which are depicted in Fig. 3-23. Hexokinase, the first enzyme in the glycolytic pathway, is subject to end-product inhibition by glucose-6-phosphate and 3-phosphoglycerate. A major regulatory step is the conversion of fructose-6-phosphate to fructose-1,6-diphosphate by the enzyme phosphofructokinase. This enzyme is very sensitive to ATP levels and is activated when the ATP level falls slightly.

Actively contracting muscles lose amino acids, particularly alanine, that presumably come from the degradation of proteins. Alanine is removed from the blood by the liver and is a principal amino acid utilized for gluconeogenesis. This pathway for providing energy is probably quantitatively important only during starvation and/or sustained strenuous exercise.

The oxidation of free fatty acids also leads to ATP formation. Most of the fatty acids utilized by skeletal muscle come from the white fat in the body, which amounts to 15 kg or more. Fatty acids from the plasma or mobilized from intracellular lipid droplets combine with carnitine, a low molecular weight base, to form carnitine acyl esters, which are soluble in the cytoplasm. The carnitine acyl esters are hydrolyzed at the mitochondrial membrane and the fatty acids degraded in the sarcosomes by β-oxidation to acetyl-CoA, which then enters the Krebs' cycle.

Fatty acids, rather than glucose, provide the major source of energy for muscle contraction[37,55] in a fasting person. This was determined by a series of ingenious experiments in which the brachial artery and deep vein to the forearm were cannulated in human volunteers.[165] Blood flow to the forearm and the arteriovenous differences of glucose, free fatty acids, O_2, CO_2, and other substances were measured. The respiratory quotient (CO_2 produced divided by O_2 utilized) was found to be 0.76, closer to that of fats (0.7) than to that of glucose (1.0). Only about 16% of the O_2 consumption of resting muscle could be attributed to glucose metabolism, whereas about 84% comes from oxidation of fatty acids. After a heavy carbohydrate-containing meal, however, glucose may be utilized to a greater extent by peripheral muscle.

Heat production

A. V. Hill[13,80-85] was the first to measure the small amount of heat produced by a muscle by means of a thermopile, a collection of bimetallic thermocouples wired to a galvanometer. At rest the basic metabolic processes in muscle result in a low, uniform production of heat. In frog sartorius muscle at 20° C the rate of heat generated at rest amounts to approximately 2×10^{-3} cal/(gm · min). Using the same figure for human muscle and a temperature coefficient for 10° C (Q_{10}) of 2.5, A. V. Hill estimated the rate of resting heat production from 30 kg of skeletal muscle in an average 70 kg human to be about 300 cal/min. This is equivalent to the consumption of 60 ml O_2 per minute, a large fraction (about 25%) of the total O_2 consumption under basal conditions.

Muscles can contract many times in the absence of O_2. In the absence of O_2 (i.e., under anaerobic conditions), isolated muscles produce only about half as much heat as they do in the presence of O_2. Glycogen stores are depleted, whereas lactic acid and other products of anaerobic metabolism accumulate in the muscle. When O_2 is readmitted to the system, the accumulated metabolic intermediates are oxidized and a corresponding amount of heat is generated.

During contraction a substantial amount of additional heat is produced over and above resting heat production.* After a stimulus is applied the additional heat production can be detected even before tension production and is associated, therefore, with the very early events in formation of the active state of the contractile elements. In the presence of oxygen the additional heat produced by muscle contraction, whether a twitch or a tetanus, can be divided into two phases; the *initial heat* and the *heat of recovery* or delayed heat. In tetanized muscle, heat production increases rapidly on stimulation, soon reaching a steady-state value of about 1.2 cal/(gm · min). After stimulation is stopped, heat production rapidly falls to about 1% initial heat rate. The excess heat production during recovery continues for 10 to 90 minutes, so that the total heat of recovery is approximately equal to the initial heat.[5,54] The heat of recovery corresponds to the metabolic reactions of oxidation of lactate and other intermediates and the resynthesis of glycogen. When muscle contracts in the absence of O_2, the initial heat is almost the same as in the presence of O_2, but the heat of recovery is almost zero. If a muscle is made to contract under anaerobic conditions and O_2 is readmitted to the system some time afterward, the delayed heat of recovery is generated after O_2 is available.

A. V. Hill showed that energy change in muscle during contraction can be described by the following relation:

$$U = A + W + \alpha\Delta x + M \qquad (4)$$

where U is the total energy change associated with contraction, A is the *activation heat* (i.e., the heat production associated with the activation of the contractile elements), W is the mechanical work performed by the muscle by lifting a load, $\alpha\Delta x$ is the *shortening heat,* and M is the *maintenance heat* of contraction.

The activation heat begins and is almost completely liberated before any tension is developed[93,118,148]; it corresponds in time to the latency relaxation mentioned previously. Thus activation heat seems to be associated with the internal work required to transform the contractile elements from the resting to the active state. Quantitatively, it is quite small, amounting to about 10 cal/gm muscle. Part of the activation heat probably is associated with a change in the elastic properties of muscle, but on the basis of indirect evidence, it is surmised that about two thirds of the activation heat is associated with the release of Ca^{2+} from the triads, its binding by

*See references 54, 56, 62, 63, 118, 139, 140, 154, 161, and 162.

troponin, and the subsequent rearrangement of the thin filament proteins. The activation heat is greatest for the first twitch after a period of rest and becomes smaller with succeeding twitches.

The maintenance heat begins at about the time tension begins and can be divided into two parts: the *labile maintenance heat* (h_A) and the *stable maintenance heat* (h_B), as depicted in Fig. 3-24. In tetanized frog muscle at 0° C the labile heat initially amounts to about 4×10^{-3} cal/(gm · sec); it decays exponentially, reaching one half the initial rate after about 1.2 sec. Thus the total labile heat generated amounts to 5 to 6×10^{-3} cal/(gm · sec).

Stable heat is generated at a constant rate during a tetanus, amounting to about 4×10^{-3} cal/(gm · sec) until fatigue sets in, when it declines abruptly. It has a very high temperature coefficient. The stable heat has quite different values in functionally different muscles; it is low when the muscle maintains tension efficiently and vice versa. Generally the maintenance heat is lower for type I slow muscles and higher for type II fast muscles. For isometric contractions at shorter than rest length, both the labile and the stable heats diminish. For stretched muscle, the labile heat is approximately constant, whereas the stable heat diminishes with stretching and is roughly proportional to the degree of overlap. The chemical equivalent of the stable maintenance heat is accounted for by phosphocreatine hydrolysis in iodoacetate-poisoned muscle; the chemical equivalent of the labile heat is not yet clearly known.

In 1923 W. O. Fenn[66] showed by heat measurements that when a muscle performs actual mechanical work by shortening, it mobilizes additional energy. Thus a contracting muscle liberates approximately a constant amount of heat and does more or less mechanical work.[92] In iodoacetate poisoned muscle, there is an additional amount of phosphocreatine hydrolysis that corresponds to the work done over and above heat liberated so that about 9.8 kcal work is done per mole phosphocreatine split. Furthermore, when a muscle performs work, the recovery heat exceeds the initial heat by an amount that is energetically roughly equal to or even slightly in excess of the work done.

In 1938 A. V. Hill[80] discovered that muscle contracting isotonically liberated an extra amount of heat that was proportional to the amount of shortening, which he called the shortening heat. The shortening heat is proportional mainly to the distance of shortening and does not depend greatly on the load, the speed of shortening, or the amount of work performed. Since mechanical work is $W = P\Delta x$, substituting this in equation 4 and rearranging the terms gives the following:

$$U = A + (P + \alpha)\Delta x + M \qquad (5)$$

Hill was struck by the resemblance of the term $(P + \alpha)$ in the heat equation and the term $(P + a)$ in the force-velocity equation (equation

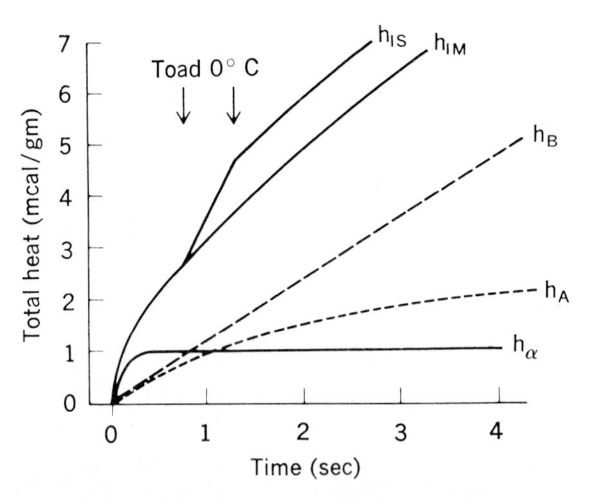

Fig. 3-24. Diagram to illustrate total heat production during maintained isometric contraction of toad muscle at 0° C, compared to total heat production of muscle permitted to shorten for short period of time (between arrows). h_{IM} is total heat production by muscle contracting isometrically. h_{IS} is total heat production by muscle permitted to shorten. Difference between these two is $\alpha\Delta x$, heat of shortening. Activation heat, A, and the components of M, the maintenance heat, are also depicted: h_A, labile maintenance heat; h_B, stable maintenance heat. (From Carlson and Wilkie.[6])

1) and was able to show a rough equivalence between the coefficient of the shortening heat α and the force-velocity constant a, although they clearly are not identical.[90]

Shortening heat is greatest for the first twitch after a period of rest and is less for subsequent twitches. Hill showed in 1964 that α increases somewhat with the load or decreases slightly with velocity and can be described approximately by the following empirical relation:

$$\alpha = 0.16\, P_0 + 0.18\, P \qquad (6)$$

where P is the actual load and P_0 is the maximum tension the muscle can produce. An interesting problem facing muscle physiologists is that no chemical equivalent of shortening heat, that is, no corresponding amount of metabolism, has been found.

Last, note should be made of *thermoelastic heat*. As mentioned previously, resting muscle has rubberlike thermoelastic properties, whereas actively contracting muscle has springlike thermoelastic properties. During the development of tension the change in elastic properties is accompanied by an absorption of heat by the muscle. As tension falls during relaxation, an equivalent amount of heat is released by the muscle owing to its elastic properties. The various kinds of muscle heat must be corrected for the thermoelastic heat. However, for a complete cycle of contraction and relaxation, the net heat produced by thermoelastic mechanisms is zero.

Properties of myosin ATPase

Now we come full circle and again consider the myosin ATPase, examining its complex mechanism for hydrolysis of ATP. Myosin ATPase does not follow Michaelis-Menten enzyme kinetics. There is a very rapid initial burst of inorganic phosphate appearance that can only be explained by a complex series of reactions. On the basis of very rapid kinetic experiments,[57,121,122,151] it has been possible to show that there are at least six steps in the hydrolysis reaction:

$$M + ATP \underset{k_{-1}}{\overset{k_1}{\rightleftharpoons}} M^* \cdot ATP \underset{k_{-2}}{\overset{k_2}{\rightleftharpoons}} M^* \cdot ADP \cdot P \underset{k_{-3}}{\overset{k_3}{\rightleftharpoons}} \qquad (7)$$

$$M \cdot ADP \cdot P \underset{k_{-4}}{\overset{k_4}{\rightleftharpoons}} M \cdot ADP + P_i \underset{k_{-5}}{\overset{k_5}{\rightleftharpoons}} M + ADP$$

where M is myosin, M* is an energy-rich form of myosin, and P_i is inorganic phosphate. The findings can be summarized as follows:

1. Myosin reacts rapidly with ATP to form a complex; the myosin is converted to an energy-rich form, as deduced from protein fluorescence studies for instance.
2. While complexed to the myosin, ATP is hydrolyzed to ADP and P_i. Reaction 2 is much more rapid than reaction 1. This step is extremely temperature sensitive.
3. In reaction 3, while the ADP and P_i are still attached to the myosin, the latter is converted to a low-energy form. This step is slow, rate limiting in the sequence of reactions, and insensitive to temperature changes.
4. Reactions 4 and 5 are rapid. In the presence of actin (i.e., under physiologic conditions) the reaction sequence proceeds as shown below.

Here it is important to note that the reaction rates of the various steps are such that the energy-rich actomyosin-ATP complex (AM* · ATP) is dissociated before the ATP in the complex is hydrolyzed and then recombines with actin before the myosin is converted from the energy-rich to the low-energy form. Thus actomyosin-ATPase has obligatory steps in which the actin and myosin are dissociated. In actomyosin gels hydrolyzing ATP under conditions that are as close to physiologic as possible, it is estimated that about half the myosin molecules are attached to actin, whereas the other half are not. The other important point to note is that the conversion of the AM* · ADP · P complex to the low-energy form is still the rate-limiting step. As a result, most of the myosin and actomyosin will be present in the form of a phosphorylated complex with a high negative charge. This fits data obtained from x-ray diffraction studies that suggest in intact muscle under conditions of great contractile activity, less than one third of the cross bridges are attached to the thin filaments at any given moment.

Last, the rate constants for all reactions except step 2 are greater for actomyosin than for myosin alone, particularly the hydrolysis of AM* · ADP · P to product, which explains how actin activates myosin ATPase activity.

$$\begin{array}{ccccc} & & M^* \cdot ATP & \rightarrow & M^* \cdot ADP \cdot P \\ & & \uparrow & & \downarrow \\ AM \cdot ATP & \rightarrow & AM^* \cdot ATP & & AM^* \cdot ADP \cdot P \rightarrow AM \cdot ADP \cdot P \rightarrow AM \cdot ADP + P_i \end{array} \qquad (8)$$

(AM · ADP returns to AM · ATP: ATP in, ADP out)

MECHANISM OF MUSCLE CONTRACTION

The mechanism of muscle contraction is a topic of great contemporary research interest. The problem is hardly solved, and any comments must be regarded as tentative. In analyzing any machine, man-made or natural, two kinds of explanations are required. The first kind of explanation concerns the mechanical arrangement of the moving parts, and the second kind of explanation concerns the way in which energy in one form is transduced to energy in another form. Thus in describing how a gasoline engine works, we first talk about the cylinders containing a moving piston connected to a drive shaft, valves also connected to the drive shaft, and the two-cycle series of events that comprise the mechanical events involved in the operation of the engine. Then we talk about a carburetor in which air and vaporized gasoline are mixed and sucked into an engine cylinder, the subsequent compression of the mixture and its ignition by a spark, followed by forceful expansion owing to the fact that the products of the chemical reaction occupy much more volume at ambient temperature and pressure than the reactants. In discussing the mechanism of muscle contraction, we must first explain the mechanical details of the machine, and, second, discern how energy transduction from chemical to mechanical energy takes place. Furthermore, whereas man-made machines are almost always large (macroscopic) devices with bulky parts, natural biologic machines are molecular machines. A single molecule of myosin reacts with a single molecule of actin to produce tension. The macroscopic mechanical properties of muscle result from summation of all the individual molecular interactions.

Concerning the mechanical events (and their relation to the biochemical extents), A. F. Huxley first proposed the two-state model in 1957.[14] According to this analysis the myosin heads and cross bridges are elastic elements with a mechanism for attaching themselves transiently to specific sites on the thin filaments[78,126,150] (Fig. 3-25). The following cyclic sequence of events is believed to take place[87,88]:

1. The cross bridges extend and attach them-

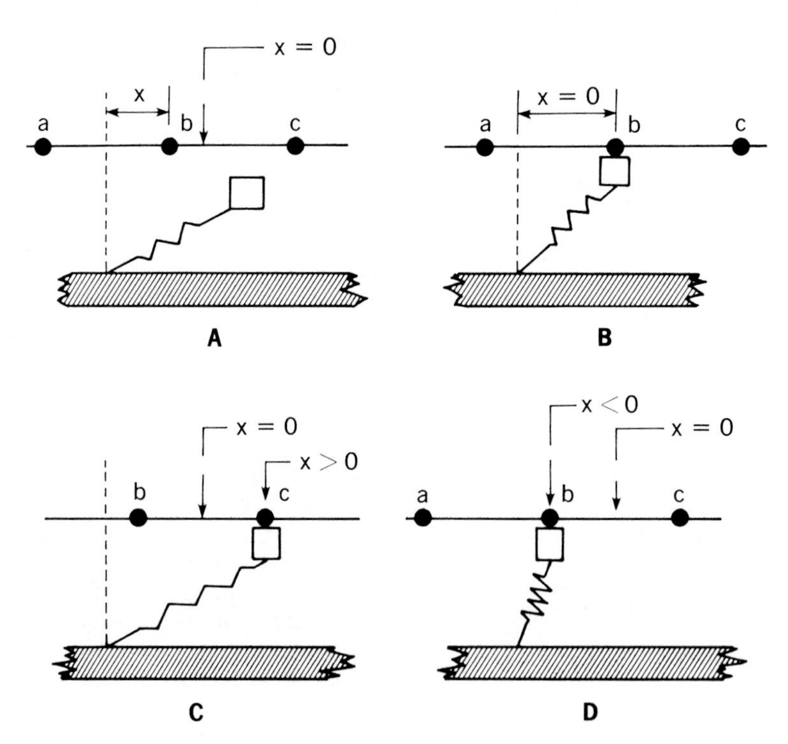

Fig. 3-25. Schematic diagram to illustrate operation of two-state mechanism for muscle contraction showing cross bridge and its thick filament plus three attachment sites (*a*, *b*, and *c*) on a thin filament. Actin-binding site on cross bridge is drawn as square. **A,** Detached cross bridge. **B,** Attached cross bridge at x = 0. Spring is exerting no force at all. **C,** Cross bridge at positive x. Cross bridge spring is extended and is exerting tension favorable to contraction. **D,** Cross bridge at negative x. Spring is compressed and is exerting force opposing contraction.

selves to specific sites (presumably actin sites) on the thin filaments. The probability that attachment will occur is $f(x)$, where x is the instantaneous distance between the equilibrium position (0) and the maximum distance for attachment h along the myofibrillar axis. Thus:

$$f(x) = \begin{cases} f_1(x/h) & \text{for } 0 < x < h \\ 0 & \text{elsewhere} \end{cases} \quad (9)$$

where f_1 is the attachment rate constant.

2. The cross bridges detach, possibly at stage 2A of the biochemical sequence of reactions in ATP hydrolysis by actomyosin described previously. The probability that detachment will occur is $g(x)$ and:

$$g(x) = \begin{cases} g_1(x/h) & \text{for } x \geq 0 \\ g_2 & \text{for } x < 0 \end{cases} \quad (10)$$

where g_1 and g_2 are the detachment rate constants.

Several simplifying assumptions have been made in formulating this model:

1. The 6 or more biochemical states of ATP hydrolysis can be represented by two states.
2. The cross bridges act as independent but equivalent units.
3. Each cross bridge behaves mechanically as though it had only a single head.
4. Changes in interfilament spacing does not affect cross-bridge performance.
5. The rate functions f_1, g_1, and g_2 do not depend on external conditions.
6. The g is *not* the inverse rate function to f, nor vice versa. The inverse rate functions to f and g may be neglected.

If we let N equal the density of cross bridges (i.e., the number of cross bridges per g) and n the fraction of cross bridges that are attached, then nN equals the density of attached cross bridges. A rate equation for cross-bridge attachment can be written as follows:

$$\frac{dn}{dt} = f(x)[1 - n(x, t)] - g(x)n(x, t) = f(x) - [f(x) + g(x)]n(x, t) \quad (11)$$

The equation can be integrated to give the following for an isometric steady state:

$$n(x) = \begin{cases} \dfrac{f(x)}{f(x) + g(x)} = \dfrac{f_1}{f_1 + g_1} & \text{for } 0 < x < h \\ 0 & \text{elsewhere} \end{cases} \quad (12)$$

For an isotonic steady state the solution for velocity v is as follows:

$$n_v(x) = \begin{cases} \dfrac{f_1}{f_1 + g_1}\left\{1 - \exp\left[-\dfrac{\phi}{v}\left(1 - \dfrac{x^2}{h^2}\right)\right]\right\} & \text{for } 0 \leq x < h \\ \dfrac{f_1}{f_1 + g_1}\left[1 - \exp\left(-\dfrac{\phi}{v}\right)\right]\exp\left(\dfrac{2g_2x}{sv}\right) & \text{for } x < 0 \\ 0 & \text{for } x \geq h \end{cases} \quad (13)$$

where s is the sarcomere length and $\phi = (h/s)(f_1 + g_1)$. By making intelligent assumptions about the values of the various parameters, the two-state attachment/detachment model predicts that about 30% of the cross bridges will be attached at any moment, a value that corresponds to the prediction from x-ray diffraction data.

Similarly, the model leads to expressions for the force developed by the cross bridges. For an isometric steady-state contraction the force is given by the following:

$$P_0 = \frac{1}{2} Nh \cdot kh \frac{f_1}{f_1 + g_1} \quad (14)$$

and for isotonic steady states the force-velocity relation predicted by the model is as follows:

$$P_v = P_0\left\{1 - \frac{v}{\phi}(1 - e^{-\phi/v})\left[1 + \frac{1}{2}\left(\frac{f_1 + g_1}{g_2}\right)^2 \frac{v}{\phi}\right]\right\} \quad (15)$$

where $k(x)$ is the stiffness of the cross-bridge spring. To simplify the model, it is usually assumed that the cross-bridge spring obeys Hooke's law, and therefore k is constant.

The two-state model that describes the molecular mechanical behavior of muscle can be used as the basis for obtaining gross muscle mechanical properties by computer simulation.[110-112] A typical twitch is depicted in Fig. 3-26, and a typical force-velocity curve computed from the two-state cycle of cross-bridge activity is depicted in Fig. 3-27. The latter result is particularly important because it means that the macroscopic force-velocity relation can be derived from the microscopic molecular properties of the interacting contractile proteins. The relations between the macroscopic force velocity parameters and the microscopic ones are, in addition to P_0, given by equation 3-14:

$$v_M = \frac{1}{2} hg_2 \quad (16)$$

$$b = v_M \frac{g_2}{f_1 + g_1} \quad (17)$$

The two-state model describing the microscopic molecular mechanical properties of muscle may be overly simplistic. It compresses the various multiple stages of ATP hydrolysis by actomyosin into two. It does not adequately de-

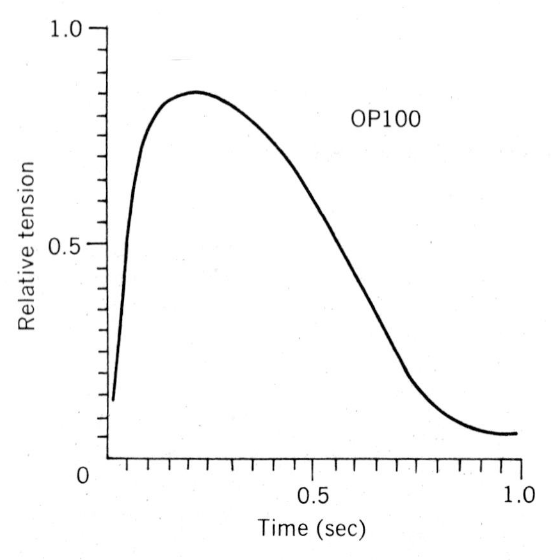

Fig. 3-26. Isometric twitch simulated by computer programmed with model based on two-state mechanism of muscle contraction. (From unpublished data of Schaar and Dowben.)

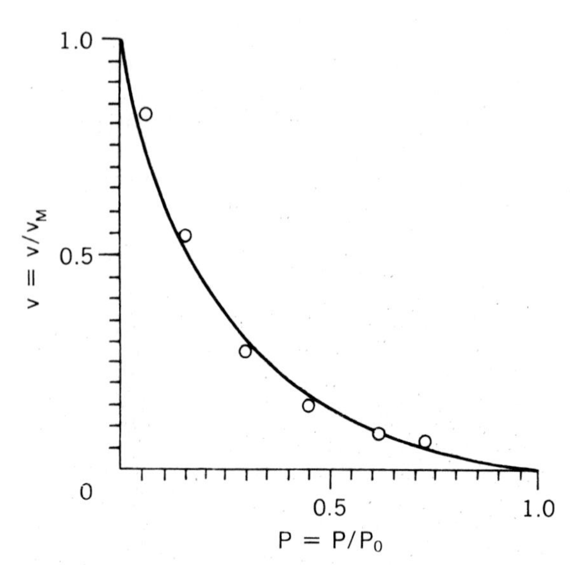

Fig. 3-27. Force-velocity curve from computer model. Curve is Hill equation for $u = a/P_0 = 0.25$. Circles are from isotonic tetanic contractions of frog sartorius muscle on which model is based. (From unpublished data of Schaar and Dowben.)

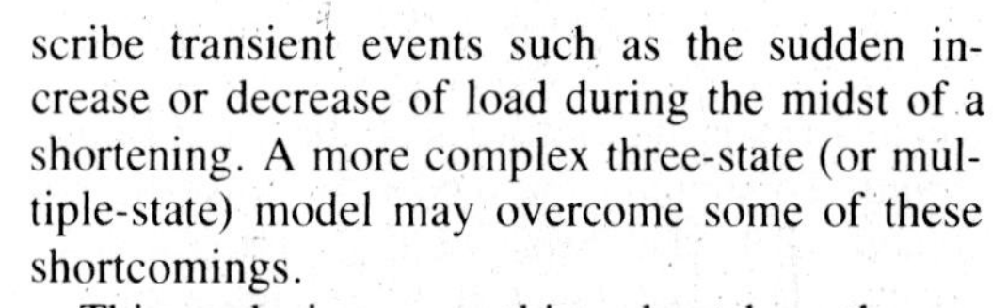

scribe transient events such as the sudden increase or decrease of load during the midst of a shortening. A more complex three-state (or multiple-state) model may overcome some of these shortcomings.

This analysis says nothing about how the energy made available by ATP hydrolysis leads to the generation of a force or tension. Concerning the mechanism of energy transduction in muscle, that is to say, how chemical energy is converted into mechanical work, our ideas are presently even more uncertain. One kind of theory holds that the myosin head undergoes a change in its three-dimensional shape (i.e., a conformational change driven by the energy released during ATP hydrolysis). The conformational change is thought to produce a mechanical deformation in the myosin head (perhaps a swiveling of the myosin head at the joint with the cross bridge), which is resolved by pushing or pulling the thin filament toward the center of the sarcomere. Thus the cross bridges undergo *repeated cycles of movement* that involve attachment of the myosin head to the thin filaments, a structural change in the myosin heads causing translational movement of the thin filaments, and, finally, detachment and relaxation of the myosin heads to a starting position farther down the thin filaments. In recent versions of this theory the myosin heads are not thought to push the thin filaments straight out but rather to rotate at the site of attachment.

Another group of theories about the mechanism of energy transduction in muscle proposes an electrostatic mechanism of force generation. ATP hydrolysis produces a negative surface charge on the myosin heads in part due to the phosphorylated intermediate. In the simpler electrostatic theories the negatively charged thick filaments are thought to repel each other, or the negatively charged myosin heads are attracted to positively charged sites on the thin filaments. Serious objections have been raised to these simple electrostatic mechanisms, and they do not easily accommodate the cyclic mechanical process described previously. A different type of electrostatic mechanism[146,164] is worthy of serious consideration, however. It has been shown that the thin filaments are highly electrically polarizable in the transverse direction. The electrostatic field of the negatively charged myosin heads polarizes the thin filaments, thereby generating a force that draws the thin filaments toward the center of the sarcomere. This mechanism is the electrostatic analog of the electromagnetic solenoid; the solenoid consists of a soft iron bar that is magnetized when an electrical current passes through a wire coil and is drawn into the center of the coil. The binding of myosin to actin in the complexes seems to be electrostatic in nature.[57,142] ATP hydrolysis is periodic, and the negative surface charge on individual myosin heads is formed and dissipates in a cyclic manner corresponding to the mechanical cycle described previously.

Some interesting experiments of Oplatka et al.[131,132] should be mentioned. These investigators first soaked muscle fibers in 50% glycerol to remove all the cytoplasm, leaving only the structural and contractile proteins. Such glycerinated muscle fibers contract when placed in a buffer containing ATP. Oplatka et al. then extracted the myosin from the glycerinated muscle fibers using Hasselbach-Schneider solution (0.5M KCl and 0.1M phosphate + pyrophosphate buffer at pH 6.4). Glycerinated muscle fibers from which the myosin has been extracted show no thick filaments on electron micrographs and do not contract on the addition of ATP. Oplatka et al. now soaked the extracted glycerinated fibers in heavy meromyosin, some being taken up by the fibers. Electron microscopy did not show thick filaments, but the I bands became more dense and the thin filaments were "decorated" with HMM arrowheads. Glycerinated muscle fibers from which myosin was extracted and which were subsequently irrigated with HMM contracted on addition of ATP and actually developed one third or more of the tension developed by ordinary glycerinated muscle fibers. These experiments show that the thick filamentous structure is not required for contraction, only the myosin heads. The thick filaments give mechanical advantage and direction to contraction in muscle.

Rigor

When muscle fibers are completely depleted of phosphorylcreatine and ATP (e.g., after death), they develop extreme rigidity, a state called *rigor*. Rigor results from fixed interactions between the myosin heads and actin molecules that are different from the ordinary interactions during contraction.[104] X-ray diffraction studies show that almost all the cross bridges are attached to thin filaments in an axial repeat distance of about 2 × 376 Å, characteristic of the arrangement of the actin monomers in the thin filaments. The cross bridges do not detach during rigor, and ATP is not hydrolyzed. The muscles are extremely stiff and resist both stretching and compression.

UNDIFFERENTIATED CONTRACTILE SYSTEMS

Cell motility

Although muscle is composed of highly differentiated cells whose principal function is to generate motion, movement is an attribute of cells in general. Movement in nonmuscle cells has many important consequences for the whole organism. One of the most universal and remarkable kinds of movement is *cytoplasmic streaming,* the intracellular flow of cytoplasm that is particularly striking in slime mold, in the giant cells of green algae such as *Nitella* or *Chara,* and in other plant cells. Cytoplasmic streaming has been easily and extensively studied in green algae. The greater part of the volume of these cells is occupied by a centrally located vacuole in which metabolic products accumulate. The inner cytoplasm flows around the vacuole in two oppositely directed helical streams, reaching velocities as great as 100 μ/sec and carrying along various particulate intracellular components. Cytoplasmic streaming requires the presence of O_2 and active metabolism.

In contrast to the turbulent movement of the inner cytoplasm, the outer cytoplasm lying just beneath the cell membrane is rather stiff and does not move. Electron microscopy reveals a system of fibers composed of filaments 50 to 70 Å in diameter lying between the outer cortical layer of still cytoplasm and the inner layer of actively streaming cytoplasm. The orientation of the fibers corresponds to the direction of the cytoplasmic streaming. The streaming is more rapid close to the fibers and more vigorous where the fiber system is more developed.

Although cytoplasmic streaming is less dramatic in animal cells, it is nevertheless widespread. It is perhaps most prominent in nerve cells, particularly in the axons, which in man may be as long as several meters. Because all protein synthesis takes place in the cell body of the neuron, it is necessary to transport essential enzymes and other proteins to the most distant extensions of the cell. It may be noted that some protein moves from adjacent glial cells into neuronal processes, but this is not sufficient to maintain axonal function. Cytoplasm moves down axons at varying rates; frequently axonal flow is lumped into a slow component, moving at the rate of 1 to 2 mm/24 hr, and a fast component, moving at the rate of 250 to 400 mm/24 hr. Axonal flow is energy dependent and apparently requires ATP hydrolysis.

The fast component of axonal flow[119,130] is conducted in the granular reticulum, an anastomosing network of membranous tubules that extend from the cell body to the ends of the axon. By the use of electron-opaque markers, it can be demonstrated that flow is bidirectional, moving toward the cell body as well as away from it. The slow component of axonal flow[90] only moves away from the cell body and requires great stability of the components because the flow amounts to about a meter per year. Transport of microtubules, microfilaments, and other structural proteins makes up the slow component.

Mitosis is another ubiquitous kind of cell movement that is apparently mediated by the fibers of the mitotic spindle. A related but different kind of motion during cell division is *cytokinesis,* the formation of the contraction band that gradually serves to separate the two daughter cells. Actinlike proteins are involved in cleavage by cytokinesis, whereas tubulin is involved in mitosis.

Nonmuscle contractile proteins

Starting from the early work of Loewy,[120] who extracted an actomyosin-like material from slime mold, several investigators have purified contractile proteins from numerous nonmuscle cells.[21] Actin is a major protein of many kinds of cells. Nonmuscle actin resembles muscle actin in many properties, including molecular weight and amino acid composition (including the presence of methylated amino acids). However, some differences are found, and actin may exist in the form of isoenzymes with distinctive tissue distributions, similar to the isoenzymes of lactate dehydrogenase and other soluble proteins.

Under many conditions the major portion of intracellular actin in cells like macrophages exists in the nonsedimentable monomer form. Nonmuscle actin does polymerize to form double helical strands, which can be visualized in many cells as microfilaments 50 to 100 Å in diameter. These filaments are identified as actin in electron micrographs because of their ability to bind specific antibodies and HMM to form "decorated" fibers with periodic arrowheads. In some cells these filaments are associated with cytoplasmic streaming, lying between the still and moving portions of the cytoplasm and oriented in the direction of flow.[128,152]

Some motile processes dependent on microfilaments are inhibited by *cytochalasin B,* a metabolite isolated from cultures of the mold *Helminthosporum.* Among the motile processes inhibited by cytochalasin B are phagocytosis, cytokinesis, cell adhesion, and changes of cell shape or cell movement. It has been shown that cytochalasin B combines with both nonmuscle and muscle actin, usually causing depolymerization of the microfilaments.

When highly purified plasma membranes are prepared from many kinds of mammalian cells, also isolated are large amounts of microfilaments that often appear to be attached to the cytoplasmic surface of the cell membrane. Actin microfilaments in association with plasma membranes are particularly prominent where there is considerable bulk transport. For instance, actin microfilaments are particularly prominent in the microvilli of intestinal epithelium. These actin microfilaments may mediate changes in cell shape, cell migration, etc.

Myosin-like proteins have also been isolated from nonmuscle cells. All these myosin-like proteins are able to bind reversibly to actin filaments and possess an actin-stimulated Mg^{2+}-activated ATPase. However, in contrast to nonmuscle actin, nonmuscle myosins vary considerably in molecular weight and other physical and enzymatic properties. Myosin filaments can be found in some nonmuscle cells, in granulocytes and platelets, for example. More frequently, however, myosin is found largely in a soluble, nonsedimentable form in nonmuscle cells.

Tropomyosin has been isolated from several kinds of nonmuscle cells, including platelets and brain cells. Nonmuscle tropomyosin has a slightly lower molecular weight (about 15% less), and the paracrystal periodicity is less than that of muscle tropomyosin. Like muscle tropomyosin, nonmuscle tropomyosin is composed of two almost completely helical subunits arranged in a superhelix.

Some nonmuscle cells such as platelets contain troponin-like proteins, and motility is regulated by changes in the intracellular Ca^{2+} concentration. In other cell types a high molecular weight protein cofactor is required for actin-stimulated myosin ATPase activity. This cofactor is involved in regulating motile activity; the motile activity in the latter type of cells is switched on or off by some cytoplasmic factor that has not yet been identified, but it clearly is not Ca. Myosin is phosphorylated in some nonmuscle cell systems, and the phosphorylation modulates cell motility. Last, most nonmuscle cells have complex mechanisms that control the intracellular distribution of contractile proteins and their assembly into filaments as well as the disaggregation into nonsedimentable monomers.

Microtubules and tubulin

In addition to proteins that resemble the contractile proteins of muscle, many nonmuscle cells contain an additional protein involved in movement, *tubulin.*[149] This globular protein has a molecular weight of approximately 110,000 daltons and is composed of two different subunits, α-tubulin and β-tubulin, about 55,000 daltons each. Tubulin contains two molecules of tightly bound GTP, one on each subunit. In the presence of certain acidic accessory protein, the *microtubule-associated proteins* (MAPs), the absence of Ca^{2+}, and the presence of Mg^{2+}, tubulin dimers aggregate to form highly characteristic structures called *microtubules*. Microtubule formation is accompanied by hydrolysis of the tubulin-bound GTP to GDP.

Microtubules are hollow, helical structures about 240 Å in diameter containing approximately 13 tubulin subunits per turn of the left-handed helix. Microtubules may reach a length of several microns. The formation of microtubules[40] occurs in two distinct phases: (1) nucleation and ring formation and (2) growth by elongation. The MAPs are important in nucleation, although other polycations will also promote nucleation, and polyanions are inhibitory. Nucleation can occur spontaneously, but it is enhanced by certain structures (site nucleation), including homologous microtubules, basal bodies (centrosomes), and chromosomes. Linear growth appears to be controlled by complex cytoplasmic factors. Reversible modification of tubulin monomers such as phosphorylation of specific serine residues may be involved in the control of microtubule formation. Several environmental factors such as high pressure, low temperature, and high pH cause disaggregation of microtubules, whereas in vitro high tubulin concentration, glycerol, and sucrose promote microtubule assembly.

A number of drugs, including colchicine, periwinkle (vinca) alkaloids, podophyllotoxin, and the antibiotics griseofulvin and maytansine, can bind to tubulin, preventing the assembly of new microtubules and frequently causing the disaggregation of existing microtubules. The effect of these drugs is particularly spectacular in dividing cells, where they interfere with spindle formation and cause arrest of mitosis in metaphase.

Cilia

Cilia and flagella are motile, hairlike extensions that project from 2 μ to 3 mm from the

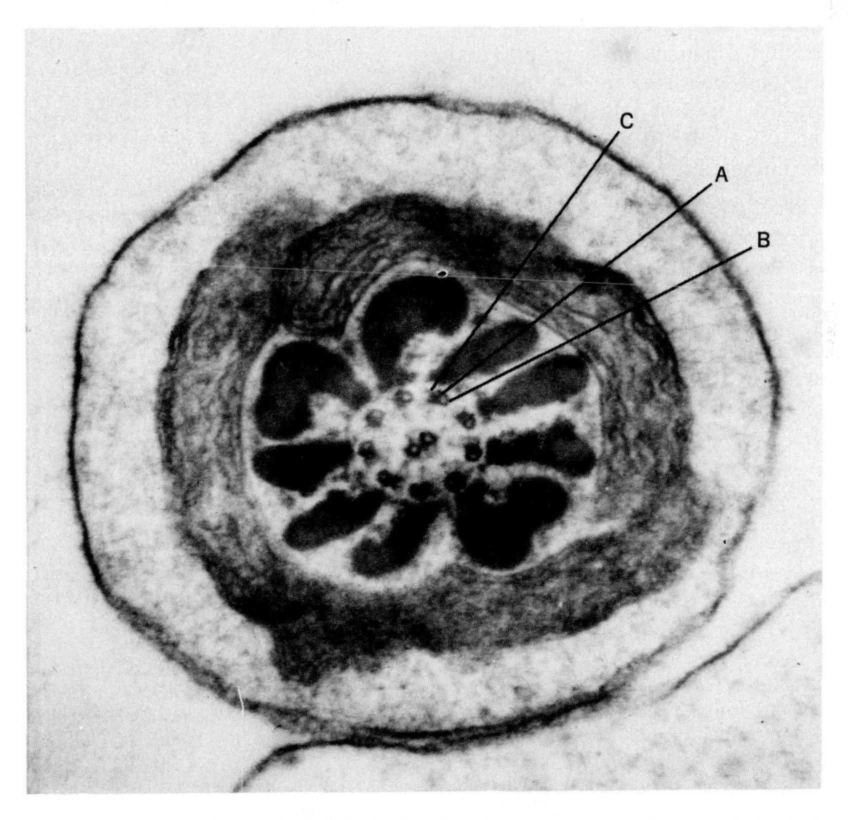

Fig. 3-28. Cross section of a sperm tail (rat) showing the arrangement of microtubules in the *axial filaments,* two central microtubules surrounded by a ring of nine sets of microtubules. The peripheral set of microtubules consists of a central one *(A),* another complete microtubule on the clockwise side *(B),* and side arms on the anticlockwise side *(C).* Protein spokes connect the central microtubules with the peripheral ones.

free surface of certain cells. The diameter of various eukaryotic cilia is remarkably uniform, about 0.2 μ. Regardless of the species of origin, eukaryotic cilia show the same internal structure, namely, a ring of nine sets of microtubules surrounding two central microtubules. This nine + two unit structure is called the *axial filament* (Fig. 3-28).

The outer nine elements of the axial filament consist of double microtubules composed of units of α-tubulin and β-tubulin associated in a heterodimer.[154] The central A microtubule contains 13 protofilament monomers per turn. The adjacent B microtubule contains 10 protofilament monomers per turn and shares the remaining 3 with the A microtubule. The peripheral microtubules are connected to the two central microtubules by protein spokes. The A microtubules also have two arms that project toward the next element in the ring. The arms are composed of a myosin-like protein called *dynein* that hydrolyzes energy-rich nucleotides. Satir[145] proposed that the dynein interacts in a cyclic fashion with the B microtubule in the next element in the ring in a manner analogous to the interaction of the myosin heads with the thin filaments in muscle. This interaction causes the beating motion of cilia and flagella. Dextrocardia is a hereditary condition that may be accompanied by abnormalities in structure of the microtubular side arms and by abnormal function of cilia.

REFERENCES

General reviews

1. Aidley, D. J.: The physiology of excitable cells, London, 1971, Cambridge University Press.
2. Aubert, X.: Le couplage energetique de la contraction musculaire, Brussels, 1956, Editions Arsciasa.
3. Bendall, J. R.: Muscles, molecules, and movement, New York, 1969, American Elsevier Publishing Co., Inc.
4. Blinks, J. R.: Calcium transients in striated muscle cells, Eur. J. Cardiol. **1:**135, 1973.
5. Buchthal, F., Kaiser, E., and Rosenfalck, P.: The rheology of the cross striated muscle fiber with particular reference to isotonic conditions, Dan. Biol. Med. **21:**1, 1951.
6. Carlson, F. D., and Wilkie, D. R.: Muscle physiology, Englewood Cliffs, N.J., 1974, Prentice-Hall, Inc.
7. Dowben, R. M.: Cell biology, New York, 1971, Harper & Row, Publishers, Inc.
8. Dreizen, P., Gershman, L. C., Trotta, P. O., and Stracher, A.: Myosin subunits and their interactions, J. Gen. Physiol. **50**(suppl.):85, 1967.
9. Ebashi, S., and Endo, M.: Calcium ion and muscle contraction, Prog. Biophys. Mol. Biol. **18:**123, 1968.
10. Ebashi, S., Endo, M., and Ohtsuki, I.: Control of muscle contraction, Q. Rev. Biophys. **2:**351, 1969.
11. Ebashi, S., Oosawa, F., Sekine, T., and Tonomura, Y., editors: Molecular biology of muscular contraction, Amsterdam, 1965, Elsevier.
12. Gergely, J., editor: Biochemistry of muscle contraction, Boston, 1964, Little, Brown & Co.
13. Hill, A. V.: Trails and trials in physiology, Baltimore, 1965, The Williams & Wilkins Co.
14. Huxley, A. F.: Muscle structure and theories of contraction, Prog. Biophys. Mol. Biol. **7:**255, 1957.
15. Huxley, H. E.: Muscle cells. In Brachet, J., and Mirsky, A. E., editors: The cell, New York, 1960, Academic Press, Inc., vol. 4.
16. Huxley, H. E.: The mechanism of muscle contraction, Science **164:**1356, 1969.
17. Jobsis, F. F.: Energy utilization and oxidative recovery metabolism in skeletal muscle. In Sanadi, D. R., editor: Current topics in bioenergetics, New York, 1969, Academic Press, Inc., vol. 3.
18. Mommaerts, W. F. H. M.: Energetics of muscular contraction, Physiol. Rev. **49:**427, 1969.
19. Podolsky, R. J.: Membrane systems in muscle cells, Symp. Soc. Exp. Biol. **22:**87, 1968.
20. Podolsky, R. J., editor: Contractility of muscle, Englewood Cliffs, N.J., 1971, Prentice-Hall, Inc.
21. Pollard, T. D., and Weihing, R. R.: Actin and myosin and cell movement, CRC Crit. Rev. Biochem. **2:**1, 1974.
22. Pringle, J. W. S.: Mechano-chemical transformation in striated muscle, Symp. Soc. Exp. Biol. **22:**67, 1968.
23. Stadtman, E. R.: Allosteric regulation of enzyme activity, Adv. Enzymol. **28:**41, 1966.
24. Szent-Gyorgyi, A.: Chemical physiology of contraction of body and heart muscle, New York, 1953, Academic Press, Inc.
25. Tiegs, O. W.: Innervation of voluntary muscle, Physiol. Rev. **33:**90, 1953.
26. Tonomura, Y., and Oosawa, F.: Molecular mechanism of contraction, Annu. Rev. Biophys. Bioeng. **1:**159, 1972.
27. White, D. C. S., and Thorson, J.: The kinetics of muscle contraction, Prog. Biophys. Mol. Biol. **27:**173, 1973.
28. Woledge, R. C.: Heat production and chemical change in muscle, Prog. Biophys. Mol. Biol. **22:**37, 1971.
29. Zierier, K. L.: Some aspects of biophysics of muscle. In Bourne, G. H., editor: Structure and function of muscle, New York, 1973. Academic Press, Inc., vol. 3.

Original papers

30. Abbott, B. C., and Wilkie, D. R.: The relation between velocity of shortening and the tension-length curve of skeletal muscle, J. Physiol. **120:**214, 1953.
31. Adrian, R. H., and Freygang, W. H.: The potassium and chloride conductance of frog muscle membrane, J. Physiol. **163:**61, 1962.
32. Adrian, R. H., Chandler, W. K., and Hodgkin, A. L.: Voltage clamp experiments in striated muscle fibres, J. Physiol. **208:**607, 1970.
33. Adrian, R. H., Constantin, I. L., and Peachey, L. D.: Radial spread of contraction in frog muscle fibres, J. Physiol. **204:**231, 1969.
34. April, E. W.: The myofilament lattice; studies on isolated fibers. IV. Lattice equilibria in striated muscle, J. Mechanochem. Cell. Motil. **3:**111, 1975.
35. Aubert, X., and Lebacq, J.: The heat of shortening during the plateau of tetanic contraction and at the end of relaxation, J. Physiol. **216:**181, 1971.
36. Bahler, A. S., Fales, J. T., and Zierler, K. L.: The dynamic properties of mammalian skeletal muscle, J. Gen. Physiol. **51:**369, 1968.
37. Baltzan, M. A., Andres, R., Cader, G., and Zierler,

K. L.: Heterogeneity of forearm metabolism with special reference to free fatty acids, J. Clin. Invest. **41:** 116, 1962.
38. Barany, M.: ATPase activity of myosin correlated with speed of muscle shortening, J. Gen. Physiol. **50:**197, 1967.
39. Bonner, R. F., and Carlson, F. D.: Structural dynamics of frog muscle during isometric contraction, J. Gen. Physiol. **65:**555, 1975.
40. Borisy, G. G., Olmsted, J. B., Marcum, J. M., and Allen, C.: Microtubule assembly in vitro, Fed. Proc. **33:**167, 1974.
41. Boyd, I. A.: The structure and innervation of the nuclear bag muscle fibre system and the nuclear chain muscle fibre system in mammalian muscle spindles, Philos. Trans. R. Soc. Lond. (Biol.) **245:**81, 1962.
42. Buller, A. J., Eccles, J. C., and Eccles, R. M.: Differentiation of fast and slow muscles in the cat hind limb, J. Physiol. **150:**399, 1960.
43. Carlson, F. D.: Kinematic studies in mechanical properties of muscle. In Remington, J. W., editor: Tissue elasticity, Washington, D.C., 1957, American Physiological Society.
44. Carlson, F. D., Hardy, D. J., and Wilkie, D. R.: Total energy production and phosphocreatine hydrolysis in the isotonic twitch, J. Gen. Physiol. **46:**851, 1963.
45. Carlson, F. D., Hardy, D., and Wilkie, D. R.: The relation between heat produced and phosphorylcreatine split during isometric contraction of frog's muscle, J. Physiol. **189:**209, 1967.
46. Chaplain, R. A., and Frommelt, B.: The energetics of muscular contraction. I. Total energy output and phosphoryl creatine splitting in isovelocity and isotonic tetani of frog sartorius, Pfluegers Arch. **334:**167, 1972.
47. Civan, M. M., and Podolsky, R. J.: Contraction kinetics of striated muscle fibres following quick changes in load, J. Physiol. **184:**511, 1966.
48. Close, R. I.: The pattern of activation in the sartorius muscle of the frog, J. Gen. Physiol. **46:**1, 1962.
49. Close, R. I.: The relations between sarcomere length and characteristics of isometric twitch contractions of frog sartorius muscle, J. Physiol. **220:**745, 1972.
50. Constantin, L. L., and Podolsky, R. J.: Depolarization of the internal membrane system in the activation of frog skeletal muscle, J. Gen. Physiol. **50:**1101, 1967.
51. Couteaux, R.: Innervation du muscle strié et organisation du sarcoplasme au niveau des terminaisons motrices. In Schapira, G., editor: Le muscle: étude de biologie et de patholgie, Paris, 1950, L'Expansion Scientifique Française.
52. Curtin, N. A., and Woledge, R. C.: Energetics of relaxation in frog muscle, J. Physiol. **238:**437, 1974.
53. Curtin, N. A., and Woledge, R. C.: Energy balance in DNFB-treated and untreated frog muscle, J. Physiol. **246:**737, 1975.
54. Curtin, N. A., Gilbert, C., Kretzchmar, K. M., and Wilkie, D. R.: The effect of the performance of work on total energy output and metabolism during muscular contraction, J. Physiol. **238:**455, 1974.
55. Dagenais, G. R., Tancredi, R. G., and Zierler, K. L.: Evidence for an intramuscular lipid pool in the human forearm, J. Clin. Invest. **50:**23a, 1971.
56. Dickinson, V. A., and Woledge, R. C.: The thermal effects of shortening in tetanic contractions of frog muscle, J. Physiol. **223:**659, 1973.
57. Eisenberg, E., and Moos, C.: The adenosine triphosphatase activity of acto-heavy meromyosin. A kinetic analysis of actin activation, Biochemistry **7:** 1486, 1968.
58. Eisenberg, R. S., and Gage, P. W.: Ionic conductances of the surface and transverse tubular membranes of frog sartorius fibers, J. Gen. Physiol. **53:**279, 1969.
59. Elliott, G. F., Lowy, J., and Worthington, C. R.: An x-ray and light diffraction study of the filament lattice of striated muscle in the living state and rigor, J. Mol. Biol. **6:**295, 1963.
60. Endo, M.: Length dependence of activation of skinned muscle fibers by calcium, Cold Spring Harbor Symp. Quant. Biol. **37:**505, 1972.
61. Engelhardt, W. A., and Ljubimowa, M. N.: Myosine and adenosine-triphosphatase, Nature **144:**668, 1939.
62. Fales, J. T.: Muscle heat production and work: effect of varying isotonic load, Am. J. Physiol. **216:**184, 1969.
63. Fales, J. T., and Zierler, K. L.: Relation between length, tension, and heat: frog sartorius muscle, brief tetani, Am. J. Physiol. **216:**70, 1969.
64. Falk, G., and Fatt, P.: Linear electrical properties of striated muscle fibres observed with intracellular electrodes, Proc. R. Soc. Lond. (Biol.) **160:**69, 1964.
65. Feinstein, B., Lindegaard, B., Nyman, E., and Wohlfart, G.: Morphologic studies of motor units in normal human muscles, Acta Anat. **23:**127, 1955.
66. Fenn, W. O.: The relation between the work performed and the energy liberated in muscular contraction, J. Physiol. **58:**373, 1923.
67. Fenn, W. O., and Marsh, B. S.: Muscular force at different speeds of shortening, J. Physiol. **85:**277, 1935.
68. Franzini-Armstrong, C.: Studies of the triad. I. Structure of the junction in frog twitch fibres, J. Cell Biol. **47:**488, 1970.
69. Franzini-Armstrong, C., and Porter, K. R.: Sarcolemnal invaginations and the T-system in fish skeletal muscle, Nature **202:**355, 1964.
70. Franzini-Armstrong, C., and Porter, K. R.: The Z disc of skeletal muscle fibrils, Z. Zellforsch. Mikrosk. Anat. **61:**661, 1964.
71. Gage, P. W., and Eisenberg, R. S.: Action potentials, and excitation-contraction coupling in frog sartorius fibers without transverse tubules, J. Gen. Physiol. **53:**298, 1969.
72. Gilbert, C., Kretzschmar, K. M., Wilkie, D. R., and Woledge, R. C.: Chemical change and energy output during muscular contraction, J. Physiol. **218:**163, 1971.
73. Gonzalez-Serratos, H.: Inward spread of activation in vertebrate muscle fibres, J. Physiol. **212:**777, 1971.
74. Gordon, A. M., Huxley, A. F., and Julian, F. J.: The variation in isometric tension with sarcomere length in vertebrate muscle fibres, J. Physiol. **184:**170, 1966.
75. Gower, D. and Kretzschmar, K. M.: Heat production and chemical change during isometric contraction of rat soleus muscle, J. Physiol. **258:**659, 1976.
76. Hanson, J. and Huxley, H. E.: The structural basis of contraction in striated muscle, Symp. Soc. Exp. Biol. **9:**228, 1955.
77. Hanson, J., and Lowy, J.: The structure of F-actin and actin filaments isolated from muscle, J. Mol. Biol. **6:**46, 1963.
78. Haselgrove, J. C. and Huxley, H. E.: X-ray evidence for radial cross-bridge movement and for the sliding filament model in actively contracting skeletal muscle, J. Mol. Biol. **77:**549, 1973.
79. Hellam, D. C. and Podolsky, R. J.: Force measure-

ments in skinned muscle fibres, J. Physiol. **200:**807, 1969.
80. Hill, A. V.: The heat of shortening and the dynamic constants of muscle, Proc. R. Soc. Lond. (Biol.) **126:** 136, 1938.
81. Hill, A. V.: The heat of activation and the heat of shortening in a muscle twitch, Proc. R. Soc. Lond. (Biol.) **136:**195, 1949.
82. Hill, A. V.: The energetics of relaxation in a muscle twitch, Proc. Roy. Soc. (Biol.) **136:**211, 1949.
83. Hill, A. V.: The abrupt transition from rest to activity in muscle, Proc. Roy. Soc. (Biol.) **136:**399, 1949.
84. Hill, A. V.: The effect of load on the heat of shortening of muscle, Proc. Roy. Soc. Lond. (Biol.) **159:**297, 1964.
85. Hill, A. V.: The variation of total heat production in a twitch with velocity of shortening, Proc. Roy. Soc. (Biol.) **159:**596, 1964.
86. Hill, D. K.: Tension due to interaction between the sliding filaments in resting striated muscle: the effect of stimulation, J. Physiol. **19:**637, 1968.
87. Hill, T. L.: Theoretical formalism for the sliding filament model of contraction of striated muscle. Pt. I, Prog. Biophys. Mol. Biol. **28:**267, 1974.
88. Hill, T. L.: Theoretical formalism for the sliding filament model of contraction of striated muscle. Pt. II, Prog. Biophys. Mol. Biol. **29:**105, 1975.
89. Hodgkin, A. L., and Horowicz, P.: The effect of sudden changes in ionic concentrations on the membrane potential of single muscle fibres, J. Physiol. **153:**370, 1960.
90. Hoffman, P. N., and Lasek, R. J.: The slow component of axonal transport, J. Cell Biol. **66:**351, 1975.
91. Homsher, E., and Rall, J. A.: Energetics of shortening muscles in twitches and tetanic contractions. A reinvestigation of Hill's concept of the shortening heat, J. Gen. Physiol. **62:**663, 1973.
92. Homsher, E., Mommaerts, W. F. H. M., and Ricchiuti, N. V.: Energetics of shortening muscles in twitches and tetani contractions. II. Force determined shortening heat, J. Gen. Physiol. **62:**677, 1973.
93. Homsher, E., Mommaerts, W. F. H. M., Ricchiuti, N. V., and Wallner, A.: Activation heat, activation metabolism and tension-related heat in frog semitendinosus muscles, J. Physiol. **202:**601, 1972.
94. Homsher, E., Rall, J. A., Wallner, A., and Ricchiuti, N. V.: Energy liberation and chemical change in frog skeletal muscle during single isometric tetanic contractions, J. Gen. Physiol. **65:**1, 1975.
95. Howell, J. N.: A lesion of the transverse tubules of skeletal muscle, J. Physiol. **201:**515, 1969.
96. Hunt, C. C., and Kuffler, S. W.: Motor innervation of skeletal muscle: multiple innervation of individual muscle fibers and motor unit function, J. Physiol. **126:**293, 1954.
97. Hutter, O. F., and Noble, D.: The chloride conductance of frog skeletal muscle, J. Physiol. **151:**89, 1960.
98. Huxley, A. F., and Niedergerke, R.: Interference microscopy of living muscle fibres, Nature **173:**971, 1954.
99. Huxley, A. F., and Simmons, R. M.: Mechanical properties of the cross bridges of frog and striated muscle, J. Physiol. **218:**59, 1971.
100. Huxley, A. F., and Simmons, R. M.: Mechanical transients and the origin of muscular force, Cold Spring Harbor Symp. Quant. Biol. **37:**669, 1972.
101. Huxley, A. F. and Taylor, R. E.: Local activation of striated muscle fibres, J. Physiol. **144:**426, 1958.
102. Huxley, H. E.: The double array of filaments in cross-striated muscle, J. Biophys. Biochem. Cytol. **3:**631, 1957.
103. Huxley, H. E.: Electron microscopic studies of the structure of natural and synthetic protein filaments from striated muscle, J. Mol. Biol. **7:**281, 1963.
104. Huxley, H. E., and Brown, W.: The low angle x-ray diagram of vertebrate striated muscle and its behavior during contraction and rigor, J. Mol. Biol. **30:**383, 1967.
105. Huxley, H. E., and Hanson, J.: Changes in the cross-striations of muscle during contraction and stretch and their structural interpretation, Nature **173:**973, 1954.
106. Jarcho, L. W., Dowben, R. M., Berman, B., and Lilienthal, J. L.: Site of origin and velocity of conduction of fibrillary potentials in denervated skeletal muscle, Am. J. Physiol. **13:**129, 1954.
107. Jewell, B. R., and Wilkie, D. R.: An analysis of the mechanical components in frog striated muscle, J. Physiol. **143:**515, 1958.
108. Jewell, B. R., and Wilkie, D. R.: The mechanical properties of relaxing muscle, J. Physiol. **153:**30, 1960.
109. Jobsis, F. F., and O'Connor, M. J.: Calcium release and reabsorption in the sartorius muscle of the toad, Biochem. Biophys. Res. Commun. **25:**246, 1966.
110. Julian, F. J.: Activation in a skeletal muscle contraction model with a modification for insect fibrillar muscle, Biophys. J. **9:**547, 1969.
111. Julian, F. J.: The effect of calcium on the force-velocity relation of briefly glycerinated frog muscle fibres, J. Physiol. **218:**117, 1971.
112. Julian, F. J., and Sollins, M. R.: Variation of muscle stiffness with force at increasing speeds for shortening, J. Gen. Physiol. **66:**287, 1975.
113. Julian, F. J., Sollins, K. R., and Sollins, M. R.: A model for the transient and steady state mechanical behavior of contracting muscle, Biophys. J. **14:**546, 1974.
114. Knappeis, G. G., and Carlsen, F.: The ultrastructure of the Z disc in skeletal muscle, J. Cell Biol. **13:**323, 1962.
115. Kuffler, S. W.: The two skeletal nerve muscle systems in frog, Arch. Exp. Pathol. Pharmakol. **220:**116, 1953.
116. Kushmerick, M. J., and Davies, R. E.: The chemical energetics of muscle contraction. II. The chemistry, efficiency and power of maximally working sartorius muscle, Proc. R. Soc. Lond. (Biol.) **174:**315, 1969.
117. Kushmerick, M. J., and Paul, R. J.: Aerobic recovery metabolism following a single isometric tetanus in frog sartorius muscle at 0° C, J. Physiol. **254:**693, 1976.
118. Kushmerick, M. J., Larson, R. E., and Davies, R. E.: The chemical energetics of muscle contraction. I. Activation heat, heat of shortening and ATP utilization for activation-relaxation processes, Proc. R. Soc. Lond. (Biol.) **174:**293, 1969.
119. Lasek, R. J.: Axonal transport and the use of intracellular markers in neuroanatomical investigations, Fed. Proc. **34:**1603, 1975.
120. Loewy, A. G.: An actomyosin-like substance from the plasmodium of a myxomycete, J. Cell. Comp. Physiol. **40:**127, 1952.
121. Lymn, R. W. and Taylor, E. W.: Transient state phosphate production in the hydrolysis of nucleoside triphosphates by myosin, Biochemistry **9:**2975, 1970.
122. Lymn, R. W., and Taylor, E. W.: Mechanism of adenosine triphosphate hydrolysis by actomyosin, Biochemistry **10:**4617, 1971.
123. Matsubara, I., and Elliott, G. F.: X-ray diffraction

studies on skinned single fibres of frog skeletal muscle, J. Physiol. **230:**62, 1972.
124. Matsumoto, Y.: Validity of the force-velocity relation for muscle contraction in the length region $l < l_0$, J. Gen. Physiol. **50:**1125, 1967.
125. Mendelson, R. A., Morales, M. F., and Botts, J.: Segmental flexibility of the S-1 moiety of myosin, Biochemistry **12:**2250, 1973.
126. Mittenthal, J. E.: A sliding filament model for skeletal muscle: dependence of isometric dynamics on temperature and sarcomere length, J. Theor. Biol. **52:**1, 1975.
127. Mommaerts, W. F. H. M., Seraydarian, K., and Marechal, G.: Work and chemical change in isotonic muscular contractions, Biochim. Biophys. Acta **57:**1, 1962.
128. Mooseker, M. S., and Tilney, L. G.: The organization of an actin filament-membrane complex; filament polarity and membrane attachment in the microvilli of intestinal epithelial cells, J. Cell Biol. **67:**725, 1975.
129. Morris, C. J., and Raybould, J. A.: Fiber type grouping and end-plate diameter in human skeletal muscle, J. Neurol. Sci. **13:**181, 1971.
130. Ochs, S.: Fast transport of materials in mammalian nerve fibers, Science **176:**252, 1972.
131. Oplatka, A., Gadasi, H., and Borejdo, J.: The contraction of "ghost" myofibrils and glycerinated muscle fibers irrigated with heavy meromyosin subfragment-1, Biochem. Biophys. Res. Commun. **58:**905, 1974.
132. Oplatka, A., et al.: Demonstration of mechanochemical coupling in systems containing actin, ATP and nonaggregating active myosin derivatives, J. Mechanochem. Cell Motil. **2:**295, 1974.
133. Peachey, L. D.: The sarcoplasmic reticulum and transverse tubules of the frog's sartorius, J. Cell Biol. **25:** 209, 1965.
134. Pepe, F. A.: The myosin filament. II. Interaction between myosin and actin filaments observed using antibody straining in fluorescent and electron microscopy, J. Mol. Biol. **27:**227, 1967.
135. Podolsky, R. J., and Constantin, L. L.: Regulation by calcium of the contraction and relaxation of muscle fibers, Fed. Proc. **23:**933, 1964.
136. Podolsky, R. J., and Nolan, A. C., and Zaveler, S. A.: Cross bridge properties derived from muscle isotonic velocity transients, Proc. Natl. Acad. Sci. USA **64:** 504, 1969.
137. Podolsky, R. J., and Teichholz, L. E.: The relation between calcium and contraction kinetics in skinned fibres, J. Physiol. **211:**19, 1970.
138. Potter, J. D., and Gergely, J.: Troponin, tropomyosin, and actin interactions in the Ca^{2+} regulation of muscle contraction, Biochemistry **13:**2697, 1974.
139. Rall, J. A., and Schottelius, B. A.: Energetics of contraction in phasic and tonic skeletal muscles of the chicken, J. Gen. Physiol. **62:**303, 1973.
140. Rall, J. A., Homsher, E., Wallner, A., and Mommaerts, W. F. H. M.: A temporal dissociation of energy liberation and high energy phosphate splitting during shortening in frog skeletal muscles, J. Gen. Physiol. **68:**13, 1976.
141. Rapoport, S. I.: Mechanical properties of the sarcolemma and myoplasm in frog muscle as a function of sarcomere length, J. Gen. Physiol. **59:**559, 1972.
142. Rizzino, A. A., Barouch, W. W., Eisenberg, E., and Moos, C.: Actin-heavy meromyosin binding: determination of binding stoichiometry from ATPase kinetic measurements, Biochemistry **9:**2402, 1970.
143. Romanul, F. C. A., and Van der Muelen, J. P.: Slow and fast muscles after cross innervation. Enzymatic and physiological changes, Arch. Neurol. **17:**387, 1967.
144. Rudel, R., and Taylor, S. R.: Striated muscle fibers; facilitation of contraction at short lengths by caffeine, Science **172:**387, 1971.
145. Satir, P.: Studies on cilia. III. Further studies on the cilium tip and a "sliding filament" model of ciliary motility, J. Cell Biol. **39:**77, 1968.
146. Schaar, P. L., and Dowben, R. M.: Energy transduction in striated muscle, Ann. N.Y. Acad. Sci. **227:** 268, 1974.
147. Schoenberg, M., and Podolsy, R. J.: Length-force relation of calcium activated muscle fibers, Science **176:** 52, 1972.
148. Smith, I. C. H.: Energetics of activation in frog and toad muscle, J. Physiol. **220:**583, 1972.
149. Snyder, J. A., and McIntosh, J. R.: Biochemistry and physiology of microtubules, Annu. Rev. Biochem. **45:**699, 1976.
150. Stein, R. B., and Wong, E. Y. M.: Analysis of models for the activation and contraction of muscle, J. Theor. Biol. **46:**307, 1974.
151. Taylor, E. W., Lymn, R. W., and Moll, G.: Myosin product complex and its effect on steady-state rate of nucleoside triphosphate hydrolysis, Biochemistry **9:** 2984, 1970.
152. Tilney, L. G., and Detmers, P.: Actin in erythrocyte ghosts and its association with spectrin, J. Cell Biol. **66:**508, 1975.
153. Walsh, T. H., and Woledge, R. C.: Heat production and chemical change in tortoise muscle, J. Physiol. **206:**457, 1970.
154. Warner, F. D.: The fine structure of the ciliary and flagellar axoneme. In Sleigh, M. A., editor: Cilia and flagella, New York, 1974, Academic Press, Inc., p. 11.
155. Weber, A., and Winicur, S.: The role of calcium in the superprecipitation of actomyosin, J. Biol. Chem. **236:**3198, 1961.
156. Weiss, P. A., and Hiscoe, H. B.: Experiments on the mechanism of nerve growth, J. Exp. Zool. **107:**315, 1948.
157. Weltman, J. K., and Dowben, R. M.: Relatedness among contractile and membrane proteins: evidence for common ancestral genes, Proc. Natl. Acad. Sci. USA **70:**3230, 1973.
158. Wilkie, D. R.: Heat work and phosphorylcreatine break-down in muscle, J. Physiol. **195:**157, 1968.
159. Winegrad, S.: Intracellular calcium movements of frog skeletal muscle during recovery from tetanus, J. Gen. Physiol. **51:**65, 1968.
160. Winegrad, S.: The intracellular site of calcium activation of contraction in frog skeletal muscle, J. Gen. Physiol. **55:**77, 1970.
161. Woledge, R. C.: The energetics of tortoise muscle, J. Physiol. **197:**685, 1968.
162. Woledge, R. C.: In vitro calorimetric studies relating to the interpretation of muscle heat experiments, Cold Spring Harbor Symp. Quant. Biol. **37:**629, 1972.
163. Yamada, T., Shimizu, H., and Suga, H.: A kinetic study of the energy storing enzyme-product complex in the hydrolysis of ATP by heavy meromyosin, Biochim. Biophys. Acta **305:**642, 1973.
164. Yu, L. C., Dowben, R. M., and Kornacker, K.: The molecular mechanism of force generation in striated muscle, Proc. Natl. Acad. Sci. **66:**1199, 1970.
165. Zierler, K. L., et al.: Muscle metabolism during exercise in man, Trans. Assoc. Am. Physicians **81:**266, 1968.

4 Vertebrate smooth muscle

JEAN M. MARSHALL

Smooth muscles, in contrast to striated muscles, are a heterogeneous group that show great diversity both in their morphologic arrangement and in their physiologic properties. In the walls of the gastrointestinal tract, blood and lymph vessels, uterus, vas deferens, and nictitating membrane, smooth muscle is arranged in sheets or layers of contiguous cells. Single smooth muscle cells, however, are found in the capsule and trabeculae of the spleen, and isolated contractile units called myoepithelial cells occur in some glands. Occasionally, smooth muscle cells are arranged in small groups analogous to skeletal muscles; small, cylindric pilomotor muscles are inserted onto the hairs in mammalian skin. The smooth muscles of the hair follicles, most blood vessels, vas deferens, and nictitating membrane are activated only by their motor nerves, whereas visceral smooth muscles exhibit spontaneous activity. The response of various smooth muscles to neurohumoral transmitter substances is not uniform; norepinephrine relaxes intestinal muscle and contracts vascular smooth muscle in most mammals. There is also diversity in the behavior of functionally comparable muscles of different species; norepinephrine causes the uterus of the rabbit to contract but inhibits the contractions of the rat uterus.

Because of these diverse morphologic and physiologic properties, it is difficult to make generalizations about the fundamental properties of smooth muscles. Experiments with smooth muscle are not easy to perform, and the results of investigations are often confusing and ambiguous. There are, nevertheless, three physiologic characteristics of smooth muscles that are of general enough occurrence to be listed here: (1) Smooth muscles are capable of slow, sustained contractions that can be maintained with a minimum expenditure of energy. (2) Their motor innervation is exclusively autonomic. (3) They all exhibit a certain degree of intrinsic "tone," that is, basal resting tension on which contractions are superimposed.

STRUCTURE AND CHEMISTRY OF SMOOTH MUSCLE CELLS

When examined with the light microscope, smooth muscle cells appear as long, spindle-shaped fibers, 2 to 5 μm in diameter and 50 to 100 μm in length, with a single nucleus situated in the middle, widest portion of the cell body. Each cell is surrounded by its plasma membrane, the sarcolemma, and there is no protoplasmic continuity between one cell and its neighbors.[2,76]

The extracellular space in most smooth muscles is two to three times greater than in striated muscle. This large extracellular space, combined with the small size of the smooth muscle cells and their large surface-volume ratio, favors rapid exchange of materials between the cells and the external medium. The extracellular space contains blood vessels, nerves, collagen and elastin networks, and a ground substance composed of glycoproteins and mucopolysaccharides.

The cytoplasm of smooth muscle cells is homogeneous when viewed under the light microscope. The contractile proteins, myofilaments, are not arranged in distinct sarcomeres, and the alternating dark and light bands characteristic of striated muscles are lacking—hence the name "*smooth* muscle." Nevertheless, electron microscopic studies reveal the presence of thick (13 to 17 nm in diameter) and thin (5 to 8 nm) filaments, believed to be the organized arrangements of myosin and actin, respectively.[76] In several smooth muscles examined in detail (pulmonary artery and portal–anterior mesenteric vein of the rabbit), rosettes consisting of about 15 thin filaments surrounding a thick filament have been observed.[76] The distance between a thin filament and its nearest thick-filament neighbor is about 1.5 nm, which would permit interaction through cross bridges between the two types of filaments. In high-resolution electron micrographs, cross-bridge–like lateral projections are sometimes visible on the thick filaments.[77] The ultrastructural organization of the thick and thin filaments and the cross bridges

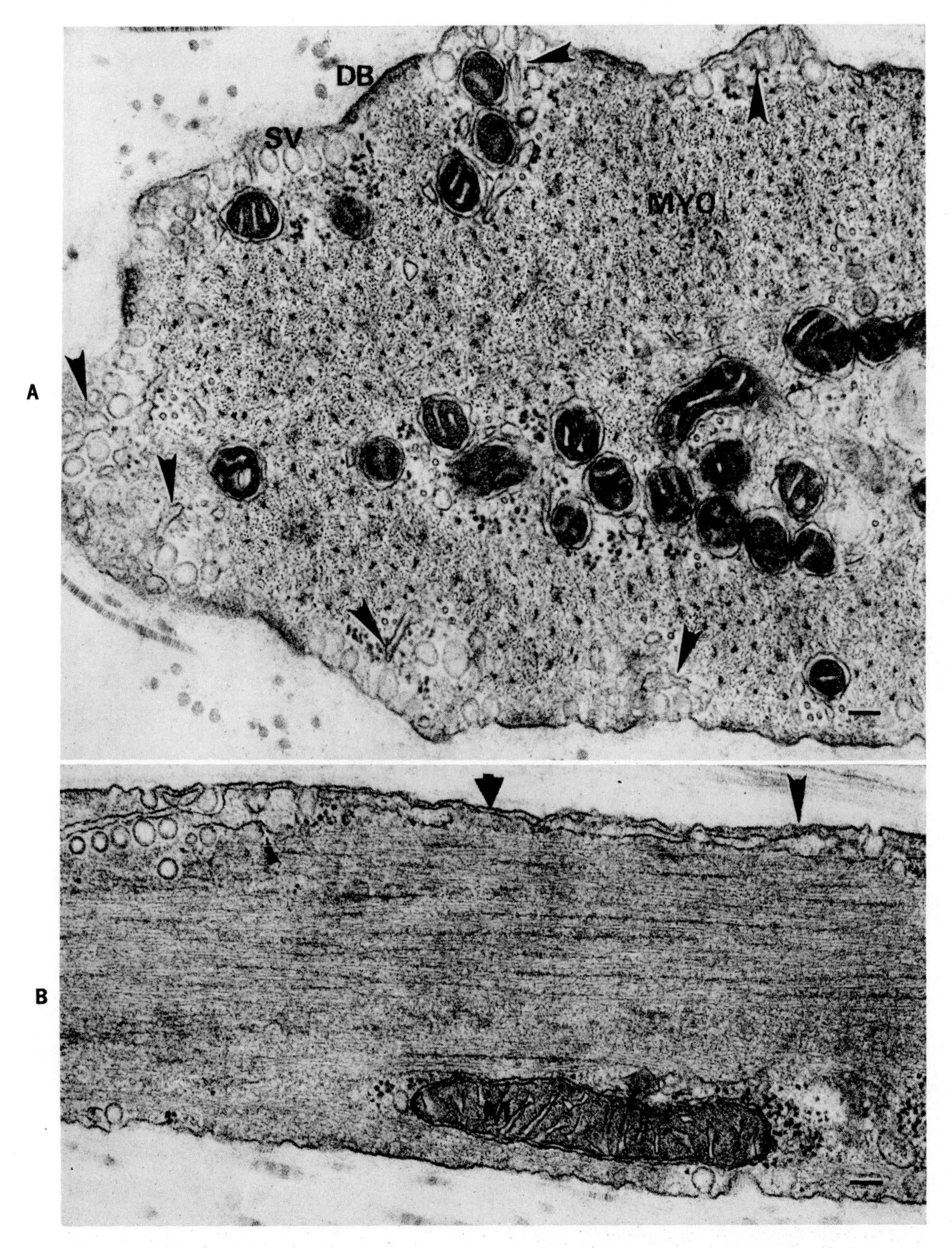

Fig. 4-1. A, Transverse section of portion of smooth muscle cell from portal anterior mesenteric vein (PAMV) of rabbit. Thick and thin myofilaments *(MYO)* are present, with regular spacing of about 70 nm between thick filaments. Sarcoplasmic reticulum (arrowheads) approaches surface vesicles *(SV)* at various points along cell membrane. Also shown is area of cell membrane occupied by dense body *(DB)*. Calibration (lower right) = 0.1 μm. (magnification ×47,000). **B,** Longitudinal section through smooth muscle cell from PAMV showing unusually long element of sarcoplasmic reticulum–cell membrane relationship (arrowheads) and dilated portion of sarcoplasmic reticulum near some surface vesicles (arrow). Thick and thin filaments are seen in cytoplasm, as is mitochondrion *(M)* close to surface vesicle. Calibration (lower right) = 0.1 μm (magnification ×47,000). (From Devine et al.[37])

on the thick filaments are consistent with a sliding filament mechanism of contraction in vertebrate smooth muscle.[8,75]

Scattered throughout the cytoplasm and at certain regions of the sarcolemma are aggregates of electron-dense material called dense bodies (Fig. 4-1). Thin filaments enter these bodies, and this relationship suggests that the dense bodies are functionally analogous to the Z lines of striated muscle.

Natural actomyosin (a mixture of tropomyosin-troponin, α-actinin, and some other proteins) has been isolated from smooth muscle,[66] although its concentration as well as its ATPase activity is much lower in smooth than in skeletal muscle.[65] Also, the concentration of high-energy phosphate compounds (ATP, phosphocreatine) is much lower in smooth than in skeletal muscle.[65]

The plasma membrane of smooth muscle cells is surrounded by an extraneous glycoprotein coat similar to the basement membrane of epithelial cells. At certain points along the surface the sarcolemma is divested of this extraneous coat, and the plasma membrane of one cell approximates that of an adjacent cell. High-resolution electron micrographs of these close contacts suggest that they are of at least two types, the gap junction or nexus and attachment plaques.[43,48] At the nexus the outer leaflets of the adjoining membranes are close together, with a gap of only about 2 nm between them (Fig. 4-2). The plasma membranes within the nexus are specialized and contain hexagonally packed particles. These structures on the opposing membranes are not in actual contact with one another but are in register, and each particle has a central channel extending through the gap from cytoplasm to cytoplasm. Attachment plaques are characterized by a thick layer of electron-dense material on the cytoplasmic aspects of the two opposing membranes. Microfilaments penetrate and merge into this material—hence the name "attachment plaque." The gap between the cells at the plaques is between 10 and 30 nm wide and filled with an inter-

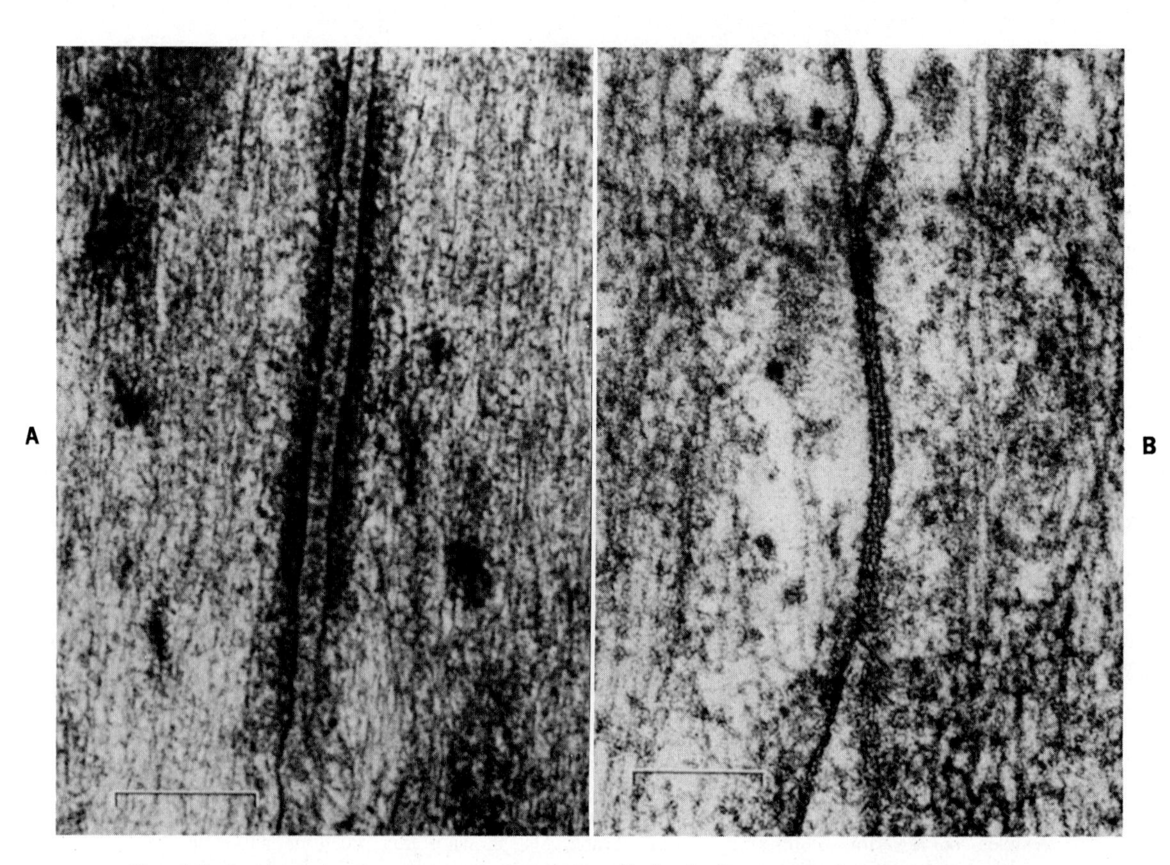

Fig. 4-2. A, Nexus between two smooth muscle cells in circular muscle from guinea pig ileum. Calibration (lower left) = 0.1 μm. (Magnification ×182,000.) **B,** Attachment plaque between two muscle cells in circular muscle from guinea pig ileum. Calibration (upper right) = 0.25 μm. (Magnification ×81,000.) (From Gabella.[43])

vening layer of electron-opaque substance (Fig. 4-2). Attachment plaques are thought to be permanent, stable structures whose function, as yet undefined, is related to contraction. The nexus, on the other hand, may be a dynamic structure, forming and dispersing within hours or even minutes.[48] It has been suggested that the nexus is the site of electrical coupling between smooth muscle cells. There are, however, smooth muscles where there is good evidence for electrical coupling but where few or no nexuses have been observed, for example, longitudinal muscles of the intestine, uterus, and vas deferens.[43] Therefore it should not be assumed that the nexus is the only site of electrical coupling between cells.

The plasma membrane of smooth muscle cells exhibits flask-shaped invaginations—the so-called surface vesicles or "caveolae." They are arranged in rows parallel to the main axis of the cell and freely communicate with the extracellular space (Fig. 4-1). Although their function is not known, they significantly increase the surface area of the cell membrane and may be analogous to the transverse tubules of striated muscle.[43,76]

All smooth muscles contain a sarcoplasmic reticulum (SR), although it is less extensive than that of skeletal muscle.[43,76] It consists of a three-dimensional network of tubules whose peripheral elements approach the plasma membrane and caveolae to within 10 to 20 nm and also communicate directly with SR located near the cell nucleus. Because of its strategic position close to the cell surface and by analogy with striated muscle, it has been suggested that the SR may play an essential role in the storage and release of Ca in smooth muscle.[43,76] The SR of smooth muscle can accumulate divalent cations, but its ability to release these ions (especially Ca) when the cell is stimulated has not yet been demonstrated.[43]

PHYSIOLOGIC CLASSIFICATION OF SMOOTH MUSCLES

Some years ago Bozler[18] proposed that smooth muscles be classified into two groups according to their physiologic properties—unitary muscles and multi-unit muscles. The unitary muscles are characterized by spontaneous activity initiated in pacemaker areas within the tissue that spreads throughout the whole muscle as if the muscle were a single unit. Multi-unit muscles do not contract spontaneously and normally are activated in more than one region by multiple motor nerves.

The smooth muscles of the gastrointestinal tract, uterus, and ureter are good examples of unitary muscles. The contractions of these muscles are not initiated by nerve impulses, but they may be coordinated and regulated by nervous mediation. In this respect the unitary or visceral muscles resemble cardiac muscle. Another property of unitary muscle is its ability to respond to stretch by developing active tension.

The multi-unit muscles include the pilomotors, the muscles of the nictitating membrane, ciliary muscles, iris, and larger blood vessels. These muscles usually respond to a single volley of nerve impulses with a contraction. Multi-unit muscles are organized roughly into "motor units"—hence their name. These "units," however, are diffuse and show considerable overlap. In general, multi-unit muscles do not react to stretch by developing tension.

This classification should not be regarded as rigid, since some smooth muscles do not fall clearly into one or the other of these two groups but combine properties of both. For example, the muscles of the vas deferens are usually activated by stimulation of their extrinsic motor nerves, but when placed in an isolated organ bath, they may become spontaneously active. Smooth muscles in certain arterioles, venules, and veins are autorhythmic; they show spontaneous activity, yet they also respond to extrinsic nerve stimulation. A single volley of nerve impulses can initiate contraction of the urinary bladder, but this muscle also can be made to contract by stretching. Although these and other exceptions exist, the classification of smooth muscles into two general categories will be followed here because it provides a convenient framework for describing physiologic properties of smooth muscle. Visceral muscle will be considered first, followed by a discussion of multi-unit muscle. The chapter ends with a consideration of the transmission of excitation from autonomic nerves to smooth muscle cells.

VISCERAL OR UNITARY SMOOTH MUSCLE

The foundations of visceral smooth muscle physiology were established by the classic experiments of Emil Bozler from 1938 to 1948.[18,19] He recorded, by means of extracellular electrodes, electrical activity in visceral smooth muscle and correlated it with the contractions of the muscle both in vivo and in vitro. From these studies he concluded that the spontaneous rhythmic contractions of the muscle were initiated and maintained by periodic electrical discharges (action potentials) from the muscle cells. These action potentials originated from pacemaker areas within the muscle and were conducted over the muscle in a manner similar to that observed in cardiac muscle. The spontaneous, rhythmic discharge of action potentials persisted in the presence of ganglion and nerve blocking agents. Hence they were myogenic, that is, they originated within the muscle cells themselves.

Recently these observations have been confirmed and extended with the aid of improved instrumentation and techniques, especially those recording electrical activity with intracellular microelectrodes or with the sucrose gap method. These methods are described in Chapters 1 and 2. Most studies utilizing these techniques have been done on intestinal and uterine muscle.[3,4,6] Thin segments of the external longitudinal muscle layers can be dissected free from the underlying muscle and ganglion cells (intestine). When placed in an isolated organ bath and bathed in oxygenated physiologic salt solution at 37° C, the strips of muscle will contract spontaneously for many hours. It is possible to penetrate the individual muscle cells with very fine glass capillary microelectrodes (tip diameter 0.5 μm) and obtain records of the transmembrane potentials from individual cells while the contractile tension developed by the entire strip is measured simultaneously with a muscle lever or force transducer. At present, most of our information about visceral smooth muscle comes from such studies.

When a microelectrode is inserted into a quiescent visceral smooth muscle cell, a potential difference of between 55 and 60 mV is recorded across the cell membrane, with the inside of the cell negative relative to the outside. Spontaneously active unitary smooth muscles are never in a true resting state comparable to that of an inactive skeletal muscle; thus the term "*resting* membrane potential" cannot really be applied to these muscles. Nevertheless, the term is used in describing autorhythmic smooth muscles to refer to the maximal level of membrane polarization attained between periods of activity.[4]* The resting membrane potential of a quiescent visceral smooth muscle cell differs from that of a skeletal muscle cell at rest in two ways: it has a lower value (Table 4-1) and is more labile. The latter

*In this chapter the membrane potential refers to the potential, measured by an *intracellular* microelectrode, relative to the extracellular fluid. The resting membrane potential will therefore have a negative sign (e.g., −55, −60 mV). Changes in resting potential (e.g., from −60 to −40 mV) will be noted as "depolarization" or "decreased negativity" shifts in the opposite direction (e.g., −60 to −80 mV) will be designated "hyperpolarization" or "increased negativity."

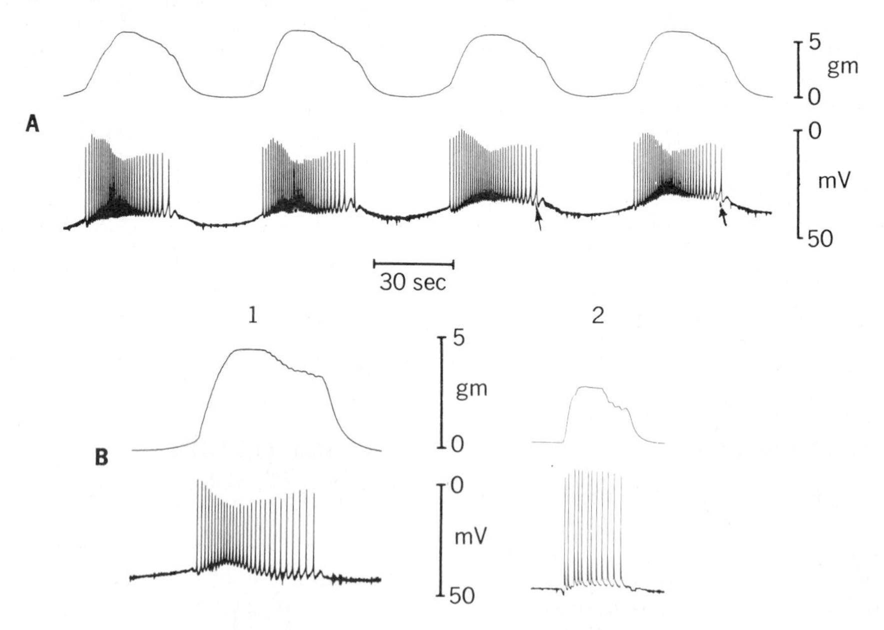

Fig. 4-3. A, Slow waves and prepotentials (arrows) in isolated segment of myometrium from parturient rat. **B,** Pacemaker cell *(1)* showing prepotentials and "follower" cell *(2)* showing conducted action potentials that rise abruptly from flat baseline and are larger than pacemaker spikes. Records taken from isolated segment of myometrium from nonpregnant rat where slow waves are not present and pacemaker activity is less organized than in **A.** Note that spike frequency and train duration are less in 2 than in record *1*, resulting in a smaller contraction in record *2*. (From Marshall.[61])

property enables visceral smooth muscle cells to generate action potentials spontaneously.

Initiation of spontaneous activity

The spontaneous generation of action potentials in most smooth muscle cells results from at least two types of fluctuations of their electrical activity: slow, rhythmic oscillations of the membrane potential and more rapid, localized depolarizations of the membrane. The slow oscillations, variously called "basic electrical rhythm," "pacesetter potential," or simply "slow waves," are especially prominent in the longitudinal muscle layer of the intestine and in the myometrium during parturition[10,69] (Fig. 4-3). Their periods may be seconds or minutes, depending on the particular tissue and its environment. In these muscles the rapid, localized depolarizations appear at the crest of the slow waves and generate the action potentials and are called pacemaker potentials or prepotentials (Fig. 4-3). The prepotentials determine the action potential frequency and the duration of the burst of action potentials and hence the force and duration of each contraction (Fig. 4-3). The slow waves set the interval between contractions and hence the contraction frequency (Fig. 4-3). It has been suggested that slow waves spread across the longitudinal muscle layer and into the underlying circular layer and serve to synchronize spike discharge and contractions over a large area of muscle.[6,69]

In some situations, for example, the mid- or nonpregnant uterus and the longitudinal muscle from the guinea pig cecum (taenia coli), prepotentials often arise spontaneously in the absence of well-organized slow waves and generate action potentials that are conducted to neighboring cells[4] (Fig. 4-3). Under such circumstances, spike discharge is frequently asynchronous in various areas of the muscle because of the multiple foci of pacemaker activity. Therefore the contractions are irregular in frequency, amplitude, and duration.

Although the cellular mechanisms responsible for the generation of slow waves and prepotentials are not entirely known, the slow waves are thought to be metabolic in nature and may involve the rhythmic activation of an electrogenic* Na pump.[6,69] The prepotential seems to be a voltage-dependent increase in membrane conductance, most probably to Ca and/or Na, which produces a transient inward current resulting in membrane depolarization.[10] The interaction between slow waves and prepotentials might be as follows: at the trough of a slow wave the electrogenic pump activity is minimal and as a result the membrane depolarizes and brings the potential to threshold for the voltage-dependent prepotential activity, which triggers spike discharge. Electrogenic pump activity then increases and the membrane repolarizes, bringing the potential below threshold for spike generation and inactivating the voltage-dependent prepotential mechanism.

Conduction of electrical activity

Action potentials and slow waves are conducted throughout the muscle in a manner not yet completely understood. Since smooth muscle cells do not have intercellular cytoplasmic continuity, how can electrical activity propagate from one cell to another? It seems most likely that transmission is electrical and is made possible by the presence of low-resistance pathways between the interiors of adjacent muscle cells.[13]

If an active cell is adjacent to an inactive one, some of the current accompanying the action potential in the former will pass through the membrane of the latter. However, if the membrane of the inactive cell has a much higher resistance than the extracellular fluid, the fraction of current traversing the membrane of the inactive, adjacent cell will be insufficient to reach threshold. In order to devise a feasible electrical theory of transmission, one must include the postulate of a low-resistance pathway between the interiors of adjacent cells. Such pathways would allow sufficient current from an active cell to pass through the membrane of a neighboring, inactive cell so that threshold for excitation would be reached.

If the cells are connected electrically through low-resistance pathways, the membrane potentials of neighboring cells should be affected when current is injected into one cell by means of a microelectrode. The results of such an experiment in intestinal muscle indicate that stimulation of one cell through an intracellular microelectrode produces localized membrane depolarization in some of the nearby cells.[4,50,80] The spatial decay of this electrotonic potential is sharp, and its time course is rapid (space constant 0.25 mm, time constant 31 msec), suggesting that the current density in the vicinity of the stimulating electrode dissipates quickly. In other words, a

---

*An electrogenic ion pump is one that transfers charge across the membrane and thereby causes a change in membrane potential. An electrogenic Na pump, when activated, transports Na^+ outward across the membrane against the Na concentration gradient without accompanying anions and without a one-to-one inward transport of K^+. As a result, there is a net outward movement of positive charge and the membrane hyperpolarizes. When the pump is turned off or its activity is reduced, there is an immediate depolarization.

significant fraction of the current injected into the cell is shunted through low-resistance paths into the cytoplasm of neighboring cells. This shunting of current through a complex three-dimensional network of cells probably explains why it is difficult to elicit a conducted action potential in many visceral muscles by localized, intracellular stimulation.[80]

On the other hand, if the stimulating electrodes are extracellular and are large relative to the dimensions of a single muscle cell, the current flow between the electrodes causes a simultaneous change in the membrane potential of many cells.[4,80] In this situation the magnitude of the electrotonic potential declines exponentially relative to distance from the stimulating electrodes, suggesting that the electrical properties of the tissue resemble those of a uniform core conductor. The externally applied current spreads longitudinally in one dimension, with space and time constants of 1.9 mm and 100 msec, respectively (intestinal muscle).[80] Since a single cell is only 5 μm wide and about 200 μm long, the characteristics of the electrotonic spread of externally applied current indicate that the functional units for conduction are bundles of interconnected cells rather than individual cells. This concept is strengthened by the observation that spike generation becomes graded in longitudinal strips of muscle less than 100 μm wide and when the diameter of the stimulating electrodes is less than 50 μm.[4] Thus in order to initiate a conducted action potential, many cells have to be depolarized at the same time. The conducted action potential is apparently capable of depolarizing many cells simultaneously,[80] which indicates that there is electrical homogeneity among groups of cells.

The regions of cell membrane where the cells are electrically connected may be the gap junctions. The number of such junctions per cell is not known, nor is their electrical resistance. Furthermore, there is no direct evidence indicating that electrical spread of excitation actually does occur at these sites.

Ionic basis of electrical activity in visceral muscle

The ionic basis of electrical activity in visceral muscle is qualitatively similar to that already described for nerve and striated muscle fibers (Chapters 1 and 2), the membrane potentials being related to the distribution of ions across the cell membrane and to the ionic permeability of the membrane.

The resting potential of many excitable cells arises from the concentration gradient for K^+ across the membrane (K^+ being more concentrated inside than outside) and from the relative impermeability of the resting membrane to ions other than K^+. If the resting potential is exclusively a K^+ diffusion potential, then (1) its magnitude should be equal to the K^+ equilibrium potential and (2) it should be linearly related to the logarithm of the K^+ concentration gradient across the membrane. In skeletal muscle, the resting potential and the K^+ equilibrium potential are almost identical, but the resting potential in smooth muscle is considerably less than the K^+ equilibrium potential (Table 4-1). This suggests that in visceral muscle other ions in addition to K^+ contribute to the resting potential. The most likely candidates are Na and Cl. In intestinal muscle, over 95% of the intracellular Na exchanges with the extracellular Na with a half-time of less than 1 min.[46] This exchange rate is many times greater than in skeletal muscle at rest. The ratio of Na to K^+ permeability is 10 to 30 times higher in smooth muscle than in skeletal muscle (Table 4-1). Cl^- also exchanges rapidly in visceral muscle, at least 85% of the tissue Cl exchanging with a half-time of 8 min.[46]

Despite these appreciable Na and Cl permeabilities, the resting potential of visceral muscle is nevertheless *predominantly* a K^+ potential. The magnitude of the resting potential is inversely related to the concentration of K^+ in the extracellular medium (Fig. 4-4). However, this relationship is linear only at concentrations of K^+ above 30 mM.

The deviation from linearity at concentrations below 30 mM is probably the result of the significant contributions of Na and Cl permeabilities to the resting potential. Furthermore, probably all visceral smooth muscles have an Na-K pump whose activity contributes to the resting potential at physiologic concentrations of extracellular K.[6,36]

In skeletal muscle and most nerve axons the action potential is produced by movements of Na^+ and K^+ across the membrane successively and respectively depolarizing and repolarizing the membrane.

A detailed quantitative analysis of the ionic currents accompanying the action potential, similar to that for skeletal muscle and nerve, is not available for smooth muscle. The primary reason for this deficiency in smooth muscle is that the voltage clamp technique, which permits one to estimate the magnitude and time course of ionic

Table 4-1. Comparison of electrolyte distribution, equilibrium potentials, resting potentials, and permeabilities in smooth and skeletal muscle in vitro

Tissue	Extracellular* (mM)			Intracellular (mM)			Equilibrium potential† (mV)			Resting potential (mV)	Permeability ($\times 10^{-8}$ cm/sec)			P_{Na}/P_K	P_{Cl}/P_K	Reference
	Na	K	Cl	Na	K	Cl	Na	K	Cl		Na	K	Cl			
Taenia coli (guinea pig)	137	5.9	134	13	164	58	+62	−88	−22	−57	1.8	11	6.7	0.16	0.61	34
Aorta (rat)	150	10	150	12	163	56	+67	−74	−26	−50	2.3	6.0	12.6	0.38	2.00	53
Sartorius (frog)	120	2.5	121	10	140	3	+63	−102	−92	−90	0.8	60	120	0.01	2.00	56

*Values for physiologic salt solutions bathing the muscle.

†Equilibrium potential is calculated as $E_{ion} = \frac{RT}{nF} \ln \frac{[ion]_o}{[ion]_i}$.

permeability changes as a function of membrane potential, is not readily applied to smooth muscle. The small size of the smooth muscle cells and their complex three-dimensional arrangement preclude the use of intracellular microelectrodes for measuring current and voltage within the same cell. Nevertheless, a qualitative analysis of the ionic mechanisms responsible for the action potential has been attempted by use of the double sucrose gap technique* or by experiments with microelectrodes where the extracellular ions are replaced or where drugs known to interfere with ion movements are applied before and during action potential generation.

The rising velocity and amplitude of the action potential in visceral muscles depend on both Na and Ca, although the relative contribution of each of these ions varies from one type of muscle to another.[6,58,80] In some intestinal muscles and the smooth muscle of the vas deferens the spike amplitude and rising velocity are unchanged when the extracellular Na concentration is reduced to one tenth its normal value. However, a similar reduction in extracellular Ca causes a significant decrease in spike amplitude and rising velocity even when Na concentration is normal.[80] In these muscles, action potentials disappear in Ca-free solutions or in solutions containing agents that prevent Ca influx across the cell membrane.[6,80] Voltage clamp experiments on intestinal muscle substantiate these results and provide additional evidence that in the intestine and vas deferens the inward current producing the depolarization of the action potential is carried primarily by Ca.[52]

*An extracellular method whereby a small segment of muscle is placed in a methyl methacrylate chamber containing three pools of saline solution separated from one another by high-resistance sucrose gaps. The muscle is situated so that the ends are in the side pools and the sucrose forms insulating cuffs around the middle portion or "node." It is essential that the node be kept as small as possible and always well below the length constant of most visceral muscles (about 1 mm). Also, the muscle bundles within the node should be uniformly oriented in a longitudinal direction. Constant current pulses are applied between the saline pool at one end of the muscle and the node while the voltage is monitored between the opposite saline pool and the node. The limitations of this method include uncertainty about the uniformity of current distribution within the node and about the homogeneity of control of the membrane potentials of all cells within the node during clamping (for critical discussion see Ramon et al.[70]) Because of these limitations, voltage control is poor during rapid transient changes in membrane potential but good during steady holding potentials. Despite these difficulties, the extracellular technique has been useful in qualitative analyses of the slow currents during spike generation.

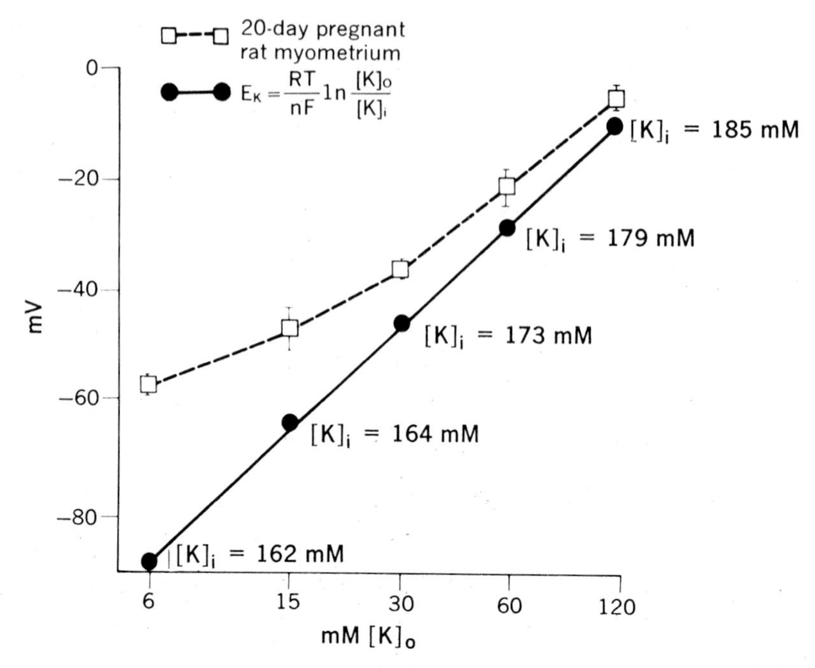

Fig. 4-4. Relation between resting membrane potential, calculated potassium equilibrium potential (E_K), and external potassium concentration ($[K]_o$) in uterine smooth muscle. Squares are measured resting potentials (mean ±SD). Circles are equilibrium potentials calculated from Nernst equation. Note that as $[K]_o$ is raised, $[K]_i$ also increases. (From Casteels and Kuriyama.[33])

On the other hand, in the smooth muscle of the uterus, urinary bladder, and ureter the rising velocity and amplitude of the action potential are significantly reduced in Na-deficient solutions (less than one tenth normal value) and the action potentials disappear in Na-free solutions.[32,80] Reduction of external Ca concentration also decreases spike amplitude and rise time despite the presence of normal amounts of Na, whereas elevation of Ca level increases these parameters of the spike.[80] Voltage clamp experiments on the myometrium have substantiated these findings and suggest that both Na and Ca are needed for spike electrogenesis.[12,54]

Thus in the intestine and vas deferens, Ca^{2+} probably carries most of the inward current accompanying the action potential under normal circumstances, but Na can substitute for Ca in Ca-deficient solutions. On the other hand, in the uterus, bladder, and ureter, Na^+ is primarily responsible for the spike, although there is an appreciable Ca-current that is unmasked in Na-deficient solutions.

The repolarization phase of the action potential is presumably associated with an increase in K permeability, although there is no direct evidence for this assumption.[6]

IONIC DISTRIBUTION IN VISCERAL MUSCLES

When evaluating the experimental findings for the distribution of electrolytes (intracellular vs extracellular) in visceral muscle, one must keep in mind that this tissue is never quiescent for long periods of time. Its functional state at the time of the electrolyte analyses is not precisely known. Therefore steady-state resting levels of ionic distributions, comparable to those in skeletal muscle at rest, cannot be determined in visceral muscle. The intracellular ionic concentrations in smooth muscle probably represent the mean values that exist somewhere between complete quiescence and some degree of activity of the muscle.

The data in Table 4-1 show that the intracellular concentrations of Na and K in smooth muscle are similar to those of skeletal muscle, whereas the internal Cl concentration is considerably higher in smooth muscle. The instability of the membrane (i.e., its tendency for spontaneous depolarization) in visceral muscle may be related to the fact that the distribution of both Na and Cl is not in equilibrium with the resting potential. A slight increase in Na or Cl permeability (or decrease in K permeability) would drive the

Table 4-2. Muscle cell parameters related to excitation-contraction coupling*†

Muscle	Cell diameter (μ)	Cell surface area (cm^2/gm‡)	Cell volume (cm^3/gm)	Actomyosin content (mg/gm)	Ca^{2+} required for coupling (mμM/gm)	Ca^{2+} influx per action potential (mμM/gm)
Sartorius (frog)	100	300§	0.81	70	91	0.1
Atria (guinea pig)	25	1,000	0.70	38	50	1.0
Taenia coli (guinea pig)	5	3,300	0.50	10	13	4.0

*Data from Bianchi.[15]
†Cell surface area calculated on the basis of cell diameter and assuming the cell is a cylinder.
‡Wet weight of muscle.
§Exclusive of transverse tubular element, which is about 2,100 cm^2/gm.

potential toward the equilibrium potentials for Na and Cl. The membrane would then depolarize, and when its potential reached threshold, action potentials would appear. One might also predict, on the basis of the information in Table 4-1, that a similar situation exists in aortic smooth muscle. However, this muscle is not spontaneously active probably because of certain unique electrical characteristics of its cell membrane (see p. 136).

Despite the appreciable resting Na permeability and the steep electrochemical gradient for this ion, the concentration of Na^+ within the smooth muscle cell remains constant and at a much lower level than in the extracellular medium (Table 4-1). Hence Na^+ must be actively ejected as fast as it enters the cell, that is, smooth muscle must have an effective Na pump. The existence of a metabolically dependent Na pump is evidenced by the fact that low temperature or metabolic inhibitors cause visceral muscles to gain Na and lose K. Rewarming the tissue or washing out the inhibitor results in an extrusion of Na and reaccumulation of K.[36] In the taenia coli, this pump has a Na to K pumping ratio of 3:2 and therefore is electrogenic with respect to Na.[2] The equilibrium potential for Cl is some 25 mV less negative than the resting potential, and therefore the high internal Cl concentration (see skeletal muscle data in Table 4-1) is maintained against a steep electrochemical gradient, necessitating an ion pump. An active Cl pump has also been postulated for smooth muscle.[34]

Excitation-contraction coupling

Contractions of visceral muscles are preceded by depolarization of the cell membrane and action potential discharge. Although the nature of the process connecting excitation to contraction in smooth muscle is not entirely known, Ca^{2+} is believed to play an important role. A rise in intracellular free Ca^{2+} concentration is considered to be the primary event in excitation-contraction coupling.[15] Smooth muscle, like striated, contracts when the intracellular Ca^{2+} rises above 10^{-7}M and contraction is maximal at 10^{-6}M. Relaxation occurs when the Ca^{2+} concentration falls below 10^{-7}M.[75] This "activator" Ca^{2+} may come from the extracellular medium, from the plasma membrane, or from the sarcoplasmic reticulum. The relative contributions of each of these sources varies in different types of smooth muscle.[37]

Since the amount of Ca^{2+} needed for maximum activation of the contractile proteins is 1.3 μM/gm actomyosin, the actomyosin content of a muscle can be used to estimate the amount of Ca^{2+} needed for excitation-contraction coupling.[15] From Table 4-2, we see that this amount is highest for skeletal and lowest for smooth muscle. In the taenia coli the Ca^{2+} influx during one action potential could supply almost one third of this activator Ca^{2+} (Table 4-2) and with repetitive spiking (which usually accompanies each contraction in most visceral muscles, as shown in Fig. 4-6) the influx of extracellular Ca^{2+} would be more than adequate to produce a maximum contraction.

Since smooth muscles develop tension more slowly than most striated muscles, the need for rapid activation of the contractile elements in the former is not as great as in the latter. It will be recalled that the sarcoplasmic reticulum is not as well developed in smooth as in skeletal muscle and smooth muscle lacks a T system. Since the smooth muscle cell has a small diameter and a large surface to volume ratio (Table 4-2), the time for inward diffusion of activator Ca^{2+} from the surface membrane to the contractile elements is thought to be well within the limits of the latency between membrane excitation and contraction.[68]

The latency between excitation of the muscle and the onset of its contraction is of interest, since it is relevant to the kinetics of Ca movements in the coupling of excitation to contraction. Recently an attempt has been made to measure this latency in single, isolated smooth muscle cells from the stomach of the toad.[41] The isometric contractions of the muscle cell in response to brief electrical field stimulation were recorded, and the minimal latency between the electrical stimulus and onset of active force was about 400 msec at 22° C. Another approach for estimating the "excitation-contraction" latency in mammalian smooth muscle is to record simultaneously the action potential and contraction from a very small segment of muscle where the propagation of excitation is rapid and uniform and the conduction distance is short. The segment is stimulated electrically at multiple points along its surface in order to minimize the effect of impulse propagation. In this situation, conduction time is only a small fraction of excitation-contraction latency. The stimulus intensity is adjusted so that each shock evokes only one action potential. The results of one such experiment are shown in Fig. 4-5, which also includes, for comparison, the excitation-contraction interval in a single fiber of frog skeletal muscle and in a short strip of mammalian ventricle. Excitation-contraction latency is short in skeletal (about 10 msec) and cardiac (about 25 msec) muscle but consid-

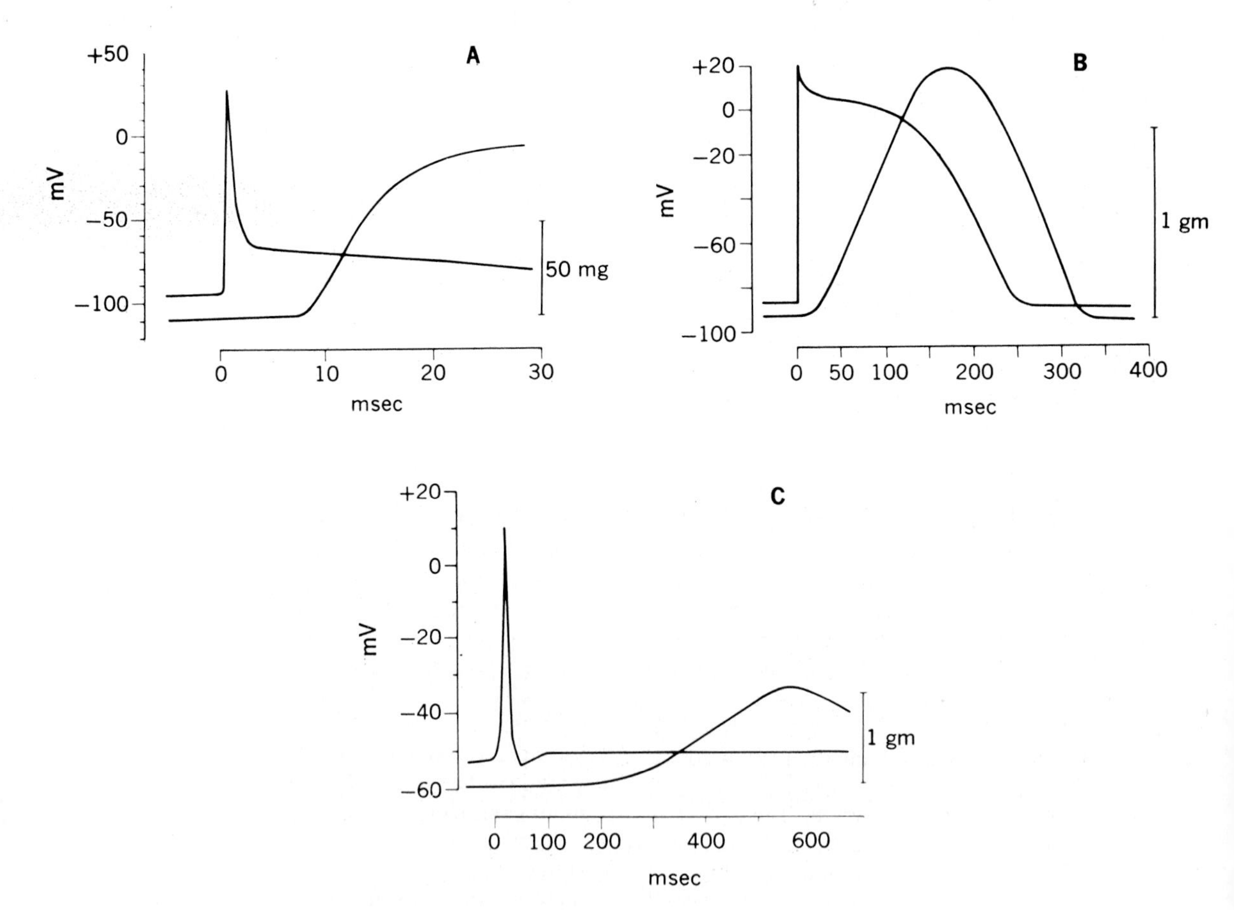

Fig. 4-5. Diagrams showing relation between transmembrane action potential and isometric tension curve. **A,** Single muscle fiber from frog semitendinosus muscle. Temperature 20° C. Excitation-contraction latency (time from onset of depolarization to initial development of tension) about 8 msec. **B,** Small segment of ventricular muscle of cat. Temperature 37° C. Excitation-contraction latency about 10 to 15 msec. **C,** Small segment of uterine muscle of rat. Temperature 37° C. Excitation-contraction latency about 200 msec. (**A** from Hodgkin and Horowicz[49]; **B** modified from Brooks et al.[20])

erably longer in smooth muscle (about 200 msec at 37° C). Since tension is developed more slowly in smooth than in striated muscle, the need for rapid activation of the contractile elements is not as imperative in smooth muscle as in striated. Smooth muscle cells have small diameters and a large surface to volume ratio, and the time for inward diffusion of Ca has been estimated to be less than one hundredth of the 200 to 400 msec excitation-contraction latency in these cells.[41,68]

The magnitude of the contraction of visceral muscle is directly related to (1) the frequency of action potential discharge from the individual muscle cells and (2) the total number of cells synchronously active. The relationship between action potential frequency and force of contraction is best illustrated in very small strips of muscle where conduction distance is short and the propagation of action potentials is uniform, that is, electrical activity of one cell is representative of all the cells in the strip. Simultaneous electrical and mechanical records from one such segment of uterine smooth muscle are shown in Fig. 4-6. A single action potential produces a small, discrete increase in muscle tension, multiple action potentials at low frequency produce partially fused contractions, and a further increase in action potential frequency results in a smooth tetanic contraction. The duration of the contraction is controlled by the duration of the action potential train. This direct correlation between action potential frequency and contraction exists, of course, only if all the individual muscle cells are active in synchrony. If the activity is asynchronous or if some cells do not participate in every contraction, the electrical record from an individual cell is not necessarily representative of all cells.

Factors influencing performance of visceral smooth muscle

Stretch. Visceral smooth muscles are sensitive to stretch and respond with a depolarization of the cell membrane. In quiescent muscles, if this depolarization reaches threshold, action potentials are generated and the muscle contracts.[4,58] If the muscle is rhythmically active when it is stretched, the resulting membrane depolarization converts the intermittent discharge of action potentials into a continuous train, producing a sustained contraction of the muscle. The magnitude of the membrane depolarization is related to the degree of the stretch. When a segment of intestinal muscle is progressively stretched in a stepwise manner, there is a graded depolarization of the cell membrane as well as a graded increase in spike frequency and tension.[4]

The stretch sensitivity is a property of the smooth muscle cell itself and does not depend on the nervous elements within the tissue. It is present in nerve- and ganglion-free preparations and is not affected by ganglion- or nerve-blocking substances.[4] Smooth muscle thus differs from the stretch-insensitive skeletal muscle fiber.

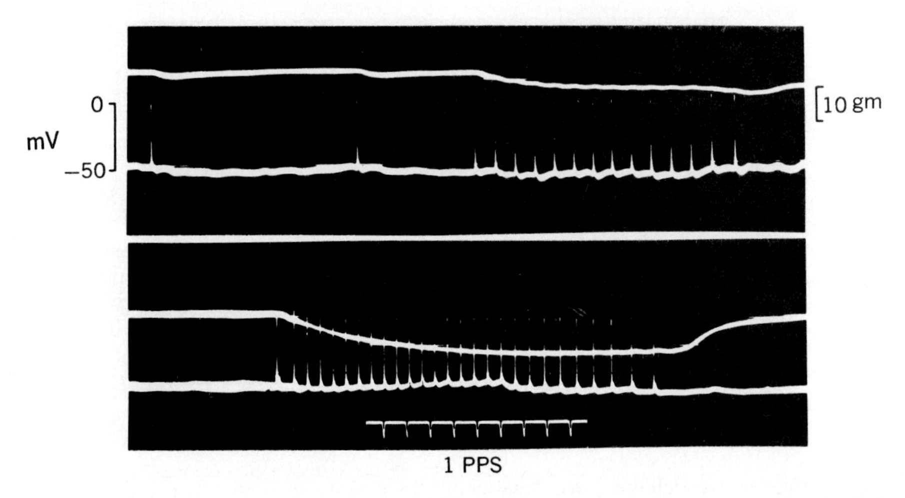

Fig. 4-6. Relation between rate of action potential discharge and force of contraction in isolated segment of estrogen-treated rat myometrium. Isometric tension record, top trace (downward deflection indicates increase in tension); transmembrane potentials, lower trace. Continuous record with microelectrode remaining within same smooth muscle cell during several successive contractions of muscle. Magnitude and duration of contractions are regulated by rate and duration of action potential discharge. (From Marshall.[60])

The sensitivity to stretch is of obvious functional importance in visceral muscles. An increase in the intraluminal pressure within an organ such as the ureter, bladder, uterus, or intestine can elicit a contraction of the smooth muscle elements within its walls. The contraction will expel or propel the contents of the lumen.

Humoral factors

Acetylcholine. Acetylcholine stimulates most visceral smooth muscles and increases the force and frequency of their contractions. In spontaneously contracting muscles the primary effect of low concentrations of acetylcholine is an increase in contraction frequency mediated by an increase in the frequency of the slow wave component of the electrical activity.[17] At higher concentrations there is a marked membrane depolarization, an increase in membrane conductance, and the initiation or acceleration of spontaneous spike discharge (Fig. 4-7).[22] These actions are prevented by atropine but not by ganglion- or nerve-blocking agents and occur in nerve-free segments of the smooth muscle in chick amnion. Thus acetylcholine acts directly on the muscle cells and not indirectly via neural elements in the tissue. Cholinergic receptors in visceral muscles are probably distributed over the entire surface of the muscle cell membrane rather than localized at a specific site such as the motor end-plate of skeletal muscle.[17]

The depolarization and increase in membrane conductance caused by acetylcholine are primarily the result of an increase in Na conductance.[22] Ca also contributes to the actions of acetylcholine, since both the depolarization and contraction are abolished in Ca-free media.[22,50,58]

Although normally acetylcholine acts initially on the excitable membrane of the smooth muscle cell, it can still evoke a contraction when the membrane is depolarized to a level where the spike electrogenesis is inactivated (e.g., by exposure of the muscle to isotonic K solutions).[39] In K-depolarized muscles the action of acetylcholine depends on the presence of Ca in the external environment and is accompanied by an increase in the movement of K^+ and Cl^- as well as Na^+ across the cell membrane.[38] It should be emphasized, however, that under physiologic conditions the cell is not bathed in isotonic K solutions, and acetylcholine acts initially on the electrical activity of the cell membrane.

Epinephrine and norepinephrine. The actions of the adrenergic amines epinephrine and norepinephrine can be either excitatory or inhibitory, depending on the type of visceral muscle and on the species of animal from which it comes. For example, in the rabbit and in the human the amines inhibit intestinal muscle and stimulate the uterus, whereas in the rat they inhibit both the intestine and the uterus. These actions are mediated by a combination of the

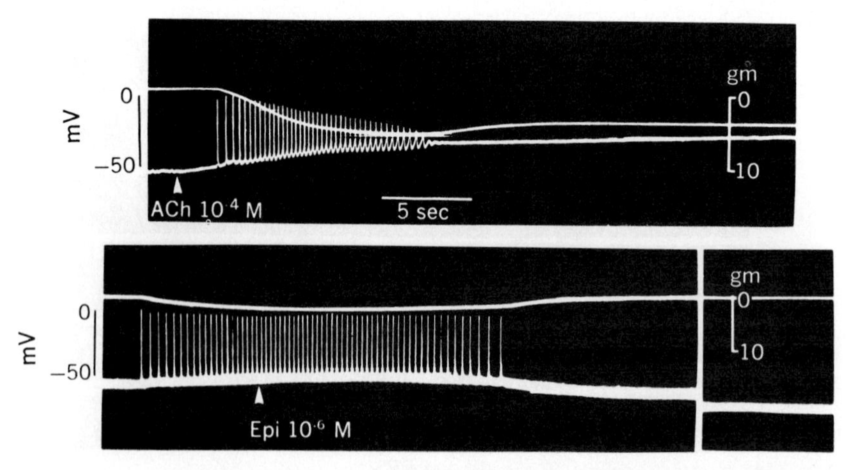

Fig. 4-7. Effect of acetylcholine, *ACh,* and epinephrine, *Epi,* on contractions (upper trace) and transmembrane potentials (lower trace) in uterine muscle of rat. In experiment shown in top frame, high concentration of acetylcholine produced sustained depolarization of cell membrane and prolonged contraction of muscle. In experiment illustrated in lower frame, epinephrine was added during spontaneous contraction of muscle. Within about 15 sec, action potentials disappeared and muscle relaxed. Records in right-hand part of frame were taken 5 min after application of epinephrine, and at this time membrane potential had increased to around −65 mV.

amines with specific receptor sites on or in the muscle cell, the so-called adrenoceptors.[21] The classification of adrenoceptors into α and β groups is now generally accepted. In smooth muscle, activation of the α-receptors usually results in excitation, whereas β-receptors mediate inhibition, with the exception of the intestine, where inhibition is mediated by both α- and β-receptors. In most smooth muscles, norepinephrine activates primarily α-receptors, whereas epinephrine activates both α- and β-receptors.

The stimulatory actions of epinephrine and norepinephrine are superficially similar to those of acetylcholine in that they are associated with depolarization, an increase in membrane conductance, accelerated spike discharge, and increased tension.[22] However, the specific mechanisms underlying these actions vary from tissue to tissue. In the myometrium the depolarization is due primarily to an increased conductance to Cl, whereas in the vas deferens depolarization is mediated by an increase in Na conductance.[22,59] In the ureter the depolarization is thought to result from a decrease in K conductance.[74] In all tissues, however, the stimulatory effects depend on the presence of Ca in the external medium.

The inhibitory actions of the adrenergic amines are characterized by a cessation of action potential discharge and a hyperpolarization of the cell membrane, that is, the membrane potential becomes more negative (Fig. 4-7). The amount of hyperpolarization depends on the magnitude of the membrane potential when the amines are applied. If a strip of intestinal muscle is stretched so that the membrane potential is reduced to about −45 mV, epinephrine increases the potential to about −60 mV. In the absence of stretch or when the muscle is quiescent, the membrane potential may already be around −60 mV, and in this situation the hyperpolarizing effects are minimal.[58]

The mechanisms underlying these inhibitory actions have been studied most extensively in the intestine and myometrium. In the intestine, inhibition is the combined result of activation of both α- and β-receptors; α-receptor activation produces membrane hyperpolarization and β-receptor activation suppresses spontaneous pacemaker discharge.[23,24] In the myometrium, both the potential change and the suppression of pacemaker activity are mediated by β-receptor activation.[57] A number of studies suggest that these actions result from an increase in membrane conductance to K[23,24,57,74] or an increase in the potential gradient for K^+ across the cell membrane and an increase in the K^+ selectivity of the membrane.[55]

Electrical stimulation of the muscle, after its spontaneous activity is abolished by β-adrenergic agents, causes a contraction of the muscle, although the magnitude of this contraction is considerably less than normal. This, plus the finding that these agents relax K-depolarized muscles without an accompanying change in membrane potential, suggests an interference with excitation-contraction coupling.[21]

Another prominent action of β-adrenergic amines is stimulation of cellular metabolism. In many tissues, including the intestine and uterus, this action is related to an increase in the level of cyclic AMP within the muscle cell.[21,62] Cyclic AMP in turn activates a protein kinase that stimulates the active binding of Ca to the cell membrane and to intracellular sites as well as the outward extrusion of Ca across the cell membrane. In this way the amount of ''activator'' Ca available to the contractile elements is reduced and the muscle relaxes.[62] The increased binding of Ca to the cell membrane is also believed to suppress the pacemaker electrical activity.[21]

Hormonal factors. Although the behavior of all smooth muscles is probably influenced by their hormonal environment, the only visceral smooth muscle that has been extensively studied with respect to the hormonal regulation of its activity is the myometrium (uterine muscle). This discussion will therefore concern itself only with this muscle, which is uniquely sensitive to the ovarian hormones estrogen and progesterone and to oxytocin, a hormone of the neurohypophysis.

The uterus from an animal in diestrus shows little spontaneous activity and is virtually insensitive both to electrical and chemical stimulation and to stretch. During estrus the uterus becomes increasingly spontaneously active and sensitive to external stimuli and to stretch.[19] A similar evolution of uterine activity can be induced in ovariectomized or immature animals by the administration of estrogen. In addition to its effects on the excitability of the myometrium, estrogen also stimulates the synthesis of actomyosin and high-energy phosphate compounds within the myometrial cell.[35] After priming with estrogen, further treatment with progesterone results in a reduction of uterine motility and of sensitivity to external stimuli and to stretch. Progesterone has no significant effects on the contractile components of the muscle cell.[65]

One of the means by which estrogen and progesterone modify myometrial activity is through alterations in the level of the resting membrane potential and in the propagation of action potentials through the myometrium.[60,61] The resting membrane potential of the myometrial cells in an immature or ovariectomized animal is only about −35 mV. At this potential the membrane is relatively inexcitable, supposedly because the membrane is in a state of "accommodation,"* and the muscle is quiescent. Estrogen treatment brings the potential into the range (about −50 mV) in which spontaneous discharge of action potentials occurs and the muscle becomes rhythmically active. Estrogen also enhances the conduction of excitation throughout the muscle, producing a well-synchronized activation of many cell groups. Thus the strong contractions typical of the estrogen-dominated uterus result from the actions of estrogen on the excitable membrane as well as on the contractile elements of the myometrial cell.

Progesterone stabilizes the membrane potential near the resting level. This effect diminishes the excitability to the muscle and produces areas of conduction block within the tissue.[4,61] Consequently, the propagation of action potentials from cell to cell is impeded. Not all regions of the muscle are active during each contraction, and the force of contraction is reduced. The weak, uncoordinated contractions typical of the progesterone-dominated uterus result from the effects of progesterone on the excitable membrane of the myometrial cell and not from an alteration in the contractile machinery.

During the course of pregnancy changes occur in the myometrium that parallel those just described for estrogen and progesterone. In the rat and rabbit myometrium the membrane potential of the muscle cells gradually becomes more negative, reaching a maximum (about −60 mV) around midpregnancy and remaining at this level until the end of term. During this period the uterus may show some spontaneous contractions, but these are localized, irregular, and weak. About 24 hr before parturition the membrane begins to depolarize and the uterus becomes progressively more active. At parturition the membrane potential is about −50 mV and uterine motility is maximal, with action potentials and waves of contraction spreading uniformly through the muscle.[33] The concentration of high-energy phosphate compounds and actomyosin within the myometrial cells also increases throughout pregnancy, reaching a maximum several days before parturition.[35]

The increased negativity (e.g., about −60 mV) of the membrane potential at midpregnancy comes at a time when the concentration of progesterone in the maternal blood is relatively higher than that of estrogen. The decline in negativity of the membrane potential, with the subsequent increase in uterine conductivity and motility at the terminal stage of pregnancy, may be the result of the rising level of estrogen in the blood that occurs during the latter part of gestation.

These changes in membrane potential and propagation of action potentials may reflect hormonally induced modifications in the transmembrane ionic concentration gradients and in the ionic permeability of the cell membrane. To explore these possibilities, Casteels and Kuriyama[33] measured the ionic distributions (intracellular vs extracellular) and permeabilities in the myometrium of the rat at various stages of pregnancy. They found no significant differences in the Na, K, or Cl concentration gradients during pregnancy. On the other hand, the K permeability gradually increased until midpregnancy and then remained high until near the end of gestation. Na permeability was relatively low throughout pregnancy but increased several days before delivery. The electrophysiologic changes in the myometrium that accompany pregnancy might therefore be explained on the basis of these permeability variations.

Oxytocin, a hormone from the neurohypophysis, is one of the most selective of all visceral smooth muscle stimulants, acting primarily on the myometrium and on the myoepithelial cells of the mammary glands.[61] In quiescent muscles, oxytocin initiates rhythmic bursts of action potentials and contractions. In spontaneously active muscles, low concentrations of oxytocin accelerate the frequency of contractions primarily by increasing the frequency of the slow waves.[11] At higher concentrations, oxytocin depolarizes the membrane and increases the spike frequency, causing more forceful contractions. Very high concentrations produce sustained depolarization and contracture. In the human, rabbit, and rat the myometrium is most sensitive to oxytocin near the end of gestation or when the uterus is primarily under the influence of estrogen.[61]

The depolarizing action of oxytocin depends on the presence of both Na^+ and Ca^{2+} in the external medium.[67] Oxytocin increases the Ca^{2+}

*A prolonged, steady depolarization of its membrane makes a cell intrinsically less excitable, presumably because of the high degree of inactivation of Na conductance. Therefore a stronger stimulus is required to increase Na conductance sufficiently to produce a regenerative response (i.e., an action potential). This decrease in excitability (increase in threshold) accompanying a sustained membrane depolarization is termed accommodation (Chapter 2).

conductance of the myometrial membrane and thereby not only accentuates rhythmic spike discharge but also augments contractile force.[64]

MULTI-UNIT SMOOTH MUSCLE

Nictitating membrane and pilomotor muscles

Of the two classic examples of multi-unit muscles, the nictitating membrane and the pilomotor fibers, the former has long been a favorite of the investigator interested in the effects of drugs on autonomic neuroeffector systems. In the cat, where the nictitating membrane is more highly developed than in most species, this organ consists of two thin sheets of smooth muscle inserted into adjacent sides of a T-shaped cartilage on the nasal portion of the orbit. The nictitating membrane is innervated exclusively by postganglionic adrenergic nerve fibers arising from cells in the superior cervical ganglion, and it can be easily exposed with both pre- and postganglionic nerve fibers intact. Normally the membrane shows no spontaneous activity but can be made to contract by stimulation of the pre- or postganglionic nerves. Several attempts have been made to record with extracellular electrodes the electrical activity of the smooth muscle cells during nerve stimulation,[73] but the records are a mixture of complex waveforms and are difficult to interpret. The structure of the nictitating membrane has been studied with the electron microscope, and it is interesting that the smooth muscle cells in this organ do not show the gap junctions so characteristic of some visceral smooth muscles.[25] Each smooth muscle cell in the nictitating membrane is completely surrounded by a basement membrane that separates the cells by distances of at least 600 Å. Furthermore, many small nerve fibers run between the muscle cells, and apparently each individual cell is reached by at least one or more axons. These morphologic characteristics substantiate the physiologic findings that the activation of the smooth muscle cells in the nictitating membrane is neurogenic and that excitation by myogenic conduction from cell to cell, so typical of visceral smooth muscle, does not occur.

Although the nictitating membrane is normally activated only through its motor nerves, it can contract in response to a variety of smooth muscle stimulants (e.g., epinephrine, norepinephrine, histamine, and acetylcholine) when these substances are either topically applied or injected into the blood supplying the organ. The finding that acetylcholine stimulates the muscle is interesting because it shows that the cells are sensitive to the cholinergic transmitter even though the muscle itself does not possess cholinergic innervation. As would be expected for a multi-unit muscle, the nictitating membrane is insensitive to stretch.[4]

The pilomotor muscles are discrete microscopic bundles of smooth muscle cells attached at one end to hair follicles and at the other end to the inner surface of the basal layer of the epidermis. Their motor innervation comes exclusively from sympathetic adrenergic fibers. When the muscles contract in response to nerve stimulation, they throw the tissue into folds and erect the hairs. An in vitro method for studying the activity of the pilomotor fibers in the skin of the cat's tail has been devised.[47] Although individual muscle bundles cannot be isolated, tubes of skin containing many such bundles can be removed from the tail. When these tubes are everted and suspended in an isolated organ bath and the nerves within the tissue are stimulated, the pilomotor muscles contract and the skin tube shortens. The shortening can be registered with a suitable muscle lever or transducer. The effects of nerve stimulation are mimicked by norepinephrine and epinephrine, but unlike the nictitating membrane, the pilomotor muscles are insensitive to acetylcholine. Nothing is known about the mechanism of the effects of nerve stimulation or of exogenously applied neurohumoral agents on the individual pilomotor muscle cells.

Vascular smooth muscle

It is difficult to study vascular smooth muscle in vivo, since many factors, both intramural and extramural, influence its performance. In the intact animal as well as in isolated vascular beds the influences of neurogenic control, humoral vasoactive substances, tissue metabolites, and vessel wall thickness cannot be divorced from those of the experimental test substances.

Because of these in vivo complications, much of the recent work on vascular smooth muscle has been done in vitro, where the environment can be more effectively controlled and where the activity of the muscle can be measured directly. Since the smooth muscle in the walls of arteries is oriented in a close spiral or circular manner, helical strips of muscle can be dissected from the wall, so that the long axis of the strip is parallel to the long axes of the individual muscle fibers.[16] Such strips, when cut from large-conduit vessels (carotid artery, aorta, pulmonary artery) or from small-diameter (200 to 300 μm) resistance vessels and placed in an isolated organ bath, remain viable for many hours. A prominent longitudinal layer of smooth muscle is found in certain veins (portal, inferior vena cava, renal, adrenal) and segments of these muscles can be easily isolated and arranged for experimental use. The results of experiments with these two types of preparations have considerably advanced our understanding of the physiology of vascular muscle over the past decade.[1,6,7,78]

The functional characteristics of vascular muscle are reasonably uniform from species to species but vary considerably within different regions of the vascular bed in any one animal. The behavior of smooth muscle from the large-conduit arteries (aorta, carotid) resembles that of typical multi-unit muscles. The large arteries are not spontaneously active and do not respond to stretch by developing active tension. Their muscle cells are separated from one another by connective tissue elements, and nexuses appear to be less frequent than in smaller arteries and veins.[7,25] The electrolyte distribution and ionic permeability of the smooth muscle of large arteries is similar to that of visceral muscle (Table 4-1). The resting potential of the muscle cells in large arteries is around −50 mV and within the range reported for visceral muscle (Table 4-1). Despite this low resting potential, the membrane potential of the large arteries is quite stable, and spontaneous action potentials are not observed, nor can they be induced by electrical stimulation.[7,63] This electrical stability is probably related to the delayed rectification that occurs when the membrane is depolarized below −40 mV.[63] The rectification is attributed to an increase in K permeability, which prevents depolarizing currents from producing an action potential. Stimulants such as epinephrine, norepinephrine, 5-hydroxytryptamine, angiotensin, and acetylcholine produce graded depolarizations of the cell membrane that result in slow, sustained contractions.[1,8,78] Thus graded depolarization rather than spike generation appears to be the normal electrical trigger for contraction in these muscles.

The behavior of small-resistance vessels and veins resembles that of visceral muscle in many respects.[8,78] For example, smooth muscle from these vessels often displays spontaneous contractions that are initiated by action potentials arising from pacemaker areas within the muscle. The muscles are also stretch sensitive, and their cells lie quite close together. Cell-to-cell junctions are a prominent morphologic feature of these muscles, suggesting a high degree of electrical continuity between cells.[80] The usual response of these muscles to stimulatory agents (e.g., histamine, vasopressin, 5-hydroxytryptamine, and angiotensin) is depolarization initiation of action potentials in quiescent muscles, or an increase in the frequency of action potential discharge in contracting muscles. The response of the small arteries, arterioles, and veins to epinephrine and norepinephrine varies from one given vascular bed to another and is determined by the relative sensitivity of the vascular smooth muscle in these different areas to α- (excitatory) and β- (inhibitory) adrenergic activation.[79,83] Excitation and vasoconstriction are the usual responses of the most vascular smooth muscles to the adrenergic amines. These excitatory effects are prevented and often reversed by α- adrenergic blocking agents. In the coronary circulation, however, the inhibitory action of epinephrine and norepinephrine predominates, resulting in a pronounced dilatation of the coronary bed when these agents are present.[83] This inhibitory action is prevented by appropriate β-adrenergic blocking agents. Venous smooth muscle is also frequently dilated by epinephrine and norepinephrine, this action being mediated by activation of β-adrenoceptors within the muscle.[8]

The predominant effect of acetylcholine on small-resistance vessels and veins is usually inhibitory, producing a marked vasodilatation.[1,78] This action is prevented by atropine, suggesting that acetylcholine is acting directly on the smooth muscle cells rather than on neural elements in the tissue. Acetylcholine has been reported to cause contraction of the carotid and main pulmonary arteries in vitro when applied in relatively high concentrations.[77] Since the contractions are prevented by ganglionic and adrenergic blocking agents, it is likely that the acetylcholine is acting indirectly via activation of adrenergic nerves within the muscle.[1,78]

The membrane depolarization accompanying the excitatory action of norepinephrine probably results from a nonspecific increase in permeability to Na^+, K^+, and Cl^-.[53] The inhibitory actions of β-adrenergic amines are characterized by membrane hyperpolarization that is K dependent and is probably mediated by an increase in cyclic AMP.[8]

In vascular as in visceral muscle, membrane potential changes are not essential for drug-induced contractions and relaxations, since these actions also occur in K-depolarized muscles.[8] However, Ca^{2+} plays a key role in excitation-contraction coupling in both normal and K-depolarized muscles. This "activator" Ca^{2+} comes from both extracellular and intracellular sources. Large arteries (aorta, pulmonary, carotid) maintain their contractile response to norepinephrine for prolonged periods in Ca^{2+}-deficient solutions. Smaller arteries (mesenteric) and veins, on the other hand, do not contract in Ca^{2+}-deficient solutions. These results correlate positively with the greater volume of sarcoplasmic reticulum in the smooth muscle of the large arteries, suggesting a greater number of intracellular binding sites for Ca^{2+} in the larger vessels.[37]

AUTONOMIC NERVE: SMOOTH MUSCLE TRANSMISSION

Electrophysiologic aspects

The basis of our knowledge about the mechanism of transmission of excitation by neurohumoral agents from motor nerves to effector cells comes from the classic electrophysiologic studies on neuromuscular transmission in skeletal muscle by Katz.[56] Recent experiments show that many of the principles that apply to skeletal neuromuscular transmission are also relevant for smooth muscle.[4,51] The subsequent discussion attempts to summarize the pertinent aspects of neuroeffector transmission in smooth muscle.

Although a number of smooth muscle–autonomic nerve preparations of visceral and vascular smooth muscle are now available,[5,9] the two "classic" and therefore most extensively studied ones are the vas deferens–hypogastric nerve preparation from the guinea pig and the nerve–intestinal muscle preparation from the rabbit. This discussion is limited primarily to these preparations. Intestinal muscle can be isolated with both parasympathetic and sympathetic nerves attached so that both excitatory and inhibitory effects can be observed.[44,45] The vas deferens, on the other hand, is innervated exclusively by the sympathetic nervous system, which in this muscle is entirely excitatory.[51]

When a microelectrode is inserted into a smooth muscle cell in the vas deferens, the resting membrane potential is found to be stable and lies between −60 and −70 mV. If the hypogastric nerve is stimulated repeatedly with pulses of submaximal intensity and about 1 msec in duration, each stimulus produces a small localized depolarization of the cell membrane, the so-called excitatory junction potential (EJP).[29] If the nerve is stimulated with sufficient rapidity, the individual potentials sum, reducing the membrane potential to about −35 mV, where an action potential is initiated and the muscle contracts (Fig. 4-8). Junction potentials are recorded in practically every cell impaled by a microelectrode, and they resemble in many respects the end-plate potentials of skeletal muscle. Adrenergic blocking agents (bretylium and phentolamine) or pretreatment of the animal with reserpine (a substance that depletes the adrenergic nerve terminals of their transmitter—norepinephrine) abolishes the EJP's and the muscle no longer responds to nerve stimulation. These findings indicate that the EJPs are produced by the release of the adrenergic transmitter norepinephrine during stimulation of the hypogastric nerve.

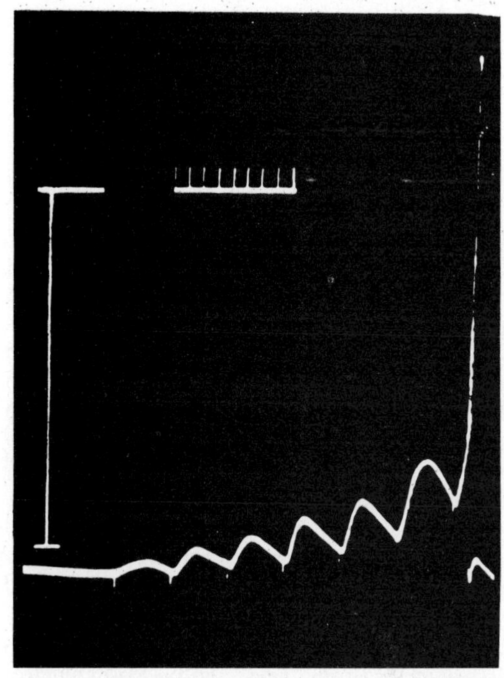

Fig. 4-8. Effect of repeated stimulation of hypogastric nerve on transmembrane potential of single smooth muscle cell in vas deferens of guinea pig. Each stimulus (indicated by small vertical "pip" on trace) is followed by slow depolarization of cell membrane, the so-called excitatory junction potential (EJP). Individual EJPs sum, bringing membrane potential to threshold, around −35 mV, where action potential is initiated.

Vertical calibration, −50 mV, with horizontal top portion of calibration equaling 0 mV. Time mark = 100 msec intervals. (From Burnstock and Holman.[29])

When the nerves are stimulated at frequencies of less than 1 pulse/sec, the individual EJPs are easily seen, since they do not sum and can be studied without the complications of action potential discharge and muscle contraction. Under these circumstances the amplitude of the EJPs is remarkably constant from one cell to another, but their latency (time from nerve stimulation to appearance of junction potential) and rate of development vary considerably. Latencies from 20 to 70 msec have been recorded from cells only 2 mm apart. Assuming that the conduction time for the nerve impulse is about 10 msec, a minimal delay of 10 msec is attributable to the transmission processes, which include the diffusion time for the transmitter from the nerve to the effector site.[29]

Since there are more muscle cells than nerve fibers, the presence of EJPs in every cell suggests that each axon of the hypogastric nerve is able to influence the membrane potentials of many different muscle cells. In order to accomplish this, the individual axons might branch extensively, so that each muscle cell receives a discrete nerve ending, or one branch of an axon might liberate its transmitter substance directly on some cells and in the vicinity of many others. In the latter case the cells would be activated by a general diffusion of transmitter. Activation by generalized diffusion would also account for the marked variations in latency and rising velocity of the individual EJPs. The concept of a generalized diffusion of transmitter was first suggested some years ago by Rosenblueth[73] to explain his finding that in many autonomic neuroeffector systems the effects of variations in strength and frequency of nerve stimulation were "quantitatively interchangeable." Similarly, the amplitude and rate of depolarization of the EJPs in the vas deferens are increased either by raising the strength of nerve stimulation, that is, by increasing the number of active fibers (at constant frequencies above 1 pulse/sec), or by increasing the frequency (at constant strength). These findings for the vas deferens also support the idea of a generalized diffusion of transmitter during nerve stimulation.

In addition to the EJPs, which occur only when the nerve is stimulated, another form of electrical activity has been recorded from the smooth muscle cells in the vas deferens. The second type is seen in the absence of nerve stimulation and consists of a random discharge of small, subthreshold potentials that resemble the "miniature" potentials arising from the postjunctional membrane of the nerve-muscle junction in skeletal muscle (Fig. 4-9).[29,30] These spontaneous potentials do not elicit a contraction of the muscle and are recorded from every cell penetrated with the microelectrode. The amplitude and frequency of the "miniatures" are unaffected by atropine but are reduced by 90% in preparations from reserpine-treated guinea pigs. They presumably reflect the response of the smooth muscle cell to the spontaneous release of norepinephrine from the adrenergic nerve fibers within the tissue.

In contrast to the miniature end-plate potentials recorded from the motor end-plate in skeletal muscle, the amplitude and time course of the spontaneous miniature potentials from the vas deferens vary over a wide range (Fig. 4-9). The largest potentials with a fast time course might be produced by the release of transmitter from nerves close to the point of impalement of the microelectrode on the effector cell membrane; the smaller, slower potentials might be caused by transmitter liberated at greater distances from the electrode. The shape of the amplitude distribution curve for the spontaneous potentials is continuous and similar to that obtained from other muscle cells having multiple innervation (e.g., crustacean muscle and "slow" fibers from frog and chick skeletal muscle).[30] In these cells the microelectrode simultaneously records the activity of many different nerve endings at various distances from the point of impalement. The amplitude distribution curves from muscles that have localized nerve endings (e.g., frog and mammalian "twitch" fibers) show a gaussian or bell-shaped distribution. In such muscles the microelectrode records from only one discrete motor end-plate.

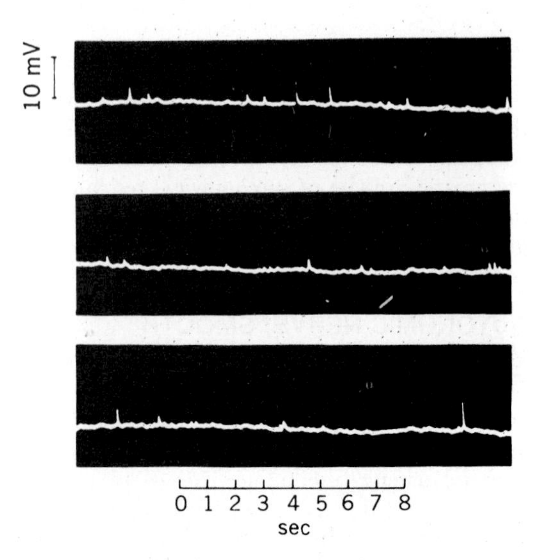

Fig. 4-9. Continuous intracellular recording from smooth muscle cell in vas deferens of guinea pig. Note spontaneous discharge of miniature junction potentials that appear in quiescent muscle cell in absence of hypogastric nerve stimulation. (From Burnstock and Holman.[29])

Katz[56] has shown that the miniature potentials at the motor end-plate in frog skeletal muscle are due to the spontaneous release of quanta of acetylcholine and that the neurally evoked end-plate potential arises from the synchronous release of many quanta of transmitter. At present, techniques are not sufficiently refined to permit recording from a single neuroeffector junction in the smooth muscle of the vas deferens. Consequently, it is impossible to determine whether the spontaneous potentials in this tissue arise in a quantal fashion (i.e., whether they are produced by the release of equal amounts of transmitter at different distances from the locus of the microelectrode). Furthermore, it is not possible to say whether the junction potentials in smooth muscle are composed solely of miniature potentials. Indirect evidence, however, indicates that they are. Pretreatment of the guinea pig with reserpine or partial denervation of the vas deferens reduces the *amplitude* of the neurally evoked potentials and the *rate* of discharge of the spontaneous miniature potentials. Furthermore, when the strength of nerve stimulation is initially low and is gradually increased, the amplitude of EJP increases in a stepwise fashion.[4,51]

EJPs have also been recorded in smooth muscle cells from the longitudinal layer of the rabbit colon in response to stimulation of the pelvic (parasympathetic) nerves.[44] Although in this situation the transmitter is acetylcholine, the EJPs have many features in common with those just described for the vas deferens. One advantage of the rabbit colon preparation is that both the right and left pelvic nerves can be isolated and stimulated. In this manner the interactions of the two nerves can be observed on the effector cell membrane. EJPs are recorded from every smooth muscle cell when either the right or left pelvic nerve is stimulated with single or repetitive shocks. If both nerves are stimulated simultaneously, there is no increase in the rate of development or in the amplitude of the individual EJPs. These findings show that each pelvic nerve has access to every smooth muscle cell in the colon, and that the effects of simultaneous stimulation of both nerves are not additive.[45] Since the colon is spontaneously active, the amplitude and time course of the EJPs elicited by nerve stimulation are usually complicated by the appearance of spontaneously generated action potentials, and it is not possible to study the EJPs as thoroughly in this muscle as in the vas deferens. The high degree of spontaneous motility also makes it technically difficult to keep the microelectrode securely inside one muscle cell for long periods of time, and it is impossible to demonstrate the existence of "miniature" spontaneous potentials in this tissue.

Stimulation of the lumbar colonic nerves abolishes the spontaneous discharge of action potentials and the muscle becomes quiescent. During inhibition the membrane potential of the smooth muscle cells remains at its normal resting level. If the colon is stretched so that the membrane potential of the muscle cells is reduced and the spike discharge becomes continuous, the inhibitory effects of nerve stimulation are accompanied by a generalized hyperpolarization of the cell membrane. The hyperpolarization is abolished by adrenergic blocking agents, suggesting that the inhibitory transmitter is norepinephrine or a mixture of epinephrine and norepinephrine.[45]

Discrete, localized inhibitory junction potentials (IJPs) have been recorded in stretched segments of smooth muscle from the tenia coli of the guinea pig.[31] In these experiments the intramural nerves (those running within the muscle coats) were activated by electrical field stimulation with pulses of short duration (less than 1 msec) at frequencies from 1 to 30 shocks/sec. Following each stimulus there was transient localized hyperpolarization of the muscle cell membrane (Fig. 4-10). Repetitive stimulation at frequencies greater than 1/sec resulted in summation of successive inhibitory potentials, cessation of action potential discharge, and relaxation of the muscle. These IJPs were unaffected by atropine but abolished by procaine (in concentrations that blocked conduction in small nerve fibers), suggesting that the IJPs were neurally mediated but that the neurotransmitter was not acetylcholine. The inhibitory potentials were not prevented by adrenergic blocking agents and were unaffected after degeneration of the sympathetic adrenergic neurons. The principal active substance released by these inhibitory nerves is a purine nucleotide, probably ATP, and therefore these nerves are called "purinergic."[27] They are found throughout the gastrointestinal tract and possibly also in the lung, trachea, uterus, bladder, esophagus, and vascular system.[27]

Morphology of neuroeffector junctions—correlation with electrophysiology

The term *"nerve ending"* as applied to an autonomic neuroeffector junction is best defined in the functional sense to mean the part or parts of the neuron from which the release of transmitter substance occurs during stimulation.[25,72] Studies with the electron microscope have greatly increased knowledge of the neuroeffector junction. Of the various smooth muscles examined, the vas deferens will be discussed here. The fine structure of the innervation of this tissue is essentially similar to that of other smooth muscles (uterus, urinary bladder, intestine, gallbladder, colon, iris),[25,28] and the electrophysiology of

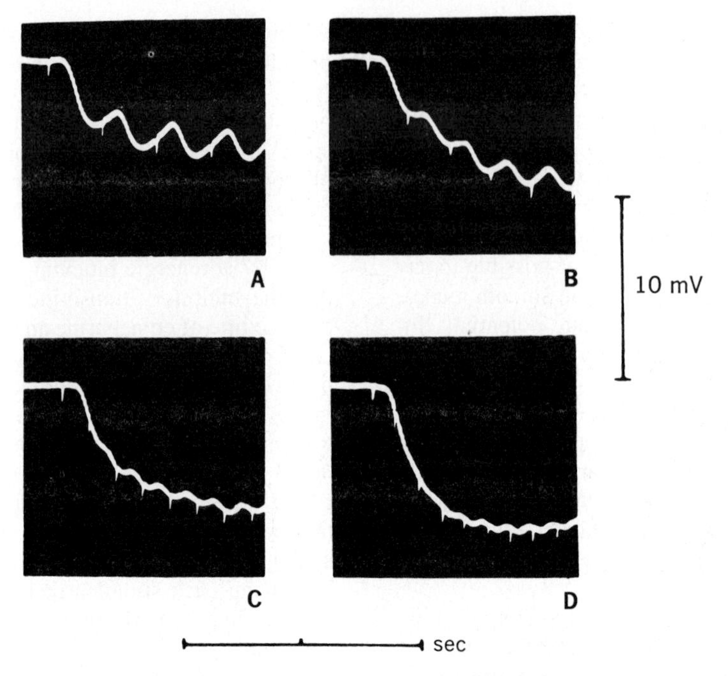

Fig. 4-10. Inhibitory junction potentials (IJPs) recorded from smooth muscle cells in taenia coli of guinea pig in response to stimulation of intrinsic nerves. Electrical records obtained with sucrose gap method. Downward deflection of trace indicates hyperpolarization of membrane. Frequencies of stimulation: **A,** 2/sec; **B,** 3/sec; **C,** 4/sec; **D,** 6/sec. (From Burnstock et al.[31])

neuroeffector transmission has been more extensively studied in the vas deferens than in these other tissues.

As the hypogastric nerves enter the vas deferens, they divide into numerous branches of nonmyelinated axons supported by a network of Schwann cells, the so-called autonomic ground plexus. Fine strands of the plexus, containing small bundles of two to eight axons, ramify within the muscular tissue, and occasionally single axons, partially or completely free of their Schwann cell sheath, come to lie within 20 to 30 nm of the smooth muscle cell membrane.[28,71] Sometimes these unsheathed axons cause indentations in the surface of the muscle cell and when viewed in longitudinal section appear as "beadlike" enlargements of the axon lying in a cleft of the muscle cell membrane.[28,71] In these instances the space separating the nerve and muscle cell membranes may be only 15 nm wide. The axon emerges from the groove on one muscle cell and proceeds to form close contacts with other muscle fibers, constituting a series of "synapses en passage." It has been suggested that the nerve action potential may release transmitter from many points along the axon—from the regions of close contact, the "synapses en passage," as well as other points in the autonomic ground plexus more distant from the muscle cells.[28,71] In this manner one axon may influence many different smooth muscle cells. This arrangement could be the morphologic basis for the electrophysiologic findings that EJPs are recorded from every smooth muscle cell in the vas deferens. The variations in latency and time course of the individual junction potentials may be caused by the release of transmitter, directly on some cells (at the "synapses en passage") and at greater distances from other cells (in the autonomic ground plexus).

In the vas deferens of the rat the innervation is so dense that every cell may be intimately associated with one or more axons.[71] In the vas deferens of the guinea pig, however, there is only one close axon-muscle contact per 100 muscle cells. These observations led Burnstock and Holman[29] to postulate that in the guinea pig some cells were activated directly by the release of transmitter, whereas others were activated indirectly, either by generalized diffusion of transmitter from axons in the autonomic ground plexus or by passive electrotonic spread of current from the directly excited cells.

Unlike the motor end-plate of skeletal muscle, the nerve and smooth muscle cell membranes are usually not specialized at their points of contact. The axons lying close to the smooth muscle cell membrane as well as those running within the autonomic ground plexus, however, are filled with vesicles and mitochondria.[28] It will be recalled that vesicles are also found within the nerve terminals at the neuromuscular junction in skeletal muscle. It has been suggested, but not proved, that acetylcholine might be stored in these vesicles, each vesicle representing a "packet" of transmitters, and that the release of such packets produces the miniature end-plate potentials. The finding of vesicles in the autonomic axons within the vas deferens and the presence of spontaneous miniature potentials in the muscle cells suggest that the transmission processes in the vas deferens and at the skeletal muscle neuromuscular junction are essentially similar.[28]

Since many smooth muscles, including the vas deferens, are innervated by both adrenergic and cholinergic nerves, the transmitter liberated at the neuroeffector sites can be either norepinephrine or acetylcholine. A crucial problem, therefore, is to distinguish between cholinergic and adrenergic neuroeffector junctions. One experimental approach is to examine the relationship between the axoplasmic vesicles and the transmitter substances and/or their precursors. Two types of vesicles, granular and agranular, have been seen in the axons of both the rat and guinea pig vas deferens.[25,28,71] The agranular vesicles are between 40 and 60 nm in diameter and resemble those found in the nerve terminals at the skeletal neuromuscular junction. The granular vesicles range in diameter from 30 to 60 nm and contain a core of electron-dense material. It is now generally agreed that the nerve profiles containing predominantly small agranular vesicles (40 to 60 nm) represent cholinergic axons and that the vesicles are the storage sites for acetylcholine, whereas adrenergic nerves are characterized by the predominance of small granular vesicles (30 to 60 nm) with a dense core.[28] Support for the latter idea comes from the finding that tritiated norepinephrine, injected intravenously, can be localized by electron microscopic autoradiography within the granular vesicles in sympathetic nerves in the rat heart and brain and in the pineal gland.[28,82] Also, drugs that deplete or enhance the storage of norepinephrine within adrenergic neurons cause parallel decreases or increases in the percentage of the small granular vesicles within these nerve terminals[28] (Fig. 4-11).

Electron microscopic studies on nerve axons in the sphincter muscle from the rabbit iris, which is believed to be innervated predominantly by cholinergic nerves, show that the axoplasmic vesicles are of the agranular type. The axons in the dilator muscle, thought to be adrenergically innervated, contain most granular vesicles.[72] Purinergic nerves are believed to contain predominantly large (100 to 200 nm) opaque vesicles.[27] However, direct evidence from specific histochemical and autoradiographic methods to substantiate this belief is not yet available.

Combined histochemical and electron microscopic techniques have been particularly useful in the identification of adrenergic and cholinergic neurons.[25,82] Catecholamines develop an intense fluorescence when treated with formaldehyde, and this reaction has been exploited to visualize adrenergic neurons in smooth muscle.[40] When freeze-dried muscles are exposed to formaldehyde and then sectioned and examined under the fluorescence microscope, areas of intense fluorescence appear as beadlike varicosities within the cell body and along the entire lengths of the terminal axons of the adrenergic nerves (Fig. 4-12, *A*). Chemical analyses of the catecholamine content of the muscles correlate positively with the size and distribution of these varicosities. Combined studies using fluorescence histochemistry and electron microscopy show that the varicosities are especially prominent where the axons containing granular vesicles come in close contact with smooth muscle cells, providing additional evidence that the adrenergic neurotransmitter is localized and probably released at these sites.[4,25]

Histochemical identification of cholinergic axons presumes that functionally cholinergic axons contain higher levels of acetylcholinesterase (AChE) than noncholinergic axons. The histochemical procedure for the localization of AChE involves incubation of the tissue with acetylthiocholine and $CuSO_4$. Tissue AChE hydrolyzes the acetylthiocholine to thiocholine and acetic acid. Thiocholine then precipitates as copper thiocholine, which in the presence of H_2S forms dark-brown deposits at the sites of AChE activity within the tissue. These sites are visible with the light microscope[25] and presumably indicate the areas of acetylcholine release and/or storage. This histochemical method has recently been modified for use in electron microscopy. The innervation of the bladder, vas deferens, and nictitating membrane has been examined with this technique.[25] Heavy deposits of AChE appear in the plasma membranes of some of the axons in these tissues, identifying them as cho-

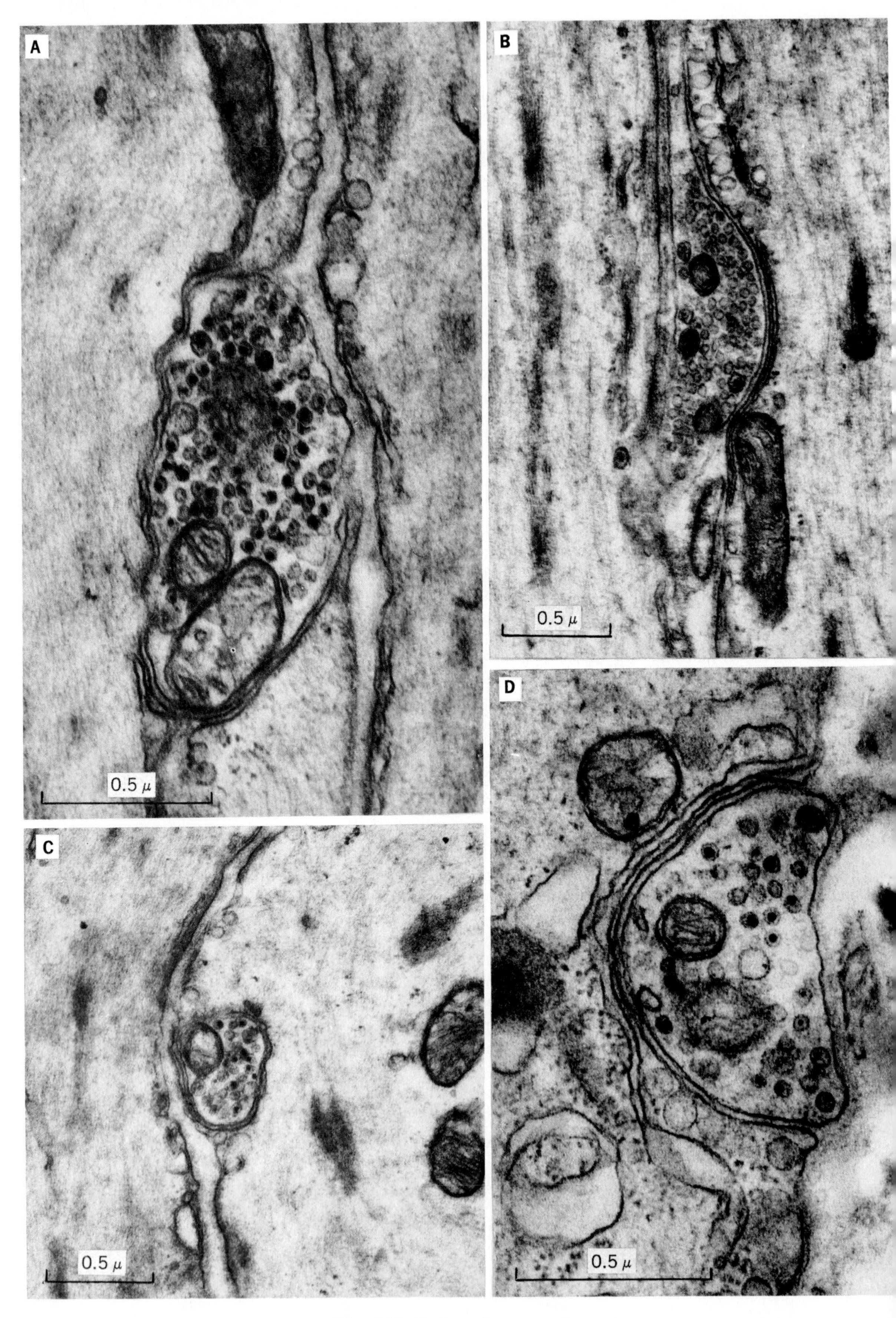

Fig. 4-11. For legend see opposite page.

linergic (Fig. 4-12, *B*). Many cholinergic axons are found in the bladder but relatively few in the vas deferens and nictitating membrane. AChE activity and norepinephrine (as indicated by autoradiography) do not occur together within the same axon (i.e., a single axon does not liberate both transmitters), but separate cholinergic and adrenergic axons sometimes run together in the same nerve bundle in many autonomically innervated muscles.[4,25]

Nerve-muscle relationships in vascular muscle

Although the characteristics of autonomic innervation of vascular muscle are similar to those of visceral muscle, the neuroeffector organization in vascular muscle is characterized by a relatively long distance between terminal axons and smooth muscle cells.[1,7,79] In general the minimum separation distance between nerve and muscle increases the vessel diameter. For example, the minimum neuromuscular distance in arterioles, small arteries, and veins is about 80 to 120 nm, whereas in medium to large arteries it is 200 to 1,000 nm.[26] In most blood vessels the autonomic ground plexus lies within the adventitia, and the terminal axons rarely penetrate beyond the outer surface of the media. This anatomic distribution, especially well documented for adrenergic nerves, varies considerably not only among different blood vessels but also within different regions of the same vessel.[7,26] In large elastic arteries, the adrenergic nerves are generally limited to the outer portion of the vessel; in muscular arteries, they penetrate for a short distance into the media; some thick-walled cutaneous veins contain many adrenergic fibers in their media, whereas veins in skeletal muscle have few nerve fibers in their smooth muscle layers. In the proximal part of the main pulmonary artery of the rabbit, adrenergic nerves terminate on the outer surface of the media, whereas in the distal portion of the same vessel the nerves extend well into the media.[7] The nerve-muscle relationships in most blood vessels become more intimate as vessel size diminishes. In terminal arterioles and precapillary sphincters and venules the ratio of axons to muscle cells is greater than in large arteries. In the smaller vessels the regions of close contact between axons and muscle cell are also more numerous.[1,7,78,79] The cholinergic innervation, although less well studied than adrenergic innervation, is also restricted to the medioadventitial junction.[7,78]

The innervation pattern of blood vessels has functional implications related to the control of these vessels by the sympathoadrenal system. Catecholamines released from adrenergic nerve terminals or diffusing into the tissue from the circulation are inactivated primarily by uptake into the nerve terminals following their action on the effector cell membrane. In densely innervated muscles (e.g., vas deferens and iris), where the nerve terminals are distributed throughout the tissue, exogenous (circulating) catecholamines are rapidly inactivated by uptake into the dense nerve plexus, and their effects are weak and brief. Therefore circulating catecholamines play a minor role in the control of such muscles. Actually, in these muscles the adrenergic nerves not only provide nervous control of the muscles but also protect them from the unwanted effects of exogenous amines.[25] Blood vessels, on the other hand, are concerned with homeostatic responses, and the asymmetric disposition of their innervation is suitable for control of the vessels by both adrenergic nerves and circulating catecholamines.[42] Stimulation of adrenergic nerves in small arteries and veins causes prompt vasoconstriction. EJPs have been recorded from smooth muscle cells in the outer region of the media, and the time course and latency of these potentials are similar to those in the vas deferens during stimulation of the hypo-

---

Fig. 4-11. Electron micrographs showing relationship of autonomic nerve axons to smooth muscle cells in vas deferens of rat. **A,** Axon, free of Schwann cell sheath, lying in deep groove in surface of smooth muscle cell. Axon contains numerous granular and agranular vesicles and mitochondria. **B,** "Beadlike" enlargement of nerve axon, packed with agranular vesicles, lying in groove on surface of smooth muscle cell. Space between nerve and muscle membranes is about 15 nm. Obliquely sectioned lower portion of axon reenters muscular cleft and appears to be turning to left to form second "bead" on adjacent muscle cell. **C,** Tiny axon almost submerged beneath surface of muscle cell on right. Axon contains granular and agranular vesicles and one mitochondrion. **D,** Axon containing fewer vesicles and two mitochondria. In this instance, separation between nerve and muscle membrane is about 25 nm. Row of pinocytotic vesicles associated with muscle cell membrane is located within lower part of neuroeffector junction. (From Richardson.[71])

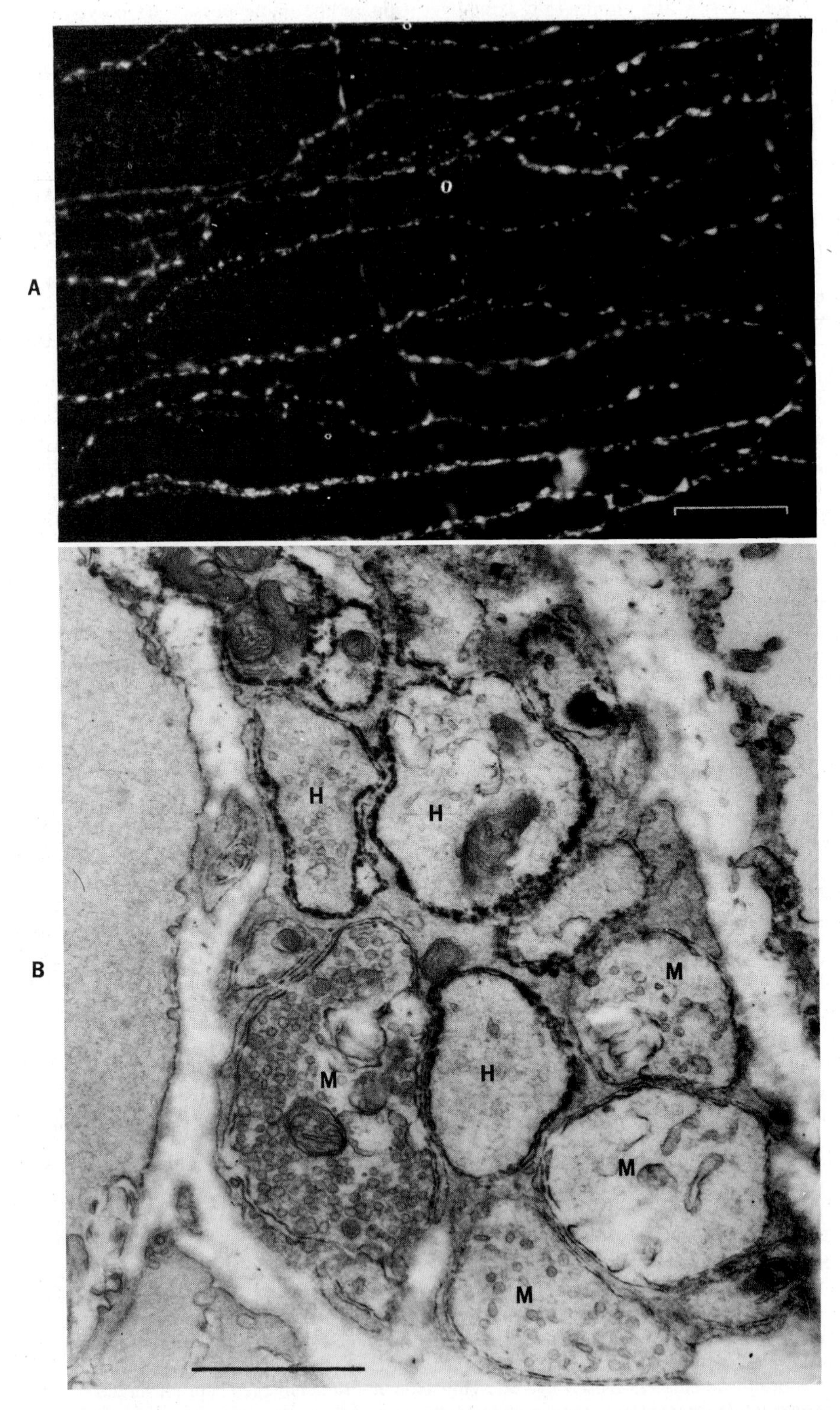

Fig. 4-12. A, Whole mount of mesenteric vein from sheep; freeze-dried preparation incubated with formaldehyde for 1 hr. Bundles of nerve fibers containing noradrenaline are identified by beadlike regions of intense fluorescence along axons. Calibration (lower right) = 50 μm. **B,** Electron micrograph of axon bundle in vas deferens of guinea pig. Acetylcholinesterase (AChE) incubation. Heavy deposits of AChE appear in plasma membranes of some axons, *H*, and lighter deposits in others, *M*. Calibration (lower left) = 1 μm. (From Burnstock.[25])

gastric nerve.[78] EJPs are also recorded from cells in the inner media, a region far from the nerve plexus, but the time course and latency of these potentials indicate that they arise by electrotonic spread from the "directly" innervated cells in the outer media.[78] Since the smooth muscle of the inner media is devoid of adrenergic nerve terminals and therefore of principal sites for catecholamine inactivation, this muscle is quite responsive to circulating epinephrine and norepinephrine. Hence the innervation pattern of these blood vessels is organized so that they are effectively controlled by both adrenergic nerves and circulating catecholamines. In large elastic arteries (aorta, carotid) where the density of innervation is less and the distance between nerve and muscle is larger than in the smaller vessels, the response to nerve stimulation has a longer latency and is much weaker and no discrete junction potentials are observed in the muscle cells.[25] Therefore these large blood vessels appear to be primarily under the control of circulating catecholamines.

In one reported electrophysiologic study of a blood vessel having cholinergic innervation (uterine artery of the guinea pig), stimulation of the cholinergic nerves or the administration of acetylcholine had no effect on the muscle cell membrane potential, although a pronounced vasodilatation occurred.[14] The dilatation might be mediated by a blockade of the propagation of electrical activity.

SUMMARY

Smooth muscles can be divided generally into two groups according to their physiologic properties: (1) unitary and (2) multi-unit. Examples of unitary muscles are those of the viscera—intestine, uterus, and ureter. These muscles are characterized by their spontaneous motility and by their sensitivity to stretch. The spontaneous rhythmicity results from two types of membrane potential fluctuations: slow waves and prepotentials. The periodicity of the slow waves is seconds or minutes, depending on the particular muscle. The prepotentials appear at the crest of the slow waves and initiate action potentials, which trigger the contractions. The slow waves set the frequency of contractions, and the prepotentials, by controlling spike frequency and duration, determine the strength and duration of each contraction. Although the spontaneous activity may be independent of nerve stimulation, it is usually modified and coordinated by the autonomic nervous system. The excitability of the muscle is also influenced by stretch, since unitary smooth muscle cells behave as stretch receptors, their membranes being depolarized by stretching. The action of muscle stimulants (e.g., acetylcholine) is also characterized by a depolarization of the cell membrane and by an increase in frequency of action potential discharge. Inhibitors of muscle activity (e.g., epinephrine in the intestine) suppress action potential discharge and "stabilize" the membrane at its normal quiescent level or at a slightly more negative (hyperpolarized) level.

Examples of multi-unit muscles are the pilomotor fibers in the skin, the ciliary muscles and the muscles of the nictitating membrane of the eye, and some vascular smooth muscles. All these muscles are normally quiescent and are activated only through their autonomic nerves, each muscle being composed of multiple "motor units." Multi-unit muscles are generally insensitive to stretch. Some of them, for example, those of the nictitating membrane and of various blood vessels, can be stimulated by the direct application of certain humoral substances (e.g., norepinephrine). The electrophysiologic aspects of such activation are not well established.

Neuroeffector transmission processes associated with neural activation in smooth muscle are qualitatively similar to those occurring at the motor end-plate in skeletal muscle. Discrete nerve endings have never been seen in smooth muscle, however, and it seems likely that during nerve stimulation the transmitter is released at various points along the nerve axon. Some of these regions lie close to the muscle cell membrane, whereas others are located farther away. Thus some smooth muscle cells are activated directly and others by a generalized diffusion of transmitter. Unlike the motor end-plate in skeletal muscle, there is no specialization of the nerve and smooth muscle cell membranes at their points of contact. However, the axons do contain vesicles, a number of which are granular and are believed to be storage sites for norepinephrine. In other axons the vesicles are predominantly agranular, and these may contain acetylcholine or possibly precursors of norepinephrine.

REFERENCES

General reviews

1. Bevan, J. A., et al., editors: Second international symposium on vascular neuroeffector mechanisms, Basel, 1976, S. Karger AG.
2. Bülbring, E., and Needham, D. M., organizers: A discussion on recent developments in vertebrate smooth muscle, Philos. Trans. R. Soc. Lond. (Biol.) **265:**7, 1973.

3. Bülbring, E., and Shuba, M. F., editors: Physiology of smooth muscle, New York, 1975, Raven Press.
4. Bülbring, E., Brading, A. F., Jones, A. W., and Tomita, T., editors: Smooth muscle, Baltimore, 1970, The Williams & Wilkins Co.
4a. Casteels, R., Godfraind, T., and Ruegg, J. C., editors: Excitation-contraction coupling in smooth muscle, New York, 1977, Elsevier North-Holland, Inc.
5. Daniel, E. E., and Paton, D. M., editors: Methods in pharmacology. Smooth muscle, New York, 1975, Plenum Press, vol. 3.
5a. Kao, C. Y.: Electrophysiological properties of uterine smooth muscle. In Wynn, R. M., editor: Biology of the uterus, New York, 1977, Plenum Press.
6. Prosser, C. L.: Smooth muscle, Annu. Rev. Physiol. **36:**503, 1974.
7. Somlyo, A. P., and Somlyo, A. V.: Vascular smooth muscle. I. Normal structure, pathology, biochemistry, and biophysics, Pharmacol. Rev. **20:**197, 1968.
8. Somlyo, A. P., and Somlyo, A. V.: Vascular smooth muscle. II. Pharmacology of normal and hypertensive vessels, Pharmacol. Rev. **22:**249, 1970.
9. Worcel, M., and Vassort, G., editors: Smooth muscle pharmacology and physiology, Paris, 1976, INSERM.

Original papers

10. Anderson, N. C.: Membrane potential oscillation in smooth muscle. In Worcel, M., and Vassort, G., editors: Smooth muscle pharmacology and physiology, Paris, 1976, INSERM.
11. Anderson, N. C., and Ramon, F.: Interaction between pacemaker electrical behavior and action potential mechanism in uterine smooth muscle. In Bülbring, E., and Shuba, M. F., editors: Physiology of smooth muscle, New York, 1975, Raven Press.
12. Anderson, N. C., Ramon, F., and Snyder, A.: Studies on calcium and sodium in uterine smooth muscle excitation under current-clamp and voltage-clamp conditions, J. Gen. Physiol. **58:**322, 1971.
13. Barr, L., and Dewey, M. M.: Electrotonus and electrical transmission in smooth muscle. In Code, C. F., editor: Handbook of physiology. Alimentary canal section, Washington, D.C., 1968, American Physiological Society, vol. 4.
14. Bell, C., and Burnstock, G.: Cholinergic vasomotor neuroeffector junctions. In Bevan, J. A., Furchgott, R. F., Maxwell, R. A., and Somlyo, A. P., editors: Physiology and pharmacology of vascular neuroeffector systems, Basel, 1971, S. Karger AG.
15. Bianchi, C. P.: Pharmacology of excitation-contraction coupling in muscle, Fed. Proc. **28:**1624, 1969.
16. Bohr, D. F., Goulet, P. L., and Taquini, A. C.: Direct tension recording from smooth muscle of resistance vessels from various organs, Angiology **12:**478, 1961.
17. Bolton, T. B.: Voltage-dependent behavior of drug-operated ion channels. In Worcel, M., and Vassort, G., editors: Smooth muscle pharmacology and physiology, Paris, 1976, INSERM.
18. Bozler, E.: Action potentials and conduction of excitation in muscle, Biol. Symp. **3:**95, 1941.
19. Bozler, E.: Conduction, automaticity and tonus of visceral smooth muscles, Experientia **4:**213, 1948.
20. Brooks, C. M., Hoffman, B. F., Suckling, E. E., and Orias, O.: Excitability of the heart, New York, 1955, Grune & Stratton, Inc.
21. Bülbring, E.: Action of catecholamines on the smooth muscle cell membrane. In Rang, H., editor: Drug receptors, Baltimore, 1973, University Park Press.
22. Bülbring, E., and Szurszewski, J. W.: The stimulant action of noradrenaline (α-action) on guinea-pig myometrium compared with that of acetylcholine, Proc. R. Soc. Lond. (Biol.) **185:**225, 1974.
23. Bülbring, E., and Tomita, T.: Increase in membrane conductance by adrenaline in the smooth muscle of the guinea-pig taenia coli, Proc. R. Soc. Lond. (Biol.) **172:**89, 1969.
24. Bülbring, E., and Tomita, T.: Suppression of spike generation by catecholamines in the smooth muscle of the guinea-pig taenia coli, Proc. R. Soc. Lond. (Biol.) **172:**103, 1969.
25. Burnstock, G.: Structure of smooth muscle and its innervation. In Bülbring, E., Brading, A. F., Jones, A. W., and Tomita, T., editors: Smooth muscle, Baltimore, 1970, The Williams & Wilkins, Co.
26. Burnstock, G.: Innervation of vascular smooth muscle: histochemistry and electron microscopy, Clin. Exp. Pharmacol. Physiol. **2**(suppl.)7, 1975.
27. Burnstock, G.: Purinergic transmission. In Iverson, L. L., Iverson, S. D., and Snyder, S. H., editors: Handbook of psychopharmacology, New York, 1975, Plenum Press, vol. 5.
28. Burnstock, G., and Costa, M.: Adrenergic neurons. Their organization, function and development in the peripheral nervous system, New York, 1975, John Wiley & Sons, Inc.
29. Burnstock, G., and Holman, M. E.: The transmission of excitation from autonomic nerve to smooth muscle, J. Physiol. **155:**115, 1961.
30. Burnstock, G., and Holman, M. E.: Spontaneous potentials at sympathetic nerve endings in smooth muscle, J. Physiol. **160:**446, 1962.
31. Burnstock, G., Campbell, G., Bennett, M., and Holman, M. E.: Inhibition of the smooth muscle of the taenia coli, Nature **200:**581, 1963.
32. Bury, V. A., and Shuba, M. F.: Transmembrane ionic currents in smooth muscle cells of ureter during excitation. In Bülbring, E., and Shuba, M. F., editors: Physiology of smooth muscle, New York, 1975, Raven Press.
33. Casteels, R., and Kuriyama, H.: Membrane potential and ionic content in pregnant and non-pregnant rat myometrium, J. Physiol. **177:**263, 1965.
34. Casteels, R., Droogmans, G., and Hendrickx, H.: Active ion transport and resting potential in smooth muscle cells, Philos. Trans. R. Soc. Lond. (Biol.) **265:**47, 1973.
35. Csapo, A.: The four direct regulatory factors of myometrial function. In Ciba Foundation: Progesterone: its regulatory effect on the myometrium, Ciba Foundation study group No. 34, London, 1969, J. & A. Churchill, Ltd.
36. Daniel, E. E., et al.: The sodium pump in smooth muscle. In Bevan, J., Furchgott, R. F., Maxwell, R. A., and Somlyo, A. P., editors: Physiology and pharmacology of vascular neuroeffector systems, Basel, 1971, S. Karger AG.
37. Devine, C. E., Somlyo, A. V., and Somlyo, A. P.: Sarcoplasmic reticulum and excitation-contraction coupling in mammalian smooth muscles, J. Cell. Biol. **52:**690, 1972.
38. Durbin, R. P., and Jenkinson, D. H.: The effect of carbachol on the permeability of depolarized smooth muscle to inorganic ions, J. Physiol. **157:**74, 1961.
39. Evans, D. H. L., Schild, H. O., and Thesleff, S.: Effects of drugs on depolarized plain muscle, J. Physiol. **143:**474, 1958.

40. Falck, B.: Observations on the possibilities of the cellular localization of monoamines by a fluorescence method, Acta Physiol. Scand. **56**(suppl. 197):1, 1962.
41. Fay, F. S.: Mechanical properties of single isolated smooth muscle cells. In Worcel, M., and Vassort, G., editors: Smooth muscle pharmacology and physiology, Paris, 1976, INSERM.
42. Folkow, B., and Neil, E.: Circulation, New York, 1971, Oxford University Press, Inc.
43. Gabella, G.: Fine structure of smooth muscle, Philos. Trans. R. Soc. Lond. (Biol.) **265:**7, 1973.
44. Gillespie, J. S.: The electrical and mechanical responses of intestinal muscle cells to stimulation of their extrinsic parasympathetic nerves, J. Physiol. **162:**76, 1962.
45. Gillespie, J. S.: Electrical activity of the colon. In Code, C. F., editor: Handbook of physiology. Alimentary canal section, Washington, D.C., 1968, American Physiological Society, vol. 4.
46. Goodford, P. J.: Distribution and exchange of electrolytes in intestinal smooth muscle. In Code, C. F., editor: Handbook of physiology. Alimentary canal section, Washington, D.C., 1968, American Physiological Society, vol. 4.
47. Hellman, K.: The isolated pilomotor muscles as an in vitro preparation, J. Physiol. **169:**603, 1963.
48. Henderson, R. M.: Cell-to-cell contacts. In Daniel, E. E., and Paton, D. M., editors: Methods in pharmacology. Smooth muscle, New York, 1975, Plenum Press, vol. 3.
49. Hodgkin, A. L., and Horowicz, P. C.: The differential action of hypertonic solutions on the twitch and action potential of a muscle fibre, J. Physiol. **136:**17P, 1957.
50. Holman, M. E.: Introduction to electrophysiology of visceral smooth muscle. In Code, C. F., editor: Handbook of physiology. Alimentary canal section, Washington, D.C., 1968, American Physiological Society, vol. 4.
51. Holman, M. E.: Junction potentials in smooth muscle. In Bülbring, E., Brading, A. F., Jones, A. W., and Tomita, T., editors: Smooth muscle, Baltimore, 1970, The Williams, & Wilkins Co.
52. Inomata, H., and Kao, C. Y.: Ionic currents in the guinea-pig taenia coli, J. Physiol. **255:**347, 1976.
53. Jones, A. W.: Reactivity of ion fluxes in rat aorta during hypertension and circulatory control, Fed. Proc. **33:**133, 1974.
54. Kao, C. Y., and McCullough, J. R.: Ionic currents in uterine smooth muscle, J. Physiol. **246:**1, 1975.
55. Kao, C. Y., Inomata, H., McCullough, J. R., and Yuan, J. C.: Voltage clamp studies of the actions of catecholamines and adrenergic blocking agents on mammalian smooth muscles. In Worcel, M., and Vassort, G., editors: Smooth muscle pharmacology and physiology, Paris, 1976, INSERM.
56. Katz, B.: Nerve, muscle and synapse, New York, 1966, McGraw-Hill Book Co.
57. Kroeger, E. A., and Marshall, J. M.: Beta-adrenergic effects on rat myometrium: mechanisms of membrane hyperpolarization, Am. J. Physiol. **225:**1339, 1973.
58. Kuriyama, H.: Effects of ions and drugs on the electrical activity of smooth muscle. In Bülbring, E., Brading, A. F., Jones, A. W., and Tomita, T., editors: Smooth muscle, Baltimore, 1970, The Williams & Wilkins Co.
59. Magaribuchi, T., Ito, Y., and Kuriyama, H.: Effects of catecholamines on the guinea-pig vas deferens in various ionic environments, Jpn. J. Physiol. **21:**691, 1971.
60. Marshall, J. M.: Regulation of activity in uterine smooth muscle, Physiol. Rev. **42**(suppl. 5):213, 1962.
61. Marshall, J. M.: Effects of neurohypophysial hormones on the myometrium. In Knobil, E., and Sawyer, W. H., editors: Handbook of physiology. Section 7: the hypothalamo-hypophysial complex, Washington, D.C., 1974, American Physiological Society, vol. 4, ch. 17.
62. Marshall, J. M., and Kroeger, E. A.: Adrenergic influences on uterine smooth muscle, Philos. Trans. R. Soc. Lond. (Biol.)**265:**135, 1973.
63. Mekata, F.: Current spread in the smooth muscle of the rabbit aorta, J. Physiol. **242:**143, 1974.
64. Mirroneau, J.: The effects of oxytocin on ionic currents underlying the rhythmic activity and contraction in uterine smooth muscle, Pfluegers Arch. **363:**113, 1976.
65. Needham, D. M., and Shoenberg, C. F.: Proteins of the contractile mechanism in vertebrate smooth muscle. In Code, C. F., editor: Handbook of physiology. Alimentary canal section, Washington, D.C., 1968, American Physiological Society, vol. 4.
66. Nonomura, Y., and Ebashi, S.: Isolation and identification of smooth muscle contractile proteins. In Daniel, E. E., and Paton, D. M., editors: Methods in pharmacology. Smooth muscle, New York, 1975, Plenum Press, vol. 3.
67. Osa, T., and Taga, F.: Effects of external Na and Ca on the mouse myometrium in relation to the effects of oxytocin and carbachol, Jpn. J. Physiol. **23:**97, 1973.
68. Peachy, L. D., and Porter, K. R.: Intracellular impulse conduction in muscle cells, Science **129:**721, 1959.
69. Prosser, C. L., Weems, W. A., and Connor, J. A.: Types of slow rhythmic activity in gastrointestinal muscles. In Bülbring, E., and Shuba, M. F., editors: Physiology of smooth muscle, New York, 1975, Raven Press.
70. Ramon, F., Anderson, N. C., Joyner, R. W., and Moore, J. W.: Axon voltage-clamp simulation. IV. A multicellular preparation, Biophys. J. **15:**55, 1975.
71. Richardson, K. C.: The fine structure of the nerve endings in smooth muscle of the rat vas deferens, J. Anat. **96:**427, 1962.
72. Richardson, K. C.: The fine structure of the albino rat iris with special reference to the identification of adrenergic and cholinergic nerves and nerve endings in its intrinsic muscles, Am. J. Anat. **114:**173, 1964.
73. Rosenblueth, A.: The transmission of nerve impulses at neuroeffector junctions and peripheral synapses, New York, 1950, John Wiley & Sons, Inc.
74. Shuba, M. F., et al.: Mechanism of excitatory and inhibitory actions of catecholamines on the membrane of smooth muscle cells. In Bülbring, E., and Shuba, M. F., editors: Physiology of smooth muscle, New York, 1975, Raven Press.
75. Somlyo, A. P.: Excitation-contraction coupling in vertebrate smooth muscle: correlation of ultrastructure with function, Physiologist **15:**338, 1972.
76. Somlyo, A. P., and Somlyo, A. V.: Ultrastructure of smooth muscle. In Daniel, E. E., and Paton, D. M., editors: Methods in pharmacology. Smooth muscle, New York, 1975, Plenum Press, vol. 3.
77. Somlyo, A. P., Devine, C. E., Somlyo, A. V., and Rice, R. V.: Filament organization in vertebrate smooth muscle, Philos. Trans. R. Soc. Lond. (Biol.) **265:**223, 1973.
78. Speden, R. N.: Excitation of vascular smooth muscle. In Bülbring, E., Brading, A. F., Jones, A. W., and Tomita, T., editors: Smooth muscle, Baltimore, 1970, The Williams & Wilkins Co.

79. Su, C., and Bevan, J. A.: Adrenergic transmitter release and distribution in blood vessels. In Bevan, J. A., Furchgott, R. F., Maxwell, R. A., and Somlyo, A. P., editors: Physiology and pharmacology of vascular neuroeffector systems, Basel, 1971, S. Karger AG.
80. Tomita, T.: Electrophysiology of mammalian smooth muscle, Prog. Biophys. Mol. Biol. **30:**185, 1975.
81. Verity, M. A.: Morphologic studies of the vascular neuroeffector apparatus. In Bevan, J. A., Furchgott, R. R., Maxwell, R. A., and Somlyo, A. P., editors: Physiology and pharmacology of vascular neuroeffector systems, Basel, 1971, S. Karger AG.
82. Wolfe, D. E., Potter, L. T., Richardson, K. C., and Axelrod, J.: Localizing tritiated norepinephrine in sympathetic axons by electron microscopic autoradiography, Science **138:**440, 1962.
83. Zuberbuhler, R. C., and Bohr, D. F.: Responses of coronary smooth muscle to catecholamines, Circ. Res. **16:**431, 1965.

II

THE BIOLOGY OF NERVE CELLS

5 Neuromuscular transmission

WILLIAM L. NASTUK

In the body, information in the form of nerve impulses flows from one point to another over elements of the nervous system. The transmission pathways are generally made up of several neurons in a chain. Thus nerve impulses not only travel along the surface membranes of individual neurons, they must also pass from cell to cell. In this chapter, we are concerned with the mechanisms involved in cell-to-cell transmission.

The region at which two neurons make contact with each other is called a synapse. At their peripheral terminations, motoneurons also make contact with muscle cells, and this region of contact is known as the neuromuscular junction. Synapses and neuromuscular junctions are characterized by unique morphologic and physiologic features. Details of these specializations will be given in this and subsequent chapters.

It is important to recognize that each neuron that enters into synaptic contact with another maintains its surface membranes intact. At synaptic regions the unbroken surface membranes of the individual neurons are separated by a 20 nm gap called the synaptic cleft. This cleft is continuous with the extracellular space and is presumed to be filled with a solution resembling extracellular fluid. At the neuromuscular junction the motoneuron and muscle fiber are also separated by a cleft, but it is larger and more irregularly shaped than the synaptic cleft.

It might be supposed that the transmission of impulses from one excitable cell to another is accomplished by the same mechanisms as those involved in axonal transmission. This idea, which has been described as the "electrical theory" of synaptic or neuromuscular transmission, was much espoused in earlier years. The theory was believed to be applicable to all synapses, but as powerful evidence accumulated, it became more and more certain that cell-to-cell transmission is commonly carried out by chemical intermediaries. Thus synaptic and neuromuscular transmission is now most generally described in terms of a "chemical theory." Nonetheless, in 1959 with modern experimental techniques, it was clearly demonstrated that at the giant motor synapse of the crayfish, electrical coupling is operative.[69] In fact, there are cases[132,133] where the coupling involves both chemical and electrical mechanisms operating in parallel, with one bearing a heavier duty than the other. One should no longer be surprised to find both chemical and electrical cell-to-cell coupling mechanisms being employed in invertebrate and vertebrate animals (see Kandel,[4] Chapter 11).

In Chapter 2 the details of the processes underlying excitation and conduction of nerve impulses along axons were presented. In an axon, adjacent regions of the surface membrane are electrically coupled to each other, since they are shunted by an electrolyte-containing fluid present in the axoplasm and outside the cell surface. For this reason a change in membrane potential in one region produced, let us say, as a result of an Na^+ inrush causes changes in membrane potential of the electrically coupled adjacent regions of the axon membrane. When two adjacent regions of axon membrane have unequal potential differences across them, electrical currents (eddy currents) flow between these regions. Such interactions occur during axonal conduction, and, thereby, resting regions of the axon that lie ahead of the oncoming nerve impulse become depolarized and liminally excited.

One can see intuitively that several conditions must be met for synaptic or neuromuscular transmission to occur via an electrical mechanism. Among them are the following: (1) The electrical coupling between the cells must be relatively tight, so that the eddy currents generated by an active presynaptic cell are forced to cross the membrane of the neighboring quiescent postsynaptic cell. Flow of this electrical current causes the membrane potential of the quiescent postsynaptic cell to shift from its resting level. Close electrical coupling between cells can be achieved by interdigitating their membranes; one example of such a morphologic arrangement is found at the intercalated discs of the ventricular myocardium. (2) A second requirement for electrical transmission is that the presynaptic membrane and the membrane contiguous with the postsynaptic membranes must be electrically excitable. In nature there are junctions where the latter requirement is not met.[17,76] For example,[25] in the frog the membranes of slow (tonus) fibers are not electrically excitable, and neuromuscular transmission is effected by a chemical mechanism. The suppression of electrical excitability is neurally controlled because action potentials can be electrically initiated in denervated slow fibers.[143] (3) The third requirement for electrical transmission is that the presynaptic elements must be capable of driving sufficient electrical current across the postsynaptic membranes so that the potential difference across these membranes falls at least to the critical level, at which an action potential is initiated. This condition may not be met for several reasons. For example, the presynaptic elements are sometimes physically small, and, because of their small membrane area, they have limited capacity to generate strong electrical current. On the other hand, the area of the postsynaptic chemoreceptive membrane can be relatively large, and hence the presynaptic elements have to be capable of displacing a large postsynaptic electrical charge in order to change the potential difference across the postsynaptic membranes (for a mathematical analysis of these aspects of electrical transmission, see Katz,[5] pp. 99-106).

Neuromuscular transmission in "twitch" fibers of vertebrate skeletal muscle involves the steps by which the motor nerve impulse leads to the initiation of a propagated action potential in the muscle fiber. This propagated muscle action potential in turn leads to changes that activate the contractile elements of the fiber (excitation-contraction coupling). A neuromuscular junction is classed as excitatory if its activation leads to a reduction of the potential difference across the postjunctional membrane. This nonpropagated localized depolarization, which is commonly known as an end-plate potential (EPP), has maximal amplitude at the postjunctional region. Because the muscle fiber has distributed electrical properties like those of an electrical cable, the EPP spreads with decrement along the muscle fiber membrane, which is contiguous with the postjunctional membrane. If and when the EPP reaches a critical magnitude, a propagated muscle action potential is initiated. Normally, conduction takes place from neuron to muscle fiber (dromic conduction), but under certain circumstances the eddy currents accompanying the muscle action potential can electrically excite the nerve terminals, and thereby antidromic conduction occurs in the motoneuron.

Although only excitatory neuromuscular junctions have been found in vertebrate skeletal muscle, it has been shown that the neuromuscular junctions of muscle fibers in invertebrate animals are not all excitatory; another type of junction classed as inhibitory is also present. Activation of an inhibitory neuromuscular junction can block the initiation of action potentials and lead to muscular relaxation. The reason for this is that arrival of a nerve impulse at an inhibitory junction ultimately causes specific increases in the ionic conductance of the postjunctional membrane that result in either an increase in the potential difference across the postjunctional membrane (hyperpolarization) or a more effective maintenance of the membrane potential near its resting value. Such changes, as has been pointed out in Chapter 2, lead to a diminution in excitability.

Peripheral inhibition has not been observed at the neuromuscular junctions of vertebrate skeletal muscle fibers. In vertebrate animals, inhibitory synaptic activity is seen in the central nervous system (CNS). The details of central inhibition are presented in Chapter 6.

For the skeletal muscle of vertebrate animals the evidence indicates that neuromuscular transmission is accomplished via a chemical intermediary, acetylcholine (ACH). This agent, which is released from the motor nerve terminal subsequent to the arrival of a nerve action potential, diffuses across the synaptic cleft and combines with a receptor protein in the postjunctional membrane. Thereupon the permeability of the postjunctional membrane changes, and transmembrane ion movements occur by which means the EPP is generated. Thus the neuromuscular transmission process involves neurosecretion on one hand and chemoreception on the other.

Because of its accessibility, much experimental work on cell-to-cell transmission has been carried out on the neuromuscular junction; as a result, more is known about the details of the transmission process for this junction than of chemically mediated synaptic transmission processes that occur at other regions of the nervous system. However, understanding gained at the neuromuscular junction provides a basis for interpreting synaptic transmission at central excitatory synapses, autonomic ganglia, and other neuroeffector junctions. Details of the transmission processes at the autonomic ganglia and for smooth muscle or gland cells innervated by autonomic neurons will be described in Chapters 4, 6, and 33.

MORPHOLOGIC FEATURES OF NEUROMUSCULAR JUNCTION[11,15,34,174]

The general arrangement of the motor innervation of striated muscle is described in Chapter 23. Each motor nerve fiber and the extrafusal muscle fibers it innervates represent a divergent system. Each branch of the myelinated motor axon approaches the muscle fiber and further divides to form a divergent array of unmyelinated terminal filaments that spread along the muscle fiber in both directions, often occupying several thousand square micrometers of its surface. The form of this terminal tree, as well as its extent, varies greatly from one vertebrate species to another and may even differ between different muscles in the same animal. The ultrastructure of the pre- and postjunctional elements, however, appears much the same for all those vertebrate forms in which it has been studied with the electron microscope.

The general features of these elements are illustrated by the diagrams in Fig. 5-1. Each unmyelinated axon terminal lies embedded in an

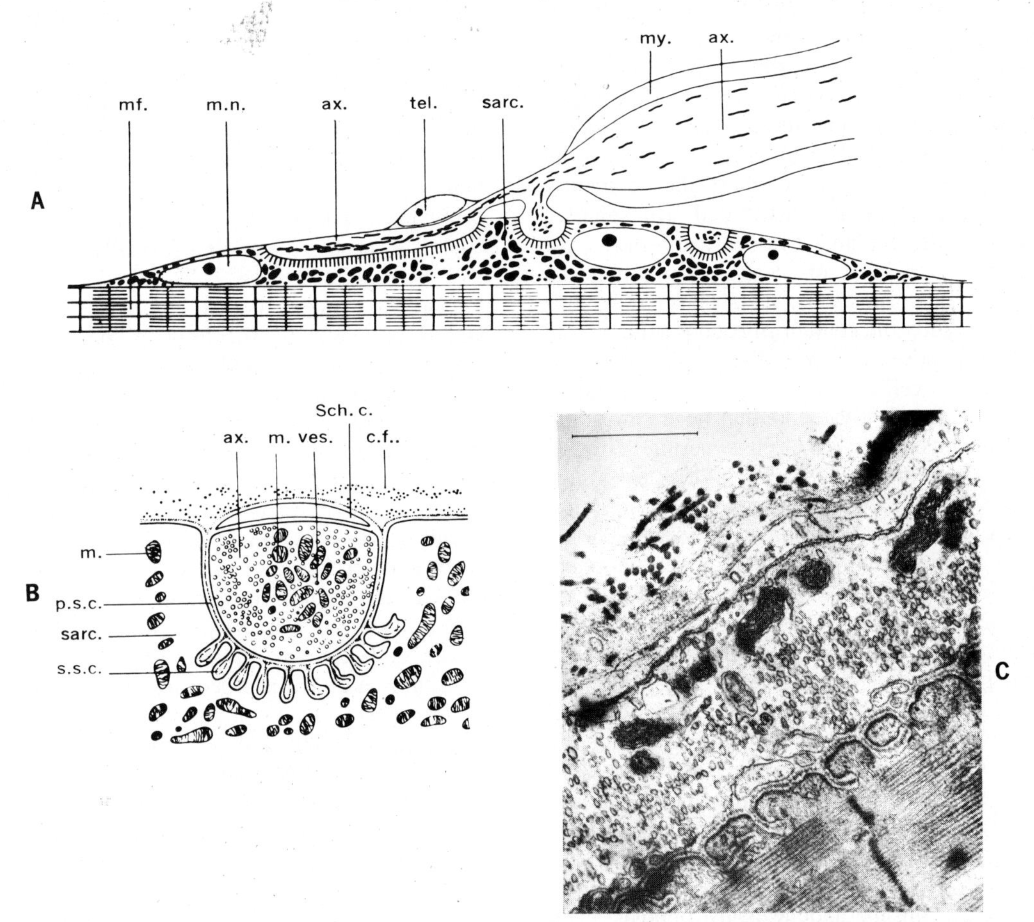

Fig. 5-1. A, Schematic drawing of typical neuromuscular junction. Terminal nerve branches lie in synaptic troughs. At axoplasmic-sarcoplasmic interface, transversely cut subneural lamellae are seen as rodlets about 1 μm in length. *ax.,* Axoplasm of motor fiber; *mf.,* myofibrils; *m.n.,* muscle cell nuclei; *my.,* myelin sheath; *sarc.,* muscle cell sarcoplasm and its mitochondria; *tel.,* thin Schwann cell that completely covers nerve endings—only nucleus is indicated. **B,** Schematic drawing of synaptic trough seen in cross section. *ax.,* Axoplasm; *c.f.,* collagen fibrils; *m.,* mitochondria; *p.s.c.,* primary synaptic cleft; *sarc.,* sarcoplasm; *Sch. c.,* Schwann cell; *s.s.c.,* secondary synaptic cleft; *ves.,* synaptic vesicles. **C,** Electron micrograph of neuromuscular junction of frog. Calibration (upper left) = 1 μm. From upper left to lower right: connective tissue with collagen fibrils, Schwann cell, motor nerve terminal filled with synaptic vesicles and mitochondria, primary synaptic cleft containing electron-dense ground substance, secondary synaptic clefts, muscle cell sarcoplasm, and myofibrils. Globular-shaped elements of irregular size located in primary synaptic cleft are fingerlike projections of Schwann cell. (**A** and **B** modified from Robertson[174] and Couteaux[33]; **C** from Birks et al.[15])

indentation in the surface of the muscle fiber. The cell membranes of nerve and muscle are each everywhere continuous and distinct and are separated by a synaptic cleft approximately 50 nm wide. The outer surface of the nerve terminal, the side away from the muscle cell, is completely covered with a Schwann cell sheath, and this is also true for the unmyelinated preterminal axon, so that the nerve cell membrane is nowhere exposed to the surrounding connective tissue.

The neuromuscular junction differs from other synapses by virtue of the remarkable morphologic specialization of the postsynaptic cell membrane. In the region of the synaptic gutter the muscle cell membrane is thrown into folds (secondary synaptic clefts) that open into the primary synaptic cleft. In some forms, these folds occur in a fairly regular arrangment, as shown for the frog in Fig. 5-1, but in others the secondary clefts are less regularly distributed. The synaptic cleft is differentiated extracellular space, for within it is a layer of electron-dense material called ground substance. This dense layer follows faithfully the contour of each postjunctional fold, and indeed it covers the entire surface of the muscle cell. At the edge of the synaptic gutter, it fuses with a similar layer spreading over the outside surface of the covering Schwann cells. From the physiologic evidence, it must be concluded that the layer of ground substance within the cleft is not a diffusion barrier.

The terminal endings of the motoneurons show the specialized features of nerve terminals (Fig. 5-1). Each terminal ending is packed with mitochondria and with many small globular bodies having structureless interiors. These bodies, called synaptic vesicles, may not be distributed randomly within the terminal. In frog muscle the motor nerve terminals along their course exhibit a succession of nodulelike enlargments that are densely packed with vesicles, whereas the intervening regions show much lower vesicle density.[134] Vesicles are frequently seen to be congregated within the axoplasm just opposite each postjunctional fold of the muscle cell membrane, and it is at this position that the axon membrane shows thickened local regions called "active zones."

The synaptic troughs usually occur in the summit of small hillocklike elevations of the muscle cell membrane, elevations caused by the accumulation of sarcoplasm, mitochondria, and many muscle cell nuclei. This region may also contain some vesicular inclusions, which are usually smaller and more variable in form and size than are the synaptic vesicles of the terminal axoplasm.

GENERAL FEATURES OF NEUROMUSCULAR TRANSMISSION

From the time of the classic experiments of Claude Bernard, performed over a century ago, it has been inferred that a process having a special chemical nature is responsible for the excitation of muscle by nerve and that this process is localized at neuromuscular junctions. Bernard observed, as have many since, that after paralysis is induced by the alkaloid curare, the muscle continues to contract in response to a directly applied electrical stimulus, and conduction in nerve fibers is unimpaired. He concluded, therefore, that curare acts on some special chemical entity believed to operate at the neuromuscular junction. With time, increasing knowledge of the mechanisms of axonal conduction generated fresh questions concerning the mechanism of neuromuscular transmission. The long-enduring problem of whether transmission is electrical or chemical in nature was first formally defined by DuBois-Reymond in 1877. To explain, *electrical transmission* means that excitation of the muscle cell is produced by the flow of ionic currents across its membrane, and these ionic currents are generated by the arrival of an action potential in the nerve terminals. *Chemical transmission* means that depolarization releases from the nerve terminal a specific substance that reacts with the muscle cell membrane to produce its excitation. It will be seen that for the vertebrate neuromuscular junction the evidence now available supports a chemical transmission mechanism and indicates that ACh is the transmitter agent.

The development of knowledge of chemical transmission at skeletal neuromuscular junctions is intimately linked with transmission in autonomic ganglia and that between postganglionic autonomic fibers and the smooth muscle or gland cells they innervate. The latter subject will be discussed in Chapters 4 and 33. Insight into the mechanism of transmission at each of these junctions was derived from the study of drugs or naturally occurring substances that, on injection or local application, mimic the effects produced by nerve impulses. The muscle postjunctional membrane has been shown to be especially sensitive to many drugs, which supports the idea that it has a chemoreceptor function. It was Langley, for example, who showed that the stimulating action of the drug nicotine is confined to the end-plate regions of muscles, and the stimulating effect is unimpaired following the degeneration of the nerve terminals produced by nerve section. This stimulating action of nicotine is blocked by curare to a degree determined by the relative concentrations of the two drugs.[113] Langley's quantitative studies of the competitive nature of the actions of nicotine and curare led him to conceive the generalization that such compounds compete for the same "receptor substance" in the postsynaptic membrane and that the loose union of the drug with this hypothetical substance leads, in the case of curare, to a block of neuromuscular transmission and, with nicotine, to excitation of the muscle cell. The idea that combination of the receptor substance with one drug inhibited its union with another was used to explain the competitive nature of their actions. During the years following Langley's work the concept of postjunctional membrane receptors has been widely used by physiologists and pharmacologists in explaining the junctional activities of various quaternary ammonium compounds and other drugs.

In the past there have been many attempts to isolate cholinergic receptors but only in recent years has the successful isolation of nicotinic cholinergic receptors been accomplished. For a literature review and interesting original contributions the reader should consult Kandel[4] and Axelson and Thesleff.[12] The nicotinic cholinergic receptor, a glycoprotein, was obtained from receptor-rich electrical organs of two electric fish, *Electrophorus* and *Torpedo*. The isolation of the receptor was greatly aided by the earlier discovery of Chang and Lee[26] of α-bungarotoxin, an active component of snake venom. This polypeptide toxin blocks neuromuscular transmission by binding tightly to the receptors of the postjunctional membrane of the skeletal muscle fibers. The toxin can be labeled with radioactive iodine (^{125}I), and when labeled toxin reacts with cholinergic receptors, it provides a marker for guiding the purification procedures. ^{125}I–α-bungarotoxin also can be used in autoradiography to determine the location and density of receptors in muscle fiber membranes. The receptor molecule, which has a molecular weight estimated to be approximately 250,000 to 400,000, is composed of several subunits. The receptor is distinct from acetylcholinesterase (AChE). As expected, it effectively binds ACh and related cholinergic agonists. It also binds Ca^{2+}, and the physiologic and pharmacologic effects of such bound Ca^{2+} are being actively investigated.

Sequence of events[155]

The train of events in neuromuscular transmission is as follows: ACh is synthesized by motoneurons and is stored in a sequestered form in the neuron terminals. When liberated from this store by a nerve impulse, ACh diffuses across the synaptic cleft and reacts with receptor molecules in the muscle cell membrane. The formation of this transmitter-receptor complex brings about an increased cation permeability of the cell, which leads to a local reduction in membrane potential, the EPP. When the EPP reaches a critical level, a conducted action potential is initiated in the adjacent muscle fiber membrane, and this action potential causes activation of the contractile elements. The ACh is either quickly destroyed by the hydrolytic action of the enzyme AChE or it is removed by diffusion, as shown below.

Acetylcholine as transmitter agent

The evidence provided by the research of recent years, particularly that obtained using biophysical methods, allows a precise description in quantitative terms of several of the sequential steps just described. It is appropriate to pause at this point, however, to indicate the background of knowledge from which these advances have been made. Direct evidence that ACh is the transmitter agent at the neuromuscular junction was obtained from the early investigations of Dale,

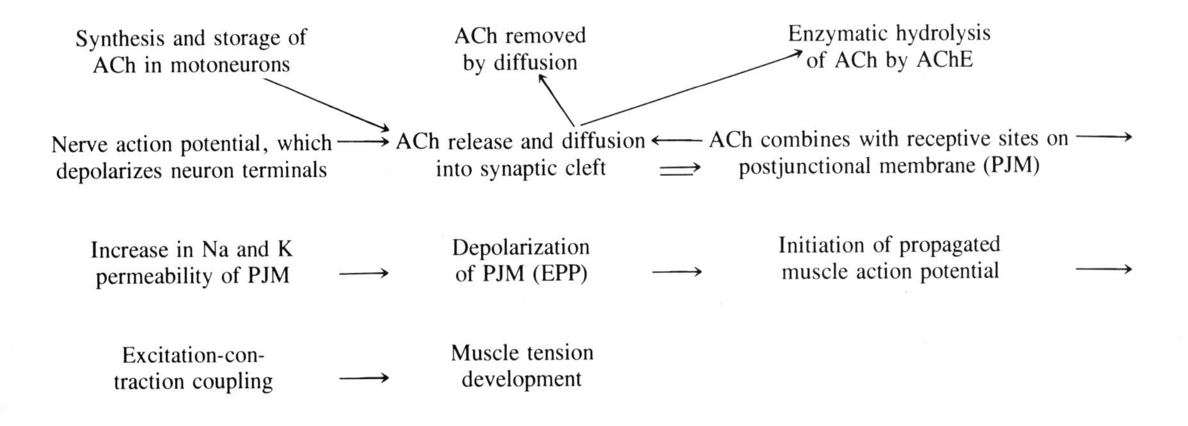

Brown, Feldberg, and their colleagues,[21,23,35,36] whose pioneering work has been expanded by contributions of many other investigators. Some of the critical evidence can be summarized as follows:

1. By section and cross union of nerves and by observation of the capacity for reinnervation after regeneration, it has been shown that motor nerves to skeletal muscles can replace and be replaced by other cholinergic nerves. It is well known that such reinnervation fails when cross unions are made between nerves that operate with different transmitter agents.

2. Stimulation of motor nerves to perfused voluntary muscle causes the liberation of ACh into the perfusate, provided that enzymatic destruction of ACh is prevented by inhibiting AChE with an appropriate drug, such as physostigmine. ACh release by nerve stimulation is essentially unaffected when muscle contraction is blocked by curare.

3. Close arterial injection of ACh causes a quick twitchlike contraction in both normal and denervated skeletal muscle. This takes the form of a brief asynchronous tetanus, and electrical recording shows that the muscle impulses originate at the region of the end-plates.[20,44,110] A powerful and now widely used iontophoretic technique for very rapid localized application of ACh and other ionized drugs, as originated by Nastuk,[151] is shown schematically in Fig. 5-2, *A*. By such iontophoresis a minute quantity of ACh can be rapidly ejected from a loaded micropipet placed at the external surface of the PJM. Fig. 5-2, *B*, shows that stepwise increase in the amounts of ACh applied causes *graded* increases in depolarization of the PJM. In the final upper trace the depolarization produced became large enough to initiate a propagated action potential. Another very significant point brought out by means of the iontophoretic technique is that when ACh is delivered to the interior of the muscle cell, neither local depolarization nor a conducted action potential results.[44] This indicates that the ACh receptor sites are present on the outside but not on the inside of the PJM. This view is further supported by the fact that when *d*-tubocurarine, a drug that competes with ACh for receptor sites, is injected into the muscle cell beneath the PJM, it has no effect on the EPP.[47]

Although ACh-sensitive receptors are abundant on the outer surface of the PJM, they are found more sparsely distributed at the extrajunctional regions.[137,138] The ACh sensitivity of these zones is comparatively low (approximately 1,000 times less), and the time course of the depolarization produced is relatively slow. Recent evidence indicates that junctional and extrajunctional receptors behave similarly,[128] although earlier work indicated that they represent two different groups.[66,67] The sensitivity of the extrajunctional regions to ACh increases over a period of many days following denervation of the muscle fiber. This denervation sensitization has been

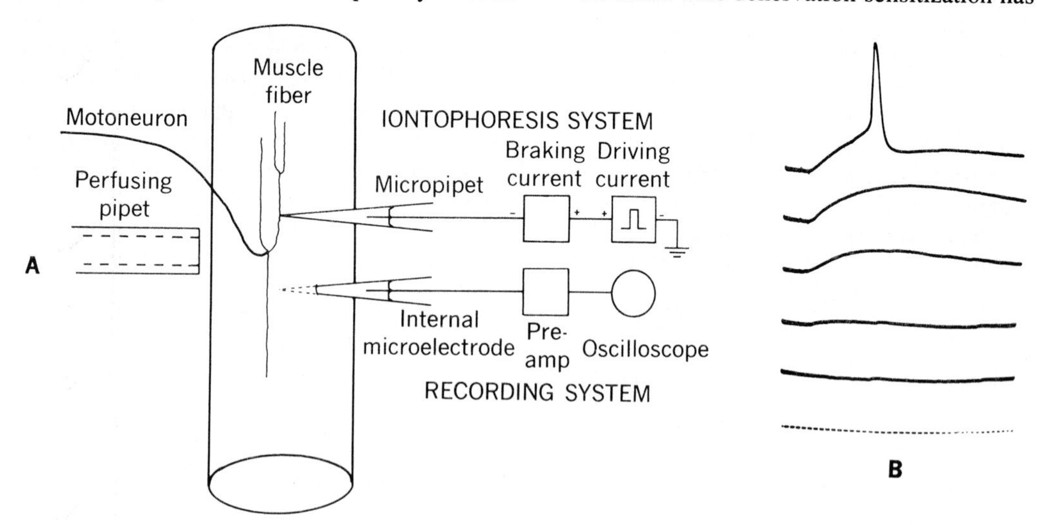

Fig. 5-2. A, Schematic diagram showing experimental arrangement used when recording changes in transmembrane potential produced during iontophoretic application of acetylcholine (ACh) ions to postjunctional membrane (PJM). Pipet to left is used when changing composition of extracellular fluid in this zone. **B,** Intracellular recording showing depolarization of PJM produced by iontophoretic application of ACh ions in successively increasing amounts. Time base markers = 1 msec. (From Nastuk.[152])

studied by many investigators over the past years. Some important aspects of this phenomenon were investigated by Miledi,[136] who utilized the advantages of the iontophoretic technique.

4. Normally a single motor nerve impulse initiates a single muscle action potential, and this, in turn, leads to the production of a brief contraction of the muscle fiber ("twitch"). These relationships are changed by the application of drugs that inhibit AChE (e.g., physostigmine, neostigmine, and edrophonium). After such a treatment a single nerve impulse causes the initiation of a train of muscle action potentials, and the muscle contraction becomes "tetanic" in character.[158] When measures are taken to prevent the initiation of muscle action potentials, as was done for the experiment illustrated in Fig. 5-3, it is seen that after cholinesterase inhibition, there is a considerable prolongation of the EPP (see Fig. 5-3, *B*) produced in response to a single nerve impulse. (See also Fatt and Katz,[62] Kordas,[108] and Nastuk and Alexander.[158]) Prolonged depolarization also arises in the wake of repetitive nerve discharge of esterase-inhibited nerve-muscle preparations. From their studies of such preparations, Katz and Miledi[100,101] concluded, after considering several interesting possibilities, that

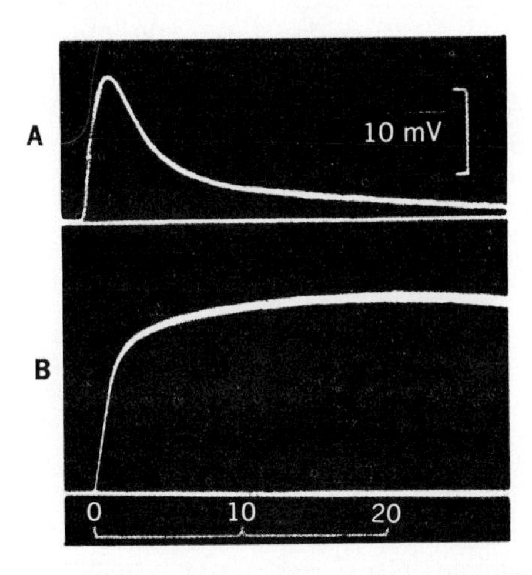

Fig. 5-3. Results of experiment illustrating effect of neostigmine, an inhibitor of acetylcholinesterase (AChE), on end-plate potential (EPP), which is recorded via intracellular electrode placed in muscle cell just beneath end-plate. **A,** EPP evoked by nerve impulse. Successful evocation of action potential in muscle cell is prevented by lowering Na^+ of external fluid (four fifths of Na^+ replaced by sucrose), thus revealing full time course of EPP. **B,** Muscle has now been treated with neostigmine bromide, 10^{-6}M; inhibition of AChE greatly prolongs EPP but alters its amplitude only slightly. Time scale in milliseconds. (From Fatt and Katz.[62])

removal of released ACh by diffusion is slowed because the ACh is bound to postsynaptic receptors.

End-plate potential[19,62,150]

The discovery by Göpfert and Schaefer[73] in 1938 and by Eccles and O'Connor[56] in 1939 that a unique electrical event is interposed between nerve impulse and muscle action potential provided a means for more detailed analyses of the mechanism of neuromuscular transmission. As mentioned previously, this intervening event, EPP, represents a transient intense local depolarization of the postjunctional region of the muscle cell. Following the arrival of the impulse in the terminals of the nerve fiber there is a synaptic delay of 0.5 to 1 msec, after which time the EPP begins its appearance.[92] Normally the EPP increases in amplitude rapidly and excites the muscle cell, and therefore much of its true time course is masked by the superposition of the initiated muscle action potential. This can be appreciated by examining the oscillograms shown in Fig. 5-4, *A* and *B*, which are typical records taken at neuromuscular junctions. Fig. 5-4, *C*, shows for comparison a record taken of a nonjunctional region. The EPP and the action potential may be separated by utilizing the drug *d*-tubocurarine, which inhibits and reduces the postsynaptic depolarizing action of ACh (*d*-tubocurarine is a pure alkaloid found in crude curare). The differences between the EPP and the muscle action potential are made clear by intracellular records made at a curare-blocked neuromuscular junction during progressive washout of the curare. As the drug is removed from the PJM, the EPP elicited by nerve stimulation increases in amplitude until it reaches the critical level at which a muscle action potential is initiated (Fig. 5-5). The EPP, a local depolarization of the PJM, is caused by a transient increase in the ionic permeability of this membrane. As we shall see, the ACh-induced portion of the EPP lasts only about 2 to 3 msec, and thus much of the falling phase of the EPP represents the passive return of the membrane to its resting potential along a time course determined by the distributed resistance (a function of extrajunctional ionic permeability) and capacitance of the muscle cell membrane. To better appreciate this the reader should refer to the discussions of local response of nerve and the cablelike properties of the cell membrane (Chapter 2). The membrane cable properties determine also the extent to which the EPP spreads, with decrement, into the adjacent nonjunctional regions of the muscle fiber. The spa-

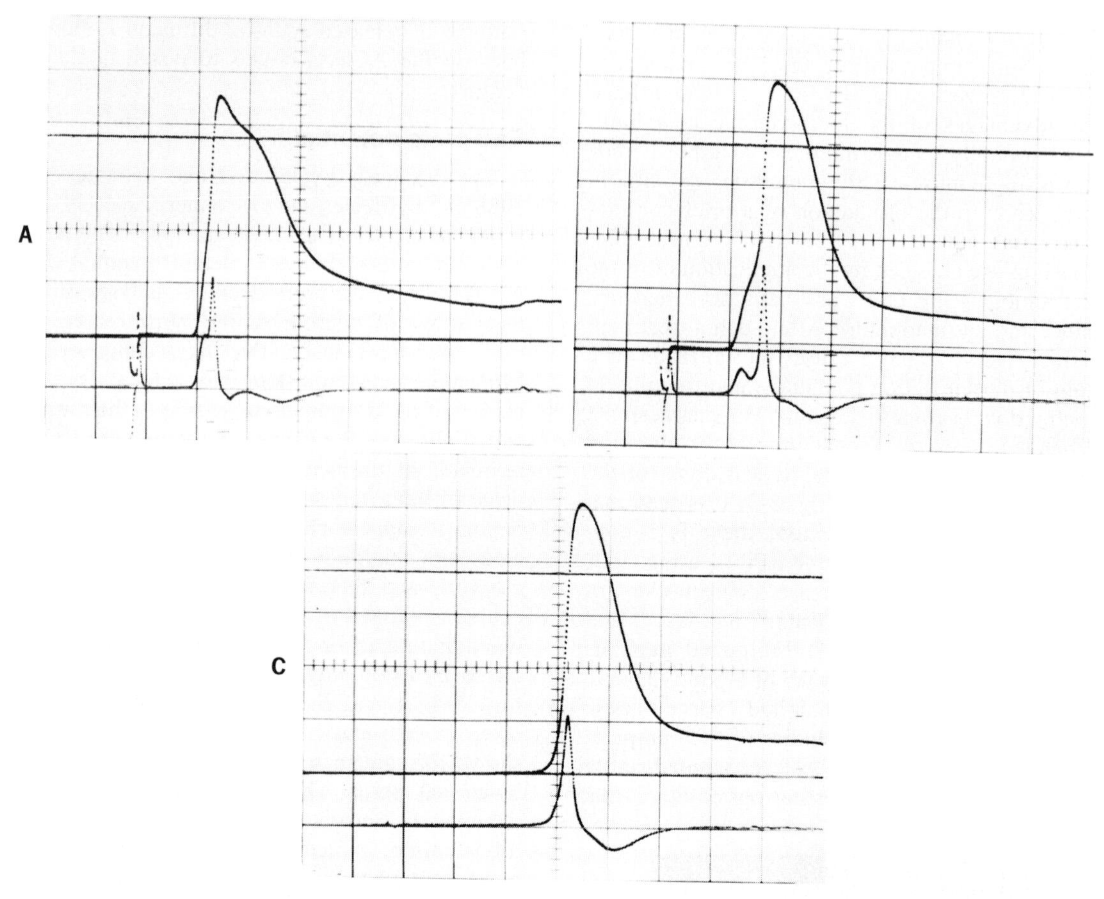

Fig. 5-4. A, Transmembrane potential record showing neurally initiated action potential obtained at neuromuscular junction on frog sartorius muscle. Resting potential = −93 mV. First phase of action potential record represents EPP, which rises rapidly to reach critical membrane potential, at which point a propagated action potential is initiated. Trace below action potential is its first derivative $\left(\frac{dVm}{dt}\right)$. First disturbances at left are stimulus artifacts. Calibration; x axis = 2 msec/block; y axis for RP and AP = 25 mV/block, y axis for $\frac{dVm}{dt}$ = 250 V/sec/block. **B,** Same as **A** except record was obtained at junction with lower transmission safety margin. Note that EPP rises more slowly than in **A. C,** Same as **A** except record was obtained at nonjunctional region. Note that no EPP is present, and action potential is larger in amplitude and falls faster than in **A.**

tial extent of the EPP is indicated by the records of Fig. 5-6, obtained at a curarized junction. It is important to recognize that when the postjunctional membrane has normal chemosensitivity (no blocking agents such as curare are present), the amount of ACh released by a nerve impulse is more than sufficient to drive the PJM potential to the critical level at which an action potential is initiated. For this reason, there is a three- to four-fold safety factor in the neuromuscular transmission process. It is this safety factor that accounts for the remarkable reliability of transmission at the neuromuscular synapse, and it explains why a one-to-one relation prevails between nerve and muscle impulses under most normal conditions.

The EPP has other properties characteristic of local responses. The first of these is that it exhibits no refractory period because if in a curarized muscle, one nerve impulse follows another by the proper interval, the depolarizations produced by them summate. If curarization is not too heavy, this summed depolarization may reach the threshold level of the muscle cell, producing in it a conducted action potential.[57] The phenomenon

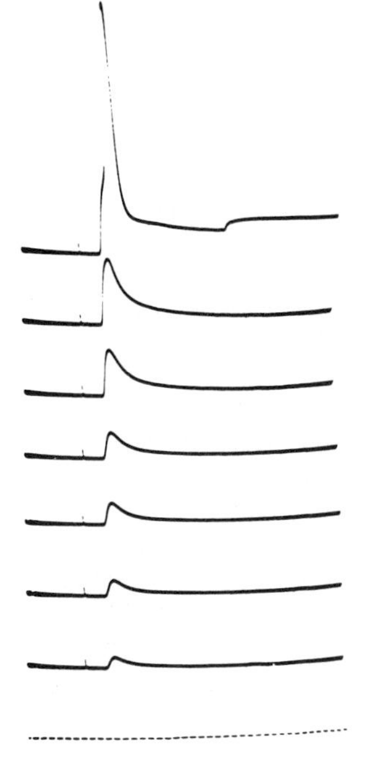

Fig. 5-5. Intracellular recording from muscle fiber showing serial group of EPPs recorded during washout of d-tubocurarine from junctional region. Motor nerve was stimulated every 3 sec, and washout began with bottom record. Time base markers = 1 msec. (From Nastuk.[152])

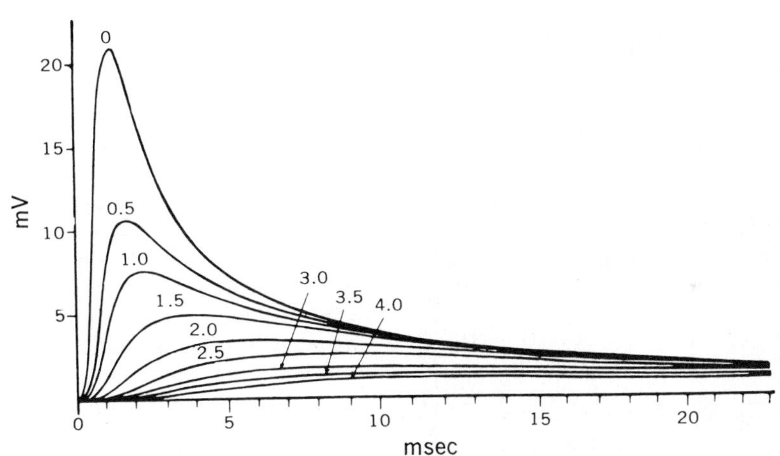

Fig. 5-6. Records obtained in response to nerve volleys taken with intracellular microelectrode located at different distances from end-plate focus of curarized frog muscle. Records have been superimposed on single time scale, taking instant of nerve stimulation as common point. Numbers indicate approximate distances in millimeters from end-plate at which each record was taken. Displacement of electrical charge from muscle cell membrane at end-plate reaches maximum about 2 msec after start of EPP. Restoration of charge follows exponential time course. Active phase of neuromuscular transmission is brief, impulsive event, and prolonged time course and spatial spread of change in membrane potential are determined by resistance and capacitance of resting muscle cell membrane. (From Fatt and Katz.[62])

is an example of *temporal summation,* a property seen in synaptic action in the CNS.

Further experimental studies that reveal the processes responsible for the production of the EPP will be described in the sections that follow. However, at this stage, it will be instructive to highlight some of the experimental evidence and arguments whereby the electrical theory of neuromuscular transmission was discarded.[111]

1. The total electrical charge transferred across the muscle cell membrane during a single EPP is larger than the estimated *total* ionic content of

the nerve terminals. Thus the EPP cannot result solely from ionic eddy currents associated with an action potential in the neuronal terminals.[62]

2. Recording of the nerve action potential in single telodendrites by carefully placed external microelectrodes shows that this action potential is separated in time from the onset of the EPP[91,92] (Fig. 5-7). This synaptic delay, which has been estimated to average 0.75 msec in the frog, is only partly accounted for by the estimated diffusion time of ACh across the synaptic cleft, and other factors related to the probability of ACh release enter in.

3. The flow of subthreshold currents across the nerve terminal membrane produces no change in the membrane potential of the muscle cell. The same result is obtained whether the local currents are produced by local electrical stimulation of the nerve terminals or by an approaching action potential that is blocked just proximal to the nerve terminal arborization. In the second case the terminal membrane supplies the eddy currents that are associated with the early rising phase of the action potential.

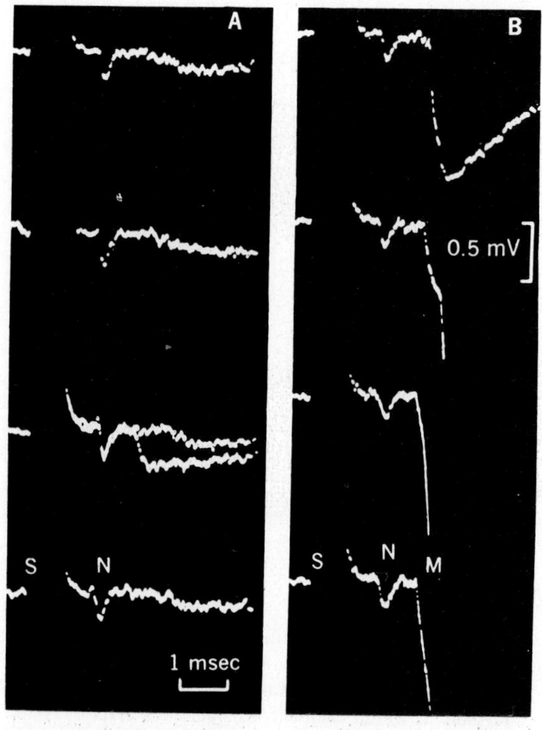

Fig. 5-7. Extracellular recording from frog sartorius muscle obtained using microelectrode containing 0.5M $CaCl_2$ placed at point of contact between terminal filament of motoneuron and its adjacent muscle fiber. Muscle was bathed in low Ca, high Mg solution to block neuromuscular transmission. Negative going potential is downward. Stimulus *(S)* to motor nerve evoked action potential *(N)* in nerve terminal followed by variable postsynaptic response *(M)*. In **A,** efflux of Ca from electrode was stopped. In **B,** efflux of Ca is permitted to occur and, thereafter, postsynaptic potential increases in amplitude. (From Katz and Miledi.[90,93,94])

4. A muscle action potential can be electrically initiated distant from the neuromuscular junction and allowed to propagate along the muscle fiber into the junctional region. By proper timing, at the moment when the muscle action potential is making the junctional transit, a nerve impulse can be arranged to arrive at the motor nerve terminals. At this stage the muscle membrane is occupied with a muscle action potential, and hence it should be absolutely refractory and beyond any significant influence by the eddy currents associated with the "colliding" nerve impulse. However, despite this prediction, experimental results show that the PJM can undergo a powerful change in permeability on arrival of the nerve impulse (Fig. 5-13).

ENZYMATIC SYNTHESIS AND HYDROLYSIS OF ACETYLCHOLINE[78,79,170]

Synthesis, distribution, and storage

For ACh to function as the transmitter agent at the neuromuscular synapse, it is necessary that this ester be released suddenly from the nerve ending and that motoneurons contain a mechanism for its synthesis, for it is clear that cholinergic neurons, when subjected to long-term repetitive stimulation, will release more ACh from their endings than the amount stored in the resting neuron terminals.[14,22,168,171] A second requirement is that released ACh be removed from its site of action at the postsynaptic membrane within 1 to 2 msec. These stipulations have been met by the discoveries that cholinergic neurons contain an enzyme (choline acetyltransferase, earlier called choline acetylase) that catalyzes the acetylation of choline and that at the postsynaptic regions there is present a second enzyme, AChE, in quantities apparently sufficient to hydrolyze the released ester rapidly. These relationships are of general significance, for ACh has been proved to be the transmitter agent at the synapses of autonomic ganglia, and substantial evidence indicates that ACh serves as a transmitter at certain synapses in the CNS as well.

ACh synthesis in vivo was first demonstrated in 1937 by Brown and Feldberg.[22] Shortly thereafter Mann et al.[129] found that respiring slices of mammalian brain perform this synthesis in vivo, and Stedman and Stedman[179] found that the synthetic mechanism survived the cell rupture produced by grinding. An important contribution was that of Nachmansohn and Machado,[148] who found in 1943 that the synthesis of ACh by brain is greatly accelerated by the addition of adenosine triphosphate (ATP) and choline and that the synthesizing enzyme contained —SH groups. Thereafter the continuing investigations of Feldberg[64] and Nachmansohn[144] and their colleagues established that (1) the synthesizing enzyme system is present in aqueous extracts of acetone-dried brain; (2) the system yields maximal activity on the addition of ATP, choline, K^+, and (if incubated anaerobically) an —SH-containing compound such as cysteine; and (3) dialysis reduces activity, which is restored when the dialysate is returned. This "activator" as described by Feldberg has been identified as coenzyme A.

It is now apparent that acetylation occurs in two steps and that choline acetyltransferase catalyzes only the second. The first is the combination of acetate with coenzyme A to form "active acetate," that is, acetylcoenzyme A (acetyl-CoA):

Step 1: ATP + CoA + Acetate $\rightleftharpoons$ Acetyl-ScoA + AMP + Pyrophosphate

Step 2: Acetyl-ScoA + Choline $\xrightleftharpoons[\text{acetyltransferase}]{\text{Choline}}$ ACh + HS-CoA

The overall reaction indicated by step 1 may occur via a series of intermediate stages. It should be emphasized that this is not the only possible source of acetyl-CoA, for it is likely that there is a common pool to which many acetyl donor systems contribute and from which various acceptor systems draw. There is no evidence that any one donor system is specifically channeled into step 2.

Distribution of choline acetyltransferase[65,78]

In higher vertebrates, choline acetyltransferase has been detected in those parts of the peripheral and autonomic nervous systems that contain cholinergic neurons. Thus in the central and peripheral regions of the nervous systems of higher vertebrates the distribution of choline acetyltransferase parallels that of ACh itself, that is, both are constituents of cholinergic neurons. The quantity of choline acetyltransferase present in any given region appears to depend on the proportion of individual cholinergic neurons present in the neuronal population and not on wide variation in the amount of this enzyme in individual neurons. In the central and peripheral nervous systems the distribution of cholinergic neurons is not uniform; for this reason certain structures, such as the caudate nucleus, contain large amounts of the enzyme, whereas others, such as the cerebellar cortex, contain practically none. Choline acetyltransferase is abundant in ventral spinal roots, but dorsal roots contain little of this enzyme and that which is found is probably confined to those sensory neurons that are cholinergic and also contain AChE.[71]

The intracellular distribution of various elements of the ACh system has been intensively investigated by Hebb,[78,79] Whittaker,[188,190] DeRobertis,[49] and many other workers. For practical reasons, much of the work is done on subcellular fractions obtained from mammalian cerebral cortex (e.g., see Marchbanks[130] and Whittaker[190]) or other tissues rich in cholinergic neurons. Although results of earlier studies indicated that choline acetyltransferase is a constituent of synaptic vesicles (see subsequent discussion), it now appears that the enzyme is largely found in the cytoplasm. However, some of the enzyme may be bound to negatively charged sites on intracellular membrane surfaces. ACh synthesized in the cytoplasm is taken up into synaptic vesicles where it is stored, protected from hydrolysis by AChE. Studies of the microphysiology of neuromuscular transmission detailed in a later section indicate that this bound packet of ACh finds its way to the nerve terminal membrane, from which it is released by the depolarization associated with the nerve impulse.

Enzymatic hydrolysis of acetylcholine[146,191,192]

Esterases of various types are widely distributed in animal tissues, and, of these, at least two show some degree of specificity for choline esters. One is found in large quantities in mammalian blood serum, and the second is present in red blood cells and in many of the conducting tissues of the nervous system. The latter, called "true" or acetylcholinesterase (AChE), shows a high but not exclusive affinity for ACh, and the rate of substrate hydrolysis decreases with increase in length of the acyl chain: acetyl > propionyl > butyrylcholine. There is a well-defined optimal substrate concentration; the rate decreases for weaker or stronger solutions, a fact that can be depicted by a bell-shaped curve that relates hydrolytic activity to substrate concentration. The serum esterase (called "pseudocholinesterase"), on the other hand, has a much weaker affinity for ACh, and the rate of hydrolysis increases both with substrate concentration and with increasing length of the acyl chain.[149]

It is now apparent that there is a reasonably good correlation between the distributions in the nervous system of AChE, ACh, and choline acetyltransferase.[24] The concentration of AChE is generally highest in those regions in which a high concentration of synapses is joined with a high content of choline acetyltransferase, a conjunction to be expected where synaptic transmission is mediated by ACh. The ventral roots, for example, contain a large amount of choline acetyltransferase, and there is a high concentration of AChE at the neuromuscular end-plates. Histochemical studies, particularly by Couteaux[33,34,37,46] and by Koelle,[102,103,105] have shown that most of the AChE is concentrated on the muscle side of the junction and is associated closely with the folds of the postjunctional membrane.[33,34,37,102-104] It is most likely that reactive sites of the enzyme are exposed on the membrane outer surface. AChE is more plentiful in the end-plate region than elsewhere along the muscle, and this distribution is at least partially explained by the existence of convolutions of the postjunctional membrane that increase its surface area at that zone. When the nerve terminals degenerate following nerve section, the concentration of AChE is reduced, but appreciable amounts remain in this region. Thus it may reasonably be assumed that an appreciable amount of the postsynaptic AChE is of presynaptic origin, and Koelle[104] has suggested that this relationship applies to cholinergic synapses of autonomic ganglia. Although biosynthesis of some neuronal constituents such as ACh takes place in neuronal terminals, one should also keep in mind the growing lines of evidence showing that many cellular constituents and organelles found in the terminals of various types of neurons may be synthesized in the cell body and transported to the axonal terminals by axonal flow. For the reader who wishes orientation on various ideas and details of such cellular dynamics, there are a number of helpful publications.[6,7,12,13]

As described in the schematic diagram on p. 155, ACh that is released into the synaptic cleft is rapidly removed by diffusion or enzymatic hydrolysis. Although removal of ACh by diffusion is more effective than is commonly realized,[55,165] enzymatically catalyzed hydrolysis of released ACh is an important mechanism for terminating the active phase of the EPP produced by the action of this transmitter. The velocity of the hydrolysis of ACh by AChE depends on many factors, and past estimates are complicated by many uncertainties. Wilson and Harrison[193] estimated the turnover number at 25° C, pH 7.0, ACh 2.5×10^{-3}M, to be 1.2×10^4 moles/sec, and they concluded that a biologically reasonable concentration of enzyme would be required to hydrolyze, in 2 msec, the ACh released during neuromuscular transmission. The turnover time for AChE has been estimated by Lawler[114] at less than 100 μsec.

The process of hydrolytic enzyme action takes place in two stages.[144] The first is the formation of an enzyme-substrate complex. The apparent dissociation constant (Michaelis' constant) of the AChE-ACh complex (1.2×10^{-4}) indicates that this complex is stable and reversible. A simplified schematic representation of a way in which this binding might occur is shown in Fig. 5-8. The positively charged N^+ of ACh is bound by coulombic forces to an anionic site on the enzyme. The existence of such a negative site has been demonstrated with the aid of competitive inhibitors and appropriate substrates. An additional binding is attributed to the nonspecific van der Waals forces. It is also postulated that a region on the active surface of the enzyme, close to the anionic site, contains a basic group that forms a covalent bond with the acyl carbon of the ester. The further rearrangements leading to the hydrolysis of the bound ester are thought to proceed through a first step involving acetylation of the enzyme and the elimination of choline and a second involving an acid enzyme complex that leads to regenerated enzyme and acetate ion. Detailed investigations of these hydrolytic mechanisms have been made by Wilson.[191,192]

Although the functional role of AChE at the

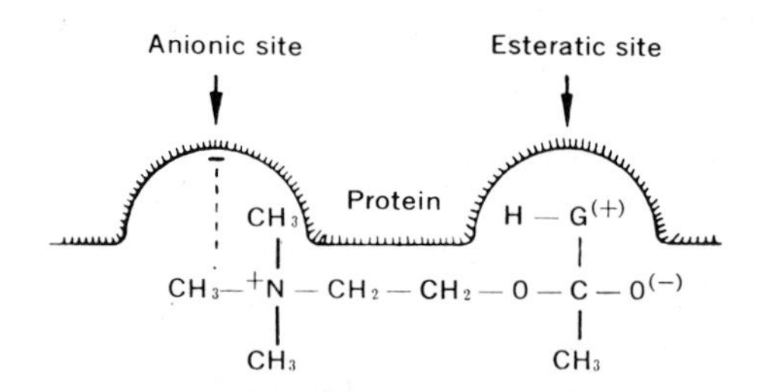

Fig. 5-8. Schematic representation of complex formed between AChE and ACh. Complex is stabilized by coulombic and van der Waals' forces at anionic site and by covalent bond formation between carbonyl carbon and basic group at esteratic site. This basic group is symbolized by *G*, and *H* represents dissociable hydrogen atom not involved in binding. Hydrolysis then follows by a process thought to occur in two consecutive steps: first, acetylation of enzyme with elimination of choline and second, regeneration of enzyme and acetic acid. (From Nachmansohn.[144])

neuromuscular junction is well established, the fact that this enzyme is closely associated with extrajunctional axonal and muscle fiber membranes raises questions as to its functional significance at such sites. For many years, Nachmansohn[144,145,147] has maintained that the release and electrogenic action of ACh are essential first steps that lead to sequentially timed increases in Na^+ and K^+ conductances, which underlie production of action potentials. This hypothesis and its corollaries have been subject to much experimental testing and scientific argument, and many investigators find them unacceptable. It is not possible here to review these arguments, many of which hinge on technical details. A few examples of the points of discussion are the following: (1) choline acetyltransferase is not universally distributed in the nervous system and some neurons do not contain ACh[24,78,79]; (2) AChE is present in the membranes of some nerve cells that are not cholinergic[103,104]; (3) tetrodotoxin, a powerful poison obtained from the puffer fish, blocks the Na conductance mechanism of axons and muscle fibers but has no effect on the response of muscle postjunctional receptors to ACh. For additional facts and viewpoints the reader will be aided by the reviews of Hebb,[78] Castillo and Katz,[45] and Katz.[88]

MICROPHYSIOLOGY OF THE NEUROMUSCULAR JUNCTION*

Before proceeding with this section the reader should recall some of the morphologic details given in Fig. 5-1. The presynaptic terminals contain large numbers of globular bodies (synaptic vesicles) that are about 50 nm in diameter and have a structureless interior. Similar-appearing bodies are occasionally seen in the muscle fiber cytoplasm underlying the postsynaptic membrane. A primary synaptic cleft 50 nm wide separates the pre- and postsynaptic membranes, except for regions where the latter is infolded (secondary synaptic clefts).

Spontaneous and activated release of acetylcholine from motor nerve terminals[18,63,75,80,115]

In 1952 Fatt and Katz discovered that the motor nerve endings are not at rest even in the absence of nerve impulses. The records of Fig. 5-9 show that small spontaneous depolarizations of the muscle cell membrane, known as miniature end-plate potentials (MEPPs), occur at the region of the end-plate and that these are rapidly attenuated as the intracellular recording site is moved away from the junctional region. Extracellular recording, which allows a more precise localization, showed that MEPPs occur at many very discrete loci about a nerve ending.[46] The combined discharge of the ending is disordered in time, for the probability that an MEPP will occur at any given instant fits that predicted for a completely random sequence (Fig. 5-10). The waveform of the MEPP resembles that of an EPP, but its amplitude is only 0.5 to 1.0 mV. The MEPP amplitude is changed only by factors that alter the responsiveness of the postsynaptic membrane, that is, it is increased in size and prolonged in time course by such anti-AChE drugs as neostigmine or eserine. The frequency of discharge, on the other hand, is changed by factors that alter the state of the nerve terminal membrane, particularly the membrane potential, for the discharge frequency is powerfully increased by depolarization and decreased by hyperpolarization of the terminal.[42,88,117] These facts led Katz and his colleagues to the conclusion that the MEPPs represent the response of the muscle PJM

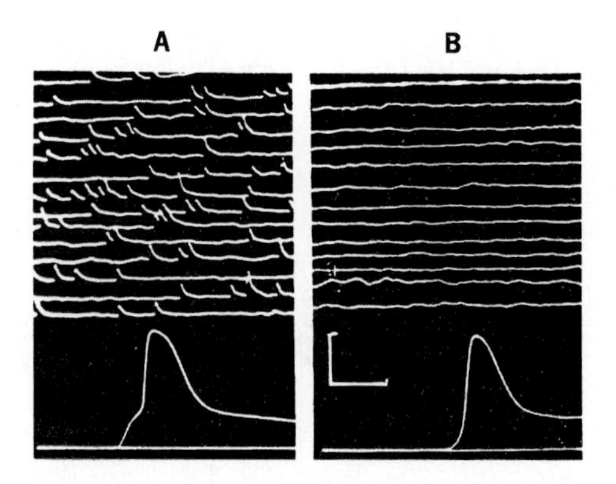

Fig. 5-9. Records illustrating fact that spontaneous miniature end-plate potentials (MEPPs) occur only at end-plate regions of muscle fiber. For **A,** microelectrode was placed inside frog muscle fiber at nerve-muscle junction. Successive lines from above downward were recorded consecutively in time. Note random occurrence of spontaneous potential changes of about the same amplitude, which occasionally superimpose and sum. Record below, obtained at much lower amplification and on faster time base, is EPP–muscle action potential complex recorded at this end-plate region. For **B,** microelectrode was placed 2 mm away from junction in same muscle fiber. Records above show amplifier noise but no MEPPs. Record below shows conducted action potential in this fiber set up by nerve impulse, with little sign of EPP. Calibrations for lower records = 50 mV and 2 msec. (From Fatt and Katz.[63])

*See references 3, 5, 46, 61, 88, and 131.

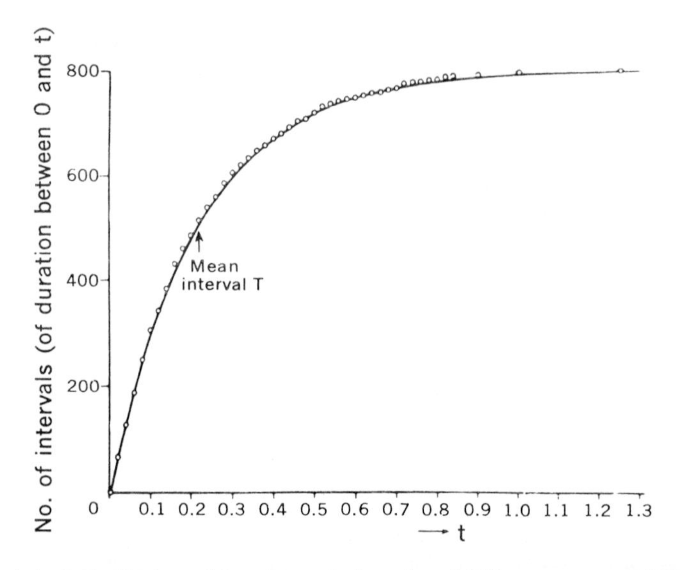

Fig. 5-10. Statistical distribution of time intervals in series of 800 spontaneous MEPPs. That sequence is random is shown by fact that distribution of intervals is asymmetric and follows simple exponential law. Probability that MEPP will occur in given interval increases with that time interval along simple exponential curve of type $p = 1 - e^{-t/T}$, in which t is interval chosen for observation and T is mean interval between successive events. (From Fatt and Katz.[63])

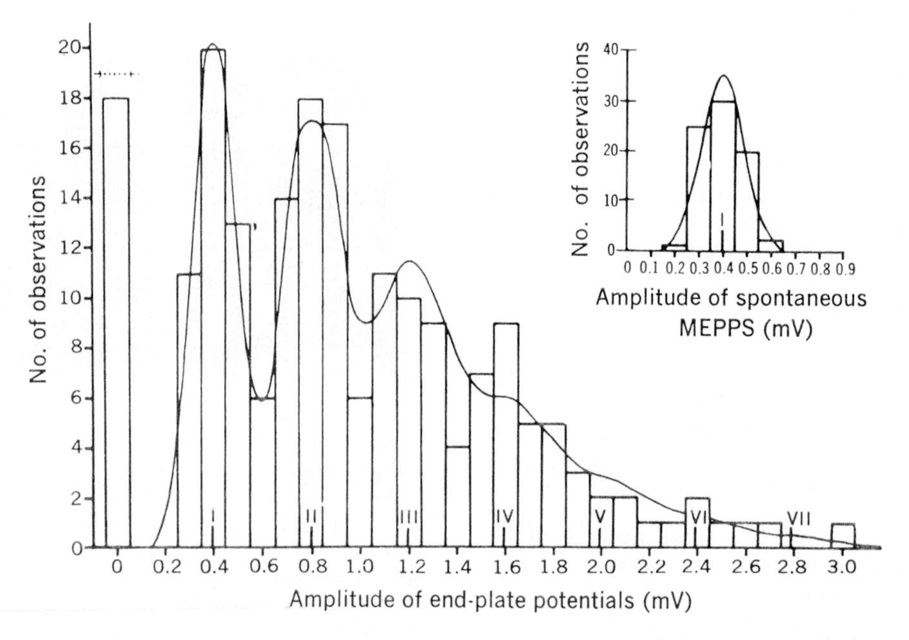

Fig. 5-11. Histograms of neurally evoked EPPs and amplitude distribution of spontaneous MEPPs (inset) in fiber of mammalian skeletal muscle in which neuromuscular transmission was blocked by increasing magnesium concentration of bathing Krebs' solution to 12.5 mM. Peaks of neurally evoked EPP amplitude distribution occur at 1, 2, 3, etc. times the mean amplitude of MEPPs. Gaussian curve is fitted to MEPP amplitude distribution and used to calculate theoretical distribution of neurally evoked EPP amplitude (continuous curve), which fits well with amplitude distribution of those experimentally observed (bar graph). Bar placed at zero indicates number of failures observed in series of trials; arrows and dotted line show number of failures expected theoretically from Poisson's law. (From Boyd and Martin.[19])

to spontaneously released packets of ACh (quanta). The amplitude of the MEPP shows statistical variations (Fig. 5-11, inset) but amplitude variations can also result from changes in postjunctional chemosensitivity or differences in the electrophysiologic characteristics of individual muscle fibers. From this and various other lines of reasoning, Katz proposed that the quantity of ACh released during each spontaneous event is relatively constant, at least over short periods of observation. Experimental evidence, as detailed later, shows that the MEPP is not generated by reaction of the postsynaptic membrane with one ACh ion; rather, each MEPP is produced by the action of one quantum of ACh, which amounts to 2,000 to 10,000 ACh ions (see Cold Spring Harbor Symposia,[1] pp. 175-186).

Relation between spontaneous release of acetylcholine and that evoked by nerve impulse[40,88,116]

The mechanism that transports the packet of ACh from the nerve terminal into the synaptic cleft is not yet completely understood, but this subject has been under intensive investigation and many theoretical ideas have appeared in the literature.[80,81] Synaptic vesicles have been isolated by differential cell fractionation techniques, and they have been shown to contain ACh (see, for example, Whittaker[189,190] for recent work and other literature citations). Thus it is now commonly believed that each quantum of ACh represents the ACh contained in one presynaptic vesicle and that an MEPP is generated when such a vesicle ruptures and discharges its contents into the synaptic cleft to reach and react with receptors in the postsynaptic membrane. Hubbard[80] has reviewed various suggested mechanisms concerning the interaction of the synaptic vesicles with sites on the inner surface of the neuronal terminal membrane. Direct electron micrographic evidence showing presynaptic rupture of synaptic vesicles has appeared in several publications (e.g., Hubbard and Kwanbunbumpen[82]). Further evidence establishing the links between synaptic vesicles and ACh release during neuromuscular transmission also comes from other sources. For example, application of black widow spider venom to the neuromuscular junction causes a marked rise and then a fall of MEPP frequency. After such treatment, electron micrographs show that presynaptic terminals contain very few synaptic vesicles.[29,120] A more physiologic and reversible depletion of synaptic vesicles can be produced by applying an isotonic K^+ propionate–Ringer's solution to depolarize neuronal terminals.[70] According to current views, when exocytosis of synaptic vesicles occurs, the synaptic vesicle membrane fuses with the nerve terminal axolemma. Thus, following the onset of intense (high K^+) depolarization of the nerve terminal, whereby exocytosis occurs very rapidly, the synaptic vesicle membrane accumulates in the neurilemma. In the mixed neurilemmal membrane whose area has increased, infoldings occur, forming large cisternlike structures that invade the nerve terminal. New synaptic vesicles, covered with a characteristically fuzzy coat, bud from these membrane infoldings of the neurilemma and move into the cytoplasm. There these new vesicles are refilled with ACh, lose their fuzzy coating, and become a part of the store of vesicles available for release. Electron micrographic evidence and further details of the just-described vesicle cycling processes are presented in Gennaro et al.[70]

It is now well established that the value of the membrane potential has a powerful influence on ACh release. As mentioned previously, depolarization of the nerve terminal produced by increasing extracellular K^+ increases the rate of release of ACh. Katz[88] showed that MEPP frequency increases logarithmically with decreases in membrane potential. At present, direct experimental measurements of transmembrane potential of motor nerve terminals is ordinarily not possible because they are so small in diameter. However, in the case of a giant synapse found in the stellate ganglion of the squid, one can, with intracellular electrodes, measure transmembrane potentials in both the pre- and postjunctional fibers. Katz and Miledi[95,97] have utilized this anatomically favorable synaptic arrangement to study the relationship between transient reduction of the membrane potential of the presynaptic fiber and the ensuing postsynaptic potential produced in the postsynaptic fiber. In their work with this synaptic preparation the action potentials were blocked by applying tetrodotoxin and tetraethylammonium to inhibit Na and K conductance increases, which are initiated by depolarization. Under these circumstances the presynaptic terminal could be depolarized to a controlled degree by applying electrical current pulses to it. Fig. 5-12, *A,* shows the experimental arrangement, and the results are given in Fig. 5-12, *B* to *D*. It is apparent in *B* that as presynaptic membrane depolarization exceeds values greater than about 35 mV, the postsynaptic response increases rapidly. As shown in *D,* the postsynaptic responses increase logarithmically with respect to presynaptic depolarization. Thus for this synapse

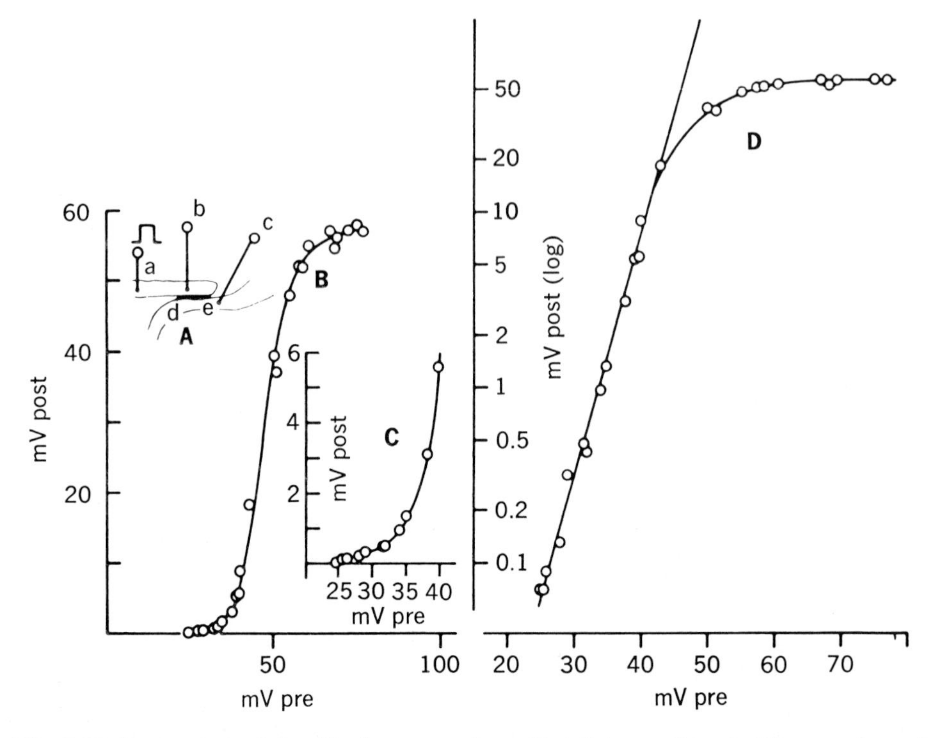

Fig. 5-12. Input-output relationship for synapse in squid stellate ganglion. **A,** Diagram of experimental arrangement showing *(a)* intracellular electrode for applying depolarizing current pulses to presynaptic terminal; *(b)* electrode for recording membrane potential of presynaptic terminal; and *(c)* electrode for recording membrane potential of postsynaptic fiber at region *(e)*. **B,** Plot showing relationship between presynaptic depolarization (abscissa) and postsynaptic response (ordinate). **C,** Initial portion of curve in **B** to show more detail. **D,** Semilogarithmic plot of input-output relationship as shown in **B.** (From Katz and Miledi.[95])

the "input-output" relationship is like that found for the neuromuscular junction. For results of further studies of this giant synapse, see Miledi[139] and Miledi and Slater.[140]

A further understanding of spontaneous and action potential–actuated release of ACh has come from study of the effects of changes in extracellular Ca^{2+} and Mg^{2+} concentrations on neuromuscular transmission.[3,39,80,86] Although such changes cause relatively small alteration in the frequency of the spontaneous MEPPs, they have powerful effects on the response to a nerve impulse. Excess of Mg^{2+} blocks neuromuscular transmission by interfering with the presynaptic release of transmitter agent. Ca^{2+} opposes the effect of Mg^{2+} and relieves the block. Reduction in Ca^{2+} concentration, on the other hand, produces neuromuscular block in a way quite similar to that caused by excess Mg.$^{2+}$ As the block produced by elevated Mg^{2+} progresses, electrical recording of the events produced by testing nerve volleys shows that the amplitude of the evoked EPP falls in a series of small steps and that just before complete block occurs, the postsynaptic response is similar in amplitude and duration to that of spontaneously occurring MEPPs. Before this stage, it is seen that the amplitudes of a series of neurally evoked EPPs vary in size by a quantal factor that is also equal to the amplitude of the miniature EPPs. When the average quantum content of the EPP is small, these amplitude fluctuations occur in a manner predictable by Poisson's law. A statistical analysis of this variation is given in Fig. 5-11. This leads to the conclusion that the MEPP is the basic unit of action at the neuromuscular junction. The depolarization of the nerve terminal accompanying the nerve impulse increases enormously the probability of quantal release. Thus the normal EPP is generated by the virtually synchronous release of quanta of transmitter whose individual spontaneous liberations give rise to the miniature potentials.

Under normal conditions the arrival of a nerve action potential at the motoneuron terminals liberates approximately 200 to 300 ACh quanta (quantal content). The quantal content can be estimated by various experimental techniques, some of which involve statistical analysis.[3,131] Additional experimental methods outlined by Hubbard et al.[3] allow one to estimate that roughly 300,000 quanta are stored in the neuronal terminals at a single neuromuscular junction. Thus the stores of preformed ACh may be depleted during high-frequency repetitive neuronal discharge, and replenishment of the stores and intracellular deployment of ACh quanta are important factors in maintaining neuromuscular transmission during sustained activity.

Various hypothetical models have been proposed[3,80] to explain how synaptic vesicles move to the inner surface of the presynaptic terminals and discharge their contents into the synaptic cleft. One of the firmly established facts is that the ACh release process depends on the extracellular $[Ca^{2+}]_o$ activity. In fact, many excitation-secretion processes involving release of neurotransmitters and hormones are Ca^{2+}-dependent.[176]

Over the past few years, many aspects of the action of Ca^{2+} and other ions on presynaptic ACh release have been studied quantitatively. By employing iontophoretic techniques (see Fig. 5-2, *A*) Katz and Miledi[90,93,94] have shown that Ca^{2+} facilitates transmitter release if applied immediately before depolarization of presynaptic terminals occurs. They have concluded that depolarization of the axon terminal membrane opens a channel for Ca^{2+}, allowing it to move to the inside of the axon membrane, where it participates in a reaction that increases the rate of transmitter release. These steps contribute a large part of the synaptic delay time. Additional aspects of the presynaptic action of Ca^{2+} come from a detailed study of the relationship between $[Ca^{2+}]_o$ and the amplitude of the EPP (Dodge and Rahamimoff[51]). From their work, these investigators concluded that the cooperative action of four Ca^{2+} ions is required for neural release of one quantum of transmitter.

The rate of ACh release from motoneuron terminals increases rapidly during the membrane depolarization that occurs when an action potential traverses the motoneuron terminals. Because the amplitude of the action potential depends on the extracellular $[Na^+]_o$, one would expect reduction in $[Na^+]_o$ to decrease the quantal content and thus to cause the EPP amplitude to diminish. This result was obtained by Colomo and Rahamimoff[30] when the extracellular $[Ca^{2+}]_o$ was normal. However, with low $[Ca^{2+}]_o$ reduction of $[Na^+]_o$ increased the quantal content and EPP amplitude. They interpreted this result on the basis that increased influx of Ca^{2+} occurs with reduced $[Na^+]_o$ because Na^+ and Ca^{2+} compete for transit across the terminal membrane. Such Ca^{2+}-Na^+ antagonism is not unique; it has been observed in other systems such as the frog heart.[121] To summarize the work of Dodge, Rahamimoff, and Colomo, three reactions are assumed to occur:

$$Ca + X \rightleftharpoons CaX$$

$$Mg + X \rightleftharpoons MgX$$

$$nNa + X \rightleftharpoons Na_nX$$

where X indicates membrane sites. CaX is effective in ACh release, but MgX and Na_nX are not.

Various lines of evidence indicate that the intracellular concentration of Ca^{2+} controls release of ACh from neuronal terminals. The intracellular level of free Ca^{2+} is determined by influx and efflux of this cation and the Ca binding and buffering action of various intracellular constituents, especially mitochondria. Thus (see subsequently) repetitive firing of the neuronal terminal would be expected to increase the intracellular $[Ca^{2+}]_i$ and raise quantal content. Also, any agent that interferes with the powerful Ca^{2+} uptake mechanism of mitochondria would be expected to cause an increase in $[Ca^{2+}]_i$ and thereby increase ACh release. Details of investigations supporting this idea are given by Rahamimoff et al.[1]

Postsynaptic response of the muscle cell[3,45,72]

After its release from the neuronal terminals, ACh must move across the synaptic cleft to reach its sites of action on the PJM. Because of the short distance involved, this transit can occur by simple diffusion in times well within the synaptic delay. Katz and Miledi[92] estimate the time from the moment of ACh release to the onset of the MEPP at less than 50 μsec. In this estimation, they made the reasonable supposition that the rise time of the MEPP represents a rise in ACh concentration at the receptor site. The release of ACh into the perfusing fluid of a stimulated eserinized muscle and the quick excitatory action of ACh when applied locally near the PJM indicate that this ion is not bound to macromolecules within the synaptic cleft but diffuses as it would in a free solution.

The ACh released from neuronal terminals reaches the postjunctional membrane, where it reacts with receptor sites and thereby causes an increase in ionic permeability. The resulting ionic fluxes cause the PJM potential to decrease (depolarization), and, if the reduction is sufficient, a muscle action potential is initiated in the passively depolarized surrounding conductile membrane of the muscle fiber. The electrogenic action of ACh on the PJM is one example of a large number of chemoelectrical transductions that are operative at the chemosensitive membranes found at many synapses, on smooth and cardiac muscle cells, on the endings of some sensory neurons, on many nonconductile cells, etc. At present, much experimental work is directed to the isolation and characterization of receptor molecules and to an understanding of the mechanisms whereby membrane ionic permeability is controlled.

Direct evidence[44] has shown that cholinergic receptor sites are found on the external but not the internal surface of the muscle fiber membrane. These sites are distributed on the PJM, but they are also found on the extrajunctional membrane.[89,137,138] Many investigators (e.g., Koelle[104]) who have utilized electron microscopic histochemical localization techniques have shown that AChE is present at both prejunctional membrane and PJM. Thus ACh liberated into the synaptic cleft can reach the PJM to react either with the receptor sites or with AChE located there, or it can be lost by diffusing out of the synaptic cleft.

The hypothesis that ACh exerts its action on the muscle cell by union with a specific receptor substance of the cell membrane derives from the original proposition of Langley, to which reference has been made. Langley's ideas were greatly elaborated by Clark. Further evidence for the existence of a cholinergic receptor has come from studies of the action of various drugs on neuromuscular transmission. It is well known that many quaternary ammonium compounds, including ACh, can depolarize the PJM when they are applied to this region by iontophoresis or by microperfusion techniques.[156] Such drugs, which are known as receptor activators, cause increases in the ionic permeability of the PJM, presumably by producing a conformational change in the receptor molecule. It is worth noting that if such activators are applied for relatively long periods of time (seconds or more), the initially produced receptor activation is lost, and PJM permeability, as well as membrane potential, returns toward control values.[56,160] This and some other pharmacologic aspects of neuromuscular transmission will be discussed in a following section.

In addition to the receptor activators, there exists a large group of drugs that are receptor inhibitors, that is, when they combine with receptor sites, no permeability change is produced, but the occupation prevents receptor sites from reacting readily with neurally released ACh. *d*-Tubocurarine, a bisquaternary compound, is a good example of such a drug; α-bungarotoxin is another. For more information the reader should consult various textbooks of pharmacology and appropriate reviews and monographs.

In recent years, many attempts have been made to isolate the molecules located in PJMs that bear ACh-receptive sites. An interesting account of the earlier history and progress in receptor isolation has been given by DeRobertis.[49] One difficulty that impeded these early efforts is that receptor molecules represent only a small fraction of the molecular constituents that make up the membranes found in the tissues from which isolation of the receptor was attempted. Furthermore, many of the various structural molecular entities exhibit nonspecific binding for ACh and related compounds. Estimates of the distribution of cholinergic receptors in PJMs[1,173,186] show that they are well packed in this region, but this chemoreceptive region represents only a small portion of the entire sarcolemma.

Receptor isolation was successful when the receptor-rich electrical organs of electric fish were used as a source. Another advance was the employment of labeled α-bungarotoxin (a constituent of snake venom that binds tightly to receptors) as a marker in guiding the choice of separation techniques.

Many investigators, including Changeux et al.[27] and Miledi et al.,[142] have used α-bungarotoxin to isolate and characterize the cholinergic receptor protein abundantly found in electrical tissue. The isolated cholinergic receptor has been characterized from results obtained by several investigators.[1] One of the problems in determining the properties of the receptor is that the molecule may undergo structural alteration during the isolation and purification procedures. Another approach to the characterization of the active sites in the receptor molecule is to determine changes in behavior that are produced when the receptor is reacted in situ with various chemical agents.[28,87] By this means, it has been found that at the active site, the cholinergic receptor contains a disulfide bond. The possibility that phosphate groups are involved in the membrane channels whose permeability to cations is controlled by the receptor molecule was suggested by Liu and Nastuk,[119] who showed that UO_2^{2+} that binds strongly to phosphate groups inhibits the response of the receptor to carbamylcholine (a stable analog of ACh). Various interesting proposals depicting the structures of the receptor and conformational changes it is believed to undergo have been given.[1]

We may now turn to the electrophysiologic phenomena that occur at the muscle PJM during neuromuscular transmission. The reader will find Ginsborg's review[72] of this subject clear and instructive.

When ACh reacts with receptors of the PJM, the ionic conductance of this structure is greatly increased, and, as a result, rapid ionic movements occur across it. These movements reduce the electrical charge on the PJM, and thus the potential difference across it falls. As mentioned earlier, this membrane depolarization is commonly known as the EPP. In the ACh-activated PJM the channels through which ions move are open as a group for a very brief period (about 2 msec) because ACh is quickly dissociated from the receptor site. The reason for this is that [ACh] in the synaptic cleft is rapidly reduced both by enzymatic hydrolysis and by diffusion. If the EPP does not reach the critical level (about −50 mV), a muscle action potential is not initiated and the PJM potential is restored to its normal resting value by a passive outward diffusion of K^+.

The action potential neurally initiated at the neuromuscular junction has a form different from that seen in action potentials recorded at nonjunctional regions. Typical recordings made at the neuromuscular junction are seen in Fig. 5-4. The first phase of the record represents the EPP, which, in this case, rises to the critical level (about −50 mV), at which a propagated action potential is initiated. Thereafter the record becomes complicated, since it represents the combined activity of the PJM and the adjacent electrically excitable nonjunctional membrane. Furthermore, the action potential record shows a reduced amplitude and longer falling phase than is seen with action potentials recorded at sites distant from the end-plate. In seeking the explanation for this reduced "overshoot" of the action potential, Fatt and Katz[62] proposed that the action of ACh on the PJM is to increase its permeability to all species of ions, and they termed their proposal the "short-circuit hypothesis." Additional support for this hypothesis was provided by "collision experiments," as shown in Fig. 5-13. From these and other results, one can see that the membrane potential of the ACh-activated PJM is driven from its resting value of −92 mV to a value in the −10 to −20 mV range (Fig. 5-14).

Although subsequent experimental work[153,154]

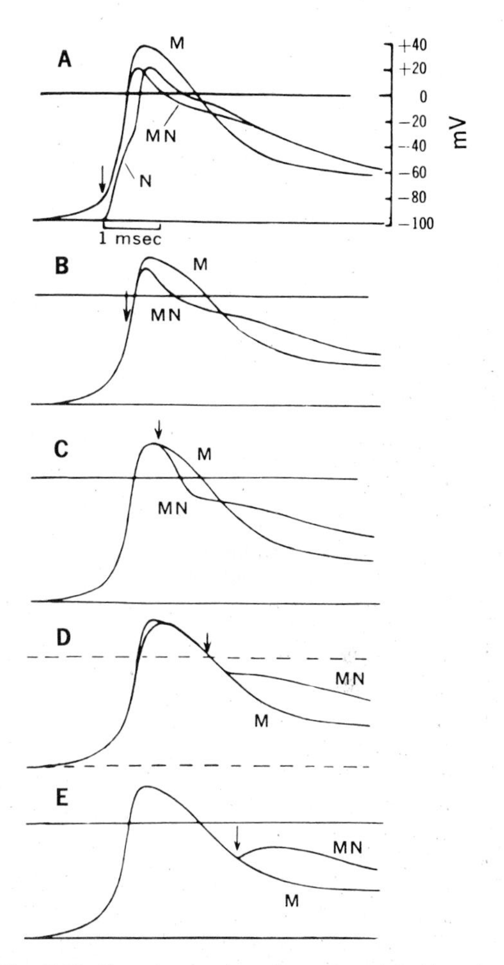

Fig. 5-13. Superimposed tracings of records from single end-plate showing interaction, at various phases, between ACh released by nerve impulse, and directly excited muscle action potential passing site of intracellular recording at end-plate *(M)*. Arrows indicate starts of responses to interjected nerve impulses *(N)*. When ACh release is timed to occur during peak of muscle action potential, as in **C,** it tends to drive membrane potential toward level around −15 mV, and thus record MN is produced. When ACh is released after membrane begins to repolarize, as in **E,** it tends to drive membrane potential back toward the −15 mV level. This −15 mV level represents "reversal potential," that is, it is potential toward which ACh-activated PJM is driven. (From del Castillo and Katz.[43])

raised some doubts about the validity of the short-circuit hypothesis, it was not discarded until the specific ionic permeability changes produced at the ACh-activated postjunctional membrane were clarified by the satisfying and well-conducted experiments of Takeuchi and Takeuchi.[183] Using a curare-blocked nerve-muscle preparation, these investigators applied a voltage

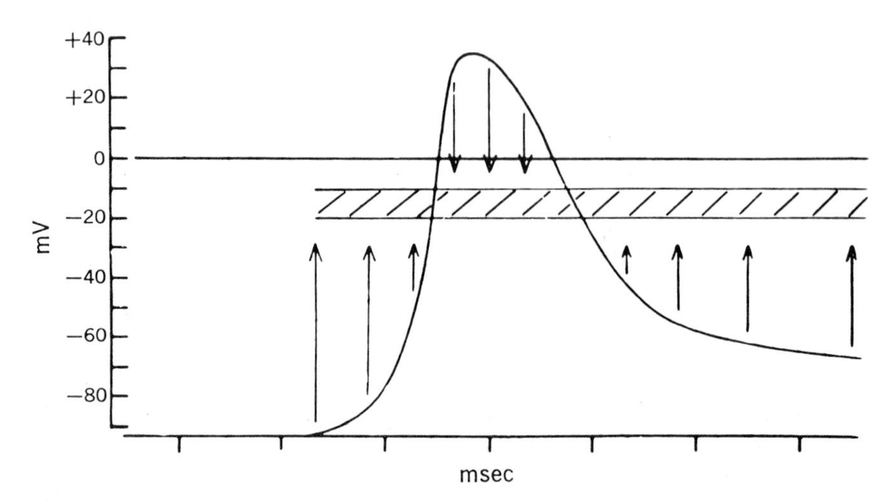

Fig. 5-14. Diagram summarizing results obtained by means of null-point experiment, illustrated in Fig. 5-13, which indicates direction of EPP produced by ACh applied at various values of membrane potential (represented by different phases of muscle action potential). Arrows indicate direction and relative magnitude of potential change due to release of ACh. The hatched area indicates approximate level at which EPP reverses sign, "reversal potential." (From del Castillo and Katz.[44])

clamp to the PJM and measured the EPP and the end-plate current (EPC) after nerve stimulation (Fig. 5-15). They found[184] (Fig. 5-16) that over a wide range of values, the EPC is proportional to the membrane potential. Extrapolation of the plot indicated that the EPC reaches zero and changes direction at the −10 to −20 mV level (reversal potential). Takeuchi and Takeuchi[184] went on to show that the value of the reversal potential was changed by variation in extracellular Na^+, but variations of extracellular Cl^- (glutamate substituted) did not affect its value. From these and other results, Takeuchi and Takeuchi proposed that ACh increases the Na^+ permeability (P_{Na}) and K^+ permeability (P_K) of the PJM more or less simultaneously, and for this reason the potential of the PJM moves to a new value, lying in the range of −10 to −20 mV. This region is roughly midway between the equilibrium potentials for K^+ and Na^+. The ACh-activated PJM is known to become permeable not only to Na^+ and K^+, but also to Ca^{2+}[185] and to small-diameter cations such as NH_4^+ and $(CH_3)_4N^+$.[69a,154] However, under in vivo conditions, only the transmembrane fluxes of K^+ and Na^+ are large enough to have substantial influence in altering membrane potential (Table 5-1).

In subsequent publications Takeuchi[181,182] further analyzed the basis for the reversal potential (called equilibrium potential in these papers). She calculated that a reversal potential of −15 mV would be produced in the ACh-activated PJM if the ratio of the increased Na and K conductances were 1.29:

$$\frac{\Delta G_{Na}}{\Delta G_K} = 1.29$$

She found that raising the extracellular $[Ca^{2+}]_o$ from 1.8 to 18 mM made the reversal potential more negative because increased $[Ca^{2+}]_o$ reduced ΔG_{Na}, leaving ΔG_K unchanged. Because *d*-tubocurarine does not affect the value of the reversal potential, the ratio of ΔG_{Na} to ΔG_K is unaffected, although the individual conductances are greatly reduced. Thus *d*-tubocurarine reduces the number of channels opened by released ACh.

A number of factors in addition to the $[Ca^{2+}]_o$ can cause variations in the reversal potential. Among the drugs, procaine can shift it in the negative direction,[122] and atropine shifts it in the positive direction.[123] These results have been interpreted on the basis of variations in ΔG_{Na}, ΔG_K, or both. Maeno[122] supposed that separate membrane channels were available for Na^+ and K^+, but this view is not generally accepted[67,72,107] and other explanations have been suggested that do not necessarily require synthesis of different receptor molecules.[67] A recent study[50] of EPC fluctuations indicates that only one type of ionic channel exists in the PJM of the frog neuromuscular junction.

The liberation of ACh from presynaptic terminals, as is now well established, occurs in quanta, each of which amounts to several thou-

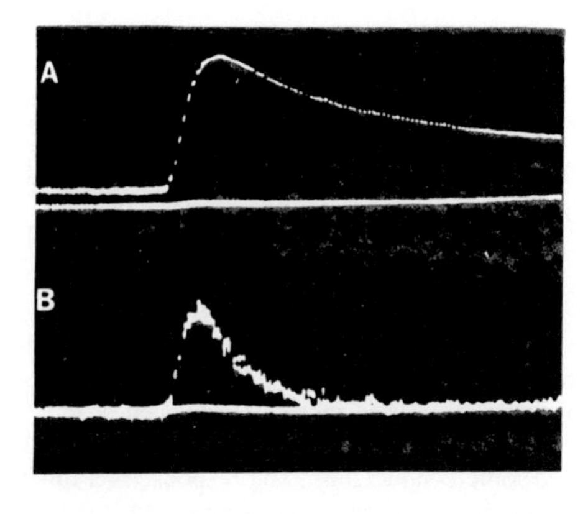

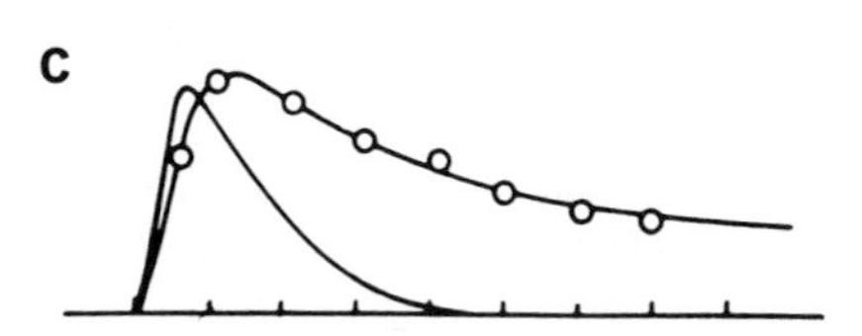

Fig. 5-15. A, EPP recorded intracellularly from d-tubocurarine–blocked muscle fiber. **B,** Lower trace is membrane potential recorded as in **A,** but with PJM voltage clamped. Upper trace is end-plate current (EPC) supplied by voltage clamp. **C,** Superimposed traces of EPP (curve with open circles) and EPC recorded from same end-plate. Open circles are values of membrane potential calculated from EPC. (From Takeuchi and Takeuchi.[183])

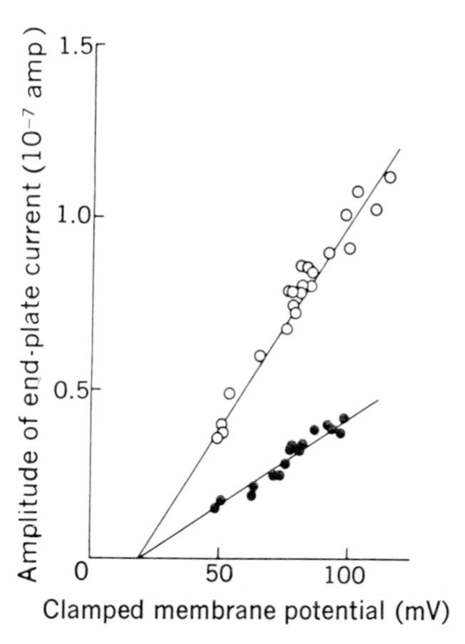

Fig. 5-16. Showing relationship between EPC measured at various values of PJM potential maintained by voltage clamping d-tubocurarine–blocked preparation. Open circles mark results with dTC 3×10^{-6} gm/ml. Closed circles mark results with dTC 4×10^{-6} gm/ml. (From Takeuchi and Takeuchi.[184])

Table 5-1. Properties of conducted action potential and EPP*

	Conducted action potential	EPP
Initiated by changes of membrane conductance	(Outward) electrical current	ACh
During rising phase	Specific increase of Na permeability	Increased permeability to Na and K
During falling phase	Specific increase of K permeability	No increase of ion permeability above resting condition (i.e., "passive" decay)
Equilibrium potential of active membrane	Na potential (approximately +50 mV, inside positive)	Reversal potential (approximately −15 mV inside negative)
Other distinguishing features	Regenerative ascent followed by refractory period	No evidence for regenerative action or refractoriness

*Modified from del Castillo and Katz.[45]

sand ACh^+ ions. The action of one quantum of ACh generates a miniature EPP. Generation of an MEPP involves the activation of a considerable number of individual cholinergic receptors, each of which controls the ionic permeability of a single membrane channel. The opening of each individual membrane channel to Na^+, K^+, and Ca^{2+} causes a very small elementary conductance increase. Katz and Miledi[99] found that the steady application of ACh to the PJM gave rise to a membrane depolarization associated with an increased membrane "noise," which they concluded was a sign of the random operation of many elementary events, each being involved in the opening and closing of single membrane channels. Their analysis of this membrane noise showed that a single elementary event produces a membrane voltage change of about 0.3 μV, about 1,000 times smaller than that of the MEPP. The conductance increase per channel, estimated to be 10^{-10} mho, is associated with a charge transfer amounting to 10^{-4} coulombs, which is equivalent to 5×10^4 univalent ions. More recently the membrane current that flows during a single channel opening has been measured directly in extrajunctional regions of denervated muscle.[163]

Not long before Katz and Miledi studied membrane noise, Kordas[106,108] discovered that the time course of the EPC produced during neuromuscular transmission was prolonged by membrane hyperpolarization. This discovery was extended by Magleby and Stevens[124,125] and by Anderson and Stevens,[10] who formulated a model to explain the voltage dependence of the channel ionic conductance control. In this model, one or two ACh ions combine with each receptor to induce a conformational change that opens a channel for Na^+, K^+, and Ca^{2+}. The channel is considered to exist in the completely open or completely closed state, and the transition between them is short. This model may be too simple, however, because there is evidence that the value of the single channel conductance, which is a measure of the open state, varies, depending on the agonist used to activate the receptor.[32] The average duration of the channel open time usually ranges from 0.5 to 2 msec, but it can be longer, depending on the membrane potential, the temperature, the nature of the cholinergic agonist, and other factors. Ionic channel characterization rests on theoretical models and experimental evidence involving noise and perturbation analysis as applied to chemically and electrically excitable membranes. Further details cannot be given here, and the reader should consult Stevens[1] and Takeuchi[4] for brief treatments and further references. DeFelice[38] has described the history and basis of fluctuation analysis as applied to neurobiology.

ADDITIONAL FACTORS IN THE CONTROL OF POSTJUNCTIONAL MEMBRANE PERMEABILITY

The literature dealing with various aspects of the action of drugs on PJM is well developed and extends over a long period of time. The mechanisms of action of many of the drugs that influence neuromuscular transmission are of interest to physiologists, but details cannot be presented here, and the reader will wish to consult textbooks and reviews, many of which are in the realm of pharmacology. However, two topics in this field warrant a brief introduction at this point.

Cooperativity[31,75]

The reaction between ACh (A) and postjunctional receptors (R) is often represented as follows:

$$[A] + [R] \rightleftharpoons [AR] \rightleftharpoons [AR]^*$$

The brackets denote concentrations. AR* is the activated form of the agonist-receptor complex in which the postjunctional ionic channel, which it controls, is in the open state. The response produced at the PJM can be measured either as depolarization or as increased ionic conductance. The response is a function of the number of receptors activated, and it is usually plotted vs the agonist concentration (such plots are often called "dose-response" curves). Reliable determinations of the relationship between response and agonist concentration are difficult to carry out for many reasons, although such measurements are frequently made, often without sufficient consideration of factors that bear on the validity and interpretation of the results.

If we wish to assume that the model depicting agonist-receptor reaction is correct, the plot of response vs agonist concentration will have a hyperbolic form. However, in many such experiments, sigmoid-shaped curves have been obtained. Some of the papers in which such results have been published have been cited by Colquhoun,[31] Hartzell et al.,[77] and Sheridan and Lester.[177] The latter two papers present further data of this kind in addition to theoretical treatments of it.

The explanation of sigmoidal response-concentration curves requires more complicated models than the one just depicted, and in these the concept of cooperativity is introduced. For example, in the following two-stage sequential model:

$$2[A] + [R] \rightleftharpoons [A] + [AR] \rightleftharpoons [A_2R]$$

the binding of the first molecule of agonist A to the receptor R facilitates the binding of the second molecule of A. One can see that such facilitation, which would be called positive cooperativity, could cause an unexpectedly large increase in response as the concentration of A is increased. For further details the reader should consult the references cited previously. The subject has been briefly introduced here in order to illustrate that the response produced by application of agonists to tissues is more complicated than one might have supposed from earlier treatments.

Desensitization

The term "desensitization" means a loss of response of a receptive membrane that occurs during continuous application of a receptor-activating agonist. For example, during the sustained application of ACh to the muscle PJM, initially there is a rapidly produced increase in membrane conductance and resulting membrane depolarization, but with time (seconds), these changes decline toward the control level. The decline in response of the PJM can be sufficient to cause a blockade of neuromuscular transmission. Receptor desensitization has long been known to physiologists and pharmacologists. Its importance is also well appreciated by anesthesiologists who block neuromuscular transmission by administering succinylcholine, thereby producing the muscle relaxation required for many surgical procedures.

Recently a review of publications concerning receptor desensitization has been published (Nastuk[157]). In the review the characteristics of desensitization, factors that influence its development, and models proposed to explain this phenomenon are outlined. The basic mechanisms underlying desensitization are not fully understood, but it seems likely that binding Ca^{2+} by both the receptor and perhaps other intrinsic molecular constituents making up the ionic channel may be involved.

SOME FURTHER ASPECTS OF NEUROMUSCULAR TRANSMISSION

Neuromuscular fatigue and facilitation[4]

Under normal conditions, conduction in the motor nerve and transmission at the neuromuscular junction are relatively indefatigable. During a sustained maximal effort, it is the contractile mechanism that fails, and it is here that fatigue first makes its appearance. This fact is demonstrated by Merton's study[135] of the functional capacity of the *adductor pollicis* in normal, waking human subjects. From his observations the following conclusions can be drawn: (1) A maximal voluntary effort of a muscle produces a tension equal to that caused by a high-frequency stimulation of all the motor nerve fibers to the muscle. (2) During intense voluntary effort a single electrical stimulation of the motor nerve evokes a normal muscle action potential, but there is no increase in tension. Such a stimulus produces synchronous impulses in every motor nerve fiber not at that instant refractory, because it is conducting an impulse of natural origin. This result provides independent evidence that the voluntary effort can activate the contractile mechanism to the full. (3) When the strength of the maintained maximal voluntary effort fails, tension cannot be restored by electrical stimulation of the motor nerve, even though such a stimulus evokes a maximal nerve volley and it appears that neuromuscular transmission is fully operative because a maximal action potential in the muscle results. This is true even when extreme fatigue of the contractile mechanism has reduced the tension of the voluntary effort to zero.

Fatigue of neuromuscular transmission can occur, however, under certain conditions. For example, if the nerve of an isolated nerve-muscle preparation, which receives no circulation, is stimulated at a low frequency for a period of several minutes, there is a slow fall to zero of the muscle tension produced by each stimulus, and the muscle may relax incompletely. At this time the motor nerve continues to conduct impulses, and direct electrical stimulation of the muscle elicits a response of maximum tension. This neuromuscular fatigue is due to a slow, steady decline in the amount of ACh released by each nerve impulse, so that gradually the EPP produced by the released ACh falls in amplitude and fails to initiate action potentials in the muscle fibers. It has been shown by del Castillo and Katz[41] that the decreased output of ACh is due to a reduction in the number of ACh quanta released and not to a drop in the amount of ACh per quantum. This is true at least over the short term of several minutes. Furthermore, the progressive decline in the number of quanta released per impulse is associated with an increase in the spontaneous release rate, as shown by the increase in MEPP frequency. It might be supposed that the repeated discharge of ACh on the PJM might depress the responsiveness of the latter, since it is known that such desensitization can occur during the sustained application of depolarizing quaternary ammonium compounds such as ACh and its analogs. However, it has been shown[166] that postjunctional desensitization apparently is not an important factor in explaining the reduction of the EPP during a train of nerve impulses.

Under ordinary physiologic conditions, during repetitive stimulation of the motoneuron, depression of neuromuscular transmission develops. This depression is attributed to a reduction in the store of ACh quanta readily available for neuronal release, and hence the quantal output of ACh per nerve impulse (quantal content) falls. If $[Ca^{2+}]_o$ in the extracellular fluid is reduced and $[Mg^{2+}]_o$ is increased, the quantal content is small, and thus each neuronal action potential places very small demand on the store of readily releasable ACh quanta. Under these conditions, repetitive neuronal discharge is not associated with depression of neuromuscular transmission. On the contrary, if amphibian[43] or mammalian[18,117] nerve-muscle preparations are bathed in low Ca^{2+}–high Mg^{2+} solutions, it is found that the EPP amplitude is *increased* during repetitive nerve stimulation. Statistical analysis indicates that this increase occurs because more and more quanta of ACh are released with each successive nerve impulse.

The physiologic characteristics of neuromuscular facilitation and the mechanisms responsible for its production have been studied by many investigators, and the publications cited here can guide the interested reader to more details. Mallart and Martin[126,127] using both Mg-blocked (low quantal content) and curare-blocked (high quantal content) preparations showed that potentiation occurs in two phases. The first phase appeared within 5 msec of the conditioning impulse and decayed exponentially with a time constant of about 35 msec. The second phase became evident about 60 to 80 msec after the conditioning impulse, rose to a peak in approximately 120 msec, and decayed with a time constant of about 250 msec. These investigators concluded that in both phases of facilitation an increased probability of ACh release occurs and that the two components can sum linearly.

Rahamimoff[172] and Katz and Miledi[96] studied the effect of variation of $[Ca^{2+}]_o$ on neuromuscular facilitation in order better to understand the mechanisms responsible for it. They systematically varied the extracellular $[Ca^{2+}]_o$ and $[Mg^{2+}]_o$ in the bathing fluid or relied on iontophoretic techniques to produce local changes in $[Ca^{2+}]_o$ at appropriate times. In both papers, it is concluded that $[Ca^{2+}]_o$ is an important factor in determining the facilitation produced by a nerve impulse. The models presented involve entry of Ca into the axon terminal during propagation of an impulse. The Ca that enters is thought to combine with specific sites on the inner surface of the membrane, thereby raising the probability of ACh release. Thus neuromuscular facilitation (which is associated with an increased probability of ACh release) would be exhibited during the time that the Ca-membrane complex persists. These workers came to grips with the problem of explaining the relatively long time course of neuromuscular facilitation. Part of the problem may be resolved by assuming that four Ca^{2+} ions cooperate in release and that the dissociation of the Ca-membrane complex follows a nonlinear rate equation. However, it is recognized by these workers and others that the mechanisms determining the extent and time course of neuromuscular facilitation are manifold and are as yet far from understood.[80,141] The problem may be better attacked by studying synapses such as those existing in the squid stellate ganglion, where it is possible to use intracellular techniques to measure and manipulate membrane potentials and to apply ions and drugs to both pre- and postsynaptic elements. For further information the reader should consult Katz and Miledi,[95,97,98] Miledi,[139] and Miledi and Slater.[140]

The preceding discussion deals with the neuromuscular facilitation that appears early in the wake of a single nerve impulse and during repetitive discharge of the motoneuron. However, after such tetanic discharge is completed, neuromuscular transmission can show immediate depression followed by a period of facilitation that can last up to several minutes. This slowly developing, long-enduring facilitation is known as posttetanic potentiation. It can be demonstrated by applying a train of stimuli to the motor nerve, following which appropriately timed test stimuli are delivered and the EPP or the contractile response is recorded. An example taken from Hutter[83] is shown in Fig. 5-17. In this case the nerve-muscle preparation was partially curarized (a large fraction of the junctions was blocked) as shown by the fall in muscle tension output (Fig. 5-17, *A* and *B*). Tetanic stimulation caused a brief neuromuscular facilitation followed by depression, which resulted in a complete neuromuscular block. However, during the posttetanic period the tension output in response to test stimuli rose nearly to the normal level. The plots in Fig. 5-17, *C* and *D*, illustrate that the degree and duration of posttetanic potentiation depend on the frequency and duration of the conditioning tetanus. Posttetanic potentiation was studied more directly by Liley[115] and Liley and North[118] who showed that during such potentiation the quantal content is increased and the frequency of spontaneously released quanta (MEPPs) is raised. More recently a number of variables affecting posttetanic potentiation were studied by Miledi and Thies,[141] Rosenthal,[175] and Weinreich.[187] They found that posttetanic potentiation was prolonged or intensified by raising extracellular $[Ca^{2+}]_o$. From this and other evidence, they suggest that the phenomenon may be related to accumulation of Ca in the neuronal terminals.

There are two other forms of neuromuscular facilitation that deserve mention. The first is that produced by stretch of the muscle.[85] If in a partly curarized nerve-muscle preparation the muscle is stretched, there is an increase in the number of muscle fibers that respond to a nerve volley (i.e. recruitment of fibers that had previously failed to contract). This facilitation results from an increase in quantal content, that is, mechanical deformation in some way causes a larger number of ACh quanta to be released by each nerve impulse and thus increases the amplitude of the EPP. Like all forms of facilitation mentioned so far, that produced by stretch is associated with an increased spontaneous release rate.

The last form of facilitation to be mentioned was first described by Orbeli many years ago. He observed that when neuromuscular transmission

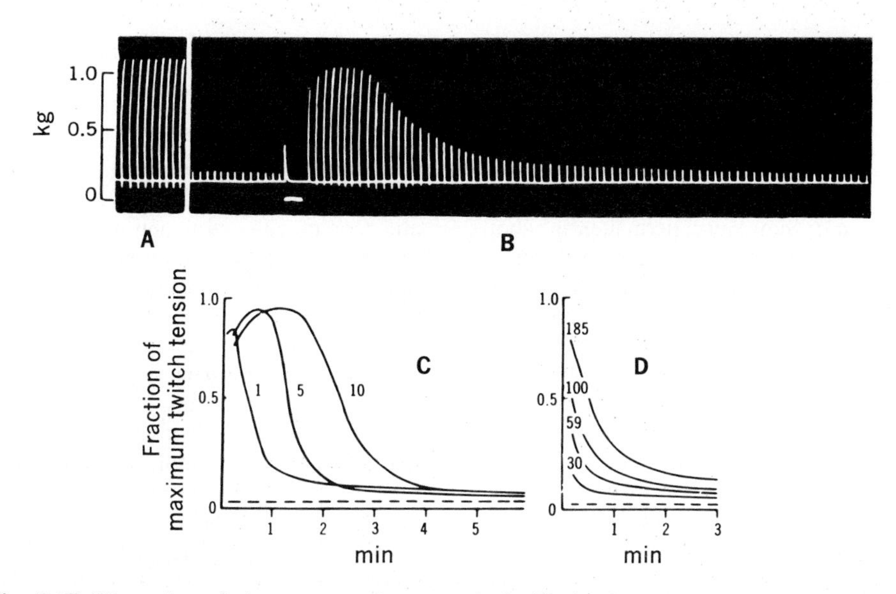

Fig. 5-17. Illustration of phenomenon of posttetanic facilitation in partially curarized muscle. **A** and **B,** Contractions of tibialis muscle of anesthetized cat produced by supramaximal single stimulus to sciatic nerve every 10 sec, recorded mechanically. **A,** Untreated preparation. **B,** After partial curarization, contractions greatly reduced in amplitude—many end-plates blocked. At horizontal signal, nerve was stimulated for 20 sec at 80/sec; thereafter stimulation at rate of once every 10 sec resumed. Curare block was completely overcome (decurarization) for about 1 min. **C** and **D,** Graphs illustrating time course of posttetanic potentiation in partially curarized tibialis muscle of cat. Experimental arrangement as above. Twitch tension, expressed as fraction of maximum twitch tension in normal preparation, plotted against time in minutes after end of tetanization of sciatic nerve, which contains efferent motor fibers innervating this muscle. **C,** Duration of potentiation depends on duration of conditioning tetanus; results plotted here following tetani of 1, 5, and 10 sec at 250 stimuli/sec. **D,** Time course of potentiation after 1 sec repetitive stimulation at different frequencies. (From Hutter.[83])

has been fatigued by prolonged, low-frequency stimulation of the motor nerve, stimulation of the sympathetic innervation of the muscle restores neuromuscular transmission and results in an increase in twitch tension. The effect is due to the recruitment of muscle fibers that in the fatigued state had failed to respond. It is produced also by epinephrine and by norepinephrine, and Hutter and Loewenstein[84] have shown that these agents or sympathetic nerve stimulation increases the amplitude of the EPP produced by the nerve impulse. They concluded that the facilitation is postjunctional because these agents were found to increase the sensitivity of the postjunctional receptors to ACh. However, in a more recent paper, Kuba and Tomita[109] showed that noradrenaline augments the probability of ACh release.

Changes in neuromuscular transmission following motor nerve section[16]

Section of the efferent nerve to a skeletal muscle results in its paralysis, a complete loss of voluntary and reflex contractions. Stimulation of the distal end of the cut nerve will continue to elicit contractions of the muscle for many hours. The first change to occur is a failure of neuromuscular transmission, and there is evidence that for any one junction this breakdown occurs rather suddenly. There is little doubt that the change is entirely due to a failure of transmitter release from presynaptic terminals, for at this stage there is no change in the sensitivity of the PJMs of the muscle cells to ACh. The mechanism of this failure is unknown. The main axons of the motor fibers will continue to conduct action potentials in response to electrical stimuli for several additional hours. It is of some importance to note that although the resting and action potentials (the latter set up by direct electrical stimulation) of denervated muscle cells are normal, there are changes in the properties of the membranes of these cells that lower the minimum current required to excite them, which accounts for the moderately increased electrical excitability of denervated skeletal muscle.[164] In poikilothermic animals the time course of these events is more prolonged with lower body temperature.

In both mammals and amphibians following denervation the spontaneously occurring MEPPs cease abruptly when transmission fails.[115,136] It has been shown for the frog that MEPPs are not, however, permanently abolished, for during the second week after denervation they reappear, although at a frequency some 100 times lower than normal. They occur at the places on the muscle cells where the remnants of the myelinated axons can be seen to terminate. These local potentials appear some weeks before the muscle cell develops hypersensitivity to ACh (see subsequently), and at this time the pharmacologic reactions are similar to those of normal muscle. It seems likely that the slowly recurring MEPPs are produced by the quantal release of ACh from outside the muscle cell, and it has been suggested that the Schwann cell, which survives degeneration of the axon, is responsible for this slow production and release of ACh quanta.[16] On reinnervation of the muscle cells by regenerating motor fibers, the spontaneous MEPPs once again occur at normal frequencies. Their occurrence during the period of denervation has not been observed in mammalian skeletal muscle.[115]

It has been known for a long time that denervated mammalian skeletal muscle develops a great increase in its sensitivity to topically applied ACh. This can be shown by injection of ACh into the arteries supplying a chronically denervated muscle. The increase in the sensitivity of a denervated effector cell is a general phenomenon (Chapter 32) exhibited by muscle, nerve, and gland cells. Denervated effector cells become increasingly sensitive to the transmitter agent secreted by the nerve fiber that innervates them. In skeletal muscle a 100- to 1,000-fold increase in chemosensitivity may develop during the weeks after denervation.

Various lines of evidence show that ACh receptors are concentrated in PJMs. The sensitivity to ACh in the PJM and surrounding membrane has been carefully mapped using improved microscopic visualization and iontophoretic technique.[54,112] In normal muscle, as one moves away from the edge of the postsynaptic gutter, the sensitivity to ACh falls rapidly. For the more distant extrajunctional membranes the sensitivity is 1,000 times less, but a significant response to ACh is still present.[89] These variations in sensitivity to ACh are correlated with the density of receptors as determined by use of radioactive-labeled α-bungarotoxin. The differential distribution of receptors in normal fibers depends on the presence of functional motor innervation. Prior to being innervated the muscle fiber is sensitive to ACh at all regions along its length, but when neuromuscular transmission becomes established, the receptor density increases in PJMs and decreases in extrajunctional regions. Following section of the motor nerve, with time the muscle fiber becomes increasingly sensitive to ACh, and the number of receptors increases in the extrajunctional regions. An increased postjunctional receptor density persists for some time after denervation (Frank et al.[1]).

The sensitization phenomenon has been investigated by the local application of small quantities of ACh, delivered iontophoretically via micropipets to the surface of denervated muscle cells.[12,136] The results indicate that there is no change in the sensitivity of localized chemosensitive sites on the PJMs of denervated muscle. This may seem surprising because it is well known that there is an increased responsiveness of denervated skeletal muscle to bulk-applied ACh. The apparent paradox is believed to be explained by an increase, in denervated muscle, in the total number of receptive sites available for combination with ACh.

The changes in chemosensitivity produced after section of the motor nerve are to some extent a result of inactivity because direct electrical stimulation of the denervated muscle causes the chemosensitivity to return toward normal levels (Lømo and Westergard[1]). The paralysis of denervated muscle cannot itself be the sole factor in increasing its chemosensitivity because Miledi[136] has shown that for muscle fibers having two or more end-plates, section of one motor fiber leads to an increased ACh sensitivity around the denervated neuromuscular junction, but this does not occur at the remaining innervated junction.

Receptors are synthesized intracellularly by muscle fibers, as is the case for other protein constituents. Newly synthesized receptors move to the surface membranes, in which they become incorporated. There the receptors remain for a period of hours to days until they are returned once more to the intracellular phase to undergo enzymatic breakdown, and certain of the breakdown products are extruded from the cell. Many details of the receptor cycling process have been worked out by Fambrough and his colleagues, and reviews presented by them and other workers should be consulted for more details.[1]

Because ACh receptors and other molecular constituents in membranes undergo a cycle involving incorporation and removal from the membrane and these molecules can diffuse along the membrane, which is a rather fluid structure, one might expect that receptors in postjunctional and extrajunctional membranes are identical. However, some evidence for differences between the receptors located in these regions has been obtained,[1] but not all this evidence is unequivocal. For example, Feltz and Mallart[67] showed that extrajunctional receptors of both normal and denervated muscles have reversal potentials of −42 mV, and this value was also obtained for receptor sites at the "postjunctional areas" of denervated muscles. Recently Mallart et al.[128] have reinvestigated this matter, and they pointed out that these values for reversal potential are in error because nonlinearities in the membrane potential–EPC relationship were not taken into account. They found the reversal potential for junctional and extrajunctional receptors to be identical, and both were in the 0 to −10 mV range.

These findings have a general significance in that the motor nerve fiber exerts an important influence on the chemosensitivity of the muscle cell it innervates. This is only one example of the so-called trophic influence of nerve. An abundant literature[52] shows that the trophic influence is exerted on many of the physiologic mechanisms of effector cells.

In mammals, denervated muscle displays very fine, diffuse, spontaneous movements. Usually the tremor is so slight in degree that it is seen well only when the bared surface of the muscle is observed in reflected light. The surface of the muscle is then observed to be "rippled by a restless agitation without either apparent rhythm or obvious center of activity."[48] This confused medley of small twitches constitutes true *fibrillation,* a condition produced by the randomly occurring twitches of individual muscle fibers. The cause of this spontaneous contraction is uncertain. It may be the result of excitation of the hypersensitive muscle cells by small quantities of circulating or locally released ACh. It is quite different from the randomly occurring contraction of entire motor units called *fasciculation,* which is commonly seen in certain diseases of the CNS (e.g., amyotrophic lateral sclerosis).

PATHOLOGIC DISTURBANCES OF NEUROMUSCULAR TRANSMISSION

It can now be appreciated that neuromuscular transmission is a complicated process whose many steps are potentially vulnerable to pathologic derangement and interference. For aid in considering the possibilities, the steps in neuromuscular transmission and a few of the factors known to influence them are given here[155]:

- Presynaptic elements
  - ACh
    - Synthesis
    - Storage
    - Mobilization
    - Release
      - Membrane depolarization (via action potential or electrotonically)
      - Ca, Mg, botulinum toxin, etc.
  - Extent of terminal arborization
- Synaptic cleft
  - Diffusion of ACh
- Postsynaptic elements
  - Extent of PJM area
  - Reaction of PJM receptor sites to ACh
    - Desensitization mechanism
  - Resting potential of muscle fiber
    - Ionic composition of extra- and intracellular fluids
  - Excitability of conductile system
  - Hydrolysis of ACh
- Contractile elements
  - Excitation-contraction coupling
  - Performance of the actomyosin system

Myasthenia gravis[68,74]

One very well-known example of a pathologic disturbance in neuromuscular transmission is that which appears in a relatively rare disease known as myasthenia gravis. If a patient with this malady is asked to carry out a muscular task such as the repeated clenching of the fist, he is found to be able to produce strong contractions at the start, but very rapidly thereafter, muscular weakness appears. This weakness may be exhibited in many of the skeletal muscles, including those involved in eye movements, limb movements, and deglutition. Some regions may be more greatly affected than others. There is a vast literature dealing with both clinical and basic research in the field of myasthenia gravis. Many other papers have been devoted to problems concerning the treatment of patients with this disease. It is impossible here to give a complete review of the numerous past developments in this field, but from the general references the reader should be guided to further details.

In myasthenic patients, recording of muscle action potentials (electromyography) from affected regions shows that during repetitive stimulation of the motor nerve, neuromuscular transmission fails. This finding is further supported by the long-known fact that the muscular strength of the myasthenic patient is appreciably improved after administration of anticholinesterase drugs such as neostigmine. Such treatment has no curative value, but many patients afflicted with the disease are able to maintain reasonable daily activities if anticholinesterase medication is continuously administered orally.

There have been many attempts to pinpoint the exact nature of the neuromuscular transmission defect of myasthenic patients. Past speculations have included the following: presence of a circulating curare-like agent that binds to postjunctional receptors, thus reducing the response to released ACh; an excessive accumulation of AChE at the neuromuscular junction; and inadequate synthesis or release of ACh from the motor nerve terminals. Until recently there existed no compelling evidence to support the first two hypotheses, but a few years ago, experimental evidence was obtained by Elmqvist et al.,[58] who interpreted it as indicating that a presynaptic defect was the major factor in causing faulty transmission. These investigators used single fiber techniques on muscle biopsied from patients with myasthenia gravis and found that MEPPs were much reduced in amplitude. In their study, postjunctional chemosensitivity was found to be unchanged and the increase in MEPP frequency produced by presynaptic depolarization was normal. On the basis of this evidence, it was concluded that ACh quanta are released in the usual number by nerve impulses, but the *size* of each quantum is small. This solely presynaptic interpretation of the transmission defect in myasthenia gravis, which is based on the reduction in MEPP amplitude,[185] was widely accepted until recently, but evidence obtained later provides alternative explanations for the small-amplitude MEPPs. One such contribution is derived from the work of Engel and Santa.[58a] Their electron micrographs show that in myasthenia gravis the primary and secondary synaptic clefts are increased in size and the secondary clefts are less numerous than usual. Under such circumstances the ACh discharged into the clefts by each quantum undergoes greater dilution, which would diminish the ACh concentration at the postsynaptic receptor sites. Evidence of possible PJM involvement was provided by Fambrough et al.,[60] who showed, using ^{125}I-labeled α-bungarotoxin and radioautography, that a reduction in ACh receptors occurs in muscle fibers of myasthenic patients. Further direct evidence of postsynaptic involvement was obtained by Albuquerque et al.,[9] who showed, using iontophoresis, that the PJMs of muscle fibers from myasthenic patients have a significantly reduced sensitivity to ACh. For these and other reasons (see subsequent discussion) the explanation of the defect in neuromuscular transmission has shifted to consideration of the PJM. However, one should not exclude presynaptic involvement because pathologic changes in the terminal arbor of the motoneuron may contribute to the neuromuscular transmission deficit.

Although it is well established that neuromuscular transmission is faulty in myasthenic patients, much remains to be learned about the pathologic mechanisms whereby the faults are produced. The possibility that the disease might have an etiology based on immunologic derangements was proposed in two independent papers by Nastuk et al.[161] and Simpson.[178] Nastuk et al. based their view on experimental work concerning the detection of cytolytic activity and altered serum complement activity in the blood of myasthenic patients and on further evidence provided by Strauss et al.,[180] who demonstrated that the serum of some myasthenic patients contains an antibody against skeletal muscle. In formulating their hypothesis, Nastuk et al. made an effort to take into account and to interpret many aspects of myasthenia gravis, including the following: (1) existence of immunologic abnormalities; (2) presence of enlarged thymus glands in many myasthenic patients, which was connected with item 1 by virtue of the demonstrations by others that the thymus gland has an important immunologic function; (3) recognition that significant histopathologic changes occur in the muscles of myasthenic patients; and (4) consideration of the possibility that the neuromuscular defects seen in myasthenia gravis might be, to some extent, irreversible or at least that some functional defects might persist long after the etiologic agents that produced them had vanished.

In 1966 a critique of the autoimmune theory and an appraisal of past and current research in myasthenia gravis was presented by Nastuk and Plescia.[159] A number of hypothetical suggestions for pathologic disease–producing mechanisms were made in the review, and the need for an animal model of myasthenia gravis was emphasized. Production of myasthenia in rabbits immunized with a crude muscle extract was reported by Plescia et al.[169] in 1968, but this result could not be duplicated consistently. In the years immediately following this work, cholinergic receptor protein was isolated from the electrical organs of certain electric fish. Patrick and Lindstrom,[167] in preparing rabbit antibodies against the cholinergic receptor, found that the immunized animals developed muscle weakness that, as further investigation revealed, was caused by a failure of neuromuscular transmission. These experiments have since been repeated by many other groups of investigators who have produced experimental myasthenia in several mammalian species and, more recently, the frog.[162]

During the past 5 years a large number of publications have defined various characteristics of experimental myasthenia gravis. Results of these publications have been briefly presented in an informative review.[53] Many immunologic aspects of myasthenia gravis have been established, and it is now clear that in this disease, antireceptor antibodies can be detected not only in the circulation but also bound to PJMs in situ.[59] These receptor-antibody complexes have also been demonstrated to bind one of the constituents of serum complement.[59] Furthermore, the morphologic changes in ultrastructure demonstrated for human myasthenic muscle also appear in the muscles of receptor-immunized animals. Such lines of evidence provide support for the hypothesis postulated earlier by Nastuk et al.[161] that an immune cytolytic process may cause alteration or even destruction of junctional membranes in myasthenia gravis. It is evident that the solution and full understanding of this disease will require the further collaborative efforts of electrophysiologists, immunologists, and others.

REFERENCES

General reviews

1. Cold Spring Harbor Symposia on Quantitative Biology; The synapse, Cold Spring Harbor, N.Y., 1975, Cold Spring Harbor Laboratory, vol. 50.
2. Eccles, J. C.: The physiology of synapses, New York, 1964, Academic Press, Inc.
3. Hubbard, J. I., Llinas, R., and Quastel, D. M. J.: Electrophysiological analysis of synaptic transmission, Baltimore, 1969, The Williams & Wilkins Co.
4. Kandel, E. R., editor: Handbook of physiology. Section 1: the nervous system, Baltimore, 1977, The Williams & Wilkins Co.
5. Katz, B.: Nerve, muscle and synapse, New York, 1966, McGraw-Hill Book Co.
6. Quarton, G. C., Melnechuk, T., and Schmitt, F. O., editors: the neurosciences. A study program, New York, 1967, The Rockefeller University Press.
7. Schmitt, F. O., editor: The neurosciences. A second study program, New York, 1970, The Rockefeller University Press.
8. Siegel, G. J., Albers, R. W., Katzman, R., and Agranoff, B. W., editors: Basic neurochemistry, ed. 2, Boston, 1976, Little, Brown & Co.

Original papers

9. Albuquerque, E., Rash, J. E., Mayer, R. F., and Satterfield, J. R.: An electrophysiological and morphological study of the neuromuscular junction in patients with myasthenia gravis, Exp. Neurol. **51**:536, 1976.
10. Anderson, C. R., and Stevens, C. F.: Voltage clamp analysis of acetylcholine produced end-plate current fluctuations at frog neuromuscular junctions, J. Physiol. **235**:665, 1973.
11. Anderson-Cedergren, E.: Ultrastructure of motor end-plate and sarcoplasmic components of mouse skeletal muscle fiber, J. Ultrastruct. Res. **1**(suppl.): 1, 1959.
12. Axelson, J., and Thesleff, S.: A study of supersensitivity in denervated mammalian muscle, J. Physiol. **147**:178, 1959.
13. Barondes, S. H., editor: Cellular dynamics of the neuron, New York, 1969, Academic Press, Inc.
14. Birks, R., and MacIntosh, F. C.: Acetylcholine metabolism in a sympathetic ganglion, Can. J. Biochem. **39**:787, 1961.
15. Birks, R., Huxley, H. E., and Katz, B.: The fine structure of the neuromuscular junction of the frog, J. Physiol. **150**:134, 1960.
16. Birks, R., Katz, B., and Miledi, R.: Physiological and structural changes at the amphibian myoneural junction, in the course of nerve degeneration, J. Physiol. **150**: 145, 1960.
17. Bishop, G. H.: Natural history of the nerve impulse, Physiol. Rev. **36**:376, 1956.
18. Boyd, I. A., and Martin, A. R.: Spontaneous subthreshold activity at mammalian neuromuscular junctions, J. Physiol. **132**:61, 1956.
19. Boyd, I. A., and Martin, A. R.: The end-plate potential in mammalian muscle, J. Physiol. **132**:74, 1956.
20. Brown, G. L.: Action potentials of normal mammalian muscle. Effects of acetylcholine and eserine, J. Physiol. **89**:220, 1937.
21. Brown, G. L.: Transmission at nerve endings by acetylcholine, Physiol. Rev. **17**:485, 1937.
22. Brown, G. L., and Feldberg, W.: The acetylcholine metabolism of a sympathetic ganglion, J. Physiol. **88**: 265, 1937.
23. Brown, G. L., Dale, H. H., and Feldberg, W.: Reactions of the normal mammalian muscle to acetylcholine and to eserine, J. Physiol. **87**:394, 1936.
24. Burgen, A. S. V., and Chipman, L. M.: Cholinesterase and succinic dehydrogenase in the central nervous system of the dog, J. Physiol. **114**:296, 1951.
25. Burke, W., and Ginsborg, B. L.: The action of the neuromuscular transmitter on the slow fibre membrane, J. Physiol. **132**:599, 1956.
26. Chang, C. C., and Lee, C.-Y.: Electrophysiological study of neuromuscular blocking action of cobra neurotoxin, Br. J. Pharmacol. **28**:172, 1966.
27. Changeux, J.-P., Kasai, M., and Lee, C.-Y. Use of a snake venom toxin to characterize the cholinergic receptor protein, Proc. Natl. Acad. Sci. USA **67**:1241, 1970.
28. Changeux, J.-P., Podleski, T., and Meunier, J.-C.: On some structural analogies between acetylcholinesterase and the macromolecular receptor of acetylcholine, J. Gen. Physiol. **54**:225s, 1969.
29. Clark, A. W., Mauro, A., Longenecker, H. E., and Hurlbut, W. P.: Effects of black widow spider venom on the frog neuromuscular junction: effects on the fine structure of the frog neuromuscular junction, Nature **225**:703, 1970.
30. Colomo, F., and Rahamimoff, R.: Interaction between sodium and calcium ions in the process of transmitter release at the neuromuscular junction, J. Physiol. **198**: 203. 1968.
31. Colquhoun, D.: Mechanisms of drug action at the voluntary muscle end-plate, Annu. Rev. Pharmacol. Toxicol. **15**:307, 1975.
32. Colquhoun, D., Dionne, V. E., Steinbach, J. H., and Stevens, C. F.: Conductance channels opened by acetylcholine-like drugs in muscle end-plates, Nature **253**:204, 1975.

33. Couteaux, R.: Localization of cholinesterases at neuromuscular junctions, Int. Rev. Cytol. **4**:335, 1955.
34. Couteaux, R.: Morphological and cytochemical observations on the postsynaptic membrane at motor end-plates and ganglionic synapses, Exp. Cell Res. **5**(suppl.):294, 1958.
35. Dale, H.: Transmission of nervous effects by acetylcholine, Harvey Lect. **32**:229, 1937.
36. Dale, H., Feldberg, W., and Vogt, M.: Release of acetylcholine at voluntary motor nerve endings, J. Physiol. **86**:353, 1936.
37. Davis, R., and Koelle, G. B.: Electron microscopic localization of acetylcholinesterase and nonspecific cholinesterase at the neuromuscular junction by the gold-thiocholine and gold-thiolacetic acid methods, J. Cell Biol. **34**:157, 1967.
38. DeFelice, L. J.: Fluctuation analysis in neurobiology, Int. Rev. Neurobiol. **20**:169, 1977.
39. del Castillo, J., and Katz, B.: The effect of magnesium on the activity of motor nerve terminals, J. Physiol. **124**:553, 1954.
40. del Castillo, J., and Katz, B.: Quantal components of the end-plate potential, J. Physiol. **124**:560, 1954.
41. del Castillo, J., and Katz, B.: Statistical factors involved in neuromuscular facilitation and depression, J. Physiol. **124**:574, 1954.
42. del Castillo, J., and Katz, B.: Changes in end-plate activity produced by presynaptic polarization, J. Physiol. **124**:586, 1954.
43. del Castillo, J., and Katz, B.: The membrane change produced by the neuromuscular transmitter, J. Physiol. **125**:546, 1954.
44. del Castillo, J., and Katz, B.: On the localization of acetylcholine receptors, J. Physiol. **128**:157, 1955.
45. del Castillo, J., and Katz, B.: Biophysical aspects of neuromuscular transmission. In Butler, J. A. V., editor: Progress in biophysics and biophysical chemistry, New York, 1956, Pergamon Press, vol. 6, p. 121.
46. del Castillo, J., and Katz, B.: Localization of active spots within the neuromuscular junction of the frog, J. Physiol. **132**:630, 1956.
47. del Castillo, J., and Katz, B.: A study of curare action with an electrical micromethod, Proc. R. Soc. Lond. (Biol.) **146**:339, 1957.
48. Denny-Brown, D. E., and Pennybacker, J. B., Fibrillation and fasciculation in voluntary muscle, Brain **61**: 311, 1938.
49. DeRobertis, E.: Molecular biology of synaptic receptors, Science **171**:963, 1971.
50. Dionne, V. E., and Ruff, R. L.: End-plate current fluctuations reveal only one channel type at frog neuromuscular junction, Nature **266**:263, 1977.
51. Dodge, F. A., Jr., and Rahamimoff, R.: Co-operative action of calcium ions in transmitter release at the neuromuscular junction, J. Physiol. **193**:419, 1967.
52. Drachman, D. B.: Neuromuscular transmission of trophic effects, Ann. N.Y. Acad. Sci. **183**:158, 1971.
53. Drachman, D. B.: Myasthenia gravis. Pt. I and II, N. Engl. J. Med. **298**:136, 186, 1978.
54. Dreyer, F., and Peper, K.: The acetylcholine sensitivity in the vicinity of the neuromuscular junction of the frog, Pflúegers Arch. **348**:273, 1974.
55. Eccles, J. C., and Jaeger, J. C.: The relationship between the mode of operation and the dimensions of the junctional regions at synapses and motor end-organs, Proc. R. Soc. Lond. (Biol.) **148**:38, 1957.
56. Eccles, J. C., and O'Connor, W. J.: The responses which nerve impulses evoke in mammalian striated muscles, J. Physiol. **97**:440, 1939.
57. Eccles, J. C., Katz, B., and Kuffler, S. W.: Nature of the "end-plate" potential in curarized muscle, J. Neurophysiol. **5**:362, 1941.
58. Elmqvist, D., Hoffmann, W. W., Kugelberg, J., and Quastel, D. M. J.: An electrophysiological investigation of neuromuscular transmission in myasthenia gravis, J. Physiol. **174**:4; 7, 1964.
58a. Engel, A. G., and Santa, T.: Histometric analysis of the ultrastructure of the neuromuscular junction in myasthemia gravis and in the myasthenic syndrome, Ann. N.Y. Acad. Sci. **183**:46, 1971.
59. Engel, A. G., Lambert, E. H., and Howard, F. M., Jr.: Immune complexes (IgG and C_3) at the motor end-plate in myasthenia gravis, Mayo Clin. Proc. **52**:267, 1977.
60. Fambrough, D. M., Drachman, D. B., and Satyamurti, S.: Neuromuscular junction in myasthenia gravis: decreased acetylcholine receptors, Science **182**:293, 1973.
61. Fatt, P.: Skeletal neuromuscular transmission. In Field, J., editor-in-chief: Handbook of physiology. Neurophysiology, Baltimore, 1959, The Williams & Wilkins Co. vol. 1, ch. 6.
62. Fatt, P., and Katz, B.: An analysis of the end-plate potential recorded with an intra-cellular electrode, J. Physiol. **115**:320, 1951.
63. Fatt, P., and Katz, B.: Spontaneous subthreshold activity at motor nerve endings, J. Physiol. **117**:109, 1952.
64. Feldberg, W.: Present views on the mode of action of acetylcholine in the central nervous system, Physiol. Rev. **25**:596, 1945.
65. Feldberg, W., and Mann, T.: Properties and distribution of the enzyme system which synthesizes acetylcholine in nervous tissue, J. Physiol. **104**:411, 1946.
66. Feltz, A., and Mallart, A.: An analysis of acetylcholine responses of junctional and extrajunctional receptors of frog muscle fibers, J. Physiol. **218**:85, 1971. (Appendix by R. Kahn and A. Le Yaounanc.)
67. Feltz, A., and Mallart, A.: Ionic permeability changes induced by some cholinergic agonists on normal and denervated frog muscles, J. Physiol. **218**:101, 1971.
68. Fields, W. S., editor: Myasthenia gravis, Ann. N.Y. Acad. Sci. **183**:3, 1971.
69. Furshpan, E. J., and Potter, D. D.: Transmission at the giant motor synapses of the crayfish, J. Physiol. **145**: 289, 1959.
69a. Furukawa, T., Furukawa, A., and Takagi, T.: Fibrillation of muscle fibers produced by ammonium ions and its relation to the spontaneous activity at the neuromuscular junction, Jpn. J. Physiol. **7**:252, 1957.
70. Gennaro, J. F., Jr., Nastuk, W. L., and Rutherford, D. T.: Reversible depletion of synaptic vesicles induced by application of high external K^+ to the frog neuromuscular junction, J. Physiol. **280**:237, 1978.
71. Giacobini, E.: Histochemical demonstration of AChE activity in isolated nerve cells, Acta Physiol. Scand. **36**:276, 1956.
72. Ginsborg, B. L.: Ion movements in junctional transmission, Pharmacol. Rev. **19**:289, 1967.
73. Göpfert, H., and Schaefer, H.: Uber den direkt und indirekt erregten Aktionsstrom und die Funktion der Motorischen End-platte, Pfluegers Arch. **239**:597, 1938.
74. Grob, D., editor: Myasthenia gravis, Ann. N.Y. Acad. Sci. **274**:1, 1976.
75. Grundfest, H.: Synaptic and ephaptic transmission. In Field, J., editor-in-chief: Handbook of physiology,

Neurophysiology, Baltimore, 1959, Waverly Press, vol. 1, ch. 5.
76. Grundfest, H.: Synaptic and ephaptic transmission. In Bourne, G. H., editor: The structure and function of nervous tissue, New York, 1969, Academic Press, Inc., vol. 2, Ch. 8, p. 463.
77. Hartzell, H. C., Kuffler, S. W., and Yoshikami, D.: Post-synaptic potentiation: interaction between quanta of acetylcholine at the skeletal neuromuscular synapse, J. Physiol. **251**:427, 1975.
78. Hebb, C. O.: Biochemical evidence for neural function of acetylcholine, Physiol. Rev. **37**:196, 1957.
79. Hebb, C. O.: Formation, storage and liberation of acetylcholine. In Koelle, G. B., subeditor: Handbuch der experimentellen Pharmakologie. Cholinesterases and anticholinesterase agents, Berlin, 1963, Springer-Verlag, vol. 15, p. 55.
80. Hubbard, J. I.: Mechanism of transmitter release. In Butler, J. A. V., and Noble, D., editors: Progress in biophysics and molecular biology, New York, 1970, Pergamon Press, Inc., vol. 21, pp. 33-124.
81. Hubbard, J. I.: Mechanism of transmitter release from nerve terminals, Ann. N.Y. Acad. Sci. **183**:131, 1971.
82. Hubbard, J. I., and Kwanbunbumpen, S.: Evidence for the vesicle hypothesis, J. Physiol. **194**:407, 1968.
83. Hutter, O. F.: Post-tetanic restoration of neuromuscular transmission blocked by d-tubocurarine, J. Physiol. **118**:216, 1952.
84. Hutter, O. F., and Loewenstein, W. R.: Nature of neuromuscular facilitation by sympathetic stimulation in the frog, J. Physiol. **130**:559, 1955.
85. Hutter, O. F., and Trautwein, W.: Neuromuscular facilitation by stretch of motor nerve-endings, J. Physiol. **133**:610, 1956.
86. Jenkinson, D. H.: The nature of the antagonism between calcium and magnesium ions at the neuromuscular junction, J. Physiol. **138**:434, 1957.
87. Karlin, A.: Chemical modification of the active site of the acetycholine receptor, J. Gen. Physiol. **54**:245s, 1969.
88. Katz, B.: The transmission of impulses from nerve to muscle, and the subcellular unit of synaptic action, Proc. R. Soc. Lond. (Biol.) **155**:455, 1962.
89. Katz, B., and Miledi, R.: Further observations on the distribution of acetylcholine-reactive sites in skeletal muscle, J. Physiol. **170**:379, 1964.
90. Katz, B., and Miledi, R.: Localization of calcium action at the nerve-muscle junction, J. Physiol. **171**: 10P, 1964.
91. Katz, B., and Miledi, R.: Propagation of electrical activity in motor nerve terminals, Proc. R. Soc. Lond. (Biol.) **161**:453, 1965.
92. Katz, B., and Miledi, R.: The measurement of synaptic delay, and the time course of acetylcholine release at the neuromuscular junction, Proc. R. Soc. Lond. (Biol.) **161**:483, 1965.
93. Katz, B., and Miledi, R.: The effect of calcium on acetylcholine release from motor nerve terminals, Proc. R. Soc. Lond. (Biol.) **161**:496, 1965.
94. Katz, B., and Miledi, R.: The timing of calcium action during neuromuscular transmission, J. Physiol. **189**: 535, 1967.
95. Katz, B., and Miledi, R.: A study of synaptic transmission in the absence of nerve impulses, J. Physiol. **192**:407, 1967.
96. Katz, B., and Miledi, R.: The role of calcium in neuromuscular facilitation, J. Physiol. **195**:481, 1968.
97. Katz, B., and Miledi, R.: Tetrodotoxin-resistant electric activity in presynaptic terminals, J. Physiol. **203**: 459, 1969.
98. Katz, B., and Miledi, R.: The effect of prolonged depolarization on synaptic transfer in the stellate ganglion of the squid, J. Physiol. **216**:503, 1971.
99. Katz, B., and Miledi, R.: The statistical nature of the acetylcholine potential and its molecular components, J. Physiol. **224**:665, 1972.
100. Katz, B., and Miledi, R.: The binding of acetylcholine to receptors and its removal from the synaptic cleft, J. Physiol. **231**:549, 1973.
101. Katz, B., and Miledi, R.: The nature of the prolonged end-plate depolarization in anti-esterase treated muscle, Proc. R. Soc. Lond. (Biol.) **192**:27, 1975.
102. Koelle, G. B.: The elimination of enzymatic diffusion artifacts in the histochemical localization of cholinesterases and a survey of their cellular distributions, J. Pharmacol. Exp. Ther. **103**:153, 1951.
103. Koelle, G. B., subeditor: Handbuch der experimentellen Pharmakologie. Cholinesterases and anticholinesterase agents, Berlin, 1963, Springer-Verlag, Inc., pp. 187-298.
104. Koelle, G. B.: Current concepts of synaptic structure and function, Ann. N.Y. Acad. Sci. **183**:5, 1971.
105. Koelle, G. B., Davis, R., and Devlin, M.: Acetyl disulfide $(CH_3COS)_2$ and bis-(thioacetoxy)aurate (I) complex, $Au(CH_3COS)_2$, histochemical substrates of unusual properties with acetylcholinesterase, J. Histochem. Cytochem. **16**:754, 1968.
106. Kordas, M.: The effect of membrane polarization on the time course of the end-plate current in frog sartorius muscle, J. Physiol. **204**:493, 1969.
107. Kordas, M.: The effect of procaine on neuromuscular transmission, J. Physiol. **209**:689, 1970.
108. Kordas, M.: An attempt at an analysis of the factors determining the time course of the end-plate current. I. The effects of prostigmine and of the ratio of Mg^{2+} to Ca^{2+}, J. Physiol. **224**:317, 1972.
109. Kuba, K., and Tomita, T.: Noradrenaline action on nerve terminal in the rat diaphragm, J. Physiol. **217**: 19, 1971.
110. Kuffler, S. W.: Specific excitability of the end-plate region in normal and denervated muscle, J. Neurophysiol. **6**:99, 1943.
111. Kuffler, S. W.: Symposium on physiology of neuromuscular junctions: electrical aspects, Fed. Proc. **7**: 437, 1948.
112. Kuffler, S. W., and Yoshikami, D.: The distribution of acetylcholine sensitivity at the post-synaptic membrane of vertebrate skeletal twitch muscles: iontophoretic mapping in the micron range, J. Physiol. **251**: 465, 1975.
113. Langley, J. N.: On the contraction of muscle, chiefly in relation to the presence of "receptive" substances. Part IV. The effect of curare and some other substances on the nicotine response of the sartorius and gastrocnemius muscle of the frog, J. Physiol. **39**:235, 1909-10.
114. Lawler, H. C.: Turnover time of acetylcholinesterase, J. Biol. Chem. **236**:2296, 1961.
115. Liley, A. W.: An investigation of spontaneous activity at the neuromuscular junction of the rat, J. Physiol. **132**:650, 1956.
116. Liley, A. W.: The quantal components of the mammalian end-plate potential, J. Physiol. **133**:571, 1956.
117. Liley, A. W.: The effects of presynaptic polarization on the spontaneous activity at the mammalian neuromuscular junction, J. Physiol. **134**:427, 1956.
118. Liley, A. W., and North, K. A. K.: An electrical in-

vestigation of effects of repetitive stimulation on mammalian neuromuscular junction, J. Neurophysiol. **16:** 509, 1953.
119. Liu, J. H., and Nastuk, W. L.: The effect of UO_2^{2+} ions on neuromuscular transmission and membrane conduction, Fed. Proc. **25:**570, 1966.
120. Longenecker, H. E., Hurlbut, W. P., Mauro, A., and Clark, A. W.: Effects of black widow spider venom on the frog neuromuscular junction: effects on end-plate potential, miniature end-plate potential and nerve terminal spike, Nature **225:**701, 1970.
121. Lüttgau, H. C., and Niedergerke, R.: The antagonism between Ca and Na ions on the frog's heart, J. Physiol. **143:**486, 1958.
122. Maeno, T.: Analysis of sodium and potassium conductances in the procaine end-plate potential, J. Physiol. **183:**592, 1966.
123. Magazanik, L. G., and Vyskocil, F.: Different action of atropine and some analogues on the end-plate potentials and induced acetylcholine potentials, Experientia **25:**618, 1969.
124. Magleby, K. L., and Stevens, C. F.: The effect of voltage on the time course of end-plate currents, J. Physiol. **223:**151, 1972.
125. Magleby, K. L., and Stevens, C. F.: A quantitative description of end-plate currents, J. Physiol. **223:**173, 1972.
126. Mallart, A., and Martin, A. R.: An analysis of facilitation and transmitter release at the neuromuscular junction of the frog, J. Physiol. **193:**679, 1967.
127. Mallart, A., and Martin, A. R.: The relation between quantum content and facilitation at the neuromuscular junction of the frog, J. Physiol. **196:**593, 1968.
128. Mallart, A., Dreyer, F., and Peper, K.: Current-voltage relation and reversal potential at junctional and extrajunctional ACh-receptors of the frog neuromuscular junction, Pfluegers Arch. **362:**43, 1976.
129. Mann, P. J. G., Tennenbaum, M., and Quastel, J. H.: Acetylcholine metabolism in the central nervous system. The effects of potassium and other cations on acetylcholine liberation, Biochem. J. **33:**822, 1939.
130. Marchbanks, R. M.: Biochemical organization of cholinergic nerve terminals in the cerebral cortex. In Barondes, S. H., editor: Cellular dynamics of the neuron, New York, 1969, Academic Press, Inc., p. 115.
131. Martin, A. R.: Quantal nature of synaptic transmission, Physiol. Rev. **46:**51, 1965.
132. Martin, A. R., and Pilar, G.: The dual mode of synaptic transmission in the avian ciliary ganglion, J. Physiol. **168:**443, 1963.
133. Martin, A. R., and Pilar, G.: Transmission through the ciliary ganglion of the chick, J. Physiol. **168:**464, 1963.
134. McMahan, U. J., Spitzer, N. C., and Peper, K.: Visual identification of nerve terminals in living isolated skeletal muscle, Proc. R. Soc. Lond. (Biol.) **181:**421, 1972.
135. Merton, P. A.: Voluntary strength and fatigue, J. Physiol. **123:**553, 1954.
136. Miledi, R.: The acetylcholine sensitivity of frog muscle fibres after complete or partial denervation, J. Physiol. **151:**1, 1960.
137. Miledi, R.: Junctional and extra-junctional acetylcholine receptors in skeletal muscle fibres, J. Physiol. **151:**24, 1960.
138. Miledi, R.: Induction of receptors. In Mongar, J. L., and de Reuck, A. V. S., editors: Enzymes and drug action, Boston, 1962, Little, Brown & Co., p. 220.
139. Miledi, R.: Spontaneous synaptic potentials and quantal release of transmitter in the stellate ganglion of the squid, J. Physiol. **192:**379, 1967.
140. Miledi, R., and Slater, C. R.: The action of calcium on neuronal synapses in the squid, J. Physiol. **184:**473, 1966.
141. Miledi, R., and Thies, R.: Tetanic and post-tetanic rise in frequency of miniature end-plate potentials in low-calcium solutions, J. Physiol. **212:**245, 1971.
142. Miledi, R., Molinoff, P., and Potter, L. T.: Isolation of the cholinergic receptor protein of Torpedo electric tissue, Nature **229:**554, 1971.
143. Miledi, R., Stefani, E., and Steinbach, A. B.: Induction of the action potential mechanism in slow muscle fibers of the frog, J. Physiol. **217:**737, 1971.
144. Nachmansohn, D.: Chemical and molecular basis of nerve activity, New York, 1960, Academic Press, Inc.
145. Nachmansohn, D.: Role of acetylcholine in neuromuscular transmission, Ann. N.Y. Acad. Sci. **135:** 136, 1966.
146. Nachmansohn, D.: Chemical control of the permeability cycle in excitable membranes during electrical activity, Ann. N.Y. Acad. Sci. **137:**877, 1966.
147. Nachmansohn, D.: Proteins in excitable membranes, Science **168:**1059, 1970.
148. Nachmansohn, D., and Machado, A. L.: The formation of acetylcholine. A new enzyme: "choline acetylase," J. Neurophysiol. **6:**397, 1943.
149. Nachmansohn, D., and Rothenberg, M. A.: Studies on cholinesterase, I. On the specificity of the enzyme in nerve tissue, J. Biol. Chem. **158:**653, 1945.
150. Nastuk, W. L.: The electrical activity of the muscle cell membrane at the neuromuscular junction, J. Cell Comp. Physiol. **42:**249, 1953.
151. Nastuk, W. L.: Membrane potential changes at a single muscle end-plate produced by transitory application of acetylcholine with an electrically controlled microjet, Fed. Proc. **12:**102, 1953.
152. Nastuk, W. L.: Neuromuscular transmission: fundamental aspects of the normal process, Am. J. Med. **19:** 663, 1955.
153. Nastuk, W. L.: Action of acetylcholine at the muscle end-plate membrane, Fed. Proc. **18:**112, 1959.
154. Nastuk, W. L.: Some ionic factors that influence the action of acetylcholine at the muscle end-plate membrane, Ann. N.Y. Acad. Sci. **81:**317, 1959.
155. Nastuk, W. L.: Fundamental aspects of neuromuscular transmission, Ann. N.Y. Acad. Sci. **135:**110, 1966.
156. Nastuk, W. L.: Activation and inactivation of muscle postjunctional receptors, Fed. Proc. **26:**1639, 1967.
157. Nastuk, W. L.: Cholinergic receptor desensitization. In Cottrell, G. A., and Usherwood, P. N. R., editors: Synapses, Glasgow, 1977, Blackie & Son, Ltd.
158. Nastuk, W. L., and Alexander, J. T.: The action of 3-hydroxy-phenyldimethylethylammonium (Tensilon) on neuromuscular transmission in the frog, J. Pharmacol. Exp. Ther. **111:**302, 1954.
159. Nastuk, W. L., and Plescia, O. J.: Current status of research on myasthenia gravis, Ann. N.Y. Acad. Sci. **135:**664, 1966.
160. Nastuk, W. L., and Gissen, A. J.: Actions of acetylcholine and other quaternary ammonium compounds at the muscle postjunctional membrane. In Paul, W. M., Daniel, E. E., Kay, C. M., and Monckton, G., editors: Muscle, London, 1965, Plenum Publishing Co., Ltd. p. 389.
161. Nastuk, W. L., Plescia, O. J., and Osserman, K. E.: Changes in serum complement activity in patients with myasthenia gravis, Proc. Soc. Exp. Biol. Med. **105:** 177, 1960.
162. Nastuk, W. L., et al.: Myasthenia in frogs immunized against cholinergic receptor protein. Am. J. Physiol.

236:C53, 1979; Am. J. Physiol. Cell Physiol. **5:**C53, 1979.
163. Neher, E., and Sakmann, B.: Single channel currents from membrane of denervated frog muscle fibers, Nature **260:**799, 1976.
164. Nicholls, J. G.: The electrical properties of denervated skeletal muscle, J. Physiol. **131:**1, 1956.
165. Ogston, A. G.: Removal of acetylcholine from a limited volume by diffusion, J. Physiol. **128:**222, 1955.
166. Otsuka, M., Endo, M., and Nonomura, Y.: Presynaptic nature of neuromuscular depression, Jpn. J. Physiol. **12:**573, 1962.
167. Patrick, J., and Lindstrom, J.: Autoimmune response to acetylcholine receptor, Science **180:**871, 1973.
168. Perry, W. L. M.: Acetylcholine release in the cat's superior cervical ganglion, J. Physiol. **119:**439, 1953.
169. Plescia, O. J., Nastuk, W. L., and Johnson, V.: Immune response in rabbits to homologous cardiac muscle antigens. In Rose, N. R., and Milgram, F., editors: International convocation on immunology, Buffalo, New York, 1968, Basel, 1968, S. Karger AG.
170. Potter, L. T.: Acetylcholine metabolism at vertebrate neuromuscular junctions. In Costa, E., and Giacobini, E., editors: Biochemistry of simple neuronal models, New York, 1970, Raven Press, p. 163.
171. Potter, L. T.: Synthesis, storage and release of [^{14}C]acetylcholine in isolated rat diaphragm muscles, J. Physiol. **206:**145, 1970.
172. Rahamimoff, R.: A dual effect of calcium ions in neuromuscular facilitation, J. Physiol. **195:**471, 1968.
173. Rang, H. P.: Acetylcholine receptors, Q. Rev. Biophys. **7:**283, 1974.
174. Robertson, J. D.: The ultrastructure of a reptilian myoneural junction, J. Biophys. Biochem. Cytol. **2:**381, 1956.
175. Rosenthal, J.: Post-tetanic potentiation at the neuromuscular junction of the frog, J. Physiol. **203:**121, 1969.
176. Rubin, R. P.: The role of calcium in the release of neurotransmitters and hormones, Pharmacol. Rev. **22:** 389, 1970.
177. Sheridan, R. E., and Lester, H. A.: Rates and equilibria at the acetylcholine receptor of Electrophorus electroplaques, J. Gen. Physiol. **70:**187, 1977.
178. Simpson, J. A.: Myasthenia gravis: a new hypothesis, Scott. Med. J. **5:**419, 1960.
179. Stedman, E., and Stedman, E.: The mechanism of the biological synthesis of acetylcholine, Biochem. J. **33:** 811, 1939.
180. Strauss, A. J. L., et al.: Immunofluorescence demonstration of a muscle binding complement-fixing serum globulin fraction in myasthenia gravis, Proc. Soc. Exp. Biol. Med. **105:**184, 1960.
181. Takeuchi, N.: Some properties of conductance changes at the endplate membrane during the action of acetylcholine, J. Physiol. **167:**128, 1963.
182. Takeuchi, N.: Effects of calcium on the conductance change of the endplate during action of the transmitter, J. Physiol. **167:**141, 1963.
183. Takeuchi, A., and Takeuchi, N.: Active phase of frog's endplate potential, J. Neurophysiol. **22:**395, 1959.
184. Takeuchi, A., and Takeuchi, N.: On the permeability of endplate membrane during the action of transmitter, J. Physiol. **154:**52, 1960.
185. Thesleff, S.: Acetylcholine utilization in myasthenia gravis, Ann. N.Y. Acad. Sci. **135:**195, 1966.
186. Waser, P. G.: Receptor localization by autoradiographic techniques, Ann. N.Y. Acad. Sci. **144:**737, 1967.
187. Weinreich, D.: Ionic mechanism of post-tetanic potentiation at the neuromuscular junction of the frog, J. Physiol. **212:**431, 1971.
188. Whittaker, V. P.: The isolation and characterization of acetylcholine-containing particles from brain, Biochem. J. **72:**694, 1959.
189. Whittaker, V. P.: Origin and function of synaptic vesicles, Ann. N.Y. Acad. Sci. **183:**21, 1971.
190. Whittaker, V. P.: Subcellular localization of neurotransmitters. In Clementi, F., and Ceccarelli, B., editors: Advances in cytopharmacology, New York, 1971, Raven Press, p. 319.
191. Wilson, I. B.: The mechanism of enzyme hydrolysis studied with acetylcholinesterase. In McElroy, W., and Glass, B., editors: A symposium of the McCollum Pratt Institute, Baltimore, 1954, The Johns Hopkins University Press.
192. Wilson, I. B.: The inhibition and reactivation of acetylcholinesterase, Ann. N.Y. Acad. Sci. **135:**177, 1966.
193. Wilson, I. B., and Harrison, M. A.: Turnover number of acetylcholinesterase, J. Biol. Chem. **236:**2292, 1961.

6 Synaptic transmission

VERNON B. MOUNTCASTLE and ANTONIO SASTRE

A synapse is a region of near contact between neurons or between a neuron and a muscle or gland cell that is specialized to allow the excitation or inhibition of one by the other. Synapses are the intercellular contacts across which one cell exerts a trophic influence on another. The neuromyal synapse is the prototype of what will be called *chemical* synapses, at which an effect on a postsynaptic element is produced by the neurosecretion from presynaptic endings of specific chemical substances that produce changes in the permeabilities of the postsynaptic membrane. Synapses possessing the basic features of the neuromuscular junction (i.e., chemical synapses) predominate in vertebrates, especially mammals. At many synapses in invertebrates and at some in vertebrates the effect of a presynaptic impulse on a postsynaptic neuron is achieved by the flow of action currents from the former across the plasma cell membrane of the latter. These are called *electrical* synapses, and they possess characteristic morphologic features by which they differ from chemical synapses. Electrical synapses occur rarely in the mammalian CNS, and for this reason, greatest attention will be given to chemically operated synapses.

The physiologic properties of synapses within each of these two major classes are remarkably uniform, so that certain general principles may be applied to synapses in different locations. These principles have emerged from three lines of research: study of the ultrastructure of synapses by electron microscopy, observations of the electrical events in nerve cells by means of intracellular recording, and identification of transmitter agents by biochemical and histochemical methods.

The action of a presynaptic impulse on a postsynaptic element is either excitatory or inhibitory. The sign of its action at a synapse is not determined by the particular transmitter released at the nerve terminal, but by the nature of the permeability change that results from its action on the postsynaptic membrane. Some transmitters may be excitatory at some synapses, inhibitory at others. A central neuron receives a mixture of inputs of opposite action and expresses by its rate of discharge the net—the integrated—influence of all. Moreover, central synaptic transfers are in many places complicated by the action of interneurons, cells intercalated between the incoming and the outgoing elements of a nuclear region; such internuncial cells may modify and control synaptic transfer through the region.

The neuron doctrine

The essence of the neuron doctrine is that each nerve cell is a structural entity bounded by its plasma cell membrane and that a functional discontinuity parallels that of structure. This generalization was established by the anatomist Ramón y Cajal. It is supported by many experimental observations, including those that follow:

1. Electron microscopic studies show a cytoplasmic discontinuity between nerve cells; any central neuron is everywhere separated from the processes of other neurons or glial cells by an intercellular cleft 15 to 20 nm wide.

2. All parts of a neuron depend metabolically on its cell body. When an axon is severed, the ensuing wallerian degeneration extends to its terminals, but usually no change occurs in the cells they contact. In certain cases, if all or nearly all the presynaptic axons ending on a cell are cut, that cell may shrink in size, an isolation atrophy occurring, for example, in the cells of the lateral geniculate body when the optic nerves are cut.

3. When the axon of a neuron is severed, there is a transient change in the distribution of the granular endoplasmic reticulum of the cell body (it may later disappear), and the nucleus moves to an eccentric position. This chromatolysis recedes after several weeks if the axon regenerates, as do those of motoneurons but not those of neurons wholly central. During chromatolysis, no change is seen in the synaptic boutons ending on the altered cell.

4. When neurons of the mammalian CNS are studied by means of intracellular recording techniques, their interiors are observed to be separated from the extracellular surround and from

the insides of adjacent neurons and glial cells by a phase possessing a high electrical resistance and capacitance compared to cytoplasm or the extracellular fluid. That phase is the cell membrane.

5. The neuroblasts of the mantle layer of the embryonic neural tube remain distinct cellular entities throughout their development.

It was the major contribution of Sherrington[128] to provide the functional counterpart of the neuron theory and to show that intercellular synaptic linkages confer on the nervous system its basic properties: the ability of a given group of neurons to vary their response to presynaptic impulses, a fractionation of activity, and facilitation and inhibition—in a word, integration. The mechanism of this integrative action is the central problem of experimental neurology.

MICROSCOPIC ANATOMY OF SYNAPTIC LINKAGES

Nerve cells are related to one another at intercellular junctions in a wide variety of ways, but the presynaptic to postsynaptic relation is commonly axodendritic, axosomatic, or axoaxonic. Synapses may also be classified in terms of the specialized nature of their membrane interfaces and intracellular organelles. A classification made by Bodian[23] using these criteria is given in the following outline.

Classification of junctions between neurons in vertebrates, based on topography and vesicular types

- I. Vesicular synapses, chemically operated
  - A. Conventional topography
    - 1. Axodendritic, predominantly type I with spherical vesicles
    - 2. Axosomatic, predominantly type II with flattened vesicles
    - 3. Axoaxonic, terminating on:
      - a. Axon hillock with predominantly flattened vesicles
      - b. Axonal shaft (e.g., at node of Ranvier); rare; vesicle type unknown
      - c. Axonal terminals with spherical vesicles
  - B. Unconventional topography
    - 1. Telotelodendritic, reciprocal (e.g., the retina)
    - 2. Dendrodendritic, reciprocal (e.g., the olfactory bulb)
    - 3. Axodendrodendritic (e.g., in sympathetic ganglia)
- II. Nonvesicular synapses, electrically operated
  - A. Conventional topography; gap junctions
    - 1. Axodendritic
    - 2. Axosomatic
  - B. Unconventional topography, gap or desmosome-like junctions
    - 1. Dendrodendritic
    - 2. Dendrosomatic
    - 3. Somasomatic
- III. Mixed junctions, chemically and electrically operated

These varieties of synapses are widely distributed in vertebrates and invertebrates; thus far electron microscopic studies have not revealed specific types of synapses unique to complex brains. It appears that the remarkable development of the nervous system in mammals is due to the great increase in the numbers of neurons and of their interconnections, not to innovations in synaptic structure.

Nerve fibers commonly branch repeatedly to terminate on a number of postsynaptic cells; a single neuron may thus have synaptic effects on many others, for its impulses are usually conducted into each branch and terminal. Reciprocally, a given neuron may receive presynaptic fibers from many neurons that may be of quite different types and reside in different locales. These two facts lead to the principles of divergence and convergence, which, it will be seen, are of considerable functional significance. Superimposed divergent and convergent arrangements occur to degrees varying from the rare 1:1 relation to those in which the pre-post and post-pre ratios number several hundred. These ratios determine the synaptic security within a neural system and thus its capacity for rapid action in the temporal domain, the degree of "local sign" or spatial specificity, etc.

Synaptic ultrastructure[69]

The general features of synaptic ultrastructure are illustrated in Fig. 6-1. At chemically operated synapses the cell membranes are separated by a uniform cleft of 200 to 300 Å that is continuous at its edges with the somewhat narrower intercellular cleft system. At the synaptic interface, both cellular membranes show regions of increased electron density that may be asymmetric; at some synapses an electron-dense material is seen within the synaptic cleft. A regular feature of chemical synapses is the presence in the presynaptic, but not the postsynaptic, location of large numbers of intracellular vesicles 20 to 80 nm in diameter. They are called synaptic vesicles because of this location and because chemical analysis of subcellular centrifugation fractions of neural tissue shows that they contain high concentrations of transmitter agents. In peripheral and central adrenergic neurons these vesicles

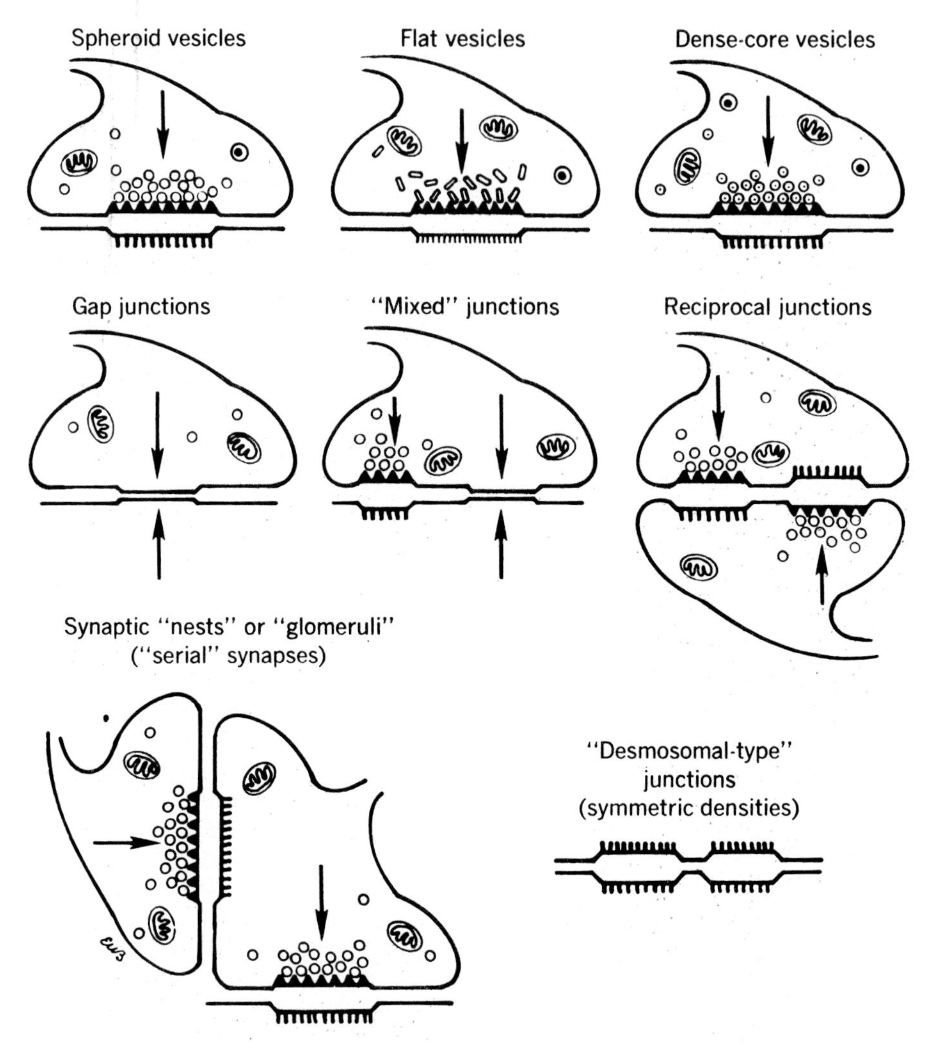

Fig. 6-1. Schematic representation of variations of synaptic structure and topography, presumably associated with functional diversity.

Upper row: "Asymmetric" synaptic contacts characteristic of most central presynaptic bulbs. Rigid cleft of about 16 nm, usually delimited by variable amounts of pre- and postsynaptic electron-dense material after osmium or phosphotungstic acid treatment, is characteristic. Dense material (not shown here) is also present in cleft. In addition to mitochondria, presynaptic bulbs contain one of several varieties of microvesicles of about 20 to 40 nm, of which three major types are shown. *Spheroid vesicles* that resist deformation in fixation process are associated with certain excitatory synapses. Vesicles that are readily flattened after aldehyde fixation have been related to a number of known inhibitory synapses *(flat vesicles)*. Vesicle populations of intermediate susceptibility to flattening are characteristic of still other types of presynaptic bulbs, including peripheral cholinergic synapses. Microvesicles with electron-dense cores are characteristic of peripheral adrenergic axon terminals, whereas central "adrenergic" endings are thought to be associated with clusters of larger *dense-core vesicles* of about 60 to 80 nm.

Middle row: Narrow clefts of about 2 nm (gap junctions) are characteristic of synaptic contacts at which electrotonic transmission has been demonstrated. These are devoid of junctional densities or clusters of microvesicles. Same synaptic interface may also exhibit separate sites with morphologic characteristics of either chemical or electrotonic junctions. Reciprocally asymmetric junctions have also been defined in same interface in olfactory bulb and retina.

Lower row: Specialized, serial synapses may couple dendrites to other dendrites, to cell bodies, or even to axons. They are sometimes referred to as "triads." The desmosomal-type junction is similar to that found in other epithelia; it shows a symmetry of interface densities, delimiting an intercellular cleft of about 20 nm. (From Bodian.[23])

may have an electron-dense core, whereas those of peripheral cholinergic and nonadrenergic central synapses have structureless, electron-translucent interiors. There is good evidence that synaptic vesicles are constructed within the neuronal cell body and moved by axonal transport to the synaptic boutons, a transport known to depend on the integrity of the microtubules (p. 216).

The regular association of certain of these ultrastructural characteristics at some synapses, and of other sets at others, has led to classifications of chemically operated synapses. The two major classes (types I and II) originally proposed by Gray include more than 80% of central synaptic boutons.[33,60] *Type I synapses are commonly axodendritic and excitatory*. They display a slightly wider synaptic cleft and greater asymmetry than do type II synapses, with a continuous thickening of the presynaptic membrane. A layer of electron-dense material lies within the synaptic cleft. The synaptic vesicles are spheroid in shape and strongly resist form distortion in any fixative. Accumulations are frequently seen at the dense projections. *Type II synapses are commonly axosomatic and inhibitory*. Here the membrane thickenings are restricted to small spots in the area of synaptic contact. Synaptic clefts are about 20 nm wide and contain no electron-dense material. The synaptic vesicles of type II synapses are oblong and flattened in aldehyde-fixed material,[22,135] a property that clearly distinguishes them from the synaptic vesicles of type I.

There are two classes of junctions between cells in the vertebrate brain and other tissues without the special characteristics of chemically operated synapses.[1] *Tight junctions* appear as five-layered structures in electron micrographs because the outer leaflets of the membranes of the two cells are fused. They commonly form continuous closed belts around the perimeters of epithelial cells, functioning to maintain structural integrity and to impede the passage of material across the epithelial sheet. At *gap junctions* between neurons and between glial cells the intercellular cleft is narrowed from 20 to 3 nm. This narrow zone is crossed by perpendicularly oriented structures that contain low impedance channels linking the interiors of the two cells; they allow ready passage of small mobile ions and even of molecules as large as polypeptides between the cells.[4] Gap junctions between neurons are electrically operated synapses. Similar junctions link astrocytes and are thought to account for the low-impedance link between them (p. 217), but these do not function as electrically operated synapses, for glial cells, unlike neurons, do not possess the capacity for impulse generation.

In summary, the neuronal and glial elements of the brain are separated by an intercellular cleft system about 20 nm wide. This cleft is slightly widened and is associated with a variety of membrane and organelle specializations at chemically operated synapses. It is narrowed to a slit at gap junctions and obliterated at tight junctions. Gap junctions are locales of decreased impedance to ion movement from one cell to the other; they function as electrically operated synapses between neurons and as low-impedance links between astrocytes.

SYNAPTIC TRANSMISSION AT INVERTEBRATE AXOAXONIC JUNCTIONS

Synapses between large fibers and cells in invertebrate nervous systems are favorable sites for the study of synaptic transmission; they can be isolated by microdissection and several microelectrodes can be placed in each synaptic element. The large volumes permit microchemical analysis of the axoplasm of single elements. They have been used in several classic studies of synaptic function. Two excitatory invertebrate synapses will be described: one is chemically and the other is electrically operated.

Chemically operated invertebrate excitatory synapse

The mantle muscles of cephalopods are innervated by giant axons synaptically linked at axoaxonic junctions to equally large presynaptic fibers with diameters as great as 500 μm (Fig. 6-2). Presynaptic impulses set up by electrical stimuli invariably evoke impulses in the postfiber, and for short periods of time the synapse can be driven in a 1 : 1 manner at rates as high as 400/sec. It then exhibits no signs of integrative action. If stimuli are delivered to the postfiber, impulses propagate backward—antidromically—to the synaptic region, but no event is observed in the prefiber. The synapse is thus directionally oriented and provides an example, as does the neuromuscular junction, of the *principle of forward conduction,* a property of all chemically operated synapses.

The three pairs of pre- and postfiber intra-axonal records in Fig. 6-2, *A* to *C,* were taken just as a high-frequency train of impulses ended in transmission failure. They illustrate several of the general properties of chemically operated synapses, which are listed in Table 6-1. First,

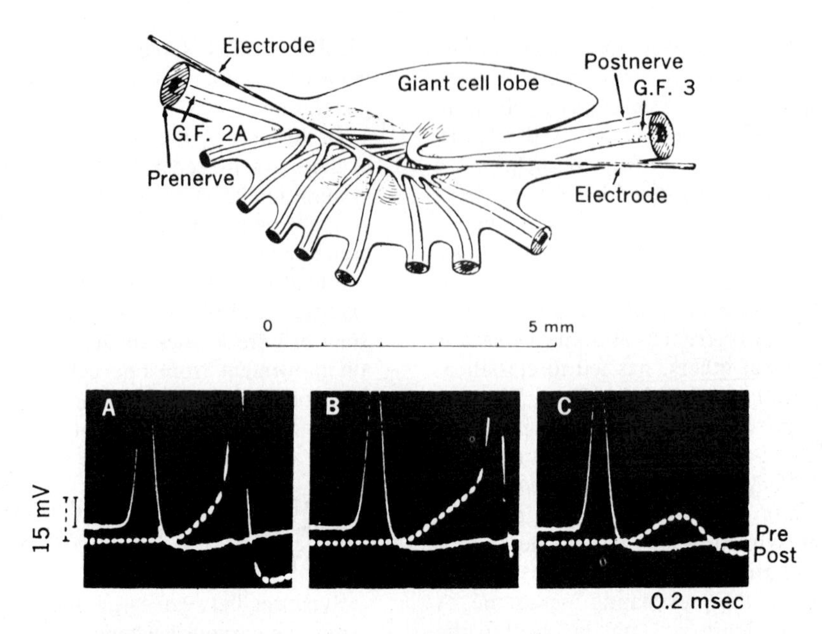

Fig. 6-2. Top: Diagrammatic representation of axoaxonic synapse between second- and third-order giant axons (*G.F. 2A* and *G.F. 3*) of the squid. Microelectrodes are shown in pre- and postfibers at their approximate angles of approach. Pairs of stimulating electrodes are also applied to two nerves during experiment but are not shown in drawing. **A** to **C,** Simultaneous records taken via intracellular microelectrodes in prefiber (continuous line) and postfiber (broken lines) of squid giant axon synapse shown above. Both electrodes are close to region of synaptic contact. Potential changes occurring at tip of each electrode are amplified and displayed on the sweeps of a double-beam cathode-ray oscilloscope. Three sets of records were taken just as transmission began to fail, during prolonged high-frequency presynaptic stimulation. Successive frames show prolongation of postsynaptic response time and, in the third, local postsynaptic response remaining after conduction of postsynaptic spike has failed. (Top from Bullock and Hagiwara[24]; bottom from Hagiwara and Tasaki.[65])

there is an irreducible period of time after arrival of the presynaptic impulse (solid line) during which no change occurs in the postfiber membrane potential (dotted line). This period is the *synaptic delay,* which for chemically operated neuroneural synapses has values of 0.5 to 1.0 msec. Second, the initial depolarizing postsynaptic event has a slower time course than the action potential, to which it leads at an inflection point. This time is the postsynaptic response time, which varies with conditions of excitability and facilitation in contrast to the nearly constant synaptic delay. As the high-frequency train continues, the initial event decreases in amplitude, then fails to depolarize the postfiber to the critical level for impulse initiation and is revealed alone (Fig. 6-2, *C*). This depolarizing event is the *excitatory postsynaptic potential,* or *EPSP*. It has properties in common with the local event at the neuromuscular junction: it is not conducted but spreads over the postfiber membrane for a distance determined by the resistance and capacity of that membrane, and its amplitude can be graded, as it is in this case reduced by rapid repetition.

The transmission failure illustrated in Fig. 6-2 occurs at the synaptic region itself, for after failure, both pre- and postfibers still conduct impulses at high frequencies. Thus one of the characteristics of synapses is a susceptibility to transmission failure, in contrast to the virtually indefatigable nature of axonal conduction. Study of this synapse after failure reveals that if two presynaptic impulses arrive at a short interval, the EPSP evoked by the second sums with that of the first; their summed depolarization may reach threshold and elicit a postfiber impulse. Under these conditions the synapse is not obligatory and displays a *temporal summation of synaptic excitation, one of the characteristics of integrative action.*

The giant synapse of the squid has been studied using multiple microelectrode recording techniques and the voltage clamp method. The results reveal that the effect of a presynaptic action potential is to drive the inside potential of the postsynaptic fiber close to the potential level of the extracellular cleft.[65] A voltage clamp of the postfiber beyond this point causes a reversal of the sign of the EPSP, strong evidence that the synapse is chemically operated. This conclusion is further supported by the observation of miniature EPSPs occurring spontaneously in the quiescent postal fiber.[102] The nearly zero equilibrium potential for the EPSP indicates that the permeability change produced in the postfiber by transmitter action is nonspecific in the sense that it is not restricted to Na^+ ions. The permeability change allows a charge movement from outside to inside (the EPSP); if large enough, this movement leads to the conducted action potential. There is some evidence that the glutamate ion is the transmitter agent at this synapse,[102] but this identification is still uncertain. The release of transmitter agent is preceded by the movement of Ca^{2+} into the presynaptic terminal during its depolarization by the presynaptic action potential.

Electrically operated excitatory synapse[58]

Electrical or electrotonic synapses are distinguished from those that are chemically operated by the differing properties listed in Table 6-1. Prominent among these is a specialized low-impedance pathway that provides an effective electrical coupling between pre- and postsynaptic cells. A further requirement for an electrically operated synapse is a proper electrical match between the two cells, for the presynaptic element must be large enough to supply ionic current adequate to depolarize the postsynaptic one. There is now a large number of synapses in invertebrates and submammalian vertebrates proved to be electrically operated and at which narrow gap junctions exist. These junctions isolate the synaptic interface from the low-impedance shunt of the intercellular cleft system

Table 6-1. Some general characteristics of chemically and electrically operated synapses

Chemically operated synapses	Electrically operated synapses
1. Interface ultrastructure and organelle content usually asymmetric	1. Ultrastructure usually symmetric, without interface or organelle specialization
2. Low-impedance shunt to intercellular space via widened synaptic cleft; no change in cell-to-cell impedance	2. Narrowed synaptic cleft, reducing shunt; low-impedance pathway between cells
3. Presynaptic action currents have minimal effect on postsynaptic membrane potential	3. Presynaptic action currents are immediate agent for synaptic transmission
4. Spontaneous quantal release of transmitter agent produces miniature PSPs in postsynaptic cell	4. No comparable event
5. Presynaptic action potential causes synchronous release of large number of quantal units of transmitter requiring Ca^{2+}	5. No such event
6. Event 5 leads to local, nonconducted, gradable, summable response of postsynaptic cell (PSP), which reverses sign with shift of postsynaptic membrane potential	6. Similar response produced by flow of presynaptic action currents across postsynaptic membrane; not reversed by membrane potential changes
7. If of sufficient amplitude, local EPSP leads to postsynaptic action potential; response of opposite sign, IPSP occurs at inhibitory synapses	7. Similar sequence from EPSP to impulse
8. Postsynaptic membrane receptor molecules combine with transmitter, leading to permeability changes and PSPs	8. No such changes
9. Pre- or postsynaptic membrane may contain hydrolysing enzyme for transmitter or other mechanism for transmitter inactivation that occurs pre- or postsynaptically	9. No such mechanism
10. Pre- and postsynaptic events are modifiable by chemical agents	10. No comparable susceptibility
11. Transsynaptic conduction is unidirectional	11. Conduction either uni- or bidirectional, commonly the latter
12. Local PSP allows integrative action by temporal and/or spatial summation	12. Similar, but many are 1:1 with minimal integrative properties
13. Transsynaptic action sensitive to temperature changes	13. Relatively insensitive to temperature changes

and provide just such a shunt linking the two synaptic cells.

The electrically operated synapse between the lateral giant fiber and the giant motor fiber of the crayfish is diagrammed in Fig. 6-3. The records displayed were obtained from microelectrodes in pre- and postfibers. They show that conduction occurs only from the pre- to the postfiber; an antidromic impulse in the postfiber produces little change in the prefiber (Fig. 6-3, *E*). Dromic conduction occurs with virtually no synaptic delay. The initial postsynaptic response is a typical EPSP (Fig. 6-3, *D*), which, if it is of sufficient amplitude, leads to an action potential in the postfiber. Its onset after no delay indicates a direct electrical spread of depolarizing current across the synaptic gap. Another feature that clearly distinguishes the two types is this: EPSPs at chemically but not at electrically operated synapses may be reversed in sign by depolarization of the postsynaptic membrane to a level beyond the equilibrium potential for transmitter action.

At this crayfish synapse (Fig. 6-3, *A*) currents may spread in the dromic direction only, a rectifying property of the synapse not yet explicable in terms of its ultrastructure. At many electrical synapses, however, conduction occurs with ease in either direction. The minimal synaptic delay, large safety factor for synaptic action, and reciprocity underlie an important function of electrotonic synapses: cells linked by them tend to discharge in synchrony. The reciprocal conduction observed at electrical synapses indicates that the law of forward conduction can be applied with certainty only to chemically operated synapses.

Ephaptic interactions between nerve cells

In the intact organism a nerve cell lies in a conducting medium, and the extrinsic ionic currents associated with its action potentials flow through the volume conductor of the extracellular cleft system. An adjacent neuron will be penetrated by those currents to a degree determined by its own membrane resistance relative to that of the cleft system, and its excitability will be changed thereby in a direction determined by the direction and density of the transmembrane current. In the nervous system the packing density of neurons and their processes is very high, and the question arises as to whether ephaptic interactions, as these nonsynaptic influences are called, are of physiologic consequence.

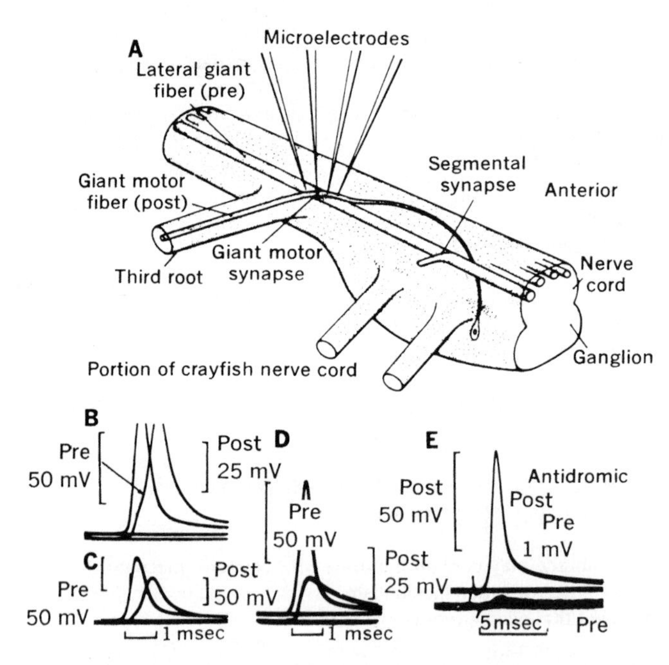

Fig. 6-3. A, Drawing of one ganglion in crayfish abdominal nerve cord showing courses of one giant motor fiber from cell body to exit. Its junction with lateral giant prefiber is shown; other synapses are not. **B** to **D,** Orthodromic nerve impulse transmission recorded as in **A,** with intraaxonal recording from prefiber (upper trace) and postfiber (lower trace). **B** and **C** at different amplifications. In **B,** arrow indicates inflection from EPSP to nerve impulse. In **D,** EPSP did not reach threshold of postfiber. Postfiber action potential shown in **E,** elicited antidromically, produced only negligible depolarization of prefiber. Minimal synaptic delay is characteristic of electric synapses. This one is unidirectional; others conduct in either direction. (From Furshpan and Potter.[58])

Such ephaptic influences occur at regions of injury to nerve trunks, and this cross excitation is frequently evoked to explain some of the aberrant sensory phenomena that may follow nerve injury. It is important to emphasize that such effects have only been observed at sites of injury and that under normal conditions they must be very small indeed. The *principle of isolated conduction remains valid;* impulses traveling along normal nerve fibers of a group do not spread laterally to travel along adjacent fibers. The compact intermingling of fibers and neurons of diverse functions in the CNS makes this almost a priori for orderly function.

Although these effects are too small to produce nonsynaptic cross excitation between axons, it is possible that they influence neuronal activity in a more subtle fashion. Terzuolo and Bullock[133] have shown, for example, that extracellular currents of a value comparable to those under discussion, although insufficient to excite adjacent neurons, may affect the frequency of discharge of those already active by influencing the excitability of the spike-generating zone in the cell soma. These ephaptic influences may play a role in causing closely adjacent neurons to discharge at the same frequency.

CENTRAL SYNAPTIC CONDUCTION

The simplest activities of CNS synapses are reflexes, in which some sensory stimulus evokes from the nervous system a discharge to muscles or glands, producing a response appropriate to the stimulus. Many of these reflexes traverse neural arcs of the spinal cord, which can be isolated from the remainder of the CNS by section, thus reducing the complexity of events in distal segments and making anesthesia unnecessary. Each spinal segment possesses reflex arcs of two (monosynaptic) and three or more (polysynaptic) neurons that can be studied simultaneously. In experiments the sensory stimulus is frequently replaced by a nerve volley elicited by an electrical stimulus, which allows precise timing. The fiber groups conducting the afferent volley and their peripheral origin can be chosen selectively. Postsynaptic discharges in motor nerves can be studied either by direct recording or by measuring the muscle contractions they produce. Moreover, the full panorama of methods centered around intracellular recording has been applied to the study of synaptic events in motoneurons and provides the base for a modern understanding of synaptic transmission in the CNS.

General properties of central synapses

The relay of impulses through central synapses, like the relay at chemically operated synapses elsewhere, is strictly unidirectional. Indeed, it has been known since the time of Bell (1811) and Magendie (1822) that the dorsal and ventral roots consist, respectively, of afferent sensory fibers entering the cord and motor fibers leaving it, and that conduction through spinal reflex arcs is strictly one way, from the former to the latter. Transmission through even the most direct of these synaptic arcs occurs with an *irreducible synaptic delay* of about 0.5 msec at the local region of synaptic contact. An illustrative case is given by the experiment shown in Fig. 6-4.[35] A microelectrode was placed inside a gastrocnemius motoneuron of a cat spinal cord, and afferent volleys to it were generated repetitively in monosynaptically related afferents. Each volley produced a local postsynaptic response after a synaptic delay of about 0.5 msec, which was identical for each of the 30 volleys. These responses are superimposed to compose the record of Fig. 6-4. That record shows also that 10 of the 30 volleys elicited EPSPs just large enough to lead to conducted impulses, which are indicated by lines going off the record. Each was initiated after a variable interval, the *postsynaptic response time*.

Normally, pools of central neurons receive trains of impulses occurring asynchronously in afferent fibers. Postsynaptic discharge may then begin after a period of time called the *nuclear delay,* which allows for a buildup through spatial and temporal summation of the excitatory action of these impulses to the threshold levels of the

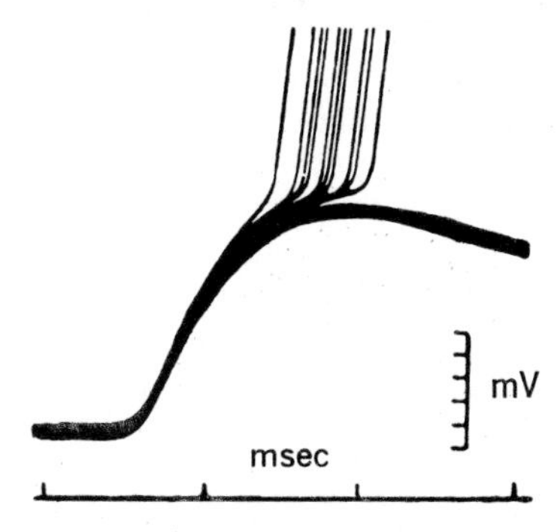

Fig. 6-4. Responses of cat gastrocnemius motoneuron evoked by repetitive stimulation of large afferents from that muscle monosynaptically linked to its motoneurons. Potential changes at tip of intracellular microelectrode displayed on superimposed successive sweeps of cathode-ray oscilloscope. Resting membrane potential of cell was −74 mV. Strength of afferent volley just critical for evoking conducted action potential, which some volleys do (action potentials are strokes off the record), while others evoke only EPSPs, which, superimposed, form heavy smooth curve. (From Coombs et al.[35])

most excitable cells in the pool. Nuclear delay is not related to any synaptic delay, and its value will vary with the number of active fibers, the state of excitability within the postsynaptic ensemble, etc. Only under the special conditions of synchrony of afferent input and maximal excitability will it approach the limiting value of synaptic delay.

Fractionation of a neuron pool and the functional results of convergence

A single incoming volley in an afferent pathway projecting on a pool of central neurons (e.g., a motor nucleus) may evoke the discharge of some but not all postsynaptic elements. They are thus divided into those of the *discharge zone* and those of the *subliminal fringe;* neurons of the latter receive subthreshold synaptic excitation. This fractionation depends on some mixture of two variables: (1) neurons of equal excitability may receive different numbers of presynaptic boutons from the active fibers or (2) neurons with equal presynaptic excitation may have different discharge thresholds (e.g., smaller neurons are more readily discharged than large ones).

With the rare exception of 1:1 convergence numbers, each neuron of a central nucleus receives presynaptic boutons from more than one presynaptic fiber and frequently from a large number; these presynaptic fibers may arise from different afferent sources. The functional results of this converging overlap are of great importance for understanding neural function. They are revealed by the simple experiment illustrated schematically in Fig. 6-5. Arrangements are made to record the postsynaptic discharge from a central nucleus, for example, to record the synaptically evoked discharge from a motor nucleus in a ventral root or motor nerve, and to stimulate two sets of dorsal root afferents, each of which contain fibers projecting directly on the cells of the motor nucleus. It will then be observed that the postsynaptic discharge evoked by simultaneous volleys in the two sets of converging fibers is less than the sum of the two discharges evoked by each separately. This deficit of discharge is called *occlusion*.

If now the number of fibers composing each volley is reduced, the postsynaptic discharge evoked by simultaneous volleys may be much greater than the sum of those evoked by separate volleys. This *spatial summation* results from the reciprocal overlap of the subliminal fringes, for the effects produced in these fringe cells by the two sets of synaptic endings have summed to

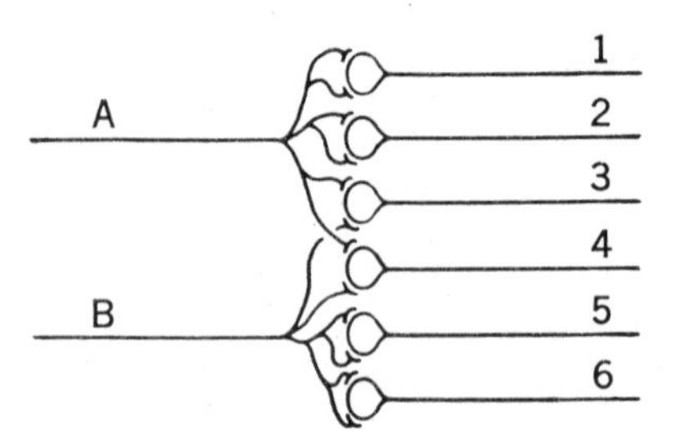

Fig. 6-5. Schematic illustration of functional results of convergence: occlusion and spatial summation. If activation of two synaptic endings is required to secure postsynaptic discharge, impulse in *A* or *B* alone will discharge two postsynaptic cells. However, simultaneous impulses in *A* and *B* will discharge all six cells. Cells *3* and *4* are discharged as result of spatial summation of presynaptic excitatory actions of impulses in *A* and *B*. However, if only one active ending is adequate to secure synaptic transmission, impulse in either *A* or *B* will each discharge four cells. Their simultaneous action can discharge only six cells. There is a discharge deficit of two cells, a phenomenon called occlusion.

bring them to threshold. In a similar way, if two weak volleys in a single group of afferent fibers, each ineffectual when alone, are delivered at a brief interval, the subliminal effects of the second sum with the persisting effects of the first to bring the postsynaptic cells to threshold depolarizations. This is an example of *temporal summation,* quite comparable to that previously described for the squid giant synapse. Spatial and temporal summation in monosynaptically related systems follow time courses identical with those of the local excitatory postsynaptic responses to be described later. Where interneurons participate in overall transmission through a central nucleus, spatial and temporal summation may be prolonged (p. 201).

SYNAPTIC REACTIONS OF NERVE CELLS[2,3]

The description of synaptic actions on central neurons will point out that both synaptic excitation and postsynaptic inhibition fit the paradigm of chemically operated synaptic actions outlined in Table 6-1. In each case the postsynaptic membrane potential changes (EPSPs and IPSPs) result from changes in ionic conductances of the postsynaptic membrane caused by chemical transmitters.

Synaptic excitation[3]

The local EPSP of a cat motoneuron, recorded intracellularly, is shown in Fig. 6-6. Such

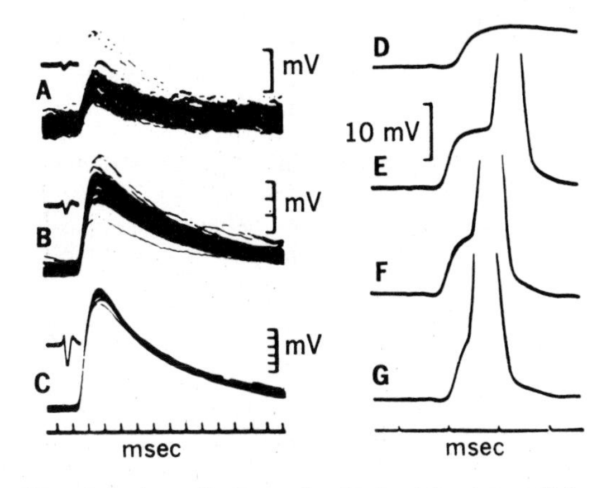

Fig. 6-6. **A** to **C,** Records obtained by intracellular microelectrode of EPSPs evoked in cat spinal motoneuron by volleys in monosynaptically related afferents. Potential changes occurring at its tip were amplified and displayed on cathode-ray oscilloscope. For each record, 20 to 30 responses were superimposed. Upward deflection indicates depolarization. Size of afferent volley increased from **A** to **C,** as indicated by inset records obtained by recording on adjacent dorsal root with another electrode. Increase in size of EPSP, indicated by decreasing amplification from **A** through **C** (note calibrations), is example of spatial summation. Impulse discharge was prevented by deep anesthesia. **D** to **G,** Recordings of potentials in spinal motoneuron generated by volleys in monosynaptically related afferents, obtained as described above. Volleys increased progressively in size from **D** through **G.** Weakest volley evokes EPSP only, whereas larger volleys evoke EPSP of sufficient size to attain discharge threshold and produce impulses written partially off records. Note progressive decrease in postsynaptic response time but invariant synaptic delay. (**A** to **C** from Coombs et al.[35]; **D** to **G** from Eccles.[48])

a motoneuron receives monosynaptically the excitatory endings of many afferent fibers originating from stretch receptors of the muscle it innervates. A single volley in these afferents generates a transient depolarization, the EPSP, which is increased in amplitude but unchanged in time course as the number of fibers conducting impulses is increased (Fig. 6-6, *A* to *C*). Active synapses generate local postsynaptic responses of similar time course, and the EPSPs shown in Fig. 6-6 result from the summation of such individual synaptic potentials. This simple experiment provides an example of the classic concept of spatial summation put forward by Sherrington. In this experiment, precautions were taken to prevent impulse discharge, thus revealing the time course of the EPSPs. Under other conditions, as increasingly larger volleys produce larger EPSPs, a point of depolarization is reached at which Na conductance across the membrane becomes regenerative, and a conducted impulse is generated (Fig. 6-6, *D* to *G*). This point, the *threshold,* differs among neurons. Further analyses have shown that the local EPSPs of central neurons possess the general properties of the local responses of chemically operated synapses previously considered at neuromuscular, axoaxonic, and ganglionic junctions (Table 6-1).

The inset in Fig. 6-7, *A*, shows that the ionic current that produces the EPSP flows inwardly beneath the active excitatory endings and thus outwardly (the depolarizing direction) through surrounding areas of cell membrane. The time course of the inwardly directed current can be calculated (if the time constant of the membrane is first determined) from the relation $dv/dt = I/C - V/\gamma m$, in which dv/dt is the slope of the recorded potential curve at any instant (t), C and I are, respectively, the capacity of the membrane and the current flowing through it, and γm is the electrical time constant of the membrane. The results of calculations for a series of times is plotted in Fig. 6-7, *A*. The active current flow is a brief event that rapidly depolarizes the membrane—the rising phase of the EPSP.

Further important facts are revealed by an application of the voltage set technique, carried out by inserting a double-barreled microelectrode into a motoneuron. Steady current is passed through one barrel, presetting the membrane potential at any desired level, and the steady membrane potential and changes in it are recorded through the second barrel. The results (Fig. 6-8) show that the equilibrium potential toward which the membrane potential is driven during synaptic action is close to zero. This suggests that the excitatory transmitter substance causes a large nonspecific change in membrane permeability, comparable to that produced by ACh at the neuromuscular junction (Chapter 5).

In summary, excitatory synaptic transmission is accomplished at the large majority of synapses in the CNS by means of chemical operations: substances are released from presynaptic terminals by the arrival of nerve impulses, and in each case the substance that is released appears to change the ionic conductance of the postsynaptic membrane in such a way as to drive it in the depolarizing direction, leading to excitation. The transmitter agents that are active at central excitatory synapses will be described in a following section.

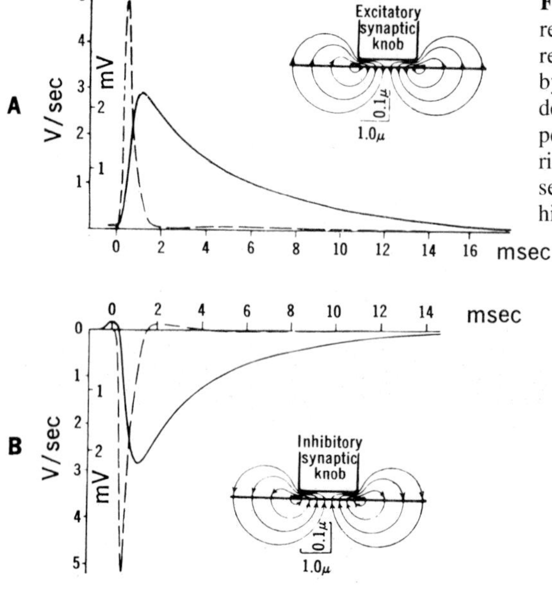

Fig. 6-7. Solid lines are mean values of EPSP and IPSP, **A** and **B**, respectively, evoked by just maximal volleys in muscle afferents respectively excitatory or inhibitory for motoneuron under study by intracellular recording. These potential changes are analyzed as described in text, after time constant of cell was determined experimentally. Excitatory and inhibitory synaptic currents so derived are plotted as broken lines on ordinate scale of volts per second. Insets show directions of ionic current at excitatory and inhibitory synapses. (From Curtis and Eccles.[40])

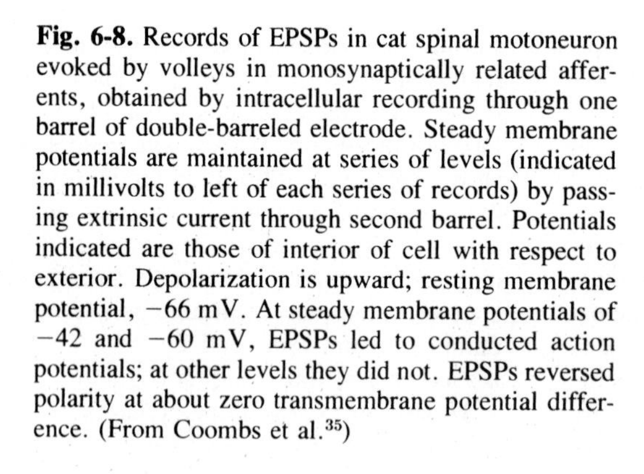

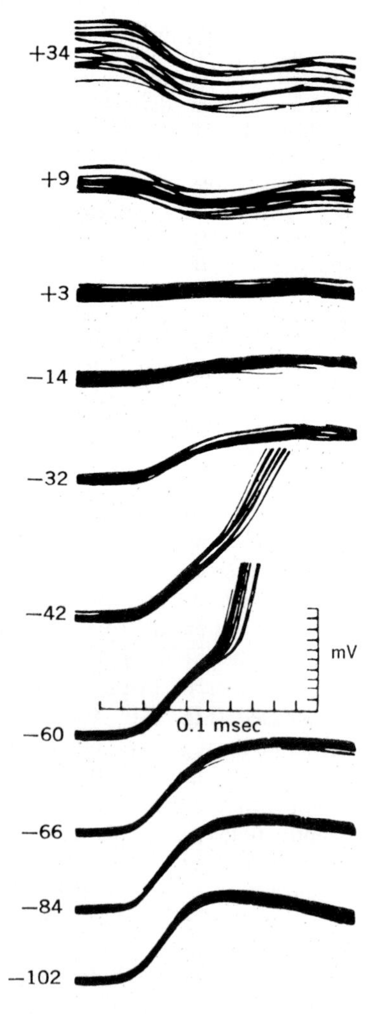

Fig. 6-8. Records of EPSPs in cat spinal motoneuron evoked by volleys in monosynaptically related afferents, obtained by intracellular recording through one barrel of double-barreled electrode. Steady membrane potentials are maintained at series of levels (indicated in millivolts to left of each series of records) by passing extrinsic current through second barrel. Potentials indicated are those of interior of cell with respect to exterior. Depolarization is upward; resting membrane potential, −66 mV. At steady membrane potentials of −42 and −60 mV, EPSPs led to conducted action potentials; at other levels they did not. EPSPs reversed polarity at about zero transmembrane potential difference. (From Coombs et al.[35])

Initiation of impulses in central neurons[2,3]

In the classic view, synaptic excitation from axon terminals to dendrites leads to an impulse that sweeps over the cell body and down the axon. It is now clear that this is not the case and that neurons integrate synaptic action by summing local events at a critical site for impulse initiation. Intracellular recording reveals that the motoneuron impulse has three components (Fig. 6-9): the first is the EPSP and the last two are all-or-nothing events. The EPSP leads, at a critical level of about 10 mV depolarization, to a conducted impulse that arises first in the initial segment of the cell and axon, the IS spike. It is only when depolarization reaches 30 mV that an impulse is set up in the cell body itself. (The dendrites of spinal motoneurons appear to be passive conductors of electrotonic [synaptic] potentials. In other neurons, dendrites may conduct action potentials.) This seemingly reverse sequence of events is due to the low threshold of the initial segment. Thus synaptically evoked transmembrane ionic currents set up in the cell body and dendrites will depolarize the initial segment to its threshold before that of the soma is reached, even though the former is spatially more distant.

Direct evidence for this sequence has been obtained by recording simultaneously from a motoneuron and from its axon in a small ventral root filament.[3] The conducted axon spike was initiated by the initial segment and under some conditions of high excitability even reached the ventral root *before* the spike was initiated within the cell body itself. The origin of impulses in the initial segment, or under some conditions even at the first node of Ranvier for cells with myelinated axons, seems to be a general property of nerve cells.

Local synaptic events[76,117]

Intracellular studies of mammalian spinal motoneurons, under conditions that eliminate afferent impulses, reveal spontaneous miniature PSPs comparable to the spontaneous quantal events at the neuromuscular junction.[21,26,86] The number of quanta released from a single synaptic bouton when it is invaded by an impulse is small in comparison with the number released at the neuromuscular junction; such a "unitary" PSP in a motoneuron is commonly composed of one to five-quanta. The ways in which such unitary PSPs initiated at different locations on the cell may affect its excitability and may interact with each other have been the subjects of studies by a number of investigators, and a highly successful quantitative model of the motoneuron has been developed by Rall.[116] Interaction between concept, model, and experiment has provided a solution for the formerly perplexing problem of how synaptic impingements occurring on the distal dendritic branches of a neuron could affect the initial segment and the soma, in which impulses are generated. Taking into account the cable properties of neurons, it is now clear that even in cells that possess very long and widely branching dendritic trees (e.g., motoneurons) the overall electrotonic length of the cell is no more than twice the membrane space constant. This means that synaptic action, even on the most distant dendrites, can cause an appreciable flow of ionic current across the membrane of the initial segment.

The records in Fig. 6-6 indicate that, under the conditions of that experiment, EPSPs generated by volleys of impulses in an increasing number of converging presynaptic fibers were of about the same time course. This is true only if the synaptic impingements are of about the same (averaged) electrotonic distance

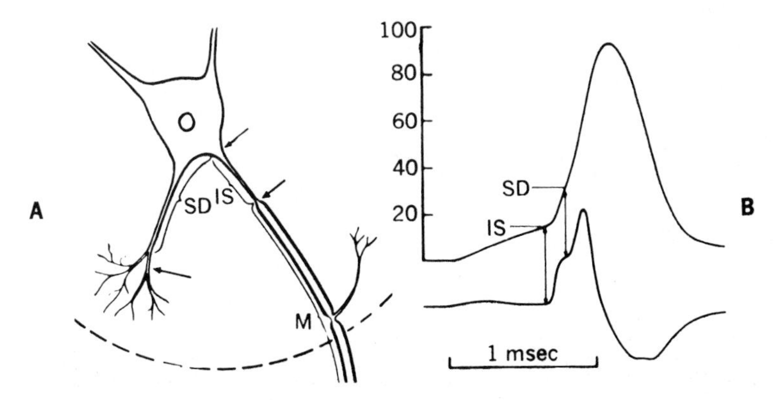

Fig. 6-9. A, Schematic drawing of motoneuron of spinal cord. Recurrent axon collateral leaves axon at point *M*. Portions of cell referred to as initial segment, *IS,* and somadendritic part, *SD,* are indicated by arrows and brackets. **B,** Intracellular recording of response of motoneuron to synchronous monosynaptically related afferent volley. Record below is taken simultaneously and electrically differentiated, which accentuates change in slope of depolarization from EPSP to *IS* spike and from latter *SD* spike. (**A** from Eccles[47]; **B** from Coombs et al.[34])

from the cell body. Obviously the closer a synaptic action to the soma, the more powerful its effect on the excitability of the cell. The more distant such a synapse, the less its effect due to the decrement in amplitude and slowing of time course caused by electrotonic conduction.

PSPs will sum linearly if they are generated at membrane locations so far apart that they do not interact with each other. When closer, the change in the membrane potential (e.g., at the region of synapse A), caused by synaptic action at synapse B, will change the driving potential at A and hence reduce the amplitude of the PSP caused by synaptic action at A. Examples of the two cases, linear and nonlinear summation, are shown in Fig. 6-10.

Recent experimental observations by Edwards et al.[50,51] have raised some doubt whether the general paradigm of chemically operated synaptic transmission can be applied directly to the monosynaptic link between I_a muscular afferents and motoneurons. The very low quantal number observed at this synapse and the fact that the fluctuation in charge transfer is in general nonquantal in nature and cannot be described by binomial or Poisson statistics, suggest that (1) transmission at any single terminal of such an afferent may be an all-or-none event and (2) the fluctuation in transmission, that is, in the amplitude of EPSPs evoked by successive impulses in a single afferent fiber, may result from the combined effects of intermittent failure of transmission at first one and then another of the several terminals of the fiber due to intermittent failure of the action potential to invade all terminals on all trials. Moreover, Edwards et al. were unable to reverse by polarizing current the EPSPs evoked at synaptic terminals on the somata of motoneurons.[52]

Synaptic inhibition

Crustacean muscle cells, unlike the striated muscle of vertebrates, receive a dual innervation, one excitatory and the other inhibitory. The two types of motor nerve fibers end in adjacent loci on the surfaces of muscle cells. Inhibitory action is exerted in two ways.[46] First, by increasing the conductance of the muscle cell membrane to Cl^- and K^+ but not to Na^+, the action of the inhibitory transmitter stabilizes membrane potential at a level between E_{Cl^-} and E_{K^+} or drives it to that level by a hyperpolarizing PSP, the IPSP. The postsynaptic membrane is then less readily depolarized by the action of the transmitter released from excitatory presynaptic terminals: this is *postsynaptic* inhibition. Second, the inhibitory axons also terminate, in axoaxonic synapses, on the terminals of the excitatory axon, and the

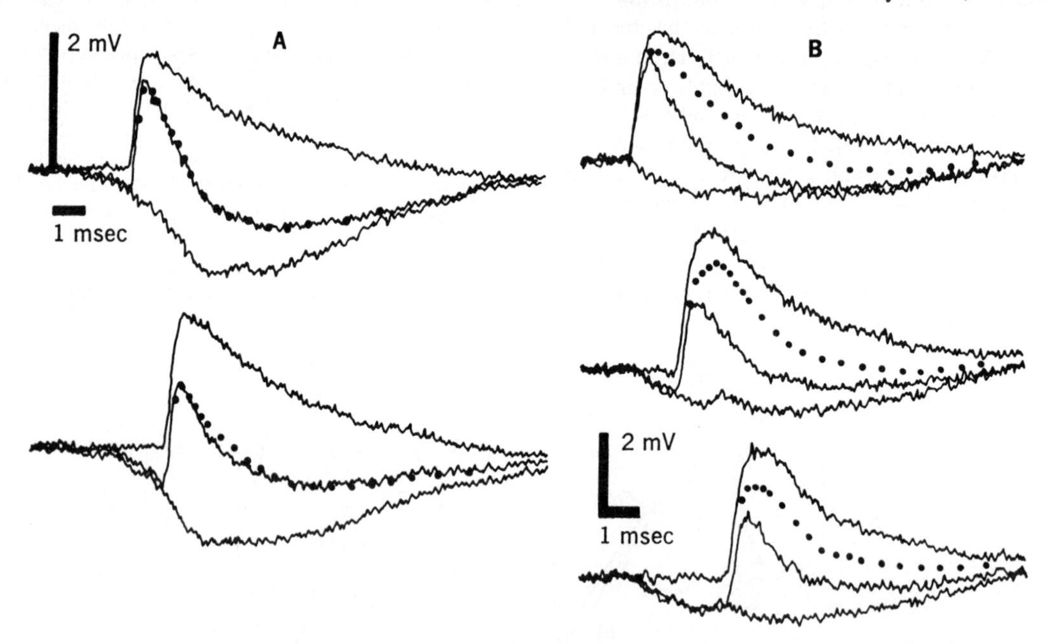

Fig. 6-10. Records of averaged PSPs recorded in spinal motoneuron of cat. Column **A** illustrates linear summation of EPSP generated by lateral gastrocnemius nerve volley (upper set in each trace), with IPSP generated by volley in nerve to antagonistic muscle (lowest trace in each set). Middle records were generated when two volleys were generated so that their synaptic actions would reach the motoneuron simultaneously; they are neatly superimposed on dots that represent algebraic summation of upper and lower records (i.e., linear summation). Opposite case is shown in column **B,** obtained from another gastrocnemius motoneuron, where summation is obviously markedly nonlinear. Most motoneurons studied in this way showed intermediate degrees of nonlinearity. (From Rall et al.[117])

action of the transmitter there is to depolarize the axon terminals. Impulses in the excitatory axon invading these partially depolarized terminals release less transmitter per impulse, for the number of quanta released is a function of the total change in membrane potential and its rate of change. The net effect is a reduction of the depolarizing action on the postsynaptic cell of impulses in excitatory presynaptic axons. This is called *presynaptic* inhibition, even though the inhibition results from an excitatory depolarization of the presynaptic terminals and occurs without any conductance or membrane potential change in the postsynaptic cell.

In vertebrates, such an antagonistic innervation of peripheral effectors occurs only in the autonomic nervous system. Yet the arrest of an active contraction or a slackening of existing tension is a common response of skeletal muscle to an afferent input. *Since vertebrate skeletal muscle receives no efferent inhibitory fibers, a reflexly evoked cessation of muscle contraction must result from a decrease in the discharge of excitatory motoneurons. It is a central event.* Inhibition is important for orderly function even in simple reflex actions: active contraction of one muscle is commonly accompanied by a decreased contraction of its antagonist, an example of reciprocal innervation that will be considered from the reflex point of view in Chapter 28. Here it is pertinent to consider the mechanisms of synaptic inhibition, which occurs in the mammalian CNS in both its postsynaptic and its presynaptic forms.

Postsynaptic inhibition

The records in Fig. 6-11 show that an afferent volley inhibitory for a motoneuron elicits in it a hyperpolarizing local response, provided the cell is at its resting membrane potential of −70 mV. This inhibitory postsynaptic potential (IPSP) is, like the EPSP, local and not conducted; it extends electrotonically and can be graded in amplitude by varying the size of the afferent inhibitory volley evoking it (Fig. 6-11, right, records A to D). The ionic current producing the IPSP is thought to flow outward across the membrane beneath an active inhibitory synaptic bouton and hence inward, the hyperpolarizing direction, across surrounding areas of cell membrane. The IPSP relaxes passively from its peak toward the resting membrane potential along a time course determined by the membrane time constant (Fig. 6-7, *B*).

The equilibrium potential for the IPSP is about −80 mV, as shown by the records of Fig. 6-11, a level midway between E_{K^+} of −90 mV and E_{Cl^-} of −70 mV. It appears, however, that the change in permeability produced by the inhibitory transmitter agent is mainly for Cl^- because changes in intracellular Cl^- concentration produced by iontophoresis causes parallel changes in the IPSP equilibrium potential. Glycine and γ-aminobutyric acid are now established as inhibitory transmitter agents in the vertebrate CNS,[108] but the former is the predominant inhibitory agent at spinal motoneurons.

Both afferent fibers and those of long CNS tracts are excitatory in nature; yet activity in them commonly leads to inhibition of some sets of neurons near their terminations. This inhibition is exerted via small, short-axoned, inhibitory interneurons that release inhibitory transmitter. The linkages of this sort are thought to account for reciprocal inhibition, for example, in the myo-

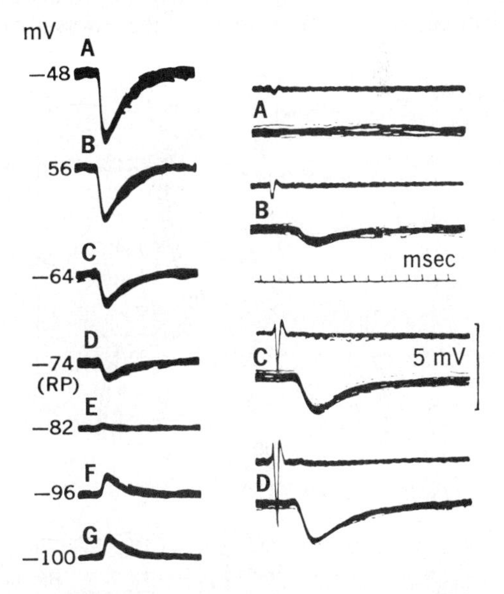

Fig. 6-11. Left: Potentials recorded intracellularly from cat spinal motoneuron via double-barreled microelectrode. Records were formed by superimposition of about 40 faint traces of cathode-ray oscilloscope; they show IPSPs evoked by volleys in nerve fibers inhibitory for cell under study. By means of steady current through one barrel of electrode, membrane potential was preset at voltage indicated on each record. *A* to *G* indicate that IPSP reversed its polarity at about −80 mV transmembrane potential difference. Right: Lower records of each pair give intracellular recorded responses of cat motoneuron to an inhibitory afferent volley, which was increased in size from *A* to *D* and is indicated by sharp downward deflections in upper record of each pair, obtained via another electrode placed on adjacent dorsal root. Records illustrate gradation of IPSP and spatial summation of inhibitory action. (From Coombs et al.[35])

tatic reflex pathways of the spinal cord; they are diagrammed in Fig. 6-12, *A* and *B*.

Direct postsynaptic inhibition is a widespread phenomenon in the CNS and has been studied with detailed microphysiologic methods in many locations. Synaptic inhibition possesses many of the general properties of synaptic excitation. It exhibits peripheral local sign and central localization, it may be summed temporally and spatially, and, like all synaptic actions, it shows in comparison with axonal conduction a greater susceptibility to fatigue, anoxia, and the action of drugs.

Presynaptic inhibition[3]

A volley of impulses in dorsal root afferents leaves in its wake an enduring depolarization of the intraspinal segments of both the active and the adjacent fibers. The depolarization extends electrotonically along the dorsal root fibers and can be recorded from them as the "dorsal root potential." It is not conducted, can be summed spatially and temporally, and displays a latency that allows time for several intraspinal synaptic delays; it thus possesses the properties of a local PSP occurring in axons. This primary afferent depolarization, or PAD, has a striking effect on the synaptic excitatory capacity of impulses invading the partially depolarized synaptic boutons, and it is thought to operate by the same mechanism of "presynaptic inhibition" as that described at the crustacean neuromuscular junction (p. 196). The terminals of some branches of afferent fibers are linked by one or more excitatory interneurons to the terminals of adjacent fibers via axoaxonic synapses. Activation of these chemically operated excitatory synapses leads to the prolonged depolarization of the axon terminals, which when occurring in dorsal root afferents is recorded as the PAD. When thus depolarized, such terminals release less transmitter when they are themselves invaded by im-

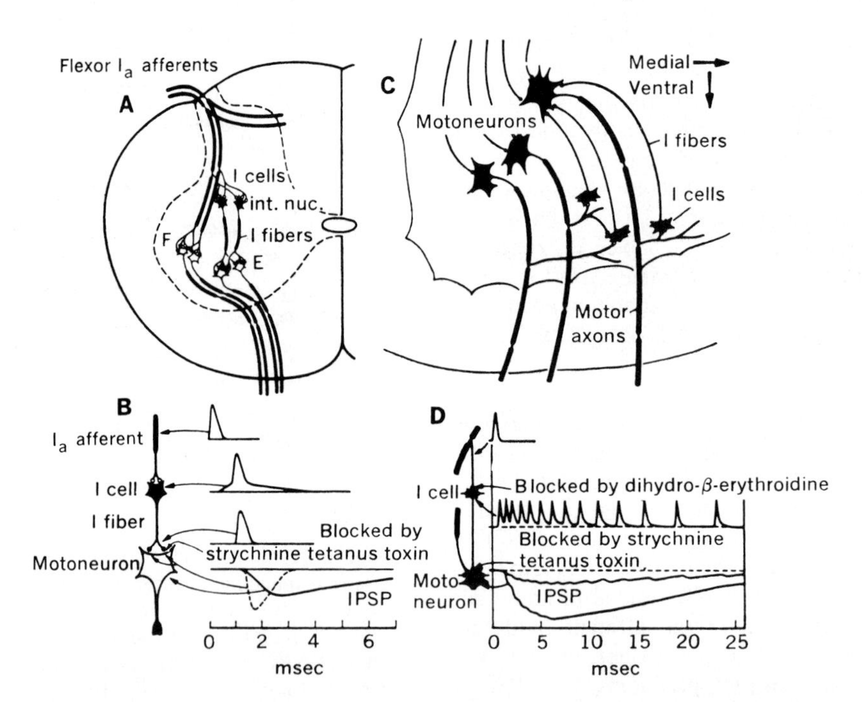

Fig. 6-12. A represents pathway for reciprocal action within spinal cord. Flexor Ia afferents are shown terminating directly on flexor motoneurons—the monosynaptic excitatory pathway of the stretch reflex. Collaterals of these first-order afferents terminate on inhibitory interneurons *(I cells)* that project on extensor motoneurons—the disynaptic inhibitory component of the stretch reflex. Events in this pathway are shown schematically in **B.** Final effect on inhibited motoneuron is shown by inhibitory synaptic current (dotted line) and resulting IPSP. **C,** Recurrent inhibitory pathway. Collaterals of motor axons project on Renshaw interneurons *(I cells),* which in turn project back on cells of same (and perhaps other) neuron pool. **D,** Postulated events in this pathway are shown diagramatically. Impulse in axon collateral evokes prolonged repetitive discharge in Renshaw interneuron, and prolonged IPSPs are evoked in motoneuron on which it impinges. Heavier rippled line shows greatly increased IPSP that results from convergence on motoneuron of many Renshaw interneurons. (From Eccles.[3])

pulses, and a degree of "inhibition" results. The occurrence of such a depolarization has been shown by direct intraterminal recording in some cases and in many others by the fact that during presynaptic inhibition the terminal segments of fibers are more readily excited by local, direct electrical stimulation.

Powerful and reciprocal presynaptic inhibitory linkages exist between primary afferent fibers and between the terminals of second- and third-order fibers in subcortical regions of sensory systems, for example, the dorsal horn of the spinal cord, between fibers terminating in the dorsal column nuclei, and between optic tract fibers in the lateral geniculate, but not at the level of the cerebral cortex. In sensory systems, presynaptic inhibition appears to be distributed in the pattern of surround inhibition. For example, a *local* skin stimulus elicits a train of impulses in fibers innervating the stimulated spot, impulses that, via the interneuronal arcs described, "inhibit" the terminals of fibers whose peripheral branches terminate in adjacent areas of skin.[122] The net effect is to limit and sharpen the profile of central neural activity set up by the stimulus. This mechanism operates as a powerful negative feedback control on a variety of primary afferent fibers and serves to limit and regulate the further central transmission of intense activity in first-order elements. How and to what degree presynaptic inhibition operates in spinal reflexes is uncertain.

Recurrent inhibition[3]

Renshaw discovered that antidromic impulses in motor axons of a ventral root inhibited neighboring motoneurons and that such a volley evoked a high-frequency discharge in interneurons of the ventral horn, which have since been called Renshaw cells. This recurrent inhibitory pathway is diagrammed in Fig. 6-12, *C;* the inhibition is postsynaptic and chemically operated. The excitatory transmitter at the junction of the motor axon collateral and the Renshaw interneuron is ACh, as it is at the peripheral neuromuscular terminations of the same motoneurons. Thus Renshaw cells are depressed by curare-like drugs, their excitation is prolonged by anticholinesterases, and they are stimulated by locally applied ACh or nicotine.[42] The recurrent inhibitory pathway is important in the organization and control of spinal reflex action; it will be discussed in Chapter 28. Recurrent inhibitory pathways are ubiquitous in the CNS, where they stabilize and limit the spread of neuronal activity.

SYNAPTIC MECHANISMS IN INTEGRATIVE ACTION

In the preceding discussions of synaptic transmission, experimental evidence obtained by studies of single cells has been used to elucidate synaptic microphysiology. It is unlikely, however, that the activity of a single neuron is of critical significance in the overall operation of the nervous system. Central actions involve large populations of neurons, and knowledge of the spatial and temporal distributions of activity in neuron pools is required for an understanding of brain function. Central neurons are bombarded by trains of impulses that vary in frequency in any given fiber and are out of phase in adjacent fibers. It is important to learn how these patterns of activity are transformed at synaptic junctions into postsynaptic patterns of discharge.

Integrative action at the cellular level

It was from his studies of spinal cord reflexes that Sherrington derived his perceptive generalization concerning the integrative action of the nervous system.[39,128] He conceived that an afferent volley excitatory for a motoneuron created in the synaptic region an enduring change, a central excitatory state, which outlasted in time the afferent input, was not itself conducted, and could sum in both space and time with other such events. Afferent volleys inhibitory for a motoneuron created the central inhibitory state with similar properties but of opposite sign.

It is possible now to explain how an individual motoneuron—or any nerve cell—integrates its total constellation of excitatory and inhibitory synaptic inputs, as diagrammed in Fig. 6-13. The key discoveries were that the impulse is generated in the initial segment of the cell and that such an impulse need not traverse the cell body itself and is probably seldom conducted into the dendrites. Thus the membrane potential of the initial segment continuously integrates the excitatory and inhibitory currents flowing across it, currents set in motion by transsynaptic actions occurring at many and variously located sites on the soma and dendrites. The level of depolarization of the initial segment determines the rate of discharge of the cell.

Transsynaptic input-output functions

It has already been noted than an afferent volley of sufficient size will discharge some cells of a neuron pool, activate others subliminally, and at least in some systems leave still other cells unaffected. The quantitative relation between the number of active presynaptic fibers

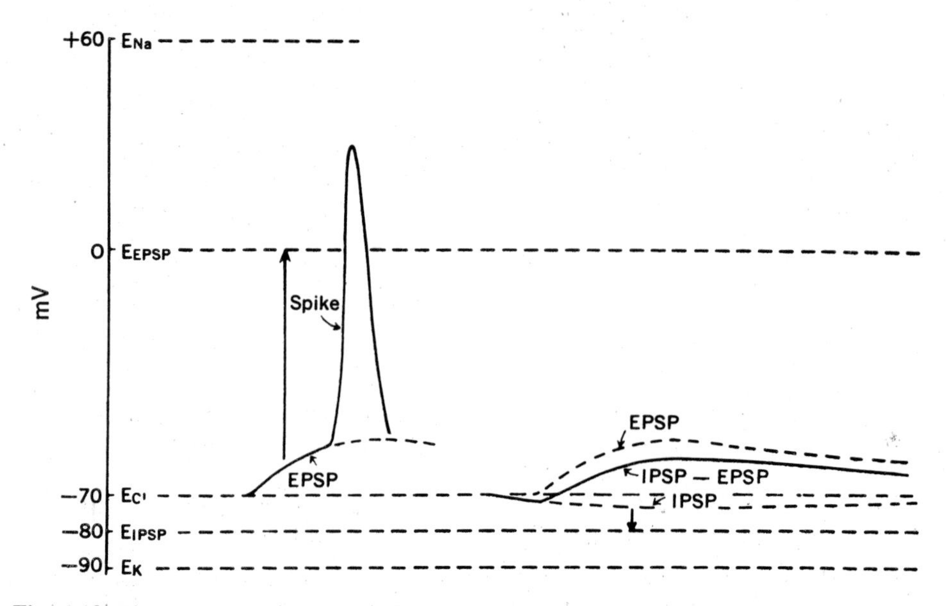

Fig. 6-13. Diagram summarizes mode of action of postsynaptic excitation and inhibition at chemically operated synapses in CNS in terms of ionic hypothesis. Equilibrium potentials for Na^+, K^+, and Cl^- and for EPSP and IPSP given as dotted lines. At left, EPSP is seen driving membrane potential in depolarizing direction, and at threshold eliciting an action potential in cell. To right, IPSP and EPSP are shown alone (dotted lines) and when they interact (net effect, continuous line). EPSP is now so depressed by simultaneous inhibitory effect that it does not reach cell threshold. Interaction of synaptic influences of opposite signs is essence of integrative action of single neurons. (From Eccles.[49])

and these fractions of a neuron pool is important, for the extent of the subliminal fringe determines the degree of summation possible. Recent evidence suggests that in the stretch reflex arc of the spinal cord all the neurons of a motor nucleus receive synaptic input (of varying degrees) from each monosynaptically related stretch afferent fiber (Chapter 28). Motoneurons are graded, that is, they enter the discharge zone, in terms of their size, as thresholds are lower for smaller cells. Synaptic actions within motoneuron nuclei thus include a very large subliminal fringe that accounts for the spatial and temporal summation between converging presynaptic inputs, for example, from two muscle synergists.

In other synaptic systems the gradation between discharge zone and subliminal fringe may be determined by the degree of convergence if cell size is the same throughout the neuron pool. Moreover, the relations between these fractions of a neuron pool vary widely. In the major sensory systems (e.g., at the dorsal column nuclei or the lateral geniculate body) the synaptic link is secure and little subliminal fringe exists; almost all cells receiving presynaptic impulses will discharge. Undoubtedly all variations of these relations exist in the nervous sytem, and for any given synaptic region they must also vary with the general level of excitability of the neuron pool, a level frequently controlled by presynaptic inputs other than the one considered.

Central facilitation and depression

It is a characteristic feature of central synaptic action that the postsynaptic discharge evoked by a given afferent input is modified by repetition. The temporal and spatial summations of local responses contribute to an overall facilitation of the responses to repetitive input; inhibitory postsynaptic responses are similarly summed. Repetition produces no known change in the capacity of the postsynaptic cell to respond with EPSPs or IPSPs. Repetition does, however, cause marked increases or decreases in the synaptic stimulating efficacy of the presynaptic elements, presumably due to a change in the amount of transmitter released per terminal per impulse; the size and direction of such a change depend on the frequency of afferent impulses. Once again the monosynaptic link between large muscle afferents and spinal motoneurons has provided a suitable system for study. Intracellular recordings show that the second of two volleys in these afferents produces a larger EPSP in the related motoneuron than does the first, a true facilitation that reaches 120% of the control value and declines gradually over 15 to 20 msec, during which time the two responses are temporally summed.[41] Both summation and facilitation thus contribute to a greater depolarization of the cell and to a greater likelihood of impulse generation. When, on the other hand, the interval between stimuli is lengthened, there is a decrease in the second response lasting for many seconds. The evidence is that these changes also have a presynaptic origin and leads to the conclusion that at short intervals the active terminals release greater amounts of transmitter by virtue of their prior activation; at longer intervals they release less. Such frequency-bound alterations also occur during repetitive stimulation. At low frequencies there is an enduring depression; at intermediate frequencies (60 to 100/sec) there is a relative facilitation. At still higher rates of synaptic drive (250/sec) a steady state of maintained depolarization is reached, suggesting that the maximal rate of overall transmitter output has been achieved.

In summary, the size of the discharge zone and the frequency of activity of its neurons for any synaptic system expresses the net effect of several factors acting in concert. Among these are (1) the general state of excitability in the neuron pool, a level conditioned by the "background" activity in all presynaptic pathways; (2) the proportion of the fibers of the presynaptic pathway that are inhibitory rather than excitatory; (3) the degree of spatial and temporal summation of the PSPs of the neurons; (4) changes during repetitive activity in the rates of production or release of transmitter at the presynaptic terminals; (5) the presence of recurrent inhibition or facilitation produced by impulses in the axon collaterals of the active neurons of this or other neuron pools; and (6) the subnormality of the neurons that discharge, a factor limiting impulse frequency.

Prolonged facilitation or posttetanic potentiation. Larrabee and Bronk[88] first discovered the prolonged facilitation phenomenon in their studies of synaptic transmission through autonomic ganglia. They found that after repetitive stimulation of the preganglionic nerve, there followed a prolonged period of facilitation of synaptic transmission. This change was found not to be due to any change in the excitability of postsynaptic cells, but to an increased release of transmitter agent from each tetanized presynaptic terminal when tested by a single impulse during the posttetanic period. The degree of facilitation is a function of the frequency and duration of the conditioning stimulation and may last for several minutes. A similar phenomenon was observed by Lloyd[96] in the monosynaptic pathway through the spinal cord, and it is thought to be a general property of synapses in the vertebrate CNS.

The action of interneurons[10]

First-order afferents and motoneurons are highly specialized cells that because of their accessibility are by far the best known central neural elements. The monosynaptic link between large muscle afferents and spinal motoneurons has been the locus of the most important studies of central synaptic transmission. The two-neuron arc is a special case, however, for most of the synaptic relay centers of the brain and spinal cord are complicated by the presence of interneuronal cells. In this restricted definition, interneurons are cells whose axons do not enter white matter and that, by virtue of recurrent feedback looping or by feed-forward connections, play important roles in controlling nuclear transmission. For example, note the interneurons of the spinal reflex pathways diagrammed in Fig. 6-12. In a more general sense, however, very large regions of the brain may be thought of as giant chains of interneurons relating to reflex arcs or to long projecting systems and influencing activity in them.

There is a general belief that those aspects of the integrative action of the nervous system that cannot be explained on the basis of the integrative mechanisms of single cells depend on the complex interactions within and between *populations* of neurons. The overall action of a neural population thus reveals ensemble properties that are generated by the population action per se and are not obvious in the action of any single element. It has not as yet been possible to set down general rules concerning these actions that will apply to all or even to many interneuronal populations. Their study, case by case, makes up a large part of central neurophysiology, which is discussed in many chapters of this book. Here it is pertinent to indicate how synaptic transmission at interneurons differs from that at the motoneuron, the prototype cell just discussed.

Interneurons form indirect, multisynaptic, divergent and reentrant, convergent loops in nuclear regions, superimposed on more direct afferent and efferent pathways.* Thus afferent ac-

*"Afferent" and "efferent" are used to indicate the major inflow to and outflow from a synaptic center in the nervous system as well as in their more traditional use to indicate the dorsal and ventral root fibers of the spinal cord.

tivity, which may reach efferent neurons directly, will also evoke activity that, delayed and amplified through interneuronal chains, will continue to reach efferent neurons during a considerable period of time. It appears that interneurons play an important role in prolonging and sustaining activity in the CNS. The simplest form of this sustaining action is known as *afterdischarge*. For example, a single volley of impulses in a cutaneous nerve elicits flexor muscle contractions that may continue for 0.1 sec or more due to a repetitive, reflexly evoked discharge of flexor motoneurons that is sustained by the repetitive and prolonged activity in the interneuronal chains linking input to output in the spinal reflex pathway.

Direct study of interneurons has been made most extensively in the spinal cord.[98] Even the interneurons of a single spinal segment are heterogeneous; some link afferent fibers to motoneurons, others are inhibitory neurons in recurrent pathways, others link long descending systems to motoneurons, and still others are intercalated between afferent fibers and the cells of origin of long ascending systems. Moreover, there are powerful interactions between these groups, a subject only recently brought under study.[98] Background or "spontaneous" activity, occurring in the absence of obvious stimulation, is a prominent feature of interneuronal activity, as it is of a very large proportion of all CNS neurons. A second prominent aspect of interneuronal activity is the tendency to discharge long trains of impulses, even in response to a single excitatory volley. Intracellular recording from interneurons reveals that EPSPs provoked by such a volley are prolonged and irregular.[72] Their action potentials, in contrast to those of motoneurons, are not followed by a period of hyperpolarization; there is no prolonged period of subnormality after impulse discharge. Thus one may suppose that the lowest threshold portion of the cell, the initial segment, will respond with impulses again and again as long as the ionic current is adequate to drive the EPSP to threshold. The combination of such events in the several elements of divergent-convergent and circular reentrant linkages of neurons will contribute to prolonged interneuronal and efferent neuronal discharge. Moreover, all those aspects of monosynaptic transmission making for facilitation (temporal and spatial summation, temporal facilitation, posttetanic potentiation) will be operative at interneuronal synapses also, all in concert tending to increase and prolong the output of the circuit, be it excitatory or inhibitory. Undoubtedly the actions of interneurons are responsible for one constant feature of central neurons, indeed of the brain as a whole: a steady, slightly oscillating state of high excitability, *a readiness to respond*.

Synaptic transmission at suprasegmental levels of the central nervous system[10]

The majority of synapses in suprasegmental parts of the mammalian CNS show the ultrastructural characteristics of chemically operated synapses. A number of studies have shown that the general paradigm of chemical transmission obtains here as at spinal motoneurons, but with significant modifications. Particularly, certain quantitative aspects of transmission differ, for cerebral neurons are smaller than motoneurons and have higher input resistances and longer time constants, factors favoring temporal and spatial summation. They seldom show afterhyperpolarization and thus may discharge high-frequency bursts or sustained high-frequency trains when under intense or sustained synaptic drive. Each cerebral nucleus or cortical area appears to contain neurons that differ greatly in their rates of adaptation to steady synaptic drive, a property thought to be determined by both intrinsic membrane properties and local synaptic circuitry. The most striking difference, however, pertains to the properties of dendrites and the mechanism of cellular integration. It has been shown for a number of cell types, particularly for the Purkinje cells of the cerebellum and the hippocampal pyramidal cells, that dendrites are not only passive electrotonic conductors. Activation of excitatory axodendritic synapses can lead to the generation of action potentials in local regions of dendritic membrane. These regions can summate across stretches of electrically inexcitable dendritic membrane and influence directly the spike-generating mechanism of the cell soma. These "booster zones" in the dendritic tree appear to operate much more efficiently in the direction of the cell body than away from it, so that the patterns of local synaptic events within the dendritic trees are not wiped out with each impulse discharge of the cell body.

Thus in many central neurons there occurs a form of cellular integration somewhat more complex than that implied by the "single trigger zone" model described for spinal motoneurons. Active dendritic responses compensate for the very long electrotonic length of the very long, thin dendritric trees of some central neurons, such as the pyramidal cells of the neocortex, so that even the most distant excitatory synapses may influence the overall discharge rate of the cell.

Inhibitory synapses in cerebral structures are predominantly axosomatic in location, so that even though their number is small, only 5% of the total, they exert a powerful influence on central neurons. IPSPs are very large and have very long time courses, and inhibition has been shown to play a crucial role in shaping the spatial patterns of neural activity in central populations, for example, in the central nuclei and cerebral cortical areas of sensory systems, described in later chapters.

SYNAPTIC TRANSMITTER AGENTS IN THE MAMMALIAN AUTONOMIC GANGLIA AND CENTRAL NERVOUS SYSTEM

The action of a neurotransmitter on a postsynaptic element, whether a neuron or an effector organ, requires that the postsynaptic membrane be receptive to the transmitter in question. It was postulated early in this century by a number of investigators[29,87] that the sensitivity of a postsynaptic element depended on the existence of specific neurotransmitter "receptors." Until the late 1960s the receptor postulate was supported by much indirect evidence, but no direct proof of the existence of a receptor had been obtained nor was its molecular structure elucidated. Early in the 1970s a number of workers used the method of affinity chromatography to show that the ACh receptor at the neuromuscular junction is a glycoprotein and an integral part of the subsynaptic plasma membrane.[53] This ACh receptor is the best characterized of a large number of neurotransmitter receptors thought to exist in the CNS and the peripheral nervous system.

Receptor classification is usually based on pharmacologic criteria of the interaction between the receptor, the endogenous neurotransmitter, and various drugs. For a given transmitter, more than one pharmacologically defined receptor subtype may exist and exhibit a characteristic tissue and regional distribution. The names given are arbitrary, often based on historical precedent. For example, the receptor responsible for the action of ACh at the skeletal neuromuscular junction is called a *nicotinic* ACh receptor because nicotine can mimic the action of ACh at this synapse. In contrast, the actions of ACh on the heart, smooth muscle of the gut, exocrine glands, and many regions of the CNS are poorly reproduced by nicotine but can be mimicked by muscarine, causing this receptor to be called *muscarinic*. Some cells (e.g., autonomic neurons) can possess both types of ACh receptors (p. 204).

Recent studies indicate that the receptor (i.e., the protein that recognizes and binds the transmitter) and the macromolecular entities that, when activated by a transmitter-receptor complex, mediate the physiologic response are distinct. The ACh receptor molecule and the ion channel that opens transiently on cholinergic stimulation appear to be different entities.[54,131] Similar results have been obtained for the β-adrenergic receptor and its effector, namely the enzyme adenylate cyclase.[120,124] The fact that the receptor and the effector are different macromolecules serves to rationalize the observation that responses triggered by the same receptor subtype may differ in different tissues. For example, activation of muscarinic receptors in the smooth muscle of the gut is excitatory, whereas the muscarinic action of ACh in the heart is inhibitory. Similarly, activation of myocardial β-adrenergic receptors is excitatory, but in the cerebellum, β-adrenergic action is inhibitory.

Identification of neurotransmitters

Identifications of central transmitters are based on techniques and concepts developed in the study of neuromuscular and ganglionic synapses, with added steps to circumvent the blood-brain barrier and to deal with the anatomic complexity of the CNS. Synaptic delays, polarized conduction, characteristic ultrastructure, and reversible PSPs all indicate that most central synapses are chemically operated. Moreover, central pharmacologic effects are frequently best understood as the action of drugs on transmitter synthesis and release or on postsynaptic receptor molecules.

ACh and norepinephrine are substances commonly considered to be central transmitters, as they are at peripheral junctions, but it is likely that a number of other substances function in a similar way. There are certain criteria to be met for establishing transmitter identity, namely, that (1) the substance and the enzymatic system for its synthesis are present in neurons, (2) the substance is released via a calcium-dependent process in adequate amounts from presynaptic neurons by impulses in them, (3) mechanisms exist for the rapid inactivation of the candidate transmitter, (4) local application of the substance elicits postsynaptic permeability changes identical to those produced by synaptic activity, and (5) drug effects potentiating or inhibiting actions of naturally or experimentally delivered transmitter are similar. Based on these criteria, it has been established that ACh and norepinephrine are transmitters at peripheral synapses.[6] All these criteria have been met to establish that ACh is the transmitter at the Renshaw synapse (p. 199), whereas evidence for the identity of transmitters elsewhere in the CNS is still somewhat indirect.

A number of substances of low molecular weight have been proposed as potential transmitters; the list of candidates has recently lengthened greatly because of the discovery of a number of short peptides that show markedly inhomogeneous distributions in the central and peripheral junctions and that are extremely potent in inducing behavioral alterations when injected into the cerebrospinal fluid of experimental animals. Some of these substances may not function as neurotransmitters in the classic mode, as ACh and norepinephrine do in the peripheral nervous system. They may serve rather to *modulate* ongoing neural activity in more subtle ways. Agents of this sort may modulate neuronal transactions in a number of different ways: (1) by changing the threshold for action potential generation, thus increasing or decreasing the efficiency of other synaptic input (p. 193); (2) by changing the voltage- or time-dependent properties of the ionic channels that generate action potentials, and these changes in turn possibly modifying the discharge frequencies of neurons or even inducing self-sustaining activity, as has been shown to be the effect of some peptides in the nervous system of *Aplysia*[15]; or (3) by modifying the amount of neurotransmitter released from presynaptic terminals by an action potential, for example, the inhibition of norepinephrine release from adrenergic nerve endings by prostaglandins of the E series.[11] Such modulator roles remain a possibility for central synapses for such "classic" neurotransmitter candidates as serotonin (5-HT) and dopamine.

Studies of the ontogeny of the synaptic machinery of neurons have led to a reexamination of some long-held working hypotheses. Dale[43] first proposed and Eccles[3] further elaborated the idea that a given neuron releases only one transmitter agent from all its terminals. Recent experiments on sympathetic ganglion neurons studied in vitro have demonstrated that immature neurons can at certain stages of development release more than one transmitter,[59] and that cellular and humoral cues can determine which neurotransmitter a given neuron will synthesize.[8,19,137] At present, there is no convincing evidence that mature neurons, in mammals, can synthesize and release more than one transmitter agent. However, the possibility that this is the case remains open, for the existence of a transmitter with a peptide modulator in the same synaptic terminal has been suggested on the basis of histochemical observations in primary afferent fibers in the rat spinal cord.[114]

GANGLIONIC TRANSMISSION

Synaptic events

An action potential in a preganglionic fiber evokes a series of synaptic events in the postganglionic neurons. The first is a depolarization (fast EPSP or fEPSP), which if of sufficient amplitude will trigger an action potential in the postsynaptic cell. The fEPSP is produced by the action of presynaptically released ACh on the nicotinic cholinergic receptors of the postganglionic neuron. This local, graded event is in all essential aspects identical to the end-plate potential (EPP) at the muscle end-plate, described in Chapter 5. Miniature EPSPs (mEPSPs) occur spontaneously and randomly in autonomic ganglion cells. The available evidence[20] supports the view that here, as in the skeletal neuromuscular junction, the mEPSPs are responses of the postsynaptic cell to spontaneously released multimolecular "packets" of ACh, believed to be stored in synaptic vesicles.

The short-latency fEPSP is usually followed by two other events of longer latency and more prolonged time course; one is hyperpolarizing (slow IPSP or sIPSP), the other depolarizing (slow EPSP or sEPSP). The tenth paravertebral sympathetic ganglion of the bullfrog is particularly well suited for the study of the electrogenesis of these slow PSPs. This ganglion contains two types of cells that can be identified by the antidromic conduction velocity of their axons: cells with axons of B-fiber conduction velocity are called B cells, and those with conduction velocities of C fibers are called C cells. The slow PSPs are generated differentially in the two cell types: an sEPSP in B cells and an sIPSP in C cells. The two cell types receive separate preganglionic input and can be activated independently: B cells from preganglionic B fibers in the sympathetic chain and C cells from C fibers in the seventh spinal nerve.[134]

Fig. 6-14, *A*, shows that the fEPSP is augmented in amplitude on hyperpolarization of the membrane, and it is first reduced in amplitude and finally reversed in polarity on depolarization. The conductance of the membrane increases during the fEPSP; the reversal potential is −14 mV; this fact taken together with evidence obtained in ion-substitution experiments is consistent with the notion that the fEPSP is produced by a transient increase in the membrane perme-

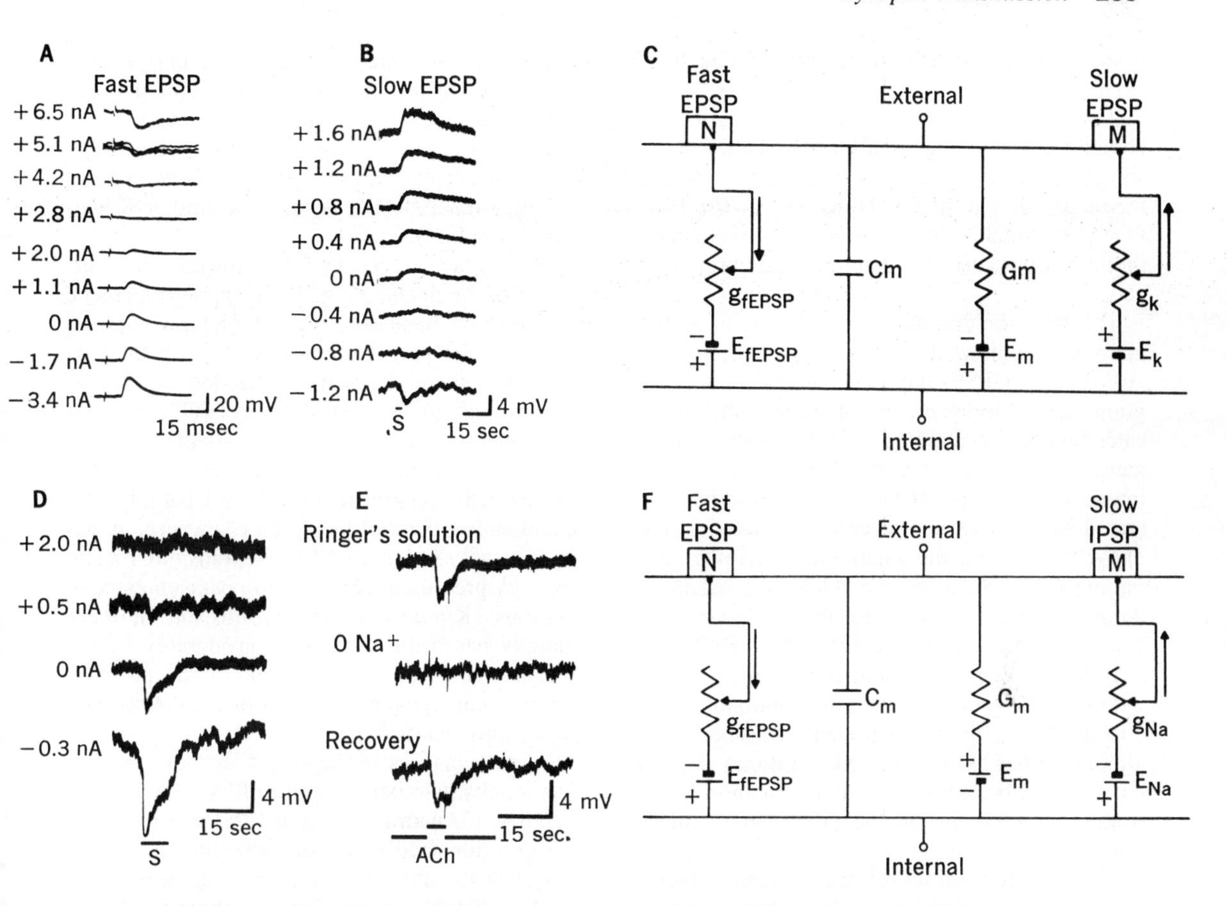

Fig. 6-14. A, Intracellular recordings of fast EPSPs (fEPSPs) from sympathetic frog neurons. Numbers at left indicate current injected into cell to change resting potential. **B,** Intracellular recordings of slow EPSPs (sEPSPs) from frog sympathetic neurons. *S* indicates onset and duration of preganglionic stimulus. Numbers at left indicate current injected into cell to change resting potential. **C,** Equivalent circuit for neuronal membrane indicating how *reduction* in g_K can lead to depolarization observed with sEPSP. **D,** Intracellular recording from frog sympathetic neurons exhibiting increase in amplitude of slow IPSP on hyperpolarization of neuronal membrane. *S* indicates period of preganglionic stimulation. **E,** Same as **D,** using iontophoretic application of ACh. Slow IPSP is abolished in Ringer's solution containing 0 Na^+. **F,** Equivalent circuit for neuronal membrane illustrating slow IPSP due to *reduction* in g_{Na}. (From Weight, F. F.: Synaptic potentials resulting from conductance decreases. In Bennett, M. V. L., editor: Synaptic transmission and neuronal interaction, New York, copyright © 1974, Raven Press, New York.)

ability to Na^+ and K^+. In contrast, the sEPSP is *increased* on depolarization and is first reduced in amplitude and eventually reversed when the membrane is driven in the opposite direction to potential levels more negative than E_K (Fig. 6-14, *B*). The membrane conductance is reduced during the sEPSP. This fact together with the results of experiments in which ACh was applied externally supports the suggestion that the sEPSP is generated by a transient *reduction* in the membrane permeability to K^+ caused by activation of a second set of ACh receptors on the B cell membrane, in this case muscarinic receptors (Fig. 6-14, *C*).

The sIPSP differs from that found in spinal motoneurons (p. 197) in that it, like the sEPSP, results from a decrease in membrane conductance. The sIPSP, whether elicited by preganglionic stimulation or by iontophoresis of ACh onto C cells, is enhanced by hyperpolarization and abolished in Na^+-free Ringer's solution (Fig. 6-14, D and E). This sIPSP is due to a muscarinic action of ACh that transiently reduces the resting Na^+ permeability, thus driving the membrane potential to a more hyperpolarized level (Fig. 6-14, *F*).[138]

Amphibian sympathetic ganglia possess catecholamine-containing small cells, believed by some[94] to be interneurons responsible for the mediation of the sIPSP. However, in the bullfrog these small cells do not synapse on sympathetic neurons; they are extra-adrenal chromaffin cells.[139]

Transmission through mammalian sympathetic ganglia is somewhat more complicated, for the sEPSPs and sIPSPs can occur in the same ganglion cell. Moreover, in mammals the small catecholamine-containing cells do synapse on the ganglionic neurons, possibly functioning as true interneurons.[99] The sIPSP is believed to be generated by a muscarinic receptor–mediated action of ACh acting on the small catecholamine-containing cell, which in turn releases a catecholamine (perhaps dopamine) that activates the ganglionic neuron (Fig. 6-15). The sIPSPs and sEPSPs can occur in mammalian ganglion cells without a detectable change in membrane conductance[93,107]; their electrogenesis is unclear, although it has been suggested that it may depend on the selective activation of one or more electrogenic ion pumps on the postsynaptic membrane.

The neurotransmitter released at neuroeffector junctions by postganglionic sympathetic neurons in mammals is norepinephrine. The only known exceptions to this rule are the vasodilator fibers innervating vessels in skeletal muscle and the axons innervating sweat glands. These neurons are anatomically part of the sympathetic system, but they release ACh as their transmitter (Chapters 33 and 34).

In contrast to the detailed information available for sympathetic ganglia, parasympathetic ganglia have been little studied, and most of their functions are poorly understood. This is because the majority of them are embedded in effector organs and are not readily accessible for experimentation. However, the available evidence suggests that the synaptic processes in parasympathetic ganglia fit the general paradigm of chemically operated synapses elsewhere. It has been established that ACh is the transmitter agent for *both* pre- and postganglionic parasympathetic neurons. Knowledge of transmission in these ganglia has been increased considerably by the recent studies of Kuffler et al. of the postganglionic parasympathetic neurons in frog and mudpuppy hearts.[67,101]

These synapses in the frog heart have the following characteristics: (1) mEPSPs occur spontaneously; (2) neurally evoked EPSPs of multiunit composition lead to action potentials at critical levels of membrane depolarization; (3) neurally evoked EPSPs reverse sign at about -12 mV

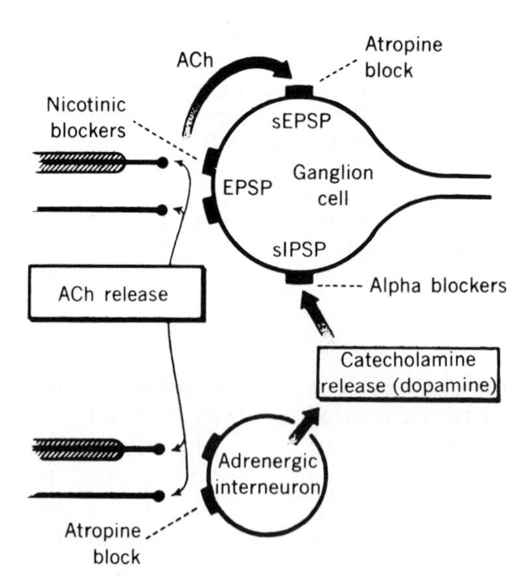

Fig. 6-15. Schematic outline of synaptic mechanisms at peripheral sympathetic ganglion of the mammalian autonomic nervous system. All preganglionic fibers are cholinergic; released ACh acts at two receptor sites, one nicotinic, leading to rapid EPSP, the other muscarinic, leading to prolonged sEPSP. Some presynaptic fibers, operating via interneuron utilizing dopamine as transmitter, exert prolonged inhibitory synaptic action on postganglionic cells. (From Libet.[92])

membrane potential, indicating the nonspecific permeability change produced by ACh; (4) all these events can be mimicked by ACh applied iontophoretically through a micropipet placed close to a synaptic area[44]; and (5) these synaptic areas, where boutons are seen, are much more sensitive to ACh than are areas of the cell surface that are free of boutons.[66]

This last observation allowed Kuffler et al.[83] to study the changes in ACh sensitivity following preganglionic nerve section. It has been known for a long time that denervated structures, whether nerve or muscle cells, become more sensitive to the transmitter substances that normally play on them. This *law of denervation* of Cannon is discussed in detail in Chapter 34. Kuffler et al. found that within a few days after denervation the region of ACh sensitivity spread to include the entire surface of the postganglionic neuron, but there was no absolute increase in sensitivity at any restricted locus.

Study of the parasympathetic neurons in the mud puppy cardiac septum revealed that here, as in sympathetic ganglia, fEPSPs and sIPSPs could be recorded from the same cell. Both the fEPSP and the sIPSP were shown to be due to the action of the same transmitter, ACh, acting on different postsynaptic receptors.[67] This is one of the few exceptions to the general rule that the action of a given transmitter on a single neuronal element will be either excitatory or inhibitory, but not both. Moreover, this system serves to illustrate the general principle that *the postsynaptic action of a transmitter will be excitatory or inhibitory depending on the nature of the ionic channels altered by the activation of the neurotransmitter receptor and not on the nature of the neurotransmitter itself.* ACh in these neurons will give rise to an fEPSP via its interaction with nicotinic receptors and will generate an sIPSP by interacting with muscarinic receptors in the same cell. Thus the nature (excitatory or inhibitory) and the kinetics (fast or slow) depend on the type of receptor activated.

It is not always possible to draw generalizations from receptor type to physiologic response. Although muscarinic responses are in general slower in onset and longer in duration than nicotinic ones (the former are measured in hundreds of milliseconds to seconds, the latter in a few milliseconds[115]), no generalization is possible with respect to the type of physiologic response. For example, muscarinic actions of ACh on frog sympathetic ganglia can give rise to sEPSPs or sIPSPs.

Central cholinergic transmission

It is now well established that ACh is the transmitter agent at neuromuscular synapses and in autonomic ganglia. Its presence within the CNS has been known since the 1930s, and its inhomogeneous distribution there suggests a specific function in some regions but not in others. The dorsal roots, optic nerves, and cerebellum contain virtually no ACh, whereas the ventral roots, spinal cord, caudate nucleus, and retina contain large amounts, as much as 7 μg/gm tissue. The distribution of the synthetic enzyme choline acetyltransferase (CAT) generally parallels that of ACh, and so does the distribution of muscarinic receptors.[130] The distribution of the degradative enzyme acetylcholinesterase (AChE) is wider than is that of ACh or CAT, being present in regions that receive no known cholinergic input. The cellular location of AChE is less restricted in central axons than at the skeletal neuromuscular junction, for it appears to be present both in the synaptic cleft and within the presynaptic axon, where it may hydrolyze any ACh not protected by vesicular storage.[31]

Chemical studies of brain tissue after homogenization and differential density-gradient centrifugation, particularly in combination with electron microscopic studies of the resulting fractions, have proved valuable in correlating chemical composition with subcellular location. In such a brain homogenate, 70% to 75% of the total ACh is contained within a fraction made up almost wholly of nerve terminals. When these "synaptosomes" are themselves disrupted, the ACh is found within a fraction containing mainly synaptic vesicles. CAT is also present in nerve endings, but its precise intraneuronal localization is still unsettled. At least a part of this enzyme activity is found either free in the cytoplasm or in microsomal particles throughout the neuron. It is synthesized within the cell body and moved to the nerve endings by axonal transport. The ACh within the synaptic vesicles is thought to be in a storage form, protected from hydrolysis until its release from the nerve endings, and replaced by the synthetic action of CAT. The activity of the released ACh is presumably terminated by hydrolysis with AChE, which is bound to membrane fragments in brain homogenates. Most of the cholinergic receptors are found in fractions enriched with synaptic membranes, suggesting a plasma membrane localization.[130]

The steady-state concentration of ACh in cholinergic neurons is remarkably constant under physiologic conditions, suggesting that the rate of synthesis is increased during periods of en-

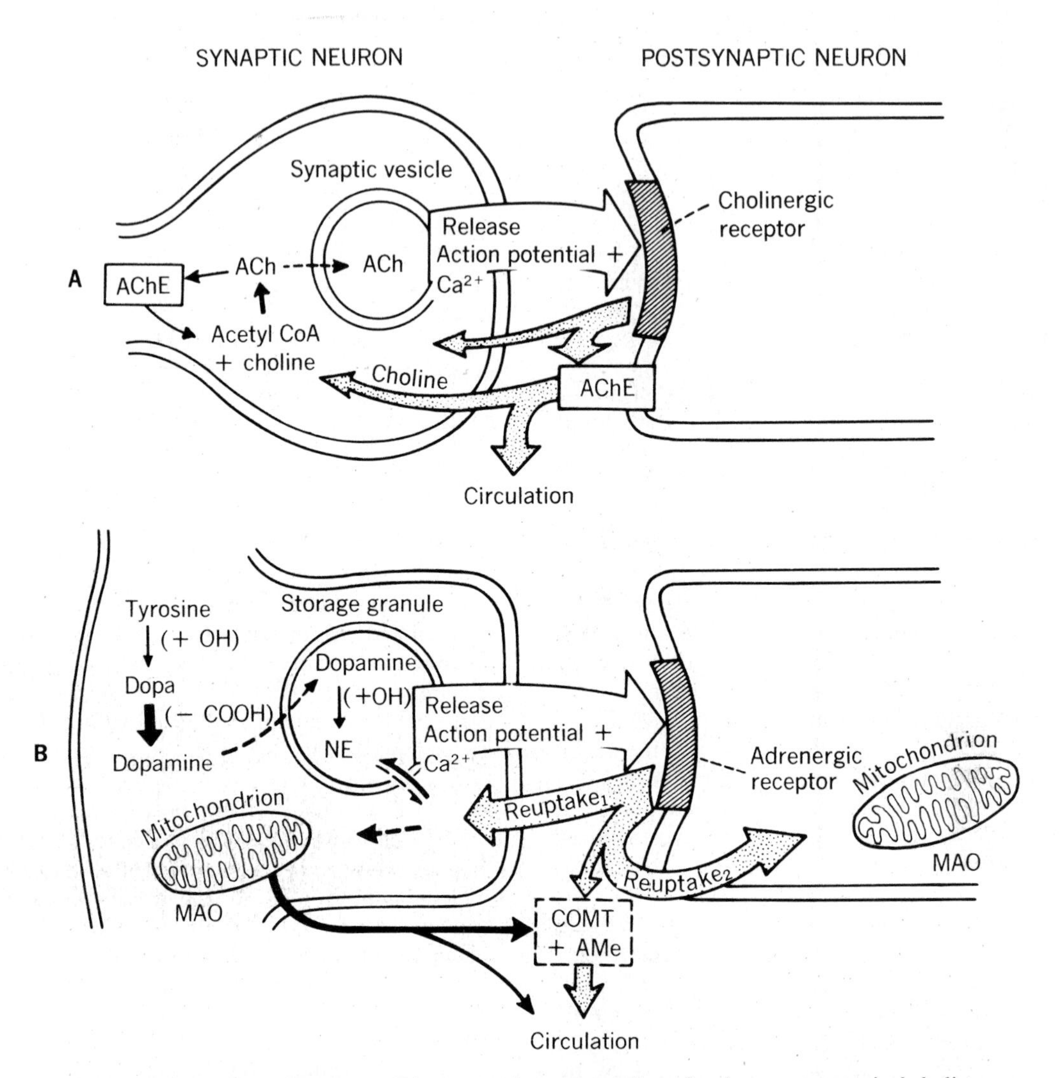

Fig. 6-16. A, Cholinergic synapse. Diagrammatic representation of major events at typical cholinergic synapse, including synthesis of ACh by choline acetyltransferase (probably in cytoplasm), storage in and release from synaptic vesicles, postsynaptic interaction, and subsequent inactivation by AChE, which probably also acts presynaptically to remove excess ACh. Mechanism of release by nerve impulses (white arrow) is not certain but depends on external Ca^{2+} and perhaps Na^{+} and probably does not involve intermediate discharge into free presynaptic cytoplasm. Black arrow represents metabolic pathway of ACh catabolism at postsynaptic neuron, with choline release into circulation and feedback of choline into presynaptic synthetic sites. **B,** Adrenergic synapse. Synthesis of NE occurs from phenylalanine via tyrosine and directly from endogenous tyrosine. First step, hydroxylation of tyrosine to form dihydroxyphenylalanine (dopa), appears to be rate-limiting (small arrow). This step and subsequent decarboxylation to form dopamine occur in cytoplasm. Dopamine is then taken into storage vesicles for final β-hydroxylation to form NE. NE that leaks into the cytoplasm is destroyed by mitochondrial monoamine oxidase (MAO). Release by nerve impulses requires Ca^{2+}. After interaction with receptor sites, NE is removed mainly by mechanism of active reuptake but partly also by enzymatic deactivation by catechol-*O*-methyltransferase (COMT), utilizing as the methyl donor *S*-adenosylmethionine (AMe). This latter enzymatic step produces *O*-methylated metabolites detectable in blood, CSF or urine. Unmetabolized catecholamines are not able to escape rapidly from brain because of blood-brain diffusion barrier.

hanced neuronal activity leading to transmitter release. The available evidence[31,68] suggests that (1) CAT activity is not the rate-limiting step; (2) the availability of acetyl-CoA derived from mitochondria, the cosubstrate for the acetylation of choline, is not limiting; and (3) the concentration of choline is at or below its K_m for CAT, which suggests that it plays a rate-limiting role. Choline obtained from the hydrolysis of ACh or elsewhere from the extracellular fluid is transported to the site of synthesis by a vigorous high-affinity, Na^+-dependent uptake process. The choline uptake process is tightly coupled to subsequent acetylation,[16,68] and most investigators believe that the rate of transport of choline is the rate-limiting step in the synthesis of ACh. Furthermore, the uptake velocity of the high-affinity system is increased by neuronal activity that leads to increased ACh release.[32,68] This is at present the favored explanation for the ability of cholinergic terminals to maintain a near-constant supply of ACh in the presence of widely fluctuating neuronal activity. The molecular mechanisms involved in activity-dependent changes in choline uptake velocity remain unclear.

Multiple-barreled micropipets allow one to apply substances through one barrel to a very local region by iontophoresis while recording through the other. The chemoreceptive properties of the synaptic membrane of Renshaw interneurons, described on p. 199, were established in this way. This method of electrical recording combined with iontophoresis has been used widely in studies of central neurons, with results that are sometimes contradictory and difficult to interpret. They do include evidence of the "nicotinic" properties of some CNS synapses.[6] In some locations, however, the excitatory responses of neurons to ACh resemble more closely a "muscarinic" action of ACh; they are blocked by atropine but not by *d*-tubocurare, they are slow in onset, and they often persist for many seconds after even brief applications. Many cells give intermediate responses that are not clearly either nicotonic or muscarinic.[18] In the cerebral cortex and caudate nucleus, inhibitory cholinergic transmission may occur.

It should be emphasized that in these iontophoretic experiments the local concentrations of ACh and the sites at which it acts are unknown; the actions observed may be pharmacologic. Nevertheless, the evidence suggests strongly that ACh is a synaptic transmitter agent in many brain regions, including the cerebral cortex, hippocampus, thalamus, and basal ganglia. What is certain is that the ACh system (Fig. 6-16) is present in the brain, that it is localized to synaptic boutons, and that some central synapses are cholinergic. It is equally certain that many others are not.

Catecholamines

A number of compounds possessing a catechol ring (i.e., a benzene ring with two adjacent hydroxyl substituents) connected via a two-carbon chain to an amine group occur in the CNS and peripheral nervous system and function as neurotransmitters, synaptic modulators, or hormones. The most important are dopamine (DA), norepinephrine (NE), and epinephrine (E) (Table 6-2). DA is a putative transmitter in the CNS, retina, and sympathetic ganglia. NE is the transmitter released by postganglionic sympathetic neurons, with only two exceptions (see p. 204 and Chapter 33), and is also believed to be a transmitter in the CNS. Epinephrine is released as a hormone (with NE) from the adrenal medulla; it may also function as the transmitter at certain synapses in the CNS. The study of catecholaminergic pathways in the CNS was greatly facilitated by the discovery that these substances, when exposed to formaldehyde vapor, form fluorescent dihydroisoquinoline derivatives, which can be visualized with conventional fluorescence microscopy.[55]

The biosynthesis of catecholamines can start from the amino acid phenylalanine, but tyrosine is the normal precursor under physiologic conditions. Tyrosine is converted to dihydroxyphenylalanine (DOPA) by the action of tyrosine hydroxylase (TH). TH is the rate-limiting step in the synthesis of catecholamines.[90] It is a pteridine-dependent enzyme that uses molecular oxygen as a cosubstrate. The apparent K_m of TH for tyrosine and its cofactor tetrahydrobiopterin vary under different conditions. Tyrosine is present in brain at concentrations well above the measured K_m; therefore the availability of tyrosine is not rate limiting. The concentration of the cofactor is unknown; its availability may be rate limiting. The availability of O_2 could also be rate limiting. Variation in the activity of TH is responsible for homeostatic regulation of transmitter availability in the face of widely varying levels of neuronal activity. There are at least two ways in which this might occur: (1) TH is feedback inhibited by the end products DA and NE, and the inhibition is noncompetitive for the substrate tyrosine and competitive with respect to the pteridine cofactor.[136] Thus an increase in neuronal activity leading to increased release of DA (or NE) from the nerve terminals will reduce the DA (or NE) available *inside* the terminals, leading to a decrease of feedback inhibition and an increase

Table 6-2. Proposed central neurotransmitters; summary of candidate substances, their structures, and metabolism

Compound	Structure	Synthesis	Inactivation
Acetylcholine (ACh)	$H_3C\,C(=O){-}O{-}(CH_2)_2\,\overset{+}{N}(CH_3)_3$	Choline acetylation	Hydrolysis by AChE
Dopamine (DA)	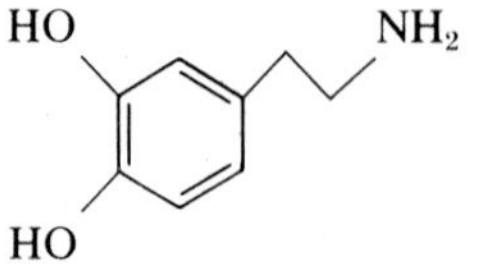 HO, HO, NH_2	Tyrosine hydroxylation, dopa decarboxylation	Reuptake, MAO, COMT
Norepinephrine (NE)	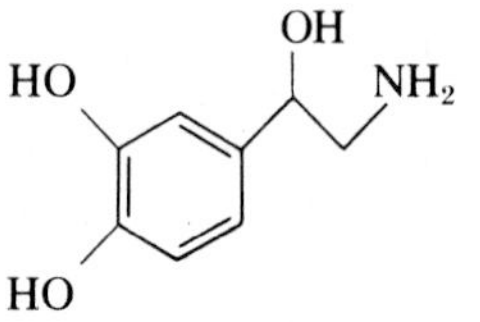 OH, HO, NH_2, HO	Dopamine-β-hydroxylation	Reuptake, MAO, COMT
Epinephrine (E)	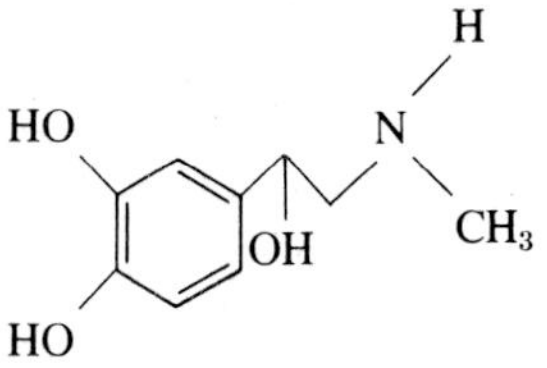 H, HO, N, OH, CH_3, HO	*N*-Methylation, transfer via PNMT	Reuptake, MAO, COMT
5-Hydroxytryptamine (5-HT), or serotonin	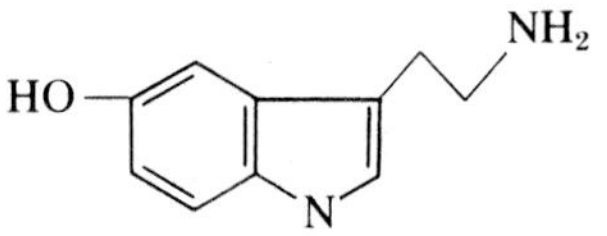 NH_2, HO, N	Tryptophan hydroxylation, 5-OH-tryptophan decarboxylation	Reuptake, MAO
Histamine	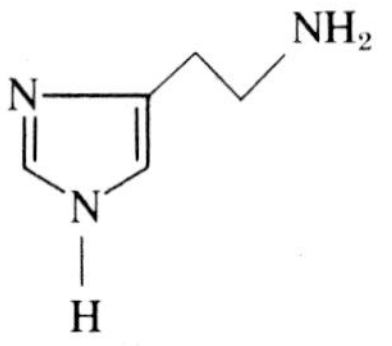 NH_2, N, N, H	Histidine decarboxylation	*N*-Methylation by histamine-*N*-methyltransferase; oxidative deamination (MAO?)
Excitatory dicarboxylic amino acids (e.g., glutamate, aspartate)	$CH_2\,(CH_2)_n\,CH\,NH_2$ with COOH, COOH; $n = 0\text{-}1$		Reuptake, decarboxylation, NH_3 fixation
Inhibitory amino acids (e.g., GABA, glycine)	$CH_2\,(CH_2)_n\,NH_2$ with COOH; $n = 0\text{-}4$	GABA by glutamate decarboxylation	GABA by reuptake, transamination, and oxidation to succinate

of TH activity. This will replenish the DA (or NE) stores. (2) An increase in impulse traffic in the nerve terminals leads to a short-term (1- to 10-minute) change in the physical properties of TH by an unknown mechanism. The altered physical state of TH is then such that it exhibits a reduction in its K_m for the pteridine cofactor, thus elevating the reaction velocity for a given amount of cofactor, and simultaneously decreases the inhibition constant (K_I) for DA and NE. Thus more DA or NE would be required to inhibit the enzyme.[140]

TH is also subject to long-term control. On severe electrical or chemical stimulation of the n. locus coeruleus or the adrenal medulla, the measured activity of TH in these structures rises with a lag time of 8 to 24 hr after stimulation and remains elevated for 1 to 2 weeks. Immunotitration studies have shown that the increased activity in the n. locus coeruleus is due to increased catalytic activity of the same number of enzyme molecules (activation), whereas the increase seen in the adrenal medulla is due to an increase in the number of TH molecules, with no apparent increase in the catalytic activity of a given enzyme molecule (induction). The physiologic significance of these forms of long-term regulation remains unclear.[91]

Most of the TH activity is found in the soluble fraction from brain homogenates. Immunocytochemical studies have shown that axonal TH is associated with microtubules, presumably for transport from the soma to the terminals. The exact location of TH in the terminals has not been established.[112]

The DOPA produced by TH is rapidly decarboxylated by the enzyme aromatic L-amino acid decarboxylase (AADC) to yield DA. AADC is a cytoplasmic pyridoxal-5′-phosphate–dependent enzyme, ubiquitously present in neuronal and nonneuronal tissues. AACD has, as its name implies, a wide substrate specificity and is present in sufficiently large quantities that DOPA does not accumulate to any extent and has never been found to be rate limiting.

In dopaminergic neurons, such as those in the pars compacta of the substantia nigra, the synthetic pathway proceeds no further. DA itself is stored in granules and is believed to serve as a neurotransmitter. In some neurons, such as those forming the n. locus coeruleus or some of the cells in the adrenal medulla, DA is taken into storage granules that contain the enzyme dopamine-β-hydroxylase (DβH). DβH is a tetrameric glycoprotein, a copper-containing, mixed-function oxidase requiring ascorbate as a cofactor. DβH catalyzes a stereospecific hydroxylation, the product *l*-NE. DβH does not exhibit absolute substrate specificity for DA and can therefore produce "false transmitters" when provided by pharmacologic manipulations with a different substrate. Part of the DβH in a vesicle is soluble and released with other vesicle contents on neural stimulation; the remaining DβH is bound to the vesicular membrane (Fig. 6-16).

NE in the vesicles is stored in a metabolically and osmotically inactive, ATP-bound form. A specific protein, chromogranin A, is also present. NE, ATP, DβH, and chromogranin A are all released via a Ca^{2+}-dependent process when a nerve impulse reaches an adrenergic terminal.

In cells of the adrenal medulla that release E and in two groups of neurons in the medulla,[70] the enzyme phenylethanolamine-*N*-methyltransferase (PNMT) is present. PNMT catalyzes the transfer of the "activated" methyl group in *S*-adenosylmethionine to the amine group in NE to yield E. The mammalian enzyme will only methylate phenylethylamines with a β-hydroxy group and has the lowest K_m for *l*-NE. This suggests that the latter is the natural substrate. PNMT is a cytoplasmic enzyme, and it is unclear how NE synthesized by vesicular DβH is made available to the cytoplasm. It is also unclear how E is then repackaged into vesicles.[104]

Degradation of catecholamines proceeds via the enzymes monoamine oxidase (MAO) and catechol-*O*-methyltransferase (COMT). Neither enzyme shows any specificity with respect to tissue distribution. MAO is bound to the external mitochondrial membrane; COMT is soluble and widely distributed, and its precise localization in neuronal tissues is unsettled. COMT uses *S*-adenosylmethionine as the methyl donor.

Unlike ACh, for which the physiologic mode of termination of synaptic action is by the degradation of ACh and the uptake system present is for a hydrolysis product, choline, MAO and COMT are involved in degradation and maintenance of a steady state over long time scales. Thus if a MAO or COMT inhibitor is applied to an adrenergic synapse, little immediate potentiation of the effects of NE is achieved. The physiologic mode of termination of action of NE is via two specialized uptake systems present in the pre- and postsynaptic membranes. These take up NE (or E) itself, rather than its degradation products.[73,74] A separate DA uptake system has also been demonstrated.[38] The two NE (E) uptake systems are labeled $uptake_1$ and $uptake_2$. The first is present in the presynaptic nerve terminal and is a high-affinity, low-capacity stereospecific system. *l*-NE is sequestered by $uptake_1$ and is not available for degradation by MAO; it is released as a transmitter. $Uptake_2$ is present in the postsynaptic target tissue and has a lower affinity for NE and a higher capacity than does $uptake_1$. It does not exhibit stereospecificity, and the catecholamines taken up by this system are then degraded by MAO and COMT. These two systems are responsible for the physiologic termination of action of transmitter released on nerve stimulation. An immediate potentiation of the response to NE is apparent on selective inhibition of $uptake_1$ (e.g., by cocaine) or $uptake_2$ (e.g., by metanephrine or corticosterone). As may be expected from their anatomic locations, the action of $uptake_1$ predominates in densely innervated structures and that of $uptake_2$ in sparsely innervated regions[74] (Fig. 6-16).

Other proposed transmitters

5-Hydroxytryptamine (serotonin). The indoleamine 5-hydroxytryptamine (5-HT) has been proposed frequently as a CNS transmitter agent. 5-HT has been found with uneven regional distribution in all vertebrate brains examined. The concentrations range from 20 to 35 ng/mg protein in the hypothalamus, preoptic nuclei, olfactory tubercle, globus pallidus, and raphe nuclei to a level of 207 ng/mg protein in the cerebral and cerebellar cortices.[119] 5-HT is retained in nerve ending fractions of brain homogenates and may be stored in granular synaptic vesicles.[28]

5-HT, like the catecholamines, will form a fluorescent product when exposed to formaldehyde vapor. Although the 5-HT fluorophore develops with much less intensity than does its catecholamine counterpart, it has been possible with this technique to map some of the major serotonergic pathways in the mammalian CNS (Fig. 6-17). These arise from nine groups of midline pontine and upper brain stem raphe nuclei: the more caudal groups project to the spinal cord and the more rostral groups emit extensive projections to the diencephalon and telencephalon. The n. raphe medianus provides most of the

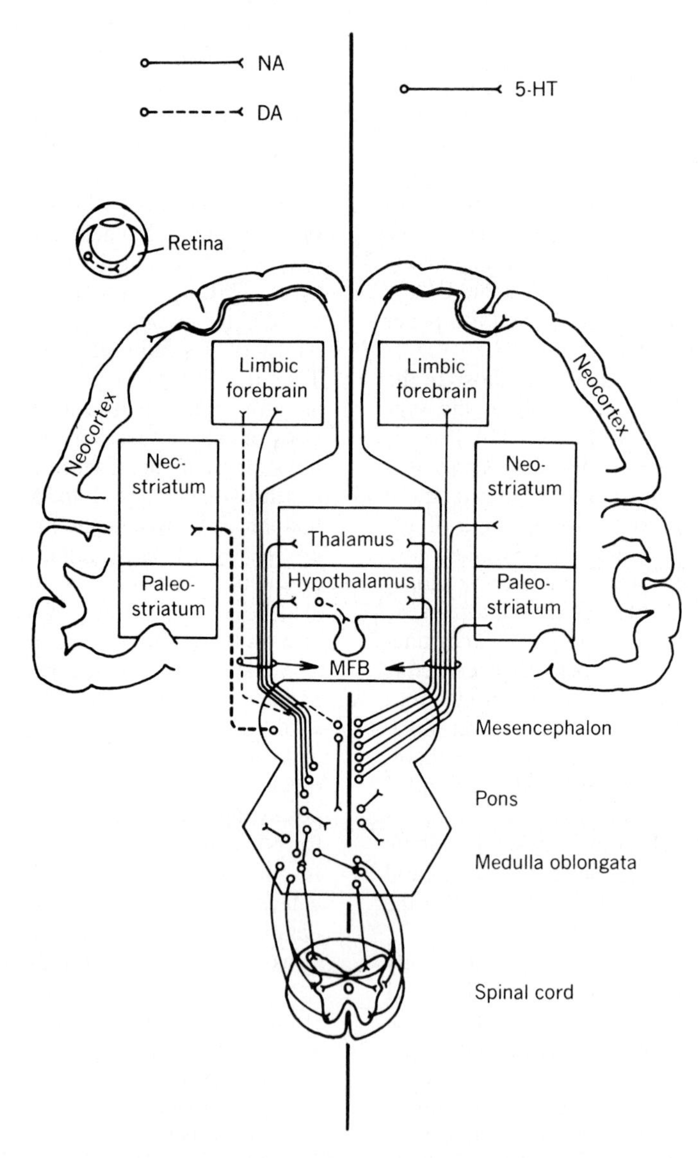

Fig. 6-17. Schematic representation of distribution of neurons containing norepinephrine (NE), dopamine (DA), or serotonin (5-hydroxytryptamine, 5-HT) in mammalian CNS as determined by histofluorescence techniques. (From Anden et al.[13])

serotonergic fibers innervating the limbic system, and the dorsal raphe projects to the neostriatum and cerebral and cerebellar cortices.[12,13]

Synthesis of 5-HT proceeds from the precursor amino acid tryptophan via tryptophan hydroxylase (TrpH). This enzyme is located exclusively in the neurons of the raphe nuclei and their processes. TrpH has been demonstrated by immunocytochemical methods to be located in these neurons in the cytoplasm and in association with the Golgi apparatus and the endoplasmic reticulum. In dendrites and axons, it is associated mainly with microtubules and granular vesicles. This suggests that the enzyme is synthesized in the cell bodies and eventually transported to the terminal varicosities of raphe neurons.[77,113]

TrpH is—like TH—a pteridine-dependent-enzyme that uses O_2 as a cosubstrate. The K_m for tryptophan when tetrahydrobiopterin is used as the cofactor is about 5×10^{-5} M. This value is more than twice the concentration of tryptophan in the brain. Thus the enzyme is normally unsaturated in situ and is itself the rate-limiting step in the synthesis of 5-HT. Tryptophan reaches the brain from the plasma via an uptake system that transports other neutral amino acids (such as tyrosine, valine, and isoleucine) as well. Plasma tryptophan is in turn derived from dietary protein intake, for mammals cannot synthesize this amino acid. Thus the amount of tryptophan in the diet will influence plasma and thus brain levels of the amino acid. Given that TrpH is normally unsaturated with respect to this substrate, an increase in the available free tryptophan should increase brain levels of 5-HT, and this has proved to be the case.[142] Furthermore, exposure of experimental animals to an atmosphere of 100% oxygen also increases their cerebral 5-HT content, which suggests that TrpH is normally unsaturated with the cosubstrate molecular oxygen. The product of the action of TrpH, 5-hydroxytryptophan, is then rapidly decarboxylated by AADC to yield 5-HT.

The main route for degradation of 5-HT is oxidation via MAO, a step then followed by a second oxidation via aldehyde dehydrogenase to 5-hydroxyindoleacetic acid (5-HIAA), the major metabolite.

The mechanisms responsible for maintaining the steady-state concentration of 5-HT are unclear. Unlike TH, neither 5-hydroxytryptophan nor 5-HT inhibit TrpH. Moreover, inhibition of MAO produces large increases in total brain 5-HT, which indicates that TrpH is not subject to feedback inhibition. A reuptake system for 5-HT can be demonstrated in brain slices,[127] but it is not certain whether oxidation, reuptake, or both are primarily responsible for terminating the action of 5-HT under physiologic conditions. A complicating feature is that the uptake systems for NE and DA will take up exogenous 5-HT, if made available in high concentration, and release it as a "false transmitter."

Iontophoretically applied 5-HT has a variety of excitatory and inhibitory effects. The absence of a single, unambiguously characterized serotonergic synapse mars the interpretation of iontophoretic experiments. The results obtained may represent pharmacologic effects rather than normal physiologic processes. A confounding problem has been encountered when peripheral antagonists of 5-HT, such as the drug LSD, have been used to clarify the actions of 5-HT in the CNS. LSD and related hallucinogens appear to mimic rather than antagonize 5-HT in the mammalian CNS. Similar problems cloud the interpretation of experiments aimed at identifying postsynaptic 5-HT receptors.[17] At present the case for 5-HT as a neurotransmitter remains incomplete.[61] The matter is of considerable importance, for there is evidence that pathways that are putatively serotonergic in action play important roles in the central mechanisms controlling pain and sleep (Chapters 10 and 13).

Amino acids

Several aliphatic amino acids are known from iontophoretic experiments to have profound effects on the state of polarization of many central neurons.[6] Among the compounds generally found to be excitatory the dicarboxylic amino acids L-glutamate and L-aspartate are the best known. They produce rapid and striking transient depolarization when applied to neurons but not when injected intracellularly. Many natural and synthetic compounds with a terminal carboxyl or sulfonyl moiety separated by one or two carbon atoms from an α-carbon bearing an amino and carboxyl group possess such activity. The natural L-enantiomers of glutamate and aspartate are widely distributed in the cytoplasm of mammalian central neurons, particularly in isolated nerve endings.[108] They can be released from the cerebral cortex during thalamic and brain stem stimulation at rates many times greater than their spontaneous efflux and are released rather specifically, compared with other amino acids, from electrically stimulated CNS slices or even from cortical synaptosomes in vitro.[108] The mechanisms responsible for the rapid termination of action of L-aspartate or L-glutamate are unclear;

active reuptake by both neuronal and nonneuronal cells is the favored hypothesis.

As can be expected for substances that are important for protein synthesis and other metabolic events, definite evidence for their proposed role as synaptic transmitters is lacking. The case for L-glutamate or L-aspartate is weakened by their nearly ubiquitous excitatory effect on neurons and the absence of a specific antagonist for their action. The lack of such a specific pharmacologic antagonist has so far prevented the critical test of whether the behavior of these putative transmitters is identical with that of the endogenous transmitter(s) responsible for excitatory responses in the cerebrum and spinal cord.

Amino butyric acid. More is known about γ-amino butyric acid (GABA) than about any other proposed mammalian central inhibitory transmitter. In the crustacean peripheral inhibitory motor fibers, GABA has been identified as the transmitter.[111] In mammalian brains, GABA shows some unique features: (1) Its synthesis is restricted to the CNS; only trace amounts are detectable in the peripheral nervous system or in other organs. (2) GABA is further restricted to the CNS by an efficient blood-brain barrier across which it does not pass. (3) GABA distribution in the brain and spinal cord is heterogeneous. (4) Its steady-state concentration and turnover rate are 10 to 100 times greater than those of other transmitters (e.g., ACh or DA).[37,108]

GABA is formed by the α-decarboxylation of glutamic acid. The synthetic enzyme glutamic acid decarboxylase (GAD) is a pyridoxal-5′–dependent neuronal enzyme with a high degree of substrate specificity for L-glutamate. Catabolism proceeds via a reversible reaction catalyzed by GABA-transaminase (GABA-T), a pyridoxal-5′–dependent enzyme with an absolute requirement for α-ketoglutarate as the amine receptor. GABA-T (unlike GAD) is not found exclusively in neurons. Although both GAD and GABA-T are associated with particular subcellular fractions, their cellular distributions are different. GAD is found in high concentrations in centrifuged fractions containing presynaptic nerve endings, whereas GABA-T is found in the main associated with cell body mitochondria; the mitochondria from presynaptic endings show little GABA-T activity. Immunocytochemical studies by Roberts et al.[100,121] on the mouse cerebellum have confirmed the differential distribution of GAD and GABA-T suggested by the fractionation studies. In electron micrographs, GAD was localized to synaptic boutons *on* neuronal somata and dendrites. GABA-T was found *inside* Purkinje and stellate cell bodies, as well as in glial cells.

Brain slices and synaptosomal preparations rapidly take up GABA from the extracellular fluid. Moreover, GABA-T inhibitors fail to potentiate synaptic inhibition or the action of iontophoretically applied GABA. These two observations suggest that termination of the action of synaptically released GABA is by reuptake rather than degradation, as is the case for NE.

With few exceptions, activation of a GABA receptor leads to synaptic inhibition. The properties of the ion channel opened by the GABA-receptor interaction have not been completely elucidated, but on neurons from Deiters' nucleus and in invertebrate preparations a transient increase in Cl^- permeability appears to be responsible for the inhibition.

Glycine. At all levels of the neuraxis neurons can be found that are sensitive to the action of GABA, and generally the action of GABA is inhibitory. In contrast, the action of glycine is limited to the spinal cord. Iontophoretically applied glycine diminishes the discharge rate and the excitability of both spinal motoneurons and interneurons. Small doses of strychnine, a convulsant known to reduce spinal postsynaptic inhibition, can selectively abolish the response to glycine. The regional distribution of glycine in the cat spinal cord suggests an association with interneurons. Transient anoxia of the lumbosacral cord, produced by clamping the thoracic aorta, preferentially destroys interneurons. When the cord is analyzed for amino acid content after such an anoxia, the only candidate inhibitory amino acid transmitter whose concentration is markedly reduced is glycine.[108] The reversal potentials for glycine-induced hyperpolarization and natural IPSPs are similar and respond in an analogous fashion to increases in intracellular K^+ and Cl^-. An increased conductance to K^+ and Cl^- of the motoneuron plasma membrane is responsible for the generation of IPSPs and the eventual reduction of the discharge rate in these cells.[108]

The most likely synthetic pathway for glycine in the CNS is from the amino acid serine by the action of serine transhydroxymethylase, a folate-dependent enzyme. As for other amino acid transmitters, it is not yet known whether decarboxylation, transamination, or reuptake is the physiologic mechanism leading to inactivation of glycine and termination of its synaptic action. A high-affinity uptake system for glycine is present in the pons, medulla, and spinal cord.[14]

Histamine. Histamine is an imidazole amine that occurs in some as yet unidentified hypothalamic neurons and is widely present in brain tissue that contains mast cells. Histamine not associated with mast cells is located in homogenate fractions containing nerve endings and has a rapid turnover. Histamine concentration in the cerebral cortex is reduced by lesions in the region of the medial forebrain bundle, which suggests the existence of a histamine-containing diencephalic tract that may project diffusely to cortical areas.[126]

Histamine is formed in a one-step decarboxylation by a specific enzyme with a K_m of about 10^{-4} M for the precursor histidine. This specific decarboxylase can be distinguished from the AADC involved in NE and 5-HT synthesis (p. 209) by the differential inhibition of the former by α-hydrazinohistidine.[62]

Degradation of histamine occurs via imidazole-*N*-methyltransferase and MAO; the latter will accept *N*-methylhistamine as a substrate.

Iontophoretic application of histamine can produce a variety of excitatory and inhibitory effects on neurons, but there is little direct physiologic evidence that histamine receptors exist in the mammalian CNS.

Neuroactive peptides

A number of apparently unrelated avenues of research have converged within the past 5 years to define the field of neuroactive peptides, and research in this field is increasing at a rapid rate. That peptides could affect neuronal activity has been known since von Euler and Gaddum discovered substance P (SP) in 1931. This substance was shown by Chang and Leeman to be an undecapeptide 40 years later. However, SP was not considered seriously as a neurotransmitter or modulator until a number of other peptides were found to be present in neuronal tissue and to affect neuronal activity and behavior. A number of these peptides, such as angiotensin II, were known for their action on peripheral vascular tissue. Others, such as thyrotropic-releasing hormone (TRH) and luteinizing hormone–releasing hormone (LH-RH), were discovered in hypothalamic extracts by Shally and Guillemin in their search for factors that stimulated the anterior pituitary gland. The discovery of two pentapeptides in the mammalian CNS that had opiate-like activity,[71] both in brain receptor assays and in in vitro smooth muscle assays, brought endocrinology and neuroscience together. It was realized that the entire sequence of one of these pentapeptides (Met^5-enkephalin) is contained within the pituitary hormone β-lipotropin. At present a growing list of neuroactive peptides is rapidly making the distinction between what is a hormone and what is a neurotransmitter vague and opening new dimensions in cellular neurochemistry. Some of these peptide transmitter candidates are listed in Table 6-3. None of those listed yet fulfills all the criteria to be met by a bona fide transmitter, but the progress of this field is so rapid that statements of this sort are often outdated by the time they appear in print!

Table 6-3. Structures of some neuroactive peptides presently being considered as potential neurotransmitters or neuromodulators*

Peptide	Structure
TRH	(Pyro)Glu—His—Pro(NH_2)
Enkephalins	Tyr—Gly—Gly—Phe—Met—OH Tyr—Gly—Gly—Phe—Leu—OH
Angiotensin II	Asp—Arg—Val—Tyr—Ile—His—Pro—Phe—OH 1 2 3 4 5 6 7 8
Vasopressin†	CyS—Tyr—Phe—Glu(NH_2)—Asp(NH_2)—CyS—Pro—Lys—Gly(NH_2) 1 2 3 4 5 6 7 8 9
Oxytocin	CyS—Tyr—Ile—Glu(NH_2)—Asp(NH_2)—CyS—Pro—Leu—Gly(NH_2)
LH-RH	(Pyro)Glu—His—Trp—Ser—Tyr—Gly—Leu—Arg—Pro—Gly(NH_2)
Substance P	Arg—Pro—Lys—Pro—Gln—Gln—Phe—Phe—Gly—Leu—Met(NH_2)
Neurotensin	(Pyro)Glu—Leu—Tyr—Glu—Asn—Lys—Pro—Arg—Arg—Pro—Tyr— Ileu—Leu—OH
Somatostatin	Ala—Gly—CyS—Lys—Asn—Phe—Phe—Trp—Lys—Thr—Phe—Thr—Ser—CyS

*From Cooper et al.[36]
†Position 8 is Lys in pigs and hippopotamuses; in other mammals, Lys is replaced by Arg.

AXONAL TRANSPORT[5,97]

The great length of neuronal processes, particularly axons, implies that a special mechanism exists for the transport of materials from one part of a neuron to another, for most of the machinery for synthesizing proteins and other large molecules is confined to the cell body and dendrites. Weiss and Hiscoe[141] discovered that axoplasmic material accumulated proximal to an axonal constriction and that axonal diameter and content decreased distal to it. When the constriction was removed, the accumulated material moved down the nerve fiber at a rate of about 1 mm/day, an event thought to occur continuously in normal axons. Knowledge of this matter has increased greatly since the introduction of the method of radioactive labeling of molecules within the cell body and tracing their movement by radioautography with light and electron microscopy.[45] A number of molecular species and intracellular organelles are known to be synthesized or constructed within the cell body and moved down the axon to its terminals. This axonal transport functions to maintain the neuron, supplying even its most distant parts with materials synthesized in the cell body. It may, in addition, supply substances that influence the morphologic, biochemical, and functional specification of postsynaptic cells, exerting a "trophic" influence of neurons on the cells they innervate.[64]

It is now clear that there are two quite different rates of axonal transport that depend on different mechanisms and are likely to serve different functions in the overall economy of the neuron. *The fast component of axonal transport* moves at a rate of about 400 mm/day, occurs in both directions between cell body and axon terminals, and is composed of phospholipids, glycoproteins, the constituents of synaptic and neurosecretory vesicles, and the enzyme systems that synthesize transmitter agents. It is thought to play an important role in the dynamic function of the nerve ending in synaptic transmission and neurosecretion and in replenishing the terminal membrane. The role of fast retrograde transport in neuronal function is less certain. It has been shown that large molecules such as the enzyme horseradish peroxidase, placed in the vicinity of nerve terminals, may be taken up into the terminals by a process of reverse pinocytosis and reach the cell body by the retrograde rapid transport system. This provides a mechanism by which substances originating in postsynaptic target cells might influence neuronal function, thus allowing a trophic influence of target tissues on the neurons that innervate them.[64]

Slow axonal transport moves at about 1 mm/day, only in the orthodromic direction, and consists of a high proportion of soluble proteins, the protein constituents of microtubules, and mitochondria. It is thought to be responsible for the continuous replacement of major constituents of the axon, including neurofilaments and neurotubules. The transport rate equals that of the regeneration of severed nerve fibers. A similar slow transport process exists in dendrites as well.[79] The mechanisms of the two types of axonal transport are unknown. There is evidence that fast transport of organelles depends on their interaction with microtubules, for their transport ceases in the presence of substances like colchicine that bind to and alter the molecular configuration of the microtubules. What controls these transport processes is unknown.

In summary, rapid axonal transport provides components of transmitter and neurosecretory systems, terminal nerve membrane, and perhaps

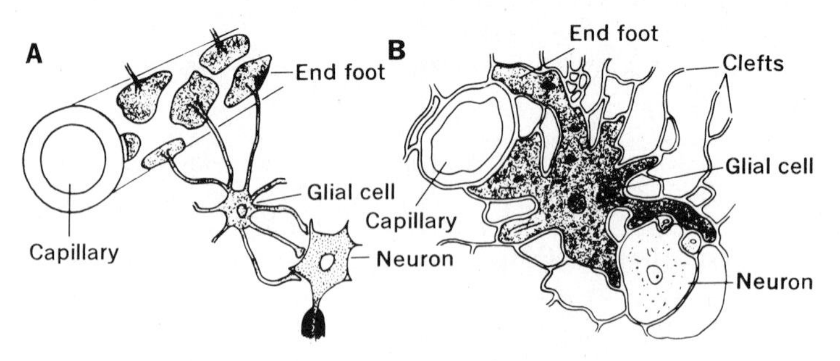

Fig. 6-18. A, Schematic drawing shows neuron-glia-capillary relation as seen with light microscope. Some "end feet" of astrocyte make contact with capillary wall and others with neuron. This relation is basis for original idea that glial cells might serve as channels for passage of substances between blood and neurons, and vice versa. **B,** Sketch shows same relation as seen with electron microscope. All cells, axons, dendrites, and astrocytes are tightly packed but separated by intercellular cleft system that is usually of 15 to 20 nm width. (From Kuffler.[7])

other substances required at nerve endings; the duality of direction allows for the transport of putatively trophic substances in either direction—for action of nerve on target cells and vice versa. The slow axonal transport is a bulk movement of the axon itself and raises the possibility that axons and their endings are continuously growing or being replaced and that these changes can be modified by the activity of the neuron.

NEUROGLIA[81,89]

Neuroglial cells of the brain greatly outnumber its neurons and occupy half its volume. They possess no axons, neither make nor receive synapses, and are linked with one another but not to neurons by gap junctions, regions of low resistance to ion flow.[82] Neuroglia possess the potential to continue to divide throughout life, a property lost by neurons early in development.[129] The space between neurons is occupied by glial cells, leaving clefts of 15 to 20 nm separating all cellular elements. These intercellular clefts make up 5% to 10% of the volume of the brain, constitute its true extracellular space, and are the avenues for the diffusion of ions and molecules within the brain and between it and the vascular system. This special transport function was previously and is thought now erroneously assigned to glial cells because of their apparent interposition between blood vessels and nerve cells (Fig. 6-18). Glial cells display the elements of actively metabolizing cells: mitochondria, endoplasmic reticulum, ribosomes, glycogen, fat, ATP, etc. They congregate in regions of the brain in which neurons are destroyed by injury or disease and play an important role in repair and scar formation.[118]

There are three types of glial cells. The origin of *microglia* is still uncertain. Undoubtedly, some are of neuroectodermal origin, while others enter the CNS from the bloodstream and function there as macrophages (e.g., in response to infection). The *oligodendroglia* and *astrocytes* are true neuroglia and, like neurons, are of ectodermal origin. The oligodendroglia form and maintain myelin sheaths around axons, as do the Schwann cells of peripheral nerves.[25] Direct evidence concerning astrocyte function was first obtained by Kuffler and his colleagues, who applied microphysiologic and microchemical methods to invertebrates, in which glial cells can be identified and observed directly.[7,81]

Experimental preparations.[82,105,106] Direct observations of glial cells were made in the CNS of the leech and in the optic nerves of amphibians[30,84,110] which have large glial cells accessible in situ. The neuron-glial relation in a central ganglion of a leech is shown in Fig. 6-19. In each packet, 50 to 60 neurons are embedded within invaginations of a single large glial cell, but all cell boundaries are distinct and separated by 15 nm clefts, except at the gap junctions that link adjacent glial cells. The cleft system opens freely to the outer capsule and capillary endothelium. Both glial cells and neurons can be studied with intracellular micropipets, and the movement of substances from outside into the ganglion can be measured by their effects, if any, on the membrane potentials of those cells or by microchemical analysis.

The optic nerve fibers of *Necturus* are similarly encased within invaginations of large glial cells.[84,110] The blood vessels from which nutrient substances

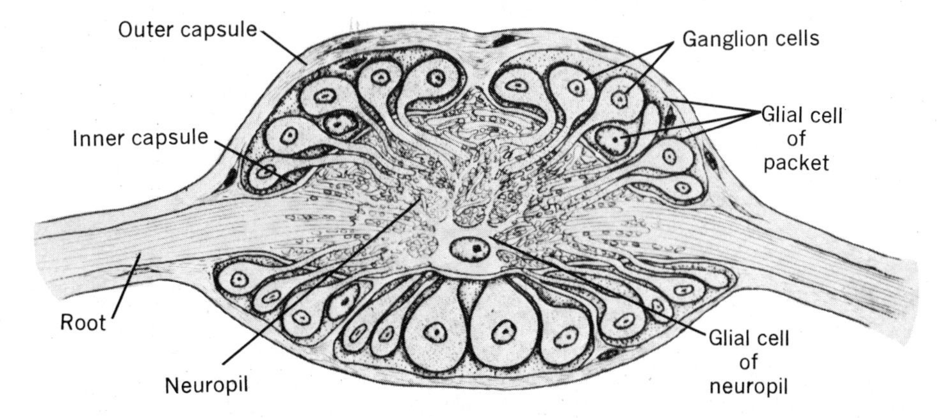

Fig. 6-19. Semidiagrammatic representation of transverse section through ganglion of leech CNS. Each peripheral compartment or packet contains numerous nerve cells, all invested by single large glial cell. Neurons send their processes through inner capsule into neuropil, which occupies central compartment and is where synaptic contacts are made. Neuropil itself is completely invaginated within embrace of other glial cells, one of which is shown on ventral aspect of central compartment. (From Coggleshall and Fawcett.[30])

reach the fibers run on the surface of the nerve, to which the intercellular cleft system opens freely. Glial cell membrane potentials can be observed during changes in the composition of the extracellular fluid and when the nerve fibers are excited electrically or by light flashes to the retina.

Physiologic properties of glial cells[7,9]

A great deal has been learned about glial cells in studies of the preparations described. The membrane potential of a glial cell varies as predicted by the Nernst equation when K_o^+ is changed (Fig. 6-20); thus the membrane potential of the glial cell is an excellent indicator of K_o^+ for it follows faithfully changes in the extracellular concentration of that ion. In contrast to neurons, the membrane potential of glial cells can be displaced passively over a wide range by transmembrane current without evoking active membrane responses (Fig. 6-20, *B*). The important conclusion is that *glial cells do not possess the capacity for generating local responses or conducted impulses, as do neurons, and do not participate directly in the rapid signal transmission functions of the nervous system.* Imposed currents do not spread directly from neurons to glia or vice versa. Moreover, for the intact ganglion of the leech, both direct observation and theoretical calculation show that (1) the intercellular cleft system is an adequate avenue for diffusion of ions between the external medium and the vicinity of neurons, (2) it provides an equally open pathway for the ionic currents of action potentials, and (3) no significant portion of those currents flows through glial cells. After destruction of their investing glial cells, neurons of the leech ganglion still generate and conduct action potentials, use oxygen, and metabolize glucose. This indicates that if glia play an important "metabolic support" role in relation to neurons, that role must operate over a longer time scale than the several hours these denuded neurons were observed.[82,105]

It is clear from their physiologic properties that glial cells play a role in clearing the intercellular spaces of the K^+ extruded from active neurons and that by virtue of the low-impedance gap junctions that link adjacent glial cells, they

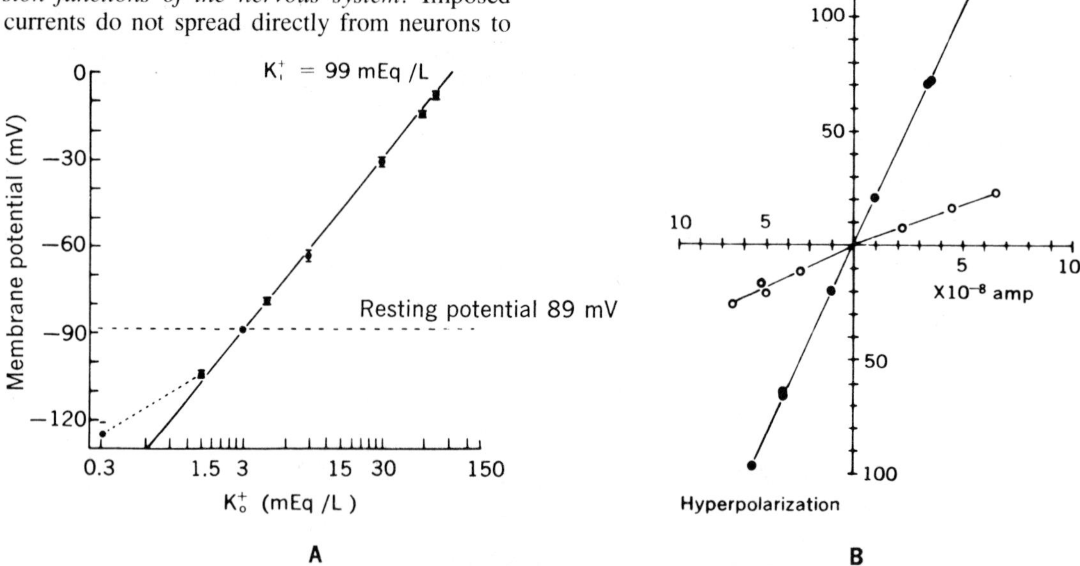

Fig. 6-20. A, Relation between glial membrane potential and K^+ concentration in bathing fluid (K_o^+) in optic nerves of *Necturus*. Forty-two measurements were made with K^+ concentrations below or above normal. Mean resting potential in Ringer's solution was 89 mV. Horizontal bars = ±SD of mean. Solid line has slope of 59 mV/10-fold change in K^+ concentration. It fits observed points except at lowest concentration tested. Membrane potential when $K_o^+ = K_i^+$; K_i^+ is therefore about 99 mEq/L. **B,** Current-voltage relation for two glial cells. Square pulses of current were passed through one intracellular electrode; membrane potential changes that resulted were recorded through second electrode in same cell. Resistance of each cell remained constant even though in the series represented by steeper curve, membrane potential was displaced over total range of 200 mV. No active membrane responses were observed, even with rapid depolarizations. Glial cells do not give regenerative responses similar to those seen in neurons. (From Kuffler et al.[84])

may function as a "spatial buffer," removing K^+ from the extracellular space during times of intense neuronal activity. Whether the associated glial depolarization serves as a signal evoking from the glial cells a response (metabolic?) of importance for the continued function of neurons is unknown.

Glial cells of the mammalian central nervous system

A class of cells presumed to be glial cells has been studied within the mammalian cerebral cortex,[27,63,78] medulla,[80] and spinal cord[132] by means of intracellular recording of membrane potentials and with K^+-selective micropipets (for review see Somjen[9]). These cells have in some studies been identified by intracellular injection of marking dyes and later study of histologic sections. Mammalian glial cells, like those of the invertebrates described previously, never show local postsynaptic responses or action potentials, but their membranes are slowly depolarized by the accumulation of K_o^+ accompanying the activity of adjacent neurons. There is good evidence that these slow depolarizations are the major source of the sustained potential shifts that can be recorded across the active cerebral cortex. Whether they contributed to the electroencephalogram is described in Chapter 9.

Neuron-glia relations

It is clear from the foregoing that glial cells are not directly involved in the signaling function of the CNS. Nevertheless, there are suggestions that an important functional relation exists between neurons and glial cells, although its nature is still unknown. For example, glia adjacent to motoneurons are greater in number the longer the axon of the motoneuron[56,57] and increase in number in parallel with increases in motoneuronal activity.[85] Schwann cells of the periphery,[109] glial cells of dorsal root ganglia[103,123] and of the CNS,[75] and glioma cells in tissue culture possess a high-affinity uptake mechanism for GABA, an established inhibitory transmitter agent in the CNS, and release it when depolarized. Thus one candidate response of glial cells to the K^+-mediated depolarization described might be to release an agent that would, at least in some parts of the nervous system, suppress the neural activity leading to the depolarization. Glial cells have uptake mechanisms for other candidate transmitter agents, such as glutamate, and for neutral amino acids, as do all cells. However, glia possess the GABA uptake mechanism in regions where neurons possess no GABA receptors. Thus whether glial cells exert a modulating influence on synaptic transmission or exchange metabolites with neurons remains uncertain.

REFERENCES

General reviews

1. Bennett, M. V. L.: Electrical transmission: a functional analysis and comparison to chemical transmission. In Kandel, E. R., editor: Handbook of physiology. Section 1: the nervous system, cellular biology of neurons, Bethesda, Md., 1977, American Physiological Society, vol. 1, pt. 1.
2. Burke, R. E., and Rudomin, P.: Spinal neurons and synapses. In Kandel, E. R., editor: Handbook of physiology. Section 1: the nervous system, cellular biology of neurons, Bethesda, Md., 1977, American Physiological Society, vol. 1, pt. 1.
3. Eccles, J. C.: The physiology of synapses, Berlin, 1964, Springer-Verlag.
4. Gilula, N. B.: Electrotonic junctions. In Schmitt, F. O., and Worden, G. F., editors: The neurosciences: fourth study program, Cambridge, Mass., 1978, The M.I.T. Press.
5. Grafstein, B.: Axonal transport: the intracellular traffic of the neuron. In Kandel, E. R., editor: Handbook of physiology. Section 1: the nervous system, cellular biology of neurons, Bethesda, Md., American Physiological Society, vol. 1, pt. 1, pp. 691-717.
6. Krnjevic, K.: Chemical nature of synaptic transmission in vertebrates, Physiol. Rev. **54:**418, 1974.
7. Kuffler, S. W.: Neuroglial cells: physiological properties and a potassium mediated effect of neuronal activity on the glial membrane potential, The Ferrier Lecture, Proc. R. Soc. Med. **168:**1, 1967.
8. Patterson, P. H.: Environmental determination of autonomic neurotransmitter functions, Annu. Rev. Neurosci. **1:**1, 1978.
9. Somjen, G. G.: Electrophysiology of neuroglia, Annu. Rev. Physiol. **37:**163, 1975.
10. Spencer, W. A.: The physiology of supraspinal neurons in mammals. In Kandel, E. R., editor: Handbook of physiology. Section 1: the nervous system, cellular biology of neurons, Bethesda, Md., 1977, American Physiological Society, vol. 1, pt. 1.
11. Westfall, T. C.: Local regulation of adrenergic transmission, Physiol. Rev. **57:**659, 1977.

Original papers

12. Aghajanian, G. K., Haigler, H. J., and Bennett, J. L.: 5-Hydroxytryptamine in brain. In Iversen, L. L., Iversen, S. D., and Snyder, S. H., editors: Handbook of psychopharmacology, New York, 1975, Plenum Press, vol. III, p. 63.
13. Anden, N. E., et al.: Ascending monoamine neurons to the telencephalon and diecephalon, Acta Physiol. Scand. **67:**313, 1966.
14. Aprison, M. H., Davidoff, R. A., and Werman, R.: Glycine: its metabolic and possible transmitter roles in nervous tissue. In Lajtha, A., editor: Handbook of neurochemistry, New York, 1970, Plenum Press, vol. 3.
15. Barker, J. L.: Peptides: roles in neuronal excitability, Physiol. Rev. **56:**435, 1976.
16. Barker, L. A., and Mittag, T. W.: Comparative studies of substrates and inhibitors of choline transport and choline acetyltransferase, J. Pharmacol. Exp. Ther. **192:**86, 1975.

17. Bennett, J. P., and Snyder, S. H.: Serotonin and lysergic acid diethylamide binding in rat brain membranes: relationship to postsynaptic serotonin receptors, Mol. Pharmacol. **12:**373, 1976.
18. Bird, S. J., and Aghajanian, G. K.: The cholinergic pharmacology of hippocampal pyramidal cells: a microiontophoretic study, Neuropharmacology **15:**273, 1976.
19. Black, I. B.: Regulation of autonomic development, Annu. Rev. Neurosci. **1:**183, 1978.
20. Blackman, J. C., Ginsberg, B. L., and Ray, C.: Synaptic transmission in the sympathetic ganglion of the frog, J. Physiol. **167:**355, 1963.
21. Blankenship, J. E., and Kuno, M.: Analysis of spontaneous subthreshold activity in spinal motoneurons of the cat, J. Neurophysiol. **31:**195, 1968.
22. Bodian, D.: An electron microscopic characterization of classes of synaptic vesicles by means of controlled aldehyde fixation, J. Cell. Biol. **44:**115, 1970.
23. Bodian, D.: Neuron junctions: a revolutionary decade, Anat. Rec. **174:**73, 1972.
24. Bullock, T. H., and Hagiwara, S.: Intracellular recording from the giant synapse of the squid, J. Gen. Physiol. **40:**565, 1957.
25. Bunge, R. P.: Glial cells and the central myelin sheath, Physiol. Rev. **48:**197, 1968.
26. Burke, R. E.: Composite nature of the monosynaptic excitatory postsynaptic potential, J. Neurophysiol. **30:**1114, 1967.
27. Castelluci, V. F., and Goldring, S.: Contribution to steady potential shifts of slow depolarization in cells presumed to be glia, Electroencephalogr. Clin. Neurophysiol. **28:**109, 1970.
28. Chan-Palay, V.: Serotonin axons in the supra- and subependymal plexuses and in the leptomeninges; their roles in local alterations of cerebrospinal fluid and vasomotor activity, Brain Res. **102:**103, 1976.
29. Clark, A. J.: Mode of action of drugs on cells, London, 1933, Edward Arnold (Publishers), Ltd.
30. Coggeshall, R. E., and Fawcett, D. W.: The fine structure of the central nervous system of the leech, *Hirudo medicinalis,* J. Neurophysiol. **27:**229, 1964.
31. Collier, B.: Biochemistry and physiology of cholinergic transmission. In Kandel, E. R., editor: Handbook of physiology. Section 1: the nervous system, Bethesda, Md., 1976, American Physiological Society, vol. 1.
32. Collier, B., and MacIntosh, F. C.: The source of choline for acetylcholine synthesis in a sympathetic ganglion, Can. J. Physiol. Pharmacol. **47:**127, 1969.
33. Colonnier, M.: Synaptic patterns on different cell types in the different laminae of the cat visual cortex, an electron microscope study, Brain Res. **9:**268, 1968.
34. Coombs, J. S., Curtis, D. R., and Eccles, J. C.: The generation of impulses in motoneurones, J. Physiol. **139:**232, 1957.
35. Coombs, J. S., Eccles, J. C., and Fatt, P.: Excitatory synaptic action in motoneurones, J. Physiol. **130:**374, 1955.
36. Cooper, J. R., Bloom, F. E., and Roth, R. H.: The biochemical basis of neuropharmacology, New York, 1978, Oxford University Press, Inc.
37. Costa, E.: Some recent advances in the biochemical pharmacology of γ-aminobutyric acid. In Usdin, E., Hamburg, D. A., and Barchas, J. C., editors: Neuroregulators and psychiatric disorders, New York, 1977, Oxford University Press, Inc., p. 372.
38. Coyle, J. T., and Snyder, S. H.: Catecholamine uptake by synaptosomes in homogenates of rat brain: stereospecificity in different areas, J. Pharmacol. Exp. Their. **170:**221, 1969.
39. Creed, R. S., et al.: Reflex activity of the spinal cord, Oxford, 1932, Clarendon Press.
40. Curtis, D. R., and Eccles, J. C.: The time courses of excitatory and inhibitory synaptic actions, J. Physiol. **145:**529, 1959.
41. Curtis, D. R., and Eccles, J. C.: Synaptic action during and after repetitive stimulation, J. Physiol. **150:**374, 1960.
42. Curtis, D. R., and Ryall, R. W.: The acetylcholine receptors of Renshaw cells, Exp. Brain Res. **2:**66, 1966.
43. Dale, H. H.: Pharmacology and nerve endings, Proc. R. Soc. Med. **28:**319, 1935.
44. Dennis, M. H., Harris, A. H., and Kuffler, S. W.: Synaptic transmission and its duplication by locally applied acetylcholine in parasympathetic neurons in the heart of the frog, Proc. R. Soc. Lond. (Biol.) **177:**509, 1971.
45. Droz, B., and LeBlond, C. P.: Axonal migration of proteins in the central nervous system and peripheral nerves as shown by radioautography, J. Comp. Neurol. **121:**325, 1963.
46. Dudel, J., and Kuffler, S. W.: Presynaptic inhibition at the crayfish neuromuscular junction, J. Physiol. **155:**543, 1961.
47. Eccles, J. C.: The central action of antidromic impulses in motor nerve fibres, Arch. Gesamte Physiol. **260:**385, 1955.
48. Eccles, J. C.: Excitatory and inhibitory synaptic action, Harvey Lect. **51:**1, 1955.
49. Eccles, J. C.: Modes of communication between nerve cells, Australian of Science Year Book, Sydney, 1963, White & Bull.
50. Edwards, F. R., Redman, S. J., and Walmsley, B.: Statistical fluctuations in charge transfer at IA synapses on spinal motoneurons, J. Physiol. **259:**665, 1976.
51. Edwards, F. R., Redman, S. J., and Walmsley, B.: Non-quantal fluctuations and transmission failures in charge transfer at I-a synapses on spinal motoneurons, J. Physiol. **259:**689, 1976.
52. Edwards, F. R., Redman, S. J., and Walmsley, B.: The effect of polarizing currents on unitary IA excitatory post-synaptic potentials evoked in spinal motoneurones, J. Physiol. **259:**705, 1976.
53. Eldefrawi, M. E., and Eldefrawi, A. T.: Acetylcholine receptors. In Cuatrecasas, P., and Greaves, M. F., editors: Receptors and recognition, London, 1977, Chapman & Hall, Ltd., vol. 4, ser. A.
54. Eldefrawi, A. T., et al.: Perhydrohistrionicotoxin: a potential ligand for the ion conductance modulator of the acetylcholine receptor, Proc. Natl. Acad. Sci. USA **74:**2172, 1977.
55. Flack, B., Hillarp, N. A., Thieme, G., and Thorpe, A.: Fluorescence of catecholamines and related compounds condensed with formaldehyde, J. Histochem. Cytochem. **10:**348, 1962.
56. Friede, R. L.: Relationship of body size, nerve cell size, axon length and glial density in the cerebellum, Proc. Natl. Acad. Sci. USA **49:**187, 1963.
57. Friede, R. L., and van Houten, W. H.: Neuronal extension and glial supply: functional significance of glia, Proc. Natl. Acad. Sci. USA **48:**817, 1962.
58. Furshpan, E. J., and Potter, D. D.: Transmission at the giant synapses of the crayfish, J. Physiol. **145:**289, 1959.
59. Furshpan, E. J., MacLeish, P. R., O'Lague, P. R., and Potter, D. D.: Chemical transmission between rat sympathetic neurons and cardiac myocytes developing in microcultures: evidence for cholinergic, adrenergic

and dual-function neurons, Proc. Natl. Acad. Sci. USA **73**:4225, 1976.

60. Gray, E. G.: Electron microscopy of excitatory and inhibitory synapses: a brief review, Prog. Brain Res. **31**:141, 1969.
61. Gershon, M. D.: Biochemistry and physiology of serotonergic transmission. In Kandel, E. R., editor: Handbook of physiology. Section 1: the nervous system, Bethesda, Md., 1977, American Physiological Society, vol. 1, p. 573.
62. Green, J. P., Johnson, C. L., and Weinstein, H.: Histamine as a neurotransmitter. In Lipton, M. E., Killam, K. G., and DiMascio, A., editors: Psychopharmacology—a generation of progress, New York, 1978, Raven Press.
63. Grossman, R. G., and Hampton, T.: Relationships of cortical glial cell depolarizations to electrocortical surface wave activity, Electroencephalogr. Clin. Neurophysiol. **28**:95, 1970.
64. Guth, L.: "Trophic" influences of nerve on muscle, Physiol. Rev. **48**:645, 1968.
65. Hagiwara, S., and Tasaki, I.: A study on the mechanism of impulse transmission across the giant synapse of the squid, J. Physiol. **143**:114, 1958.
66. Harris, A. J., Kuffler, S. W., and Dennis, M. J.: Differential chemosensitivity of synaptic and extrasynaptic areas on the neuronal surface membrane in parasympathetic neurons of the frog, tested by microapplication of acetycholine, Proc. R. Soc. Lond. (Biol.) **177**:541, 1971.
67. Hartzell, H. C., Kuffler, S. W., Stickgold, R., and Yoshikami, D.: Synaptic excitation and inhibition resulting from direct action of acetylcholine on two types of chemoreceptors on individual amphibian parasympathetic neurones, J. Physiol. **271**:817, 1977.
68. Haubrich, D. R., and Chippendale, T. J.: Regulation of acetylcholine synthesis in nervous tissue, Life Sci. **20**:1465, 1977.
69. Heuser, J. E., and Reese, T. S.: Structure of the synapse. In Kandel, E. R., editor: Handbook of physiology. Section 1: the nervous system, cellular biology of neurons, Bethesda, Md., 1977, American Physiological Society, vol. 1, pt. 1.
70. Hokfelt, T., Fuxe, K., Goldstein, M., and Johansson, O.: Immunohistochemical evidence for the existence of adrenaline neurons in the rat brain, Brain Res. **66**:235, 1974.
71. Hughes, J. H., et al.: Identification of two related pentapeptides from the brain with potent opiate agonist activity, Nature **258**:577, 1975.
72. Hunt, C. C., and Kuno, M.: Properties of spinal interneurones, J. Physiol. **147**:346, 1959.
73. Iversen, L. L.: Role of transmitter uptake mechanisms in synaptic transmission, Br. J. Pharmacol. **41**:571, 1971.
74. Iversen, L. L.: Biochemical aspects of synaptic modulation. In Schmitt, F. O., and Worden, F. G., editors: The neurosciences: third study program, Cambridge, Mass., 1974, The M.I.T. Press.
75. Iversen, L. L., and Kelly, J. S.: Uptake and metabolism of gamma-aminobutyric acid by neurons and glial cells, Biochem. Pharmacol. **24**:933, 1975.
76. Jack, J. J. B., Miller, S., Porter, R., and Redman, S. J.: The time course of minimal excitatory postsynaptic potentials evoked in spinal motoneurones by group Ia fibres, J. Physiol. **215**:353, 1971.
77. Joh, T. H., Shikimi, T., Pickel, V. M., and Reis, D. J.: Brain tryptophan hydroxylase: purification of, production of, antibodies to, and cellular and ultrastructural localization in serotonergic neurons of rat midbrain, Proc. Natl. Acad. Sci. USA **72**:3375, 1975.
78. Karahashi, Y., and Goldring, S.: Intracellular potentials from "idle" cells in cerebral cortex of cat, Electroencephalogr. Clin. Neurophysiol. **20**:600, 1966.
79. Kreutzberg, G. W., Schubert, P., Toth, L., and Rieske, E.: Intradendritic transport to postsynaptic sites, Brain Res. **62**:399, 1973.
80. Krnjevic, K., and Morris, M. W.: Extracellular K^+ activity and slow potential changes in spinal cord and medulla, Can. J. Physiol. Pharmacol. **50**:1214, 1972.
81. Kuffler, S. W., and Nicholls, J. G.: The physiology of neuroglial cells, Ergeb. Physiol. **57**:1, 1966.
82. Kuffler, S. W., and Potter, D. D.: Glia in the leech central nervous system. Physiological properties and the neuron-glia relationship, J. Neurophysiol. **27**:290, 1964.
83. Kuffler, S. W., Dennis, M. J., and Harris, A. J.: The development of chemosensitivity in extrasynaptic areas of the neuronal surface after denervation of parasympathetic ganglion cells in the heart of the frog, Proc. R. Soc. Lond. (Biol.) **177**:555, 1971.
84. Kuffler, S. W., Nicholls, J. G., and Orkand, R.: Physiological properties of glial cells in the central nervous system of amphibia, J. Neurophysiol. **29**:768, 1966.
85. Kuhlenkampf, H.: Verhalten der neuroglia in den Vorderhornern des Ruckenmarkes der weissen Maus unter dem Reiz physiolopischen Tatigkeit, Z. Anat. Entwickl. Gesch. **116**:304, 1952.
86. Kuno, M.: Quantum aspects of central and ganglionic synaptic transmission, Physiol. Rev. **51**:647, 1971.
87. Langley, J. N.: On the contraction of muscle, chiefly in relation to "receptive" substances. IV. The effect of curare and some other substances on the nicotine response of the sartorius and gastrocnemius muscle of the frog, J. Physiol. **39**:235, 1909-1910.
88. Larrabee, M. G., and Bronk, D. W.: Prolonged facilitation of synaptic excitation in sympathetic ganglia, J. Neurophysiol. **10**:139, 1947.
89. Lasansky, A.: Nervous function at the cellular level: glia, Annu. Rev. Physiol. **33**:241, 1971.
90. Levitt, M., Spector, S., Sjoerdsma, A., and Udenfriend, S.: Elucidation of the rate-limiting step in norepinephrine biosynthesis in the perfused guinea-pig heart, J. Pharmacol. Exp. Ther. **148**:1, 1965.
91. Lewander, T., Joh, T. H., and Reis, D. J.: Tyrosine hydroxylase: delayed activation in central noradrenergic neurons and induction in adrenal medulla elicited by stimulation of central cholinergic receptors, J. Pharmacol. Exp. Ther. **200**:523, 1977.
92. Libet, B.: Generation of slow inhibitory and excitatory postsynaptic potentials, Fed. Proc. **29**:1945, 1970.
93. Libet, B.: The role SIF cells play in ganglionic transmission. In Costa, E., and Gessa, G. L., editors. Advances in biochemical psychopharmacology, New York, 1977, Raven Press.
94. Libet, B., and Kobayashi, H.: Adrenergic mediation of slow inhibitory postsynaptic potential in sympathetic ganglia of the frog, J. Neurophysiol. **37**:805, 1974.
95. Llinas, R., and Nicholson, C. O.: Calcium role in depolarization-secretion coupling: an aequorin study in squid giant synapse, Proc. Natl. Acad. Sci. USA **72**:187, 1975.
96. Lloyd, D. P. C.: Post-tetanic potentiation of response in monosynaptic pathways of the spinal cord, J. Gen. Physiol. **33**:147, 1949.

97. Lubinska, L.: On axoplasmic flow, Int. Rev. Neurobiol. **17**:241, 1975.
98. Lundberg, A.: Convergence of excitatory and inhibitory action on interneurons in the spinal cord. In Brazier, M. A. B., editor: The interneuron, Los Angeles, 1969, University of California Press.
99. Matthews, M. R., and Raisman, G.: The ultrastructure and somatic efferent synapses of small granule-containing cells in the superior cervical ganglion, J. Anat. **105**:255, 1969.
100. McLaughlin, B. J., et al.: The fine structural localization of glutamate decarboxylase in synaptic terminals of rodent cerebellum, Brain Res. **76**:377, 1974.
101. McMahan, U. J., and Kuffler, S. W.: Visual identification of synaptic boutons on living ganglion cells and of varicosities in postganglionic axons in the heart of the frog, Proc. R. Soc. Lond. (Biol.) **177**:485, 1971.
102. Miledi, R.: Spontaneous synaptic potentials and quantal release of transmitter in the stellate ganglion of the squid, J. Physiol. **192**:379, 1967.
103. Minchin, M. D., and Iversen, L. L.: Release of (3H) gamma-aminobutyric acid from glial cells in rat dorsal root ganglia, J. Neurochem. **23**:533, 1974.
104. Musacchio, J. M.: Enzymes involved in the biosynthesis and degradation of catecholamines. In Iversen, L. L., Iversen, S. D., and Snyder, S. H., editors: Handbook of psychopharmacology, New York, 1975, Plenum Press, vol. 3, p. 1.
105. Nicholls, J. G., and Kuffler, S. W.: Extracellular space as a pathway for exchange between blood and neurons in central nervous system of leech: the ionic composition of glial cells and neurons, J. Neurophysiol. **27**:645, 1964.
106. Nicholls, G. H., and Kuffler, S. W.: Na and K content of glial cells and neurons determined by flame photometry in the central nervous system of the leech, J. Neurophysiol. **28**:519, 1965.
107. Nishi, S.: Ganglionic transmission. In Hubbard, J. I., editor: The peripheral nervous system, New York, 1974, Plenum Press.
108. Obata, K.: Biochemistry and physiology of amino acid transmitters. In Kandel, E. R., editor: Handbook of physiology. Section 1: the nervous system, cellular biology of neurons, Bethesda, Md., 1977, American Physiological Society, vol. 1, pt. 1, p. 625.
109. Orkand, P. M., and Kravitz, E. A.: Localization of the sites of gamma-aminobutyric acid (GABA) uptake in lobster nerve-muscle preparations, J. Cell. Biol. **49**:75, 1971.
110. Orkand, R. K., Nicholls, J. G., and Kuffler, S. W.: The effect of nerve impulses on the membrane potential of glial cells in the central nervous system of amphibia, J. Neurophysiol. **29**:788, 1966.
111. Otsuka, M., Iversen, L. L., Hall, W. W., and Kravitz, E. A.: Release of gamma-aminobutyric acid from inhibitory nerves of lobster, Proc. Natl. Acad. Sci. USA **56**:1110, 1966.
112. Pickel, V. M., Joh, T. H., and Reis, D. J.: Ultrastructural localization of tyrosine hydroxylase in noradrenergic neurons of brain, Proc. Natl. Acad. Sci. USA **72**:659, 1975.
113. Pickel, V. M., Joh, T. H., and Reis, D. J.: Monoamine synthesizing enzyme in central dopaminergic, noradrenergic and serotonergic neurons, J. Histochem. Cytochem. **24**:792, 1976.
114. Pickel, V. M., Reis, D. J., and Leeman, S. E.: Ultrastructural localization of substance P in neurons of rat spinal cord, Brain Res. **122**:534, 1977.
115. Purves, R. D.: Function of muscarinic and nicotinic acetylcholine receptors, Nature **261**:149, 1976.
116. Rall, W.: Core conductor theory and cable properties of neurons. In Kandel, E. R., editor: Handbook of physiology. Section 1: the nervous system, cellular biology of neurons, Bethesda, Md., 1977, American Physiological Society, vol. 1, pt. 1.
117. Rall, W., et al.: Dendritic location of synapses and possible mechanisms for the monosynaptic EPSP in motoneurons, J. Neurophysiol. **30**:1169, 1967.
118. Ramon y Cajal, S.: Neuron theory or reticular theory? Objective evidence for the anatomical unity of nerve cells (translated by M. U. Purkiss and C. A. Fox), Madrid, 1954, Consejo Superior de Investigaciones Cientificas.
119. Saavedra, J. M.: Distribution of serotonin and synthesizing enzymes in discrete areas of the brain, Fed. Proc. **36**:2134, 1977.
120. Sahyoun, N., Hollenberg, M. D., Bennett, V., and Cuatrecasas, P.: Topographic separation of adenylate cyclase and hormone receptors in the plasma membrane of toad erythrocyte ghosts, Proc. Natl. Acad. Sci. USA **74**:2806, 1977.
121. Saito, K., et al.: Immunohistochemical localization of glutamate decarboxylase in rat cerebellum, Proc. Natl. Acad. Sci. USA **71**:269, 1974.
122. Schmidt, R. F., Senges, J., and Zimmerman, M.: Determination of the peripheral receptive field and excitability measurements of the central terminals of single mechanoreceptive afferents, Exp. Brain Res. **3**:220, 1967.
123. Schon, F., and Kelly, J. S.: Autoradiographic localization of (^{3}H)GABA and (^{3}H)glutamate over satellite glial cells, Brain Res. **66**:275, 1974.
124. Schramm, M., Orly, J., Eimerl, S., and Korner, M.: Coupling of hormone receptors to adenylate cyclase of different cells by cell fusion, Nature **268**:310, 1977.
125. Schrier, B. K., and Thompson, E. J.: On the role of the glial cells in the mammalian nervous system, J. Biol. Chem. **249**:1769, 1974.
126. Schwartz, J. C.: Histamine as a transmitter in brain, Life Sci. **17**:503, 1975.
127. Shaskan, E. A., and Snyder, S. H.: Kinetics of serotonin uptake into slices from different regions of rat brain, J. Pharmacol. Exp. Ther. **175**:404, 1970.
128. Sherrington, C. S.: The integrative action of the nervous system, New Haven, Conn., 1906, Yale University Press.
129. Sidman, R. L., and Rakic, P.: Neuronal migration, with special reference to developing human brain: a review, Brain Res. **62**:1, 1973.
130. Snyder, S. H., Chang, K. J., Kuhar, M. H., and Yamamura, H. I.: Biochemical identification of the mammalian muscarinic cholinergic receptor, Fed. Proc. **34**:1915, 1975.
131. Sobel, A., Heidmann, T., Hofler, J., and Changeux, J.-P.: Distinct protein components from torpedo marmorata membranes carry the acetylcholine receptor site branes carry the acetylcholine receptor site and the binding site for local anesthetics and histrionicotoxin, Proc. Natl. Acad. Sci. USA **75**:510, 1978.
132. Somjen, G.: Evoked sustained focal potentials and membrane potentials of neurons and of unresponsive cells of the spinal cord, J. Neurophysiol. **33**:562, 1970.
133. Terzuolo, D., and Bullock, T. H.: Measurement of voltage gradient across a neuron adequate to modulate its firing, Proc. Natl. Acad. Sci. USA **42**:687, 1956.
134. Tosaka, T., Chichibu, S., and Libet, B.: Intracellular

analysis of slow inhibitory and excitatory postsynaptic potentials in sympathetic ganglia of the frog, J. Neurophysiol. **31**:396, 1968.

135. Uchizono, K.: Characteristics of excitatory and inhibitory synapses in the central nervous system of the cat, Nature **207**:642, 1965.
136. Udenfriend, S., and Dairman, W.: Regulation of norepinephrine synthesis. In Weber, G., editor: Advances in enzyme regulation, Oxford, Eng., 1971, Pergamon Press, Ltd., vol. 9.
137. Varnon, S. S., and Bunge, R. P.: Trophic mechanisms in the peripheral nervous system, Annu. Rev. Neurosci. **1**:327, 1978.
138. Weight, F. F.: Synaptic potentials resulting from conductance decreases. In Bennett, M. V. L., editor: Synaptic transmission and neuronal interaction, New York, 1974, Raven Press, p. 141.
139. Weight, F. F., and Weitsen, H. A.: Identification of small intensely fluorescent (SIF) cells as chromaffin cells in the bullfrog sympathetic ganglia, Brain Res. **128**:213, 1977.
140. Weiner, N., Lee, F. L., Waymire, J. C., and Posiuviata, M.: Regulation of tyrosine hydroxylase activity in adrenergic nervous tissue. In Wolstenholme, G. E. W., and Fitzsimmons, D. W., editors: Aromatic amino acids in the brain. CIBA Foundation Symposium No. 22 (new ser.), Amsterdam, 1974, Elsevier/North Holland Biomedical Press.
141. Weiss, P., and Hiscoe, H. B.: Experiments on the mechanism of nerve growth, J. Exp. Zool. **107**:315, 1948.
142. Wurtman, R. J., and Fernstrom, J. D.: Nutrition and the brain. In Schmitt, F. O., and Worden, F. G., editors: The neurosciences: third study program, Cambridge, Mass., 1974, The M.I.T. Press.

III

PRINCIPLES OF SYSTEM THEORY AS APPLIED TO PHYSIOLOGY

7 Systems and models

JAMES C. HOUK

The life of a multicellular organism such as man depends as much on the ability of cells, tissues, and organs to function in a cooperative manner as it does on the integrity of the individual cells. The performance of essential body functions is the beautiful result of a complex interaction between cells of different types, each specialized to provide small, but unique, contributions to the whole process that constitutes life. For example, cells of all types rely on the blood as a source of chemical nutrients and a sink for the disposal of metabolites. The suitability of the blood in the performance of this source and sink function is hardly an accident—it derives from the coordinated contributions of numerous cell types. Cells specialized for contraction circulate the blood; cells specialized for absorption procure a supply of nutrients; cells specialized for storage conserve nutrients in times of abundance and release them in times of need. A major problem in physiology is to understand how individual actions such as these are integrated into larger functions that then contribute to the survival of the organism. *System theory* is a science that deals specifically with complex interacting processes, and the purpose of this part of the book is to summarize those of its principles that are particularly helpful in understanding the physiology of the body.

This chapter introduces basic concepts in system theory and explains the rationale behind several of the methods used to construct models of systems. This material provides background for topics covered in many chapters of this book; in particular, it provides an introduction to the following chapter, where the principles of regulation and control are discussed and applied to homeostasis and other control problems in physiology.

ORIENTATION TO THE SYSTEM APPROACH

The system concept is a very general one. A system can be defined as any assemblage of objects united by processes of physical interaction that relate measured quantities one to another. The measurements of the relevant quantities are usually expressed as sets of values that are either continuous or discrete in time; these are variously called *signals, variables,* or time *functions*. System theory deals primarily with the relations between such variables, with much less emphasis on the actual physical nature of the quantities they describe.

In the study of the relations between variables, one usually comes to view some of the variables as being determined independently, by causes extrinsic to the system, and others as being dependent on these, although this distinction is sometimes arbitrary. The independent variables are called *inputs,* the dependent ones are called *outputs,* and the system is defined to include the particular collection of physical processes that intervenes to relate outputs to inputs. In this manner a small portion of the universe is parceled out to be contained within the boundaries of the system, and the remainder constitutes the environment (Fig. 7-1).

Another approach is to define the boundaries of the system first and then to identify relevant inputs and outputs. Outputs are quantities that derive from the *state* (or condition) of the system, whereas inputs are quantities that influence its state. Usually only a subset of the actual inputs and outputs of a system are designated as relevant ones. An input is considered relevant if it has an appreciable influence on the state of the system. Relevant outputs are selected as ones being particularly important or interesting, based on independent considerations.

The characterization problem

Once the inputs, outputs, and boundaries of a system have been defined, the next step in system analysis is to discover the relations between the inputs and outputs. The two basic approaches to this characterization problem are an empirical one, the experimental study of input-output relations, and a theoretical one, the derivation of the relations from first principles, for example, from

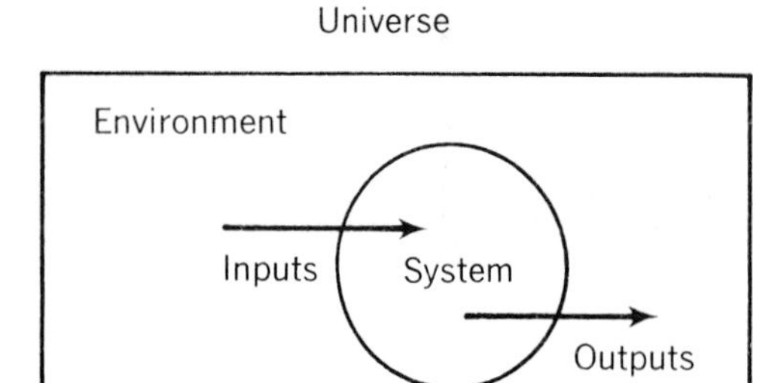

Fig. 7-1. System is pictured as small portion of universe selected for study. Inputs from environment influence state of system; outputs derive from its state.

basic laws of thermodynamics. Of course the origin of the latter is also empirical. In either case, one obtains a model of the system, expressed by one or more mathematical equations or by a set of graphs. A simple example of a mathematical model of a system is the Fick equation (Chapter 1, equation 3). Here one might consider the inputs to be the concentrations of an uncharged molecule on either side of a membrane, which is the system, and the single output to be the transmembrane flux of this molecule.

A key concept in system theory is the notion of an operator or transfer function. These equivalent terms arise from an abstraction—a view of a system as a process that operates on input variables to transform them into output variables. This notion provides a convenient way of thinking about a mathematical model of a system. The model is an equation that details the steps in the conversion process. Inputs are added, subtracted, multiplied, divided, integrated, and differentiated to convert them into outputs. The equation specifies which of these elementary mathematical operations are required together with the particular order in which they must be performed. The emphasis is on the mathematical operations rather than on the actual physics responsible for the operations. In contrast, contemporary physiology places more emphasis on the biophysical processes than on the mathematical relations used to describe them. However, most physiologists do not contest the advantage of using mathematics rather than words to describe a well-understood process, since equations are usually both more precise and more concise than are verbal descriptions.

Equations can also be written in a more abstract manner. In this case a whole sequence of elementary mathematical operations is represented by a single symbol, the operator or transfer function mentioned earlier. For example, a general equation characterizing any system is as follows:

$$y = S[x] \tag{1}$$

Here it is assumed that the input (x) is a vector consisting of a set of input functions ($x_1, x_2, x_3, \ldots x_m$) and, similarly, the output (y) is a vector consisting of a set of output functions ($y_1, y_2, y_3, \ldots y_n$). S is an operator that symbolizes whatever sequence of elementary mathematical operations is required to transform the inputs into outputs. The use of complex operators and vector notation greatly simplifies the appearance of an equation. However, a definition must be given that details the precise meaning of the operator symbol and the system variables. For example, we might designate x as a two-dimensional input that represents the molecular concentrations on either side of a membrane, assign y to the molecular flux, and let S represent subtraction of the two inputs and multiplication by a constant (K) that includes the diffusion coefficient and membrane dimensions. This defines the Fick equation:

$$y = K(x_1 - x_2) \tag{2}$$

Both equations 1 and 2 are now suitable mathematical characterizations of a membrane system providing passive transport of a molecular species.

In this particular example, not much space is saved by writing equation 1 instead of equation 2, but the savings can be appreciable when the system is more complex. Actually, the important reason for using operator notation goes beyond saving space; it is to encourage abstract thinking about a system as being characterized by a single complex operation.

Prediction problems

The mathematical characterization of a system can be used to predict the time course of outputs of a system in response to known input functions. The block diagram of Fig. 7-2, *A*, portrays this application schematically for the general system specified by equation 1. The basic idea is to operate on the input functions (x), which are given, with the system model (S), which is known. This yields the corresponding output functions (y). Some of the specific techniques that are used to implement this procedure will be discussed later.

Sometimes the outputs of a system are known and the problem is to deduce the inputs that produced them, which is the important problem in diagnosis. Here it is useful to introduce the con-

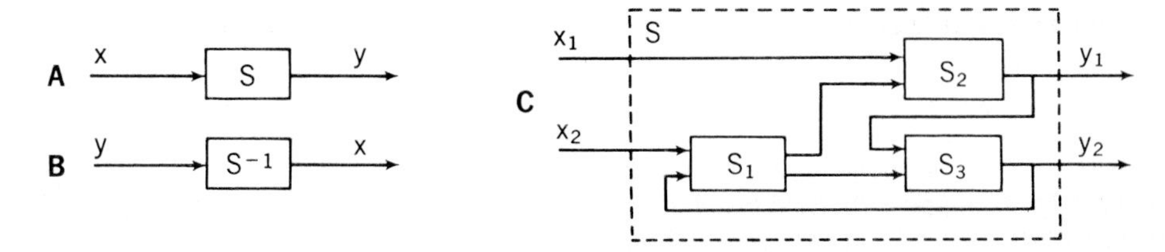

Fig. 7-2. Problems addressed by system theory. **A,** System model *S* can be used to predict response *y* to any input *x*. **B,** Inverse of system model (if a unique one exists) can be used to calculate input *x* that caused some measured output *y*. **C,** Models of subsystems *(S_1, S_2, S_3)* can be combined to obtain model of overall system *S (synthesis);* alternatively, one may attempt to specify subsystems that underlie measured properties of overall system *(analysis)*.

cept of inverse operations. The inverse operation S, written S^{-1}, is an operation that undoes the original one. Thus if S describes the conversion of an input (x) into an output (y), S^{-1} describes a process that converts y back into x (Fig. 7-2, *B*). As a simple example, division is an operation that is the inverse of multiplication. However, inversion of an operator is often much more complex than this, and in some cases a unique inverse does not exist. For example, a system that squares the input to yield the output has no unique inverse. Knowing the output (y) indicates only that the magnitude of the input is the square root of y; its sign may be either plus or minus. Another problem arises when a system receives multiple inputs, in which case any given output can be produced by several different combinations of these inputs. Here the observer needs to measure internal variables of the system in order to determine the particular input combination responsible for the observed output.

Synthesis and analysis

Problems of synthesis and analysis are among the most important issues to which system theory is addressed. These concern the study of relations between relations, that is, the manner in which the input-output relations of subsystems combine to determine the overall input-output relations of a larger system. A physiologic example of an analysis problem is that of determining the factors responsible for the overall pumping characteristics of the heart, and an example of a synthesis problem is that of deriving the characteristics of blood pressure regulation from the characteristics of the heart and other components of the cardiovascular system. The solutions to such problems are especially important, since they often lead to a new degree of understanding of the systems in question, such as a knowledge of which factors are crucial in determining the limitations of overall performance.

A block diagram illustrating the analysis-synthesis problem is given in Fig. 7-2, *C*. The properties of the overall system are designated by the operator (S) and those of the component subsystems by the subscripted operators. The synthesis problem is to specify S in terms of S_1, S_2, and S_3. In special cases, such problems can be solved in general, yielding an equation for S in terms of the component operators with few or no assumptions as to their specific characteristics. In Chapter 8 the principles of negative feedback will be demonstrated with a solution of this type. The analysis problem, that of determining the subsystems constituting an overall system (S), is much more difficult. In general there will be a variety of combinations of subsystems, each of which gives rise to the same overall system properties. The discovery of which combination is correct requires additional knowledge of the internal workings of the system.

SYSTEM CHARACTERIZATION (MODEL BUILDING)

Certain concepts, techniques, and terminology that pertain to the problems of obtaining and using mathematical characterizations of systems are widely used in physiology, and the student needs to have some understanding of them. Here the treatment is concise, and the interested reader is referred to the texts on these topics listed in the bibliography for more complete expositions.

Block diagrams are schematic summaries that portray signal flow in a system along with selected information concerning the properties and boundaries of systems or subsystems. They are often used in conjunction with physical diagrams to aid in the definition of a system together with the relevant variables. Block diagrams are also used to define the relationships between subsystems and the overall system. When block diagrams are sufficiently detailed, they can serve as complete mathematical characterizations of a system.

As a simple example, consider the system portrayed by the physical diagram of Fig. 7-3, *A*, which consists of a cell suspended in a reservoir of nutrient medium. Suppose that the problem is to characterize the dependence of the internal concentration (c_2) of some uncharged molecular substance on the external concentration (c_1) and the rate (v) at which it is utilized by cellular metabolism. The overall block diagram of this cell reservoir system, shown in Fig. 7-3, *B*, identifies the important input and output variables. Further contemplation of the problem leads to the more detailed diagram of Fig. 7-3, *C*, which decomposes the overall system into three well-understood processes, while also introducing two important internal variables—the flow rate (f) of the substance across the cellular membrane and the amount (n) that is accumulated inside the cell. Each of these processes is characterized by a simple equation. The transport process is described by the Fick equation:

$$f = K(c_1 - c_2) \tag{3}$$

where K is the transport coefficient defined earlier (equation 2). Intracellular accumulation of the substance is described by a mathematical statement of the conservation of matter:

$$n = \int (f - v)dt \tag{4}$$

The amount (n) in the cell is a cumulative (integration) result of the difference between inflow (f) and utilization rate (v). The final output of the system is determined by the third process, in which internal concentration is specified by the dilution principle:

$$c_2 = \frac{1}{\theta} n \tag{5}$$

where θ is the intracellular volume.

The detailed block diagram shown in Fig. 7-3, *D*, can be constructed from these equations and the block diagram of Fig. 7-3, *C*. Here the system is decomposed into a sequence of five elementary mathematical operations, thus providing a detailed mathematical characterization of the cell reservoir system suitable for computer simulation.

Differential equations constitute another commonly used format for a system model. The particular equation can be selected empirically, or it can be derived from a set of equations known to describe the properties of the relevant subsystems (the synthesis problem). As a simple example of this latter approach, equations 3 and 5 describing membrane transport and the dilution principle can be substituted into equation 4, the conservation expression, to obtain a differential equation that characterizes the cell reservoir system of Fig. 7-3:

$$\theta c_2 = \int [K(c_1 - c_2) - v]dt \tag{6}$$

Differentiation of both sides of equation 6 yields:

$$\theta \dot{c}_2 = K(c_1 - c_2) - v \tag{7}$$

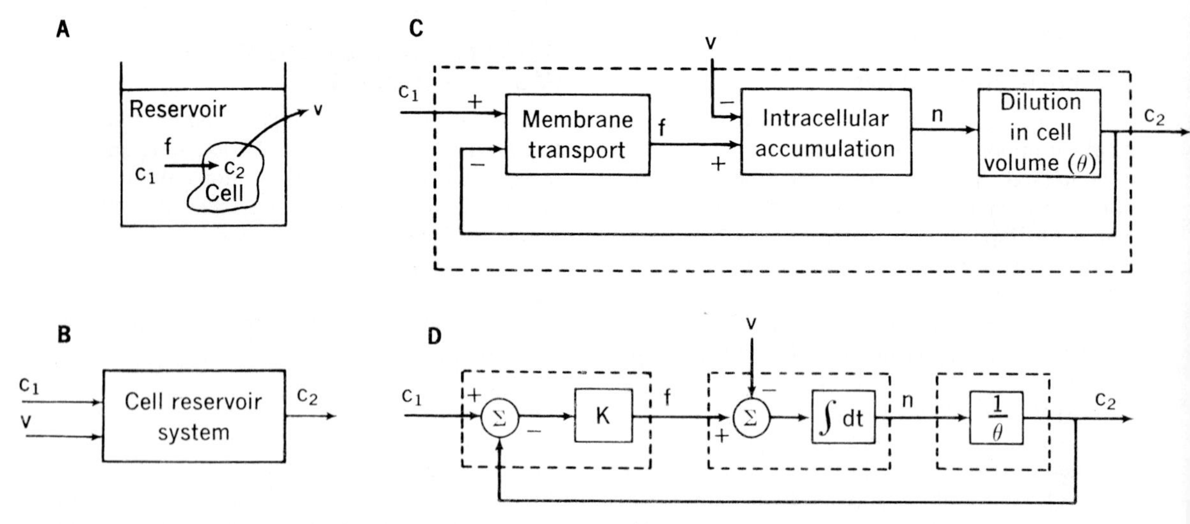

Fig. 7-3. Different diagrams of cell reservoir system. **A,** Physical diagram. **B,** Overall block diagram. **C,** Block diagram decomposing system into three biophysical processes. **D,** Block diagram decomposing system into five elementary mathematical operations. Variables: C_1, reservoir concentration; c_2, cellular concentration; *f*, transport rate; *n*, amount accumulated intracellularly; *v*, utilization rate. Parameters: *K*, transport coefficient; θ, cytoplasmic volume.

The convention $\dot{c}_2$ to represent the time derivative of the variable c_2 will be used throughout this chapter. Usually the terms in the equation are arranged to place the output (dependent variable), in this case c_2, on the left and the inputs (independent variables) on the right:

$$\theta\dot{c}_2 + Kc_2 = Kc_1 - v \tag{8}$$

The following example illustrates how a system equation of this type can be used to solve relevant problems.

Example 1: Time course of a hormonal action

Consider a suspension of cells of a type known to utilize a certain substrate at a rate controlled by the concentration of a hormone in the bathing medium. The problem is to determine the time course of this hormonal action, and let us assume that the substrate utilization rate is inaccessible to direct measure and must be inferred from measurements of intracellular concentration. An assumption that the cells are homogeneous allows the cell reservoir model (Fig. 7-3 and equation 8) to be applied to this problem.

The first step is to evaluate θ and K, which are characteristic parameters of the system. Suppose that the net intracellular volume (θ) is estimated from the volume of cells settling on centrifugation, whereas the transport coefficient (K) is to be derived from a control experiment in which the cells are suddenly (at time zero) transferred from a medium having initial substrate concentration C_{1_i} to a medium having final substrate concentration C_{1_f}. In both media the hormone is absent and the utilization rate is assumed zero; thus equation 8 becomes:

$$\frac{\theta}{K}\dot{c}_2 + c_2 = c_1 \tag{9}$$

This first-order (contains no time derivatives higher than the first) linear differential equation has a solution of the form:

$$c_2(t) = A + Be^{-t/T} \tag{10}$$

for $t \geq 0$, where $T = \theta/K$. The constants A and B are evaluated from initial and final conditions, both of which involve an equilibrium with $c_2 = c_1$, since transport is passive. Thus $c_2(0) = A + B = C_{1_i}$ and $c_2(\infty) = A = C_{1_f}$; from which $B = C_{1_i} - C_{1_f}$; finally, substitution into equation 10 yields:

$$c_2(t) = C_{1_f} + (C_{1_i} - C_{1_f})e^{-t/T} \tag{11}$$

The solid curve in Fig. 7-4, *A*, is a graph of this solution, the predicted change in internal concentration in response to a stepwise change (dashed lines) in external concentration. Internal concentration decays from the initial to the final value with an exponential time course, the speed of which is determined by the parameter T, called the *time constant*. T is the time taken for the decay to be 63% complete, and it serves as a convenient index for the time scale in Fig. 7-4. As a general rule, time constants present in responses derive from the characteristic parameters of the system, in this case the ratio between the volume and the transport coefficient (θ/K in equation 9). Thus the experimental measurement of T from records obtained in the control experiment allows K to be evaluated from the ratio θ/T.

Now consider the effect of adding hormone to the bathing medium. The subsequent stimulation of substrate utilization will lead to a reduction in intracellular substrate concentration, but the change in c_2 will not occur as rapidly as the change in v. This is because the reduction in concentration must await the progressive depletion of intracellular substrate molecules, which is slowed by the time constant of the system. Despite this, the actual time course of the hormonal stimulation of utilization rate can be estimated from the observed $c_2(t)$, if an appropriate correction is made. Solution of equation 8 for v specifies what that correction must be:

$$v(t) = -K[c_2(t) + T\dot{c}_2(t) - C_1] \tag{12}$$

The initial negative sign denotes that v and c_2 are reciprocally related—a fall in c_2 results from a rise in v. The addition of the first derivative, $\dot{c}_2(t)$, weighted by T, is the factor that compensates for the slowness of $c_2(t)$ as an indicator of $v(t)$. The upper case symbol (C_1) is used to indicate that reservoir substrate concentration is held constant.

Fig. 7-4, *B*, provides a hypothetical example in which the solid curve represents an observed time course of c_2 in response to a sudden addition of hormone to the bathing medium. The dashed curve was calculated from equation 12 to predict the time course of hormonal action in controlling substrate utilization.

In summary, this example illustrates how system parameters and internal variables that are inaccessible to direct measurement can be deduced from other

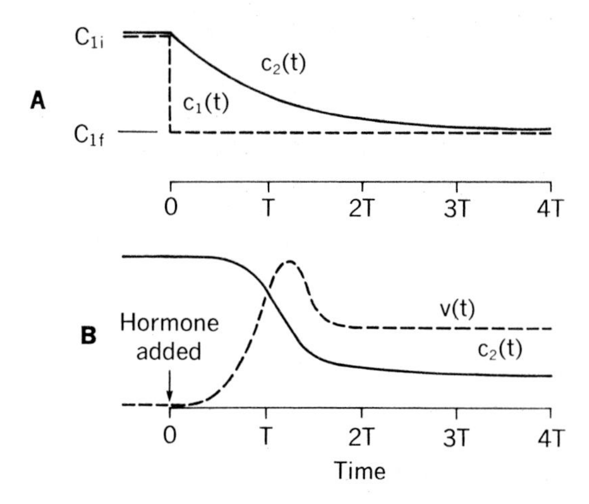

Fig. 7-4. Responses $c_2(t)$ of cell reservoir system to, **A,** step change in reservoir concentration $c_1(t)$ and, **B,** change in utilization rate $v(t)$ provoked by addition of hormone at time 0. C_{1i}, Initial reservoir concentration; C_{1f}, final reservoir concentration; *T*, time constant.

more easily measured responses. It is important to stress that results obtained in this manner are only as valid as the model from which they are calculated.

Special features of linear systems

Linearity is an ideal property created to facilitate mathematical analysis. Although real systems are never linear, they may approximate linear behavior, and important insight concerning actual system behavior can frequently be obtained from an analysis based on the assumption of linearity. Linear models are particularly useful because they obey the principle of *superposition* (or additivity) and the corollary principle of *proportional scaling* (or homogeneity). The latter guarantees that larger input functions will yield outputs that are larger by precisely the same proportion. Thus there is no need to analyze responses to different amplitudes of input. The principle of superposition guarantees that the response to a given input, or set of inputs, can be obtained by adding up (superimposing) the responses to the members of any decomposition of the inputs into component parts. For example, the response of the linear model of the cell reservoir system (Fig. 7-3, *D*) to a simultaneous change in reservoir concentration (c_1) and utilization rate (v) can be obtained by summing the responses to each individually. Alternatively, each of the input functions can be decomposed further into any number of components in order to facilitate analysis.

One widely used decomposition of an input signal is in terms of a sequence of brief pulses called impulses or Dirac delta functions. As shown in Fig. 7-5, *B*, the impulses are imagined to occur at regular intervals and have amplitudes that reflect the amplitude of the input during the corresponding interval. Presume that the response to an impulse of some standard amplitude, called an impulse response, is known from a prior measurement (e.g., Fig. 7-5, *C*). The response to the present input then can be calculated by summing appropriately scaled and delayed impulse responses, as illustrated in Fig. 7-5, *D*. This example is convenient for illustrative purposes, but it fails to show that the actual decomposition of the input is usually in terms of an infinite series of infinitesimally spaced impulses. Thus the summation portrayed in Fig. 7-5, *D* is replaced by an integration according to an equation that is called the *convolution integral:*

$$y(t) = \int_{-\infty}^{t} x(t')h(t - t')dt' \qquad (13)$$

The function h(t) represents the standard impulse response and $x(t')h(t - t')$ is a scaled and delayed version. (The argument $t - t'$ signifies delay by an amount t'.) The output at time t is calculated by adding up (integrating) the output components resulting from impulses at all preceding times.*

The convolution approach is extremely powerful due to the fact that the input-output behavior of a linear system is fully characterized by a single function, the impulse response. Once an impulse response has been measured with a single observation, or derived from differential

*The formulations of the convolution integral and of Fourier integrals given here presume time invariance (to be discussed later) as well as linearity. However, alternative formulations are available for use when the system is time varying.

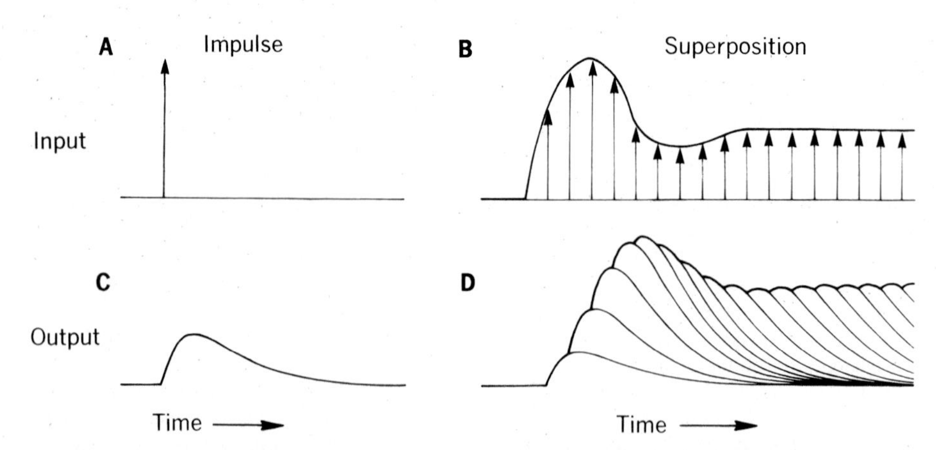

Fig. 7-5. Schematic explanation of superposition integral. **A,** Impulse function. **B,** Decomposition of input waveform into series of impulses. **C,** Response to single impulse. **D,** Response to input waveform in **B** obtained by superimposing (adding up) responses to constituent impulses.

equations describing the system, the response to any input can be calculated with the convolution integral.

Another decomposition of an input signal that is widely used is one in terms of a set of sinusoidal functions. This decomposition is accomplished by applying an equation called the *Fourier integral:*

$$X(\omega) = \int_{-\infty}^{\infty} x(t)e^{-j\omega t}\,dt \qquad (14)$$

This integral is said to perform a Fourier transformation, that is, the conversion of a time function x(t) into a frequency function $X(\omega)$. In effect the equation decomposes an input signal x(t) into an infinite set of sinusoidal functions, the amplitudes and phases of which are described by the function $X(\omega)$. The argument ω is the sinusoidal frequency in radians per second ($\omega = 2\pi f$ where f is the frequency in cycles per second or Hertz).

The exponential function $e^{\pm j\omega t}$, used in equation 14 and later in equation 15, is simply a compact representation for a sinusoidal function. The term $j = \sqrt{-1}$ signifies the imaginary part of a complex number. Although complex number theory is required for effective use of Fourier integrals, here the purpose is only to indicate the rationale behind such equations. The reader is encouraged to think of the Fourier transformation abstractly, simply as a special mathematical operation that decomposes a time function into sinusoidal components. The next important question concerns the manner in which a system will respond to these sinusoidal components.

A linear system, or, more precisely, a linear model of a system, is unique in that it will always respond to a sinusoidal input with a sinusoidal output having the same frequency as the input. However, the output sinusoid may have a different amplitude and phase than the input, as illustrated in Fig. 7-6. The response to a particular frequency of input is thus characterized by a gain (or attenuation) factor and a phase shift. When different frequencies of input are applied, different gain factors and phase shifts are obtained; data of this sort are usually summarized by a set of graphs referred to collectively as a *Bode plot,* or *frequency response.* The ordinates of these graphs are the logarithm of the gain and a linear representation of the phase shift, and the common abscissa is the logarithm of the frequency. An example is shown in Fig. 7-7.

The input-output properties of a linear system are completely characterized by a Bode plot, although this presumes that an adequate range of frequencies was used to obtain the graphs. The output of the system in response to an input x(t), having a transform $X(\omega)$, can be calculated from an equation called the *inverse Fourier integral:*

$$y(t) = \frac{1}{2\pi}\int_{-\infty}^{\infty} H(\omega)X(\omega)e^{j\omega t}d\omega \qquad (15)$$

where $H(\omega)$ is a function that symbolizes the gain and phase data contained in a Bode plot. In essence, equation 15 calculates the output to each individual frequency of input as the product $H(\omega)X(\omega)$, converts these to sinusoidal time functions with the multiplier $e^{j\omega t}/2\pi$, and summates the individual output components by the process of integration. Thus the technique of sinusoidal analysis, like the convolution approach, is based on the principle of superposition.

The use of *Laplace transformations* represents a simple modification of the Fourier transformation approach, wherein the complex argument $s = \sigma + j\omega$ is used instead of the purely imaginary $j\omega$. This has certain mathematical advantages, but, in fact, there is little difference between the two techniques. The transfer function H(s) obtained with the Laplace approach is the same as the one $H(\omega)$ obtained from the Fourier approach when the substitution $s = j\omega$ is performed.

It may appear that the convolution and Fourier transformation approaches represent separate and unrelated methods for linear system characterization. Actually, they are intimately related, since the function $H(\omega)$ summarizing a Bode plot is equivalent to the Fourier transformation of the

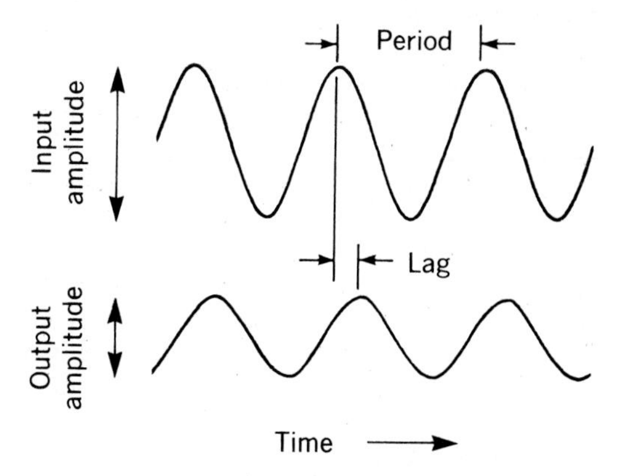

Fig. 7-6. Response of linear system to sinusoidal input. In this case, output sinusoid is attenuated (gain less than 1), and phase lag is introduced. Frequency of sinusoid is 1/period.

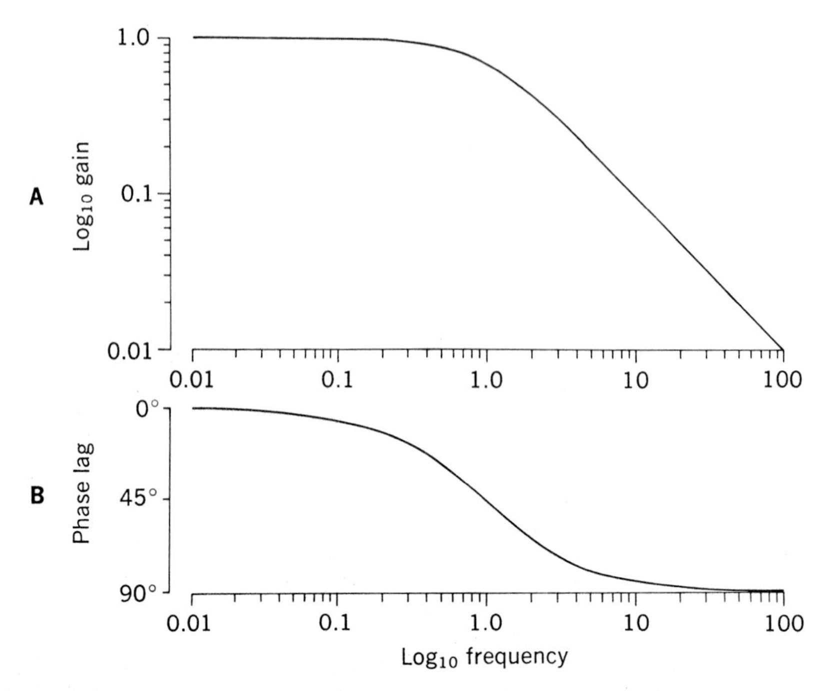

Fig. 7-7. Bode plot (or frequency response) summarizing sinusoidal responses of system that attenuates high-frequency inputs. System equation is $T\dot{y} + y = x$, where $T = \frac{1}{2}\pi$.

impulse response h(t). This equivalence derives from the fact that the Fourier transformation of a standard impulse consists of an infinite set of sinusoids, each having the same standard amplitude and a zero phase shift. Thus applying an impulse to a system is like applying simultaneously all the sinusoidal functions that are required to obtain a Bode plot. The Fourier transformation of these simultaneous sinusoidal responses extracts the gain and phase information corresponding to each frequency. An operator (e.g., H) is a noncommital way of referring to a system characterization, since it can symbolize either the frequency response, $H(\omega)$ together with the inverse Fourier integral or the impulse response h(t), together with the convolution integral.

Given that an impulse response reveals the entire frequency response of a system in a single measurement, one may question the utility of ever determining frequency response by the more tedious method of applying individual sinusoids. The two major advantages of the latter procedure are practical ones—first, more precise measurements usually can be obtained by applying one frequency at a time and, second, sinusoids are gentle inputs in comparison with an impulse, since the latter is equivalent to an abrupt shock and may disrupt normal system performance or even damage the system. However, these arguments are not entirely compelling, since the added precision may not be required and the potentially disruptive effect of an impulse can be circumvented by using a milder transient input. For example, one can derive the impulse response of a system from the response to a step input function, the rationale being as follows. A step function is the time integral of an impulse. Thus the response obtained is the time integral of the impulse response. The latter can be obtained by differentiating the step response, since differentiation is an operation inverse to integration.

The example just given invokes another important property of a linear system (provided it is also time invariant), which is that the order of operations may be interchanged freely without altering the result. The procedure of inverting the order of operations is called *commutation*. Fig. 7-8 details the logic involved in the application of this rule to the previous example. In *A* the differential operator (a linear operation) is used simply to indicate the definition of an impulse as the time derivative of a step function, and the operator (H) symbolizes the properties of a linear system. The commutation of these two operations in *B* demonstrates that the impulse response can be obtained by differentiating the step response.

In summary, the key properties of a linear model are superposition, proportional scaling,

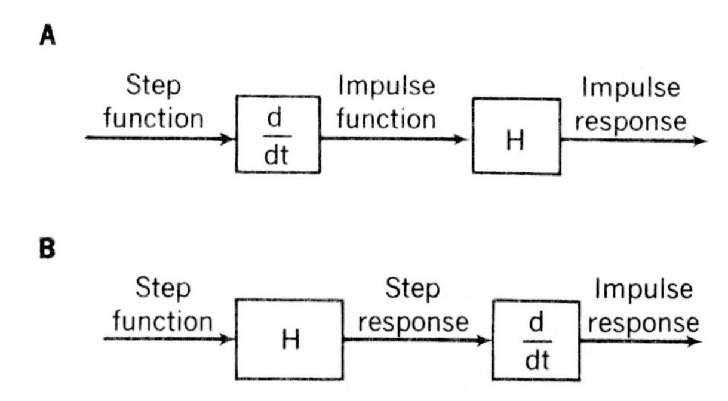

Fig. 7-8. Application of principle of commutation. Here, operation of differentiation is commuted with system operator *H* to demonstrate that impulse response can be obtained by differentiating step response.

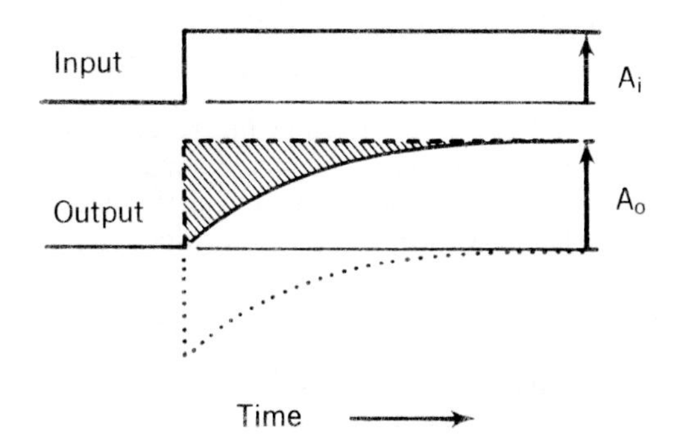

Fig. 7-9. Static (dashed) and dynamic (dotted) components of transient response (solid output trace). Steady-state gain is the ratio A_o/A_i.

and commutability. The superposition property is the basis of the two most commonly used techniques in systems analysis—the convolution approach, which decomposes input signals into a series of impulses, and the Fourier (or Laplace) transformational approach, which decomposes inputs into a set of sinusoids. The discussion given here has emphasized the rationale behind these techniques; their application to the characterization of dynamic properties of systems is introduced in the next section and further illustrated in example 2.

Dynamic and static properties

Systems are generally dynamic, which means that the outputs are functions of both present and past values of the inputs. If the outputs can be specified adequately based only on present values of inputs, the system is said to be static. The properties of static systems are characterized by algebraic relations, whereas differential or integral operations (or a delayer) are required to characterize dynamic systems.

Fig. 7-9 illustrates an example of a response of a dynamic system to a step input. The entire response (solid trace), called a *transient response,* consists of a transient phase, during which the output continues to change even though the input remains constant, and a *steady-state response,* to which the output eventually settles. A transient response is often subdivided into static and dynamic components, respectively illustrated by the dashed and dotted traces in Fig. 7-9. The static component of response is defined as the output of a static model of the system, a model in which the steady state is achieved immediately rather than progressively over time. The *dynamic component* is the remainder, usually a transient that decays toward zero when the input is held constant. The exception occurs in a system that performs pure integration, since the response of an integrator to a constant input is an output that grows larger continuously. There is no static model of an integrator; it is purely dynamic.

The relative importance of dynamic and static components of response depends on the rate of change of an input. A system may achieve a steady state continuously if the input changes slowly, in which case the response will reveal static properties. With rapid fluctuations of an input the response is usually dominated by dynamic properties. It is for this reason that the frequency response of a system generally reveals static properties at low sinusoidal frequencies and dynamic properties at high frequencies. For example, the Bode plot of Fig. 7-7 characterizes a system with a marked drop-off in gain at high frequencies, and this suggests that the response to an abrupt input such as a step would be sluggish. Fig. 7-9, which happens to illustrate the step response of a similar system, confirms this prediction. In contrast, a system having an increased gain at high frequencies would respond to a step input with overshoot (example 2, p. 237).

Gain is a term denoting the amplitude ratio between output and input. If a system is both static and linear, its gain is a constant multiplier, but this is a unique situation. The introduction of dynamics makes gain a function of frequency (as seen earlier), nonlinearity makes it a function of amplitude, and time variance makes it a function of time. When a system is both dynamic and linear, the gain at low sinusoidal frequencies characterizes the static properties of the system. This multiplier is called the *steady-state gain,* since it is also defined by the ratio of output to input in the steady state (e.g., A_o/A_i in Fig. 7-9).

Time variance and nonlinearity

If the parameters of a system (e.g., the transport coefficient K and the cytoplasmic volume θ in Fig. 7-3) do not change with time, the system is said to be time invariant (or stationary), a property that has been tacitly assumed until now. Frequently, however, these parameters are themselves variables, as is the case, for example, if a hormone released by some source extrinsic to a cell causes changes in membrane permeability. Fig. 7-10, *A,* illustrates the manner in which this time-varying system can be characterized with a block diagram. Here k represents the transport coefficient formerly designated as K—the lower case symbol is used to emphasize that this quantity is now a variable, a new input to the system, rather than a constant multiplier. Note also the difference between the summation symbol and the multiplying symbol.

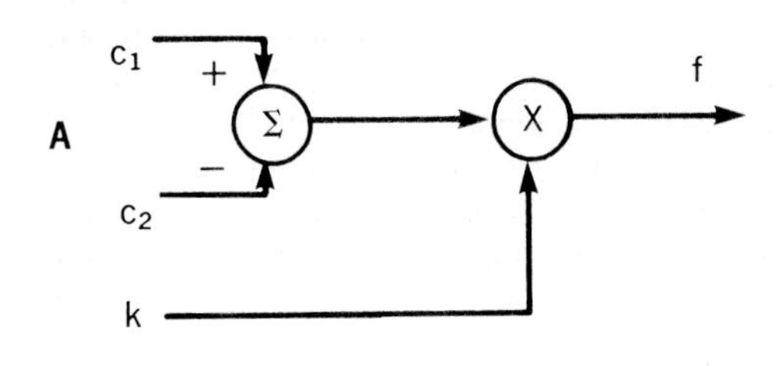

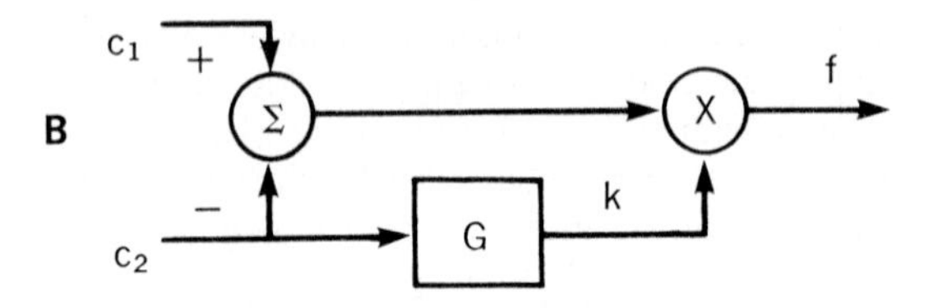

Fig. 7-10. Examples of time-varying and nonlinear systems. If *k* were constant, this transport system would be linear and time invariant. In **A,** *k* is an independent variable, making system time varying. In **B,** *k* changes whenever c_2 changes, making system nonlinear.

If the permeability of the cell membrane were a function of the internal concentration (c_2) rather than an independent variable, the system would no longer be classified as time varying, but it would be nonlinear. Fig. 7-10 characterizes this problem with a block diagram in which the dependence of the transport coefficient on c_2 is designated by the operator G. This operator could be a constant multiplier or it could be a more complex function; let us presume the former. The nonlinearity of the system is demonstrated by the equation relating output to inputs, which can be written from an inspection of Fig. 7-10.

$$f = Gc_2(c_1 - c_2) = G(c_1c_2 - c_2^2) \quad (16)$$

It is clear, for example, that the response to a step increase in internal concentration will not scale in proportion to the input amplitude, and it can also be shown that superposition does not hold. Lacking these defining properties of linearity, the system is nonlinear by default.

Continuity in space and time

Frequently the variables associated with a system are continuous functions of both space and time, in which case the system is said to be *distributed* (or continuous in space) and *continuous* (in time). For example, the concentration of a substance in a tissue generally varies with the particular site as well as with time. A precise characterization of this situation requires partial differential equations, the solutions to which may be complex. Another approach is to approximate the continuous spatial variations with a set of discrete ones by lumping the parameters of the medium into a small number of compartments. Such *lumped-parameter approximations* are widely used in system theory and in physiology, since they are simpler to deal with and, in most cases, yield results that are sufficiently accurate. Thus one usually treats the intracellular and interstitial spaces as two lumped compartments, the concentration in each being the spatial average throughout that compartment. Variations in time are usually treated in a continuous manner, although they can also be approximated by difference equations that lump time into epochs, in which case the characterization is said to be discrete (in time).

The treatment of systems as being lumped or discrete is quite natural in some instances. For example, the movement of an inertial mass is a lumped problem, since the mass is a rigid body and the motions of one point (its center of gravity) accurately characterize the motions of the entire mass. The operations performed by a digital computer are naturally characterized as discrete events in time. Action potentials of nerve cells are also discrete events, but it is often advantageous to deal with the frequency at which they occur as a continuous variable.

Value of approximations

Generally speaking, systems are distributed, continuous, nonlinear, dynamic, and time varying, whereas the properties lumped, discrete, linear, static, and time invariant are mutually independent abstractions created to facilitate analysis. Fortunately, problems can still be solved when systems do not conform to these simplifying abstractions, but this may entail the use of so-called brute force techniques such as *computer simulation* or *numeric analysis*. Solutions obtained with these techniques usually lack generality; for example, it may be necessary to compute from the start the response to an input that is only slightly different from another for which a solution has just been obtained. This, of course, greatly interferes with one's ability to grasp the general significance of the operations performed by a system. For this reason, it is often preferable to accept simplifying assumptions, provided it is realized that the insights to which they may lead are only approximations of the behavior of the real world. In other cases the essential features of a system may derive from nonlinear, time-varying, or distributed properties, and their neglect may lead to quite erroneous conclusions. The proper choice requires good judgment and careful analysis.

A good example of a problem in which computer simulation is mandated by the complexity of the model is discussed in Chapter 3 (Figs. 3-26 and 3-27).

Example 2: Phasic responsiveness of sensory receptors

A characteristic feature of many receptor organs is that they show greater sensitivity to rapidly changing inputs than to slower ones, a dynamic property referred to as phasic responsiveness. In some cases, phasic responsiveness is well modeled by a linear differential equation, but in others, nonlinear equations are required. Examples will be given of each.

Equation 17 provides one of the simplest linear models of a system having phasic responsiveness:

$$\dot{r} + ar = b\dot{x} + cx \qquad (17)$$

The variables r and x represent the discharge rate of the receptor and the stimulus intensity, respectively, and the coefficients a, b, and c are constant parameters. The response of this model to a step change in stimulus intensity is:

$$r(t) = AK(1 + Be^{-t/T}) \qquad (18)$$

for $t \geq 0$. (In most cases the intermediate steps in the derivations will not be provided, since the purpose of this chapter is to teach concepts rather than techniques.) The definitions of the new constants are as follows: A is stimulus intensity, $K = c/a$ is the static gain of the receptor, $B = ab/c$ is an overshoot factor that is a measure of phasic sensitivity, and $T = 1/a$ is the time constant of the exponential decay. A plot of the step response (Fig. 7-11) illustrates the initial overshoot at the onset of the stimulus and the exponential decay to the steady state, the amplitude of which equals that of the input, since K was set equal to 1.

How can one go about determining if this model adequately describes the dynamic properties of a receptor? The first step is to adjust the parameters of the model to obtain the best fit to a given response. If the fit is still poor, a more complex model must be chosen. Next, one uses the model (now without altering the parameters) to predict responses to other stimuli, which are then compared with actual responses of the receptor. One way of doing this is to differentiate the step response to obtain the impulse response (see Fig. 7-8). The convolution integral (equation 13) then is used to compute the responses to new input functions. The ramp and sinusoidal responses shown in Fig. 7-11, *B* to *D*, were obtained in this manner. In effect, what is being carried out is a test of the superposition hypothesis. If there is good agreement with actual data, superposition holds, and one concludes that the receptor behaves linearly. Poor agreement probably indicates nonlinear behavior, but it may indicate important time-varying or distributed properties.

Several general points are illustrated by the responses to the various input functions in Fig. 7-11, *A* to *D*. First, note that the amount of overshoot is greatest in response to the step, less with the fast ramp, and least with the slow ramp. This illustrates the point made earlier, that rapidly changing inputs bring out the dynamic properties of a system, whereas slowly changing ones show mostly static properties. Second, note that the exponential component that appears in both the rising and falling phases of the ramp response has the same time constant as does the exponential decay in the step response. This illustrates a general property of linear systems—time constants are characteristic features of the system and merely take on alternative weightings when different input functions are used. The same point is illustrated by the transient phase of adjustment to the sinusoidal input, which shows an exponential term with the same time constant. It is worth emphasizing that the gain and phase parameters that one uses in constructing a Bode plot are measured only after this transient phase dies out. In the example of a sinusoidal response provided the frequency is sufficiently high to illustrate the phase advance and enhanced amplification characteristic of phasic responsiveness. At a lower frequency there

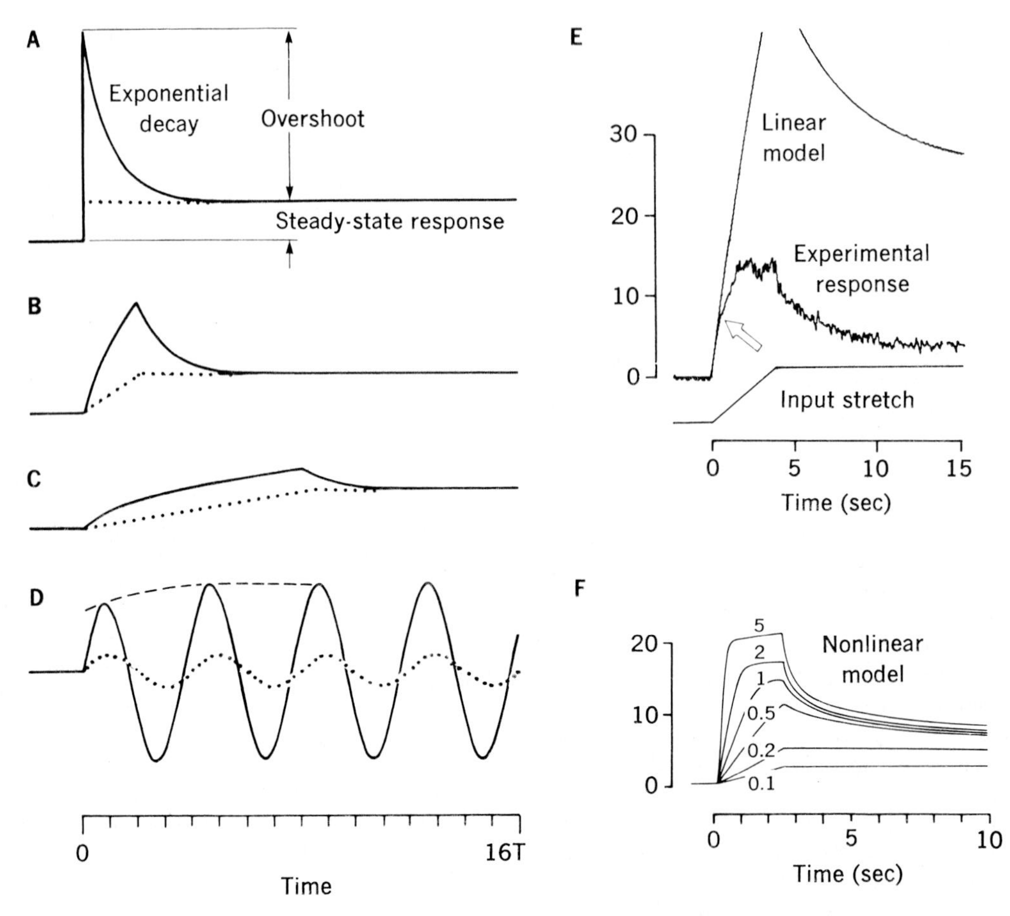

Fig. 7-11. Examples of systems showing phasic sensitivity. **A** to **D** are responses (solid traces) to different input waveforms (dotted traces) of a system characterized by equation 17. Steady-state gain K = c/a was set equal to 1, and time scale is relative in terms of time constant T = 1/a. **E** compares response (increment in discharge rate) of linear model of spindle receptor (primary ending) with experimental response to ramp stretch of intermediate amplitude.[14] Linear model, based on small-amplitude sinusoidal stretches, predicts experimental response accurately up to the point indicated by arrow; beyond this, receptor behavior becomes markedly nonlinear. **F** shows responses of nonlinear model (equation 19) of spindle receptor to ramp stretches having plateau amplitudes in the range 0.1 to 5 mm as indicated.[15] Model response to 2 mm stretch captures salient nonlinear features of experimental response shown in **E.**

would be no phase advance and the amplification would equal the steady-state gain, which in this case is K = 1.

Response properties of Golgi tendon organs are approximated by linear models of the type just illustrated, although additional terms are required to characterize adequately responses to both slow and fast inputs.[16] A variation on the same approach has been applied to muscle spindle receptors.[14,17,19] In this case, sinusoidal inputs were used to obtain the initial characterization of the receptor, in terms of a Bode plot, and the receptor was then tested for linearity with transient inputs. Reasonable agreement was obtained only when inputs were restricted to very small amplitudes. The failure that was shown between the prediction of the linear model and the actual response with an input of intermediate amplitude is illustrated in Fig. 7-11, *E*.

At present, no one has devised a general model of the spindle, that is, one capable of predicting its responses to a variety of inputs. The reason for this difficulty is that the spindle shows several complex nonlinear features. A tentative model[15] that captures some of these features is given by equation 19, and examples of its responses to ramps of several amplitudes are shown in Fig. 7-11, *F*.

$$\dot{r} + a(r - bx)^5 = c\dot{x} \qquad (19)$$

Since the different inputs (amounts of stretch) are simply scaled versions of each other, the fact that the

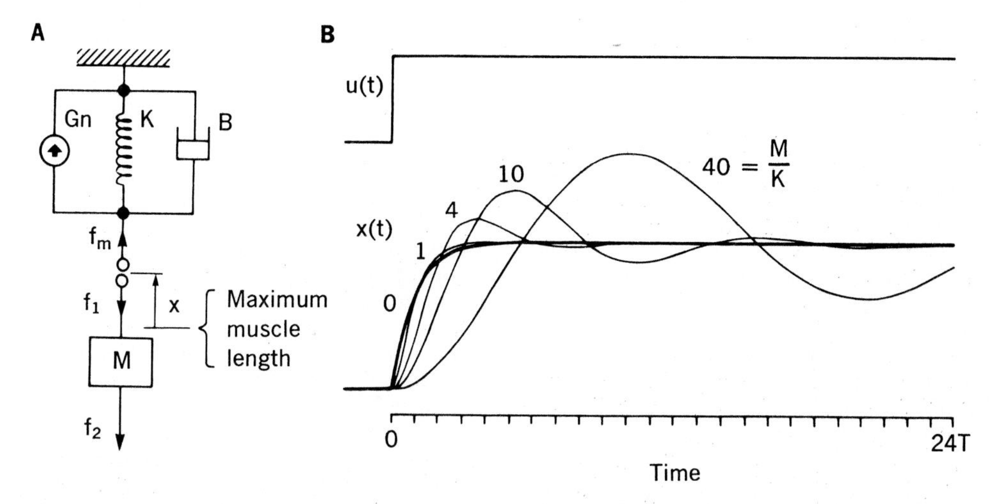

Fig. 7-12. Model of muscle attached to inertial load. **A,** *M* is inertial mass, *B* is viscous coefficient representing slope of force-velocity curve, *K* is spring constant representing slope of length-tension curve, *Gn* represents output of force generator (*n* is number of muscle fibers recruited), *x* is amount of muscle shortening, and *f*'s represent different forces. **B** shows family of step responses (muscle shortening) of model calculated by assuming different values of mass in range 0 to 40 times the spring constant *K*. Viscous coefficient *B* was set equal to 2*K*, and time scale is in terms of time constant T = B/K.

outputs fail to scale (instead, they change their entire time course) is immediate demonstration of nonlinear behavior. In addition, note the different saturation tendencies for the dynamic and static phases of response. This feature gives rise to the impression that the dynamic phase during the ramp is responsive to the first derivative of the input (i.e., to stretch velocity). However, actual tests indicate a very weak dependence on velocity. When physiologic systems are as nonlinear as this one, linear approximations are not very useful as characterizations, except under restricted conditions.

Example 3: Muscle mechanics and movement

Consider the problem of determining how different loading conditions influence the movement produced by a muscle in response to a neural command. Here it is instructive to start with the results of classic studies of muscle mechanics, which demonstrate that the developed force depends on muscle length and the velocity of shortening. (Chapter 3 discusses muscle mechanics; Fig. 3-12 shows the dependence on length [length-tension curve]; Fig. 3-15 shows the dependence on velocity [force-velocity curve]; and Fig. 3-16 captures the dual dependence of force on both length and velocity.) The controlling signals sent in motor axons act by modulating these basic mechanical properties. Stronger contractions are commanded either by a recruitment of additional muscle fibers (in groups called motor units) or by increasing the rate at which action potentials are sent to already recruited fibers (Chapter 26). The purposes of this example are well served by considering recruitment as the sole mechanism for neural control. The reader is referred to other sources for alternative treatments of this problem.[12,18,20]

Equation 20 outlines the manner in which muscle force (f_m) depends on the number of muscle fibers recruited (n) and on the length (x) and velocity ($\dot{x}$) of the muscle:

$$f_m = \frac{n}{n_m} F(x,\dot{x}) \qquad (20)$$

F is a nonlinear function that represents the mechanical properties of a fully activated muscle ($n = n_m$), and the ratio n/n_m specifies the fraction of muscle fibers recruited. Since the fibers are functionally in parallel, the approximate effect of recruitment is simply to multiply (scale) the mechanical dependencies. Even with these assumptions the relation is quite nonlinear and difficult to handle analytically (see Fig. 3-16). A nonlinear relation usually can be subjected to linearization,[24] and this procedure may yield a reasonable description of the salient features of a system, within certain bounds that must then be explored carefully.

Equation 21 is a linearized version of equation 20, and Fig. 7-12, *A*, provides a schematic interpretation of each term:

$$f_m = Gn - Kx - B\dot{x} \qquad (21)$$

x is defined as the amount of muscle shortening from a maximal value. The new parameters are defined as follows: G is a gain factor that converts the neural signal into a force, specifically the force developed at maximal length (x = 0) and at zero velocity; K is the slope of the length-tension curve, which is analogous to a spring constant; and B is the slope of the force-velocity curve, which is analogous to a viscous coefficient. The main shortcoming of this equation is its failure to capture the dependence of K and B on the number of muscle fibers recruited (the multiplicative dependence mentioned earlier). In addition, it neglects the nonlinearity of the length-tension and force-velocity relations, the series elasticity of muscle, and the delayed time course of the activation process (Chapter 3). Nevertheless, the equation is adequate to illustrate the basic control features of muscle.

Muscles act on a variety of mechanical loads in real situations, and the resulting movements can be predicted by combining equation 21 with another equation that specifies the properties of the load. Here it is assumed that the load consists of an inertial mass (M) and an external force (f_2), as shown in Fig. 7-12, *A*. The resulting force on the muscle (f_1) is characterized by Newton's law as follows:

$$f_1 = M\ddot{x} + f_2 \qquad (22)$$

where $\ddot{x}$ represents the acceleration (second derivative of x). External force disturbances combine with gravity to determine the value and time course of f_2.

The equality of f_m and f_1 allows equations 21 and 22 to be combined to yield:

$$M\ddot{x} + B\dot{x} + Kx = Gn - f_2 \qquad (23)$$

The conceptual significance of this step is important to emphasize. Equation 21 describes muscle properties, and equation 22 describes load properties. When the load and the muscle are connected physically, neither equation alone determines either the force of the muscle (or of the load) or the length of the muscle (or position of the load). Instead the values of both variables are determined by the simultaneous solution of the two equations, that is, by the properties of the system as a whole.

Another point worth emphasis is that the mass (M), viscosity (B), and spring constant (K) each serve as *parameters* of the equation, whereas recruitment (Gn) and external force (f_2) appear as *input functions*. Furthermore, one can consider the two latter terms each to be components of a single composite input function, $u = Gn - f_2$. This is convenient for computational purposes, since the solutions to the equation (the predicted movements) depend only on the net balance between the neural input and external force rather than on their individual values. The functional equivalence between increases in Gn and decreases in f_2 means also that a given movement can be produced with either neural signals or external disturbances serving as inputs.

Finally, it should be noted that the equation derived here is one encountered frequently in system theory. It is the equation for any linear, second-order system.

Fig. 7-12, *B*, shows computed responses of this model to a step change in the composite input function (increased neural command or decreased load force). The different curves were obtained by assuming different values for the inertial mass. When M = 0, the second-order term ($M\ddot{x}$ in equation 23) disappears; the system then becomes first order and is characterized by a single exponential component of length change (time constant T = B/K). Small values of M introduce a second exponential term, which has the effect of slowing the initial onset of length change. Yet larger values give rise to overshoot and oscillation.

Application of these results to a particular muscle system requires initially a measurement of the characteristic parameters of the muscle (K and B). An example to which the student is referred is Robinson's[20] study of the eye muscles, in which the parameters were estimated and the predicted effect of added inertia was verified experimentally. The actual model used by Robinson is more complex than the one analyzed here, but the salient features are, in fact, well captured by the simpler model.[21]

In summary, the present example together with the previous one illustrate the two approaches commonly used to characterize a system. In example 2, the basic form of the model (i.e., the type of equation) was chosen empirically, whereas in example 3 (and also in example 1), it was derived from more fundamental information concerning the properties of the elements of the system. Regardless of the method used to obtain the model, its parameters must then be evaluated experimentally. The derivation of the model from more fundamental information has the advantage that it serves to integrate physiologic knowledge at different levels of study. A more complex example of this to which the student is referred concerns a model of the heart based on its geometry and the contractile properties of cardiac muscle.[23]

GENERAL APPROACH TO MODELING TRANSPORT AND STORAGE PROCESSES

Models of physiologic systems can be formulated at several levels of knowledge, ranging from the interactions of molecules to the interactions of organ systems, or even organisms. Whereas the goal of physiology is to relate all these levels to each other, it is usually advantageous to formulate any given problem at some intermediate level not too far removed from the level at which one must arrive to answer the problem. It is also wise not to include more detail than is necessary to provide a valid solution. In this section these points are illustrated while introducing a general approach to the modeling of transport and storage mechanisms.

Processes in which a substance, or energy, is stored at a given locale or is transported from one locale to another are ubiquitous in physiologic (and physical) systems. The input-output properties of such processes are usually characterized by relations between variables expressing the amount of a substance (or of energy), its flow rate, and the forces that propel the flow. The general manner in which the variables amount, flow, and force relate to each other is indicated by the following equations:

Transport:

$$\text{Flow} = G[\text{force gradient}] \tag{24}$$

Storage:

$$\text{Amount} = C[\text{force}] \tag{25}$$

Conservation:

$$\text{Amount} = \int (\Sigma \text{ flows})\, dt \tag{26}$$

Table 7-1 lists some examples of processes that comply with these general relations.

The transport equation specifies that the flow rate of a substance, or energy, from one locale to another depends on a gradient in force between the two locales (e.g., a gradient in concentration propels diffusion, whereas a gradient in temperature propels heat flow). The operator G, specifying the conversion of force gradient into flow, represents the conductance of the process (the ease with which flow occurs). This relation may also be expressed in an inverse fashion:

$$\text{Force gradient} = R[\text{flow}] \tag{27}$$

where $R = G^{-1}$ represents the resistance of the transport process. Sometimes the conductance of a process is approximately constant (e.g., the diffusion transport coefficient K in equation 3 is approximately constant over the usual range of interest), but in general, it is some nonlinear function. Thus the resistance of a vascular bed is a complex function that specifies the relation between blood flow and the force that propels it, which is a gradient in blood pressure (Chapter 39). This function is nonlinear due to the fact that the diameter of a blood vessel is not a constant; diameter increases with blood pressure and resistance decreases with the fourth power of diameter. The resistance of a blood vessel is also time varying due to the fact that the contractile activity of vascular smooth muscle in the walls of the vessel is controlled by sympathetic nerve discharge. Increased activity decreases the diameter of the vessel, causing an increased resistance to flow.

The storage equation specifies that the amount stored is related to the force at which it is stored by the capacitance of the process. For example, the volume of blood (amount) stored in a blood vessel is related to blood pressure (force) according to the distensibility of the blood vessel. Like conductance, capacitance is often a nonlinear and time-varying operator, as it is in the case of a blood vessel. However, in some processes, capacitance is linear and time invariant. For example, solutes are stored in solvents according to the dilution principle:

$$\text{Number of moles} = \theta(\text{concentration}) \tag{28}$$

The volume θ of cytoplasm, the capacitance of this process, is ordinarily considered to be a constant.

The conservation equation expresses the principle of the conservation of matter (or energy). At any given locale an amount changes whenever the inflows and outflows at that locale do not balance. Thus the volume of blood in a vessel accumulates when the inflow exceeds the outflow, in analogy with the accumulation of a molecular species in the cell reservoir system described earlier (Fig. 7-3). The conservation relation is always a linear, dynamic, and time-invariant operation. Many problems in physiology can be formulated under the assumption that the only dynamic process is this one. However, if conductance and storage processes are examined in detail, one usually finds that they are also dynamic. For example, the Fick equation represents a steady-state characterization of a process that is actually dynamic. However, the dynamics of membrane diffusion can usually be ignored,

Table 7-1. Amount, flow, and force variables in several biophysical systems that obey the general relationships specified in equations 24 to 26

Process	Amount	Flow	Force
Diffusion of uncharged molecules	Number of moles	Rate of diffusion	Concentration
Diffusion of charged molecules	Charge (moles)	Current (flux)	Voltage (electrochemical potential)
Enzyme-catalyzed reaction	Number of moles converted	Reaction rate	Concentration
Fluid system	Volume	Flow	Pressure
Thermal system	Heat content	Heat flow	Temperature

since the steady state is achieved very rapidly in comparison with the dynamics of accumulation.

Equations 25 and 26 deal with the storage and conservation of amounts. For some processes, one needs also to consider the storage and conservation of momentum, in which case two additional equations relate flows and forces to momentum:

Storage of momentum:

$$\text{Momentum} = I[\text{flow}] \quad (29)$$

Conservation of momentum:

$$\text{Momentum} = \int (\Sigma \text{ forces}) \quad (30)$$

The operator I represents the inertia of the material being transported. These relations are sometimes required in models of fluid processes (whenever kinetic energy becomes appreciable in comparison with potential energy).

Physiologic systems often can be analyzed as a set of biophysical processes, each of which obeys one of the general relations just described. In this way a system model can be constructed by an appropriate interconnection of the relevant component processes, the outputs from some of the processes serving as inputs to others. The cell reservoir system discussed earlier (Fig. 7-3) is a simple example that the student should review before going on to the example given here. The interested reader is also referred to several texts[1,6,9] that develop these concepts of analogous systems in some detail.

Example 4: Thermal circuit of the body

Biochemical reactions proceed at rates controlled by local mechanisms of enzyme expression, by levels of circulating hormones, and by action potentials in motor nerves, as in the case of muscular contraction. The various reactions all produce heat as a by-product of metabolism, and the reaction rates are influenced, in turn, by the changes in temperature that result from heat production. This influence of temperature on reaction rate is a disturbance of metabolic control processes, particularly since increased reaction rates cause increases in temperature, which then increase reaction rates further in a regenerative cycle. Changes in the temperature of the environment (ambient temperature) are another source of disturbances, since they also act to change body temperature. In mammals and birds these disturbances in temperature are minimized by the actions of a thermoregulatory system that controls heat production and heat loss. The controlled portion of this system is constituted by the thermal properties of body tissues, to which may be applied the laws that govern the storage and flow of heat (Chapter 59). This thermal system is used here as an example of a complex controlled system that can be understood and reasonably well characterized in terms of a few component processes that follow the general relations between amount, flow, and force. In thermal systems these variables become heat content (Q), heat flow ($\dot{Q}$), and temperature (T) (Table 7-1).

The diagrams in Fig. 7-13 illustrate three ways of viewing the thermal circuit of the body. *A* is a simplified physical diagram that lumps thermal properties into two compartments: the skin, which provides the major thermal interface with the environment, and the remainder, called the core, where most of the metabolic heat is produced and stored. Published models frequently incorporate a larger number of compartments,[13,22] but two will be adequate here. *B* provides a more detailed description of this model in which storage, transport, and conservation principles are included. The frequent convention of using electrical symbols to represent these processes is explained in the next paragraph. *C* emphasizes the control aspects of this problem by focusing on the important inputs and outputs. The important output variables are the temperatures of the two compartments. The inputs that influence these temperatures are of two types: disturbances and forcing functions. The former cause undesired changes in temperature, whereas the latter are the controls that correct these errors.

A capacitor symbol is used in Fig. 7-13, *B*, to represent the heat capacitance of the body (C), that is, the storage process of equation 25. Heat is stored like electrical charge; when more is stored, the temperature (voltage) rises. The capacitance for heat storage in the skin is small and will be neglected. The symbol for a resistor represents the transport process of equations 24 and 27. Heat normally flows down a temperature gradient from core to skin (through R_{cs}) to the ambient environment (through R_{sa}). Reverse flow can also occur when the gradient is in the opposite direction. The other symbols represent two sources of heat flow —metabolism ($\dot{Q}_m$) and sweating ($\dot{Q}_s$)—and a source of temperature—the ambient temperature (T_a) of the environment. Sources are used to represent independent variables (i.e., ones controlled by processes external to the functional boundaries of the system). Note that different symbols are used to distinguish sources of flow and sources of force (temperature). The principle of the conservation of energy (equation 26) is represented by the branching of the lines connecting the different components. For example, the portion of the heat produced by metabolism ($\dot{Q}_m$) that does not flow immediately from core to skin (through R_{cs}) must flow instead into the capacitor, where it accumulates as heat content, producing a rise in T_c. The neglect of skin capacitance means that heat flow from core to skin must at all times equal the sum of the flow produced by sweating and the flow through the resistance (R_{sa}), representing the physical processes of conduction, convection, and radiation.*

*The model neglects heat loss associated with evaporation from the respiratory passages, which can be modeled by a flow source connected between the core and the environment. It is particularly important to include this feature in studies of carnivores, since respiratory loss is controlled via the mechanism of panting to provide a major forcing function for the control of core temperature.

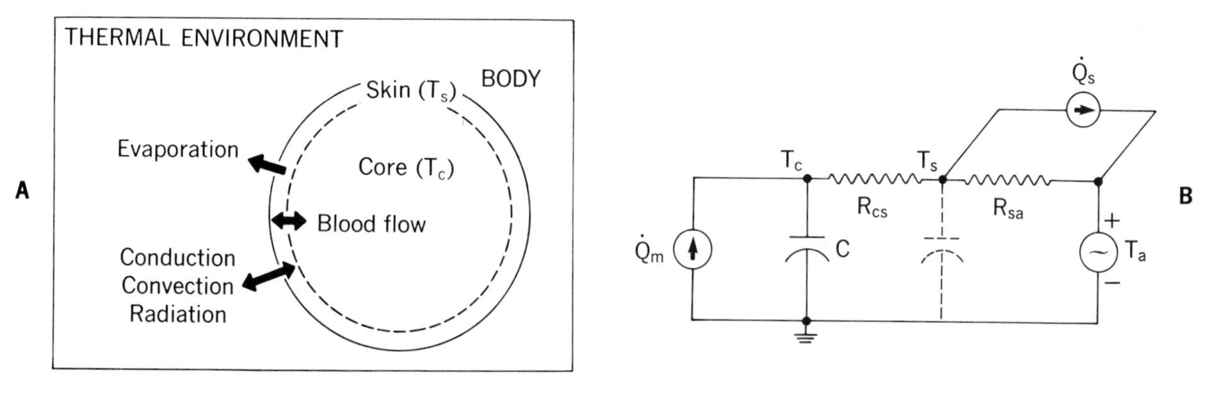

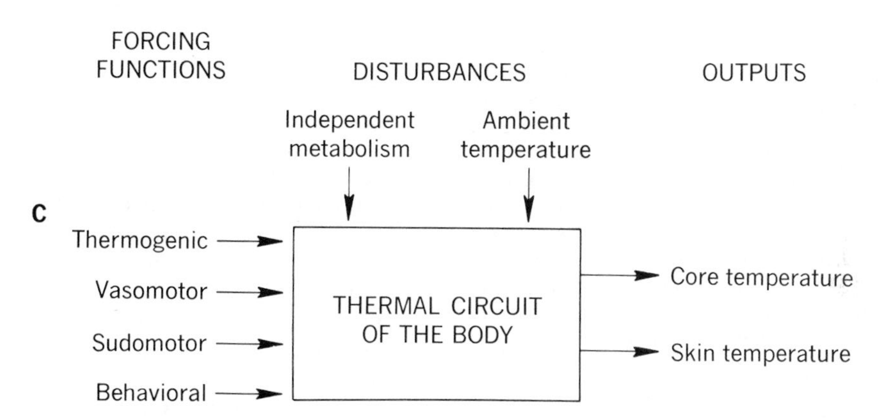

Fig. 7-13. Thermal model of the body. **A,** Simple physical model lumping thermal properties into skin and core compartments. **B,** Thermal circuit. $\dot{Q}_m$, Rate of metabolic heat production; T_c, core temperature; C, thermal capacitance of core; R_{cs}, thermal conductance between core and skin (related to skin blood flow); T_s, skin temperature; R_{sq}, thermal conductance between skin and environment (conduction, convection, and radiation); $\dot{Q}_s$, evaporative heat loss due to sweating; T_a, ambient temperature. **C,** Overall block diagram indicating forcing function and disturbance inputs and the two output temperatures.

To gain familiarity with circuit models the student is urged to use Fig. 7-13, *B,* as an aid in outlining the sequence of events that would follow an abrupt increase in metabolic heat production, as during exercise. Another good exercise is to consider the effect of a depression in ambient temperature.

In both the suggested examples the initial changes are mediated strictly by the thermal parameters of the body, leading to changes in core and skin temperature. We have yet to consider the compensatory actions of thermoregulation, which are mediated by the forcing functions shown in Fig. 7-13, *C.* The manner in which these actions can be incorporated into the thermal circuit (Fig. 7-13, *B*) is as follows. Thermogenic inputs increase the rate of metabolic heat production ($\dot{Q}_m$) via the mechanism of shivering, an important compensatory mechanism during cold exposure. Vasomotor inputs control the flow rate of blood between the core of the body and the skin, this flow being a highly effective medium for heat exchange between the two compartments. In terms of the thermal circuit the resistance R_{cs} is modulated, thus promoting (increased blood flow) or restricting (decreased blood flow) the loss of heat from the core to the skin. Sudomotor inputs produce sweating, an effective mechanism of heat loss, provided the sweat evaporates. The interesting feature of this control is that it operates quite independently of a temperature gradient, which is the reason for using a flow source ($\dot{Q}_s$) to characterize it. This independence is important, since it allows the body to pump heat into the environment against a temperature gradient. The behavioral forcing function represents the action of putting on or taking off clothes. The former increases the resistance between the skin and the environment (R_{sa}), whereas the latter reduces it.

Models of this type are very useful to the student of physiology, since they provide a concise summary of the most important features of a system. Furthermore, the student can use the model as a conceptual aid in thinking through the responses to various disturbances.

As an example, consider the thermal consequence of being in an environment in which ambient temperature is elevated above normal body temperature. Here, as in most other cases, it is advantageous to consider first the direct effects of the disturbance on the system in the absence of regulatory actions. The thermal circuit (Fig. 7-13, *B*) indicates that the elevation of ambient temperature reverses the normally positive gradients in temperature from core to skin to environment, thus causing heat to flow passively from the environment into the body. Consequently, metabolic heat is not eliminated but combines with passive flow to be stored in the body capacitance (C) as an increase in core temperature. Core temperature will rise progressively until it exceeds ambient temperature by an amount sufficient to restore the passive flow of metabolic heat to the environment. One concludes that in the steady state any increase in ambient temperature would produce an equal increase in core temperature, in the absence of regulatory actions.

The vasomotor and sudomotor forcing functions are activated by the increased body temperature via a feedback mechanism discussed in Chapters 8 and 59; their actions may now be superimposed on the passive response. Usually it is advantageous to consider each regulatory action separately before making a conclusion concerning their combined actions. The vasomotor action provokes the following chain of events: dilation of skin blood vessels, leading to increased skin blood flow, leading to reduced resistance to heat flow between core and skin; thus R_{cs} decreases. Ordinarily this would facilitate greater heat loss from core to skin; however, since the gradient $T_c - T_s$ is reversed by the high ambient temperature, the decrease in R_{cs} promotes greater heat gain instead. Although the vasomotor mechanism appears to produce an inappropriate response in isolation, the response becomes appropriate when the sudomotor action is added.

Increased sweating promotes evaporative heat loss from the skin, which lowers skin temperature. Since this restores the positive temperature gradient between core and skin, the vasomotor controlled reduction in R_{cs} now becomes effective in promoting heat loss from the core. Note, however, that there remains a positive temperature gradient between the environment and the skin through which the body will passively absorb heat via the mechanisms of conduction, convection, and radiation (R_{sa}). In the steady state, heat loss due to sweating ($\dot{Q}_s$) must counterbalance the flow from the core plus the passive gain of heat from the environment.

In summary, one concludes that the key factor in the ability of the body to withstand a warm environment is the maintenance of a cool skin temperature via the sudomotor mechanism. If it were not for this, metabolic heat would accumulate and core temperature would rise. More generally, this example illustrates both the complexity of the interaction among processes in body systems and the utility of a model in coping with these complexities.

Summary

System theory deals with two basic topics—the relations between inputs and outputs and the relations between whole systems and their component processes. These are, of course, also two fundamental concerns in physiology. What system theory provides is a systematic and quantitative approach to these problems, one that stresses the mathematical nature of the relations rather than the physical nature of the variables. Since physiologic systems typically involve complex interactions between large numbers of biophysical processes, both the formal approach to the characterization of input-output relations, in terms of operators or transfer functions, and the systematic way of dealing with interactions between processes are of great value in the study of physiology.

The aim of this introduction has been to provide a general understanding of key concepts and some of the methodology of system theory. Even without advanced study of the topic, the student should find this perspective helpful in future explorations of specific physiologic systems. I also hope that some students will be encouraged to construct their own models of physiologic systems, either in the form of block diagrams or in the form of circuit analogs, such as the one in example 4. These models are valuable study aids in the integration of the factual material of physiology into a working knowledge of living systems.

One of the major applications of system theory concerns the understanding and design of systems that control other systems, and this topic is taken up in the next chapter.

REFERENCES

General reviews

1. Blesser, W. B.: A systems approach to biomedicine, New York, 1969, McGraw-Hill Book Co.
2. Buckley, W.: Modern systems research for the behavioral scientist, Chicago, 1968, Aldine Publishing Co.
3. Grodins, F. S.: Control theory and biological systems, New York, 1963, Columbia University Press.
4. Huggins, W. H., and Entwisle, D. R.: Introductory systems and design, Waltham, Mass., 1968, Blaisdell Publishing Company.
5. Milhorn, H. T., Jr.: The application of control theory to physiological systems, Philadelphia, 1966, W. B. Saunders Co.
6. Milsum, J. H.: Biological control systems analysis, New York, 1966, McGraw-Hill Book Co.
7. Padulo, L., and Arbib, A. A.: System theory—a unified state-space approach to continuous and discrete systems, Philadelphia, 1974, W. B. Saunders Co.
8. Riggs, D. S.: The mathematical approach to physiological problems, Cambridge, Mass., 1963, The M.I.T. Press.
9. Riggs, D. S.: Control theory and physiological feedback mechanisms, Baltimore, 1970, The Williams & Wilkins Co.
10. Rosen, R., editor: Foundations of mathematical biology. Supercellular systems, New York, 1973, Academic Press, Inc., vol. 3.
11. Simon, W.: Mathematical techniques for physiology and medicine, New York, 1972, Academic Press, Inc.

Original papers

12. Bawa, P., Mannard, A., and Stein, R. B.: Predictions and experimental tests of a visco-elastic muscle model using elastic and inertial loads, Biol. Cybern. **22:**139-145, 1976.
13. Cornew, R. W., Houk, J. C., and Stark, L.: Fine control in the human temperature regulation system, J. Theor. Biol. **16:**406-426, 1967.
14. Hasan, Z., and Houk, J. C.: Analysis of response properties of deefferented mammalian spindle receptors based on frequency response, J. Neurophysiol. **38:**663-672, 1975.
15. Hasan, Z. and Houk, J. C.: The transition in the sensitivity of spindle receptors that occurs when the muscle is stretched more than a fraction of a millimeter, J. Neurophysiol. **38:**673-689, 1975.
16. Houk, J. C., and Simon, W.: Responses of Golgi tendon organs to forces applied to muscle tendon, J. Neurophysiol. **30:**1466-1481, 1967.
17. Matthews, P. B. C., and Stein, R. B.: The sensitivity of muscle spindle afferents to small sinusoidal changes in length, J. Physiol. **200:**723-745, 1969.
18. Partridge, L. D.: Signal-handling characteristics of load-moving skeletal muscle, Am. J. Physiol. **210:**1178-1191, 1966.
19. Poppele, R. E.: Systems approach to the study of muscle spindles. In Stein, R. B., et al., editors: Control of posture and locomotion, New York, 1973, Plenum Press, pp. 127-146.
20. Robinson, D. A.: The mechanics of human saccadic eye movement, J. Physiol. **174:**245-264, 1964.
21. Robinson, D. A.: Models of the saccadic eye movement control system, Kybernetik **14:**71-83, 1973.
22. Stolwijk, J. A. J., and Hardy, J. D.: Temperature regulation in man—a theoretical study, Pfluegers Arch. **291:** 129-162, 1966.
23. Suga, H., and Sagawa, K.: Mathematical interrelationship between instantaneous ventricular pressure-volume ratio and myocardial force-velocity relation, Ann. Biomed. Eng. **1:**160-181, 1972.
24. Thaler, G. J., and Pastel, M. P.: Analysis and design of nonlinear feedback control systems, New York, 1962, McGraw-Hill Book Co., p. 16.

8 Homeostasis and control principles

JAMES C. HOUK

One of the great events in the history of physiology occurred in 1878 when Claude Bernard enunciated the general principle of the constancy of the "milieu intérieur," based on many detailed observations indicating that the internal environment of the body is closely regulated.[13] This regulated environment is characterized by a constancy of certain biophysical variables, for example, temperature, blood pressure, the concentrations of many substances in body fluids, the volumes of fluids in body compartments, and the density of red cells in the blood. Bernard recognized that the maintenance of an internal environment independent of the external environment provides animals and man a great deal of freedom in choosing a habitat, as contrasted with single-celled organisms that are confined to an environment suitable for their survival. He stated his case forcefully when he wrote, "It is the fixity of the internal environment which is the condition of free and independent life, all the vital mechanisms, however varied they may be, have only one object, that of preserving constant the conditions of life in the internal environment."

In the period 1910 to 1932 Walter Cannon made many important contributions to the study of the mechanisms involved in regulating the internal environment of the body, and he coined the term "homeostasis" to refer to this regulation.[15] Homeostasis, when literally translated, means "similar condition," which Cannon believed was appropriate, since the conditions are not actually fixed. Rather, the internal variables are held within narrow ranges by the actions of homeostatic mechanisms. This subject of automatic regulatory action is one of the important topics discussed in this chapter, and it will be a major theme in many other sections of this book.

A constancy of the physicochemical environment is, of course, only a first step in providing for the integrative activities of the cells and tissues. These activities must adapt to changing requirements, and, in fact, they are altered in a dynamic manner as we make transitions from one state to another (e.g., from resting to running or from fasting to eating). Different tissues must respond in different ways to each situation, and these responses must be coordinated.

Because of the physical separation of the tissues and organs, mechanisms for communication over distances are required for the integration and adaptation of bodily functions. The communication systems of the body are nervous and endocrine; neural signals are transmitted over long distances by the propagation of action potentials in peripheral nerve fibers, and endocrine signals are transmitted over short distances by diffusion and over longer distances by way of the circulation. These control signals act on organs and tissues, which are themselves characterized by a complex of biophysical interactions. For example, the contraction of a muscle is controlled by the neural signal sent to it in motor axons, but the resultant movement depends on the biochemical state of the muscle, the mass of the load that is being lifted (example 3 in Chapter 7), the mechanics of the skeleton, and the forces that are being produced simultaneously by other muscles. Similarly, the control of blood flow and pressure in one set of blood vessels interacts with the fluid mechanical state of other parts of the vasculature. The control processes of the body must deal with all these interacting variables simultaneously. The purpose of this chapter is to describe the basic principles and strategies used by the body in achieving this control.

ORIENTATION TO CONTROL THEORY

A system is said to be controlled when its inputs are manipulated in a manner that causes its outputs to vary in some prescribed way or to remain constant at prescribed values. This statement introduces the inescapable link to teleology

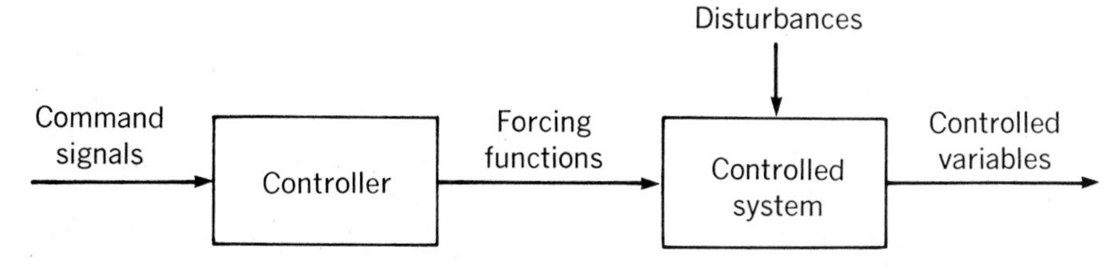

Fig. 8-1. Basic signals and components of a control system. Command signals designating desired performance are operated on by controller to produce forcing functions. The latter, introduced as inputs to controlled system, produce desired changes in controlled variables. Uncontrolled inputs called disturbances produce undesired changes in controlled variables.

that confronts us when we seek to understand the control processes of the human body. These processes quite obviously have functions,[31] and a major problem in analysis is to specify the functions as precisely as possible.

Fig. 8-1 introduces a conventional view of the control problem and provides definitions of terms used in this chapter, some of which have already been mentioned in example 4 of Chapter 7. The purpose of the overall system is to control the values of certain variables within prescribed limits. These *controlled variables* are a subset of the outputs of what is called the *controlled system*. The latter is typically constituted by an interacting set of physical (or biophysical) processes, the state of which is subject to the influences of two categories of input: *disturbances* and *forcing functions*. Disturbances are uncontrolled inputs that cause undesired changes of the state of the controlled system and, hence, of controlled variables. Forcing functions instead are inputs that can be manipulated to cause (or force) desired changes in state. Forcing functions are also outputs, those from another system called the controller. The function of the controller is to generate appropriate forcing functions based on its inputs, called *references, set points,* or *command signals,* which designate the desired behavior of the overall system.

Another category of input to the controller, not shown in Fig. 8-1, derives from sensors that monitor the states of the controlled system and other variables. Information from sensors is required when disturbances are sufficiently large to produce unacceptable changes in controlled variables, and control processes designed primarily to compensate for these disturbances are called *regulators*. Since most physiologic control systems are subjected to large disturbance inputs, regulator function will be a major topic in this chapter. However, before advancing to specific physiologic examples, it is helpful to contrast the differences between the control problems faced by engineers and by physiologists, since this will illustrate some of the inherent problems in the application of control theory to physiology.

Typically an engineer is asked to design a system for controlling certain physical quantities (the controlled variables) in a precisely defined way. Given this clear statement of the function of the system, he may then follow a more or less systematic sequence of steps. First, he selects components for the controlled system, the physical devices that will be used to manipulate the controlled variables. Second, he derives a mathematical model that characterizes the responses of the controlled system to its forcing functions and to potential disturbances. Third, he then uses this model to determine the operations that must be performed by the controller to achieve the required performance. This is the most interesting problem for control theorists, who have developed a variety of approaches to the problem of selecting mathematical operations that are suitable or even optimal. Fourth, and finally, he constructs the controller from electronic components (or writes a digital computer program that implements the controller operations), connects it to the controlled system, and tests out the overall performance of the system to determine if the design requirements have been met. If so, the system is ready for delivery, and, if not, he must return to one of the earlier steps to revise his solution.

The problems faced by a physiologist differ from those of the engineer in several important respects. First, the function of a physiologic control system usually is known only in some vague fashion initially—its precise function must be discovered by experiment. Second, steps in engineering analysis that require the synthesis of a larger system from subsystems are replaced by ones requiring analysis, the experimental discovery of the important subsystems that interact to make up the controlled system and the controller as well as the discovery of which neural or hormonal signals are important forcing functions. Analysis is inherently more difficult than synthesis, since there are numerous ways in which a system subjected to analysis might be organized; in contrast, unique solutions usually exist for problems of synthesis. A third difference is that the systematic steps in engineering designs are replaced by a different, and less rigid, sequence. Ideally the analysis of a physiologic system begins with measurements of its overall characteristics (the last step in engineering design), from which one hopes to specify the function in more or less precise terms. One then attempts to identify the forcing functions and study the characteristics of the controlled system, culminating in a model of the controlled system. Finally, one attempts to characterize the operations performed by the controller and discover its mechanisms. In practice, analysis usually proceeds on all fronts simultaneously, since different investigators specialize in each of the steps outlined and since any individual step usually cannot be completed until some of the results obtained from the other steps become available. For example, the precise function of the system may not become completely clear until the operations performed by the controller are known. Thus the systematic approach followed by the engineer contrasts with the parallel approach required of physiologists.

REGULATORY STRATEGIES FOR HOMEOSTASIS

The overall homeostasis of the body is achieved by the simultaneous operation of a group of physiologic regulators, each of which functions to maintain one or more variables relatively constant. These *regulated variables* are outputs of the different controlled systems of the body, such as the muscle and thermal circuit discussed in the previous chapter. A controlled system thus constitutes one of the component processes that make up a regulatory system. The other components are a controller that generates the forcing functions that act on the controlled system and a set of sensors that provide the information on which the actions of the controller are based. The sensors can be situated to detect potential disturbances or to detect the effects the disturbances have on regulated variables (Fig. 8-2). The former configuration is called *feedforward* and the latter is called *feedback*. Each of these two control strategies has particular advantages and disadvantages.

Disturbances generally occur in advance of the disturbing effects they provoke. Thus the detection of disturbances by the sensors in a feedforward regulator provides predictive information about impending changes in regulated variables. The controller must then use this information to calculate the effects the measured disturbances are likely to have on the regulated variables and the forcing functions required to counteract these effects. In order to do this the controller of a feedforward regulator must, in essence, contain a model of how the controlled system behaves. The precision of regulation clearly depends on the accuracy of this model. Since the body's controlled systems are frequently nonlinear and time varying, the models contained within the brain must be either highly complex or else simplified representations of these complex

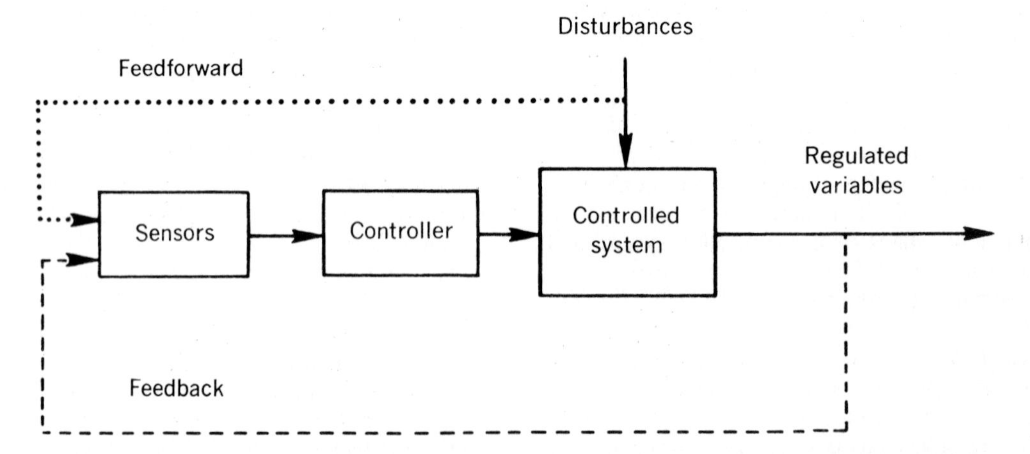

Fig. 8-2. Feedforward and feedback configurations for regulation. Function of a regulator is to diminish effects of disturbances on regulated variables. Regulatory actions are based on outputs of sensors that detect either potential disturbances, in feedforward system, or effect disturbances have on regulated variables, in feedback system.

systems. The requirement of complex models for precise regulation is one of the major disadvantages of feedforward control. Another disadvantage derives from the fact that many variables must be monitored to specify all the potential disturbances and to provide an up-to-date model of a controlled system, since its properties usually change with time. Sensors capable of monitoring all these variables may not be available, in which case a feedforward regulator is bound to make errors.

One of the intriguing examples of feedforward regulation in body homeostasis concerns the classic conditioned reflex. Pavlov,[28] who was one of the first to investigate this phenomenon, demonstrated that practically any sensory cue (the conditioned stimulus) can be caused to trigger a given physiologic response, provided the former is presented in association with a stimulus (the unconditioned stimulus) that normally elicits the latter. Thus the controller for salivary secretion, which acts to keep the mouth appropriately moist, responds to the sounding of a bell after a period of training during which bell ringing is followed by the introduction of food into the mouth, the latter being a normal stimulus for salivation. In this example the conditioned reflex provides a feedforward mechanism that moistens the mouth in preparation for food that is about to arrive. It is clear that errors in performance (too much moisture) are bound to occur the first time the bell is not followed by the presentation of food. The predictive value of the bell is good only as long as the environmental contingencies remain unchanged.

Regulation based on the feedback configuration is free of the major disadvantages of feedforward control, but it has other disadvantages. The fact that feedback control is based on the current values of regulated variables eliminates the possibility of completely erroneous regulatory actions. Furthermore, the variables that must be monitored are only the regulated ones, and the calculations performed by a feedback controller can be made extremely simple. There are three major disadvantages of feedback regulation: (1) speed—the regulatory actions must await the consequences of the disturbances; (2) stability—due to the presence of a closed loop of control, excessive corrective action can be propagated around the loop in a nonending cycle of oscillation; (3) error—the correction depends on the existence of an error; thus correction is not complete (except in certain ideal situations mentioned later).

Although feedback and feedforward are the two basic strategies for controlling moment-to-moment regulatory actions, a third strategy, *adaptive control,* functions over longer time periods. Adaptive control is loosely defined as any control that changes to meet changing needs. All regulatory systems satisfy this definition. However, in the more strict sense of the term, an adaptive modification is distinguished as a beneficial change in the moment-to-moment properties of a system that occurs over a long time period, one longer than that required for individual responses.[1,3] Biologic examples include the hypertrophy of a muscle that occurs as a long-term response to physical training, the acquisition of a new conditioned reflex, or the learning of a new and better response through operant reinforcement.

An adaptive control system has, in addition to the component processes already described, a special subsystem that evaluates some measure of the quality of the responses to stimuli, either as they occur naturally or as they are evoked by internally generated test signals. This measure of quality is then used as a basis for adjusting the parameters or structure of the main regulatory system. Thus a well-designed adaptive system should continue to improve its performance based on past experience and readily adjust to new situations. For this reason, adaptive control is equated with learning. Needless to say, the general theory of adaptive systems is presently at a rather primitive stage, although progress is being made.[5,7]

The homeostatic systems of the body employ all these strategies for regulation in combinations that appear to utilize the respective advantages of each. In the following sections, particular attention will be given to negative feedback systems, simply because these are best understood at the present time.

NEGATIVE FEEDBACK CONCEPTS

The essential feature of a negative feedback system is the provision for a closed loop of control through which any disturbance in output is opposed. Fig. 8-3 shows how this can be accomplished with the use of a few simple components. A sensor is situated to detect one of the outputs of the controlled system, the *regulated variable*. The output of the sensor, which is proportional to the actual value of the regulated variable, is subtracted from a *reference* signal (or *set point*), which represents a desired value of the regulated variable, to form an *error signal*. This step is called *error detection*. The error signal is then amplified and sent as a forcing func-

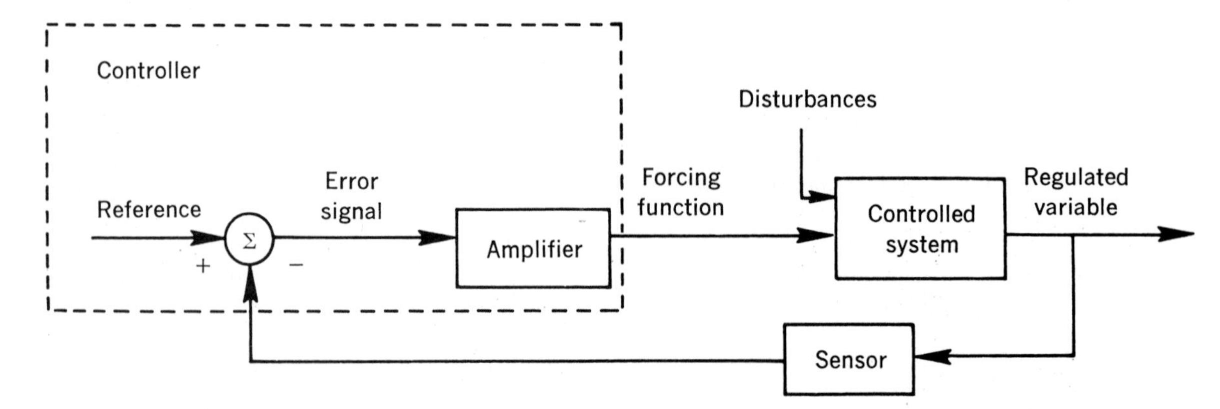

Fig. 8-3. Diagram to illustrate potential simplicity of feedback controller. Appropriate forcing function can be generated by two elementary operations: error detection (subtraction) and amplification.

tion to the controlled system. Whenever a disturbance acts to modify the regulated variable, an error will be detected, amplified, and supplied to the controlled system, thereby forcing the regulated variable back toward the reference value. For example, a disturbance that depresses slightly the value of the regulated variable will result in a small, positive error signal, which, when amplified and delivered to the controlled system, will act to elevate (and hence restore) the value of the regulated variable.

Loop gain and error reduction

The extent to which negative feedback reduces the errors caused by disturbances is uniquely determined by a single parameter (or function) called the loop gain of the system. The meaning of loop gain can be understood if one considers the output of a general type of feedback regulator (Fig. 8-4) to consist of two components: (1) a disturbance component (y_d) that represents the uncompensated response to a disturbance, obtained when feedback is absent, and (2) a compensatory component (y_c) that represents the portion of the output attributable to feedback. Loop gain can then be defined as the relationship of the compensatory component to the net response:

$$y_c = G[y] \tag{1}$$

where the operator G is the loop gain. The idea here is that y, the input to the feedback loop, is amplified to yield the compensatory response y_c. The loop gain may be a constant (in a linear static system), but in general, it depends on frequency (dynamic system), and it may also depend on amplitude (nonlinear system) and time (time-varying system).

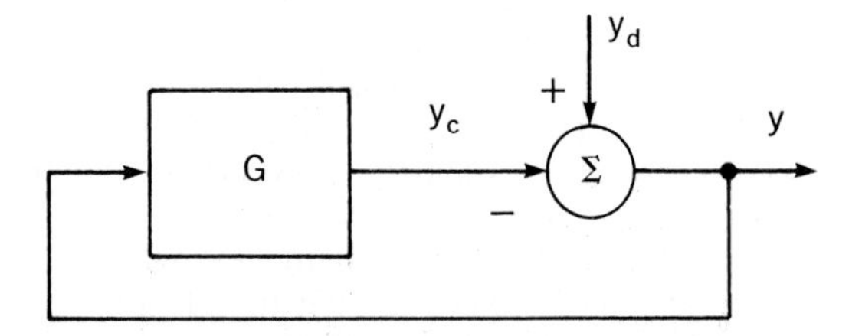

Fig. 8-4. Generalized diagram of negative feedback system. Regulated variable y is represented as difference between alteration y_d, produced by disturbance, and compensatory response y_c, produced by feedback path having loop gain G.

The precise relation between error reduction and loop gain is easily derived as follows:

$$y = y_d - y_c \tag{2}$$

$$y = y_d - G[y] \tag{3}$$

$$(1 + G)[y] = y_d \tag{4}$$

The quantity (1 + G) is itself an operator, and if its inverse exists, the attenuation operator A can be defined.

$$y = A[y_d] \tag{5}$$

where:

$$A = (1 + G)^{-1} \tag{6}$$

The significance of this result is the following. The variable y_d measures the severity of a disturbance in terms of the alteration in the controlled variable it would provoke without feedback. Any response to a disturbance, and hence to y_d, can be considered an error. The operator A defines the extent to which feedback diminishes this error; error reduction thus depends in an inverse manner on the magnitude of the loop

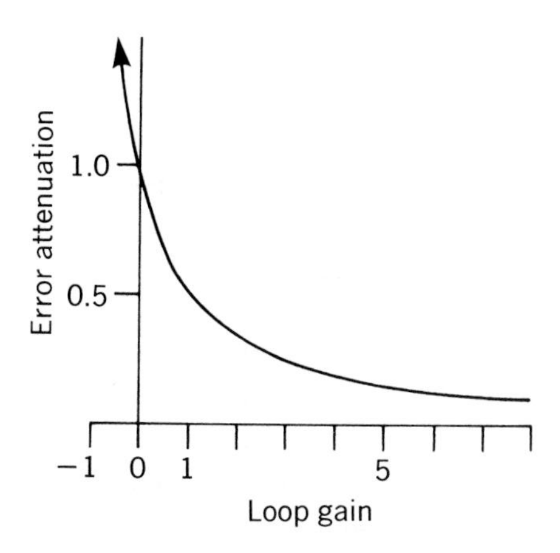

Fig. 8-5. Relation between error attenuation and loop gain. Error attenuation (A in equation 5) is measure of efficacy with which feedback reduces changes in regulated variable produced by disturbance. In linear, static case shown here, $A = y/y_d = 1/(1 + G)$ where y, y_d, and G are defined in Fig. 8-4.

gain. Fig. 8-5 shows the precise dependence of error attenuation on gain for a linear static system (the steady-state dependence of a linear dynamic system). Larger loop gains result in greater attenuation, as expected, but the dependence is nonlinear. The greatest improvements occur as the loop gain increases from 0 to 1, and the improvements are progressively less with further increases in gain. What constitutes good attenuation is, of course, relative, but a loop gain of 9 yields a 10:1 reduction in error. Negative values of loop gain represent positive feedback, and it is apparent that this increases the error rather than decreasing it.

Example 1: Error reduction in the thermoregulatory system

Alterations in metabolic activity or changes in ambient temperature were discussed in Chapter 7 (see Fig. 7-13) as inputs to the thermal circuit that tend to disturb body temperature. Disturbed values are detected by peripheral temperature receptors and by thermosensitive central neurons that register core temperature (Fig. 7-6). These thermosensors connect to efferent neural pathways, particularly via the hypothalamus, controlling the various forcing functions that act on the thermal circuit to regulate body temperature (Chapter 59).

The forcing functions were detailed earlier, in example 4 of Chapter 7. One of them, the vasomotor signal, is particularly important in compensating for small errors in core temperature in a region of regulation that provokes neither shivering nor sweating. One can appreciate the negative feedback nature of this regulatory loop through the central nervous system (CNS) by tracing the compensatory actions. Increases in core temperature in response to alterations in ambient temperature provoke vasodilation of skin arterioles, which increases the heat conductance (or decreases the resistance, R_{cs}) from the core to the skin, thus promoting heat loss and a restoration of core temperature. Conversely, decreases in core temperature provoke vasoconstriction, decreased conduction, and reduction of heat loss. Thus the disturbance in temperature will be opposed by a regulatory action, the defining feature of negative feedback.

The compensatory action does not completely restore core temperature, although error reduction is appreciable. Changes in ambient temperature of several degrees cause changes in core temperature of only a few tenths of a degree.[21] Error attenuation has also been calculated from an independent estimate of the loop gain of the system[16]; the steady-state value of 9 (note that loop gain is dimensionless) predicts, from equation 6, an error attenuation of $^1/_{10}$, in general agreement with the observed errors.

The error-reducing power of thermoregulation is augmented considerably when other compensatory mechanisms are added to the vasomotor mechanism, as occurs under conditions of more severe thermal stress. Consider the thermoregulatory performance during exercise. Fig. 59-3 (p. 1419) indicates that a 20-fold increase in metabolic heat production, associated with muscle contraction during exercise, results in a 1.5° C error in core temperature. In order to appreciate the significance of this value, one must obtain an estimate of the error in core temperature that would be expected in the absence of negative feedback, which is the value of y_d in equation 5. Since the thermal circuit in Fig. 7-13, *C*, becomes a linear model when regulatory changes in thermal conductance are prevented, one can apply the principle of proportional scaling. Thus the 20-fold increase in metabolic heat production should produce a 20-fold increase in the temperature gradient from the core to the environment. Ordinarily this gradient is about 10° C, assuming a 27° C ambient temperature; exercise should increase the gradient to 200° C, which amounts to a core temperature of 210° C and an hypothetical error of 210 − 37 = 173° C. Since the observed error is only 1.5° C, the attenuation afforded by thermoregulation amounts to 0.009, which is indeed remarkable performance. A value of loop gain in excess of 100 is obtained by substitution into equation 6.

Stability

In practice the benefits of high gain must be weighed against the possibility that the system may become unstable. Any real system is limited by a finite response time, and problems arise when the correction is delayed sufficiently as no longer to be appropriate. Then the correction itself becomes a disturbance that in the presence of high gain may be propagated around the feedback loop to produce an output that continues in the absence of a corresponding input. This unstable output may be either an oscillation or a

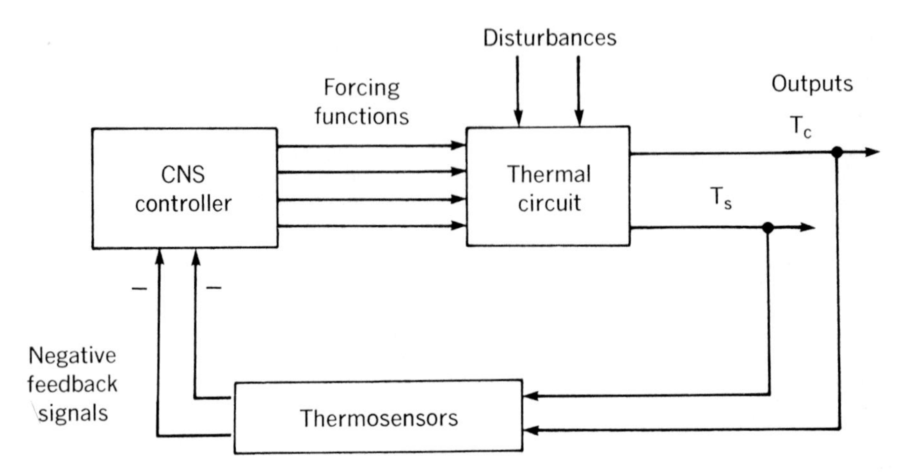

Fig. 8-6. Block diagram of the thermoregulatory system. Thermal circuit (see Fig. 7-13, *C*) is controlled system. Changes in its outputs, temperature of core *(T_c)* and of skin *(T_s),* are sensed by thermosensors to yield feedback signals. The latter are converted into appropriate forcing functions by neural pathways that constitute CNS controller.

maintained offset that drives the system into saturation. In control system design, one strives to achieve a compromise between the beneficial effects of high gain and a tendency toward instability.

Physiologic control systems are generally quite stable, although they may become unstable when disease processes alter the properties of the system. For example, Cheyne-Stokes breathing is a cyclic variation in ventilatory rate that may result when disease processes lengthen the delay between changes in blood gas concentration and the corresponding compensatory responses (Chapter 75). This additional delay in the feedback pathway makes the compensatory response arrive late and persist beyond the time at which it is appropriate; the resultant overcompensation in turn leads to a depression of respiratory drive in a cycle that may continue indefinitely.

Rate and integral feedback

One of the disadvantages of negative feedback mentioned earlier is that the compensatory actions may be excessively slow, since they must await the development of the error that drives them. This problem can be alleviated by introducing feedback sensitive to the rate of change of the regulated variable. In engineering design this is accomplished either by differentiating the output of the sensor or by installing another sensor that is itself sensitive to rate of change. Usually rate feedback is added to normal proportional feedback in a combination that yields satisfactory results. Fig. 8-7, *A,* will be used to explain how this can improve performance.

Consider that a disturbance causes the controlled variable y to deviate with the time course shown in the upper trace. The rate of change of this signal (the time derivative shown in the middle trace) jumps to a large value at the onset of the disturbance response (time t_1) before the error fully develops. This is the reason why rate signals are said to be predictive. Correspondingly, the use of a rate signal as an input to the feedback loop provides a mechanism for generating a strong compensatory action in advance of the full error. Although rate feedback may arrest the development of an error, it is incapable of generating the steady compensatory action required to oppose a steady disturbance. This is because the rate of change of y falls to zero whenever y becomes constant (time t_2). However, if rate and proportional signals are added to form a composite feedback signal, one can obtain the advantages of each. Rate feedback speeds the compensatory action, whereas proportional feedback provides steady-state compensation. In physiologic regulators this combining of rate and proportional sensitivity frequently occurs within the sensor itself, as exemplified by the phasic responsiveness of muscle receptors discussed earlier (example 2, Chapter 7).

Rate feedback may also be helpful in improving the stability of a regulator, since its predictive aspect can provide some compensation for the time lags that lead to oscillation. However, too much rate feedback can cause instability just as too much proportional feedback can, and the efficacy of rate feedback in improving stability depends in a complex manner on the overall

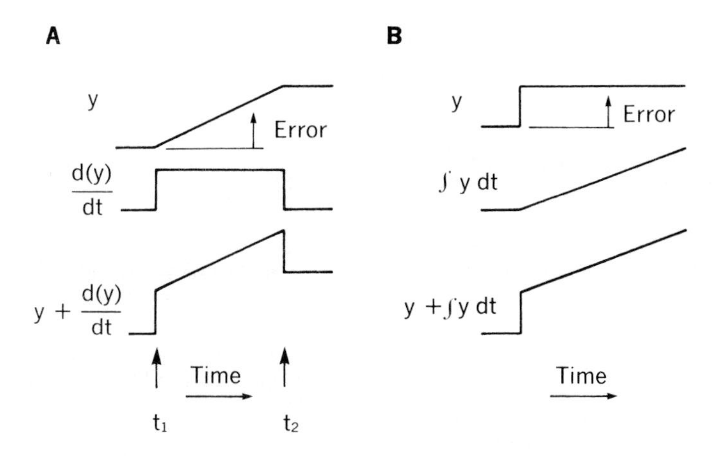

Fig. 8-7. Diagram to illustrate advantages of rate and integral feedback. **A,** Rate signal $\frac{d(y)}{dt}$ provides means for speeding detection of developing error in regulated variable *(y)*. **B,** Integral signal $\int y\, dt$ provides means for detecting small maintained error in regulated variable.

properties of the system. Stein[34] has shown that rate feedback in the neuromuscular control system actually decreases the margin of stability; however, it quickens the response time.

Equation 5 and Fig. 8-5 show that the only way to eliminate completely an error response to a disturbance is to increase the loop gain to infinite values. Ordinarily this cannot be done because of instability, but a way around this is to introduce integral feedback, as illustrated in Fig. 8-7, *B*. The upper trace shows a presumed disturbance response, in this case a step change in the regulated variable y. The middle trace shows the time integral of this signal, which grows progressively at longer time intervals. Thus a maintained error in y will eventually lead to an infinite output from the integrator, even if the error is small. If the integral of y is used as a feedback signal, the compensatory response will eventually become infinitely large, demonstrating an infinite loop gain that, in theory, would eliminate all error in the steady state. The sluggishness of the compensatory response evident from the middle trace in Fig. 8-7, *B*, can be overcome by including proportional feedback (and perhaps also rate feedback) in addition to integral feedback (lower trace).

One practical problem with this scheme is that the feedback mechanisms saturate before the compensatory action becomes infinite, and another is the difficulty of building a perfect integrator. Integral control is approached in some homeostatic systems in which the compensatory actions control flow variables, since the resultant changes in amounts are proportional to the integrals of flows (equation 26 in Chapter 7). A good example is the control of the fluid volume of the body, in which case an important forcing function is the level of antidiuretic hormone, which controls the rate of urine formation.

Capricious nature of set point in physiologic regulators

The reference, or set point, of a regulator is determined by signals that are applied as additive (or subtractive) inputs to the feedback loop. In the example of Fig. 8-3 the reference is a single signal applied to a localized summing node, the error detector. However, any additive signal to the loop will bias the value that a regulated variable assumes and therefore will contribute to the set point of the regulator. The controller for a physiologic regulator is usually a succession of neuronal stages that convert the feedback signals sent from sensors into appropriate forcing functions. In general, each neuron at each stage will receive other inputs as well as feedback signals. Consequently, the reference for a physiologic regulator is a distributed rather than a localized signal, made up of the combined effects of additive signals sent to each neuron that participates in the feedback loop. It follows that there is no localized error detector either; instead, error detection is an operation that is distributed among each of the neurons in the feedback chain.

It is impractical to measure the reference of a physiologic regulator directly, due to its distributed nature. Consequently, one is forced to use an operational definition. One possibility is to designate the set point as the value the regulated variable assumes in the absence of all disturbances. However, often it is impossible or nonsensical to eliminate disturbances. For example, the two major categories of disturbance to thermoregulation are independent metabolism and ambient temperature (see Fig. 7-13, *C*). Setting either metabolism or temperature to zero causes death rather than defining the set point. As an alternative to zero, one might assign the basal metabolic rate to be the "no disturbance" value, but one is still left with the arbitrary decision as to which ambient temperature constitutes no disturbance. Different choices lead to different set points, although all are in the vicinity of 37° C. The practical resolution of this problem is to choose some temperature in the normal range (usually 37.0° C) as an arbitrarily defined set point and to treat departures from this value as errors.[16]

The student should realize that an inability to define a unique reference presents no substantial problems in practice, even though it may be perplexing conceptually. The essential function of feedback—the minimization of *changes* in regulated variables—can be measured and appreciated without invoking the concept of a unique set point. This is also consonant with Cannon's definition of homeostasis as a confinement of internal variables of the body to narrow limits of variation rather than an absolute constancy. The physician who must assess the absolute values of regulated variables such as temperature and blood pressure simply compares the measured values with a "normal" range.

Example 2: Regulation of blood pressure and flow

The circulation of blood functions as a mass transport system that carries metabolites and hormones over long distances from one body tissue to another (Chapters 35 to 44). Whereas the cardiovascular system is complicated by a vast network of blood vessels, by a heart of complex structure, and by several regulatory mechanisms operating both locally and globally, the salient features relevant to the control of the circulation can be captured by relatively simple models, ones not so complex as to inhibit intuitive understanding by the student. The model presented here ignores the pulmonary circulation and deals only with steady-state values of mean pressures and flows. The interested reader is referred to other sources for descriptions of more detailed models.[18,19,33]

A precise statement of the function of the cardiovascular control system, based on experimental observations of its overall performance, is that it serves to maintain arterial blood pressure constant while simultaneously delivering blood flow to individual tissues and organs in accordance with their metabolic activity (see Guyton[20]). The appreciation of how this is accomplished begins with an understanding of the functional properties of the controlled system, the heart, and the vasculature. The student may wish to review the discussion of generalized transport and storage processes in Chapter 7 and the example there describing the thermal circuit before proceeding with the following description, which employs similar concepts.

Cardiovascular circuit. Fig. 8-8, *A*, is a simple physical diagram of the circulation (excluding the pulmonary portion) and Fig. 8-8, *B*, shows a corresponding circuit analog that serves to define the basic relations between blood pressure and flow. The vasculature is represented by a capacitor and two resistors. The former symbolizes the ability of blood vessels to store blood under pressure as a result of the distensibility of the vessel walls.

$$\text{Pressure} = C^{-1}[\text{blood volume}] \tag{7}$$

Although all blood vessels show this property, the most significant contribution occurs in the venules and veins; the present model places all the vascular capacitance at this site. Flow in the remainder of the vasculature is thus assumed to be governed only by the property of resistance, the relation between pressure gradient and flow:

$$\text{Pressure gradient} = R[\text{flow}] \tag{8}$$

or:

$$\text{Flow} = R^{-1}[\text{pressure gradient}] \tag{9}$$

The circuit model of the heart is based on empirical studies of isolated cardiac pumping behavior. The flow source on the output side derives from the observation that cardiac output (f_a) is essentially independent of mean arterial pressure (P_a) over a broad range that includes normal arterial pressure. The pressure source on the input side derives from the observation that the pressure (P_h, related to mean atrial pressure) at which the heart accepts venous return (f_v) remains nearly constant over a broad range that includes normal circulatory flow rates. The actual value of P_h is slightly above atmospheric; the small increase that occurs as either P_a or f_v increase can be neglected in this model of the whole circulation, even though it is an important factor in analytic models of cardiac function.

The third important feature of the cardiac pump is that its outflow is equal to its inflow ($f_a = f_v$), except during brief transients. Since the heart has a very limited capacity to store blood, whatever amount enters from the venous vasculature must also be pumped out into the arterial vasculature. Normally the heart supplies whatever energy is required to keep outflow equal to inflow via the Starling mechanism. However, this feature is taxed if either arterial pressure or venous return become excessive or if the heart suffers damage. Under these overload conditions the heart becomes grossly distended and P_h increases. Overload can be included in the model by introducing

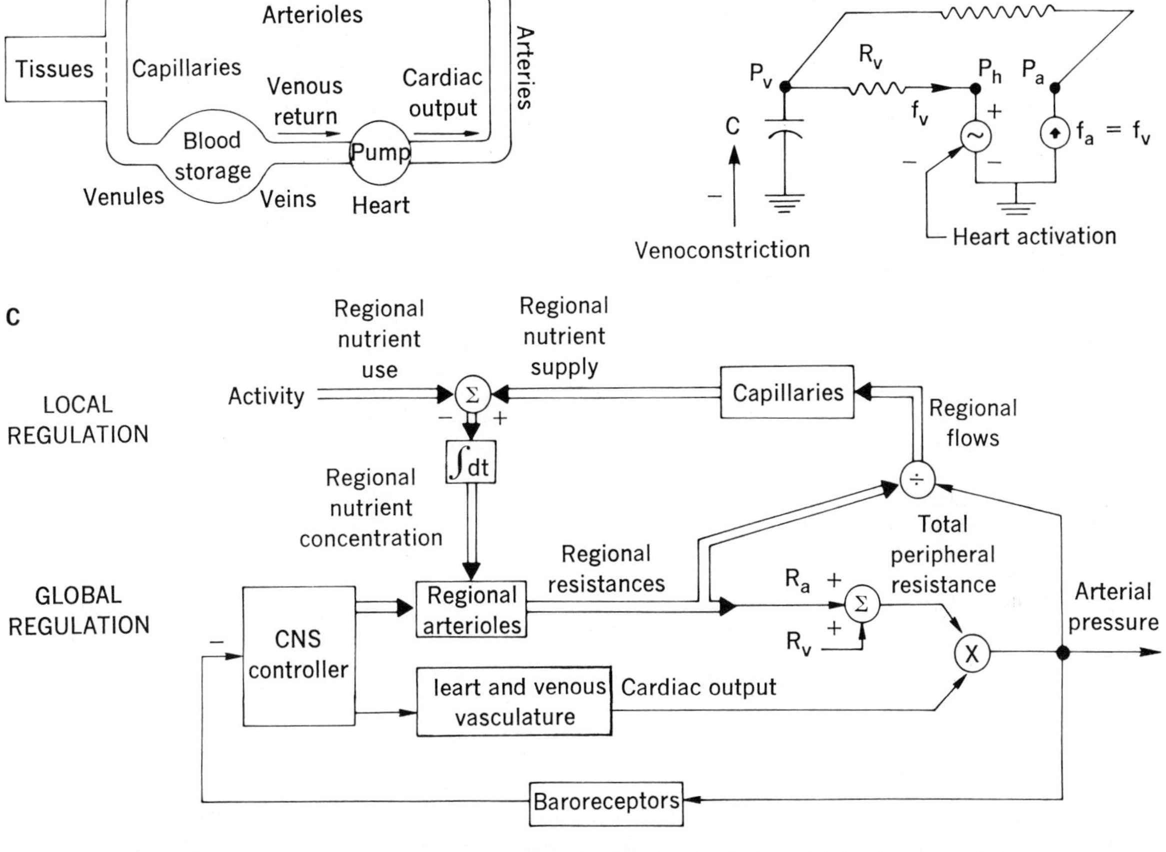

Fig. 8-8. Model of the cardiovascular system. **A,** Simple physical model that ignores pulmonary vasculature. **B,** Circuit model of heart and vasculature. *C,* (venous) capacitance for blood storage; P_v, venous pressure; R_v, venous resistance; f_v, venous return; P_h, pressure at which heart accepts venous return; f_a, cardiac output; P_a, arterial pressure; R_a, arterial resistance. **C,** Block diagram showing local and global regulatory loops controlling blood flow and arterial pressure, respectively.

a dependence of the pressure source P_h on P_a and f_v, but this is neglected here, since the normal dependence is weak.

One application of this circuit model is its use in resolving a question that often confuses students—does pressure cause flow or does flow cause pressure at different sites in the vasculature? The resistance property of blood vessels is indeterminate on this issue, since either pressure gradient (equation 8) or flow (equation 9) can be considered as the independent variable; one must turn to other circuit components to determine which variables are properly designated as the independent or causal ones. Since amounts do not change instantaneously, but only progressively over time (see equation 26 in Chapter 7), they are natural choices. In the present example the only amount is the volume of blood (v) stored in C, which in turn determines venous pressure (P_v) according to the following equation:

$$P_v = C^{-1}[v] \tag{10}$$

The pressure source (P_h) represents another independent variable, provided the heart does not become overloaded. The independence of P_v and P_h leads to the conclusion that the flow through the venous resistance (R_v) should be considered the dependent variable in the following equation:

$$f_v = R_v^{-1}[P_v - P_h] \tag{11}$$

Although it is correct to treat pressure gradient as the cause of flow through the venous resistance, the same is not true for the arterial resistance. This is because cardiac output (f_a) equals (and is determined by) venous return (f_v) independently of events on the arterial side of the circulation. It therefore serves, along with P_v, as an independent variable in the following equation:

$$P_a = R_a[f_a] + P_v \tag{12}$$

where the dependent variable is arterial pressure.

Forcing functions. Cardiovascular regulation is achieved by the actions on vascular smooth muscle and on cardiac muscle of neural signals sent in efferent autonomic nerve fibers and of neurohumeral signals transported by the circulating blood. The contraction of the smooth muscle produced by these forcing functions causes vasoconstriction, which changes the resistance and capacitance properties of the vessels. Resistance is increased by any reduction in the diameter of the vessels, and capacitance is decreased by a stiffening of the vessel walls. Arterioconstriction and venoconstriction are treated separately, since the predominant effect on the arterial vessels is an increase in resistance (R_a), whereas the predominant effect on the venous vessels is a decrease in capacitance (C). The autonomic actions on cardiac muscle fibers cause an increased heart rate and strength of contraction. This results in a more complete emptying of the heart that in turn reduces the pressure at which the heart receives blood. Thus the cardiac actions are well modeled by a decrease in the pressure source (P_h). Fig. 8-8, *B*, indicates the actions of the three forcing functions as providing inputs to R_a, C, and P_h.

The next step is to use the cardiovascular circuit to trace the consequences of these controls. The increase in R_a (arterioconstriction) does not affect the output of the flow source (f_a), but it does increase the pressure gradient across R_a, thus increasing arterial pressure (equation 12). Venoconstriction decreases C, and, since the volume of blood does not change, the result is an increase in venous pressure (equation 10). The immediate upstream effect is a small increase in P_a, (equation 12), but the major effect is a downstream one that is best considered along with the effect of heart activation. An increase in P_v (venoconstriction) together with a decrease in P_h (via heart activation) combine to intensify the pressure gradient ($P_v - P_h$) that propels venous return (equation 11). The resultant increase in f_v is matched by a corresponding increase in f_a, and this elevation in cardiac output (resulting from venoconstriction and heart activation) multiplies with the elevated resistance (resulting from arterioconstriction) to produce an increase in arterial pressure.

Thus increased levels of the three forcing functions act in combination to raise cardiac output and arterial pressure. Conversely, decreased levels of arterioconstriction, venoconstriction, and heart activation combine to lower cardiac output and arterial pressure; the student is encouraged to trace each of these latter actions to gain familiarity with the natural flow of causality that emerges from the cardiovascular circuit.

Although causality is clear in the simple model presented here, it is not always as clear in the actual circulation, nor would it be in a more complex model of the circulation. The value of a well-designed simple model is that it helps one to focus quickly on the most significant factors that contribute to any given response. The student who uses such a model should then supplement the conclusions derived from it with his more detailed knowledge of the complete physiologic system.

Global regulation of arterial pressure. Baroreceptors situated in the walls of blood vessels function as sensors of blood pressure that provide negative feedback to a CNS controller located in the medullary portion of the brain stem. The forcing functions discussed previously are the outputs of this controller. The overall system, shown in the bottom half of Fig. 8-8, *C*, is one that acts globally to regulate arterial blood pressure; the local regulatory system, shown in the upper half of Fig. 8-8, *C*, should be ignored for the moment.

An increase in blood pressure inhibits the signals serving as forcing functions, whereas decreased pressure excites them. The resultant compensatory responses are summarized in Fig. 8-8, *C*, as actions of two categories—those that influence blood pressure by first modifying cardiac output (venoconstriction and heart activation) and the one that influences pressure without appreciably affecting flow (arterioconstriction). The latter controls R_a, and this is added to R_v to obtain the total peripheral resistance of the vasculature. The product of peripheral resistance and cardiac output determines arterial pressure. The student should verify that compensatory responses mediated by the neural feedback loop are appropriate to oppose changes in arterial blood pressure.

The steady-state loop gain of this system has been assessed in dogs by comparing the disturbance in blood pressure produced by blood loss before and after blocking the regulatory loops.[32] The values obtained are in the range 3 to 4, considerably less than the values of the loop gain of thermoregulation mentioned earlier. Although a value of 4 does afford an appreciable attenuation of errors (Fig. 8-5), one wonders why the gain is not higher. The reason may relate to stability problems, since damped oscillations in blood pressure called Mayer waves, suggesting a low margin of stability, have been observed by several investigators. Perhaps the forces of evolution have increased the loop gain as high as is compatible with stable performance.

The set point of the cardiovascular regulator can be estimated by measuring blood pressure under resting conditions, which yields mean values in the vicinity of 100 mm Hg. Here one assumes that the current blood volume of the subject, determined by the actions of other homeostatic systems, is the "ideal" one that characterizes the no disturbance state.

Local regulation of blood flow. Superimposed on these pressure-regulating actions are local regulatory responses in each of the tissues of the body. The smooth muscle in the walls of arterioles is responsive to alterations in local metabolites that occur whenever nutrients are depleted or by-products accumulate. Regional vasculatures differ in their respective sensitivities to different chemical substances (Chapter 44) but share the same functional feature of dilating when nutrients are depleted or by-products accumulate and of constricting when the reverse occurs. Fig. 8-8, *C*, illustrates the situation in the case of nutrients. The double arrows are used to indicate that the variables are multidimensional—one dimension applying to each tissue in the body. Any imbalance between nutrient supply and use in a particular vasculature will lead to an accumulation or depletion of the nutrient substance there, provoking a local regulatory response in the arterioles.

Fig. 8-8, *C*, can be used to trace the complex of responses to an alteration in the activity state of a particular tissue. For example, an increase in activity tends to deplete the regional nutrients. Decreased nutrient concentration provokes local vasodilation, thus decreasing the resistance of the local arterioles, which has both local and global consequences. Locally it results in greater blood flow, which tends to restore the nutrient concentration, whereas globally it decreases total peripheral resistance, which tends to lower arterial blood pressure. The latter constitutes a disturbance to pressure regulation, provoking a global regulatory response consisting of vasoconstriction and stimulation of the heart. Vasoconstriction of arterioles tends to restore the peripheral resistance, but this action must compete with the local regulatory responses in each of the regional vasculatures. The actions on the heart and venous vasculature instead serve to increase cardiac output, which accomplishes two things: It tends to restore arterial blood pressure and it also provides the extra blood flow required to meet the metabolic needs of the active tissue without sacrificing blood flow to other tissues.

This example can be summarized by noting that the cardiovascular regulator provides a mechanism for keeping the pressure up while the local regulatory mechanisms drain off as much blood as the regional tissues require. The local responses resolve local needs but in so doing disturb overall pressure regulation. These disturbances in turn force the homeostatic system to meet the real needs of the body, which are not a constant blood pressure but rather an adequate flow of blood to the tissues that need it.

INTEGRATED HOMEOSTATIC ACTIONS

Several important principles regarding the strategies employed in maintaining homeostasis emerge when one compares the regulatory responses of several organ systems. The purpose of this section is to provide some general statements and examples concerning these integrated homeostatic actions.

Activity states

Changes in the activity states of the body are common events that must be supported by adjustments in the flows of substances and energy to and from the active organs and tissues, as illustrated by the previous example. This requires new patterns of blood flow, alterations in respiratory gas exchange, adjustments in heat loss, and many other responses, most of which involve changes in flow variables. However, physiologic regulators generally control force variables (blood pressure, concentrations of substances, and temperature) rather than flow variables. Controlled increases in these force variables might be used to meet the needs for increased flow, but this is not the strategy employed. Instead, homeostatic systems maintain the force variables approximately constant, whereas flows are controlled independently, frequently by local mechanisms in the active tissues themselves (example 2). This strategy ensures that flow is not wasted, since it is delivered preferentially to those locales that need it.

Feedforward regulation

The physiologic mechanisms for feedforward regulation are not well understood at the present time, although it appears likely that this control strategy is an important one in the maintenance of homeostasis. This statement is based partly on the fact that regulated variables may not show the errors in response to disturbances that ordinarily are required to initiate and maintain compensatory feedback actions. For example, the pronounced metabolic vasodilation of muscle arterioles during exercise is a disturbance to the cardiovascular regulator that would be expected to reduce arterial pressure by 10 to 20 mm Hg, based on the assumption that the only compensatory mechanism is negative feedback via baroreceptors and that the loop gain of the system is approximately 4, as discussed earlier. Instead, what is observed is no reduction in arterial pressure, or even a slight rise. Furthermore, there are major cardiovascular changes that occur just before or at the very onset of exercise; these must be attributed to feedforward control, since there is no feedback stimulus to drive them. The situation is similar for the respiratory system, where minute volume increases abruptly at the onset of exercise prior to any changes in blood gas concentrations, and normal blood concentrations of carbon dioxide and oxygen are well maintained during the ensuing exercise, when muscles greatly increase the production of carbon dioxide and the utilization of oxygen.

These features and others suggest that the CNS has effective mechanisms for feedforward homeostatic regulation and that these mechanisms are particularly important in the coordination of transitions in activity state. However, an understanding of the neural processes that achieve these controls awaits the results of future studies.

Homeostatic interactions

Interaction between homeostatic systems arises whenever more than one regulator sends forcing functions to the same organ or tissue. In some cases the different sources of forcing function call for the same response of a target tissue, which presents no problem, but in other cases the requisite responses are opposite in direction, which gives rise to competition.

During exercise, for example, the cardiovascular regulator calls for generalized vasoconstriction that includes the skin vasculature, and this acts to divert skin blood flow to the dilated vasculature of the muscles participating in the exercise. At the same time the increased metabolism of skeletal muscle generates heat, causing body temperature to rise, and the thermoregulator responds by signaling the skin blood vessels to dilate. The thermoregulatory-induced vasodilation of skin blood vessels increases the heat conductance from core to skin, thus promoting heat loss, but this action also lowers peripheral resistance and disturbs cardiovascular regulation. The resulting competition presents no serious problem if the mechanisms for increasing cardiac output are not saturated, since the net result is an increase of blood flow to both muscle and skin (see Table 40-3, p. 1042).

However, during heavy exercise the competition between cardiovascular and thermoregulatory needs becomes critical, since there is a physical limit to the maximal value of cardiac output. The muscles need more blood flow to power contraction, whereas the skin needs more blood flow to conduct heat to the environment. What actually happens is that skin blood flow is sacrificed, which allows additional muscular output while permitting core temperature to rise. Clearly this is a transient situation that has to be terminated (by exhaustion) before body temperature becomes excessive.

One can summarize this discussion by stating that cardiovascular regulatory actions override those of thermoregulation during maximal exercise, whereas during more modest exercise the thermoregulatory actions are satisfied by causing a disturbance to cardiovascular function, one that is appropriately compensated.

As a second example of competition, consider food deprivation in a cold environment. In this case, temperature is maintained by thermogenesis even though this depletes valuable energy reserves, and in the extreme the animal is driven to starvation. Here, thermoregulatory actions override the systems regulating energy conservation. Similarly, there are many other examples of competitive interaction between the homeostatic systems of the body, and the student should explore these as each organ system is encountered.

Pathologic disturbances to homeostasis

Disease states of the body are the result of complex sequences of interactions between pathologic defects and normal physiologic mechanisms. The planning of effective treatment requires that the physician understand the intricacies in these patterns of interaction, for it is only in this manner that he can safely and effectively reverse the etiology and return the patient to a normal physiologic state. The complexity of pathologic states often confuses the student, who may have difficulty in deciding where to begin his conceptual analysis of these complex systems. Here some general guidelines are provided and illustrated with practical examples. The point is also made that normal homeostatic mechanisms may compensate well for some pathologic situations, but they may act poorly in others and actually worsen the state. The physician must, of course, be prepared to block homeostatic actions whenever they are detrimental to recovery.

The etiology of a disease state typically consists of a sequence of transitions from the normal state of homeostasis to states that are progressively more pathologic. Although these transitions are actually continuous, conceptual analysis is greatly facilitated by dealing with discrete stages, as will be illustrated. The problem of diagnosis is, of course, the inverse of this, since one starts with the disease state and attempts to trace backward in time to the initial disturbance that provoked it.

Example 3: Homeostatic adjustments to hemorrhage

Blood loss, or hemorrhage, is a disturbance to cardiovascular function that is compensated by a spectrum of regulatory actions. As a general rule, it is wise to trace the full effects a disturbance would have on the controlled system in the absence of regulatory actions and then consider the regulatory actions, one at a time. The relevant controlled system in this case consists of the biophysical processes making up the circulation. The model discussed earlier (Fig. 8-8, *B*) points to the capacitance properties of the veins as the starting point in tracing the effects of blood loss on the unregulated circulation. A reduction in blood volume causes a reduced storage pressure in the veins, and this reduced venous pressure diminishes venous return, cardiac output, and arterial pressure. The fall in blood flow provokes local regulatory actions, and the fall in pressure provokes global ones. Consider the local actions first.

Reduced flow to any given tissue results in a depletion of nutrients that stimulates arteriolar vasodilation by way of the local regulatory loops shown in Fig. 8-8, *C*. The function of the dilation is to restore local blood flow, but the mechanism is rather ineffective in the absence of pressure regulation. Vasodilation will reduce the total peripheral resistance and simply cause a further fall in arterial pressure, without a significant restoration of cardiac output. The tendency for arterial pressure to fall further can be considered as an additional stimulus to the homeostatic system regulating blood pressure.

The neural regulatory actions restore pressure by either increasing venous return and cardiac output or increasing peripheral resistance. The former action meets the requirement for a restoration of blood flow, whereas the latter does not. Instead, it diverts the limited flow from tissues such as skin and viscera that have weak local regulatory responses to ones less capable of withstanding a reduction in blood flow, such as the heart and CNS, that have strong local regulation. Thus the compensatory actions of the pressure regulator can be expected to provide only a partial restoration of cardiac output. Note also that the pressure regulator does nothing to restore blood volume.

The next level of homeostatic compensation appended to the previous actions concerns the mechanisms that restore blood volume. One is the bulk movement of interstitial fluid into the vascular compartment that is promoted by a reduction in capillary hydrostatic pressure (Chapter 43). The reduced venous pressure and arteriolar constriction mentioned earlier are separate factors contributing to the reduced capillary pressure and the movement of fluid. Finally, there is another homeostatic system that regulates the volume of body fluids through physiologic actions on the rate of urine formation and behavioral actions controlling thirst. Volume regulation is slower in action than is arterial pressure regulation, and it may fail to compensate adequately when the hemorrhage is rapid and severe.

When blood loss is modest, the combined compensatory actions just described function in a coordinated manner to provide an adequate blood supply to critical tissues while blood volume is restored. Note that none of the mechanisms acting in isolation would be adequate—the efficacy of homeostasis rests on their integrated actions. Modest hemorrhage thus may result in no critical failure in cardiovascular function. However, when a severe deficit in blood volume persists for some time, the reduced blood flow to the skin and viscera may lead to anoxia of the vascular muscle there, followed by a collapse of peripheral resistance and a marked fall in blood pressure. The pathology then spreads to other tissues and may become irreversible. Present treatment is based on action before these irreversible changes begin. Blood volume is restored by transfusion, and vasodilator substances are given to relieve the anoxia of the vascular smooth muscle. The "normal" homeostatic action of vasoconstriction must be interfered with to promote recovery.

Example 4: Fever

Sometimes pathologic disturbances act at early stages of homeostatic systems rather than on the controlled system, as in the previous example. This produces a particularly devastating situation, since negative feedback does not compensate well for disturbances in controller or sensor function. As an example, consider fever, which is provoked by direct actions of pyrogens, released by bacteria, on the thermosensitive neurons serving as sensors for thermoregulation. The actions increase the discharge rates of cold-sensitive neurons just as a decrease in temperature normally would do. The thermoregulatory system, being unable to distinguish between a decrease in temperature and a pyrogenic action, responds by initiating all the "normal" physiologic mechanisms for heat conservation. Core temperature rises and is then maintained constant at an abnormally high level. This illustrates the general point that a disturbance that is applied at an early state in a feedback loop has an effect analogous to a modification in the set point of the system. Efforts to cool the patient are resisted by vasoconstriction and shivering thermogenesis, since they are interpreted as disturbances tending to reduce temperature below its disturbed reference level.

Multiple sensors and the nature of regulated variables

It is important to identify regulated variables, since this is one way of specifying the function of a feedback system. For cases in which only a single variable is fed back, it follows that this variable (or a variable derived directly from it) is the regulated one, since the best a negative feedback system can do is to minimize disturbances in the monitored variable. However, physiologic regulators receive feedback from large numbers of sensors, each of which may monitor a different variable, depending on the site of the sensor and its special properties. For example, temperature receptors and thermosensitive neurons are located at sites throughout the body, and temperature changes at these various sites are adequate stimuli for thermoregulatory responses.

Given that more than one variable is available as feedback to a controller, the next question concerns the manner in which this additional information is used. The possibilities vary along a continuum between two extremes in strategy. One extreme is to use the information to control each of the monitored variables independently and simultaneously. The other is to combine the information about several variables to form a single *regulated property*. A first step in distinguishing between the alternative strategies is to determine the number of degrees of freedom available in the control of forcing functions. If there is but a single degree of freedom in this control, there cannot be more than one regulated property. As the number of degrees of freedom increases, so does the potentiality for the simultaneous control of several variables or properties.

The physiologic mechanisms of temperature

regulation appear to be organized to regulate a single property, since there is little suggestion of more than a single degree of freedom in control (however, the behavioral mechanisms may add extra flexibility[14]). As described earlier, there are three principal categories of forcing function for thermoregulation: the vasomotor mechanism that is used alone in the region of fine control, the sudomotor mechanism controlling evaporative heat loss that is added under conditions of heat stress, and the thermogenic mechanism that is added under conditions of cold stress. Although these mechanisms could be controlled independently, the evidence suggests that they are evoked as coordinated responses in a stereotypic pattern, regardless of which thermosensors are stimulated. Apparently the central controller for thermoregulation combines the thermosignals from all parts of the body to create a single regulated property, and recent research efforts have been devoted to the discovery of the relative contributions of different body temperatures.[14] One finding is that in man and other large animals the dependence on core temperature is much greater than that on skin temperature, but the reverse is true for small animals.

An interesting consequence of having both skin and core temperature contribute to the regulated property occurs during exposure to a cold environment. Heat loss from the skin is promoted by a low ambient temperature. The subsequent reduction in skin temperature promotes heat loss from the core, but it also provokes a thermoregulatory response, one consisting of vasoconstriction of skin blood vessels and shivering thermogenesis. Skin vasoconstriction also has two effects that are important to trace: a reduction of the thermal conductance from core to skin, which prevents excessive heat loss from the core, and a further fall in skin temperature, which reinforces the original thermoregulatory responses. The reinforcing nature of the vasoconstrictor response, which represents a positive feedback component of the overall response, is often sufficiently powerful to drive the shivering mechanism in the absence of any reduction in core temperature; in fact, Fig. 59-29 illustrates the results of an experiment in which core temperature actually rose during exposure to a cold environment.

The inclusion of skin temperature in the thermal reference is apparently advantageous (presuming, for the moment, that the real goal is to maintain core temperature) when disturbances are restricted to changes in ambient temperature. Skin temperature reflects these changes and provides an "early warning" mechanism for thermoregulation. However, when the disturbance is an increase in metabolism, the situation is less clear, since increases in core temperature are tolerated, whereas skin temperature may actually fall (as if skin temperature were the real regulated variable). These puzzles concerning biologic design are circumvented if one simply concludes that the thermoregulatory system functions to minimize disturbances in a thermal property made up of a weighted combination of temperatures throughout the body.

Example 5: Regulation of the stiffness of skeletal muscle

The performance of skeletal muscles during movement and posture is regulated by feedback from muscle stretch receptors (Chapter 28). Fig. 8-9, *A*, is a simplified block diagram of this system in which the reflex pathways that serve as neural feedback loops have been divided into two basic categories: a loop from muscle spindle receptors that monitors and regulates muscle length and a loop from Golgi tendon organs that monitors and regulates muscle force. The sensors of this system (spindles and tendon organs) and the controlled system (muscle and load) provided the topics for examples 2 and 3 of Chapter 7, and the student may find it helpful to review these before proceeding with the present example.

One of the interesting features of this regulatory system is the dual feedback from sensors monitoring two different mechanical variables. This suggests that some regulated property composed of muscle length and force components may be maintained constant by the compensatory actions, rather than either length or force variables individually. The alternative, that both length and force are regulated simultaneously, can be excluded, since it is physically impossible to keep both constant when disturbances cause changes in mechanical load. (Length can be maintained only by resisting the new load with a new force and, conversely, force can be maintained only by failing to resist the stretch imposed by an added load.) Experimental evidence and theoretical considerations both suggest that the regulated property is stiffness, which is the ratio of force change to length change.[27]

The student will recall from an earlier example that the inherent mechanical properties of the muscle present a certain stiffness in the absence of any neural feedback (the spring constant K in Fig. 7-12, *A*). However, the stiffness presented by the muscle is quite variable, since it depends in a highly nonlinear manner on muscle length, velocity, and level of recruitment (example 3 in Chapter 7). Neural feedback apparently functions to minimize these variations and, in so doing, it maintains the net stiffness presented by the neuromuscular system relatively constant. Fig. 8-9, *B*, shows an example in which the mechanical stiffness of the muscle failed during the application of a stretch and neural regulatory actions promptly compensated for this failure.

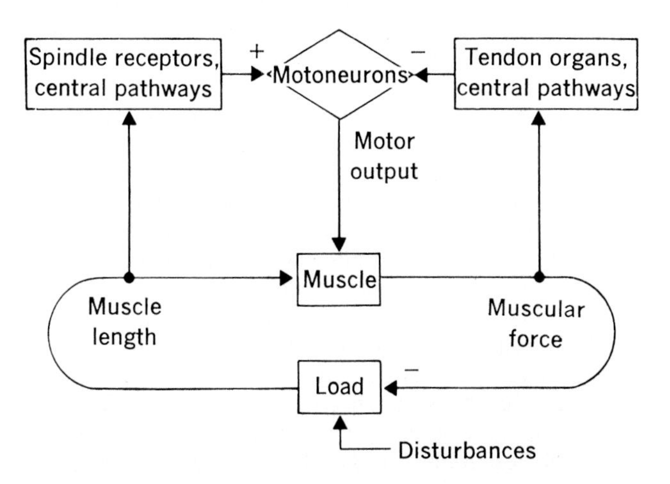

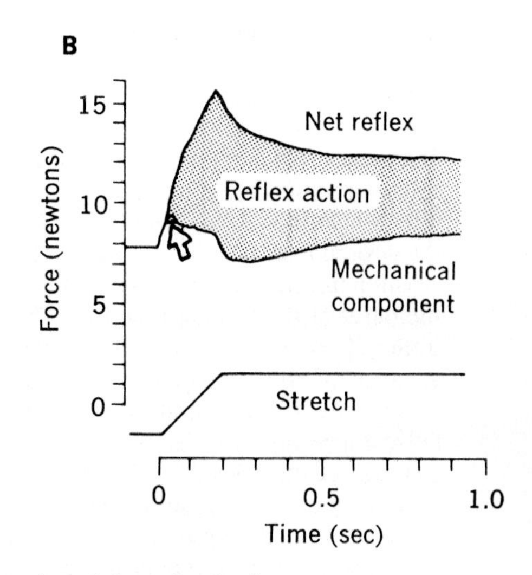

Fig. 8-9. Regulation of stiffness resulting from combined length and force feedback to motoneurons. **A,** Block diagram showing mechanical loop of interaction between muscle and load and two neural pathways conveying length and force feedback. **B,** Example of stiffness regulation.[27] Trace labeled mechanical component represents response to stretch when neural feedback is inoperative; trace labeled net reflex is response of intact system; difference represents reflex action, that is, effect of neural feedback. Failure in muscular stiffness that occurs after small amount of stretch (arrow) is promptly compensated by reflex action.

The maintenance of a constant stiffness appears to have two biologic advantages. One is that it provides a springlike interface between the body and its mechanical environment, thus absorbing the impacts of external disturbances in analogy with the suspension system of a car. A second advantage may be the constancy itself, since this would simplify the task of higher neural centers that compute movement commands. These commands must be adjusted to be appropriate for controlling the properties of the system to which they are sent, and, if these properties are unvarying, the adjustments can be made routinely.

THE CONTROL PROBLEM

Previous sections of this chapter have emphasized problems in regulation, the goal being to maintain controlled variables constant, whereas the purpose of this section is to discuss contemporary approaches to the control problem, where the goal is to cause the controlled variables to vary in some prescribed manner. Fig. 8-10 illustrates the control problem and a trivial (and usually unrealizable) solution to it, one based on the feedforward principle. *A* shows a controlled system (S) and a controller (C). The latter must generate a forcing function (x) that causes a controlled variable (y) to vary with time in the manner specified by an arbitrary command signal (w). The trivial solution is to make C equivalent to the inverse of a model of S, i.e., $C = S^{-1}$, as shown in *B*. This done, the general result will be y = w, as shown in *C*. The realistic limitations to this solution are well illustrated by considering a biologic example, a problem concerning the control of eye movements.

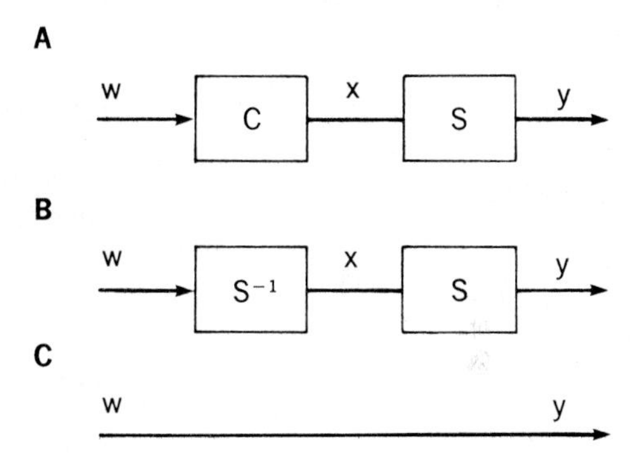

Fig. 8-10. Ideal solution to control problem. **A,** Command signal *w* to controller *C* produces forcing function *x* that is applied to controlled system *S* to change controlled variable *y*. **B,** Controller is assumed to have properties that are the inverse (S^{-1}) of those of controlled system. **C,** This yields ideal performance y = w.

The function of the eye movement system appears to be one of aiming the eyes at interesting targets and maintaining the gaze on targets that move. The controlled system, discussed in example 3 of Chapter 7, consists of the mechanical properties of the eye muscles, the eyeball, and the orbit. Fig. 8-11, *A*, summarizes its dynamic characteristics by showing a step response, the movement y produced in response to the step-forcing function x sent by ocular motoneurons. Note that there is a time delay due to conduction in motor axons and muscle activation processes, after which the movement sluggishly approaches the desired final position. Note also that there is a reaction time delay between target movement and the onset of the forcing function. The ideal controller ($C = S^{-1}$) (1) sends a forcing function in advance of the desired eye movement and (2) makes this signal infinitely large initially (impulse function) in order to overcome the sluggishness of the response (Fig. 8-11, *B*).

Neither of these requirements can be met in practice. The first assumes knowledge of a target's position before it appears, which defies the law of causality. (The inverse of a time delay exists, but it is unrealizable.) The second requirement is also unrealizable, since infinitely large signals will saturate any real system.

Fig. 8-11, *C*, shows the practical solution provided by the oculomotor control system. The reaction time and conduction delays are not compensated, but the sluggishness of the position change is diminished by the generation of what is called a pulse-step type of forcing function.[30] The pulse component acts to accelerate the initial trajectory of the eye movement, whereas the step component holds the new position.

The reaction time is most interesting to investigators of brain function. It represents the time taken by the biologic controller to generate a new forcing function following the appearance of a novel target. At present, little is known about the actual nature of this process, although many current models presume that it begins with a calculation of an error signal representing the difference in position between the target and the fovea.[29] This type of controller exemplifies a classic design, the servomechanism, employed in the majority of technologic control systems in use today. In the remainder of this section a brief summary will be given of this and several other contemporary approaches to controller design.

A *servomechanism* is a control system that operates on the principle of negative feedback, discussed earlier with respect to regulator function (Fig. 8-3). The major difference here is that the

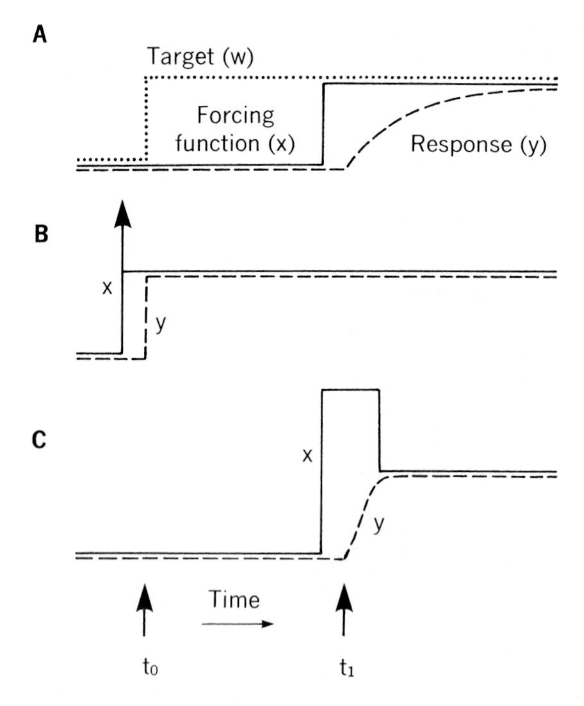

Fig. 8-11. Eye-movement control example. **A,** Forcing function is assumed to be step change in oculomotor output occurring a reaction time after movement of visual target. Response is eye movement with exponential time course eventually settling on target position. **B,** Unrealizable forcing function required to achieve ideal control, as in Fig. 8-10; upward arrow represents impulse function, generated in advance of target motion. **C,** Actual forcing function is of pulse-step form. Pulse portion serves to reduce time required for eye to achieve target position, but there is no compensation for reaction time delay. The variables w, x, and y are as in Fig. 8-10.

reference signal, previously assumed constant, is now allowed to vary as a function of time and, correspondingly, is called a command signal. A derivation similar to equations 1 through 6 can be used to demonstrate that the controlled variable y will respond to the command signal w according to the following:

$$y = SC(1 + FSC)^{-1}[w] \quad (13)$$

where S, C, and F are operators representing the respective properties of the controlled system, the controller, and the sensor. Assuming a faithful sensor (with properties F = 1) and a high loop gain (FSC $\gg$ 1), equation 13 can be approximated by the following:

$$y \cong SC(SC)^{-1}[w] = w \quad (14)$$

One concludes that ideal control (y = w) can be approached if it is possible to raise the loop gain to very high values. In practice, this requirement may be difficult or impossible to achieve due to the problems of instability in feedback loops discussed earlier. This certainly is a relevant limitation in eye movement control due to the long latency in the control pathway associated with a reaction time.

Sampled-data control systems are servomechanisms that use feedback in an intermittent manner as a result of the introduction of a discontinuous element called a sampler. For example, the controlled variable may be sampled at discrete increments in time, in which case controller operation during the intersample period is based on the value of the previous sample(s). Although this control strategy suffers the obvious disadvantage of throwing away information (the values of the controlled variable during the intersample period), it presents some advantages that may be significant biologically. First, it opens the possibility of time-sharing operations, in analogy with those performed on large computers. Although it is certain that the brain performs many operations simultaneously (unlike most contemporary digital computers) and, therefore, does not require time sharing, there is clear evidence that some operations interfere with others, which may necessitate time scheduling, if not actual time sharing.

Second, and perhaps more significant, there is the advantage that sample-data control can lead to improved stability in a feedback system. If the sample rate is sufficiently low, the system will settle to the output state before it is time to take a subsequent sample and update the forcing function. A system operating in this manner can be designed to have a near-perfect steady-state response (y = w) and yet remain stable. The major disadvantage is that the speed of response to a novel stimulus is limited by the sampling interval. There is excellent evidence that human motor control involves intermittent operations (see Navas and Stark[26]), although they are probably not strictly analogous to those of a sample-data control system (see Robinson[29]).

Model reference control systems are of several types, but all have in common the use of a component that represents a model of the actual controlled system (see Houk[22] and Truxal[35]). The forcing function is sent to both the model S_2 and the actual system S_1, as shown in Fig. 8-12, *A*, and an error signal ($y_2 - y_1$) is created by subtracting the actual output from the model output. Different uses may then be put to this error signal as shown by options 1, 2, and 3 in Fig. 8-12, *A*. The basic idea of model reference comparison was originated by Helmholtz to explain the absence of a sensation of world movement during voluntary eye movement and later applied to a variety of biologic problems by von Holst and Mittlestaedt, MacKay, and others.[24] Some of the engineering designs based on this principle are discussed here.

In a conditional feedback system (option 1) the error is processed (e.g., amplification) and summed with the original forcing function to provide a modified forcing function that is sent to the controlled system. This configuration is designed to cancel feedback under conditions in which the controlled system responds precisely, as does the model, which is made to represent ideal performance. When disturbances interfere with ideal performance, an error signal is produced, amplified, and applied to the controlled system so as to reduce the error. The efficacy of error reduction depends on the loop gain of the negative feedback pathway in the same manner discussed previously, and the system is also subject to instability if there is too much delay in the loop or if the gain is too high. The major advantage of the conditional feedback configuration is that feedforward and feedback features can be designed separately, the former by altering the dynamic properties of the controller (attempting to make C approach S_1^{-1}) and the latter by altering gain and dynamic properties of the feedback loop. This might be advantageous in biologic design (evolution) as well.

Additional theoretical advantages result if the processed error signal is also fed back to the model system (option 2 in Fig. 8-12, *A*). This creates a positive feedback loop that, if appropriately tuned, can actually result in improved sta-

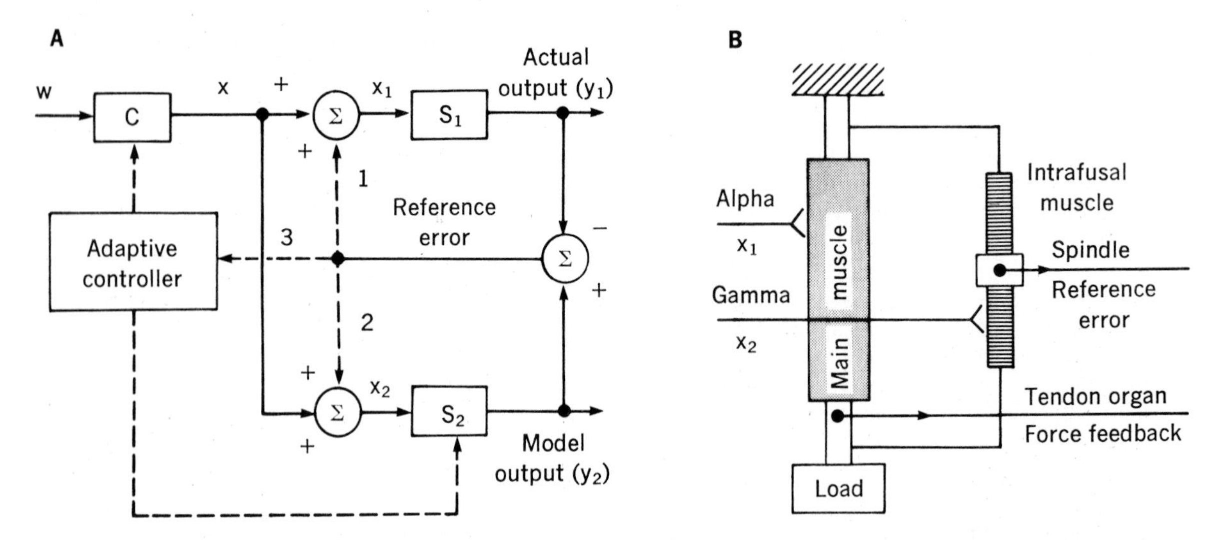

Fig. 8-12. Model reference control configurations. **A,** S_2 is model of actual controlled system S_1. Reference error detects differences between model's ideal output (y_2) and actual output (y_1). Options *1, 2,* and *3* show different ways in which reference error signal can then be used. **B,** Diagram to illustrate how spindle receptor can be viewed as model reference detector. Controlled system S_1 is main muscle and its load, intrafusal muscle is assumed to be model system S_2, and receptor is situated to detect differences between degree of main and of intrafusal muscle shortening.

bility along with greater attenuation of errors. In fact, errors may be reduced to zero if the gain of the positive feedback loop is precisely 1, and this gives rise to what has been called a "zero-sensitivity" system, since it is insensitive to disturbances.[35] The third option in Fig. 8-12, *A*, involves an adaptive controller that uses the error signal as a basis for modifying the properties of the main controller C and/or the model S_2.

Example 6: The muscle spindle as a model reference sensor

The special structure of the muscle spindle receptor (Chapter 25) is one that suggests the possibility of model reference operation. This structure and its relationship to the main muscle are shown schematically in Fig. 8-12, *B*. The spindle consists of miniature muscle fibers (called *intrafusal*) that have a central sensory zone where spindle afferent fibers terminate. The sensory zone displays less contractile activity than the poles; thus when the intrafusal muscle is activated, the sensory zone is stretched, causing afferent discharge. Furthermore, the whole structure lies functionally in parallel with the main muscle; thus when the muscle lengthens, this also stretches the sensory zone, causing afferent discharge.

The dual responsiveness to intrafusal contraction and muscle stretch provides the basis for a model reference calculation, if it is assumed that the intrafusal muscle represents a model of the controlled system, the main muscle and its load. In this view the forcing function to the controlled system (x_1 in Fig. 8-12, *A*) is conveyed by action potentials in the large α-motor fibers innervating the main muscle, whereas the forcing function to the model (x_2) is conveyed by signals in the small γ motor fibers that innervate intrafusal muscle. Movements are produced by sending signals in both α- and γ-motor axons. If the subsequent shortening of the main muscle is equal to the shortening of the intrafusal muscle, the stretch applied to the sensory zone should not change, nor should spindle discharge. This represents the null condition described earlier, in which actual output equals model output, and the reference error in Fig. 8-12, *A* is zero. The student should work through the situation in which a large load interferes with the shortening of the main muscle, causing a reference error in the form of increased discharge of spindle afferents.

The further development of this analogy with model reference control systems requires that spindle afferents project appropriately to complete the feedback loops. The well-documented excitatory projections to α-motoneurons complete the loop required for the conditional feedback configuration (option 1 in Fig. 8-12, *A*). Matthews[25] has supported this analogy and pointed out that the resultant reflex would reinforce contractions that are heavily loaded, a phenomenon he refers to as "servo assistance." However, no one has yet investigated the specific loading conditions under which the null condition prevails (i.e., the situation in which the controlled system behaves precisely as the model). There are probably also feedback connections with γ-motoneurons (option 2), but they appear to be weak. The extent to which they meet the unity gain condition required for zero sensitivity is not presently known. Spindle afferents also project to the brain, where reference errors may be used to compute adaptive modifications (option 3), although even less is known about this possibility.

One shortcoming of this analogy is its failure to take into account the force signals from tendon organs (Fig. 8-12, *B*), which, as discussed in example 5, provoke inhibition of the main motoneurons. However, this added projection requires only that one modify the model reference theories, which assume that muscle length is the regulated variable, to include the idea that it is actually a property, the stiffness of the system, that is maintained by regulatory actions. The error in stiffness would be computed at the level of α-motoneurons by combining the reference error, which is a length error, with the inhibitory signal from tendon organs.

Optimal control theory

At the beginning of this section, it was pointed out that ideal control ($C = S^{-1}$) usually cannot be achieved due to time delays and limits for the maximal values of forcing functions. A tacit assumption in this statement is that ideal control involves a perfect replication of the trajectory of a target. However, part of the design policy may be to limit the amplitudes of signals or to prevent energy consumption from becoming excessive, even though this may necessitate a relaxation of the requirement for a speedy response to rapidly changing inputs.

Classic control theory deals with these problems of realistic limits to the values of variables and realistic criteria for the evaluation of performance in a trial and error manner. A particular controller is assumed, and the consequences of this assumption are tested by calculating the overall properties of the system or responses to typical inputs. If the consequences are unacceptable, a new assumption for the controller is tried, and this process is continued until the performance is judged acceptable and realistic.

Optimal control theory provides a more systematic approach to these problems. The criteria for acceptable performance are stated as a mathematical equation, a *criterion function,* and realistic limits of the variables are stated in the form of other equations called *constraints*. The problem then becomes one of minimizing (or maximizing) the criterion function subject to the constraints, and this problem is solved by the calculus of variations or by a computer algorithm (dynamic programming). The essence of this approach concerns the use of the criterion function, which may embody a variety of design policies. For example, the policy may be to minimize the time required for the control system to follow a given class of inputs, it may be to minimize energy expenditure, it may be to maximize the value of one or more output variables, it may be to minimize the average error in response to a class of input functions, or it may be some combination of such criteria.

Although the optimal approach is quite general, its utility is limited by the fact that only certain forms of criterion functions can be evaluated using the calculus of variations. Dynamic programming[5] overcomes many of these difficulties at the expense of computational time, which may become excessive even when the calculations are done on large computers, and there remain categories of important problems for which algorithms that are generally applicable have not been discovered. Thus although much progress has been made in theoretical approaches to the problem of optimal controller design, there remain many important problems that have not been solved.

Function generators

A successful solution to an optimal control problem yields the time course of the forcing functions required to optimize performance. One is still left with the problem of constructing a function generator that will produce these signals in response to a simpler command signal, for example, a triggering pulse or a signal depicting the trajectory of a target. This stage of the problem presents no major difficulty in engineering, since a variety of physical devices are available to do the job. However, the discovery of the mechanisms by which neurons interact to generate forcing functions remains a significant problem in biology today. Recently, important progress has been made in understanding the organization of the pattern generators for locomotion, although the specific neural mechanisms have not yet been defined.[17] Much less is known about the mechanisms that generate signals controlling willed movements, although some intriguing hypotheses have been proposed.[23]

SUMMARY

A fundamental goal in physiology is to understand the mechanisms by which living systems control their internal processes and transactions with the surrounding environment. Control theory provides a general language and some guiding principles helpful in approaching these problems.

Many internal variables of the body are maintained remarkably constant (homeostasis) due to the actions of physiologic control systems that function as regulators. The values of the regulated variables are generally influenced by uncontrollable inputs, called disturbances, and by forcing functions sent from a controller; regulation is achieved when the latter counterbalance the former.

The two basic strategies for computing the required forcing functions are feedback and feedforward. The controller of a feedback system may be no more complex than a pathway through which changes in regulated variables are sensed, amplified, and applied to the controlled system as forcing functions. The sign must be arranged so that any change in a regulated variable results in forcing functions that oppose that change (i.e., *negative* feedback). The efficacy of regulation is increased by higher gain up to the point at which this causes overcompensation and instability. Feedback compensation is incomplete, since it depends on an error as the stimulus for the compensatory response, and it may be slow, since compensation must await the development of the error. The controller of a feedforward regulator anticipates errors in regulated variables by sensing disturbances directly; this allows compensatory responses to be generated before the development, or even in the absence, of errors in the regulated variables. However, the required computations are more complex than in the case of feedback, and failures may result in quite erroneous responses that would remain undetected in the absence of feedback. Regulators in living systems appear to employ both principles in a balanced combination, although at present most is known about feedback mechanisms.

In other cases, control systems function to coordinate specific changes in output variables, rather than to keep them constant. Servomechanisms employ negative feedback for this purpose; as in the case of regulator function, high gain improves performance, but may lead to instability. The introduction of sampling can improve stability, at the expense of speed, and it also opens possibilities for time-sharing operations. Several interesting control configurations are described that derive from the concept of model reference error signals, the latter representing the difference between the actual response and the response of a model that represents ideal performance. The chapter ends with a description of the optimal control approach to controller design.

REFERENCES

General reviews

1. Adolph, E. F.: Physiological regulations, Tempe, Ariz., 1943, Jaques Cattell Press.
2. Anderson, B. D. O., and Moore, J. B.: Linear optimal control, Englewood Cliffs, N.J., 1971, Prentice-Hall, Inc.
3. Arbib, M. A.: The metaphorical brain—an introduction to cybernetics as artificial intelligence and brain theory, New York, 1972, Wiley-Interscience.
4. Baylis, L. E.: Living control systems, San Francisco, 1966, W. H. Freeman & Co., Publishers.
5. Bellman, R. E.: Adaptive control processes: a guided tour, Princeton, N.J., 1961, Princeton University Press.
6. Elgerd, O. I.: Control systems theory, New York, 1967, McGraw-Hill Book Co.
7. Mishkin, E., and Braun, L., Jr.: Adaptive control systems, New York, 1961, McGraw-Hill Book Co.
8. Stark, L.: Neurological control systems: studies in bioengineering, New York, 1968, Plenum Press.
9. Tou, J. T.: Digital and sampled data control systems, New York, 1959, McGraw-Hill Book Co.
10. Wiener, N.: Cybernetics or control and communication in the animal and the machine, New York, 1961, The M.I.T. Press and John Wiley & Sons, Inc.
11. Yamamoto, W. S., and Brobeck, J. R.: Physiological controls and regulations, bicentennial volume, Philadelphia, 1965, W. B. Saunders Co.

Original papers

12. Adolph, E. F.: Physiological adaptations: hypertrophies and superfunctions, Am. Sci. **60:**608-617, 1972.
13. Bernard, C.: Leçons sur les phénomènes de la vie communs aux animaux et aux végétaux, Paris, 1878, Editions J-B Baillière.
14. Cabanac, C.: Temperature regulation, Annu. Rev. Physiol. **37:**415-439, 1975.
15. Cannon, W. B.: The wisdom of the body, New York, 1932, W. W. Norton & Co., Inc.
16. Cornew, R. W., Houk, J. C., and Stark, L.: Fine control in the human temperature regulation system, J. Theor. Biol. **16:**406-426, 1967.
17. Grillner, S.: Locomotion in vertebrates: central mechanisms and reflex interaction, Physiol. Rev. **55:**247-304, 1975.
18. Grodins, F. S., and Buoncristioni, J. F.: General formulation of the cardiovascular control problem—mathematical models of the mechanical system. In Reeve, E. B., and Guyton, A. C., editors: Physical bases of circulatory transport: regulation and exchange, Philadelphia, 1967, W. B. Saunders Co., pp. 61-75.
19. Guyton, A. C.: Circulatory physiology: cardiac output and its regulation, Philadelphia, 1963, W. B. Saunders Co.
20. Guyton, A. C.: Regulation of cardiac output, N. Engl. J. Med. **277:**805-812, 1967.
21. Hardy, J. D., and DuBois, E. F.: Basal metabolism, radiation, convection and vaporization at temperatures of 22 to 35° C, J. Nutr. **15:**477-512, 1938.
22. Houk, J. C.: The phylogeny of muscular control configurations. In Drischel, H., and Dettmar, P., editors: Biocybernetics, Stuttgart, 1972, Gustav Fischer Verlag, vol. 4, pp. 125-155.
23. Kornhuber, H. H.: Motor functions of the cerebellum and basal ganglia, Kybernetik **8:**157-162, 1971.
24. MacKay, D. M., and Mittelstaedt, H.: Visual stability and motor control (reafference revisited). In Keidel, W. D., Handler, W., and Spreng, M., editors: Cybernetics and bionics, Munich, 1974, R. Oldenbourg Verlag GmbH, pp. 71-80.
25. Matthews, P. B. C.: Mammalian muscle receptors and their central actions, Baltimore, 1972, The Williams & Wilkins Co., pp. 566-574.
26. Navas, F., and Stark, L.: Sampling or intermittency in hand control system dynamics, Biophys. J. **8:**252-302, 1968.

27. Nichols, T. R., and Houk, J. C.: Improvement in linearity and regulation of stiffness that results from actions of stretch reflex, J. Neurophysiol. **39:**119-142, 1976.
28. Pavlov, I. P.: Work of the digestive glands, London, 1910, Charles Griffin & Co., Ltd.
29. Robinson, D. A.: Models of the saccadic eye movement control system, Kybernetic **14:**71-83, 1973.
30. Robinson, D. A.: Oculomotor control signals. In Lennerstrand, G., and Bach-y-rita, P., editors: Basic mechanisms of ocular motility and their clinical implications, Oxford, England, 1975, Pergamon Press, Ltd., pp. 337-374.
31. Rosenblueth, A., Wiener, N., and Bigelow, J.: Behavior, purpose and teleology, Philos. Sci. **10:**18-24, 1943.
32. Sagawa, K.: Relative roles of the rate sensitive and proportional control elements of the carotid sinus during mild hemorrhage. In Baroreceptors and hypertension, Oxford, 1967, Pergamon Press, pp. 97-105.
33. Sagawa, K.: The use of control theory and systems analysis in cardiovascular dynamics. In Cardiovascular fluid dynamics, London, pp. 115-171, 1972, Academic Press, Inc. (London), Ltd.
34. Stein, R. B.: The peripheral control of movement, Physiol. Rev. **54:**215-243, 1974.
35. Truxal, J. G.: Control systems—some unusual design problems. In Michkin, E., and Braun, L., editors: Adaptive control systems, New York, 1961, McGraw-Hill Book Co., Chapter 4.

IV

GENERAL PHYSIOLOGY OF THE FOREBRAIN

9

GIAN F. POGGIO and VERNON B. MOUNTCASTLE

Functional organization of thalamus and cortex

In previous chapters the mechanisms of axonal conduction and synaptic transmission have been considered in some detail. The ground was thus laid for study of the central nervous system, which is introduced in this and the following chapter with a discussion of the general aspects of thalamocortical structure and function, and of those systems of the forebrain concerned with controlling levels of excitability, and thus awareness. A nervous system is a collection of cells specialized to convey signals in rapid tempo and with great fidelity. Its forte is the transmission in neural code of information concerning the state of the body's internal measuring devices, of the immediate external environment, and of more distant objects and events in the external world and the dispatch of command signals to effectors for corrective action and—at will, independent of input—for the initiation of action on the environment. Between these input and output channels there intervenes the working brain, an information-processing, data-storing, decision-making machine: the executant organ of behavior. The complexity of that behavior sets animals endowed with nervous systems apart from other forms o life, and in animal phylogeny, increasing complexity of behavior is in general paralleled by increasing size and complexity of brain. The functions of brains fall into two classes: homeostatic regulation and the initiation of action, although many functions show properties of both.

In the domain of homeostatic regulation, for example, are mechanisms that control body temperature, maintain appropriate levels of blood pressure and pulmonary ventilation, regulate steady and cyclic endocrine function, and preserve or restore an erect body position in the face of the collapsing force of gravity. They are in general invariant and automatic in that they occur without conscious volition: they are reflexes, stereotyped items of behavior destined for execution on appropriate signal unless consciously withheld, a restraint possible only for those involving somatic musculature. In the domain of initiation of action lie those items of exploratory and appetitive behavior that apparently begin independently of any external evoking stimulus. Although these may serve regulatory ends, at certain levels of complexity they appear to be truly spontaneous and mark that independence and freedom from environmental contingencies characteristic of animals with large brains. Along this spectrum they merge with those activities that are free of both stimulus from and action on the environment—learning, thinking, remembering—actions known only by introspective evaluation and, at choice, public description.

Study of brains from small to large reveals no striking phylogenetic innovations in cellular morphology or synaptic mechanisms.[2] What is noteworthy along this scale is the increasing number of neuronal units, the extensive elaboration of the number and complexity of their interconnections, and the flowering of interconnected systems and subsystems. Thus the emerging intricacies of brain function will find explanation, it is believed, in the emergent functional capacities of large populations of neural elements, properties not immediately and completely deducible from those of single cells, as far as they are presently known. Thus attention is first directed to the general properties of the largest accumulation of cells, the forebrain, and to the systems and subsystems that exert general, not specific, control of its function.

ANATOMIC ORGANIZATION OF THALAMOCORTICAL SYSTEMS

The thalamic nuclear mass includes a variety of cellular aggregates with different anatomic organization and functional significance. Some of these nuclei or groups of nuclei are associated with systems that subserve the processing of signals from the external world; other thalamic regions are interconnected with neural structures, both cortical and subcortical, that are known to play important roles in somatic motor and autonomic functions. Finally, a large portion of the dorsal thalamus, especially developed in man and primates, is part of the telencephalic systems operating in perceptual and integrative higher functions. Thus the thalamus and cerebral cortex, with their complex interconnections, constitute an anatomic substratum on which rest our perceptions of the external world and of ourselves within it, as well as our actions and motivational responses.

Development of the thalamus

At early stages of development the paired thalamic primordium lies in the diencephalic wall of the third ventricle, separated from the more ventral hypothalamus by the sulcus hypothalamicus. Within the thalamic plate, three areas differentiate: the dorsal area is termed the *epithalamus;* the middle and larger, the *dorsal thalamus;* and the most ventral, the *ventral thalamus* (Fig. 9-1). Within each of these areas, clusters of cells differentiate at various times during ontogenesis, the dorsal thalamus being still largely undifferentiated when the nuclei of the epithalamus and ventral thalamus have already developed completely.

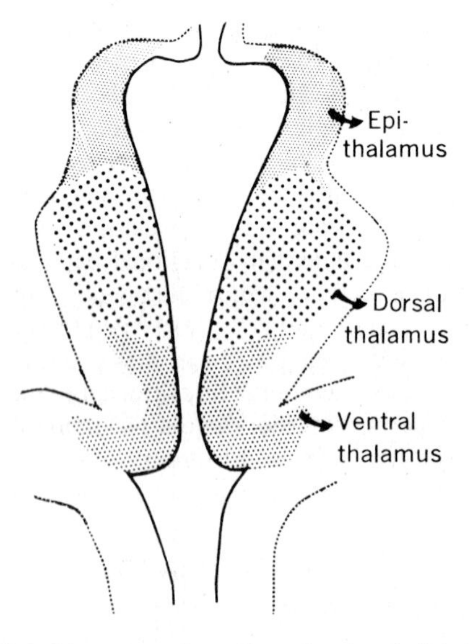

Fig. 9-1. Diagrammatic outline of oral end of thalamic plate from an 18 mm rabbit embryo. The three subdivisions are shown. (From Ajmone Marsan.[1])

General classification of thalamic nuclei and their telencephalic projections

In the adult the thalamic mass may be subdivided on the basis of topographic relations by differences in the cyto- and myeloarchitectural characteristics of different parts, as well as by the fiber connections between thalamic parts and the remainder of the central nervous system (CNS). Such a division is given in Table 9-1, a classifi-

Table 9-1. Thalamic subdivisions and nuclear classification

Subdivision	Classification of nuclei
Epithalamus	N. habenularis (Hb) N. paraventricularis (Pv) Regio pretectalis (PT)
Dorsal thalamus	Anteromedial N. anteroventralis (AV) N. anterodorsalis (AD) N. anteromedialis (AM) N. lateralis dorsalis (LD) N. medialis dorsalis (MD) Nuclei of the midline Intralaminar N. paracentralis (Pc) N. centralis lateralis (Cl) N. parafascicularis (Pf) N. centrum medianum (CM) Ventrolateral N. ventralis anterior (VA) N. ventralis lateralis (VL) N. ventralis posterior: ventroposterolateralis (VPL) ventroposteromedialis (VPM, VPMpc) ventroposteroinferior (VPI) Caudal N. geniculatus lateralis dorsalis (GLd) N. geniculatus medialis (GM) Posterior complex: N. suprageniculatus (Sg) N. limitans (Li) N. posterior (Po) Dorsolateral N. lateralis posterior (LP) Pulvinar complex: pulvinar anterior (PA) pulvinar medialis (PM) pulvinar lateralis (PL) pulvinar inferior (PI)
Ventral thalamus	N. reticularis (R) N. geniculatus lateralis ventralis (GLv)

cation derived from the studies of Walker[85] and Olszewski[56] on the monkey thalamus, from the studies of Rose and Woolsey[73] on a representative series of mammals, and from the studies of Van Buren and Borke[9] and of Hassler[6] on the human thalamus. The thalamotelencephalic organization in subhuman primates is not thought to differ significantly from that of man, and the description that follows deals in the main with that organization in the monkey. It is thought to be typical of the primate pattern.

Differences exist between the three primordial thalamic areas in their connections with the telencephalon. Neurons of the epithalamus and ventral thalamus do not project to the cerebral cortex or striatum; their axons terminate elsewhere. Neurons of most nuclei of the dorsal thalamus, on the other hand, send their axons to the cerebral cortex and make up the thalamocortical projections. A small number of nuclei projects mainly on the caudate and putamen, forming the thalamostriate projections.

Epithalamus

Cellular groupings of epithalamic origin are the habenular and the paraventricular nuclei and the nuclei of the pretectum, no one of which sends axons to the corpus striatum or the cerebral cortex. The habenula is thought to be related to the olfactory system, for its main afferent connections are through the stria medullaris from the septal nuclei, from parts of the amygdala, and perhaps from the hypothalamus as well. Its main efferent pathways terminate in the n. interpeduncularis of the midbrain (tractus habenulo-interpeduncularis), which, in turn, projects on midbrain structures. The connections of the n. paraventricularis are with the hypothalamus, whereas the pretectal nuclei are associated chiefly with the visual system, receiving inputs from retina, superior colliculus, and visual cortex and projecting to the dorsal thalamus (pulvinar and posterior group of nuclei).

Ventral thalamus

The relations between the two nuclei of ventral thalamic origin and the telencephalon are incompletely understood. The ventral lateral geniculate nucleus remains intact after removal of the endbrain, and, in the cat, it is reciprocally connected with the superior colliculus and pretectum. The reticular nucleus is a sheath of cells that surrounds the anterior and lateral aspects of the dorsal thalamus, intercalated between the external medullary lamina and the internal capsule. It is transversed by all thalamocortical and corticothalamic fibers, which produce its reticulated appearance. The prevailing orientation of the dendritic fields of reticular cells is perpendicular to the course of the traversing fibers, from which they receive extensive innervation via axon collaterals. Reticular cell axons project into the dorsal thalamus, where they branch very widely and synapse with neurons of thalamic nuclei. Thus the reticular nucleus is optimally placed to regulate the activity of thalamic cells by feeding back on them some integral of their own activity and that of the corticothalamic systems projecting on them.[27,52,76]

Dorsal thalamus

The dorsal thalamus is the largest portion of the thalamic mass and a large number of nuclei originate in it (Table 9-1 and Fig. 9-2). The internal medullary lamina separates the thalamus into medial and lateral (ventral and dorsal) nuclear regions and bifurcates dorsorostrally to enclose the anterior nuclei. Several cellular groupings differentiate in and about the internal medullary lamina and together make up the intralaminar nuclei. Two major structures, the medial and lateral geniculate bodies, develop early in fetal life in the caudal thalamus. A transitional zone in the posterior diencephalon that includes several nuclei of varied cellular morphology constitutes the posterior nuclear complex.

On the basis of morphologic characteristics, afferent and efferent connections, and functional role, several major thalamocortical systems may be recognized. In what follows a brief summary of their anatomic substrate is given.

Thalamic nuclei associated with efferent control mechanisms. The anterior nuclei, n. lateralis dorsalis, n. medialis dorsalis, and nuclei of the midline may be regarded as parts of neural systems controlling visceral efferent and endocrine mechanisms and are thought to play roles in the CNS mechanisms controlling emotional behavior as well.

The anterior nuclear group (AV, AD, and AM) receives input from the hippocampus both directly via the fornix and relayed through the mamillary body (mamillothalamic tract). It projects to the cortex of the cingulate gyrus on the medial aspect of the hemisphere (areas 23, 24, and 32).[72,89] The n. lateralis dorsalis (LD), which is located posterior to and in continuity with the anterior nuclear group,[85] also receives afferent fibers through the fornix. Its efferent projections in primates are not known; they are possibly to the cortex of the cingulate gyrus posteriorly to those of the anterior nuclei.

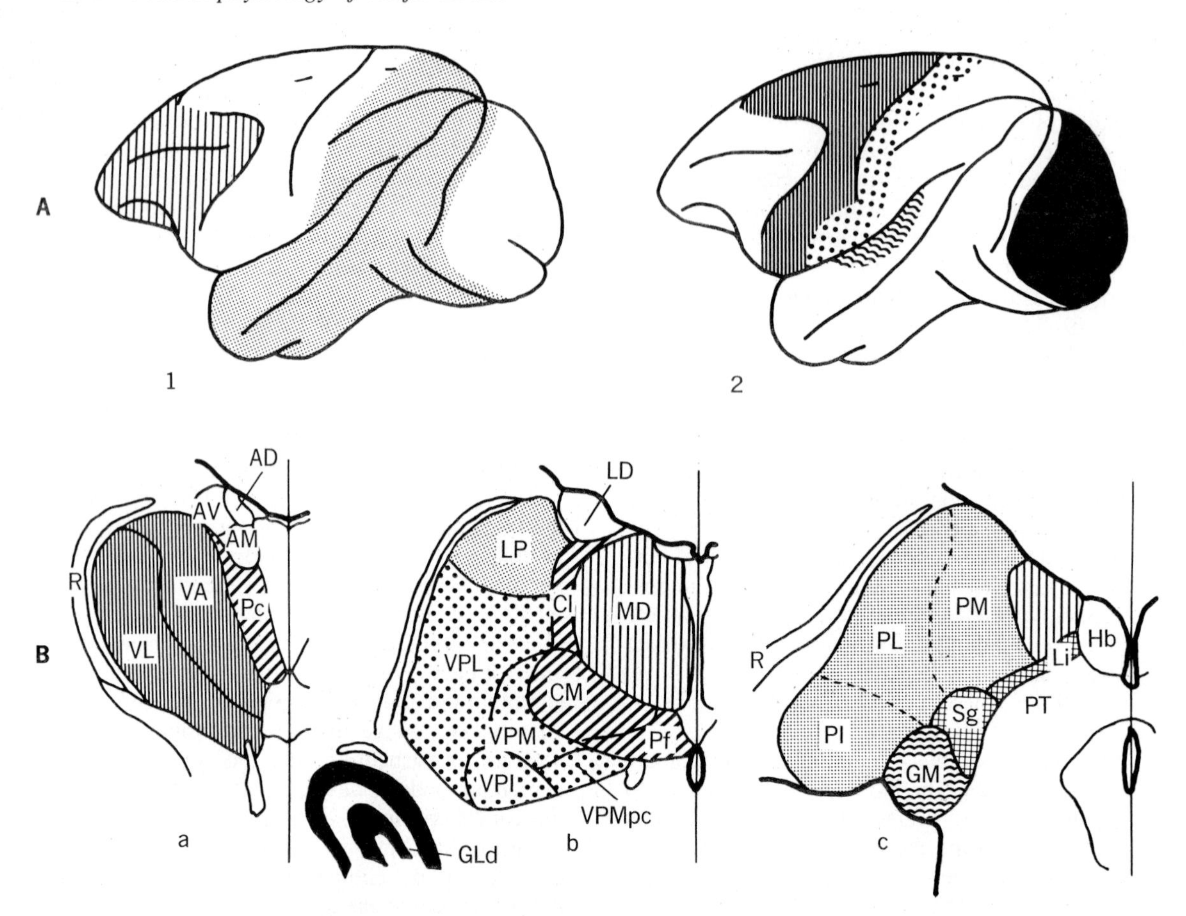

Fig. 9-2. Schematic diagram of anatomic organization of thalamocortical systems. Drawings in **B** outline nuclear configuration of monkey thalamus and those in **A** lateral surface of ipsilateral cerebral hemisphere. Areas similarly marked indicate thalamic nuclear groups and cortical regions to which they project. Outlines of thalamus are of coronal sections in frontocaudal sequence from *a* to *c* and are derived from plates of stereotaxic atlas of macaque thalamic by Olszewski.[56] Abbreviations used are listed in Table 9-1. Epithalamic and midline nuclei as well as anterior nuclei and n. lateralis dorsalis are not marked with symbols. Projections of latter two nuclear groups to cingulate gyrus on medial surface of hemisphere are not illustrated. **A,** *1,* shows cortical regions that receive connections from n. medialis dorsalis and from lateralis posterior-pulvinar complex (groups I and IV in text). In **A,** *2,* are indicated projection areas of motor relay nuclei (group II), in front of central sulcus, and behind it those of primary sensory nuclei (group III).

The larger nuclear complex of the medial thalamic mass, the n. medialis dorsalis (MD) receives connections from the hypothalamus and amygdala[54] and sends a massive projection to the entire homotypical frontal cortex, rostral to areas 6 and 32.[14] The phylogenetically older midline nuclei, poorly defined in primates, appear to have connections with the hypothalamus and sparse projections to localized regions of the cortex of the frontal pole and of the medial surface of the hemisphere.[38]

Thalamic nuclei for the control of movement and posture. The intralaminar nuclei, n. ventralis anterior and n. ventralis lateralis, are components of the thalamic nuclear mass that are heavily interconnected with telencephalic structure subserving motor functions: the intralaminar with the neostriatum and the ventralis (VA and VL) with the precentral and frontal cortex. Afferent fibers to the rostral intralaminar nuclei, the paracentralis (Pc) and the centralis lateralis (Cl), are from the medial spinothalamic pathway (paleospinothalamic) as well as from the cerebellothalamic tract.[50] The more caudal nuclei, the centrum medianum (CM) and the parafascicularis (Pf), receive only scanty ascending connections, and their afferent input is chiefly from the cerebral cortex (area 4 to CM, area 6 to

Pf) and the medial segment of the globus pallidus, which sends a dense projection to CM.[49] The main efferent connections of the intralaminar nuclei are with the neostriatum (Pc and Cl with the head of the caudate; CM and Pf with the putamen).[64] These projection systems are highly organized topographically both in anteroposterior and mediolateral planes. In addition to the heavy striate projection, the intralaminar nuclei send a sparse and diffuse projection to the cerebral cortex of the frontal, parietal, and limbic regions but have no connections with posterior parietal and occipital cortex.[38]

The n. ventralis lateralis (VL) is the motor relay nucleus of the thalamus, and its connectional organization is similar to that of the sensory relay nuclei (see subsequent discussion). The major input to VL is from the contralateral cerebellum, mainly from the dentate nucleus, and from the red nucleus via the cerebellorubrothalamic tract. It also receives fibers from the globus pallidus and the intralaminar nuclei. The efferent projection of VL is to the heterotypical agranular cortex of precentral areas 4 and 6, from which it receives reciprocal connections.[14] The VL-precentral system is topographically organized in a detailed and orderly manner.

The n. ventralis anterior (VA) receives fibers from the cerebellum,[50] from the medial segment of the globus pallidus, and from the substantia nigra; its efferent connections appear to be with other thalamic nuclei, as well as with large portions of the cortex of the frontal lobe, including area 6.[15,28,50] The medial part of VA sends fibers to the orbitofrontal cortex, and it may be regarded as part of the thalamus associated with efferent control mechanisms[1,28,77] (see previous discussion).

Thalamic nuclei associated with sensory pathways. Each sensory system originates in a peripheral sheet of receptors and ascends through the nervous system to reach the cerebral cortex; it is composed of several neuron populations synaptically connected in the direction from periphery to centers. With the exception of the olfactory, all sensory systems reach directly one or another specific region of the dorsal thalamus, the axons of whose cells project, in turn, on "sensory" areas of the cerebral cortex.

PRIMARY SENSORY NUCLEI. The n. ventralis posterior, n. geniculatus lateralis dorsalis, and n. geniculatus medialis are the primary sensory nuclei. The larger portion of the n. ventralis posterior is represented by the nuclei ventroposterolateralis (VPL) and ventroposteromedialis (VPM), often referred to together as the ventrobasal complex. These nuclei are the site of termination of the ascending fibers of the somatic afferent systems (medial lemniscus, spinothalamic, trigeminothalamic) and thus the thalamic transfer region for somatic sensibility. They project to the heterotypical koniocortex of the postcentral gyrus (areas 3, 1, and 2) and to the cortex of the second somatic area, from both of which they receive reciprocal connections. The system is topographically organized with the body surface orderly represented both within the thalamus and in the somatosensory cortex (SI and SII)[41] (Chapter 12).

The more ventral VPL receives scattered afferents from vestibular nuclei[25,45] and projects to a region of cortex within SI where responses to vestibular stimulation can be evoked.[34] Two smaller subdivisions of the n. ventralis posterior are parts of other sensory pathways. The n. ventroposteromedialis parvocellularis (VPMpc) appears to be associated with the gustatory afferent system, whereas the n. ventroposteroinferior (VPI) is probably another thalamic relay region for the vestibular system.[25,32] Both nuclei project to the cortex of the insula, and the "taste" nucleus (VPMpc) also projects to SI within the area of tongue representation.[24,70]

The n. geniculatus lateralis dorsalis (GLd), a laminated structure in the caudal thalamus, is the thalamic transfer region for the visual afferent system. Fibers from the ipsilateral temporal and contralateral nasal hemiretinas (contralateral hemifield of view) terminate in each GLd, whose neurons, in turn, project to the primary visual cortex (VI), a zone of heterotypical koniocortex in the occipital lobe called the area striata, or area 17. The geniculostriate pathway is retinotopically organized to replicate in orderly fashion the visual field it subserves (Chapter 18). In addition to the reciprocal and topologic connections from the visual cortex, the dorsal lateral geniculate nucleus also receives a projection from the superficial layers of the superior colliculus and possibly fibers from the brain stem reticular formation.

The medial geniculate nucleus (GM) is the thalamic transfer region for the auditory system. The large afferent bundle to GM is the brachium of the inferior colliculus, which contains "auditory" fibers of different origin (inferior colliculus, n. of lateral lemniscus, and superior olivary complex). Separate divisions of GM project to different cortical areas, which include the area of heterotypical koniocortex (area 41) in the depth of the sylvian fissure (primary auditory cortex, AI) and surrounding regions in the superior temporal gyrus (AII). As for the other sensory systems, the auditory thalamocortical connections are reciprocal and topographically organized and reflect the spatial distribution of the peripheral receptive elements in the organ of Corti and thus furnish a neural replication of the scale of audible tones (Chapter 15).

POSTERIOR NUCLEAR COMPLEX. The heterogeneous cellular group that includes the n. suprageniculatus, n. limitans, and n. posterior receives inputs from subcortical structures associated with different sensory systems. Within its boundaries terminate fibers from the deep layers of the superior colliculus (visual) and from the medial geniculate body (auditory) as well as ascending lemniscal and spinothalamic fibers (somatosensory). Of particular interest are the physiologic findings of characteristic response properties of the neurons of the posterior complex, which suggest that this thalamic region may play a major role in central pain mechanisms.[60,62] The cortical projections of the posterior complex are with the insula (Sg and Li) and adjacent regions in the depth of the sylvian fissure (Po).[24]

Thalamic nuclei operating in integrative, perceptual mechanisms. The n. lateralis posterior and pulvinar complex constitutes a large component of the thalamic mass, the pulvinar in particular, and attain a development and internal differentiation characteristic of the thalamus of primates, especially man. The development of these thalamic nuclei parallels that of the cortical areas with which they are connected, that is, the parieto-occipitotemporal homotypical cortex intercalated between the primary projection fields of the somatic, visual, and auditory systems.

The afferent connections to the dorsolateral nuclear complex are not fully known. These nuclei do not receive large ascending fibers of primary sensory pathways. On the other hand, the various subdivisions of the pulvinar (PI, PM, PA, and PL) and the n. lateralis posterior (LP) receive fibers from the superior colliculus (superficial layers to PI and deep layers to PM, PA, and LP) as well as from the pretectal nuclei (to PL).[17] The efferent projections of LP are to the posterior parietal cortex areas 5 and 7.[57,85] The projections of the pulvinar are to a wide extent of neocortex and appear to be precisely organized among the various pulvinar subdivisions. In summary,[18,24,82,85] the medial nucleus (PM) sends fibers mainly to the cortex of the temporal lobe (areas 20, 21, and 22); the anterior (PA) and the laterocentral pulvinar regions are connected with the posterior parietal cortex (areas 5 and 7), whereas the more lateral (PL) and inferior (PI) pulvinar subdivisions project to prestriate cortex areas 18 and 19. Reciprocal corticopulvinar connections exist for most of these systems.[83,85] The pulvinar-prestriate connections appear to be topographically organized with reference to the contralateral visual field.[16]

ANATOMIC ORGANIZATION OF CEREBRAL CORTEX[3,4,68]

The cerebral cortex is a convoluted and laminated sheet of neurons that develops from the primitive pallial outpouching of the telencephalon and evolves most extensively in phylogeny, reaching its greatest development in primates. In man it covers some 2,500 cm^2 of surface, is from 2.5 to 4.0 mm thick, occupies about 600 cm^3 volume, contains several billion neurons, and has a large but unknown number of glial cells. Areas of cortex receive afferent fibers from subcortical structures, in particular from the dorsal thalamus, from cortical areas of the same hemisphere, and from usually homologous areas of the opposite hemisphere. Cortical neurons may project their axons intracortically locally, to cortical areas of the same hemisphere through the white matter (association fibers), via the great commissures to the opposite hemisphere (commissural fiber), or to subcortical cellular masses even as far away as the spinal cord (projection fibers). In concert with subcortical nuclear regions and afferent and efferent systems the cerebral cortex receives and analyzes sensory information, stores a record of experience in memory, programs and governs the execution of movements, regulates homeostatic processes, and is the essential but not exclusive neural substratum of those complex aspects of brain function indicated by such words as thinking, remembering, calculating, planning, and judging, Although no understanding of these events is possible now at the level of mechanism, the available evidence supports the view that the cerebral cortex is their prime seat of action. Yet in these functions the cortex interacts with virtually all other regions of the CNS. *Total behavior is the result of the total function of the nervous system.*

Cortical cell types

Cortical neurons are not all alike, and many classifications have been made based on the size and form of cell bodies, the length and distribution of their dendritic trees, and the destinations and degree of branching of their axons.[68] There are three general cell classes, and within the first two many subtypes have been identified.[4]

Pyramidal cells. Pyramidal cell bodies are commonly triangular or trapezoidal in silhouette, with base downward and apex directed toward the cortical surface. Their complex dendritic trees usually consist of (1) a basilar dendritic arborization that ramifies in the immediate locale of the cell body, largely horizontally, and (2) an apical dendrite up to 2 mm in length that ascends from the cell body through overlying cellular layers, frequently reaching and branching terminally within the outermost layer. Both the apical trunk and its numerous branches may be covered with the specialized postsynaptic protrusions called spines.[78] Pyramidal cells are frequently classified in terms of axonal destination. Many emerge from the cortex as association, commissural, or projection fibers, frequently sending recurrent collateral branches back on the cellular regions from which they sprang. Axons of some pyramidal cells (the cells of Martinotti) turn back toward the cortical surface, never leaving the gray matter, to end via their many branches on the dendrites of other cells. Many subtypes have been described; some of them occur only in restricted cortical areas. Pyramidal cell bodies vary greatly in size, from axial dimensions of $15 \times 10\ \mu$ up to $120 \times 90\ \mu$ or more for the giant pyramids of the motor cortex, which are called Betz cells after their discoverer.

Stellate or granule cells. Stellate or granule cells differ remarkably from pyramidal cells. Their cell bodies are small, and dendrites spring from them in all directions to ramify in the immediate vicinity of the cell of origin. The axon may arise from a large dendrite and commonly divides repeatedly to terminate on the cell bodies and dendrites of immediately adjacent cells. The axons of other granular cells turn upward to end in superficial layers or, uncommonly, may leave the cortex.

Fusiform cells. Fusiform cells have spindle-shaped cell bodies, and branching dendritic trees may arise from both ends of the spindle. The axons usually project from the cortex after emitting recurrent collaterals; less commonly, they turn toward the cortical surface to terminate in more superficial cortical layers.

Layered distribution of cortical cells

Cortical cells are not randomly distributed along the axis normal to the cortical surface; this orderliness affects both the distribution of cell types and their packing density. Relative segregation by depth produces a stratification; each stratum is called a cortical layer. Over the greater portion of the cerebral hemispheres the cortex displays six layers. This stratification is apparent in the sixth month of fetal life in the human being and is well developed at birth. Cortex that possesses six layers in adult life or that possessed six layers in ontogeny is termed isocortex or, more commonly, neocortex. Neocortex occupies some 90% to 95% of the cortical surface. The cortex lying medial to the rhinal sulcus never possesses six layers, and the degree of its stratification varies from place to place. It is termed allocortex and is characterized in places by an external layer of myelinated fibers; it includes cortex of the hippocampus and rhinencephalon.

The cortical layers are usually named and defined as follows, from the pial surface inward:

Molecular or plexiform layer (I). The molecular layer contains the terminal branches of the apical dendrites of pyramidal cells of layers II, III, V, and VI and some of the axonal terminals of cortical cells with ascending axons (Martinotti cells) and of unspecific cortical afferents. It contains a few neurons, the horizontal cells of Cajal.

External granular layer (II). Layer II contains a large number of tightly packed, small pyramidal cells. Their basilar dendrites ramify within this layer, their rather short apical ones within the overlying molecular layer. Only a few of their axons project from the cortex, the large majority branch repeatedly and terminate about the cells of layers V and VI. Afferent axons to this layer are those of Martinotti cells, of granular cells of layer IV, of association fibers, and of recurrent collaterals of axons of the pyramidal cells of deeper layers.

Pyramidal cell layer (III). The pyramidal cell layer is a continuation of layer II without a sharp line of demarcation but possessing larger pyramidal cells. Their apical dendrites ramify upward, yielding branches in layer II and terminating in layer I; their basilar dendritic field is expanded horizontally in layer III. The cells of the lower edge of layer III receive specific thalamic afferents, and synapses are made throughout the layer from association axons, axons of granular cells of layer IV, and axons of Martinotti cells of layers V and VI. Some efferents from layer III end in layers V and VI, but many leave the cortex as association or commissural fibers.

Granular cell or internal granular layer (IV). Layer IV is packed with small stellate cells that give the name "granular" to those regions of cortex in which this layer is best developed, e.g., primary sensory receiving areas, which are also called koniocortex, and the granular cortex of some regions of the frontal lobe. The dendrites of the granule cells arborize locally within layer IV and receive a very heavy synaptic impingement from the axon terminals of the specific thalamocortical afferents. The granular cell axons rarely leave the cortex; some ascend to terminate in layers I to III, whereas others descend to end on cells of layers V and VI. The motor cortex of the precentral gyrus is called agranular because of the virtual absence of this layer. Layer IV also contains a number of pyramids with apical dendrites that reach to layer I.

Ganglion cell or giant pyramidal layer (V). Layer V is sharply demarcated from layer IV. The large pyramids have apical dendritic shafts that reach layer I, where they end in complicated brushes; the basilar dendrites and the collaterals of the ascending shafts are distributed exclusively in layer V. The efferent axons are usually projection fibers and with rare exceptions have branches and recurrent collaterals that ascend to terminate in layers II, III, and even I. Cells with ascending axons are found in this as in all other cellular layers. The pyramidal cells of layer V vary greatly in size in different areas of the cortex.

Fusiform or multiform cell layer (VI). The fusiform layer contains many spindle cells. Dendrites may arborize from either or both ends of the cells, and larger ones may ascend to layer I without branching. Smaller spindles have dendritic fields confined to layout of the cortex, mainly to the dorsal thalamus, and give off branches and recurrent collaterals before leaving.

Summary of cortical afferents[36,37,47]

1. Thalamocortical projection fibers, the so-called specific thalamic afferents (e.g., lateral geniculate to visual cortex), ascend still myelinated through layers V and VI and divide repeatedly to form extensive terminal plexi in layer IV and the lower part of layer III. Synapses are made in great numbers on the dendrites of the granular cells of layer IV and also on the apical dendrites of pyramids of layers V and VI as they pass through layers III and IV.
2. A second type of thalamocortical afferent projects to more than one cortical area, giving off collaterals to all layers in passing through the cortex and ending primarily in layer I. These fibers also are thalamic in origin.
3. Association and commissural fibers may give collaterals to cells of layer VI on entering the cortex, but their main field of termination is in layers I to IV, especially in layers II and III.

Summary of cortical efferents[39,40,42,48]

1. The pyramidal cells of layers II and III are the cells of origin of the commissural and the ipsilateral corticocortical efferent fibers.
2. The pyramidal cells of layer V are the cells of origin of corticofugal axons projecting to the basal ganglia, brain stem, and spinal cord.
3. The pyramidal and fusiform cells of layer VI are the cells of origin of corticothalamic fibers.

With rare exceptions, all efferent axons leaving the cerebral cortex are axons of pyramidal cells, and all pyramidal cells emit efferent axons. Almost all pyramidal cell axons emit recurrent collateral branches as they pass down through the cortical layers, particularly into layer V. The ratio of pyramidal cells to stellate cells in the neocortex is about 2:1, a ratio that is virtually unchanged over the series of mammals.

FUNCTIONAL IMPLICATIONS OF THE CELLULAR CONNECTIVITY OF THE CORTEX: COLUMNAR ORGANIZATION

It should now be clear that the weight of synaptic relations in the cerebral cortex is in a vertical direction, normal to the pial surface, and that pathways and synapses that might allow horizontal intragriseal spread of activity are few in number and concentrated in layers I, IV, and V. From the functional point of view the cortex must be regarded as made up of complex chains of interneurons that do not differ in principle of organization from such chains elsewhere. These chains are interposed between input and output over a vertical column of cells extending across all cellular layers. Initial input to such a chain, (e.g., over the specific thalamic afferents) gains a powerful and rapid access via some granular cells of layer IV downward to the efferent neural elements of layers V and VI and via others upward to the more superficial pyramids of layers II and III. Access to the efferent elements of the cortex may in the first place occur fairly directly, over one or a few synapses, if excitability and convergence are high. Second, a quick translation of activity to the outer cellular layers of the cortex will occur, activating arcs of neurons superimposed on the initially activated oligosynaptic neurons and delivering repetitive trains of impulses by recurrent and reentrant activation to the efferent cortical neurons. These, in turn, will influence the activity of those superimposed chains via recurrent collaterals, which activity may be either excitatory or inhibitory, and which is distributed in a patterned, not a random, spatial array. It is likely, for example, that the efferents from each column, via recurrent inhibition, will depress the activity of neurons in adjacent columns, thus contributing further to the vertical segregation of cortical neuronal activity. The neurons of the superimposed reentrant chains, lying mainly in layers II, III, and IV, will also receive input from other cortical areas via association and commissural fibers and from the intralaminar nuclei of the thalamus. What is usually referred to as "an integration of neural activity" will result: the influences reaching these groups of neurons will be titrated in such a way that the resulting output is some net result of all. The meaning of this titration in terms of "what cortex does" to its input to produce its output is unknown. *It is the central problem of cortical physiology.*

Physiologic evidence that a vertical column of cells such as that just referred to indeed serves as an elementary functional unit in the primary sensory cortices will be given in Chapters 12 and 18.

VARIATIONS IN THE LAYERED DISTRIBUTION OF CORTICAL CELLS: CYTOARCHITECTURE

Even a casual inspection of cortical sections reveals that (1) the layers vary in packing density and in thickness from place to place, (2) in some areas certain layers are greatly reduced or even absent, (3) special cell types appear in some areas and are not present in others, and (4) the numbers of afferent and efferent fibers and their segregation at some cortical depths (e.g., the line of Gennari of the visual cortex) are not the same from one place to another. These observations led to the idea that the cerebral cortex might be divided into a number of fields, within each of which these morphologic characters are uniform, and each differing significantly from its neighbors. On this cytoarchitectural basis a large number of areas have been identified by different investigators, and maps of the cerebral cortex have been constructed (Fig. 9-3). Common morpho-

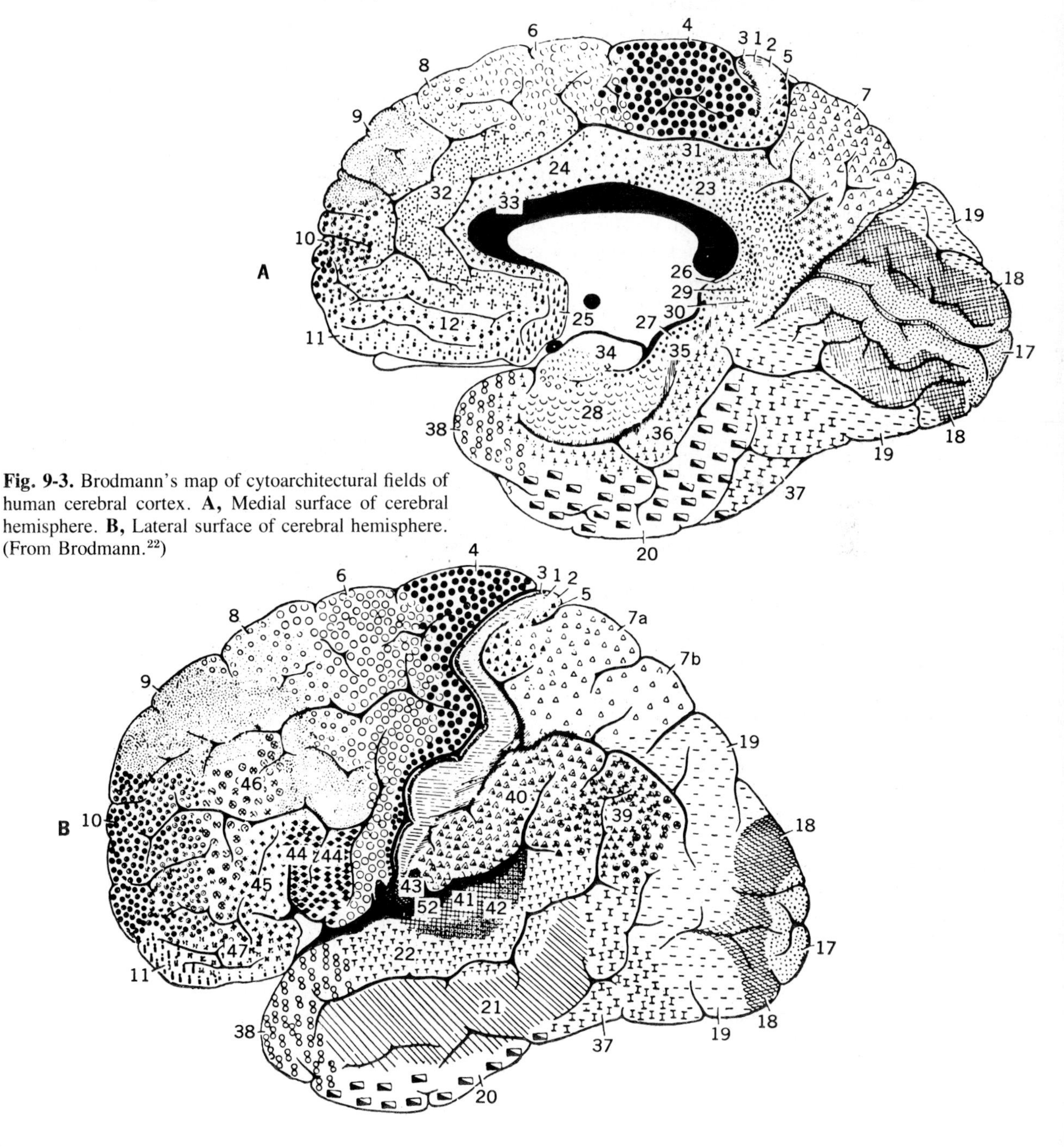

Fig. 9-3. Brodmann's map of cytoarchitectural fields of human cerebral cortex. **A,** Medial surface of cerebral hemisphere. **B,** Lateral surface of cerebral hemisphere. (From Brodmann.[22])

logic features, on the other hand, make it possible to outline larger regions of cortex, each including many cytoarchitechtonic areas. Two basically different morphologic types of cerebral cortex have been described by von Economo[84]: the homotypical and the heterotypical. The *homotypical cortex* is characterized by six well-developed cellular layers that can be always identified even though there exists variation in thickness and development from region to region. Homotypical cortex is found in the anterior frontal lobe and in temporal, occipital, and parietal convolutions; it is characteristic of those regions of cerebral cortex often referred to as "association cortex." The cortex of "sensory" and "motor" regions, on the other hand, is characteristically *heterotypical cortex,* so called because in it the six cellular layers cannot be clearly identified. The heterotypical cortex of areas that receive primary sensory systems has richly developed granular layers II and IV and relatively poorly developed layers III and V. This type of cortex is called granular cortex or koniocortex (*konis,* dust). The heterotypical cortex of motor areas, on the other hand, has poorly developed granular layers II and IV, whereas layers III and V are well developed. This type of cortex is called agranular cortex; it is found in precentral areas 4 and 6 and in certain other parts of the frontal lobe.

Study of serial sections of the cortex reveals to every observer that in some places the juncture between fields is obvious and sharp and can be marked on a photomicrograph with a line. This is true, for example, of the juncture between the striate cortex (area 17) and the prestriate cortex (area 18). At other conjunctions the change is not so abrupt but occurs gradually over a distance of 1 to 3 mm, a region labeled a transition zone; for example, area 3a in the depths of the central sulcus is an area of transition from the granular cortex of the postcentral gyrus to the agranular motor cortex of the precentral gyrus. All observers agree on facts such as these. When parcellations are attempted in such areas as the granular frontal cortex (areas 9 to 12) or the temporal lobe, however, observers disagree. Their disagreement varies from the statement that in such a large area as the granular frontal cortex no separate fields exist at all to the position that a very large number of separate fields can be identified. For the student concerned with the complexities of the cortical maps that have resulted from cyto- and myeloarchitectonic studies, a working definition of what may constitute a cortical field may be useful.

Definition of a cortical field

1. A cortical field is a region within which the cell types and packing densities of the cortical layers may be uniform. Within many fields, however, variants of morphology occur. Rose[71] put forward the idea that these variants should be grouped together, a natural series constituting a field. Thus a cortical field as defined by other criteria may display continuing change in structural detail. Changes that vary systematically are designated the gradients of a field; the line dividing two fields is placed at that point at which these gradients, which may be occurring in all layers as one moves from point to point across the field, are changing most rapidly. The important conclusion is that two adjacent fields that have definite overall differences may yet have no sharp border between them.

2. A cortical field may receive a thalamocortical projection from a particular thalamic nucleus. When the extent of that projection coincides with the cytoarchitectonic definition of a field determined independently, that definition is greatly strengthened (e.g., the projection of the lateral geniculate body on area 17 and the projection of each of the anterior thalamic nuclei on precisely defined architectonic fields of the cingular gyrus). In other areas the thalamocortical projection field is not so precisely defined, nor is the cytoarchitectural definition; in general the more precise the one, the sharper the other.

3. The extent of a field defined by physiologic means is frequently found to coincide with that determined by the two methods previously described. This is particularly true of the primary cortical areas to which the great afferent systems project.

When a cortical area that is a candidate for identification as a separate field meets all three criteria, identification cannot be doubted.

Cortical maps and functional localization

Fig. 9-3 displays maps of the human cerebral cortex that were published by Brodmann.[23] Studies of the human and other primate brains had been made earlier by Brodmann,[22] by Campbell,[26] and by others and have been repeated since.[84] The Brodmann maps are presented here because his system of numeration is the most widely used in both clinical and experimental neurology.

One possible implication of the cytoarchitectonic differences is that cortical areas function differently and deal with separate functions *because* of these cellular differences. This seems unlikely: there is nothing peculiarly *motor* about the cellular arrangements in the *motor* cortex. An alternative hypothesis is that cortical areas function differently and deal with particular functions because of differences in their afferent and efferent connections. It may be that cytoarchitectural differences are themselves due to variations in the number and destinations of afferent fibers growing into the uniform fetal neocortex and in the numbers and lengths of the efferent axons growing out. A corollary of this hypothesis that function depends on connection is that what cortex does with its input to produce its output may be, at least in principle, identical from one neocortical field to another.

Thus the problem is posed of what is meant by "localization of function" within the cerebral cortex. It does not mean that the neocortex is made up of a number of distinct organs, each dealing with a particular function or "quality" and all fitted together in a complex mosaic. What it does mean is that *some particular areas, by virtue of their connections, are involved with one particular aspect of function*. For example, stimulation and ablation experiments as well as clinical observations support the statement that the precentral motor cortex is intimately involved in the control of movement and the regulation of posture. What is *not* meant is that motor function resides there and there alone, and the partial recoveries of motor function that follow its removal indicate clearly the part played in this function by other regions of the brain. On the other hand, the prepotent executive role of the motor cortex in movement is obvious.

GENERAL PHYSIOLOGY OF THE CEREBRAL CORTEX

Cellular properties and synaptic transmission

The subject of synaptic transmission in the CNS was considered in Chapter 8. The electrical phenomena that occur in different parts of nerve cells were used to support a hypothesis of their integrative action. The techniques of intracellular recording have been applied to both neocortical cells[61,79] and the pyramidal cells of the hippocampus.[44] Results indicate that the general concepts elucidated for spinal motoneurons apply with some modifications to neurons located in more rostral portions of the nervous system.

Cortical cells, like other neurons, are capable of three types of electrical activity: (1) the regenerative change in Na^+ conductance that occurs at a critical level of depolarization and leads to impulse conduction, (2) depolarizing (excitatory) local postsynaptic responses, and (3) polarizing (inhibitory) local postsynaptic responses. There is some evidence that, as for spinal motoneurons, all parts of cortical neurons are not equally capable of local and regenerative processes. The low threshold of the initial segment, or axon hillock region, for regenerative action, coupled with the propensity of soma and dendrites to support under some conditions only local PSPs, affords an explanation for the integrative action of cortical cells. The frequency of impulses discharged from the initial segment and down the axon is a running and probably linear function of the net ionic current flowing across its membrane, the intensity and direction of which are determined by the number, sign, and location of the local PSPs occurring elsewhere on the cell membrane. It is important to emphasize certain properties of PSPs, particularly with reference to subjects discussed later. They are local, nonconducted changes in membrane potential produced by local changes in ionic permeability; thus they affect other regions of the cell (e.g., the initial segment) only by electrotonic extension. They may be algebraically summed both in time and space over the cell surface.

When an afferent volley to the cerebral cortex is either purely excitatory or (after interneuronal relay) purely inhibitory for a cell under study, that cell responds with PSPs that are depolarizing or (hyper) polarizing. Such a purity of synaptic input rarely occurs, so that the PSPs observed are commonly diphasic, a combination of sequentially timed excitatory and inhibitory synaptic responses (Fig. 9-4). The PSPs of cortical cells differ from those of spinal motoneurons in time course, which is longer than can be explained by the membrane time constant. The prolongation of PSPs of cortical cells beyond that predicted by their membrane time constants may be due to a persistence of the transmitter agents released by single presynaptic impulses or, what is more likely, to the repetitive activity of cortical cells, particularly of the excitatory and inhibitory interneurons (i.e., the granular cells) that relay afferent input to the large pyramidal cells.

The membrane potentials of cortical neurons are not constant but oscillate more or less continuously near the threshold level for spike generation. This continuous variation and the accompanying impulse discharges are frequently referred to as *spontaneous*, a word when used in central neurophysiology or electroencephalography, designates neural activity in the absence of any overt or intentionally provoked afferent input. Spontaneous activity is not thought to be generated by, or not wholly by, an intrinsic cellular pacemaker but by a continuous and ongoing synaptic input. The result is that large numbers of cells in the cerebrum are active at rates of discharge of 10 to 30/sec. Thus neuronal transactions take on a second degree of freedom: either an increase or a decrease in rate of discharge may serve as a positive signal.

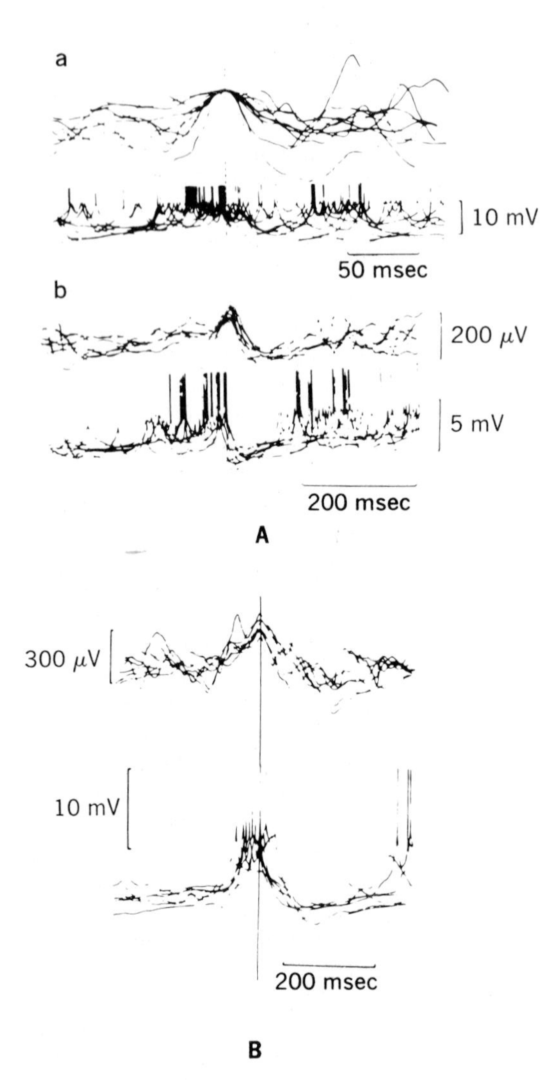

Fig. 9-4. In all records shown, upper trace is electrical activity recorded at surface of cat cortex with "gross electrode," and lower traces are recorded from micropipet tip within cortical cell.

A, Drawings of EEG waves of two different types are superimposed and oriented in time by summits of negative waves and correlated with simultaneous intracellular events. In *a,* mainly surface-negative waves are shown to correlate with EPSPs and associated impulse discharges of cortical cells. In *b,* negative-positive waves recorded at surface are shown to correlate well with EPSPs and impulse discharge and an IPSP following in sequence. **B,** Records of mainly negative waves of surface-recorded EEG (upper traces) just at onset of increased electrical activity produced by the drug Metrazol, which later led to convulsions. Lower traces show correlation with these surface-negative waves of EPSPs and synaptic events plus impulse discharges of subjacent cortical cell. (From Creutzfeldt et al.[31])

Cortical neurons tend to discharge repetitive trains of impulses even when the excitatory presynaptic volley consists of but a single impulse per presynaptic fiber. This lends further credence to the idea that the action potential is initiated at a site separate and distal in the cell from the PSP loci and that the action potential does not invade those loci. On recovery from the first spike discharge the spike-generator membrane is again influenced by the persisting currents set up by the persisting synaptic effect and so on until those currents fall below threshold.

Observations made on cortical cells are consistent with the assumption that synaptic transmission in the cerebral cortex is "chemical" in nature (Chapter 8), that is, the depolarization of synaptic endings by nerve impulses causes them to release substances that diffuse across the synaptic cleft to combine with specific receptor substances in the subsynaptic membrane. That combination leads to changes in the ionic permeability of the postsynaptic cell and the resulting PSPs. Candidate synaptic transmitters have been described in Chapter 8.

Cortical pyramidal cells and the origin of surface-recorded electrical activity

When records of electrical potential differences are made between one electrode resting on the cortical surface and a second placed a considerable distance away, the array represents a special case of recording at the boundary of a large conductile medium containing active elements, explicable only in part on the basis of the classic volume conductor theory. It is likely that conducted action potentials in axons contribute little to cortical surface records, for insofar as they occur asynchronously in time in large numbers of axons that run in many directions relative to the surface, their net influence on an electrode at that surface will be zero. An exception is the special case in which large numbers of thalamocortical axons are activated simultaneously by electrical stimulation of thalamic nuclei or their afferent pathways. It will be shown later that surface records obtained under other circumstances signal principally the net effect of local PSPs of cortical cells. These may be of either sign and may occur immediately beneath the electrode or at some distance from it; a potential change recorded at the surface is the measure of the net IR between the surface site and the distant electrode, produced by the extracellular current flows associated with local PSPs. It is obvious, however, that if all the cell bodies and dendrites of cortical cells were randomly arranged in the cortical matrix, the net influence of their synaptic currents would be zero. *Any* electrical change recorded at the surface must be due to the orderly and symmetric arrangement of some class of cells within the cortex.

The cortical pyramidal cells seem the most likely candidates. Their long apical dendrites are arranged parallel to one another and normal to the cortical surface. Potential changes in one part of such a cell relative to other parts create "open" fields of current flow that can be detected at the surface of the cortex or indeed at the surface of the head. The granular cells, on the other hand, are unlikely to contribute substantially to surface records. Their spatially restricted dendritic trees are arranged radially around their cell bodies, so that charge differences between dendrites and somata will produce "closed" fields of current flow that will add to zero when viewed from the relatively great distance of the cortical surface.

The influence on the surface record of a PSP depends on its sign, orientation, and location. Each may be regarded as creating a radially oriented dipole. Thus continuing synaptic input creates a series of potential dipoles and resulting current flows that are staggered but overlapped in space and time. Surface potentials of any form can be generated by one population of presynaptic fibers and the cells on which they terminate, depending on the proportion that are inhibitory or excitatory, whether predominantly axodendritic or axosomatic, the level of postsynaptic cells in the cortex, etc.

A general and empirical description of the oscillatory changes in electrical potential that can be recorded from the surface of the brain or from the scalp is given in the following section. The question of the relation of these waves to cellular events, introduced previously with general statements, has recently come under direct study. Comparisons have been made between surface-recorded electrical activity and cellular events recorded via intracellular microelectrodes.[31,33,61,79] Convincing evidence has been adduced that the frequency, signs, and amplitudes of the surface-recorded slow waves reflect the net effect of the PSPs of subjacent cortical cells. The records shown in Fig. 9-4, taken from the work of Creutzfeldt et al.,[31] illustrate some of these correlations. The results obtained may be summarized as follows:

1. Excitatory (depolarizing) PSPs in neural elements located close to the cortical surface generally induce negativity in the surface record; when deep within the cortex, they generally induce positivity in the surface record. Inhibitory (polarizing) PSPs generally induce opposite effects in the surface record.
2. The correlation between the cortical surface wave and the PSP sequence in any single cortical element cannot always be predicted. The result depends on simultaneous events, which may be of either sign or of the two in either sequence, in many neurons located at various depths.
3. Some correlations are very strong, however, especially under certain well-defined circumstances.
   a. The primary evoked potential, the positive-negative wave recorded on the cortical surface in response to sensory, afferent nerve, or thalamic relay nuclear stimulation, is correlated with initially depolarizing synaptic electrogenesis in the cortical depths, with succeeding polarizing PSPs in these same elements, and with slower depolarizing PSPs in more superficial cortical elements. Delayed or prolonged PSPs lead to more complicated "secondary" waves in the evoked potential.
   b. The surface-negative recruiting response (Chapter 10) recorded on the cortical surface in response to stimulation of the generalized thalamocortical system is associated with summated depolarizing PSPs in cortical cells with the same time course. The waxing and waning of the evoked cortical waves during slow iterative stimulation (5 to 7/sec), which is characteristic of the recruiting responses, is paralleled by a waxing and waning of the cellular EPSPs as well as the size of the thalamocortical volley (p. 300).
   c. During the spontaneous waves of the surface record, whether occurring in the anesthetized or the sleeping animal, or when synchronized by a convulsive agent, there is a close positive correlation between cortical surface negativity and cellular depolarization, during which a high-frequency discharge of impulses may occur in the cortical cell. The implication is that the depolarizing PSPs predominate throughout all the layers of the cortex.

To date a detailed analysis of this sort has not been made in the waking and alerted state, which is characterized by higher frequency, asynchronous, lower amplitude oscillations in the surface record. The asynchrony of the cortical waves in the waking state is produced by asynchronous synaptic events in the subjacent cortical cells.

Synchronous activity in large numbers of neural elements temporarily raises the K^+ concentration in the extracellular clefts within the brain, producing purely passive changes in the resting membrane potential of glial cells. The close electrical coupling between glial cells (Chapter 8) indicates that these electrical changes may have widespread effects and may contribute to slow potential changes recorded by electrodes placed on the surface of the head or on the scalp.

ONGOING ELECTRICAL ACTIVITY OF THE BRAIN: THE ELECTROENCEPHALOGRAM

Oscillations in electrical potential occur almost continuously between any two electrodes placed on the surface of the head or on the cerebral cortex itself, and the records are termed, respectively, the electroencephalogram (EEG) and the electrocorticogram. These oscillations differ in frequency and amplitude from place to place and in different states of awareness. They persist in altered form during excitement, drowsiness, sleep, coma, anesthesia, epileptic attacks and through severe changes in blood gas or cerebral metabolite concentrations. They never cease short of massive cerebral catastrophe or impending or actual death. In theory, these wavelike potential changes might serve as direct and measurable indices of brain activity. Although this objective is not yet realized, the association between certain brain-wave patterns and some states of altered brain function is so constant and so readily recognized that the study of brain waves has become an important diagnostic tool in clinical medicine.

The spontaneous electrical activity of the brain was discovered by Caton in 1875, although the phenomenon was observed independently at about the same time by others.[21] However, it was Hans Berger who, between 1929 and 1938, showed that electrical activity could be recorded from the scalp surface of human beings, developed methods for making such recordings, and described several varieties of the activity that commonly occur.[19] Adrian et al.[12,13] confirmed and extended these observations. Berger also studied alterations in the EEG produced in certain disease states, and he is thus rightly regarded as the founder of the medical specialty of electroencephalography.

EEG recordings are ordinarily made from a large number of electrodes positioned on the surface of the head. Three types of electrode connections are used: (1) between each of a pair, (2) between each as a monopolar lead against a "distant" electrode, commonly placed on the ear, and (3) between each as a monopolar lead and the average of all. Although the same electrical events are recorded in each of the three ways, they will appear in a different format in each. The potential changes that occur are amplified by high-gain, differential, capacity-coupled amplifiers. The output signals are usually displayed by ink-writing oscillographs (writing on moving paper), a method that limits the frequency response to the range between 0.5/sec and 80 to 100/sec. Alternatively the output signals may be displayed by an inertialess system for preservation of high-frequency response, the cathode-ray oscilloscope, or recorded on magnetic tape for later analysis. In ordinary practice, records are analyzed by direct inspection and measurement. More sophisticated analyses are measurement of the frequency-power spectrum and of auto- and cross-correlation functions[10,43,69] and by display of potential-space-time three-dimensional plots. Recordings are made directly from the cortical surface in human beings during neurosurgical procedures or from surface and depth electrodes that may be implanted during such procedures and left on or in the brain even for extended periods of time. The development of these techniques has opened a wide and hopefully fruitful field of investigation in which correlations between behavior and the slow-wave electrical events occurring in the brain are attempted.

The potential waves recorded may vary in frequency from 1 to over 50/sec and when led from the head surface are 50 to 200 μV in amplitude. In a normal individual at rest with eyes closed in a quiet room the dominant rhythm varies from 8 to 13/sec and is seen at greatest amplitude in the parietal and occipital regions; the amplitude may slowly wax and wane. This is the *alpha* or Berger rhythm, which is illustrated by the records in Fig. 9-5. The normal alpha rhythm varies in amplitude and spatial distribution from one individual to another, and occasionally an individual thought to possess a normal and a normally functioning brain may never show an alpha rhythm. In any one person, however, the dominant alpha frequency is remarkably constant from time to time, scarcely varying by as much as 1 Hz. The alpha rhythm is thought to originate from an alert but relatively "unoccupied" brain. On sensory stimulation, especially with light, or a conscious effort for vision or purposeful mental activity, it is quickly replaced by a higher frequency (13 to 25 Hz), lower voltage pattern (Fig. 9-5). This *beta* rhythm is commonly referred to as the "activated" or desynchronized pattern, although it may also appear in other states.[11] (Chapter 10). Slower waves than the alpha occur rarely in normal individuals, and when they do occur in wak-

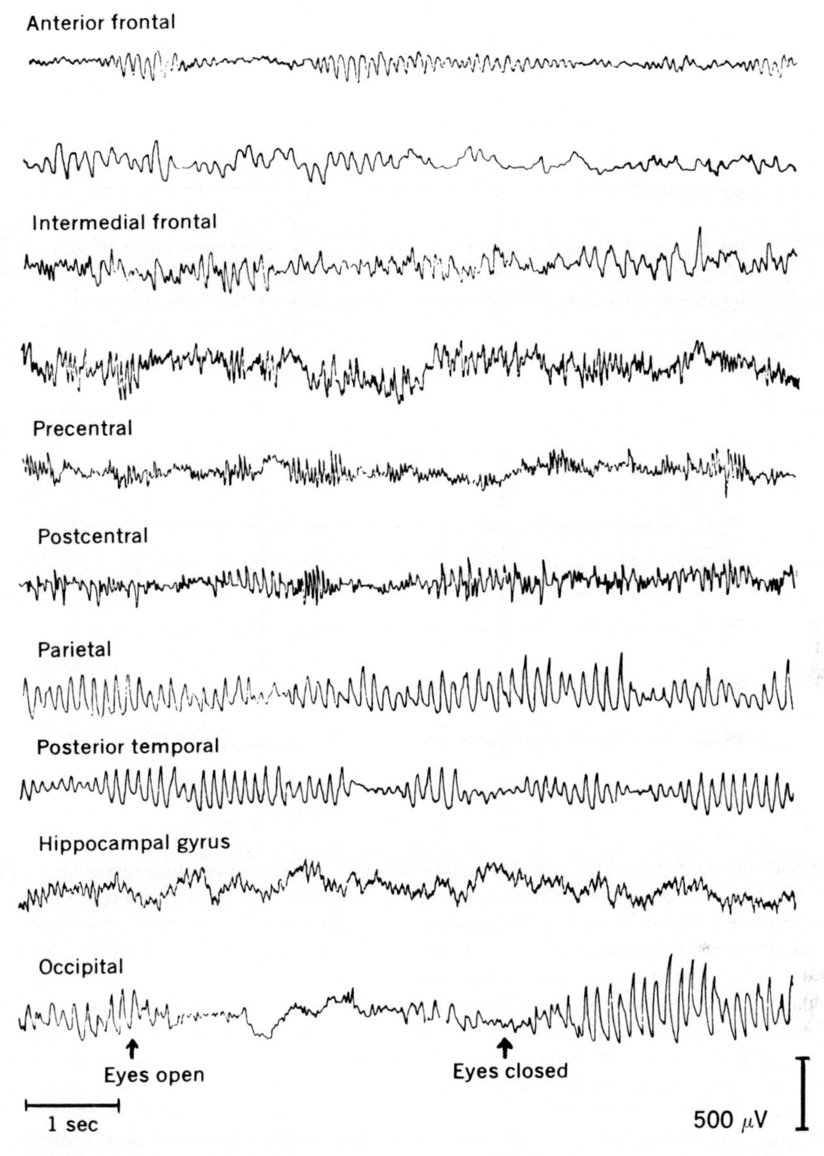

Fig. 9-5. Spontaneous electrical activity, or "resting rhythms," from different cortical areas in man. Sample tracings were taken directly from exposed cortex with silver-chloride, cotton-wick bipolar electrodes. Alpha rhythms are most prominent from entire parietal, posterior temporal, and occipital regions, with exception of postcentral gyrus itself. More rapid activity is present in anterior regions, with relatively pure beta rhythm in precentral gyrus. Note blocking of alpha rhythm in occipital region when eyes are open. (From Penfield and Jasper.[8])

ing subjects other than newborn infants, they usually indicate disease or injury to the brain. These are the *theta* (3 to 7 Hz) and the *delta* (0.5 to 3.5 Hz) rhythms. The changes in the EEG associated with drowsiness and sleep will be considered in Chapter 10. The desynchronization of the alpha rhythm ("alpha blocking") by light[35] can be made contingent on stimuli delivered over other sensory pathways (Fig. 9-6).[87] If light is preceded by a noise and the pair are then associated for a number of trials, the alpha blocking will be produced by the noise delivered alone, even though the noise when delivered previously in isolation had been presented so many times as to lose its own capacity for alpha blocking, a phenomenon termed habituation. Whether this contingent alpha blocking fits the paradigm of classic pavlovian conditioning is unsettled,[51] but it can be shown to be similar to it on a number of counts.

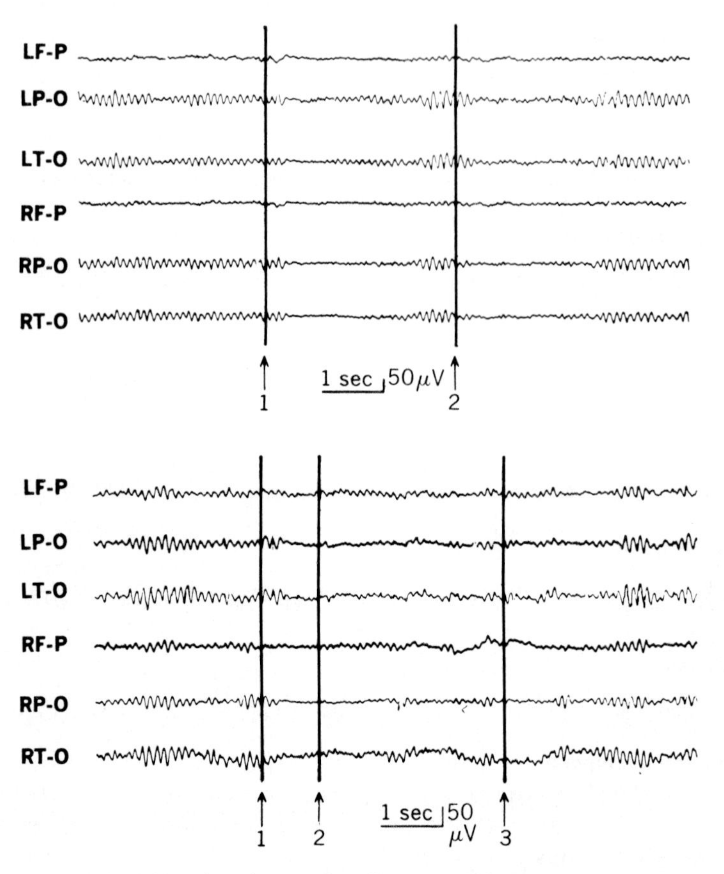

Fig. 9-6. Contingent alpha blocking, frequently called "conditioned cerebral response," in human being. Records are EEGs recorded from scalp of subject resting in quiet, semidarkened room. Upper set illustrates blocking of alpha rhythm, recorded under these conditions, with presentation of light stimulus of 3 sec duration (on at arrow *1* and off at arrow *2*). Sound stimulus (500 Hz, 50 dB above threshold, 4 sec duration) was then slowly repeated until *its* alpha blocking effect disappeared, a phenomenon called habituation. Paired sound and light stimuli were then presented, the light appearing about 1 sec after onset of sound, the two then continuing together for 3 sec. Lower records, obtained after several presentations of paired stimuli, show that tone to which habituation had occurred now produces alpha blocking before onset of light. Whether this "contingent" acquisition of alpha blocking capacity by tone (to which habituation had occurred), by virtue of its association with light, should be regarded as true conditioning in the pavlovian sense is disputed. Paired leads are *LF-P,* left frontoparietal; *LP-O,* left parieto-occipital; *LT-O,* left temporo-occipital; *RF-P,* right frontoparietal, etc. (From Wells.[87])

Considerable attention has been given to the ontogenetic development of the EEG. Records of newborn infants are characterized by continuous, irregular, asymmetric waves from all areas; no regular rhythms appear. Faster activity appears intermittently during the first few weeks and months of life, becoming persistent after 1 year of age. The further development to the adult pattern occurs gradually over the early years to adolescence.

It will be clear from the preceding section that slow waves recorded from the brain, such as those of the EEG, are the compounded result of the local postsynaptic responses of cortical cells.[33] A major conclusion is that the synchronization of cortical potential waves, their rhythmicity, and their spatial progression over the cortical surface result from the postsynaptic effects of ordered patterns of impulses in presynaptic fibers rather than from intrinsic rhythmic properties of cortical cells. The question of whether the spontaneous waves themselves exert some intercellular influence remains open. It has been shown that rhythmically firing neurons are extraordinarily sensitive to voltage gradients in their extracellular surround,[81] but whether the extracellular voltage gradients associated with a local PSP of one cell are sufficiently intense to affect the excitability of adjacent cells is unknown. Such a mechanism would provide an alternative, but not mutually exclusive, explanation (other than that of synchronous presynaptic input) for the fact that large masses of neurons tend to beat in unison, particularly when idling. There is no evidence that slow waves per se constitute a separate signaling mechanism within the brain.

METHOD OF SINGLE-UNIT ANALYSIS AND ITS APPLICATION TO THE STUDY OF BRAIN FUNCTION

It is only by empirical correlation, not by elucidation of mechanism, that the study of brain waves has proved useful in clinical medicine. In a later section and in later chapters it will emerge that one slow-wave event, the potential change evoked by sensory stimulation, has proved a valuable tool for mapping the central topography of sensory systems. The complexities of slow waves and the uncertain knowledge of their relation to cellular events has in the past limited their use in the study of mechanism, for only in special cases can the patterns of activity in a population of neurons be deduced with certainty from the amplitudes, forms, durations, or signs of the slow potential changes recorded from them. Yet it is just those neuron firing patterns that must be defined before the function of large aggregates of nerve cells is understood. In the first place, it must be known how a single neuron, as a representative of a class of neurons in a given locale, performs during the execution of various functions by the neural populations of which it is a member. Equally important are the relations between the activity of such a neuron and that of its neighbors. Experimental objectives such as these can now be reached by use of the method of single-unit analysis and its corollary, the derived reconstruction of population events. This is possible because central nerve cells, which cannot be isolated anatomically, may sometimes be isolated electrically.

The method is applied in various forms in neurophysiologic experiments. The way in which certain aspects of sensory stimuli are encoded in terms of afferent impulses in peripheral nerve fibers is described in Chapter 11, largely on the basis of the study of nerve fibers isolated by microdissection. Observation of cellular events by intracellular recording has led to the present understanding of neuromuscular and synaptic transmission, and this method has been applied successfully at the level of the cerebral cortex. When the experimental objectives are those just given, the method of recording via microelectrodes whose tips lie in an extracellular position is especially appropriate, for single neurons can then be observed for long periods of time and a considerable number of cells of a given population can be studied *seriatim* under standard, well-controlled conditions.

It is now clear that using these methods the neural replication of sensory events can be determined in great detail (Chapters 12, 15, and 16). Most importantly, the study of many cells of a given locale allows a post hoc reconstruction of the total profile of activity in all the neurons of a population that are influenced by a peripheral stimulus. Findings are then correlated closely with histologic identification of recording sites. Thus attention of investigators is directed toward the dynamic and time-dependent aspects of the activity of central sensory neurons, that is, how they signal such things as changes in stimulus positions or intensity, stimulus shape, quality (e.g., color), temporal cadence, etc. Comparison of the response properties of populations of neurons at different levels of a sensory system (e.g., of the retinal ganglion cells, the cells of the lateral geniculate, and those of the visual cortex) establishes the transformations and abstractions of neural activity that occur between periphery and center. Significant signals are thought to derive from either increases or decreases in the ongoing spontaneous rate of activity present in many but not all central sensory neurons. There is another possibility—that at one and the same rate of discharge two different signals might be made by virtue of differences in the temporal order with which impulses occur. The possibility that pulse-interval modulation constitutes a significant signaling mechanism has attracted intensive study in recent years. It has so far been shown to be a signaling device only in certain special cases. Indeed, it is unlikely on a priori grounds that the precise sequential timing of impulses is important in those neural systems in which a considerable convergence of presynaptic fibers on postsynaptic neurons occurs at each level of the system. The unique temporal sequence of impulses in any one presynaptic fiber will, after the transformation imposed by the PSPs of the postsynaptic cell, scarcely be identifiable in the temporal pattern of the train of impulses emitted by that cell.

In other systems, notably the great afferent

sensory pathways in which convergence is restricted and the security of transmission is very high, the rhythmic pattern in a synchronously active group of first-order sensory fibers is transmitted with great fidelity, at least through the first stage of cortical activation. Therefore there is the possibility that in these systems the temporal order in which impulses occur is of critical importance for information transmission, over and above the general level of activity. In either case the study of this "internal structure of the neural message" is an elegant and frequently a most revealing method of data analysis.[63]

Statistical nature of neural activity[30,53,58,59]

The study of central neurons has confirmed what is intuitively predictable from inspection of brain-wave recordings—that neural activity is a variable, a statistical affair that for quantitative evaluation must be treated from the probabilistic point of view. The spontaneous discharges of central neurons occur irregularly in time. Impulses do not appear randomly, however, for the time spacing is subject to certain constraints such as refractoriness and afferent or recurrent inhibition. This uncertainty exists even at the level of first-order fibers, although their discharge sequences under a steady sensory stimulus are much less variable than are those of central sensory neurons. It is apparent that the asynchronous convergence of trains of impulses in first-order fibers, even though quasi-periodic, will evoke from the postsynaptic element on which they impinge trains of impulses that show a much wider dispersion of impulse intervals, and so on at successive relays. In some locales these discharge sequences are also influenced by cyclic variations in the likelihood of discharge, an influence undoubtedly exerted by other converging systems.[63]

A certain variability pertains also to the responses of the first- (and the nth-) order neurons activated by even brief sensory stimuli, as shown by the variations in the responses of mechanoreceptive afferents from the skin of the monkey's hand to mechanical stimuli (Fig. 12-13). Inspection of this graph reveals that threshold of response is itself a statistical matter, as is the value of the response to a stimulus of a given strength. Thus a central detecting and interpretive mechanism, itself composed of neurons, must identify a neural signal as such against a background of ongoing neural activity and discriminate between two signals (i.e., between two trains of impulses) as being different, when each will from one trial to the next show considerable fluctuation, even though each population of responses is evoked by a repeated stimulus of constant strength. The neurophysiologist uses a variety of methods for averaging responses to estimate true means and for distinguishing signals from background, but it should be emphasized that the central detecting apparatus must frequently decide whether a stimulus has occurred at all or whether one stimulus is stronger than another on the basis of a single brief period of altered neural activity. How this matter may be treated from the standpoint of statistical decision theory is considered in some detail in Chapter 20. Whether brains function in this way is unknown, but some such mechanism seems a prerequisite for orderly function.

PRIMARY EVOKED POTENTIAL: CORRELATION WITH SYNAPTIC ORGANIZATION AND CELLULAR ACTIVITY[66]

The term "evoked potential" identifies the electrical change that may be recorded in some part of the brain in response to the deliberate stimulation of sense organs, the afferent fibers of peripheral nerves, or some point on the sensory pathway leading from periphery to center. It refers to the evoked slow-wave events rather than the cellular discharges that may be associated with it. Although most frequently observed in sensory systems, the electrical changes evoked in one part of the brain by stimulation of a distant part with which it is linked fall into the same category. This provides one method for identifying intracerebral connections, for example, those that link homologous areas of the two cerebral hemispheres via the corpus callosum. Evoked potentials differ from spontaneous brain waves in several ways. They usually appear in definite time relations to the stimuli that evoke them. They commonly are observed in only a local part of the brain (e.g., in a sensory area on stimulation of the appropriate receptors), although this restricted distribution depends on the level of excitability of the brain and the particular component of the waveform to which reference is made. Under certain experimental conditions, sensory evoked potentials may be used to map the central representation of sensory systems, for they are more or less predictable and reproducible both from time to time and from one animal to another.

The hypothesis now seems well established that the evoked slow waves recorded in or on the surface of the cortex or within subcortical nuclei are the net result of the extracellular current flows generated by the local postsynaptic responses of neurons. Action potentials in axons are discernible only in records made from the cortical surface when large numbers of thalamocortical fibers are activated synchronously, as by electrical stimulation of thalamic relay nuclei.

Typical evoked potentials recorded from the postcentral gyrus of a monkey, evoked by brief mechanical stimuli to the skin, are shown in Fig. 9-7. The positive-negative waveform recorded on the surface (inset) could result from one or a combination of the following: (1) depolarization of cell somata deep within the cortex, followed by or overlapped with depolarization of membranes lying immediately beneath the electrode (i.e., the apical dendrites); (2) polarization of dendrites followed by their depolarization; (3) depolarization of deep-lying somata followed by their polarization; etc. Some further evidence, not completely conclusive, is obtained by observing the changes in the evoked potential as a recording microelectrode is passed through the cortex in a direction normal to its surface. As the records in Fig. 9-7 show, the initial positive component dwindles quickly after penetration,

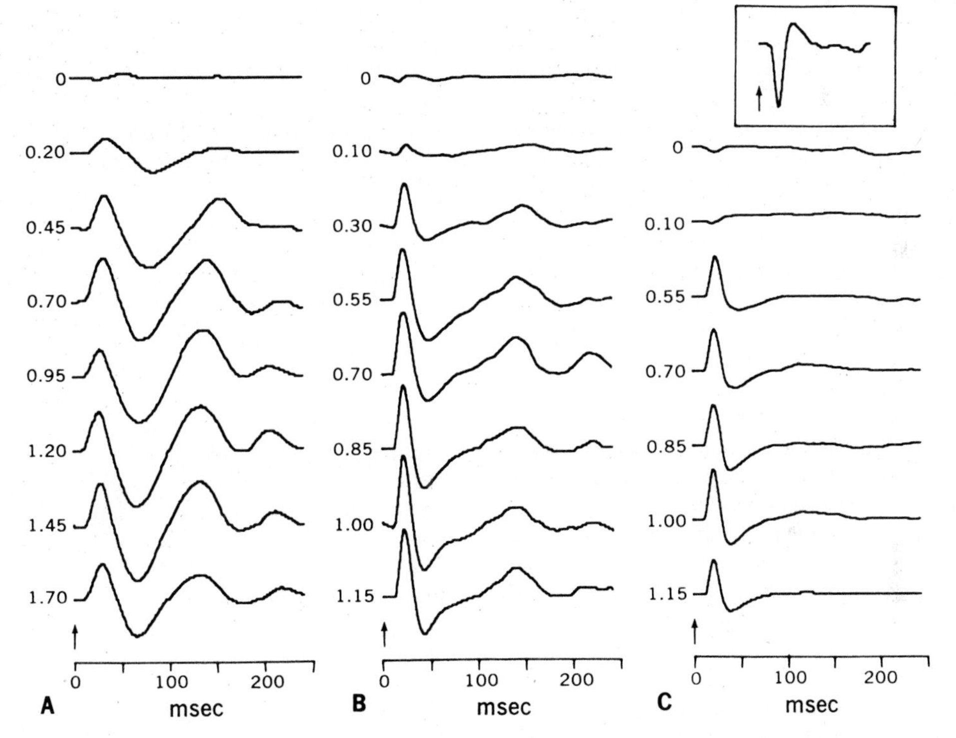

Fig. 9-7. Cortical potentials from postcentral gyrus of monkey evoked by brief mechanical stimulus to skin; positivity at electrode tip downward.

In inset at upper right, a typical evoked potential recorded on cortical surface with 2 mm, ball-tipped platinum wire electrode, a "gross electrode," is shown. Three columns of superimposed tracings (**A** to **C**) show evoked potentials recorded with microelectrode passed through cortex in direction normal to its surface. Numbers to left of each series indicate depth (in millimeters below cortical surface) at which responses were obtained. Each tracing represents computed average of 16 successive responses. Vertical arrows mark onset time of 10 msec mechanical stimulus to skin. Stimulus was delivered repetitively at rate of ½ sec; skin indentation was about 1 mm. In all instances, stimulus was applied at peripheral locus from which maximal cortical activation was obtained, located on first digit of contralateral hand.

In **A** are plotted results of microelectrode penetration into cortex of postcentral gyrus of unanesthetized monkey. Subsequently, 10 mg/kg of sodium pentobarbital was administered intravenously to animal, and another cortical penetration was made close to the first. Recorded evoked potentials are shown in **B.** With electrode deep within cortex (1.15 mm), second and equal dose of anesthetic was given, and evoked responses were observed at various depths as microelectrode was withdrawn. These later results are shown in **C.** Sensitivity of recording amplifier was 4 times greater in **A** than in **B** and **C.**

whereas the negative component grows remarkably and remains of great amplitude throughout the traverse of the cellular layers.

Under these conditions the afferent barrage of impulses has produced, either mono- or oligosynaptically, a set of standing dipoles distributed throughout the cellular layers of the cortex, dipoles oriented with positive ends toward the cortical surface. The vertical orientation of synaptic relations within the cortex fits this suggestion. In this particular case the site of microelectrode penetration was at that cortical locus most intensely activated by the peripheral stimulus. The initial and predominantly negative configuration of the slow-wave evoked responses recorded within the cortex is taken to indicate the predominance of an early excitatory inflow to the cells in that region. The presence of PSPs, here recorded extracellularly as slow waves, indicates that the anesthetic agent has not blocked synaptic transmission within the cortex. It has, however, exerted a powerful effect on the spike-generator mechanism of the cortical neurons, for under deep anesthesia only a few cells, relative to the total number present, discharge action potentials.

When the same experiment is repeated in the absence of anesthesia (Fig. 9-7, *A*), the slow-wave evoked responses are more complex, and averaging methods are commonly used to bring out the signal—the evoked potential—from the ongoing EEG. The sequential reversal of the sign of the initial phase of the evoked potential is identical to that observed in anesthetized animals, and the initial wave is frequently followed by "late" or secondary waves. The secondary waves, even when evoked by sensory stimuli, are commonly not confined to the relevant cortical sensory area and may be very widely distributed over the surface of the hemisphere. When recorded outside the primary receiving area, they may be initially and wholly negative in sign, an event interpreted to indicate an intense depolarization of the apical dendrites. This fact, plus their wide distribution, suggests that they are the sign of collateral activation by the afferent inflow of the generalized thalamocortical system, which is described in Chapter 10.

The technique of deriving the evoked potential from the ongoing EEG waves by the method of averaging makes it possible to study evoked potentials in human beings, using recordings from the surface of the scalp with appropriate precautions to eliminate artifact due to evoked reflex activation of muscle.[20] Investigations have now been undertaken to correlate these electrical changes with the behavioral events of sensation and perception as well as their alterations in patients with disease or injury of the nervous system.[88]

Correlation of the evoked potential with the impulse discharge of neurons

Thus far, efforts to correlate the events in single cortical cells with the spontaneous ongoing EEG recorded on the surface have not been very successful, and more statistical studies are required to determine the relation. This is so because the waves of the EEG are the complex result of all the membrane conductance changes in neurons within the "view" of the recording electrode (Fig. 9-4). More precise correlations have been observed between the surface-recorded evoked potential and the EPSP-IPSP sequence of membrane potentials produced in cortical cells by stimulation of thalamic relay nuclei.[31] Further, a very precise correlation can sometimes be shown to exist between the evoked potential recorded with a microelectrode within the cortex and close to the cell, and the impulse discharges of that cell. However, such a parallelism is not always observed, for some cells may be inhibited when the form of the local intracortical potential predicts excitation.

What is certain is that from the moment a recording microelectrode penetrates layer II of a sensory cortex, down across all the cellular layers of the cortex, an appropriately placed sensory stimulus evokes at the microelectrode an initially negative wave. This is indicative of intense postsynaptic depolarization, and associated with it is the impulse activity of cells in all layers of the cortex.

Volume conductor theory

F. J. Brinley, Jr.

The most direct method of recording an action potential from an excitable tissue is by placing a microelectrode inside the cell. However, in some cases (e.g., clinical electroencephalography and electrocardiography), intracellular microelectrode recording of transmembrane potentials is not feasible, and it is necessary to record electrical potentials with electrodes placed some distance from the actual site at which the signal is generated. Although the potentials observed in the two recording situations appear dissimilar, they are generated by the same events and are related mathematically. The analysis that describes this relation is called the volume conductor theory.

As an illustration of the sort of reasoning involved, consider the relation between electrical signals recorded by electrodes located in two different positions as an action potential propagates along a nerve. The two recording arrangements to be compared are (1) an intracellular micropipet and (2) an external electrode located some distance from the axon surface. For mathematical simplicity the fiber is assumed to be unmyelinated, of uniform diameter, infinitely long, and contained in an infinite, electrically homogeneous conducting volume.

Since the action potential occupies a finite period of time and propagates along the nerve with a finite velocity, at any given instant the action potential occupies a discrete length of nerve. The length of fiber (L) so involved is given by L = Vt, where V is the velocity of propagation (assumed constant) and t is the time required for a complete action potential. The assumption of a constant velocity of propagation implies that the temporal variation of potential at a fixed point is equivalent to a spatial variation at a fixed time. Experimentally it may be easier to use a fixed electrode and measure the time course of an action potential as it passes by, but conceptually it is simpler to consider the action potential as "frozen in time" and study the spatial variation in potential as an exploring electrode is moved along the nerve.

The theory can be considered conveniently in three steps: (1) The cable theory relates the transverse membrane potential at a particular point to a transverse current density at the same point. (2) Ohm's law relates a flow of current (i.e., current density) in a region to the strength of the electrical field in that region. (3) Coulomb's law relates the potential observed at a point in space to a charge distribution.

1. It can be shown from the cable theory (p. 52) that the transmembrane current flow (amperes per cubic centimeters) is proportional to the second derivative of the transmembrane potential (V_m) at that point (Fig. 9-8):

$$i_m(x = x_i) = \left.\frac{\partial^2 V_m}{\partial x^2}\right|_{x = x_i} \cdot \frac{1}{R_o + R_i} \qquad (1)$$

V_m can represent either the total observed potential across the membrane or only the electrotonic potential (i.e., the numeric value of the deviation of the actual potential from the resting potential, which is assumed to be temporally and spatially invariant). Since the transverse membrane current is proportional to the second derivative of the potential, a constant resting potential will not contribute to i_m. R_o and R_i are the external (tissue fluid) and internal (axoplasmic) electrical resistances, respectively (ohms per centimeter); x represents distance along the nerve (in centimeters).

2. The magnitude of the current density at a particular point on the surface of the nerve will be proportional to the electrical field at that point (Ohm's law):

$$i_m(x = x_1) \propto E(x = x_1) \qquad (2)$$

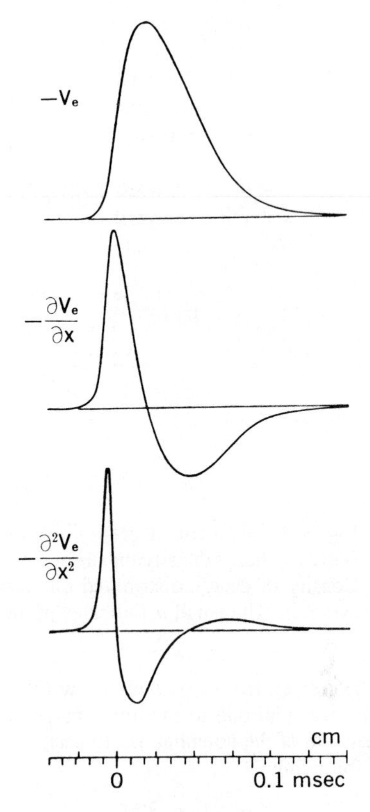

Fig. 9-8. Graphic differentiation of transmembrane action potential, $-V_e$, upper curve, to give membrane current, lower curve. Time and distance scales are given at bottom of figure. (From Lorente de Nó.[46])

This electrical field can be considered to result from the presence of electrical charges on the surface of the fiber. Therefore the spatial variation of transmembrane potential along a nerve fiber may be regarded as determining a charge distribution on the outside of the fiber. It should be understood that the assumption of net surface electrical charge made here is a mathematical artifice to aid the derivation. Such charges do not exist physically and should not be confused with the separated ion charges on either side of the membrane that are responsible for the resting membrane potential. A distribution of fictitious ion charges such as that described here is frequently called a "virtual charge" distribution. The magnitude of this assumed charge, at a specified point, will be proportional to the second derivative of the transmembrane potential at that point. Since the transmembrane current (which is also proportional to the second derivative) can be either positive or negative, the charge can have either sign. Because the membrane potential is radially symmetric (i.e., varies only along the length of the nerve and not around the circumference) the charge distribution will likewise be radially symmetric (Fig. 9-9).

3. The electrical potential at a point (P) not on the surface of the fiber, due to this array of charges, can be

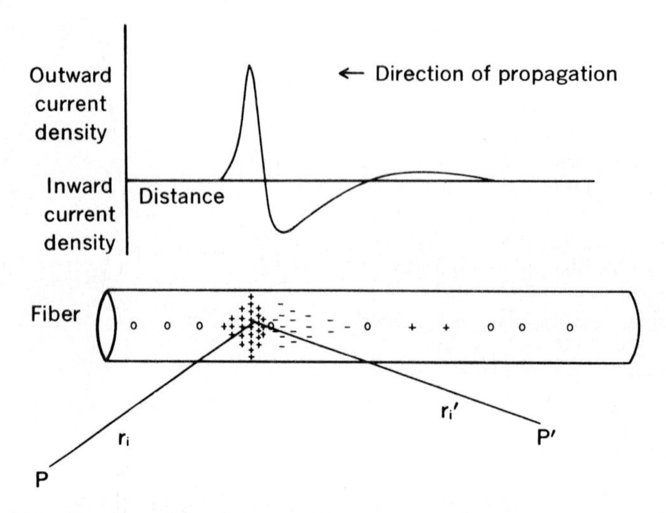

Fig. 9-9. Schematic diagram illustrating relation between membrane current density and conceptual (virtual) charge density on surface of nerve fiber used to calculate external field by Coulomb's law. Density of charge distributed on surface of fiber is indicated qualitatively by number of + or − symbols. The small *o* indicates no net surface charge.

simply calculated from Coulomb's law ($V_p = q/r_p$). Since the potential due to multiple charges is simply the linear sum of the potential due to each charge, one has the following:

$$V_p = \sum_n \frac{q(n)}{r_p(n)} \qquad (3)$$

where V_p is the potential at point P and r_p is the distance of the nth charge q to point P. The summation extends over all charges.

In the foregoing statements it has been assumed that the diameter of the fiber is very small compared to the distance to the field point (P). Even if the field point is assumed to be so close to the surface of the fiber that the charge must be considered as distributed on a circular surface rather than concentrated at a point, no new physical principles are involved in the analysis. Coulomb's law can still be applied. However, the mathematical formulation is somewhat more complicated.

The cable theory used to derive equation 1 assumes that there is no radial component of current in either the external or internal medium, that is, the longitudinal currents flow only in the regions immediately adjacent to the nerve membrane and do not circulate to any extent in the medium distant from the membrane. Although this is approximately true for an isolated fiber immersed in oil or suspended in a moist chamber where the current in the external medium is constrained to flow in the narrow path of conducting fluid outside the nerve, it is not generally true for the nerve in situ. Clark and Plonsey[29] have presented a formal mathematical solution for a long unmyelinated nerve fiber in a conducting medium conducting an action potential along its surface. The formal solution contains two terms. The first term, which is proportional to the membrane current, is equivalent to the solution of equation 1. The second term is a complicated integral, which can be considered as a correction term indicating the deviation of the simple cable theory from the true physical situation. Clark and Plonsey have evaluated this integral with the aid of a computer. The calculations indicate that the radial current (and hence radial component of the field) in the external medium may be substantial. According to their analysis, an action potential recorded in the external medium remains triphasic, as in the simple analysis given here, but the amplitude and phase relationship of the potential recorded in the volume may be quite different from that predicted from simple cable theory.

Applications of volume conductor theory

The computational complexities of applying equation 3 to each element of a large population of excitable cells with different spatial orientations conducting asynchronously preclude a quantitative calculation of the configuration of extracellular potentials recorded in a volume conductor, even if the basic assumptions such as homogeneity and infinitely large size of the volume conductor were to be met. For this reason, in practice, electrophysiologists and clinicians alike rely on semiempirical correlations between function of an excitable tissue and configuration of the observed potentials. Nevertheless, a few useful qualitative generalizations can be made on the basis of the simple analysis just presented and illustrated with three specific examples.

Electrocardiogram (ECG). The general configuration of the ECG as recorded with either limb or unipolar leads is obtained from the second derivative of the transmembrane action potential of the myocardial muscle cells. As can be seen from Fig. 9-9, when a conducting fiber is oriented so that the action potential is propagating toward the sensing electrode (e.g.,

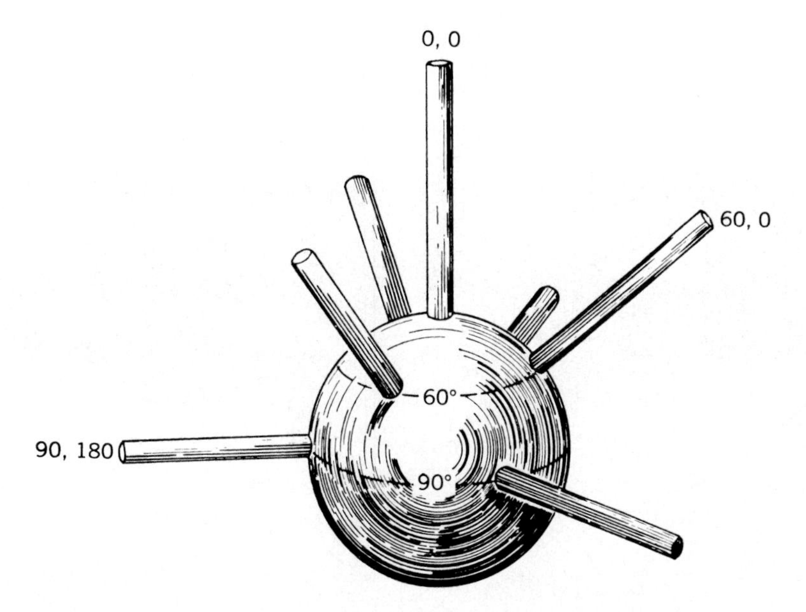

Fig. 9-10. Diagram illustrating relative orientation of seven "dendrites," but not their lengths, around spherical "soma" used to calculate potential field shown in Fig. 9-11. Numbers indicate angular coordinates of the three dendrites shown in Fig. 9-11. (From Rall.[67])

point P), the observed deflection will be positive because the electrode is relatively nearer the positive than the negative charge. When the action potential is propagating away from the electrode (e.g., point P′), the observed deflection will be negative because the electrode is now closer to negative charge. A practical example can be seen in the precordial leads of the ECG. The small initial positive R wave seen in the right precordial leads represents progressive depolarization of the right ventricle muscle fibers moving in the general direction of the right sternal border. This small R wave is quickly superceded by the larger negative S wave that represents activation of the much larger left ventricular mass. Propagation in the left ventricle is directed toward the left axilla and away from the sternum; hence the deflection is negative. The potential configuration is approximately reversed in the left precordial leads, and a large initial R wave representing left ventricular depolarization is recorded.

Extracellular potentials recorded within substance of the brain by small microelectrodes. Diagrams similar to Fig. 9-9 are sometimes used to locate regions of "active" or "passive" membrane along a length of nerve or on the surface of an excitable cell. Negative potentials are presumed to represent sinks of current toward which current flows in the extracellular field. Since the current flow across the membrane in this sink region is inward, the membrane is considered to be actively depolarized (i.e., undergoing regenerative Na and K conductance changes). Positive potentials are presumed to indicate sources of current flow, implying that the membrane nearest the sensing electrode is passively supplying current to a more remote, active region where the ionic current is inward.[75] Although this interpretation is approximately true for a very simple case (e.g., an isolated single nerve fiber), it is not valid generally for the complicated arrangements of conducting elements found in the CNS.

The theoretical extracellular fields of some relatively simple arrangements of dendrites and somata have been calculated by Rall.[67] One such orientation of dendrites is illustrated in Fig. 9-10. Fig. 9-11 shows the extracellular field resulting from this particular orientation during the peak of an action potential generated in the soma (i.e., the soma is considered to be active and all the dendrites passive). The extracellular field along the dendrites, away from their immediate surface, is negative out to a distance of about eight soma radii, although these elements are not undergoing a regenerative depolarization. The potential field shown in Fig. 9-11 has been calculated assuming an infinite, homogeneous, isotropic extracellular space. Since practically all the extraneuronal volume in mammalian brain is occupied by glial cells, the true extracellular space is probably rather small. This circumstance could cause considerable quantitative modification of the shape of the potential surfaces illustrated in Fig. 9-11. However, Nelson and Frank[55] have confirmed experimentally the theoretical conclusions for the spinal motoneuron; the extracellular field actually is predominantly negative and is recordable (i.e., is greater than a few microvolts) to at least 500 to 700 μ from the soma.

Fig. 9-12 shows the theoretical relations between the entire time course of an intracellular action potential arising from the soma of a hypothetical cell with passive, symmetrically oriented dendrites and extracellular potentials recorded at varying distances from

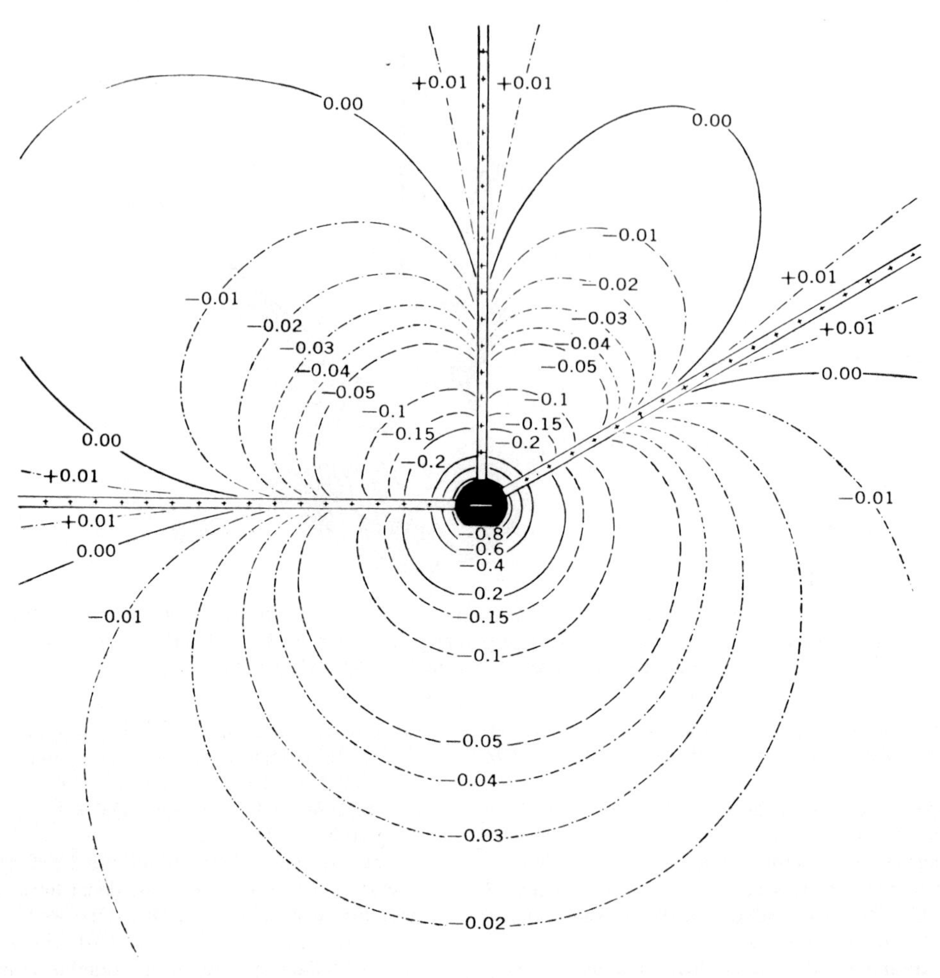

Fig. 9-11. Isopotential contours for dendrite-soma model illustrated in Fig. 9-10. Plane of figure is that defined by three labeled dendrites in Fig. 9-10. Zero potential contour is shown as solid line. Field corresponds to that which occurs at peak of action potential generated in neuron with active soma and passive dendrites. Figures are in millivolts. (From Rall.[67])

the soma. In actual experiments, Terzuolo and Araki[80] used two parallel microelectrodes joined together and recorded intra- and extracellular potentials simultaneously from spinal motoneurons. Since the tip separations were about 3 to 20 μ, the pipet recording the extracellular potential must necessarily have been reasonably close to the cell soma. In this recording circumstance the extracellular potential sequences following antidromic activation were usually positive-negative-positive instead of simply negative-positive, as shown in Fig. 9-12. The first additional positive wave in these triphasic responses represents current flow from the initially inactive soma-dendritic complex into the depolarized initial segment.

In some cases, relatively large, positive, extracellular action potentials can be recorded from neurons in both brain and spinal cord.[55] It is thought that these positive potentials result when the electrode tip is pushed against the cell membrane, causing sufficient local damage to render the spot under the electrode, but not the adjacent areas, inexcitable. The amplitude of the positive extracellular potentials, up to several millivolts, is considerably larger than that of the usual negative extracellular spikes, which are no more than a few hundred microvolts. In the case of the positive spikes the difference in amplitude presumably results because the tip of the microelectrode is pressed against the surface of the cell or perhaps partially plugged with torn glial cell membranes, creating a higher resistance for current flow. The resultant IR drop is therefore greater than in the case of a microtip well out in the extracellular space.

Electrical potentials recorded from surface of the brain. A variety of multiphasic electrical potentials can be recorded from the cortical surface in response to diverse stimuli to the nervous system, peripheral as well as central. These electrical responses result from current flow in the extracellular space caused by

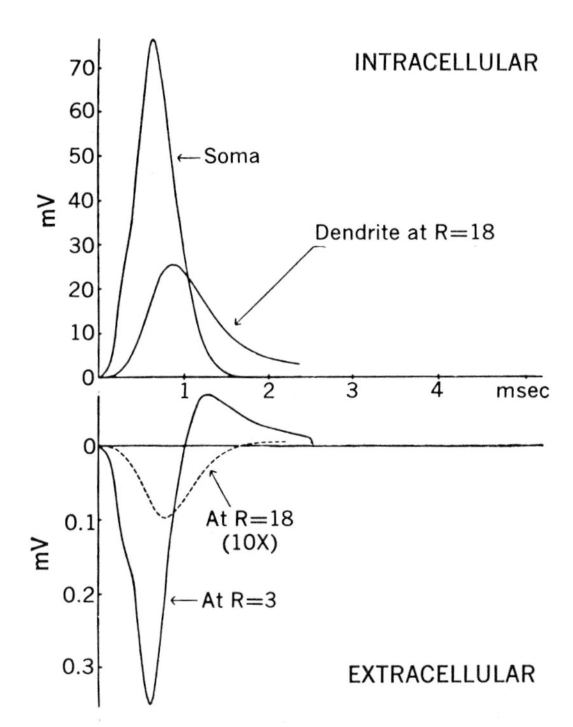

Fig. 9-12. Time course of calculated intra- and extracellular potentials, assuming that action potentials arise in soma (as by intracellular stimulation) and that electrically passive dendrites are symmetrically arranged around soma. Curve labeled "dendrite" in upper record shows delay and attenuation of soma action potential as it spreads electronically along dendrites. (From Rall.[67])

action potentials or synaptic potentials at the surface of many individual neurons. The observed surface potentials bear only a general relationship to the membrane currents generated by a single neuron or part of a neuron for three reasons: (1) the relatively large surface electrodes commonly used that permit recording from many units, (2) the complex orientation of cortical elements, and (3) asynchronous variation in membrane potential of the units. However, the principles presented in the initial paragraphs of this section can be used to correlate the sign of the observed surface potential with the extracellular currents flowing within the brain substance.

As an example of the application of volume conductor theory to the evoked activity recorded from the cortical surface, consider a large electrode resting on the surface of the exposed brain. A second electrode is placed some distance away on an inactive portion of the cortex or skull. This second or "indifferent" electrode is placed so far away from the site of electrical activity that its potential does not vary significantly with time. The potential at this point can therefore be taken as defining a "zero of potential," against which fluctuations of potential at the first recording electrode can be compared.

If cells or cell elements in the immediate vicinity of the recording electrode are acting as sources of membrane current (i.e., current is flowing out of the region of the recording electrode), the potential will become positive with respect to an indifferent electrode.* If the units are acting as electrical sinks, the current will flow into the region and the brain substance under the recording electrode will become negative. This analysis identifies only the general direction and intensity of net current flow in the region near the recording electrode. It provides no information concerning the distribution, number, or identity of the cellular elements acting as sources or sinks within the region, nor does it identify the regions elsewhere that are supplying current to or receiving current from the region under the electrode. If the electrical activity generated within the brain by the original stimulus is complex, the evoked surface potentials may also reflect this complexity and consist of multiphasic waves whose component deflections have varying durations and/or polarities.

Since the magnitude of the potential is equal to the IR drop between the tip of the recording electrode and the indifferent electrode, the actual size of the potential will depend to some extent on the amount of fluid shunt on the surface of the brain and the firmness with which the electrode rests on the cortical surface as well as on the location and intensity of the intracortical sources and sinks. The size of the potential will not, however, depend on the location of the indifferent electrode as long as it is a great distance away from the recording site.

REFERENCES
General reviews

1. Ajmone Marsan, C.: The thalamus. Data on its functional anatomy and on some aspects of thalamo-cortical integration, Arch. Ital. Biol. **103:**847, 1965.
2. Bullock, T. H., and Horridge, G. A.: Structure and function of the nervous systems of invertebrates, San Francisco, 1965, W. H. Freeman & Co., Publishers.
3. Chow, K. L., and Leiman, A. L.: The structural and functional organization of the neocortex, Neurosci. Res. Symp. Summ. **5:**149, 1971.
4. Colonnier, M.: The structural design of the neocortex. In Eccles, J. C., editor: The brain and conscious experience, Berlin, 1966, Springer Verlag.
5. Glaser, G. H.: The normal electroencephalogram and its reactivity. In Glaser, G. H., editor: EEG and behavior, New York, 1963, Basic Books, Inc., Publishers.
6. Hassler, R.: Anatomy of the thalamus. In Schaltenbrand, G., and Bailey, P., editors: Introduction to stereo-

*These statements are only approximately true, in view of Rall's demonstration that negative potentials (i.e., with respect to an indifferent electrode) can be observed in the vicinity of a passive membrane acting as a source (Fig. 9-12). However, the cell bodies of the majority of cortical neurons lie 100 μm or more beneath the cortical surface. The nerve elements in immediate proximity to an electrode on the arachnoid surface are therefore axons or dendrites, each relatively distant from its own cell body. In such cases the conventional notions of source or sink apply fairly well.

taxis with an atlas of the human brain, New York, 1959, Grune & Stratton, Inc., vol. 1.
7. MacKay, D. M.: Evoked brain potentials as indicators of sensory information processing, Neurosci. Res. Symp. Summ. **4**:397, 1970.
8. Penfield, W., and Jasper, H. H.: Epilepsy and the functional anatomy of the human brain, Boston, 1954, Little, Brown & Co.
9. Van Buren, J. M., and Borke, R. C.: Variations and connections of the human thalamus, New York, 1972, Springer Verlag New York, Inc., vol. 1.
10. Walter, D. O., and Brazier, M. A. B.: Advances in EEG analysis, Electroencephalogr. Clin. Neurophysiol. **27**(suppl.): 1, 1969.

Original papers

11. Adey, R. W.: Spectral analysis of EEG data from animals and man during alerting, orienting and discriminative responses. In Neurophysiology of attention, London, 1968, Butterworth & Co. (Publishers), Ltd.
12. Adrian, E. D., and Matthews, B. H. C.: The Berger rhythm: potential changes from the occipital lobes in man, Brain **57**:355, 1934.
13. Adrian, E. D., and Yamagiwa, K.: The origin of the Berger rhythm, Brain **58**:323, 1935.
14. Akert, K.: Comparative anatomy of frontal cortex and thalamofrontal connections. In Warren, J. M., and Akert, K., editors: The frontal granular cortex and behavior, New York, 1964, McGraw-Hill Book Co.
15. Angevine, J. B., Locke, S., and Yakovlev, P. I.: Thalamocortical projection of the ventral anterior nucleus in man, Arch. Neurol. **7**:518, 1962.
16. Benevento, L. A., and Davis, B.: Topographical projections of the prestriate cortex to the pulvinar nuclei in the macaque monkey: an autoradiographic study, Exp. Brain Res. **30**:405, 1977.
17. Benevento, L. A., and Fallon, J. M.: The ascending projections of the superior colliculus in the rhesus monkey (Macaca mulatta), J. Comp. Neurol. **160**:339, 1975.
18. Benevento, L. A., and Rezak, M.: The cortical projections of the inferior pulvinar and adjacent lateral pulvinar in the rhesus monkey (macaca mulatta): an autoradiographic study, Brain Res. **108**:1, 1976.
19. Berger, H.: Das Elektrenkephalogramm des Menschen, Acta Nova Leopold. **6**:173, 1938.
20. Bickford, R. G., Jacobsen, J. L., and Cody, D. T.: Nature of average evoked potentials to sound and other stimuli in man, Ann. N.Y. Acad. Sci. **112**:204, 1964.
21. Brazier, M. A. B.: A history of the electrical activity of the brain. The first half-century, New York, 1961, Macmillan Publishing Co., Inc.
22. Brodmann, K.: Vergleichende Lokalisationslehre der Grosshirnrinde in ihren prinzipien dargestellt auf Grund des Zellenbaues, Leipzig, 1909, Johann Ambrosius Barth.
23. Brodmann, K.: Feinere Anatomie des Grosshirns. In Handbuch der Neurologie: allgemeine Neurologie, Berlin, 1910, Springer Verlag, vol. 1.
24. Burton, H., and Jones, E. G.: The posterior thalamic region and its cortical projection in New World and Old World monkeys, J. Comp. Neurol. **168**(2):249, 1976.
25. Büttner, U., Büttner-Ennever, J. A., and Henn, V.: The vestibular thalamus: neurophysiological and anatomical studies in the monkey. In Hood, J. D., editor: Vestibular mechanisms in health and disease, London, 1978, Academic Press, Inc. (London), Ltd., chap. 7, pp. 80-85.
26. Campbell, A. W.: Histological studies on the localization of cerebral function, Cambridge, England, 1905, University Press.
27. Carman, J. B., Cowan, W. M., and Powell, T. P. S.: Cortical connexions of the thalamic reticular nucleus, J. Anat. **98**:587, 1964.
28. Carmel, P. W.: Efferent projections of the ventral anterior nucleus of the thalamus in the monkey, Am. J. Anat. **128**:159, 1970.
29. Clark, J., and Plonsey, R.: A mathematical evaluation of the core conductor model, Biophys. J. **6**:95, 1966.
30. Cowan, J. D.: Statistical mechanics of neural networks. In Gerstenhaber, M., editor: Mathematical questions in biology: proceedings of the second annual symposium, Providence, 1969, American Mathematical Society.
31. Creutzfeldt, O. D., Watanabe, S., and Lux, H. D.: Relations between EEG phenomena and potentials of single cortical cells (parts I and II), Electroencephalogr. Clin. Neurophysiol. **20**:1, 1966.
32. Deecke, L., Schwarz, D. W., and Frederickson, J. M.: Nucleus ventroposterior inferior (VPI) as the vestibular thalamic relay in the rhesus monkey, Exp. Brain Res. **20**:88, 1974.
33. Elul, R.: Brain waves: intracellular recording and statistical analysis help clarify their physiological significance. In Rochester conference on data acquisition and processing in biology and medicine, Oxford, 1966, Pergamon Press, Ltd., vol. 5.
34. Frederickson, J. M., Figge, U., Scheid, P., and Kornhuber, H. H.: Vestibular nerve projection to the cerebral cortex of the rhesus monkey, Exp. Brain Res. **2**:318, 1966.
35. Goldstein, S.: Phase coherence of the alpha rhythm during photic blocking. Electroencephalogr. Clin. Neurophysiol. **29**:127, 1970.
36. Hubel, D. H., and Wiesel, T. N.: Laminar and columnar distribution of geniculo-cortical fibers in the macaque monkey, J. Comp. Neurol. **146**:421, 1972.
37. Jones, E. G.: Lamination and differential distribution of thalamic afferents within the sensory-motor cortex of the squirrel monkey, J. Comp. Neurol. **160**:167, 1975.
38. Jones, E. G., and Levitt, R. Y.: Retrograde axonal transport and the demonstration of non-specific projections to the cerebral cortex and striatum for thalamic intralaminar nuclei of the rat, cat and monkey, J. Comp. Neurol. **154**:349, 1974.
39. Jones, E. G., and Powell, T. P. S.: Connexions of the somatic sensory cortex of the rhesus monkey. I. Ipsilateral cortical connexions, Brain **92**:477, 1969.
40. Jones, E. G., and Powell, T. P. S.: Connexions of the somatic sensory cortex of the rhesus monkey. II. Contralateral cortical connexions, Brain **92**:717, 1969.
41. Jones, E. G., and Powell, T. P. S.: Connexions of the somatic sensory cortex of the rhesus monkey. III. Thalamic connexions, Brain **93**:37, 1970.
42. Jones, E. G., and Wise, S. P.: Size, laminar and columnar distribution of efferent cells in the sensory-motor cortex of monkeys, J. Comp. Neurol. **175**:391, 1977.
43. Joseph, J. P., Remond, A., Rieger, H., and Lesevre, N.: The alpha average. II. Quantitative model and the proposition of a theoretical model, Electroencephalogr. Clin. Neurophysiol. **26**:350, 1969.
44. Kandel, E. R., Spencer, W. A., and Brinley, F. J., Jr.:

Electrophysiology of hippocampal neurons (parts I to IV), J. Neurophysiol. **24**:225, 1961.
45. Liedgren, S. R., et al.: Representation of vestibular afferents in somatosensory thalamic nuclei of the squirrel monkey (Saimiri sciureus), J. Neurophysiol. **39**:601, 1976.
46. Lorente de Nó, R.: A study of nerve physiology, Stud. Rockefeller Institute Med. Res. **132**:1, 1947.
47. Lund, J.: Organization of neurons in the visual cortex, area 17, of the monkey (Macaca mulatta), J. Comp. Neurol. **147**:455, 1973.
48. Lund, J. S., et al.: The origin of efferent pathways from the primary visual cortex, area 17, of the macaque monkey as shown by retrograde transport of horseradish peroxidase, J. Comp. Neurol. **164**:287, 1975.
49. Mehler, W. R.: Further notes on the center median nucleus of Luys. In Purpura, D. P., and Yahr, M. D., editors: The thalamus, New York, 1966, Columbia University Press.
50. Mehler, W. R.: Idea of a new anatomy of the thalamus, J. Psychiatr. Res. **8**:203, 1971.
51. Milstein, V.: Contingent alpha blocking: conditioning or sensitization? Electroencephalogr. Clin. Neurophysiol. **18**:272, 1965.
52. Minderhoud, J. M.: An anatomical study of the efferent connections of the thalamic reticular nucleus, Exp. Brain Res. **12**:435, 1971.
53. Moore, G. P., Perkel, D. H., and Segundo, J. P.: Statistical analysis and functional interpretation of neuronal spike data, Annu. Rev. Physiol. **28**:493, 1966.
54. Nauta, W. J. H.: Neural associations of the amygdaloid complex of the monkey, Brain **85**:505, 1962.
55. Nelson, P. G., and Frank, D.: Extracellular potential fields of single motoneurons, J. Neurophysiol. **27**:914, 1964.
56. Olszewski, J.: The thalamus of the Macaca mulatta. An atlas for use with the stereotaxic instrument, Basel, 1952, S. Karger AG.
57. Pearson, R. C. A., Brodal, P., and Powell, T. P. S.: The projection of the thalamus upon the parietal lobe in the monkey, Brain Res. **144**:143, 1978.
58. Perkel, D. H., Gerstein, G. L., and Moore, G. P.: Neuronal spike trains and stochastic point processes. I. The single spike train, Biophys. J. **7**:419, 1967.
59. Perkel, D. H., Gerstein, G. L., and Moore, G. P.: Neuronal spike trains and stochastic point processes. II. Simultaneous spike trains, Biophys. J. **7**:440, 1967.
60. Perl, E. R., and Whitlock, D. G.: Somatic stimuli exciting spinothalamic projection neurons in cat and monkey, Exp. Neurol. **3**:256, 1961.
61. Phillips, C. G.: Intracellular records from Betz cells of cat, Q. J. Exp. Physiol. **41**:58, 1956.
62. Poggio, G. F., and Mountcastle, V. B.: A study of the functional contributions of the lemniscal and spinothalamic systems to somatic sensibility. Central nervous mechanisms in pain, Johns Hopkins Med. J. **106**:266, 1960.
63. Poggio, G. F., and Viernstein, L. J.: Time series analysis of the discharge sequences of thalamic somatic sensory neurons, J. Neurophysiol. **27**:517, 1964.
64. Powell, T. P. S., and Cowan, W. M.: A study of thalamo-striate relations in the monkey, Brain **79**:364, 1956.
65. Powell, T. P. S., and Cowan, W. M.: The interpretation of the degenerative changes in the intralaminar nuclei of the thalamus, J. Neurol. Neurosurg. Psychiatry **30**:140, 1967.
66. Purpura, D. P.: Nature of electrocortical potentials and synaptic organization in cerebral and cerebellar cortex, Int. Rev. Neurobiol. **1**:48, 1959.
67. Rall, W.: Electrophysiology of a dendritic neuron model, Biophys. J. **2**:145, 1962.
68. Ramón y Cajal, S.: Histologie du systeme nerveus, Paris, 1909-1911, Maloine, vol. 2.
69. Remond, A., et al.: The alpha average. I. Methodology and description, Electroencephalogr. Clin. Neurophysiol. **26**:245, 1969.
70. Roberts, T. S., and Akert, K.: Insular and opercular cortex and its thalamic projection in Macaca mulatta, Schweiz. Arch. Neurol. Neurochir. Psychiatr. **92**:1, 1963.
71. Rose, J. E.: The cellular structure of the auditory region of the cat, J. Comp. Neurol. **91**:409, 1949.
72. Rose, J. E., and Woolsey, C. N.: Structure and relations of limbic cortex and anterior thalamic nuclei in rabbit and cat, J. Comp. Neurol. **89**:279, 1948.
73. Rose, J. E., and Woolsey, C. N.: Organization of the mammalian thalamus and its relationships to the cerebral cortex, Electroencephalogr. Clin. Neurophysiol. **1**:391, 1949.
74. Rose, J. E., and Woolsey, C. N.: Cortical connections and functional organization of the thalamic auditory system of the cat. In Harlow, H. F., and Woolsey, C. N., editors: Biological and biochemical bases of behavior, Madison, 1958, University of Wisconsin Press.
75. Rosenthal, F.: Relationships between positive-negative extracellular potentials and intracellular potentials in pyramidal tract neurons, Electroencephalogr. Clin. Neurophysiol. **30**:38, 1971.
76. Scheibel, M. E., and Scheibel, A. B.: The organization of the nucleus reticularis thalami: a Golgi study, Brain Res. **1**:43, 1966.
77. Scheibel, M. E., and Scheibel, A. B.: The organization of the ventral anterior nucleus of the thalamus: a Golgi study, Brain Res. **1**:250, 1966.
78. Scheibel, M. E., and Scheibel, A. B.: On the nature of dendritic spines—report of a workshop, Comm. Behav. Biol. **1**:231, 1968.
79. Stefanis, C., and Jasper, H.: Intracellular microelectrode studies of antidromic responses in cortical pyramidal tract neurons, J. Neurophysiol. **27**:828, 1964.
80. Terzuolo, C. A., and Araki, T.: An analysis of intra- versus extracellular potential changes associated with activity of single spinal motoneurons, Ann. N.Y. Acad. Sci. **94**:547, 1961.
81. Terzuolo, V. A., and Bullock, T. H.: Measurement of imposed voltage gradient adequate to modulate neuronal firing, Proc. Natl. Acad. Sci. USA **42**:687, 1956.
82. Trojanowski, J. Q., and Jacobson, S.: Areal and laminar distribution of some pulvinar cortical efferents in rhesus monkey, J. Comp. Neurol. **169**:371, 1976.
83. Trojanowski, J. Q., and Jacobson, S.: The morphology and laminar distribution of cortico-pulvinar neurons in the rhesus monkey, Exp. Brain Res. **28**:51, 1977.
84. von Economo, C.: The cytoarchitectonics of the human cerebral cortex (translated by S. Parker), Oxford, 1929, Humphrey-Milford.
85. Walker, A. E.: The primate thalamus, Chicago, 1938, University of Chicago Press.
86. Walter, W. G.: Intrinsic rhythms of the brain. In Magoun, H. W., editor: Handbook of physiology, Neurophysiology section, Baltimore, 1959, The Williams & Wilkins Co.
87. Wells, C. E.: Alpha wave responsiveness to light in man.

In Glaser, G. H., editor: EEG and behavior, New York, 1963, Basic Books, Inc., Publishers.

88. Whipple, H. E., and Katzman, R., editors: Sensory evoked response in man, Ann. N.Y. Acad. Sci. **112:** entire issue, 1964.

89. Yakovlev, P. I., Locke, S., and Angevine, J. B., Jr.: The limbus of the cerebral hemisphere, limbic nuclei of the thalamus, and the cingulum bundle. In Purpura, D. R., and Yahr, M. D., editors: The thalamus, New York, 1966, Columbia University Press.

10

VERNON B. MOUNTCASTLE

Sleep, wakefulness, and the conscious state

INTRINSIC REGULATORY MECHANISMS OF THE BRAIN

In treating the problem of consciousness, neurobiologists assume that they deal with a certain aspect of the functional organization of brains that will eventually be defined in terms of neural mechanisms. Consciousness is an aspect of brain function for which, even from the behavioral point of view, there is no ready, brief definition, but several of its observable attributes are known with reasonable certainty:

1. *Consciousness is a neural phenomenon.* Organisms without nervous systems never display the publicly observable attributes of consciousness.

2. *Consciousness exists in other animals as well as in man.*[12] Evidence exists that consciousness appears in animals *pari passu* with the development of a complex nervous system as the master regulatory mechanism of the animal. But a nervous system per se is not enough. The attributes of consciousness do not appear when nervous systems containing relatively few neurons control behavior by linking afferent input via "releasers" to neural mechanisms governing innate behavioral patterns. It is difficult to draw any line separating those animals that are conscious from those that are not, and it is perhaps better to regard the different species as distributed along a continuum of an increasing degree and complexity of consciousness. The presence of the conscious control of action may, as an operational definition, be assumed when an organism displays the capacity for choice of action, the ability to set one goal aside in favor of another, the power to withhold action or reaction. Certainly a high order of consciousness is involved in anticipatory planning for action, in modifying action once initiated in terms of then-current events, and in the preparation of alternative stratagems to deal with abstract conceptualizations of events that may be encountered. All these latter compose a property of brains called intelligence.

What are some of the publicly observable aspects of behavior in man and in animals that suggest the presence of consciousness?

a. The *act of attention* and the capacity to shift attention selectively.
b. *Manipulation of abstract ideas,* preeminently a conscious process characteristic of human beings—the representation of abstract or general ideas by words or other symbols. No animal approaches this level, but evidence for elaborate means of communication between individual animals of many species suggests that this ability is present in higher animals in a rudimentary form.
c. *Capacity for expectancy* is further indicated by the use of tools by animals in the wild, by the organization of troops for hunting or food gathering, by the posting of sentinels, etc.
d. *Self-awareness and the recognition of other selves,* evidenced by the social and familial behavior of animals, by their organized play and elaborately imitative behavior, and by their ability to copy novel acts or utterances. This involves also the ability of some animals to profit by their fellows' experience.
e. *Esthetic and ethical values,* although rarely observed, certainly exist in animals. Thus the rescue operations of dolphins to aid their comrades in respiratory distress indicate concern for the welfare of others.

3. *Consciousness is a variable quantity in a*

given individual; human beings are from time to time more or less conscious. The state of introspective clarity and awareness of the external world varies in the same person over a range from excitement through normal alertness, to drowsiness, to light sleep, and then to deep sleep. Consciousness has the property of lability, and movement along this scale is relatively free: the cyclic variation from sleep to wakefulness is common experience. Perhaps the temporarily or permanently irreversible loss of consciousness called coma should be set apart from this continuum, as should the anesthetic state, the loss of consciousness associated with some epileptic attacks, and the alteration of consciousness produced by certain drugs. These all give further evidence that the level of consciousness depends on the state of brain activity.

4. *Consciousness presupposes perception,* a continual updating of the central neural reflection of the state of events in the external world. Leaving aside such special states as hypnosis, the available evidence indicates that a reduction or a lack of variety in afferent input leads to a reduction in the level of consciousness. Whether total reduction of input leads to total loss of consciousness is unsettled, perhaps because complex nervous systems may replicate afferent input by imagery, thus maintaining the conscious state.

5. *Consciousness presupposes memory,* a continuous storage of information about internal and external events, a record against which anticipatory plans for action are laid.

The neural mechanisms involved in these actions that give evidence for consciousness are poorly understood. Considerable information is available concerning systems of the brain that control and maintain levels of excitability, however, and some reason attaches to the proposition that the *overall level of consciousness depends on the level of excitability of the brain.* For this reason, attention will now be given to those intrinsic systems that regulate the excitability—the readiness for action—of central neuronal populations, particularly those of the cerebral cortex.

INTRINSIC REGULATORY MECHANISMS OF THE FOREBRAIN: THE GENERALIZED THALAMOCORTICAL SYSTEM

Fundamental to orderly function of the brain is the presence within it of systems of neurons that, by virtue of their wide distribution and terminal synaptic organization, serve to regulate neuronal excitability, exerting this influence from the level of the cerebral cortex to that of spinal motoneurons. This system occupies part of the tegmentum of the medulla, pons, and mesencephalon, projecting upward on telencephalic distributors, one of which is the generalized thalamocortical system. A descending projection, the reticulospinal system, serves to funnel excitatory and inhibitory influences from the forebrain on the spinal segmental apparatus. The *ascending reticular activating system* lies parallel with and receives collateral input from the great afferent systems and, driven thus indiscriminately, tunes appropriately the ongoing levels of excitability of cells of the cerebral cortex, the basal ganglia, and other large gray structures of the forebrain: it sets the stage for action. It is their indiscriminate activation by a variety of afferents, a property of their cells both individually and collectively considered, and the wide distribution of their effects that are indicated by the term "generalized." In contrast, the so-called *specific* afferent systems that they parallel are precisely organized for the transfer of sensory information, whereas the ascending reticular systems are engaged more generally in arousal and alerting functions. Both are essential for normal function; neither may be categorized as higher or lower than the other, for "without the former the animal is blind, deaf, or anesthetic; without the latter it cannot be aroused from a sleep-like state."[33] Almost all areas of the cortex receive projection fibers of the two types. The term "specific," in this context, indicates not only the thalamocortical stages of the great afferent systems but all others that project on the cortex in a precise manner (e.g., the projection of the ventralis lateralis on the motor cortex or of the medialis dorsalis on the frontal granular cortex). Although any given cortical area, with rare exception, receives overlapping thalamic projections from a specific and generalized system, the two engage intracortical synaptic organizations in different ways and thus exert different effects on cortical neurons.

Generalized thalamocortical system and the recruiting phenomenon[105]

Morison and Dempsey discovered in 1942 that electrical stimulation within the nuclei of the generalized thalamocortical system evokes changes in the electrical activity of large areas of the cerebral hemispheres. Further analysis of this phenomenon, particularly by Jasper, suggests that it receives activating influences from the brain stem reticular formation, in turn exerting control over neuronal excitability and thus the ongoing electrical activity of the cortex.

The electrical phenomena are illustrated by the records in Fig. 10-1. In an anesthetized animal, electrical stimulation of nuclei of the generalized thalamocortical system evokes slow waves over a large part of the cerebral cortex, of greater amplitude in frontal areas than elsewhere. The slow waves are usually and predominantly surface negative in sign. They show several remarkable characteristics. First, the amplitude of the evoked response grows from the first response, which may be very small or absent, to the second, to the third, etc. This property is called recruiting. Second, the amplitude of the response tends to cycle through recurrently waxing and waning phases. Third, recruiting responses are of long latency, frequently greater than 25 msec, as compared with the latency of 1 to 2 msec for cortical responses to stimulation of specific thalamic nuclei. Fourth, the recruiting waves can be evoked only by stimulation within a frequency range of 6 to 12/sec; indeed, they are best seen when stimuli are delivered at frequencies close to or slightly higher than the dominant frequency of the normal alpha rhythm. In waking animals with implanted recording and stimulating electrodes, as in anesthetized animals, recruiting responses are evoked by stimulation of the nuclei of the central commissural system. They are most readily produced in quiet, relaxed animals and may then, with continued stimulation, lead to the onset of sleep. Recruiting waves are reduced in

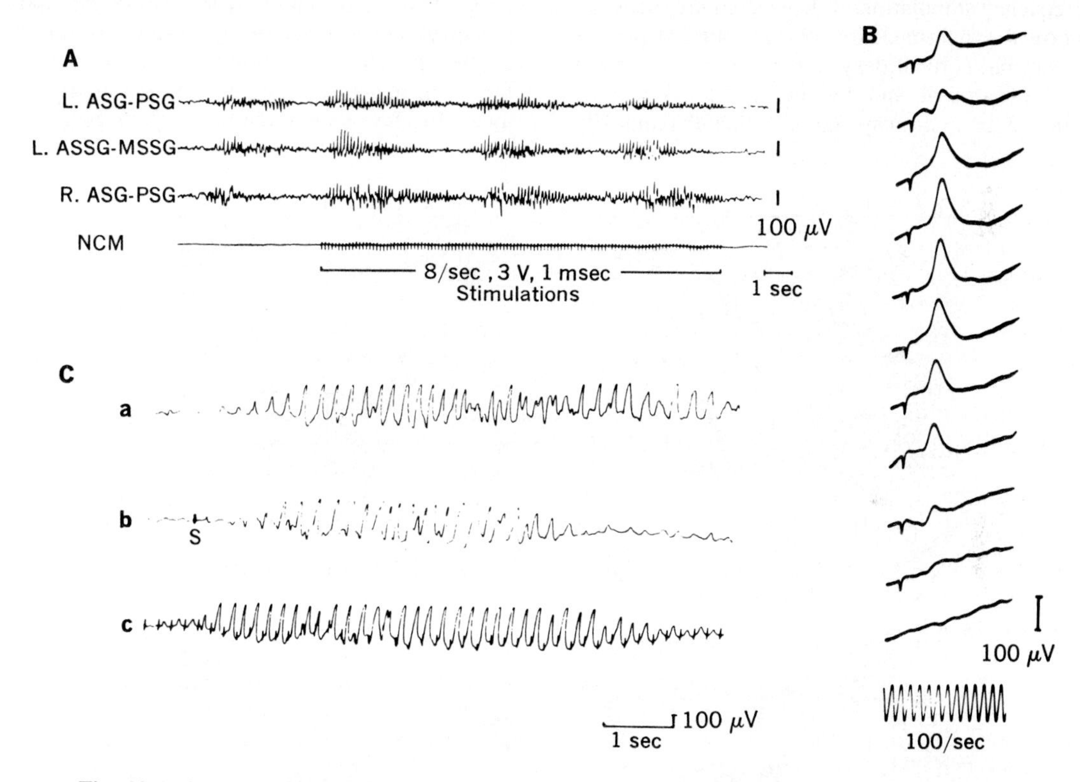

Fig. 10-1. A, Recruiting responses recorded from cerebral cortex of cat anesthetized with sodium pentobarbital, evoked by stimulation of n. centralis medialis at rate of 8/sec. Stimulation of central commissural system evokes responses bilaterally when stimulated locus is near midline of thalamus, as here. Note waxing and waning of recruiting responses during train of stimuli. Spontaneous spindle burst (''barbiturate'' spindle) preceded period of stimulation. Bipolar recordings from *L. ASG-PSG,* left anterior sigmoid gyrus vs left posterior sigmoid gyrus; *L. ASSG-MSSG,* left anterior suprasylvian gyrus vs middle suprasylvian gyrus; and *R. ASG-PSG,* right anterior sigmoid gyrus vs right posterior sigmoid gyrus. *NCM,* trace indicating stimuli. **B,** Recruiting responses recorded via implanted recording electrodes from anterior sigmoid gyrus of waking cat in response to 8/sec stimulation of n. centralis lateralis via implanted stimulating electrodes. Monopolar tracings, negativity upward. Records read from below upward, one trace for each stimulus. Note waxing and waning of response similar to recruiting responses obtained in anesthetized animals. **C,** Records obtained from cortex of cat under pentobarbital anesthesia. *a,* Spontaneous or ''barbiturate'' spindle; *b,* similar spindle tripped by single stimulus, *S,* delivered to n. centralis medialis; *c,* typical recruiting sequences evoked at some locus by stimulation at 5/sec. (**A** and **B** from Yamaguchi et al.[113]; **C** from Jasper.[57])

amplitude in alert animals with desynchronized EEGs, for example, in the arousal produced by sensory or brain stem reticular stimulation.[109,110,113] In animals lightly asleep or under barbiturate narcosis, the slow waves of the EEG are frequently interrupted by spontaneous runs of waves at or slightly above the alpha frequency ("sleep spindles"; Fig. 10-5), and in this state a single stimulus delivered to a nucleus of the generalized thalamocortical system may trip such a spindle (Fig. 10-1). These facts, taken in concert, first suggested the role of this system in regulating (i.e., in pacesetting) the rhythmic electrical activity of the cortex, particularly the alpha rhythm. It is important to note that high-frequency stimulation of the system produces not cortical synchronization or recruiting waves or sleep, but cortical desynchronization resembling that of arousal and the alert state. When so elicited in a drowsy animal, this thalamically induced arousal differs from that evoked by natural sensory or electrical reticular stimulation in that it may not outlast the period of stimulation and, depending on the thalamic nucleus stimulated, it may be restricted to one hemisphere.

The areas of the cat thalamus from which cortical recruiting responses can be evoked are shown in Fig. 10-2, and the system has a similar distribution in the primate thalamus. Stimulation at one point in an intralaminar nucleus produces recruiting responses that spread throughout the system, perhaps over multisynaptic chains of neurons collaterally activated, forward through its major outflow funnel, the n. ventralis anterior and its adjacent cap of n. reticularis. This is a region from which the most powerful cortical recruiting responses can be elicited by electrical stimulation at shortest latency. In general, strong stimulation at any one locus in the system tends to elicit recruiting

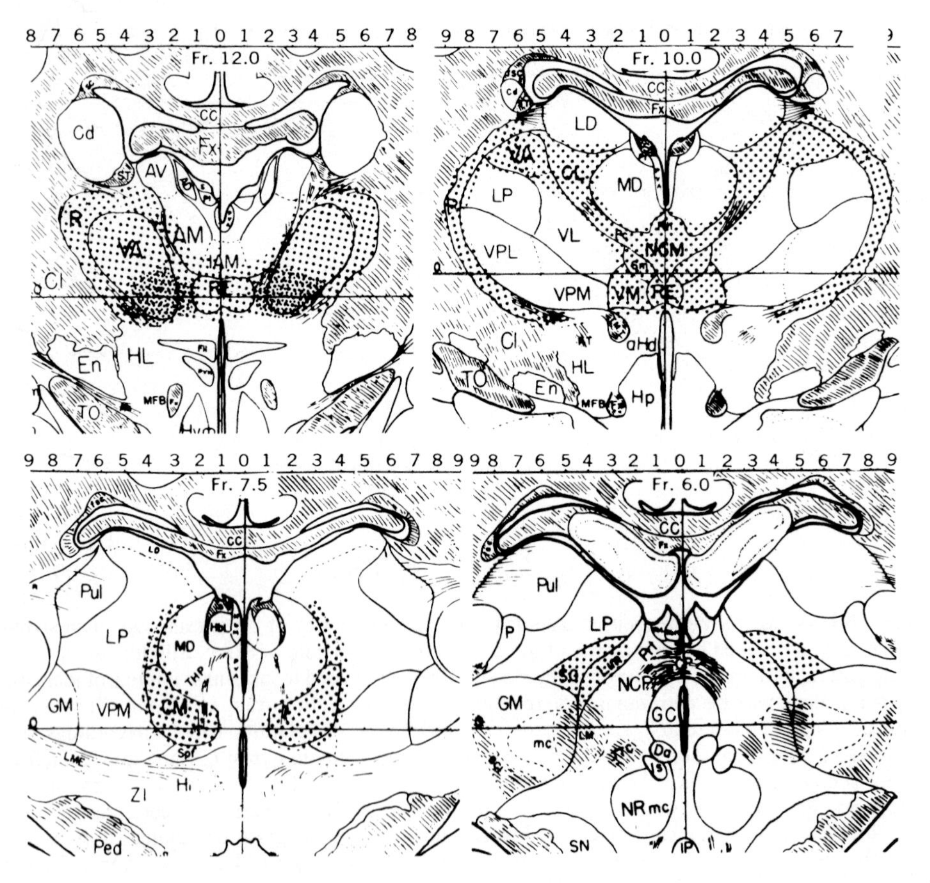

Fig. 10-2. Drawings of cross sections of cat thalamus; four frontal planes from anterior (Fr. 12.0) to posterior (Fr. 6.0). Nuclei of origin of generalized thalamocortical system are indicated by stippling. Areas identified are those within which electrical stimuli at 6 to 12/sec will evoke recruiting responses in cerebral cortex. System has similar distribution in primate brain but is of relatively smaller size there in comparison to greatly enlarged specific relay and associational nuclei. For nuclear designations, see p. 274. (From Jasper.[57])

activity widely distributed in both thalamus and cortex, and stimulation of midline nuclei does so bilaterally. Nevertheless, there is a pattern of topographic organization within it, for threshold stimulation of the dorsolateral portions evokes responses preferentially in parieto-occipital regions of the cerebral cortex, whereas medioventral portions activate primarily the cortex of the frontal pole and the medial wall of the hemisphere. There is both anatomic and physiologic evidence that nuclei of the generalized system project intrathalamically,[46] perhaps via axonal collaterals, on nuclei that are related to "association" cortex as well as on cells of the specific sensory nuclei. These latter are not, however, essential links in the thalamocortical pathways of the recruiting system, for intralaminar stimulation continues to evoke widespread cortical recruiting responses after their destruction, at which time the responses are especially prominent in the primary receiving areas of the cortex.[53]

The nuclei of the central commissural system project on the neocortex in a loose anteroposterior topographic arrangement. The fact that local, slowly repeated stimulation at any one point in the system produces recruiting responses that spread throughout the system and produces powerfully synchronizing effects on other thalamic nuclei as well can be explained by the hypothesis that the system is connected by short intrathalamic axons or by axon collaterals of thalamocortical axons with nearly all thalamic nuclei. In addition, or alternatively, later evidence suggests that the n. reticularis may serve as this distributor of reentrant activity. Its cells receive collaterals of the thalamocortical axons of both specific and generalized nuclei (and of corticothalamic axons as well) and distribute their axons very widely back on the thalamus itself.[101]

The generalized thalamocortical system receives, in addition to input from the ascending brain stem reticular system, the paleospinothalamic tract. Thus cells of this system receive afferent input both from the cerebral cortex and from all major afferent systems, either directly via the paleospinothalamic system or indirectly via the brain stem reticular system. This global convergence affords a suitable mechanism for titration of input and regulation of the level of excitability of cortical neurons. Such a convergence is achieved with the loss of specific information about driving stimuli and thus should not be regarded as a higher order integrative action, a conclusion that fits the ancient phylogenetic history of the system.

Augmenting phenomenon. Dempsey and Morison[45] first observed that the cortical response evoked by electrical stimulation of the thalamic relay nucleus for a sensory system differs from that evoked by stimulation of the appropriate receptors. When such a thalamic stimulus is repeated at rates of 6 to 12/sec, the cortical response, particularly its second (the negative) component, grows rapidly in amplitude over the first few responses of the train (Fig. 10-3). The aug-

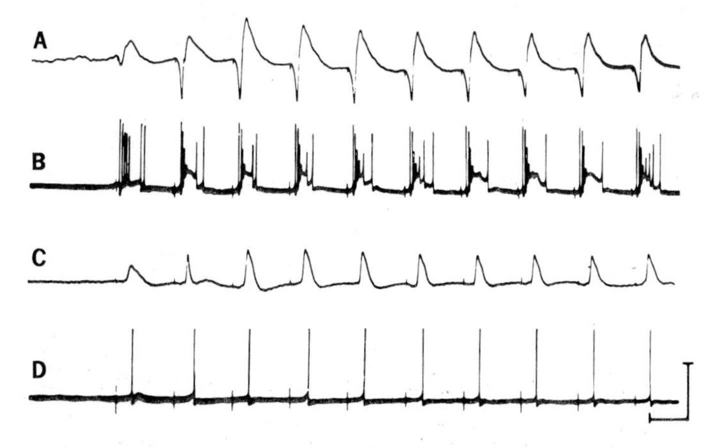

Fig. 10-3. Responses evoked by low-frequency stimulation, 7/sec, of specific relay nucleus of thalamus (VL) and recorded, **A,** from surface of motor cortex and, **B,** via intracellular electrode placed in cortical cell of origin of pyramidal tract axon, a PT cell. **A** shows typical augmenting increment of response. **B** shows intense synaptic depolarization of cell, with "inactivation by excessive depolarization," for there is failure of impulse discharge at peak of depolarization. Lower two records were obtained directly thereafter, but this time during 7/sec stimulation within central commissural system. **C** shows typical waxing and waning of amplitude of almost purely negative wave, recruiting phenomenon recorded at surface of amplitude of almost purely negative wave, recruiting phenomenon recorded at surface of cortex. **D** shows that transsynaptic relay on cortical cell is much less powerful than for input from specific relay nucleus, for it evokes only slow and weak depolarization that leads to single impulse per response. All records negative up. Calibration: horizontal bar 0.1 sec for all records; vertical bar 50 mV for records **B** and **D.** Augmenting and recruiting responses in **A** and **C** are 0.5 to 1.5 mV in amplitude. (From Purpura et al.[93])

menting phenomenon is not evoked by repetitive stimulation of a sensory system at any point peripheral to the thalamus. The augmented negative component is distributed over a somewhat wider area than the limited cortical focus in which the primary evoked potential appears. The suggestion is that the electrical stimulation of the thalamic nucleus engages a second set of neural elements that impinges on cell bodies and dendrites of the more superficial layers of the cortex. Since the generalized thalamic nuclei apparently project collaterals of their axons on the cells of the specific relay nuclei, the phenomenon of augmentation is probably due to the combined excitation of specific and generalized thalamocortical systems.

Synaptic engagement of cortical neurons by thalamocortical systems

Specific and generalized thalamocortical systems engage the neuronal chains of the cortex in different ways. Volleys in generalized afferents produce an intense depolarization of the dendrites and cell bodies in the outer two or three cortical layers, which accounts for the fact that recruiting responses are negative in sign when recorded by a surface electrode and positive when recorded by a micorelectrode deep in the cortex. Specific volleys set up by sensory stimuli or by electrical stimulation of a thalamic relay nucleus, on the other hand, evoke an initially positive potential change when recorded at the cortical surface (p. 288). This fits with the notion that these afferents initially produce an intense depolarization of neurons of the middle layers of the cortex, although rapid and powerful synaptic relays lead to EPSPs in cells of all layers. These differences are further revealed by studying the afferent discharge from the cortex produced by volleys of the two types, discharges observed by recording either from axons of the pyramidal tracts[93] or intracellularly from the neurons from which they arise (called PT cells).[93] Synchronous volleys in specific afferents lead to a depolarization of PT cells so intense as to inactivate spike generation in the midst of the high-frequency burst of impulses (Fig. 10-3). Volleys in generalized afferents, on the other hand, produced much weaker EPSPs in PT cells and elicited but a single impulse, as shown in Fig. 10-3. The two systems converge on the final common path of the pyramidal tract neurons over interneuronal chains that differ, a fact predictable from their different intracortical terminal synaptic distribution.

The sequence of local PSPs of thalamic and cortical neurons evoked by stimulation of a generalized thalamic nucleus gives some indication of how this system may phase cortical activity and the waves of the EEG.[89,90,92] Such stimuli evoke short EPSPs followed by large and prolonged IPSPs in cells of other generalized as well as specific thalamic nuclei. For the latter this phasic inhibition is so powerful as to block completely afferent thalamocortical transmission. Thus slow rhythmic discharge in the central commissural system will bring into synchronous oscillation widely distributed populations of thalamic neurons (and thus of the cortical neurons on which they project) between brief excitation and a powerful and prolonged inhibition. It is this mechanism that is referred to as the "thalamic pacemaker," thought to possess this rhythmic tendency inherently, a tendency displayed especially when the system is free of ascending excitatory drive. This accounts for the spindling cortical waves characteristic of light sleep or barbiturate anesthesia and perhaps for the alpha rhythm of relaxed, quiet wakefulness. The results of studies using intracellular recording now allow a tentative understanding of this shift from synchronization to desynchronization of cortical activity as the generalized system shifts from idling to activation.[91] The prolonged inhibitory PSPs produced by generalized volleys are replaced by sustained depolarization. This is particularly obvious when the system comes under the drive of the ascending reticular activating system, which controls and indeed initiates this change during the shift from sleep to wakefulness. It is thus a major subject for discussion in the section that follows.

SLEEP AND WAKEFULNESS

The nature of sleep[5,17]

The term "sleep" describes two behavioral states that differ from alert wakefulness by a readily reversible loss of reactivity to environmental events, a reversibility that differentiates sleep from other altered states of consciousness (e.g., anesthesia or coma). The two sleep states are characterized by quite different alterations in the functional state of the brain. In the first, called slow-wave sleep (SWS), the brain waves are slow and synchronous as compared to waking; cardiovascular, respiratory, and autonomic system levels are somewhat reduced but steady; and subjects awakened from SWS seldom report dreaming. In the second sleep state, variously called activated, paradoxical, or desynchronized sleep (DS), the brain waves are of low voltage and variable frequency, resembling those in the waking state, but the threshold for arousal is much higher than in SWS; there are marked irregularities in the blood pressure, heart rate, respiration, and autonomic activity; outbursts of rapid eye movements (REMs) occur; and subjects awakened from DS commonly report dreaming.

Sleep in mammals is a special example of a more general phenomenon; in their activity, all plants and animals to some degree show a periodicity, cyclic variations timed by internal clocks of uncertain nature. In simple animals, rudimentary periodicities may survive destruction of the CNS, but *in mammals, sleep is a neural phenomenon.*

Biologic clocks and the 24 hr sleep-wakefulness cycle[1,11]

The rhythmic alterations referred to are ubiquitous both in the kinds of organisms and in the kinds of functions considered. When the period of oscillation approximates the period of the earth's rotation, they are called *circadian rhythms.* The available evidence supports the proposition that the "clocks" governing these rhythms are endogenous and innate and that they are usually self-sustaining and undamped. Whether such oscillations in activity appear at levels of organization simpler than that of single cells is unknown. When free-running, circadian rhythms are remarkably precise, for the observed standard errors may be no more than ± 2 min per nearly 24 hr cycle. Over a very wide range, circadian rhythms are nearly independent of temperature and are remarkably insensitive to chemical perturbations. They are, however, sensitive to the intensity of light, and the nearly 24 hr circadian rhythms of many types may be *entrained* to an exact 24 hr rhythm by 12:12 hr light-dark alternations. A free-running cycle ending before the superimposed dawn or dusk resets by stepwise delay, one ending afterward by stepwise advance. The phase of some free-running or nonentrained circadian rhythms can be shifted by a single brief alteration or stimulus, particularly in the light regimen.

Nocturnal mammals such as the rat show cyclic variations between rest and activity, with a period of 60 to 90 min associated with increased motor activity, feeding, grooming, gastric and intestinal activity, etc. Under a 12:12 hr light-darkness regimen or in the natural habitat these periods of activity begin promptly at dusk and cease at dawn. If, however, the animal is never exposed to light or if the eyes are enucleated or the optic nerves cut, this cyclic activity drifts gradually with reference to the 24 hr day, following a period slightly less or greater than 24 hr. This indicates that under normal circumstances the circadian rhythm is entrained to an exact 24 hr period by a light-darkness cycle of that duration. Once set free by light deprivation, this cycle is astonishingly free of atmospheric or climatic conditions, runs independently in a number of similarly deprived animals housed under identical conditions, and is unaffected by a host of experimental manipulations. Removal of the endocrine glands, including the gonads and pituitary, pineal, thyroid, and adrenal glands in different animals, produces no change. The cycle survives periods of anoxia or of epileptic convulsions and is impervious to periods of deep anesthesia, taking up at the appropriate point in the cycle once consciousness is regained. Alcohol intoxication, the administration of tranquilizing drugs, and decreases or increases in activity of the autonomic nervous system (produced by drugs) are all equally without influence. Of lesions of the forebrain, only those that injure the hypothalamus produce an effect, but this effect is profound: the precisely timed cycle dissolves into a random time series of brief periods of rest and activity. Although it is clear that this internal clock depends on the hypothalamus for its existence, the nature of the clock and the underlying mechanisms of timing are unknown. Examples of such cyclic changes occur also in the severity of some diseases and in the presence or absence of symptoms, and these often show periods greater than 1 day. The role of these periodicities in the pathogenesis of disease is still uncertain.[94]

Like the rat, the newborn infant shows rest-activity cycles of about a 90 min duration (Fig. 10-4). When sleeping, the troughs of these cycles are occupied by SWS and the peaks by DS. During the first year of life these cycles become entrained to the ordinary daily rhythm of human existence, although the 90 min undulations persist, associated during sleep with the periodic appearance of episodes of DS (Fig. 10-12). This entrainment to a 24 hr cycle may be due in part to familial and social pressures; it is associated with a gradual decrease in the need to sleep and increased activity during periods of wakefulness. However, it is of interest to note that a similar 24 hr cycle is inherent and innate in man and other mammals.[95] Efforts to entrain human activities to periods of considerably shorter or longer lengths of time have so far been relatively unsuccessful, although more extensive study is required before the matter can be regarded as settled.

Physiologic variables and the 24 hr sleep-wakefulness cycle[6]

Together with the gradual entrainment of the sleep-wakefulness cycle in the human infant, usually established by the end of the second year of life, cyclic variations appear in other func-

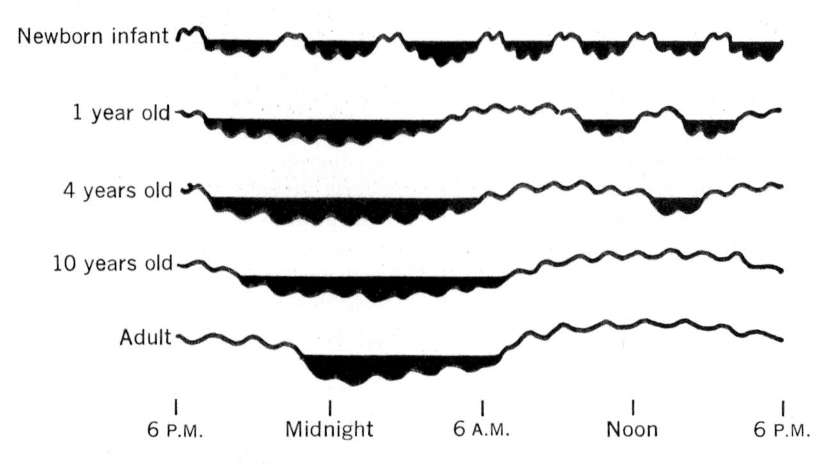

Fig. 10-4. Schematic representations of gradual transition from polycyclic alternations between sleep and wakefulness of newborn infant to monocyclic pattern in adult. Black areas indicate sleep. Secondary undulations indicate rest-activity periodicity, which gradually lengthens from 50 to 60 min in infants to 80 to 90 min in adults. (From Kleitman.[7])

Table 10-1*

	Sleep	Fatigued wakefulness	Normal rested
Number of cases	6	6	11
Mean arterial pressure (mm Hg)	89.70	94.20	86.50
Cerebral blood flow (ml/100 gm/min)	65.00	59.20	54.80
Cerebral O_2 consumption (ml/100 gm/min)	3.42	3.52	3.34
Hemoglobin concentration (gm/100 ml)	14.36	14.30	14.57
Arterial O_2 content (vol%)	19.43	19.42	19.44
Arterial CO_2 tension (mm Hg)	46.30	46.00	41.30

*Data from Mangold et al.[69]

tions. During SWS there is a mild reduction in alveolar ventilation accompanied by a slight rise in mixed venous and alveolar CO_2. The heart slows, and there is a moderate drop in blood pressure. The minimal changes in the concentrations of many blood constituents are accounted for by the equally slight dilution that follows the movement of extravascular water into the blood on assumption of the recumbent position. A diurnal cyclic variation in hypothalamic-pituitary function is evidenced by the activity of one of its target organs, the adrenal cortex: there is a decreased blood level and urinary excretion of ketosteroids and an increase in the eosinophil count in the blood. The morning eosinopenia is less marked in blind than in normal individuals. There is a cyclic variation in the rate of release of growth hormone, that rate being greatest during SWS.[83,107] Water excretion by the kidney is lowered during the night. Perhaps the most constant cyclic variation is in the body temperature, which drops by as much as 2° F during the night. It has been surmised at one time or another that each of these changes stood in a causal relation to sleep, in particular that the slowing of the heart and mild drop in blood pressure would result in decreased blood flow through the brain, that is, that sleep is produced by a cerebral anemia. That this is not the case has been shown by direct measurement (Table 10-1).[69]

Changes in central nervous activity during sleep. The remarkable changes in CNS activity during sleep are not limited to those regulating the level of consciousness. With sleep, there is an elevation of thresholds for many reflexes, and in very deep sleep (SWS), certain pathologic reflexes may appear, even in healthy individuals. There is simultaneously a reduction in muscle tone, and the limbs may be completely flaccid. Movements occur during any lighter sleep, and the degree of motility varies cyclically with a period of 60 to 90 min, although motility patterns differ from one individual to another. Striking changes occur in the musculature of the eye: the pupil is narrowly constricted by an increase in frequency of discharge of neurons of the Edinger-Westphal nucleus. Although the pupils are narrow, the light reflex is present. The eyes

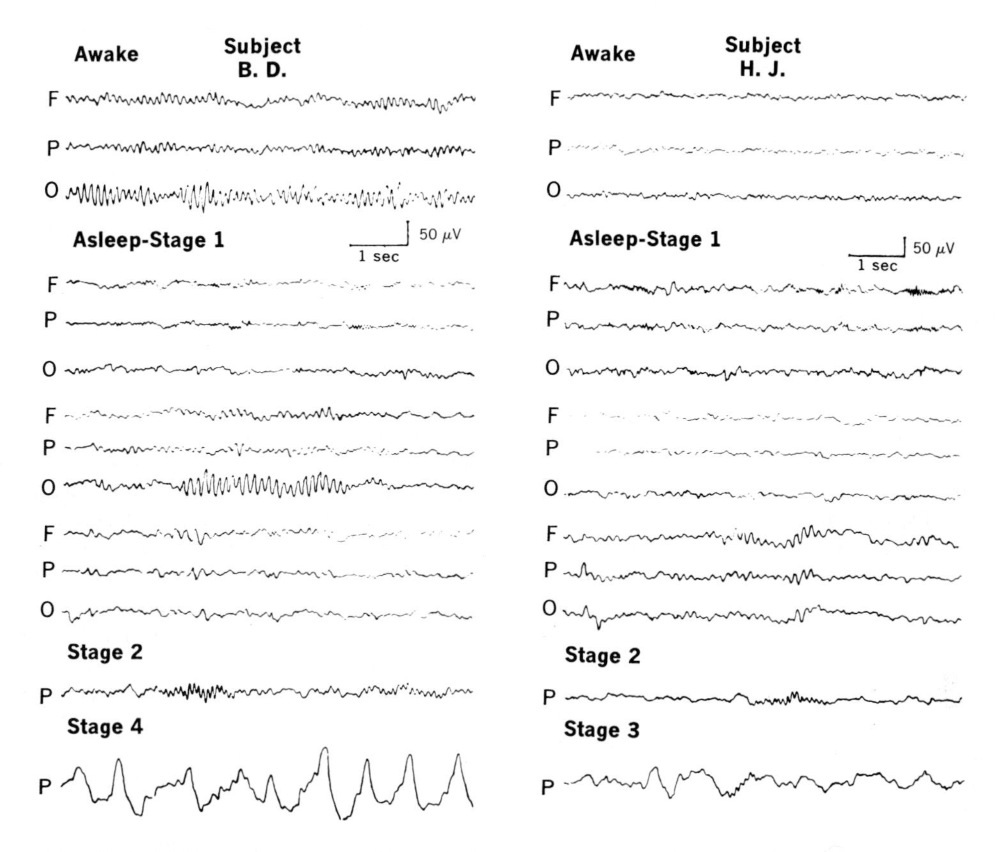

Fig. 10-5. EEG records of two subjects during various stages of sleep to illustrate classification given below. *F*, Frontal, *P*, parietal, and *O*, occipital locations of leads. (From Dement and Kleitman.[44])

most frequently assume a position of upward divergence. A decrease in peripheral sympathetic activity is evidenced by increased electrical resistance of the palmar skin; there is, however, a tonic contraction of the vesical and rectal sphincters.

Changes in the EEG characteristic of different stages of sleep have proved to be a useful tool for experimental studies of sleep. Various schemes of classifying EEGs in man and other primates[7,29,48,61] and correlating them with sleep have been proposed, among which is the commonly used one of Dement and Kleitman,[44] given in the following outline and Fig. 10-5. Automatic methods are now commonly used for analyzing and classifying EEG records obtained during sleep.[63]

Stage 1: Record in which EEG alpha waves are slowed very slightly; seen in quiet relaxed wakefulness or as drowsiness supervenes.

Stage 2: Record with spontaneous "sleep spindles," runs of a few seconds' duration consisting of regular 14 to 15/sec waves superimposed on a low-voltage background with admixture of 3 to 6/sec waves; a state from which the sleeper is readily aroused.

Stage 3: Record with some sleep spindles but now on a background of slower 1 to 2/sec delta waves; associated with sleep of intermediate depth.

Stage 4: Record consisting entirely of high-voltage slow delta waves without spindling, associated with a phase of sleep in which the threshold for awakening is greatly elevated.

Cyclic variations in the EEG pattern occur during normal sleep, with a period of 80 to 120 min, the pattern changing on the peak of each cycle to a modified or "emergent" stage 1. This phase is associated with the presence of REMs, an acceleration of the heart and respiratory rates, and occasionally by gross body movements occurring against a background of decreased muscle tone (Fig. 10-12). This emergent stage 1 differs from the descent through stage 1 on going to sleep in that the threshold for arousal by an auditory stimulus is elevated 5 to 10 times. By awakening sleepers during various phases of the cycle, Dement and Kleitman[44] discovered that the *emer-*

gent stage 1 with REMs is commonly associated with dreaming. This finding has permitted study of dream content, of the ontogenetic development of dreaming, of the time scale in dreaming, of the effect of presleep events and conditions on dream content, etc.[2] This stage of sleep in man is probably not completely identical to the special state of deep sleep with EEG desynchronization in animals, which will be described in a later section.

Activation or arousal of the EEG in human beings is said to occur when the high-voltage slow waves of sleep are replaced by the low-voltage fast activity usually seen in wakefulness. This transition is commonly abrupt and direct, but in more gradual awakenings the transitional intervening stages of sleep spindles, modified alpha rhythm, etc., may be seen in reverse sequence from that of the transition from waking to sleep. In experimental work in animals (most commonly cat or monkey), three states are usually specified, although intervening states do occur: (1) low-voltage fast activity with widened pupils of wakefulness, (2) high-voltage slow activity with pupillary constriction of moderately deep sleep, and (3) the special state of deep sleep with EEG desynchronization referred to previously as paradoxical or desynchronized sleep (DS). Activation of the EEG is frequently equated with the transition from sleep to wakefulness when the behavioral correlates cannot be evaluated, although the two are occasionally dissociated in otherwise normal individuals, may be dissociated by the action of certain drugs such as atropine, and are characteristically dissociated in DS in animals.

CENTRAL NEURAL MECHANISMS REGULATING SLEEP AND WAKEFULNESS

Two important discoveries generated modern concepts concerning the neural mechanisms involved in sleep and wakefulness. The first of these was made by Hess,[18,54] who discovered that slow, rhythmic stimulation (8/sec), via implanted electrodes, of a diencephalic region encompassing the thalamic nuclei of the generalized thalamocortical system in waking cats produced, after several minutes of stimulation, all the behavioral signs of normal sleep, that is, the sequence of postures, etc. that are normally seen in a cat going to sleep and in actual sleep itself. Hess also discovered that stimulation of a more ventral and posterior region, including the posterior hypothalamus and the gray matter at the mesodiencephalic junction, produced all the behavioral signs of awakening. These observations led to the concept of sleep as an active process; in other words, activation by some means of what is now termed the generalized thalamocortical system (the "recruiting" system) influences the excitability and the rhythmic patterns of activity of cerebral cells in such a way that sleep ensues. This hypothesis has been modified in the light of subsequent findings, however, but as will be apparent in the following discussion, the synchronizing mechanisms do play an important part in a unified concept of the cerebral mechanisms in sleep.

In a novel and important experiment, Bremer discovered in 1935 that when the neuraxis of a cat is transected at the level of the first cervical segment, with artificial respiration and precautions for maintenance of blood pressure (Bremer's *encéphale isolé* preparation) the animal shows the EEG and pupillary signs of normal sleep-wakefulness cycles. In contrast, when the transection is made at the mesencephalic level, just caudal to the motor nuclei of the third cranial nerve (Bremer's *cerveau isolé*), there ensued a permanent condition resembling sleep.[31]

Bremer's discovery led to the concept of sleep as a passive process, as a deactivation phenomenon, that is, wakefulness is an active state maintained by afferent input to the brain, and sleep ensues when that input is removed, as in the cerveau isolé cat, or falls below a certain critical level, as in normal sleeping. In the cerveau isolé preparation, olfactory input to the brain remains, but strong olfactory stimuli produce only a transient activation that does not outlast the stimulus. Visual pathways from retina to cortex are also intact, but visual stimuli do not evoke widespread activation of the EEG in the cerveau isolé animal with high mesencephalic transection, as they do in the intact animal. Although Bremer tentatively concluded that deafferentation per se is sufficient to induce sleep, this last observation concerning visual stimuli indicates that some neural mechanism in addition to the direct sensory pathways is required for the maintenance of wakefulness. Indeed, all the investigations of the last decades have given results that support the point of view that this additional mechanism involves collateral activation of neural populations occupying the reticular formation of the brain stem, an important concept that derives largely from the work of Moruzzi[78] and Moruzzi and Magoun.[79] This neural system, by virtue of its upward projection on the cerebrum, is thought to control the level of excitability of the forebrain and thus the state of consciousness.

A number of clinical observations support the idea that brain stem mechanisms play an essential role in regulating the state of consciousness. Large areas can be removed from any part of one cerebral hemisphere, symmetric parts can be removed from both, or indeed one entire hemisphere can be removed without loss of consciousness or disruption of normal sleep-wakefulness patterns. Lesions of the brain stem, however, frequently produce alterations in the state of consciousness, as do certain forms of encephalitis that cause the death of neurons in the periaqueductal gray matter of the mesencephalon.[111] Patients with this latter type of lesion frequently exhibit altered sleep patterns, and some pass into prolonged somnolence as the disease progresses. In early stages of the disease these same patients may display a persistent insomnia, thought to be due to the irritative effects of the acute disease process. These clinical observations, together with the discoveries of Bremer and Moruzzi and Magoun, direct attention to the detailed anatomy and physiology of the reticular formation of the brain stem.

Brain stem reticular formation: anatomic considerations

The term "brain stem" designates the medulla, pons, and mesencephalon, a region extending from the level of the obex to the mesodiencephalic junction. The term "reticular formation (RF) of the brain stem" identifies those areas of the brain stem that are characterized by aggregations of cells of different types and sizes, interspersed with a wealth of nerve fibers traveling in many directions, and excludes more definitely circumscribed groups of cells (e.g., the sensory and motor nuclei of the cranial nerves, the red nucleus, the substantia nigra, and the inferior olive). The RF is frequently referred to as being diffuse in its organization. This generalization is not without some ground, although almost every anatomic investigation of recent years has shown that the RF is not only complexly built but that regions of it differ considerably from one another with regard to their cytoarchitecture and their afferent and efferent connections with other regions of the brain.[32,82,97] (For a detailed description, see Berman.[28]) The RF consists of a variety of cellular fields that grade almost imperceptibly from one to another but within which a large number of nuclei are differentiated. In many physiologic investigations, precise anatomic identification of regions stimulated or ablated or from which recordings of electrical activity were made was not attempted. For this reason it will be necessary in the discussion that follows to use such terms as the *mesencephalic* or the *pontile* RF. The large majority of studies have been carried out in the cat. However, it should be emphasized that there are important changes in this region across the evolutionary series from rabbit to man. For example, the giant neurons that are so conspicuous in the RF of rabbit and cat are far less numerous in that of the primate brain stem. The region emits systems of long axons projecting both caudally and rostrally from the brain stem, but a prominent characteristic is the profuse interconnections of its parts, both by the collaterals of projecting axons and by those which ramify wholly within the brain stem. A large part of what is known of the function of this region has resulted from experiments in which electrical stimuli are delivered to the RF and observations made of the bodily changes that result. It is not surprising, then, in view of the profuse anatomic interconnections of regions of the RF, that the simultaneous engagement by such electrical stimuli of cells in the locale and of fibers of passage gave results that led to the idea that the region is only "diffusely" organized, for with strong stimuli it is possible to produce such diverse physiologic phenomena as changes in the EEG, in reflex excitability of the spinal cord, in respiratory rhythm, and in peripheral blood pressure. How to untangle these complex and interconnecting control systems is a problem of major importance for which no ready solution is apparent.

Major efferent systems of the noncerebellar reticular formation

Reticulospinal connections. Cells of origin of the reticulospinal system are concentrated in, although not exclusive to, two areas of the medial or "effector" regions of the RF. The first is the entire n. gigantocellularis of the medulla, together with adjacent portions of n. reticularis ventralis and n. reticularis lateralis. In the pons the reticulospinal neurons lie in the n. reticularis pontis oralis and part of the n.r.p. caudalis. Reticulospinal fibers course caudally in the ventrolateral funiculus of the spinal cord; those from the pons course ipsilaterally; and those from the medulla are bilaterally distributed. Few reticulospinal fibers descend below the thoracic level, and the system is thought to influence more distal spinal segments via the propriospinal system.

Ascending connections.[80] Long ascending connections originate only from the medial two thirds of the RF, from a medullary and a pontile region, which in each case are shifted caudally from areas of origin of the reticulospinal systems (Fig. 10-6). The bulk of these fibers pass through the mesencephalon in or near the central tegmental tract. The system divides at the caudal border of the diencephalon. A major division projects on the generalized thalamocortical system (i.e., on nuclei of the central commissural system and perhaps also on the reticular nucleus of the thalamus). It should be pointed out also that elements of the paleospinothalamic tract pass in the same region and make the same terminations. A second component projects on the subthalamic body of Luys, the zona incerta, and the entopeduncular nucleus and contributes fibers to the fields of Forel. Another major system originates wholly from the mesencephalic RF and projects very widely on the mammillary nuclei, the peri-

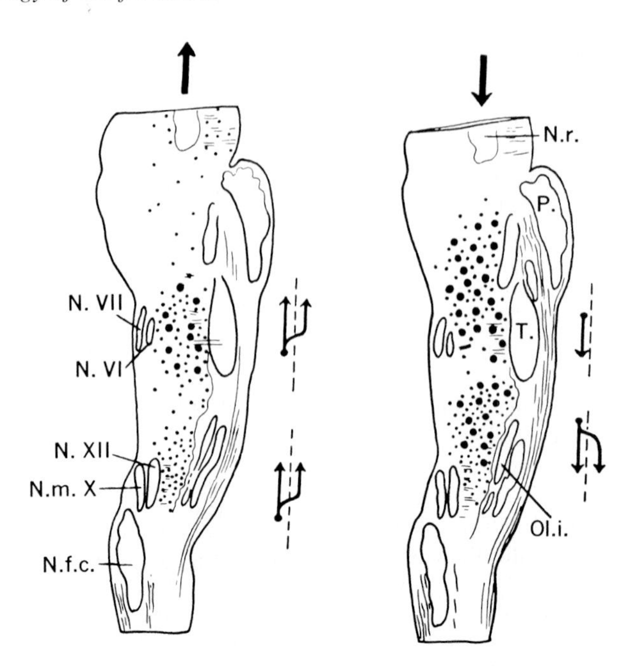

Fig. 10-6. Distribution of cells of RF that send long axons to spinal cord are shown on right, projected on parasagittal section; large dots indicate large cells. Medullary RF projects bilaterally, pontine RF homolaterally. Similarly, on left, distribution of cells having long axons that extend above mesencephalon. Both medullary pontile areas project bilaterally. Distributions for two projections are not identical, and in both medulla and pons heaviest concentrations of reticulospinal neurons lie oral to those for cells with ascending axons. What is not illustrated is that some cells emit axons that divide and send one branch to spinal cord and one orally, above mesencephalon. (From Brodal.[32])

ventricular and lateral hypothalamic areas, the septa nuclei, the preoptic region, and the basal ganglia.

In the regions of origin of the long projecting systems, more than one half of the neurons emit such long axons; indeed, many emit axons that divide into an ascending branch that reaches the cerebrum and a descending branch that reaches the spinal cord, branches that emit numerous collaterals to reticular cells as they pass.[100]

In addition to these efferent systems projecting outside the brain stem, there are profuse projections from reticular regions on the motor and sensory nuclei of the cranial nerves as well as on cells of the superior colliculus.

Major afferent systems to the noncerebellar reticular formation

Fibers of cortical origin course through the corticobulbar system to terminate in the brain stem, some in the pontile and medullary RF, and in particular in the nuclei that emit long reticulospinal fibers. These descending elements originate from the frontal cortex, particularly from the motor cortex and less certainly from the cortex of the parietal and temporal lobes and from the medial wall of the hemisphere. The total number of corticofugal fibers reaching the RF is not large, and fibers from widely separate cortical areas project on the same RF locales. Other fibers of cerebral origin project from the basal ganglia, the subthalamus, and the hypothalamus on the mesencephalic RF.

Fibers of spinal origin reach the reticular formation in several ways. Collaterals of ascending fibers of the anterolateral system project on the small-celled, more lateral component of the medial two thirds of the RF, which also receives input of vestibular and auditory origin and projects on the more medial effector region of the RF. However, large numbers of direct spinoreticular fibers reach this same effector region directly, and their terminations are concentrated in those regions that emit long ascending reticulocerebral fibers.

• • •

In summary, the RF can be regarded as encompassing several subsystems. Some portions relay input of spinal and cerebral cortical origin to the cerebellum. The more medial or effector portions of the RF subserve both a cerebroreticulospinal and a spinoreticulocerebral system, and in both pons and medulla the relay zones for the two are not coincident, those of the former being orally displaced from those for the latter (Fig. 10-6). The mesencephalic RF is reciprocally connected with the hypothalamus, septum, basal ganglia, and preoptic region.

It is therefore obvious that the RF is not organized

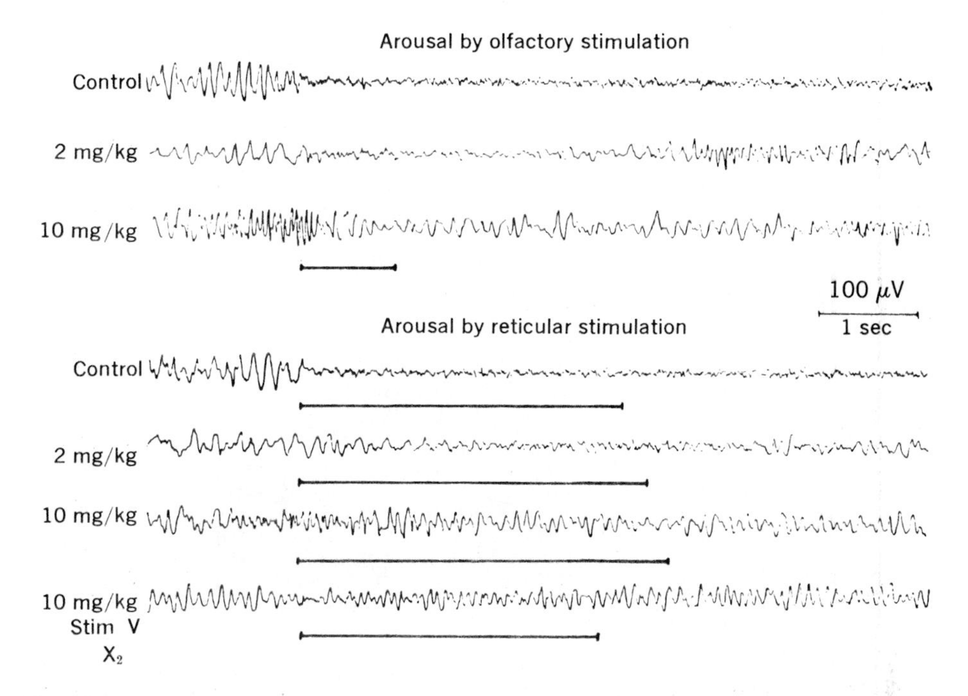

Fig. 10-7. EEG of rabbit showing, for two traces labeled "control," arousal or activation by sensory stimulus, above, and by high-frequency electrical stimulation of reticular activating system, below. In each case, effect is abolished with increasing doses of anesthetic (sodium pentobarbital). (From Arduini and Arduini.[20])

in a wholly "diffuse" fashion. It is just as important to emphasize, however, the extensive overlap between the areas subserving these systems and the profuse synaptic interconnections between them. Thus whether the RF is to be considered as "diffusely" or "specifically" organized is a redundant question: it is both.

Sleep as a passive process: reticular deactivation[77,97]

In 1949 Moruzzi and Magoun[79] discovered that rapid stimulation (50 to 200/sec) of the brain stem produced activation of the EEG, an effect evoked by stimulation of the central core of the brain stem in a region extending upward from the bulbar RF to the mesodiencephalic junction, the dorsal hypothalamus, and the ventral thalamus, thus including the dynamogenic area of Hess.[54] This phenomenon is illustrated by the records in Fig. 10-7, and the relevant anatomic information is summarized in Fig. 10-8. In many features the activation produced by RF stimulation resembles the arousal produced by natural stimulation. The discovery led to a host of succeeding investigations, the results of which support the idea that *withdrawal of the energizing action of the reticular activating system plays an important role in the genesis of sleep.*

When the RF is stimulated via implanted electrodes in sleeping animals, behavioral awakening and EEG desynchronization result. This is also true in animals after section of the long ascending sensory systems in the mesencephalon (i.e., the lemnisci) but does not occur after lesions of the mesencephalic RF. Indeed, after extensive lesions of the mesencephalic RF, animals may be comatose for many days and unresponsive to any stimuli.[49] If they survive, they may show good recovery of sensory and motor functions but display various and sometimes prolonged periods of somnolence, with marked refractoriness for arousal, which, when evokable, may not outlast the arousing stimuli. In contrast, animals surviving transection of the long ascending and descending tracts of the midbrain, but with no RF lesion, show no alterations of the sleep-wakefulness cycle, are readily aroused, and then show activated EEGs, although they are profoundly deficient in the sensory-motor spheres.

These influences of the RF on the activity of the forebrain are mediated via two upwardly projecting pathways. The first pathway projects on the central commissural nuclei, thus influencing excitability of the entire cortex by entraining the elements of the generalized thalamocortical system. That a second pathway exists is suggested by the finding that arousal can still be produced

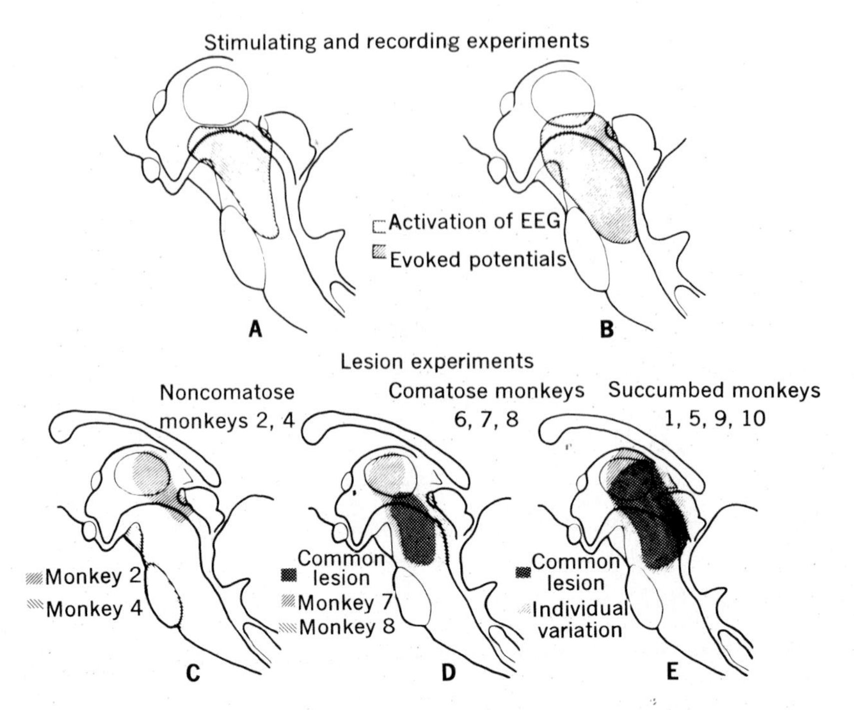

Fig. 10-8. Diagrams summarizing studies of brain stem RF by electrical recording, by electrical stimulation, and by observation of animals after reticular lesions. Reticular regions involved are shown projected on midsagittal drawings of monkey brain stem. **A,** Area within which high-frequency electrical stimulation produced EEG activation and arousal of sleeping animals. **B,** Area of brain stem from which evoked potentials were recorded in response to a variety of peripheral stimuli, with considerable convergence. Areas shown in **A** and **B** are nearly coincident. Contrast of **C** and **D** indicates that, when lesions involve central core of pontile and mesencephalic RF, permanent loss of consciousness may ensue, certainly when lesions are complete, but that more peripheral lesions that may sever long ascending and descending tracts will not affect consciousness or disturb sleep-waking cycles. **D** indicates that many large reticular lesions are incompatible with life. (From French and Magoun.[49])

by RF stimulation after massive lesions of the generalized thalamocortical system that leave the ventral thalamus intact. This second pathway is postulated to pass through the ventral thalamus to the internal capsule.

Electrophysiologic studies have shown that the RF receives input from all afferent systems, including the olfactory and visceral. Studies of single RF cells, however, indicate that under some experimental conditions more than half are insensitive to any sensory stimuli. Of those that are sensitive the convergence is wide but rarely global. Where a wide convergence does exist, the range of response of the reticular neuron provides no neural signal that is unique for each source, nor for the spatial, intensive, or temporal aspects of each sensory input. The point is that, whereas the ascending system is driven by sensory input, it should not be regarded as a sensory system, fitting the usual concept of that term no better than do the spinal motoneurons.[102,103] Although behavioral and EEG arousal and reticular activation can be produced by any sufficiently intense sensory input, the somatic sensory input is the most potent in this regard. Indeed, the periods of activation that are interspersed with long periods of EEG synchronization and ocular signs of sleep in the encéphale isolé cat are largely sustained by the remaining somatic sensory input, for almost continuous sleep follows transection of the trigeminal nerves.

Corticoreticular interactions. There is both anatomic and electrophysiologic evidence that neuronal systems of cortical origin project directly on the brain stem RF, influencing the transmission of ascending reticular activity as well as that of the reticulospinal system. Electrical stimulation of one of many effective cortical areas produces generalized bilateral desynchronization of the EEG in the encéphale isolé preparation, an effect still present after section of the corpus callosum. Behavioral arousal and

EEG desynchronization are produced by stimulation via implanted electrodes in otherwise normal, sleeping animals, but cortical activation by such cortical stimulation is no longer evoked after a high mesencephalic transection. These observations indicate that the corticoreticulocortical systems may function as reentrant circuits (i.e., as a control system), regulating what Bremer has called the *tonus cérébrale*. Such a reentrant controlling system would require negative feedback (inhibition) or perhaps a combination of inhibition and excitation, but the effect of cortical efferents on reticular neurons has been shown by intracellular recording to be predominantly excitatory.[65] This is, however, not incompatible with the idea that cortex and RF compose a homeostatic mechanism for setting and controlling the excitability of the former,[41] for, as will be apparent from the following section (p. 314), ascending reticular influences are not only desynchronizing and arousing but contribute positively to cortical synchronization and sleep as well.

It has been known for a long time that the regions of the brain stem now referred to generally as the RF exert controlling influences over a wide range of bodily functions, particularly in what might be termed the vegetative sphere. Its controlling function in respiration and circulation is described in some detail in Chapters 41 and 62, and it appears likely that this region is also concerned in the control of many autonomic and endocrine functions (e.g., the hypothalamo-pituitary mechanisms). How all these functions are coordinated by this system in a way that leads to constancy and not chaos is an important subject for future study. Some concepts in this regard have been put forward by Dell,[41] particularly from the standpoints of homeostasis and cybernetic theory.

Bremer's observation[31] that EEG activation in the encéphale isolé preparation evoked by sound no longer occurred following bilateral removal of the auditory areas of the cortex raises the possibility that the cortex plays a role via corticoreticular systems in selective or differentiated arousing responses to different stimuli. When, for example, an animal is repeatedly aroused by a sound of one tone and the stimulus is repeated, the behavioral and EEG arousals gradually dwindle until the stimulus is no longer effective, an example of a general phenomenon called habituation. Such an animal is, however, quickly and completely aroused by another tone differing from the first by only 30 to 40 Hz when the base tone is 200/sec, although there is some spread of the habituation to tones closer to the basic habituating one.[19] Behavioral correlates of this phenomenon are obvious: the awakening of a sleeper when his name is softly called, although he is unaffected by other names called loudly; the awakening of a mother to the cry of her child, although much louder noises have no effect, etc. The integrity of auditory cortex is deemed essential for such differentiated arousals.

Activity of cortical neurons during sleep.[55] Studies of single cortical neurons in unanesthetized animals have shown that during the transition from wakefulness to drowsiness to SWS, the large majority of cortical neurons slow their rate of discharge, whereas such large cells as those from which pyramidal tract fibers originate may accelerate. Although a slow fall in frequency does occur in the net activity of the cortical neuronal population, there does not appear to be a universal inhibition of cortical neurons in SWS. In the transition from SWS to DS, almost all cortical neurons increase their discharge rates to levels even higher than those of quiet wakefulness,[70] although still below those of alert activity. What is more striking than rate changes between these various states is the change in the temporal pattern of discharge.[81] During wakefulness, cortical neurons discharge at intervals that compose a time series neither random nor perfectly periodic. As sleep ensues, discharges appear in short, high-frequency trains with long silent periods intervening. During the spindling phase of sleep these clustered discharges are likely to occur in synchrony with the spindle waves, and closely adjacent neurons tend to discharge together. The reverse transition, from sleep to wakefulness, is marked by disappearance of the clusters and differential changes in the discharge frequencies of different neurons but only a moderate increase in overall activity. During brisk arousal and during purposeful activity, that overall rate increases very greatly. Cells of origin of the pyramidal tract are no exception to this pattern.

This change in the temporal patterns is thought to reflect similar changes in that of neurons of the generalized thalamocortical system, the thalamic pacemaker. In wakefulness these cells discharge rapidly and asynchronously, producing a smoothly maintained depolarizing pressure on cortical neurons and thus on their quasi-periodic discharge sequences. Wih decrease in ascending activating influences at the onset of sleep and a simultaneous increase in ascending synchroniz-influences (to be described later) the cells of this system tend to discharge in unison at the rate of 3 to 5/sec, characteristic of cortical slow waves in sleep. Each such thalamocortical volley evokes EPSP-IPSP sequences, which are the basis of the slow waves observed, and entrain cortical neurons to discharge high-frequency bursts in response to each thalamocortical volley.

Sleep as an active process: synchronizing mechanisms of the lower brain stem[8]

Largely due to the work of the Italian school of physiology led by Moruzzi, it is clear that the brain stem activating mechanisms just described are paralleled by others that exert a reciprocal effect, tending to synchronize the EEG and produce behavioral sleep by an *active* process. The initial discovery was made by Batini et al.,[23-25] who found that after complete transection of the brain stem rostral to the medulla—their midpontine, pretrigeminal preparation—animals display a relative insomnia characterized by activated EEGs and the ocular signs of behavioral wakefulness. For example, when observed continuously (up to 9 days), the EEGs of such animals were activated for 78 ± 10% of the time, as compared with 37 ± 12% for normal cats, when both groups were observed under the same conditions and without any intentional stimulation. That a state of true wakefulness existed in these animals was further evidenced by the finding that orienting reflexes were present and true conditional responses could be established.[15,16] The fact that prolonged periods of somnolence appear in the encéphale isolé preparation suggested that the synchronizing structures separated from the forebrain by the midpontine transection must lie in the lower brain stem and that they must be tonically active. That this mechanism is localized to the region of the rostral solitary tract and the adjacent n. reticularis ventralis is shown by the facts that (1) low-frequency (10/sec) stimulation of this region produces EEG synchronization, whereas high frequencies produce strong arousal[64]; (2) single neurons of the region, observed in free-moving, unanesthetized animals, begin to discharge 1 to 2 min before the onset of sleep and are relatively inactive during wakefulness[36]; (3) following small bilateral lesions in this region, animals show increased spontaneous activation and that produced by stimulation of the reticular activating system is prolonged[30]; (4) local cooling of the bulbar RF in sleeping cats produces arousal[27]; and (5) differential anesthetization of bulbar or mesencephalic regions produces differential EEG effects, an experiment illustrated in Fig. 10-9.

The proximity of the bulbar synchronizing region to that known to receive afferents influenced by cardiovascular events raises the question of whether the two are identical. Indeed, it has been known for a long time that high pressure within a vascularly isolated but neurally innervated carotid sinus will produce behavioral sleep. The weight of evidence, however, suggests that the synchronizing reticular neurons and those played on by cardiovascular afferents are not the same and that the soporific effect of the latter depends on inhibition of the reticular

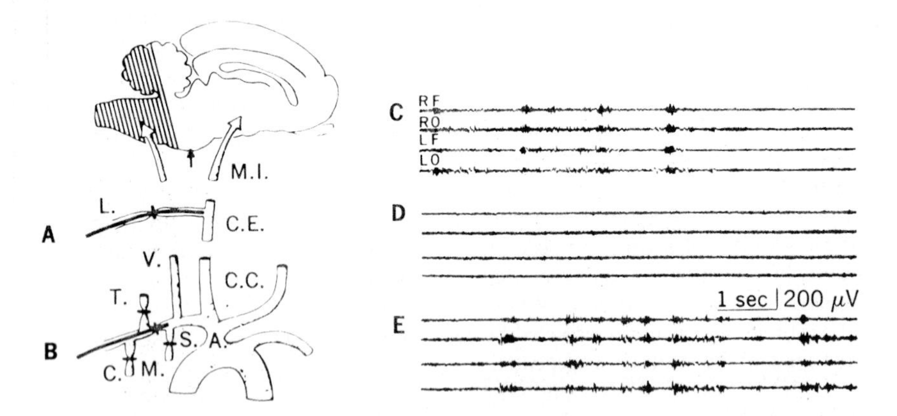

Fig. 10-9. A and **B,** Procedure for separate perfusion of medulla and caudal pons via vertebral circulation, *A,* or all brain rostral to midpontine region via carotid, *B.* Clip placed on basilar artery at arrow. **C** to **E,** EEG records of cat prepared as indicated in diagram but also with transection of spinal cord at C_1, an encéphale isolé preparation. Four traces are shown: *RF,* right frontal; *RO,* right occipital; *LF,* left frontal; *LO,* left occipital. Five-second intervals between **C** and **D** and between **D** and **E.** At beginning, EEG shows typical spindling activity of drowsing or lightly sleeping encéphale isolé preparation. During black signal in **C,** 0.3 μg sodium pentobarbital was injected into vertebral artery. *Anesthetization of medulla produces activation of forebrain,* presumably by anesthetization of medullary synchronizing mechanisms. (From Magni et al.[66])

activating system. How pressoreceptors from the abdominal cavity (a full stomach!) exert their sleep-inducing effect is unknown, but it is probably caused by excitation of the bulbar synchronizing region and/or inhibition of the reticular activating system.

The capacity of somatic sensory afferents to induce sleep when activated by low-frequency stimulation has been clarified by ingenious experiments of Pompeiano and Swett.[85-87] These investigators used stimulating and recording electrodes implanted on peripheral nerves and studied the effect of stimulation in unanesthetized, freely moving animals. They found that low-frequency stimulation of cutaneous mechanoreceptive afferents (group II) at rates of 10/sec or lower elicited the onset of natural sleep with EEG synchronization (Fig. 10-10). More rapid excitation of these same afferents or excitation of nociceptive afferents (group III) at any frequency produced prompt arousal in sleeping cats. They also showed that group II afferents preferentially activate bulbar reticular neurons and that group III afferents activate those of the pons and mesencephalon, all via pathways in the anterolateral columns of the spinal cord. Thus an explanation is provided for the everyday observation that sleep may be induced by slowly repeated, monotonous stimuli and for the habituation of the orienting reaction and the induction of sleep described by Pavlov.

The hypothesis that besides the reticular activating system there is an antagonistic group of synchronizing and therefore sleep-inducing structures in the lower brain stem is supported by evidence that is partly indirect but nevertheless strong. How this system may exert its effects and how it interacts with the reticular activating system are important questions yet unanswered. An attractively simple explanation is that each acts in an antagonistic way on the thalamic pacemaker, the final outcome—sleep or wakefulness

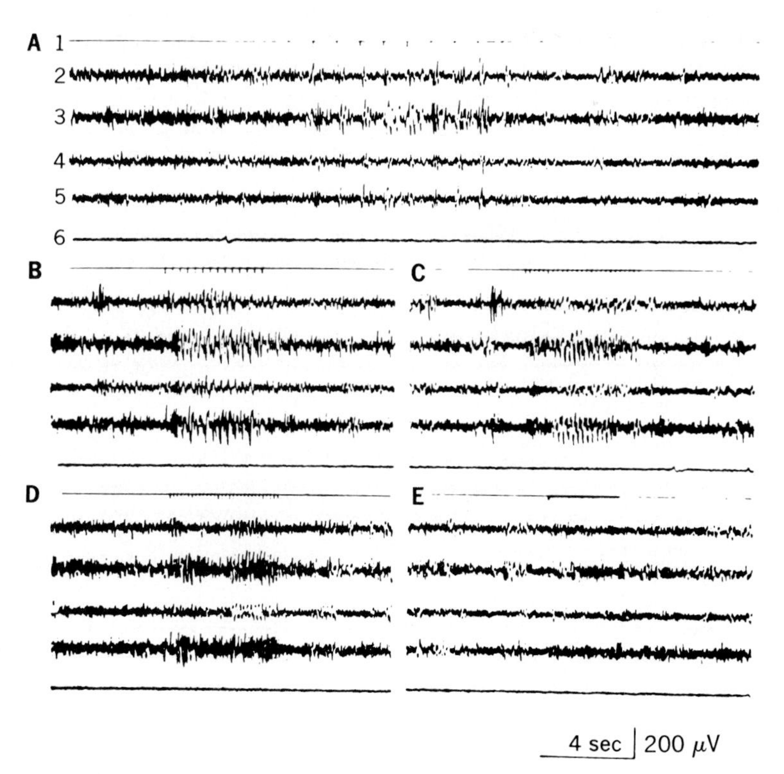

Fig. 10-10. Synchronization of EEG produced by stimulation of right superficial radial nerve at group II strength in intact unanesthetized cat. Cerebral recording and nerve-stimulating electrodes implanted 17 days earlier. Traces: *1,* stimulus marker; *2,* left parieto-occipital lead; *3,* left temporo-parietal lead; *4,* right parieto-occipital lead; *5,* right temporo-occipital lead; *6,* left neck electromyogram. **A,** Spindles tripped by each stimulus at 1/sec. **B** to **D,** Stimulation at 3, 5, and 6/sec, respectively, produced synchronization of the EEG waves, more or less at stimulus frequency. **E,** Stimulation at 16/sec produced no clear synchronization. Stimulation at higher frequencies produced clear-cut activation patterns and arousal. (From Pompeiano and Swett.[85])

—depending on the relative degree of activity in each. However, there is evidence that the synchronizing region may operate by inhibition of the reticular activating system, to which it is reciprocally linked, and both are influenced by corticoreticular systems. Thus the final state of awareness must certainly depend on the result of these more complex interactions, which remain to be defined more precisely.

Basal forebrain system

It has been known for a long time that lesions of the neural structures just anterior to and above the optic chiasm or of the neocortex of the orbital surface of the frontal lobe lead to hyperactivity and insomnia.[73] In an extended series of experiments, Clemente and Sterman and their colleagues[37,38] have shown that electrical stimulation of these basal forebrain regions via chronically implanted electrodes will elicit all the behavioral, electroencephalographic, and reflex signs of synchronized sleep. The effect is best produced by stimuli of low frequency but is not changed by rapid stimuli (up to 150/sec). In acute experiments with electrodes implanted in both the basal forebrain region and the reticular activating system of the brain stem, these investigators were able to demonstrate an interaction of the opposite behavioral and electroencephalographic effects produced by stimulation of these two regions. Moreover, it was possible to establish classic pavlovian conditioning and differential conditioning by pairing electrical stimulation of the basal forebrain region with previously neutral sounds of different frequencies. The results of such an experiment are illustrated in Fig. 10-11; they provide further evidence for a functional relation between the basal forebrain and the cerebral cortex. It is thought that the orbital cortex and the subcortical, preoptic areas give origin to a multisynaptic system of short-axoned cells that exerts an inhibitory influence over the locus of origin of the ascending reticular activating system in the brain stem. However, the role this system plays in regulating the periodic cycles of sleep and wakefulness is not yet fully understood.

SLEEP WITH DESYNCHRONIZED EEG AND DREAMING[5]

The *synchronized* or *slow-wave sleep* (SWS) of mammals is interrupted periodically by transitions to a different state that is characterized by desynchronization of the EEG and a powerful, active, descending inhibition of the segmental motor apparatus. Episodes of *desynchronized*

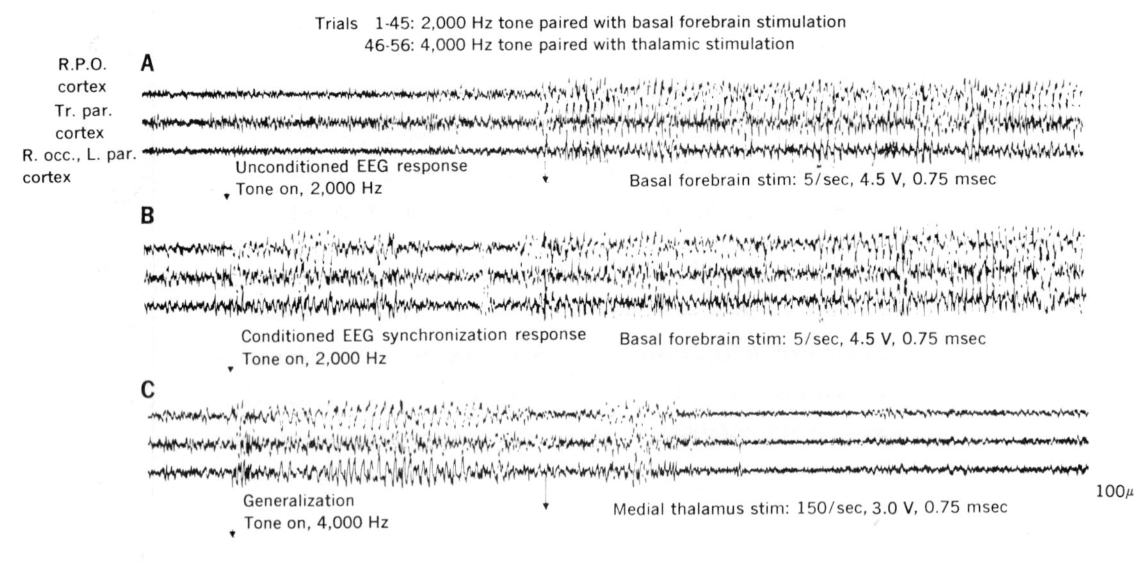

Fig. 10-11. EEG records from cat show development and generalization of conditioned EEG synchronization. **A,** Synchronization elicited by slowly repetitive stimulation of basal forebrain system. Early in series of conditioning trials, presentation of 2,000 Hz tone produced no change in ongoing EEG. **B,** After additional pairings, tone evokes shift to slow-wave, high-voltage EEG activity. **C,** Later presentation of 4,000 Hz tone evokes similar change in EEG records, an example of generalization of conditioned rseponse. **C** also shows that when electrical stimuli are delivered to basal forebrain system, there is (slightly delayed) *desynchronization* of EEG, an example of the fact that at many locations within the brain at which low-frequency stimulation produces synchronization, high-frequency stimulation has opposite effect. (From Clemente et al.[39])

sleep (DS) are regularly preceded by periods of SWS, reappear at intervals of about 90 min (the length of the basic mammalian rest-activity cycle,[106] Fig. 10-4, and last from 5 to 30 min or more, tending to lengthen as the night's sleep progresses, as shown in Fig. 10-12. DS is dual in nature; its tonic phase, characterized by a low-voltage, variable-frequency EEG and muscle atonia, is frequently broken by phasic outbursts of the rapid conjugate movements of the eyes called REMs and by a phasic deepening of the muscle atonia and superimposed muscle twitches and sometimes general and even violent movements. The REMs are preceded and accompanied by sharply phasic neural activity originating in the pontile RF, which is projected via polysynaptic pathways to the lateral geniculate bodies and from thence to the visual cortex. This activity is seen in the cat electrocorticogram as series of high-voltage waves called ponto-geniculo-occipital (PGO) spikes. They have a likely counterpart in the saw-toothed waveforms sometimes seen in the human EEG preceding and during the REM episodes of DS. It is important to emphasize that during this phase of sleep there is a dissociation between the EEG pattern and the behavioral state, for the desynchronized EEG of DS resembles, at least superficially, that of the waking state, whereas the threshold for behavioral arousal is greatly elevated over that of SWS.

The discovery by Aserinsky and Kleitman[21] and Dement and Kleitman[44] that human beings commonly report dreaming if awakened during DS but rarely do so if awakened during SWS led to greatly increased study of sleep as a biologic phenomenon. In particular, these studies concerned the cyclic nature of sleep and the periodic relation between SWS and DS, the need for sleep and the effects of sleep deprivation, dreaming, and such pathologic sleeplike states as narcolepsy.[98] The periodicity of DS is illustrated

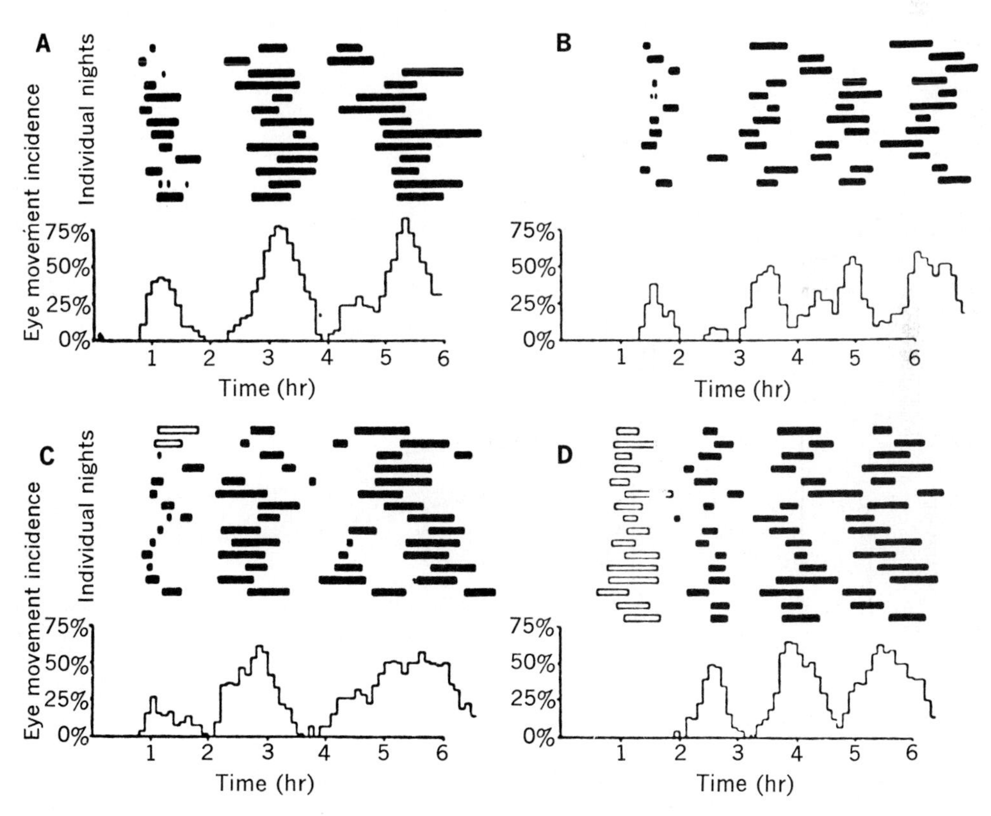

Fig. 10-12. Rhythmic occurrence of desynchronized sleep with REMs, an emergent stage 1 EEG, and dreaming for four subjects studied for a number of nights. Each bar represents single period of eye movements; each row of bars a single night's sleep. Composite histograms of incidence of REMs for several nights' sleep of each subject are placed under the series of bars. Subjects **A** and **D** cycled somewhat more regularly than **B** and **C.** Open bars indicate "cycles" when expected REMs did not appear, and during these times a persisting stage 2 was seen in EEG. (From Dement and Kleitman.[44])

by the study of four human subjects summarized in Fig. 10-12. It is now well documented by a number of studies[68,84,98] that DS is most prevalent at birth and decreases both in actual amount and as a percent of total sleep during ontogeny (Fig. 10-13). This has suggested a hypothesis concerning the functional meaning of DS—that the widespread activation of the forebrain that accompanies DS is important for the maturation of the developing nervous system and in the maintenance of synaptic connectivity in the adult, although there is thus far no direct evidence to support this idea.

Physiologic changes during desynchronized sleep

During SWS there is a progressive decline in average heart and respiratory rates and an early, although slight, fall in blood pressure. During DS there are slight increases in the average levels of these measures but, more characteristically, a marked increase in their short-term variability.[26,67,104] Thus the phasic periods of DS may be accompanied by sharp rises or falls in blood pressure and by wide variations in heart rate and respiratory pattern. These changes are thought to precipitate the catastrophic nocturnal events common in patients with cardiovascular and respiratory disease.[5] Other episodic changes occur during DS: there is an increased blood flow through and O_2 consumption by the brain, penile erection, and both renal and endocrinologic signs of CNS activation of both the anterior and posterior lobes of the pituitary gland.[68] All these events occur commonly but irregularly and thus are thought to be results and not causes of the episodes of DS.

Neural mechanisms in desynchronized sleep[3]

Experiments have shown that cats no longer display periods of DS following bilateral lesions of a particular part of the pontile RF and the n. reticularis pontis, oralis, and caudalis.[35] Cats with such lesions continue to show SWS in normal periodicity. It was Jouvet[60] who also showed that all aspects of DS that are mediated through effector systems of the brain stem and spinal cord

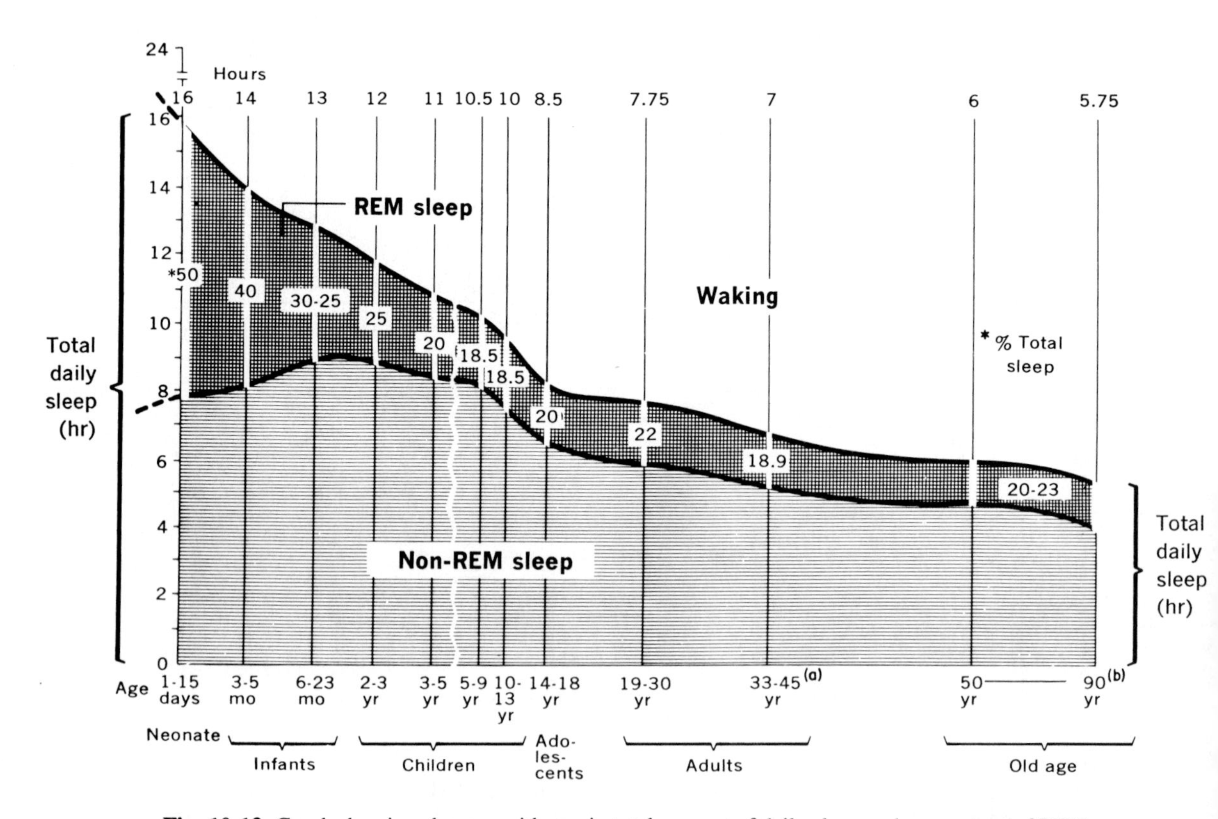

Fig. 10-13. Graph showing changes with age in total amount of daily sleep and percentage of REM (activated) sleep. There is a sharp drop in amount of REM sleep after early years, falling from 8 hr at birth to less than 1 hr in old age. Change in amount of non-REM sleep is much less marked, falling from 8 hr to about 5 hr over life-span. (From Roffwarg et al.[96])

occur regularly in animals that survive into the chronic state after total transection of the mesencephalon.* The desynchronization of the EEG, observed at cortical, mesencephalic, and diencephalic levels, is accompanied by a moderate increase in the discharge rate of cortical neurons. Each period of DS is preceded by the appearance of large, slow electrical waves in the pontile RF, and McCarley and Hobson[71] found that at this time the neurons of the reticular nuclei just referred to *increase their rates of discharge by factors of 5 to 10.* Simultaneously, large, slow electrical waves appear in the hippocampus, but the role of this and other limbic structures in DS is still uncertain.[34] During the tonic phase of DS, spinal motoneurons are subjected to a powerful postsynaptic inhibition, which Pompeiano[10] showed to be exerted by descending activity in reticulospinal systems of the ventral quadrants of the spinal cord.

In an important series of experiments, Pompeiano and his colleagues have shown that many of the events that characterized the *phasic* component of DS are initiated by activity that begins in the vestibular nuclei.[10] During outbursts of REMs these vestibular neurons discharge high-frequency bursts of impulses. Bilateral destruction of the medial and descending vestibular nuclei in the cat has the following effects on the phasic component of DS: (1) it eliminates REMs and the phasic increases in pupil diameter and heart rate that accompany them, leaving the desynchronized EEG unaffected; (2) it abolishes the phasic increment in motor inhibition, shown to be due to a presynaptic inhibition of the terminals of segmental afferents, and (3) it eliminates the excitatory pyramidal tract discharges that, breaking through even the phasically increased segmental inhibition, produce the muscular twitches and jerks that accompany the REMs.[76]

The powerful motor suppression exerted during DS has its counterpart on the sensory side, for there is a parallel suppression of afferent transmission, for example, in the dorsal column–medial lemniscal component of the somatic afferent system.[52]

Thus the state of deep sleep with desynchronized EEG is characterized by special and somewhat different sets of neural actions during its tonic and phasic periods. The tonic phase is initiated by an outburst of activity that begins in the pontile RF and seemingly engages the nervous system from the level of the cerebral cortex to that of the segmental efferents. The timing mechanism that controls its appearance is entirely unknown and the further events that trigger the superimposed phasic outbursts of REMs and associated phenomena are not understood. Among the several hypotheses that have been put forward concerning these control mechanisms, that postulating special biochemical mechanisms has received the most attention and is discussed in a following section.

*Such animals, usually cats and dogs, can be maintained for many months following high mesencephalic transection if an island of hypothalamus is left intact over the pituitary gland with an intact blood supply. This assures normal regulation of water balance. The animals are poikilothermic and must be kept in a temperature-regulated environment, with careful attention to food intake, etc.[22]

The need for sleep

Although its physiologic role is unknown, there is a clear biologic need for sleep and for each of its two distinct states, SWS and DS.[43] If, for example, a healthy human being is selectively deprived of DS by systematic awakening when it appears or by action of a drug that selectively affects DS, a rebound in DS occurs to levels higher than normal when undisturbed sleep is once again permitted. This increase in DS occurs without affecting SWS; moreover, selective deprivation of SWS produces just the reciprocal rebound effects during the recovery period.[42] If subjects are deprived of all sleep, the recovery period is marked at first by an excess of SWS, and the rebound in DS is delayed until the second or third recovery night.

Thus sleep is an active state of the nervous system that appears in two qualitatively quite different forms that depend on two sequentially linked but distinct patterns of neural activity. It is as if the circadian appearance of sleep and the periodic alternation between SWS and DS are controlled by some clocklike timer that reaches a set point and then, given compatible environmental circumstances and drive, initiates first SWS and then periodically thereafter episodes of DS. The long time course of these cycles and their slow recovery when disrupted by deprivation suggest that these periods are controlled by regulating humoral factors that operate on a longer time scale than do synaptic transmitter actions and other short-term neural effects.

BIOCHEMICAL REGULATORY MECHANISMS IN SLEEP[4,13,59,108]

Recent studies of the candidate biochemical mechanisms regulating sleep suggest that the induction and maintenance of the stages of sleep and waking are controlled, at least in part, by biogenic amines, specifically by the relative concentrations of serotonin (5-hydroxytryptamine, or 5-HT) and norepinephrine (NE), *or their related catabolic products,* at certain pertinent sites of action within the brain. The importance of these amines has been emphasized by the recent discovery, using the histochemical fluorescence technique, that neurons of the nuclei of the median raphe of the brain stem contain large amounts of 5-HT and that they project their small axons upward very widely to the diencephalon and cerebral cortex.[40,50,51] Similarly, it has been shown that neurons of the brain stem containing large amounts of NE are heavily concentrated in the lateral pontile RF, particularly in the locus coeruleus. Lesions of either of these areas lead to a marked reduction in the brain content of the relevant amine.

The evidence that the 5-HT system plays a role in sleep regulation is as follows: (1) Depletion of 5-HT in the brain or blockage of its synthesis by appropriate drugs (e.g., *p*-chlorophenylalanine) produces a sudden decrease in both SWS and DS that lasts 1 to 2 days and is followed by a slow recovery.[88,112] (2) Lesions of the raphe nuclei in cats lead to effective 5-HT depletion and to a severe and longer lasting insomnia[58] that is quickly relieved by administration of 5-hydroxytryptophan, a precursor of 5-HT that, unlike 5-HT, crosses the blood-brain barrier. (3) Close arterial injection of 5-HT into the brain stem circulation leads to a brief arousal followed by a prolonged hypersynchrony of the EEG.[99] (4) Surgical destruction of the area postrema of the medulla prevents this synchronizing effect of intra-arterial 5-HT, whereas local application of 5-HT to the area postrema on the medullary surface leads to immediate hypersynchrony.[61] It is postulated that 5-HT receptors of the area postrema activate the dendrites of neurons that lie in the n. tractus solitarius or its immediate vicinity,[75] part of the medullary synchronizing center (p. 314). (5) Administration of the drug reserpine depletes the brain of both 5-HT and NE and produces insomnia in both cats and men.[56] If reserpine is followed by 5-hydroxytryptophan, the precursor of 5-HT, SWS is immediately restored, but DS is not. In the cat, if reserpine is followed by a precursor of NE, dihydroxyphenylalanine (dopa), DS but not SWS is restored. Such a clear-cut effect is not seen in man. (6) Complete bilateral destruction of the locus coeruleus in cats causes a complete suppression of DS and has no effect on SWS.

There appears to be a functional link between SWS and DS, for after destruction of the raphe system or depletion of brain 5-HT by a drug blocking its synthesis, the amounts of both SWS and DS are related to the remaining levels of brain 5-HT. DS does not reappear until returning SWS reaches about 15% of normal. This leads to the idea that the 5-HT mechanism of SWS somehow "acts as a priming mechanism for triggering paradoxical sleep" (i.e., DS).[58] Nevertheless, there is evidence that the NE-containing neurons of the pontile RF, and particularly those of the locus coeruleus, play an important role in the regulation of DS. (1) Bilateral selective destruction of the locus coeruleus in the cat produces a selective suppression of DS and a fall of NE levels in rostral parts of the brain. (2) Drugs that selectively inhibit NE synthesis suppress DS. However, Williams[13] has shown that humans maintained on a synthetic diet low in phenylalanine (NE precursors) display a reversible syndrome that includes a disruption of sleep patterns, a marked decline in DS, but impaired memory and reduced physical activity as well.

It is obvious from the foregoing that a combination of biochemical and behavioral research on sleep has uncovered a host of new facts of great potential significance for understanding the mechanisms within the brain that regulate its own excitability and program its cyclic oscillation between sleep and wakefulness. Jouvet,[4] Williams,[13] and others suggest that it is the rate of turnover of 5-HT, or of one of its catabolic products, that is monitored by a clocklike mechanism within the brain and regulates the sleep cycles.

Is there a more general hypnogenic factor? The idea that a sleep-inducing substance might be produced in other tissues, or indeed within the brain itself, long antedates the biochemical investigations just described and continues to receive some experimental support. There are reports, for example, that a dialysate of venous blood draining the brain of a sleeping donor animal will induce sleep in an alert animal, whereas such a dialysate from an alert donor has no such effect.[74] The hypothetical sleep-inducing substance has not been identified. Other investigators report that in a cross-circulation experiment in dogs, the induction of sleep in a donor animal by electrical stimulation of the medial thalamus (in the manner of Hess, p. 308) induced EEG hypersynchrony within the recipient dog within 20 to 30 sec.[62] The candidate hypnogenic substance has not been identified.

Pappenheimer and his colleagues, taking a novel approach, have developed a method for collecting the cerebrospinal fluid of animals (they used goats) in the chronic condition.[47] They found that the intraventricular injection of small amounts of CSF from sleep-deprived animals induced EEG and behavioral signs of sleep in well-rested recipient animals. Identification studies of this substance are not yet complete. The sleep-promoting factor has a molecular weight of less than 500. Its actions are not duplicated by the intraventricular injection of 5-HT or several other candidate hypnogenic substances.

REFERENCES

General reviews

1. Cold Spring Harbor symposia on quantitative biology. Biological clocks, Baltimore, 1960, The Waverly Press, Inc.
2. Hartmann, E., editor: Sleep and dreaming, Int. Psychiatry Clin. **7:**entire issue, 1970.
3. Jouvet, M.: Neurophysiology of the states of sleep, Physiol. Rev. **47:**117, 1967.
4. Jouvet, M.: Biogenic amines and the states of sleep, Science **163:**32, 1969.
5. Kales, A.: Sleep: physiology and pathology, Philadelphia, 1969, J. B. Lippincott Co.
6. Kety, S. S., Evarts, E. V., and Williams, H. L., editors: Sleep and altered states of consciousness, Baltimore, 1967, The Williams & Wilkins Co.
7. Kleitman, N.: Sleep and wakefulness, ed. 2, Chicago, 1963, University of Chicago Press.
8. Moruzzi, G.: Active processes in the brain stem during sleep, Harvey Lect. **58:**233, 1962-1963.
9. Moruzzi, G.: Sleep and instinctive behavior, Arch. Ital. Biol. **107:**175, 1969.
10. Pompeiano, O.: The neurophysiological mechanisms of the postural and motor events during desynchronized sleep, Res. Publ. Assoc. Res. Nerv. Ment. Dis. **45:** 351, 1967.
11. Richter, C. P.: Biological clocks in medicine and psychiatry, Springfield, Ill., 1965, Charles C Thomas, Publisher.
12. Thorpe, W. H.: Ethology and consciousness. In Eccles, J. C., editor: Brain and conscious experience, Berlin, 1966, Springer Verlag.
13. Williams, H. L.: The new biology of sleep, J. Psychiatr. Res. **8:**445, 1971.
14. Witkin, H., and Lewis, H.: Experimental studies of dreaming, New York, 1967, Random House, Inc.

Original papers

15. Affani, A., Marchiafava, P. L., and Zernicki, B.: Orientation reactions in the midpontine pretrigeminal cat, Arch. Ital. Biol. **100:**297, 1962.
16. Affani, A., Marchiafava, P. L., and Zernicki, B.: Conditioning in the midpontine pretrigeminal cat, Arch. Ital. Biol. **100:**305, 1962.
17. Akert, K., Bally, C., and Schadé, J. P., editors: Sleep mechanisms, Prog. Brain Res. **18:**entire issue, 1965.
18. Akert, K., Koella, W. P., and Hess, R., Jr.: Sleep produced by electrical stimulation of the thalamus, Am. J. Physiol. **168:**268, 1952.
19. Apelbaum, J., Silva, E. E., Fruck, P., and Segundo, J. P.: Specificity and biasing of arousal reaction habituation, Electroencephalogr. Clin. Neurophysiol. **12:** 829, 1960.
20. Arduini, A., and Arduini, M. G.: Effect of drugs and metabolic alterations on brain stem arousal mechanism, J. Pharmacol. Exp. Ther. **110:**76, 1954.
21. Aserinsky, E., and Kleitman, N.: Regularly occurring periods of eye motility, and concomitant phenomena, during sleep, Science **118:**273, 1953.
22. Bard, P., Woods, J. W., and Bleier, R.: The effects of cooling, heating, and pyrogen on chronically decerebrate cats, Comm. Behav. Biol. **5A:**31, 1970.
23. Batini, C., Palestini, M., Rossi, G. F., and Zanchetti, A.: EEG activation patterns in the midpontine pretrigeminal cat following sensory deafferentation, Arch. Ital. Biol. **97:**26, 1959.
24. Batini, C., et al.: Effects of complete pontine transections on the sleep-wakefulness rhythm, the midpontine pretrigeminal preparation, Arch. Ital. Biol. **97:**1, 1959.
25. Batini, C., et al.: Neural mechanisms underlying the enduring EEG and behavioral activation in the midpontine pretrigeminal cat, Arch. Ital. Biol. **97:**13, 1959.
26. Berger, R. J.: Physiological characteristics of sleep. In Kales, A., editor: Sleep: physiology and pathology, Philadelphia, 1969, J. B. Lippincott Co.
27. Berlucchi, G., Maffei, L., Moruzzi, G., and Strata, P.: EEG and behavioral effects elicited by cooling of medulla and pons, Arch. Ital. Biol. **102:**372, 1964.
28. Berman, A. L.: The brain stem of the cat. A cytoarchitectonic atlas with stereotaxic coordinates, Madison, 1968, University of Wisconsin Press.
29. Bert, J., et al.: A comparative sleep study of two cercopithecinae, Electroencephalogr. Clin. Neurophysiol. **28:**32, 1970.
30. Bonvallet, M., and Allen, M. B., Jr.: Prolonged spontaneous and evoked reticular activation following discrete bulbar lesions, Electroencephalogr. Clin. Neurophysiol. **15:**969, 1963.
31. Bremer, F.: The neurophysiological problem of sleep. In Adrian, E. A., Jasper, H. H., and Bremer, F., editors: Brain mechanisms and consciousness, Springfield, Ill., 1954, Charles C Thomas, Publisher.
32. Brodal, A.: The reticular formation of the brain stem. Anatomical aspects and functional correlations, Edinburgh, 1957, Oliver & Boyd, Ltd.
33. Bullock, T. H., and Horridge, G. A.: Structure and function in the nervous systems of invertebrates, San Francisco, 1965, W. H. Freeman & Co., Publishers, chap. 1 and 5.
34. Cadillac, J., Passouant-Fontaine, T., and Passouant, P.: Modifications de l'activité de l'hippocampe suivant les divers stades du sommeil spontane chez le chat, Rev. Neurol. Psychiatr. **105:**171, 1961.
35. Carli, G., and Zanchetti, A.: A study of pontine lesions suppressing deep sleep in the cat, Arch. Ital. Biol. **103:**751, 1965.
36. Caspers, H.: Die Veränderungen der corticalen Gleichspannung und ihre Beziehungen zur senso-motorischen Aktivitat (Verhalten) bei weckreizungen am freibeweglichen Tier, Proceedings of the twenty-second congress, Leiden, 1962, vol. 1, part 1, p. 443.
37. Clemente, C. D.: Forebrain mechanisms related to internal inhibition and sleep, Cond. Reflex **3:**145, 1968.
38. Clemente, C. D., and Sterman, M. B.: Basal forebrain mechanisms for internal inhibition and sleep, Res. Publ. Assoc. Res. Nerv. Ment. Dis. **45:**127, 1967.
39. Clemente, C. D., Sterman, M. B., and Wyrwicka, W.: Forebrain inhibitory mechanisms: conditioning of basal forebrain induced synchronization and sleep, Exp. Neurol. **7:**404, 1963.
40. Dahlstrom, A., and Fuxe, K.: Evidence for the existence of monoamine-containing neurons in the central nervous system. I. Demonstration of monoamines in the cell bodies of brainstem neurons, Acta Physiol. Scand. **62**(suppl. 232):1, 1964.
41. Dell, P.: Reticular homeostasis and critical reactivity. In Moruzzi, G., Fessard, A., and Jasper, H. H., edi-

tors: Brain mechanisms, Amsterdam, 1963, Elsevier Publishing Co.
42. Dement, W.: Effect of dream deprivation, Science **131:** 1705, 1960.
43. Dement, W.: The biological role of REM sleep (circa 1968). In Kales, A., editor: Sleep, physiology and pathology, Philadelphia, 1969, J. B. Lippincott Co.
44. Dement, W., and Kleitman, N.: Cyclic variations of EEG during sleep and their relations to eye movements, body motility and dreaming, Electroencephalogr. Clin. Neurophysiol. **9:**673, 1957.
45. Dempsey, E. W., and Morison, R. S.: The electrical activity of thalamocortical relay systems, Am. J. Physiol. **138:**283, 1942.
46. Desiraju, T., and Purpura, D. P.: Organization of specific-nonspecific thalamic internuclear synaptic pathways, Prog. Brain Res. **21:**169, 1970.
47. Fencl, V., Koski, G., and Pappenheimer, J. R.: Factors in cerebrospinal fluid from goats that affect sleep and activity in rats, J. Physiol. **216:**565, 1971.
48. Freemon, F. R., McNew, J. J., and Adey, W. R.: Chimpanzee sleep stages, Electroencephalogr. Clin. Neurophysiol. **31:**485, 1971.
49. French, J. D., and Magoun, H. W.: Effects of chronic lesions in central cephalic brain stem of monkeys, Arch. Neurol. Psychiatry **68:**591, 1952.
50. Fuxe, K.: Evidence for the existence of monoamine neurons in the central nervous system. IV. Distribution of monoamine nerve terminals in the central nervous system, Acta Physiol. Scand. **64**(suppl. 247): 37, 1965.
51. Fuxe, K., Hokfelt, T., and Ungerstedt, V.: Localization of indolealkylamines in CNS, Adv. Pharmacol. **6A:**235, 1968.
52. Ghelarducci, B., Pisa, M., and Pompeiano, O.: Transformation of somatic afferent volleys across the prethalamic and thalamic components of the lemniscal system during the rapid eye movements of sleep, Electroencephalogr. Clin. Neurophysiol. **29:**348, 1970.
53. Hanbury, J., and Jasper, H. H.: Independence of diffuse thalamocortical projection system shown by specific nuclear destruction, J. Neurophysiol. **16:**252, 1953.
54. Hess, W. R.: The functional organization of the diencephalon, New York, 1957, Grune & Stratton, Inc.
55. Hobson, J. A., and McCarley, R. W.: Cortical unit activity in sleep and waking, Electroencephalogr. Clin. Neurophysiol. **30:**97, 1971.
56. Hoffman, J. S., and Domino, E. F.: Comparative effects of reserpine on the sleep cycle of man and cat, J. Pharmacol. Exp. Ther. **170:**190, 1969.
57. Jasper, H. H.: Unspecific thalamocortical relations. In Magoun, H. W., editor: Handbook of physiology. Neurophysiology section, Baltimore, 1960, The Williams & Wilkins Co., vol. 2.
58. Jouvet, M.: Insomnia and decrease of cerebral 5-HT after destruction of the raphe system in the cat, Adv. Pharmacol. **6B:**265, 1968.
59. Jouvet, M.: Serotonin and sleep. In Blum, J. J., editor: Biogenic amines as physiological regulators, Englewood Cliffs, N.J., 1969, Prentice-Hall, Inc.
60. Jouvet, M., and Jouvet, D.: A study of the neurophysiological mechanisms of dreaming, Electroencephalogr. Clin. Neurophysiol. **24**(suppl.):133, 1963.
61. Koella, W. P.: Serotonin and sleep, Exp. Med. Surg. **27:**157, 1969.
62. Kornmüller, A. E., Lux, H. D., Winkle, K., and Klee, M.: Neurohumoral ausfeloste Schlafzustande an Tieren mit gekreuztem Kreislauf unter der Kontrolle von EEG-Ableitungen, Naturwissenschaften **48:**503, 1961.
63. Larsen, L. E., and Walter, D. O.: On automatic methods of sleep staging by EEG spectra, Electroencephalogr. Clin. Neurophysiol. **28:**459, 1970.
64. Magnes, J., Moruzzi, G., and Pompeiano, O.: Synchronization of the EEG produced by low-frequency electrical stimulation of the region of the solitary tract, Arch. Ital. Biol. **99:**33, 1961.
65. Magni, F., and Willis, W. D.: Cortical control of brain stem reticular neurons, Arch. Ital. Biol. **102:**418, 1964.
66. Magni, F., Moruzzi, G., Rossi, G. F., and Zanchetti, A.: EEG arousal following inactivation of the lower brain stem by selective injection of barbiturate into the vertebral circulation, Arch. Ital. Biol. **97:**33, 1959.
67. Mancia, G., Baccelli, G., Adams, D. B., and Zanchetti, A.: Vasomotor regulation during sleep in the cat, Am. J. Physiol. **220:**1086, 1971.
68. Mandell, A. J., and Mandell, M. P.: Biochemical aspects of rapid eye movement sleep, Am. J. Psychiatry **122:**391, 1965.
69. Mangold, R., et al.: The effects of sleep and lack of sleep on cerebral circulation and metabolism of normal young men, J. Clin. Invest. **34:**1092, 1955.
70. McCarley, R. W., and Hobson, J. A.: Cortical unit activity in desynchronized sleep, Science **167:**901, 1970.
71. McCarley, R. W., and Hobson, J. A.: Single neuron activity in cat gigantocellular tegmental field: selectivity of discharge in desynchronized sleep, Science **174:**1250, 1971.
72. McGinty, D. J.: Encephalization and the neural control of sleep. In Sterman, M. B., McGinty, D. J., and Adinolfi, A. M., editors: Brain development and behavior, New York, 1971, Academic Press, Inc.
73. McGinty, D. J., and Sterman, M. B.: Sleep suppression after basal forebrain lesions in the cat, Science **160:**1253, 1968.
74. Monnier, M., and Hosli, L.: Humoral transmission of sleep and wakefulness. II. Hemodialysis of a sleep inducing humor during stimulation of the thalamic somnogenic area, Pfluegers Arch. **282:**60, 1965.
75. Morest, D. K.: Experimental study of the projections of the nucleus of the tractus solitarius and the area postrema in the cat, J. Comp. Neurol. **130:**277, 1967.
76. Morrison, A. R., and Pompeiano, O.: Vestibular influences during sleep, Arch. Ital. Biol. **108:**154, 1970.
77. Moruzzi, G.: the physiological properties of the brain stem reticular formation. In Adrian, E. D., Jasper, H. H., and Bremer, F., editors: Brain mechanisms and consciousness, Springfield, Ill., 1954, Charles C Thomas, Publisher.
78. Moruzzi, G.: Reticular influences on the EEG, Electroencephalogr. Clin. Neurophysiol. **16:**1, 1964.
79. Moruzzi, G., and Magoun, H. W.: Brain stem reticular formation and activation of the EEG, Electroencephalogr. Clin. Neurophysiol. **1:**455, 1949.
80. Nauta, W. H. J., and Kuypers, H. G. J. M.: Some ascending pathways in brain stem reticular formation. In Jasper, H. H., et al., editors: Reticular formation of the brain, Boston, 1958, Little, Brown & Co.
81. Noda, H., and Adey, W. R.: Firing variability in cat association cortex during sleep and wakefulness, Brain Res. **18:**513, 1970.
82. Olszewski, J., and Baxter, D.: Cytoarchitecture of the human brain stem, Philadelphia, 1954, J. B. Lippincott, Co.
83. Parker, D. C., Sassin, J. F., and Mace, J. W.: Human

growth hormone release during sleep; electroencephalographic correlation, J. Clin. Endocrinol. Metab. **29:** 871, 1969.
84. Parmelee, A. H., et al.: Maturation of EEG activity during sleep in premature infants, Electroencephalogr. Clin. Neurophysiol. **24:**319, 1968.
85. Pompeiano, O., and Swett, J. E.: EEG and behavioral manifestations of sleep induced by cutaneous nerve stimulation in normal cats, Arch. Ital. Biol. **100:**311, 1962.
86. Pompeiano, O., and Swett, J. E.: Identification of cutaneous and muscular afferent fibers producing EEG synchronization or arousal in normal cats, Arch. Ital. Biol. **100:**343, 1962.
87. Pompeiano, O., and Swett, J. E.: Action of graded cutaneous and muscular afferent volleys on brain stem units in the decerebrate, cerebellectomized cat, Arch. Ital. Biol. **101:**552, 1963.
88. Pujol, J. F., et al.: The central metabolism of serotonin in the cat during insomnia. A neurophysiological and biochemical study after administration of p-chlorophenylalanine or destruction of the raphe system, Brain Res. **29:**195, 1971.
89. Purpura, D. P., and Cohen, B.: Intracellular recording from thalamic neurons during recruiting responses, J. Neurophysiol. **25:**621, 1962.
90. Purpura, D. P., and Shofer, R. J.: Intracellular recording from thalamic neurons during reticulocortical activation, J. Neurophysiol. **26:**494, 1963.
91. Purpura, D. P., McMurtry, J. G., and Maekawa, K.: Synaptic events in ventrolateral thalamic neurons during suppression of recruiting responses by brain stem reticular formation, Brain Res. **1:**63, 1966.
92. Purpura, D. P., Scarff, T., and McMurtry, J. G.: Intracellular study of internuclear inhibition in ventrolateral thalamic neurons, J. Neurophysiol. **28:**487, 1965.
93. Purpura, D. P., Shofer, R. J., and Musgrave, F. S.: Cortical intracellular potentials during augmenting and recruiting responses. II. Patterns of synaptic activities in pyramidal and nonpyramidal tract neurons, J. Neurophysiol. **27:**133, 1964.
94. Richter, C. P.: Sleep and activity: their relation to the 24-hour clock, Res. Publ. Assoc. Res. Nerv. Ment. Dis. **45:**8, 1967.
95. Richter, C. P.: Inborn nature of the rat's 24-hour clock, J. Comp. Physiol. Psychol. **75:**1, 1971.
96. Roffwarg, H. P., Muzio, J. N., and Dement, W. C.: Ontogenetic development of the human sleep-dream cycle, Science **152:**604, 1966.
97. Rossi, G. F., and Zanchetti, A.: The brain stem reticular formation. Anatomy and physiology, Arch. Ital. Biol. **95:**199, 1957.
98. Roth, B., Bruhova, S., and Lehovsky, M.: REM sleep and NREM sleep in narcolepsy and hypersomnia, Electroencephalogr. Clin. Neurophysiol. **26:**176, 1969.
99. Roth, G. I., Walton, P. L., and Yamamoto, W. S.: Area postrema: abrupt EEG synchronization following close intra-arterial perfusion with serotonin, Brain Res. **23:**223, 1970.
100. Scheibel, M. E., and Scheibel, A. B.: Structural substrates for integrative patterns in the brain stem reticular core. In Jasper, H. H., et al., editors: Reticular formation of the brain, Boston, 1958, Little, Brown & Co.
101. Scheibel, M. E., and Scheibel, A. B.: The organization of the nucleus reticularis thalami: a Golgi study, Brain Res. **1:**43, 1966.
102. Segundo, J. P., Takenaka, T., and Encabo, H.: Electrophysiology of bulbar reticular neurons, J. Neurophysiol. **30:**1194, 1967.
103. Segundo, J. P., Takenaka, T., and Encabo, H.: Somatic sensory properties of bulbar reticular neurons, J. Neurophysiol. **30:**1221, 1967.
104. Snyder, F., Hobson, J. A., Morrison, D. F., and Goldfrank, F.: Changes in respiration, heart rate, and systolic blood pressure in human sleep, J. Appl. Physiol. **19:**417, 1964.
105. Steriade, M.: Ascending control of thalamic and cortical responsiveness, Int. Rev. Neurobiol. **12:**87, 1970.
106. Sterman, M. B., and Hoppensbrouwers, T.: The development of sleep-waking and rest-activity patterns from fetus to adult in man. In Sterman, M. B., McGinty, D. J., and Adinolfi, A. M., editors: Brain development and behavior, New York, 1971, Academic Press, Inc.
107. Takahashi, Y., Kipnis, D. M., and Daughaday, W. H.: Growth hormone secretion during sleep, J. Clin. Invest. **47:**2079, 1968.
108. Torda, C.: Biochemical and bioelectric processes related to sleep, paradoxical sleep, and arousal, Psychol. Rep. **24:**807, 1969.
109. Veslasco, M., and Lindsley, D. B.: Effect of thalamocortical activation on recruiting responses, Acta Neurol. Lat. Am. **14:**188, 1968.
110. Velasco, M., Weinberger, N. M., and Lindlsey, D. B.: Effect of thalamocortical activation on recruiting responses. I. Reticular stimulation, Acta Neurol. Lat. Am. **14:**99, 1968.
111. von Economo, C.: Sleep as a problem of localization, J. Nerv. Ment. Dis. **71:**249, 1930.
112. Weitzman, E. D., Rapport, M. M., McGregor, P., and Jocoby, J.: Sleep patterns of the monkey and brain serotonin concentration: effect of p-chlorophenylalanine, Science **160:**1361, 1968.
113. Yamaguchi, N., Ling, G. M., and Marczynski, T. J.: Recruiting responses observed during wakefulness and sleep unanesthetized chronic cats, Electroencephalogr. Clin. Neurophysiol. **17:**246, 1965.

V

CENTRAL NERVOUS MECHANISMS IN SENSATION

11

VERNON B. MOUNTCASTLE

Sensory receptors and neural encoding: introduction to sensory processes

Sensation is that mode of mental functioning referring to immediate stimulation of the organism, including hearing, smelling, and seeing; specifically, it is the direct behavioral experience evoked by immediate stimulation of the sense organs. Sensation is no longer strictly differentiated from perception, but the latter term is used frequently to designate a more complete behavioral experience that may involve the combination of different sensations and their conjunction with past experience in apprehending and understanding the objects and facts we encounter in everyday life. The study of sensation begins with the subject of sensory transduction, that is, how the peripheral endings of afferent nerve fibers convert impinging energy into local excitation and how this in turn leads to trains of afferent nerve impulses. This process of peripheral encoding determines what information the nervous system receives about the quality, locus, intensity, and temporal pattern of stimuli that elicit sensations.[3,55] The goal is to learn how this initial neural replication of events in the external world is further elaborated and transformed within the brain and how it leads to perception, sensory discrimination, storage in memory, and, at choice, motor response or public description.

It is everyday human experience that different stimuli elicit differing sensory experiences that can be classified and named. Electromagnetic waves 400 to 760 nm in length elicit experiences called visual, and those elicited by lights of different wavelengths are described differently, for they are seen as different colors. Objects examined tactually are described as hard or soft, as warm or cool, and as having certain spatial contours or temporal patterns. Substances are readily differentiated by the way they taste or smell. It seems to be given directly to conscious experience that stimuli of a given kind elicit sensory experiences that, within certain limits of resolution and discrimination, can be defined with precision and recognized whenever encountered. Each of these readily distinguished classes is termed a sensory modality, a class of sensations connected along a qualitative continuum. Those evoked by light compose the visual modality, those by sound the auditory. General somatic sensibility is made up of several distinct modalities: mechanoreception (touch-pressure), warmth, coolness, pain, and sense of position and movements of the limbs (kinesthesia) can each be identified separately. *These depend in the first instance on afferent activity in certain specific sets of primary afferent nerve fibers.* Other aspects of mechanoreception are equally distinct but depend on spatial or temporal patterns of activity in the mechanoreceptive afferents. Stereognosis, the sense of form and contour, depends on the spatial pattern of activity in the entire population of mechanoreceptive afferents engaged by the stimulus. The sense of flutter-vibration depends on the appearance of periodic activity in two particular sets of mechanoreceptive afferents that, when active in other temporal patterns, elicit only the sense of mechanical contact. Stereognosis and flutter-vibration are derived modes of the mechanoreceptive sense. How we identify the quality of a stimulus is a major problem in the neurophysiology of sensation and will influence treatment of the subject from the level of first-order fibers to that of total behavior.

Humans locate stimuli in space with considerable accuracy, for example, the position of a mechanical stimulus on a finger pad can be reproduced to within 1 to 2 mm, and with foveal vision the position of a light can be replicated to within a degree of arc. The problem of how a stimulus is located in physical space reduces to the question of how a brain identifies the locations of peaks of activity in central neural fields, perhaps in the primary sensory cortical areas. This remains a major unsolved problem in central nervous physiology.

Human subjects identify accurately the intensity as well as the place and quality of sensory stimuli, but do poorly when rating stimuli of the same quality but of different magnitudes along physical scales, such as those for heat, light, or pressure. We discriminate well between two such stimuli if they are presented close together in space and in time. Finally, humans possess a refined ability to identify or to discriminate between stimuli that differ in spatial configuration or temporal pattern. We excel in sensing spatial and temporal transients, but not steady states.

A nervous system faces the external world indirectly via the afferent input reaching it over first-order sensory nerve fibers. It derives from that input an ongoing, continually changing, and nearly up-to-date picture of the external environment. Our perceptual images of the external world are to some extent abstractions of physical reality. They are determined in the first instance by the selective transformations that occur at the level of first-order fibers and then by others imposed subsequently in the long chain of neural events leading to perception, including those determined by genetic and experiential sets. It is the aim of this and the following chapters to explain as far as is presently possible the neural mechanisms of these transformations. Initially I shall show how sensory capacities are determined by the transducer action of peripheral nerve endings and, as is sometimes the case, the specialized nonneural receptor cells associated with them. Those serving what is termed somesthesis will be treated in this chapter and the relevant central neural mechanisms in that which follows.

DEFINITIONS AND PRINCIPLES

1. The terms *"sensory receptor," "ending,"* and *"organ"* are used interchangeably by sensory physiologists. In many cases the true sensory endings, the transducers, are the peripheral terminals of the afferent nerve fibers themselves. In others the nerve terminals are linked to specialized nonneural transducer cells, such as the hair cells of the cochlea. These transducer cells may determine the sensitivity and dynamic range of the nerve endings.

2. Sensory receptors possess *thresholds;* of stimuli of the same quality but different magnitudes, some never and others always excite the nerve endings. Transitional between these are stimuli increasingly likely to excite. Threshold is therefore a variable quantity and is usually defined by a statistical estimator. The threshold for human sensation in the limiting case approaches that of the relevant sets of first-order fibers.

3. A *sensory unit* is a single primary afferent nerve fiber, all of its peripheral branches and central terminals, and any associated nonneural transducer cells as well.

4. A *peripheral receptive field* is that spatial area within which a stimulus of sufficient magnitude and proper quality will evoke a discharge of impulses in a sensory unit, for example, the area of skin in which a mechanical stimulus will excite a cutaneous mechanoreceptive afferent fiber.

5. The peripheral branches of adjacent sensory units are successively intertwined. This overlap shifts gradually across a sensory surface such as the skin. It is unlikely, therefore, that any stimulus normally encountered that is above behavioral threshold ever engages only one single fiber. This *principle of partially shifted overlap* is important for understanding central neural mechanisms in spatial discrimination.

6. The peripheral receptive fields of primary afferent fibers that terminate within a given sensory surface may vary greatly in size, usually inversely with the number of innervating nerve fibers per unit area. The volume of central nervous tissue given to the representation of various parts of a sensory field varies directly with this *peripheral innervation density,* and both vary with sensory acuity.

7. Sensory fibers differ in the manner in which they respond to a continuing, steady stimulus (i.e., in the rate of *adaptation*). Some are detectors of transients only, for they discharge a few impulses on the application and at the removal of a stimulus, but none during its steady application. Other afferents respond to stimulus onset with a high-frequency discharge determined by the rate of stimulus application and its final magnitude and continue to discharge during stimulus application at a frequency determined by that magnitude.

8. Different sets of nerve fibers, when active, elicit different sensations, and any one set elicits the same sansation no matter how it is activated. This *modality specificity* is thought to be determined both by the different central connections of different sets of afferent fibers and by the differential sensitivity of their peripheral endings to

different forms of impinging energy. Although the mechanism of this qualitative specificity is still largely unknown, it provides the basis for a useful classification of first-order sensory nerve fibers, given in Table 11-1, which includes most of the types known to exist in mammals.

THE TRANSDUCER PROCESS

The primary event in the process of sensing is the conversion of stimulus energy to a depolarization of sensory endings and via that event the encoding of stimulus features in an afferent impulse message. The terminal nerve membrane is itself the primary transducer in many sensory nerve fibers, including many mechanoreceptive afferents. It is an area of nerve membrane specialized to react to a particular class of stimuli with a physicochemical change, perhaps in molecular configuration, leading to alterations in membrane permeability and via the sequence of events diagrammed later to centripetally conducted nerve impulses. Many primary transducer fibers end freely among the cells of the tissues they innervate, but the terminals of others are related to one or more specialized nonneural receptor cells or are encapsulated by small organs of such cells. In the case of an encapsulated primary transducer nerve terminal, like that in the pacinian corpuscle, the sensory organ influences the sensitivity and the dynamic range of the nerve ending, but it does not determine the form of energy to which the ending is differentially sensitive. For other mechanoreceptive afferents the linkage between the nonneural receptor cells and the nerve ending has the ultrastructural characteristics of a chemically operated synapse, the specialized nonneural cell being presynaptic. This suggests, but by no means proves, that the primary transduction occurs at the nonneural receptor cell, which is then surmised to release a transmitter substance that excites the nerve terminal as described in Chapter 8 for chemically operated synapses.[19] In such a case the nonneural cell may determine the specific sensitivity of the nerve terminal, but this is still uncertain. Examples of this second type of transducer arrangement are the linkages between auditory nerve fibers and the hair cells of the cochlea[16] and those between a certain set of afferents and specialized Merkel cells in the skin. The sensitivity of some mechanoreceptors of this type (e.g., those in the cochlea) is several orders of magnitude greater

Table 11-1. Classification of some first-order afferents

Incident stimulus	Intermediate mechanism	Examples of receptor types and function served
Mechanical force	Unknown; possibilities are: 1. Change in static properties of nerve ending (e.g., in capacitance, resistance, etc.) 2. Intermediate release of specific chemical agent and chemoreception at nerve ending	Mechanoreceptors, serving: 1. Touch-pressure in skin and subcutaneous tissues; both organized and free nerve endings 2. Position sense and kinesthesia: mechanoreceptors of joints and vestibular receptors of inner ear 3. Mechanoreceptors of cochlea, serving hearing 4. Stretch receptors of muscle and tendon 5. Visceral pressure receptors: carotid and right atrium
Light	Photochemical transduction, leading to excitation of nerve endings by a synaptic mechanism	Photoreceptors of eye, serving vision
Heat	Unknown (by regulation of chemical reaction that influences state of nerve ending)	Thermoreceptors, separately for: 1. Warmth 2. Cold
Substances in solution	Uncertain; probably excitation of receptor cell or nerve ending by specific chemical combination, leading to change in permeability	Chemoreceptors, separately for: 1. Taste 2. Smell Osmoreceptors Carotid body receptors
Extremes of mechanical force, heat or its absence, presence of certain chemicals	Incipient or actual destruction of tissue cells (release of substance exciting nerve ending)	Nociceptors, serving pain

than is that of other mechanoreceptive afferents ending freely among nonspecialized tissue cells in some species.

The sequence of events leading from stimulus to nerve impulse may be diagrammed as follows:

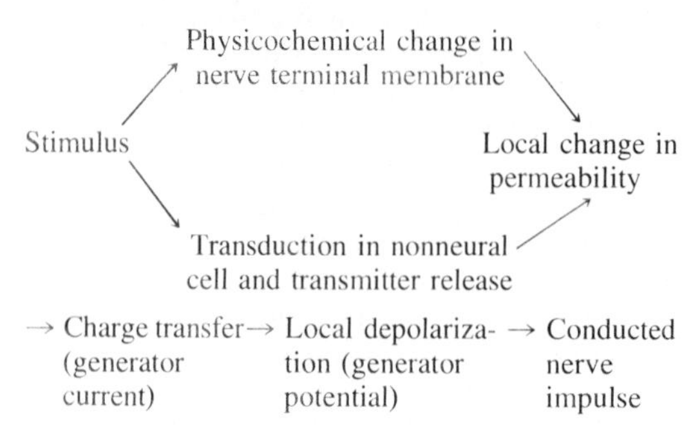

The large majority of mechanoreceptors are either quickly or slowly adapting; those with intermediate rates of adaptation are rare. Quickly adapting receptors signal stimulus velocity; others may signal higher derivatives of stimulus position. The pattern of activity in a population of such afferents innervating a sensory sheet such as the skin may signal the direction of stimulus movement, but they provide little information about stimulus magnitude. In some cases, they signal stimulus frequency with great accuracy. Slowly adapting receptors, on the other hand, play a part in signaling transient changes and the steady magnitudes of stimuli, but not stimulus frequency.

Transducer mechanism of a quickly adapting receptor: the pacinian corpuscle[45]

Mammalian mechanoreceptive nerve endings are commonly so embedded in the tissues they innervate that they cannot be isolated for direct study. The pacinian corpuscle is one exception, for it is found in the mesentery, from which individual corpuscles and their axons can be isolated either in situ with normal or perfused circulation or after removal for study in vitro. This receptor organ is also present in the dermis, in subcutaneous and intramuscular connective tissue, and in the periosteum. The corpuscle is an ellipsoidal body made up of a number of concentric lamellae of connective tissue cells (Fig. 11-1). The myelinated nerve fiber enters at one end, a final node of Ranvier occurs within the corpuscle, the myelin and Schwann sheaths disappear, and the bare, nearly straight nerve terminal occupies the center of the inner core of the corpuscle. The inner core lamellae are hemiconcentric; they are the thin flat extensions of cells whose nuclei lie in the intermediate zone of the capsule. These cells are not linked to the nerve terminal membrane in any special way; the terminal is regarded as the primary transducer element. The intralaminar spaces are filled with extracellular fluid and contain collagenous fibers. The corpuscle is turgid and scarcely compressible, so that a mechanical stimulus to its surface can only affect the nerve ending by a differential displacement of corpuscular elements.[31,45]

A pacinian corpuscle and its innervating axon can be isolated for study in the manner shown in Fig. 11-1, *A*. The upper records in Fig. 11-1, *B*, show that weak mechanical stimuli elicit a local change in membrane potential that, if of sufficient amplitude, leads to regenerative depolarization at the first node of Ranvier and a conducted action potential in the stem axon. This local event is the generator potential, a change caused by the local transmembrane flow of ionic current. It has the following properties: (1) It is generated in the nerve terminal itself, not in the cells of the sensory organ. (2) It can be graded in amplitude by stimuli of different magnitudes. (3) It is local, spreads electrotonically over the membrane, and is not conducted. (4) It can be summed, both spatially and temporally, which means that generator response set up by two weak stimuli delivered simultaneously to two spots or sequentially to the same spot may add to depolarize the nodal membrane to the threshold level. Thus the generator event has many properties in common with the end-plate potential of muscle cells and with the local excitatory postsynaptic response of nerve cells.

The equilibrium potential of the generator event is about zero,[46] which suggests that the permeability change initiated by mechanical stimulation is not restricted to Na^+; indeed, removal of Na^+ from the external solution reduces the generator potential to about one third its control level, but does not eliminate it.[62] The Na^+ conductance mechanism of the stem axon and the nerve terminal appear to differ somewhat, for tetrodotoxin in low concentration poisons the regenerative Na^+ conductance mechanism of the stem axon and eliminates the action potential, but affects the generator potential only slightly.[48] The local concentration of K^+ is 2 to 3 times higher in the intracapsular fluid surrounding the nerve terminal than in blood plasma, but what role K^+, Ca^{2+}, or other small ions may play in the production of that portion of the generator current not carried by Na^+ is unknown. It is possible that the high concentration of K^+ around the nerve terminal contributes to its sensitivity.[30]

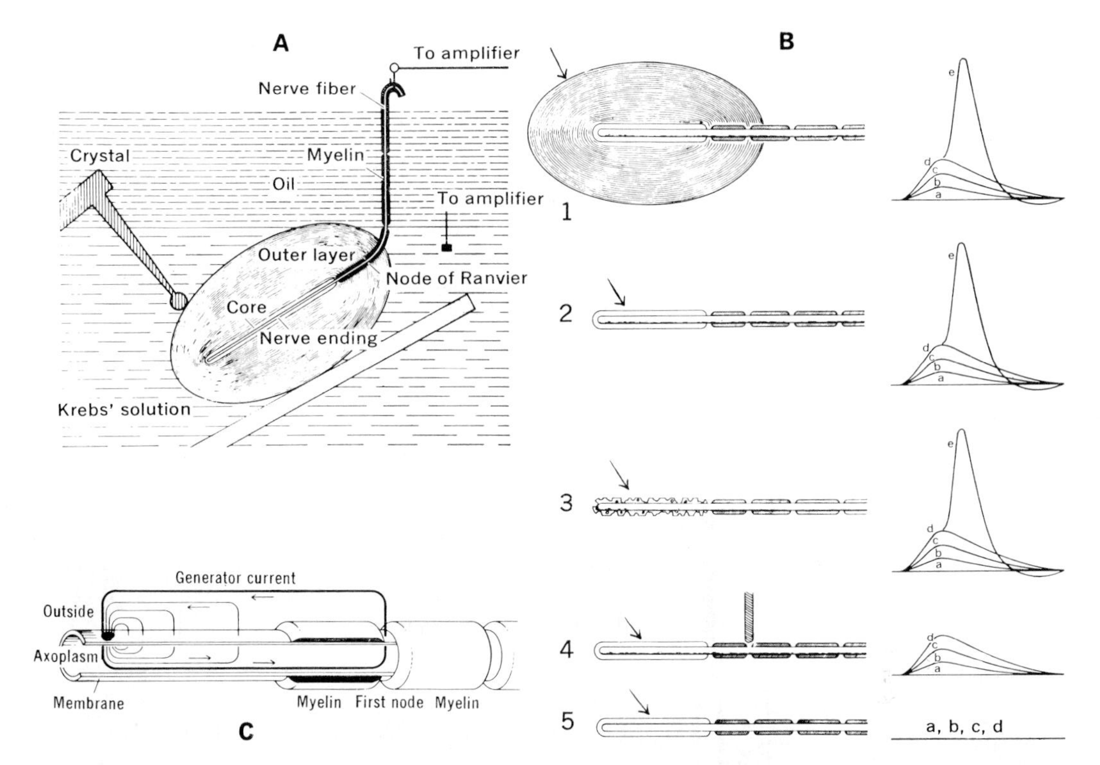

Fig. 11-1. **A** illustrates method of stimulation of pacinian corpuscle after it and its axon have been dissected free of cat mesentery. Recordings of generator potential and action potentials were made with one electrode on axon and the other in volume conductor surrounding corpuscle. Recording electrode is occasionally placed closer to corpuscle for better observation of generator potentials. Glass stylus attached to Rochelle salt crystal moves when crystal is activated by short electrical pulses. **B,** *1,* Records *a* to *d* show increasing generator potential produced by successively stronger stimuli that for *e* produced generator potential that reached firing level for axon, resulting in conducted action potential. Similar sequence for *2* is unchanged after removal of all outer lamellae of corpuscle or, *3,* after bits of inner core have been removed. In *4,* pressure at node of Ranvier blocks production of action potential without affecting generator process. Records in *5* illustrate that mechanical stimulation to decapsulated core produces no response when axon has been caused to degenerate by section several days earlier. **C** indicates concept of local nature of change in membrane permeability and resulting flow of generator current reaching first node of Ranvier. (Modified from Loewenstein.[42a])

Over a considerable range the relation between stimulus strength and the amplitude of the generator response is linear[48] (Fig. 11-2).

There is no evidence that a special chemical event intervenes between the mechanical stimulus and the resulting changes in permeability of the nerve terminal membrane: the terminal is the primary transducer.[67] The membrane of the stem axon is by comparison to that of the terminal virtually insensitive to mechanical strain. The capacity of the nerve terminal membrane to respond to even minute mechanical deformations with generator events is an example of the specialization of function of a sensory nerve ending.

The threshold level of the nodal membrane for the initiation of action potentials in the stem axon is nearly constant over a temperature range of 12° to 40° C.[32] Charge transfer through the receptor membrane of the pacinian nerve terminal is, on the other hand, markedly affected by temperature changes; the amplitude and rate of rise of the generator potential increase with temperature with Q_{10}s of 2.0 to 2.5. Nevertheless, Lowenstein[43] showed that no generator changes and no conducted action potentials are produced in a pacinian afferent by even very large and rapid changes in temperature. Thus although the charge transfer process is sensitively dependent on temperature, the ending is a highly specific mechanoreceptor.

It seems likely that the sensory nerve terminal

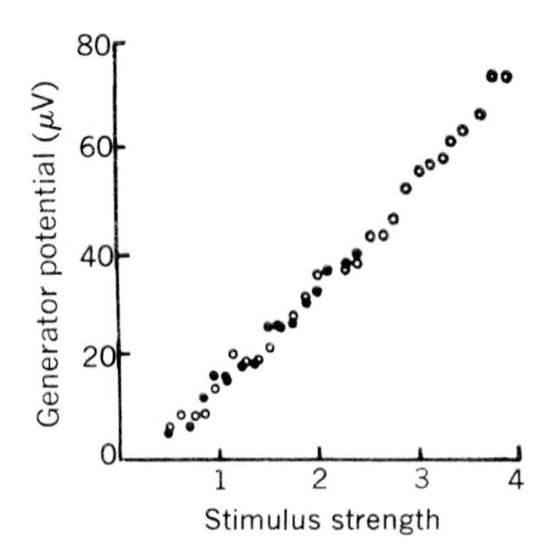

Fig. 11-2. Relation between generator potentials and strengths of mechanical stimuli evoking them for pacinian corpuscle before (filled circles) and after (open circles) block of action potential generation by application of tetrodotoxin, a toxic substance obtained from puffer fish. Relation is approximately linear over considerable range of stimulus strengths. (From Loewenstein et al.[48])

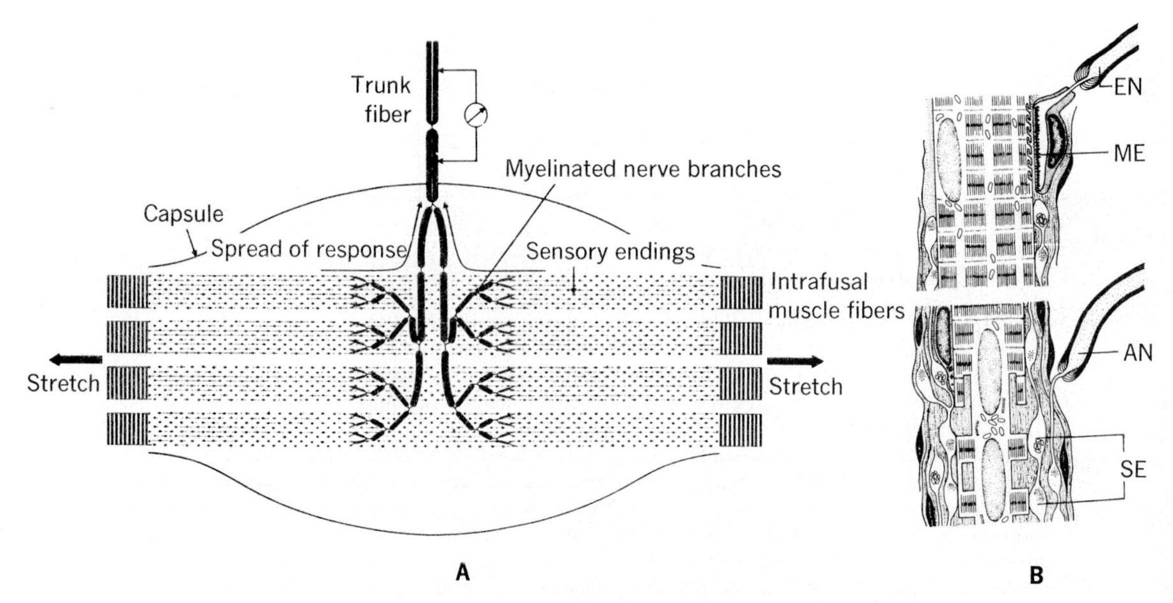

Fig. 11-3. A, Schematic diagram of branching of afferent axon within frog muscle spindle on assumption of six orders of dichotomous branching. In this model there are eight parallel rows of 50 end-bulbs on each intrafusal fiber. This gives a conservative estimate of number of end-bulbs in even small muscle spindle. **B,** Imaginary drawing of frog muscle spindle illustrating some aspects of its longitudinal structure in different zones. *AN,* Afferent nerve fiber; *SE,* sensory nerve endings; *EN,* efferent motor nerve fiber; *ME,* motor nerve endings. (**A** from Ottoson and Shepard[54]; **B** from Karlsson et al.[36])

of the pacinian corpuscle can support both the local, nonconducted generator event and the regenerative Na^+ conductance of the action potential, for dromic impulses can, under certain circumstances, arise in the terminal segment, and antidromic ones invade it.[52] It appears, however, that under normal circumstances, impulses are initiated at the first node of Ranvier. This would occur if the depolarization threshold for the action potential is lower there than in the terminal membrane. In this case the first node would function much as does the initial segment of the motoneuron (Chapter 6). All available facts suggest that the generator process occurs in local patches of nerve terminal membrane that are not electrically excitable, that is, they cannot support a regenerative change in Na^+ conductance, and that these areas are interspersed with areas of membrane that can. The nerve terminal and its stem axon illustrate the two types of electrogenesis of which nerve cells are capable. One is all or none and explosively conductile, and the other is local, gradable, summable, and not conducted but rather electrotonically extended.

Transducer mechanism of a slowly adapting receptor: the muscle spindle[4]

The peripheral terminals of many other afferent nerve fibers respond to long-lasting stimuli with persisting generator responses that elicit repetitive discharges of nerve impulses in their stem axons. Such a fiber may discharge initially at a frequency sensitively determined by the rate of stimulus application. Following this *onset transient,* the discharge declines to a more or less steady state determined by stimulus magnitude. Obviously the generator processes at the terminals of quickly and slowly adapting afferents differ in some important way. Information concerning those in slowly adapting receptors has come in large part from study of the stretch receptors in muscle.

Muscle spindles are complex sensory organs that vary greatly in structure between species. They commonly contain several specialized muscle cells, called intrafusal fibers, and in many species, including mammals, contain two different muscle cell types that receive different efferent and afferent innervations. The role of these complex receptor organs in the regulation of tone and movement is described in Chapter 22. The simpler spindle organ of frog muscle has been a favorite object for the study of sensory transduction, for it can be isolated by dissection and maintained in a controlled ionic environment, the spindle stretched for exact distances at controlled rates, and records made of the electrical signs of both local and conducted neural events. Fig. 11-3, *A,* shows a frog spindle with several intrafusal muscle fibers, and its single large afferent fiber branching many times before terminating on the surfaces of the muscle cells in a series of bulbar swellings connected by thin axon segments. These latter are shown in Fig. 11-3, *B,* made from an electronmicrograph. The bulbous swellings are closely applied to the surface membranes of the muscle cells, but there are none of the special characteristics of a synaptic relation between the two.

It has been known since the experiments of Adrian and Zotterman[7] that stretch of the frog spindle evokes action potentials in the afferent nerve and that this repetitive discharge declines very slowly in frequency with maintained stretch. It was Katz[37] who discovered events interposed between the mechanical deformation of the stretch and the conducted nerve impulses in the stem axon, namely, a local depolarization of the nerve endings and brief impulselike activity thought to be elaborated in the small myelinated segments of the nerve within the cap-

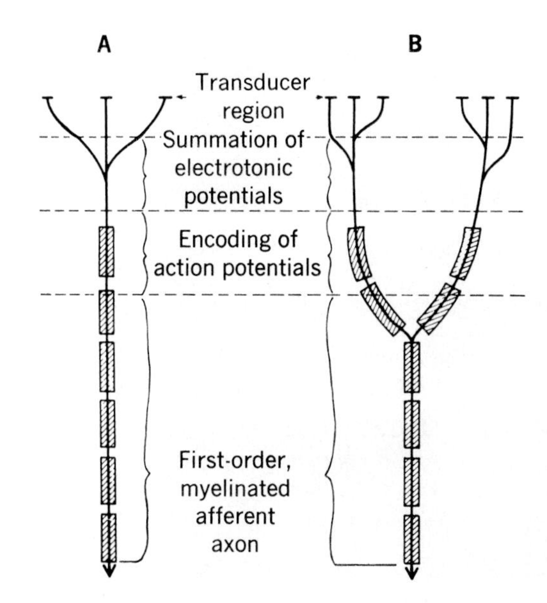

Fig. 11-4. Schematic diagrams illustrating alternative innervation patterns of afferent fibers. **A,** Innervation with single functional encoder. **B,** Innervation with two separate encoders. Regular discharges commonly observed in afferent fibers with two (or more) encoding sites is explained on the assumption that action potential initiated at any one encoding site will invade all others, thus "resetting" them. (From Eagles and Purple.[15]).

sule.[53,54] The generator or receptor potential spreads electrotonically along the axon membrane to a trigger or encoding site in the parent axon, thought to be at the first node of Ranvier (Fig. 11-4). The frequency of discharge is a linear function of the degree of stretch, both during the dynamic phase of muscle lengthening and during steady extensions of the muscle at different lengths (Fig. 11-5, *B*). Thus the length of the spindle at any instant is signaled by the instantaneous frequency (the interspike interval) and the velocity of the applied stretch by the rate of increase of that frequency. The generator, revealed by the blocking action of local anesthetic on the action potential mechanism (Fig. 11-6), is graded by the velocity and amount of stretch as well. Its dynamic and static phases are related in a linear way to the velocity and amount of stretch.[4] When small stretches are delivered, brief and sometimes abortive action potentials may be recorded from a spindle; they are thought to arise in the short unmyelinated segments of the axonal branches within the spindle. These impulses are confluent at the node of Ranvier. With greater degrees of stretch the more intense generator potentials appear to govern the prepotential–action potential sequence at the

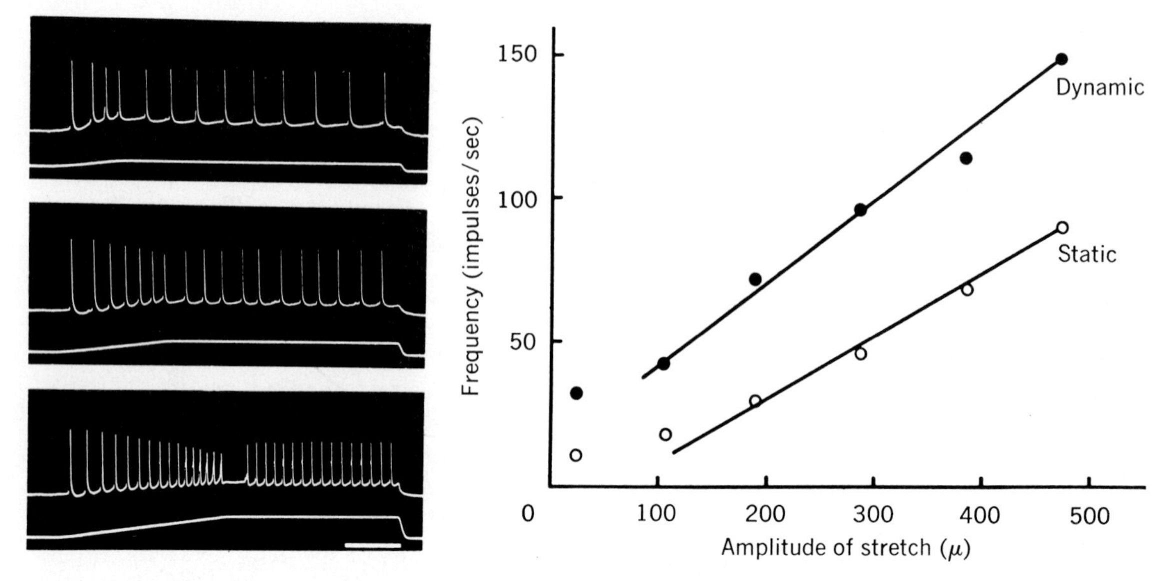

Fig. 11-5. Left, Action potentials recorded from afferent nerve fiber innervating frog muscle spindle, evoked by linearly increasing stretches at rate of 5 mm/sec, to three different amplitudes, as shown on lower traces. Graphs to right show that there is linear increase in peak frequency of discharge with amplitude of stretch for both dynamic and static phases of discharge. With greater velocities of stretching, slope of relation for dynamic discharge is steeper. (From Ottoson and Shepard.[4])

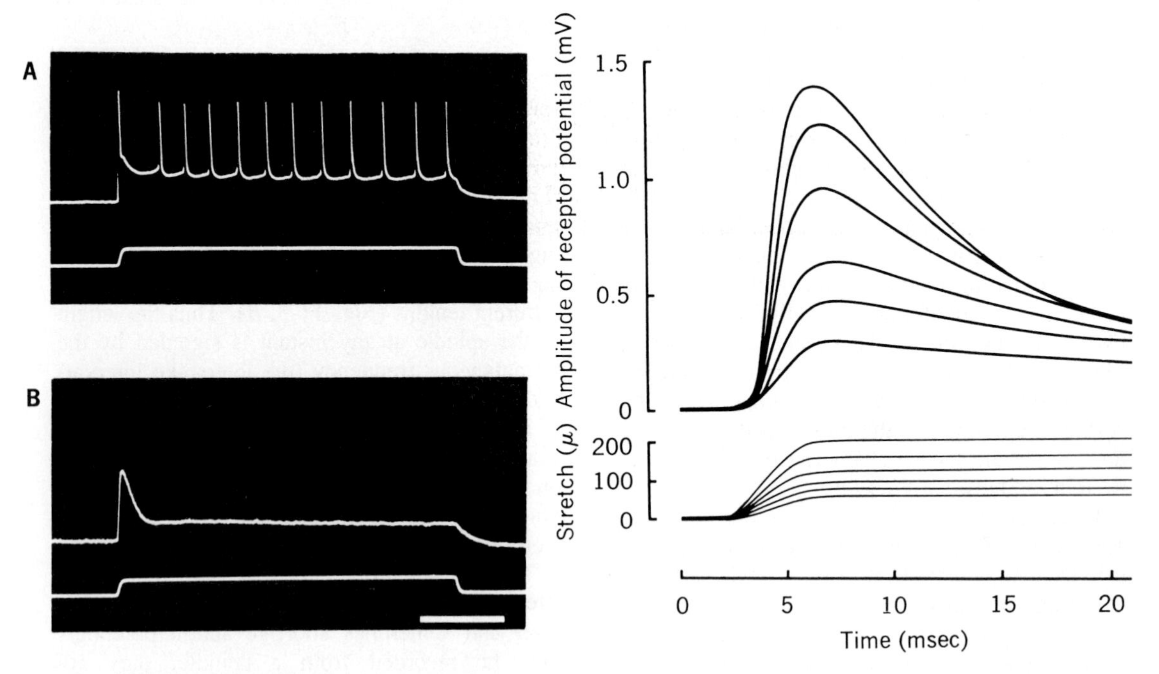

Fig. 11-6. Left, Response of isolated frog muscle spindle to stretch of 400 msec duration before, **A,** and after, **B,** treatment of preparation with local anesthetic at concentration that differentially blocks impulse initiation, leaving receptor potential isolated, as shown in **B.** Receptor potential shows initial dynamic phase developed during period of stretch, and a following steadily maintained depolarization of ending during steady stretch. Time bar in lower left of **B,** 50 msec. Right, Upper records show dynamic phases of receptor potential elicited by stretches of spindle to different lengths after impulse initiation has been blocked with local anesthetic. Over a considerable range, amplitude of dynamic phase of receptor potential is linear function of degree of stretch, with slope determined by velocity of stretch. (From Ottoson and Shepard.[4])

node of Ranvier more directly. The small prepotentials linking the maintained generator potential and each action potential can be seen in Figs. 11-4 and 11-5. Many of these features of the physiology of the muscle spindle have been observed also in dissected, isolated spindles from mammalian muscle.[2,25]

The mechanism of mechanoelectrical conversion in the muscle spindle is unknown. The most likely explanation is that the nerve endings are deformed directly, leading to a conductance change and the flow of depolarizing current across the terminal membrane. That current is carried largely but not exclusively by Na^+, for 25% to 30% of an evoked generator potential persists after prolonged immersion of a spindle in an Na^+-free solution. The permeability change produced by the mechanical event is thus not restricted to Na^+, although the role of other small ions is uncertain. The Na^+ channels of the terminal must differ from those of the stem axon, for they are unaffected by tetrodotoxin, which abolished the conducted action potentials in the stem axon by block of the Na^+ channels at concentrations that leave the generator potential virtually unaffected.

How is it that the afferent trains of impulses in the parent axon of the spindle are so regular, even though there may be two or indeed many local encoding sites? The simple situation with one encoding site is shown in Fig. 11-6, *A*, where summation of the electrotonically conducted generator potential occurs at the first node of Ranvier. Analysis of cases with two or more encoding sites (Fig. 11-6, *B*) shows that the mean rate will be increased (for a given stimulus) and will indeed be very regular in the parent axon on the assumption of a "simultaneous reset" mechanism,[15] that is, an action potential at one encoding site invades antidromically and resets all other encoders. Some such mechanism must operate to produce the regular discharges in the parent axon, for it is well known that if a number of independent trains of regular events of different frequency are combined with only the interaction of refractoriness, the resulting train will approach a random time series, even with as few as three confluent trains.

Sensory receptor adaptation and nerve accommodation

The dynamic (frequency) ranges of mechanoreceptors are set by the mechanical properties of tissue about the nerve endings and by the accommodation of the axonal membrane to generator current. The lamellated outer portion of the pacinian corpuscle, for example, acts as a mechanical high-pass filter composed of both viscous and elastic elements.[24,44] High-frequency compressional forces are transmitted directly to the central core, with a time course reflected in that of the receptor potential. The steady component of an applied compression is transmitted to the nerve terminal with a marked attenuation. On impulsive release of steady compression this stored force is transmitted from stretched elastic to viscous elements, producing a second dynamic event, the "off" generator response shown on the left in Fig. 11-7. After removal of this mechanical filter by microdissection, steady compression of the decapsulated central core elicits a maintained generator potential (Fig. 11-7, right). Nevertheless, a quasi-steady generator potential of even 5 times threshold evokes only one or two impulses at the first node of Ranvier. This quickly accommodating property of the nodal membrane and the high-pass mechanical filtering action of the corpuscle itself are complementary and together account for the fact that pacinian afferents are selectively sensitive to sinusoidal mechanical stimuli with a best frequency at about 250 Hz, at which stimuli of only 0.1 to 0.2 μm peak-to-peak amplitude elicit one impulse in phase with each sine wave cycle.[61,66]

In slowly adapting receptors such as the muscle spindle, no mechanical filter exists and maintained generator potentials are produced by steady stretch. There is, however, a marked adaptation from the onset transient to that steady level (Fig. 11-5, *B*), which is due in part to the fact that the tension within the spindle itself declines along such a time course during the early period after a quick stretch[26,27] and in part to changes in the local ionic environment of the

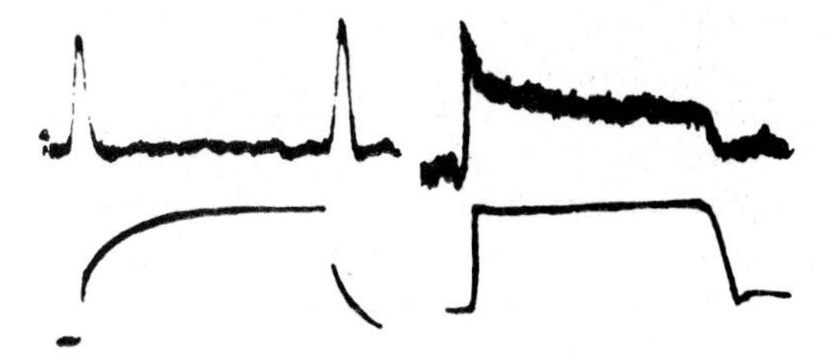

Fig. 11-7. Records to left illustrate generator potential (upper trace) evoked in nerve ending of intact pacinian corpuscle at application and release of mechanical compression pulse approximately 50 msec in duration, indicated by lower trace. Records to right were obtained similarly, but after removal of outer lamellae of corpuscle. Generator on-response in second case declines slowly, but is maintained throughout duration of stimulus. (From Loewenstein and Mendelsohn.[47])

nerve terminal during stretch.[28] It is not due to accommodation of the nodal membrane of the stem axon, for passage of a steady current through the node produces an almost instantaneous rise to a regular discharge rate, then maintained.[42] Mechanoreceptors of different types differ in the relative importance of mechanical and membrane properties in determining their rates of adaptation to steady stimuli.

In summary, for many receptors a local event intervenes between a stimulus to a sensory receptor and the discharge of nerve impulses in its stem axon. The local event is a transfer of ionic charge across the nerve membrane, a depolarization called the generator potential. This generator event is not conducted but invades adjacent regions of the parent axon by electrotonic extension. It can be summed both temporally and spatially, can be graded by stimuli of different magnitudes, shows no refractory period, and when the depolarization in spike-generating membrane reaches threshold, conducted action potentials result, sometimes via the intermediation of local prepotentials. The differences in the rates of adaptation, seen most notably in mechanoreceptors, are due both to the special properties of the nonneural elements of the receptor organs, which in some cases function as high-pass filters, and in varying proportions to the fact that the spike-generating membrane of the stem axons of different receptors differ greatly in the rates at which they accommodate to generator current.

COMMUNICATION AND CODING IN THE NERVOUS SYSTEM

The cellular elements of the nervous system are interconnected in divergent-convergent arrays of different complexity. The ratio numbers of these relations determine to some extent the functional properties of a neural subsystem within the brain, for example, its synaptic security and thus its frequency-carrying capacity. The business of this network is communication; it receives, may transform, stores, retrieves, and transmits information—a word used here in its general, not Shannon, sense. This is achieved by successive processes of encoding, transmission, readout, and reencoding. Something akin to computation occurs in each nuclear region, where trains of impulses arriving over several or indeed many channels combine with the intrinsic properties of the recipient neurons and the local neural network to determine the nature of the output: integrative action occurs. Thus the problem of neural coding includes both the coding of sensory events into impulse patterns in afferent nerve fibers[3] and the transmission and modification of neural signals between different populations of cells within the nervous system. Where these are even a few millimeters apart, impulse codes must intervene, but in local neuronal circuits, communication occurs in a number of ways without the intermediation of conducted action potentials.[5]

Code of the labeled line

The most firmly established of all neural codes is one independent of the mode of transmission of information, for it depends only on which particular set of fibers or neurons is active and not on the temporal pattern of the signals. Once the identity of the active line is known, the information stored within the system tells the meaning of the activity. An example is the modality specificity of sensory afferents, but the concept is generalized to apply to any neuron anywhere, sensory or not. It includes also the proposition that the locus of activity within the brain carries information. For example, increases in activity within the visual cortex lead solely to perceptions of light and no other; the locus of that activity within the visual cortex determines the sector of the visual field to which the sensation of light is referred.

It is important to emphasize that the information carried by a given neural line may be embedded in more than one code. Activity of any sort within a labeled set of mode-specific afferents always evokes a qualitatively specific sensory experience, but different impulse frequencies in those specific fibers may signal different intensities of that sensation.

Impulse codes[55,64]

No single impulse code is universal, but the number of candidate codes is not large. The identification of what is to be regarded as a neural impulse code is an empirical problem: by what criteria can the validity of a code be established? First, the neural impulse pattern must be shown to occur within the brain or its afferent connections under natural conditions, for example, be evoked by natural sensory stimuli. Second, it must be shown that there exists a set of neurons whose activity changes in response to the coded signals received over connecting neural channels. In practice this is frequently studied by recording the activity of central neurons activated by sensory stimuli or in relation to naturally occurring movements, or by measuring a change in behavior taken to infer a change in the activity of some sets of central neurons. For example, if a uniquely identifiable sensory experience is evoked only when a certain temporal pattern of a stimulus is impressed on a certain set of afferent fibers and that temporal pattern is identified in the afferent volleys and in the central neural activity evoked by those stimuli, the temporal pattern may be accepted as a naturally occurring neural code.

Rate or frequency codes. Rate or frequency codes include the common frequency code just referred to, that is, the coding of the magnitude of a sensory stimulus or the level of a presynaptic input by the frequency of the output neural discharge (for an example, see Fig. 12-13). It is assumed that this frequency is then averaged over some short period of time compatible with the integrating time constants of target neurons. Thus this code seems suitable for the slowly but not the quickly adapting first-order afferents and similarly for the sets of central neurons on which the two classes of fibers project. Frequency modulation of a quasi-steady rate of discharge is one variant of the basic frequency code.[68] In multistable systems the rate code may provide an all-or-none signal, with frequency above or below a certain level setting in one or the other system state.

Sequential order codes.[55] The time occupied by a nerve impulse is usually a small fraction of the time between impulses, so that the impulse train can be regarded as a point process. Thus some of the information transmitted may reside in the sequential order and the lengths of the time intervals between impulses. Obviously the carrying capacity of a line is increased if the system can encode and read out different micropatterns at each mean rate. This aspect of neural coding has been studied intensively, but no code of this general nature has been established with certainty, except for the limiting cases. These are the simple "go" command signals, for example, of a motoneuron, and the signal of the frequency of a vibrating stimulus by the periodic activity entrained by it in primary afferent fibers.[66]

Coincidence gating codes. Coincidence gating codes, and their reciprocals produced by inhibition, depend on the axiom that a particular coincidence or locked sequential pattern in two or n converging elements, or, in a more general sense, the temporal relations between impulses in adjacent lines, will transfer specific information by virtue of that temporal relation. An elegant example of this type of code has been discovered by Rose et al.[60] for some central neurons of the auditory system of the cat. This cell population is thought to signal the position of sound in space, using for that purpose the differences in latency of excitation/inhibition signals arriving from the two ears—at the microsecond level—to code in the population pattern of discharge the azimuth of the stimulating sound (Chapter 14).

Distributed or ensemble codes.[18] Distributed or ensemble codes are those in which information is transmitted by the profile of activity in a set of afferent fibers or the neurons of a central field, including both the distribution of levels of activity and the times of occurrence of impulses in members of such a neural population. For example, although the *frequency* of a vibrating stimulus delivered to the palmar skin of the hand is coded in the periodic pattern of discharge in a certain set of mechanoreceptive afferents,[66] the *amplitude* of the vibration is coded by the total activity in the population of afferents engaged by the stimulus.[35] It is likely that the spatial contours of stimuli are coded in a similar way in the distributed profile of activity in neural populations.

Nonimpulse codes

Nerve cells interact in many ways that do not involve spike electrogenesis, particularly in local networks of cells with short axons where transmission distances are short.[57] Changes in membrane potential in one part of the cell can influence by electrotonic extension chemically operated synapses in other parts. Cells functioning in this way without conducted action potentials have been identified in the retina[13] and in the olfactory bulb.[65] It is surmised that some of the short axon cells of the cerebral cortex may function similarly. Moreover cells in some parts of the central nervous system (CNS) are linked via electrically operated synapses (Chapter 8) through which electrotonically transmitted changes in one cell can influence the excitability of the other. Lastly, where neural processes are tightly packed, as in the neuropil of local circuits, purely ephaptic interactions between cell processes may be of importance in signal transmission and thus serve as codes.

Slow electrical events. Both the PSPs and the action potentials of central neurons are associated with extracellular current flows in brain tissue. The potential differences between loci caused by these current flows can be recorded locally or from the surface of the brain or head; in the latter case the record is called an electroencephalogram (EEG) (Chapter 7). The components of the currents generated by active neurons penetrate the membranes of others, may condition their excitability, and thus to some degree function as signals. A great deal of study has been devoted to the proposition that those summed potential changes recorded from masses of brain tissue are themselves carriers of information, particularly in the study of conditioning, learning, and memory.[34] Whether the EEG is indeed an active signaling agent or the summed reflection of the activity of large populations of neurons controlled in other ways is still an open question, but the latter alternative seems more likely.

Molecular transfer as a signaling mechanism. The rapid bidirectional movement of large molecules not only in axons but also in dendrites and cell bodies of neurons makes possible the transport of substances from any part of the cell to any of its sites of contact with other cells. The evidence is clear that molecules such as peptides pass readily from one neuron to another or to target effector cells, both at synapses and at less specialized cellular interfaces. This molecular cell-to-cell transport is a powerful means of information transmission and undoubtedly is the basis for the trophic relation between neurons and between neurons and effector cells described in Chapter 8.

PRIMARY AFFERENT FIBERS, RECEPTORS PROPERTIES, AND SOMATIC SENSATION

Five different qualities of sensation can be evoked by stimualtion of body tissues: touch-pressure, warmth, cold, pain, and the sense of position and movement of the joints. The first four are best developed in the skin, although not completely restricted to it. Each is served by a specific set of cutaneous afferent fibers. The fifth depends on activity in afferents innervating the ligaments and capsules of the joints, with the added possibility that muscle afferents may also contribute to this sense of kinesthesis. More complex sensations are synthesized within the CNS system from the activity in different sets of nerve fibers, each of which, when acting alone, evokes a single primary sense quality. The perception of the size and shape of an object grasped with the hand is a case in point. Other varieties of somesthesis depend on specific quantitative or temporal aspects of the activity in a homogenous set of fibers, such as the sense of vibration evoked by tuned, periodic activity in the pacinian afferents. Other temporal patterns in these very same afferents may elicit only a sense of mechanical contact. Humans identify the elementary sense qualities when appropriately simple and local stimuli occur, but more commonly we compose a wide range of sensory experiences by a central synthesis of afferent input in several sets of fibers.

The existence of these elementary qualities of somatic sensibility is proved in a simple experiment. Blix discovered that there are in the skin small local regions of differential sensitivity for each of the four elementary qualities of somatic sensiblity, interspersed in a manner that varies from place to place and between individuals. It is particularly clear in regions of low innervation density. For example, on the hairy skin of the forearm one can identify touch, warmth, cold, and pain spots using appropriately restricted mechanical or electrical stimuli. Indeed, no matter how such a spot is stimulated, as long as the stimulus is local to it, only one elementary sensory experience is elicited, although its derived variations may be evoked by temporal or intensive changes in the stimulus (e.g., touch to flutter by increasing frequency or itch to pain by increasing intensity).[8] The temporal or intensive variations of a purely local stimulus do not produce shifts in the modality of sensation, such as from touch to warmth or cold to touch.

Fiber diameter and modality type

Afferent fibers of different sizes do not occur in equal numbers in peripheral nerves.[9,17] The distribution of fiber sizes in one human nerve is shown by the histogram of Fig. 11-8, together with the compound action potential to be expected in such a nerve after conduction for several centimeters from a site at which all its myelineated fibers were excited synchronously by a brief electrical stimulus. Fig. 11-8 shows the large number of myelinated A-beta and A-delta fibers in the nerve, but does not reveal that cutaneous nerves also contain large numbers of unmyelinated C fibers, of which 85% to 90% are dorsal root afferents and the remainder sympathetic postganglionic efferents. The ratio of C fibers to A fibers varies from 5:1 in nerves innervating proximal parts to 1:1 in nerves innervating face and hand.[58] Those body regions in which sensation is most acute receive the highest proportion of innervation to myelinated fibers.

Two facts of importance with regard to peripheral neural mechanisms in somesthesis have now been established beyond reasonable doubt. The first is that although fiber classes of different sizes serve different sensory functions, there is sometimes a complete overlap in the fiber size distributions of some classes with diffrent function. Size/conduction velocity does not define uniquely a modality-specific class of fibers. The

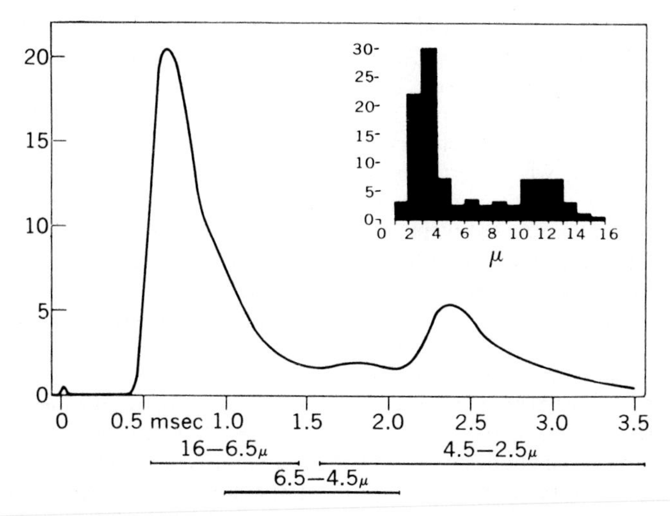

Fig. 11-8. Form of compound action potential in human sensory nerve (medial cutaneous) calculated from distributions of fiber diameters shown in inset histogram, which plots numbers of fibers for each diameter.[51] Data were recorded as fiber diameters but converted to axon diameters for calculation of reconstruction. Ordinate: relative amplitude of expected compound action potential. Abscissa: time scale; latency is that expected after 4 cm of conduction; micron scale for fiber, not axon, diameters. Construction made on basis of conversion factor of velocity in meters per second equals 9.2 times axon diameter in microns. Largest afferent fibers of peripheral nerves, group I, are not present in cutaneous nerves. (From Gasser.[17])

second is that each single afferent fiber is modality specific, even though fibers of different specificity may be of the same size. Evidence supporting these statements has come from several lines of investigations, which will be considered here.

Sensations evoked by electrical stimulation of human nerves. Electrical stimulation of nerve trunks brings to action nerve fibers in a descending order of size as stimulus strength is increased. When this is done with direct monitoring of the afferent volleys evoked by the stimuli,[22] it is clear that those stimuli that excite only the beta fibers elicit in waking human subjects only sensations of mechanical events. The different types of mechanoreceptive experiences cannot be differentiated by electrical stimulation of nerve trunks, for there is a nearly complete overlap of the sizes of the fibers serving them. When the stimulus strength is increased to include the A-delta fibers in the afferent volley, subjects report the added sensation of fast pricking pain; when the stimulus is sufficiently strong to activate the C fibers, a sense of burning pain is felt, with a delay caused by the slow conduction in C fibers—hence the terms "slow" or "second" pain. Cool and warmth are known from other evidence to be served by A-delta and C fibers, respectively, but they are seldom identified separately in nerve trunk stimulation experiments, undoubtedly because of the dominating nature of the accompanying painful experiences. Thus this technique allows only a crude differentiation between evoked sensations and reveals little about the function of fiber groups of the same size but with quite different functions.

Remaining sensory capacities in humans after differential block of different fiber groups in peripheral nerves. When local anesthetics in low concentrations are infiltrated into and around a cutaneous nerve, it is possible to show by electrical stimulation on one side of the block and eletrical recording on the other that C fibers are blocked first and then, after a delay, the A fibers are blocked in an ascending order of size. The first sensation lost along with the C fiber component of the afferent volley is second pain, followed then as the block progresses by warm, pricking pain, and cool as the A-delta fibers succomb. Touch-pressure sensibility survives until the very largest myelinated fibers are blocked. This sequence and the association between fiber group blockade and the sensations lost have been confirmed during direct recording with electrodes inserted into the peripheral nerves of waking subjects.[69]

Block of a peripheral nerve by local asphyxia, as by local pressure, blocks A fibers first and then, after a delay of 30 to 45 minutes, the C fibers are blocked,[41] an order roughly the reciprocal of that produced by local anesthetic. Most components of touch-pressure are lost with block of the beta fibers. A vague sense of contact may remain after the beta block; it, pricking pain, and the low-threshold temperature sensibilities are lost as the A fiber block progresses to include the A-delta fibers.[40] At this time, when the C fibers appear to be conducting normally through the region of block, the remaining sensibility in the skin innervated by the partially blocked nerve is limited to second pain and the noxious extremes of heat and cold.

The results of these two sets of experiments suggest that those fiber groups that can be grossly separated by blocking serve different sensory modalities. Unequivocal evidence for this modality specificity has come from studies in which individual sensory nerve fibers have been observed directly.

Functional specificity of single afferent nerve fibers and sensory innervation of peripheral tissues

This and the following several chapters rely heavily on evidence obtained in studies of sensory and motor systems using the methods of single-unit analysis described in Chapter 9. Investigations of this sort have been made of peripheral nerve fibers in many species, including man.[23] Work in humans has accelerated since Hagbarth and Vallbo[21] discovered that the electrical signs of impulses in axons of peripheral nerves of humans could be recorded by passing a tungsten microelectrode directly through the skin and into a nerve trunk (Fig. 11-9). Under these circumstances in man, and with a number of equivalent methods in monkeys, information has accumulated concerning the conduction velocities, general functional properties, peripheral receptive fields, and differential sensitivities of each of the several classes of first-order afferents in peripheral nerves. The evidence has steadily mounted that regardless of the fiber class or receptor type considered, and no matter what the peripheral tissue innervated, the first-order sensory fibers of the somatic afferent systems of mammals are modality specific: they are differentially sensitive to one or another form of impinging energy. It is now possible to describe the properties of the mechanoreceptive afferents that innervate the glabrous and hairy skin in primates, although much work needs to be done to define precisely the parametric relations between stimulus and response and particularly to discover the mechanisms of peripheral transduction.

Mechanoreceptive innervation of glabrous skin in primates. The hands of monkeys and humans are similar in structure, innervation, and capacity as sensing agents, and the properties of the first-order cutaneous afferents innervating them appear to be identical.[20,33,38,51,66] Three

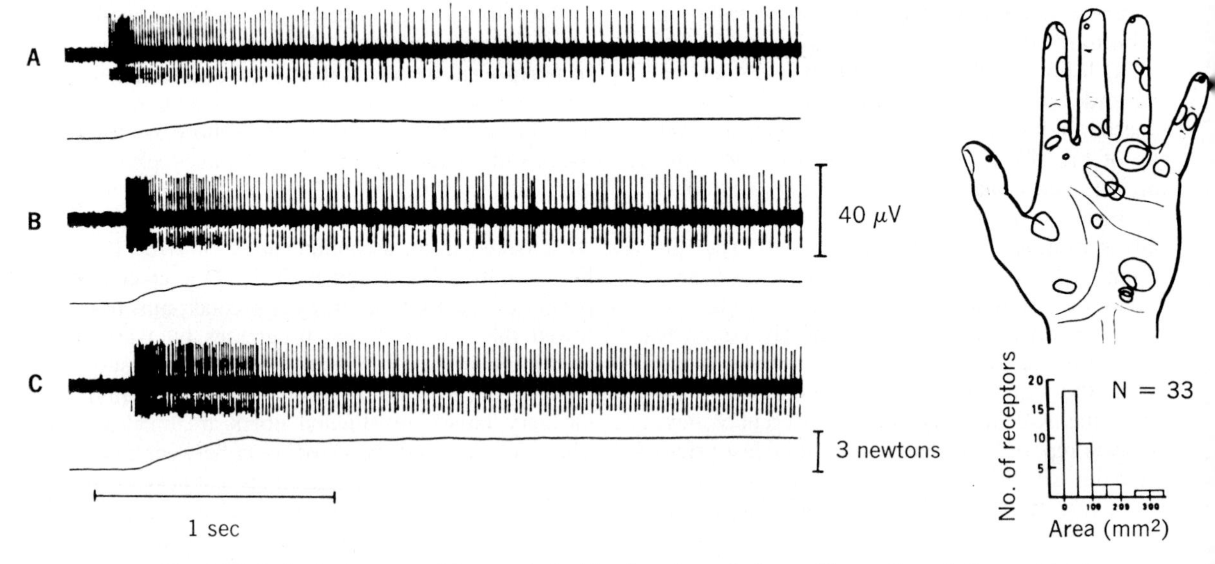

Fig. 11-9. Left, Action potentials recorded with microelectrode from afferent fiber of ulnar nerve of waking human subject. Fiber innervated receptive field on volar surface of fifth finger, middle phalanx. It adapted slowly to stimulating forces shown by lower traces of each set. Right, Drawings of peripheral receptive fields of a number of first-order afferent nerve fibers of median and ulnar nerves innervating glabrous skin of human hand. All adapted slowly to steadily maintained pressure delivered to receptive fields. (From Knibestol and Vallbo.[39])

classes of large mechanoreceptive afferents innervate the glabrous skin (Table 11-2). The Meissner afferents discharge a brief burst of impulses on a step indentation of the skin; the discharge adapts quickly to steady pressure and recurs in a brief burst on removal of the stimulus. They are selectively sensitive to low-frequency sinusoids, are best in the range of 30 to 40 Hz, and are thought to serve the submodality of flutter (Chapter 12). Pacinian corpuscles are actually located just beneath the skin, as shown in Fig. 11-10; they occur in many deeper tissues as well. They are, however, so sensitive to mechanical stimuli delivered to the surface of the skin that they are conveniently classed as quickly adapting cutaneous mechanoreceptors. The pacinian corpuscles are selectively sensitive to high-frequency mechanical sinusoids, are best at 250 Hz, and are thought to serve the submodality of vibration (Chapter 12). The Meissner and pacinian afferents are velocity detectors, and the pacinian fibers may be sensitive to even higher derivatives of position. The Merkel afferents respond to a step identation of the skin with an

Table 11-2. Peripheral afferent nerve fibers in primates and their roles in sensation

Type of skin	Fiber class, size, and conduction velocity	Peripheral termination	Differential sensitivity and adaptive property†	Sensory function
Glabrous	A-beta,* 6-12 μ, 35-75 m/sec	Merkel's cells	SA mechanoreceptors	Velocity and position detectors; sense of *touch-pressure*
		Meissner's corpuscles	QA mechanoreceptors	Velocity and perhaps instantaneous position; sense of *contact* and *flutter;* best at 30-40 Hz
		Pacinian corpuscles	QA mechanoreceptors	Velocity and perhaps higher derivatives of position; sense of *contact* and *vibration;* best at 250-300 Hz
	A-delta, 1-5 μ, 5-30 m/sec	Bare nerve endings	SA thermoreceptors	Sense of *cooling*
		Bare nerve endings	SA nociceptors	Sense of *pricking pain* and, at low frequencies, *tickle*
	C fibers, 0.2-1.5 μ, 0.5-2 m/sec	Bare nerve endings	SA thermoreceptors	Sense of *warming*
		Bare nerve endings	SA nociceptors	Sense of *burning pain* and, at low frequencies, *itch*
Hairy	A-beta,* 6-12 μ, 35-75 m/sec	Hair follicle apparatus	QA mechanoreceptor	Velocity detectors; sense of *contact* and *flutter;* best at 30-40 Hz
		Bare nerve endings, "field" receptors	QA mechanoreceptor	Velocity detectors; sense of *contact*
		Pinkus domes, Merkel cells	SA "type I" mechanoreceptor	Stimulation in human elicits no sensation
		Ruffini organs	SA "type II" mechanoreceptor	Velocity and position detectors; sense of *touch-pressure*
		Pacinian corpuscles	QA mechanoreceptor	Velocity and perhaps higher derivatives of position; sense of *vibration;* best at 250-300 Hz
	A-delta, 1-5 μ, 5-30 m/sec	Bare nerve endings	QA mechanoreceptor	Velocity; sense of *contact*
		Bare nerve endings	SA thermoreceptor	Sense of *cooling*
		Bare nerve endings	SA nociceptors	Sense of *pricking pain* and, at low frequencies, *tickle*
	C fibers, 0.2-1.5 μ, 0.5-2 m/sec	Bare nerve endings	SA thermoreceptors	Sense of *warming*
		Bare nerve endings	SA nociceptors	Sense of *burning pain* and, at low frequencies, *itch*
		Bare nerve endings	SA mechanoreceptors	Occur rarely in monkey skin and not present in human skin

*In earlier classifications these fibers were labeled A-alpha. They are, however, similar in size to the second group of afferent fibers in muscle nerves. Thus they are termed A-beta fibers here, which allows a correlation between the systems of classification as follows: group I = A-alpha; group II = A-beta; group III = A-delta; and group IV = C fibers.

†SA = slowly adapting; QA = quickly adapting.

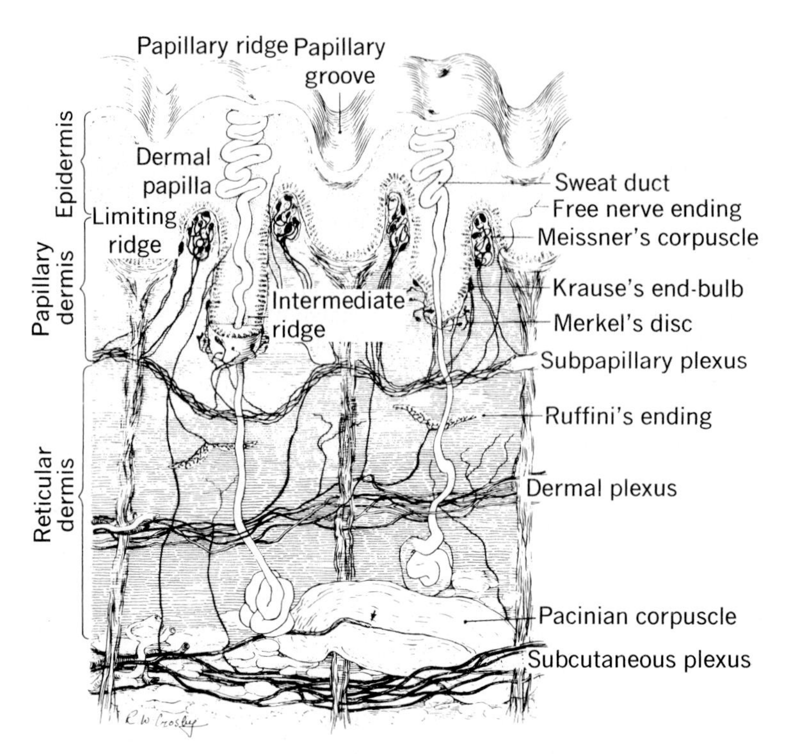

Fig. 11-10. Section of glabrous skin cut across papillary bridges. There are three major networks of nerves, corresponding generally to vascular plexuses: subcutaneous, dermal, and subpapillary. Deeper plexus contains more and larger myelinated fibers. Any single axon may travel in one plexus to its locus of termination or, less commonly, move from one plexus to another. Limiting ridge is tightly adherent to dermal collagen, fixed to underlying bone, and contains relatively few nerve endings. Intermediary ridge tends to lie free in papillary dermis and is relatively mobile; its inner surface is rich in nerve endings with tapered or expanded tips, Merkel's discs. Pacinian corpuscles lie in deep dermis or subcutaneous tissue. Meissner's corpuscles occupy uppermost reaches of papillary dermis. Other, less common, organized endings such as those of Ruffini or Krause may appear at any place in dermis. Tapered-tip or free endings occur throughout all layers of dermis and may cross basement membrane to terminate in lower reaches of epidermis. (Courtesy Dr. M. E. Jabalay.)

onset transient discharge, whose frequency is a function of the rate of indentation and then declines along a double exponential time course to a more or less steady rate of discharge determined by the degree of skin indentation (i.e., by stimulus magnitude). The majority of the "adaptation" appears to be due to changes in force at the stimulator-skin interface due to viscous flow in the tissue. These afferents are detectors of both velocity and steady positions. Records of the activity of such a fiber and a map of the receptive fields of a number of them innervating the human hand are shown in Fig. 11-9. The parametric relation of the discharges in these fibers to the stimuli that excite them are given in Chapter 12.

The activity of these three sets of mechanoreceptive afferents illustrates several of the neural codes described in an earlier section: (1) the simple on-off signal of a quickly adapting afferent that an event has occurred; (2) its periodic code of stimulus frequency; and (3) the rate or frequency code of the slowly adapting afferent that signals stimulus amplitude. Moreover, it is clear that the codes for stimulus movement, direction, speed, and contour must reside in the ensemble of activity in the population of mechanoreceptors engaged by the stimulus.

Small myelinated (A-delta) and C fibers sensitive to weak, noninjurious mechanical stimuli are common in the innervation of the furry skin of some mammals; they do not occur in that of the glabrous skin of the human or monkey hand.[14,56,70] The sense of touch-pressure and all its varieties in the hands of primates appears to depend almost exclusively on afferent input in larger myelinated fibers. When only delta and C fibers are intact and functioning, one can perceive pain and temperature but only the vaguest sense of mechanical contact with the hand.

Mechanoreceptive innervation of the hairy skin of primates. Large myelinated fibers that terminate in the hair follicle apparatus are mechanoreceptors exquisitely sensitive to movement of the hairs. They are thought to provide the relevant afferent input for the sense of flutter evoked from the hairy skin by low-frequency mechanical sinusoids (10 to 40 Hz) and in the sensing of mechanical contact and movement. The pacinian corpuscles lie just beneath the dermis, as well as in deeper tissues, and their afferents run in mixed nerves. Like the pacinian afferents innervating the glabrous skin, they provide the afferent signals for the sense of vibration, best at 250 to 300 Hz.[49] The large myelinated fibers that terminate in the skin between the hairs, presumably in bare endings, are also detectors of transient motion. Two types of slowy adapting afferents that innervate the hairy skin and terminate there in complex receptor organs have been identified in many mammals and in the hairy skin of monkeys and humans.

The first type (I) terminates in the Pinkus endorgan, a small (200 to 300 μ diameter) translucent elevation of the skin found most commonly on the trunk and abdomen, and in the proximal limbs. Its afferent fiber divides, and each branch terminates in a close (synaptic ?) relation with a Merkel cell of the basilar layer of the epidermis. Curiously, mechanical stimuli restricted to this sensory organ elicit no sensory experiences in man, even at stimulus intensities readily appreciated when delivered to the skin alongside.[49] It is therefore likely that the second slowly adapting afferent (type II), which terminates in the skin in a Ruffini ending, plays a role in the sense of touch-pressure in the hairy skin.[12,49]

The thermoreceptors and nociceptors of the A-delta and C fiber classes that inntevate the hairy skin are similar in properties to those innervating the glabrous skin. One difference is clear: no A-delta or C fiber afferents sensitive to weak, noninjurious mechanical stimuli innervate the glabrous skin. Afferents of this type occur in the innervation of the hairy skin of the monkey, but are more plentiful in the furred skin of carnivores and other species (for classifications, see Burgess and Perl[1]). It appears, therefore, that with the remarkable increase in the capacity of the hand as a sensing organ, there has been a very great increase in its innervation density and a shift in its pattern of innervation toward large, rapidly conducting myelinated afferent fibers. Moreover, there is a shift from monkey to man toward an increase in the percentage of the mechanoreceptive afferent fibers that are slowly adapting.[33,38]

Mechanoreceptive innervation of deep tissues other than muscle and joints in primates. It has been known for a long time that when a local patch of skin is anesthetized without affecting the underlying tissue, as by cocaine iontophoresis, the threshold for detection of pressure is increased several orders of magnitude. There is little direct evidence available concerning the mechanoreceptive innervation of the fascia and other deep connective tissue, but this finding indicates that it is sparse and that individual afferents have high thresholds. We are, in contrast, exquisitely sensitive to vibratory stimuli delivered to the deep tissues, particularly to bones and interosseous membranes, where pacinian corpuscles and their afferents are found in large numbers. Fascia and other deep structures are apparently provided with an ample nociceptive innervation (Chapter 13), but little is known of thermal sensitivities in these locations.

The *concept of population* is important for understanding CNS function. Consider, for example, the events that occur when even a local stimulus is delivered to the skin of the finger. It will fall at different positions in the overlapping receptive fields of a number of different fibers. Thus the degree of activity will grade from maximum for those engaged at their field centers to minimum for those engaged only marginally. The stimulus will activate a population of fibers, and that activity will be graded across this "neural space" in a manner that reflects the size and shape of the stimulus. Obviously, localization of a stimulus, identification of two points on the skin as two and not one, and recognition of more complex contours cannot be explained on the basis of activity in single afferent nerve fibers. A central detecting apparatus must read profiles of neural activity.

The relations between first-order afferent input, central nervous mechanisms, and sensory performance in the mechanoreceptive senses of touch-pressure, flutter-vibration, and kinesthesis will be discussed in Chapter 12.

In summary, there is a correlation between the sensations that are mediated and the diameters of the first-order sensory fibers mediating them. The A-beta fibers serve a number of sensations that can be grouped together as mechanoreceptive (touch-pressure, flutter-vibration, and position sensibility), whereas some cruder components of contact sensibility may be mediated by A-delta–sized fibers for some parts of the body, but not the hand. Pricking pain is mediated by fibers in the A-delta range, as is the sense of cool. Burning pain is mediated by C fibers, and those serving warmth are of C fiber conduction velocity. These statements deal with central tendencies, but the results of a large number of investigations of the properties of single afferent fibers, in both animal and human subjects, have now established beyond reasonable doubt that afferent fibers are highly specific in nature, that is, they are differentially sensitive to one or another form of energy. They are modality specific, labeled lines; the mechanism of this remarkable specificity is unknown. When active, each group elicits one elementary sensory experience or contributes that quality to combine with others in the compounding of complex sensations.

Morphology of receptors and its relation to sensation

The modality specificity of afferent nerve fibers raises the question of whether this depends on the special properties of the multicellular, nonneural receptor organs that in many cases encapsulate their peripheral terminals.[29] The discovery of sensory spots in the skin and the results of differential blocking experiments led morphologists to seek a structurally unique receptor organ to account for the differential sensitivity of different nerve fibers. A large number of different end-organs were identified, each thought to serve in this way a single sensory quality. It is now clear that the plethora of such identifications can be accounted for by the vagaries of the histologic methods used. Evidence accumulates that the organized endings of mammalian skin—the hair follicle apparatus, Merkel's discs and Meissner's corpuscles of the glabrous skin, Ruffini's endings, and pacinian corpuscles—are linked to mechanoreceptive afferents. Their function is to condition the dynamic response characteristics of these fibers rather than their modality specificity.

Anatomic studies suggest that the many varieties of sensory endings found in the skin of mammals, and in subcutaneous and deep tissues as well, can be lossely classified into three groups.[50] These are the bare nerve endings, endings with expanded tips, and encapsulated endings. In the epidermis and the papillary dermis of the human palm, for example, the three sets of nerve endings terminate in relation to certain skin structures (Fig. 11-10). Free nerve terminals end among the connective tissue cells of the papillary dermis as well as between the dermal cells; indeed, they penetrate almost to the stratum corneum. They are densely distributed throughout all layers of the skin. The expanded tip endings are closely related to the specialized epidermal cells of Merkel found only in the lowest layer of the stratum germinativum of the intermediate ridges. The encapsulated endings, the Meissner's corpuscles, are found in the dermal papillae, and in young persons nearly every cross section of a papillary ridge shows a pair. Each Meissner's cell receives the terminals of two to five large myelinated fibers, and each fiber branches to send terminals to corpuscles distributed over a receptive field of about 2 mm^2. The dermis contains encapsulated groups of endings with expanded tips in the Ruffini organs and those in pacinian corpuscles. These latter occur in large numbers in the deep tissues of the hand.

The form, number, and distribution of sensory receptor organs vary with age, region of the body, and even occupation. Meissner's corpuscles of the glabrous skin, for example, are rarer at birth than in childhood, more plentiful on hands than on feet, and decrease with advancing years, when many atypical and transitional forms occur.

In summary, the peripheral endings of myelinated fibers innervating the body tissues are differentially sensitive to one form of impinging energy, whether or not their terminals are encapsulated by sensory end-organs. Fibers with such organs are mechanoreceptors; the end-organs appear to determine the dynamic sensitivities of the nerve endings. A-delta and C fibers terminate as free endings in peripheral tissues, but these are no less specific with regard to modality, for certain sets are differentially sensitive to cooling, others to warming, and still others to stimuli that tend to damage tissue cells; these latter are nociceptors.

TROPHIC INTERACTIONS BETWEEN AFFERENT NERVE FIBERS AND CELLS OF THE TISSUES THEY INNERVATE[6,71]

There is a rapidly accumulating body of evidence that neurons influence one another or the peripheral target tissues they innervate by mechanisms other than the direct synaptic transmission of excitation or inhibition (Chapter 8). The influence of a central neuron on peripheral cells is an intrinsic property of the nerve cell and is independent of synaptic influence on it from other central neurons. This controlling, or trophic, influence depends on protein synthesis within the cell body and is undoubtedly exerted by axoplasmic transport of proteins or polypeptides from cell body to nerve terminal, from whence the influence is exerted on target cells by mechanisms still unknown but that certainly stimulate gene expression in that target cell. The trophic interactions between afferent neurons and the peripheral tissues they innervate appear to vary greatly for different types of sensory fibers. When gustatory afferents that innervate the tongue are sectioned, for example, the taste buds of the lingual mucosa degenerate rapidly. They are re-formed as the regenerating terminals of the gustatory afferents approach the mucosa, and in mammals, no other type of afferent can exert this effect on the mucosal cells (Chapter 18).

There appears to be a gradient between this type of total trophic dependence and total receptor organ independence among the various types of cutaneous afferents in mammals. The Merkel cells of the Pinkus dome receptor organ, for example, degenerate when denervated, and with time the entire organ disappears. That organ is re-formed and Merkel cells reappear if regenerating afferents approach the skin, but only in the specific locations in which those organs existed previously. In this case the receptor organ cells are trophically dependent on the neuron, but, conversely, the nerve fiber can evoke the formation of a specialized organ only from certain sets of skin cells. Moreover, afferent fibers of a "foreign" sensory nerve, sprouting into the denervated region, can elicit receptor organ formation, but only at those same specific loci.[10,11] Muscle spindle organs are somewhat more independent, for they degenerate only if denervated during the first few days of life in mammals. Later, transection of spindle afferent nerve fibers may produce some atrophic changes in the spindles, but they remain intact.

Pacinian corpuscles, in contrast, appear to be much more independent of the influence of their afferent nerve fibers. They do not degenerate when denervated and can be reinnervated by afferent nerve fibers of a foreign nerve known to contain no pacinian afferents.[63] A transducer mechanism appears on the terminal membrane of this regenerating foreign nerve fiber when it enters the denervated corpuscle, a reverse trophic influence of tissue cells on the nerve fiber.

Little is known about the trophic dependence of the specialized receptor organs of the glabrous skin of the primate hand on their innervation. When the hairy skin of the forearm is transplanted to the hand and reinnervated by afferent fibers that previously innervated glabrous skin, certain changes occur. The hair follicles and sweat glands disappear; the dermal ridges characteristic of glabrous skin do not appear, but the deeper layers of the epidermis assume the characteristics of volar digital skin. However, no Meissner's corpuscles are formed in the reinnervated skin graft.[59]

REFERENCES

General reviews

1. Burgess, P. R., and Perl, E. R.: Cutaneous mechanoreceptors and nociceptors. In Iggo, A., editor: Handbook of sensory physiology. Somatosensory system, New York, 1973, Springer-Verlag New York, Inc., vol. 2.
2. Hunt, C. C.: The physiology of muscle receptors. In Hunt, C. C., editor: Handbook of sensory physiology. Muscle receptors, New York, 1974, Springer-Verlag New York, Inc., vol. 3, pt. 2.
3. Mountcastle, V. B.: The problem of sensing and the neural coding of sensory events. In Quarton, G. C., Melnechuk, T., and Schmitt, F. O., editors: The neurosciences. A study program, New York, 1967, The Rockefeller University Press.
4. Ottoson, D., and Shepherd, G. M.: Transducer properties and integrative mechanisms of the frog's muscle spindle. In Loewenstein, W. R., editor: Handbook of sensory physiology. Principles of receptor physiology, New York, 1971, Springer-Verlag New York, Inc., vol. 1.
5. Schmitt, F. O., Dev, P., and Smith, B. H.: Electrotonic processing of information by brain cells, Science **193:**114-120, 1976.
6. Werner, J. K.: Trophic influence of nerves on the development and maintenance of sensory receptors, Am. J. Phys. Med. **53:**127-142, 1974.

Original papers

7. Adrian, E. D., and Zotterman, Y.: The impulses produced by sensory nerve-endings. Part 2. The response of a single end-organ, J. Physiol. **61:**151-171, 1926.
8. Bishop, G. H.: Neural mechanisms of cutaneous sense, Physiol. Rev. **26:**77-102, 1946.
9. Bishop, G. H.: The relation of nerve fiber size to modality of sensation. In Montagna, W., editor: Cutaneous innervation, Oxford, 1960, Pergamon Press, Ltd., pp. 88-98.
10. Burgess, P. R., and Horch, K. W.: Specific regeneration of cutaneous fibers in the cat, J. Neurophysiol. **36:**101-114, 1973.
11. Burgess, P. R., English, K. B., Horch, K. W., and Stensaas, L. J.: Patterning in the regeneration of type I cutaneous receptors, J. Physiol. **236:**57-82, 1974.
12. Chambers, M. R., Andres, K. H., von Duering, N., and Iggo, A.: The structure and function of slowly adapting type II mechanoreceptor in hairy skin, Q. J. Exp. Physiol. **57:**417-445, 1972.
13. Dowling, J. E.: Synaptic arrangements in the vertebrate retina: the photoreceptor synapse. In Bennett, M. V. L., editor: Synaptic transmission and neuronal interaction, New York, 1974, Raven Press, pp. 87-103.
14. Dykes, R. W.: Mechanisms of thermal sensibility in the primate—thermoreceptive afferent fibers in the glabrous skin of the monkey. Doctoral dissertation, Baltimore, 1970, The Johns Hopkins University Press.
15. Eagles, J. P., and Purple, R. L.: Afferent fibers with multiple encoding sites, Brain Res. **77:**187-193, 1974.
16. Flock, A.: Sensory transduction in hair cells. In Loewenstein, W. R., editor: Handbook of sensory physiology. Principles of receptor physiology, New

York, 1971, Springer-Verlag New York, Inc., vol. 1, pp. 397-441.
17. Gasser, H. S.: Conduction in nerves in relation to fiber types, Res. Publ. Assoc. Res. Nerv. Ment. Dis. **15:** 35, 1935.
18. Gerstein, G. L.: Functional association of neurons: detection and interpretation. In Schmitt, F. O., editor: The neurosciences, second study program, New York, 1970, The Rockefeller University Press, pp. 648-661.
19. Grundfest, H.: The general electrophysiology of input membrane in electrogenic excitable cells. In Loewenstein, W. R., editor: Handbook of sensory physiology. Principles of receptor physiology, New York, 1971, Springer-Verlag New York, Inc., vol. 1, pp. 136-166.
20. Gybels, J., and van Hees, J.: Unit activity from mechanoreceptors in human peripheral nerve during intensity discrimination of touch, International Congress Ser. No. 253, Amsterdam, 1971, Excerpta Medica, pp. 198-206.
21. Hagbarth, K. E., and Vallbo, A. B.: Mechanoreceptor activity recorded percutaneously with semimicroelectrodes in human peripheral nerves, Acta Physiol. Scand. **69:**121-122, 1967.
22. Hallin, R. G., and Torebjork, H. E.: Electrically induced A and C fibre responses in intact human skin nerves, Exp. Brain Res. **16:**309-320, 1973.
23. Hensel, H., and Boman, K. K. A.: Afferent impulses in cutaneous sensory nerves in human subjects, J. Neurophysiol. **23:**564-578, 1960.
24. Hubbard, S. J.: A study of rapid mechanical events in a mechanoreceptor, J. Physiol. **141:**198-218, 1958.
25. Hunt, C. C., and Ottoson, D.: Impulse activity and receptor potential of primary and secondary endings of isolated mammalian muscle spindles, J. Physiol. **252:** 259-282, 1975.
26. Husmark, I., and Ottoson, D.: Relation between tension and sensory response of the isolated frog muscle spindle during stretch, Acta Physiol. Scand. **79:**321-334, 1970.
27. Husmark, I., and Ottoson, D.: The contribution of mechanical factors to the early adaptation of the spindle response, J. Physiol. **212:**577-592, 1971.
28. Husmark, I., and Ottoson, D.: Is the adaptation of the muscle spindle of ionic origin? Acta Physiol. Scand. **81:**138-140, 1971.
29. Iggo, A.: Is the physiology of cutaneous receptors determined by morphology? Prog. Brain Res. **43:**15-31, 1976.
30. Ilyinsky, O. B., Krasnikova, T. L., Akoev, G. N., and Elman, S. I.: Functional organization of mechanoreceptors, Prog. Brain Res. **43:**195-203, 1976.
31. Ilyinsky, O. B., Volkova, N. K., Cherepnov, V. L., and Krylov, B. V.: Morphofunctional properties of Pacinian corpuscles, Prog. Brain Res. **43:**173-185, 1976.
32. Ishiko, N., and Loewenstein, W. R.: Effects of temperature on the generator and action potentials of a sense organ, J. Gen. Physiol. **45:**105-124, 1961.
33. Jarvilehto, T., Hamalainen, H., and Laurinen, P.: Characteristics of single mechanoreceptive fibres innervating hairy skin of the human hand, Exp. Brain Res. **25:** 45-61, 1976.
34. John, E. R.: Switchborad versus statistical theories of learning and memory, Science **177:**850-864, 1972.
35. Johnson, K. O.: Reconstruction of population response to a vibratory stimulus in quickly adapting mechanoreceptive afferent fiber population innervating glabrous skin of the monkey, J. Neurophysiol. **37:**48-72, 1974.
36. Karlsson, U., Anderson-Cedergren, E., and Ottoson, D.: Cellular organization of the frog muscle spindle as revealed by serial sections for electromicroscopy, J. Ultrastruct. Res. **14:**1-35, 1966.
37. Katz, B.: Depolarization of sensory terminals and the initiation of impulses in the muscle spindle, J. Physiol. **111:**261-282, 1950.
38. Knibestol, M.: Stimulus-response functions of slowly adapting mechanoreceptors in the human glabrous skin area, J. Physiol. **245:**63-80, 1975.
39. Knibestol, M., and Vallbo, A. B.: Single unit analysis of mechanoreceptor activity from the human glabrous skin, Acta Physiol. Scand. **80:**178, 1970.
40. Landau, W., and Bishop, G. H.: Pain from dermal, periosteal, and fascial endings and from inflammation, Arch. Neurol. Psychiatry **69:**490-504, 1958.
41. Lewis, T., Pickering, G. W., and Rothschild, P.: Centripetal paralysis arising out of arrested bloodflow to the limbs, Heart **16:**1, 1931.
42. Lippold, O. C. J., Nicholls, J. G., and Redfearn, J. W. T.: Electrical and mechanical factors in the adaptation of a mammalian muscle spindle, J. Physiol. **153:**209-217, 1960.
42a. Loewenstein, W. R.: Biological transducers, Sci. Am. **203:**98, 1960.
43. Loewenstein, W. R.: On the "specificity" of a sensory receptor, J. Neurophysiol. **24:**150-158, 1961.
44. Loewenstein, W. R.: Rate sensitivity of a biological transducer, Ann. N.Y. Acad. Sci. **156:**892-900, 1969.
45. Loewenstein, W. R.: Mechano-electric transduction in the Pacinian corpuscle. Initiation of sensory impulses in mechanoreceptors. In Loewenstein, W. R., editor: Handbook of sensory physiology. Principles of receptor physiology, New York, 1971, Springer-Verlag New York, Inc., vol. 1.
46. Loewenstein, W. R., and Ishiko, N.: Effects of polarization of the receptor membrane and of the first Ranvier node in a sense organ, J. Gen. Physiol. **43:** 981, 1960.
47. Loewenstein, W. R., and Mendelson, M.: Components of receptor adaptation in a Pacinian corpuscle, J. Physiol. **177:**377, 1965.
48. Loewenstein, W. R., Terzuolo, C. A., and Washizu, Y.: Separation of transducer and impulse-generating processes in sensory receptors, Science **142:**1180, 1963.
49. Merzenich, M. M., and Harrington, T.: The sense of flutter-vibration evoked by stimulation of the hairy skin of primates: comparison of human sensory capacity with the responses of mechanoreceptive afferents innervating the hairy skin of monkeys, Exp. Brain Res. **9:**236-260, 1969.
50. Miller, M. R., Ralston, H. J., and Kasahara, M.: The pattern of cutaneous innervation of the human hand, foot and breast. In Montagna, W., editor: Cutaneous innervation, London, 1960, Pergamon Press, Ltd., Chap. 1, pp. 1-47.
51. Mountcastle, V. B., Talbot, W. H., and Kornhuber, H. H.: The neural transformation of mechanical stimuli delivered to the monkey's hand. In de Reuck, A. V. S., and Knight, J., editors: Touch, heat and pain. A Ciba Foundation symposium, London, 1966, Churchill Livingstone.
52. Oseki, M., and Sato, M.: Initiation of impulses at the non-myelinated nerve terminal in Pacinian corpuscles, J. Physiol. **170:**167-185, 1964.
53. Ottoson, D., and Shepard, G. M.: Steps in impulse generation in the isolated muscle spindle, Acta Physiol. Scand. **79:**423-430, 1970.
54. Ottoson, D., and Shepard, G. M.: Synchronization of activity in afferent nerve branches within the frog's

muscle spindle, Acta Physiol. Scand. **80:**492-501, 1970.
55. Perkel, D. H., and Bullock, T. H.: Neural coding. In Schmitt, F. O., Melnechuk, T., Quarton, G. C., and Adelman, G., editors: Neurosciences research symposium summaries, Boston, 1969, The M.I.T. Press.
56. Perl, E. R.: Myelinated afferent fibers innervating the primate skin and their response to noxious stimuli, J. Physiol. **197:**593-615, 1968.
57. Rakic, P.: Local circuit neurons, Neurosci. Res. Program Bull. **13**(3):291-446, 1975.
58. Ranson, S. W., Droegemueller, W. H., Davenport, H. K., and Fisher, C.: Number, size, and myelination of the sensory fibers in the cerebrospinal nerves. Res. Publ. Assoc. Res. Nerv. Ment. Dis. **15:**3, 1935.
59. Ridley, A.: A biopsy study of the innervation of forearm skin grafted to the finger tip, Brain **93:**547-554, 1970.
60. Rose, J. E., Gross, N. G., Geisler, C. D., and Hind, J. E.: Some neural mechanisms in the interior colliculus of the cat which may be relevant to localization of a sound source, J. Neurophysiol. **29:**288-314, 1966.
61. Sato, M.: Response of Pacinian corpuscles to sinusoidal vibration, J. Physiol. **159:**391-409, 1961.
62. Sato, M., Ozeki, M., and Nishi, K.: Changes produced by sodium-free condition in the receptor potential of the non-myelinated terminal in pacinian corpuscles, Jpn. J. Physiol. **18:**232-237, 1968.
63. Schiff, J., and Loewenstein, W. R.: Development of a receptor on a foreign nerve fiber in a pacinian corpuscle, Science **177:**712-715, 1972.
64. Segundo, J. P.: Communication and coding by nerve cells. In Schmitt, F. O., editor: The neurosciences, second study program, New York, 1970, The Rockefeller University Press, pp. 569-587.
65. Shepherd, G. M.: Synaptic organization of the mammalian olfactory bulb, Physiol. Rev. **52:**864-917, 1972.
66. Talbot, W. H., Darian-Smith, I., Kornhuber, H. H., and Mountcastle, V. B.: The sense of flutter-vibration: comparison of the human capacity with response patterns of mechanoreceptive afferents from the monkey hand, J. Neurophysiol. **31:**301-334, 1968.
67. Teorell, T.: A biophysical analysis of mechano-electrical transduction. In Loewenstein, W. R., editor: Handbook of sensory physiology. Principles of receptor physiology, New York, 1971, Springer-Verlag New York, Inc.
68. Tersuolo, C. A.: Data transmission by spike trains. In Schmitt, F. O., editor: The neurosciences, second study program, New York, 1970, The Rockefeller University Press, pp. 661-671.
69. Torebjork, H. E., and Hallin, R. G.: Perceptual changes accompanying controlled preferential blocking of A and C fibre responses in intact human skin nerves, Exp. Brain Res. **16:**321-332, 1973.
70. van Hees, J., and Gybels, J. M.: Pain related to single afferent C. fibers from human skin, Brain Res. **48:**397-400, 1972.
71. Zelena, J.: Development, degeneration, and regeneration of receptor organs, Prog. Brain Res. **13:**175-211, 1964.

12

VERNON B. MOUNTCASTLE

Neural mechanisms in somesthesis

The term "somesthesis," like the synonymous phrase "somatic sensibility," designates sensations aroused by stimulation of the bodily tissues other than those of sight, hearing, smell, and taste. Normal observers identify several primary qualities in somesthesis: mechanoreception (touch-pressure), warmth, coolness, pain, and the sense of position and movement of the joints. Other somatic sensations appear unique, but are evoked by quantitative or dynamic variations of stimuli in one or another of the primary modes. Indeed, in ordinary life our somesthetic experiences are subtle mixtures of those regarded as primary; such blends yield readily recognized and easily remembered sensations.

A nervous system receives abstracted and coded information about the external world over its first-order sensory nerve fibers. The distorted image that results is determined by first-order transformations as well as others imposed in the chain of central neural events leading to perception. That image is conditioned by the complexion of current input with previously received and stored information; perceptions depend in part on previous experience and current expectations. Our understanding of these mechanisms is less certain for central than for peripheral events. It is the aim here to describe those for somatic sensibility as far as we presently know them.

The idea of a "sensory system" within the brain is a common-sense abstraction useful to the neurologic experimentalist or clinician; it allows the classification of events and observations and the assignment of causes. A sensory system is composed of the afferent pathways, subcortical nuclei, and cortical areas that receive their major afferent input via one sensory portal and that *can be shown on other grounds to play a role in the sensation considered.* The observer's segregation of sensory experiences into easily differentiated classes suggests a parallel degree of anatomic segregation of the relevant neural systems.

The somatic afferent system is dual in nature. One major part begins with the myelinated fibers of the dorsal roots larger than 4 to 6 μm that project either directly via the ipsilateral dorsal column (DC) or indirectly via the ipsilateral dorsolateral column on the dorsal column nuclei (DCN) and adjacent sensory nuclei of the brain stem. These latter, in turn, project on the ventrobasal thalamic nuclear complex, whose cells send their axons to the postcentral gyrus of the cerebral cortex. Within the core of this system there is a precise replication of the body form, a preservation of the modality specificity of the myelinated first-order fibers, and a degree of synaptic security well suited for neural action in rapid tempo. It is precisely organized to serve the discriminative aspects of mechanoreceptive sensibility, and the term *"lemniscal"* will be used to designate this system and the functional properties of its neurons.

A second component of the somatic system originates at the segmental level and is composed of the smaller myelinated and the unmyelinated fibers of the dorsal roots. These project on interneurons of the dorsal horn and serve both local and multisegmental reflexes. The ascending components originate from neurons of the dorsal horn, and their axons run cranially for the most part in the contralateral anterolateral column. This *anterolateral* system has three subsystems: the spinobulbar, paleospinothalamic, and neospinothalamic. It serves a general and poorly defined form of mechanoreception and the senses of pain and temperature. These two major divisions of the somatic afferent system are thus divergent through prethalamic pathways. Parts of them converge again to a certain extent at thalamocortical levels of the brain. How they work synergistically in signaling sensory events is a major problem in understanding the neural mechanisms in somesthesis. A similar duality characterizes the trigeminal component of the system.

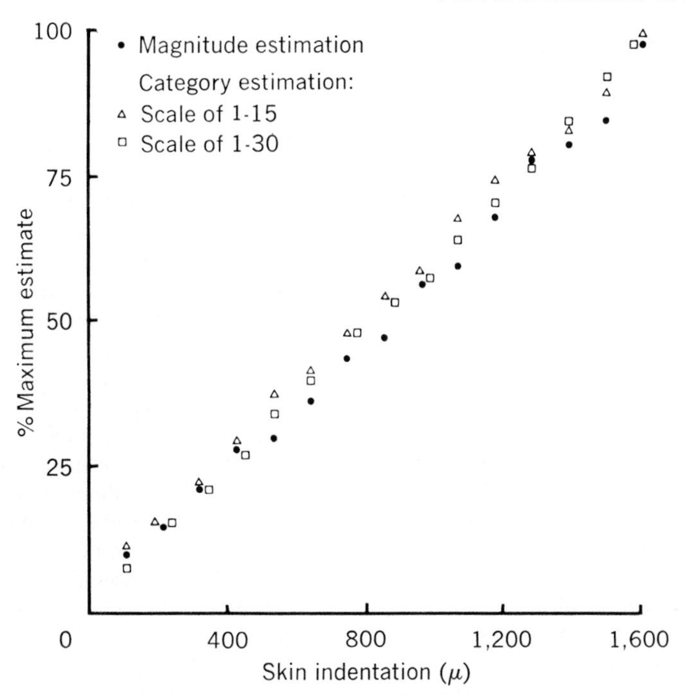

Fig. 12-1. Human rating of touch-pressure stimuli. Subjects were asked to estimate subjective magnitude of pressure sensations elicited by stimuli of different amplitudes in three ways: (1) category estimation with rating scale of 1 to 15; (2) the same, with scale of 1 to 30; and (3) in procedure in which subjects used numbers of their own choice based on their perceived ratio of one pressure intensity to another, method of "magnitude estimation." Ten naive subjects were used in each procedure. Stimuli were step identations of 900 msec duration, delivered every 8 sec, in pseudorandom sequence to distal pad of finger. Rounded probe tip, 2 mm in diameter. Estimates expressed as average of percent of maximum judgments for each subject. Results linear for each of three methods. (From LaMotte.[57])

Human capacity in somesthesis[6]

The human capacity for sensing in any domain is defined by the ability to detect stimuli when they are present and to identify, scale, and differentiate between stimuli of the same quality that differ in place, amplitude, and spatial and temporal patterns. These psychophysical measurements are often made simultaneously with recordings of the peripheral or central neural events evoked by sensory stimuli. The advantages and some limitations of this method for the study of sensation will be apparent in this and following chapters.

Touch-pressure

The threshold for touch-pressure, like that of any sensation, is a variable quantity; it is determined by measuring the probability that a subject will detect each of a series of stimuli of the same quality but of different amplitudes. The 0.5 point on such a psychometric function is taken as threshold. Those for touch-pressure are lowest in regions of the body like the hands, where innervation density is high, and highest on parts like the back and legs, where innervation density is low. Sensory threshold is determined in part by spatial summation, but may also be influenced by differences in the thresholds of individual primary afferent fibers that innervate different body parts. Thresholds are lowered, and sensory acuity is greatest when the observer moves the receptor sheet, such as his fingertip, over a test object in an oscillatory manner. This may be due in part to the sharpening of the profiles of neural activity in the central reaches of the somatic system by the movement, whether active or passive. In addition, the "motor act of sensing" may itself contribute to sensory capacity, perhaps by centrally reentrant pathways, but there is no direct evidence that this is so. Mechanoreceptive thresholds are about equal on the two sides of the body and in both right- and left-handed individuals.

Intensity discrimination and rating. For brief mechanical stimuli delivered to the fingertip the human observer's estimate of stimulus intensity is a nearly linear function of the degree of skin indentation or of applied force (Fig.

12-1). The just detectable increment varies from 3% to 10% of the base comparison for various forms of mechanoreceptive sensibility over the midrange of stimulus intensities. This *Weber fraction* rises for both weak and more intense base stimuli.

Spatial and temporal discrimination. An important feature of human somesthesis is the capacity to identify static and dynamically changing spatial arrays of stimuli delivered to the skin surface. At the level of mechanism the question is how a central neural apparatus specifies a locus of increased activity in a neural field and discriminates between different and often rapidly changing contours of activity within it. Two simple tests of this capacity often used in clinical examinations are accuracy in reproducing the location of a point stimulated and identification of two points stimulated as two and not one. Thresholds for two-point discrimination vary from about 2 to 3 mm on the fingertips to 20 mm on the forearm to 40 mm on the back (Fig. 12-2). Threshold is lowest for mechanical stimuli of intermediate intensity and may be lowered further if the two points are stimulated successively.

The classical two-point test commonly used clinically measures a form of tactile discrimination, but not the capacity for spatial resolution. Phillips et al.[88] have found that subjects working in the forced choice procedure apparently use cues based on differences in the overall amount of neural activity evoked in each test, so that they can discriminate one point from two even when the two are at zero separation (Fig. 12-3, *A*). These authors measured the spatial resolving capacity of humans for stimuli delivered passively in three more experiments, whose results are also shown in Fig. 12-3. The detection of a gap in a smooth surface and the discrimination of a vertical from a horizontal grating both showed thresholds of about 0.8 mm. In the fourth experiment, subjects were asked to identify raised letters of different vertical dimension passively pressed against the skin of the forefinger. The results show (Fig. 12-3, *D*) that the maximal information transmission of 4.7 bits/stimulus, for the 26-letter alphabet, was approached with letter heights of 7.8 mm, with an average detection threshold of 75% correct responses of 3.6 mm letter height. For this size of letter, the two-dimensional Fourier transform for

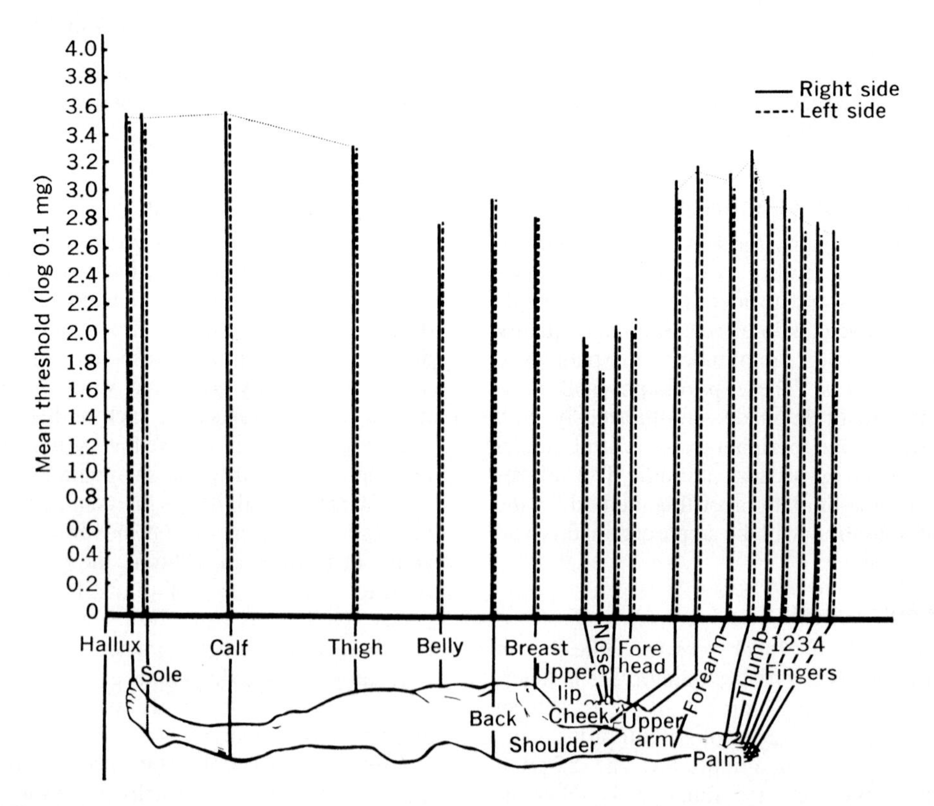

Fig. 12-2. Pressure sensitivity thresholds for different regions of male body surface. (From Weinstein.[121a])

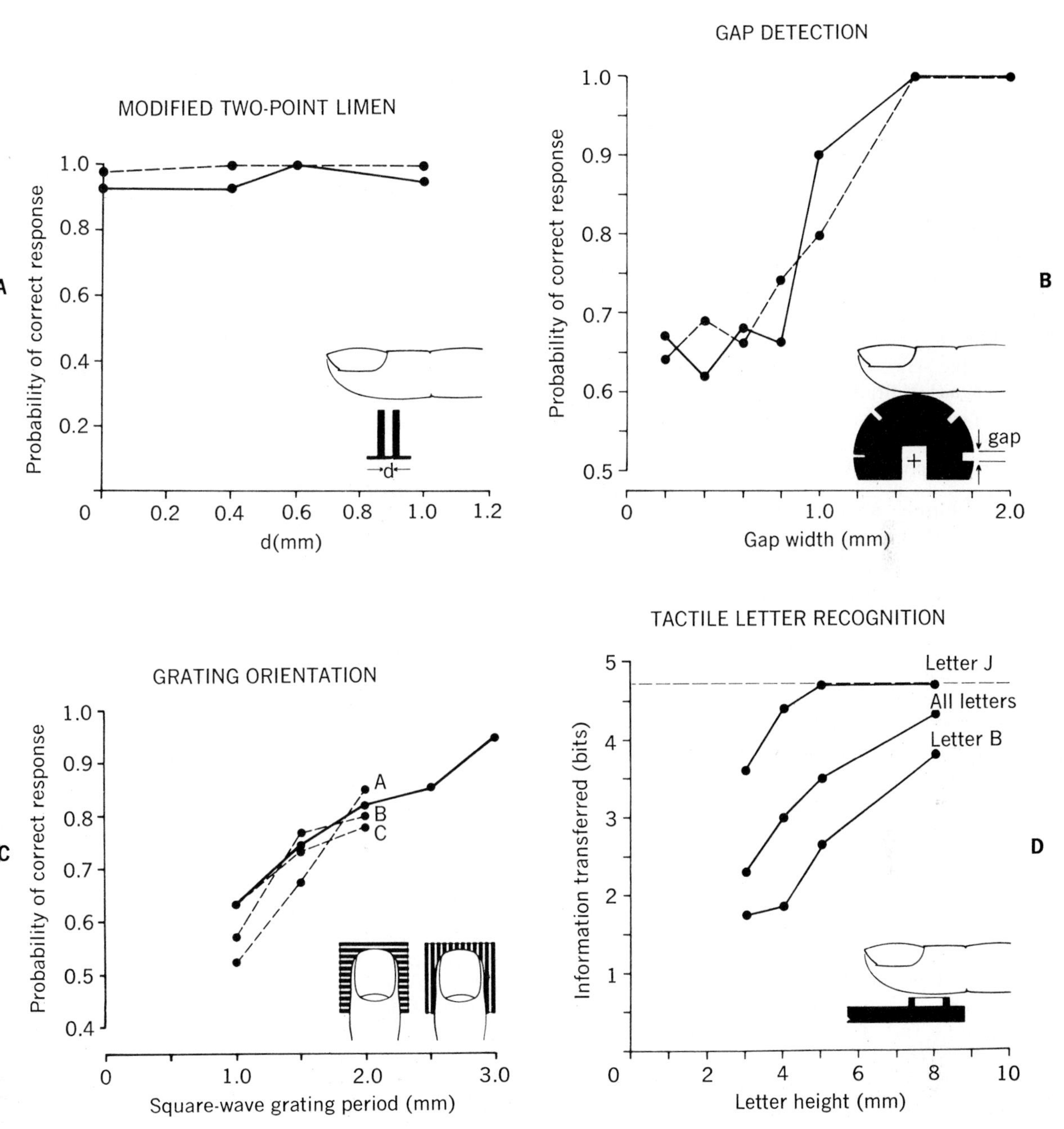

Fig. 12-3. A, Results for two subjects working in forced-choice procedure, identifying second of two successively delivered stimuli as one or two points. Discrimination persisted until zero separation; presumably then discrimination is between size of two-point stimulus vs that of single point. **B,** Performance of two subjects in identifying on which of two occasions surface pressed against skin of finger contained gap. Limen is about 0.8 mm, at 75% detection point. **C,** Subject's performance in identifying orientation of second grating. Heavy line, 900 μm steady indentation. Dashed lines, 900 μm indentation with superimposed vibration; at *A*, 200 Hz, 10 μm amplitude; at *B*, 40 Hz, 300 μm; at *C*, 200 Hz, 35 μm. Vibration had no effect on discrimination limen. **D,** Subject's performance in letter recognition task; 16 presentations were made of each letter at each letter height. Results presented in terms of information transmitted, described in text. No lateral movements of finger were permitted in any test, and all were delivered passively, except for vibration test, **C.** (Courtesy Dr. Kenneth O. Johnson.)

many letters shows a dominant peak corresponding to a spatial period of approximately 1.7 mm and thus a peak:trough dimension of 0.85 mm. Thus on three separate tests the spatial resolving power of the human observer is about 0.8 mm for stimuli presented passively and with no lateral movement of finger or stimulus.

Examples of the importance of serial order in somesthesia are the ability to recognize letters or numerals written on the skin and the performance of the blind in reading Braille type. Highly skilled persons read about 100 words (600 letters)/min, approximately the rate at which the sighted read aloud. The very high information flow suggested (60 bits/sec) is greatly reduced by the serial dependencies and redundancies of language. Braille reading, like other complex somesthetic discriminations, depends on the detection of varying blends of spatial, intensive, and temporal cues.

Definition: functional properties of central sensory neurons

The term "functional property" is used to indicate those characteristics of a neuron by which it can be classified with reference to others, beyond those cellular properties of membrane potential, energy metabolism, synaptic mechanism, etc., which are in one variant or another common to all central neurons. It is thought that for central cells of a sensory system these properties determine the immediate aspects of sensation, that is, whether information is transmitted permitting recognition of the quality of a stimulus and its location, the rating of and discrimination between stimuli ordered along intensive or extensive continua, and the identification of stimulus rhythms. These properties are of two classes: static and dynamic.

Static properties are those set by the synaptic linkages between first-order fibers and central neurons and the transducer properties of the former. They change little over wide changes in forebrain excitability, or the influence of converging systems, or the effect of anesthetic agents. Static properties are those of place and modality. The term "modality" commonly refers to the quality of a sensation experienced and described by human observers. It is used here in a special way to indicate the nature of the peripheral stimulus that will activate a sensory neuron, peripheral or central.

Dynamic properties include those that depend on the quantitative aspects of synaptic transmission. They are thus influenced readily by the action of converging or reentrant systems that impinge on sensory synaptic relays and are sensitive to the actions of anesthetics or other drugs and to changes in overall excitability of the brain (e.g., between sleep and wakefulness).

Obviously, certain properties of central neurons are both static and dynamic (e.g., afferent inhibition).

The application of the method of single-unit analysis to the study of sensory systems depends on the careful identification and measurement of static and dynamic properties of large numbers of neurons at each level of a system. The course of events throughout a sensory system during and following a peripheral stimulus can then be reconstructed and explanations of sensory behavior sought in terms of neural mechanics. Much of the discussion that follows concerns the results obtained in experiments of this sort.

PRETHALAMIC COMPONENTS OF SOMATIC AFFERENT SYSTEM

Lemniscal system[5]

The term "lemniscal" is used first in its anatomic meaning to identify those components of the somatic afferent system that transit the brain stem in the medial lemniscus to terminate in the dorsal thalamus. More generally, it is used to designate a set of properties and a mode of neural processing. The distinguishing feature of the lemniscal system is that information concerning the location, form, quality, and dynamic attributes of stimuli that impinge on the body is encoded in neural transforms that are transmitted with considerable precision across the nuclear relays of the system to the cerebral cortex. This major characteristic depends on the facts that (1) the peripheral sheets of receptors are represented centrally in precise and detailed spatial patterns, with less divergence than in the anterolateral system; (2) central neurons of the system most commonly receive direct or relayed afferent input that originates in first-order fibers of only one modality type; and (3) the synaptic relations of the system ensure a capacity for a high level of dynamic activity. These general properties define a *lemniscal neuron,* a cell that subtends a restricted receptive field, is activated by a single form of physical stimulus, and possesses powerful synaptic transmitting capacities. Each nuclear relay zone of the system receives, in addition, descending fibers from the forebrain; these descending systems can control or modulate afferent synaptic transmission in the system, but the nature of those controls and their meaning for sensation are still obscure (p. 380).

Dorsal ascending systems of the spinal cord and somatic sensory nuclei of the brain stem

The ascending systems of the dorsal half of the spinal cord that feed into the medial lemniscus are the DCs via the DCN and an ascending component of the dorsolateral column via nucleus Z, which lies at the caudal end of the brain stem, just anterior and lateral to the DCN. The dorsal column is composed of ascending myelinated stem axons of fibers of the dorsal roots larger than 3 to 4 μm and of second-order ascending axons of neurons of the spinal cord.[38,103,104] The zones of the peripheral tissue innervated by fibers of adjacent dorsal roots overlap by as much as a full dermatome. The shifts in the position on the body surface of the peripheral receptive fields (RFs) of consecutively adjacent dorsal rootlets trace an intermittently recursive path, moving smoothly through the sequence for a single root, but jumping back a step on transition from one root to another. Just cephalad to the entry of each dorsal root the DC contains a full representation of its myelinated fibers (<3 to 4 μm) arranged in a narrow band applied laterally to similar bands from more caudal segments.

Two rearrangements occur as the lumbosacral dorsal root contingents ascend the DC.[123,128,130] The first is topographic, for the fibers of adjacent dorsal roots are shuffled in such a way that the intermittently recursive sequence of the dorsal roots' RFs is changed to a continuous one. First-order fibers with contiguous RFs of the same or adjacent dorsal roots are themselves contiguous at the highest levels of the DC.[129] In contrast, fibers that enter distantly separate dorsal roots are not contiguous in the DC, even though their RFs may be adjacent peripherally. The fibers originating from hand or foot are each clustered together at the most cephalad levels of the DC, separating the representations of the preaxial and postaxial sides of arm and leg, respectively. The general result is a representation composed by dermatomal alignment distorted by differences in peripheral innervation densities and as nearly complete topologically as is possible in the mapping of a three-dimensional surface onto a two-dimensional plane.[7]

The second rearrangement within the dorsal ascending systems concerns modality. Each dorsal rootlet of the lumbosacral segments contains a full sample of the fibers of the modality types of Table 12-1, although proportions differ from one rootlet to another. After a few segments' travel in the cephalad direction, a root's contingent in the DC is much reduced and is then composed largely of quickly adapting primary afferents innervating the skin and deep tissues. The majority of the large-diameter muscle afferents and of the slowly adapting cutaneous afferents have turned out of the DC to synapse on neurons of the dorsal horn, including such separate nuclei as Clarke's column. This may account for the fact that only 25% of the myelinated fibers of the dorsal roots (<3 to 4 μm) project stem axons directly to the DCN in the cat; this figure is likely to be higher in primates.

Several second-order systems arise from dorsal horn cells and ascend from the lumbosacral regions of the cord in its dorsolateral columns. These include the spinocerebellar and spinovestibular systems and two specific somatic sensory systems. The first is the lateral cervical system, which is of significant size in carnivores but much reduced if present at all in man (p. 355). The second is composed mainly of axons of cells activated by large-diameter joint and muscle afferents, travels upward in the most medial part of the dorsolateral column, and terminates in nucleus Z in the lower brain stem.[14,60,65,81] This nucleus lies just cephalad and lateral to the DCN. The axons of its cells decussate and ascend the brain stem with the contralateral medial lemniscus, terminating in the anterior portion of the ventrobasal complex of the thalamus.[43] It is unknown to what degree a similar re-sorting occurs in the ascending projections of fibers of the cervical roots. Certainly the topographic reshuffling has occurred for afferents from arm and hand as from leg and foot at the zone of entry into the DCN. Some reshuffling of modality types has occurred, also, for example, the powerful spino–lateral cuneate–cerebellar system and the rostral spinocerebellar tract.[83] It is unknown whether a re-sorting has occurred among the deep and slowly adapting afferents from the arm, as from the leg, that are destined to project via relays to the cerebral cortex. So far only a few investigations of these matters have been made in the primate, and more information is badly needed.

The available information suggests that each mapping unit of the DC projects on a long, thin, dorsoventrally oriented sheath of DCN cells and that these projections together compose the quasi-topologic transform of the body form. The modality spectrum contained in the dorsal roots has to a great extent been reconstituted in the brain stem sensory nuclei by a combination of afferents in the two components of the dorsal ascending systems. From caudal to rostral within each segmental lamina of the DCN there is a shift from a predominance of cutaneous mechanoreceptive afferents posterodorsally to those activated by mechanical stimulation of deep tissues or by joint rotation anteroventrally. Quickly adapting cutaneous afferents project in a highly specific topographic pattern on the clustered groups of cells ("nests") of the dorsal and caudal parts of the DCN, and almost all these cells send their axons into the medial lemniscus. Cells of the more reticulated region of the cephalad and anterior parts of the DCN receive three known inputs: (1) second-order axons of the DCN arising from dorsal horn cells, which subtend large RFs and are of unknown function; (2) second-order axons from deep tissues, mainly muscle,[101,102] which presumably reach this portion of the DCN via the dorsolateral column; and (3) descending axons of cerebral origin. Only a portion of the cells of this part of the DCN send their axons into the medial lemniscus; the remainder are surmised to function as local interneurons. A striking feature of synaptic relations within the DCN, as at higher levels of the lemniscal system, is that modality specificity is as precise for central as for the first-order elements listed in Table 12-1. The implication is that each DCN cell receives its presynaptic input only from first- or second-order cells of a single modality type. Within this mode-specific pattern there is a restricted spatial convergence and divergence within each modality type.

The more dynamic aspects of synaptic processing in the DCN will be discussed in a later section.

In summary, the re-sorting of afferents in and between dorsal ascending systems of the spinal cord establishes that both topography and modality are mapping principles in the somatic afferent system and that they preset the pattern of representation throughout the system. The mapping element at the level of entry to the brain stem sensory nuclei is a group of first- and second-order neurons representing a segment of a dorsal root input. These groups are arranged in a pattern allowing a maximally continuous representation of the body, although incomplete topologically. Moreover, there is a differential displacement within this pattern of neurons serving different modes of mechanoreception. This general representation is projected on the somatic sensory relay nuclei of the thalamus and from there on somatic sensory area I of the cerebral cortex.[29] These topographic and modality representations and the synaptic mechanisms of the DCN create a system projecting upward via the medial lemniscus that possesses the three salient features of a "lemniscal" system: detailed and specific topographic representation, modality specificity of elements with a minimal conver-

Table 12-1. First-order fibers feeding the lemniscal system via the dorsal ascending pathways of the spinal cord

Source	Type	Mechanoreceptive submode
Hairy skin	Quickly adapting fibers sensitive to hair movement (movement detectors)	Touch-pressure (and flutter component of flutter-vibration)
	Slowly adapting fibers innervating tactile organs (detectors of transients and of steady states; types I and II)	Touch-pressure
	Slowly adapting fibers innervating distributed fields in skin surface, few in number (ending unknown)	Touch-pressure
Glabrous skin	Quickly adapting fibers innervating dermal ridges, probably Meissner's corpuscles (movement detectors)	Touch-pressure (and flutter component of flutter-vibration)
	Slowly adapting fibers innervating dermal ridges, probably Merkel's disc (detectors of movements and steady states)	Touch-pressure
	Slowly adapting fibers innervating distributed receptive fields, few in number (ending unknown)	Touch-pressure
Dermis and deep tissues, other than muscle	Fibers ending in pacinian corpuscles (detectors of high-frequency transients)	Vibratory sensibility
	Slowly adapting fibers ending in fascia and periosteum (Ruffini cell–like)	Touch-pressure
	Fibers ending in joing capsules and ligaments (Ruffini cell–like)	Position sense and kinesthesia
Muscle	Spindle afferents, Ia and II	Kinesthesia (?)
	Tendon organs afferents, group Ib	Sense of effort (?)

gence between different types, and synaptic security of a high order. It is important to emphasize that the DCN are integrating centers as well as relay nuclei.[42]

Lateral cervical system[5,16]

The second-order axons of the lateral cervical system arise from neurons of layers III, IV, and V of the dorsal horn and ascend in the most medial corner of the dorsolateral column to terminate in the lateral cervical nucleus, a group of cells located just lateral to the dorsal horn of the first and second cervical segments. The axons of these cells cross the cord to join the contralateral medial lemniscus and, with it, terminate within the ventrobasal complex of the dorsal thalamus, which projects on the somatic sensory areas of the cerebral cortex. The cells of this system are almost all sensitive to light mechanical stimulation of the ipsilateral skin, but a few are activated by noxious stimuli and many receive input from both myelinated and unmyelinated fibers of the dorsal roots. At least that is true in the cat, the animal in which most studies of this system have been made. The lateral cervical nucleus in the cat is more than a simple relay, for synaptic transmission within it is modulated in complex ways by descending control systems. The lateral spinocervical system has been identified in carnivores, ungulates, and cetaceans, but appears much reduced in monkeys and rudimentary if present at all in man.[120] In those species possessing it this system may account for the persistence of certain aspects of mechanoreceptive sensibility after lesions of the DC. What role, if any, it may play in human somesthesis is unknown.

Anterolateral system[5,67]

The anterolateral system is phylogenetically older than the DC system, it is less precisely organized topographically, and its cells of origin are activated by a broader spectrum of dorsal root afferents, including A-delta and C fibers as well as larger myelinated cutaneous afferents.[32] Its axons arise from cells distributed throughout several layers of the dorsal horn, including the marginal zone; the majority cross via the anterior commissure and turn cephalad; a small number may ascend ipsilaterally. Fibers from each segment are applied medioventrally to those from caudal regions. All the fibers turning cephalad in the anterolateral columns do not reach supraspinal levels; those that do not are part of the propriospinal system, and those that do compose the anterolateral system. The latter has several major components, each with different supraspinal projection targets:

1. The *spinobulbar component* is a major afferent pathway to the brain stem and mesencephalic reticular formation, from which activity is relayed upward via the reticulothalamic systems to terminate in the intralaminar nuclei of the thalamus.

2. The *paleospinothalamic component* projects directly from the cord to the n. centralis lateralis of the intralaminar group.[67] It contains A-delta and C fibers and converges with the relayed spinobulbar projections on common thalamic targets.

3. The *neospinothalamic component* projects on the posterior nuclear group and the ventrobasal complex of the thalamus, reaching the latter in a topographic pattern generally merging with that of the medial lemniscus, although its terminations are concentrated more posteriorly in the ventrobasal complex. The neospinothalamic tract grows rapidly in phylogeny. It is scarcely discernible in rodents and carnivores, but increases in size rapidly in primates. In man the tract contains 1,500 to 2,000 myelinated fibers up to 4 to 6 μm in diameter. Whether it also contains C fibers is unknown. It is the only topographically organized component of the anterolateral system.

The important role of the anterolateral system in pain and temperature sensibility is considered in Chapter 13, together with the problem of synaptic processing in the dorsal horn.

Trigeminal system[2]

The system serving facial sensation recapitulates the dual system just described, although it differs in detail. In many mammals, facial sensation is important in environmental exploration, a fact reflected in the dominance of the head and face in central replications of the body form (Fig. 12-8). Although in man this tactile exploratory function is greatly reduced, sensory input from intraoral structures is important in deglutition and speech. Indeed, for many tests of somesthetic capacity the tongue is the most sensitive area of the body; for example, two-point discrimination on its tip is 3 times better than on the tips of the fingers.

First-order fibers

Nerve fibers innervating the skin of the face are of the same modality types as those of the body: mechano-, thermo-, and nociceptive. Their cell bodies are distributed in a topographically orderly way in the semilunar ganglion. The centrally directed axonal branches are as a group partially rotated in the trigeminal root, so that at entry to the pons there is an inverted dorsoventral representation of the face, a pattern retained throughout their central course and in the second-order trigeminal cellular regions of the medulla on which they project. On entering, most but not all of the myelinated fibers divide into short ascending and long descending branches, the former terminating within the main sensory nucleus, the latter making up the bulk of the trigeminal spinal tract. The level of termination of these descending fibers is not related to their peripheral origin, for the topographic representation of the face is complete to the full caudal extension of the tract and the spinal nucleus of the trigeminal. In addition, some cutaneous afferents from the head enter the brain stem via the seventh, ninth, and tenth cranial nerves, join the trigeminal tract, and descend to the n. caudalis.

The trigeminal tract contains a lower proportion of unmyelinated fibers than do comparable spinal nerves; many of those that are present arise from regions of the facial and buccal surfaces that are highly sensitive to thermal and noxious stimuli. These C fibers pass without bifurcation into the spinal tract and terminate almost wholly within the n. caudalis. Not all C fibers of the trigeminal tract are of external origin, however, for many arise from neurons of the n. caudalis and constitute a trigeminal linkage system that is perhaps comparable to Lissauer's tract of the spinal cord.[112]

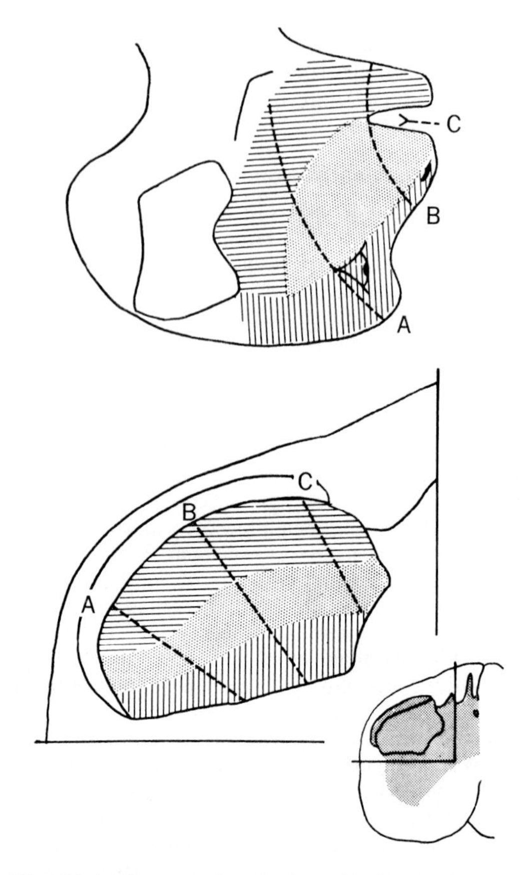

Fig. 12-4. Diagram of projection of ipsilateral face and oral structures onto transverse plane through trigeminal nuclear complex. Nucleus caudalis is represented, but similar projection occurs at all rostrocaudal levels. Skin fields of separate trigeminal divisions are differentiated by shading, correlated to their representation within nucleus. Representation is inverted, with mandibular division represented dorsally, etc. Radial projection of primary fibers also results in ordered mediolateral pattern, with most oral structures represented most medially, as in sequence *C, B, A*. (From Darian-Smith.[2])

Organization of the brain stem trigeminal nuclei

The main sensory and spinal trigeminal nuclei are structurally distinct and are related to the ascending and descending branches of the entering trigeminal tract fibers, respectively. The spinal nucleus is further separable into three cytoarchitecturally distinct nuclei: oralis, interpolaris, and caudalis. The main sensory nucleus and the n. oralis appear functionally analogous to the DCN and the lemniscal system.[82] The pattern of representation within them is inverted, with perioral and intraoral areas represented most medially (Fig. 12-4). Most of the cells of these nuclei send their axons to the contralateral ventrobasal complex of the thalamus, ascending closely applied to the dorsomedial aspect of the medial lemniscus. Some cells of the main sensory nucleus project to the ipsilateral thalamus, constituting the dorsal trigeminothalamic tract, and account for the small ipsilateral face representation in lemniscal patterns at thalamic and cortical levels.

The n. caudalis resembles cytoarchitecturally the dorsal horn of the spinal cord, and its cells project their axons, some contra- and some ipsilaterally, to the thalamic targets of the anterolateral system, the intra-laminar and posterior nuclear groups.[110] Its similarity to the anterolateral system in function is further strengthened by the fact that section of the the trigeminal spinal tract at the level of the obex and thus cephalad to the n. oralis produces in both man and animals loss of facial sensations of pain and temperature but results in only trivial changes in mechanoreceptive sensibility.

Neurons of the main sensory nucleus and of n. oralis possess static properties quite similar to those of the cells of the DCN. They respond to only one form of mechanical stimulation and subtend small and continuous receptive fields that together compose the topologically intact pattern shown in Fig. 12-4. The n. interpolaris resembles cytoarchitecturally the external cuneate nucleus and, like the latter, many but not all of its cells project on the cerebellar cortex; neither is thought to contribute directly to somesthesia.

In summary, the medial lemniscal analog within the trigeminal system includes the main sensory nucleus and n. oralis, although the latter differs from the medial lemniscus system in sending collateral projections to the RF. N. caudalis, on the other hand, resembles the anterolateral system in its cytoarchitecture and its cephalad connections. The ipsilateral face is represented in toto in both components of the trigeminal system. This central pattern is a somewhat distorted image of the peripheral body form, reflecting the heavy peripheral innervation of peri- and intraoral areas. Any single locus on the ipsilateral face or inside the mouth is represented centrally by a column of cells extending the whole length of the trigeminal complex, with a differential distribution for modality within the column.

THALAMOCORTICAL COMPONENTS OF SOMATIC AFFERENT SYSTEM

The ventral thalamic mass receives in an anteroposterior sequence three ascending systems, the dentatothalamic projection from the cerebellum, the medial lemniscus, and the neospinothalamic tract. At the mesodiencephalic junction the medial lemniscus contains the latter two systems in a single topographic pattern, together with the trigeminothalamic projections, and terminates in a region of distinctive architecture and functional properties, the *ventrobasal (VB) complex*.[73,122] There are three salient features of this region: a detailed topographic representation of the contralateral body form, the highly specific static properties of its neurons with regard to place and modality, and the great synaptic security between incoming elements of the medial lemniscus and the thalamocortical cells.[90]

The topographic details in the monkey are shown in the figurine map of Fig. 12-5; a similar pattern has been observed directly by recording in humans. The distorted pattern is set at entry to the DCN by the place sorting of the DC and reflects the fact that the central replication is related to peripheral innervation density rather than to body geometry. From locus to locus across the thalamic pattern there is a gradually shifting overlap of the peripheral zones represented. The afferents from any single spinal segment project on a thin curving lamella of thalamic tissue, concave medially.

VB neurons are highly specific for place and modality. They are activated by stimuli delivered within discrete, contralateral, continuous receptive fields and are preferentially responsive to but one stimulus quality. The classification of first-order fibers given in Table 12-1 applies with equal certainty to DCN, VB, and postcentral neurons, implying that DCN neurons receive presynaptic terminals from fibers of only one of the classes of Table 12-1, and that, in turn, each VB neuron is linked to but one class of DCN neurons, etc. It is not known what factors are operative in ontogenesis to produce such a precise ordering of relations in the specific sensory systems.

Within the topographic pattern there is a differential distribution of the neurons of different modality classes. Those related to cutaneous afferents are concentrated posteriorly, whereas those activated from deep fascia, periosteum, and joint afferents are shifted anteriorly. Still further anteriorly is a zone thought to receive relayed activity originating in low-threshold fibers of muscle nerves. This differential distribution becomes more marked in the transition from cat to monkey to man.[28,50,116]

Studies of human thalamus[28,50,116]

Observations on the somatic sensory nuclei of the thalamus have been made in human beings, the opportunity being afforded by the discovery that local destruction of a small area of the ventrolateral thalamic mass, in front of the VB complex, reduces both the tremor and the increased muscle tone of some patients with Parkinson's disease. Thalamic explorations are carried out under local anesthesia, and recording-stimulating electrodes are guided into the thalamic target by a stereotaxic instrument that ensures an exact orientation of the electrode traverse with reference to identifiable landmarks. In humans the ventrolateral thalamic mass is even more differentiated than in the monkey, so that in its ventral portions four nuclei can be defined.[47]

N. ventralis caudalis (V.c.). The V.c. receives medial lemniscal and neospinothalamic input and projects on the postcentral gyrus. Electrical stimulation elicits mechanoreceptive paresthesias referred to local regions of the contralateral body surface, and neurons of the region are activated from small contralateral cutaneous receptive fields (Fig. 12-6). These may be surrounded by inhibitory zones in the pattern of afferent inhibition described later. Taken together, the receptive fields of these cells compose a precise topographic representation of the contralateral body surface.

N. ventralis intermedius (V. im.). The V.im. ap-

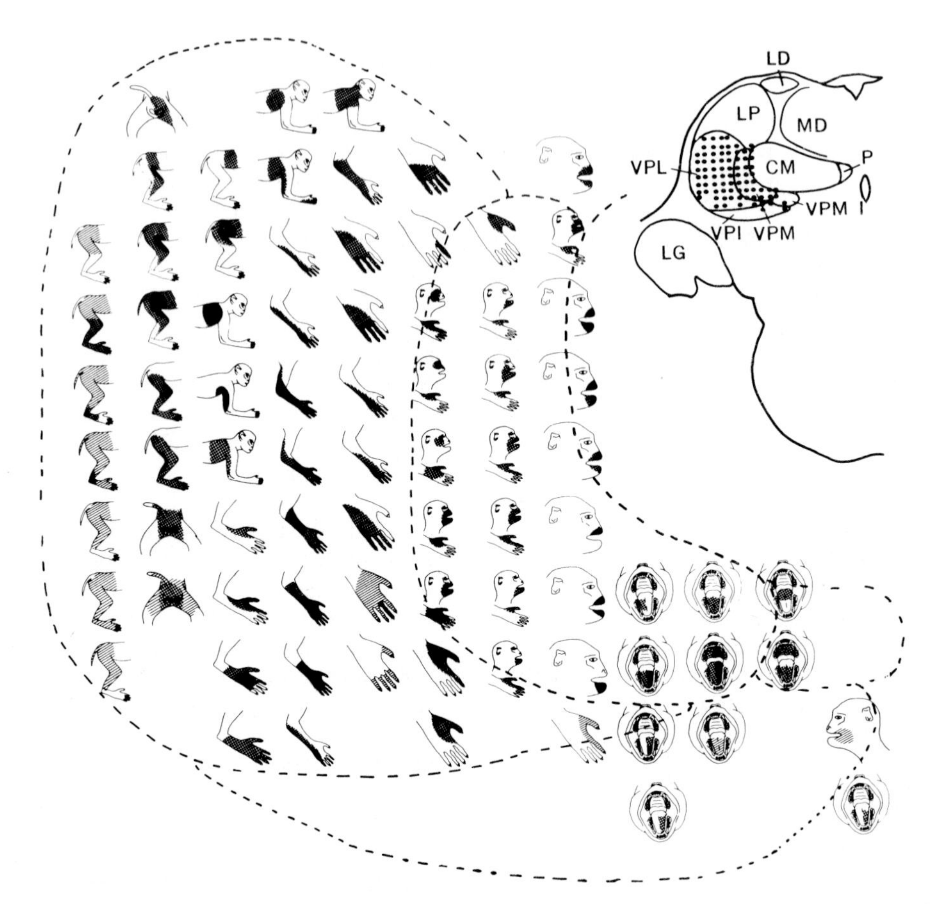

Fig. 12-5. Representation of cutaneous tactile sensibility in one frontal plane of thalamus of monkey, as determined by evoked potential technique. Inset drawing prepared from frontal section of brain in plane of electrode penetrations; dots indicate positive points, and each figurine drawing is arranged accordingly. Tactile stimulation of skin of areas marked on figurines evoked responses at points indicated. Gradation in intensity of projection, from most to least, indicated by solid shading and cross-hatching. With exception of ipsilateral intraoral and perioral regions, all responses were obtained only from stimulation of contralateral side of body and head. *VPL, VPM,* and *VPI* indicate divisions of ventral posterior nuclei, also called ventrobasal complex. *CM,* Centre médian; *P,* parafascicularis; *MD,* medialis dorsalis; *LP,* lateralis posterior; *LD,* lateralis dorsalis; and *LG,* lateral geniculate body. (From Mountcastle and Henneman.[73])

parently receives only selections of the lemniscal and no spinothalamic input, for its cells are activated by the passive rotation of the contralateral limbs at their joints and not by skin stimulation. These cells project on the postcentral gyrus. The region is homologous to the anterodorsal parts of VB in the monkey.[90] A puzzling report is that electrical stimulation here produces no sensory experience in waking humans, but more observations are needed on this point.

N. ventralis oralis posterior (V.o.p.). The V.o.p. receives the dentatothalamic system from the cerebellum and projects on the precentral motor cortex, area 4. Some of its cells are activated by muscle stretch, but none by skin stimulation; others discharge only during voluntary movement, and they may discharge in synchrony with the 4 to 5/sec tremor of parkinsonians. Stimulation produces an acceleration of voluntary movements but no sensory experience. Lesions here are sometimes effective in relieving the tremor of Parkinson's disease.

N. ventralis oralis anterior (V.o.a.). The V.o.a. receives the pallidothalamic system, a major output from the basal ganglia, and projects on Brodmann's area 6. Electrical stimulation here increases muscle tone and slows voluntary movement but elicits no sensory experience, and it is here that small lesions may relieve the hypertonia of Parkinson's disease.

Neither electrical stimulation nor destruction of the loci in which lesions modify the signs of Parkinson's disease produce alterations in somatic sensation. The implication is that the thalamic relay nuclei for the cerebellothalamocortical systems play important roles in the forebrain regulation of posture, tone, and movement but no essential role in somesthesia.

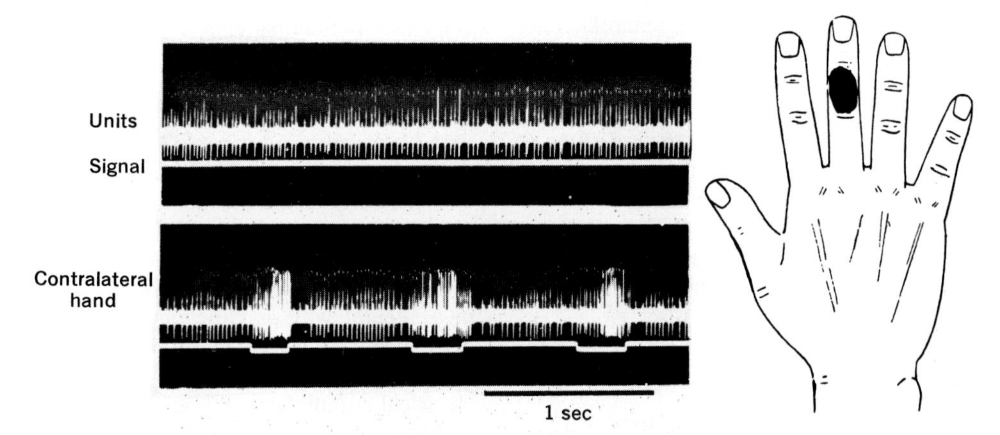

Fig. 12-6. Records, obtained during microelectrode exploration of thalamus in waking human being undergoing stereotactic thalamotomy for parkinsonism, of impulse discharges of single cell located in posterior portion of ventral thalamic group, homolog of ventrobasal complex in monkey. Cell was activated by light tactile stimulation of area of skin on dorsum of contralateral middle finger, shown in black. Upper records: spontaneous activity. Lower records: acceleration of spontaneous activity produced by three light strokes across receptive field, indicated by signal line. (From Jasper and Bertrand.[50])

Thalamic areas activated by anterolateral system and functional properties of anterolateral neurons

The spinobulbar and paleospinothalamic systems funnel somatic afferent inflow to the ascending reticular and generalized thalamocortical systems and thus play an important role both in regulating levels of excitability in the forebrain (Chapter 10) and in pain and temperature sensibilities (Chapter 13). Elements of these systems possess static and dynamic properties quite different from those characteristic of lemniscal neurons.

In carnivores the neospinothalamic component of the anterolateral system contains only a small number of fibers. In primates and especially in man this newer anterolateral tract increases greatly in the number and the size of its fibers, and a major projection on the VB complex appears. After section of the dorsal ascending systems of the spinal cord in monkeys, stimulation of cutaneous nerves elicits the response of some cells of the VB complex and of the postcentral gyrus in a predominantly contralateral projection. Little is known, however, of the details of the neospinothalamic-VB projection in primates save that it is in topographic register with that of the medial lemniscus and contains cutaneous mechanoreceptive elements but none driven by joint rotation.[86,125] It is not known whether the two sets of fibers preempt exclusive sets of VB cells or converge commonly on them, or what the dynamic properties of the synaptic linkages are. The neospinothalamic system may account for the rudimentary capacity for mechanoreceptive sensibility that remains in monkeys and men after lesions of the dorsal systems.

The neospinothalamic system also projects on the posterior nuclear group (PO), an area of thalamus intercalated between the VB complex anteriorly, the geniculate bodies laterally and posteriorly, and the pulvinar above.[89] PO cells may be activated from large receptive fields that may be contralateral or ipsilateral or both, may be discontinuous, and may in the limiting case cover nearly the entire body surface. There is no regular topographic representation of the body form in this region, in contrast to the detailed representation in the VB complex. Unlike the mode-specific cells of the lemniscal system, the large majority of PO cells are polyvalent with respect to the adequate stimuli that activate them, some being responsive to light mechanical stimuli in one part of their receptive fields, only to noxious stimuli delivered to other parts, and to sound and light as well. The majority of PO cells, whether polyvalent or not, are activated by stimuli destructive of tissue.[89]

• • •

In summary, the lemniscal and anterolateral projections on the thalamus differ strikingly. The former is organized as a large-fibered, rapid-tempo system, with the functional properties necessary for the discriminative aspects of somatic sensibility: identification of place, contour, and quality of mechanical stimuli, the sensing of the position and movement of the limbs, the resolution of serial order, etc. The anterolateral system serves a variety of more general functions. Its phyletically ancient projection, either directly or via the ascending brain-stem systems, serves an energizing, not an elaborative, function by driving the generalized thalamocortical system. Its most recent development, the neospinothalamic projections on the VB complex, reveals rudimentary lemniscal properties. The role of the anterolateral system in pain sensation is discussed in Chapter 13.

SOMATIC SENSORY AREAS OF THE CEREBRAL CORTEX

Somatic sensory area I: postcentral gyrus[93]

Lesions of the postcentral gyrus (somatic sensory area I, SI) produce defects in somatic sensibility in humans and other primates (p. 381), and electrical stimulation within this region in conscious human patients elicits somatic sensory experiences referred to local regions on the contralateral side of the body. Indeed, electrical stimulation of the paracentral cortex combined with observations of the movements induced and the sensations evoked, in patients under local anesthesia, is of great localizing value in neurosurgical procedures.[85] Its use has resulted in general maps of the body representation in the postcentral gyrus, which are frequently represented in a general way as caricature drawings of the body form. It is possible under some circumstances to elicit somatic sensory experiences by electrical stimulation of a number of other cortical areas, but it is in the postcentral area that such experiences can be provoked most consistently at lowest threshold, be referred so regularly to local contralateral parts, and, when composed together in spatial order, reveal a regular topographic pattern. That *general* pattern has

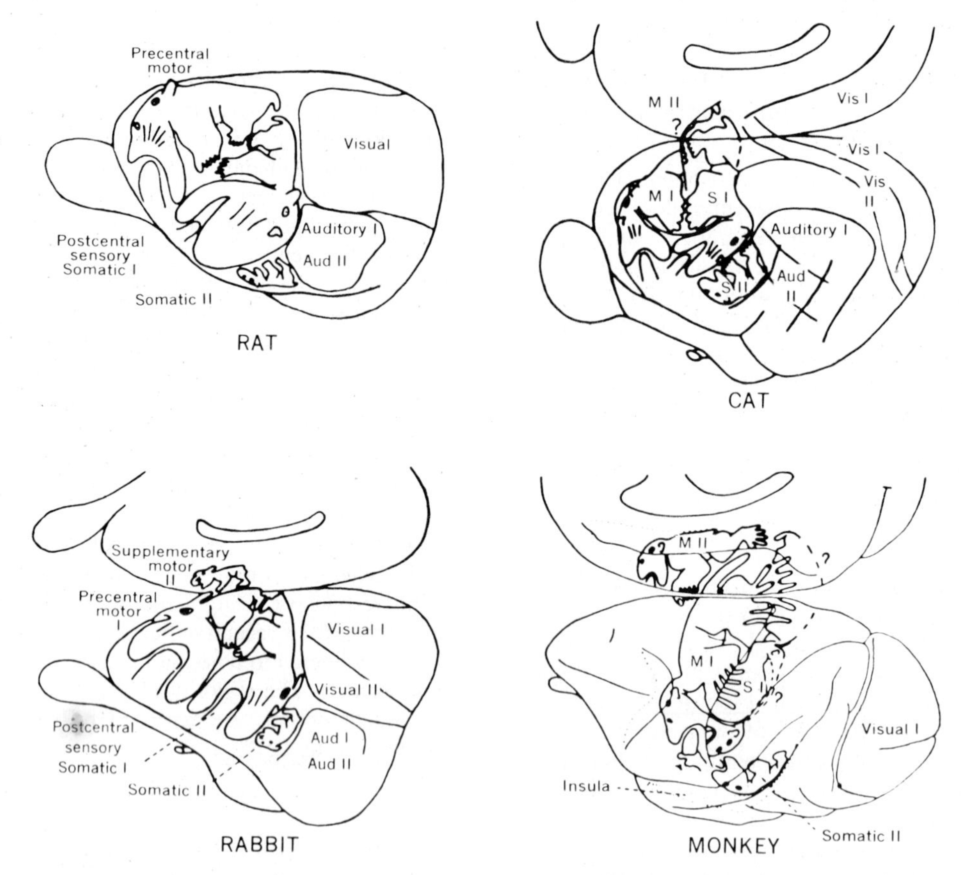

Fig. 12-7. Diagrams of cortices of rat, rabbit, cat, and monkey *(Macaca mulatta)* showing locations and general plans of organization of precentral motor *(MI)*, supplementary motor *(MII)*, postcentral sensory *(SI)*, and second somatic sensory *(SII)* areas. Relations to auditory and visual cortices are shown, except for monkey auditory area, which lies hidden on lower bank of sylvian fissure. For rabbit, cat, and monkey, medial walls of hemispheres are swung upward in drawings to occupy same planes as lateral surfaces. (From Woolsey.[131])

been confirmed and further detailed by application of the evoked potential method in humans.

It is largely through the studies of Woolsey that we know the general plans of representation of the sensory and motor systems in the cerebral cortices of experimental animals. The maps of Fig. 12-7 are general diagrams of the motor and sensory representations in the cerebral cortices of rat, rabbit, cat, and monkey determined by the methods of electrical stimulation (motor) and evoked potentials (sensory). These methods do not reveal some details described subsequently, but show the pattern of representation of the body parts in the somatic sensory areas.[131,132]

The finer detail of the plan of representation of the body parts in the postcentral gyrus of monkeys has now been studied by the method of recording with microelectrodes the electrical signs of the activity of single cortical neurons.[7,77,78,94] In these experiments, large number of microelectrode penetrations are made at close intervals and the receptive field and modality type of each neuron identified. The representation indicated by the cartoons of Fig. 12-7 has been confirmed insofar as the general relation between the body parts is concerned. However, the recent and extensive studies of Merzenich and Kaas and their colleagues,[70,84,114] in both New and Old World monkeys, have revealed a detailed mode of representation not hitherto suspected: there is a complete representation of the body in each of the three cytoarchitectural zones of the postcentral somatic area, areas 3b, 1, and 2, and there is some evidence to suggest that a fourth separate pattern exists in the transitional area 3a, although this may be an extension of the pattern of the precentral motor cortex. The representation patterns in areas 3b and 1 of an Old World monkey, *Macaca fascicularis,* are shown in Fig. 12-8. The representations in 3b and 1, as well as that in 2 (not illustrated), are organized in parallel with feet and tail medially and face laterally. Common body parts are represented at the boundaries between areas, whereas the representations of more distal parts are roughly but not exactly mirror images.

What is not illustrated in Fig. 12-8 is the fact, described subsequently, that the different submodalities of somatic sensibility are differentially represented in each of these cytoarchitectural fields. The general conclusion from these recent experiments is one of great importance: the x-y defining parameters for the larger processing units of the postcentral gyrus are place (mediolaterally) and modality (anteroposteriorly).

Lemniscal neurons from the level of the first-order fibers to postcentral cells are each preferentially activated by a particular type of mechanical stimulation of peripheral tissues. The modality classes of postcentral cells replicate those of the myelinated first-order fibers, listed in Table 12-1. These classes are distributed in a differential manner in the postcentral gyrus, a separation presaged by the modality segregation that occurs in ascending spinal pathways. Almost all the neurons of the most posterior zone, area 2, are activated by the rotation of joints or mechanical stimulation of periosteum or fascia, whereas the majority of those of area 3b are linked to slowly adapting and those of area 1 to quickly adapting cutaneous afferents.[84,94,126,127] The neurons of area 3a, a region considered to be either transitional between motor and sensory cortex or a part of the motor field,[52] are activated by stretch afferents from muscle.[64,87]

Columnar organization of somatic sensory cortex[4,72,94]

A description was given in Chapter 9 of the columnar organization of the neocortex, namely, that the basic functional unit of the neocortex is a vertically oriented group of cells extending across all the cellular layers and heavily interconnected in the vertical direction, sparsely so horizontally. Such a unit is capable of complex input-output operations, independently of extensive spread of neural activity tangentially in the gray matter. The identification parameters of such a "macrocolumn" of SI are the static ones of place and modality, as described previously. These variables are set congruently by the segregated transsynaptic projections to the cerebral cortex of activity in small sets of first-order sensory nerve fibers having closely overlapping receptive fields and a common sensory transducer capacity, that is, modality (Fig. 12-9). There is preliminary evidence that more dynamic aspects of intracortical processing are also arranged in a columnar manner. Thus the somatic sensory cortex illustrates one of the salient features of the columnar mode of organization: it allows the mapping—the "representation"—of a number of variables within the two-dimensional matrix of the cerebral cortex, with preservation of an orderly topology.

The columnar organization for place and modality has now been observed in each of the cytoarchitectural divisions of SI in a large number of animal species and under a variety of experimental conditions, including those of the waking,

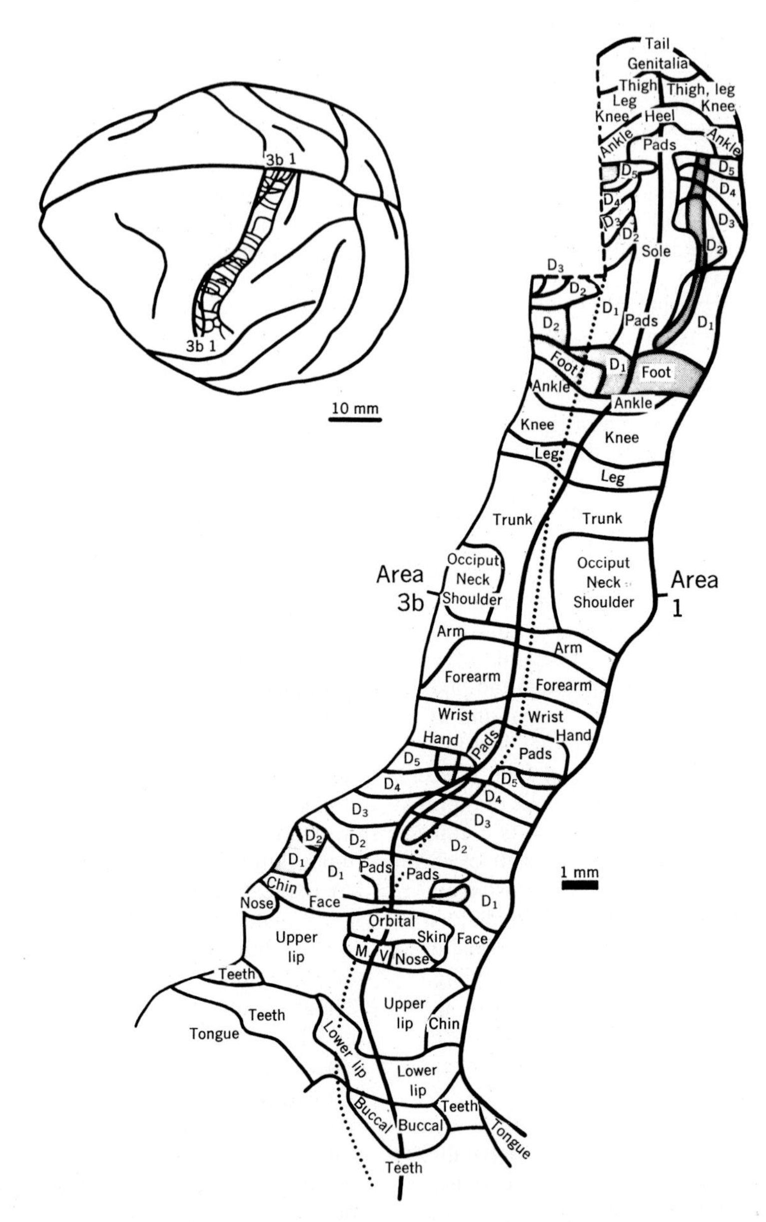

Fig. 12-8. Detailed map of representation of body parts in contralateral postcentral gyrus, somatic sensory area I. Inset shows drawing of lateral surface of hemisphere of old-world monkey, *Macaca fascicularis*. Larger map was constructed from results obtained in microelectrode mapping experiments in which many penetrations were made normal to cortical surface, close to each other. In each penetration, receptive fields and modality properties of cortical neurons were recorded. There is an independent map of contralateral body in cytoarchitectural areas 3b, 1, and 2 (the last not shown in this drawing). Fourth may exist in area 3a, just in front of area 3b; alternatively, map of area 3a may be congruent with that of motor cortex. Representation does *not* compose topologically intact map of body, nor can it be described as representation in terms of dermatomes. Its salient features are representation of body areas commonly found but not always continuously, and segregation of cortical neurons in four cytoarchitectural areas 3a, 3b, 1, and 2 by modality types. Defining parameters in x-y dimensions, for processing units of somatic sensory cortex, are therefore those of place and modality. (Courtesy Dr. J. H. Koos.[114])

Kaas

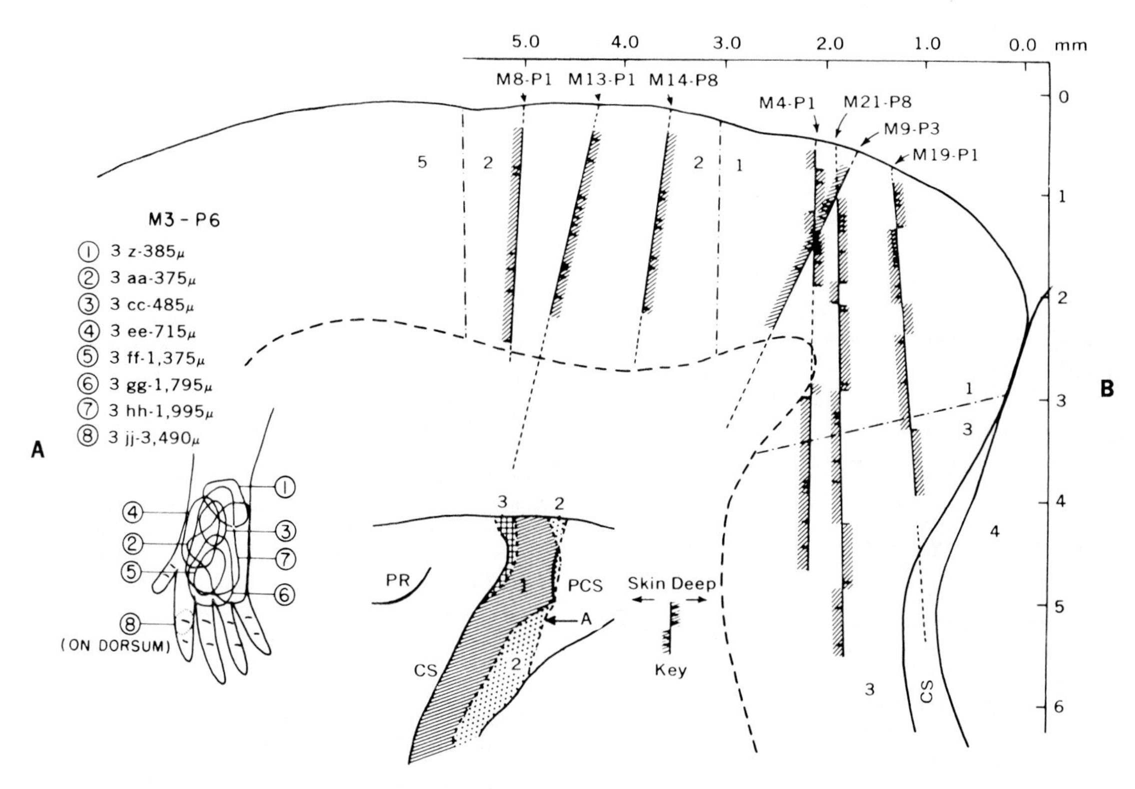

Fig. 12-9. Reconstruction from serial sections of seven microelectrode penetrations made in postcentral gyrus of macaque at level marked *A* on inset drawing of postcentral gyrus. Reconstructed tracts shown as if all were occurring on outline drawing of gyrus in cross section, with central sulcus, *CS,* now placed to right. Along each penetration, cross-hatching and horizontal lines to left indicate, respectively, a multi-unit record observed and an "isolated" neuron studied that was activated by stimulation of skin. Similar markings to right for those activated by stimulation of joint, periosteum, and fascial receptors. When penetrations are made normal to cortical surface and pass through cortex parallel to vertical columns of cells, all those encountered are of same modality type, are activated from superimposed peripheral fields (as shown for penetration *M3-P6* on inset, upper left), and respond with minimal differences in latency. Where penetrations pass at angles across columns, en bloc reversal of modality types are observed, for electrode tip is then passing successively from one cell column to another. (From Powell and Mountcastle.[94])

behaving monkey, and in the second somatic area (SII) as well.[130]

Somatic sensory area II[7]

Adrian and Woolsey discovered that the body surface is represented in a second cortical area (SII) that occupies the parietal cortex of the superior bank of the sylvian fissure in primates, including man. The body regions are represented in SII, as in SI, in the sequence of dermatomes, although the degree of overlap is greater in SII. A distinctive feature of SII is that both the *contralateral and the ipsilateral body halves are mapped to SII in a single superimposed image of the body.* The receptive fields of SII neurons consist of pairs of matched and symmetric body areas. The two parts of a field are disjoint if they occupy the apices of the limbs; for more proximal body parts the ipsilateral and contralateral receptive fields may be continuous across the midline.

Tabulation of the static properties of neurons of SII with regard to cortical location and depth reveals a columnar organization as precise as that of SI.[7] On other counts there are marked differences. First, no neurons of SII are activated by joint rotation. Second, SII neurons are less securely linked to peripheral input than are those of SI. Neurons of SII appear particularly sensitive to the direction of movement of a mechanical stimulus, and for a given neuron the optimal direction for excitation may be reciprocal in the two (the contra- and ipsilateral) parts of the receptive field. Whether SII plays a special role in processing for the special feature of stimulus movement is unknown.

The properties of the neurons of the transitional cortex linking SII with the auditory areas resemble those of the posterior nuclear group of the thalamus: these neurons show a polysensory convergence; many may be activated only by noxious mechanical stimulation; and they are commonly related to very large and often asymmetric receptive fields. In the cat, neurons of this type are distributed in columns that overlap the posterior border of SII[24]; in the monkey, they appear segregated in a well-circumscribed and homogeneous area.[130]

Precentral motor cortex. The precentral motor cortex area of the primate CNS is one of the major sources of the descending outflows via the pyramidal and extrapyramidal systems that terminate in the basal ganglia, brain stem, and spinal cord, described in chapters dealing with the central control of movement (Part VII). The region receives projections from the cerebellothalamocortical and the pallidothalamocortical systems, avenues allowing modulations of cortical efferent output by activity of the cerebellum and basal ganglia, respectively—the great reentrant circuits of the motor system. The pre- and postcentral gyri are interconnected by corticocortical fibers thought to provide an integration of sensory and motor activities at the cortical level.

The precentral motor cortex plays no role in any primary quality of somesthesis, although it receives a relatively direct projection of activity generated in peripheral mechanical receptors, thus forming afferent-efferent linkages at the cortical level, thought to play an important role in motor regulation (Chapter 32). The degree and tightness of this coupling differs markedly in different species. In marsupials the sensory and motor cortices are coextensive within a single topographic pattern.[61] In carnivores the cells of the motor cortex may be activated by stimulation of skin and deep tissues, particularly by stimuli that lead to movement.[15] Experiments employing microstimulation reveal a columnar arrangement within the cat motor cortex, with a tight coupling between input and output, that is, peripheral stimulation excites the cells in cortical loci, at which electrical stimulation elicits movements of the limbs in the direction of the stimuli delivered.[10]

It appears that the cutaneous projection to area 4 is much reduced in monkeys, perhaps limited to afferent signals from the volar surface of the hand, thought to play a role in tactually guided movement. The majority (71%) of area 4 neurons activated by passively delivered peripheral stimuli are related to joint afferents,[62,63] an afferent input thought to provide a feedback signal of joint position to the highest level of motor control. A smaller proportion (16%) is related to afferents from muscle, and there is some evidence that this muscle projection arises predominantly from the secondary spindle afferents, although no class of muscle afferents has been excluded.[49,92]

Studies in locally anesthetized humans undergoing intracranial surgical procedures for the treatment of epilepsy suggest that the direct mechanoreceptive input to the motor cortex is further reduced in humans, for potentials evoked there by stimulation of deep tissues could be recorded in some but not all subjects.[39] Recordings made from single neurons in the hand area of the human motor cortex, however, yielded results resembling those in the monkey, but the population of neurons studied so far is necessarily small.[40]

In summary, there appears to be an important difference between man and experimental animals, so that the afferent projections to the hand area of the motor cortex in humans is much reduced from that in monkeys and is exclusively kinesthetic in nature. It appears that the function of processing of sensory input from the periphery is important for the central regulation of movement and is a prominent feature of the motor cortex of the cat, much less so in that of the monkey, and much reduced in man.

Thalamo- and corticocortical connectivity in somatic afferent system

Much of the foregoing description of the somatic sensory thalamic and cortical regions was synthesized from classic anatomy and from the results of electrophysiologic experiments. Important new advances have recently been made in detailing the neural connections between somatic areas of the thalamus and cortex using a battery of new anatomic tracing methods that depend on the bidirectional axonal transport of identifiable molecules (Chapter 9). These studies include those of the intra- and interhemispheric connections between somatic sensory and other cortical areas, which together are thought to compose the anatomic substratum for the perceptual elaboration of somesthetic experience, as well as those allowing somatic sensory guidance of skilled movements.

The VB thalamic complex projects in an organized manner to all the cytoarchitectural divisions of SI,[56] thus accounting for the lemniscal properties of SI neurons. The differential distribution of modality types in the brain stem sensory nuclei and in the VB complex is reflected in a similar differential distribution in SI. The VB complex projects also on SII, but it is uncertain if the dual projection of this nuclear complex arises from a single or separate cell populations; neither explains the ipsilateral projection to SII. The medial portion of the posterior nuclear complex projects on the transitional cortical area linking SII to the auditory cortex, referred to previously, which may be one cortical target in the central pain pathway.[21] Both SI and SII project backward to the ipsilateral VB complex, in a strictly ordered topographic fashion, so that only those local zones or cortex receiving fibers from a local zone of VB complex project back on it. Centrifugal fibers therefore respect both the topographic and modality properties of small groups of cells in thalamic relay nuclei. Both SI and SII project to the restricted part of the posterior nuclear group, which receives ascending spinal and lemniscal fibers, but with little sign of topographic organization.

SI projects via the corpus callosum on somatotopically homologous areas of SI and SII of the opposite hemisphere.[3,53,54] The SI projection derives only from the body, head, and proximal limb areas of the cortical projection pattern. The apices of the limbs, those regions most highly developed for somesthetic function (in the monkey, *both* the hand and the foot), are free of callosal connections. The precentral motor cortex, area 4, is reciprocally connected with area 2 of the postcentral gyrus, as is area 3a with area 1. In addition, there is a "stepwise outward progression from the main sensory areas within (i.e., into) the parieto-temporal and frontal lobes, with an interlocking of each new parieto-temporal and frontal step."[55] The three cytoarchitectural divisions of SI project backward on area 5, which is reciprocally linked with area 6 of the frontal lobe. Area 5 in turn projects backward on area 7, and area 7 is reciprocally connected with area 46 of the frontal lobe; areas 46 and 6 are linked to each other. In these successive steplike projections, general somatopy is successively reduced, and there occurs an increase in intrasystem convergence. The third-order projection areas in turn project on the areas of the frontal, orbitofrontal, and temporal cortices and in those steps converge with similarly elaborated projections of the visual and auditory systems. With this intersystem convergence there appear projections from the convergent areas on those regions of the brain thought to be concerned with memory (the entorhinal cortex and the hippocampus) and with affective behavior (the limbic cortex). The detailed connections of the somatic sensory cortex have been greatly elaborated by recent studies of Powell and Jones and their colleagues; the interested student should consult Jones et al.[52] and Shanks et al.[108,109] The results obtained by them are consonant with the classic hierarchical model of brain organization.

Recapitulation: summary and comparison of lemniscal and anterolateral systems

The meaning for somesthetic function of the duplicate representation of the somatic afferent system in the cerebral cortex and of the several prethalamic systems projecting on the forebrain is still to a considerable degree uncertain. The lemniscal system, by virtue of the properties of its neurons, appears designed to serve the discriminative forms of somesthesis, for it can present at the cortical level neural transforms signaling the senses of touch-pressure, kinesthesia, and flutter-vibration. This conclusion fits with clinical and experimental observations, for lesions of this system produce severe deficiencies in these quantitative aspects of somatic sensibility.

The anterolateral system seems to serve more general aspects of sensation and to transmit information about certain qualitatively distinct sensations, namely, those of a thermal and painful nature (Chapter 13). It provides much less precise information concerning the place, spatial pattern, and temporal cadence of stimuli. Its wide projection via the spinobulbar and spinoreticular systems on the ascending reticular and generalized thalamocortical systems indicates its potent role in arousal and in the control of forebrain excitability. It is not yet possible to define precisely the contributions of the first and second cortical somatic areas to function. To what extent each contributes and how they must work synergistically in the cortical elaboration of afferent activity leading to the perception of somatic sensory events in all the shades and degrees of which human beings are capable is an important problem for future study. Contributions toward its solution have come from the study of humans and animals after lesions of afferent pathways or of the cerebral cortex, a body of information detailed in a subsequent section.

LEMNISCAL MECHANISMS IN TACTILE, VIBRATORY, AND KINESTHETIC SENSATION

The static and dynamic functional attributes of lemniscal neurons will be used to describe, as far as presently possible, the central neural processing mechanisms in the tactile, vibratory, and kinesthetic forms of mechanoreceptive sensibility. The system transmits a precise neural transform of peripheral events for initial cortical processing, but little is known of intracortical mechanisms beyond that first stage. It is hoped that study of the initial cortical transform of afferent signals and their comparison with both the overt reactions of human subjects to sensory stimuli and their subjective descriptions of the experiences those stimuli evoke will lead to testable hypotheses concerning the neural mechanisms intervening between system input and output.

Neural mechanisms in tactile sensibility

The tactile sense depends on information about the place, intensity, and temporal and spatial patterns of neural activity evoked by mechanical stimulation of the skin. Identification of spatial pattern depends on the spatial distribution of the intensity of a stimulus and detection of the direction and speed of stimulus movement across the skin on the sequential spatial and temporal translations of activity across neural populations.

More subtle aspects of the tactile sense depend on combinations of these three: place, intensity, and temporal order.

The lemniscal system is highly organized to preserve the local sign of a peripheral stimulus at successively more central levels, yet the divergence within even this highly specific system is such that the capacity for localization or identification of the spatial contours of stimuli cannot be explained in terms of isolated neural lines. It may be understood, at least in part, in terms of the contoured profiles of neural activity evoked at the peripheral and each central level of the system and especially by mechanisms for sharpening and preserving those activity profiles.

A suggestion of the shape of the distributed neural activity evoked in the population of first-order fibers engaged by a spatially distributed mechanical stimulus is gained by experiments of the sort illustrated and described in Fig. 12-10. Here, use is made of the supposition that the single neural element under study has occupied at each step of the serial experiment a series of different positions in an active neural population, positions occupied by adjacent neurons when the stimulus is delivered to a single locus. The contour of such a graph can then be interpreted to represent the spatial and temporal "contours" of the neural activity in central populations. *It indicates the transform in neural space of the intensity, contour, and location of the peripheral stimulus.* The cascaded divergence of the lemniscal system means that the population of neurons activated by such a stimulus has a considerable spatial extent in its central projections that would limit severely the capacity for spatial discriminations. Something more is needed to limit lateral spread and to sharpen profile edges; that something is afferent inhibition.

This phenomenon is illustrated in Fig. 12-11. This postcentral neuron of a monkey responded to light pressure applied to its excitatory receptive field, and both its spontaneous and evoked activities were inhibited by light pressure delivered to a region of skin surrounding the excitatory receptive fields. The graph plots the inter-

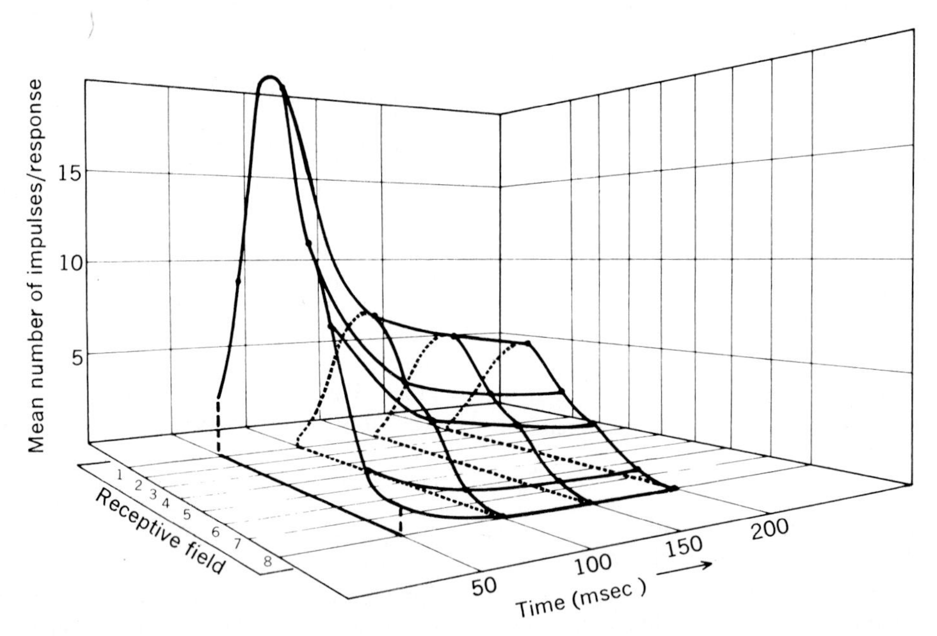

Fig. 12-10. Results of receptive field study in monkey of mechanoreceptive myelinated afferent of median nerve (slowly adapting type) that innervated peripheral receptive field extending across seven to eight dermal ridges of distal pad of thumb. A line bisecting receptive field forms left horizontal axis. Numbered dermal ridges occur at about 300 μm intervals; full field measured about 2.4 mm in diameter. Right horizontal axis is time, and vertical axis is mean number of impulses, occurring in each successive 50 msec periods, evoked by abrupt indentation of skin of supramaximal strength. Stimulating probe tip was 0.5 mm in diameter, machined to a one-third spherical surface. Contoured lines can be thought to represent distribution of activity in entire population of first-order afferents activated by stimulus, on the assumption that as stimulus is moved from place to place across receptive field from edge to center to edge again, fiber under observation occupies similar series of positions in populations of fibers activated at each stimulus position. Contours show high-frequency onset transient and decline to quasi-steady rate of discharge. (From Mountcastle et al.[79])

action of the two opposite influences on the response of the neuron. This form of spatially ordered afferent inhibition is a central neural event and has been observed at each level of the lemniscal system. Afferent inhibition, by limiting and shaping neural profiles, may contribute to two-point discrimination and hence to all spatial discriminations. The phenomenon is a general one in mammalian sensory systems; its role in the CNS mechanisms in hearing and vision is described in Chapters 15 and 18.

Recent studies of synaptic stations of the lemniscal system have shown that the mechanisms of afferent inhibition are the presynaptic and postsynaptic mechanisms already familiar in segmental reflex pathways. The presynaptic form is prevalent at the DCN,[8] the postsynaptic, in thalamus and cortex. A scheme of these relations applicable to any level of the system is given in Fig. 12-12. It is a general principle that inhibition is exerted by local, intranuclear mechanisms (i.e., no primary afferent neuron and no neurons with long axons linking DCN-thalamus-cortex in either direction exert inhibition by direct monosynaptic action). The shaping effect of inhibition might occur in either or all of three ways:

1. *Feed-forward mechanisms,* via branches of axons entering a nuclear region and terminating (a) on excitatory interneurons that in turn end on the terminals of neighboring entering axons, thus exerting presynaptic inhibition (p. 198), or (b) on inhibitory interneurons that terminate on the cell bodies of adjacent relay neurons, exerting postsynaptic inhibition (p. 196).

2. *Local feedback mechanisms,* via recurrent collaterals of axons leaving a relay nucleus, inhibiting the relay of activity in the belt of surrounding neurons, by either of the mechanisms listed under 1.

3. *Reflected feedback mechanisms,* via descending elements from more cephalad regions of the forebrain that, impinging on more inferior relay regions from above, may shape the spatial extent and form of the relayed activity, again by either of the mechanisms given under 1. There is evidence that such reflected feedbacks may also be positive in nature. The possibility of positive feedback to the center of an active nuclear relay zone, combined with negative feedback to the surround, would at the same time limit the spatial spread of that discharge zone and facilitate the relay of impulses through its center.

Intensity functions and linearity of transmission in lemniscal system

The neural scaling of stimulus intensity is particularly important for tactile sensibility, for which the degrees of freedom are only place, intensity, and temporal pattern. The slowly adapting, myelinated, mechanoreceptive afferents of large caliber that innervate the glabrous skin of the hand are, unlike quickly adapting

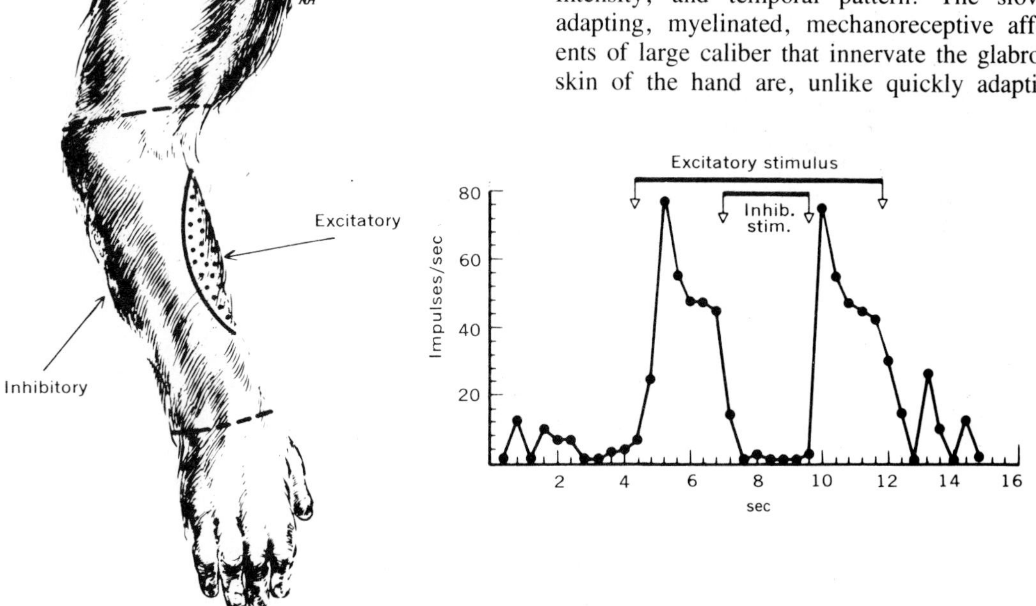

Fig. 12-11. Illustration of interaction of excitatory and inhibitory effects on neuron of postcentral gyrus of monkey that was produced by stimuli delivered to its peripheral receptive field on contralateral forearm. Cell was excited by stimuli delivered to field on preaxial side of arm and inhibited by stimuli delivered anywhere within large surrounding area (only dorsal half is shown). Graph plots impulse frequency vs time during excitatory-inhibitory interactions. Application of excitatory stimulus evoked high-frequency onset transient discharge that declined toward steady plateau until interrupted by application of inhibitory stimulus. On removal of latter, sequence was repeated in response to continuing excitatory stimulus. (From Mountcastle and Powell.[78])

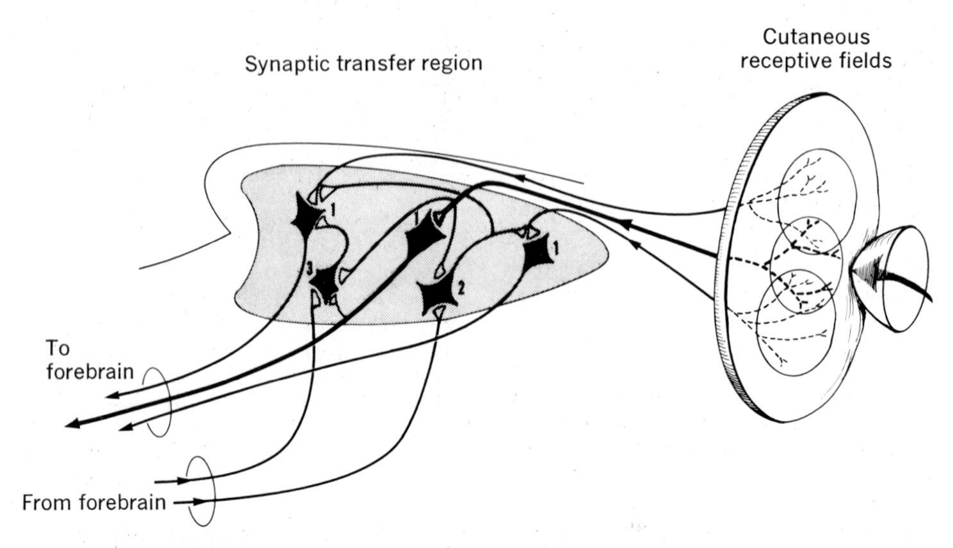

Fig. 12-12. Types of synaptic linkage subserving afferent inhibition in synaptic transfer region within lemniscal system. Input axons may terminate on primary relay neurons, *1,* on interneurons, *2,* with axoaxonal synaptic contact on neighboring entering fiber terminals, or on interneurons, *3,* which themselves possess polarizing synaptic terminals on adjacent relay neurons. Type 2 cells subserve presynaptic inhibitory mechanism and type 3 cells are interneurons in postsynaptic inhibitory pathway. Both presynaptic and postsynaptic inhibitory interneurons may be discharged via recurrent collaterals of axons of relay cells and via descending pathways from forebrain. Not all such interneurons, however, will have excitatory inputs from each of these sources. Relative inputs from these different sources will vary with different interneurons in nucleus and in different nuclei.

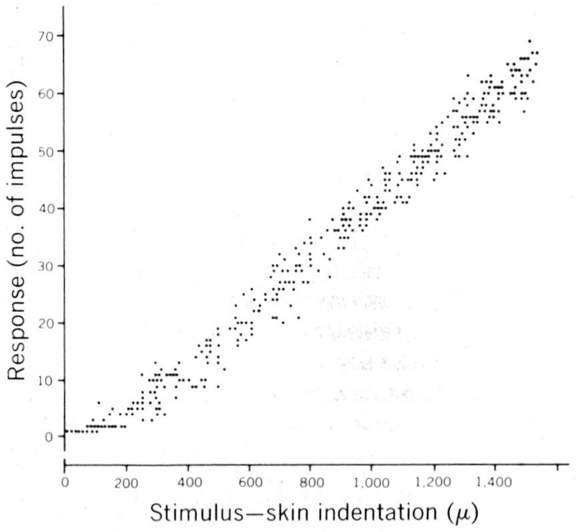

Fig. 12-13. Stimulus-response relation for slowly adapting myelinate fiber innervating glabrous skin of monkey hand. Stimuli were step indentations of skin of 600 msec duration, delivered in random order of intensity at rate of 12/min. Stimulator tip 2 mm in diameter, machined to one-third spherical surface. Number of impulses evoked by each stimulus is plotted as function of measured intensity of that stimulus. (From Mountcastle et al.[79])

afferents, sensitive both to the rate of a stimulus application and to its intensity (Chapter 11). Together with the central lemniscal neurons to which they are linked, these fibers can account for the intensity discriminations that humans make with their hands. A stimulus-response relation for such an afferent is shown in Fig. 12-13.[79] It is linear, and the degree of variability between responses is small and independent of response level. Thus for this afferent the initial transformation across the skin is linear, on the assumption that the operative neural code is that of frequency (p. 337). Experiments of this same type carried out for neurons of the VB complex of the thalamus and the postcentral gyrus in the monkey have yielded similar results. The implication is that transmission across the intervening nuclear relays imposes no further transformation on the first-order input with regard to intensity. The quantitative relation between the peripheral stimulus measured on some physical scale and the neural activity scaled in frequency is identical at all levels.

The perceptive and interpretive mechanisms of the brain intervene between the postcentral gyrus and a behavioral output evoked by such a mechanical stimulus. Although little is known of those mechanisms, the output can be measured. The result of such an experiment in a human subject is given in Fig. 12-1. In this experiment the series of stimuli delivered was exactly similar to the series used in the study of monkey first-order fibers. The response variable was, however, the human subject's estimate of the magnitude of each stimulus. The result indicates that whatever the intervening transformations may be, they are linear in sum. Findings for other sets of afferents in this and other systems are similar and lead to the general conclusion that *the relation of the human to the external environment, with regard to the variable of stimulus intensity, is set by the transfer properties of the peripheral terminals of first-order fibers.*

Harrington and Merzenich[46] have tested this generalization by study of the hairy skin of monkey and man. They found that if the stimulus probe tip is moved only a few centimeters from the palmar to the hairy skin of the adjacent forearm, the subjective magnitude estimation function for human observers changes from a linear to a negatively accelerating one, best described by a power function with an exponent of about 0.5. They then confirmed the fact that the stimulus-response function for large, myelinated, slowly adapting mechanoreceptive afferents innervating the hairy skin of the monkey's arm is also negatively accelerating and also best described by a power function with an exponent of about 0.5.

Sense of flutter-vibration and its neural mechanisms[58,74,115]

A sinusoidal mechanical stimulus applied to the body surface evokes a sensation humans identify as qualitatively different from other mechanical stimuli. This sensation of vibration consists of two different but related qualities. The first is a sense of local cutaneous flutter evoked by low-frequency oscillations in the range of 5 to 40 Hz. The second is the more diffuse and penetrating sense of hum evoked by sinusoids in the range of 60 to 400 Hz. Frequencies above 500 to 600 Hz are felt as stationary. There are two transitions in the sensory experiences evoked by vibratory stimuli of gradually increasing amplitude. The first is the absolute threshold, that level at which a sensory event is first detected but at which the frequency of the stimulus cannot be identified nor frequency discriminations made. The second occurs with stimuli some 7 to 8 dB greater in amplitude, at which levels the sensory qualities of flutter or vibration can be identified and frequency discriminations made. The difference between these two thresholds is called the *atonal interval*. A similar interval occurs in the sense of hearing (Chapter 15). Some results of psychophysical studies of flutter-vibration in humans and monkeys are given in Fig. 12-14. Psychometric functions obtained with stimuli at 30 Hz are shown in *A*. Repetition of this threshold measurement over the frequency range from 2 to 400 Hz yields the frequency-threshold curves shown in *B*. It is obvious that humans and monkeys possess thresholds and frequency ranges that are almost identical. For both, thresholds are much lower for high than for low frequencies and are lowest at about 250 Hz. Humans and monkeys discriminate between vibratory stimuli of the same frequency but different amplitudes by about 10%, and they discriminate between those of the same subjective amplitude but different cycle length (frequency) by about 5% to 10%. These Weber fractions are constant over a wide range, but trend upward at low and high values of the comparison stimuli. Finally, the intensities of vibratory stimuli of different amplitudes are rated by humans (in subjective magnitude estimation experiments) in a nearly linear relation to actual stimulus amplitudes. It has not yet been possible to carry out such rating experiments in monkeys.

The sense of flutter-vibration is of general interest with regard to central neural mechanisms because the periodic nature of the stimuli and the neural activity evoked by them at all levels of the somatic system allow study of a neural code depending on temporal order, as well as the dynamic frequency range of the system of lemniscal neurons linking periphery and cortex. The explanatory power of the results of experiments in monkeys, for understanding human sensation, is greatly strengthened by the observation that the sensory performance of monkeys equals that of man, at least for the modes of somesthesis tested. Direct correlations can be made between the sensory performance of monkeys and the

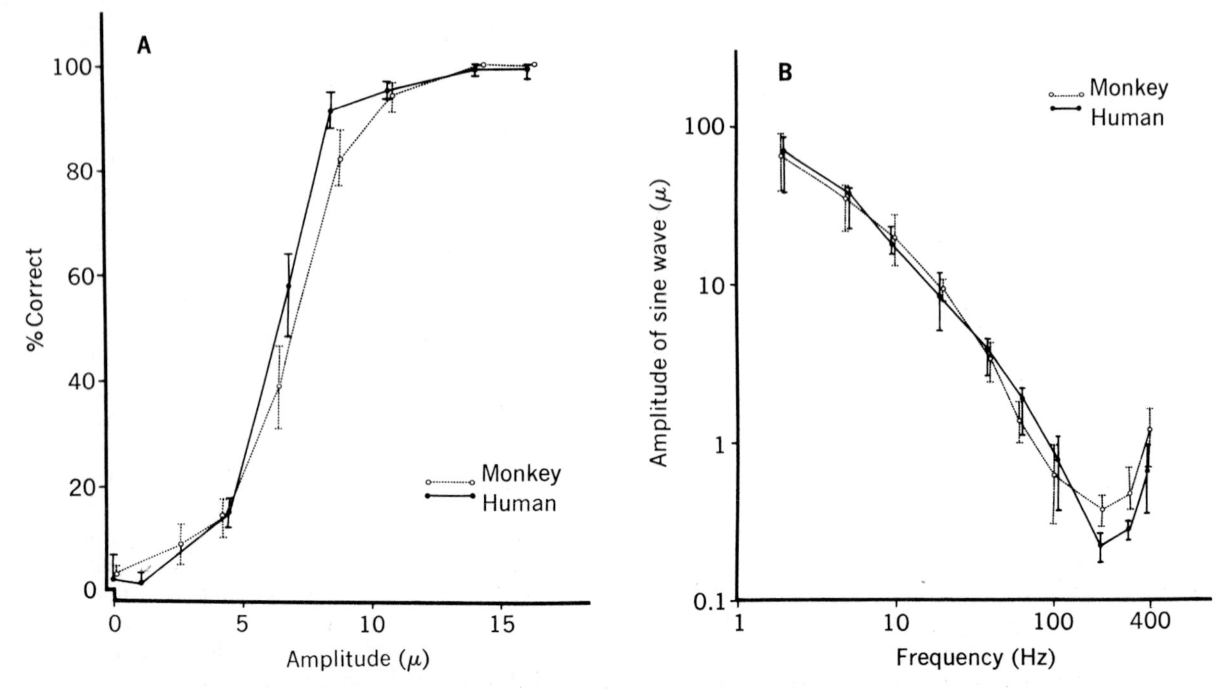

Fig. 12-14. A, Average psychometric functions for groups of human and monkey subjects asked to detect presense of 40 Hz mechanical sinusoid delivered to glabrous skin of their hands. Vertical lines = 2 × SE of means. **B,** Frequency-threshold functions for groups of human and monkey subjects asked to detect the presence of mechanical sinusoids of different frequencies, delivered to the glabrous skin of their hands. Vertical lines = 2 × SE of means. (From Mountcastle et al.[74])

neural events evoked in their nervous systems by the same sets of stimuli. Frequently the two sets of observations can be made in waking monkeys as they execute the relevant sensory tasks. It seems reasonable to conclude that the neural events observed in the monkey are comparable to those evoked in man by the same stimuli delivered under comparable behavioral conditions.

Analysis of the functional properties of the myelinated mechanoreceptive afferent fibers innervating the glabrous skin of the monkey's hand has provided evidence that this dual sense of flutter vibration is served by two distinct sets of afferent fibers. The first terminates in the Meissner corpuscles of the dermal ridges, adapts quickly to steady stimuli, and is differentially sensitive to low and not high frequencies, best at 30 to 40 Hz. The second terminates in the pacinian corpuscles of the subcutaneous tissues, is also quickly adapting, and is especially sensitive to high and not low frequencies, best at about 250 Hz. The local iontophoresis of cocaine into the superficial layers of the glabrous skin anesthetizes the terminals of the Meissner afferents without affecting the pacinian corpuscles and elevates thresholds in the low-frequency but not the high-frequency range. Fig. 12-15 illustrates in the form of interval histograms the patterns of discharge in a fiber of each class, activated by a vibratory stimulus at each fiber's best frequency, at a number of stimulus amplitudes. In each case the lowest histograms display analyses of the responses evoked at the *absolute thresholds*. They show that nerve impulses occurred at the cycle length only rarely and more commonly at interval multiples of the cycle length (i.e., the frequency of the stimulus was not directly preserved in the neural signal). However, for each fiber, stimuli at somewhat greater amplitudes evoked a perfect entrainment of the neural discharge locked at one impulse per stimulus cycle. This latter amplitude level is defined as the *tuning threshold,* for at this level the stimulus frequency is perfectly coded in the afferent neural discharge. Repetition of this experiment at a number of frequencies yields U-shaped absolute threshold and tuning curves for fibers of each class. The tuning curves are displaced upward on the amplitude axis from the absolute threshold curves by 6 to 8 dB. The conclusions are that the dual sense of flutter-vibration depends on two sets of primary afferent fibers, that a sensory

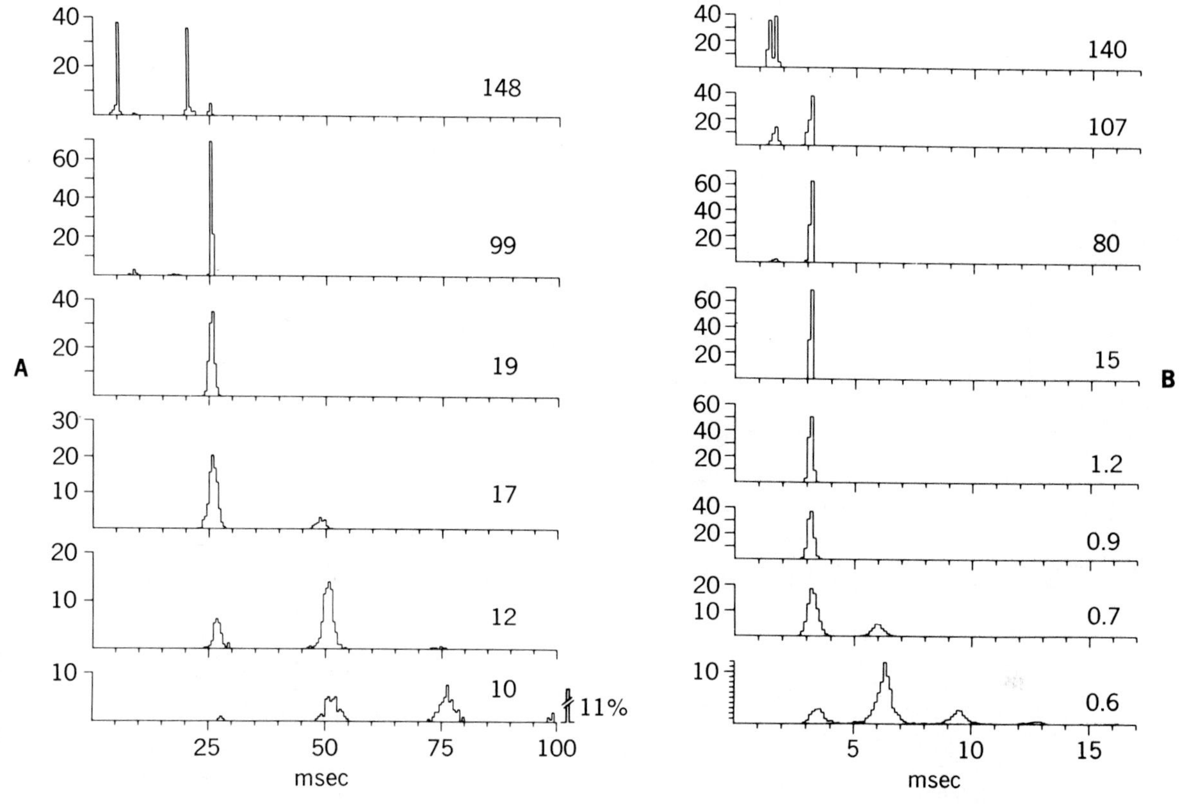

Fig. 12-15. Interval histograms for responses of Meissner afferent, **A,** and pacinian afferent, **B,** each dissected from median nerve of deeply anesthetized monkey for recording. Mechanical sine waves were delivered to center of receptive fields innervated by fibers, on volar surface of hand, at 40 Hz for Meissner afferent and 300 Hz for pacinian afferent. Lowest histograms were made from responses evoked by stimuli at absolute thresholds, 10 μm for Meissner and 0.6 μm for pacinian. They show that impulses occurred both at, and at interval multiples of, cycle lengths of stimuli. At stimulus amplitudes of 19 μm for Meissner and 0.9 μm for pacinian, fibers tuned perfectly, so that each interval between nerve impulses was locked at cycle length of stimulus. Doubling occurred for each fiber with very strong stimuli. (From Talbot et al.[115])

event is detected on the appearance of any activity in a small number of fibers of one or the other class, and that the capacity to identify frequencies and to discriminate between them depends on the appearance of periodically tuned neural activity at the input to the system. The psychophysical atonal interval described previously is determined by a similar ''neural atonal interval'' in the activity of first-order fibers.

The charts of Fig. 12-16 show the tuning points for populations of Meissner *(A)* and pacinian *(B)* afferents innervating the monkey hand, superimposed on the average behavioral frequency-threshold function for monkeys. The absolute thresholds for these fibers are shifted downward about 8 dB from the tuning points illustrated in Fig. 12-16. It is not possible to know at which of the two peripheral thresholds a monkey is responding in a detection task, but it is clear from Fig. 12-16 that the monkey's capacity cannot be accounted for in terms of the response properties of either set of first-order fibers alone, but requires both.

The importance of periodically tuned activity for sensory tasks more complex than detections is apparent when the monkey or human subject must discriminate between stimuli set at subjectively equal strengths, but of different frequencies.[58] Psychometric functions for frequency discriminations by a monkey subject are shown in Fig. 12-17; the DL for frequency was about 3 Hz at a base frequency of 30 Hz when stimuli were intense. Capacity deteriorated as stimulus intensity decreased, and at 16.5 μm significant discrimination could not be made. This amplitude was about 8 dB above the detection threshold at the base frequency, which indicates that this particular subject detected the presence of a

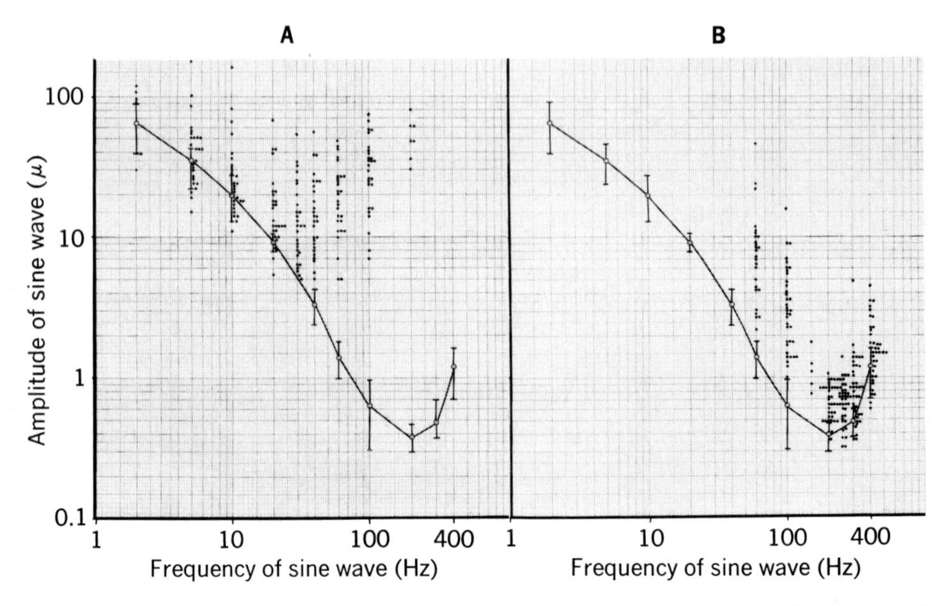

Fig. 12-16. Solid line in each graph is average of frequency threshold functions for six monkey subjects. Vertical lines at each tested frequency indicate ± 1 SE of means. **A,** Each dot indicates tuning point for quickly adapting, large myelinated mechanoreceptive afferent fiber innervating glabrous skin of monkey hand. Fibers were isolated for study by microdissection of median nerves; response of each was determined for a number of sine wave amplitudes at each of several frequencies. Thresholds of monkeys for perception of vibration in low-frequency range from 5 to 40 Hz could be accounted for by hypothesis that what is required for determination that mechanical stimulus is oscillating and not steady is appearance of tuned discharge in some small number of fibers of this class. **B,** Similar results for a number of pacinian afferent fibers terminating in hand. Similar statement can be made concerning appearance of tuned discharges in pacinian fibers in frequency range of 60 to 400 Hz and concerning sensation of vibration. (From Mountcastle et al.[74])

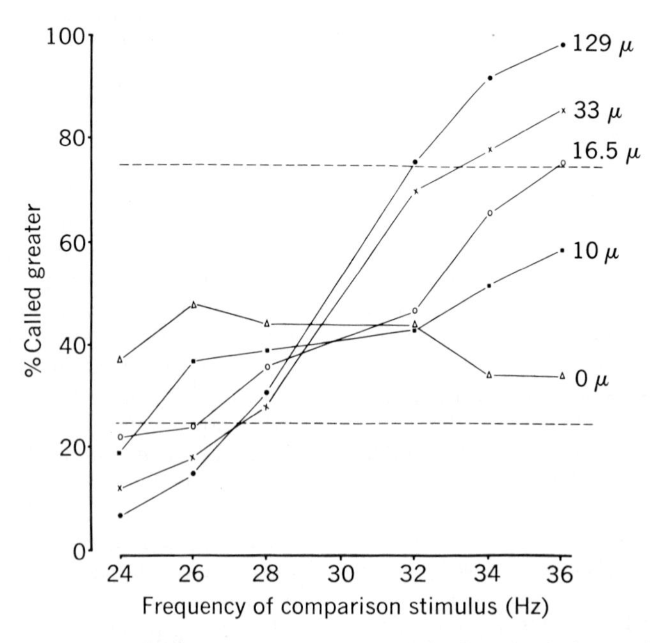

Fig. 12-17. Psychometric functions for frequency discriminations made by monkey subject, illustrating deterioration in performance as amplitudes of stimuli were decreased toward detection threshold, which was 5 μm. Figure alongside each function is amplitude value of 30 Hz standard stimulus, to which all comparisons of stimuli at test frequencies were matched in subjective intensity for each run. Results define atonal interval, in this case, amplitude range from threshold at 5 μm to about 16.5 μm. (From LaMotte and Mountcastle.[58])

stimulus at or close to the absolute threshold of peripheral fibers, but could make frequency discriminations only when stimulus amplitudes exceeded the tuning thresholds for some small set of Meissner afferents.

The idea that periodicity is the peripheral and central neural code for flutter-vibration has been tested further in studies of the responses of postcentral neurons to mechanical sinusoids delivered to the hands of unanesthetized monkeys.[80] The analysis of the records of such a neuron, shown in Fig. 12-18, reveals the periodic nature of the discharge of a cortical neuron linked to Meissner afferents, a periodicity that appears at about the average monkey threshold and increases gradually with increases in stimulus amplitude. More recent experiments in which sensory detection and cortical neuronal activity were observed simultaneously in waking, behaving monkeys[22] reveal strikingly parallel increases in periodicity and the likelihood of stimulus detection, but no causal relation has yet been established.

It is obvious from the fact that Meissner afferents have parallel tuning curves that frequency discrimination in the low-frequency range cannot be made on the basis of which fibers are active. It is unlikely that such discrimination can be made on the basis of the differences in the total number of impulses evoked by two discriminable stimuli, for the cycle to cycle jitter in the responses or cortical neurons is such that the overall average frequency of discharge in the two cases might be identical over the short stimulus durations required for discrimination. All the available evidence suggests that it is the dominant interval in the periodically entrained neural activity that is the relevant signal and that the differential recognition of frequency of an oscillating mechanical stimulus in the range of flutter depends on a CNS mechanism for discriminating between different dominant periods in the activity of postcentral neurons.

The human subject's estimate of the intensity of a low-frequency mechanical sinusoid delivered to the hand is a nearly linear function of stimulus amplitude. Yet there is no obvious linearity

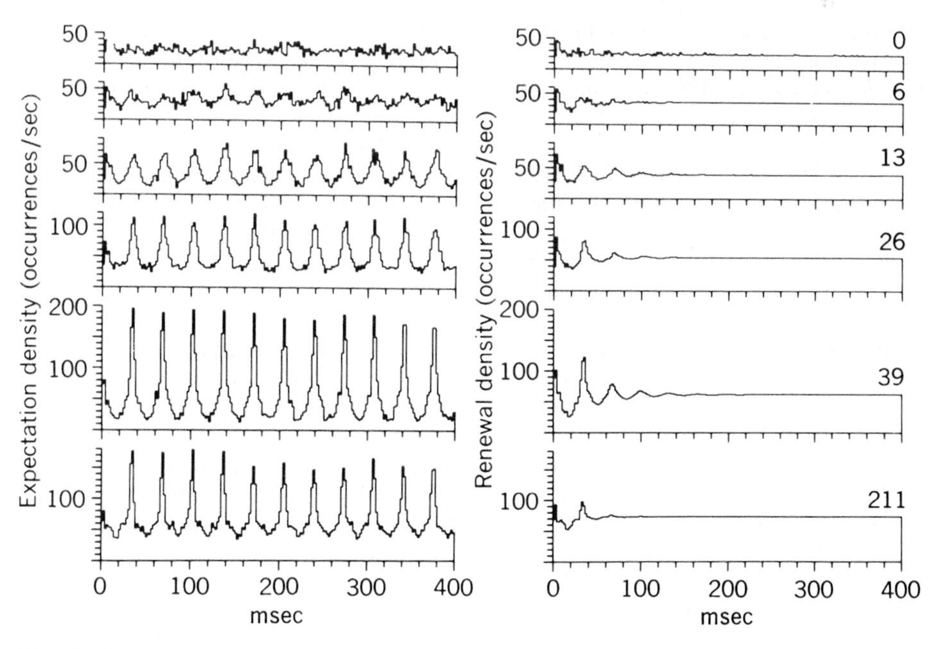

Fig. 12-18. Expectation density and renewal density histograms for responses of neuron of postcentral gyrus of unanesthetized monkey linked to quickly adapting mechanoreceptive afferents innervating glabrous skin of contralateral hand. Driving peripheral stimulus at 30 Hz, delivered at a number of different amplitudes, indicated in micrometers to far right for each pair of histograms. Expectation density analysis resembles autocorrelation function and shows likelihood of occurrence of impulse in each small sequential period of time. Histograms show strong replication of driving stimulus in rhythmic discharge of cortical neuron, suggested at stimulus amplitude of 6 μm and clear and strong at 12 μm. Monkey detection threshold at 30 Hz is 6 to 8 μm. Renewal density analysis is essentially repetition of expectation density after random shuffling of impulse sequence. Results show importance of *temporal order in which impulses occur* for this particular neuronal code. (From Mountcastle et al.[80])

Table 12-2. Peripheral and cortical neuronal coding mechanisms for sense of flutter*

Psychophysical event	Peripheral neural code	Central neural code
Identification: flutter or vibration?	• Place code: which set of peripheral fibers is active	• Place code: which set of cortical neurons shows an increment in activity
Detection of flutter	• Place plus frequency code: appearance of any activity in Meissner afferents	• Place plus frequency code: increment in activity in set of postcentral cortical neurons on which Meissner afferents project
Frequency discrimination	• Place plus temporal order code: appearance of tuned discharges in some small number of Meissner afferents	• Place plus temporal order code: increase above a certain minimal level of cyclic entrainment of activity of that set of postcentral cortical neurons exclusively activated by Meissner afferents
Subjective magnitude estimation	• Place and spatial distribution code: linear increase in size of population of Meissner afferents activated by stimulus	? Place and spatial distribution code: linear growth in size of population of cortical cells in which increments of activity occur
Amplitude discrimination	? Place and spatial distribution code: differences in size of active population of Meissner afferents in two cases discriminated	? Place and spatial distribution code: differences in size of population of crotical cells with incremented activity in two cases

*Codes preceded by a bullet are established with reasonable certainty; those preceded by a question mark are considered to be reasonable inferences from the information available. (From LaMotte and Mountcastle.[58])

in the response of the Meissner afferent to account for this, for the increment in neural activity occurs in a discontinuous, although monotonic, fashion with increases in stimulus amplitude; above the tuning point there is a wide range of increasing intensity that produces no further change in response until very intense stimuli elicit double discharges per cycle (Fig. 12-15). Johnson[51] was the first to show that the spatial recruitment of Meissner afferents to the active population occurs in a linear fashion with increments in stimulus amplitude. This linearity holds for the number of active fibers as well as for the total activity within that population. Thus the peripheral code for the intensity of a vibrating stimulus is a spatial one, namely, the overall size of the active population of neuronal elements engaged by the stimulus.

Table 12-2 presents an attempt to specify the peripheral and central codes thought to operate in the sense of flutter for the detection of a mechanical sinusoid in that frequency range, in frequency and amplitude discriminations, and in subjective magnitude estimations. Fig. 12-19 summarizes the detection and discrimination capacities of humans and monkeys and relates them to the neural atonal interval.

Senses of position and movement and their neural mechanisms

Humans possess an acute sense of the position of the body in the gravitational field and of the relative positions and movements of the body parts, a complex sense that depends on an integration of afferent information reaching the CNS over the visual, vestibular, and somatic afferent systems. This integration appears to be a special function of the areas of the parietal lobe between the primary sensory areas of the three systems involved. Closure of any one of these portals may distort but not destroy the concept of the body form and position. Congenitally blind individuals are not disoriented in space, and their drawings of themselves suggest that they conceive of body form in terms of its peripheral innervation density rather than actual size and shape. The manual recognition of the shape and size of objects is known as *stereognosis*, a complex sense thought to depend on a combination of tactile and joint afferent input.

Steady positions of the limbs are sensed and remembered with an accuracy greatest at proximal and least at distal joints.[17,95] A static position of one shoulder can be matched at the other with an error of only 2 to 3 degrees, even after an interval of several minutes. Change in the angle of the knee joint can be detected even when the movement producing the change is so slow that no movement is sensed.[48] Thresholds for detection of passive movements of the joints against relaxed muscles are functions of the rates of movement and may be as slow as 0.2 degrees for a passive abduction of the hop at 2 degrees/sec.[45] The capacity to detect the direction and

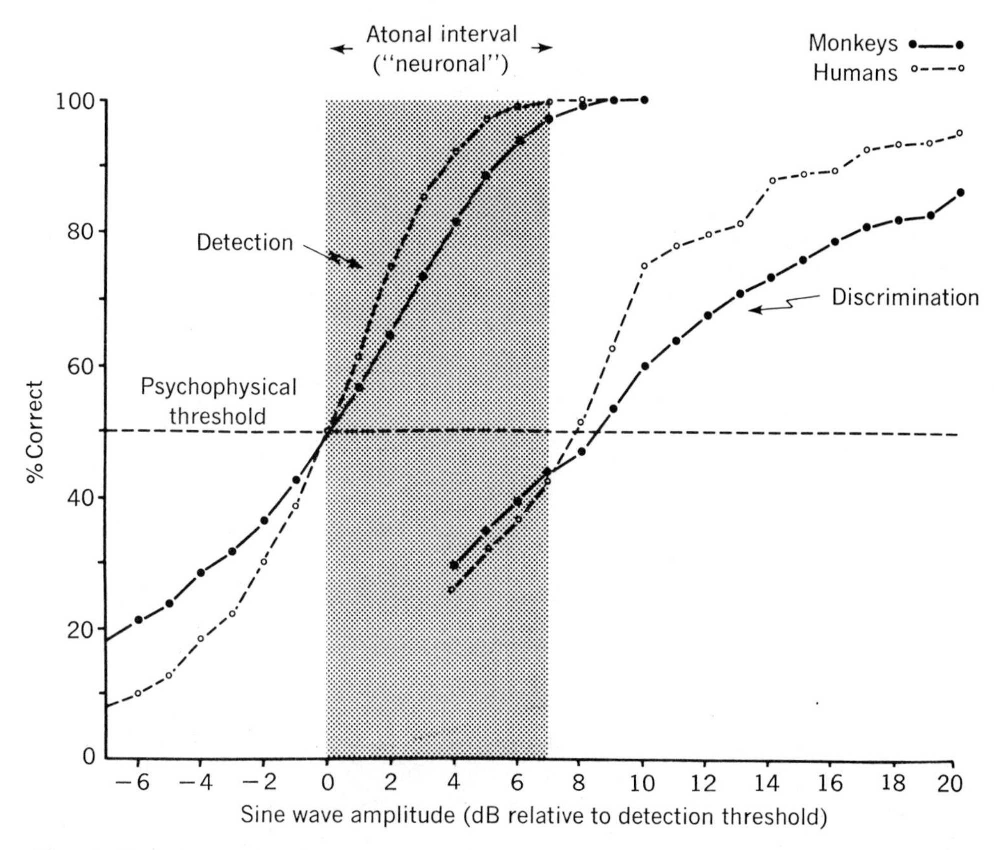

Fig. 12-19. Relation of detection and frequency discrimination thresholds for monkey and human subjects to neural "atonal interval" for peripheral afferent fibers for sense of flutter. Curves to left are psychometric functions for humans and monkeys for detection of 30 Hz stimuli as function of their amplitudes. Curves to right relate net correct score for frequency discrimination to stimulus amplitude. Shaded area is range of stimulus amplitude between threshold and tuning levels for group of Meissner afferents innervating glabrous skin of monkey's hand. Horizontal dotted line defines psychophysical threshold for both detection and discrimination tasks, placed arbitrarily halfway between pure chance and perfect performance. All amplitudes related to detection threshold, set to 0 dB for scaling purposes. (From LaMotte and Mountcastle.[58])

rate of movement is improved if passive movements are imposed during active contractions of the muscles operating across the joint moved, especially when movements are executed by the voluntary contractions of those muscles.

These simple observations suggest that the somatic sensory component of the sense of position and movement is itself complex and depends on the signals in sets of peripheral afferent fibers with different but overlapping functional properties. There appear to be five aspects of this sensibility that can be identified:

1. The sense of steady joint angles
2. The sense of the imposition of passive movements on the limbs at their joints, with muscles relaxed
3. The sense of movement produced by active muscular contraction (kinesthesis)
4. The sense of the tension exerted by contracting muscles
5. The sense of effort

There is now considerable evidence that both joint afferents and an unidentified set(s) of muscle afferents contribute to the sense of movement and position. The degree to which muscle afferents play such a role has been studied after selective elimination of the joint afferents, either by local anesthesia[17,33,95,98,99] or after replacement of a joint with a prosthesis.[26,45] Study of the contribution of joint afferent input without accompanying muscle afferent discharge has been accomplished by Gandevia and McCloskey,[33] who took advantage of the fact that in a certain posture the muscles operating at the distal interphalangeal joint of the middle finger are disengaged. An example of their results is given in Fig. 12-20. The graph above shows the elevation of the threshold for detection of movement at the distal interphalangeal joint of the middle finger, caused by muscular disengagement. A more severe and highly variable change followed differential elimination of the skin and joint afferents by local anesthesia, leaving the muscular afferents undisturbed (Fig. 12-20, *B*). The conclusion suggested is that the normal performance shown by filled circles in both graphs depends on a central integration of afferent signals in two or more quite different sets of peripheral afferent fibers. A number of similar studies indicate that the defect in the movement sense produced by local anesthesia of joints may be compensated to some degree by some subjects if they exert active opposition to the passively imposed movement

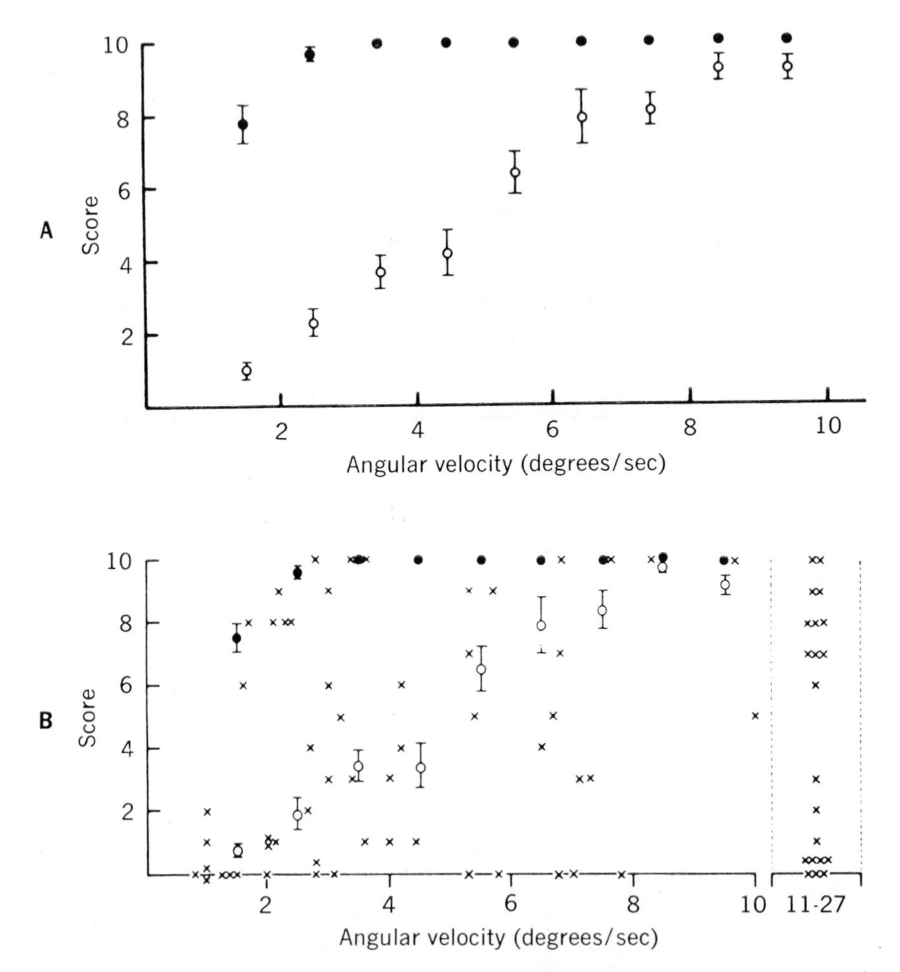

Fig. 12-20. A, Averaged scores for performance of 12 subjects asked to detect direction of movement of distal interphalangeal joint of middle finger at different angular velocities; perfect score = 10. Filled circles, normal conditions; open circles, after positioning finger so as to disengage muscles operating across joint. Performance at slow speeds was enhanced when both muscle and joint afferents were functioning. In each case, performance improved as angular velocity increased, although total angular displacement remained constant at 10 degrees. **B,** Results obtained under same general conditions for seven subjects, before and after test finger was anesthetized by digital nerve block, removing joint and cutaneous afferent contribution to position sense. Mean scores obtained with finger unanesthetized are shown for normal circumstances (filled circles) and after disengagement of muscles operating across joint (open circles). Crosses, individual scores made by subjects after finger anesthetized, that is, when muscle afferents alone could contribute to position sense. Performance of subjects using muscle sense was variable. No single set of afferents can alone account for normal performance. (From Gandevia and McCloskey.[33])

by muscle contraction. The sense of kinesthesia may be virtually normal when the joints are anesthetized, at least in some subjects.[99]

The afferents from skin have properties that might provide signals that a movement is beginning, but no more specific information. There is some evidence that they exert a more general facilitatory effect on the central nervous mechanisms in the senses of position and movement.

The afferents from joints contribute importantly to this sensibility. Slowly adapting afferent fibers terminate in Ruffini cell–like end-organs in the capsular tissue; others end in Golgi organs in the capsular tissue; and still others end in Golgi organs on the ligaments. The discharge of the former is affected to some extent by the action of muscles whose tendons cross the joint and may tense its capsule; that of the latter is unaffected by muscle action.[44] Fibers ending in paciniform corpuscles occur rarely. All these fibers are of the beta or group II size; they travel centrally in both muscle and cutaneous nerves—a "pure" nerve of either type probably does not exist. The response properties of joint afferents have been studied by a number of investigators[9,12,20] and summarized by Skoglund.[111] Only a small percentage are quickly adapting. The slowly adapting joint afferents respond to movement of the joint into their "excitatory angles" with a high-frequency discharge (Fig. 12-21) whose rate of change is a sensitive indicator of the direction and speed of movement. If the movement ends within the excitatory angle of the fiber, the frequency declines over several seconds to a lower and more or less steady frequency that may persist undiminished for minutes or hours. This continuing discharge is a variable determined by joint angle. Movements from a more to a less excitatory position within the excitatory angle produce a transient decrease in discharge, with recovery to a lower steady state—with a hysteresis due at least in part to long-term adaptation. The excitatory angles of joint afferents are of two types. A few are "double-ended" and placed at different positions along the range of joint movement; a much larger number are "single-ended" and discharge maximally at either full extension or full flexion. Thus the entire population of afferents supplying a given joint codes its steady position and the direction, rate, and extent of its movements.

Presently, which sets of muscle afferents may contribute to kinesthesia is unknown. The spindle afferents provide a sensitive measure of the velocity of muscle elongation and the secondary spindle afferents a signal of instantaneous length (Chapter 25). The discharge from spindle receptors is influenced by the level of activity of the

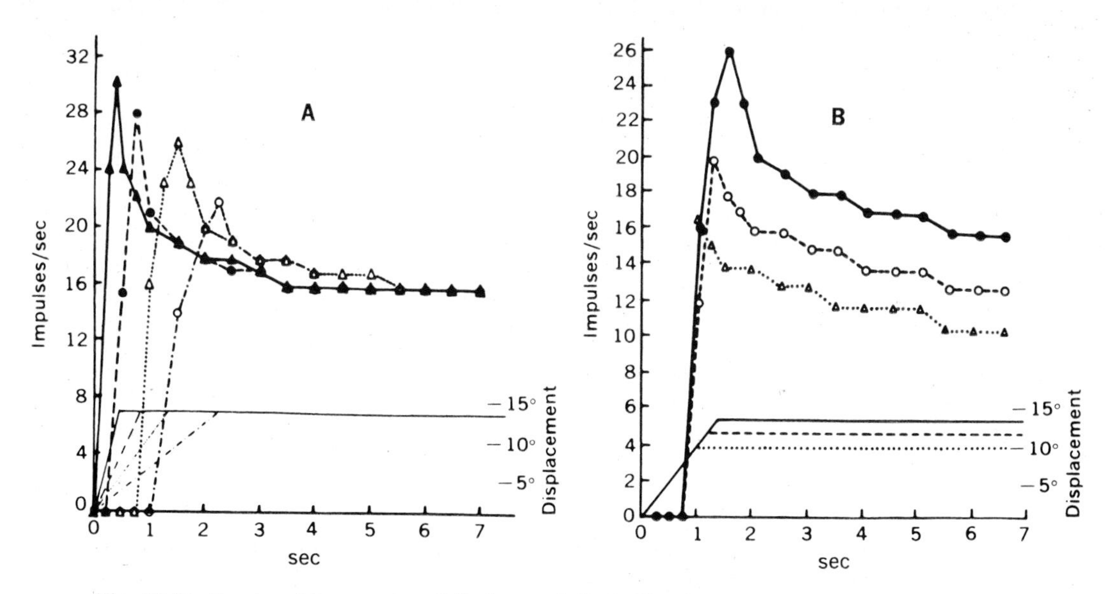

Fig. 12-21. Graphs of frequencies of discharge of single fiber innervating knee joint of cat. **A,** Response to flexion of leg through angle of 14 degrees at four different rates: closed triangles, 35 degrees/sec; closed circles, 17 degrees/sec; open triangles, 10 degrees/sec; open circles, 6 degrees/sec. Displacements are indicated by lines below. Steady impulse frequency with leg in final position is same in each case, but onset transient is function of velocity of movement. **B,** Frequencies of discharge of such a fiber during flexion of leg at knee at rate of 10 degrees/sec through three different angles, as indicated. (From Boyd and Roberts.[12])

fusimotor neurons that innervate them, so that the rates of spindle discharge may differ from one time to another for the same positions or movements. Thus on this hypothesis the CNS would need to sense and titrate the spindle afferent discharge and the level of activity in fusimotor neurons, derive from this an absolute measure of muscle length, and from that derive a transform into joint angle. The hypothesis remains to be explored.

Central neural mechanisms.[76,77] Afferents from the joints enter the DC and in the ascending reshuffle (p. 353) are sorted in such a way that quickly adapting afferents project directly to the DCN, whereas slowly adapting ones exit from the DC, and, after synaptic relay in the dorsal horn, second-order elements project upward via the dorsolateral columns to terminate in the more cephalad regions of the DCN and adjacent regions.[7] These latter are linked to VB thalamic neurons with similar properties.[76] There is a heavy projection of thalamic joint neurons to the postcentral gyrus, mainly to area 2.[77] This linked system of neural elements presents signals of the position and the velocity and direction of joint movement[133] for central processing.

The graph of Fig. 12-22 plots the discharge frequency of a thalamic joint neuron activated by movements of the contralateral knee, carried in steps through the excitatory angle of the cell. Each movement elicited a high-frequency transient discharge that quickly subsided to a quasi-steady rate determined by joint angle. The excitatory angle for this neuron was about 75 degrees in extent, with maximal excitation in full flexion. The relation between joint angle and discharge rate is negatively accelerating and described by a power function with an exponent of about 0.5 to 0.6, which resembles that characterizing the human subjective estimation of the degree of joint displacement.[45] Fitted functions for the responses of a number of thalamic joint neurons are shown in Fig. 12-23, where the data are presented on normalized scales to allow a synthetic reconstruction of the events in the population of thalamic cells as a joint is moved from one position to another through its full range. A majority of cells subtend angles greater than half the total range of joint rotation, and all the curves are "single-ended." The cortical joint neurons subtend similar broad, single-ended excitatory angles.[77] The first-order afferents from, for example, the knee joint of the monkey, are related to excitatory angles much narrower in extent, and a minority are located at different positions in the range of movement and are "double-ended." Thus thalamic and cortical neurons, by varying their rates of discharge in the manner shown in Fig. 12-23, provide running spatial integrals of the activity in a considerable number

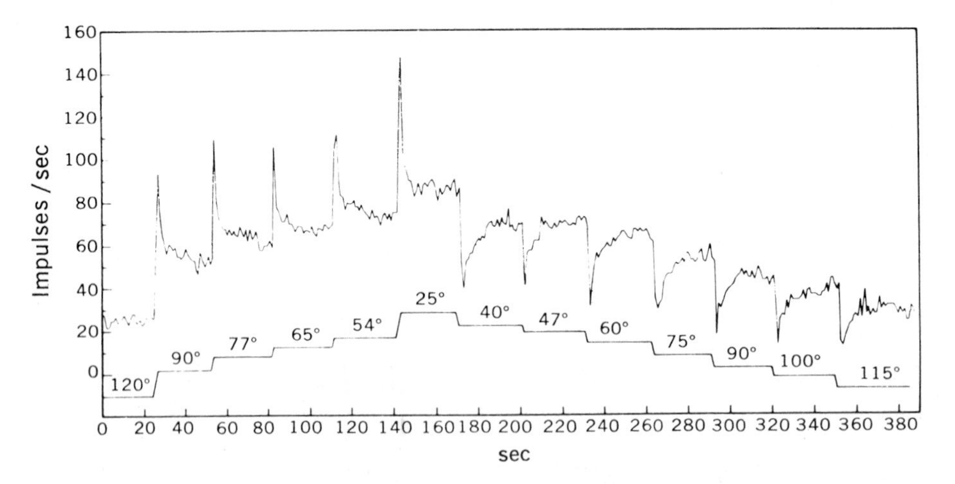

Fig. 12-22. Results of excitatory angle study of ventrobasal thalamic neuron activation by flexion of contralateral knee in unanesthetized monkey. Knee was rotated in short steps from position outside excitatory angle (from 120-degree joint angle) to full flexion (25-degree joint angle) and back again. Each movement in flexor direction produced transient discharge that declined over a few seconds to more or less steady discharge rate determined by joint angle. Movements in extensor direction produced, reciprocally, off-transients and recovery to plateau determined, once again, by joint position, but with obvious hysteresis. Plateau of discharge obtained on first limb of study used to plot excitatory angle functions such as those shown in Fig. 12-23. (From Mountcastle et al.[76])

of first-order fibers that must converge on them across the intervening synaptic relays. In a sense a combined spatial and intensive pattern of representation of joint position within the first-order neural population is converted at the levels of thalamus and cortex to a pattern varying only intensively in the domain of discharge frequency. At these higher levels the datum of the *position* of a given neuron in the neural field provides information concerning the joint moved; the degree of joint displacement is signaled by impulse frequency. It is common in recordings at thalamic and cortical levels to see pairs of adjacent neurons activated reciprocally by opposite movements of a joint.

Much less is known about the CNS projection of muscle afferents and the role those projections play in kinesthesis. It is clear from the description on p. 353 that the afferent fibers from muscle project into the dorsal ascending systems of the spinal cord and on the sensory-motor areas of the cortex, notably to area 3a, a region transitional between motor and sensory cortex.[49,87] A small class of neurons of area 5 is also activated by muscle afferents.[18,75,105] It is not known which classes of muscle afferents project in this way, nor how the information conveyed over this component of the somatic afferent system may be integrated at the cortical levels with that initiated peripherally in the joint afferents.

Perception of tension exerted by muscle contraction and central sense of effort. Humans possess an acute sense of the force exerted by their muscles.[97,99] The graph of Fig. 12-24 shows that the capacity to discriminate the force exerted between thumb and finger in compressing springs of different stiffness is little affected when the afferents from skin and joints of the compressing fingers are blocked by local anesthesia. The implication is that the relevant afferent information comes from the contracting muscles themselves. Among muscle receptors the Golgi tendon organs appear the best suited by their properties to provide that information, but nothing is known of their ascending central projections into the somatic afferent system.

It has been known for a long time that the perceived heaviness of lifted objects is increased for fatigued muscles or for those weakened by lesions of the central motor system without sensory defect, and this has been confirmed by the study of subjects with partial curarization of lifting muscles.[34,35] It seems likely that, in addition to the sense of tension that depends on peripheral

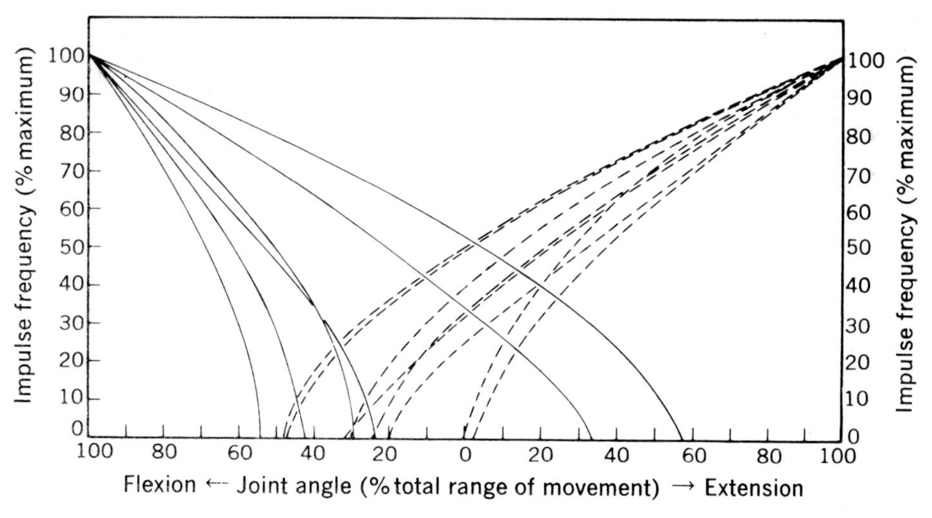

Fig. 12-23. Plots of excitatory angle functions for 14 ventrobasal thalamic neurons related to hinge joints of contralateral limbs studied in unanesthetized monkeys in manner of experiment illustrated in Fig. 12-22. Abscissal scale is normalized for different joints by changing each to percent of maximal movement possible in either direction from midposition of joint. Ordinal scale is normalized for each neuron by converting maximal discharge rate to 100. Majority of neurons subtend excitatory angles greater than one half the total range of movement of joint to which they are related; maximally excitatory positions are always at either full flexion or full extension, never both: curves are monotonic. Considered as a population, neurons will discharge reciprocally as joint moves back and forth through middle range of its movement, and such a reciprocal action of cortical and thalamic joint neurons is regularly observed. (From Mountcastle et al.[76])

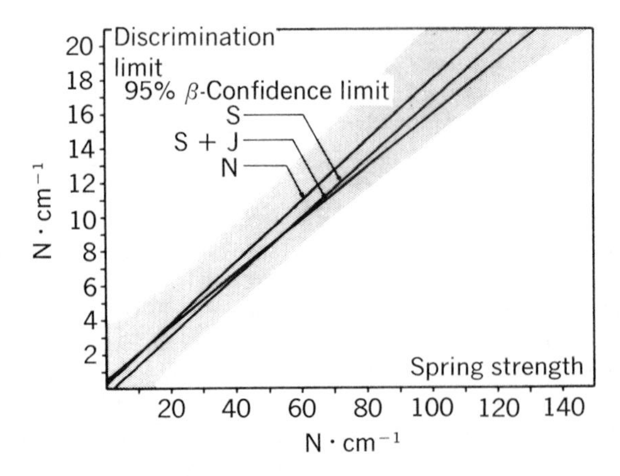

Fig. 12-24. Force discrimination. Graph showing capacity of human subjects to discriminate between forces required to compress different springs as function of spring strength. *N,* With normal conditions; *S,* with anesthesia of skin of fingers; *S + J,* anesthesia of skin and joints of fingers. Stippled area, 95% confidence limits for regression without anesthesia. Anesthesia of skin and joints caused only small increase in discrimination limens for force. (From Roland and Ladegaard-Pedersen.[100])

afferent signals, humans are able to sense the "central effort" exerted in a motor act independently of peripheral input.[68,69] Presumably this entails a centrally reentrant circuit that provides for some unknown perceptual mechanism information concerning the central motor commands for peripheral action. This is the general idea sometimes referred to as the "corollary discharge" or "efference copy." There is no direct experimental evidence concerning its nature.

It has recently been observed that vibratory stimuli (100 to 200 Hz, 0.5 to 1.5 mm peak to peak amplitude) delivered to the skin over a tendon such as the biceps elicits an illusion of joint movement.[30,41,66] This is thought to support the proposition that muscle afferents contribute to the perception of the position and movements of the limbs. Questions concerning the role of the muscle afferents are raised by a number of other observations, namely, that electrically induced trains of impulses confined to the group I afferents from muscle (1) evoke no sensations and, in particular, no illusion of movement; (2) do not elicit behavioral responses in waking animals, other than segmental reflexes; (3) cannot serve as conditioning stimuli; and (4) will neither arouse sleeping animals nor desynchronize their EEGs.[91,113]

Our understanding of the neural mechanisms of position sense and kinesthesis is still very limited. The role of particular sets of muscle afferents must be defined, and their common or differential CNS projections determined. How can signals from muscles be used to serve such a precise and quantitative sense when their discharge at any given position or at any given rate of movement may vary greatly from time to time as functions of the level of fusimotor activity? Finally, a matter of more general interest is how sensory information originating in several groups of different sets of peripheral fibers can be integrated to result in what appears subjectively to be a uniform sensibility.

DESCENDING SYSTEMS INFLUENCING AFFERENT TRANSMISSION IN SOMESTHETIC PATHWAYS[119]

The somatic afferent system contains descending components, corticofugal neurons that terminate, either directly or after an intervening subcortical relay, on synaptic regions of the sensory pathways. It is an old idea that these descending elements play a role in the selective attention to one sensory portal to the exclusion of others. What experimental evidence there is suggests that this is not the case, and it seems more likely that this action occurs at a higher level, so that the primary representations of peripheral events are continually present at the first stage of cortical processing and available for inspection at will. Descending systems may contribute to the shaping and funneling of the ascending neural activity evoked by sensory stimuli, perhaps by reinforcing or modifying the afferent surround inhibition that functions to improve resolution in the neural replication of stimulus contours, adding contrast to stimulus display. Finally, these descending systems may play a role in facilitating the particular zones of afferent pathways about to be activated by active movement of receptor sheets. On this idea, activity in these systems may be regarded as parallel and corollary to the central commands for movement and to play a role in that stimulus search so characteristic of primate behavior.[25,36]

Anatomic studies

It has been shown by anatomic methods that there are heavy descending projections to the spinal and bulbar relay zones of the somatic system and to the trigeminal component as well. In primates, those terminating in the DCN and the dorsal horn of the spinal gray matter arise mainly in the postcentral gyrus, with less dense projections from the motor cortex and the posterior parietal areas. The projection is topographically organized to complete a feedback loop, from hindleg cortical area to hindleg portion of the DCN, etc. The projections from one hemisphere are distributed bilaterally, but the heaviest projection is to the contralateral side. The large majority of the projections are direct from cortex to DCN; a smaller component relays in the reticular formation of the brain stem. The cortical projection to the dorsal horn of the spinal gray matter is also topographically organized (i.e., from hindleg area to the dorsal horns of the lumbar cord, etc.). The majority of these fibers reach their destinations via the pyramidal tract. The VB complex of the thalamus receives corticothalamic fibers from both SI and SII, so that cortical and thalamic patterns are linked in a reciprocal somatotopy.

The action of descending systems has been studied by recording the changes in synaptic transmission in subcortical zones of the somatic system produced by (1) electrical stimulation of the cortical region emitting descending systems and (2) reversible blockade of those systems by cortical cooling. These methods allow study of topographic relations and identification of the signs of the descending effect and its local synaptic mechanisms, but do not reveal how the system is brought to action normally. Towe[119] has made an extensive survey of the effect of cortical stimulation on cells of the DCN. Of 1,431 neurons studied, 416 were affected by stimulation of the topographically related region of pre- or postcentral gyrus: 117 were excited or facilitated, 253 were inhibited, and 46 showed mixed effects. The descending inhibitory effects were exerted both pre- and postsynaptically by local interneuronal action, as in the cat[8]; the effects were transmitted mainly via the pyramidal system. A somewhat similar pattern of corticofugal influences on neurons of the trigeminal complex has been observed in the cat.[27] Fetz[31] has made a detailed study of the effects of pyramidal tract volleys on the neurons of the dorsal horn in the cat. The corticospinal effect changed from predominantly inhibitory in Rexed's lamina IV to predominantly excitatory in lamina VI. Thus descending systems can influence or regulate synaptic relay into both local reflex and ascending sensory pathways that originate in the dorsal horn.

Cooling of the first somatic cortical area results in a net increment in synaptic transfer from the VB complex to SI, but not to SII, whereas cooling SII results in increased responses (to lemniscal volleys) of both SI-directed and SII-directed thalamocortical neurons.[19] This finding suggests a special role for the second somatic area: it may exert a controlling influence on transmission in the lemniscal component of the somatic afferent system.

In summary, efferent systems originate from the sensory and motor areas of the cortex and project on the subcortical relay nuclei of both the lemniscal and anterolateral systems. The projections are mainly direct via the pyramidal system, but in part indirect via the reticular formation of the brain stem. The function of these systems is not yet understood, but it is undoubtedly more complex than an on-off valving of input. Heuristic hypotheses are that they (1) shape the total population replication of sensory stimuli, lending contrast to the spatially distributed patterns of neural activity, and/or (2) tune the somatic afferent system during active tactile exploration.

DEFECTS IN SOMESTHESIS PRODUCED BY NERVOUS SYSTEM LESIONS IN MAN AND OTHER PRIMATES

Lesions of the lemniscal system produce defects in the discriminative aspects of somesthesis. What is lost or degraded by such lesions is the elegance and precision that normally characterize somesthetic functions, particularly those that require analysis of temporal order and changing spatial patterns of afferent neural activity. In normal individuals this information permits precise localization of stimuli; the identification of quality, relative and absolute intensities, and temporal patterns; the recognition of spatial extent and form; and the position of the body and its parts and their movements. These will be considered in that which follows, together with the effects produced by lesions at levels of neural organization that, although concerned with somesthesis, cannot be specified as lemniscal or anterolateral or indeed even as exclusively somesthetic in function. These effects are complex disturbances of appreciation and awareness of body form, spatial orientation, and the relation of the body to extrapersonal space—functions that obviously depend on integration of information in several sensory spheres.

Lesions of the lemniscal system, in certain locations in certain patients, produce not only defects but abnormalities of sensation, pathologic events that are attributed to the released and unbridled action of the anterolateral system on the forebrain. These latter, together with the effects produced by lesions of the anterolateral system per se, will be considered in Chapter 13 on pain mechanisms.

The study of sensory defects in humans is important for an understanding of sensation. A verbal description of a sensory experience has a unique value in itself and allows the use of objective measures of the quantitative aspects of sensory experiences.

In animals with lesions placed in sensory pathways, measures of sensory capacity and of discrimination are made in training experiments such as that illustrated in Fig. 12-25. It is important to emphasize three aspects of these experimental methods that occasionally make their results difficult to interpret.

1. In both man and animal, more is tested than the sensory experience itself. The human report depends on introspection, immediate recall, and verbal expression; the animal report depends on the capacities to learn and to respond with an overt, directed motor act.

2. In such experiments, *one observes the remaining capacity for sensory function; one infers from any deficiency the function of the lesioned part*. The function of a part (e.g., a cortical area) is not confined to its intrinsic neural mechanisms,

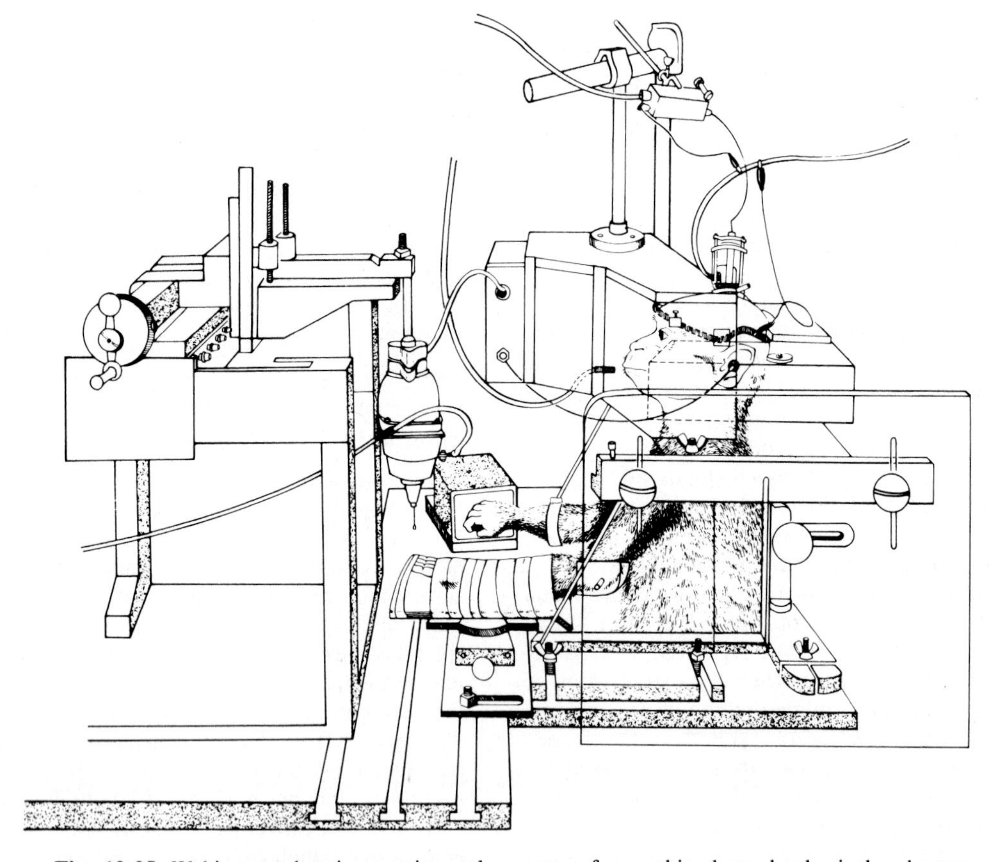

Fig. 12-25. Waking monkey in experimental apparatus for combined psychophysical and neurophysiologic experiments. Mechanical stimulator carried on rack and pinion devices at left delivers stimuli of variable intensity, duration, and temporal pattern to exposed glabrous skin of immobilized left hand. Animal manipulates signal key with his right hand to initiate trials, indicate detection, etc. Head is fixed in periods of study that are repeated from day to day. Tube delivers liquid reward to monkey's mouth. Drawings above indicate microelectrode chamber implanted over right postcentral gyrus and its attached microdriver, input preamplifier, etc. In this way, simultaneous studies can be made of animal's response to mechanical stimuli delivered to left hand and responses of cells of contralateral sensory cortex activated by same stimuli.

but includes all its complex relations with other parts. It is not likely that the defects in the remaining capacity will simply reflect the function of the part removed.

3. In both animal and man, particularly in young individuals, the lesioned nervous system possesses a remarkable power to *compensate* for structural loss by an unknown process. With increasing time there is frequently a considerable recovery of functional capacity, and the long-term residual deficiency following a CNS lesion is commonly much less than that manifested in the days immediately following the lesion.

In traditional neurology and psychology a clear distinction was drawn between sensation and perception. The first term described the "pure" and "simple" experiencing of the primary sensory qualities, a process to be explained directly in terms of the functional properties of receptors, afferent nerve fibers, and central sensory systems up to some poorly defined "intermediate" level of complexity. Perception, on the other hand, subsumed the total behavioral experience of apprehending an object, more particularly the experiencing of the complex and holistic aspects of sensory stimuli, the patterns, selectivity, contrasts, and similarities between stimuli, and the apprehension of serial order. Somesthetic *sensations* were thought to depend on the primary sensory pathways and to be elaborated in the postcentral gyrus. Somesthetic *perceptions* were thought to depend on these and, in addition, a further neural elaboration and integration in the posterior parietal association areas.

Studies of recent years have cast doubt on this strict sensation-perception dichotomy and indicate a sensation-perception continuum of increasing complexity. Indeed, the presumed "simple" sensory experiences may themselves be complex and may not be governed by isomorphic relations between stimulus qualities and those experiences. It appears likely that the progressively more elaborate processing of sensory input, its comparison with information stored in immediate and long-term memory, its evolution into efferent motor activity or into that undefined central neural reflection subject to introspective analysis and verbal description, and its recombination into neural activities that lead to total sensory experiences all involve not only the postcentral gyrus but other cortical areas as well, in addition to their interactions with subcortical referrants. This process of progressively more elaborate action is undoubtedly a continuum, and lesions at one level or another will affect function differentially at one or more points along it. Examples of these variations follow.

Defects in somatic sensibility produced by lesions of afferent pathways[1]

Lesions of the myelinated afferents entering the lemniscal system may occur in peripheral nerves or dorsal roots or in their central ascending trajectories into the dorsal systems of the spinal cord. Loss of the group I and II fibers of peripheral nerves, which occurs in certain demyelinating diseases, or a *combined lesion* of the dorsal column and the adjacent most medial part of the dorsolateral column, in primates, results in profound alterations in somatic sensibility on the ipsilateral side of the body. In the dorsal system syndrome there is a loss of those aspects of tactile sensibility characterized by and requiring identification of the directional or temporally changing or spatially distributed aspects of stimuli, as well as deficiencies in tactually guided manual exploration. Position sense and kinesthesis are markedly defective, and there is a decrease in the accuracy of projected movements of the limbs toward targets in surrounding space. The subject no longer recognizes the shape of objects explored tactually, a defect called astereognosis, and there are defects in vibratory sensibility and the ability to judge lifted weights. Combined and complete lesions of both dorsal and dorsolateral columns occur rarely, so that many patients with lesions of the dorsal half of the spinal cord show some fragment of the total pattern of defects described. The ability to recognize a mechanical stimulus to peripheral tissue remains when only the anterior and anterolateral columns of the cord are intact, but the anterolateral system cannot serve the discriminative aspects of somesthesis so characteristic of dorsal system function. Pain and temperature sensibilities remain after dorsal lesions, although certain abnormalities in the pain sense appear (Chapter 13).

Isolated lesions of either the DC or the dorsolateral column are rarely described and verified anatomically in humans. Knowledge of the sensory losses produced by such lesions has come from study of monkeys trained to make detections and discriminations in behavioral tasks and tested before and after lesion. The modality rearrangement from DC to dorsolateral column that occurs shortly after lumbar dorsal root fibers enter the DC (p. 353) predicts the losses produced in the leg and foot by lesions made in the thoracic region. Isolated transection of the DC at this level leaves intact those aspects of sensibility that depend on deep and particularly on slowly adapting afferents (e.g., position sense and kinesthesis). It affects severely those that depend on signals in quickly adapting afferents, such as the discrimination of the direction of stimulus movement,[121] or those that require active tactile exploration.[11] The addition to the transection of the most medial part of the dorsolateral columns reveals the full syndrome of the dorsal system lesions described previously. These losses of function appear immediately after tract section, but may lessen somewhat during long periods of retraining, so that a monkey may regain function to a degree determined by the difficulty of the task and perhaps also by differences between subjects. This change with training shows the ability of the damaged nervous system to learn to achieve a behavioral end, although slowly and inaccurately. In the present case, one may surmise that the animal, deprived by tract section of clear signals of the relevant stimulus parameters tested for, now learns to derive cues from signals transmitted over other pathways, even though these latter cues are by comparison only faintly correlated with the stimulus parameters.

It is still unknown whether the modality shuffle from DC to dorsolateral column occurs for afferents of the cervical and upper thoracic areas as it does for lumbar dorsal roots. Lesions restricted to the cuneate fasciculus at the C1 to C2 level result in a defect in discriminative aspects of tactile sensibility in the arm. The movement performance of the arm remained relatively normal in these experiments, including independent finger movements, which suggests that afferents serving position sense and kinesthesis were undisturbed by the lesion.[13] In other experiments of a similar nature, it was reported that such a lesion in the cervical region caused a profound disturbance in the projection of the arm toward targets in extrapersonal space,[37] which suggests that the lesions in the latter series might have encroached on the medial aspect of the dorsolateral columns.

Immediately after section of the cervical and upper thoracic dorsal roots the anesthetic arm of a monkey hangs as if paralyzed at the side, virtually devoid of spontaneous movements. With time and especially in young animals, spontaneous but poorly controlled movements begin. Taub et al.[118] and Taub and Berman[117] have shown that with intensive and prolonged training such an animal can learn to project arm and hand accurately into space toward an unseen target, to squeeze a rubber bulb to a certain level of pressure for reward, etc. Great care was taken by these investigators to eliminate associated peripheral cues from which such an animal might obtain information to associate with movements of the insentient arm, for example, afferent signals produced by associated movements of normally innervated body parts. The implication of this important result is that there exists a centrally reentrant signal paralleling or corollary to the neural commands for movement and that with prolonged training a monkey can learn which particular pattern of command signals leads to reward and repeat that central pattern from memory.

Hemisection of the spinal cord produces a constellation of defects termed the Brown-Séquard syndrome. It consists of (1) loss of the discriminative aspects of somesthesia on the ipsilateral side, a DC defect; (2) loss of pain and temperature sensibility on the contralateral side, an anterolateral system defect; and (3) partial paralysis of movement and changes in muscle tone on the ipsilateral side, due to interruption of descending motor control systems, including the pyramidal tract.

Lesions of the DCN or the medial lemniscus itself occur rarely in humans and are difficult to make with precision in monkeys. The available facts indicate that lesions in either place produce changes resembling the "dorsal systems" syndrome described previously, but no studies of such animals have been made using modern methods of behavioral testing of sensory capacities.

Similarly, naturally occurring lesions rarely destroy the thalamic VB complex entirely, without other damage. Smaller lesions within it have been made during stereotaxic explorations of the human thalamus, aimed at alleviating intractable pain. They produce a profound contralateral mechanoreceptive loss. Very large lesions are sometimes produced by vascular occlusions or hemorrhages, destroying both VB and adjacent structures; they may produce, in addition to somesthetic defects, the constellation of elevated threshold for pricking pain associated with severe spontaneous pain of central origin, called the "thalamic syndrome" (Chapter 13).

Defects in somesthesis produced by cerebral cortical lesions

Removal of the first somatic area in monkeys produces sensory defects on the contralateral side of the body, and partial removals do so in topographically related body parts. Such an animal regains the capacity to respond to contralateral mechanical stimuli, but thresholds remain elevated 2 to 3 times. A defect in position sense and kinesthesis is obvious as a lack of awareness of passive movements of the contralateral limbs and the frequent arrest of movements, always reluctantly begun, in awkward positions maintained for long periods. There is a loss of the discriminative aspects of somesthesis, a gross impairment in tactile, shape, size, and roughness discrimination that is never relearned.[106,107]

This suggests that the capacity for somesthetic function of a brain without SI depends primarily on the difficulty of the sensory task attempted. A recent study was aimed at specifying which tasks among a group with differing complexities depend absolutely on the processing mechanisms of the postcentral gyrus.[59] Monkeys were trained in a series of somesthetic tasks of increasing difficulty in the somesthetic modality of flutter-vibration (1) to detect the presence of such a stimulus; (2) to discriminate between the different qualitative natures of flutter and vibration; (3) to discriminate between mechanical sine wave stimuli of different amplitudes but the same frequency; (4) to discriminate between different frequencies of the same subjective intensities; and (5) to categorize stimuli of different frequencies. Animals were trained before and tested for long periods after cortical removals, in the arrangement illustrated in Fig. 12-25. No deficiency in any test was observed in the hand ipsilateral to a cortical lesion. With regard to flutter-vibration, postcentral removals elevated contralateral thresholds 2 to 3 times for simple detection and impaired permanently, but not completely, the capacity to discriminate between flutter and vibration and to make amplitude discriminations. The degree of partial recovery of function during prolonged postoperative periods of testing varied from one animal to another and appeared to depend in part on the duration and intensity of preoperative training. What is permanently lost is the capacity to make frequency discriminations, as shown in Fig. 12-26, which illustrates the random performance in this task of an animal after SI removal. The implications are that the first somatic area is necessary for the normal performance in all the aspects of flutter-vibration tested and that its neural processing mechanisms are necessary for any performance above chance when the task requires discrimination between neuronal periodicities (i.e., discriminations between temporal orders). Undoubtedly even more complex tasks that require recognition of differences in temporal order or changing spatial arrays in afferent neural activity depend equally on the processing mechanisms of the first somatic area. No unique function for the second somatic area has been revealed by the lesion method of study.

Complete removals of the postcentral gyrus in humans and/or lesions of the fiber systems linking it to the dorsal thalamus produce severe defects in somesthesis.[96,98,99] There are permanent deficits in the senses of passive position and movements of the limbs, in the sense of tension or force, and in kinesthesis. There are equally severe impairments in those aspects of the cutaneous mechanoreceptive sense that depend on temporally dynamic or spatially distributed patterns of afferent neural activity (e.g., there is a marked astereognosis). The sense of mechanical contact is preserved, although with elevated threshold. Pain and temperature sensibilities are preserved, although the threshold for pricking pain may be elevated and spontaneous pain of central origin resembling that of the thalamic syndrome may appear (Chapter 13).

Roland has recently studied the defects in somesthesis that occurred in a large series of pa-

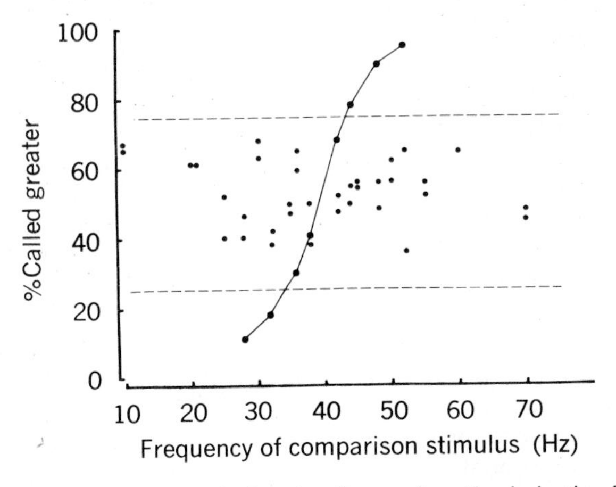

Fig. 12-26. Solid line plots psychometric function for monkey discriminating between oscillating mechanical stimuli at 40 Hz and others of different frequency, but equal subjective amplitude, before operation. Scattered points indicate monkey's performance in same task tested over many weeks after removal of entire postcentral gyrus contralateral to the hand tested. There was complete loss of capacity to make significant discrimination. (From LaMotte and Mountcastle.[59])

tients.[100] Ninety-four with verified and well-defined lesions of the cerebral hemispheres were selected from a much larger number undergoing operations and cortical removals for cerebral disease. The exposed gyri and sulci were identified in every patient and the region of cortex removed mapped. Only those patients sustaining direct damage to or undercutting of the postcentral gyrus showed impairment in primary somesthetic modes or in more complex ones like stereognosis and kinesthesis. This finding does not imply that more posteriorly located association areas of the cerebral cortex do not play a role in somesthesis; lesions there may produce complex perceptual disorders without primary somesthetic defect (Chapter 22). Roland has studied in this same series the defects in position sense and kinesthesis that are produced by smaller lesions restricted to different anteroposterior parts of the hand area[98] and has measured in an elegant and quantitative manner the defect produced by cortical lesions and whether those defects may be further accentuated by local anesthesia of the skin and joints of the fingers. His results are important in relation to the evidence given earlier that different somesthetic modes are represented in different cytoarchitectural areas of the postcentral gyrus. They may be summarized as follows:

1. Patients with defects in the sense of the tension or the force exerted by muscular contraction had lesions involving the central sulcus or its connections. It is uncertain whether the impairment seen in these cases was due to destruction of a projection area for afferents signaling tension from the periphery (e.g., the Golgi tendon organs of muscle), to the associated paresis, or to both.

2. Patients with severe impairment of position sense and kinesthesis that persisted with or without anesthesia of the skin and joints of the fingers had extensive lesions of the postcentral gyrus and severe deficits in other aspects of somesthesis as well.

3. A third group of patients retained a considerable capacity for position sense and kinesthesis that survived anesthesia of the skin alone, but was lost completely on anesthesia of the joints. These patients had lesions centered around the postcentral, not the central, sulcus, presumably in area 5 or at the area 2–area 5 border.

The tentative conclusions from these important new findings by Roland conform nicely with evidence presented earlier, namely, that (1) some undefined set(s) of muscle afferents project to areas 3a and 4, perhaps including both spindle and tendon organ afferents; (2) cutaneous afferents dominate the projection to areas 3b and 1; (3) joint afferents dominate the projection to area 2; and, (4) another muscle afferent projection, perhaps from spindles, reaches area 5 and/or the area 2–area 5 border region.

The effects of lesions of the posterior parietal association areas on the role of interhemispheric mechanisms in somesthesis are described in Chapter 22.

REFERENCES

General reviews

1. Beck, C. H. M.: Dual dorsal columns: a review, Can. J. Neurol. Sci. **3**:1, 1976.
2. Darian-Smith, I.: The trigeminal system. In Iggo, A., editor: Handbook of sensory physiology. Somatosensory system, New York, 1973, Springer-Verlag New York, Inc., vol. 2.
3. Jones, E. G., and Powell, T. P. S.: Anatomical organization of the somatosensory cortex. In Iggo, A., editor: Handbook of sensory physiology. Somatosensory system, New York, 1973, Springer-Verlag New York, Inc., vol. 2.
4. Mountcastle, V. B.: An organizing principle for cerebral function: the unit module and the distributed system. In Edelman, G. M., and Mountcastle, V. B.; The mindful brain: cortical organization and the group-selective theory of higher brain function, Cambridge, Mass., 1978, The M.I.T. Press.
5. Perl, E. R., and Boivie, J. G.: Neural substrates of somatic sensation. In Hunt, C. C., editor: MTP international review of science. Physiology, Series 1, Neurophysiology, Baltimore, 1975, University Park Press, vol. 3.
6. Stevens, S. S.: Psychophysics. In Stevens, G., editor: Introduction to its perceptual, neural and social prospects, New York, 1975, John Wiley & Sons, Inc.
7. Werner, G., and Whitsel, B. L.: The somatic sensory cortex: functional organization. In Iggo, A., editor: Handbook of sensory physiology. Somatosensory system, New York, 1973, Springer-Verlag New York, Inc., vol. 2.

Original papers

8. Anderson, P., Eccles, J. C., Schmidt, R. F., and Yokota, T.: Depolarization of presynaptic fibers in the cuneate nucleus, J. Neurophysiol. **27**:92, 1964.
9. Andrew, B. L., and Dodt, E.: The deployment of sensory nerve endings at the knee joint of the cat, Acta Physiol. Scand. **28**:287, 1953.
10. Asanuma, H.: Recent developments in the study of the columnar arrangement of neurons within the motor cortex, Physiol. Rev. **55**:143, 1975.
11. Azulay, A., and Schwartz, A. S.: The role of the dorsal funiculus of the primate in tactile discrimination, Exp. Neurol. **46**:315, 1975.
12. Boyd, I. A., and Roberts, T. D. M.: Proprioceptive discharges from stretch-receptors in the knee joint of the cat, J. Physiol. **122**:38, 1953.
13. Brinkman, J., Bush, B. M., and Porter, R.: Deficient influences of peripheral stimuli on precentral neurones in monkeys with dorsal column lesions, J. Physiol. **276**:27, 1978.
14. Brodal, A., and Pompeiano, O.: The vestibular nuclei in the cat, J. Anat. **91**:438, 1957.

15. Brooks, V. B., and Stoney, S. D., Jr.: Motor mechanisms: the role of the pyramidal system in motor control, Annu. Rev. Physiol. **33**:337, 1971.
16. Brown, A. G.: Ascending and long spinal pathways, dorsal columns, spinocervical tract and spinothalamic tract. In Iggo, A., editor: Handbook of sensory physiology. Somatosensory system, New York, 1973, Springer-Verlag New York, Inc., vol. 2, pp. 315-338.
17. Browne, K., Lee, J., and Ring, P. A.: The sensation of passive movement at the metatarso-phalangeal joint of the great toe in man, J. Physiol. **126**:448, 1954.
18. Burchfiel, J. L., and Duffy, F. H.: Muscle afferent input to cells in primate somatosensory cortex, Brain Res. **45**:241, 1972.
19. Burchfiel, J. L., and Duffy, F. H.: Corticofugal influences upon cat thalamic ventrobasal complex, Brain Res. **70**:395, 1974.
20. Burgess, P. R., and Clark, F. J.: Characteristics of knee joint receptors in the cat, J. Physiol. **203**:317, 1969.
21. Burton, H., and Jones, E. G.: The posterior thalamic region and its cortical projection in the New World and Old World monkeys, J. Comp. Neurol. **168**:249, 1976.
22. Carli, G., LaMotte, R. H., and Mountcastle, V. B.: A comparison of sensory behavior and the activity of postcentral cortical neurons, observed simultaneously, elicited by oscillatory mechanical stimuli delivered to the contralateral hand in monkeys, Proceedings of the Twenty-Fifth International Congress of Physiology, 1971.
23. Carmon, A.: Stimulus contrast in tactile resolution, Percept. Psychophysics **3**:241, 1968.
24. Carreras, M., and Andersson, S. A.: Functional properties of neurons of the anterior ectosylvian gyrus of the cat, J. Neurophysiol. **26**:100, 1963.
25. Coulter, J. D.: Sensory transmission through lemniscal pathway during voluntary movement in the cat, J. Neurophysiol. **37**:831, 1974.
26. Cross, M. J., and McCloskey, D. I.: Position sense following surgical removal of joints in man, Brain res. **55**:443, 1973.
27. Darian-Smith, I., and Yokota, T.: Corticofugal effects on different neuron types within the cat's brain stem activated by tactile stimulation on the face, J. Neurophysiol. **29**:185, 1966.
28. Donaldson, I. M. L.: The properties of some human thalamic units. Some new observations and a critical review of the localization of thalamic nuclei, Brain **96**: 419, 1973.
29. Dreyer, D. A., Schneider, R. J., Metz, C. B., and Whitsel, B. L.: Differential contributions of spinal pathways to body representation in postcentral gyrus of *Macaca mulatta*, J. Neurophysiol. **37**:119, 1974.
30. Eklund, G.: Position sense and state of contraction; the effects of vibration, J. Neurol. Neurosurg. Psychiatry **35**:606, 1972.
31. Fetz, E. E.: Pyramidal tract effects on interneurons in the cat lumbar dorsal horn, J. Neurophysiol. **31**:69, 1968.
32. Foreman, R. D., Applebaum, A. E., Beall, J. E., Trevino, D. L., and Willis, W. D.: Responses of primate spinothalamic tract neurons to electrical stimulation of hindlimb peripheral nerves, J. Neurophysiol. **38**:132, 1975.
33. Gandevia, S. C., and McCloskey, D. I.: Joint sense, muscle sense, and their combination as position sense, measured at the distal interphalangeal joint of the middle finger, J. Physiol. **260**:387, 1976.
34. Gandevia, S. C., and McCloskey, D. I.: Changes in motor commands, as shown by changes in perceived heaviness, during partial curarization and peripheral anesthesia in man, J. Physiol. **272**:673, 1977.
35. Gandevia, S. C., and McCloskey, D. I.: Sensations of heaviness, Brain **100**:345, 1977.
36. Gibson, J. J.: Observations on active touch, Psychol. Rev. **69**:477, 1962.
37. Gilman, S., and Denny-Brown, D.: Disorders of movement and behavior following dorsal column lesions, Brain **89**:397, 1966.
38. Gless, P., and Soler, J.: Fibre content of the posterior column and synaptic connection of nucleus gracilis, J. Zellforsch. **36**:381, 1951.
39. Goldring, S., Aras, E., and Weber, P. C.: Comparative study of sensory input to motor cortex in animals and man, EEG Clin. Neurophysiol. **29**:537, 1970.
40. Goldring, S., and Ratchenson, R.: Human motor cortex: sensory input data from single neuron recordings, Science **175**:1493, 1972.
41. Goodwin, G. M., McCloskey, D. I., and Matthews, P. B. D.: The contribution of muscle afferents to kinaesthesia shown by vibration induced illusions of movement and by the effects of paralyzing joint afferents, Brain **95**:705, 1972.
42. Gordon, G.: The concept of relay nuclei. In Iggo, A., editor: Handbook of sensory physiology. Somatosensory system, New York, 1973, Springer-Verlag New York, Inc., vol. 2.
43. Grant, G., Boivie, J., and Silfvenius, H.: Course and termination of fibres from the nucleus Z of the medulla oblongata. An experimental light microscopical study in the cat, Brain Res. **55**:55, 1973.
44. Grigg, P.: Mechanical factors influencing response of joint afferent neurons from cat knee, J. Neurophysiol. **38**:1473, 1975.
45. Grigg, P., Finerman, G. A., and Riley, L. H.: Joint position sense after total hip replacement, J. Bone Joint Surg. **55-A**:1016, 1973.
46. Harrington, T., and Merzenich, M. M.: Neural coding in the sense of touch: human sensations of skin indentation compared with the responses of slowly adapting mechanoreceptive afferents innervating the hairy skin of monkeys, Exp. Brain Res. **10**:251, 1970.
47. Hassler, R.: Thalamic regulation of muscle tone and speed of movement. In Purpura, D. B., and Yahr, M. D., editors: The thalamus, New York, 1966, Columbia University Press, pp. 419-436.
48. Horch, K. W., Clark, F. J., and Burgess, P. R.: Awareness of knee joint angle under static conditions, J. Neurophysiol. **38**:1436, 1975.
49. Hore, J., Preston, J. B., Durkovic, R. G., and Cheney, P. D.: Responses of cortical neurons (areas 3a and 4) to ramp stretch of hindlimb muscles in the baboon, J. Neurophysiol. **39**:484, 1976.
50. Jasper, H. H., and Bertrand, G.: Thalamic units involved in somatic sensation and voluntary and involuntary movements in man. In Purpura, D. P., and Yahr, M. D., editors: The thalamus, New York, 1966, Columbia University Press.
51. Johnson, K. O.: Reconstruction of population response to a vibratory stimulus in quickly adapting mechanoreceptive afferent population innervating the glabrous skin of the monkey, J. Neurophysiol. **37**: 48, 1974.
52. Jones, E. G., Coulter, J. D., and Hendry, S. H. C.: Intracortical connectivity of architectonic fields in the

somatic sensory, motor and parietal cortex of monkeys. (In press.)

53. Jones, E. G., Coulter, J. D., and Wise, S. P.: Commissural connectivity in the sensory-motor cortex of monkeys. (In press.)
54. Jones, E. G., and Powell, T. P. S.: Connections of the somatic sensory cortex of the rhesus monkey. II. Contralateral connections, Brain **92:**504, 1969.
55. Jones, E. G., and Powell, T. P. S.: An anatomical study of converging sensory pathways within the cerebral cortex of the monkey, Brain **93:**793, 1970.
56. Jones, E. G., and Powell, T. P. S.: Connections of the somatic sensory cortex of the rhesus monkey. III. Thalamic connections, Brain **93:**37, 1970.
57. LaMotte, R. H.: Psychophysical and neurophysiological studies in tactile sensibility. In Hollies, R. S., and Goldman, R. F., editors: Clothing comfort: interaction of thermal, ventilation, construction and assessment factors, Ann Arbor, Mich., 1977, Ann Arbor Science Publishers, Inc., pp. 83-105.
58. LaMotte, R. H., and Mountcastle, V. B.: Capacities of humans and monkeys to discriminate between vibratory stimuli of different frequency and amplitude: a correlation between neural events and psychophysical measurements, J. Neurophysiol. **38:**539, 1975.
59. LaMotte, R. H., and Mountcastle, V. B.: Disorders of somesthesis after lesions of the parietal lobe, J. Neurophysiol. **42:**400, 1979.
60. Landgren, S., and Silfvenius, H.: Nucleus Z, the medullary relay in the projection path to the cerebral cortex of group I muscle afferents from the cat's hind limb, J. Physiol. **218:**551, 1971.
61. Lende, R. A.: A comparative approach to the neocortex: localization in monotremes, marsupials, and insectivores, Ann. N.Y. Acad. Sci. **167:**262, 1969.
62. Lenon, R. N., Hanby, J. A., and Porter, R.: Relationship between the activity of precentral neurones during active and passive movements in conscious monkeys, Proc. R. Soc. Lond. (Biol.) **194:**341, 1976.
63. Lenon, R. N., and Porter, R.: Afferent input to movement-related precentral neurones in conscious monkeys, Proc. R. Soc. Lond. (Biol.) **194:**313, 1976.
64. Lucier, G. E., Ruegg, D. C., and Wiesendanger, M.: Responses of neurones in motor cortex and in area 3A to controlled stretches of forelimb muscles in cebus monkeys, J. Physiol. **251:**833, 1975.
65. Magherini, P. C., Pompeiano, O., and Sequin, J. J.: Responses of nucleus Z neurons to vibration of hindlimb extensor muscles in the decerebrate cat, Arch. Ital. Biol. **113:**150, 1975.
66. Matthews, P. B. C.: Muscle afferents and kinesthesia, Br. Med. Bull. **33:**137, 1977.
67. Mehler, W. R.: Some observations on secondary ascending afferent systems in the central nervous system. In Knighton, R. S., and Dumke, P. R., editors: Pain, Boston, 1966, Little, Brown & Co.
68. Merton, P. A.: Human position sense and the sense of effort, Symp. Soc. Exp. Biol. **18:**387, 1964.
69. Merton, P. A.: The sense of effort. In Porter, R., editor: Breathing: Hering-Breuer Centenary Symposium, Edinburgh, 1970, Churchill Livingstone, pp. 207-211.
70. Merzenich, M. M., Kaas, J. H., Sur, M., and Lin, C. S.: Double representation of the body surface within cytoarchitectonic areas 3b and 1 in "SI" in the owl monkey (Aotus trivirgatus), J. Comp. Neurol. **181:** 41, 1978.
71. Millar, J.: Flexion-extension sensitivity of elbow joint afferents in cat, Exp. Brain Res. **24:**209, 1975.
72. Mountcastle, V. B.: Modality and topographic properties of single neurons of cat's somatic sensory cortex, J. Neurophysiol. **20:**408, 1957.
73. Mountcastle, V. B., and Henneman, E.: The representation of tactile sensibility in the thalamus of the monkey, J. Comp. Neurol. **97:**409, 1952.
74. Mountcastle, V. B., LaMotte, R. H., and Carli, G.: Detection thresholds for stimuli in humans and monkeys: comparison with threshold events in mechanoreceptive afferent nerve fibers innervating the monkey hand, J. Neurophysiol. **35:**122, 1972.
75. Mountcastle, V. B., Lynch, J. C., Georgopoulos, A., Sakata, H., and Acuna, C.: Posterior parietal association cortex of the monkey: command functions for operations within extrapersonal space, J. Neurophysiol. **38:**908, 1975.
76. Mountcastle, V. B., Poggio, G. F., and Werner, G.: The relation of thalamic cell response to peripheral stimuli varied over an intensive continuum, J. Neurophysiol. **26:**807, 1963.
77. Mountcastle, V. B., and Powell, T. P. S.: Central nervous mechanisms subserving position sense and kinesthesis, Johns Hopkins Med. J. **105:**173, 1959.
78. Mountcastle, V. B., and Powell, T. P. S.: Neural mechanisms subserving cutaneous sensibility, with special reference to the role of afferent inhibition in sensory perception and discrimination, Johns Hopkins Med. J. **105:**201, 1959.
79. Mountcastle, V. B., Talbot, W. H., and Kornhuber, H. H.: The neural transformation of mechanical stimuli delivered to the monkey's hand. In de Reuck, A. V. S., and Knight, J., editors: Ciba Foundation Symposium on Touch, Heat and Pain, London, 1966, J. & A. Churchill.
80. Mountcastle, V. B., Talbot, W. H., Sakata, H., and Hyvarinen, J.: Cortical neuronal mechanisms in flutter-vibration studied in unanesthetized monkeys. Neuronal periodicity and frequency discrimination, J. Neurophysiol. **32:**452, 1969.
81. Nijensohn, D. E., and Kerr, F. W. L.: The ascending projections of the dorsolateral funiculus of the spinal cord in the primate, J. Comp. Neurol. **161:**459, 1975.
82. Olszewski, J.: On the anatomical and function organization of the spinal trigeminal nucleus, J. Comp. Neurol. **92:**401, 1950.
83. Oscarson, O.: Three ascending tracts activated from Group I afferents in forelimb nerves of the cat, Prog. Brain Res. **12:**180, 1964.
84. Paul, R. L., Goodman, H., and Merzenich, M. M.: Alterations in mechanoreceptor input to Brodmann's areas 1 and 3 of the postcentral hand area of *Macaca mulatta* after nerve section and regeneration, Brain Res. **39:** 1, 1972.
85. Penfield, W., and Jasper, H.: Epilepsy and the functional anatomy of the human brain, Boston, 1954, Little, Brown & Co.
86. Perl, E., and Whitlock, D. G.: Somatic stimuli exciting spinothalamic projections to thalamic neurons in cat and monkey, Exp. Neurol. **3:**256, 1961.
87. Phillips, C. G., Powell, T. P. S., and Wiesendanger, M.: Projection from low threshold muscle afferents of hand and forearm to area 3a of baboon's cortex, J. Physiol. **210:**59p, 1970.
88. Phillips, J. R., Johnson, K. O., and Darian-Smith, I.: Tactile spatial resolution. (In press.)
89. Poggio, G. F., and Mountcastle, V. B.: A study of the functional contributions of the lemniscal and spinothalamic systems to somatic sensibility. Central nervous mechanisms in pain, Johns Hopkins Med. J. **106:**266, 1960.

90. Poggio, G. F., and Mountcastle, V. B.: The functional properties of ventrobasal thalamic neurons studied in unanesthetized monkeys, J. Neurophysiol. **26:**775, 1963.
91. Pompeiano, O., and Swett, J. E.: Identification of cutaneous and muscular afferent fibers producing EEG synchronization or arousal in normal cats, Arch. Ital. Biol. **100:**343, 1962.
92. Porter, R.: Influences of movement detectors on pyramidal tract neurons in primates, Ann. Rev. Physiol. **38:**121, 1976.
93. Powell, T. P. S.: The somatic sensory cortex, Br. Med. Bull. **33**(2):129, 1977.
94. Powell, T. P. S., and Mountcastle, V. B.: Some aspects of the functional organization of the cortex of the postcentral gyrus of the monkey: a correlation of findings obtained in a single unit analysis with cytoarchitecture, Johns Hopkins Med. J. **105:**133, 1959.
95. Provins, K. A.: The effect of peripheral nerve block on the appreciation and execution of finger movements, J. Physiol. **143:**55, 1958.
96. Roland, P. E.: Tactile manual agnosia, Dan. Med. Bull. **19:**1, 1972.
97. Roland, P. E.: Do muscular receptors in man evoke sensations of tension and kinaesthesia? Brain Res. **99:**162, 1975.
98. Roland, P. E.: Asterognosis: tactile recognition after localized hemispheric lesions in man, Arch. Neurol. **33:**543, 1976.
99. Roland, P. E.: Sensory feed-back of tension and kinaesthesia to the cerebral cortex of man, 1978. (In press.)
100. Roland, P. E., and Ladegaard-Pedersen, H.: A quantitative analysis of sensations of tension and of kinesthesia in man. Evidence for a peripherally originating muscular sense and for a sense of effort, Brain **100:** 671, 1977.
101. Rosen, I.: Localization in caudal brain stem and cervical spinal cord of neurons activated from forelimb group I afferents in the cat, Brain Res. **16:**55, 1969.
102. Rosen, I., and Sjolund, B.: Organization of group I activated cells in the main and external cuneate nuclei of the cat: identification of muscle receptors, Exp. Brain Res. **16:**221, 1973.
103. Rustioni, A.: Non-primary afferents to the nucleus gracilis from the lumbar cord of the cat, Brain Res. **51:**81, 1973.
104. Rustioni, A.: Non-primary afferents to the cuneate nucleus in the brachial dorsal funiculus of the cat, Brain Res. **75:**247, 1974.
105. Sakata, H., Takaoka, Y., Kawarasaki, A., and Shibutani, H.: Somatosensory properties of neurons in the superior parietal cortex (area 5) of the rhesus monkey, Brain Res. **64:**85, 1973.
106. Semmes, J.: Somesthetic effects of damage to the central nervous system. In Iggo, A., editor: Handbook of sensory physiology. Somatosensory system, New York, 1973, Springer-Verlag New York, Inc., vol. 2.
107. Semmes, J., Porter, L., and Randolph, M. C.: Further studies of anterior postcentral lesions in monkeys, Cortex **10**(1):55, 1974.
108. Shanks, M. F., Pearson, R. C. A., and Powell, T. P. S.: The intrinsic connections of the primary somatic sensory cortex of the monkey, Proc. R. Soc. Lond. (Biol.) **200:**95, 1978.
109. Shanks, M. F., Rockel, A. J., and Powell, T. P. S.: The commissural fibre connections of the primary somatic sensory cortex, Brain Res. **98:**166, 1975.
110. Shende, M. C., Stewart, D. H., Jr., and King, R. B.: Projections from the trigeminal nucleus caudalis in the squirrel monkey, Exp. Neurol. **20:**655, 1968.
111. Skoglund, S.: Joint receptors and kinesthesis. In Iggo, A., editor: Handbook of sensory physiology. Somatosensory system, New York, 1973, Springer-Verlag New York, Inc., vol. 2, pp. 111-137.
112. Stewart, W. A., and King, R. N.: Fiber projections from the nucleus caudalis of the trigeminal nucleus, J. Comp. Neurol. **121:**271, 1963.
113. Swett, J. E., and Bourassa, C. M.: Comparison of sensory discrimination thresholds with muscle and cutaneous nerve volleys in the cat, J. Neurophysiol. **30:**530, 1967.
114. Sur, M., Nelson, R. J., and Kaas, J. H.: Postcentral somatosensory cortex in macaque monkeys: multiple body representations and neuron properties, Neurosci. Abstracts **4:**559, 1978.
115. Talbot, W. H., Darian-Smith, I., Kornhuber, H. H., and Mountcastle, V. B.: The sense of flutter-vibration: comparison of the human capacity with response patterns of mechanoreceptive afferents from the monkey hand, J. Neurophysiol. **31:**301, 1968.
116. Tasker, R. B.: Thalamotomy for pain: lesion localization by detailed thalamic mapping, Can. J. Surg. **12:** 62, 1969.
117. Taub, E., and Berman, A. J.: Movement and learning in the absence of sensory feedback. In Freedman, S. J., editor: The neurophysiology of spatially oriented behavior, Homewood, Ill., 1968, Dorsey Press.
118. Taub, E., Goldberg, E. A., and Taub, P.: Deafferentation in monkeys: pointing at a target without visual feedback, Exp. Neurol. **46**(1):178, 1975.
119. Towe, A. L.: Somatosensory cortex, descending influences on ascending systems. In Iggo, A., editor: Handbook of sensory physiology. Somatosensory system, New York, 1973, Springer-Verlag New York, Inc., 1973, vol. 2, pp. 701-718.
120. Truex, R. C., Taylor, M. J., Smythe, M. Q., and Gildenberg, P. L.: The lateral cervical nucleus of cat, dog, and man, J. Comp. Neurol. **139:**93, 1970.
121. Vierck, C. J., Jr.: Tactile movement detection and discrimination following dorsal column lesions in monkeys, Exp. Brain Res. **20:**331, 1974.
121a. Weinstein, S.: Intensive and extensive aspects of tactile sensitivity as a function of body part, sex and laterality. In Kenshalo, D. R., editor: The skin senses, Springfield, Ill., 1968, Charles C Thomas, Publisher.
122. Welker, W. I.: Principles of organization of the ventrobasal complex in mammals, Brain Behav. Evol. **7:** 253, 1973.
123. Werner, G., and Whitsel, B. L.: The topology of dermatomal projection in the medial lemniscal system, J. Physiol. **192:**123, 1967.
124. Werner, G., and Whitsel, B. L.: Topology of the body representation in somatosensory I of primates, J. Neurophysiol. **31:**856, 1968.
125. Whitlock, D. G., and Perl, E.: Thalamic projections of spinothalamic pathways in monkey, Exp. Neurol. **3:**240, 1961.
126. Whitsel, B. L., and Dreyer, D. A.: Comparison of single unit data obtained from the different topographic subdivisions of the postcentral gyrus of the macaque: implications for the organization of somatosensory projection pathways, Exp. Brain Res. **1:**415, 1976.
127. Whitsel, B. L., Dreyer, D. A., and Roppolo, J. R.: Determinants of body representation in postcentral

gyrus of macaques, J. Neurophysiol. **34:**1018, 1971.
128. Whitsel, B. L., Petrocelli, L. M., Ha, H., and Dreyer, D.: Modality representation and fibre sorting in the fasciculus gracilis of squirrel monkeys, Exp. Neurol. **29:**227, 1970.
129. Whitsel, B. L., Petrocelli, L. M., and Sapiro, G.: Modality representation in the lumbar and cervical fasciculus gracilis of squirrel monkeys, Brain Res. **15:**67, 1969.
130. Whitsel, B. L., Petrocelli, L. M., and Werner, G.: Symmetry and connectivity in the map of the body surface in somatosensory area II of primates, J. Neurophysiol. **32:**170, 1969.
131. Woolsey, C. N.: Organization of somatic sensory and motor areas of the cerebral cortex. In Harlow, H. F., and Woolsey, C. N., editors: Biological and biochemical bases of behavior, Madison, 1958, University of Wisconsin Press.
132. Woolsey, C. N., Marshall, W. H., and Bard, P.: Representation of cutaneous tactile sensibility in the cerebral cortex of the monkey as indicated by evoked potentials, Johns Hopkins Med. J. **70:**399, 1942.
133. Yin, T. C. T., and Williams, W. J.: Dynamic response and transfer characteristics of joint neurons in somatosensory thalamus of the cat, J. Neurophysiol. **39:**582, 1976.

13

VERNON B. MOUNTCASTLE

Pain and temperature sensibilities

Pain is that sensory experience evoked by stimuli that injure or threaten to destroy tissue, defined introspectively by every man as that which hurts. Pain is nevertheless such a universal experience of everyday life that verbal descriptions of it provide recognizable signals from one person to another concerning its presence, nature, intensity, duration, location, reference, and temporal course. Indeed, descriptions of particular pains are telltale signs to the inquiring physician of the nature and extent of the pathologic processes that arouse them.

Pain is composed, first, of a separate and distinct sensation and, second, of the individual's *reaction to pain,* with accompanying emotional overtones, widespread reflected activity in both somatic and autonomic effectors, and volitional efforts of avoidance or escape. This reaction, akin to suffering and related to the sometimes life-threatening nature of the experience, differs widely among persons, being influenced by age, sex, and race; by the nature, duration, and intensity of the pain; and above all by the personality of the sufferer. Pain, moreover, is commonly produced by stimuli that at weaker intensities evoke other somatic sensations, for example, warmth, cold, or mechanical contacts.

The weight of evidence indicates that pain is a separate and identifiable sensation evoked by afferent activity in a specific set of first-order nerve fibers. Uncertainty as to the central nervous pathways activated by impulses in nociceptive afferents is not a sign that pain involves any operations of the nervous system unknown in the neural mechanisms of other sensory phenomena.

HUMAN PERCEPTION OF PAIN

There are at least three qualities of pain on which most observers agree. The first is the bright, pricking pain so readily evoked by a brisk needle jab of the skin. It is accurately localized, subsides quickly, and is so much less potent in evoking that emotional overtone common with other pains that it is readily studied in a quantitative manner. It is first or "fast" pain and depends on activation of a certain set of delta fibers in peripheral nerves.

The second is burning pain, a sensation of slower onset, greater persistence, and less certain location that may continue as an "afterimage" for many seconds after removal of the provoking stimulus. It is this pain that is so difficult to endure and that readily evokes the cardiovascular and respiratory reflexes characteristic of pain. This second or "slow" pain is produced by activation of a certain set of unmyelinated fibers (C fibers) in peripheral nerves.

Visceral and somatic deep structures, when noxiously stimulated, give rise to pain most commonly described as aching, sometimes with an additional quality of burning. This aching pain is important for the physician, for it commonly signals life-threatening disturbances of vital organs. It is just this pain that is most difficult to define precisely, for the location of the structures concerned makes experimental observation difficult, and naturally occurring pains of visceral origin are frequently referred to sites distant from the actual location of this disturbing process (p. 414).

Adequate stimuli for pain and the human threshold for perception of pain

In contrast to other sensations, the application of a pain-provoking stimulus commonly produces local tissue changes such as erythema of the skin, and continued noxious stimulation alters the tissue and its function. This leads to the suggestion that *the adequate stimulus for pain is the rate of destruction of a tissue innervated by pain fibers,* including changes short of cell rupture. The importance of rate of change in reaching the pain threshold is shown by the strength-duration curve in Fig. 13-1.

Pain can be evoked by energy in any one of

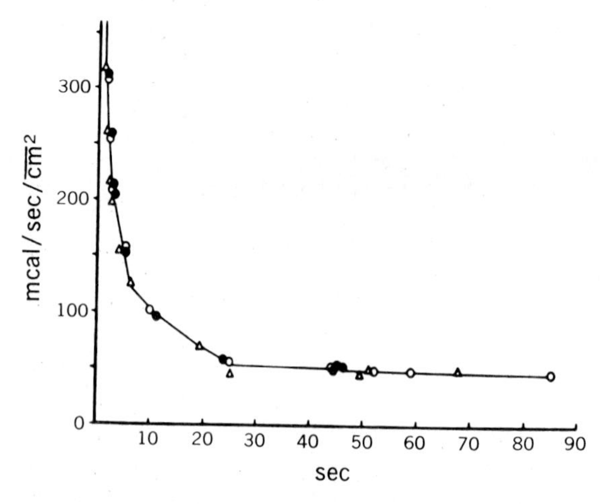

Fig. 13-1. Relation between duration of thermal irradiation and its intensity, at threshold for pricking pain, in three normal subjects. (From Hardy et al.[64])

four forms: electrical, mechanical, chemical, or thermal. Electrical stimulation of nerves is useful in determining which peripheral nerve fibers will elicit pain when active and in determining certain temporal aspects of pain. Pain is readily elicited by pressure on the skin. Mechanical stimuli have been used successfully in the study of visceral pain (e.g., by distention of a hollow viscus with inflatable balloons). Chemical agents have been used in attempts to identify neurohumoral agents that may excite pain endings (p. 401).

Of the many methods used experimentally to produce pain, that of thermal radiation has been of the greatest value. A dolorimeter designed by Hardy et al.[64] consists of a heat source (a lamp) and a lens-condensing system that produces an even distribution of heat at the skin surface. The area exposed and the intensity and duration of stimuli are controlled, and radiometric calibration allows the designation of stimuli in terms of heat transfer as milligram-calories per

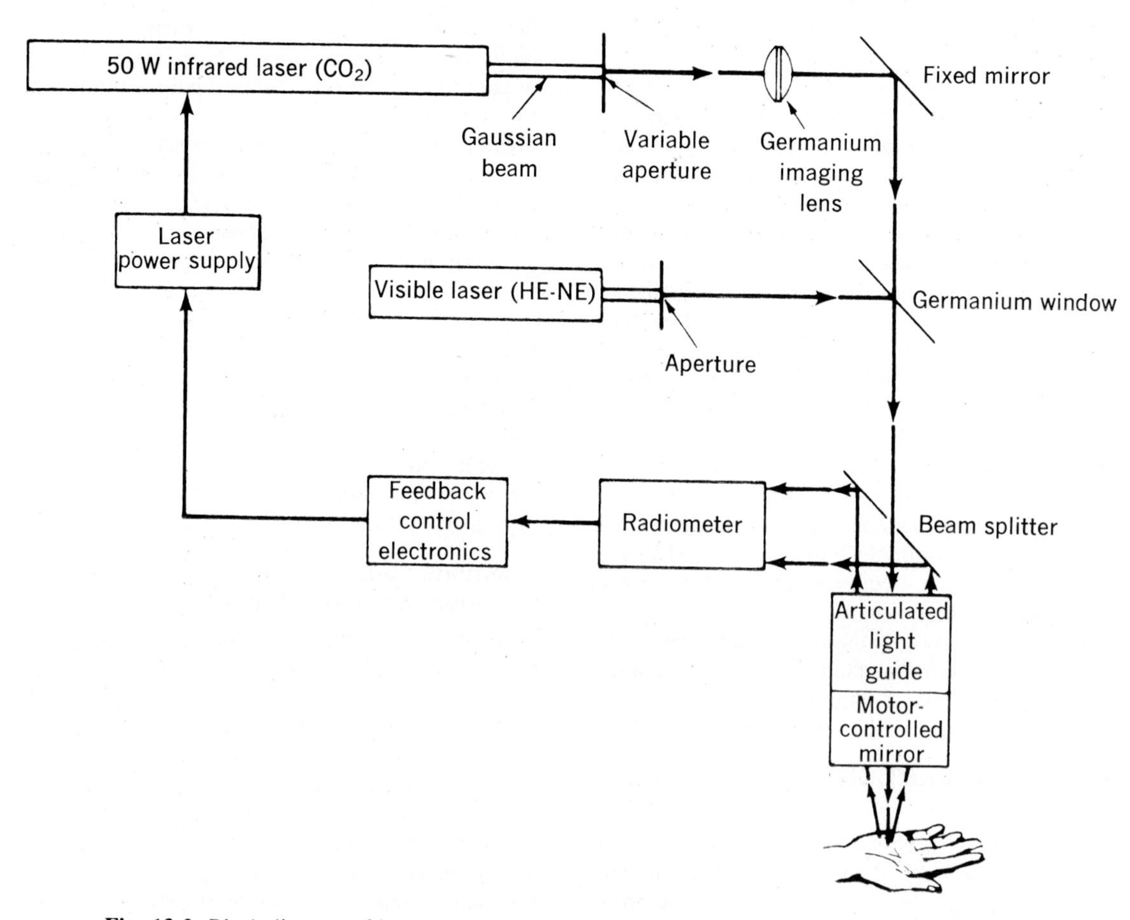

Fig. 13-2. Block diagram of laser-powered radiant heat stimulator used in study of pain in human subjects and of nociceptive afferents in monkeys. (From Meyer et al.[110])

second per square centimeter of skin (mcal/sec/cm²). Blackening the skin surface raises heat transfer to better than 95%, makes it independent of natural skin pigmentation, and prevents penetration to deeper layers of the skin. The skin temperature is measured radiometrically or with a small bead thermister. Successful use of the method requires that the subject understand the end point sought. For many studies Hardy et al. chose as the end point a small, distinct stab of pain occurring just at the end of a 3 sec exposure to thermal radiation.

More recently, a thermal stimulator has been designed with a laser as heat source and a closed loop control of temperature via an intrinsic radiometer.[110] With it, constant temperature steps, rather than rates of energy transfer, can be delivered to the skin over a range of 0.1° to 25° C above base temperature, with rise times of 30° C/sec, and regulated with a range of 0.1° C (Fig. 13-2).

The histogram in Fig. 13-3 shows the distribution of pricking pain thresholds measured in terms of the rate of heat transfer. Stimuli were delivered to the skin of the forehead in a large number of instructed but untrained individuals who varied in age (14 to 70 years), sex, race, and occupation. The mean value is 206 ± SD = 21 mcal/sec/cm². The skin of the forehead was chosen for stimulation in these experiments because in ordinary room environments its temperature is nearly constant at about 34° C. The graph in Fig. 13-4 shows that skin temperature is an important variable in such measurements; the threshold for thermally evoked pricking pain is the rate of heat transfer to the skin that exceeds the rate of heat loss by an amount just sufficient to drive the skin temperature to about 45° C within any given period of time. Indeed, pain occurs spontaneously at skin temperatures of about 44° to 45° C, regardless of previous thermal history. The pricking pain threshold measured by thermal radiation varies from place to place on the body surface, but Hardy et al.[64] found that for every locus tested the final skin temperature at threshold fell within a degree or so of 45° C.

If a skin temperature of about 45° C is maintained for a number of hours, it will lead to epidermal necrosis. The difference in time scale is important, for pain and the associated avoidance reactions are elicited quickly at 45° C, so that irreversible tissue damage does not occur.

The constancy of the pain threshold in terms of skin temperature raises the question of whether the threshold for the reflex avoidance reactions to pain is similar. Hardy[63] studied a patient who had survived for many years a complete transection of the spinal cord in the thoracic region. The

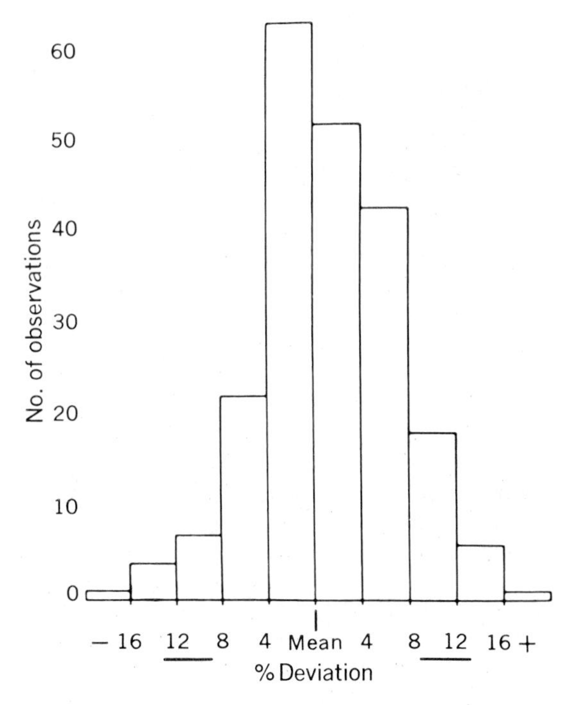

Fig. 13-3. Distribution of pain thresholds measured by thermal radiation method in large population of individuals. Threshold is taken as that intensity of thermal radiation delivered to blackened skin of forehead that produces trace of pricking pain at end of 3 sec period of exposure. It is expressed as rate of heat transfer: millicalories per second per square centimeter; for this population the mean is 206 ± SD = 21 mcal/sec/cm² and corresponds to skin temperature at end of 3 sec period of about 45° C. (From Hardy et al.[64])

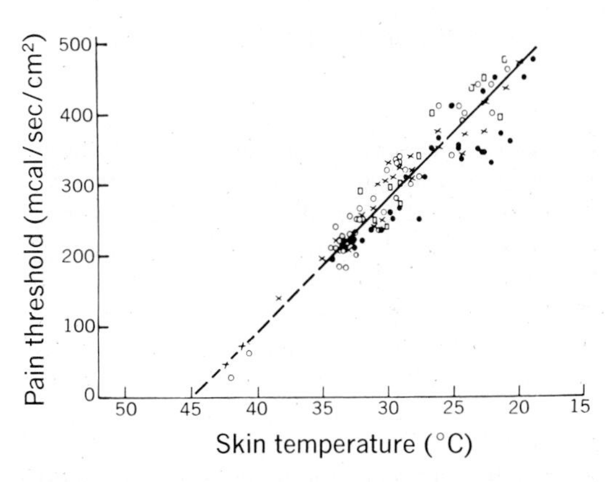

Fig. 13-4. Relation of initial temperature of skin of forehead (site of stimulation) to rate of heat transfer required to elicit threshold pricking pain, studied in three normal subjects. Linear relation intercepts abscissa at about 45° C, a skin temperature that, if continued, will produce irreversible skin damage. (From Hardy et al.[64])

pricking pain threshold on the normally innervated skin of the dorsum of the hand was 43.6° ± 1.1° C, well within the normal range. The stimulus was then shifted to the skin of the dorsum of the foot, a region linked only with the isolated segments of the spinal cord, where the reflex threshold was 44.1° ± 0.75° C, a value not distinguishable from the pain threshold in the same individual. The suggestion is that the two thresholds may be identical, although that for reflex withdrawal, normally under descending control, may vary from time to time in the normal individual.

Burning pain is a cutaneous sensation qualitatively different from pricking pain, and evidence adduced from blocking experiments indicates that it is served by unmyelinated or C fibers. It is this burning pain that is so prominent a feature of pain produced by disease or injury arising from the skin and skin deprived of its myelinated fiber innervation. The threshold for burning pain, when evoked by the method of thermal radiation, is some 20 to 40 mcal/sec/cm² *below* that for pricking pain, and the variation among individuals is somewhat greater.

Aching pain arises from deeper tissues, in particular the viscera, the periosteum, and the joints and the tissues that surround them. It can be evoked by thermal radiation sufficiently intense to raise the temperature of deeper tissues; its threshold is in the range of 300 to 320 mcal/sec/cm². Pains that arise from the viscera and from the body cavities and the special quality of pain called headache will be discussed in later sections.

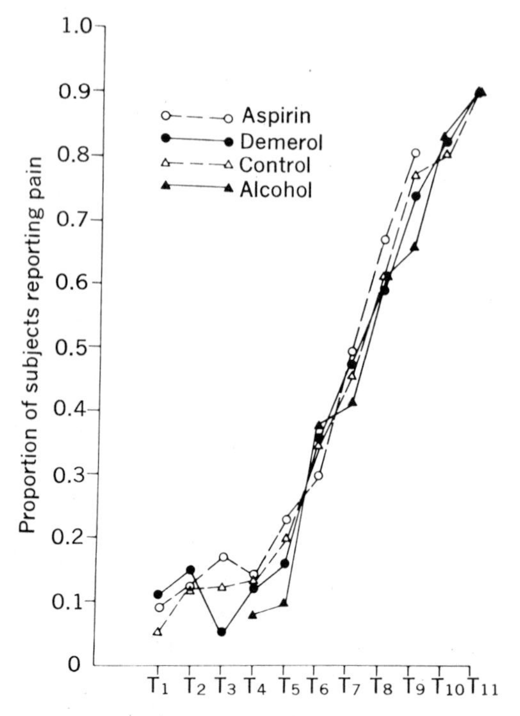

Fig. 13-5. Distribution functions of pain thresholds for a number of subjects after administration of no drug, after analgesic aspirin, after narcotic Demerol, and after ethyl alcohol. Functions plot percentage of positive identifications of pain when stimuli of each intensity were delivered for a number of times. Stimuli were thermal irradiations and are plotted here as temperature of irradiated skin reached at end of 3 sec period of exposure. T_1 = 39.1° C, T_2 = 40.0° C, T_3 = 40.7° C, T_4 = 41.4° C, T_5 = 42.2° C, T_6 = 43.0° C, T_7 = 43.7° C, T_8 = 44.5° C, T_9 = 45.3° C, T_{10} = 46.0° C, T_{11} = 46.8° C. (From Chapman et al.[35])

Factors influencing the pain threshold

Under the conditions just described the threshold for pain induced by thermal radiation is not affected by age, sex, fatigue, or the minor alterations of mood that all regard as normal. It is affected by the local conditions of the area stimulated, particularly by those that affect the transfer of heat to or its dissipation from the skin: its absorbent capacity, resting temperature, wetness or dryness, etc. The threshold of the skin is also influenced by its immediate past history: the number, intensities, and lengths of previous stimuli and the interval of time since the last. These latter factors are related to the appearance first of suppression and then of primary hyperalgesia (p. 401), which is accompanied by a lowering of threshold for pain.

Is there adaptation to painful stimuli?

When painful stimuli at or just above threshold are delivered one after another to the same skin location, the pain evoked by each may gradually become less intense and after a few stimuli may not be painful at all. An intense noxious stimulus may reduce the intensity of the pain produced by a weaker one delivered shortly thereafter to the same spot. This phenomenon is termed suppression (Fig. 13-7). A similar suppression is observed in the successive responses of first-order nociceptive afferent fibers (p. 398). Humans may adapt to continuous noxious stimuli that are just at the threshold level of intensity, that is, up to a 47° C skin temperature. However, continuing noxious stimuli only slightly more intense produce continuing pain, and, in contrast to the adaptation that occurs for just suprathreshold stimuli, may with time produce pain of increasing intensity, with the onset of primary hyperalgesia.

An experiment bearing on this point is illustrated in Fig. 13-6. Five trained observers were

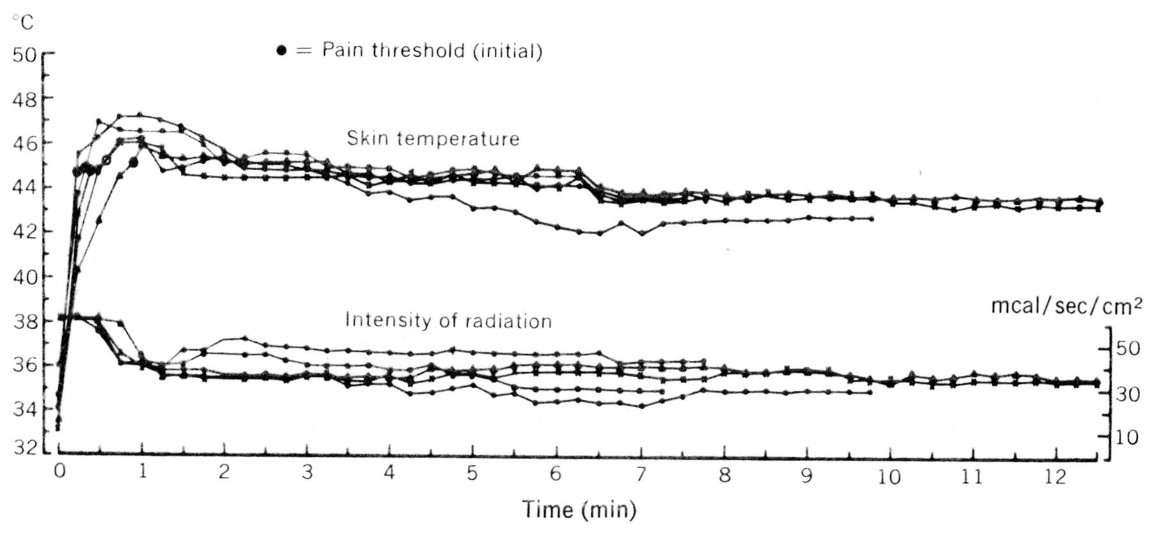

Fig. 13-6. Results of experiment aimed at testing degree of adaptation to painful stimuli, using thermal radiation method, in five subjects. Each controlled for himself intensity of thermal radiation just required to produce continued pricking pain. Intensities they chose are plotted below, skin temperatures above. Rather than adaptation, skin temperature required just to maintain a sensation of pain actually dropped about 1° C during exposure, a change thought to be due to onset of primary hyperalgesia. (From Greene and Hardy.[60])

asked to maintain by their own control of an unseen dial an intensity of thermal radiation that just continued to evoke pricking pain.[60] Note that the intensities of radiation that they chose are similar in strength to the rheobase of the strength-duration curve in Fig. 13-1. Rather than adaptation, the curves show that there was a gradual lowering of the skin temperature required for the pain threshold, a change attributed to the onset of *primary hyperalgesia*.

Is there spatial or temporal summation of pain?

The term "spatial summation" as applied to sensations refers to two related ideas. The first is the possibility that a stimulus that is subthreshold for sensation when applied to a small area will evoke sensations when applied over a larger area of the receptor surface. Presumably this is due to the additive central effects of the activation of an increasing number of primary sensory fibers. The second is that if the stimulus is initially supraliminal when applied to a small area it will, when applied to larger areas, elicit more intense sensations. Spatial summation is a classic attribute of vision (Ricco's law) and occurs for the cutaneous senses of touch, warmth, and cold. It occurs to only a limited extent for pricking, burning, or aching pain. For example, comparison of the senses of warmth and pain, elicited in the two cases by thermal radiation of different intensities, revealed that for warmth there was more than a 200-fold increase in threshold as the area of skin stimulated was reduced from 200 to 0.2 cm^2; the pricking pain threshold remained constant over the range of skin areas from 0.5 to 28 cm^2, the largest area tested.[64]

A sharp distinction should be made between the lack of spatial summation in the threshold and in the intensity of pain sensation and its importance in certain reactions to pain. Spatial summation is a functional property of the segmental reflex mechanisms of the spinal cord, for example, and most powerfully so for the flexion reflexes evoked by noxious stimuli. Moreover, the total experience of pain, in contrast to the sensation of pain, is undoubtedly influenced by the spatial extent of pain-provoking injury or disease: the life-threatening effect of an extensive as contrasted to a local burn of the body surface, for example, may certainly influence in some subjects the total pain experience associated with the injury.

Temporal summation, on the other hand, plays an important role in pain sensation; indeed, it is required for sensing certain pains. Collins et al.[41] studied the effect in conscious human subjects of afferent volleys in C fibers evoked by electrical stimulation of peripheral nerves after block of myelinated fibers. A single C fiber volley, proved to have entered the nervous system, evoked no sensation whatsoever! At frequencies of 3/sec the subjects always experienced pain, and with continued stimulation at this or higher rates the sensation grew quickly to one of unbearable pain.

Intensity functions and the scaling of pain

Pain is a unique sensation in many ways, not least because of the narrow range that separates stimuli of threshold value from those that evoke a pain experience of maximal intensity: it is only 2:1. Using the heat transfer method described earlier, Adair et al.[19] have studied the relation of the subjective estimation of the intensity of a pain experience to the physical values of provoking stimuli. They found this relation to be linear over the first two thirds of the total range of pain experience. LaMotte and Campbell,[94] using the stimulator diagrammed in Fig. 13-2, have studied the way human subjects categorize the magnitude of warmth and pain sensations evoked by stimuli that ranged in temperature from 40° to 50° C. They constructed from these category estimates a scale of subjective thermal intensity including, in a continuum, both warmth and pain (Fig. 13-7). It is a monotonically increasing function with a slight positive acceleration over the range of temperatures studied. Fig. 13-7 also illustrates by the difference between the function marked by filled circles and that by open circles the suppressive effect that occurs when noxious stimuli are repeatedly delivered to the same skin location at frequent intervals.

Pain scales established in this way have proved valuable for the study of pain mechanisms, particularly for the correlation of psychophysical measurements of pain produced by stimuli and the neural events evoked in both monkeys and men by those same stimuli. Those scales have not yet proved of equal value for ranking the severity of pain suffered by patients with injury or disease largely, it is believed, because in this setting the sensation of pain and the total pain experience of it cannot be separated subjectively. Many ordinal scales have been constructed for the rating of the

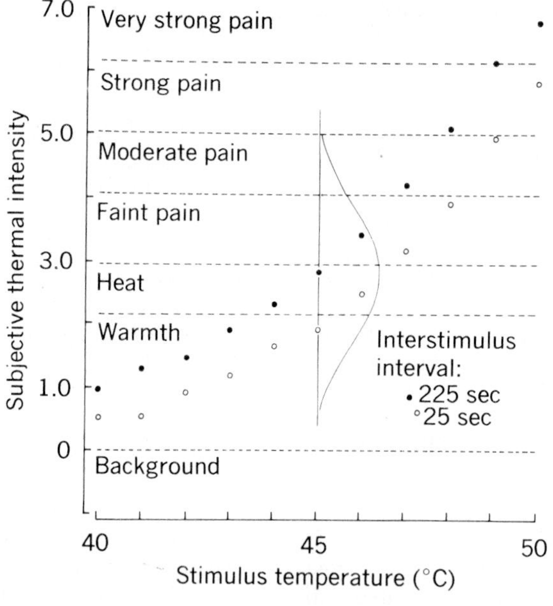

Fig. 13-7. Scale of subjective intensity of radiant thermal stimuli. Subjects categorized magnitude of warmth and painful sensations evoked by radiant heat delivered to glabrous skin of hand. Numeric values represent responses to stimuli on equal-interval scale. Closed circles represent scale values obtained when radiant heat stimuli were delivered to nine separate loci (8 mm in diameter) on thenar eminence, in rotating sequence. Mean interstimulus interval (ISI) was then 225 sec. Open circles give scale values using ISI of 25 sec, with stimuli delivered to single locus on thenar eminence. Dotted lines mark scale values for boundaries between each category. Each stimulus scale value is mean of normal distribution of category responses obtained for a given stimulus temperature and ISI. Distance along ordinate is provided in standard deviation units. Distributions of responses to 45° C stimulus delivered with ISI = 225 sec is illustrated and is representative of other distributions. Magnitude of warm and painful sensations depends not only on stimulus temperature, but also on rate of stimulus delivery. More rapid delivery suppresses thermal intensity. Warmth and pain sensations form subjective continuum, and subjective magnitude is regular function of stimulus intensity. (From LaMotte and Campbell.[94])

severity of the total pain experience ("clinical"), but none has proved wholly satisfactory. As a result, some have chosen to rate the severity of pain in terms of the amounts of analgesic agents required to relieve pain by a certain degree, as estimated by patient and observers. Used with the "double-blind" method, this has proved useful in rating analgesic drugs when assessed against the placebo effect. It does not provide a scale for the intensity of pain. The principal action of analgesic drugs appears to be on the reaction to pain, not its threshold (Fig. 13-5).

Signal detection theory and pain[2]

The psychophysical methods of limits and of constant stimuli were used in the studies of pain chosen for description and illustration previously. The results obtained with these methods vary greatly between different studies, particularly those of clinical pain. Much of this variability can be attributed to differences in response bias (i.e., the subject's willingness to report pain) rather than to differences in sensory capacity. In recent years, signal detection theory has been used to isolate and measure independently these two aspects of the pain experience: (1) the *sensory variable* (d'), the subject's capacity to discriminate among stimuli that differ along a physical continuum, and (2) the psychological component (L_x), which specifies the subject's report criterion, the location along the intensive continuum above which the subject declares stimuli painful and below which he does not. The method is described in Chapter 21. The sensory variable, unlike the traditional threshold measurement, remains relatively unchanged under a variety of stimulus and behavioral conditions. It is the perceived magnitude or criterion level (L_x) that varies so greatly between stoical individuals and their opposites and is shifted upward by placebo, suggestion, hypnosis, and acupuncture.[38] Some narcotics and analgesics apparently influence both sensory capacity and criterion level, others only the latter.

More generally, studies using signal detection theory have given explicit descriptions of these two separable aspects of painful experiences and suggest that the neural mechanisms responsible for each, although interdependent and partially congruent, are to some degree and at some level separate. The sensory capacity is little affected by the affective overtones that frequently accompany some pains, particularly those caused by disease states, whereas the response criterion is sensitive to them, and changes in it reflect changes in the overall subjective experience of pain.

HUMAN REACTION TO PAIN AND THE TOTAL PAIN EXPERIENCE

It is the thesis of the foregoing account that human observers can identify the sensation of pain, examine it introspectively, and describe it publicly despite accompanying reactions to pain. These reactions are of three types: local, reflex, and behavioral. Local reactions will be discussed in a following section. Reflex actions are provoked by noxious stimuli and may occur whether or not those stimuli also evoke the sensation of pain, although they commonly do.

Somatic reflex reactions. The reflex contraction of flexor muscles evoked by noxious stimulation is part of the withdrawal response to pain. It shows spatial summation, in contrast to pain sensation, and in experimental animals or in paraplegic man, its threshold may equal that for human pain sensation. In normal man, flexion withdrawal may be suppressed completely or it may occur with the anticipation or threat of pain. Its threshold is thus variable and indefinable, in contrast to that of pain sensation.

Continuing noxious stimulation, particularly that arising in deep somatic or visceral tissues, commonly elicits a steadily maintained reflex contraction, usually of adjacent but sometimes of distant muscles. Splinting of the muscles of the abdominal wall, for example, may be the first sign of intra-abdominal disease. It may appear at a stage when the frequency of discharge in pain fibers innervating the area of tissue injury is too low to evoke the sensation of pain. With continued contraction the muscles may themselves become painful, a process greatly intensified and accelerated if they are deprived of blood supply or their venous drainage is blocked. It is thought that this muscle tenderness or "soreness" is due to the release of pain-producing substances from the continually contracting and thus partially ischemic muscle cells. This pain of muscle origin frequently becomes more severe than the pain that initiates it, and it may dominate the clinical picture. It should not be confused with referred pain (p. 414), although referred pain may accompany the initiating process.

Autonomic reflex reactions. Reflex reactions engaging autonomic effectors commonly accompany the somatic motor responses to noxious stimulation, particularly when the pain provoked is severe and of sudden and unexpected onset. They are parts of the more generalized mobilization for defensive or aggressive reaction to attack, or to the threat of it, which depend for integration on executant neural mechanisms of the forebrain, particularly of the hypothalamus (Chapter 34). They may include cardiac acceleration and peripheral vasoconstriction, a rise in blood pressure, dilatation of the pupils, and secretion from the sweat glands and the adrenal medulla—all signs of intense activity in sympathetic efferent nerve fibers. All pains do not evoke similar reflex patterns, however, for visceral pain of sudden onset (testicular crushing, pain of chemical peritonitis, etc.) may provoke vasodilatation and a drastic fall in blood pressure, together with a decrease in somatic motor tone. In the intact mammal these autonomic effects are welded into the total behavioral pattern, even though some are basically spinal reflexes. In chronic spinal man, for example, with a high thoracic transection, distention of the bladder may elicit intense vasoconstriction, sweating, and piloerection in the parts of the body innervated by the isolated spinal segments.

The phasic sympathetic discharge to the sweat glands evoked by a painful stimulus is accompanied by a change in the electrical resistance of the skin. Although this change is not wholly due to the excretion of sweat, it is a ready measure of the integrity of the sympathetic efferent pathways: it is called the galvanic skin reflex (GSR). How completely such a segmental reflex may be dominated by descending influences of forebrain origin is shown by the fact that the GSR adapts or "habituates" to repeated stimuli, but more especially by the fact that its presence and amplitude vary greatly from one individual to another and from time to time in a single person. It appears to reflect in part the affective state of the subject, his attitude toward the situation in which he receives the painful stimulus, and the meaning he attaches to such stimuli. Thus many of the reflex reactions to pain cannot be separated from the overall emotional attitude of the subject toward pain. It is the meld of both with the sensation itself that constitutes the total pain experience.

PERIPHERAL NEURAL MECHANISMS IN PAIN

Peripheral nerve fibers and pain[1]

Cutaneous pain is served by two specific sets of peripheral nerve fibers, each differentially sensitive to destructive stimuli. A specific set of A-delta fibers serves fast, pricking, or "first" pain, and an equally specific set of C fibers serves slow, burning, or "second" pain. The latency difference is due in part to the difference in conduction velocities of the two sets. The qualitative differences in these two aspects of pain sensation are partially due to differences in the functional properties of the two sets of fibers and in the transducer mechanisms at their peripheral endings, but perhaps more importantly to differences in their central connections.

A new method for recording from single fibers in the peripheral nerves of humans[164] has provided (1) confirmation of these general statements and (2) new information concerning the peripheral neural mechanisms in pain. It has long been known that pressure blocks conduction in nerve fibers in an order from large to small and that local anesthesia blocks in a mixed but generally reverse direction. These observations have now been repeated in combination with single-fiber recording in the nerves of human subjects, and the results correlated with their verbal descriptions of the sensations evoked by a variety of cutaneous stimuli as block began and progressed. The double nature of pain persists as long as fibers of the A-delta class and smaller continue to conduct through a partial pressure block. The discriminative aspects of mechanoreceptive sensibility are lost with block of the A-beta fibers, leaving pain of both types and warmth and cool intact, as well as a vaguely described and imprecise form of contact sensibility. Pricking pain and cool disappear when conduction in delta fibers fails; warmth and burning pain persist as long as C fibers continue to conduct through the block.[30,57,102,159,160] The remaining second pain is often exaggerated, a central release phenomenon discussed on p. 419. A reserve sequence of loss of the two forms of pain occurs with the onset and progression of nerve block by local anesthetic.

The specific, differential sensitivity of the A-delta and C fiber nociceptive afferents to destructive stimuli has now been established beyond any reasonable doubt in a large number of investigations in experimental animals.[1,120] Studies of

Table 13-1. Distribution of nociceptive fibers of the A-delta and C fibers among different groups in terms of the qualitative nature of the most adequate stimulus for each group*

Group	A-delta fibers		C fibers	
	No.	%	No.	%
Mechanical	160	46	19	10
Mechanothermal	177	52	164	86
Mechanical + heat	79		75	
Mechanical + cold	35		13	
Mechanical + heat + cold	63		76	
Thermal	8	2	8	4
Heat	7		5	
Cold	0		0	
Heat + cold	1		3	
TOTAL	345	100	191	100

*From Georgopoulos.[58]

the innervation of the monkey hand are of particular interest, for the glabrous skins of the monkey and human hands appear to be identical in histologic structure, neural innervation, and sensory capacity (Chapter 12). Table 13-1 lists the proportions of nociceptive afferents of different types identified by Georgopoulos in a population study of the innervation of the monkey's hand.[58] Many are sensitive only to mechanical stimuli that are destructive of tissue. The cumulative distribution plots of Fig. 13-8 show that although there is a considerable overlap between

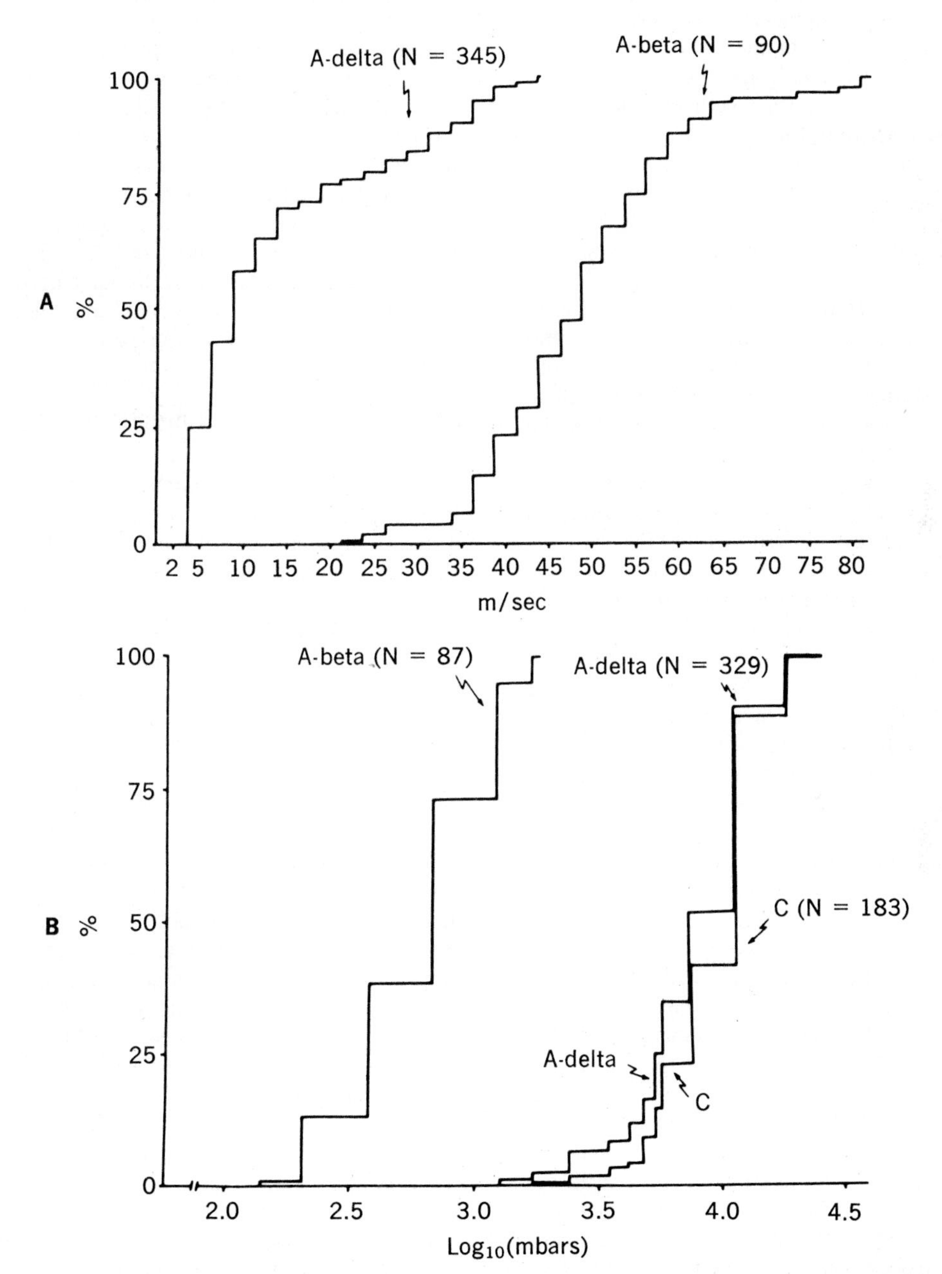

Fig. 13-8. A, Accumulative distributions of conduction velocities of A-beta and A-delta fibers innervating palmar and digital skin of monkey's hand. There is overlap in range of 25 to 45 m/sec. **B,** Accumulative distributions of A-beta, A-delta, and C fibers innervating palmar and digital skin of monkey's hand with regard to mechanical thresholds. Those for A-delta and C fibers are in noxious range, and there is no overlap with thresholds of A-beta fibers. Thus A-beta and A-delta fibers of same conduction velocities, in range of overlap, nevertheless differ as to functional properties and mechanical thresholds. (From Georgopoulos.[58])

the distributions of the conduction velocities of A-beta and A-delta mechanoreceptive afferents, and thus their sizes, there is a clear separation in the distribution of their mechanical thresholds, and the threshold distributions of A-delta and C fibers are identical. Table 13-1 also demonstrates a fact originally emphasized by Perl, that the majority of both A-delta and C fiber nociceptive afferents are polymodal, that is, their specific sensitivity is to the destructive nature of stimuli, not to the particular form of energy producing the destruction. A number of studies have been made of the stimulus-response relation for nociceptive afferents, with the general result that over the relatively narrow range from threshold to maximal, the relation is linear or only slightly positively accelerating.[59,94] For example, Georgopoulos found that the average exponent for the best-fitting power function for the population he studied was 1.23 ± 0.28 SD, with a median value of 1.1.[59] Thus the portion of the category function of Fig. 13-7 that begins with "faint pain" and progresses to the right can be accounted for by the stimulus-response function of the relevant sets of nociceptive first-order afferent fibers.

Perl discovered that when noxious stimuli are delivered repeatedly to a local skin area, a spontaneous discharge may appear in nociceptive afferents previously silent, along with a lowering of their thresholds and an increase in their responses to supramaximal stimuli.[88,121,123] This sensitization may occur in nociceptive afferents innervating skin some distance from that damaged and depends on the axon reflex mechanism.[51] It is thought to account in part for the persisting pain that accompanies tissue injury or inflammation. Under some circumstances, however, sensitization is not a prominent feature, and the common time-dependent effect of repeated stimuli is a marked suppression of response.[58,94,132] Which of these changes will occur appears to be determined by the degree of tissue damage.

It has recently been discovered that the ventral roots of cats, monkeys, and humans contain significant numbers of unmyelinated afferent fibers, neurons whose cell bodies lie in the dorsal root ganglia, but whose centrally directed axons enter the spinal cord in the ventral roots.[23,40] This constitutes an exception to the Bell-Magendie law. The majority of these ventral root unmyelinated fibers innervate the viscera of the pelvis and have high mechanical thresholds.[39] Whether they contribute to pain sensibility is uncertain.

In summary, the two aspects of pain are served by two distinct and specific sets of nociceptive afferents, one of A-delta and the other of C fibers. Some of these are activated by a particular form of noxious stimulation, but the majority are polymodal in nature, that is, they are specifically sensitive to the destruction of tissue cells, not to the form of energy causing that destruction. The stimulus-response function of nociceptive afferents is nearly linear over the narrow (only 2:1) range of pain sensibility, and this can account for the nearly linear human estimation of the intensity of noxious stimuli over that same range. Pain is a specific sensation with its own particular sets of specific peripheral afferent fibers. This conclusion does not imply that other sets of afferent fibers, when discharging in concert with nociceptive afferents, may not contribute to the total perceptual experience of which pain may be only a part.

Peripheral transducer mechanisms in pain

It is not known how noxious stimuli are transduced at the peripheral terminals of nociceptive afferents. It is possible but unlikely that the endings are activated directly by noxious stimuli, thus initiating a sequence of permeability change, local response, and conducted nerve impulses, in the mode of the mechanoreceptive afferents described in Chapter 11. There is evidence to suggest that an intervening step involves the release of one or more "pain producing substance(s)," either from damaged tissue cells and nerve terminals or from the nerve terminals themselves, and that interaction of such a substance with receptor molecules of the nerve terminal membrane leads to excitation, much in the manner of chemoreceptors and resembling the process of excitation at chemically operated synapses. Attempts to identify this substance make up an important part of research in the field of pain, for if the chain of biochemical events could be identified, it might be possible to design a drug to block it, thus interfering with the pain process at the level of its inception, perhaps with no major effect on the CNS. Several older lines of evidence suggest that such a chemoreceptive process is important in the initiation of pain.

The triple response. Shortly after injury to the skin there appears in the injured area intense vasodilatation that leads to local edema and formation of a wheal a few millimeters in diameter, soon surrounded by a less intense but much wider (several centimeters) vasodilatory flushing of the skin. Local reddening, wheal formation, and surrounding flare make up the triple response of

Lewis.[10] Within the injured area and in part of the surrounding region of flare, a pinprick produces more intense pain than before injury and the pain threshold is lowered; stimuli not previously painful become so, and these changes may persist for days. This heightened sensibility is termed *primary hyperalgesia*. The hypersensitivity to pinpricks that also develops in a wide region outside the flare may persist for only a few hours; within this region the threshold for pain is not lowered. This is termed the region of *secondary hyperalgesia*. The two hyperalgesias are thought to differ in causal mechanism.

After acute section of a peripheral nerve the triple response persists as long as the peripheral axons. After their degeneration (5 to 6 days), only local reddening and wheal formation follow local injury. The full triple response is produced by cutaneous injury after section of the dorsal roots central to the ganglion, for then the peripheral axons do not degenerate. The vasodilatation seen as flare is therefore an *axon reflex;* impulses traveling centrally from the region of injury are conducted antidromically over other branches of the nociceptive afferents, invade their endings, and elicit the vasomotor changes. Indeed, some branches of dorsal root C fibers end in close relation to small peripheral blood vessels, although not directly innervating them. The implication is that antidromic impulses invading C fiber terminals release from those terminals a vasodilator substance. Autonomic efferents are not involved, for the triple response survives complete sympathectomy.

This phenomenon is closely related to the vasodilatation produced by electrically induced antidromic volleys in dorsal root C fibers, discussed in some detail in the following section. The mechanism is thought to be identical for the two cases of cutaneous vasodilatation and to be important for understanding the chemoreceptive nature of nociceptive afferents. Foerster,[52] for example, showed that in man, if an adjacent root on either side of a transected dorsal root were intact, stimulation of the peripheral portion of the transected root resulted in cutaneous vasodilatation in its peripheral field of distribution and in sensations of pain. The adjacent roots, of course, have overlapping peripheral fields. The proposition that antidromic invasion of C fiber terminals either releases or causes the formation in the intercellular fluid of a substance having the properties of relaxing arteriolar smooth muscle *and* activating the endings of nociceptive afferents (thus accounting for Foerster's observation) fits with the idea that the normal excitation of nociceptive afferents after cellular injury is similar in nature.

Neurohumoral mediators and pain. The identity of a pain-producing substance (PPS) activated or released by cellular injury has proved elusive during nearly 50 years of search. Many toxins[33] and other naturally occurring molecules do produce pain on intradermal injection[78]; among them are histamine, K^+, 5-hydroxytryptamine (5-HT), acetylcholine (ACh), adenosine triphosphate (ATP)[28], and the polypeptides bradykinin and substance P (SP). No one of these or any other has been proved unequivocally to be the only PPS active under normal conditions. Chapman et al.[37] provided further evidence that such a substance is active normally in producing pain in injured tissues. They perfused the subcutaneous space in man and examined the perfusate for PPSs under a variety of conditions. They found that when the skin above the perfused area was untouched, the perfusate collected was no more potent in producing pain than is NaCl in physiologic concentrations. When, however, the skin overlying the perfused region was injured to a degree producing the triple response, the perfusate contained a polypeptide that caused vasodilatation and pain on intradermal injection. This polypeptide had many of the biologic and biochemical properties of the nonapeptide bradykinin (Arg-Pro-Pro-Gly-Phe-Ser-Pro-Phe-Arg), one of the most potent of naturally occurring PPSs.[24] It was also released into the subcutaneous perfusate during stimulation of the distal end of a transected dorsal root innervating the skin of the region perfused or when that skin received a severe thermal injury. Bradykinin is present in blister fluid and in other painful tissue exudates.[78] These observations led to the hypothesis that when cells are injured, they release into the interstitial fluid one or more proteolytic enzymes and that the resulting hydrolysis of the interstitial fluid proteins yields a number of polypeptides. Among them is bradykinin, which binds to receptor molecules in the membrane of the nociceptive afferent terminals, leading to local response and nerve impulses in those fibers. More recent experiments have shown that both bradykinin and 5-HT, injected intra-arterially, are potent stimuli for the terminals of the unmyelinated afferents innervating muscle, including the nociceptive afferents.[55,67] Bradykinin is not so selective in its effect on cutaneous afferents, for both the C and the A-delta nociceptive afferents and the A beta slowly adapting mechanoreceptive afferents are activated by it.[26] Moreover, one important observation made by Chapman et al.[36] does not fit with this scheme: a PPS was *not* released into a subcutaneous perfusate on injury to the overlying skin, if that skin had been previously denervated, with degeneration of all peripheral nerve fibers.

If it is true that a PPS is released from injured tissue only when that tissue is innervated by afferent fibers, presumably nociceptive afferents, the possibility is raised that nociceptive afferent terminals themselves contain such a substance that is released from them on tissue injury and depolarizes both the same and neighboring nerve terminals. The most likely candidate for such a substance is SP. This naturally occurring undecapeptide was first isolated from brain tissue by von Euler and Gaddum,[165] and its structure was determined by Chang et al.[34] It is found in the dorsal horn of the spinal cord and, when applied iontophoretically in this region, selectively depolarizes both second-order cells of the dorsal horn activated by noxious stimulation of the skin and central terminals of those nociceptive afferents.[135] After ligation of a dorsal root the level of SP drops markedly in the dorsal horn but increases in the distal segment of the ligated root. Hokfelt et al. have used an immunofluorescent method to localize SP more precisely.[68,69,70] It is present in highest concentration in the substantia gelatinosa and Lissauer's tract, in small-diameter afferent fibers, and in the sets of small dorsal root ganglion cells thought to emit unmyelinated and small myelinated fibers. SP has now been further localized to large vesicles within the central terminals of those fibers, terminals that also contain smaller vesicles.[42,126] This last observation raises the possibility that these fibers operate via some other transmitter and that SP acts as a "modulator" of synaptic transmission. SP is also present in the peripheral portions of small-diameter fibers that terminate in the dental pulp[116] and skin.[68] These same studies have revealed another set of small dorsal root ganglion cells that contain somatostatin, the 14–amino acid polypeptide that, synthesized and released by hypothalamic neurons, suppresses release of growth hormone by cells of the anterior pituitary gland. When somatostatin is iontophoretically applied in the dorsal horn, it suppresses the activity of neurons. If there exists a set of primary afferent fibers that operate with somatostatin as a synaptic transmitter or modulator, it is the first instance in which primary afferents have been shown to operate directly as inhibitory fibers; all others known are excitatory in nature.

In summary, the specific transducer mechanism that operates in peripheral tissues to link tissue injury to nerve impulses in nociceptive afferents is not known with certainty, but it is likely to be chemically mediated. The identity of the "transmitter" PPS(s) is still uncertain, and the origin, mechanism of release, and mode of action are equally obscure. The weight of evidence suggests that this peripheral PPS is one or more of several polypeptides shown to be powerful in this regard.

To return to the mechanisms of cutaneous hyperalgesia, the sensory changes at the local site of injury and in the flaring surround are classed together as primary hyperalgesia, for in either case the presence of PPSs lowers threshold and causes a stimulus of a certain strength to evoke more discharges in more afferent fibers than it normally would—the hyperalgesic phenomenon. In the region of local injury, the substance is released from the injured cells; in the region of flare, it is released by the action of antidromic impulses in C fibers. However, no such peripheral mechanism can account for the secondary hyperalgesia of the nonerythematous surround, for local changes do not occur there and the area is too wide to be reached by axon reflexes. It has been suggested that secondary hyperalgesia results from a heightened level of excitability in interneuron pools receiving afferent input from both the injured and the secondarily hyperalgesic regions. This would account for hyperalgesia without change in the pain threshold. No direct evidence relating to this idea is yet available.

CENTRAL NERVOUS MECHANISMS IN PAIN

The question of the CNS mechanisms in pain is undoubtedly one of the most difficult in brain physiology. Ideas concerning those mechanisms may be inferred from anatomic facts and from alterations in pain sensibility that occur in patients with disease or surgical lesions of the nervous system. Much has been learned by electrical stimulation of the brain in animals and in man and from electrophysiologic experiments in animals.

Two facts serve as starting points. First, pain is evoked by impulses in either of two sets of specific nociceptive afferent fibers, described previously, or by electrical stimulation of the ascending systems of the ventral quadrant of the spinal cord, into which activity in those sets of primary fibers flows via the processing circuits of the dorsal horn. Second, a full surgical transection of those ascending systems in the ventral quadrant of the cord causes a marked elevation of pain and temperature thresholds (as well as of itch) on the contralateral side, beginning a few segments below the transection (Fig. 13-9).

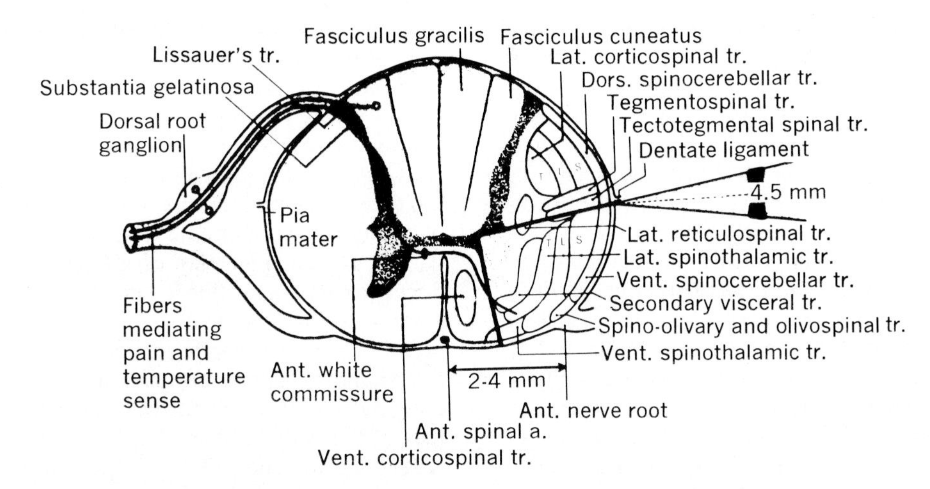

Fig. 13-9. Drawing of cross section of human spinal cord at level of third thoracic segment to illustrate some cord pathways and extent of section required for adequate anterolateral cordotomy. *T, L,* and *S* refer to thoracic, lumbar, and sacral, respectively, and illustrate relatively crude topographic organization in anterolateral system. Visceral afferents probably do not run in separate tract, but lie most medial because in large part they enter cord over thoracic dorsal roots. (From Taren and Kahn.[158])

Those elevations may amount to 50% or more and may last for a long time. Neither the peripheral fiber size classes nor the ascending pathways are devoted exclusively to signals of pain and temperature. The nociceptive sets of A-delta and C fibers are modality specific, but other sets of these same classes of fibers serve other modes in an equally specific way (e.g., the low-threshold mechanoreceptive afferents particularly sensitive to slowly moving stimuli). Moreover, a ventral quadrant section like that outlined in Fig. 13-9 is followed by a moderate elevation of the cutaneous mechanoreceptive threshold on the contralateral side of the body that is as definite, although not as striking, as the accompanying hypalgesia.

Local segmental mechanisms[6,7]

Cytoarchitectural[139] and ultrastructural[133] studies in the cat have revealed that the dorsal horn may be divided into six horizontal laminae on the basis of differences in the sizes and distributions of neurons, dendritic orientations, and synaptic organization within each. Lamina I is the most dorsal and contains the large marginal cells of Waldeyer; lamina II and the upper part of lamina III correspond to the substantia gelatinosa (SG); and the lower part of lamina III and laminae IV to VI make up the nucleus proprius of the dorsal horn and contain large neurons. Later physiologic studies have shown that this organization has some functional significance. No comparable cytoarchitectural study has yet been made in any primate, but investigators commonly use the laminar designations of Rexed[139] to refer to putatively homologous zones of the primate dorsal horn. I shall do the same in that which follows, with the reservation that the dorsal horn of the primate is unlikely to be a simple replication of that of the cat.

The cells of origin of the contralateral ascending axons of the anterolateral system (spinothalamic tract axons and cells [STT]) have been defined by (1) their retrograde degeneration after contralateral anterolateral cordotomy in man[91]; (2) their antidromic excitation by electrical stimulation of the contralateral thalamus[162] or anterolateral funiculus[90,130,172] in the monkey; and (3) the filling of cell bodies by identifiable tracers reaching them by retrograde axonal transport after injection of those tracer substances about STT axonal terminations in the contralateral thalamus in the monkey.[79,161] The results show that STT cells are widely distributed in the spinal gray matter, including the marginal neurons of lamina I, but probably no small neurons of the SG; a large number of neurons in laminae IV to VI; and a smaller number in even deeper layers. A major problem in understanding central neural mechanisms in pain concerns the segmental processing mechanisms of the dorsal horn that intervene between afferent activity in primary nociceptive fibers and the patterns of impulses in ascending systems of the spinal cord that lead to painful or thermal sensations.

It has been known for a long time and recently

confirmed in a number of studies in primates that nociceptive afferents of the A-delta and C fiber groups enter the spinal cord over the lateral division of each dorsal root[92,148,150] and terminate in the gray matter of the dorsal horn after a short longitudinal course in the extrinsic sector of Lissauer's tract; some fibers may reach the SG directly. That this projection is nociceptive in function is shown, in addition, by the fact that multiple shallow incisions of Lissauer's tract produce a zone of hypalgesia in the related ipsilateral dermatomes in both animals and humans.[74,134] Kumazawa and Perl[89] discovered in electrophysiologic experiments that the two sets of primary afferents have different projection targets within the dorsal horn (Figs. 13-10 and 13-11), and this difference has been confirmed in anatomic experiments.[92,148] Each marginal neuron of lamina I of the monkey is selectively activated by impulses in one coherent class of A-delta–sized primary afferent fibers, that is, the noxious mechanoreceptive, the cooling thermoreceptive, or (if the neurons in the lateral extension of the neck of the dorsal horn are included) that class of low-threshold mechanoreceptive afferents sensitive to slowly moving mechanical stimuli, fibers that may serve the sense of itch-tickle. Excitatory convergences between these types are rare, inhibitory convergences more common. Lamina I of the dorsal horn should not, however, be regarded as a simple relay of afferent activity in dorsal root A-delta fibers into the anterolateral system, for the marginal neurons receive several other classes of synaptic inputs.[112] SG neurons, for example, supply terminals to the somata of marginal neurons and to the presynaptic terminals of the A-delta fibers that terminate on them (Figs. 13-10 and 13-11). The terminals of aminergic descending elements reach the marginal neurons as well. Thus this relay through

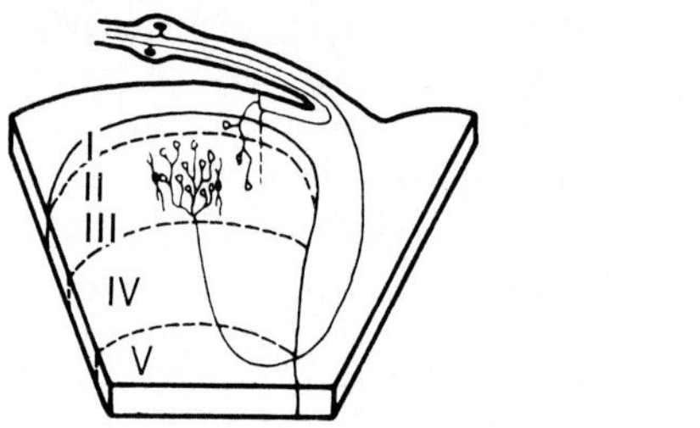

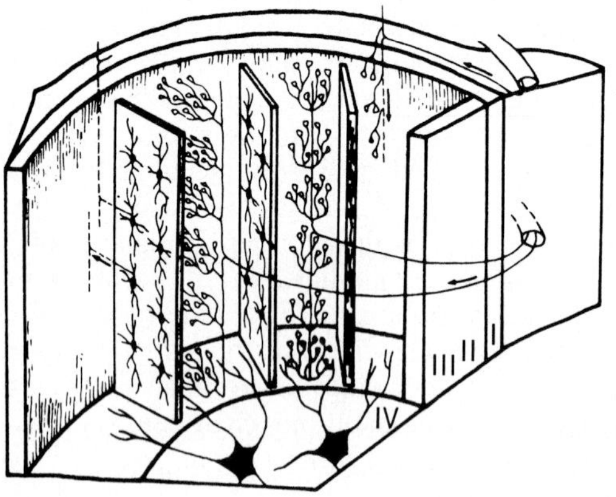

Fig. 13-10. A, Distribution of large and small primary afferent fibers of dorsal root into medial and lateral components and laminae of termination of those fibers in dorsal horn. Dorsal horn is depicted in **B** as if viewed from its ventral aspect; note very large number of terminals of large fibers of medial division of dorsal root, and compartmentalization of cells of substantia gelatinosa into modules. (From Kerr and Casey.[7])

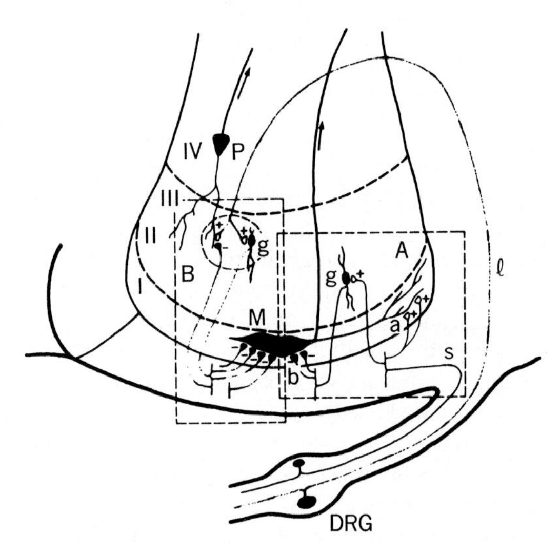

Fig. 13-11. Some of circuitry of dorsal horn. In circuit *A,* fine myelinated afferent of dorsal root is shown terminating on marginal neuron *(M).* Circuits are shown terminating on dendrites of *M* to emphasize that they are probably excitatory in nature. Gelatinosa cell *(g)* is shown sending afferents to terminate on cell body of *M,* suggesting that it is inhibitory; it may be enkephalinergic inhibitory interneuron. Circuit *B* indicates that large myelinated fibers of medial division that are not nociceptive terminate as far dorsally in dorsal horn as lower part of substantia gelatinosa, layer III. They are shown to synapse on dorsally projecting dendrites of large neurons of IV *(P),* which project to suprasegmental levels. They also synapse on gelatinosal cells that project on *M* cells of marginal zone and are thought to be inhibitory in nature. Projection of C fibers into gelatinosa is not shown. All gelatinosal cells are not inhibitory, for C fiber input to region evokes ascending activity in spinothalamic system. (From Kerr and Casey.[7])

lamina I from A-delta fibers to marginal neurons is open to regulation by both local segmental mechanisms and by descending control systems.

Cells of the substantia gelatinosa are predominantly activated by afferent impulses in C fibers. Kumazawa and Perl[89] accomplished the difficult feat of recording directly with microelectrodes the action potentials of SG neurons in the monkey. Any single cell of SG is activated by either the polymodal nociceptive afferents or the low-threshold mechanoreceptive afferents of the C group that are sensitive to slowly moving stimuli, fibers common in the innervation of hairy but not glabrous skin. Many SG neurons receive convergent input from A-delta and C fibers of the same functional class. SG axons only rarely project for long distances; for example, only an occasional SG cell can be activated antidromically by stimulation of the anterolateral system or its thalamic terminations,[7,90,162] and very few are labeled by retrograde axonal transport from the contralateral thalamus.[79] Yet it is clear that impulses in afferent nociceptive C fibers are relayed into the anterolateral systems projecting toward the forebrain and that they evoked the sensation of severe pain even when they are delivered without accompanying or preceding activity in larger fibers. Many SG axons enter Lissauer's tract, project within it for short distances, and then reenter lamina I to terminate on its marginal neurons, on SG neurons of II or III, or on the synaptic boutons of other axons terminating in those regions. This is one reason for the proposition referred to previously and illustrated in Fig. 13-10, that the SG may be a part of a local controlling or regulating mechanism for the relay between A-delta fibers to the marginal neurons, as well as a relay zone for C fiber activity.

The drawing of Fig. 13-11 shows that the larger myelinated fibers of the medial division of the dorsal root emit segmental axon collaterals before projecting into the dorsal columns. These collaterals pass ventrally, medial to the dorsal horn, then turn laterally and upward again in gentle curves to terminate in longitudinally extended axonal ramifications that alternate with rows of SG neurons to produce the modular arrangement of the dorsal horn shown in Fig. 13-11.[81,138,144] The bulbous terminations of these fibers form the central elements of the glomeruli of this region and make synaptic contacts with the dorsally directed dendrites of the large projection neurons of the n. proprius (IV to VI), as well as with what are surmised to be fine dendrites of SG neurons and axonal terminals of unknown origin. Some uncertainty exists concerning (1) how far dorsally the terminals of the large afferents penetrate into the SG and (2) the details of the synaptic relations within the glomeruli.

These connections undoubtedly form a major pathway between the large myelinated afferents and the STT neurons of laminae IV to VI, which in the monkey contain more than one half of all the spinal neurons that project to the contralateral thalamus via the anterolateral system,[79] together with many more neurons with different projections. The functional properties of the STT neurons of the nucleus proprius have been studied intensively in the monkey.* They include one set activated only by light mechanical stimulation of the skin via large myelinated fibers, another responsive to both light mechanical and noxious mechanical and thermal stimuli via convergent large and small fiber projections, and a third sensitive only to noxious stimuli and activated by impulses in A-delta and C fibers. What mechanisms are interposed between the zone of A-delta terminations in I and of C fibers largely in SG and the STT neurons of laminae IV to VI are largely unknown. It has been shown that convergent interaction between afferent input in large and small fibers occurs at this segmental level, that these may be excitatory or inhibitory, and that they may operate via post- or presynaptic mechanisms. Several theories concerning the CNS mechanisms in pain are based on these interactions. The latter may account for the well-known facts that (1) activity in large myelinated fibers may suppress pain arising in the peripheral region innervated by them as well as the responses of STT cells to impulses in nociceptive afferents,[54] which is the basis of several methods aimed at controlling pain, and (2) spontaneous pain may arise in a peripheral area denervated of its large-fiber innervation.

Ascending nociceptive systems of the spinal cord and their forebrain projections

There can be little doubt that the ascending systems important for the sensation of pain, itch, and temperature are located in the ventral half of the spinal cord. The major pathways travel in the anterolateral quadrant, as suggested by the marked contralateral hypalgesia that follows anterolateral cordotomy in man. Further, local electrical stimulation in the anterolateral quadrant in man, via penetrating microelectrodes, elicits sensations of pain and temperature,[105,156]

*See references 53, 81, 130, 131, 166, 171.

but surprisingly no mechanoreceptive sensations, despite the facts that (1) mechanoreceptive afferents project into the anterolateral system,[22] (2) anterolateral cordotomy produces a mild contralateral elevation of touch thresholds, and (3) a rudimentary contact sensibility survives transection of systems in the dorsal half of the cord. Sweet et al.[156] found in a series of 200 such local anterolateral quadrant stimulations that sensations of pain were evoked at 54% of the stimulated locations, feelings of heat at 37%, and feelings of cold at the remaining 9%. The sensations were referred to the contralateral side in 82%, ipsilaterally only in 12%, and to both sides in 6%. It is well known that the ipsilateral projection is usually small, but may vary greatly in some individuals and in the very rare case may be the dominant projection.

It is therefore clear that as far as the peripheral nerve fiber and ascending spinal components of the nervous system are concerned, *it is neural action in certain specific sets of peripheral nerve fibers and ascending central axons that provoke pain*. No special "spatiotemporal" pattern of impulses in fibers otherwise unspecific for pain is required or sufficient. The pattern imposed by electrical stimulation of either peripheral fibers or ascending tracts does not occur under any natural circumstances, yet it evokes pain. Variation of the pattern or frequency of electrical stimulation may change the intensity or the temporal cadence but not the quality of the sensory experience elicited. Patterns of impulses in a variety of other afferents and of other central systems may contribute to, regulate, or suppress the total pain experience, but the pain per se, if evoked, is independent of that activity with regard to its quality.

Anterolateral cordotomy results in a profuse and widely distributed antegrade degeneration, revealing ascending projections to (1) the central reticular core of the brain stem and mesencephalon; (2) an element of the posterior nuclear group (PO) and the ventrobasal (VB) complex of the thalamus, sometimes called the neospinothalamic tract or system; and (3) the nuclei of the intralaminar nulear group of the thalamus, notably n. centralis lateralis, a projection sometimes called the paleospinothalamic tract or system (Fig. 13-12).[107] An anterolateral cordotomy

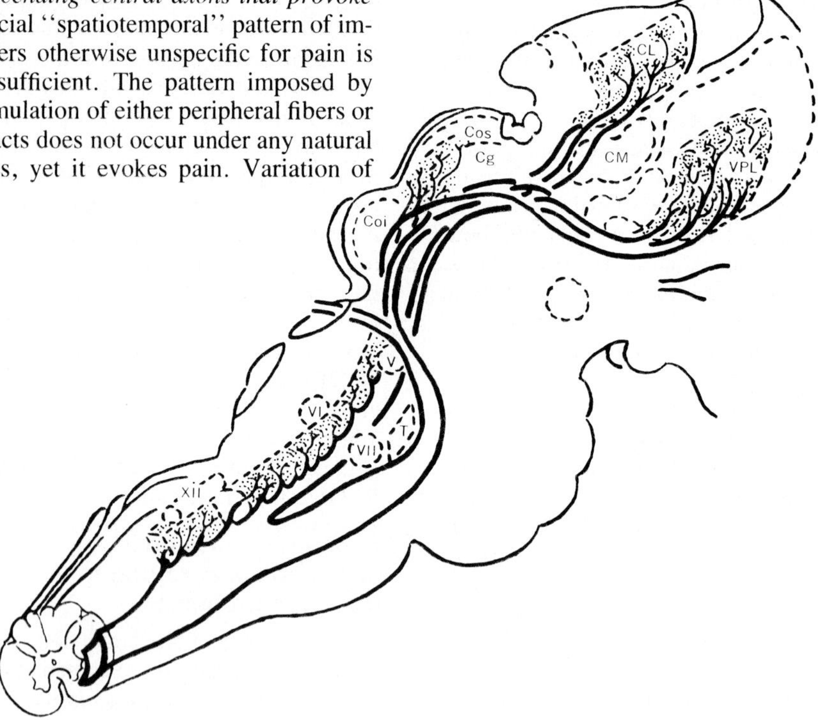

Fig. 13-12. Zones of terminal degeneration of ascending fiber systems transected by full anterolateral cordotomy in human. Fiber systems of anterolateral fasciculus are shown by heavy outlines; their projection regions are stippled. These include various medullary and pontine reticular nuclei, regions of periaqueductal gray matter *(Cg)* and tectum of the superior colliculus *(Cos)*, and ultimately nucleus ventralis posterolateralis *(VPL)*, part of ventrobasal nuclear complex. System also projects on nucleus centralis lateralis *(CL)*, one of intralaminar nuclei. Contralaterally directed lateral spinothalamic track is only one portion of this entire system; it projects on mesencephalic tegmentum and thalamic targets shown. Ventral spinothalamic tract is not shown, and it engages same targets as does lateral spinothalamic tract. (Courtesy Dr. William R. Mehler.)

transects systems arising ipsilaterally, as well as those originating on the contralateral side of the spinal cord that are known to constitute the essential pathway for pain in man. Kerr has been able to isolate this contralateral projection by cutting the anterior commissure at several adjacent spinal segments and tracing the resulting antegrade degeneration.[6-8,81,83] The contralaterally directed system contains small myelinated fibers and no C fibers.[97] It projects on (1) PO and VB of the thalamus; (2) the intralaminar nuclei, particularly n. centralis lateralis; and (3) the periaqueductal gray matter and the n. cuneiformis of the mesencephalon. This last nucleus projects upward on the intralaminar nuclei, the subthalamus, and the dorsal hypothalamus.[48] An important observation is that in the monkey, and presumably in man, no ascending elements of contralateral origin terminate in the reticular formation of the medulla or pons. Those projections, shown in Fig. 13-12, must therefore arise at the segmental level from the ipsilateral side.[7] These observations have led Kerr to the generalization that the spinoreticular projections of the anterolateral quadrant function predominantly or exclusively to elicit somatic and autonomic reactions to noxious stimuli, not in evoking painful sensations. Axonal transport studies of the contralateral projections to the mesencephalon and intralaminar nuclei suggest that that to the latter derives largely from the marginal cells of lamina I and that to PO and VPL largely from the STT neurons of laminae IV to VI. This segregation cannot be absolute, however, for there is electrophysiologic evidence that some single STT axons may divide to innervate both VPL and the mesencephalic reticular formation.[131] Kerr has also brought new anatomic evidence that a ventral spinothalamic tract exists as an entity separate from the lateral tract in the monkey and presumably in man and that its axons project to the same targets as do those of the lateral tract.[82]

The anterolateral projections that arise ipsilaterally are the cephalad extensions of the intersegmental systems of the cord commonly referred to as the propriospinal system, or Bechterew's "ground bundle" of fibers, which closely surrounds the gray columns of the cord. In animals such as carnivores, activity set up in nociceptive afferents may reach the forebrain via this multichain, polysynaptic system.

Forebrain mechanisms and pain

It has been known for a long time that lesions in some locations in the CNS, notably in thalamus or cortex, can lead to spontaneous pain, the "pain of central origin" discussed on p. 419. Spontaneous pain may also appear after selective loss of the larger fibers in peripheral nerves, with the smaller fibers intact. These observations lead to the general hypothesis that there exists within the brain a system that controls and regulates the transmission of nociceptive afferent input into ascending systems, as well as of those forebrain mechanisms involved in the overall perceptual experience of pain. The fact that spontaneous pain may occur after differential loss of the larger peripheral afferent fibers suggests that the central regulating system depends for a part of its tonic drive on continuing activity in the large, mechanoreceptive afferent fibers of peripheral nerves. The general hypothesis of a central regulation of the neural mechanisms in pain was greatly strengthened by the discovery that electrical stimulation of several loci within the CNS will suppress pain, both in humans and in experimental animals, and by the equally important discovery that there exists an endogenous opiate system within the brain. These central regulatory mechanisms will be discussed in a following section. Here I describe what is known about the regions of thalamus and cortex activated *directly* by the ascending nociceptive systems of the spinal cord, regions thought to be important for the perception of pain.

The thalamic and mesencephalic targets of the contralaterally directed lateral and medial spinothalamic tracts (LSTT and MSTT) were described on p. 406. To recapitulate, they include (1) the VB complex; (2) portions of the PO; (3) the intralaminar group, mainly n. centralis lateralis; and, (4) the periaqueductal gray matter and n. cuneiformis of the mesencephalon, which regions in turn project upward on the intralaminar nuclei, the subthalamic region, and the dorsal hypothalamic area.

The ventrobasal complex and the postcentral gyrus. The VB complex receives mechanoreceptive inflow from both the medial lemniscus and the spinothalamic tracts. Several thousand cells of the VB complex,[98,122,128] the postcentral gyrus,[111,168] and the second somatic area[169] have now been studied in electrophysiologic experiments in monkeys in a variety of conditions. Only rarely has a neuron been observed to be differentially sensitive to noxious stimulation of peripheral tissues. A few cells have been found to be sensitive to thermoreceptive input.[87,129] These findings suggest that the VB complex receives the specific mechanoreceptive and thermoreceptive STT axons; whether it also receives nociceptive input is still uncertain, for these regions of thalamus and cortex have not yet been studied electrophysiologically in waking, behaving monkeys trained to detect and react to noxious stimuli.

Electrical stimulation of the VB complex in waking humans elicits mechanoreceptive but only rarely nociceptive sensations ("paresthesias") referred to the contralateral side of the body; much more commonly this stimulation suppresses ongoing pain sensations. Lesions of the VB complex may produce a transient contralateral hypalgesia, soon followed by the onset of severe pain of central origin—the thalamic syndrome described on p. 419. Stimulation of the postcentral gyrus in waking humans elicits contralaterally referred paresthesias that are only occasionally described as painful.[16] Lesions or surgical removals of the postcentral gyrus cause an immediate hypalgesia of the contralateral side accompanied by loss of some of the discriminative aspects of somatic sensibility. In occasional patients the hypalgesia may persist for long periods of time. In the majority of cases, however, the relief of pain and the hypalgesia are soon followed by hyperpathia. This result of lesions of either the VB complex or the postcentral gyrus indicates that this system is not a major component of the central pain pathway, in the usual sense, but that lesions in it remove a potent tonic drive on which the central system regulating pain described subsequently depends; hyperpathia may be considered a "release" phenomenon.

The posterior nucleus and the retroinsular cortex. The medial portion of the posterior nucleus of the thalamus (POm, one component of the posterior nuclear complex) is a projection target of STT axons and itself projects on a local cortical target, the retroinsular field that bounds the posterior border of the second somatic area, deep within the sylvian fissure.[31] Many of its neurons are differentially sensitive to noxious stimuli,[122,127] and others show by their properties convergent mechanoreceptive and nociceptive input. The neurons of its projection target, the retroinsular cortex, possess similar properties.[32,169] On the electrophysiologic evidence this system might be regarded as one part of a central pain pathway. Electrical stimulation of the homologous area of the human thalamus evokes severe pain in waking subjects, which is referred to the contralateral side. Electrolytic lesions in this thalamic region result in an initial contralateral analgesia to pinprick and relief from intractable pain of peripheral origin. This relief may be short lived, however, and is frequently followed in days or weeks by the reappearance of the pain for which surgical therapy was initiated, possibly proceeding to a state of excruciating hyperpathia. A lesion of the parietal operculum, which contains the second somatic and the retroinsular cortex in man, may result in disturbances of pain sensation without appreciable loss of discriminative aspects of somatic sensibility.[27] Talairach et al.[157] found that electrical stimulation of the white matter just beneath this region evokes sensations that are often painful and that destruction of these fiber systems may alleviate a severe pain of pathologic origin on the contralateral side of the body.

Mesencephalon and the intralaminar system of the thalamus. Stimulation of the spinothalamic or trigeminothalamic tracts at the level of entry to the diencephalon elicits severe contralateral pain, as does stimulation of the periaqueductal gray matter and the mesencephalic tegmentum (n. cuneiformis), which are both projection targets of the STT.[113] To be effective in relieving pain, lesions placed at the mesodiencephalic junction must transect both the laterally ascending components of the STTs and the ascending projections from the mesencephalic tegmentum. The desired extension of such a lesion medially into the tegmentum is limited by unwanted effects on vital functions. Stimulation of the intralaminar nuclei of the thalamus, which are targets of both direct STT projection and the ascending projection of the mesencephalic tegmentum, elicits not severe pain but feelings of anxiety in waking humans. The effects are described as unpleasant but not overtly painful and are occasionally accompanied by changes in the conscious state. Lesions in this region are reported to relieve chronic pain in some cases; such lesions usually include n. centralis lateralis, but commonly extend to other thalamic nuclei as well.

Other areas of the cerebral cortex. Disturbances in pain sensibility may occur as one of the many alterations in function produced by lesions of the *posterior parietal areas* described in Chapter 22. Lesions of the dominant hemisphere, particularly when they include the marginal gyrus, produce what Schilder termed an asymbolia for pain.[16,143] A patient with such a lesion may retain a normal threshold to pain but no longer appreciate its destructive significance (e.g., he may no longer withdraw from threatening gestures). These patients also show the disturbances of body schema termed amorphosynthesis by Denny-Brown, associated with agnosias of various degrees and types. The observations indicate that there is a further higher order processing in the parietal lobe of activity evoked by noxious stimuli and that pain plays a role in sensory and perceptive phenomena at the highest level.

When the severe and intractable pain of advancing disease is not relieved by attempted interruption of afferent systems concerned with pain, more drastic procedures are occasionally undertaken in attempts to alleviate suffering during the terminal months or years of life. Among these, lesions of the mediodorsal nucleus of the thalamus or transection of the fibers linking it to areas 9 to 12 of the frontal lobe on which it projects ("frontal leukotomy") are often undertaken. Bilateral lesions of the mediodorsal nucleus or bilateral frontal leukotomy may diminish the anguish of constant pain, but such lesions also produce drastic changes in personality and intellectual capacities. The reactions of these patients to individual noxious stimuli may even be exaggerated, but they state that although their pain persists as before, it is no longer so disturbing and they may require little or no pain-relieving medication. With time there is usually some regression of the disorders in personality and intellectual functions caused by these lesions, but with it a recurrence of severe suffering from pain. Such interventions are considered by neurosurgeons as a last resort in the sequence of procedures aimed at the relief of intractable pain. The results do not indicate a projection of a pain system on the mediodorsal nucleus and thence on the frontal lobe.

In summary, the ascending spinothalamic tracts project on and engage a number of telencephalic systems, no single one of which can be regarded as *the* central pain tract. Undoubtedly, each plays an important part in the overall central mechanisms in pain. New information is badly needed concerning which STT components project into which system, the functional properties of the neurons in each, and the relation of their activity to the painful experience. This information could now be obtained in electrophysiologic experiments in waking, behaving monkeys trained to detect and to rate painful stimuli. Neurosurgical attempts during the last decades to alleviate severe and persistent chronic painful conditions by making lesions at thalamic and cortical loci, putatively parts of the "pain system," have been generally disappointing, both as therapeutic measures and in revealing physiologic mechanisms. Each lesion that relieves pain temporarily is often followed by hyperpathia of central origin. This suggests that at higher levels of the CNS the "sensory system" for pain is inextricably mixed with those involved in the central regulation of pain and that a lesion of the system successful in interrupting the sensory function will almost certainly interfere with the controlling function as well.[65]

CENTRAL MECHANISMS CONTROLLING PAIN

Two recent discoveries have established that the brain contains an elaborate system for the control and regulation of nociceptive afferent input. The first is that electrical stimulation of discretely distributed loci in the brain elicits analgesia that is frequently powerful and may outlast the period of stimulation by minutes or hours. The second is that the brain contains an analgesic system in which synaptic transmission at some locations is mediated by endogenous morphinelike substances, the enkephalins, naturally occurring ligands for opiate receptor molecules located in the membranes of groups of discretely distributed central neurons. These two "systems" are largely but not completely congruent at brain stem and thalamic levels. Under natural circumstances this controlling system depends in part for background drive on continuing input over the somatic afferent system. Thus lesions of somatic nerves or the central pathways into which those nerves project are surmised to reduce activity in the controlling system and thus release the "spontaneous pain of central origin" that commonly follows lesions of the somatic afferent system in man.

Analgesia produced by electrical stimulation

It has been known for a long time[76] that electrical stimulation of peripheral tissues, actually the afferent nerve fibers that innervate them, causes a reduction in pain sensitivity in those tissues, an analgesia. In recent decades the method of electrical stimulation of peripheral nerves, of afferent pathways of the spinal cord, and of the brain itself has been explored as a mode of therapy for painful states in humans[11,141] and as an experimental tool in the study of central neural pain mechanisms in animals.[12] The therapeutic attempt was made possible by human stereotaxic instruments that allow passage of stimulating electrodes to many selected brain areas with considerable accuracy and with minimal damage to overlying structures. Effective loci in the somatic afferent system for stimulation-produced analgesia include the skin itself, peripheral nerve trunks,[11,154] dorsal columns,[100,155] ventral columns,[11,95] VB thalamic complex,[106] and internal capsule.[50] Potent central loci that are not parts of the somatic afferent system include the periventricular and periaqueductal gray matter of thalamus and mesencephalon[3,71,142] and certain of the midline nuclei of the brain stem, particularly the n. raphe magnus.[3,117,140] Stimulation of loci in the limbic system may affect the overall reaction to pain, but usually without objective analgesia.

The analgesia produced by electrical stimulation has the following characteristics: (1) It may be swift and complete, even allowing surgical procedures in animals.[140] (2) Stimulation of afferent pathways is usually not effective in man unless it elicits referred paresthesias. (3) The more peripheral in the somatic afferent system the locus stimulated, the more likely that analgesia will be restricted to the tissue innervated by or represented at that locus. This is not always the case, for in some instances, stimulation of one afferent nerve trunk elicits analgesia in a region including both its own field of distribution and that of adjacent nerves, obviously via a central mechanism for the latter. (4) The more central the locus stimulated, the more likely that the period of analgesia will outlast the period of stimulation, sometimes for hours, and that the analgesia will be distributed over a large part of and in the limit of the entire body. This is often the case for stimulations of the periventricular and periaqueductal gray matter and the raphe nuclei. (5) The analgesia may be largely, but not completely, reversed or prevented by administration of naloxone, a morphine and enkephalin antagonist.[20,118] (6) The analgesia elicited by central stimulation may be partially reversed or prevented by serotonin depletion or blockade.

Mechanisms of stimulation-produced analgesia. There is evidence that several mechanisms may play roles in stimulation-produced analgesia. (1) Electrical stimulation of peripheral nerves or afferent pathways may produce a direct block of conduction in nociceptive afferent fibers or ascending STT axons (e.g., by electrotonic polarization). It has been suggested that stimulation of the dorsal columns may produce analgesia in part because the stimulating current flows to and blocks ascending axons of the ventral quadrants; direct electrical stimulation of the anterior surface of the cord appears to be equally or even more effective than is stimulation of its posterior surface.[95,100] (2) Afferent impulses in large myelinated afferents of peripheral nerves and dorsal columns inhibit the STT neurons directly.[54] (3) Electrical stimulation of afferent nerves or at central brain locations drives the central regulating system, *thus producing analgesia by a descending blockade of synaptic transmission in the nociceptive afferent pathway at the segmental level.* (4) Stimulation of some limbic or other forebrain structures may influence the overall affective reaction to pain by other and unknown mechanisms.

Further experiments have shown that electrical stimulation of n. raphe magnus and the immediate surround suppresses the responses of STT neurons to nociceptive afferent input at the segmental level.[170] Raphe-spinal fibers project to those laminae of the dorsal horn that contain the cells of origin of the spinothalamic tracts at all segmental levels, and transection of the raphe-spinal tract in the dorsolateral column greatly reduces the analgesia produced by electrical stimulation of the brain. A component of the raphe system with similar properties projects on the spinal trigeminal nucleus. The n. raphe magnus is considered a principal outflow funnel of the descending system controlling the nociceptive afferent system. The system operates with serotonin as the transmitter agent for the long raphe-spinal fibers, for both stimulation-produced analgesia and the decerebrate inhibition of segmental flexor reflexes are markedly reduced after serotonin depletion or blockade.[21,49,108]

Analgesia produced by local brain injection of opiates[18]

Local injections of small amounts of morphine into the dorsal horn of the spinal cord produces a powerful blockade of synaptic transmission from nociceptive afferents to the STT neurons of the spinal cord and to the interneurons of the substantia gelatinosa. A general analgesia is produced by local microinjection of opiates into brain stem locations, particularly into the periaqueductal and periventricular gray matter, the sites at which electrical stimulation is most effective in producing analgesia.[124,173] In general the distribution of loci at which the two agents are effective are similar and replicate the distribution maps of the enkephalin-opiate receptor system shown in Fig. 13-13 and described subsequently. These effects of morphine are reversed or blocked by naloxone, and the general analgesia produced by local brain stem injection is blocked or greatly reduced by section of the dorsolateral columns, which contain the descending raphe-spinal system described previously, or by lesions of n. raphe magnus itself. These observations reinforce the conclusion that this nucleus and the neurons immediately surrounding it constitute a major outflow funnel of the descending CNS mechanism controlling pain. The precise

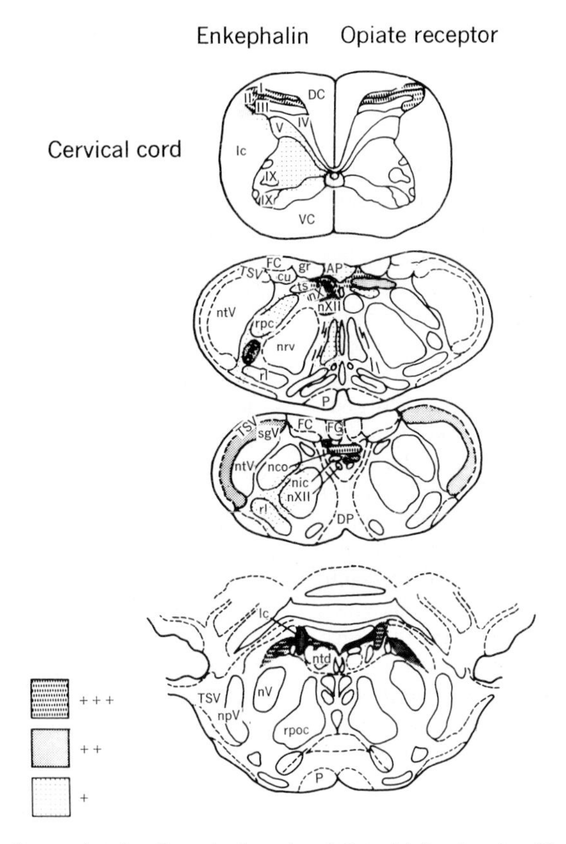

Fig. 13-13. Brain of rat at levels of cervical cord and three higher levels of brain stem to illustrate distributions of enkephalins *(left)* and opiate receptors *(right)*. Key indicates differences in concentrations. *AP*, Area postrema; *cu*, n. cuneatus; *DC*, dorsal column; *DP*, pyramidal decussation; *FC*, fasciculus cuneatus; *FG*, fasciculus gracilis; *gr*, n. gracilis; *lc*, locus ceruleus; *nco*, n. commissuralis; *nic*, intercalatus; *npV*, principal nucleus of trigeminal; *nrv*, n. reticularis pars ventralis; *ntd*, dorsal tegmental nucleus of Gudden; *ts*, n. tractus solitarius; *ntV*, spinal nucleus of trigeminal; *nX*, dorsal nerve of vagus; *nXII*, hypoglossal nucleus; *P*, corticospinal tract; *rl*, n. reticularis parvocellularis; *rpoc*, n. reticularis pontis caudalis; *sgV*, substantia gelatinosa of trigeminal; *ts*, tractus solitarius; *TSV*, spinal tract trigeminal; *VC*, ventral columns. (From Uhl, G. R., Childers, S. R., and Snyder, S. H.: Opioid peptides and the opiate receptor. In Ganong, W. F., and Martini, L., editors: Frontiers in endocrinology, New York, copyright © 1978, Raven Press, New York.)

and restricted distribution of the opiate-sensitive loci within the brain and the extremely minute amounts of opiate required to produce general analgesia when locally injected into the brain stem are consonant with two general ideas. Namely, morphine acts by binding specifically with opiate receptors that occur naturally within the brain, and neurons with those receptors are part of the essential machinery by which the CNS regulates the central processing of nociceptive afferent input.

Opiate receptors and their natural ligands, the enkephalins[5,13,14,15]

Several of the pharmacologic properties of opiates such as morphine or etorphine suggest that they act by binding to specific receptor molecules in the membranes of cells, particularly neurons. Opiates produce their analgesic and other effects when present in low concentration: they are stereospecific, and their effects can be reversed by antagonistic drugs such as naloxone that have similar steric structures but appear to block rather than "activate" the postsynaptic receptors. Opiate receptors have been located in the membranes of central neurons, their naturally occurring ligands identified, and the binding affinities of both narcotics and those natural ligands, the enkephalins, determined. Opiate receptors and enkephalins are present in the CNS of every vertebrate and no single invertebrate of the many examined,[125,145] and their concentrations and distributions change little over the series of mammals from mouse to man, so that the distribution map for the rat shown in Fig. 13-13 resembles in general that of a primate.

The discovery of opiate receptors in the brain suggested at once that opiate-like molecules occur there naturally and function as synaptic transmitters or synaptic-modulating agents. Two pentapeptides with opioid properties have been isolated from brain and identified as met-enkephalin (H-Tyr-Gly-Gly-Phe-Met-OH) and leu-enkephalin (H-Tyr-Gly-Gly-Phe-Lue-OH).[72,119] These molecules possess a conformational structure at the tyrosine end that resembles that of a portion of the morphine molecule and that of several other narcotics, which suggests common noncovalent interactions with a geometrically complementary site — one presumed to be a part of the opiate receptor molecule. Enkephalins compete with narcotics for opiate receptor binding stereospecifically and equal narcotics in potency in producing analgesia when injected into those brain stem and thalamic structures in which opiate receptors are concentrated and in which electrical stimulation also induces analgesia. Enkephalins are rapidly destroyed by proteolytic enzymes and thus are not effective when injected intravenously or into the cerebral ventricles.

Regional distribution and cellular location of the enkephalin-opiate receptor system.[14,15] It is a current general hypothesis that enkephalins are transmitters in specific neural systems of the brain and that these transmitters and their opiate receptors are likely to be found in the brain structures involved in each physiologic process modulated by the administration of opiate drugs. Enkephalins and opiate receptors are found in the synaptosomal fraction after selective centrifugation of brain homogenates. A synaptosome is a pinched-off nerve ending and its contents of vesicles and cytoplasm, together with the tightly attached subsynaptic area of the membrane of the postsynaptic cell. On further separation of the synapsomal fraction the opiate receptors are found bound to the cellular membranes, presumably the postsynaptic ones, and not to the vesicles. Immunofluorescent studies show that the enkephalins are localized to nerve fibers and their endings, it is believed in the synaptic vesicles.[114,163] This method of immunofluorescent microscopy has also been used to work out the regional distribution of the enkephalins in the brain.[146,147] The distribution of the opiate receptors has been studied by measuring the stereospecific binding of radioactively labeled ligands in homogenates of dissected brain parts and more precisely by autoradiography of brain sections exposed to radioactively labeled ligands. The general mammalian pattern of distribution is shown in Fig. 13-13. At the spinal level the transmitter system is concentrated most densely in the dorsal horn, particularly in laminae I to III (and in the "SG" of the spinal nucleus of the trigeminal nerve). The opiate receptor concentration in the dorsal horn is greatly reduced after transection and degeneration of the dorsal root fibers,[93] but enkephalin levels are not affected by either this procedure or by a more cephalad transection of the spinal cord. The conclusion is that lamina I and the substantia gelatinosa (laminae II and III) of the dorsal horn contain enkephalinergic interneurons and that a large proportion of these terminate on the presynaptic terminals of some dorsal root afferents, presumed to be the nociceptive afferents because of their spinal terminations.

At brain stem levels the system is found in the vagal nuclei and in associated nuclei and tracts, which is consonant with the powerful effect of narcotics on vagal reflexes and respiratory regulation. In the present context, it is of greatest interest that the transmitter-receptor system is concentrated in the periaqueductal and periventricular gray matter of mesencephalon and thalamus, for it is just in these regions that both electrical stimulation and local injections of narcotics are most effective in inducing a general analgesia. In the forebrain the enkephalin-opiate receptor system is found in high concentrations in the hypothalamus and in limbic structures such as the amygdaloid nuclear complex, a distribution thought to be related to the powerful effect of narcotics on emotional behavior and their capacity to evoke inner feelings of pleasure and well-being. High concentrations occur in the caudate nucleus and globus pallidus, for which no correlation is obvious except that narcotics produce a remarkable slowness in the initiation of movements. The neocortex contains low concentrations of enkephalins or opiate receptors.

Mode of action of enkephalins. Local inotophoresis of morphine agonists or enkephalins into regions of the CNS containing narcotic receptors leads to suppression of the activity of neurons in the immediate vicinity, observed at the level of the spinal cord,[47,96] brain stem,[29,66] and forebrain,[56] and in tissue culture.[101] These effects are reversed by naloxone. It is likely that they are mediated by the control of Na^+ conductance exerted by the narcotic agonist-receptor complex. At the level of the spinal cord the presynaptic inhibition of the terminals of dorsal root fibers[93,101] appears to depend on a depolarization of the terminals caused by an increase in Na^+

conductance. Elsewhere the mechanisms of inhibition may be postsynaptic, in which case the narcotic agonist-receptor complex is thought to decrease Na^+ conductance and thus antagonize the depolarizing action of excitatory transmitter released from the adjacent terminals of other cells.[9] How the suppression of neurons in and near the potent analgesic sites of the brain stem leads to activation of the descending pain-controlling system is not yet clear, but direct recording has revealed that systemic morphine produces an increase in the activity of cells in the n. raphe magnus.[115] This suggests that the action of the enkephalinergic inhibitory interneurons is to suppress tonically active inhibitory neurons that project on the n. raphe magnus, thus releasing the raphe-spinal system from a tonic inhibitory control; it is the mechanisms of disinhibition.

On a more general level, narcotic agonists and enkephalins affect the rates of synthesis of the cyclic nucleotides in neurons with narcotic receptors.[9] The enzyme adenylate cyclase is suppressed and thus the rate of synthesis of cyclic adenosine phosphate decreased; the enzyme guanylate cyclase is stimulated and the rate of synthesis of cyclic guanosine monophosphate accelerated. How this shift in the balance between these two "intracellular messengers" affects the cellular expression of some effects of narcotics is not clear. The subject has now been studied in tissue culture of a hybrid neuroblastoma-glial cell with narcotic receptors by Niremburg and by Klee.[9] During chronic exposure of the cells to narcotic agonists or to enkephalin, the synthesis of cyclic AMP gradually recovers to the normal level. If the narcotic agonists are then suddenly removed from the culture medium, the levels of cyclic AMP rise far above normal and then gradually subside. This cell system appears to provide a model of tolerance and withdrawal characteristic of narcotic addiction.

The endorphins

The endorphins[4,9] are short-chain neuropeptides derived from B-lipotropin,[73] a lipolytic peptide of 91 amino acid residues found in the pituitary gland, particularly in its pars intermedia, and in the brain as well, but in very low concentrations. The amino acid sequence 61 . . . 65 of B-lipotropin replicates that of met-enkephalin, and all fragments containing this sequence and larger than enkephalin itself (i.e., fragments 61 . . . 65 to 6 . . . 91) are called endorphins. These substances have a high affinity for opiate receptors of brain and pituitary gland and, when injected at appropriate places in the brain or into the ventricular system, elicit analgesia that on a molar basis is many times more powerful than that of morphine itself.[99] This effect is blocked by naloxone. The cerebral distribution of the endorphins is, in general, congruent with that of the opiate receptors, with disproportionately high concentrations in caudate and globus pallidus, amygdala, and hypothalamus. They are always located within neurons, particularly in nerve terminals. Some endorphins induce muscular rigidity and a catatonic state in addition to analgesia, and, when injected into the ventricular system, changes in temperature regulation may also be induced. It has been suggested that systems with endorphins as synaptic transmitters or modulators are important in homeostatic regulations and that disturbances in the synthesis or action of these substances could lead to complex alterations in behavior.[61]

PAIN ARISING FROM VISCERAL AND DEEP SOMATIC STRUCTURES

The pain that arises from the viscera is of great importance to the physician, for knowledge of its mechanisms and particularly of the phenomenon of reference is essential for the diagnosis and location of disease processes in the thoracic and abdominal cavities. These depend on visceral afferent nerve fibers whose cell bodies lie in dorsal root ganglia or in the ganglia of certain cranial nerves. Despite the fact that their axons course peripherally through autonomic nerves, they are in every way analogous to afferent fibers of somatic nerves.

Visceral pain

In considering the nociceptive innervation of the viscera, it is important to differentiate those fibers and pathways that are both afferent and sensory from those in which impulses evoke regulatory reflexes of various sorts but elicit no sensory experience. Aside from a sensory innervation of the upper trachea and esophagus, the vagus nerve, for example, contains no afferent *sensory* fibers. Stimulation of the vagus at any point below the recurrent laryngeal nerve evokes no sensation in conscious human subjects and no reflexes characteristic of pain in animals. Yet about 80% of vagal fibers are afferent and an even larger percent are unmyelinated. They are mechano- and chemoreceptors concerned with reflex regulation of gastric motility and secretion. Nociceptive afferents from the lower urinary and reproductive tracts reach the CNS via the pelvic parasympathetic nerves. Otherwise those innervating the thoracic and abdominal viscera run in sympathetic pathways, the cardiac and splanchnic nerves. These contain large numbers of unmyelinated and delta-sized myelinated afferent fibers. The splanchnic nerves contain in addition a few hundred larger myelinated fibers that end in the pacinian corpuscles of the mesentery. Their function is uncertain, but it is known that these receptors are exquisitely sensitive to vibratory stimuli. The parietal pleura and peritoneum and the outer borders of the diaphragm are innervated by branches of the thoracic and lumbar spinal nerves, and the center of the diaphragm is innervated by afferents of the phrenic nerve. Thus both visceral and somatic afferents may signal noxious processes that involve both the viscera and the body wall.

The parietal pleura and peritoneum are exquisitely sensitive, and light mechanical stimuli to them evoke pain in conscious humans. Under normal conditions, excessive distention of the gut or bladder evokes pain, as does strong contraction of their muscular walls, especially when working against obstructions to movement of their contents. Indeed, increased tension in the gut wall appears to be the only *normal* adequate stimulus evoking pain. The exposed normal intestinal mucosa is insensitive to mildly noxious stimuli. However, if the mucosa is hyperemic or inflamed, even light mechanical stimuli or dilute solutions of acid or alkali suffice to evoke the deep aching pain so characteristic of intra-abdominal disease. This is frequently attributed to a sparse innervation, but it seems more likely that the inflammatory process changes the threshold of nociceptive afferents, perhaps by the polypeptide mechanisms described on p. 401.

Referred pain

Pain that arises in the viscera of the thoracic and abdominal cavities may be felt at the site of primary stimulation, although poorly localized. Frequently such pain may be perceived as if occurring at a distant site on the skin surface innervated from the same spinal segment as the visceral locus of origin. Referred pain may appear either simultaneously with or indeed without pain appreciated as arising at the site of noxious stimulation. To a lesser extent this is also true of the deep somatic pain arising from muscles, joints, and periosteum. Hyperalgesia of the second type may occur in the region of reference (p. 401). A knowledge of the reference patterns produced by visceral disturbances is necessary for the accurate diagnosis and localization of disease processes; they are detailed in clinical treatises.

It has been observed that when the intensity of the disease process initiating a visceral pain is low or just beginning, the pain may disappear when the site of reference or the afferent nerves innervating it are locally anesthetized. When, however, the pain-provoking process increases in severity, this may not be the case, and the pain may be referred to an area of skin even when the latter is completely anesthetic. The first observation suggests that the low-frequency sustained afferent discharge in some cutaneous nociceptive afferents, not normally sufficient to maintain a suprathreshold frequency in ascending anterolateral elements, sums with convergent input from visceral nociceptive afferents to do so. The convergence and summation could of course occur at a higher level of the anterolateral system. Why the sensation of pain then evoked is interpreted as arising from the skin and not the viscera is unknown. It is frequently stated that this false reference is due to the fact that any given segment receives many times more somatic than visceral afferents and that during life experience reference is "learned" for the dominant input. There is no experimental evidence for such a statement.

The second observation, that local anesthesia of the area of reference may not affect that reference when visceral nociceptive input is intense, suggests that this latter may then alone drive anterolateral elements at suprathreshold frequencies. The problem of the mechanism of reference is the same.

In some cases the mechanisms of referred pain may be even more complex. If, as seems likely, nociceptive afferents from deep and superficial tissues are related to one another in the presynaptic inhibitory mode described in Chapter 8, then afferent impulses in one may at a certain frequency elicit antidromic impulses in the other. This antidromic activity in C fibers innervating the skin would be expected to produce there a vasodilatation and primary hyperalgesia (p. 401); indeed, this is exactly what is seen in some loci of referred pain. The possibility is open that referred pain, like hyperalgesia, may be produced by two quite different mechanisms, one peripheral and the other central.

Deep somatic pain

Pain of a particularly severe and aching quality is evoked by injury to deep structures of the body other than the viscera—from the muscles, fascia, joints, periosteum, and tendons. These deep structures are innervated by both A-delta and C fiber nociceptive afferents.[85] Pressure, cutting, and heat—any stimulus destructive of tissue cells—are adequate stimuli for evoking pain from deep structures. The most potent stimulus for evoking pain from muscle is sustained or repetitive contraction; the pain is especially severe if the working muscle is deprived of its blood supply. This is thought to be due to the release of algesic substances from the anoxic contracting cells, probably one or more of the bradykinin-like polypeptides.

Visceral and deep somatic pains commonly evoke powerful reflex contraction of skeletal muscle (e.g., the continuous splinting contraction of the abdominal musculature provoked by intra-abdominal pain or the cocontraction of all muscles acting to fixate an injured joint). Muscles steadily contracting over long periods of time become sources of painful afferent input, probably due to the release of proteolytic enzymes from the continually contracting muscle cells and the pain-provoking action of the resulting polypeptides. This muscle "soreness" may persist for hours or days after the original source of pain has disappeared.

The central nervous projections of visceral and deep somatic nociceptive afferents are thought to be identical with those previously described for the pain input from the skin.

HEADACHE MECHANISMS[17]

The most common deep somatic pain is headache, perhaps the most frequent complaint made by patients to their physicians. Regardless of their source, headaches are remarkably similar; the pain is of a deep, aching, diffuse nature, quick to arouse reflex contractions of extracranial muscles of the head, which commonly then become additional sources of nociceptive input themselves. Pain is frequently referred to a region of the head distant from the site of its initiation. Headache as a pain is similar in quality to that evoked from deep somatic tissues elsewhere in the body or that arising from extracranial deep tissues of the head itself (e.g., the teeth or the orbital tissues).

Pain from within the cranium does not arise from the brain itself but from its supporting tissues, except when pain pathways are stimulated directly or when an increase in their activity occurs following brain lesion, the hyperpathia of central origin. Neurons and glia are not themselves innervated by nociceptive afferents. The intracranial structures that *are* so innervated have been identified by direct electrical or mechanical stimulation during neurosurgical procedures performed on conscious patients after local anesthesia of extracranial tissues.[136] They are (1) the great venous sinuses and their large venous tributaries on the surface of the brain; (2) parts of the dura, particularly at the base of the skull; and (3) the cerebral arteries, particularly the middle meningeal and the great cerebral vessels close to their origin. The cranium itself, the brain parenchyma, most of the dura and the pia-arachnoid, the ependyma, and the choroid plexuses are insensitive. Thus pain may be evoked from within the head by traction on large veins or the venous sinuses, traction or dilatation of the middle meningeal artery, traction on the large arteries at the base, pulsatile distention of those arteries, or inflammatory processes about any pain-sensitive structure. Pain may also result from an irritating pressure on an afferent nerve (e.g., the root of the trigeminal in its intracranial course). Headaches of intracranial origin are explained in terms of these mechanisms. Pain originating on the upper surface of the tentorium or anywhere in the cranial vault above it is referred anterior to a frontal intra-aural plane and is mediated via nerve V. Pain originating in the posterior fossa is referred behind this plane and is mediated via nerves IX and X and the upper two or three cervical nerves.

Headache associated with changes in intracranial pressure. In the adjustments made to changes in posture the CSF follows closely the venous pressure measured at the same level of the intracraniospinal system. In the erect position, both reach negative values, whereas pressure in the lumbar sac rises. The difference in specific gravity between the brain and the fluid in which it is suspended leaves a net weight of about 40 gm, which is carried by the intracranial supporting structures, principally the large veins and venous sinuses, and the tentorium. When CSF is removed, this net suspended weight evokes pain by stretch or distortion of the suspending tissues. It is of course greatly increased when CSF is replaced by air. The pain evoked by drainage is at first referred to the vertex and front of the head but may then increase in severity and spread widely. It is relieved by restoration of the CSF and at least partially by assumption of a recumbent position.

Increased intracranial pressure per se does not produce headache. Brain tumors cause increased intracranial pressure when they block the free flow of CSF through the ventricular system of the brain; they produce pain by direct mechanical traction on or distortion of pain-sensitive structures within the head.

Headache produced by distention of cerebral arteries. Severe headache is produced by direct mechanical distention of cerebral vessels resulting from excessive pulsation. Such an event follows the intravenous injection of histamine. As shown in Fig. 13-14, histamine injection first

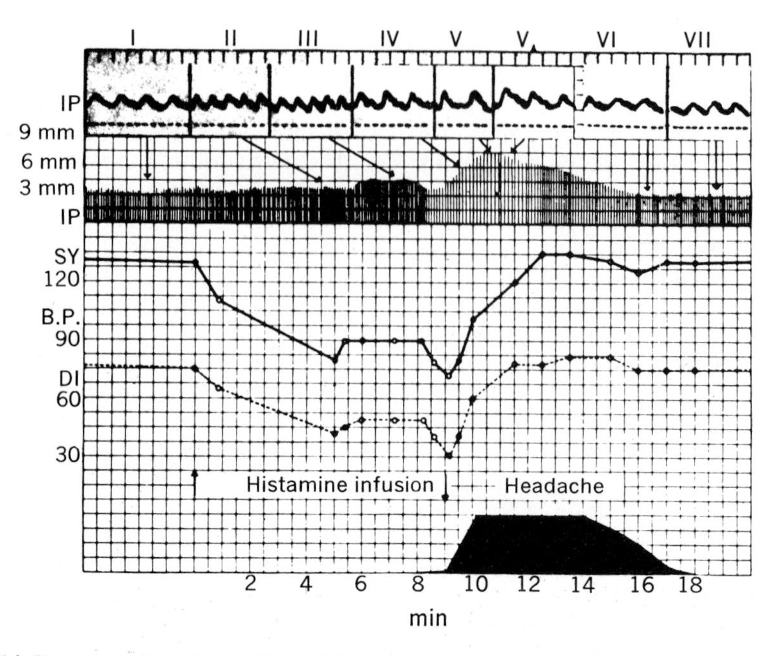

Fig. 13-14. Representation of several physiologic variables in human subject when potent vasodilator, histamine phosphate, was infused continuously for 9 min at rate of 0.1 mg/min. *SY* and *DI,* Systolic and diastolic blood pressures. *IP,* Record of intracranial pulsations, parts of which are displayed on expanded time scale in top record. Systemic blood vessels recover tone more rapidly than do intracranial ones; resulting increased intracranial pulsations produced severe headache. Both increased pulsations and headache can be prevented by *increasing* intracranial pressure. (From Wolff.[17])

produces a drop in systemic blood pressure due to a generalized vasodilatation. Recovery proceeds more rapidly in extracranial than in intracranial vessels, so that the recovering blood pressure then distends the passive cerebral vessels with each systole of the heart, producing a severe throbbing headache that subsides with the recovery of intracranial vascular tone and decreased intracranial pulsation. The extracranial and meningeal arteries are thought to play only a minor role in the headache produced by histamine. This experimentally induced pulsation headache can be abolished by increases in the intracranial pressure. Headaches associated with fever and those produced by the inhalation of vasodilator substances are thought to be of this pulsation type.

Headache arising from extracranial tissues of the head. Of all headaches that occur, those just described are relatively rare. Much more common are those that arise from extracranial tissues. Diseases of the nose, the paranasal sinuses, and the orbital contents, for example, frequently announce their presence by producing headache, and it is these that elicit powerful and pain-provoking contraction of the temporalis muscle and the muscles of the neck. Knowledge of their origin and particularly their patterns of reference is important for accurate diagnosis; they are detailed in monographs[17] and in clinical texts. Many headache syndromes of particular interest arise from the extracranial vessels, particularly from the temporal arteries, which are exquisitely sensitive to manipulation or distention. Intense headache is produced by inflammation of the walls of these vessels. More common are those severe headaches grouped under the generic term "migraine," a periodically recurring headache that is frequently unilateral in onset but that may become generalized and is sometimes associated with a more generalized vascular disorder. The pain is produced by excessive pulsation of the temporal arteries, and an attack can frequently be aborted by a vasoconstrictor agent (Fig. 13-15). If a migraine attack is allowed to proceed, edema fluid collects in the extracranial perivascular tissues. This fluid contains proteolytic enzymes and pain-provoking polypeptides of the neurokinin-bradykinin type (p. 401).[37,136] Thus the mechanism of headache production in migraine is dual in nature: an initial vasodilatation and increased pulsatile distention of the arteries, followed by the extravasation of pain-producing polypeptides.

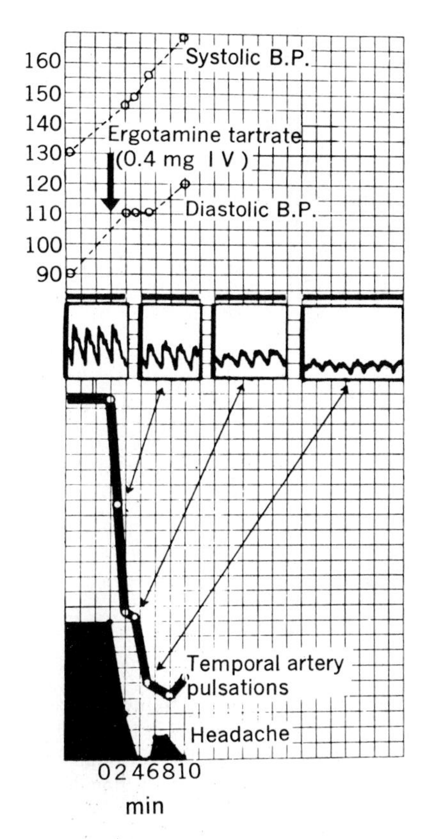

Fig. 13-15. Records of course of events when headache of migraine type is interrupted by administration of potent vasoconstrictor. Decreased amplitude of pulsations of temporal artery (middle line of records and plot below) is associated with remarkable decrease in intensity of headache, although systemic blood pressure rises. (From Wolff.[17])

Headache is frequently associated with the continued elevation of blood pressure called "essential" hypertension. The headache, like that of migraine, is due to excessive pulsation of the extracranial arteries. It occurs most commonly when smooth muscle tone is low and pressure is high, but it may occur if the former is low enough when the latter is normal.

DISORDERS OF THE PERCEPTION OF PAIN AND PAIN OF CENTRAL ORIGIN

Many disorders of the peripheral nerves and of the CNS lead to a condition of spontaneous pain or to pain that is produced by trivial, normally innocuous stimuli. These pains are among the most severe known to man, may come to dominate completely the behavior of the sufferer, and present extraordinarily difficult therapeutic problems. Of these, only a few that cast light on normal pain mechanisms will be discussed.

Painful states produced by disorders of peripheral nerves

Nerve section. Head made a major contribution to sensory physiology by describing in detail the changes in sensation that follow transection of a cutaneous nerve innervating the distal parts of an extremity.[65] When a peripheral nerve innervating the arm and hand is sectioned, as Head did in his own arm, the changes in sensation produced are not uniform throughout the area affected (Fig. 13-16). Between the zone of normal sensation proximally and the small area of complete anesthesia distally, there are three zones of altered sensibility, characterized by the fact that normally neutral stimuli produced unpleasant sensations when applied to them. In the first, bordering skin with normal sensation, an extremely unpleasant, poorly localized sensation is produced by a moving cotton wisp, by light pinprick, or by cold. An extremely light touch or warmth is not felt. Distal to this, in the second area, pinpricks, heat, cold, and pressure are all identified as painful. Last, next to the zone of total anesthesia, the skin is insensitive, but deep pressure evokes a deep throbbing pain that radiates up the wrist and arm. It is important to emphasize, as Denny-Brown[46] has done, that these changes appear *immediately* after nerve transection and do not depend on differential rates of nerve regeneration, although the picture may be altered during regeneration. The present interpretation placed on these findings is that the overlap in the fields of distribution of peripheral nerves differs for different fiber groups, being least for those myelinated afferents above approximately 6 μ in size that project into the lemniscal system and greatest for the small myelinated delta fibers and the C fibers innervating the skin and deep somatic tissues that project into the anterolateral system. Section of a nerve therefore exposes a zone of peripheral tissue, both skin and subcutaneous, which is represented centrally only in the anterolateral system. The lack of the discriminative capacities in these zones, the diffuse and radiating character of sensations evoked from them, and above all the hyperpathia that appears immediately after nerve section are all consistent with the view that under normal circumstances the transmission of activity into and up the anterolateral system is controlled by simultaneous activity in the lemniscal system. It is apparent that this control may be exerted at every level of the somatic system from dorsal horn to cerebral cortex, for the release phenomena described may follow lesions at any level.

Causalgia.[153] The classic syndrome of causalgia was described in men surviving gunshot wounds of the extremities by Mitchell in 1864; it occurs in less than 5% of such wounds that damage the nerves, most commonly when the median and sciatic nerves are partially injured but not transected; injury to arteries is not a necessary concomitant. A severe, burning hyperpathia appears in the distribution of the nerve soon after injury. The hand or foot is first warm and dry and is later cool and sweats profusely. When severe, the pain is so great that the patient cannot bear the slightest

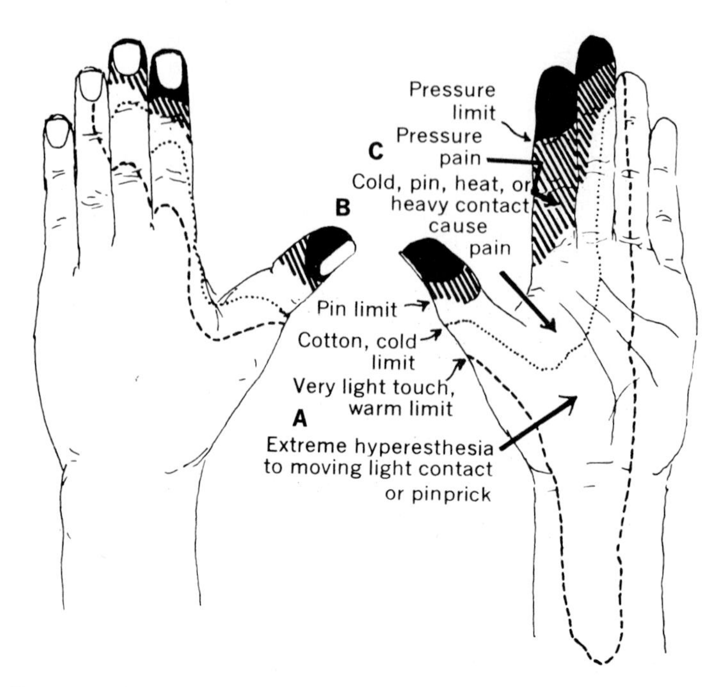

Fig. 13-16. Representation of sensory findings in right hand of 39-year-old patient following transection of medial nerve, with no regeneration. Areas *A* and *B* exhibit hyperesthesia, it is thought, because these regions are now represented centrally only in anterolateral system and not in lemniscal system: symptoms are release phenomena. Excessive pain on deep pressure in zone *C* is thought to be similar in origin. (From Denny-Brown.[46])

touch, even of clothing or puffs of air. The pain is greatly accentuated during emotional disturbances or with any increased peripheral sympathetic activity. Stimulation of sympathetic efferents to the extremity greatly accentuates the pain,[167] and sympathectomy is a specific and almost uniformly successful treatment. It is therefore not a hyperpathia of the type that follows complete section of a nerve described previously. It is frequently hypothesized that injury to the nerve damages both sympathetic efferents and nociceptive afferents in such a way that the extrinsic currents accompanying impulses in the former elicit afferent impulses in the latter at the site of injury, a theory that fits well the exacerbation of the pain during increased sympathetic activity and its relief by sympathectomy. There is, however, no direct evidence that this is so in causalgia, and it is possible that a zone of nerve anoxia is produced by vascular injury within the nerve, so that afferent impulses in C fibers normally of subthreshold frequency now elicit at the site of injury prolonged and repetitive trains of impulses that enter the CNS, a well-known property of anoxic nerve. According to this theory, sympathectomy would be effective by eliminating vasomotor efferents to the vasa nervorum, thus improving local blood supply.

It has recently been shown that the pain of causalgia, like several other types of pain arising peripherally, may be suppressed by electrical stimulation of the large myelinated afferent fibers innervating the region in which the pain arises.[109] It is surmised that in this case, as in others, the activity in large myelinated afferents impinges on and increases the controlling action of the central regulating systems described on p. 409.

Trigeminal neuralgia.[16] One of the most severe of the neuralgias occurs in the distribution of the fifth cranial nerve. It is characterized by brief, often serially repetitive attacks of severe, lightning-like pains, limited unilaterally to the distribution of the trigeminal nerve, interspersed with intervals free of pain. When severe, the attacks occur rapidly and without any obvious provoking stimulus; more commonly they are elicited by any light mechanical stimulus delivered to what may be very small trigger zones on the upper lip or the nasolabial folds. Cold, warmth, or noxious stimuli do not evoke attacks; it is activity in the smallest myelinated and C fiber mechanoreceptive afferents that is most potent in doing so. Between attacks no abnormalities of facial sensation can be demonstrated; indeed, only rarely is any neurologic deficit apparent, and hyperesthesia of the face does not occur. The condition is relieved by transection or block of peripheral branches of the trigeminal or, more permanently, by section of its root between the gasserian ganglion and the brain stem. Presumably the resulting anesthesia is effective by eliminating the background inflow of impulses on which, it is inferred, the spontaneous attacks depend and by preventing provocation by trigger stimuli. Sjöqvist[149] has shown that transection of the descending trigeminal tract at the level of the obex effectively relieves the condition and produces a dissociation of facial sensation so that pain and temperature sensibilities disappear, with only a slight reduction in tactile acuity. This is strong evidence that the n. caudalis of the trigeminal complex is essential for the transmission of pain from the face. The pathophysiology of trigeminal neuralgia is not known with certainty. Frequently it appears to be produced by an irritative compression of the gasserian ganglion or the root by vessels, dural strictures, etc.; at least temporary relief is afforded in many cases by a "decompression" of the ganglion and root. According to this theory, damage to the fibers of the root or ganglion allows a "cross-talk" between mechanoreceptive and nociceptive fibers, similar to one explanation of the causalgia-like pain of the extremities that may follow nerve injury. Indeed, demyelination of trigeminal fibers in the root and ganglion has been described. Other investigators believe that the condition is produced by a disorder of synaptic mechanisms within the n. caudalis itself. The evidence supporting these conflicting points of view has been summarized.[80,84]

Pain of central origin: the thalamic syndrome

Pain in the absence of overt noxious stimulation may be produced by acute "irritative" disease processes involving the central pain pathways or may occur when central neurons concerned with pain discharge at high frequencies during epileptic attacks. There is, however, another form produced by stable, nonirritative lesions of the CNS, called central pain, defined as spontaneous pain, with a characteristic overreaction to external stimuli. The prototype is the thalamic syndrome originally described by Dejerine and Roussy,[45] which may follow vascular lesions that destroy parts of the ventral or lateral nuclear complexes of the dorsal thalamus. It is characterized by a contralateral, severe, and persistent pain, frequently so intolerable as to be uncontrolled by narcotic agents. The threshold for pain sensation may be normal or elevated, but in either case once threshold is reached by a stimulus there is an explosive onset of pain that may spread to regions far removed from the offending stimulus. In severe cases, such explosive episodes are produced by normally neutral stimuli. When the lesion involves the VB complex, there is some loss of discriminative cutaneous sensibility, which may be absolute, and position sense and thus a variable degree of astereognosis. These general features of the central pain syndrome may appear with lesions at any level of the somatic afferent system. They may occur following the degeneration of the myelinated fibers of peripheral nerves caused by vitamin deficiencies or tabes dorsalis and are produced commonly by lesions at the thalamic level, only rarely by lesions of the DCs, and occasionally by lesions of the parietal cortex itself. The appearance of spontaneous pain of central origin caused by destructive but nonirritative lesions is interpreted as a release phenomenon similar in principle to that occurring after section of a peripheral nerve in the zone between normal and totally anesthetic tissue. Such lesions interrupt afferent inflow over myelinated fibers and the translation of that activity over dorsal cord systems to thalamus and cortex. This activity appears to play an important role in maintaining the tonic activation of the central systems regulating transmission in nociceptive afferent pathways, described on p. 409. How this interaction or tonic activation occurs is still uncertain, but it appears to take place at every level from that of segmental afferent inflow to that of the cerebral cortex.

ITCH AND TICKLE[77,78]

Itching is that sensation, recognized by all persons with normal innervation of the skin, that evokes a desire to scratch. It may vary from a barely perceptible annoyance to a sensation so intense that it dominates behavior as totally as does severe pain. Tickle is the itching component of the sensation evoked by a light moving mechanical stimulus. Both itch and tickle are followed by a strong sensory "afterimage" that may last for seconds after removal of the stimulus. Examination of the skin with a fine mechanical stimulus or by electrical stimuli delivered via an intracutaneous microelectrode reveals a punctate distribution of itch spots in the skin; they appear to be identical with the pain spots. Itch differs from pain in that it can be evoked only from the skin, the palpebral conjunctiva, and portions of the skin and mucous membranes of the nose, but not from any deep tissue of the body.

Several lines of evidence indicate that impulses evoking itch are conducted in C fibers: (1) the reaction time to itch evoked by electrical stimulation of itch spots varies from 1 to 3 sec, with body location; (2) itch disappears pari passu with burning pain with the onset of the block of cutaneous nerves produced by local anesthetic when the discriminative aspects of somesthesia are still intact; and (3) itch persists along with burning pain when nerve trunks are blocked by pressure at a time when only C fibers are still conducting. In common with second pain, itch is exaggerated, diffuse, and poorly localized in skin deprived of A fiber innervation.

Itch spots identified by stimulation have, like pain spots, been shown to be regions of increased density of bare nerve terminals of unmyelinated and the thinnest myelinated fibers. Those evoking itch appear to terminate more superficially in the skin than do those of nociceptive afferents, for itch is not evoked by stimuli directed beneath the epidermodermal border.

Local mechanical stimuli or the intradermal injection or natural release of chemicals are adequate itching stimuli (Fig. 13-17). Of these, histamine is a powerful pruritogenic agent. It evokes itch with a long latency (5 to 10 sec) compared to that of mechanical stimuli. When pruritic, it invariably produces wheal and flare in the skin. It can be diluted to a point at which it no longer evokes itching, but it then still produces wheal and flare. Spicules of cowhage (the common itching powder) produce itching by virtue of their content of a proteolytic enzyme, mucanain, and it has been shown that proteinases from a wide variety of sources produce itching on intradermal injection, even in a very low concentration.[25] Whether they act directly or by formation of polypeptides is uncertain; the latter seems likely, for polypeptides of the bradykinin type are themselves pruritogenic in very low concentrations. Proteinases or the polypeptides they produce may act independently of histamine release, for they still produce itching in (1) histamine-insensitive subjects and (2) regions of skin previously depleted of histamine. Some proteolytic enzymes, like trypsin, are not pruritogenic under these conditions and presumably do act by histamine release.

A region of local itch is surrounded by an area of itchy skin in the hyperalgesic paradigm, presumably produced by the axon reflex mechanism. Painful stimuli, on the other hand, inhibit itching sensation: scratching must usually be painful to be effective.

The central neural mechanisms responsible for itching are similar to those described for pain, and only very rarely have reports of their dissociation by central lesions been made. However, some central lesions (e.g., tumors of the spinal gray or of the pons) may early in their course produce intolerable itching, presumably by an excitatory irritation of central neurons concerned with itching sensation.

From much of the previous discussion, it might be concluded that itch is a form of pain served by the same peripheral C fibers and central neural mechanisms. Indeed, this hypothesis has frequently been put forward, with the statement that itch is produced by low-frequency discharges in pain fibers. There are several facts that make this unlikely:

1. Increasing frequency of electrical stimulation of an itch spot, without increasing intensity, may increase the intensity of the itch without evoking pain.
2. The reflex effects evoked by noxious and by itching stimuli are totally different; that of the first is flexion withdrawal, that of the second, the temporally and spatially organized patterns of the scratch reflex.
3. The two can be dissociated peripherally. Immersion of the skin in water at 41° C quickly abolishes itch but intensifies pain. Skin stripped of its epidermis is insensitive to any itch-provoking stimulus but exquisitely sensitive to noxious ones.

A more likely hypothesis is that itch is served peripherally by a set of highly sensitive mechanoreceptive C fibers.

THERMAL SENSIBILITY

Temperature sensibility is composed of two separate and distinct qualities, warmth and cool, which are served by two separate and differentially sensitive sets of first-order afferent fibers.

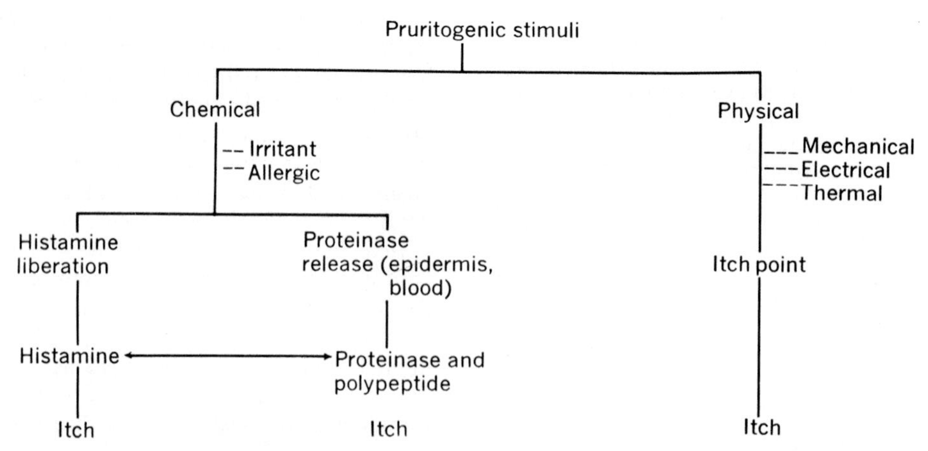

Fig. 13-17. Scheme illustrating ways in which itching is thought to be produced in skin: (1) by direct physical excitation of nerve endings or (2) by release of histamine and/or proteolytic enzymes from damaged cells. (From Arthur and Shelley.[25])

Sensitivity to temperature change is not uniform over the skin and mucous membranes, but distributed in a spotlike fashion; sensitivity is greatest where spots are most dense. In general, cold spots are more numerous than warm spots by ratios of 4:1 to 10:1, and both are more common on the face and hands than elsewhere on the body. The ratio and hence relative sensitivity vary markedly; for example, the forehead is very sensitive to cold but relatively insensitive to warmth, and some regions are completely insensitive to the latter. The sense of temperature differs from pain in two ways: over a certain range of temperature, adaptation to a state of sensory neutrality occurs, and spatial summation is a salient feature of warmth and cool. The important role of the peripheral thermoreceptive afferents in the regulation of body temperature is described in Chapter 59.

Using the method of subjective magnitude estimations, Stevens and Stevens[152] have shown that the relations between the human estimates of the intensities of thermal stimuli and their temperatures are adequately described by power functions, as shown in Fig. 13-18. This is true, however, only when the skin temperature at the level of thermal neutrality is entered as a subtractive constant, thus creating a ratio scale for the independent variable in terms of physiologic meaning as well as physical rating. A major factor in the detection of temperature changes and the scaling of temperature sensations is the area of skin stimulated, for there is a continuing spatial summation of the input as the exposed area is increased, especially for warmth. Stevens and Mark[151] have shown that for any areal extent stimulated, the degree of apparent warmth does indeed grow as a power function of the intensity level, but the smaller the area, the greater the exponent of that function. The exponent increases from about 1.0 for large areas to 1.6 for small ones, as in the experiment illustrated in Fig. 13-18. These families of power functions for warmth extrapolate to a point of convergence at about the level of heat intensity that leads to tissue damage and pain. Intensity and area trade one for the other to evoke the same sensations of warmth over a considerable range. When the entire body is exposed in a climate chamber, the region of thermal indifference is greatly narrowed, and even at rates of in-

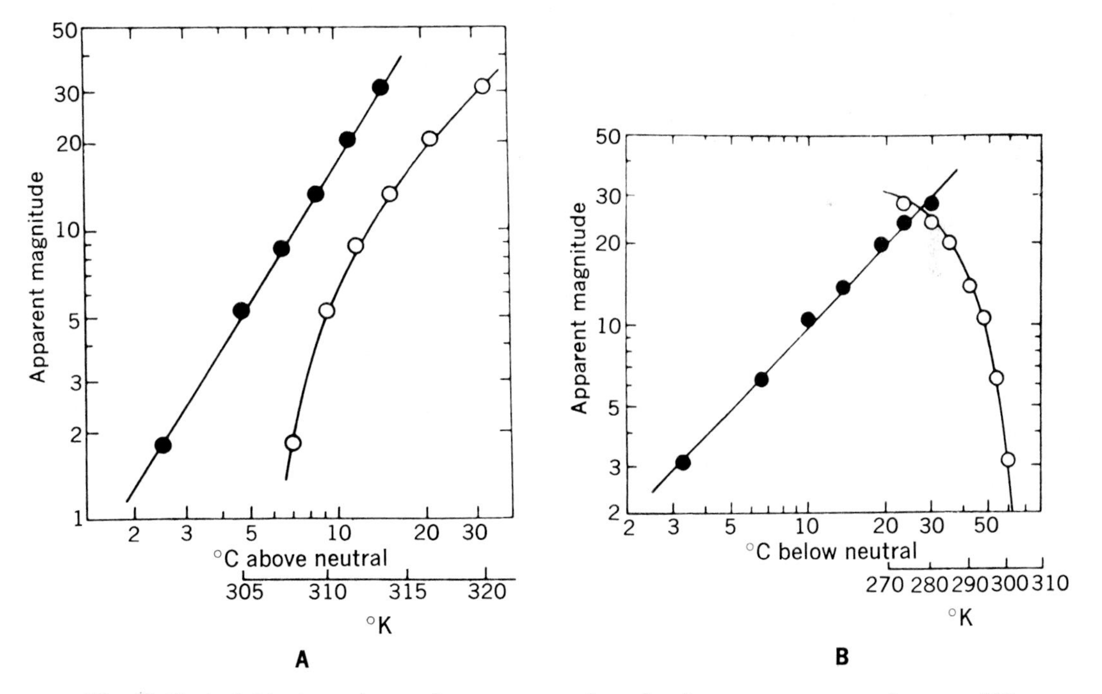

Fig. 13-18. A, Subjective estimate of apparent warmth; each point represents geometric mean of 36 estimates made by 12 observers of warmth of metallic cylinders, at various temperatures, applied to skin of forearm. Lower abscissa, for unfilled points, is log scale of absolute temperature (in degrees Kelvin). Upper abscissa, for filled points, is log scale of difference in temperature (degrees centigrade) between stimulus and assumed "physiologic zero" of 32.5° C. Line connecting filled points has slope of 1.6. **B,** Results of similar experiment using cold stimuli and an assumed "physiologic zero" of 31.0° C. Slope of line connecting filled points is about 1.0. (From Stevens and Stevens.[152])

crease in skin temperature as low as 0.001° C/sec, the warmth threshold is reached at a skin temperature slightly less than 35° C.[104]

The subjective magnitude estimation function for the sense of cool is nearly linear (Fig. 13-18). This sense shows another property of interest. Removal of a cool stimulus is followed by a persisting, gradually fading feeling of cool lasting for many seconds. This is due to the fact that the cool afferents continue to discharge as long as the temperature at their terminals is below a certain level, regardless of the direction of the thermal gradient, which is reversed when the stimulus is removed from the skin.

Relation between temperature change in skin and thermal sensations

A constant thermal sensation persists indefinitely when the temperature of the skin is below 20° or above 40° C. Adaptation occurs in the range between, so that the thermal sensation produced by an imposed temperature step gradually fades to one of thermal neutrality. The period of persisting sensation long outlasts the time required for the intracutaneous temperature to reach a constant value. It is clear from this fact as well as from the persistence of a constant sensation when skin temperature is below 20° or above 40° C that the adequate stimulus is the absolute skin temperature, not a transcutaneous temperature gradient.

Within the range of 20° to 40° C, for uniform rates of change of different slopes, adaptation produces a shift in threshold that depends on the initial skin temperature and the slope of the imposed change. Very rapid rates of temperature change in either direction evoke a thermal sensation when the skin temperature has shifted by only 1° to 2° C. Very gradual rates of change, however, produce very large shifts in skin temperature before any report of thermal sensation is made. Outside the range of 20° to 40° C the rate of change required to evoke sensation is zero: these are the zones of constant thermal sensation.

Peripheral thermoreceptive nerve fibers

Nerve blocking experiments using pressure or local anesthetic agents show that the two thermal senses disappear together, simultaneously with pricking pain. When all myelinated fibers are blocked, leaving only C fibers conducting, human subjects are sensitive only to those extremes of heat and cold nociceptive afferents. This suggests that both warm and cold thermoreceptive afferents in man are members of the thinly myelinated delta group of fibers. These terminate in the skin as bare nerve endings, scarcely discriminable by light microscopy from the similar endings of C fibers.

Knowledge of the functional properties of cold and warm fibers has been greatly advanced by Darian-Smith and his colleagues,[43,44,75] particularly because their studies were made on fibers innervating the glabrous skin of the hands of monkeys. The palmar skin and its innervation appear identical in man and monkey. A quantitative comparison was made between the electrophysiologic studies of nerve fibers innervating the monkey hand and psychophysical measures of human subjects' capacity to discriminate between thermal stimuli delivered to their hands. The results are summarized as follows:

1. Cold fibers are unresponsive to mechanical deformation of the skin. They constitute about one third of the A-delta fibers innervating the monkey hand and have a mean conduction velocity of about 14 m/sec; each terminates in a spotlike receptive field (< 1 mm diameter) in the glabrous skin. These afferents respond with a high-frequency onset transient to the onset of a cooling pulse that declines to a more or less steady rate of discharge for pulses up to a few seconds in duration. There is a remarkably small variability in their responses to slowly repeated cooling pulses of the same intensity. The stimulus-response function is linear over a range of 8° to 10° C downward from neutral skin temperature. It is only this input in cold fibers that can account for the human capacity to differentiate the intensity of cooling pulses. Although many large-fibered, slowly adapting mechanoreceptive afferents innervating the glabrous skin are excited by cooling pulses, they are more than an order of magnitude less sensitive to cold than are the cold afferents themselves. *Such a differential sensitivity is the essence of modality specificity.*

2. Warm fibers are unresponsive to mechanical deformation of the skin, but are exquisitely sensitive to warm pulses. Their conduction velocities have been measured in the range of 0.5 to 2.5 m/sec in monkey and in man.[44,62,86,94] They are C fibers, which conforms with the observation that in man the perception of warmth survives pressure block of all myelinated fibers.[103] The response of warm fibers differs from that of cold fibers, for there is no onset transient discharge at the onset of an abrupt warming pulse; the frequency of discharge gradually rises over the first second after pulse onset to a steady level set by the intensity of that pulse. The variability of responses to trains of slowly repeated pulses of equal intensity is low; the stimulus-response function is linear. It is only these warm fibers that can account for the human subject's capacity to

discriminate between warming pulses of slightly different intensities.

3. The human discriminable increment between pairs of warming or pairs of cooling pulses is fully accounted for by the differences in the neural discharge evoked by members of the pairs. This increment is identical regardless of the overall level of activity during the two pulses, which may of course be quite different, depending on the general intensity level of the two pulses. The form of the Weber function, which relates fractional increment in cool or warmth that can be discriminated to intensity level at which the discrimination is made, is set by the characteristics of populations of thermoreceptive afferents and not by subsequent central neutral events.

REFERENCES

General reviews

1. Burgess, P. R., and Perl, E. R.: Cutaneous mechanoreceptors and nociceptors. In Iggo, A., editor: Handbook of sensory physiology. Somatosensory system, New York, 1973, Springer-Verlag New York, Inc.
2. Clark, W. C.: Pain sensitivity and the report of pain: an introduction to sensory decision theory, Anesthesiology **40:**272, 1974.
3. Fields, H. L., and Basbaum, A. I.: Brainstem control of spinal pain transmission neurons, Annu. Rev. Physiol. **40:**217, 1978.
4. Goldstein, A.: Opioid peptides (endorphins) in pituitary and brain, Science **193:**1081, 1976.
5. Iverson, L. L., Nicoll, R. A., and Vale, W. W.: Neurobiology of peptides, Neurosci. Res. Program Bull. **16:** 211, 1978.
6. Kerr, F. W. L.: Segmental circuitry and ascending pathways of the nociceptive system. In Beers, R. S., and Bassett, E. G., editors: Mechanisms of pain and analgesic compounds, eleventh Miles Symposium, New York, 1979, Raven Press. (In press.)
7. Kerr, F. W. L., and Casey, K.: Pain, Neurosci. Res. Program Bull. **16:**1, 1978.
8. Kerr, F. W. L., and Wilson, P. R.: Pain, Annu. Rev. Neurosci. **1:**83, 1978.
9. Klee, W. A.: Endogenous opiate peptides. In Gainer, H., editor: Peptides in neurobiology, New York, 1977, Plenum Publishing Corp.
10. Lewis, T.: Pain, New York, 1942, Macmillan Publishing Co., Inc.
11. Long, D. M., and Hagfors, N.: Electrical stimulation in the nervous system: the current status of electrical stimulation of the nervous system for the relief of pain, Pain **1:**109, 1975.
12. Mayer, D. J., and Price, D. D.: Central nervous system mechanisms of analgesia, Pain **2:**379, 1976.
13. Snyder, S. H.: Opiate receptors and internal opiates, Sci. Am. **236:**44, 1977.
14. Synder, S. H.: Opioid peptides in the brain. In Schmitt, F. O., and Worden, F. G., editors: The neurosciences: fourth study program, Cambridge, Mass., 1978, The M.I.T. Press.
15. Uhl, G. R., Childers, S. R., and Snyder, S. H.: Opioid peptides and the opiate receptor. In Ganong, W. F., and Martini, L., editors: Frontiers in endocrinology, New York, 1978, Raven Press, vol. 5.
16. White, J. C., and Sweet, W. W.: Pain and the neurosurgeon: a forty-year experience, Springfield, Ill., 1969, Charles C Thomas, Publisher.
17. Wolff, H. G.: Headache and other pain, New York, 1963, Oxford University Press, Inc.
18. Yaksh, T. L., and Rudy, T. A.: Narcotic analgetics: CNS sites and mechanisms of action as revealed by intracerebral injection techniques, Pain **4:**299, 1978.

Original papers

19. Adair, E. R., Stevens, J. C., and Marks, L. E.: Thermally induced pain, the Dol scale, and the psychophysical power law, Am. J. Psychol. **81:**147, 1968.
20. Adams, J. E.: Naloxone reversal of analgesia produced by brain stimulation in the human, Pain **2:**161, 1976.
21. Akil, H., and Liebeskind, J. C.: Monoaminergic mechanisms of stimulation-produced analgesia, Brain Res. **94:**279, 1975.
22. Applebaum, A. E., Beall, J. E., Foreman, R. D., and Willis, W. D.: Organization and receptive fields of primate spinothalamic tract neurons, J. Neurophysiol. **38:** 572, 1975.
23. Applebaum, M. L., et al.: Unmyelinated fibers in the sacral 3 and caudal 1 ventral roots of the cat, J. Physiol. **256:**557, 1976.
24. Armstrong, D.: Pain. In Erdos, E. G., editor: Bradykinin, kallidin, and kallikrein, Handbook Exp. Pharmacol. **25:**434, 1970, Heidelberg, 1970, Springer-Verlag.
25. Arthur, R. P., and Shelley, W. B.: The peripheral mechanism of itch in man. In Wolstenholme, G. E. W., and O'Connor, M., editors: Pain and itch; nervous mechanisms, Boston, 1959, Little, Brown & Co.
26. Beck, P. W., and Handwerker, H. O.: Bradykinin and serotonin effects on various types of cutaneous nerve fibers, Pfluegers Arch. **347:**209, 1974.
27. Biemond, A.: The conduction of pain above the level of the thalamus opticus, Arch. Neurol. **75:**231, 1956.
28. Bleehen, T., and Keele, C. A.: Observations on the algogenic actions of adenosine compounds on the human blister base preparation, Pain **3:**367, 1977.
29. Bradley, P. B., Briggs, I., Gayton, R. J., and Lambert, L. A.: Effects of microiontophoretically applied methionine-enkephalin on single neurones in rat brain, Nature **261:**425-426, 1976.
30. Burke, D., MacKenzie, R. A., Skuse, N. F., and Lethlean, A. K.: Cutaneous afferent activity in median and radial nerve fascicles: a microelectrode study, J. Neurol. Neurosurg. Psychiatry **38:**855, 1975.
31. Burton, H., and Jones, E. G.: The posterior thalamic region and its cortical projection in new world and old world monkeys, J. Comp. Neurol. **168:**249, 1976.
32. Carreras, M., and Andersson, S. A.: Functional properties of neurons of the anterior ectosylvian gyrus of the cat, J. Neurophysiol. **26:**100, 1963.
33. Chal, L. A., and Kirk, E. J.: Toxins which produce pain, Pain **1:**3, 1975.
34. Chang, M. M., and Leeman, S. E.: The isolation of a sialogogic peptide from bovine hypothalamic tissue and its characterization as substance P, J. Biol. Chem. **245:**4784, 1970.
35. Chapman, L. F., Dingman, H. F., and Ginzberg, S. P.: Failure of analgesic agents to alter the absolute sensory threshold for the simple detection of pain, Brain **88:** 1011, 1965.
36. Chapman, L. F., Ramos, A. O., Goodell, H., and Wolff, H. G.: Neurohumoral features of afferent fibers in man. Their role in vasodilatation, inflammation, and pain, Arch. Neurol. **4:**617, 1961.

37. Chapman, L. F., et al.: A humoral agent implicated in vascular headache of the migraine type, Arch. Neurol. **3:**223, 1960.
38. Clark, W. C., Hall, W., and Yang, J.: Changes in pain discriminability and pain report criterion after acupunctural or transcutaneous electrical stimulation. In Bonica, J. J., and Albe-Fessard, D., editors: Advances in pain research and therapy, New York, 1976, Raven Press.
39. Clifton, G. L., Coggeshall, R. E., Vance, W. H., and Willis, W. D.: Receptive fields of unmyelinated ventral root afferent fibres in the cat, J. Physiol. **256:**573, 1976.
40. Coggeshall, R. E., et al.: Unmyelinated axons in human ventral roots, a possible explanation for the failure of dorsal rhizotomy to relieve pain, Brain **98:**157, 1975.
41. Collins, W. F., Nulsen, F. E., and Shealy, C. N.: Electrophysiological studies of peripheral and central pathways conducting pain. In Knighton, R. S., and Dumke, P. R., editors: Pain, Boston, 1966, Little, Brown & Co.
42. Cuello, A. C., Jessell, T. M., Kanazawa, I., and Iverson, L. L.: Substance P.: localization in synaptic vesicles in rat central nervous system, J. Neurochem. **29:** 747, 1977.
43. Darian-Smith, I., Johnson, K. O., and Dykes, R.: "Cold" fiber population innervating palmar and digital skin of the monkey: responses to cooling pulses, J. Neurophysiol. **36:**325, 1973.
44. Darian-Smith, I., Johnson, K. O., and LaMotte, C.: Warming and cooling the skin: peripheral neural determinants of perceived changes in skin temperature. In Kornhuber, H. H., editor: The somatic sensory system, Stuttgart, 1975, Georg Thieme Verlag.
45. Dejerine, J., and Roussy: The thalamic syndrome, Arch. Neurol. **20:**560, 1969.
46. Denny-Brown, D.: The release of deep pain by nerve injury, Brain **88:**725, 1965.
47. Duggan, A. W., Hall, J. G., and Headley, P. M.: Morphine enkephalin and the substantia gelatinosa, Nature **264:**456, 1976.
48. Edwards, S. B., and De Olmos, J. S.: Autoradiographic studies of the projections of the midbrain reticular formation: ascending projections of nucleus cuneiformis, J. Comp. Neurol. **165:**417, 1976.
49. Engberg, I., Lundberg, A., and Ryall, R. W.: Is the tonic decerebrate inhibition of reflex paths mediated by monoaminergic pathways? Acta Physiol. Scand. **72:** 123, 1968.
50. Fields, H. L., and Adams, J. E.: Pain after cortical injury relieved by electrical stimulation of the internal capsule, Brain **97:**169, 1974.
51. Fitzgerald, M.: The sensitization of cutaneous nociceptors by spread from a nearby injury and its blockade by local anaesthesia, J. Physiol. **278:**44p, 1978.
52. Foerster, O.: In Bumke, O., and Foerster, O., editors: Handbuch der Neurologie, Berlin, 1936, Springer-Verlag, vol. 6.
53. Foreman, R. D., et al.: Responses of primate spinothalamic tract neurons to electrical stimulation of hindlimb peripheral nerves, J. Neurophysiol. **38:**132, 1975.
54. Foreman, R. D., et al.: Effects of dorsal column stimulation on primate spinothalamic tract neurons, J. Neurophysiol. **39:**534, 1976.
55. Franz, M., and Mense, S.: Muscle receptors with group IV afferent fibres responding to application of bradykinin, Brain Res. **92:**369, 1975.
56. Frederickson, R. C. A., and Norris, F. H.: Enkephalin-induced depression of single neurons in brain areas with opiate receptors—antagonism by naloxone, Science **194:**440, 1976.
57. Fruhstorfer, H., Zenz, M., Nolte, H., and Hensel, H.: Dissociated loss of cold and warm sensibility during regional anaesthesia, Pfluegers Arch. **349:**73, 1974.
58. Georgopoulos, A. P.: Functional properties of primary afferent units probably related to pain mechanisms in primate glabrous skin, J. Neurophysiol. **39:**71, 1976.
59. Georgopoulos, A. P.: Stimulus-response relations in high-threshold mechanothermal fibers innervating primate glabrous skin, Brain Res. **128:**547, 1977.
60. Greene, L. C., and Hardy, J. H.: Adaptation of thermal pain in the skin, J. Appl. Physiol. **17:**693, 1962.
61. Guillemin, R., et al.: Characterization of the endorphins, novel hypothalamic and neurohypophyseal peptides with opiate-like activity. Evidence that they induce profound behavioral changes, Psychoneuroendocrinology **2:**59, 1977.
62. Hallin, R. G., and Torebjork, E.: Receptors with C-fibres responding specifically to warmth in human skin, Proc. Int. Un. Physiol. Sci. **27:**301, 1977.
63. Hardy, J. D.: Threshold of pain and reflex contraction as related to noxious stimuli, J. Appl. Physiol. **5:**525, 1953.
64. Hardy, J. D., Wolff, H. G., and Goodell, H.: Pain sensations and reactions, Baltimore, 1952, The Williams & Wilkins Co.
65. Henson, R. A.: Henry Head's work on sensation, Brain **84:**535, 1961.
66. Hill, R. G., Pepper, C. M., and Mitchell, J. F.: Depression of nociceptive and other neurones in the brain by iontophoretically applied met-enkephalin, Nature **262:**604, 1976.
67. Hiss, E., and Mense, S.: Evidence for the existence of different receptor sites for algesic agents at the endings of muscular group IV afferent units, Pfluegers Arch. **362:**141, 1976.
68. Hokfelt, T., Kellerth, J.-O., Nilsson, G., and Pernow, B.: Experimental immunohistochemical studies on the localization and distribution of substance P in cat primary sensory neurons, Brain Res. **100:**235, 1975.
69. Hokfelt, T., et al.: Immunohistochemical evidence for separate populations of somatostatic-containing and substance P–containing primary afferent neurons in the rat, Neuroscience **1:**131-136, 1976.
70. Hokfelt, T., et al.: Immunohistochemical analysis of peptide pathways possibly related to pain and analgesia: enkephalin and substance P, Proc. Natl. Acad. Sci. USA **74:**3081, 1977.
71. Hosobuchi, Y., Adams, J. E., and Linchitz, R.: Pain relief by electrical stimulation of the central gray matter in humans and its reversal by naloxone, Science **197:** 183, 1977.
72. Hughes, J.: Isolation of an endogenous compound from the brain with pharmacological properties similar to morphine, Brain Res. **88:**295, 1975.
73. Hughes, J., et al.: Identification of two related pentapeptides from the brain with potent opiate agonist activity, Nature **258:**577, 1975.
74. Hyndman, O. R., and Wolkin, J.: Anterior cordotomy. Further observations on physiological results and optimum manner of performance, Arch. Neurol. Psychiatry **50:**129, 1943.
75. Johnson, K. O., Darian-Smith, I., and LaMotte, C.: Peripheral neural determinants of temperature dis-

crimination in man: a correlative study of responses to cooling skin, J. Neurophysiol. **36**:347, 1973.
76. Kane, K., and Taub, A.: A history of local electrical analgesia, Pain **1**:125, 1975.
77. Keele, C. A.: Chemical causes of pain and itch, Annu. Rev. Med. **21**:67, 1970.
78. Keele, C. A., and Armstrong, D.: Substances producing pain and itch, London, 1964, Edward Arnold (Publishers), Ltd.
79. Kenshalo, D. R., and Willis, W. D.: Laminar distribution of spinothalamic cells in the primate lumbosacral spinal cord, Anat. Rec. **190**:443, 1978.
80. Kerr, F. W. L.: Evidence for a peripheral etiology of trigeminal neuralgia, J. Neurosurg. **26**:168, 1967.
81. Kerr, F. W. L.: Neuroanatomical substrates of nociception in the spinal cord, Pain **1**:325, 1975.
82. Kerr, F. W. L.: The ventral spinothalamic tract and other ascending systems of the ventral funiculus of the spinal cord, J. Comp. Neurol. **159**:335, 1975.
83. Kerr, F. W. L., and Lippman, H. H.: The primate spinothalamic tract as demonstrated by anterolateral cordotomy and commissural myelotomy. In Bonica, J. J., editor: Advances in neurology. International symposium on pain, New York, 1974, Raven Press, vol. 4, pp. 147-156.
84. King, R. B.: Evidence for a central etiology of tic douloureux, J. Neurosurg. **26**:175, 1967.
85. Kniffki, K.-D., Mense, S., and Schmit, R. F.: Responses of group IV afferent units from skeletal muscle to stretch, contraction and chemical stimulation, Exp. Brain Res. **31**:511, 1978.
86. Konietzny, F., and Hensel, H.: Warm fiber activity in human skin nerves, Pfluegers Arch. **359**:265, 1975.
87. Kreisman, N. R., and Zimmerman, I. D.: Representation of information about skin temperature in the discharge of single cortical neurons, Brain Res. **55**:343, 1973.
88. Kumazawa, T., and Perl, E. R.: Primate cutaneous sensory units with unmyelinated (C) afferent fibers, J. Neurophysiol. **40**:1325, 1977.
89. Kumazawa, T., and Perl, E. R.: Excitation of marginal and substantia gelatinosa neurons in the primate spinal cord: indications of their place in dorsal horn functional organization, J. Comp. Neurol. **177**:417, 1978.
90. Kumazawa, T., Perl, E. R., Burgess, P. R., and Whitehorn, D.: Ascending projections from marginal zone (lamina I) neurons of the spinal dorsal horn, J. Comp. Neurol. **162**:1, 1975.
91. Kuru, M.: Sensory paths in the spinal cord and brain stem of man, Osaka, 1949, Sogensha.
92. LaMotte, C.: Distribution of the tract of Lissauer and the dorsal root fibers in the primate cord, J. Comp. Neurol. **172**:529, 1977.
93. LaMotte, C., Pert, C. B., and Synder, S. H.: Opiate receptor binding in primate spinal cord: distribution and changes after dorsal root section, Brain Res. **112**:407, 1976.
94. LaMotte, R. J., and Campbell, J. N.: Comparison of responses of warm and nociceptive C-fiber afferents in monkey with human judgements of thermal pain, J. Neurophysiol. **41**:509, 1978.
95. Larson, S. J., et al.: A comparison between anterior and posterior spinal implant systems, Surg. Neurol. **4**:180, 1975.
96. Le Bars, D., Menetrey, D., Conseiller, C., and Besson, J. M.: Depressive effects of morphine upon lamina V cells activities in the dorsal horn of the spinal cat, Brain Res. **98**:261, 1975.
97. Lippman, H. H., and Kerr, F. W. L.: Light and electron microscopic study of crossed ascending pathways in the anterolateral funiculus in monkey, Brain Res. **40**:496, 1972.
98. Loe, P. R., Whitsel, B. L., Dreyer, D. A., and Metz, C. B.: Body representation in ventrobasal thalamus of macaque: a single-unit analysis, J. Neurophysiol. **40**: 1339, 1977.
99. Loh, H. H., Tseng, L. F., Wei, E., and Li, C. H.: β-Endorphin is a potent analgesic agent, Proc. Natl. Acad. Sci. USA **73**:2894, 1976.
100. Long, D. M., and Erickson, D. E.: Stimulation of the posterior columns of the spinal cord for relief of intractable pain, Surg. Neurol. **1**:134, 1975.
101. MacDonald, R. L., and Nelson, P. G.: Specific opiate–induced depression of transmitter release from dorsal root ganglion cells in culture, Science **199**:1449, 1978.
102. MacKenzie, R. A., Burke, D., Skuse, N. F., and Lethlean, A. K.: Fibre function and perception during cutaneous nerve block, J. Neurol. Neurosurg. Psychiatry **38**(9):865, 1975.
103. MacKenzie, R. A., Burke, D., Skuse, N. F., and Lethlean, A. K.: Fibre function and perception during cutaneous nerve block, Proc. Aust. Assoc. Neurol. **12**:65, 1975.
104. Marechaux, E. W., and Shafer, K. E.: Über Temperaturempfindungen bei Einwirkung von temperaturreizen verschiedener Steilheir auf den ganzen Körper, Arch. Ges. Physiol. **251**:765, 1949.
105. Mayer, D. J., Price, D. D., and Becker, D. P.: Neurophysiological characterization of the anterolateral spinal cord neurons contributing to pain perception in man, Pain **1**:51, 1975.
106. Mazars, G. J.: Intermittent stimulation of nucleus ventralis posterolateralis for intractable pain, Surg. Neurol. **4**:93, 1975.
107. Mehler, W. R.: The anatomy of the so-called "pain tract" in man: an analysis of the course and distribution of the ascending fibers of the fasciculus anterolateralis. In French, J. D., and Porter, R. W., editors: Basic research in paraplegia, Springfield, Ill., 1962, Charles C Thomas, Publisher.
108. Messing, R. B., and Lytle, L. D.: Serotonin-containing neurons: their possible role in pain and analgesia, Pain **4**:1, 1977.
109. Meyer, G. A., and Fields, H. L.: Causalgia treated by selective large fibre stimulation of peripheral nerve, Brain **95**:163, 1972.
110. Meyer, R. A., Walker, R. E., and Mountcastle, V. B.: A laser stimulator for the study of cutaneous thermal and pain sensations, IEEE Trans. Biomed. Eng. **23**:54, 1976.
111. Mountcastle, V. B., and Powell, T. P. S.: Neural mechanisms subserving cutaneous sensibility, with special reference to the role of afferent inhibition in sensory perception and discrimination, Johns Hopkins Med. J. **105**:201, 1959.
112. Narotzky, R. A., and Kerr, F. W. L.: Marginal neurons of the spinal cord, Brain Res. **139**:1, 1978.
113. Nashold, B. S., Jr., Wilson, W. P., and Slaughter, D. G.: Sensations evoked by stimulation in the midbrain of man, J. Neurosurg. **30**:14, 1969.
114. Neale, J. H., Barker, J. L., Uhl, G. R., and Solomon, S. H.: Enkephalin-containing neurons visualized in spinal cord tissue cultures, Science **201**:467, 1978.
115. Oleson, T. D., Twombly, D. A., and Liebeskind, J. C.: Effects of pain-attenuating brain stimulation and mor-

phine on electrical activity in the raphe nuclei of the awake rat, Pain **4:**211, 1978.

116. Olgard, L., Hokfelt, T., Nilsson, G., and Pernow, B.: Localization of substance P-like immunoreactivity in nerves in the tooth pulp, Pain **4:**153, 1977.
117. Oliveras, J. L., Redjemi, R., Guilbaud, G., and Besson, J. M.: Analgesia produced by electrical stimulation of the inferior centralis nucleus of the raphe in the cat, Pain **1:**139, 1978.
118. Oliveras, J. L., et al.: Opiate antagonist, naloxone, strongly reduces analgesia induced by stimulation of a raphe nucleus (centralis inferior), Brain Res. **120:**221, 1977.
119. Pasternak, G. W., Goodman, R., and Snyder, S. H.: An endogenous morphine-like factor in mammalian brain, Life Sci. **16:**1765, 1975.
120. Perl, E. R.: Myelinated afferent fibres innervating the primate skin and their response to noxious stimuli, J. Physiol. **197:**593, 1968.
121. Perl, E. R.: Sensitization of nociceptors and its relation to sensation. In Bonica, J. J., and Albe-Fessard, D. G., editors: Advances in pain research and therapy. Proceedings of the First World Congress on Pain, New York, 1976, Raven Press, vol. 1, pp. 17-28.
122. Perl, E. R., and Whitlock, D. G.: Somatic stimuli exciting spinothalamic projections to thalamic neurons in cat and monkey, Exp. Neurol. **3:**256, 1961.
123. Perl, E. R., Kumazawa, T., Lynn, B., and Kenins, P.: Sensitization of high threshold receptors with unmyelinated (C) afferent fibers, Prog. Brain Res. **43:**263, 1976.
124. Pert, A.: Analgesia produced by morphine microinjections in the primate brain, Neurosci. Res. Program Bull. **13:**87, 1975.
125. Pert, C. B., Aposhian, D., and Synder, S. H.: Phylogenetic distribution of opiate receptor binding, Brain Res. **75:**356, 1974.
126. Pickel, D. J., Reis, D. J., and Leeman, S. E.: Ultrastructural localization of substance P in neurons of rat spinal cord, Brain Res. **122:**535, 1977.
127. Poggio, G. F., and Mountcastle, V. B.: A study of the functional contributions of the lemniscal and spinothalamic systems to somatic sensibility. Central nervous mechanisms in pain, Johns Hopkins Med. J. **106:**266, 1960.
128. Poggio, G. F., and Mountcastle, V. B.: The functional properties of ventrobasal thalamic neurons studied in unanesthetized animals, J. Neurophysiol. **26:**775, 1963.
129. Poulos, D. A.: Central processing of peripheral temperature information. In Kornhuber, H. H., and Thieme, G., editors: The somatosensory system, Littleton, Mass., 1975, PSG Publishing Co., pp. 79-93.
130. Price, D. D., and Mayer, D. J.: Neurophysiological characterization of the anterolateral quadrant *neurons* subserving pain in *M. mulatta,* Pain **1:**59, 1975.
131. Price, D. D., Hayes, R. L., Ruda, M., and Dubner, R.: Spatial and temporal transformations of input to spinothalamic tract neurons and their relation to somatic sensations, J. Neurophysiol. **41:**933, 1978.
132. Price, D. D., Hu, J. W., Dubner, R., and Gracely, R. H.: Peripheral suppression of first pain and central summation of second pain evoked by noxious heat pulses, Pain **3:**57, 1977.
133. Ralston, H. J., III: The fine structure of neurons in the dorsal horn of the cat spinal cord, J. Comp. Neurol. **132:**275, 1968.
134. Rand, R. W.: Further observations on Lissauer tractolysis, Neurochirurgia **3:**151, 1960.
135. Randic, M., and Miletic, V.: Effect of substance P in cat dorsal horn neurones activated by noxious stimuli, Brain Res. **128:**164, 1977.
136. Ray, B. S., and Wolff, H. G.: Experimental studies on headache. Pain sensitive structures of the head and their significance in headache, Arch. Surg. **41:**813, 1940.
137. Rethelyi, M.: Preterminal and terminal arborizations in the substantia gelatinosa of the cat's spinal cord, J. Comp. Neurol. **172:**511, 1977.
138. Rethelyi, M., and Capwoski, J. J.: The terminal arborization pattern of primary afferent fibers in the substantia gelatinosa of the spinal cord in the cat, J. Physiol. **73:**269, 1977.
139. Rexed, B.: The cytoarchitectonic atlas of the spinal cord of the cat, J. Comp. Neurol. **100:**297, 1954.
140. Reynolds, D. V.: Surgery in the rat during electrical analgesia induced by focal brain stimulation, Science **164:**444, 1969.
141. Richardson, D. E.: Brain stimulation for pain control, IEEE Trans. Biomed. Eng. **23:**304, 1976.
142. Richardson, D. E., and Akil, H.: Pain reduction by electrical brain stimulation in man. Part 2. Chronic self-administration in the periventricular gray matter, J. Neurosurg. **47;**184, 1977.
143. Rubins, J. L., and Friedman, E. D.: Asymbolia for pain, Arch. Neurol. **6:**554, 1948.
144. Scheibel, M. E., and Scheibel, A. B.: Terminal axon patterns in cat spinal cord. II. The dorsal horn, Brain Res. **9:**32, 1968.
145. Simantov, R., Goodman, R., Aposhian, D., and Synder, S. H.: Phylogenetic distribution of a morphine-like peptide, "enkephalin," Brain Res. **14:**204, 1976.
146. Simantov, R., Kuhar, M. J., Pasternak, G. W., and Synder, S. H.: The regional distribution of a morphine-like factor enkephalin in monkey brain, Brain Res. **106:**189, 1976.
147. Simantov, R., Kuhar, M. J., Uhl, G. R., and Snyder, S. H.: Opioid peptide enkephalin: immunohistochemical mapping in rat central nervous system, Proc. Natl. Acad. Sci. USA **74:**2167, 1977.
148. Sindou, M., Quoex, D., and Baleydier, C.: Fiber organization at the posterior spinal cord-rootlet junction in man, J. Comp. Neurol. **153:**15, 1974.
149. Sjöqvist, O.: Studies on pain conduction in the trigeminal nerve. A contribution to the surgical treatment of facial pain, Acta Psychiatr. Scand. **27**(suppl.):1, 1938.
150. Snyder, R.: Organization of the dorsal root entry zone in cats and monkeys, J. Comp. Neurol. **174:**47, 1977.
151. Stevens, J. C., and Marks, L. E.: Spatial summation and the dynamics of warmth sensation, Percept. Psychophysics **9:**391, 1971.
152. Stevens, J. C., and Stevens, S. S.: Warmth and cold: dynamics of sensory intensity, J. Exp. Psychol. **60:**183, 1960.
153. Sunderland, S.: Pain mechanisms in causalgia, J. Neurol. Neurosurg. Psychiatry **39:**471, 1976.
154. Sweet, W. H.: Control of pain by direct electrical stimulation of peripheral nerves, Clin. Neurosurg. **23:** 103, 1976.
155. Sweet, W. H., and Wepsic, J. G.: Stimulation of the posterior columns of the spinal cord for pain control: indications, technique, and results, Clin. Neurosrug. **21:** 278, 1974.
156. Sweet, W. H., White, J. C., Selverston, B., and Nilges, R. G.: Sensory responses from anterior roots and from surface and interior of spinal cord in man, Trans. Am. Neurol. Assoc. **75:**165, 1950.
157. Talairach, J., Tournax, P., and Bancaud, J.: Chirugie

parietale de la douleur, Acta Neurochir. **8:**153, 1960.
158. Taren, J. A., and Kahn, E. A.: Thoracic anterolateral cordotomy. In Knighton, R. S., and Dumke, P. R., editors: Pain, Boston, 1966, Little, Brown & Co.
159. Torebjörk, H. E.: Afferent C units responding to mechanical, thermal and chemical stimuli in human non-glabrous skin, Acta Physiol. Scand. **92:**374, 1974.
160. Torebjörk, H. E., and Hallin, R. G.: Perceptual changes accompanying controlled preferential blocking of A and C fibre responses in intact human skin nerves, Exp. Brain Res. **16:**321, 1973.
161. Trevino, D. L., and Carstens, E.: Confirmation of the location of spinothalamic neurons in the cat and monkey by the retrograde transport of horseradish peroxidase, Brain Res. **98:**177, 1975.
162. Trevino, D. L., Coulter, J. D., and Willis, W. D.: Location of cells of origin of spinothalamic tract in lumbar enlargement of the monkey, J. Neurophysiol. **36:**750, 1973.
163. Uhl, G. F., et al.: Immunohistochemical mapping of enkephalin containing cell bodies, fibers and nerve terminals in the brainstem of the rat, Brain Res. 1979. (In press.)
164. Vallbo, A. B., and Hagbarth, K.-E.: Activity from skin mechanoreceptors recorded percutaneously in awake human subjects, Exp. Neurol. **21:**270, 1968.
165. von Euler, U. S., and Gaddum, J. H.: An unidentified depressor substance in certain tissue extracts, J. Physiol. **72:**74, 1931.
166. Wagman, E. H., and Price, D. B.: Responses of dorsal horn cells of *M. mulatta* to cutaneous and sural nerve A and C fiber stimuli, J. Neurophysiol. **32:**803, 1969.
167. Walker, A. E., and Nulsen, F.: Electrical stimulation of the upper thoracic portion of the sympathetic chain in man, Arch. Neurol. Psychiatry **59:**559, 1948.
168. Whitsel, B. L., Dreyer, D. A., and Roppolo, J. R.: Determinants of body representation in postcentral gyrus of macaques, J. Neurophysiol. **34:**1018, 1971.
169. Whitsel, B. L., Petrucelli, L. M., and Werner, G.: Symmetry and connectivity in the map of the body surface in somatosensory area II of primates, J. Neurophysiol. **32:**170, 1969.
170. Willis, W. D., Haber, L. H., and Martin, R. F.: Inhibition of spinothalamic tract cells and interneurons by brainstem stimulation in the monkey, J. Neurophysiol. **40:**968, 1977.
171. Willis, W. D., Maunz, R. A., Foreman, R. D., and Coulter, J. D.: Static and dynamic responses of spinothalamic tract neurons to mechanical stimuli, J. Neurophysiol. **38:**587, 1975.
172. Willis, W. D., Trevino, D. L., Coulter, J. D., and Maunz, R. A.: Responses of primate spinothalamic tract neurons to natural stimulation of hindlimb, J. Neurophysiol. **37:**358, 1974.
173. Yeung, J. C., Yaksh, T. L., and Rudy, T. A.: Concurrent mapping of brain sites for sensitivity to the direct application of morphine and focal electrical stimulation in the production of antinociception in the rat, Pain **4:** 23, 1977.

14 The auditory periphery

MOÏSE H. GOLDSTEIN, Jr.

The vestibular and auditory structures of the vertebrate inner ear and the lateral line organ found in fish and some amphibia derive from the same mesectodermal anlage and can be grouped together on morphologic grounds. Although differences in the various sensory structures of the inner ear and lateral line exist, they all have certain features in common: specialized sensory cells, called hair cells, are embedded in an epithelium of supporting cells; cilia protrude from the tops of the hair cells; and the hair cells are innervated by neurons (primary nerve fibers), which form the eighth cranial nerve. Our main interest in this chapter is in the events leading to excitation of cochlear hair cells and the resulting neural activity. Before considering the physiology of hair cells, we will briefly indicate the relevant gross anatomy and consider mechanical processes in audition.

Gross anatomy of the labyrinth

The orientation of the bony labyrinth in the compact petrous part of the temporal bone and its relationship to the middle ear bones, the tympanic membrane, the eustachian tube (tuba pharyngotympanica), and other structures are shown in Fig. 14-1. The membranous labyrinth consists of thin-walled sacs and ducts filled with a clear fluid, endolymph. Fig. 14-2 illustrates the innervation of the inner ear structures by ramifications of the two branches of the eighth nerve. The vestibular organs of the inner ear are surrounded by endolymph, and for these organs the perilymph, a clear fluid similar chemically to cerebrospinal fluid (CSF), serves as a cushion between the membranous labyrinth and the walls of the bony labyrinth. On the other hand, the cochlea is situated so that it is sensitive to the acoustic energy imparted to the perilymph by the stapes.

MECHANICAL PROCESSES IN AUDITION

If the spatial and temporal pattern of evoked neural events in the auditory nerve could be identified for a specified sound wave, a great deal could be understood concerning the neural coding of acoustic signals by the auditory periphery. The modification of the acoustic stimulus by the outer and middle ears and the spatiotemporal pattern of motion of the cochlear partition are the principal topics of this section. Neural coding, with emphasis on the current knowledge of coding in the fibers of the auditory nerves, will be discussed later.

Physics of sound[16]

The movements of the cone of a loudspeaker produce compression and rarefaction of the air particles in front of it. These variations of pressure and displacements of particles are propagated by reason of the elastic nature of the medium. The situation for a sinusoidal plane wave is illustrated in Fig. 14-3. The distribution at an instant in time of air particles in a sound wave traveling to the right is shown at the top of Fig. 14-3. The two variables, pressure and particle velocity, are, respectively, the differential pressure due to the acoustic disturbance and the average particle velocity that excludes the random brownian motion of gas molecules. Note the region indicated by the vertical line in the top portion of Fig. 14-3. Particles just to the left and just to the right are moving toward this region, as indicated by the arrows. A compression of air particles and an increase of pressure result, so that a short time later the particles are distributed as shown in the bottom portion of Fig. 14-3, and the pressure and particle velocity are as indicated below. By continuing this sort of reasoning, it can be seen that the sound wave will travel, although the individual air particles simply move back and forth.*

Near a sound generator the propagating waves may have a curved front, but as they move away from the source, they approximate plane waves. Mathematical analysis of plane waves indicates that they will travel undistorted at a constant velocity ($c = \sqrt{\beta/\rho}$). β is the bulk modulus of

*For a more detailed treatment, see Zwislocki.[121]

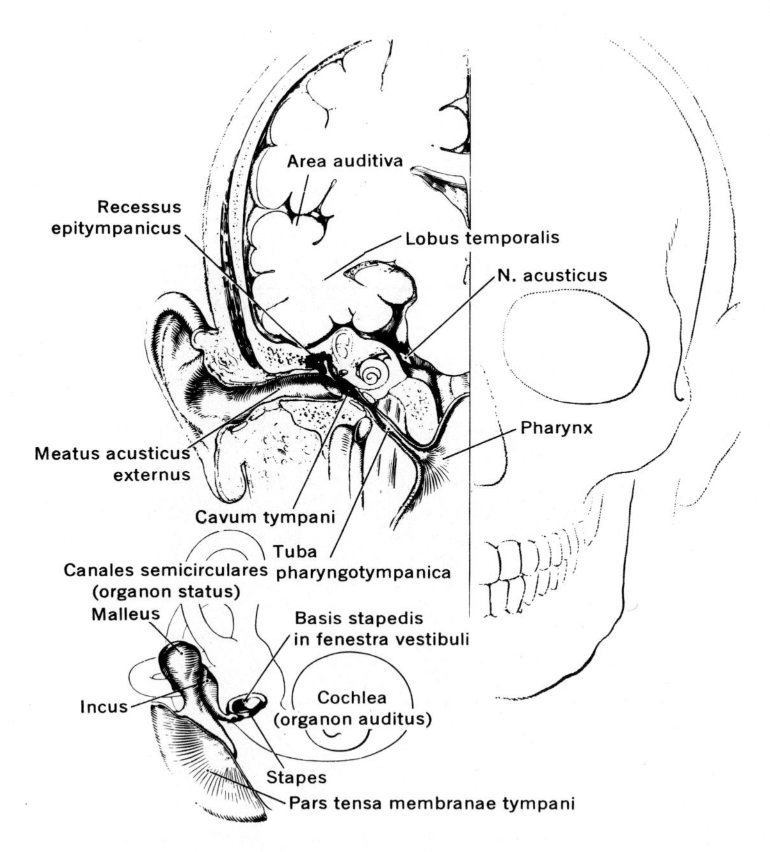

Fig. 14-1. Orientation of external, middle, and inner ear structures. (Courtesy B. Melloni.)

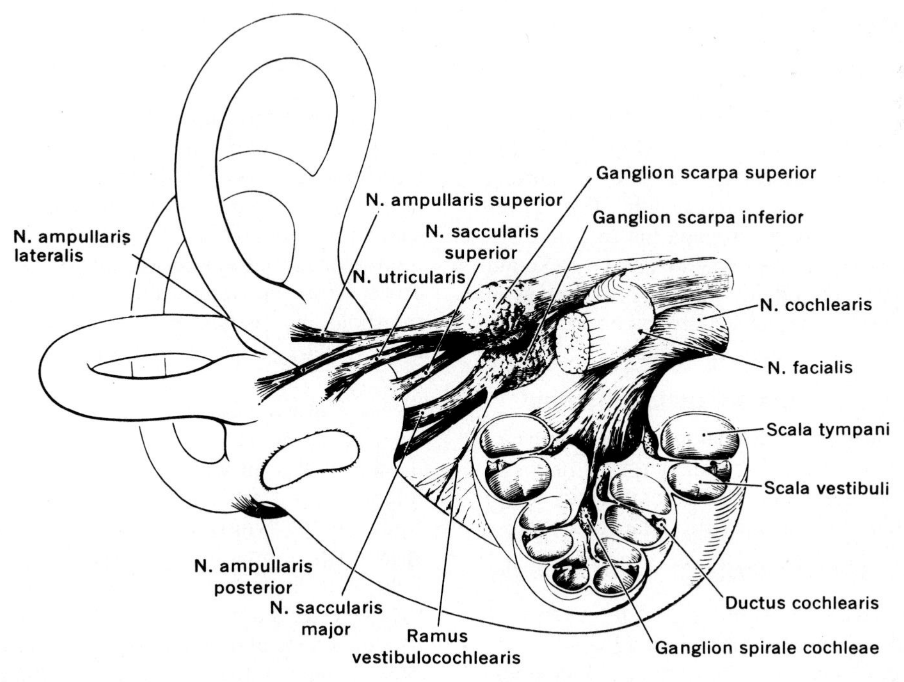

Fig. 14-2. Innervation of inner ear end-organs, also showing cross section of cochlea. (Courtesy B. Melloni.)

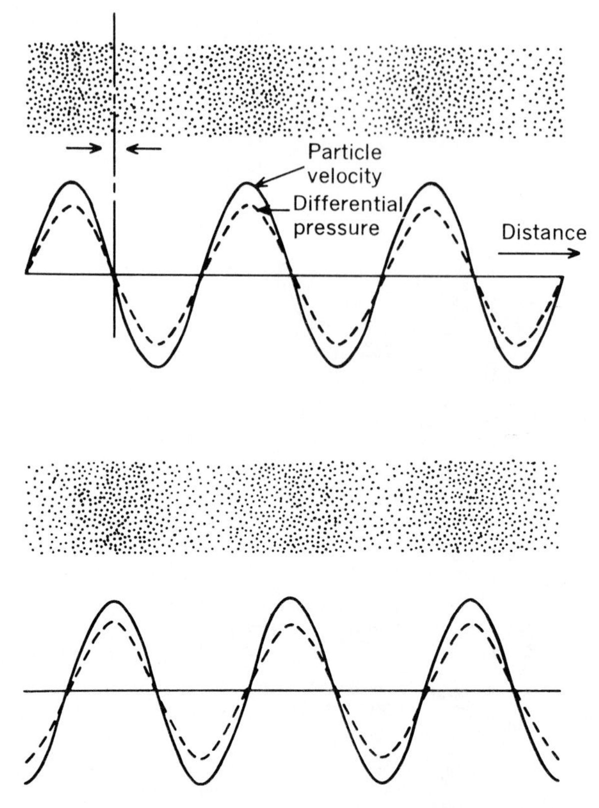

Fig. 14-3. Fluctuation of particle velocity and differential pressure as a function of distance for traveling wave at two different instants. (Modified from Pierce and David.[78])

elasticity, and ρ is the density of the medium.[17] For sinusoidal traveling waves (Fig. 14-3) the distance between peaks of pressure or particle velocity is called the wavelength. The wavelength (λ), velocity of propagation (c), and frequency in Hertz (f) are related by $c = \lambda f$. Sound waves will travel in fluids and solids as well as in gases; however, unlike electromagnetic waves, they will not propagate in a vacuum.

Specification of the acoustic stimulus

The auditory system operates over a remarkable range. In discussing the energy or intensity levels of acoustic stimuli the common measure is a relative one, the decibel (dB). The decibel measure is equal to $10 \log_{10} \frac{E}{E_R}$, or 10 times the logarithm (base 10) of the ratio of the energy of the signal to some reference energy. If, as is usual, we measure pressure rather than energy, the corresponding measure is $20 \log_{10} \frac{P}{P_R}$ (energy varies as the square of pressure). The reference pressure most often used in audition is 0.0002 dynes/cm^2, which is close to the human threshold at 1,000 Hz. Intensity measures relative to this reference are given in decibels as the sound pressure level (SPL). For stimuli in the region of maximum auditory sensitivity (2,000 to 4,000 Hz) the range of intensities between threshold and discomfort is about 120 dB, or an energy range of 1 million million and a pressure range of 1 million.

There are a number of places at which the acoustic stimulus can be specified. In experimental studies for which accurate specification is important, it is preferable to specify the pressure signal at the tympanic membrane. Other measures such as pressure in a free (i.e., echo-free) field or at the entrance to the ear canal require that the effects of sound diffraction caused by the listener's head and pinna and the effect of the ear canal be taken into account. These effects are important above 1 kHz. The main effect of the ear canal is to introduce a broadly tuned resonance at around 3,000 Hz.[112]

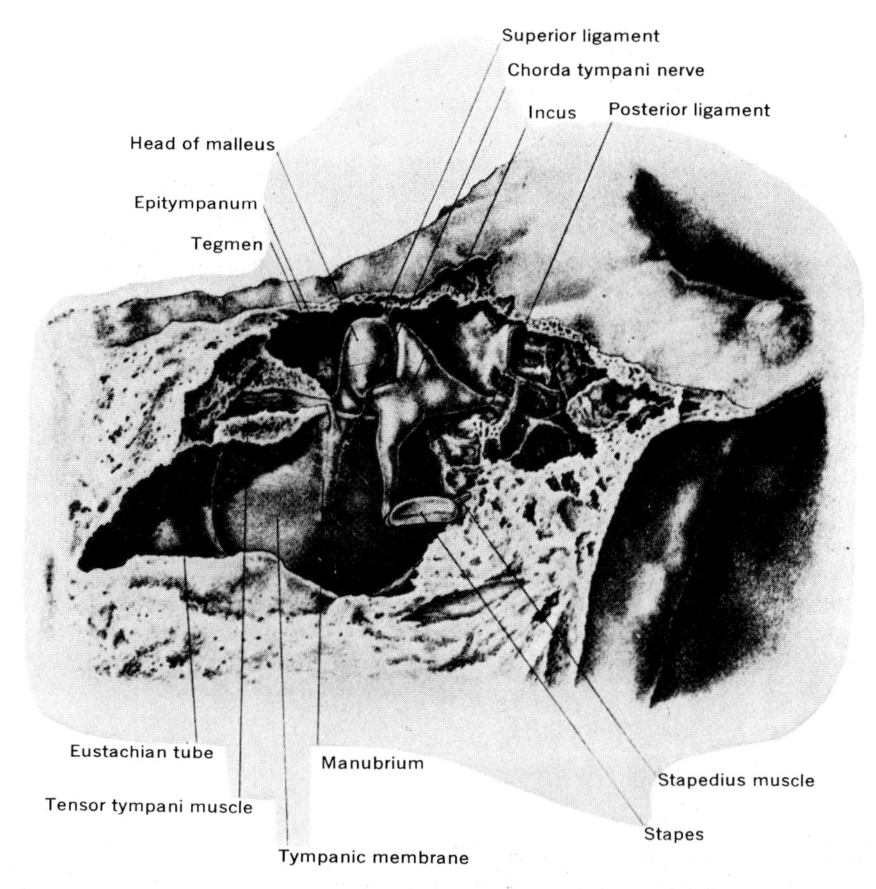

Fig. 14-4. Middle ear on right side, view from within. For view from front, see Fig. 14-1. (From Deaver: Surgical anatomy of the human body, Philadelphia, 1926, The Blakiston Co.)

Transmission of sound energy through the middle ear*

The physical properties of a medium that are relevant to the propagation of sound can be characterized by its impedance. Acoustic impedance is defined for sinusoidal variations of pressure and particle velocity; its magnitude equals the ratio of the amplitude of the pressure variation to the amplitude of the variation of volume velocity. Volume velocity is defined as the product of particle velocity and the area through which the sound is propagated. In general, impedance has both magnitude and phase. A sound wave is not completely transmitted through a boundary between media of different impedances. Instead, part of the incident wave is reflected. The amount of reflection depends on the mismatch of impedances. When one is swimming with ears submerged, it is hard to hear airborne sounds. The impedance mismatch between air and water is such that only 0.1% of the incident energy is transmitted through the boundary—the remainder is reflected.

There is an impedance mismatch problem in transmitting the energy of airborne sound to the cochlea, which has a much higher impedance. When both impedances are resistive, it can be shown that the ratio of transmitted energy (E_t) to incident energy (E_i) is as follows:

$$\frac{E_t}{E_i} = \frac{4r}{(r + 1)^2}$$

in which r is the ratio of impedances. The ratio for the cat was experimentally determined to be 10^4.[109] From this equation, it is clear that the impedance mismatch would cause a loss of 34 dB.

The middle ear structures are illustrated in Figs. 14-1 and 14-4. One of the functions of the tympanic membrane and the delicate middle ear bones that provide the mechanical linking between the tympanic membrane and the stapes is to effect an impedance match. There are three factors that cause pressure at the tympanic membrane to be lower than at the footplate of the stapes and volume velocity to be higher.[108] First,

*See references 12, 52, 58, 71, 110, and 121.

the major factor is the ratio of the area of the tympanic membrane to that of the stapes footplate. The second factor is the lever system that consists of the malleus and incus. The third factor is the leverage that results from the curvature of the tympanic membrane and the manner in which it vibrates.[58] The curved membrane mechanism was first formulated by von Helmholtz[10] more than a century ago. Until recently, von Helmholtz' scheme was rejected in favor of the idea that at low frequencies the tympanic membrane moves as a stiff, hinged plate.[9] However, using the technique of time-averaged laser holography, Khanna and Tonndorf[58] have demonstrated the correctness of von Helmholtz' formulation. The pattern of vibration they obtained is shown in Fig. 14-5. Their calculations indicate an almost complete impedance match in cats, so that the 34 dB that would be lost due to mismatch of impedances is recovered.

Such an analysis of middle ear action is valid only for frequencies below about 3,000 Hz, at which the motion of the tympanic membrane is frequency independent[108] and the ossicles move as a rigid body.[52] The analysis also assumes that the stapes footplate motion is pistonlike.[52] Even in this frequency range the analysis gives an overly simplified picture, since the mechanical and acoustic properties of the middle ear structures and cavities and of the cochlear windows affect the transmission of acoustic energy to the cochlea.[120]

At frequencies above approximately 3,000 Hz the vibratory pattern of the eardrum breaks up into quasi-independent subpatterns.[58] Also at higher frequencies the elasticities in the joints of the middle ear chain must be taken into account.[52]

Otosclerosis is a disease in which the footplate of the stapes is immobilized in the oval window by an abnormal bone growth. Modern practice is to replace the stapes. An older method, the fenestration operation, consisted of creating a new window in the ampullar end of the horizontal semicircular canal. Then the transmission of sound to the cochlea would be via the round window instead of the oval window and would be without the aforementioned impedance transfor-

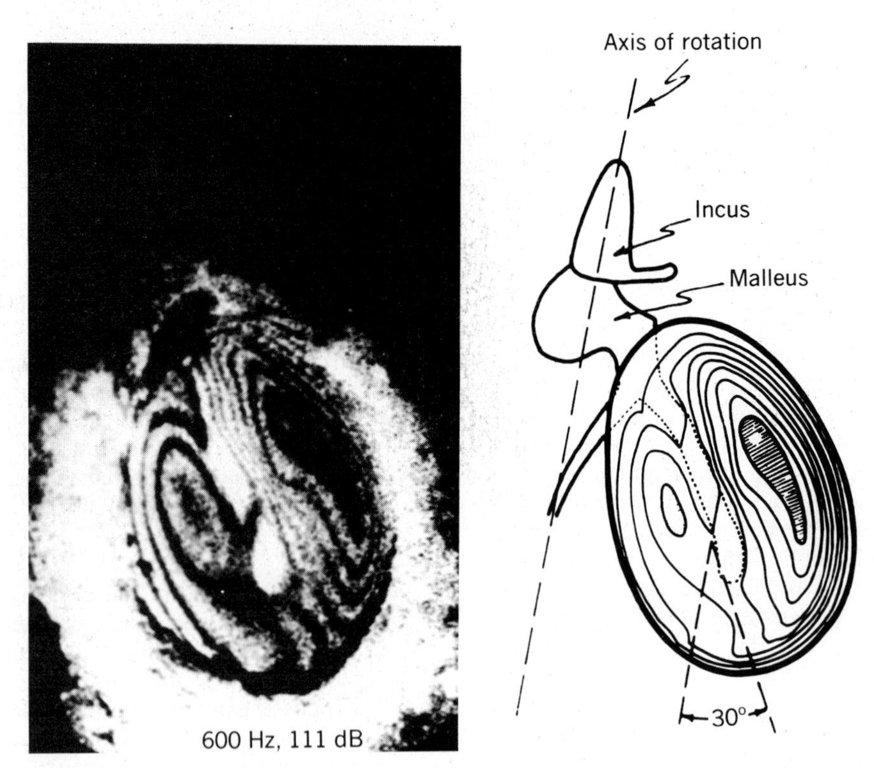

Fig. 14-5. A, Photograph of time-averaged hologram of left tympanic membrane of cat (frequency, 600 Hz; pressure, 111 dB SPL). **B,** Schematic drawing of hologram. Dark and bright fringes in photograph indicate isoamplitude deflection. Vibration of posterior (right) peak is 14.6×10^{-5} cm; vibration of anterior peak is 7.52×10^{-5} cm. Method detects motions as small as 1.2×10^{-5} cm. Note that dark fringes visible along manubrium run parallel to ossicular axis, indicating that manubrium rotates about that axis. (From Tonndorf, J., and Khanna, S.: Ann. Otol. Rhinol. Laryngol. **79:**743, 1970.)

mation. The dynamic range of audition, normally more than 120 dB, is such that an acceptable level of hearing was achieved, although with reduced sensitivity.

The middle ear apparatus has two muscles: the tensor tympani muscle, which is inserted on the manubrium of the malleus, and the stapedius muscle, which is inserted on the neck of the stapes. The former is innervated by a branch of the trigeminal cranial nerve, the latter by the facial nerve. The situation of the middle ear muscles is shown in Fig. 14-4. Excitation of the tensor tympani muscle pulls the manubrium of the malleus inward, resulting in a movement of the drum. Excitation of the stapedius muscle moves the stapes, but does not result in a significant movement of the drum. When the mobility and transmission properties of the middle ear mechanism are considered, the two muscles are seen to act as synergists, with the effect of the separate contraction of each of the muscles of the same order of magnitude. The action of the muscles causes an upward shift in the principal resonance of the middle ear mechanism, resulting in a loss in transmission for the lower frequencies.[71,115]

The middle ear muscles are reflexly activated by intense sound. In man the threshold of activation is more than 70 dB above the threshold of hearing,[21,70] and contraction increases with increasing stimulus intensity over a range of about an additional 30 dB. The reflex in one ear can be excited by stimulation of that ear, of the opposing ear, or of both. Sensitivity of the reflex is greatest for bilateral, least for contralateral, and intermediate for ipsilateral stimulation.[69] The latency of the reflex is 140 to 160 msec, so its effect on brief sounds such as pistol shots is minimal. The acoustic middle ear reflex is frequently considered as a mechanism to protect the cochlea from loud damaging sounds (e.g., jet aircraft, rock music). These sorts of sounds result from the advance of man's technology, so the evolutionary pressure for such a mechanism is obscure. The muscles are also observed to contract in the absence of loud evoking sounds. For example, they are activated during REM sleep[15] and just before vocalizations or chewing. The latter observations have led to suggestions that the middle ear muscle mechanism evolved to permit the cochlea to be maximally sensitive to environmental sounds and minimally sensitive to self-produced sounds.[9,19,91]

The eustachian tube, which connects the tympanic cavity with the pharynx, is normally closed but opens during swallowing, yawning, chewing, and sneezing. It functions to equalize the static pressure on the two sides of the eardrum. An imbalance of static pressure reduces the transmission of the middle ear structures,[51] an effect easily observed during the landing of an aircraft as the cabin pressure increases above the level maintained during flight.

COCHLEAR MECHANICS

Fig. 14-6 shows the guinea pig cochlea in cross section. Most mammalian cochleae are quite similar morphologically. The human

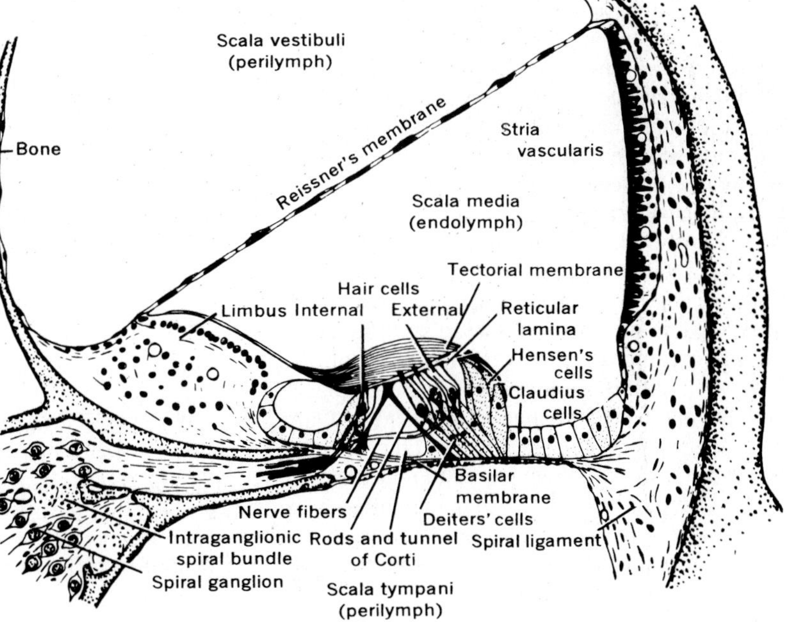

Fig. 14-6. Cross section of cochlear partition of guinea pig. (From Davis.[2])

cochlea is a spiral canal of about two and a half turns, 35 mm in length. The cochlear duct runs the length of the cochlea. It separates the scala vestibuli and the scala tympani, which communicate at the apical end of the cochlea by a small opening called the helicotrema. The oval window is the seat of the footplate of the stapes and thus the point of excitation of the normal cochlea. The round window is a flexible membrane at the basal end of the scala tympani. The relationship of the structures to the middle ear and exernal meatus is shown in Fig. 14-1.

The cochlear partition contains the hair cells, the receptor cells of audition. Mechanical excitation of these cells leads to excitation of primary fibers of the cochlear nerve. The detailed structure of the end-organ will be considered later; at present the cochlear partition can be considered as a structure with mechanical properties that change continuously from the basal to the apical end of the cochlea, providing the basis for a spectral analysis of the acoustic stimulus. The basilar membrane plays the major role in producing these changes in mechanical characteristics of the cochlear partition.

By direct observation of the motion and mechanical properties of cochlear structures in the temporal bones of fresh cadavers and in animal preparations, von Békésy indicated how cochlear partition vibration provides a basis for frequency analysis of the acoustic stimulus. The reader is urged to consult the collection of his papers[9] for a detailed description of the experiments.

In one series of experiments, von Békésy measured the elastic properties of the basilar membrane. Under a microscope, glass threads were pressed at right angles on the surface of the basilar membrane and the resulting pattern of deformation was observed. The circular patterns observed indicated the same elastic properties in the longitudinal and transverse directions. Furthermore, when fine cuts were made in the basilar membrane, the cut surface did not draw apart. Thus the basilar membrane is not under tension; it can best be likened to a gelatinous sheet covered by a thin, homogeneous layer of fibers.[9] Since the hairs would bend at a certain maximum pressure, this method could be used to measure the static compliance of the structures of the cochlear partition. Compliance is the displacement per unit force, the inverse of stiffness. Only the basilar membrane exhibited elastic properties that changed along the length of the cochlea.

The compliance of the basilar membrane was also measured by attaching a fluid-filled tube to the pierced round window and blocking the helicotrema with a mixture of agar and gelatin, producing a static differential pressure between the scala tympani and the scala vestibuli.[9] Under these conditions the displacement of the cochlear partition is proportional to its compliance. The results indicated an increase in compliance by a factor of more than 100 from near the oval window to the helicotrema. This physical characteristic to a large extent determines the pattern of vibration of the cochlear partition.

A most difficult task and one crucial to an understanding of cochlear dynamics is the observation of the actual pattern of motion of the cochlear partition. von Békésy observed cochlear vibrations under a binocular microscope. Stroboscopic illumination was used to determine the amplitude and phase of sinusoidal motion. Small particles of silver were scattered on the cochlear partition to make it more clearly visible and to allow accurate measurement.

von Békésy reconstructed the pattern of vibration for sinusoidal stimulation of a given frequency. The results indicated a traveling wave moving from base to apex, shown in Fig. 14-7 for a stimulus of 200 Hz. The actual form of displacement is shown at a number of sequential instants of time within a cycle, the dark line occurring last. The vertical scale is greatly magnified, since even for the high pressures von Békésy used (about 140 dB SPL), the maximum amplitude of vibration of the cochlear partition was only about 3×10^{-3} mm. Note that the velocity of the traveling wave of cochlear displacement decreases from base to apex. The baseline crossings in Fig. 14-7 illustrate this well, since the time increments between successive waveforms were equal. The traveling wave of cochlear partition displacement took several milliseconds to go from base to apex, whereas

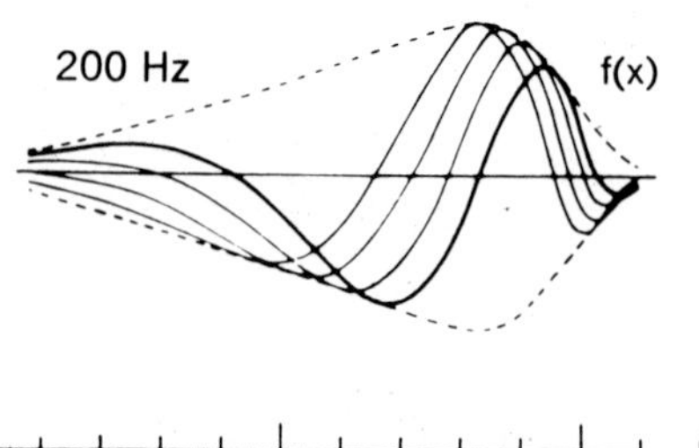

Fig. 14-7. Traveling wave along basilar membrane produced by 200 Hz stimulus. Displacements are shown at a number of sequential instants within a cycle, the dark curve occurring last. (From von Békésy.[9])

the acoustic pressure wave traveled the length of the cochlea in about 25 μsec.

For sinusoidal stimulation the movement at any given position was sinusoidal at the stimulus frequency. The envelope of the pattern of vibration (indicated by the dotted lines in Fig. 14-7) showed a rather broad maximum at about 28 mm. Fig. 14-8 shows the envelope of vibration for a sequence of increasing frequencies. (The dotted lines are extrapolations later verified by direct observation.) The point of maximum vibration moves toward the base with increasing frequency, and for any given frequency the pattern of vibration drops off more sharply toward the apex than toward the base. Note that relative

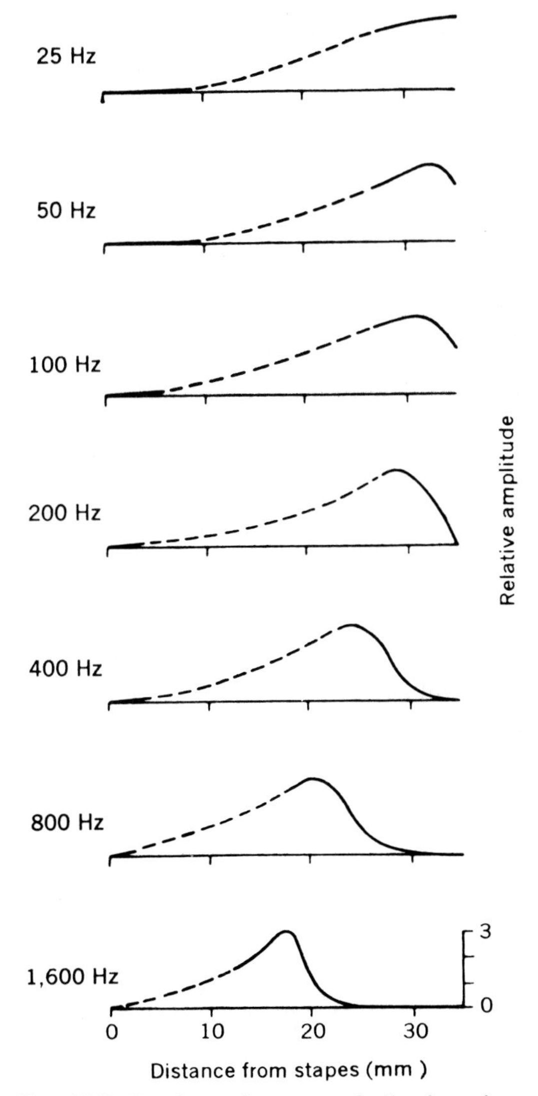

Fig. 14-8. Envelope of patterns of vibration of cochlear partition of cadaver specimen for various frequencies. (From von Békésy.[9])

rather than absolute amplitude is plotted (i.e., the curves are drawn to have equal maxima).

Using light microscope techniques, von Békésy obtained a remarkably complete picture of basilar membrane motion in response to sinusoidal stimuli. Because of the limitation of his method, he had to use stimulation intensities over 100 dB SPL. Also, most of his measurements were on cadaver temporal bones. We know now that significant changes occur shortly after death.[82] It seems that experimental findings by investigators using more sensitive techniques and working with living preparations will result in modification and sharpening up of the picture presented by von Békésy.

One method that has been used successfully to measure basilar membrane motion is the Mössbauer technique.[57,81] The Mössbauer effect is a Doppler phenomenon at the nuclear level, permitting the measurement of small velocities. With this technique, measurements of basilar membrane motion have been made using SPLs of 70 to 100 dB, 40 to 60 dB lower than possible with light microscope techniques.

Fig. 14-9 shows the placement of a Mössbauer source on the basilar membrane. After the source was positioned the scali tympani was refilled with physiologic saline solution. Control experiments showed that neither the opening in the scali tympani nor the loading of the basilar membrane with the source, which weighed about 0.1 μg, affected basilar membrane movement.[81]

A second Mössbauer source was placed on the umbo of the tympanic membrane to measure displacement of the malleus. The ratio of basilar membrane displacement to malleus displacement as a function of stimulus frequency gives a mechanical tuning curve. Such tuning curves for three stimulus SPLs are shown in Fig. 14-10. The tuning curve has a definite peak. On the low-frequency side of the peak there is a region of 24 dB/octave slope followed by a region of 6 dB/octave slope. On the high-frequency side the slope is about 100 dB/octave. Only a small region at the base of the cochlea could be studied due to anatomic constraints. It was observed that moving the source to a more basal position resulted in a tuning curve with higher peak frequency.[81,82]

Rhode's results were in general agreement with von Békésy's except on two important features of basilar membrane motion. First, the peaks of the tuning curves obtained with the Mössbauer technique were sharper than the peaks of von Békésy's tuning curves. The second difference involves *linearity*. As we study the trans-

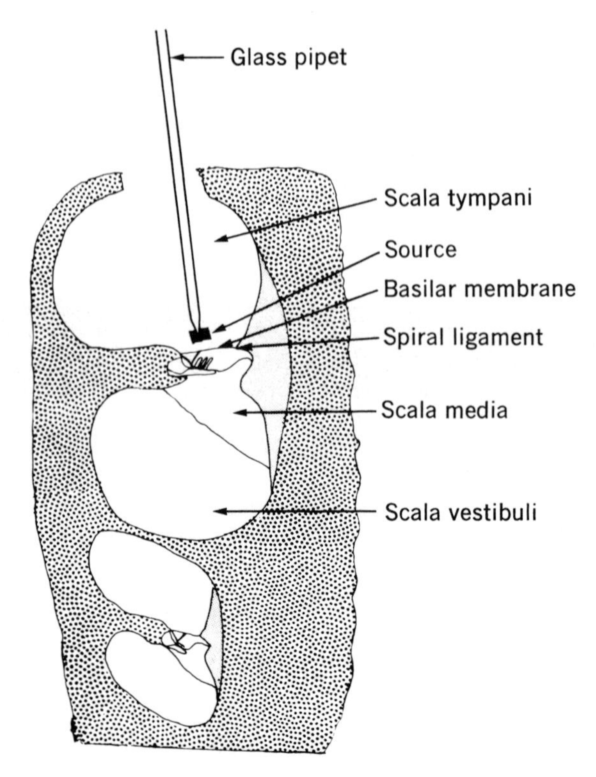

Fig. 14-9. Cross section of squirrel monkey cochlea showing placement of a Mössbauer source with the aid of a section pipet. (From Rhode.[81])

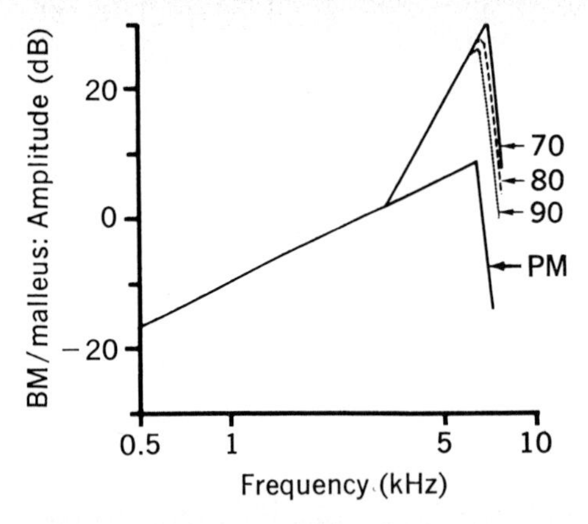

Fig. 14-10. Schematized transfer ratio for basilar membrane (BM) to umbo (malleus) displacement. Ratio is plotted for three SPLs: 70, 80, and 90 dB. *PM*, Postmortem transfer ratio for these intensities. (Modified from Rhode and Robles.[82])

mission of signals through a system like the auditory periphery, it is useful to know if a given part of the system is linear. This is because if it is linear, we can draw broad inferences from measurements with relatively few signals, for example, measurements of responses to sinusoids or to clicks allow prediction of responses of the given part of the system to any arbitrary signal.

A linear system obeys the rules of *scaling* and *superposition*. Scaling means that if the input is made larger or smaller, the response is correspondingly larger or smaller (e.g., if the input is made twice as large, the response is twice as large). Superposition means that if an input signal is constructed by superposing (adding) a number of component signals, the response is the superposition of the responses to these component signals.

Fig. 14-10 shows mechanical tuning curves for stimuli of three intensities: 70, 80, and 90 dB SPL. Since the tuning curve is a ratio of basilar membrane displacement (response) to umbo displacement (input), if the system is linear, the scaling property requires that the curves exactly overlay each other. Note that for frequencies that result in the peak in the mechanical tuning curve the linear relationship breaks down.[81]

The nonlinearity seen by Rhode has been confirmed by many repetitions of the experiment; responses to clicks show a nonlinearity in keeping with that for tones,[82] and control experiments were performed. Yet other researchers have failed to see similar nonlinear phenomena in

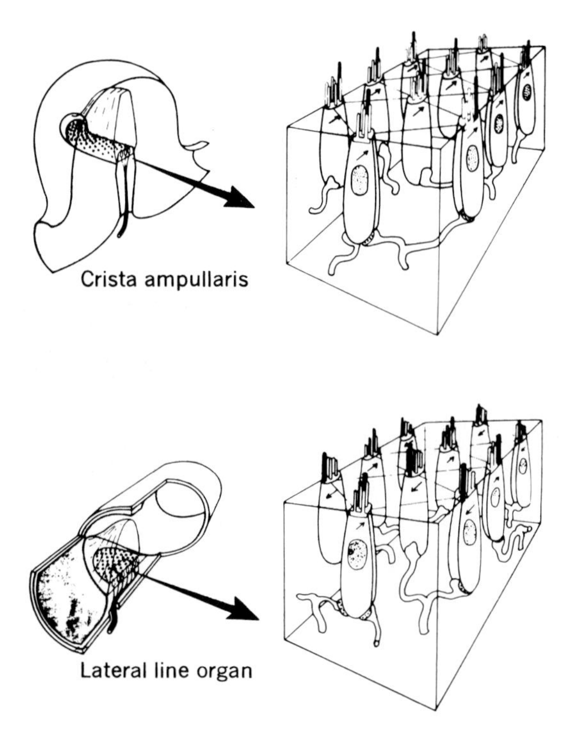

Fig. 14-11. Diagram of crista ampullaris and lateral line organ with enlarged areas of sensory epithelia. Kinocilium is painted black. Cupula overlying epithelium is omitted in drawings at higher magnification. (From Flock and Wersäll.[43])

studies using the Mössbauer technique[57] or a capacitance probe[118] in which guinea pigs were the experimental animals. There may be a species difference, but in any case the discrepancy is troublesome, and I hope that future work will make clear its cause.

The tuning curve marked PM in Fig. 14-10 illustrates the postmortem tuning curves obtained in the squirrel monkey.[82] Note that (1) there is a decrease in the frequency of the peak of the curve; (2) the peak ratio of response to input decreases by more than 10 dB postmortem; and (3) on the low-frequency side of the peak the region of 24 dB/octave slope disappears, leaving only a 6 dB/octave slope. It was also observed that within 1 hour of death the vibration of the basilar membrane for all frequencies is linear. The postmortem changes begin in minutes, implicating some metabolic factor. It is of interest that the postmortem results resemble the findings of von Békésy. Since most of von Békésy's studies used cadaver temporal bones, it seems likely that postmortem changes account for the discrepancy between his results[9] and results of recent studies,[57,81,118] which show considerably sharper peaks in the tuning curves.

Physiology of hair cells[43,113,114]

Fig. 14-6 shows the mammalian cochlear partition in cross section. The first step in transduction of the mechanical vibration of the cochlear partition to a pattern of neural impulses in the eighth nerve presumably involves some deformation or stressing of the hair cells. It seems likely that mechanisms involved in the transduction of mechanical stimulation to neural activity are similar in hair cells of the cochlea, the vestibular organs, and the lateral line organ. More is known about the physiology of the lateral line and vestibular hair cells than of the cochlear hair cells, so we shall briefly leave the cochlea.

The crista ampullaris and the lateral line organ exhibit striking differences in their electrophysiologic responses to sinusoidal stimulation, although their gross morphologic features are similar (Fig. 14-11). Both organs exhibit a receptor potential called the microphonic potential, which is probably related to depolarization and hyperpolarization of the hair cells. In the crista ampullaris the frequency of the microphonic potential is that of a sinusoidal stimulus; however, in the lateral line organ, it is twice the stimulus frequency.

Fig. 14-12 shows schematically the ultrastructure of the sensory epithelium of the lateral line organ. This figure labels many parts of hair cells and related structures.[40] A difference in the orientation of the hair cells in the receptors as seen by the electron microscope explains the difference in the microphonic potentials.[43] In Fig. 14-11 the orientation of the sensory hair bundles is indicated by the arrows on each hair cell: in the crista ampullaris, all the hair cells have the same orientation; in the lateral line organ, adjacent hair cells are polarized in opposite directions. Fig. 14-13 shows how such an arrangement could account for the difference in microphonic potentials. Intracellular recordings obtained from hair cells of the lateral line[53] provide direct support for the concept of morphologic polarization.

The first step in transduction of the mechanical vibration of the cochlear partition to a pattern of neural impulses in the eighth nerve presumably involves some deformation or stressing of the hair cells. The ends of Deiters' cells (see Fig. 14-6 for terminology) that face the scala media form a stiff open network, the reticular lamina. The hair-bearing ends of the sensory cells are firmly held in the openings of this lamina. The opposite ends of the hair cells rest in cuplike supports that are also formed by Deiters' cells and receive the nerve endings of the afferent and efferent auditory fibers. The rods of Corti form a

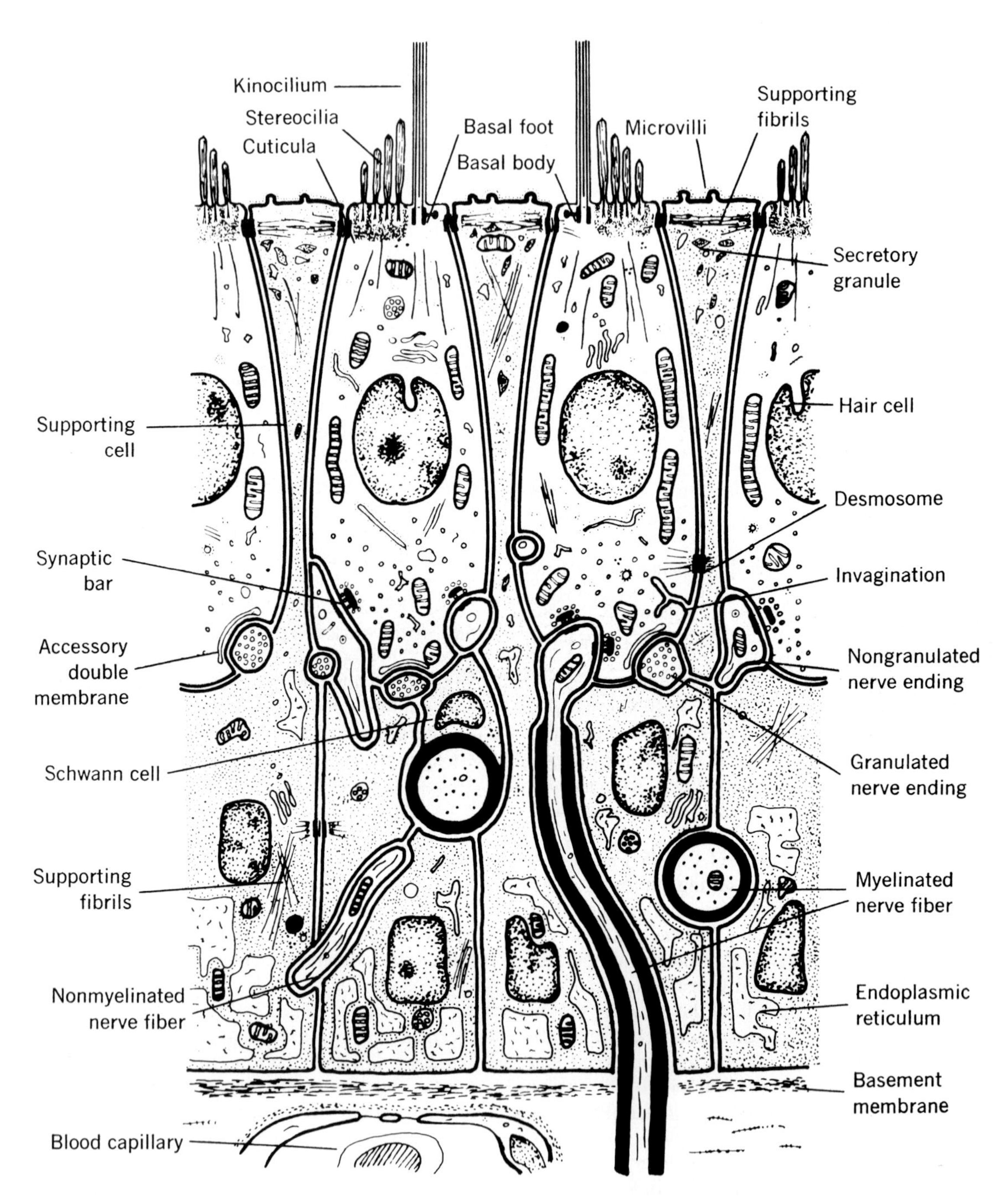

Fig. 14-12. Schematic drawing showing ultrastructure of sensory epithelium of lateral line canal organ of *Lota vulgaris*. (From Flock.[40])

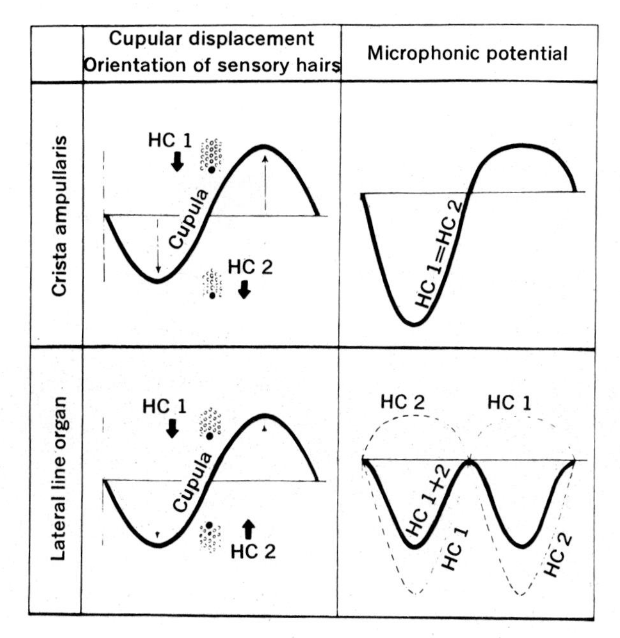

Fig. 14-13. Illustration of hypothesis accounting for difference in microphonic potentials from crista ampullaris and lateral line organ. In crista, all hair cells are polarized in same direction, as indicated by *HC 1* and *HC 2*. Cupular displacement in this direction is excitatory for both cells and is followed by decrease in potential (depolarization), whereas displacement in other direction is accompanied by increase in potential (hyperpolarization), as shown by right curve representing responses from *HC 1* and *HC 2*. In lateral line organ, *HC 1* and *HC 2* are polarized in opposite directions. Potential changes induced by cupular displacement contributed by *HC 1* will follow course indicated by dotted curve marked *HC 1* in right figure, whereas potential changes contributed by *HC 2* will follow course indicated by curve *HC 2*. Recorded microphonic potential represents sum of these two partially canceling waveforms, curve *HC 1* + *HC 2*, and will consequently show a frequency double that of cupular displacement. (From Flock and Wersäll.[43])

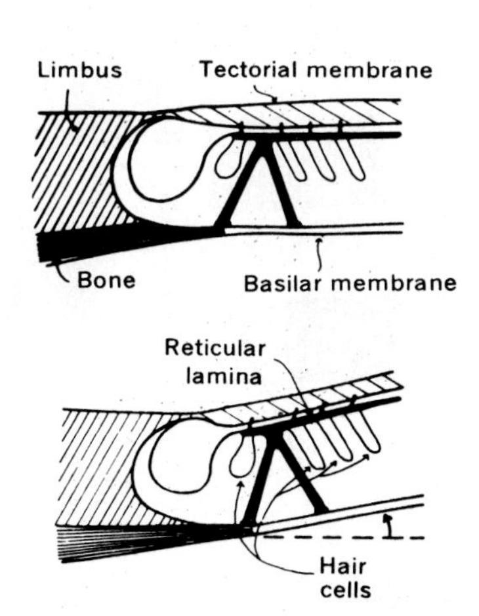

Fig. 14-14. Movement of organ of Corti and tectorial membrane, based on descriptions by von Békésy.[9] Shearing action between two stiff structures (tectorial membrane and reticular lamina) bends hairs of hair cells. (From Davis.[2])

stiff triangular supporting structure for the reticular lamina. The flask-shaped inner hair cells are in a single row along the inner edge of the reticular lamina; in humans, they number about 3,500. The external hair cells are cylindric in shape and are arranged in three rows on the opposite (outer) side of the tunnel of Corti. They number about 20,000. These relationships are shown schematically in Figs. 14-14 and 14-17.

The upper and lower sketches of Fig. 14-14 show one way in which the cilia of the hair cells could be deformed. The reader should understand that the sketches are highly schematic (for one thing, the motion of the basilar membrane is vastly exaggerated) and the notion that it is in this fashion that the first step in transduction takes place is highly speculative. It does seem likely that the cilia are quite stiff and remain straight even as a shearing force makes them pivot.[41] There is good evidence that the tips of the tallest cilia of the outer hair cells are imbedded in the tectorial membrane,[63] but reports concerning the embedding of the hairs of the inner sensory cells are not in agreement.[54,63,98]

Exactly how these forces on the hair cells lead to excitation of the fibers of the eighth nerve is not known. One approach to a better understanding of the mechanical-neural transduction has been through studies of the electrical potentials of the cochlea.

COCHLEAR POTENTIALS

A number of potentials may be recorded from the cochlea by gross electrodes placed inside the cochlea through small holes drilled in the bony wall or by an external electrode placed near the round window.

Resting potentials[9,27,68,106]

The DC or steady potentials of the unstimulated cochlea (the resting potentials) have been studied most commonly in the guinea pig, an animal in which the four and a half turns of the cochlea are readily explored. The scala tympani (see Fig. 14-6 for terminology) and scala vestibuli are filled with perilymph, which has the chemical composition of CSF. The scala media is filled with endolymph, which differs markedly in chemical composition, having high K and low Na concentrations. High K concentration is also found in the tectorial membrane and the space between it and the reticular lamina, indicating that the sensory hairs of the cochlear hair cells are exposed to the ionic environment provided by endolymph,[42] whereas the cell bodies are in perilymph. Endolymph probably is secreted by the stria vascularis and is reabsorbed in the endolymphatic sac.

As an electrode is advanced from the scala tympani through the basilar membrane, the potential drops 80 to 90 mV when the electrode enters the organ of Corti. In this region the potential fluctuates considerably, and it seems likely that the large negative potentials are intracellular potentials from cells in the organ of Corti, including hair cells, although no systematic study of them has been made.[22] As the electrode passes into the scala media, the potential rises to a value that is 80 to 90 mV positive with respect to the scala tympani. This is called the endocochlear potential (EP). Further penetration through Reissner's membrane into the scala vestibuli returns the potential close to that of the scala tympani. The source of the EP is the stria vascularis,[26,106] and the potential depends on the animal's metabolic state.[68]

Cochlear potentials in response to sound[25,26]

A number of cochlear potentials are evoked by acoustic stimulation. Differential recording from the scala vestibuli and scala tympani yields the cochlear microphonic (CM) potential, which is related to the instantaneous pattern of vibration of the cochlear partition in the vicinity of the position of the electrodes, and two summating potentials (SP+ and SP−), which follow the envelope of the acoustic stimulus.[25]

Fig. 14-15 shows the relationship of CM amplitude to the input sound pressure level.[2] Growth is linear up to about 80 dB SPL, but not for higher input levels.[66] Another test of linearity is the faithfulness with which the CM reflects the input waveform, in this instance a sinusoid. We expect the CM to be periodic with the period of the input, but to have *distortion components*, most likely at integral multiples of the fundamental frequency. At moderate intensities, distortion components of CM potentials were found at the region of the cochlea stimulated most by the input sinusoid; at high intensities the locus of distortion components is determined by their own frequency and the spectral filtering properties of the cochlear partition.[23] One aspect of nonlinearity in the CM potential is illustrated by the input-output curve in Fig. 14-15. For high-frequency tones the waveform of CM potentials is almost undistorted for intensities that far exceed the linear region of the input-output relationship (insets, Fig. 14-15). It is important to remember

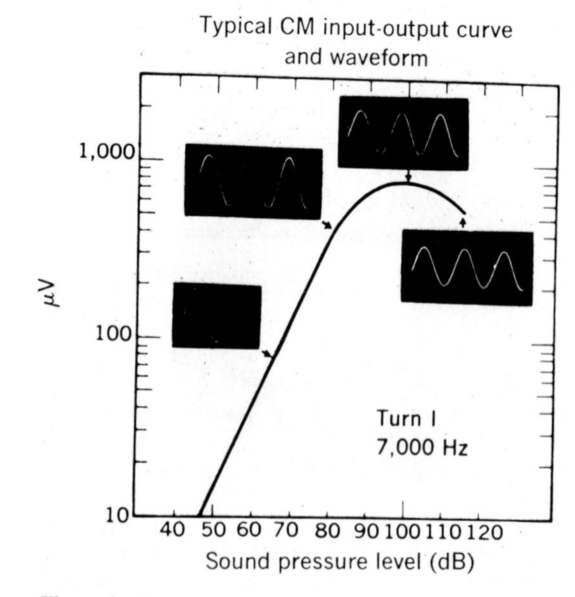

Fig. 14-15. Input-output curve for cochlear microphonic (CM) response of first turn of guinea pig cochlea to 7,000 Hz tone bursts. Insets show waveform of differentially recorded CM. Note absence of peak limiting, even at highest sound intensity. (From Davis.[2])

that the CM potential is a weighted summation of hair cell potentials from a region of the cochlea and is not equivalent to hair cell potentials from a single longitudinal position. In recent experiments, potentials from single inner hair cells in the guinea pig cochlea were recorded intracellularly.[86] A positive potential, probably the source of SP+, was studied in detail. The tuning curve for this potential was found to be quite sharp.

By recording CM potentials at a number of positions along the cochlea, it is possible to reconstruct the "traveling wave" of this potential.[107] The traveling wave reflected by CM in the guinea pig cochlea (Fig. 14-16) is similar to that of the human cochlear partition produced by a sudden displacement of the stapes. von Békésy found that it takes somewhat over 3 msec for the wave in response to a brief transient stimulus to traverse the length of the human cochlea; in the guinea pig, this time is nearer 2 msec.

Receptor mechanisms

Fig. 14-17, presents a surface view of the tops of the inner and outer hair cells, showing the orientation of the hairs, which is constant throughout the cochlea.[34,44] Outer hair cells of the cat have about 100 stereocilia; inner hair cells have about 50.[100] The morphologic polarization of both outer and inner hair cells is in a radial direction, with the stereocilia increasing in length toward the outer edge of each hair cell, and with a spot in which the cuticle is missing near the outer edge. Some species[44] (but not the cat[100]) typically have a basal body near the outer edge. Cochlear hair cells of all adult mammals lack kinocilia.

If the hair cells of the cochlea were analogous to those of the lateral line and vestibular organs, we would expect that when the cilia were moved radially, outward, depolarization of the hair cells would result, with subsequent excitation of afferent fibers. It is very difficult to test this schema directly. Experiments in which the basilar membrane was deflected and held for more than 10 msec[97] or statically[65] toward either the scala vestibuli or the scala tympani indicate that the majority of the afferent auditory nerve fibers fire more for displacement toward the scala tympani than for displacement toward the scala vestibuli. These results seem anomalous, but until we know the relative motion of the basilar membrane (actually the reticular laminar) and the tectorial membrane, we cannot say how anomalous they are.

Summarizing, at present we have limited knowledge of cochlear receptor mechanisms, and much of the knowledge we have is inferential.

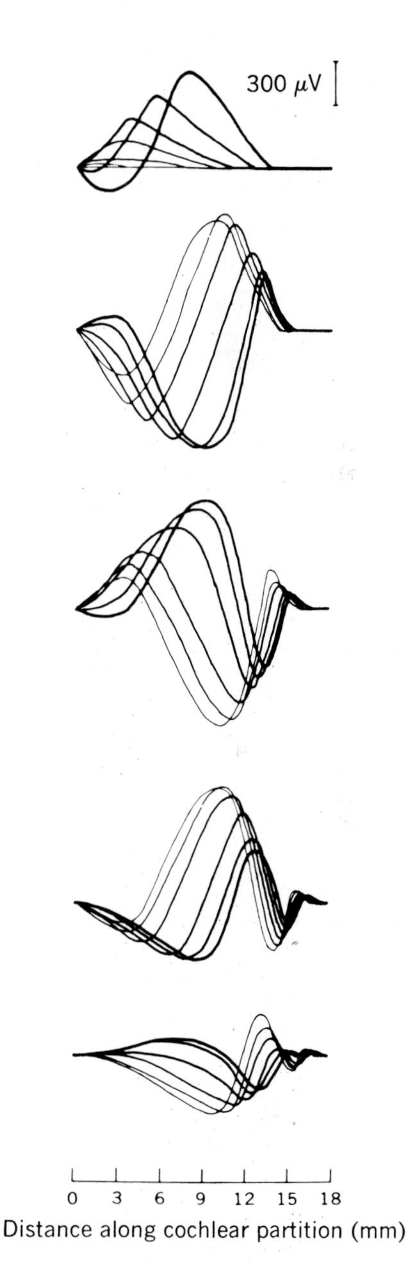

Fig. 14-16. Instantaneous CM voltage as a function of distance along cochlear partition at intervals of 0.1 msec. Thicker lines represent more recent times. Progression from top to bottom set is a continuous sequence. Between sets go from thickest line in set above to thinnest line in set below. Acoustic signal is diphasic transient, with about 1 msec between peaks. (From Teas et al.[107])

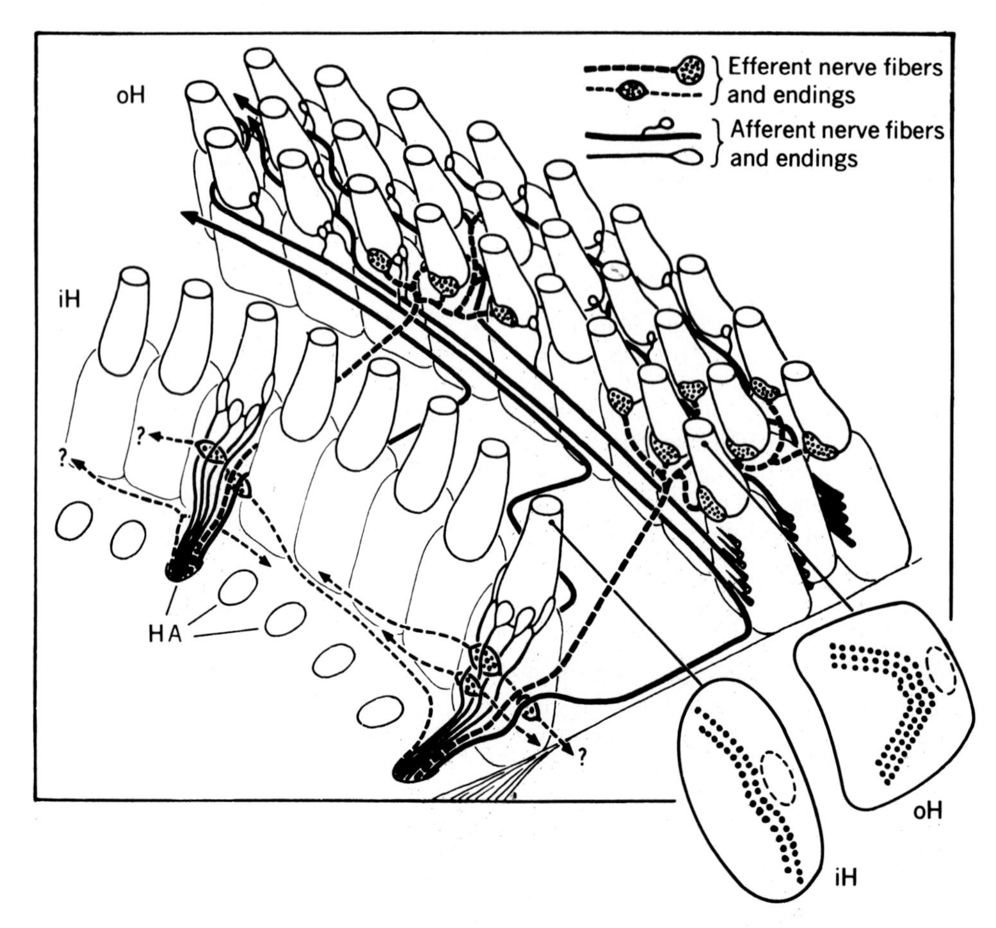

Fig. 14-17. Schema of innervation pattern of organ of Corti in cat. At lower right of illustration is top view of one inner hair cell, *iH*, and one outer hair cell, *oH*, showing stereocilia (dots) and place where cuticle is missing (dashed oval). Only representative examples of nerve fibers arriving to organ of Corti through two habenular openings, *HA*, are shown. Afferent fibers are shown as continuous lines; efferent fibers are dashed. (See text for quantitative information.) Full spiral basalward extension of afferent fibers from outer hair cell (outer spiral fibers) cannot be shown because of limited space. Indicated nerve fibers and endings do not correspond to their actual numbers. (Schema from Spoendlin[99a]; tops of hair cells after Flock et al.,[44] Engström et al.,[34] and Spoendlin.[100])

The relevant physiologic questions are as follows: (1) How does the basilar membrane move in response to acoustic stimuli? (2) What forces on the hair cell cilia and bending of these cilia result from that motion? (3) What changes (e.g., in intracellular potential and membrane permeability of the hair cells) follow? (4) What changes lead to release of a chemical transmitter? (5) What leads to impulse generation in the fibers of the auditory nerve? As we have seen, von Békésy's classic studies of the first question are presently being extended by researchers using modern techniques for measuring small motions. Direct experimental approach to the second question seems beyond present technology, so that we are forced to use models and to make inferences. Progress is being made in answering the third and fourth questions with recent studies of intracellular electrical responses from cochlear hair cells,[72,86] opening the way for significant advances. It is worth noting that the afferent chemical transmitter in hair cells has yet to be identified. At present there is a lot of information about the fifth question, impulse activity in cochlear nerve fibers, as we shall see further along in this chapter. In fact, much of the present thinking about receptor mechanisms is inferred from the picture of sound-evoked activity in the fibers of the auditory nerve of normal preparations and of animals with noise- or drug-induced cochlear damage.

Whole-nerve action potential

None of the potentials discussed thus far is a direct sign of nerve impulses in the auditory fibers. If, instead of recording differentially

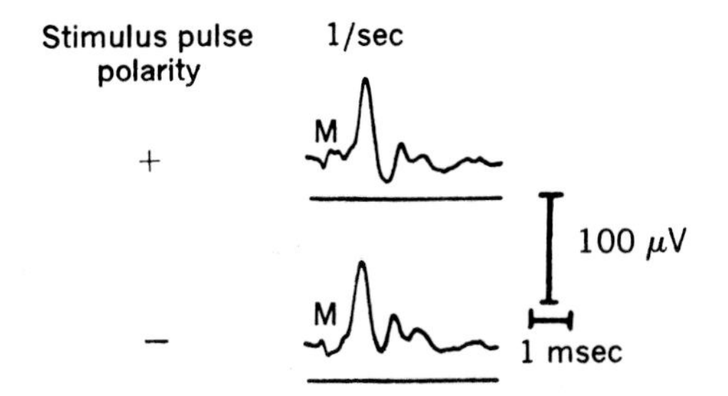

Fig. 14-18. Responses to rarefaction (+) and condensation (−) clicks at repetition rate of 1/sec. Only microphonic *M* reverses with reversal of stimulus polarity. Upward deflection of electrical signal indicates negative potential at electrode located near round window. (From Peake et al.[76])

across the cochlear partition, as in the study of CM potentials, the potentials from the scala vestibuli and scala tympani are summed and referred to a reference electrode on the neck, the whole-nerve action potential of nerve evoked by sound may be obtained almost free of CM potentials and SP.

Action potentials (APs) may also be obtained by recording between a point near the round window and a distance reference electrode. In this case a microphonic component is also obtained. Fig. 14-18 illustrates round window responses to rarefaction and to condensation clicks. Quite similar potentials may be recorded from humans by an electrode that is passed through the tympanic membrane and placed on the promontory.[79] Other recording sites are the earlobe[96] or ear canal.[20] In all these cases a distance reference (e.g., the nose) is used. All these signals share with the round window potential the feature that APs and microphonic responses to brief acoustic transients (e.g., clicks) may be separated because the microphonic has shorter latency than the AP. The neural and microphonic signals recorded transtympanically are small, and those recorded from canal or earlobe are very small, so that averaging of many responses is necessary. These recordings are finding increasing clinical use. Their clinical application is called electrocochleography.[6]

Fig. 14-18 shows that the microphonic potential reverses polarity with the stimulus, but the neural responses do not. The microphonic potential recorded in this manner is a spatial average of CM potentials, weighted so that the basal portion is predominant. The neural potential depends on a synchronous discharge of auditory fibers.[50] Since it is the fibers innervating the basal turn that fire synchronously to a brief transient stimulus,[107] the neural component shown in Fig. 14-18 also disproportionately weights electrical activity from the basal portion of the cochlea. The two negative peaks (N_1 and N_2) of the AP have been studied in detail. The latency between the microphonic and AP decreases with increasing stimulus intensity. The AP follows repetitive stimuli to rates of approximately 3,000/sec.[76]

The most direct method of investigating the patterns of neural activity in the eighth nerve is to record impulses from individual fibers, an experimental method now commonly used. The results obtained are the subject of the next section.

CODING OF ACOUSTIC STIMULI IN FIBERS OF THE EIGHTH NERVE*

There are about 30,000 afferent neurons in the human auditory nerve. The cell bodies are arranged in a spiral ganglion that parallels the organ of Corti but is within the bony modiolus. The portion of such a neuron peripheral to the cell body loses its myelination only when it goes through the habenula perforata into the organ of Corti. A schema of the innervation pattern of the hair cells in the cat is shown in Fig. 14-17.[99a] There are about 55,000 afferent neurons, approximately 95%[100] of which innervate inner hair cells in the radial fashion shown. The other afferent fibers cross the tunnel of Corti at the bottom, spiral basalward in regular rows between Deiters' cells, and finally terminate on the outer hair cells.[100,101] The afferent neurons are bipolar, and the myelinated portion functions as the axon. The centrally directed portions of the fibers pass through the hollow core of the cochlea to form the cochlear portion of the eighth nerve. They enter the cranial cavity through the internal auditory meatus and end on the cells of the cochlear nucleus of the medulla.

Each radial afferent fiber seems to make a single synapse with one inner hair cell. Each inner hair cell has terminations from 10 to 20 radial afferent fibers. On the other hand, each spiral afferent fiber innervates about 10 outer hair cells, and outer hair cells have about 4 afferent terminals. The spiral afferent fibers run along the cochlea for a distance of 0.7 to 1 mm, sending out terminal branches to the outer hair cells from approximately the final 0.2 mm of their length.[100,101] There is a small (0.5%) subpopulation of afferent neurons with very large axons that run in a spiral direction apicalward and basalward for a total distance of about 10 inner hair cells, making synaptic contact with every inner hair cell along their course.[101,102] The cell

*See references 4, 36, 74, 77, 85, and 105.

bodies of these neurons and of the afferent neurons innervating the outer hair cells are of one type; the cell bodies of the afferent fibers that make a single synapse with an inner hair cell are of another type. The two types show morphologic differences and different degeneration behavior following cochlear nerve section.[101,102] It is not yet clear to what extent the pattern of innervation in other mammals corresponds to that seen in the cat.

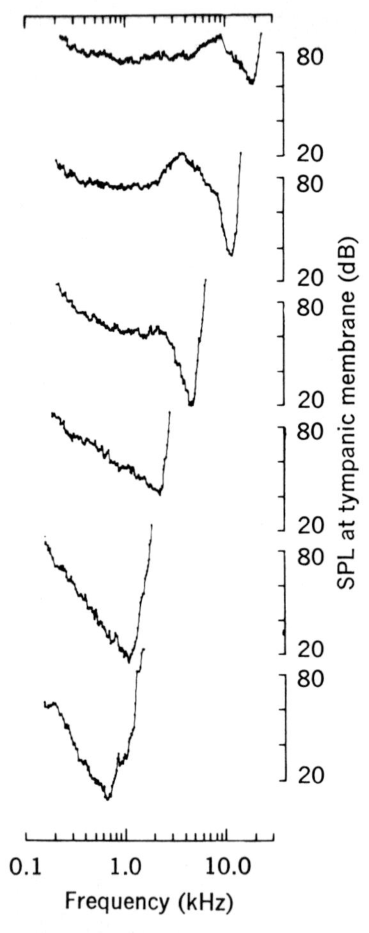

Fig. 14-19. Tuning curves for six auditory nerve units of different characteristic frequency (CF). Tone bursts (50 msec duration, 2.5 msec rise-fall time) were presented at 10/sec, and discharge rate during each burst was compared with the discharge rate during intervening intervals. At any one tone frequency, level (SPL) of successive bursts was changed until a criterion difference in discharge rate of 1 spike/tone burst was reached. Procedure was then repeated at successively lower tone frequencies until desired frequency range was covered. All curves shown were obtained from one cat. (From Kiang and Moxon.[59])

It is important to know how the temporal-spatial pattern of vibration of the cochlea is encoded as a temporal-spatial pattern of nerve impulses in cochlear nerve fibers, which are the initial portion of the intricately interconnected structures of the auditory pathway. Coding in the primary fibers should indicate those properties of the acoustic signal that can possibly be sensed by the organism.

Spontaneous activity[4,61,67]

When advancing a microelectrode into the auditory nerve of an experimental animal, it is usual to present an acoustic stimulus. Trains of clicks or noise bursts are commonly used, since they excite the cochlea broadly. Study proceeds after the nerve impulses of a single fiber are isolated. The first question is whether a fiber exhibits "spontaneous" activity, that is, whether impulses are present in the absence of acoustic stimulation. Most (if not all) primary auditory fibers of mammals exhibit spontaneous activity. The average rate of this discharge may be above 100/sec, although it is usually considerably lower. In cats, over 10% of the fibers have spontaneous rates of less than 0.5 spike/sec.[67] The impulses occur in a random time sequence closely modeled by a modified Poisson process. In this model the occurrence of an impulse is equally likely at any time and does not depend on previous events; it is modified only to account for a brief refractory period after the occurrence of an impulse.

Tuning curves

Auditory nerve fibers are differentially sensitive to sounds of different frequency. This property is shown in Fig. 14-19. These curves of threshold as a function of tone frequency are called tuning curves. All the curves have a V-shaped minimum. The frequency of this minimum is called the characteristic frequency (CF). Note that the high-CF units have sharp "tips" near CF and a broad "tail" at lower frequencies.[59] As fibers leave the cochlea, they maintain their relative positions in the spiraling bundle,[90] so that one finds an orderly progression of characteristic frequencies as an electrode is advanced through the cochlear nerve.[4]

Tuning curves of single neural units are very important in auditory neurophysiology. They are in some ways analogous to receptor fields in vision. For cochlear nerve afferent fibers, all tonal stimuli within the tuning curve will yield suprathreshold responses; all tonal stimuli outside are subthreshold. Shortly, the nature of suprathreshold responses will be considered, but first more consideration should be given to the tuning curves themselves.

There are a number of ways experimentally to manipulate cochlear function. These include hypoxia, death, dosage with otoxic drugs, and exposure to high-intensity sounds. It has recently been shown that during hypoxia[35] the sharp tips in the tuning curves may be more or less eliminated. A similar result was obtained in animals exposed to ototoxic drugs or high-intensity noise.[60] The hypoxia technique has the advantage of being reversible. Destruction of cochlear elements such as inner hair cells, outer hair cells, supporting cells, and nerve fibers by noise exposure or ototoxic drugs can be traced by histology.

To what extent is the neural tuning of primary (i.e., cochlear afferent) units related to mechanical tuning in the cochlea?[47] A salient postmortem change in mechanical tuning was the elimination of the sharp peak.[82] Considering this and the change observed in neural tuning during hypoxia, the conclusion can be made that mechanical and neural tuning are closely related. However, it takes more detailed consideration than space permits to explore questions as to whether the mechanical tuning is sharp enough to account for the very sharp neural curves; to what extent nonlinearities seen in neural coding have their basis in the mechanical response; etc. Right now the relevant data come from many laboratories and several different animal species, and the answers are not yet in. No doubt the future will bring clarification.

Intensity functions

As stated previously, the tuning curve tells us which sinusoidal signals would have excited a unit (i.e., would have been above threshold) and which would have been ineffective. Next, we will consider the strength of responses to suprathreshold sinusoids (tones). Results are summarized in Fig. 14-20. The curves show unit response in spikes per second as a function of stimulus level in decibels SPL. Such curves are called intensity functions. Results from two units with characteristic frequencies (CFs) of 1.3 and 4 kHz are illustrated. The parameter on the curves is tone frequency in kilohertz.[88]

First, let us consider the intensity functions for CF tones. For both units, rate increases rapidly over a range of 20 to 30 dB above threshold. For higher levels the response rate of the unit illustrated on the left saturates; for the unit illustrated on the right there is a bend in the intensity function, but rate continues to increase. These two examples illustrate the range of behaviors observed.

For tone frequencies above CF the slope of the intensity function is a decreasing function of frequency (illustrated by the dashed curves). For tone frequencies below CF the intensity functions are similar to those for tones at CF except the curves are shifted to the right (illustrated by the solid curves) These relationships were found to be quite regular and were fit by a model.[88]

Responses to two-tone stimuli

Earlier (p. 436) the defining properties of linear systems were given. An important question

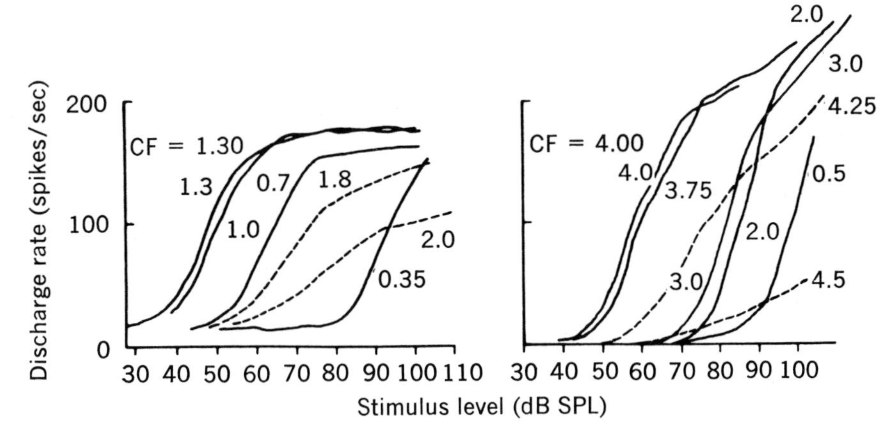

Fig. 14-20. Intensity functions for two cat auditory nerve fibers. Stimuli were presented in 400 msec bursts once every 1.5 sec, and average discharge rate was computed. Parameter on curves is tone frequency. Note fiber CFs. (From Sachs and Abbas.[88])

is to what extent the auditory periphery has linear properties. To start, to what extent is the response to two tones presented simultaneously the superposition of the responses to the tones presented separately?

When two-tone stimuli are presented, a suppression of activity in response to one (excitatory) tone can result when the second (inhibitory) tone is presented.[87,89] This phenomenon is illustrated schematically in Fig. 14-21. The dashed, V-shaped curve is the tuning curve of a unit. A pure tone within this curve increases firing rate by 20% or more above the spontaneous rate. When an excitatory stimulus of frequency and intensity indicated by the triangle was added to it a second tone with frequency-intensity parameters in the shaded regions, the rate drops from that in response to the excitatory tone alone by 20% or more. All primary units show two-tone suppression.

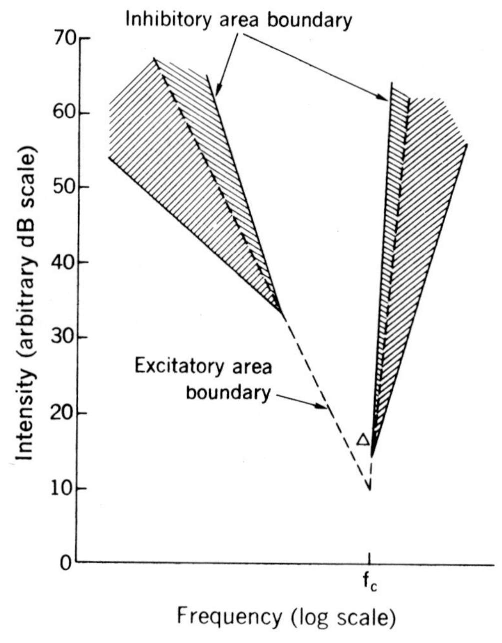

Fig. 14-21. Idealized inhibitory and response areas for typical auditory nerve fiber. Tuning curve as in dashed lines. Second tone in shaded areas suppresses response to excitatory tone (triangle). Regions for which the second tone presented alone would result in excitation are shaded differently than regions for which the second tone alone would not excite the fiber. (From Sachs.[87])

Since a tone outside the tuning curve can affect response to a tone within the curve, by this test the auditory periphery does not have linear properties. Many other studies of single-unit activity in auditory neurons[48,49,83] have shown nonlinearities in the auditory periphery even for low–stimulus intensity levels.

Temporal coding

The responses of primary auditory fibers to continuous stimuli are of a tonic nature. At the onset of a noise or tonal stimulus that excites the unit the discharge rate is relatively high; subsequently it adapts to a more or less constant value.[4,119] For wide-band noise or high-frequency tone stimulation the cadence of nerve impulses is not regular but has a random pattern, similar to the spontaneous activity but with a higher average rate.

For sinusoidal stimuli, low-frequency fibers of the auditory nerve are more likely to discharge in one part of the stimulus cycle than in others. Fig. 14-22 illustrates the phenomenon. Note that although spike occurrence was quite random, when spikes did occur, it was more or less at the same stimulus phase. This phase-locking appears in recordings from monkeys for frequencies as high as 4,000 to 5,000 Hz.[83] The interspike interval histogram in Fig. 14-23 illustrates the stochastic nature of the responses. Some intervals are clustered about the period of the pure tone, some about twice this value, etc. Such a distribution is observed even with low frequencies, for which the period is considerably longer than the neural refractory time. This is not a simple frequency division in which the cell discharges on every cycle and then at higher frequencies on every second cycle of the stimulus, then on every third, etc. (the simplest statement of the volley theory of Wever[11]). Instead, the phase-locking occurs in a probabilistic way.

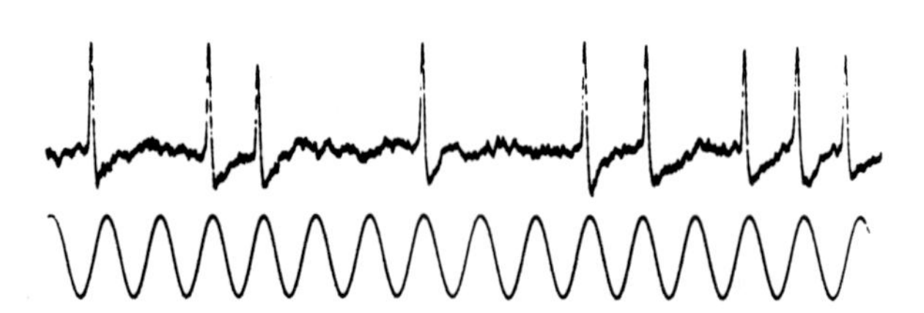

Fig. 14-22. Continuous record of response of guinea pig auditory nerve fiber to 300 Hz tone. Bottom trace shows electrical signal to earphone. (From Evans.[36])

Kiang et al.[4,61] discovered a remarkable ability of primary fibers to code the temporal patterns of vibration of the cochlear partition in response to clicks. The result is illustrated in Fig. 14-24. Many hundreds of stimulus presentations were used to obtain the poststimulus time histogram of unit discharges. This statistical display gives the distribution of latencies of the unit impulses relative to the onset of the click. Note that the poststimulus time histogram in Fig. 14-24, *B*, has several evenly spaced peaks, Examination of the individual records in Fig. 14-24, *A*, shows that in any given stimulus presentation the fiber does not discharge at the time of each and all of the peaks. These are, however, the times when the probability of discharge is greatest. The multipeaked pattern was observed for units with characteristic frequencies up to about 5,000 Hz (Fig. 14-24, *B* to *D*). Furthermore, the space between peaks was found to be almost exactly equal to the inverse of the characteristic frequency of the unit under study.

Population studies of the auditory nerve units[62,77]

The single-unit studies cited previously consisted of more or less extended study of a limited number of neurons in each experiment. Recent work in which a population of units representative of the auditory nerve is studied in each experiment shows great promise. It takes a population of hundreds of units to represent well the activity in the auditory nerve. Ideally, one would record simultaneously from the units; however, the technology is not so advanced, and units are studies individually in sequence. This means that each unit should be studied with a small set of stimuli, the same set being used for all units. There must also be some control for the stability of the preparation. Implicit is the assumption that the discharge patterns of units in the population are to a reasonable approximation independent. Fortunately, for the auditory nerve this seems to be the case.[56]

Pfeiffer and Kim[77] did the germinal population studies of cochlear nerve fiber responses. For each unit studied, they determined CF; this parameter allows one to order individual units according to the place on the cochlea that has a peak of vibration at the unit's CF. Some examples will illustrate the power of this approach.

Fig. 14-25 shows an analysis of the population response to a 20 dB SPL, 600 Hz sinusoidal stimulus.[77] In both *A* and *B*, each data point is from a different unit and is placed along the abscissa according to the unit's CF. Fig. 14-25, *A*,

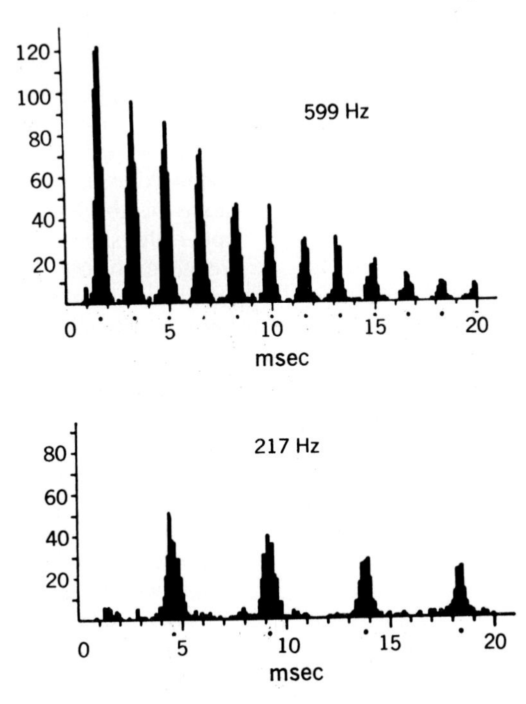

Fig. 14-23. Interval histogram for responses to pure tones. Abscissa: interval in msec. Ordinate: number of intervals in discharges to a 100 dB SPL tone of 20 sec duration. Tone frequencies: 599 Hz, near best frequency of unit, and 217 Hz. Intervals cluster around integral multiples of period of tone indicated by dots below ordinate scale. (From Rose et al.[83])

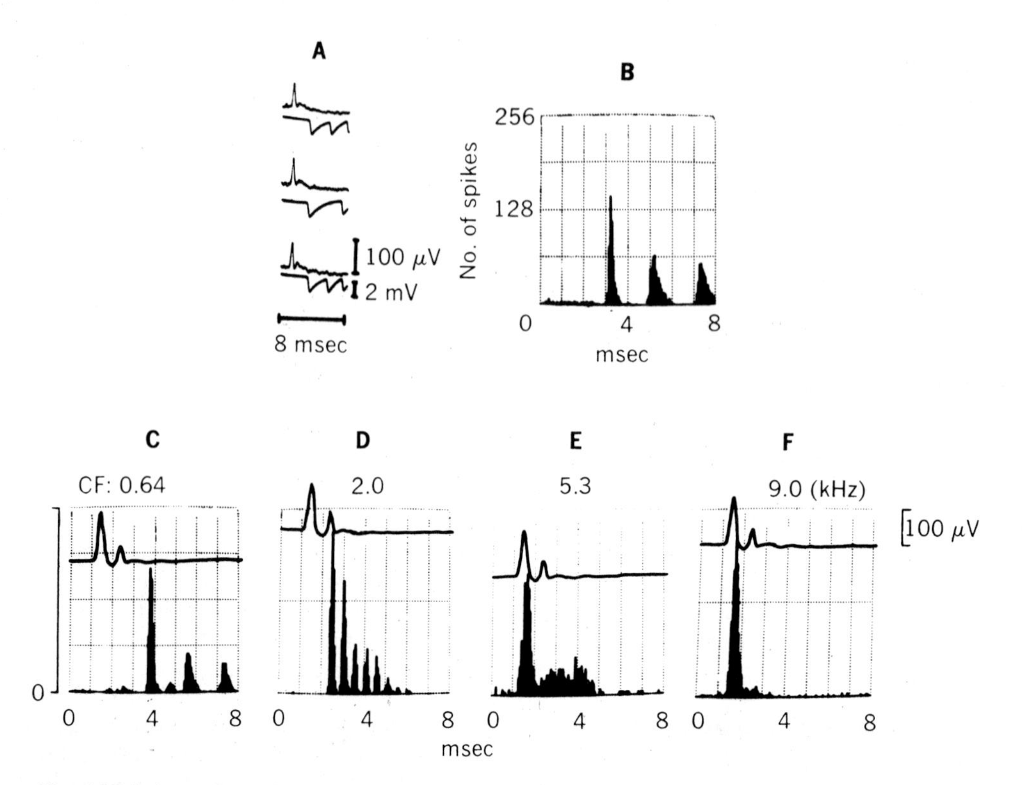

Fig. 14-24. A and **B,** Responses of single cat cochlear fiber to click stimuli. **A,** Responses to three successive click presentations. Start of each trace was synchronized with electrical pulse to earphone. Upper traces: record from electrode at round window showing click-evoked AP response. Lower traces: spike discharges from fiber. Negativity upward. **B,** Poststimulus time (PST) histogram showing averaged temporal pattern of activity in response to 600 clicks. CF of fiber, 0.54 kHz. **C** to **F,** PST histograms for four more cat cochlear fibers of differing CF compared to AP response recorded from round window. CFs in kilohertz indicated above plots. Linear ordinate scale. Full scale: **C,** 256; **D,** 128; **E,** 64; and **F,** 128 spikes. Data samples 1 min each; clicks presented at 10/sec in each case. (From Kiang et al.[4])

shows the normalized fundamental component of the phase-locked response to the sinusoid. Fig. 14-25, *B*, shows the relative phase of the response of each unit. From these two records the traveling wave of cochlear nerve fiber activity was determined as shown in Fig. 14-25, *C*. It is interesting to compare this traveling wave with that of the basilar membrane shown in Fig. 14-7. Note that the abscissa in Fig. 14-25 starts at zero and that of Fig. 14-7 does not. Considering differences in species and stimulus frequencies the traveling waves are remarkably similar.

Unit responses to two-tone stimulation were discussed previously. The population approach contributes significantly to our understanding of neural coding on tone pairs. In the experiment whose results are shown in Fig. 14-26, tones of 2,100 Hz and 2,700 Hz were presented simultaneously.[62] A Fourier analysis was made of the resulting phase-locked single-unit activity. In Fig. 14-26, responses of the individual units are displayed as in Fig. 14-25, as data points placed along the abscissa according to the unit CF. The upper plot shows the responses to the difference frequency of 600 Hz. This frequency would be heard as a combination tone.[10] The middle and lower plots show the responses at the frequencies of the two tones.

Of special interest are the responses to the difference frequency. Two peaks are in that plot, one at the 600 Hz place and one at the place of responses to the two-tone pair. It is worth noting that the stimulus level was 34 dB SPL, below the level of a quiet conversation. Obviously the nonlinear system behavior reflected in the activity of auditory nerve fibers is substantial even near threshold.

Form and function

An important question that is still quite open is the relationship of the anatomic arrangement of auditory afferent neurons and their physiology.

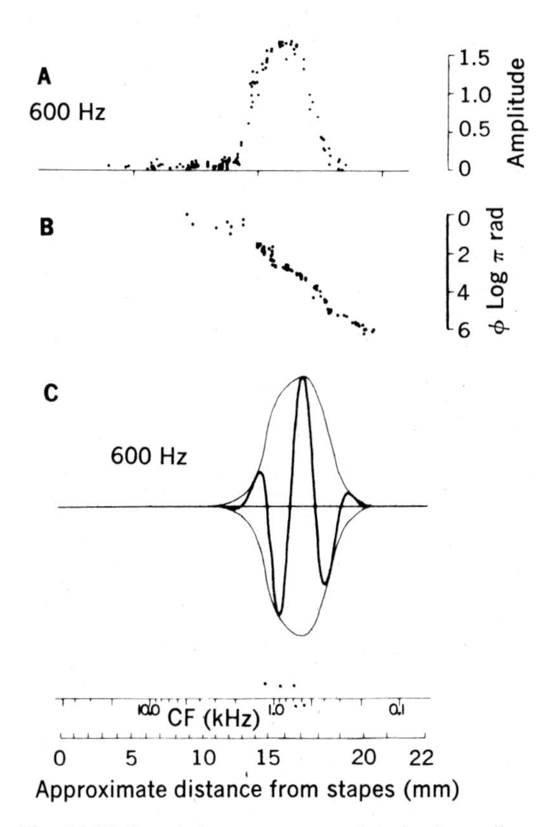

Fig. 14-25. Population responses and derived traveling wave of cochlear nerve fiber activity in response to stimulation by 600 Hz sinusoid. Intensity level of stimulus, 20 dB SPL. **A,** Normalized amplitudes of Fourier coefficients of fundamental components of period histogram. (Period histogram gives statistical display of timing of nerve spikes relative to period of the stimulus.) **B,** Phase of fundamental component of period histogram relative to sound pressure at eardrum. **C,** Traveling wave of cochlear nerve fiber activity obtained from data of **A** and **B.** Data were averaged visually to obtain envelopes of response (thin lines) and averaged phase. These data were used to determine amplitudes of traveling wave (heavy lines). (From Pfeiffer and Kim.[77])

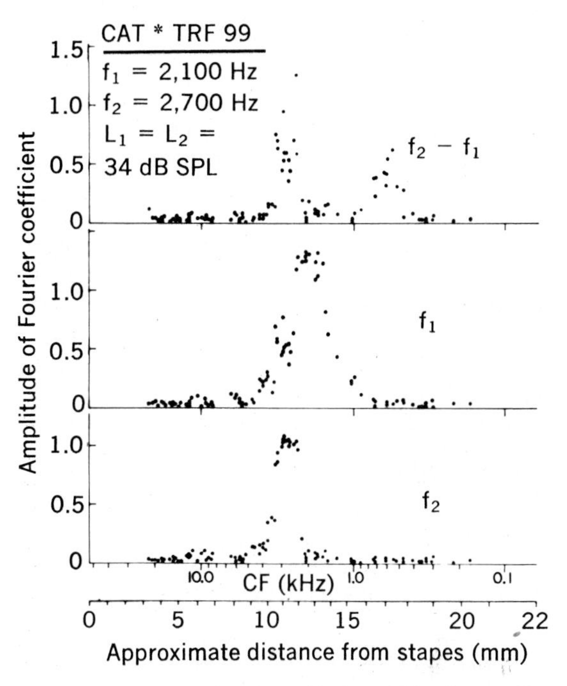

Fig. 14-26. Normalized amplitudes of Fourier coefficients of $(f_2 - f_1)$, f_1, and f_2, components of period histogram of single cochlear nerve fiber responses to two tones, f_1 and f_2, plotted vs fiber characteristic frequency *(CF)* and distance from stapes. Individual dots in each plot represent different nerve fibers obtained from same cat. Stimulus level of two tones was 34 dB SPL. (From Kim, D. O., and Molnar, C. E.: Cochlear mechanics: measurements and models. In Tower, D. B., editor: The nervous system. Human communication and its disorders, New York, Copyright © 1975, Raven Press, New York.)

In this case, Spoendlin and Kiang have done intensive studies of, respectively, the anatomy and physiology of afferent neurons in cats. On anatomic grounds it might be expected that there would be two main neural populations (corresponding to fibers innervating the inner and outer hair cells), one outnumbering the other by about 20 to 1.[101,102] To date, no such dichotomy has been shown in the physiologic population. However, the work of several laboratories[24,60,97] clearly indicates that functional properties of cochlear afferent fibers are influenced by inner *and* outer hair cells. The anatomy shows no fibers connected to both inner and outer hair cells, but there are opportunities for interaction, for example, near the habenula where the fibers are unmyelinated and fibers to inner and outer hair cells are enveloped by a special type of satellite cell.[100,102]

In addition to the afferent fibers considered in this section, there are other neural systems innervating the cochlea. The autonomic nervous system sends nerve endings to the labyrinthine artery and its larger branches and also forms a dense terminal plexus in the area of the habenula perforata.[103] Another more extensively studied efferent system innervating the cochlea and intimately connected with the afferent fiber system is the olivocochlear afferent system. Anatomic and physiologic studies of this system are the topics of the next section.

OLIVOCOCHLEAR EFFERENT SYSTEM*

The auditory pathway includes descending as well as ascending tracts throughout its course. Two of these descending conduction systems project into the cochlea; they are the crossed and uncrossed olivocochlear fibers. A recent study[111] that used retrograde horseradish peroxidase tracer indicates that in the cat about 1,800 efferent fibers project to each cochlea; the major part of this projection seems to be ipsilateral (uncrossed). There is also, at least in rodents, a reticulocochlear pathway of efferent fibers ending in the cochlea.[84]

Morphologic considerations

By making selective lesions and observing patterns of terminal degeneration, anatomists have determined the manner in which the efferent fibers innervate the cochlea.[55,64,93,101] Results of findings in the cat are summarized in Fig. 14-17. Note that efferent innervation of the outer hair cells is radial and innervation of the inner hair cells is spiral. Also, the inner hair cell efferent synapses appear to be presynaptic on the afferent dendrites, and the outer hair cell efferent synapses are presynaptic on the hair cells.

A number of electron microscopic studies of the nerve endings on the hair cells have been made. One type of synaptic pattern found for guinea pig outer hair cells is shown in Fig. 14-27. The pattern for cat outer hair cells is similar. The first row of outer hair cells has the most abundant supply of efferent (type 2) synapses found in the cochlea. In the upper (more apical) turns, they gradually disappear from the second and third

*See references 28, 38, 80, 104, 111, and 117.

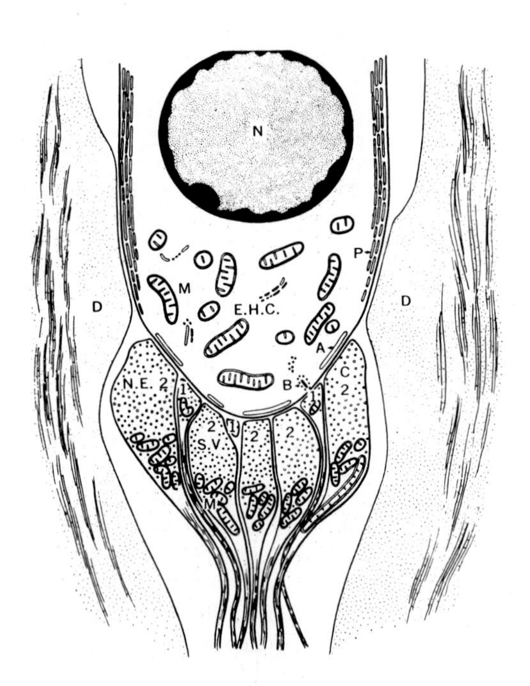

Fig. 14-27. Schematic drawing of outer hair cell from third cochlear turn of guinea pig. This pattern of synapses is typical for hair cells in inner row (all turns) and more basal portions of second and third rows. *A*, Accessory membrane; *B*, synaptic bar; *C*, region of vesicle concentration; *D*, Deiters' cell; *E.H.C.*, external hair cell; *M*, mitochondrion; *N*, nucleus; *1*, type 1 nerve ending; *N.E. 2*, type 2 nerve ending; *P*, vesicular peripheral membranes; *S.V.*, synaptic vesicles. (From Smith and Sjöstrand.[94])

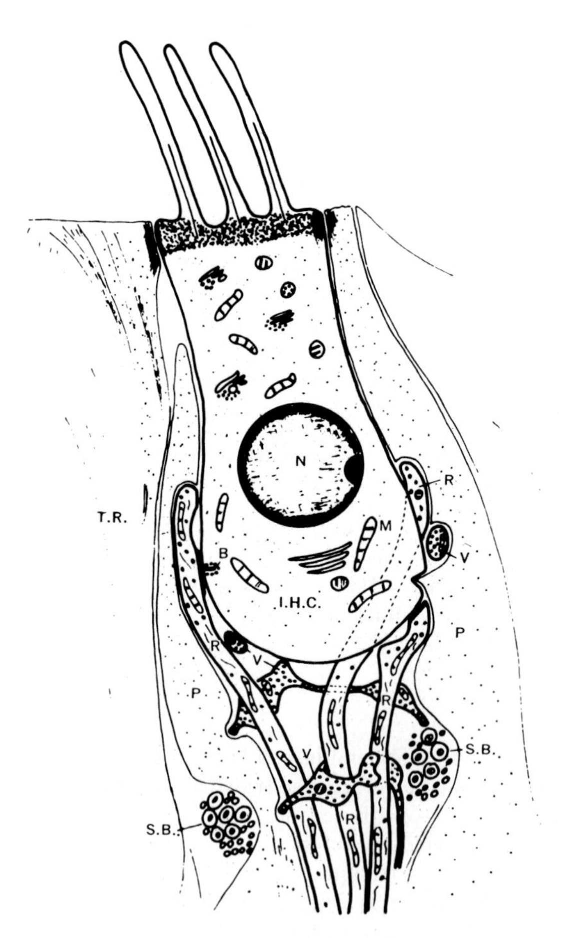

Fig. 14-28. Schematic drawing showing innervation of internal hair cell of guinea pig cochlea. *B*, Synaptic bar; *I.H.C.*, internal hair cell; *M*, mitochondrion; *N*, nucleus; *P*, phalangeal cell; *R*, radial nerve fiber; *S.B.*, spiral nerve bundle; *T.R.*, tunnel rod; *V*, vesiculated nerve with enlargement. (From Smith.[92])

row of hair cells.[100] The number of efferent endings on the outer hair cells has been estimated at 40,000.[99] Obviously there is a great deal of branching of the fibers of the olivocochlear bundle.[75]

The innervation of a typical internal hair cell is shown schematically in Fig. 14-28. The elongated terminals belong to the radial afferent fibers that terminate on the internal hair cells. The small, heavily vesiculated nerve endings rarely terminate on the hair cell, usually making synaptic contact with the afferent dendrites by means of vesiculated enlargements. These efferent endings belong to the small nerve fibers of the inner spiral bundle.

Serial reconstructions of electron micrographs indicate that for essentially all afferent synapses on inner hair cells there is a synaptic bar in the hair cell near the region of synaptic contact. In cats the synaptic bars are not consistently found near the afferent terminals on the outer hair cells.[99]

Electrophysiologic studies

This discussion of the electrophysiology of the olivocochlear pathways will have two main parts. First, the nerve discharge activity evoked by sound in the olivocochlear fibers will be described. Second, the effect of electrical stimulation of the olivocochlear bundle on activity of the auditory nerve fibers will be considered.

Fex[38] was able to record from individual fibers of the crossed olivocochlear bundle of cats at the vestibular-cochlear anastomosis. In order to place the electrode in the olivocochlear bundle under visual control, it was necessary to destroy the cochlea on that side. Thus acoustic stimulation of only the contralateral ear was possible, although the afferent fibers of the ipsilateral ear could be stimulated electrically.

Fig. 14-29 illustrates the pattern of response of an efferent fiber to a sinusoidal signal. The firing pattern in Fig. 14-29, *A*, is quite regular in contrast to the pattern of response of an afferent fiber, shown in Fig. 14-29, *B*. Rates of discharge of less than 50 spikes/sec, even with strong stimulation, and an absence of firing with no stimulation were also characteristic of these fibers. The tuning properties of the crossed efferent fibers indicated that axons going to the basal turn of the cochlea generally responded to higher frequencies than axons going to the apical turns. Thus excitation of a region of one cochlea would result in efferent activity at roughly the same region of the contralateral cochlea. Thresholds of excitation by sound were usually above 40 dB SPL, although a few fibers with lower thresholds were observed.

The crossed efferent fibers were also activated by electrical stimulation of the ipsilateral auditory nerve. When sound was presented to the contralateral ear during this electrical stimulation, responsiveness was increased, indicating a convergence of inputs from the two ears on the cells of origin of the crossed efferent pathway. The properties of the uncrossed efferent fibers are similar to those of the crossed fibers, with the following exceptions[39]: (1) many showed resting activity that with a few exceptions could be inhibited by a tonal stimulus presented to the contralateral ear; (2) inhibition of response to one tonal signal was produced by another tone of a different frequency; and (3) when activated by sound, the uncrossed efferents in the basal fascicle of the anastomosis generally responded to higher tone frequencies than did fibers in the apical fascicle, suggesting that afferents and uncrossed efferents from homotopic cochlear points in opposite cochleae are connected.

In another series of experiments, Fex studied the effects of electrical stimulation of the crossed olivocochlear bundle on single-unit activity of primary afferent fibers. The inhibition of the response of an afferent fiber excited by a tone is shown in Fig. 14-30. Here the crossed efferent fibers were stimulated by a tetanus of repetitive

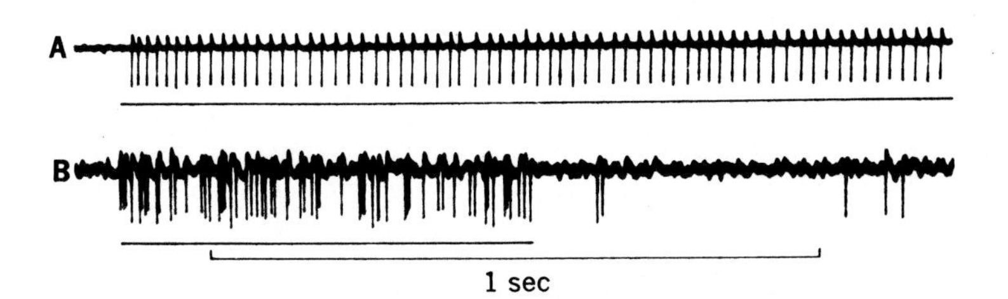

Fig. 14-29. Responses to tones from single efferent fiber, **A,** and from primary auditory fiber, **B.** Note regular firing pattern in **A** in contrast to irregular pattern in **B.** Records are from different experiments. Duration of sound stimuli is indicated by horizontal bars under records. (From Fex.[38])

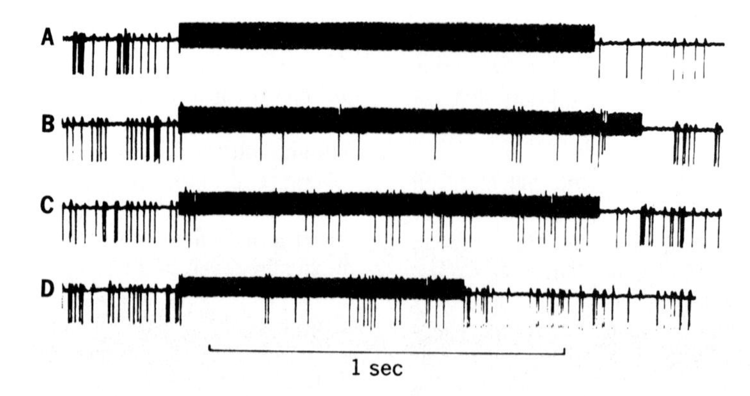

Fig. 14-30. Effect of tetanic electrical stimulation of crossed efferents on tone-evoked activity in primary auditory neuron. Primary afferent is stimulated with a tone of its best frequency, 950 Hz. In **A,** sound pressure was kept at constant level between 5 and 10 dB relative to threshold of fiber. Between each record, sound pressure was increased 5 dB. Tetanic bursts were applied to efferents. Note total inhibition in **A** and some poststimulatory inhibition after cessation of stimulation. Note that in **B, C,** and **D** there was dissipation of inhibition from efferent stimulation after about a quarter of a second. (From Fex.[38])

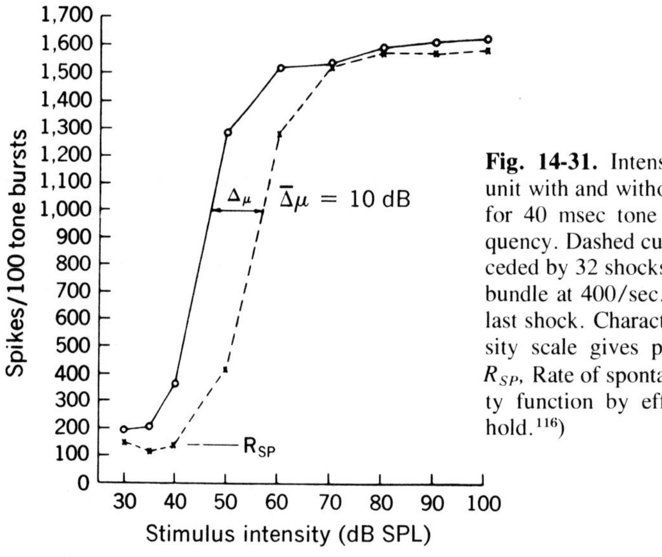

Fig. 14-31. Intensity functions for primary afferent unit with and without efferent stimulation. Solid curve for 40 msec tone bursts at unit's characteristic frequency. Dashed curves for same acoustic stimulus preceded by 32 shocks presented to crossed olivocochlear bundle at 400/sec. Tone burst starts at 10 msec after last shock. Characteristic frequency, 16.5 kHz. Intensity scale gives peak-to-peak sound pressure level. R_{SP}, Rate of spontaneous activity; $\Delta\mu$, shift of intensity function by efferent stimulation. (From Wiederhold.[116])

electrical shocks (rates from 250 to 425/sec were found most effective) delivered by electrodes placed in the fiber tract as it crosses the floor of the fourth ventricle. The effect of such activation of the efferent pathway was always inhibitory.

These results have been extended by Weiderhold[116,117] who demonstrated that in nearly all fibers in the cat the effect of shocks to the crossed efferent bundle on the intensity function of a primary afferent fiber is to shift the curve to the right. The result is shown in Fig. 14-31. The average amount of shift in the region of rapid growth of the intensity function was 10 dB in this case. The effect of efferent bundle stimulation and thus the shift varied from fiber to fiber, and values ranging from 1 to 25 dB were obtained. Shifts were greatest for units with characteristic frequencies in the range of 6 to 10 kHz.

Earlier, Galambos[45] found that the neural potential recorded from near the round window (action potential) may be reduced or abolished by tetanic stimulation of the crossed efferent fibers. Close observation reveals not only that the neural potential is abolished by stimulation of the efferent fibers but also that the microphonic component is increased in magnitude.[37] This effect, although small, is quite consistent and also can be demonstrated with tonal acoustic stimulation. The parameters of the tetanus most effective for reducing the neural potential are also most effective for increasing the size of the microphonic potential. Furthermore, the time course of the two effects is similar, as is their dissipation after cessation of the tetanus.[28] It has been demonstrated that the electrical stimulation of the uncrossed olivocochlear bundle will also reduce the size of the click-evoked action potential.[29] The effect is smaller than that observed with stimulation of the crossed bundle,[29,95] and there is no increase in the CM potential.[95]

All the phenomena just described that result from electrical stimulation of the olivocochlear fibers are diminished or disappear following intravenous injection of strychnine sulfate or the related alkaloid brucine.[30] A similar block of inhibition following administration of these drugs is seen for postsynaptic inhibition in the cat's spinal cord,[17] a phenomenon that suggests that the mechanism of inhibition in the two systems is similar (i.e., through release of a potassium and/or chloride permeability–increasing transmitter that has the effect of "clamping" the nerve membrane at a subthreshold potential). A recent report[31] indicates that for the crossed olivocochlear pathway the transmitter release leads to increased conductance to chloride.

Function of efferent fibers in audition

Note that in Fig. 14-31 electrical stimulation of the crossed olivocochlear bundle has very little effect on the response of the primary afferent fiber for intensities greater than 70 dB SPL. The result is typical and in keeping with the finding that the intensity function is shifted to the right, since intensity functions of most afferent fibers saturate at levels approximately 30 dB above threshold.[4] Similarly, with click stimulation there is a greater reduction of action potentials by olivocochlear bundle stimulation for low-intensity clicks than for high-intensity clicks.[28,95]

Noting such results, Dewson[32] hypothesized that the function of the crossed olivocochlear bundle is to raise the threshold of auditory nerve fibers, making them insensitive to masking noise but still responsive to signals of moderate or high intensity. He demonstrated, in cats, that the attenuation of the click-evoked action potential by noise masking or by electrical stimulation of the olivocochlear bundle did not summate when the conditions were delivered simultaneously.[32] A more dramatic result has been shown in guinea pigs.[73] In this case, action potential responses to moderately intense clicks were almost entirely masked by the addition of high-rate, low-intensity clicks. When the crossed olivocochlear bundle was stimulated electrically, the responses were "unmasked." Support for Dewson's hypothesis came from behavioral tests of rhesus monkeys before and after surgical section of the crossed olivocochlear bundles.[33] Discrimination of moderately intense vowel sounds presented in background noise was poorer following section of the bundles.

There has also been a suggestion, based on behavioral findings in squirrel monkeys, that the olivocochlear efferent system has a role in frequency discrimination.[18] Unfortunately in these studies the control discriminations were poor, although they did become even worse after transection of the olivocochlear bundle.

It is interesting to note that although the action potential is reduced by electrical stimulation of the olivochoclear bundles, attempts to demonstrate changes that could be attributed to the olivocochlear pathways in action potential recordings from cats in long-term studies have been without positive results. When the cats were asleep, awake, placid, or interestedly observing a mouse, no changes assignable to the efferent pathways could be observed in the characteristic N_1, N_2 neural-evoked potential.[14,46] Perhaps in view of the hypothesis that the efferent system acts to reduce masking by background noise, future experiments of this nature should employ masking and masked stimuli.

CONCLUSION

In this chapter, four aspects of the periphery of the auditory system have been considered: (1) the mechanical events by which an acoustic signal is transformed to a spatiotemporal pattern of vibration of the cochlear partition; (2) the mechanisms by which this vibration leads to excitation of the fibers of the cochlear branch of the eighth nerve; (3) the spatiotemporal patterns of nerve fiber discharges in the eighth nerve (that have been called coding), and (4) the olivocochlear efferent system.

In all these areas there are interesting questions yet to be answered. Perhaps intracellular studies will lead to further understanding of the mechanism of excitation of the auditory fibers. The efferent system is so often cited in speculations concerning the physiologic basis of behavior that it would be most helpful to have further experimental demonstrations of its functional significance. The coding of acoustic stimuli by the fibers of the eighth nerve cannot yet be described in general terms for a wide class of stimuli. Some fundamental nonlinearities have been brought to light, but the work to date is just a beginning.

The eighth nerve serves as the gateway to more central structures that are interconnected and anatomically organized in a most intricate manner. Within the central structures the patterns of activity of the eighth nerve fibers are processed and reprocessed in ways that are only beginning to be indicated by experimental investigation. The representation of acoustic signals in the CNS is one of the topics of the next chapter.

REFERENCES

General reviews

1. Dallos, P.: The auditory periphery, New York, 1973, Academic Press, Inc.
2. Davis, H.: Excitation of auditory receptors. In Magoun, H. W., editor, Neurophysiology section: Handbook of physiology, Baltimore, 1959, The Williams & Wilkins Co., vol. 1.
3. Keidel, W. D., and Neff, W.D., editors: Handbook of sensory physiology. Auditory system, Heidelberg, 1975, Springer-Verlag, vol. 5.
4. Kiang, N. Y-S., Watanabe, T., Thomas, E. C., and Clark, L. F.: Discharge patterns of single fibers in the cat's auditory nerve, Cambridge, Mass., 1965, The M.I.T. Press.
5. Møller, A. R., editor: Basic mechanisms in hearing, New York, 1973, Academic Press, Inc.
6. Ruben, R. J., Elberling, C., and Salomon, G., editors: Electrocochleography, Baltimore, 1976, University Park Press.
7. Tobias, J. V., editor: Foundations of modern auditory theory, New York, 1970 and 1972, Academic Press, Inc., vols. 1 and 2.
8. Tower, D. B., editor in chief: The nervous system. In Eagles, E. L., volume editor: Human communication and its disorders, New York, 1975, Raven Press, vol. 3.
9. von Békésy, G.: Experiments in hearing (research articles from 1928 to 1958), New York, 1960, McGraw-Hill Book Co.
10. von Helmholtz, H. L. F.: Die Lehre von den Tonempfindungen als physiologische Grundlage für die Theoric der Musik, ed. I, Brunswick, Germany, 1863, Friedr Vieweg & Sohn Verlagsgesellschaft. (Translated and adapted by Ellis, A. J.: Sensations of tone, New York, 1954, Dover Publications, Inc.)
11. Wever, E. G.: Theory of hearing, New York, 1949, John Wiley & Sons, Inc.
12. Wever, E. G., and Lawrence, M.: Physiological acoustics, Princeton, 1954, Princeton University Press.
13. Zwicker, E., and Terhardt, E., editors: Facts and models in hearing, New York, 1974, Springer-Verlag New York, Inc.

Original papers

14. Baust, W., Berlucchi, G., and Moruzzi, G.: Changes in auditory input during arousal in cats with tenotomized middle ear muscles, Arch, Ital. Biol. **102:**675, 1964.
15. Baust, W., Berlucchi, G., and Moruzzi, G.: Changes in the auditory input in wakefulness and during synchronized and desynchronized stages of sleep, Arch. Ital. Biol. **102**:657, 1964.
16. Beranek, L. L.: Acoustics, New York, 1954, McGraw-Hill Book Co.
17. Bradley, K., Easton, D.M., and Eccles, J. C.: An investigation of primary or direct inhibition, J. Physiol. **122:**474, 1953.
18. Capps, M. J., and Ades, H.W.: Auditory frequency discrimination after transection of the olivocochlear bundle in squirrel monkeys, Exp. Neurol. **21:**147, 1968.
19. Carmel, P. W., and Starr, A.: Acoustic and non-acoustic factors modifying middle-ear muscle activity in waking cats, J. Neurophysiol. **26:**598, 1963.
20. Cullen, J. K., Ellis, M. S., Berlin, C. I., and Lousteau, R. J.: Human acoustic nerve action potential recordings from the tympanic membrane without anesthesia, Acta Otolaryngol. **74:**15, 1972.
21. Dallos, P.: Dynamics of the acoustic reflex: phenomenological aspects, J. Acoust. Soc. Am. **36:**2175, 1964.
22. Dallos, P.: On the negative potential within the organ of Corti, J. Acoust. Soc. Am. **44:**818, 1968.
23. Dallos, P., Shoney, Z. G., Worthington, D. W., and Cheatham, M. A.: Cochlear distortion: effect of direct-current polarization, Science **164:**449, 1969.
24. Dallos, P., et al.: Cochlear inner and outer hair cells: functional differences, Science **177:**356, 1972.
25. Davis, H.: Biophysics and physiology of the inner ear, Physiol. Rev. **37:**1, 1957.
26. Davis, H.: Some principles of sensory receptor action, Physiol. Rev. **41:**391, 1961.
27. Davis, H., et al.: Modification of the cochlear potentials produced by streptomycin poisoning and by extensive venous obstruction, Laryngoscope **68:**596, 1958.
28. Desmedt, J. E.: Auditory-evoked potentials from cochlea to cortex as influenced by activation of the efferent olivo-cochlear bundle, J. Acoust. Soc. Am. **34:**1478, 1962.
29. Desmedt, J. E., and LaGrutta, V.: Function of the uncrossed olivo-cochlear fibers in the cat, Nature **200:**472, 1963.
30. Desmedt, J. E., and Monaco, P.: Suppression par la strychnine de l'effect inhibiteur contrifuge exerce par le faisceau olivo-cochleaire, Arch. Int. Pharmacodyn. Ther. **129:**244, 1960.
31. Desmedt, J. E., and Robertson, D.: Ionic mechanism of the efferent olivocochlear inhibition studied by cochlear perfusion in the cat, J. Physiol. **247:**407, 1975.
32. Dewson, J. H., III: Efferent olivocochlear bundle: some relationships to noise masking and to stimulus attenuation, J. Nuerophysiol. **30:**817, 1967.
33. Dewson, J. H., III: Efferent olivocochlear bundle: some relationships to stimulus discrimination in noise, J. Neurophysiol. **31:**122, 1968.
34. Engström, H., Ades, H. W., and Hawkins, J. E.: II. Structure and functions of the sensory hairs of the inner ear, J. Acoust. Soc. Am. **34:**1356, 1962.
35. Evans, E. F.: The effects of hypoxia on the tuning of single cochlear nerve fibers, J. Physiol. **238:**65, 1973.
36. Evans, E. F.: Cochlear nerve and cochlear nucleus. In Kiedel, W. D., and Neff, W. N., editors: Handbook of sensory physiology, Heidelberg, 1975, Springer-Verlag, vol. 5, part 2.
37. Fex, J.: Augmentation of the cochlear microphonics by stimulation of efferent fibres to cochlea, Acta Otolaryngol. **50:**540, 1959.
38. Fex, J.: Auditory activity in centrifugal and centripetal cochlear fibers in cat. A study of a feedback system, Acta Physiol. Scand. **189**(suppl.):1, 1962.
39. Fex, J.: Auditory activity in uncrossed centrifugal

cochlear fibers in cat, Acta Physiol. Scand. **64:**43, 1965.
40. Flock, Å.: Transducing mechanisms in the lateral line canal organ receptors, Symp. Quant. Biol. **30:**133, 1965.
41. Flock, Å.: Studies on the sensory hairs of receptor cells in the inner ear, Acta Otolaryngol. **83:**85, 1977.
42. Flock, Å.: Electron probe determination of relative ion distribution in the inner ear, Acta Otolaryngol. **83:**239, 1977.
43. Flock, Å., and Wersäll, J.: A study of the orientation of the sensory hairs of the receptor cells in the lateral line organ of fish, with special references to the function of the receptors, J. Cell Biol. **15:**19, 1962.
44. Flock, Å, Kimura, R. S., Lundquist, P. G., and Wersäll, J.: Morphological basis of directional sensitivity of the outer hair cells of the organ of Corti, J. Acoust. Soc. Am. **34:**1351, 1962.
45. Galambos, R.: Suppression of auditory nerve activity by stimulation of efferent fibers to cochlea, J. Neurophysiol. **19:**424, 1956.
46. Galambos, R.: Studies of the auditory system with the implanted electrodes. In Rasmussen, G. L., and Windle, W. F., editors: Neural mechanisms of the auditory and vestibular system, Springfield, Ill., 1960, Charles C Thomas, Publisher.
47. Geisler, C. D., Rhode, W. S., and Kennedy, D. T.: Responses to tonal stimuli of single auditory nerve fibers and their relationship to basilar membrane motion in the squirrel monkey, J. Neurophysiol. **37:**1156, 1974.
48. Goblick, T. J., and Pfeiffer, R. R.: Time domain measurements of cochlear nonlinearities using combination click stimuli, J. Acoust. Soc. Am. **46:**924, 1969.
49. Goldstein, J. L., and Kiang, N. Y-S.: Neural correlates of the aural combination tone $2f_1$-f_2, Proc. IEEE **56:** 981, 1968.
50. Goldstein, M. H., and Kiang, N. Y-S.: Synchrony of neural activity in electric responses evoked by transient acoustic stimuli, J. Acoust. Soc. Am. **30:**107, 1958.
51. Guinan, J. J., Jr., and Peake, W. T.: Motion of the middle ear bones, Quarterly Progress Report No. 74, Cambridge, Mass., 1964, The M.I.T. Press.
52. Guinan, J. J., Jr., and Peake, W. T.: Middle-ear characteristics of anesthetized cats, J. Acoust. Soc. Am. **41:**1237, 1967.
53. Harris, G. G., Frishkopf, L. S., and Flock, Å.: Receptor potentials from hair cells of the lateral line, Science **167:**76, 1970.
54. Hoshino, T.: Attachment of inner sensory cell hair to tectorial membrane, Ann. Otol. Rhinol. Laryngol. **38:**11, 1976.
55. Iruato, S.: Efferent fibers to the sensory cells of Corti's organ, Exp. Cell Res. **27:**162, 1962.
56. Johnson, D. H., and Kiang, N. Y-S.: Analysis of discharges recorded simultaneously from pairs of auditory nerve fibers, Biophys. J. **16:**719, 1976.
57. Johnstone, B. M., Taylor, K. J., and Boyle, A. J.: Mechanics of the guinea pig cochlea, J. Acoust. Soc. Am. **47:**504, 1970.
58. Khanna, S. M., and Tonndorf, J.: Tympanic membrane vibrations in cats studied by time-averaged holography, J. Acoust. Soc. Am. **51:**1904, 1972.
59. Kiang, N. Y-S., and Moxon, E. C.: Tails of tuning curves of auditory-nerve fibers, J. Acoust. Soc. Am. **55:**620, 1974.
60. Kiang, N. Y-S., Liberman, M. C., and Levine, R. A.: Auditory-nerve activity in cats exposed to ototoxic drugs and high-intensity sounds, Ann. Otol. Rhinol. Laryngol. **85:**752, 1976.
61. Kiang, N. Y-S., Watanabe, T., Thomas, E. D., and Clark, L. F.: Stimulus coding in the cat's auditory nerve, Ann. Otol. Rhinol. Laryngol. **71:**1009, 1962.
62. Kim, D. O., and Molnar, C. E.: Cochlear mechanics: measurements and models. In Tower, D. B., editor: The nervous system. Human communication and its disorders, New York, 1975, Raven Press, vol. 3.
63. Kimura, R. S.: Hairs of the cochlear sensory cells and their attachment to the tectorial membrane, Acta Otolaryngol. **61:**55, 1965.
64. Kimura, R. S., and Wersall, J.: Termination of the olivocochlear bundle in relation to the outer hair cells of the organ of Corti in guinea pig, Acta Otolaryngol. **55:**11, 1962.
65. Konishi, T., and Nielson, D. W.: The temporal relationship between motion of the basilar membrane and initiation of nerve impulses in auditory nerve fibers, J. Acoust. Soc. Am. **53:**325, 1973.
66. Laszlo, C. A., Gannon, R. P., and Milsum, J. H.: Measurement of the cochlear potentials of the guinea pig at constant sound-pressure level at the eardrum. I. Cochlear-microphonic applitude and phase, J. Acoust. Soc. Am. **47:**1063, 1970.
67. Liberman, M. C.: Auditory-nerve response from cats raised in a low-noise chamber, J. Acoust. Soc. Am. **63:**442, 1978.
68. Misrahy, G. A., De Jonge, B. R., Shinberger, E. W., and Arnold, J. E.: Effects of localized hypoxia on the electrophysiological activity of cochlea of the guinea pig, J. Acoust. Soc. Am. **30:**705, 1958.
69. Møller, A. R.: Acoustic reflex in man, J. Acoust. Soc. Am. **34:**1524, 1962.
70. Møller, A. R.: The sensitivity of contraction of the tympanic muscles in man, Ann. Otol. Rhinol. Laryngol. **71:**86, 1962.
71. Møller, A. R.: An experimental study of the acoustic impdeance of the middle ear and its transmission properties, Acta Otolaryngol. **60:**129, 1965.
72. Mulroy, M. J., Altman, D. W., Weiss, T. F., and Peake, W. T.: Intracellular electric responses to sound in a vertebrate cochlea, Nature **249:**482, 1974.
73. Nieder, P. C., and Nieder, I.: Crossed olivocochlear bundle: electric stimulation enhances masked neural responses to loud clicks, Brain Res. **21:**135, 1970.
74. Nomoto, M., Suga, N., and Katsuki, Y.: Discharge pattern and inhibition of primary auditory nerve fibers in the monkey, J. Neurophysiol. **27:**768, 1964.
75. Nomura, Y., and Schuknecht, H. F.: The efferent fibers in the cochlea, Ann. Otol. Rhinol. Laryngol. **74:**289, 1965.
76. Peake, W. T., Goldstein, M. H., and Kiang, N. Y-S.: Responses of the auditory nerve to repetitive acoustic stimuli, J. Acoust. Soc. Am. **34:**562, 1962.
77. Pfeiffer, R. R., and Kim, D. O.: Cochlear nerve fiber responses: distribution along the cochlear partition, J. Acoust. Soc. Am. **58:**867, 1975.
78. Pierce, J. R., and David, E. E.: Man's world of sound, New York, 1958, Doubleday & Co., Inc.
79. Portmann, M., and Aran, J. M.: Electrocochleography, Laryngoscope **81:**899, 1971.
80. Rasmussen, G. L.: The olivary peduncle and other fiber projections of the superior olivary complex, J. Comp. Neurol. **84:**141, 1946.
81. Rhode, W. S.: Observations of the vibration of the basilar membrane in squirrel monkeys using the Mössbauer technique, J. Acoust. Soc. Am. **49:**1218, 1971.

82. Rhode, W. S., and Robles, L.: Evidence from Mössbauer experiments for nonlinear vibration in the cochlear, J. Acoust. Soc. Am. **55:**588, 1974.
83. Rose, J. E., Brugge, J. F., Anderson, D. J., and Hind, J. E.: Phase-locked response to low frequency tones in single auditory nerve fibers of the squirrel monkey, J. Neurophysiol. **30:**771, 1967.
84. Rossi, G., and Cortesina, G.: Research on the efferent innervation of the inner ear, J. Laryngol. Otol. **77:**202, 1963.
85. Rupert, A., Moushegian, G., and Galambos, R.: Unit responses to sound from the auditory nerve of the cat, J. Neurophysiol. **26:**449, 1963.
86. Russell, I. J., and Sellick, P. M.: Tuning properties of cochlear hair cells, Nature **267:**858, 1977.
87. Sachs, M. B.: Stimulus-response relation for auditory nerve fibers: two-tone stimuli, J. Acoust. Soc. Am. **45:**1025, 1969.
88. Sachs, M. B., and Abbas, P. J.: Rate versus level functions for auditory-nerve fibers in cats: tone-burst stimuli, J. Acoust. Soc. Am. **56:**1835, 1974.
89. Sachs, M. B., and Kiang, N. Y-S.: Two-tone inhibition in auditory-nerve fibers, J. Acoust. Soc. Am. **43:**1120, 1968.
90. Sando, I.: The anatomical interrelationships of the cochlear nerve fibers, Acta Otolaryngol. **59:**417, 1965.
91. Simmons, F. B.: Perceptual theories of middle ear muscle function, Ann. Otol. Rhinol. Laryngol. **73:**724, 1964.
92. Smith, C. A.: The innervation pattern of the cochlea: the internal hair cell, Ann. Otol. Rhinol. Laryngol. **70:**504, 1961.
93. Smith, C. A., and Rasmussen, G. L.: Recent observations on the olivocochlear bundle, Ann. Otol. Rhinol. Laryngol. **72:**489, 1963.
94. Smith, C. A., and Sjöstrand, F. S.: Structure of the nerve endings in the external hair cells of the guinea pig cochlea as studied by serial sections, J. Ultrastruct. Res. **5:**523, 1961.
95. Sohmer, H.: A comparison of the efferent effects of the homolateral and contralateral olivo-cochlear bundles, Acta Otolaryngol. **62:**74, 1966.
96. Sohmer, H., and Feinmesser, M.: Routine use of electrocochleography (cochlear audiometry) on human subjects, Audiology **12:**167, 1973.
97. Sokolich, W. G., Hammernik, R. P., Zwislocki, J. J., and Schmiedt, R. D.: Inferred response polarities of cochlear hair cells, J. Acoust. Soc. Am. **59:**963, 1976.
98. Soudijn, E. R., et al.: Scanning electron microscopic study of the organ of Corti in normal and sound-damaged guinea pigs, Ann. Otol. Rhinol. Laryngol. **85**(suppl. 29): 1976, 1-58, No. 4, part 2.
99. Spoendlin, H.: The organization of the cochlear receptor, Basel, 1966, S. Karger.
99a. Spoendlin, H.: The innervation pattern of the organ of Corti, J. Laryngol. Otol. **81:**717, 1967.
100. Spoendlin, H.: Structural basis of peripheral frequency analysis. In Plomb, B., and Smörenburg, G. F., editors: Frequency analysis and periodicity detection in hearing, Leiden, The Netherlands, 1970, A. W. Sijthoff International Publishing Co.
101. Spoendlin, H.: Degeneration behavior of the cochlear nerve, Arch. Klin. Exp. Ohr.-Nas. U. Kehlk. Helilk. **200:**275, 1971.
102. Spoendlin, H.: Innervation densities of the cochlea, Acta Otolaryngol. **73:**235, 1972.
103. Spoendlin, H., and Lichtensteiger, W.: The adrenergic innervation of the labyrinth, Acta Otolaryngol. **61:**423, 1966.
104. Starr, A.: Influence of motor activity in click-evoked responses in the auditory pathway of waking cats, Exp. Neurol. **10:**191, 1964.
105. Tasaki, I.: Nerve impulses in individual autitory nerve fibers of guinea pig, J. Neurophysiol. **17:**97, 1954.
106. Tasaki, I., and Spyropolous, C. S.: Stria vascularis as a source of endocochlear potential, J. Neurophysiol. **22:**149, 1959.
107. Teas, D. C., Eldredge, D. H., and Davis, H.: Cochlear responses to acoustic transients: an interpretation of whole-nerve action potentials, J. Acoust. Soc. Am. **34:**1438, 1962.
108. Tonndorf, J., and Khanna, S.: The role of the tympanic membrane in middle ear transmission, Ann. Otol. Rhinol. Laryngol. **79:**743, 1970.
109. Tonndorf, J., Khanna, S. M., and Fingerhood, B. J.: The input impedance of the inner ear in cats, Ann. Otol. Rhinol. Laryngol. **75:**752, 1966.
110. von Békésy, G., and Rosenblith, W. A.: The mechanical properties of the ear. In Stevens, S.S., editor: Handbook of experimental psychology, New York, 1951, John Wiley & Sons, Inc.
111. Warr, W. B.: Olivocochlear and vestibular efferent neurons of the feline brain stem: their location, morphology and number determined by retrograde axonal transport and acetylocholinesterase histochemistry, J. Comp. Neurol. **161:**159, 1974.
112. Weiner, F. M., and Ross, D.: The pressure distribution in the auditory canal in the progressive sound field, J. Acoust. Soc. Am. **18:**401, 1946.
113. Wersäll, J., and Flock, Å.: Physiological aspects of the structure of vestibular end organs, Acta Otolaryngol. **192**(Suppl.):85, 1964.
114. Wersäll, J., Flock, Å., and Lundquist, P. G.: Structural basis for directional sensitivity in cochlear and vestibular sensory receptors, Symp. Quant. Biol. **33:**115, 1965.
115. Wever, E. D., and Vernon, J. A.: The effect of the tympanic muscle reflexes upon sound transmission, Acta Otolaryngol. **45:**433, 1955.
116. Wiederhold, M. L.: Variations in the effects of electric stimulation of the crossed olivocochlear bundle on cat single auditory-nerve fiber responses to tone bursts, J. Acoust. Soc. Am. **48:**966, 1970.
117. Wiederhold, M. L., and Kiang, N. Y-S.: Effects of electric stimulation of the crossed olivocochlear bundle on single auditory-nerve fibers in the cat. J. Acoust. Soc. Am. **48:**950, 1970.
118. Wilson, J. P., and Johnstone, J. R.: Capacitive probe measures of basilar membrane vibration. In Symposium on Hearing Theory, IPO, Eindhoven, The Netherlands, 1972, Eindhovensche Drukkerij, p. 172.
119. Young, E. D., and Sachs, M. B.: Recovery of single auditory-nerve fibers from sound exposure, J. Acoust. Soc. Am. **50:**94, 1971.
120. Zwislocki, J.: Analysis of the middle-ear function. I. Input impedance, J. Acoust. Soc. Am. **34:**1514, 1962.
121. Zwislocki, J.: Analysis of some auditory characteristics. In Luce, R. D., Bush, R. R., and Galanter, E., editors: Handbook of mathematical psychology, New York, 1965, John Wiley & Sons, Inc., vol. 3.

15

VERNON B. MOUNTCASTLE

Central nervous mechanisms in hearing

The mechanical transducer action of the ear implies that the frequency of a pure tone determines the profile of active elements in the distributed array of eighth nerve fibers. Other stimulus properties determine the overall frequencies, durations, temporal modulation, and internal timing in the trains of impulses in those active fibers. From this time-dependent distribution, humans with normal hearing make accurate discriminations for spectral pitch and loudness (the perceptual derivatives of frequency and intensity), locate the positions of sounds in space, and identify the time and intensity modulations of complex sounds (e.g., the virtual pitch of complex tones, consonance, and musical intervals).

The transduction of sound made in the cochlea and the neural transforms of it reaching the central nervous system (CNS) are not themselves sufficient to explain the human capacity to hear. This is obvious from the fact that even a pure tone of low intensity sets up an active region that extends for a considerable length along the cochlear partition. Thus there cannot be isolated zones of activity in the neural field of eighth nerve fibers for each discriminable tone. In this, as in other afferent systems, two stimuli placed close together on a peripheral receptor sheet elicit two partially shifted but overlapping profiles of activity in the relevant sets of afferent nerve fibers and in the central neural populations to which the latter are linked. Two sounds that differ in frequency by a just discernible interval are equivalent to two profiles of disturbance slightly separated in space along the cochlear partition. It is surmised that the difference in the two distributions of neural activity evoked by them provides a detectable cue for a central neural apparatus capable of reading the two neural profiles to a fine degree. A major problem in auditory physiology is to understand how that central mechanism operates so well to differentiate between sets of neural activity that may differ so slightly.

A description is first given of the human capacity to hear. A review of the anatomy and physiology of the central auditory system follows, in which evidence from both anatomic description and electrophysiologic experiments is blended to elucidate the central representation of the cochlear partition, and thus of frequency, at various levels of the system. Studies of the remaining capacity of animals and man to hear after lesions of the central auditory pathway are used to indicate the differing functional capacities of which those various levels are capable. A description will then be given of the dynamic pattern of neural activity evoked in the central auditory system by sounds, which might provide the basis for discriminations of pitch, loudness, lateralization, etc. Some discussion will be devoted to the subject of attention to a sound stimulus as a prototype of the more general problem of how one may attend to one sensory input to the relative exclusion of others.

THE HUMAN CAPACITY TO HEAR[7,53]

The measurement of the sensory attributes of auditory stimuli and of their relation to the physical characteristics of sounds is the business of auditory psychophysics. A major problem in this endeavor is to determine along which scales subjective sensory experience should be rated. The physical scale of the stimulus is most commonly used, but the total sensory experience is determined not only by the physical characteristics of the stimulus, but also by its peripheral transduction and by the further transformation imposed on primary afferent activity by the CNS. The human capacity to hear relative to the physical dimensions of sounds heard is described to establish what is to be understood in terms of the central neural mechanisms in hearing.

Sensitivity

The lower group of curves of Fig. 15-1 covers the range of audibility curves determined for human observers in a number of different ways.[53]

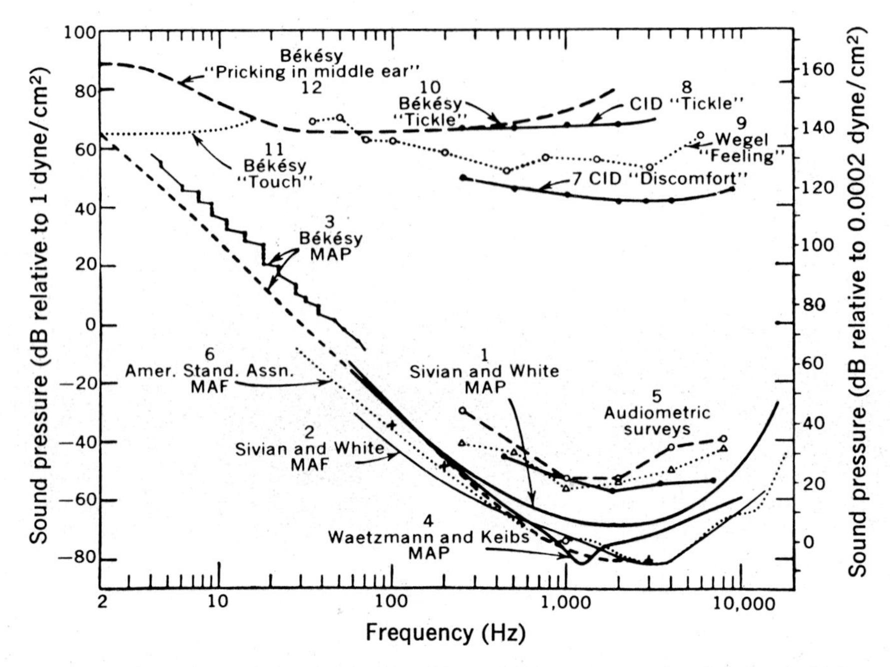

Fig. 15-1. Determinations of threshold of hearing at different frequencies and threshold for nonauditory sensations produced by sound: definition of auditory area. Curves *1* to *6* represent various measurements of threshold audibility curve; audiometric surveys may include individuals with impaired hearing. *MAP* = minimum audible pressure at eardrum; *MAF* = minimum audible pressure in free sound field measured in space occupied by observer's head during test. Curves *7* to *12* represent different measurements of threshold of nonauditory sensations produced by sounds and which are uncomfortable. (Curves collected from various sources by Licklider.[53])

The lowest thresholds for humans are between 1,000 and 3,000 Hz, where they are 100 to 10,000 times lower than at the lower and upper ends of the sensitivity curve, respectively. The threshold for hearing in this best frequency range is generally about 0.0002 dyne/cm^2 sound pressure in young adults free of ear or central nervous disease. The two ordinates of Fig. 15-1 indicate the two ways in which sound stimuli are commonly scaled for intensity. A reversed scale often used in human audiometry expresses hearing loss as the difference in decibels between threshold for normal individuals and that of a particular subject tested at a series of frequencies. Very intense sounds may evoke discomfort and/or mechanical sensations; these levels are indicated by the upper set of curves in Fig. 15-1. The area between the two sets of curves in the figure is the dynamic range of human hearing and is sometimes called the auditory area.

Auditory sensations are also produced when vibrations are delivered to the bones of the head, and this is the basis of certain types of hearing aids. The effect is produced by the bone conduction of audiofrequency vibrations directly to the cochlear partition.[91] The thresholds for hearing by bone conduction are about 50 to 60 dB above those for hearing by air conduction; this low sensitivity results from the gross impedance mismatch between the skull and the surrounding air.

Hearing with two ears is significantly better than with one and at moderate sound pressure levels approaches perfect summation. The neural inputs from the two ears thus sum for both loudness and for subliminal stimuli.[37] With advancing age there is commonly a loss of hearing, especially for frequencies higher than 3,000 to 4,000 Hz. The threshold for hearing is limited by the properties of the ear and of the CNS, not by the physical characteristics of the air. A range of about 10 dB separates the human threshold at best frequency from that level of sensitivity at which the random thermal agitation of gas molecules would become audible.

A question of theoretical interest is how that physiologic process set up by a just audible sound differs from that evoked by a sound just weak enough to be inaudible. This matter is treated in Chapter 20 from the standpoint that what is threshold is a continuous variable, and it is described in the context of signal detection theory. Some investigators[7] propose that the difference

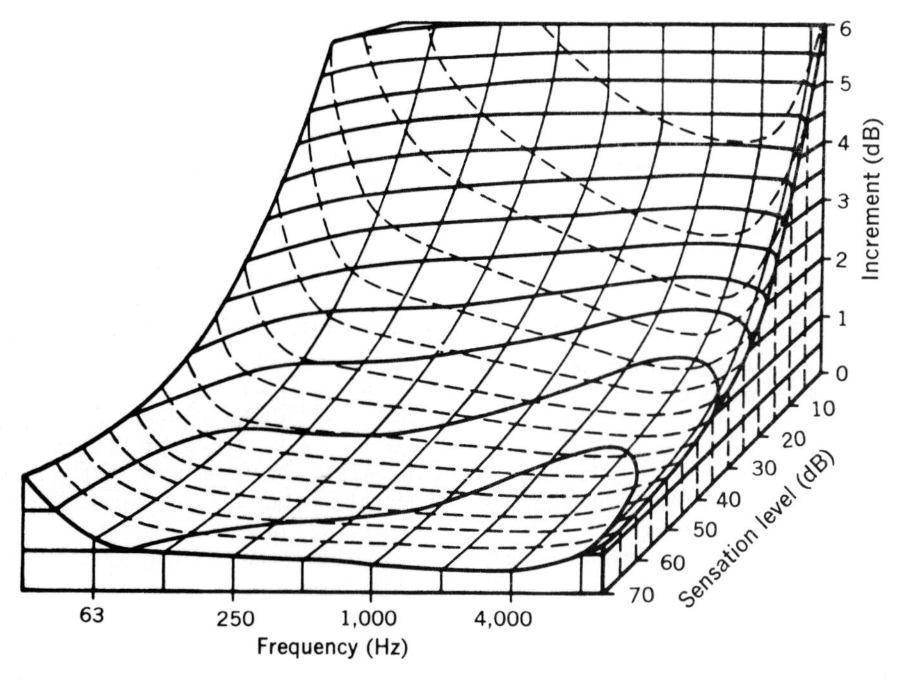

Fig. 15-2. Three-dimensional graph showing differential intensity threshold as function of frequency and intensity of standard tone. Threshold is given as difference in decibels between standard intensity and standard plus increment. (Based on data by Riesz, 1928; from Licklider.[53])

between the two sets of neural activity is a single "neural quantum," an elementary increase not further defined in their psychophysical experiments, but regarded by them as a step change in neural activity that raises the total to a supraliminal level. The two ideas are compatible, for neural activity must certainly increase by some quantal jump, in the limit by one additional impulse discharge by one or several neurons. The decision theory model may describe the changing level along the neural quantal scale at which a central detecting apparatus accepts the decision—*yes,* a stimulus has been delivered, vs *no,* it has not. It is this decision level that fluctuates and that is influenced by factors other than stimulus intensity.

Difference thresholds and loudness functions (scaling)[7,97]

The difference threshold or the difference limen (DL) is the measure of how small an increase in intensity of a sound an average listener can detect as an increase in loudness. The DL for pure tones varies with both the frequency and the intensity of the sound when its loudness is compared to that of a standard stimulus. Observers can detect changes as small as 1 dB difference in sound intensity for tones well above threshold and in the best frequency range for human hearing. If the intensity of the comparison stimulus is lowered, however, the detectable increment as a fraction of the strength of the comparison stimulus rises rapidly. If the comparison intensity is raised, the DL drops rapidly, reaching less than 0.2 dB at 100 dB sound pressure levels (SPLs), as shown in Fig. 15-2.[97] Thus the DL for loudness is not constant over the intensive continuum, nor is it a constant fraction of the comparison stimulus. These facts lead to the nonlinear Weber function for sound intensity shown in Fig. 20-7, p. 611.

Loudness is the intensive attribute of an auditory sensation in terms of which sounds may be ordered on a scale from soft to loud. Such a scale cannot be constructed by adding the number of DLs over the intensity range of audible sounds. A suitable scale can be made by asking observers to assign numbers to sounds of different intensities (magnitude estimations) or to adjust the intensity of a sound to produce a loudness that matches a series of numbers (magnitude production). When this is done, the results show that a loudness ratio of about 2 : 1 is produced by a pair of stimuli that differ in intensity by about 10 dB, that is, by a ratio of 3 : 1. This relation has been shown to hold over a wide range of audible intensities,[46] and it follows that the loudness (L) of a sound can be approximated by a power function of the intensity of the sound (I), as follows.[88]

$$L = kI^{0.3}$$

where I is measured in units of energy and k is a constant of proportionality. The exponent is therefore 0.6 when sound is scaled in units of pressure, as was done for the graph of Fig. 15-3. Some alinearity in the loudness functions (in log-log relation) is usually observed for soft sounds in the intensity region up to 40 dB above threshold. This deviation for weak tones indicates that a proper loudness function must take into account the loudness required of a tone that is just at threshold. Thus a loudness function most accurate for all intensities up to at least 100 dB is as follows:

$$L = k(I^a - I_0^a)$$

where I_0 is the threshold intensity and a is the exponent of the power function.

The accepted scale for loudness commonly used is the *sone* scale; a sone is defined as the loudness produced by a 1,000 Hz tone 40 dB above 0.0002 dyne/cm².

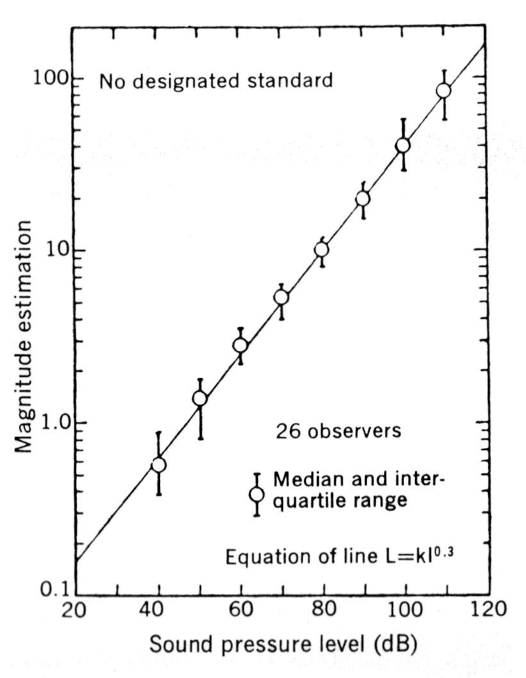

Fig. 15-3. Loudness function for human observers determined by method of magnitude estimation. Each subject was asked to assign a number to each of a series of tones in accordance with his estimate of its intensity. Number scales for different observers were normalized and averaged data plotted as shown. For tones above 40 dB relative to 0.0002 dyne/cm², loudness function is almost perfectly described by power function of form shown, plotted here in log-log coordinates. Exponent $n = 0.3$ is in energy units. (From Stevens.[88])

Pitch of a pure tone and pitch discrimination

Pitch is the subjective attribute of hearing that is in part determined by the frequency of the sound heard. The pitch experience is evoked by sounds of sufficient intensity over a frequency range from 15,000 to about 20,000 Hz. Very weak tones of any intensity evoke no experience of pitch. This *atonal interval* covers the first 3 to 5 dB above threshold on the intensity scale. Above that, further changes in intensity produce changes in subjective pitch of no more than ±2% of stimulus frequency.[19] The precise identification of pitch requires tonal stimuli of 15 to 20 msec duration, a time that does not change with frequency. Thus the lower audible tones can be identified after only one or a few cycles are heard, very high ones only after many.

Loudness and pitch differ in a fundamental way.[7] Loudness is an attribute scaled along a prothetic continuum: an increase in loudness produces increases in the rates of activity in a centered and slowly expanding population of neurons. Pitch, on the other hand, is an attribute scaled along a metathetic, or extensive, continuum: successive changes in the frequency of sound can be presumed to result in successive shifts in neural space of the activated sets of cells within the accessible neural populations. The cochlear partition is mapped onto the sensory projection areas of the auditory system in such a way as to preserve ordinal relations, which must also be preserved in whatever other neural elaborations serve perception.

The validity of the identification of loudness and pitch as different types of sensation is shown by the fact that although the DL for the former is a curvilinear function of intensity, that of the latter is uniform over a considerable range of the auditory area. The DL for frequency is 2 to 3 Hz over an intensity range from 20 to 70 dB and for frequencies up to about 3,000 Hz. The DL rises very rapidly for weak sounds or for higher frequencies. Obviously these relations will lead to a nonlinear Weber function for pitch.

Localization of sound in space[47]

The ability of human beings to locate the position of a pure tone in space varies with the frequency of the tone and the location tested. The minimal detectable angle for a tone of 750 Hz varies from 1 degree at 0 degree azimuth (dead ahead) to 7 degrees at 75 degree azimuth.[61] It rises for frequencies of 2 to 4 kHz and falls to a minimum at about 6 kHz. It is always less for complex sounds or noise than for pure tones.

There are two possible cues when the head is fixed, because both the amplitude and the phase of a plane sound wave emanating from an eccentrically placed sound source differ at the two ears. Phase differences are significant and amplitude differences are negligible at low frequencies (up to about 2,500 to 3,000 Hz); at higher frequencies the reverse holds, for the sound shield of the head is for them more effective in increasing the interaural difference in amplitude, whereas the possible time differences due to phase shifts are quite small. The sensing of the inclination (elevation) of a sound source is poor when the head is fixed. In normal circumstances, it is greatly improved by the kinesthetic cues from neck muscles and the changes in level and phase at the two ears produced by head movement. Localization of sounds in the vertical plane and front-back discriminations are enhanced by the external ear; when it is absent or fixed flat to the head, discriminations in the vertical plane fall to chance levels.

The time and intensity cues can be separated by delivering two tones or short clicks to the ears by earphones, which allows independent control of the amplitude of each and the time interval between them. When the sounds delivered are identical in amplitude and synchronous in time, the subject describes a perceptual image of a sound located exactly in the center of the head. If the sound in one ear is more intense than that in the other, or leads in time, the sound image shifts toward that ear to a degree determined by the time difference. There is evidence that the lateralization of sounds in this manner is for the subject the same task as localizing sounds in external space. A lag in time can be compensated by an increase in the amplitude of the lagging sound, and vice versa; that is, a time-intensity trading relation exists. The ratio varies with different tasks, sound intensities, and frequencies, but an average figure of 20 to 30 μsec/dB fits many observations.[42,48] The mechanisms for sound localization and of time intensity trade are neural, not cochlear. They are described on p. 471.

Masking and the critical bands[82]

The pure tone threshold may be raised by the simultaneous presence of a second auditory stimulus, a phenomenon called masking. The degree of masking depends on the frequency components and the power of the masking stimulus, for as its frequency spectrum is widened around the frequency of the testing tone, at constant power, a width is reached at which there is no further increase in the degree of masking of the test tone. This width is known as the critical band. When a masking noise at the critical band is increased in intensity, with both test and masking sounds delivered to the same ear, the degree of masking increases linearly in proportion to the power of the masking noise, up to about 60 to 80 dB above threshold.[38,39] Above that level, masking noise segments acquire the capacity to mask testing tones of quite different frequency. This is due to a combination of nonlinear distortion in the cochlea itself and the attenuation of the test tone produced by the contraction of the middle ear muscles, evoked by the intense masking noise.

The critical band width for masking a pure tone is an exponential function of the distance along the basilar membrane, beginning at the helicotrema, that is, band width is a nearly linear function of frequency. The band width is equivalent in its frequency range to about 1 mm distance along the basilar membrane, and this may be regarded as the area over which the cochlea integrates the power of signals. The *masked threshold* is defined as that intensity of a testing tone required to produce a detectable change in the afferent signals produced by the critical band and the masked stimulus when the two are summed.

The attributes of loudness, pitch, and location in space are the simplest characteristics of audible sounds. Many other aspects of the sounds we hear in ordinary life (e.g., their roughness, timbre, and sharpness) are multidimensional in nature. They have been treated from the overall theoretical point of view,[90] but are still only vaguely understood in terms of neural mechanisms.

ANATOMY AND TOPOGRAPHY OF THE AUDITORY SYSTEM

A general problem for all sensory systems is to determine which afferent pathways, nuclear complexes, and cortical fields compose it. The definition of a sensory system is in a certain sense an abstraction, for the processing of sensory input and its integration with ongoing activity in other systems may involve very large parts of the CNS. Certainly some brain regions are more directly involved in the sensory mechanisms of audition than are others, particularly when the concept of what is sensory is separated from what is afferent but not directly sensory. The following criteria will be used: those regions that (1) can be shown to receive a major share of their input either directly from the eighth nerve or via subsequent central relays of eighth nerve activity and (2) can also be shown on other grounds to play a role in audition, for example, by behavioral testing of the capacity to hear after lesions, will be classified as auditory structures. This definition includes a group of auditory regions arrayed from the eighth nerve to the cerebral cortex, but excludes regions of the brain that may receive auditory input as one of several inputs, even though from the behavioral point of view such a region may play an important role in phenomena evoked by auditory stimuli.

A schematic outline of the auditory system is shown in Fig. 15-4 and some details of its brain stem components in Fig. 15-5. All eighth nerve fibers terminate within the cochlear nuclei (CN), most by branching to innervate each of its three divisions: the dorsal (DCN), posteroventral (PVN), and anteroventral (AVN) nuclei. No fibers of the eighth nerve project to any "higher" structures within the system. The ascending projections from the CN are many (the striae of Held and von Monakow, the trapezoid body), but can be grouped into the dorsal and ventral pathways. The dorsal pathway arises in the DCN, may make connections with the superior olivary complexes of each side, and projects strongly to the contralateral inferior colliculus (IC) via the lateral lemniscus, with a collateral projection to the nucleus of the lateral lemniscus. The ventral pathway is more complex, arises from several different nuclear groups in the AVN and PCN, and projects via the trapezoid body to the various subgroups of the superior olivary complexes of each side; some of its fibers enter the lateral lemniscus and reach the IC. The cellular components of the olivary complex project via the lateral lemnisci to the IC of each side. The IC projects on the medial geniculate body of the dorsal thalamus, which in turn projects on the auditory areas of the cerebral cortex.

A striking characteristic of the auditory system is that it is disposed in a combined series and parallel arrangement, in which some efferents from a nucleus may end in the next immediate nucleus, whereas others bypass it to terminate in a more central one. Moreover, efferent fibers descend from one level to the next or the next but one in the same bypass arrangement as that of the afferent system; they provide a neural substratum for an efferent control of afferent input at every level (Fig. 15-4, *B*). One final stage in this controlling system is the efferent olivocochlear bundle, described in Chapter 14. The ascending and descending connections of the system are in some cases reciprocal, in others not. An important characteristic of the afferent system is that the cochlea is represented more than once at each of its levels. Throughout there is a central core of a highly specific and localized representation of the cochlear end-organ and thus of sound frequencies. Surrounding this central core region are one, two, or even more additional representations of the cochlea, which may differ in functional characteristics from the central one. Final-

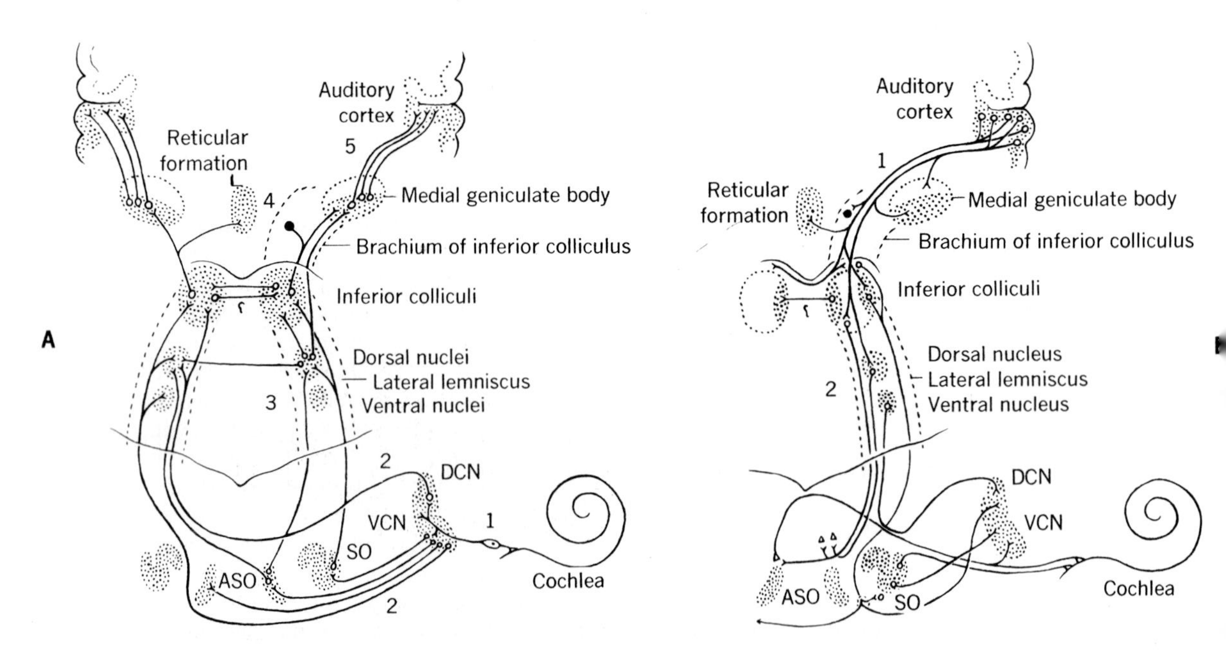

Fig. 15-4. Outline drawings of the auditory system. **A,** Ascending components. **B,** Descending components. *DCN,* Dorsal cochlear nucleus; *VCN,* ventral cochlear nucleus; *SO,* superior olive; *ASO,* accessory superior olive. (From Gacek.[30])

ly there is a collateral offshoot from the auditory system into the reticular formation of the brain stem and on the pontile, medullary, and spinal motor systems, providing in the first case avenues for influencing general forebrain excitability and in the second for more or less automatic auditory reflex control of certain motor mechanisms.

Cochlear nuclei[3]

It is possible using electrophysiologic methods to determine the frequency tuning curves of the single neurons encountered in microelectrode passages through central auditory structures. Such a curve is constructed by determining the intensity of a pure tone just sufficient to evoke an increase in the discharge of the cell under study for a number of tones of different frequencies. Determination of the tuning curves of cells observed seriatim in many microelectrode passages allows one to reconstruct the representation of frequency, and thus of the cochlear partition, in the structure under study. Studies of the CN using this method have revealed, as the triplicate projection of each eighth nerve fiber predicts,[77] that the full range of frequencies transduced by the cochlea is represented in each of the three divisions of the CN complex. High frequencies are represented dorsally and low frequencies ventrally in each division; the cochlear partition is represented unwound with basal end up and apex down. Sample tuning curves for CN fibers *(A)* and CN cells *(B)* in the cat are shown in Fig. 15-6. The range of sound frequencies to which a single neuron is sensitive depends on sound intensity, and at just suprathreshold intensity the range is very narrow around a "best" or "critical" frequency. Moreover, for many neurons the tuning curves are slightly asymmetric, sloping more rapidly on the high-frequency than on the low-frequency side of the critical frequency, an appearance exaggerated by the logarithmic method of plotting. Finally, the tuning curves of CN cells are no broader than are those of the eighth nerve fibers, despite the fact that the ratio between these first- and second-order elements is mildly divergent at 1:3.

A powerful factor tending to sharpen the tuning curves of some CN neurons is *afferent inhibition*. In the auditory as in the visual and somesthetic systems, afferent inhibition is topographically disposed in such a way that delivery of a stimulus that excites a certain number of cells in a neural field will suppress the activity of the surrounding cells, thus tending to limit the

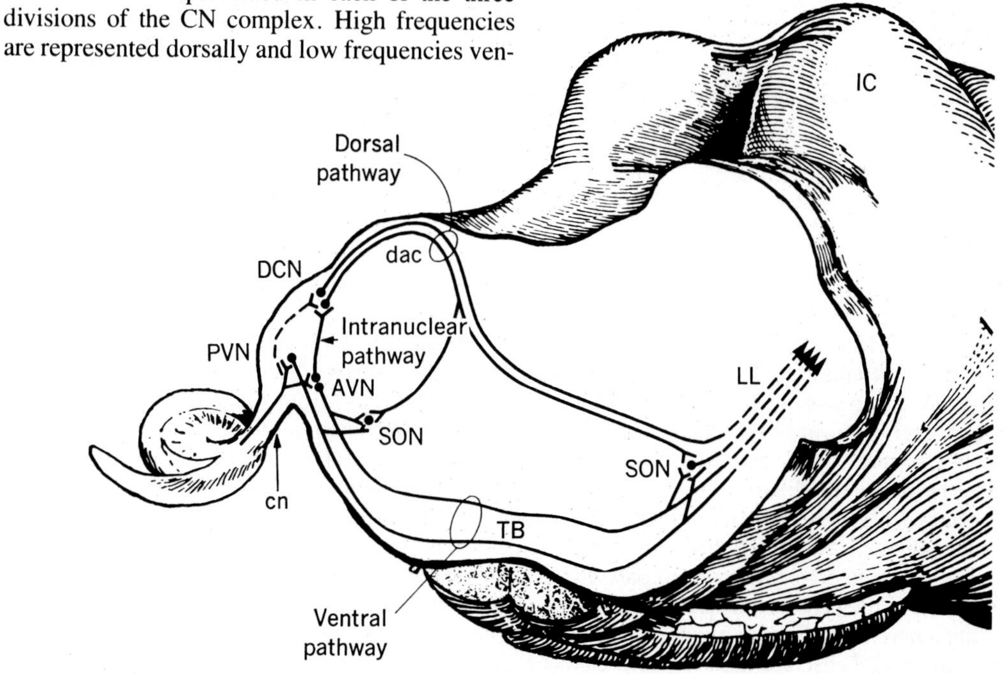

Fig. 15-5. Diagrammatic transverse section at level of cochlear nuclei of brain stem, illustrating dorsal and ventral second-order pathways of auditory system. *DCN, AVN,* and *PVN,* dorsal, anteroventral, and posteroventral divisions of cochlear nuclear complex; *dac,* dorsal acoustic striae; *SON,* superior olivary complex, including nucleus of trapezoid body; *TB,* trapezoid body; *LL,* lateral lemniscus; *IC,* inferior colliculus; *cn,* cochlear nerve. Second-order pathways are simplified. (From Evans and Nelson.[29])

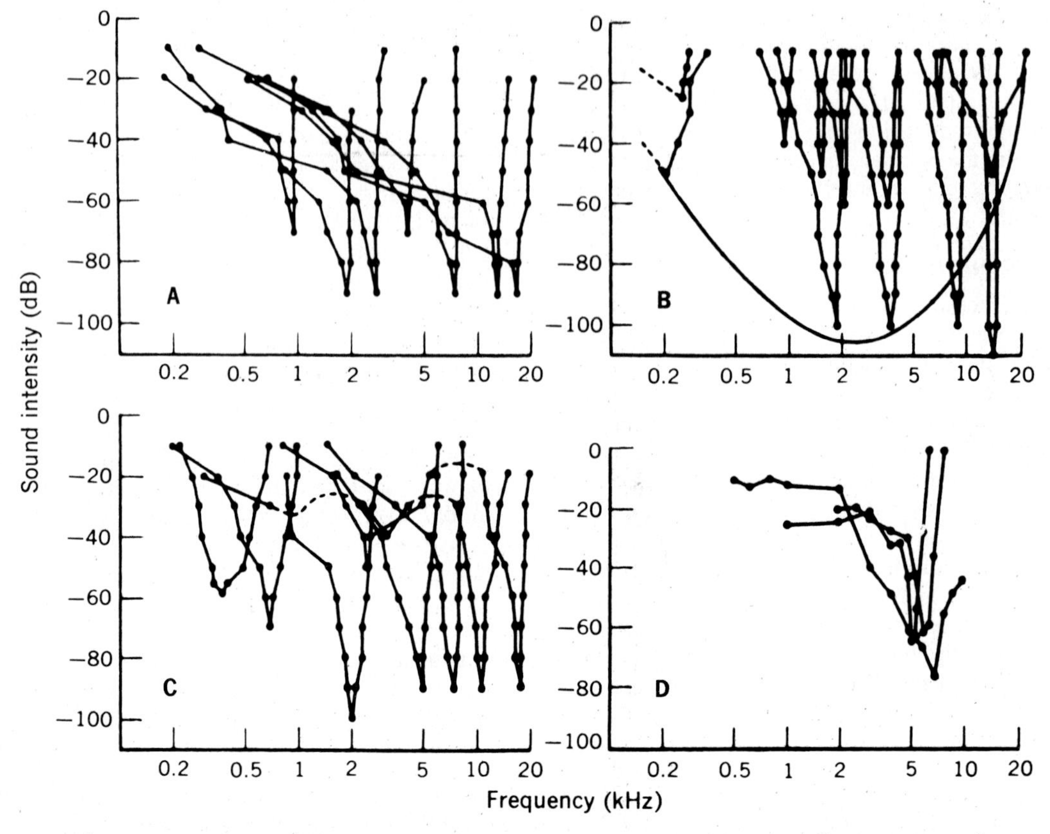

Fig. 15-6. Tuning curves for neurons located at various levels of auditory system of cat. **A,** Cochlear nerve fibers. **B,** Inferior colliculus. **C,** Neurons of the region of the trapezoid body. **D,** Medial geniculate body. Curve for each neuron plots, in decibels below a reference level, intensity of tones of different frequencies required to just excite it. Later, more extensive studies have shown that tuning curves of medial geniculate cells are no broader than are those of neurons at lower levels of auditory system. (From Katsuki.[49])

size and sharpen the edges of the active zone in the neural field. The phenomenon is illustrated by the graph of Fig. 15-7. In this case a single CN neuron was held under observation and a series of tones of different frequencies and intensities delivered as test stimuli. It is seen that tones either higher or lower in frequency than the best frequency suppressed the activity of the cell. In this case, as successive tones were delivered, the cell occupied a series of different positions in each successive neural population engaged, varying from the excited center to its suppressed surround. On the reciprocal interpretation, we conclude that a pure tone stimulus excites a population of neurons in a discrete locale of the CN and suppresses those that surround it. Evans and Nelson[28] have found, in their studies of unanesthetized cats, that afferent inhibition is a common property of cells of the DCN, but occurs rarely for cells of the two ventral divisions of the CN complex. The inhibition of cells of the DCN is exerted, at least in part, via an intranuclear inhibitory pathway that arises in the AVN and projects on the DCN.[29]

The neuronal processing mechanisms of the DCN and the two ventral divisions of the CN complex differ in other ways as well,[5] but the meaning for function of the triple representation of the cochlea at this first central stage of the auditory system is still unclear. Not only is surround inhibition a much more common property of DCN than of PVN or AVN cells, but many of them are *only* inhibited by sounds,[28] and others are inhibited by intense tones that at weak and intermediate intensities excite them. Otherwise, cells of the three divisions display tuning curves that are virtually identical in shape and sharpness.[73] Their thresholds at critical frequency are as low as those of cochlear nerve fibers in the cat and, considered as an ensemble, match the behavioral threshold for hearing in those animals. A puzzling aspect of the properties of CN cells,

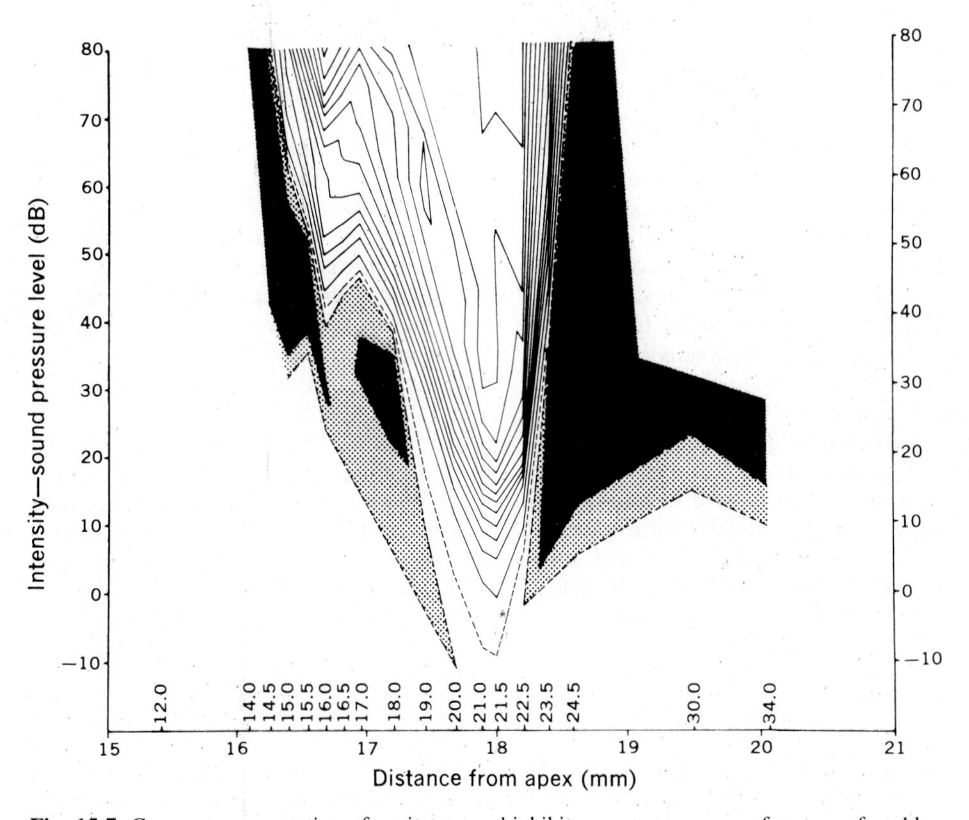

Fig. 15-7. Contour representation of excitatory and inhibitory response areas of neuron of cochlear nuclear complex of cat. Each solid contour line connects all frequency-intensity combinations that evoked same number of nerve impulses. Dashed line connects points at which number of impulses evoked exceeded mean of spontaneous rate of discharge by 1 SD; first solid line above dashed line is contour for total count of 200, and subsequent lines above represent counts that increase by counts of 100 impulses. Crosshatched areas indicate regions in which stimuli inhibited spontaneous activity by at least 3 SD. By making a reciprocal interpretation, this map for single neuron studied at series of different frequencies can be thought to represent distribution of activity in cochlear neural field produced by intense tone at single frequency: a center of very intense activity surrounded by inhibition. Abscissae: upper, frequency of testing tones in kilocycles; lower, estimated positions along basilar membrane for best response to given frequency. Ordinate: intensity of testing tones in decibels relative to 0.0002 dyne/cm^2. (From Greenwood and Maruyama.[40])

as of first-order fibers, is that, regarding stimulus intensity, they display a limited dynamic range (usually no greater than 40 to 50 dB), compared to the very broad range over which sounds of different intensity can be rated (p. 460). A number of studies have been made of the time-dependent aspects of the discharge of CN cells in response to tones, clicks, and more complex sound stimuli.[5,32,33,52] A large variety of cells with different response patterns has been observed, but it is not yet possible to correlate these properties directly with nuclear location, neuronal cell type, or behavioral aspects of hearing. Although for low frequencies of sound, CN fibers and some CN cells may follow the sound waves beat for beat, there is little other evidence to support the "volley" theory of pitch discrimination. Study of neurons of more central parts of the auditory system suggests that frequency following does not occur there for other than the very lowest frequencies of the audible sound spectrum. To a first approximation, cells of the two ventral divisions of the CN resemble in their discharge properties primary afferent fibers, whereas cells of the DCN show the results of what is thought to be more complex "neural integration." On the basis of these and other facts, Evans[3] has suggested that the ventral pathway of Fig. 15-5 is predominantly concerned with processing neural activity related to temporal information, sound localization, and reflex actions, whereas the DCN may represent the first stage in "higher level" processing of auditory input. The latter remains to be defined more precisely.

Auditory regions of brain stem and thalamus[2,4]

The precise representation of the cochlear partition within the CN is preserved in one or more of the nuclei of each more central complex of the auditory system. The DCN projects via the dorsal acoustic stria and the lateral lemnisci, contralaterally to the ventral and bilaterally to the dorsal nuclei of the lateral lemnisci. Few axons of the DCN cells reach the IC or the dorsal thalamus directly. The nuclei of the lateral lemnisci are tonotopically organized,[9] their cells are as sharply tuned to frequency as are those of the DCN, and neurons of its dorsal component are sensitive to interaural intensity differences.[13] Axons of cells of the ventral CN project via the ventral pathways on (1) the ipsilateral lateral superior olive and the preolivary nuclei, (2) the trapezoid body to the contralateral nucleus of that body, and (3) the lateral dendrites of cells of the ipsilateral and medial dendrites of cells of the contralateral medial superior olivary nuclei (Figs. 15-4 and 15-5). Thus each neuron of the medial superior olive receives input from each ear and projects its axon upward via the lateral lemniscus. Cells with this pattern of synaptic input are thought to provide signals of the spatial location of sounds (p. 471). Each of the olivary nuclei is tonotopically organized, and their neurons are sharply tuned to frequency.[4] The cells of the preolivary nuclei make connections with the reticular formation, with cranial motor nerve nuclei, and by descending pathways in the ventral columns with the spinal segmental apparatus. These descending connections are thought to serve reflexes evoked by sound that orient head and body toward its source.

The large majority of the neurons of the superior olivary–trapezoid group project via the lateral lemniscus to terminate in cells of its nuclei; more do so contralaterally than ipsilaterally. Few of these axons reach the IC directly, and probably none reach the medial geniculate body. The cells of the dorsal and ventral nuclei of the lateral lemniscus project in turn on those of the IC, although some make direct and many make collateral connections with neurons of the adjacent reticular formation.

The IC is a major component of the afferent and efferent auditory pathways. The lamination of its central nucleus and the patterning of the ascending axons as they enter it suggest a precise tonotopic representation,[64] and electrophysiologic studies reveal a precise replication of the audible sound frequency spectrum in both the central and the external nuclei.[10,59,78] The tuning curves of cells of the central nucleus are as narrow as those of the cochlear nerve fibers, a restriction that depends in part on collateral afferent inhibition (Fig. 15-6). Many collicular neurons are sensitive to the timing and/or intensity of sounds delivered to the two ears, like neurons of the medial superior olive, and are thus likely to play a role in the location of sounds in space.[79]

The medial geniculate body is an extrinsic nucleus of the dorsal thalamus. It receives auditory fibers from the IC and descending ones from the auditory cortex, on which it projects.[2] It has three divisions, identified in both cat and monkey. The ventral division is organized in a laminar manner,[65] it contains a precise tonotopic representation,[8,41] and its neurons have narrow tuning curves. The dorsal division is similarly organized and, in addition to its input from the IC, receives a fine-fibered ascending lateral tegmental tract of unknown origin. The dorsal and ventral divisions make up the *pars principalis* of some nomenclatures. The medial or *magnocellular* division is a region of quite different cytoarchitecture and receives converging inputs from both the auditory and the somatic sensory systems. The central core of the system at the thalamic level is the ventral division. It emits an essential thalamocortical projection to the auditory cortex and receives from it a precisely ordered reciprocal corticothalamic projection. Both the central core and the regions that surround it project on a number of other auditory fields in a sustaining fashion.[75]

Auditory areas of the cerebral cortex

The peripheral auditory apparatus executes a spectral analysis of sound in such a way that frequency is specified by the position of the activated elements within the total array of eighth nerve fibers. This *place* representation of frequency is preserved through a central core of the subcortical components of the auditory system and in several of its cortical projection areas. The most intensively studied of these is the first or "primary" auditory area, which is surrounded by a number of other fields also dominated by auditory inflow.

The primary auditory region (AI) has been defined in terms of its distinctive cytoarchitecture, its thalamic relations, and by the evoked potential method, described in Chapter 9.[72,74,75] In man, it occupies the transverse temporal gyrus of Heschl, Brodmann's areas 41 and 42,[17] as shown in Fig. 15-8, *A*. Stimulation of this area in waking humans elicits auditory illusions that are almost always referred to the contralateral

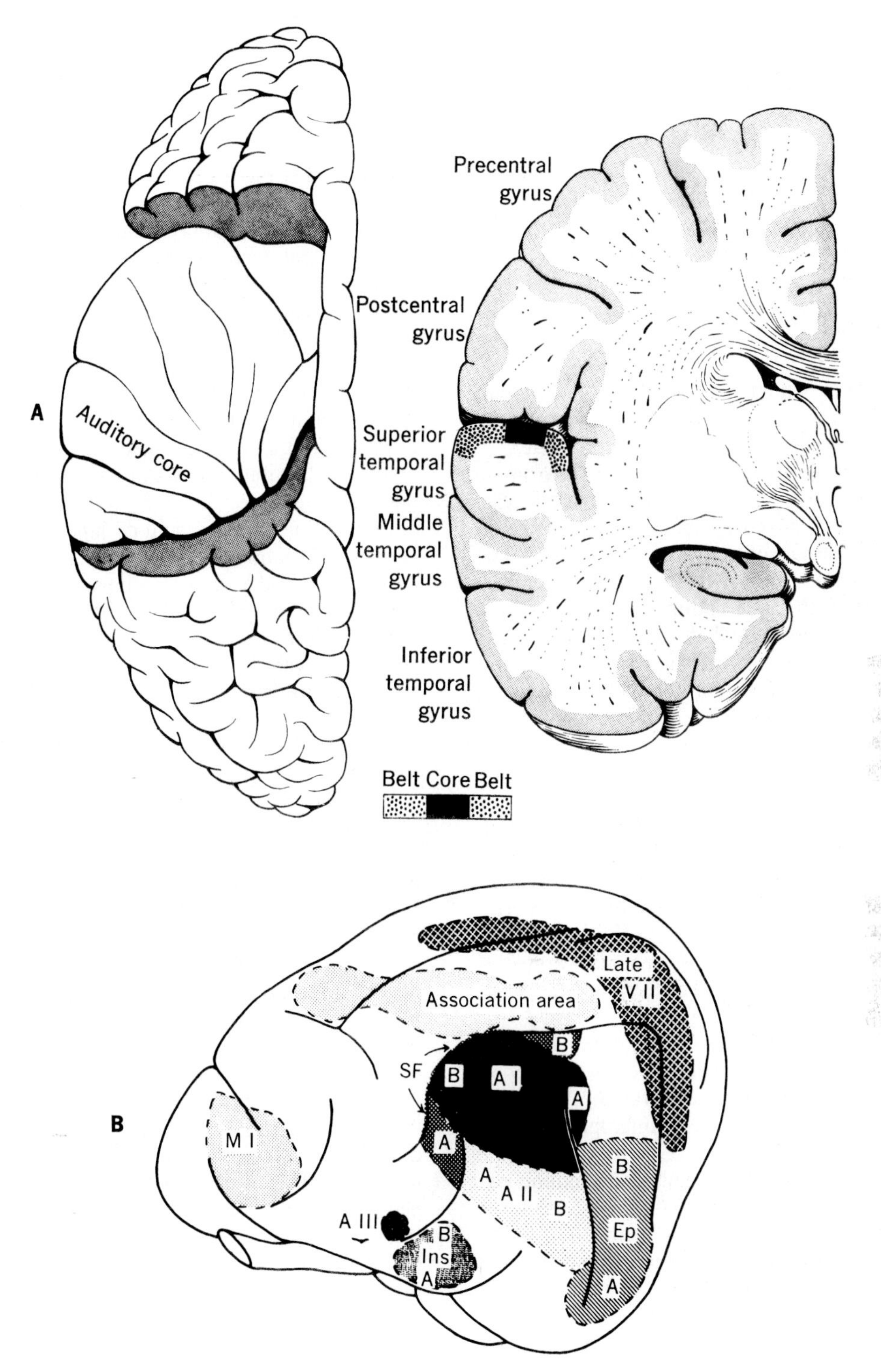

Fig. 15-8. A, Drawings to show location of the human auditory cortex in first transverse gyrus of temporal lobe, the "auditory core." On left, parietal lobe has been cut away to reveal superior temporal plane, seen from above. On right, frontal section passing through auditory cortex showing core area buried in sylvian fissure, surrounded by a belt area. **B,** Summary diagram showing areas of cat cerebral cortex that can be activated by auditory input under variety of experimental conditions. In each area, A = apex and B = base of cochlea (i.e., low and high frequencies, respectively). Central core is auditory I, *A I,* partially surrounded by auditory II, *A II,* ectosylvian posterior area, *Ep,* and suprasylvian sulcus, *SF.* Both insular, *Ins,* and infratemporal fields lateral to *A II,* not indicated in this drawing, receive patterned input from cochlea. Small area marked *A III* also activated by sound, but whether in detailed pattern is unknown. Association and motor cortices, *M I,* respond to sound under special conditions and in all-areas or no-areas fashion, as described in text: they are thought to be activated via generalized thalamocortical system. *Late* indicates visual area II, *V II,* of cortex, within which responses to brief auditory stimuli occur with 100 msec delay under chloralose anesthesia. (**A** from Neff et al.[6]; **B** from Woolsey.[94])

ear.[69] The most extensive studies of AI have been made in cats, for in these animals the area is exposed on the lateral surface of the hemisphere. Fig. 15-8, *B*, shows the auditory areas identified in the cat by the evoked potential method and summarizes the extensive studies of this area by Woolsey.[94] This drawing shows that AI is nearly surrounded by at least three other areas activated by sound. AI and the surrounding belt of auditory fields have also been identified in the monkey.[68]

Studies in which the method of single-unit analysis has been used have revealed a number of important facts concerning the auditory cortex:

1. The majority of auditory cortical neurons have narrow tuning curves that are frequently flanked by inhibitory sidebands in the format of surround inhibition. The thresholds at best frequencies are comparable to those of auditory nerve fibers and thus to behavioral thresholds. Some cells are more complexly tuned (e.g., to frequency or amplitude modulation of sound stimuli), but the role of these neurons that appear to make higher order abstractions of stimulus properties in central auditory mechanisms is still unknown.
2. There is a precise tonotopic organization within AI in both cat and monkey and presumably in man. Frequencies are represented from high to low and from anterior to posterior, with partially shifted overlap. Each frequency is represented in a mediolateral band of cortex in both cat[60] and monkey.[58]
3. The auditory cortex is organized in a columnar fashion, as described in Chapter 9. The defining parameter for the columnar processing units in the anteroposterior dimension is frequency. Within each isofrequency strip the mediolateral defining parameter is composed of those aspects of sound stimuli and their neural transforms that signal sound location in space (see later discussion).[45]

It should be emphasized that the projection of the auditory system on the cerebral cortex is bilateral: the two cochleae are represented in registered overlay. However, thresholds for activation of auditory cortical neurons are 5 to 20 dB higher for ipsilateral than for contralateral stimuli in the cat.[44]

Auditory fields surrounding auditory area I. The map of Fig. 15-8, *A*, shows that AI is nearly surrounded by a cortical belt composed of three fields, within each of which there is a spatial representation of the cochlea. The tonotopic representation in the anterior auditory field, for example, is as precise as is that in AI,[50] and it is known that these surrounding fields may sustain behavior that depends on complex discriminations between auditory inputs in the absence of the central core area, as described in a later section. Removal of AI, or even small lesions within it, causes a degeneration of the ventral division of the pars principalis of the medial geniculate body. Small lesions within any one of the belt areas produces only insignificant degeneration in the medial geniculate body, whereas removal of all auditory fields results in degeneration of all the subdivisions of that thalamic nuclear complex.

In addition to the auditory areas described, certain association areas of the parietotemporal cortex may under some circumstances be activated by auditory as well as by other sensory stimuli. Little is known of the higher order integration of various sensory inputs presumed to take place in these cortical association areas.

Columnar organization of the auditory cortex

It is clear from the foregoing that the frequency of stimulating sounds is represented in the auditory cortex in an orderly manner in a series of isofrequency bands running across AI. The koniocortex of this region is the most markedly columniated of any heterotypical cortex; its radial arrangement of cells is obvious in sections cut normal to the cortical surface. Studies of the physiologic properties of single neurons of AI have revealed that cells encountered as a microelectrode passed down such a vertical column are tuned to a nearly identical frequency and that their tuning curves have sharp roll-offs in both directions (Fig. 15-6) due at least in part to lateral inhibition. The overlap of the tuning curves of neurons in adjacent isofrequency strips is sufficient to account for the partially shifted overlap of frequency representation observed in surface mapping experiments such as those from which the drawings of Fig. 15-8 were made.

It has now been shown in both cat[45] and monkey[12] that other properties of auditory stimuli are mapped along each isofrequency strip in a direction orthogonal to that of the change in frequency. In the high-frequency region, it is the properties of aural dominance and binaural interaction that vary along the long dimension of an isofrequency band. Neurons located along the vertical axis of the cortex, in a given locus within such a band, display similar properties of frequency sensitivity and binaural interaction. Binaural summation may produce a suppression of response as compared with monaural stimulation; in all such cases the contralateral ear is dominant. On the other hand, binaural stimulation may produce summation, and in this case, either ear may dominate. The suppression and summation columns are arranged in continuous bands placed more or less orthogonally to the isofrequency contours. Within each summation set there is a further columniation for which the identification parameter is ear dominance. It is well known that the interaural differences in intensity provide cues for localization in space of high-frequency sounds (i.e., those above about 3,000 Hz). The type of interaural interaction differs in the low-frequency regions, for neurons in these isofrequency strips appear to be most sensitive to interaural delay, and the property of characteristic delay (p. 475) is mapped across the isofrequency contours more or less orthogonally to the parameter of frequency. Thus it appears that there is a precise mapping onto the columns of the auditory cortex of those properties of sounds that allow their localization in the contralateral half-field of space. Those dynamic properties will be described in a later section.

REMAINING CAPACITY FOR HEARING AFTER BRAIN LESIONS[6]

The bilateral cerebral projections of the auditory pathways account for the fact that only bilateral lesions of central auditory structures of the forebrain produce obvious defects in hearing. These lesions occur only rarely in man, but a number have been identified for study before death, and anatomic studies have been made afterward. It is clear from them that bilateral, complete destruction of the cortical auditory areas or section of the auditory radiations results in near-total deafness.[18] This is evidence for the progressive phyletic encephalization of the auditory system, for such a remarkable deficiency in auditory function does not follow even bilateral and complete lesions in cats or monkeys. Patients with unilateral lesions of the auditory cortex show an elevated hearing threshold and defects in loudness and temporal order discriminations in the contralateral ear and a disorganization of the contralateral spatial field, with a decrease in the capacity to localize sounds within it.[81]

A major field of research is the study of auditory capacities of animals before and after surgical destruction of a part of the auditory system. Those capacities can be measured in nonhuman primates by combining operant conditioning methods with psychophysical ones.[84] Although many variations are used, the experimental protocol is generally as described here.

Monkeys are trained to sit quietly in restraining chairs with heads fixed and earphones applied. They learn to initiate a behavioral trial by depressing one key. After a variable delay a pure tone is delivered; a response on a second key during the presentation of the tone is rewarded with food or liquid. The intensity of the tone is varied and, from the results obtained, a psychometric function can be constructed by plotting the percent of correct detections as a function of tone intensity. Threshold is taken as the point at which 50% of the tones are detected. DLs for sound intensity and frequency can be determined in somewhat similar protocols. For tests of the capacity to localize sounds, an animal may be required to identify or move toward the position of a source of sound, to be rewarded for correct choice. A recent advance has been to combine this behavioral method with electrophysiologic ones by recording simultaneously slow-wave potentials from the surface of the head or the electrical signs of the activity of single neurons of the auditory system via implanted microelectrodes. Results obtained in these ways will be noted in a later section.

An example of the results of studies of this sort made by Stebbins[85] are shown in Fig. 15-9. They show that the macaque's hearing range extends about one octave above that of man, to nearly 45 kHz. Similar threshold functions have been determined using the reaction time paradigm.[71] At frequencies below 8 kHz the monkey's absolute sensitivity is comparable to that of man to changes in either frequency or intensity. Measurements of this sort are made before and after brain lesions, with the hope that correlation between changes in sensory performance and the location and extent of the brain lesion will indicate the role played in audition by the part destroyed. Postoperatively the animals are tested for retention of the learned act and the capacity for relearning it, if lost. The method is a geographic or anatomic one and is not expected to disclose underlying dynamic neuronal mechanisms. Some findings obtained with its use are summarized in the following paragraphs.

Stimulus detection and intensity discrimination[6]

After removal of all the cortical auditory areas, cats[67] and monkeys[24] retain or learn again (1) to make a conditional response to the appearance of a sound and (2) to discriminate between two sounds of different intensities. Thresholds are usually elevated very little or not at all. If lesions of the IC are added to those of the cortex or if the brachium of the IC is transected bilaterally, the

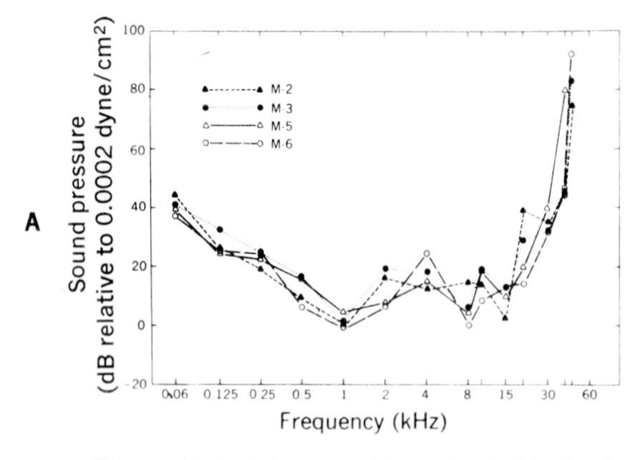

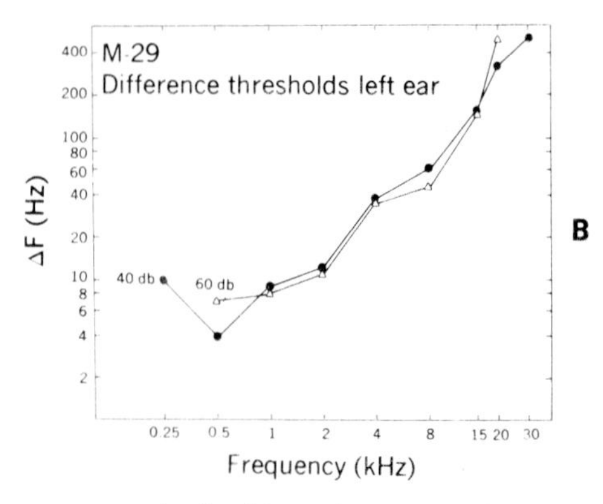

Fig. 15-9. A, Normal auditory thresholds for four macaques, animals able to hear sound frequencies as high as 45 kHz, at least one octave above upper limit for humans. Sensitivity functions are otherwise quite similar to those for humans, except for slight elevation at about 4 kHz. Compare with human sensitivity functions shown in Fig. 15-1. **B,** Frequency difference thresholds (ΔF) for one macaque, shown as function of base frequency, at intensity levels 40 and 60 dB above absolute threshold. Macaque's capacity to discriminate between sounds of different frequencies is only slightly less than that of humans. (**A** from Stebbins et al.,[86] copyright by American Association for the Advancement of Science; **B** from Stebbins.[84])

DLs may be increased by 7 to 10 dB. Thus it appears that in these animals the tonotopically organized thalamocortical component of the auditory system is not necessary for sound detection and intensity discrimination, measured in conditioning paradigms, but that when intact, it ensures a low threshold and a low DL in these two tests, respectively. This is true also in human patients with even large bilateral lesions of the temporal cortex, but those patients in whom anatomic studies have shown lesions to be complete are described as behaviorally deaf. Obviously, further studies are needed of the auditory capacity of humans with lesions of both temporal lobes, to be correlated with detailed histologic studies of their brains after death.

Complete, bilateral transection of the ascending auditory pathways at the level of the lateral lemnisci in cats produces animals that can detect sounds only if they are 90 to 100 dB above preoperative thresholds.[6] Sounds of this intensity are likely to activate somatic sensory mechanoreceptors, as well as the cochlea. The conclusion suggested is that the auditory centers of the pons and medulla do not provide neural mechanisms necessary for a conditional response to auditory stimuli, although they are sufficient for mediating descending reflex mechanisms activated by sound.

Simple frequency discrimination

Removal of all tonotopically organized cortical areas in the cat eliminates a learned avoidance response that depends on correct discrimination between tones of different frequency. Such animals are capable of relearning frequency discriminations even when the task requires the recognition of stimulus frequency as well as the detection of frequency change.[21] Bilateral section of the brachium of the IC in cats is followed by loss of and incapacity in relearn frequency discriminations.[36] It is not yet clear why transection of the ascending input to the thalamocortical components of the auditory system produces a more severe deficit in frequency discrimination than does complete removal of auditory cortical areas, which is, of course, followed by retrograde degeneration of auditory thalamic regions.

The results of similar experiments in monkeys have produced quite different results, for in them a learned response to tones of different frequencies is lost completely. The deficit is severe and permanent; thresholds for even the simplest frequency discrimination remain elevated 15 to 20 times preoperative levels, even with prolonged training.[54] It seems likely that in man an even more severe deficit exists, for patients with proved total bilateral lesions of auditory cortical areas are described (and describe themselves) as deaf. However, such complete bilateral lesions occur only rarely, and patients with them have not yet been studied in a quantitative manner.

Discrimination of tonal pattern

Cats can be trained to discriminate between sequences of tones of the same frequencies and intensities but of different order (e.g., ABA vs BAB). Such a test requires identification of the tone sequences and their retention in short-term memory for comparison. This capacity is lost and cannot be relearned after bilateral removal of all auditory cortical areas,[25] even though such animals can be retrained to detect tonal onset and to discriminate between different frequencies and between different frequency modulation sequences. Further study has shown that the central core of the auditory cortex, AI, is essential but not sufficient for this discriminatory behavior. Lesions of the insular and temporal auditory fields lateral to AI and AII lead to as severe and permanent deficiency in tonal pattern discriminations as that following removal of all auditory fields.[6] Moreover, there is some evidence that removals of insulotemporal cortex in cats produces severe deficits in temporal patterns of changes in sound intensity as well as of frequency and in making discriminations between different temporal patterns of visual and somatic sensory stimuli.[20]

A similar loss in the capacity to recognize auditory patterns occurs in monkeys after removal of homologous auditory association areas of the superior temporal gyrus of the temporal lobe.[87] The loss is particularly severe if the task requires discrimination between different speech sounds.[23] These results illustrate an important difference in function between auditory cortical areas. The central core region suffices for simple identification of frequencies and discrimination between them. Auditory tasks of greater complexity that require processing of information about temporal pattern and other complex qualities depend, in addition, on the auditory association fields of the temporal lobe. Even more complex disorders of auditory perception appear in humans with lesions of these regions (Chapter 22).

Localization and lateralization of sound in space[1]

The task of localizing in space the source of a sound differs in the head-fixed and head-free conditions. Normally the human or animal makes scanning head movements to elicit differences in the quality and intensity of the sound, whether listening with one or two ears, and fixation of the head causes a severe deterioration of performance. In either condition the subject with normal hearing uses differences in time of onset, phase, and intensity of sounds arriving at the two ears to derive information about the spatial location of sound sources. This suggests that the CNS structures essential for sound localization[57] are those that receive convergent input from the two ears, and studies of animals with brain lesions confirm that this is the case.

Transection of the trapezoid body, but of none of the higher commissures of the auditory system, produces a severe deficit in sound-localizing capacity, as do lesions of the superior olivary complex.[16,56,62] These findings are consonant with the fact that the auditory pathways converge on the superior olivary complex, as well as with electrophysiologic observations of a convergent excitatory-inhibitory projection from the two ears on cells of the superior olive (p. 466). Lesions of the IC or of its brachium produce a similar deficit.[89] Bilateral removal of the auditory cortical areas in cats and monkeys results in as severe a deficit as do lesions of afferent projection pathways, particularly when the animal is required to move through space to approach the source of the sound.[43,55,66] Unilateral lesions in both humans and experimental animals produce defects in the location of sound in the contralateral half-field of space.

In summary, animals and humans function at many levels of complexity with regard to audition, and in some forms when the sensory system is reduced by ablation of the auditory cortical areas, the remainder may suffice to process information about sound stimuli of simple configuration and thus provide cues for differential conditional response. Cats and monkeys, after complete bilateral removal of all auditory cortical areas, may still respond to the onset of sounds and to changes in the intensity or frequency of tones. It is highly unlikely that this is true in man, but the evidence is presently not decisive on this point. The cortical auditory areas are essential, in all mammals studied, for tasks in the sphere of audition that require for execution any of the following: (1) a short-term memory storage and the immediate or delayed recall of information about auditory stimuli; (2) the integration of neural events for short periods of time; (3) the identification of the duration of stimuli. Lesions at the lowest level of bilateral convergence in the system, the superior olivary complex and the commissure of the trapezoid body, lead to as severe deficits in the localization of sound as do removals of the cortical auditory areas.

For carnivores, auditory stimuli are obviously potent guides for behavior and gain priority in attention, in contrast to the visual domination among sensory inflow so characteristic of the life of primates. It appears likely that the progressive corticalization of the specific sensory systems in the latter means that in them the capacity for function of the subcortical components of the auditory system in the absence of cortex is much reduced over that in carnivores. Certainly this is the implication of reports of deafness in those few human subjects in whom postmortem studies have revealed a total destruction of the auditory cortical areas, but more quantitative studies of hearing in humans with temporal lobe lesions are required to elaborate this point.

DYNAMIC PROPERTIES OF AUDITORY NEURONS

In what has gone before I have dealt with the static properties of neurons of the auditory system, those determined by system connectivity and local synaptic microstructure. These properties persist unchanged no matter what the changes in the temporal patterns of neural activity. Notable among them are the frequency-tuning curves of individual neurons and the spatial representation of the cochlear partition at each level of the auditory system. Dynamic properties are, on the other hand, the temporal, spatial, and intensive aspects of neural activity. Study of these latter properties is aimed at understanding the dynamic and time-dependent operations of the nervous system. The major objective is to discover how the auditory performance of an organism can be explained in terms of these neural mechanisms.

The method used is that of single-unit analysis, described in Chapter 9. With this method the activity of central auditory neurons can be monitored in either anesthetized or in waking, behaving animals under a number of conditions, including those in which animals execute auditory detections and discriminations. The aim is to study serially a large and representative sample of neurons in a given part of the auditory system (e.g., the medial geniculate nucleus or the auditory cortex), so that a post hoc reconstruction can be made of the neural events occurring in that part in the response to auditory stimuli and correlations made between that activity and the stimuli that evoke it, as well as with concurrent behavioral events. There are formidable problems in how to measure and scale neuronal activity and how to make those correlations. The method most often used is that of analyzing the sequential time order in which neural impulses occur, combined with auto- and cross-correlation measures developed for this purpose.[26,63] The results obrained with these methods in the last decade have provided new insights into many of the central neural mechanisms in audition. Instances will be described in that which follows.

Spontaneous activity

Auditory neurons from the level of the eighth nerve to the cerebral cortex are spontaneously active in the absence of auditory stimuli. The origin of this activity at the cochlear end of the eighth nerve fibers is unknown. For central neurons, it is undoubtedly the result of both the spontaneous end-organ input and the influence of other systems on the auditory neurons. The central core regulating systems (Chapter 10) influence the degree of activity at all levels of the brain. How this central influence interacts with that of the spontaneous discharges in auditory nerve fibers is not clear, but it is known that the cells of some portions of the CN are still spontaneously active after destruction of the ipsilateral cochlea.[51] A third major influence that may condition spontaneous activity is that of the descending components of the auditory system, about which more will be said in a later section.

Given that spontaneous activity will vary with the general level of brain excitability, an important question is how spontaneous activity interacts with that evoked by auditory stimuli: are the two simply additive or is their interaction multiplicative? For neurons of the IC, both types of interaction have been observed[15]: for some cells there is simple addition of the two, for others there is a constant background-increment ratio. The form of those interactions at every level of the system is important for hypotheses about how a cortical discriminative mechanism may operate; it deserves further intensive study.

The interval sequences of the spontaneous activity of auditory neurons occur in different forms. For neurons at lower levels of the system, histograms of the intervals between nerve impulses are commonly poissonian in form, with a "dead time" to allow for neuronal refractoriness: the sequences approach randomness.[70] For other neurons the intervals cluster in quasi-gaussian distributions around the mean. At higher levels of the system the interval distributions observed are more variable, change with the state of alertness of the animal, and in sleep or other depressed states such as anesthesia are frequently bimodal in form. The meaning for function of the ubiquitous spontaneous activity in this, as in other central systems, is obscure.

Intensity functions of auditory neurons

The rate of discharge of auditory neurons, at all levels of the system, commonly increases rapidly along an S-shaped curve[11] when discharge rate is plotted against a log scale of stimulus intensity (Fig. 15-10, *A*). The peak

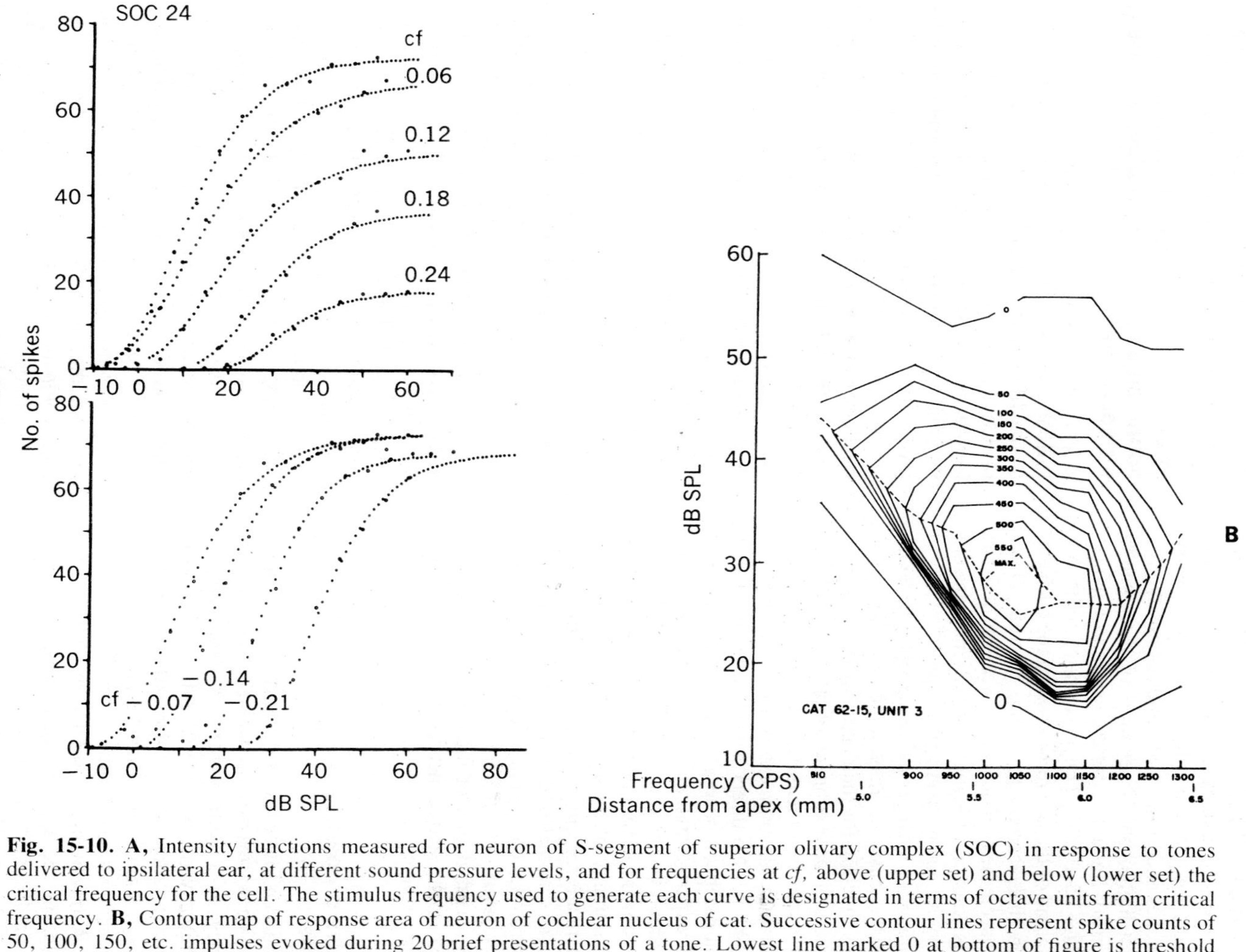

Fig. 15-10. A, Intensity functions measured for neuron of S-segment of superior olivary complex (SOC) in response to tones delivered to ipsilateral ear, at different sound pressure levels, and for frequencies at *cf*, above (upper set) and below (lower set) the critical frequency for the cell. The stimulus frequency used to generate each curve is designated in terms of octave units from critical frequency. **B,** Contour map of response area of neuron of cochlear nucleus of cat. Successive contour lines represent spike counts of 50, 100, 150, etc. impulses evoked during 20 brief presentations of a tone. Lowest line marked 0 at bottom of figure is threshold tuning curve for excitation of cell. Thus as intensity level of a tone effective for exciting cell was increased, there was first an increase in discharge of cell, and then, with stronger stimuli, a decrease: intensity-response function is not monotonic. (**A** from Boudreau and Tsuchtitani[11]; **B** from Greenwood and Maruyama.[40])

frequencies reached are higher for cells of lower levels of the system than for neurons of the auditory cortex. In all cases the dynamic range over which single neurons code stimulus intensity is 30 to 50 dB, which contrasts with the total range of 100+ dB over which humans can rate the intensities of sounds, as shown in Fig. 15-3. Moreover, for many auditory neurons at every level further increases in stimulus intensity produce not higher but lower rates of discharge[96]: the intensity-frequency functions are not monotonic (Fig. 15-10, *B*). This is surmised to be due to the spread of the active zone along the cochlear partition that recruits to action afferent fibers with adjacent best frequencies. Activity in those fibers is known to evoke central mechanisms inhibitory for the neuron under study.[35,40]

In an earlier section, evidence was presented that on the basis of magnitude estimation the loudness function for man approximates a power function with an exponent of about 0.6, with stimulus intensity measured in pressure units, and that that function extends over a range of about 100 dB. No direct neural correlate of that function has been observed in the auditory system of experimental animals. Increasingly intense stimuli produce increases in discharge for neurons tuned to stimulus frequency; with further increases, many of those neurons are inhibited, but others are recruited to the discharging population. It is surmised that a central mechanism must generate the overall input-output power function descriptive of human auditory behavior by some combination of intensive and extensive changes in the active neural field. How those population events are coded and interpreted is a mystery.

Adaptive properties of auditory neurons

Neurons at lower levels of the auditory system respond to a steadily maintained tone of best frequency in a wide variety of temporal patterns.[28] Many discharge a high-frequency transient at stimulus onset, after which the rate declines to a much lower level but may then persist for long periods of time as the stimulus continues. A much smaller proportion of cells, for example, in the CN complex, responds only with a brief burst to tonal onset; remains silent, although the stimulus continues; and may discharge a second burst when the stimulus is turned off. The sequence is thought to be due to a temporally mixed input complexed of both excitatory and inhibitory influences evoked by the stimulus. At higher levels of the auditory system a progressively higher proportion of cells responds to transients and fewer steadily to maintained stimuli. Like other sensory systems the auditory seems organized to detect and accentuate preferentially *changes* in the environment, the steady nature of which is signaled by only a portion of the available neural elements.

Phase-locking of auditory neurons to sound stimuli and the problem of coding for sound frequency

The most likely way in which the nervous system codes sound frequency is by place (i.e., by the differences in the spatial locations of zones of activity set up in central auditory structures by sounds of different frequency). There is a differential spatial mapping for frequency at every level from the cochlear nerve to the auditory cortex. Alternatively, it is possible that the discrimination for pitch depends on the discharge of neural elements in synchrony with each sound wave, thus providing a temporal code for frequency. Such a "volley theory" is untenable if it demands a discharge in any central neuron for each cycle of a sound wave, for the overall rate of discharge of those cells is seldom higher than 100 to 200/sec. However, study of the temporal sequences of discharge of auditory nerve fibers[76] and neurons at low levels of the auditory system[5,52,80,93] has revealed a period-time code. It has been shown that for those fibers and neurons with best frequencies below 2,000 to 3,000 Hz, neuronal discharges occur in a preferred relation to the phase of the stimulating sound wave. The periods of time between successive impulses in single elements are always integral multiples of the wavelength of the stimulating sound wave, even though the overall neuronal discharge rate is some small fraction of sound frequency. Given that adjacent elements responding to a tone respond to different successive cycles of the sound wave, it would be possible for a central detecting network to decode sound frequency by integrating the information carried in a number of converging elements. How such a detecting network might achieve that integration is unknown, for the time constants of the membranes of central neurons are long compared to the sound wavelengths. For sound frequencies above 2,000 to 3,000 Hz, this phase-locking does not occur, and for all higher audible frequencies, place remains the best possible candidate code for sound frequency. The periodic discharge evoked by sounds of low frequency may play a role in the phenomenon of periodicity pitch.[76,83]

Neural mechanisms for sound localization[1,27]

Experimental animals and human beings after unilateral loss of the auditory cortical areas show disturbances of sound localization in the contralateral half-field of space that may increase DLs from 3 to as much as 15 degrees. Humans with loss of hearing in one ear show a profound loss of the capacity to localize sounds when their heads are fixed in space. This suggests that both binaural input and the central interaction between those inputs are required for localization. The first binaural cue for localizing sound is the difference in timing (and thus of phase) between the arrival of a sound at the two ears, which might operate for low-frequency sounds, below 3,000 Hz. This difference amounts to about 29 μsec/cm of interaural distance. The second cue is the difference in the intensity of a sound at the two ears and is likely to be important only for those sounds (above 3,000 Hz) for which the head acts as an effective sound shadow.

The first location of the auditory system at which binaural interaction occurs is in the superior olivary complex. Each neuron of the medial superior olive receives a contralateral relayed input on its medially directed dendrite and an ipsilateral relayed input on its lateral dendrite. For most of these cells the former is excitatory and the latter inhibitory. A number of investigators have shown that the discharge rate of olivary neurons with low best frequencies (below 3,000 Hz) is strongly affected by variations in *interaural phase delay* and is maximal for each neuron at a "characteristic delay" that is different for different neurons. In contrast, neurons with higher best frequencies that are excited by input from the contralateral and inhibited by input from the ipsilateral ear are *sensitive to interaural intensity differences* and relatively insensitive to the overall intensity level.[34] The interaural phase delay for low-frequency neurons and intensity differences for high-frequency neurons will produce a net increase in the level of activity in the contralateral nucleus and a net decrease in activity on the ipsilateral side. The result is a weighting of ascending activity toward the contralateral side of the cerebral cortex for contralateral stimuli, thus providing a cue for lateralization. Within the total population of ascending elements of the superior olivary complex there is a spatial mapping for phase delay and intensity differences, which provide cues for localization within the contralateral half-field (Fig. 15-11).

These differential cues in the output of the superior olive are reflected in the response properties of neurons of higher levels of the auditory system on which they project, the nucleus of the lateral lemniscus,[13] IC,[79] the medial geniculate body[2] and the cerebral cortex.[12,14,45] For example, as the phase relation between the waves of a pure tone delivered to the two ears at equal intensity is changed, neurons of the IC with best frequencies below 3,000 Hz inscribe periodic alterations in discharge rate, there being a certain "characteristic delay" of the ipsilateral behind the contralateral, at which the net excitatory ef-

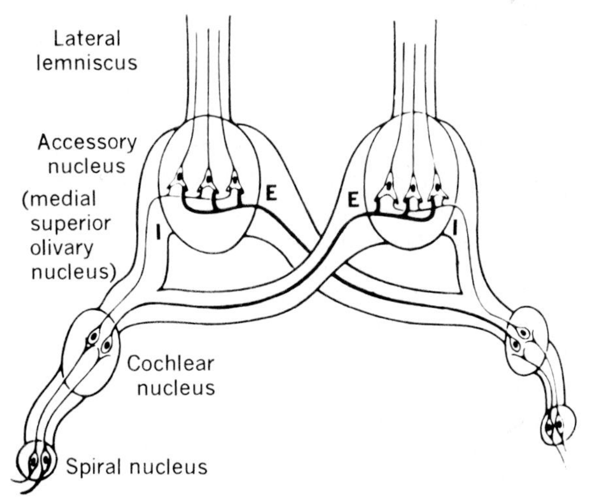

Fig. 15-11. Outline of features of von Békésy–van Bergeijk model of role neurons of medial superior olive may play in localization of sound in space. Neurons of cochlear nuclei predominantly inhibit cells of ipsilateral accessory nucleus and excite those of contralateral one. Thus if neural input from one ear is more intense than that from the other or leads it in time, ascending neural activity will be weighted toward contralateral side of auditory system. (From van Bergeijk.[92])

fect of the two stimuli on the cell is maximal (Fig. 15-12).[79] This characteristic delay is a constant feature for any given neuron, invariant over the frequency range to which the cell is sensitive and over a considerable range of overall intensity; it varies systematically for cells of the population. If it is assumed that maximal rate of discharge is an important signal, each neuron can be regarded as subtending a hyperboloid in the contralateral sound field, as illustrated in Fig. 15-12. For neurons with very short characteristic delays the arms will diverge and at simultaneity will scribe a line in the sagittal plane. With longer delays the hyperboloid becomes more acute, and in the limit the arms fuse to form a line perpendicular to the sagittal plane, along which sound localization should be most acute. For other neurons, confusion between the locus in the anterior and posterior quadrants might arise for sounds placed at equal angles of incidence to the sagittal plane, and indeed this is just what is found in human subjects. Similar observations have been made for neurons of the auditory cortex. The hypothesis under consideration is that such auditory neurons subtend different *receptive fields* within the contralateral space and that the spatial field for sound is mapped onto a spatially distributed population of neurons in much the same way that spatial position is mapped in other sensory systems. A group of neurons with differing characteristic delays can be regarded as subtending a nested series of hyperboloids such as that in Fig. 15-12. It is an interesting example of a time-space transform in the CNS.

DESCENDING COMPONENTS OF THE AUDITORY SYSTEM AND THE PROBLEM OF ATTENTION TO AUDITORY STIMULI[31,95]

In everyday life states of selective attentiveness are easily observed aspects of human behavior: we may attend to a particular sensory channel to the relative exclusion of others. This selectivity in the adaptive behavior of organisms appears to operate at all psychologic levels, whether observed introspectively by man or in behavioral studies of men and animals, and particularly for sensation. If, for example, a subject wearing earphones is required to repeat words delivered to one ear, he may ignore a different set of words delivered to the other until items of special significance for him (e.g., his name) are inserted in the unattended set. This suggests at once that exclusion from recognition of unattended stimuli is a phenomenon occurring at a

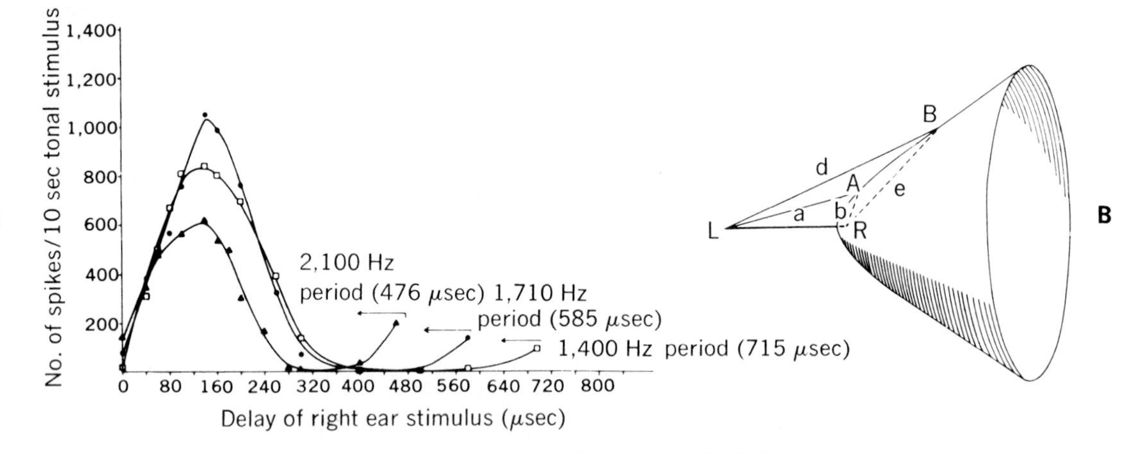

Fig. 15-12. A, Periodic discharge curves generated by neuron of inferior colliculus of cat at three frequencies of sound stimulation when tone delivered to right ear was successively delayed in time (i.e., in phase) relative to that in left ear. Characteristic delay at which excitatory effects sum to maximal for this neuron is about 140 μsec. With further delay, response is completely suppressed. Characteristic delay is independent of frequency of stimulating tone for any given neuron. **B,** Drawing of hyperboloid to indicate schematically possible locations in space of sound source that would result in interaural time differences matching characteristic delay for a particular neuron of contralateral inferior colliculus. *L* and *R* represent location of ears. Sound sources at *A* or *B* will produce the same time delay at *L* relative to *R* ($a - b = d - e$), which describes a hyperbola locating all sound sources in one plane that produce the same time delay. Rotation of hyperbola around *L-R* axis locates all possible sound sources giving this same interaural time difference, neglecting sound shadows, etc. For neurons with different characteristic delays a nest of such hyperboloids would indicate all possible sound sources. (From Rose et al.[79])

high functional level of the CNS and that neural signals concerning the unattended input are immediately available when *attention* is directed toward them. Moreover, experiment has shown that information concerning unattended stimuli is stored in the memory for short periods of time and can be recalled, with decreasing accuracy over the first 10 sec or so after delivery over the unattended channel. Even when attention is narrowly focused on one sensory channel, it seems necessary to conclude that information concerning other peripheral stimuli is admitted to the CNS and evaluated in terms of significance for survival.

The presentation of a novel stimulus evokes in animals a reaction of orientation, interest, and investigation. If the stimulus is slowly repeated and if it possesses for the animal no other interest value than that is has occurred, attention quickly wanes and the orienting reaction disappears after 10 to 15 stimuli have been presented: this phenomenon is termed habituation. The continued slow repetition of an uninteresting stimulus has a powerful soporific effect. Interest in this phenomenon directs attention toward the descending components of sensory systems.

The efferent or descending components of the auditory system are schematized in Fig. 15-5. They take origin from the auditory cortex, particularly from the infratemporal fields, and course with relays through the more medial components of the medial geniculate complex and the nucleus of the lateral lemniscus to terminate in the CN. It is not known whether the descending system impinges also on the olivary origin of the olivocochlear bundle, which is known to exert a suppressing or inhibiting effect on the cochlear hair cells (p. 450). Desmedt[22] has shown that electrical stimulation of the descending system causes a reduction in the slow-wave response evoked in the CN by sounds, without necessarily diminishing eighth nerve input. Thus the descending system can exert an influence on afferent transmission at the level of first synaptic relay independently of olivocochlear action. The latter, however, does exert a powerful control over the cochlear (and hence the CNS) response to sound. In Desmedt's experiments, he was able to show that repetitive stimulation of the olivocochlear bundle reduces eighth nerve response to a click by an amount equivalent to an 18 to 25 dB reduction in sound intensity. An interesting observation made also by Desmedt is that reduction in the eighth nerve volley produced a linearly proportional decrease in the evoked response at each central level of the auditory system. The implication is that, whatever transforms occur in the stimulus-response relation of the auditory system, they are largely cochlear in location and that thereafter transmission is linear, at least as measured by the slow-wave evoked potentials. The neurons of the olivocochlear bundle may be activated by sound; possess tuning curves not noticeably different from those of auditory neurons elsewhere; and, when active, discharge impulses at very regular rates. Little is known about what central neural mechanisms control this system.

The middle ear muscles provide a third mechanism for controlling auditory input, as detailed in the previous chapter. Their reflex contractions decrease sensitivity of the ear to sounds of the lower audible frequencies, and thus they may function as a protective device. The direct activation of the muscles during vocalization, however, suggests a more important function: to minimize sensitivity to sounds that we ourselves produce.

What function can then be attributed to the descending component of the auditory system? No evidence presently exists, but an hypothesis worthy of intensive study is that, far from "blocking" or "valving" sensory input in a nonselective manner, this system plays an important role in shaping sensory input and thus in discrimination.

REFERENCES

General reviews

1. Erulkar, S. D.: Comparative aspects of spatial localization of sound, Physiol. Rev. **52**:237, 1972.
2. Erulkar, S. D.: Physiological studies of the inferior colliculus and medial geniculate complex. In Keidel, W. D, and Neff, W. D., editors: Handbook of sensory physiology. Auditory system, Berlin, 1975, Springer-Verlag, vol. 5, part 2, p. 145.
3. Evans, E. F.: Cochlear nerve and cochlear nucleus. In Keidel, W. D., and Neff, W. D., editors: Handbook of sensory physiology. Auditory system, Berlin, 1975, Springer-Verlag, vol. 5, part 2, p. 2.
4. Goldberg, J. M.: Physiological studies of auditory nuclei of the pons. In Keidel, W. D., and Neff, W. D., editors: Handbook of sensory physiology. Auditory system, Berlin, 1975, Springer-Verlag, vol. 5, part 2, p. 109.
5. Møller, A. R.: Coding of sounds in lower levels of the auditory system, Qt. Rev. Biophys. **5**:59, 1972.
6. Neff, W. D., Diamond, I. T., and Casseday, J. H.: Behavioral studies of auditory discrimination: central nervous system. In Keidel, W. D., and Neff, W. D., editors: Handbook of sensory physiology. Auditory system, Berlin, 1975, Springer-Verlag, vol. 5, part 2, p. 307.
7. Stevens, S. S.: In Stevens, G., editor: Psychophysics. Introduction to its perceptual, neural and social prospects. New York, 1975, John Wiley & Sons.

Original papers

8. Aitkin, L. M., and Webster, W. R.: Medial geniculate body of the cat: organization and responses to tonal stimuli of neurons in ventral division, J. Neurophysiol. **35**:365, 1972.
9. Aitkin, L. M., Anderson, D. J., and Brugge, J. F.: Tonotopic organization and discharge characteristics of single neurons in nuclei of the lateral lemniscus of the cat, J. Neurophysiol. **33**:421, 1970.
10. Aitkin, L. M., Webster, W. R., Veale, J. L. and Crosby, D. C.: Inferior colliculus. I. Comparison of response properties of neurons in central, pericentral, and external nuclei of adult cat, J. Neurophysiol. **38**:1196, 1975.
11. Boudreau, J. C., and Tsuchitani, C.: Cat superior olive S-segment cell discharge to tonal stimulation, Contrib. Sens. Physiol. **4**:143, 1970.
12. Brugge, J. F., and Merzenich, M. M.: Responses of neurons in auditory cortex of the macaque monkey to monaural and binaural stimulation, J. Neurophysiol. **36**:1138, 1973.
13. Brugge, J. F., Anderson, D. J., and Aitkin, L. M.: Responses of neurons in the dorsal nucleus of the lateral lemniscus of cat to binaural tonal stimulation, J. Neurophysiol. **33**:441, 1970.
14. Brugge, J. F., Dubrovsky, N. A., Aitkin, L. M., and Anderson, D. J.: Sensitivity of single neurons in auditory cortex of cat to binaural tonal stimulation: effects of varying interaural time and intensity, J. Neurophysiol. **32**:1005, 1969.
15. Bures, J., and Buresova, O.: Relationship between spontaneous and evoked unit activity in the inferior colliculus of rat, J. Neurophysiol. **28**:641, 1965.
16. Casseday, J. H., and Neff, W. D.: Auditory localization: role of auditory pathways in brain stem of the cat, J. Neurophysiol. **38**:842, 1975.
17. Celesia, G. G.: Organization of auditory cortical areas in man, Brain **99**:403, 1976.
18. Clark, W. E. L., and Russell, W. R.: Cortical deafness without aphasia, Brain **61**:375, 1938.
19. Cohen, A.: Further investigation of the effects of intensity upon the pitch of pure tones, J. Acoust. Soc. Am. **33**:1363, 1961.
20. Colavita, F. B., Szeligo, F. V., and Zimmer, S. D.: Temporal pattern discrimination in cats with insular-temporal lesions, Brain Res. **79**:153, 1974.
21. Cranford, J. L., Igarashi, M., and Stramler, J. H.: Effect of auditory neocortex ablation on pitch perception in the cat, J. Neurophysiol. **39**:143, 1976.
22. Desmedt, J. E.: Physiological studies of the efferent recurrent auditory system. In Keidel, W. D., and Neff, W. D., editors: Handbook of sensory physiology. Auditory system, Berlin, 1975, Springer-Verlag, vol. 5, part 2, p. 219.
23. Dewson, J. H., III, Cowey, A., and Weiskrantz, L.: Disruptions of auditory sequence discrimination by unilateral and bilateral cortical ablations of superior temporal gyrus in the monkey, Exp. Neurol. **28**:529, 1970.
24. Dewson, J. H., III, Pribram, K. H., and Lynch, J. C.: Effects of ablations of temporal cortex upon speech sound discrimination in the monkey, Exp. Neurol. **24**:579, 1969.
25. Diamond, I. T., and Neff, W. D.: Ablation of temporal cortex and discrimination of auditory patterns, J. Neurophysiol. **20**:300, 1957.
26. Dickson, J. W., and Gerstein, G. L.: Interactions between neurons in auditory cortex of the cat, J. Neurophysiol. **37**:1239, 1974.
27. Evans, E. F.: Neural processes for the detection of acoustic patterns and for sound localization. In Schmitt, F. O., and Worden, F. G., editors: The neurosciences. Third study program, Cambridge, 1974, The M.I.T. Press, p. 131.
28. Evans, E. F., and Nelson, P. G.: The responses of single neurones in the cochlear nucleus of the cat as a function of their location and the anesthetic state, Exp. Brain Res. **17**:402, 1973.
29. Evans, E. F., and Nelson, P. G.: On the functional relationship between the dorsal and ventral divisions of the cochlear nucleus of the cat, Exp. Brain Res. **17**:428, 1973.
30. Gacek, R. R.: Neuroanatomy of the auditory system. In Tobias, J. V., editor: Foundations of modern auditory theory, New York, 1972, Academic Press, Inc., vol. 2, p. 239.
31. Garner, W. R.: Attention: the processing of multiple sources of information. In Carterette, E. C., and Friedman, M. P., editors: Handbook of perception. Psychophysical judgement and measurement, New York, 1974, Academic Press, Inc., vol. 2, chap. 2, p.23.
32. Godfrey, D. A., Kiang, N. Y.-S., and Norris, B. E.: Single unit activity in the posteroventral cochlear nucleus of the cat, J. Comp. Neurol. **162**:247, 1975.
33. Godfrey, D. A., Kiang, N. Y.-S., and Norris, B. E.: Single unit activity in the dorsal cochlear nucleus of the cat, J. Comp. Neurol. **162**:269, 1975.
34. Goldberg, J. M., and Brown, P. B.: The response of binaural neurons of dog superior olivary complex to dichotic tonal stimuli: some physiological mechanisms of sound localization, J. Neurophysiol. **32**:613, 1969.
35. Goldberg, J., and Greenwood, D. D.: Response of neurons of the dorsal and posteroventral cochlear nuclei of the cat to acoustic stimuli of long duration, J. Neurophysiol. **29**:72, 1966.
36. Goldberg, J. M., and Neff, W. D.: Frequency discrimination after bilateral section of the brachium of the inferior colliculus, J. Comp. Neurol. **116**:265, 1961.
37. Green, D. M., and Yost, W. A.: Binaural analysis. In Keidel, W. D., and Neff, W. D., editors: Handbook of sensory physiology. Auditory system, Berlin, 1975, Springer-Verlag, vol. 5, part 2, p. 461.
38. Greenwood, D. D.: Auditory masking and the critical band, J. Acoust. Soc. Am. **33**:484, 1961.
39. Greenwood, D. D.: Critical bandwidths and the frequency coordinates of the basilar membrane, J. Acoust. Soc. Am. **33**:1344, 1961.
40. Greenwood, D. D., and Maruyama, N.: Excitatory and inhibitory response areas of auditory neurons in the cochlear nucleus, J. Neurophysiol. **28**:863, 1965.
41. Gross, N. B., Lifschitz, W. S., and Anderson, D. J.: The tonotopic organization of the auditory thalamus of the squirrel monkey (Saimiri sciureus), Brain Res. **65**:323, 1974.
42. Hafter, E. R., and Jeffress, L. A.: Two-image lateralization of tones and clicks, J. Acoust. Soc. Am. **44**:563, 1968.
43. Heffner, H., and Masterton, B.: Contribution of auditory cortex to sound localization in the monkey (Macaca mulatta), J. Neurophysiol. **38**:1340, 1975.
44. Hind, J. E.: Unit activity in the auditory cortex. In Rasmussen, G. L., and Windle, W. F., editors: Neural mechanisms of the auditory and vestibular systems, Springfield, Ill., 1960, Charles C Thomas, Publisher, chap. 14, p. 201.
45. Imig, T. J., and Adrian, H. O.: Binaural columns in the

primary field (AI) of cat auditory cortex, Brain Res. **138:**241, 1977.
46. Irwin, R. J., and Corballis, M. C.: On the general form of Stevens' law for loudness and softness, Percept. Psychophysics **3:**137, 1968.
47. Jeffress, L. A.: Localization of sound. In Keidel, W. D., and Neff, W. D., editors: Handbook of sensory physiology. Auditory system, Berlin, 1975, Springer-Verlag, vol. 5, part 2, p. 449.
48. Jeffress, L. A., and McFadden, D.: Differences of interaural phase and level in detection and lateralization, J. Acoust. Soc. Am. **49:**1169, 1971.
49. Katsuki, Y.: Neural mechanisms of auditory sensations in cats. In Rosenblith, W. A., editor: Sensory communication, Cambridge, Mass., 1961, The M.I.T. Press, p. 561.
50. Knight, P. L.: Representation of the cochlea within the anterior auditory field (AAF) of the cat, Brain Res. **130:**447, 1977.
51. Koerber, K. C., Pfeiffer, R. R., Warr, W. B., and Kiang, N. Y.-S.: Spontaneous spike discharges from single units in the cochlear nucleus after destruction of the cochlea, Exp. Neurol. **16:**119, 1966.
52. Lavine, R. A.: Phase-locking in response of single neurons in cochlear nuclear complex of the cat to low-frequency tonal stimuli, J. Neurophysiol. **34:**467, 1971.
53. Licklider, J. C. R: Basic correlates of the auditory stimulus. In Stevens, S. S., editor: Handbook of experimental psychology, New York, 1951, John Wiley & Sons, Inc., chap. 25, p. 985.
54. Massopust, L. C., Jr., Wolin, L. R., and Frost, V.: Frequency discrimination thresholds following auditory cortex ablations in the monkey, J. Aud. Res. **11:**227, 1971.
55. Masterton, R. B., and Diamong, I. T.: Effects of auditory cortex ablation on discrimination of small binaural time differences, J. Neurophysiol. **27:**15, 1964.
56. Masterton, B., Jane, J. A., and Diamond, I. T.: Role of brainstem auditory structures in sound localization. I. Trapezoid body, superior olive, and lateral lemniscus, J. Neurophysiol. **30:**341, 1967.
57. Masterton, B., Thompson, G. C., Bechtold, J. K., and Robards, M. J.: Neuroanatomical basis of binaural phase-difference analysis for sound localization: a comparative study, J. Comp. Physiol. Psychol. **89:**379, 1975.
58. Merzenich, M. M., and Brugge, J. F.: Representation of the cochlear partition on the superior temporal plane of the macaque monkey, Brain Res. **50:**275, 1973.
59. Merzenich, M. M., and Reid, M. D.: Representation of the cochlea within the inferior colliculus of the cat, Brain Res. **77:**397, 1974.
60. Merzenich, M. M., Knight, P. L., and Roth, G. L.: Representation of cochlea within primary auditory cortex in the cat, J. Neurophysiol. **38:**231, 1975.
61. Mills, A. W.: On the minimal audible angle, J. Acous. Soc. Am. **30:**237, 1958.
62. Moore, C. N., Casseday, J. H., and Neff, W. D.: Sound localization: the role of the commissural pathways of the auditory system of the brain, Brain Res. **82:**13, 1974.
63. Moore, G. P., Perkel, D. H., and Segundo, J. P.: Statistical analysis and functional interpretation of neuronal spike data, Annu. Rev. Physiol. **28:**493, 1966.
64. Morest, D. K.: The laminar structure of the inferior colliculus of the cat, Anat. Rec. **148:**314, 1964.
65. Morest, D. K.: The laminar structure of the medial geniculate body of the cat, J. Anat. **99:**143, 1965.
66. Neff, W. D., and Casseday, J. H.: Effects of unilateral ablation of auditory cortex on monoaural cat's ability to localize sound, J. Neurophysiol. **40:**44, 1977.
67. Osterreich, R. E., Strominger, N. L., and Neff, W. D.: Neural structures mediating differential sound intensity discrimination in the cat, Brain Res. **27:**251, 1971.
68. Pandya, N. D., and Sanides, F.: Architectonic parcellation of the temporal operculum in rhesus monkey and its projection pattern, Z. Anat. Entwickl. Gesch. **139:** 127, 1973.
69. Penfield, W., and Rasmussen, T.: The cerebral cortex of man. A clinical study of localization of function, New York, 1950, Macmillan Publishing Co., Inc.
70. Pfeiffer, R. R., and Kiang, N. Y.-S.: Spike discharge patterns of spontaneous and continuously stimulated activity in the cochlear nucleus of anesthetized cats, Biophys. J. **5:**301, 1965.
71. Pfingst, B. E., Hienz, R., and Miller, J.: Reaction-time procedure for measurement of hearing. II. Threshold functions, J. Acoust. Soc. Am. **57:**431, 1975.
72. Rose, J. E.: The cellular structure of the auditory region of the cat, J. Comp. Neurol. **91:**409, 1949.
73. Rose, J. E.: Organization of frequency sensitive neurons in the cochlear nuclear complex of the cat. In Rasmussen, G. L., and Windle, W. F., editors: Neural mechanisms of the auditory and vestibular systems, Springfield, Ill., 1960, Charles C Thomas, Publisher, chap. 9, p. 116.
74. Rose, J. E., and Woolsey, C. N.: The relations of thalamic connections, cellular structure and evocable electrical activity in the auditory region of the cat, J. Comp. Neurol. **91:**441, 1949.
75. Rose, J. E., and Woolsey, C. N.: Cortical connections and functional organization of the thalamic auditory system of the cat. In Harlow, H. F., and Woolsey, C. N., editors: Biological and biochemical bases of behavior, Madison, Wis., 1958, University of Wisconsin Press.
76. Rose, J. E., Brugge, J. F., Anderson, D. J., and Hind, J. E.: Some possible neural correlates of combination tones, J. Neurophysiol. **32:**402, 1969.
77. Rose, J. E., Galambos, R., and Hughes, J. R.: Microelectrode studies of the cochlear nuclei of the cat. Johns Hopkins Med. J. **104:**211, 1959.
78. Rose, J. E., Greenwood, D. D., Goldberg, J. M., and Hind, J. E.: Some discharge characteristics of single neurons in the inferior colliculus of the cat. I. Tonotopical organization, relation of spike-counts to tone intensity, and firing patterns of single elements, J. Neurophysiol. **26:**294, 1963.
79. Rose, J. E., Gross, N. B., Geisler, C. D., and Hind, J. E.: Some neural mechanisms in the inferior colliculus of the cat which may be relevant to localization of a sound source, J. Neurophysiol. **29:**288, 1966.
80. Rose, J. E., Kitzes, L. M., Gibson, M. M., and Hind, J. E.: Observations on phase-sensitive neurons of anteroventral cochlear nucleus of the cat: Non-linearity of cochlear output, J. Neurophysiol. **37:**218, 1974.
81. Sanchez-Longo, L. P., and Forster, F. M.: Clinical significance of impairment of sound localization, Neurology **8:**119, 1958.
82. Scharf, B.: Critical bands. In Tobias, J. V., editor;

Foundations of modern auditory theory, New York, 1970, Academic Press, Inc., vol. 1, p. 159.
83. Small, A. M.: Periodicity pitch. In Tobias, J. V., editor: Foundations of modern auditory theory, New York, 1970, Academic Press, Inc., vol. 1, p. 3.
84. Stebbins, W. C., editor: Animal psychophysics; the design and conduct of sensory experiments, New York, 1970, Appleton-Century-Crofts.
85. Stebbins, W. C.: Hearing. In Schrier, A., et al., editors: Behavior of nonhuman primates, New York, 1970, Academic Press, Inc., vol. 3.
86. Stebbins, W. C., Green, S., and Miller, F. L.: Auditory sensitivity of the monkey, Science **153:**1646, 1966.
87. Stepien, L. S., Cordeau, J. P., and Rasmussen, T.: The effect of temporal lobe and hippocampal lesions on auditory and visual recent memory in monkeys, Brain **83:**470, 1960.
88. Stevens, S. S.: The measurement of loudness, J. Acoust. Soc. Am. **27:**815, 1955.
89. Strominger, N. L., and Oesterreich, R. E.: Localization of sound after section of the brachium of the inferior colliculus, J. Comp. Neurol. **138:**1, 1970.
90. Terhardt, E.: Pitch, consonance, and harmony, J. Acoust. Soc. Am. **55:**1061, 1974.
91. Tonndorf, J.: Bone conduction. In Tobias, J. V., editor: Foundations of modern auditory theory, New York, 1972, Academic Press, Inc., vol. 2, p. 195.
92. van Bergeijk, W. A.: Variation on a theme by Bekesy: a model of binaural interaction, J. Acoust. Soc. Am. **34:**1431, 1962.
93. van Gisbergen, J. A. M., Grashuis, J . L., Johannesma, P. I. M., and Vendrik, A. J. H.: Neurons in the cochlear nucleus investigated with tone and noise stimuli, Exp. Brain Res. **23:**387, 1975.
94. Woolsey, C. N.: Organization of cortical auditory system: A review and synthesis. In Rasmussen, G. L., and Windle, W. F., editors: Neural mechanisms of the auditory and vestibular systems, Springfield, Ill., 1960, Charles C. Thomas, Publisher, chap. 12, p. 165.
95. Worden, F. G.: Attention and auditory electrophysiology. In Stellar, E., and Sprague, J. M., editors: Progress in physiological psychology, New York, 1966, Academic Press, Inc., vol. 1, p. 45.
96. Young, E. D., and Brownell, W. E.: Responses to tones and noise of single cells in dorsal cochlear nucleus of unanesthetized cats, J. Neurophysiol. **39:**282, 1976.
97. Zwicker, E.: Scaling. In Keidel, W. D., and Neff, W. D., editors: Handbook of sensory physiology. Auditory system, Berlin, 1975, Springer-Verlag, vol. 5, part 2, p. 401.

16

GERALD WESTHEIMER

The eye

INCLUDING CENTRAL NERVOUS SYSTEM CONTROL OF EYE MOVEMENTS

The human eyeball (Fig. 16-1) approximates in shape a sphere with a diameter of just less than an inch. The function of the organ is to produce an optical image of the world on light-sensitive cells. The method used for this in the vertebrate eye (but not, for example, in the insect eye) is the one we are familiar with in cameras: a transparent lens system producing an inverted image on a screen, the retina. The light-transducing properties and neural characteristics of the retina are discussed in detail in Chapter 17 and the central mechanisms of the visual process in Chapter 18. Here some of the physiologic aspects of the eye's protective mechanism and of the processes involved in retaining the eye's shape and transparency will be discussed. We then go on to consider the eye's primary function of optical image formation, including some of its simpler anomalies and methods of their correction.

PROTECTIVE MECHANISMS

The eye, unlike the ear, has lids to close it off from the environment. The process of closing the lids, mediated by a relaxation of the levator palpabrae muscle, supplied by nerve III, coupled with a contraction of the orbicularis oculi muscle, supplied by nerve VII, occurs in the following circumstances:

1. *As a protective reflex* whose afferent arc is the sensory system of the external eye, chiefly of the cornea, and nerve V. Protective lid closure also occurs with certain visual stimuli such as sudden bright lights (dazzle reflex) or rapidly approaching objects (menace reflex).

2. *During sleep,* when there is a tonic contraction of the orbicularis.

3. *During spontaneous blinking,* which takes place throughout waking life once every few seconds, the rate depending on many factors. A blink is bilateral, lasts about a quarter of a second, and the eyeballs move up during a blink, as indeed they do also during sleep. Blinking serves to keep the corneal surface clear of mucus and the lacrimal fluid spread evenly over it.

4. *In voluntary lid closure,* where it can be utilized to effectually disrupt the visual process. About 1% of the incident light, evenly distributed over the whole visual field, passes through the closed eyelid with average skin pigmentation.

On the margins of the eyelids there are *eyelashes,* or cilia. Totaling about 200 in each eye, they each have an average life of a few months. The follicle of each cilium is innervated by mechanoreceptive nerve endings that discharge impulses when the cilium is bent. Activation of these receptors evokes a blink reflex when a foreign body touches the lid aperture.

The upper outer portion of each orbit contains the *lacrimal glands*. They are innervated by the midbrain autonomic outflow traveling first along nerve VII, synapsing in the sphenopalatine ganglion, and reaching the glands via the mandibular division of nerve V and the zygomatic nerve. Secretion of tears, as well as being of psychic origin, is produced by activation of receptors of the lids and conjunctivae.

The normal secretion of tears amounts to less than 1 ml/day. The tear fluid has a pH of about 7.4 and osmotic pressure equivalent to about 0.9% NaCl. Reports of higher tonicity (up to 1.4% NaCl equivalent) can probably be accounted for by evaporation. An important constitutent of tears is *lysozyme,* a mucolytic (hydrolyzing mucopolysaccharides) enzyme with bactericidal action. The tear fluid serves (1) to provide lubrication for lid movements, (2) to provide a good optical surface for the cornea by forming a thin film over it, and (3) as an emergency mechanism to wash away noxious agents.

Most of the tear fluid normally formed is lost

by evaporation. The normal drainage channels are the lacrimal canaliculi, sac, and duct leading into the interior nasal meatus.

OPTIC MEDIA

In its passage through the eye, light successively traverses the cornea, the aqueous humor, the crystalline lens, and the vitreous humor (Fig. 16-1). The cornea is anatomically continuous with and histologically similar to the sclera. Together they form the outer shell of the eyeball, which contains the lens, and the gel-like vitreous humor, whose internal pressure and hence shape are maintained by the appropriate balance between the rates of formation and elimination of aqueous humor.

Cornea

The cornea is about 11 mm in diameter and has a thickness of 1 mm near its junction with the sclera and 0.5 mm near its center. Its outer surface, the epithelium, is continuous with the epithelial layers of the conjunctiva. Most of the cornea consists of the so-called stroma, a lamellar structure of submicroscopic collagen fibrils in an organized arrangement that makes it transparent to light. There are normally no blood vessels in the cornea; as a consequence, it derives its oxygen supply by diffusion from the air and from surrounding structures. This is also the pathway of CO_2 loss. The supply of glucose and transfer to lactic acid seem to involve diffusion from the aqueous and other sites.

The transparency of the cornea depends on the maintenance of normal intraocular pressure and many other factors as well. Severe elevation of the intraocular pressure, lack of proper osmotic equilibrium with the surrounding fluids, changes in temperature, and improperly fitted contact lenses are some of the factors that may cause edema of the cornea, with a resulting increase in the light scattered by this structure. The earliest signs of this are halos seen around bright lights, but later visual acuity is lowered. The effect is usually reversible. A seriously injured cornea is repaired by scar tissue, which is usually not transparent. Because the cornea is avascular, corneal grafts may be uncomplicated by immunologic reactions.

The cornea is supplied by sensory fibers from nerve V. They lose their myelination shortly after entering the cornea and terminate as free nerve endings in the epithelium and the stroma. Weak mechanical stimuli to the cornea evoke tactile sensations. Stronger ones are painful. Sensory localization is poor on the cornea, and there is some doubt whether heat or cold can be sensed there. If the corneal nerves are cut (e.g., by an

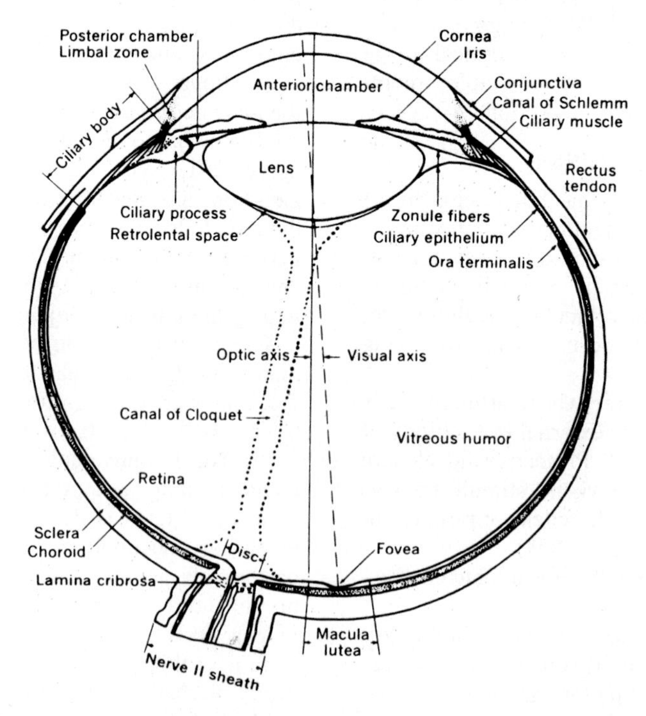

Fig. 16-1. Horizontal section of right human eyeball. (From Walls.[21])

incision for removal of cataract), they regenerate within a few weeks.

Lens

The crystalline lens also is an avascular structure. It is suspended by fibers, the zonule of Zinn (Fig. 16-2). The lens consists of an elastic membrane, the capsule, that encloses a single layer of epithelium and a complicated system of transparent fibers derived from it. New fibers are continually laid down, but the volume of the lens changes little throughout life. In youth the lens substance is malleable, which allows the changes in shape called accommodation to occur, but with age it becomes harder, has a higher solid content, and usually loses some of its transparency. An eye with an opaque lens is said to have a cataract, but there are many degrees of loss of transparency, all resulting in some loss of definition in the retinal image. Cataracts may be caused by physical trauma, radiation, or chemical factors such as the elevated glucose concentration of the aqueous in a patient with uncontrolled diabetes.

The lens uses O_2 and glucose and produces lactic acid. Ascorbic acid seems to be an important agent in lens metabolism. All these substances are carried in the aqueous humor and diffuse across the lens capsule, inside which a potential of about −70 mV is maintained, a result of the differential concentration of Na^+ and K^+ actively maintained between the inside of the lens fibers and the anterior chamber.

Vitreous

Little is known about the distinctive features of the vitreous. Its gel-like physical consistency is thought to be the result of a network of submicroscopic fibers containing within its spaces large molecules of hyaluronic acid; the enzyme hyaluronidase produces an increased fluidity of the vitreous.

Aqueous

The composition of the aqueous humor closely approximates that of protein-free plasma (Table 16-1). The mechanism of formation and drainage of the aqueous is important not only because the aqueous is the principal carrier of metabolites for the lens and the cornea but also because it regulates the intraocular pressure, on which the maintenance of the eye's shape and

Table 16-1. Values of two distribution ratios*

	Concentration in aqueous / Concentration in plasma	Concentration in dialysate / Concentration in plasma
Na	0.96	0.945
K	0.955	0.96
Mg	0.78	0.80
Ca	0.58	0.65
Cl	1.015	1.04
HCO_3	1.26	1.04
Urea	0.87	1.00

*From Davson.[11]

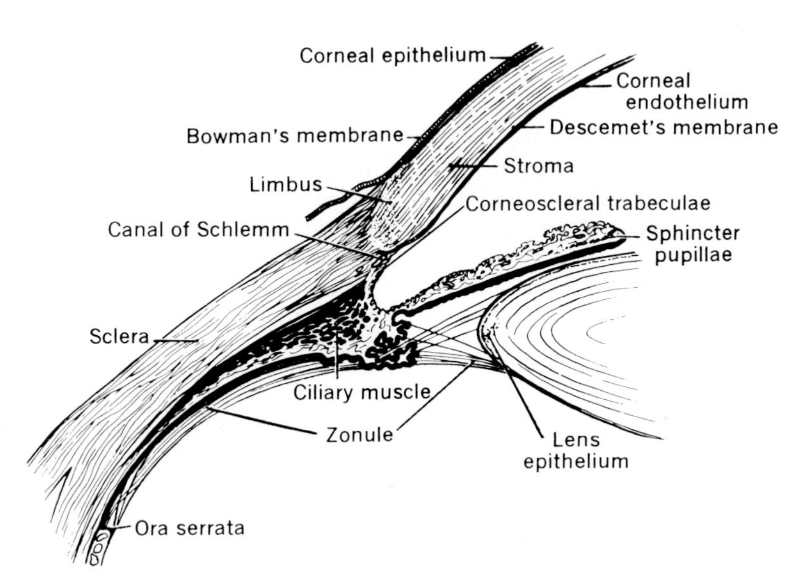

Fig. 16-2. Section showing details of anterior segment of human eye. (From Davson.[12])

Table 16-2. Optical constants of a typical normal eye

Surface	Radius of curvature (mm)	Refractive index		Distance from anterior surface of cornea (mm)	Refractive power (D)
		Anterior	Posterior		
Anterior cornea	7.8	1.000 (air)	1.376	0	+48.2
Posterior cornea	6.8	1.376	1.336 (aqueous)	0.5	− 5.9
Anterior lens	10.0*	1.336	1.386†	3.6*	+ 5.0†
Posterior lens	−6.0	1.386†	1.336 (vitreous)	7.2	+ 8.3†
Retina				24.0	

*During maximum accommodation the anterior surface of the lens has a radius of curvature of 5 mm and its anterior surface is moved forward to be nearly 3 mm behind the anterior surface of the cornea. Partial accommodation will produce values between these values and those given in the table.

†The index of refraction of the lens varies from 1.386 near each surface to 1.406 in the center. The indicated refractive power is for the lens surfaces only. The gradient of refractive index within the lens produces additional refractive power.

transparency depends. The aqueous fluid is formed in the processes of the ciliary body. Comparison of the constituents of aqueous and blood plasma shows that it is not formed by simple filtration; for example, bicarbonate ions are in considerably higher concentration in aqueous than in plasma. The mechanism of secretion of aqueous is not fully understood, but it is known that carbonic anhydrase inhibitors such as acetazolamide (Diamox) reduce the rate of aqueous formation. This drug is of considerable assistance in controlling glaucoma, a condition in which the intraocular pressure is elevated because of interference with normal aqueous drainage. Since the aqueous bathes the lens, the iris, and the posterior surface of the cornea and is in contact with the vitreous, diffusion at the surface of these structures will affect its composition.

After the aqueous is formed, it flows forward between the lens and the iris and in this way into the anterior chamber. The temperature gradient in the anterior chamber (the cornea is nearer the external environment and hence cooler) causes a thermal circulation of the aqueous, the layers nearer the posterior surface of the cornea rising and those nearer the lens falling. Drainage channels for the aqueous are located near the junction between the sclera and the cornea (Fig. 16-2). A trabecular meshwork leads into the canal of Schlemm, which drains into the venous system of the eye.

Intraocular pressure

Normally the intraocular pressure is about 20 mm Hg above atmospheric pressure. It varies with respiration and with the pulse, and there is also a diurnal cycle. These changes are usually no more than a few millimeters of mercury. Intravenous injection of adrenaline in the intact animal causes a sudden rise in the intraocular pressure. External pressure such as that caused by contraction of the lid and the extraocular musculature may cause a marked increase. Prolonged contraction of the sphincter of the iris opens the drainage channels and may therefore lower intraocular pressure in eyes in which these are obstructed; the lowering due to reduction in the rate of formation of aqueous produced by acetazolamide has already been referred to.

A good estimate of the intraocular pressure in an intact eye may be obtained by an instrument called the tonometer, which measures the resistance that is encountered in producing a small indentation of the cornea by a plunger. Since this depends to some extent on the rigidity of the cornea, these instruments must be calibrated on cadavers.

OPTICAL IMAGE FORMATION IN THE EYE

To produce an image of the outside world on the retina the eye has a refracting apparatus consisting of the cornea and the crystalline lens. The optical properties of refracting surface are determined by their radius of curvature and the index of refraction of the media that they separate, and the characteristics of the optical apparatus as a whole depend on those of the surfaces and their separation.

The details for each of the individual surfaces in a typical normal eye are given in Table 16-2. Included in the table is the specification of the refracting power of each surface. The unit in which refracting power is measured is the diopter, the reciprocal of the focal distance in meters. It is seen from the table that the anterior surface of the cornea is the major refracting surface of the eye—hence any irregularities in it are of major significance in image formation in the eye.

The total refracting power of the eye cannot be obtained by simple addition of the values in the last column of Table 16-2 for two reasons: first, the separation of the surfaces affects their cumulative refractive effect, and second, the crystalline lens behaves in quite an unusual way. The refractive index of the lens is not uniform but varies between 1.386 near its surfaces to 1.406 at its center. A structure with such a refractive index gradient produces an optical effect corresponding to a much higher refractive index than is actually found. It has been estimated that if the human lens were filled with a medium of uniform refractive index, this would have to be 1.416 to produce the same refractive power as that of the actual lens. This phenomenon is even more marked in the fish eye, for here the cornea, being immersed in water, has little influence on the light rays, and most of the refraction necessary to bring an image onto the retina is provided by the lens. Some fish lenses have a gradient of refractive index reaching values as high as 1.5 in the center, and their power is equivalent to homogeneous lenses filled with medium that has a refractive index of 1.7.

Although human eyes show considerable individual differences, it is useful to devise an optical model for the typical eye, with the aid of which a number of important calculations can be done very simply and with surprisingly good general validity. These optical models are called schematic eyes, and by far the least complicated of these is the *reduced eye*. In it a single surface has been substituted for a cornea and lens, just as it is helpful at times to substitute a single thin lens for a complicated many-component optical system. In the reduced eye the single surface, which separates the vitreous from air, obviously has to have a different curvature from that of the real cornea, since it alone must focus rays on the retina. The position of this single surface is also not quite the same as that of the cornea; rather it occupies the place at which a single lens or surface would have the same effect as that of the whole optical system of the eye—the principal plane. A more detailed comparison between the reduced eye and a typical real eye may be made from Fig. 16-3.

The optical constants of the reduced eye have been chosen so that in its most salient features it operates as an image-forming mechanism most like the real eye. There are some important differences—the reduced eye cannot accommodate in the way the real eye does—but the optical properties are sufficiently similar and so very much simpler that it is an ideal model on which to illustrate the major optical effects and defects occurring in a normal human eye.

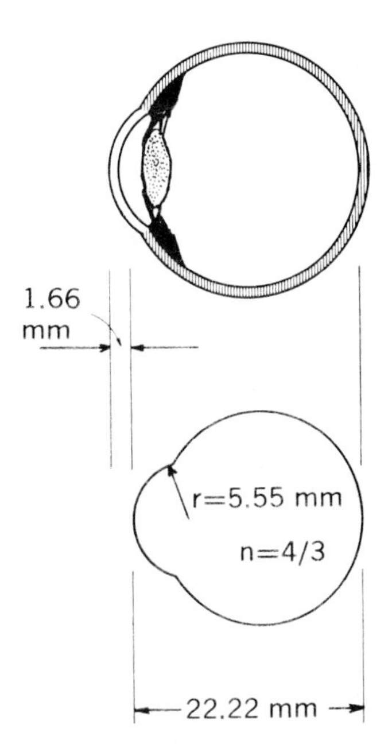

Fig. 16-3. Schematic diagram of optic media of eye (upper) and of "reduced" eye (lower). Reduced eye consists of single surface of 5.55 mm radius of curvature separating air from water, and its image-forming properties are similar to those of typical human eye.

Refractive errors

A normal eye with accommodation relaxed will produce an image of an infinitely far object in its focal plane, which is nearly 24 mm behind the corneal vertex and 22.22 mm behind the vertex of the reduced eye. If the retina is situated in this plane, a sharp image will be formed on the receptors and the optical prerequisite for clear vision is met. This matching of focal length of the optical part of the eye and axial length is the condition of *emmetropia*. It must be remembered that this merely satisfies the prerequisite for clear retinal imagery; whether, in fact, the patient will *see* clearly depends on the integrity of the photochemical or neurophysiologic stages of the visual process, the subject of later chapters.

What is the effect of a mismatch between focal length of the optics of the eye and the eye's axial length? A typical example of this is myopia—a condition in which the image of a distant object is formed not on the retina but in front of it (Fig. 16-4). This may occur because the particular eye has an optical refracting system more powerful (shorter focal length) than usual but has a normal axial length or, more commonly, because the focal length is normal but the eye is too long. In

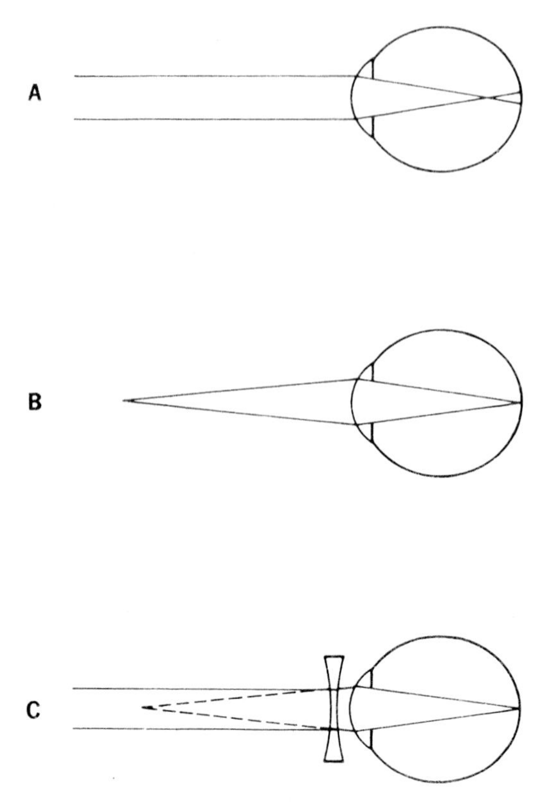

Fig. 16-4. Myopic eye. **A,** Rays from point object at infinity form an image in vitreous and are intercepted as a blur patch on retina. Size of blur patch for given state of out-of-focusness depends on pupil size. **B,** In the same eye, close-up target is imaged on retina. Reciprocal of object distance in meters gives amount of myopia in diopters. **C,** Myopia may be corrected by a lens of such diverging power that it makes rays from target at infinity appear to come from far point of myopic eye.

any case the rays from a point object at infinity come together in a point focus in the vitreous and will then diverge again. When the rays reach the retina, the cross section of the bundle will form not a point but a patch, the so-called blur patch, whose size depends on the extent of out-of-focusness and the diameter of the pupillary aperture. Objects at a great distance, as a consequence, will appear blurred. The same laws of optical image formation, according to which the image of a distant object will be formed in the focal point and hence in the myopic eye in front of the retina, allow us to determine the position of a target so that its image in such an eye *is* formed on the retina. This point in object space is called the *far point*. The reciprocal of the distance, in meters, between the eye and the far point is the amount of myopia in diopters. For example, a reduced eye whose focal point is at a distance of 22.22 mm behind the surface but whose retina is 23.4 mm behind the surface will have its far point ⅓ m in front of the eye and will therefore be 3 D myopic. For an axial length of 25.6 the figures will be ⅛ m and 8 D, respectively.

The only way to produce a clear retinal image in such an eye when the object is at a long distance is to place a lens in front of the eye, which changes the vergence of the bundle of rays emanating from this object in such a way that when the rays emerge from the lens, they appear to be coming from the patient's far point. There is thus a simple relationship between the far point distance and the focal length of the correcting lens. In myopia the latter has to be a negative lens, since a parallel incident bundle of rays must emerge from it as a diverging bundle to impinge on the eye as if it were coming from the far point. In turn this bundle is then imaged on the retina by the optical system of the eye. There is thus optical conjugacy between the distant object and the patient's far point with respect to the correcting lens, and the far point and the retina with respect to the eye's optics. This concept of optical conjugacy permits ready handling of such a problem as finding the lens power necessary to correct an eye when the lens is placed farther away than usual—the lens now has to have the power necessary to make the object point conjugate to the far point with respect to the new lens position.

In any case a myopic eye associated with the lens that brings a distant object conjugate to its far point is equivalent to an emmetropic eye: in each case the retina receives a clear image when the target is far away.

When the axial length of an eye is too short relative to its focal length, the retina will intercept the bundle of rays from a distant object before it comes to a focus. An eye with this kind of defect is said to be *hyperopic* (Fig. 16-5). A good way of looking at this problem is to note that there is a deficit of refracting power in the optical system of the eye. Bringing the target in from infinity does not help as it does in myopia—it merely moves the image even farther behind the retina. What this eye needs is additional refracting power. This can be provided by a positive lens placed in front of the eye, which gives the bundle of rays from a distant object some convergence and this, added to the refracting power of the eye, will produce a clear image on the retina.

The eye is usually not perfectly symmetric,

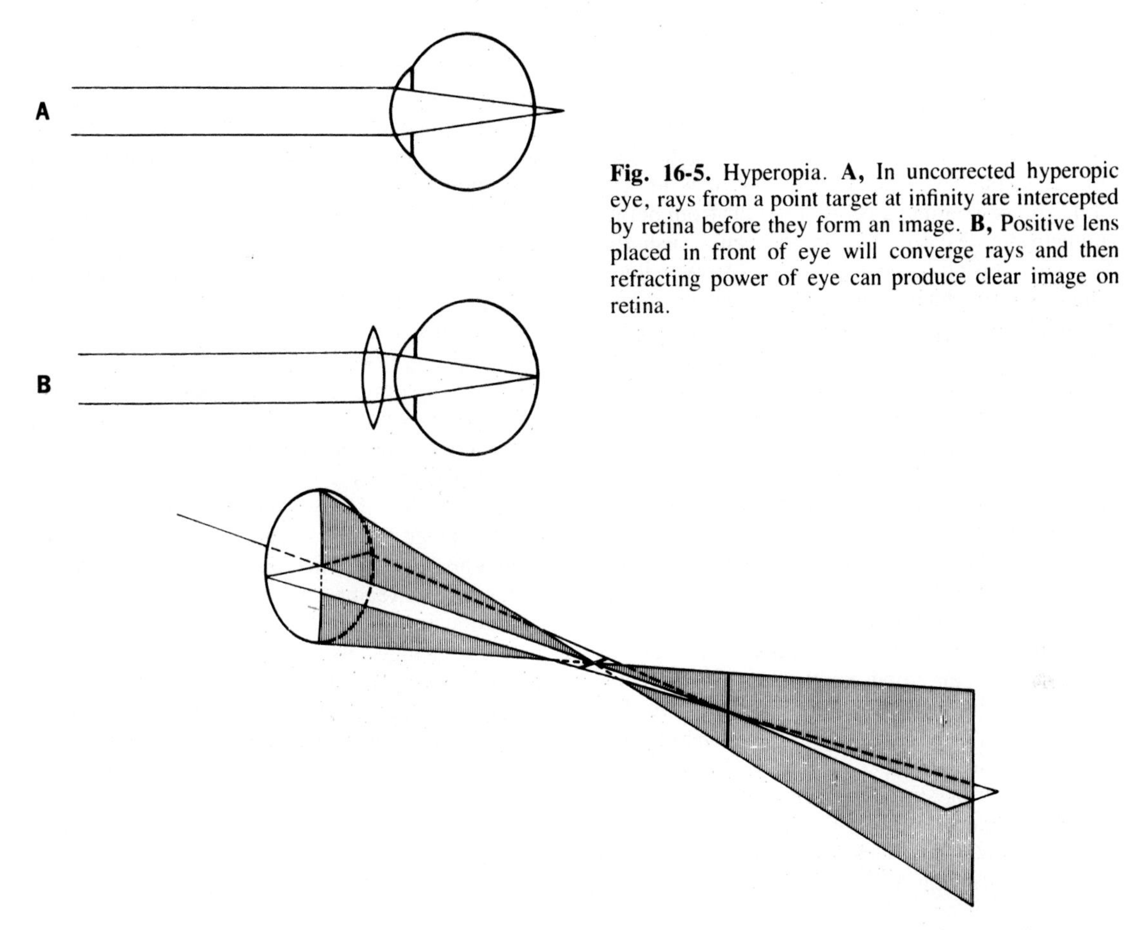

Fig. 16-5. Hyperopia. **A,** In uncorrected hyperopic eye, rays from a point target at infinity are intercepted by retina before they form an image. **B,** Positive lens placed in front of eye will converge rays and then refracting power of eye can produce clear image on retina.

Fig. 16-6. Astigmatic imagery. When optical system has unequal refracting power in two meridians at right angles to each other, bundle of rays from object at infinity will undergo refraction to produce configuration illustrated in this figure. Cross section of bundle in image space is never a point, but an ellipse in general and a line in two special positions.

but as a first approximation one may postulate the existence of an anteroposterior axis of symmetry, which is also the optical axis of the eye (i.e., the line on which the centers of curvature of the various surfaces lie). A plane section of the eye containing this anteroposterior axis is called a meridian of the eye. Thus the horizontal (0 to 180 degrees) meridian of the eye bisects it into an upper and a lower hemisphere, and the vertical (90 to 270 degrees) meridian bisects it into the nasal and temporal hemispheres.

An important optical anomaly is one in which the curvature of an optical surface, usually that of the cornea, is not the same in all meridians. For example, a cornea may have a higher curvature in the vertical than in the horizontal meridian. As a result, the sheet of rays lying in the vertical meridian will be refracted more than that in the horizontal meridian, and the point object will give rise in image space to a complicated bundle (Fig. 16-6) whose cross section is never a point, but either an ellipse or, in one special situation, a vertical line and, in another, a horizontal line. This condition is called *astigmatism*. The uncorrected astigmatic eye, depending on the position of the retina, may have generally blurred vision or a particular kind of imagery in which line targets in just one meridian appear sharp. The correction of an astigmatic eye is achieved by means of a lens that also has different curvatures in its meridians, so that its astigmatism is exactly complementary to that of the eye. When this lens has no power in one principal meridian, it is called a cylindric lens; when it has power, but differing in the two principal meridians, it is a spherocylindric or toric lens.

Myopia, hyperopia, or astigmatism is called a *refractive error,* or *ametropia*. It occurs when the optical refracting apparatus of the eye is incapable of producing a sharp image of a distant object or when it does so but the retina is not situated in the correct position. Such conditions are quite common. There is a strong genetic influence and the evidence that environmental factors contribute materially to its origin is uncertain. The newborn eye is often not far from being emmetropic. Myopia, when it occurs, usually develops during the teens and is irreversible. There is little change in the refractive error during adult life, but when lenticular changes commence with age, there may be associated anomalies of refraction.

For most purposes an ametropic eye with the appropriate spectacle correction functions like an emmetropic eye. The lens may be worn in frames in front of the eyes or, in suitable patients, they may be made in the form of thin shells worn directly on the cornea—contact lenses. There is always a thin layer of tear fluid or specially prepared buffer solution between the cornea and the contact lens, a fact that makes the optical properties of the total contact lens correction somewhat complicated. Contact lenses have the advantage not only of being less conspicuous than spectacles but also of effectively eliminating the anterior surface of the cornea as a refracting surface, an invaluable aid to clear retinal imagery when this surface is irregular or deformed by scarring or disease.

Accommodation

The diopter, the reciprocal measure of distance in meters, which is used to scale the degree of ametropia and the correcting lens power of an eye, is also of value in expressing the amount of accommodation exerted by an eye. The following discussion applies alike to emmetropic eyes and eyes made artificially emmetropic by spectacles or contact lenses. Such eyes have a target at optical infinity imaged on the retina when the ciliary muscle is relaxed. In order to image a close-up target on the retina, it is necessary to increase the optical refracting power of the eye. This occurs when the crystalline lens is allowed to assume a more biconvex (i.e., more highly refracting) shape as a result of contraction of the ciliary muscle. If the distance of the target that has its image formed on the retina when the eye is in a given state of accommodation is measured in meters, its reciprocal is the amount of accommodation in diopters. Thus an emmetropic eye that has changed its refractive power to bring a target at ¼ m into focus is exerting 4D accommodation. Due to the progressive sclerosis of the lens substance the amplitude of accommodation decreases with age (i.e., the nearest point to which an eye can accommodate recedes). This is illustrated in Fig. 16-7 both in terms of this near point of accommodation (in meters) and also of its reciprocal, the accommodative am-

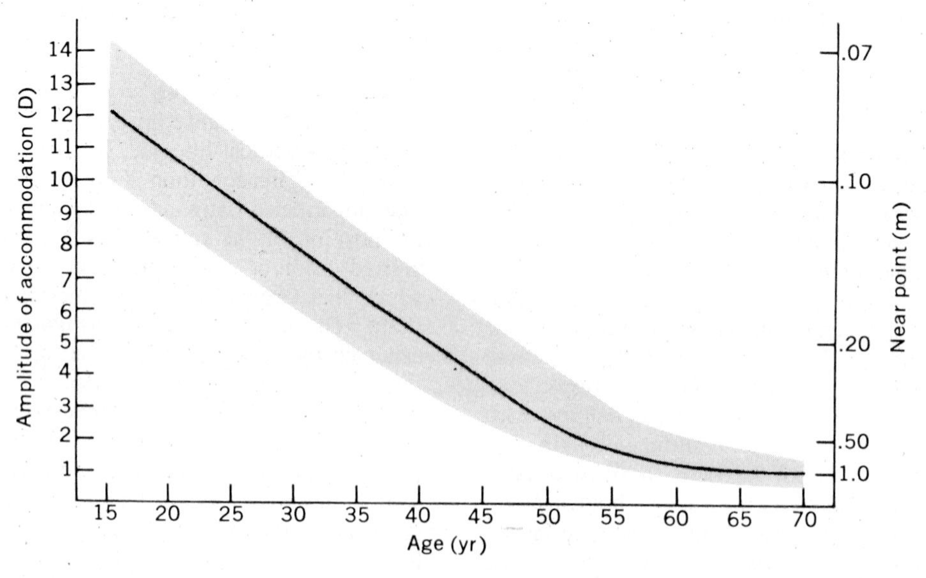

Fig. 16-7. Changes in accommodation with age. Graph shows typical value and range of accommodative amplitude accepted as normal for a given age. Ordinates at left, diopters, and at right, nearest point of clear vision for an emmetrope.

plitude (in diopters). When accommodation is no longer adequate to achieve or maintain clear focus on the desired close-up target (e.g., reading material), it may be supplemented by positive lenses that can supply the necessary additional refractive power. Such a patient is said to be *presbyopic* and can be helped by bifocal or trifocal lenses that provide the refractive correction for distant objects, if necessary, and also the additional power ("the add") for one or two specified near viewing distances.

Insofar as accommodation adds refractive power to the eye, it can also be of help to the uncorrected hyperopic eye: clear vision of objects at optical infinity can be achieved by an effort of accommodation in diopters equal to its amount of hyperopia, also in diopters.

THE RETINAL IMAGE AND VISUAL ACUITY

The image, that is, the replica of the outside environment, formed by the optical system of the eye on the retina is two dimensional. The geometric relationship between the outside world and the retinal image is based on a projection system, using the center of the eye's entrance pupil as the center of projection. An object sends out light energy in many directions, but only a small proportion—that which passes through the pupil—contributes to the formation of the retinal image. The most representative light "ray" therefore is the one passing through the center of the eye's entrance pupil—the chief ray. Since each light ray in object space has as its conjugate a light ray in the image space, the most representative ray for the purposes of defining the retinal position of the image (clear or blurred) is the image-sided chief ray. It passes from the center of the eye's exit pupil to the geometric image or, if a blur patch is present on the retina rather than a clear image, to the center of the blur patch. If it is desired to state where on the retina a target in the outside world is being imaged, one consequently draws a straight line from the object to the center of the eye's entrance pupil and then finds the intersection of its image-sided correlate with the retina. In any given eye there is a single and unique relationship between these two, and it suffices therefore to state object position and dimensions in terms of angles subtended at the center of the entrance pupil.

If the detailed calculations are carried out in a typical emmetropic human eye, it is found that an object subtending an angle of 1 degree at the eye will form an image on the retina that is 0.3 mm in size. One can thus look at the process of image formation in the eye as a mapping of directions in object space (determined by the line between the object point and the center of the eye's entrance pupil) and position occupied by the image of the object point. Along each direction only one radial distance will be in focus, that for which the eye is accommodated. This explains the common practice of interchanging retinal distances (in millimeters) and angular object subtense (in degrees or radians); for example, one describes the width of the optic nerve head as 5 degrees rather than 1.5 mm.

Field of fixation and visual field

The most important target direction for any eye is the one that corresponds to the center of the fovea, for here the anatomic and functional organization of the retina is most favorable for good resolution. Normally the two eyes will be positioned to image the object of regard in the centers of the foveas. This process of fixation is the major function of the whole oculomotor system, a fact attested to by the relative absence of eye movements in species possessing no foveas. The object fixated is called the fixation point, and the line joining the fixation point and the center of the eye's entrance pupil is called the primary line of sight.

If the head is held still, the subject is asked to fixate a small object, and this object is moved about, it is possible to map out the *field of fixation*. It is a measure of the capacity of the extraocular muscles to move the eyes and it usually extends about 45 degrees in all directions, often more in a downward direction. When there is paralysis of an extraocular muscle or a mechanical obstruction in the orbit, the field of fixation may be restricted.

Consider, on the other hand, the situation when a stationary target is presented to an eye and the eye steadily fixates it. This target now represents the center of the fovea, and all other retinal locations may be specified by other points on a sphere, with the center of the entrance pupil as its center. On this sphere, one may map out the extent of all those regions from which a light sensation may be elicited. This is called the *visual field*—it measures the regions of peripheral retina of the stationary eye that respond to light, as distinct from the field of fixation, which measures the range of target positions that can be foveally fixated with the moving eye.

Visual acuity

The analysis of the capacity to discriminate small differences in spatial configuration of targets involves a study of the quality of the replica available at the retinal level, the anatomic and functional characteristics of the retina, and the central nervous system (CNS).

If the retinal image were a point-by-point replica of the target, resolution would in theory be unlimited. However, even a perfect optical instrument does not form a point image, but spreads the light coming from a point object into a diffraction pattern whose size is inversely related to the pupil aperture. When the pupil of the eye is small, the spread of the focused image of a point, the so-called point-spread function, is entirely that given by diffraction theory. As the pupil increases in diameter beyond about 2 mm, the point-spread function no longer conforms to the diffraction limitation (which would demand its continued narrowing), but stays near the shape shown in Fig. 16-8, only to widen again when the pupil diameter exceeds 5 mm. The failure of the eye to perform as an ideal optical instrument when its pupil is larger than 2 mm is due to the aberrations inherent in its optical system. The eye has some correction for spherical aberration in its cornea, which is aspherical, but none for chromatic aberrations. These aberrations become progressively more important as the pupil is enlarged. But the photopic visual system has a built-in compensation device in its receptors: light impinging obliquely on the retinal cones is less effective in stimulating the transducing process than light coming in head-on (Stiles-Crawford effect). Rods, which are intended to be highly efficient light gatherers, have little Stiles-Crawford effect.

The best optical performance of the eye, then, is found when the pupil is in the range of 2 to 5 mm, but this is only true when the eye is in good focus. We have seen that out-of-focusness results in the imaging of a point as a blur patch whose size is directly proportional to the pupil diameter, so that out-of-focus imaging is better the smaller the pupil, often down to very narrow pinholes.

Fig. 16-8 shows that a point object such as a star will even under the best possible conditions spread its light over an appreciable retinal region. When two point sources such as two stars are close together, each will produce its own light spread, but the two light distributions will summate: what a retinal receptor responds to is the sum of all the light reaching it from all sources and over its whole acceptance area. The shape of the summed light distribution from two closely adjacent point sources differs insignificantly from that of a single, brighter target, and it is therefore impossible to make the distinction between a single and a double light source. If the two point sources are now separated, the intensity configuration on the retina will show two distinct humps with a trough between them. The condition for resolution has now been reached, provided, however, that the receptor mosaic has a "grain" so that the trough and its flanking humps each fall on individual mosaic units, and that the physiologic pathway from each of these units is capable of transmitting independent signals. This example illustrates the factors involved in the resolution of spatial details or, as it is called, visual acuity: optical (point-spread function),

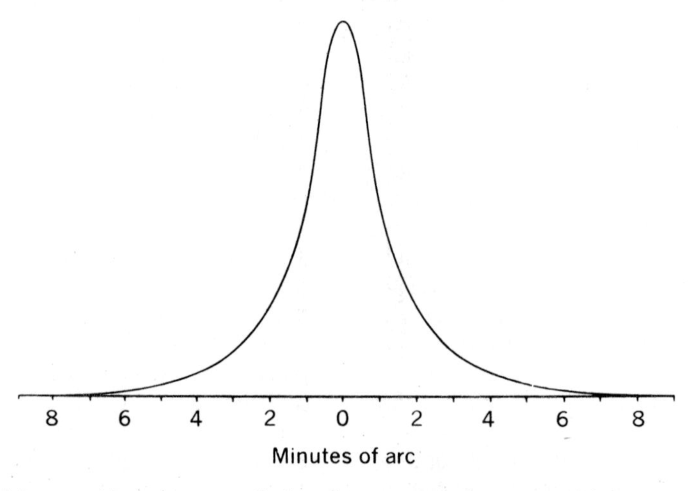

Fig. 16-8. Light spread in human eye. Optics of eye, even in best state of focus, spread light from line source over finite distance. This is in part due to diffraction and in part due to aberrations. Spread function, illustrated in this diagram, provides physical limitation to resolution in eye. (From Westheimer.[24])

anatomic (receptor mosaic), and functional (independent pathways for transmitting signals). In an eye in good focus, in the center of the fovea, when the light stimulus lasts sufficiently long and is sufficiently intense, these three factors operate optimally and have evolved to about the same limit of performance. It follows that the total resolution performance suffers when there is an abnormality in any of its components.

Tests of visual acuity involve the discrimination between spatial configurations that lead to differing sensations only if details of a given angular subtense are resolved. Examples are the Landolt C's, which are seen as C's of the correct orientation and not O's if the subject can resolve the gap. The scale for these letters is gap size in minutes of arc, usually the gap is one-fifth the overall size of the letter. Snellen letters are based on the same principle; the letter is 5 times the width of a limb.

It has been accepted for at least a century that normal visual acuity is 1 minute of arc, that is, separation of two stars by 1 minute of arc leads to the sensation of two stars, and by less than 1 minute of arc to the sensation of one star. On the same basis, if the gap in a Landolt C is 1 minute of arc or larger, the C can be distinguished from an O. This, together with the convention of carrying out these tests with objects at 6 m or 20 feet, is the origin for designating a resolving capacity of 1 minute as 20/20 visual acuity; that is, a letter with a feature subtending 1 minute of arc at 20 feet can then be resolved at 20 feet. If an eye requires a 2 minutes of arc gap to recognize a letter, this corresponds to a letter whose gap would subtend 1 minute of arc at 40 feet; hence the designation of such a visual acuity as 20/40.

Under good conditions, most normal eyes have a resolving capacity somewhat better than 20/20; 20/15 is not unusual. And with attention to all factors and a repetitive target like a grating, good observers can resolve down to nearly ½ minute of arc—which for a 2 mm pupil represents just about the limit of resolution for even an ideal optical system like the eye's.

The visual system is capable of even finer spatial discrimination when resolution is not a factor. For example, two clearly demarcated features can be detected as nonaligned when one is displaced by just a few seconds of arc. This kind of hyperacuity, involving discrimination of retinal distances much smaller than the size of the individual receptive elements, depends critically on the separation of the features; it is testimony to the extraordinary finesse of neural processing of optical signals.

OCULAR MUSCULATURE

There are three muscular systems associated with the eye. The first controls the state of focus and permits a clear retinal image to be formed for a wide range of object distances. The second controls the aperture of the eye. Both of these are situated within the eye and are called intraocular. They are made up of smooth muscles innervated by the autonomic system. The third system functions to control the direction in which the eye is pointing. The muscles are attached to the outside of the eye and hence this system is called the extraocular motor system.

Ciliary muscle and accommodation

The ciliary muscle is in its action a sphincter muscle and is innervated by the parasympathetic bulbar outflow via nerve III and the short ciliary nerves.[25] The crystalline lens is suspended by the fibers of the zonule of Zinn, which run from the ciliary body to the lens. In the unaccommodated state the ciliary muscle is relaxed and the zonular fibers taut. The result is that the anterior surface of the lens is flatter than the posterior surface. Contraction of the ciliary muscle makes the ciliary processes bulge inward toward the anteroposterior axis of the eye. The effect is a slackening of the tension in the zonular fibers, now leaving unopposed the elastic forces in the lens capsule that tend to mold the malleable lens substance into a different shape: the middle of the anterior surface is more highly curved and bulges into the anterior chamber (Fig. 16-9). This process is called accommodation and depends on (1) the contraction of the ciliary muscle and (2) the capacity of the lens substance to assume a more biconvex shape when the tension on the zonule is released. Failure of accommodation to occur when neural impulses arrive at the ciliary muscle may be due to muscular or lenticular factors. The ciliary muscle is a smooth muscle innervated by the parasympathetic system, and it may be paralyzed by atropine, homatropine, or similar drugs that block the myoneural junction to acetylcholine. (It may also be stimulated by drugs such as eserine that inactivate cholinesterase.) Although such an effect is usually produced by local application of drugs, occasionally it may be the result of systemic administration.

A normal contraction of the ciliary muscle may also fail to evoke a change in lens shape if the lens substance is unable to yield to the physical molding force of the lens capsule. We have seen

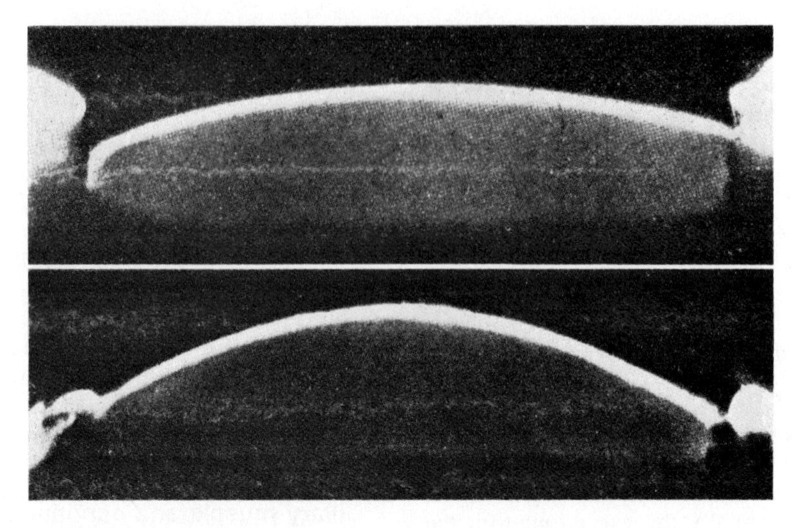

Fig. 16-9. Slit-lamp section showing anterior surface of normal human crystalline lens in unaccommodated (upper) and accommodated (lower) states. (From Fincham.[15])

that the lens substance gets harder with age. This process is a progressive one, but it becomes noticeable at an age of 40 years and more when only a small response in the lens shape results even on maximal contraction of the ciliary muscle. In the fifties a stage is usually reached when the lens cannot change shape at all, even when completely freed from the zonule of Zinn.

Accommodation serves as a focusing device in the eye, a response evoked by the presence of a blurred image on the retina. The reaction time of this response is about ⅓ sec, and its execution occupies another ½ sec (Fig. 16-10). The two eyes respond equally and simultaneously.

During steady viewing of a close-up target there is an unsteadiness of the ciliary muscle response. The resulting oscillations of accommodation are also binocular and have a frequency of about 2 Hz. They have been interpreted as the inevitable consequence of the accommodation system acting as a servomechanism. When a person is confronted with an empty visual space, as in complete darkness or when looking at a vast expanse of sky, one might expect the ciliary muscle to be relaxed. In fact, there is a slight contraction, focusing the eye not at infinity but at about 1 m. This is called night or empty field myopia.

The neural pathway subserving accommodation is coupled with that for convergence of the two eyes, for the effort to produce accommodation usually results in some degree of convergence, even when one eye is covered—the so-called accommodation-convergence synkinesis. It exists even in people who never had binocular vision (e.g., when there is uniocular congenital cataract). A similar synkinesis exists between accommodation and pupil constriction.

Iris musculature

The aperture of the iris, the pupil, is controlled by the state of contraction of the two muscles in the iris: the sphincter and the dilator. The sphincter is a parasympathetically innervated muscle; its innervation and drug susceptibility closely resemble those of the ciliary muscle, but unlike the latter it is innervated by neurons in the ciliary ganglion.[26] The dilator's innervation comes from the thoracic sympathetic outflow.

Contraction of the sphincter alone or in association with relaxation of the dilator causes constriction of the pupil, or miosis. Contraction of the dilator and relaxation of the sphincter (the sphincter muscle is stronger than the dilator) causes dilation of the pupil, or mydriasis. Drugs producing miosis (Fig. 16-11), so-called miotics, are either those interfering with the normal mode of operation of cholinesterase (e.g., eserine) or those competing with acetylcholine for the receptor sites of the sphincter muscle (e.g., pilocarpine). There are two major kinds of drugs producing mydriasis, so-called mydriatics: the parasympatholytic drugs such as atropine and its derivatives and the sympathomimetic drugs (various catecholamines, such as phenylephrine, and the amphetamines).

The size of the pupil at any time reflects the relative state of contraction of its two muscles. There is some doubt whether the two major effector substances have a reciprocal pharmacologic action on the two muscles or whether one

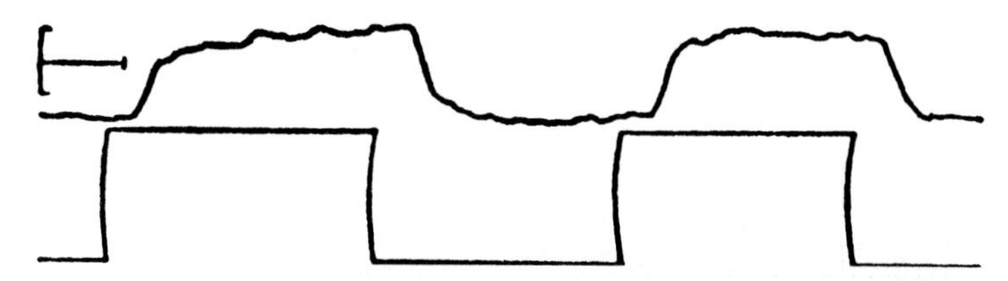

Fig. 16-10. Accommodation responses to step focus changes in normal human eye. Top line: accommodation (length of horizontal line, 1 sec; height of arc, 1 D). Upward movement represents far-to-near accommodation. Bottom line: stimulus signal (same scale). (From Campbell and Westheimer.[10])

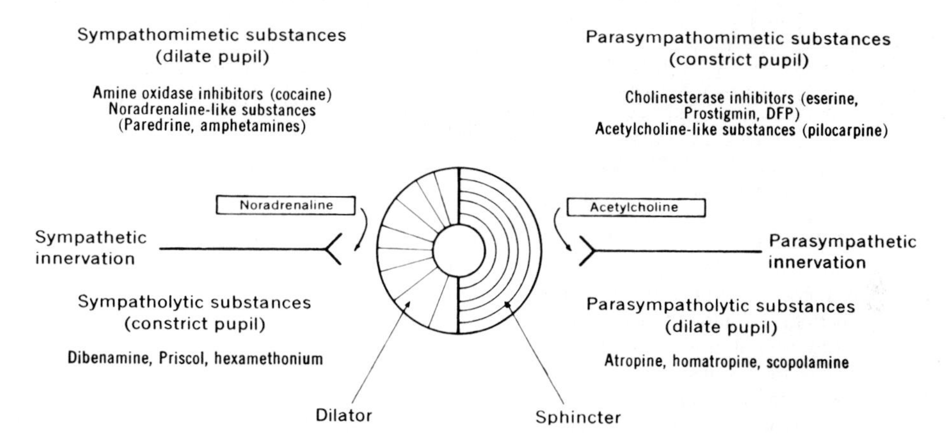

Fig. 16-11. Pharmacology of iris musculature.

muscle, the sphincter, is the more powerful and the pupil is predominately affected by its influence. States such as fright that are characterized by high sympathetic activity produce mydriasis, as does pain. Morphine causes a pronounced constriction of the pupil, an effect probably produced by the central nervous action of this drug.

Constriction of the pupil, effected largely by the parasympathetic activity via nerve III and the sphincter muscle, is a reflex for which the stimulus is light. The pathway for this pupillary light reflex includes afferents from the retina to the pretectal area in the midbrain, internuncial fibers to the Edinger-Westphal portion of nerve III nucleus, preganglionic efferents synapsing in the ciliary ganglion, and postganglionic fibers that reach the sphincter of the iris through the short ciliary fibers. Illumination of one eye causes pupillary constriction of the same eye (direct reflex) and the other eye (consensual reflex). The pupillary light reflex has a reaction time of about 0.2 sec (Fig. 16-12). The pupil constricts also when a person regards a close-up object (near reflex), a response associated with accommodation. An interesting neurologic sign is the Argyll Robertson pupil, commonly encountered in tertiary syphilis, where the pupil response to light is lost but that to near objects is retained.

The iris musculature exerts only a moderate control over the quantity of light admitted to the eye. The pupil diameter of the normal human eye ranges from 8 mm in total darkness to 2 mm under very bright conditions. The maximal possible change in area is thus 16 times, which is far too small to maintain a constant incident light flux in the total range of brightness over which the eye operates.

The pupils undergo a characteristic series of changes with the depth of anesthesia. They are dilated in the beginning stages and also in the deepest stage. In intermediate stages of surgical anesthesia as well as during sleep the pupils are constricted. Curiously, accommodation follows a similar pattern.[25]

Under normal conditions the pupil exhibits small oscillations called physiologic hippus. These changes occur synchronously in the two eyes and may reach an amplitude of 1 mm or more.

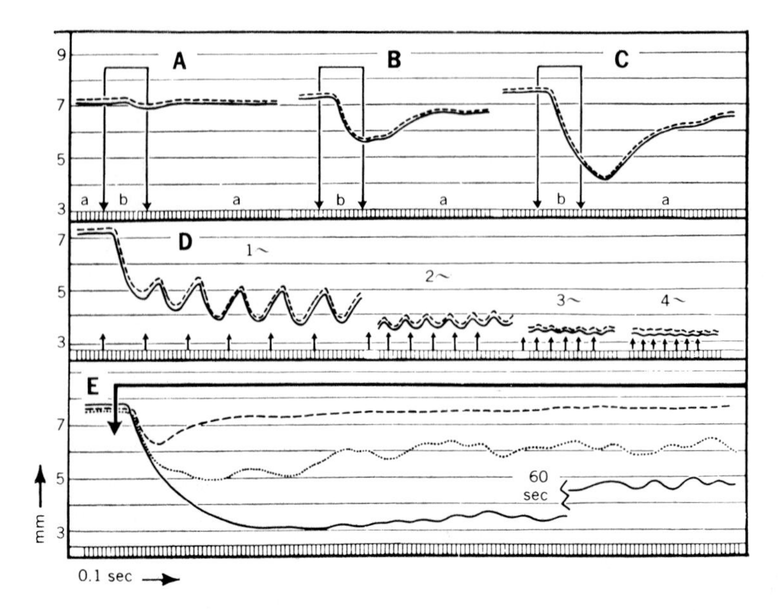

Fig. 16-12. Pupillary reactions to light in normal man. Pupillary diameter (in millimeters) against time (in 0.1 sec units). Solid lines represent right pupil's reactions; broken lines, left pupil's simultaneous responses (except in **E**). First row (**A** to **C**): At *a,* subject's eyes were in darkness. During 1 sec periods, *b* (outlined by dougle arrows), white light flashes were presented. Light intensity was varied by neutral gray filters. In **A,** light was 1, in **B** it was 4, and in **C** it was 8 $\log_{10}$ units brighter than subject's absolute visual threshold. With increasing stimulus intensity, pupillary reflexes increased in extent and speed and latent period shortened. Reactions were equal on both sides, although only right eye was stimulated while left eye remained in darkness. Second row **(D):** At moments marked by small arrows, short bright light flashes (5 msec duration, same brightness as in **C**) were presented to subject's right eye at rates of 1 to 4/sec. With increasing rate of stimulation, pupillary oscillations become smaller and mean pupillary diameter decreases. Third row **(E):** Pupillary movements elicited by prolonged light stimulation (duration framed by arrow). Only right pupil's movements are shown. Intensities were 8 (solid line), 5 (dotted line), and 2 (broken line) $\log_{10}$ units above subject's visual threshold. After initial contraction, pupil dilated partially, more so when light was dim than when it was bright. Pupillary oscillations appeared that were faster in bright than in dim light. (From Lowenstein and Loewenfeld.[20])

Extraocular muscle system

The muscular apparatus that moves the eye in the orbit serves to stabilize the eye with respect to the surrounding environment, compensating for changes in head position; these positional changes depend in large part on vestibular input. In primates and especially in man there has evolved a specialized region of the retina called the fovea within which spatial resolution is markedly better than elsewhere on the retina. The extraocular muscles function to "point" the eye toward a desired target so that its image is located on the fovea.

The eye moves within the orbit like a ball in a socket joint; no lateral movement occurs, that is, one point within the eye remains fixed in the orbit during movement. The center of rotation of the eye is about 13 mm behind the apex of the cornea. There is one pair of muscles corresponding to each degree of freedom for movement of such an apparatus, six muscles in all (Fig. 16-13). They originate in the wall of the orbit and are wrapped around the eye for some distance before inserting thereon. Horizontal movements are produced by reciprocal contraction and relaxation of the medial and lateral recti. Vertical and rotational (about the anteroposterior axis of the eye) movements are produced by similar reciprocal contractions of the superior and inferior recti coordinated with actions of the two oblique muscles. The muscles are striated, are relatively powerful for their load, and have fast contraction times; their motor units have a low ratio of muscle-to-nerve fibers. There are stretch-sensitive receptors within the extraocular muscles, but their central reflex or sensory functions are unknown: they do not evoke the stretch reflexes common to other muscles. The levator palpebrae elevates the upper lid and frequently operates in conjunction with the superior rectus.

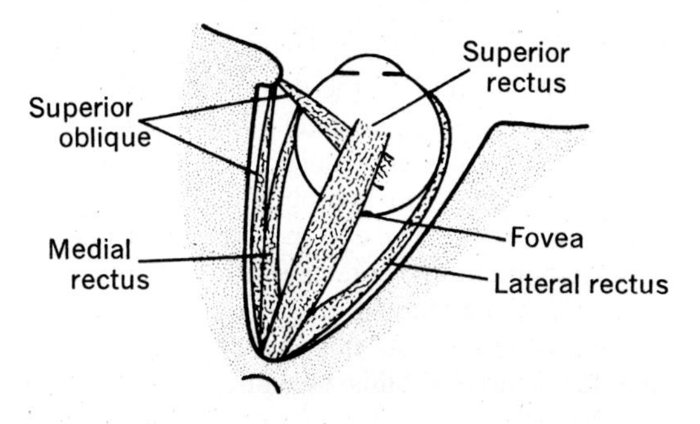

Fig. 16-13. Extraocular muscles of right eyeball as viewed from above. Inferior oblique runs parallel to final portion of superior oblique but is situated below eyeball. Inferior rectus runs parallel to superior rectus but is situated below eyeball. Eyeball is tightly packed into orbit with orbital fat, and content of orbit is held back by lids. As a consequence, when muscles contract, they will rotate eyeball around axis through center of rotation normal to plane containing points of origin and insertion of muscles and center of rotation.

The neuromuscular junctions in these muscles are particularly susceptible to the action of curare-like agents, and the earliest signs of neuromuscular disorders such as myasthenia gravis (Chapter 5) may appear in them as drooping of the eyelid (ptosis) or divergence or convergence of the eyes themselves. In some species the orbit contains smooth muscle innervated by autonomic nerves, which on contraction protrudes or retracts the eye relative to the orbit, but, if present at all, the muscles are rudimentary in man.

One of the major organizational principles of primate ocular motility is that normally both eyes stay and move in parallel. This so-called conjugacy holds in all but one set of circumstances (the exception being when a target is close and needs to be converged on). As simple and obvious as this arrangement appears, it is in fact based on a complicated neural organization because the two orbits are mirror symmetric and each contains 6 muscles in a complex mechanical layout. Moreover, each superior oblique muscle is innervated by nerve IV, whose nucleus is on the other side! Each lateral rectus muscle is innervated by nerve VI and the remaining muscles by nerve III. The operation of the 12 muscles in the two orbits as a single functional unit is thus a remarkable central neural achievement.

During deep sleep or anesthesia, no impulses reach the ocular muscles along their nerves and the eye assumes its anatomic position of rest. During the waking state or during activated sleep (Chapter 10), all the muscles are in a graded, reciprocal state of contraction and the eye position is determined by the net balance of all the forces acting on the eyeball.

Two aspects of oculomotor behavior thus need explanation: the neural basis of the maintenance of steady gaze and the signal routing leading to the execution of movements of the eyes. The absence of changes in load and effect of gravity make this a special muscular system, and one need not be surprised that little help can be obtained from studying limb or trunk movement. In fact, the opposite may be true: the relative compactness of the oculomotor apparatus and its close physical relationship with the brain stem mechanisms subserving vital functions such as breathing and alertness make some defects in it of more direct diagnostic import for many kinds of neural dysfunction.

One of the more striking features of oculomotor function is a remarkable stereotypy of the executed movements, which betrays the existence of just a few modes of operation of the underlying neural control circuits.

Because the motor apparatus is very responsive and the load is small and constant, actual eye positions and velocities represent with high fidelity the "intent" of the CNS. In fact, it is so usual for a CNS command to have been executed faithfully, that no special facility exists to monitor eye position—the oculomotor apparatus operates "open loop": the sending of signals through the motor nerves is regarded as equivalent to the movement having been executed. This situation gives rise to a wholly different way of estimating the relation between the eye and the outside world than is the case with the limbs. It has been called "command copy" and means that whenever a CNS motor instruction to move the eye has been issued, a copy of this command is sent to the sensory centers, where it can be compared with the movement of the retinal image. If the movement of the retinal image matches the commanded movement of the eyes, the outside world is perceived as having remained stationary. Such a system, which does not depend on sensors (e.g., muscle spindles) is, however, susceptible to perceptual errors (e.g. when a part of the visual world moves and another remains fixed) and to severe difficulties in lower motor paralysis (e.g., when one or more muscles do not follow the command through peripheral nerve or neuromuscular junction block).

Three basic types of ocular movements can be observed in normal alert primates. In two, the saccades and the smooth or gliding movements, the eyes move conjugately, and in the third, the vergence movements, the two eyes move relative to each other. At this stage of our knowledge of the nerve control circuits, it is reasonable to assume that these represent the only three basic movement patterns and the problem of neural control can be broken down into the mechanism by which these movement patterns are generated and, separately, the way they are called into play by sensory and higher cortical signals. In addition to the saccadic, smooth, and vergence movements, there may occasionally be seen other types of movements. They are usually indicative of CNS pathology (e.g., see-saw nystagmus) or improper sensory development (the pendular nystagmus of albinism), or they occur when alertness fails (drifts, often binocularly disjunctive, during onset of sleep and anesthesia).

In a description of the neural mechanisms generating eye movements the best approach is to work in the peripheral-to-central direction. There are of the order of 3,000 motoneurons leading to each of the 12 extraocular muscles; this total outflow of about 40,000 fibers represents about 10% of all motoneurons in the body. The simplest behavioral situation is one in which an alert person has both eyes looking straight ahead, that is, both eyes are in their primary position. Although no special movement impulses are at play, the motoneurons carry a steady impulse rate in the range of 10 to 100/sec, each neuron firing at a rate somewhat different from the next. Only when alertness fails will the rates reduce to zero. This means that in an awake person the eye at rest does not constitute a passive situation; the eye position is actively maintained by the appropriate level of impulses in all muscles.

Eye positions other than the primary are represented by proportionally different impulse rates in the different eye muscles. The origin of this maintained impulse rate in the oculomotoneurons of an alert primate is not known. It does not arise within the motoneurons themselves, which have been shown to be in general similar to other motoneurons, that is, they can initiate impulses only on synaptic input. An attractive hypothesis would be a regenerative circuit of the nature of a positive feedback via axonal recurrent collaterals. However, such collaterals do not exist in the case of oculomotoneurons; therefore the source of the continuous synaptic input has to be looked for elsewhere.

Another hypothesis to account for the maintained impulse rate is to postulate that after each eye movement, the impulse burst that initiates the movement is retained in a supranuclear regenerating neural circuit. This circuit would have to have the property of holding the impulse rates at the level demanded by the control signal. However, the steady impulse rates in the oculomotoneurons exist even in reduced oculomotor systems in primates with major brain stem lesions who cannot make any eye movements at all. In these cases the impulse rates start up on awakening and are maintained at a steady level until drowsiness sets in. At this stage of our knowledge the source of this synaptic input, which must be prototypical for maintained posture in general, has to be placed in a hypothetical pacemaker system dependent on alertness. The close anatomic relationship between the oculomotor apparatus in the brain stem and the reticular formation, to which physiologic substrates of sleep and alertness have been traditionally assigned, makes this supposition plausible. Neural pacemaker circuits have been widely studied in the invertebrate nervous system.

Once steady maintenance of eye posture is understood to be the manifestation of a given set of steady impulse rates in the oculomotoneurons, eye *movements* can be viewed as manifestations of central signals to change these impulse rates. The most ubiquitous eye movement in man and primates is the saccade, which is already present at birth. It is a sudden, binocularly conjugate, preprogrammed displacement. It lasts 20 to 50 msec, depending on the extent of the movement. The trajectory is very fast, showing velocities of up to 600 degrees/sec, and often there are overshoots. The saccade is the consequence of simultaneous changes in impulse rates in oculomotoneurons from the set representing the initial eye position to that representing the final eye position. These changes are accompanied by quite marked transients (Fig. 16-14). Neurons that change their rate downward often cease firing briefly altogether, and neurons that change upward may show brief bursts of impulse rates of even up to 1,000 Hz in the monkey. (The monkey has been used extensively for a single-cell analysis of this system. Except where perceptual factors are concerned, the macaque's oculomotor apparatus is similar to man's, although the velocities of movement can be higher by a factor of up to 2).

Several classes of brain stem neurons have been observed whose firing patterns change in a manner closely related to those of the oculomotoneurons in time spans closely associated with saccades.[13,19] Some change their firing just before (4 to 5 msec) the beginning of the saccadic transients in the motoneurons and can be thought of as initiating the neural sequence of events end-

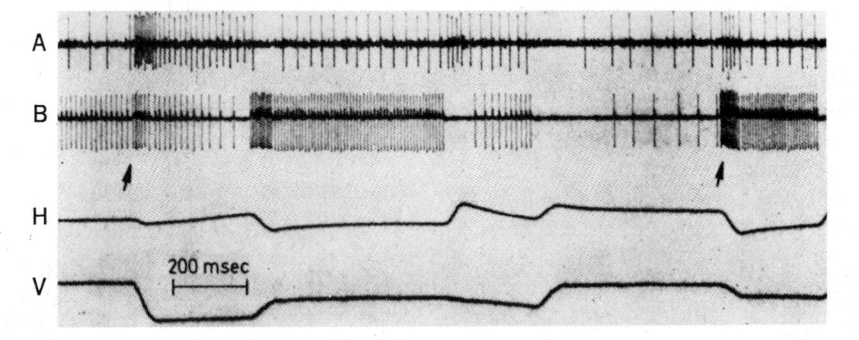

Fig. 16-14. Simultaneous record of impulses in two ocular motoneurons in an alert macaque during spontaneous saccadic and smooth eye movements. *A,* Left inferior oblique motoneuron; *B,* right abducens motoneuron; *H,* horizontal eye position (↑ left); *V,* vertical eye position (↑ down). Record shows position coding of motoneuron impulses and saccadic transients. (From Eckmiller et al.[14])

ing in the abrupt movement of both eyes called saccade. Others definitely follow saccades and may be carrying corollary information to other CNS regions that an eye movement has just occurred.

Just as the saccadic movements have their origin in sudden changes, albeit with heavy transients, in motoneuron impulses, so do the smooth conjugate movements. Their code in premotoneuron cell firing is still obscure.

Yet there is strong evidence that the penultimate site of origin of eye movement signals (the ultimate being in the motoneurons) is in the region of the brain stem adjacent to the motor nuclei of the eye muscles, lying in general between the top of the midbrain and the vestibular nuclei and intimately connected with the reticular formation. Unfortunately, the anatomic layout is not even remotely regular; the various levels seem to be intertwined with each other and with other systems, a fact that makes anatomic localization of single cells no definitive guide to their place in the sequence of neurons ending on the motoneurons. Nevertheless in addition to the widespread presence of eye movement–related signals in neurons there, we have the following evidence implicating the brain stem region outlined previously in the initiation of eye movements:

1. *Anatomy.* Injection of horseradish peroxidase into the nerve III nucleus of cats shows that neurons in the reticular formation, colliculus, and vestibular complex send axons there.[17]
2. *Stimulation.* Insertion of electrodes into the reticular formation and electrical stimulation in the alert monkey causes bilaterally conjugate slow or fast eye movements, depending on the mode of stimulation.[9] In some places, stimulation causes the inhibition of all saccades, no matter what their origin.
3. *Lesions.* Lesions in the reticular formation cause a characteristic syndrome of gaze paralysis, where no lateral saccades can be made with either eye.[9]

The difference between the activity specific to the individual eye muscles and that assigned by unit recording, stimulation, and lesion localization to the next higher, that is, supranuclear, stage is that here we generally have a binocularly coordinated system, and this implies the existence between the two levels of a switching circuit that distributes the excitation in appropriate measure to the neurons of the 12 individual muscles. It is innate (saccades are almost perfectly conjugate at birth), and if it changes at all during life, it does so very ineffectually, because peripheral injury to a nerve or muscle leaves relatively permanent defects in conjugacy.

Stimulation and lesion making in animals, and pathologic studies in human patients, have also implicated several other neural regions as playing a role in eye movement control. In addition to the vestibular system, which will be described in detail subsequently, they are the superior colliculi, the cerebellum, and various thalamic and cortical loci. However, it seems that most eye movement patterns are not critically dependent on these regions, because their ablation leaves the ability to make these movements intact, although the mode of their utilization for visual tasks is greatly changed. Usually, stimulation causes saccades, whose amplitude and direction critically depend on the place. The colliculus has the additional interesting feature that a place whose stimulation causes a certain saccade also harbors neurons that can be visually activated and whose receptive field is situated in such a place in the visual field that the stimulated saccades will bring the fovea there. This system has properties needed for a primitive visual fixation mechanism. One region in area 8 of the frontal cortex, the frontal eye field, has a particularly low threshold for stimulation.[9] Contralateral saccades are induced here, with a latency of 10 to 15 msec. It is significant that saccades can be induced by stimulation of the frontal eye fields or colliculi only if the animal is awake. This fact, together with the absence of direct fibers from these regions to the oculomotoneurons and the binocularly conjugate nature of these movements, again emphasizes the significance of the brain stem distributing network and the crucial role of the alertness-dependent pacemaker probably located there.

The vergence eye movement system deserves a special note. Although operating via the same muscles, it is slower than most saccadic and sawtooth nystagmic movements, and it may still be present when signals for other movements do not reach the medial rectus motoneurons. Reciprocal interchange of signals between medial and lateral rectus muscles takes place via the prominent fiber bundle called the medial longitudinal fasciculus, situated just ventral to the fourth ventricle. Its section causes an inability of the ipsilateral eye to turn in (adduct) for saccadic or vestibular stimuli; convergence stimuli can, however, still induce adduction of the eye.

OCULOMOTOR RESPONSES

Ideally one would like to be able to describe in physiologic terms the operation of the whole sequence that starts with a physical stimulus to a sense organ and ends with a motor act. Among oculomotor acts, there are, to be sure, also movements that are not immediately predicated on a sensory stimulus change, the so-called voluntary movements (e.g., during reading), but these may not unreasonably be left out of consideration in the present context.

Of interest are eye movement responses (1) to changes in head position and (2) to changes in the visual world. The first will be described in some detail as vestibulo-ocular responses, the second more sketchily as visual eye movement responses. Neither are simple reflexes; the second, in fact, demonstrates a considerable involvement of higher centers. In addition, there are complex responses in which coordinated head and eye movements occur jointly. They can be initiated by stimulation of collicular and brain stem regions, but whenever such a head movement occurs, it in turn triggers a vestibulo-ocular response, which is then superimposed on all other eye movement. This shows the primitive, basic nature of the vestibulo-ocular connection, which is also of the greatest evolutionary antiquity.

Vestibulo-ocular responses[18]

Changes in rotational head velocity trigger signals in the peripheral vestibular mechanism that are initially proportional to head acceleration, but decay slowly with a time constant of about 5 sec. The effects can take place in any of the 3 rotational degrees of head movements, because the sensory apparatus has separate mechanisms in three mutually perpendicular planes. For the sake of simplicity the following description is restricted to head rotation around an axis that is normal to the plane of the horizontal semicircular canals whose responses are coupled to horizontal eye movements. In man the canals are embedded in the head in a plane elevated at an angle 30 degrees to the horizontal; this means tests specific to the horizontal canals have to be carried out with a head stance that restricts rotational stimulation to that plane. Fig. 16-15 illustrates the relationship between changes in head acceleration, velocity, and position, position of the cupula, impulses in the vestibular afferents, and the various parameters of eye velocity, position, and gaze. The most instructive situation is one in which a person's head or whole body suddenly starts rotating with constant velocity to the right. This corresponds to a step stimulus in the realm of head velocity ($\dot{H}$) and an impulse function in that of head acceleration ($\ddot{H}$). In accordance with Steinhausen's model of semicircular canal and cupula function (Chapter 27), the consequence is a sudden cupular deflection (right cupula to the left and left cupula to the right) that decays slowly. Impulses in the vestibular afferents contained in nerve VIII, which in a resting head have a constant high rate (even during sleep!), are now modulated in the manner shown: the right increases and the left decreases. Synaptic connections in the vestibular nucleus give rise to impulses to the oculomotor neurons designed to give a smooth eye velocity of both eyes to the left (i.e., opposing the direction of the head acceleration). The velocity is at first fairly accurately proportional to the head velocity, but then decays to zero with a time constant of 15 to 20 seconds. Since the eyes would soon reach the limit of their gaze, a series of oppositely directed saccades is interspersed. The sequence of smooth movements to the right and saccades to the left is called a vestibular nystagmus to the left (a nys-

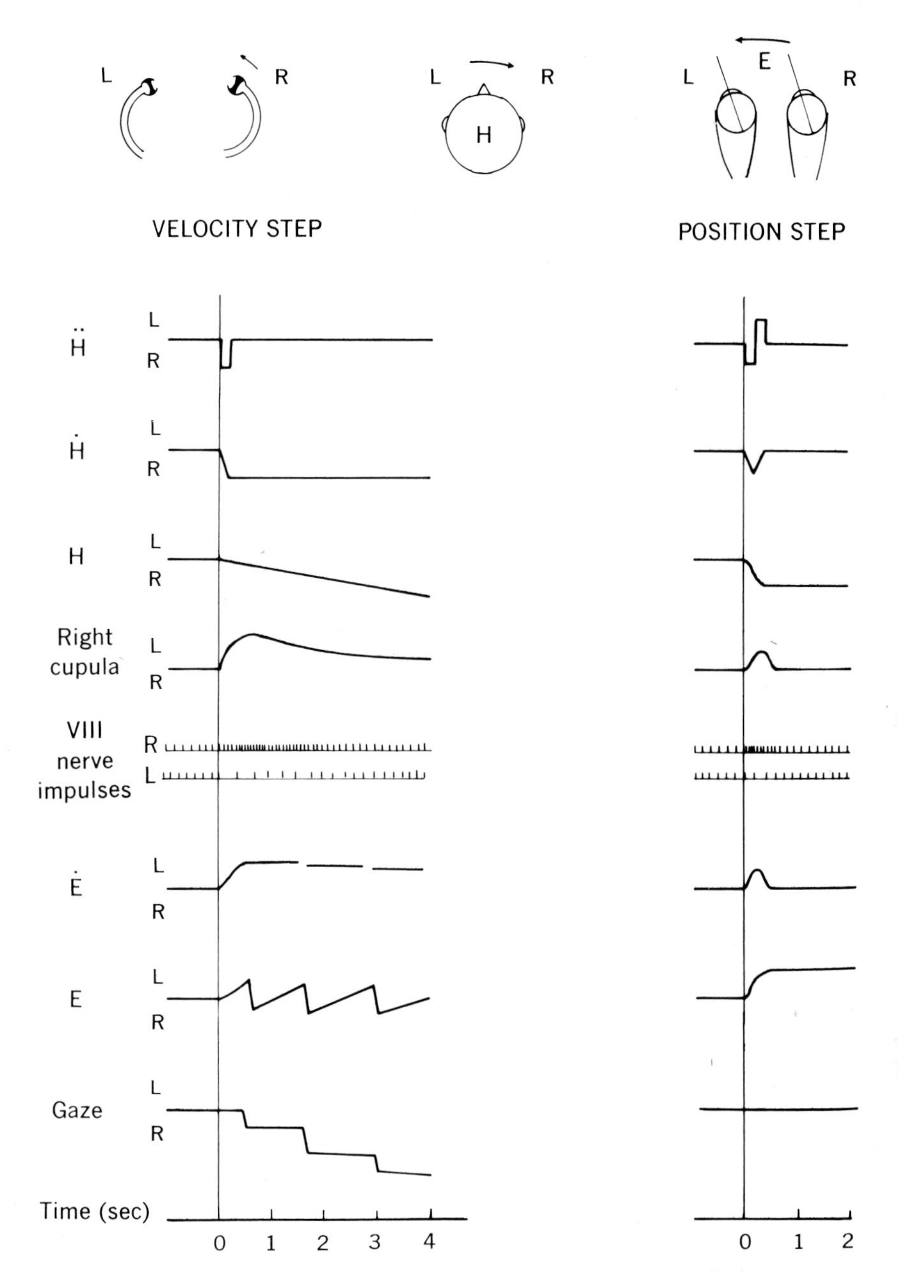

Fig. 16-15. Two typical examples of vestibulo-ocular responses and their physical and physiologic component stages. Individual traces are schematic. *Top from left to right:* Top view of horizontal semicircular canals, head, and eyes. *Left:* Velocity step (i.e., sudden acceleration from rest to rotation with constant velocity to right). *Right:* Position step (i.e., quick turn to right). *Traces from top to bottom:* $\ddot{H}$, rotational acceleration of the head; $\dot{H}$, rotational velocity of head; *H,* position of head. *Position of right cupula* with respect to its resting position. After displacement due to acceleration pulse, cupula returns to its resting position with time constant of about 5 sec. Left cupula undergoes mirror-symmetric change. *Discharge rate of eighth nerve fibers* from horizontal semicircular canals is constant when head remains steady. Excursion of cupula toward midline increases firing rate; excursion away from midline decreases it. $\dot{E}$, Velocity of smooth component of vestibulo-ocular responses is related to cupular deflection but, in unhabituated person, returns to zero with slower time constant. *E,* Eye position shows also saccades that are interspersed with slow movements. Total response is called vestibular nystagmus and has its direction named after direction of saccades. Acceleration pulse to right shown here causes right vestibular nystagmus. Stopping constant-velocity rotation to right would correspond to acceleration pulse to left, and response would be left vestibular nystagmus (i.e., smooth movements to right, saccades to left). *Gaze,* that is, difference between head position and eye position corresponding to position of eye in space, without superimposed visually guided movements. Right side of figure illustrates that vestibulo-ocular response well serves intent of system to keep gaze (and hence retinal image of fixed object) stabilized during fast head maneuvers.

tagmus takes its name from the direction of the fast component), and it clearly works to retain the images of the outside world stationary on the retina. When, in response to some pathologic condition, the saccadic apparatus is inoperative, the eyes move to the extreme position and stay there. This shows that vestibular nystagmus is composed of two phases: the smooth vestibular component and the opposing fast component, which are saccades. These saccades are closely coupled to the vestibular signals because they occur in the dark also and even are apparent in the discharges of neurons in the vestibular nuclei, but they tend to disappear during drowsiness. The smooth vestibular eye movements are sustained longer than other eye movements during the onset of sleep or anesthesia, but during deep sleep and lower planes of anesthesia, they are absent.

Fig. 16-15 also shows the situation with respect to all these variables during a single head turn. It is seen that a remarkably accurate stabilization of the outside world on the retina occurs, and this is clearly the feature for which this system has evolved. At a beginning level of study the vestibulo-ocular reflex shows the kind of regularity that invites the application of control system analysis. For example, the left column of Fig. 16-15 lends itself to the statement that a step stimulus of head velocity leads to smooth eye movements whose velocity decays with a time constant of τ seconds; the smooth component of the eye movements at the outset has a velocity ($\dot{E}$) that is related to the step in head velocity ($\dot{H}$) by the dimensionless quantity gain $G = \dot{E}/\dot{H}$. The two parameters, τ and G, are relatively easily measured, and there is reasonable concordance with the system's properties when the latter are determined by a sinusoidal analysis that gives gain and phase relationship between input (i.e., head velocity) sinusoidal stimuli and output (i.e., eye velocity) sinusoidal responses as a function of frequency of the sinusoids.

This kind of analysis, in its attempts to characterize the properties of the reflex in numeric terms, can be used to demonstrate the time variance of the responses. There are some significant adaptive changes in vestibulo-ocular responses, and these, because they allow a rather direct approach to the structural and functional substrates of neural plasticity, are in the forefront of neurobiologic thinking.

A given head velocity input may induce different vestibular eye movement responses depending on the preceding utilization of the system. This may occur in several levels of complexity:

1. *Alertness changes*. With decreasing alertness, for example, onset of sleep or anesthesia, the gain of the system reduces, dropping even to zero in deep sleep or deep anesthesia.

2. *Habituation*. After repeated stimulation, especially if visual stimuli are present to help visual fixation, the time constant (τ) is reduced.[17] Figure skaters and ballet dancers have a time constant that may be only one fourth that of a normal subject. In animals there is no habituation when the stimuli are given during sleep, suggesting that the habituation takes place centrally. This phenomenon is probably associated with reduction in the activity of a brain stem neural circuit that normally extends the time course of eye movement responses well beyond those of cupular responses (compare line cupula with line $\dot{E}$ in Fig. 16-15).

3. *Adaptive gain change to compensate externally imposed visual conditions*. When optical devices such as magnifying or minifying lenses, or reversing prisms[16] are worn continuously, the subject constantly needs to modify the amplitude of his vestibulo-ocular responses to head rotation in order for stabilization of the visual world to occur. Within 1 or 2 weeks, vestibulo-ocular responses even in the dark (i.e., in the absence of the visual stimuli that induce the compensation) show that an amplitude of responses has changed from their original values toward those needed for visual stability with the devices. In other words, the needed change in gain (eye velocity/head velocity) of the response has been institutionalized. There is some evidence that the cerebellum is involved in these adaptive changes, although there is some doubt whether the adaptation itself occurs in the cerebellum or whether the known inhibitory influences of the cerebellum on the vestibular nuclei facilitate brain stem changes.

The major significance of these findings on habituation and adaptive gain changes is that even as simple a motor act as the vestibulo-ocular response involves a complex interplay of neuronal influences. The minimum pathway between the vestibular periphery and the oculomotoneurons has three synapses (vestibular hair cells → vestibular afferents → vestibular nuclei → oculomotoneurons), and there is clearly enough synaptic interplay to allow major response modification as a result of recent experience, as well as, of course, changes in affect and alertness.

Whereas head rotation is the adequate stimulus, vestibulo-ocular responses can also be elicited by caloric stimulation of the ear. When a subject's head is tilted back 60 degrees, the horizontal canals are in a vertical plane. Irrigation of the outer auditory meatus with warm (45° C) water will induce an upward convection current in the semicircular canals, tending to push the cupula inward. This is equivalent to the effect on this cupula of a head acceleration in an ipsilateral direction and will set up an ipsilateral vestibulo-ocular nystagmus that can be quite rapid and may last for several minutes. Cold water irrigation has an effect in the opposite direction. The caloric test has clinical value because, unlike a head rotation, it can be used to determine the functional integrity of each labyrinth separately.

When for some reason such as disease or accident, nerve VIII on one side suddenly ceases to conduct impulses, the input from the other side becomes unbalanced, giving correspondent vestibular nystagmus. This tends to disappear within a few weeks. That central compensation has taken place is shown by the fact that sudden loss of the other labyrinth will then set up vestibular nystagmus in the direction associated with the labyrinth lost first.

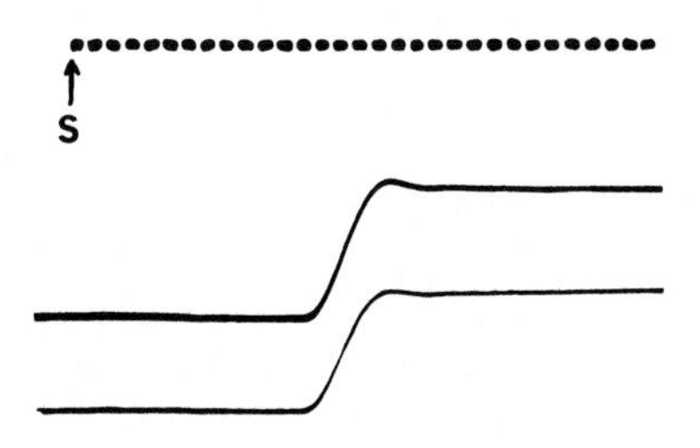

Fig. 16-16. Time course of 20-degree human saccadic eye movement. Time trace interrupted every 0.01 sec. Stimulus presented at *S*. (From Westheimer.[22])

Visual eye movement responses

In contrast to the relatively simple brain stem reflex pathways responsible for vestibular nystagmus, the majority of eye movements are more complex, and they are influenced by neural action in many locales of the nervous system. The predominant afferent influence on eye position and movement is that of the retina itself, an activity relayed centrally via the optic nerve and tract and the geniculocortical system.

The following is a brief survey of the relationship between visual target movements and the eye movements they elicit. If a saccadic movement is initiated as a consequence of target displacement, it follows the latter with a reaction time that varies between ⅛ and ¼ sec. Saccadic movements do not, however, occur only in response to target displacement; they occur continually during environmental scanning. During reading or looking at a picture the eye movements are only of the saccadic type, following each other repetitively at intervals of about ¼ sec. Smooth pursuit movements, on the other hand, occur only when there is a moving target in the visual field. The eyes then move smoothly with good binocular coordination, and the velocity of the tracking movement matches that of the target as long as the latter does not exceed about 30 to 40 degrees/sec.

The two types of eye movements are occasionally employed together (e.g., in optokinetic nystagmus). If a subject watches a continuously moving target such as a train, the eyes fix a point on the target and move with the angular velocity required to keep the image of that point stationary on the fovea. When they reach the limit of the field of eye movements, a quick saccadic move-

Fig. 16-17. Photograph of human eye movement in response to target moving with constant angular velocity of 5 degrees/sec. Heavy trace, stimulus; light trace, eye; vertical lines, time in 0.01 sec. There is a reaction time, and then eye makes up accumulated steady-state error with saccadic movement and tracks target with constant-velocity "smooth" movement. All features of these types of eye movements are conjugate, that is, parallel in the two eyes. (From Westheimer.[23])

ment occurs in the opposite direction, bringing the eyes to a new point on the moving target, and the maneuver is repeated.

Small, spontaneous saccades and other movements occur when the eyes may otherwise appear at rest. These small oscillations are of interest because some theories of visual acuity are based on the supposition that the small scanning oscillations convert a steady light stimulus into a time-varying one to which the sensory receptive mechanism of the retina is much more sensitive than it is to steady illumination. It is now thought unlikely that these oscillations aid in resolution, but it has been shown that if the visual image is stabilized on the retina by special optical techniques, a curious "washing out" of contrast occurs, even though resolution per se may not be diminished.

Disjunctive eye movements

In addition to the binocularly parallel or conjugate movements so far described, animals with binocular vision have a further mechanism that allows the two foveas to be directed to a close-up target—the disjunctive movements. They are slower than saccadic movements but have a similar latency. Not only can the two eyes converge but they can also diverge somewhat, and there is a facility for small vertical and cyclorotational vergences as well. Exact registration of the images of a given target on corresponding points in the retina of each of the two eyes is a necessary prerequisite to good binocular vision with its process of accurate spatial localization, stereoscopy—hence the need for a mechanism to allow this to take place when the target is closer than infinity and to make small adjustments when, owing to anatomic or other factors, the two eyes are not quite accurately aligned.

When the images of a single target on the two retinas give rise to a single coordinated sensory impression, they are said to be fused. When fusion is prevented by covering one eye, it is possible to measure the natural alignment of the two eyes. A slight misalignment, but one that can be overcome by the normal process of convergence (or vertical vergence if the misalignment is vertical), is called a *heterophoria*. A patient in whom normal vergence movements are not adequate to maintain fusion has strabismus, or *squint*. The relationship between accommodation and convergence has already been referred to. It depends on a neural linkage between the midbrain centers subserving the two functions.

REFERENCES

General reviews

1. Duke-Elder, W. S.: Textbook of ophthalmology, St. Louis, 1942, The C. V. Mosby Co., vol. 1.
2. Emsley, H. H.: Visual optics, ed. 5, London, 1955, Hatton Press, Ltd., vol. 1.
3. Helmholtz, H.: Treatise on physiological optics, New York, 1962, Dover Publications, Inc., vol. 1-2.
4. Maurice, D. M.: The cornea and sclera. In Davson, H., editor: The eye, ed. 2, New York, 1969, Academic Press, Inc., vol. 1.
5. Moses, R. A.: Adler's physiology of the eye, clinical application, ed. 5, St. Louis, 1970, The C. V. Mosby Co.
6. Sorsby, A., et al.: Emmetropia and its aberrations, Medical Research Council Special Report No. 293, London, 1957, Her Majesty's Stationery Office.
7. Westheimer, G.: Visual acuity and spatial modulation thresholds. In Jameson, D., and Hurwich, L. M., editors: Handbook of sensory physiology, section 4, Berlin, 1972, Springer-Verlag, vol. 7.

Original papers

8. Alpern, M.: Movements of the eyes; also Accommodation. In Davson, H., editor: The eye, ed. 2, New York, 1969, Academic Press, Inc., vol. 3.
9. Bender, M. B., editor: The oculomotor system, New York, 1964, Harper & Row, Publishers.
10. Campbell, F. W., and Westheimer, G.: Dynamics of accommodation responses of the human eye, J. Physiol. **151:**285, 1960.
11. Davson, H.: Physiology of the ocular and cerebrospinal fluids, London, 1956, J. & A. Churchill, Ltd.
12. Davson, H.: The intra-ocular fluids; also The intraocular pressure. In Davson, H., editor: The eye, ed. 2, New York, 1969, Academic Press, Inc., vol. 1.
13. Dichgans, J., and Bizzi, E., editors: Cerebral control of eye movement and motion perception, Basel, 1972, S. Karger AG. (Vol. 82 of Bibliotheca Ophthalmologica.)
14. Eckmiller, R., Blair, S. M., and Westheimer, G.: Oculomotor neuronal correlations shown by simultaneous unit recordings in alert monkeys, Exp. Brain Res. **21:** 241-250, 1974.
15. Fincham, E. F.: Mechanism of accommodation, Br. J. Ophthalmol. **8**(suppl.):5, 1937.
16. Gonshor, A., and Jones, G. M.: Extreme vestibulo-ocular adaptation induced by prolonged optical reversal of vision, J. Physiol. **256:**381-414, 1976.
17. Graybiel, A. M., and Hartwieg, E. A.: Some afferent connections of the oculomotor complex of the cat, Brain Res. **81:**543-551, 1975.
18. Kornhuber, H. H., editor: Handbook of sensory physiology. Vestibular system, Berlin, 1974, Springer-Verlag, vol. 6, parts 1-2.
19. Lennerstrand, G., and Bach-y-Rita, P.: Basic mechanisms of ocular motility and their clinical implications, Oxford, 1975, Pergamon Press, Ltd.
20. Lowenstein, O., and Loewenfeld, I. E.: The pupil. In Davson, H., editor: The eye, ed. 2, New York, 1969, Academic Press, Inc., vol. 3.
21. Walls, G. L.: The vertebrate eye, Bloomfield Hills, Mich., 1942, Cranbrook Institute of Science.
22. Westheimer, G.: Mechanism of saccadic eye movements, Arch. Ophthalmol. **52:**710, 1954.
23. Westheimer, G.: Eye movement responses to horizontally moving-visual stimulus, Arch. Ophthalmol. **52:** 932, 1954.

24. Westheimer, G.: Optical properties of vertebrate eyes. In Fuortes, M. G. F., editor: Handbook of sensory physiology, section 2, Berlin, 1972, Springer-Verlag, vol. 7.
25. Westheimer, G., and Blair, S. M.: Accommodation of the eye during sleep and anesthesia, Vision Res. **13:** 1035, 1973.
26. Westheimer, G., and Blair, S. M.: The parasympathetic pathways to internal eye muscles, Invest. Ophthalmol. **12:**193, 1973.

17

KENNETH T. BROWN

Physiology of the retina

In Chapter 16 it is shown that the major optical function of the eye is to gather light rays from the primary stimulus, which is the external world, and focus them into a secondary stimulus, which is an image on the retina. This image is a two-dimensional representation of the external world, reduced in size to fit the retina, and it is the stimulus that leads to spatially patterned excitation of the photoreceptors. The ultimate end products of the visual system are sensations that represent the external world with considerable accuracy and that can therefore guide adaptive behavior in relation to the external environment. Since these sensations are part of conscious experience, they fall within the subject matter of psychology. Thus one of the overall tasks of visual physiology is to bridge the gap between physics and psychology by explaining how light, as a phenomenon of physics, is translated into light as a conscious experience. Our visual experiences contain much information and attain considerable complexity, so the problems of visual physiology are correspondingly complex. In this chapter, only retinal aspects of visual physiology will be considered; more central levels of the visual system will be taken up in Chapter 18. Emphasis will be placed on evidence that applies most clearly to man, since the human retina is of primary interest in medicine.

FUNCTIONAL ANATOMY

It is helpful to look first at some major features of retinal structure, because retinal histology tells much about how the retina functions. Most knowledge of retinal structure, which may be applied with confidence to the human eye, has been obtained from macaque monkeys. These monkey retinas are very similar to those of humans but are more readily obtained under good conditions for histologic study.

Cells and their synaptic relations in the peripheral retina

The main types of retinal cells have been revealed by Golgi preparations,[1,8] and details of their synaptic contacts have been studied by electron microscopy.[5,25,81,101] Fig. 17-1 summarizes the results of such work in the macaque monkey. The retina contains two main classes of receptors: the *rods* and the *cones*. Each type of receptor possesses both outer and inner segments, which are joined by a connecting cilium. In retinal anatomy the terms "outer" and "inner" designate relative distances of structures from the center of the eye. In the peripheral retina, both the outer and inner segments of rods are slender and roughly cylindric; hence their appearance is rodlike. In peripheral cones the outer segments are rather conical in form, and the inner segments are thicker than those of rods. The outer segments of both cones and rods contain visual photopigment that absorbs the light, leading to excitation. These photoreceptors are best regarded as neurons with outer and inner segments that are highly specialized. The axon terminals of the photoreceptors exhibit expansions; the rod expansion is roughly spheric and is thus termed a rod spherule, whereas the cone expansion spreads more widely and is called a cone pedicle.

The second-order neurons, in the direct line of transmission, are bipolar cells. The ganglion cells are thus third-order neurons, and their axons lie close to the retinal surface in the optic fiber layer. These axons proceed directly to the optic disc, where they leave the eye and course through the optic nerve and tract to terminate on fourth-order neurons of the lateral geniculate body. The axons of these fourth-order cells complete the line of transmission to the visual cortex.

As shown in Fig. 17-1, bipolar cells have been classified into three main types. The *midget bipolar cell* makes multiple dendritic synapses with only one cone, whereas multiple axonal synapses are made with a *midget ganglion cell*. Since the midget ganglion cell does not synapse directly with any other bipolar cell, it appears to provide the third stage of a "private line" leading from a single cone to the brain.[8] The cone synapse with a midget bipolar cell features presynaptic vesicles clustered about a presynaptic ribbon, so this

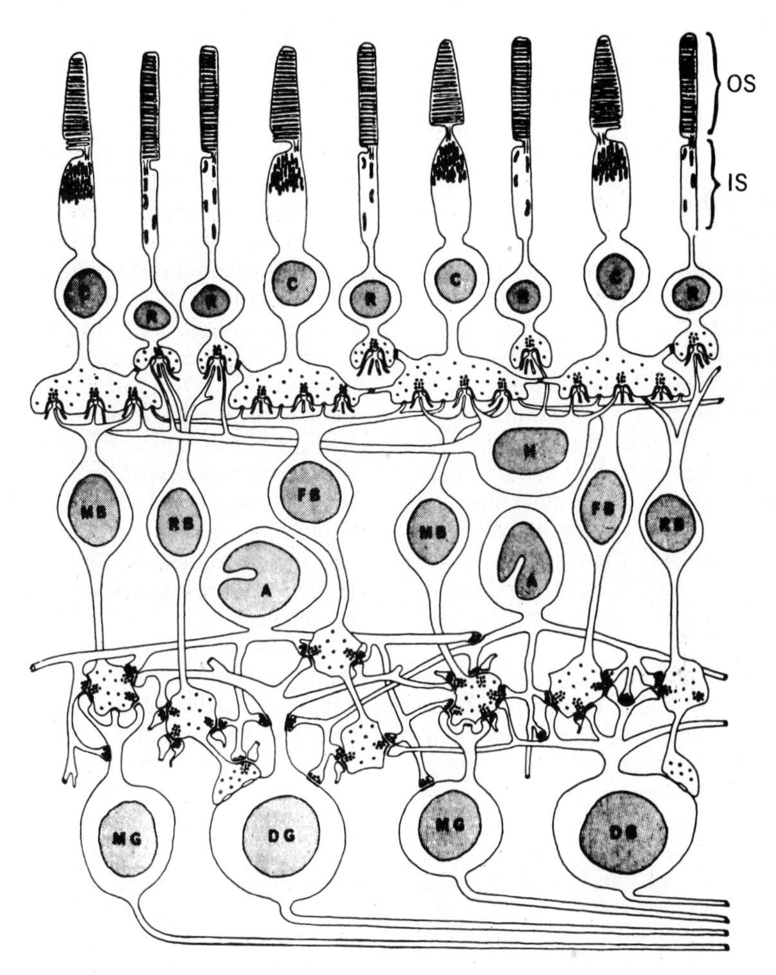

Fig. 17-1. Summary diagram of macaque monkey retina. Abbreviations: *OS,* outer segments; *IS,* inner segments; *R,* rod; *C,* cone; *MB,* midget bipolar; *RB,* rod bipolar; *FB,* flat bipolar; *H,* horizontal cell; *A,* amacrine cell; *MG,* midget ganglion cell; *DG,* diffuse ganglion cell. (From Dowling and Boycott.[5])

synapse appears to be chemical by anatomic criteria. The *rod bipolar cell* makes dendritic synapses only with rods. The fine structure of these synapses is similar to that of a synapse between a cone and midget bipolar cell, but the rod bipolar cell contacts many rods and thus represents converging rod pathways. The *flat bipolar cell* contacts many cones by means of widely spreading processes; these contacts are superficial indentations that lack typical features of either chemical or electrical synapses.[5] The *diffuse ganglion cell* receives axonal contacts from all three types of bipolar cells. Hence this cell represents a considerable convergence of rod pathways and is also a stage at which cone and rod inputs are combined.

The anatomic situation is greatly complicated by lateral connections at several levels of the retina. Fig. 17-1 shows cone-rod contacts and cone-cone contacts between the axon terminals of adjacent receptors.[5,101] Also, the retina contains two cell types that appear specialized for mediating lateral interactions. Recent work shows that the macaque monkey retina has a single type of *horizontal cell,* which makes short dendritic connections only with cones, whereas a long axon proceeds laterally and makes contact only with rods.[25,81] Thus Fig. 17-1 is probably incorrect in showing that horizontal cell dendrites, and also axons, contact both types of receptors. More proximally, the *amacrine cell* exhibits widely spreading processes that contact bipolar cells, ganglion cells, and other amacrine cells. These lateral connections confer to the retina an anatomic complexity rivaling that of the brain. This is probably related to the embryology of the retina, which is formed as an outpouching of the brain and may be regarded as a sample of brain tissue. The retina is a unique case of receptors that are intimately connected with brain tissue at a peripheral level of the nervous system.

Retinal layers and their circulatory supply

Fig. 17-2 is a photomicrograph that shows some additional features of retinal structure. The precise stratification of the retina into various layers is illustrated, and the major layers and membranes are labeled. Note that the outer nuclear layer, containing the cell bodies of receptors, has more cells than the inner nuclear layer. This occurs despite the fact that the inner nuclear layer contains cell types other than bipolars, illustrating a convergence of pathways from receptors to bipolars. There are still fewer cells in the ganglion cell layer, illustrating an additional convergence of pathways to the ganglion cells.

The circulatory supply of the retina has two main divisions, the so-called retinal circulation and the choroidal circulation. The retinal circulation is formed from the central retinal artery, which enters the eye at the optic disc. This artery branches into smaller arteries that course over the retinal surface; the collecting veins return over the retinal surface, and all join to form the central retinal vein that leaves the eye at the optic disc. The major blood vessels of the retinal circulation tend to run in artery-vein pairs; one such pair is shown in Fig. 17-2. The capillaries of the retinal circulation occur in strata at several retinal levels, and the deepest capillary plexus occurs with consistent precision at the outer margin of the inner nuclear layer.[9] Major vessels of the choroidal circulation penetrate the sclera to one side of the optic disc and form a rich system of large vessels in the choroid[9]; some of these may be seen in Figs. 17-2 and 17-3. An exceedingly rich capillary plexus, called the choriocapillaris, is derived from the choroidal circulation. As shown in Fig. 17-2, this lies immediately against the basement membrane of the pigment epithelium, which is called Bruch's membrane. Note that the entire deeper half of the retina, from Bruch's membrane to the synapses between photoreceptors and second-order neurons, is devoid of blood vessels. Thus there are no capillaries in direct contact with any portion of the photoreceptors.

Because of the dual circulatory supply of the retina, it long seemed likely that cells of the ganglionic and inner nuclear layers are served primarily by the retinal circulation, whereas the receptors are supported partly from the retinal and partly from the choroidal circulation. This has been confirmed by studies in which needles were inserted into the monkey eye and used as channels for recording microelectrodes and other devices.[3,33] A steel rod was introduced through one needle, and its rounded end was pressed on

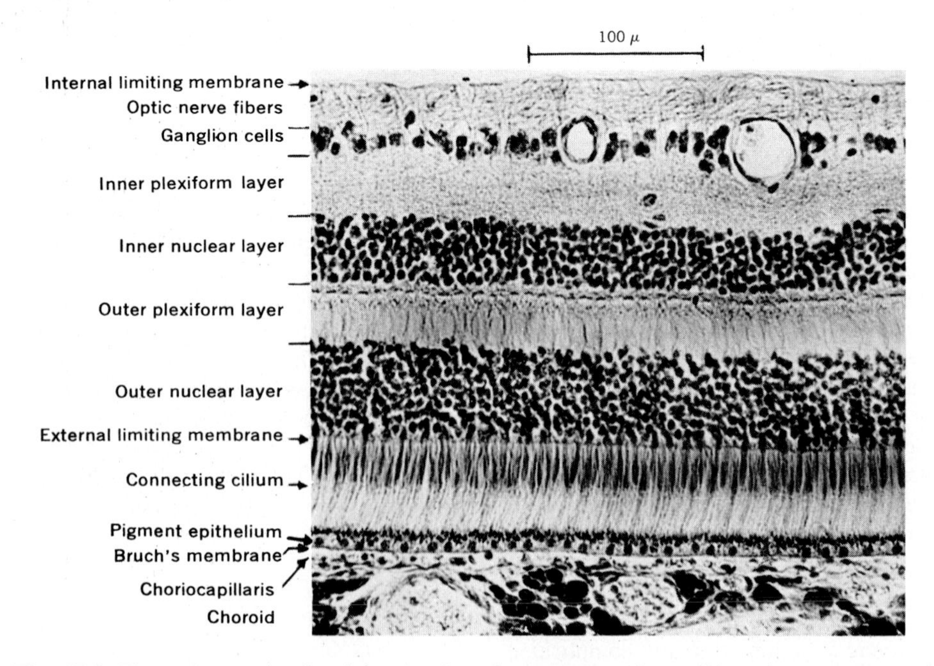

Fig. 17-2. Photomicrograph of peripheral retina of macaque monkey, with major structures and layers labeled. (From Brown.[2])

the optic disc to occlude the retinal circulation without occluding the choroidal circulation. Within a few minutes after this procedure the electrical responses of ganglion cells and most cells of the inner nuclear layer were abolished, but receptor potentials remained for many hours. Thus the photoreceptors derive an important portion of their metabolic requirements from the choriocapillaris.

Retinal supporting tissue

The retina is mechanically delicate and bound together mainly by a type of glial cell called the Müller cell. These cells are oriented perpendicularly to the retinal surface and send processes in both directions from their somata in the inner nuclear layer. The processes directed toward the vitreous humor form expansions when they reach the inner surface of the retina; these fuse together to form the internal limiting membrane. The processes passing the other direction terminate at the bases of receptor inner segments. There they form the external limiting membrane, which is not a true membrane at all but an aggregation of the distal tips of Müller cells. The Müller cells bind the retina together over the greater part of its thickness, but there is no true connection between the receptors and pigment epithelium. Hence retinal detachments resulting from severe blows, or other pathologic effects, usually occur between photoreceptor outer segments and the pigment epithelium.

Optical effects of retinal tissue

Light must pass through the entire retina before it is absorbed by photopigment in the receptor outer segments. This probably limits visual acuity in the peripheral retina. Such a thin sheet of nervous tissue is almost transparent, but its optical effects cannot be neglected, and the stimulating light must also pass through the retinal blood vessels.

Light not absorbed in receptor outer segments passes on to the pigment epithelium or choroid. If this light were reflected back, it would degrade the retinal image. However, this light is largely absorbed by melanin, a black pigment contained in granules in the pigment epithelium and choroid, as shown in Figs. 17-2 and 17-3. By greatly reducing reflected light, the optical quality of the image at the photoreceptors is thus improved.

Anatomic specialization of the fovea

Fig. 17-3 shows a histologic section through the fovea, a small central part of the retina that is used whenever we look directly at an object; it occupies the subjective center of the visual field. This is the retinal area of highest visual acuity, and the fovea has a number of anatomic specializations that favor visual acuity. Note that the pigmentation of the pigment epithelium is stronger in the fovea than in the peripheral retina; reflected light is thus especially well suppressed in the fovea. The central fovea contains only

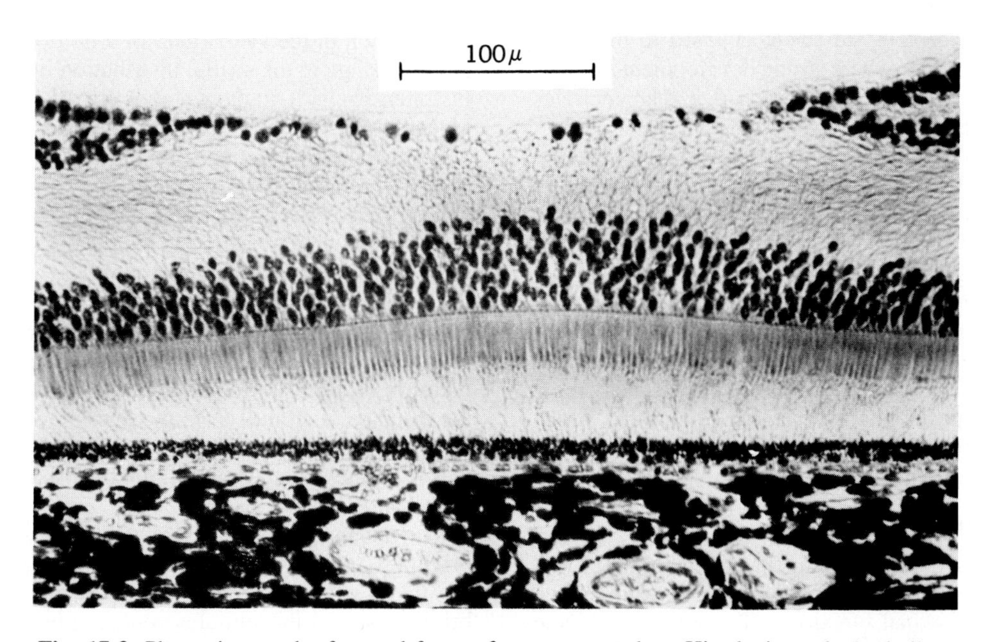

Fig. 17-3. Photomicrograph of central fovea of macaque monkey. Histologic methods similar to those used for Fig. 17-2. Very thin outer segments of foveal cones are poorly preserved. (From Brown et al.[3])

cones. The outer and inner segments of foveal cones are rather elongated and cylindric by comparison with peripheral cones. Hence foveal receptors look superficially like rods but may be recognized as cones by the shape of the axon terminal, the cone pedicle. The foveal cones are very slender; this increases the fineness of the "grain" of the foveal receptor surface, thus enhancing visual acuity. In the very center of the fovea the receptors are especially slender, having a diameter in humans of 1 to 1.5 μm.[8] The increase of receptor density in the central fovea results in a thickening of the outer nuclear layer, as shown in Fig. 17-3. The length of the inner segment does not increase in the central fovea, but in humans the length of the outer segment approximately doubles.[8] Since the outer segment contains the photopigment, this increased length probably increases the photopigment content of these receptors. Thus the slenderness of receptors in the central fovea favors visual acuity, and the increased length of the outer segment reduces the loss of sensitivity that would otherwise occur.

In the central fovea the ganglionic and inner nuclear layers are almost completely absent, and there are no blood vessels of the retinal circulation. Thus retinal impediments in the path of light to the photopigment are greatly reduced in this area. The axons of foveal photoreceptors turn laterally and make synaptic contact with bipolar cells in the parafoveal region. In Fig. 17-3 the inner nuclear and ganglionic layers may be seen forming on either side of the foveal center. Thus the second- and third-order neurons that serve foveal receptors are swept outward to the parafoveal region, where strong development of these layers creates a parafoveal ridge that encircles the fovea.

Convergent pathways from receptor to ganglion cell, as exhibited by the rod receptor system, probably limit visual acuity. In contrast, the cone pathway through the midget bipolar cell and midget ganglion cell should favor visual acuity. According to Polyak,[8,9] the midget bipolar and midget ganglion cells are the most typical cell types in their respective cell layers of the parafovea. Hence foveal cones appear to take considerable advantage of this type of cone pathway, which probably plays a significant role in the high acuity of the fovea.

Duplicity theory

In 1866 the anatomist Max Schultze[109] first advanced the theory that rods and cones serve different retinal functions. His view has become known as the duplicity theory, which states that in the human retina the cones function at the high intensities of daylight vision, conferring advantages of high visual acuity and color vision, whereas the rods have the greater sensitivity required for night vision but do not mediate color and cannot resolve fine details. His theory was developed largely by comparing receptor structure with known visual abilities in many vertebrate species. Thus he noted that the retinas of diurnal birds (e.g., the falcon) are rich in cones, whereas the retinas of nocturnal species (e.g., the owl and bat) are rich in rods. He also compared anatomic and functional differences between the fovea and periphery of the human retina. The duplicity theory is now so well supported and so universally accepted that it is more doctrine than theory, and it is probably the most useful single principle in visual physiology. In view of contemporary evidence, however, the principle applies better to the *rod system* and the *cone system*, rather than to the rods and cones alone. This is because some of the functional differences between rod and cone systems are not due entirely to differences between rods and cones themselves but result partly from synaptic connections in the rod and cone pathways. The division of the retina into rod and cone receptor systems is of such profound importance that it affects all major retinal functions, as shown in the following section.

ABILITIES OF THE VISUAL SYSTEM

Resolution of stimuli in space: visual acuity

Detection in the visual field of a border, or an abrupt change in the spatial distribution of stimulus intensity, is a fundamental ability of the eye. Under certain conditions the eye can discriminate very fine details in the stimulus pattern. This ability is referred to as visual acuity, which has already been discussed from certain points of view and which may be measured by determining the fineness of stimulus detail that can be seen. Thus measurements of visual acuity indicate the resolving power of the eye in two-dimensional space. Stereoscopic acuity, by comparison, is a measure of ability to discriminate differences in the distance of stimuli from the eye. Since depth perception depends partly on integration of signals from the two eyes, it will not be discussed in this chapter.

The manner in which visual acuity varies with brightness of the stimulus and its distance from the fovea is shown in Fig. 17-4. At the lowest brightness level the stimulus could not be seen in the fovea. It was first detected when presented at

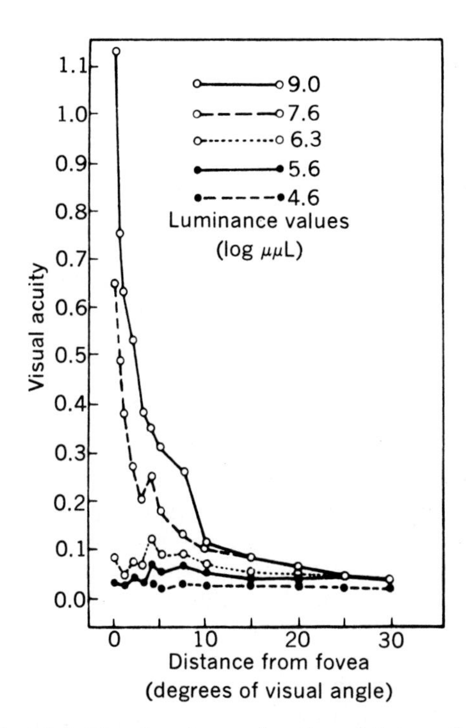

Fig. 17-4. Visual acuity as a function of distance from fovea at five different luminance levels of test stimulus. Test stimulus was the gap in a ring known as a Landolt C, and visual acuity is expressed in arbitrary units on a relative scale. Measurements were made along horizontal meridian of retina on temporal side. For definition of the unit of luminance, the lambert (L), see *The Science of Color*.[4, p.231] (From Mandelbaum and Sloan.[89])

about 4 degrees from the fovea, and there was no distinct change of visual acuity as the stimulus was moved further toward the periphery. This illustrates foveal blindness at low levels of illumination; astronomers have long known that dim stars, which cannot be seen in direct vision, can often be seen by looking slightly to one side. At the highest brightness level, on the other hand, acuity was maximum in the fovea and decreased rapidly as the stimulus was displaced to about 10 degrees from the fovea; beyond that point, the decrease of acuity was more gradual. Thus Fig. 17-4 illustrates how the duplicity theory applies to visual acuity.

Note in Fig. 17-4 that although stimulus intensity has a marked effect on acuity in the pure-cone fovea, it has very little effect in the predominantly rod area at the extreme periphery of the retina. The improvement of visual acuity with stimulus intensity, which occurs most strongly in the fovea, is probably due to a finer ''grain'' in the *cone response pattern* that results from increased intensity. Let us assume for simplicity that there is no background illumination. For any given receptor in an illuminated stimulus area the probability of absorbing light and reacting will increase with stimulus intensity. Thus an increased stimulus intensity will increase the proportion of receptors that respond within the stimulus area. As this occurs, the ''grain'' of the receptor response pattern becomes finer and more closely represents the details of the stimulus itself. Of course, this increased detail in the cone response pattern could not be effective unless each cone possessed a pathway to the brain that approximated a private line. Such a pathway also seems necessary to account for the fact that under light-adapted conditions, acuity increases as the fovea is approached, in a manner that corresponds roughly with increased cone density.[8] Convergent pathways from the rods, on the other hand, are probably responsible for the acuity of rod vision being relatively fixed, at a lower level than that of cones, with little improvement resulting from either increased stimulus intensity or increased rod density. Although the rod density increases toward the periphery and reaches a maximum at about 20 degrees from the human fovea,[89] the lowest curve of Fig. 17-4 shows that there are no corresponding changes of visual acuity. Although detailed pathways from receptor to brain are not yet worked out, it seems established that there is much less convergence for peripheral cones than for rods and that convergence is minimal in the case of foveal cones. Cell counts have shown that in the predominantly rod retina of the guinea pig there are 220,000 rods, 50,000 cones, and 7,000 ganglion cells; in a pure-cone squirrel retina there are 200,000 cones, no rods, and 90,000 ganglion cells.[119] Thus the ratio of rods to ganglion cells in the guinea pig has a minimum value of about 31:1, but the ratio of cones to ganglion cells in the squirrel retina is only slightly greater than 2:1. It appears that the functional duplicity of the human retina, as it relates to visual acuity, results largely from this basis. This illustrates why the duplicity theory is best applied to rod and cone systems rather than to rods and cones alone.

Resolution of stimuli in time: critical fusion frequency

The eye can also resolve stimuli that are separated in time. The limit of this ability is indicated by measuring the minimum frequency at which repetitive stimuli appear to fuse together into a continuous stimulus. This minimum frequency of subjective fusion is called the critical fusion frequency (CFF). As visual acuity is a measure of the spatial-resolving power of the eye, CFF is an analogous measure of the time-resolving power of the eye. The resolution of stimuli in time is limited because the response to a given stimulus does not cease exactly when the stimulus ceases but persists for a time thereafter. The phenomenon of subjective fusion has useful applications; for example, it provides the physiologic basis for motion pictures. The visual continuity of "movies" is attained by flashing successive frames of the film at a frequency in excess of CFF; motion pictures were called "flickers" during the period when the rate at which the frames were flashed was not sufficiently high.

Fig. 17-5 shows CFF as a function of both retinal illumination (intensity of the flickering stimulus) and retinal location. In the fovea, CFF rises rapidly with stimulus intensity and then falls slightly after attaining a maximum value. The foveal CFF can exceed 50 Hz, so the time-resolving ability of the eye is well developed. In the curves at both 5 and 20 degrees from the fovea (Fig. 17-5), CFF first increases with stimulus intensity and reaches a plateau, which is followed by a second and more rapid rise of the curve to a higher level. In the fovea the first limb of these curves is not found, but the second limb is present and maximally developed. This indicates that in the peripheral retina the first limb of each curve represents rod function, whereas the second and higher limb represents cone function. This interpretation is supported by many lines of evidence. Hence the rod and cone receptor systems also affect the time-resolving power of the visual system. The fact that cone CFF can go much higher than that of rods appears to be explained by the different time courses of cone and rod receptor potentials, as shown later in this chapter.

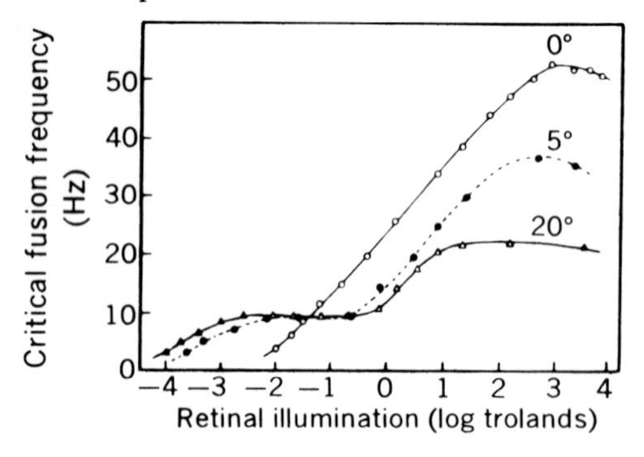

Fig. 17-5. Effect of retinal illumination (stimulus intensity) on critical fusion frequency in fovea (0 degree) and at 5 and 20 degrees above fovea. For definition of the unit of retinal illumination, the troland, see *The Science of Color*.[4,p.232] (From Hecht and Verrijp.[68])

Spectral sensitivity

The sensitivity of the eye to a flash of light varies with the wavelength of the stimulus flash. If the wavelength of the light is varied, and the relative energy required to produce a sensation of constant brightness is determined, the reciprocals of these energies will give a curve called a *luminosity function*. Such a function shows the relative effectiveness (luminosity) of various wavelengths of light and hence the spectral sensitivity of the eye. The scotopic and photopic luminosity functions are obtained under dark-adapted and light-adapted conditions, respectively. Such functions are shown in Fig. 17-6. Note that these are relative functions in which the maximum luminosity of each curve has been set equal to one; this permits convenient comparison of the two curves. If expressed on an absolute basis, luminosity is greater in the scotopic case.

Fig. 17-6 shows that the eye responds best to a limited band of wavelengths called the *visible spectrum,* extending from slightly below 400 nm to somewhat above 700 nm. This places its sensitivity in the wavelength range between ultraviolet and infrared radiation. Wavelengths shorter than 400 nm are normally absorbed by the lens of the eye.[120] In the lensless *(aphakic)* eye the visible spectrum extends downward to at least 315 nm, and the ultraviolet light in this extended spectrum is not injurious to the retina.[120] Wavelengths shorter than 315 nm tend to be injurious to tissue; they are absorbed by the cornea

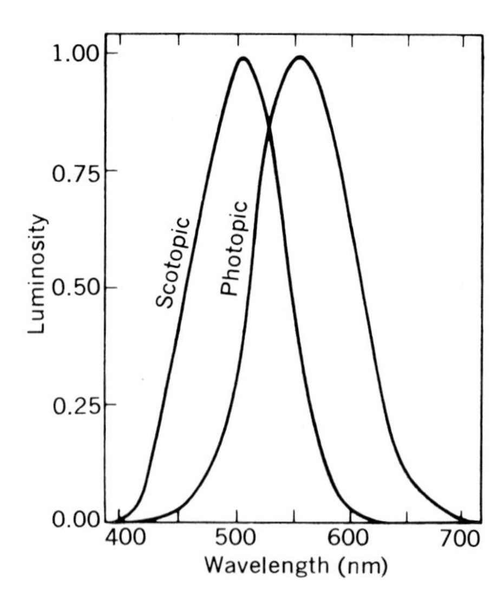

Fig. 17-6. Relative scotopic and photopic luminosity functions of normal human eyes. These functions were adopted as standards by International Commission on Illumination in 1951 (scotopic) and 1924 (photopic). (From *The Science of Color*.[4])

and do not reach the retina.[120] The upper limit of the visible spectrum is determined by the sensitivity of the retina itself, and very low levels of sensitivity to infrared radiation extend to approximately 1,000 to 1,050 nm.[62]

Fig. 17-6 shows that the photopic luminosity function is shifted toward longer wavelengths than the scotopic function. This shift of spectral sensitivity is called the *Purkinje shift* because it was first recognized by Johannes Purkinje in 1825. Maximum sensitivity of the scotopic function is at 507 nm, whereas maximum sensitivity of the photopic function is at 555 nm. The scotopic curve has been shown to represent rod function, whereas the photopic curve depends on the cones, so the Purkinje shift is another aspect of the duplicity theory.

Rod photopigment. The form of the scotopic luminosity function has been explained by studying the visual photopigment in the outer segment of human rods. This has been done best by macerating a human retina, a process that fragments the outer segments. Particles of the rod outer segments were separated from the retina by centrifugation. The absorption spectrum of the suspension of rod particles was then determined and plotted as one of the curves in Fig. 17-7. For comparison, scotopic sensitivity was determined by two separate methods. In one method the scotopic luminosity function was determined and corrected for light absorption in the ocular media in order to represent sensitivity to light arriving at the retina. In the other method the scotopic luminosity function was determined in an aphakic eye; the principal colored structure of the ocular media, the yellow lens, had been removed in a cataract operation. Because of these corrections for light absorption in the ocular media, maximum scotopic sensitivity in Fig. 17-7 is at 500 nm instead of 507 nm. Furthermore, rod sensitivity is expressed as reciprocals of the number of light quanta at each wavelength rather than as reciprocals of light energy; this is required for accurate comparison of sensitivity data with absorption data. A single curve fits all three sets of measurements in Fig. 17-7, thus establishing that the scotopic luminosity curve results from the absorption spectrum of the photopigment in rod outer segments.

If rod outer segments are treated with digitonin, the photopigment may be extracted and studied in solution. The photopigment from human rods is called *rhodopsin;* it is similar to the rod pigment of other mammals. The results given

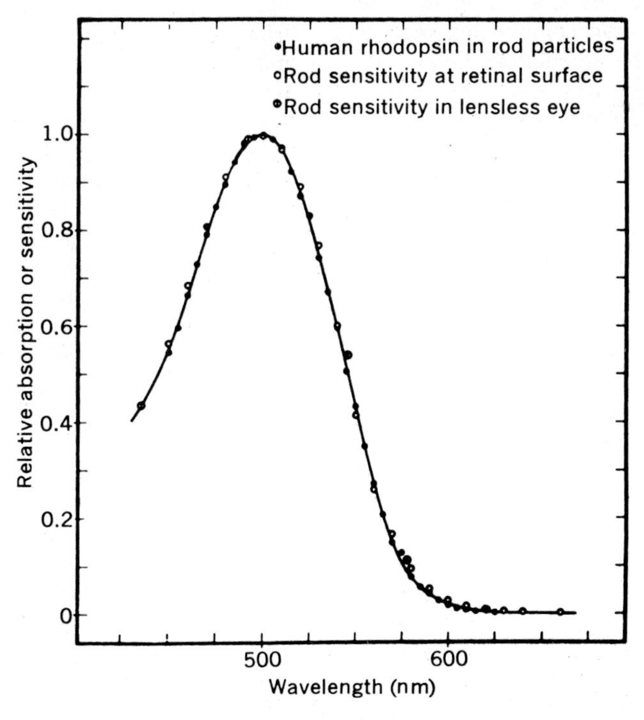

Fig. 17-7. Relative absorption spectrum of human rhodopsin, as measured in rod particles, compared with human rod sensitivity determined by two methods. See text for explanation. (From Wald and Brown.[122])

in Fig. 17-7 show that equal numbers of quanta absorbed by rhodopsin are equally effective throughout the visible spectrum in initiating excitation. When passing from maximum sensitivity at 500 nm to longer or shorter wavelengths, the sensitivity falls to half at a wavelength at which twice the number of quanta are required to be incident on the retina. However, Fig. 17-7 shows that relative absorption of these quanta by the rhodopsin correspondingly falls to half, so that the number of quanta absorbed is the same as at maximum sensitivity. Thus equal numbers of absorbed quanta produce a constant brightness, since constant brightness is the criterion in determining a luminosity curve. This means that a quantum of light *absorbed* by rhodopsin is equally effective at any wavelength in the visible spectrum, although the energy content of a single quantum is inversely proportional to wavelength. Fig. 17-7 also shows that all quanta are not absorbed with equal ease, the relative sensitivity of the dark-adapted retina to different wavelengths of light being governed by the relative effectiveness of rhodopsin in trapping quanta at different wavelengths.

Cone photopigments and color vision. The study of cone photopigments has always presented greater difficulties than the study of rods, partly because cone pigments are difficult to extract, and partly because color vision requires the wavelength of the stimulating light to be coded in the receptor response. The Young-Helmholtz theory of color vision assumed three different photopigments with absorption maxima in the blue, green, and red portions of the spectrum, respectively. In subsequent work, it was generally assumed that each of these photopigments is present in only one of three major classes of cones. This theory was long supported by indirect evidence, but final confirmation required the determination of absorption spectra on the pigments contained in single cones.

The necessary technique of microspectrophotometry was first developed in Japan in 1957 by Hanaoka and Fujimoto.[66] Intact outer segments were obtained from frog or carp retinas, placed on a microscope slide, and spectrophotometry was performed only on the beam of light passing through a single outer segment. The effects of pigments other than photopigments were eliminated by taking difference spectra. In this procedure an absorption spectrum is obtained, a strong light is then used to partially or wholly bleach the photopigment, and another absorption spectrum is obtained. The difference between the two absorption spectra is the difference spectrum. It excludes the effects of any photostable pigments in the measuring beam and therefore represents the absorption characteristics only of the photopigment bleached by light. Outer segments of frog rods were found to have difference spectra resembling that of rhodopsin, with maximum absorption at around 500 nm. Cone outer segments from the carp retina gave difference spectra that fell into several groups, each with its characteristic absorption maximum. Studies by Marks,[90] who examined 113 cone outer segments of the goldfish retina, showed that the difference spectra fell into three separate groups with absorption maxima that were respectively in the blue, green, and red portions of the spectrum. This confirms the Young-Helmholtz theory for the goldfish retina and gives background data from a species in which many cones may be examined. Similar experiments were then reported on human and monkey retinas.[11,39,91] In these studies the retinas were removed and placed on a slide with the receptors facing upward. Light was passed axially through the receptors from below, in the normal direction, and spectrophotometry was performed on the light passing through single cones of the parafovea where the cones are especially large. Although the number of receptors studied was small because of technical difficulties, the results indicate that the distribution of photopigments in single cones of human and monkey retinas is similar to that in goldfish. Fig. 17-8 shows the results of one of these studies in which the difference spectra of the pigments in single cones fell into three distinct groups. The peak absorptions of the three groups were at approximately 445, 535, and 570 nm. The first two of these are called blue and green pigments, which are the colors of the wavelengths maximally absorbed. The third pigment absorbs best in the yellow part of the spectrum, but it is called the red pigment because of historical precedent and the fact that it absorbs red light better than any of the other pigments. Note that there is considerable overlap between the difference spectra of the three cone classes. Thus a given wavelength will stimulate at least two, and in some cases all three, types of cones. This work reveals a response code for color at the receptor level, since each wavelength of light will elicit a unique ratio of responses among the three fundamental cone classes. In contrast, single rods of the human parafovea all appear to contain the same visual pigment, rhodopsin.[39]

Since the scotopic luminosity function is determined by the absorption spectrum of rhodopsin in rods, the photopic luminosity function may be expected to result from cone photopigments. There are at least three different cone pigments, and the simplest possibility is that the summed absorption spectrum of all the cone pigments will fit the photopic luminosity function. The relative amounts of the three cone pigments are critical in determining the summed absorption spectrum, so results are best obtained on a sample of retina containing the normal proportions. This has been done by applying the technique of microspectrophotometry to the human retina, using a light beam 0.2 mm in diameter and confined to the pure-cone portion of the fovea.[38] The difference spectrum of a representative sample of foveal cone pigments was obtained in this way. The foveal photopic luminosity function was then obtained and expressed on a quantum basis. The resulting pair of curves matched satisfactorily. Thus the photopic luminosity curve represents cone function, and its shape appears determined by the relative absorption of different wavelengths of light by the combined effects of all the cone pigments.

The Young-Helmholtz theory has long provided the most accepted basis for explaining many of the facts of color mixing. The dominant *hues* (color sensations) resulting from stimulation by narrow-wavelength bands of the visible spectrum are four in number: blue, green, yellow, and red. But it is well established that any color sensation that can be experienced may be elicited by an appropriate mixture of three pure primary colors from the blue, green, and red portions of the spectrum. Thus the sensation of yellow may be produced by mixing pure red and green stimuli; and the sensation of purple, which is not elicited by any pure wavelength, may be produced by mixing red and blue. These empirical rules of color mixing hold only when the colors are additive, as when mixing pure spectral bands of light. By comparison, a paint gains color because its pigment absorbs all light except that which is reflected; hence the mixing of paints is referred to as subtractive color mixture. The empirical *rules* of color mixing are different in the two cases, but the *principles* are the same if one considers only the light that reaches the observer's eye. If all three spectral primaries are mixed in appropriate proportions, cancellation of color can be achieved. Variation of stimulus intensity will then produce the entire achromatic gray scale. These are some of the major rules of color mixing; they may be explained by assuming that all color sensations (plus achromatic sensations) result from an appropriate ratio of stimulation of three kinds of receptors, each responding strongly to one of the primary colors. Although the receptor code for color now seems estab-

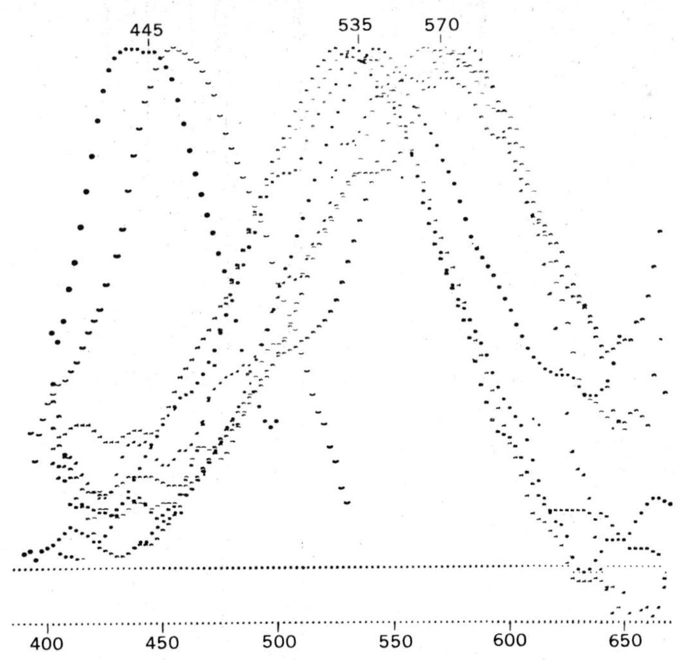

Fig. 17-8. Difference spectra of visual pigments in 10 individual cones of macaque and human retinas. Curves recorded from macaque monkey cones are represented by numbers, whereas those from human cones are shown by open parentheses. Maximum absorption of all curves has been adjusted to same level. (From Marks et al.[91])

lished, the photoreceptors are only the first stage in our understanding of color vision. There is evidence that the response code for color changes markedly at later stages of the visual system and that the later type of response code influences color sensations. Hence these later stages are likewise important for understanding color vision and will be taken up in Chapter 18.

Color blindness. The three cone pigments also provide a basis for understanding most aspects of color blindness.[11] Normal color vision is said to be *trichromatic,* since all the hues seen by a normal individual may be produced by using three primary colors. Trichromatic color vision also includes *protanomalous* and *deuteranomalous* types of color defects. If red and green primary colors are mixed to match a spectral yellow, the protanomalous subject will require more red than normal, whereas the deuteranomalous subject will require more green than normal. Thus defects in the cone pigment systems are not all or none but can occur in varying degrees. Dichromatic vision represents a more severe color defect, in which all the hues that can be seen may be produced by mixing only two primary colors. This classification includes three subcategories (*protanopia, deuteranopia,* and *tritanopia*), depending on whether the nonfunctional pigment system is the first (red), second (green), or third (blue). The *monochromatic* subject may be assumed to be totally color blind, seeing the world as it appears in black and white photographs, since the sensation produced by any wavelength of light may be matched by varying the intensity of only one primary color. The typical totally color blind subject is a *rod monochromat,* who possesses little or no cone function.

As shown in Fig. 17-8, either red or green spectral bands will stimulate both the red and green pigment systems. Thus the sensations of both red and green may be expected to depend on a particular ratio of response in the two types of cones, and loss of either pigment system should cause color deficits in both red and green portions of the spectrum. Correspondingly, both the protanope and deuteranope confuse red and green and are said to be *red-green* blind. The color confusions of color-blind subjects are the basis for all of the common tests of color vision.

Although color blindness may result from acquired retinal disorders, including vitamin A deficiency,[123] most cases are inherited and cannot be corrected. Red-green blindness is inherited as a simple sex-linked recessive trait. It is passed from a man to all his daughters, who are color-normal carriers if the man's wife is not also a carrier, and it reappears in half his daughters' sons. Some degree of this type of defect is estimated to be present in about 8% of men and less than 0.5% of women.[12] Tritanopia is much more rare, and total color blindness is likewise very rare.

Absolute sensitivity: dark adaptation

The sensitivity of the eye to a flash of light cannot be represented by any fixed value. It changes dramatically with light and dark adaptation, the effects of which are known from common experience. When passing from bright light into a darkened place (e.g., a movie theater), it is sometimes difficult at first even to see which seats are empty, but after a few minutes the surroundings can be seen quite well. On leaving the theater, one is dazzled by the bright light, but the eyes quickly become less sensitive and the discomfort disappears. The increased sensitivity of the eye in the dark is called dark adaptation, and its converse is light adaptation.

Fig. 17-9 shows dark-adaptation functions, which represent absolute threshold as a function

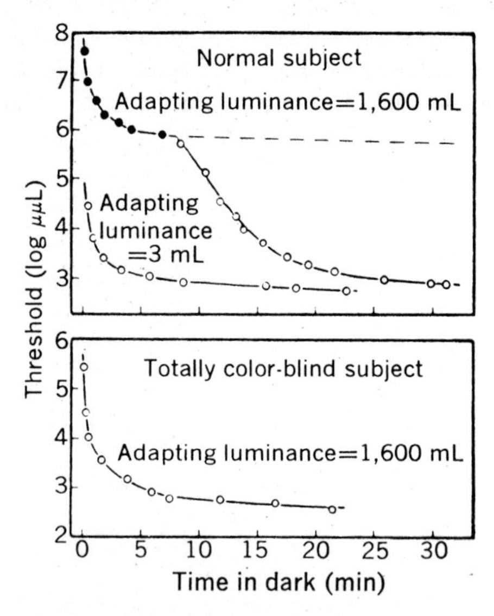

Fig. 17-9. Absolute thresholds as a function of time in the dark. Results are shown from normal subject after both high and low levels of light adaptation, and from totally color-blind subject following high level of light adaptation. Thresholds were measured with circular test field subtending 3 degrees of visual angle, presented 7 degrees from fovea on temporal side of retina. In results from normal subject, dashed line was not part of original experiment; it is added here to show approximate time course of later part of dark adaptation when only cones are stimulated. (From Hecht et al.[69])

of time in the dark. The *absolute visual threshold* is the minimum intensity of light that evokes a light sensation in the absence of background illumination. It is absolute only in the sense of being the threshold in complete darkness, and it falls during dark adaptation. It may be contrasted with the *incremental threshold,* the stimulus intensity that must be added to a background illumination in order to be seen against that background, a subject dealt with later in this chapter. In Fig. 17-9 the results in the normal subject, following strong light adaptation, illustrate the classic dark-adaptation function for the peripheral retina. This curve has two separate limbs. The first limb shows a rapid drop in threshold to an approximate plateau. The onset of the second limb occurs after about 8 min in the dark and may be seen as a second rapid decrease in threshold. Dark adaptation is almost complete after 30 min, but the threshold continues to fall slowly for more than 1 hr.

It has been demonstrated by several methods that the first limb of this classic dark-adaptation curve represents cone function, whereas the second limb represents rod function. The test stimulus in the experiment of Fig. 17-9 was violet in color. The solid black dots represent thresholds at which this violet color was seen, and this occurred only during the first limb of the curve. The open circles indicate thresholds at which the stimulus was colorless, and this was the case during the second limb of the curve. Pure-cone dark-adaptation curves have been obtained primarily by two methods. In one case the dark-adaptation function is measured by a small test stimulus confined to the fovea. This test stimulus may also consist only of red light, which is shown in Fig. 17-6 to activate cones and have minimal effects on rods. Dark-adaptation functions under these conditions show an initial portion that coincides with the cone limb of the curve illustrated; a second limb of the curve does not occur, and the later part of the foveal curve follows the approximate time course shown by the dashed line in Fig. 17-9.[67] The second method employs subjects with severe congenital cases of night blindness *(nyctalopia),* in whom rod function appears entirely absent. When a dark-adaptation function is obtained in any portion of the retinas of these subjects, the results are similar to those obtained from the foveas of normal subjects.[67]

The second limb of the classic dark-adaptation curve has also been isolated by two major methods. When the light-adapting luminance (Fig. 17-9) was decreased from 1,600 to 3 millilamberts (mL), the dark-adaptation curve of the normal subject was quite different from that following the high adapting luminance. In this case the color of the test stimulus was not seen at any time during dark adaptation. The threshold was already relatively low at the beginning of dark adaptation, and it dropped much more rapidly than following the high adapting luminance. Thus if light adaptation is not sufficiently strong to make cone function predominant, only the rod limb of the dark-adaptation curve is found in the periphery of the normal retina. The other method is to use subjects with rare total color blindness (rod monochromats) who show little or no evidence of cone function. The lower part of Fig. 17-9 shows results from the peripheral retina of such a subject following the same high level of light adaptation that was used for the normal subject. The dark-adaptation curve of the rod monochromat does not exhibit a cone limb, and sensitivity falls quickly to about the same level that was attained only after a considerably longer time in the normal subject.[69,107] Thus dark-adaptation studies reveal another aspect of the duplicity theory. Whereas the cones are capable of a small amount of dark adaptation, the rods can dark adapt to a much greater extent.

Note in Fig. 17-9 that when exposure to strong light was followed by 30 min of dark adaptation, the sensitivity of the normal peripheral retina increased by about 5 log units. Sensitivity can increase by 1 additional log unit during the first 0.5 sec of dark adaptation, prior to the first measurement of Fig. 17-9.[16] Thus dark adaptation can increase retinal sensitivity by at least 6 log units (1 million times).

The combined results of Figs. 17-6 and 17-9 explain the fact that red light permits one to perform tasks involving high visual acuity while maintaining readiness for rapid dark adaptation. The red light stimulates primarily cones, which can engage in photopic tasks such as reading, and has little light-adapting effect on the rods. On entering the darkness, thresholds are already relatively low and the eye rapidly dark adapts. This is why red goggles are used in situations requiring readiness for dark-adapted requirements (e.g., by military night pilots on defensive alert).

The ability of the retina to alternate between cone and rod dominance of visual functions is at the heart of the duplicity theory. One may now ask why the cones are dominant in the light-adapted state and early dark adaptation, whereas the rods become dominant in the more fully dark-adapted state. It is also important to understand the processes that underlie the increased sensitivity of both cones and rods during dark adaptation.

There is now evidence in the macaque monkey retina, as discussed later in this chapter, that under photopic conditions the rod signals are suppressed via a cone-initiated pathway. This may explain why rods make little or no contribution to visual functions under photopic conditions.

A technique of reflection densitometry has been developed for measuring photopigment densities in the living human eye.[40] This technique has been used to study both cone and rod pigments, including bleaching of the pigments by a strong light and the time course of regeneration in the dark.[40,108] The cone pigment in the fovea of the protanope regenerates more rapidly than rhodopsin.[108] Hence one reason why cone thresholds are lower than rod thresholds during the early part of dark adaptation is the more rapid regeneration of cone pigments.

As light-adapting intensity is increased, the break between the cone and rod limbs of the dark-adaptation curve occurs at a progressively later time during dark adaptation. But regardless of when the break occurs, it appears when rhodopsin is about 90% regenerated.[106] Hence rod thresholds become lower than cone thresholds when rhodopsin has regenerated to about 90% of its fully dark-adapted concentration. This suggests that rhodopsin concentration is critical in the changeover from cone to rod dominance of visual sensitivity during dark adaptation.

We may next ask why the fully dark-adapted sensitivity of the rod system is greater than that of the cone system. Most dark adaptation functions, such as those shown in Fig. 17-9, are obtained with relatively large stimulus spots. The threshold to a large stimulus spot will be determined not only by the threshold of each individual receptor, but by the extent to which area summation occurs. There is little or no convergence of cone pathways, so increasing the size of the stimulus spot will have little effect on the cone threshold. But if the activities of many receptors converge on a common site, as indicated by the anatomy of rod pathways, increased size of the stimulus spot will be quite effective in decreasing the rod threshold. If dark-adaptation functions are obtained with very small stimulus spots, the difference between final thresholds in the fovea and periphery is much reduced.[15] Thus when the test stimulus is fairly large, as in Fig. 17-9, one reason for the final rod threshold being lower than that of cones is that the rods can take greater advantage of the large stimulus area.

PHOTORECEPTOR STRUCTURE

Inner and outer segments

The fine structure of the connecting zone between inner and outer segments is similar in cones and rods and is shown schematically in Fig. 17-10. The connection itself is a true cilium. In cross section, it shows an outer membrane and nine pairs of peripheral fibrils; this is the usual structure of nonmotile cilia throughout the animal kingdom. The intracellular fibrils of the connecting cilium arise from a basal body (centriole) located at the outer end of the inner segment. These fibrils penetrate into the outer segment and traverse a portion of its length. De Robertis has shown that the entire outer segment is derived from a cilium. In morphogenetic studies of mice sacrificed at various times after birth, sequential stages were demonstrated in differentiation of the outer segment from a primitive cilium.[50] Hence the vertebrate photoreceptor is a ciliated cell in which the main portion of the cilium is highly differentiated and specialized, with the primitive ciliary characteristics maintained only at the connection between inner and outer segments.

Fig. 17-10 also shows that in longitudinal section the rod outer segment contains a tightly packed stack of discs. Each disc appears to be a closed membranous saccule with a flattened form similar to that of a red blood corpuscle. These saccules are formed in both rods and cones by infolding of the outer segment membrane, as shown schematically in Fig. 17-11. At the base of the rod outer segment of a frog, saccules are shown to be formed by infolding along an incision passing around the outer segment for a portion of its circumference. Proceeding distally from the base of the rod outer segment, this incision becomes shorter and quickly disappears. Thus saccules in the greater part of rod outer segments have no direct connection with the outer membrane, as depicted also in Fig. 17-10. In comparison, the outer segment of a frog cone typically tapers and is much shorter than the rod outer segment, and its saccules are continuous with the outer membrane by a zone of infolding that extends the entire length of the outer segment. Thus rod and cone outer segments follow a similar structural plan but show distinct variations from that plan. Recently the outer segments of living frog eyes were infiltrated with procion yellow, a highly diffusible fluorescent dye, which clearly confirmed that cone saccules have patent openings to the extracellular medium throughout the outer segment, whereas rod saccules have such patent openings only at the

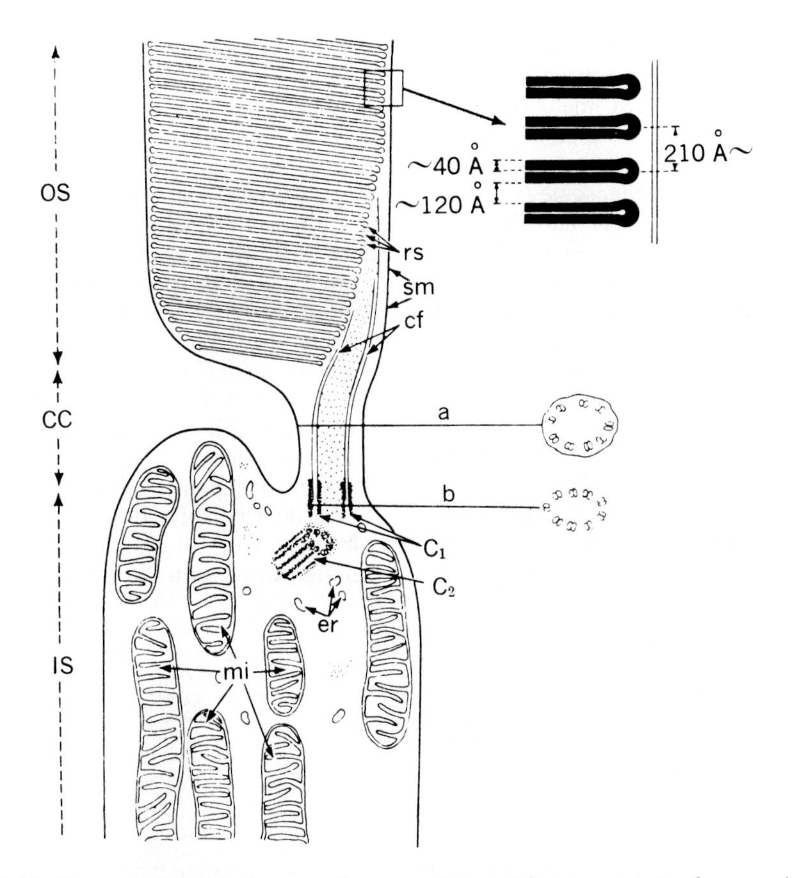

Fig. 17-10. Diagram of connecting zone between outer and inner segments of mammalian rod, based on electron microscopy. Main structures shown include base of outer segment, *OS;* connecting cilium, *CC;* and outer portion of inner segment, *IS*. C_1 is major centriole, or basal body, of connecting cilium; C_2 is secondary centriole. Other structures are labeled as follows: *cf,* ciliary filaments; *rs,* rod saccules; *sm,* surface membrane; *mi,* mitochondria; *er,* endoplasmic reticulum. At right are cross sections through, *a,* connecting cilium and, *b,* basal body. Portion of outer segment is also shown at greater magnification. (From De Robertis.[50])

base of the outer segment.[87] Studies of lanthanum infiltration into glutaraldehyde-fixed retinas indicate that in monkeys also it is only the cone saccules that have openings to the outside along the entire length of the outer segment.[44]

Electron microscopy has produced evidence that the visual photopigment is contained in the saccule membranes. After extraction of rhodopsin by digitonin an electron-dense component of the saccule membranes in rod outer segments is absent, and it is more difficult to demonstrate after light adaptation.[57] Rhodopsin has a molecular weight of 27,000 to 28,000 in several mammals,[70] and biochemical evidence also indicates that it is a major component of saccule membranes. Thus infolding of the outer membrane to produce the saccules seems to have the functional significance of greatly increasing the amount of membrane available to contain photopigment. This increases the photopigment content of the outer segment, improves its light-trapping efficiency, and hence increases the light sensitivity of both rods and cones.

Renewal of outer segments

Young[13,131] has shown that rod outer segments have a remarkable renewal mechanism. When radioactive protein precursor was injected into rhesus monkeys, rats, mice, and frogs, it quickly appeared in rod outer segments in the sequence of stages shown in Fig. 17-12. In stage 1, shown at left, the black dots in the inner segment represent radioactive protein revealed by a high-resolution autoradiograph. This radioactive protein migrated through the connecting cilium (stage 2) and formed a dense narrow band at the base of the outer segment (stage 3). This band then migrated along the axis of the outer segment (stage 4) and reached the distal end of the outer segment (stage 5). Finally, the distal tip of the outer segment that contained the radioactive band was phagocytized by the pigment epithelium cell, within which it appeared as a cytoplasmic inclusion called a phagosome (stage 6). Since the protein is incorporated into the outer segment in a narrow band, saccules are forming constantly at the base of the outer segment, after which they move along the outer segment, with the oldest material at the distal tips being phagocytized and digested by the pigment epithelium cells. In rhesus monkey the radioactive band traverses the outer segment in only 9 to 13 days, so the turnover is quite rapid.[131] When visual pigment was extracted from the frog retina while the radioactive band was traversing the outer segment, 80% to 85% of the total radioactivity was recovered.[65] Thus the main protein synthesized from radioactive precursors is incorporated into the visual pigment itself.

Phagocytosis of the tips of rod outer segments does not occur at a constant rate under normal conditions. When albino rats or frogs were kept on a diurnal light cycle, little phagocytosis by the pigment epithelium took place in the dark; a large synchronous burst of phagocytosis then occurred 1 to 2 hours after onset of the light, and during the remainder of the light period the resulting phagosomes were being digested by the pigment epithelium.[18,88] In frog this cycle appears controlled only by the light itself,[18] but in the albino rat it has become established as a circadian rhythm that can continue for at least 3 days in continuous darkness.[88]

An injection of radioactive protein precursor also results in diffuse labeling at scattered sites throughout the outer segment. In rods this scat-

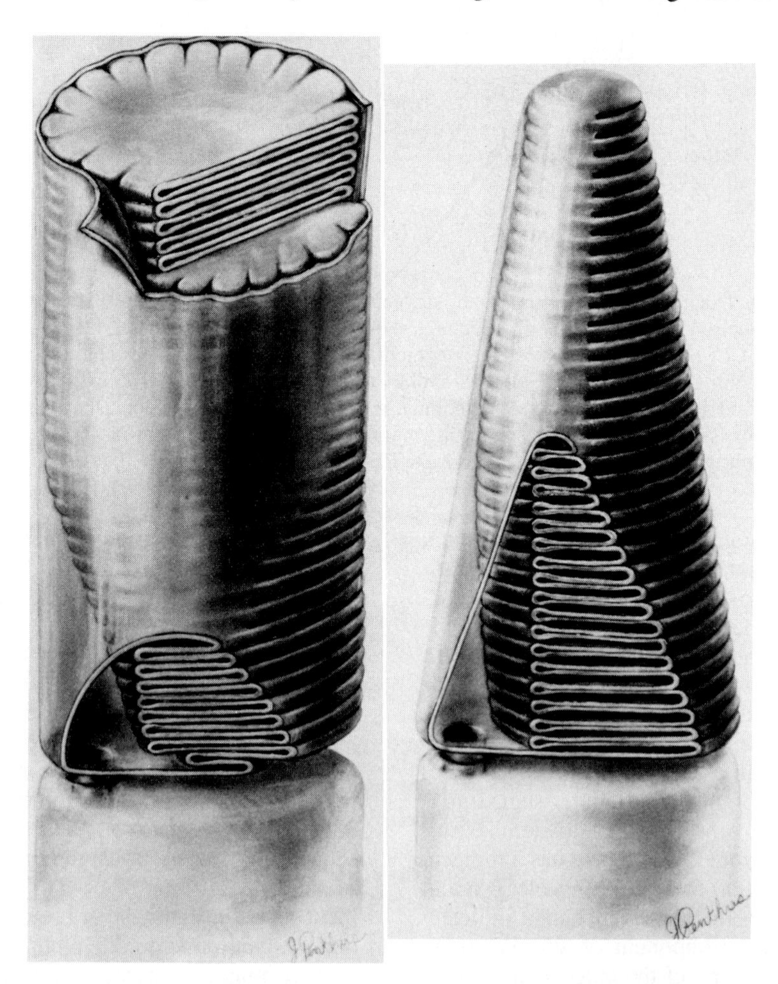

Fig. 17-11. Schematic structure of outer segments of frog rod (left) and cone (right), based on electron microscopy. Drawings show entire cone outer segment, but only base of rod outer segment because of its greater length with no significant change of structure beyond point shown. (Courtesy Dr. R. W. Young.)

tered renewal of individual protein molecules may be of minor importance. In cones, however, it is the only renewal mechanism that has been demonstrated, since the bandlike renewal has not been seen in the cones of any species studied.[13]

The pigment epithelium sends out villous processes that interdigitate between rod outer segments. In contrast, in both cat and human a specialized type of contact has now been demonstrated between the pigment epithelium and the outer segments of cones; the villous processes flatten into leaflets, several of which wrap closely about the tip of each cone outer segment.[112,114] This arrangement may facilitate metabolic exchange between the pigment epithelium and cone outer segments. In the human retina, it has also been shown now that the tips of cone outer segments are phagocytized, following which phagosomes appear in the pigment epithelial processes that contact the cone outer segments.[75,114] This indicates that the cones must also renew their outer segments efficiently, but the mechanism is less well understood than in rods.

These results answer the important question of how photoreceptors remain functional throughout life, although their high sensitivity renders them

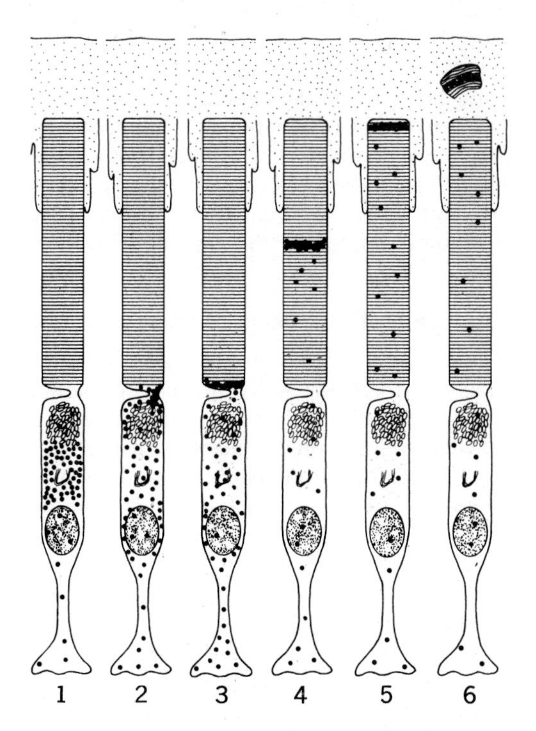

Fig. 17-12. Continuous renewal of base of rod outer segment and removal of distal tip of outer segment by pigment epithelium cells, as revealed by high-resolution autoradiography. See text for explanation. (Courtesy Dr. R. W. Young.)

vulnerable to damage by normal intensities of the very light stimuli they are designed to receive.[93] Being nerve cells, they cannot renew themselves by division. Instead, they have evolved mechanisms for renewing the specific portion of the cell that absorbs light and is thus most subject to damage.

The mechanisms of outer segment renewal must also be important in recovery from damage by a variety of other factors. For example, after experimental retinal detachment in the rhesus monkey the outer segments of both cones and rods degenerated; following surgical reattachment of the retina, both cone and rod outer segments regenerated and regained their normal appearance.[83]

This type of work has likewise suggested an explanation of human retinitis pigmentosa, a blinding disease in which the postmortem histology shows a marked accumulation of cellular debris between the receptors and pigment epithelium.[71] Such an effect would be expected if phagocytic activity of the pigment epithelium cells was markedly defective or absent. As rod outer segments continued to be formed, debris would accumulate and eventually block the normal exchange of materials between the receptors and choriocapillaris. This suggestion is greatly strengthened by the study of a hereditary retinal degeneration in rats in which the postmortem histology is similar, and for which detailed studies have confirmed the suggested etiology of the condition.[24,71,72]

PHOTOCHEMISTRY OF VISUAL PIGMENTS

Effects of light on rhodopsin

The photochemistry of rhodopsin has been determined primarily by extracting it from the outer segments of rods and studying it in isolation.[7,76] In common with all visual pigments studied to date, rhodopsin is a *protein* bearing a light-absorbing *chromophore,* to which it owes its color and sensitivity to light. Its general structure and the major effects of light absorption are shown in Fig. 17-13. The protein part of the molecule is called *opsin.* The chromophore is called *retinal*$_1$, an abbreviation of retinaldehyde$_1$, a name resulting from the chromophore being an aldehyde of vitamin A_1. In the resting state this chromophore is in the 11-*cis* form, which fits closely into the opsin portion of the molecule. The absorption of light isomerizes the chromophore, which straightens into the all-*trans* configuration, and this stage is called prelumirhodopsin. The reaction may be stopped at this stage

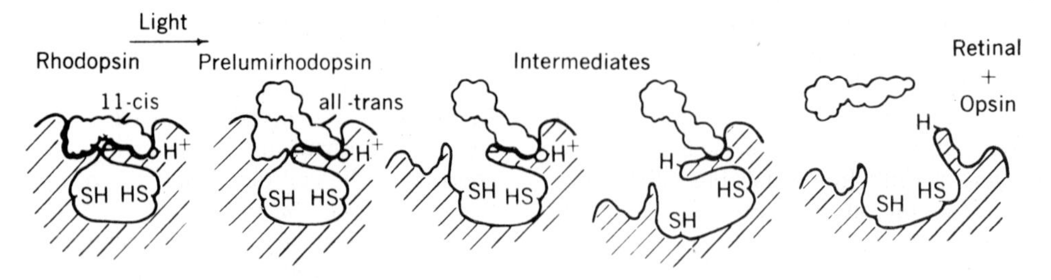

Fig. 17-13. Action of light on rhodopsin. See text for explanation. (Modified from Wald and Brown.[11])

if the temperature is below about −140° C. But prelumirhodopsin is highly unstable, presumably because the chromophore no longer fits closely to the opsin. With progressive warming the opsin opens up in several stages, which are designated intermediates in Fig. 17-13. Above about −140° C the first intermediate, called lumirhodopsin, is formed. Above about −40° C, another stage, metarhodopsin I, occurs, and metarhodopsin II is formed when the temperature exceeds about −15° C. Finally, at temperatures above 0° C and in the presence of water, the molecule is hydrolyzed into its two main fractions of all-*trans* retinal and opsin. Low temperatures are convenient for studying this sequence of events, which can thus be stopped at any stage. The same sequence of events occurs under physiologic conditions,[7] but the sequence occurs very rapidly, and the unstable stages between rhodopsin and its hydrolysis are more difficult to detect because they are short lived. Note that in this sequence of events the only action of light is to isomerize retinal from the 11-*cis* to the all-*trans* form.[121] All the succeeding reactions are energy yielding and proceed spontaneously under normal physiologic conditions. Even the isomerization of retinal is an energy-yielding reaction,[121] so the only energy required of light is that which triggers an energy-yielding reaction. This is a typical characteristic of highly sensitive cellular reactions in which the stimulus is only required to trigger the release of a preloaded source of potential energy, so that there is an energy amplification at the initial stage of the reaction.

Resynthesis of rhodopsin

Following its hydrolysis from the opsin, retinal is reduced by enzymatic action to vitamin A, with which the retinal comes into equilibrium. Some of this vitamin A appears to be stored in the pigment epithelium, at least if the light is sufficiently strong and prolonged for vitamin A to be formed from retinal in large quantities.[51] Since the photopigment molecule is broken down after light absorption, a process often referred to as bleaching, it must regenerate in the dark to prepare for another response to light. Following its hydrolysis from the opsin, retinal is in the all-*trans* form; the first step required for resynthesis of rhodopsin is conversion of this retinal back to the 11-*cis* form.[121] This can take place in the dark through the action of an enzyme, and this reaction consumes energy. Combination of the 11-*cis* retinal with opsin to form rhodopsin then occurs spontaneously. This reaction promotes the re-formation of retinal from vitamin A in two ways.[121] First, this spontaneous reaction yields energy that is used for the oxidation of vitamin A to retinal. Second, by removing retinal from one side of the reaction the equilibrium is altered to favor conversion of vitamin A to retinal. When all the opsin present has recombined with retinal, there can be no further energy-yielding reactions to support further accumulation of retinal. Thus the action is limited by the amount of available opsin.[121]

Structure and reactions of cone pigments

Cone pigments have proved more difficult to extract in sufficient quantity for study in solution. However, one pigment called iodopsin has been extracted from the predominantly cone retina of the chicken. The chromophore of this pigment is also retinal$_1$, but the opsin is different from that of rhodopsin.[7,121] The breakdown and synthesis of iodopsin appear to occur in basically the same manner as in rhodopsin.[7]

Spectrophotometry of the human fovea has detected the presence of pigments that absorb strongly in the red and green portions of the spectrum, respectively,[38] and may be called red and green pigments for convenience. Iodopsin appears to correspond to the red pigment of the human fovea.[38] When foveal pigments were bleached by a strong light, addition of 11-*cis* retinal$_1$ resulted in resynthesis of both pigments.[38] Thus both the red and green pigments of the human retina contain the same chromophore as rhodopsin. This makes it probable that all human photopigments possess the same chromophore, $retinal_1$, and differ among themselves only in the opsin component. The fact that retinal is formed from vitamin A explains why an experimental vitamin A deficiency in humans results in higher dark-adapted thresholds for both cones and rods, and why normal cone and rod thresholds are recovered following vitamin A administration.[123]

Early receptor potential

An intense brief flash of light elicits from the retina an extremely rapid electrical response. This was first found by microelectrode studies in the macaque monkey retina, and the response was shown to have maximum amplitude when the electrode was near the distal tips of the photoreceptors.[29] It was designated the *early receptor potential* (ERP) because of its receptor origin and the absence of any detectable latency.[29] It is typically biphasic in form,[30] and the two phases have been designated R_1 and R_2. As shown in Fig. 17-14, this response may also be recorded between wick electrodes on opposite sides of an isolated retina.

The ERP shows little or no sensitivity to ionic composition of the bathing medium and hence cannot be generated by a transmembrane ion flux.[96] It is, however, dependent on the integrity of the visual photopigment[29] and on normal orientation of the photopigment molecules in the saccule membranes.[47] In a pure-rod eye the spectral sensitivities of both R_1 and R_2 match well to the absorption spectrum of rhodopsin.[98] A variety of methods have now shown that the response is a direct electrical consequence of light-induced changes in the form of the photopigment molecule.[97] Apparently the altered molecular form shifts the position of electrical charges on the molecule, thus generating a transient electrical response. The potential developed by a single molecule must be extremely small, but the intense flash required to elicit the ERP causes synchronous events in a great many molecules, yielding a measurable response. With lowered temperature, R_2 may be abolished selectively, thus isolating R_1, which has been recorded at temperatures as low as −35° C.[99] It appears likely that R_1 results from a molecular step in the conversion of prelumirhodopsin to metarhodopsin I, whereas R_2 results from a later step that is more sensitive to lowered temperature.[99] These results indicate that the ERP is a convenient

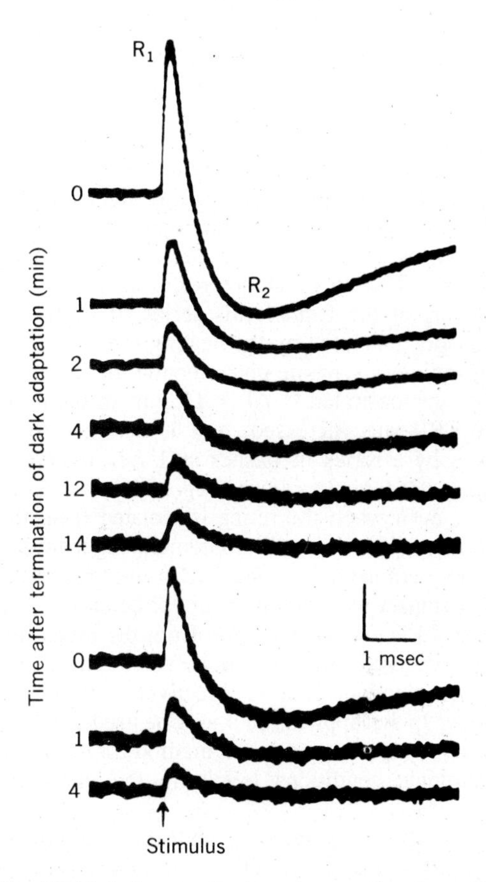

Fig. 17-14. Influence of adaptation on first phase, R_1, and second phase, R_2, of early receptor potential of isolated toad retina. Response recorded between wick electrodes on opposite sides of retina; upward deflection indicates positivity of vitreous side of retina, as in ERG recording. Stimulus was intense 0.7 msec flash covering retina, which was initially dark adapted. Retina was then light adapted by 15 stimulus flashes at 1 min intervals, dark adapted again for 15 min, and finally light adapted by another 5 flashes at 1 min intervals. Voltage calibration indicates 100 μV for top three records and 50 μV for all other records. (From Brown.[2])

method for in vivo detection of rapid light-induced changes of molecular form in the photopigment, and it is being applied in this type of investigation.

As expected from the manner of its generation, the amplitude of the ERP is linearly proportional to the number of photopigment molecules that are activated by absorbing quanta from the light flash.[45] With the flash intensity constant, the number of quanta absorbed will increase linearly with concentration of the photopigment. Under these conditions the amplitude of the ERP is linear with photopigment concentration, thus giving a long-needed method for the rapid and convenient measurement of photopigment concentration. In Fig. 17-14 this method is applied to show the bleaching of photopigment by light and its regeneration in the dark. Beginning with an isolated and dark-adapted toad retina, a series of 15 flashes at 1 min intervals quickly light adapted the retina and reduced the photopigment concentration, as shown by the reductions in both R_1 and R_2. The small stable response finally remaining at the end of this series of flashes is probably generated by the heating effect of light absorption, a separate phenomenon that has also been demonstrated.[63] After 15 min in darkness the ERP was increased, and it was abolished again by a series of flashes at 1 min intervals. Hence toad photopigment regenerates in darkness, even when the retina is isolated from the pigment epithelium. Photopigment regeneration is more efficient, however, when the toad retina is in contact with the pigment epithelium.[27]

An ERP may be recorded from the intact human eye by a corneal electrode and a remote reference electrode that is effectively behind the eye.[129] In this application it may be used diagnostically, as it is a convenient method for detecting pathologic conditions involving the photopigment.[23]

An ERP is generated by both rods and cones, but the signal generated by a single cone is larger than that generated by a single rod.[3] This effect is so great that although cones contain only about 10% of the total photopigment in human and macaque monkey retinas, the ERP from both species is contributed mainly by cones.[59,60] This is probably a consequence of all the cone saccules being open to the extracellular medium, so that electrical events in the cone photopigments are more readily recorded by extracellular electrodes.

Functions of photochemical cycle

The breakdown of visual pigment by light and its regeneration in the dark is one mechanism by which the retina adapts its sensitivity to the prevailing illumination. During light adaptation the reduced photopigment concentration decreases the light-trapping efficiency of the outer segment, thus lowering sensitivity; converse effects occur during dark adaptation. This increases the range of light intensities to which the retina can respond, which is perhaps the main functional advantage underlying the evolution of such an elaborate photochemical cycle. Under physiologic conditions, rhodopsin in the squid retina isomerizes, but it does not bleach or hydrolyze into its two major fractions following light absorption.[77] Thus the more elaborate photochemical cycle of vertebrates may represent a higher evolutionary development in the service of light and dark adaptation.

Absorption of light by the photopigment also triggers the process of excitation. Initial changes in the molecular form of rhodopsin are exceedingly rapid, but hydrolysis appears to be too slow to account for the rapidity of light sensations.[10] Hence it is generally assumed that at least one of the early molecular events shown in Fig. 17-13 plays a direct role in photoreceptor excitation. Another approach is to record the electrical activity of the photoreceptor and determine the characteristics that excitatory events must have in order to initiate the electrical activity. This approach seems required to select among the candidates for excitatory molecular events in the photopigment, and retinal neurophysiology has now made a start on this problem.

MAMMALIAN ELECTRORETINOGRAMS

The electroretinogram (ERG) is a relatively slow and complex electrical response of the retina to illumination; it was first recorded by Holmgren in 1865. The most conventional recording method is to place an electrode in contact with the cornea while the reference electrode is at a remote location that is effectively behind the eye. Recent analytic work in mammals has been done primarily by an electrode introduced through a needle into the vitreous humor of an experimental animal. If a microelectrode is used, it may be introduced into the retina itself to record the ERG from various retinal levels.[35-37]

Fig. 17-15 is an ERG recorded from the cat retina by an active electrode in the vitreous humor and a reference electrode in the orbit be-

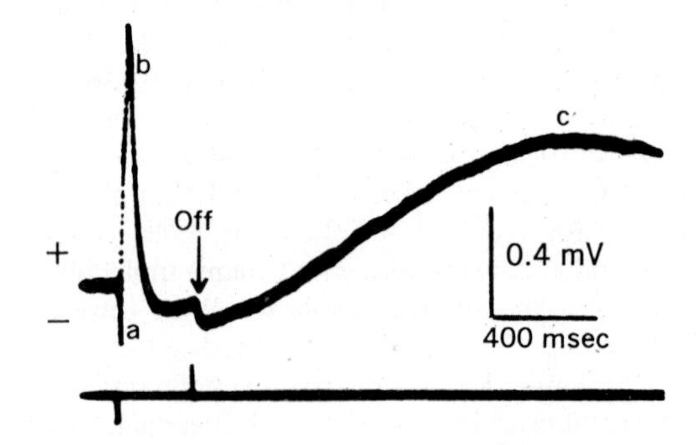

Fig. 17-15. ERG of cat, with major deflections, conventionally labeled. Eye was light adapted and retina was stimulated by a large and intense light spot. Onset and termination of stimulus shown in lower record. See text for further explanation. (From Brown.[2])

hind the eye. Positivity of the active electrode is shown as an upward deflection, which is the convention for ERG work and also for intracellular recording. That convention will be followed for all electrical responses illustrated in this chapter. The cat ERG is typical of that elicited from a predominantly rod retina. Shortly after onset of the stimulus there is a negative a-wave, followed by a larger positive b-wave. The positive c-wave is also initiated by onset of the stimulus, but the c-wave rises so slowly that with short stimuli its peak occurs long after the stimulus. On termination of the stimulus, the off-response is a simple negative deflection.

Electroretinogram components and their origins

It is obvious from Fig. 17-15 that the ERG is not a unitary response but consists of several components. Since the ERG may be recorded by gross electrodes, each component must consist of the summed activity of many cells of a given class that are responding in synchrony to the light stimulus. Hence the ERG offers the unique advantages of a readily recorded response that reveals the electrical activity of certain classes of retinal cells. The basic problems have always been to analyze this complex response into its components and to identify the type of cell generating each component. Solution of these problems is required so that the ERG may be used to maximum advantage in studying retinal physiology and diagnosing retinal disorders.

A combination of several methods has shown that the mammalian ERG consists of four major components. These are (1) the *c-wave,* (2) an ionically generated *receptor potential,* (3) *the b-wave,* and (4) a *dc component,* so called because its time course resembles that of a direct current pulse.

The c-wave is generated by the pigment epithelium, as shown most clearly by intracellular recording of the c-wave from pigment epithelium cells of the cat retina.[113] The c-wave is initiated, however, by light absorption in outer segments of the photoreceptors.[113] In frog the use of special electrodes that record potassium concentration has now shown that following a light flash there is a transient reduction of extracellular potassium concentration that closely follows the time course of the c-wave.[95] This effect is maximal in the photoreceptor layer. It thus appears that light-activation of the photoreceptors leads to a brief local lowering of the extracellular potassium concentration, which results in a transient hyperpolarization of pigment epithelium cells.

The physiologic significance of the c-wave is not clear, but it cannot be in the direct chain of nervous activity that leads to visual sensations. Thus Fig. 17-16 shows the general form of the ERG in the absence of the c-wave, and how this response can be accounted for by summation of the three major neural components that have been identified. Such an analysis is shown for the ERG from a predominantly rod retina and also for the ERG of a pure-cone eye. The macaque monkey and human are intermediate cases, since their retinas contain both rods and cones.

The neural components of the ERG, and their origins, have been revealed most clearly by studies with penetrating microelectrodes.[2,36,37] An intraretinal electrode can record a local ERG that is only from a small retinal area surrounding the electrode.[36] Since the recorded response is only from the area penetrated, the amplitude of each component should be maximum at the etinal level where it is generated. Electrode depth has been determined by physiologic methods and confirmed by electrode marking.[32,37] Such studies in the cat retina showed that the a-wave attains maximum amplitude near the distal tips of the photoreceptors, indicating that the a-wave results from the rising phase of a receptor potential.[37] In contrast, maximum amplitudes of both the b-wave and the dc component were in the inner nuclear layer.[37] The b-wave and dc component were then separated by two methods. Lidocaine (Xylocaine), a local anesthetic, abolishes the b-wave without affecting the dc component.[37] Reduction of stimulus intensity can isolate the dc component, thus permitting its shape to be determined.[36]

The type of cell that generates the dc component has not been identified. Intracellular recording in Müller cells of *Necturus* has indicated, however, that the b-wave is generated across the membranes of these glial cells.[92] Analytic work on glial cells indicates that they respond only to ionic composition of the extracellular fluid, as when the potassium concentration is altered by the activity of nearby nerve cells.[85] Hence generation of the b-wave by the Müller cells is probably secondary to the activity of nerve cells of the inner nuclear layer. In any event the b-wave is closely comparable to ganglion cell activity as an indicator of retinal sensitivity,[52] and the b-wave is especially useful for this purpose because it is so readily recorded.

Foveal recording in the macaque monkey shows a great reduction of both the dc component and b-wave, corresponding to almost complete absence of the inner nuclear layer, whereas

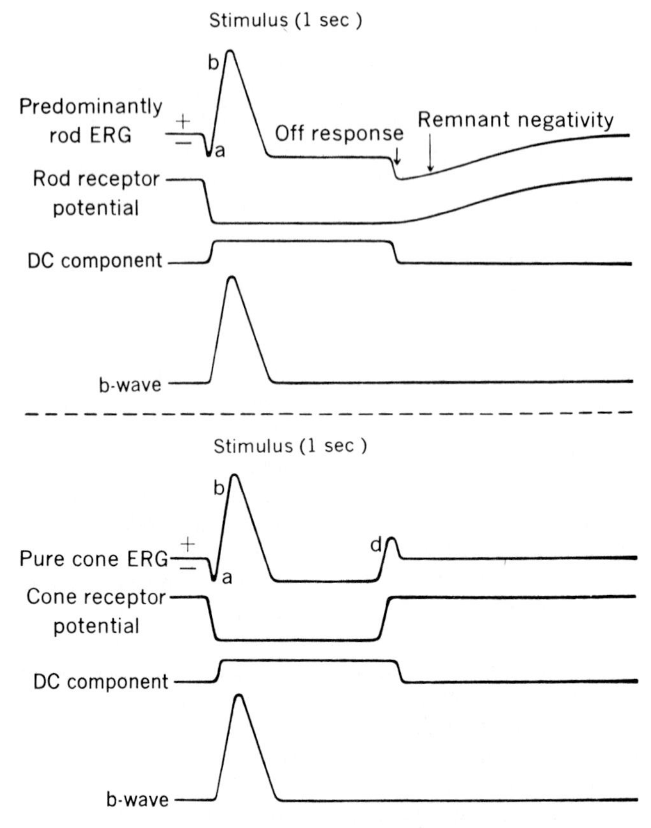

Fig. 17-16. Schematic component analyses of mammalian ERGs from predominantly rod and from pure-cone eyes. In both rod and cone analyses, c-wave from pigment epithelium is omitted, and remaining ERG is shown to be accounted for by algebraic summation of the three major neural components that have been identified. (From Brown.[2])

the a-wave from the fovea is larger than from the peripheral retina.[2] When the retinal circulation of the macaque monkey is mechanically clamped at the optic disc, the dc component and b-wave are quickly abolished, but the receptor potential survives for many hours, being well supported by the choroidal circulation.[2,33] The c-wave also survives but may be abolished in the macaque monkey by light adaptation, thus isolating the receptor potential. The receptor origin of these isolated responses has been demonstrated by a variety of criteria,[2,3,33] and these isolated responses have revealed the time courses of both cone and rod receptor potentials.[33,34,125]

Analysis of the rod ERG (Fig. 17-16) shows that the initial rise of the rod receptor potential contributes the rising phase of the a-wave, which is abruptly terminated by onset of the large b-wave and dc component, both of which are of opposite polarity from the receptor potential. These components have longer latencies than the receptor potential, as they are generated by cells of the inner nuclear layer. After the b-wave has run its course, there is usually a steady level of negative potential during the remainder of the stimulus, apparently because the negative receptor potential is larger than the positive dc component. Immediately after the stimulus, rapid decay of the dc component gives the negative deflection called the off-response. This is followed by a slowly decaying phase, usually referred to as "remnant negativity,"[6] which results from the very slow decay of the rod receptor potential.

The form of the cone ERG is very similar to that of the rod ERG until the stimulus terminates. At that time the cone ERG typically shows a d-wave, consisting of a positive rising phase and a negative falling phase, following which there is no remnant negativity. The cone ERG appears to consist of the same neural components as the rod ERG, but the cone receptor potential decays rapidly when the stimulus terminates.[2,33,125] In the cone ERG, termination of the stimulus is followed first by decay of the cone receptor poten-

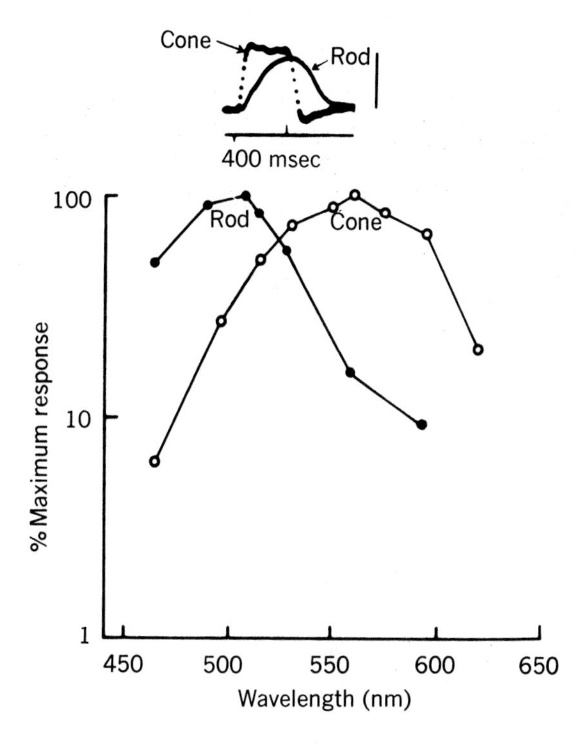

Fig. 17-17. Top, superimposed recordings of isolated cone and rod receptor potentials of macaque monkey. After clamping retinal circulation, these responses were recorded from microelectrode inserted beside fovea to about depth of inner segments. Cone response was elicited by 560 nm stimulus spot that covered mainly pure-cone fovea, with light entering eye only through center of pupil. Rod response was elicited by 508 nm stimulus of equal quantum content, delivered in annulus that stimulated primarily the parafovea, with light entering eye only through periphery of fully dilated pupil. Each response obtained by averaging about 100 records. The two responses are shown at similar amplitude to facilitate comparison of their time courses. Voltage calibration represents 80 μV for cone response and 200 μV for rod response. Spectral response curves are plotted below for both cone and rod responses. Ordinate of these curves is percentage of maximum response on a logarithmic scale. (From Whitten and Brown.[125])

tial, giving the positive rising phase of the d-wave. The falling phase of the d-wave then results from decay of the dc component, which is opposite in polarity and which decays later because it is generated in the inner nuclear layer.[2] Since the cone receptor potential has decayed by the time the d-wave is completed, remnant negativity is not found in the cone ERG. Thus the main differences in form between rod and cone ERGs result from the different decay rates of rod and cone receptor potentials.

ROD AND CONE RECEPTOR POTENTIALS

The early receptor potential does not usually appear in the ERG, as it requires higher stimulus intensities than are used in most ERG work. When the early receptor potential does appear, the peaks of both R_1 and R_2 precede the a-wave.[29,30] Thus the slower cone and rod receptor potentials, the leading edges of which create the a-wave, are often referred to as late receptor potentials. Here they will be called simply receptor potentials, since they are generated by altering ionic fluxes, in common with other types of receptor potentials. Although the early receptor potential is an inevitable consequence of light absorption by the photoreceptor, there is strong evidence that it is not a causal step in the direct chain of excitatory events.[97] On the other hand, the slower ionic receptor potentials appear to be transmitted along photoreceptors and to initiate the activity of second-order cells.[31] Hence these receptor potentials are especially important to the understanding of visual functions.

Different time courses of receptor potentials from monkey cones and rods

The time courses of cone and rod receptor potentials were first revealed in macaque monkey retinas,[33,34] in which the photoreceptors have thus far proved too small for intracellular recording. Fig. 17-17 shows receptor potentials of the macaque monkey that were isolated from other ERG components by methods already described, then recorded locally by inserting a microelectrode beside the fovea to about the depth of the inner segments.[125] Although receptor potentials are negative in polarity when recorded by conventional ERG methods, they are positive when recorded intraretinally at the inner segments, because of the altered electrode locations.[2,3] Cone and rod responses were recorded selectively by using predominantly foveal or parafoveal stimulation. Also, stimulus rays were directed along the axis of foveal cone outer segments, but approached the outer segments at a considerable off-axis angle for the rod case; this latter maneuver reduces the effectiveness of stimulation for cones, but not for rods (the Stiles-Crawford effect). Essentially pure cone and pure rod responses were thereby obtained, as shown in Fig. 17-17 by the cone response exhibiting peak amplitude at about 560 nm, whereas the rod response peaked at about 508 nm. The illustrated cone and rod responses were elicited by equal quantum stimuli in the intensity range where both rods and cones are activated. Note that the cone response exhibits a shorter latency, more rapid rise, and more rapid decay than the rod response. In the respective ranges of stimulus intensity that influence rod and cone responses (see Fig. 17-21), both rod and cone responses exhibit faster time courses at higher stimulus intensities. But in the macaque monkey, it was shown that variation of stimulus intensity could not yield a rod receptor potential with such a rapid time course as that of the cone receptor potential.[125] Hence these findings reveal a third major aspect of the duplicity theory. Cones and rods differ not only in their anatomy and in their contained photopigments, but also in the time courses of their electrical responses.

Intracellular receptor potentials from turtle cones and rods

Much information has now been obtained by intracellular recording from photoreceptors, as confirmed by injecting marker dyes into the cell through the recording electrode, followed by histologic identification of the marked cell. This work has been conducted in lower vertebrates with especially large photoreceptors, such as *Necturus* and turtles. For example, the snapping turtle *(Chelydra serpentina)* has an abundance of both rods and cones that have large enough inner segments for intracellular work. The records of Fig. 17-18 were obtained by impaling both a rod and a nearby cone by two separate microelectrodes. The responses of both receptors were then recorded simultaneously in response to a large stimulus flash that covered the retinal area around the electrodes. Several points are revealed that have proved typical of intracellular responses of vertebrate photoreceptors. First, the intracellular response to a light flash is negative in polarity. Hence the photoreceptor is *hyperpolarized* by light stimulation, unlike all other types of vertebrate receptors that have yielded to intracellular recording. Second, neither the cone nor rod response shows any sign of impulse activity, the responses consisting only of slow potentials. Third, the time course of the cone response is more rapid in all respects than the time course of the rod response.

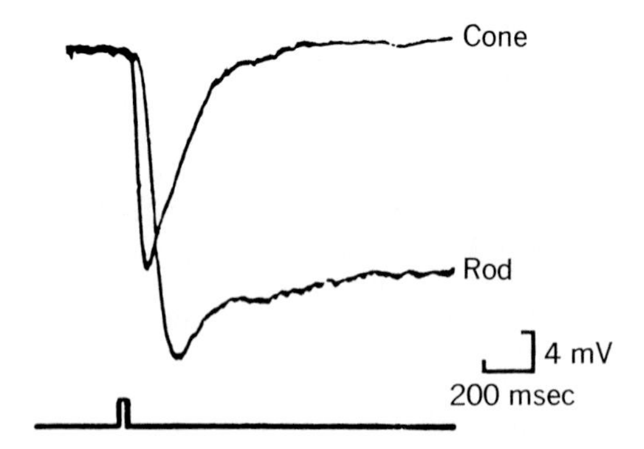

Fig. 17-18. Responses from a rod and a nearby red-sensitive cone of snapping turtle, recorded intracellularly by two independent microelectrodes and elicited simultaneously by same light flash. Stimulus was a 20 msec flash of 633 nm light delivering 720 quanta/μm^2 at receptor surface and large enough to cover retinal area that activated these receptors. (From Copenhagen and Owen.[49])

Effects of stimulus intensity on rod and cone receptor potentials

Fig. 17-19, *A*, shows the effects of stimulus intensity on the amplitude and time course of the monkey's rod and cone receptor potentials. These records were obtained from the peripheral retina of a macaque monkey after clamping the retinal circulation, using an extracellular microelectrode near the distal tips of the photoreceptors. In comparison, the records of Fig. 17-19, *B*, were recorded intracellularly from the inner segment of a cone in the red-eared turtle *(Pseudemys scripta elegans)*, which has a predominantly cone retina. In both experiments shown in Fig. 17-19 the responses were evoked by a 20 μsec flash of white light covering the area around the electrode, and superimposed responses are shown to flashes in which intensity was varied by steps of 0.6 log unit. The monkey retina responded to the lowest stimulus intensities with only a rod receptor potential, which rose and decayed very slowly. With increased intensity the amplitude of this rod response increased, and in the middle record a superimposed cone response appeared with rapid rise and decay. This cone response then increased in its rate of rise as its amplitude increased to the maximum, beyond which still higher stimulus intensities increased the duration of the cone response prior to the beginning of its decay. The intracellular records show, in greater detail, similar effects of stimulus intensity on responses from the single turtle cone. Hence with closely comparable stimulus conditions, the effects of stimulus intensity on the amplitude and time course of the cone response were very similar for the monkey and turtle.

Fig. 17-20 shows receptor potentials recorded intracellularly from a turtle rod as a function of stimulus intensity. This rod showed a resting

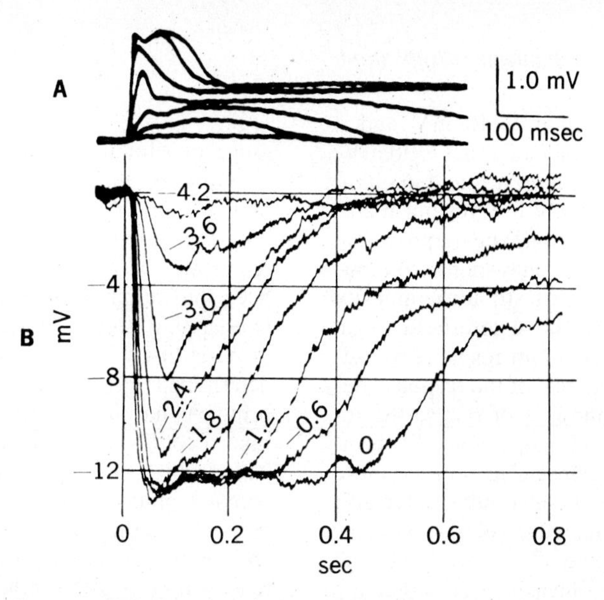

Fig. 17-19. A, Isolated extracellular receptor potentials from peripheral retina of macaque monkey, recorded by microelectrode near distal tips of receptors. Responses evoked by 20 μsec flash covering large area around microelectrode. Superimposed responses are shown to maximum flash intensity and to flash intensities that decreased by steps of 0.6 log unit. **B,** Intracellular receptor potentials from single turtle cone, evoked by 20 μsec flashes covering area of 160 μm diameter around impaled cone. Here also, superimposed responses are shown to maximum flash intensity (0) and to flash intensities reduced by steps of 0.6 log unit. (**A** from Brown and Murakami[31]; **B** from Baylor and Fuortes.[20])

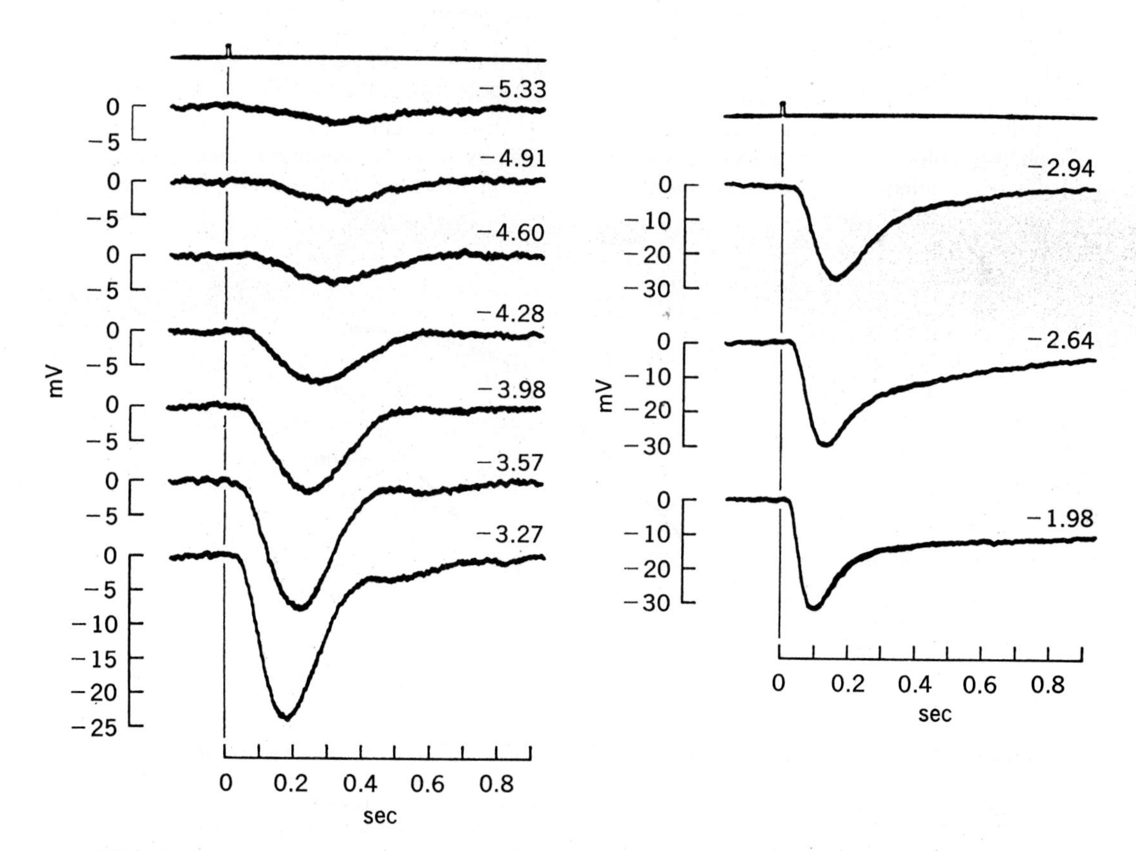

Fig. 17-20. Intracellular responses as a function of stimulus intensity, recorded from a rod in retina of red-eared turtle. Stimulus was 10 msec flash of white light delivered to circular area 300 μm in diameter that was centered on receptor. Stimulus intensity for each response is given in log units of attenuation below maximum stimulus intensity available. (From Baylor and Hodgkin.[21])

membrane potential of about −40 mV and a maximum response amplitude of about −30 mV. These values are about the largest found in either cones or rods. The relatively low resting membrane potential seems an intrinsic requirement for a cell that hyperpolarizes in response to light. Fig. 17-20 shows that when stimulus intensity was increased from just above threshold to an intensity that yielded maximum response amplitude, the latency to initial rise of the response decreased markedly and the rate of rise to the response peak became much more rapid. With increased intensity the response also decayed more rapidly up to a relatively high stimulus intensity, beyond which further increases of intensity resulted in a very slow decay. This greatly slowed decay at high stimulus intensities is called the *rod aftereffect*. It also appears in the monkey's rod receptor potential, as shown in Fig. 17-19, *A*, and is responsible for decay of the rod receptor potential being especially slow at the high stimulus intensities normally used to evoke an ERG. There is a comparable *cone aftereffect* at high photopic intensities,[127] as illustrated in Fig. 17-19 for both monkey and turtle.

Figs. 17-17 through 17-20 illustrate that extracellular and intracellular methods can yield very similar results when studying the amplitude and time course of receptor potentials. They also indicate that certain aspects of intracellular work in the photoreceptors of lower vertebrates can be applied with considerable confidence to monkey, and hence to human, retinas.

Fig. 17-19, *A*, shows that a relatively low range of stimulus intensity affects only the rod response of the macaque monkey, whereas a higher range of stimulus intensity seems to alter only the cone receptor potential. In the monkey retina the rod and cone receptor potentials have been shown to respond to different but somewhat overlapping ranges of stimulus intensity,[3,31] indicating that this aspect of the duplicity theory is determined largely at the receptor level. This principle has been well illustrated by intracellular recording in *Necturus*, as shown in Fig. 17-21, in which the amplitudes of both rod and cone receptor potentials have been plotted as a function of stimulus intensity. Note that the rod response increases rapidly over a relatively low intensity range, within which only the higher intensities begin to influence the amplitude of the cone response. At the intensity level where rod responses cease to increase rapidly with stimulus intensity, the cones take over and exhibit a rapid increase of amplitude in a higher range of stimulus intensity. Thus rods and cones share the stimulus intensity range, with both types of receptors coding stimulus intensity in terms of response amplitude.

Rod and cone receptor potentials as determinants of CFF

The time courses of rod and cone receptor potentials must be important determinants of all visual functions that involve the dimension of time. As discussed earlier in this chapter, the

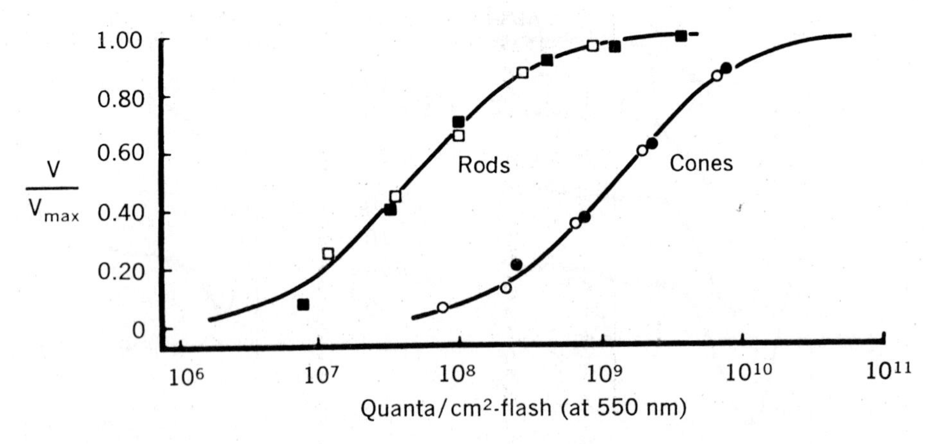

Fig. 17-21. Response amplitudes of two rods and two cones of *Necturus* retina, recorded intracellularly and plotted as a function of stimulus intensity. Stimulus was brief flash of 550 nm light covering large retinal area, and stimulus intensity is plotted as quanta per square centimeter delivered by flash at retinal surface. Response amplitude is plotted as V/V_{max}, where V is peak voltage of a given response and V_{max} is peak voltage of largest responses obtained from rods or cones, respectively. (From Fain and Dowling[55]; Copyright © 1973 by the American Association for the Advancement of Science.)

ability of the visual system to resolve stimuli that are separated in time is measured by the CFF. The main results of Fig. 17-5, which shows the human CFF as a function of stimulus intensity, now seem explicable from the time courses of receptor potentials. For example, the CFF of cones can extend much higher than that of rods. Since the cone receptor potential rises and decays considerably more rapidly than the rod receptor potential, of course cone responses are better suited to resolve stimuli that are closely spaced in time. For both the rod and cone limbs of the CFF curves in Fig. 17-5, CFF increases markedly with stimulus intensity. Figs. 17-19 and 17-20 show that increased stimulus intensity speeds the rise to the peak of both rod and cone receptor potentials, and this should improve resolution in the time domain. At very high photopic intensities, however, Fig. 17-5 shows the cone CFF to decrease somewhat. This is probably a consequence of the cone aftereffect. The cone aftereffect is probably also the basis for positive afterimages; these are images that persist for a short time after the stimulus, and they occur only at very high photopic intensities.

Intensity discrimination by foveal cones

Under natural conditions a major task of cone vision is to detect an increment or decrement of stimulation against a prevailing steady background illumination. Fig. 17-22 shows a study of cone intensity discrimination that was conducted by recording the isolated receptor potential from the monkey's pure cone fovea. Each curve shows amplitude of the cone receptor potential as a function of intensity of the incremental test flash. One such curve was obtained in darkness, and other curves were obtained against background illuminations covering a range of 6 log units. The results fit well to the theoretical curves drawn. The basic response curve in darkness is well fit by the following expression[26]:

$$R_t = \frac{I_t^n}{I_t^n + K_r^n}$$

In this equation, R_t is response amplitude to the test flash, and I_t is intensity of the test flash in trolands, whereas the constant K_r is the number of trolands required to elicit a cone receptor potential of half-maximum amplitude. This equation is a power law at low to moderate intensities, in agreement with the effects of stimulus intensity on sensation magnitudes as determined psychophysically in a variety of sensory modalities.[116] In the case of monkey cones the exponent of this power law has a value of about 0.70.[26] As background intensity increases, the response curves in Fig. 17-22 move to the right, indicating that higher stimulus intensities are re-

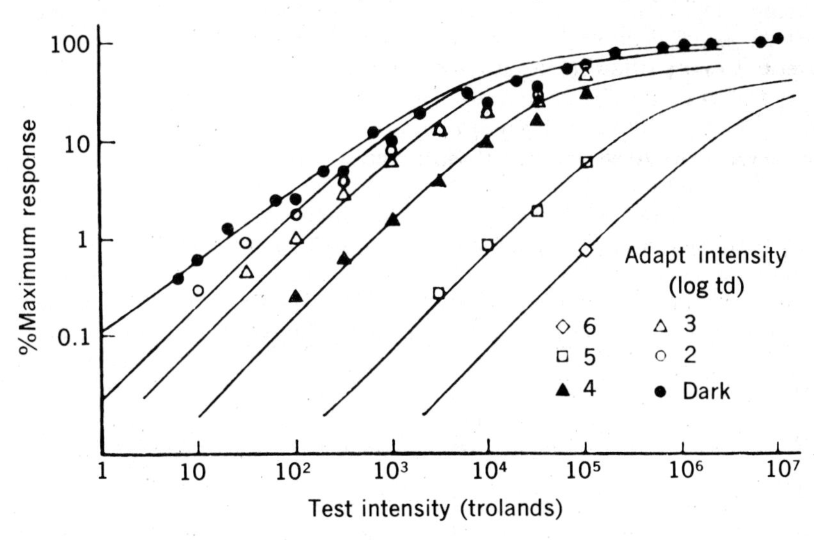

Fig. 17-22. Amplitude of foveal cone receptor potential as function of stimulus intensity. After clamping retinal circulation, potentials were recorded by extracellular microelectrode near distal tips of foveal receptors. Pure-cone origin of receptor potential confirmed by spectral response function. Response amplitude measured at termination of stimulus and expressed as percentage of maximum response amplitude. Test stimuli were presented in complete darkness and against a variety of adapting background intensities. Both background and test flashes had a retinal diameter of 1.1 mm and were centered on fovea. Stimulus was 150 msec flash of light at 580 nm. Data are mean values from six animals, and curves through data points are theoretical. (From Boynton and Whitten[26]; Copyright © 1970 by the American Association for the Advancement of Science.)

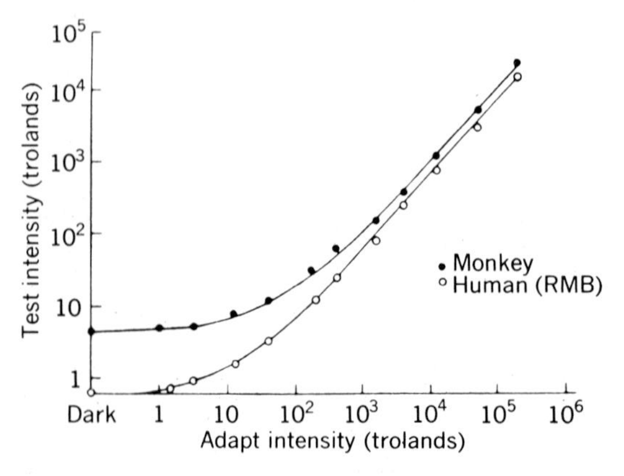

Fig. 17-23. Incremental thresholds as function of steady adapting intensity for human observer and for receptor potential of macaque monkey. Both curves obtained with same optical stimulator, using yellow adapting field 1.1 mm in diameter and centered on fovea. Test stimulus was 150 msec flash of yellow light. For human, test stimulus was confined to fovea, and thresholds were obtained by adjusting test stimulus to minimum detectable intensity. In monkey, pure-cone foveal receptor potential was recorded as in Fig. 17-22 and evoked by 1.1 mm test flash centered on fovea. Responses were averaged for detection of small responses, and 10 μV amplitude of receptor potential was taken as threshold. (From Boynton and Whitten[26]; Copyright © 1970 by the American Association for the Advancement of Science.)

quired to elicit an incremental receptor potential of constant amplitude.

In Fig. 17-23 the intensity discrimination of the human fovea is compared directly with detection of incremental intensities by cone receptor potentials of the monkey fovea. The psychophysical function in humans shows the effect of steady adapting intensity on the incremental test intensity required to elicit a threshold sensation. This function rises at first slowly and then more rapidly. The monkey results show the effect of steady adapting intensity on the incremental test intensity required to elicit a very small cone receptor potential of constant amplitude. The similarity of the two curves indicates that the effect of background intensity on incremental sensitivity is determined largely at the receptors, and the reasons for this have been analyzed.[26]

At greater than approximately 10^3 trolands of adapting intensity the two curves of Fig. 17-23 have similar slopes, indicating that in this range the psychophysical effects of adapting intensity are strongly determined by the receptors. In this high range of intensities, it might be expected that a background intensity would be reached that would elicit a cone receptor potential of maximum amplitude, causing the incremental threshold function to rise vertically as the threshold became infinitely high. Fig. 17-23 shows that this does not happen; instead, cones are able to give incremental brightness sensations against indefinitely high background illuminations. The basis for this important aspect of cone function now seems clear.[26] It is only above approximately 10^3 trolands that the bleaching of cone photopigment by a steady background illumination becomes significant. Above this level the cone receptor potential elicited by a steady background illumination is only about half the amplitude elicited by a brief flash of equal intensity. This is because the long duration of the steady background causes photopigment bleaching, thus reducing receptor sensitivity and the amplitude of cone receptor potential elicited by a steady background. In the intensity range in which this effect occurs, each doubling of background intensity will roughly halve the photopigment concentration, so that the number of quanta absorbed per unit time remains constant, and the cone receptor potential elicited by the background illumination is unchanged. Thus photopigment bleaching serves to hold the cone receptor potential elicited by intense backgrounds to only about half the maximum cone response. By adapting to the background intensity in this manner, the cone response elicited by the background is kept in the center of the cone's response range, so that the cone receptor potential can either increase or decrease in response to incremental or decremental stimuli. Of course the incremental threshold rises in this same range of background intensity, as shown in Fig. 17-23, because progressively less photopigment is available to absorb the incremental light flash. Thus bleaching of photopigment reduces sensitivity to incremental light flashes, but serves to extend greatly the upper limit of background illuminations against which cones are capable of intensity discrimination. This mechanism is not available to rods, apparently because rod receptor potentials reach maximum amplitude at background intensities that do not cause significant bleaching of photopigment.[26] Thus with increased background intensity the incremental threshold of rods rises sharply to infinity at the upper end of the rod response range.[14] This does not impair intensity discrimination because at these background intensities the cones, endowed with a remarkable mechanism of intensity discrimination up to extremely high levels of background intensity, have already taken over.

Adaptation and intensity discrimination have also been studied intracellularly in both rods and cones of *Necturus*.[94] Note in Fig. 17-21 that at a given level of adaptation, both rods and cones of *Necturus* exhibit significant changes of response amplitude over only about 3 log units of stimulus intensity. Of course, intensity discrimination should be optimal in the central portions of these curves, where the slopes are greatest. Intracellular work has shown that during light adaptation the cone response curve remains similar in form, but it moves to the right along the intensity axis.[94] Thus light adaptation adjusts the cone response range to center on the intensity of the background illumination, so that the cones are always optimally sensitive to increments or decrements of intensity relative to the prevailing background illumination.

RECEPTOR EXCITATION

The outlines of the subject of receptor excitation are now defined, and some of the critical steps are known. The vertebrate photoreceptor is unique among receptors and nerve cells studied to date, in that steady current flows along the photoreceptor in the absence of stimulation; this is called *dark current*. Penn and Hagins[100] first showed that a steady dark current flows extracellularly from the inner to the outer segment. This subject was greatly elaborated by Zuckerman,[132] who analyzed the extracellular voltage and resistance gradients along the axis of frog photoreceptors, from which the transmembrane dark current was determined for various positions along the receptor axis. His results are shown schematically in Fig. 17-24. The dark current was found to be generated primarily by an electrogenic pump in the inner segment. By definition, an electrogenic pump extrudes sodium more rapidly than it brings potassium into the cell. It thus acts as a continuous source of current flowing outward from the inner segment. The electrogenic nature of the pump was demonstrated by rapid reduction of the dark current after ouabain treatment, and by highly sensitive increases of dark current in response to raised extracellular potassium concentration.[132] In Fig. 17-24 the electrogenic pump is shown near the outer portion of the inner segment, where net outward current was most intense.[132] This is also a site of intense metabolic activity, as indicated in Fig. 17-10 by the high concentration of mitochondria, which consistently cluster in the outer portion of the inner segment. Ion substitution experiments demonstrated that the current flowing outward from the inner segment was carried mainly by sodium.[132] Fig. 17-24 shows that this outward sodium current was found to return by

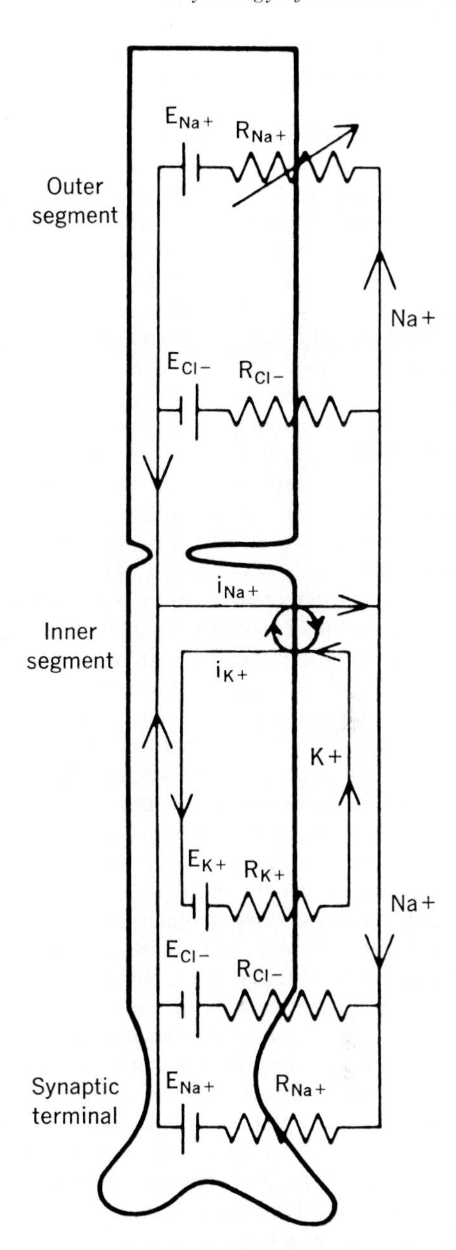

Fig. 17-24. Summary diagram of dark current in vertebrate rod photoreceptors. (Modified from Zuckerman.[132])

two loops, one of which flows distally to reenter the cell through the outer segment, whereas the other flows proximally to reenter through the synaptic terminal. After poisoning the electrogenic pump with ouabain, the proximal current loop was abolished and the distal loop considerably weakened. The remaining current in the distal loop resulted from a passive mechanism in which the sodium permeability of the outer segment was greater than that of the inner segment[132]; apparently it was only this passive component of the distal loop that was observed by Penn and Hagins in isolated slices of the rat retina.[100]

The dark current is important because it appears that light-induced modulation of the dark current generates the receptor potential. Korenbrot and Cone[82] studied the ionic permeability of the outer segments of frog rods by isolating the outer segments and measuring their osmotically determined changes of length in a variety of solutions. The outer segments proved impermeable to potassium, consistent with Zuckerman's results shown in Fig. 17-24, where the potassium component of dark current is shown confined to the inner segment. More importantly, Korenbrot and Cone showed that light reduced the sodium permeability of outer segments, but had no effect on membrane permeability for a variety of other ions. As may be seen in Fig. 17-24, a reduced sodium permeability of the outer segment would reduce the amount of sodium reentering the cell through the distal loop. If the electrogenic pump extrudes current from the inner segment at a constant rate during light stimulation, for which there is some evidence,[132] a reduced sodium current in the distal loop must be accompanied by an increased sodium current in the proximal loop. It thus appears that the vertebrate photoreceptor acts as a current divider, with light shifting some of the dark current in the distal loop to renter the cell through its synaptic terminal. A reduced sodium entry through the outer segment membrane will have a hyperpolarizing effect on the membrane potential. Since light presumably does not alter sodium permeability of the synaptic terminal, the increased sodium current in the proximal loop will likewise have a hyperpolarizing effect. It thus appears that these two effects account for hyperpolarization of the photoreceptor membrane by light, as recorded intracellularly. Similarly, the light-induced reduction of current in the distal loop, and the resulting increased current in the proximal loop, will both cause the distal end of the photoreceptor to become extracellularly more positive, relative to the proximal end. The polarity of the extracellular receptor potential is thus readily understood, since this potential is always positive when recorded between a microelectrode in the distal part of the receptor layer and a reference electrode in the vitreous humor (see Figs. 17-17 and 17-19, *A*). It is also evident that the demonstrated similarity of intra- and extracellular receptor potentials, in time course and in many other respects, results from their common origin in light-induced modulation of the dark current.

In vertebrate photoreceptors the transduction problem now concerns the mechanism by which light reduces the sodium conductance of the outer segment membrane. As shown in Figs. 17-10 and 17-11, the saccules throughout most of the rod outer segment have no connection with the membrane that bounds the outer segment. Since the rod photopigment is contained predominantly in the saccule membranes, the action of light on that photopigment must be transferred in some way to the bounding membrane. A general hypothesis for how that occurs is now being followed by many investigators. This hypothesis assumes that a chemical transmitter is concentrated within the saccules. Evidence has already been presented that the visual photopigment is a structural component of the saccule membrane and that the conformation of the photopigment is altered when a quantum of light is absorbed. It is assumed that this alteration of molecular structure opens a channel to release from the saccule some of its contained transmitter, which then diffuses to the outer segment membrane, where it decreases sodium permeability by partially or entirely blocking sodium channels. Yoshikami and Hagins have shown that increased extracellular calcium ions can reduce dark current flowing in the distal loop,[130] and there is some evidence that increased intracellular calcium ion has similar effects.[64] Also, intracellular recording in the outer segments of toad rods has shown that increased extracellular calcium mimics light adaptation by hyperpolarizing the cell, reducing the amplitude of the light response, and speeding the time course of the light response.[28] Hence calcium is now a candidate for the internal transmitter.[64] There are indications, however, that the present simple form of the calcium transmitter hypothesis may not be adequate. There is a strong question, for example, based on reasonable assumptions of how high the calcium concentration could be within the saccule, whether absorption of a single quantum of light could release the amount of calcium that is theoretically required to mimic the action of a single absorbed quantum of light on the dark current.[46]

In applying the calcium transmitter hypothesis to the cone case, it must be remembered that cone saccules maintain their continuity with the other membrane and in fact consist of deep invaginations of that membrane. The calcium hypothesis assumes that the extracellular calcium ion concentration is higher than the intracellular concentration; when a quantum of light is absorbed by photopigment in a cone invagination, calcium is thus assumed to enter the cell and act on the inside of the bounding membrane to reduce its sodium permeability.[64]

Clearly, much further work is required on this subject. In particular, both the general hypothesis and the identity of the transmitter need to be established. If the chemical transmitter hypothesis proves correct, receptor potentials should have much in common with postsynaptic potentials. This would provide a basis for understanding many characteristics of the electrical activity of vertebrate photoreceptors.

SYNAPTIC TRANSMISSION AT RECEPTOR TERMINALS

Both anatomic and physiologic evidence suggest that transmission between the photoreceptors and second-order cells is mediated by chemical synapses. The receptor terminals exhibit vesicles similar to those found in the presynaptic terminals of well-demonstrated chemical synapses.[5] In other types of receptors and nerve cells that have been studied, depolarization of the presynaptic membrane leads to transmitter release. If vertebrate photoreceptors follow this principle, then transmitter must be released continuously in darkness, so that hyperpolarization of the photoreceptor by light can reduce transmitter release and thus mediate postsynaptic effects. Cells postsynaptic to receptors include both bipolars and horizontal cells. Because of their larger size, the horizontal cells have been more accessible to intracellular recording. These cells, like the photoreceptors, typically exhibit low resting membrane potentials and are hyperpolarized by light. Dowling and Ripps[53] have shown in a fish retina that magnesium ions, which block transmitter release at a variety of presynaptic sites, mimic the action of light on the horizontal cell. Increased extracellular magnesium caused the horizontal cell to hyperpolarize from a resting potential of about −30 mV to a value of about −60 mV, without significantly affecting the light responses of photoreceptors. Another presynaptic blocking agent, cobalt, also mimics the effects of light on horizontal cells.[80] Although a depolarizing type of transmitter appears to be released by photoreceptors, it has not been identified; candidates include sodium aspartate and sodium glutamate, both of which rapidly depolarize horizontal cells with little accompanying effect on receptor potentials.[41]

In summary, it appears that dark current maintains the photoreceptor in a relatively depolarized state that is accompanied by the continuous release of a depolarizing chemical transmitter. Since light absorption reduces dark current, it hyperpolarizes the receptor, thus reducing transmitter release and also hyperpolarizing second-order cells. Dark current is thus central to the unconventional aspects of signal generation and synaptic transmission in vertebrate photoreceptors. Although a functional significance of this highly specialized mechanism may be expected, it has not yet been identified.

RECEPTOR INTERACTIONS

Until recently it was generally thought, despite anatomic indications to the contrary, that photoreceptors acted as independent light detectors whose signals did not interact with each other. It has now been shown that receptors interact in specific ways that appear to have important functional consequences.

Signal sharing among photoreceptors

One type of interaction is a system of signal sharing that is well developed in the rod case, as demonstrated in both snapping turtle and toad.[49,56] In this type of interaction, each rod's response is dissipated partly into other rods. Hence with a stimulus covering many receptors, the response of each rod is generated partly by its own signal-generating mechanism, but a significant portion of the response is also received from other rods. The existence of such signal sharing is demonstrated in Fig. 17-25, which shows intracellular responses from a rod of the

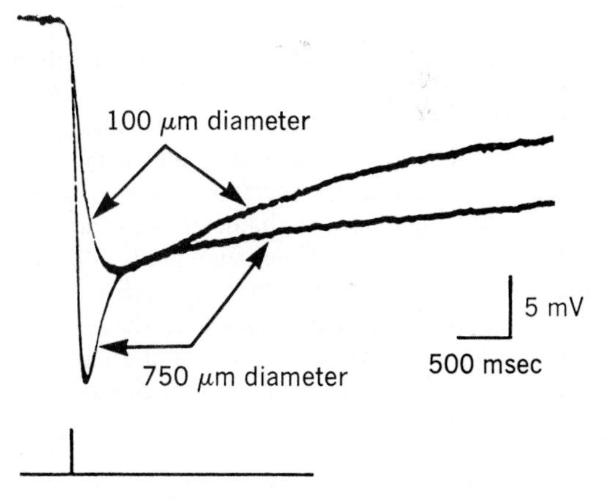

Fig. 17-25. Summative interaction between rods, demonstrated by intracellular recording in rod of snapping turtle. Spot of light 100 μm in diameter was first centered and focused on impaled receptor, and 20 msec flash of 514 nm light was used to evoke a response, stimulus intensity having been adjusted to about 0.5 log units greater than required to evoke maximum response amplitude with this type of stimulus. Spot size was then increased to 750 μm, which increased response amplitude and slowed decay of response. (From Copenhagen and Owen.[49])

snapping turtle. A stimulus of 100 μm diameter was first centered on the rod, and the stimulus intensity was increased until the response no longer increased with higher stimulus intensity. At that point the rod must have been giving the maximum response that it could generate on its own. When the intensity was held constant at that level and the stimulus diameter was then increased to 750 μm, the initial phase of the response increased markedly in amplitude. Also, the response decay was slowed, an apparent example of the rod aftereffect. Since the impaled rod responded more strongly to the larger stimulus, this increased response must have resulted from the stimulation of other rods that were more than 50 μm away from the impaled rod. In the snapping turtle, Copenhagen and Owen have also studied this phenomenon by impaling two nearby rods with two independent microelectrodes.[48] When either rod was hyperpolarized by passing current, a hyperpolarization was also seen in the other rod. Depolarizing currents produced qualitatively similar effects. Hence rods are coupled in a manner that permits reciprocal signal sharing, a kind of coupling that typically requires electrical synapses based on gap junctions.[101] In the toad retina, gap junctions have been identified between the inner segments of adjacent rods,[56] whereas in the snapping turtle, the location of rod-rod junctions has not been determined.

A similar system of signal sharing occurs between cones,[19,21,49] but for cones the retinal area over which the sharing occurs is smaller than for rods of the same species. This has been well demonstrated in the snapping turtle.[49] During intracellular recording a focused stimulus spot was centered on the impaled receptor, and the stimulus spot was increased in diameter until the sensitivity of the cell no longer increased with further increase of spot diameter. The cell's sensitivity became maximum at a stimulus diameter of about 125 μm in the cone case and at about 300 μm in the rod case. In the cone case this type of coupling is probably based on gap junctions that have been identified between the synaptic terminals of cones in several species, including the turtle and the macaque monkey.[101]

This system of signal sharing appears to occur primarily between receptors of a given class. Thus rods share signals mainly with other rods. In cones the signal sharing is even more specific, as demonstrated in the predominantly cone retina of the red-eared turtle. In that case, it has been shown that the red, green, and blue cones each share signals only with cones of the same class.[19] In this way, signal sharing can occur in cones without disturbing the color coding of cone signals.

The functional advantage of signal sharing appears to have been revealed by Simon et al.[111] They noted that cones exhibiting little or no signal sharing with other cones have relatively high levels of random noise in the membrane potential during intracellular recording in the dark. The origin of this dark noise in the cell's membrane potential is not clear, but it probably results from the random opening and closing of sodium channels in the outer segment. In any event this noisiness of the membrane potential probably results in synaptic noise, defined as random variations in the release of transmitter from the photoreceptor's terminal. This synaptic noise probably limits visual sensitivity, in which the problem of the bipolar cell is to discriminate a light-induced release of transmitter from random releases of transmitter in the dark. Since the membrane noise in any given receptor is random, the simultaneous pooling of signals from a number of receptors would result in the membrane noise being reduced, as in the case of signal averaging. In accordance with this hypothesis, the receptors that share signals over the largest retinal areas exhibit the lowest levels of dark membrane noise, and this relationship has been shown to be systematic.[86] Thus signal sharing reduces membrane noise in the dark, and this should improve the signal-to-noise ratio for synaptic transmission, thus enhancing the ability of the visual system to detect weak stimuli. On the other hand, signal sharing appears to be a disadvantage to visual acuity. The rods are specialized for high sensitivity, but have low acuity, and these functional characteristics of rods may result partly from the sharing of rod signals over a relatively large retinal area. Cones are specialized for high acuity in the photopic intensity range, where less sensitivity is required, and these cone characteristics seem to correspond with the sharing of cone signals over a smaller retinal area.

Inhibitory interaction between cones

In recordings from single cones of the red-eared turtle, an inhibitory type of interaction has also been observed.[22,110] This occurs over a longer distance than the signal-sharing type of interaction between cones. Also, the inhibitory interaction exhibits a distinct delay. These features result from the horizontal cell acting as an interneuron in mediating the inhibitory interaction. Fig. 17-26 shows records obtained while

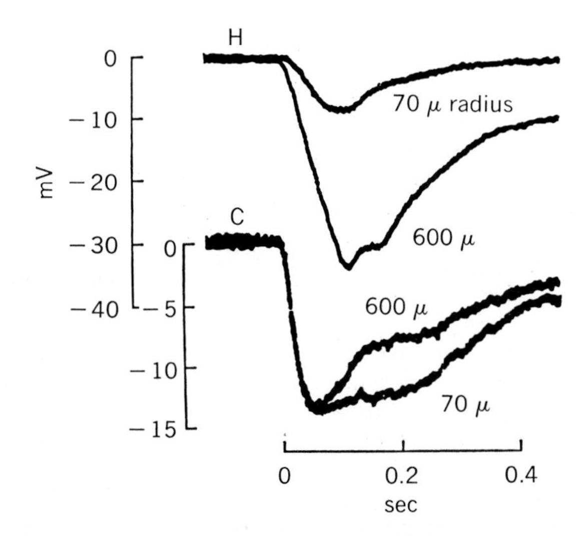

Fig. 17-26. Simultaneous intracellular recordings from a cone (C) and a nearby horizontal cell (H) in red-eared turtle. Small stimulus of 70 μm radius was centered on cone and used to elicit simultaneous responses from both cells. Then stimulus was enlarged to radius of 600 μm, and this procedure was repeated. Small stimulus gave cone response of maximum amplitude; larger stimulus gave cone response of similar amplitude, followed by delayed depolarization relative to control response, illustrating delayed inhibitory effect on cone receptor potential. (From Baylor et al.[22])

stimulating spots were focused on a small retinal area, within which two independent electrodes recorded intracellularly from a cone and a nearby horizontal cell. When the radius of the stimulus was increased from 70 to 600 μm, the amplitude of the cone response was unaffected, but the amplitude of the horizontal cell's response increased markedly. This is in line with other evidence that the horizontal cell summates receptor activity over a much larger retinal area than that over which cones exhibit signal sharing. Furthermore, the larger spot caused a delayed decrease of the cone response, and this delayed depolarization from the cone's control response became maximum just after the peak of the horizontal cell's response. In other experiments it was shown that when a horizontal cell was hyperpolarized by passing current through an intracellular electrode, this induced depolarization in the intracellular recording from a nearby cone.[22] Thus horizontal cells can mediate inhibitory lateral interaction between receptors. In the red-eared turtle, it has been shown that this type of interaction is mediated reciprocally between red and green cones.[110] Thus a red stimulus will cause a delayed reduction in the response of green cones and vice versa. This appears to be a mechanism for intensifying color sensations. In color mixing, red and green are complementary colors that, when mixed in appropriate proportions, will give only the sensation of gray. Because of the mechanism just described, a pure red stimulus will not only activate red receptors, but will also inhibit green receptors, thus enhancing the purity of the resulting sensation of red. Since the mechanism is reciprocal, it also operates well for a pure green stimulus.

It is not clear whether the described mechanism for enhancing the purity of color sensations exists in primates. But these results show that horizontal cells can act as inhibitory interneurons mediating interactions between receptors, and this basic role of horizontal cells is likely to be consistent among vertebrate species.

Cone-activated suppression of rod receptor potentials in the primate retina

The duplicity theory has always contained an important question concerning the functional status of rods at photopic intensities. This question is rendered more crucial by the rod aftereffect, which occurs at stimulus intensities above that required to elicit a rod response of maximum amplitude. If there were no mechanism for suppressing rod responses at photopic stimulus intensities, this rod aftereffect would be expected to strongly compromise the time-resolving ability of the visual system. In that event the high CFFs characteristic of photopic vision would not seem possible. Also, in that event the low visual acuity of the rod system would be expected to compromise the visual acuity that can be achieved by the cone system.

In the macaque monkey there is now evidence that when special care is taken to maintain normal physiologic conditions, the activation of cones at photopic intensities results in the rod receptor potential being markedly reduced or abolished.[126] This is indicated by disappearance of the slowly decaying rod receptor potential at photopic stimulus intensities. The horizontal cell probably acts as an inhibitory interneuron in this pathway. As pointed out earlier in this chapter, present evidence indicates that the macaque monkey has a single type of horizontal cell, from which short dendrites contact only cones, whereas a long axon proceeds laterally and contacts only rods.[25,81] If this horizontal cell conducts in the expected direction and acts as an inhibitory interneuron, it appears anatomically well suited to mediate a cone-activated suppression of rod receptor potentials.

Intracellular work in photoreceptors of lower vertebrates has thus far not revealed a cone-activated suppression of rod responses. Hence this mechanism may have evolved only in the higher vertebrates. This would not be surprising in view of marked species differences in the types of horizontal cells found and the connections that they make. Lower vertebrates, such as fish, have at least four anatomically different kinds of horizontal cells.[115] Hence the single type of horizontal cell in the macaque monkey indicates an evolutionary trend toward simplification of horizontal cells to serve fewer functions, which may safely be assumed to be critical functions for the primate retina.

Although the underlying mechanism is not established, suppression of the monkey's rod responses at photopic stimulus intensities appears to have the important consequence of clearing the entire pathway through the retina for the handling of only cone signals. The cones are thus permitted to realize their full capabilities for high CFF and visual acuity, unimpeded by the limitations that rod signals would otherwise impose on those functions. In accordance with this view, it has also been shown in the macaque monkey that rod signals fail to appear in the responses of ganglion cells when cones are activated simultaneously.[61] In total color blindness, where cone functions are absent or minimal, the described photopic suppression of rod responses should not occur. Hence the photophobia and abnormal dazzle by bright lights, which are consistent symptoms of total color blindness, are explicable on this basis. Also, work in the macaque monkey has shown that the pathway for photopic suppression of rod responses is extremely sensitive to functional interruption by anoxia and barbiturates.[126] This suggests an explanation for why photophobia is also such a common transient symptom following the excessive use of a variety of drugs, such as alcohol.

PROCESSING OF SIGNALS THROUGH THE RETINA

The type of signal generated by retinal cells becomes modified as the signal progresses through the retinal pathway. Also, an important concept in retinal signal processing is the *receptive field* of a cell. This is defined as the retinal receptor area within which light stimuli influence the activity of the cell whose signals are being recorded. Each major type of retinal cell has now been studied by intracellular recording in a variety of vertebrates. For each cell type, these studies agree well concerning the kind of signal generated and the general organization of the receptive field. The results are well illustrated by work in *Necturus,* where all classes of cells are unusually large and amenable to intracellular recording.[124] Fig. 17-27 shows recordings obtained by Werblin and Dowling from each class of cell, with certain cells of each class being identified by electrophoretic injection of Niagara blue dye, followed by histologic study to determine the location and form of the cell. Each type of cell was stimulated by a light spot 100 μm in diameter that was centered on the impaled cell and also by a concentric annulus having an inside diameter of 500 μm.

In Fig. 17-27 the receptor responded to the light spot with a hyperpolarizing slow potential, as shown also in previous figures. Although the receptors themselves have receptive fields, resulting from receptor interactions such as signal sharing, these receptive fields tend to be small. Hence the annular stimulus gave only a very small hyperpolarizing response, and this probably resulted mainly from stray light that was scattered from the stimulus onto the impaled receptor.

The horizontal cell also gave only a hyperpolarizing slow potential, but exhibited a larger receptive field than the receptor. This was indicated by the response of the horizontal cell to the annular stimulus being about 60% as large as this cell's response to the light spot. The larger receptive field of the horizontal cell undoubtedly results from the demonstrated convergence of many receptors to form synapses with a given horizontal cell.

The bipolar cell likewise responded only with slow potentials, but the response polarity depended on the part of the receptive field stimulated. For the bipolar cell illustrated in Fig. 17-27, a centered spot always gave a hyperpolarizing response. The annulus alone was ineffective as a stimulus. But when the centered spot was used as a continuous stimulus to maintain the cell's membrane potential in a relatively hyperpolarized state, as was done for the annular response in Fig. 17-27, the annulus gave a depolarizing response. The receptive field of the bipolar cell was thus concentrically organized into two antagonistic zones, so that stimulation of the peripheral part of the cell's receptive field reduced the response to central illumination. In response to a centered spot, about half the bipolar cells were hyperpolarized, as shown, whereas the other half exhibited depolarizing responses. In either case the response to continuous illumination of the center of the receptive field was reduced by stimulating the periphery of the receptive field with an annulus. In goldfish the center-surround organization in the receptive fields of bipolar cells is even more pronounced, since the annulus alone consistently gives a response opposite in polarity from the response to a centered stimulus.[79]

The amacrine cell represents a transition from sustained to transient slow potentials. Also, the amacrine cell responds only with depolarizations, which can serve as generator potentials that initiate regenerative nerve impulses. Thus in Fig. 17-27 the amacrine cell responded with transient depolarizations when the stimulus was turned either on or off. In response to the centered spot, the "on" response was relatively large and at its peak gave rise to several nerve impulses, whereas the "off" response was only a small transient slow potential. In contrast, the annulus elicited relatively large transient depolarizations at both "on" and "off," but only the "off" response was large enough to initiate one or two nerve impulses. Thus amacrine cells can also give different responses to stimulation of the central and peripheral portions of their receptive fields, but for amacrine cells the difference is between a predominantly "on" or "off" response. Some amacrine cells in *Necturus*, and all amacrine cells studied thus far in goldfish,[79] have failed to show this type of center-surround organization in their receptive fields. These cells have yielded "on" and "off" responses of similar magnitude in response to either centered spots or annular stimuli. The anatomy of amacrine cells, as shown in Fig. 17-1, suggests that they mediate lateral interac-

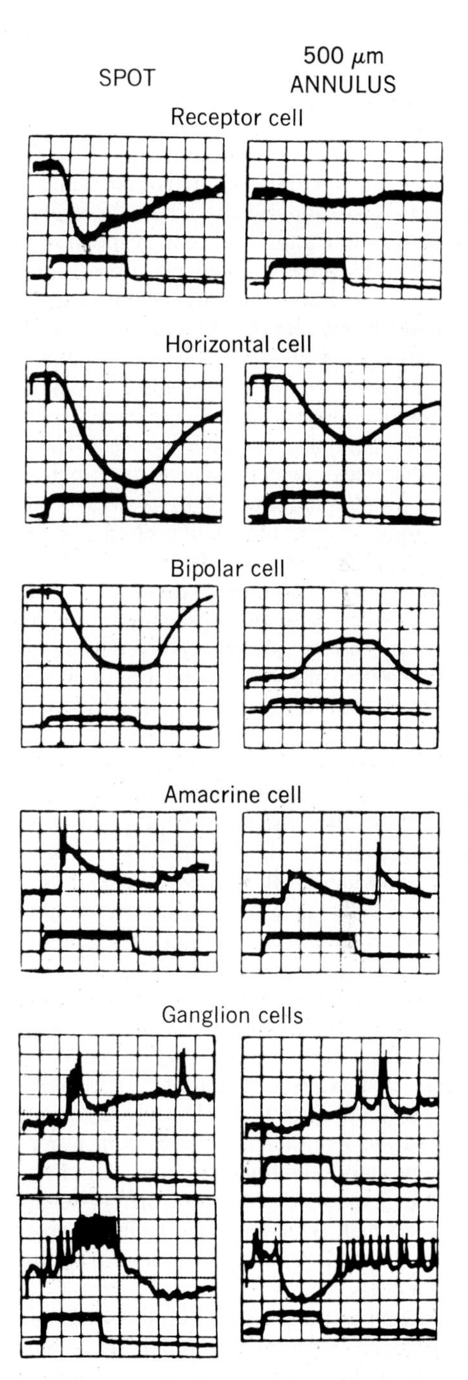

Fig. 17-27. Intracellular recordings from main types of cells in retina of *Necturus*. For each cell, responses were elicited both by centered spot 100 μm in diameter and by annulus with inner diameter of 500 μm. Single grid division represents 1 mV for the receptor, 5 mV for the amacrine cell, and 2 mV for all other cases, whereas single division represents 200 msec on time scale. (From Werblin and Dowling.[124])

tions at the level of the inner plexiform layer. Although their detailed functions are little known as yet, they appear to signal especially when the stimulus is turned on or off, and they may thus be especially important in the detection of moving stimuli.

In ganglion cells the discharge of nerve impulses is the signal transmitted to the brain; hence the pattern of impulse discharges is of great interest. Certain ganglion cells respond to light stimuli with *transient* impulse discharges, whereas others yield *sustained* impulse discharges that continue throughout the stimulus. Both cell types are shown in Fig. 17-27, where the upper responses are from a cell of the transient type, whereas the lower responses are from a sustained type of ganglion cell. In response to a central spot the transient cell gave a brief but strong "on" discharge followed by a weaker "off" discharge, whereas the annulus elicited almost no "on" discharge but a relatively strong "off" discharge. The sustained type of cell showed a similar center-surround organization of its receptive field. In this case the centered stimulus depolarized the cell, accompanied by a strong and sustained "on" discharge, whereas following the stimulus the cell became hyperpolarized and there was no "off" discharge. In response to the annulus there was a distinct hyperpolarization and no impulse discharge during the stimulus, but following the stimulus there was a sustained "off" discharge. It is thus evident that most ganglion cells have an antagonistic center-surround organization of their receptive fields. Stimulation of either portion of the receptive field will not only elicit a response characteristic of that part of the receptive field, but it will also change the membrane potential to oppose the response characteristic of the other part of the receptive field. This holds for all such ganglion cells in the light-adapted state, with a centered stimulus spot eliciting an "on" discharge in about half these cells and an "off" discharge in the other half. This has been well demonstrated in a number of vertebrate species, including cat and monkey.[78,84]

Since bipolar and ganglion cells exhibit similar organization of their receptive fields, it is likely that this antagonistic center-surround type of organization is determined at the level of the bipolar cells and passed on directly to the ganglion cells. For example, consider a bipolar cell in which the receptive field has a depolarizing center and a hyperpolarizing periphery. If these signals were passed on to a ganglion cell without inverting response polarity, the receptive field of that ganglion cell would have an "on" center and an "off" periphery.

Much work has demonstrated, and Fig. 17-27 illustrates, that in early stages of the pathway through the retina the cells generate only graded slow potentials. Hence the transmission of signals through these early cells and the mixing of signals when cells at one stage provide synaptic inputs to the next stage are accomplished entirely by slow potentials. This raises the question of what functional advantage may have endowed this arrangement with survival value during its evolution. It is theoretically efficient for the early stages of signal processing to be accomplished without impulse activity. For example, this avoids the necessity for changing an amplitude-modulated receptor potential to an impulse frequency code for purposes of transmission, then changing back to an amplitude-modulated code in the postsynaptic potential of the first synapse. Since transmission over any appreciable distance requires impulses, the elimination of impulses requires neurons to be quite short. The retina has achieved this arrangement by evolving an outpouching of the brain to meet the receptors. There is a definite advantage in the brain tissue going out to meet the receptors rather than vice versa; the retina can thus move independently of the brain for scanning the visual scene. In the retinal pathway, it is not until the ganglion cells that impulse transmission becomes required because of the length of ganglion cell axons. Hence it is not surprising that the ganglion cells are the first ones in this pathway to transmit signals based exclusively on nerve impulses. Since the retina is formed embryologically as an outpost of the brain, the observation of extensive signal processing in the retina on the basis of slow potentials alone suggests that this may also be an important principle in other parts of the brain that have not yet yielded to systematic intracellular work.

EFFECTS OF ADAPTATION AT GANGLION CELL LEVEL

When the retina is dark adapted, the type of discharge elicited by stimulating the peripheral part of the ganglion cell's receptive field changes to match that of the center.[17] Thus an on-center cell, which gives off-discharges from the periphery in the light-adapted state, gives on-discharges from the periphery after dark adaptation. The same principle applies to off-center cells. Hence the central and peripheral portions of the receptive field have summative effects in the dark-adapted state. This change of receptive field organization shows that dark adaptation changes the functional pathways from the receptors to the ganglion cell. This must occur because stimulation of the same receptors in the peripheral portion of the receptive field of a ganglion cell can elicit an off-discharge in one case but an on-discharge in another case.

The receptive field organization of ganglion cells in the light-adapted state has a theoretical advantage for the detection of a border in the retinal stimulus. Although a detailed explanation will not be developed here, it may be assumed that detection of a border requires different discharge rates for ganglion cells that are close together but on opposite sides of the border. It appears that the effect of mutual inhibition between the central and peripheral parts of the receptive field is to *enhance* this difference between discharge rates of ganglion cells on opposite sides of the border, thus improving the ability of the retina to detect borders. After dark adaptation the ganglion cell can integrate the effects of stimuli throughout its receptive field, thereby improving its ability to detect weak light stimuli.

In summary, the light-adapted receptive field seems to be organized for the detection of borders and fine details of the stimulus at the expense of sensitivity, whereas the dark-adapted receptive field seems to be organized for improved light detection at the expense of detailed vision. This indicates that although the anatomy of the retina is complex, it is highly organized in the service of visual requirements. There is even a flexibility of the functional pathways through the retina, which can change to meet the demands of different conditions.

CLASSIFICATION OF GANGLION CELLS

Based primarily on work in the cat retina, ganglion cells have now become classified by their functional and anatomic properties into three main categories: W, X, and Y cells.[54,117,118] The axonal conduction velocity increases from W to X to Y cells.[118] All X and Y cells have roughly circular receptive fields with antagonistic center-surround organization, as already described; the X cells give sustained responses that continue throughout the stimulus, whereas the Y cells give transient responses.[43] Apparently the X cells give continuous information about stimulus intensity during the stimulus, whereas the Y cells signal mainly when the stimulus goes on or off. Some W cells also give sustained or transient responses with antagonistic center-surround organization.[117] But some W cells have receptive field centers that respond when the stimulus is turned either on or off, whereas others are directionally selective (responding best to a spot moved across the receptive field in a particular direction); still others are suppressed by contrast, their maintained firing rates being decreased by placing either a bright or dark spot over the center of the receptive field.[117] In the cat, both the W and the X cells are most densely represented in the area centralis, which corresponds to the primate fovea.[58] In contrast, the retinal density of Y cells is maximal 1.5 to 3.0 degrees from the center of the area centralis and decreases both centrally and toward the peripheral retina.[58] For each type of ganglion cell the size of the receptive field increases with distance from the area centralis[117]; at any given retinal location, however, X cells have the smallest receptive fields, whereas W and Y cells have larger receptive fields of similar size.[117]

The three categories of ganglion cells also have distinctive central projections in the cat. The X and Y cells both project to the A-laminae of the cat's lateral geniculate nucleus (LGN), and separate subsets of LGN cells relay information from the X and Y cells to the visual cortex.[74] The superior colliculus (SC), which is intimately concerned with the initiation of eye movements, also receives a branch from the axon of the Y cell.[58] The W cells project to the C-laminae of the LGN, from which relay cells carry the information to the visual cortex.[128] The W cell likewise sends an axon branch to the SC.[73] To summarize, all three types of ganglion cells have parallel and separate pathways that pass to the LGN, from which their messages are relayed to the visual cortex. Only the W and Y cells have axon branches passing to the SC, and in cat the W cells provide about 90% of the retinal input to the SC.[73]

Although the functional significance of this ganglion cell classification is not yet clear, it seems likely that the X cells bear primary responsibility for high visual acuity. Because of their projections to the SC, the W cells in particular (and the Y cells as well) may be involved with the control of eye movements. Detailed understanding of the functional aspects of ganglion cell classification awaits further work on this subject.

Since ganglion cells send the final retinal message to the brain, the coding of their impulse messages is of special interest. Stimulus intensity appears to be coded in terms of impulse frequency, although it is not known whether impulse frequency provides all the information used by the brain in sensing stimulus intensity. The ganglion cells must also carry a code for stimulus color, and this code seems established. Since the best work on that subject has been conducted at more central levels of the visual system, the coding of color information will be presented in Chapter 18.

EFFECTS OF LIGHT DEPRIVATION

The effects of severely reduced visual stimulation are of interest because of the information provided about the normal role that stimulation plays in the maturation and functioning of the visual system. The eye is unique for this type of study because light stimulation may be prevented by raising animals in darkness or fitting opaque covers over the eyes. In other sense organs the total absence of stimulation is difficult or impossible to ensure without destruction of the sense organ, and the ensuing tissue degeneration complicates interpretation of the results.

Systematic experimental work in this field was first conducted by Riesen and co-workers[42,102-105] in chimpanzees. Animals were placed in total darkness at various times after birth and for varying periods. Visual abilities were determined by training animals to respond differentially to paired stimuli that tested color discrimination, visual acuity, and pattern vision. Visual reflexes were also tested, and certain animals were sacrificed at appropriate times for histologic study. Immediately after light deprivation, visual abilities were markedly affected. These effects included difficulty in fixating objects and in following moving stimuli. Visual acuity was also restricted. But the most marked effects were in the discrimination of patterns. This was shown most dramatically by obvious behavior; for example, although normal animals quickly recognize a feeding bottle by sight, as shown by vocalizations and other anticipatory responses, the light-deprived animals did not respond until actually touched by the feeding bottle. Effects were more severe when the duration of light deprivation was increased or when it was introduced earlier in infancy. Complete recovery occurred following total darkness from birth to 7 months of age. But when kept in total darkness from birth to 16 months or more, the effects were severe and irreversible. Aside from pupillary reactions, startle responses to turning on a light, and primitive pursuit movements of the eyes to follow a gross visual stimulus, such animals were blind. Histologic study of their retinas showed almost complete degeneration of the ganglion cell layer.[42]

If each day chimpanzees were given 90 min of light that was supplied to the immobilized animal through a diffusing plastic dome, retinal degeneration was prevented, but the animals were still highly deficient in pattern vision.[103,104] A control animal given 90 min of normal light experience each day was scarcely distinguishable from animals raised under completely normal conditions. Thus specific experience with forms seems necessary to develop normal pattern vision.

The effects of form deprivation are also seen in congenital cataract patients after the cataracts are removed and corrective lenses are supplied. Triangles and squares are initially distinguished only by *counting* the number of corners, and a long learning period is required before recognition of even such simple forms is immediate. If the cataracts are removed by the end of the first year, reasonably good vision can be attained, but if they are removed too late, the effects of form deprivation are serious and cannot be overcome. Thus the relatively complex ability to discriminate patterns, which is accomplished primarily at central levels of the visual system, requires some type of learning process that must be started early in life.

The role of prior experience in appreciating even more subtle visual qualities is well expressed by Thoreau's statement that "beauty is in the eye of the beholder." To those who study vision, his words are doubly meaningful. For the eye itself is an object of infinitely ordered beauty, which even dimly viewed is scarcely forgotten. Like all true beauty it grows with deeper understanding, for which the following references are recommended.

REFERENCES
General reviews

1. Boycott, B. B., and Dowling, J. E.: Organization of the primate retina: light microscopy, Philos. Trans. R. Soc. Lond. **255B:**109, 1969.
2. Brown, K. T.: The electroretinogram: its components and their origins, Vision Res. **8:**633, 1968.
3. Brown, K. T., Watanabe, K., and Murakami, M.: The early and late receptor potentials of monkey cones and rods, Symp. Quant. Biol. **30:**457, 1965.
4. Committee on Colorimetry, Optical Society of America: The science of color, New York, 1953, Thomas Y. Crowell Co.
5. Dowling, J. E., and Boycott, B. B.: Organization of the primate retina: electron microscopy, Proc. R. Soc. Lond. (Biol.) **166B:**80, 1966.
6. Granit, R.: Sensory mechanisms of the retina, New York, 1963, Hafner Publishing Co.
7. Hubbard, R., Bownds, D., and Yoshizawa, T.: The chemistry of visual photoreception, Symp. Quant. Biol. **30:**301, 1965.

8. Polyak, S.: The retina, Chicago, 1941, University of Chicago Press.
9. Polyak, S.: The vertebrate visual system, Chicago, 1957, University of Chicago Press.
10. Wald, G.: Molecular basis of visual excitation, Science **162:**230, 1968.
11. Wald, G., and Brown, P. K.: Human color vision and color blindness, Symp. Quant. Biol. **30:**345, 1965.
12. Wright, W. D.: Researches on normal and defective colour vision, London, 1946, Henry Kimpton.
13. Young, R. W.: Visual cells and the concept of renewal, Invest. Ophthalmol. **15:**700, 1976.

Original papers

14. Aguilar, M., and Stiles, W. S.: Saturation of the rod mechanism of the retina at high levels of stimulation, Opt. Acta **1:**59, 1954.
15. Arden, G. B., and Weale, R. A.: Nervous mechanisms and dark-adaptation, J. Physiol. **125:**417, 1954.
16. Baker, H. D.: Initial stages of dark and light adaptation, J. Opt. Soc. Am. **53:**98, 1963.
17. Barlow, H. B., Fitzhugh, R., and Kuffler, S. W.: Change of organization in the receptive fields of the cat's retina during dark adaptation, J. Physiol. **137:** 338, 1957.
18. Basinger, S., Hoffman, R., and Matthes, M.: Photoreceptor shedding is initiated by light in the frog retina, Science **194:**1074, 1976.
19. Baylor, D. A.: Lateral interaction between vertebrate photoreceptors, Fed. Proc. **33:**1074, 1974.
20. Baylor, D. A., and Fuortes, M. G. F.: Electrical responses of single cones in the retina of the turtle, J. Physiol. **207:**77, 1970.
21. Baylor, D. A., and Hodgkin, A. L.: Detection and resolution of visual stimuli by turtle photoreceptors, J. Physiol. **234:**163, 1973.
22. Baylor, D. A., Fuortes, M. G. F., and O'Bryan, P. M.: Receptive fields of cones in the retina of the turtle, J. Physiol. **214:**265, 1971.
23. Berson, E. L., and Goldstein, E. B.: The early receptor potential in sex-linked retinitis pigmentosa, Invest. Ophthalmol. **9:**58, 1970.
24. Bok, D., and Hall, M. O.: The etiology of retinal dystrophy in RCS rats, Invest. Ophthalmol. **8:**649, 1969.
25. Boycott, B. B., and Kolb, H.: The horizontal cells of the rhesus monkey retina, J. Comp. Neurol. **148:**115, 1973.
26. Boynton, R. M., and Whitten, D. N.: Visual adaptation in monkey cones: recordings of late receptor potentials, Science **170:**1423, 1970.
27. Brown, K. T.: An early potential evoked by light from the pigment epithelium-choroid complex of the eye of the toad, Nature **207:**1249, 1965.
28. Brown, K. T., and Flaming, D. G.: Opposing effects of calcium and barium in vertebrate rod photoreceptors, Proc. Natl. Acad. Sci. U.S.A. **75:**1587, 1978.
29. Brown, K. T., and Murakami, M.: A new receptor potential of the monkey retina with no detectable latency, Nature **201:**626, 1964.
30. Brown, K. T., and Murakami, M.: The biphasic form of the early receptor potential of the monkey retina, Nature **204:**739, 1964.
31. Brown, K. T., and Murakami, M.: Delayed decay of the late receptor potential of monkey cones as a function of stimulus intensity, Vision Res. **7:**179, 1967.
32. Brown, K. T., and Tasaki, K.: Localization of electrical activity in the cat retina by an electrode marking method, J. Physiol. **158:**281, 1961.
33. Brown, K. T., and Watanabe, K.: Isolation and identification of a receptor potential from the pure cone fovea of the monkey retina, Nature **193:**958, 1962.
34. Brown, K. T., and Watanabe, K.: Rod receptor potential from the retina of the night monkey, Nature **196:**547, 1962.
35. Brown, K. T., and Wiesel, T. N.: Intraretinal recording with micropipette electrodes in the intact cat eye, J. Physiol. **149:**537, 1959.
36. Brown, K. T., and Wiesel, T. N.: Analysis of the intraretinal electroretinogram in the intact cat eye, J. Physiol. **158:**229, 1961.
37. Brown, K. T., and Wiesel, T. N.: Localization of origins of electroretinogram components by intraretinal recording in the intact cat eye, J. Physiol. **158:**257, 1961.
38. Brown, P. K., and Wald, G.: Visual pigments in human and monkey retinas, Nature **200:**37, 1963.
39. Brown, P. K., and Wald, G.: Visual pigments in single rods and cones of the human retina, Science **144:**45, 1964.
40. Campbell, F. W., and Rushton, W. A. H.: Measurement of the scotopic pigment in the living human eye, J. Physiol. **130:**131, 1955.
41. Cervetto, L., and MacNichol, E. F., Jr.: Inactivation of horizontal cells in turtle retina by glutamate and aspartate, Science **178:**767, 1972.
42. Chow, K. L., Riesen, A. H., and Newell, F. W.: Degeneration of retinal ganglion cells in infant chimpanzees reared in darkness, J. Comp. Neurol. **107:**27, 1957.
43. Cleland, B. G., Dubin, M. W., and Levick, W. R.: Sustained and transient neurones in the cat's retina and lateral geniculate nucleus, J. Physiol. **217:**473, 1971.
44. Cohen, A. I.: Further studies on the question of the patency of saccules in outer segments of vertebrate photoreceptors, Vision Res. **10:**445, 1970.
45. Cone, R. A.: The early receptor potential of the vertebrate eye, Symp. Quant. Biol. **30:**483, 1965.
46. Cone, R. A.: The electrophysiology of rhodopsin, Invest. Ophthalmol. (suppl.) **16:**94, 1977.
47. Cone, R. A., and Brown, P .K.: Dependence of the early receptor potential on the orientation of rhodopsin, Science **156:**536, 1967.
48. Copenhagen, D. R., and Owen, W. G.: Coupling between rod photoreceptors in a vertebrate retina, Nature **260:**57, 1976.
49. Copenhagen, D. R., and Owen, W. G.: Functional characteristics of lateral interactions between rods in the retina of the snapping turtle, J. Physiol. **259:**251, 1976.
50. De Robertis, E.: Some observations on the ultrastructure and morphogenesis of photoreceptors, J. Gen. Physiol. **43**(suppl.):1, 1960.
51. Dowling, J. E.: Chemistry of visual adaptation in the rat, Nature **188:**114, 1960.
52. Dowling, J. E., and Ripps, H.: Visual adaptation in the retina of the skate, J. Gen. Physiol. **56:**491, 1970.
53. Dowling, J. E., and Ripps, H.: Effect of magnesium on horizontal cell activity in the skate retina, Nature **242:** 101, 1973.
54. Enroth-Cugell, C., and Robson, J. G.: The contrast sensitivity of retinal ganglion cells of the cat, J. Physiol. **187:**517, 1966.
55. Fain, G. L., and Dowling, J. E.: Intracellular record-

ings from single rods and cones in the mudpuppy retina, Science **180:**1178, 1973.
56. Fain, G. L., Gold, G. H., and Dowling, J. E.: Receptor coupling in the toad retina, Symp. Quant. Biol. **40:**547, 1976.
57. Fernandez-Moran, H.: The fine structure of vertebrate and invertebrate photoreceptors as revealed by low-temperature electron microscopy. In Smelser, G. K., editor: The structure of the eye, New York, 1961, Academic Press, Inc.
58. Fukuda, Y., and Stone, J.: Retinal distribution and central projections of Y-, X-, and W-cells of the cat's retina, J. Neurophysiol. **37:**749, 1974.
59. Goldstein, E. B.: Contribution of cones to the early receptor potential in the rhesus monkey, Nature **222:** 1273, 1969.
60. Goldstein, E. B., and Berson, E. L.: Cone dominance of the human early receptor potential, Nature **222:** 1272, 1969.
61. Gouras, P., and Link, K.: Rod and cone interaction in dark-adapted monkey ganglion cells, J. Physiol. **184:** 499, 1966.
62. Griffin, D. R., Hubbard, R., and Wald, G.: The sensitivity of the human eye to infra-red radiation, J. Opt. Soc. Am. **37:**546, 1947.
63. Hagins, W. A., and McGaughy, R. E.: Molecular and thermal origins of fast photoelectric effects in the squid retina, Science **157:**813, 1967.
64. Hagins, W. A., and Yoshikami, S.: A role for Ca^{2+} in excitation of retinal rods and cones, Exp. Eye Res. **18:**299, 1974.
65. Hall, M. O., Bok, D., and Bacharach, A. D. E.: Biosynthesis and assembly of the rod outer segment membrane system. Formation and fate of visual pigment in the frog retina, J. Mol. Biol. **45:**397, 1969.
66. Hanaoka, T., and Fujimoto, K.: Absorption spectrum of a single cone in carp retina, Jpn. J. Physiol. **7:** 276, 1957.
67. Hecht, S.: Rods, cones, and the chemical basis of vision, Physiol. Rev. **17:**239, 1937.
68. Hecht, S., and Verrijp, C. D.: Intermittent stimulation by light, J. Gen. Physiol. **17:**251, 1933-1934.
69. Hecht, S., et al.: The visual functions of the complete colorblind, J. Gen. Physiol. **31:**459, 1948.
70. Heller, J.: Comparative study of a membrane protein. Characterization of bovine, rat, and frog visual pigments$_{500}$, Biochemistry **8:**675, 1969.
71. Herron, W. L., Riegel, B. W., and Rubin, M. L.: Outer segment production and removal in the degenerating retina of the dystrophic rat, Invest. Ophthalmol. **10:**54, 1971.
72. Herron, W. L., Riegel, B. W., Myers, O. E., and Rubin, M. L.: Retinal dystrophy in the rat—a pigment epithelial disease, Invest. Ophthalmol. **8:**595, 1969.
73. Hoffman, K.-P.: Conduction velocity in pathways from retina to superior colliculus in the cat: a correlation with receptive-field properties, J. Neurophysiol. **36:** 409, 1973.
74. Hoffman, K.-P., Stone, J., and Sherman, M.: Relay of receptive-field properties in dorsal lateral geniculate nucleus of the cat, J. Neurophysiol. **35:**518, 1972.
75. Hogan, M. J., Wood, I., and Steinberg, R. H.: Phagocytosis by pigment epithelium of human retinal cones, Nature **252:**305, 1974.
76. Hubbard, R., and Kropf, A.: Molecular aspects of visual excitation, Ann. N.Y. Acad. Sci. **81:**388, 1959.
77. Hubbard, R., and St. George, R. C. C.: The rhodopsin system of the squid, J. Gen. Physiol. **41:**501, 1957.
78. Hubel, D. H., and Wiesel, T. N.: Receptive fields of optic nerve fibres in the spider monkey, J. Physiol. **154:**572, 1960.
79. Kaneko, A.: Physiological and morphological identification of horizontal, bipolar and amacrine cells in goldfish retina, J. Physiol. **207:**623, 1970.
80. Kaneko, A., and Shimazaki, H.: Synaptic transmission from photoreceptors to bipolar and horizontal cells in the carp retina, Symp. Quant. Biol. **40:**537, 1976.
81. Kolb, H.: Organization of the outer plexiform layer of the primate retina: electron microscopy of Golgi-impregnated cells, Philos. Trans. R. Soc. Lond. **285B:** 261, 1970.
82. Korenbrot, J. I., and Cone, R. A.: Dark ionic flux and the effects of light in isolated rod outer segments, J. Gen. Physiol. **60:**20, 1972.
83. Kroll, A. J., and Machemer, R.: Experimental retinal detachment and reattachment in the rhesus monkey, Am. J. Ophthalmol. **68:**58, 1969.
84. Kuffler, S. W.: Discharge patterns and functional organization of mammalian retina, J. Neurophysiol. **16:** 37, 1953.
85. Kuffler, S. W., and Potter, D. D.: Glia in the leech central nervous system: physiological properties and neuron-glia relationship, J. Neurophysiol. **27:**290, 1964.
86. Lamb, T. D., and Simon, E. J.: The relation between intercellular coupling and electrical noise in turtle photoreceptors, J. Physiol. **263:**257, 1976.
87. Laties, A. M., and Liebman, P. A.: Cones of living amphibian eye: selective staining, Science **168:**1475, 1970.
88. LaVail, M. M.: Rod outer segment disk shedding in rat retina: relationship to cyclic lighting, Science **194:** 1071, 1976.
89. Mandelbaum, J., and Sloan, L. L.: Peripheral visual acuity, Am. J. Ophthalmol. **30:**581, 1947.
90. Marks, W. B.: Visual pigments of single goldfish cones, J. Physiol. **178:**14, 1965.
91. Marks, W. B., Dobelle, W. H., and MacNichol, E. F., Jr.: Visual pigments of single primate cones, Science **143:**1181, 1964.
92. Miller, R. F., and Dowling, J. E.: Intracellular responses of the Müller (glial) cells of mudpuppy retina: their relation to b-wave of the electroretinogram, J. Neurophysiol. **33:**323, 1970.
93. Noell, W. K., and Albrecht, R.: Irreversible effects of visible light on the retina: role of vitamin A, Science **172:**76, 1971.
94. Normann, R. A., and Werblin, F. S.: Control of retinal sensitivity. I. Light and dark adaptation of vertebrate rods and cones, J. Gen. Physiol. **63:**37, 1974.
95. Oakley, B., and Green, D. G.: Correlation of light-induced changes in retinal extracellular potassium concentration with c-wave of the electroretinogram, J. Neurophysiol. **39:**1117, 1976.
96. Pak, W. L.: Some properties of the early electrical response in the vertebrate retina, Symp. Quant. Biol. **30:**493, 1965.
97. Pak, W. L.: Rapid photoresponses in the retina and their relevance to vision research, Photochem. Photobiol. **8:**495, 1968.
98. Pak, W. L., and Cone, R. A.: Isolation and identification of the initial peak of the early receptor potential, Nature **204:**836, 1964.
99. Pak, W. L., and Ebrey, T. G.: Visual receptor po-

tential observed at sub-zero temperatures, Nature **205:** 484, 1965.
100. Penn, R. D., and Hagins, W. A.: Signal transmission along retinal rods and the origin of the electroretinographic a-wave, Nature **223:**201, 1969.
101. Raviola, E., and Gilula, N. B.: Gap junctions between photoreceptor cells in the vertebrate retina, Proc. Natl. Acad. Sci. U.S.A. **70:**1677, 1973.
102. Riesen, A. H.: Arrested vision, Sci. Am. **183:**16, 1950.
103. Riesen, A. H.: Plasticity of behavior: psychological aspects. In Harlow, H. F., and Woolsey, C. N., editors: Biological and biochemical bases of behavior, Madison, Wisc., 1958, University of Wisconcin Press.
104. Riesen, A. H.: Effects of stimulus deprivation on the development and atrophy of the visual sensory system, Am. J. Orthopsychiatry **30:**23, 1960.
105. Riesen, A. H.: Effects of early deprivation of photic stimulation. In Osler, S. F., and Cooke, R. E., editors: The biological basis of mental retardation, Baltimore, 1965, The Johns Hopkins University Press.
106. Rushton, W. A. H.: Dark-adaptation and the regeneration of rhodopsin, J. Physiol. **156:**166, 1961.
107. Rushton, W. A. H.: Rhodopsin measurement and dark-adaptation in a subject deficient in cone vision, J. Physiol. **156:**193, 1961.
108. Rushton, W. A. H.: Cone pigment kinetics in the protanope, J. Physiol. **168:**374, 1963.
109. Schultze, M.: Zur anatomie und physiologie der retina, Arch. Mikr. Anat. **2:**175, 1866.
110. Simon, E. J.: Feedback loop between cones and horizontal cells in the turtle retina, Fed. Proc. **33:**1078, 1974.
111. Simon, E. J., Lamb, T. D., and Hodgkin, A. L.: Spontaneous voltage fluctuations in retinal cones and bipolar cells, Nature **256:**661, 1975.
112. Steinberg, R. H., and Wood, I.: Pigment epithelial cell ensheathment of cone outer segments in the retina of the domestic cat, Proc. R. Soc. Lond. **187B:**461, 1974.
113. Steinberg, R. H., Schmidt, R., and Brown, K. T.: Intracellular responses to light from cat pigment epithelium: origin of the electroretinogram c-wave, Nature **227:**728, 1970.
114. Steinberg, R. H., Wood, I., and Hogan, M. J.: Pigment epithelial ensheathment and phagocytosis of extrafoveal cones in human retina, Philos. Trans. R. Soc. Lond. **277B:**459, 1977.
115. Stell, W. K., and Lightfoot, D. O.: Color-specific interconnections of cones and horizontal cells in the retina of the goldfish, J. Comp. Neurol. **159:**473, 1975.
116. Stevens, S. S.: On the psychophysical law, Psychol. Rev. **64:**153, 1957.
117. Stone, J., and Fukuda, Y.: Properties of cat retinal ganglion cells: a comparison of W-cells with X- and Y-cells, J. Neurophysiol. **37:**722, 1974.
118. Stone, J., and Hoffman, K.-P.: Very slow-conducting ganglion cells in the cat's retina: a major, new functional type? Brain Res. **43:**610, 1972.
119. Vilter, V.: Histologie et activité electrique de la rétine d'un Mammifère strictement diurene, le Spermophile (Citellus citellus), Compt. Rend. Soc. Biol. **148:**1768, 1954.
120. Wald, G.: Alleged effects of the near ultraviolet on human vision, J. Opt. Soc. Am. **42:**171, 1952.
121. Wald, G.: The photoreceptor process in vision. In Magoun, H. W., editor, Neurophysiology section: Handbook of physiology, Baltimore. 1959, The Williams & Wilkins Co., vol. 1.
122. Wald, G., and Brown, P. K.: Human rhodopsin, Science **127:**222, 1958.
123. Wald, G., Jeghers, H., and Arminio, J.: An experiment in human dietary night-blindness, Am. J. Physiol. **123:**732, 1938.
124. Werblin, F. S., and Dowling, J. E.: Organization of the retina of the mudpuppy, Necturus Maculosus. II. Intracellular recording, J. Neurophysiol. **32:**339, 1969.
125. Whitten, D. N., and Brown, K. T.: The time courses of late receptor potentials from monkey cones and rods, Vision Res. **13:**107, 1973.
126. Whitten, D. N., and Brown, K. T.: Photopic suppression of monkey's rod receptor potential, apparently by a cone-initiated lateral inhibition, Vision Res. **13:** 1629, 1973.
127. Whitten, D. N., and Brown, K. T.: Slowed decay of the monkey's cone receptor potential by intense stimuli, and protection from this effect by light adaptation, Vision Res. **13:**1659, 1973.
128. Wilson, P. D., and Stone, J.: Evidence of W-cell input to the cat's visual cortex via the C laminae of the lateral geniculate nucleus, Brain Res. **92:**472, 1975.
129. Yonemura, D., Kawasaki, K., and Hasui, I.: The early receptor potential in the human ERG, Acta Soc. Ophthalmol. Jpn. **70:**120, 1966.
130. Yoshikami, S., and Hagins, W. A.: Light, calcium, and the photocurrent of rods and cones, Biophys. J. **11:**47a, 1971.
131. Young, R. W.: The renewal of rod and cone outer segments in the rhesus monkey, J. Cell Biol. **49:** 303, 1971.
132. Zuckermann, R.: Ionic analysis of photoreceptor membrane currents, J. Physiol. **235:**333, 1973.

18 Central neural mechanisms in vision

GIAN F. POGGIO

One form of experience of the physical world of objects is furnished by the capacity man and other creatures possess to receive and interpret information carried by a particular type of radiant energy called light. This capacity for *vision* is provided by a specialized biologic system that accomplishes physicophysiologic transformations and neural decision processes leading to an integration of the information received into perceptual experience.

In previous chapters the optical properties of the receiving apparatus, the eye, and the mechanisms for the formation of images on the retinal surface have been discussed. Reference has also been made to the motor component of the system by which the position of the eye can be controlled either voluntarily or automatically and directed in space. The transducer function of the retinal receptors, the rods and cones, has been described as well as the anatomic and functional relation of these receptors to the retinal ganglion cells, the neural elements in which the neurally coded visual message originates.

This chapter deals with the portion of the visual system that extends from the retinal ganglion cells to the primary receiving area of the cerebral cortex, and beyond it, with those central systems that appear to be implicated in the elaboration of the visual experience. The evidence presented is obtained, whenever possible, from experimental investigations in primates and from studies in man. Findings from other animals, the cat in particular, are described when they are the only available ones and it is thought they may contribute to a better understanding of the central mechanisms in vision.

ORGANIZATION OF THE VISUAL AFFERENT SYSTEM

The axons of the retinal ganglion cells, the third-order neurons of the visual pathway, form the innermost layer of the retina (Chapter 17) and converge at the optic papilla into a bundle of fibers named the optic nerve, the chiasma, and the optic tract in different portions along its centripetal course. In primates the great majority of these fibers terminate on cells of the lateral geniculate nucleus.

A second group of optic fibers passes to the midbrain and terminates in the superior colliculus and pretectal region. The axons of geniculate neurons form the visual radiation and project fully on the striate cortex of the occipital lobe. Beyond the area striata, visual information is distributed to other areas of the cortex both adjacent and distant, as well as subcortically to the pulvinar of the dorsal thalamus, the tectum, and the lateral geniculate nucleus. The second, extrageniculate visual pathways progress from the retina through the superior colliculus to the posterior thalamus, the pulvinar in particular; fibers from the superior colliculus descend to lower midbrain structures associated with the control of eye movement and visual orientation mechanisms.

Like other major afferent systems, the visual system is topographically organized: within most of its components, the retinal surface is spatially represented in an orderly manner, and patterns of retinal activity correspond topologically to patterns of activity in central nervous structures.

Topographically the retina may be subdivided with respect to its specialized functional center, the fovea, into four quadrants. A vertical straight line, the vertical meridian, passing through the center of the fovea separates the retinal surface into a temporal and a nasal half. Because of the eccentric position of the fovea, these two halves are unequal in extent, the nasal half being larger than the temporal half (Fig. 16-1). A horizontal straight line, the horizontal meridian, divides each half into dorsal and ventral quadrants. Any number of retinal sectors may be defined by straight lines radiating from the center of the fovea.

Along the centroperipheral radial direction, two concentric principal retinal regions or zones are commonly recognized: a *central* and a *peripheral* region. Further zonal subdivisions may be

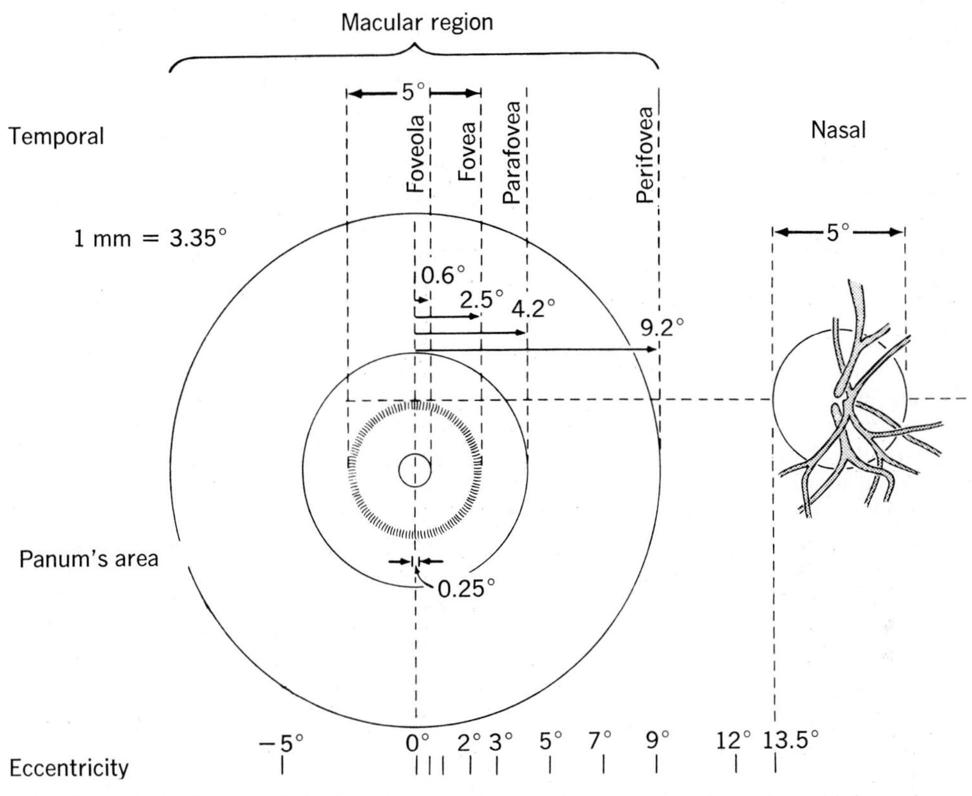

Fig. 18-1. Scale diagram of fundus of human right eye including macular region with its various concentric subdivisions and papilla of optic nerve. Macular region is the more central area of retina, 5.5 mm in diameter (1 mm = 3.35 degrees); its outer boundary corresponds to place where retinal ganglion cells are reduced to single layer. It must be distinguished from macula lutea, a smaller, poorly defined area surrounding fovea and characterized by yellow pigment in ganglion cells and bipolar cells. Also indicated is extent (0.25 degrees) of Panum's area, the region of space proximal and distal to fixation point over which objects are seen binocularly as single. (Modified from Bishop.[1])

made by any number of circles concentric to the fovea. Anatomically the division between central and peripheral retina rests on the different accumulation of ganglion cells in the two regions. The central retina contains small cells that are densely packed in six to seven rows at the edge of the foveal depression and reduce to one row at the outer margin of the area. In the peripheral retina the ganglion cells form a continuous, occasionally broken, single layer.[6] The central area of the retina measures 5 to 6 mm in diameter, which corresponds to about 15 to 20 degrees of visual angle. In clinical literature the term "central area" generally designates the more restricted *macular* region occupied by the fovea and its vicinity, which is characterized by yellow pigmentation (macula lutea) and measures about 3 mm in diameter (Fig. 18-1).

A similar topographic subdivision can be made of the visual field, that is, of that portion of space which the retina subtends when the eye is held stationary (Fig. 18-7). Because of the transparent lens system, there exists, of course, an inverse spatial relation between the position of any point in the field and its image on the retinal surface.

RETINOGENICULOSTRIATE PATHWAY

In their intraretinal portion the axons of ganglion cells of the peripheral retina run toward the papilla of the optic nerve in an approximately radial course, whereas axons from the central retina, a large number of which are of small caliber, follow a straight or slightly arched horizontal path. At the papilla the macular fibers are located laterally and are flanked above and below by the fibers from the dorsal and ventral temporal retina, respectively, whereas the fibers from the nasal retina are located medially. On leaving the eyeball the fibers acquire their myelin sheaths and run backward in the optic nerve.

At the level of the chiasm the fibers from the nasal half of the retina cross the midline and enter the contralateral optic tract; the fibers from the temporal half pass into the ipsilateral tract. This is true for macular fibers as well as for fibers from the peripheral retina, and the vertical meridian separates the portion of retina with a crossed projection from that with an uncrossed projection. Accordingly, the suprachiasmatic portion of the afferent visual system on each side of the brain carries a representation from each of the two retinas and subserves the contralateral hemifield of vision. Topographically the fibers from the temporal retina maintain their relative position on the outside of the chiasm, whereas the crossing nasal fibers occupy the central portion. Crossing macular fibers are situated posteriorly and dorsally and spread over a large extent of the central chiasm.[7,63]

On leaving the chiasm, crossed and uncrossed fibers from the two homonymous dorsal retinal quadrants move to occupy the medial dorsal portion of the optic tract and those from the ventral quadrants its lateral ventral aspect. The macular fibers come to lie dorsocentrally in the tract and mix diffusely in medial and lateral areas containing the peripheral projections. Within the limits of this gross topographic arrangement the fibers from the two eyes are fairly evenly distributed and intermingled in an apparently random fashion.[25,63]

Thus there exists in all portions of the retinogeniculate projection a topographic organization of the optic fibers with respect to their retinal origin, but this organization is not a precise one. On the other hand, as the optic fibers reach the lateral geniculate nucleus, they are sorted in accordance with their origin and terminate within this structure in a highly detailed topographic order.

Lateral geniculate nucleus

The cells of the lateral geniculate nucleus (LGN), a thalamic nucleus, are arranged in layers, or laminae, separated by interlaminar bundles of fibers. In primates, six major cellular laminae may be recognized, numbered 1 to 6 beginning at the hilus and continuing toward the dorsal aspect of the nucleus. Cells in ventral laminae 1 and 2 *(magnocellular laminae)* are larger and more uniform in size and shape than the cells in the dorsal laminae 3 to 6 *(parvocellular laminae)*. In each lamina, two types of neurons are found: a large neuron, thought to be a geniculostriate projection cell (P cell), and a small neuron forming about 10% of the total neuron population, considered to be an interneuron (I cell).[81] Synaptic junctions are almost as densely distributed between cell laminae as within them. The large retinogeniculate axon terminals are confined to the cellular layers, whereas two other types of axons, both intra- and extrageniculate, form synaptic contacts in all parts of the nucleus.[55,81]

The laminar differentiation of the LGN is accompanied by a characteristic sorting of the incoming fibers from the two eyes: crossed and uncrossed projections terminate in a segregated fashion in different cell layers. The fibers from the contralateral nasal hemiretina reach laminae 1, 4, and 6, whereas the fibers from the ipsilateral temporal retina distribute to laminae 2, 3, and 5 (Fig. 18-2). The topographic representation of the retina within the LGN is precisely organized. The replication of the horizontal meridian coincides with the median dorsoventral axis of the nucleus along its entire anterposterior extent. Anatomically corresponding areas of the two hemiretinas project—each to the respective laminae—onto successive radial sectors on either side of the median axis, the dorsal retina medially and the ventral retina laterally. Retinal zones of increasing eccentricity project serially along the posteroanterior axis of the LGN with the fovea posteriorly and the more peripheral zones anteriorly. The central visual field, to about 15 degrees of eccentricity, is represented by all six cell laminae, whereas the rest of the binocular field is reflected in four layers only (two parvocellular and two magnocellular). The monocular crescent is represented by one parvocellular and one magnocellular layer.[84] The termination of the projection fibers is sharply circumscribed, and minute lesions in the retina are followed by transneuronal degeneration of discrete clusters of cells in the appropriate laminae of the LGN. The projection locales from the two eyes in the respective laminae lie in exact register, so that any small area in the contralateral field of view can be shown as a straight dorsoventral radial column through the nucleus (Fig. 18-2).

The axons of the principal geniculate cells form the visual radiation and terminate wholly in the striate cortex of the ipsilateral occipital lobe, the primary visual cortex. On leaving the confines of the LGN the visual radiation fibers pass through the posterior part of the internal capsule between the somatic and auditory radiations, spread out, and sweep backward to form an orderly stratified lamina along the lateral wall and the posterior horn of the lateral ventricle. The fibers arising from the median posterior portion of the geniculate nucleus (representing central retina) follow a fairly straight anteroposterior course and are flanked dorsally by the fibers that originate from the medial part of the nucleus (homolateral dorsal retinal quadrants).

The fibers that arise from the lateral part of the LGN (homolateral ventral retinal quadrants) loop downward and forward into the temporal lobe to sweep around the lateral horn of the ventricle (temporal loop of Meyer) before turning backward to join the other projection fibers and form the ventral portion of the visual radiation. If a small lesion is made in the area striata of a monkey, the ensuing retrograde cellular degeneration in the LGN is localized to a sharply defined region. Circumscribed lesions in the optic radiation, on the other hand, lead to much less clear-cut foci of degeneration within the geniculate nucleus. Thus the orderly topographic representation of the retina at the LGN becomes somewhat smeared in the optic radiation, and it is reconstituted again in an exquisitely precise manner at the primary visual cortex.[25] The retrograde degeneration affects all six geniculate laminae along a given narrow sector of the nucleus. This indicates that cell groups in adjoining laminae that receive afferents from corresponding points in right and left retinas project close together to the striate area, thus providing the basis for a complete retinotopic map in the visual cortex as well as the initial condition for the convergence of visual information from the two eyes on single neurons.

Striate cortex

In man and other primates the striate cortex, area 17 of Brodmann, possesses distinctive mor-

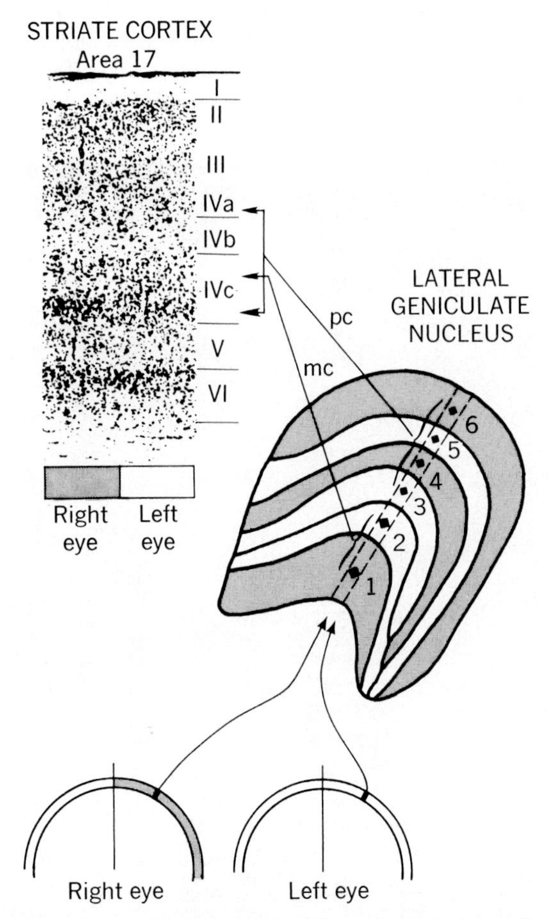

Fig. 18-2. Retinogeniculostriate pathway, left side. Clusters of retinal ganglion cells in homologous parts of two hemiretinas project to radial column of cells in lateral geniculate nucleus. Principal horizontal meridian is represented along median sector of lateral geniculate nucleus: dorsal retinas project medially, ventral retinas laterally. Fibers from contralateral eye end in geniculate laminae 1, 4, and 6 and those from ipsilateral eye in laminae 2, 3, and 5. Striate projection from parvocellular lateral geniculate nucleus (laminae 3 to 6) *(pc)* ends chiefly in lower layer IVc and also in IVa. Magnocellular projection (laminae 1 and 2) *(mc)* reach upper part of layer IVc. Ipsilateral and contralateral eye inputs end segregated in layer IV in alternating patches 0.4 mm wide. (See Fig. 18-3.)

phologic characteristics that separate it sharply from the surrounding area 18. Approximately at the middle of the thickness of the cortex there is a zone composed mainly of a band of medullated fibers. This zone, usually visible macroscopically, is named the "stria of Gennari" (layer IVb), and it runs between narrow layer IVa, often difficult to separate from the more superficial layers, and layer IVc, the inner granular layer. Geniculate projection fibers terminate within layer IV; the bulk of the afferents from parvocellular geniculate laminae (3 to 6) end in the lower, denser part of layer IVc, and a smaller contingent reach layer IVa. Fibers from the magnocellular geniculate laminae (1 and 2) terminate in the upper part of layer IVc (Fig. 18-2). The large majority of geniculate terminals make synaptic contacts with dendritic spines of pyramidal cells and likely also of stellate cells. Other terminals end directly on dendritic shafts and a very few on stellate cell bodies.[51]

At the sites of termination of the geniculate projections, inputs from right and left eyes are still segregated. Fibers from geniculate laminae 1, 4, and 6 (contralateral eye) and from laminae 2, 3, and 5 (ipsilateral eye) terminate in layer IV in alternating patches about 0.4 mm wide (Fig. 18-3). Functionally, these monocular patches extend vertically to superficial and deep cortical layers to form alternating right-eye, left-eye slabs, or columns, of ocular dominance. Cells within a column are activated exclusively (layer IVc) or preferentially by stimulation of the same eye.[69]

The geniculate projection is restricted to area 17, of which it constitutes nearly the total input. The only other known afferents originate in the pulvinar and terminate chiefly in layer I.[17,92] Few and sparse callosal connections are restricted to a very narrow zone of striate cortex bordering on area 18.[50] Efferent projections from striate cortex distribute to other cortical areas and to

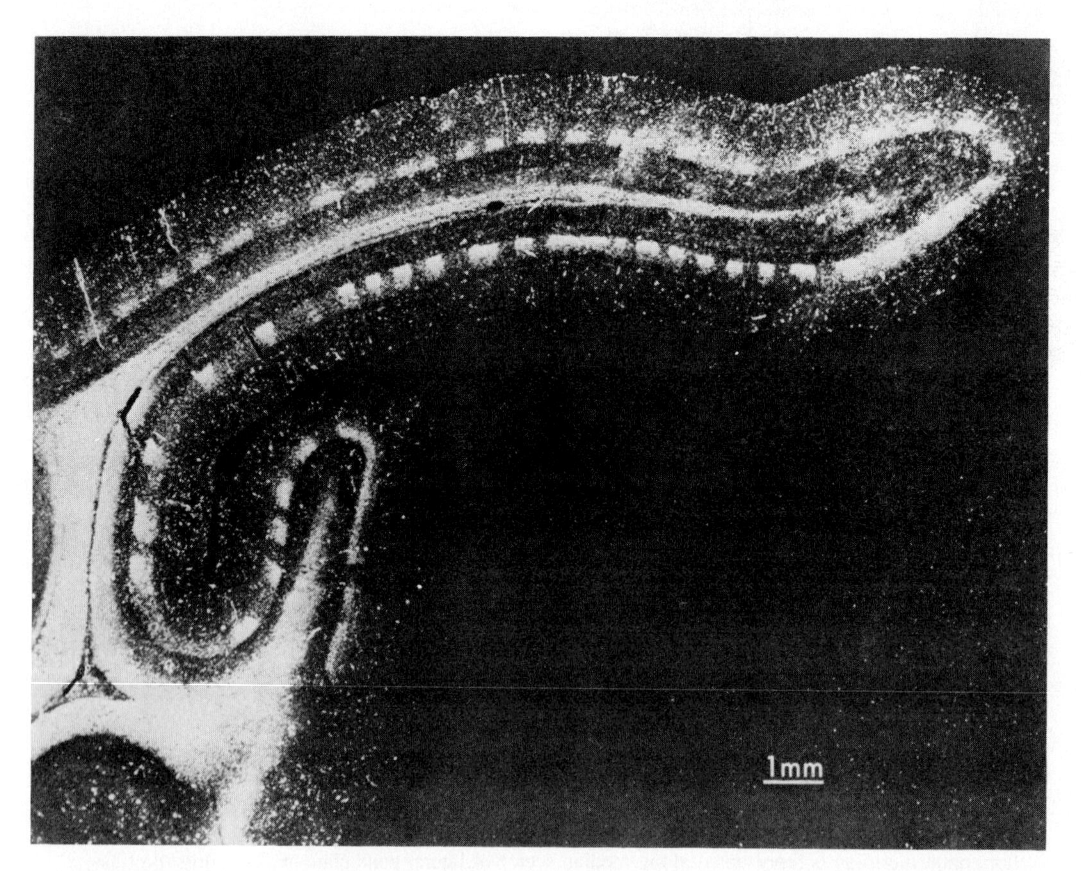

Fig. 18-3. Distribution of geniculate terminals in layer IV of striate cortex. Dark-field autoradiograph of area 17 of adult macaque following monocular injection of radioactive fucsine-proline mixture 2 weeks before. Photograph shows part of exposed surface of striate cortex and buried part immediately beneath; some of buried part has fallen away during sectioning. Labeled terminal stands out clearly in layer IVc as white patches (about 0.4 mm wide) separated by gaps of similar width and corresponding to eye that had not been injected. (From Hubel and Wiesel.[69])

subcortical structures as well. No efferent projections originate from layers receiving geniculate fibers (layers IVa and IVc) or from layer I—the input layers. Axons of pyramidal cells in superficial layers, II and III in particular, reach the ipsilateral prestriate cortex, whereas those in deeper layers terminate subcortically; layer V projects to the inferior pulvinar and to the superior colliculus and pretectum, layer VI to the lateral geniculate nucleus.[82]

The retinotopic geography of the striate cortex of man has been determined largely by clinical studies of war wounds and of surgical ablations of the occipital lobe.* These observations have outlined the spatial representation of the retina on the cortical surface but have provided only limited information on its detailed organization. More accurate maps of the visual cortex have been obtained experimentally with electrophysiologic techniques in animals.[23,37,121]

The topographic organization of the primary visual cortex appears to be similar in nonhuman primates and in man. In man the area striata is almost totally confined to the superior and inferior lips of the calcarine fissure on the medial aspect of the occipital lobe, and its posterior boundary is located at about the occipital pole (Fig. 18-4). In the monkey, on the other hand, a large portion of the striate cortex extends over the lateral surface of the hemisphere (Fig. 18-5). The cortical replication of the horizontal retinal meridian runs longitudinally along the depth of the calcarine fissure and around the occipital pole onto the lateral surface of the brain. It divides the primary visual cortex into upper and lower portions, where dorsal and ventral retinal quadrants are represented, respectively. The replication of the vertical retinal meridian coin-

*See references 7, 8, 56, 62, 117, and 118.

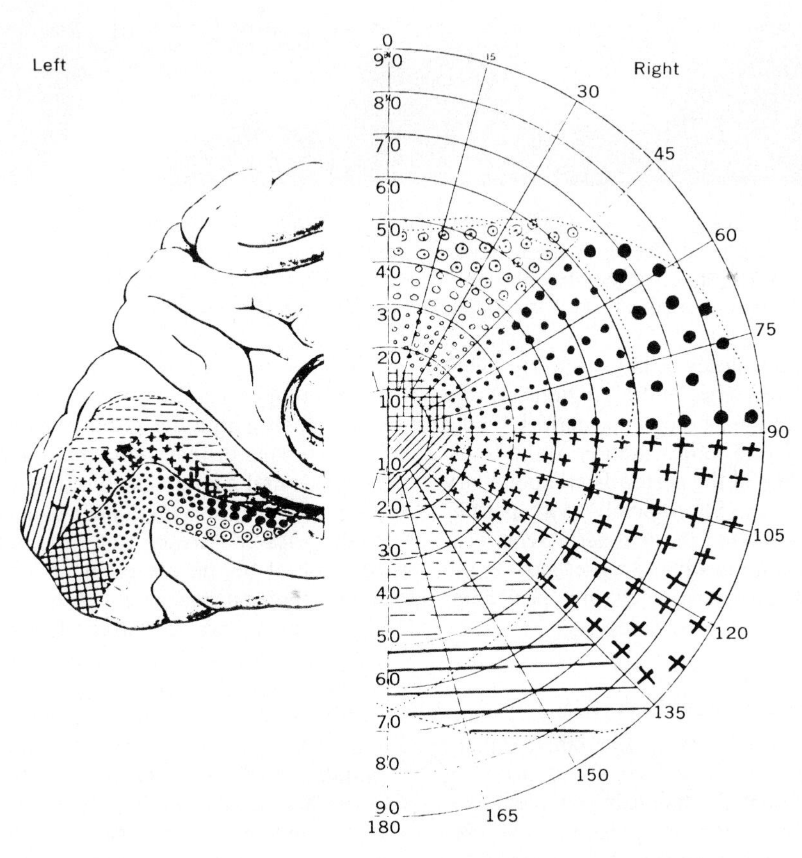

Fig. 18-4. Cortical representation of contralateral half of field of vision according to Holmes.[62] Striate area of left hemisphere of brain is shown with calcarine fissure widely opened. Areas of right half of visual field are marked corresponding to their areas of cortical representation. Macular region of retina is represented posteriorly in brain and is relatively large; peripheral retina is represented anteriorly and is relatively small. Horizontal meridian is projected along depth of calcarine fissure and vertical meridian along upper and lower margins of striate cortex. (From Duke-Elder.[47])

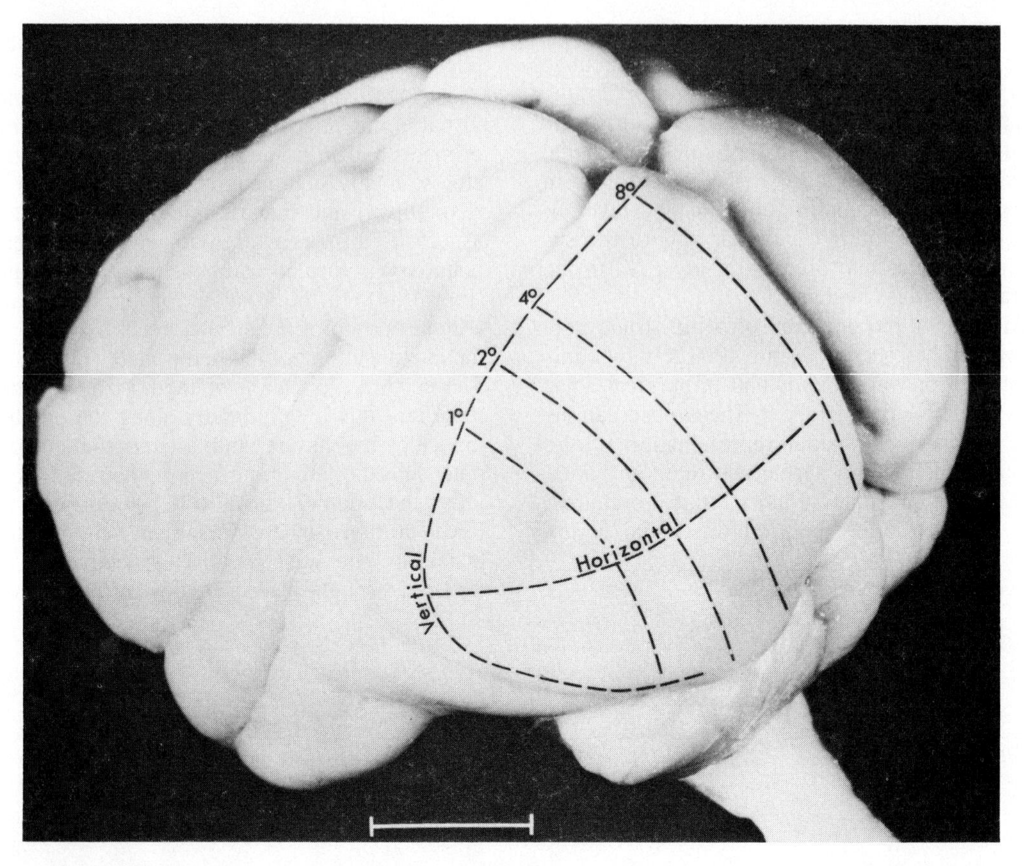

Fig. 18-5. Schematic outlines of representation of central 8 degrees of contralateral half of visual field on lateral striate cortex of rhesus monkey. Left hemisphere, posterolateral view. Numbers indicate degrees of eccentricity from central fovea, which is represented most laterally in cortex, about junction of horizontal and vertical principal meridians. Calibration = 1 cm.

cides with the boundaries of the striate area with prestriate area 18. The central regions of the homolateral hemiretinas project posteriorly in the hemisphere, with the representation of the fovea at about the middle of the posterolateral boundary of the primary area. Successively more peripheral zones are arranged in an orderly sequence, the most peripheral ones being represented at the anterior end of the calcarine fissure (peripheral nasal retina of the opposite side).

FUNCTIONAL ASPECTS OF TOPOGRAPHIC ORGANIZATION

The primary visual system, like other primary sensory systems, is organized to replicate in neural space the spatial relationships of the physical world of objects that we perceive. The precise topographic arrangement of the retinogeniculocortical pathway represents the more general aspect of the organization subserving the maintenance of spatial relations. Thus contiguous areas of the retinal surface are represented contiguously in the LGN and in the cerebral cortex: the topologic relations of the external world are maintained within the visual system.

As in other primary afferent systems, the number of central neural elements assigned to various parts of the sensory periphery is not proportional to the physical size of that part, but it is an expression of the degree of innervation density of the receptive surface (i.e., of the number of sensory elements per unit area). The sensory visual receptors, the rods and cones, are connected through a complicated intraretinal network to the ganglion cells (Chapter 17). Each ganglion cell thus defines a retinal receptive unit, that is, the population of receptors distributed over an area of retinal surface that is connected to the brain by a single fiber. In considering the functional relation between retina and central structure, the ganglion cell may be taken as the representative member of the peripheral surface.

When one compares the topographic anatomy of the ganglion cell layer in the retina with that

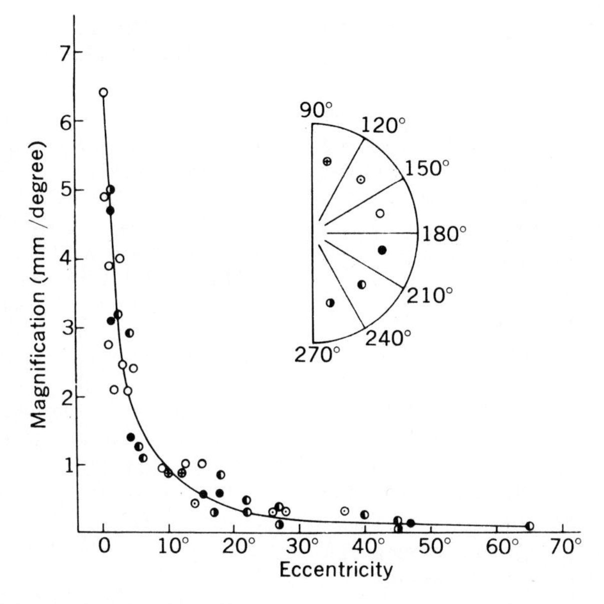

Fig. 18-6. Plot of "magnification factor" vs degrees of retinal eccentricity in macaque (see text). Data for various radii explored have been grouped in six sectors; decrease in magnification from center to periphery appears to be equal in all directions. (From Daniel and Whitteridge.[37])

of the striate cortex, an important difference is apparent. Although the retina ganglion cells are distributed along a centroperipheral density gradient,[7,123] the cell density of various parts of the striate cortex appears remarkably uniform.[28] As an approximate generalization, the area relation between retina and striate cortex may be visualized by imagining the population of ganglion cells spread out in a uniform layer and this layer superimposed on the receiving cortical surface. The representation of the central retina, where the density of ganglion cells is higher, occupies an area of cortex larger than the area covered by the peripheral retina. A numeric estimate of this relation may be obtained by measuring the extent of cortical surface from which a neural response can be evoked by photic stimulation of a discrete area of the retina. For obvious optical reasons the amount of retinal surface devoted to 1 degree of visual field is constant. In the cortex the corresponding amount of surface varies, as shown in Fig. 18-6, where the "magnification factor" (millimeters of cortex per degree of visual field) is plotted with respect to retinal eccentricity.[37,69]

The following facts suggest that the magnification factor represents an important anatomic basis for the visual capacity for two-dimensional resolution.[30,37] In the macaque the diminution in magnification factor from the foveal projection area to the peripheral projection area resembles the reduction in visual acuity in man from the fovea to the periphery (Fig. 17-4). Macaque and squirrel monkeys possess very similar visual acuity and similar magnification factors (6 mm) for the central 1 degree of visual field.[29] Finally, in the squirrel monkey the cortical representation of the extrafoveal retina is much more compressed than in the macaque, and indeed, removal of the foveal projection area in the striate cortex leads to greater impairment of visual acuity in the squirrel monkey than in the macaque.[30]

EFFECTS OF STRIATE CORTEX LESIONS IN PRIMATES

In man and other primates, injury to the occipital cortex produces a permanent loss of vision in the relevant portion of the visual field. In the monkey, however, the visual deficit caused by lesions of the striate area appears to be less severe than that observed in man. Klüver[76] has shown that the visual behavior of monkeys after bilateral removal of the primary visual cortex depends chiefly on the ability they retain to use differences in the total amount of light entering the eye as a factor for discrimination. No form or color vision is left. Recent experimental evidence indicates that after ablation of the striate cortex the monkey retains a certain capacity for visual discrimination controlled by the spatial location of the stimulus, for it can be trained to detect, orient itself toward, and reach out for localized "events" of certain kinds, especially moving events.[125]

After partial damage of the striate cortex the loss of function is limited to a portion of the visual field, the size, shape, and position of the defect being dictated by the topographic arrangement of the primary projection to the cortex. Because of their specificity of place and modality, visual field defects are of great clinical significance for the topical diagnosis of lesions involving any portion of the visual pathway; they will be summarized in the next section. In the monkey, partial lesions of the area striata are followed by partial defects in the visual field, with the expected topographic characteristics. In the defective portion of the field, however, the animal still possesses a certain degree of visual capacity. Although the recognition of objects that lie within the area of the defect is lost, the animal can respond to flashes of light within it: the deficit is a greatly reduced sensitivity rather than total blindness.[32] In addition to the visual field defect, partial striate lesions that include the foveal area impair visual acuity and, transiently, visually guided reaching. Such lesions have little or no effect on postoperative learning or retention of visual discrimination habits, whereas this capacity is severely impaired by lesions of cortical areas beyond the striate cortex (p. 556).[33,129] The lack of discrimination deficit after foveal striate lesions may be attributed to the ability of man and monkey to use extrafoveal parts of the retina by fixating eccentrically.[8,32] The orderly topographic representation of the retinal surface in the striate cortex and the absence of any extensive spatial convergence within it make it possible for visual information received at the extrafoveal region of the retina to be processed—largely unaffected by the lesion—by the extrafoveal cortex and transmitted forward to other cortical areas for further elaboration.

There appear to be no basic differences in the organization of the primary visual system in man and monkey. In both, removal of the striate cortex leads to complete degeneration of the dorsal LGN. The residual visual capacity of the destriated monkey is not surprising, for we know that neural activity evoked by light is transmitted from the retina not only to the geniculostriate system but also to regions of the midbrain, which may provide the basis for a response to light. The functional significance of the retinomesencephalic projection diminishes in phylogeny, and its role in vision seems to be inversely related to the extent to which the geniculostriate system has developed.

THE VISUAL FIELD AND ITS ABNORMALITIES

The visual field is defined as the bounded portion of space that is visible at any one time while the eye is fixating a stationary point in it. The image of any point in the field falls spatially inverted on the retina. Accordingly, the visual field may be topographically subdivided, like the retinal surface, into four quadrants by vertical and horizontal straight lines crossing at the center of the fixation point as well as into concentric central and peripheral regions.

Accurate mapping of the visual field in humans may be done by means of a simple instrument, the perimeter, which allows the presentation of a visual stimulus at any point in the subject's field of view. In this way, functional defects may be outlined and charted. Perimetric mapping has also furnished important information on several other aspects of the functional organization of the visual system such as acuity, sensitivity, extent of the color field, and response to movement.

Because of the normally recessed position of the eye in the orbit, the portion of space viewed by the retina with the eye fixed on a point straight ahead is limited in its extreme extent by the nose, eyebrows, and cheek bones. The *monocular visual fields* (Fig. 18-7) are slightly irregular ovals that extend from the fixation point, approximately 60 degrees nasally and above, to about 70 degrees below and 90 degrees temporally. The papilla of the optic nerve produces the physiologic *blind spot* in the temporal half of the field of vision, a vertical oval area, located about 16 degrees from the point of fixation and 2.5 degrees below it (Fig. 18-1). The *binocular visual field* is the combination of the partially overlapping right and left monocular fields. Its central part represents the portion of space visible simultaneously with the two eyes directed toward a common fixation point; it is roughly circular in shape and has a diameter of about 120 degrees. On either side of this paired portion there extends the nonoverlapping part of each monocular field, the so-called temporal crescent, which corresponds to the most peripheral nasal retina.

It was previously mentioned that the study of the visual field defects has helped us to understand the anatomic organization of the visual afferent system in man, in particular its topographic arrangement. Clearly the characteristics of the visual field defect can provide most valuable information for the clinical diagnosis of the site, extent, and in some cases also the nature of a pathologic process involving the visual projection system.

Lesions at any level of the visual pathway may produce functional defects in the area of the visual field represented in the damaged part of the system. The intensity of the defect may range from a slight reduction of visual acuity to total loss of function within the affected area. The size, shape, and position of the defect depend on the location and extent of the lesion within the visual pathway (Fig. 18-8).

Visual field defects are fundamentally different

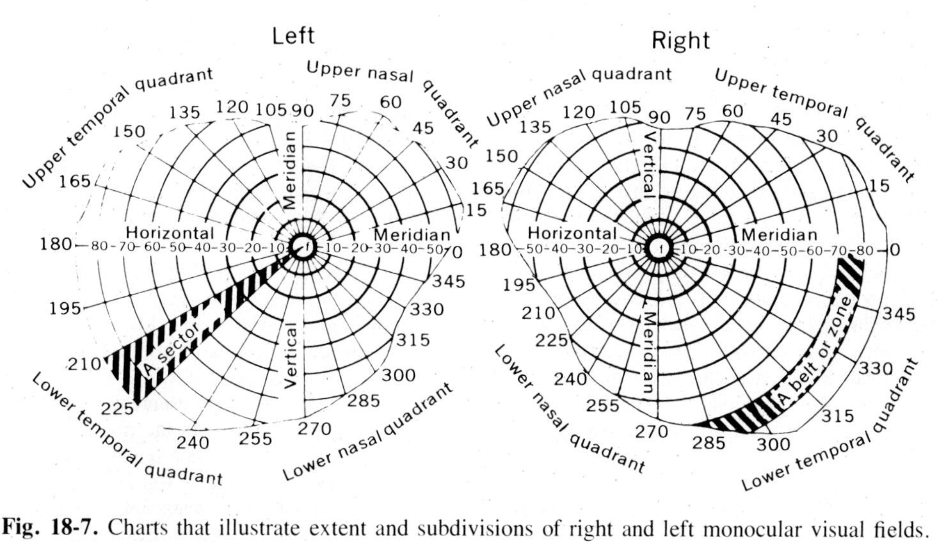

Fig. 18-7. Charts that illustrate extent and subdivisions of right and left monocular visual fields. (From Polyak.[7])

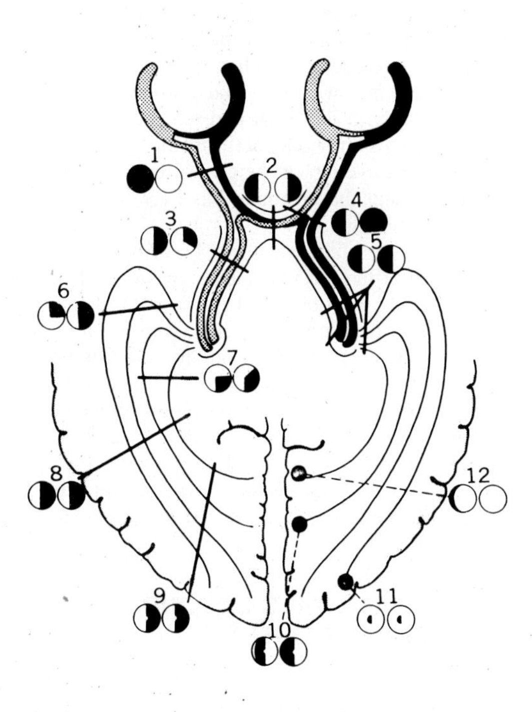

Fig. 18-8. Visual field defects produced by interruption of visual pathways at various sites. *1,* Optic nerve: blindness in eye on side of lesion, with normal vision in contralateral eye. *2,* Sagittal lesion of chiasm: bitemporal hemianopsia. *3,* Optic tract: contralateral incongruous homonymous hemianopsia. *4,* Optic nerve–chiasma junction: blindness on side of lesion with contralateral temporal hemianopsia. *5,* Posterior optic tract, lateral geniculate nucleus, and posterior limb of internal capsule: contralateral homonymous hemianopsia. *6,* Optic radiation, temporal loop: contralateral incongruous homonymous hemianopsia or superior quadrantic hemianopsia. *7,* Optic radiation, medial fibers: contralateral incongruous homonymous inferior quadrantic hemianopsia. *8,* Optic radiation in parietal lobe: contralateral homonymous hemianopsia. *9,* Optic radiation in posterior parietal lobe and occipital lobe: contralateral homonymous hemianopsia with macular sparing. *10,* Midportion of calcarine cortex: contralateral homonymous hemianopsia with macular sparing and sparing of contralateral temporal crescent. *11,* Pole of occipital lobe: contralateral homonymous hemianoptic scotomas. *12,* Anterior tip of calcarine fissure: contralateral loss of temporal crescent with otherwise normal visual fields. (From Harrington.[58])

when the interrupting lesion is located in front of or behind the optic chiasm:

1. Unilateral damage to the portion of the visual pathway peripheral to the decussation of the optic fibers gives rise to a functional defect limited to the field of vision of the eye on the affected side. Complete severance of the optic nerve is, of course, followed by total loss of function in the corresponding eye. Partial lesions determine a variety of defects that correspond to the area of the field represented by the fibers that are interrupted. This is particularly evident for damage to the optic fibers in their intraretinal course, where, because of their convergent arrangement toward the optic papilla, lesions similar in all respects produce an extensive field defect when situated near the optic nerve head and a small one when located in the retinal periphery. When two unilateral lesions exist, two visual defects are produced,

one in each monocular field of view. In general the two defects are different in size, shape, and other characteristics, a condition termed "incongruity," which may also be produced by unilateral lesions of more proximal portions of the afferent pathway.

2. Interruption of the visual pathway anywhere from the chiasm to and including the area striata results in field defects in which the essential feature is some degree of *hemianopsia,* that is, loss of function that extends over one half of the visual field. Hemianopsic defects are termed homonymous or heteronymous, depending on whether the corresponding or the noncorresponding halves of the two respective monocular visual fields are affected. As determined by the anatomic arrangement of the visual pathway, heteronymous hemianopsia is characteristic of lesions of the chiasm and homonymous hemianopsia of unilateral damage to the suprachiasmal, intrahemispheric portion of the system (i.e., optic tract, geniculate nucleus, visual radiation, and striate cortex).

Pathologic processes at the level of the chiasm usually affect the intermediate, bridgelike portion of this structure. Consequently, they interrupt the two sets of decussating fibers (i.e., the fibers originating from the nasal half of each retina) and give rise to the characteristic visual defect in the temporal half of the two monocular fields of view—bitemporal hemianopsia. Very infrequent is the occurrence of the reciprocal condition, the interruption of the uncrossed fibers and consequent binasal hemianopsia, which can only be explained on the basis of bilateral lesions.

Lesions in the suprachiasmal portion of the visual pathway produce homonymous hemianopsia. Each cerebral hemisphere relates to the contralateral halves, nasal and temporal, of both monocular fields of view, hence to the contralateral half of the binocular visual field. Thus right homonymous hemianopsia results from lesions of the optic tract, geniculate nucleus, visual radiation, and striate cortex of the left side, and left homonymous hemianopsia results from damage to the same structures in the right hemisphere. Total hemianopsia implies, of course, total interruption of the visual pathway behind the chiasm on one side. Hemispheric lesions most often involve the upper portion of the visual system, the visual radiation, and the visual cortex. The large spread of the optic pathway in the posterior portion of the hemisphere offers a possibility for partial lesions and consequently limited defects, always hemianopsic in character, in the contralateral field of view. These partial defects may take a great variety of forms and involve only a small area (scotoma), a segment of a quadrant, an entire quadrant or more. Unilateral field defects from lesions to the suprachiasmal visual pathway may occur as the result of the interruption of the unpaired central projection of the most peripheral nasal fibers of the ventrolateral hemiretina (temporal crescentic defects).

A common occurrence in cases of homonymous hemianopsia is the preservation of visual function in a portion of the field of view of each eye around the fixation point. Perimetric mapping shows that in these cases the border between functioning and nonfunctioning halves of the visual field does not pass through the point of fixation but curves around it. In most cases the preserved central vision subtends 2 to 5 degrees, but it may vary from 1 degree or less from fixation to a large portion of the affected half-field. This remarkable phenomenon, termed *central* or *macular sparing,* may occur with lesions of any portion of the suprachiasmal visual pathway, but more commonly with lesions of the posterior portion.

Some forms of central sparing may be accounted for on the basis of a lesser vulnerability of central as compared to peripheral fields.[8] The representation of the macula is very large and occupies a great proportion of the visual pathway. The occipital pole, where the macula is represented, receives a double vascular supply, from both the middle and the posterior cerebral arteries.[7] These features make it possible for even large hemispheric lesions, particularly those of a vascular nature, to leave intact the macular projection and thus to spare an island of central vision.

Recent anatomic studies in the macaque monkey[22] have shown the existence of a bilateral central projection from a 1-degree vertical strip along the midline of the retina. Within this strip, ipsilaterally and contralaterally projecting ganglion cells are intermixed. At the fovea the strip splits and curves as 0.5-degree bands along both the nasal and temporal rims of the foveal pit. This arrangement results in a 1- to 3-degree nasotemporal overlap of the central field of view and could represent the anatomic basis for the phenomenon of macular sparing.

In a number of cases, it can be demonstrated that the sparing of the macula is apparent rather than real. In these cases, central vision is extended to the blind half of the field by employing in fixation not the foveal area, but an area of retina on the functioning side (pseudofovea). The development of eccentric fixation may be facilitated by the physiologic movement of the eye scanning the image of the object on which attention is directed. If a more complete picture is obtained from an area outside the fovea, this area becomes the center of attention and an automatic fixation is shifted to it.[8,62]

VISUAL PATHWAYS BEYOND THE STRIATE CORTEX

The devastating functional sequences of the destruction of the geniculostriate pathway clearly indicate that in primates the primary projection system plays a fundamental role in vision, possibly an essential one in man. Moreover, it has long been recognized that visual disturbances may result from lesions of cortical areas outside the area striata. These disturbances reflect, in general, an inability of the affected individual to utilize correctly the visual information provided by a healthy primary projection system.

In humans, they include difficulties in the recognition and identification of objects, patterns, or color and visual disorientation in space and are often associated with other disorders, particularly of speech, reading, and writing.

These clinical observations, strongly supported by experimental investigations in monkeys, have led to the notion that from the striate cortex visual information is brought to converge and interact within other cortical areas in which a further elaboration of the visual message occurs, an elaboration that is essential for the total perceptual experience.

The cortical progression of the visual pathway beyond the striate cortex is an orderly one, and its organization in the monkey has been outlined in recent years both anatomically* and functionally.[28,31,87,108] In brief, the visual projection extends forward from the striate cortex (area 17) to the prestriate cortex (areas 18 and 19) and from the latter to the inferior convexity of the temporal lobe, the inferotemporal cortex (areas 20 and 21). Moreover, areas 17 and 18 project to a small region in the posterior bank of the superior temporal sulcus (STS).[131,132] From the visual areas, fibers reach the frontal cortex: the striate and prestriate areas project to area 8 (the frontal eye field); the inferotemporal cortex also projects to area 8, as well as to other premotor and prefrontal areas (Fig. 18-9). Homologous regions of the two sides, with the notable exception of the area striata, are connected through commissural fibers: the frontal and prestriate cortex via the corpus callosum and the inferotemporal cortex chiefly via the anterior commissure. In addition, cortical visual areas are part of complex corticosubcortical circuits whose functional significance is little understood. These connections will be mentioned in the following sections.

Prestriate cortex

In man and other primates the primary visual area is surrounded by the prestriate cortex, which Brodmann subdivided in two concentric areas: area 18, adjoining the striate cortex, and area 19, rostral to the former and extending into the parietal and temporal lobes. Morphologically the transition between these two areas is difficult to determine, and in recent years the terms "area 18" and "area 19" have been used with different meanings by different investigators, often without reference to cytoarchitecture. On the basis of anatomic connections and neuronal properties, Zeki[131-133] has outlined several independent areas in the prestriate cortex of the macaque. Surrounding the striate cortex, or V1, there are two concentrically adjacent areas, V2 and V3, largely buried in sulci, and each receives a topographically organized projection from V1, with the lower visual field represented dorsally and the upper visual field ventrally. The representation of the fovea occupies a large extent of prestriate cortex between upper and lower field representations. From areas V2 and V3 projection fibers pass forward to the prelunate sulcus (V4), where a detailed retinotopic organization is less evident. Intracortical connections from V4 progress to an area of prestriate cortex in the posterior bank of the superior temporal sulcus (STS), where independent projections from V1 and V2 also converge. At the prestriate cortex the visual pathways of the two sides subserving right and left contralateral hemifields are brought together by commissural fibers. Prestriate callosal connections are complexly organized and exist chiefly, if not exclusively, between areas of representation of the vertical meridian and of the fovea. A few, sparse callosal fibers also exist in striate cortex and only in the vicinity of the striate-prestriate border.[34,133] The prestriate cortex, that in macaque receives no input from the lateral geniculate nucleus, has major connections with the thalamus and tectum. Topographically organized reciprocal connections exist between prestriate areas, the inferior pulvinar (which receives connections from the superior colliculus), and the adjacent lateral pulvinar. Other prestriate connections occur with the lateroposterior thalamic complex and the reticular nucleus. The prestriate-collicular projection, like all corticotectal visual projections, is retinotopically organized.[15,17,92]

The specific role of the prestriate cortex in visual functions is not fully understood. As previously described, a monkey who has had the striate cortex removed bilaterally retains some elementary form of visually induced behavior. Under this condition the animal is still able to perform efficiently in a familiar surrounding; visual recognition of still objects is lost, but moving objects attract gross movements of the eyes and hand and may evoke prehensile reactions; visual placing reactions may still be observed.[44,76,125] Additional removal of the prestriate cortex, area 18 in particular, results in a loss of all visual reactions, leaving only pupillary reflexes and blink to bright flashes. Bilateral removal of prestriate cortex alone is followed by disturbances in spatial judgment and confusion of moving objects, while prehensile and placing

*See references 34, 73, 74, 93, and 130.

reactions remain intact.[44] Moreover, in the total absence of this region the monkey is not able to learn, or retain without deficit, pattern discrimination tasks. With subtotal ablations this deficit is particularly severe only when the lesion involves the prestriate area receiving a projection from that part of the striate cortex linked to the central retina.[31] Indeed, foveal prestriate cortex is crucial for pattern discrimination whereas, as already indicated, foveal striate cortex is not. Electrical stimulation of prestriate cortex produces an orderly pattern of conjugate eye movements related to the visual field representation in this area.[124]

These findings indicate that the prestriate cortex is the fundamental link over which information regarding visual stimuli progresses from striate cortex to other cortical regions of the same and opposite hemisphere. The prestriate cortex plays a major role in regulating visuospatial adjustments, and it may serve visual functions in the absence of input from the primary visual projection. It may be then assumed that the contribution of areas 18 and 19 to the organization of visual behavior depends *also* on the connections these areas entertain with subcortical structures, which in turn receive visual information directly or indirectly via retinal afferents that do not synapse in the LGN. The cortical pathway, however, is essential for higher visual functions, for bilateral destruction of the pulvinar does not affect pattern discrimination learning,[24] whereas cortical lesions do.

Inferotemporal cortex

In 1938 Klüver and Bucy described a complex syndrome of behavioral alterations in monkeys that follows bilateral removal of the temporal lobe. They observed that "while the monkey showed no gross defect in the ability to discriminate visually, she seemed to have lost entirely the ability to recognize and detect the meaning of objects on the basis of optic criteria alone."[77] Many neuropsychologic investigations have since shown that the neocortex of the inferior convexity of the temporal lobe serves visual functions exclusively and that it is of crucial importance in visual discrimination learning.*

The inferotemporal cortex includes the middle and inferior temporal gyri and corresponds closely to areas 20 and 21 of Brodmann. Visual information reaches the temporal lobe chiefly via the striate-prestriate cortical pathway. The first evidence of the importance of this projection was provided by the experiments of Mishkin,[87] in which both occipital and temporal cortices were removed unilaterally in monkeys trained to perform a visual discrimination task. When the two lesions were on the same side of the brain, the animal relearned the problem almost immediately. When, on the other hand, the occipital cortex was removed in one hemisphere and the temporal cortex in the other, the monkey required a large number of trials to relearn the discrimination. Finally, when the corpus callosum was also sectioned in animals with "crossed" lesions, relearning was very difficult if possible at all. Interruption of callosal fibers eliminated the link between the intact occipital lobe on one side and the temporal lobe on the other. Recent anatomic investigations have detailed the steps in the progression of the visual pathway to the inferotemporal cortex (Fig. 18-9). Fibers from the prestriate cortex reach the cortex of the inferior temporal gyrus (area 20) and possibly also the posterior part of the middle temporal gyrus. Area 20 in turn projects to the middle temporal gyrus (area 21).[74,93] From the inferotemporal cortex, fibers pass over to cortical areas of the frontal lobe, to the tip of the temporal lobe, and to a part of the cortex buried in the STS. Subcortically, the inferotemporal cortex is connected with a number of structures, among them the amygdala, pulvinar, superior colliculus, and pretectum.[73,74] There is no detailed topographic representation of the field of vision in the temporal lobe, and subtotal lesions of the prestriate cortex are followed by degeneration that spreads over large areas of the inferotemporal cortex.

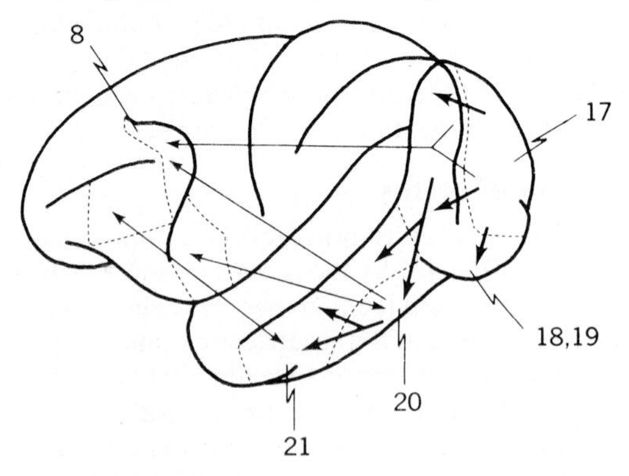

Fig. 18-9. Schematic diagram of progression of visual pathways beyond striate cortex. Forward connections exist from primary visual cortex (area 17) to prestriate cortex (areas 18 and 19) and from the latter to inferotemporal cortex (areas 20 and 21). Areas 17, 18, 19, and 20 also project to area 8 on lateral aspect of frontal lobe. Reciprocal connections exist between inferotemporal cortex and other premotor and prefrontal regions. (Refer to Jones and Powell[74] for details.)

*See references 31, 33, 54, 87, and 129.

When the inferotemporal cortex is removed bilaterally, the behavior of the monkey appears normal, with no detectable visual field defects, impairment of visual acuity,[33] or impairment of threshold for brightness discrimination.[49] On the other hand, certain learning capacities are severely disturbed: the animal is unable to perform a variety of visual choice tasks even though it had learned them preoperatively. The deficit produced by inferotemporal ablations is exclusively visual: no somesthetic or auditory defects are present. Its nature is not clearly understood; it cannot be attributed to "sensory" changes, to the animal's inability to comprehend the testing situation, or to the complete disappearance of specific memory traces corresponding to specific visual habits. What is known is that the inferotemporal impairment in discrimination learning does not depend critically on the dimension in which the visual cues differ (e.g., size, brightness, hue, or pattern), but rather on the nature of the discrimination problem. A task that requires a large number of trials for a normal animal to learn is usually impossible for a monkey with inferotemporal lesions, whereas a simple task may be learned correctly.[99] The impairment is most pronounced for visual discrimination involving alternatives,[49] particularly for concurrent discrimination (i.e., parallel learning of several simple discriminations).[31] There is some evidence that certain visual defects that occur in man after temporal lobe damage are similar to those observed in monkeys following inferotemporal lesions.

The foregoing neurologic observations on the functional significance of cortical regions beyond the primary receiving area provide evidence that these regions are essential for certain complex dimensions of vision. On the other hand, the notion that the area striata is a "receptive" area serving relatively simple functions and the "associative" prestriate and inferotemporal cortices are privileged integrative centers and sites of higher functions is not tenable in its extreme form. It is suggested that the total replica of visual events projected in neural code from the eye to the brain undergoes a series of successive transformations, each essential to that following, along the central visual system. The perceptual process develops out of the continuum of these transformations.

Frontal cortex

It has long been known that in man and monkey electrical stimulation of an area on the lateral aspect of the frontal lobe (the frontal eye field), with lowest threshold at Brodmann's area 8, elicits horizontal and oblique deviations of the eyes to the contralateral side, the precise direction of which depends on the locus of stimulation. These movements are often accompanied by turning of the head toward the opposite side. Adversive eye deviations can also be produced by stimulation of several other brain regions, most readily from prestriate and striate cortex, again in a topographically organized pattern.[102,124]

In humans, lesions involving the lateral aspect of the frontal lobe interfere with conjugate eye movements, producing paralysis of gaze or forced contralateral deviation of eyes and head. These disturbances are usually transitory. In monkeys the two main effects observed after unilateral frontal eye field lesions are ipsilateral deviation of eyes and head and a neglect of contralateral visual stimuli, the result of a contralateral relative visual field defect, a hemiamblyopia.[80] Complete recovery from these impairments is commonly observed within a few weeks.

Anatomically, the frontal eye field receives projections from ipsilateral occipital and temporal visual areas (Fig. 18-9) and is also connected with the posterior parietal cortex (area 7). Subcortically, there are important reciprocal connections with the dorsomedial nucleus of the thalamus, as well as connections with other thalamic nuclei, the basal ganglia, and the superior colliculus/pretectal complex. The fibers to the colliculus terminate in superficial and intermediate layers where the afferents from the retina and the projections from occipital cortex are also found. There is no evidence of a direct projection from area 8 to the oculomotor nuclei.[79]

Electrophysiologic studies have shown that there are two major types of neurons in the frontal eye field of the alert monkey.[20,89] Some neurons respond unspecifically with a short burst of impulses to visual stimuli appearing anywhere over large portions of the animal's field of view. Neurons of the second type discharge in association with movements of the eyes and head: some cells are active during saccadic eye movements, others when the eyes are slowly drifting or held stationary at a specific position, and still others when the head is turned in a particular direction. Most important, in nearly all instances the discharge of the eye movement–related neurons occurred *after* the beginning of the movement. Moreover, saccadic eye movements evoked by electrical stimulation of the frontal eye field are not abolished by ablation of the superior colliculus, whereas those evoked by stimulation of visual occipital cortex are.[106]

These findings suggest that frontal eye field

neurons do not participate in the initiation of oculomotor action but rather monitor oculomotor activity (e.g., eye shifts and relative positions in orbit). The frontal eye fields do not possess specific "motor" or "sensory" functions and may be regarded as an interactive site of neural events operating in the control of eye movements and more generally in the complex coordination of motor behavior in the visual sphere.

FUNCTIONAL PROPERTIES OF VISUAL NEURONS

In its course from the retina to the cerebral cortex the visual message, translated in neural code, undergoes a series of transformations that are reflected in the activity of the neurons of the system. These transformations depend on the anatomic connections within and between the neuronal aggregates; on their mode of action, which can be either excitatory or inhibitory; and finally on the qualitative and quantitative nature of the interaction between them. A single visual neuron, as representative of the population of cells of which it is a member, reflects to a certain degree the ability of the visual system to perform those discriminating and abstracting functions on which the recognition and interpretation of the visual stimulus is based.

Visual neurons at all levels of the system may discharge impulses in the absence of any specific retinal stimulation. This maintained activity is generally irregular and changeable; it exhibits a great variety of patterns and occasionally rhythmic fluctuations. Like all types of neural activity, it is greatly influenced by anesthetic agents, particularly barbiturates, and by the state of alertness of the animal. Variations in the rate and pattern of discharge occur spontaneously and concomitantly with the changes in the EEG and behavioral signs that are characteristic of sleep and wakefulness.[115] Against this background of activity, changes in illumination over the retinal surface, with either white or color light, modify the frequency and pattern of firing of visual neurons, whose impulse discharges may increase, decrease, or be completely suppressed. These visual responses may be described as *excitatory* or *inhibitory,* with no implication as to how they are determined in the neural network. They are also termed *on, off,* and *on-off* depending on whether the neuron yields a burst of impulses to dark-to-light transitions, to light-to-dark ones, or to both conditions. Some cells respond with *transient* impulse bursts, others with *sustained* discharges lasting for the duration of the stimulus.

Visual neurons, like other sensory neurons, do not relate anatomically and functionally to the entire peripheral receptive surface, but only to a discrete region of it. The *receptive field* of a neuron of the visual system may be defined as that area in the visual field—or corresponding area of retinal surface—within which an adequate stimulus influences the activity of that neuron.[103]

Visual receptive fields may be composed of two or more subfields with different and often opposite functional properties, arranged concentrically (center-surround organization of retinal ganglion cells and geniculate neurons), or side-by-side (simple cortical cells), or in a spatially homogeneous organization (complex cortical neurons). There is evidence that under certain experimental conditions the responses of visual neurons may be modified or influenced by stimulation of retinal areas outside the receptive field.[78,83,85]

In summary, the activity of visual neurons in response to a photic stimulus depends on the intensity, the spectral characteristics, and the spatial and temporal configuration of the pattern of illumination falling on their receptive fields. The central replication of the visual message is determined not only by the physical dimensions of the stimulus and by the intrinsic properties and state of excitability of the neural network implicated, but also by the spatial relationships of the excitatory and inhibitory components within the receptive field of each element of the population of neurons involved.

SPATIOTEMPORAL ORGANIZATION

Lateral geniculate nucleus

In the light-adapted state the receptive field of geniculate neurons, like that of retinal ganglion cells (Chapter 17), is composed of a central region, either excitatory (on-center neurons) or inhibitory (off-center neurons), surrounded by an annular region giving the opposite response. Illumination of part or all of the excitatory region increases the firing rate of the neurons (either transiently or for the duration of the stimulus), whereas illumination of the inhibitory regions reduces or suppresses it and in general evokes a discharge at *off* (Fig. 18-10). This form of center-surround organization of the receptive field, circular in shape or with various degrees of radial symmetry, obtains for the very large majority of geniculate and retinal neurons of the cat and the monkey.* At all levels of the (afferent) visual system the *size* of the receptive field center is an

*See references 46, 64, 105, 110, and 127.

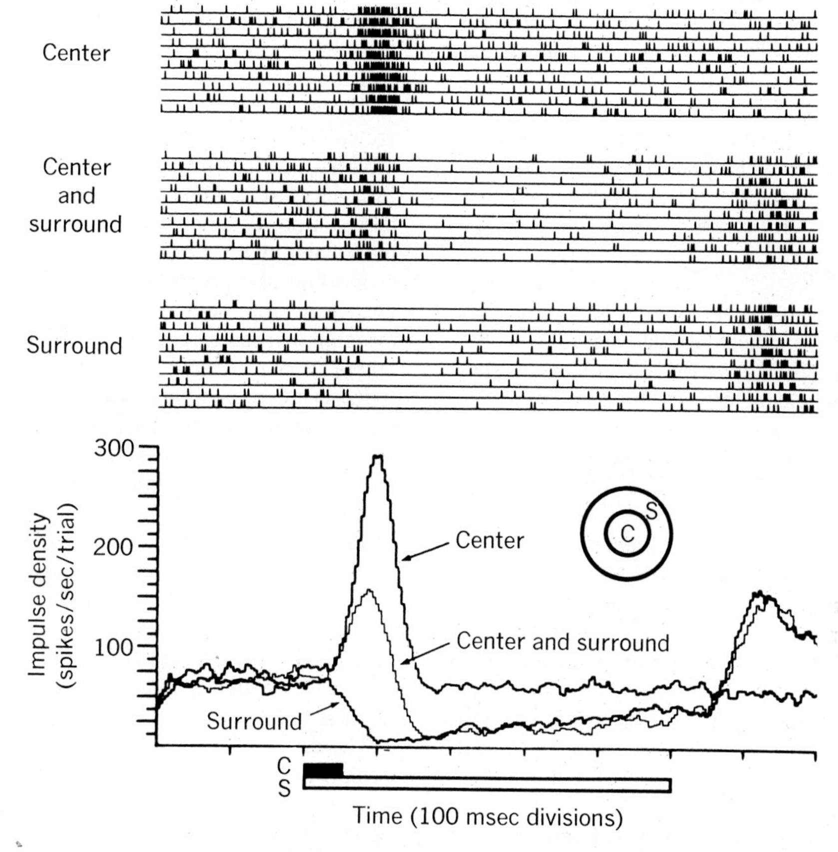

Fig. 18-10. Response of on-center cat geniculate neuron to discrete flash stimulation of center *(C)* and surround *(S)* regions of its concentric receptive field with spot and annulus, respectively. Upper part of figure shows replicas of sequences of neural impulses before, during, and after illumination of the two regions separately and together. In lower part of figure, corresponding time histogram of impulses, constructed over 50 stimulus repetitions, are plotted superimposed. Times of presentation of stimuli are shown at bottom of figure: center spot, 0.6 degrees in diameter, 50 msec flash duration; surround annulus, 1 to 2.7 degrees inner-outer diameters, 500 msec flash duration. Stimulation of center region evokes on-response; stimulation of surround evokes suppression of activity and off-response following cessation of flash. Simultaneous stimulation of center and surround elicits reduced excitatory response that reflects antagonistic interaction between the two regions; on termination of brief center stimulus, response continues with characteristics similar to those of response evoked by illumination of surround alone. (From Poggio et al.[97])

increasing function of retinal eccentricity: fields located near the fovea have smaller centers than do fields in the retinal periphery. In the cat's retina, field centers range from 0.5 to 8 degrees of arc in diameter, although larger ones have been found occasionally. In the monkey, LGN field centers as small as 2 minutes of arc and up to 1 degree were observed[127] for neurons related to the central visual field. These findings suggest that the centroperipheral gradient of visual acuity in man "may well be related to variations in receptive field center size similar to those found in the monkey and cat."[64]

Within each region of the receptive field there exists a direct relation between the size of the stimulus and the magnitude of the neural response (spatial summation). Each region, however, is not equally sensitive over its entire extent: the center generally gives the strongest response when stimulated near its geometric center, the surround near its internal boundary.

Between the two regions of the receptive field there exists mutually antagonistic interaction, so that on simultaneous stimulation of center and surround the two effects tend to cancel; in the limit, no response at all will be evoked. Usually the net result will be an on-response, an off-response, or an on-off–response, depending on the relative effectiveness of the two antagonistic components; in general the response will reflect the characteristics of the center region (Fig. 18-10). There is evidence that excitatory and inhibitory mechanisms do not occupy spatially separated regions, but overlap over the entire extent of the receptive field. One mechanism, however, is more sensitive over the center region of the field and the opponent mechanism over the surround.[10,97,105]

Striate cortex

In contrast with their stereotyped properties in retina and geniculate nucleus, visual neurons in the striate cortex possess different and specific functional characteristics. First, at this level, inputs from the two eyes converge and interact on single neurons, hence constitute the beginning of binocular vision. Second, a selectivity for certain spatiotemporal stimulus parameters develops: optimal responses are obtained with elongated straight contrast patterns (bright or dark bars, edges, etc.), and little or no response occurs with diffuse illumination; for any one neuron, certain patterns are more effective than others. Most characteristically, the *orientation* of the contrast border over the receptive field and often its *direction* of motion are critical and specific parameters for a cortical cell. These characteristics are illustrated in Fig. 18-11 for a neuron in monkey striate cortex responding best to optimally oriented black bar moving from left to right in a direction perpendicular to the optimal orientation.

These functional properties of striate neurons are chiefly the results of intracortical network inhibitory mechanisms,[35,112,114] and indeed, most cells in cortical layer IVc, the principal site of termination of the geniculate projection, are monocular and not orientation sensitive. A small proportion of cells, both monocular and binocular, responding equally well to stimuli at all orientations, are also found in other cortical layers. Some of these neurons have functionally uniform receptive fields; others display properties not essentially different from those of geniculate cells *(nonoriented neurons)*.

The majority of striate neurons are binocular and selective for stimulus orientation and, often, motion direction. These may be divided into two main types, termed "simple" and "complex" by Hubel and Wiesel[65-67] and, more generally, "S-type" and "CX-type" by Schiller et al.[112] The classifying criteria for simple and complex cells rest on the spatial organization of the receptive fields.

Simple cells (S cells) have receptive fields composed of one or more spatially distinct and adjacent subfields. Each subfield is sensitive to light increments (on) or to light decrements (off)

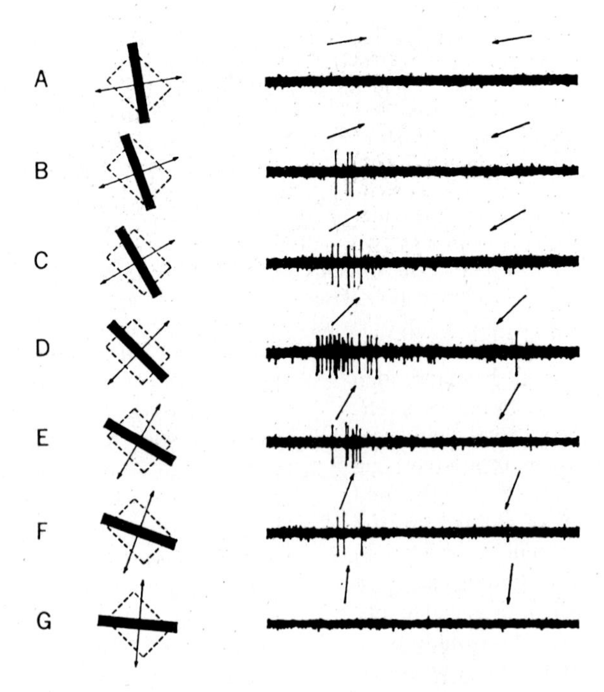

Fig. 18-11. Responses of a complex cell in right striate cortex (layer IVa) macaque to various orientations of moving black bar. Receptive field in left eye indicated by interrupted rectangles, approximately ⅜ × ⅜ degrees in size. Duration of each record, 2 sec. Arrows indicate direction of stimulus motion. (From Hubel and Wiesel.[67])

and to one or both directions of motion perpendicular to field's orientation. Most commonly, S-type cells display directional preference, and a large proportion of them are strictly unidirectional (Fig. 18-12). On the basis of the subfield properties and arrangements, a variety of S cell types have been described in area 17 of the cat[19] and of the macaque monkey.[112] The seven S-type cells recognized by Schiller et al.[112] are shown in Fig. 18-13.

Complex cells (CX cells) have orientation-sensitive, spatially uniform receptive fields: no separate subfields may be identified within them, and responses are evoked by light increments (on) or light decrements (off) from anywhere in the field. A majority of CX cells are direction selective, although not as strongly as S-type cells (Fig. 18-14).

A third type of cortical visual neurons, the *"hypercomplex" neurons,* has been recognized by Hubel and Wiesel[66,67] in cats and monkeys. In addition to a selectivity for stimulus orientation, these cells display a selectivity for stimulus length: for optimal responses a properly oriented stimulus must be limited (stopped) on one or both ends along the axis of receptive field organization. Hubel and Wiesel have considered the hypercomplex neuron as an independent functional type of cortical cell with unique operational significance. Schiller et al.,[112] on the other hand, have not found stimulus end stopping to be sufficiently strong and specific in the monkey striate cortex to warrant the identification of a separate functional type of cortical neuron. Possibly the hypercomplex attribute is the expression of a nonspecific general inhibitory network in the more superficial layers of the striate cortex.

On the basis of receptive field characteristics and intracortical locations of neurons with different spatial properties, Hubel and Wiesel[65,66] proposed a hierarchic model of organization of the striate cortex based on successive convergences of sets of neurons with similar properties on other cortical cells thought to subserve a higher order integrative function than the former. The first transformation would occur at the striate neurons receiving the geniculate input. In the following stages, S cells converge onto CX cells and CX cells onto hypercomplex ones. In contrast with this

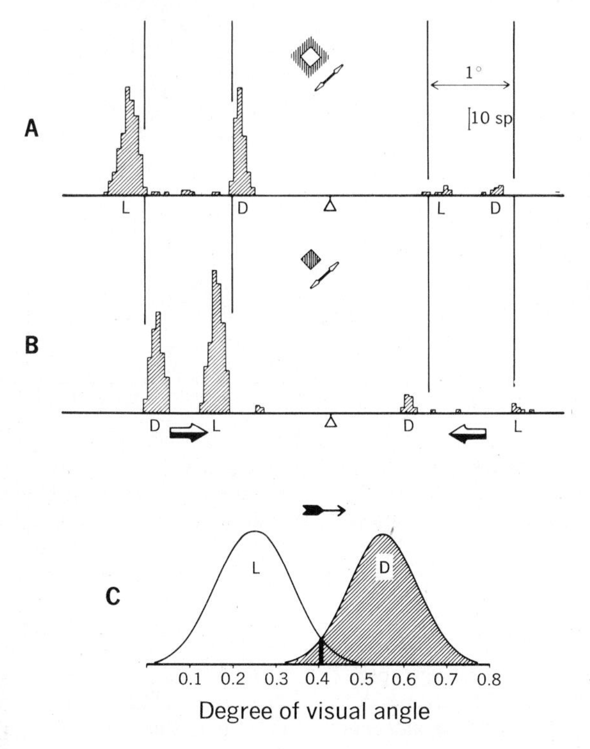

Fig. 18-12. Simple (S-type) cell in striate cortex of rhesus monkey. Responses are shown as time histograms of impulses. Arrows indicate direction of movement across receptive field. Cell sensitive both to light edge *(L)* and to dark edge *(D)*. Receptive field composed of two separate subfields responding to opposite contrast and same unidirectional movement. **A,** Response to 1-degree light square. **B,** Response to 1-degree dark square. **C,** Schematic drawing of receptive field showing size, separation, and directionality of the two subfields. *sp,* Spikes. (From Schiller et al.[112])

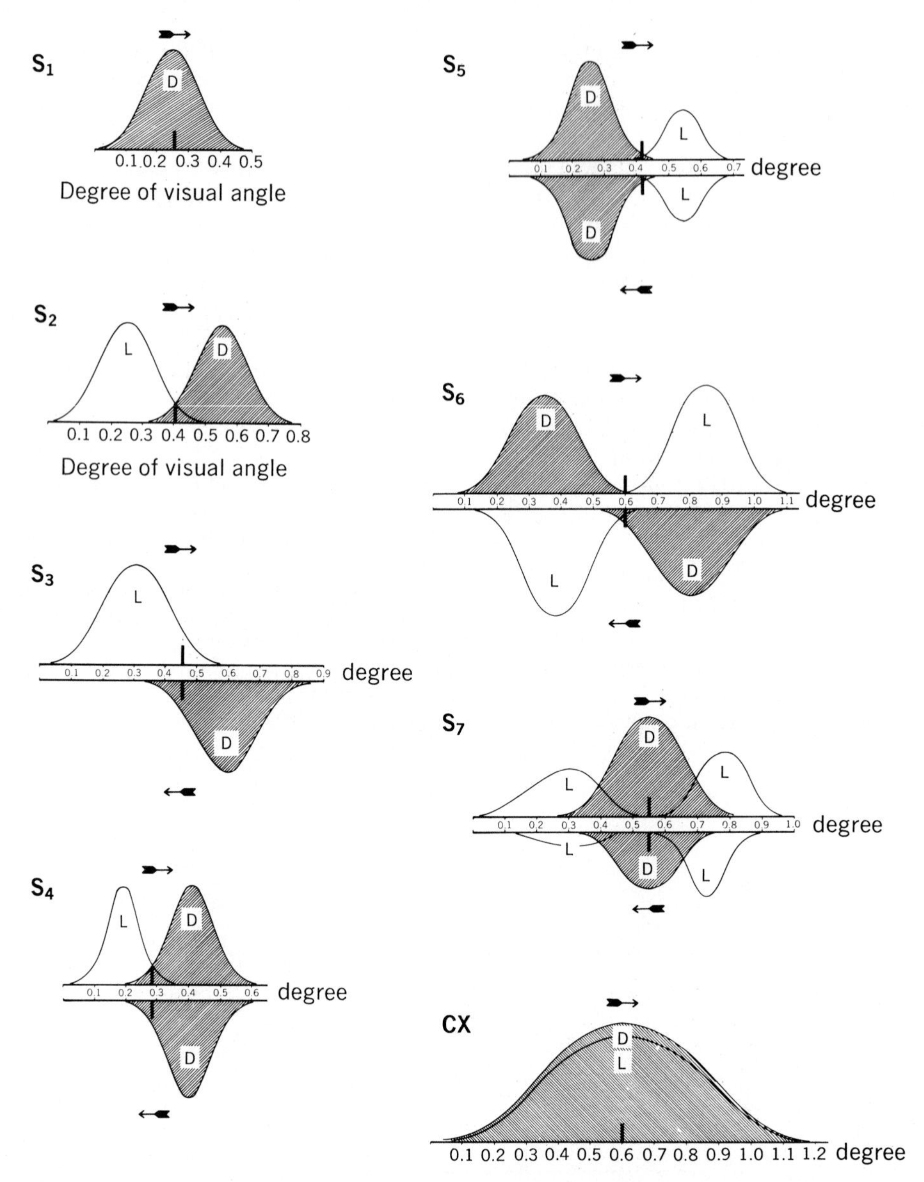

Fig. 18-13. Schematic drawing illustrating seven S-type cells and one CX-type cell observed in striate cortex of macaque monkey. Each drawing shows approximate size and spatiotemporal response properties of receptive field. *D,* Dark edge–sensitive regions; *L,* light edge–sensitive regions. Arrows indicate direction of effective stimulus motion across receptive field. (From Schiller et al.[112])

serial processing stands the parallel processing scheme that in recent years has received some strong experimental support. For example, S and CX cells may receive direct geniculate afferents[61]; S cells respond preferentially to slower stimulus speed than CX cells do[90,94]; S cells are insensitive to movements of randomly textured patterns, to which CX cells respond strongly.[57] The conclusion that can be derived from the available evidence is that (1) the spatiotemporal properties of cortical visual neurons depend on more complicated mechanisms than either a serial convergence from neuron to neuron or a simple parallel processing[116] and (2) a comprehensive scheme to explain the complexities of cortical organization has yet to be developed.

Prestriate and inferotemporal cortex

The spatiotemporal properties of neurons in area 18, or V2, are not essentially different from

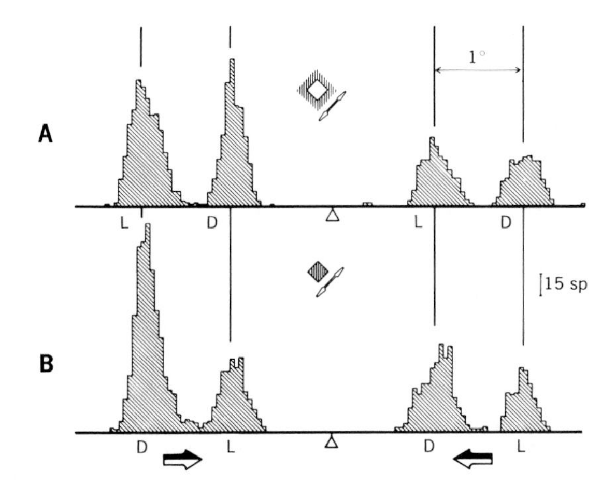

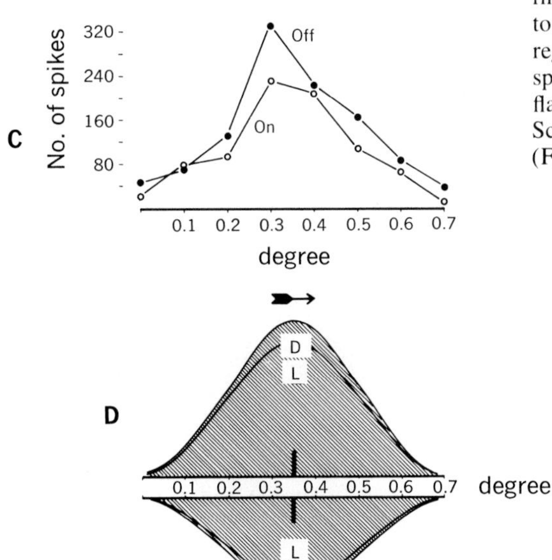

Fig. 18-14. Complex (CX-type) cell in striate cortex of rhesus monkey. Unit is bidirectional and responds both to light *(L)* and dark *(D)* edges throughout activating region. **A,** Response to moving light square. **B,** Response to moving dark square. **C,** Responses to square flashed at different locations over receptive field. **D,** Schematic diagram of receptive field. *sp,* Spikes. (From Schiller et al.[112])

those of striate neurons. This area is retinotopically organized, and its cells are orientation/direction sensitive with simple or complex characteristics. Receptive fields are larger, however, than those in area 17 at the same corresponding eccentricities, and many neurons respond to stimulus velocities higher than those to which striate neurons are sensitive.[9] Nearly all cells are binocular and, like area 17 neurons, have properties suggesting they may subserve stereoscopic mechanisms.[68,95]

A small prestriate area in the posterior bank of the superior temporal sulcus has been shown to contain cells specialized to signal motion, frequently irrespective of the spatial pattern of the stimulus, and to be directionally selective.[131] Next to this area, another small STS region exists in which nearly all neurons are color coded (p. 568).[133]

In inferotemporal cortex, where there is no detailed topographic representation of the field of view, all neurons have very large receptive fields, usually more than 10 × 10 degrees. Most fields are binocular and bilateral and extend across the vertical meridian into both hemifields, almost always including the fovea.[104]

CHROMATIC PROPERTIES

Lateral geniculate nucleus

The response properties of neurons in the LGN to color stimuli have been investigated in some detail in the macaque, an animal that behavioral tests have shown to possess color vision very similar to that of man. Throughout the primary visual system of primates, from retina to cerebral cortex, two major types of neuronal spectral sensitivity have been identified. Some neurons are color opponent: they respond with an increase in firing rate to a certain region of the visible spectrum and with a decrease to other regions; white light is usually not a very effective stimulus. Other cells are spectrally broad band: they are uniformly excited or inhibited by light of all wavelengths and by white light. Cells of the two categories are thought to subserve different visual functions: the former to carry information about the color of the light, the latter about its brightness.

Color-opponent cells. In the macaque LGN, color-opponent cells are found in the parvocellular layers, where they form more than 75% of the neuron population. On the basis of their response to diffuse flash stimulation of the retina with monochromatic light, De Valois et al.[42] have recognized four classes of opponent cells (Fig. 18-15). Some cells are excited at long wavelengths and inhibited at short wavelengths: (1) red excitatory–green inhibitory (+R−G cells) and (2) yellow excitatory–blue inhibitory (+Y−B cells). Other cells have the opposite chromatic organization: (3) green excitatory–red inhibitory and (4) blue excitatory–yellow inhibitory. R-G cells are found more frequently than Y-B cells.

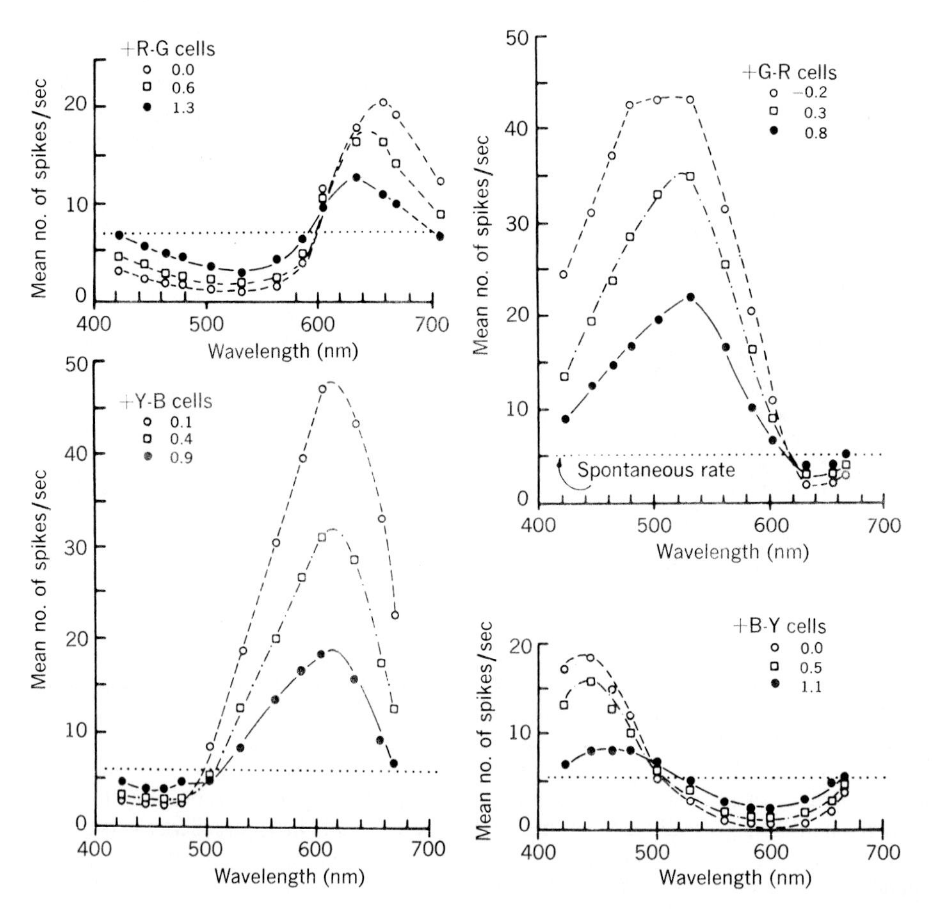

Fig. 18-15. Average firing rate of samples of the four classes of spectrally opponent cells described by De Valois et al.[42] in macaque geniculate nucleus. Diffuse flash stimulation with monochromatic lights of equal energy. *+R−G,* Red excitatory–green inhibitory; *+G−R,* green excitatory–red inhibitory; *+Y−B,* yellow excitatory–blue inhibitory; *+B−Y,* blue excitatory–yellow inhibitory. For each class of cells, spectral response curves to three levels of light intensity are shown; numbers next to each curve represent log attenuation relative to maximum available. Open symbols and vertical lines at each point enclose 1 SE of mean. Dotted horizontal line indicates mean firing rate in absence of stimulation for each type. (From De Valois.[40])

The color names used to define these neurons do not refer to retinal photopigments, but rather to the color that we see at those wavelengths that in any one cell produce excitation and inhibition and that correspond to the four subjective primary colors in the spectrum. As the graphs in Fig. 18-15 show, neither the shape of the spectral curves nor the wavelength of maximal effects that obtain for the four classes of geniculate opponent neurons are similar to the absorption spectra of any of the photopigments found in cone receptors of the retina of primates (Fig. 17-8). On the other hand, cone receptors subserve color vision, and geniculate cells receive an input from them.[127] Obviously, spectrally opponent cells are functionally connected to at least two receptor types, usually one excitatory and the other inhibitory.

The spectral sensitivities of the inputs may be determined by means of chromatic adaptation experiments. An example is shown in Fig. 18-16, where the response of a geniculate neuron is plotted as a function of wavelength for three different conditions of adaptation. Under dark adaptation (no bleach) the neuron responds to wavelengths in the middle spectral range with an increase of its impulse frequency over the spontaneous rate (the zero point on the ordinate scale) and to longer wavelengths with a decrease in frequency (+G−R cell). This response may be separated into excitatory and inhibitory components by selective bleaching of retinal pigments with intense adapting lights (chromatic adaptation). When the system of retinal receptors with red-absorbing pigments is selectively adapted out with a red light (680 nm bleach), only excitatory responses can be evoked that reflect the excitatory input to the cell from the middle wavelength–sensitive system. Similarly, an isolated inhibitory input from the long wavelength system is revealed under conditions of green adaptation (510 nm bleach). The spectral response curves for the two antagonistic components are broad and overlap over a large extent; the points of maximal effect have shifted from their positions under dark adaptation and correspond closely to the points of maximum absorption of the middle and long wavelength photopigments (Fig. 17-8).

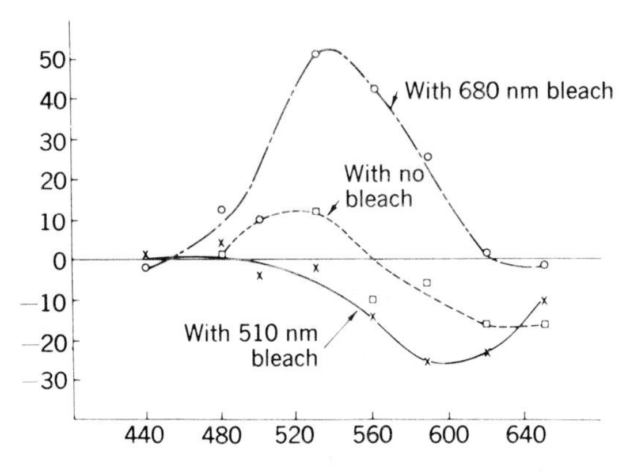

Fig. 18-16. Spectrally opponent cell of lateral geniculate nucleus of macaque (green excitatory, red inhibitory). Plot of response of neuron to diffuse retinal stimulation with various monochromatic lights during dark adaptation and during chromatic adaptation from two different spectral regions (680 nm and 510 nm). Abscissa: wavelength in nanometers. Ordinate: number of impulses per second during light stimulus with respect to spontaneous rate taken as zero for this plot. (From De Valois.[39])

These chromatic adaptation studies indicate that the response of opponent geniculate neurons is the result of neural interaction (algebraic summation[40]) between pairs of functionally antagonistic inputs from cone receptor systems. A single cell receives one input—either excitatory of inhibitory—from the cone receptors sensitive to long wavelengths (maximum absorption at 570 nm) and an opponent input from either the middle or short wavelength receptors (535 and 445 nm). Long and middle wavelength combinations produce the two reciprocal classes of R-G opponent neurons and long and short wavelength combinations the Y-B cells.

Studies of properties of lateral geniculate neurons with discrete visual stimuli[46,110,127] have shown that there are two types of organization with respect to the spatial distribution of the color-opponent systems within the neuron's receptive field (Fig. 18-17). The large majority of geniculate cells in parvocellular laminae have a concentric receptive field with typical center-surround organization to white light (type I of Wiesel and Hubel[127]). Monochromatic light stimulation reveals that the two spatial subfields are spectrally opponent: excitatory or inhibitory responses are obtained by stimulation of the field's center with lights in a given spectral range and respectively opposite responses by stimulation of the surround with another band of wavelengths. Most type I geniculate cells have a red/green sensitivity, the more common receptive field being "red on-center/green off-surround." Blue-sensitive cells are much fewer and are found predominantly in the ventral pair (3 and 4) of the parvocellular laminae.[111] A minority of less than 10% of color-opponent neurons are type II[127]: for these neurons the spatial distributions of the two spectral inputs are identical, and

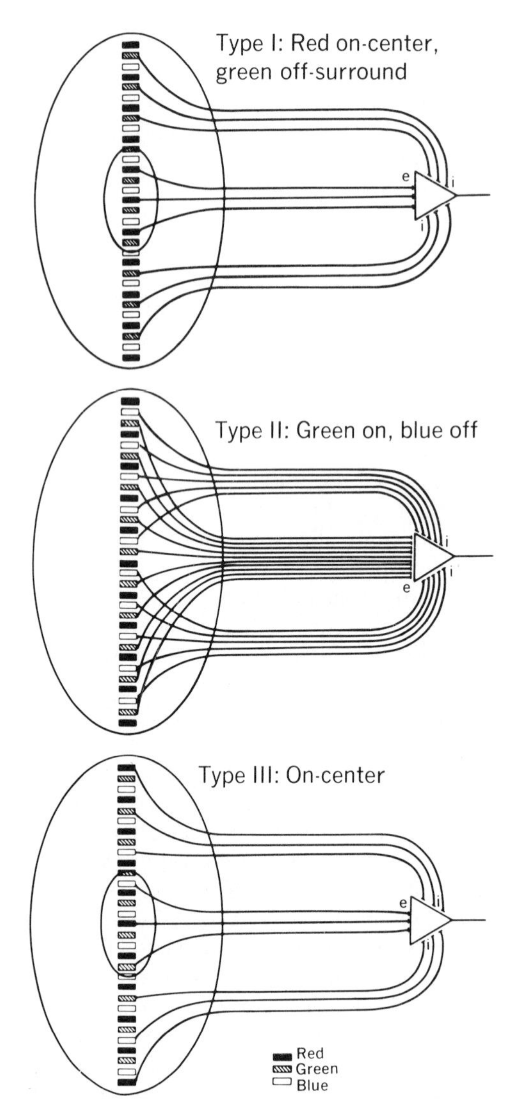

Fig. 18-17. Schematic diagrams to illustrate contribution of cones to type I, II, and III geniculate neurons, as proposed by Wiesel and Hubel. Three types of cones are indicated with different symbols, and for simplicity, receptors are shown only along one line through receptive field. Cones project to geniculate cells via intervening synapses, not indicated in drawings, and activation of receptors leads to excitation, *e*, of geniculate neuron or to inhibition, *i*. Example of type I cell: excitatory input from red-sensitive cones in field center and inhibitory input from green-sensitive cones in periphery. Three cone types from center are arbitrarily shown as being present in ratio of 1 : 1 : 1, and their ratio is the same in periphery. Example of type II cell: excitatory input from green-sensitive cones and inhibitory input from blue-sensitive cones. Relative contribution of two cone types is same in all parts of receptive field. Example of type III cell: input from all three types of cones, excitatory in nature from center and inhibitory from periphery. Proportion of three types of receptors is the same for two antagonistic regions of receptive field. (Redrawn from Wiesel and Hubel.[127])

no center-surround organization exists within the receptive field. These cells respond with excitation or inhibition to stimulation over the entire receptive field at some wavelengths and in the opponent fashion at other wavelengths. White light stimulation is relatively ineffective.

In summary, geniculate neurons of the macaque monkey that are differentially sensitive to wavelengths receive functionally opponent inputs from paired sets of the three cone receptor systems. In most instances, opponent systems are spatially segregated in center and surround regions of the receptive field of these neurons. There is evidence that intraretinal mechanisms may account for the sorting of the excitatory and inhibitory components and that color-opponent neural coding originates at the ganglion cell of the retina and is maintained and elaborated at higher levels of the system.[38]

Broad-band cells. Other LGN neurons are not differentially sensitive to wavelength. These cells are found throughout the geniculate nucleus, although in different proportions and with different characteristics in dorsal and ventral laminae. In the parvocellular laminae, broad-band cells are a minority (20% to 25%), whereas they constitute nearly the entire neuron population of the magnocellular laminae. These cells (type III of Wiesel and Hubel[127]) (Fig. 18-17) have concentric receptive fields, the center and surround regions of which have identical spectral sensitivities, and from each of them the same kind of response, either excitatory or inhibitory, is obtained at all wavelengths and with white light.

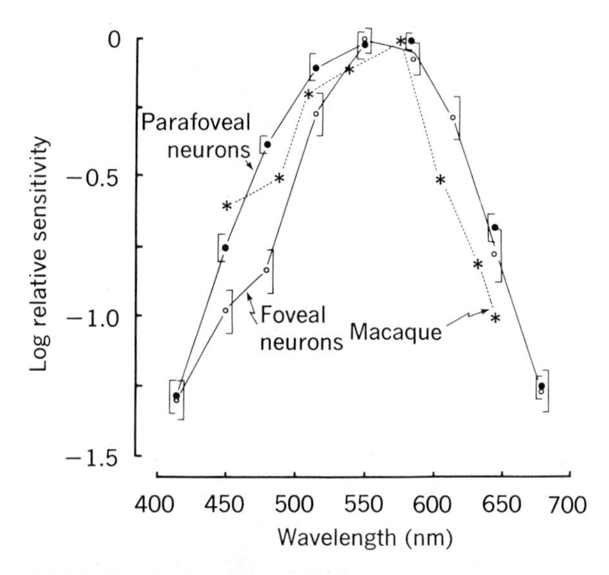

Fig. 18-18. Average spectral sensitivity curves for broad-band neurons in foveal (0 to 2 degrees of eccentricity) and parafoveal (4 to 7 degrees) striate cortex of the macaque plotted together with photopic spectral sensitivity curve for same animal determined behaviorally by De Valois et al.[43] Neuronal curves were derived from response threshold determinations at various wavelengths with stimuli of optimal spatiotemporal characteristics. All neurons were orientation sensitive, simple or complex, and were located in various cytoarchitectural layers. (From Poggio et al.[98])

Broad-band neurons are functionally connected to all types of cone receptors as well as to rod receptors. Broad-band cells are maximally responsive to the middle part of the spectrum, but will give exactly the same response to lights of any wavelength if they are appropriately adjusted in luminance. Indeed, the spectral sensitivity of broad-band neurons in the LGN (as well as in the retina and striate cortex) matches the luminosity function of the animal, as measured by behavioral tests of visual discrimination.[4,43] These findings suggest that broad-band neurons are operating primarily in the transmission of information about the luminous intensity of the stimulus. Their properties will be discussed further in a following section.

Striate cortex

Contrary to expectations based on the heavy afferent projection to striate cortex from color-opponent parvocellular LGN, only a minority of neurons in area 17 of the macaque appears to be color coded. It is possible that the experimental design of a number of studies on the chromatic properties of cortical cells was not appropriate for revealing unequivocal chromatic responses, and that, consequently, incorrect quantitative estimates have been made. On the other hand, several qualitatively different types of color-coded cells have been observed in striate cortex next to neurons that respond at all wavelengths of the visible spectrum, the broad-band neurons. The proportion of color cells appears to increase from nonfoveal to foveal striate cortex, where color vision is more highly developed.[53,98]

The broad-band neurons, both in foveal and in parafoveal cortex, have spectral sensitivities similar to the overall spectral sensitivity of the macaque (Fig. 18-18) and are operationally unaffected by the chromatic dimension of the stimulus. Of the color-coded striate neurons the majority have nonoriented concentric receptive fields or simple (S) fields. Some of these cells have color-opponent organization similar to that of the type I geniculate cells of Wiesel and Hubel.[127] Other cells display a more elaborate dual-opponent organization. These cells are opponent for both color and space; within their receptive fields the central region has one opponent color system, usually red/green, and the surround or flanks the opposite opponent-color properties. Dual-opponent neurons are generally insensitive to white light and respond best to simultaneous contrast conditions when one color is presented over the field center while the functionally opponent color is presented over the surround (Fig. 18-19). Other striate cortex neurons, particularly neurons with complex (CX) spatial properties, are color selective and respond preferentially or exclusively to a restricted band of

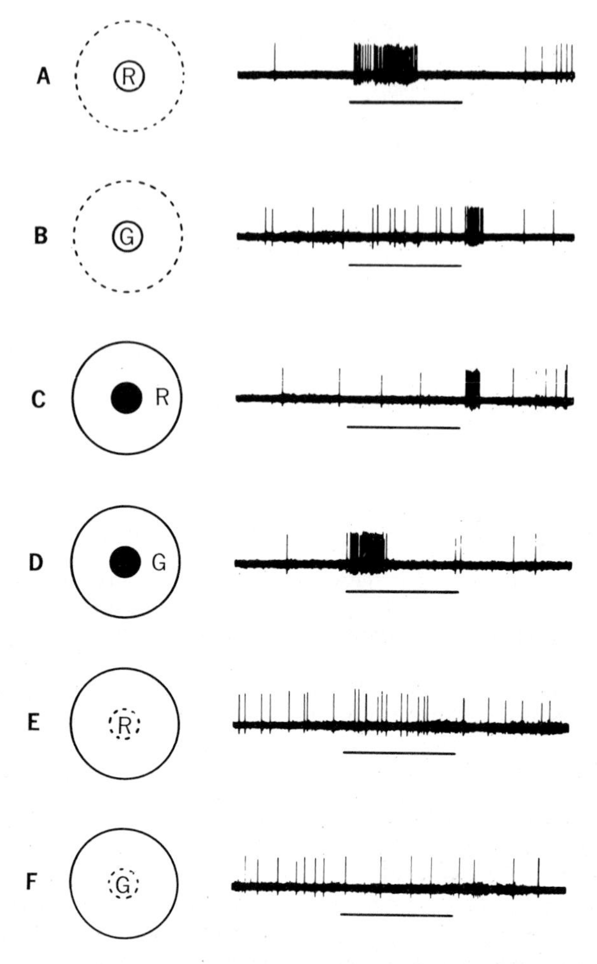

Fig. 18-19. Responses to monochromatic stimuli of light-adapted cell with concentric receptive field in striate cortex of macaque. Field center was 0.5 degree in diameter; total diameter of field was 8 degrees. Stimuli were a 0.5-degree spot, a 0.5-degree ID–8 degree OD annulus, and an 8-degree spot. **A,** Centered red spot produced on-discharge. **B,** Green spot produced off-discharge. **C,** Red annulus evoked off-response. **D,** Green annulus evoked on-response. **E** and **F,** Larger monochromatic spots were without effect. The 1 sec stimulus duration is indicated by the black bar below each oscilloscope trace. Duration of each sweep, 3 sec; *R,* red, 630 nm; *G,* green, 500 nm. (From Michael.[86])

wavelengths toward one or the other end of the spectrum. Among them, red-selective cells are more common. White light is an effective stimulus for most cells of this type (Fig. 18-20).

It is likely that dichromatic mechanisms similar to those observed at the geniculate level[40,41,86,127] are the basis for the functional organization of many color-coded cortical cells. On the other hand, there is evidence that all three cone mechanisms can intersect on a single striate neuron, not only on neurons with broad-band spectral characteristics, but also on neurons with evident color properties.[53]

Prestriate cortex

Beyond the striate cortex, color-specific neurons have been found in defined cortical regions, where they appear to be particularly numerous and give a "color-specific" identity to that patch of cortex. Zeki[133] has described two regions of color-coded neurons in prestriate cortex: in the prelunate gyrus, specifically in V4, and in the lateral part of the visual area in the posterior bank of the superior temporal sulcus (STS). These neurons are binocular, usually not orientation/direction sensitive, and color selective for certain wavelengths often at either the long or the short end of the visible spectrum.

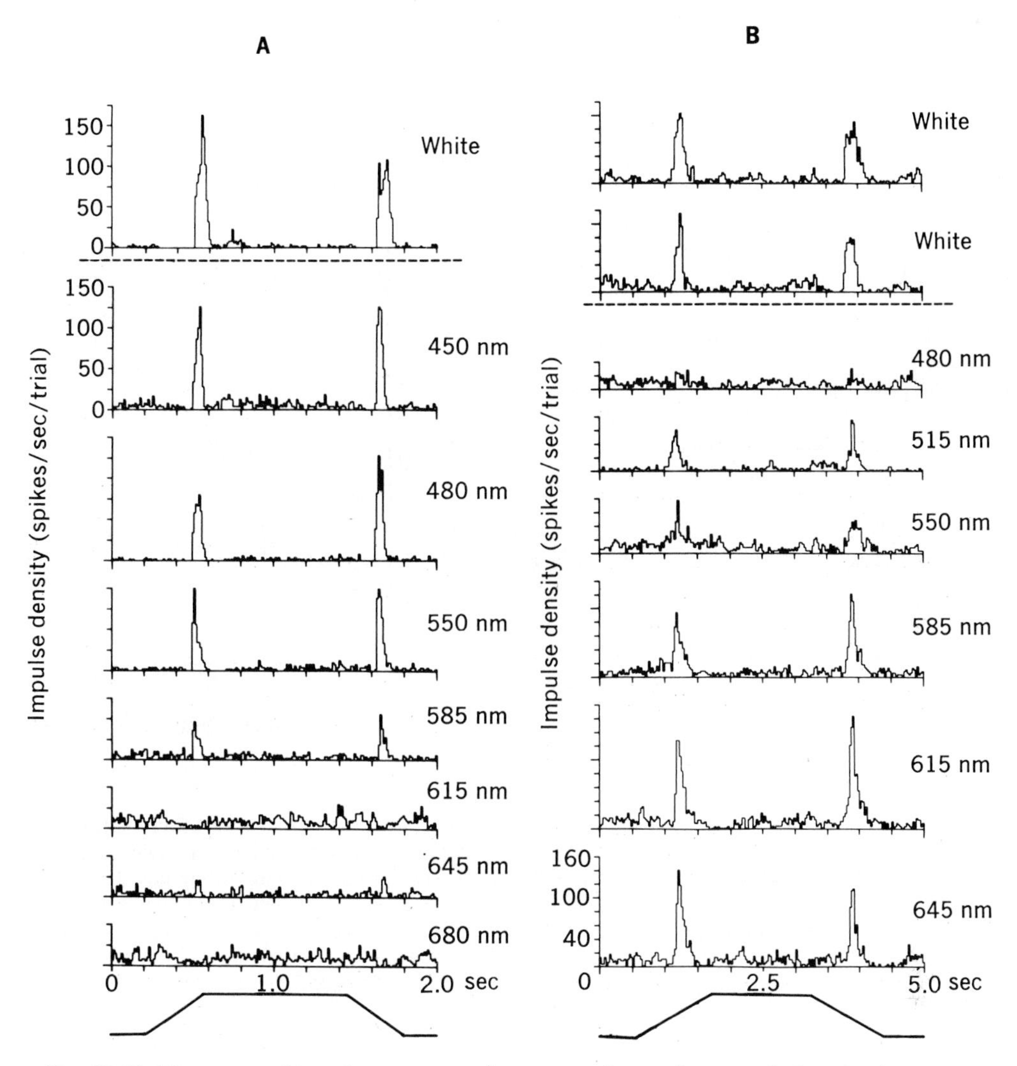

Fig. 18-20. Histograms of impulse responses of two spectrally tuned neurons in foveal striate cortex of the macaque. Stimuli of different wavelength equated in luminance. For each neuron, spatiotemporal stimulus pattern set for optimal response to white light (top histograms). Time course of stimuli moving across receptive field shown at bottom. **A,** Blue/green-tuned neuron (layer IVb; nonoriented uniform receptive field). **B,** Red-tuned neuron (layer V; complex receptive field). (From Poggio et al.[98])

INTENSITY CODING

All other parameters of stimulation being constant, the change in the pattern and frequency of firings of central visual neurons in response to a flash of light is a function of the intensity of the flash. Fig. 18-21 shows examples of the responses evoked from a broad-band neuron in striate cortex with monochromatic lights at a series of intensities. As the luminous intensity of the stimulus increases, the neural response also increases irrespective of wavelength, and over a portion of the effective range, it stands in linear relation to the logarithm of the stimulus intensity.

The general features of the neural replication of the intensive dimension of the visual stimulus are indicated by studies of Jacobs[72] of the LGN of the squirrel monkey, in which the behavior of broad-band neurons was analyzed. The plots in Fig. 18-22 show the results obtained at four different adapting luminances for a series of stepped increments and decrements around each of them, over a total range of ± 0.6 logarithmic unit. When the retina is diffusely illuminated with

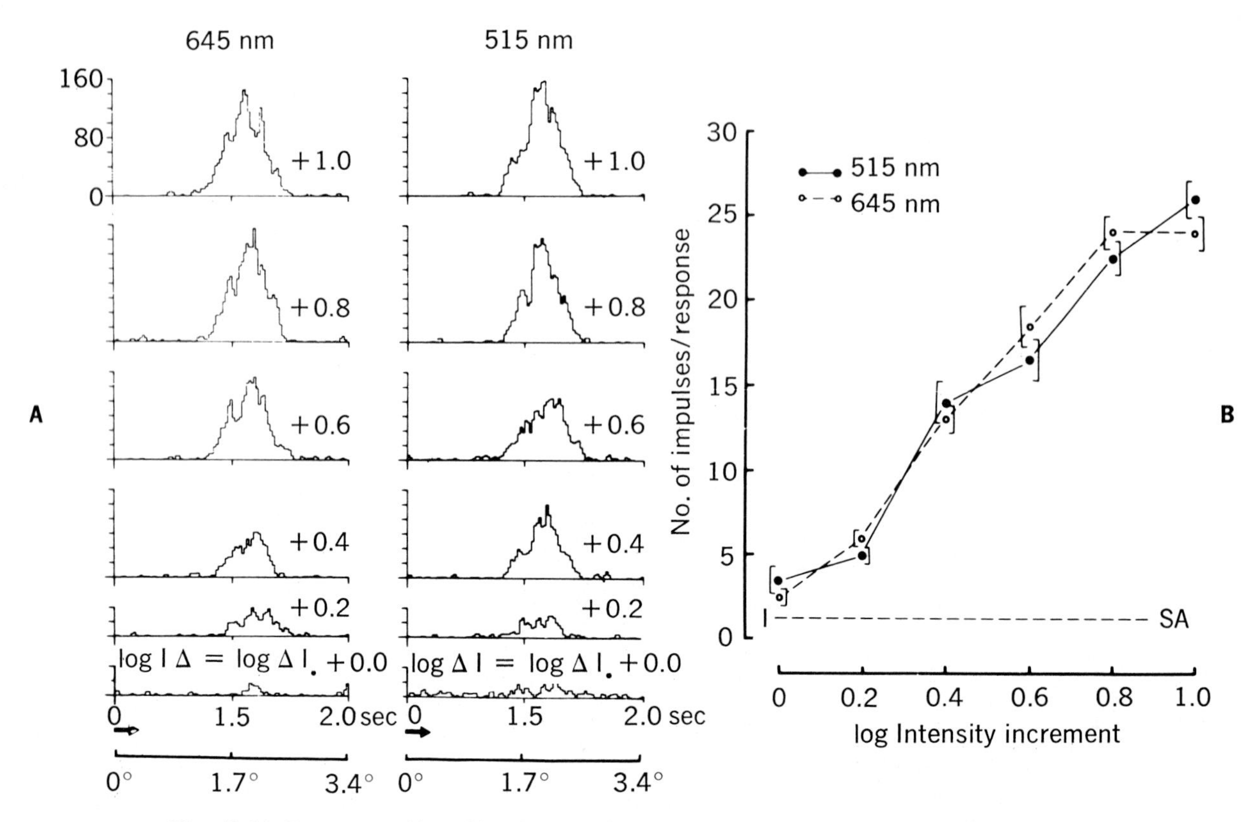

Fig. 18-21. Responses of broad-band neuron in macaque striate cortex as a function of stimulus intensity. Unit unidirectionally sensitive. **A,** Histograms of average responses of two series of intensities with stimuli of monochromatic lights at 645 and 515 nm moving in optimal direction. Lowest histogram of each series shows responses to threshold intensity, and successive histograms upward show responses to stimuli of increasing intensities (0.2 log unit increments). **B,** Average number of impulses per response at various stimulus intensities. Mean spontaneous activity *(SA)* shown by horizontal dashed line. Brackets indicate ±1 SE. (From Poggio et al.[98])

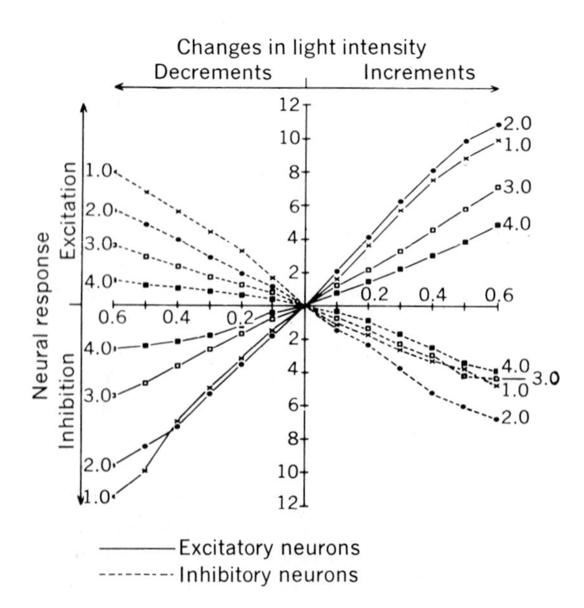

Fig. 18-22. Graphs plotting number of impulses discharged by excitatory and inhibitory broad-band squirrel monkey geniculate neurons in response to increments and decrements in light intensity about different levels of adaptation. Four adaptation luminances from which stimulus shifts were made are indicated at both ends of each curve as log attenuations. Abscissa: size of stimulus shift in relative log units. Ordinate: mean change in frequency of firing (impulses/sec) above and below mean spontaneous discharge rate. (Replotted from data by Jacobs.[72])

flashes of light whose luminance is greater than the luminance level at which the eye is adapted, broad-band geniculate neurons respond with either an increase (excitatory neurons) or a decrease (inhibitory neurons) of their firing frequency. Conversely, when the intensity of the test stimulus is less than that of the adapting level, a decreased impulse rate of excitatory neurons and an increased activity of inhibitory neurons are produced.

The intensity range over which graded unsustained changes in neural response are obtained from these neurons is a limited one, perhaps just over 1 log unit of intensity about the adaptation intensity, and the magnitude of the changes is influenced by the level of retinal adaptation. As shown by the slopes of the curves of Fig. 18-22, the effective range of changes in neural activity is greater at higher than at lower adapting luminances. The operating range of LGN neurons observed in these experiments is considerably less than that of the human visual system, whose effective range is greater than 10 log units of light intensity. There is evidence, however, from studies in the awake monkey for the existence of neurons in the LGN,[12] and in the striate cortex as well,[13] that are capable of graded responses over several log units of light intensity, of the order of 4 to 5 log units, approaching the overall operating range of the visual system. It may be that there are neurons effectively sensitive over different intensity ranges, some limited and other wider, that may subserve different aspects of intensity analysis of the visual patterns.

X AND Y SYSTEMS

During the past several years, studies of the anatomic and functional properties of visual neurons of the cat have led to a classification of these neurons in three main types generically named W, X, and Y.[27,48,120] The criteria on which the distinction is based are outlined in Chapter 17. Recent experimental observations have provided evidence that similar differential properties may also be discerned in the visual system of the monkey.[38,46,109,110]

In the macaque, LGN cells of the two types are anatomically segregated in different layers. One group of cells (X-like) receive their excitatory inout over slowly conducting optic fibers. These cells give sustained on- or off-responses to stationary patterns and are not sensitive to rapidly moving stimuli. X cells occur in the parvocellular laminae (3 to 6) of the LGN, and the large majority of them have color-opponent properties. Another group of geniculate cells (Y like) receive their input over rapidly conducting retinal axons. Stationary stimuli over their receptive fields evoke transient responses, and fast-moving stimuli drive these cells very effectively. Geniculate Y cells are found in the magnocellular laminae (1 to 2) and commonly give mixed on-off–responses; they show no color opponency and generally have a broad-band spectral sensitivity.

The evidence for the extension of the X and Y systems into visual cortex is as yet incomplete. Both slowly and fast-conducting projection fibers to the striate cortex have been identified, the former being the axons of X geniculate cells, the latter of Y cells.[110] Also, cortical neurons with X- and Y-like response properties have been described in cats and monkeys,[96,112,120] but the distribution and organization of these systems in the visual cortex is not known.

BINOCULAR INTERACTION

The large majority of neurons in the monkey visual cortex are binocular, that is, each of these neurons receives inputs from the right and the left eyes. Along the primary visual pathways up to the cells in input layer IV of area 17, the two eyes are functionally segregated. In other layers of the striate cortex, and in visual prestriate and inferotemporal cortical areas, single neuron binocularity is the norm.

For many binocular neurons, two receptive fields, one in each eye, can be precisely defined, although for some neurons, one of the fields may be difficult to characterize. The spatiotemporal organizations of the two fields are frequently very similar, but may also have different characteristics. For example, the spatial arrangement of the subfields of S cells (see Fig. 18-14) is not necessarily the same in the two eyes; directional selectivity may be more pronounced or only present for one eye, or it may be in diametrically opposite directions in the two eyes.[95,131] Orientation selectivity is usually the same, but the sharpness of orientation tuning may differ for right and left fields. The amplitudes of the two monocular responses are often unequal, and the eye from which the larger excitatory response is evoked is referred to as the "dominant" eye. Several degrees of eye dominance have been recognized, from total dominance (monocular neurons) to binocular equilibrium.[69] In brief, the response properties of binocular neurons to monocular stimulation of right and left eyes may be essentially identical (ocular balance) or may differ in strength, quality, or both (ocular imbalance).[95]

An additional and most important asymmetry between right- and left-eye receptive fields of cortical neurons resides in their relative spatial position, which, for many neurons, is not exactly the same. Thus there are binocular cortical neurons whose receptive fields, or their component subfields, occupy "corresponding" positions in the two retinas and others whose fields occupy different or "disparate" positions. The significance of this phenomenon for the neural mechanisms of stereoscopic vision will be discussed in a later section.

During normal binocular vision the response behavior of a binocular neuron is dynamically determined by the spatiotemporal characteristics of the retinal images and of the neuron's receptive fields in the two eyes actively directed in space. Response facilitation or summation follows stimulation of functionally homologous regions of the two receptive fields, simultaneously or in appropriate temporal sequence, and response suppression or inhibition is obtained by stimulation of functionally opponent regions. Although the influences of the "dominant" eye are often evident, the binocular response may also be determined by input from the "weak" eye.

FUNCTIONAL ORGANIZATION OF THE STRIATE CORTEX

The general architectural plan of the cerebral cortex (Chapter 9) is based on systems of discrete vertical "columns" perpendicular to the cytoarchitectonic layers from cortical surface to white matter. The columns are regarded as information-processing modules and are identified by common functional properties of the constituent neurons. For some columnar systems the anatomic substrate of the column has also been recognized.

In the striate cortex of cat and monkey, two vertical column systems have been identified and their architectural organization elegantly reconstructed in a series of electrophysiologic and anatomic studies by Hubel and Wiesel and their collaborators.[69,71] The elements of these two systems are the *ocular dominance column* and the *orientation column*. The origin of the column is in input layer IVc, and from there the column extends and develops vertically above and below that layer.

The ocular dominance column system is set up by the arrangement of the geniculostriate pathway: in the macaque monkey, LGN projection fibers functionally related to the right or left eye have segregated terminal sites, about 400 μm wide, in layer IVc of the striate cortex (Fig. 18-3). Nearly all cells in this layer are monocular. Above and below layer IVc the vast majority of cells receive inputs from both eyes, but the excitatory dominance of one or the other eye extends to all cortical layers. Thus throughout the striate cortex there exists an orderly alternation of vertical slabs or columns of neurons each preferentially excited by the right or left eye. Within each eye dominance column, however, the inputs from the two eyes converge on single cells, their interaction leading to the specific binocular response properties of these cortical neurons.

The system of orientation columns is finer and more detailed. These vertical elements, estimated width 50 μm, are precisely arranged so that the full 180 degrees of possible orientations are represented in sequences of columns across the cortex, the orientation sensitivity shifting some 10 degrees from one column to the next.

Hubel and Wiesel[69] have designated as a "hypercolumn" a complete set of adjacent columns of each system: either a set of columns representing the right and left eyes or a set of columns subserving 180 degrees of orientation. A schematic block of cortex containing hypercolumns of each type is shown in Fig. 18-23. For simplici-

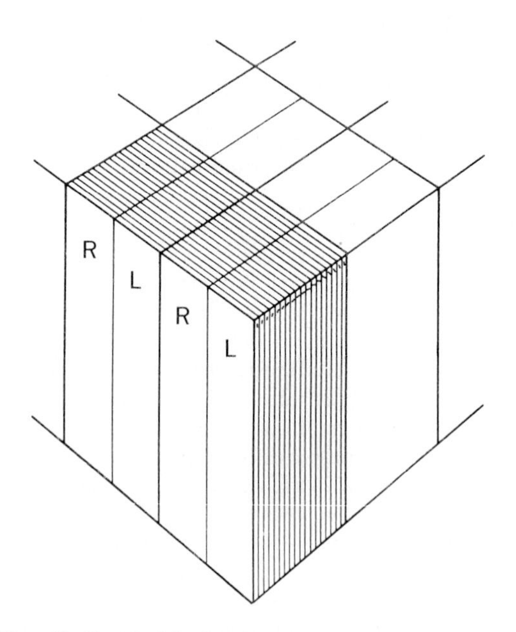

Fig. 18-23. Model of striate cortex to show roughly the dimensions of ocular dominance slabs *(L, R)* in relation to orientation slabs and cortical thickness. Thinner lines separate individual columns; thicker lines demarcate hypercolumns, two pairs of ocular dominance columns, and two sets of orientation columns. (From Hubel and Wiesel.[69])

ty the two vertical systems are drawn at 90 degrees to each other, although no evidence exists as to whether this is their actual arrangement or whether the systems intersect at some other angle. Dominance hypercolumns and orientation hypercolumns have similar and constant widths; a right-eye/left-eye pair of dominance columns occupies about 1 mm across the cortex, and so does the set of orientation columns covering 180 degrees. Thus throughout the binocular part of the striate cortex there exist essentially identical units, each capable of analyzing a region of visual field for ". . . light-dark contours in all orientation and with both eyes."[69] Because of the much larger "cortical magnification" of central than peripheral retina (Fig. 18-6) a greater number of functional units is available for the detailed analysis of the central visual field than for the coarser resolution of the periphery.

Within the visuotopic cortical map, many stimulus variables other than orientation and binocularity must be analyzed. How this is done is not known at this time, but the evidence just described suggests that the information may be processed by discrete and vertically organized cortical modules.

BINOCULAR VISION AND DISCRIMINATION OF DEPTH

Psychophysics

During the course of evolution there occurs a shift in the position of the two eyes from a symmetric location on opposite sides of the head to a frontal binocular position. This allows the development in primates of exquisite manual skills executed under visual guidance. Most important, in binocular vision there emerges a new and vivid perception of depth not possible in monocular vision.*

Binocular visual space comprises a solid angle determined by the overlapping part common to the two monocular fields. Within this binocular field the two eyes are stimulated from a single point in space at the same time, but not necessarily at the same position on each retina. Because of the horizontal separation between the two eyes, the two-dimensional images of a tridimensional object occupy slightly different horizontal positions on right and left retinas (binocular parallax). This horizontal shift between the two images, or *retinal image disparity,* is the essential cue for binocular depth discrimination or stereopsis.

When the two eyes are directed simultaneously to a given fixation point, the images of that point fall on the foveas of left and right retinas (Fig. 18-24—point F and its retinal projections f_1 and f_2). The two images are combined, or "fused," by central neural mechanisms, and the point is seen as single and localized "in one and the same visual direction, no matter whether the stimulus reaches the retinal elements in one eye alone, its corresponding partner in the other eye alone, or both simultaneously."[23] By definition, retinal elements that give rise to a localization in the same subjective visual direction when stimulated are said to be *corresponding retinal elements* in that one may be suppressed without affecting the result.

For a given position of the eyes there exists in space a locus of points, about the fixation point, whose images fall on corresponding retinal elements and thus are seen as single and localized in the same visual direction. This locus of points is called the *horopter*. Its simplest geometric con-

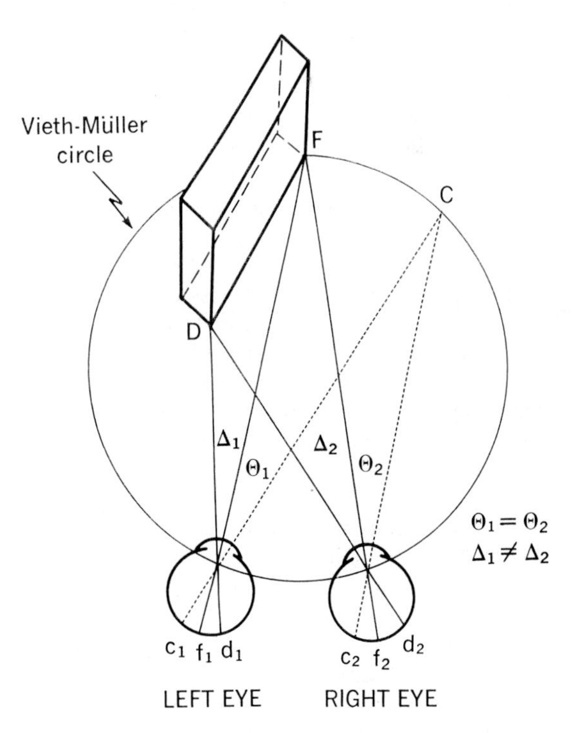

Fig. 18-24. Diagram showing geometric construct of corresponding (c_1, c_2) and disparate (d_1, d_2) retinal images and Vieth-Müller horopter circle.

*When we see the world with one eye, the sense of the third dimension can be added to the visual image by indirect cues such as movements of the head to produce successive images of the same object on different retinal elements (monocular parallax), knowledge of the angular subtense of the object, and shadow effects on distant objects by nearer ones. The process by which depth is perceived in monocular vision is quite different from that operating in binocular depth perception. Its nature is not as yet fully understood.

struct is a circle that passes through the fixation point and the optical centers of the two eyes: the Vieth-Müller circle (Fig. 18-24—point C and its retinal projections c_1 and c_2: $\Theta_1 = \Theta_2$). Experimental determinations, however, have shown that the horopter does not coincide with the circle.[5] For a fixation distance of approximately 1 m the empirical horopter is concave toward the observer and lies between the Vieth-Müller circle and the frontoparallel plane through the fixation point. With increasing viewing distances the shape of the horopter changes from concave to convex, but there is evidence that the set of corresponding retinal elements remains the same.[75]

Points in the visual world farther or nearer to the observer than the special position just described project optically to different horizontal positions in the two eyes (Fig. 18-24—point D and its retinal projections d_1 and d_2: $\Delta_1 = \Delta_2$). The retinal elements stimulated are noncorresponding or disparate retinal elements, and the images of that point, being localized in separate directions, may be seen as double. Horizontal retinal disparity is the cue for binocular depth discrimination. Vertical disparities may also occur, determined by such factors as slight differences in the elevation of the eyes or magnification differences between the two eyes for close objects. Vertical disparity does not contribute to the stereoscopic depth experience.

When we fixate on some point in space, we are not ordinarily aware that most of the objects in the visual world can be seen as double. In fact, there exists only a restricted spatial region in front of and behind the horopter plane within which binocular single vision is obligatory. Clearly, optical correspondence of retinal images is not a prerequisite for singleness of vision, and Panum made the suggestion that each point on the retina should be regarded as corresponding not with a point but with an area in the retina of the other eye (see Boring[21]). As long as the retinal image disparities are within the limits of this area, the images in the two eyes are combined in the brain and seen as a fused, single image in depth. This retinal area, and by extension the corresponding region in space, is called *Panum's fusional area:* it is smallest at the center of the fovea and increases toward the retinal periphery. Outside the limits of Panum's area, vision is double (Fig. 18-25).

The experience of depth in binocular vision is a vivid and independent sensation (much as the sensation of color and brightness) by which the tridimensional quality of the visual world

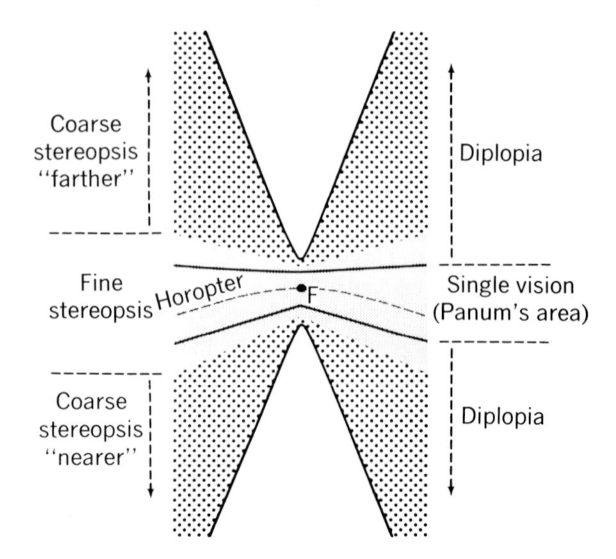

Fig. 18-25. Schematic representation of relative position of horopter and of overlapping regions of binocular single vision and of fine stereopsis as well as of regions of double vision and coarse stereopsis. *F*, Point of fixation.

emerges in a striking manner. The stereoscopic experience is one of relative depth only: one object is perceived as farther or nearer than another in an ordinal scale of depth. Stereopsis alone does not provide for the sensation of absolute distance of objects from the observer.

With small retinal disparities (e.g., those occurring within Panum's area as well as in a limited region in front of and behind it) the experience of stereoscopic depth is evident and compelling, and within this range the subjective depth is correlated with the magnitude of the disparity.[5] Clearly, fusion of disparate images (single vision) is not a prerequisite for binocular depth perception, and it is still possible to localize objects correctly even though their images have retinal disparities so large that the objects may appear double (diplopia). On either side of this narrow region of *fine stereopsis,* depth may still be experienced but only as a less specific and purely qualitative sensation of "nearer" and "farther," a *coarse stereopsis.*[5,18,100] With increasingly large image disparities the sense of depth gradually fades and eventually disappears (Fig. 18-25).

Stereoscopic acuity, that is, the ability to discriminate differences in depth, is comparable in sensitivity to the finest monocular visual acuity, for example, the ability to recognize two line segments as not aligned if displaced by only a few seconds of arc (Chapter 16). Considering that binocular depth perception involves the ocu-

Fig. 18-26. Random-dot stereo pair. When viewed monocularly, two fields appear as random sets of dots without recognizable features. However, when stereoscopically combined (e.g., by crossing eyes), there is a vivid depth impression of a diamond suspended over the background. (From Julesz.[3])

lomotor cooperation of the two eyes, this fact is indeed proof of the remarkable capacities of the visual apparatus. Stereoscopic acuity is maximal in the vicinity of the horopter and declines rapidly with departure from it. At the limit of stereoscopic acuity, it is possible to discriminate a difference in depth as small as 1.5 mm between two objects placed a distance of 1 m from the eyes.

The extensive investigations by Julesz[3] have provided conclusive evidence that retinal image disparity is the unique visual determinant of binocular depth perception. Julesz has used computer-generated random dot patterns as stereo pairs, an example of which is given in Fig. 18-26. The right and left images are identical random dot patterns, except for certain areas that, although identical, are shifted relative to each other in the horizontal direction by an integral number of dots. When viewed monocularly, the two figures appear as homogeneously random without recognizable features. When binocularly combined, there emerges a vivid depth impression of a pattern suspended above the background.

These results indicate that stereoscopic depth may be experienced without any monocular or binocular cues to depth except horizontal retinal image disparity and that the neural integration of dissimilarities in the images seen by the two eyes is established without monocular recognition of any pattern or form. Morever, these findings suggest that the neural mechanisms operate on the basis of a point-to-point analysis, since line contours of any appreciable length are not required for the stereoscopic experience.

Neural mechanisms

Our understanding of the central neural mechanisms operating in binocular vision rests on considerations of the anatomic organization of the visual afferent system and of the functional properties of its constituent neurons. The decussation of the optic fibers at the chiasm brings together in the same hemisphere the central projections of the two homolateral hemiretinas, and within this pathway the representations of topographically corresponding retinal areas are kept closely associated in a precisely organized retinotopic pattern. The inputs from the two eyes pass through the LGN with minimal or no interaction and come together for the first time at the striate cortex. It is at the cortical level that during the past several years the neurophysiologic mechanisms of binocular vision have been actively investigated with the method of single-unit analysis.*

Cortical visual neurons are characteristically sensitive to spatial patterns and respond best to contrast stimuli of particular size, orientation, direction of motion, etc. For most neurons there exist two receptive fields, one in each eye, with the same or somewhat different organization (p. 560). At any given site in the retinotopic cortical map there are neurons whose fields have exactly corresponding positions in the two eyes and

*See references 11, 26, 68, 75, 91, and 95.

neurons whose fields have slightly different positions in the horizontal and vertical directions *(receptive field disparity)*. Whereas vertical disparities do not contribute to stereopsis, horizontal receptive field disparities most likely do by providing the conditions for the simultaneous stimulation of a binocular neuron's two receptive fields by the same object at a specific position in depth.

The significance of positional receptive field disparity in the neural mechanisms underlaying binocular depth discrimination may be appreciated from Fig. 18-27. The schematic drawing shows a horizontal plane cutting through the two eyes and including the region of space in front of them. Three pairs of binocular receptive fields, one for each of the three representative cortical neurons (A, B, and C), are outlined in right and left retina, and each is marked with a different pattern. To simplify the description, neurons are chosen whose left-eye members of the receptive field pairs (a, b, c) all have the same monocular visual direction and thus fall on each other onto the retinal surface. The members of the receptive field pairs in the right eye, on the other hand, are not in register and occupy different horizontal positions (a′, b′, c′). The left-eye and the right-eye receptive fields of neurons B (b and b′) occupy exactly corresponding retinal locations (disparity = 0), and thus the two fields superimpose in space at the horopter plane. The right and left fields of neurons A and C, on the other hand, have different separations. Accordingly, receptive fields a and a′ superimpose in front of it. The two fields of neuron A are said to have divergent disparity and those of neuron C convergent disparity.

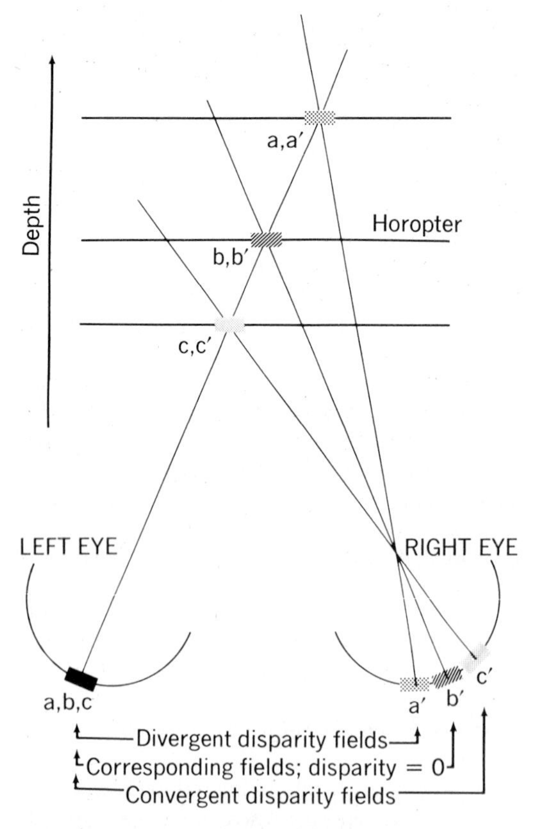

Fig. 18-27. Diagram illustrating how horizontal receptive field disparity might form basis of neural mechanisms for binocular depth discrimination. (See text.)

Although not explicitly shown in Fig. 18-27, it is clear that in front and behind the plane of receptive field superposition there exists for each binocular neuron a series of planes of receptive fields overlap. The range of depths over which overlap occurs is the range of spatial binocular interaction. The binocular responses of a cortical neuron over this range constitute the neuron's stereoscopic depth response profile, whose characteristics are thought to reflect the neuron's functional significance in mechanisms of depth perception.

Direct evidence for depth-sensitive neurons in the visual cortex, both in striate and prestriate cortex, has been obtained in the alert and visually active macaque.[95] In these animals, known to possess excellent stereopsis, the majority of cells in areas 17 and 18 subserving central vision respond differentially to stimuli positioned at various depths, in front of, behind, or at the plane of fixation.

The most common type of depth-sensitive cortical neuron is the *tuned excitatory* (Fig. 18-28). Cells of this type give similar responses to monocular stimulation of either eye and display binocular facilitation over a narrow excitatory region about the fixation distance and binocular suppression at closer or farther distances. Less frequently observed is the type termed *tuned inhibitory*. These cells give strong responses to monocular stimulation of one of the two eyes, the "dominant" eye. Binocular stimulation evokes similar strong responses at all distances, except at distances about the fixation distance, where the interaction between the inputs from the two eyes results in a reduction or suppression of the cell's activity.

Two other binocular neuron populations reciprocally organized for depth occur in monkey visual cortex: the *far* and the *near* neurons (Figs. 18-29 and 18-30). These cells, named for the region of space over which excitatory responses are obtained, usually display strong eye dominance.

Binocularly, they respond in a reciprocal manner to stimuli positioned nearer than and to stimuli positioned farther than the fixation distance and often over a relatively large range of distances (i.e., 1 degree of convergent or divergent disparity).

For the region of central vision within 2 degrees of fixation, about half of the depth-tuned neurons are maximally influenced binocularly by stimuli at the fixation distance (disparity = 0). The other tuned cells instead respond best to stimuli at other distances, either in front of or behind fixation over a narrow range of depths (±0.25 degree of disparity). The depth-tuned neurons may provide the substrate for central fusion of slightly disparate retinal images and ensure a single cortical replica of the narrow region of space in depth that includes the current plane of fixation and its immediate nearer and farther neighborhood (Panum's fusional region of single vision). The sharpness of tuning and the depth distribution of peak sensitivities possessed

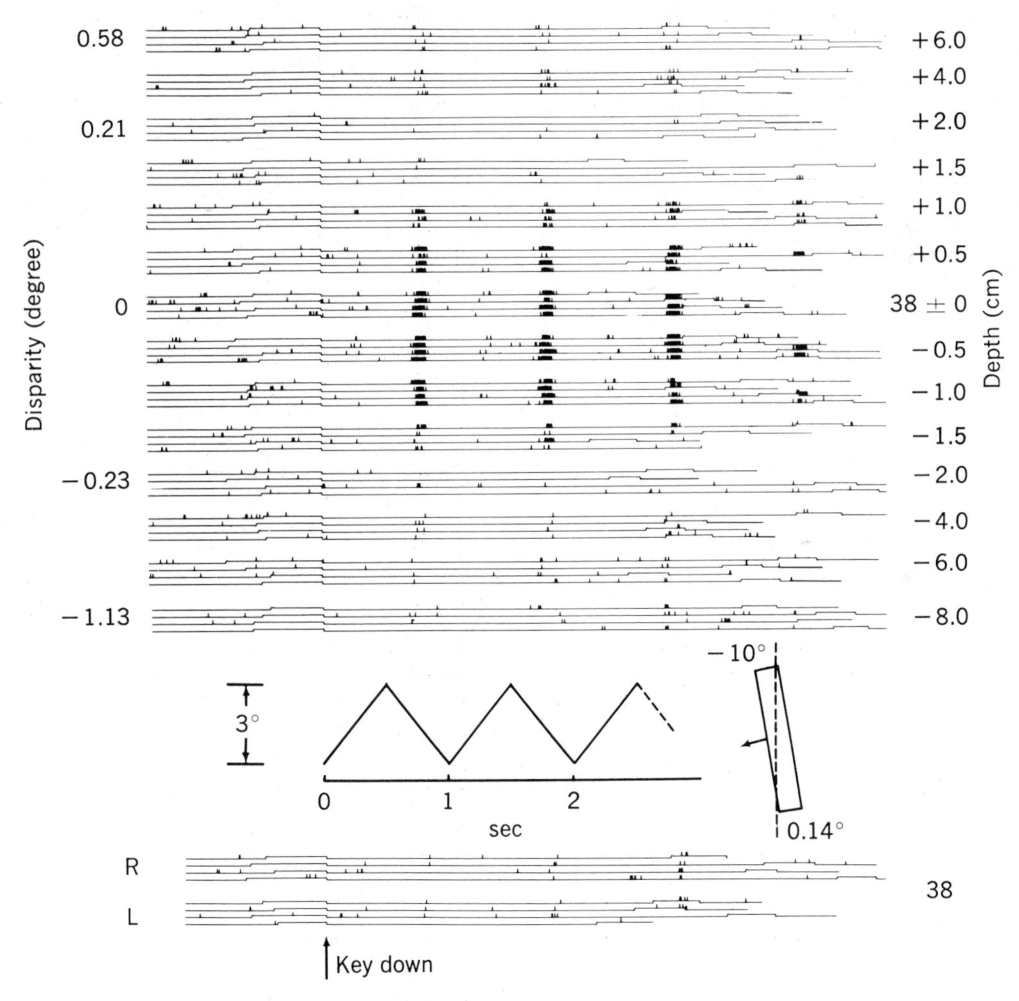

Fig. 18-28. Replicas of impulse discharges of depth-tuned excitatory neuron (area 17, unidirectional). Neural activity recorded in alert macaque during periods of behaviorally controlled binocular fixation of gaze beginning at "*key down.*" Amplitude and time course of stimulation, as well as orientation and width of bright bar stimulus, are shown in figure between depth series of binocular responses (above) and two monocular responses (below). Small arrow indicates initial direction of movement. Depth is given at right as deviation (in centimeters) from fixation distance (38 cm); corresponding horizontal retinal image disparities (in degrees) are shown at left. Right eye *(R)* and left eye *(L)* monocular responses are poor and about equal. Binocular facilitation occurs over a range of approximately 2.5 cm; inhibitory sidebands nearer and farther are evident. (From Poggio and Fischer.[95])

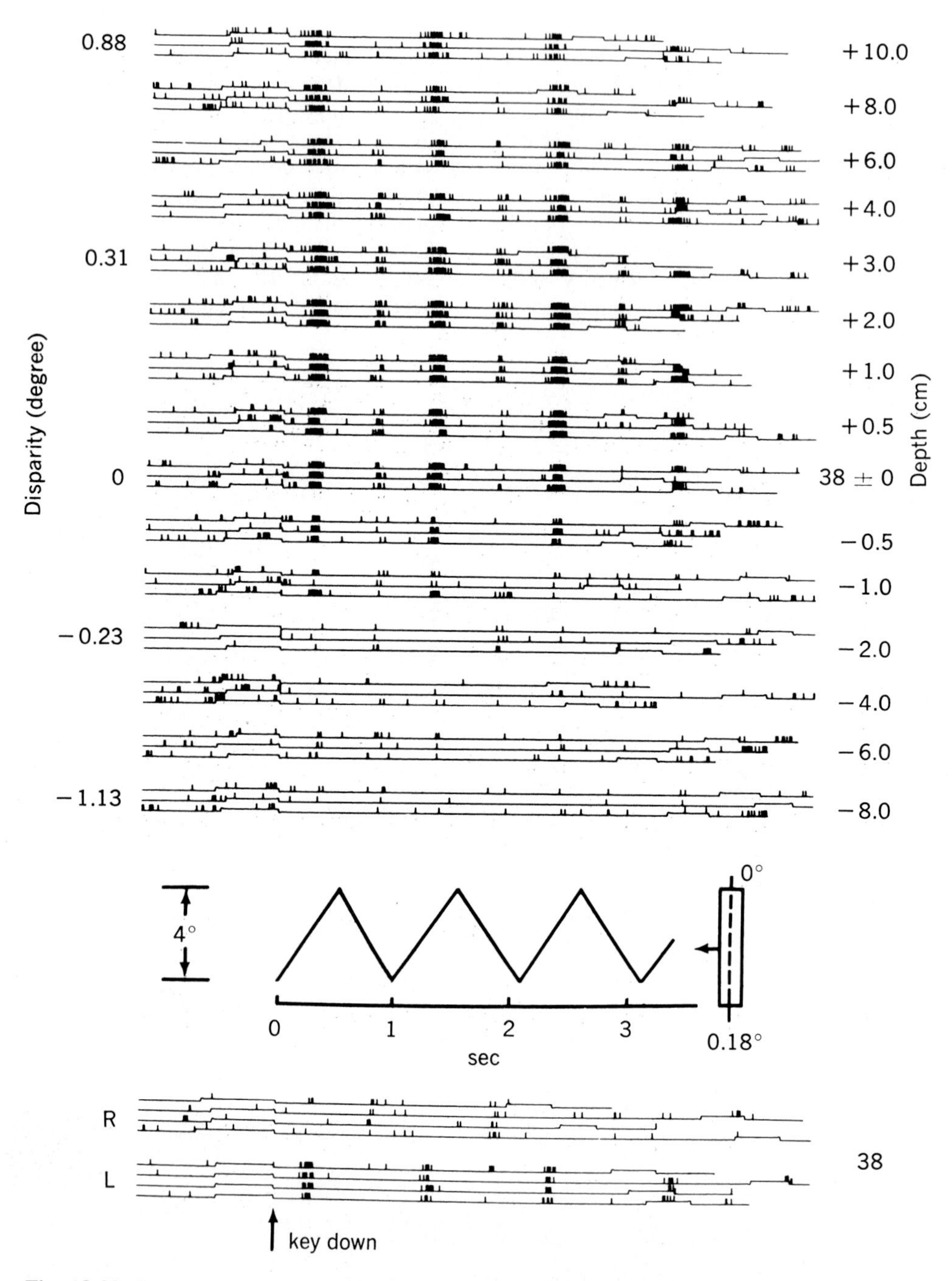

Fig. 18-29. Responses of a far neuron at different depths (area 18) in alert and behaving monkey. Labels and conventions as in Fig. 18-28. Left eye monocular dominance and opposite directional selectivity in two eyes. Binocularly, neuron responded best to stimuli about and behind plane of fixation; at distances nearer than fixation, responses were suppressed over wide range. (From Poggio and Fischer.[95])

by the tuned excitatory cells could provide the basis for neuronal mechanisms leading to the three-dimensional perception of objects with high stereoacuity (fine stereopsis).

The other two types of depth cells observed in monkey's visual cortex (far/near) respond differentially over larger depth ranges to binocular stimuli nearer than and farther than the fixation distance. These neurons may be part of a more primitive stereo system of two reciprocally organized binocular neuron populations whose activity may provide information leading to

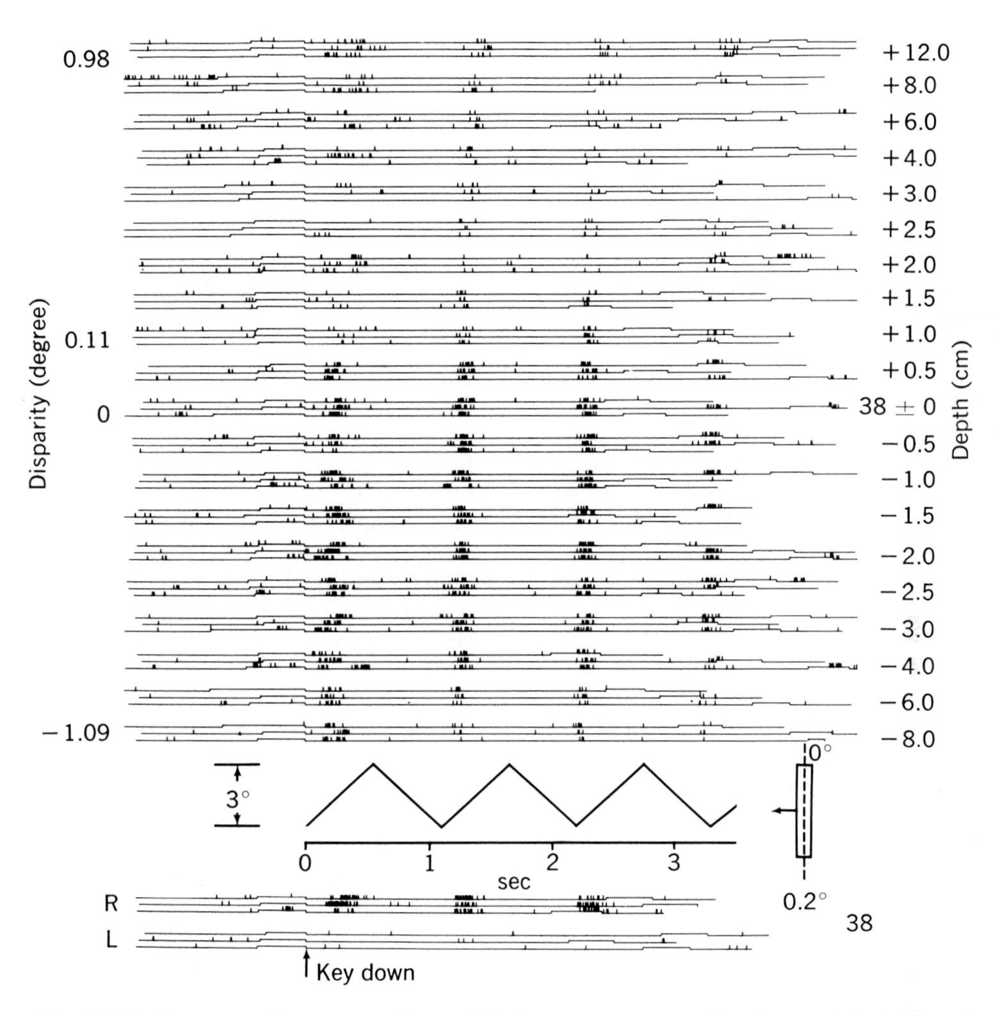

Fig. 18-30. Responses of near neuron in area 18 of macaque cortex to stimuli presented at different depths. Labels and conventions as in Fig. 18-28. Unidirectional cell, right eye monocular dominance. Binocularly, excitatory responses to stimuli at and nearer than fixation distance, and suppression of activity over range of farther distances. (From Poggio and Fischer.[95])

qualitative depth estimates in the presence of double vision (coarse stereopsis).

The experimental evidence provided by single-neuron studies suggests that under conditions of normal binocular vision, a tridimensional object in space, imaged two dimensionally in slightly different positions onto the two retinas and translated in neural code, is "reconstructed" in the visual cortex by the profile of activity in the populations of neurons topographically related to that part of the visual field in which the object is located. Because of the scatter in the horizontal separation of the receptive fields in the two eyes, different neurons are activated by objects at different distances. Moreover, functionally distinctive types of cells subserve the two aspects of depth perception recognized psychophysically: one type of neurons processes information leading to the single vision and high stereoacuity in the vicinity of the fixation plane; the other type contributes to the diplopic estimates of "nearer" and "farther." Finally, the proposition may be entertained that disparity-sensitive neurons may be elements of cortical mechanisms for the control of disjunctive eye movements (Chapter 16). Because of the wide range of depth sensitivity, near and far neurons may participate in the initiation of vergence movements, whereas tuned excitatory neurons have properties that are appropriate for the completion of vergence and maintenance of binocular fixation.[126]

RETINOMESENCEPHALIC PROJECTION AND ROLE OF THE MIDBRAIN IN VISION

Visual information is forwarded from the retina to central structures along two separate pathways: the retinogeniculate projection described previously and the retinomesencephalic projection. Fibers of retinal origin terminate bilaterally in the midbrain, chiefly in the superior colliculus and pretectal region. Input to the superior colliculus is also provided by a major projection from the ipsilateral visual cortex. From the tectum, descending connections exist to the reticular formation and other structures of the brain stem, and ascending projections extend to the thalamus, the LGN, and, in particular, the pulvinar, which is in turn reciprocally connected with the prestriate and inferotemporal cortices.[16,59]

The primary visual projections and its forward extension, on the one hand, and the extrageniculate visual pathways, on the other, constitute two interconnected systems that are thought to subserve different visual functions. The former provides the neural mechanisms for the capacity to discriminate and recognize objects on the basis of their shape and motion; the latter subserves the ability to orient visually to these objects, shifting the eyes to bring and to maintain the central retina (fovea) on these objects and to relate movements of head and body to their location in space.[60,122] The extent of functional dissociation between the two systems is not clearly defined. Although there is little doubt that in higher mammals the midbrain structures play a fundamental role in visual orientation and the visual cortex in form discrimination, there is also evidence that in monkeys, lesions of the superior colliculus severely mar pattern vision[45] and that in the cat, the "midbrain-pulvinar-cortical pathways provide the first stage in simple, coarse form perception and discrimination."[119]

Anatomic and electrophysiologic findings have shown that in the tectum of all vertebrates there exists an orderly topographic representation of the visual field. Homonymous halves of the field of view of the two eyes are represented superimposed in the contralateral colliculus. Upper and lower field quadrants are represented in medial and lateral parts of the tectum, respectively, and are separated by the replication of the horizontal meridian. The central region of the field is represented rostrally and the periphery caudally. As in the geniculostriate pathways, the map of the visual field is distorted, with the central 20 degrees occupying about two thirds of the colliculus and the remaining 70 degrees being compressed in the posterior third. Axons of retinal ganglion cells terminate in the superficial gray layer of the colliculus. In the monkey the projection from peripheral retina is dense and more continuous, whereas that from central retina is thin and sparse.[70,128] The majority of retinotectal cells have nonconcentric receptive fields, slowly conducting axons, and broad-band spectral sensitivity.[109] The projection from the ipsilateral cortex has an uniformly dense distribution throughout the superficial layers of the colliculus, and within this region the fibers from the frontal eye field (area 8) also terminate.[79] Retinotectal and striatotectal projections are topographically corresponding.

The functional importance of the superior colliculus in the control of eye and head movements, and more generally as the center of visuomotor coordination, has been shown by observations of the effects of lesions on the animal's behavior, by the results of stimulation experiments, and by analysis of the properties of collicular neurons.

After unilateral removal of the colliculus the monkey can look up and down or to either side and reach out to a moving object with either hand, but it maintains an expressionless facies and appears unconcerned with visual events in the hemifield contralateral to the lesion. In that field the animal cannot fixate but can only look approximately in the direction of an object. Bilateral destruction of the superior colliculus results in profound disturbances of vision and general behavior. The monkey stares fixidly into space, shows no orientation to visual events, and spends most of the time silently in a corner of the cage. Some recovery of vision for moving objects may occur, but no vision for still objects is ever regained. Bilateral collicular removal and striate cortex ablation have very similar effects on the monkey visual behavior.[45]

Single-unit analysis of collicular neurons in the alert monkey has revealed the existence of two major types: neurons that respond to visual stimuli only and neurons that discharge in relation to eye movements and generally to visual input as well. Neurons responding exclusively to retinal stimulation are located in superficial layers of the colliculus. Usually activated equally well from both eyes, these neurons respond best to small, luminous moving patterns and display little selectivity for shape, orientation, and direction of motion of stimuli. Receptive field centers are less than 1 degree in diameter at the fovea and 10 to 30 degrees or larger in the periphery.[52,107] These cells are sensitive to nonspecific "events" appearing in the field of vision.

Cells in the deeper collicular layers discharge in association with saccadic eye movements of specific directions and sizes. Cells in the right colliculus are associated with saccades toward the left and neurons in the left colliculus with movements to the right. In all instances the neural discharge *precedes* the eye movement by 50 to 200 msec; these cells are likely to operate in the initiation of oculomotor action. Moreover, the majority of these neurons are also influenced by visual stimuli, and the receptive field of each neuron is typically located in that area of the visual field to which the eyes move as the result of the associated saccade.[107,111]

Electrical stimulation of the colliculus in unrestrained animals evokes orienting movements of eyes, head, and body.[14] In particular, collicular stimulation in the alert monkey evokes conjugate saccades that, irrespective of where the eye is in the orbit, causes it to move to that region of the visual field topographically represented in the stimulated part of the colliculus (Fig. 18-31).[36,101,111]

These experimental findings have elucidated important aspects of the organization of the tectum in visuomotor activity. In the intermediate and deep layers of the colliculus, most neurons are functionally associated with eye movements and discharge *before* the initiation of specific saccades. Within these layers there exists an orderly eye movement map that is visuotopically in register with the sensory map present in the superficial layers. Each superior colliculus processes topographically organized input from the contralateral hemifield of vision, and its neurons send signals to the motoneurons for the eye muscles as well as to those for the head and trunk that may activate a sequence of orienting movements. These signals, however, do not reach the acting motoneurons directly. The descending connections from deep tectal layers terminate within a region of the brain stem known to pro-

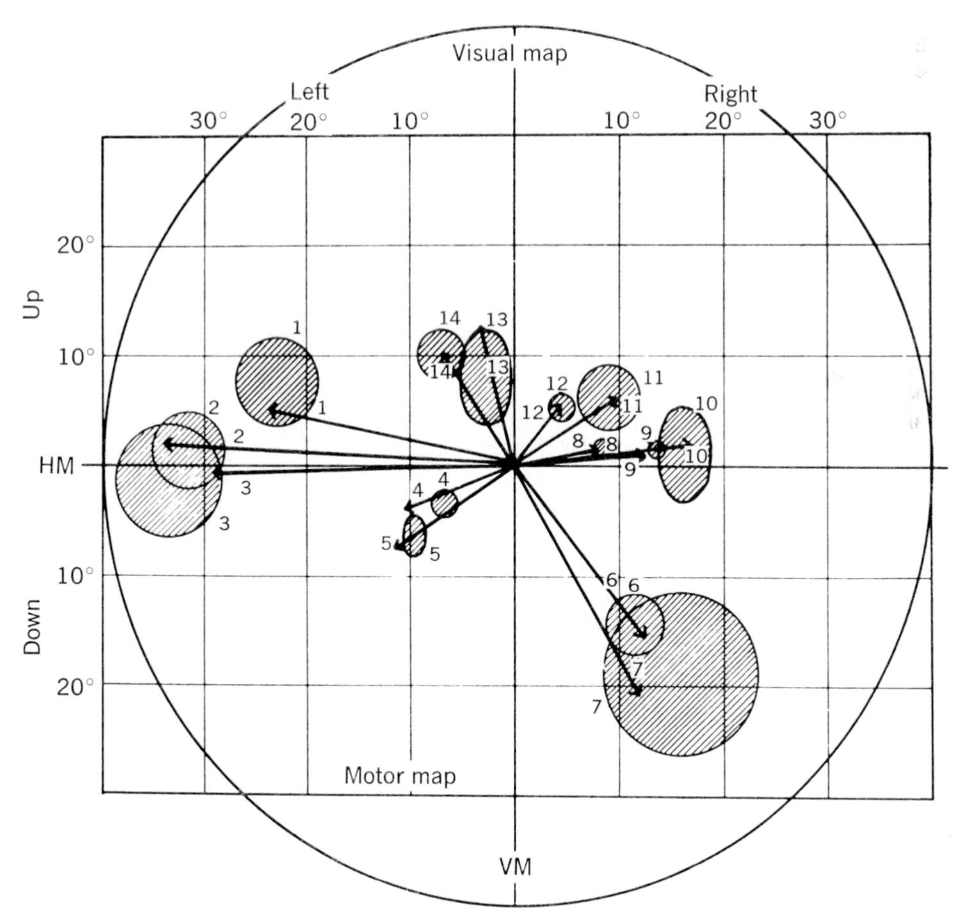

Fig. 18-31. Recording and stimulation in superficial layers of superior colliculus of rhesus monkey. *Visual map* with receptive fields of 14 cells is superimposed on *motor map* with its arrows representing electrically elicited saccades at each of 14 sites. Length of each arrow represents mean length of 8 to 14 stimulation-elicited saccades; direction of each arrow represents mean direction of saccades. *HM*, Horizontal meridian; *VM*, vertical meridian. (From Schiller and Stryker.[111])

ject to the cerebellum and spinal cord gray matter, but no fibers of collicular origin are known to reach oculomotor nuclei or motor nuclei for head and neck movement.[59]

In man and other primates, fixation of gaze is largely reflex in nature and depends on the striate cortex.[44] The visual cortex projects to the colliculus in an orderly topographic fashion, and electrical stimulation of the occipital cortex evokes eye and head movements similar to those elicited by tectal stimulation. The striatotectal projection arises from pyramidal cells in layer V that have receptive fields of complex properties, relatively large size, and poor orientation selectivity.[112] In layers V and VI of the foveal striate cortex, many neurons are "depth tuned" and respond selectively to binocular stimuli at or close to the plane of fixation.[95] Striate cortex ablation or cooling has little effect on the visual responses of neurons in superficial layers of the colliculus but disrupts those of deeper neurons; eye-movement cells still discharge prior to saccades, but they no longer have visual receptive fields.[113] Saccadic eye movements evoked by electrical stimulation of visual occipital cortex are abolished by collicular ablation.[106] On the other hand, neither striate cortex nor colliculus is essential for visual detection or visually guided saccades in the alert monkey.[88] These findings strongly suggest that in normal vision, cortex and tectum are functionally complementary. The corticotectal and retinomesencephalic projections interact at the colliculus to operate in the mechanisms for fixation of gaze, in the reflex shifts of the eye toward an object attracting the attention of the animal, and also in voluntary visuospatial adjustments.

REFERENCES

General reviews

1. Bishop, P. O.: Binocular vision. In Moses, R. A., editor: Adler's physiology of the eye: clinical application, ed. 6, St. Louis, 1975, The C. V. Mosby Co., p. 558.
2. Carterette, E. C., and Friedman, M. P.: Handbook of perception, New York, 1975, Academic Press, Inc., vol. 5.
3. Julesz, B.: Foundations of cyclopean perception, Chicago, 1971, University of Chicago Press, pp. xiv and 406.
4. LeGrand, Y.: Light, colour and vision, London, 1957, Chapman & Hall Ltd.
5. Ogle, K. N.: The optical space sense. In Davson, H., editor: The eye, New York, 1969, Academic Press, Inc., vol. 4.
6. Polyak, S. L.: The retina, Chicago, 1941, University of Chicago Press.
7. Polyak, S. L.: The vertebrate visual system, Chicago, 1957, University of Chicago Press.
8. Teuber, H. L., Battersby, W. S., and Bender, M. B.: Visual field defects after penetrating missile wounds of the brain, Cambridge, 1960, Harvard University Press.

Original papers

9. Baizer, J. S., Robinson, D. L., and Dow, B. M.: Visual responses of area 18 neurons in awake, behaving monkey, J. Neurophysiol. **40:**1024, 1977.
10. Baker, F. H., Riva Sanseverino, E., Lamarre, Y., and Poggio, G. F.: Excitatory responses of geniculate neurons of the cat, J. Neurophysiol. **32:**916, 1969.
11. Barlow, H. B. C., Blakemore, C., and Pettigrew, J. D.: The neural mechanisms of binocular depth discrimination, J. Physiol. **193:**327, 1967.
12. Barlow, R. B., Jr., Snodderly, D. M., and Swadlow, H. A.: Intensity coding in primate visual system, Exp. Brain Res. **31:**163, 1978.
13. Bartlett, J. R., and Doty, R. W., Sr.: Responses of units in striate cortex of squirrel monkeys to visual and electrical stimuli, J. Neurophysiol. **37:**621, 1974.
14. Bender, M. B., and Shanzer, S.: Oculomotor pathways defined by electric stimulation and lesions in the brainstem of monkey. In Bender, M. B., editor: The oculomotor system, New York, 1964, Harper and Row, Publishers, Inc., pp. 81-104.
15. Benevento, L. A., and Davis, B.: Topographical projections of the prestriate cortex to the pulvinar nuclei in the macaque monkey: an autoradiographic study, Exp. Brain Res. **30:**405, 1977.
16. Benevento, L. A., and Fallon, J. H.: The ascending projections of the superior colliculus in the rhesus monkey *(Macaca mulatta),* J. Comp. Neurol. **160:**339, 1975.
17. Benevento, L. A., and Rezak, M.: The cortical projections of the inferior pulvinar and adjacent lateral pulvinar in the rhesus monkey *(Macaca mulatta):* an autoradiographic study, Brain Res. **108:**1, 1976.
18. Bishop, P. O., and Henry, G. H.: Spatial vision, Annu. Rev. Psychol. **22:**119, 1971.
19. Bishop, P. O., Coombs, J. S., and Henry, G. H.: Responses to visual contours: spatiotemporal aspects of excitation in the receptive fields of simple striate neurones, J. Physiol. (Lond.) **219:**625, 1971.
20. Bizzi, E., and Schiller, P. H.: Single unit activity in the frontal eye fields of unanesthetized monkeys during eye and head movements, Exp. Brain Res. **10:**151, 1970.
21. Boring, E. G.: Sensation and perception in the history of experimental psychology, New York, 1942, Appleton-Century-Crofts.
22. Bunt, A. H., Minckler, D. S., and Johanson, G. W.: Demonstration of bilateral projection of the central retina of the monkey with horseradish peroxidase neuronography, J. Comp. Neurol. **171**(4):619, 1977.
23. Burian, H.: Sensorial retinal relationship in concomitant strabismus, Trans. Am. Ophthalmol. Soc. **43:**373, 1945.
24. Chow, K. L.: Anatomical and electrographical analysis of temporal neocortex in relation to visual discrimination learning in monkeys. In Delafresnave, J. F., editor: Brain mechanisms and learning, Oxford, 1961, Blackwell Scientific Publications Ltd.
25. Clark Le Gros, W. E.: The sorting principle in sensory analysis as illustrated by the visual pathways, Ann. R. Coll. Surg. Engl. **30:**299, 1962.
26. Clarke, P. G. H., Donaldson, I. M. L., and Whitteridge, D.: Binocular visual mechanisms in cortical areas I and II of the sheep, J. Physiol. (Lond.) **256:**509, 1976.

27. Cleland, B. G., Dubin, M. W., and Levick, W. R.: Sustained and transient neurones in the cat's retina and lateral geniculate nucleus, J. Physiol. (Lond.) **217:**473, 1971.
28. Cowey, A.: Projection of the retina on the striate and prestriate cortex in the squirrel monkey, *Saimiri sciureus,* J. Neurophysiol. **27:**366, 1964.
29. Cowey, A., and Ellis, C. M.: Visual acuity of rhesus and squirrel monkeys, J. Comp. Physiol. Psychol. **64:**80, 1967.
30. Cowey, A., and Ellis, C. M.: The cortical representation of the retina in squirrel and rhesus monkeys and its relation to visual acuity, Exp. Neurol. **24:**374, 1969.
31. Cowey, A., and Gross, C. G.: Effects of foveal prestriate and inferotemporal lesions on visual discrimination by rhesus monkeys, Exp. Brain Res. **11:**128, 1970.
32. Cowey, A., and Weiskrantz, L.: A perimetric study of visual field defects in monkeys, Q. J. Exp. Psychol. **15:**91, 1963.
33. Cowey, A., and Weiskrantz, L.: A comparison of the effects of inferotemporal and striate cortex lesions on the visual behaviour of rhesus monkeys, Q. J. Exp. Psychol. **19:**246, 1967.
34. Cragg, B. G.: The topography of the afferent projections in the circumstriate visual cortex of the monkey studied by the Nauta method, Vision Res. **9:**733, 1969.
35. Creutzfeldt, O. D., Kuhnt, U., and Benevento, L. A.: An intracellular analysis of visula cortical neurones to moving stimuli: responses in a co-operative neural network, Exp. Brain Res. **21:**251, 1974.
36. Cynader, M., and Berman, N.: Receptive-field organization of monkey superior colliculus, J. Neurophysiol. **35:**187, 1972.
37. Daniel, P. M., and Whitteridge, D.: The representation of the visul field on the cerebral cortex in monkeys, J. Physiol. **159:**203, 1961.
38. De Monasterio, F. M., and Gouras, P.: Functional properties of ganglion cells of the rhesus monkey retina, J. Physiol. (Lond.) **251:**167, 1975.
39. De Valois, R. L.: Behavioral electrophysiological studies of primate vision. In Neff, W. D., editor: Contributions to sensory physiology, New York, 1965, Academic Press, Inc.
40. De Valois, R. L.: Physiological basis of color vision. In Tagungsbericht Farbtagung, Color 69, Stockholm, 1969, pp. 29-47.
41. De Valois, R. L., and De Valois, K. K.: Neural coding of color. In Carterette, E. C., and Friedman, M. P.: Handbook of perception, New York, 1975, Academic Press, Inc., vol. 5, pp. 117-166.
42. De Valois, R. L., Abramov, I., and Jacobs, G. H.: Analysis of response patterns of LGN cells, J. Opt. Soc. Am. **56:**966, 1966.
43. De Valois, R. L., et al.: Psychophysical studies of monkey vision. I. Macaque luminosity and color vision tests, Vision Res. **14:**53, 1974.
44. Denny-Brown, D., and Chambers, R. A.: Physiological aspects of visual perception. I. Functional aspects of visual cortex, Arch. Neurol. **33:**219, 1976.
45. Denny-Brown, D., and Fischer, E. G.: Physiological aspects of visual perception, II. The subcortical visual direction of behavior, Arch. Neurol. **33:**228, 1976.
46. Dreher, B., Fukada, Y., and Rodieck, R. W.: Identification classification and anatomical segregation of cells with X-like and Y-like properties in the lateral geniculate nucleus of old-world primates, J. Physiol. (Lond.) **258:**433, 1976.
47. Duke-Elder, S.: A textbook of ophthalmology, St. Louis, 1952, The C. V. Mosby Co., vol. 1.
48. Enroth-Cugell, C., and Robson, J. G.: The contrast sensitivity of retinal ganglion cells of the cat, J. Physiol. (Lond.) **187:**517, 1966.
49. Ettlinger, G.: Visual discrimination with a single manipulandum following temporal ablations in the monkey, Q. J. Exp. Psychol. **3:**164, 1959.
50. Fisken, R. A., Garey, L. J., and Powell, T. P. S.: The intrinsic, association and commissural connections of area 17 of the visual cortex, Philos. Trans. R. Soc. Lond. (Biol.) **272**(919):487, 1975.
51. Garey, L. J., and Powell, T. P. S.: An experimental study of the termination of the lateral geniculo-cortical pathway in the cat and monkey, Proc. R. Soc. Lond. (Biol.) **179:**41, 1971.
52. Goldberg, M. E., and Wurtz, R. H.: Activity of superior colliculus in behaving monkey. I. Visual receptive fields of single neurons, J. Neurophysiol. **35:**542, 1972.
53. Gouras, P.: Opponent-colour cells in different layers of foveal striate cortex, J. Physiol. (Lond.) **238:**583, 1974.
54. Gross, C.: Inferotemporal cortex and vision. In Stellar, E., and Sprague J. M., editors: Progress in physiology psychology, New York, 1973, Academic Press, Inc., vol. 5, pp. 77-123.
55. Guillery, R. W., and Colonnier, M.: Synaptic patterns in the dorsal lateral geniculate nucleus of the monkey, Z. Zellforsch. **103:**90, 1970.
56. Halstead, W. C., Walker, A. E., and Bucy, P. C.: Sparing and nonsparing of "macular" vision associated with occipital lobectomy in man, Arch. Ophthalmol. **24:**948, 1940.
57. Hammond, P., and MacKay, D. M.: Differential responsiveness of simple and complex cells in cat striate cortex to visual texture, Exp. Brain Res. **30:**275, 1977.
58. Harrington, D. O.: The visual fields: a textbook and atlas of clinical perimetry, ed. 4, St. Louis, 1976, The C. V. Mosby Co.
59. Harting, J. K.: Descending pathways from the superior colliculus: an autoradiographic analysis in the rhesus moneky *(Macaca mulatta),* J. Comp. Neurol. **173**(3):583, 1977.
60. Held, R.: Dissociation of visual functions by deprivation and rearrangement, Psychol. Forsch. **31:**338, 1968.
61. Hoffmann, K. P., and Stone, J.: Conduction velocity of afferents to cat visual cortex: a correlation with cortical receptive field properties, Brain Res. **32:**460, 1971.
62. Holmes, G.: The organization of the visual cortex in man, Proc. R. Soc. Lond. (Biol.) **132:**349, 1945.
63. Hoyt, W. F., and Luis, O.: The primate chiasm: details of visual fiber organization studied by silver impregnation techniques, Arch. Ophthalmol. **70:**69, 1963.
64. Hubel, D. H., and Wiesel, T. N.: Receptive fields of optic nerve fibers in the spider monkey, J. Physiol. (Lond.) **154:**572, 1960.
65. Hubel, D. H., and Wiesel, T. N.: Receptive fields, binocular interaction and functional architecture in the cat's visual cortex, J. Physiol. (Lond.) **160:**106, 1962.
66. Hubel, D. H., and Wiesel, T. N.: Receptive fields and functional architecture in two non-striate visual areas (18 and 19) of the cat, J. Neurophysiol. **28:**229, 1965.
67. Hubel, D. H., and Wiesel, T. N.: Receptive fields and functional architecture of monkey striate cortex, J. Physiol. (Lond.) **195:**215, 1968.
68. Hubel, D. N., and Wiesel, T. N.: Cells sensitive to

binocular depth in area 18 of the macaque monkey cortex, Nature **225:**41, 1970.
69. Hubel, D. H., and Wiesel, T. N.: Functional architecture of macaque monkey visual cortex, Proc. R. Soc. Lond. (Biol.) **198:**1, 1977.
70. Hubel, D. H., LeVay, S., and Wiesel, T. N.: Mode of termination of retinotectal fibers in macaque monkey: an autoradiographic study, Brain Res. **96:**25, 1975.
71. Hubel, D. H., Wiesel, T. N., and Stryker, M. P.: Anatomical demonstration of orientation columns in macaque monkey, J. Comp. Neurol. **177:**361, 1978.
72. Jacobs, G. H.: Effects of adaptation on the lateral geniculate response to light increment and decrement, J. Opt. Soc. Am. **55:**1535, 1965.
73. Jones, E. G.: The anatomy of extrageniculostriate visual mechanisms. In Schmitt, F. O., and Worden, F. G. editors: The neurosciences, third study program, Cambridge, Mass., 1974, The M.I.T. Press, pp. 215-227.
74. Jones, E. G., and Powell, T. P. S.: An anatomical study of converging sensory pathways within the cerebral cortex of the monkey, Brain **93:**793, 1970.
75. Joshua, D. E., and Bishop, P. O.: Binocular single vision and depth discrimination: receptive field disparities for central and peripheral vision and binocular interaction on peripheral single units in cat striate cortex, Exp. Brain Res. **10:**389, 1970.
76. Klüver, H.: Functional significance of the geniculostriate system, Biol. Symp. **7:**253, 1942.
77. Klüver, H., and Bucy, P. C.: An analysis of certain effects of bilateral temporal lobectomy in the rhesus monkey, with special reference to "psychic blindness," J. Psychol. **5:**33, 1938.
78. Krüger, J., Fischer, B., and Barth, R.: The shift-effect in retinal ganglion cells of the rhesus monkey, Exp. Brain Res. **23:**443, 1975.
79. Künzle, H., and Akert, K.: Efferent connections of cortical area 8 (frontal eye field) in *Macaca fascicularis:* a reinvestigation using the autoradiographic technique, J. Comp. Neurol. **173:**147, 1977.
80. Latto, R., and Cowey, A.: Visual field defects after frontal eye-field lesions in monkeys, Brain Res. **30:**1, 1971.
81. LeVay, S.: On the neurons and synapses of the lateral geniculate nucleus of the monkey, and the effects of eye enucleation, Z. Zellforsch. **113:**396, 1971.
82. Lund, J. S., et al.: The origin of efferent pathways from the primary visual cortex, area 17, of the macaque monkey as shown by retrograde transport of horseradish peroxidase, J. Comp. Neurol. **164:**287, 1975.
83. Maffei, L., and Fiorentini, A.: The unresponsive regions of visual cortical receptive fields, Vision Res. **16:**1131, 1976.
84. Malpeli, J. G., and Baker, F. H.: The representation of the visual field in the lateral geniculate nucleus of *Macaca mulatta,* J. Comp. Neurol. **161:**560, 1975.
85. McIlwain, J. T.: Receptive fields of optic tract axons and lateral geniculate cells: peripheral extent and barbiturate sensitivity, J. Neurophysiol. **27:**1154, 1964.
86. Michael, C. R.: Color vision mechanisms in monkey striate cortex: dual-opponent cells with concentric receptive fields, J. Neurophysiol. **41**(3):572, 1978.
87. Mishkin, M.: Visual mechanisms beyond the striate cortex. In Russel, R. W., editor: Frontiers in physiological psychology, New York, 1966, Academic Press, Inc., pp. 93-119.
88. Mohler, C. W., and Wurtz, R. H.: Role of striate cortex and superior colliculus in visual guidance of saccadic eye movement in monkeys, J. Neurophysiol. **40:**74, 1977.
89. Mohler, C. W., Goldberg, M. E., and Wurtz, R. H.: Visual receptive fields of frontal eye field neurons, Brain Res. **61:**385, 1973.
90. Movshon, J. A.: The velocity tuning of single units in cat striate cortex, J. Physiol. (Lond.) **249:**445, 1975.
91. Nikara, T., Bishop, P. O., and Pettigrew, J. D.: Analysis of retinal correspondence by studying receptive fields of binocular single units in cat striate cortex, Exp. Brain Res. **6:**353, 1968.
92. Ogren, M. P., and Hendrickson, A. E.: The distribution of pulvinar terminals in visual areas 17 and 18 of the monkey, Brain Res. **137:**343, 1977.
93. Pandya, D. N., and Kuypers, H. G. J. M.: Corticocortical connections in the rhesus monkey, Brain Res. **13:**13, 1969.
94. Pettigrew, J. D., Nikara, T., and Bishop, P. O.: Responses to moving slits by single units in cat striate cortex, Exp. Brain Res. **6:**373, 1968.
95. Poggio, G. F., and Fischer, B.: Binocular interaction and depth sensitivity in striate and prestriate cortex of behaving rhesus monkey, J. Neurophysiol. **40:**1392, 1977.
96. Poggio, G. F., Doty, R. W., Jr., and Talbot, W. H.: Foveal striate cortex of behaving monkey: single-neuron responses to square-wave gratings during fixation of gaze, J. Neurophysiol. **40:**1369, 1977.
97. Poggio, G. F., Baker, F. H., Lamarre, Y., and Riva Sanseverino, E.: Afferent inhibition at input to visual cortex in the cat, J. Neurophysiol. **32:**892, 1969.
98. Poggio, G. F., et al.: Spatial and chromatic properties of neurons subserving foveal and parafoveal vision in rhesus monkey, Brain Res. **100:**25, 1975.
99. Pribam, K. H., and Mishkin, M.: Simultaneous and successive visual discrimination by monkeys with inferotemporal lesions, J. Comp. Physiol. Psychol. **48:**198, 1955.
100. Richards, W.: Visual space perception. In Carterette, E. C., and Friedman, M. P.: Handbook of perception, New York, 1975, Academic Press, Inc., vol. 5, pp. 351-386.
101. Robinson, D. A.: Eye movements evoked by collicular stimulation in the alert monkey, Vision Res. **12:**1795, 1972.
102. Robinson, D. A., and Fuchs, A. F.: Eye movements evoked by stimulation of frontal eye fields, J. Neurophysiol. **32:**637, 1969.
103. Robson, J. G.: Receptive fields: neural representation of the spatial and intensive attributes of the visual image. In Carterette, E. C., and Friedman, M. P.: Handbook of perception, New York, 1975, Academic Press, Inc., vol. 5, pp. 81-116.
104. Rocha-Miranda, C. E., Bender, D. B., Gross, C. G., and Mishkin, M.: Visual activation of neurons in inferotemporal cortex depends on striate cortex and forebrain commissures, J. Neurophysiol. **38:**475, 1975.
105. Rodieck, R. W., and Stone, J.: Analysis of receptive fields of cat retinal ganglion cells, J. Neurophysiol. **28:**833, 1965.
106. Schiller, P. H.: The effect of superior colliculus ablation on saccades elicited by cortical stimulation, Brain Res. **122:**154, 1977.
107. Schiller, P. H., and Koerner, F.: Discharge characteristics of single units in superior colliculus of the alert rhesus monkey, J. Neurophysiol. **34:**920, 1971.
108. Schiller, P. H., and Malpeli, J. G.: The effect of striate

cortex cooling on area 18 cells in the monkey, Brain Res. **126:**366, 1977.
109. Schiller, P. H., and Malpeli, J. G.: Properties and tectal projections of monkey retinal ganglion cells, J. Neurophysiol. **40:**428, 1977.
110. Schiller, P. H., and Malpeli, J. G.: Functional specificity of lateral geniculate nucleus laminae of the rhesus monkey, J. Neurophysiol. **41:**788, 1978.
111. Schiller, P. H., and Stryker, M.: Single-unit recording and stimulation in superior colliculus of the alert rhesus monkey, J. Neurophysiol. **35:**915, 1972.
112. Schiller, P. H., Finlay, B. L., and Volman, S. F.: Quantitative studies of single-cell properties in monkey striate cortex, J. Neurophysiol. **39:**1288, 1976.
113. Schiller, P. H., Stryker, M., Cynader, M., and Berman, N.: Response characteristics of single cells in monkey superior colliculus following ablation or cooling of visual cortex, J. Neurophysiol. **37:**181, 1974.
114. Sillito, A. M.: Inhibitory processes underlying the directional specificity of simple, complex and hypercomplex cells in the cat's visual cortex, J. Physiol. (Lond.) **271:**699, 1977.
115. Singer, W.: Control of thalamic transmission by corticofugal and ascending reticular pathways in the visual system, Physiol. Rev. **57:**386, 1977.
116. Singer, W., Tretter, F., and Cynader, M.: Organization of cat striate cortex: a correlation of receptive-field properties with afferent and efferent connections, J. Neurophysiol. **38:**1080, 1975.
117. Spalding, J. M. K.: Wounds of the visual pathways. I. The visual radiation, J. Neurol. Neurosurg. Psychiatry **15:**99, 1952.
118. Spalding, J. M. K.: Wounds of the visual pathways. II. The striate cortex. J. Neurol. Neurosurg. Psychiatry **15:**169, 1952.
119. Sprague, J. M., Levy, J., DiBerardino, A., and Berlucci, G.: Visual cortical areas mediating form discrimination in the cat, J. Comp. Neurol. **172:**441, 1977.
120. Stone, J., and Dreher, B.: Projection of X- and Y-cells of the cat's lateral geniculate nucleus to areas 17 and 18 of visual cortex, J. Neurophysiol. **36:**551, 1973.
121. Talbot, S. A., and Marshall, W. H.: Physiological studies on neural mechanisms of visual localization and discrimination, Am. J. Ophthalmol. **24:**1255, 1941.
122. Trevarthen, C. B.: Two mechanisms of vision in primates, Psychol. Forsch. **31:**299, 1968.
123. Van Buren, J. M.: The retinal ganglion cell layer, Springfield, Ill., 1963, Charles C Thomas, Publisher.
124. Wagman, I. H.: Eye movements induced by electric stimulation of cerebrum in monkeys and their relationship to bodily movements. In Bender, M. D., editor: The oculomotor system, New York, 1964, Harper & Row, Publishers, pp. 18-39.
125. Weiskrantz, L., Cowey, A., and Passingham, C.: Spatial response to brief stimuli by monkeys with striate cortex ablations, Brain **100:**655, 1977.
126. Westheimer, G., and Mitchell, D. E.: The sensory stimulus for disjunctive eye movements, Vision Res. **9:**749, 1969.
127. Wiesel, T. N., and Hubel, D. H.: Spatial and chromatic interactions in the lateral geniculate body of the rhesus monkey, J. Neurophysiol. **29:**1115, 1966.
128. Wilson, M. E., and Toyne, M. J.: Retino-tectal and cortico-tectal projections in *Macaca mulatta,* Brain Res. **24:**395, 1970.
129. Wilson, W. A., and Mishkin, M.: Comparison of the effects of inferotemporal and lateral occipital lesions on visually guided behavior in monkeys, J. Comp. Physiol. Psychol. **52:**10, 1959.
130. Zeki, S. M.: Representation of central visual fields in prestriate cortex of monkey, Brain Res. **14:**271, 1969.
131. Zeki, S. M.: The functional organization of projections from striate to prestriate visual cortex in the rhesus monkey, Symp. Quant. Biol. **40:**591, 1976.
132. Zeki, S. M.: The projections to the superior temporal sulcus from areas 17 and 18 in the rhesus monkey, Proc. R. Soc. Lond. (Biol.) **193:**199, 1976.
133. Zeki, S. M.: Colour coding in the superior temporal sulcus of rhesus monkey visual cortex, Proc. R. Soc. Lond. (Biol.) **197:**195, 1977.

19

LLOYD M. BEIDLER

The chemical senses: gustation and olfaction

The olfactory and gustatory systems monitor the chemical environment in which we live. When early man searched for food, his taste helped him to discriminate the bitter poisons from the sweet, high-caloric foods. Many other animals also utilize taste to regulate food intake, and novel food is always regarded with suspicion. If the animal becomes sick, it correlates the sickness with any recently eaten novel food and avoids this food or any with a similar taste for many weeks thereafter.

It has been suggested that man's craving for sweetness is a rather strong biologic drive. This sweet preference is observed at birth, and studies with pregnant women show that at 4 months of gestation the human fetus will increase its swallowing rate if saccharin is injected into the mother's amniotic fluid.[17,23,24] The strong drive for sweetness may account for the difficulty in regulating caloric intake by obese persons.

The olfactory system increases the breadth of flavor experience and enhances the pleasure of food. In addition, odors play a great role in the reproductive behavior of most animals. Special odors (pheromones) released by female insects attract the males and lead to copulation. Special glandular odors play a similar role with many mammals.[26] The importance of olfaction in man's reproductive behavior is still not well understood.

The olfactory and gustatory systems can detect and respond to many thousands of different chemicals over wide ranges of concentration. A better understanding of the interactions between the chemical stimuli and receptors may help us to also understand similar interactions with many other cells of the body.

GUSTATION

Taste structures

Taste buds, the organs of taste, are widely distributed throughout the oral cavity, although they are most concentrated on the tongue. These taste buds are associated with specialized papillae, namely, fungiform, foliate, and circumvallate. Hundreds of mushroomlike fungiform papillae are located on the anterior two thirds of the dorsal tongue surface (Fig. 19-1). Each contains one to five taste buds on its dorsal surface that are innervated by the chorda tympani nerves. Several large circumvallate papillae on the posterior part of the tongue contain thousands of taste buds in the trenches that surround each papilla. They are innervated by the glossopharyngeal nerves. The foliate papillae, which are often more pronounced in mammals other than man, contain thousands of taste buds in their grooves. The papillae are on each side of the tongue near its posterior region and are innervated by both the chorda tympani and the glossopharyngeal nerves.

Since the fungiform taste buds are readily exposed, they respond quickly to substances taken into the mouth, whereas these same substances must diffuse into the circumvallate trenches before stimulating these taste buds. Movement of the tongue squeezes the taste stimuli out of the trenches, and secretions of von Ebner's glands at the trench bottom help the flushing action. Both processes take time, which explains why many taste sensations experienced at the back of the tongue often linger for several minutes. The circumvallate taste buds are very sensitive to bitter stimuli as compared to the buds of the fungiform papillae, which respond well to sweet and salty substances.

The epithelium covering the dorsal surface of the tongue is not well penetrated by chemicals applied to its surface. A sheet of epithelium can be stripped from the dorsal tongue surface of the rate after collagenase injection. Isotope-labeled stimuli placed on its surface can be recovered on the other side at various times and their concentration measured with a scintillation counter. The ease of penetration is related to the substance's

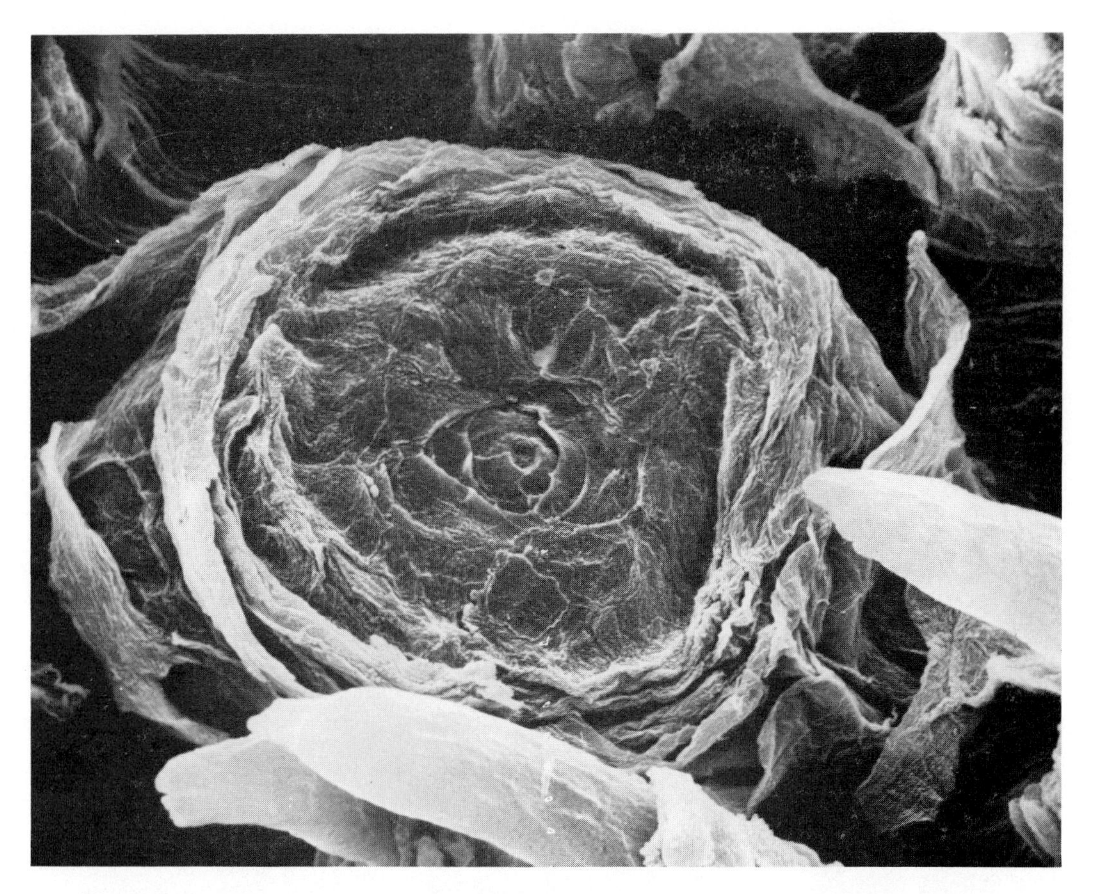

Fig. 19-1. Scanning electron photomicrograph of rat fungiform papilla showing taste pore of taste bud beneath it. (Magnification ×1200.)

relative solubility in lipid and aqueous media.

Each taste bud is a flask-shaped collection of 40 to 50 cells (Fig. 19-2). Microvilli extend from the apices of the taste cells (Fig. 19-3) and are in direct contact with the saliva covering the tongue.[62] It is thought that the taste molecules interact with these microvilli. The taste cells are innervated at their base. Each taste nerve fiber divides many times before approaching the base of the taste bud. In fact, a single fiber of the rat chorda tympani nerve bundle may innervate several taste buds and many cells within each taste bud.[9] Thus the electrical activity recorded from such a fiber represents the activity from many taste cells.

The integrity of the taste bud depends on its innervation. If the taste nerve is cut, the taste buds it normally innervates disappear within several days.[40,64] The taste buds reappear after the taste nerve again regenerates into the papillae, although an intact central connection of the taste nerve is not necessary for this trophic effect on the taste buds. The nature of the trophic influence is still not well understood, although it is commonly assumed that it is regulated by specific molecules manufactured in the cell body of the taste nerve. Tritiated leucine injected into the rat nodose ganglion is quickly utilized to form labeled peptides and proteins that move down the axoplasm toward the epiglottal taste buds, where they accumulate within 10 to 20 hours.

The individual cells within the taste bud are continually replaced by daughter cells of dividing epithelial cells surrounding the taste bud. These cells enter the taste bud, differentiate, function as normal taste cells for about 8 days, degenerate, and then are again replaced.[10,21] Thus although the integrity of the taste bud is always maintained, the individual cells comprising the taste bud have limited lives and are in constant flux. These dynamics require that the taste nerve endings within the taste bud also continuously move and innervate the new taste cells. Continuous labeling of newly dividing cells with tritiated thymidine for a period of over a month indicated that all the cells of the taste bud are replaced. Some electron microscopists classify taste bud cells into three or four types, based

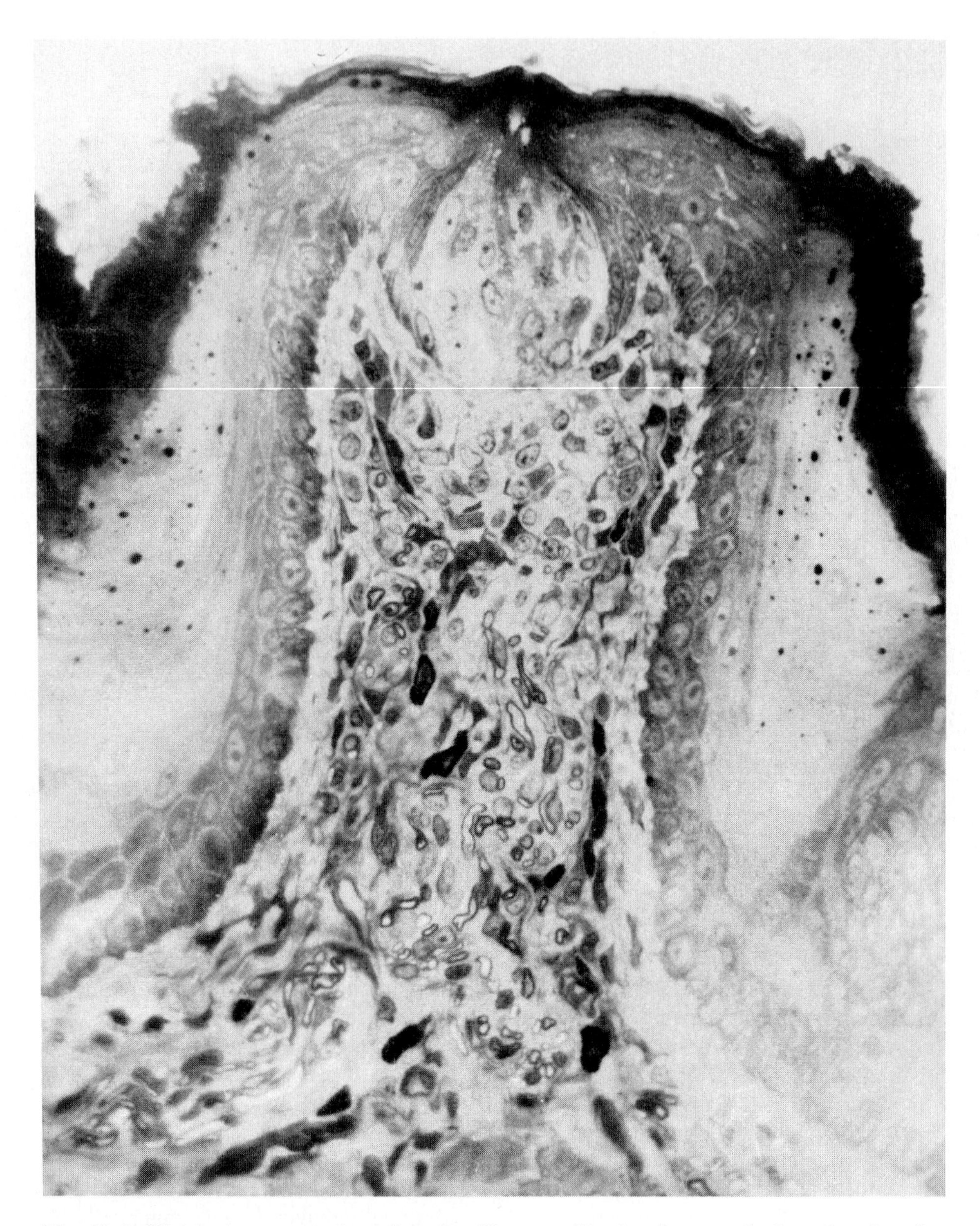

Fig. 19-2. Histologic cross section of rat fungiform papilla showing taste bud on dorsal surface with pore to surface. (Magnification ×512.)

on their electron opacity and subcellular structure.[62] Assigning functional differences to these types, however, is difficult.[34]

Gustatory pathways

The facial, glossopharyngeal, and vagus nerves innervate a majority, if not all, of the primate taste buds. The cell bodies of their taste fibers are located in the geniculate, superior petrosal, and nodose ganglia, respectively (Fig. 19-4).

The afferent taste fibers travel to the medulla and terminate in the ipsilateral nucleus of the tractus solitarius (NTS). Neural responses have been recorded in the NTS of mammals when taste substances were applied to the tongue, although responses to thermal and mechanical stimuli were often recorded at the same site.[41,53,70]

The thalamic taste relay has been identified in the monkey, cat, and rat as being in the ventromedial division of the ventral posterior thalamus.

Fig. 19-3. Electron photomicrograph of apical end of taste bud (rabbit foliate) showing numerous taste microvilli extending into taste pore. (From Murray and Murray.[62])

Fig. 19-4. Diagrammatic sketch of major taste pathways from tongue.

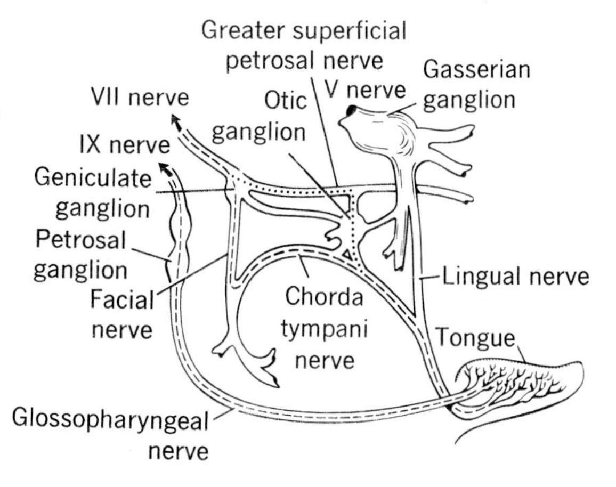

Single-unit recording has shown that most taste neurons respond also to thermal, but not mechanical, stimuli.[12,28,49]

The taste fibers leave the thalamic relay and pass via the internal capsule to terminate either in the parietal cortex or in the anterior opercular-insular cortex overlying the claustrum.[13,14] These two taste nerve projections are close to those of the somatic sensory system.

Norgren recently found a nucleus in the dorsal pons of the mammalian brain stem, interposed between the medullary and thalamic taste nuclei.[63] This nucleus is the source of a bifurcated pathway. One portion goes to the classic thalamocortical projections. The other passes below the thalamus and provides some projections to the lateral hypothalamus and a larger projection to the central nucleus of the amygdala. Stimulation or ablation of the amygdala affects food intake.[69] Hypothalamic lesions produce aphagia and adipsia.

Neural coding

Early researchers classified tastes into four distinct groups: sour, bitter, salty, and sweet. More complex tastes were thought to be but mixtures of these four. Today these four primary qualities are still assumed, although it is often considered that this assumption is an oversimplification if applied to the tastes of foods normally eaten. The question remains, however, of how the nerves signal information concerning the type of substance placed on the tongue. Perhaps there are specific taste buds that are sensitive to salty substances, others to bitter, or sour, or sweet. Such a simple scheme was proved wrong when Pfaffmann first recorded from single taste nerve fibers.[68] He, and many others that followed, showed that a single taste fiber can respond to a number of different types of stimuli and that no two fibers possess the same response profile.[29,31,65] This was elegantly demonstrated by Ogawa et al.,[65] who recorded from 50 single rat taste fibers in response to each of four stimuli representing the four primary taste qualities (Fig. 19-5). It was concluded that the coding for a given taste quality was afforded by a specific pattern of neural activity as the result of the relative excitation of a large population of taste fibers of varying specificity and sensitivity. This rather complex concept was restudied many years later by Frank,[31] who determined the sensitivities of 79 hamster chorda tympani fibers. She agreed; a single taste fiber can respond to a wide variety of taste stimuli. However, if the fibers are grouped according to the primary taste quality to which they respond best, then she noted that any one fiber has a similar response profile to those of others in the same group (Fig. 19-6). She concluded, therefore, that taste fibers can be grouped according to the four primary taste qualities.

It is often desirable to consider the natural food of an animal, determine the components that are water soluble, and use these chemicals in the studies of neural coding, without regard to any previously described primary taste qualities. This approach was used by Boudreau and Alev[16] when recording from single units of the cat geniculate ganglion, which contains the cell bodies of the chorda tympani taste fibers. Using a large

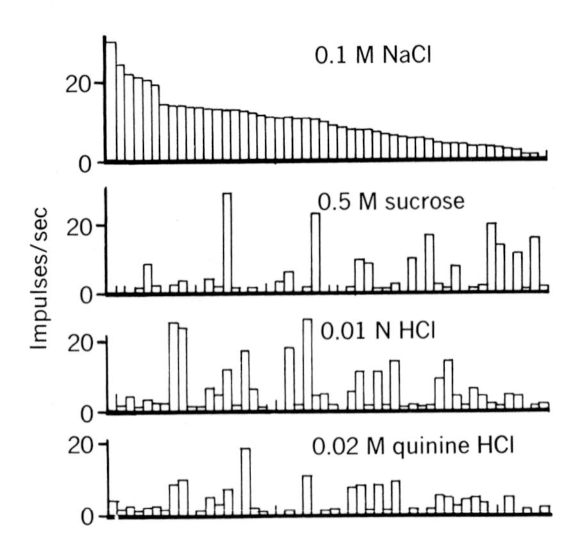

Fig. 19-5. Responses of 50 single rat taste fibers (chorda tympani) to stimuli corresponding to salty, sweet, sour, and bitter qualities. (Redrawn from Ogawa et al.[65])

number of diverse taste stimuli, they found they could subdivide the geniculate cells into three different independent groups according to spontaneous activity, latency to electrical stimulation, and response to the chemical stimuli chosen. It has long been appreciated that the patterns of nerve impulses of taste units vary considerably with the nature of the chemical stimulus, and these patterns have been studied in detail.[58] Their importance in neural coding is still not ascertained, however.

Some of the complexity of neural coding may be due to the multiple innervation of several taste cells, some residing in different fungiform papillae, by a single taste nerve fiber. Miller stimulated two papillae individually while recording from a single chorda tympani fiber that innervated both.[57] The total response was additive for most stimuli applied to both papillae, but inhibitory interaction was noted for a few.

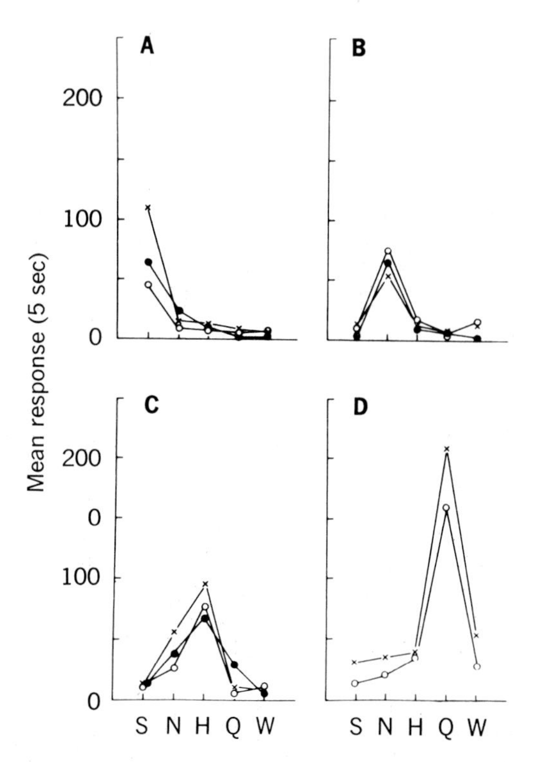

Fig. 19-6. Mean frequency of nerve impulses recorded from single taste nerve fibers in response to stimuli corresponding to four primary taste qualities. *S*, 0.1M sucrose; *N*, 0.03M NaCl; *H*, 0.003M HCl; *Q*, 0.001M quinine hydrochloride; *W*, water; *x*, circumvallate; open circles, foliate; solid circles, fungiform. Responses are grouped according to stimulus to which they respond best. (From Pfaffmann et al.[69a])

Transduction

The taste cells respond quickly to a variety of chemicals over a large concentration range. The integrity of the cells is maintained after several hours of continued flushing of the tongue with water or high salt concentrations. This indicates that the cell interior does not equilibrate with the solution bathing the tongue. It is suggested that the chemical stimuli adsorb to the surface of the taste cell microvilli that project into the saliva. The rapid reversibility of the taste response also suggests that the binding is weak, as would be expected if hydrogen bonding or electrostatic forces were involved instead of covalent bonding.

If the taste substances are bound to the microvilli, how does this lead to nerve excitation? A first thought is that stimulus binding leads to a membrane conformation change, which results in a change of ionic permeability of microvillus membrane and a corresponding change in membrane potential. Indeed, Sato and Beidler[72] found that the transmembrane potential of rat cells depends on Na^+ influxes from the extracellular spaces, but emphasized that these Na^+ influxes do not originate from the NaCl stimulus applied to the tongue surface. One may conclude that there is some type of messenger that correlates a stimulus binding at the microvilli to a change in membrane permeability at the base of the cell, where innervation occurs. Whether the messenger is in the form of a specific ion or molecule, a change in membrane properties, or an electrical event is not known.

If the microvilli membranes are protein-lipid structures, as are other cell membranes, there should be many sites where salts, acids, and even more complex organic molecules can bind. Since the binding forces are so weak, the binding properties at a given site depend on the environment in which the site exists as well as the nature of the site itself. Any small variations in membranes from cell to cell should influence the binding. Evidence for this is obtained by recording from a series of single taste fibers in response to a series of inorganic salts. No two fibers have the same response profile.[29] Similarly, if one records from the total chorda tympani that innervates a population of taste buds, the total response to Na^+ vs K^+ varies according to species.[11] Rodents respond better to Na^+ and carnivores to K^+. Species differences are also observed when recording responses to more complex molecules such as sugars and bitters. Microelectrodes inserted into single cells of the taste bud of the frog, rat, or hamster can record the changes in resting potential of the cell as well as the change in transmembrane resistance. Ozeki showed that membrane conductance increased and the cell depolarized in response to NaCl, sucrose, and HCl, whereas the conductance decreased and the cell hyperpolarized in response to quinine.[67]

A quantitative analysis of taste cell and taste nerve responses to a variety of chemical stimuli increases our understanding of the transduction process. The data support the concept that the first event is the binding of the stimulus molecule to receptor membrane sites. If it is assumed that the taste cell membrane has a finite number of equivalent but independent sites for stimulus adsorption, the simple mass action law can be applied to obtain the following equation:

$$R = \frac{CKR_s}{1 + CK}$$

where R is the magnitude of response to a given stimulus, C is the stimulus concentration, R_s is the maximum response at high concentration, and K is the binding constant.[6] Responses to chemicals that evoke pure taste qualities (e.g., NaCl and sucrose) can be described by this equation. More complex stimuli that evoke mixed tastes require the additional assumption that several different receptor sites are involved.[7] The cation is most important for stimulation, but the anion also plays a role.

If the chemical stimulus is adsorbed to a particular macromolecule of the taste cell membrane, it should be possible to isolate the macromolecule and study its adsorption properties. Dastoli and Price[22] isolated from the tongue epithelium a protein that binds sugars and other sweet substances in a manner predicted by the previous taste equation. Subsequent researchers have increased the yield of taste cells in the tongue fraction utilized in biochemical studies, improved methods for characterization, and extended the number of stimuli bound to the receptor protein.*

Binding of ions and molecules to the plasma membrane is a property common to most body cells, as is the change in resting potential when ions are added to the surrounding fluid. Taste cells differ in that a change in their transmembrane potential leads to nerve excitation. Also, the taste cells can respond to a wide variety of chemicals of different concentrations without irreversible damage.

*See references 19, 43, 46, 50, and 52.

Function-structure correlations

Sweet. Application of the taste equation to experimental data led to the conclusion that most common chemical stimuli are bound to the taste cell membrane with an energy of 1 to 2 kcal/mole.[6] Since some organic molecules can form a hydrogen bond to membranes with this binding force, Shallenberger and Acree[73] examined a number of sweet stimuli to see how much hydrogen bonding could be used in taste stimulation. They concluded that all sweet stimuli must have an electronegative atom (A) to which a hydrogen atom is attached by a single covalent bond and another electronegative center or atom (B) within a 3 Å distance.[73] The sweet-tasting compound can then bind by two hydrogen bonds to two similar atoms or to centers of the taste bud receptor site:

$$\text{Taste bud receptor site}\ \left\{\begin{array}{l} \text{—A—H-----------B—} \\ \text{—B-----------H—A—} \end{array}\right\}\ \text{Sweet-tasting}$$

The ε-amino group of lysine in the taste receptor protein could provide the A—H structure, and the carbonyl atom of the peptide bond could provide the B structure. The A—H, B unit is a necessary but insufficient requirement for predicting the sweet taste of a molecule. The molecular conformation, including steric factors, is also important.

One should not forget that specific binding sites are located in a protein-lipid membrane. The importance of the lipid component was emphasized by Deutsch and Hansch,[25] who related relative sweetness of a series of sweetners to their partition coefficient or relative solubility in a polar and a nonpolar solvent. Lipid solubility may be related to the need of the sweet stimulus to penetrate into the membrane before site interaction. Shallenberger and Lindley[74] also considered the fact that a lipophilic element in a stimulus molecule may help to orient the molecule, so that the A—H, B hydrogen bonding can take place. They designated a hydrophobic site (γ) and indicated the relationships shown in Fig. 19-7.

In the preceding discussion, it was assumed that there exists but one kind of site to which sweet molecules can bind. This may not necessarily be correct, since two different receptor sites have been found in the flesh fly.[75] The furanose and pyranose forms of sugars interact at two independent sites.

The sweetness of dipeptides has been studied extensively since it was first demonstrated that L-aspartyl-L-phenylalanine methyl ester (aspartame) is extremely sweet.[4,18,56] Tastes of some peptides are shown here:

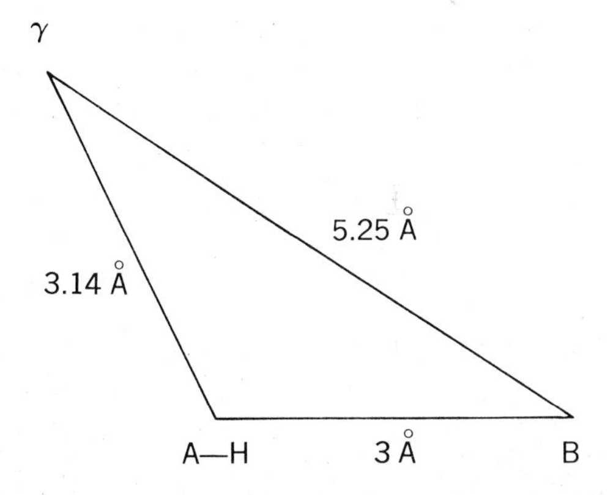

Fig. 19-7. Lipophilic point (γ) is located near centers of hydrogen bonding, *A—H* and *B*. (From Shallenberger and Lindley.[74])

Sour
- Gly-L-Asp
- Gly-L-Glu
- L-Ser-L-Asp
- L-Pro-Gly-Gly-L-Glu

Bitter
- Gly-L-Try
- L-Ala-L-Phe
- L-Leu-Gly
- L-Lys-L-Ala

Little or no taste
- Gly-Gly
- L-Ala-Gly
- L-Ala-L-Ala
- L-Phe-L-Phe

Bitter. Many sweet compounds can be made bitter by small changes in molecular structure. No single structural component can be related to all bitter substances. Kubota and Kubo[45] have shown that certain bitter diterpenes also contain an A—H, B system in which the two units are separated by about 1.5 Å. However, no distinct interorbital distance can yet be established for the bitterness of most compounds.[15]

Different individuals show a marked difference in threshold to a group of bitter molecules containing the N—C=S group.[30] This difference depends on a single autosomal allelic gene pair that presumably controls specific bitter sites of the receptor membrane. The so-called tasters represent two genotypes with the heterozygous Tt or the homozygous TT genetic constitution, whereas the nontasters have the recessive homozygous tt gene.[44]

Sour. The presence of the hydrogen ion is a necessary requirement for a sour-tasting substance, although other factors play a role in the determination of the intensity of sourness at a given pH.[8] Not all acids are equally sour at equi-pH and equinormality, as shown in Table 19-1. Two factors appear to be of particular importance: the ionic strength of the solution and the electrical charge of the receptor membrane. The latter becomes more positive as an increasing number of hydrogen ions are adsorbed, which in turn partially repels the addition of more hydrogen ions. This can be overcome by simultaneous adsorption of the negative anions of the acid. Thus the ability of a particular anion of an acid to bind to the membrane helps to determine the number of hydrogen ions bound at a given pH and the resultant degree of sourness.

Table 19-1. Equal sour response*

Acid†	Acid (mM)	$[H^+]$ (mM)	pH
Sulfuric	2.2	4.20	2.38
Oxalic	3.3	3.50	2.46
Hydrochloric	5.0	5.00	2.30
Citric	5.5	1.97	2,71
Tartaric	5.9	2.38	2.62
Nitric	5.9	5.90	2.23
Maleic	6.4	4.82	2.32
Dichloroacetic	9.0	7.70	2.11
Succinic	10.0	0.824	3.08
Malic	10.0	1.91	2.72
Monochloroacetic	10.4	3.13	2.50
Glutaric	11.0	0.73	3.14
Formic	11.6	1.32	2.88
Adipic	14.0	0.74	3.13
Glycolic	15.0	1.45	2.84
Lactic	15.6	1.41	2.85
Mandelic	25.0	3.11	2.51
Acetic	64.0	1.06	3.00
Propionic	130.0	1.32	2.88
Butyric	150.0	1.44	2.84

*From Beidler.[8]
†These acid concentrations produce a summated electrophysiologic response equal in magnitude to that produced by 5 mM HCl applied to the tongue of the rat.

An increase in the ionic strength of an acid solution results in an increase in the amount of acid bound to wool protein. In like manner, addition of a salt to an acid solution may increase the sourness, even though the pH may be higher. This has been shown to be true in taste buds of such diverse species as man, rat, and fish.

Salty. The taste of NaCl is pure salty except near threshold and at very high concentration. Most other salts produced mixed tastes. The stimulating efficiency of salts depends primarily on the cation, although the anion is also important. The ability of a taste receptor cell to bind a given cation varies from cell to cell, and thus

Table 19-2. Possible taste stimuli

Protein	Molecular weight	Molar threshold	Animal
Monellin	10,700	10^{-7}	Man
Thaumatin	21,000	10^{-7}	Man
Miraculin	44,000	10^{-7}	Man
Serum albumin	65,000	10^{-7}	Marine snail

the response varies from one taste fiber to the next. The overall response of a population of taste fibers also varies from one species to another. These observations have led to the concept that cation binding depends on the relative size of the hydration shell surrounding the cation, which is partially determined by the binding strength of the receptor site.[7]

Proteins as stimuli. In 1968 Kurihara and Beidler isolated a protein (molecular weight 44,000) that elicited sweetness at low pH and had no taste at moderate and high pH.[46,47] This molecule, called miraculin, has a low threshold, binds strongly, and can elicit sweetness up to 1 hour after initial application. This discovery led to the consideration of other high molecular weight compounds as possible taste stimuli in a variety of species (Table 19-2).[39,59,60,81]

OLFACTION

Olfactory structures

An odor enters the nares, passes up the moist passageway of the nose, and sweeps past the 5 cm² olfactory area. The concentration of the odorant may decrease due to adsorption and absorption before it enters the olfactory area.[76] The amount of odorant and its flow rate are regulated by the force of the sniff and the vasodilatation or constriction of the nasal blood vessels.

The olfactory mucosa consists primarily of olfactory receptors, supporting cells, and basal cells (Fig. 19-8). This tissue is covered by a variable thickness of moving mucus in which the cilia of the olfactory receptors are embedded. There are about 100 million olfactory receptors in the rabbit and a smaller number in the human. Each receptor is a bipolar neuron with 50 to 150 μ cilia extending from its apex into the mucus (Fig. 19-9). The basal end of the receptor constricts to form a 0.2 μ unmyelinated axon. Hundreds of these axons are bundled together by a Schwann sheath[33] (Fig. 19-10). The primary olfactory nerve bundle, which contains the axons of all the receptors, passes through the thin and bony cribriform plate before entering the olfactory bulb. The olfactory nerve bundle is very short in man, but can extend up to 30 to 50 cm in the long-nosed garfish.[27] This fish is particularly suitable for the study of axoplasmic transport, since labeled leucine placed in the nares is absorbed by the olfactory receptors and utilized in the cell bodies to manufacture proteins, which then are transported toward the olfactory bulb at a speed of about 200 mm/day.[38]

The olfactory mucosa is well protected by its location deep in the nose. However, permanent loss of olfaction can occur during accidental whiplash if the thin cribriform plate is broken and the olfactory axons cut. Temporary and sometimes complete loss is also noted if man is exposed to certain solvents for prolonged periods. Fortunately, under certain circumstances, the olfactory receptors can be replaced. Graziadei[35] has shown, using autoradiographic studies with tritiated thymidine, that the basal cells can divide and form new olfactory receptors, including their axons, which reinnervate the cells of the olfactory bulb!

The olfactory receptors are often separated from each other by supporting epithelial cells. Their apex terminates with fine, short microvilli that also extend into the mucus. The prevalence of endoplasmic reticulum and the numerous granules suggest that these cells may have a secretory function. Much of the yellow pigmented secretion of the olfactory area is produced by Bowman's glands, found deep in the olfactory epithelium.

Olfactory pathways

The millions of axons of the olfactory receptors enter the olfactory bulb, where they converge to form glomeruli. Each glomerulus contains the endings of about 25,000 olfactory receptor axons and makes contact with about 25 mitral cells, the second-order neuron of the olfactory system.[1] Reciprocal inhibition of mitral cell activity is effected by horizontal interneurons that synapse with dendrites of neighboring mitral cells. Centrifugal fibers also synapse in this region of the olfactory bulb.

The axons of the mitral cells, the number of which is about one thousandth the number of primary olfactory nerves, enter the olfactory tract and travel to the olfactory tubercle, the prepiriform cortex, and the periamygdaloid area.[51,82] Olfactory messages are then relayed by projections to the thalamus and hypothalamus, where they play an important role in the control of feeding and reproductive behavior.[71] Centrifugal fibers may also enter the olfactory tract

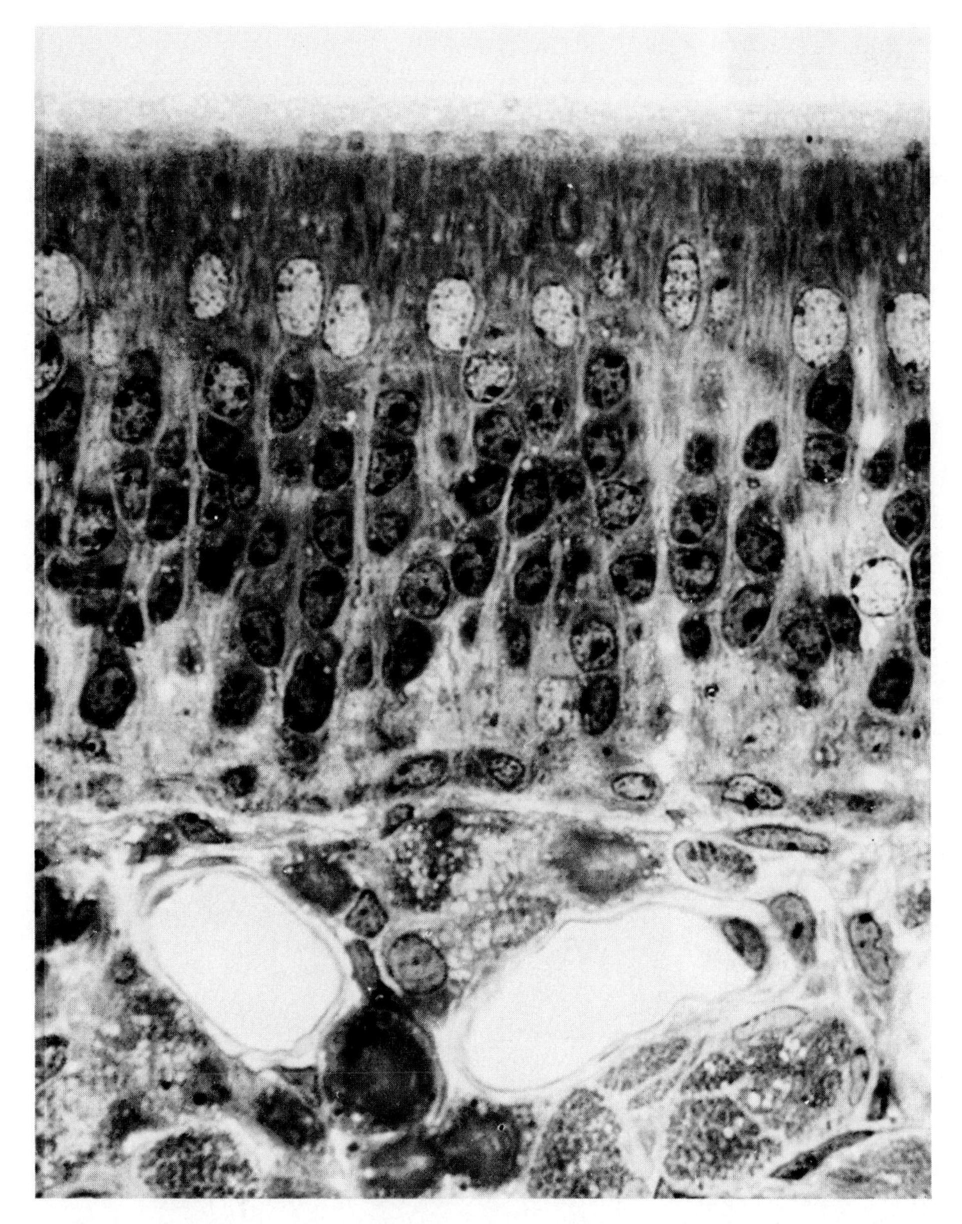

Fig. 19-8. Histologic cross section of mouse olfactory epithelium. Note dark nuclei of receptor cells and light nuclei of supporting cells. (Magnification ×1460.) (Courtesy Dr. P. P. C. Graziadei.)

and exert a predominantly inhibitory influence on the olfactory bulb.

Neural coding

Man can smell odors of thousands of different chemicals, including those newly synthesized and first introduced to the living world. Odor memory is often quite extraordinary, with certain highly specific odors initiating the recall of a specific incident that occurred in childhood. Perfumers and psychologists often group chemicals according to their odor quality. One such assembly included camphoraceous, musky, floral, pepperminty, etheral, pungent, and putrid.[2] These have been referred to as primary odors.

Are there specific olfactory receptors for each of the seven primary odors? Electrophysiologists attempted to answer this question by recording simple unit activity in the olfactory epithelium. This is a difficult task, and the olfactory c fibers respond with a frequency usually

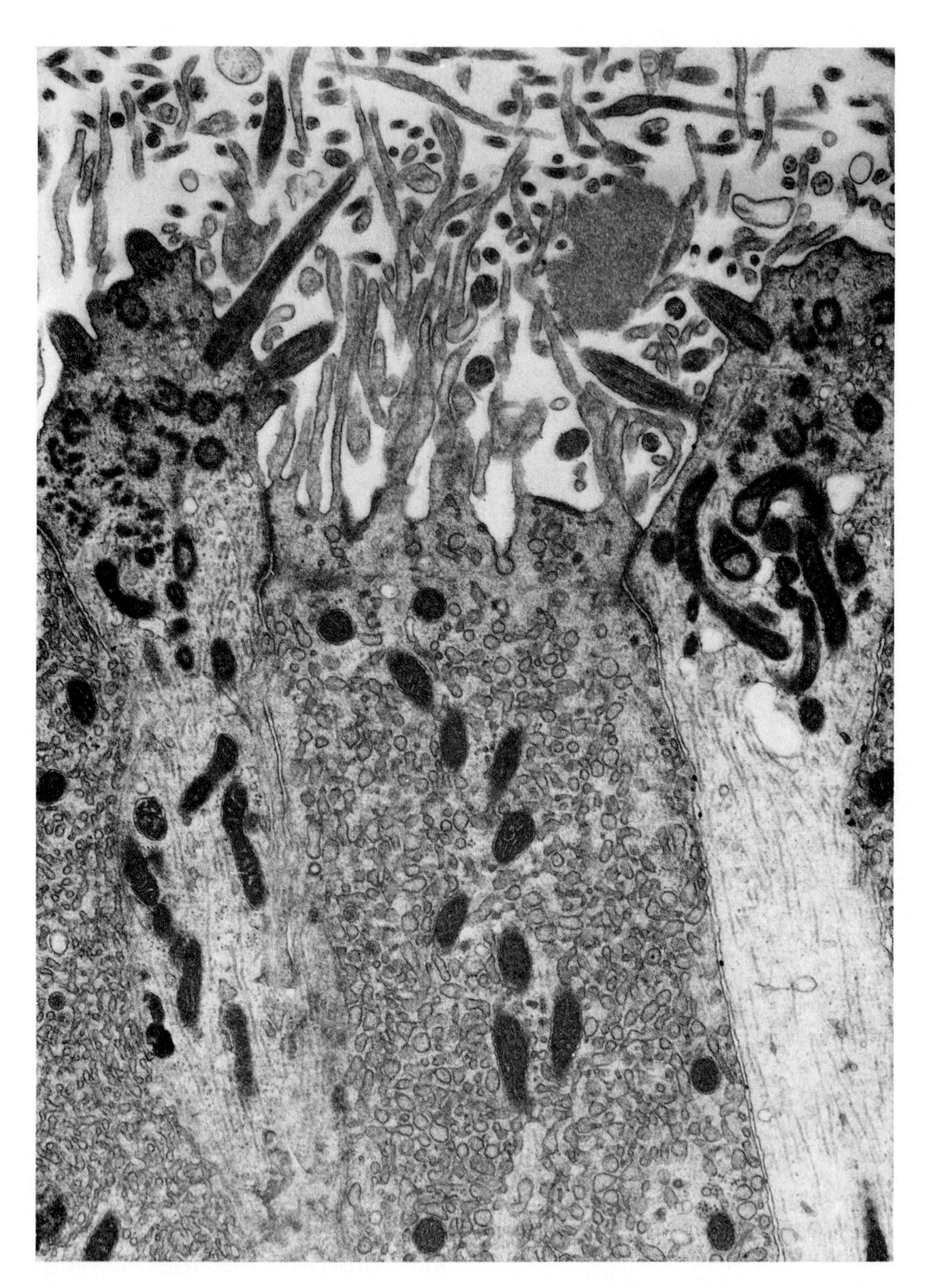

Fig. 19-9. Electron photomicrograph of cross section of dorsal surface of mouse olfactory epithelium. Note apices of two olfactory receptors with cilia projecting into mucus above. Supporting cells are vesicular and project microvilli. (Magnification ×30,750.) (Courtesy Dr. P. P. C. Graziadei.)

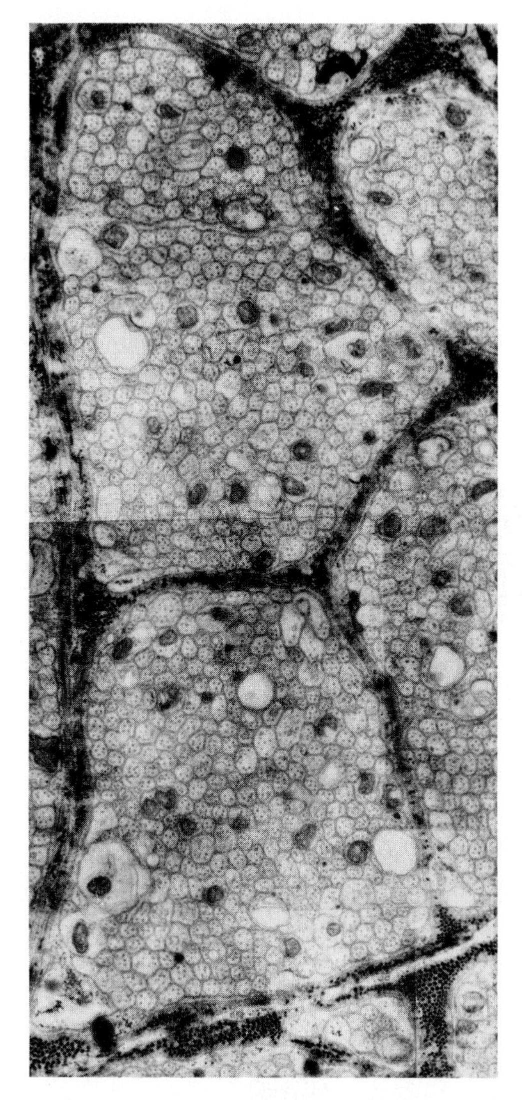

Fig. 19-10. Electron photomicrograph of portion of garfish olfactory nerve, showing two Schwann domains, each containing about 400 olfactory axons. (Magnification ×12,000.) (From Gross and Beidler.[37])

below 10/sec but with a possible maximum 2 to 3 times greater. If a battery of 15 different odorous molecules are used, the low spontaneous activity does not change for about 15% to 20% of the receptors. Of those that do respond, no two respond in the same manner, with some only responding to one of the test odors and very few responding to all[54] (Fig. 19-11). Attempts to group receptor responses to correspond to the seven primary odor categories has not been successful. Recently, the concept of seven primary odors has been questioned.

Most responsive receptors are excited by odors, but some may be inhibited by a few selected odors. Mitral cell recording is easier, and a similar battery of odors can be used. The mitral cells have a higher spontaneous frequency that can either be increased or decreased by odor stimulation of the receptor input (Fig. 19-12). Again, however, grouping like-responding mitral cells is difficult, and each mitral cell appears to have a response profile different from any other.[54]

One may conclude that the olfactory receptors are not highly tuned to chemicals of one category, whether based on molecular structure of the stimulus or the sensory quality of the sensation. Perhaps a pattern of the relative responses of a large number of receptors is necessary for the brain to encode odor quality.

Another unusual approach is that of Mozell.[61] He suggests that the responsiveness of receptors varies over the sheet of olfactory mucosa, so that an odor passing over the mucosa will leave its own "fingerprint" of activity analogous to a chromatogram. The relative sensitivity to odors varies with the place on the olfactory epithelium.

Transduction

The cilia of the olfactory receptors project into the mucus covering the olfactory epithelium and are the first portion of the receptor to interact with the odorant. Removal of the cilia decreases, but does not eliminate, the ability of a receptor to respond to odors.[78] Transport of an odor to the olfactory epithelium requires an odorant vapor pressure high enough so that a reasonable number of molecules enter the gaseous phase on entering the nares. Penetration into the mucus requires an odorant aqueous solubility, and entrance into the plasma membrane of the cilium requires a reasonable lipid solubility. Thus vapor pressure is unimportant. Amino acids, for example, are good stimuli for fish olfaction.[20]

Some insect olfactory receptors can respond to but one odorant molecule, and one molecule of certain odors may be sufficient to excite one human olfactory receptor as well.[76] This would imply that the olfactory sense is the most sensitive of all our sensory capabilities. It is thought that adsorption of the odorant molecule by a cilium produces a conformational change in its plasma membrane. This would lead to a change in ionic permeability (possibly that of Ca^{++}) that, in turn, would produce the generator current necessary to excite the axonal end of the receptor cell. Since millions of such receptors are closely packed into a small area of epithelium, the individual gen-

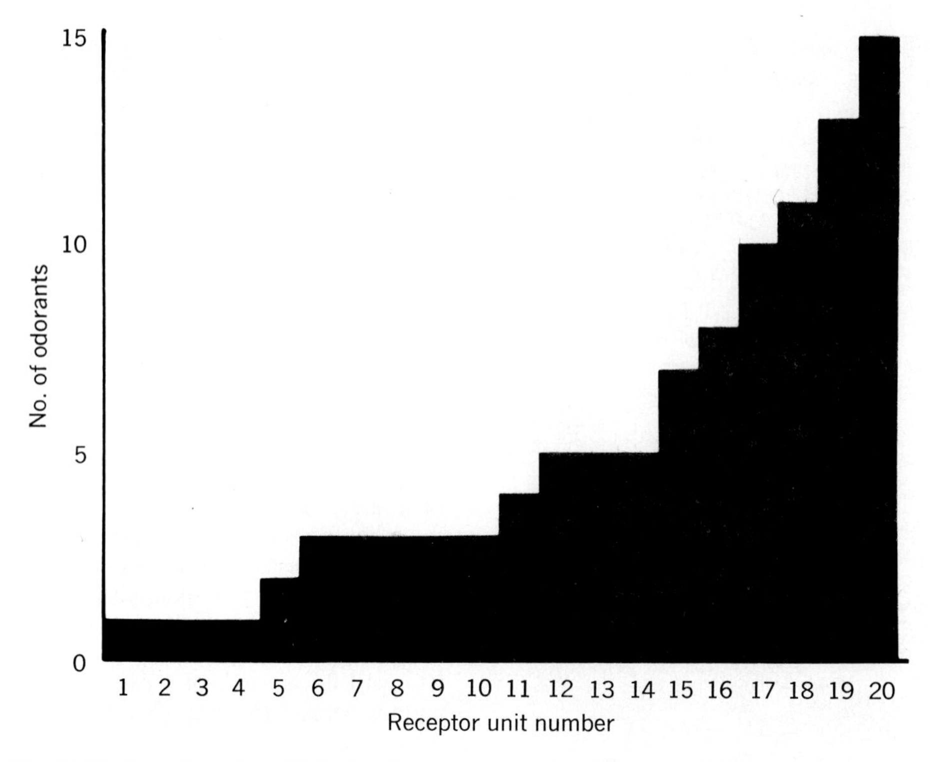

Fig. 19-11. Recordings from 20 single olfactory receptors in response to 15 different odorant stimuli show that each receptor differs in its response profile. (Courtesy Dr. D. Mathews.)

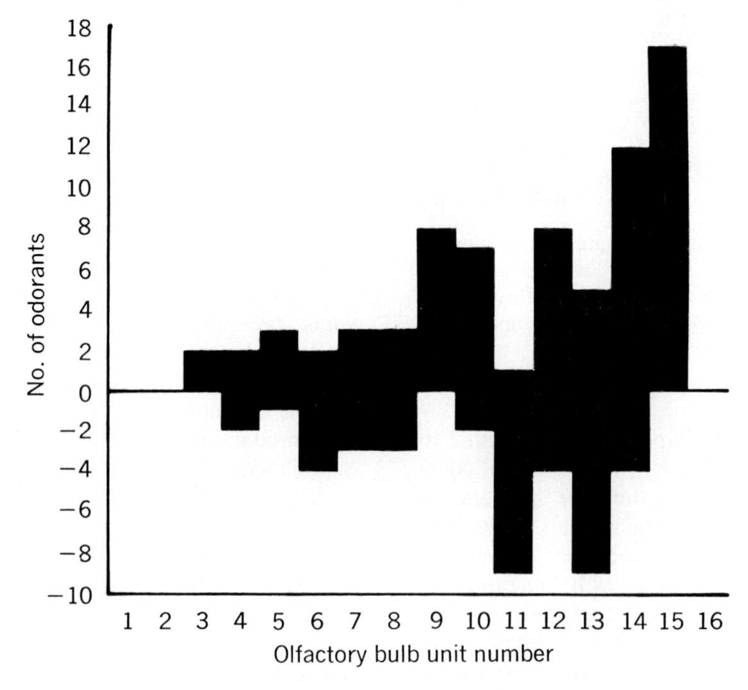

Fig. 19-12. Recordings from 16 mitral cells in response to 18 odorant stimuli indicate that response may be excitatory or inhibitory, and each cell has a different response profile. (Courtesy Dr. D. Mathews.)

erator currents may add, so that a rather large potential difference is generated across the olfactory epithelium. Ottoson[66] studied these potentials in detail and called them electro-olfactograms (EOGs). The EOG is a monophasic negative potential recorded from the surface of the olfactory epithelium when odor is applied (Fig. 19-13). It has a latency of 100 msec or less, rises to a height that is odor-concentration dependent, and declines exponentially after the odor is removed. Potentials other than generator potentials may be recorded with surface electrodes, since supporting cells and Bowman's glands are also located in the olfactory epithelium, and the secretion of both may be influenced by odor application. Since the EOG is maximum at the epithelial surface, current flow enters the cilia, passes down through the receptor cell, and goes out near the axon hillock, which excites the axon and initiates nerve action potentials. Decreasing the mucosal shunting by applying high sucrose concentrations increases the magnitude of the recorded EOG.[80]

Function-structure correlations

Early researchers tried to classify odorants according to their odor quality. It was thought that these would represent primary odors and that specific receptors exist that respond to odors within one of the classifications. Microelectrode studies failed to find specific olfactory receptors, and the quest turned to specific anosmia. Some individuals are anosmic to certain odors. By grouping such odors according to the type of anosmia, a system of primary odors might develop.[3] However, the number of groupings became so large that the concept of primary odors became of little interest.

One can argue that all molecular structures should have a similarity if they elicit similar odor qualities. On this basis, many structure-function studies were initiated. The two most important parameters are the overall molecular shape and the presence of functional groups within this shape.[5] The exact nature of the functional group may not change the odor quality much. For example, nitrobenzene and cyanobenzene have the odor of bitter almonds (Fig. 19-14). The same compounds with a 3-methoxy and 4-hydroxy substitution have a vanillin odor (Fig. 19-15). The position of a substitution may also be important (Fig. 19-16). Many times a similarity in odor does not correlate with a similarity in molecular structure. For example, there are macrocyclic musks, steroid musks, nitrogen musks, indane musks, tetraline musks, and benzene musks. Whether all these molecules can react

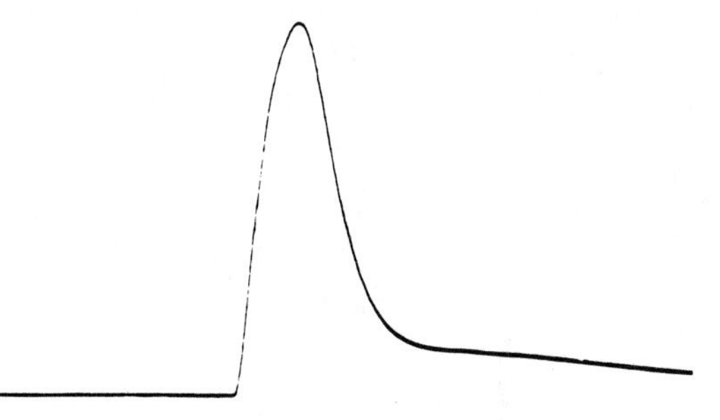

Fig. 19-13. EOG recorded from surface of box turtle olfactory epithelium in response to puff of geraniol odor. Peak amplitude, 8 mV. (Courtesy Dr. D. Tucker.)

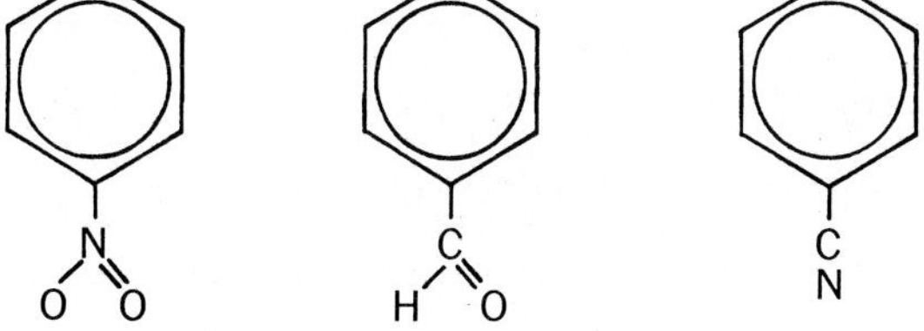

Fig. 19-14. Bitter almond odor.

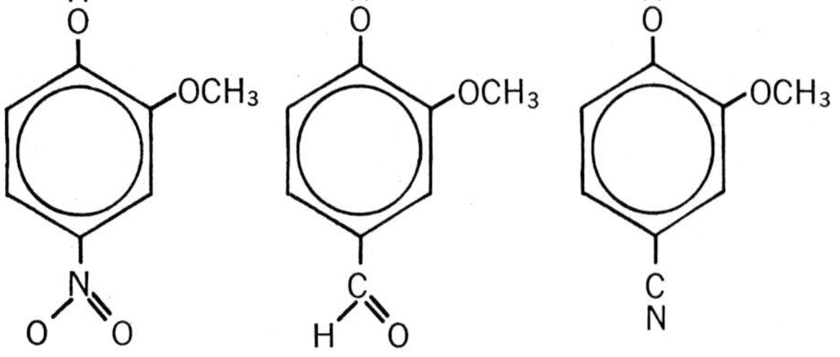

Fig. 19-15. Vanillin odor.

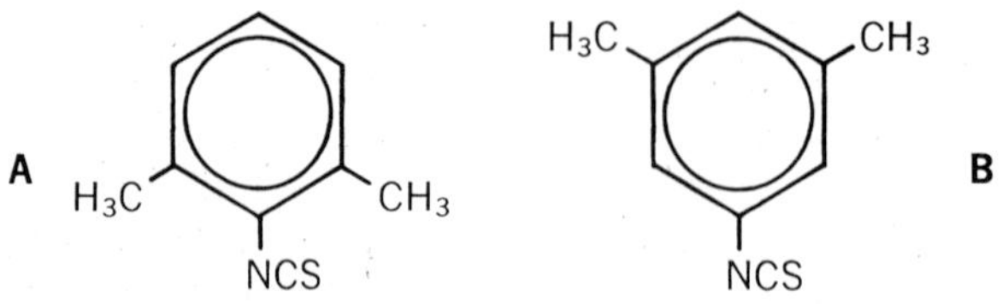

Fig. 19-16. A, Sweet odor. **B,** Pungent odor.

on the same receptor site of the olfactory plasma membrane is not known. More likely there are many different kinds of receptor sites associated with the same olfactory receptor cell. This situation is very similar to that which exists with the taste receptor.

The odor concentration is never known at the surface of the human olfactory epithelium. This limits analysis of structure-function data. Studies of fish olfactory receptors do not have this limitation. Quantitative studies of fish olfaction show responses to some amino acids well below 10^{-7} M.[20,42,77] The trout receptor site is viewed as a cationic subsite and a anionic subsite that can interact with ionized α-amino and α-carboxyl groups of the amino acid stimulus.[42] These two subsites are near another subsite that can accommodate the α-hydrogen atom of the L-isomer, but that excludes the D-isomer. Interaction with the α-carboxyl group is not necessary for catfish taste receptors. However, both olfactory and gustatory receptors of fish can respond to certain amino acids at very low concentrations.[20]

REFERENCES

1. Allison, A. C., and Warwick, P. T. T.: Quantitative observations on the olfactory system of the rabbit, Brain **72:**186, 1949.
2. Amoore, J. E.: The stereochemical theory of olfaction. I. Identification of seven primary odors, Proc. Sci. Sec. Toilet Goods Assoc. **37**(suppl.):1, 1962.
3. Amoore, J. E.: Specific anosmia and the concept of primary odors, Chem. Senses Flav. **2:**267, 1977.
4. Ariyoshi, Y.: The structure-taste relationships of aspartyl dipeptide esters, Agri. Biol. Chem. **40:**983, 1976.
5. Beets, M. G. J.: Odor and molecular constitution, Am. Perfum., June 1, 1961.
6. Beidler, L. M.: A theory of taste stimulation, J. Gen. Physiol. **38:**133, 1954.
7. Beidler, L. M.: Taste receptor stimulation, Prog. Biophys. Mol. Biol. **12:**109, 1962.
8. Beidler, L. M.: Anion influences on taste receptor response. In Hayashi, T., editor: Olfaction and taste II, Oxford, 1967, Pergamon Press, Ltd.
9. Beidler, L. M.: Innervation of rat fungiform papilla. In Pfaffmann, C., editor: Olfaction and taste III, New York, 1969, The Rockefeller University Press.
10. Beidler, L. M., and Smallman, R.: Renewal of cells within taste buds, J. Cell Biol. **27:**263, 1965.
11. Beidler, L. M., Fishman, I. Y., and Hardiman, C. W.: Species differences in taste responses, Am. J. Physiol. **181:**235, 1955.
12. Benjamin, R. M.: Some thalamic and cortical mechanisms of taste. In Zotterman, Y., editor: Olfaction and taste I, Oxford, 1963, Pergamon Press, Ltd.
13. Benjamin, R. M., and Burton, H.: Projection of taste nerve afferents to anterior opercular insular cortex in squirrel monkey (Saimiri sciureus), Brain Res. **7:**221, 1968.
14. Benjamin, R. M., Emmers, R., and Blomquist, A. J.: Projection of tongue nerve afferents to somatic sensory area I in squirrel monkey (Saimiri sciureus), Brain Res. **7:**208, 1968.
15. Birch, G. G., and Lee, C. K.: Structural functions and taste in the sugar series: the structural basis of bitterness in sugar analogues, J. Food Sci. **41:**1403, 1976.
16. Boudreau, J. C., and Alev, N.: Classification of chemoresponsive tongue units of the cat geniculate ganglion, Brain Res. **54:**157, 1973.
17. Bradley, R. M., and Stern, I. B.: The development of the human taste bud during the foetal period, J. Anat. **101:**743, 1967.
18. Brussel, L. B. P., Peer, H. G., and Von der Heijden, A.: Structure-taste relationship of some sweet-tasting dipeptide esters, Z. Lebensm. Unters. Forsch. **159:**337, 1975.
19. Cagan, R. H.: Biochemical studies of taste sensation. I. Binding of ^{14}C-labeled sugars to bovine taste papillae, Biochim. Biophys. Acta **252:**199, 1971.
20. Caprio, J.: Electrophysiological distinctions between the taste and smell of amino acids in catfish, Nature **266:**850, 1977.
21. Conger, A., and Wells, M. A.: Radiation and aging effect on taste structure and function, Radiat. Res. **37:**31, 1969.
22. Dastoli, F. R., and Price, S.: Sweet sensitive protein from bovine taste buds: isolation and assay, Science **154:**905, 1966.
23. De Snoo, K.: Das trinkende kind im uterus, Monatschr. Geburtshilfe **105:**88, 1937.
24. Desor, J. A., Maller, O., and Greene, L. S.: Preference for sweet in humans: infants, children, and adults. In Weiffenbach, J. M., editor: Taste and development. The genesis of sweet preference, Bethesda, 1977, Department of Health, Education and Welfare.
25. Deutsch, E. W., and Hansch, C.: Dependence of relative sweetness on hydrophobic bonding, Nature **211:**75, 1966.
26. Doty, R. L.: Mammalian olfaction, reproductive processes, and behavior, New York, 1976, Academic Press, Inc.
27. Easton, D. M.: Garfish olfactory nerve: easily accessible source of numerous long, homogeneous, nonmyelinated axons, Science **172:**952, 1971.
28. Emmers, R.: Separate relays of tactile pressure, thermal and gustatory modalities in the cat thalamus, Proc. Soc. Exp. Biol. Med. **121:**527, 1966.

29. Fishman, I. Y.: Single fiber gustatory impulses in rat and hamster, J. Cell. Comp. Physiol. **49:**319, 1957.
30. Fox, A. L.: The relationship between chemical constitution and taste, Proc. Natl. Acad. Sci. U.S.A. **18:** 115, 1932.
31. Frank, M.: An analysis of hamster afferent taste nerve response functions, J. Gen. Physiol. **61:**588, 1973.
32. Frank, M.: Personal communication, 1978.
33. Gasser, H.: Olfactory nerve fibers, J. Gen. Physiol. **39:**473, 1956.
34. Graziadei, P. P. C.: The ultrastructure of vertebrate taste buds. In Pfaffmann, C., editor: Olfaction and taste III, New York, 1969, The Rockefeller University Press.
35. Graziadei, P. P. C.: Personal communication, 1978.
36. Graziadei, P. P. C., and MontiGraziadei, G. A.: Continuous nerve cell renewal in the olfactory system. In Jacobson, M., editor: Handbook of sensory physiology, Berlin, 1978, Springer-Verlag, vol. 9.
37. Gross, G. W., and Beidler, L. M.: Fast axonal transport in the c-fibers of the garfish olfactory nerve, J. Neurobiol. **4:**413, 1973.
38. Gross, G. W., and Beidler, L. M.: A quantitative analysis of isotope concentration profiles and rapid transport velocities in the c-fibers of the garfish olfactory nerve, J. Neurobiol. **6:**213, 1975.
39. Gurin, S., and Carr, W. E.: Chemoreception in Nassarius obsoletus: the role of specific stimulatory proteins, Science **174:**293, 1971.
40. Guth, L.: Taste buds on the cat's circumvallate papillae after reinnervation by glossopharyngeal, vagus, and hypoglossal nerves, Anat. Rec. **130:**25, 1958.
41. Halpern, B. P., and Nelson, L. M.: Bulbar gustatory response to anterior and to posterior tongue stimulation in the rat, Am. J. Physiol. **209:**105, 1965.
42. Hara, T. J.: Further studies on the structure-activity relationships of amino acids in fish olfaction, Comp. Biochem. Physiol. (A) **56:**559, 1977.
43. Hiji, Y., and Sato, M.: Isolation of the sugar-binding protein from rat taste buds, Nature N. Biol. **244:**91, 1973.
44. Kalmus, H.: Genetics of taste. In Beidler, L. M., editor: Handbook of sensory physiology. II. Chemical senses—taste, New York, 1971, Springer-Verlag New York, Inc., vol. 4.
45. Kubota, I., and Kubo, I.: Bitterness and chemical structure, Nature **223:**97, 1956.
46. Kurihara, Y.: Effect of taste stimuli on the extraction of lipids from bovine taste papillae, Biochim. Biophys. Acta **306:**478, 1973.
47. Kurihara, K., and Beidler, L. M.: Taste-modifying protein from miracle fruit, Science **161:**1241, 1968.
48. Kurihara, K., and Beidler, L. M.: Mechanism of the action of taste-modifying protein, Nature **222:**1176, 1969.
49. Landgren, S.: Thalamic neurons responding to tactile stimulation of cat's tongue, Acta Physiol. Scand. **48:** 238, 1960.
50. Lo, C. H.: The plasma membranes of bovine circumvallate papillae isolation and partial characterization, Biochem. Biophys. Acta **291:**650, 1973.
51. Lohman, A. H. M., and Lammers, H. G.: On the structure and fibre connections on the olfactory centres in mammals. In Zotterman, Y., editor: Sensory mechanisms, Amsterdam, 1967, Elsevier Publishing Co.
52. Lum, C. K. L., and Henkin, R. I.: Sugar binding to purified fractions from bovine taste buds and epithelial tissue, Biochim. Biophys. Acta **421:**380, 1976.
53. Makous, W., Nord, S., Oakley, B., and Pfaffmann, C.: The gustatory relay in the medulla. In Zotterman, Y., editor: Olfaction and taste I, Oxford, 1963, Pergamon Press, Ltd.
54. Mathews, D. F.: Response patterns of single neurons in the tortoise olfactory epithelium and olfactory bulb, J. Gen. Physiol. **60:**166, 1972.
55. Mathews, D. F.: Personal communication, 1978.
56. Mazur, R. H., Schlatter, J. M., and Goldkamp, A. H.: Structure-taste relationships of some dipeptides, J. Am. Chem. Soc. **91:**2684, 1969.
57. Miller, I. J.: Peripheral interactions among single papilla inputs to gustatory nerve fibers, J. Gen. Physiol. **57:** 1, 1971.
58. Mistretta, C. M.: A quantitative analysis of rat chorda tympani fiber discharge patterns. In Schneider, C., editor: Olfaction and taste IV, Stuttgart, 1972, Wissenschaftliche Verlagsgesellschaft mbH.
59. Morris, J. A., and Cagan, R .L.: Purification of monellin, the sweet principle of Dioscoreophyllum cumminsii, Biochim. Biophys. Acta **261:**114, 1972.
60. Morris, J. A., Martenson, R., Deibler, G., and Cagan, R. H.: Characterization of monellin, a protein that tastes sweet, J. Biol. Chem. **248:**534, 1973.
61. Mozell, M. M.: The spatiotemporal analysis of odorants at the level of the olfactory receptor sheet, J. Gen. Physiol. **50:**25, 1966.
62. Murray, R. G., and Murray, A.: Fine structure of taste buds of rabbit foliate papillae, J. Ultrastruct. Res. **19:** 327, 1967.
63. Norgren, R., and Leonard, C.: Taste pathways in rat brainstem, Science **173:**1136, 1971.
64. Oakley, B.: Reformation of taste buds by crossed sensory nerves in the rat's tongue, Acta Physiol. Scand. **79:**88, 1970.
65. Ogawa, H., Sato, M., and Yamashita, S.: Multiple sensitivity of chorda tymapni fibers of the rat and hamster to gustatory and thermal stimuli, J. Physiol. **199:**223, 1968.
66. Ottoson, D.: Electrical signs of olfactory transducer action. In Wolstenholme, G. E. W., and Knight, J., editors: Ciba symposium on taste and smell in vertebrates, London, 1970, J. & A. Churchill, Ltd.
67. Ozeki, M., and Sato, M.: Responses of gustatory cells in the tongue of rat to stimuli representing four taste qualities, Comp. Biochem. Physio. (A) **41:**391, 1972.
68. Pfaffmann, C.: Gustatory afferent impulses, J. Cell. Comp. Physiol. **17:**243, 1941.
69. Pfaffmann, C.: Biological and behavioral substrates of the sweet tooth. In Weiffenbach, J. M., editor: Taste and development: the genesis of sweet preference, Bethesda, 1977, Department of Health, Education, and Welfare Publication No. 20 (NIH) 77-1068.
69a. Pfaffman, C., Frank, M., and Norgren, R.: Neural mechanisms and behavioral aspects of taste, Annu. Rev. Psychol. **30:**296, 1979.
70. Pfaffmann, C., Erickson, R. P., Frommer, G. P., and Halpern, B. P.: Gustatory discharges in the rat medulla and thalamus. In Rosenblith, W. B., editor: Sensory communication, Cambridge, Mass., 1961, The M. I. T. Press.
71. Powell, T. P. S., Cowan, W. M., and Raisman, G.: The central olfactory connections, J. Anat. **99:**791, 1965.
72. Sato, T., and Beidler, L. M.: Membrane resistance change of frog taste cells in response to water and NaCl, J. Gen. Physiol. **66:**735, 1975.
73. Shallenberger, R. S., and Acree, T. E.: Molecular theory of sweet taste, Nature **216:**480, 1967.

74. Shallenberger, R. S., and Lindley, M. G.: A lipophilic-hydrophobic attribute and component in the stereochemistry of sweetness, Food Chem. **2:**145, 1977.
75. Shimada, I., Shiraishi, A., Kijima, H., and Morita, H.: Separation of two receptor sites in a single labellar sugar receptor of the fleshfly by treatment with *p*-chloromercuribenzoate, J. Insect Physiol. **20:**605, 1974.
76. Stuiver, M.: The biophysics of the sense of smell, Ph.D. dissertation, Groningen University, 1958.
77. Sutterlin, A. M., and Sutterlin, N.: Electrical responses of the olfactory epithelium of Atlantic salmon (Salmo salar), J. Fish. Res. Board Can. **28:**565, 1971.
78. Tucker, D.: Olfactory cilia are not required for receptor function, Fed. Proc. **26:**544, 1967.
79. Tucker, D.: Personal communication, 1978.
80. Tucker, D., and Shibuya, T.: A physiologic and pharmacologic study of olfactory receptors. In cold Spring Harbor Symposia on Quantitative Biology: Sensory receptors, New York, 1965, Cold Spring Harbor Laboratory of Quantitative Biology, vol. 30.
81. Van der Wel, H., and Loeve, K.: Isolation and characterization of thaumatin I and III, the sweet-tasting proteins from Thaumatococcus danielli Benth, Eur. J. Biochem. **31:**221, 1972.
82. White, L. E.: Olfactory bulb projections of the rat, Anat. Rec. **152:**465, 1965.

VI

SOME ASPECTS OF HIGHER NERVOUS FUNCTION

20

GERHARD WERNER

The study of sensation in physiology

PSYCHOPHYSICAL AND NEUROPHYSIOLOGIC CORRELATIONS

Research in psychology and work in physiology closely interrelate in the study of sensation. Sensory processes can be explored at various levels of analysis—at the introspective level, with the subject providing verbal reports of his sensory experience; at the level of externally observable behavior, such as adjusting a certain stimulus under the subject's control to a criterion value; and, finally, in terms of neural processes engendered by sensory stimuli. Each of these levels of analysis will be illustrated with examples intended to demonstrate how sensation is studied and in what form sensory performance in psychophysical and behavioral tasks can be correlated with spatial and temporal patterns of neural activity. The underlying notion is that events in the physical environment of a given organism find a representation in its nervous system; the goals are to discover the relation between sensory performance and spatiotemporal patterns of neural activity and to account for the former in terms of the latter.

Representation and coding in the nervous system

In general the representation of physical events involves the transformation of input signals into output messages by means of transducers. In this process of transformation, any physical resemblance between the input signal and the output may be discarded, as is the case when chemical or physical stimuli impinging on the body's receptors generate trains of nerve impulses in sensory pathways. Another illustration is the representation of verbal or numeric messages as magnetized points on a metallic tape, as in computer technology. Thus a representation may be quite abstract, reflecting only some of the qualitative properties or quantitative relations in the input signal and excluding others. Yet the transformation of the input signal to its representation retains some correspondence between the two, to the extent that certain events (or combinations of events) of the repertoire of input signals are unambiguously matched by certain features of the output.

The nature of the correspondence between signals and their representation may be of different forms. In the simplest case there is a well-defined quantitative relation between input into and output from the transducer, for example, between sound pressure acting on a microphone and the electrical current thus generated. In such cases the rules of correspondence between input and output are of the form of a continuous function, expressed in mathematical terms. The study of stimulus-response relations in psychophysics and in some areas of quantitative neurophysiology follow this paradigm.

In other cases, it is more appropriate to view the representation of a signal as a discrete mapping from the domain of input signals into the domain of output symbols; this is a form of coding.

To illustrate the concept of coding, consider the set of events that can occur at the input as consisting of the elements $V = \{v_1, v_2, v_3, v_4\}$; let $A = \{o, 1\}$ be the elements of which the encoded representation of events of V can be composed. An example of a code could be the following mappings (M) from V to A:

$$M(v_1) = 1$$
$$M(v_2) = 011$$
$$M(v_3) = 010$$
$$M(v_4) = 00$$

A code such as this can be thought of as a simple kind of automaton, that is, a device that accepts symbols from the set V and produces symbols composed of the elements of A according to predetermined rules.

Clearly there can be a sequence of consecutive encoding schemes interposed between the stimulus domain and an interpretive device so that the output of one encoder serves as the input to an encoder of a higher level of abstraction. This is a distinct possibility in the central nervous system (CNS) wherever stimulus-evoked afferent neural activity is transmitted through a succession of sensory relay stations and cortical receiving areas, the neurons of which tend to signal stimulus configurations of progressively increasing complexity.

A simple illustration of this possibility is the interpretation by Hubel and Wiesel[26] of some of their own experimental findings with the receptive field organization of neurons in the lateral geniculate body and of the neurons with "simple" receptive fields in cortical area 17 (Chapter 17). For these cortical neurons, specifically oriented lines tend to replace circular light spots as the optimal stimuli. Fig. 20-1 illustrates a possible scheme that is capable of generating a "code" of a higher order abstraction (consisting of line segments as elements of the code alphabet) from a lower order representation of light stimuli in the form of neurons with circular receptive fields.

These introductory remarks on the concepts of coding and representation are intended to set the stage for an appreciation of the experimental strategy employed in the investigations of the nature of correspondences between neural and sensory events.

In the attempts to recognize in neural activity the features that represent the properties of stimuli, several experimental approaches in neurophysiology have proved to be informative. On the one hand, electrical responses in populations of neurons, recorded as "evoked potentials" with relatively gross electrodes, can be taken as indicative of the location and magnitude of a neural response to a stimulus. On the other hand, the stimulus response of individual neurons can be observed with microelectrodes. In this case, it is possible to analyze the neural response in terms of a mean discharge rate and also in terms of the sequential ordering of the neuron discharge, in the expectation that the latter might bear some relation to certain stimulus attributes.

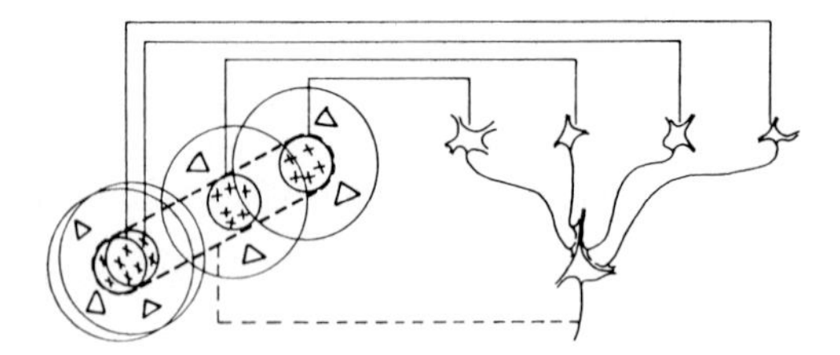

Fig. 20-1. Possible scheme for explaining organization of simple receptive fields. Large number of lateral geniculate cells (four are shown at right) have receptive fields with "on" centers arranged along straight line on retina. All project on single cortical cell, and synapses are supposedly excitatory. Receptive field of cortical cell will then have elongated "on" center (indicated by interrupted lines in receptive field diagram to left). (From Hubel and Wiesel.[26])

Applications of these approaches will be illustrated later in this chapter, after an introductory discussion of the principles that underlie the evaluation of sensory functions in psychophysics.

PSYCHOPHYSICS: INTROSPECTIVE AND BEHAVIORAL METHODS USED TO MEASURE SENSATION

G. T. Fechner,[2] who is generally considered to be the founder of psychophysics as it is currently understood, developed the ideas that sensations, differing in degree, could be assigned numeric values and that their relative magnitudes would stand in a certain mathematical relation to the magnitudes of the corresponding physical stimuli.

Fechner set out to determine how "sensation intensity" varies with physical stimulus magnitude. The method was essentially that used earlier by Weber: he would set a variable stimulus (S_v) equal to a standard stimulus (S_s) and then change S_v by small steps until the observer reported that he felt a just noticeable difference (jnd). The method was therefore based on introspection, a procedure that uses one and the same person as both subject and instrument of measurement. With early empiricism the belief grew that introspection as a form of immediate experience is incorrigible and, by definition, not subject to error; it provides privileged access to private experience that can be shared with others through the medium of language. However, in the absence of objective referents, the words of the language of introspection can never be unambiguously defined, except perhaps in the restricted context of a well-defined psychophysical test situation consisting of observation of a single physical event in isolation.

Criticisms such as this and the lack of reliability of the introspective method were some of the forces that contributed to the rise of the school of behaviorists in the early 1910s. J. B. Watson,[13] as did others, found that animals can learn to respond differentially to different stimuli. In the language of the then-current sensationalistic psychology this implied that they could discriminate, say, one tonal frequency from another, that is, that their behavior could be viewed as indicating differences in sensory experiences. Consequently, there would be no longer any need for considering subjective sensory experience as an independent experimental variable in psychologic studies; rather, psychophysics could be brought into harmony with the principle of the scientific method by examining correlations between two sets of objectively observable phenomena: physical stimuli and overt behavior. Actually, Watson showed that it was possible to describe all psychologic facts established at that time in terms of behavioral manifestations only, with no need to bring consciousness or sensation into the picture. By 1930 the mainstream of behaviorism was fused with and largely superseded by the dominant philosophic movements of that period, namely, logical positivism and operationism. The principal repercussions of this were twofold. In the first place, Stevens[42] could rightly point out the possibility of measuring subjective sensation magnitude by requiring no more than that the subject make some observable behavioral response; this was not to deny the existence of a subjective response but to replace it, for the purpose of measurement, by a behavioral expression. Accordingly, sensation is not judged from introspective reports; instead, its measurement is based on overt behavioral acts. This concept of operationist psychophysics makes it possible to study sensation in animal experiments by essentially the same procedures as are applied in man, except that differential reinforcement techniques for control of animal behavior take the place of verbal instruction to human subjects. The common feature is the type of stimulus control technique that was developed by Békésy for human audiometry studies.[47]

The principle of this procedure is that the intensity of a certain stimulus increases continuously as long as a signal button is pressed and decreases automatically when this button is released. Through control of the button the subject is able to let the stimulus intensity fluctuate between just above and just below threshold. By recording these intensity fluctuations, a difference limen for intensity and an absolute intensity threshold can be determined without any verbal (introspective) report from the subject.

In an analogous situation, experimental animals can be made to confront two response keys (A and B) and a small lighted stimulus patch. In a training period (which is, in effect, the counterpart to the verbal instruction given a human subject) the light patch changes in a random sequence between illumination and darkness: pressing key A blanks out the stimulus patch and pressing key B is followed by food reward if the light patch is dark. In this manner the animal acquires the behavior of pressing key A when light is visible and pressing key B when the light patch is dark. In addition, appropriate switching circuitry reduces the luminance of the stimulus if key A is pressed, whereas pressing key B increases luminance. As a result, the stimulus is kept oscillating about threshold. This procedure has been successfully employed to determine, for instance, the scotopic spectral sensitivity in various species.[18]

The extent to which behavioral and neural data can jointly be brought to bear on studies of sensation is illustrated with an example of recent work on vision. As an important attribute of human vision, stereopsis has been studied for many years as the sensation of relative visual depth that results from the dissimilarities of the images seen by the two eyes. To avoid interference by various familiarity cues and other uncontrollable circumstances, Julesz[30] devised a novel technique for random dot stereo images. An example of such a stereo pair is shown in Fig. 20-2. When viewed monocularly, both fields of Fig. 20-2 give a homogeneous random impression without any apparent features, but when viewed stereoscopically, a center square is vividly perceived in front of its surround.

Fig. 20-3 illustrates how Fig. 20-2 and similar random-dot stereo images are generated. The description of the procedure is here reproduced in Julesz's own words:

The equally probable randomly selected black and white picture elements which are contained in corresponding areas in the left and right fields are labeled in three categories: (1) Those contained in corresponding

Fig. 20-2. Basic random stereo pair. When two fields are viewed stereoscopically, center square appears in front of background. (From Julesz, B.: Binocular depth perception without familiarity cues, Science **145**:356, copyright © 1964 by the American Association for the Advancement of Science.)

1	0	1	0	1	0	0	1	0
1	0	X	A	A	B	B	0	0
0	0	Y	B	A	B	A	1	1
0	1	0	0	1	1	1	0	1
1	1	A	B	A	B	A	0	0
0	0	B	A	B	A	B	1	0
1	1	0	1	0	1	1	0	0
1	0	A	A	B	A	X	0	1
1	1	B	B	A	B	X	1	0
0	1	0	0	0	1	1	1	1

1	0	1	0	1	0	0	1	0
1	0	A	A	B	B	Y	0	0
0	0	B	A	B	A	X	1	1
0	1	0	0	1	1	1	0	1
1	1	B	A	B	A	B	0	0
0	0	A	B	A	B	A	1	0
1	1	0	1	0	1	1	0	0
1	0	Y	A	A	B	A	0	1
1	1	Y	B	B	A	B	1	0
0	1	0	0	0	1	1	1	1

Fig. 20-3. Illustration of method by which stereo pair in Fig. 20-2 was generated. (From Julesz, B.: Binocular depth perception without familiarity cues, Science **145:**356, copyright © 1964 by the American Association for the Advancement of Science.)

areas with 0 disparity (which when viewed stereoscopically are perceived as the surround) are labeled 0 or 1. (2) Those contained in corresponding areas with non 0 disparity (which when viewed stereoscopically are perceived in front of or behind the surround) are labeled A or B. (3) Those contained in areas which have no corresponding areas in the other field (that is, project on only one retina and thus have no disparity) are labeled X and Y. The 0 and 1 picture elements are identical in corresponding positions on the two fields. The positions of the A and B picture elements belonging to corresponding areas in the fields are also identical, but are shifted horizontally as if they were a solid sheet. Because of this shift some of the picture elements of the surround are uncovered and must be assigned new brightness values (X and Y). Since these areas lack disparity, they can be regarded as undetermined in depth. Fig. 20-3 contains three rectangles in the left and right fields, composed of A and B picture elements. Each field contains an upper, middle and lower rectangle which can be regarded as corresponding left and right "projections" of a rectangular planar surface located in depth when viewed from different angles. The projections of the upper rectangle (that is the corresponding upper rectangles in the left and right fields) are horizontally shifted relative to each other in the nasal direction by one picture element, the corresponding lower rectangles are shifted in the temporal direction to the same extent, while the corresponding middle rectangles have a one-picture-element periodicity and may be regarded as being shifted in either direction. The low density of picture elements and the large disparities would prevent stereopsis in a pattern corresponding to Fig. 20-2. In order to achieve stereopsis the number of picture elements would have to be decreased considerably. For this reason a computer is used.*

By combining the random-dot pattern technique with standard operant conditioning procedures, Bough[19] succeeded in determining that, like man, macaques have stereoscopic vision. The animals were required to distinguish random-dot patterns that, if viewed binocularly, would seem to have an inner square displaced behind the plane of the surround, from random-dot patterns in which no inner square was distinguishable during binocular vision. In 200 trials the two animals in this study reached correct performance levels of 90% and 98%, respectively.

In neurophysiologic studies with macaques, Hubel and Wiesel[28] ascertained that some 40% of the neurons in area 18 function as "binocular depth cells." These neurons usually respond most vigorously if both eyes are stimulated simultaneously. In some instances the responses are elicited by stimulation of exactly corresponding retinal regions; in other instances there is a disparity in the positions of the two receptive fields. This suggests that neurons in area 18 play a role in the elaboration of the stereoscopic depth mechanism that was shown to exist in this species in the behavioral tests of Bough.[19]

This example has been described in some detail, for it illustrates with particular clarity how a *qualitative* comparison between neurophysiologic and behavioral observations can contribute to the recognition of the neural processes that play a part in a particular sensory function. In the following sections, similar comparisons between sensory and neural activity will be discussed, with emphasis on *quantitative relations*.

Quantitative correlations between sensation and neural activity

The neurophysiologic line of inquiry essentially parallels the traditional approach of the psychophysicist, except that the latter tried to correlate stimulus values with sensation magnitudes, whereas the neurophysiologist attempts to correlate stimulus values with differences in the neural responses. A distinct stimulus attribute with well-defined physical properties is selected, and the magnitude of this attribute is varied over a range of values; for example, in the case of tactile stimuli restricted to a point of the skin, the displacement of the skin (measured in units of length), the force acting on the skin, or the impulse of a rapidly applied stimulus of the cutaneous surface (measured as the product of mass times impact velocity of the stimulating agent) could constitute suitable scales for the quantification of the stimulus. Similarly, the pitch and loudness of tones are stimulus attributes that can be directly measured on appropriate scales.

The next step involves measuring the stimulus-evoked neural activity and seeking a corre-

*Julesz, B.: Binocular depth perception without familiarity cues, Science **145:**356, 1964.

spondence between the stimuli of different values on their measurement scale and the neural responses they evoke.

This raises some difficult questions. What is an appropriate way to "measure" neural activity? And what is a suitable rule for assigning numeric values to neural activity so that differences in magnitude of neural activity can be represented as relations of greater and smaller and, under more exacting conditions, as numeric differences on a continuous and monotonic scale?

These questions are not unique to the measurement of neural activity: they underlie all measurement. But most conventional measurement scales (e.g., weight or length) are embedded in some conceptual framework from which they derive their justification. In employing these conventional scales of measurement, we do not ordinarily reflect on what Wiener referred to: "Things do not, in general, run around with their measures stamped on them like the capacity of a freight car: it requires a certain amount of investigation to discover what their measures are."[56]

In the absence of a suitable conceptual framework to guide the selection of an appropriate scale for measuring neural activity, the procedure is purely experimental and pragmatic: it is necessary to vary the stimulus, to "measure" the neural activity in various ways, and to discover the scale of neural activity measurement that makes the functional relation between the stimulus and the magnitude of the neural response tally with that between the stimulus and a psychophysical measurement of sensation. It is then possible to test further whether these relationships also tally under a variety of conditions, for example, variations of the attentive set, or under the influence of pharmacologic agents with known effects on behavior and subjective experience in man.

The ultimate goal is to find a particular means of characterizing neural responses in quantitative terms (i.e., a scale to measure neural responses) so that the functional relation between stimulus and neural response is of the same form as the relation of stimulus to "sensation," as measured in an independent psychophysical or behavioral experiment. It may then be said that this particular functional relation between stimulus and neural response qualifies as the "neural code" of the stimulus attribute under study.

The selection of distinct stimulus attributes for study reflects the idea that sensation can be viewed as being composed of individual and mutually independent attributes, of which classic psychophysics defined four: quality, intensity, duration, and extension. The independence of these attributes is fictitious, for there cannot be an intensity in sensation except as an intensity *of* a quality, and this argument can be applied mutatis mutandis to the relation of the other attributes as well. Nevertheless, for experimental purposes the concept of isolated attributes (notably the stimulus attribute of intensity) can be useful and informative, as long as one remains mindful of its restricted validity.

INTENSITY ATTRIBUTE IN PSYCHOPHYSICS AND NEUROPHYSIOLOGY

There are four types of questions that can be asked concerning the magnitude of a sensation.[3] The question *"Is anything there?"* leads to the *detection problem;* the *problem of recognition* includes the systematic attempts to answer the question *"What is it?";* we cannot meaningfully and consistently answer the question *"How much of it is there?"* unless we give a systematic account of the *scaling problem;* and finally the question that, historically, led to the foundation of quantitative psychophysics and its original preoccupation with the *discrimination problem: "Is this different from that?"*

Each of these questions has its counterpart in quantitative neurophysiology. In observing neural activity, we wish to know whether the response to stimulus A differs from that to stimulus B and, if so, in what quantitative manner. This reflects the close interrelation of discrimination and scaling. When we wish to detect from ongoing neural activity whether a stimulus has been applied and, if so, which one of a given set of alternative possibilities it was, we seek to determine the way in which the CNS could conceivably perform the task of stimulus detection and recognition.

Discrimination problem

Investigations of sensory discrimination were begun early in the 19th century by the physiologist E. H. Weber. Weber's procedure was based on what is now known as "comparative unidimensional judgments": one particular stimulus of a certain intensity (the standard stimulus) is applied in alternation with one of a number of other stimuli (the comparison stimulus) that are of the same type but that differ in physical stimulus magnitude. A set of standard and comparison stimuli could, for instance, be a series of tones of equal pitch and different loudness or tactile stimuli differing in the force applied to the skin. The subject is instructed to note whether the second stimulus in each pair is stronger or weaker than the standard stimulus, which is applied as the first stimulus in each pair. When this is done a large number of times for each stimulus pair, the probability with which the comparison stimulus in each pair is judged stronger or weaker than the standard can be estimated. Fig. 20-4 presents experimental data on loudness discrimination, and Fig. 20-5 displays schematically the typical properties of these data. Note that at the point with the ordinate $p = 0.5$, comparison and standard are judged to be of equal intensity; this point of subjective intensity frequently differs from objective equality. The difference generally depends on the time interval between the presentation of comparison and standard stimuli and tends to decrease as this interval becomes larger.

The interquartile range of this discrimination function is that interval on the abscissa that is delineated by the probabilities 0.25 and 0.75, respectively, of judging the comparison stimulus stronger than the standard (Fig. 20-6). By convention, this difference in stimulus intensity at the 0.25 and 0.75 point is divided by two and then measures the jnd; in the example in Fig. 20-6 the jnd is 0.9 stimulus units. Obviously, good discrimination capacity means a small jnd and vice versa.

The discrimination function depicted in Fig. 20-6 is designated the "psychometric function." It is informative with respect to the nature of stimulus discrimination in the neighborhood of the standard stimulus chosen for this particular series of measurements. One can now ask how this function varies with changes in the magnitude of the standard stimulus? This Weber determined, and he came to the empirical conclusion that the jnd varies with S in the form $\Delta S = K \cdot S$, where $K > 0$ is called the Weber fraction. This relation is known as Weber's law. It has become customary to plot such data as $\Delta S/S$ vs S, in which case Weber's law describes a line parallel to the abscissa.

Later investigators extended such discrimination measurements over a wider range of stimulus intensities than were applied by Weber. Their results revealed a considerable departure from the linear relation of Weber's law for most sense modalities tested, notably for low intensities of the standard stimulus (Fig. 20-7). Following a recommendation of Luce and Edwards,[6] it is now general usage to designate the empirical relation $\Delta S/S = f(S)$ as the Weber function, of which Weber's law would be the special case of linearity over the entire range of S.

Weber functions for tactile discrimination depart in monkeys as they do in man from the linearity of Weber's law.[40] This may now be used for comparison with observations in neurophysiologic studies. However, although the criterion for the stimulus discrimination in psychophysical measurements is the jnd, what might constitute a comparable, liminal increment in neural activity is not at all certain. Therefore the objective of such comparisons between psychophysical and neurophysiologic discrimination studies is to find, essentially by trial and error, units and a scale for measuring neural activity that would bring the Weber function of neural responses in a certain modality into correspondence with that

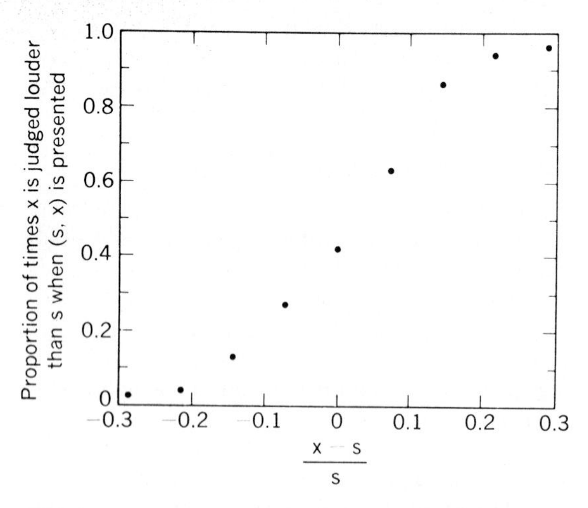

Fig. 20-4. Loudness discrimination when pairs of stimuli, x, s, are presented: s = standard stimulus (50 dB or 0.0002 dynes/cm^2); x = comparison stimulus. Each data point is based on 105 observations collected over three sessions. Note that abscissa is plotted on linear dimension scale with $x - s/s$. (Unpublished data by Galanter; from Luce et al.[7])

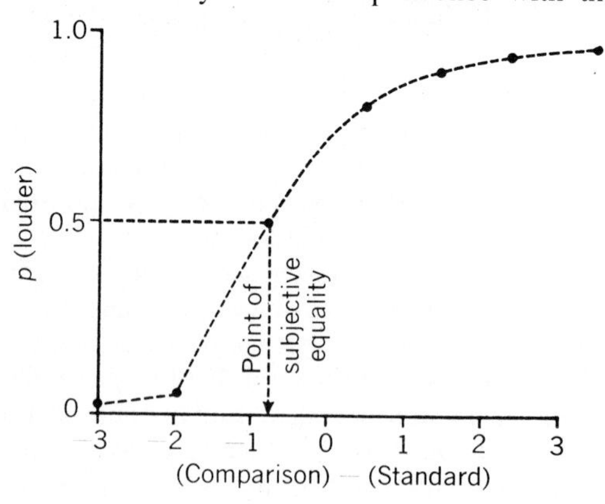

Fig. 20-5. Schematic representation of series of measurements of stimulus discrimination obtained with method of comparative unidirectional judgment. Ordinate plots proportion of times comparison stimulus is judged more intense as probability; abscissa plots difference between standard and comparison stimulus intensity. (From Galanter.[3])

of its psychophysical Weber function. This approach was applied to neural activity evoked by tactile stimuli in first-order cutaneous afferents[53]: it was found that correspondence between neural and psychophysical Weber functions could be attained if the critical neural response increment (ΔR) for "recognition" of a stimulus increment was assumed to be a constant number of impulses, equal at any response level for any standard stimulus. This assumption about the nature of the discriminable ΔR is parsimonious, and it is conceivable that other, more complex assumptions concerning ΔR could lead to the same correspondence. On the other hand, the assumption that a constant fractional increment (ΔR/R = C) in neural activity is discernible over the entire stimulus range does *not* lead to a correspondence of neural and psychophysical Weber functions (Fig. 20-8). This supports the suggestion that the primate nervous system may measure equal increments of neural activity, rather than equal fractional increases, in order to generate the type of Weber functions established in psychophysical experiments.

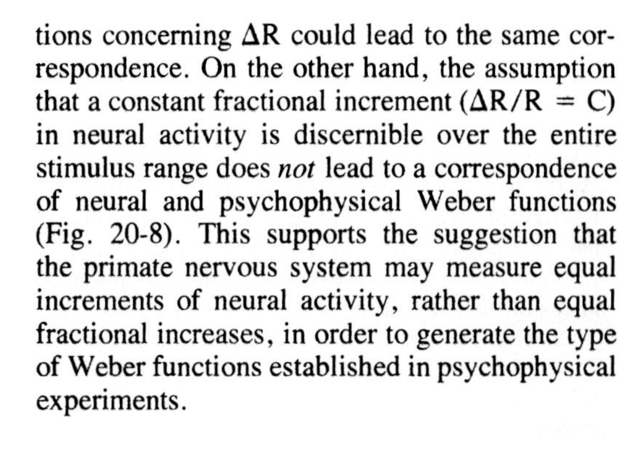

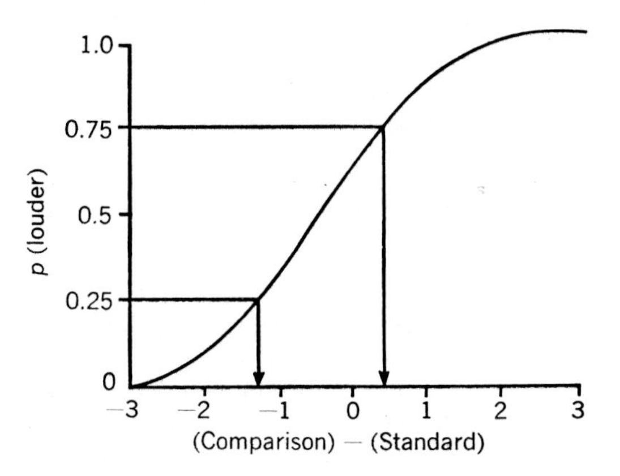

Fig. 20-6. Illustration of definition of "just noticeable difference" (jnd). See text for discussion. (From Galanter.[3])

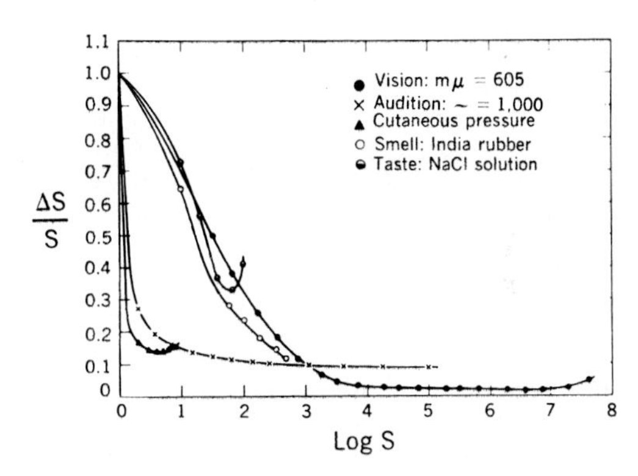

Fig. 20-7. Plots of Weber functions obtained for five different sense modalities: on ordinate, Weber fractions; on abscissa, intensity of base stimulus in logarithmic units. Weber fractions are normalized to unity at threshold. Note that if Weber's law were valid, individual graphs would assume form of straight lines parallel to abscissa. (Based on data by Holway and Pratt; from Luce et al.[7])

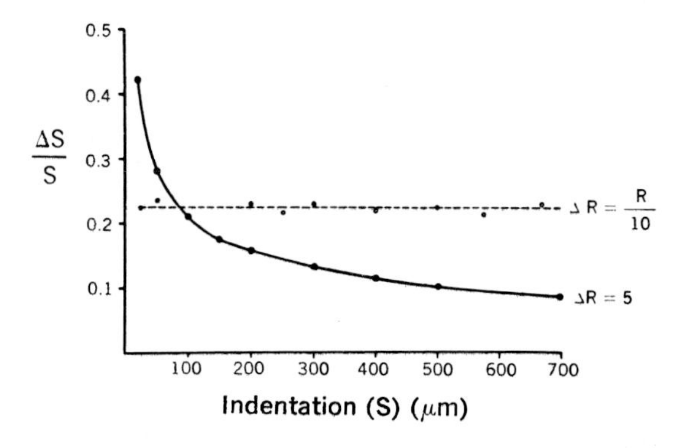

Fig. 20-8. Graphs of two Weber functions, ΔS/S = f(S), computed from stimulus-response curve of first-order cutaneous afferent nerve fiber responsive to mechanoreceptor stimulation. Stimulus-response curve for skin indentations of 1 sec duration was $R = 7.004 \times S^{.444}$ (*R* = number of nerve discharges in 1 sec; *S* = skin indentation in micrometers). Solid points connected by curve were obtained by assuming that least discriminable response increment, ΔR, is 5 impulses/response at any intensity of standard stimulus, *S*. Open circles connected by dashed line were obtained by assuming that ΔR is constant fraction (i.e., 1/10) at response at any intensity of standard stimulus. Note that only curve ΔR = 5 is of the same form as psychophysically determined Weber functions. (From Werner and Mountcastle.[53])

Although concerned with an extensive rather than an intensive stimulus property, size and distance are aspects of the discrimination problem. Since Weber's work on cutaneous sensibility, it has been customary to use the minimal distance at which two points touching the skin can be recognized as separate stimuli as a measure of sensitivity for spatial stimulus attributes. This sensitivity differs in a characteristic manner at different body sites, for example, it is maximal at the apices of the extremities and diminishes with a proximodistal gradient toward the trunk.[50] The comparison of two-point discrimination thresholds with the capability to recognize size differences of disks gently placed on the skin revealed a far superior performance in the latter test.[46] Thus the cutaneous sense appears much more suited to perform size than distance discriminations. The general implication is that the choice of the test to characterize the capability of a sensory system is critical: in general the task should match the role this sensory system plays in the life of the species under natural conditions, for these conditions shaped the design of the nervous system in evolution.

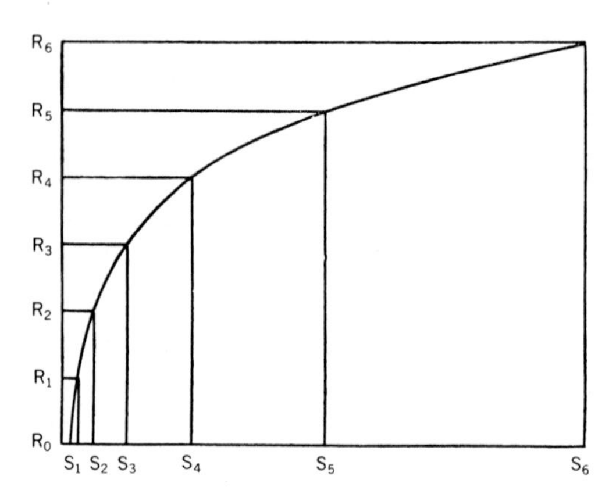

Fig. 20-9. Schematic representation of psychophysical scale proposed by Fechner (see text). Equidistant segments on ordinate are sensation jnds; abscissa plots stimulus magnitude and indicates stimulus increments (stimulus jnds) that correspond to different sensation jnds at different points along sensation continuum.

Scaling problem

As outlined on p. 606, Fechner's idea was this: for many of the primary physical stimulus attributes (e.g., loudness, brightness, weight) there would exist corresponding dimensions of sensory experience; changes in the magnitude of the one would be related to changes in the magnitude of the other in the form of a single-valued, monotonic, and everywhere differentiable (smooth) function. This requires a distinction between two kinds of jnds: one that measures the just discriminable increment of the physical stimulus magnitude as defined earlier by Weber and a second, distinctly different, that is measured in the units of the appropriate sensation continuum. Fechner defined the latter jnd (the sensation jnd) as the unit on the sensation scale and assumed all sensation jnds to be subjectively equal over the entire sensation continuum. The implication of this is shown in Fig. 20-9: two stimuli separated by one (stimulus) jnd at the low end of the intensity scale give rise to the same subjective difference in sensation intensity as two stimuli at the upper end of the intensity scale that are also one stimulus jnd apart. Accepting the validity of Weber's law ($\Delta S/S = K$), this definition of the sensation jnd (ΔR = constant) permitted Fechner to formulate the simple mathematical relationship $\Delta S/S = A \cdot \Delta R$, where A is, by the definition of the magnitudes involved, a constant. Next Fechner applied a principle often used in mathematical physics: what is a valid relation for small differences remains valid in the limit as the differences approach infinitesimally small values. Accordingly, the previous equation can be written in the form $dS/S = A \cdot dR$, which on integration yields Fechner's law in the form $R = A \cdot \log S/S_0$; S_0 is a constant of integration, to be interpreted as the threshold stimulus strength below which there would be no corresponding sensation elicited. This seemed to accomplish Fechner's objective to bringing physical stimulus intensity and sensation magnitude into a unique mathematical relationship of the same general form for all sensation continuums.

For more than 50 years, Fechner's law remained almost entirely unrivaled and his assumptions unchallenged. Finally, criticisms arose. First, evidence accumulated to indicate that Weber's law is not in all cases entirely valid (Fig. 20-7). This leads to an internal logical inconsistency in the derivation of Fechner's law, for Luce and Edwards[6] proved that, except for the special case of Weber's law, the assumption of the equality of the sensation jnds and the transition, in the limit, to the differential equation are mutually contradictory. A second source of dissatisfaction relates to Fechner's "indirect" method of scaling, in which the unit of measurement is derived from the discriminatory capacity and its statistical fluctuations.

To circumvent this latter objection, two alternative methods for psychophysical scaling were proposed and are presently in use. These are, first, the scaling procedures that require the observer to partition a given segment of a certain sensation continuum into a predetermined number of subjectively equal intervals and, second, those procedures that require the observer to make a direct judgment of apparent sensation

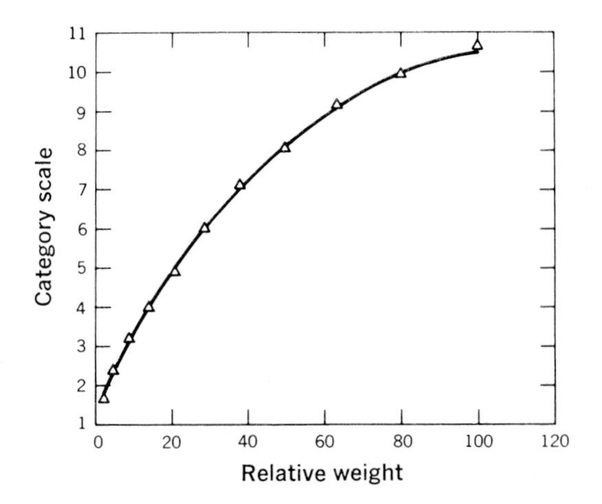

Fig. 20-10. Category scale for apparent weight. Lifted weights cover range from 43 to 292 gm. Weights were logarithmically placed within that range. (Modified from data by Stevens and Galanter.[44])

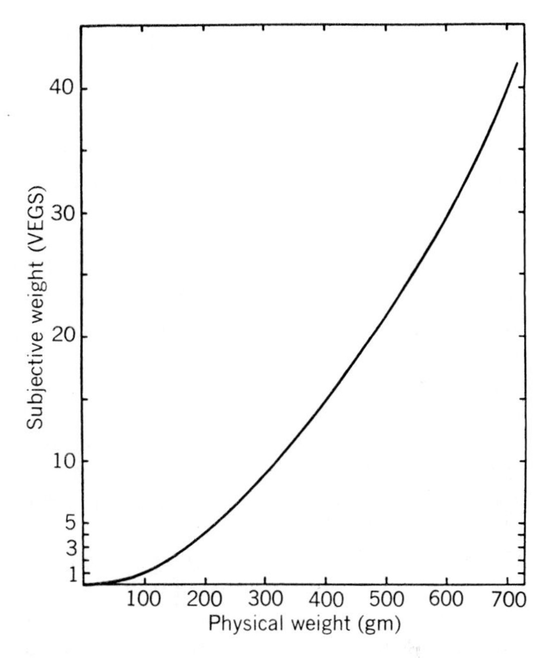

Fig. 20-11. VEG scale of subjective weight. Curve shows how subjective weight varies as function of physical weight. One VEG is weight experienced by lifting 100 gm. (From Harper and Stevens.[24])

magnitudes. In the former case the general protocol of the experiment is as follows: the subject is told the designation of a set of categories (e.g., the first 11 integers); he is then exposed to the weakest and the strongest stimuli that delineate the segment of the stimulus range that will be used in the final test, and he is told to place the weakest stimulus in category 1 and the strongest in category 11. The task is now to distribute the subsequently presented stimuli among the 11 categories in such a manner that the sensation intervals between categories are perceived as subjectively equal. The simplest analysis of the final data consists of calculating the mean category assignment for the different stimulus intensities used in the test; these mean values are the sensation scale values of the respective stimulus intensities. In Fig. 20-10 the ordinate is marked off in 11 equal intervals that correspond to the subjectively equidistant sensation categories. The abscissa plots the relative intensity of the stimuli used (in this case, weights). The category scale for apparent weight results directly from this.

The second group of psychophysical scaling methods, aimed at a direct estimation of the apparent sensation magnitude, is based on the following general protocol: a certain stimulus is presented, and the subject is asked to adjust a variable stimulus to a value that is subjectively either half or twice as large. Using weight as an example, we define as a unit of subjective weight the apparent weight perceived when 100 gm is lifted; Harper and Stevens[24] have designated this unit of subjective weight as 1 VEG. The subject now determines in a series of trials that a weight of, say, 72 gm is subjectively half as heavy as the standard of 100 gm. Therefore a subjective scale value of 0.5 is assigned to correspond to 72 gm physical weight. Next, we may find that 140 gm is judged twice as heavy as 100 gm: accordingly, a subjective scale value of 2 VEG is assigned to correspond to the physical weight of 140 gm. The systematic continuation of this procedure leads to the scale shown in Fig. 20-11. In this case, fractionations of sensation magnitude elicited by a standard stimulus were the operations that led to the scale construction.

Of course, subjects do not invariably assign the same number (or category) to a particular stimulus. The variation in the values assigned to the same stimulus in repeated trials may be considerable: the standard deviation of the estimated magnitude may be as much as 20% to 40% of the mean value. Most of the data published in the psychophysical literature are based on "averages" obtained in trials with different subjects and are intended to represent a "typical" scale.

The comparison of the psychophysical scales obtained with the three different methods discussed so far (i.e., the fechnerian method, category estimation, and magnitude estimation) discloses some important general relations. First, the resulting scales often do not correspond with one another. This is shown in Fig. 20-12; the departure of the curves for magnitude and category scales from the psychometric function depicting the fechnerian scale is indicative of the failure of the former two scales to fulfill Fechner's assumption of the equality of the sensation jnds. On the other hand, there are certain sensation continua for which scales obtained by the three different scaling procedures are indeed linearly related. Stevens and Galanter[44] have shown that sense continua can be divided into two classes, depending on whether or not the relation between magnitude and category scales is linear. They have designated as class I (or prothetic) continua those for which the relation of the two scales always deviates from linearity in a manner illustrated in Fig. 20-13. These examples, depicted schematically, indicate (and other instances corroborate) that this relation applies to the intensity dimension in sensation. It has been suggested that it is characteristic for those sensation attributes for which increase in stimulus intensity is accompanied by some neural additive process of excitation. Sense continua of the second type (class II, or metathetic continua), for which magnitude and category scales may be linearly related, include pitch, position, and others for which there is reason to believe that the stimulus-evoked neural activity shifts from one locale to another, as the stimulus parameter in question varies over a series of values.

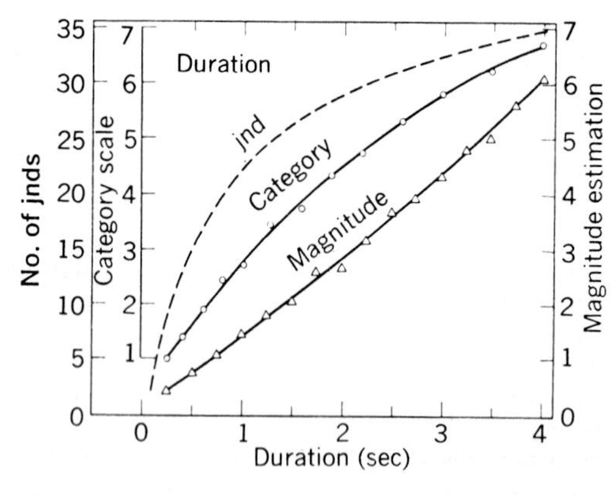

Fig. 20-12. Three kinds of psychological measures of apparent duration. Triangles represent mean magnitude estimations by 12 observers who judged apparent durations of white noises. Circles represent mean category judgments by 16 observers on scale from 1 to 7. Two end stimuli (0.25 and 4 sec) were presented at outset to indicate range, and each observer twice judged each duration on 7-point scale. Dashed curve is discriminability scale obtained by counting off jnds. (Modified from Stevens.[11])

The second generally valid conclusion that has emerged from the studies on psychophysical scaling concerns a surprising uniformity of the scales obtained with the magnitude estimation procedure for a large number of different sense continua. In the cases listed in Table 20-1, and in others as well, Stevens[11] and co-workers have shown that the sensation magnitude relates to the physical stimulus magnitude in the form of a power function $R = K \cdot S^n$, with the exponent n being a numeric parameter, characteristic for the individual sense continua.

It is important to note that the general validity of this "power law" depends on the appropriate selection of the physical stimulus scale. It was pointed out earlier that magnitude estimation techniques involve essentially a fractionation of sense magnitudes; therefore these procedures lead to a numeric scale for sensation that is a "ratio scale," that is, a scale that reflects quantitative relations of the measured property faithfully only if no algebraic operations other than multiplication or division are performed on the measured values. The addition or subtraction of a constant leads to loss of information. Therefore the origin of a ratio scale (its zero value) must be an invariably fixed reference point, like a constant of nature. In psychophysical measurements the threshold of a stimulus subserves this role of a fixed reference point; it is then possible to measure stimuli in terms of a ratio scale of *distance* from threshold. This becomes particularly important at low values of sensation magnitude. As shown in Fig. 20-14, when apparent temperature is scaled by magnitude estimation and the results are plotted against the absolute temperature scale (with both scales in logarithmic units), the data fall on a concave curve, but when plotted in terms of degrees above threshold intensity for warmth sensation, the data plot on a straight line. Thus the data fit a power function of the general form:

$$R = k(S - S_0)^n$$

where S_0 is the threshold stimulus intensity.[41]

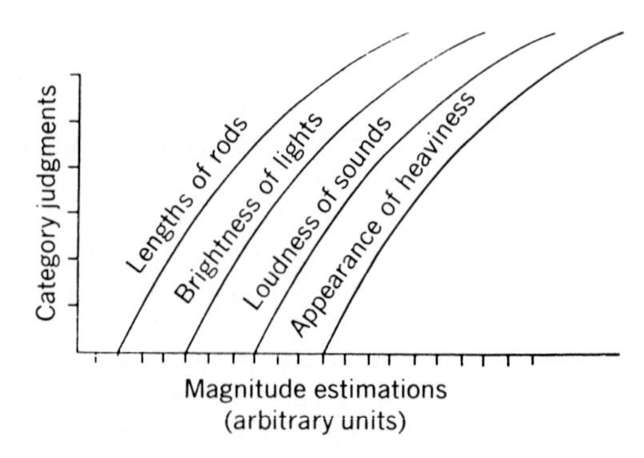

Fig. 20-13. Relation between magnitude and category judgments. (From Galanter.[3])

The third important consideration pertains to the influence that the selection of the stimulus values and the temporal sequence in which the stimuli of different intensities are applied has on the sensation estimate of the subjects. Although category scales of one and the same subject vary with these parameters, it is a notable property of magnitude scales that they are invariant with respect to changes in the stimulus ensemble. As a result, it has been suggested that magnitude scales reflect the events on which sensory judgments are based more directly than psychophysical measurements involving category scales.

Scaling of stimulus and response in quantitative neurophysiologic studies. Mountcastle et al.[33] have conducted an investigation in which the discharges of individual neurons responding to steady position and movement of joints were recorded in the ventrobasal thalamic nuclei of macaques. The experimental records permitted evaluation of the quantitative relation between the angular position of a joint and the mean discharge rate of a thalamic neuron that was responsive to differences in position of that same joint.

For an analysis of the precise functional relation between these variables the principle of converting measurements to ratio scales was applied, that is, the numeric values of the variables were referenced to their natural zero values as scale origin. This required expressing the angular position as increment over the joint angle that is the threshold position for the neuron under study (Θ_T) and plotting as response the increment of mean discharge rate over the spontaneously ongoing activity (c) that persists while the joint is outside the excitatory angle of the neuron under study. The straight-line relation between the variables displayed in Fig. 20-15 indicates that the relation between joint position and mean neuronal discharge rate, after appropriate scaling of the variables, can be described by a power function.

The essential point of this study is the demonstration of a lawful, quantitative relation between the appropriately scaled stimulus intensity (provided, in this case, by the angular position of

Table 20-1. Representative exponents of power functions relating psychophysical magnitude to stimulus magnitude on prothetic continua*

Continuum	Exponent	Stimulus conditions
Loudness	0.60	Binaural
Loudness	0.54	Monaural
Brightness	0.33	5° target—dark-adapted eye
Brightness	0.50	Point source—dark-adapted eye
Lightness	1.20	Reflectance of gray papers
Smell	0.55	Coffee odor
Smell	0.60	Heptane
Taste	0.80	Saccharine
Taste	1.30	Sucrose
Taste	1.30	Salt
Temperature	1.00	Cold—on arm
Temperature	1.60	Warmth—on arm
Vibration	0.95	60 Hz—on finger
Vibration	0.60	250 Hz—on finger
Duration	1.10	White noise
Repetition rate	1.00	Light, sound, touch, and shocks
Finger span	1.30	Thickness of wood blocks
Pressure on palm	1.10	Static force on skin
Heaviness	1.45	Lifted weights
Force of handgrip	1.70	Precision hand dynamometer
Autophonic level	1.10	Sound pressure of vocalization
Electrical shock	3.50	60 Hz—through fingers

*From Stevens.[43]

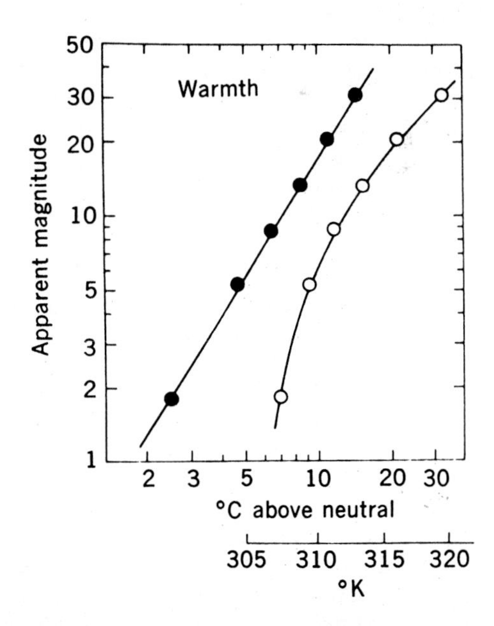

Fig. 20-14. Magnitude estimation of apparent warmth. Each point is geometric mean of 36 estimates (12 observers). Upper abscissa, for filled points, is log scale of difference in temperature (Celsius) between stimulus and "physiologic zero." Lower abscissa, for unfilled points, is log scale of absolute temperature (Kelvin). (From Stevens and Stevens.[41])

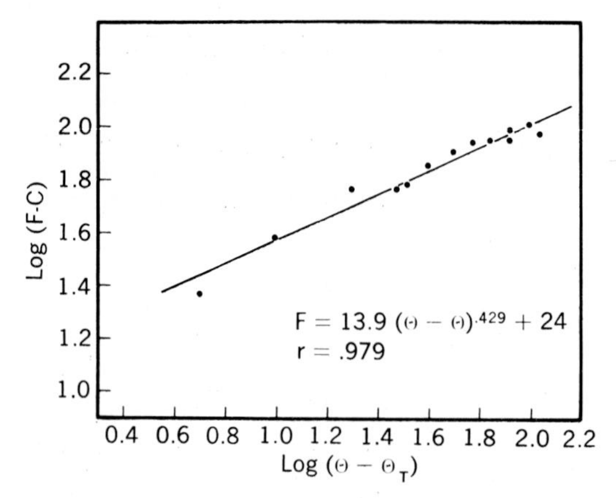

Fig. 20-15. Plot of mean rate of discharge (impulses per second) vs angle for ventrobasal thalamic neuron, driven by extension of contralateral knee joint. Abscissa marks in logarithmic units departure of angular position of joint, Θ, from threshold position for excitation of neuron, Θ_T. Ordinate plots on logarithmic scale the increment of discharge rate above spontaneous activity, *C*. Thus both dependent and independent variables are expressed in reference to absolute zero point (Θ and *C*, respectively) and are ratio scales. Equation of functional relation between dependent and independent variables is written as inset (value of 24 represents spontaneous activity in impulses per second; *r* = Pearson). (From Mountcastle et al.[33])

joints) and the mean rate of thalamic cell discharges—a stimulus-response relation that is of the same general form as that which describes the functional dependence of the subjective sensation magnitude in man on the intensity of stimuli in a large number of different sensation continua.

In this study the neural activity was measured in the third-order sensory relay station of the medial lemniscal pathway, that is, at a level of the somatosensory system where extensive convergence of neural activity from peripheral receptors of the joint under study has occurred. The neural activity elicited in first-order cutaneous nerve fibers by graded indentation of slowly adapting cutaneous touch corpuscles was the subject of a similar experimental study[53]: in this case, too, the stimulus-response relation is best described by a power function, as in the human observer's estimate of stimulus magnitudes.[29]

One implication becomes immediately apparent: the serially superposed neural transforms leading from the site of neural activity recorded in these studies to the final behavioral (verbal) response must also be power functions, at least with regard to the intensity attribute.

Other more qualitative arguments can be adduced to strengthen this contention. One is that the sensation of passive movements of joints in man depends only on the function of the receptors in the joint capsule; muscle and tendon receptors are not involved.[37] The other argument is provided by the finding of Hensel and Bowman[25] that a single impulse in a single or in each of a very few cutaneous afferent fibers is sufficient to evoke a conscious sensation in man.

Detection problem

Historically the detection problem developed out of the question of the absolute threshold in sensation. Fechner's concepts and methods incorporated the idea of a lower limit of sensitivity, thought to be characteristic of sensory systems. He and his successors were aware of the inherent instability of this sensation threshold: a certain stimulus of fixed intensity would at some presentations elicit the response "Yes, I detect it," and at others the response "No, I don't detect it." It was also recognized from the beginning that the observer's attitude affected this threshold estimate. In this form the idea of a sensory threshold persisted as an essentially fictitious and sometimes controversial concept in psychophysics until the more recent availability of methods that permitted the isolation and quantification of the factors determining the limits of stimulus detection at any one time and their dependence on certain experimental variables.

One important step in this development was a change in the experimental design: instead of determining the threshold by varying the stimulus, the stimulus was kept at a fixed value and the question became how often the presentation of this stimulus elicited the correct response of detection, and how often the stimulus was reported to have occurred although it was actually not applied ("false alarm"). Accordingly, two kinds of errors can be made in this situation: failures to detect a signal when it is there and reports of the perception of a signal when it is not there. Detection is concerned with the factors that influence the relation between these two types of errors, and their implications for sensory mechanisms in general. With this the emphasis shifted to the problem of specifying and controlling the criteria the observer may use in making perceptual judgments.

The typical experimental situation is this[12]: sensory events occurring in a fixed interval of time are observed by the subject, who must then decide whether the observation interval contained only background interferences or a signal (stimulus) as well. The background interference is thought to be random; it may either by introduced by the experimenter or be inherent in the sensory process. It is commonly designated as

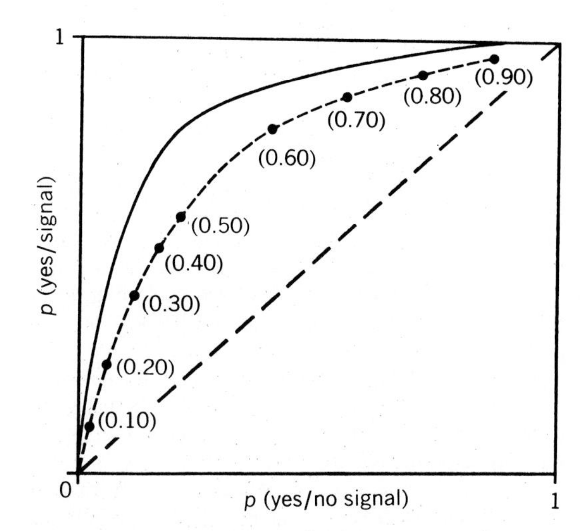

Fig. 20-16. Likelihood of correct signal detection—*p*(yes/signal)—and of false alarm—*p*(yes/no signal)—in relation to probability of signal presentation and signal intensity. Dashed curve with dots represents this relation for signal of medium intensity; numbers in parentheses are respective probabilities of signal presentation. Solid curve depicts this relation for signal of higher intensity. (From Galanter.[3])

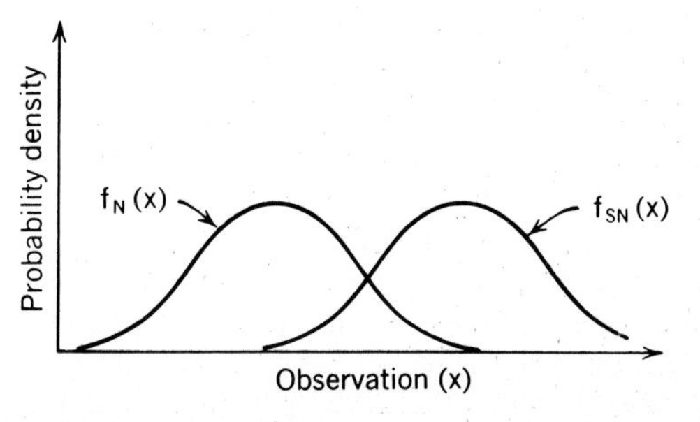

Fig. 20-17. Probability density functions of signal and signal plus noise (see text). (From Swets et al.[12])

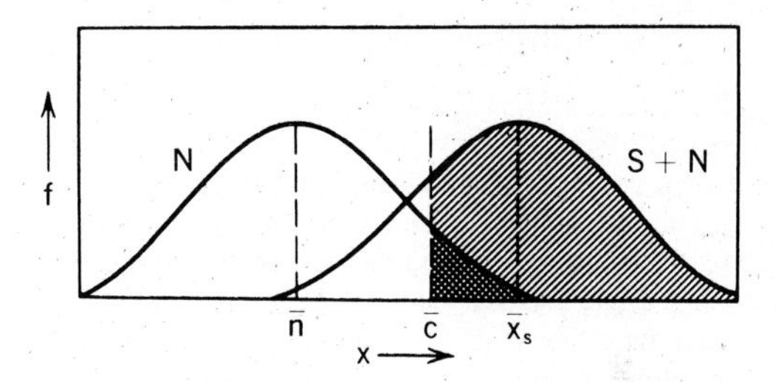

Fig. 20-18. Probability density of neural activity: *x*, for noise alone and for noise plus signal; *c* is criterion. (From Eijkman and Vendrik.[21])

noise (N). In the alternative case the sensory event is judged to consist of noise on which a signal is superimposed (SN); for example, the signal may be a flash of light in a known location on a uniformly illuminated background or a brief sound. The trials are repeated a certain number of times, under identical conditions, with observation periods *with* and *without* signals randomly intermingled. The outcome of a series of trials is expressed as the probabilities of "false alarm" and "correct detection."

One of the experimental variables found to affect the subject's performance under such conditions was prior information regarding the probability of signal presentation in the observation periods, as shown schematically in Fig. 20-16. When the subject is told that the signal will be present in a large percentage of the observation periods, the false alarm rate is very high, approaching almost 100%. On the other hand, with prior information as to a low relative rate of signal presentation, the subject often fails to detect the signal. Similarly, it is possible to modify the subject's performance in accordance with rewards offered for correct recognition and costs incurred with false alarms.

Although it is not necessary for the quantitative evaluation of the psychophysical data, it may be helpful to think of the sensory events that form the basis for the perceptual judgment as some measure of neural activity, perhaps the number of discharges in a specific period of time in a certain locale in the CNS. This enables one to visualize in hypothetical form the probability that a certain sensory event of magnitude (x) will occur when noise alone is present, $f_N(x)$, and the probability of occurrence of the same sensory event when a signal is also present, $f_{SN}(x)$ (Fig. 20-17).

With this general model in mind, one may formulate the factors governing signal detection in terms of a decision between two alternative hypotheses: the subject must decide whether a certain observation (x) is more likely to be a member of one of two partially overlapping probability density functions. This can be done by setting a criterion cutoff value for x (=c) so that for any $x > c$ the observation will be placed in the category SN, and for any $x < c$, in the category N (Fig. 20-18). Depending on the position of c along the x-axis, the ratio of errors in correct detection and false alarms will vary: a moment's reflection will make it clear that the area to the left of c under the distribution curve SN represents the likelihood of stimulus presentations that the observer labels "noise," whereas the area to the right of c under N (crosshatched in Fig. 20-18) represents the likelihood of false alarm. Accordingly, the shaded area under SN is the likelihood of correct decision. Thus the fundamental quantities in the evaluation of the performance of the subject are the two areas under the distribution curves in Fig. 20-18 that are marked by shading and cross-hatching. These areas under the respective probability density curves of the hypothetical model correspond to the experimentally measured correct decision and false alarm probabilities. The hypothetical criterion value enables one to visualize one means by which the actual perceptual judgment could be under the control of a variety of factors (e.g., anticipation and motivation). In this form the hypothesis of signal detection is accessible to experimental neurophysiologic tests.

Mountcastle et al.[32] and Werner and Mountcastle[52] designed a statistical estimator for the discrimination between two trains of neuron discharges evoked, for example, in thalamic joint neurons by two slightly different positions of the relevant joint within the excitatory angle for that neuron. The statistical method chosen was the sequential probability ratio test; its application to the specific problem is illustrated in Fig. 20-19. Beginning with the first discharge interval of the impulse train to be tested, a ratio is established between the probabilities with which that interval occurs in each of the parent populations between which a decision is to be attempted. The ratios for successive intervals are similarly established, and these ratios are successively multiplied. A decision is reached when the consecutive product thus formed reaches either of two levels that are preset by the degree of statistical validity desired; one level indicates that the impulse train examined should be considered as belonging to the interval distribution $f_0(x)$ (and the respective joint position) and vice versa. The difference in joint positions that could be detected from comparisons of the respective neuronal impulse trains with this test were of the same order that can be discriminated in psychophysical experiments in man.

The potential relevance of statistical tests of this kind for sensory functions of the CNS is perhaps in part related to the startling phenomenon that the relative variability of discharge intervals in impulse trains of central sensory neurons is practically constant over a large range of mean discharge rates. The possible implication of this for the discrimination of neural activity evoked by sensory stimuli of different intensities is schematically illustrated in Fig. 20-20; on the

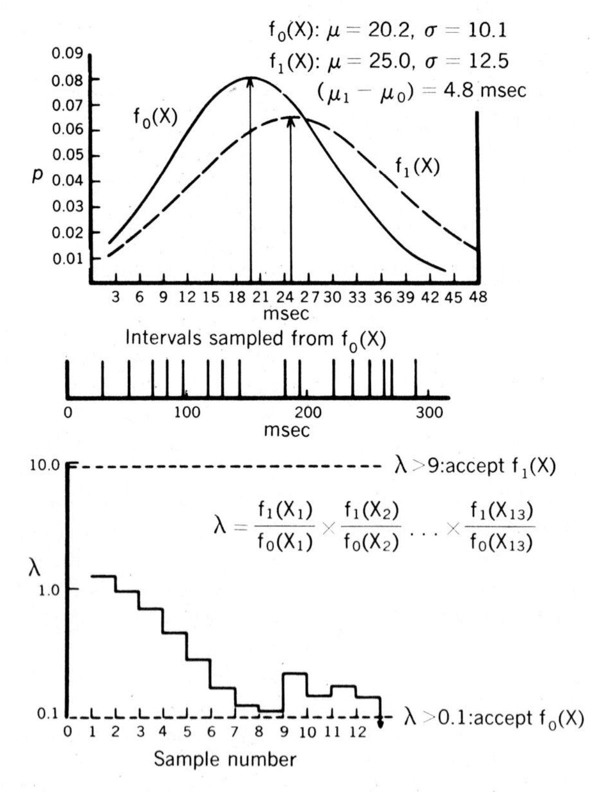

Fig. 20-19. Illustration of use of sequential probability ratio test (see text). Above are two probability density functions between which discrimination is sought, with actual mean value, μ, for each as observed in study of thalamic neuron under slightly different intensities of sensory drive. Next is sample of sequence of intervals belonging to $f_0(X)$. Graph below plots successive changes in probability ratio, λ, determined as shown by inset. (From Mountcastle et al.[32])

left a curve plots the relation between the mean discharge rate of a thalamic joint neuron and the intensity of the exciting stimulus, the joint position (this is the power law relation described earlier). From this, we read off the mean discharge interval for the neural activity at two pairs of positions along the continuum: the members of each pair are separated by equal increments in stimulus intensity, a $\Delta\Theta$ of 10 degrees. In the first case, illustrated by the next column to the right, we calculate the interval distributions about each of the four means, assuming normality, and in this case a constant value of the standard deviation (=30 msec), for each of the four distributions. The overlap of the distributions reflects pictorially what was also confirmed by the application of the sequential probability ratio test: that position discrimination would, under these circumstances, become increasingly difficult as the test increment is moved to the right along the stimulus continuum. This is due to the negatively accelerating nature of the power law relation between mean response frequency and joint angle. The task of position discrimination is quite different, however, if the constant relation between mean and standard deviation, observed experimentally, is taken into account. This situation is illustrated to the right in Fig. 20-20 for the case of a constant coefficient of variation (CV) of 50%. The relatively much smaller overlap of the pair of distributions at the upper end of the stimulus response curve is apparent and preserves discriminability. Thus the diminution of the spread of the interval distribution with increasing mean frequency offsets, at least to some extent, the negatively accelerating nature of the stimulus-response relation.

Recognition problem

When the task is one of identifying a particular stimulus with a single name or number, perceptual judgments are probably made by comparing the presented stimulus with some subjective standard. To set this clearly aside from discriminatory judgments of whether one stimulus is greater than, less than, or the same as another stimulus, the former type of judgment is often designated as "stimulus rating" or "absolute judgment." Stevens has estimated that the young human listener can distinguish 350,000 different tones when they are presented in pairs for *discrimination*. But if tones of the same frequency range (100 to 8,000 Hz) are presented for *individual* identification, there are no more than five to seven different pitches that the untrained individual never confuses.[36] Garner[4] and Miller[8] have shown that the precise number of stimulus categories that can be distinguished by absolute judgment can be calculated from experimental data by applying concepts and algorithms of information theory; thus the recognition problem relates directly to some quantitative considera-

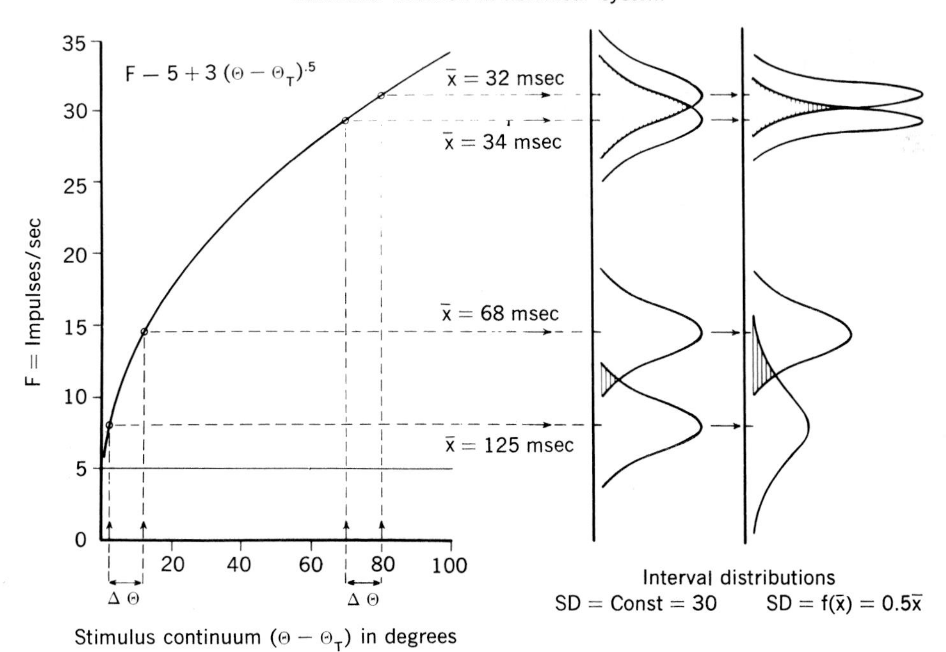

Fig. 20-20. Schematic presentation of manner in which constancy of relative variability (coefficient of variation) of discharge intervals during steady-state activity of individual neurons could affect statistical decision between different intensities of sensory drive (see text). (From Werner and Mountcastle.[52])

tions of information transmission in sensory systems.

The relation between stimuli and the subject attempting identification of these same stimuli is thought to be analogous to that between a source emitting messages and a receiver of these messages. Information exists in a message only if there is a priori some uncertainty about what the message will be; the *amount* of information transmitted is equal to the amount by which the uncertainty in the receiver is reduced by the message received. Uncertainty, like information, is measured in *bits,* where 1 bit is the uncertainty involved in an event with two possible and a priori equally likely outcomes. The measure that satisfies the condition that uncertainties of independent events are additive is logarithmic, and it has become common practice to use logarithms to the base 2 for this measure.

For the actual design and evaluation of experiments, consider a certain number of stimuli chosen from a set of different, discrete stimulus intensities. Depending on the number of different intensities in the set and on the relative frequency, p(S), with which stimuli of each intensity occur in the group, there exists some uncertainty that any one given stimulus of the group is of a certain specific magnitude. This stimulus uncertainty can be computed as follows:

$$H(S) = -\Sigma\, p(S) \cdot \log_2 p(S)$$

The absolute value of these expressions is plotted on the abscissae of the graphs presented later in this section. For a given set of stimuli, defined by its uncertainty value, the experimental data can be displayed in the form of a matrix with the joint occurrences of a certain stimulus and a certain response as entries. Table 20-2 depicts the absolute number of paired occurrences expressed as probabilities.

The matrix of Table 20-2 represents a bivariate distribution of the relative number of instances (normalized to probabilities) at which the stimulus S_i was associated with the response R_j. A "joint uncertainty" may be calculated for this bivariate distribution in a manner analogous to that used for the computation of the stimulus uncertainty:

$$H(S,R) = -\Sigma\, P(S,R) \cdot \log P(S,R)$$

This result can then be compared with the uncertainty in a reference matrix containing as entries those values that would result if stimuli and responses were not lawfully associated with one another but occurred in a strictly random association only. The entries to this latter reference matrix are calculated as the products of the respective marginal values of R and S, and its uncertainty is maximal (H_{max}). The difference by which the uncertainty value of the experimentally determined matrix is decreased from the theoretical maximal value determines the actual amount of information transmitted in the psychophysical experiment. This value is plotted on the ordinate of the graphs in Figs. 20-21 and 20-23.

The information transmitted in experiments involving absolute judgments varies in a characteristic fashion with the number of different stimulus categories used (and, accordingly, with the stimulus uncertainty). An example is shown in Fig. 20-21. In this particular case, information transmission equals the stimulus uncertainty up to a value of the latter of about 2.3 bits; at higher values of stimulus uncertainty (i.e., when a larger number of different stimulus categories are presented), information transmission reaches a plateau value of about 2.5 bits. The interpretation of this is that, regardless of the number of categories into which a given stimulus range is divided, the observer can single out no more than about five or six (i.e., the antilogarithm of 2.3 with the base 2) values for the stimulus that he will never confuse with another.

Fig. 20-21 reflects the characteristic feature of any imperfect information-transmitting system, namely, an upper limit of information transmission that is a characteristic magnitude of the system and code under

Table 20-2. Matrix of probabilities of joint occurrences of fixed value of R (R_j) and fixed value of S (S_i)*†

		Response categories (R_j)				
		1	**2**	**3**	**4**	**Marginal sums**
Stimulus categories (S_i)	**1**	0.08	0.02	0	0	0.10
	2	0.17	0.20	0.03	0	0.40
	3	0	0.03	0.18	0.09	0.30
	4	0	0	0.04	0.16	0.20
	Marginal sums	0.25	0.25	0.25	0.25	

*From Garner.[4]

†Internal entries are joint probabilities $p(S_iR_j)$. Marginal sums give values of p(R) and p(S), respectively.

study. This is sometimes also defined as channel capacity.

The startling result of psychophysical studies of this kind is that the maximum amount of information transmission in a large number of sense continua, notably intensive ones, is on the order of about 2.5. This is the basis for the statement that the number of stimuli that can be identified by absolute judgment is far below the number of discriminable stimulus pairs.

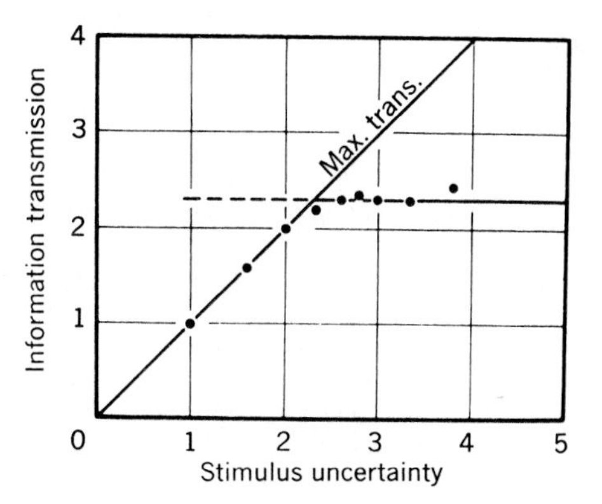

Fig. 20-21. Information transmission for absolute judgments of auditory pitch. (Based on data from Pollack[36]; from Garner.[4])

The procedure outlined for the information evaluation of absolute psychophysical judgments can also be applied to a quantitative evaluation of stimulus-evoked neural activity. In this case a measure of the stimulus-evoked neural response (e.g., the number of discharges evoked) forms the entries in the (S-R) matrix. Fig. 20-22 gives such a display, constructed from the activity recorded from a mechanoreceptive afferent nerve fiber activated by graded mechanical stimulation of its cutaneous receptive field. The numeric evaluation of data such as those in Fig. 20-22 indicated that the maximal information transmission with regard to skin indentation is, on the average, 2.5 bits for slowly adapting touch corpuscles of the hairy skin and 3.0 bits for receptors of the glabrous skin (Fig. 20-23). These results correspond to those cited earlier for information transmission in psychophysical judgments. The implications are that the limiting link in information transmission across the skin, insofar as in-

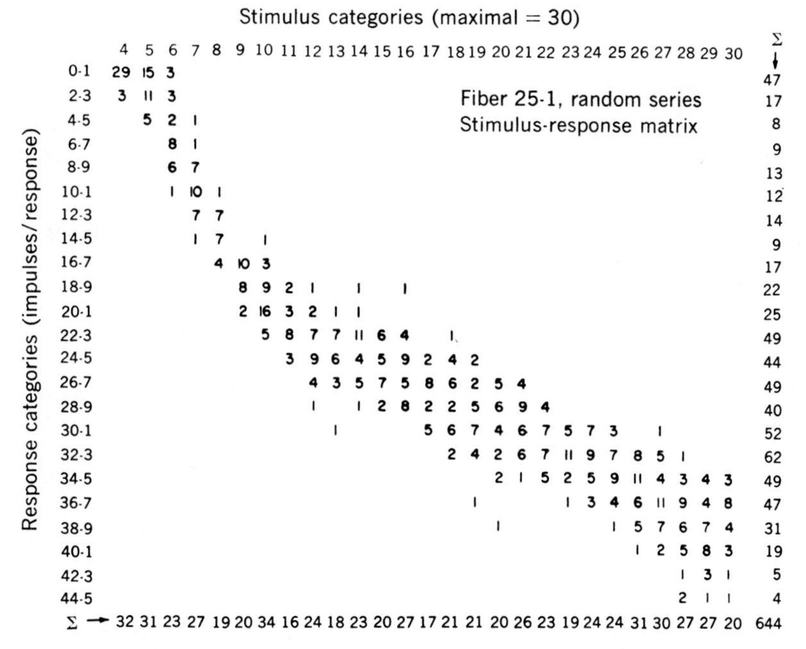

Stimulus categories (maximal = 30)

Fiber 25-1, random series
Stimulus-response matrix

Response categories (impulses/response)	4	5	6	7	8	9	10	11	12	13	14	15	16	17	18	19	20	21	22	23	24	25	26	27	28	29	30	Σ ↓
0-1	29	15	3																									47
2-3	3	11	3																									17
4-5		5	2	1																								8
6-7			8	1																								9
8-9			6	7																								13
10-1			1	10	1																							12
12-3				7	7																							14
14-5				1	7		1																					9
16-7					4	10	3																					17
18-9						8	9	2	1		1		1															22
20-1						2	16	3	2	1	1																	25
22-3							5	8	7	7	11	6	4		1													49
24-5								3	9	6	4	5	9	2	4	2												44
26-7									4	3	5	7	5	8	6	2	5	4										49
28-9									1		1	2	8	2	2	5	6	9	4									40
30-1										1				5	6	7	4	6	7	5	7	3		1				52
32-3															2	4	2	6	7	11	9	7	8	5	1			62
34-5																	2	1	5	2	5	9	11	4	3	4	3	49
36-7																1				1	3	4	6	11	9	4	8	47
38-9																	1					1	5	7	6	7	4	31
40-1																							1	2	5	8	3	19
42-3																									1	3	1	5
44-5																									2	1	1	4
Σ →	32	31	23	27	19	20	34	16	24	18	23	20	27	17	21	21	20	26	23	19	24	24	31	30	27	27	20	644

Fig. 20-22. Stimulus-response matrix for mechanoreceptive fiber activated by light mechanical stimuli delivered to its corpuscular ending. Afferent fiber was isolated for study by dissection of saphenous nerve in thigh. Maximal movement of stimulus probe was 690 μ. This was divided into 30 equal steps. Threshold lay between steps 4 and 5 throughout study; stimulus duration was 500 msec. Stimuli were delivered once every 3 sec. Stimuli of different intensities were delivered in random order until 644 were given. Numbers of stimuli of different intensities are indicated by lower horizontal row of numbers. Responses to each stimulus are categorized by number of impulses in steps of two impulses. (From Werner and Mountcastle.[53])

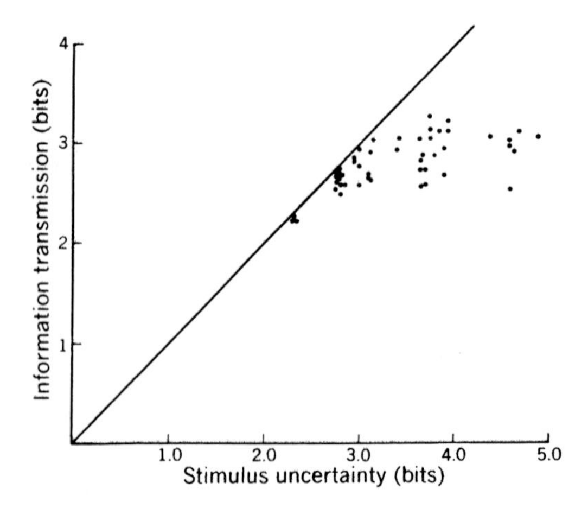

Fig. 20-23. Information transmission calculated from a series of stimulus-response data obtained for each of 17 mechanoreceptive fibers ending in palm of hand (macaque). Stimuli were applied at rate of 6 or 12/min; observation times were 500 msec or longer but not exceeding 1 sec. (From Werner and Mountcastle.[54])

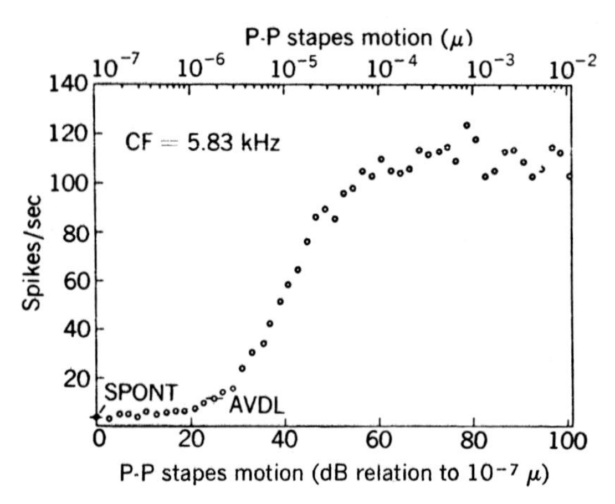

Fig. 20-24. Discharge rate of single auditory nerve fiber as function of stimulus level. Stimulus was continuous tone at characteristic frequency of fiber (5.83 kHz). Horizontal scale is expressed in terms of stapes displacement. Each point on graph represents 1 min run. There were no rest periods between runs, which were made for a series of ascending levels in 2 dB steps. Filled circle, *SPONT*, represents level of spontaneous discharge; point *AVDL* represents discharge rate at "threshold," defined as audiovisual detection level. Stimulus at *AVDL* corresponds to sound-pressure level of 2 dB with reference to 0.0002 dyne/cm². (From Kiang.[5])

tensity is concerned, is at the level of the first-order fiber and that certain central mechanisms preserve all, or nearly all, the information they receive.[54]

Code for stimulus intensity

The examples chosen to illustrate instances of the neural representation of the intensity attribute of stimuli did not require more than measuring the mean rate of discharges or the average number of responses of individual neural elements in order to obtain the correspondence between neural and psychophysical stimulus-response relationships. Thus, depending on the test situation, mean discharge rate and average response numbers can be considered in these examples to be a neural code for stimulus intensity, at least at the levels of the afferent pathways where these measurements were taken.

A more complex situation appears to exist with the representation of loudness by discharges of auditory nerve fibers, as illustrated in Fig. 20-24. The discharge rate in these fibers reaches a plateau within some 40 dB of threshold. Accordingly, the dynamic response range of individual fibers is narrow. Moreover, thresholds of different afferent fibers do not differ markedly. Yet psychophysical loudness estimates vary monotonically over a stimulus range of 100 dB or more without attaining a saturation level. This discrepancy suggests that the neural code for loudness is complex and possibly related to the temporal spacing of discharges in a population of afferent fibers or affected, in some not yet understood manner, by the activity in the olivocochlear efferent system.[5]

TIME STRUCTURE OF DISCHARGES AS A NEURAL CODE

In Chapter 14, evidence was presented to indicate that the discharges of auditory nerve fibers and of neurons of the cochlear nuclear complex and the inferior colliculus may be time locked to a segment of the cycle of a sinusoidal auditory stimulus. Accordingly, these neural elements reflect the frequency of a persisting tone by producing discharges at intervals grouped around the value of the tone period and its integral multiples, at least at tone frequencies up to the order of 2,000 to 5,000 Hz, depending on the animal species. At these low frequencies, phase-locking of discharges in auditory nerve fibers reflects the tone frequency unambiguously, for it is not at all affected by stimulus duration or strengths.[38]

This leads to the idea that the auditory system represents tone frequency by a period-time code, at least at low frequencies and at its input level from the periphery. In pursuit of this notion, Rose et al.[39] obtained some further evidence from the use of complex periodic sound stimuli.

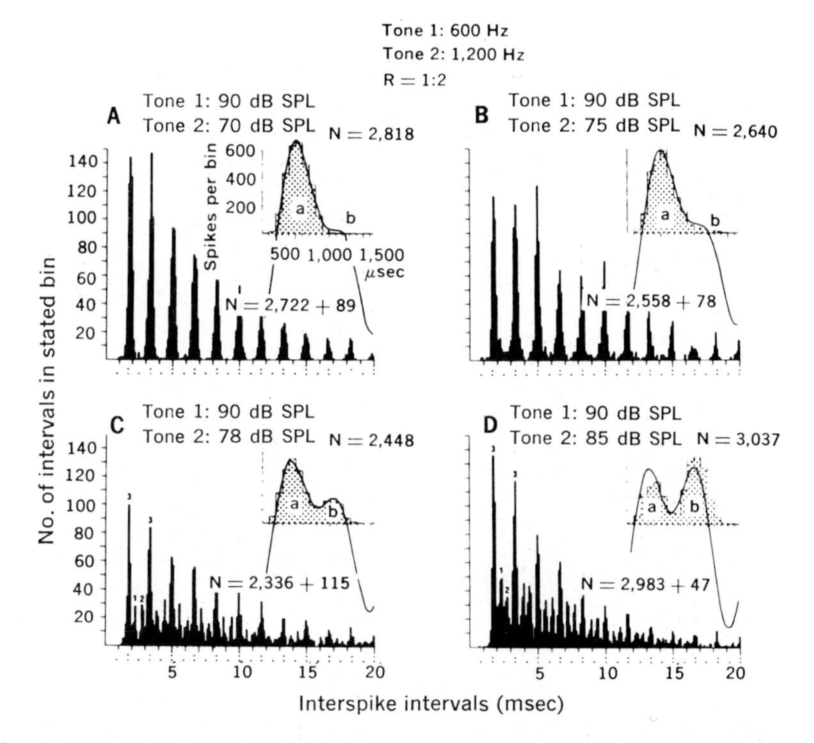

Fig. 20-25. Relation between interspike intervals and distribution of spikes in period histograms. Tone of 600 Hz was locked to tone of 1,200 Hz in ratio of 1:2. **A** to **D,** Interspike interval and period histograms obtained when lower frequency tone was at 90 dB SPL, while strength of higher frequency tone varied as indicated. Each sample based on responses to two tonal presentations. Duration of each presentation: 10 sec. Interspike-interval histograms: abscissa, time in milliseconds; each bin = 100 µsec. Dots below abscissa indicate integral multiples of periods of both primary tones. Upper row of dots indicates integral multiples of period of 600 Hz tone; second row of dots indicates integral multiples of period of 1,200 Hz tone. Ordinate: number of interspike intervals in each bin. Number of intervals, *N,* is given by two numbers; the first indicates number of plotted intervals; the second is number of intervals that exceeded value of abscissa. Period histograms: abscissa, period of complex wave in microseconds; each bin = 100 µsec. Ordinate: number of spikes in each bin. Period of complex sound: 1,666 µsec. *N* = number of spikes in sample. Each period histogram is fitted by curve that is the sum of two sine functions. Different subpopulations in period histograms are identified in alphabetical order from left to right. (From Rose et al.[39])

Fig. 20-25 illustrates, in addition to period histograms and the appropriate fitted waveforms, distributions of interspike intervals for four stimulus conditions, as indicated. In Fig. 20-25, *A,* approximately 98% of all intervals are aa and bb intervals grouped around integral multiples of the period of the lower frequency tone. This indicates that the fiber transmitted information pertaining almost exclusively to the lower primary tone of 600 Hz. Increase of the loudness of tone 2 in Fig. 20-25, *B,* leads to the appearance of some ab and ba intervals, although the lower tone is still by far dominant. In data shown in Fig. 20-25, *C,* all three possible interval values appear in substantial numbers. Approximately 90% are ab intervals with a mean duration of 563 µsex; ba intervals and ab intervals occur with equal frequency and their main duration is 1,103 µsec; the remaining 62% are aa and bb intervals, with the mean duration of 1,666 µsec, or an integral multiple of this value. The pattern of interspike intervals in Fig. 20-25, *C,* consists of triads that will repeat themselves every 1,666 µsec. The members of the first four triads are identified as modes 1, 2, and 3 in the histogram.

From these experimental findings stems the evidence that discharges can recur periodically at frequencies for which there is no spectral component in the acoustic stimulus. Instead, these latter discharges reflect the peaks of unidirectional elevations of the waveform of the stimulus. In man the same auditory stimuli elicit the sensation of combination tones, that is, tones that are heard but for which there is no spectral component in the acoustic stimulus. Thus it appears that combination tones can be a consequence of the way in which tone frequency is encoded in the ear. Moreover, the occurrence of periodicities in the auditory nerve discharge pattern and the sensation of tones, both unrelated to spectral components in the auditory input, strengthens the contention that the temporal pattern of neural discharges in auditory nerve fibers is indeed a significant determinant of the perceived tone, at least at low tone frequencies.

The situation is unresolved with regard to tone frequencies in excess of 5,000 Hz when refractoriness of the auditory nerve fibers prevents the occurrence of time intervals between consecutive discharges shorter than some 700 to 800 μsec. In this case the discharges accumulate at integral multiples of the period, which are markedly longer than the refractory period. Some questions also remain as to the mechanism in the central auditory pathways that is capable of interpreting the time-period code of the afferent nerve fibers and as to the manner in which the time-period code may be transformed into a different central representation. A possible candidate for an alternative form of tone frequency coding is the tonotopic organization of the subcortical structures of the auditory system (Chapter 15).

This complex of questions is, to some extent, similar to those raised by the investigations of Mountcastle and co-workers[34,45] regarding the cutaneous sense of flutter vibration (Chapter 12). The conclusion of these studies was that the perception of regular oscillatory movements of the skin depends on the appearance of periodic trains of nerve impulses in primary afferent nerve fibers, with the periodicity in the impulse train reflecting the stimulus frequency. Two classes of afferent fibers were distinguished: those originating from the glabrous skin, which were most sensitive to frequencies of 5 to 40 Hz, and those originating from deep cutaneous tissue, which responded in a frequency range of 60 to 300 Hz.

In the projection to somatosensory area I of the postcentral gyrus the two classes of afferents remain essentially distinct and activate different cortical neurons; one class of quickly adapting cortical neurons readily follows the sinusoidal mechanical stimulus of the skin over a frequency range of 5 to 80 Hz, with a strong periodic recurrence of impulses at intervals close to the cycle length of the vibratory stimulus. Thus information on stimulus frequency remains preserved in the form of the temporal discharge pattern of these cortical neurons. This suggests that the capability of discriminating frequencies of vibratory stimuli in behavioral and psychophysical tests may be attributable to a central neural mechanism that can detect differences between the period lengths in impulse trains of this class of neurons.

The situation is different in another class of cortical neurons, which receives its afferent input from the pacinian corpuscles located in the deeper cutaneous layers. Vibratory stimuli with a frequency of 80 to 400 Hz, which entrain periodic discharges in the afferents originating from pacinian corpuscles, are reflected only by an increase of discharge rates in these cortical neurons. But there is no relation between the magnitude of the increase in firing rate and the stimulus frequency, nor is there any periodicity in the discharges that would reflect the stimulus frequency. Yet the human observer is capable of discriminating between different frequencies of vibratory stimuli, irrespective of whether the stimuli engage, according to their frequency, the quickly adapting system of afferents from the glabrous skin or the pacinian elements.

The general implication is that a certain stimulus attribute, which by virtue of the choice of the selected physical measure (e.g., frequency, as in the case under consideration) can be represented on a continuous and monotonic scale, need not be processed by the nervous system in an equally continuous and homogeneous fashion. Instead, different ranges of a particular attribute may be represented in the nervous system by means of entirely different types of codes. In the extreme a mode of neural representation may be adopted that is entirely unlike the scales of sensory experiences and the conventional measures of stimulus properties.

The work of Gesteland et al.[22] with the olfactory system is an example. Neural activity in olfactory nerve fibers may be augmented or reduced by odors, and every olfactory cell is affected by many different odors, some excitatory, some inhibitory. If tested with a group of different odors seriatim, it appears that every olfactory nerve fiber ranks these odors differently according to the degree of excitation (or inhibition) each of them elicits. The chances of encountering two nerve fibers with characteristically different responses to one pair of odors selected at random from the group would be small, but with many fibers engaged in the response, each of them ranking the odors differently according to the degree of excitation or inhibition they produce, it would be possible for the entire population as a whole to distinguish many different odors. In this way an entire population of neural elements generates a code, the effectiveness of which cannot be appreciated when the responses of individual elements are examined in isolation.

SENSE MODALITY AND TEMPOROSPATIAL PATTERNS OF NEURAL DISCHARGES

Since the time of Helmholtz it has been customary to categorize sensations according to modalities, that is, classes of sensations that form qualitative continua. Thus tone perception is a single modality, since hearing encompasses a continuous series of tones with no qualitative gaps. Unlike hearing or color vision, the cutaneous sense consists of several classes of sensations with different introspective qualities (Chapter 11).

Two contrasting points of view have been proposed to explain the multiplicity of modalities of the cutaneous sense in neuroanatomic and neurophysiologic terms. One viewpoint, which originated with Johannes Müller, states in current terms that experienced sense quality is determined by the central connections of the activated nerve fibers, irrespective of the physical nature of the stimulus that elicited this activity. Some of the ramifications of this view are that afferent nerve fibers with different diameters are essentially specific for different sense modalities and that the different modalities of cutaneous sensation have their own peripheral receptors that are preferentially sensitive to a particular form of stimulus energy (e.g., heat or mechanical deformation).

The second viewpoint originated with Nafe[35] and found active proponents in Weddell[49] and Sinclair.[10] Its principal claim is that the complex spatially and temporally dispersed pattern of neural activity is the determining factor in the experience of sensory quality. Thus activity in a given group of fibers or neurons could at one point in time contribute to the experience of touch and at another point contribute to the experience of pain, cold, or warmth.[10]

More recently, it has been proposed that these contrasting viewpoints (commonly designated as the "specific modality" and "pattern" theories, respectively) are not necessarily mutually exclusive. Recognition of receptor and neural pathway specialization for the transduction and transmission of particular types of cutaneous stimuli does not preclude that differences in temporal impulse spacings in individual afferents and differences in the relative distribution of afferent activity in a spectrum of nerve fibers could also be an important aspect of the central neural representation of a stimulus. The implication is, for instance, that a discharge pattern that consists of a rapid rise followed by a slow decline of discharge rate could be transmitted and perceived differently than a discharge pattern in the same afferent that rises slowly and declines rapidly.[48] Whether there are neurons in the CNS that are specifically triggered by one and not another temporal pattern of afferent input is still uncertain. The possibility exists, however, in view of several studies in invertebrates, since certain postsynaptic neurons responded differently when stimulated with impulse trains of different temporal spacing, although the mean stimulus rate was held constant.

STIMULUS FEATURE DETECTION BY NEURONS

Our central theme thus far in this chapter has been the proposition that mean rate and periodicity of neural discharges represent stimulus intensity and frequency in afferent neural pathways. The argument revolved around the demonstrations of correspondence between these parameters of neural discharges and measures of sensation, the latter obtained largely in man. In the conversion of physical stimulus attributes to neural activity, each physical magnitude value is made to correspond to an appropriate neural magnitude value. The relation between the physical and the neural scales may in some instances be a linear and in others a nonlinear function. This general concept was shown to satisfy certain kinds of experimental data, but it also became apparent that this principle of stimulus representation does not exhaust the scope of stimulus encoding by the CNS.

As an alternative, coding schemes that reduce redundancy in sensory stimuli have attracted attention since the mid-1950s.[1,14] The underlying notion was that evolutionary adaptation of the organisms to certain types of redundancies, which are always present in the environment, would have occurred. The guiding principle in these considerations was Shannon's concept of "optimal codes," which match the statistical structure of regularities in the available repertoire of messages.

As far as it was known at that time, there were only two mechanisms available to the nervous system to reduce redundancy of information in the stimulus. In the temporal domain these were mechanisms specifically sensitive to onset and cessation of a stimulus, and in the spatial domain there was lateral inhibition. The potential significance of such redundancy-reducing codes consists of economy in signal transmission, because these codes exploit lawful regularities in the stimulus source. For instance, the duration of a stationary stimulus is uniquely defined by the moments of its beginning and end, and thus there is no need for generating neuronal impulses in the interval between these points in time.

The concept of the "stimulus feature" can be considered a generalization of this principle. This concept becomes applicable whenever a neuronal discharge signals with relatively high selectivity the occurrence of an input state that contains in its specifications the concomitant occurrence of certain regularities in the stimulus space, in addition to being specific for a certain place on the receptor sheet and stimulus modality. For instance, because matter is cohesive, objects can be fully characterized in terms of their boundaries; hence boundaries, that is, edges, corners, and angles, become the "features" in terms of which the spatial layout of a stimulus object can be unambiguously represented.

Such features may be purely of a spatial nature, consisting of stationary contours or patterns, or they may combine spatial with temporal information in the form of a stimulus motion. The economy consists, then, of limiting the characterization of a shape to the signaling of its boundaries or of emphasizing change over stationarity.[14]

Carried to its logical consequence, this principle implies that the CNS takes the information available in the proximal stimulus on receptor sheets out of its original context and imposes a classification into disjunctive entities. The general principle appears to be that the number of neurons available to signal stimulus information increases with progression along the afferent pathway but that any given neuron is less frequently activated as the constraints of its stimulus feature become more severe.[16]

The neurophysiologic reality of feature detection is by now amply substantiated. A characteristic example is provided by a class of neurons in the auditory cortex that responds most effectively to an appropriate phase difference between binaurally applied tone stimuli, with that phase difference itself being a function of the frequency of the tone employed, whereas another class of neurons responds preferentially to the intensity difference of binaural tones.[20] Other examples are the neurons with "simple," "complex," and "hypercomplex" receptive fields in the visual cortex[27] and the relative selectivity with which neurons in somatic sensory area I respond to cutaneous stimuli moving across the receptive field of the skin (Fig. 20-26).[55]

A recent observation of single neural responses in the inferotemporal cortex of macaques underscores the nature of the problem the investigator faces when searching for the "optimal" stimulus for a certain neuron. In some instances, neurons responded most strongly when the animal was presented with displays of relatively complex geometric figures. This may indicate that factors other than the stimulus configuration per se (e.g., the significance of the stimulus for the animal's behavioral repertoire) entered into determining the magnitude of the neural response.[23]

The common denominator of these experimental findings, and others as well,[31] is that individual neurons can be shown to be "triggered" to discharge by a relatively specific spatial or temporospatial stimulus configuration; there is some gradation of the density of discharge with variation of the stimulus properties, but there is generally a clearly defined stimulus context (e.g., shape or motion of a stimulus in the neuron's receptive field) that elicits a maximal response. Thus neurons in a population can be classified according to their "best" stimuli. These "best" stimuli are contexts of spatial simultaneity (in the case of shape) or temporal succession (in the case of stimulus motion) of excitation in the neuron's peripheral receptive field. A less than maximal response indicates only departure from the optimal stimulus, not the direction of departure or which component of the context was altered. The occurrence of the maximal discharge in a particular neuron can be thought of as the

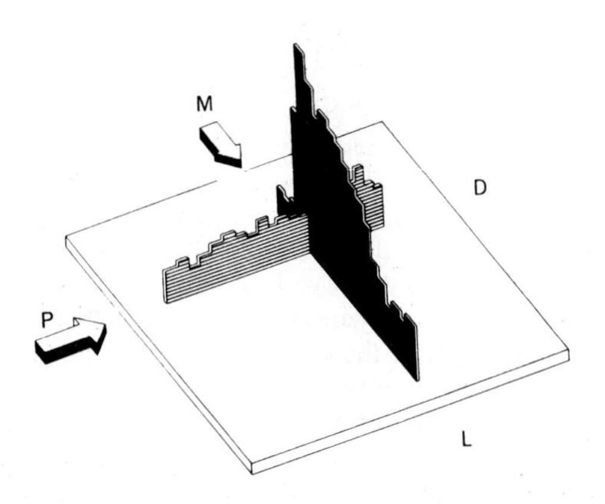

Fig. 20-26. Responses of neuron in somatosensory area I (postcentral gyrus) of unanesthetized macaque to movements of fine brush across cutaneous receptive field of neuron, which was located on dorsum of contralateral foot. Two directions of stimulus movement are shown: from medial to lateral (black, $M \rightarrow L$) and from proximal to distal (striped, $P \rightarrow D$). Height of bars composing vertical displays signifies number of discharges in each of consecutive 50 msec bins of duration of stimulus motion across receptive field, averaged over 25 consecutive and identical stimuli and expressed as discharges per second. Stimuli moved at constant velocity of 30 mm/sec. Spontaneous activity of neuron was 19 ($\pm$2.0) impulses/sec, exactly the height of striped plane at beginning of stimulus motion near border of receptive field. Intersection of vertical planes reflecting responses to stimulus movement in two orthogonal directions shows that an area in receptive field that responds most vigorously when stimulus moves across it from medial to lateral does not respond at all when stimulus moves across it from proximal to distal. (Based on unpublished experiments of Whitsel, Petrucelli, and Werner.)

"code" of the particular stimulus context that this neuron represents.

In the visual and the somatosensory cortical receiving areas in which this form of stimulus-feature encoding has been investigated in some detail,[27,55] it has been found to be associated with a functional architecture that places neurons with common response properties in juxtaposition. In the vertical dimension of the cortex, neurons with near-identical receptive fields are assembled to cell columns; in the horizontal dimension, neurons with similar optimal stimuli are assembled to cortical laminae, each lamina differing from the other in the complexity of stimulus features its neurons represent. Thus in a first approximation, it appears that a sensing surface (retina or cutaneous body surfaces) is mapped many times over; each map represents a particular stimulus feature, and all maps are in register with one another with regard to place on the peripheral receptor sheet.

One of the most intriguing problem areas in the study of sensation in neurophysiology is understanding the mechanisms and the sequence of events that intervene between the fragmentation of stimulus attributes and features as "seen" by individual cortical neurons and the perception of objects with permanent identities that move in a perceptual space along continuous trajectories. These problems range from clarifying in neurophysiologic experiments the synaptic mechanisms enabling feature detection to establishing the perceptual relevance of feature-sensitive neurons by psychologic means.

This latter issue comes into sharper focus if we ask which place and what role feature-signaling neurons in the nervous system could play in a mechanistic account of behavioral and perceptual functions and in a psychologic theory of information processing. To illustrate contrasting possibilities of their involvement, we may consider features as a hierarchy of elements in terms of which entire scenes of sensory input are analyzed and described[17]; alternatively, we may consider that feature elements stand at the interface between perception and action in the sense that they trigger or "release" behavioral acts in an automaton-like fashion.[51] A discussion of this issue must also take cognizance of the fact that perception of organisms is for the most part action oriented, both in the sense of motor acts enabling the acquisition of snapshots of scenes following each other serially in time and also with regard to purposeful actions to be carried out with the objects of perception. Hence the analogy is perhaps more with the program control of a robot than it is with the computer program that labels and categorizes patterns. But even disregarding for the moment the motor component in perceptual activity, there is need for postulating additional operational principles and neural mechanisms in the attempt to account for the perceptual achievements that entail selective attention and cognitive mechanisms on the part of the perceiving organism.[9]

REFERENCES

General reviews

1. Attneave, F., and Benson, B.: Spatial coding of tactual stimulation, J. Exp. Psychol. **81:**216-222, 1969.
2. Fechner, G. T.: Elements of psychophysics (translated by Adler, H. E., translator, and Howes, D. H., and Boring, E. G., editors), New York, 1966, Holt, Rinehart & Winston.
3. Galanter, E.: Contemporary psychophysics. In Brown, R., Galanter, E., Hess, E. H., and Mandler, G., editors: New directions in psychology, New York, 1962, Holt, Rinehart & Winston.
4. Garner, W. R.: Uncertainty and structure as psychological concepts, New York, 1962, John Wiley & Sons, Inc.
5. Kiang, N. Y. S.: A survey of recent developments in the study of auditory physiology, Ann. Otol. Rhinol. Laryngol. **77:**656, 1968.
6. Luce, R. D., and Edwards, W.: The derivation of subjective scales from just noticeable differences, Psychol. Rev. **65:**222, 1958.
7. Luce, R. D., Bush, R. R., and Galanter, E.: Handbook of mathematical psychology, New York, 1963, John Wiley & Sons, Inc., vol. 1.
8. Miller, G. A.: The magical number seven, plus or minus two; some limits on our capacity for processing information, Psychol. Rev. **63:**81, 1956.
9. Neisser, U.: Cognition and reality, San Francisco, 1976, W. H. Freeman & Co. Publishers.
10. Sinclair, D.: Cutaneous sensation, Oxford, 1967, Oxford University Press.
11. Stevens, S. S.: On the psychophysical law, Psychol. Rev. **64:**153, 1957.
12. Swets, J. A., Tanner, W. P., and Birdsall, T. G.: Decision processes in perception, Psychol. Rev. **68:**301, 1961.
13. Watson, J. B.: Psychology from the standpoint of a behaviorist, Philadelphia, 1919, J. B. Lippincott Co.
14. Werner, G., editor: Feature extraction by neurons and behavior, Cambridge, Mass., 1975, The M. I. T. Press.

Original papers

15. Barlow, H. B.: Sensory mechanisms, the reduction of redundancy and intelligence. Symposium on the mechanization of thought processes at the National Physical Laboratory, London, H.M. Stationery Office, Symp. No. 10, 535-539, 1959.
16. Barlow, H. B.: Trigger features, adaptation and economy of impulses. In Leibovic, K. N., editor: Information processing in the nervous system, Berlin, 1969, Springer-Verlag.
17. Barlow, H. B., Narasimhan, R., and Rosenfeld, A.: Visual pattern analysis in machines and animals, Science **177:**567-575, 1972.
18. Blough, D. S., and Schrier, A. M.: Scotopic spectral sensitivity in the monkey, Science **139:**493, 1963.

19. Bough, E. W.: Stereoscopic vision in the macaque monkey: a behavioral demonstration, Science **225:**42, 1970.
20. Brugge, J. F., Dubrovsky, N. A., Aitkin, L. M., and Anderson, D. J.: Sensitivity of single neurons in auditory cortex of cat to binaural tonal stimulation—effects of varying interaural time and intensity, J. Neurophysiol. **35:**1005, 1969.
21. Eijkman, E., and Vendrick, A. J.: Detection theory applied to the absolute sensitivity of sensory systems, Biophys. J. **3:**65, 1963.
22. Gesteland, R. C., Lettvin, J. Y., and Pitts, W. H.: Chemical transmission in the nose of the frog, J. Physiol. **181:**525, 1965.
23. Gross, C. G., Bender, D. B., and Rocha-Miranda, C. E.: Visual receptive fields of neurons in inferotemporal cortex of the monkey, Science **166:**1303, 1969.
24. Harper, R. S., and Stevens, S. S.: A psychological scale of weight and a formula for its derivation, Am. J. Psychol. **61:**343, 1948.
25. Hensel, H., and Bowman, K. K. A.: Afferent impulses in cutaneous sensory nerves in human subjects, J. Neurophysiol. **23:**564, 1960.
26. Hubel, D. H., and Wiesel, T. N.: Receptive fields, binocular interaction and functional architecture in the cat's visual cortex, J. Physiol. **160:**106, 1962.
27. Hubel, D. H., and Wiesel, T. N.: Receptive fields and functional architecture of monkey striate cortex, J. Physiol. **195:**215, 1968.
28. Hubel, D. H., and Wiesel, T. N.: Cells sensitive to binocular depth in area 18 of the macaque monkey cortex, Science **225:**41, 1970.
29. Jones, F. N.: Some subjective magnitude functions for touch. In Hawkes, G. R., editor: Symposium on cutaneous sensibility, Report No. 424, Fort Knox, 1960, U.S. Army Medical Research Laboratory.
30. Julesz, B.: Binocular depth perception without familiarity cues, Science **145:**356, 1964.
31. Maturana, H. R., Lettvin, J. Y., McCulloch, W. S., and Pitts, W. H.: Anatomy and physiology of vision in the frog, J. Gen. Physiol. **43:**129, 1960.
32. Mountcastle, V. B., Poggio, G. F., and Werner, G.: The neural transformation of the sensory stimulus at the cortical input level of the somatic afferent system. In Gerard, R. W., and Duyff, J. W., editors: Information processing in nervous system, Proceedings of the International Union of Physiological Sciences, New York, 1962, vol. 3.
33. Mountcastle, V. B., Poggio, G. F., and Werner, G.: The relation of thalamic cell response to peripheral stimuli varied over an intensive continuum, J. Neurophysiol. **26:**807, 1963.
34. Mountcastle, V. B., Talbot, W. H., Sakata, H., and Hyvarinen, J.: Cortical neuronal mechanisms in flutter vibration studied in unanesthetized monkeys. Neuronal periodicity and frequency discrimination, J. Neurophysiol. **32:**452, 1969.
35. Nafe, J. P.: The psychology of felt experience, Am. J. Psychol. **39:**213, 1957.
36. Pollack, I.: The information of elementary auditory displays, J. Acoust. Soc. Am. **24:**745, 1952.
37. Provins, K. A.: The effect of peripheral nerve block on the appreciation and execution of finger movements, J. Physiol. **143:**55, 1958.
38. Rose, J. E., Brugge, J. F., Anderson, D. J., and Hind, J. E.: Phase-locked response to low frequency tones in single auditory nerve fibers of the squirrel monkey, J. Neurophysiol. **30:**769, 1967.
39. Rose, J. E., Brugge, J. F., Anderson, D. J., and Hind, J. E.: Some possible neural correlates of combination tones, J. Neurophysiol. **32:**386, 1969.
40. Ruch, T. C., Fulton, J. F., and German, W. H.: Sensory discrimination in monkey, chimpanzee and man after lesions of the parietal lobe, Arch. Neurol. **39:**919, 1938.
41. Stevens, J. C., and Stevens, S. S.: Warmth and cold dynamics of sensory intensity, J. Exp. Psychol. **60:**183, 1960.
42. Stevens, S. S.: The operational basis of psychology, Am. J. Psychol. **47:**323, 1935.
43. Stevens, S. S.: The psychophysics of sensory function. In Rosenblith, W. A., editor: Sensory communication, Cambridge, Mass., 1961, The M.I.T. Press.
44. Stevens, S. S., and Galanter, E. H.: Ratio scales and category scales for a dozen perceptual continua, J. Exp. Psychol. **54:**377, 1957.
45. Talbot, W. H., Darian-Smith, I., Kornhuber, H. H., and Mountcastle, V. B.: The sense of flutter vibration—comparison of the human capacity with response patterns of mechanoreceptive afferents from the monkey hands, J. Neurophysiol. **31:**301, 1968.
46. Vierck, C. J., and Jones, M. B.: Size discrimination on the skin, Science **158:**488, 1969.
47. von Bekesy, G.: A new audiometer, Acta Otolaryngol. **35:**411, 1947.
48. Wall, P. D., and Cronly-Dillon, J. R.: Pain, itch and vibration, Arch. Neurol. **2:**365, 1960.
49. Weddell, G.: The anatomy of cutaneous sensibility, Br. Med. Bull. **3:**167, 195, 1945.
50. Weinstein, S.: Intensive and extensive aspects of tactile sensitivity as a function of body parts, sex and laterality. In Kenshalo, D. R., editor: The skin senses, Springfield, Ill., 1968, Charles C Thomas, Publisher.
51. Werner, G.: Neural information processing with stimulus feature extractors. In Werner, G., editor: Feature extraction in neurons and behavior, Cambridge, Mass., 1975, The M.I.T. Press.
52. Werner, G., and Mountcastle, V. B.: The variability of central neural activity in a sensory system, and its implications for the central reflection of sensory events, J. Neurophysiol. **26:**958, 1963.
53. Werner, G., and Mountcastle, V. B.: Neural activity in mechanoreceptive cutaneous afferents—stimulus response relations, Weber functions and information transfer, J. Neurophysiol. **28:**359, 1965.
54. Werner, G., and Mountcastle, V. B.: Quantitative relations between mechanical stimuli to the skin and neural responses evoked by them. In Kenshalo, D. R., editor: The skin senses, Springfield, Ill., 1968, Charles C Thomas, Publisher.
55. Whitsel, B. L., Roppolo, T. R., and Werner, G.: Cortical information processing of stimulus motion on primate skin, J. Neurophysiol. **35:**691, 1973.
56. Wiener, N.: A new theory of measurement—a study in the logic of mathematics, Proc. Lond. Math. Soc. **19:** 181, 1920.

21

GERHARD WERNER

Higher functions of the nervous system

INTRODUCTION: THE ASSOCIATION CORTEX

The sensory deficits after partial or total ablation of primary cortical receiving areas are in large measure predictable on the basis of the modality specificity and the topologic organization of the projections they receive from the periphery via the specific thalamic relay nuclei. These "extrinsic" neural systems are set apart from the "intrinsic" systems that, in contrast, consist of the cortical projections from thalamic nuclei without any major extrathalamic or extratelencephalic input. Traditionally, these latter cortical areas are known as "association cortex"; a designation that reflects the idea of 19th century psychophysics that perception is based on "associating" the elementary units of sensation to more complex ideas, patterns, and relations. In this tradition the "association cortex" was thought to be the neural structure in which the modality-, place-, and intensity-specific inputs from primary cortical receiving areas would interact and eventually initiate an integrated motor output by way of transcortical action.

To the extent to which this dichotomy between sensation and perception in psychology was discredited under the influence of Gestalt psychology, phenomenology, and certain trends in operational behaviorism (Chapter 20), the considerable evidence that has accumulated during the past two decades from neurobehavioral studies and clinical neurology has demanded a different conceptualization of the function of the association cortex. These current ideas are no longer compatible with a role of "associating" diverse sensory inputs and serving as a link between sensory and motor cortex. Instead, the range of functions in which the "intrinsic" cortical areas play a role is now thought to encompass the spatial structure of perception, associative learning, and behavior and to involve transactions at a symbolic level in which "meaning" of stimulus context and "intentions" of behavioral acts appear to play some role.

The primate cortical areas in question are located within the parietopreoccipital convexity (the classic sensory association areas) and include, in addition, the prefrontal area occupying the anterior pole of the frontal lobe (the classic frontal association area). These areas of homotypical neocortex have enlarged dramatically in mammalian evolution, particularly in primates (Chapter 7). There exists some regional specificity among them for different sense modalities and thus different forms of perception, particularly in those areas close to the primary sensory and motor areas. Thus the parieto-occipital cortex is related in function to the somesthetic sphere of perception and behavior, the middle temporal region to auditory function, and the inferior temporal region to vision. Many of the functions of the homotypical "association" cortex transcend modality-specific behavior and involve the properties of dealing with the spatial arrangement of objects and events in relation to the body form and position, the identification of the meaning of objects and events for the organism in terms of its internal drive states, the direction of attention to and action toward external events, and the temporal sequencing of behavior.

POSTERIOR PARIETAL AREA

Parietal lobe syndrome[34,41]

A patient with a gross lesion of the parietal lobe presents obvious and striking abnormalities of behavior that are not attributable to defects in the elementary aspects of sensation. The manifestations differ according to the hemisphere involved. Hemispheric specialization was first recognized in connection with language function, which is attributed to the left hemisphere in right-handed subjects; hence we have the notion of hemispheric dominance. However, lesions of the parietal lobe have also produced evidence of hemispheric specialization for functions other than language.

Damage of the parietal lobe in the right (minor) hemisphere of right-handed subjects leads to difficulties in recognizing spatial relationships in the contralateral half of body and extracorporeal space: the patient may act as though the limbs on the side opposite the lesion did not exist; he may not use the affected hand, yet deny abnormality in it; or he may claim that a contralateral limb belongs to another person. He may disregard events taking place in the side of the environment contralateral to the lesion, and he will tend to give naive and irrational explanations of these abberations. He may deny the existence of the contralateral body half altogether or deny illness present in it. The patient may fail to complete that portion of drawings or to correctly construct (e.g., with building blocks) the side of a scene contralateral to his lesion. The latter symptom is known as constructional apraxia.[42]

Denny-Brown and Chambers[36] characterized this class of disorders as amorphosynthesis and interpreted it as a physiologic deficit similar to the behavioral disorders observed in monkeys after parietal cortex ablation. The parietal syndrome in the monkey is contralateral and symmetrically equivalent on the two sides. Such an animal shows a prominent *neglect* resembling that described for humans: there is neglect of the contralateral limbs, paucity of spontaneous movements, and errors in reaching with the contralateral hand toward targets in either half of the immediate extrapersonal space. A catatonic persistence of unusual limb positions occurs, and with bilateral lesions there is a marked reduction in all exploratory behavior. Perceptual rivalry and extinction (see subsequent discussion) occurs in the visual, auditory, and somesthetic domains. There is a release of the tactile avoiding reaction—indeed of all avoiding reactions—so that animals with parietal lobe lesions show an excessive response to even mildly noxious stimuli delivered anywhere on the contralateral side, and the instinctive grasp reaction is largely abolished. This latter consists of an orienting movement of the hand elicited by contact with any place on it that brings the hand into proper orientation for grasping, which follows.

The common denominator of these abnormalities and the clinical signs of amorphosynthesis are thought to exist in a deficit of sterotaxic exploratory behavior and orientation in space; both factors lead to a fragmentation of the perceptual process. At least to some extent, this deficiency may be the result of a more fundamental process: a frequent manifestation of parietal lobe damage consists of the lack of appreciation of a stimulus on the affected side of the body when an equivalent stimulus is simultaneously presented to the unaffected side. This phenomenon is known as *extinction:* the patient may recognize only the stimulus delivered to the side opposite the normal hemisphere, or, when the two stimuli are simultaneously delivered to the side opposite the diseased hemisphere, the patient may recognize only the proximal and disregard the distal stimulus, particularly if the former has been applied to the face *(proximal dominance).*[3] Extinction may also occur between stimuli of different modalities; for example, a visual stimulus opposite the normal hemisphere can obliterate perception of a somatic stimulus. Thus there arises some *perceptual rivalry* between simultaneously applied stimuli that normal subjects can differentiate and interpret in terms of their relation to one another without difficulty. Accordingly, some of the behavioral defects associated with right parietal lobe lesions may not only be due to neglect of the left, but also to excessive preoccupation with the right side of body and extracorporeal space; constructional apraxia from right-sided lesions or confusion of garments or in putting them on (apraxia of dressing) is in this sense an excessive distraction by mostly right-sided parts. The same can be found in the types of dyscalculia and dysgraphia resulting from right-sided lesions, in which preoccupation with small parts of numerals and letters distracts and prevents normal perception.

Disorders caused by parietal lobe lesions in the left (dominant) hemisphere differ from those attributable to the minor hemisphere by characteristically involving *both* sides of body and extracorporeal space.[41] These disorders, which involve disturbances related to classification and naming and to manipulating the symbol of a particular class of percepts, are traditionally subsumed under the term *"agnosias."* They are described in Chapter 22.

The distinction between left and right parietal lobe function does not imply that the dominant hemisphere lacks the capability for morphosynthesis; rather, it appears that this function is overlaid, and its defects obscured, by a process of higher abstraction that encompasses the perception of the entire body and extracorporeal space in a unitary, indivisible manner that is one important aspect of cerebral dominance; purely contralateral disturbances of morphosynthesis are therefore rarely seen after parietal lesions in the dominant hemisphere. Denny-Brown[35] suggests that the condition of asymbolia for pain is an instance of bilateral amorphosynthesis and not an agnosia, for such patients can feel pain and discuss it, although they do not exhibit the normal behavioral response to painful stimuli.

Body image. A particularly dramatic demonstration of the important, although not exclusive, role of the parietal lobes in the spatial organization of perception and behavior is associated with pathologic disturbances of the mental image that a person possesses about his own body and its physical attributes. Before the turn of the century, Bonnier called attention to some striking distortions shown by some of his patients in the attitude toward their own bodies: he described individuals who actually felt that their entire body had vanished (aschematia). Slightly later, A. Pick designated as autotopagnosia the disturbances in orientation on one's own body surface (e.g., the inability to distinguish left and right). Pick suggested that each individual develops a spatial image of his body from information supplied by the sensory systems. The idea of this body image may be so persistent that a person may generate a phantom sensation after losing a leg or an arm. Phantom limbs are to be expected in the majority of amputees, unless the limb is lost very early in life.[33]

Henry Head developed the concept of body image into a central theme of his neurologic thinking[58]: he concluded that each person constructs in the course of development a model of himself that becomes the frame of reference against which all body movements and postures are judged. This spatial scheme is not confined to the anatomic limits of the body; it may also incorporate instruments held in the hand or objects otherwise attached to the body.

Although frequently associated with parietal lobe damage, body-image pathology may also be associated with damage to other brain areas.[4] Some of the best examples of spatial alterations of body image are to be found in states of drug intoxication. Moreover, schizophrenic patients can show almost the same range of abnormal body-image phenomena.[19]

Haptic sensitivity: tactile apprehension of object quality and shape

Identification of objects through the sense of touch presupposes the combined and coordinated contribution of two modalities: cutaneous touch and position sense. Together, these two sources contain the information needed to specify the layout of the surfaces of an object being manipulated. J. J. Gibson[6] argued that the perception of the layout of surfaces *is* the perception of space and that the concurrent sensory influx from skin and articular angle and motion conveys stimulus information in its own right, namely, that of haptic touch. This implies a functional unity between sensory information originating within the body space (i.e., sensory events signaling joint position) and sensory information related to physical stimuli impinging on the body from without (i.e., cutaneous sensory events).

The sequence of events that Gibson noted when an observer is required to discriminate the shapes illustrated in Fig. 21-1 is typically the following:

(1) He curves his fingers around its face, using all fingers and fitting them into the cavities; (2) he moves his fingers in a way that can only

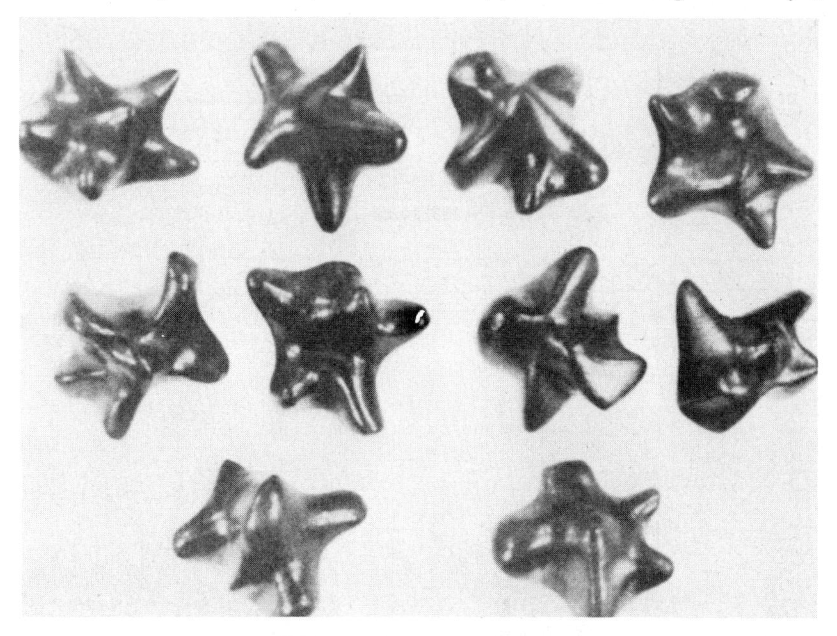

Fig. 21-1. Sculptured objects for studying haptic perception. (From Gibson.[6])

be called exploratory, since the movements do not seem to become stereotyped, or to occur in any fixed sequence, or even to be clearly repeated; (3) he uses oppositions of thumb and finger, but different fingers; he rubs with one or more fingers and occasionally he seems to trace a curvature with a single finger. The activity seems to be aimed principally at obtaining a set of touch-postures, the movement as such being incidental to this aim. Introspection bears out the hypothesis that the phenomenal shape of the subject does emerge from such a series of covariant transformations. No subject ever tried to run his finger over the whole array of curves in a systematic manner such as that of the scanning beam of a television tube.*

Indeed, the eye and the hand sample different properties of the object. In a haptic task, left-right reversals and up-down changes of the position of test figures are felt to be very dissimilar; visually the same changes produce little sense of differences. Conversely, changes in the curvature of contours elicit a sharp sense of difference visually and little sense of difference haptically.[39]

Psychologic studies of the blind confirm that the haptic sense contributes to the subject's idea or notion of surrounding space. Révész[18] considers the blind person a "pure haptic." The blind are conscious of the space generated by the surfaces that are accessible to haptic exploration. The particular role of the hand in the generation of this haptic space was recognized some 80 years ago by Féré[38] who wrote, "The hand merits the notice of physiologists and psychologists who have up to now rather neglected it: for the hand is both an agent and an interpreter in the growth of the spirit."

*Gibson, J. J.: The senses considered as perceptual systems, Boston, 1966, Houghton Mifflin Co.

Supramodal mechanism in stereognosis. When the inability to recognize the form of objects by touch, known as astereognosis, occurs in the absence of any overt signs of sensory deficits as measured by the conventional tests of somatic sensation (i.e., two-point discrimination, joint position sense, etc.), the concept of tactual agnosia has traditionally been invoked and associated with damage to the posterior parietal area. Although this aspect of the posterior parietal lobe function is well established, there is now also evidence that other brain lesions that spare this and primary sensory and motor areas can also produce some impairment of shape discrimination.[20]

In the chronic state after parietal injury, when the dramatic manifestations of apraxias and agnosias for the body and for surrounding space recede, more subtle tests of spatial orientation are required to bring deficiencies into prominence. One such task, employed by Semmes et al.,[20] consisted of a series of maps, reproduced in Fig. 21-2, that are presented to the subjects to test their capacity for route finding.

The following description is taken from Teuber's account of the test.

The person to be tested is led into a large room where nine dots are marked on the floor, all equi-

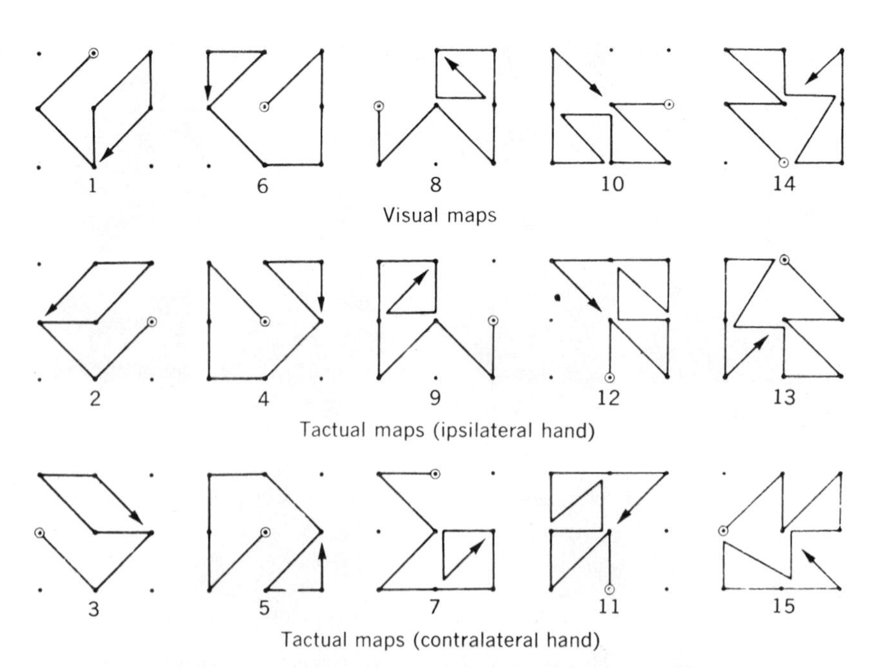

Fig. 21-2. Maps used in tests of route finding. (From Teuber.[70])

distant from each other in a square array. These dots are represented on each of the maps with a line connecting them, indicating where in the room the person tested has to start and which path he is to follow by active locomotion, from dot to dot, until he reaches the goal symbolized by the arrowhead. All maps are carried by the patient, one map at a time; he is not permitted to reorient the map as he looks about. However, one edge of the map, as well as the appropriate wall of the room, have been marked to indicate north. Five of the maps [top row, Fig. 21-2] are presented visually, being line drawings on a square of cardboard. The remaining maps are made of raised tacks with cords running between them and these are carried on a tray attached to the patient's chest who palpates them under a black cloth.*

The patients who have difficulties with this task show equally inferior performance for the visual and the tactual-kinesthetic (or haptic) modes of presentation. Yet this general difficulty in route finding by means of maps is specifically dependent on the parietal lobes, since only patients with parietal lesions show a significant deficit. This deficiency cannot be classified as an *agnosia* in the usual sense, as it implies disorders limited to "higher functions" *within* a given sensory modality. Instead, these observations suggest that one component of the parietal lobe syndrome consists of an inability to evaluate sensory information in terms of its spatial organization and that the contribution of the parietal lobe to the structuring of sensory information into a spatial framework is not limited to the somesthetic senses. The integrity of some structure in this region may be necessary for the utilization of maps and schema as devices to represent space internally.[69]

Although not demonstrated by the route-finding test, there is other evidence to indicate that defects in the apprehension of spatial data or topographic concepts are preferentially associated with lesions in the right parietal lobe.[42]

Associative learning of tasks in somesthesis

There is firm evidence that the posterior parietal lobe lesions in macaques produce well-delineated deficiencies in the acquisition and performance of tasks involving discrimination of tactual stimuli, deficiencies of a nature that excludes the involvement of the ataxic component of the parietal lobe syndrome described earlier.

The general method used in these studies involves one or the other variant of the so-called Wisconsin General Test Apparatus (WGTA), described subsequently. As a rule, the subjects are first given trials on a large number of problems similar to that which will later constitute the test selected for the definitive study; this is to develop the monkey's ability to "learn to learn," based on the learning-set procedure of Harlow.[9]

There are two different aspects to the role of the posterior parietal lobe in associative learning of somesthetic stimulus discriminations. In the first place, in tests involving the discrimination of roughness of textures (sandpapers of different grades), the deficit induced by posterior parietal ablations consists of a decrement of the capacity to resolve fine differences between stimuli, whereas the performance remains unimpaired when the difference between the stimuli is great.[74] Thus the posterior parietal lobe determines, as it were, a "set point" for the degree of resolution in stimulus discrimination.

The second aspect concerns the selection of the strategy for handling multiple-object, problem-solving tasks. The test procedure in the WGTA is as follows:

The animals are initially confronted with two objects placed over two holes, on a board containing 12 holes in all (with a peanut under one of the objects). An opaque screen is lowered between the monkey and the objects as soon as the monkey has displaced one of the objects from its hole (a trial). When the screen is lowered, separating the monkey from the 12 hole board, the objects are moved (according to a random number table) to two different holes on the board. The screen is then raised and the animal again confronted with the problem. The peanut remains under the same object until the animal finds the peanut five consecutive times (criterion). After the monkey reaches criterion performance, the peanut is shifted to the second object and testing continues (discrimination reversal). After an animal again reaches criterion performance a third object is added. Each of the three objects in turn becomes the positive cue; testing proceeds as before until the animal reaches criterion performance with each of the objects positive, in turn. Then a fourth object is added and the entire procedure repeated. As the animal progresses, the number of objects is increased serially through a total of 12 (Fig. 21-3).*

The experimenter is interested in analyzing the monkey's performance in terms of recurrent and persistent trends. If a consistent pattern of choices can be seen, one tends to interpret this as a reflection of a "hypothesis" the monkey may have adopted regarding the consistencies in

*Teuber, H. L.: In Eccles, J. C., editor: Brain and conscious experience, New York, 1966, Springer-Verlag New York, Inc.

*Pribram, K. H.: Behavioral Sci. **4**:245, 1959.

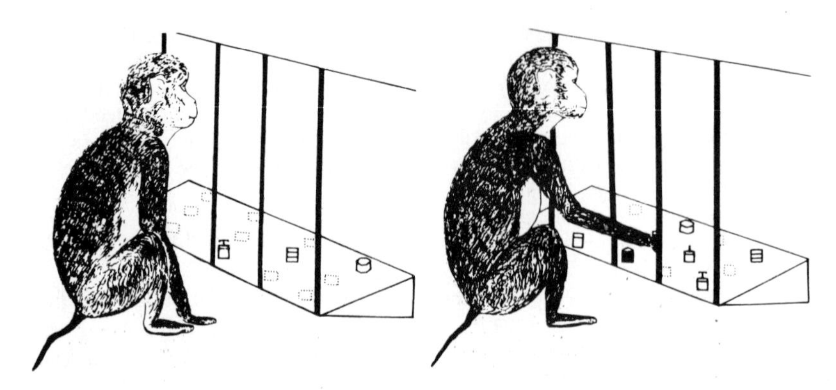

Fig. 21-3. Diagrams of multiple-object, problem-solving examples of three- and seven-object situations. Food wells are indicated by dashed circles, each of which is assigned a number. Placement of each object over food well was shifted from trial to trial according to random-number table. Record was kept of object moved by monkey on each trial, only one move being allowed per trial. Trials were separated by lowering an opaque screen to hide objects from monkey as they were repositioned. (From Pribram.[17])

the environment he faces in order to guide his behavior toward optimization of reward. Some typical questions the experimenter wishes to answer are as follows: Given a particular stimulus situation on consecutive trials, what pattern of responses does a monkey show? Does he choose the object on the right side consistently? Or does he alternate?

The problem just described is particularly interesting because for optimal performance it requires two different strategies that should be followed in alternation: (1) During search, move on successive trials each of the objects until the peanut is found. (2) After search, select the object under which the peanut had been found on the preceding trial (i.e., lose-shift, win-stay). These strategies should be appropriately alternated to obtain the maximal number of rewards.

The result of this experiment by Pribram[60] was that animals with posterior *intrinsic sector* ablations show a striking reduction in searching, that is, they will sample fewer objects for the reward. This deficit appears specific for the parietal lesions; the deficit found after frontal intrinsic sector ablations is of an entirely different nature (p. 637).

Based on these and similar observations, Pribram[60] formulated a generalization that has significant implications for an understanding of the operational principles of the posterior intrinsic cortex. The essential nature of the posterior parietal lobe deficit becomes apparent when one compares the failure in the multiple-choice discriminative task with a behavioral situation in which normal animals and those with parietal lobe lesions perform indistinguishably: a monkey with drastic resection of the posterior parietal lobe is able to catch a flying gnat in midair and perform similar isolated acts with undiminished skill and precision; his reaction is entirely normal in simple "go–no go" situations that merely require the decision of whether or not to execute a certain behavior. In contrast, when presented with a series of stimuli for selection and required to choose different responses according to preceding outcomes of the test, the animal with a parietal lobe lesion fails.

Accordingly, it appears that the posterior intrinsic cortical mechanism enables the subject to deal with more complex units of behavior requiring a selection from "contextual alternatives" into which the sensory information can be partitioned. It is as if a set of environmental circumstances were divided into different mutually exclusive classes, as if one of these classes were selected as the relevant environmental state (its "image"), and as if a plan were adopted that would control an appropriate sequence of actions to be performed.[12]

INFEROTEMPORAL CORTEX

The recognition of the role of the temporal lobe in vision originated with an observation of Klüver and Bucy[50]: one of the most startling defects following bilateral temporal lobectomy in macaques was their inability to recognize objects by vision alone ("psychic blindness"). Discrimination of two-dimensional patterns is the most severely impaired visual ability, but there is also some deficiency in the discrimination of object quality in terms of color, size, and brightness. Although widespread within vision, the defect does not extend beyond it. Subsequent studies led to a progressively more precise delineation of a circumscribed region in the temporal lobe, ablation of which abolished the monkey's ability to discriminate between pairs of visual patterns that he was successful in discriminating preoperatively. The area in question is the inferior temporal convexity, comprising the middle and inferior temporal convolutions.[14,46]

On gross observation the monkey with an inferotemporal lesion is indistinguishable from a normal animal; the visual discrimination impairment becomes evident only in more formal training situations and becomes more pronounced as the difficulty of the discrimination tasks is increased. The defect consists primarily of a retardation of the rate at which a particular visual discrimination can be acquired. This raises the question of whether the trained monkey with an inferotemporal lesion responds to the same "distinctive stimulus features" as the normal monkey or whether he perhaps utilizes cues provided by the stimulus in a different manner.

Experimental evidence supports this latter idea[31]: if complex stimuli are presented (e.g., panels depicting a series of parallel bars that are alternately black and colored, with both angular orientation of the bars and color hue as variable stimulus attributes), the monkey with an inferotemporal lesion performs, at least at a certain stage of training, as though he had only learned at most to respond to "colored bars" on the panel and not to the more narrowly defined stimulus of "blue vertical bar," as would the normal monkey. These and similar findings suggest that the nature of the deficit of the inferotemporal monkey is not merely attributable to a deficit in perception, learning, or retention.

On theoretical grounds, one would predict that the discrimination performance of normal subjects would improve with stimulus patterns of increasing redundancy; concomitantly, retention would diminish. Normally there is a trade-off between attending to redundant stimulus features to optimize discrimination on the one hand and focusing on a particular aspect of the stimulus to optimize retention on the other.

Characteristically, monkeys with inferotemporal lesions perform with stimuli of different redundancy as if they were attempting to process *all* the available stimulus information: it seems that they try to attend to all the choices, correct or incorrect, and attempt to remember also the redundant features of the stimuli.[73] These results fit with the idea that the inferotemporal cortex exerts some efferent control over the input from the primary visual system, with the effect of reducing redundancy in visual input.

Inferotemporal neurons have a high degree of specificity and individuality in their responses to visual stimuli; length, width, and shape of the stimulus, as well as contrast with background, color, and direction and speed of movement affect the response. The receptive fields are relatively large, may be unilateral or bilateral, and invariably include the fovea. Characteristically, the responses diminish with repeated stimulus presentation.[40]

The anatomic connections of inferotemporal cortex are complex and place it in a strategic position for visual information processing. On the one hand, the inferotemporal cortex has two-way connections to the optic tectum, which accommodates the primary central visual mechanisms in lower vertebrates; in the primate, this pathway involves also the pulvinar. On the other hand, there are reciprocal connections to the phylogenetically more recent geniculostriate visual system via the circumstriate belt of cortical tissue. Consequently, the inferotemporal cortex receives information related to visual orienting processes (via the superior colliculus and pulvinar) and also the information extracted from the visual scene by the "feature'detecting" neurons of visual cortex. There is evidence for a defect in visuomotor integration after inferotemporal lesion: the patterns of eye movements during visual discrimination are altered. This may reflect a primary defect of scanning and information gathering in the visual scene as the cause of the visual deficit, or it may be the result of some perceptual or mnemonic defect that interferes with search or selection of cues.[29a]

Visual deficits after temporal lobe damage in man. In contrast to the language disturbances caused by temporal lobe damage in the dominant hemisphere, a constellation of deficits in nonverbal task performance arises after cortical excision in the right temporal lobe; some of these deficits appear closely related to the visual deficits of monkeys with inferotemporal lesions.[55] Among these visual deficits are disturbances in tachistoscopic recognition and dot numerosity,[49] in classification of pictures of faces in which normal contour lines have been eliminated[51] and in the detection of small differences in complex patterns.[53] The recall of geometric drawings, but not of stories or word pairs, is also impaired.[56]

MIDDLE TEMPORAL AREA: AUDITORY DISCRIMINATION

The evidence for modality-specific auditory discrimination deficits after temporal lobe lesions is less definite than for tactile and visual deficits with regard to their respective intrinsic cortical areas. To some extent this is related to the fact that monkeys, the preferred species for studies of this general type, have great difficulty in acquiring behavior guided by auditory stimuli. However, Penfield and Kristiansen's observations[16] indicate the existence of a circumscribed area in the temporal lobe of man that, if stimulated electrically during brain surgery, elicits the patient's recall of past auditory experiences (e.g., the specific rendition of a piece of music he may have heard years ago). As during temporal lobe epileptic discharges, the patient feels "distant" and removed from the reenacted experience.

The experiments of Diamond and Neff,[37] carried out in cats, demonstrated that bilateral removal of a cortical area corresponding to primate temporal neocortex produces a specific deficit that is not sensory. Ablation of the superior temporal convolution is followed by a lasting deficit in the discrimination of tonal patterns in the absence of any obvious frequency or intensity discrimination deficit. In addition, the cat's ability to localize the source of the sound is disrupted: it is as though the sound image produced by a pair of clocks had lost the attribute of locus in an extracorporeal space. Thus location of a sound source in space is no longer a component of the animal's perception of the external environment.[52]

Auditory perception deficiencies in man. In its most dramatic form the auditory deficit entails a failure to comprehend spoken language (receptive aphasia). Most patients with such a condition also have severe expression disorders; they produce a rush of speechlike but incomprehensible sounds, often without appreciating their own disability. An important and not yet entirely resolved question is whether there is in the human brain a regional specialization for processes underlying perception and comprehension of speech patterns and other nonspeech sound patterns (Chapter 22).

The evidence suggestive of lateralization of specific auditory functions is derived from the responses to dichotic tasks involving the separate presentation of two different strings of signals (e.g., the numbers 234 and 357 or other verbal material), one to each ear. The normal listener will first deal with the input to one ear (i.e., 234) while holding the other ear's input in short-term memory and then reproduce the input to the other ear (i.e., 357). When the stimuli are presented in strict synchrony , there is a superiority in performance on the side opposite the hemisphere that mediates speech (normally, the left); this asymmetry is established early in life.[28] A preference for the stimulus delivered to the left ear exists for nonverbal material (e.g., brief melodies). As one would expect from this asymmetry, patients with long-standing right temporal lobe lesions perform in the dichotic number test as normal subjects do, whereas left temporal lobe lesions lower the performance score.[13] The situation is reversed for dichotic presentation of melodies.

Information processing in perception

The preceding sections were presented in order to demonstrate that behavior controlled by perceptual processes appears to involve internal models of reality that permit the structuring of sense experiences into spatial order and temporal sequence, but current sensory information never mediates behavior directly.[7] This conception is at variance with the more traditional stimulus-response reflex arc concept that ties current action rather closely to current sensory input.

The term "model" is used here in the sense proposed by Minsky: "To an observer B, an object A* is a *model* of an object A to the extent that B can use A* to answer questions about A."[57] This implies that the model subserves a dual function and consists, as it were, of two compartments: one that contains the required knowledge that the organism has about itself and its world (like a data file on a computer tape or disk) and one that contains the machinery for addressing questions and decoding answers. Although bipartite, this constitutes one information-processing system.

Information-processing theories of perceptual processes in problem solving, including those employing computer simulation as their means of formalization and analysis, have in recent years generated valuable insights. It is now generally agreed that problem solving can be represented as searching through a large set of possibilities until a satisfactory solution is found. For instance, a characteristic way in which people respond to a set of symbol sequences is to adopt some hypotheses as to the regularities in the sequence; look for evidence to see whether the hypothesis holds true; if needed, search for additional evidence; and eventually adopt the hypothesis or test another alternative. In this sense, patterns in sets of elements are "conceived" and in some sense generated in the perceiver's mind by some iterated process of devising a plan, testing its consequences, and either acception or rejecting it.[12,66]

In solving problems that require actions as outcome, this process of "pattern conception" becomes an integral part of the problem-solving process itself. Evidence for the character of the initial perceptual activity comes largely from situations where problems are presented to subjects in visual form. The evidence takes two forms: (1) a record of the subject's eye movements during the first few seconds after problems (e.g., a chess position) are presented to them, showing the succession of fixations during this time, and (2) the subject's ability to retain information about complex visual displays after brief exposure.

With regard to the second point, the important feature is that configurations in which the subject recognizes some "meaning" can be retained in short-term memory as one "chunk"[11]: a chess master can reproduce a position on the board almost without error, placing all or almost all the pieces correctly, but with random boards his performance drops to that of an amateur. With regard to the "perception" of the chess position while selecting the next move, the observations indicate that the succession of fixations involves not only the positions of the figures actually on the board but also the connections between present and future positions after a hypothetical move. In this sense the eye movements reflect thought processes.[26] Accordingly, there are two processes in operation that focus on the "map" provided by the chess board: (1) the recording of information into "meaningful" chunks, where meaning is related to the subject's past knowledge and familiarity with the situation, and (2) a goal-directed exploration of alternative changes to be brought about in the position on the board with these alternative changes actually carried out, one by one, by the appropriate eye movement. Both processes are ingredients of the computer programs that successfully mimic the chess player's performance.[21]

This last example has been discussed in some detail because it offers an analogy or perhaps a prototype for the kind of information processing the association cortex is part of: one could conceive of the primary cortical receiving areas, which are topologic maps of their respective receptor sheets, as fulfilling the role of the chess board in the preceding example. The figures on the chess board would in this view represent the fractionated and fragmented stimulus-feature representations in the specific sensory systems (discussed in Chapter 20). The role of the modality-specific association cortex would be to provide the programs and strategies that operate on the "data structure" in the primary cortical receiving areas, to partition the fractionated stimulus feature to meaningful "chunks" in accordance with past experience, to test on these same maps various ways of combining "chunks" to objects of perception with "meaning" in the light of past experience, and to test the outcomes of plans for future action.

FRONTAL INTRINSIC SECTOR

One of the most striking difficulties of the macaque with a defective frontal lobe is manifested in behavioral situations involving delays between presentation of a stimulus and the opportunity to respond to it. The tasks involving such delays were originally devised to demonstrate that animals were capable of symbolic processes in the sense that an "idea" would be required to solve a problem for which the clues to a solution are given prior to, and not at the time of, opportunity for solution.[45]

The delayed reaction test is conducted in the following form or some variant of it: a monkey is allowed to view through bars a piece of food being deposited beneath one of two or more cups on a sliding tray. An opaque door is then lowered in front of the animal for a chosen interval. At the end of this interval the tray is pushed forward to the cage, the door is raised, and the animal is permitted to reach the cups. The animal is allowed to select one cup, the reward being obtained if the proper cup is selected. With training, a normal monkey makes successful choices after delays as long as 90 sec between seeing the food and making the selection.

The delayed alternation test is similar. In this case the animal is required to make alternate right (R) and left (L) turns and to remember which turn comes next in a sequence; there is a delay imposed between completion of one turn and the initiation of the next in the sequence. A special case is that of double alternation, in which the correct choices are RRLLRRLL.

In the studies of Jacobsen[47] the delayed response task proved to be a selective index of primate frontal lobe injury: after bilateral orbital-frontal lesions, delays as short as 5 sec between seeing the reward and selecting its location reduce success in the choice to a matter of chance; the animal is simply at a loss in choosing the container that conceals the food.

The interpretation of this phenomenon has been the subject of much discussion: Jacobsen[47] favored the idea that an impairment of recall, manifested as a defect in an "immediate-memory" process, could account for the deficit. Others emphasized increased distractibility. It was also suggested that locomotor hyperactivity occurring after frontal surgery in the monkey might interfere with the utilization of mnemonic devices (e.g., spatial cues) adopted to help bridge the interval between stimulus exposure and response opportunity. None of these suggestions appeared to adequately explain the behavior deficit.

Additional and more refined experiments by Pribram,[61] and Pribram et al.[64] provided evidence for a more precise characterization of the consequences of frontal intrinsic sector ablation. First, there are several factors that can be eliminated as elements contributing to the behavior deficit. The deficit is not due to a deficiency of function in one particular sense modality; it is independent of whether visually discriminable or visually indistinguishable cues are to be relied on for problem solving, and impairment of proprioception is not responsible. Second, there is some more positive evidence as to what constitutes the basic defect: in the multiple-object discrimination problem (p. 638), which requires alternation at appropriate times between a systematic search strategy and a strategy to persevere once a successful solution is obtained, the monkey with a defective frontal lobe fails to take the cue that indicates the point at which to switch strategies.

When a monkey chooses the positive cue, five times in a row, he attains criterion. From this moment on, his best strategy would be to return to the original search (i.e., to move on successive trials each of the objects until the peanut is found again). The time for changing from the previous strategy (win-stay) to search is signaled to him by the fact that a response to the previously rewarded object is no longer rewarded. The animal with a defective frontal lobe does not take this cue; instead, it perseveres with the object that was previously but is no longer rewarded.

It appears as though the deficit in an animal with a frontal lobe lesion implies an inability to cope with situations in which an element of remoteness and ambiguity is introduced into the relation between stimulus and success of response. A particularly clear-cut experimental situation devised by Pribram et al.[64] affirmed this idea. The experimental protocol in question was as follows:

> Two objects, a small tobacco tin and a flat ash tray, served as cues. All subjects were given 30 trials a day and were initially rewarded only when they chose the tobacco tin. When the tobacco tin had been chosen for ten consecutive responses, the reward (a peanut) was placed under the ash tray until ten consecutive correct responses were again obtained. Another reversal was then instituted. Reversals were continued to the "ten correct" criterion until 500 trials were accomplished. The procedure was then changed so that reversals were given after an animal had reached a criterion of only five consecutive correct responses; the reversals to the five correct criterion were continued until another 500 trials were completed. After this, the monkeys were run to criteria "four correct," "three correct," and "two correct," in that order.*

The prominent feature of this situation is that the number of consecutive successful discriminations required to meet criterion keeps changing; the animal with a defective frontal lobe is unable to cope with this problem.

Behavior guided by stimulus contingencies that are no longer present when the response is to take place implies some temporal extension of the reaction to the stimulus presentations that can bridge the period to the response. What is the nature of this process? As stated before, monkeys with frontal ablations routinely fail the delayed alternation task, but this result changed radically when Pribram[63] used the following stimulus schedule: between each pair of R-L presentations, a 15 sec delay was interposed, so that the temporal pattern of the task had this appearance:

R-L—R-L—R-L—R-L, etc.

This change in stimulus program enabled the animal with a defective frontal lobe to perform the delayed alternation task as well as normal animals. It was critical that the 15 sec alternation occurred between pairs of R-L sequences and not between each R-L alternation. Thus external "pacing" provided by the environment can compensate for the failure of the mechanism that enables the normal animal to overcome the temporal separation between stimulus presentation and response opportunity. The conjectures from this observation are that the role of the frontal intrinsic sector consists of imposing some temporal organization on the stream of stimuli impinging on the organism and that the proper division or "chunking" of the stimulus stream is an important factor in guiding the organism's behavior when the relation between stimulus and response is fraught with ambiguous expectations.

A suggested analogy is the parsing of sentences and its relation to the meaning a string of words may take on. The well-known example by Chomsky illustrates this point for the word string "they are flying planes," which, according to the segmentation in components, may have entirely different meanings: "(they) (are flying) (planes)" or "(they) (are) (flying planes)."

The point of view derived from these observations is that the "short-term memory," which intervenes between the presentation of stimuli and the opportunity for response, involves active working processes of input coding and programming and not merely the deposition of a memory trace of some sort, which fades according to some law of decay.

SENSORY-MOTOR COORDINATION

The example of the relation between eye movements, perception, and planning an activity that was discussed earlier emphasizes the intimate interrelations between sensory and motor processes in the information transactions of the CNS. These and similar phenomena have been interpreted as indicating a mechanism whereby a self-produced movement involves not only the classic efferent discharge to the musculature of the effector organ but also a concomitant (corollary) discharge from motor to sensory areas within the CNS. This corollary discharge is assumed to be present in all voluntary movements but absent in purely passive ones; it presets the sensory systems for the changes in the input that are anticipated consequences of the intended movement.[72]

The significance of the interrelations between sensory and motor systems has become apparent

*Pribram, K. H., et al.: A progress report on the neurological processes disturbed by frontal lesions in primates. In Warren, J. M., and Akert, K., editors: The frontal granular cortex and behavior, New York, 1964, McGraw-Hill Book Co.

from experiments on restricted rearing (with motor rather than sensory deprivation), in which a normal adult was fitted with distorting (prismatic) spectacles. The prisms imposed tilts and curvatures on contours and displaced the visual scene. Held and Freedman[43] have shown that far-reaching adaptation to the optically induced distortions and displacements can be achieved, provided the subject actively moves about while viewing his visual environment through the distorting prisms. In contrast, when the subject is moved about in a wheelchair while wearing the distorting spectacles, he fails to produce any adaptation. This proves that self-initiated movement is a necessary prerequisite for perceptual adaptation under conditions of deranged sensory input. It is thought that the hypothetical corollary discharges associated with the self-produced movement permit the development of this perceptual adaptation.

Held and Hein[44] performed an additional experiment that seems to indicate that movement plays a similar role in the initial acquisition of sensory-motor coordination during early development, as it does in adaptation to rearranged sensory input. A kitten whose visual exposure is limited from birth to the inside of the drum shown in Fig. 21-4 will exhibit essentially normal form and depth perception when tested after several months, provided that he moves himself about actively (kitten *A* in Fig. 21-4). On the other hand, the same exposure under passive conditions, for example, within a gondola (kitten *P* in Fig. 21-4), in which a "passive" kitten is carried about by an "active" littermate, precludes the normal development of visual motor coordination. The interpretation is that the passively moved kitten lacks the opportunity to correlate its self-produced motor output with the corresponding changes of visual input.

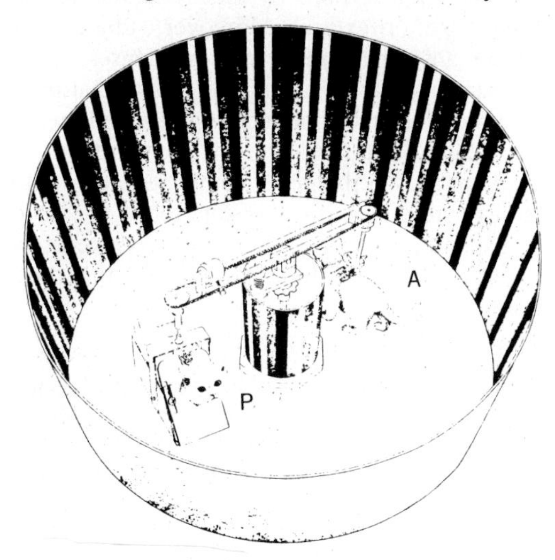

Fig. 21-4. Active and passive movements of kittens were compared using this apparatus. Active kitten walked about more or less freely; its gross movements were transmitted to passive kitten by chain and bar. Passive kitten, carried in gondola, received essentially same visual stimulation as active kitten because of unvarying pattern on wall and on center post. Active kittens developed normal sensory-motor coordination; passive kittens failed to do so until several days after being freed. (From Held, R.: Plasticity in sensory-motor systems. Copyright © 1965 by Scientific American, Inc. All rights reserved.)

There is evidence that profound and lasting disorders of the postulated corollary discharge mechanism are related to lesions of the frontal lobe or the basal ganglia in man.[24] A more direct demonstration exists for the monkey; Bossom[30] has shown severe deficiencies in adaptation to distorting prisms in monkeys with bifrontal lobectomies or bilateral lesions in the head of the caudate nucleus.

INTERHEMISPHERIC INTEGRATIONS

This topic directs attention immediately to the large forebrain commissures, of which the corpus callosum is the most prominent structure. Evidence that has been forthcoming since 1950 dealt at first with the role of the corpus callosum in the interhemispheric transfer of visual discrimination learning and later with more general aspects of interhemispheric relations.[22]

Sectioning the optic chiasm in a mammal with crossed optic fibers leaves the major part of the visual field intact, but each eye then projects only to its homolateral hemisphere (Fig. 21-5). Cats with sectioned chiasms, trained to discriminate visual patterns through one eye while the second eye is covered with a mask, can also execute the same task through the other, previously covered eye. Thus there is intercommunication between the two brain halves. However, this cross-availability of information between the two hemispheres can be abolished by sectioning the corpus callosum. Accordingly, with both optic chiasm and corpus callosum sectioned, the cat can learn a different, even opposite, discrimination task with each eye without mutual interference. This functional independence of the surgically separated hemispheres with respect to discrimination learning has since been amply substantiated.

There are some exceptions to this general rule. For instance, easy brightness discriminations learned with one eye do transfer to the other hemisphere in split-brain preparations.[54] Moreover, split-brain cats and monkeys can discriminate brightness between one

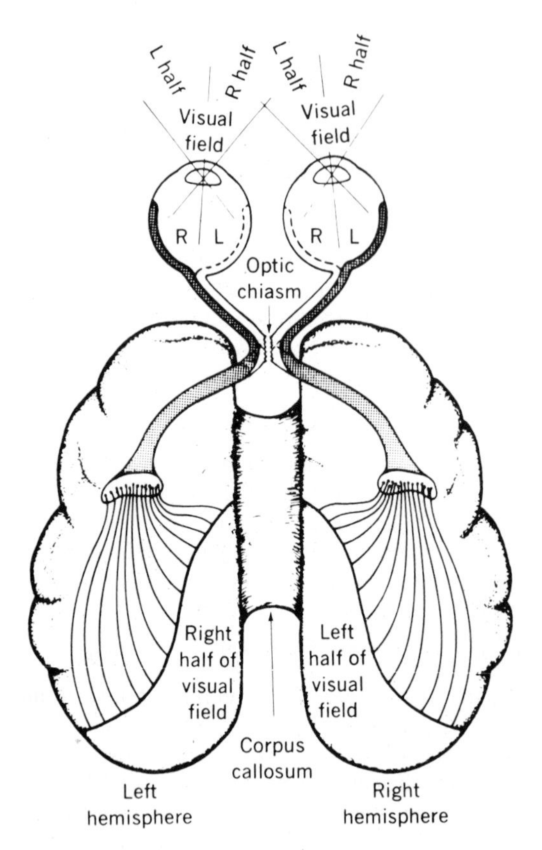

Fig. 21-5. Sketch indicating effects of midsagittal division of optic chiasm. (From Sperry, R. W.: Hemisphere deconnection and unity in conscious awareness, Am. Psychol. **23:**723, copyright © 1968 by the American Psychological Association; reproduced by permission.)

stimulus presented to one eye and another stimulus presented to the second eye.[65] Thus certain cross comparisons are still achieved between the two divided hemispheres: it appears that discrimination involving visual identification of objects is a highly corticalized process that requires the corpus callosum for interhemispheric transfer; on the other hand, discriminations between spatial distributions of otherwise identical stimulus components (e.g., five-pointed star vs six-pointed star) transfer from one side to the other even in the absence of the corpus callosum, presumably via brain stem pathways.[71]

Corpus callosum section has been carried out in humans in an attempt to cope with otherwise unmanageable epileptic seizures. These patients have undergone extensive tests of their perceptual and cognitive abilities, and many interesting findings were uncovered, notably with regard to the subtleties of the differentiation between the major and minor (subordinate) hemispheres. Right-handed patients whose left hemisphere is equipped with the expressive mechanism for speech and writing and with the main centers for comprehension and organization of language can communicate verbally about the visual experiences of the right half of the visual field and about the somesthetic experiences of the right body half. The right hemisphere in aphasic and agraphic patients who are unable to express themselves verbally can be shown to perform inter- and cross-modal generalizations of perceptual or mnemonic material. For example, after a picture of some object (e.g., a cigarette) has been presented to the minor hemisphere through the left visual field, the subject can retrieve the item pictured from a collection of objects using blind touch with the left hand, which is mediated through the right hemisphere. Unlike the normal person, however, the commissurotomy patient is obliged to use the corresponding hand (the left hand, in this case) for retrieval and fails when he is required to search with the right hand. Using the right hand the subject recognizes and calls off the names of each object, but the right hand, or its hemisphere, does not know what it is looking for. The minor hemisphere can also perform certain generalizations and abstractions (e.g., directing the sorting of objects into groups by touch on the basis of shape, size, or texture). Moreover, subjects give correct answers with manual responses of the left hand in tests requiring the determination of sums or products of numerals perceived stereographically, but they are unable to report the same answers verbally. In some cases the minor hemisphere is found to be superior to the major hemisphere, notably in tasks that involve drawing spatial relationships and performing block design tests.[23]

The general implication is that patients with neurologic lesions have various strategies at their command to compensate for their defects. Therefore it is very difficult to obtain a conclusive neurologic description of a patient with brain damage.

Another factor is the difference in symptoms caused by a certain lesion, depending on the stage of neural maturation at which the defect occurred. A case of agenesis of the corpus callosum, studied by Sperry,[68] illustrates the high degree of functional reorganization and compensation that is possible with congenital defects in the CNS. This patient was able to read rapidly words that fell partly in one visual half-field and partly in the other, a task at which patients with surgical separation of the hemispheres in later life fail entirely. Interestingly, persons with congenital corpus callosum defects are very often ambidextrous, suggesting that speech is represented bilaterally. Compensation is not equally effective for all tasks and in all sense modalities; information of tactile, positional, and kinesthetic patterns acquired by one hand cannot be utilized as fully to guide motor action of the opposite hand as it is in normal subjects. This deficit is particularly obvious in manual maze tests, in which each hemisphere has to acquire, essentially independently, the skill to solve the problem.

MECHANISMS OF LEARNING AND MEMORY

The process of learning is concerned with establishing a relation between a present and a past event; furthermore, when successive stimuli can be compared and a response that is contingent on that comparison can be made, we speak of memory. The implication is that the original event must leave some change (a trace or engram) in the nervous system that permits the present event to be related to it.

Research on memory has focused on the properties and nature of this hypothetical trace. However, with the exception of some measurements of human recall, it is for all practical purposes impossible to study memory in isolation without also involving associative learning, notably in animal experiments applying procedures to elucidate the character of this trace.

Weiskrantz has analyzed this complex situation lucidly: "In order to know whether an event can be stored and retrieved (more strictly, can produce a change within the organism that permits a subsequent repetition of the event to be related to the original), we must have a response that we can measure. We obtain such a response by associating our event with another that is known to control behavior in a predictable manner, i.e., a reward or punishment."[25] Learning, then, is bound up with the study of memory as a matter of practical necessity.

Natural history of the memory trace. In the typical experimental situation the subject initially undergoes one of a series of *exposure trials* for learning; for example, an exposure trial may consist of the application of a visual or an auditory stimulus, followed by an aversive stimulus that the subject can avoid by a specific act or instrumental performance (like pressing a bar). On the subsequent *retention trials* the subject is expected to "remember" the association between the neutral and aversive stimuli. The overt evidence for the establishment of this association is provided by the subject's act to prevent the aversive stimulus from occurring.

When an electroconvulsive shock (ECS) is administered immediately after the exposure trial, the performance at the retention trial is usually impaired; however, if the ECS is administered after some longer interval (e.g., some 6 hours after the exposure trials), the impairment, if any, is unappreciable.[32] Various other interferences such as brain concussion and anesthetics have the same retrograde amnesic effect. These and similar findings have led to the idea that a gradual change occurs in the trace initially set up by the exposure trial, in the sense of a *consolidation*.

The original theory of consolidation was that information is initially stored as some form of persistent, reverberating neural activity that is gradually converted into structural changes at synapses. In more general terms the implication is that there are at least two kinds of traces: one that is short-lasting and susceptible to disrupting treatments and one that is more stable and long lasting. There is a suggestive parallel between the distinction between these types of traces and the concepts of a more immediate short-term and more indirect long-term memory, based on observations in man.

Storage versus accessibility. The retrograde amnesia that occurs after ECS does not necessarily imply that this treatment interferes with consolidation. One may argue that the disturbance affects the subsequent availability of the trace for retrieval. The question is: How permanent is the apparent retention impairment? Experimental as well as clinical evidence[75] indicates improvement of retention with time after the disruptive treatment; for example, subjects who had no retention 24 hr after ECS may have some residual when tested 48 hr later. This suggests that the interfering trauma does not cause total trace destruction, but that the retrieval of traces may be temporarily interfered with.

Neurophysiologic mechanisms related to learning. A first approach to the study of neural mechanisms that may be involved in the formation of the memory trace is to study the electroencephalographic (EEG) changes during the acquisition of new behavior.

In the nonattentive, relaxed state the EEG exhibits the characteristic 8 to 13 Hz oscillations of the alpha rhythm (Chapter 8). Any novel and unexpected stimulus results in an arrest or blocking of this alpha activity, and the electrical record assumes the form of low-voltage, fast activity. There are concomitant somatic and autonomic signs: eye movements toward the stimulus source, respiratory irregularities, change in electrical skin resistance, and an increased muscle tone. The sum total of these changes is the *orienting* response, of which the blockade of the alpha rhythm is the EEG sign. If the stimulus has no behavioral significance to the animal and is repeated a number of times, the orienting response quickly subsides; *habituation* has occurred. This habituation is quite specific for the repeatedly applied stimulus. Certain kinds of stimuli lead more readily to habituation than others. For instance, visual stimuli may require hundreds of presentations before the alpha blockade over the occipital area fails to take place, whereas the cortical arousal with auditory stimuli fades after some 20 to 30 stimulus presentations.

An important new aspect was introduced in 1935 when Durup and Fessard, in experiments with habituation to visual stimuli, accidentally discovered that the click of the shutter of a camera used for intermittent photography of the oscilloscope record also elicited alpha block after some time, even when the light stimulus was withheld. The click had not previously produced a change of electrocortical activity and had apparently acquired the ability of altering the occipital alpha rhythm as a result of its temporal association with the visual stimulus. When the click was subsequently presented without association with the light stimulus, it soon lost the acquired capacity of altering the occipital rhythm.

These experiments, and others performed later in different animal species and man, clearly demonstrated that the occipital alpha rhythm of the EEG can be depressed or blocked by sound, a previously ineffective stimulus, if it has been repeatedly paired with light as an unconditioned stimulus. Prior to this finding, it was generally accepted that responses that could be conditioned in the classic pavlovian sense were of three kinds: (1) glandular, smooth-muscle, and blood pressure responses, (2) involuntary responses in striate muscle (e.g., eye movements), and (3) potential voluntary responses (e.g., withdrawal movements or locomotion). To these a fourth group was added—electrical activity of cortical centers themselves.

The work of Adey[1] and co-investigators introduced a new perspective to this line of investigation: the emphasis in these studies shifted to attempts to discern, in the simultaneously recorded electrical activity of various cortical and subcortical structures, consistent mutual relations that would characterize the participation of and interaction between different neuronal populations in the learning process. Before learning started the electrical waves in the hippocampus were found to lead those in the entorhinal cortex; when the cat learned to reach the food reward in the T maze, this phase relation was reversed.[27] One possible interpretation of this phenomenon is the assumption that the information acquired during the learning process was stored in the entorhinal cortex and served subsequently as a "standard" against which sensory input was compared to decide which action should be taken. Observations such as these may be taken to support Penfield's suggestion[59] that the hippocampus functions in consolidation and recall of memory in conjunction with other neuronal systems, rather than solely as an independent and isolated respository of memory traces.

The idea of the participation of a multiplicity of neural structures in the information transactions that underlie learning is particularly well illustrated by an experiment of John and Killam[48]; this experiment also directs attention to differences in the involvement of different neural structures as learning progresses. The conditioning stimulus was a light flashing 10 times/sec, called a tracer-conditioned stimulus because it allowed the investigators to trace the signal in the brain wherever it would appear in the form of evoked potentials at 10 Hz or multiples therof. The cat had to learn to jump from one compartment of a double-grid box to the other to avoid shock whenever the conditioned stimulus was presented. Later a differential stimulus consisting of a light flashing 7 times/sec was introduced. No shock followed the presentation of this stimulus.

The conditioning stimulus initially gave rise to photic driving at many points in the brain, which faded out with repeated presentation as evidence for habituation. As soon as the conditioning started (when shock was presented after the conditioning stimulus), the tracer potentials reappeared at all recording sites, including the visual and auditory cortices, the reticular formation, the nucleus ventralis anterior of the thalamus, the fornix, the septum, the hippocampus, and the amygdala. As the cat learned to avoid shocks, the tracer potentials disappeared again, except from the visual cortex, hippocampus, and midbrain reticular formation. Later still, when the cat had completely learned the avoidance response, the tracer potentials also disappeared from the hippocampus but reappeared in the nucleus ventralis anterior of the thalamus.

During differential training, it was found that the 7/sec flicker first generated a 10/sec evoked response in the visual cortex and that the cat would make its usual avoidance response. As the cat began to discriminate behaviorally, however, the cortical responses also became differentiated, and the 7/sec flashes evoked 7 Hz waves. During the intermediate phase the cat usually made an error when the frequency of the evoked potential was not that of the presented stimulus. In other words, the behavior corresponded to the cortical activity rather than to the objective conditioning stimulus.

Human memory processes: a conceptual scheme. Contemporary memory research in normal individuals has characterized several stages in human memory processes (Table 21-1).[5,15,67] In the first place, there is evidence for a *sensory memory* of limited capacity that holds raw data of sensory events for a period of a few hundred milliseconds for attention, scanning, and further processing. This first stage in the processing of sensory information prior to further storage imposes in some sense a discontinuous, quantal character on the continuous stream of incoming sense impressions, much like a sample-and-hold device does in electronic data-gathering systems: the continuous flow of sensory signals becomes "packaged" in discrete units, each unit being either selected for transfer into another storage medium or discarded. Information is removed from sensory memory either by spontaneous decay or by a process of "erasure" in response to incoming new information.

The successful transfer from this short-lived store requires that it be attended to and recoded into a more stable form. This may occur in two different forms.[15] First, a verbal label may be attached to the content of the sensory memory, which is then thought to be transferred to another limited-capacity storage system, the *primary memory*. Although not accurately determined, there is a suggestion that its information capacity matches the limit of information capacity in sensory systems. Forgetting occurs in primary memory as new items enter and displace old ones. In the absence of such interference the information storage extends over several seconds.

An alternate route, if verbal recoding does not take place, consists of transfer from sensory to *secondary memory*, a large and more permanent storage system. Forgetting in secondary memory appears to require "unlearning" and interference by either previously (proactive inhibition) or subsequently (retroactive inhibition) learned material. The secondary memory can also serve as a recipient for information displaced from primary memory. The transfer mechanisms in this case are associated with rehearsal, which consists of recycling of material through primary memory. The probability of successful transfer appears to increase with the amount of time an item spends in primary memory: thus the success of transfer increases with the number of rehearsals associated with transfer cycles.

Present evidence suggests that "meaning" is a significant factor in the organization of secondary memory, in contrast to the operating mode of the primary memory. This difference is illustrated by the kinds of errors made on recall from the one as opposed to the other store[29]: in primary memory most errors result from phonetic confusion; items that sound alike (e.g., Bs and Vs) are substituted for one another. Errors in secondary memory, on the other hand, involve items with similar meaning. Thus part of the recoding in secondary memory is semantic, but other types of relationships (e.g., temporal and spatial contiguity) also play a role.

Finally, there are memory traces of extreme durability, ready access, and high resistance to brain damage and disease. This is the store for

Table 21-1. Human memory processes*

	Storage system			
	Sensory memory	**Primary memory**	**Secondary memory**	**Tertiary memory**
Capacity	Limited by amount transmitted by receptor	The 7 ± 2 of memory span	Very large (no adequate estimate)	Very large (no adequate estimate)
Duration	Fractions of a second	Several seconds	Several minutes to several years	May be permanent
Entry into storage	Automatic with perception	Verbal recoding	Rehearsal	Overlearning
Organization	Reflects physical stimulus	Temporal sequence	Semantic and relational	?
Accessibility of traces	Limited only by speed of readout	Very rapid access	Relatively slow	Very rapid access
Types of information	Sensory	Verbal (at least)	All	All
Types of forgetting	Decay and erasure	New information replaces old	Interference: retroactive and proactive	May be none

*From Ervin and Anders.[5]

which Ervin and Anders[5] proposed the designation of *tertiary memory;* it is thought to be the repository of such highly overlearned material as one's own name or how to read and write—material and skills outlasting retrograde amnesias that may affect other, less stable components of memory.

Remembering or reconstructing. The attempts to distinguish between the processes of remembering and reconstructing are a current theme in memory research and are based on techniques of inferring the properties of memory from the types of errors made by subjects. Bartlett[2] developed a theory of remembering that was based on the observation that when subjects tried to recall stories he had asked them to learn, their version was shorter, the phraseology updated, and the entire tale more coherent and consequential. Moreover, the subjects were often unaware that they were substituting rather than remembering, and often the very part that was created anew was the part the subject was most certain about. Thus there was some invention or, in Bartlett's term, *"constructive rendering."*

Bartlett was led to propose that recall from past experience is seldom merely reduplicative; instead, we reconstruct and schematize the remembered material in accordance with past experiences, circumstances, and expectations.[2] Thus memory and recall have more nearly the features of dynamic encoding and decoding processes than the properties one would expect from an immutable trace or engram.[62]

Current trends in thought on human information processing suggest that in order for material to be stored within secondary memory, it must be assimilated into an existing organization of rules and schemata and that retrieval from memory is a generative process in which these same rules and schemata reenact, as it were, the events we attempt to remember.[15]

REFERENCES

General reviews

1. Adey, W. R.: Spontaneous electrical brain rhythm accompanying learned responses. In Schmidt, F. O., editor: The neurosciences. Second study program, New York, 1970, The Rockefeller University Press.
2. Bartlett, F. C.: Remembering, London, 1932, Cambridge University Press.
3. Bender, M. B.: Disorders in perception, Springfield, Ill., 1952, Charles C Thomas, Publisher.
4. Critchley, M.: The parietal lobe, Baltimore, 1953, The Williams & Wilkins Co.
5. Ervin, F. R., and Anders, T. R.: Neural and pathological memory, data and conceptual scheme. In Schmitt, F. O., editor: The neurosciences. Second study program, New York, 1967. The Rockefeller University Press.
6. Gibson, J. J.: The senses considered as perceptual systems, Boston, 1966, Houghton Mifflin Co.
7. Gregory, R. L.: On how so little information controls so much behaviour. In Waddington, C. H., editor: Towards a theoretical biology, Chicago, 1969, Aldine-Atherton, Inc., vol. 2.
8. Gross, C. G.: Visual functions of inferotemporal cortex. In Jung, R., editor: Handbook of sensory physiology, Heidelberg, 1973, Springer Verlag, vol. 7.
9. Harlow, H. F.: Learning set and error factory theory. In Koch, S., editor: Psychology—a study of a science, New York, 1959, McGraw-Hill Book Co., vol. 2.
10. Held, R.: Plasticity in sensory-motor systems, Sci. Am. **213:**84, 1965.
11. Miller, G. A.: The magical number seven, plus or minus two; some limits on our capacity for processing information, Psychol. Rev. **63:**81, 1956.
12. Miller, G. A., Galanter, E., and Pribram, K. H.: Plans and the structure of behavior, New York, 1960, Holt, Rinehart & Winston, Inc.
13. Milner, B., and Teuber, H. L.: Alterations of perception and memory in man: reflections on method. In Weiskrantz, L., editor: Analysis of behavioral change, New York, 1968, Harper & Row, Publishers.
14. Mishkin, M.: Visual mechanisms beyond the striate cortex. In Russell, R. W., editor: Frontiers in physiological psychology, New York, 1966, Academic Press, Inc.
15. Norman, D. A.: Memory and attention: an introduction to human information processing, New York, 1969, John Wiley & Sons, Inc.
16. Penfield, W., and Kristiansen, K.: Epileptic seizure patterns, Springfield, Ill., 1951, Charles C Thomas, Publisher.
17. Pribram, K. H.: The physiology of remembering. In Bogoch, S., editor: The future of brain science, New York, 1969, Plenum Press.
18. Révész, G.: Psychology and art of the blind, New York, 1950, Longmans, Green, & Co., Inc.
19. Schilder, P.: The image and appearance of the human body, New York, 1964, John Wiley & Sons, Inc.
20. Semmes, J., Weinstein, S., Ghent, L., and Teuber, H. L.: Somatosensory changes after penetrating brain wounds in man, Cambridge, Mass., 1960, Harvard University Press.
21. Simon, H., and Barenfeld, M.: Information-processing analysis of perceptual processes in problem solving, Psychol. Rev. **76:**473, 1969.
22. Sperry, R. W.: Some general aspects of interhemispheric integration. In Mountcastle, V. B., editor: Interhemispheric relations and cerebral dominance, Baltimore, 1962, The Johns Hopkins University Press.
23. Sperry, R. W.: Hemisphere deconnection and unity in conscious awareness, Am. Psychol. **23:**723, 1968.
24. Teuber, H. L.: The riddle of frontal lobe function in man. In Warren, J. M., and Akert, K., editors: The frontal granular cortex and behavior, New York, 1964, McGraw-Hill Book Co.
25. Weiskrantz, L.: Memory. In Weiskrantz, L., editor: Behavioral change, New York, 1968, Harper & Row, Publishers.
26. Yarbus, A. L.: Eye movements and vision, New York, 1967, Plenum Press.

Original papers

27. Adey, R.: Studies of hippocampal electrical activity in approach learning. In Delafresnaye, J. F., editor: Brain mechanisms and learning, Springfield, Ill., 1961, Charles C Thomas, Publisher.

28. Averbach, E., and Coriell, A. S.: Short term memory in vision, Bell Syst. Tech. J. **40**:309, 1961.
29. Baddeley, A. D., and Dale, M. C. A.: The effects of semantic similarity on retroactive interference in long and short term memory, J. Verb. Learn. Behav. **5**:417, 1966.
29a. Bagshaw, H., Mackworth, N. H., and Pribram, K. H.: The effect of inferotemporal cortex ablations on eye movements during discrimination training, Int. J. Neurosci. **1**:153, 1970.
30. Bossom, J.: The effect of brain lesions on prism-adaptation in monkey, Psychosom. Sci. **2**:45, 1965.
31. Butter, C. M., Mishkin, M., and Rosvold, H. E.: Stimulus generalization in monkeys with inferotemporal and lateral occipital lesions. In Mostofsky, D. I., editor: Stimulus generalization, Stanford, 1965, Stanford University Press.
32. Chorover, S. L., and Schiller, P. H.: Short term retrograde amnesia in rats, J. Comp. Physiol. Psychol. **59**:73, 1965.
33. Critchley, M.: The body-image in neurology, Lancet **1**: 335, 1950.
34. Critchley, M.: Tactile thought, with special reference to the blind, Brain **76**:19, 1953.
35. Denny-Brown, D.: Discussion. In Mountcastle, V. B., editor: Interhemispheric relations and cerebral dominance, Baltimore, 1962, The Johns Hopkins University Press.
36. Denny-Brown, D., and Chambers, R. A.: The parietal lobe and behavior, Res. Pub. Assoc. Res. Nerv. Ment. Dis. **36**:35, 1958.
37. Diamond, I. T., and Neff, W. D.: Ablation of temporal cortex and discrimination of auditory cortex, J. Neurophysiol. **20**:300, 1957.
38. Féré, C.: Le main, la préhension et le toucher, Rev. Phil. **41**:621, 1896.
39. Goodnow, J.: Eye and hand: Differential sampling of form and orientation properties, Neuropsychologia **7**: 365, 1969.
40. Gross, C. G., Schiller, P. H., Wells, C., and Gerstein, G. L.: Single-unit activity in temporal association cortex of the monkey, J. Neurophysiol. **30**:833, 1967.
41. Hecaén, H.: Brain mechanisms suggested by studies of parietal lobes. In Darley, F. L., editor: Brain mechanisms underlying speech and language, New York, 1967, Grune & Stratton, Inc.
42. Hecaén, H., Penfield, W., Bertrand, C., and Malmo, R.: The syndrome of apractognesia due to lesions of the minor cerebral hemisphere, Arch. Neurol. Psychiatry **75**:400, 1956.
43. Held, R., and Freedman, J.: Plasticity in human sensorimotor control, Science **142**:455, 1963.
44. Held, R., and Hein, A.: Movement produced by stimulation in the development of visually guided behavior, J. Comp. Physiol. Psychol. **56**:872, 1963.
45. Hunter, W. S.: The delayed reaction in animals and children, Behav. Monogr. **2**:1, 1913.
46. Iwai, E., and Mishkin, M.: Further evidence on the locus of the visual area in the temporal lobe of the monkey, Exp. Neurol. **25**:585, 1969.
47. Jacobsen, C. F.: Studies of cerebral function in primates. I. The function of the frontal association areas in monkeys, Comp. Psychol. Monog. **13**:3, 1936.
48. John, E. R., and Killam, K. F.: Electrophysiological correlates of avoidance conditioning in the cat, J. Pharmacol. Exp. Ther. **125**:252, 1959.
49. Kimura, D.: Right temporal lobe damage, Arch. Neurol. **8**:264, 1963.
50. Klüver, H., and Bucy, P. C.: An analysis of certain effects of bilateral temporal lobectomy in the rhesus monkey with special reference to "psychic blindness," J. Psychol. **5**:33, 1938.
51. Lansdell, H. C.: Effect of extent of temporal lobe ablation on two lateralized deficits, Physiol. Behav. **3**:271, 1968.
52. Masterton, R. B., and Diamond, I. T.: Effects of auditory cortex ablation on discrimination of small binaural time differences, J. Neurophysiol. **27**:15, 1964.
53. Meier, M. J., and French, L. O.: Lateralized deficits in complex visual discrimination and bilateral transfer of reminiscence following unilateral temporal lobectomy, Neuropsychologia **3**:261, 1965.
54. Meikle, T. H., Jr., and Sechzer, J. A.: Interocular transfer of brightness discrimination in "split-brain" cats, Science **132**:734, 1960.
55. Milner, B.: Intellectual function of the temporal lobes, Psychol. Bull. **51**:42, 1954.
56. Milner, B.: Visual recognition and recall after right temporal lobe excision in man, Neuropsychologia **6**:191, 1968.
57. Minsky, M. L.: Matter, mind and models. In Minsky, M., editor: Semantic information processing, Cambridge, Mass., 1968, The M.I.T. Press.
58. Oldfield, R. C., and Zangwill, O. L.: Head's concept of the schema and its application in contemporary British psychology, Br. J. Psychol. **32**:18, 1942.
59. Penfield, W.: Functional localization in temporal and deep sylvian areas, Res. Publ. Assoc. Res. Nerv. Ment. Dis. **36**:210, 1958.
60. Pribram, K. H.: On the neurology of thinking, Behav. Sci. **4**:245, 1959.
61. Pribram, K. H.: A further experimental analysis of the behavioral deficit that follows injury to the primate frontal cortex, Exp. Neurol. **3**:432, 1961.
62. Pribram, K. H.: The amnestic syndromes: disturbances in coding? In Talland, G. A., and Waugh, N. C., editors: The pathology of memory, New York, 1969, Academic Press, Inc.
63. Pribram, K. H.: The primate frontal cortex, Neuropsychologia **7**:259, 1969.
64. Pribram, K. H., Ahumada, A., Hartog, J., and Ross, L.: A progress report on the neurological processes disturbed by frontal lesions in primates. In Warren, J. M., and Akert, K., editors: The frontal granular cortex and behavior, New York, 1964, McGraw-Hill Book Co.
65. Robinson, J. S., and Voneida, T. J.: Interocular perceptual integration in cats with optic chiasma and corpus callosum sectioned, Am. Psychol. **16**:447, 1961.
66. Shipstone, E. I.: Some variables affecting pattern conception, Psychol. Monogr. **74**:1, 1960.
67. Sperling, G.: Successive approximation to a model for short term memory. In Sanders, A. F., editor: Attention and performance, Amsterdam, 1967, North Holland Publishing Co.
68. Sperry, R. W.: Cerebral dominance in perception. In Young, F. A., and Lindsley, D. B., editors: Early experience and visual information processing in perceptual and reading disorders, Washington, D.C., 1970, National Academy of Sciences.
69. Teuber, H. L.: Space perception and its disturbances after brain injury in man, Neuropsychologia **1**:147, 1963.
70. Teuber, H. L.: Alterations of perception after brain injury. In Eccles, J. C., editor: Brain and conscious experience, New York, 1966, Springer-Verlag New York, Inc.

71. Trevarthen, C. B.: Two mechanisms of vision in primates, Psychol. Forsch. **31:**299, 1968.
72. von Holst, E., and Mittelstaedt, H.: Das Reafferenzprinzip (Wechselwirkungen zwischen Zentralnervensystem und Peripherie), Naturwissenschaften **37:**464, 1950.
73. Wilson, M., and Kaufman, H. M.: Effect of inferotemporal lesions upon processing of visual information in monkeys, J. Comp. Physiol. Psychol. **69:**44, 1969.
74. Wilson, M., Stamm, J. S., and Pribram, K. H.: Deficits in roughness discrimination after posterior parietal lesions in monkeys, J. Comp. Physiol. Psychol. **53:** 535, 1960.
75. Zinkin, S., and Miller, A. J.: Recovery of memory after amnesia induced by electroconvulsive shock, Science **155:**102, 1967.

22

NORMAN GESCHWIND

Some special functions of the human brain

DOMINANCE, LANGUAGE, APRAXIA, MEMORY, AND ATTENTION

This chapter deals with a loosely defined group of brain functions usually called "higher functions." They generally include capacities characteristic of humans that are poorly developed or entirely absent in subhuman forms. The use of language is included, whether in the spoken form, a nearly universal attribute of humans, or in the written form, which has only in the last century spread widely through many but by no means all human societies. The higher functions include the capacity for memory storage, highly selective attention, complex motor learning, and a number of special talents, for example, musical or drawing skills, mathematical ability, or the rapid recognition of faces. Cerebral dominance refers to the specialization of each hemisphere for certain functions, such as musical skills or language, and is generally classed as a higher function.

The term "higher cortical functions" is not used, since it suggests an often-held but untenable theory that the higher functions depend exclusively on the cerebral cortex. This view has sometimes led to the neglect of the more complex capacities of subcortical structures. It was assumed for many years, for example, that the memory disorders of the Korsakoff syndrome, to be discussed later, result from neocortical damage, but we now know that the causative lesions are usually in nonneocortical structures.

Most of the observations described here were made in studies of human beings with brain lesions. Although recent work suggests that chimpanzees may have some language ability,[12,40] there is so far no evidence of linguistic ability in any other nonhuman species. There is no information about the parts of the brain involved in language, except in humans. It is equally true that for many other higher functions humans are the exclusive or nearly exclusive source of information. Human musical, drawing, or mathematical abilities appear to differ so sharply from the abilities of any other animal that no reasonable animal experiment concerning them has been devised, at least up to the present time. There are even some "elementary" activities in which humans hold a special, although not unique, position. Humans, for example, are the only mammals who routinely assume the bipedal upright position.

About 50 or 60 years ago, much of the then-current neurophysiology was based on observations made in humans. The earliest maps of the projections of the visual fields onto the cortex, for example, were made after study of humans with visual cortical lesions, after correlation of the functional disorders with postmortem study of the lesioned brains. In the last half-century, however, our understanding of visual physiology has advanced far beyond that derived from such clinical studies, largely due to sophisticated recording methods used in studies of the brains of subhuman forms. There appear to be no similar animal models for study of many of the higher functions of the brain, and it is still necessary to rely on the experiments of nature on the human brain for information. This is often true even in areas in which there have been intensive studies of nonhuman species. For example, study of frontal lobe function in monkeys has contributed little to understanding the function of the frontal lobe in humans. Most studies of the frontal lobe in monkeys have emphasized the "delayed response" deficit produced by frontal lesions.[50]

Yet there is no clear equivalent of this disorder among the defects seen in humans after frontal lobe damage. Many other examples could be cited, and we are presently in the position of depending almost completely on clinical observations to understand the neural mechanisms of higher functions. The human brain is, of course, rarely directly accessible in life for physiologic study, and the discovery of usable animal models would allow an increase in our knowledge of those mechanisms.

A full coverage of all aspects of the higher functions of the human brain would deal in detail with issues that are primarily of psychologic or linguistic interest. I will in general not deal with these matters, despite their great interest and importance. The major purpose of the present chapter is to present our knowledge of the anatomic bases and the presumed physiologic mechanisms of those aspects of brain function that are uniquely human, or nearly so.

CEREBRAL DOMINANCE

The human brain differs markedly with regard to cerebral dominance from the brains of other mammals. This term means the predominance of one side of the brain for certain functions.[42,47,53]

Handedness

The most obvious example of cerebral dominance is handedness. The great majority of humans are right-handed, and left-handers comprise only 7% to 10% of the population. Some of the differences in incidence depend on the precise definitions of right- and left-handedness. We are forced to adopt some arbitrary definitions, since we do not understand the causes of handedness. The term means the preferential use of one hand in most or all unimanual activities. Most right-handed individuals are fully right dominant. Thus they write, hold a knife, throw a ball, flip a coin, brush the teeth, comb the hair, or hold a hammer with the right hand. The right leg is also preferred for such activities as kicking a ball or stamping on an insect. Left-handers are on the average less likely to be so purely left-limb dominant and often carry out a certain number of tasks with the right hand. Whether this is a sign of less manual laterality in the left-hander or an effect of social pressure is unknown. Thus although the instances are rarer today than in the past, individuals are occasionally forced to write with their right hands against their innate predilection. The canons of table manners and the rarity of certain special utensils (e.g., special scissors or baseball gloves) often force left-handers to carry out many acts with the right hand. It is therefore difficult to separate the degree of the innate tendency to sinistrality from the effects of social pressure. Although some authors separate the left-handed, the ambidextrous, and the right-handed, others prefer to use a simpler classification: right-handers, those who carry out nearly all actions with the right hand, and nonright-handers, those who show even moderate degrees of preference for the left hand.

An obvious question is whether left-handedness is inherited. Unfortunately even this simple question cannot be answered with certainty, because a significant percentage of left-handers are "pathologic left-handers," that is, persons who are genetically right-handed but who, because of an early injury to the left hemisphere, came to use the right hand. Among identical twins, about 12% of the pairs are of opposite handedness.[42] It is well-known that twin births are more difficult and more likely to be associated with brain damage. The left-handed member of these nonconcordant twin pairs is most likely the one who suffered injury.

Right-handers are overwhelmingly left hemisphere dominant for speech, and they develop language disorders after left hemisphere damage, but almost never after right-sided lesions. The situation in left-handers is more complex. It appears from the observations of Gloning et al.[20] that left-handers develop language disorders after damage to either hemisphere, but that they recover better from right hemisphere than from left hemisphere damage. The left hemisphere thus appears to be more important for langauge in a majority of left-handers, but the right and left sides of the brains of left-handers differ less in their language abilities than do those of right-handers.

The clinical syndromes produced by brain lesions appear to differ between left- and right-handers,[23] which suggests that their brains differ, also. This is one example of an important principle: there are individual differences in brains, so that no single physiologic account will describe everyone. It is not clear whether there are male-female differences in human brains, although such differences have been described in other species.[41] Women are, on the average, more often right-handed than are men, but it is not clear whether this is the effect of innate endowment or cultural pressures.

Other types of dominance

Although handedness is the most obvious form of cerebral dominance, there are many functions that depend more heavily on one hemisphere than on the other. In the 1860s, Broca first called attention to the fact that damage to restricted areas of the brain led to language disorders. One of his great contributions was to point out that such patients nearly all suffered from left hemisphere lesions. The importance of the left hemisphere for language is, in fact, the most striking example of dominance. Of 100 consecutive patients with language disorder resulting from a brain lesions, about 97 will have a left hemisphere lesion and only 3 a right-sided lesion.[53]

Left dominance for language does not mean that the right hemisphere has no language capacity at all. Patients with total destruction of the left hemisphere often show certain language capacities mediated by the right side of the brain.[45] We cannot be certain of the degree of language competence the right hemisphere possesses, and it probably varies from one individual to another.

In contrast, musical abilities appear to depend more heavily on the right than on the left hemisphere. The difference between musical and language competence is readily demonstrated by a method devised by Wada and Rasmussen.[49] Sodium amytal (a barbiturate anesthetic) is injected into one internal carotid artery. The anesthetic temporarily inactivates one cerebral hemisphere. This method has been used by neurosurgeons for determining which hemisphere is dominant for speech before brain operations. In one study, Bogen and Gordon[3] compared the effects of injections into the left and right internal carotid arteries. The left-sided injection led to a paralysis of the right side of the body and to a gross impairment in the ability to produce language, although the patients could still sing on key. When the right side was injected at another time, the left side of the body was paralyzed. The patients could then speak correctly but could no longer sing on key. This is a surprising result, for one might expect that the brain would use the same structures for producing sounds, whether they were the sounds of language or those of music. However, the right hemisphere seems to contain the neural mechanisms for music, the left those for language.

The Wada test also brings out another surprising form of dominance, for after anesthesia of the left hemisphere a patient appears depressed and may weep, whereas after anesthesia of the right side the patient is often inappropriately cheerful. Gainotti[19] has documented the same findings in a large group of patients with lesions in the hemispheres. It thus appears that the two halves of the brain have different emotional responses, with the right side being more important for the emotions of sadness, fear, or horror. Other studies have shown that the right hemisphere is more important for the recognition of emotion in other people and for recognizing the emotional tone of speech.

The right hemisphere is also dominant for certain aspects of the visuospatial sense. Patients with large right hemisphere lesions have difficulty in copying designs and make errors particularly in the outside contours. Patients with left hemisphere lesions also make poor copies of designs, but are particularly likely to make errors on the internal details. Obviously, any copy made by a normal person depends on cooperation of the two hemispheres, but we have little idea at this time how this integration is achieved.

Anatomic asymmetries of the brain

The simplest explanation for hemispheric dominance would be a structural difference between the two sides. Heschl and Flechsig described asymmetries in the structure of the cerebral hemispheres, and in the 1920s and 1930s Pfeifer[39] called attention to differences between the upper surfaces of the temporal lobes. This difference is illustrated in Fig. 22-1. Geschwind and Levitsky[17] measured the area of the planum temporale, that is, the region bounded anteriorly by the sulcus of Heschl and posteriorly by the posterior border of the sylvian fossa. Working with 100 adult brains, they found this area to be larger on the left in 65, equal on the two sides in 24, and larger on the left in 11 ($p < 0.001$). In absolute terms the length of the outer border of the left planum was 3.6 ± 0.9 cm, that of the right 2.7 ± 1.2 cm ($p < 0.001$). Hence the left planum was on the average nearly a full centimeter longer and, in relative terms, nearly one-third larger than the right. In many instances the left planum is several times larger than its fellow on the right.

A second asymmetry in this region is in the number of transverse gyri. On the left, there is usually only one, whereas on the right, two anterior transverse gyri are frequently present.[6] The asymmetry of the planum temporale was found by Wada in the brains of fetuses and newborns. Chi et al.[7] showed that the first clear signs of this lateral difference is present at 31 weeks of gestation.

More recently, Galaburda et al.[11] have measured the size of the cytoarchitectural fields in each hemisphere. The primary auditory cortex, area 41, is approximately the same size on the two sides. The major increase in the size of the

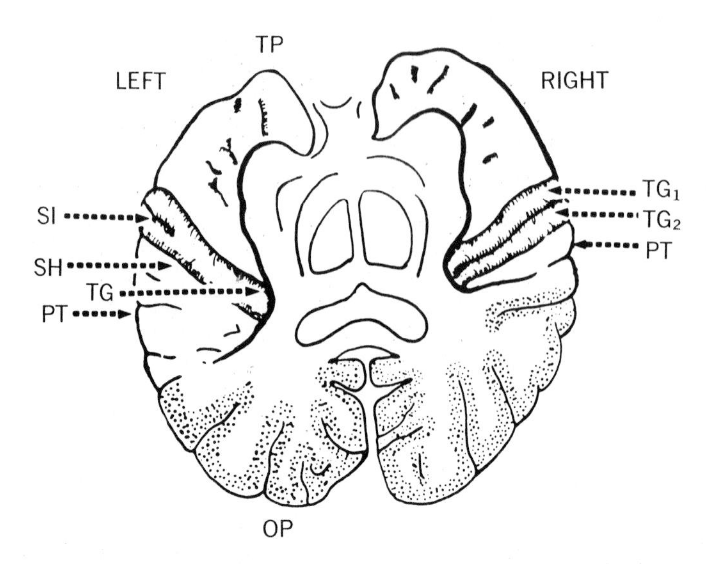

Fig. 22-1. Upper surfaces of two temporal lobes are exposed by knife cuts that follow sylvian fissures and then cut through brain. *TP*, Temporal pole; *OP*, occipital pole; *TG*, transverse gyrus of Heschl, containing primary auditory cortex. Two transverse gyri *(TG_1, TG_2)* are more common on right. *SI*, Sulcus intermedius; *SH*, sulcus of Heschl (posterior boundary of TG); *PT*, planum temporale. Note larger left planum.

planum on the right is due to the disproportionate size of area 22. In one brain with considerable asymmetry, this area was several times larger on the left than on the right. We shall see later that this region corresponds closely to Wernicke's area, one of the major regions of the cortex involved in language functions; it is reasonable to speculate that these asymmetries are of functional significance.

Other asymmetries have also been observed but not yet related to cytoarchitectonic areas. Thus LeMay and Culebras[28] have shown from the study of arteriograms carried out in life (arteriograms are x-ray films made after a radiopaque material is injected into the arteries of the brain for the purpose of diagnosis) and from the study of brains after death that the right sylvian fissure is typically angled upward more sharply than is the left. The impression of the sylvian fissure can often be seen on the inner surface of the skull, and this asymmetry has been demonstrated in the fossil skulls of a Neanderthal man (over 30,000 years old) and of Peking man (about 300,000 years old). Moreover, the same asymmetry exists in the brains of great apes, especially in those of orangutans, but not in the brains of monkeys.[29,52] In the brains of most right-handed humans the right frontal region and the left parieto-occipital regions are wider than the corresponding areas on the opposite side. The skull in these areas shows asymmetries corresponding to those of the underlying brain.

Table 22-1. Comparative widths of left and parieto-occipital regions in right- and left-handers, as measured in computerized x-ray films of the skull

Group	Left wider	Equal size	Right wider
Right-handers	64%	20%	16%
Left-handers	22%	32%	46%

The use of radiologic methods has made it possible to correlate asymmetries with handedness. On the basis of these studies,[11] it is possible to state some generalities concerning these side differences: (1) There are marked individual differences in the extent of the asymmetry from brain to brain. At present, one can only speculate that these correspond to individual differences in specific abilities. (2) Among left-handers there are more brains not showing a significant asymmetry than among right-handers. (3) There is a higher percentage of brains with reverse asymmetry among left-handers. Table 22-1 contains the data demonstrating these principles for one particular asymmetry. Note that in the left-handers there is a greater percentage of brains without asymmetry (32% as compared to 20% for the right-handers). Also note that the ratio of the percentages of asymmetric brains on the two sides is lower for the left-handers (46:20) than for the right-handers (64:16). It seems likely that the number of asymmetries in the human brain will be very large and even conceivable that each individual has a distinctive pattern. Observation that structural asymmetries exist in the brain of the higher primates makes it possible that experimental studies of certain aspects of cerebral dominance may be carried out.

LANGUAGE AREAS OF THE BRAIN

It is important to recall the contrast between speech and language. *Language* refers to the vocabulary and syntactic rules independent of the mode of production or comprehension. *Speech* refers to the actual production of spoken language. A disorder of language refers to a defect in either the production or comprehension of vocabulary or syntax. Such disorders are called *aphasias*. A disorder of speech is one in which the sounds or rhythm are badly produced despite correct grammar and choice of words. Despite this distinction, "speech area" is often used synonymously with "language area." *Vocalization* is the production of sound without linguistic content.

Most of what we know about the central neural mechanisms controlling language functions has come from the study, in life, of patients with language disorders and, after death, of the lesions in their brains. Many difficulties attend such studies, for brain lesions of many types are frequently multiple, may produce disordered function "at a distance" by altering intracranial pressure relations, and may in early stages cause local or neighborhood changes in function that reverse with time. The most useful studies involve patients who sustain and survive local occlusion of cerebral blood vessels and who can then be studied for months or years after this event in a stable state and whose brains can after death be subjected to detailed histologic study.

Certain neurosurgical patients also provide important data, particularly those who undergo precisely circumscribed and identified cortical excisions for the control of epilepsy. Studies of these patients have been carried out in many clinics, particularly by Milner and her associates at the Montreal Neurological Institute, and they have provided particularly well-documented information about the function of the cortex. A major difficulty is that many of these patients have epileptic lesions dating from childhood, and it is well-known that lesions to speech areas early in life are compensated for far better than when such lesions occur in adult life. Moreover, there is some evidence that epilepsy of long standing may itself lead to altered cerebral physiology. Some studies have been made in which electroencephalograms (EEGs) were made in patients speaking or listening to spoken language, and the recordings were processed with computer techniques. Although some useful results have been obtained, this method is still in its early stage of development.[8]

A further factor complicating the study of the effects of brain lesions on higher functions is that even in patients with fixed and delimited lesions, changes in function occur gradually with time, frequently over many years. We do not know all the mechanisms of these "plastic" changes in brain function, although they include sprouting, transsynaptic, and transneuronal degeneration, and supersensitivity of denervation.[14]

Thus there is not available at the present time an exact method for the study of the brain mechanisms in higher functions in humans. Nevertheless a large body of knowledge concerning this subject has accumulated. It is likely that many of the difficulties in these studies, such as the plastic properties of brain function, will stimulate more fundamental studies of the biologic mechanisms involved so that they may be controlled or stimulated in such a way as to promote restoration of brain function to the fullest possible extent.

Speech areas mapped by stimulation methods

The method of the electrical stimulation of the cerebral cortex in patients under local anesthesia for the purpose of therapeutic operations on the brain was developed by Foerster[10] and Penfield. Fig. 22-2 represents a composite of the results obtained by Penfield and Roberts.[38] Stimulation in the classic speech areas produces no overt response if delivered when the patient is silent. If the patient is speaking, stimulation in these areas causes an arrest of speech or leads to errors in speech. Vocalization may be produced by stimulation of the lower part of the precentral gyrus in either hemisphere. Errors in language (i.e., errors of grammar or word choice) are produced only by stimulation of the left hemisphere, with the exception that stimulation of the supplementary motor area on the medial surface of either hemisphere can lead to errors of language. Errors in language, meaning the production of incorrect language, are usually called *aphasic errors*. The supplementary motor area is the only one of the speech areas in which stimulation may lead to actual speech. When evoked, it is typically the repetition of a syllable or word, or even a phrase: for example, "ba-ba-ba-" or "slap me, slap me."

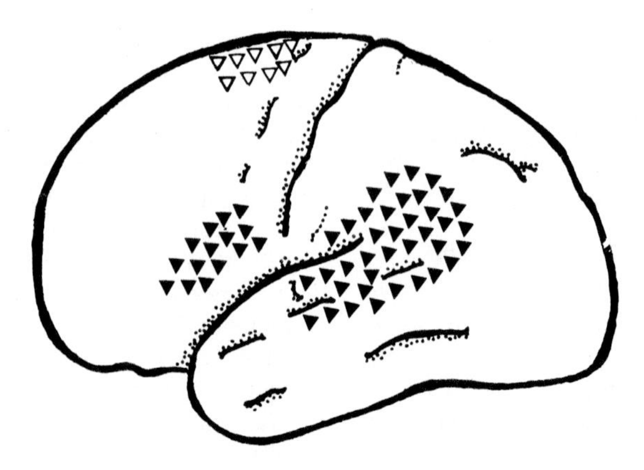

Fig. 22-2. Speech areas of left hemisphere as determined by electrical stimulation. Areas indicated by the triangles are those in which electrical stimulation leads to *aphasic* errors, that is, errors in use of language. Solid triangles show language areas along borders of sylvian fissure. Open triangles show language area on *medial* surface of hemisphere (supplementary motor area). (Modified from Penfield and Roberts.[38])

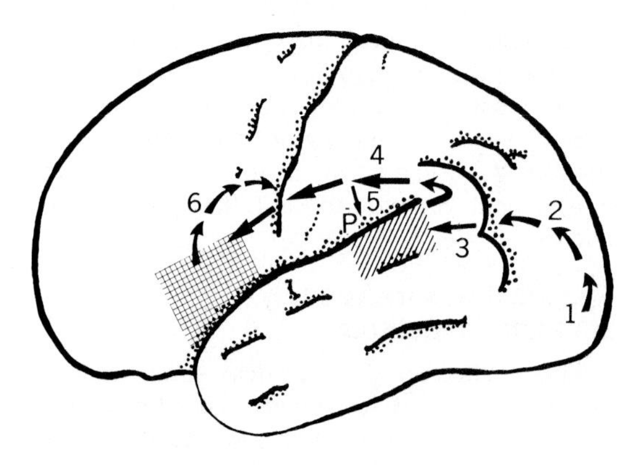

Fig. 22-3. Language areas and their connections as determined by postmortem studies of lesions in brains of aphasic patients. Diagonal lines indicate Wernicke's area. Cross-hatched region is Broca's area. Arrows indicate presumed connections involved in naming a seen object. Visual image *(1)* is projected *(2)* to angular gyrus region *(3)*. This arouses learned auditory form of word in Wernicke's area, which is then transmitted via arcuate fasciculus *(4)* to Broca's area, which contains programs that control *(6)* cortical motor region involved in speech in precentral gyrus. Arrow *(5)* indicates that there is probably an additional synapse in parietal operculum *(P)*. Effects of lesions at different sites on this route are described in text.

The diagram of Penfield and Roberts corresponds closely to that produced by a different method, namely, by mapping the cortical sites in which damage leads to disturbance in language production or comprehension (Fig. 22-3). The one major difference is the absence from the classic lesion diagram of the supplementary motor area. Although lesions in this region can lead to disorders of language, they do not persist in severe form and usually disappear within a few weeks.

Stimulation in the pulvinar of the dorsal thalamus also leads to speech arrest or to errors in language. Ojemann et al.[37] observed that stimulation in the pulvinar leads to speech errors and/or to language errors and that stimulation in this region may facilitate language learning. There are now several cases in which left thalamic lesions led to language disorder. In all the cases described so far the patient recovered from the aphasia completely or nearly so. We do not understand the role of these two areas, the supplementary motor area and the pulvinar, lesions of which produce only transitory aphasias. They may play a role in the original acquisition of language or in recovery from damage to other speech areas, but there is no direct evidence to support these conjectures. *Permanent* aphasias occur only after lesions in the regions shown in Fig. 22-3, but not all lesions in these regions lead to permanent aphasia.

Stimulation in the nonhuman primate of areas of the cortex that appear homologous to the speech areas of the human brain does not evoke vocal responses. Vocalization, including species-specific cries, is elicited by stimulation of subcortical limbic structures. Similarly, vocalization is rarely obtained by stimulation of the human speech areas, and changes are usually observed only if the subject is speaking. A similar experiment is, of course, impossible in the nonhuman primate. There has not been a systematic study of the effects of stimulation of the deep limbic structures in humans.

Classic language areas of the cerebral cortex

The classic language zones are the areas lying along the sylvian fissure (on the left side in most persons) in which stimulation leads to aphasic errors and lesions lead in many cases to permanent disorders in the production or comprehension of language. The anterior speech area is called *Broca's area.* It is located in the third frontal gyrus (Fig. 22-3). The more anterior part of this gyrus is called the pars triangularis and the most posterior part the ''foot of the third frontal convolution.'' Electrical stimulation in any part of this region leads to speech arrest or to aphasic errors, but it is uncertain whether the entire anterior speech area has a similar function or whether it is composed of subareas with differing functions. Moreover, it is possible that the anterior speech area is even larger than the region mapped by electrical stimulation.

Thus we can conclude that there is an anterior speech area and that damage to this region leads to an aphasia. Recent observations by Mohr[35] make it likely that small lesions within this region are followed by a good degree of recovery and that only large lesions produce a lasting aphasia of the classic Broca type.

Although the diagram of Penfield and Roberts shows a continuous posterior speech region, it seems almost certain that this area contains more than one functional unit. The area is not homogeneous in cytoarchitecture. Moreover, lesions below the sylvian fissure usually produce a clinical picture that differs from that produced by lesions above it. The term ''Wernicke's area'' should be confined to the speech area of the temporal lobe below the sylvian fissure; this designation also fits with Wernicke's original description.

Organization of the language areas

In this section a description of the functions of the language areas and their patterns of interconnection is presented. A simple theory is offered that fits well with current clinical and anatomic knowledge, although little is known of the dynamic neural mechanisms in the coding and decoding of language.

The anterior speech area lies in front of the lower end of the motor cortex, just in front of a zone referred to as the face area, that part of the precentral-motor cortex in which electrical stimulation leads to (1) movements of the skeletal muscles innervated by the cranial nerves, except for the extraocular muscles, and (2) movements of the muscles of respiration. Stimulation here leads to movements of the jaw, face, and laryngeal, pharyngeal, and respiratory muscles. The face area thus ''innervates'' all the muscles involved in speaking. Since Broca's area lies immediately anterior to the face area, one might conjecture that it is the source of central ''programs'' for controlling the activity of the face area in the production of speech. I shall define this function more precisely later.

Wernicke's area lies in the posterior part of the first temporal gyrus on the convexity of the brain and extends onto the upper surface of the temporal lobe. Its cytoarchitecture is transitional between that of the primary auditory cortex and that of homotypical association cortex. Except for the congenitally deaf and the rare person raised by deaf, nonspeaking parents, humans learn language via the auditory route. We can postulate that Wernicke's area contains the neural mechanisms for the recognition of speech patterns relayed to it from the left primary auditory cortex.

Let us surmise that as the auditory patterns of language are learned, they are relayed from Wernicke's area to Broca's area. These two regions are connected by the *arcuate fasciculus.* This bundle arises in Wernicke's area, passes around the posterior end of the sylvian fissure into the lower parietal lobe, and runs forward to Broca's area. This region, through its mechanisms for controlling the speech musculature, converts the auditory inputs from Wernicke's area into the motor patterns for speech. Consider now in a simple paradigm the functions of that part of the left angular gyrus involved in language. Suppose a child hears the word ''banana'' spoken by an adult. He learns that the sound ''banana'' is associated with a particular class of visual images. The left angular gyrus region contains the rules for this association. The sight of a banana arouses a neural replication of the image, and this pattern of neural activity (or some more abstract code of it) reaches the angular gyrus. (The word ''image'' is a shorthand used to describe the neural coding of the stimulus or its transformations.) This region, in turn, evokes the sound-form ''banana'' in Wernicke's area. Conversely, when the child hears the word ''banana,'' neural mechanisms of Wernicke's area recognize this as one of the patterns of the language. This sound pattern is then transmitted to the angular gyrus, which, in turn, leads to arousal of the visual image, presumably in some portion of the visual association cortex.

We can now consider the process of vocal repetition. The neural replicate of a spoken word passes through the primary auditory area and is relayed to Wernicke's area, where the pattern is "recognized" and forwarded to Broca's area. This cortical region contains the rules for controlling the face area, which generates the proper pattern of signals to the speech musculature, so that the sound is reproduced.

Now consider the process of *naming* a seen object. The neural image of the object is transmitted from the visual region to the angular gyrus, which arouses a corresponding auditory pattern in Wernicke's area, which, in turn, relays the auditory form to Broca's area, and this leads to the arousal of the correct articulatory form.

This simple model can also be applied to an analysis of the comprehension of spoken words. The neural replicate of the heard word reaches Wernicke's area, whence the appropriate pattern is relayed to the angular gyrus, which, in turn, arouses the corresponding visual, auditory, and/or somesthetic images. The process of learning to read can be conceptualized as follows. In general, reading skill is acquired after the child has acquired comprehension of auditory language. The neural replicate of the seen word is eventually transmitted to the angular gyrus, which, in turn, arouses the auditory form of the word in Wernicke's area.

It is much more difficult to devise a model for the process of writing the name of a seen object. It is possible that a seen object arouses the auditory form of the word, which is conveyed to the angular gyrus. This region, in turn, arouses the written form, which then, by a route that is not known, reaches the motor regions, which generate the motor commands for the written form.

Thus the simple model used deals with many of the phenomena observed when different parts of the speech areas are damaged. It does not deal with grammatic structure, and the analysis of the act of writing is incomplete. It assumes a simple serial pathway, yet there is some evidence that alternative routes, which are not well known, exist in parallel with this serial system. Finally, it is an anatomic model that deals with structure at the gross level and makes no assumptions or predictions about the intrinsic neural mechanisms involved in the operations postulated.

Lesions of Broca's area

After large lesions of Broca's area, spoken language is slow and effortful and the sounds poorly produced. There are, however, certain other effects not predicted by the general outline just given.

First, in general, such a patient sings correctly, in most cases without words.[51] The easy production of melody contrasts with the effortful utterance of words and suggests that the central neural mechanisms for the production of musical sounds are located separately from those for language production, probably in the right hemisphere.

Second, the patient does not have equal difficulty with the production of all words. He tends to produce names or attributes of objects but has greater difficulty in producing the small grammatic words (e.g., "if," "is," or "he"), so that his speech takes on a telegraphic quality. This suggests that some part of Broca's area is important for the grammatic structure of an utterance. Moreover, the patient is usually not mute, although mutism may persist for a few days after the cerebral injury. This indicates that there is some alternative route for producing the remaining ungrammatic phrases. It appears that in some cases the right hemisphere may be responsible for this abnormal language production.

Third, the patient writes incorrectly, which suggests that the process of writing is in some sense "downstream" from the formulation of language in Broca's area, although the exact route is unknown. Other aspects of the patient's performance are consistent with this general outline. He understands written and spoken language far better than he speaks it, and this corresponds with the intactness of Wernicke's area and of its connections to the angular gyrus. On the other hand, some persons with Broca's aphasia have difficulty with special aspects of language comprehension. The patient often performs poorly in understanding the grammatic aspects, just as he produces them badly. Thus he may understand "source of illumination" or "apparatus for amplifying the voice" but may fail with phrases in which comprehension depends on grammatic features, for example, "Is my brother's wife a man or a woman?" or "If the lion was killed by the tiger, which animal died?"

Lesions of Wernicke's area

The scheme just outlined predicts that a lesion of Wernicke's area will produce defects in speaking, writing, repetition, and comprehension of written and spoken language. The most striking feature of the person with Wernicke's aphasia is the quality of the spoken language, which, in a typical case, contrasts with that produced by a patient with a lesion of Broca's area. The latter speaks in a slow manner with great effort and produces the sounds of language poorly. In contrast, the patient with Wernicke's aphasia may produce speech that is more rapid than normal, effortless, with well-preserved rhythm and melody, and with full and correct production of English phonemes. The patient with Broca's aphasia produces speech that is agrammatic, often with a telegraphic quality. The individual with Wernicke's aphasia constructs phrases with a normal grammatic skeleton. Thus neither prosody, nor articulatory quality, nor grammatic structure are primarily affected in Wernicke's aphasia. The salient feature is the failure to find the correct word. The patient may simply omit it: "I went to the . . . they brought me to the place where . . . it wasn't the one that. . . ." Here the patient fails after several attempts to supply the word "hospital" or a near equivalent. Sometimes there are obvious pauses, giving the speech an air of hesitancy that at first might suggest the output of a patient with Broca's aphasia. At other times the incomplete grammatic phrases follow one another in a way that, without closer inspection, may give the impression of a grammatic jumble. The patient may circumlocute ("I saw the one who takes care of curing") or use words almost empty of precise meaning: "He was the one who does what you do with it." In other cases an incorrect word is substituted for the correct one; this is called a paraphasia. The two words are sometimes related by meaning ("My hand hurts," uttered as the patient points to a bruise on his foot) and sometimes totally unrelated ("This is an Argentinian rifle," said as the patient shows the examiner a 5-cent piece). Such substitutions of one word for another are called verbal paraphasias, whereas the term *"literal,"* or preferably *"phonemic,"* paraphasia is used for the substitution of a well-articulated but inappropriate phoneme in a word (e.g., "lork" for "fork"). Finally, words may be produced that are apparently purely random collections of sounds (e.g., "flieber" or "sodent"). These are termed *neologisms*.

The patient with Wernicke's aphasia is the prototype of the fluent aphasic, whereas one with Broca's aphasia is the prototype of the nonfluent aphasic. One explanation of the mechanism of fluent aphasias is that after damage to Wernicke's area, Broca's area runs on in an unbridled manner, carrying out its own overlearned activities, that is, correct pronunciation, prosody, and grammatic structure. Many observations are compatible with this idea. Fluent forms of aphasia occur rarely in children, which fits with the idea that it takes years of training for Broca's area to run on freely. Furthermore, Isserlin,[25] observing wounded in World War I, pointed out that fluent forms of aphasia were less common in young soldiers but increasingly common with greater age. At no age do all posterior lesions lead to grossly fluent aphasia, suggesting that in some individuals Broca's area never acquires this ability to run on unbridled. The production of a fluent aphasia by an anterior lesion is, if it ever occurs, an exceedingly rare event.

The general scheme of the functional linkages of the areas of the brain related to language must be modified somewhat. If Broca's area functioned totally in isolation, one would expect the output to be a meaningless string of grammatic and well-articulated utterances. In most cases, however, the output is organized in some way (e.g., the grossly paraphasic speech may describe the events that brought the patient to the hospital). This suggests that an input, although a defective one, reaches Broca's area and that there must be alternative neural pathways to it. The information transmitted over these alternative routes apparently does not contain the precise specification of the desired words. It is, of course, possible that these alternative systems are in the right hemisphere.

The patient with Wernicke's aphasia speaks and repeats incorrectly and writes aphasically. There is an incomprehension of spoken and written language, since Wernicke's area is a necessary way station in the process of comprehension.

Lesions of parietal operculum

Lesions of the parietal operculum leave Wernicke's and Broca's areas intact but disconnect the two from each other. When Broca's area is disconnected in this way, it may run on in an uninhibited fashion. We would expect that the patient with a lesion of the parietal operculum would also suffer from a fluent aphasia, and this is usually the case. The aphasia occurring after a lesion in this location is usually referred to as *conduction aphasia*. The condition differs from that following a lesion in Wernicke's area in that the former patient produces more phonemic and fewer verbal paraphasias than does the latter. This may be a clue to one of the functions of the synaptic relay of the arcuate fasciculus in the parietal operculum. The conduction aphasic comprehends spoken and written language well. Since Wernicke's area is intact, it can arouse associations in the posterior portions of the parietal lobe and in the occipital lobe, although it is disconnected from Broca's area. Such a patient also repeats badly, as would be expected, since the lesion cuts the pathway from Wernicke's area to Broca's area. It is usually the small function words rather than the long ones that are badly repeated. Thus "He was here" is generally more difficult than "Washington," and "Was he here?" still more difficult. In contrast, number words are often repeated better than any other class. This suggests the presence of an alternative route for repetition available for some types of words and not for others. The criterion that favors repeatability may be related to the ability to code the word in some concrete form. Thus one can conceive that "Washington" may be pictured (as a city or person) but "is" cannot. Some conduction aphasics produce paraphrases of the words they hear.

Kleist observed that some patients with lesions in Wernicke's area will show exactly the syndrome of conduction aphasia.[27] The explanation is that in some cases there is a recovery from the comprehension defect of Wernicke's aphasia, so that the patient is left with a fluent aphasia with poor repetition capacity.

Isolation of the speech area

Isolation of the speech area occurs more rarely than any syndrome discussed previously; it illuminates the cerebral mechanisms of language. Geschwind et al.[18] described a patient observed over a period of 9 years, until death, whose brain was studied in detail in serial sections. The patient was a young woman who had suffered carbon monoxide poisoning. She produced spontaneously only occasional expletives. On the other hand, she repeated almost any random English sentence correctly. Even more striking was the completion of many well-known phrases. Thus if the examiner said "Roses are red," the patient would say, "Roses are red, violets are blue" and continue on to give the full rhyme. In addition, if the examiner preceded a phrase by "Say" or "Repeat," the patient would repeat the phrase exactly as a normal person would, dropping the instruction.

One feature of this grossly impaired patient was the ability to carry on verbal learning. If a song that had not existed before her illness was played to her several times, she would eventually come to sing along with the record. After a few more trials, it was observed that if the record was stopped as soon as the patient had begun to sing, she would continue to produce the words and music correctly through to the end. She was thus able to learn the words of several songs.

The patient was otherwise grossly deteriorated. She showed no evidence of comprehension of langauge, nor did she ever communicate spontaneously in any fashion. She could not feed or dress herself or use her unparalyzed upper limbs appropriately. It was not certain that she recognized any of those around her, except perhaps her father.

The brain sections showed that the entire speech circuit was intact: Wernicke's area, the arcuate fasciculus, and Broca's area. The auditory pathways and the face area of the percentral gyrus and their efferent connections were also intact. There was a large lesion shaped like a reversed "C" that destroyed both cortex and white matter and almost encircled the speech regions. The patient thus had preserved the full circuit needed for repetition: the auditory pathway, Heschl's gyrus, Wernicke's area, the parietal operculum, Broca's area, the precentral face area, and the corticobulbar tracts. The preserved capacities for repetition could all depend on activities within this circuit. Thus the commands "Repeat" or "Say" can be conceived as instructions to Broca's area. ("In response to 'Repeat' say only the words following this word.") The well-known sequences that were completed could have been stored in the preserved auditory association cortices. The failure to comprehend normally can be seen as the incapacity of the heard words reaching Wernicke's area to arouse associations anywhere in the brain except in Broca's area. Finally, we must ask how this "isolated speech area" could carry on verbal learning. The serial sections revealed that the hippocampus and the central speech areas were preserved on both sides. We shall see later that the hippocampus is necessary for the laying down of memories.

Alexia without agraphia

The patient with alexia without agraphia suffers from the acute onset of a right hemianopia and the sudden loss of the ability to read. He is unable to understand written and spoken language, although he speaks and writes correctly. His inability to read is not the result of an elementary disorder of vision, since he can correctly copy words he fails to comprehend and name correctly objects he sees. The failure of some patients to read even such a simple word as "cat" contrasts with the capacity to describe photographs of complex scenes. In some patients the ability to read numbers, even six-digit numbers, is retained despite the incapacity to read words.

This syndrome is usually produced by an occlusion of the left posterior cerebral artery with infarction of the left visual cortex and destruction of the splenium (the posterior end) of the corpus callosum. Destruction of the left calcarine cortex produces a right hemianopia. Written words can be seen only when presented in the left visual field and projected to the right visual cortex; but for a word "perceived" in the right visual cortex to be expressed verbally, its neural transform must be transmitted to the left, the language-dominant hemisphere, via the splenium of the corpus callosum. That transmission is prevented by the lesion described.

The syndrome does not occur with a left hemianopia, except in some left-handers. A lesion confined to the left visual cortex does not produce the syndrome if the splenium is intact. Moreover, a lesion of the splenium with preservation of both occipital lobes produces a syndrome in which a patient can read in the right visual field but not in the left. This was first described by Trescher and Ford,[48] who studied a patient in whom the splenium had been sectioned to allow surgical removal of a tumor of the third ventricle.

If this explanation is correct, how can such a patient name objects or read numbers? These capacities also should require that information be transmitted from the intact left field via the right to the left hemisphere. One explanation may be that this information concerning these spared classes of visual stimuli is relayed across commissures that remain intact, such as the more anterior portions of the corpus callosum or the anterior commissure, which was shown in experiments on chimpanzees by Black and Myers[2] to transmit visual information between the hemispheres.

Pure word deafness

Patients with pure word deafness fail to understand or repeat spoken language, although their hearing is normal, or nearly so, and all other aspects of language are intact. The syndrome is produced by destruction of the left auditory radiation, which connects the left medial geniculate body to the primary auditory cortex in Heschl's gyrus.[21] As a result, the neural transform of heard words can be relayed only to the right temporal lobe. In order to be comprehended as language, they would have to be transmitted across the corpus callosum to the left temporal lobe. However, the lesion destroys both the left auditory radiation and the corresponding callosal fibers from the right side to the left auditory region. As a result, Wernicke's area can receive no auditory input. In other cases of this syndrome there are bilateral lesions. The mechanism in these cases is still unclear.

Dichotic listening

The phenomenon of dichotic listening casts further light on the transmission of auditory verbal material between the hemispheres. This method was devised by Broadbent[5] and used by Kimura[26] in studying humans with brain lesions. The subject wears earphones. A series of three words is presented to the right ear, and simultaneously a series of three other words is presented to the left ear. The normal subject reports first the entire series delivered to the right and then those to the left ear. After temporal lobe excisions for epilepsy there is a drop in the accuracy of reporting. Following right temporal excisions the patient continues to report correctly the words presented to the right ear, but makes errors in repeating those presented to the left ear.

In patients with lesions of the corpus callosum there is a very distinctive alteration of performance.[46] When words are presented to only one ear and white noise to the other, words are reported correctly. When different words are presented simultaneously to the right and left ears, none of the words are reported from the left ear, whereas those presented to the right are correctly reported. Under these conditions the neural transform of a word presented to the left ear can reach the left temporal lobe only via ipsilateral components of the auditory pathway. Evidently this route is capable of transmitting words accurately to the left temporal lobe, since when words are presented to the left ear and noise to the right, the patient reports them correctly. But when words are presented simultaneously to both ears, those from the left ear are not reported. This suggests that when the left auditory region receives words simultaneously input from the right ear via the contralateral auditory pathway and from the left via the ipsilateral pathway, the latter is suppressed. In normal persons, both messages are reported. The words heard by the left ear would be suppressed if they were projected only via the ipsilateral pathway to the left temporal lobe. The message from the left ear must therefore, under normal conditions, have an alternate route: it travels via the contralateral pathway from the left ear to the right auditory cortical region and from there across the corpus callosum to the left auditory region. The message from the right ear leads to suppression of a left ear message that travels ipsilaterally, but does not suppress that delivered from the other side via the commissure.

Recovery from aphasia

Destruction of one of the primary speech regions may lead to permanent aphasia, yet there are cases in which the aphasic disorder improves considerably, and rarely it may disappear. We do not understand the mechanism of this recovery. We do know that left hemisphere lesions produce aphasia in childhood, but that recovery is very good if the lesions occur before about 9 years of age. The recovery rate is lower in adult life, but does not fall to zero at any age. Luria[31] found that about 30% of patients with penetrating wounds in the primary speech areas showed good recovery. All the factors that determine the degree of recovery in adults are not known. Size of the lesion is, of course, one factor (i.e., recovery is poorer with larger lesions). Left-handed patients are more likely to recover than are right-handers. Right-handers with family histories of left-handedness in parents, siblings, or children show better recovery than do those without such histories. These observations fit with the fact that the left-hander has less marked cerebral dominance than does the right-hander.

It is often argued that the recovery of language functions after large lesions of the left hemisphere is the result of "taking over" by the right hemisphere. This could mean that the right side *relearns* language after the storehouse on the left is destroyed. Another interpretation is that left dominance for language does not imply a unique storehouse of learning on the left side, but rather that learning occurred on both sides but that the right hemisphere was somehow suppressed by the left. Following destruction of the left side the latent storehouse on the right would, after an interval, become available. This would occur commonly before the age of 9 years and in a minority of adult cases. We shall see later that there is good evidence for the view that there are instances in which learned information is present in the brain but unavailable for use. The phenomenon of "shrinking retrograde amnesia" (p. 661) shows that such phenomena occur. Obviously the possibility that recovery from aphasia depends on activation of a latent but inactive store of language learning in the right hemisphere opens a possibility for a physiologically based mode of therapy. If such a storehouse does not exist on the right side of the brain, the prospects for recovery from an aphasia in adult right-handers is poor.

Evoked potential studies of higher functions

There have been no studies of the postulated neural mechanisms of language in humans employing the electrophysiologic methods routinely used in research in animals. There have, however, been some preliminary studies of the EEG changes in relation to language, and it is likely that with improved methods, this will become an important way for studying the neural mechanisms of this and other higher functions. For example, McAdam and Whitaker[32] observed that just before speech a potential change occurs in the left lower frontal region in the vicinity of Broca's area. Later studies showed that heard words are likely to evoke a left posterior temporal potential change (i.e., in Wernicke's area). The interested student should refer to Desmedt,[8] who reviews knowledge of these methods and the results obtained with them. Another promising method is one in which cerebral blood flow is measured in different parts of the brain using radioactive tracers. Striking local changes in blood flow have been observed during different activities, such as speaking. Undoubtedly, this method will provide information about the location of cortical areas involved in higher functions, as well as a means of determining which areas play a role in the recovery of function.

THE APRAXIAS

The word "apraxia" was first used by Liepmann at the turn of the century to describe the failure of a patient to carry out correctly certain movements in response to stimuli that normally elicit them, in the absence of weakness, other motor disorders, or sensory loss, and with an intact comprehension of language. The term has since been applied loosely to a variety of disorders of the execution of movement, for example, failure to copy designs correctly ("constructional apraxia"), failure to put clothing on correctly ("dressing apraxia"), and gait disorders with frontal lobe lesions ("frontal ataxia of gait"). We will use the term, however, only in the sense just defined. Liepmann distinguished two major types of apraxia: *ideomotor* and *ideational*. The definition refers primarily to ideomotor apraxia. Ideational apraxia was used by Liepmann to describe failures in carrying out sequences of acts, although individual movements were correct. The term "ideational apraxia" is sometimes used to refer to the incorrect handling of objects, but in the discussion that follows this disorder will be dealt with as one of the ideomotor apraxias.

Consider a specific example. A patient with a lesion of the anterior four fifths of the corpus callosum showed the following behavior. When asked to carry out (in response to verbal command) certain movements with the right arm (e.g., "salute"), the patient did so correctly. When asked to perform the same movement with the left arm, he either did nothing or made an incorrect movement. If the examiner himself made the correct movement, the patient could imitate it correctly with his left arm. The patient's ability to perform correctly with the right arm in response to verbal command rules out incomprehension, inattention, or lack of cooperation. His ability to imitate correctly with the left arm indicates that the failure to respond properly with the left arm to verbal command was not the result of an elementary sensory or motor disorder. The diagram of Fig. 22-4 illustrates the pathways involved when movements are carried out to verbal command. A proper response is made with the right arm if the message is transmitted from Wernicke's area to some more anterior region, such as the left premotor area, and from there relayed to the left precentral motor cortex. A correct response to verbal command with the left arm depends on transmission of the command from the left to the right hemisphere. Two possible routes are shown in the diagram. The posterior one is probably not the route normally used, since right posterior lesions do not lead to apraxia of the left arm. The route followed is therefore probably the anterior one marked in the diagram.

A patient with a lesion in midcallosum correctly carries out movements of the right arm to verbal command, but fails to respond correctly with the left arm, since the command cannot be relayed across the corpus callosum. When the movement is carried out by the examiner, the patient's right hemisphere is able to see the movement and imitate it; the left hand performs correctly in imitation. Movements of the cranial musculature are executed correctly on command,

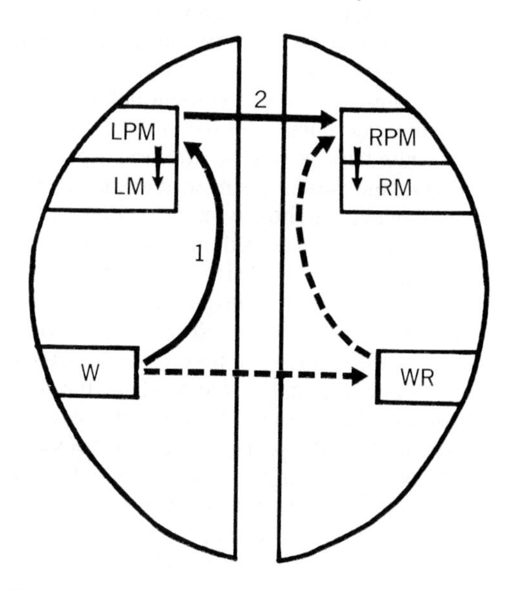

Fig. 22-4. Diagrammatic representation of pathways involved in carrying out movements to verbal command. Command to be carried out with right limbs reaching Wernicke's area *(W)* is relayed *(1)* to left premotor region *(LPM)* and from there to left motor cortex *(LM)*. Command to be carried out with left limbs follows route from *W* to *LPM* and from there is relayed via corpus callosum *(2)* to right premotor region *(RPM)* and thence to right motor cortex *(RM)*. Alternative route indicated by dotted lines, from Wernicke's area to its homolog on the right *(WR)* and thence to right premotor and motor regions, appears not to be the one normally used for carrying out movements with left arm to verbal command. Text describes effects of lesions at different sites.

undoubtedly due to the fact that the face area of the motor cortex of each side has a considerable control of the cranial musculature on both sides.

Movements of the extraocular muscles and of the axial musculature are performed well to verbal command.

Lesions in the left posterior frontal lobe are usually large and produce a right hemiplegia along with a Broca's aphasia. Such a patient shows apraxia of the left limbs to verbal command, since the lesion destroys the origin of the callosal fibers in the cortex. Such a patient differs from one with a midline callosal lesion in that he is also apraxic in the cranial musculature: the left face area is damaged, and the verbal command cannot reach the right face area because of the lesion of the callosal fibers at their origin. If the examiner carries out the facial movement, the patient may imitate perfectly, since the right hemisphere, on seeing the correct movement, can innervate both sides of the cranial musculature. This apraxia is called *bucco-facial apraxia* and commonly accompanies Broca's aphasia. It is the most common of the syndromes that result from lesions of the corpus callosum.

A lesion in the left parietal operculum produces conduction aphasia. Although verbal commands reach Wernicke's area and are understood, they cannot be relayed to either the left or the right premotor region. The patient may be apraxic in the face and the extremities of both sides in response to verbal command. Patients with Broca's aphasia or conduction aphasia do, however, carry out movements involving the extraocular muscles and those of the trunk in response to verbal command.

The geographic scheme presented accounts for some of the observed phenomena, but is incomplete. In the examples given the patients failed to make certain movements to verbal command, but did so on imitation. This is not the case in all apraxic patients. Goodglass and Kaplan[22] showed that about 35% of patients apraxic to verbal command imitated correctly, but 65% continued to perform poorly even after the examiner demonstrated the movement. Nevertheless, the majority of these succeed in the presence of an actual object. A small group of patients perform poorly even with the actual object. How are we to interpret these failures? We can approach an explanation by considering the likelihood that in right-handers the left hemisphere contains a greater storehouse of motor learning than does the right and that imitation requires access to such a store. This has two implications for the understanding of left-handedness. In right-handers the stores of language and motor learning are in the same hemisphere. In left-handers the major store of *motor* learning is in the right hemisphere, but there is also a store on the left that becomes accessible after damage to the right side. This explanation is compatible with the general rule that left-handers show apraxia less often than right-handers, and with the principle stated earlier, that is, that dominance effects are less marked in left-handers.

Eye movements and axial movements are spared to a greater or lesser degree in all cases of apraxia, regardless of lesion site. The diagram of Fig. 22-4 shows the precentral motor cortex as the only source of movement commands. This is not fully correct, since eye movements are not produced by stimulation of this region but can be elicited by electrical stimulation of several other areas. These so-called "eye fields" are present in each of the four lobes of the hemisphere, and stimulation within them elicits both eye movements and adversive movements, which involve the axial musculature. One cannot, however, elicit discrete movements of the muscles of the distal extremities by stimulating these eye fields. In the temporal lobe there is a large area of such "nonpyramidal" motor cortex that corresponds to Wernicke's area. Thus a motor command for eye movements can be elicited via the nonpyramidal system descending from this region, along pathways not shown in the diagram of Fig. 22-4. However, if discrete distal movements are required, the command must reach the cortical origins of the pyramidal system. The lesions shown in the diagram cut off access to the pyramidal system. This interpretation of the preservation of axial movements enables us to understand another feature of the apraxic patient. If Wernicke's area had no access to motor pathways other than the pyramidal tract, a lesion of the corpus callosum should lead to a failure of the patient to carry out any movement of the left hand to verbal command. Yet in many cases the patient carries out an incorrect movement, one that appears not to require pyramidal mechanisms in its execution. He may, however, execute the correct movement in the presence of an object. Thus one can cause the patient to use first a nonpyramidal system arising from the left hemisphere (when a verbal command is given) and, with a nonverbal stimulus, cause a switch to the pyramidal system projecting from the right hemisphere. Brinkman and Kuypers[4] have shown that in the monkey with appropriate lesions one can cause a shift from one motor system to another by controlling the stimulus. These shifts are easier to detect in the human patient because of the large number of learned movements that can be tested.

MEMORY DISORDERS

In this section, I describe briefly the so-called amnestic or Korsakoff syndrome. This occurs frequently after head injury and in thiamine deficiency, a disorder almost confined to alcoholics, at least in the United States. Rarely, the syndrome follows surgical lesions or vascular infarctions. The unusual clinical features of this syndrome were not predictable from anything known from the study of human memory, which illustrates the fact that study of the damaged human nervous system often reveals aspects of psychologic function otherwise totally unsuspected. The general demeanor of such a patient is in most cases quite normal. He appears fully alert, awake, cooperative, and attentive. All language functions are normal, and the patient may perform well on tests of abstraction, such as the interpretation of proverbs. The only major disorder is in certain memory functions, but not in all. Immediate memory is intact; by this is meant the ability to repeat without delay what is said by the examiner. The patient will repeat a string of numbers as well as do normal persons. In contrast, the patient's performance in remembering after even a short interval of time (a few minutes) is greatly impaired; he may be unable to recall a single one of a list of seven items or numbers after 2 or 3 minutes' delay. This deficit in the acquisition of new information is sometimes called *anterograde amnesia*. Such a patient may not know the day of the week or the date, the names of those around him (except for those well known from the past), or the place in which he is located. Although he may spend the day watching television, he may be unable to recount anything about current events.

This memory deficit is not confined to the anterograde amnesia. There is nearly always a deficit in remembering events that occurred before the onset of his illness, called *retrograde amnesia*. This varies in degree from patient to patient, even in those with comparable deficits in anterograde memorization. In some cases there is a gradient in the retrograde deficit, which is most severe for more recent months or years. The oldest memories are the best preserved. Some patients, however, show almost no gradient and fail even on questions from the most remote past; for example, the patient's father's first name cannot be recalled. The amnesias so common in films and fiction, in which the subject loses all personal memory, occur only in psychiatric disorders or in malingerers. The patient with Korsakoff's syndrome retains knowledge of his language, of calculations, of prayers, and of other well-learned material acquired early in life.

A simple explanation of retrograde amnesia is that a memory when first laid down is for a time unstable. In this period the memory trace is readily disrupted by many disturbances, including head injury. Once the memory is "consolidated," it is immune to these disruptions. Since retrograde amnesias are usually very long, it would appear at first that the period of consolidation could last for many years. Some doubts are cast on this hypothesis by the description of patients in whom the retrograde amnesia appears to "shrink."[1,43] Some patients who, after head injury, display a full Korsakoff's syndrome with its typical anterograde and retrograde amnesia may gradually recover. Over a period of time the patient's ability to acquire new information returns. He may be left with an island of amnesia for a period of time before his injury, and in the final recovered state this period is usually very short. In such a patient the recovery of memories that were apparently lost means that those memory traces were not destroyed by the injury; the failure must have been in the capacity to retrieve them. On the other hand, the permanent retrograde amnesia that remains is compatible with the idea of consolidation, but its time course must in those cases have been brief.

The lesions in cases of Korsakoff's syndrome involve portions of the limbic system. In some cases, parts of this system are destroyed (e.g., in alcoholic Korsakoff's syndrome), whereas in others, they appear to be temporarily incapacitated (e.g., head injury). The process of immediate recall clearly does not but the ability to recall new events does depend on these limbic structures. Further, once information is stored, its retrieval depends for many years on these same portions of the brain. Eventually, however, a memory can be retrieved without the mediation of the limbic system.

Anatomic structures involved in memory

It was assumed for many years that the Korsakoff syndrome appeared only after cortical lesions. However, a number of patients were then studied who had undergone bilateral surgical removal of the medial temporal lobes; the lesions involved the amygdala, hippocampus, and parahippocampal gyri.[44] These patients develop a severe and persistent amnestic syndrome. Patients with vascular lesions in these same structures, on both sides of the brain, develop a similar amnestic syndrome. It is not known whether any single one of these structures is critical for normal memory. In addition, studies of the brains of patients with alcoholic Korsakoff's syndrome have shown lesions in noncortical structures, especially in the mamillary bodies, the thalamus, and the periaqueductal region of the midbrain. Mamillary body lesions are regarded by many investigators as the critical ones, and this idea is supported by the fact that a major efferent pathway of the hippocampus, the fornix, projects to the mamillary bodies.

Why are the limbic structures involved in the process of storing memories and retrieving them? Memories are not stored in these structures but elsewhere; otherwise, memories would never be available after the destruction of these areas. Some studies in animals have shown that electrical changes occur in the hippocampus during learning, as well as in one of the targets of its efferent fiber systems, the medial septal nuclei. The majority of the functions in which the limbic system plays a role are of immediate importance to the survival of self or of the species: eating, drinking, aggression, and sex. It is possible that the limbic structures involved in memory are designed to select for storage and retrieval events of major importance for these basic biologic functions. Lesions in the limbic system or its immediate connections may account for another common feature of the early Korsakoff's syndrome: such a patient rarely shows a normal concern for his deficit, and some are blatantly euphoric.

Our understanding of the mechanism of this syndrome is too incomplete to explain the degree of recovery that frequently occurs. A patient with Korsakoff's syndrome may show a slowly increasing ability to store new information. The euphoria present at onset usually disappears, and such a patient may at a later stage show a more appropriate emotional response to his condition.

Dominance effects in memory

The most complete studies of laterality effects of lesions on memory have been carried out by Milner,[34] who studied patients undergoing temporal lobe surgery for epilepsy. It was found that after left temporal excisions, especially those including the medial temporal structures, verbal memories were likely to be impaired, and after right temporal removal, memory for certain nonverbal materials (e.g., arbitrary visual patterns or melodies) was likely to be impaired. Each hemisphere thus appears to depend to a major extent on its own medial temporal region. After destruction of one temporal lobe the storing of memories appropriate to the function of that hemisphere is deficient. After some months the deficit decreases, but Milner's studies show that some lateralized disability may persist indefinitely.

Other features of the Korsakoff syndrome

It is clear that there is more than one memory system, for even in the most severe case of Korsakoff syndrome the patient is not aphasic. Language is surely learned, and one might expect stored linguistic information to be lost or irretrievable if only one memory system existed. Similarly, the Korsakoff patient does not forget how to use eating utensils or how to dress himself. It may be that these abilities are preserved because they are learned early in childhood and practiced heavily. No case of Korsakoff's syndrome in a young child has been studied, so this idea has not been tested.

In one alternative view there are many systems involved in the storage and retrieval of information. A major one certainly exists in the medial temporal lobe and its connections, but others are located elsewhere. It is thus possible that certain language and motor functions are learned without involvement of the medial temporal structures, but little is known of them.

ATTENTION DISORDERS

Disorders of attention are among the most common of the impairments of higher functions in humans, and failures of attention are a major cause of learning difficulty in many otherwise normal individuals. The complexity of the process of attention is shown in a simple experiment carried out by Broadbent[5] that demonstrates what is easily observed in everyday life. The subject received different messages through earphones. The left delivered one recorded speech while the right transmitted a different one. If the subject was instructed to attend to one message, he was able afterward to describe the message delivered to that ear, but not the message to the other. The experiment was then repeated in an identical manner, except that at one point the subject's name was delivered to the unattended ear. The subject afterward recalled nothing from the unattended ear except the fact that his name was spoken. This demonstrates that the information from the unattended ear was not "cut off"; it must have been received and stored, for the unexpected appearance of the subject's name was recognized and remembered. It appears that certain important stimuli may cause a shift of directed attention.

Attention systems have two related functions. They maintain a central orientation to one stimulus or class of stimuli, and at the same time they monitor in a scanning mode inputs coming over other sensory systems. The attention system must operate under complex rules that specify the value of the stimulus that will divert attention. Stimuli with an emergency or survival value will commonly evoke such a shift, but others will not. Thus the brain mechanisms for attention maintain the organism in a narrow balance between an excessive distractibility and an excessively restricted attention. Clinical disorders of attention vary from a failure of any stimulus to attract attention (as in patients with akinetic mutism, caused by a lesion of the mesencephalic reticular formation) to cases in which a random and haphazard shift of attention occurs frequently, as in the delirious patient. Here I shall deal primarily with disorders of attention that result from hemispheric lesions.

Clinical picture of confusional states

The disorders of attention that occur in humans with brain lesions or disease are so varied they cannot be classified in an orderly way. There are some features common to nearly all. What is most striking is the disorganization of the patient's thought processes. He may respond well to some questions and fail on equally simple ones. He may begin to speak and suddenly shift the subject or stop talking. In many cases the patient is fully oriented and knows the answers to questions, but fails to reply correctly because of erratic attention. Most normal persons have experienced the same state, for example, when during marked drowsiness they attempt to do some complex mental arithmetic or to follow an involved lecture.

Bilateral disorders of attention

Transient failures of attention are common in everyday life. Clinical attention disorders result commonly from metabolic or toxic causes (e.g., from intoxication by alcohol or other drugs) and occur in hepatic, renal, or respiratory disease, although each disease process may lead to a disorder with distinctive features. We do not know the precise nature of the disorder of central nervous system function in any of these toxic or metabolic states.

A number of focal hemispheric lesions lead to

confusional states. Infarction of the undersurface of the occipital lobe on either side may cause such a state.[24] This particular confusional state may be caused by an extension of the lesion into the hippocampal region, but it should not be confused with the memory disorder caused by hippocampal lesions described previously.

The confusional states caused by unilateral lesions of the convexity of the hemispheres are more common. There is in them a marked dominance effect, for confusion occurs rarely after acute lesions of the left hemisphere (for a description of amorphosynthesis caused by such lesions, see Chapter 21). On the other hand, Mesulam et al.[33] found that acute lesions of the right hemisphere may lead to such states. The available evidence indicates that either right frontal or right parietal lesions may result in a general or global confusion. In some cases, this may be the only clear clinical sign produced by such a lesion, and such a patient may easily be mistaken for one with a metabolic or toxic disorder rather than a focal brain lesion.

The preferential production of bilateral disorders of attention by right-sided lesions suggests a right hemisphere dominance for the general attention process. Other observations fit with this idea. Thus Lewis et al.[30] showed a great activation of the electroencephalogram of the right hemisphere during certain vigilance tasks that was abolished by the ingestion of alcohol. Dimond[9] has studied the capacities of each hemisphere to maintain vigilance in patients after section of the corpus callosum and anterior commissure for epilepsy. He found that the isolated right hemisphere performs significantly better than the left in a vigilance task. These observations support the proposition that some portion of the right hemisphere takes the lead in monitoring all inputs to both hemispheres, that is, in the scanning function described previously.

The unilateral disorder of attention that is part of the parietal lobe syndrome is described in Chapter 21.

SYNDROMES OF THE CORPUS CALLOSUM

Lesions of the corpus callosum have been useful for studies of many aspects of cerebral dominance. Thus they provided evidence for the dominance of one hemisphere in motor learning.[15] There has been a revival of interest in this subject in recent decades, largely due to the demonstration by Myers and Sperry[36] of the callosal syndromes in animals, the description of the callosal syndrome in humans by Geschwind and Kaplan,[16] and the extensive and detailed studies by Sperry and his colleagues of humans after complete surgical section of the cerebral commissures.[13] Many aspects of these subjects have already been discussed here and in Chapter 21.

DIRECTIONS OF FUTURE RESEARCH

I have described in this chapter only a small sample of the disorders of the higher functions that appear in humans after lesions of the brain. The question arises of how a further understanding of these complex functions of the human brain is to be achieved. One might continue to collect and describe syndromes and study the lesions responsible for them. The results are of great usefulness to the clinical neurologist. One might study these syndromes from the psychologic or linguistic point of view, an approach that has been successful in the hands of such investigators as Benton, Hécaen, Milner, De Renzi, Vignolo, Zangwill, and many others. It appears likely, however, that a further understanding of dynamic neural mechanisms is required before effective therapy can be devised.

Studies of simpler nervous systems will provide information important for that understanding, but it is probable that an advanced nervous system like that of the human exhibits principles of organization that must be investigated directly. The advent of new methods, such as computerized x-ray studies of the brain and refined methods of studying evoked potentials or local metabolic changes, will undoubtedly supply sources for major advances.

However, a detailed study of neurologic mechanisms will at least for the foreseeable future be possible only in nonhuman primates. It may be necessary for the more complex aspects of human behavior to be translated into simpler but still relevant behavioral paradigms in which nonhuman primates can be trained and thus be available for direct study. Consider, for example, the complex function of the frontal lobe in humans. It has been argued that this region contains neural mechanisms essential for foresight and socialization. It is obviously difficult or impossible to translate that concept into a combined behavioral-electrophysiologic experiment in the monkey. An alternative hypothesis is that the frontal lobe is responsible for the acquisition of complex responses to limbic stimuli (e.g., to anger, fear, sexual excitement, or hunger). This type of learning may underlie the socialization of the child, and it is at least possible that subhuman primates learn in a similar way the rules of their group for reacting to limbic stimuli. Although the translation to the primate training laboratory strains analogy, it is possible that it is only through such attempts that the neural mechanism of the truly "higher" functions can be brought under direct experimental observation.

REFERENCES

1. Benson, D. F., and Geschwind, N.: Shrinking retrograde amnesia, J. Neurol. Neurosurg. Psychiatry **30:** 539-544, 1967.
2. Black, P., and Myers, R. E.: A neurological investigation of eye-hand control in the chimpanzee. In Ettlinger, E. G., editor: Functions of the corpus callosum, London, 1965, J. & A. Churchill, Ltd., pp. 47-59.
3. Bogen, J. E., and Gordon, H. W.: Musical tests for functional lateralization with intracarotid amobarbitol, Nature **230:**524-525, 1971.
4. Brinkman, J., and Kuypers, H. G. J. M.: Cerebral control of contralateral and ipsilateral arm, hand, and finger movements in the split-brain rhesus monkey, Brain **96:**653-674, 1973.
5. Broadbent, D. E.: The role of auditory localization in attention and memory span, J. Exp. Psychol. **47:**191-196, 1954.
6. Campain, R., and Minckler, J.: A note on the gross configuration of the human auditory cortex, Brain Lang. **3:**318-323, 1976.
7. Chi, J. G., Dooling, E., and Gilles, F. H.: Gyral development of the human brain, Ann. Neurol. **1:**86-93, 1977.
8. Desmedt, J.: Language and hemispheric specialization in man: cerebral event-related potentials, Basel, 1977, S. Karger.
9. Dimond, S. J.: Depletion of attentional capacity after total commissurectomy in man, Brain **99:**347-356, 1976.
10. Foerster, O.: Motorische Felder und Bahnen. In Bumke, O., and Foerster, O., editors: Handbuch der Neurologie, Berlin, 1936, Springer-Verlag, vol. 6.
11. Galaburda, A., LeMay, M., Kemper, T., and Geschwind, N.: Right-left asymmetries in the brain, Science **199:**852-856, 1978.
12. Gardner, R. A., and Gardner, B. T.: Teaching sign language to a chimpanzee, Science **165:**664-672, 1969.
13. Gazzaniga, M. S.: The bisected brain, New York, 1970, Appleton-Century-Crofts.
14. Geschwind, N.: Late changes in the nervous system: an overview. In Stein, D. G., Rosen, J. J., and Butters, N., editors: Plasticity and recovery of function in the central nervous system, New York, 1974, Academic Press, Inc.
15. Geschwind, N.: The apraxias: neural mechanisms of disorders of learned movement, Am. Sci. **63:**188-195, 1975.
16. Geschwind, N., and Kaplan, E.: A human cerebral deconnection syndrome, Neurology **12:**675-685, 1962.
17. Geschwind, N., and Levitsky, W.: Human brain: left-right asymmetries in temporal speech region, *Science* **161:**186-187, 1968.
18. Geschwind, N., Quadfasel, F. A., and Segarra, J. M.: Isolation of the speech area, Neuropsychologia **6:**327-340, 1968.
19. Gainotti, G.: Emotional behavior and hemispheric side of the lesion, Cortex **8:**41-55, 1972.
20. Gloning, I., Gloning, K., Haub, G., and Quatember, R.: Comparison of verbal behavior in right-handed and non right-handed patients with anatomically verified lesion of one hemisphere, Cortex **5:**43-52, 1969.
21. Goldstein, K.: Language and language disturbances, New York, 1948, Grune & Stratton, Inc., pp. 220-221.
22. Goodglass, H., and Kaplan, E.: Disturbance of gesture and pantomime in aphasia, Brain **86:**703-720, 1963.
23. Hécaen, H.: Aphasic, apraxic and agnosic syndromes in right and left hemisphere lesions. In Vinken, P. J., and Bruyn, G. W., editors: Handbook of clinical neurology: disorders of speech, perception, and symbolic behavior, Amsterdam, 1969, North Holland Publishing Co., vol. 4.
24. Horenstein, S., Chamberlain, W., and Conomy, J.: Infarction of the fusiform and calcarine regions: agitated delirium and hemianopia, Trans. Am. Neurol. Assoc. **92:**85-89, 1967.
25. Isserlin, M.: Aphasie. In Bumke, O., and Foerster, O., editors: Handbuch der Neurologie, Berlin, 1936, Springer-Verlag, vol. 6, p. 671.
26. Kimura, D.: Cerebral dominance and the perception of verbal stimuli, Can. J. Psychol. **15:**166-171, 1961.
27. Kleist, K.: Sensory aphasia and amusia, Elmsford, N.Y., 1962, Pergamon Press, Inc.
28. LeMay, M., and Culebras, A.: Human brain-morphologic differences in the hemispheres demonstrable by carotid arteriography, N. Engl. J. Med. **287:**168-170, 1972.
29. LeMay, M., and Geschwind, N.: Hemispheric differences in the brains of great apes, Brain Behav. Evol. **11:**48-52, 1975.
30. Lewis, E. G., Dustman, R. E., and Beck, E. C.: The effects of alcohol on visual and somato-sensory evoked responses, Electroencephalogr. Clin. Neurophysiol. **28:** 202-205, 1970.
31. Luria, A.: Traumatic aphasia, The Hague, 1970, Mouton.
32. McAdam, D. W., and Whitaker, H. A.: Language production: electroencephalographic localization in the normal human brain, Science **172:**499-502, 1971.
33. Mesulam, M-M., Waxman, S. G., Geschwind, N., and Sabin, T. D.: Acute confusional states with right middle cerebral artery infarctions, J. Neurol. Neurosurg. Psychiatry **39:**84-89, 1976.
34. Milner, B.: Laterality effects in audition. In Mountcastle, V. B., editor: Interhemispheric relations and cerebral dominance, Baltimore, 1962, The Johns Hopkins University Press.
35. Mohr, J.: Broca's area and Broca's aphasia. In Whitaker, H., editor: Studies in neurolinguistics, New York, 1976, Academic Press, Inc.
36. Myers, R. E., and Sperry, R. W.: Interocular transfer of a visual form discrimination habit in cats after section of the optic chiasm and corpus callosum, Anat. Rec. **115:** 351-352, 1953.
37. Ojemann, G. A., Blick, K. I., and Ward, A. A., Jr.: Improvement and disturbance of short-term verbal memory with human ventrolateral thalamic stimulation, Brain **94:**225-240, 1971.
38. Penfield, W., and Roberts, L.: Speech and brain mechanisms, Princeton, N.J., 1959, Princeton University Press.
39. Pfeifer, R. A.: Pathologie der Hörstrahlung und der corticalen Hörsphäre. In Bumke, O., and Foerster, O., editors: Handbuch der Neurologie, Berlin, 1936, Springer-Verlag, vol. 6.
40. Premack, D.: On the assessment of language competence in the chimpanzee. In Schrier, A. M., and Stollmitz, F., editors: Behavior of nonhuman primates, New York, 1971, Academic Press, Inc.
41. Raisman, G., and Field, P. M.: Sexual dimorphism in

the neuropil of the preoptic area of the rat and its dependence on neonatal androgen, Brain Res. **54:**1-29, 1973.

42. Roberts, L.: Aphasia, apraxia, agnosia in abnormal states of cerebral dominance. In Vinken, P. J., and Bruyn, G. W., editors: Handbook of clinical neurology: disorders of speech, perception, and symbolic behavior, Amsterdam, 1969, North Holland Publishing Co., vol. 4.
43. Russell, W. R., and Nathan, P.: Traumatic amnesia, Brain **69:**280-300, 1946.
44. Scoville, W. B., and Milner, B.: Loss of recent memory after bilateral hippocampal lesion, J. Neurol. Neurosurg. Psychiatry **20:**11-21, 1957.
45. Smith, A.: Speech and other functions after left (dominant) hemispherectomy, J. Neurol. Neurosurg. Psychiatry **29:**467, 1966.
46. Sparks, R., and Geschwind, N.: Dichotic listening in man after section of neocortical commissures, Cortex **4:**3-16, 1968.
47. Subirana, A.: Handedness and cerebral dominance. In Vinken, P. J., and Bruyn, G. W., editors: Handbook of clinical neurology: disorders of speech, perception, and symbolic behavior, Amsterdam, 1969, North Holland Publishing Co., vol. 4.
48. Trescher, J. H., and Ford, F. R.: Colloid cyst of the third ventricle, Arch. Neurol. Psychiatry **37:**959-973, 1937.
49. Wada, J., and Rasmussen, T.: Intracarotid injection of Sodium Amytal for the lateralization of cerebral speech dominance, J. Neurosurg. **17:**266-282, 1960.
50. Warren, J. M., and Akert, K.: The frontal granular cortex and behavior, New York, 1964, McGraw-Hill Book Co.
51. Yamadori, A., Osumi, Y., Masuhara, S., and Okubo, M.: Preservation of singing in Broca's aphasia, J. Neurol. Neurosurg. Psychiatry **40:**221-224, 1977.
52. Yeni-Komshian, G. H., and Benson, D. A.: Anatomical study of cerebral asymmetry in the temporal lobe of humans, chimpanzees and rhesus monkeys, Science **192:** 387-389, 1975.
53. Zangwill, O. L.: Cerebral dominance and its relation to psychological function, Springfield, Ill., 1960, Charles C Thomas, Publisher.

VII

NEURAL CONTROL OF POSTURE AND MOVEMENT

23

ELWOOD HENNEMAN

Organization of the motor systems

A PREVIEW

The motor systems of the brain exist to translate thought, sensation, and emotion into movement. At present the initial steps in this process lie beyond analysis. We do not know how voluntary movements are engendered, nor where the "orders" come from. Most of the information that is available concerns the circuits that execute these shadowy commands.

Movement is the end product of a number of control systems that interact extensively. Their complexity demands that we proceed logically by (1) defining the nature of movement in terms of muscles and joints, (2) presenting an outline of the motor systems so that the relation of the parts to the whole is apparent from the outset, and (3) explaining how "control" is achieved. This introductory chapter is designed to meet the first two of these needs.

THE NATURE OF MOVEMENT: MUSCLES AND JOINTS

In most forms of skeletal movement the motion occurs at joints, where two or more bones come together to form a nearly frictionless pivot. Muscles are arranged so that their ends are attached on opposite sides of the joint, which may act as a fulcrum (Fig. 23-1). Since individual muscles can pull but not push, at least two muscles on opposite aspects of the joint are generally required to provide a full range of movement in both directions. At some joints there are several pairs of antagonistic muscles disposed so as to produce abduction, adduction, and rotation as well as flexion and extension.

There is a great deal of interaction between the various muscles in a limb. One reason for this is that muscular forces are exerted as much on the point of origin as on the insertion. Movement of one member of a joint therefore requires fixation of the other member by muscles at the next joint. In movements of the forearm, for example, the muscles of the shoulder generally contract to fix the humerus. If powerful forces are involved, fixation may be required at joints some distance away. A series of interconnections has developed in the spinal cord to link the motoneurons of muscles that commonly work together. The nature of the connections depends on the kind of mechanical interactions involved. The most closely related muscles are the direct antagonists such as the biceps and triceps (Fig. 23-1), for when either one of the pair shortens, the opposing muscle is necessarily lengthened. Direct antagonists are of great importance because precise control of movements frequently involves application of braking action. Each muscle acts as a brake for its antagonist, serving to bring rapid movements to a quick, smooth stop. Recordings of the electrical activity of the human biceps and triceps muscles, for example, reveal that during a contraction of the biceps there is usually some activity in the triceps and that toward the end of a rapid contraction, this activity increases sharply.

In general, muscles can be divided into two groups, depending on their action in the body. One group plays a special role in upright posture, standing, and locomotion. These are the muscles that normally oppose the force of gravity. Special reflexes and postural reactions involving them are highly developed in vertebrates. As a matter of tradition, physiologists refer to all antigravity muscles as "extensors" and to their antagonists as "flexors," regardless of their particular actions at joints. The flexor digitorum longus in the leg, for example, is regarded as an extensor muscle because it assists in raising the weight of the body upward against gravity. Another example

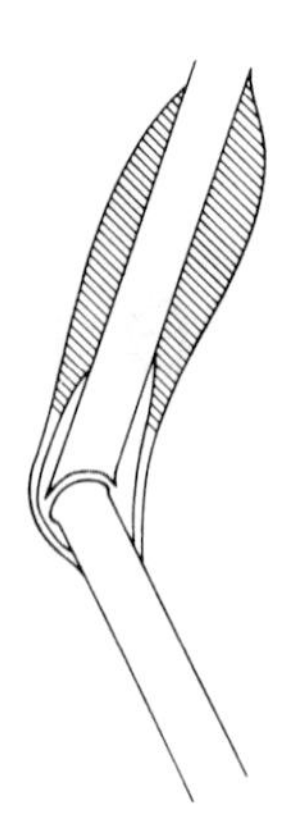

Fig. 23-1. Schematic representation of typical joint showing pair of antagonistic muscles that flex and extend it.

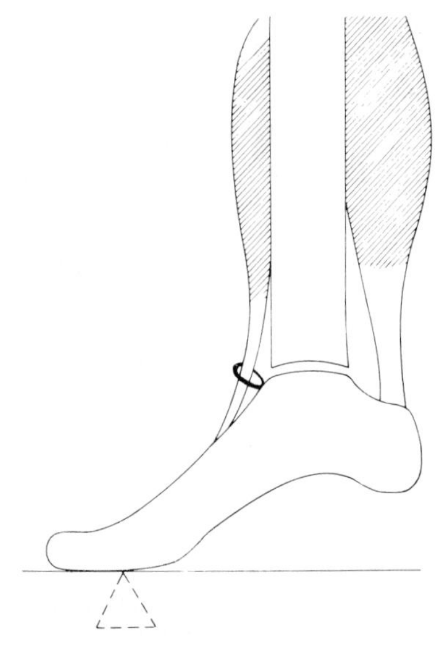

Fig. 23-2. Antigravity action of gastrocnemius muscle that raises heel and weight of body transmitted through tibia.

of a muscle that plays a prominent antigravity role is the gastrocnemius, which is represented diagrammatically in Fig. 23-2. Taking origin from the back of the tibia, it inserts on the calcaneus. Contraction of this muscle raises the heel and by lever action lifts the tibia, with the ball of the foot acting as the fulcrum. In addition to the obvious antigravity muscles, there are others, not usually regarded as extensors, that oppose gravity in less direct ways. The adductors of the thigh, for example, qualify as extensors because adduction includes a definite antigravity vector. Other muscles that are not concerned with posture or stance at all (e.g., those that raise the lower jaw) also oppose gravity and are classified as extensors. In pronounced antigravity reactions (e.g., decerebrate rigidity), all these muscles are actively contracted at the same time, a vivid illustration of their common role.

The other group of muscles, which is antagonistic to the extensors, has in common the function of withdrawing the body reflexly from painful stimuli. Any muscle that takes part in flexor withdrawal reflexes is classified as a flexor regardless of its joint action. The tibialis anterior, which opposes the gastrocnemius, is a flexor at the ankle; the extensor digitorum longus, despite its anatomic name, is a physiologic flexor of the toes and participates in most withdrawal reactions of the foot.

Finally, it should be emphasized that even simple movements are deceptively complex in their underlying mechanics. The number of muscles involved and the interactions between them greatly complicate the problem of neural control. In contrast to most engineering control systems, there is not one process or action to be regulated, but many simultaneously. As we shall see, the task of coordinating them is so formidable that it requires an elaborate center containing, according to recent estimates, a total of 10^{11} neurons!

OUTLINE OF THE MOTOR SYSTEMS

In this section an attempt has been made to isolate the parts of the nervous system that are primarily motor in function and to describe their roles in motor activity as concisely as possible. The aim is to provide a functional outline that will permit the reader of subsequent chapters to fit each new circuit into its proper relation with the whole system.

Fig. 23-3 is a block diagram of the motor systems, greatly simplified for the sake of initial orientation. Each block represents a major subdivision of the nervous system. The flow of signals between blocks is indicated by arrows (unshaded for sensory connections, shaded for motor and nonsensory connections). To begin with, it should be emphasized that all the various motor pathways shown in this schematic nervous system ultimately converge on a series of simple circuits that link each muscle with the spinal cord. As Fig. 23-4 indicates, the sensory neurons carrying impulses from a given muscle are connected with the motoneurons, which transmit impulses back to it. A closed loop that regulates the activity of each muscle individually is thus formed. This is the basic

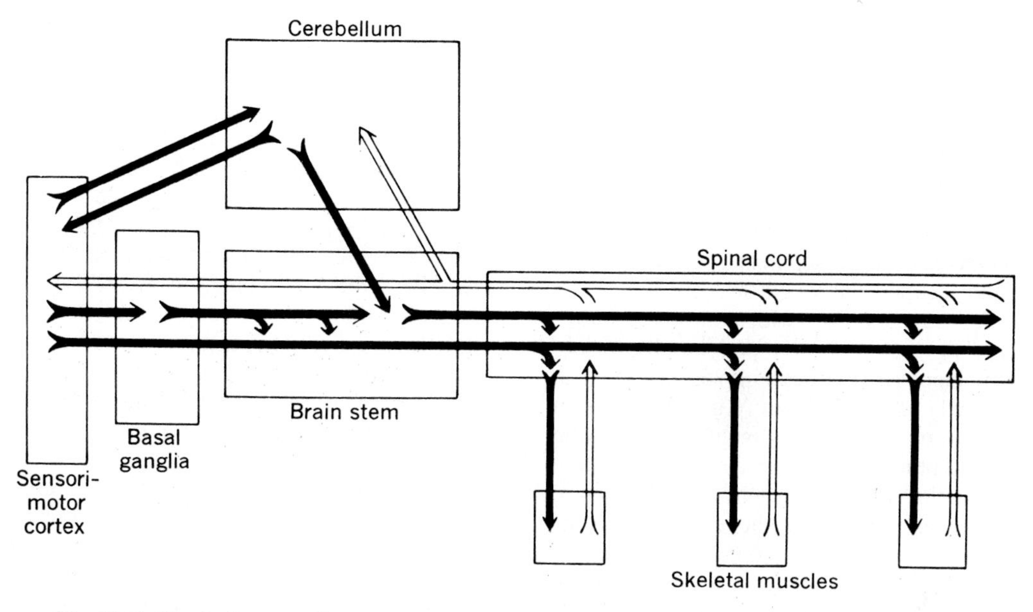

Fig. 23-3. Block diagram of motor systems. Each block represents a major subdivision of the nervous system. Unshaded arrows indicate sensory connections; shaded arrows represent motor and nonsensory connections.

control mechanism of the motor systems, for no movement, reflex or voluntary, can occur except through the agency of this circuit. Although these circuits may function autonomously in simple reflexes, their activity is largely controlled by centers at higher levels. From these centers, descending tracts run the entire length of the spinal cord, giving off fibers at all levels. One such tract, the corticospinal, originates in the cerebral cortex and transmits signals from the sensorimotor cortex directly to the spinal level. All the other tracts that descend into the spinal cord arise in the brain stem. The brain stem is a prespinal integrating center of great complexity that receives signals from all higher centers and processes them for transmission to the spinal cord. Superimposed on the upper end of the brain stem are the basal ganglia, a heterogeneous group of nuclei that receive signals from the sensorimotor cortex and discharge into the brain stem. At the highest level is the sensorimotor cortex, which presides over the entire motor system. Interconnected with all levels and functioning as an overall coordinator of motor activities is the cerebellum, a highly organized center with an extensive cortex of its own. The brief account of the motor systems that follows this introduction will begin at the spinal level and will take up the other subdivisions of the nervous system in ascending order.

The closed loop that links each muscle with the spinal cord is shown schematically in Fig. 23-4. As the figure indicates, some of the sensory fibers from the muscle establish a direct connection with the motoneurons, whereas others do so indirectly through internuncial cells. In the latter type of circuit the loop in-

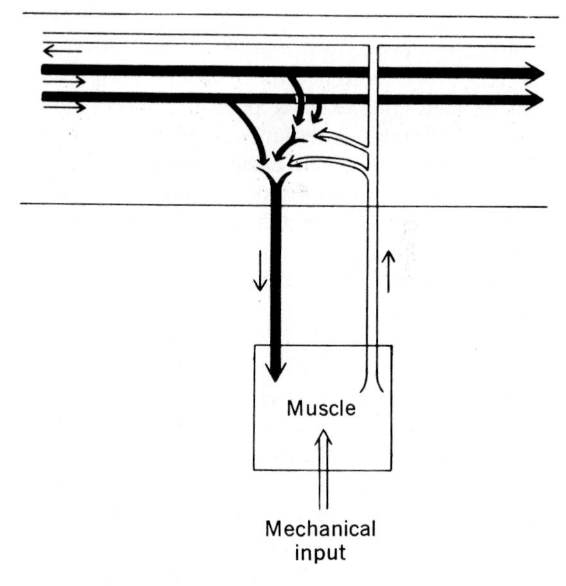

Fig. 23-4. Schematic representation of closed loop that links each muscle with spinal cord. Shaded arrow to muscle represents motoneuron. Sensory and motor connections to it are indicated by unshaded and shaded arrows, respectively.

cludes the following parts: the cell bodies of the motoneurons in the spinal cord, their axons running out in the ventral roots, the neuromuscular junctions, the muscle itself, the sensory receptors with their fibers entering the spinal cord through the dorsal roots and their terminations in the spinal cord, and finally one or more internuncial cells sending axons to the motoneurons. These segmental circuits automatically regulate the length and tension of the muscle in accordance with various requirements. They are responsive peripherally to mechanical input (e.g., stretching of the muscle) and centrally to neural input (e.g., signals from higher centers). Mechanical input stimulates the stretch receptors, provoking circuit action that elicits a reflex response of the muscle; neural input initiates activity in the central parts of the loop that also travels around the entire circuit. Signals from higher centers impinge chiefly on the internuncial cells and, to a lesser extent, directly on the motoneurons themselves.

It has been emphasized that muscles interact extensively. Much of the interaction is controlled automatically by spinal circuits. The sensory feedback from muscles is not restricted to their own motoneurons; rather, it spreads out through collaterals of primary sensory neurons and through internuncial circuits to reach the motoneurons of all closely related muscles and, to a decreasing extent, those of more distant muscles. A stretch or a contraction of one muscle affects its own motoneurons most strongly, those of its direct antagonist somewhat less, and those of other synergists and antagonists around it still less. Thus every primary loop is part of a larger feedback network serving a group of muscles.

It is a general rule of the nervous system that sensory impulses arising in a primary sensory neuron are distributed by way of collaterals, projection tracts, and secondary relays to widely separated parts of the nervous system. As Figs. 23-3 and 23-4 indicate, signals arising in skeletal muscles not only serve in local segmental circuits but are also transmitted up the spinal cord to higher levels of the nervous system. The spinal circuits involve very little delay and ensure rapid responses when speed is essential and time-consuming processing of sensory data is not required. While this immediate response is occurring, the same signals are being transmitted to higher centers for more elaborate analysis of their information content and for combination with signals from other types of receptors. After variable delays, signals from the higher centers are relayed back to segmental levels. Thus "long loops" involving higher centers and more delay help to regulate the activity of the spinal circuit.

Superimposed on the orderly, segmental organization of the spinal cord is the more complex brain stem. Its structure provides no easily read clues to its functions. Its central role in motor activities and in immediate control over the spinal cord is obvious, however, for *all the descending tracts to the spinal cord except one, the corticospinal tract, arise in the brain stem.* As indicated in Fig. 23-3, these descending tracts terminate at all levels of the spinal cord. They can thus control or regulate activity at many spinal levels simultaneously.

It is not yet possible to specify the function of the brain stem centers and their exact relations to the segmental circuits, but it is clear that they extend motor capacity far beyond the stereotyped regulatory behavior of the spinal animal. Animals with a brain stem (but without higher centers) are capable of integrated activities such as standing, walking, and postural adjustments. These activities require the participation of various righting reflexes and antigravity mechanisms, they demand precise control of equilibrium, and they involve coordination of muscles throughout the body.

Running the length of the brain stem is a core of gray matter called the reticular formation (RF), which is centrally involved in motor activities. It receives connections from the spinal cord, the cerebral cortex and basal ganglia, and the cerebellum. It functions as an integrating center, adding, subtracting, and combining the influences of all these centers. Two subsystems, which have not been adequately defined anatomically, have been identified within the RF. A reticular inhibitory center in the medulla gives origin to a reticulospinal tract that transmits impulses that are predominantly inhibitory to extensor motoneurons. A reticular facilitatory system located in the higher portions of the brain stem transmits impulses that are predominantly facilitatory to extensor motoneurons. The functions of these two systems are far from clear, but it appears that they serve as common efferent pathways through which the entire extrapyramidal system exerts its influence on stretch reflex circuits.

Closely related to the reticular systems in their postural activities are the elaborate sensing devices in the vestibule of the ear and their ramifications in the nervous system. Through its connections with the RF and through the vestibulospinal tract, this system regulates muscle in response to changes in the position of the head in space and in response to all accelerations that the body experiences.

At the next higher level of the motor system are the basal ganglia, a group of nuclei that receive descending connections from the sensorimotor cortex and project to the RF of the brain stem. The fact that they occupy a very large volume of the brain suggests that their motor functions are of major importance. The details of their structure and connections, however, provide few clues as to their precise roles. Certain pathologic lesions confined to the basal ganglia result in characteristic motor disturbances without sensory deficits or mental impairment. Although these syndromes are often dramatic and very disabling, they offer surprisingly little insight into the normal function of these systems. The basal ganglia, with the descending fibers they receive from the cerebral cortex, and with the RF and the reticulospinal tracts, comprise the bulk of the "extrapyramidal" motor system.

At the highest level of the nervous system there are several cortical areas that are partially or wholly devoted to motor functions. They include the area just anterior to the central sulcus, generally known as the "motor cortex," and the area just posterior to the central sulcus, which is a "receiving area" of the somatic sensory system. Together these areas are often referred to as the sensorimotor cortex, a term that properly emphasizes the important role of sensory systems in motor control. In each of the subdivisions of the sensorimotor cortex the parts of the body are represented in orderly fashion. Electrical stimulation within any of these areas results in movements of the particular muscle groups represented under the electrodes. The effects of stimulation are due to activation of two different descending systems, which are indicated in Fig. 23-3. One of them consists of corticifugal fibers from all subdivisions of the sensorimotor cortex. These fibers come together and form the pyramidal or corticospinal tract, which runs without interruption from the cerebral cortex to all levels of the spinal cord. Most of these long fibers end on internuncial cells that form part of the feedback loop from muscle to motoneurons. The organization of the pyramidal tract suggests that it is designed for precise control of individual muscle groups. The other corticifugal system is the extrapyramidal system, which sends fibers to some of the basal ganglia and to the RF of the brain stem. The pyramidal and extrapyramidal systems, although dissimilar in their organization, work together harmoniously.

Acting as a coordinating center for motor systems as a whole is a large, highly organized outgrowth of the brain stem known as the cerebellum. It receives signals relayed from receptors in muscles, tendons, joints, and skin as well as from visual, auditory, and vestibular end-organs. This great volume of sensory input serves purely motor functions, chiefly regulatory in effect. Signals from the cerebral cortex and from other motor regions also reach the cerebellum in great numbers. All this enormous influx is analyzed and integrated in an elaborate but orderly cortical network. From the cerebellum, efferent fibers pass to the thalamus, red nucleus, vestibular nuclei, and RF. Through these connections the cerebellum influences motor centers from the cerebral cortex down to the spinal motoneurons, coordinating the activity of motor circuits at all levels of the central nervous system.

24 Skeletal muscle

ELWOOD HENNEMAN

THE SERVANT OF THE NERVOUS SYSTEM

The physicochemical mechanisms that enable muscle to shorten, lengthen, and produce tension were described in Chapter 3. How these mechanisms are organized into functional units under the control of the nervous system is our present concern and is the first step toward an understanding of movement and posture.

Skeletal muscles and the motoneurons that control them are the products of evolution. Survival placed a premium on speed of movement to capture prey or escape predators, on the capacity to resist fatigue, on a favorable ratio between the weight and strength of muscle, and, perhaps above all, on efficient use of energy. Muscle is a tissue that uses large quantities of energy, hence there has been considerable selection pressure for efficiency. A. V. Hill[30] realized that some qualities could only be enhanced at the expense of others. He pointed out that "every muscle or group of muscles will show qualitatively the sort of properties that a very intelligent engineer, knowing all the facts, would have designed for them in order to meet, within wide limits, the requirements of their owner. These properties must provide a compromise between various needs." The nature of this compromise will become apparent when the properties of the individual motor units in a muscle are described.

To be a useful servant of the nervous system, a muscle must be susceptible to precise control over a wide range of lengths, tensions, speeds, and loads. This requirement entails great difficulties for the systems that control muscles and probably accounts for much of their complexity. A particular input to the motoneurons that supply a muscle will produce contractile effects that differ with the initial state of the muscle, its prior history, and other factors such as the activity of its synergists and antagonists. The nervous system must somehow be informed of these variables and take them into account in fashioning inputs to motoneurons.

Certain properties tend to be mutually exclusive in a single muscle fiber. For example, rapid contraction and economy in the utilization of energy are not found in the same fiber. Evidently a single, uniform type of fiber was not adequate to embody all the characteristics needed in a muscle. Three types of fibers differing morphologically, chemically, and functionally evolved to meet the varied requirements of mammals. The great majority of mammalian muscles consist of a mixture of these three types.

TYPES OF FIBERS IN MAMMALIAN SKELETAL MUSCLE*

In 1874 Ranvier[51] observed that mammalian skeletal muscles differed in color and in the microscopic appearance of their fibers. Later it was noted that fibers of small diameter were darker and more granular than larger fibers in the same muscle. Small fibers predominated in red muscles, large fibers in white muscles. Accordingly, they were referred to as "red" and "white" fibers. Differences between fibers were not conspicuous in routine stains and received little attention until histochemical methods were developed for localizing enzymatic activity in sections. The activity of succinic dehydrogenase reflected the mitochondrial content and oxidative activity of a fiber, whereas the degree of myosin- or myofibrillar-ATPase activity indicated differences in the contractile apparatus.[43,47] The concentration of myoglobin was demonstrated by means of its peroxidase activity, and the sites of glycogen synthesis were revealed by radioautographs.[14]

*A more comprehensive account of this topic may be found in a review article by Gauthier.[22]

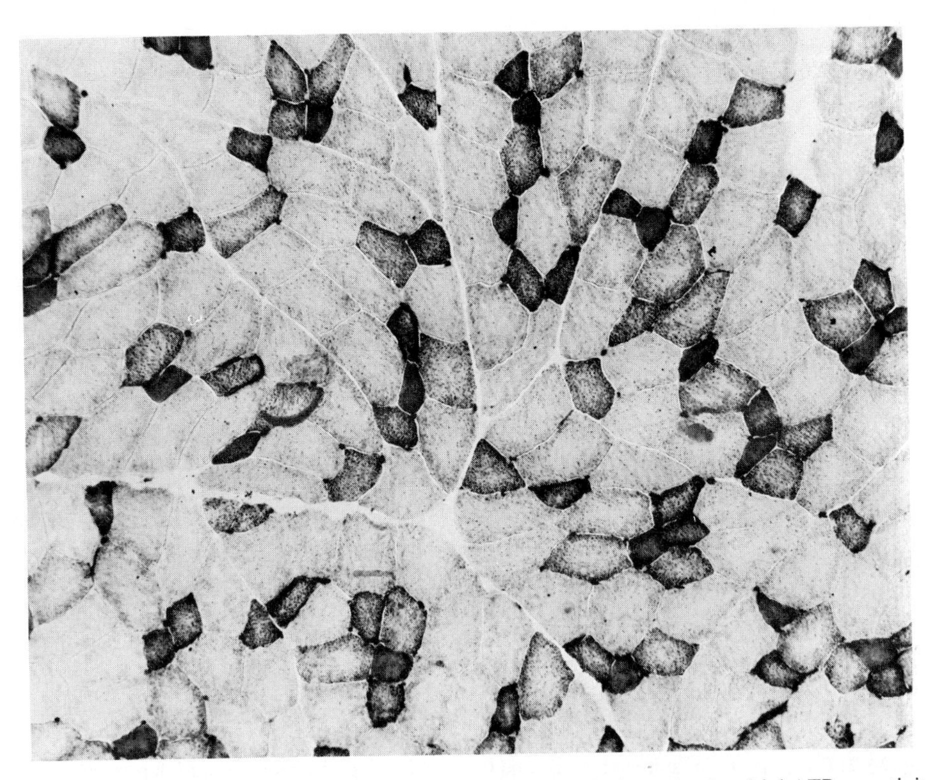

Fig. 24-1. Cross section of m. gastrocnemius muscle of cat showing mitochondrial ATPase activity of three types of fibers. Large, pale fibers are type A; small, dark fibers are type C; and fibers of intermediate size and density are type B. (From Henneman and Olson.[27])

Classification of fiber types has been controversial because of the different criteria employed. For example, types I and II have been used to designate the low and high levels, respectively, of phosphorylase or myofibrillar ATPase activity. When a variety of cytochemical procedures, including the localization of succinic dehydrogenase and mitochondrial ATPase, were compared in serial sections, where the same fibers could be followed, three types of fibers could be recognized.[57] These were designated A, B, and C, or "white," "intermediate," and "red," respectively. These two equivalent terminologies are based on ultrastructural as well as histochemical criteria, which are consistent with each other.

Figs. 24-1, 24-2, and 24-5 illustrate the appearance of the three types of fibers in a heterogeneous muscle (medial gastrocnemius of the cat) when treated to demonstrate mitochondrial adenosine triphosphatase (ATPase). Figs. 24-3 and 24-4 show the results of applying the same technique to a homogeneous muscle (soleus) of the cat, which consists of just one type of fiber. This technique causes the mitochondria to stand out as black dots of various sizes and darkens the cytoplasm of different fibers to a variable degree. It also produces an intense reaction in the capillaries around the muscle fibers. The appearance and distribution of these features under low magnification are illustrated in Fig. 24-1. The large, pale-staining fibers, containing relatively few mitochondria, which predominate in all parts of the muscle, are designated type A. Scattered among them are small fibers that look nearly solid black. They are called type C. A third type of fiber (B), intermediate in size and in the intensity of its ATPase reaction, has the majority of its mitochondria situated at the periphery of the cell, most densely near the capillaries (examples are in Figs. 24-2 and 24-4). The three types of fibers are shown at higher magnification in Fig. 24-2. Figs. 24-2 to 24-4 illustrate the presence of the very dark capillaries often located at the interstitial angles between fibers. The number of capillaries is greatest around the small, dark C fibers. In longitudinal sections (Fig. 24-5), it is apparent that capillaries may run for long distances along the periphery of B or C fibers. The number of capillaries is somewhat less around B than C fibers and is least around A

Fig. 24-2. Characteristic features of A, B, and C fibers in m. gastrocnemius at high power: type A, large, with few mitochondria; type B, intermediate in size and enzymatic activity, with subsarcolemmal distribution of large mitochondria; type C, small, with marked background activity and numerous small mitochondria. ATPase section. (From Henneman and Olson.[27])

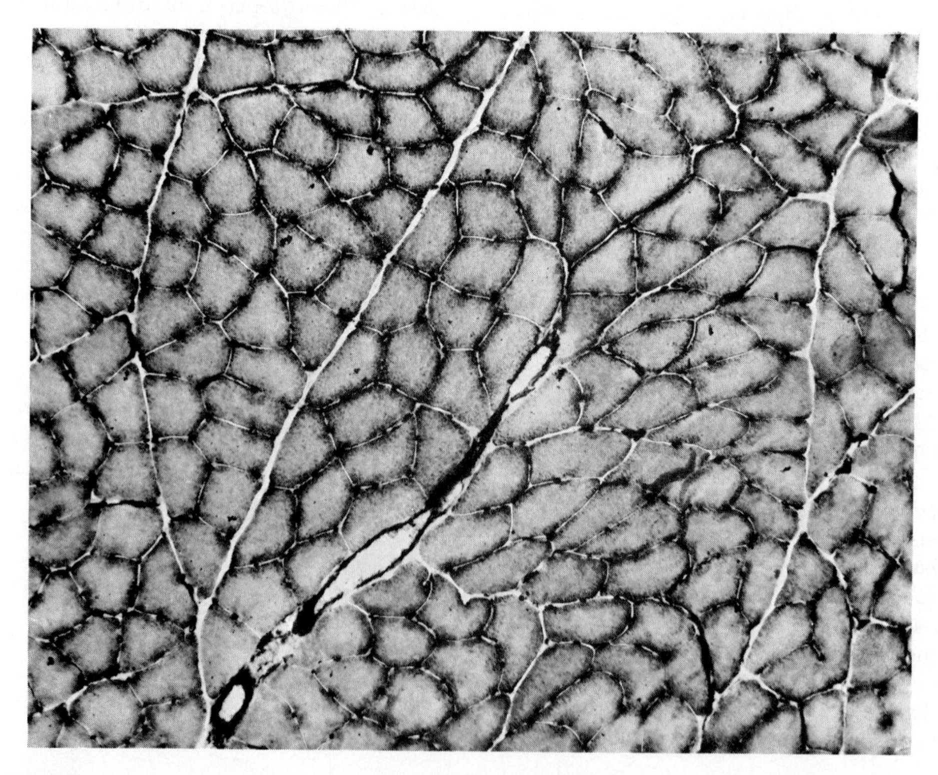

Fig. 24-3. Cross section of soleus muscle of cat showing uniformity of ATPase activity and fiber size. Note intense ATPase activity of capillaries around each fiber. (From Henneman and Olson.[27])

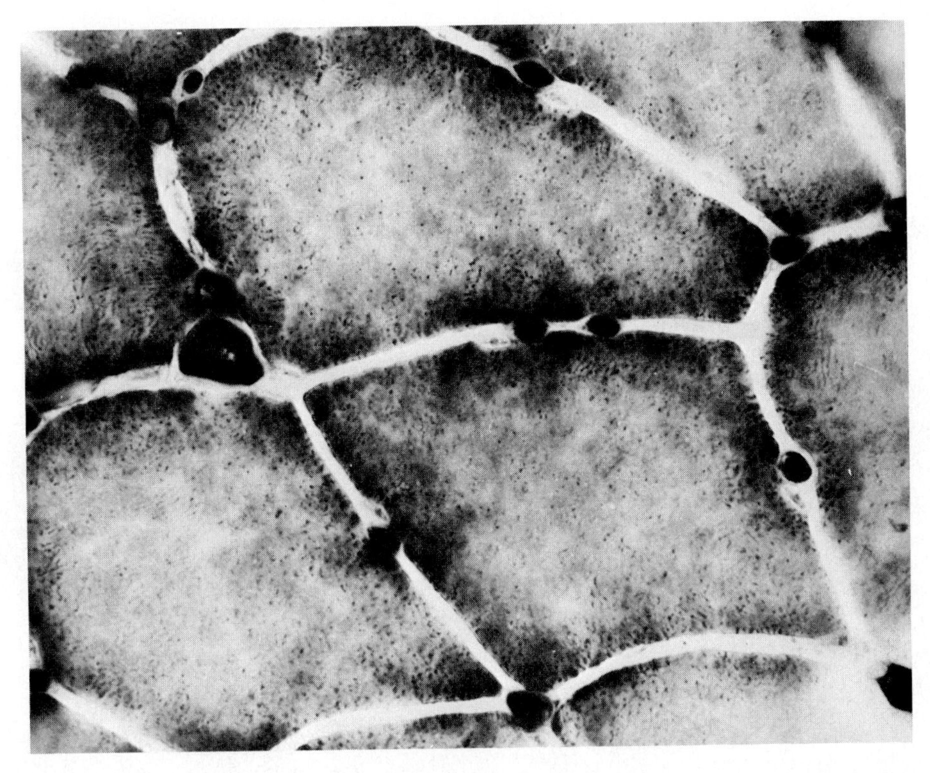

Fig. 24-4. Characteristic features of soleus fibers at high power. Mitochondrial density is greatest at periphery of fibers and especially near the numerous capillaries. ATPase section. (From Henneman and Olson.[27])

fibers. Unless they are adjacent to B or C fibers, it is clear in cross sections that A fibers may have no neighboring capillaries at all.

Recognition of the three types of fibers does not depend on the particular histochemical method used. Stein and Padykula[57] found that, although each technique emphasized different characteristics, the same fibers could be identified in serial sections by the distribution of either succinic dehydrogenase, esterase, glycogen, or mitochondrial ATPase. The same muscle may vary in its histochemical composition in different animals, depending on their size, speed of movement, and metabolic rate. Using 36 different mammalian species, Gauthier and Padykula[23] found that small animals with a high metabolic rate and fast breathing rate had diaphragms composed largely of small, red fibers rich in mitochondria. The diaphragms of large mammals, such as the cow, with slow, phasic respiration, consisted chiefly of large, white fibers with a low mitochondrial content. Mammals of intermediate size, such as the cat, had various mixtures of the three types of fibers that were most appropriate for their particular patterns of respiratory activity.

Red, white, and intermediate fibers can also be identified by their ultrastructural characteristics.[19,20,21,61] In red fibers, such as the one labeled RF in Fig. 24-6, *A,* large mitochondria (MT_1) with closely packed cristae are abundant, particularly beneath the sarcolemma. These peripheral aggregations correspond to sites of mitochondrial enzymatic activity visible with the light microscope. Similar large mitochondria form longitudinal rows between myofibrils, providing them with a nearby source of energy. Elliptic profiles of mitochondria are aligned on both sides of the Z bands, which are much thicker in red fibers than in the other types. These profiles are sections through mitochondria that encircle the myofibrils at the I bands. In intermediate fibers, such as the one labeled IF in Fig. 24-6, *A,* the mitochondria in subsarcolemmal aggregations and between the myofibrils are smaller and the cristae are less densely packed. The Z bands are thinner than in red fibers. In white fibers (Fig. 24-6, *B*), mitochondria are scarce even in the subsarcolemmal region and are usually absent between myofibrils. Z bands are narrower in white fibers than in the other two types. The sarcoplasmic

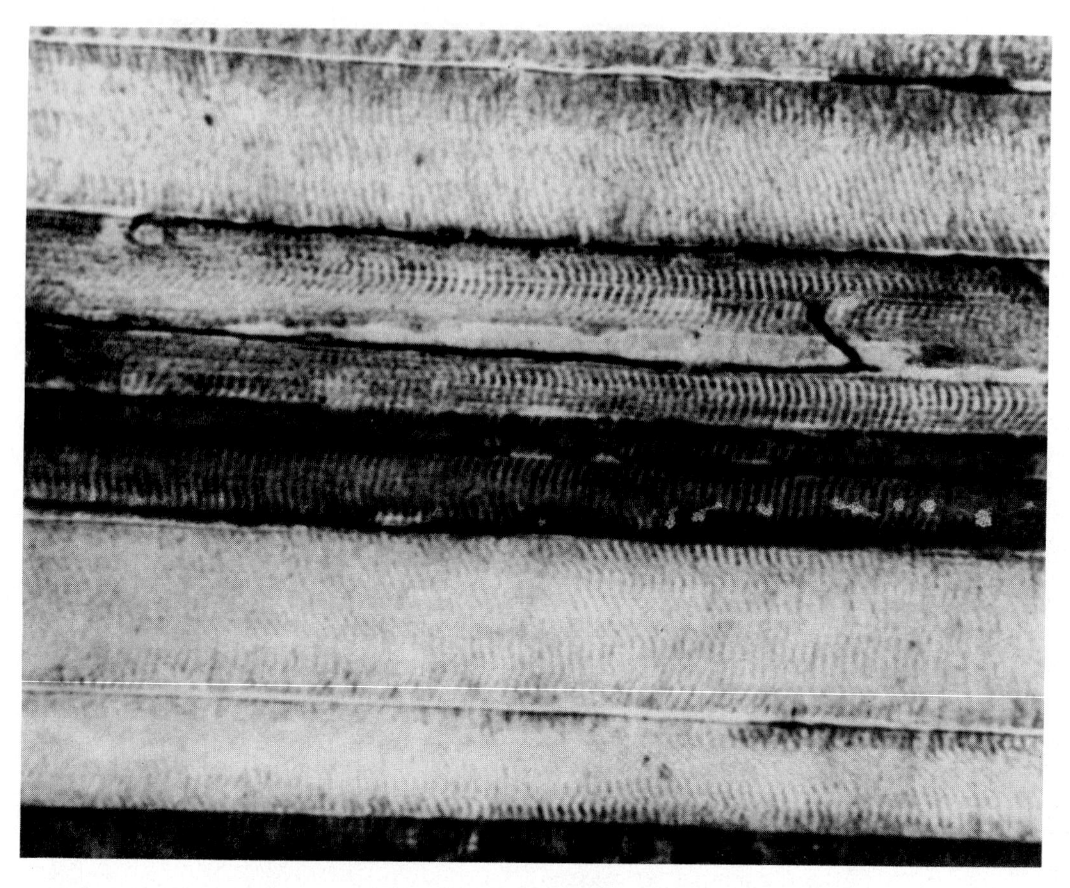

Fig. 24-5. Longitudinal section of extensor digitorum longus of cat. Each fiber exhibits uniform ATPase activity throughout its visible length. Note dark blood vessels running adjacent to muscle fibers. Large, pale A fibers have relatively poorer blood supply than smaller, darker fibers. (From Olson.[45])

reticulum also differs in the three types of fibers. Red fibers have an elaborate network of narrow tubules at the H band. A more compact arrangement of broad, parallel tubules occurs at the same site in white fibers. Red fibers appear to have a greater membranous surface area at the H band than do white fibers. Table 24-1 summarizes some of these features and includes data on the motor end-plates of each type of fiber. The table shows that the mitochondrial content, the concentration of myoglobin, and the richness of the capillary network are all inversely proportional to the diameters of their muscle fibers. This suggests that red fibers are equipped for a high level of metabolic activity, a conclusion reached by Coimbra[14] in his radioautographic study of glycogen synthesis in muscle.

The great majority of muscles in mammals consist of a mixture of the three types of fibers described previously, although the proportions of A, B, and C fibers vary widely in different muscles. All such heterogeneous muscles are known as pale muscles because their gross appearance is, in general, lighter in color than that of "red" muscles, such as the soleus. In cats and in man the soleus muscle consists of B fibers throughout and therefore looks homogeneous under the microscope, as shown in Figs. 24-3 and 24-4. These two illustrations show the numerous small mitochondria more densely distributed at the periphery of the cell near each capillary. The proximity of many mitochondria to the blood supply in these B fibers and the smaller size of C fibers serve to reduce the distances that oxygen and other substances, which are transported to the enzyme-containing mitochondria, must travel to maintain the aerobic metabolism characteristic of red fibers. The larger size of A fibers and the scarcity of capil-

Fig. 24-6. Ultrastructure of red *(RF)*, intermediate *(IF)*, and white muscle fibers. **A,** Subsarcolemmal mitochondria (Mt_1) are wider and more extensive in *RF* than in adjacent *IF*. Interfibrillary mitochondria are also more conspicuous in *RF* (not apparent in this picture). Mitochondria are larger and cristae are more closely packed in *RF* than *IF* and Z bands are thinner in *IF* than *RF*. **B,** Nuclear region *(N)* of white fiber, showing less extensive subsarcolemmal mitochondria than in *RF* and *IF*. Interfibrillary rows of mitochondria are rare. Pairs of filamentous mitochondria (Mt_3) at I band are predominant form. Z bands are narrowest in white fiber. (From Gauthier.[21])

Table 24-1. Cytologic characteristics of fiber types in mammalian skeletal muscle*

Fiber type	Fiber diameter (μm)†	Vascular supply	Mito-chondria	Myoglobin content	Z band width (Å)‡	Motor end-plate: Axonal vesicles	Motor end-plate: Junctional folds	
Red	27,45,47	High	Abundant	High	634	Moderate	Few	Short
Intermediate	34,52,47	Intermediate	Moderate	Intermediate	433	Moderate	Few	Long
White	44,59,69	Low	Few	Low	339	Abundant	Many	Long

*From Gauthier.[22]
†Average for rat diaphragm and semitendinosus red and white regions, respectively.
‡Average for rat diaphragm.

laries, mitochondria, and myoglobin correlate with their largely anaerobic metabolism. In these fibers, glycogen supplies the energy required for contraction.

In subsequent sections of this chapter we shall see how these histologic and histochemical features are reflected in the properties of the motor units.

CONTRACTILE PROPERTIES OF WHOLE MUSCLES AND THEIR GREAT VARIABILITY

Early investigators were unaware of the great variability in the individual fibers of a muscle and generally treated muscle as a rather homogeneous entity. Since the studies of Ranvier in the last century, it has been known that muscles differ widely in their speed of contraction, fatigability, and response to different rates of stimulation and that these differences are correlated to some extent with their gross paleness or redness. Fig. 24-7 illustrates the great range of contraction speeds found in three different muscles of the cat. Curves a, b, and c are called *twitch contractions*. They are the briefest contractions a muscle can undergo. They occur in response to a single, brief, electrical stimulation of the muscle nerve that activates all the fibers in the muscle almost simultaneously. The extraocular muscles that rotate the eyeball require only about 7.5 msec to reach their maximal tension in a twitch. The soleus muscle, a slow antigravity muscle in the hind limb, requires about 100 msec. The contraction speeds of most pale muscles fall somewhere between these extremes. Speed of contraction is clearly related to the functional role of the muscle. Movements of the eye require great speed. Movements of the limbs require much less speed, but greater endurance. Each muscle in the body is adapted for its special role and if that role changes, the properties of the muscle alter accordingly. Muscles are among the most flexible and adaptable organs in the body, and the mechanisms of this adaptability are currently the subject of much investigation. In emphasizing the varied roles of many muscles, it is often forgotten that even very specialized muscles such as those designed primarily for great speed must also be capable of some degree of maintained activity without fatigue. It is therefore not surprising that such muscles usually consist of a mixture of three types of fibers to provide a broader range of contractile properties.

In observations on whole muscles the details of motor unit activity are obscured, and a very inadequate view of the varied capacities of the muscles results. The characteristics of the individual motor units in a muscle differ greatly, endowing the muscle with a broad spectrum of contractile properties. The distribution of these properties confers on the muscle functional capacities that it would not otherwise possess. These properties must be described in some detail in order to explain how control over muscle is achieved.

PROPERTIES OF INDIVIDUAL MOTOR UNITS

The concept of the "motor unit" was introduced by Sherrington[56] to describe a single alpha motoneuron plus the group of muscle fibers innervated by it. The muscle fibers are not arranged in a compact group, but are scattered widely through a three-dimensional territory. A nerve impulse, set up in the motoneuron, is conducted down its axon and into all its branches, terminals, and end-plates, where it initiates impulses in all the muscle fibers of the unit. The motor unit is, thus, a functional entity.

In the 1960s a technique was developed that made it possible to activate a single motor unit in an otherwise quiescent muscle and to study its properties in isolation.[8,40] The axon of a

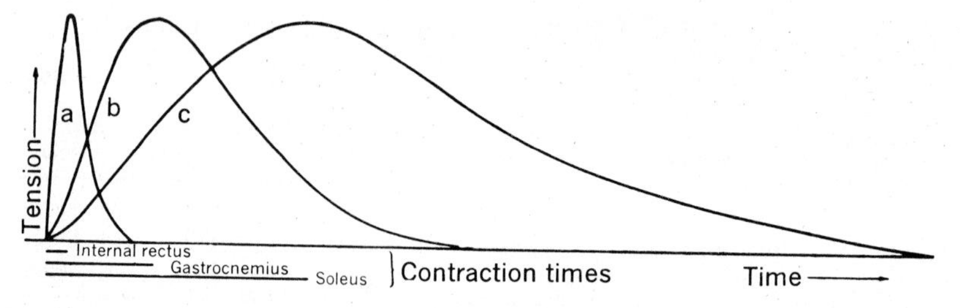

Fig. 24-7. Twitch contractions of three cat muscles, arranged to show the great differences in their speeds. Curve *a* represents internal rectus, curve *b* represents gastrocnemius, and curve *c* represents soleus. (From Cooper and Eccles.[15])

single motor unit was separated from others supplying the same muscle by repeatedly subdividing ventral root filaments until a thin strand was found that contained only one axon supplying the muscle under investigation. The conduction velocity of this axon was obtained by measuring the distance between stimulating electrodes on the muscle nerve and recording electrodes on the ventral root filament and the time required for the impulse to be conducted along this length of nerve. The twitch contraction of this motor unit was recorded by stimulating the ventral root filament with single shocks and recording the tension with a force transducer attached to the muscle's tendon. The "contraction time" from the onset of the twitch to the peak of tension was taken as a measure of the contraction speed of its muscle fibers. By stimulating the ventral root filament at various rates, the maximal tetanic tension was recorded and the susceptibility of the unit to neuromuscular and contractile fatigue was evaluated. More than 100 motor units from the medial gastrocnemius, a pale muscle, were studied[63] and compared with a similar group from the soleus, a red muscle.[40] These two muscles insert on the same tendon, and both extend the ankle, yet their properties differ.

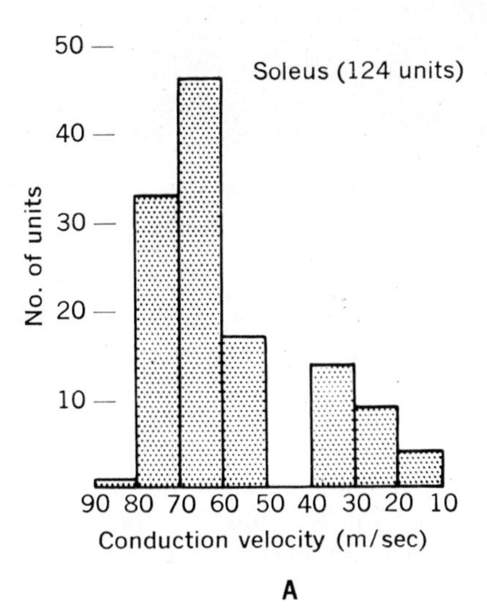

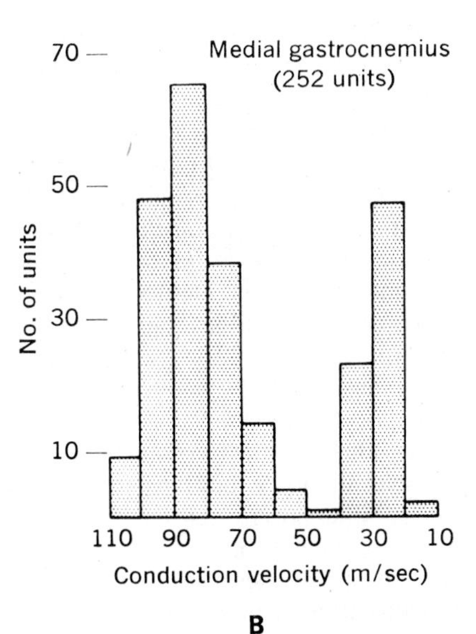

Fig. 24-8. A, Conduction velocities of 124 soleus axons, grouped in decades. **B,** Conduction velocities of 252 axons of m. gastrocnemius. (**A** from McPhedran et al.[40]; **B** from Wuerker et al.[63])

One of the differences between gastrocnemius and soleus motor units is the size of the nerve fibers innervating them. As Fig. 24-8 indicates, the conduction velocities of soleus axons (except for those supplying muscle spindles) fall chiefly between 50 and 80 m/sec, with a peak distribution between 60 and 70 m/sec. Gastrocnemius axons range from about 60 to 110 m/sec, with a peak between 80 and 90 m/sec. Conduction velocity is a direct function of axonal diameter,[33] hence the greater conduction velocities of gastrocnemius axons are due to their larger diameters. Since the diameter of an axon is directly related to the size of its cell body,[50] it may be concluded that the motoneurons supplying the medial gastrocnemius are larger than those innervating the soleus. The presence of large, rapidly conducting motor fibers in the nerve to the gastrocnemius is correlated with the abundance of large, pale, type A muscle fibers in this muscle that are absent in the soleus. As Fig. 24-8 indicates, muscle nerves also contain a group of smaller, more slowly conducting (10 to 40 m/sec) *gamma* fibers, which innervate the tiny muscle fibers inside muscle spindles.

When single motor units are stimulated at various rates and their tensions are recorded, as in Fig. 24-9, there are striking differences in the maximal tetanic tensions they develop. In the soleus muscle, the maximal tensions vary from 3 to 40 gm. In the gastrocnemius, they range from 0.5 gm to about 120 gm, a more than 200-fold difference! The distributions of maximal tetanic tensions for soleus and gastrocnemius muscles are plotted in Fig. 24-10. In both histograms the largest number of motor units occurs in the 0 to 10 gm range. In general there is a progressive decrease in the number of motor units in each higher decade. These histograms are typical of all muscles that have

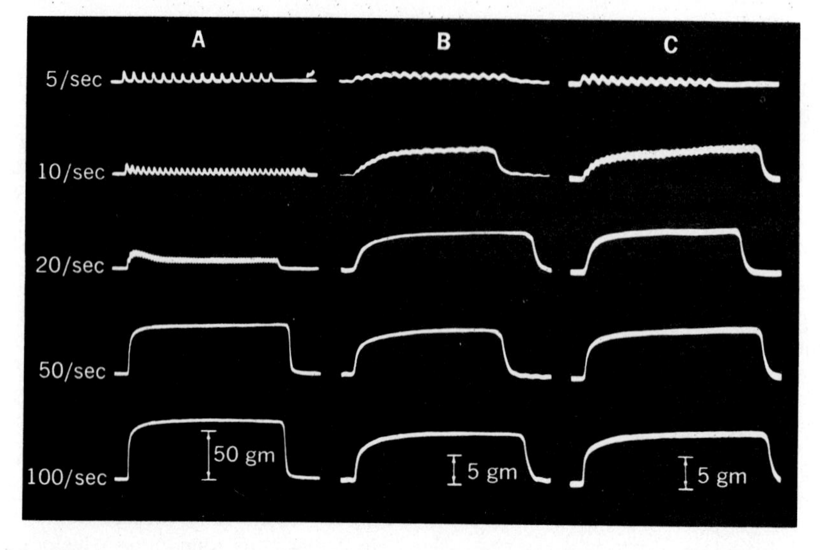

Fig. 24-9. Tetanic tensions developed by three units (**A, B,** and **C**) stimulated at 5, 10, 20, 50, and 100/sec. Units **A** and **B** from m. gastrocnemius; unit **C** from soleus. (From Wuerker et al.[63])

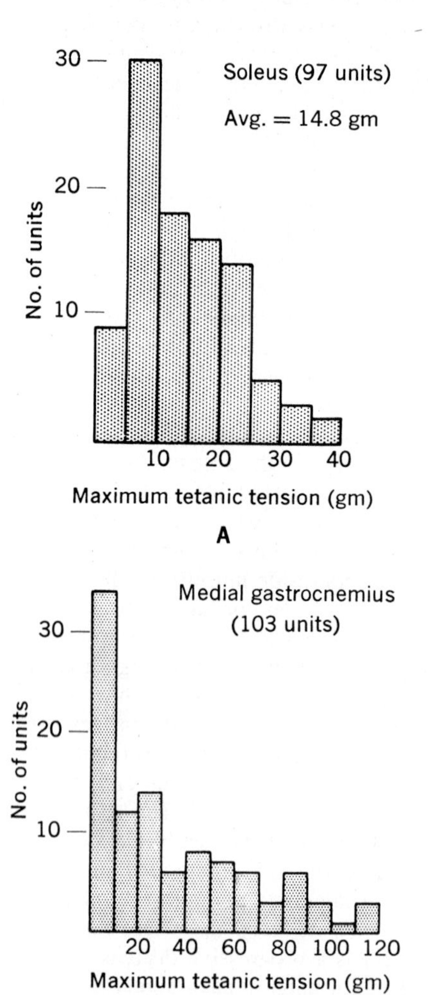

been studied in this way, including such specialized muscles as the tensor tympani[60] and those that move the eye. The functional significance of this distribution pattern will be explained in detail later. For the present, it is sufficient to point out that the abundance of small units makes possible precise, finely graded control of muscle tension in the lower ranges, whereas the presence of very large units provides for the addition of large increments of tension during maximal efforts.

By counting all the muscle fibers in the soleus and dividing by the total number of motor fibers in its nerve, Clarke[13] calculated that each motoneuron innervated 120 muscle fibers. Correcting his estimate for the small gamma fibers to muscle spindles, which constitute about 33% of the total, an average innervation ratio of 180:1 is obtained. This would apply to the average motor unit that develops 14.8 gm of tetanic tension in soleus. By extrapolation, the smallest motor unit (3.2 gm) and the largest unit (40.4 gm) in this sample would consist of 39 and 491 muscle fibers, respectively. Since the cat's soleus muscle consists entirely of B fibers, these figures are reasonably accurate. This approach cannot be applied to the gastrocnemius, however, because it has three types of fibers with different properties.

Fig. 24-10. Distribution of maximal tetanic tensions of motor units of soleus, **A,** and m. gastrocnemius, **B.** (**A** from McPhedran et al.[40]; **B** from Wuerker et al.[63])

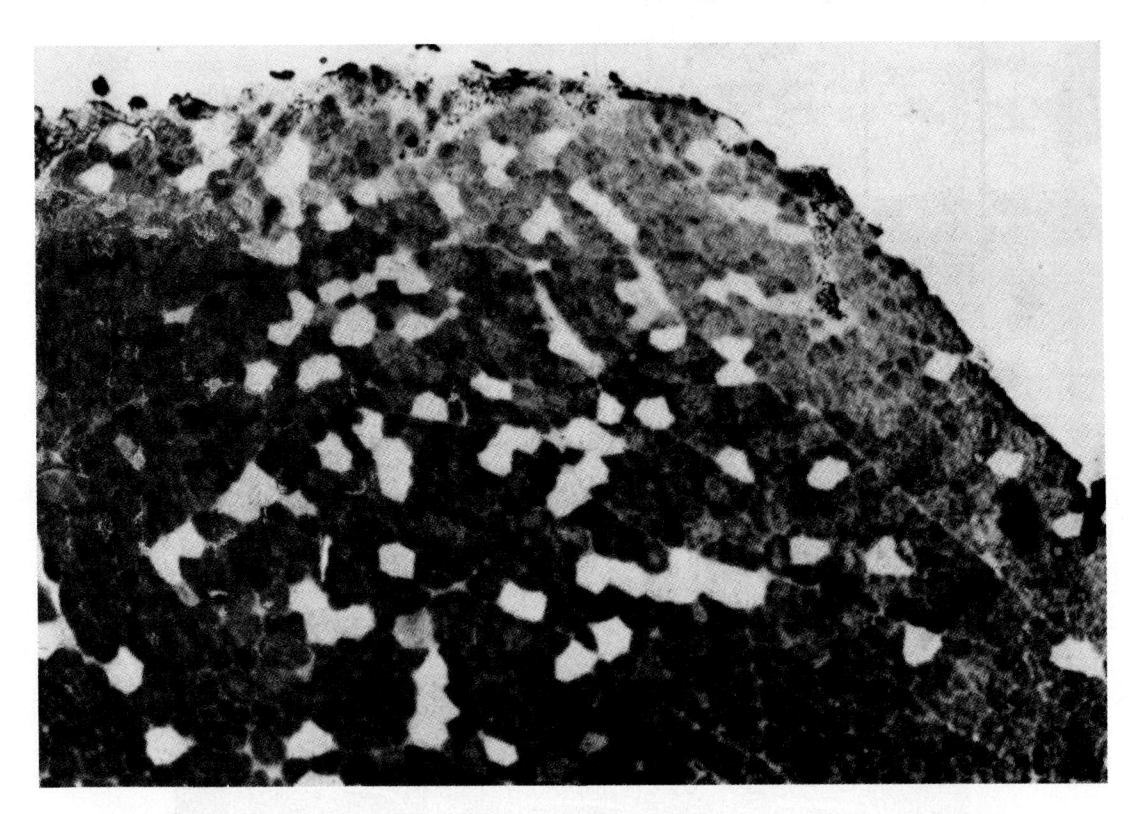

Fig. 24-11. Cross section showing distribution of fibers of a single motor unit; fibers are PAS negative. Section taken from distal part of rat's anterior tibial muscle, where unit is smaller than in middle of muscle. Magnification about ×40. (From Edström and Kugelberg.[17])

A more direct method of determining the number of muscle fibers in a unit is to deplete their supply of glycogen by prolonged stimulation of the axon at slow rates.[17] The depleted fibers can readily be identified by their pale or white appearance (Fig. 24-11) in histologic sections stained with the periodic acid–Schiff (PAS) technique. Using this method, Burke and Tsairis[11] estimated that there were about 750 fibers in a large motor unit of the cat's gastrocnemius. Thus one motoneuron can trophically maintain hundreds of muscle fibers!

Glycogen depletion by prolonged stimulation at a slow rate has also been used to identify the fibers of a motor unit, so that their histochemical reactions could be compared. Serial sections, treated with a variety of histochemical techniques, indicate that all the fibers of a motor unit are homogeneous. Histochemical homogeneity is not sufficient, however, to prove functional homogeneity. Further proof comes from a consideration of the contraction times of motor units, which are shown in Fig. 24-12. The soleus muscle of the cat, which appears completely homogeneous in its histochemical reactions, has units with contraction times ranging from 58 to 193 msec. If motor units had random samples of muscle fibers with various speeds of contraction, their contraction times would differ very little. The fact that individual units differ so greatly in speed of contraction indicates that each unit has a complement of fibers that contract at the same speed. As noted previously,[11] a large unit in the cat's medial gastrocnemius muscle would consist of about 750 fibers. A sample that large would include a significant number of slow fibers if the distribution were random. Large units, however, are not average in their speed of contraction. They are all fast. Neither the time courses of their twitch nor their fusion characteristics suggest that they contain a mixture of slow fibers. Moreover, very slow units do not include any fast fibers, for they develop their maximal tetanic tensions at low rates of stimulation. In short, there is good evidence that motor units are homogeneous functionally as well as histochemically.

The contraction speed of a unit determines how rapidly it must be excited to produce its

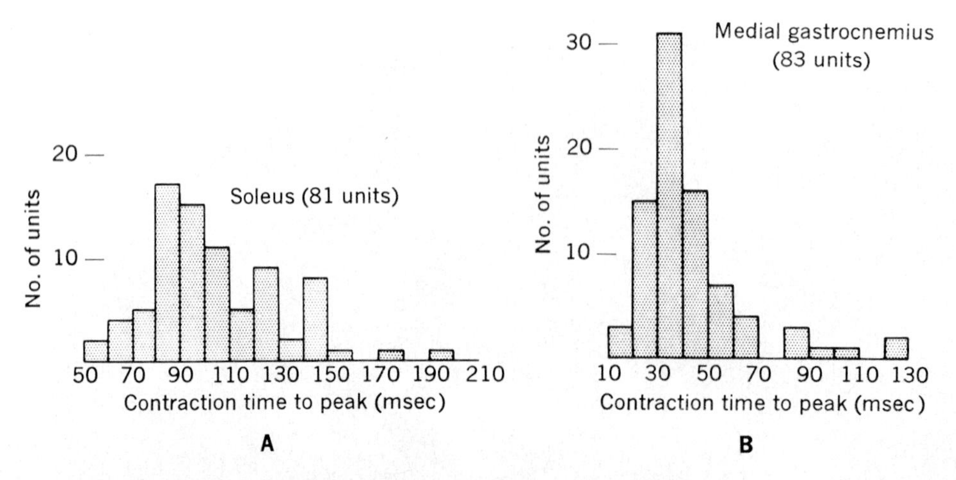

Fig. 24-12. Distribution of contraction times of soleus, **A**, and m. gastrocnemius, **B**, motor units. (**A** from McPhedran et al.[40]; **B** from Wuerker et al.[63])

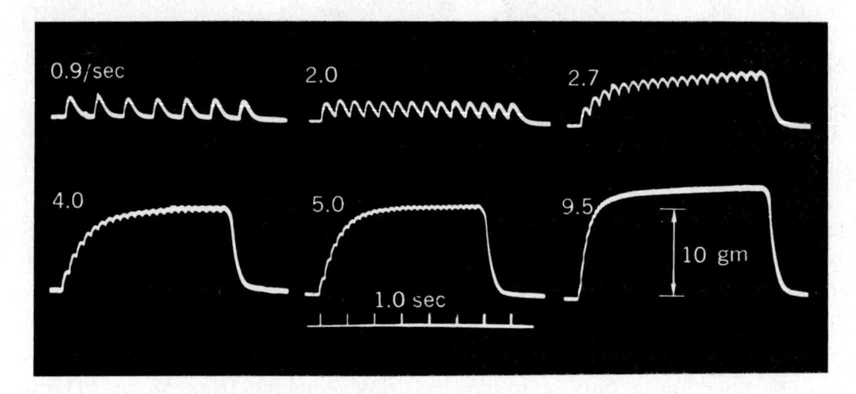

Fig. 24-13. Oscillographic records of tensions developed by a soleus motor unit with a very long contraction time (193 msec) when stimulated at rates of 0.9 to 9.5/sec. (From McPhedran et al.[40])

maximal output. In Fig. 24-9 the large motor unit, whose tension records appear in column A, developed relatively little tension when stimulated at 5 to 20 times/sec, whereas the units in columns B and C developed a considerable proportion of their maximal tension at these rates. The frequency at which consecutive twitches overlap one another and summation of their tension occurs depends on the total time required for each twitch contraction and relaxation. Fig. 24-13 illustrates the process of mechanical summation in a soleus motor unit with a very slow contraction.

Not only do the tetanic tensions developed by motor units differ widely, but also the capacities to maintain these tensions. Susceptibility to one type of fatigue is illustrated in Fig. 24-14. The effects of stimulation at 50 and 100/sec on the muscle action potentials (upper records in each pair) and the tensions of two motor units are reproduced from an experiment on the medial gastrocnemius. With stimulation at 50/sec, neither unit A nor unit B showed any reduction in tension during a 5 sec tetanus, although the action potentials of unit B diminished slightly. At 100/sec, however, the action potentials of unit B decreased to a very low level during the tetanus, indicating progressive failure of transmission at the neuromuscular junctions. This failure resulted in a parallel decline in tension. Neuromuscular fatigue frequently occurred in large motor units at lower frequencies and shorter durations of stimulation than in small units. Large units, in fact, sometimes showed early failure of junctional transmission at frequencies below those necessary to develop maximal tension. If neuromuscular fatigue occurs to the same extent under normal conditions, it must seriously limit intensive use of large motor units.

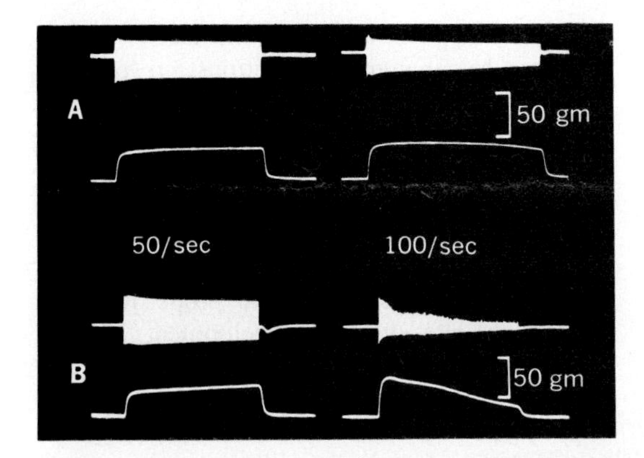

Fig. 24-14. Effect of repetitive stimulation at 50 and 100/sec on the electromyogram and tension of two motor units of m. gastrocnemius. (From Wuerker et al.[63])

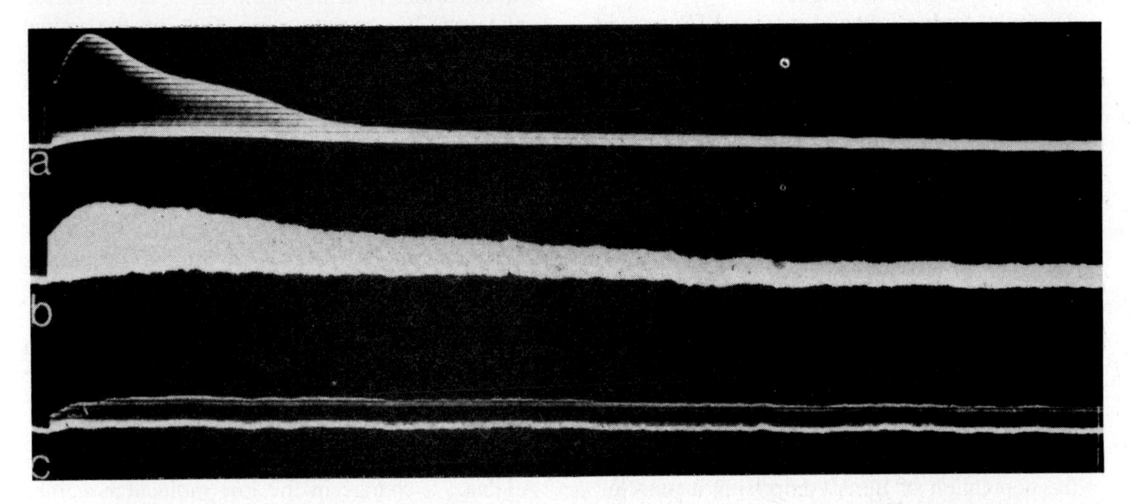

Fig. 24-15. Records showing effect of repetitive contractions at 10/sec for 10 min on three types of motor units. *A, B,* and *C* units in *a, b,* and *c,* respectively, show 90% decline, 50% decline, and 100% increase in twitch tensions. Initial twitch tensions were 1.25, 1.0, and 0.5 gm, respectively. Note that the type C unit showed no fatigue, but increased its twitch contraction. (From Edström and Kugelberg.[17])

Kugelberg and Edström[36] showed that fatigue of the contractile mechanisms could be produced experimentally and distinguished from junctional failure by stimulation at slow (2 to 10/sec) rates for more prolonged periods. The ability to identify the fibers belonging to individual motor units also made it possible for them to recognize three types of units corresponding to fiber types A, B, and C of prior description. They found a striking relationship between contractile fatigability and motor unit type, as illustrated in Fig. 24-15. When stimulated at 10/sec, the twitch tensions of type A units in the rat's anterior tibial muscle declined to 10% or less of their original values. Most of the decline occurred within 3 min (i.e., during the first 1,800 contractions). The responses of type B units declined more slowly and seldom failed completely. The twitches of type C units increased by as much as 100% initially and thereafter showed no fatigue. The mean twitch tensions of A, B, and C units were 1.8 gm, 0.75 gm, and 0.4 gm, respectively. These large differences in fatigability have important functional implications. The lower rates of natural discharges are within the 5 to 10/sec range and are found in tonic contractions for maintenance of posture. Type A units are not capable of prolonged activity at these low frequencies, whereas type C units are well adapted for it.

Data drawn from about 400 motor units in several different muscles of the cat's hindlimb indicate that the largest units contract most rapidly. This correlates well with their high levels of myosin ATPase. These units fatigue rapidly because they have few mitochondria and low levels of the oxidative enzymes that seem to determine endurance. Small units fall into two groups, a fast and a slow contracting type.[26,63] Both types are less susceptible to fatigue than are the large units. Slowly contracting small units have low levels of myosin ATPase, but abundant mitochondria and oxidative enzymes, which correlate with their almost complete resistance to fatigue. The rapidly contracting type have uniformly higher levels of myosin ATPase, but somewhat lower levels of oxidative enzymes, correlating with their partial susceptibility to fatigue.

There is evidence that myosin ATPase and mitochondrial enzymes are independently variable,[35] permitting a wide range of speed and endurance. Extraocular muscles, which have the fastest contractions of any muscles in the body and the most rapidly firing motoneurons (600 to 800/sec), have three types of fibers.[43] All three types are small in diameter and high in myosin ATPase and mitochondrial enzyme content.

It is apparent from this account that the chief characteristic of motor units is their diversity. Although the muscle fibers in a single motor unit are uniform in their histochemical reactions and contractile properties, there is a surprisingly wide range of properties represented in the population of units comprising a muscle. It is natural to ask how the central nervous system (CNS) makes use of this diversity. Does it select from among the various motor units just those it requires for a certain task? Does it mobilize just the large, powerful, rapidly contracting units to supply the speed and power needed by a high jumper to clear a 7-foot bar? May it activate only small, slower units to provide the delicacy and precision a watchmaker needs? The answers are not obvious. They require information about the motoneurons that control motor units. For the present, it is enough to emphasize that the sizes, speeds, tensions, and fatigabilities of motor units are distributed systematically in every skeletal muscle, but in a uniquely different pattern in each muscle.

ADAPTIVE PROPERTIES OF SKELETAL MUSCLE*

The picture presented so far has perhaps overemphasized the static organization of muscle and left the reader without a proper appreciation of the fact that muscle is highly adaptive to changing conditions, responding throughout life to the demands placed on it. The rapid increase in strength and endurance caused by systematic training are examples of this adaptive capacity. Many of the effects of training occur in the muscle tissue itself.

In most species, slow and fast muscles do not acquire their characteristic properties until a relatively late stage of development, usually postnatally. This process of differentiation appears to depend on the establishment of the adult pattern of motoneuron activity. If the motoneurons of a newborn kitten are rendered quiescent by operative isolation of the spinal cord shortly after birth, muscles that would normally have become slow develop as fast muscles.[9] Even in the adult animal, muscles continue to adapt to the pattern of motor impulses they are receiving. A fast muscle, subjected to a prolonged pattern of low-frequency activation by means of an implantable stimulator, undergoes changes in all the properties that normally distinguish it from a slow muscle. For example, there is an increase in capillary density, a switch from a glycolytic to an oxidative enzyme profile, a decrease in the rate and total capacity for uptake of calcium ions by sarcoplasmic reticulum, a decrease in the specific activity and alkali stability of myosin ATPase, a change in the low molecular weight subunits (light chains) associated with the globular head of the myosin molecule, and alterations in the myosin heavy chains as well. These alterations are presumably responsible for the marked reduction in contractile speed, which was the first effect of prolonged stimulation to be described.[55]

That a tissue as highly differentiated as skeletal muscle retains into adult life the capacity for such profound changes is a surprising discovery, which was first revealed by the experiments of Buller et al.,[10] in which the motor nerves to fast and slow muscles were cut and cross-anastomosed. After reinnervation the fast muscle became slower and the slow muscle faster. Subsequent studies by these and other investigators[49,53,65] revealed that many histochemical and biochemical changes occurred in the reinnervated muscles that were consistent with their slowing or speeding up. Fast, pale muscles took on a redder, more vascular appearance, they developed more mitochondria and myoglobin, and their metabolism became more aerobic as they slowed up, whereas an opposite series of changes occurred in slow,

*Based on Salmons and Sreter.[54]

red muscles that became faster. After reinnervation of a homogeneous muscle, such as the cat's soleus muscle, by the nerve to a mixed muscle, three types of muscle fibers appeared, arranged for the most part in clumps or islands of similarly staining cells. Although Buller et al. considered the possibility that the changes in contractile speed might be due to the altered patterns of impulse activity reaching the muscles, they proposed instead that chemical trophic factors carried by the motor nerves were responsible for the transformation of the muscles.

These alternatives were put in a new perspective by the recent work of Salmons and Sreter,[54] in which 20 weeks of low-frequency (i.e., 10/sec) stimulation of the nerve to the extensor digitorum longus (EDL), a fast muscle, were carried out, so that this muscle could be compared with the unstimulated EDL and soleus of the contralateral limb. In response to the prolonged stimulation, the EDL and several other similarly stimulated fast muscles became slower and even more resistant to fatigue than the normal soleus! The biochemical results were equally striking. The light chains of myosin in the EDL were entirely replaced by others characteristic of slow muscles, and there was some evidence that the heavy chains had also been replaced.

However, it could still be argued that changes in impulse activity and cross union achieve similar end results by different mechanisms. To analyze the situation further, the effects of long-term stimulation were superimposed on those of cross innervation.[54] Motor nerves to the tibialis anterior and EDL, both fast muscles, were cut, freed, and joined to the distal end of the soleus nerve. Eight weeks after cross union, stimulators were implanted in each of the animals so that the cross-innervated soleus muscle received a continuous 10/sec train of motor impulses via its new innervation. After a further 8 weeks the animals were examined terminally.

According to the activity pattern hypothesis, the usual consequences of cross innervation from a "fast" nerve should be absent in the muscles receiving a steady stream of impulses, for despite the new innervation, the impulse activity reaching these muscles would be essentially unchanged. This was borne out. In unstimulated cross-innervated muscles the usual increase in speed of slow muscles occurred. In contrast, the *cross-innervated* and *stimulated* soleus muscle contracted and relaxed even more slowly than its counterpart on the control side! Gel electrophoretograms showed myosin light chains characteristic of fast muscle in the cross-reinnervated soleus muscles. But similar cross-reinnervated soleus muscles that had also received stimulation had light chain patterns indistinguishable from those of control soleus muscles.

Since it is highly unlikely that different mechanisms could produce results so much alike in all details, it was concluded that the effects of impulse activity and those of cross innervation are mediated by the same basic mechanisms. The available evidence leads to two conclusions: (1) in the absence of a change in the pattern of motor activity reaching a muscle, cross reinnervation has no significant effect or (2) an alteration in the flow of impulses to a muscle, unaccompanied by a change in the nerve supplying it, can produce effects equaling or surpassing those brought about by nerve cross union.

These recent observations make it unlikely that the physiologic and biochemical differences between fast and slow muscles are due to chemotrophic differences in their motor nerves, but still leave other possible roles for substances carried to muscles by axoplasmic flow. It now seems probable that gene expression in skeletal muscle is subject to the regulation of one or more substances whose intracellular concentrations alter with the level of contractile or metabolic activity.

Experiments carried out by Lømo et al.[39] on denervated, curarized muscles, in which neural factors could not contribute to the results, provide additional evidence regarding the importance of the activity pattern in the muscle itself. In one group of rats the soleus muscle was stimulated through large, platinum plates for 0.5 sec periods at a rate of 100/sec, once every 25 sec, thus producing a mean contraction frequency of 2/sec. Another group received 10/sec stimuli for periods of 10 sec, repeated every 50 sec. The mean frequency of contraction for this group was also 2/sec. The brief periods of 100/sec stimulation were intended to resemble the phasic activity of fast muscles, whereas the longer periods of 10/sec stimulation were similar to the tonic activity of slow muscles. After 36 days of stimulation in vivo the isometric contractions of these muscles were compared with curarized normal muscles and muscles denervated for 39 days. The mean isometric twitch contraction time (T_c) of muscles stimulated at a rate of 100/sec was 16 msec, as compared with 35 msec in normal and 41 msec in denervated, unstimulated muscles. In contrast, denervated muscles stimulated at 10/sec responded almost like normal muscles, with a mean T_c of 33 msec. The twitch-tetanus ratio was 0.12 for the group stimulated at 100/sec, as compared with 0.33 for the group contracting at 10/sec. The rapid loss of mass usually occurring in denervated muscles was largely prevented regardless of the pattern of stimulation. Muscles stimulated at 100/sec were much more resistant to fatigue during prolonged intermittent stimulation than normal fast muscle. The ability of a muscle to resist fatigue is apparently related more to its overall activity, which was similar in both groups, than to impulse frequency within bursts of activity. Barnard et al.[6] have also shown that endurance and contraction speed may be independently controlled in experiments showing that exercise may increase endurance without affecting contraction speed.

It may be concluded that the contractile properties of soleus muscles respond to the pattern of muscle activity in the absence of any neural trophic factor. Intermittent, brief bursts of activity at high frequency induce a number of properties characteristic of fast muscles. Under the influence of more prolonged low frequency activity the muscle remains slow.

In addition to the dramatic effects of motoneurons on muscles, there is also evidence that motoneurons respond adaptively to influences emanating from muscles. The best known example is the effect of denervated muscle fibers on nearby nerve fibers, which put out new branches and enlarge their field of innervation. Recent experiments indicate that denervated fibers in a muscle may also exert an influence on the cell bodies of the intact motoneurons supplying that muscle. Huizar et al.[32] partially denervated the cat's soleus muscle by cutting some of the fibers in the soleus nerve. Three weeks later, intracellular recordings from soleus motoneurons whose axons were uncut showed significantly shorter after-hyperpolarization (AHP) potentials than those in normal controls. Other electrophysiologic properties were apparently unaffected. The more extensive the denervation, the greater was its effect on AHP. According to Kuno et al.[38] soleus motoneurons show no significant changes in conduction velocity or AHP 29 to 46 days after section of the lumbosacral dorsal roots. Alterations in the properties of intact soleus motoneurons would, therefore, not result from the sensory deprivation associated with partial denervation of the muscle.

An interesting aspect of this study is that the decrease in AHP in soleus motoneurons was associated with a shortening of contraction times in the intact motor units 1 to 2 weeks later. It is generally agreed that the size and duration of AHP are factors limiting the frequency of motoneuron discharge. Since AHP drives membrane potentials farther below their firing thresholds, it decreases the excitability of motoneurons. A decrease in AHP, therefore, allows the motoneurons to fire more rapidly than before denervation. The effects of the increased firing rate are consistent with the previous conclusion that the activity pattern of motoneurons determines the contraction times of their muscle fibers. In the absence of the slow rates of discharge in soleus motoneurons with normal AHPs the contraction times of their muscle fibers become shorter.

Thus motoneurons and muscles are strongly linked to each other by adaptive mechanisms. These mechanisms come into play in a variety of neuromuscular diseases, and the differences between primary disease processes and the adaptive responses to them are beginning to be recognized and used diagnostically by clinical neurologists.

PROPERTIES OF MOTONEURONS THAT DETERMINE ACTIVITY PATTERNS AND INFLUENCE MUSCLE FIBERS

In the preceding section, it was concluded that the pattern of contractile activity of a muscle determines many of its properties. Normally, this pattern is impressed on the muscle by impulses from its motoneurons. The wide range of contractile speed, strength, and endurance in the different units of a muscle presumably results from an equally wide range of firing behavior in its motoneurons. Although the factors responsible for the discharge patterns of individual motoneurons are not fully understood in biophysical terms, some have been identified and studied.

One of the most important determinants of firing pattern is motoneuron size. Experiments described in detail in Chapter 25 show that the excitability of a motoneuron is inversely proportional to its size. The smaller a cell is, the less surface area there is available for the flow of synaptic currents across its membrane and the greater is its total input resistance to such flow. By Ohm's law, depolarizing synaptic currents flowing across greater resistances produce larger excitatory postsynaptic potentials (EPSPs) in motoneurons. If other factors are equal, a smaller excitatory input is, therefore, required to discharge a small cell than a large one. This may explain why small motoneurons are more frequently fired during normal activity, although distribution of excitatory and inhibitory input may play an important role, also. Small motoneurons have thin axons that can subdivide into a limited number of terminals, forming small motor units. In homogeneous muscles, such as the soleus, small motoneurons fire slowly and their motor units contract slowly. It is apparent that the combination of frequent discharge and slow firing rates causes a pattern of contractile activity that results in slowly contracting muscle fibers. Larger motoneurons have larger, more rapidly conducting axons that subdivide into numerous terminals, forming larger motor units. These cells are less excitable than small motoneurons and require larger excitatory inputs to be fired. When they are discharged, they tend to fire more rapidly. Their activity pattern is typically one of less frequent bursts of activity at higher rates than small cells, hence they engender more rapid contraction times in their muscle fibers. The larger gastrocnemius motoneurons fire at still faster rates, but only for a short time. Their activity patterns consist of brief, infrequent bursts of activity at high rates. Their muscle fibers contract very rapidly, producing large tensions, but, due to the relatively small amount of total contractile activity they undergo, these units do not develop much endurance.

The influence of motoneuron size is apparent in a variety of studies relating axonal conduction velocities (CVs) to the properties of their motor units. In the first deep lumbrical muscle of the cat, which consists of only 4 to 10 motor units, Laporte and his colleagues showed that the maximal tetanic tensions and contraction times of the units were related directly to the CVs of their axons.[8] Similar findings were obtained in a study of 43 motor units of a somewhat larger muscle, the first superficial lumbrical.[1] Fig. 24-16 illustrates how strongly the contractile properties of motor units, as well as the amplitudes of their muscle action potentials, are correlated with the CVs of their axons.

The first large limb muscle that was investigated thoroughly was the cat soleus muscle,[40] a homogeneous muscle consisting entirely of intermediate (B) fibers. Fig. 24-17 illustrates that the maximum tetanic tensions of its motor units vary directly with the CVs of their axons, as in the small muscles of the foot. Although the slopes of the regression lines differed from one experiment to another, correlations similar to those in Fig. 24-17 were found in all 10 of the experiments, comprising a total of 97 units. A recent reexamination of the 1965 data also showed a correlation between CV and twitch contraction time in every experiment.

A study from another laboratory, in which data on 30 soleus units in six preparations were pooled, did not show these correlations.[42] Prompted by this apparent discrepancy, McPhedran et al.[40] replotted the data on all 97 soleus units on one graph (not illustrated). No correlation between CV and tetanic tension was then apparent. Bagust[2] has recently demonstrated that pooling tends to obscure relations between contractile variables. Correlations similar to those originally reported were found whenever large numbers of soleus units were sampled from single muscles. "In no case was the correlation coefficient (r) obtained from [pooled] data significant at the 10% level, whereas the values obtained from individual experiments . . . never fell below this value."[2]

Further evidence concerning the significance of motoneuron size and axonal CV is compiled in Tables 24-2 and 24-3. In Table 24-2 the mean tensions of cat soleus units are grouped according to their CVs, and the contraction times of the same group of units are listed. Corresponding data for the medial gastrocnemius appear in Table 24-3. Both tables indicate that *on the average* slowly conducting axons (i.e., small fibers) innervate motor units that produce small tensions and contract slowly. Large fibers supply units that produce larger tensions and contract more rapidly. In the case of the soleus the mean tensions increase quite uniformly (considering that these are pooled data) from the slowest to the fastest group of fibers. The correlations for the gastrocnemius are equally impressive up to about 100 m/sec, but the last two groups with six and two units each are slightly below the peak of mean tension. All these data indicate that the contractile properties of a motor unit are correlated with the CV of its axon, that is, the size of its motoneuron.

Closer inspection of Tables 24-2 and 24-3 indicates, however, that the contractile properties of a motor unit are not determined exclusively by the size of its motoneuron. Comparison of the contraction times of soleus and gastrocnemius units with CVs in the three groups between 70.0 and 84.9 m/sec reveals that in each group the

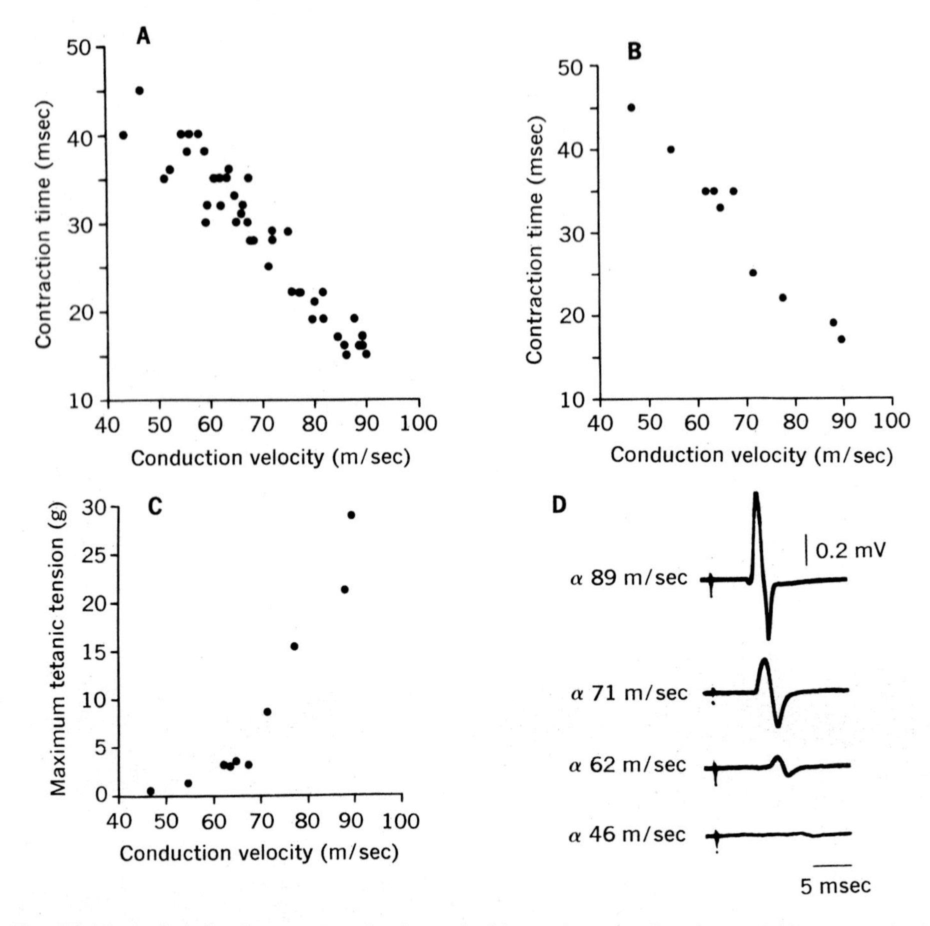

Fig. 24-16. A, Relation between conduction velocities and contraction times of 43 motor units in first superficial lumbrical muscles of six cats. **B,** Same as **A,** but for 10 motor units in one cat, showing more linear relationship with less scatter. **C,** Relation between maximal tetanic tensions and conduction velocities in same experiment. **D,** Muscle action potentials of four single motor units recorded from surface of same muscle. (From Appelberg and Emonet-Dénand.[1])

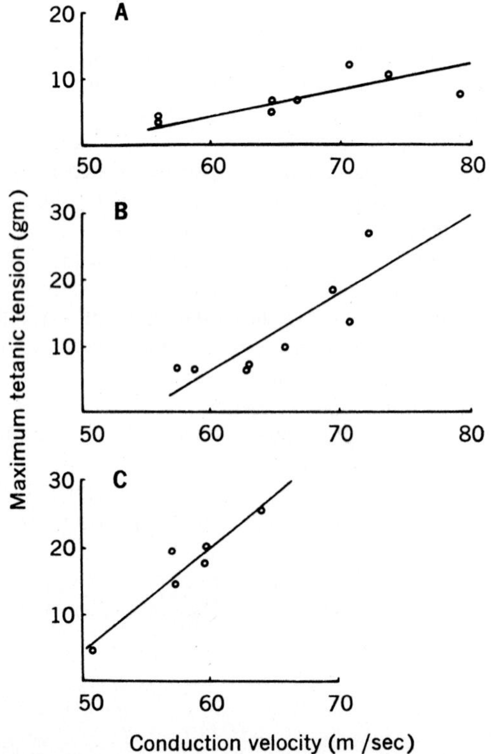

Fig. 24-17. Plots of relation between maximum tetanic tension of individual motor units and conduction velocity of their axons in three different experiments. (From McPhedran et al.[40])

Table 24-2. Mean tensions and contraction times of soleus motor units grouped according to their conduction velocities*

	Conduction velocity (m/sec)						
	50.0-54.9	55.0-59.9	60.0-64.9	65.0-69.9	70.0-74.9	75.0-79.9	80.0-84.9
No. of units	1	14	12	24	21	9	1
Mean tensions (gm)	4.7	9.5	14.9	16.2	15.0	17.0	20.6
Contraction times (msec)	177	109	115	104	101	85	90

*From Henneman and Olson.[27]

Table 24-3. Mean tensions and contraction times of m. gastrocnemius motor units grouped according to their conduction velocities*

	Conduction velocity (m/sec)										
	55.0-59.9	60.0-64.9	65.0-69.9	70.0-74.9	75.0-79.9	80.0-84.9	85.0-89.9	90.0-94.9	95.0-99.9	100.0-104.9	105.0-109.9
No. of units	1	1	4	8	15	17	17	21	15	6	2
Mean tensions (gm)	0.5	3.7	5.4	8.5	17.4	35.2	47.8	45.3	49.4	43.5	44.4
Contraction times (msec)†			129 (1)	72.1 (4)	54.8 (7)	49.0 (14)	39.1 (15)	40.2 (20)	40.3 (13)	34.2 (5)	25.5 (2)

*From Henneman and Olson.[27]
†Numbers in parentheses indicate number of units on which contraction times were obtained.

gastrocnemius units contracted more rapidly. As previously noted, there is a significant linear relationship between the axonal CVs of soleus units and both their contraction times and their AHPs.[32] Three weeks after partial denervation of the soleus nerve there was a decrease in the duration of AHPs without a change in the mean conduction velocities of the intact fibers. Conduction velocity was no longer correlated with contraction time, but AHP was still significantly correlated. Although the implications of these recent experiments are not entirely clear, they suggest that the contraction speed of muscle fibers is determined, at least in part, by some factor related to the duration of AHP. The difference in the contraction times of the three groups of soleus and gastrocnemius motor units referred to above may, therefore, be due to differences in the AHPs of their motoneurons.

In a previous section of this chapter, it was noted that any increase in contractile activity caused by electrical stimulation of a muscle or its nerve resulted in decreased susceptibility to fatigue. The activity pattern of normal muscles apparently determines their endurance in a similar way. Soleus motor units are relatively nonfatigable because their motoneurons are relatively small and excitable and, consequently, have a high level of daily activity. Gastrocnemius motor units that have small motoneurons are also resistant to fatigue.[63] In a systematic study of fatigability in this muscle, Reinking et al.[52] recently found that units with contraction times longer than 45 msec were essentially nonfatigable (24 out of 26 units) and small (25 out of 26 units). Gastrocnemius units that contracted more rapidly were relatively resistant to fatigue only if they had small motoneurons. Those with large motoneurons had little endurance.

Recent observations indicate that another factor of major importance may influence the properties of motor units. In the past, it has been assumed that a pool is a homogeneous population of cells differing only in size. Now there is evidence[26] that there is more than one type of motoneuron in the pool supplying a heterogeneous muscle, perhaps one type for each of the three kinds of muscle fibers. To date the only characteristic distinguishing different types of motoneurons is their rate of firing. Plantaris motoneurons of the same size, for example, may fire at rates differing by 2 or 3:1. There is also evidence that the firing rates of different types of motoneurons may be affected differentially by inhibitory inputs. The firing pattern of each type of motoneuron probably differs sufficiently to account for considerable variation in the contractile properties of muscle fibers, but it is possible that other, so far undefined characteristics of each motoneuron type may also influence its muscle fibers. It is likely that the relation of axonal CV to contractile properties may differ for each type of motoneuron; in fact, this possibility was suggested in earlier studies.[46,63] If so, this would help to explain the scatter in experimental data and clarify its significance. One group[24,58] impressed with the scatter in experimental data, has challenged the evidence relating CV and contractile properties. Another group, however, has found that despite the scatter, maximum tetanic tensions and contraction times are correlated with CV at the 0.01% level in pooled data and even more strongly in single experiments on adult cats.[5] This group has also confirmed the validity of the relationship during postnatal development in kittens[4] and in self-reinnervated fast and slow twitch muscles.[3] If there are three separate sets of relations between CV and contractile properties and they are intermingled in the data on heterogeneous muscles, it is not surprising that unanimity has not yet been reached on this question.

NEUROMUSCULAR JUNCTIONS ON RED, WHITE, AND INTERMEDIATE FIBERS

The fine structure of neuromuscular junctions and the mechanism of transmission were described in Chapter 5. Variations on this general theme have recently been demonstrated by Padykula and Gauthier[48] at the junctions on red, white, and intermediate muscle fibers. Their observations add to the body of evidence indicating that there are three basic types of fibers and motor units in most mammalian muscles. The new findings are correlated with differences in susceptibility to neuromuscular fatigue, which is the chief feature of myasthenia gravis, certain drug intoxications, and a secondary effect of other illnesses.

Fig. 24-18 illustrates the ultrastructure of junctions on red fibers, *A,* and white fibers, *B.* The most obvious differences are the total sizes of the junctions and the amounts of surface area they present for discharge and reception of transmitter. The axonal terminals on red fibers are small and elliptic. The areas of contact they make on a single muscle fiber are relatively discrete and separate. The junctional folds, which the sarcoplasmic membrane of the muscle forms, are relatively shallow, sparse, and irregular in their arrangement. Axoplasmic vesicles are mod-

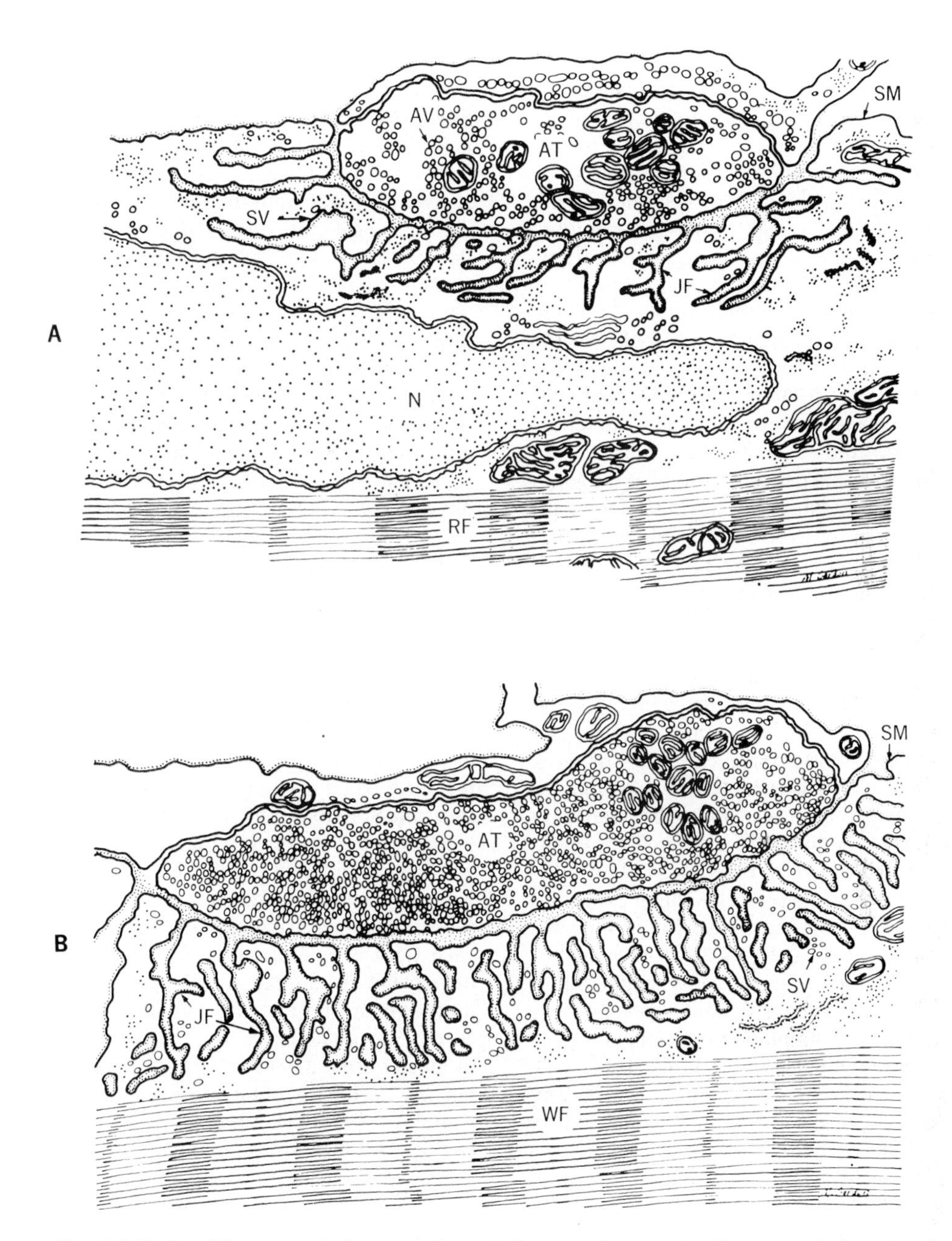

Fig. 24-18. Semidiagrammatic interpretations of electron micrographs of motor end-plates on, **A,** a red fiber *(RF)* and, **B,** a white fiber *(WF)*. Axonal terminals *(AT)* of motor fibers contain axoplasmic vesicles *(AV)* and mitochondria. Terminals are separated from surface of muscle cell by a space that contains basal lamina. In this region, sarcoplasmic membrane *(SM)* is thrown into junctional folds *(JF)*, and there are numerous mitochondria. Basal lamina occupies primary synaptic cleft separating relatively smooth plasma membrane of axon terminal from sarcolemma. Single basal lamina follows contours of sarcolemma into secondary synaptic clefts, where it can be seen lining each junctional fold. *AV* are moderate in number, and sarcoplasmic vesicles *(SV)* are sparse in RF. Both types of vesicles are more numerous in WF. (Based on Padykula and Gauthier.[48])

erate in number, and sarcoplasmic vesicles are sparse.

Axonal terminals on white fibers are longer and flatter and have a much greater area for release of acetylcholine. The postjunctional surface formed by the branching, closely spaced junctional folds provides a vast receptive area for the transmitter. At adjacent contacts on the same muscle fiber the junctional folds may merge with one another, so that synaptic contiguity is more continuous and widespread. Axoplasmic and sarcoplasmic vesicles are numerous.

Intermediate fibers receive relatively large axonal terminals. Their junctional folds are the most widely spaced and deepest of the three types. They are relatively straight and unbranched. Axoplasmic vesicles are less numerous than in white fibers. Sarcoplasmic vesicles are conspicuous near the deeper portions of the junctional folds. According to Padykula and Gauthier,[48] junctions on intermediate fibers are not just a gradation between the types on red and white fibers, but are a distinct third type (Table 24-1).

On the presynaptic side of the myoneural junction the dimensions and form of the axonal terminals are correlated with the caliber of the parent axon.[44] On the postsynaptic side the distinctive features of the three types of junctions are correlated with the diameters of their muscle fibers. The larger a muscle fiber, the smaller is its input resistance[34] and the greater is the input required to depolarize it to its firing threshold. Large white fibers, therefore, require extensive synaptic surfaces. The greater density of synaptic vesicles in their axonal terminals is presumably associated with the release of more acetylcholine over a wider area. The resulting end-plate potential is sufficient in amplitude and extent to set up a propagated impulse in the large, low-resistance muscle fiber. Smaller junctions with fewer vesicles are adequate to ensure transmission to the more easily excited red fibers.

Junctions on white fibers are able to transmit impulses at a high enough rate to produce a fused tetanus in the motor unit initially, but those endings fatigue rapidly, as illustrated in Fig. 24-18, *B*, and transmission may cease completely in a few seconds. The significance of these variations in fatigability is not known. The contractile mechanism in white fibers fatigues rapidly and recovers slowly. Perhaps junctional failure, which precedes contractile fatigue but recovers much sooner, protects the fiber from the more serious consequences of the more prolonged fatigue.

CLASSIFICATION OF MOTOR UNITS

In view of the homogeneity of individual motor units, it is reasonable to assume that there are three kinds of units in mixed muscles, corresponding to the A, B, and C types of muscle fibers that can be distinguished histochemically and ultrastructurally. When the axonal conduction velocities, tetanic tensions, and contraction speeds of motor units in several hindlimb muscles of the cat were analyzed and compared with the histochemistry of the same muscles,[27,46] many of the units could readily be classified as type A, B, or C. However, it was sometimes difficult to distinguish among the three kinds of units with certainty, because the properties of each type were distributed over a wide range and there was considerable overlap between them. Although the ABC classification was obviously sound, it appeared that additional or different physiologic criteria were needed for definitive identification of all units.

The most comprehensive classification of motor units devised so far is that of Burke et al.,[12] who used combinations of physiologic properties to distinguish three groups of units and then compared their results with the histochemical profiles of the same units. After a systematic examination of physiologic criteria on which to base a classification, they found two properties that enabled them to separate gastrocnemius units into unambiguous and nonoverlapping groups. One of these properties, sensitivity to fatigue, was measured quantitatively by tetanizing motoneurons intracellularly for 330 msec once each second and measuring the contractile response to each train of stimuli. Fig. 24-19, *A*, illustrates the course of fatigue in three types of motor units, which Burke et al. designated FF, FR, and S, as elaborated later. The ratio of the tension produced after 2 min of this stimulation to the tension during the first tetanus was called the "fatigue index." The three-dimensional diagram in Fig. 24-19, *B*, shows that the "fatigue index" separated 79 of the 81 units in these experiments into two groups, one with an index of less than 0.25 and the other with an index above 0.75. Only two exceptions were apparent. The second criterion used to separate groups of units was the shape of the unfused tetanus during the first 1 to 2 sec of stimulation. The units denoted by open circles in Fig. 24-19, *B*, all had unfused contractions that rose to maximum within about 250 msec and then showed a slow decline or "sag" in amplitude. The upper two tension records in column b show this pattern of response. In contrast, the units with stippled circles all showed a

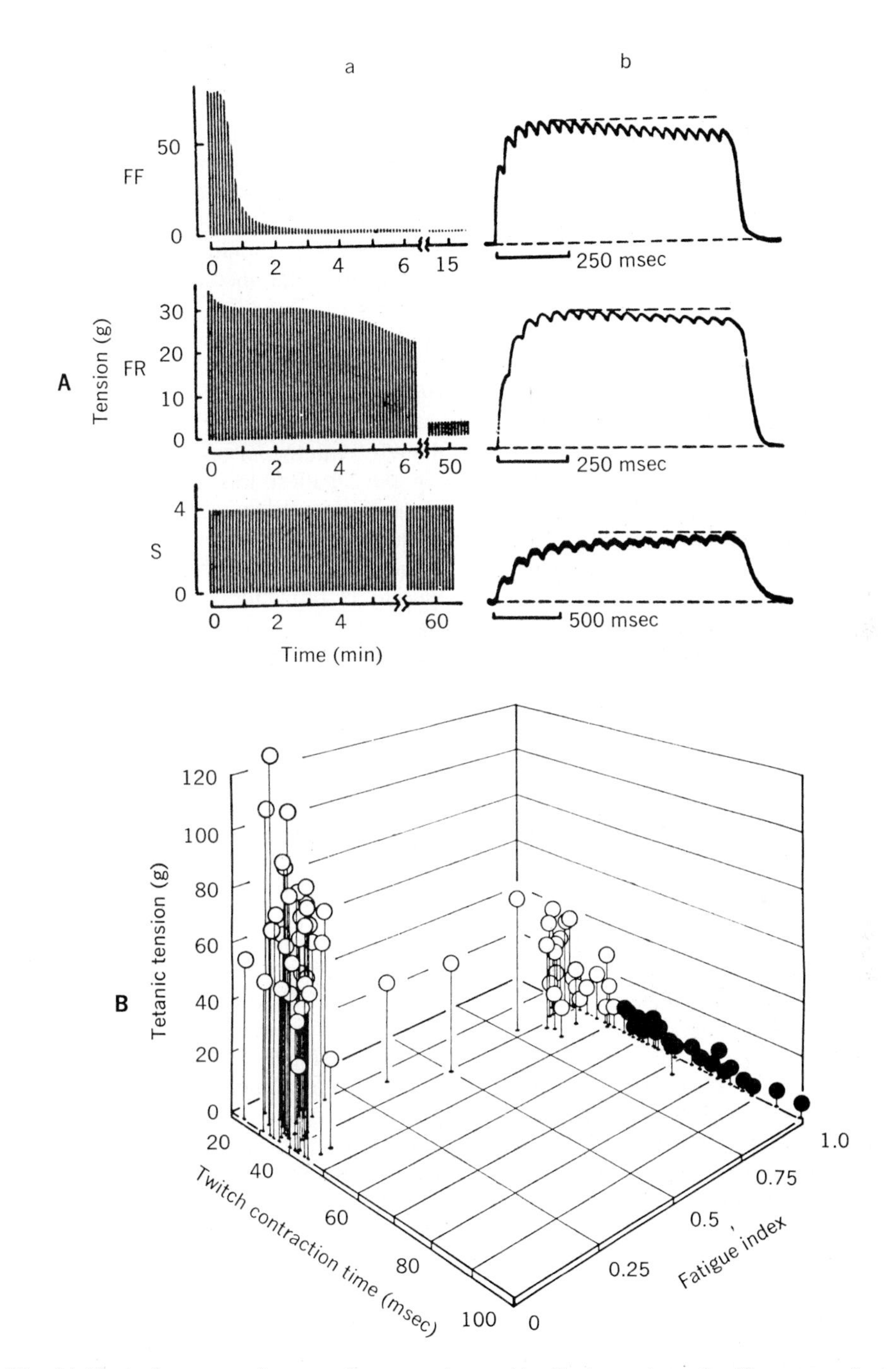

Fig. 24-19. A, Summary of contractile properties used by Burke et al. to classify motor units in m. gastrocnemius muscle of cat. Column *a:* Graphs of tensions produced by three different types of motor units during 40/sec tetani, each lasting 330 msec and repeated every 1 sec for times shown on abscissa. Note rapid decline in tension within 2 min in *FF* motor unit, greater resistance to fatigue in *FR* unit, and complete resistance to fatigue in *S* unit. Column *b:* Records of unfused tetanic responses in same three units, showing slight decline (sag) in tension in late portions of the responses in *FF* and *FR* units, and increase in tension in *S* unit. **B,** Three-dimensional diagram summarizing contractile properties of 81 m. gastrocnemius motor units. Units with sag represented by open circles (*FF* and *FR* units); units without sag denoted by stippled circles (*S* units). Note two unclassified units with fatigue indexes between 0.25 and 0.75. (From Burke et al.[12])

gradual increase in tension throughout the first 1 to 2 sec and no tendency to sag. The lowest of the three records under b illustrates this pattern.

After separating three groups of units by these criteria, Burke et al. noted that other properties of motor units, such as twitch contraction times and maximal tetanic tensions, were obviously related to their three groupings. The cluster of units in the left-hand corner of the three-dimensional diagram, which had fatigue indices below 0.25 and obvious "sag," also contracted most rapidly and produced the largest tetanic contractions. This group, corresponding to type A units, was designated as fast contracting, fast fatiguing (FF). Another group of rapidly contracting units with fatigue indices greater than 0.75 was designated as fast contracting, fatigue resistant (FR). It produced tetanic tensions intermediate between the FF units and a third group (stippled circles) designated type S. This S group was characterized by uniformly small tetanic tensions, slow contractions, and extreme resistance to fatigue. The distribution of twitch contraction times in the FR and S groups was apparently continuous, but there was no overlap between the two groups.

After examining the contractile properties of each unit, Burke et al. continued stimulating the motoneurons intracellularly at a slow rate in order to deplete the glycogen in their muscle fibers by the technique of Kugelberg and Edström.[36] The muscle fibers identified in this way were then stained for a variety of other substrates. It was found in these studies that the repetitive stimulation used to deplete glycogen did not alter fiber-staining characteristics in the other histochemical reactions. "Each physiological unit type had a unique histochemical profile and no exceptions were found to the exact match between physiological and histochemical profiles."[12] Despite this exact match, Burke et al. hesitated to use the terminology of histochemistry for their classification of motor units.

The three-dimensional diagram in Fig. 24-19 suggests that large motor units are more numerous than they really are. This is because large motoneurons are much easier than small ones to penetrate and "hold" for intracellular stimulation. In fact, small motoneurons and small motor units greatly outnumber large units, as the following section will make clear.

PRINCIPLES UNDERLYING THE ORGANIZATION OF MUSCLE

The combination within one motor unit of a particular contraction speed, a certain capacity for producing tension, a measurable susceptibility to fatigue, and a specific histochemical profile is not a random one. There are well-defined principles that specify how different properties are combined and distributed.

Fig. 24-20 summarizes some of the information that has been assembled in the form of a histogram, which illustrates the distribution of properties in a typical population of motor units. The histogram is a smoothed version of the one obtained from the medial gastrocnemius muscle (Fig. 24-10), in which the number of motor units in each size range is plotted against the maximal tetanic tensions they produce. Similarly constructed histograms for different limb muscles[40,46,63] vary in shape, but all are skewed to the left, with the largest number of units producing the smallest individual tensions. Thus the number of units in each size range varies systematically with the sizes of the units in that range. The actual distribution of unit sizes in a muscle is probably a more perfect regression curve than any obtained experimentally. In a study of the flexor digitorum longus, Olson and Swett[46] grouped the 83 units developing less than 10 gm each on an expanded scale of intervals of 1 gm. Even within this restricted range, there was a clear tendency for the number of units in each 1 gm interval to decrease in number as their tensions increased.

Fig. 24-21 suggests that there is an anatomic basis for the distribution of the tetanic tensions and the other properties listed in Fig. 24-20. The histogram in Fig. 24-21 is a rescaled version of one constructed by Hodes et al.[31] from measurements of the cell volumes of all the motoneurons in the seventh cervical segment of the rhesus monkey. Comparison of the two figures indicates that the large number of motor units producing small tensions is innervated by a correspondingly large population of small motoneurons and that the few motor units generating large tensions are supplied by a relatively small population of large cells. This conclusion is in harmony with the observations of a number of investigators,[1-5,8,40,46] showing that the axonal diameters of motor units are correlated directly with their tetanic tensions. The numerous units at the left in Figs. 24-20 and 24-21, for example, have the thinnest axons. Throughout the nervous system there is a general rule, first noted by Ramon y Cajal,[50] stating that the different parts of individual neurons are all constructed on the same scale. In general, thin axons come from small cell bodies with dendrites of proportionate length and size. Recently, motoneurons that had first been studied physiologically with intracellular electrodes were filled with a dye to mark them and were then examined histo-

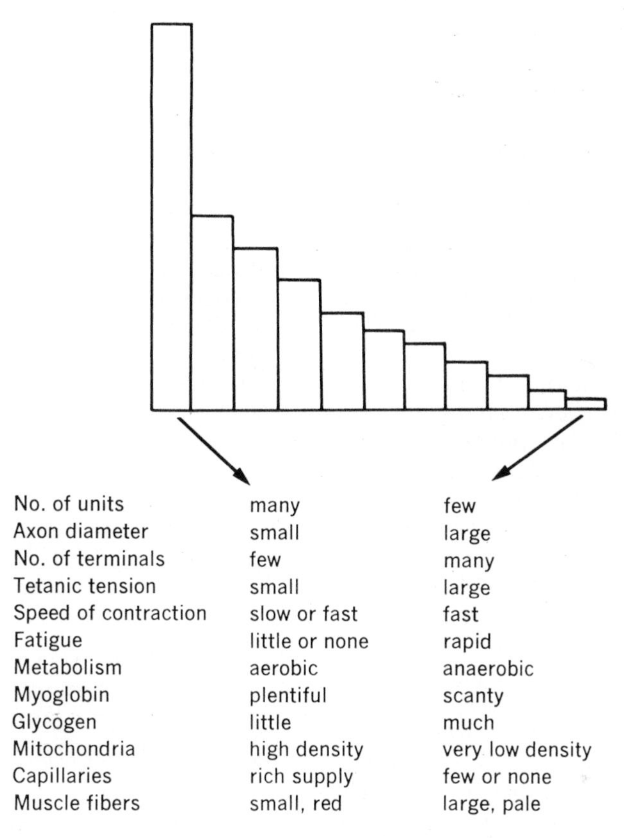

Fig. 24-20. Histogram illustrating numeric distribution of motor unit properties according to their sizes (i.e., maximal tetanic tensions). Characteristics correlated with motor unit size are shown for the two ends of the histogram. An approximately continuous gradation between these limits is assumed.

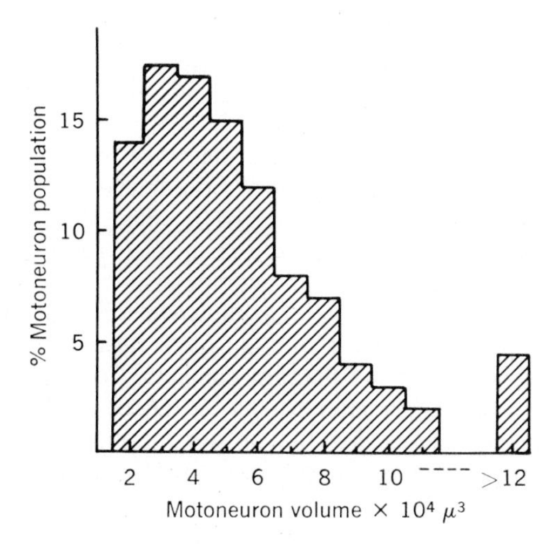

Fig. 24-21. Sizes of motoneurons in seventh cervical segment of spinal cord of normal rhesus monkey. Motoneurons with cell volumes greater than 120,000 μm^3 (4.5% of total population) include cells up to 237,000 μm^3. These have been put into a single group to simplify the histogram. An estimate of the possible errors in these data, which apply chiefly to smaller cells in the ventral horn, is given in original paper. (Redrawn and rescaled from Hodes et al.[31])

logically. Using the conduction velocity of the axons as a measure of their diameter, Barrett and Crill[7] found that axonal diameter correlated closely with the size of the cell body and the extent of its dendritic tree. In addition, as Ramon y Cajal[50] noted, the diameter of an axon determines the number of terminals it gives off. Eccles and Sherrington[16] devoted considerable attention to the branching of motor fibers and proposed that since division of large fibers begins far proximal to the muscle ". . . the large fiber prepares early to form an extensive motor unit." Thus the size of a motoneuron is an important element in determining the capacity of its motor unit to produce tension. However, other factors, such as the type of muscle fibers it supplies, also play a role.

Fig. 24-20 indicates that many other features of motor units—morphologic, physiologic, and biochemical—also correlate closely with motoneuron size. When it was found that both the excitability and the inhibitability[29] of motoneurons were size dependent, as described in Chapter 26, it became apparent that the set of interrelated properties shown in Fig. 24-20 was just part of a more comprehensive neuromuscular organization. The influence of motoneuron size on this constellation of neural and muscular properties came to be known as the "size principle."[27,28]

Fig. 24-20 brings out the important correlations between the size of a motor unit and its other properties. All large units are rapidly contracting and composed of pale muscle fibers that are poorly supplied with mitochondria and capillaries. These units produce large tensions, but even when stimulated at slow rates, they are the first to show contractile fatigue, as Fig. 24-15 illustrates. This susceptibility to fatigue is not well understood. It may be related to the lack of an abundant blood supply and of sufficient mitochondria for the oxidative enzymes necessary for aerobic metabolism. Instead, type A fibers utilize glycogen, which is capable of supplying energy anaerobically at a rapid rate, but only for a brief time. They are quickly depleted of their glycogen, as shown in the periodic acid–Schiff reaction. After a short period of activity the tension output of a large motor unit drops to a low level, as in Fig. 24-15, and may take a long time to recover. These units also show rapid fatigue at their neuromuscular junctions (Fig. 24-14).

It must be emphasized that Fig. 24-20 is a distribution plot of motor unit sizes and the properties that correlate with size. Fig. 24-19 is a different type of diagram, designed for the purposes of classification of motor units rather than for their numeric distribution. Both figures indicate that large units contract rapidly and are very susceptible to fatigue. The smallest units depend most on oxidative pathways for their energy. Their muscle fibers are richly supplied with capillaries, and their small diameter serves to reduce the distance over which oxygen and other metabolites must be transported from capillaries to mitochondria. The distribution of myoglobin is very similar to that of mitochondria in muscle. Fibers rich in one are always rich in the other. The reason for this association, as Wittenberg[62] has demonstrated in a series of elegant papers, is that myoglobin facilitates oxygen diffusion from capillaries to mitochondria. Myoglobin also contributes to the red appearance of muscle fibers. The intermediate type of motor unit is, in general, intermediate in size, staining properties, contractile properties, and metabolism. Its energy requirements are met by a combination of aerobic and anaerobic mechanisms.

The functional significance of the distribution pattern in Fig. 24-20 will become apparent when neural control of muscle tension is described in a subsequent chapter.

METABOLIC CHANGES ASSOCIATED WITH COMPENSATORY GROWTH AND ATROPHY OF DIFFERENT TYPES OF MOTOR UNITS

Some of the adaptive properties of muscle were described in an earlier section of this chapter. It is well known that the size and strength of a muscle respond adaptively to the demands placed on it. Increased contraction, especially if it is isometric, results in compensatory hypertrophy, whereas disuse leads to atrophy. These adaptations must have had considerable evolutionary value. Hypertrophy may help the organism to improve physical performances vital to survival or to compensate for disease. Atrophy may, under some circumstances, contribute to survival by reducing metabolically expensive but unnecessary muscle bulk.

It has been pointed out that small red motor units are equipped to work more intensively and with less fatigue than large white units. The mean daily activity in a muscle is, in fact, related to the proportions of red, intermediate, and white fibers in it. Goldberg[25] has shown that the amount of protein synthesis in a muscle is related to the proportion of red fibers it contains. To do this, he used alpha aminoisobutyric (AIB) acid, an amino acid analog that is not metabolized by muscle, but is transported into muscle cells in the same way that natural amino acids are handled. The top part of Table 24-4 shows the relation between the redness of different rat muscles and their capacity to take up AIB. The differences in their distribution ratios presumably reflect differences in the rates of incorporation of amino acids into protein. The middle part of Table 24-4 supports this conclusion. Small amounts of [^{14}C]leucine were injected into rats. Samples of each muscle were analyzed for radioactivity 24 hr later. The table shows that muscle with a high proportion of red fibers incorporated more ^{14}C into both soluble and fibrillar proteins than did white fibers. Thus ^{14}C incorporation was greater in the muscles showing greater uptake of AIB. Since RNA content is essentially a measure of the amount of machinery available for protein synthesis, the RNA content of the same group of muscles was determined. The bottom part of Table 24-4 shows that red muscles have a greater RNA content per milligram than do white muscles. Thus the muscles highest in RNA are also those capable of making the most protein. The muscles of the hypophysectomized rats in these experiments remained approximately constant in size; hence protein synthesis and degradation must have been approximately equal. Consequently, it may be concluded that the

Table 24-4*

Uptake of [^{14}C]aminoisobutyric acid by red and white skeletal muscles

	Distribution ratio	Proportion of dark fibers
Soleus	3.4 ± 0.24	Predominantly dark
Diaphragm	2.4 ± 0.19	Predominantly dark
Medial gastrocnemius	2.1 ± 0.13	Predominantly dark
Flexor digitorum longus	1.9 ± 0.10	Intermediate
Plantaris	1.7 ± 0.09	Intermediate
Tibialis anterior	1.5 ± 0.18	Intermediate
Peroneus longus	1.4 ± 0.08	Predominantly white
Extensor digitorum longus	1.4 ± 0.07	Predominantly white
Lateral gastrocnemius	1.4 ± 0.06	Predominantly white
Semitendinosus	1.2 ± 0.09	Predominantly white

Muscles are arranged according to average content of dark fibers. Values given are the ratios of concentration of [^{14}C]aminoisobutyric acid in total muscle water to that in plasma 4 hr after intravenous injection. Each is the mean ±SE of seven muscles.

Incorporation of [U-^{14}C]leucine into proteins by red and white skeletal muscles

	Soluble fraction (counts/min/mg of fresh muscle)	Myofibrillar fraction (counts/min/mg of fresh muscle)
Diaphragm	4.4 ± 0.3	6.3 ± 0.3
Soleus	3.7 ± 0.2	5.6 ± 0.2
Medial gastrocnemius	3.3 ± 0.2	5.3 ± 0.4
Plantaris	3.0 ± 0.1	4.8 ± 0.2
Extensor digitorium longus	3.0 ± 0.1	4.5 ± 0.2
Lateral gastrocnemius	2.3 ± 0.2	3.7 ± 0.2

Values are the mean ±SE of at least seven observations. The myofibrillar fraction includes some nuclear and stromal material, although at least 90% of the radioactivity can be extracted as myofibrillar proteins.

RNA content of red and white skeletal muscles

	RNA (μg/mg muscle wet weight)
Soleus	7.1 ± 0.26
Medial gastrocnemius	5.8 ± 0.20
Diaphragm	5.6 ± 0.18
Plantaris	4.6 ± 0.14
Extensor digitorium longus	4.0 ± 0.18
Lateral gastrocnemius	3.8 ± 0.09

Values are the mean ± SE of 10 observations.

*From Goldberg.[25]

amount of protein usually catabolized by a muscle is proportional to its redness. The average half-lives of muscle proteins must vary in different muscles and must be shorter in tonic red muscles than phasic white ones. It must be emphasized that the rate of protein metabolism can change rapidly in response to functional demands. These observations suggest a clear association between physiologic activity and amount of protein metabolism. This matter will be discussed again in the chapter dealing with motoneurons. No good explanation for it is available, however.

IMPLICATIONS OF NEW FINDINGS FOR HUMAN MUSCLE DISEASES

The development of histochemical techniques and other experimental approaches used in animal research have had a major impact on the study of muscle disease in man, of which several hundred varieties are recognized. Human muscle has three types of fibers, which are normally arranged in a random mosaic pattern similar to that seen in the cat and rat. The percentages of the three fiber types vary from one muscle to another. A recent study of histochemical types and fiber sizes in normal human muscle[18] provides information useful for the interpretation of biochemical results obtained from human biopsy material.

Electromyographic (EMG) recordings indicate that many different motor units can be detected at any given site. This is due to the extensive intermingling of fibers from different units. In studies in which the twitch contractions associated with individual EMG potentials were extracted by an averaging technique,[41] units with contraction times ranging from 30 to 100 msec have been identified in the first dorsal interosseous muscle of the human hand. The twitch tensions in this muscle varied from 0.1 to 10.0 gm. The larger motor units had shorter contraction times than the smaller. A similar study by another group[59] on the same muscle showed that the largest, most rapidly contracting units were the most susceptible to fatigue, as in the animal experiments.

A variety of alterations in the normal appearance of muscle has been noted in disease states. In some diseases the use of appropriate histochemical techniques shows that the normal mosaic appearance has been replaced by one in which numerous fibers of the same type are clumped together, as illustrated in Fig. 24-22. These islands of similarly staining muscle fibers are known as "fiber type groupings." The grouping is presumably a result of local denervation of muscle fibers by some disease process. The denervated muscle fibers elicit sprouting from a neighboring intact nerve fiber that reinnervates all or most of the fibers in a particular region. As in the case of the cross innervation studies previously mentioned, the reinnervated fibers take on the properties dictated by their new nerve supply. Red fibers may be converted to white or vice versa. These clumps of similar fibers are concentrated portions of a single motor unit expanded in size by the sprouting. The number of fibers in a single unit often increases markedly due to this sprouting, the highest value observed being seven times greater than the normal mean number.[37] Due to the high density of such fibers,

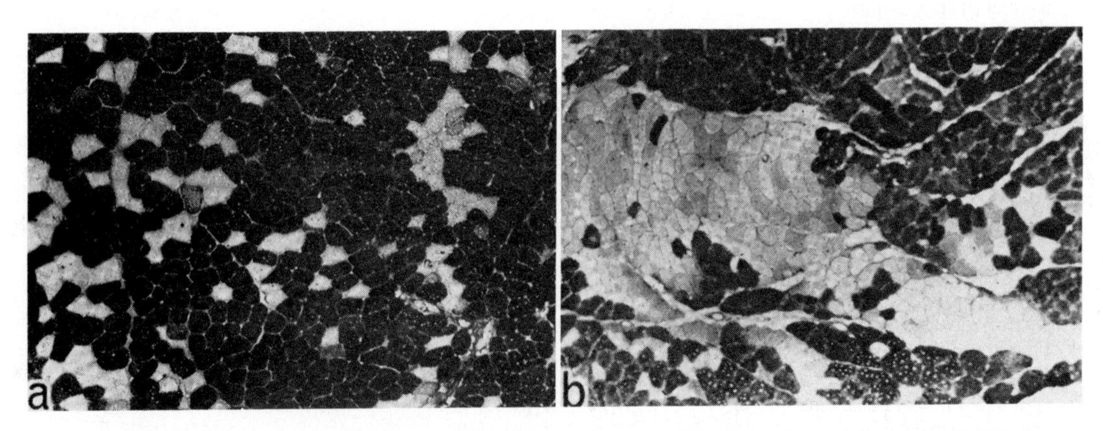

Fig. 24-22. A, Anterior tibial muscle of rat, reinnervated following section of L5 ventral root, showing some "grouping" of muscle fibers. **B,** Severe denervation and reinnervation after L4 ventral root section. Section shows part of one motor unit of 551 fibers with well-marked "grouping" of its fibers. (From Kugelberg et al.[37])

the EMG potentials they yield are often considerably larger than normal.[64]

Histochemical typing of muscle fibers has made it possible to identify a number of muscle diseases that apparently attack only one kind of fiber and spare the others. Some of these type-specific diseases undoubtedly originate in the muscle itself, whose susceptibility may depend on the biochemistry of the fiber type. But since the integrity of muscle fibers depends on their innervation to such a large extent, it is to be expected that some "muscle" diseases are, in fact, neurogenic in their origin. For a good experimental analysis of the variety of changes occurring in denervated and reinnervated muscle fibers of different types, Kugelberg et al.[37] may be consulted.

REFERENCES

1. Appelberg, B., and Emonet-Dénand, F.: Motor units of the first superficial lumbrical muscle of the cat, J. Neurophysiol. **30:**154, 1967.
2. Bagust, J.: Relationships between motor nerve conduction velocities and motor unit contraction characteristics in a slow twitch muscle of the cat, J. Physiol. **238:** 269, 1974.
3. Bagust, J., and Lewis, D. M.: Isometric contractions of motor units in self-reinnervated fast and slow twitch muscles of the cat, J. Physiol. **237:**91, 1974.
4. Bagust, J., Lewis, D. M., and Westerman, R. A.: The properties of motor units in a fast and a slow twitch muscle during post-natal development in the kitten, J. Physiol. **237:**75, 1974.
5. Bagust, J., et al.: Isometric contractions of motor units in a fast twitch muscle of the cat, J. Physiol. **231:**87, 1973.
6. Barnard, R. J., Edgerton, V. R., and Peter, J. B.: Effect of exercise on skeletal muscle. II. Contractile properties, J. Appl. Physiol. **28:**767, 1970.
7. Barrett, J. N., and Crill, W. E.: Specific membrane resistivity of dye-injected cat motoneurons, Brain Res. **28:**556, 1971.
8. Bessou, P., Emonet-Dénand, F., and Laporte, Y.: Relation entre la vitesse de conduction des fibres nerveuses motrices et le temps de contraction de leurs unites motrices, C.R. Acad. Sci. **256:**5625, 1963.
9. Buller, A. J., Eccles, J. C., and Eccles, R. M.: Differentiation of fast and slow muscles in the cat hind limb, J. Physiol. **150:**399, 1960.
10. Buller, A. J., Eccles, J. C., and Eccles, R. M.: Interactions between motoneurones and muscles in respect of the characteristic speeds of their responses, J. Physiol. **150:**417, 1960.
11. Burke, R. E., and Tsairis, P.: Anatomy and innervation ratios in motor units of cat gastrocnemius, J. Physiol. **234:**749, 1973.
12. Burke, R. E., Levine, D. N., Tsairis, P., and Zajac, F. E., III: Physiological types and histochemical profiles in motor units of the cat gastrocnemius, J. Physiol. **234:**723, 1973.
13. Clark, D. A.: Muscle counts of motor units: a study in innervation ratios, Am. J. Physiol. **96:**296, 1931.
14. Coimbra, A.: Radioautographic studies of glycogen synthesis in the striated muscle of rat tongue, Am. J. Anat. **124:**361, 1969.
15. Cooper, S., and Eccles, J. C.: Isometric responses of mammalian muscles, J. Physiol. **69:**377, 1930.
16. Eccles, J. C., and Sherrington, C. S.: Numbers and contraction-values of individual motor-units examined in some muscles of the limb, Proc. R. Soc. Lond. (Biol.) **106:**326, 1930.
17. Edström, L., and Kugelberg, E.: Histochemical composition, distribution of fibres and fatigability of single motor units, J. Neurol. Neurosurg. Psychiatry **31:**424, 1968.
18. Edström, L., and Nyström, B.: Histochemical types and sizes of fibres in normal human muscles, Acta Neurol. Scand. **45:**257, 1969.
19. Eisenberg, B. R., and Kuda, A. M.: Discrimination between fiber populations in mammalian skeletal muscle by using ultrastructural parameters, J. Ultrastruct. Res. **54:**76, 1976.
20. Fardeau, M.: Charactéristiques cytochimiques et ultrastructurales des différents types de fibres musculaires squelettiques extrafusales (chez l'homme et quelques mammifères), Ann. Anat. Pathol. **18:**7, 1973.
21. Gauthier, G. F.: On the relationship of ultrastructural and cytochemical features to color in mammalian skeletal muscles, Z. Zellforsch. **95:**462, 1969.
22. Gauthier, G. F.: The structural and cytochemical heterogeneity of mammalian skeletal muscle fibers. In Podolsky, R. J., editor: Contractility of muscle cells and related processes, Englewood Cliffs, N.J., 1971, Prentice-Hall, Inc.
23. Gauthier, G. F., and Padykula, H. A.: Cytological studies of fiber types in skeletal muscle: a comparative study of the mammalian diaphragm, J. Cell. Biol. **28:**333, 1966.
24. Gerlach, R. L., Stauffer, E. K., Goslow, G. E., Jr., and Stuart, D. G.: Relation between nerve axon size and muscle unit size and speed in motor units of cat hind limb muscles, J. Electromyogr. Clin. Neurophysiol. **16:**177, 1976.
25. Goldberg, A. L.: Protein synthesis in tonic and phasic skeletal muscles, Nature **216:**1219, 1967.
26. Harris, D. A., and Henneman, E.: Identification of two species of alpha motoneurons in cat's plantaris pool, J. Neurophysiol. **40:**16, 1977.
27. Henneman, E., and Olson, C. B.: Relations between structure and function in the design of skeletal muscles, J. Neurophysiol. **28:**581, 1965.
28. Henneman, E., Somjen, G., and Carpenter, D. O.: Functional significance of cell size in spinal motoneurons, J. Neurophysiol. **28:**560, 1965.
29. Henneman, E., Somjen, G., and Carpenter, D. O.: Excitability and inhibitability of motoneurons of different sizes, J. Neurophysiol. **28:**599, 1965.
30. Hill, A. V.: The design of muscles, Br. Med. Bull. **12:**165, 1956.
31. Hodes, R., Peacock, S. M., Jr., and Bodian, D.: Selective destruction of large motoneurons by poliomyelitis virus. II. Size of motoneurons in the spinal cord of rhesus monkeys, J. Neuropathol. Exp. Neurol. **8:**400, 1949.
32. Huizar, P., Kuno, M., Kudo, N., and Miyata, Y.: Reaction of intact spinal motoneurones to partial denervation of the muscle, J. Physiol. **265:**175, 1977.
33. Hursh, J. B.: Conduction velocity and diameter of nerve fibers, Am. J. Physiol. **127:**131, 1939.
34. Katz, B., and Thesleff, S.: On the factors which determine the amplitude of the 'miniature end-plate' potential, J. Physiol. **137:**267, 1957.
35. Kugelberg, E.: Histochemical composition, contrac-

tion speed and fatiguability of rat soleus motor units, J. Neurol. Sci. **20:**177, 1973.
36. Kugelberg, E., and Edström, L.: Differential histochemical effects of muscle contractions on phosphorylase and glycogen in various types of fibres: relation to fatigue, J. Neurol. Neurosurg. Psychiatry **31:**415, 1968.
37. Kugelberg, E., Edström, L., and Abbruzzese, M.: Mapping of motor units in experimentally reinnervated rat muscle, J. Neurol. Neurosurg. Psychiatry **33:**319, 1970.
38. Kuno, M., Miyata, Y., and Munoz-Martinez, E. J.: Differential reaction of fast and slow α-motoneurones to axotomy, J. Physiol. **240:**725, 1974.
39. Lømo, T., Westgaard, R. H., and Dahl, H. A.: Contractile properties of muscle: control by pattern of muscle activity in the rat, Proc. R. Soc. Lond. (Biol.) **187:**99, 1974.
40. McPhedran, A. M., Wuerker, R. B., and Henneman, E.: Properties of motor units in a homogeneous red muscle (soleus) of the cat, J. Neurophysiol. **28:**71, 1965.
41. Milner-Brown, H. S., Stein, R. B., and Yemm, R.: The orderly recruitment of human motor units during voluntary isometric contractions, J. Physiol. **230:**359, 1973.
42. Mosher, C. G., Gerlach, R. L., and Stuart, D. G.: Soleus and anterior tibial motor units of the cat, Brain Res. **44:**1, 1972.
43. Nachmias, V. T., and Padykula, H. A.: A histochemical study of normal and denervated red and white muscles of the rat, J. Biophys. Biochem. Cytol. **4:**47, 1958.
44. Nystrom, B.: Postnatal development of motor nerve terminals in "slow-red" and "fast-white" cat muscles, Acta Neurol. Scand. **44:**363, 1968.
45. Olson, C. B.: A functional and histochemical characterization of different types of motor units in skeletal muscles of the cat, Ph.D. thesis, Harvard University, Cambridge, Mass., 1965.
46. Olson, C. B., and Swett, C. P., Jr.: A functional and histochemical characterization of motor units in a heterogeneous muscle (flexor digitorum longus) of the cat, J. Comp. Neurol. **128:**475, 1966.
47. Padykula, H. A.: The localization of succinic dehydrogenase in tissue sections of the rat, Am. J. Anat. **91:**107, 1952.
48. Padykula, H. A., and Gauthier, G. F.: The ultrastructure of the neuromuscular junctions of mammalian red, white, and intermediate skeletal muscle fibers, J. Cell Biol. **46:**27, 1970.
49. Prewett, M. A., and Salafsky, B.: Effect of cross innervation on biochemical characteristics of skeletal muscle, Am. J. Physiol. **213:**295, 1967.
50. Ramón y Cajal, S.: Histologie du système nerveux de l'homme et des vertébrés, Paris, 1909, Librairie Maloine, vol. 1.
51. Ranvier, L.: De quelque faits relatifs à l'histologie et à la physiologie des muscles striés, Arch. Physiol. Norm. Pathol. **6:**1, 1874.
52. Reinking, R. M., Stephens, J. A., and Stuart, D. G.: The motor units of cat medial gastrocnemius: problem of their categorization on the basis of mechanical properties, Exp. Brain Res. **23:**301, 1975.
53. Romanul, F. C. A., and Van Der Meulen, J. P.: Slow and fast muscles after cross innervation, Arch. Neurol. **17:**387, 1967.
54. Salmons, S., and Sreter, F. A.: Significance of impulse activity in the transformation of skeletal muscle type, Nature **263:**30, 1976.
55. Salmons, S., and Vrbova, G.: The influence of activity on some contractile characteristics of mammalian fast and slow muscles, J. Physiol. (Lond.) **201:**535, 1969.
56. Sherrington, C. S.: Remarks on some aspects of reflex inhibition, Proc. R. Soc. Lond. (Biol.) **97:**519, 1925.
57. Stein, J. M., and Padykula, H. A.: Histochemical classification of individual skeletal muscle fibers of the rat, Am. J. Anat. **110:**103, 1962.
58. Stephens, J. A., and Stuart, D. G.: The motor units of cat medial gastrocnemius: speed-size relations and their significance for the recruitment order of motor units, Brain Res. **91:**177, 1975.
59. Stephens, J. A., and Usherwood, T. P.: The fatigability of human motor units, J. Physiol. **250:**37, 1975.
60. Teig, E.: Tension and contraction time of motor units of the middle ear muscles in the cat, Acta Physiol. Scand. **84:**11, 1972.
61. Tomanek, R. J.: Ultrastructural differentiation of skeletal muscle fibers and their diversity, J. Ultrastruct. Res. **55:**212, 1976.
62. Wittenberg, J. B.: Myoglobin-facilitated oxygen diffusion: role of myoglobin in oxygen entry into muscle, Physiol. Rev. **50:**559, 1970.
63. Wuerker, R. B., McPhedran, A. M., and Henneman, E.: Properties of motor units in a heterogeneous pale muscle (m. gastrocnemius) of the cat, J. Neurophysiol. **28:**85, 1965.
64. Yahr, M. D., Herz, E., Moldaver, J., and Grundfest, H.: Electromyographic patterns in reinnervated muscle, Arch. Neurol. Psychiatry **63:**728, 1950.
65. Yellin, H.: Neural regulation of enzymes in muscle fibers of red and white muscle, Exp. Neurol. **19:**92, 1967.

25

DALE A. HARRIS and ELWOOD HENNEMAN

Feedback signals from muscle and their efferent control

Even relatively simple movements involve the coordination of many separate muscles and may require precise regulation of their lengths and tensions. Centers in the brain that control movement may have to operate during a wide range of rapidly changing circumstances. For example, an individual may walk across a smooth, flat surface while he is well rested; at other times he may walk or run up a steep, irregular slope under conditions of increasing muscular fatigue. The motor centers cannot send down control signals to the motoneurons that are appropriate for the constantly varying task without receiving information from many sources. The eyes can supply information about the terrain over which the individual moves, and the vestibular system can help to maintain his balance. There must also be provision for a more or less continuous flow of information from the muscles that are carrying out the movements. This flow of information, which is called "feedback," may provide for adjustment of the control system in accordance with the changing lengths and tensions of the muscles. There are three types of stretch receptors in skeletal muscle. During active use of the triceps surae muscles of the cat there may be as many as 50,000 impulses/sec coming from these receptors into the spinal cord. Even during complete relaxation of the muscle there are still a few thousand impulses per second being transmitted from the spinal cord to the receptors via the gamma fiber system.

This chapter will describe the muscle receptor apparatus in detail and examine the signals that are sent to the central nervous system (CNS). It will also describe the specialized efferent system through which the CNS influences the signals sent back to it. The reader who wishes details beyond the scope of this account should consult general reviews, in particular those of Matthews[5] and Barker et al.[2]

STRUCTURE AND INNERVATION OF MUSCLE SPINDLES AND TENDON ORGANS[2,5,7,20]

Two of the three types of stretch receptors are located within a highly specialized receptive organ called a muscle spindle. Spindles are macroscopic in size, fusiform in shape, and widely scattered throughout the fleshy parts of the muscle. They are usually attached at both ends to the ordinary or "extrafusal" muscle fibers, as illustrated in Fig. 25-1. As indicated in Table 25-1, their number per gram of muscle tissue varies widely. Small distal muscles such as the interossei of the cat forepaw that require delicate control have a far greater density of spindles (119/gm) than large, powerful muscles, for example, the lateral gastrocnemius (5/gm). Examination of Table 25-1 suggests that spindle density may be inversely related to mean innervation ratio, which is reasonable if both are related to fineness of muscular control.

Each spindle consists of a connective tissue sheath containing 2 to 12 thin muscle fibers known as *intrafusal* fibers. Mammalian spindles contain two types of intrafusal fibers (Fig. 25-2). The longer and larger of these, which are less than half the diameter of extrafusal fibers, contain numerous large nuclei closely packed in a central bag; hence they are called *nuclear bag fibers*. The shorter and thinner fibers, which are about half the length and diameter of nuclear bag fibers, contain a single row of central nuclei resembling a chain and are known as *nuclear chain fibers*. There are usually two of the larger fibers and four of the smaller type in each spindle, but the numbers of both may vary. It

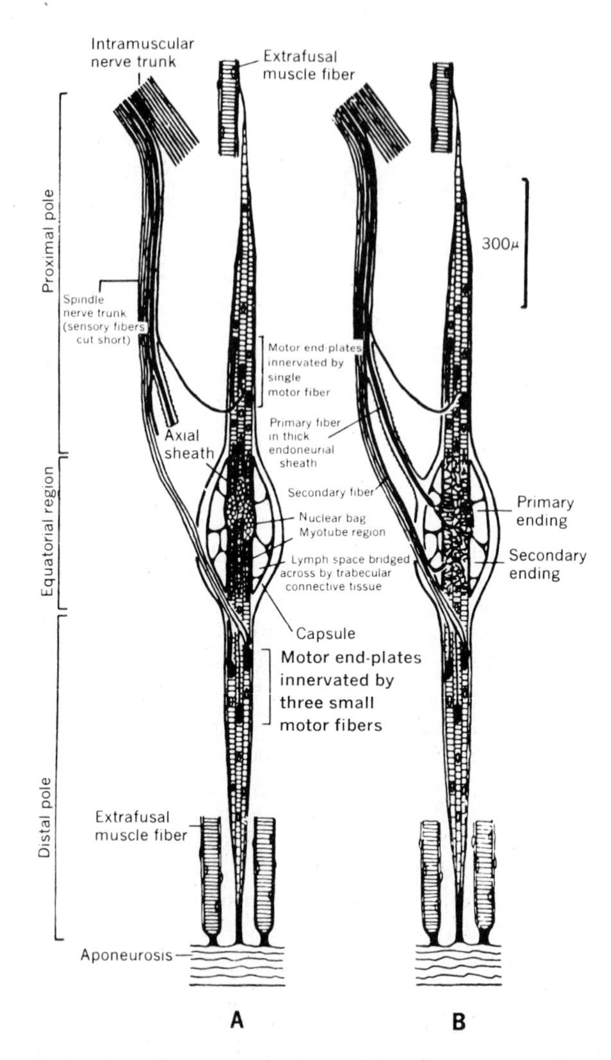

Fig. 25-1. Schematic drawing of structure and innervation of typical spindle organ from mammalian muscle. **A,** Motor innervation only. **B,** Both sensory and motor innervation. (From Barker.[14])

Table 25-1. Number and density of spindle capsules*

Muscle	Mean weight (gm)	Spindle capsule (content)		No. of spindle capsules/gm
		Range	Mean and SD	
Lateral gastrocnemius	7.61	21 (25 to 45)	35 ± 7	5
Mesial gastrocnemius	7.34	35 (46 to 80)	62 ± 9	9
Rectus femoris†	8.36	56 (77 to 132)	104 ± 14	12
Tibialis anterior†	4.57	38 (52 to 89)	71 ± 9	15
Semitendinosus	6.41	62 (80 to 141)	114 ± 14	18
Soleus	2.49	31 (40 to 70)	56 ± 7	23
Flexor digitorum longus lateral	3.25	34 (58 to 91)	75 ± 8	23
Tibialis posterior	0.78	19 (21 to 39)	31 ± 4	39
Flexor digitorum longus mesial	1.06	24 (36 to 59)	48 ± 6	45
Fifth interosseous (foot)‡	0.33	12 (22 to 33)	29	88
Fifth interosseous (hand)†	0.21	11 (21 to 31)	25 ± 2	119

*From Chin et al.[24]
†Data from Barker and Chin.[15]
‡Four muscles counted; data from Ip.[39]

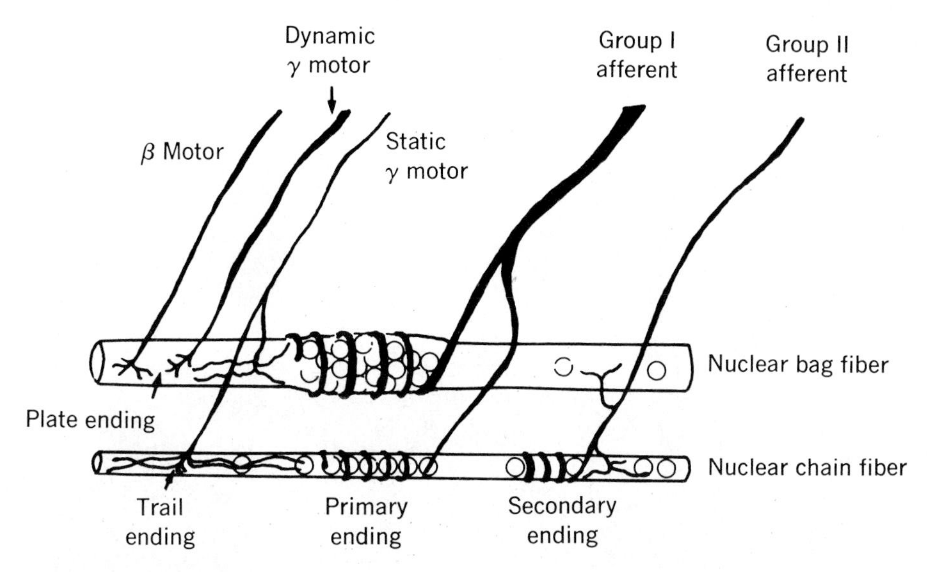

Fig. 25-2. Types of intrafusal fibers in mammalian spindles and their innervation.

should be noted that recent evidence suggests that the true situation is more complicated than is presented here. It appears that bag fibers may not constitute a homogeneous population, but rather can be divided into two physiologically and histochemically distinct types.[16,21,51]

Each spindle is innervated by axons called gamma fibers because of their diameter (1 to 8 μm). They are also referred to as "fusimotor" fibers because they innervate intrafusal muscle fibers exclusively. Although estimates as high as 25 gamma fibers per spindle have been made, more usual estimates are from 7 to 10. Gamma fibers have two morphologic types of motor endings, although it is believed that all the terminations of any given gamma fiber are of a single type. One type of ending is the discrete motor end-plate (Fig. 25-3, *B*), which lies chiefly at each pole of the spindle. The other type of termination is the "trail ending," a more diffuse ending that may extend from the central portion of the spindle all the way to the poles (Fig. 25-3, *A*). Although both kinds of termination have been observed on both bag and chain intrafusal muscle fibers,[17] it is clear that the plate endings are chiefly found on the nuclear bag fibers and rarely observed on the chain fibers. The distribution of the trail-type endings is more controversial. According to the original view, trail endings are found predominantly on nuclear chain fibers. However, according to more recent evidence, trail endings are frequently found on bag fibers as well.*

*See references 18, 19, 22, 23, and 43.

Some spindles are also supplied by skeletofusimotor (beta) fibers. Separate branches of a given beta fiber innervate both intrafusal and extrafusal muscle fibers. Although in lower vertebrates innervation of both intrafusal and extrafusal muscle fibers by the same motoneuron is the rule rather than the exception, it was for many years thought that this never occurred in mammalian muscle. It has now been demonstrated that beta or skeletofusimotor innervation occurs in a wide variety of mammalian muscles, and recent evidence suggests that in some muscles it may be relatively common.[7,27] Nonetheless, it is clear that the vast majority of mammalian intrafusal and extrafusal muscle fibers are innervated by distinct gamma and alpha motoneurons, respectively. Beta fibers terminate on spindles in plate-type endings. The efferent innervation of spindles is illustrated schematically in Fig. 25-2.

At the center of each spindle there is an expanded, fluid-filled capsule (Figs. 25-1 and 25-3, *A*); inside, the terminals of sensory fibers are attached to the intrafusal fibers. There are two types of sensory endings in most spindles: primary endings derived from group Ia (12 to 20 μm) nerve fibers and secondary endings coming from group II (4 to 12 μm) fibers. Each spindle has one primary afferent that subdivides and sends spirals around the central equatorial region of all intrafusal fibers, both bag and chain, within the spindle (Figs. 25-2 and 25-3, *A*). A given spindle may have 0 to 5 secondary afferent endings that terminate predominantly on nuclear chain fibers. The terminals lie to either side of the

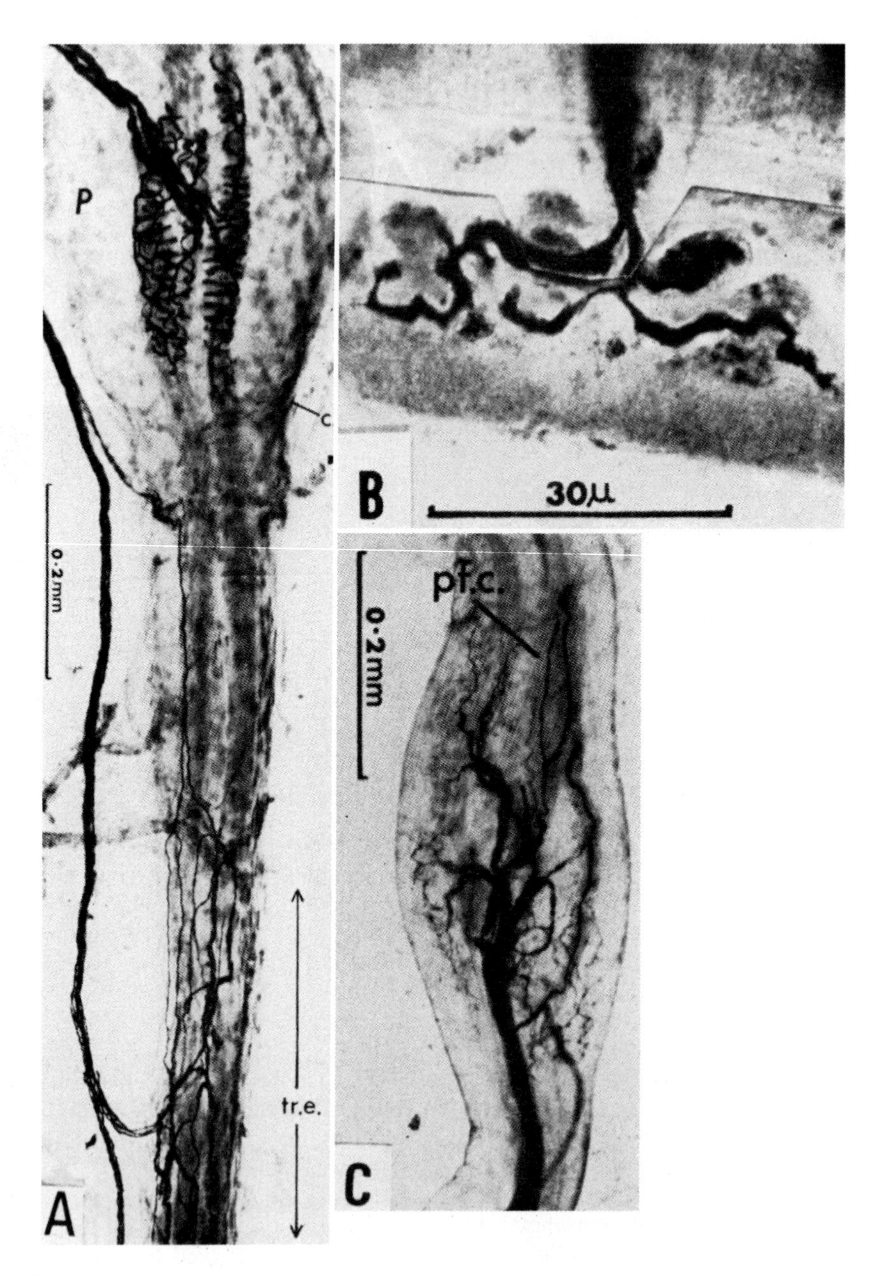

Fig. 25-3. A, Normal spindle from peroneus brevis of cat showing capsule *(c)*, primary ending *(P)*, and trail ending *(tr.e.)*. **B,** Surface view of plate ending in interosseous spindle of cat. **C,** Tendon organ from extensor digitorum longus innervated by one 10 μm group Ib fiber. Paciniform corpuscle *(pf.c.)* is visible underneath receptor capsule. (From Barker.[1])

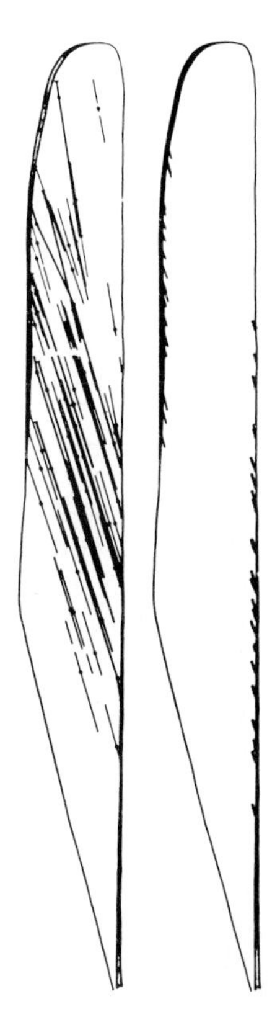

Fig. 25-4. Distribution of spindles (left) and tendon organs (right) in m. gastrocnemius of cat, as if projected onto imaginary midsagittal plane. (From Swett and Eldred.[56])

primary endings on the striated portion of the intrafusal fibers, as shown in Fig. 25-2.

The third type of stretch receptor found in mammalian muscle is the so-called tendon organ of Golgi. In contrast with spindles that lie in parallel with the muscle fibers, tendon organs are located at the musculotendinous junction and hence are in series with the contractile elements. Fig. 25-4 shows the contrasting distribution of spindles and tendon organs in the medial gastrocnemius of the cat. In general, tendon organs are less numerous than spindles; the ratio varies from about 1:1 to 1:3, tendon organs being relatively more numerous in slowly contracting muscles. A tendon organ is formed around a bundle of small tendon fascicles, immediately adjacent to the musculotendinous junction (Fig. 25-5). The receptor, which is about 500 μm long, is enclosed in a delicate capsule that blends with the connective tissue of the muscle. Its innervation is derived from a group Ib nerve fiber that divides into myelinated branches inside the capsule (Fig. 25-3, *C*) and then into sprays of unmyelinated fibers with clasplike granular swellings that are applied to the surface of the tendon fascicles. Distortion of these terminals results in a generator potential that gives rise to nerve impulses.

SIGNALS FROM SPINDLE RECEPTORS

The activity of individual stretch receptors can be recorded most readily from dorsal root filaments that have been subdivided until they contain the afferent fiber from a single active receptor. By stimulating the muscle nerve distally and recording centrally the conduction velocity of the afferent fiber coming from this ending can be measured and its diameter calculated. The tension produced by passive stretch or contraction can be recorded simultaneously with a myograph attached to the tendon of the muscle.

The responses obtained from fibers innervating primary and secondary receptors in spindles can be distinguished from those of tendon organs by their greater sensitivity to passive stretch and their characteristic pause in firing during a twitch contraction. Both these features are apparent in Fig. 25-6, which shows side by side the responses of a primary receptor *(A)* and a tendon organ *(B)*. The upper traces show the rhythmic discharge of the primary receptor in response to a tension of 10 gm and the absence of discharge of the tendon organ under 20 gm of tension. The lower traces illustrate the pause in the discharge of the primary ending and the onset of firing in the tendon organ during a twitch contraction of the muscle. This striking contrast in behavior is due largely to the locations of the receptors. Spindles are placed in parallel with the contracting extrafusal muscle fibers, whereas tendon organs lie in series with them. When a muscle is passively stretched, the intrafusal fibers in the spindle are also stretched and the endings attached to them are deformed. Tendon organs are less affected by passive stretch because they are located on inelastic tendons that lengthen very little in comparison with muscle.

In recent years, considerable effort has been made to distinguish between the signals sent to the spinal cord by the two types of spindle receptors. It now appears that they respond in much the same way to static stretch, but respond quite differently during the dynamic phase of lengthening or shortening. Fig. 25-7 illustrates the differences very clearly. The upper recording from a primary ending shows a uniform, rapid rate of discharge during the period of actual extension, a slower rate during the period of static stretch, and a cessation of discharge during re-

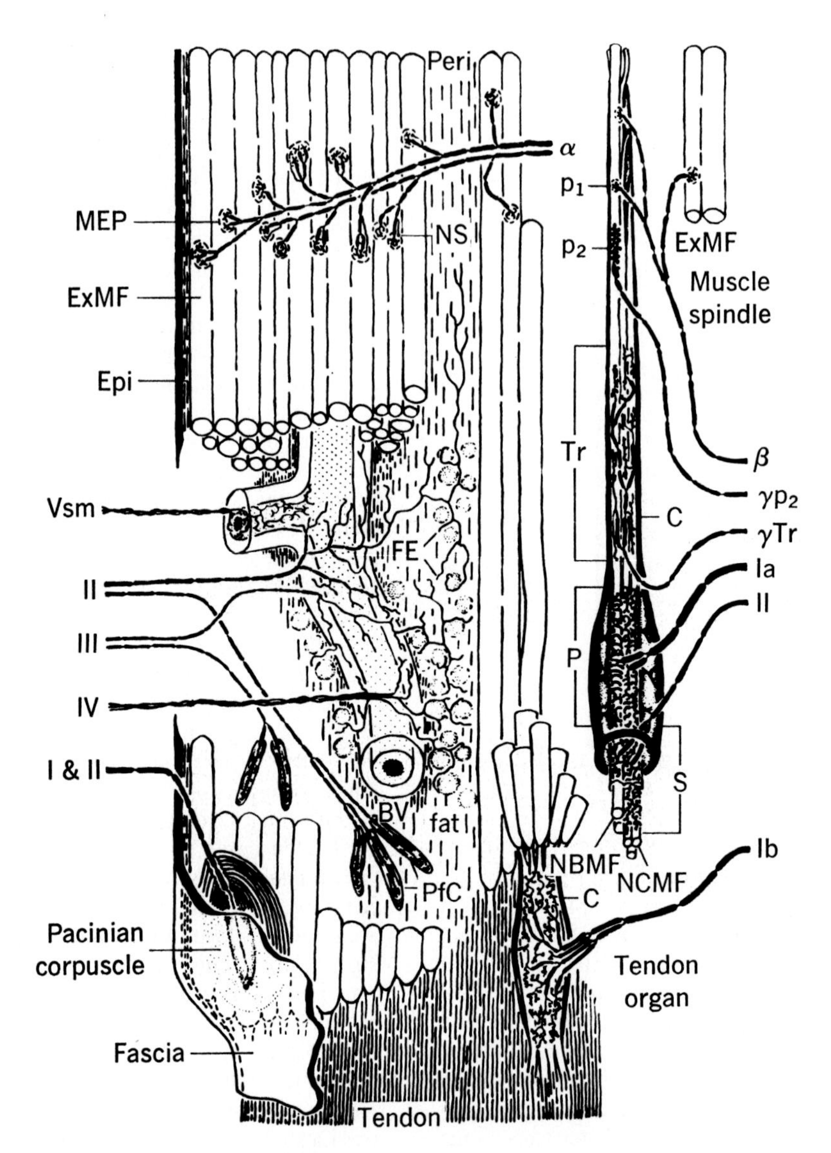

Fig. 25-5. Schema of muscle spindle showing innervation of mammalian skeletal muscle based on a study of cat hindlimb muscles. Those nerve fibers shown on right of diagram are exclusively concerned with muscle innervation; those on left also take part in innervation of other tissues. Roman numerals refer to the groups of myelinated (I, II, III) and nonmyelinated (IV) sensory fibers; Greek letters refer to motor fibers. Spindle pole is cut to about half its length, extracapsular portion being omitted. *BV,* Blood vessel; *C,* capsule, *Epi,* epimysium; *ExMF,* extrafusal muscle fibers; *NBMF,* nuclear-bag muscle fiber; *NS,* nodal sprout; *MEP,* motor end-plate; *P,* primary ending; P_1, P_2, two types of intrafusal end-plates; *Peri,* perimysium; *PfC,* paciniform corpuscle; *S,* secondary ending; *Tr,* trail ending; *Vsm,* vasomotor fibers; *FE,* free endings; *NCMF,* nuclear chain muscle fiber. (From Barker et al.[2])

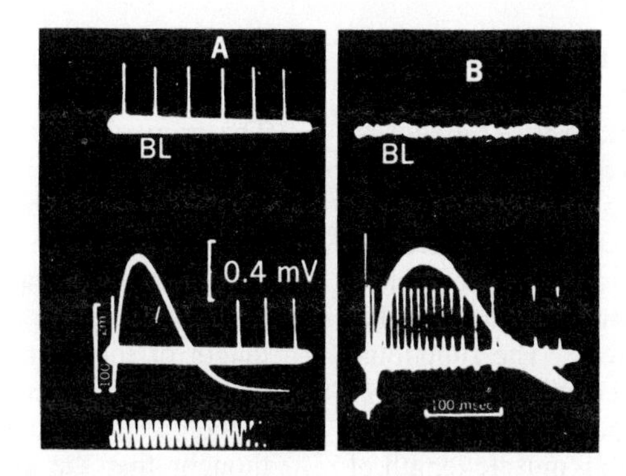

Fig. 25-6. A, Electrical impulses in single afferent fiber from soleus muscle of cat; nerve fiber innervating primary spindle receptor was isolated by microdissection of dorsal root. Initial tension of 10 gm causes baseline discharge *(BL)*. For lower records, maximal stimulus for alpha motor fibers delivered to soleus nerve produces twitch contraction, shown by curving line of strain gauge recording. During twitch contraction, afferent discharge ceases completely. Time line, 50 Hz. **B,** Single tendon organ afferent from cat soleus. Muscle tension of 20 gm caused no baseline discharge. Twitch contraction of muscle caused outburst of nerve impulses, shown in lower records. (From Kuffler and Hunt.[41])

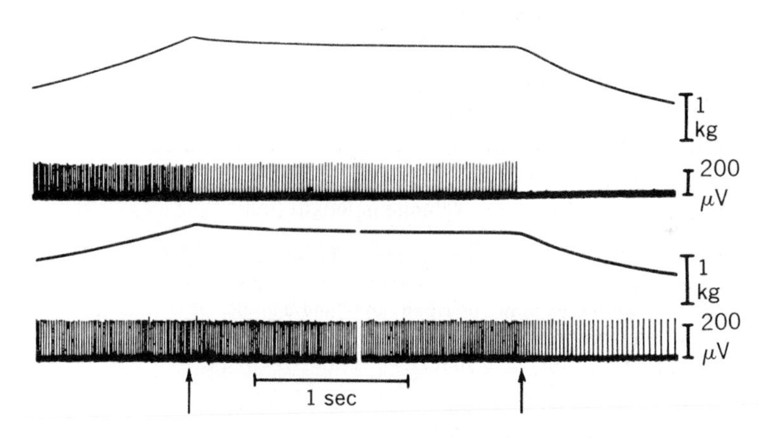

Fig. 25-7. Records obtained during stretch and release of primary ending (above) and secondary ending (below). Rate of stretch is 2.3 mm/sec in both cases. (From Harvey and Matthews.[33])

lease of stretch. The lower recording from a secondary ending shows a gradual increase in rate of discharge during the phase of extension to a rate that is maintained during static stretch. From comparisons of this type, it appears that the rate of discharge of a secondary ending is chiefly a function of the length of the muscle and is only slightly related to the rate of change in length. The discharge frequency of primary endings is related to length and velocity of stretching. Gamma innervation complicates these characteristics, as will be described later.

For small stretches, primary receptors respond linearly and with high sensitivity.[45,53] However, when the stretch exceeds 100 to 200 μm the response becomes significantly less sensitive and highly nonlinear.[34,36] Fig. 25-8 illustrates the response of a primary receptor as a function of the magnitude of passive stretch of the muscle. For very small stretches (less than 100 to 200 μm), the slope is quite steep, indicating high sensitivity. As the stretch is increased, the slope suddenly flattens out. This nonlinearity represents an abrupt decrease in receptor sensitivity. Moreover, the high sensitivity of the receptor to small changes in length can reset itself after a large stretch, as illustrated in Fig. 25-9. At a succession of muscle lengths the receptor was allowed to reach a steady state before a constant-velocity (1 mm/sec), 2 mm ramp stretch was applied. At each successive muscle length the receptor is seen to reset itself, preventing the saturation that would have occurred otherwise. The functional significance of the high-sensitivity region is unclear, since most physiologically important movements produce changes in muscle length of more than 200 μm. However, it is clear that the CNS must deal with a highly nonlinear response in the case of the primary spindle receptor. In contrast, secondary endings are less sensitive than primary endings at all magnitudes of stretch and also show less conspicuous nonlinearities. The primary endings have been found to be extraordinarily responsive to low-amplitude, high-frequency vibration and can be selectively activated by such stimuli.[44]

As discussed earlier, primary and secondary endings are found at different locations along the length of the intrafusal fibers (Fig. 25-2). It is likely that this accounts for the major differences in their response characteristics. Specifically, it is thought that the equatorial regions, where the primary endings are located, are relatively more elastic and less viscous than the more polar regions, where one finds secondary endings. Numerous models and experiments have shown that

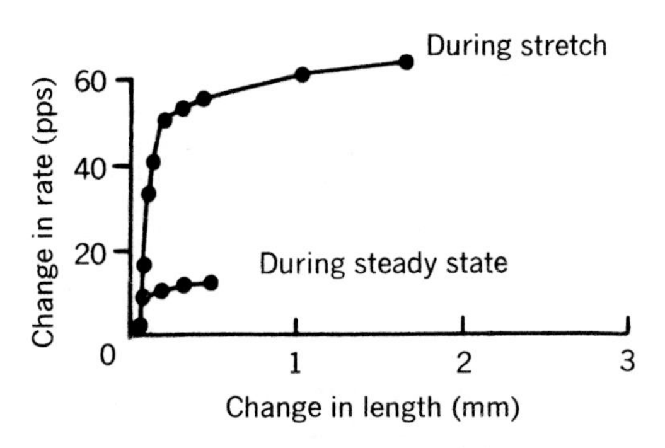

Fig. 25-8. Nonlinear response of primary receptor. Change in rate is difference between discharge rate at time of measurement and steady-state rate prior to stretch. As indicated, measurements of rate were made both during dynamic phase of stretching and also after reaching new steady state (static response). Departure from straight lines passing through origin indicates nonlinearity. Spindle has been deefferented. (Modified from Houk et al.[36])

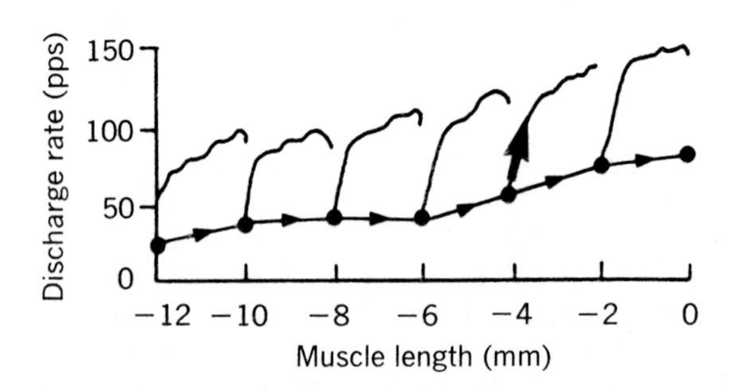

Fig. 25-9. Ability of primary receptor to reset itself to state of high sensitivity. Muscle length is measured with respect to maximum physiologic length (i.e., maximum length = 0). Filled points denote average discharge rates in the steady state. At each steady-state operating point, a constant velocity (1 mm/sec), 2 mm ramp stretch was applied and response of the ending plotted during dynamic phase of stretch as a function of instantaneous muscle length (e.g., large arrowhead). Operating points were achieved successively as indicated by small arrowheads. At each successive steady-state muscle length, receptor resets itself, preventing saturation that would otherwise occur. (Modified from Houk et al.[36])

this difference in viscoelastic properties could account for the velocity sensitivity found in the primary endings and its relative absence in the secondary endings[29,54,55]; however, contrary opinions also exist.[2,38,50]

The transition from high to low sensitivity that occurs so dramatically in the primary endings may also be due to the mechanical properties of intrafusal muscle.[5,34] It is thought that the polar, striated regions of intrafusal fibers offer resistance to stretch because of resting bonds or cross bridges between actin and myosin. As a consequence, disproportionate stretching may occur in the sparsely striated equatorial region where the primary endings are located. High sensitivity suddenly gives way to a much lower sensitivity when the stretch becomes large enough to rupture the resting bonds. The previously mentioned ability of the primary receptor to reset itself to a state of high sensitivity would correspond to a re-forming of the actin-myosin resting bonds during the steady state.

That primary and secondary endings send different messages to the CNS is well-established. Presumably the CNS recognizes and makes use of the different information conveyed by the two types of endings. However, at present there is no hard evidence to support this idea. It is known that the reflex actions of the two endings are distinct; however, this is attributable to detailed differences in the involved circuitry and not to differences in discharge patterns of the endings themselves.

EFFERENT CONTROL OF SPINDLE RECEPTORS

Muscle spindles are designed to enable the CNS to modify or control the activity of the receptors they contain. This influence is exerted chiefly by way of gamma efferents, but also to some extent by beta efferents, which, when stimulated, cause contraction of intrafusal muscle. Contraction of intrafusal fibers produces an insignificant increase in total muscle tension; thus the physiologic effect of such contraction is to deform and excite the afferent endings within the spindle. The nuclear bag fibers consist of a central noncontractile region situated between two contractile polar regions, each with its own motor innervation (Fig. 25-1). Shortening of the contractile ends of the fiber stretches the central portion where the primary receptors are located and causes them to discharge. Because of this configuration, passive stretch of the muscle and contraction of intrafusal fibers both result in stretch of the primary receptor; thus the response of their endings is determined by the additive effect of these two kinds of stimuli. Fig. 25-10 illustrates the separate and combined effects of stretch and gamma activity on a primary receptor. The three tracings in *A* show the discharges elicited by loads of 0, 5, and 20 gm. The additional effects

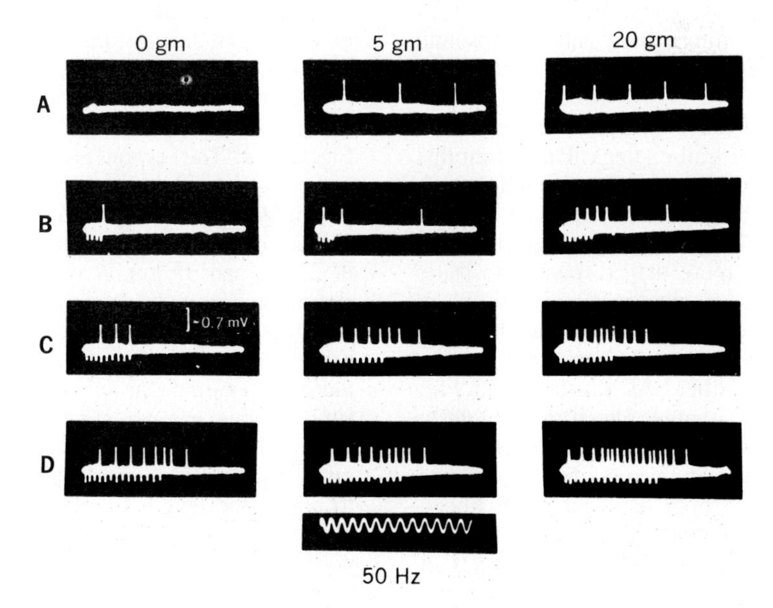

Fig. 25-10. Discharges of primary spindle receptor in soleus muscle recorded from dorsal root filament. Records obtained at 0, 5, and 20 gm of passive stretch. **A,** Baseline discharge. Stimulation of single efferent gamma fiber to soleus is indicated by small downward deflections (4 to 6 stimuli in **B,** 9 to 11 stimuli in **C,** and 14 to 16 stimuli in **D**). (From Kuffler et al.[42])

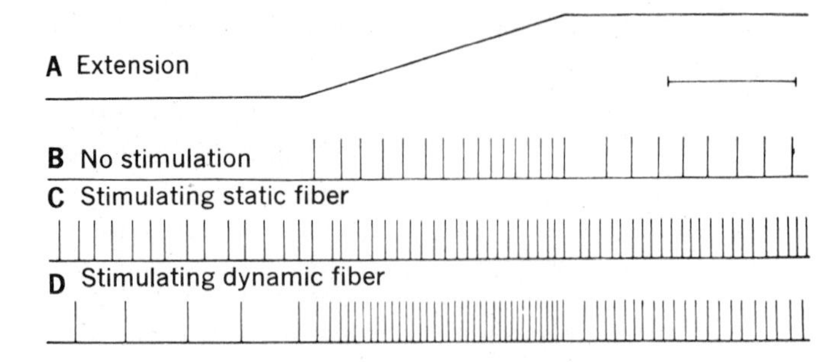

Fig. 25-11. Effects of stimulating single fusimotor fibers on response of a primary ending to stretch of 6 mm at 30 mm/sec. Throughout **C,** single static fiber was stimulated at 70/sec. Throughout **D,** single dynamic fiber was stimulated at 70/sec. Action potentials drawn from computed data. (Time bar: 0.1 sec.) (From Crowe and Matthews.[25])

of gamma stimulation are shown by the records in *B, C,* and *D*. Gamma fibers supplying this spindle were separated from the alpha fibers in the ventral root and stimulated electrically (4 to 6 pulses in B, 9 to 11 in C, and 14 to 16 in D). It is clear from this example that the rate of discharge of primary endings, and secondary endings as well, is directly related to muscle length only when there is no modulation of gamma activity. This condition seldom occurs because gamma motoneurons are continually discharging, even to resting muscle. It is not known whether the CNS has some means of allowing for the effects of gamma activity and of extracting the true length of the muscle from the combination of various inputs. Quite possibly the CNS does not need information about absolute muscle length in order to exert appropriate control.

As noted previously, gamma fibers terminate as plate endings or trail endings. Paralleling this anatomic difference, two types of functional effects have also been established. One type of gamma fiber influences afferent responses to phasic stretch far more than responses to static stretch and hence is called a "dynamic" fiber. The other type of gamma fiber increases the spindle response to static stretch and is thus called a "static" fiber. As illustrated in Fig. 25-2, the dynamic gamma fibers are thought to terminate in plate endings solely on nuclear bag fibers, whereas the static gamma fibers terminate in trail endings on both bag and chain fibers. Presumably, this anatomic arrangement plays a major role in producing the distinct effects of the two kinds of gamma efferents. Fig. 25-11 illustrates the effects of these two types of gamma efferents on a primary receptor. Stimulation of dynamic gamma fibers during phasic stretch results in marked acceleration of primary endings, but usually has no effect on secondary endings.[13] At constant length, stimulation of dynamic fibers generally has only a slight, if any, effect on primary endings and no effect on secondary endings. Stimulation of a static gamma fiber results in marked acceleration of both primary and secondary endings if they are at constant length, but causes minor and variable effects during phasic stretch.

By recording from alpha and gamma motoneurons supplying the same muscle, it has been shown that gamma activity usually occurs in parallel with alpha activity in many patterns of contraction.[37] As a consequence of this alpha-gamma coactivation, intrafusal fibers shorten as the muscle itself shortens. If intrafusal activation does not occur, the spindle receptors cease firing or slow down during a muscle contraction, as we have already seen. However, intrafusal activation by gamma motoneurons can maintain spindle firing during muscle contraction, as shown in Fig. 25-12. The responses of a primary receptor are shown with and without accompanying gamma support. The records in *A* show the discharges of a primary receptor stretched by loads of 2, 15, and 35 gm. *B* illustrates the effect of stimulating a single gamma fiber to this spindle (9 stimuli at 100/sec). *D* shows the pause in firing that occurs when the muscle contracts and there is no gamma stimulation. *C* illustrates how simultaneous gamma stimulation prevents this cessation of firing. This illustration suggests that one possible function of the gamma system is to adjust the length of intrafusal fibers so that spindle receptors always operate on a sensitive portion of their response scale. Also, during a powerful contraction with considerable shortening, it may be advantageous for the spindle receptors to continue firing in order to reinforce the power of the contraction reflexly. This is referred to as the "servoassisted" method of producing and controlling movement.[8]

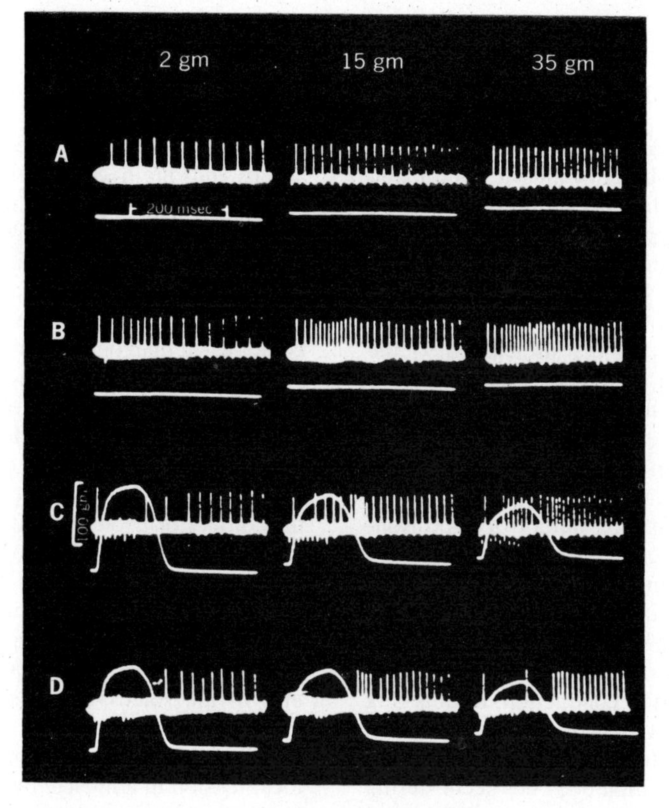

Fig. 25-12. Effect of contraction of muscle and of stimulation of its gamma motor innervation on sensory discharge from a spindle receptor. Recording from single afferent nerve fiber from flexor digitorum longus of cat, isolated by microdissection of dorsal root. Second trace of each pair is strain gauge recording of muscle tension, initially 2, 15, and 35 gm. **A,** Baseline discharge. **B,** Stimulation of gamma motor fibers (isolated by microdissection of ventral roots); nine stimuli at 10 msec intervals delivered at beginning of each record. No muscle contraction was produced, although increase in discharge rate occurred. **C,** Stimulation of alpha and gamma motor fibers produces, at higher tensions, afferent discharges during twitch contraction. **D,** Stimulation of only alpha motor fibers causes twitch contraction and cessation of afferent discharge. (From Kuffler and Hunt.[41])

Although alpha-gamma coactivation is clearly an important aspect of the organization of the motor system, it is equally clear that alpha motoneurons, dynamic gamma motoneurons, and static gamma motoneurons can be controlled independently by the CNS. Ablating the anterior lobe of the cerebellum in a decerebrate cat results in an increase in alpha discharge and a decrease in gamma discharge.[30] This illustrates clearly independent influences on the alpha and gamma systems. Independent control of static and dynamic gamma motoneurons has been demonstrated by stimulation in the region of the red nucleus, which produces a predominantly dynamic activation.[11] There are numerous other sites within the CNS that, when stimulated, produce reasonably distinct dynamic or static activation.[12] There is some evidence that alpha-gamma coactivation may in some cases involve predominantly static gamma fibers with the dynamic fibers relatively uninfluenced.

SPINDLE AFFERENT RECORDINGS IN MAN

Observing the response of muscle afferents during voluntary or involuntary motor acts in the unanesthetized animal has been exceedingly difficult. This has been accomplished in man by delicately inserting recording electrodes through the skin directly into the muscle nerve. Once inside the nerve, single spindle afferents have been successfully isolated by making very small adjustments of the electrode position. It has been found that the impulse frequency in these afferents increases during voluntary isometric contraction of the muscle in which the isolated unit is located.[57,58] As discussed earlier, this could occur only during coactivation of the alpha and gamma efferents. Supporting this conclusion is the finding that the discharge from the spindles, during muscle contraction, is reduced after a partial lidocaine nerve block affecting primarily gamma efferents. Comparison of the temporal relationship between onset of muscle contraction and onset of spindle discharges during voluntary, fast, alternating finger and foot movements suggests that in the flexor muscles studied, the contractions are initiated by the direct alpha route.[31] After a brief delay of 20 to 50 msec (due primarily to the latency of the stretch reflex) the contraction is further influenced reflexly via the coactivated gamma loop through the spindles.

The technique for studying spindle responses in man has also been applied to patients having involuntary alternating muscle contractions.[32] The contractions in resting tremor of parkinsonism are apparently organized according to the principle of alpha-gamma coactivation similar to that found in voluntary contraction of normal subjects. On the other hand, studies of spastic patients with clonus indicate that the spindle afferents are silent during contraction, discharging only during the relaxation phase. The resulting oscillation of the stretch reflex, which because of alpha-gamma coactivation does not occur during voluntary alternating contractions in the normal subject nor during the resting tremor of parkinsonism, is thought to maintain the involuntary alternating contractions of clonus.

These studies of spindle afferent responses in man have necessarily been done under rather restricted conditions. Thus one must exercise care not to overgeneralize the results.

SIGNALS FROM TENDON ORGANS

As discussed earlier, changes in muscle length do not affect tendon organs to any degree because they are located on inelastic tendon fascicles that lengthen very little in comparison with muscle. Thus these receptors provide almost no information about length. However, if active contraction occurs when the muscle is lengthened and its ends are fixed, the shortening of the contractile part of the muscle necessarily lengthens the noncontractile region, where tendon organs are located, and vigorous firing results. The rate of discharge is related to the tension that develops; however, this relationship may be more complicated than one of simple proportionality.[40] A particular tension may be developed at various muscle lengths. By virtue of their location, tendon organs measure this tension regardless of length. By systematically testing motor units in the soleus muscle, it has been found that activation of any of about 15 different units will each cause firing of a given tendon organ.[35] Fig. 25-13 illustrates the discharge of a tendon organ produced by stimulation of a single motor unit. Histologic studies have shown that 3 to 25 muscle fibers insert on each tendon organ. Hence a motor unit in the soleus that has an average of 180 fibers contributes only 1 or 2 of them to a tendon organ. Apparently a tendon organ samples the local tension in a muscle by monitoring a minute fraction of the muscle fibers from a number of motor units. The response of a tendon organ is determined by the number of active motor units inserting in it and the tension developed by each of the contributing fibers. Thus any given tendon organ reflects local conditions within the muscle rather than average tension of the entire muscle. This distinction may have special functional significance, but if so, it is not yet apparent from experiments.

The characteristic response of tendon organs to tension may play a possible role in the relief of

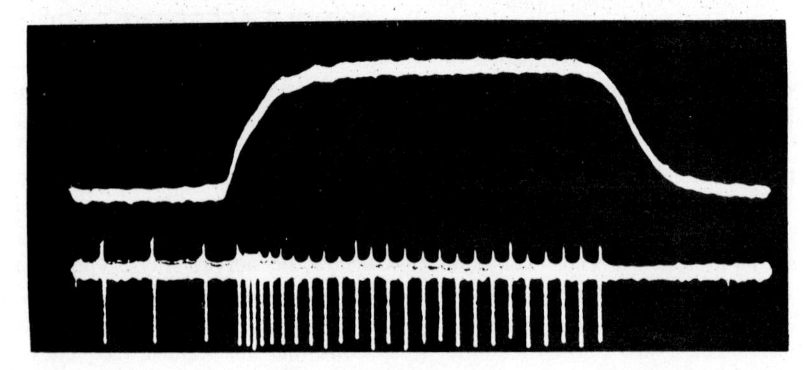

Fig. 25-13. Response of a single tendon organ in soleus (lower tracing) to tetanic stimulation of ventral root filament containing only one soleus axon. Tension developed by contraction of this single motor unit was 18 gm (upper tracing). (From unpublished records of Houk and Henneman.)

certain (but by no means all) types of muscle cramps. If an individual makes a maximum effort to contract a fully shortened muscle (a rigorously flexed biceps, for example), a painful cramp will often develop. The activation of pain receptors causes an even more vigorous contraction of the muscle by way of reflex pathways, which leads to even more intense pain.[48] At the same time, because the muscle is fully shortened, little or no force is produced on the tendon despite the vigorous contraction. Thus the tendon organs are silent. The cramping can be relieved by stretching the contracting muscle out to a long length so that force is generated on the tendon organs. This activation of tendon organs results in strong reflex inhibition that acts to relax the muscle and thus relieve cramping.[26]

Although this example demonstrates the operation of tendon organs in human subjects, it undoubtedly is a trivial example in terms of overall physiologic influence. Since tendon organs supply the CNS with information about muscle force, one would expect that their overall function is almost surely crucial in the moment-to-moment control of muscular activity.

CENTRAL CONNECTIONS AND FUNCTION OF MUSCLE RECEPTORS[3-5,8,46]

The function of muscle receptors is to observe the actions of the muscle and send certain information concerning that activity back to the CNS. As seen from the previous discussion, there is a reasonably complex array of information transmitted. The tendon organs supply information about muscle tension. The primary and secondary spindle receptors signal incremental changes in the configuration of the intrafusal muscle fibers, and the primary receptors are also responsive to the rate of change. If the gamma efferents are silent or are discharging at a constant rate, intrafusal fibers behave congruently with the extrafusal fibers, so that the spindle receptors can be thought of as responding to the changes in length of the muscle as a whole. In the presence of gamma efferent modulation the response of the muscle afferents is more complicated, although to a first approximation they can be thought of as transmitting information pertaining to the difference between the lengths of the contractile portions of the extrafusal and intrafusal muscle fibers.

More pertinent questions are what does the CNS do with this information from the muscle afferents once it is received? What effect do the muscle afferents have on the behavior of the animal? The answers to these questions have been amazingly elusive. Cortical projections from muscle afferents have been identified in the cat, monkey, and baboon.[9,10,49,52] The implications of these projections are all but unknown. The only glimpse we have into the matter comes from recent evidence showing that vigorous, phase-locked activation of the primary receptors (accomplished by vibration of the muscle tendon) will disrupt the normal accuracy of joint-angle perception in human subjects.[28] Prior to this demonstration, it was generally accepted that perception of joint angle was accomplished entirely by information provided by afferents innervating the joints themselves. The role primary spindle endings play in this affair under more physiologic conditions is controversial at present.

The muscle afferents also project onto the cerebellum.[6] Muscle afferents have been found to influence cells in the ventral, dorsal, cuneate, and rostral spinocerebellar tracts. The precise use to which the cerebellum puts the information received from muscle afferents is not known. However, it has been well established that the cerebellum plays a role in the fine control of movement. Thus it is not surprising that information from the periphery is needed.

The muscle afferents are included in a variety of reflex actions involving the spinal cord and brain stem. These reflex pathways are reminiscent of the feedback control systems used by design engineers. Such systems have a strong theoretical foundation and lend themselves to mathematical modeling. Because of this, along with the fortuitous circumstance that the spinal cord is reasonably accessible experimentally, the reflex actions of muscle afferents have been far more rigorously studied and analyzed than have their effects on higher centers of the CNS. The details of these reflex connections are the subject of another chapter in this text. Numerous discussions of the possible role of reflexes in the control of movement and posture exist in the literature.[3-5,8,47]

Summarizing, it appears that reflexes involving the muscle stretch receptors probably have the following functional roles: (1) Reflexes involving both spindle receptors and Golgi tendon organs modulate the stiffness with which a muscle resists changes in length. (2) Reflexes involving the primary spindle receptors serve to damp the neuromuscular system, thus lessening its natural tendency to oscillate. (3) Reflexes involving both spindle receptors and Golgi tendon organs partly compensate for variable behavior (due to fatigue, for example) of the neuromuscular system and for certain response nonlinearities resulting from the inherent properties of muscle. (4) The gamma motoneurons act to modulate the gain or sensitivity of the spindle reflex pathways, thereby offering a mechanism for adjusting the stiffness of the muscular system and the degree of damping. During alpha-gamma coactivation the gamma system may also be thought of as reflexly reinforcing the power of muscular contraction. (5) During locomotion the basic stepping movement of each limb is controlled by a central pattern generator in the spinal cord; interlimb coordination is achieved by interaction between the generators for each limb. The output from these generators can be modified to some extent by muscle stretch receptor and other peripheral receptor reflexes; however, the details of this influence are not well understood.

ACKNOWLEDGMENT

We wish to thank Professor Yves Laporte for reading an earlier draft of this chapter and for the numerous helpful suggestions he made.

REFERENCES

General reviews

1. Barker, D.: The innervation of mammalian skeletal muscle. In de Reuck, A. V. S., and Knight, J., editors: Ciba Foundation symposium on myotatic, kinesthetic and vestibular mechanisms, London, 1967, J. & A. Churchill, Ltd.
2. Barker, D., Hunt, C. C., and McIntyre, A. K.: Muscle receptors. In Hunt, C. C., editor: Handbook of sensory physiology, Berlin, 1974, Springer-Verlag.
3. Granit, R.: The functional role of the muscle spindles—facts and hypotheses, Brain **98:**531, 1975.
4. Grillner, S.: Locomotion in vertebrates: central mechanisms and reflex interaction, Physiol. Rev. **55:**247, 1975.
5. Matthews, P. B. C.: Mammalian muscle receptors and their central actions, London, 1972, Edward Arnold (Publishers), Ltd.
6. Oscarsson, O.: Functional organization of the spino- and cuneo-cerebellar tracts, Physiol. Rev. **45:**495, 1965.
7. Proceedings of the Anatomical Society of Great Britain and Ireland: Symposium on muscle spindles, J. Anat. **119:**183, 1975.
8. Stein, R. B.: Peripheral control of movement, Physiol. Rev. **54:**215, 1974.

Original papers

9. Albe-Fessard, D., and Liebeskind, J.: Origine des messages somatosensitifs activant les cellules du cortex moteur chez le singe, Exp. Brain Res. **1:**127, 1966.
10. Anderson, S. A., Landgren, S., and Wolsk, D.: The thalamic relay and cortical projection of group I muscle afferents from the forelimb of the cat, J. Physiol. **183:** 576, 1966.
11. Appelberg, B.: The effect of electrical stimulation in nucleus ruber on the response to stretch in primary and secondary muscle spindle afferents, Acta Physiol. Scand. **56:**140, 1962.
12. Appelberg, B., and Emonet-Dénand, F.: Central control of static and dynamic sensitivities of muscle spindle primary endings, Acta Physiol. Scand. **63:**487, 1965.
13. Appelberg, B., Bessou, P., and Laporte, Y.: Action of static and dynamic fusimotor fibres on secondary endings of cat's spindles, J. Physiol. **185:**160, 1966.
14. Barker, D.: The innervation of the muscle spindle, Q.J. Microsc. Sci. **89:**143, 1948.
15. Barker, D., and Chin, N. K.: The number and distribution of muscle-spindles in certain muscles in the cat, J. Anat. **94:**473, 1960.
16. Barker, D., and Stacey, M. J.: Rabbit intrafusal muscle fibers, J. Physiol. **210:**70P, 1970.
17. Barker, D., Stacey, M. J., and Adal, M. N.: Fusimotor innervation in the cat, Philos. Trans. R. Soc. Lond. (Biol.) **258:**315, 1970.
18. Barker, D., et al.: Morphological identification and intrafusal distribution of the endings of static fusimotor axons in the cat, J. Physiol. **230:**405, 1973.
19. Barker, D., et al.: Distribution of static and dynamic gamma axons to cat intrafusal muscle fibers, J. Anat. **119:**199, 1975.
20. Boyd, I. A.: The structure and innervation of the nuclear bag muscle fibre system in mammalian muscle spindles, Philos. Trans. R. Soc. Lond. (Biol.) **245:**81, 1962.
21. Boyd, I. A., Gladden, M. H., McWilliam, P. N., and Ward, J.: 'Static' and 'dynamic' nuclear bag fibres in isolated cat muscle spindles, J. Physiol. **250:**11P, 1975.
22. Brown, M. C., and Butler, R. G.: Depletion of intrafusal muscle fibre glycogen by stimulation of fusimotor fibres, J. Physiol. **229:**25P, 1973.
23. Brown, M. C., and Butler, R. G.: Studies on the site of termination of static and dynamic fusimotor fibres within muscle spindles of the tenuissimus muscle of the cat, J. Physiol. **233:**553, 1973.
24. Chin, N. K., Cope, M., and Pang, M.: Number and distribution of spindle capsules in seven muscles of the cat. In Barker, D., editor: Symposium on muscle receptors, Hong Kong, 1962, Hong Kong University Press.
25. Crowe, A., and Matthews, P. B.: The effects of stimulation of static and dynamic fusiform fibres on the response to stretching of the primary endings of muscle spindles, J. Physiol. **174:**109, 1964.
26. deVries, H. A.: Physiology of exercise, ed. 2, Dubuque, Iowa, 1974, William C. Brown Co., Publishers.
27. Emonet-Dénand, F., Jami, L., and Laporte, Y.: Skeletofusimotor axons in hind-limb muscles of the cat, J. Physiol. **249:**153, 1974.
28. Goodwin, G. M., McCloskey, D. I., and Matthews, P. B. C.: The contribution of muscle afferents to kinesthesia shown by vibration induced illusion of movement and by the effects of paralysing joint afferents, Brain **95:**705, 1972.
29. Gottlieb, G. L., Agarwal, G. C., and Stark, L.: Studies in postural control systems. Part III. A muscle spindle model, IEEE Trans. Systems Sci. Cybern. **SSC-6:**127, 1970.
30. Granit, R., Holmgren, B., and Merton, P. A.: Two routes for excitation of muscle and their subservience to the cerebellum, J. Physiol. **130:**213, 1955.
31. Hagbarth, K. E., Wallin, G., and Löfstedt, L.: Muscle spindle activity in man during voluntary fast alternating movements, J. Neurol. Neurosurg. Psychiatry **38:**625, 1975.
32. Hagbarth, K. E., Wallin, G., Löfstedt, L., and Aquilonius, S. M.: Muscle spindle activity in alternating tremor of Parkinsonism and in clonus, J. Neurol. Neurosurg. Psychiatry **38:**636, 1975.

33. Harvey, R. J., and Matthews, P. B.: The response of the de-efferented muscle spindle endings in the cat's soleus to slow extension of the muscle, J. Physiol. **157:** 370, 1961.
34. Hasan, Z., and Houk, J. C.: Nonlinear behavior of primary spindle receptors in response to small, slow ramp stretches, Brain Res. **44:**680, 1972.
35. Houk, J., and Henneman, E.: Responses of golgi tendon organs to active contractions of the soleus muscle of the cat, J. Neurophysiol. **30:**466, 1967.
36. Houk, J. C., Harris, D. A., and Hasan, Z.: Non-linear behavior of spindle receptors. In Stein, R. B., Pearson, K. G., Smith, R. S., and Redford, J. B., editors: Control of posture and locomotion, New York, 1973, Plenum Publishing Corp.
37. Hunt, C. C.: Muscle stretch receptors; peripheral mechanisms and reflex function, Symp. Quant. Biol. **17:**113, 1952.
38. Husmark, I., and Ottoson, D.: Is the adaptation of the muscle spindle of ionic origin? Acta Physiol. Scand. **81:**138, 1971.
39. Ip, M. C.: The number and variety of proprioceptors in certain muscles of the cat, M.Sc. thesis, University of Hong Kong, 1962.
40. Jami, L., and Petit, J.: Frequency of tendon organ discharges elicited by the contraction of motor units in cat leg muscles, J. Physiol. **261:**633, 1976.
41. Kuffler, S. W., and Hunt, C. C.: Mammalian small-nerve fibers: system for efferent nervous regulation of muscle spindle discharge, Res. Pub. Assoc. Res. Nerv. Ment. Dis. **30:**24, 1952.
42. Kuffler, S. W., Hunt, C. C., and Quilliam, J. P.: Function of medullated small-nerve fibers in mammalian ventral roots: efferent muscle spindle innervation, J. Neurophysiol. **14:**29, 1951.
43. Laporte, Y., and Emonet-Dénand, F.: Evidence for common innervation of bag and chain muscle fibres in cat spindles. In Stein, R. B., Pearson, K. G., Smith, R. S., and Redford, J. B., editors: Control of posture and locomotion, New York, 1973, Plenum Publishing Corp.
44. Matthews, P. B. C.: The reflex excitation of the soleus muscle of the decerebrate cat caused by vibration applied to its tendon, J. Physiol. **184:**450, 1966.
45. Matthews, P. B. C., and Stein, R. B.: The sensitivity of muscle spindle afferents to small sinusoidal changes of length, J. Physiol. **200:**723, 1969.
46. Merton, P. A.: Speculations on the servo-control of movement. In Malcolm, J. L., Gray, J. A. B., and Wolstenholme, G. E. W., editors: The spinal cord, Boston, 1953, Little, Brown & Co.
47. Nichols, T. R., and Houk, J. C.: Improvements in linearity and regulation of stiffness that results from actions of stretch reflex, J. Neurophysiol. **39:**119, 1976.
48. Norris, F. H., Jr., Gasteiger, E. L., and Chatfield, P. O.: An electromyographic study of induced and spontaneous muscle cramps, Electroencephalogr. Clin. Neurophysiol. **9:**139, 1957.
49. Oscarsson, O., Rosén, I., and Sulg, I.: Organization of neurones in the cat cerebral cortex that are influenced from group I muscle afferents, J. Physiol. **183:**189, 1966.
50. Ottoson, D., and Shepherd, G. M.: Length changes within isolated frog muscle spindle during and after stretching, J. Physiol. (Lond.) **207:**747, 1970.
51. Ovalle, W. K., and Smith, R. S.: Histochemical identification of three types of intrafusal muscle fibers in the cat and monkey based on the myosin ATPase reaction, Can. J. Physiol. Pharmacol. **50:**195, 1972.
52. Phillips, C. G., Powell, T. P. S., and Wiesendanger, M.: Projection from low-threshold muscle afferents of hand and forearm to area 3a of baboon's cortex, J. Physiol. **217:**419, 1971.
53. Poppele, R. E., and Bowman, R. J.: Quantitative description of linear behavior of mammalian muscle spindles, J. Neurophysiol. **33:**59, 1970.
54. Rudjord, T.: A second order mechanical model of muscle spindle primary endings, Kybernetik **6:**205, 1970.
55. Rudjord, T.: A mechanical model of the secondary endings of mammalian muscle spindles, Kybernetik **7:**122, 1970.
56. Swett, J. E., and Eldred, E.: Distribution and numbers of stretch receptors in medial gastrocnemius and soleus muscles of the cat, Anat. Rec. **137:**453, 1960.
57. Vallbo, A. B.: Discharge patterns in human muscle spindle afferents during isometric voluntary contraction, Acta Physiol. Scand. **80:**552, 1970.
58. Vallbo, A. B.: Muscle spindle response at the onset of isometric voluntary contractions in man. Time difference between fusimotor and skeletomotor effects, J. Physiol. **218:**405, 1971.

26

ELWOOD HENNEMAN

Organization of the motoneuron pool

THE SIZE PRINCIPLE

Sherrington called motoneurons "the final common path" to muscle because all the activity in the central nervous system (CNS) that influences movement converges ultimately on motoneurons. A vast superstructure of higher centers is required to formulate the commands that are conveyed to the relatively small populations of cells that control particular muscles. Each group of cells translates a numerically large and complex set of signals, excitatory and inhibitory, into a smaller output that follows simple, logical rules. We have seen that skeletal muscles are highly organized entities with collective properties that serve a wide range of needs. Similarly, the motoneurons that control them are not just an aggregation of uniform cells, but a functional ensemble whose properties are perfectly matched with those of the muscles.

CONCEPT OF THE MOTONEURON POOL

The group of motoneurons innervating a particular muscle is called a "pool."[13] With ordinary staining techniques the members of a pool are not readily identifiable among the other motoneurons scattered throughout the ventral horn. By means of a recently developed technique, however, it is possible to label the motoneurons supplying a particular muscle so that they stand out clearly from the other cells. Horseradish peroxidase (HRP), injected into a muscle, is taken up by the motor axons in the muscle and carried rapidly by retrograde axoplasmic flow back to their cell bodies, where it appears, after appropriate staining, as brown, intracytoplasmic granules. If the intramuscular injection is well dispersed, so that all the motor terminals are exposed to it, a high percentage of the motoneurons is found to contain the HRP reaction product. Fig. 26-1 illustrates the locations of labeled cells in an experiment by Burke et al.[8] in which HRP was injected into the left medial gastrocnemius (MG) muscle and the right soleus (SOL) muscle. Reconstructions of the histologic material show that each pool is arranged as a column of cells running up and down the lateral part of the ventral horn. Throughout most of their courses the two pools overlap to a great extent. Both alpha motoneurons, innervating the ordinary large muscle fibers, and gamma motoneurons, supplying the small intrafusal muscle fibers, were labeled with HRP. The latter were distributed randomly throughout both cell columns. The size of the alpha motoneurons in the SOL pool was more restricted and had a smaller mean value than the cells in the MG pool. The two pools contained about the same numbers of alpha motoneurons with average diameters less than 50 μm, but the MG pool contained many more cells with diameters greater than 50 μm. The distribution of cell sizes was largely random, but the rostral third of the MG column contained a higher ratio of large to small alpha cells than did the caudal two thirds.

For many years, it has been assumed that the motoneurons supplying a muscle form a homogeneous population of cells differing only in size and in certain size-dependent properties. Recently, however, it has been found that there may be two or more types of alpha motoneurons differing in properties that are independent of size.[22,23] Just as there are distinct types of muscle fibers in a single mammalian muscle, there may be distinct types of motoneurons that innervate them and are responsible for their properties.

The axons of the cells in different pools leave the spinal cord in one, two, or three consecutive ventral roots. Within these roots, they are inter-

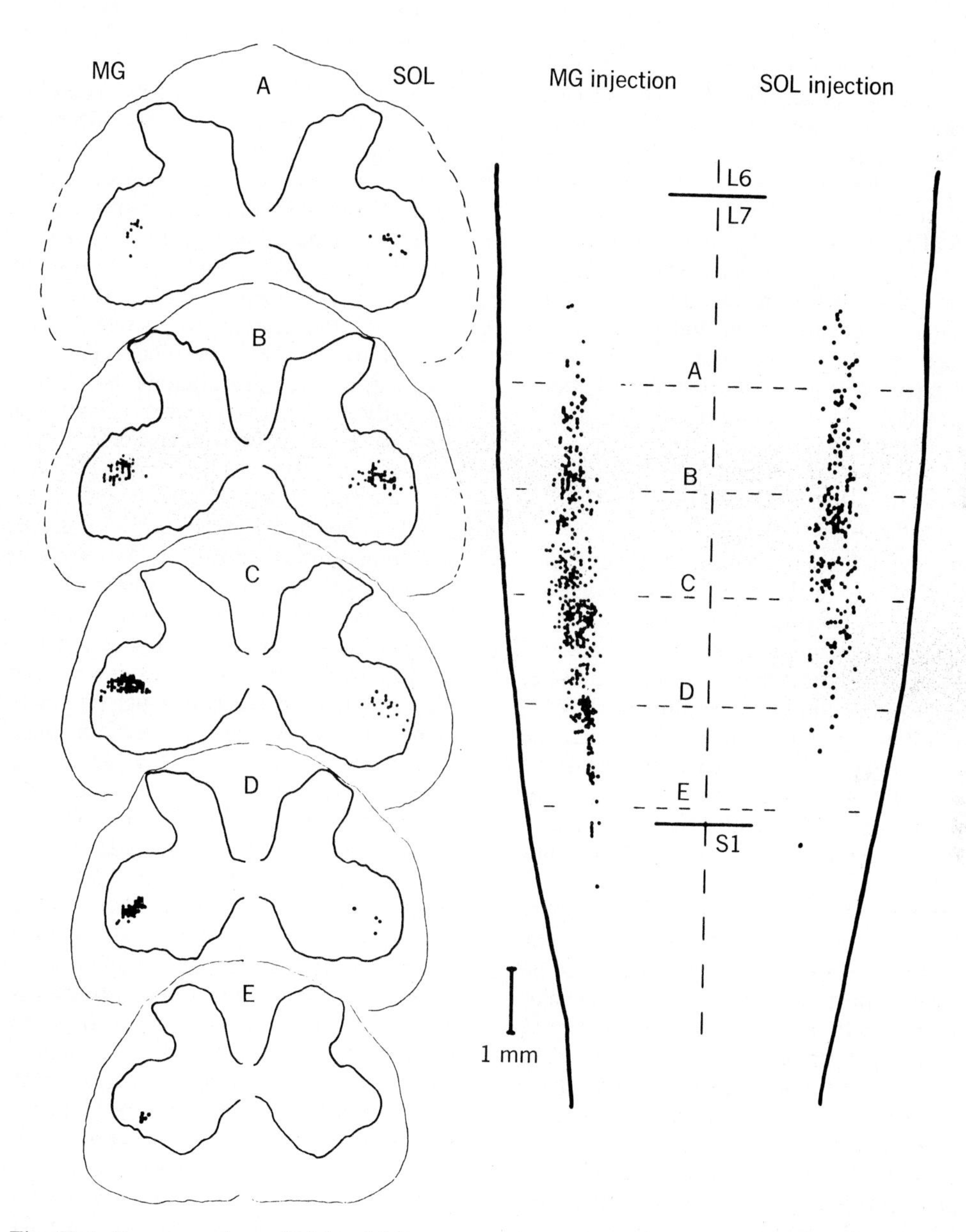

Fig. 26-1. Reconstructions of MG and SOL nuclei from serial sagittal sections of cat's spinal cord. Right: dorsal view of spinal cord outline (white matter–pia boundary in heavy lines) on which are superimposed positions (dots) of MG (left hemicord) and SOL motoneuron cell bodies (right hemicord). Boundaries between L6, L7, and S1 segments (identified by dorsal root entry zones) are indicated by heavy lines; midline denoted by longitudinal dashed line. Dashed lines across cord denote levels (labeled *A* to *E*) at which reconstructions were made. Left: reconstructions at levels *A* to *E* showing white matter–pia boundary in light lines and gray–white matter boundary in heavier lines. Neurons indicated at each level are cells located within 300 μm rostral or caudal to that cross section. (From Burke et al.[8])

mingled with axons to other muscles in the hindlimb. Eventually the axons to a particular muscle are reassembled in the muscle nerve, which contains both the sensory and the motor fibers of that muscle.

It can hardly be overemphasized that a motoneuron pool is designed for a single purpose—to produce precise mechanical effects in its muscle. It must do so by suitably timed combinations of the separate contractions of its individual motor units. These units differ so widely in their contractile properties that it is difficult to see how they can be activated to produce the exact combination of total speed and power required from the muscle. To operate under a wide variety of conditions places stringent demands on the design of the motoneuron pool and its inputs. A knowledge of pool organization and its operations is essential to an understanding of motor systems and is becoming useful in diagnostic testing of the nervous system.

FUNCTIONAL SIGNIFICANCE OF THE SIZE OF MOTONEURONS

If the motor fibers in a muscle nerve are examined histologically (after degeneration of the sensory fibers),[17] it is found that two groups can be distinguished: (1) *alpha* fibers, with diameters from about 9 to 20 μm, depending on the muscle, and (2) *gamma* fibers, ranging from about 1 to 8 μm. As Ramón y Cajal[45] first pointed out, the diameter of a nerve fiber is related directly to the size of its cell body. The largest motoneurons have surface areas of 79,000 to 250,000 μm^2,[1] and they may have as many as 10,000 synaptic knobs on their soma and dendrites. The smallest alpha cells have surface areas of 10,000 to 15,000 μm^2 and correspondingly fewer synapses. In Chapter 25, it was shown that large-diameter axons innervate motor units that develop large tensions and small-diameter axons supply motor units that produce small tensions. The sizes of the cells in a pool are apparently related to the number and size of muscle fibers they innervate. Since a motoneuron is responsible for trophically maintaining all the muscle fibers it innervates (cut off from its motor nerve, the muscle atrophies), it is not surprising that those with large trophic responsibilities are larger than those with few dependents.

The relation between the size of a motoneuron and its electrical properties did not become apparent until recently. With the introduction of the oscilloscope, it was found that conduction in peripheral nerve fibers differs systematically with their diameter. Gasser[20] showed that the amplitude of a nerve impulse recorded externally from a peripheral nerve is directly related to the diameter of the fiber transmitting it. Thus when impulses of several different amplitudes are recorded from the same nerve filament, the largest impulse signifies the firing of the largest fiber and the smaller impulses signify the discharges of correspondingly smaller fibers. Identification of single impulses is facilitated by recording from thin filaments of ventral roots, which can be teased apart more easily than peripheral nerves because they contain less connective tissue. In this way, samples of the output of a pool of motoneurons can be investigated using a variety of natural or artificial stimuli to elicit responses.

A convenient means of evoking a reflex discharge of motoneurons is to stretch an extensor (antigravity) muscle in the hind limb of a decerebrate cat. In such preparations the stretch reflex is greatly exaggerated, and the slightest degree of passive stretch results in the discharge of motoneurons and the development of muscular tension. The reflex is highly specific, the response being limited to the motoneurons of the muscle that is stretched or its closest synergists.

Fig. 26-2 illustrates the results of stretching the triceps surae and recording from a filament of the seventh lumbar ventral root. The tension applied to the tendon of the muscle is measured by the separation of the two upper beams in each frame, and the ventral root discharge appears in the lowest tracing. With the muscle completely relaxed (line 1) the only activity recorded was a steady stream of impulses of very low amplitude. These small impulses were recorded from gamma motoneurons whose axons are of smaller diameter than those of the alpha fibers also present in the filament. With slight stretch (line 2), two discharges of a much larger alpha motor fiber were recorded. With further stretch the discharges of this unit increased in rate and regularity. With greater stretch the responses of a second unit generating larger impulses appeared in line 4. Further stretch accelerated firing rates, but no additional units were recruited. Release of stretch (lines 6 to 9) caused the larger unit to cease firing before the smaller unit.

In Fig. 26-3 a similar series of tracings is reproduced to illustrate recruitment in a filament containing the axons of five triceps motoneurons. With increasing stretch, progressively larger impulses appeared. The recruitment order was 1, 2, 3, 5, 4; the order of dropout was 5, 4, 3, 2, 1.

Fig. 26-4 shows the distribution of stretch thresholds for 208 motoneurons. The tensions required to elicit maintained rhythmic firing

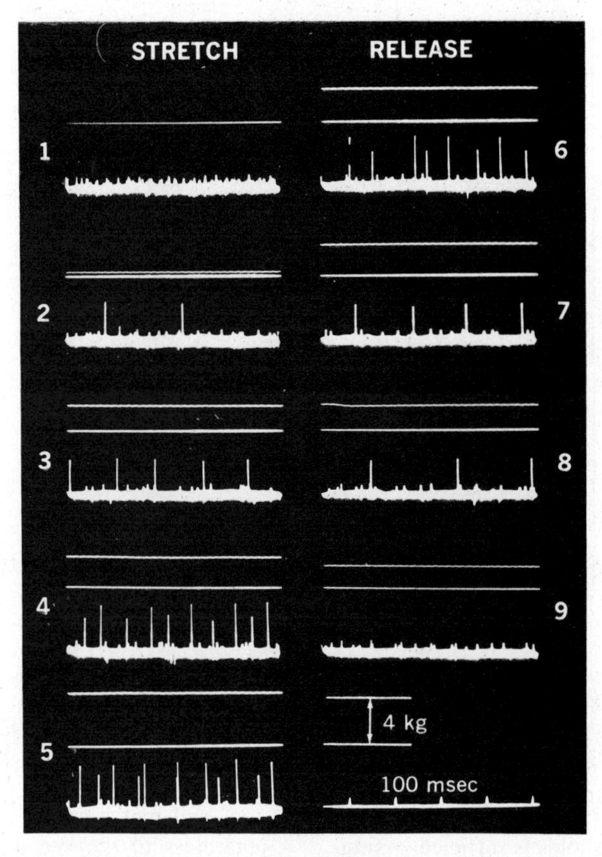

Fig. 26-2. Stretch-evoked responses of two alpha motoneurons, recorded from filament of seventh lumbar ventral root. Amount of tension applied and developed reflexly is indicated by separation of two upper beams in each frame. Several seconds elapsed between successive frames while muscle was stretched, *1* to *5*, and released, *6* to *9*. (From Henneman et al.[29])

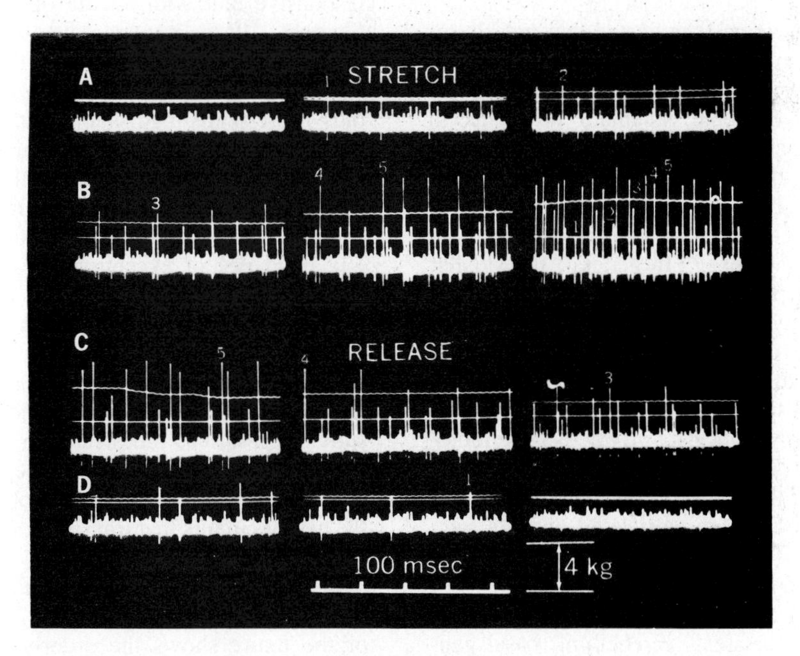

Fig. 26-3. Stretch-evoked responses of five alpha motoneurons recorded from filament of first sacral ventral root during stretch, **A** and **B,** and release, **C** and **D,** of triceps surae muscle. Small numerals above action potentials indicate rank of units according to size. (From Henneman et al.[29])

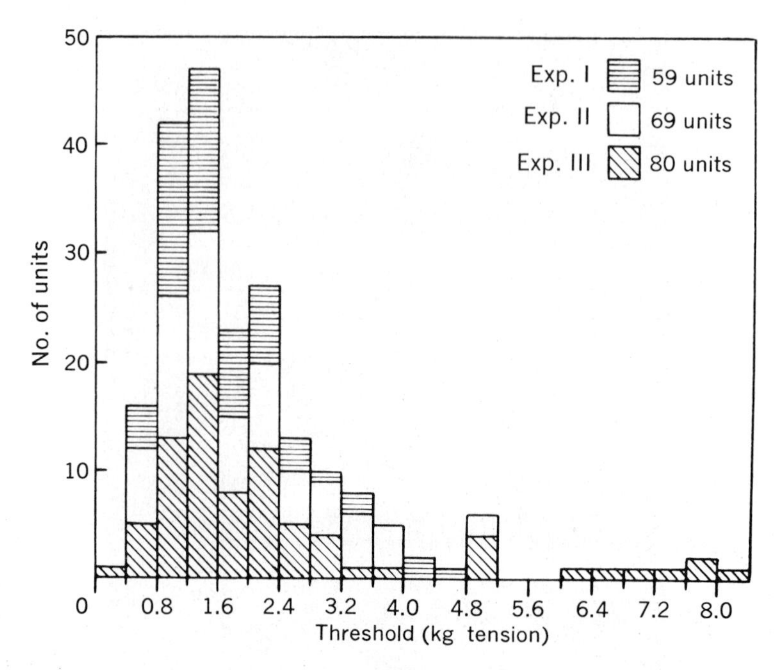

Fig. 26-4. Frequency distribution of thresholds of 208 tonically responding motoneurons. Data obtained in three experiments as indicated. Abscissa: threshold in kilograms of tension applied to deefferented triceps surae muscle. Ordinate: number of units whose thresholds fall between values indicated on abscissa. (From Henneman et al.[29])

ranged from less than 0.4 kg to more than 8 kg. The distribution of thresholds is sufficiently similar to the distribution of maximal tensions produced by motor units in the preceding chapter to indicate that the size of a motor unit and the stretch threshold of its motoneuron are closely related.

These observations[29] indicated that orderly recruitment of motoneurons must depend on differences in the excitabilities of the motoneurons themselves or on some systematic difference in the input from stretch receptors, resulting in more effective stimulation of small cells. To distinguish between these two possibilities a variety of spinal reflexes was used to discharge flexor or extensor motoneurons.[28] In general, the smaller of any two responding units was discharged at a lower intensity of stimulation regardless of whether the excitatory stimuli arose ipsilaterally or contralaterally, physiologically or electrically, whether the responses were elicited monosynaptically or polysynaptically, or whether the motoneurons were flexor or extensor. These results suggested that the susceptibility of a motoneuron to discharge is strongly correlated with its size. However, input was not excluded as an important factor in this correlation.

Trains of electrical stimuli were also applied to brain stem motor areas, cerebellum, basal ganglia, and motor cortex of the cat.[48] In general, regardless of the site of stimulation, motoneurons whose axons yielded the smallest impulses in ventral root filaments were discharged at the lowest thresholds, and other units were recruited in order of increasing size. By comparing each responsive unit with all the others in the same filament, 396 pairs of motoneurons were examined. In 86% of comparisons the smaller unit had the lower threshold, in 4% the two thresholds were indistinguishable, and in 10% the larger unit had the lower threshold. The use of electrical stimulation with widely spaced electrodes must have resulted in simultaneous discharges from several pools of motoneurons. The low incidence of exceptions to orderly recruitment suggests that most of the descending activity reaches motoneuron pools through a few penultimate internuncial circuits[37] that transform this activity into highly ordered patterns of input. In conjunction with previous findings, these results suggest that the size of a motoneuron is strongly correlated with its susceptibility to discharge regardless of the source of excitation and the neural circuits that transmit it to the motoneurons.

Susceptibility to inhibition is also a function of cell size, as illustrated in Fig. 26-5. Portion *A* of the figure shows the orderly recruitment of

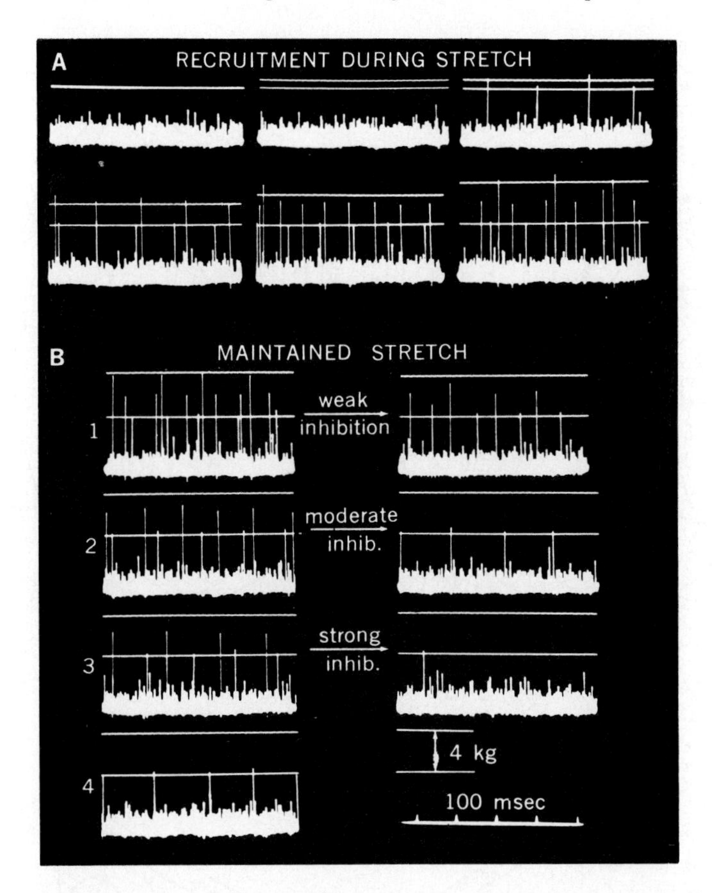

Fig. 26-5. A, Orderly recruitment of three triceps motoneurons of different sizes in response to increasing degrees of stretch. **B,** Orderly inhibition of same three units during constant stretch of 4 kg. Records *1* to *4* on left show control responses to stretch before, between, and after each of three inhibitory stimulations. Largest unit was silenced first (line *1*), intermediate unit next (line *2*), and smallest unit last (line *3*). (From Henneman et al.[28])

three triceps motoneurons with increasing stretch. After the normal pattern of recruitment has been established, a 4 kg stretch, sufficient to elicit tonic firing of all three motoneurons, was applied by elongating the muscle to a fixed length. This stretch was maintained while the effects of inhibition, shown in part *B,* were recorded. The tracings labeled 1 to 4 on the left side of Fig. 26-5, *B,* show the responses of these neurons before any inhibitory stimulation (1) and after weak (2), moderate (3), and strong (4) stimulation. Those on the right were obtained during inhibitory stimuli lasting throughout the 500 msec sweep. Weak inhibition silenced the largest unit. Inhibition of moderate intensity suppressed the unit of intermediate size as well, leaving the smallest unit discharging until a still stronger inhibition eliminated all responses to stretch. During each of the brief periods after inhibition there was a partial but incomplete recovery of the original pattern of response.

Results of this kind were obtained in a variety of experiments. In general, the larger the unit, the more susceptible it is to inhibition. Regardless of the existing level of excitatory drive, regardless of whether the motoneurons were responding rhythmically to stretch or monosynaptically to synchronous volleys, and regardless of whether the inhibition was "direct," internuncially mediated, autogenetic, or recurrent, the inhibitability of each cell was strictly size dependent. *It was concluded that, whereas the excitability of motoneurons is an inverse function of cell size, their inhibitability is a direct function of cell size.*

The foregoing observations indicate that each motoneuron functions as an integrating device. The net effects of cell size, excitatory input, and inhibition on the responses of three different motoneurons are represented quantitatively by

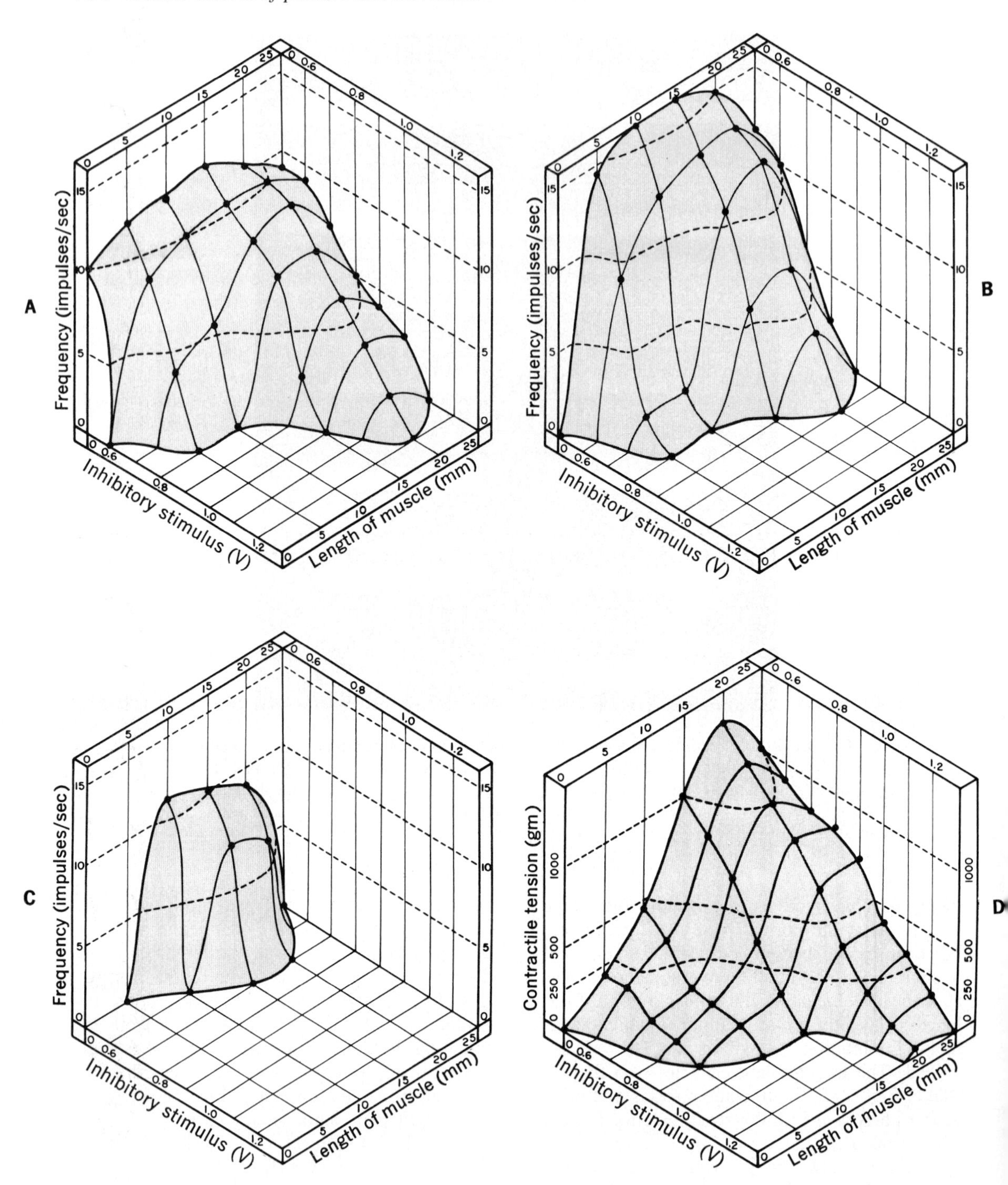

Fig. 26-6. Graphic representation of discharge frequency of three motoneurons of different sizes, **A, B,** and **C,** of triceps surae and of contractile tension developed by the muscle itself, **D,** in response to varying degrees of excitation and inhibition. X axis is intensity of inhibitory stimulation applied to ipsilateral deep peroneal nerve. Y axis is frequency of discharge in **A, B,** and **C** and contractile tension in **D.** Z axis is stretch of triceps surae in millimeters. Data for all four graphs obtained simultaneously. Each plotted point represents mean of two successive determinations. Note that unit **A** (the smallest) fired spontaneously without stretch, whereas unit **B** began to discharge between 0 and 5 mm of stretch and unit **C** (the largest) between 5 and 10 mm. (From Henneman et al.[28])

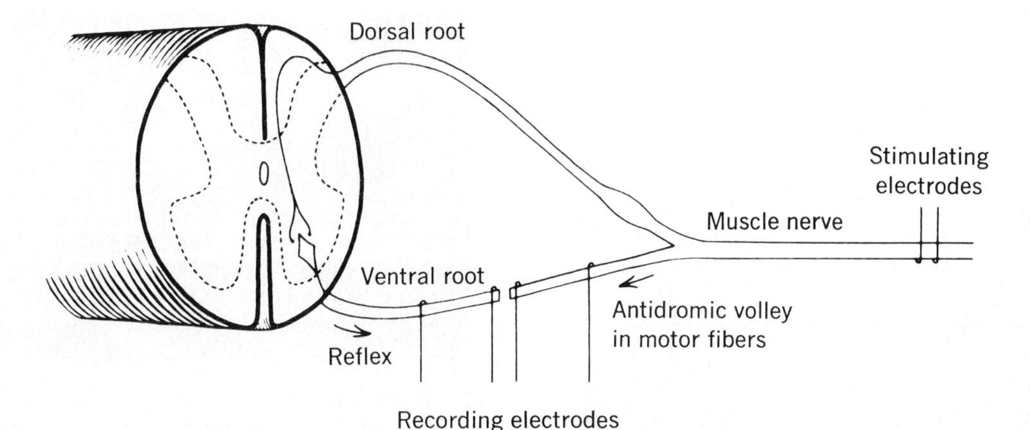

Fig. 26-7. Scheme of experiment used to measure maximal monosynaptic discharge of a motoneuron pool. See text. (From Clamann et al.[12])

means of three-dimensional graphs in Fig. 26-6. In parts *A*, *B*, and *C* the rate of discharge of each of three different motoneurons is shown on the vertical axis. The cells were subjected to various mixtures of excitation and inhibition, whose intensities can be read on the axes showing inhibition and excitation. The responses of the smallest and most excitable unit, which was spontaneously active with the muscle relaxed, are plotted in *A*. The stretch threshold of the intermediate unit *(B)* was between 0 and 5 mm extension, and that of the largest and least excitable unit was between 5 and 10 mm. At all levels of stretch the intensity of inhibitory stimulation required to silence a unit was always the greatest for the most excitable cell *(A)*. In part *D* the responses of all the motor units are combined in the contractile response of the whole muscle. The graph in *D* has the same general features as those in *A*, *B*, and *C*, although the tension of the whole muscle replaces frequency of firing on the vertical axis.

QUANTITATIVE METHOD OF MEASURING OUTPUT OF MOTONEURON POOL[12]

Although there was a striking degree of orderliness in both the neural and the contractile properties of motor units, there was a need for a more rigorous method of defining the motoneuron pool and its operations in quantitative terms. A method was devised for measuring the maximal output of a motoneuron pool and, from this, the percent output during any monosynaptic reflex. The method is illustrated in Fig. 26-7. The combined nerves to the medial gastrocnemius (MG) muscle and the lateral gastrocnemius and soleus (LG-S) muscles, which together make up the triceps surae muscle, were stimulated once a second with brief shocks that set up antidromic volleys in all the alpha motor fibers of these nerves and dromic volleys in all the group Ia fibers. The antidromic volley was recorded in the distal half of the seventh lumbar (L7) or first sacral (S1) ventral root and displayed on one trace of a two-beam oscilloscope; the dromic volley elicited a monosynaptic reflex in the proximal half of the same ventral root, which was displayed on the other trace. The baselines of these two traces are superimposed in Fig. 26-8. The left potential in each pair of responses is the antidromic volley. Its amplitude is a measure of 100% discharge of MG and LG-S alpha motor fibers as recorded from this ventral root. The right potential in each pair is the monosynaptic reflex. Its amplitude is determined by the number of MG and LG-S motoneurons discharged during the reflex. Prior to a 12 sec conditioning tetanus to the muscle nerves at 500/sec, the amplitude of the reflex was less than half that of the antidromic volley (upper row of responses in Fig. 26-8). After the conditioning tetanus (TET) the reflex amplitude increased due to posttetanic potentiation (PTP)[36] and reached temporary equality with the antidromic volley (left side of second row). Thereafter the reflex response gradually declined back to its original amplitude. At the peak of PTP, all the MG and LG-S motoneurons were evidently discharged reflexly, as nearly as could be determined with this technique. A more accurate measure of an action potential is the area obtained by integrating it electronically. This technique was used in the experiments to be described subsequently. The time integral had an

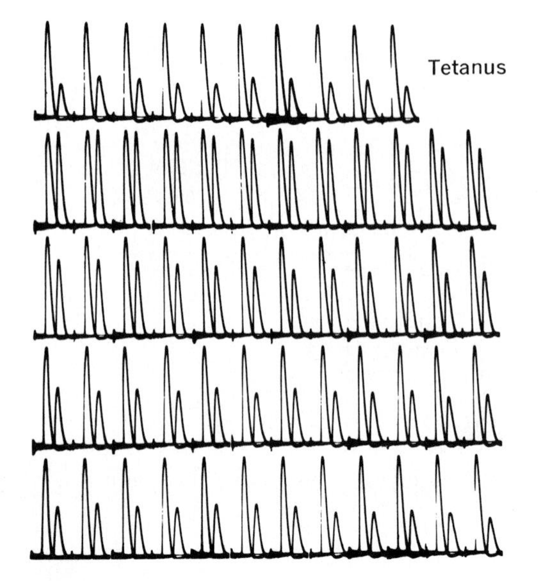

Fig. 26-8. Pairs of responses recorded as in Fig. 26-7 from proximal and distal halves of S1 ventral root in response to single-shock stimulation of combined nerves to MG and LG-S. Left and right deflections of each pair are antidromic and reflex responses, respectively, displayed on superimposed traces of a two-beam oscilloscope at equal gains. First 10 pairs in upper row recorded before 12 sec tetanus (500/sec) to MG and LG-S nerves. Subsequent pairs recorded at 2 sec intervals after tetanus. See text. (From Clamann et al.[12])

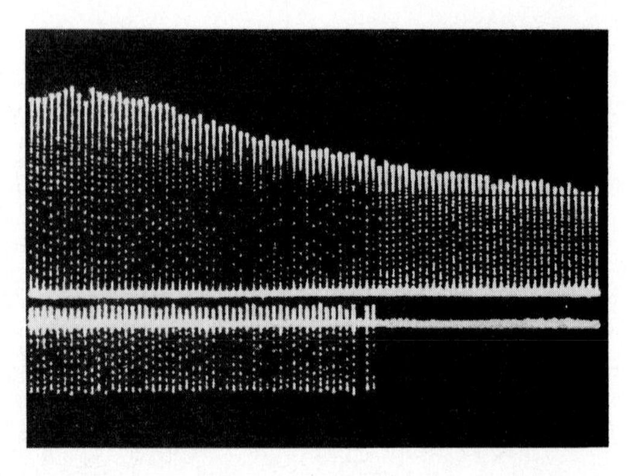

Fig. 26-9. Simultaneous recordings of a series of monosynaptic reflexes of triceps surae pool (above) and a single triceps motoneuron (below), showing critical firing level of the latter. Upper trace: time integrals of monosynaptic reflexes recorded at 1/sec from proximal half of first sacral ventral root. Height of each vertical line measures size of population response. Lower trace: monosynaptic reflexes of a single triceps motoneuron recorded from a small filament of seventh lumbar ventral root. Records were made during decline in PTP following brief train of conditioning shocks applied to combined MG and LG-SOL nerves at 500/sec. (From Henneman et al.[30])

absolute accuracy of 1% to 2% and varied by less than 0.2% on repeated trials. Thus any monosynaptic reflex could be described quantitatively, using a scale of output running from 0% to 100%.

RANK ORDER OF MOTONEURONS WITHIN A POOL[30]

With the technique just described the monosynaptic reflexes of an entire pool of triceps motoneurons were recorded once a second from an appropriate ventral root of a decerebrate cat. Simultaneously the monosynaptic reflexes of one motoneuron from the same pool were recorded from a small filament of ventral root. An example of the oscillographic records obtained in these experiments is reproduced in Fig. 26-9. The series of vertical lines in the upper tracing represents the time integrals of successive monosynaptic reflexes recorded at a rate of 1/sec. The height of each vertical line measures the response of all the motoneurons discharged by the afferent volley, that is, the percent discharge of the pool. Just before this series of responses was recorded, a brief train of conditioning shocks was applied to the combined MG and LG-S nerves at 500/sec, which resulted in posttetanic potentiation of subsequent monosynaptic reflexes. The responses in Fig. 26-9 were recorded during the decline in this PTP. The vertical lines in the lower trace are the corresponding monosynaptic reflexes of a single triceps motoneuron. Fig. 26-9 shows that as long as the response of the triceps population exceeded a certain level, the single unit from the same pool was discharged in every monosynaptic reflex. When the population response declined below this critical level, the unit ceased to respond. The first failure of the unit to discharge was followed by two successive responses before the unit ceased to fire completely. Careful examination of the record reveals that the first failure was associated with a smaller population response than the next two responses were. The "critical firing level" (CFL) was, thus, between the levels associated with the first failure and the smaller of the next two population responses. The difference between these levels was less than 1% of the maximal output of the triceps pool, which is approximately the limit of accuracy of this technique.

The records in Fig. 26-9 suggest that the CFL is very sharply defined. When the PTP was repeated several times and sufficient observations

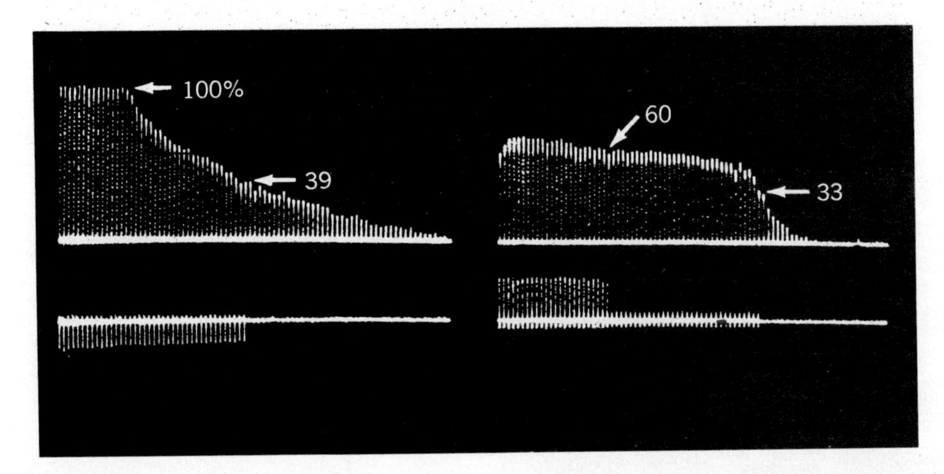

Fig. 26-10. Critical firing levels of three triceps motoneurons. Single unit in lower left tracing had a sharp threshold at 39%. Two units in lower right tracing had thresholds of 60% and 33%, respectively. Plateau in upper left record indicates 100% discharge level of triceps pool. (From Henneman et al.[30])

were available within the critical range, however, it was usually apparent that there was a slight degree of uncertainty in the behavior of the motoneuron, that is, there was a narrow range of population responses over which a unit might or might not respond monosynaptically. This is evident in Fig. 26-10, which reproduces the responses of three triceps units with relatively sharp "thresholds." The plateau at the beginning of the upper trace on the left side of the figure indicates the 100% level of the pool discharge during the early period of PTP. During the decline in the population response, each of the three units in Fig. 26-10 abruptly stopped responding at a different critical level and did not discharge again in that series. Careful examination of the record on the right side of Fig. 26-10 reveals that the unit that ceased firing at approximately the 60% level actually discharged once at 58.6% and failed to respond several times just above the 60% level. The range over which the unit's response was unpredictable (1.7% in this case) was called the "uncertain range." The CFL was defined as the arithmetic center of the uncertain range.

In this series of experiments, 203 triceps units were examined. Their CFLs ranged from 0% to 98.7%. The mean uncertain range for the entire group (2.1%) did not vary systematically with the CFL.

No technique was available for measuring CFLs directly during repetitive firing; instead, it was possible to make use of the order of recruitment in pairs of motoneurons as a measure of their susceptibility to repetitive discharge and to compare this with the CFLs established in monosynaptic tests. Fig. 26-11 reproduces recordings taken from two different pairs of plantaris motoneurons in the same preparation. To elicit the responses in traces b and c, stimuli were applied to the plantaris nerve at a rate of 300/sec. A sample of these stimuli was led to the oscilloscope. Their intensity (i.e., voltage) is indicated by the level of the traces labeled "a" with reference to the baselines labeled "d." As the intensity was first increased and then decreased manually, the two units commenced and ceased their firing at different times, thus signaling their different thresholds to repetitive discharge. In both pairs of units the motoneuron with the lower CFL (shown at the end of the trace) began to discharge sooner and ceased firing later. In the lower pair of units, differing in CFL by 3.1%, the onsets of firing were farther apart, and the last discharges were more separated in time than those of the upper pair of units, which differed by only 1.7%.

Sixty-two pairs of motoneurons were compared as illustrated in Fig. 26-11. In 57 of these pairs the unit with the lower CFL in monosynaptic tests was invariably the more susceptible to repetitive firing. In 5 pairs, which differed in CFLs by only 1.7%, 1.1%, 2.0%, 2.1%, and 5.0%, the unit with the lower CFL was slightly less susceptible to repetitive firing. The minor disagreements between the two methods of ranking motoneurons suggest that errors, which are usually not greater than ±1.0%, occur in estimating CFL. Since the repetitive stimulation used in these experiments activated several afferent systems in addition to the group I_A fibers and presumably elicited descending activity from higher centers as well as recurrent inhibition from

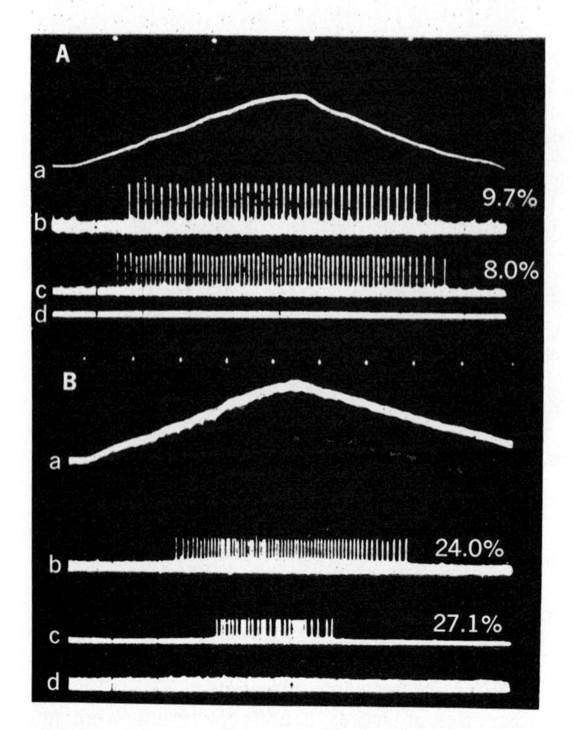

Fig. 26-11. Susceptibilities to repetitive discharge of two different pairs of plantaris motoneurons with known critical firing levels as indicated. Electrical stimulation of plantaris nerve at 300/sec in both cases. Stimulus intensity indicated by level of trace *a* with reference to baseline *d*. Responses of plantaris units recorded on traces *b* and *c*. Time marks at 1 sec intervals for both **A** and **B.** (From Henneman et al.[30])

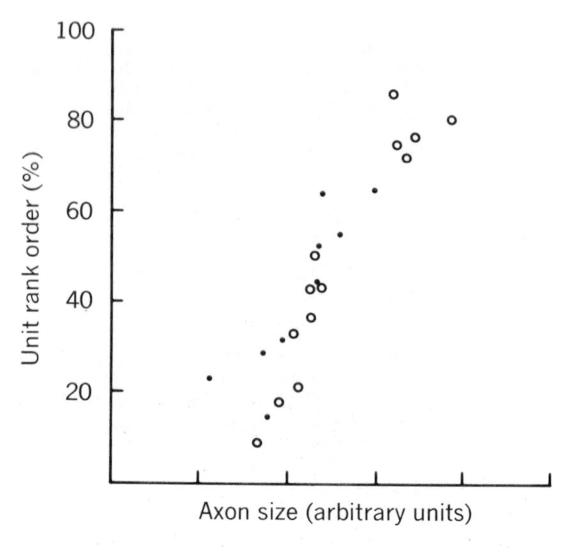

Fig. 26-12. Relation between axon size and CFL (rank order) of 21 plantaris motoneurons isolated in a single experiment. See text. (From Clamann and Henneman.[10])

the motoneuron discharges, the results obtained have a general validity that goes beyond the conditions of these particular experiments. They indicate that the rank order established with monosynaptic tests correlates very closely, if not always exactly, with the rank order obtained with repetitive discharges.

RELATION OF CRITICAL FIRING LEVEL TO AXON DIAMETER AND MOTONEURON SIZE

Defining CFL and rank order more precisely and over a wider range than ever before made it important to relate them to axonal diameter and motoneuron size by direct methods. An electrical technique, which is fully described by Clamann and Henneman,[10] was devised to do this. Three procedures were carried out: (1) Reflex responses of single motor fibers were recorded monophasically from ventral root filaments. A resistor was placed in shunt across the recording electrodes, and its value was varied until the action potentials were reduced by half. The resistance of the nerve filament was then equal to that of the shunt. Dividing the voltage of the action potentials by the resistance of the filament gave the axonal action current. (2) In experiments in which impulses were conducted antidromically over long distances to yield accurate measurements of conduction velocity, it was shown that the axonal current of an impulse varied as the square of its conduction velocity. (3) After the sizes of the impulses had been normalized in accordance with the resistances of their ventral root filaments, a direct correlation was found between impulse size and CFL. Since both CFL and axon diameter were related to impulse size, they were related to each other as illustrated in Fig. 26-12.

The linearity of the data points in Fig. 26-12 indicates that CFL is a function of a single, continuous variable, which is probably cell size. If CFL is very precisely and linearly related to cell size, it must be concluded that the distribution of input to the pool and any other factors that contribute to the relationship are also size dependent. It is appropriate, therefore, to ask whether the scatter in the data points of Fig. 26-12 indicates a degree of variability or nonlinearity in this relationship. It does not necessarily have this implication, because the variations observed when a series of observations was repeated several times were quite sufficient to account for all the scatter. Thus the relationship between CFL and axonal diameter is probably more linear than the data in Fig. 26-12 indicate. A little scatter might be anticipated, since variability in recruitment order has been observed in direct comparisons of motoneurons whose CFLs differed by less than 2.5%.[30] Barrett and Crill[1] have demonstrated a close, but not perfect, relationship between input impedance and conduction velocity and between soma size and conduction velocity, which might allow for some scatter. Their methods and assumptions, however, are probably not free of error either.

EFFECTS OF INHIBITORY INPUTS ON CRITICAL FIRING LEVEL AND RANK ORDER OF MOTONEURONS[11]

Although in the preceding experiments the motoneuron pool was subject to a variety of excitatory and inhibitory influences, the specific inputs used to elicit monosynaptic reflexes and repetitive firing were predominantly excitatory to the pool. Consequently, there was a question whether the CFLs and rank orders obtained in such experiments would be affected by the deliberate application of various types of inhibition. In order to answer this question, experiments were carried out with six different sources of inhibition.

The oscillographic record reproduced in Fig. 26-13 illustrates the effects of a potent inhibitory input from the lateral popliteal nerve on the monosynaptic reflexes of a pool of plantaris motoneurons and a single unit from the same population. The upper trace records the time integrals of the reflex discharges of the plantaris population, elicited at 2 sec intervals during the declining phase of a posttetanic potentiation. The lower trace, recorded simultaneously, shows the responses of a single plantaris unit with a CFL of 61%. Alternate reflexes of the plantaris pool were preceded by a 200 msec train of pulses applied to the lateral popliteal nerve. The parameters of this pulse train were adjusted to cause a 20% average inhibition of the maximum monosynaptic reflex of the plantaris pool. As the monosynaptic reflexes declined from their peak, the absolute amount of inhibition resulting from the peroneal input remained approximately constant. The single unit in the lower tracing discharged without exception as long as the corresponding population responses reached the level indicated by the horizontal line labeled CFL. This line was drawn at a level equal to the size of the smallest, uninhibited population response that was accompanied by a unit discharge and just above the largest uninhibited response without a unit discharge. Close examination of the record reveals that this line passes

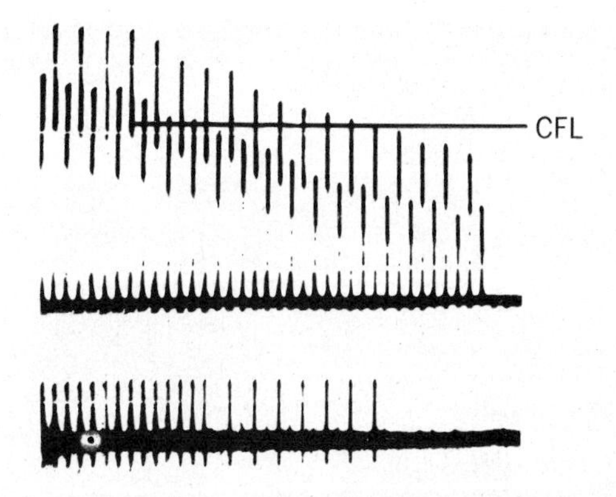

Fig. 26-13. Critical firing level (CFL) of plantaris motoneuron with and without inhibitory input from lateral popliteal nerve. Upper tracing: time integrals of monosynaptic reflexes of plantaris population elicited at 2 sec intervals during declining phase of posttetanic potentiation. Lower tracing: simultaneous responses of plantaris unit with CFL of 61%. Alternate reflexes preceded by 200 msec train of pulses to lateral popliteal nerve. Horizontal line indicating CFL drawn as described in text. (From Clamann et al.[11])

between the peak of the smallest inhibited response accompanied by a unit discharge and the top of the largest inhibited response without a unit discharge. The CFL of this unit was, thus, unchanged by the presence of a potent inhibitory input. The effects of inhibition on 32 triceps surae units with CFLs ranging from 3% to 87% were examined as previously by stimulating antagonistic nerves in the same hind limb or inhibitory sites in the medial reticular formation. The general conclusion was that inhibition had no significant effect on CFL in these experiments.

Experiments were also carried out to study the effects of inhibition on the repetitive firing of motoneurons. As a general rule, when any two motoneurons belonging to the same pool were compared directly for susceptibility to inhibition, the results could be predicted from prior measurements of their CFLs. This was always the case when the CFLs differed by more than 2.5% and was usually true when the CFLs were closer together. The oscillographic records reproduced in Fig. 26-14 illustrate this point. They show the effects of recurrent inhibition *(A)* and lateral popliteal inhibition *(B)* on the monosynaptic reflexes (traces 1 and 2) and the repetitive firing (traces 3 and 4) of a pair of plantaris motoneurons. In each of the four comparisons shown in this figure the upper unit had a slightly higher CFL, that is, 18.6% in traces 1 and 3 and 18.4% in traces 2 and 4. Each underlined monosynaptic reflex shown in *A*, 1, and *A*, 2, was preceded and

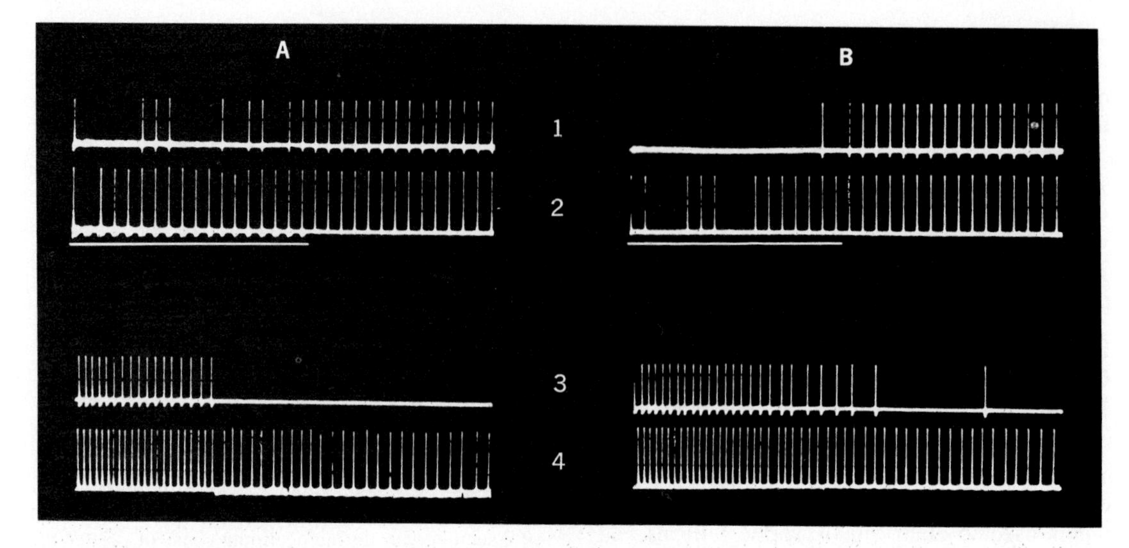

Fig. 26-14. Effects of recurrent inhibition, **A,** and lateral popliteal inhibition, **B,** on monosynaptic reflexes (traces *1* and *2*) and repetitive firing (traces *3* and *4*) of a pair of plantaris motoneurons with CFLs of 18.6% (upper) and 18.4% (lower). In traces *1* and *2*, underlined reflexes were preceded and accompanied in **A** by a 100 msec train of shocks (100/sec) to the proximal portion of L7 ventral root and in **B** by similar stimulation of the lateral popliteal nerve. In traces *3* and *4*, repetitive discharge of same units was inhibited in **A** by means of recurrent inhibition, indicated by thickened baseline in **A,** *4,* and in **B** by a progressive increase in intensity of 100/sec shocks to the lateral popliteal nerve. (From Clamann et al.[11])

accompanied by a 100 msec train of shocks (100 /sec) applied to the proximal portion of the seventh lumbar ventral root. The antidromic volleys set up by this stimulation resulted in recurrent inhibition mediated by Renshaw cells. Although 9 of the 18 conditioned reflexes in trace 1 were inhibited by this input, only 1 of those in trace 2 was silenced. In the absence of recurrent inhibition, both units responded every time.

In *A,* 3, and *A,* 4, the same pair of units is seen responding rhythmically to repetitive stimulation of the plantaris nerve. When continuous recurrent inhibition was begun, as indicated by the thicker baseline in *A,* 4, the upper unit ceased firing and the lower unit slowed to about half its original rate of discharge. In part *B* of Fig. 26-14, the same pair of units was subjected to an inhibitory input elicited by stimulating the lateral popliteal nerve at 100/sec. The underlined monosynaptic reflexes in *B,* 1, and *B,* 2, were somewhat more susceptible to this inhibition than to the recurrent inhibition. Of the 16 conditioned reflexes, 15 in trace 1 and 4 in trace 2 were silenced. In *B,* 3, and *B,* 4, the same pair of units, firing rhythmically in response to 100/sec stimulation of the plantaris nerve, was subjected to an inhibitory input whose intensity was gradually increased as the traces moved across the oscilloscope. The upper unit was finally suppressed by this input, whereas the lower one was only slowed in rate. Thus all four comparisons in Fig. 26-14 reveal that the lower unit, which had a slightly lower CFL, was more resistant to inhibition regardless of the mode of firing or the type of inhibition.

In a wide variety of experimental situations the activity of individual motoneurons is determined by the relative magnitudes of the excitatory and inhibitory inputs, as one would expect. The quantitative relation between the two types of input was examined more closely by measuring the intensities of the stimuli used to produce the inputs. The record reproduced in Fig. 26-15 is an example of the experimental data obtained. Two plantaris units, whose CFLs had been measured previously, were set into rhythmic discharge by 100/sec stimulation of the plantaris nerve. A sample of the excitatory stimuli was led to the oscilloscope. The stimulus strength of this input, indicated by the level of trace E with reference to baseline B, was constant during each test period, but varied in different tests. The lateral popliteal nerve was stimulated at 100/sec to produce a competing inhibition. The stimulus strength of this input, indicated by the trace labeled I, was gradually increased manually as the beam moved across the screen. The intensity of the inhibition required to silence each plantaris

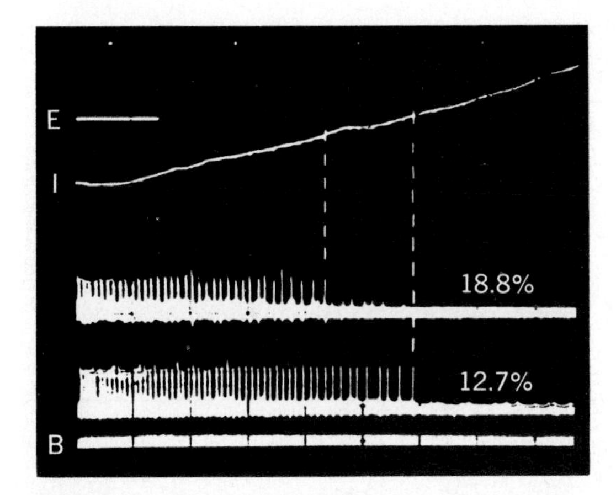

Fig. 26-15. Records illustrating quantitative relationships between intensity of excitatory stimulus causing two motoneurons with CFLs of 18.8% and 12.7% to discharge repetitively and intensity of inhibitory stimulus required to silence them. Stimulus strength of excitatory input (100/sec stimulation of plantaris nerve), indicated by level of line *E* with reference to baseline *B*, was constant during each test, but varied in different tests. Stimulus strength of inhibitory input (100/sec stimulation of lateral popliteal nerve), indicated by line *I*, was increased as trace moved across screen. Intensities of inhibition required to silence each unit were measured at time of their last discharges, as indicated by dashed lines. Time signals at top of record are 1 sec intervals. (From Clamann et al.[11])

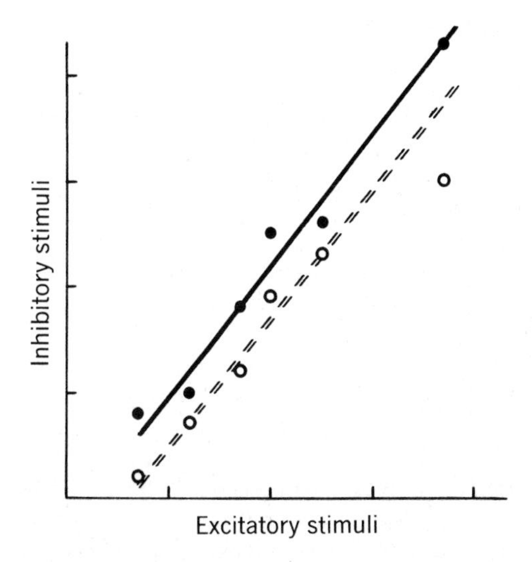

Fig. 26-16. Graph showing linear relation between intensity of excitatory stimuli causing each of two motoneurons to discharge repetitively and strength of inhibitory stimulus required to silence them. Data obtained as illustrated in Fig. 26-15. Those for lower unit, with CFL of 12.7%, are plotted with filled circles and a solid-line curve; those for upper unit, with CFL of 18.8%, are represented by open circles and a dashed curve. (From Clamann et al.[11])

unit was determined by measuring the level of trace I at the time of their last discharges, as indicated by the dashed lines. Similar measurements were made at six different levels of excitatory input. The results are graphed in Fig. 26-16. The data for the lower unit, with a CFL of 12.7%, are plotted with filled circles and a solid-line curve; those for the upper unit, with a CFL of 18.8%, are represented by open circles and a dashed curve. The plot shows that every increase in the excitatory stimuli required a proportionate increase in the inhibitory stimuli required to silence the two plantaris units. For each of the six excitatory inputs the corresponding inhibitory input was greater for the unit with the lower CFL. Furthermore, the difference between the two inhibitory inputs was approximately the same at all levels of excitation.

It must be emphasized that the stimulus intensities plotted in Fig. 26-16 do not necessarily correspond to the magnitudes of the neural inputs that result from the stimulations. Until the actual neural inputs are measured (no technique is currently available for this purpose) the quantitative validity of these results remains to be verified. Still, the findings provide further evidence for the tentative conclusion that inhibition produces effects that are similar to a simple reduction in excitation. They also suggest that the competition between excitation and inhibition in these experiments is a simple algebraic process.

The effects of inhibition on 133 pairs of plantaris motoneurons are compiled in Table 26-1. In 123 pairs (92%) the rank order for repetitive discharge during inhibition was the same as for monosynaptic reflexes. The differences in the CFLs for this group of units ranged from 0.2% to 72.1%, and their mean was 15.9%. In the 10 remaining pairs, with differences in CFLs from 0.4% to 9.9% and a mean of 3.2%, the rank order was reversed for repetitive firing.

Six different inhibitory inputs were used in these experiments, as indicated in the first column of the table. Monosynaptic reflexes were compared in a total of 170 tests and repetitive firing in 196 tests. The number of monosynaptic comparisons in which inhibition reversed the original rank order is shown in the third column. Only three reversals (1.8%) of this type occurred. In two of them the differences in CFLs were 1.0% and 0.4%. The number of comparisons in which inhibition reversed the rank order established for repetitive discharge is shown in the fifth column. Four reversals (2%) of this type oc-

Table 26-1. Effects of inhibition on pairs of motoneurons (recruitment order of pairs)*

Source of inhibitory input	Monosynaptic reflexes		Repetitive discharge	
	No. of tests	No. of reversals	No. of tests	No. of reversals
Lateral popliteal n.	36	0	38	1
Posterior biceps–semitendinosus n.	36	1	37	0
Sural n.	7	0	8	0
L7 ventral root	29	1	49	0
Contralateral S1 dorsal root	32	1	34	2
Brain stem	30	0	30	1
TOTALS	170	3	196	4

*From Clamann et al.[11]

curred. The mean difference in CFLs in these four pairs of units was 0.8%.

The six inputs used in these experiments were selected to provide contrasting types of inhibition for testing: postsynaptic and presynaptic, spinal and supraspinal, unilateral and contralateral, recurrent and nonrecurrent, cutaneous and muscular, and mixed and pure. Although these inhibitory systems and the effects they exert differ from one another in many significant respects, their effects on the rank order of motoneurons in both monosynaptic and repetitive tests were indistinguishable. As Table 26-1 indicates, none of the inputs caused a significant number of reversals. The small percentage of reversals in Table 26-1 all occurred in pairs that were nearly indistinguishable in CFL.

RECRUITMENT OF MOTOR UNITS IN MAN

Despite the limitations involved in working with human subjects, observations on motor units in man have extended the possibilities of investigation in important ways. A human subject can voluntarily produce delicately graded contractions, hold a desired tension, or make either rapid or slow contractions at will, permitting the investigator to study recruitment under a wide variety of normal conditions. Furthermore, the results are not open to the criticism that recruitment by size is the result of the use of a limited or unphysiologic input to a nervous system incapable of a more varied or flexible response. Recordings must be taken from muscles instead of nerves, hence the location of the electrodes may shift during the period of observation due to movements of active muscle fibers. This adds to the apparent number of reversed recruitments, but it is not a serious problem with isometric movements of small force and amplitude. In a previous study[42] of cat muscles the feasibility of using muscle recordings for the study of recruitment was demonstrated and the interpretation of motor unit potentials discussed.

Milner-Brown et al.[41] used an averaging technique to extract the twitch contractions of single motor units of the first dorsal interosseous muscle of the hand from the other contractile activity in the muscle and related the twitch tensions produced by single motor units to the levels of voluntary force at which they were recruited. Fig. 26-17 illustrates that the 0.1 to 10.0 gm twitch tensions varied nearly linearly as a function of the voluntary force at which they were recruited. As in the animal experiments, there was an approximately exponential relation between the numbers of motor units isolated in each size range and the tensions they produced. Larger motor units tended to contract more rapidly than smaller units. Milner-Brown et al. concluded that motor units in human muscles are recruited in orderly progression according to the size of the contractions they produce.

Recently, Freund et al.[19] reported that the tensions at which motor units in the first dorsal interosseous muscle were recruited were related directly to their axonal conduction velocities. Thus the rank order of human motor units in voluntary movements is correlated with motoneuron size, as it is in animals.

As already noted, it is the mechanical events in muscle that should be the center of attention. The motoneuron pool must be organized to achieve precise timing of these events in single motor units relative to each other and to bring about optimal combination of them for all types of movement. In an important paper, Büdingen and Freund[5] have systematically studied the relation between recruitment and the mechanical events in muscle. They began by studying voluntary contractions in the forearm with different rates of rise of tension but equal amplitudes, as

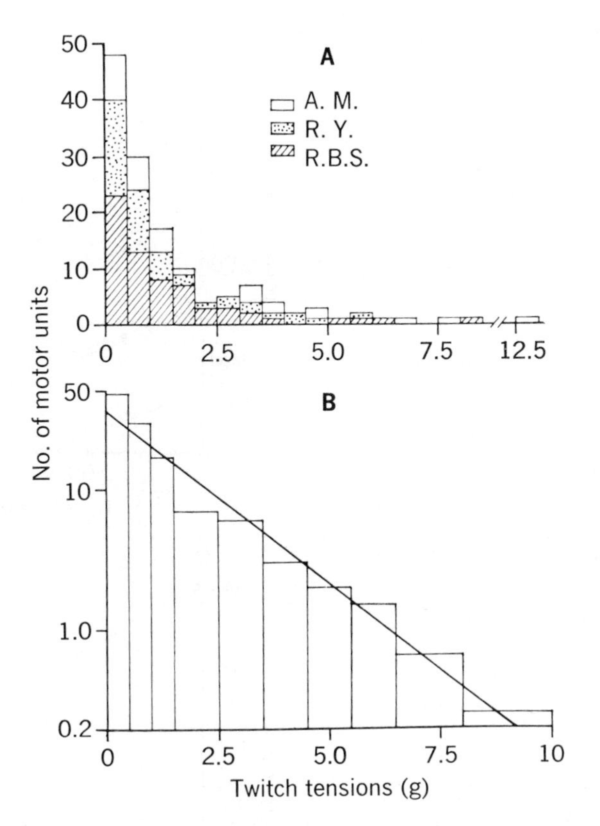

Fig. 26-17. Number of motor units, plotted on a linear scale, **A**, and on a logarithmic scale, **B**, having twitch tensions indicated. Numbers from each of three subjects are indicated. Distributions are similar for all three subjects. Computed best-fitting line on semilogarithmic plot of **B** indicates an approximately exponential relation between number of motor units and twitch tension. (From Milner-Brown et al.[41])

illustrated in Fig. 26-18, *A*. As others had demonstrated,[50] the threshold force of recruitment decreased with increasing rate of rise of tension for all motor units (note progressive reduction in threshold from d to a). Plots of the threshold forces vs the rates of rise of tension had approximately the same slopes in seven different units of the same muscle (Fig. 26-18, *C*). This monotonic decrease in threshold was consistently observed in all units tested. No evidence was found to indicate that large and small units were preferentially activated by slow (50 gm/sec) or fast (2,000 gm/sec) contractions.

Due to the general decrease in all thresholds that occurred with increasing rates of tension development, the total time between the recruitment of low and high threshold units diminished considerably. This is apparent in Fig. 26-18, *B*, which plots the time between the beginning of the voluntary muscle contraction and the onset of firing in three units of the same muscle against the rate of rise of tension. Both scales are logarithmic. With rapid contractions (between 300 and 3,000 gm/sec), recruitment was greatly compressed in time. The delay between the firing onset of low- and high-threshold units decreased from 2,500 msec during the slowest contractions to less than 100 msec during the fastest. The parallelism of the curves in part *B*, nevertheless, indicates that, although the high threshold units started firing much sooner in rapid contractions, they were never preferentially activated.

During a change in total muscle force, that force is different at the onset of firing and the subsequent twitch contraction. The relation of the first twitch contraction of a unit to the total muscle force is the mechanical counterpart of electrical recruitment. In Fig. 26-18, *D*, change in total muscle force is plotted against rate of rise of tension in y_2 for the mean contraction time of the extensor indicis (60 msec). In addition, the average decrease of the threshold force of recruitment of 12 units of this muscle during increasing rates of rise of tension is plotted in y_1. The slopes of the units with different thresholds were similar, as shown in part *C*, and can be represented

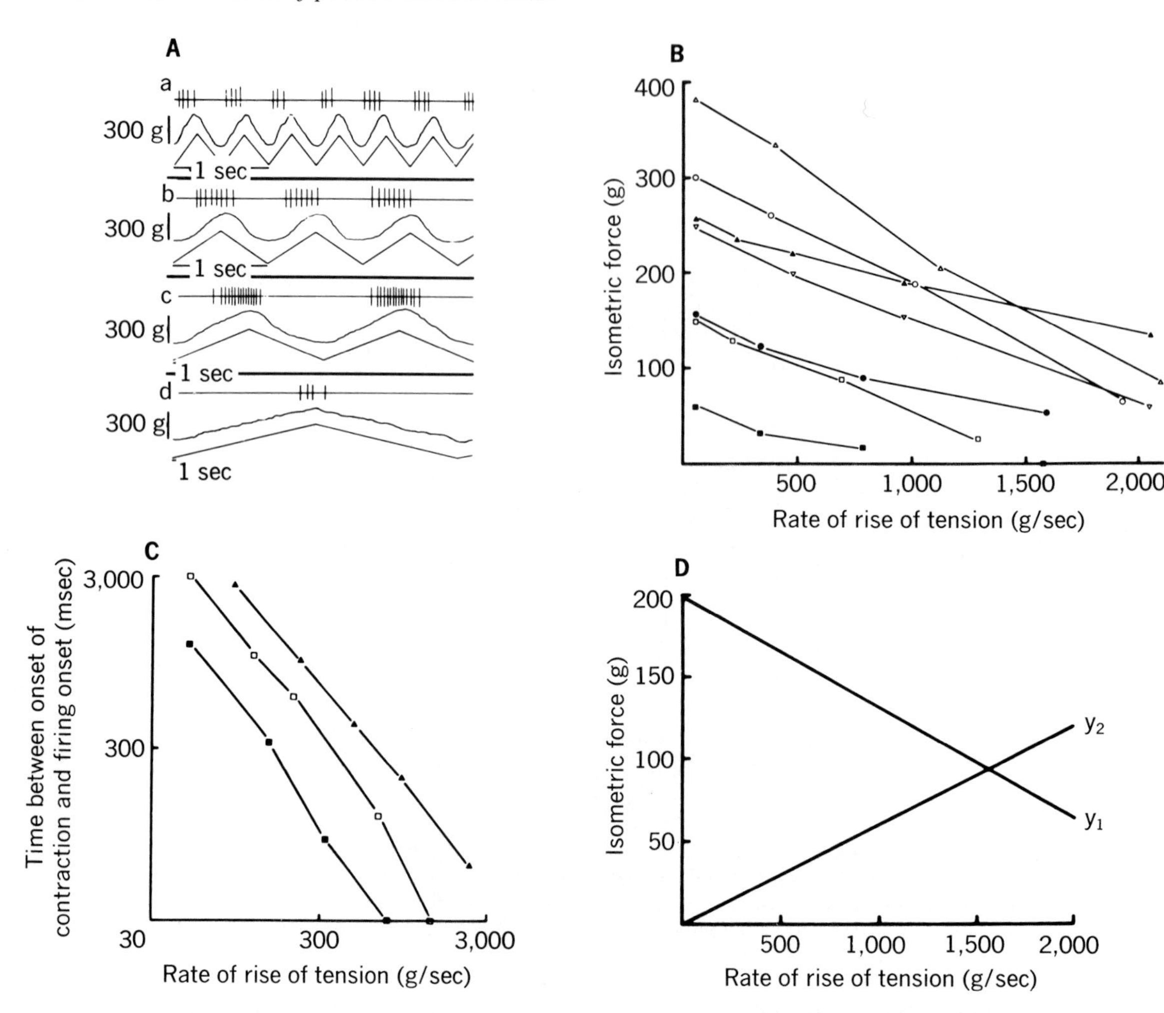

Fig. 26-18. A, Firing pattern of high-threshold motor unit recorded from extensor indicis proprius muscle during isometric contractions at different rates of rise of tension (*a* to *d*). Each trace shows from top to bottom the spike record, the isometric force, and the tracking signal, which was visually displayed to subject. Time scale in lower two records is half that of upper two records. During increasingly faster contractions, unit started firing at successively lower force levels. **B,** Plots of threshold forces of recruitment vs rates of rise of tension for seven motor units in same extensor indicis proprius muscle. Records obtained from three subjects, indicated by solid square, open circle, and solid triangle. **C,** Plots of time between beginning of voluntary muscle contraction and onset of firing (ordinate) in three units of same muscle vs rate of rise of tension (abscissa). Double logarithmic scale. The faster the contraction, the earlier the recruitment. Units are shown with same symbols as in **B. D,** Regression line of change of threshold force of recruitment on rate of rise of isometric tension ($y_1 = -0.0678 \times +198.66$) calculated from 12 motor units of extensor indicis. Increase of muscle force that occurs within mean contraction time of muscle is shown for comparison ($y_2 = \times 0.060$). See text. (Modified from Büdingen and Freund.[5])

by the single regression line y_1. The slopes of y_2 and y_1 are almost equal, but of opposite direction. Hence the sum of the two variables is approximately constant over the range examined in these experiments. *The decrease of the threshold with increasing rate of rise of tension seems to cancel the increase of force during the contraction time, so that the mechanical recruitment of a motor unit occurs at approximately the same force level regardless of the rate of rise of tension.*

Since the order of recruitment does not change with the rate of rise of tension, the proportion of the total muscle force generated by recruitment and by modulation of firing rate of the units activated prior to a particular unit must remain constant. This emphasizes how precise the relation is between the recruitment of units and the activity

Table 26-2. Motor unit threshold data*

Trial	Force (kg)	Units 1	2	3	4	5	Trial	Force (kg)	Units 1	2	3	4	5
1	0.07	+	−	−	−	−	27	2.27	+	+	+	−	−
2	0.08	+	−	−	−	−	28	2.42	+	+	+	−	−
3	0.15	+	−	−	−	−	29	2.72	+	+	+	+	−
4	0.30	+	−	−	−	−	30	2.75	+	+	+	−	−
5	0.30	+	−	−	−	−	31	3.18	+	+	+	+	−
6	0.31	+	−	−	−	−	32	3.20	+	+	+	−	−
7	0.31	+	−	−	−	−	33	3.33	+	+	+	+	−
8	0.37	+	−	−	−	−	34	3.48	+	+	+	+	−
9	0.38	+	−	−	−	−	35	3.51	+	+	+	+	−
10	0.45	+	+	−	−	−	36	3.78	+	+	+	+	−
11	0.46	+	+	−	−	−	37	4.54	+	+	+	+	+
12	0.53	+	−	−	−	−	38	4.84	+	+	+	+	+
13	0.75	+	+	−	−	−	39	4.99	+	+	+	+	+
14	0.90	+	+	−	−	−	40	5.15	+	+	+	+	+
15	0.90	+	+	−	−	−	41	5.30	+	+	+	+	+
16	0.91	+	+	−	−	−	42	5.45	+	+	+	+	+
17	1.06	+	+	−	−	−	43	5.75	+	+	+	+	+
18	1.06	+	+	−	−	−	44	6.96	+	+	+	+	+
19	1.06	+	+	−	−	−	45	8.03	+	+	+	+	+
20	1.14	+	+	+	−	−	46	8.10	+	+	+	+	+
21	1.21	+	+	+	−	−	47	8.18	+	+	+	+	+
22	1.36	+	+	−	−	−	48	8.63	+	+	+	+	+
23	1.38	+	+	+	−	−	49	9.54	+	+	+	+	+
24	1.51	+	+	−	−	−	50	9.69	+	+	+	+	+
25	1.81	+	+	+	−	−	51	10.90	+	+	+	+	+
26	1.82	+	+	+	−	−	52	12.70	+	+	+	+	+

MOTOR UNIT THRESHOLD

Unit	Ballistic force threshold (kg) Range	Mean	Ramp force threshold (kg) Range	Mean ± SD
1	0-0.07	0.035	0.14-0.48	0.34 ± 0.15
2	0.38-0.75	0.56	3.50-4.10	3.70 ± 0.28
3	1.06-1.81	1.43	3.60-4.40	4.14 ± 0.36
4	2.72-3.33	3.02	11.98-13.15	12.66 ± 0.57
5	3.78-4.54	4.16	13.73-16.41	15.46 ± 0.93

*From Desmedt and Godaux.[15]

of those already recruited. The findings demonstrate that the activation of the units in a motoneuron pool is organized to achieve a definite mechanical result. This result is a stable relation between the force output of the individual motor unit at the time of its mechanical recruitment and the force output of the whole muscle. These observations help to explain why the size and excitability of a motor unit are *necessarily* related to its contraction time, rate of firing, and recruitment order.

As described previously, recruitment is "orderly" in contractions of up to 2,500 gm/sec. Desmedt and Godaux[15] have developed a technique for investigating even faster "ballistic" contractions, which are completed in too short a time to be controlled by sensory feedback signals from the activated muscles. The commands in ballistic movements are presumably "preprogrammed" and must be dispatched to the segmental circuits and motoneurons before the contraction begins. To study these movements, subjects were asked to carry out 50 or more ballistic contractions of the tibialis anterior muscle at intervals of several seconds. Peak forces ranging from 0.05 to 12.0 kg were produced in less than 0.15 sec each. Many trials were carried out to secure a complete series of 50 specimens covering this range of forces. During each contraction the action potentials of the discharging units at the recording site were photographed along with the total contractile force. The 50 or more contractions were then ranked in order of increasing peak force and the corresponding activity of each unit was tabulated, as in Table 26-2. As the upper part of the table indicates, unit 1 fired in all 52 trials. Units 2 to 5 failed to respond below a certain peak force, but discharged with increasing probability in slightly stronger contractions. The ballistic threshold was estimated by taking the mean between the maximum peak force for which a unit never discharged and the minimum force for which it always fired. The lower part of Table 26-2 gives the thresholds of the same five units for ballistic and slow ramp contractions. The two sets of thresholds were compared by the rank correlation method,[26] which yielded a very high correlation coefficient of 0.95.

Four muscles (masseter, first dorsal interosseous, tibialis, and soleus) were investigated with similar results.[16] Since the pyramidal tract is strongly involved in ballistic movements, it was suggested that pyramidal signals impinge on the motoneuron pool in a way similar to the pattern of segmental activation by I_A spindle afferent fibers.

Further evidence was provided by studying motor units in the first dorsal interosseous muscle with the spike-triggered averaging technique that allows the investigator to extract the tension myogram of individual twitch contractions. It was clear with this technique that the more rapidly contracting units were not preferentially recruited in ballistic contractions. This is consistent with the earlier conclusion of a rather fixed recruitment order. This part of the study also showed that the motor units recruited at the initial stage of a small ballistic movement produced smaller forces than the higher threshold units.

Finally, evidence was presented that vibration-induced inhibition of motoneurons silences them in exactly the reverse order from the one in which they were recruited, as originally described in animal experiments.[16]

Thus a considerable number of the observations described in the early part of this chapter have been confirmed in experiments on human subjects, where anesthesia, surgery, and other experimental factors could play no role in the results.

Following an initial report by Harrison and Mortenson,[24] Basmajian and coworkers[2,3,4] have stated on numerous occasions that when human volunteers are provided with visual and auditory feedback from their active motor units, they easily learn within 10 or 15 min to control the activity of *any single* motor unit within a given muscle. Since these reports disagreed with the great majority of observations on animals and human subjects, an attempt was made to determine whether human subjects could indeed learn to exercise selective control over their motor units. Subjects were asked to discharge units in the extensor indicis proprius muscle by dorsiflexing the index finger. After establishing the normal order of recruitment for two motor units that were clearly distinguishable on the oscilloscope, subjects were instructed either to reverse that normal order or to silence the activity of unit 1 without silencing unit 2. These were considered to be simple but crucial tests of the claims for selective control.[27] All nine subjects were able, without difficulty, to isolate a single motor unit and to control its rate of firing from most recording sites. Slight voluntary increases in muscle contraction usually resulted in an increase in the firing rate of the first motor unit and the recruitment of a second, somewhat larger action potential, which followed the same pattern as the first. When the subject was asked to reduce tension in the muscle, the second unit was the first to cease firing, and there was a progressive decrease in the frequency of the first motor unit, the activity of which finally stopped when the muscle was at complete rest. This pattern of recruitment could be easily reproduced in all subjects tested without special training.

In six of the nine subjects, no changes in the order of recruitment were observed despite 2 hours of training with audiovisual feedback. In each experiment, recordings were made from various sites within the extensor indicis proprius muscle, and the subject was encouraged to explore maneuvers that might lead to alterations. At each new site, a minimum of 20 to 30 min was spent attempting to alter the normal order. In not a single instance out of hundreds of attempts at different recording sites was any one of these six subjects able to recruit two units in their normal order and then turn off unit 1 without silencing unit 2. It should be emphasized that even under "isotonic" conditions, the most minimal movements of the index finger, scarcely visible to the observers, were associated with activity in both units. The "critical firing levels"[30] of these pairs of motoneurons probably differed only slightly, and yet no reversals in recruitment order were observed.

The results at most recording sites in the three remaining subjects were similar to those just described. In each of these subjects, however, there was one site at which some variability in recruitment order was observed. Although one unit was recruited first and dropped out last in the great majority of tests, the unit that usually was recruited second occasionally was the first to respond and could then be activated repetitively for some seconds without any activity in the first unit. These changes in recruitment order seemed to occur randomly. None of the three subjects could on demand activate either motor unit at will or alternate their activity in a facile manner.

These studies agree with numerous reports that human subjects can easily isolate single motor units and control their frequency by simply changing the contractile force voluntarily. This "isolation" is only apparent, however, and selective control could only be demonstrated for the lowest threshold unit within recording range. There was no evidence of true voluntary or conscious control over the order of recruitment. Although switching in the recruitment order of motor units whose tension thresholds were very close was noted occasionally, these results again demonstrate that, on the whole, the pattern of recruitment in human subjects is relatively fixed.

SIZE PRINCIPLE IN OTHER SPECIES

Recruitment of motoneurons in order of increasing size by excitatory inputs of progressive intensity is not limited to mammals. It has been observed in a wide variety of animals as phylogenetically distinct as cats[29] and lobsters.[14] Since its first description in the cat,[25] it has been reported in the natural motor output to crab leg muscles[9] and eyestalk muscles,[31] cockroaches,[43] locusts,[32] dragonflies,[39] mantids,[46] water bugs,[51] newts,[49] toads,[35] fish,[38] chickens,[18] and a number of mammals, including man. Davis[14] has given a particularly good description of orderly recruitment in the motoneurons controlling the swimmerets of lobsters and has pointed out that size determines the order in which the entire population of swimmeret motoneurons is activated, rather than just the smaller groups of cells innervating single muscles. He has also pointed out that a number of other properties are correlated with size, adding further support to the use of the more inclusive term "size principle" to describe the ramifications of size. The widespread distribution of the size principle suggests that some rules governing the output of neuronal networks are common to both simple and complex nervous systems.[14] There is, moreover, suggestive evidence that the size principle may apply not only to motoneurons, but also to other neural cells. Orderly recruitment of interneurons is apparent in acridid auditory interneurons[47]; the sensitivity of pacemaker cells is said to be inversely related to their size,[7] and the size of sensory neurons in general may play a role in determining their excitability.[6,53] The variety of neurons in which size influences essential properties offers interesting clues to the biophysical mechanism of the size principle. However, this topic is too uncertain and speculative for extended treatment in a textbook.

PRINCIPLES UNDERLYING THE OPERATION OF MOTONEURON POOLS

It is apparent from the experimental observations that a motoneuron pool functions as a collective entity. Although its components are separate cells with few interconnections to bind them together, it is obvious that all the parts are functionally interrelated, so that they work together harmoniously. As we shall see in the next chapter, the collective action of the pool is due largely to the manner in which its inputs are distributed.

A pool is designed to work incrementally by adding progressively larger units to the total output or by subtracting progressively smaller units. These incremental operations must be controlled in accordance with the mechanical properties of the individual motor units so that the end result is mechanically precise and temporally appropriate. The essence of pool function is the collective action of cells whose motor units differ systematically in their contractile properties.

In early studies,[28,29,38,48] it was shown that the neural energy required to discharge a motoneuron, the energy it transmits and releases in the muscle, its mean rate of firing, and even its rate of protein[44] synthesis all correlate with its size. A pool is a hierarchy of cells organized according to size and some of the correlates of size. In functional terms the rank order in this hierarchy is defined in terms of excitability. The lowest ranking cell (No. 1) is the most susceptible to discharge; the highest ranking cell is the least susceptible. Whenever a particular motoneuron responds monosynaptically, all the lower ranking cells in the pool discharge with it. Thus all the monosynaptic outputs (O) have the following general form:

$$O_T = O_1 + O_2 + O_3 + \ldots O_x$$

where O_T is the total monosynaptic output and O_x is the highest ranking cell discharged.

This equation is a concise formulation of a basic *law of combination* that emerges from the experimental data. It specifies how the activities of motoneurons are combined in monosynaptic outputs, during repetitive firing, and during various mixtures of excitation and inhibition. Fig. 26-19 illustrates some of the features of pool organization schematically. Each vertical line represents a motoneuron, whose relative size is indicated by its amplitude. At the left end of the dia-

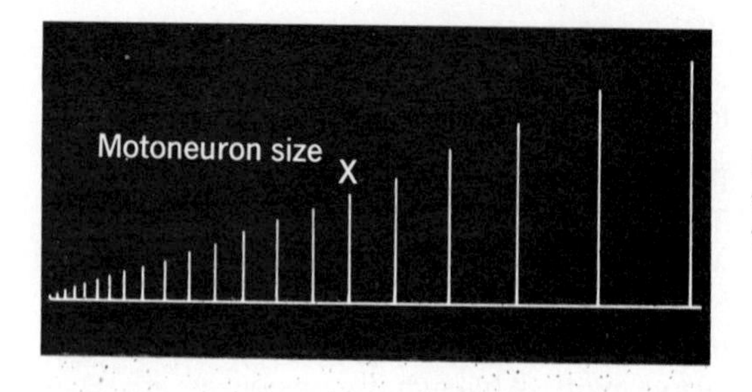

Fig. 26-19. Schematic diagram to illustrate some features of the organization of a motoneuron pool. See text.

Table 26-3. Percent contribution of single motor units to the tension developed by all smaller units of m. gastrocnemius

Grouping of units	<10 gm	10.0-19.9	20.0-29.9	30.0-39.9	40.0-49.9	50.0-59.9	60.0-69.9	70.0-79.9	80.0-89.9	90.0-99.9	100.0-109.9	110.0-119.9
Total tensions (gm)	156	152	387	252	358	323	443	519	587	278	107	345
Cumulative totals (gm)	156	308	691	947	1304	1628	2072	2590	3178	3456	3563	3908
Contribution of one unit divided by cumulative totals (%)	6.4	6.5	4.3	4.2	3.8	3.7	3.4	3.1	2.8	2.9	3.1	3.1
Contribution of one unit divided by cumulative totals (estimated for whole muscle) (%)	2.1	2.2	1.4	1.4	1.3	1.2	1.1	1.0	0.9	1.0	1.0	1.0

gram the motoneurons are small, numerous, and highly susceptible to discharge, and they differ minimally in size. At the right end the units are large, few in number, and relatively insusceptible to discharge, and they vary greatly in size. In addition to these differences in the motoneuron pool itself, there are systematic variations in the contractile properties of the motor units these cells innervate that help to ensure optimum function of the entire neuromuscular system for a wide range of speeds, forces, and loads.

It is obvious that the large number of small motor units varying slightly in their sizes could endow this system with extremely fine control at the low or vernier end of the scale. However, is there loss of fine control at the high end of the scale when the units that are recruited may contribute increments of tension up to 120 gm? Since recruitment of a new unit of any size occurs after all the smaller units in that muscle are already active, it is easy to estimate what percent of the total tension is contributed by the last addition to it. To do this, all the unit tensions are grouped in decades. The top line of Table 26-3 gives the sums of the unit tensions in each decade. The second line contains the total cumulative tensions, that is, the sum of all unit tensions in each column plus the totals of all the columns to the left. The percent of tension contributed by new units to the tensions already developed by all units of smaller size is shown in the third line. For example, the figure 6.4 in the left-hand column was obtained by dividing 10 gm (the next largest unit) by 156.3 gm (the cumulative total of all units below 10 gm). If all increments of tension are viewed as percentages of the preexisting muscle tension in this manner, it is clear that there is no loss of fine control as the total tension approaches a maximum.

A number of factors combine to make the actual control of tension more flexible and precise than the scheme just outlined would achieve. There are about 3 times as many motor units in an average MG pool as there were in our experimental sample of it. Each addition to the total tension would, therefore, represent an even smaller percentage of the total, as shown in the last line of Table 26-3. With more units the spectrum of sizes would be more nearly continuous. Finally, after units are recruited, they fire more and more rapidly with increased input as further recruitment goes on. This effect, which is called *rate modulation,* is apparent in Fig. 26-2 and 26-3. It accounts for a varying proportion of the total tension in different muscles.[34,40]

Grading of output in muscle is reminiscent of the "Weber fractions" in the field of sensory discrimination. A precise mathematical relationship for the motor system comparable to Weber's rule of the ". . . just noticeable difference of sensation . . ." is not implied here, but the analogy is clear: the smallest increment that can be added to the force exerted by a muscle becomes greater as the force of contraction itself increases.

HOW THE CENTRAL NERVOUS SYSTEM USES THE MOTONEURON POOL

No definitive information as to how the CNS uses the motoneuron pool can be given at present, but it is surely pertinent to define the functional implications of the precise and nearly invariant rank order that has been demonstrated for the higher centers where commands originate. In order to consider this in broad perspective, imagine that no rank order exists. The CNS has at its disposal motor units with a wide range of contractile properties. How should it select those it needs for a particular task, and how should it combine them to produce the muscle tension it desires? It is conceivable that the CNS might use the 300 motoneurons in the MG pool like the keys of a large adding machine, picking any combination of cells to yield the proper total. This notion is implicit in suggestions that motoneurons can be activated in a selective manner. However, this proposition would require 300 separately activatable inputs, so that each of the 300 motoneurons could be fired selectively, a very demanding requirement. It would also necessitate that circuits do something much more difficult, namely, that they calculate what combinations of active units would produce the correct total tensions. This would require formidable circuitry and considerable time. Consider the number (N) of possible combinations that might occur. For a pool consisting of 300 cells, such as the MG pool, N is calculated as follows:

$$N = \sum_{k=1}^{300} \frac{300!}{k!\,(300 - k)!}$$

In this example N is $>10^{90}$

Feedback signals from muscle could not replace circuits to precalculate the total tension. They would be too slow and, in any case, they would be needed to determine the size of the feedback correction. The solution that has evolved is the rank-ordered pool, which relieves the CNS of the necessity for selective activation of motoneurons and provides a simple rule for their combination.

EVIDENCE OF NEUROMUSCULAR PATHOLOGY IN ABNORMAL RECRUITMENT PATTERNS IN MAN

Normally it is difficult to correlate the electrical activity recorded from a patient's muscle with the histochemistry of the muscle fibers generating it because the different types of fibers are so intermingled. It is possible to do so, however, in the nearly homogeneous populations of fibers in patients with chronic, low-grade motor neuropathies and in some other diseases. The findings in these conditions are consistent with the experimental results described earlier in this chapter and serve to reinforce the conclusions drawn from them.

When an abnormal pattern of recruitment is recorded from a muscle, biopsy is almost certain to reveal pathology. Electromyography (EMG) of the most superficial parts of the muscle, followed by open biopsy of 2 to 5 mm of muscle around the tip of the electrodes, may lead to a diagnosis of long-standing neuropathy even in muscles that appear normal on gross examination. In a survey of 110 selected patients, Warmoltz and Engel[52] discovered a small group of cases with the following characteristics: (1) initial activation on minimal to mild voluntary contraction; (2) well-maintained rhythmic firing with little effort at frequencies of 6 to 10/sec; (3) sustained discharge through all ranges from mild to maximal forces; and (4) acceleration of firing rates with stronger contractions to a maximum of 18 to 20/sec. In these cases, open biopsy disclosed that 95% to 99% of the muscle fibers were small and red.

Motor units recorded from another group of patients had completely different characteristics: (1) sudden or vigorous contraction required for activation; (2) either (a) discharge in brief bursts lasting 0.5 to 5.0 sec (with no capacity to sustain them longer), during which they fired steadily at 10 to 25/sec or spluttered irregularly, or (b) discharge only once or twice; (3) peak firing frequencies reaching 16 to 50/sec at the height of bursts accompanying maximal contractions; and (4) more rapid initial and peak discharge frequencies with increasingly vigorous contractions. Biopsies from these patients contained 96% to 100% pale (type A) fibers. In brief, the normal recruitment pattern, with its wide range of potential sizes and frequencies, was lacking in both these groups.

It is reasonable to assume that the motoneurons supplying one type of muscle fiber in these cases were attacked by disease and the fibers supplied by them were denervated. Motoneurons innervating other types of fibers were apparently not susceptible to the same pathologic process and remained relatively intact. These cells presumably put out collaterals that supplied the denervated muscle fibers and in time converted them to the type characteristic of their own activity pattern, as in cross innervation experiments.

There are several puzzling features of these studies. The findings suggest that there are at least two distinct types of motoneurons supplying heterogeneous muscles, with susceptibilities to disease that may differ significantly. The extreme uniformity of the muscle population that eventually results in these cases is surprising. It may be more apparent than real, since it is unlikely that the motoneurons themselves are as homogeneous as the appearance of the muscle fibers suggests. The converted fibers appear to be normal as judged by their electrical properties. In five cases showing selective widespread, moderately severe atrophy of large, pale fibers and four cases with milder involvement, the amplitude and duration of the low-threshold unitary potentials appeared normal, suggesting that they were derived from normal muscle fibers not then involved in a disease process. In seven of these nine cases the only recognizable electromyographic abnormality was failure to achieve a full "interference pattern" of large and small spikes with maximal effort. These studies seem to require the existence of two types of motoneurons. Variations in motoneuron size and excitability may account for the differences in susceptibility to disease that apparently exist. However, the recent work of Harris and Henneman[23] on motoneurons with similar CFLs but widely differing firing rates suggests another possible explanation, namely, that there are at least two populations of motoneurons with size-independent properties sufficiently different to account for different susceptibility to disease.

No account of the organization of the motoneuron pool would be complete without a description of how excitatory and inhibitory inputs are distributed to the members of the pool and how they affect them in health and disease. This subject will be discussed in the following chapter.

REFERENCES

1. Barrett, J. N., and Crill, W. E.: Specific membrane properties of cat motoneurons, J. Physiol. (Lond.) **239:** 301, 1974.
2. Basmajian, J. V.: Control and training of individual motor units, Science **141:**440, 1963.
3. Basmajian, J. V.: Control of individual motor units, Am. J. Phys. Med. **46:**480, 1967.
4. Basmajian, J. V.: Electromyography comes of age, Science **176:**603, 1972.
5. Büdingen, H. J., and Freund, H.-J.: The relationship between the rate of rise of isometric tension and motor unit recruitment in a human forearm muscle, Pfluegers Arch. **362:**61, 1976.
6. Bullock, T. H.: Comparative aspects of some biological transducers, Fed. Proc. **12:**666-672, 1953.
7. Bullock, T. H., and Horridge, G. A.: Structure and function in the nervous systems of invertebrates, San Francisco, 1965, W. H. Freeman & Co., Publishers.
8. Burke, R. E., et al.: Anatomy of medial gastrocnemius and soleus motor nuclei in cat spinal cord, J. Neurophysiol. **40:**667, 1977.
9. Bush, B. M. H.: Proprioceptive reflexes in the legs of *Carcinus maenas* (L.), J. Exp. Biol. **39:**89, 1962.
10. Clamann, H. P., and Henneman, E.: Electrical measurement of axon diameter and its use in relating motoneuron size to critical firing level, J. Neurophysiol. **39:**844, 1976.
11. Clamann, H. P., Gillies, J. D., and Henneman, E.: Effects of inhibitory inputs on critical firing level and rank order of motoneurons, J. Neurophysiol. **37:**1350, 1974.
12. Clamann, H. P., Gillies, J. D., Skinner, R. D., and Henneman, E.: Quantitative measures of output of a motoneuron pool during monosynaptic reflexes, J. Neurophysiol. **37:**1328, 1974.
13. Creed, R. S., et al.: Reflex activity of the spinal cord, Oxford, 1932, Oxford University Press.
14. Davis, W. J.: Functional significance of motoneuron size and soma position in swimmeret system of the lobster, J. Neurophysiol. **34:**274, 1971.
15. Desmedt, J. E., and Godaux, E.: Ballistic contractions in man: characteristic recruitment pattern of single motor units of the tibialis anterior muscle, J. Physiol. **264:**673, 1977.
16. Desmedt, J. E., and Godaux, E.: Critical evaluation of the size principle of human motoneurons recruitment in ballistic movements and in vibration-induced inhibition or potentiation, Ann. Neurol. **1:**504, 1977.
17. Eccles, J. C., and Sherrington, C. S.: Numbers and contraction values of individual motor-units examined in some muscles of the limb, Proc. R. Soc. Lond. (Biol.) **106:**326, 1930.
18. Fedde, M. R., DeWet, P. D., and Kitchell, R. L.: Motor unit recruitment pattern and tonic activity in respiratory muscles of Gallus domesticus, J. Neurophysiol. **32:**995, 1969.
19. Freund, H.-J., Büdingen, J. H., and Dietz, V.: Activity of single motor units from human forearm muscles during voluntary isometric contractions, J. Neurophysiol. **38:**933, 1975.
20. Gasser, H.: The classification of nerve fibers, Ohio J. Sci. **41:**145, 1941.
21. Gelfan, S.: Neurone and synapse populations in the spinal cord: indication of role in total integration, Nature **198:**162, 1963.
22. Harris, D. A., and Henneman, E.: Effects of inhibition on firing rates of homonymous motoneurons of different types, Neurosci. Abstr. **3:**502, 1977.
23. Harris, D. A., and Henneman, E.: Identification of two species of alpha motoneurons in cat's plantaris pool, J. Neurophysiol. **40:**16, 1977.
24. Harrison, V. F., and Mortenson, O. A.: Identification and voluntary control of single motor unit activity in the tibialis anterior muscle, Anat. Rec. **144:**109, 1962.
25. Henneman, E.: Relation between size of neurons and their susceptibility to discharge, Science **126:**1345-1346, 1957.
26. Henneman, E.: Functional significance of cell size in spinal motoneurons, J. Neurophysiol. **28:**560-580, 1965.
27. Henneman, E., Shahani, B. T., and Young, R. R.: Voluntary control of human motor units. In Shahani, M., editor: The motor system: neurophysiology and muscle mechanisms, Amsterdam, 1976, Elsevier/North Holland Biomedical Press.
28. Henneman, E., Somjen, G., and Carpenter, D. O.: Ex-

citability and inhibitability of motoneurons of different sizes, J. Neurophysiol. **28:**500, 1965.
29. Henneman, E., Somjen, G., and Carpenter, D. O.: Functional significance of cell size in spinal motoneurons, J. Neurophysiol. **28:**560, 1965.
30. Henneman, E., Clamann, H. P., Gillies, J. D., and Skinner, R. D.: Rank-order of motoneurons within a pool: law of combination, J. Neurophysiol. **37:**1338, 1974.
31. Horridge, G. A., and Burrows, M.: Tonic and phasic systems in parallel in the eyecup responses of the crab *Carcinus,* J. Exp. Biol. **49:**269, 1968.
32. Hoyle, G.: Exploration of neuronal mechanisms underlying behavior in insects. In Reiss, R. F., editor: Neural theory and modeling, Stanford, Calif., 1964, Stanford University Press.
33. Hughes, G. M., and Ballintijn, C. M.: Electromyography of the respiratory muscles and gill water flow in the dragonet, J. Exp. Biol. **49:**583, 1968.
34. Kernell, D.: Recruitment, rate modulation and the tonic stretch reflex. In Homma, S., editor: Progress in brain research: understanding the stretch reflex, Amsterdam, 1976, Elsevier/North-Holland Biomedical Press, vol. 44.
35. Kobayashi, Y., Oshima, K., and Tasaki, I.: Analysis of afferent and efferent systems in the muscle nerve of the toad and cat, J. Physiol. (Lond.) **117:**152, 1952.
36. Lloyd, D. P. C.: Post-tetanic potentiation of response in monosynaptic reflex pathways of the spinal cord, J. Gen. Physiol. **33:**147, 1949.
37. Lundberg, A., Norrsell, U., and Voorhoeve, P.: Pyramidal effects on lumbo-sacral interneurons activated by somatic afferents, Acta Physiol. Scand. **56:**220-229, 1962.
38. McPhedran, A. M., Wuerker, R. B., and Henneman, E.: Properties of motor units in a homogeneous red muscle (soleus) of the cat, J. Neurophysiol. **28:**71, 1965.
39. Mill, P. J.: Neural patterns associated with ventilatory movements in dragonfly larvae, J. Exp. Biol. **52:**167, 1970.
40. Milner-Brown, H. S., Stein, R. B., and Yemm, R.: Changes in firing rate of human motor units during linearly changing voluntary contractions, J. Physiol. (Lond.) **230:**371-390, 1973.
41. Milner-Brown, H. S., Stein, R. B., and Yemm, R.: The orderly recruitment of human motor units during voluntary isometric contractions, J. Physiol. **230:**359, 1973.
42. Olson, C. B., Carpenter, D. O., and Henneman, E.: Orderly recruitment of muscle action potentials, Arch. Neurol. **19:**591-597, 1963.
43. Pearson, K. G., and Iles, J. F.: Discharge patterns of coxal levator and depressor motoneurones of the cockroach, *Periplaneta americana,* J. Exp. Biol. **52:**139, 1970.
44. Peterson, R. P.: Cell size and rate of protein synthesis in ventral horn neurones, Science **153:**1413, 1966.
45. Ramón y Cajal, S.: Histologie du systeme nerveux de l'homme et des vertebres, Paris, 1909, Librairie Maloine.
46. Roeder, K. D., Tozian, L., and Weiant, E. A.: Endogenous nerve activity and behavior in the mantis and cockroach, J. Insect Physiol. **4:**45, 1960.
47. Rowell, C. H. F., and McKay, J. M.: An Acridid auditory interneuron. I. Functional connexions and response to single sounds, J. Exp. Biol. **51:**231, 1969.
48. Somjen, G., Carpenter, D. O., and Henneman, E.: Responses of motoneurons of different sizes to graded stimulation of supraspinal centers of the brain, J. Neurophysiol. **28:**958, 1965.
49. Szekely, G., Czeh, G., and Vörös, G.: The activity pattern of limb muscles in freely moving normal and deafferented newts, Exp. Brain Res. **9:**53, 1969.
50. Tanji, J., and Kato, M.: Recruitment of motor units in voluntary contraction of a finger muscle in man, Exp. Neurol. **40:**759, 1973.
51. Walcott, B., and Burrows, M.: The ultrastructure and physiology of the abdominal airguide retractor muscles in the giant water bug, *Lethocerus,* J. Insect Physiol. **15:**1855, 1969.
52. Warmoltz, J. R., and Engel, W. K.: Open-biopsy electromyography. I. Correlation of motor unit behavior with histochemical muscle fiber type in human limb muscle, Arch. Neurol. **27:**512, 1972.
53. Yatsuki, Y., Yoshino, S., and Chen, J.: Action current of the single lateral-line nerve of the fish. II. On the discharge due to stimulation, Jpn. J. Physiol. **1:**179-194, 1951.

27

LORNE M. MENDELL and ELWOOD HENNEMAN

Input to motoneuron pools and its effects

The surface of each motoneuron in a pool is covered by 5,000 to 10,000 synaptic knobs, or *boutons,* that exert excitatory or inhibitory effects. The majority of these endings are terminals of interneurons located in the same or nearby segments of the spinal cord. Thus a great deal of the input to motoneurons is processed at the spinal level and only affects them indirectly via interneurons. Much of the mixing and blending of inputs from different sources, as well as the spatial and temporal distribution of the signals, is carried out by these interneuronal circuits. In some cases, interneurons probably function like valves, permitting or preventing access of certain signals to motoneurons. The rest of the endings on motoneurons are terminals of fibers that originate in more distant parts of the nervous system, such as the cerebral cortex, brain stem, and peripheral structures. In general, these long projections end directly on motoneurons, on interneurons, or on both.

The vast number of terminals on each motoneuron suggests that input is very finely graded and that each ending has little effect on the output. This is true, but it is worth noting that in a decerebrate preparation with tonic facilitation of extensor pools a motoneuron can be discharged by a volley of impulses in Ia fibers that contribute only a small percentage of the total number of terminals on a cell. The total number of endings on a cell is also a reflection of the number of different inputs it receives and the size of each projection. Table 27-5 summarizes information on some of the chief inputs that reach spinal motoneurons directly or fairly directly from outside the spinal cord. It is incomplete in many respects because of lack of information, particularly with regard to interneuronal systems. Despite the number and variety of terminals each cell receives, we know from the account in Chapter 26 that motoneurons in a pool, (i.e., supplying a single muscle) have a unique type of collective action. The activity of each cell is related to the total output of the pool. This suggests that the inputs to each cell in a pool are similar qualitatively if not quantitatively.

With such a rich set of afferent connections available, a major question is how these inputs are selected and combined to produce appropriate effects on the pool as a whole and on its individual members. We must assume that interneuronal circuits carry out a great deal of the premotoneuronal integration, but we know almost nothing about the mechanisms that control the combination of inputs. It is safe to assume that although impulses reaching a motoneuron can only excite or inhibit it, activity in each of the different systems that project to it has a different corollary significance because each is part of a different motor pattern in the brain. For example, vestibular inputs to a particular pool are usually just a small part of a complex pattern of signals to many muscles, calling for automatic adjustments of the body to changing positions of the head in space. At present we cannot deal with these complexities. We can only describe how certain direct and more easily accessible inputs are organized and how they affect the pool. Much of the discussion will be oriented around the problem of how input contributes to orderly recruitment of motoneurons and will assume a knowledge of Chapter 26.

It is obvious that orderly recruitment according to the size of a motor unit or the tension it produces could not occur unless there were precise rules governing all aspects of input to motoneurons and the pool as a whole. These rules must specify the distribution, density, location, and synaptic efficacy of the terminals of each afferent system. Furthermore, the inputs to members of a pool must be correlated in some manner with their cell sizes and surface areas. How this correlation is achieved is not understood, but various suggestions have been made. It has been proposed,[4,21,22,64] for example, that afferent activity is distributed selectively to individual cells and that its density and location account chiefly for recruitment order. This proposal would require an arrangement of inputs that differed qualitatively or quantitatively for each motoneuron in a pool in accordance with its size and type. A simpler explanation would be preferable, if it could account for the observations described in the preceding chapter.

Experiments on human subjects have added greatly to the range of activities in which recruitment has been studied. In tasks requiring human subjects to use different combinations of motor systems to produce slow, ramplike contractions or fast, ballistic movements, recruitment occurred in the same orderly sequence[10,11] and seldom, if ever, could be altered voluntarily.[26] It must be inferred that, in normal circumstances, different inputs are combined in a manner that always provides each cell in a pool with a balance of excitation and inhibition that does not disrupt the rank order established with inputs from its muscle. The effects of various types of inhibition described in the preceding chapter suggest, in fact, that inhibitory inputs are specifically designed to preserve the rank order observed in monosynaptic reflexes and repetitive discharges elicited from the homonymous muscle. The capacity of the nervous system to combine excitatory and inhibitory inputs without altering the relative susceptibilities of motoneurons to discharge is not understood, but it places stringent requirements on any hypothesis that may be advanced regarding the organization of input to a pool. A number of proposals[6,61,64] have been put forward, but they do not meet all the necessary requirements. An experimentally based theory of input consistent with all available facts is not yet feasible.

BACKGROUND FOR CURRENT STUDIES

Analysis of input and its role in the organization of the motoneuron pool began with Sherrington's studies of spinal reflexes (Chapter 28). From them he inferred that each afferent fiber terminates on a localized group of cells in a different portion of the pool. Depending on the level of *background* excitation reaching the pool, Sherrington believed that each input could result in a separate *discharge zone* with a surrounding *subliminal fringe* of cells that were excited but not discharged. In 1943 Lloyd[39] analyzed reflex action in relation to the pattern and peripheral source of afferent stimulation (Chapter 28). Later Eccles used the technique of intracellular recording to study the net effects of volleys of impulses from different muscles on various species of motoneurons.[14] In each of these latter cases, input was produced by an electrical shock to a muscle nerve and consisted of a synchronous volley of impulses conducted by an indeterminate number of afferent fibers. In more recent studies, techniques for examining the effects elicited in motoneurons by impulses in single afferent fibers have been introduced.[33] They permit an estimate of the number of motoneurons contacted by the terminals of single fibers, as well as a measure of the potency of these connections. An impulse in a single fiber produces an "individual" excitatory postsynaptic potential (EPSP), whose time course is not complicated by temporal dispersion, as it is for the aggregate EPSP produced by electrically evoked volleys traveling in fibers of different conduction velocities. The time course of the EPSP elicited by impulses in one fiber reflects more closely the location of the active terminals on the soma and dendrites, as will be described subsequently. This simplification of the afferent volley has enabled investigators to measure variations in the amplitudes of individual EPSPs and to make statistical studies of the process underlying release of transmitter substance.[35,51]

A number of different techniques have been developed for studying EPSPs evoked by the action of single Ia fibers. One approach has been to restrict the input to a single Ia fiber by dissection of a peripheral nerve branch.[33] In recent years a new technique, known as spike-triggered averaging (STA), has been introduced by Mendell and Henneman[50] to study the distribution and effects of impulses in single afferent fibers. This technique has yielded considerable information about the location, density, and distribution of the terminals of single Ia fibers within pools of motoneurons. Because more is known about the organization of this system than any other, it will be used as a model in the following account.

Input arising in group Ia fibers from muscle spindles

The most readily accessible and easily studied of all the inputs to motoneurons are those transmitted by Ia afferents. Group Ia fibers innervating the primary endings in muscle spindles establish direct, monosynaptic connections with motoneurons innervating the same muscle (homonymous) and synergistic muscles (heteronymous); hence they offer great advantages for

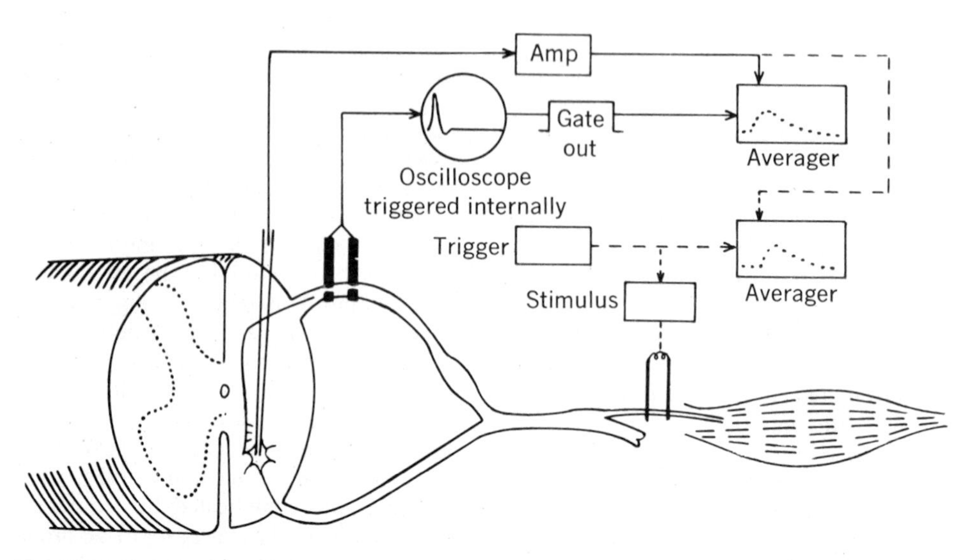

Fig. 27-1. Schematic diagram of the two methods of triggering the averaging (or summating) computer. Averager is represented twice: below, it is triggered at same time as electrical stimulus to muscle nerve (electrical stimulation of afferent fibers); above, it is triggered by stretch-evoked impulses in dorsal root filament (natural stimulation of afferent fibers). Signal from microelectrode is amplified and then led to averager. For details see text. (From Mendell and Henneman.[50])

electrophysiologic investigations. The technique used in studying the EPSPs evoked by impulses in these fibers is illustrated in Fig. 27-1. Recordings were made from an uninterrupted filament of dorsal root containing a single Ia fiber from the muscle under study, and its action potentials were used to trigger a summating computer. This device received the signals recorded by a microelectrode inside a motoneuron supplying that muscle. Each time the computer was triggered by a Ia impulse, it added the potentials occurring in the motoneuron for the subsequent 20 msec to a succession of "time bins" in its memory and displayed the total potential in each bin on an oscillographic trace. Each synaptic potential elicited by an impulse in this Ia fiber occurred after the same short conduction delay and was summed in the same bins. Because these synaptic potentials all had the same shape and sign, they appeared to grow progressively larger with repetition of the afferent impulse. Many other synaptic potentials in the same motoneuron were also summed in the computer, but since they varied randomly in shape, sign, and delay with respect to the trigger pulse, they were, in effect, averaged out of the record. Such a device is often called an averaging computer.

The presence or absence of a monosynaptic EPSP in a motoneuron was used to determine whether the single afferent fiber under study sent terminals to that particular cell. Fig. 27-2 illustrates a summated EPSP, too small to detect in single sweeps, that was extracted from the background activity by repeating it many times. Discharge of the Ia fiber was elicited by muscle stretch. Traces *A, B,* and *C* represent summated EPSPs developed after 100, 500, and 1,000 repetitions respectively. They demonstrate that the amplitude of the response grows linearly with increasing numbers of repetitions. No response was recorded (trace *D*) after the electrode had been withdrawn to a point just outside the cell, indicating that the response was generated across the membrane (i.e., was a true postsynaptic potential). EPSPs elicited by single Ia impulses, or summated in this way, are called "individual EPSPs."

In these experiments on extensor motoneurons[50] it was found that each Ia fiber projected to almost all (93%) of the motoneurons of its muscle. Individual EPSPs were recorded from 114 of the 122 homonymous motoneurons studied. Of the 22 afferent fibers isolated, 14 projected to every motoneuron impaled. The other 8 projected to all but 1 motoneuron, or possibly the EPSPs elicited were too small to detect.

Thus each of the 60 Ia fibers of the cat's medial gastrocnemius muscle sends terminals to essentially all its 300 motoneurons. These same fibers also make direct connections with about 60% of the heteronymous (lateral gastrocnemius and soleus) motoneurons,[35,50,57] with one or more

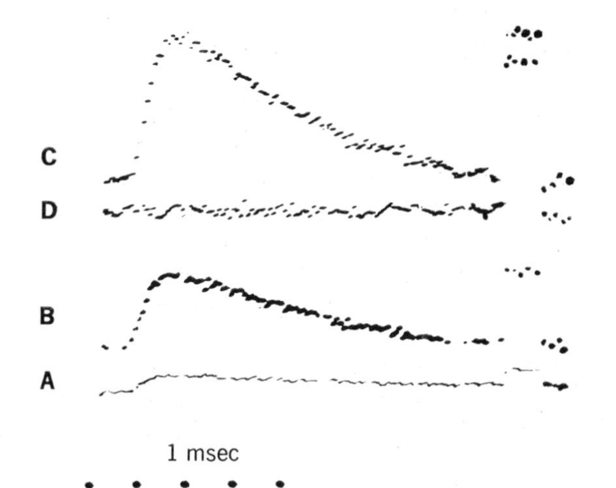

Fig. 27-2. Summated EPSPs produced by repetition of many individual EPSPs in gastrocnemius motoneuron. **A** to **C,** Results of summing 100, 500, and 1,000 sweeps, respectively. **D,** Sum of 1,000 sweeps recorded after microelectrode had been withdrawn to a point just outside motoneuron, as indicated by sudden disappearance of resting membrane potential. Calibrating pulse of 200 μV was injected at end of each sweep and summated in same manner as signal. (From Mendell and Henneman.[50])

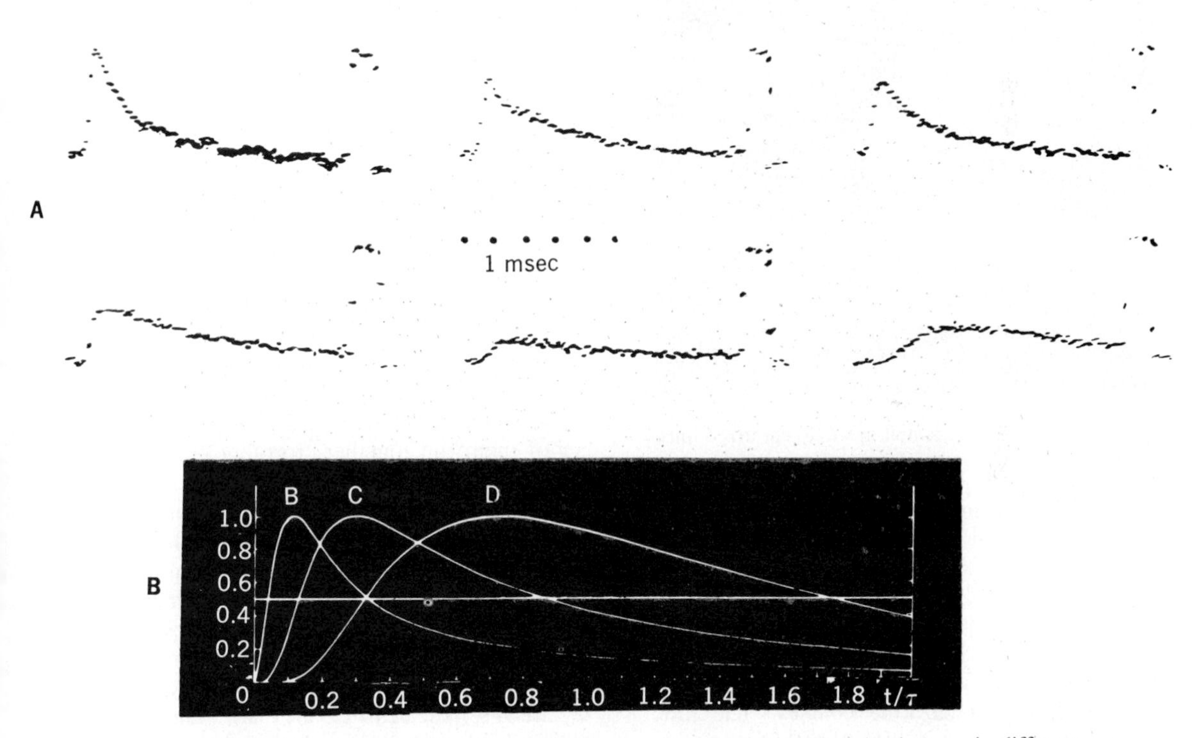

Fig. 27-3. A, Summated EPSPs evoked by impulses in same Ia fiber terminating on six different motoneurons. Calibration pulse is 200 μV in each case. **B,** Computed EPSPs for input distributed to three different parts of motoneuron. Three curves represent EPSPs produced by inputs limited to soma *(B),* proximal dendrites *(C),* and distal dendrites *(D).* (Reproduced from Fig. 2 of Rall[54] with the permission of the author; from Mendell and Henneman.[50])

groups of spinal interneurons, and send branches through the dorsal columns to end on cells of Clarke's column. Results similar to these were obtained in experiments on Ia fibers from flexor muscles and their motoneurons.[52] Occasional Ia fibers have been found, however, whose projections to motoneurons may be less extensive.[52,57]

The time course and amplitude of individual EPSPs varied over a wide range. Examples of this variation are reproduced in Fig. 27-3, *A*. The six EPSPs shown there were evoked by impulses in the same afferent fiber terminating on six dif-

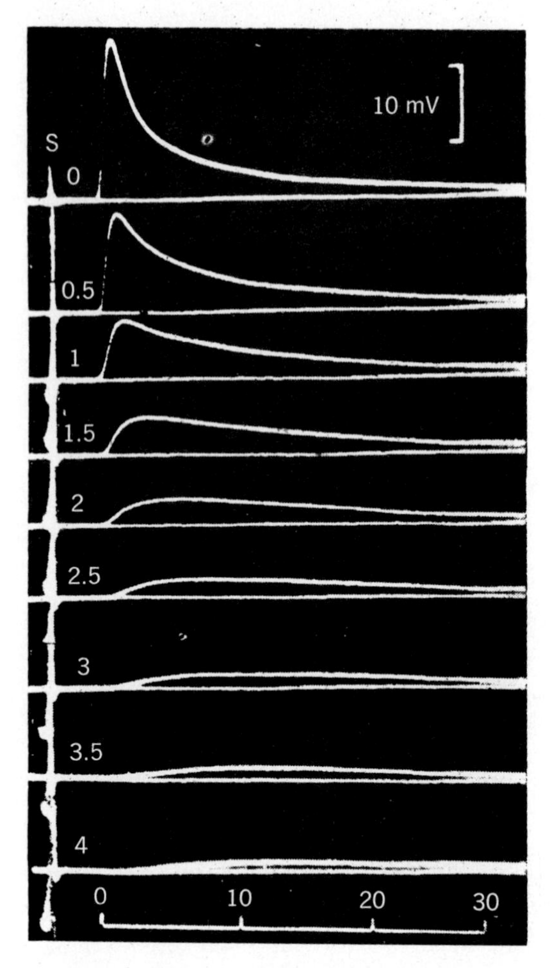

Fig. 27-4. End-plate potential in single curarized muscle fiber. Position of microelectrode was altered in successive 0.5 mm steps. Numbers give distance from focus in millimeters × 0.97. *S,* Stimulus artifact. Time in milliseconds. (From Fatt and Katz.[20])

ferent motoneurons. There was no apparent tendency in this or in other experiments for the EPSPs elicited by activity in a particular afferent fiber to have similar time courses. The time courses of individual EPSPs, however, varied in a systematic way. EPSPs with fast-rising phases generally had fast-falling phases as well. Slow-rising phases were usually accompanied by slow-falling phases.

In shape and time course, these individual EPSPs recorded from motoneurons were similar to those recorded by Fatt and Katz[20] at different distances from the end-plates of frog muscle fibers. As illustrated in Fig. 27-4, they found that the greater the distance from end-plate to recording site, the slower the rise and fall of the EPSPs. Analogously, the shapes of individual EPSPs of motoneurons suggested that they were generated at different distances from the recording electrode in the soma. This suggestion was reinforced by the studies of Rall,[54,55] who constructed a model motoneuron based on anatomic and physiologic data. With a digital computer Rall applied simulated inputs to different parts of the model cell. Curves *B, C,* and *D* in the lower part of Fig. 27-3 represent computed EPSPs limited to the soma, proximal dendrites, and distal dendrites, respectively.

Prior to this analysis there was no anatomic evidence indicating that the endings of a Ia fiber are concentrated in a single limited area, as Rall's theoretical inputs were. The close resemblance of the experimental EPSPs to the computed EPSPs therefore suggested that individual EPSPs were caused by Ia inputs concentrated within relatively small areas of motoneuronal surface.

Scheibel and Scheibel[56] recently published a Golgi study of the primary afferent collaterals in the cat's spinal cord that provides an anatomic basis for some of these findings. It is now agreed that,[56,60] contrary to earlier views, the soma and dendrites of motoneurons are disposed in a predominantly longitudinal column running for a considerable distance in the rostrocaudal axis of the spinal cord. This orientation is clearly evident in the motoneurons labeled f and g in Fig. 27-5. Each Ia fiber in the dorsal columns gives off a series of primary afferent collaterals at regular intervals. These collaterals drop ventrally with almost plumb line precision. A variable number of them run together, forming a *microbundle*. These microbundles pass perpendicularly through the dendrites of the motoneurons. Their terminals ramify almost exclusively in the mediolateral plane of the spinal cord. The terminal field of each collateral is virtually two dimensional (10 to 50 μm in thickness) and makes contact with a limited extent of the longitudinally disposed motoneurons. The simple time course of the EPSPs indicates that they are generated by terminals of a single Ia collateral that are limited to one part of the motoneuron surface. However, several examples of compound EPSPs have been encountered. Their significance is not clear.

The afferent fibers isolated in different experiments varied widely in conduction velocity (diameter). Since the number of terminals given off by a Ia fiber depends on the diameter of the parent fiber in motor axons,[13,49] the EPSPs elicited by impulses in Ia fibers may also vary in amplitude with fiber diameter. In this first series of experiments,[50] these variables were found to be directly related. The larger the Ia fiber, the

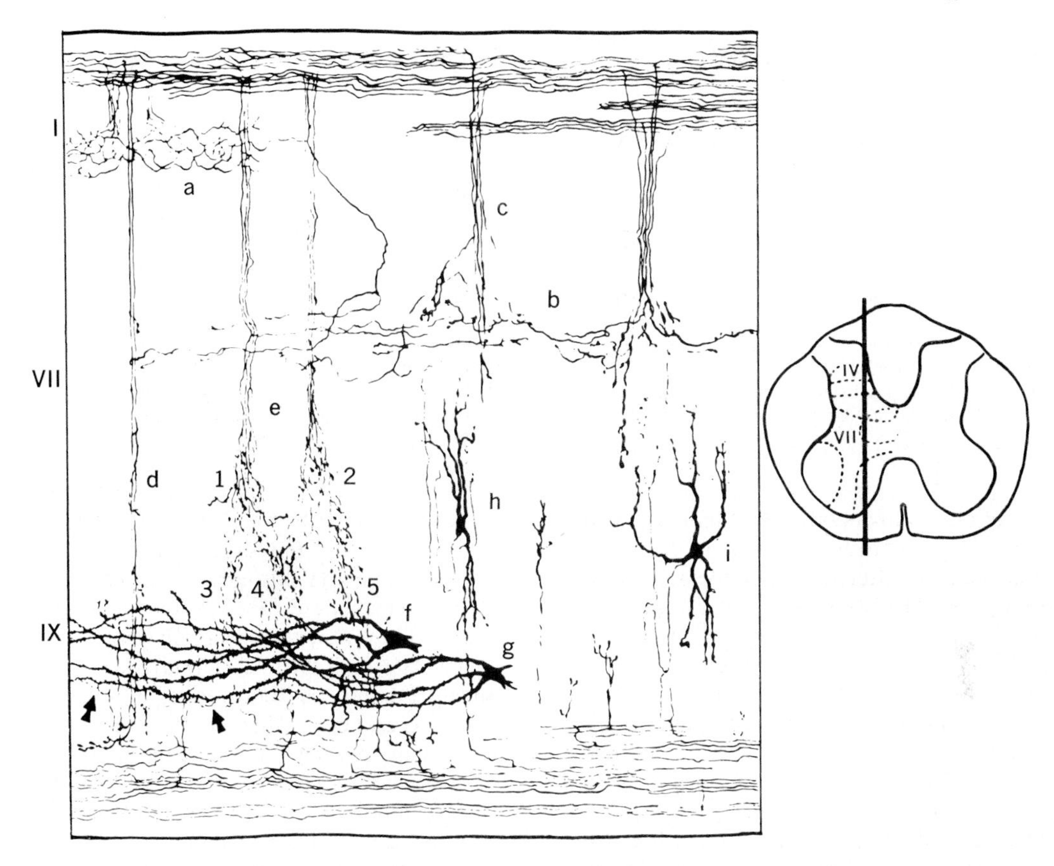

Fig. 27-5. Sagittal section through lumbrosacral cord of 40-day-old cat showing selected details of presynaptic and postsynaptic neuropil. *a,* Longitudinal neuropil of dorsal horn; *b,* elements of longitudinal neuropil of intermediate nucleus generated by short primary afferent collaterals such as those in microbundle *c; d* and *e,* microbundles made up of long primary afferent collaterals, which in the case of *e* are shown to generate primary mix *(1,2)* and secondary mix *(3* to *5)* neuropil fields; *f* and *g,* motoneuron somata in staggered position with overlapping systems of sagittaly running dendrites; *h,* interneuron of lamina VII with typical single dendritic system; *i,* interneuron of lamina VII with multiple dendrite systems. Arrows point to terminating axon collaterals from propriospinal fibers of ventral column that run parallel to motoneuron dendrites and apparently establish multiple synaptic contacts. Rapid Golgi modification. (Magnification ×200.) (Modified from Scheibel and Scheibel.[56])

bigger the EPSPs its impulses evoked. Later in this chapter it will be shown that the group II fibers from the secondary endings in muscle spindles also make direct, monosynaptic connections with homonymous motoneurons. It is perhaps significant that impulses in the much smaller group II fibers (6 to 12 μm in diameter) elicit much smaller EPSPs than Ia impulses.

Ia PROJECTIONS FROM SYNERGISTIC MUSCLES TO MOTONEURONS

Scott and Mendell[57] extended the scope of these studies by investigating the projections of the Ia fibers connecting each of the three muscles of the triceps surae group with their own motoneurons and those of their closest synergists. This group includes the medial (MG) and lateral (LG) gastrocnemius muscles and the soleus (SOL). Lloyd had shown[41] in studies on monosynaptic reflexes that the motoneurons of muscles that work together closely receive extensive Ia projections from each other. These three muscles have slightly different origins, but they all have the same insertion (the Achilles tendon). Although their actions are not identical, the major function of each is extension of the foot.

Impulses in single Ia fibers of these three muscles were the inputs in experiments with spike-triggered averaging. The connections found for each of the nine possible combinations of afferents and motoneurons is shown in Table 27-1.

Table 27-1. Projection frequency*†

Afferent	Motoneuron MG	LG	SOL
MG	78 (89)	54 (69)	63 (24)
LG	49 (49)	71 (84)	68 (19)
SOL	42 (46)	74 (61)	100 (18)

*From Scott and Mendell.[57]
†Values are percentages. Numbers in parentheses are total numbers of afferent-motoneuron combinations for each group.

Table 27-2. Mean EPSP amplitude*†

Afferent	Motoneuron MG	LG	SOL
MG	95 (69)	59 (37)	79 (15)
LG	76 (24)	141 (60)	84 (13)
SOL	50 (19)	65 (45)	143 (18)

*From Scott and Mendell.[57]
†Values are in microvolts. Numbers in parentheses are numbers of afferent-motoneuron combinations studied for each group.

For example, Ia fibers from the MG muscle projected to MG motoneurons in 78% of the 89 cells investigated, Ia afferents from the LG muscle projected to LG cells in 71% of 84 cases, and Ia fibers from SOL sent terminals to 100% of 18 SOL cells. These are all extensor (antigravity) muscles. In a similar study[52] on a flexor muscle (semitendinosus), its Ia fibers sent terminals to 82% of its motoneurons and 38% of the motoneurons supplying synergistic muscles (posterior biceps).

The mean amplitudes of the EPSPs found for each combination of Ia afferent and motoneuron species are shown in Table 27-2. Individual EPSPs recorded in homonymous motoneurons were larger than those recorded in heteronymous motoneurons. This "homonymous effect" is not sufficient, however, to explain the amplitudes of the EPSPs in Table 27-2. For example, impulses in both MG and LG afferents produced larger mean EPSPs in SOL motoneurons than in the other type of heteronymous motoneurons. Impulses in SOL afferents produced a larger mean EPSP (143 μV) in SOL motoneurons than any other in the table. The only combination in which there were no failures in projection were in SOL afferents and motoneurons. Thus individual EPSPs are, in general, larger if they are evoked in SOL motoneurons than in MG or LG cells. The findings suggest that the larger the mean amplitude of individual EPSPs in any projection, the less chance there was of failing to detect a small EPSP and the greater the projection percentage. In the projection of SOL afferents to SOL motoneurons, the combined effects of homonymous synapses and small SOL motoneurons with large input resistances apparently increased the mean EPSP amplitude to such an extent that all the EPSPs were detected and a 100% projection was found. This hypothesis, and evidence to be described later in this chapter, suggest that Ia fibers from MG and LG may also project to all homonymous motoneurons and that the lower percentages obtained in some cases may be due to failure to detect small EPSPs.

The data in Tables 27-1 and 27-2 were obtained from motoneurons of unequal sizes in different experiments. This would tend to obscure rather than produce the relationships just described. In 33 cases it was possible to minimize these factors by comparing EPSPs from a homonymous and a heteronymous afferent in the same motoneuron, thus equalizing many variables that might affect EPSPs. The ratios of the responses in these 33 cells indicated that homonymous EPSPs were larger than heteronymous responses when compared in the same motoneuron. Histograms of the EPSP amplitudes for each of the nine groups in Table 27-2 showed that the major difference in their sizes was the tendency for EPSPs larger than 150 μV to be generated in homonymous motoneurons. Small EPSPs were distributed about equally.

The Rall model[54] suggests that the size of an individual EPSP may be influenced by the location of the active terminals on the cell body. In order to test the possibility that individual EPSPs are larger in homonymous motoneurons because they may be located closer to the cell body, "shape index" curves (half width of EPSP vs rise time) were constructed for both classes of responses. Fig. 27-4 indicated that EPSPs generated on distal parts of dendrites should have longer half widths and rise times than those evoked more proximally, if other factors were equal. The data suggest that differences in location of the terminals on homonymous and heteronymous motoneurons cannot account for the amplitude differences.

The higher frequency of Ia projections to homonymous motoneurons and the larger size of the EPSPs they elicit apparently form the basis for the well-known observation that stretch reflexes are best developed in the muscle that has been stretched.[37] Studies on ventral root reflexes[40] and intracellular recordings[14] have previously revealed clear subliminal effects in heteronymous motoneurons. The observations cited previously provide more detailed evidence regarding the role of input in linking the activity of synergistic muscles.

Patterns of group Ia connections from hind limb muscles to motoneurons

In previous sections of this chapter the chief emphasis has been placed on the Ia connections of a muscle with its own motoneurons (the homonymous projection), and with its close synergists (the heteronymous projection). Here a brief review of data is given, summarizing some of the more extensive Ia connections from hind limb muscles to various species of motoneurons supplying the same limb. Since the muscles in a limb work together in a variety of harmonious movements, it is not surprising to find that the primary afferent fibers from one muscle establish direct monosynaptic connections with the motoneurons of muscles that cooperate with that muscle frequently.

In an earlier study, carried out without the benefit of averaging techniques, Eccles et al.[16] assembled data on the Ia projections of nine hind limb muscles to the motoneurons supplying them. The data were drawn from intracellular recordings of aggregate EPSPs (using nerve volleys) from approximately 500 motoneurons in the cat. Table 27-3 summarizes the monosynaptic excitatory actions of these nine species of Ia fibers so as to bring out the quantitative differences in their projections to the nine pools of motoneurons. Two criteria were employed to assess these projections: the frequency of occurrence of these aggregate EPSPs and their mean size. Vertical columns labeled DP, Per, MG, etc. at the top indicate muscle nerves stimulated with single shocks maximal for Ia fibers. Horizontal rows give fractions indicating numbers of motoneurons responding with recognizable Ia EPSPs over the total number so investigated. For example, the upper left entries show that afferent nerve volleys from DP elicited EPSPs in 140 of 140 DP motoneurons and that their mean amplitude was 3.58 mV in cats given pentobarbital. The fractions indicate whether enough Ia fibers of a muscle send terminals to a given species of motoneuron to evoke an observable EPSP, but they do not reveal how many of the Ia fibers did so in the positive cases. The DP volley produced an EPSP in 84 of 139 Per motoneurons, but the mean amplitude was only 0.80 mV. This suggests that few, or in some cases, no Ia fibers from DP established connections with Per cells. Perhaps only a certain type of cell in the Per pool receives DP terminals or the technique was not sufficiently sensitive to detect existing connections without averaging. The synergic pairs, MG + LG-S and Pl + FDB are enclosed in boxes in Table 27-3. The former pair is a good example of two synergists (ankle extensors) pulling in parallel; they exert large excitatory effects on each other. The latter pair of muscles are synergists pulling in series and acting together in plantar flexion of the digits. The Ia connections are numerous from FDB to Pl motoneurons, but less so in the opposite direction. It can be seen that entries on the diagonal from upper left to lower right contain data on the responses of motoneurons to Ia volleys from their own muscle. The EPSPs were largest in these homonymous cases and were observed in every motoneuron.

Table 27-4 provides a more detailed survey of the motoneurons supplying the pretibial flexor muscles of the cat. It confirms the impression given in Table 27-3 that the three species of peroneus motoneurons form a synergic group linked by effective monosynaptic action. These groups are enclosed in a box, as are TA + EDLB.

These tables do not provide complete or detailed data on all the Ia linkages underlying the synergic actions of muscles, but they illustrate how these inputs may contribute to such interactions. The tables also reflect the number of possible combinations of muscle inputs a motoneuron may respond to under various circumstances.

Transmission at the terminals of Ia fibers

Although electrical coupling has been suggested as a possible mechanism of transmission at Ia synapses, their fine structure does not favor this possibility. Until recently the available evidence indicated that Ia EPSPs in motoneurons were produced by postsynaptic conductance changes caused by chemical transmitters. The synaptic delay of about 0.3 msec is consistent with chemical mediation.[12,50] Nonlinear summation of Ia EPSPs has been reported,[3,34] suggesting that some chemical sites are close enough to permit mutual interactions.

Synaptic potentials associated with transmembrane ionic currents should have "equilibrium potentials"

Table 27-3. Group Ia connections from hind limb muscles to nine species of motoneurons*†

	DP	Per	MG	LGS	Pl	FDB	FDHL	PBST	SMAB
DP	140/140	80/139	0/139	0/139	0/138	5/80	12/140	0/153	0/153
mV	3.58	0.76	0.00	0.00	0.00	0.02	0.03	0.00	0.00
Per	84/139	139/139	53/141	18/142	6/142	16/64	19/140	0/139	0/139
mV	0.80	4.65	0.30	0.06	0.02	0.10	0.06	0	0
MG	0/32	0/29	47/47	48/48	16/45	0/20	5/38	0/29	0/21
mV	0.00	0.00	4.61	1.96	0.18	0.00	0.05	0.00	0.00
LG-S	0/27	1/22	36/36	62/62	52/62	4/33	10/63	0/56	0/46
mV	0.00	0.05	2.24	4.73	0.94	0.03	0.03	0.00	0.00
Pl	0/32	0/32	2/36	11/60	60/60	44/44	48/53	0/61	0/61
mV	0.00	0.00	0.00	0.06	4.62	1.64	0.61	0.00	0.00
FDB	0/14	1/14	1/14		8/8	23/23	5/14	0/14	0/14
mV	0.00	0.09	0.07		0.45	4.46	0.15	0.00	0.00
FDHL	0/61	0/61	1/61	0/42	28/60	24/91	66/66	0/42	0/42
mV	0.00	0.00	0.01	0.00	0.23	0.10	5.43	0.00	0.00
PBST	1/55	0/55	0/55	0/55	0/55	0/55	0/55	55/55	24/53
mV	0.00	0.00	0.00	0.00	0.00	0.00	0.00	2.57	0.26
SMAB	0/54	0/54	12/54	1/54	1/54	0/54	0/54	49/54	54/54
mV	0.00	0.00	0.08	0.00	0.00	0.00	0.00	1.02	1.71

*From Eccles et al.[16]

†Abbreviations: DP, pretibial flexor muscles—tibialis, anterior, extensor digitorum longus plus extensor digitorum brevis, which are supplied collectively by the deep peroneal nerve, hence the abbreviation DP; Per, peroneus muscles—longus, brevis, and tertius; MG, medical gastrocnemius; LG-S, lateral gastrocnemius plus soleus; Pl, plantaris; FDB, flexor digitorum brevis; FDHL, flexor digitorum and hallucis longus; PBST, knee flexors—posterior biceps plus semitendinosus; SMAB, hip extensors—semimembranosus plus anterior biceps. Results assembled from intracellular recordings from approximately 500 ipsilateral hind limb motoneurons. Full explanation in text.

and should be reversible if the equilibrium potential is exceeded.[12] Although several investigators have reported reversals of aggregate EPSPs, attempts to reverse individual EPSPs have not yet been successful. This test of the chemical transmission model indicates that current hypotheses may have to be revised. The available data are not adequate for any firm conclusions, but suggest that Ia EPSPs are generated by increases in the permeability to both sodium and potassium ions.

It is generally believed that the EPSPs produced by activation of single group Ia afferents consist of smaller "unit" EPSPs that are of the all-or-none type. The most extensive studies of this "quantal transmission" process in the mammalian CNS are the analyses of EPSPs elicited in spinal motoneurons by stimulation of single Ia fibers.[33-35] The conclusions of these studies are that transmission at these junctions occurs in quantal steps, the release process being subject to Poisson statistics. However, there have been difficulties with these experiments due to noise generated in the membrane and in the recording apparatus. A series

Table 27-4. Group Ia connections from hind limb muscles to motoneurons supplying pretibial flexor muscles*†

	TA	EDLB	PerL	PerB	PerT	MG	LG-S	PI	FDB	FDHL
TA	40/40	37/37	0/38	0/32	0/33	0/39	0/38	0/37	0/20	8/39
mV	2.89	1.74	0.00	0.00	0.00	0.00	0.00	0.00	0.00	0.06
EDLB	67/100	100/100	50/98	54/99	35/76	0/100	0/101	0/101	5/60	4/101
mV	0.51	2.65	0.29	0.52	0.21	0.00	0.00	0.00	0.03	0.01
PerL	11/31	33/33	42/42	34/36	28/30	0/42	0/42	0/42	1/25	0/42
mV	0.10	1.68	2.89	1.53	0.49	0.00	0.00	0.00	0.00	0.00
PerB	0/54	12/46	51/54	54/54	46/48	40/53	11/54	5/54	15/28	12/52
mV	0.00	0.10	1.41	3.06	1.11	0.48	0.11	0.03	0.22	0.13
PerT	0/12	11/12	12/12	12/12	12/12	7/12	5/12	1/12	0/11	3/12
mV	0.00	0.80	1.16	1.37	1.07	0.24	0.16	0.04	0.00	0.08

*From Eccles et al.[16]

†As in Table 27-3, but giving a more detailed survey of the motoneurons supplying the pretibial muscles. Additional abbreviations: TA, tibialis anterior; EDLB, extensor digitorum longus and brevis; PerL, peroneus longus; PerB, peroneus brevis; PerT, peroneus tertius.

of studies by Edwards et al.[17-19] has recently been carried out to reexamine this problem, using several methods to remove the contributions of the noise. The details cannot be given here, but it was concluded[18] that the fluctuations in charge transmission at Ia synapses are nonquantal, nor are they described by binomial or Poisson statistics. It was suggested instead that transmission is of the all-or-none type at each bouton and that the fluctuations resulted from the combined effects of failures in transmission at axonal branch points that occur as the Ia fiber arborizes to supply many motoneurons. The number of synaptic terminals from a single Ia fiber is believed by some investigators to be small and variable and that, in general, there are only one or two boutons on each motoneuron. In the studies by Edwards et al., some EPSPs fluctuated between two different amplitudes and time courses. Analysis[18] showed that charge transmission always occurred at one synaptic location but not always at a second location. Unfortunately, failure of the impulse to propagate into the terminals could not be clearly distinguished from failure of the terminals to release transmitter after adequate depolarization. The investigators were unable to cause reversals of single EPSPs despite injection of 150 nA depolarizing currents. Hyperpolarizing currents caused little, if any, increase in peak amplitude of EPSPs. No final conclusions can be drawn from these studies, but Edwards et al. believe that transmission does not occur in quantal steps. The model that best fits these latest results is one where charge is transferred in all-or-none fashion.[18]

Group Ia inhibition of antagonistic motoneurons via interneurons

The failure of inhibitory inputs from six different sources to alter significantly the recruitment order of homonymous motoneurons was described in the previous chapter and summarized in Table 26-1. Thus many inhibitory inputs are distributed in such a way that they do not interfere with the rank order established with excitatory inputs from muscle. Their arrangement may, in fact, be designed to preserve this order. It is not clear how this necessarily wide and systematic distribution of inhibition is achieved. Clamann et al.[8] described evidence suggesting that the density of active inhibitory endings on a motoneuron is proportional to that of its excitatory endings. Jankowska and Roberts[29] report that there may be a functional similarity between the widespread distribution of excitatory Ia terminals to homonymous motoneurons and the inhibitory terminals of the interneurons, which they also activate. Interneurons that receive collaterals of Ia fibers from the quadriceps (Q) muscle were caused to fire repetitively by applying glutamate to them iontophoretically. Their discharges were recorded extracellularly and used to trigger an averager. The postsynaptic potentials evoked by these impulses were recorded intracellularly from the motoneurons of antagonists (i.e., the posterior biceps and semitendinosus [PBST] muscles). Single impulses from the interneurons elicited individual inhibitory postsynaptic potentials (IPSPs) averaging 52 μV in amplitude. The time course of the unitary IPSPs indicated that the terminals make contact predominantly with the soma or the proximal parts of the dendrites of the motoneurons.

This technique did not allow systematic study of the projection of any single interneuron to a population of motoneurons, as has been done with Ia fibers. Comparison of the maximum composite IPSP with the mean individual IPSP, however, suggested that 10 to 20 of these interneurons converge on a PBST motoneuron. Single interneurons apparently send terminals to about 20% of the PBST motoneurons. Jankowska and Roberts[29] point out:

> If each Ia afferent fiber from Q terminates on several of these interneurons as they do on motoneurons (Mendell and Henneman, 1971), which in turn project randomly to about 20% of the PBST motoneurons, then each Ia fiber from Q might influence most of the population of PBST motoneurons. The emerging picture of the connexions between the Ia afferents and the motoneurons of the antagonist muscles would thus closely resemble that of the direct connections of the Ia afferents on the homonymous motoneurons described by Mendell and Henneman.

Stuart and his colleagues[62] have also examined Ia inhibition in antagonistic muscles. Fig. 27-6, *A,* shows the averaged response, recorded at the root entry zone, to a small, brief stretch of the MG muscle, which activates many Ia fibers synchronously. Intracellular averaging from an MG motoneuron *(B)* revealed a large EPSP elicited by this input after 512 sweeps. This response commenced 0.4 msec after the arrival of the Ia impulse at the spinal cord. The extracellular field response *(C)* to this large stimulus was only 4.5% of the EPSP amplitude. Recordings from an antagonistic motoneuron 6.5 mm away (tibialis anterior or extensor digitorum longus) showed a clear IPSP *(D)* with a longer latency (1.5 msec) similar to effects previously demonstrated.[62] Similar hyperpolarizing responses have been observed using the spike-triggered averaging technique by averaging the effects of single Ia impulses in antagonistic cells (principally tibialis anterior and extensor digitorum motoneurons) for 2,000 or more sweeps. The latency (1.2 to 2.4 msec) of these responses was consistent with a disynaptic linkage. These amplitudes (1 to 11 μV) are underestimates of true values because of the large percentage of trials in which the Ia impulse triggering the averager did not discharge the intervening interneuron.

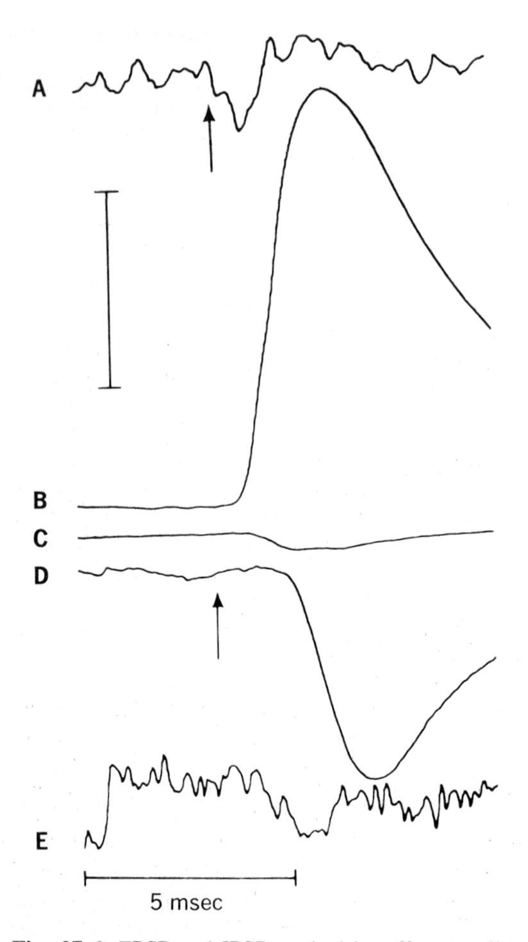

Fig. 27-6. EPSP and IPSP evoked by afferent volley due to triangular ramp stretch of MG muscle (5 msec duration and 50 μm amplitude) applied at 50 to 150 *g* initial tension. **A,** Average of 512 afferent volleys, recorded from root entry point. Arrow indicates entry of earliest components of volley. **B,** EPSP in MG motoneuron (average of 512 repetitions) with 0.4 msec latency. **C,** Extracellular control recording for **B. D,** IPSP in tibialis anterior or extensor digitorum longus motoneuron (512 repetitions) with latency of 1.5 msec. **E,** Extracellularly recorded control for **D** at 16 times its sensitivity. Calibrations = **A,** 125 μV; **B** and **C,** 2 mV; **D,** 250 μV; **E,** 15.6 μV. (From Watt et al.[62])

Group Ib input from Golgi tendon organs

The spike-triggered averaging method has also been used to study the projections from tendon organs in the MG muscle to its own motoneurons and to those of other nearby muscles. Tendon organs respond primarily to tension produced by the active contraction of motor units directly in series with the receptor, as illustrated in Fig. 25-13. Previous studies[15,36] indicated that activity in this projection was mediated by a circuit with one interneuron (i.e., a disynaptic afferent pathway). Stuart and his colleagues[62] obtained data on the projections of Ib afferent fibers to motoneurons of various species. In numerous homonymous and heteronymous motoneurons, IPSPs were seen after prolonged averaging (4,000 to 16,000 sweeps). The responses were somewhat jagged and irregular despite extensive averaging, an indication that only occasional Ib impulses triggered the discharge of the inhibitory interneuron that elicited the IPSP. Nevertheless, these potentials were more clearly defined than EPSPs evoked chiefly in direct and indirect antagonists. IPSP latencies were spread over a wide range (0.6 to 6.0 msec), and they were generally a little larger (5.2 ± 3.5 μV) than the EPSPs. These results, obtained with physiologic excitation of receptors, confirm earlier, less direct methods in showing that Golgi tendon organs have disynaptic inhibitory effects on homonymous and heteronymous motoneurons. They also confirm the existence of excitatory connections to motoneurons of antagonists.

Group II afferents from muscle spindles

Some group II afferent fibers have their peripheral terminals in the secondary endings of muscle spindles, which respond primarily to muscle length rather than to rate of change of length, as illustrated in Fig. 25-7. Matthews[47] demonstrated that impulses from these endings exert excitatory effects on homonymous motoneurons, thus contributing to the stretch reflex. There are one to five secondary endings per spindle. Activity arising in them probably exerts effects on motoneurons through several reflex pathways. It supplies essential feedback information about the length of muscles and may serve other important control functions.

Kirkwood and Sears[32] examined the effects of single group II impulses, using natural stimulation of the stretch receptors and the spike-triggered averaging technique previously described. The short central latency of the EPSPs they recorded indicated that impulses from spindle secondary endings monosynaptically excite homonymous alpha motoneurons. Individual group II fibers were found to terminate on more than one motoneuron, but the full extent of their distribution was not established. These results provided direct confirmation of Matthews' findings and showed that some group II impulses exert monosynaptic effects on motoneurons.

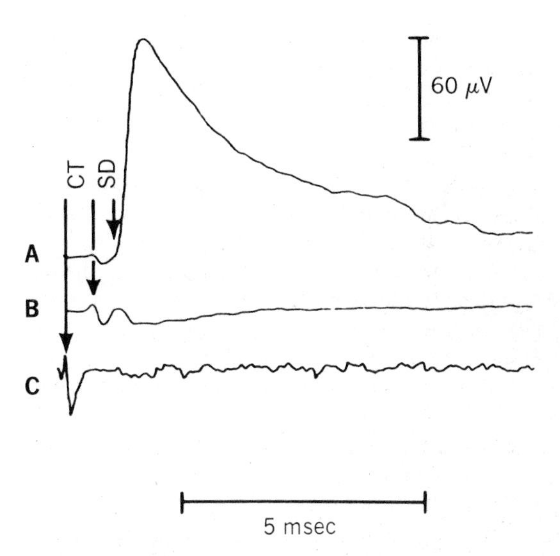

Fig. 27-7. EPSP produced in MG motoneuron by impulses in group II fiber. **A,** Average of 1,024 intracellularly recorded EPSPs. **B,** Corresponding average with microelectrode just outside motoneuron. **C,** Averaged record of group II impulses recorded from dorsal root. Intervals *CT* and *SD* indicate central conduction time and synaptic delay, respectively. (From Stauffer et al.[59])

Stuart and his colleagues[59] extended these studies by examining the projections of group II afferents from the MG muscle to more than 100 motoneurons supplying various muscles in cats' hind limbs. EPSPs were found in many motoneurons, predominantly of the triceps surae muscles. The arrival of the group II impulse at the spinal cord and its invasion of the presynaptic terminals were studied whenever possible in the averaged records, as illustrated in Fig. 27-7. Those EPSPs with a latency ≤1.4 msec were judged to be monosynaptic in confirmation of the findings of Kirkwood and Sears. The mean amplitude and rise time in MG and LG-S cells were 30.1 μV and 1.0 msec, respectively. The percentage of group II fibers projecting directly to homonymous MG and heteronymous LG-S motoneurons was apparently much lower than for Ia fibers. In this admittedly small sample of group II fibers, the projection probability was 52% for 31 MG cells and 26% for 27 LG-S cells. The mean amplitude of monosynaptic group II EPSPs was also much less than for Ia EPSPs (37.9 μV for MG cells and 12.3 μV for LG-S cells).

EPSPs with latencies greater than 1.4 msec were also observed. They were judged to be mediated by di- or trisynaptic circuits. They were smaller and had longer rise times than the monosynaptic EPSPs. In addition, a few inhibitory responses were found with latencies and rise times appropriate for di- and trisynaptic connections. Their mean amplitude was 4.6 μV. According to Stuart and his colleagues, the distribution of group II EPSPs and IPSPs was generally consistent with their exerting stretch reflex effects similar to those of Ia afferents. These observations add further confirmation to the earlier findings of Matthews[47,48] that group II endings add to the excitatory influx in the stretch reflex.

Other inputs from muscle

In addition to the fibers conveying impulses from stretch receptors to motoneurons, muscle nerves contain sensory fibers ranging in diameter from group Ia to unmyelinated C fibers. According to Stacey,[58] "at least two thirds of all the sensory fibers in muscle nerves have not as yet been unequivocally allocated endings within the muscle."

Afferent fibers with a wide range of diameters have free nerve endings. With the electron microscope, free endings have been demonstrated arising from fibers of groups II, III, and IV. A few pacinian corpuscles are supplied by fibers of group I and II origin. Cell bodies in dorsal root ganglia give rise to nonmyelinated C fibers.

The adequate stimuli for these endings are not well understood. Many C fibers are activated by subjecting the muscle to 5 to 10 gm of mechanical pressure. Prolonged contraction in conjunction with occlusion of the blood supply produces discharge of C fibers and pain. Some C fibers respond to very light touch or slight changes in temperature. Pacinian corpuscles respond to touch-pressure, whereas many free nerve endings apparently act as nociceptors.

Electrical stimulation of muscle nerves produces a varied, nonphysiologic input. Many of the afferents below group I size affect motoneurons through polysynaptic circuits, causing a net inhibition of those sup-

Table 27-5. Summary of chief inputs to motoneurons from outside spinal cord

Pathway	Origin or receptor	Fibers	To	Circuit
Dorsal roots	Primary spindle receptor	Ia (12-20 μm)	Alpha motoneuron: (1) homonymous, (2) heteronymous	Monosynaptic Monosynaptic
		Ia collaterals	Antagonist motoneurons	Disynaptic
	Secondary spindle receptors	Group II	Alpha motoneurons	Mono- di-, and trisynaptic
	Tendon organs	Group Ib (12-20 μm)	Alpha motoneurons	Di- and trisynaptic
	Joint receptors	1-20 μm in diameter	Alpha motoneurons	Polysynaptic
	Skin endings	Wide range, diameters of A-β, A-δ, C	Alpha motoneurons	Polysynaptic
	Subcutaneous endings	Wide range of diameters	Alpha motoneurons	Polysynaptic
Pyramidal or corticospinal tract	Precentral 60%: (1) area 4; (2) supplementary area	2%-8%, 10-20 μm; 8%, 4-10 μm; 90%, 0-4 μm	Direct to alpha motoneurons in primates and to interneurons	Monosynaptic Polysynaptic
	Postcentral 40%: (1) areas 3, 1, and 2; (2) second motor area	Via lateral and ventral corticospinal tracts	Interneurons in base of dorsal horn and intermediate zone	Polysynaptic
Rubrospinal tract	Red nucleus	To all levels of spinal cord	Chiefly to interneurons, a few to motoneurons	Polysynaptic
Vestibulospinal tract	Vestibulocollic cells	In lateral division: Deiters nucleus	Neck extensor motoneurons	Monosynaptic
		In medial division: (1) medial nucleus, (2) descending nucleus, (3) Deiters nucleus	In medial longitudinal fasciculus to all levels	Monosynaptic
Reticulospinal tract	Pontine reticular formation	Ipsilateral to all levels of spinal cord	Alpha motoneurons at all levels of spinal cord	Monosynaptic
	Bulbar reticular formation	To all levels of spinal cord	Probably to interneurons	Polysynaptic

plying extensor muscles and excitation of flexors. Nonphysiologic inputs, understandably, have caused confusion regarding the roles of some stretch receptors.

Central nervous system regulation of input to motoneurons

Table 27-5 summarizes the main inputs to motoneurons, for reference purposes. These projections are numerous and diverse. The signals they transmit represent the constantly changing and sometimes opposing influences that are generated in different parts of the nervous system. Movement is controlled by combining and coordinating the activity in these systems. The general principles of this process are not understood, but some of the mechanisms that the nervous system uses have been identified and will be discussed briefly in this section.

Motoneurons receive multiple inputs, each with numerous components functioning in parallel. This has obvious advantages. It is possible for many parts of the nervous system to have direct or fairly direct access to motoneurons without long delays, and the activity in each system can be graded precisely. The 5,000 to 10,000 endings on each motoneuron permit many different combinations of input. Potential disadvantages of multiple inputs are not hard to imagine, however. The most formidable problem for the nervous system would seem to be that of producing an appropriate net effect on each motoneuron from the many possible combinations of input that impinge on it.

Obviously, each input system cannot function in isolation from all others. Therefore it is necessary for the nervous system to regulate all inputs so that the mixture reaching the motoneuron is appropriate. There is a considerable body of evidence that signals destined for motoneurons are subject to modulation of different kinds at several levels of the nervous system and that the effects they produce are contingent rather than fixed. For example, the amplitude of individual EPSPs elicited in motoneurons by single impulses in Ia fibers varies in different trials with the background activity in the spinal cord[33] and may be greatly increased by transection of the cord a few hours before recording.[53] Effects such as these and the changes produced by posttetanic potentiation or presynaptic inhibition indicate that central control of input plays a major, possibly a dominant role, in routing signals to motoneurons and in regulating their synaptic effects.

Since knowledge of the mechanisms involved in controlling input and of their functional roles in integrating different inputs is limited, a necessarily brief account will be given here of this extremely important aspect of motor control. Afferent signals to motoneurons are under central control at at least three different sites:

1. *At their source*. One of the best examples of this type of central regulation occurs in muscle spindles. The sensitivity of spindle receptors to stretch is under the control of A-gamma motoneurons that supply the intrafusal muscle fibers in the spindles. Impulses in these small motoneurons cause contraction of intrafusal fibers, which may greatly enhance the response to muscle stretch or even initiate discharges from receptors when no firing is occurring. Many control centers in the nervous system receive sensory input that regulates their activity. Cells in the motor cortex, for example, receive signals from joint receptors activated by the movements they produce and from receptors in the muscles contracting during the movement.[1]

2. *At the terminals of afferent fibers*. After Ia fibers penetrate the gray matter, they give rise to terminal arborizations that subdivide into progressively smaller branches and finally end in synaptic boutons. It is not known whether impulses normally propagate into all Ia terminals and boutons or whether invasion of terminals varies with their prior activity and the degree of activity in systems that regulate the excitability of the terminals by means of presynaptic endings. Transmission in terminals or the effectiveness of transmitter release can be influenced in several ways.

A large increase in the effectiveness of transmission at Ia synapses on motoneurons occurs following tetanization of the afferent fibers at rates of more than 100/sec. Monosynaptic reflexes are greatly potentiated for periods of seconds or minutes, depending on the duration and rate of tetanization.[42] The mechanism underlying this effect is not well understood. Posttetanic potentiation has been demonstrated at many junctions in the nervous system, but its role in normal functioning is not clear.

Presynaptic inhibition (Chapter 6) is a mechanism that operates to reduce the effectiveness of synaptic transmission in many parts of the nervous system. Electron microscopic studies have revealed fine "presynaptic" endings on the terminals of Ia fibers. Activation of these endings tends to depolarize the terminal membrane (primary afferent depolarization [PAD]). This may result in reduced invasion of terminals by Ia impulses or simply in a reduction of transmitter release. Presynaptic inhibition decreases the size of EPSPs in motoneurons and thereby decreases reflex response. The organization of presynaptic systems is poorly understood, but they have an obvious role in regulating the effectiveness of inputs to motoneurons.

The most dramatic evidence that input from Ia fibers is subject to powerful controlling influences comes from recent experiments,[53] in which the spinal cord of anesthetized cats was transected at the fifth lumbar or thirteenth thoracic segment. Individual EPSPs in MG motoneurons became significantly larger following these transections than in preparations with intact spinal cords.[50] Increases in EPSPs were observed both on a short-term (within several hours after transection) and on a long-term (over several weeks) basis, but the chronic effect alone seems to require that the transection be within two segments of the region studied. The time course of these effects suggests two separate processes, since the immediate effect diminished within 5 days after transection and the long-term effect did not begin until about 2 weeks later. These increases in mean amplitude were marked by a virtual disappearance of small EPSPs (<100 μV) and also by the appearance of very large (400 to 1,200 μV), brief EPSPs not previously recorded. The mechanism underlying this enhancement could be pre- (i.e., removal of a tonic presynaptic inhibition) or postsynaptic, although the data suggest no generalized changes in motoneuron resistance.

In addition, single Ia impulses evoked EPSPs in a larger percentage (98% to 100%) of homonymous motoneurons than normally even minutes after transection, too soon for growth of new terminals. Since EPSPs smaller than 15 μV could not be detected ordinarily, this increase may have been due, in some cases, to larger mean EPSP amplitudes, with fewer small responses being overlooked. Occasionally, however, Ia fibers were found to project to a higher percentage of motoneurons after transection, even though the EPSPs were of normal amplitude. *It now seems almost certain that Ia fibers in limb muscles normally send terminals to virtually all homonymous motoneurons with access to some being blocked in the intact spinal cord.* The variability of individual EPSP amplitudes and Ia fiber connectivity under different experimental conditions should serve as a cautionary note in the interpretation of experiments that purport to measure synaptic "strength" or "efficacy." These assessments obviously depend on a number of factors in addition to the input system itself.

3. *At their terminations on interneurons.* The great majority of the neurons in the spinal cord are interneurons (Chapter 28). They receive signals from primary sensory neurons and from most of the fiber tracts descending from various supraspinal control centers. The majority of the boutons on motoneurons come from the axons of interneurons. *These facts indicate that the most important control and coordination of input signals occurs in interneurons and the circuits they form.*

The presence of interneurons in systems projecting to motoneurons makes possible a variety of interactions between them and permits various forms of processing of incoming signals before they reach the motoneurons. Only the barest understanding of interneuronal circuits and the roles they play exists, but some of their simpler functions are sufficiently well established to be described briefly under several general headings.

a. *Convergence.* There are probably no "private" interneuronal pathways in the spinal cord. Most interneurons are points of convergence in "party lines" that receive input from several sources. Lloyd,[38] for example, demonstrated convergence of impulses from the pyramidal tract and from primary sensory neurons at interneuronal levels by showing that pyramidal signals facilitated disynaptic reflexes (three neuron arcs) in the absence of any influence on monosynaptic reflexes. Since this early example there have been many other direct demonstrations of convergence, and Lundberg[43] has suggested that most interneurons that receive signals from primary afferent fibers to motoneurons also receive signals from other systems. Not only do different species of primary afferent fibers converge on single interneurons,[45] but Lundberg and his colleagues have demonstrated supraspinal control of polysynaptic reflex pathways to motoneurons by the corticospinal, rubrospinal, reticulospinal, and vestibulospinal systems.[44] Convergence suggests that input from particular sources can be modulated from complete cutoff to potent amplification. There is a strong implication that combinations of inputs have special functional significance, that is, that certain inputs are only important if they occur in conjunction with others.

b. *Reversal of synaptic effect.* Primary afferent fibers are apparently all excitatory in their action in mammals. To cause inhibition, they must discharge inhibitory interneurons in the spinal cord. For example, impulses in Ia fibers not only excite homonymous motoneurons, but also inhibit those of antagonists. This inhibition is produced via collaterals of the Ia fibers going to so-called Ia interneurons, which send inhibitory axons to the antagonistic motoneurons. These particular interneurons also receive inputs from descending tracts and from other primary afferent systems. Thus reciprocal Ia inhibition of antagonists is under strict control, and, depending on circumstances, may or may not be operative.[28]

c. *Patterning of input in space and time.* Interneurons may serve to distribute afferent signals in patterns that are under central control. They may amplify the intensity of incoming activity or prolong its duration. For example, an afferent volley from a small peripheral area lasting only a few milliseconds may be transformed by interneuronal circuits into widespread effects lasting 1,000 to 2,000 msec. This occurs typically in flexor reflexes when a brief noxious stimulus produces a widespread and prolonged withdrawal of an entire limb by activating all its flexor muscles.

These are only a few examples of functions performed by interneurons. It seems likely that the majority of supraspinal control is exerted through various types of interneuronal circuits and that better understanding of how inputs are combined and correlated will require a far more complete analysis of these circuits than is now available.

Although the functional roles of the mechanisms described in this section are poorly understood, the number and variety of such mechanisms point to the absolute necessity for central control over input systems.

Factors influencing susceptibility of motoneurons to discharge

The discharges of motoneurons are controlled by the combination of excitatory and inhibitory signals they receive and by their responsiveness to these inputs. Inherent properties of motoneurons, such as the electrical resistance between the inside and the outside (the "input resistance"), evidently play a major role in determining the size of the potentials elicited by afferent impulses. It is also apparent that the number of active endings, as well as their sizes, locations, and densities, must contribute collectively to the total response of each cell. The relative importance of pre- and postsynaptic factors, however, is difficult to evaluate in quantitative terms.

In 1957 Henneman[25] reported that the sizes of individual motoneurons discharging in flexor reflexes were directly related to the intensity of the internuncial input they were receiving. The smaller a cell was, the more readily it was discharged synaptically and the longer it continued firing to a waning input. These original observations were extended (Chapter 26) to show that susceptibility to discharge by excitatory inputs or to silencing by inhibition was strongly correlated with motoneuron size. This correlation applies to many of the pre- and postsynaptic elements that contribute to synaptic efficacy. In the following paragraphs, we will describe first some of the inherent properties of motoneurons that influence their excitability and inhibitability and then discuss the input itself. Further details and additional references are available in the recent review of spinal neurons and synapses by Burke and Rudomin.[5]

In 1957 Katz and Thesleff[30] showed that miniature end-plate potentials (MEPPs; see Chapter 5) recorded intracellularly near the myoneural junctions of frog muscle fibers varied in amplitude by more than 10:1. The number of quanta of acetylcholine (ACh) in the "packets" released spontaneously was calculated. They were approximately constant and, thus, not responsible for the variations in MEPPs. The size of the MEPPs, however, was significantly correlated with the input resistance of the muscle fibers. The thinner the muscle fiber, the smaller its surface area, the greater its input resistance, and the larger its MEPP. This finding with geometrically simple, unisynaptic muscle fibers, where many of the variables that might play a role in the case of motoneurons are absent, suggests that input resistance may be a predominant factor in synaptic efficacy elsewhere. Kuno[33] found an apparently similar relationship in motoneurons. The size of EPSPs presumably evoked by release of one quantum of transmitter (i.e., a "unit EPSP" occurring at a single synaptic terminal) was bigger in motoneurons with slow axonal conduction velocities (small cells with high input resistances) than in large ones. Previously, Eccles et al.[14] had found that a volley of impulses in all the Ia fibers in a muscle nerve produced larger aggregate EPSPs in small cells than in large ones. A similar finding was made in the systematic studies of Burke,[4] who demonstrated that small motoneurons generate larger aggregate EPSPs than do large ones. The principles applying to unit EPSPs may therefore hold for large numbers of simultaneous inputs.

In 1966 Kernell[31] published an illuminating study of factors related to input resistance. Fig. 27-8 illustrates his findings. Input resistances were determined by several methods in six different pools of motoneurons. They ranged from 0.5 to 8.1 megohms, and all yielded the same relationship to conduction velocity. When plotted against the reciprocal square of axonal conduction velocity, they formed an approximately

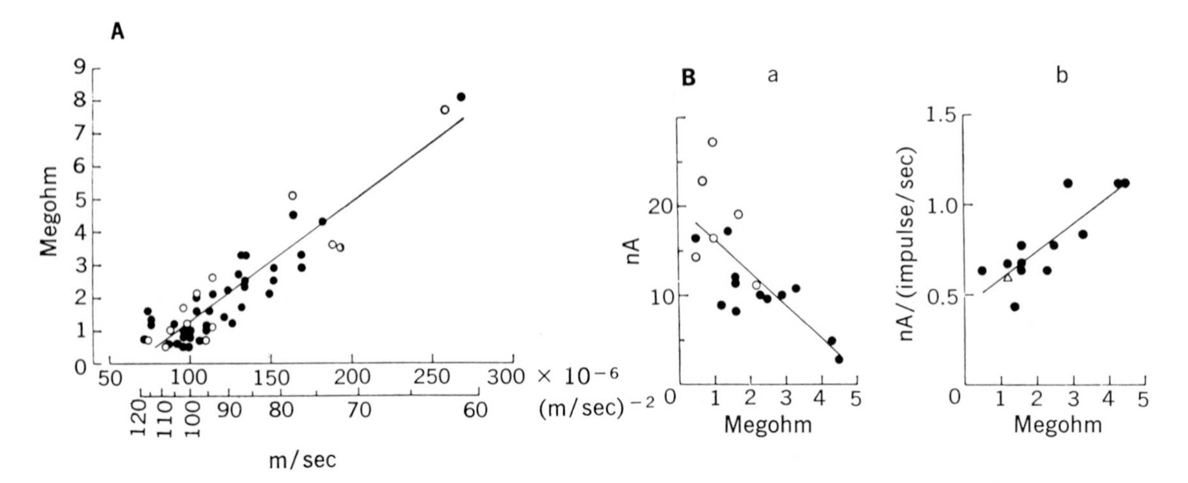

Fig. 27-8. A, Relation between input resistance and conduction velocity in 50 motoneurons. Abscissa: lower scale, conduction velocity; upper scale, square of reciprocal of conduction velocity. Ordinate: input resistance. Input resistance was measured by "direct" methods (open circles) and by "spike" method (filled circles), respectively. Regression line was calculated by method of least squares. In **B,** *a,* threshold for steady firing (1 nA = 10^{-9} amp) plotted against input resistance for 18 motoneurons. Correlation coefficient was −0.72 ($p < 0.001$). Spike size averaged 89 ± 9 mV and membrane potential 71 ± 4 mV. *b,* Increment of current strength needed per unit increase of firing rate plotted against input resistance for 13 motoneurons. Correlation coefficient was +0.84 ($p < 0.001$). Spike size averaged 89 ± 10 mV and membrane potential 73 ± 3 mV. Cells represented in both *a* and *b* are indicated by filled circles. Regression lines calculated by method of least squares. (From Kernell[31]; copyright © 1966 by the American Association for the Advancement of Science.)

linear curve, with a correlation coefficient of +0.92. Eighteen of the cells in Kernell's study were made to fire repetitively by injecting long-lasting currents intracellularly. The current strength required for repetitive discharge is plotted against input resistance in Fig. 27-8, *A*. The inverse relationship between them indicates that input resistance is important in determining the current required (synaptic as well as injected) for a steady repetitive discharge. In *B* the increment in current needed to increase the firing rate by one impulse per second is plotted. Apparently, more current is required for a unit increase in the firing rate of cells with high resistances than for those with low resistances. Thus the firing rate of the cells with high resistances was relatively unaffected by variations in current strength. Factors causing long after-hyperpolarizations in motoneurons may counteract increases in their firing rates. The amplitude and duration of after-hyperpolarization in a motoneuron are examples of properties highly correlated with cell size, as Huizar et al. have shown.[27]

In accompanying anatomic studies, Kernell[31] also showed that there is an approximately direct proportionality between the total cross-sectional area of all the visible dendritic trunks and the surface area of the soma. The extensive surfaces of large cells provide more paths for ionic current flow, which accounts for their lower resistance. Several studies[2,46,63] with new techniques agree that the size of the cell body of a motoneuron determines the extent of its dendritic tree and its total surface area. Kernell's data indicate that the ratio of dendritic to somatic conductance is the same order of magnitude for small and large cells. The input resistance of a motoneuron is thus proportional to the reciprocal of its somatic surface area, which, in turn, is related to the size of its axon.

Although input resistance may be one of the most important factors in producing orderly recruitment, a number of other properties of motoneurons, about which we know little, may contribute significantly to synaptic efficacy. Differences in the types of motoneurons in a pool have been identified[23,24] from their firing rates, although little is known about their other characteristics. Properties such as specific membrane resistance per unit area, density of sodium gates, and cell geometry may also differ and may be correlated with either cell size or type. In a recent study[7] the homonymous EPSPs elicited by stimulation of the MG nerve and the IPSPs produced by excitation of its antagonists, tibialis anterior and flexor digitorum longus, were recorded in MG motoneurons, and the same cells were stimulated intracellularly to identify the muscle unit type by its contractile characteristics. There was a significant correlation between the size of the monosynaptic EPSPs as well as the disynaptic IPSPs and the type of muscle unit. It may be that the input resistance of motoneurons varies with their type as well as their size or that other type-specific properties of motoneurons account for differences in the sizes of their EPSPs and IPSPs. In any case the correlation between the sizes of EPSPs and IPSPs in the same cell suggests that the cause lies within the motoneurons rather than in their inputs.

There is not much detailed information regarding other monosynaptic inputs to motoneurons. Clough et al.[9] reported that cells supplying muscles in the baboon's hand developed larger EPSPs in response to direct corticomotor input than the aggregate EPSPs elicited by a volley of impulses in all the Ia fibers of their own muscle. Moreover, those cells with large Ia EPSPs tended to produce larger corticomotor EPSPs than did cells with small Ia EPSPs. These observations suggest that, although some inputs exert greater effects than others, the properties of a motoneuron determine its responsiveness relative to that of all other cells in the pool.

In conclusion, it should be stressed that although the properties of motoneurons may be of great importance, the input they receive must be organized to promote rather than hinder orderly recruitment. It is in this context that the widespread projection from Ia fibers to motoneurons must be evaluated. It would permit but not ensure an "equalized" input to all members of a pool. However, many essential details of this projection are not understood. For example, it is not yet known whether all the motoneurons in a pool receive an equal density of Ia inputs or whether the density is greater on small cells. Experiments with single Ia fibers reveal no evidence of preferential convergence onto small motoneurons, but we still have little or no information about the number of boutons given off by single Ia fibers to small and large cells. For this reason, we cannot evaluate the finding that individual EPSPs (produced by the action of a single afferent fiber) and unit EPSPs (produced by the action of a single quantum of neurotransmitter) are larger in small than in large motoneurons. Whatever the explanation for these findings, it seems likely that they contribute to the result that aggregate EPSPs are larger in small than in large cells and, more importantly, to the greater excitability of small motoneurons.

REFERENCES

1. Asanuma, H., Stoney, S. D., Jr., and Abzug, C.: Relationship between afferent input and motor outflow in cat motosensory cortex, J. Neurophysiol. **31**:670-681, 1968.
2. Barrett, J. N., and Crill, W. E.: Specific membrane resistance of dye-injected cat motoneurons, Brain Res. **28**:556-561, 1971.
3. Burke, R. E.: Composite nature of the monosynaptic excitatory postsynaptic potential, J. Neurophysiol. **30**: 1114-1137, 1967.
4. Burke, R. E.: Group Ia synaptic input to fast and slow twitch motor units of cat triceps surae, J. Physiol. **196**: 605-630, 1968.
5. Burke, R. E., and Rudomin, P.: Spinal neurons and synapses. In Handbook of physiology: the nervous system I, Washington, D.C., 1977, American Physiological Society, pp. 877-944.
6. Burke, R. E., Jankowska, E., and Ten Bruggencate, G.: A comparison of peripheral and rubrospinal synaptic input to slow and fast twitch motor units of triceps surae, J. Physiol. **207**:709-732, 1970.
7. Burke, R. E., Rymer, W. Z., and Walsh, J., Jr.: Relative strength of synaptic input from short-latency pathways to motor units of defined type in cat medial gastrocnemius, J. Neurophysiol. **39**:447-458, 1976.
8. Clamann, H. P., Gillies, J. D., and Henneman, E.: Effects of inhibitory inputs on critical firing level and rank order of motoneurons, J. Neurophysiol. **37**:1350, 1974.
9. Clough, J. F. M., Kernell, D., and Phillips, C. G.: The distribution of monosynaptic excitation from the pyramidal tract and from primary spindle afferents to motoneurons of the baboon's hand and forearm, J. Physiol. **198**:145-166, 1968.
10. Desmedt, J. E., and Godaux, E.: Ballistic contractions in man: characteristic recruitment pattern of single motor units of the tibialis anterior muscle, J. Physiol. **264**:1-21, 1977.
11. Desmedt, J. E., and Godaux, E.: Fast motor units are not preferentially activated in rapid voluntary contractions in man, Nature **267**:717-719, 1977.
12. Eccles, J. C.: The physiology of synapses, New York, 1964, Academic Press, Inc.
13. Eccles, J. C., and Sherrington, C. S.: Numbers and contraction-values of individual motor units examined in some muscles of the limb, Proc. R. Soc. Lond. (Biol.) **106**:326-357, 1930.
14. Eccles, J. C., Eccles, R. M., and Lundberg, A.: The convergence of monosynaptic excitatory afferents on to many afferent species of alpha motoneurons, J. Physiol. **137**:22-50, 1957.
15. Eccles, J. C., Eccles, R. M., and Lundberg, A.: Synaptic actions on motoneurones caused by impulses in Golgi tendon organ afferents, J. Physiol. **138**:227-252, 1957.
16. Eccles, J. C., Eccles, R. M., and Shealy, C. N.: An investigation into the effect of degenerating primary afferent fibers on the monosynaptic innervation of motoneurons, J. Neurophysiol. **25**:544-558, 1962.
17. Edwards, F. R., Redman, S. J., and Walmsley, B.: Statistical fluctuations in charge transfer at Ia synapses on spinal motoneurones, J. Physiol. **259**:665-688, 1976.

18. Edwards, F. R., Redman, S. J., and Walmsley, B.: Non-quantal fluctuations and transmission failures in charge transfer at Ia-synapses on spinal motoneurones, J. Physiol. **259:**689-704, 1976.
19. Edwards, F. R., Redman, S. J., and Walmsley, B.: The effect of polarizing currents on unitary Ia excitatory post-synaptic potentials evoked in spinal motoneurones, J. Physiol. **259:**705-723, 1976.
20. Fatt, P., and Katz, B.: An analysis of the end-plate potential recorded with an intra-cellular electrode, J. Physiol. **115:**320, 1951.
21. Granit, R., and Burke, R. E.: The control of movement and posture, Brain Res. **53:**1-28, 1973.
22. Grimby, L., and Hannerz, J.: Firing rate and recruitment order of toe extensor motor units in different modes of voluntary contraction, J. Physiol. **264:**865-879, 1977.
23. Harris, D. A., and Henneman, E.: Different species of alpha motoneurons in the same pool: further evidence from the effects of inhibition on their firing rates, J. Neurophysiol. **42:**000, 1979. (In press.)
24. Harris, D. A., and Henneman, E.: Identification of two species of alpha motoneurons in cat's plantaris pool, J. Neurophysiol. **40:**16-25, 1977.
25. Henneman, E.: Relation between size of neurons and their susceptibility to discharge, Science **126:**1345-1347, 1957.
26. Henneman, E., Shahani, B. T., and Young, R. R.: Voluntary control of human motor units. In Shahani, M., editor: The motor system: neurophysiology and muscle mechanisms, Amsterdam, 1976, Elsevier North Holland Biomedical Press.
27. Huizar, P., Kuno, M., and Miyata, Y.: Differentiation of motoneurons and skeletal muscles in kittens, J. Physiol. **252:**465-479, 1975.
28. Hultborn, H.: Transmission in the pathway of reciprocal I_A inhibition to motoneurones and its control during the tonic stretch reflex. In Homma, S., editor: Progress in brain research: understanding the stretch reflex, Amsterdam, 1976, Elsevier/North Holland Biomedical Press, vol. 44, pp. 235-255.
29. Jankowska, E., and Roberts, W. J.: An electrophysiological demonstration of the axonal projections of single spinal interneurones in the cat, J. Physiol. **222:**597-622, 1972.
30. Katz, B., and Thesleff, S.: On the factors which determine the amplitude of the "miniature end-plate potential," J. Physiol. **137:**267-278, 1957.
31. Kernell, D.: Input resistance, electrical excitability, and size of ventral horn cells in cat spinal cord, Science **152:** 1637-1640, 1966.
32. Kirkwood, P. A., and Sears, T. A.: Monosynaptic excitation of motoneurones from muscle spindle secondary endings of intercostal and triceps surae muscles in the cat, J. Physiol. **245:**64-66P, 1974.
33. Kuno, M.: Quantal components of the excitatory postsynaptic potential in spinal motoneurones, J. Physiol. **175:**81-99, 1964.
34. Kuno, M., and Miyahara, J. T.: Non-linear summation of unit synaptic potentials in spinal motoneurones of the cat, J. Physiol. **201:**465-477, 1969.
35. Kuno, M., and Miyahara, J. T.: Analysis of synaptic efficacy in spinal motoneurones from 'quantum' aspects, J. Physiol. **201:**479-493, 1969.
36. Laporte, Y., and Lloyd, D. P. C.: Nature and significance of the reflex connections established by large afferent fibers of muscle origin, Am. J. Physiol. **169:**609-621, 1952.
37. Liddell, E. G. T., and Sherrington, C.: Reflexes in response to stretch (myotatic reflexes), Proc. R. Soc. Lond. (Biol.) **96:**212-242, 1924.
38. Lloyd, D. P. C.: The spinal mechanism of the pyramidal system in cats, J. Neurophysiol. **4:**525-546, 1941.
39. Lloyd, D. P. C.: Reflex action in relation to pattern and peripheral source of afferent stimulation, J. Neurophysiol. **6:**111-120, 1943.
40. Lloyd, D. P. C.: Facilitation and inhibition of spinal motoneurones, J. Neurophysiol. **9:**421-438, 1946.
41. Lloyd, D. P. C.: Integrative pattern of excitation and inhibition in two-neuron reflex arcs, J. Neurophysiol. **9:**439-444, 1946.
42. Lloyd, D. P. C.: Post-tetanic potentiation of response in monosynaptic reflex pathways of the spinal cord, J. Gen. Physiol. **33:**147-170, 1949.
43. Lundberg, A.: Supraspinal control of transmission in reflex paths to motoneurones and primary afferents. In Eccles, J. C., and Schadé, J. P., editors: Progress in brain research: physiology of spinal neurons, Amsterdam, 1964, Elsevier/North Holland Biomedical Press, vol. 12, pp. 197-221.
44. Lundberg, A.: Control of spinal mechanisms from the brain. In Tower, D. B., editor: The nervous system: the basic neurosciences, New York, 1975, Raven Press, vol. 1.
45. Lundberg, A., Malmgren, K., and Schomberg, E. D.: Convergence from Ib, cutaneous and joint afferents in reflex pathways to motoneurones, Brain Res. **87:**81-84, 1975.
46. Lux, H. D., Schubert, P., and Kreutzberg, G. W.: Direct matching of morphological and electrophysiological data in cat spinal motoneurones. In Anderson, P., and Jansen, J. K. S., editors: Excitatory synaptic mechanisms, Oslo, 1970, Universitetsforlaget, pp. 189-198.
47. Matthews, P. B. C.: Evidence that the secondary as well as the primary endings of the muscle spindles may be responsible for the tonic stretch reflex of the decerebrate cat, J. Physiol. **204:**365-393, 1969.
48. Matthews, P. B. C.: Mammalian muscle receptors and their central actions, Baltimore, 1972, The Williams & Wilkins Co.
49. McPhedran, A. M., Wuerker, R. B., and Henneman, E.: Properties of motor units in homogeneous red muscle (soleus) of the cat, J. Neurophysiol. **28:**71, 1965.
50. Mendell, L. M., and Henneman, E.: Terminals of single Ia fibers: location density and distribution within a pool of 300 homonymous motoneurons, J. Neurophysiol. **34:** 171-187, 1971.
51. Mendell, L. M., and Weiner, R.: Analysis of pairs of individual Ia-EPSPs in single motoneurones, J. Physiol. (Lond.) **255:**81-104, 1976.
52. Nelson, S. G., and Mendell, L. M.: Projection of single semitendinosus Ia afferent fibres to homonymous and heteronymous motoneurons, J. Neurophysiol. **41:**778, 1978.
53. Nelson, S. G., Collatos, T. C., Niechaj, A., and Mendell, L. M.: Immediate increase in Ia-motoneuron synaptic transmission caudal to spinal cord transection, J. Neurophysiol. **42:**655, 1979.
54. Rall, W.: Distinguishing theoretical synaptic potentials computed for different soma-dendritic distributions of synaptic input, J. Neurophysiol. **30:**1138-1168, 1967.
55. Rall, W., et al.: Dendritic location of synapses and possible mechanisms for the monosynaptic EPSP in motoneurons, J. Neurophysiol. **30:**1169-1193, 1967.
56. Scheibel, M. E., and Scheibel, A. B.: Terminal patterns in cat spinal cord. III. Primary afferent collaterals, Brain Res. **13:**417-443, 1969.

57. Scott, J. G., and Mendell, L. M.: Individual EPSPs produced by single triceps surae Ia afferent fibres in homonymous and heteronymous motoneurons, J. Neurophysiol. **39:**679-692, 1976.
58. Stacey, M. J.: Free nerve endings in skeletal muscle of the cat, J. Anat. **105:**231-254, 1969.
59. Stauffer, E. K., et al.: Analysis of muscle receptor connections by spike triggered averaging. 2. Spindle group II afferents, J. Neurophysiol. **39:**1393-1402, 1976.
60. Sterling, P., and Kuypers, H. G. J. M.: Anatomical organization of the brachial spinal cord of the cat. II. The motoneuron plexus, Brain Res. **4:**16-32, 1967.
61. Traub, R. D.: Motoneurons of different geometry and the Size Principle, Biol. Cybern. **25:**163-176, 1977.
62. Watt, D. G. D., et al.: Analysis of muscle receptor connections of spike-triggered averaging. 1. Spindle primary and tendon organ afferents, J. Neurophysiol. **39:**1375-1392, 1976.
63. Webber, C. L., and Pleschka, K.: Structural and functional characteristics of individual phrenic motoneurons, Pfluegers Arch. **364:**113-121, 1976.
64. Zucker, R. S.: Theoretical implications of the size principle of motoneurone recruitment, J. Theor. Biol. **38:** 587-596, 1973.

28

ELWOOD HENNEMAN

Organization of the spinal cord and its reflexes

ORGANIZATION OF THE SPINAL CORD

The purpose of this section is to provide a brief introduction to the spinal cord that stresses functional organization without undue involvement in specific circuits and reflexes. Attention is therefore directed to a typical segment of the spinal cord and the signals entering and leaving it. In the first portion of this section the distribution of incoming sensory impulses is reviewed; in the second part the processing of these signals by internuncial cells is discussed; and in the third part the relation of descending motor systems transmitting control signals from higher centers to the spinal mechanism is considered. The section that follows supplies detailed information on specific spinal mechanisms that are of special importance in the control of movement.

The lumbosacral part of the spinal cord and the hind limb structures that it innervates have been used extensively for physiologic experiments. In the seventh lumbar segment of the dog there are, by actual count, about 375,000 cell bodies.[17,18] Their disposition is indicated in Fig. 28-1, which shows the numbers of large and small cells in the dorsal horn, intermediate region, and ventral horn. Small cells greatly outnumber large cells in all parts of the spinal cord, as they do generally throughout the CNS. In the ventral root itself large fibers predominate. These are the axons of motoneurons that innervate skeletal muscles. There are about 12,000 sensory fibers in each dorsal root and about 6,000 motor fibers in each ventral root of this particular segment. The existence of 375,000 cells between the input and output suggests that the signals coming into this segment via dorsal roots and descending tracts are being subjected to extensive processing before they reach the motoneurons.

SENSORY INPUT

All the sensory input to the spinal cord arrives via the dorsal roots, except for a group of C fibers of uncertain function that enter via the ventral roots. The entering fibers range in size from about 22 μm to about 0.2 μm. Unmyelinated C fibers of 1 μm or less make up more than half the total. Fibers of large diameter from stretch receptors in muscle constitute a small percentage of the input. For reference purposes a classification of afferent fibers according to fiber diameter is supplied in Table 28-1.

The central end of a dorsal root breaks up into a series of rootlets, each of which separates into a medial and a lateral division. The fibers of the medial division, which are relatively large, pass medially over the dorsal horn into the posterior columns. The fine fibers of the lateral division pass directly into the tract of Lissauer at the apex of the dorsal horn. Shortly after entering the cord each fiber divides into an ascending and a descending branch. Some fibers ascend all the way to the gracile and cuneate nuclei. Others run a short distance up or down and end on cells in the posterior columns. Although they take various courses, all entering fibers apparently project to higher centers. In addition, with no known exceptions, all afferent fibers establish connections with the central gray matter of the spinal cord. Thus it appears to be a basic principle that all sensory fibers serve a double function. None project to higher centers without giving off collaterals to spinal centers and none serve a purely spinal function.

The consequences of this arrangement may be observed in a chronic "spinal" cat. In such a preparation, with no interfering input from higher

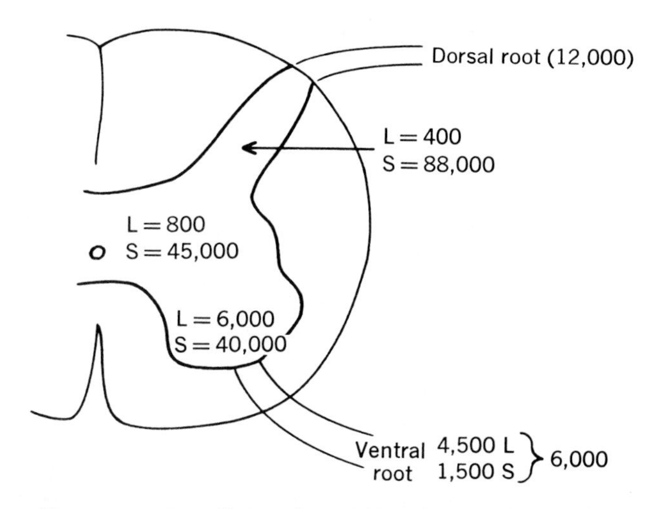

Fig. 28-1. Numbers of large, *L,* and small, *S,* cells and fibers in seventh lumbar segment of dog's spinal cord. Neurons with diameters of less than 34 μm (corresponding to surface area of 920 μm^2) were classified as small. Classification of fiber size based on position of trough between two peaks in histogram (approximately 8 μm). (Based on data from Gelfan and Tarlov.[18])

centers, it is possible to show that every type of sensory stimulus (touch, pressure, pain, heat, cold, joint movement, muscle stretch) will elicit a reflex response of some kind. This is an important observation because it indicates that all sensory signals have a spinal route to motoneurons.

The obvious inference is that immediate and delayed use is made of incoming sensory signals. At the spinal level, sensory signals serve urgent or pressing needs by eliciting rather stereotyped motor responses with very brief delays. The same signals are also transmitted to higher centers for more elaborate processing of their information content. After variable delays the effects of these signals, blended with other afferent impulses, may be felt again, greatly modified, in the output of integrating centers such as the cerebellum.

As illustrated in Fig. 28-2, the local or segmental terminations of dorsal root fibers are of two types. (1) The majority of them end on internuncial neurons situated in different parts of the central gray matter. Fibers that innervate skin terminate chiefly on cells located in the dorsal horn of the spinal cord (Fig. 28-2, *c*). Fibers from muscle are distributed to various other internuncial nuclei, a prominent example being the intermediate nucleus (Fig. 28-2, *g*). Incoming signals in these fibers affect motoneurons through the agency of one or several intervening cells that may transform the original signals. (2) A small percentage of dorsal root fibers, those coming from annulospiral endings in muscle spindles, send collaterals directly to motoneurons in the ventral horn (Fig. 28-2, *e*). These collaterals complete a two-neuron reflex arc from muscle receptor to motoneuron to muscle fibers, a so-called monosynaptic reflex pathway.

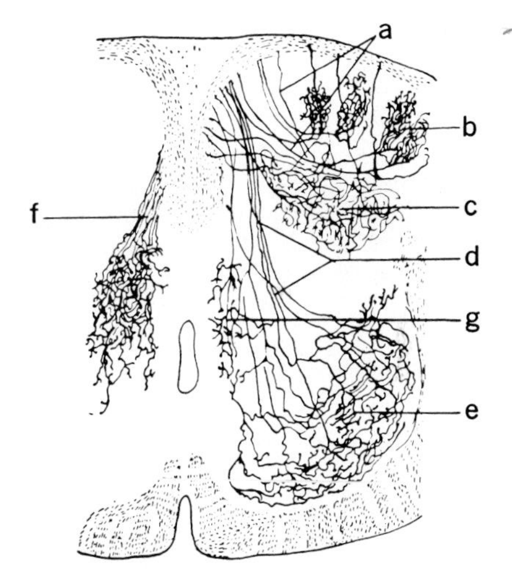

Fig. 28-2. Sketch of cross section of spinal cord indicating various terminations of reflex collaterals of primary dorsal root afferents. Some, *a,* reach neurons of substantia gelatinosa, *b,* whereas others end on cells of nucleus proprius of dorsal horn, *c*. Direct reflexomotor collaterals, *d,* reach motoneurons at *e* without an intervening synapse. Other collaterals, shown for convenience on second side at *f,* end on cells of intermediate nucleus of Ramón y Cajal, *g*. (From Ramón y Cajal.[45])

Table 28-1. Classification of dorsal root fibers

Classification of nerve fibers	Range of fiber size and velocity	Peripheral origin	Receptor organs	Effective stimulus	Type of synaptic relay	Central destination	Reflex action	Other properties
Group Ia (A-alpha)	12 to 20 μm 70 to 120 m/sec	Muscle	Annulospiral spindle endings (A-2 of Mathews)	Stretch—low threshold	Two-neuron arc "monosynaptic"	1. Motoneurons of muscle of origin	Direct excitation	Myotatic reflex; receptor discharge may cease during contraction unless small nerve motor fibers to spindles are active
						2. Motoneurons of synergists in myotatic unit	Direct excitation	
						3. Motoneurons of antagonists in myotatic unit	Disynaptic* inhibition	Reciprocal component of myotatic reflex action
Group Ib (A-alpha)	12 to 20 μm 70 to 120 m/sec	Muscle	Tendon organs of Golgi (B of Matthews)	Active contraction of muscle	Three-neuron arc "disynaptic"	1. Motoneurons of muscle of origin and its synergists in myotatic unit†	Inhibition	Receptor discharge increases during contraction and is not affected by small-nerve motor activity
						2. Motoneurons of antagonists in myotatic unit	Excitation	

Group II (A-beta and gamma)	5 to 12 μm 30 to 70 m/sec	1. Extensor muscles	Flower spray of spindle (A-1 of Matthews)	Stretch—low threshold	Mono- and polysynaptic	Excitation of homonymous and heteronymous motoneurons	Contributes to myotatic reflex
		2. Flexor muscles	Flower spray of spindle (A-1 of Matthews)	Stretch—low threshold	Mono- and polysynaptic	Excitation of flexors and inhibition of extensors throughout limb	Flexion withdrawal reflex; contralateral component is crossed extensor reflex
		3. Skin‡	Touch-pressure receptors	Mechanical deformation of skin	Multineuron arc		
Group III (A-delta)	2 to 5 μm 12 to 30 m/sec	1. Muscle	Unknown, pain receptors (?)	Destructive (?)	Multineuron arc	From either muscle or skin these afferents produce excitation of flexors and inhibition of extensors throughout limb	Flexion withdrawal reflex; contralateral component is crossed extensor reflex
		2. Skin‡	Pain—fast (?), cold, heat	Destructive (?) temperature change	Multineuron arc		
Group IV (C fibers)	0.5 to 1 μm 0.5 to 2 m/sec	From muscle and skin	Pain—slow	Destructive	Multineuron arc		

*There is evidence suggesting that an interneuron is interposed in the direct inhibitory pathway (Eccles et al.[12])
†These afferents make some connections with motoneurons of muscles at distant joints as well, but these are yet poorly understood.
‡For an earlier classification of skin nerves see p. 000. At present the terms "a" and "group I" are used synonymously to designate large afferent fibers from muscle.

INTERNUNCIAL TRANSACTIONS

Situated between the incoming and the outgoing fibers of the spinal cord are the internuncial cells or interneurons, which transmit signals from the dorsal root fibers to the spinal motoneurons. They constitute the great majority of neurons in the spinal cord, outnumbering motoneurons 30:1.[18] They do not merely intervene passively to link up afferent and efferent fibers, but serve to transform incoming signals into new and different patterns. In part the transformation may be due to the characteristics of the interneuron and in part it may result from the way that groups of internuncial cells are connected with each other to form circuits with special properties.

Interneurons are a heterogeneous group of cells that are either collected into several nuclei, as shown in Fig. 28-2, or dispersed without obvious grouping. With few exceptions they are rather small cells, with a mean diameter of only 16 μm, as compared with 48 μm for motoneurons. The surface area of the average motoneuron accommodates some 5,500 synaptic knobs, whereas the interneuron has only about 640 knobs/cell.[17] Although it is possible to record from interneurons with microelectrodes and to ascertain what afferent signals excite or inhibit them, it is not easy to determine where their axons terminate or how other neurons are affected by them. Information about interneurons is therefore limited mainly to observations of their behavior in response to various inputs.

Properties of internuncial cells[22]

The properties of interneurons do not differ radically from those of other neurons. Excitation is accompanied by membrane depolarization, and inhibition is frequently associated with hyperpolarization. In recording with a microelectrode it is usually easy to distinguish an interneuron from a motor neuron by its tendency to fire "spontaneously" (i.e., without applied stimulation, and by its characteristic repetitive response to a single afferent volley (Fig. 28-3). The frequency of its discharge may be as high as 1,500/sec in short trains,[54] whereas that of spinal motoneurons seldom exceeds 100/sec. When excited by intracellular pulses of current through a microelectrode, interneurons tend to fire repetitively to a much greater extent than do motoneurons. This capacity to discharge repetitively is apparently related to a lack of prolonged subnormality following discharge and to minimal accommodation. The cause of prolonged firing is usually prolonged excitation due to the arrival of temporally dispersed presynaptic impulses. There is no evidence for a prolonged duration of excitatory transmitter action.

Functions of internuncial cells

Internuncial cells perform a wide variety of specific functions that we are only beginning to

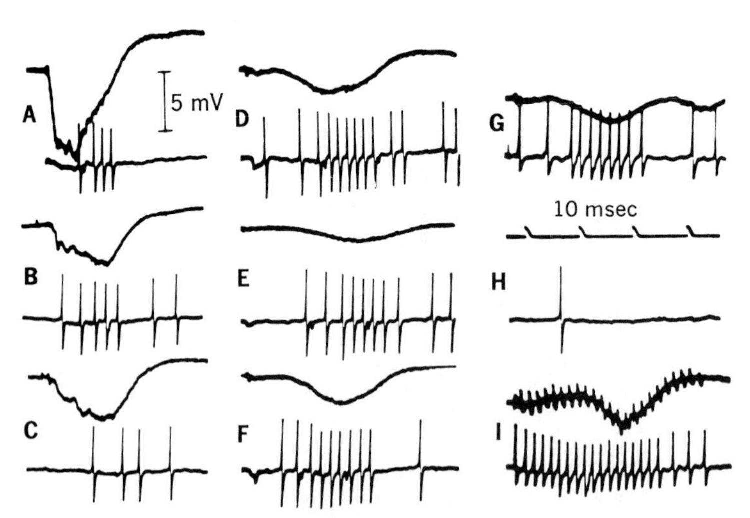

Fig. 28-3. Responses of interneuron in intermediate region of spinal cord to afferent volleys in various nerves. Upper traces are incoming volley recorded from L7 dorsal root entry zone. Lower traces are responses of interneuron recorded with extracellular microelectrode. Stimulus applied to superficial peroneal, **A;** sural, **B;** saphenous, **C;** flexor digitorum longus, **D;** plantaris, **E;** gastrocnemius, **F;** and biceps-semitendinosus, **G. H** is spontaneous response. **I** is stimulation of biceps-semitendinosus nerve at 680/sec. (From Eccles et al.[12])

appreciate. It is clear that some of these functions cannot be carried out by single cells but require a network or circuit of interconnected cells. To date, such circuits have not been demonstrated in the spinal cord, although they exist elsewhere in the nervous system.

Internuncial cells perform several types of functions:

1. They may serve as amplifiers of three different types. (a) Amplification of the intensity of an incoming signal may be achieved either by *cascading* of cells (Fig. 28-4, *M*) or through an *increase in the rate* of discharge of individual cells. In the first instance the number of active cells may be greatly increased; in the second the frequency of firing in a primary afferent fiber may be multiplied by a factor of 10 to 100. One internuncial circuit might combine both effects. (b) Amplification in time may be accomplished by means of a closed loop of interneurons (Fig. 28-4, *C*). Each of the cells in such a loop excites others, resulting in a self-perpetuating discharge. Circuits of this type may account for the fact that an incoming volley lasting only 1 msec can elicit a response of motoneurons lasting 1,000 msec. (c) Amplification in space is brought about by interneurons that distribute an incoming volley to widely separated groups of motoneurons. Ramón y Cajal recognized this type of circuit on purely anatomic grounds (Fig. 28-5, *A*).

2. Internuncial cells may serve as valves that pass or prevent transmission of afferent impulses to motoneurons. There is evidence from microelectrode studies that some interneurons are affected not only by incoming volleys from muscle but also by activity in descending systems such

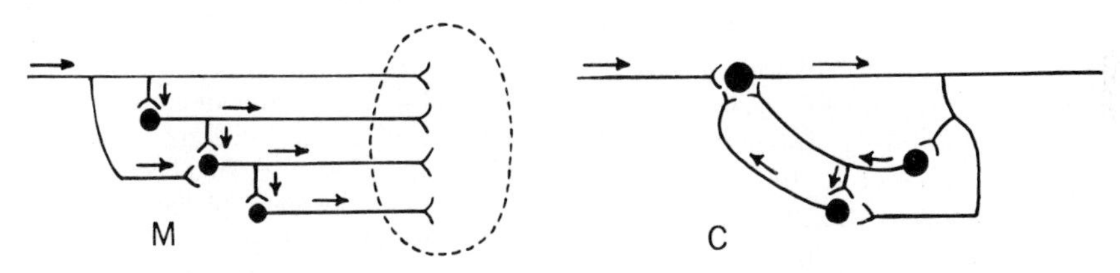

Fig. 28-4. Diagrams of two types of chains formed by internuncial neurons: *M*, multiple chain, and *C*, closed chain, which are self-reexciting. (From Lorente de Nó.[41])

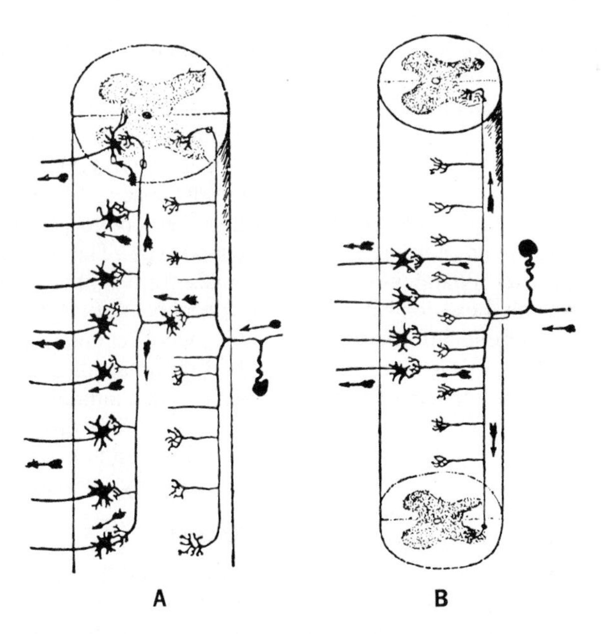

Fig. 28-5. A, Diagrammatic representation of diffuse reflex mechanism of spinal cord showing polysynaptic relay from primary dorsal root afferent, via interneuron, to motoneurons of *several* spinal cord segments. **B,** Diagrammatic representation of circumscribed reflex mechanism showing monosynaptic relay from primary dorsal root afferents to motoneurons of a *few* segments. (From Ramón y Cajal.[45])

as the corticospinal tract. Depending on the control exerted by a higher center, activity arising in tendon organs or secondary spindle receptors may or may not reach the motoneuron. In feedback terminology, this function might be referred to as "variable gain control."

3. Internuncial cells may function as signal inverters, changing an incoming excitatory signal into an inhibitory signal. For example, it is well known that impulses in group Ia fibers excite their own motoneurons and inhibit those of direct antagonists. Since release of different transmitters at different endings of the same neuron is unlikely, it has been proposed[9] that the effect on antagonistic motor neurons is mediated not directly but via interneurons that "convert" excitation to inhibition.

4. Internuncial cells probably serve as final common pathways for either excitation or inhibition. Since a number of inputs of different types may each fire a particular interneuron, these cells should be regarded as elements that are common to many afferent pathways. By taking the sum of all the excitatory and inhibitory effects impinging on it and emitting signals that are the result of all of them, an interneuron may greatly simplify the input to the motoneuron. In addition, by substituting one set of terminals for many, an internuncial cell may serve to promote neuronal economy.

The functions just described are probably the simplest and most easily recognized of a great many types of data processing by internuncial cells. In time, subtler and more complex types will certainly be uncovered.

MOTOR INPUT

In order to understand how higher centers of the brain exert their control over muscles it is necessary to know the exact terminations of the various fiber tracts that transmit these control signals to spinal levels. The required information, however, is not yet available. From both anatomic and physiologic studies it appears that the majority of descending fibers end on internuncial cells, but that some fibers pass directly to motor neurons. In the cat, for example, anatomic and electrophysiologic studies are in agreement that corticospinal fibers end chiefly or exclusively on internuncial cells. Lloyd's diagram (Fig. 28-6) summarizes the results of his investigation of this problem, showing that corticospinal fibers terminate on interneurons in the dorsal horn immediately adjacent to the fiber tract and upon interneurons in the intermediate region of the gray matter. His original studies[36] and several done subsequently[27] indicate that some of these interneurons also receive input from stretch receptors in muscle. It thus appears that in the cat the corticospinal system exerts its effects through the feedback circuits that control the length and tension of muscle. Much of the control exerted via other descending systems is probably of this general type. The precise point at which control signals from higher centers converge with local feedback control systems may be of great significance. In monkey and man some corticospinal fibers terminate directly on motoneurons.[1,21] This suggests that animals with highly developed forebrains require a direct, unconditional means of controlling motoneurons that is not subject to regulation by internuncial valves.

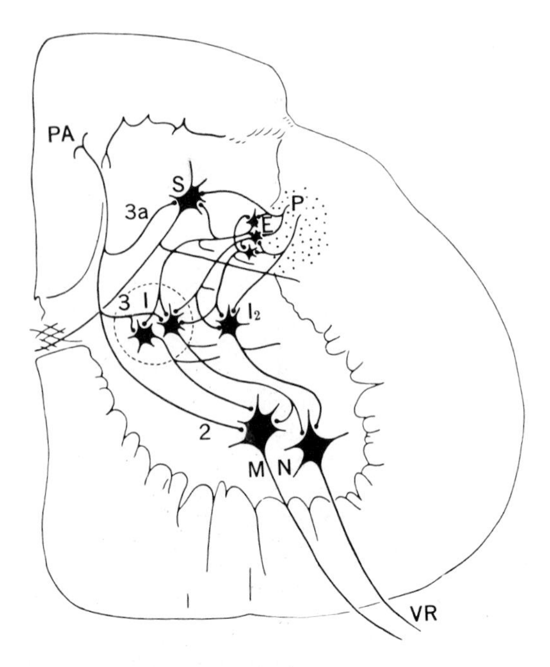

Fig. 28-6. Termination of corticospinal fibers on interneurons of dorsal horn and intermediate region. *P,* Corticospinal tract; *E,* small cells of external basilar region; *I,* intermediate nucleus of Ramón y Cajal; I_2, other neurons of intermediate region; *MN,* motoneurons; *PA,* primary afferent collaterals; *S,* solitary cells of dorsal horn; *VR,* ventral root; *2, 3,* and *3a,* terminal collaterals of primary afferent system. (From Lloyd.[36])

OUTPUT

The neurons that transmit signals from the spinal cord to muscle have cell bodies located in the ventral horn of the cord. Their axons leave the spinal cord in the ventral roots and run without interruption to skeletal muscles. In the seventh lumbar ventral root of the dog there are about 6,000 fibers. This group of cells receives signals from skin, muscles, joints, cerebellum, vestibular apparatus, brain stem, basal ganglia, and cerebral cortex. These sources of input are mentioned only to stress the great complexity of the total influx to motoneurons. As this complexity evolved, an output system capable of responding with sufficient flexibility and subtlety to the vast input had to develop with it. A homogeneous population of motoneurons with similar properties could not meet the demands. Instead, a highly organized population of cells evolved whose individual characteristics differ widely and systematically. The range and distribution of these characteristics confer on the population *collective* properties that its members do not possess individually. This subject was discussed in some detail in Chapter 26.

Gamma motor (fusimotor) neurons[22,43]

About one third of the fibers in the ventral roots innervate intrafusal fibers in muscle spindles. These fibers range from about 1 to 8 μ in diameter. Their cell bodies are scattered among the alpha motoneurons in the ventral horn of the spinal cord.

Gamma motoneurons differ from alpha motoneurons in several respects: they tend to discharge spontaneously, they often discharge at higher frequencies, and their response is more frequently repetitive than that of alpha motoneurons. They apparently lack monosynaptic excitatory connections from primary afferent fibers.

The "spontaneous" activity of gamma motoneurons is a striking phenomenon. It goes on in resting or anesthetized animals even when no alpha discharges are occurring and is never normally absent in the nerves supplying ordinary skeletal muscles. It appears to be a natural consequence of the small size of gamma motoneurons and their greater excitability. The afferent inflow from receptors in resting muscles and from other sources is evidently sufficient to maintain activity in a considerable number of gamma neurons at all times.

Various forms of natural or artificial stimulation will elicit reflex discharge in gamma motoneurons. The pattern of discharge often but not invariably parallels that of the alpha fibers innervating the same muscle. Several investigators have reported that gamma motoneurons are discharged by reflex or suprasegmental stimuli that are insufficient to discharge alpha motoneurons. Again this would appear to be due at least in part to the greater excitability of small neurons. From a control standpoint it is appropriate that the gamma system, which is an essential part of the servomechanism regulating muscle length, should be composed throughout of smaller and therefore more sensitive elements than the alpha system that it serves.

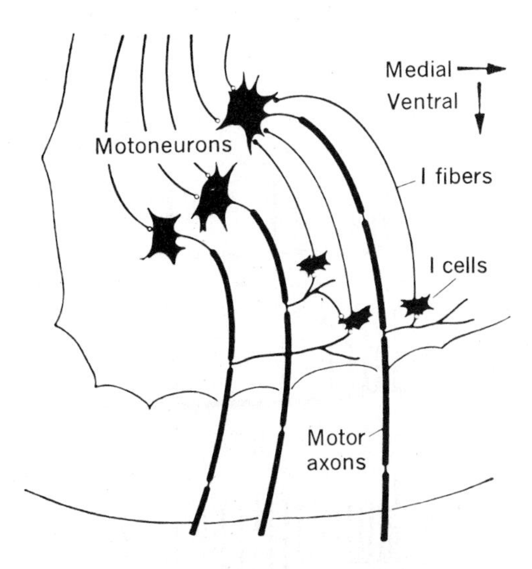

Fig. 28-7. Schematic representation of three motoneurons showing their recurrent collaterals ending on Renshaw internuncial cells, *I*, which in turn send their axons to motoneurons. (From Eccles et al.[13])

Recurrent collaterals of motor neurons

As Ramón y Cajal first demonstrated, the axons of many spinal motoneurons give off branches that turn back into the ventral horn. These "recurrent collaterals" end on small interneurons known as Renshaw cells (Fig. 28-7). A number of recurrent collaterals converge on each Renshaw cell. When ventral roots are stimulated electrically, the antidromic volley in the recurrent collaterals causes a burst of high-frequency discharges from each Renshaw cell, as illustrated in Fig. 28-8. The axons of Renshaw cells terminate on motoneurons, forming a special type of feedback circuit whose significance is not yet established. Activity of Renshaw cells is usually associated with inhibition of motoneurons in nearby portions of the spinal cord, but under some conditions certain adjacent motoneurons may be facilitated. The exact distribution of these effects is uncertain. Control of the recurrent collateral system is quite complex. Renshaw cells are excited and inhibited by input from skin and muscles and by descending impulses from higher centers. Despite extensive investigation, the functional role of the recurrent collateral system remains obscure.[46]

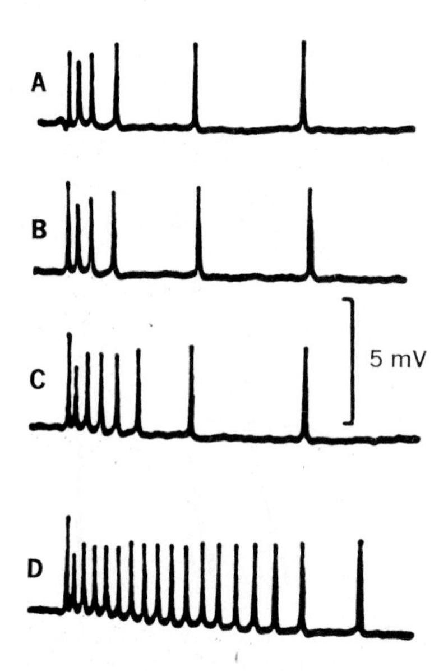

Fig. 28-8. Responses of Renshaw cell to single antidromic volleys in nerves to plantaris, **A;** soleus, **B;** medial gastrocnemius, **C;** and lateral gastrocnemius, **D.** (From Eccles et al.[13])

SPINAL REFLEXES[50]

REFLEXES OF MUSCULAR ORIGIN

The best introduction to the modern studies of muscle reflexes is an examination of the response of a single muscle to passive stretch. In such an experiment the three types of feedback circuits involved in response to stretch are all intact and capable of contributing and interacting in a normal manner. The central connections and reflex effects of each of these circuits will be described separately in later sections.

Stretch reflexes

The first significant progress in the study of stretch reflexes did not come until 1924 when Liddell and Sherrington published their famous paper.[30] Sherrington, who was professor of physiology at Oxford University, had been interested for some time in the phenomenon of decerebrate rigidity, a condition in which there is a rigid extension of the four limbs of an animal due to a transection of the brain stem. The condition has been called a caricature of standing because the limbs are thrust out stiffly as they are in the standing position. Sherrington was intrigued by the fact that there was considerable resistance to passive movement of the extended limbs and that this resistance was so marked in the muscles used in standing (i.e., the antigravity muscles). He therefore began his study by using the fully isolated quadriceps muscle of a decerebrate cat. All the nerves to the hind limb were severed except that innervating the quadriceps muscle. The pelvis was securely fixed to an experimental table and the patellar tendon was attached to a rigid myograph.

When the quadriceps muscle was stretched by lowering the table slightly, a relatively large tension was developed in response to a few millimeters of extension, as shown in Fig. 28-9. This was far more tension than the elasticity of the muscle and tendon could possibly develop as a result of passive stretch. Sherrington concluded that the tension was the result of active contraction of the muscle and was largely reflex in origin because it depended on the integrity of the reflex arc. After cutting the nerve to the quadriceps muscle, stretch caused only a slight increase in the measured tension, which was due to the elasticity of the muscle. Similarly, interrupting any other part of the reflex pathway or damaging the spinal cord abolished the reflex tension. Finally, he noted that stimulation of a sensory nerve in the same leg, which normally elicits a potent flexion reflex, also abolished the tension.

This demonstration that stretching an extensor muscle causes it to contract reflexly became at once the most important fact in reflex physiology. It led to a long series of investigations in

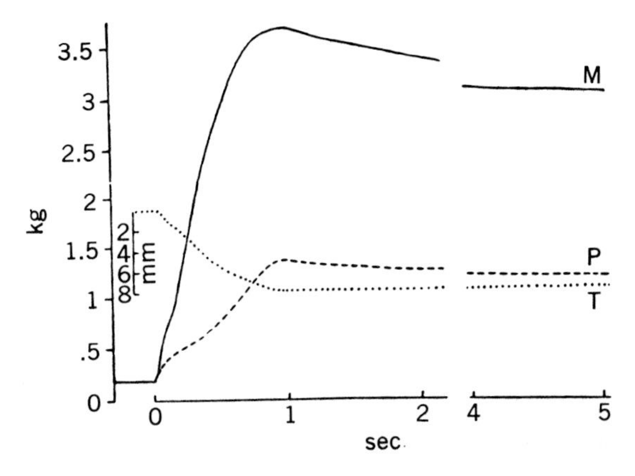

Fig. 28-9. Myotatic reflex of quadriceps muscle of decerebrate cat. *M,* Myographic record of tension change in muscle caused by its stretch, indicated by dotted line, *T*. *P* records purely elastic tension caused by exactly similar stretch of muscle after section of muscle nerve. Tension difference *(M − P)* is that produced by reflex action. (From Liddell and Sherrington.[30])

Sherrington's laboratory at Oxford University. Out of these studies came a number of important principles.

1. All skeletal muscles exhibit stretch reflexes in some degree.

2. Stretch is the adequate stimulus, but the sensitivity of the receptors is so great that jarring or vibration will also discharge them. Ordinarily gravity or the shortening of an opposing muscle causes the stretch. In a sensitive preparation a stretch of less than 1 mm is sufficient to elicit a reflex contraction.

3. The response is specific. The muscle that is stretched is the muscle that contracts. This specificity is so pronounced that stretching one head of a two-headed muscle will ordinarily elicit a contraction only in the stretched portion.

4. The reflex has a very brief latency.

5. The reflex contraction does not outlast the stretch. There is no "afterdischarge" of the motoneurons as there is in some reflexes.

6. The reflex is spinal. It remains after the cord is separated from higher centers.

7. The reflex is best developed in extensor muscles. All skeletal muscles respond to stretch with a reflex contraction, but as a rule only certain muscles show a maintained contraction (i.e., one that lasts as long as the stretch continues). These antigravity muscles are called "extensors" by physiologists, regardless of whether they flex or extend a joint. In general, flexor muscles respond to a sudden stretch with a brief phasic contraction, but they do not ordinarily exhibit a maintained contraction.

It should be emphasized that the response to stretch in the decerebrate cat, although useful for demonstrating the existence of an important mechanism, is a greatly exaggerated reflex that is never seen in normal circumstances. It is probably the result of unopposed activity in the length control system, reinforced by powerful facilitation from brain stem centers. Elongation of the muscle results in discharge of spindle receptors, which causes vigorous firing of motoneurons, but the muscle is prevented from shortening and the motoneurons continue to fire as long as the stretch is maintained. An important factor in bringing about the exaggerated response to stretch is a marked increase in the intensity of firing of gamma motoneurons. This causes continual discharge of the spindle receptors in extensor muscles, regardless of whether they are stretched or not. Passive stretch combined with gamma activity causes intense firing of spindle receptors, which results in powerful, well-maintained contractions.

Reciprocal innervation in stretch reflexes

With similar techniques it can also be shown that stretch-evoked impulses from a muscle exert an inhibitory effect on the motoneurons of antagonistic muscles. An example of this reflex inhibition is given in Fig. 28-10. The experiment is a modification of the procedure illustrated in Fig. 28-9. A stretch of the quadriceps muscle elicits a reflex contraction of that muscle that is partially inhibited by stretching the semitendinosus muscles (S) and almost completely inhibited by an additional stretch of the biceps femoris (B). Both of these latter muscles are flexors, acting in direct opposition to the quadriceps at the knee joint.

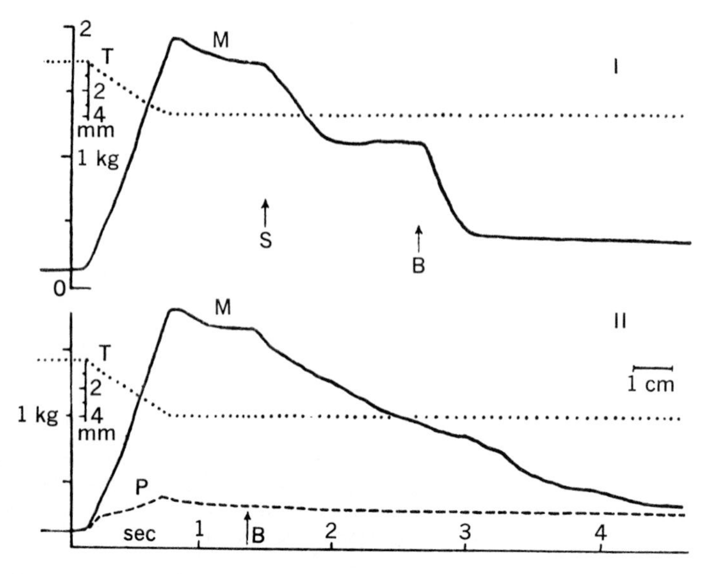

Fig. 28-10. I, Reciprocal inhibition between antagonistic muscles acting at same joint. *M,* Reflex response of quadriceps muscle of decerebrate cat produced by stretch indicated by dotted line, *T,* and recorded myographically. Stretch of semitendinosus muscle beginning at *S* produced partial inhibition of quadriceps reflex tension. Stretch of biceps begun at *B* produced complete inhibition. Inhibitory action on quadriceps motoneurons produced by stretch of two muscles antagonistic to it summed to produce complete effect. **II,** Record of quadriceps stretch reflex obtained as in **I.** Gradually increasing stretch of biceps, beginning at *B,* produced gradually increasing inhibition of quadriceps stretch reflex. Inhibitory effect can be graded by grading intensity of peripheral stimulus producing it. *P,* Passive elastic tension of quadriceps produced by stretching it after section of its nerve. (From Liddell and Sherrington.[31])

The lower records in Fig. 28-10 illustrate the progressive inhibition of quadriceps caused by a progressive stretch of biceps. The basis for this grading of inhibition was described in Chapter 26, where it was shown that the largest motoneurons were the first to be silenced and the smallest motoneurons were the last. Since muscles are arranged as antagonistic pairs, reciprocal effects occur constantly.

Length control system

Following Sherrington's work the first of the three feedback systems from muscle to be investigated in detail was that involving the primary spindle receptors. It was found by Lloyd that electrical stimulation of the nerve to a muscle results in a reflex discharge that returns to that same nerve but not to other nerves. The reflex response observed in these experiments had the same distribution as the stretch reflex itself, which suggested that it was simply the electrical manifestation of the stretch reflex. The very short latency of the reflex indicated that it might be transmitted by the two-neuron arc that Ramón y Cajal's early studies had revealed. By carefully accounting for all the time elapsing between the sudden stretch of a muscle and the appearance of a reflex discharge in the ventral roots, Lloyd[34] established the existence of a two-neuron or monosynaptic circuit that mediated stretch reflexes (Fig. 28-11). A variety of anatomic and physiologic studies indicated that the receptor involved in this reflex was the primary ending in the muscle spindle, which is innervated by a single afferent fiber of large diameter (12 to 20 μm) called a Ia fiber. These fibers give off collaterals that run directly to motoneurons in the ventral horn. The efferent limb of the reflex arc consists of the alpha motoneurons innervating the same muscle. The monosynaptic reflex elicited by a single shock to a muscle nerve is the electrical analog of the so-called tendon jerk, which is simply the phasic component of the stretch reflex. When the tendon of a muscle is tapped sharply, the whole muscle is stretched slightly and a volley of impulses goes up from the primary endings. These impulses arise simultaneously and are conducted synchronously to the motoneurons. As a consequence, a synchronous discharge of motoneurons occurs and evokes a twitch contraction of the muscle. This contraction is referred to as a knee jerk, ankle jerk, or

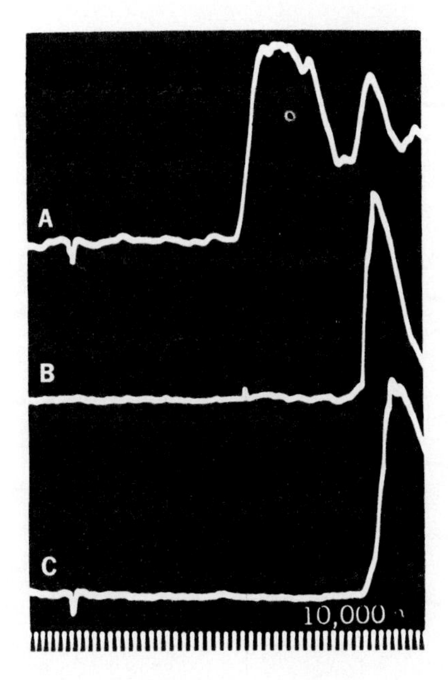

Fig. 28-11. Myotatic reflex may be transmitted through central pathway containing only one synaptic relay; it is monosynaptic. In each record, first small deflection indicates instant of stimulation. **A,** Electrical signs of *afferent* discharge evoked by brief stretch of gastrocnemius muscle of acutely spinal cat, recorded from first sacral dorsal root at a given point. **B,** Electrical signs of segmental monosynaptic reflex discharge evoked by stimulation at same point on first sacral dorsal root and recorded at given point on first sacral ventral root; initial discharge of segmental reflex is monosynaptically relayed. **C,** Electrical signs of reflex discharge into first sacral ventral root evoked by brief stretch of gastrocnemius muscle. Sum of latencies in **A** and **B** approximates that of **C,** which shows that reflex response evoked by stretch was conducted through monosynaptic arcs. Time line identical for all records. (From Lloyd.[34])

jaw jerk, depending on the muscles involved. It should be stressed that although a tendon is frequently tapped to elicit the reflex, the effective afferent volley does not arise in tendon organs but in muscle spindles.

Monosynaptic relations between synergists and antagonists[38]

As a rule there are no monosynaptic excitatory connections between muscles that act upon different joints. Between two muscles or two heads of a single muscle acting synergistically at the same joint, however, there are direct excitatory connections that link their actions closely. Whenever group Ia impulses from a particular muscle are exciting the motoneurons of that muscle, collaterals of the Ia fibers are transmitting similar but less intense effects to the motoneurons of direct synergists. Fig. 28-12 illustrates the time course of this monosynaptic facilitation. In order to appreciate the experiment from which Fig. 28-12 is derived, it should be understood that a volley in Ia fibers arriving at a pool of motoneurons ordinarily discharges only a fraction of the cells. The remainder are excited subliminally but do not fire unless some additional excitatory input is supplied. The amplitude of the monosynaptic reflex evoked by a "test" volley indicates the number of motoneurons fired. Any increase or decrease in the amplitude of this reflex occurring as a result of a previous subliminal "conditioning" volley indicates the reflex effect of that volley in a quantitative fashion. The oscillographic tracings in the upper part of Fig. 28-12 show the facilitation of monosynaptic reflexes of one portion of the biceps femoris muscle when the test volley in the nerve to that part of the muscle is preceded at varying intervals by a conditioning Ia volley in the nerve to another part of the muscle. Curve B in the lower part of Fig. 28-12 indicates the time course of this facilitation. A similar curve (A) is obtained when

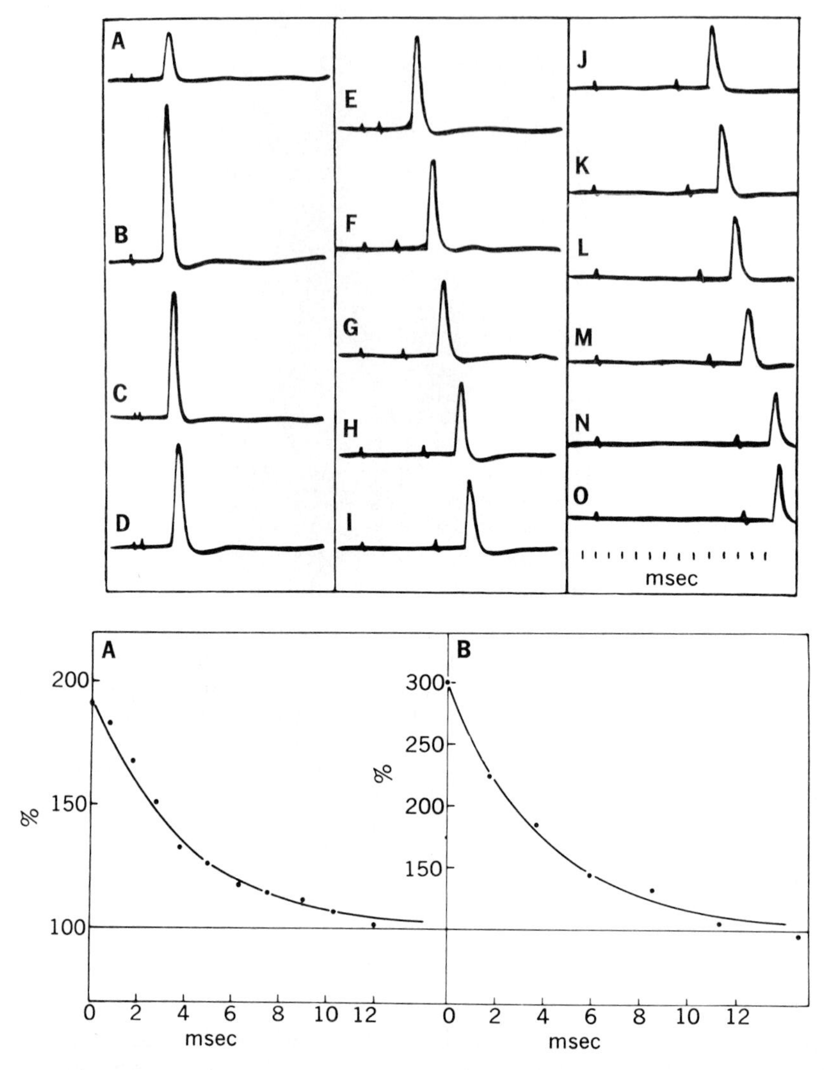

Fig. 28-12. Time course of monosynaptic facilitation. Oscillographic tracings in upper portion of figure are monosynaptic reflexes recorded from ventral root in response to stimulation of two branches of nerve to biceps femoris. Test reflex elicited by stimulation of one branch is shown in frames **A** and **O.** Frames **B** to **N** show facilitated responses obtained when test volley was preceded by a conditioning volley in the other nerve. Curves **A** and **B** in lower portion of figure plot amplitude of conditioned reflex as percentage of test reflex (ordinates) against interval between conditioning and test shocks. Curve **A** obtained from experiment on medial and lateral gastrocnemius; curve **B** obtained from experiment on biceps femoris illustrated above. (Tracings from Lloyd[37]; graphs from Lloyd.[39])

the conditioning and test volleys are set up in the nerves of synergistic extensor muscles. These curves show that the subthreshold Ia excitatory effect of the conditioning volley lasts about 12 msec, decaying exponentially, and that temporal summation of the effects of activity in different fibers plays an important role in synaptic transmission.

Similarly, impulses in group Ia afferent fibers from a given muscle inhibit the motoneurons of direct antagonists. This inhibition is illustrated by the records and curves in Fig. 28-13. The experiments were similar to those that formed the basis of Fig. 28-12. Although the time course of this inhibition appears to be approximately similar to that of monosynaptic excitation, investigation has shown that the "inhibitory" collaterals of Ia fibers end on internuncial cells that in turn cause the inhibition (Chapter 27).

The muscles that act together or in opposition

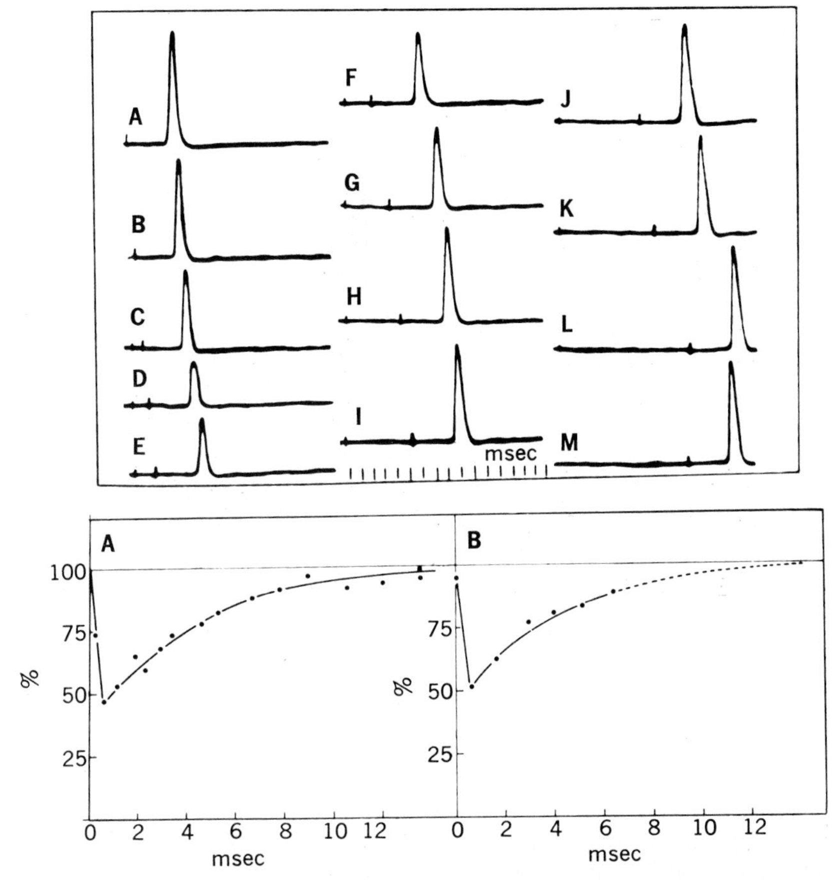

Fig. 28-13. Time course of "direct inhibition." Tracings in upper portion of figure are monosynaptic reflexes recorded in response to stimulation of nerve to gastrocnemius. Unconditioned test reflex shown in frames **A** and **M.** Frames **B** to **L** show inhibition of test reflexes by a conditioning volley in peroneal nerve (a flexor) at increasing intervals. In lower portion of figure, curve **A** shows time course of inhibition of tibialis anterior motoneurons by group Ia volleys in nerves to triceps surae. Curve **B** shows time course of inhibition of triceps motoneurons by group Ia volleys in nerves of ankle and knee flexors as illustrated above. (Tracings from Lloyd[37]; graphs from Lloyd.[39])

at a given joint are called a *myotatic unit.* The members of such a unit are mechanically interdependent, for any change in the length or tension of a particular muscle has a direct effect on its synergists and antagonists. The neural interconnections are the functional expression of this relationship.

Tension control system

The feedback system associated with the tendon organs of Golgi is sometimes called the group Ib system because the afferent fibers belong to group I and are only slightly smaller in average diameter than those from primary spindle receptors. It is well established that these afferent fibers do not end on motoneurons but on internuncial cells that project to motoneurons, completing a disynaptic reflex arc.[28] The evidence for this conclusion comes from experiments similar to those illustrated in Figs. 28-12 and 28-13, in which volleys from two muscle nerves are allowed to interact. If the first or "conditioning" volley is produced by a weak shock to a muscle nerve, it sets up afferent impulses that are limited to group Ia fibers. The effect of such a volley on the motoneurons of a synergist is illustrated by the facilitation curves in Fig. 28-12. If the conditioning shock is strengthened slightly, group Ib fibers are added to the group Ia volley. The effect of this addition is to alter the smooth facilitation curve, as shown in Fig. 28-14. Instead of curve A, curve B is obtained. Within 0.5 to 0.6 msec after the excitatory effect of the group Ia volley begins, a sudden

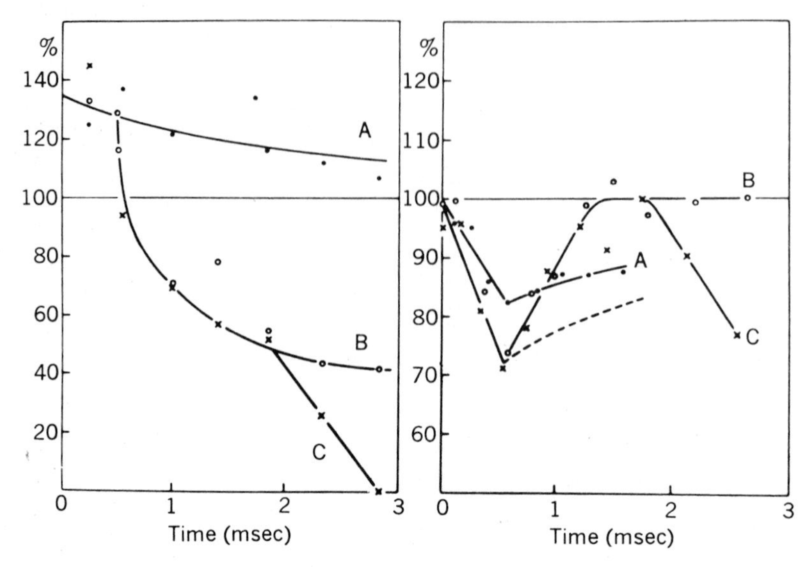

Fig. 28-14. Monosynaptic, disynaptic, and polysynaptic effects of volleys in fibers of groups Ia, Ib, and II. Left: curve *A*, facilitation of plantaris (extensor) monosynaptic reflexes by group Ia volleys from its synergist, flexor digitorum longus. Curve *B*, inhibition of same reflexes by volleys in fibers of group Ia + Ib. Curve *C*, further inhibition of same reflexes by volleys in fibers of groups Ia + Ib + II. Right: curve *A*, inhibition of plantaris monosynaptic reflexes by volleys in group Ia fibers of its antagonist, extensor digitorum longus. Curve *B*, excitatory effect of volleys in fibers of group Ib, combined with inhibitory effect of group Ia. Curve *C*, further inhibition of same reflexes by volleys in fibers of groups Ia + Ib + II. (From Laporte and Lloyd.[28])

decrease in the amplitude of test monosynaptic reflexes occurs, indicating the arrival of an inhibitory volley at the motoneurons. The inhibitory volley is more potent in its effect than the Ia excitatory volley, for their combined effects cause a net inhibition. The delay of 0.5 to 0.6 msec in the onset of inhibition signifies the existence of a single extra synapse in the afferent circuit and indicates the presence of an internuncial cell.

The effect of Ib volleys is inhibitory to the motoneurons of the same muscle and to those of direct synergists, as illustrated in Fig. 28-14, left. Motoneurons of direct antagonists, however, are facilitated by volleys in Ib fibers. As illustrated in Fig. 28-14, right, the Ia effect is inhibitory (curve A) and the Ib effect is excitatory (curve B). Thus, as in the case of the Ia system, Ib effects are distributed according to the principle of reciprocal innervation.

Following Lloyd's suggestion,[28] it has been customary to refer to the group Ib effect as the "inverse myotatic reflex" because it is opposed to the myotatic reflex. If the Ib system were actually myotatic, as is the Ia system, this terminology would be appropriate. As we have already indicated, however, the adequate stimulus for tendon organs is tension produced by active contraction of the muscle. Changes in length are ineffective. It is not logical to refer to a tension-regulating system as the inverse of a length-regulating system merely because the synaptic effects of the two are opposed; hence the term "inverse myotatic reflex" should be abandoned.

Group II reflex effects

Group II fibers from muscle apparently originate in a number of different types of receptors. This has led to confusion in the analysis of the reflex effects of activity in these fibers. One set of endings originates in endings not located in muscle spindles and not sensitive to stretch. The other type of group II fibers arises in the secondary endings of muscle spindles and is sensitive to stretch. In a number of respects the latter fibers resemble the Ia fibers arising in the primary endings of muscle spindles.

The group II fibers whose endings are not sensitive to stretch will be discussed first. If the conditioning effects of an afferent volley in all group II fibers in muscle nerves are examined in the manner previously described for Ia and Ib volleys, a third reflex effect can be distinguished. When the group II volley is added to the Ia and Ib volleys by increasing the intensity of the conditioning shock, the effect is not noticeable in the conditioning curve until 2 msec after the onset of the Ia effect. This delay is partly due to the slower conduction velocity in group II fibers, but chiefly to the presence of two or more internuncial cells in the reflex circuit. When a muscle nerve containing both types of group II fibers is stimulated with brief electrical shocks, the motor nerves of physiologic extensors are usually excited, whereas those of extensors are inhibited. These effects are illustrated by the curves labeled C in Fig. 28-14. This pattern of response was sometimes referred to as a "flexor reflex," because it resembled ordinary flexor reflexes in the distribution of excitatory and inhibitory effects. As subsequent experiments showed, this flexor pattern was not due to activity in the stretch-sensitive group II fibers.

The functional significance of the group II spindle receptors did not become apparent until more physiologic experiments were carried out. In experiments later confirmed by others, Matthews[44] showed that the effects of stretch-sensitive group II fibers were excitatory to their own (homonymous) motoneurons. In essence, the experiments compared the reflexes elicited by graded stretches with those elicited by vibrating the same muscle at various frequencies. Since secondary endings are excited by stretch but not by vibration, the larger tensions obtained during stretch presumably resulted from group II impulses. The two reflex responses did not occlude each other in the manner that would be expected if both depended solely on Ia fibers. Since this demonstration of group II excitation by Matthews, other groups have shown by spike-triggered averaging techniques that group II impulses elicit monosynaptic and probably polysynaptic EPSPs in homonymous motoneurons. The evidence for this is cited in Chapter 27. It is, therefore, generally accepted that group II impulses from secondary spindle endings contribute to the stretch reflex. The tendon jerk is probably mediated largely by Ia impulses, whereas the group II impulses from spindles probably make their greatest contribution to the maintained or tonic stretch reflex.

Significance of the three types of circuits

The full significance of the circuits found in the Ia, Ib, and group II systems is not yet established, but it has recently become apparent that transmission in circuits containing interneurons is influenced by descending activity from higher centers and by afferent impulses in cutaneous nerves.[27] In the decerebrate preparation, for example, an afferent volley in all group II fibers elicits little or no reflex discharge of flexor motoneurons. Impulses descending from the brain stem evidently inhibit the interneurons that transmit excitatory impulses to flexor motoneurons. After spinalization, these interneurons are apparently released from descending inhibition, for the previously ineffective afferent volley now elicits a large reflex response. Afferent impulses in certain fibers of cutaneous nerves may cause a similar suppression of some group II reflexes. Since depression of reflex transmission can be produced without a corresponding depression in the flexor motoneurons themselves, the internuncial cell is presumably the site at which transmission is regulated in these instances.

Impulses that impinge directly on motoneurons necessarily elicit regulatory activity in all three types of feedback control systems. Signals that impinge on internuncial cells, however, may reinforce or suppress activity in any one of the three systems selectively. It is not yet understood why the length-regulating system is direct and therefore free from this selective control, whereas the other systems are subject to it.

It is now apparent that stretch of a muscle produces effects that are not fixed and invariant but that vary according to the conditions of the experiment. The exaggerated reflex response to stretch in decerebrate preparations is thus due in part to the virtual shutdown of the control systems that are inhibitory to extensor motoneurons. In the "spinal" animal these circuits are apparently "open," for passive stretch may result in net inhibition of extensor motoneurons[20] rather than net excitation. It must be concluded that under normal circumstances descending motor systems actively regulate incoming stretch evoked impulses to produce the most suitable reflex effect under the prevailing conditions.

REFLEXES OF CUTANEOUS ORIGIN

The sensory receptors in skin and subcutaneous tissues respond to touch, pressure, heat, cold, and tissue damage. The signals from all of these receptors exert reflex effects on spinal motoneurons via internuncial cells. They provide feedback that serves to orient the animal to its immediate environment and protect it from injury. The dominant pattern of response to cutaneous stimulation of a limb is ipsilateral flexion and contralateral extension. This suggests that cutaneous activity of all kinds, not only that arising in response to injury, may elicit aversive responses that tend to withdraw the limb from a source of injury. Reflexes of cutaneous origin are not wholly flexor, however, as will be pointed out a little later.

As noted previously, the primary afferent fibers from skin send collaterals to internuncial cells in the dorsal horn. A microelectrode inserted in this region encounters many cells that respond vigorously to various kinds of cutaneous stimulation. Inhibitory as well as excitatory effects are common. The responses of motoneurons are frequently very similar to those of certain interneurons in their time course, which indicates that the patterning of motor responses may be quite a direct consequence of internuncial activity.

Flexor reflexes

The most thoroughly investigated of the cutaneous reflexes is the so-called flexor reflex, consisting of contraction of physiologic flexor muscles and relaxation of physiologic extensors. There are two kinds of flexor reflexes: those resulting from innocuous stimulation of the skin and those resulting from potentially painful and injurious stimuli. The former type consists of a weak contraction of one or more flexor muscles with little actual withdrawal of the limb. The latter type consists of a widespread contraction of flexor muscles throughout the limb that causes an abrupt withdrawal of the injured part from the source of damage.

Stimulation of almost any nerve in a limb will cause a flexor reflex. As indicated in Table 28-2, the reflex tension developed by a particular flexor muscle varies widely with the nerve stimulated. It is clear that some mechanism in the spinal cord distributes the afferent impulses to all the flexor muscles of the limb. This distribution is carried out partly by the branching of the primary afferent fibers but largely by internuncial cells. Ramón y Cajal referred to this as the "diffuse reflex mechanism" in contrast to the "circumscribed mechanism" (Fig. 28-5). The distribution of afferent impulses, although diffuse, is not entirely nonspecific. Depending on the nerve stimulated, the ipsilateral limb will take up different final positions. Table 28-3 shows how the pattern of reflex flexion varies with the nerve stimulated. Hence the reflex exhibits local sign to some extent.

Comparison of the discharges evoked in the

Table 28-2. Maximal reflex tensions developed by tibialis anterior muscle in response to tetanic stimulation of various ipsilateral hind limb nerves (m. tibialis anticus — maximum motor tension 2,160 gm)*

Afferent nerve stimulated	Tension of maximal reflex tetanus (gm)	Reflex tension expressed as percentage of maximal motor tetanus
Internal saphenous	800	32
Superficial obturator	165	6.7
Deep obturator	400	16
Nerve to quadriceps and sartorius	1,190	44
Musculocutaneous branch of peroneal	1,700	69
External plantar	1,240	50
Internal plantar	1,330	54
Small sciatic	680	28
Hamstring	565	23
Nerve of sural triceps	300 (rather low)	12
Total	8,370	

*From Creed et al.[6]

Table 28-3. Patterns of reflex tension resulting from stimulation of three hind limb nerves*

Nerve	Reflex tensions (gm)		
	Hip flexor (tensor fasciae femoris)	Knee flexor (semitendinosus)	Ankle flexor (tibialis anticus)
Internal saphenous	100	56	87
Popliteal	3 or less	42	100
Peroneal distal to tibialis anticus nerve	14	100	69

*From Creed et al.[6]

ventral root by stimulation of a muscle nerve and a skin nerve brings out the differences in their central transmission very clearly. The response to an afferent volley in group Ia fibers is a synchronous discharge of minimal delay, indicating transmission over a two-neuron pathway (Fig. 28-15, *A*). The response to a volley in a skin nerve (Fig. 28-15, *B*) consists of a more prolonged series of discharges with no monosynaptic component. The earliest deflections are the result of transmission in a reflex arc of four or more neurons; the later deflections are those that have been delayed by passage through additional interneurons. If a peripheral nerve is stimulated with shocks of increasing strength, the reflex discharge recorded from a nerve to a flexor muscle increases progressively in duration as smaller afferent fibers are added to the afferent volley. The response to a single strong shock exciting delta and C fibers may last a full second or more. If the recording electrodes are placed on a very fine muscle nerve so that individual impulses can be distinguished, it can be seen that the prolonged discharge consists of repetitive discharges of the motoneurons of small size. This prolongation of reflex effect serves the purpose of keeping a flexed limb withdrawn from a painful stimulus so that restimulation does not occur.

Crossed extension

As a ''spinal'' animal withdraws his limb in response to a noxious stimulus, the contralateral limb is simultaneously extended. The function of this reflex is to support the weight of the body when the opposite limb is lifted. Crossed extension is ''grafted on'' to the flexor reflex at the internuncial level. Collaterals of the interneurons that excite ipsilateral flexor motoneurons cross to the other side of the spinal cord and excite extensor motoneurons. At the same time, contralateral flexor muscles are relaxed due to reciprocal inhibition.

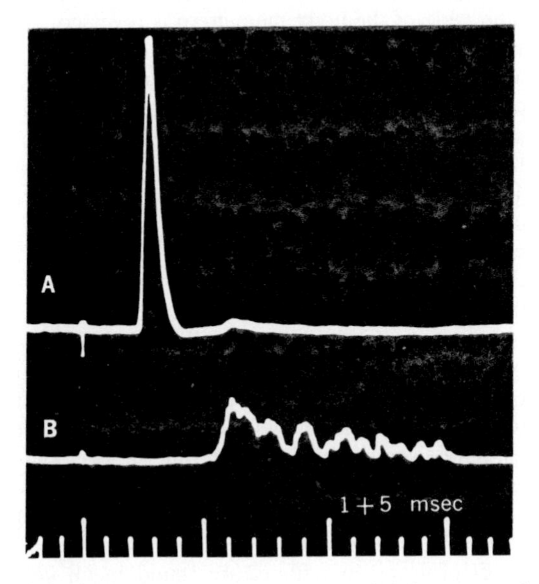

Fig. 28-15. Reflex discharges recorded from ventral root in response to stimulation of, **A,** nerves to gastrocnemius muscle and, **B,** sural nerve (a skin nerve). (From Lloyd.[35])

Reflexes associated with specific regions

Cutaneous stimulation of certain regions may elicit responses that do not fall into the flexor pattern but serve more specific purposes. Light pressure applied to the pads of the forepaw of a chronic spinal dog, for example, causes reflex extension of the whole limb.[7] This serves to reinforce the stretch reflexes and other supporting reactions brought into play when a foot is placed in contact with the ground. It is important to note that even the slightest pinch of the same toe pads will produce flexor withdrawal rather than extensor thrust. These two very different reactions serve to illustrate that the nature of a reflex action is determined by the quality of the stimulus as well as its locus. Another example of specific reflex action is the bilateral extension of the hind limbs that results from stimulation of the skin of the inner aspects of the thighs. The significance of this reflex in copulation is obvious.

REFLEXES ORIGINATING IN JOINTS

Since the position of a joint specifies the length of the muscles acting across it, the signals from joints should be of great importance in regulating movement at the spinal level, just as they unquestionably are at higher levels, where they lead to a conscious sense of position. Relatively little is known, however, about reflexes originating in joints. Several types of sensory receptors are found in and around joints, the most important of which are Ruffini's endings and tendon organs. They are located in all the ligaments and in the fibrous capsule. Every position of a joint involves stretch of some ligaments or capsular bands and relaxation of others. As might be expected, it has been found experimentally[3] that there is no position at which all joint receptors are silent. In general, each ending discharges at its maximal steady rate when the joint is at some particular angle and less rapidly when the angle is increased or decreased. Different receptors respond over different portions of the total range of movement. The pattern of firing in all the receptors signals the exact position of the joint to

Fig. 28-16. Reflex figures produced in cat with neuraxis sectioned just above first cervical segment (at calamus scriptorius). Noxious stimulation at points marked by arrows; *e*, leg extended in response; *f*, leg flexed in response. (After Sherrington.[49])

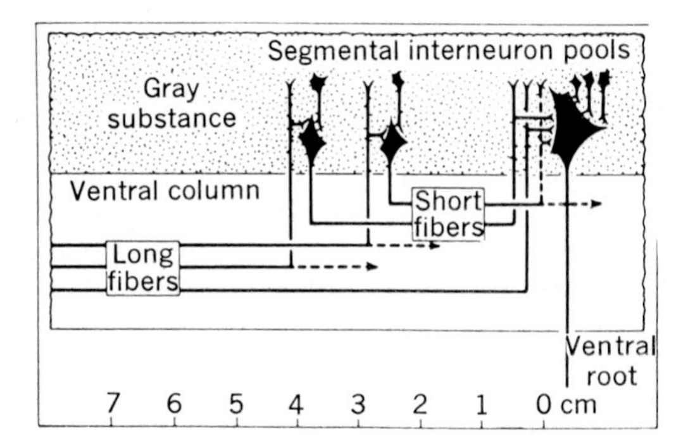

Fig. 28-17. Schema illustrating pathways connecting descending fibers of ventrolateral columns of spinal cord, the bulbospinal and long spinal systems, with motoneurons. They end in greatest concentration about cell bodies of short propriospinal neurons, located in segmental interneuron pools. Activity is relayed through them to motoneurons some 3 to 4 cm farther caudally. A direct pathway exists, but motoneurons are controlled by it only when activity is intense and motoneuron excitability high. (From Lloyd.[32])

higher centers. These signals unquestionably exert reflex effects as well, for stimulation of articular nerves elicits discharges in ventral roots.[2,16] The various patterns of signals resulting from each of the possible positions of a joint probably influence the motoneurons of the muscles, which move that joint in a systematic manner, although this has not been demonstrated experimentally.

Long spinal reflexes and the propriospinal system

In animals with four weight-bearing limbs the activity of the forelimbs and hind limbs is coordinated by so-called long spinal reflexes.[51] This can readily be observed in preparations with spinal transections in the upper cervical region. A noxious stimulus applied to any extremity results in reflex responses in all four limbs, as illustrated in Fig. 28-16. Forelimb flexion is accompanied by extension of the ipsilateral hind limb. On the other side of the body the pattern is reversed. As a result of these actions, the animal tends to move away from the stimulus. A considerable amount of normal locomotor activity in four-footed animals is probably patterned in this way. It is believed that the stimuli responsible for long spinal activity arise in the secondary spindle receptors of muscle and in cutaneous end-organs innervated by fibers of groups II, III, and IV.[33,40] Group I activity from muscles apparently has little effect on distant motoneurons.

Long spinal reflexes are mediated by *propriospinal* neurons whose cell bodies lie in the central gray matter and send their axons into the adjacent white matter, where they run up or down for variable distances before reentering the gray matter (Fig. 28-17). There is no evidence that nervous activity is ever conducted more than a minimal distance within the gray matter itself. Short propriospinal fibers are found in all columns of the white matter; long fibers are apparently found only in the lateral and ventral columns (Fig. 28-18).

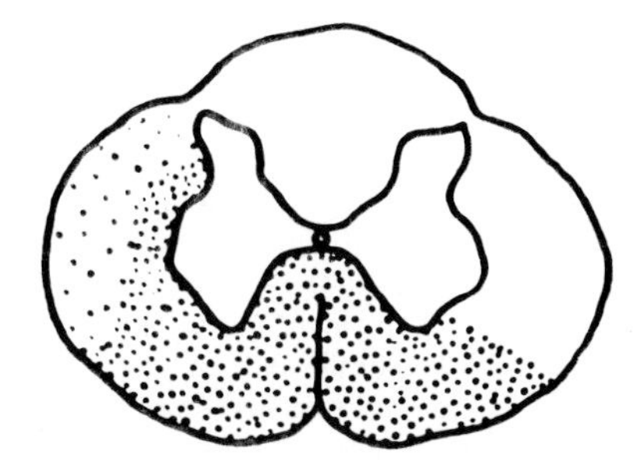

Fig. 28-18. Distribution of propriospinal neurons concerned with long spinal reflexes. Ventral columns are involved bilaterally, lateral columns unilaterally. (From Lloyd.[33])

Suggestions for review and further study

Ramón y Cajal's classic account[45] of the fine structure of the spinal cord with its numerous illustrations is invaluable. An excellent summary of the studies of the Sherrington school, containing much that has not been included in these chapters, is given by Creed et al.[6] Two monographs by Eccles[9,10] offer the most complete account of the modern work on synaptic transmission of the spinal cord, and finally, a critical review by Hunt and Perl[22] will serve as a convenient point of departure for a survey of the recent literature on spinal reflexes.

Effects of spinal transection

Vernon B. Mountcastle

SPINAL SHOCK

When the spinal cord is completely severed, two functional disasters are at once evident: (1) all voluntary motion in body parts innervated by the isolated spinal segments is permanently lost and (2) all sensation from those parts, which depends on the integrity of ascending spinal pathways, is abolished. A third conspicuous sign is an immediate spinal areflexia, a state by usage termed spinal shock.

Spinal shock is a transient condition of decreased synaptic excitability of neurons lying aboral to a transverse section of the spinal cord.[29] It varies in depth and duration with the degree of cerebral dominance of the spinal mechanisms in the species considered, that is, with the degree of encephalization. Thus in the frog the depression of reflex excitability after high spinal section lasts only a minute or so. In carnivores the duration of reflex depression is measured in hours; in the monkey it may extend over many days or weeks, whereas in anthropoid apes and in man the course of recovery of reflex excitability may extend over many months.

Spinal neurons bear on their cell bodies and dendrites hundreds of terminal end-feet, many of them endings of axons descending from supraspinal portions of the nervous system. Under normal circumstances, these deliver low-frequency trains of impulses to spinal neurons, so that at any given instant many local postsynaptic responses (subliminal depolarizations) are occurring at widely scattered sites on the postsynaptic cell. Although these may not be dense enough to provoke complete depolarization and firing of the cell, they serve to maintain it in a slightly oscillating state of high excitability, ready to respond to any spatially or temporally more concentrated presynaptic inflow. When that portion of the "background tone" taking origin in more cephalad regions is removed by spinal section, the resting excitability of the spinal cell is, for a time, greatly reduced: this is spinal shock.

Spinal shock cannot be due to any form of irritation at the cut surface of the cord, for a second transection made some days later, below the first, is followed by only a trivial reflex depression in the spinal remnant. Nor is it the result of the drop in blood pressure that follows any but the lowest transection, for oral portions of the nervous system are equally exposed to the decreased blood flow, yet display no decreased excitability, and spinal shock still appears when the hypotension is prevented by previous administration of vasoconstrictor drugs.

The phenomena of spinal shock are confined to regions aboral to the section. With the exception of the slight increase in extensor tonus in the forelegs following thoracic or lumbar section (Schiff-Sherrington phenomenon), no change in function is evident in the cephalad portions of the nervous system. Indeed, a cat or monkey immediately after cord section seems blithely unaware of this catastrophe.

Temporal course of recovery of reflex excitability

Somatic reflexes. In carnivores the period of complete areflexia is so short that it may not be observed if a long-acting anesthetic agent is used for the operation of spinal transection. Within minutes a tightening of quadriceps muscle is palpable on tapping the patellar tendon. Soon, visible knee jerks appear, and shortly thereafter, feeble flexion of the leg is produced by nociceptive stimulation of the foot. With the passage of hours the flexion withdrawal reflex gains in strength, the threshold of the reflex is lowered, and the peripheral receptive zone from which it can be evoked enlarges. Simultaneously with flexor contraction, extensors are inhibited. Cross extension reflexes and long spinal reflexes such as the scratch reflex may not reappear for days, and in general the time of recovery of a reflex action depends on the number and complexity of the synaptic relays in its central pathway. The time course of this recovery varies greatly from one individual to another and may be adversely affected by intercurrent infections, nutritional deficiencies, or other debilitations. How these conditions affect reflex excitability is unknown.

With further passage of time, reflex excitability

continues to rise, now to abnormally high levels. Sustained myotatic reflexes appear, and positive supporting reactions are evokable; in such chronic spinal animals, these reactions may suffice to support the weight of the animal for 2 or 3 min ("spinal standing") before collapse occurs. When such animals are suspended in air, the hindlegs may execute alternating flexions and extensions ("spinal stepping").

A clear example of the capacity of the isolated cord to combine elementary reflexes into movement patterns having useful purposes is given by the response of the spinal cat to immersion of the feet in water.[42] When the paw of a normal cat is immersed in water, it is withdrawn and shaken vigorously. In the intact animal, the response is independent of the temperature of either the water or of the animal, and it is evident that a tactile stimulus evokes the reflex. In the chronic spinal animal, no response is evoked by immersion of the foot in water at body temperature. However, when the animal is tested with either cold or hot water, a sharp flexion withdrawal of the foot occurs, followed by vigorous shaking of the paw. This shaking succeeds in ridding the limb of a considerable amount of water. The response to cold water is evoked by stimulation of cold receptors and is related to the temperature differences between body and water. Hot water is an effective stimulus only when sufficiently hot to excite pain receptors, and its action is independent of body temperature. In either case the total sensory input in the intact animal signals the quality of wetness, which produces a functionally appropriate response from the spinal reflex mechanism.

Mass flexion reflex. Relatively mild noxious stimulation produces widespread contraction of the flexor musculature of the limb and of the abdominal wall; evacuation of the bladder or rectum may occur at the same time but not necessarily as a part of this mass flexion reflex. It should not be supposed that the irradiation of this reflex spreads along neural channels other than those already completely formed, for there is no evidence to suggest that axons may grow to make new synaptic connections in the isolated spinal remnant. On the other hand, it is as though synaptic connections, normally sparse and tenuous, now assume a supraliminal potency.

Some light is thrown on this problem by a consideration of the phenomenon of the hypersensitivity of denervated neurons. Cannon and Rosenblueth[5] had originally shown that viscera innervated by autonomic nerves become, some days after denervation, abnormally sensitive to the synaptic transmitter agent normally active at those nerve endings. This hypersensitivity is so great that the acceleration of the denervated heart serves as a most sensitive measure of minute quantities of epinephrine in the blood. Later, Cannon and Haimovici[4] demonstrated that the motoneurons below a cord semisection become hypersensitive to acetylcholine, compared to those on the normal side of the cord, as well as to dorsal root afferent impulses and to excitant drugs. Stavraky[52] has extended these observations on central neurons, but little is known of the basic mechanisms concerned. The phenomenon is of considerable interest relative to altered states of excitability in diseases or injuries of the nervous system.

The course of reflex recovery after spinal transection in monkeys[48] resembles that in carnivores but is displaced on an extended time scale. The depth of spinal shock is profound; the body parts below the transection are flaccid and motionless. The most severe nociceptive stimulation of the skin and direct electrical stimulation of a large afferent nerve are equally ineffective in producing a motor response. In some individuals the knee jerk is the first reflex to appear, occasionally within the first hour, but it is frequently seen only at the end of the first week. In many cases the flexion-adduction of the hallux in response to scratching the sole of the foot (the plantar reflex) is the first response obtainable, and it is only rarely delayed in appearance beyond the fourth day. If adequate care is given to bladder and bowel function, to the avoidance of pressure necrosis of peripheral nerves and skin, and to the nutritional state of the animal, the subsequent course of recovery of the somatic reflexes much resembles that described in carnivores. Insufficient numbers of animals have been studied in the chronic spinal state to know whether complex extensor reflexes such as those responsible for "spinal standing" recover full reflex excitability in monkeys, but it is reasonable to suppose that they do so, for they have been observed in humans with chronic paraplegia.

Spinal shock following partial transections. Studies of cats with partial transverse lesions of the spinal cord[15] have shown that it is the reticulospinal and vestibulospinal tracts that, when severed, produce the phenomena of spinal shock, suggesting that their nuclei of origin are principal sources of descending facilitation playing on the spinal neurons. In monkeys, however, these seem less important, for isolated section of the lateral columns of the cord (the corticospinal tracts) produces nearly as severe spinal shock as does cord transection. This species difference seems due to the greatly increased direct control of the spinal mechanisms exerted by the pyramidal system in the monkey as compared with the cat. Considerable effort has been made by investigators using electrical recording techniques to determine whether interneurons or motoneurons are the more severely affected by the depression of spinal shock. In the primate the motoneurons are more severely affected than are the interneurons of the dorsal horn, whereas in carnivores the interneurons are involved and the motoneurons suffer only a slight change in excitability.

Visceral reflexes. Spinal transection also results in widespread alterations in visceral functions.

BLADDER. Immediately after section of the spinal cord there is complete atony of the smooth muscle of the bladder wall. At the same time there is an increase in constrictor tone in the sphincter muscle, presumably due to loss of inhibitory influence of central origin. As a result, urine accumulates until intravesicular pressure is sufficient to overcome sphincter resistance. Even then only small driblets of urine escape—the so-called overflow incontinence. With recovery of the somatic reflexes, tone returns to the bladder muscles and reflex emptying of the bladder occurs, produced by simultaneous contraction of its smooth muscle walls and, at least to a certain extent, relaxation of tone in the sphincter. Nevertheless, a considerable residuum of urine remains after each reflex emptying. More complete emptying is promoted if this residuum is removed each time by catheterization. Reflex emptying is greatly facilitated or at times even provoked but cutaneous stimuli, either tactile or nociceptive, delivered to abdomen, perineum, or lower extremities. The development of the reflex bladder action differs little among carnivores, monkey, and man except in its time course; in the first two the cord bladder is usually established in the first week; in the latter, only 25 to 30 days after cord section.

INTESTINE. Little is known of the function of the intestine after cord section. In dogs a loose diarrhea may develop during the first week, but this is transient and it appears that in general the processes of digestion and absorption proceed normally. Greater difficulty is met in the evacuation of waste products from the intestinal canal. Normally the presence of fecal material in the lower bowel and rectum, passively stretching the muscles of its wall, produces their active contraction and peristaltic action, combined with inhibitory relaxation of the sphincter tone, and defecation results. That this mechanism is inherent in the nerve plexus of the bowel wall is shown by the fact that such "reflex" defecation recovers after complete removal of the spinal cord. However, the presence of the superimposed reflex arcs of the cord brings a progressive character and greater fusion to rectal contractions, which results in a more massive and complete reflex response. This mechanism is depressed during the period of spinal shock, when the sphincter ani relaxes only slightly in response to passive distention of the rectum, and retention of rectal contents may occur. With recovery of reflex excitability, reflex defecation occurs; it is greatly facilitated by tactile stimulation of the skin area of the sacral segments or by manual dilatation of the sphincter muscle.

VASOMOTOR REFLEXES. Reflex actions on the peripheral vessels and other effectors innervated by the autonomic nervous system that are mediated by spinal arcs are considered in Chapter 33. Here it is pertinent to state that they, in company with reflex arcs debouching on the striated musculature, are profoundly affected during the period of spinal shock. The loss of background vasoconstrictor tone precipitates a hypotension that persists for some time, and during the early hours after cord section, even intense stimulation of an afferent nerve trunk will not produce a rise in blood pressure. With time, however, the tonic discharges return, reflex actions can once again be elicited, and an almost normal level of blood pressure is maintained in the chronic spinal animal. Spinal neurons innervating peripheral effectors concerned in the control of body temperature are, of course, permanently disconnected from descending influences originating in thermoregulatory centers, and animals with transection in the cervical region are almost wholly poikilothermic (Chapter 59). Those with lower lesions show proportionately greater degrees of temperature control, depending on the level of transection.

SEXUAL REFLEXES. The postural adjustments and reflex actions operating in the act of copulation are organized at supraspinal levels, but the reflex pathways are inherently spinal. In the spinal male dog, tactile stimulation of the penis produces erection and frequently ejaculation and a flexion of the back and hindquarters that tends to thrust the penis forward. The posture is best seen with the animal prone, belly and thighs contacting the table surface. In the spinal bitch, contractions of uterus and vaginal orifice result from stimulation of the sexual skin. Little is known of the relation of cord function to the endocrine regulation of the organs of reproduction. However, the menstrual cycle of female monkeys is little disturbed by spinal section,[53] and in several instances, impregnation, gestation, delivery, and nursing, all in relatively normal fashion, have been recorded in chronic spinal female dogs.

FUNCTIONAL CAPACITY OF THE ISOLATED HUMAN SPINAL CORD

Complete transection of the spinal cord in man due to injury or disease is a rare event in civilian life. Unfortunately such injuries caused by missiles of high velocity are not uncommon in warfare, and some of these paraplegic individuals have survived in the chronic state. Considerable success has been achieved in rehabilitating them to a certain degree for useful and comfortable lives. Descriptions of the ingenious therapeutic methods by which their survival is assured must be sought in clinical texts and papers, particularly those of Munro,[44a] Kuhn,[26] Freeman,[14] and others. Here it is more pertinent to describe the reflex activity of which the isolated remnant of spinal cord is capable.

The most informative of the early reports concerning the reflex function of the human cord is that of Theodor Kocher, published in 1896,[23] detailing a comprehensive study of 15 patients suffering a sudden cord transection, the completeness of which was subsequently verified at autopsy. Kocher's patients lived for only short periods. The classic descriptions of spinal man were provided in 1917 by Riddoch[47] and by Head and Riddoch,[19] based on studies of soldiers injured in World War I. Their detailed descriptions should be read by all those pursuing the subject. However, the periods of observation available to these clinical neurologists were frequently cut short by the deaths of patients, and their observations were made difficult to interpret by the presence of severe debilitation and frequently of infection of the urinary tract. Advanced methods of medical care permitted surgeons of World War II to study spinal man for months and years without the complications of intercurrent disease. Particular reference is made to the painstaking studies of R. A. Kuhn,[24,25] from which much of the following description has been taken. His report gains great validity also from the fact that *in every case the existence of a complete transection and its segmental level were established directly by surgical exploration of the spinal canal.* In all, 29 men were studied, their transections ranging in level from the second to the twelfth thoracic segment of the cord. Many of these men survived transection for several years. It should be emphasized that the time course of recovery of reflex activity is variable, as is the sequence of appearance of various reflexes. The sequence described below is only the usual one, and wide variations from it are to be expected in individual cases.

Stages of spinal shock. Immediately following transection there is complete loss of motor and sensory function below the level of injury, a complete areflexia is observed, and if the transection is high, a degree of hypotension exists. The duration of the complete areflexia varies greatly from one case to another; in a few, some reflex activity is elicitable within 24 hr of injury, but more commonly, none is observed for a period of 2 to 6 weeks. If complete absence of reflex actions continues for many months, as was the case in 4 of Kuhn's 29 patients, it is likely that the cord itself has been destroyed, either by direct injury or by interference with its blood supply. Paralysis of the bladder, manifested by retention of urine, is invariably complete. If the transection is high, there is also complete paralysis of the mechanism of defecation, with distention of the bowel. There are many suggestions in the reports available that in some cases certain reflexes, particularly those evoked by stimulation of genital regions, may never completely disappear. There is no record, however, of a case studied from the moment of transection; even the earliest accounts begin several hours after the injury had been sustained. In general, it can be said that the reappearance of any reflex that has been completely abolished by cord section is a sign of emerging cord activity and recovery from spinal shock.

Period of minimal reflex activity. As the profound reflex depression begins to lift, noxious stimulation of the soles of the feet elicits reflex responses. Most commonly, the first movements seen are tremulous twitchings of the toes and a brief flexion or extension of the hallux in response to plantar stimulation. At about this time, contraction of the sphincter ani appears in response to noxious plantar stimulation or to tactile stimulation of the perianal skin. The genital reflexes may also appear for the first time. This period of the onset of activity may last from 2 weeks to several months, and in the rare patient, it may represent the end stage of recovery. When this occurs, there is reason to believe that the isolated remnant has sustained severe injury.

Development of flexor activity. With further recovery the tremulous toe movements, the first elicitable, change to a typical Babinski pattern (i.e., dorsiflexion of the hallux and fanning of the toes in response to plantar stimulation), flexor withdrawal movements of the foot and ankle are evokable, and with further passage of time, flexion of the knee and the hip appear. Such a *mass flexion reflex* could be elicited in 22 of the 25 men in Kuhn's series who survived cord section for long periods of time and who manifested any reflex activity. In some instances, the mass flexion was accompanied by the crossed extension reflex, and in many instances, extension occurred during the relaxation phase from flexion. The foot, especially its plantar surface, is the reflexogenous zone *par excellence,* and here purely tactile stimuli suffice to elicit widespread flexion, which shows again that afferent fibers from tactile receptors of the skin project on the reflex mechanisms of the spinal cord as well as into its great ascending tracts. With the passage of months, it is possible to elicit mass flexion or components of it from a wider surface of the leg and, at low threshold, from the genital zone.

Development of extensor activity. In a few individuals the stage of mass flexor activity is the final one reached in recovery. However, in 18 of the 22 reflexly active patients in Kuhn's group who survived cord section for 2 or more years, a stage was reached in which the predominant reflex activity was extensor in nature, the so-called extensor spasms.* The movements of extensor muscles become evident as early as 6 months after injury and may continue for many years. The tendon reflexes are hyperactive and clonus occurs; sustained stretch reflexes are at times observed. The inverse myotatic reflex can be elicited (the lengthening reaction, or "clasp-knife" phenomenon) and with it, increased extension of the contralateral leg — Phillippson's reflex. In many individuals, strong positive supporting reflexes appear, and a few are able to

*It is apparent, therefore, that the clinical dictum that the presence of extensor spasms indicates a partial cord transection is valid only if the patient is seen within 6 months after injury.

stand for a short period of time without mechanical support (spinal standing). After development of extensor reflex activity, reflex flexion is still easily elicited by noxious plantar stimulation, so that the increase in excitability apparently affects all spinal reflex arcs.

It should be noted that the stimulus of greatest potency for strong extensor activity is a sudden, brief stretch of the flexor muscles, particularly the flexors of the thighs. It is difficult to understand this action as other than the inverse myotatic reflex, arising from flexor muscle tendon organs.

GENITAL REFLEXES. The genital zone is one from which mass flexion can be elicited by noxious stimulation at low threshold. In those patients showing any reflex activity the local genital reactions are invariably present. These are penile erection, contraction of the bulbocavernosus and sphincter ani muscles, and a delayed slow contraction of the scrotal dartos. The optimal stimulus is gentle tactile stimulation of the penile frenulum. Only rarely is penile erection accompanied by ejaculation of seminal fluid. No observations are available concerning genital reflexes in chronic spinal women.

REFLEX ACTIVITY OF BLADDER AND BOWEL. The recovery of reflex emptying of these viscera differs in no significant way from that occurring in spinal animals. However, it should be noted that reflex evacuation of the bladder, which may occasionally follow stimulation of the lower limbs, is not a part of the mass flexion reflex.[8] On the contrary, noxious stimulation most commonly produces an increase in sphincter contraction.

SUBJECTIVE SENSATIONS. Of the 29 men studied by Kuhn, 17 reported some subjective sensation referable to portions of the body below the level of the transection. Most commonly this was described as a dull, burning sensation in the buttocks, perineum, or lower abdomen. It was elicited by long-continued pressure on the ischial tuberosities, by stimulation of the anal or urethral canal, or by distention of the bladder. It seems likely that the afferents concerned travel upward over the splanchnic nerves to enter the cord over the dorsal roots rostral to the level of transection.

SUDOMOTOR ACTIVITY. Outbursts of sweating commonly occur in spinal men. In favorable cases, paroxysmal sweating can be elicited by almost any afferent inflow to the segment of cord remaining—scratching the foot, distention of the bladder, etc. Little is known of the central reflex pathways concerned.

• • •

In summary, it should be emphasized that study of men surviving cord transection for many months and years reveals that the human spinal cord contains all the inherent capacity for reflex action of which the cord of lower mammals is capable. The increasingly prolonged period of reflex recovery after cord section that is observed as one ascends from less to more complex nervous systems is paralleled by the increasing dominance of spinal mechanisms by those resident at suprasegmental levels of the brain.

REFERENCES

1. Bernhard, C. G., Bohm, E., and Petersen, I.: Investigations on the organization of the corticospinal system in monkeys, Acta Physiol. Scand. **29**(suppl. 106):79, 1953.
2. Beswick, F. B., Blockey, N. J., and Evanson, J. M.: Some effects of the stimulation of articular nerves, J. Physiol. **128:**83P, 1955.
3. Boyd, I. A., and Roberts, T. D. M.: Proprioceptive discharges from stretch-receptors in the knee-joint of the cat, J. Physiol. **122:**38, 1953.
4. Cannon, W. B., and Haimovici, H.: The sensitization of motoneurones by partial "denervation," Am. J. Physiol. **126:**731, 1939.
5. Cannon, W. B., and Rosenblueth, A.: The supersensitivity of denervated structures, New York, 1949, The Macmillan Co.
6. Creed, R. S., et al.: Reflex activity of the spinal cord, London, 1932, Oxford University Press, Inc.
7. Denny-Brown, D. E., and Liddell, E. G. T.: Extensor reflexes in the forelimb, J. Physiol. **65:**305, 1928.
8. Dusser de Barenne, J. G., and Koskoff, Y. D.: Further observations on flexor rigidity in the hindlegs of the spinal cat, Am. J. Physiol. **107:**441, 1934.
9. Eccles, J. C.: The physiology of nerve cells, Baltimore, 1957, The Johns Hopkins Press.
10. Eccles, J. C.: The physiology of nerve cells, New York, 1964, Academic Press, Inc.
11. Eccles, J. C., Eccles, R. M., and Magni, F.: Monosynaptic excitatory action on motoneurones regenerated to antagonistic muscles, J. Physiol. **154:**68, 1960.
12. Eccles, J. C., Fatt, P., and Landgren, S.: Central pathways for direct inhibitory action of impulses in largest afferent nerve fibres to muscle, J. Neurophysiol. **19:**75, 1956.
13. Eccles, J. C., Eccles, R. M., Iggo, A., and Lundberg, A.: Electrophysiological investigations on Renshaw cells, J. Physiol. **159:**461, 1961.
14. Freeman, L. W.: Treatment of paraplegia resulting from trauma to spinal cord, J.A.M.A. **140:**949, 1015, 1949; **141:**275, 1949.
15. Fulton, J. F., Liddell, E. G. T., and Rioch, D. M.: The influence of experimental lesions of the spinal cord upon the knee-jerk. I. Acute lesions, Brain **53:**311, 1930.
16. Gardner, E.: Reflex muscular responses to stimulation of articular nerves in the cat, Am. J. Physiol. **161:**133, 1950.
17. Gelfan, S.: Neurone and synapse populations in the spinal cord: indication of role in total integration, Nature **198:**162, 1963.
18. Gelfan, S., and Tarlov, I. M.: Altered neuron population in L_7 segment of dogs with experimental hind-limb rigidity, Am. J. Physiol. **205:**606, 1963.
19. Head, H., and Riddoch, G.: The automatic bladder,

excessive sweating and some other reflex conditions, in gross injuries of the spinal cord, Brain **40:**188, 1917.
20. Henneman, E.: Excitability changes in monosynaptic reflex pathways of muscles subjected to static stretch, Trans. Am. Neurol. Assoc. **76:**194, 1951.
21. Hoff, E. C., and Hoff, H. E.: Spinal terminations of the projection fibers from the motor cortex of primates, Brain **57:**454, 1934.
22. Hunt, C. C., and Perl, E. R.: Spinal reflex mechanisms concerned with skeletal muscles, Physiol. Rev. **40:**538, 1960.
23. Kocher, T.: Die Verletzungen der Virbersäule zugleich als Beitrag zur Physiologie des menschlichen Rüchenmarcks, Mit. Grenzgeb. Med. Chir. **1:**415, 1896.
24. Kuhn, R. A.: Functional capacity of the isolated human spinal cord, Brain **73:**1, 1950.
25. Kuhn, R. A., and Macht, M. B.: Some manifestations of reflex activity in spinal man with particular reference to the occurrence of extensor spasm, Bull. Johns Hopkins Hosp. **84:**43, 1949.
26. Kuhn, W. G., Jr.: The care and rehabilitation of patients with injuries of the spinal cord and cauda equina, J. Neurosurg. **4:**40, 1947.
27. Kuno, M., and Perl, E. R.: Alteration of spinal reflexes by interaction with suprasegmental and dorsal root activity, J. Physiol. **149:**374, 1959.
28. Laporte, Y., and Lloyd, D. P. C.: Nature and significance of the reflex connections established by large afferent fibers of muscular origin, Am. J. Physiol. **169:** 609, 1952.
29. Liddell, E. G. T.: Spinal shock and some features in isolation-alteration of the spinal cord in cats, Brain **57:**386, 1934.
30. Liddell, E. G. T., and Sherrington, C. S.: Reflexes in response to stretch (myotatic reflexes), Proc. R. Soc. Lond. (Biol.) **96:**212, 1924.
31. Liddell, E. G. T., and Sherrington, C. S.: Reflexes in response to stretch (myotatic reflexes), Proc. R. Socl. Lond. (Biol.) **97:**267, 1925.
32. Lloyd, D. P. C.: Activity in neurons of bulbospinal correlation systems, J. Neurophysiol. **4:**115, 1941.
33. Lloyd, D. P. C.: Mediation of descending long spinal reflex activity, J. Neurophysiol. **5:**435, 1942.
34. Lloyd, D. P. C.: Conduction and synaptic transmission in the reflex response to stretch in spinal cats, J. Neurophysiol. **6:**317, 1943.
35. Lloyd, D. P. C.: Reflex action in relation to pattern and peripheral source of afferent stimulation, J. Neurophysiol. **6:**111, 1943.
36. Lloyd, D. P. C.: The spinal mechanisms of the pyramidal system in cats, J. Neurophysiol. **4:**525, 1941.
37. Lloyd, D. P. C.: Facilitation and inhibition of spinal motoneurons, J. Neurophysiol. **9:**421, 1946.
38. Lloyd, D. P. C.: Integrative pattern of excitation and inhibition in two-neuron reflex arcs, J. Neurophysiol. **9:** 439, 1946.
39. Lloyd, D. P. C.: On reflex actions of muscular origin, Res. Publ. Res. Assoc. Nerv. Ment. Dis. **30:**48, 1950.
40. Lloyd, D. P. C., and McIntyre, A. K.: Analysis of forelimb-hindlimb reflex activity in acutely decapitate cats, J. Neurophysiol. **11:**455, 1948.
41. Lorente de Nó, R.: Analysis of chains of internuncial neurons, J. Neurophysiol. **1:**207, 1938.
42. Macht, M. B., and Kuhn, R. A.: Responses to thermal stimuli mediated through the isolated spinal cord, Arch. Neurol. Psychiatry **59:**754, 1948.
43. Matthews, P. B. C.: Muscle spindles and their motor control, Physiol. Rev. **44:**219, 1964.
44. Matthews, P. B. C.: Evidence that the secondary as well as the primary endings of the muscle spindles may be responsible for the tonic stretch reflex of the decerebrate cat, J. Physiol. **204:**365-393, 1969.
44a. Munro, D.: The rehabilitation of patients totally paralyzed below the waist, with special reference to making them ambulatory and capable of earning their own living. II. Control of urination, New Engl. J. Med. **234:** 207, 1946.
45. Ramón y Cajal, S.: Histologie du système nerveux de l'homme et des vertébrés, Paris, 1909, Librairie Maloine.
46. Renshaw, B.: Central effects of centripetal impulses in axons of spinal ventral roots, J. Neurophysiol. **9:**191, 1946.
47. Riddoch, G.: The reflex functions of the completely divided spinal cord in man compared with those associated with less severe lesions, Brain **40:**264, 1917.
48. Sahs, A. L., and Fulton, J. F.: Somatic and autonomic reflexes in spinal monkeys, J. Neurophysiol. **3:**258, 1940.
49. Sherrington, C. S.: Decerebrate rigidity and reflex coordination of movements, J. Physiol. **22:**319, 1897-1898.
50. Sherrington, C. S.: The integrative action of the nervous system, New Haven, 1906, Yale University Press.
51. Sherrington, C. S., and Laslett, E. R.: Observations on some spinal reflexes and the interconnections of spinal segments, J. Physiol. **29:**58, 1903.
52. Stavraky, G. W.: The action of adrenaline on spinal neurons sensitized by partial isolation, Am. J. Physiol. **150:** 37, 1947.
53. van Wagenen, G.: Uterine bleeding of monkeys in relation to neural and vascular processes. I. Spinal transection and menstruation, Am. J. Physiol. **105:**473, 1933.
54. Woodbury, J. W., and Patton, H. D.: Electrical activity of single spinal cord elements, Symp. Quant. Biol. **17:** 185, 1952.

29

ELWOOD HENNEMAN

Motor functions of the brain stem and basal ganglia

The majority of the motor tracts that descend into the spinal cord originate in the brain stem. The areas from which these tracts arise are integrative centers receiving signals from the cerebral cortex, the cerebellum, and the basal ganglia as well as from the spinal cord. In this chapter the organization of the motor areas of the brain stem will be discussed, and the nature of their influence on spinal mechanisms will be described. An account of the capacities and abnormalities of animals in the decerebrate and decorticate states will also be given to indicate the contributions of the brain stem to standing, walking, and postural adjustments. The latter part of the chapter will be devoted to the basal ganglia, a group of nuclear masses that play a major role in motor activity and are closely related to the brain stem centers. In this section, recent findings on the role of some of the biogenic amines will be summarized to illustrate the importance of central transmitters in motor function.

Reticular formation of the brain stem

The brain stem contains a central core of nerve cells and fibers running more or less continuously through the mesencephalon, pons, and medulla, as indicated in Fig. 29-1. Its upper end extends into the diencephalon, and its lower end extends into the spinal cord. In Weigert-stained sections, it shows up as a meshwork or reticulum of interlacing fiber bundles. Lying between and among the fibers are cells of different types and sizes. In some areas the cells are gathered together more densely as nuclei that have names. This loose aggregation of cells and fibers is called the reticular formation (RF). It lies in the central part of the brain stem. Circumscribed cell groups such as the red nucleus, the superior olive, and the cranial nerve nuclei are not ordinarily included within it.

Although the RF is not to be regarded as a simple functional entity, physiologic studies indicate that it exerts powerful facilitory and inhibitory influences on all types of motor activities at all levels of the nervous system. The medial two thirds contains many large cells and gives rise to long ascending and descending fibers. The effector functions of the RF are mediated chiefly by this medial two thirds, whereas the lateral one third evidently acts as an association area for the medial portion, influencing it through short, medially directed axons. Within the medial portion there is some segregation of cells into groups; some exert their main action on the spinal cord and others act chiefly on more rostral portions of the brain, as illustrated in Fig. 29-2.

As Scheibel and Scheibel[41] have shown with Golgi stains (Fig. 29-3), a considerable number of reticular cells give off an axon that has a long ascending as well as a long descending branch. The latter was often traced down into the spinal cord and the former up into the thalamus. Thus one cell, by means of collaterals at many levels, may influence a vast number of cells above and below it in the neuraxis. All morphologic evidence, in fact, suggests that there must be a significant correlation between the effects exerted by the RF in the rostral and caudal directions.

Golgi-stained sections of the brain stem in the parasagittal plane (Fig. 29-4) reveal that the dendrites of many reticular cells are oriented in a plane perpendicular to the long axis of the brain stem and that the collaterals of fibers that send impulses to them are flattened in the same plane (inset, Fig. 29-4).

These features suggest an arrangement for the

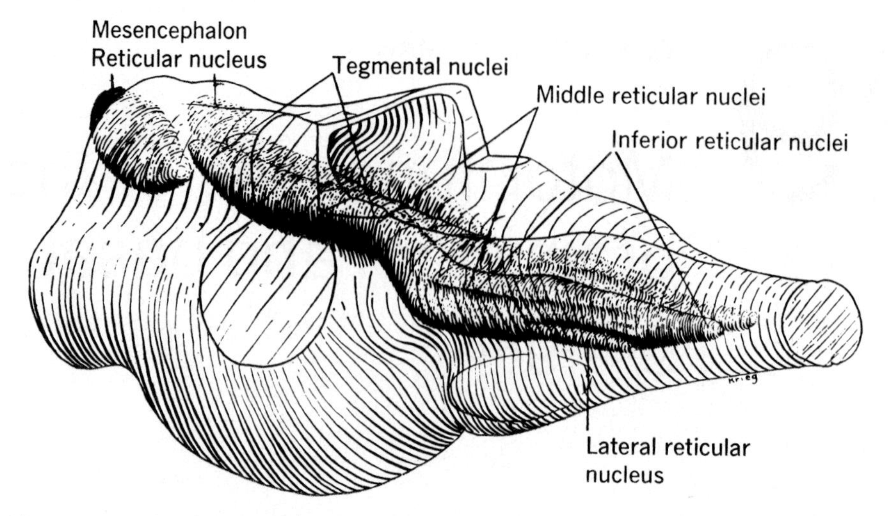

Fig. 29-1. Three-dimensional sketch illustrating position of reticular formation in brain stem. (From Krieg, W. J. S.: Functional neuroanatomy, Philadelphia, copyright © 1942 by McGraw-Hill Book Co. Used with permission of McGraw-Hill Book Co.)

collection and distribution of signals from diverse regions of the brain (i.e., an integrating mechanism). Afferent connections come from many sources. Among the most important are those from the spinal cord. Contrary to early reports, a relatively small portion of the spinal activation of the RF comes from collaterals of secondary sensory fibers. There is a massive influx of direct spinoreticular fibers that ascend in the ventrolateral funiculus; the majority are distributed to the medial two thirds of the RF. A second major afferent influx comes from the cerebellum, particularly the fastigial nucleus. These fibers are distributed chiefly to the medullary portion of the RF and supply both rostrally and caudally projecting areas. In addition, there are afferents from the *lateral hypothalamus* and the *pallidum* to the mesencephalic RF, from the *tectum* to the same regions, and from the sensorimotor areas of the cerebral cortex to several levels of the brain stem that give off reticulospinal fibers.

Reticulospinal tracts are derived from small and large cells at all levels of the medullary and pontine RF. Two maximal areas of origin can be distinguished, one from the pons and the other from the medulla (Fig. 29-2). Most of the medullary fibers come from the *nucleus reticularis gigantocellularis* and are crossed as well as uncrossed. They run bilaterally in the lateral funiculus of the cord. The pontine fibers, which are uncrossed, come from the entire *nucleus reticularis pontis caudalis* and the caudal part of the *nucleus reticularis pontis oralis*. They descend in the ventral funiculus of the spinal cord. As shown in

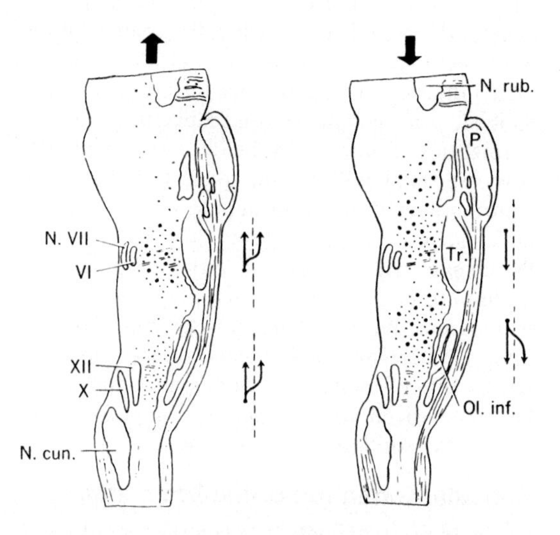

Fig. 29-2. Location of cells in reticular formation (RF) sending long axons to spinal cord (right); location of cells with axons ascending beyond mesencephalon (left). Drawn from parasagittal sections of cat brain stem. Large dots represent giant cells. Note that ascending and descending fibers take origin from almost the entire longitudinal extent of RF, but that they differ somewhat in their regions of maximal origin. Arrows at right of each drawing indicate that pontine reticulospinal fibers descend homolaterally, whereas all other fibers are crossed as well as uncrossed. (From Brodal.[9])

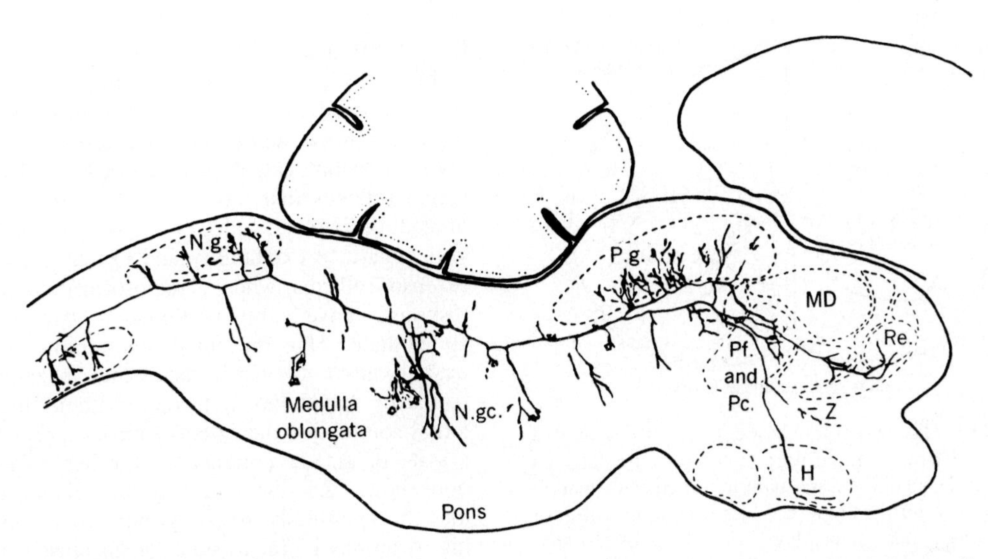

Fig. 29-3. Drawing of Golgi-stained sagittal section of brain stem of 2-day-old rat. Single large cell in reticular formation (RF) is shown. Its axon divides into ascending and descending branches. Latter gives off collaterals to adjacent parts of RF; to nucleus gracilis, *N.gc.;* and to ventral horn in spinal cord. Ascending branch gives off collaterals to RF; periaqueductal gray, *P.g.;* several thalamic nuclei; hypothalamus, *H;* and zona incerta, *Z.* Other areas identified in drawing are nucleus reuniens, *Re.*, nucleus medialis dorsalis, *MD*, nucleus parafascicularis, *Pf.*, and nucleus paracentralis, *Pc.* (From Scheibel and Scheibel.[41])

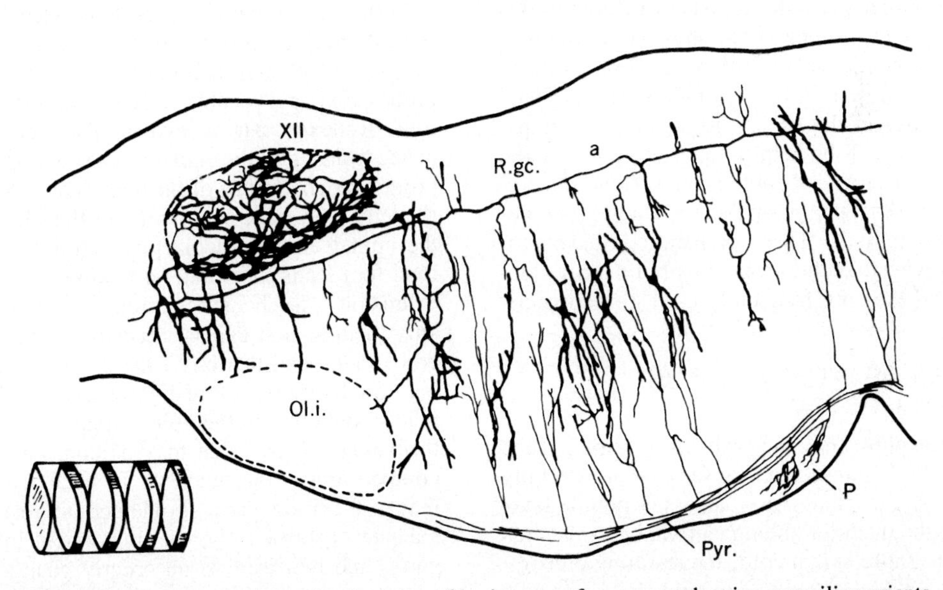

Fig. 29-4. Drawing of Golgi-stained section of brain stem of young rat showing prevailing orientation of dendrites of reticular cells in a plane perpendicular to long axis of brain stem. Collaterals of fibers that send impulses to reticular cells are flattened in same plane. *Pyr.*, Pyramidal tract; *XII*, hypoglossal nucleus; *Ol.i.*, inferior olive; *P*, pons; *R.gc.*, nucleus reticularis gigantocellularis; *a*, long axon of a reticular cell. Inset at left shows how reticular formation may be represented as series of neuropil segments. (From Scheibel and Scheibel.[41])

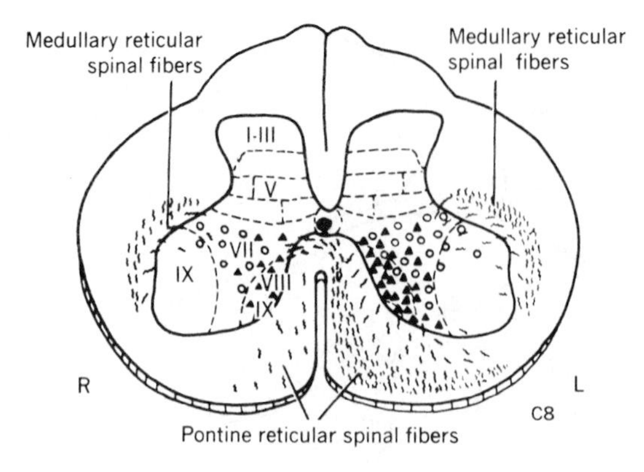

Fig. 29-5. Diagram of transverse section of spinal cord of cat at C8 showing position in cord of reticulospinal fibers from medullary and pontine reticular formation and sites of termination. ▲ = Sites of termination of pontine reticular spinal fibers. ○ = Sites of termination of medullary reticular spinal fibers. (From Nyberg-Hansen.[36])

Fig. 29-5, the pontine reticulospinal fibers terminate more ventrally than the medullary fibers. The former end in laminae VII and VIII, whereas the latter end chiefly in lamina VII. The reticulospinal tracts may be regarded as the final common pathways for a variety of extrapyramidal centers at higher levels. The signals from these centers converge in the RF, and the resultant discharge is directed to the spinal cord. Later in this chapter, some recent electrophysiologic findings on the synaptic actions of reticulospinal fibers will be described. Before taking up the functional organization of the RF and its spinal projections, however, it will be helpful to describe briefly the motor capacities of bulbospinal (low decerebrate), mesencephalic (high decerebrate), and decorticate animals. These preparations have provided invaluable clues regarding brain stem function, and they are frequently used in the experimental analysis of motor systems.

LEVELS OF INTEGRATION IN THE BRAIN STEM*

Bulbospinal (low decerebrate) animal. If the brain stem is transected near the caudal extremity of the mesencephalon or at an upper pontile level so that the medulla oblongata remains connected with the entire spinal cord, the resulting preparation is capable of more complex behavior than is the spinal animal. In contrast to spinal transection, this truncation of the lower brain stem is not followed by any depression of simple spinal reflexes. Immediately after separation of the bulbar region from the suprabulbar structures, one may elicit in rabbit, cat, dog, and monkey all those spinal reflexes that have been described in the preceding chapters. Indeed there is enhancement, with regard to excitability and execution, of all extensor reflexes, whereas reciprocally the flexor responses have a higher threshold than in the spinal state. Thus the bulbospinal animal vigorously exhibits the very reflexes that are most affected by spinal shock. In fact, without any external stimulation the extensor muscles remain in a state of steady contraction. Further examination shows that the entire system of musculature that posturally resists gravity in the standing position is in a state of increased tonic activity. This is spoken of as *decerebrate rigidity*.

In the acute bulbospinal preparation the abnormal distribution of tone in the muscles of the body continues regardless of the position of the animal. If placed upright, it remains standing, provided it is balanced, but a normal relation to the force of gravity is not at all necessary for the continuance of the rigidity. The extension of the legs and the retraction of the head persist, even becoming stronger, when the preparation is placed on its back.

Further examination of the bulbospinal animal reveals the presence of several capacities for response that are either absent from the acute spinal preparation or evocable only on strong stimulation. Reflexes can be evoked that activate the neck, trunk, and the four limbs as a whole, thus producing an attitude of the body that is definitely related to the place and quality of the stimulus. In the spinal animal, nociceptive stimulation of a hind foot leads to ipsilateral flexion with crossed extension. Such stimulation, if sufficiently strong, may also evoke extension of the ipsilateral foreleg and flexion of the contralateral foreleg. Such groups of responses are much more readily obtained in the bulbospinal animal. They form *reflex figures* that may be considered to be compensatory, inasmuch as in each the movement of certain parts is adapted to restore the balance disturbed by the movement of other parts, and the result is an orderly change in the position of the body as a whole, which is significant in meeting the exigency that has given rise to the reflex response. The reflex figure shown in Fig. 29-6, *D*, might be brought into play, for example, when a cat steps on some object that hurts its hind foot. The foot is lifted from

*This section, which was prepared by Philip Bard as part of Chapter 76 in the twelfth edition of this book, is included here with only slight modifications.

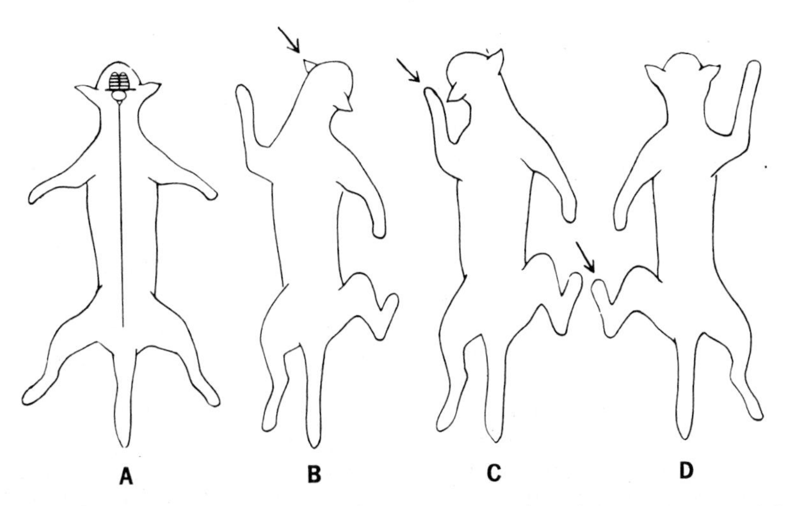

Fig. 29-6. Reflex figures. **A,** Position of cat in decerebrate rigidity. **B, C,** and **D,** respectively, are reflex figures resulting from stimulating left pinna, left forefoot, and left hind foot. (After Sherrington.[43])

the ground and the weight of the hindquarters is thus thrown on the contralateral hind leg, which extends to support this weight. At the same time the cat must prepare to move away so that the stimulus will not be encountered again, and for this act the extension of the crossed hind leg and of the ipsilateral forefoot, with their backward thrust, tend to throw the body forward and support it while the flexion of the stimulated hind leg and the crossed foreleg move these limbs forward preparatory to supporting the body at the next step. The reflex figure is thus seen to be an integral mechanism out of which is built the functional act of stepping away from a stimulus that endangers the hind foot.

Despite an ability to support its weight and to prepare attitudinally for progression, the bulbospinal cat or dog cannot fully right itself, assume a sitting or standing position, run, walk, or jump. That these deficits are due to the fact that the neural mechanisms essential for these more complex acts are situated rostrally to the bulbar region and not to some kind of "shock" has been shown by experiments in which bulbospinal animals have been kept alive for considerable lengths of time.

Bazett and Penfield[7] were the first to maintain decerebrate animals for more than the period of the usual "acute" experiment. Seven of their bulbospinal cats lived more than 1 week, and of these, two survived slightly longer than 2 weeks. In every case decerebrate rigidity persisted, although it was subject to some variations. No effective righting occurred and, of course, locomotion was not possible. Bard and Macht[5] studied a number of bulbospinal ("pontile") cats over somewhat longer survival periods (up to 40 days). None ever spontaneously righted head or body but, when placed on either side, lay with the under limbs rigidly extended and the upper ones either flaccid or slightly flexed. This posture or attitude is the first and most fundamental overt manifestation of the "body righting reaction acting on the body." On strong stimulation (e.g., of tail) the upper forelimb is flexed strongly at the elbow and retracted at the shoulder; the lower forelimb is extended and then quickly flexed so as to bring it under the chest at the same time the head is raised and rotated. Thus the under paw acts as a fulcrum while the upper limb is flexed with claws sunk in the underlying surface. In this way the shoulders and chest are pulled into an upright position. In chronic mesencephalic cats, but never in the bulbospinal preparations, the same series of acts occurs in the hind legs and leads to the assumption of a crouching or standing position. Further, in mesencephalic animals the righting occurs spontaneously.

Mesencephalic (high decerebrate) animal. When the brain stem is transected at a mesencephalic level, the preparation can scarcely be distinguished from the bulbospinal animal unless it is kept alive and in good condition for at least a week to 10 days. As indicated previously, Bard and Macht observed that in the cat, effective righting of the body and typical quadrupedal walking depend on mesencephalic mechanisms. Adequate but not wholly normal standing and locomotion are possible after a truncation that excludes all brain tissue lying rostral to the caudal extremities of the red nuclei. The capacity to right, stand, and walk develops slowly and is not fully gained until at least 10 days after decerebration. In the low mesencephalic animal, these acts rarely, if ever, occur spontaneously. Spontaneous standing and walking are exhibited by cats with brain stems truncated just above the level of exit of the third pair of cranial nerves. Animals with transections passing just caudal to the mamillary bodies exhibit locomotor abilities of the same general quality as those exhibited by decorticate cats. They are even able to run and to climb. Such defects in standing and walking as they exhibit are due to the absence of placing reactions and to the high threshold of proprioceptive corrective reactions of the legs (only imperfect hopping and proprioceptive placing reactions are present).

After decerebration the hypertonia of the antigravity muscles (decerebrate rigidity) remains in evidence throughout the survival of the animal, but it is seen only when the chronic preparation is not engaged in some phasic activity or crouching. At any time when the animal is placed on its back (and fails to execute phasic movements of the legs), it is found that the legs are rigidly extended and resist passive flexion as strongly as in the case of the freshly decerebrated animal. High mesencephalic cats in the chronic state rarely lie on their sides, and if not standing or walking, they assume a crouching position on chest and belly with all limbs flexed. If undisturbed in this attitude, they may close the eyes and allow the head to droop; indeed, they appear to sleep, and on being subjected to a gentle tap or a slight sound, they show all the overt signs of waking.

Certain postural responses of mesencephalic cats in the chronic state appear to be executed in a manner that is wholly or nearly normal. Tactile stimulation of the anal region results in the assumption of a defecatory posture; there are dorsiflexion of the tail, flexion and protraction of the hind limbs, extension and retraction of the forelimbs, and ventriflexion of the pelvis. This reflex posture is evocable in bulbospinal as well as in mesencephalic animals, but in the former, it is achieved in the lateral position without righting. If the truncation is at or above the level of exit of the third nerves, the cat invariably assumes the normal feline position before defecating. After administration of estrogenic material, high mesencephalic female cats will spontaneously assume the posture characteristic of feline estrus (crouching on chest and forearms with pelvis and tail elevated), and this attitude together with treading movements of the hind legs, typical of estrual behavior, can be induced by genital stimulation. But other more elaborate items of estrual behavior cannot be induced in these animals.

Decorticate animal. When the ablation of cerebral tissue is limited to the most recently acquired part of the cerebrum, its mantle or cortex, the deficiencies are less than those produced by mesencephalic truncation of the brain. In both cases, modes of behavior dependent on past experience are lost; all that has been learned disappears and new types of responses can be induced only by special methods of training. Conditioned reflexes can be established in decorticate cats and dogs, but the process requires special methods. There is no evidence available at present to indicate that wholly decorticate carnivores ever develop learned responses under laboratory conditions in which the normal animal readily acquires them. This combination of a loss of what are ordinarily regarded as signs of higher nervous functions with retention of many complex activities was first established experimentally in 1892 by Goltz, who succeeded in keeping a dog from which he had removed all cerebral cortex alive for many months. Goltz's achievement has since been repeated, and descriptions of a number of decorticate cats and dogs are available (for references see Bard and Rioch[6]). Because it is able to regulate its body temperature in an almost normal fashion, the decorticate mammal is much easier to maintain over long postoperative periods than is the decerebrate preparation.

Following complete decortication a cat or dog can at once right itself, stand, and walk. In view of the fact that the high mesencephalic animal does not engage in these acts until a postoperative period of from 7 to 30 days has elapsed, it seems likely that subcortical influences originating rostral to the midbrain facilitate the mesencephalic and bulbar mechanisms essential for these acts. The chief postural defects of decorticate animals are due to the loss of placing reactions and to defective hopping reactions, responses that are described later in this chapter. When animals without cortex are held suspended, they exhibit, as long as they are quiet, a marked extensor rigidity, and there can be no doubt that this *decorticate rigidity* is closely allied to *decerebrate rigidity*. Indeed, as we shall see, decerebrate rigidity is in part due to the exclusion of influences of cerebral cortical origin. The contraction of antigravity muscles observed in the acutely decerebrate animal is maintained only because the preparation is incapable of the phasic activities that the decorticate animal shows from the beginning and that emerge in the chronically decerebrate animal after some time.

DECEREBRATE RIGIDITY AND SPASTICITY

The contraction of the antigravity muscles in decerebrate rigidity is reflex in origin, for it can be abolished by section of appropriate nerves or elimination of certain reflex stimuli. In the case of the hind legs the extensor hypertonia may be reduced by section of the dorsal roots that carry the afferent impulses from the extensor muscles themselves. It may also be abolished by eliminating any pull on them. Section of nerves from other muscles in the leg or from the skin does not affect the rigidity at all.[43] These facts indicate that the myotatic reflexes of the extensors constitute an important factor in the production of their rigidity. It is not, however, the only factor, for it has been found that some time after their deafferentation the hind legs of a cat develop typical extensor hypertonia when the animal is decerebrated.[37] Under such circumstances the rigidity has its reflex origin in the neck and labyrinthine proprioceptors (see discussion of tonic neck and labyrinthine reflexes, p. 801). The rigidity of the forelegs is not reduced by deafferentation, and it is clear the proprioceptors of ear and neck constitute its essential source.[37] In any case, it is definite that after decerebration the myotatic reflexes in extensor muscles have extremely low thresholds.

Spasticity is a term that has had its widest use in the language and literature of clinical neurology. It is employed to describe a combination of symptoms—*muscle hypertonia* (increased resistance to lengthening), *hyperactive tendon reflexes,* and *clonus* (repetitive contractions in response to a suddenly applied but sustained stretch of a muscle). Spasticity is well and classically exemplified by the state of the arm and leg of a patient who has become hemiplegic as the result of severe injury or destruction of a certain portion of the frontal lobe of the cerebral hemisphere opposite the affected extremities or, more commonly, as the result of interruption of the extrapyramidal motor projection of that lobe at the level of the internal capsule. The leg is stiffly extended and resists passive flexion; the arm is held flexed and resists passive extension. Tendon reflexes are hyperactive, and clonus is evoked if the stretch is prolonged. Again it can be said that the hypertonia is chiefly encountered in antigravity muscles, for in man the flexors of the arm are the muscles that resist the force of gravity. This state of hypertonia in the arm flexors and leg extensors can be regarded as *decorticate rigidity* in man, for it is induced by a lesion that interrupts descending impulses of cortical origin, but it is to be distinguished from the decorticate rigidity of carnivores, in which the extensor hypertonia is seen in the upper as well as in the lower limbs. Actual decerebration in man, produced, for example, by a massive hemorrhage at a mesencephalic level, is followed by extreme extensor rigidity of all four extremities.

Central nervous mechanisms involved in production of spasticity and decerebrate rigidity

In his original description of decerebrate rigidity, Sherrington stated that it followed transection of the brain stem through the caudal part of the diencephalon or at any mesencephalic level. It is common practice to produce it by sectioning the brain stem at an intercollicular level. The typical rigidity then appears on removal of the anesthetic. It survives successive transections until these reach the level of entry of the eighth pair of cranial nerves. Cutting through the bulb in this region or at any level below it produces the postural and reflex condition characteristic of the spinal state. Sherrington clearly recognized that decerebrate rigidity is due to the interruption of descending pathways that originate from parts of the brain situated rostral to the mesencephalon. The previously noted fact that the rigidity persists for periods of time more than sufficient to ensure degeneration of all descending fibers cut by the transection demonstrates conclusively that the hypertonia is not due to any "irritation" set up by the mechanical insult of truncating the brain stem.

Until quite recently, it was generally accepted that decerebrate rigidity and the spasticity characteristic of hemiplegia are release phenomena in the jacksonian sense, that is, they are conditions that result simply from removal of inhibitory influences of suprabulbar origin. Already sufficient evidence has been given to show that the greater portion of the inhibitory influence has its origin above the mesencephalon. Experimental analyses, particularly those carried out by Magoun and collaborators (see especially Magoun,[30] Magoun and Rhines,[32] and Lindsley[25]) have demonstrated that *decerebrate rigidity and spasticity are due not only to the removal of inhibitory influences acting on the spinal stretch reflex but also to the maintained activity of supraspinal influences that facilitate these reflexes.* These important discoveries have implications that transcend the subject under immediate discussion, for the myotatic reflex is by no means the only spinal response subject to these opposing influences.

Vestibulospinal facilitatory influence. It has long been recognized that cutting through the medulla oblongata just below the level of the vestibular nuclei destroys the rigidity of the decerebrate cat and changes the status of the preparation to that of the spinal animal. This and other evidence suggested that in the carnivore, decerebrate rigidity depends on the integrity of one or more of the vestibular nuclei, more especially Deiter's nucleus, and that it is due to impulses descending by way of the uncrossed vestibulospinal tracts. This suggestion received strong support when Fulton et al.[20] showed that if a localized lesion is made in the region of Deiter's nucleus so that degeneration of the corresponding vestibulospinal tract occurs, subsequent decerebration fails to produce rigidity on the side of the injury and degeneration, although hyperextension and increased tendon reflexes appear in the legs of the side with intact vestibulospinal tracts.

The role of the vestibulospinal system has been further elucidated by Bach and Magoun.[1] They confirmed the observation that vestibular nuclear lesions abolish rigidity on the injured side and that this effect is chiefly if not wholly due to destruction of Deiter's nucleus. Significantly, they found that the result does not depend on any inadvertent injury of the adjacent caudal portion of the brain stem RF (Fig. 29-7), from which facilitation of spinal reflexes (including the myotatic reflexes) can so readily be obtained.

As indicated previously, an extensor hypertonia, comparable to that of decerebrate rigidity together with overactive myotatic reflexes and clonus (in short, spasticity), is produced by decortication. Ablation of the pericruciate cortex in cat or dog suffices to induce this condition in the contralateral limbs.[3,54] Destruction of the vestibular nuclei has much less effect on this condition than on decerebrate rigidity; the extensor rigidity is reduced, but the hyperactive tendon reflexes and clonus are essentially unaffected.[1] If in a carnivore pericruciate decortication is combined with removal of the anterior lobe of the cerebellum (Fig. 29-7), there ensues an antigravity posture even more pronounced than that which follows decerebration (cerebellum intact), and stretch reflexes show extremely low thresholds[47]; the animal is rendered immobile by the continued, excessive, and unbalanced activity of the central antigravity mechanism. In such preparations, Schreiner et al.[42] found that lesions limited to Deiter's nucleus were without appreciable effect; but when such lesions were superimposed on a transection of the tegmentum, spasticity was virtually abolished, the stretch reflexes being more depressed than after tegmental injury alone. As indicated before and as will be explained more fully later, there is in addition to the vestibulospinal system another lower brain stem facilitatory system, the reticulospinal. In the low decerebrate animal, destruction of Deiter's nucleus abolishes decerebrate rigidity because, as originally suggested by Bach and Magoun,[1] the truncation of the brain stem has already largely removed the source of the reticulospinal facilita-

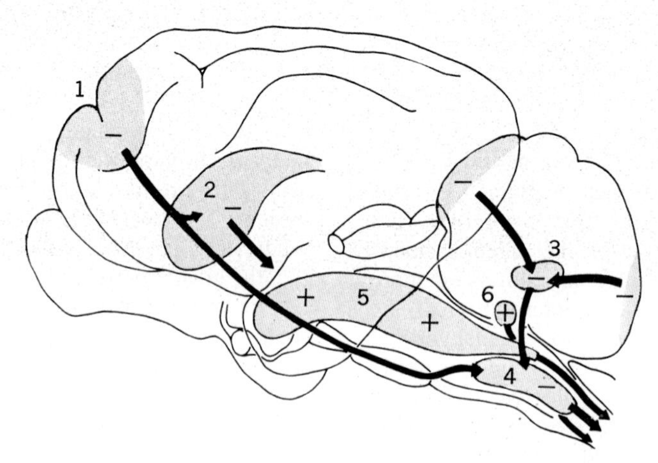

Fig. 29-7. Reconstruction of cat brain showing suppressor or inhibitory and facilitatory systems concerned in spasticity. Suppressor pathways are, *1*, corticobulboreticular, originating in pericruciate cortex; *2*, caudatospinal (probably caudatoreticulospinal); *3*, cerebelloreticular (originating in anterior lobe and paramedian lobules and relaying in nucleus fastigii); and *4*, reticulospinal. Facilitatory pathways are, *5*, reticulospinal and, *6*, vestibulospinal. (From Lindsley et al.[26])

tion. On the other hand, the spasticity of the decorticate animal or that of the animal with pericruciate areas and anterior cerebellar lobe removed is but little affected by unilateral or bilateral destruction of Deiter's nucleus, for in these cases the extensive brain stem facilitatory mechanism is still wholly intact and discharges to the cord over reticulospinal paths.

Reticulospinal facilitatory influence. In 1946 Rhines and Magoun[39] reported the very significant finding that, in cat and monkey, stimulation anywhere within a large brain stem area (approximately the same as that marked 5 in Fig. 29-7) augments the response to stimulation of the cerebral motor cortex and also markedly facilitates the knee jerk. It was made clear that these effects are produced by impulses that descend to spinal neurons from the point of stimulation and that their final common spinal path is made up of reticulospinal fibers that course chiefly in the middle portion of the lateral funiculus. The facilitatory area comprises elements of the dorsal diencephalon, hypothalamus, subthalamus, central gray matter and tegmentum of the mesencephalon, pontile tegmentum, and a large part of the bulbar RF.[25,30,32] Evidence has been obtained that the facilitatory effects can be evoked by elements at all levels of the area and that they are exerted bilaterally, the crossing taking place in the brain stem as well as in the cord.[35]

The disclosure by Rhines and Magoun that facilitation of a myotatic reflex (the knee jerk) can be elicited by stimulation of the brain stem reticulospinal facilitatory system indicated that this system may be of importance in the production of decerebrate rigidity and spasticity. As already related, Bach and Magoun[1] have shown that after destruction or interruption of the vestibulospinal system, cortically or reflexly induced movements are still markedly facilitated by stimulation of basal diencephalon, brain stem tegmentum, or RF. They also found that stimulation within the rostral portion of this extensive facilitatory area is rendered ineffective by destruction of its caudal extremity. Subsequently, Sprague et al.[48] demonstrated in decerebrate cats that stimulation of the portion of the facilitatory RF that remained in the truncated brain stem increased the already exaggerated stretch reflexes both ipsilaterally and contralaterally. This effect is well illustrated in Fig. 29-8, which shows that the experimental excitation of this reticulospinal mechanism augments and prolongs the stretch reflexes of various muscles. Also, what may be termed the intrinsic maintenance of the facilitatory process is prolonged (Fig. 29-8, *F*). The facilitation of the stretch reflex of a flexor muscle shown in Fig. 29-8, *B*, is greater than was usually found in flexors. The general application of this reticulospinal effect is indicated by the facilitation of the crossed extension reflex (Fig. 29-8, *E*). In this context, it will be recalled that movements induced by stimulation of the motor cortex or of the pyramidal tract after decortication are similarly facilitated, the effect being produced at a spinal level.[39]

The influence of this brain stem facilitatory system on extensor tone has been especially well demonstrated in experiments on cats rendered chronically spastic by ablation of the pericruciate region of the cortex. The effect has been described by Magoun and Rhines as follows:

> The animal's symptoms of spasticity were markedly reduced by the anesthesia, under which such experiments are performed, and its knee jerk, evoked recurrently, resembled a normal animal's under the same conditions. During stimulation of the brain stem facilitatory mechanism, however, this reflex not only gained in force and amplitude, as occurs in a normal animal, but a background of hypertonus appeared in the quadriceps and increased with each succeeding reflex, and a repetitive component or clonus developed and became so perseverative and extreme that the last reflex could not be detected above it.*

Here the spastic state, which had been eliminated by anesthesia, was not only restored but actually augmented by direct stimulation of the brain stem facilitatory system. Because inhibitory influences normally arising from the pericruciate area had been removed, the effect of the stimulation was greater than can be obtained under the same conditions in the normal animal. The central inhibitory influences that oppose the spastic or decerebrate state are several; these important mechanisms are considered in later discussions.

The experimental evidence just presented makes it very clear indeed that decerebrate rigidity and spasticity depend on the facilitation of spinal stretch reflexes by impulses that descend from a brain stem facilitatory mechanism. The fact that this continues to discharge powerfully after removal not only of forebrain but also of cerebellum raises the question of how this activity is maintained.

Cerebral, cortical, and cerebellar facilitory influences. There is abundant evidence that facilitation arises from the cerebellum and acts via the

*From Magoun, H. W., and Rhines, R.: Spasticity: the stretch reflex and extrapyramidal systems, Springfield, Ill., 1947, Charles C Thomas, Publisher.

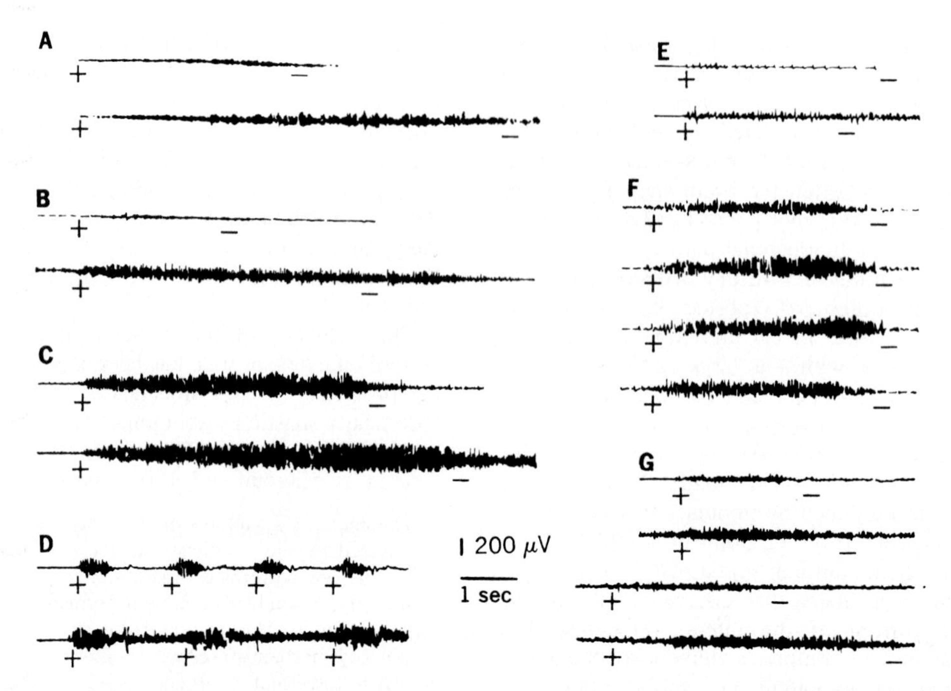

Fig. 29-8. Electromyographic evidence of reticulospinal facilitation of stretch reflexes in decerebrate cat. In such a preparation, stretch reflexes are already exaggerated. In each pair of records, **A** to **E,** upper was obtained before and lower was obtained during stimulation of facilitatory brain stem reticular formation (RF). **A** to **C,** Records of responses of quadriceps **(A),** hamstrings **(B),** and gastrocnemius **(C)** evoked by stretch (indicated between + and −). **D,** Record of gastrocnemius contractions in positive supporting reaction, upward pressure on footpad being exerted at each +. **E,** Record of gastrocnemius responses in crossed extensor reflex, with contralateral toes pinched between + and −. **F** (from above down), Gastrocnemius stretch reflex before, during, 50 sec after, and 100 sec after stimulation of facilitatory RF. **G** (from above down), Contralateral quadriceps stretch reflex before and during stimulation and ipsilateral quadriceps stretch reflex before and during stimulation of the facilitatory RF on one side of brain stem. (From Sprague et al.[48])

reticulospinal facilitory mechanism. This is doubtless more prominent in the primates than in the carnivores, but in all instances, it accompanies an inhibitory cerebellar influence. Another facilitory influence is transmitted to the spinal cord from the cerebral cortex by way of the pyramidal tracts: the discharge occurs without any specific or special stimulation (Adrian and Moruzzi), is subliminal in the sense that it does not produce any overt movements, and therefore probably plays a role in determining the activity of spinal neurons. Since there is evidence that pyramidal section alone, in the absence of any other central lesion, leads to some degree of hypotonia,[51] it is reasonable to assume that these subliminal pyramidal impulses, detectable only by electrical recording, act to facilitate the stretch reflexes. Obviously, they could not contribute to decerebrate rigidity, but it is possible that they may be involved in certain cases of spasticity.

Inhibitory influences of bulbar RF and other supraspinal central mechanisms. As already pointed out, the removal of one or another cerebral or cerebellar area leads to a lasting spastic or rigid state. This can only mean that normally these areas act to hold in check or inhibit central mechanisms essential for the development of spasticity or decerebrate rigidity. The several origins of such inhibitory action and the pathways over which it is exerted are described in some detail in Chapters 31 and 32. Here it is only necessary to keep in mind that this influence originates in a number of places, the most important (Fig. 29-7) being (1) certainly one and probably several of the cerebral motor areas, (2) the cortex of the anterior lobe and of the paramedian lobules of the cerebellum, and (3) the striatum (caudate nucleus and putamen). The cerebral influences are all mediated by way of extrapyramidal pathways; again it must be emphasized that the pyramidal tracts do not exert any inhibitory action on spinal stretch reflexes and that the

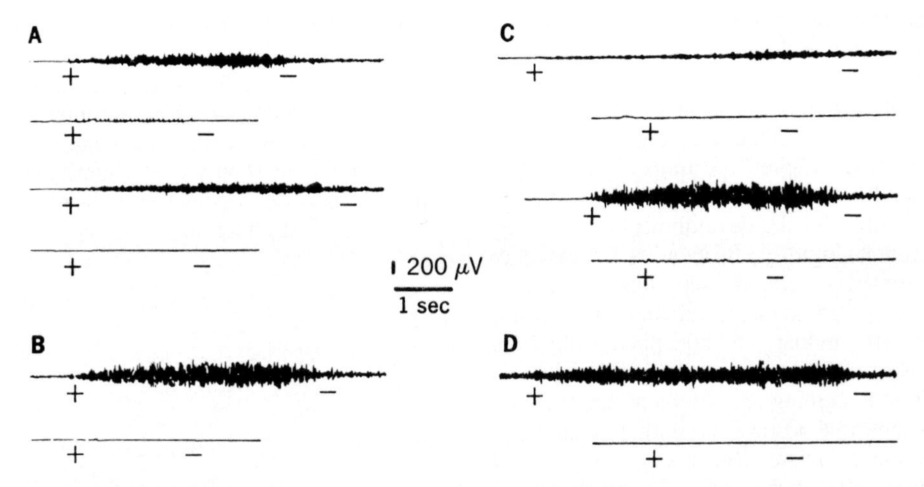

Fig. 29-9. Electromyographic evidence for inhibition of stretch reflexes in decerebrate cat by electrical stimulation of bulbar reticulospinal system. Top tracing in each pair of records was taken without stimulation; bottom tracing during stimulation. Stretch applied between + and −. **A** (from above down), Ipsilateral quadriceps stretch reflex before and during stimulation and contralateral quadriceps stretch reflex before and during stimulation of reticular formation (RF) on one side. **B,** Gastrocnemius stretch reflex before and during stimulation at level of trapezoid body. **C** (from above down), Quadriceps stretch reflex before and during stimulation and gastrocnemius stretch reflex before and during stimulation. **D,** Gastrocnemius stretch reflex before and during stimulation of medial RF overlying inferior olive. (From Sprague et al.[48])

spasticity that accompanies paralysis of voluntary movement in the primate is due to an extrapyramidal lesion.[49-51]

A most significant development in our knowledge of the central mechanisms involved in the control of muscle tone was the discovery by Magoun (see Magoun and Rhines[31]) of a powerful inhibitory mechanism in the bulbar RF. The locus of this mechanism is indicated in Fig. 29-7, where it is designated by the number 4; it occupies the ventromedial part of the bulbar RF and appears to extend as far forward as the level of the trapezoid body. The threshold of its neural elements to direct electrical stimulation is very low, and the inhibitory effects produced are evident against almost any background of motor activity. Thus in both cat and monkey, localized weak stimulation completely abolishes cortically induced movement and all spinal reflexes (multineuronal flexor reflexes as well as monosynaptic stretch reflexes). In the decerebrate animal, such stimulation causes complete loss of tone in the rigidly extended legs and renders the stretch reflexes of extensor muscles inelicitable (Fig. 29-9). These effects are bilateral, but the ipsilateral reflexes are affected to a greater extent than the contralateral. Because of its proximity to the vital respiratory and circulatory centers of the bulb, it is difficult if not impossible to determine the effects of destruction of this inhibitory area or region. It should also be understood that lesions placed here are likely to destroy neurons or fibers of the facilitory system (Fig. 29-7, 5). The marked inhibitory effects are exerted at spinal levels and are mediated by reticulospinal pathways coursing in the anterolateral white matter of the cord.[35]

It appears that most, if not all, of the suppressor or inhibitory areas of the cerebrum and the cerebellum act through the bulbar inhibitory mechanism. This is indicated in Fig. 29-7. There is also evidence that the inhibitory influence of the cerebellar cortex (anterior lobe and paramedian lobules) activates this system via a projection from the fastigial nucleus (Chapter 31). It may be supposed that inhibition of spinal activity of striatal origin is also mediated through this reticulospinal channel. The activity of the bulbar inhibitory mechanism depends on these and doubtless other inflows from higher parts of the brain. When these are eliminated, by high decerebration and decerebellation, for example, bulbospinal inhibitory discharge ceases and can only be evoked by local electrical stimulation of the reticular inhibitory area. Thus this bulbospinal mechanism for inhibition seems incapable of intrinsic activity.

"Release of function" and "influx" as dual causes of decerebrate rigidity and spasticity. The fact that injury or removal of certain parts of the brain that are known to exert inhibitory effects on stretch reflexes produces spasticity or decerebrate rigidity can properly be interpreted in terms of Hughlings Jackson's concept of "release of function." Thus we may conclude that normally these "higher" portions of the nervous system hold in check the "lower" central mechanisms essential for the development of spasticity or decerebrate rigidity. In view of the evidence just presented, it can be said that normally a number of cerebral and cerebellar mechanisms activate the bulbar reticulospinal inhibitory mechanism and thus prevent the spinal stretch reflexes from becoming so prominent that they impose an abnormal posture and limitation of movement on the organism. But this is only part, not more than half, of the story. To finish it, one must invoke, as Magoun and colleagues[25-32,42] have so clearly indicated, another jacksonian concept—that of "influx." We have seen that facilitory influences are essential for the production of decerebrate rigidity and spasticity. Here again the final common path is reticulospinal (corticospinal facilitation over the pyramidal tracts being of relatively minor significance). Without this "influx" the spinal neurons involved in the stretch reflex would be deprived of a most essential source of facilitation. Thus decerebrate rigidity and spasticity must be regarded as the results of an imbalance between the activities of two systems that exert opposite effects on spinal motoneurons.

NATURE OF CONTROL EXERTED BY RETICULOSPINAL TRACTS

Although the studies described thus far reveal the existence of powerful facilitory and inhibitory systems in the brain stem, they do not provide many clues regarding the basic principles of motor control embodied in these systems. A careful analysis of how and where reticulospinal influences are exerted was required to get at these principles. Lundberg and colleagues have carried out a series of studies on the reticulospinal tracts that reveal the extent of the control they exercise over spinal circuits. As illustrated in Fig. 29-10, electrical stimulation of the brain stem inhibitory center as described by Magoun and Rhines elicits large, postsynaptic inhibitory potentials (IPSPs) in both flexor and extensor motoneurons.[22] These IPSPs are produced by stimuli that also cause a collapse of rigidity in decerebrate preparations. They are evoked by activity in fibers that descend in the ventral quadrant of the spinal cord, that is, in the portion of the spinal cord occupied by the ventral reticulospinal tract described by anatomists (Fig. 29-5). The synaptic delays in this circuit suggest that the reticulospinal fibers in question do not end directly on motoneurons but on interneurons that send their axons to the motoneurons. Since the IPSPs are consistently of the same magnitude in flexors and extensors and are evoked by the same strength of stimulation at the same brain stem

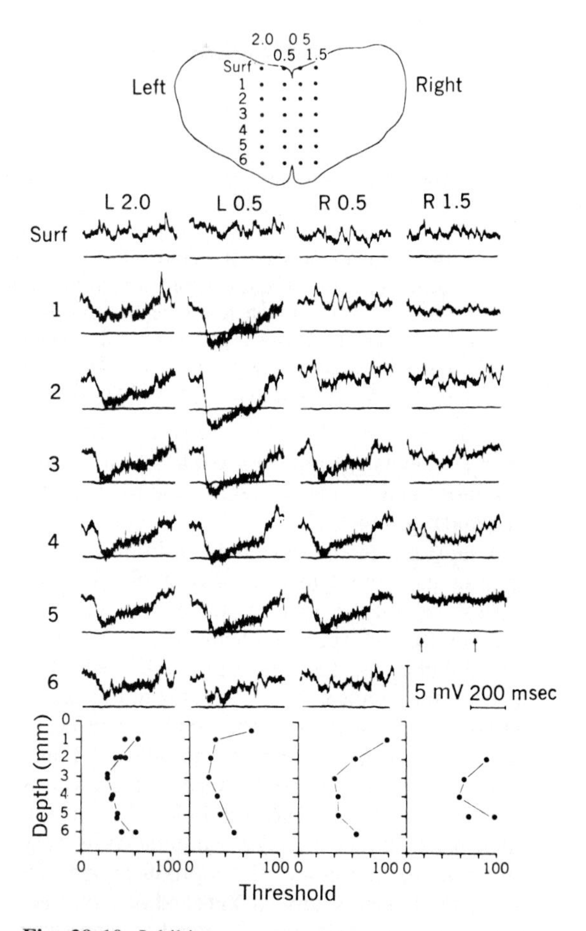

Fig. 29-10. Inhibitory postsynaptic potentials elicited in a posterior biceps or semitendinosus motoneuron by stimulation of brain stem at a series of sites 3 mm rostral to obex. Sites are shown in drawing at top. Distances (in millimeters) from midline and floor of fourth ventricle are indicated above and to left. Potentials recorded in response to these stimuli are reproduced in central portion of figure; in each pair of records, upper traces are intracellular records and lower traces are recorded from dorsal root entry zone. Stimulus intensity was constant throughout, except when measuring the thresholds for IPSPs at different depths. Thresholds for IPSPs are plotted in diagrams below potential records in arbitrary units on abscissa. Arrows mark onset and end of stimulation. (From Jankowska et al.[22])

sites, there can be little doubt about the existence of a reticular center that produces *generalized* inhibition of motoneurons as originally described by Magoun and Rhines. After the original discovery of this inhibitory area, other investigators reported that stimulation elicited mixed excitatory and inhibitory effects on motoneurons instead of pure inhibition. Although mixed effects may be elicited from some areas, Lundberg believes that at least one portion of this brain stem region can evoke pure inhibition of motoneurons when stimulated electrically.

In addition to this indirect inhibition of motoneurons, the ventral reticulospinal tract also inhibits interneuronal transmission from certain sensory fibers to motoneurons. Transmission from cutaneous and high-threshold muscle fibers is definitely suppressed, but it is not clear whether transmission from Ia and Ib afferents is influenced. Many of the descending pathways that influence interneuronal transmission in reflex paths have parallel effects on transmission to motoneurons and to primary afferent terminals.[12] This ventral reticulospinal tract is no exception, for it apparently is responsible for inhibition of transmission from Ia and Ib afferents and from flexor reflex afferents to primary afferent terminals.

Another group of inhibitory effects is mediated by a *dorsal* reticulospinal system originating in approximately the same part of the brain stem.[19] After transection of the ventral quadrant of the spinal cord to eliminate effects mediated by the ventral reticulospinal tract, stimulation of the brain stem inhibitory center no longer elicits IPSPs in motoneurons. It is still possible, however, to suppress the excitatory and inhibitory synaptic actions of flexor reflex afferents and Ib afferents on motoneurons. Fig. 29-11 illustrates this effect very clearly. The monosynaptic reflex of gastrocnemius-soleus motoneurons recorded in *A* is largely inhibited by a preceding volley in high-threshold afferents from the anterior biceps–semimembranous group. A train of five stimuli in the brain stem (*C* and *D*) effectively removes most of this inhibition *(D)* without affecting the test response *(C)* directly. The specificity of this inhibition is shown by the absence of any effect on postsynaptic potentials from Ia afferents (compare *A* and *C*) or on IPSPs from Renshaw cells (not illustrated). The inhibitory effect is exerted on interneurons; there is no increase in con-

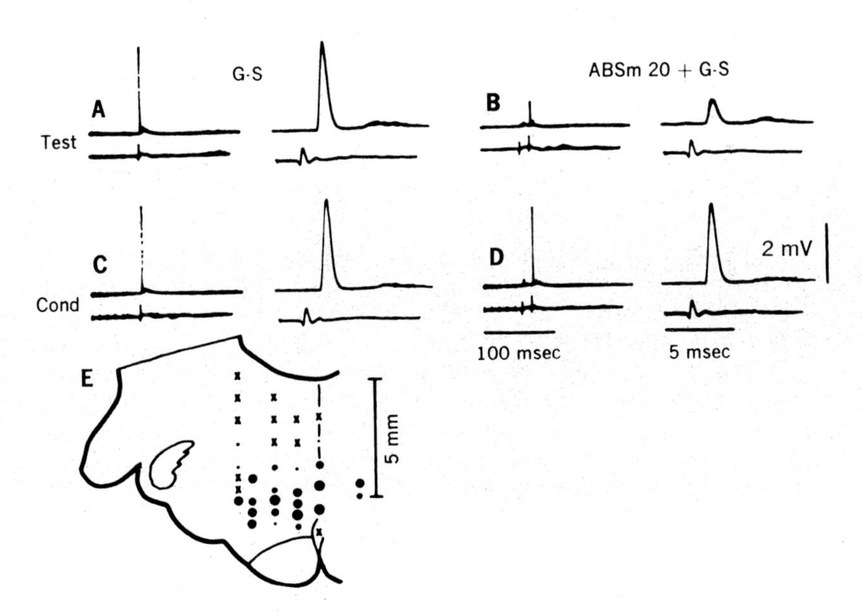

Fig. 29-11. Reticulospinal inhibition of transmission in reflex pathway. **A,** Monosynaptic reflex elicited by stimulating nerve to gastrocnemius and soleus muscles and recorded at two sweep speeds on ventral root (upper trace). **B,** Same reflex inhibited by preceding volley in high-threshold afferents in nerve to anterior biceps and semimembranosus muscles. **C** and **D,** Same as **A** and **B,** respectively, but conditioned by train of five stimuli in brain stem, which removes most of the reflex inhibition. **E,** Diagram of transverse section of brain stem 6 mm above obex. Filled circles of different sizes indicate points of stimulation and magnitude of reticulospinal inhibition. X = points stimulated without effect on reflex transmission. Lower traces in each set of records are recorded from dorsal root entry zone in L7. (From Engberg et al.[19])

ductance in the motoneurons themselves. The same dorsal reticulospinal system also inhibits transmission from flexor reflex afferents (but not from Ia and Ib afferents) to primary afferent terminals and to ascending spinal pathways. The inhibitory actions observed in these experiments were transmitted by fibers in the dorsal part of the lateral funiculus of the spinal cord. They cannot be definitely identified with any established descending tract. It is possible that there is no continuous fiber tract in this region and that the effects are relayed in propriospinal fibers; hence this pathway is referred to as the dorsal reticulospinal *system*.

An analysis of the mechanism by which interneuronal transmission is inhibited by this reticulospinal system was carried out by recording intra- and extracellularly from interneurons in the dorsal horn and intermediary regions of the spinal cord.[18] It was found that reticular stimulation evoked IPSPs in a few of these interneurons, but not in the majority of them. There was, however, an effective depression of the EPSPs and IPSPs evoked in interneurons by flexor reflex afferents. It is probable that the dorsal reticulospinal system directly inhibits only a few cells early in the chain of interneurons and that later cells in the chain are thus silenced indirectly without having any IPSPs evoked in them. The inhibition may, in fact, be exerted on the first-order interneurons that receive monosynaptic effects from primary afferents. This must be so in the Ib pathways to motoneurons, which are largely disynaptic. Four of the five interneurons (out of 78 investigated) showing IPSPs with brain stem stimulation also had monosynaptic EPSPs from cutaneous afferents. The profound inhibition of reflex transmission exerted by the reticulospinal system may thus be due to selective inhibition of a few first-order interneurons that activate many second-order interneurons. These experiments reinforce the conclusion that the reticulospinal system is primarily inhibitory. Transmission in excitatory and inhibitory paths from primary afferents to interneurons was inhibited to an equal extent, and no trace of an EPSP was evoked in any of the 78 intracellularly recorded interneurons.

A number of diverse inhibitory effects from the RF on the spinal cord have been described. In addition to those already mentioned, it has been demonstrated that reticular stimulation may produce large primary afferent depolarization that results in presynaptic inhibition of transmission from some afferent systems. These different types of inhibition do not necessarily operate in parallel. They must, in fact, be differentially active some of the time. In the decerebrate state the ventral reticulospinal pathways are probably not tonically active, whereas the dorsal reticulospinal system is tonically active, being partly responsible for decerebrate inhibition of reflex transmission.

These brain stem inhibitory centers and their spinal effects have been described in considerable detail to give specific examples of the variety and subtlety of supraspinal control. This control is exerted on sensory input, on motor output, and on the intervening processes in internuncial cells. The initial effect of these descending fibers is apparently directed exclusively to interneurons.

Very little can be said about the pathways that are responsible for reticulospinal facilitatory effects. Their spinal terminations have not yet been established anatomically. It may be assumed that facilitatory actions are transmitted over more than one route and influence several different spinal mechanisms, as in the case of reticulospinal inhibition, but experimental evidence concerning the nature of these effects is lacking.

Other descending systems

The preceding description of the reticulospinal tracts indicates that supraspinal centers exert their influences on the spinal cord in several different ways. Similar studies have been carried out on the rubrospinal and vestibulospinal systems. They will not be described because, as in the case of the reticulospinal tracts, the functional significance of the observations is not yet apparent. It is clear that descending systems control or regulate all the intrinsic motor mechanisms of the spinal cord. Sensory input is regulated by means of presynaptic endings on primary afferent terminals. Transmission between sensory afferents and motoneurons is controlled by excitatory and inhibitory endings on internuncial cells. This control may be exerted on first-order internuncial cells, as in the case of the dorsal reticulospinal system, or on some later stage in transmission. Output from the spinal cord is regulated to some extent by descending fibers that end directly on motoneurons, as in the case of some corticospinal fibers. The majority of descending signals, however, produce their effects on motoneurons indirectly by way of vestibulospinal cells. Renshaw cells, the special internuncial cells that receive input from the recurrent collaterals of motoneurons and project back to motoneurons, are among those that are subject to descending control. It is evident that all the spinal mechanisms involved in movement are under the control of higher centers. The next major step in the analysis of descending control will be to devise methods for identifying the precise aspects of movements controlled by each of these descending systems.

ATTITUDINAL AND POSTURAL REACTIONS

Attitudinal reflexes. We have seen that the hyperextension of the extremities of a decerebrate animal may be modified by elicitation, through nociceptive stimulation, of one or another reflex figure (Fig. 29-6). Magnus[28,29] and de Kleijn showed that a similarly harmonious relation of the different parts of the body may be produced by passively changing the position of the head. If the head is forcibly ventroflexed, the postural contraction of the extensor muscles of the forelegs is inhibited and the forequarters sink, while at the same time the postural contraction of the extensors of the hind legs increases, raising the hindquarters. The animal assumes the proper attitude for looking under a shelf or down a hole. On the other hand, if the head is passively dorsiflexed, the forelegs extend further and the hind legs flex. The attitude is now that of a cat looking up at a shelf and ready to spring. If the head is turned to one side, say the right, the displacement of the center of gravity is compensated for by the increased extension of the right legs, while the left are flexed, ready, as it were, to take the initial step. These are the *tonic neck reflexes*. They originate in neck proprioceptors, chiefly if not entirely in the upper joints of the neck,[33] and they last as long as the head retains the new position. *Tonic labyrinthine reflexes,* elicited by change of the position of the head in space, affect all four legs similarly. The influence of the labyrinth in the decerebrate preparation can be studied separately after excluding the neck reflexes by section of the upper cervical dorsal roots or by fixating the head and body in a plaster cast. Extensor tonus of labyrinthine origin is maximal when the animal is supine with the angle of the mouth 45 degrees above the horizontal; it is minimal when the animal is prone with the angle of the mouth 45 degrees below the horizontal. When the head is brought into other positions by rotation of the body around its transverse or longitudinal axis, intermediate degrees of extensor tonus result. The tonic neck and labyrinthine reflexes sum algebraically when both are elicitable. These reactions may be observed in normal animals, in which they may be evoked by active as well as passive movements of the head. They are easily elicited in monkeys after removal of the frontal lobes and in children with severe damage of the higher motor mechanisms of the brain.

These postural reactions are brought about by motor discharges of slow rate and they operate through the same spinal mechanism as that which manages the stretch or myotatic reflexes. It was shown by Denny-Brown[15] that the tonic neck and labyrinthine reflexes influence the stretch reflex just as they influence the posture of the limb. They involve the slow extensor muscles far more than the rapid ones; when the response is small, it is seen only in the former group. It is plain that the stretch reflexes, which as we have seen are essentially spinal in origin, can be modified by impulses originating in the labyrinth or in proprioceptors of the neck. When in a decerebrate animal the afferents from all limb muscles are interrupted by section of the appropriate dorsal roots, these and other sources of excitation may maintain slow rates of motor discharge to the postural muscles. It is for this reason that extensor rigidity reappears in deafferented limbs some time after decerebration.

Normal standing.[38] Decerebrate rigidity has been described as "reflex standing." The standing of the *bulbospinal* decerebrate preparation is, however, a mere caricature of normal standing. Comparison shows that the two conditions differ in several important respects with regard to this fundamental postural achievement.[38] An enumeration of these differences will reveal something of the character of normal standing and the extent to which suprabulbar mechanisms contribute to normal postural activity.

1. The distribution of tone in the musculature is quite different in the two cases. In the acutely decerebrate animal, the postural contraction of the extensors is accompanied by diminution of the tone of the flexors, whereas in normal standing, all the muscles around the joints of the limb are strongly contracted so as to fix the limb in the form of a rigid pillar.

2. If the bulbospinal decerebrate animal is placed on its side or back, it remains in that position, never showing the slightest tendency to right itself. Normal animals succeed in maintaining themselves right side up under the greatest variety of disturbing conditions.

3. The acutely decerebrate animal keeps its legs rigidly extended regardless of its position. It never places its feet in the correct position for standing. In contrast, the normal animal actively

places its feet in such a way that a normal stance can be assumed.

4. If a decerebrate animal is placed in a standing position, the slightest shifting of its body to one side causes it to topple over. By executing correcting movements of the legs, however, the normal animal succeeds in maintaining the standing posture during horizontal displacement of the body.

5. When made to stand on an inclined surface, the normal animal adjusts the distribution of tone in its muscles in such a way as to buttress itself against the effect of the consequent shift in its center of gravity. No trace of such a response is seen in a decerebrate preparation subjected to these conditions.

These deficiencies of decerebrate standing are the result of the absence of certain specific postural reflexes. We shall next consider the more important of these reactions.

Supporting reactions. Limbs that are freely movable when they are used in stepping or as instruments for prehension, scratching, fighting, etc. are transformed during standing into rigid pillars that give the impression of being solid columns for the support of the body. Experiments by Magnus[29] and collaborators (Schoen, Pritchard, and Rademaker) have shown that this is accomplished by a series of local static reflexes. The development of the pillarlike state is called the "*positive supporting reaction.*" It is brought about in part by an extensor response to the exteroceptive stimulus evoked by the contact of the feet with the ground *(magnet reaction)*. It is more especially the result of the proprioceptive stimuli set up in the flexor muscles of the distal joints (digits, wrist, ankle) when these are stretched by the pressure of the foot on the ground. A complex stretch reflex is thereby set up that involves not only the stretched muscles themselves but also the co-contraction of extensors and flexors, abductors and adductors, around each joint, so as to fix the entire limb for standing. In addition, the muscles of the back take part, and their participation is seen especially clearly when the whole reaction has been exaggerated by decerebellation. The resolution of this response is not wholly due to cessation of the stimulus, for active processes set up by stretching the extensors of the distal joints take part in loosening the limb. This is the *negative supporting reaction*. In the decerebrate animal, these reactions are not present; instead there is only the overactive stretch reflex of the extensors. After removal of the cerebral cortex the supporting reactions are disturbed profoundly but are not entirely absent (Rademaker).

Righting reactions. The ability to stay right side up is a universal property of animal organisms. Magnus[28,29] and de Kleijn have shown how in the higher animals, including ourselves, this capacity depends on a group of specific righting reflexes. These responses can best be studied in decorticate animals in which their reflex nature is quite apparent. If such an animal is held up by the pelvis, the head is kept in its normal position as the body is turned from one side to the other or allowed to hang down. The compensatory movements of the head are *labyrinthine righting reactions,* which are set up by stimulation of the otolithic apparatus. The responding muscles are those of the neck. These righting movements fail to occur in animals without labyrinths, and they are abolished by removing the otoliths from their maculas. As long as a decorticate or blinded animal without labyrinths is held in the air, its head takes a passive position imposed by gravity, but the moment it is placed on its side on the ground, the head is righted into the normal position. The response is due to asymmetric stimulation of the body surface, as can be shown by the fact that placing a weighted board on the uppermost side causes the head to go back into the lateral position. This is a *body righting reflex acting on the head.* When the head has been righted by either of the reflexes acting on it, the resultant twisting of the neck proprioceptively excites a *neck righting reflex* that first causes the thorax, then the lumbar region, and finally the pelvis to follow the head into the normal upright position. *Body righting reflexes acting on the body,* elicited by asymmetric stimulation of the body, may, however, cause the body to right itself even when the head is held in the lateral position. Finally, in the higher mammals (cats, dogs, monkeys), righting reflexes may be initiated through visual stimuli. The *optical righting reflexes* result in the orientation of the head and are capable of bringing this about in the absence of labyrinthine or body stimulation. They depend on the cerebral cortex. The centers for the nonvisual righting reflexes lie chiefly in the medulla and mesencephalon, but chronically decorticate cats show a very marked deficiency of the labyrinthine righting reaction.

Placing reactions. A primary requirement for normal standing is that the feet should be placed in the proper position. The normal animal accomplishes this through the agency of a group of placing reactions that have been described by Rademaker[38] and by Bard.[3,4] When the animal is lowered toward a supporting surface, visual stimuli cause the forelegs to be put down in such a way that without further adjustment they support the body in standing. With vision excluded, various exteroceptive and proprioceptive stimuli resulting from contact, position, or movement elicit other placing reactions that serve the same purpose—to bring the legs from any nonsupporting pose into a standing position.

Five nonvisual placing reactions may be distinguished. A brief description of them will serve to show how rich is the equipment for the attainment of this one postural result.

1. If a cat is held in the air with the legs free and dependent and with the head held up (so that it cannot see its forefeet or any object below and in front), the slightest contact of any aspect of either pair of feet with the edge of a table results in an immediate and accurate placing of the feet, soles down, on the table close to its edge (Fig. 29-12).
2. If the forelegs of a cat suspended in the air are held down and the chin is brought in contact with the edge of a table, both forefeet on being released are instantly raised and placed beside the jaws. Usually this is followed by extension, so that a standing position is quickly assumed. If a blinded animal is used, the forefeet, even though not held down, remain hanging until the chin touches the table.
3. If the forelegs or hind legs of a cat that is standing, sitting, or crouching on a table are thrust over the edge, they are immediately lifted so that the feet quickly regain their original positions on the table.
4. If any leg of a standing cat is passively abducted without being held, it is at once adducted and lowered so as to restore the foot to its normal standing position.
5. Although each of the foregoing reactions may be adequately studied in animals with vision intact, this final one can be evoked in pure form only after blindfolding, enucleating the eyes, or removing the visual cortex. The animal held in the air with the forelegs free is moved toward some solid object. As soon as the tips of the vibrassae of one or both sides touch the object, both forefeet are accurately placed on it. Unless the influence of the eyes is excluded, a visual placing reaction of the forelegs will be evoked under these circumstances.

Hopping reactions. These are essentially corrective movements of the legs that serve to maintain a standing posture under conditions involving displacement of the body in the horizontal plane. They are demonstrated by holding the animal so that it stands on one leg. Then on movement of the body forward, backward, or to either side, the leg hops in the direction of the displacement so that the foot is kept directly under shoulder or hip. Rademaker[38] showed that a disappearance of the supporting tone of the leg is an integral part of each hopping reaction. With the leg in the median standing position the positive supporting reaction is strong, and the leg is acting effectively as a rigid pillar, but with any displacement of the body that induces a deviation from the median position the supporting tone diminishes, the foot is raised, transposed in the direction of the displacement, and put down again to give a median support for the body in its new position. It is probable that these reactions are caused by the stretching of one or another group of muscles, that is, that they are myotatic in origin.

Central control of placing and hopping reactions, an example of strict localization of function.[3,4] When in a cat the entire cortex of one cerebral hemisphere is removed, the nonvisual placing reactions of the contralateral legs completely disappear and do not return during long survival periods (6 months to 6 years).* The hopping reactions are permanently depressed; some cannot be elicited, and others are retarded, weak, and so slow that the movement is not repeated rapidly enough to keep pace with even slow displacements of the body. Unilateral decortication does not affect the reactions of the legs ipsilateral to the removal. The control is entirely contralateral. After bilateral decortication the loss and deficits are bilaterally equal. An ablation of the frontal pole of one cortex of the extent shown in Fig. 29-12, *B*, produces as great a deficit of the reactions of the opposite legs as does complete unilateral decortication. The cortical area necessary for the elicitation of these responses includes most of the cortical points, stimulation of which causes movements of the contralateral legs. Removal of any other cortical area has no effect on the reactions. Indeed, one may remove all cortex except the frontal pole of one hemisphere (Fig. 29-12, *D*) without disturbing the reactions in the legs contralateral to the remnant. The fact that a remnant of rostrally situated cortex is able to manage the placing and hopping reactions in normal fashion shows conclusively that the cortical control of these responses is strictly localized and functionally independent of all other cortical areas.

This is an instance of the strictest kind of localization of function in the cerebral cortex. In general, it may be said that the term *"localization of function"* has been used in a rather loose sense. Since evidence indicates that different functions have quite different degrees of dependence on specific cortical areas, localization may be regarded as a relative matter. To

*A proprioceptive correcting or placing response allied to the hopping reactions remains and may become quite prominent. It consists in a placing of the foot, clumsily and often slowly, when in attempting to elicit placing reaction (1), the leg is retroflexed. This is not to be confused with the tactile placing reaction.

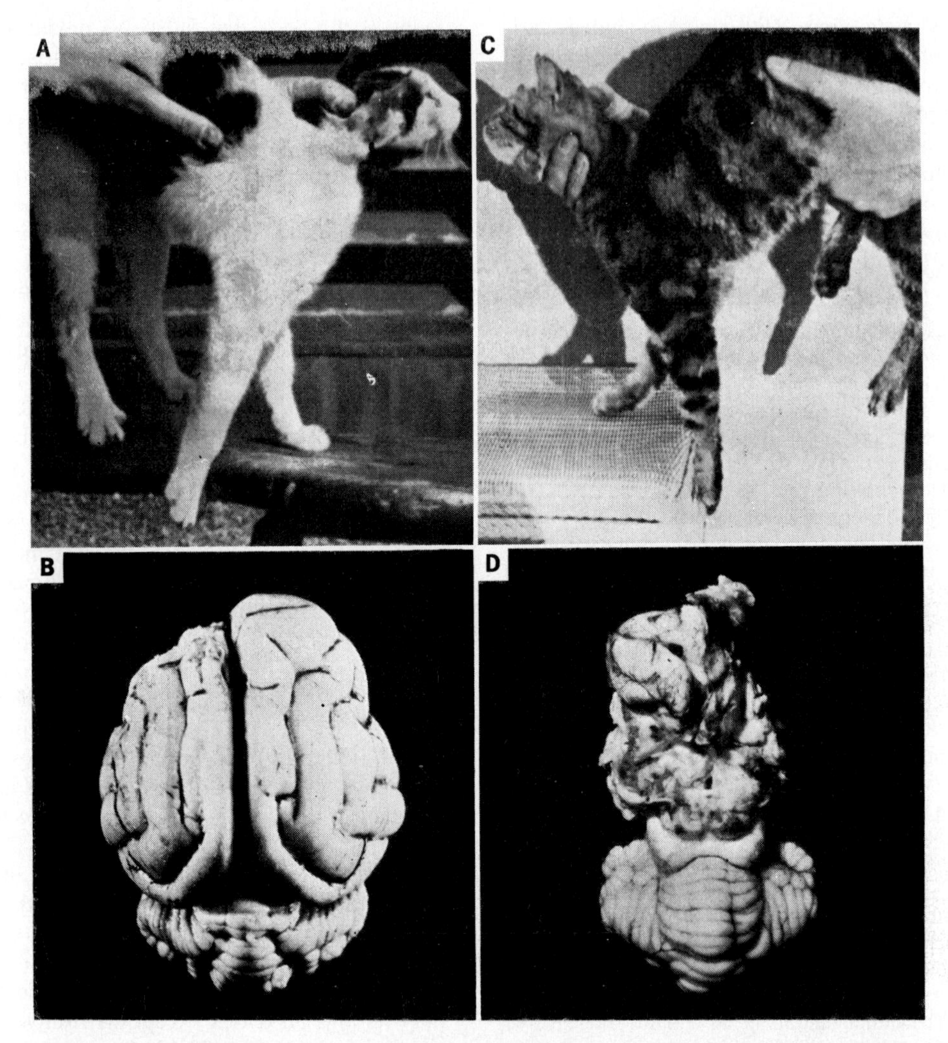

Fig. 29-12. Photographs that indicate nature and localized cortical control of placing reactions of cat. **A,** Picture of animal whose brain is shown below in **B.** Removal of small area of cortex at frontal pole of left hemisphere has permanently abolished placing reaction of right foreleg that occurs when dorsum of foot is lightly touched to edge of supporting surface (top of stool). As can be seen in **A,** ablation has not affected response of ipsilateral foreleg. Defect is as great as that shown by animals from which entire cortex of one side has been removed. **C,** Cat that had been subjected to removal (in two stages) of all cortex except frontal area, which was ablated in other animal. Its brain is shown in **D.** Note that cat has placed the right foot. This and other tests show that small remnant of left cortex was capable of managing in normal fashion the nonvisual placing reactions and the hopping reactions of opposite (right) foreleg. (Adapted from Bard.[3])

apply a rigorous definition, however, it may be said that a cortical function is localized only when a small area contains all the tissue essentially concerned in the cortical control of that function. To demonstrate such localization, it is not enough to show that the function is regularly and permanently abolished by extirpation of a restricted area, for remaining areas may act through it. To conclude that a cortical function has its sole residence in a restricted area, it is necessary to show that the function is not affected by removal of all cortex except this area. As can be seen, this has been accomplished in the case of the cortical management of the placing and hopping reactions.

It is in accord with our general conception of cortical functions that these cortically managed reactions have to do with the finer adjustments of postures that are developed by subcortical levels of integration.

From the point of view of comparative neurophysiology, it is of interest that in the monkey all hopping reactions as well as the placing reactions are completely abolished by a cortical removal (ablation of motor cortex).[4] In the reptile, such of these reactions as are present depend on central mechanisms lying below the telencephalon. In the rat and rabbit, reactions quite similar to those of the cat occur, but they are less well developed, and they are definitely less affected by decortication. Thus in the ascending scale of vertebrate quadrupeds the central control of these particular postural responses becomes more and more "corticalized." In the monkey the control can be analyzed in terms of the cortical sensory and motor components.[4]

BASAL GANGLIA

The basal ganglia are a group of nuclei in the upper brain stem and forebrain that have motor functions of great importance, as attested by the variety and severity of the abnormalities that appear when they are involved in naturally occurring diseases. Attempts to produce counterparts of these abnormalities with localized lesions in laboratory animals have been largely unsuccessful. Experimental studies of the basal ganglia, using the techniques of electrical stimulation and stereotaxic lesions as summarized by Jung and Hassler,[23] have in fact produced such a bewildering array of results that few reliable conclusions can be drawn. Under these circumstances, it seems appropriate to limit this section to a brief description of the basal ganglia and their principal connections, an account of some recent electrophysiologic studies that offer promise, a description of the major syndromes of the basal ganglia as seen by clinical neurologists, and a summary of recent findings on the biogenic amines that are involved in synaptic transmission in these nuclei.

Anatomic aspects

The basal ganglia are among the most primitive portions of the forebrain. They are relatively large and well developed in birds, which may have almost no cortex, and in reptiles, which have little cortex. In mammals there is considerable reorganization of the basal ganglia; this is presumably related to the appearance of the cerebral cortex, which is extensively interrelated with the basal ganglia.

Authorities differ somewhat regarding the composition of the basal ganglia. They are agreed, however, that the *caudate nucleus* and *putamen* (known together as the *striatum*) and the *globus pallidus* are, as their size suggests, the most important parts of the basal ganglia. Three other structures, the *subthalamic body,* the *red nucleus,* and the *substantia nigra,* all located in the mesencephalon, are usually included because of their close relations with the striatum and globus pallidus.

The globus pallidus may be regarded as the principal efferent mechanism for the striatum (caudate and putamen), from which it receives a great many fibers. The principal motor outflow of the basal ganglia is by way of the pallidal efferents known as the *ansa lenticularis.* This fiber bundle distributes signals to the thalamus, hypothalamus, zona incerta, subthalamic body, red nucleus, substantia nigra, and RF of the brain stem.

There are many fibers passing directly from the cerebral cortex to the basal ganglia. These cortical extrapyramidal fibers are distributed in large numbers to the caudate and putamen as well as the globus pallidus. Many of them were unrecognized for a long time because they are unmyelinated. In addition, many fibers in the internal capsule give off collaterals to the basal ganglia, as first pointed out by Ramón y Cajal.

Activity of neurons in globus pallidus at rest and during movement

Standard neurophysiologic techniques have not provided many clues regarding the functions of the basal ganglia. Electrical stimulation and the production of discrete lesions have yielded confusing and often contradictory results. Recordings of single-unit activity, carried out in anesthetized or paralyzed animals, have offered little insight into the normal functioning of the basal ganglia in the moving animal. Recently, however, DeLong[14] has recorded from single neurons in the globus pallidus of unanesthetized monkeys at rest and during the production of trained, voluntary movements of the forelimb or hind limb. The results of this study supply reliable information on the activity of neurons that constitute an important part of the motor outflow of the basal ganglia. In primates the globus pallidus is a composite structure, consisting of an internal and an external segment that are separated by a fiber bundle called the internal medullary lamina. A microelectrode was inserted obliquely from the surface of the brain so that it passed successively through the cerebral cortex, putamen, globus pallidus externus, globus pallidus internus, and the *substantia innominata* lying ventral to the pallidum. In this study, attention was directed chiefly to neurons in the globus pallidus.

With the monkey at rest the discharge patterns of neurons in the two segments of the globus pallidus were distinctively different. As shown in

Fig. 29-13, the external segment contains two types of units. Record *A* shows a pattern of activity characterized by recurrent periods of high-frequency discharge (HFD) separated by intervals of silence lasting up to several seconds. Record *B* shows a pattern consisting of low-frequency discharges interspersed with more rapid bursts of 5 to 20 impulses (LFD-B). These two types of neurons were randomly distributed in the external segment, 85% being high-frequency units and 15% low-frequency units with more rapid bursts. In the internal segment, only one type of unit could be distinguished. As illustrated in record *C*, this type discharges continuously at relatively high frequencies, without long intervals of silence but with frequent fluctuations in rate.

In monkeys that had been trained to carry out rapidly alternating movements of one arm, numerous units in both segments of the globus pallidus were found to discharge in consistent temporal relation to the various phases of the movements. A clear-cut correlation was shown by 66 out of 346 units in the external segment and 39 out of 205 units in the internal segment. Of the 66 units in the external segment, 54 were of the "HFD" type and 12 were of the "LFD-B" type just described. An example of a high-frequency discharge unit in the external segment is reproduced in Fig. 29-14. Record *A* illustrates the resting discharge pattern. During push-pull *(B)* and side-to-side *(C)* movements of the contralateral arm the patterns of discharge bore a consistent relationship to the phases of the movements. During ipsilateral arm movements (*D* and *E*), however, there was no consistent relationship. Contralateral movements were associated with 88% of movement-related units in the ex-

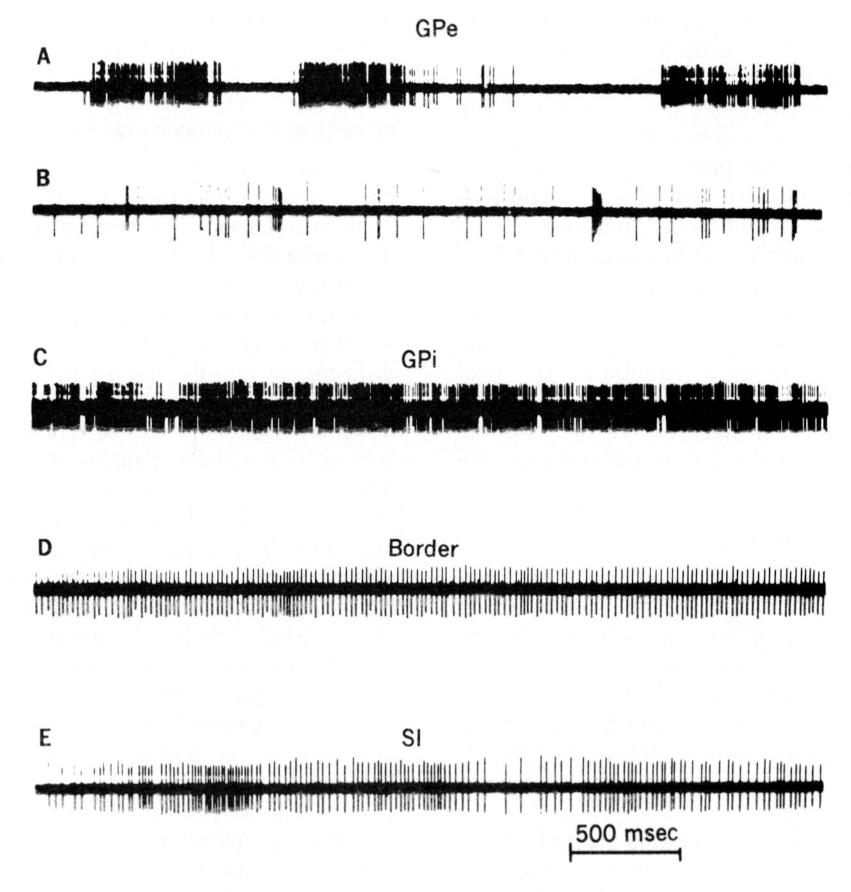

Fig. 29-13. Single-unit discharges recorded from globus pallidus of unanesthetized monkey during rest. **A,** High-frequency discharge with recurrent periods of silence seen in 85% of units in globus pallidus externa, *GPe*. **B,** Low-frequency discharge with bursting seen in 15% of units in *GPe*. **C,** Continuous, high-frequency discharge without long intervals of silence, seen in globus pallidus interna, *GPi*. **D,** Units located along borders of two pallidal segments, exhibited fourth type of discharge pattern, similar to that of neurons in **E,** substantia innominata, *SI*. (From DeLong.[14])

ternal segment and 82% of those in the internal segment of the globus pallidus; the remaining units were correlated almost equally with movements on both sides of the body. Some neurons discharged most intensely during flexion and others during extension. Increases and decreases in frequency were seen in relation to different phases of the movement cycles. In general, activity patterns were related to either arm or leg movements, but not to both.

With a wider range of movements involving more of the muscles in the limbs, or perhaps more combinations of muscles, a higher percentage of units related to movement would probably have been found. DeLong's studies indicate a definite relationship between the activity of pallidal neurons and voluntary movements of the limbs, but they do not rule out a relationship to reflex activity. It is to be hoped that the precise nature of the relationship established in these studies can be analyzed further and that a correlation with some highly specific parameter of movement can be established.

Effects of diseases of basal ganglia on movement and posture

The most useful body of information we have regarding the functions of the basal ganglia is that provided by naturally occurring diseases that attack these structures. The signs and symptoms of these diseases indicate that the basal ganglia are concerned primarily, perhaps exclusively, with motor functions. Although sensory loss or mental changes may occur in conjunction with motor abnormalities, as in Huntington's chorea, they are attributable to pathology in the cerebral cortex, thalamus, or internal capsule rather than to lesions of the basal ganglia. The principal motor effects of basal ganglia disease are akinesia, alterations in muscle tone, and several types of involuntary movement that are typical of lesions in particular nuclei.

Akinesia is believed by some neurologists to be the cardinal "deficit" symptom of extrapyramidal disease. It manifests itself as a disinclination to use the affected part of the body in a normal manner. The patient with ad-

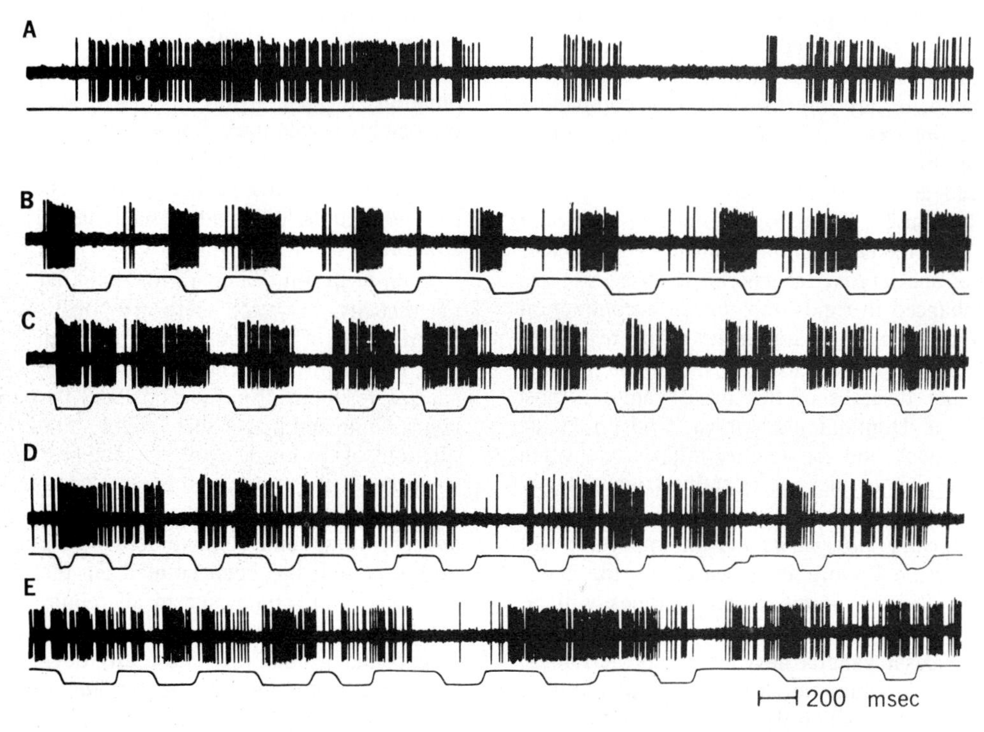

Fig. 29-14. Activity of a high-frequency discharge unit in globus pallidus externa recorded from unanesthetized monkey. **A,** During rest. **B,** During alternating push-pull. **C,** During side-to-side movements of contralateral arm. **D,** During push-pull. **E,** During side-to-side movements of ipsilateral arm. Line below record of unit activity represents position of rod that monkey is grasping. For push-pull movements, up = pull and down = push; for side-to-side movements, up = extension and down = flexion. (From DeLong.[14])

vanced *paralysis agitans,* for example, exhibits a starched or masklike expression, an unblinking, reptilian stare, a tendency to sit motionless for long intervals, and loss of certain movements normally associated with particular activities. In looking to one side, he moves his eyes but not his head; in walking, he fails to swing his arms and he turns around in a stiff, en bloc fashion; and in rising from a chair, he does not pull his feet back under him and does not use his arms to push himself up. There is a general poverty of movement, which despite the name of the disease is not due to true paralysis, for the patient can activate his immobile muscles. Rigidity is not the cause of the akinesia, for stereotaxic lesions may abolish all rigidity without relieving akinesia. The most profound akinesia is produced by lesions in the substantia nigra; the extent to which lesions elsewhere in the basal ganglia may have a similar effect is not settled.

Alterations of muscle tone characterize most diseases of the basal ganglia. Most frequently seen is an increase in the tone of opposing muscle groups that is called *rigidity*. The muscles in the involved part are firm and tense due to a continuous discharge of their motoneurons, which can easily be verified by electromyographic studies. Resistance is encountered to passive movement in any direction and is accompanied by an increased discharge of motor units. Rigidity may be widespread but seems most pronounced at the larger joints due to the greater mass of muscle there. Small muscles anywhere may be affected, including those of the face, tongue, and larynx. In general, tendon jerks are not enhanced in rigid limbs because contraction of opposing muscles limits the reflex response. Reduction in muscle tone or hypotonia is seen in certain diseases of the basal ganglia, most notably in Huntington's chorea. The limbs are usually slack and the tendon reflexes are often pendular due to lack of restraint from antagonists.

Involuntary movements are an important and characteristic feature of diseases of the basal ganglia. Several of the common types will be described here to illustrate the strong influence that the basal ganglia have on motor activities. Chorea is a dramatic example of involuntary movement over which the patient has little or no control. It is characterized by rapid, somewhat jerky or spasmodic movements involving one limb, one side of the body, or all parts of the body. It is difficult to give an adequate description of them because they are so variable and irregular. Early in the development of the disease they consist of twitches or brief, abrupt contractions of one muscle or a small group of muscles, like fragments of a purposeful act. Later, as these contractions become more frequent, they appear to flow into one another to produce an elaborate sequence of meaningless actions. Normal movements are possible, but they may have choreic features incorporated into or grafted onto them. In Huntington's chorea the cause of the abnormal activity appears to be atrophy of the caudate nucleus and putamen, resulting in "release" of some other center that discharges in a continuous, random fashion.

Athetosis is characterized by slowly spreading contractions of closely related muscles that result in sinuous, writhing movements of the extremities. The muscles of the fingers and wrist are most commonly affected, but in severe cases the forearm, arm, face, tongue, and throat may be involved. The disturbance often appears to start in the distal part of a limb and spread to the proximal parts. The individual contractions are generally slower and of longer duration than those of chorea. As in chorea, they tend to become confluent and lead to bizarre postures. In the fingers, alternating flexion and extension is often seen; in the forearm, extension and pronation tend to alternate with flexion and supination. In general, an athetotic patient cannot prevent his involuntary movements and may have great difficulty in keeping his fingers, toes, or tongue in any fixed position. As in chorea, the pathology underlying athetosis is usually in the striatum.

Torsion spasm is a condition, apparently allied to athetosis, in which there are powerful, tonic contractions of the muscles of the trunk, neck, and proximal segments of the limbs. These spasms result in grotesque postures that are uncontrollable and irresistible. Highly abnormal retractions of the head, extreme lordosis, unnatural rotations of the trunk, and bizarre contortions of an entire limb may be seen. Often the spasms appear on voluntary movement but are absent during rest. It has been pointed out that chorea, athetosis, and torsion spasm all require cortical mechanisms for their expression and that cortical or capsular lesions that cause any degree of paralysis tend to abolish the involuntary movements.

Tremor consists of rhythmic, to-and-fro movements caused by alternating contractions of agonists and antagonists. The rate is about 3 to 6/sec and remains fairly constant, but the range of movement may vary widely. As a rule, tremor is most pronounced in the distal parts of the limbs, but it may involve the trunk, head, face, tongue, or jaw. It is frequently associated with rigidity, as in paralysis agitans. There are many types of tremor; only a few of them are the result of lesions of the basal ganglia. Of these latter, one of the most important is the *static tremor* or *tremor at rest* of paralysis agitans. It usually occurs in the hands but may occasionally involve the tongue or the jaw. Characteristically it is seen when the limb is at rest. It is usually abolished during willed movements, so that it interferes very little with voluntary activity. In paralysis agitans and postencephalitic Parkinson's syndrome the lesions are predominantly in the substantia nigra.

Basal ganglia and biogenic amines

Karel V. S. Toll

More rewarding than the results of physiologic studies have been the recent biochemical, pharmacologic, and therapeutic approaches that have resulted in extraordinary advances in knowledge about the biogenic amines and the role played by these compounds and by their inhibitors and potentiators in the normal function of the basal ganglia. There is now little doubt that some monoamines serve as neurotransmitters in the CNS[10] (Chapter 6). The major pathways for the metabolism of the amines involved in this discussion are as follows:

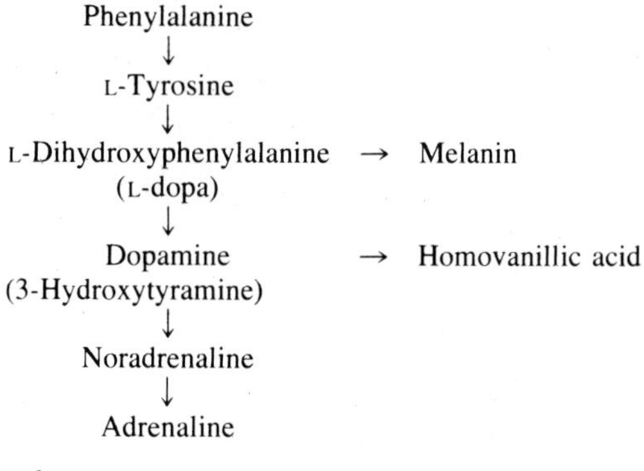

and:

L-Tryptophan
↓
L-5-Hydroxytryptophan
↓
5-Hydroxyindolacetic acid
↓
5-Hydroxytryptamine
(5-HT or serotonin)

In 1953 Twarog and Page[52] found serotonin in the brain, and in 1954 Vogt[53] established the distribution pattern of noradrenaline in various portions of the brain both normally and after drug administration. In the same year lysergic acid diethylamide was found to have a blocking effect on serotonin, and in 1955 it was demonstrated that serotonin was released from its tissue store by reserpine.[44] Indeed, the introduction in the early 1950s of the first tranquilizing drugs resulted in major advances when it was shown that the rauwolfia alkaloids (of which reserpine is a prototype) resulted in the release of norepinephrine and other amines and that following large doses of these compounds a state resembling parkinsonism develops in man and in animals. This extrapyramidal symptomatology is due to the depletion of dopamine and can be reversed by early discontinuation of the tranquilizer or by treatment with antiparkinsonian medication. The parkinsonian syndrome produced by the phenothiazine and butyrophenone derivatives seems to be due not to dopamine depletion per se but rather to blocking of dopamine receptors.[11]

Subsequently,[17] it was found that Parkinson's disease is characterized by a loss of melanin and by a deficiency of dopamine in the substantia nigra–striatal system. There is a decrease in the CSF concentration of homovanillic acid (HVA), the principal breakdown product of dopamine metabolism, lower levels of HVA being correlated with more severe degrees of akinesia. The CSF levels of 5-hydroxyindolacetic acid, a metabolite of serotonin, are also significantly decreased in patients with parkinsonism.[40]

Dopamine is found chiefly in the caudate nucleus, putamen, and substantia nigra. The main dopamine-carrying neuronal pathway arises in the substantia nigra. Axons from this nucleus run in the ventral portion of the crus cerebri and in the internal capsule to the neostriatum where dopamine, its concentration increasing as the fibers approach the target organ, is stored in synaptic vesicles located in the nerve terminals. On neuronal stimulation the transmitter molecules are released into the synapse and combine with receptor sites to affect postsynaptic neurons. Inactivation of transmitters occurs through enzyme action (monoamine oxidase in the case of serotonin; MAO or catechol-O-methyltransferase for the others) or by recapture by the membrane pump. Synthesis of new transmitter occurs within the neuron system as previously indicated.

The noradrenaline-carrying system is more widely distributed than the dopamine system, extending to most parts of the CNS. The serotonin-carrying pathways are also very extensive.

The basic pathology of advanced idiopathic Parkinson's disease, as indicated, consists of degeneration of the melanin-containing cells of the substantia nigra and of the nigrastriatal pathway. Since dopamine itself cannot pass the blood-brain barrier but its metabolic precursor dopa can, this compound was a logical choice for treatment.[2,8] Although the earlier optimism has had to be modified somewhat because of therapeutic failures and side effects, DOPA has been responsible for gratifying clinical results: two thirds of the patients who maintain treatment show an improvement of 50% or more.[34]

The abnormal involuntary movements that are a striking side effect of dopa treatment throw further light on the balance of activity in the basal ganglia. These dyskinesias (choreic, athetotic, or dystonic) resemble the symptoms of other diseases of the basal ganglia. Just as some tranquilizers produce parkinsonism as a side effect, L-dopa produces a picture of choreoathetosis. The mechanism by which L-dopa produces these side effects has not yet been established, although various possibilities have been suggested, including denervation hypersensitivity.

Efforts to enhance the effect of L-dopa by using dopa decarboxylase inhibitors have also proved fruitful. Although the exact mechanisms of their action are not understood, they make adequate treatment with much smaller doses of L-dopa possible.[8] Dopa decarboxylase depends for its functioning on the presence of pyridoxine,[27] but when given to L-dopa–treated parkinsonian patients in an effort to enhance the formation of dopamine from L-dopa, it was found to have a detrimental effect.[16] This puzzling result is prevented if the patients receive not only L-dopa but also a dopa decarboxylase inhibitor, a phenomenon that suggests that the reversal of the dopa effect produced by pyridoxine is due to increasing the dopa decarboxylase activity outside the CNS.[23]

A third neurotransmitter of great importance in the basal ganglia is acetylcholine (ACh). This substance, its synthesizing enzyme, choline acetylase, and its inactivating enzyme, acetylcholinesterase (AChE), are present in high concentrations. With new techniques designed to trace cholinergic neurons in the rat brain, Shute and Lewis[45,46] have demonstrated cholinergic pathways between the substantia nigra, the pallidum, and the caudate-putamen complex, and others have shown that electrical or chemical stimulation of the caudate nucleus causes release of ACh into the CSF. Microinjection of cholinergic drugs into the caudate nucleus results in a 20/sec tremor in rats[13] and systemically administered physostigmine produces, in addition to tremor, a rigidity that is electrophysiologically similar to that induced by reserpine. This rigidity is apparently supraspinal in origin, since it is abolished by destruction of the striatum. Physostigmine also exaggerates parkinsonian symptoms in patients not maintained on medication. This exacerbation can be reversed by anticholinergic drugs. These adverse effects of physostigmine do not occur in normal subjects, nor do they occur in patients on an optimal L-dopa dosage.

It should be noted that patients with extrapyramidal symptoms that are due to cortical lesions do not improve after administration of L-dopa, as all known dopaminergic pathways are subcortical; patients also generally do better on combined dosages of L-dopa and anticholinergic medication than on L-dopa alone. This observation lends further support to the belief that both cholinergic and dopaminergic neurons participate in the regulation of activity in the basal ganglia.

From information on the physiology and neurochemistry of acetylcholine and dopamine in the basal ganglia, it is evident that these two "neurohumors" influence the activity of some of the neuronal units within these areas in an antagonistic way. When tested on single caudate cells, acetylcholine is usually excitatory and dopamine inhibitory. It can therefore be assumed that under physiological conditions, a delicate functional equilibrium exists within the striatum between the excitatory cholinergic and the inhibitory dopaminergic mechanisms. The existence of such an equilibrium is postulated to explain the normal functioning of the striatum as a higher control center for extrapyramidal motor activity.*

*From Hornykiewicz, O.: Neurochemical pathology of Parkinson's disease. In McDowell, F. H., and Markham, C. H., editors: Recent advances in Parkinson's disease, Philadelphia, 1971, F. A. Davis Co.

REFERENCES

1. Bach, L. M. N., and Magoun, H. W.: The vestibular nuclei as an excitatory mechanism for the cord, J. Neurophysiol. **10**:331, 1947.
2. Barbeau, A.: The pathogenesis of Parkinson's disease: a new hypothesis, Can. Med. Assoc. J. **87**:802, 1962.
3. Bard, P.: Studies on the cerebral cortex. I. Localized control of placing and hopping reactions in the cat and their normal management by small cortical remnants, Arch. Neurol. Psychiatry **30**:40, 1933.
4. Bard, P.: Studies on the cortical representation of somatic sensibility, Harvey Lect. **33**:143, 1938.
5. Bard, P., and Macht, M. B.: The behaviour of chronically decerebrate cats. In Ciba Foundation symposium

on the neurological basis of behaviour, Boston, 1958, Little, Brown & Co.
6. Bard, P., and Rioch, D. McK.: A study of four cats deprived of neocortex and additional portions of the forebrain, Bull. Johns Hopkins Hosp. **60:**73, 1937.
7. Bazett, H. C., and Penfield, W. G.: A study of the Sherrington decerebrate animal in the chronic as well as the acute condition, Brain **45:**185, 1922.
8. Birkmayer, W., and Hornykiewicz, O.: Der L-Dioxyphenylalanin effekt bei der Parkinsonakinese, Wien Klin. Wochenschr. **73:**787, 1961.
9. Brodal, A.: Anatomical aspects of the reticular formation of the pons and medulla oblongata, Progr. Neurobiol. **1:**240, 1956.
10. Carlsson, A.: Basic concepts underlying recent developments in the field of Parkinson's disease. In McDowell, F. H., and Markham, C. H., editors: Recent advances in Parkinson's disease, Philadelphia, 1971, F. A. Davis Co.
11. Carlsson, A., and Lindquist, M.: Effects of chlorpromazine and haloperidol on formation of 3-hydroxytyramine and normetanephrine in the mouse brain, Acta Pharmacol. Toxicol. **20:**140, 1963.
12. Carpenter, D., Engberg, I., and Lundberg, A.: Primary afferent depolarization evoked from the brain stem and the cerebellum, Arch. Ital. Biol. **104:**78, 1966.
13. Connor, J. D., Rossi, G. V., and Baker, W. W.: Analysis of the tremor induced by injection of cholinergic agents into the caudate nucleus, Int. J. Neuropharmacol. **5:**207, 1966.
14. DeLong, M. R.: Activity of pallidal neurons during movement, J. Neurophysiol. **34:**414, 1971.
15. Denny-Brown, D.: On the nature of postural reflexes, Proc. R. Soc. Lond. (Biol.) **104:**252, 1929.
16. Duvoisin, R., Yahr, M., and Cote, L.: Pyridoxine reversal of L-dopa in parkinsonism, Trans. Am. Neurol. Assoc. **94:**81, 1969.
17. Ehringer, H., and Hornykiewicz, O.: Verteilung von noradrenalin und dopamin im gehirn des menschen und ihr verhalten bei erkrankungen des extrapyramidalen systems, Wien Klin. Wochenschr. **38:**1236, 1960.
18. Engberg, I., Lundberg, A., and Ryall, R. W.: Reticulospinal inhibition of interneurons, J. Physiol. **194:**225, 1968.
19. Engberg, I., Lundberg, A., and Ryall, R. W.: Reticulospinal inhibition of transmission in reflex pathways, J. Physiol. **194:**201, 1968.
20. Fulton, J. F., Liddell, E. G. T., and Rioch, D. McK.: The influence of unilateral destruction of the vestibular nuclei upon posture and the knee jerk, Brain **53:**327, 1930.
21. Hornykiewicz, O.: Neurochemical pathology of Parkinson's disease. In McDowell, F. H., and Markham, C. H., editors: Recent advances in Parkinson's disease, Philadelphia, 1971, F. A. Davis Co.
22. Jankowska, E., Lund, S., Lundberg, A., and Pompeiano, O.: Inhibitory effects evoked through ventral reticulospinal pathways, Arch. Ital. Biol. **106:**124, 1968.
23. Jung, R., and Hassler, R.: The extrapyramidal motor system. In Field, J., Magoun, H. W., and Hall, V. E., editors, Neurophysiology section: Handbook of physiology, Baltimore, 1960, The Williams & Wilkins Co., vol. 2.
24. Krieg, W. J. S.: Functional neuroanatomy, Philadelphia, 1942, McGraw-Hill Book Co.
25. Lindsley, D. B.: Brain stem influences on spinal motor activity, Res. Publ. Res. Assoc. Nerv. Ment. Dis. **30:**174, 1952.
26. Lindsley, D. B., Schreiner, L. H., and Magoun, H. W.: An electromyographic study of spasticity, J. Neurophysiol. **12:**197, 1949.
27. Lovenberg, W., Weissbach, H., and Udenfriend, S.: Aromatic L-amino acid decarboxylase, J. Biol. Chem. **237:**89, 1962.
28. Magnus, R.: Korperstellung, Berlin, 1924, Julius Springer.
29. Magnus, R.: Studies in physiology of posture, Lancet **211:**531, 1926.
30. Magoun, H. W.: Caudal and cephalic influences of brain stem reticular formation, Physiol. Rev. **30:**459, 1950.
31. Magoun, H. W., and Rhines, R.: An inhibitory mechanism in the bulbar reticular formation, J. Neurophysiol. **9:**165, 1946.
32. Magoun, H. W., and Rhines, R.: Spasticity: the stretch reflex and extrapyramidal systems, Springfield, Ill., 1947, Charles C Thomas, Publisher.
33. McCouch, G. P., Deering, I. D., and Ling, T. H.: Location of receptors for tonic neck reflexes, J. Neurophysiol. **14:**191, 1951.
34. McDowell, F. H., et al.: The clinical use of levodopa in the treatment of Parkinson's disease. In McDowell, F. H., and Markham, C. H., editors: Recent advances in Parkinson's disease, Philadelphia, 1971, F. A. Davis Co.
35. Niemer, W. T., and Magoun, H. W.: Reticulospinal tracts influencing motor activity, J. Comp. Neurol. **87:**367, 1947.
36. Nyberg-Hansen, R.: Sites and modes of termination of reticulospinal fibers in the cat. An experimental study with silver impregnation methods, J. Comp. Neurol. **124:**71, 1965.
37. Pollock, L. J., and Davis, L.: Studies in decerebration. VI. The effect of deafferentation upon decerebrate rigidity, Am. J. Physiol. **98:**47, 1931.
38. Rademaker, G. G. J.: Das Stehen, Berlin, 1931, Julius Springer.
39. Rhines, R., and Magoun, H. W.: Brain stem facilitation of cortical motor response, J. Neurophysiol. **9:**219, 1946.
40. Rinne, U. K., and Sonninen, V.: Acid monoamine metabolites in the cerebrospinal fluid of patients with Parkinson's disease, Neurology **22:**62, 1972.
41. Scheibel, M. E., and Scheibel, A. B.: Structural substrates for integrative patterns in the brain stem reticular core. In Jasper, H. H., et al., editors: Henry Ford Hospital symposium on the reticular formation of the brain, Boston, 1958, Little, Brown & Co.
42. Schreiner, L. H., Lindsley, D. B., and Magoun, H. W.: Role of brain stem facilitatory systems in maintenance of spasticity, J. Neurophysiol. **12:**207, 1949.
43. Sherrington, C. S.: Decerebrate rigidity, J. Physiol. **22:**319, 1897-1898.
44. Shore, P. A., Silver, S. L., and Brodie, B. B.: Interaction of reserpine, serotonin and lysergic acid diethylamide in brain, Science **122:**284, 1955.
45. Shute, C. C. D., and Lewis, P. R.: Cholinesterase-containing systems of the brain of the rat, Nature **199:**1160, 1963.
46. Shute, C. C. D., and Lewis, P. R.: The ascending cholinergic reticular system: neocortical, olfactory and subcortical projections, Brain **90:**497, 1967.
47. Snider, R. S., and Woolsey, C. N.: Extensor rigidity in cats produced by simultaneous ablation of the anterior lobe of the cerebellum and the pericruciate areas of the cerebral hemispheres, Am. J. Physiol. **133:**454, 1941.

48. Sprague, J. M., Schreiner, L. H., Lindsley, D. B., and Magoun, H. W.: Reticulo-spinal influences on stretch reflexes, J. Neurophysiol. **11:**501, 1948.
49. Tower, S. S.: The dissociation of cortical excitation from cortical inhibition by pyramid section, and the syndrome of that lesion in the cat, Brain **58:**238, 1935.
50. Tower, S. S.: Extrapyramidal action from the cat's cerebral cortex: motor and inhibitory, Brain **59:**408, 1936.
51. Tower, S. S.: Pyramidal lesion in the monkey, Brain **63:**36, 1940.
52. Twarog, B. M., and Page, I. H.: Serotonin content of some mammalian tissues and urine and a method for its determination, Am. J. Physiol. **175:**157, 1953.
53. Vogt, M.: The concentration of sympathin in different parts of the central nervous system under normal conditions and after the administration of drugs, J. Physiol. **123:**451, 1954.
54. Woolsey, C. N.: Postural relations of the frontal and motor cortex of the dog, Brain **56:**353, 1933.

30

VICTOR J. WILSON and BARRY W. PETERSON

The role of the vestibular system in posture and movement

The vestibular apparatus was likened by Sherrington[19] to a proprioceptor that acts largely as the "equilibrator of the head." In other words, it acts to stabilize the position of the head in space. A closely related function is to maintain the visual image by stabilizing the eyes in space during head movements.

In man and other mammals, these motor actions, as well as a variety of sensations, are the result of activation of a group of receptors consisting of fluid-filled semicircular canals and otolith organs lying deep in the temporal bone. Collectively, these receptors form the vestibular labyrinth. The semicircular canals are mainly stimulated by angular acceleration, whereas the otolith organs, the utricle and saccule, are stimulated by transient linear acceleration and by changes in head position with respect to the linear acceleration of gravity. Such transient or maintained stimuli evoke phasic and tonic vestibulo-ocular and vestibulospinal reflexes.

This chapter is concerned with the vestibulospinal reflexes acting on the head and limbs that are evoked by stimulation of the semicircular canals and otolith organs and with the way they interact with other reflexes in the proper maintenance of posture. Vestibulo-ocular reflexes will be touched on only briefly and are covered further in Chapter 16.

Various aspects of the vestibular system—peripheral, central, psychophysical, and clinical—have been considered in recent books, reviews, and symposia, which should be consulted for detailed treatment of some of the matters discussed in this chapter and of others that have been omitted.*

*See references 2, 4, 13, 16, 17, and 22.

PERIPHERAL VESTIBULAR MECHANISMS

Structure of the vestibular labyrinth

The vestibular labyrinth consists of a system of thin-walled sacs and ducts (the membranous labyrinth) that lies in passageways within the temporal bone (the bony labyrinth). The membranous labyrinth is filled with a potassium-rich fluid called endolymph. The space between the membranous labyrinth and the bone is filled with perilymph, which has a composition similar to that of cerebrospinal fluid.[87,109]

The structure of the membranous labyrinth in guinea pig and man is shown in Figs. 30-1 and 30-2. The figures also show the relationship between the labyrinth and the cochlea. In mammals the vestibular labyrinth consists of two saclike structures, the sacculus and utriculus, and three semicircular canals, the horizontal, anterior, and posterior. Each canal emerges from the utriculus, runs through a semicircular channel in the temporal bone, and returns to the utriculus. The planes of the three canals are approximately orthogonal to each other. At one end of each canal, where it joins the utriculus, is the canal ampulla, which contains the semicircular canal receptor, the crista ampullaris. As shown in Fig. 30-3, the crista is a mound of tissue whose upper surface contains ciliated sensory hair cells. The cilia of these cells extend upward into a gelatinous structure, the cupula, which extends across and closes off the lumen of the canal.[36,111] Fluid movements in the canal, produced by movement of the head, cause the cupula to be deflected sideways, bending the cilia of the hair cells and thereby providing the stimulus that leads to changes in activity of semicircular canal afferent fibers (see subsequent discussion).

The utriculus and sacculus each contain a patch of sensory hair cells called a macula. The

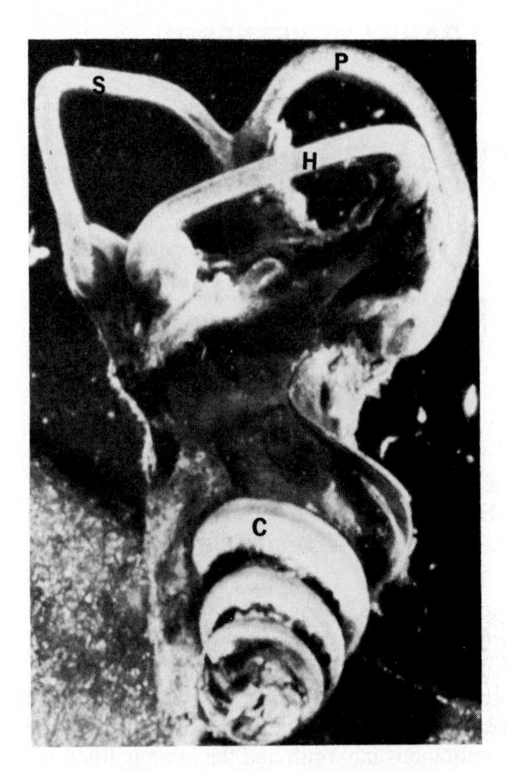

Fig. 30-1. Membranous labyrinth of guinea pig showing superior, *S*, posterior, *P*, and horizontal, *H*, semicircular canals and cochlea, *C*. (From Engstrom et al.[5])

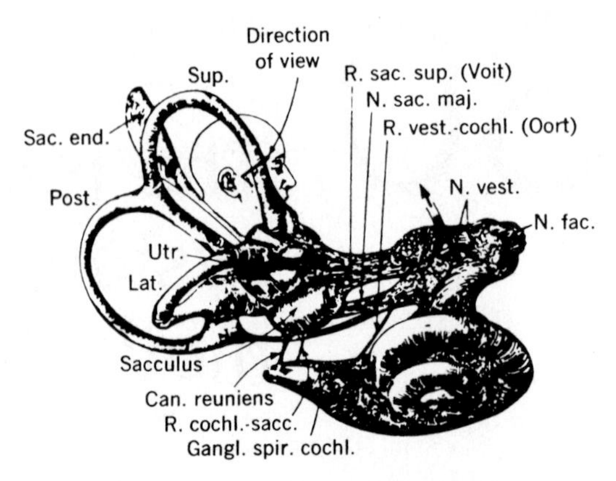

Fig. 30-2. Human vestibular apparatus showing superior, *Sup.*, posterior, *Post.*, and horizontal, *Lat.*, canals and sacculus and utriculus, *Utr.* (From Hardy.[64])

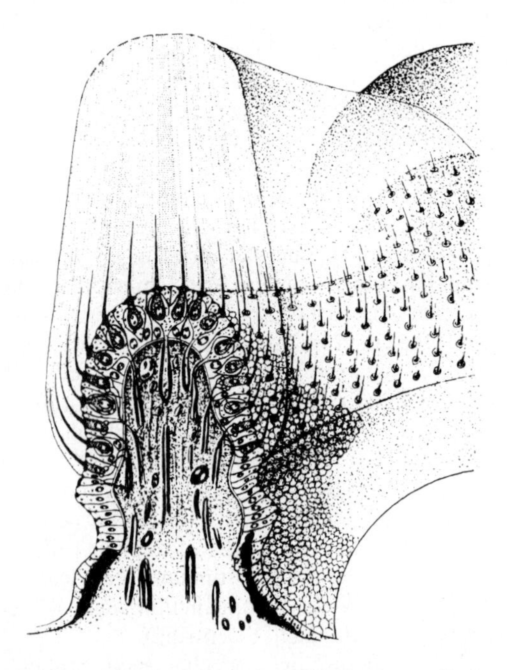

Fig. 30-3. Schematic drawing of crista ampullaris. (From Wersäll.[124])

utricular macula is approximately horizontal when the head is in its normal, upright position, whereas the saccular macula is approximately vertical.[33,43] As shown in Fig. 30-4, the cilia of macular hair cells are embedded in the otolith membrane, a gelatinous structure that contains a large number of hexagonal prisms of calcium carbonate (otoconia).[70] Because the otoconia have a density of 2.94,[5] which is greater than that of the surrounding endolymph, the otolith membrane will be displaced by the force of gravity or other linear accelerations. Such displacement bends the cilia of the underlying hair cells, causing a change in the activity of afferent fibers innervating the macula.

The labyrinthine receptors are innervated by vestibular afferent neurons, typical bipolar sensory neurons with their cell bodies in Scarpa's ganglion. From the ganglion, peripheral axon processes of the sensory neurons project to the labyrinth, whereas their central axonal processes enter the brain stem via the eighth cranial nerve and terminate in the vestibular nuclei and cerebellum (see subsequent discussion). The vestibular nerve also contains a small number of efferent axons that emanate from neurons lying close to the vestibular nuclei in the brain stem and terminate within the sensory epithelium of the ampullae and maculae.[8,12,50,52]

Sensory hair cells and mechanical transduction

The transduction of mechanical stimuli into activity of vestibular afferent fibers is a two-stage process that begins with a lateral deflection of the cilia of vestibular hair cells. As shown in Fig. 30-5, there are two types of hair cells: type I,

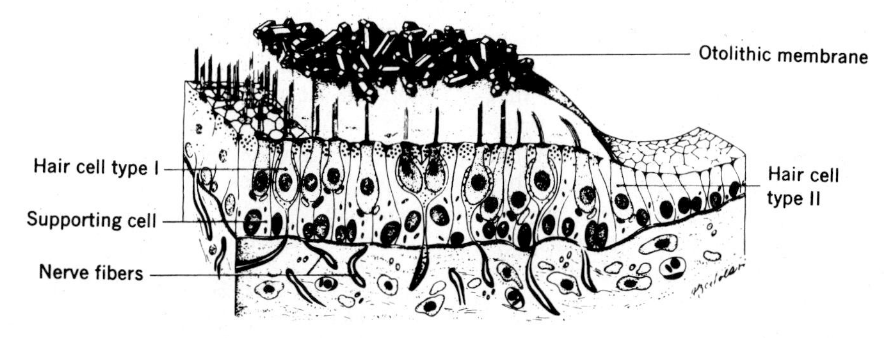

Fig. 30-4. Schematic drawing of macular epithelium. (From Iurato.[70])

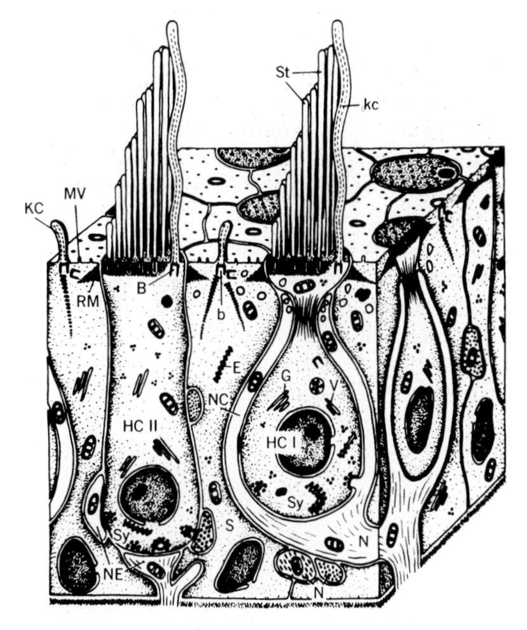

Fig. 30-5. Schematic drawing of area from vestibular epithelium, with hair cells of type I, *HC I,* and type II, *HC II.* Typical arrangement of stereocilia, *St;* kinocilia, *KC;* and modified kinocilia, *kc,* with their basal bodies and roots, *b,* in supporting cells, *S. N,* Nerve fibers; *NC,* nerve chalice; *NE,* nerve endings; *Sy,* synaptic structures; *G,* Golgi membranes; *RM,* multivesicular reticular membrane; *MV,* microvilli. (From Spoendlin.[20])

found only in mammals, and type II, found in all vertebrates. Each type of hair cell has a single kinocilium, with a structure similar to motile cilia found in other parts of the body, and 60 to 100 stereocilia.[63,108,125] The kinocilium is always located at the edge of the ciliary bundle. Studies correlating vestibular sensory responses to the orientation of the hair bundle have revealed that shearing forces that bend the cilia toward the kinocilium produce a depolarization of the hair cell and increased activity of the vestibular afferent fibers that innervate that cell, whereas bending of the cilia away from the kinocilium hyperpolarizes the hair cell and leads to a slowing of the discharge of vestibular afferent fibers.[47,76,124] Sideways bending of the cilia or forces perpendicular to the surface of the sensory epithelium have little or no effect on afferent discharge.[44,121]

Although several theories have been advanced,[6,14,32,56] little is known about the mechanism that underlies the first stage of the transduction process in which bending of cilia produces depolarizing or hyperpolarizing generator potentials in hair cells.[117] There is also no direct physiologic evidence about the mechanism of the second stage of the transduction process, in which potential changes in the hair cells are transformed into changes in the discharge of vestibular afferent fibers. Anatomic studies, which have revealed the presence of typical synaptic structures within the hair cells,[124,125] suggest that the latter transduction process may involve chemical synaptic transmission.

Because of the relationship between hair cell response and the direction of bending of the sensory cilia (i.e., toward or away from the kinocilium), the orientation of the ciliary bundles plays a major role in determining the response of different labyrinthine receptors to mechanical stimuli. In the semicircular canal ampullae, where the effective mechanical stimulus is lateral movement of the cupula in response to angular acceleration of the head, all the hair cells within the crista of an individual canal are oriented in the same direction.[76,124-126] Thus all the afferent fibers supplying a single canal will be excited by angular acceleration of the head in one direction and depressed by acceleration in the opposite direction.

The patterns of ciliary bundle orientations in the utricular and saccular maculae are much more complex than those in the ampullae, as illustrated in Fig. 30-6.[20,108,125,126] In each macula the hair cells are oriented with their directions of maximal excitation, indicated by arrows in the figure, in a fan or wavelike arrangement. This arrangement is interrupted at a dividing line, about which the ciliary bundle orientations reverse. Thus the hair cells within each macula display a wide divergence in their directional sensitivity. For example, in the left utricular macula, which is normally in a horizontal position, afferent fibers supplying hair cells in the inner zone will be excited to various degrees when the head is tilted to the left, whereas afferent fibers supplying hair cells in the outer zone will be depressed.

Because the saccular macula is oriented vertically, afferents supplying hair cells throughout much of the anteroinferior area will be strongly excited when the head is in a normal position, whereas afferents supplying hair cells in the superoposterior area will be depressed. Geometric considerations indicate that relatively little change in the activity of saccular afferent fibers will occur with small movements of the head about its normal upright position. On the other hand, saccular afferent fibers will be maximally sensitive to movements that occur when the head is lying on its side, whereas utricular afferents will be relatively insensitive when the head is in this position. Thus by acting together the utricular and saccular maculae are capable of signaling the position of the head with respect to gravity over a wide range of positions. A similar analysis indicates that the two maculae will be sensitive to a wide range of nongravitational linear accelerations.

ROLE OF THE SEMICIRCULAR CANALS IN VESTIBULAR REFLEXES AND SENSATIONS

Even a small deviation in the angular position of the head produces a significant change in the direction of gaze. Animals like man that rely heavily on vision therefore need accurate phasic vestibular reflexes to stabilize the angular position of the eyes and head in order to preserve the stability of gaze necessary for clear vision. The receptors responsible for these reflexes are the semicircular canals, which also contribute to sensations of body position.

Because they are inertial receptors, the semicircular canals are activated by angular *acceleration* of the head. The vestibular reflexes elicited by activating the semicircular canals, however, must provide an accurate shift in the angular *position* of the eyes or head to compensate for

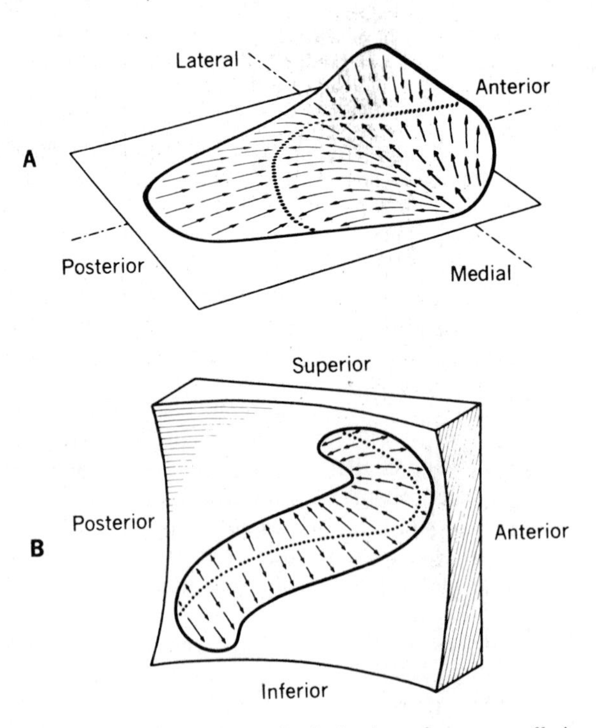

Fig. 30-6. Schematic drawing of directions of polarization of sensory cells in maculae. **A,** Utricle. **B,** Saccule. (From Spoendlin.[20])

angular displacements produced by outside forces. The angular acceleration stimulus to which the semicircular canals respond must therefore undergo a modification equivalent to a mathematical double integration to yield a reflex output proportional to angular position. For a sinusoidally modulated angular acceleration stimulus, which is often used in experimental studies, this integration manifests itself as a shift in the phase of the response. The first step of integration leads to a signal proportional to angular velocity that lags 90 degrees behind the angular acceleration stimulus. The second step of integration converts this to an angular position signal that lags 180 degrees behind the stimulus. Experiments tracing the processing of angular acceleration signals in semicircular canal reflex pathways have shown that part of the required integration is performed by the canals themselves, part by central nervous pathways, and part by the mechanics of the effector systems.

Response properties of semicircular canal receptors

Transformation of angular acceleration into neural signals by the semicircular canal apparatus can be viewed as a two-stage process. The first stage is a mechanical process in which inertial movements of the endolymph within the canal and the resulting deflection of the cupula transform angular acceleration into shearing forces that bend the cilia of the sensory hair cells in the canal crista. The second stage is the transformation of these shearing forces into changes in the activity of semicircular canal afferent fibers. The applied stimulus is modified at each stage in the process, so that the resulting neural signal differs in several ways from the applied angular acceleration. To analyze the operation of the semicircular canals, investigators have employed a combination of physical modeling and measurement of neural responses to determine how the applied stimulus is modified at each stage in the transduction process.

Because of the relatively simple structure of the semicircular canals, it has been possible to develop a physical model, usually referred to as the torsion-pendulum model, that predicts the mechanical behavior of the cupula-endolymph system.* In this model the semicircular canal is considered to be a fluid-filled toroidal ring closed at one point by the cupula, which is free to swing but restrained by springlike forces of the hair cell cilia at its base. When an angular acceleration is applied to the head, the fluid tends to remain stationary because of its inertia, thus causing motion of the canal relative to the fluid and a deflection of the cupula. This motion will be opposed by frictional forces of the fluid in the small canal lumen and by the springlike restoring force of the cupula. Thus the resulting displacement of the cupula can be expressed by the following equation:

$$\theta \frac{d^2\xi(t)}{dt^2} + \pi \frac{d\xi(t)}{dt} + \Delta\xi(t) = \theta\alpha(t) \qquad (1)$$

where $\xi(t)$ is the angular displacement of the cupula, θ the effective moment of inertia of the endolymph, π the viscous damping couple, Δ the elastic restoring couple, and $\alpha(t)$ the applied angular acceleration. This equation predicts how the applied angular acceleration will be modified by the canal mechanics. It is thus the starting point for interpreting experimental observations of the neural activity of semicircular canal afferent fibers.

Responses of semicircular canal afferent fibers have been studied in many species.* In mammals, recordings are usually made from fibers in the intracranial portion of the eighth cranial nerve. Although fibers from an individual semicircular canal are mingled with fibers from other receptors in the nerve, they are easily identified, since they respond selectively to angular acceleration in the plane of that canal. As predicted from the uniform orientation of hair cells in each canal crista, all fibers of an individual canal are excited by angular acceleration in one direction and depressed by acceleration in the opposite direction. The size and dynamic nature of the response varies from fiber to fiber, however. Goldberg and Fernandez, in a careful study of semicircular canal responses,[42,53,54] found that the response properties of fibers were related to the regularity of their ongoing discharge. They thus classified fibers as regular or irregular according to their discharge pattern and studied the responses of fibers in each class separately.

Fig. 30-7 shows the response of regular and irregular afferent fibers to a period of constant angular acceleration, indicated by the sloped portion of the angular velocity curves. For such a constant acceleration, equation 1 can be solved to give the following:

$$F(t) = S\alpha(1 - e^{-t/\tau}) \qquad (2)$$

where F(t) is the firing rate as a function of time, S the sensitivity of the fiber, α the angular

*See references 15, 61, 62, 111, and 118.

*See references 42, 53, 54, 62, 75, and 102.

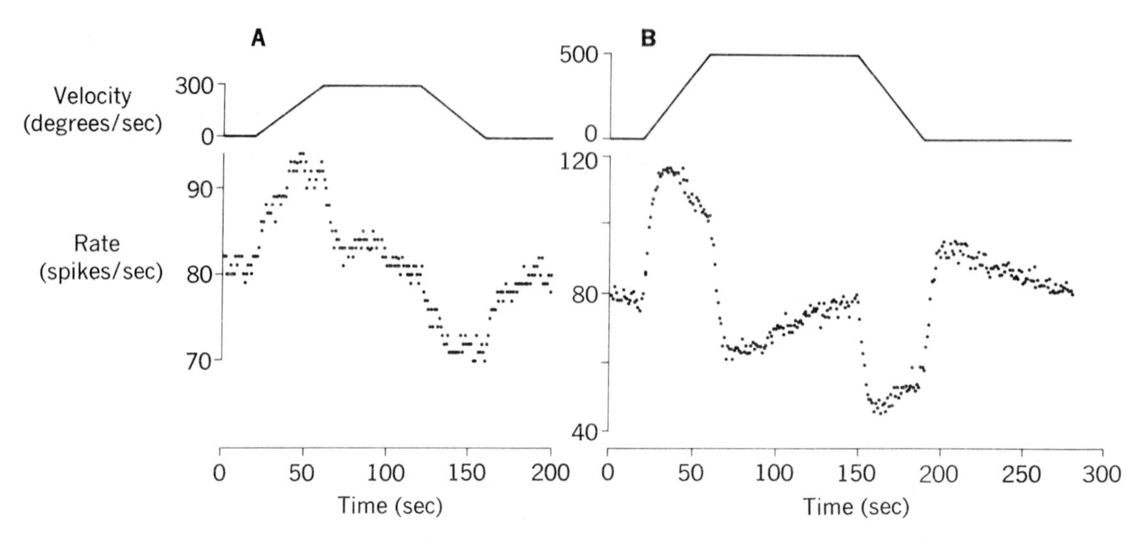

Fig. 30-7. Responses of semicircular canal afferent fibers to constant angular accelerations. Upper traces show head angular velocity, lower traces the responses of a regularly firing, *A*, and irregularly firing, *B*, afferent fiber. (From Wilson and Peterson[22]; data courtesy Dr. J. M. Goldberg.)

acceleration, and τ the time constant of the exponential rise (approximately equal to π/Δ—see equation 1). The torsion pendulum model thus predicts a steady exponential rise or decline in firing rate during constant angular acceleration. The response of the regular fiber in Fig. 30-7, *A*, conforms to this predicted pattern. The irregular fiber in Fig. 30-7, *B*, exhibits more complex behavior. After rising during the initial part of the acceleration, the firing rate declines again as the acceleration is maintained. This deviation from the prediction of the torsion-pendulum model appears to represent an adaptation of the sensory hair cells or afferent fibers to maintained stimulation. Such adaptation is more pronounced in irregular fibers.

The durations of the angular accelerations in Fig. 30-7 are much longer than those experienced during normal movements. For briefer periods of acceleration, the exponential of equation 2 can be approximated by a linear response, giving the following:

$$F(t) = S\alpha t = S\omega \qquad (3)$$

where S is the sensitivity, α the angular acceleration, and ω the angular velocity. For such brief accelerations, adaptation will also be negligible, so that both regular and irregular fibers will have similar responses. Thus under normal conditions the canals perform a single integration of the input angular acceleration to yield a response related to the angular velocity of the head.

Another approach to evaluating the responses of semicircular canal afferents is to study their response to sinusoidally modulated angular acceleration. Fig. 30-8, *A* and *C*, shows the responses of a regular and irregular afferent fiber to a sinusoidal stimulus at the frequency of 0.1 Hz. In each case the response is a sinusoidally modulated firing rate whose peak lags 70 to 80 degrees behind the applied acceleration. For such sinusoidal stimuli the torsion-pendulum model predicts that the amplitude and phase of the response should vary with the frequency of the stimulus. The predicted behavior is plotted as solid curves in the plots of Fig. 30-8, *B* and *D*. At the lower frequencies the responses of the fibers are in good agreement with the model. At higher frequencies the responses have a greater magnitude and smaller phase lag than those predicted. This difference can be explained if the hair cells are sensitive to the velocity of cupula displacement as well as to cupula position.

As explained earlier, integration of a sinusoidally modulated signal yields another sinusoid whose peak lags 90 degrees behind that of the input. The plots in Fig. 30-8 indicate that the responses of semicircular canal fibers approach this 90-degree phase lag over an input frequency range of 0.1 to 1.0 Hz. Since this range includes most naturally occurring head movements, sinusoidal analysis leads to the same conclusion as that derived by applying constant accelerations: during normal head movements the semicircular

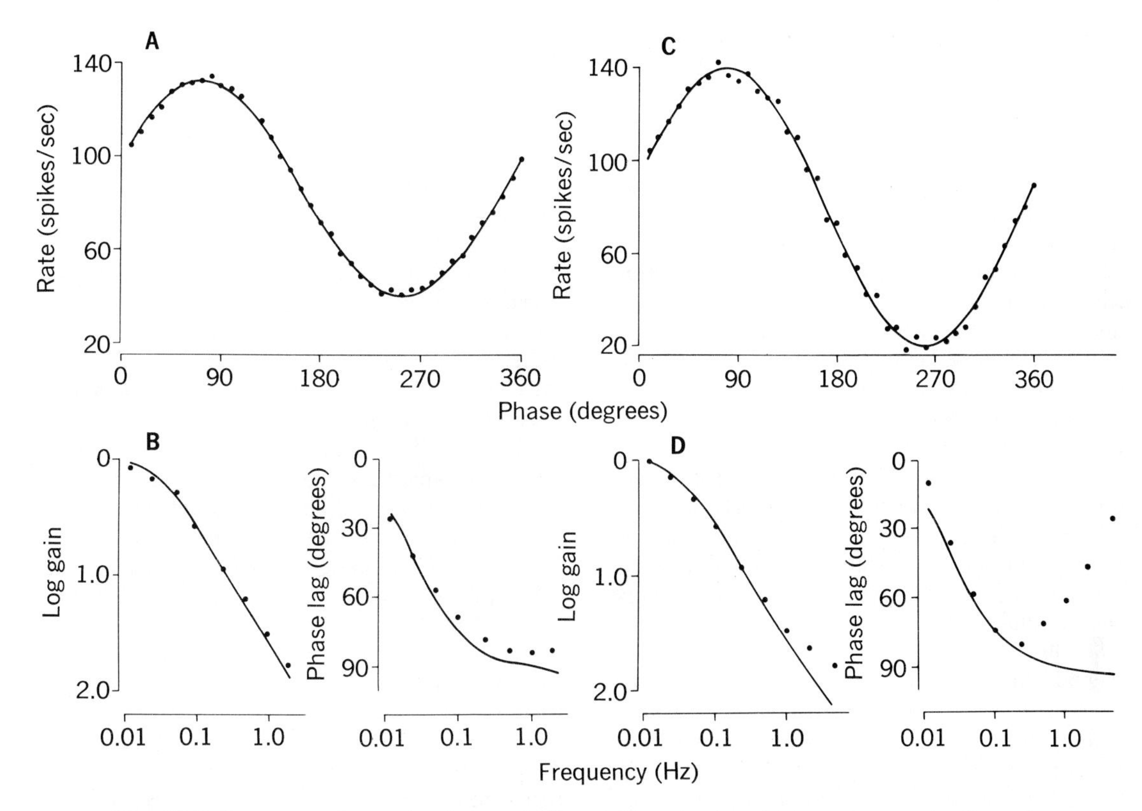

Fig. 30-8. Responses of semicircular canal afferent fibers to sinusoidal angular accelerations. **A,** Discharge rate of regularly firing fiber at various points during sinusoidal acceleration cycle. Maximum acceleration in excitatory direction occurs at 0 and 360 degrees. Solid line is sine wave with same frequency as applied angular acceleration that best fits experimental points. **B,** Bode plot of regularly firing fiber's response showing relative gain and phase as function of stimulus frequency. Solid lines plot predicted behavior of torsion-pendulum model. **C** and **D,** Plots similar to **A** and **B** for response of irregularly firing fiber. (From Wilson and Peterson[22]; data courtesy Dr. J. M. Goldberg.)

canals effectively integrate the applied acceleration to yield a signal related to angular velocity.

Reflexes elicited by semicircular canal stimulation

Activation of semicircular canal receptors can influence the activity of many muscles throughout the body. The most powerful and direct reflexes are those acting on extraocular and neck muscles. These reflexes are organized to compensate for changes in head position to keep the head stable and maintain a constant direction of gaze. For instance, if a man sitting in a rotating chair is accelerated to the right, the resulting reflexes will turn the eyes and head to the left through an angle approximately equal and opposite to the angular deviation of the chair. If the stimulus continues sufficiently long to drive the eyes toward the limit of their movement, nystagmus results: the eyes exhibit a rhythmic movement in which the slow compensatory deviation is interrupted by a quick resetting phase in the opposite direction. There may also be nystagmus of the head.[92] Similar reflexes of extraocular and neck muscles are evoked by angular motion about other axes, but analysis of the reflexes in these cases is more complex because macular receptors are also activated. The net result is nevertheless the same: stabilization of gaze and head position by compensatory reflex contractions of extraocular and neck muscles.

In the absence of vision or other positional cues, semicircular canal reflexes do a remarkably accurate job of maintaining stable gaze. Under these conditions, monkeys rotated about a vertical axis with their heads fixed to the rotating chair generate eye movements that lag the sinusoidally modulated angular acceleration by 160 degrees (i.e., within 20 degrees of the 180 degrees required for perfect compensation) and that have amplitudes approaching those of the applied angular rotation.[107] The question that then arises is, how is the semicircular canal signal that lags the applied acceleration by 70 to 80 degrees at most converted to a movement output that lags by 160 degrees or more? By comparing the firing pattern of oculomotor neurons with the resulting eye movement, Skavenski and Robinson[107] were able to show that a substantial part of the additional lag is produced by the mechanics of the extraocular muscle system when the modulation frequency is above 1 Hz. At lower frequencies, however, this factor becomes less important, and most of the phase lag is produced by mechanisms within the central nervous system.

There is an important difference between vestibulo-ocular reflexes and vestibular reflexes of the neck (vestibulocollic reflexes): the former are open-loop reflexes in which the labyrinth does not sense the resulting eye movement and the latter are closed-loop reflexes in which the labyrinth does sense the resulting head movement and is influenced by it.[92] Despite the difference, when the vestibulocollic reflex is studied experimentally in the open-loop mode, with the head restrained, the dynamic properties of vestibulocollic pathways resemble those of vestibulo-ocular pathways. If sinusoidal angular acceleration is applied to cats whose horizontal canal is in the plane of rotation, there is reciprocal excitation and inhibition, bilaterally, of the motor units in neck extensor and lateral flexor muscles. This vestibulocollic reflex is similar to the one that has been shown to be functional in man, where it makes a significant contribution to stabilization of the visual image.[92] In the cat there is, at low frequencies of stimulation, a considerable phase lag in the response of these motor units with respect to input acceleration (Fig. 30-9).[40] The mean phase lag of the responses of single motor units is 140 degrees at 0.05 to 0.15 Hz.[40] These reflex responses must also involve a central mechanism that produces the equivalent of one step of integration.

Semicircular canal reflexes acting on limb muscles, which also have extensive central processing and integration,[28] are much more variable than those acting on extraocular or neck muscles. This is understandable because the initial limb and body position must be taken into account in determining the appropriate response to an angular displacement of the head. Because of this variability, it is not possible to specify a precise pattern for canal-limb reflexes. As described by Roberts,[17] however, the reflexes appear to be organized on the same principles as neck and extraocular reflexes. They are directed to counter the applied change in head position,

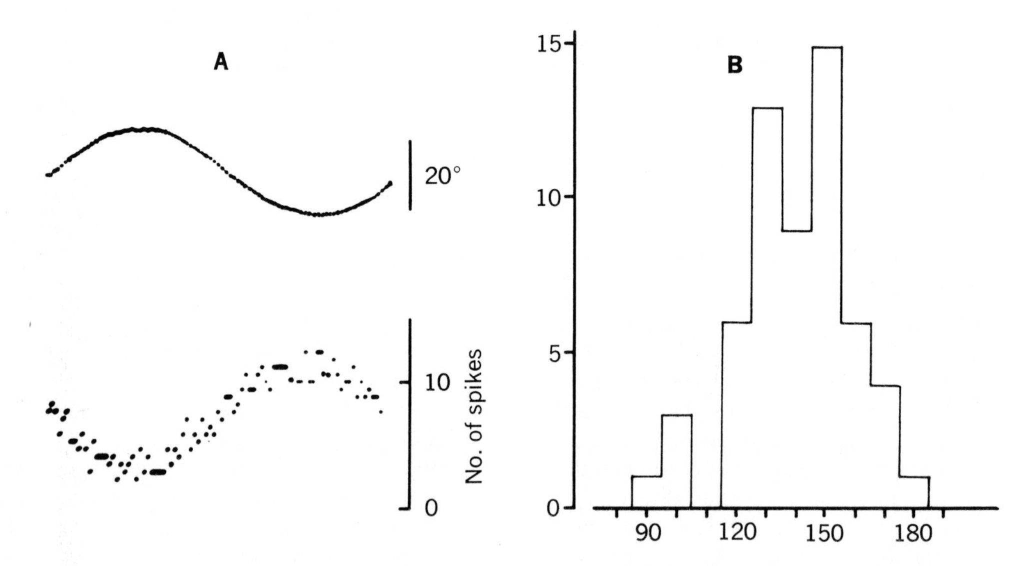

Fig. 30-9. Phase lag of motor units of dorsal neck muscles responding to horizontal angular acceleration. **A,** Response, recorded electromyographically, of motor unit in right neck muscles to stimulation at 0.1 Hz. Upper sine curve shows position of turntable (upward displacement represents rotation to right). Lower curve shows firing of motor unit; spikes were counted over 100 msec periods, and computer-averaged response represents mean value measured from 20 sweeps. Ordinate for lower curve, number of spikes. **B,** Phase lag in degrees (abscissa) for 58 units in response to 0.1 Hz stimulation. Mean = 140 ± 18.6 degrees (SD). Ordinate, number of units. (From Wilson and Peterson.[22]; unpublished data of K. Ezure, S. Sasaki, Y. Uchino, and V. J. Wilson.)

thereby maintaining stability of the head in space. For a four-legged animal, for instance, rolling to the right should produce increased activation of limb extensor muscles on the right side and increased activation of flexor muscles on the left side. In the same situation once the head has been rotated on the neck by canal-neck reflexes, reflexes arising from the change in the angle of the neck will come into play to reinforce the extension of the right limbs and flexion of the left.

Sensations elicited by semicircular canal stimulation

The semicircular canals participate in producing sensations of rotary motion. Under proper experimental conditions, human observers are able to detect angular accelerations as small as 0.1°/sec^2 in the absence of visual, auditory, or other nonlabyrinthine sensory cues.[9] The lowest sensory thresholds are obtained when the observer is allowed to view a visual target that moves with him. Under these conditions, angular acceleration evokes sensations that the target is moving in space even in the absence of movement of the image on the retina (oculogyral illusion).[31]

With proper training and instruction, observers are able to make accurate estimates of the velocity of angular rotation.[9] These estimates are consistent with the pattern of response of semicircular canal afferent fibers described earlier. Thus when angular movements are kept within the range of normally occurring head movements, the estimates correspond quite closely to the actual angular velocity. On the other hand, if the angular velocity is kept constant for a prolonged period, the observer will experience a decline in the subjective angular velocity, followed by a period of apparent rotation in the opposite direction if the movement is suddenly stopped. Both of these effects are predicted by the torsion-pendulum model of semicircular canal function. In addition, observers report a decline in sensation during maintained angular acceleration that is not predicted by the model but that may be related to the adaptation observed in some semicircular canal afferent fibers.

When observers instructed to report their angular velocity are subjected to sinusoidal angular acceleration, their sensations lag behind the applied acceleration by an amount that depends on the stimulus frequency.[118,120] The phase lag is quite similar to that shown for afferent fibers in Fig. 30-8 and approaches the 90 degrees that characterizes an angular velocity signal when movements are in the moderate frequency range typical of normal movements. In this range, observers are also able to provide accurate estimates of their angular position. Thus the sensory system must in some circumstances include a mechanism similar to that which integrates the angular velocity signal carried by semicircular canal afferents to produce reflex outputs related to angular displacement.

ROLE OF THE OTOLITH ORGANS IN VESTIBULAR REFLEXES AND SENSATIONS

The weighted otolith membranes of the utriculus and sacculus make these receptors sensitive to linear acceleration of the head. Such linear acceleration may arise either from movements of the head or from the vertical linear acceleration of gravity. Because of their sensitivity to the pull of gravity, the otolith receptors are able to provide neural signals related to the position of the head with respect to the earth. These signals give rise to static vestibular reflexes and to sensations of head orientation. The role of the phasic neural signals that result from the response of the otolith receptors to linear movement of the head is less well understood, but it appears that these signals contribute to phasic vestibular reflexes.

Response properties of otolith receptors

Unlike the semicircular canals, where the mechanics of the receptor apparatus play a major role in determining the responses of the afferent nerve fibers, the mechanics of the otolith membranes in the utriculus and sacculus contribute relatively little to the pattern of afferent response.[45] For movements within the range of normal head movements the otolith apparatus faithfully translates the total linear acceleration (due to movement and gravity) of the head into a shearing force on the cilia of utricular or saccular hair cells.

The data shown in Fig. 30-10 illustrate the response of utricular and saccular afferent fibers in the monkey to changes in the linear acceleration of gravity caused by changes in head position.[46] Essentially similar responses have been observed by other investigators.* As expected from the approximately horizontal orientation of the utricular macula,[33] the discharge rate of the utricular afferent fiber (Fig. 30-10, *A*) is at the midpoint of its range when the head is in a hori-

---

*See references 73, 74, 78, 114, and 119.

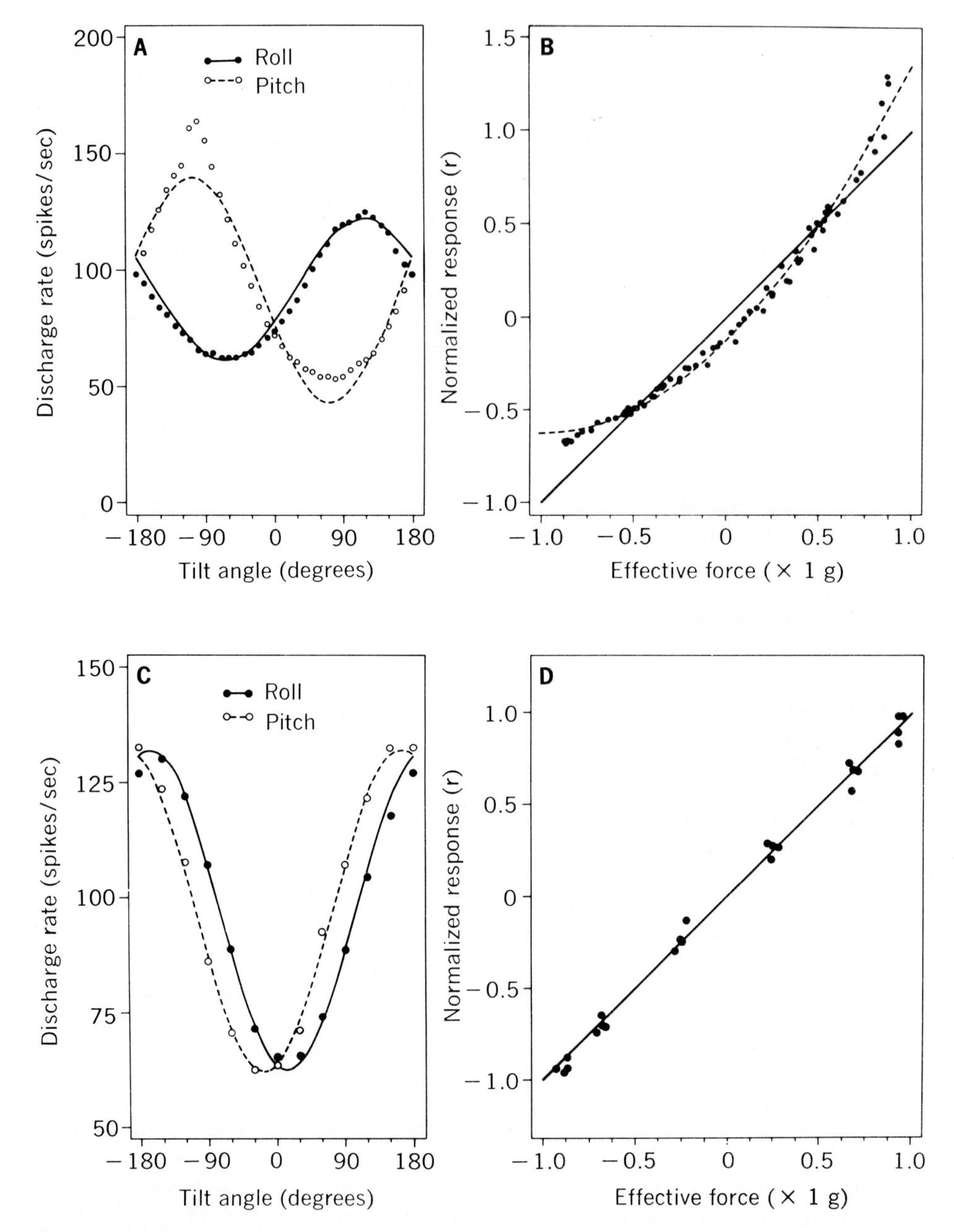

Fig. 30-10. Responses of macular afferent fibers to changes in head position. **A,** Plot of discharge rate of utricular afferent fiber to tilting head sideways (roll) or forward-backward (pitch). Curves are best-fitting trigonometric functions. **B,** Plot of relative response of fiber shown in **A** vs calculated shearing force produced by action of gravity on utricular otolith. **C** and **D,** Similar plots for response of saccular afferent fiber. (Modified from Fernandez et al.[46])

zontal position. Discharge of the saccular afferent fiber (Fig. 30-10, *C*), in contrast, is at the midpoint of its range when the head is lying on its side or pitched 90 degrees forward, as predicted from the approximately vertical position of the saccular macula. Fig. 30-10, *B* and *D*, shows that the discharge of both fibers is directly related to the force exerted on the otolith membrane by the linear acceleration of gravity as anticipated because of the simple mechanics of the otolith system. This relationship is only observed, however, when the applied linear acceleration produces shearing forces along the direction of the vector corresponding to the fiber's peak sensitivity (this vector is presumably related to the orientation of the macular hair cells with which the afferent fiber establishes synaptic contacts). Shearing forces directed 90 degrees to the best direction produce a small increase in discharge rate regardless of the sign (positive or negative) of the linear acceleration. Forces applied normal to the macular surface have no effect on afferent discharge.[44] As a result of the wide variation of macular hair cell orientations shown in Fig. 30-6, individual afferent fibers vary widely in their directional selectivity.[44] Thus afferent fibers from otolith organs provide the central nervous system with a wide variety of signals related to head position, the overall pattern of which should be unique for each orientation of the head with respect to gravity.

Although semicircular canal and macular receptors are mechanically different, both rely on sensory hair cells to convert shearing forces into changes in afferent neural discharge. Thus the hair cell transduction process should be similar in the canal and otolith systems. Careful measurements of the dynamic properties of responses in otolith afferent fibers[43-45] have shown that this is indeed the case. Utricular and saccular fibers exhibit both the adaptation to maintained shearing forces and the sensitivity to rate of change of shearing force that have been observed in the semicircular canal system. These two factors combine to make otolith afferents more sensitive to changing than to maintained linear acceleration. As in the canal system, these dynamic properties are more pronounced in fibers with an irregular discharge pattern than in those with a regular, rhythmic discharge.

Reflexes elicited by activation of otolith organs

Signals from the otolith organs are involved in both static and phasic vestibular reflexes. The static vestibular reflexes are more easily studied because stimuli that evoke them are less likely to activate other receptors. Under normal conditions, static vestibular reflexes are the result of a maintained change in the position of the head with respect to gravity. The reflexes give rise to an activation of extraocular and somatic muscles that is organized to counter the divergence of the head from its normal position and to attempt to restore it to that position.[17] When a four-legged animal is tilted to the right, for instance, static vestibular reflexes cause the eyes to counterroll to the left, activate neck muscles that turn the head to the left, and produce increased extension of the right limbs and flexion of the left limbs. If the animal is pitched forward, static vestibular reflexes cause an upward rotation of the eyes and head, forelimb extension, and hindlimb flexion (Fig. 30-11). As will be explained subsequently, static vestibular reflexes acting on the limbs normally interact with changes in limb muscle activity that are the result of neck reflexes activated by the vestibular-induced changes in head position.

Analysis of phasic reflex responses to linear acceleration is complicated by the fact that the same deflection of the otolith membranes may occur under very different conditions. If the head is maintained in a forward-tilted position, gravity will pull the otolith membranes forward. The same forward force will result if a standing subject sways backward because the center of gravity of the body will fall at the acceleration of gravity, causing the head, which is above the center of gravity, to experience an acceleration greater than that of gravity.[88] It thus appears that the reflex response in such cases must be determined by combining information from the otolith organs with information from semicircular canal and nonlabyrinthine receptors. It is thus difficult to speak of a discrete pattern of vestibular reflex response to linear acceleration. There are, however, situations involving vertical linear acceleration where the reflex response is clearer. An animal falling under the influence of gravity experiences a strong change in vertical linear acceleration and responds with an increase in extensor muscle tone that prepares it for landing.[85,123] This response to falling appears to depend in part on phasic vestibular reflexes elicited by activation of afferent fibers from the sacculus.

Sensations mediated by the otolith organs

An important sensory function of the otolith organs is their contribution to the perception of orientation of the head with respect to the vertical. Under conditions of normal gravity a blindfolded subject who is tilted about a horizontal axis is able to provide an accurate estimate of the angle at which his head is inclined when that angle is less than 30 degrees. At larger angles, systematic errors occur, first in the direction of overestimation, then underestimation.[57] Unfortunately, interpretation of the results of such experiments is complicated by the fact that tilting activates somesthetic and kinesthetic systems in addition to otolith receptors. Input from these other systems presumably accounts for the ability of patients with defective labyrinths to sense body tilt. The accuracy of the tilt sensations reported by such patients is considerably less than the accuracy attained by normal subjects, however, which indicates that labyrinthine afferents make an important contribution to the perception of body orientation. A similar conclusion arises from the results of tilting studies performed under water, where cues from other systems are eliminated. Although their perceptions were less accurate, subjects in these studies were still able to determine their position with respect to gravitational vertical.[9]

More complex techniques have been devised to test the perceptions that result from activation of otolith organs. When subjects are spun in a centrifuge, linear centripetal acceleration combines with the acceleration of gravity to produce a resultant linear acceleration that is inclined from the vertical. The resulting reorientation of spatial signals provided by otolith and other receptors leads subjects viewing a horizontal or vertical line to report that it is inclined by an angle that corresponds to the deviation between the resulting linear acceleration and true vertical (oculogravic illusion).[57] The threshold linear acceleration necessary to elicit this illusion is on the order of 0.03 g. Even weaker linear accelerations have been detected by subjects on horizontally moving sleds.[9] As in tilting experiments, procedures that minimize nonlabyrinthine inputs usually lead to an increase in the threshold stimulus necessary to produce sensations. It thus appears that the otolith organ system is only one of several receptor systems that provide information about linear acceleration of the body.

INTERACTION OF VESTIBULAR REFLEXES WITH ACTIVITY OF OTHER ORIGIN

Numerous reflexes, of which those originating in the labyrinth comprise only a part, interact to maintain proper posture. There are extensive opportunities for such interaction at various levels of the central nervous system, for example, in the brain stem, cerebellum, and spinal cord.

There is continual interplay between activity arising in proprioceptors and cutaneous receptors and activity originating in the labyrinth. Perhaps the clearest example is in the interaction with neck reflexes, discovered by Magnus and his colleagues, which is evoked by deformation of receptors of neck joints.[81] It has already been made clear that the strongest vestibulospinal labyrinthine reflex effects are exerted on the head, causing rotation of the head on the neck. As a result of this rotation, there is reflex eye movement (cervico-ocular reflex); for example, if the body is rotated with the head immobile, the eyes deviate in the direction of body rotation.[112] There are also important reflex effects on limb muscles, some of which are illustrated in Fig. 30-11.[17] If the head is kept level so that the labyrinth is not stimulated (second column), the hindlimbs are flexed when the neck is dorsiflexed (b), and extended when it is ventriflexed (h); the opposite happens to the forelimbs. The static otolith reflexes evoked by stimulation of the labyrinth only are shown by the second row of Fig. 30-11. Downward tilt causes extension of the forelimbs and flexion of the hindlimbs (f) and upward tilt vice versa (d). The four corner

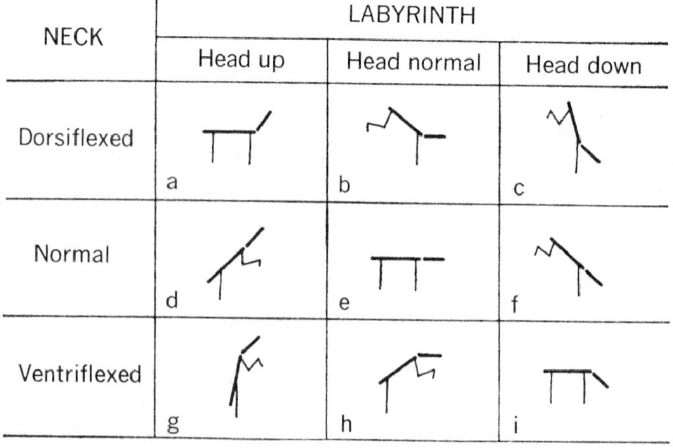

NECK	LABYRINTH		
	Head up	Head normal	Head down
Dorsiflexed	a	b	c
Normal	d	e	f
Ventriflexed	g	h	i

Fig. 30-11. Scheme of combined effects on limbs produced by positional reflexes from neck and labyrinth. Right lateral views. (From Roberts.[17])

boxes show the reflexes that result when both labyrinth and neck receptors are brought into play. In some situations, the two sets of reflexes cancel each other (a, i); in others, they reinforce each other (c, g). Roberts[72] has suggested that in general the interaction of vestibular and neck reflexes "contributes to the stability of the trunk allowing the head to move freely on the body without affecting this stability."

Visual signals also play an important role in maintaining postural equilibrium and stability of gaze. In animals without functioning labyrinths, visuomotor reflexes are sufficient to compensate almost entirely for the absence of labyrinthine reflexes, whereas in normal animals there is extensive interaction between the visual and vestibular systems. One form of interaction involves the modification of ongoing labyrinthine reflexes by visual signals. Some of this modification may take place in the vestibular nuclei.[122] If an animal that is being rotated on a turntable is presented with a visual target that moves with the turntable, vestibulo-ocular reflexes elicited by turntable rotation will be suppressed. Conversely, if the visual target is made to move in a direction opposite that of the turntable, vestibulo-ocular reflexes will be enhanced.[69]

Another form of interaction involves long-term modification of labyrinthine reflexes. If a subject is made to view the world through reversing prisms or magnifying lenses, the compensatory eye movements produced by vestibulo-ocular reflexes when the head is turned will no longer be appropriate to stabilize visual images on the retina. If this abnormal relationship between head movement and movement of retinal images is maintained for a sufficient period of time, visual signals will cause a modification in the size or direction of the vestibulo-ocular reflex so that image stability is regained.[55] This modification is different from the modification of ongoing reflexes in that it can be observed even when the adapted subject is rotated in the dark.

The visual and vestibular systems also cooperate to produce the coordinated eye and head movements that are elicited when a subject shifts his gaze from one visual target to another. The eye and head movements are both initiated at approximately the same time after the appearance of the second target, but because of its greater inertia the head is only beginning to move by the time the eyes have completed a saccadic movement that brings the target onto the fovea (Fig. 30-12, *A*). Then as the head continues to move to its final position, compensatory eye movements in a direction opposite that of the head movement keep the image of the target stationary on the retina.[30] Although the head movement would be expected to activate both semicircular canal and neck receptors, the compensatory eye movements observed in normal monkeys appear to be due almost entirely to vestibulo-ocular reflexes elicited by semicircular canal signals with little or no contribution from neck reflexes.[35] There is some evidence, however, that in humans vestibulo-ocular and neck reflexes might play a more equal role in generating the compensatory eye movements.[82,112]

COMPENSATION FOR LOSS OF VESTIBULAR FUNCTION

It is well known that loss of labyrinthine function can be compensated for very effectively.[18] There is a lasting deficit in the maintenance of postural equilibrium after loss of one labyrinth, but careful testing is required to reveal it.[7] Recovery from the initial postural abnormalities that follow such loss is due in part to the fact that, after compensation, the remaining labyrinth can to some degree carry out the functions previously performed by both. Bilateral loss of function leads to more marked deficits, but after the initial stages, even these are hard to detect under conditions of everyday life. One important exception is that the lack of vestibulo-ocular reflexes prevents full visual fixation during movement. This aside, it is difficult to distinguish between normal subjects and others without labyrinths as long as the patients are not blindfolded and are walking on an even surface.[101] We may conclude that the deficits due to withdrawal of the normal vestibular contribution to the reflexes of posture and movement can be significantly remedied by vision and somatic proprioception.

A specific example of the role of proprioception in compensation is seen in the regulation of gaze. As described in the previous section, in the intact monkey, stabilization of the eyes is carried out entirely by vestibulo-ocular reflexes. After bilateral labyrinthectomy, ocular stabilization is disturbed (Fig. 30-12, *B*), but under the conditions of this experiment, it recovers almost completely in a month or so (Fig. 30-12, *C*); this recovery is due in part to centrally preprogrammed compensatory eye movements that develop within a few days of the operation and in part to an increase in the gain of cervico-ocular reflexes.[35] Whereas the contribution of these reflexes is negligible in the intact monkey, it accounts for approximately 30% of the compensatory movement during passive head movement and even more during active head movement.

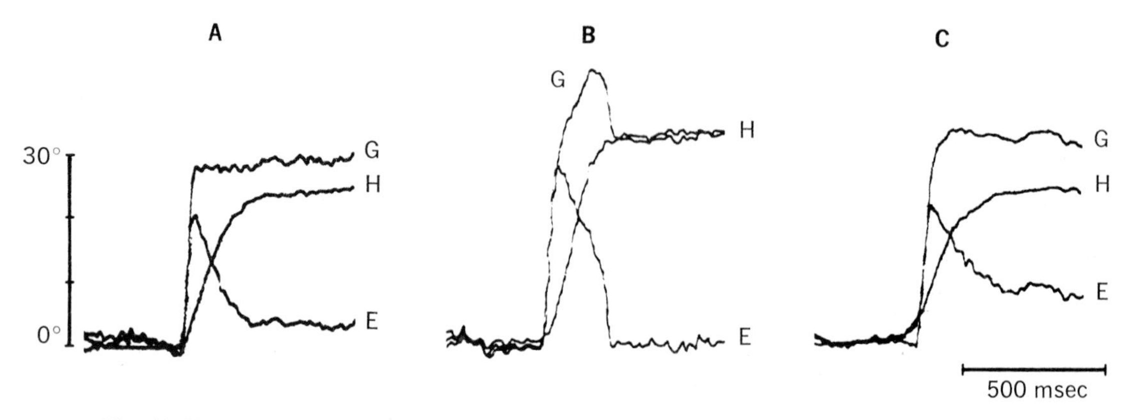

Fig. 30-12. Eye-head coordination in intact and labyrinthectomized monkey. *E*, Eye movements; *H*, head movements. Gaze movements, *G*, represent sum of *E* plus *H*. **A,** Eye-head coordination in intact monkey. **B,** Coordination 40 days after bilateral labyrinthectomy. Note gaze overshoot and corrective saccade. **C,** Coordination 120 days after labyrinthectomy. Note similarities to **A.** (From Dichgans et al.[35])

CENTRAL VESTIBULAR MECHANISMS

Anatomic organization of the vestibular nuclei and their spinal projection

There is considerable information about some of the central pathways involved in the production of the vestibular reflexes that have been discussed in previous sections. Primary sensory neurons from the semicircular canals and maculae directly transmit information mainly to two locations in the central nervous system. One is the vestibular region of the cerebellum (vestibulocerebellum), so defined because it receives vestibular afferent fibers. The principal target of vestibular afferents is the vestibular nuclear complex in the pontomedullary brain stem, consisting mainly of four vestibular nuclei—the superior, lateral (Deiters'), medial, and descending. All vestibular reflexes are executed by pathways that begin with these nuclei. Anatomic studies show that the vestibular projection to the spinal cord is from the lateral, medial, and descending nuclei and that to the extraocular motor nuclei is mainly from the superior and medial nuclei.

Vestibular nucleus neurons give rise to two vestibulospinal tracts (VST), the lateral (LVST) and medial (MVST), as shown in Fig. 30-13. The former, the classic VST, originates in the lateral nucleus and descends ipsilaterally as far as the lumbosacral cord.[1] LVST axons terminate medially in the ventral horn,[90] where it is believed that they have an excitatory action on all their target neurons.

The MVST arises in the medial, descending, and lateral nuclei.[1,26] Its fibers descend bilaterally in the medial longitudinal fasciculus and dwindle in numbers above the cervical enlargement, although some reach the thoracic spinal cord.[89] MVST fibers also terminate medially in the ventral horn; some are excitatory, others inhibitory.

Fig. 30-13 summarizes the origin and extent of the VSTs in cat and rabbit. It also shows the reticulospinal tracts (RSTs), which descend to all cord levels from the medial pontomedullary reticular formation. There are extensive interconnections between vestibular nuclei and reticular formation[1,95] and the RSTs must be included among the pathways transmitting vestibular information to the spinal cord.

Processing of labyrinthine input in the vestibular nuclei

Vestibular afferent input is unevenly distributed within the vestibular nuclei.[1,21,133] A high incidence of direct excitation has been found in Deiters' neurons projecting to the cervical spinal cord in the LVST and in the whole population of MVST neurons, most of which project to the cervical spinal cord.[21] From this and from the much lower incidence of direct excitation of more caudally projecting neurons, it follows that the rostral segments of the spinal cord, controlling neck, trunk, and forelimb muscles, are likely to be influenced more strongly by labyrinthine reflexes than are more caudal segments controlling the hindlimbs.

Not only is vestibular afferent input as a whole unevenly distributed in the nuclei, but there is also some separation between regions of termination of afferents from different receptors in the labyrinth.[51,110] The distribution of terminals of macular and canal afferents suggests that the

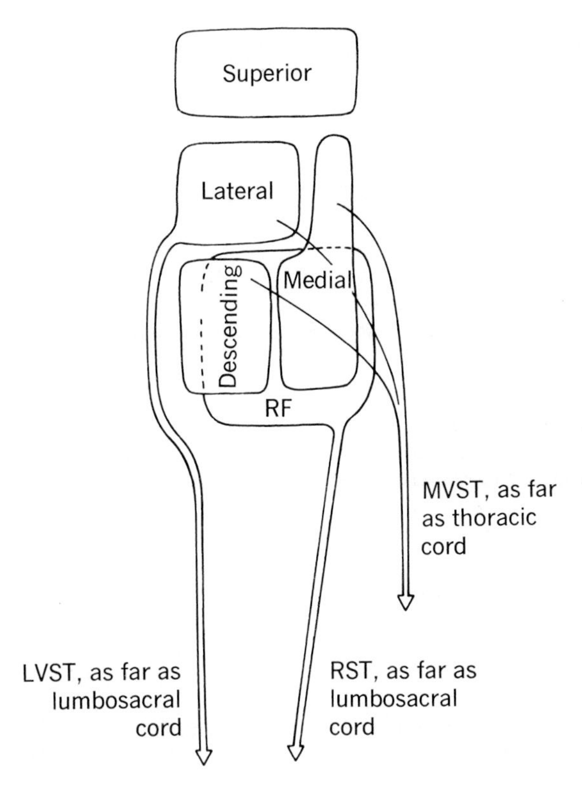

Fig. 30-13. Diagram of origin of lateral and medial vestibulospinal tracts (*LVST* and *MVST*) in vestibular nuclei and of their extent in spinal cord. Also shown are reticulospinal tracts *(RST)*, which originate in pontomedullary reticular formation *(RF)*. (From Wilson.[127])

LVST will be particularly involved in further transmission of macular signals, the MVST in transmission of canal signals. Experimental results on the canal system are in some agreement with this notion, but the division of function between the tracts is not absolute.

The regions of termination of afferents from different canals overlap, but there is nevertheless selectivity of input to second-order neurons: vestibular nucleus neurons generally respond monosynaptically to electrical stimulation of one ipsilateral canal nerve only.[80,128] Natural stimulation also reveals such specific responses,[34,83] but in addition, there is considerable convergence of canal inputs as well as convergence of utricular and canal signals.[34] In other words, in some cases vestibular afferents from different receptors have preferential access to different second-order neurons, but there are extensive opportunities for interaction of afferent signals, often by nonmonosynaptic pathways.

Responses to natural stimulation. Central vestibular neurons are classified according to their responses to angular acceleration or to static tilt. For example, second-order neurons in the horizontal canal system that are excited by angular acceleration to the ipsilateral side (ipsilateral acceleration) and inhibited by angular acceleration to the contralateral side (contralateral acceleration) are type I neurons.[37,103] The response of a type I neuron in an alert monkey to sinusoidal angular acceleration is illustrated in Fig. 30-14. There is increasing modulation of the resting discharge with increasing acceleration; at 0.93 Hz, corresponding to 670 degrees/s^2, there is intense activity during ipsilateral acceleration and complete inhibition during contralateral acceleration. There are other neuron types showing different behavior: type II neurons are inhibited and excited by ipsi- and contralateral acceleration, respectively; type III and IV neurons, less common, are excited and inhibited, respectively, by acceleration in both directions.[37] It is likely, but only partly documented, that a number of vestibulospinal neurons are type I.

Neurons responding to stimulation of the utricular macula by changes in the direction of linear acceleration due to gravity (lateral tilt) have also been subdivided[38]: alpha and beta neurons are excited by tilting the ipsi- or contra-

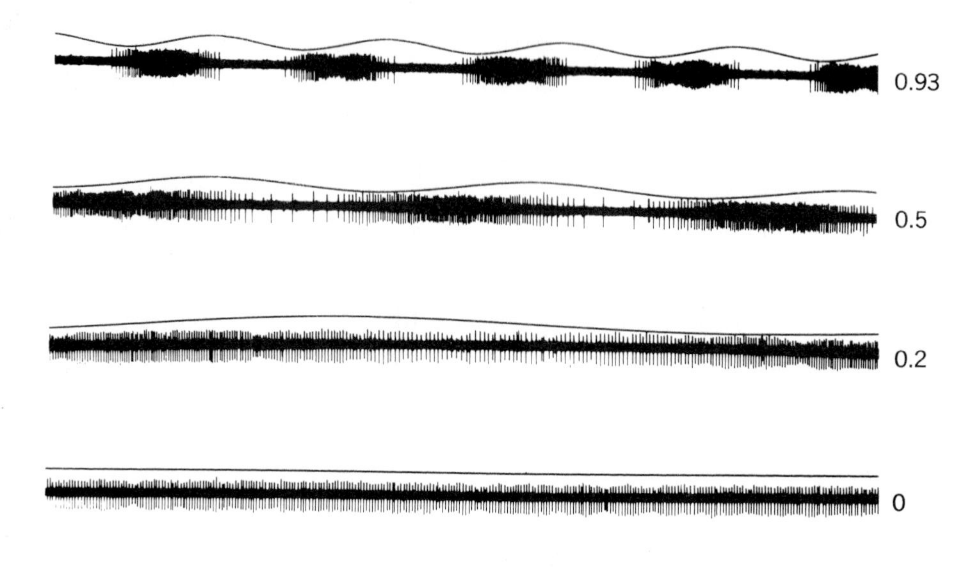

Fig. 30-14. Discharge of type I neuron in alert monkey in response to different frequencies of horizontal head acceleration. In each sweep, upper trace shows head acceleration (downward sine wave minimum is peak in ipsilateral horizontal acceleration); lower tracing shows extracellular recording of unit activity. Time mark, 1 sec. Peak acceleration is from 31.5 degrees/sec^2 at 0.2 Hz to 670 degrees/sec^2 at 0.93 Hz. (Modified from Fuchs and Kimm.[48])

lateral side down; gamma and delta neurons are excited or inhibited, respectively, by tilt in either direction. Many LVST neurons projecting to the cervical cord typically have an alpha response pattern, whereas those projecting to the lumbar cord have, on the average, a gamma pattern.[94]

Primary vestibular afferents do not terminate in the reticular formation, but reticular neurons are influenced by natural vestibular stimulation by multisynaptic pathways and have response patterns similar to those just described for vestibular neurons.[39] Among reticular neurons influenced by stimulation of the labyrinth are many RST neurons,[96] showing that the RST also acts as a relay of labyrinthine activity to the spinal cord.

Dynamic analysis of the horizontal canal system shows that information is transmitted from primary afferents to second-order neurons with relatively little change.[48,83,84,106] Comparison of Fig. 30-15, which illustrates the responses of cat second-order neurons to sinusoidal angular acceleration, with Fig. 30-8 shows that the phase lag of second-order neurons with respect to head acceleration is similar to that of first-order neurons (afferent fibers): over a considerable range of frequencies the response of second-order canal neurons is closely related to the angular velocity of head rotation. Second-order neurons have a higher sensitivity than afferent fibers, probably mainly because of the enhancing action of the commissural inhibitory pathway (see subsequent discussion).

Responses to electrical stimulation and commissural inhibition. Stimulation of the vestibular nerve evokes a monosynaptic (N_1) field potential in the vestibular nuclei, on top of which responses of second-order neurons are superimposed, as in Fig. 30-16; there may also be disynaptic (N_2) potential, with superimposed neuron activity, reflecting disynaptic activation.[99,103] Type I neurons are typically excited, mono- or disynaptically, by stimulation of the ipsilateral nerve; type II neurons are inhibited through the action of local inhibitory circuits. In addition, Shimazu and Precht[104] showed that type I neurons excited by stimulation of the *ipsilateral* nerve are often inhibited by stimulation of the *contralateral* nerve (Fig. 30-16). This ''commissural'' inhibition, mediated by short latency pathways linking the bilateral vestibular nuclei, may affect only canal-driven second-order neurons; it does not appear to influence lateral nucleus neurons whose activity is modified by static tilt.[105]

First- and second-order vestibular neurons are tonically active, and the latter exert a continual inhibitory action on contralateral second-order

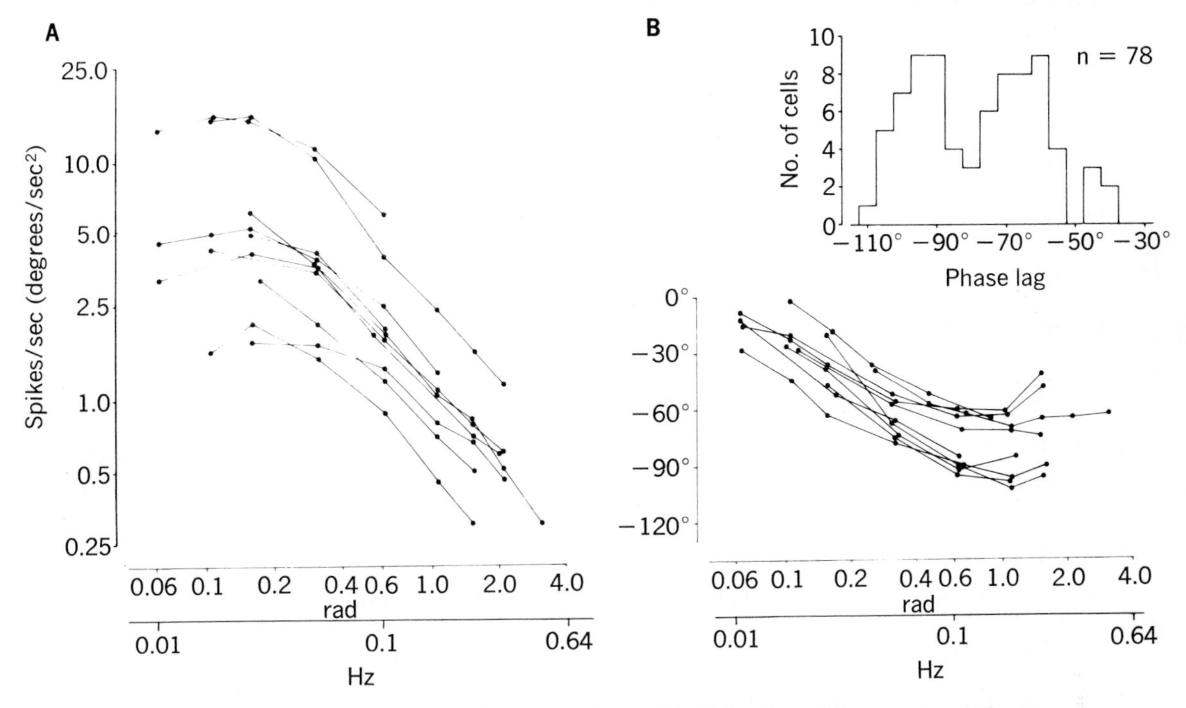

Fig. 30-15. Bode diagram of 10 neurons in vestibular nuclei. Gain, **A,** and phase angle relative to head angular acceleration, **B,** are plotted against stimulus frequency, shown in radians and hertz. Inset diagram is histogram of phase lag at frequency of 0.1 Hz for 78 neurons. (From Shinoda and Yoshida.[106])

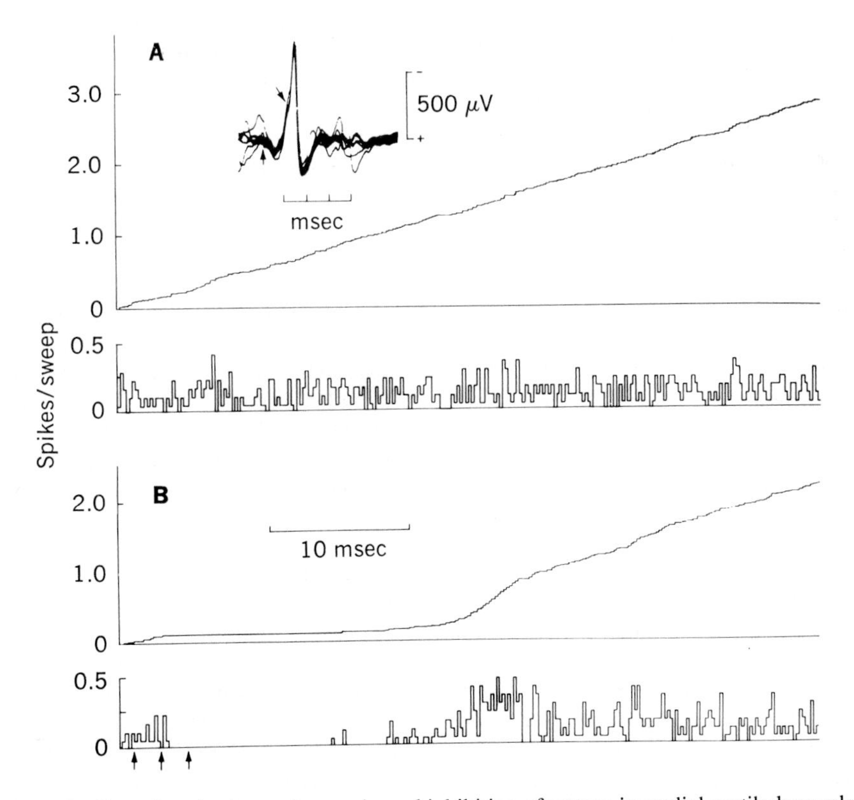

Fig. 30-16. Ipsilateral excitation and commissural inhibition of neuron in medial vestibular nucleus. Inset in **A** shows monosynaptic response, recorded extracellularly, to stimulation of ipsilateral vestibular nerve (at upward arrow). Action potential, which begins at slanted arrow, is on top of monosynaptic (N_1) field potential. **A,** Spontaneous activity of neuron shown both as PST histogram (below) and as integral of discharge (above). **B,** As in **A** but three stimuli (at three upward arrows) were delivered to contralateral vestibular nerve. (Unpublished data of S. Rapoport, A. Susswein, Y. Uchino, and V. J. Wilson.)

neurons in the canal system. Head movements that increase activity in one set of canal afferents and decrease activity in the complementary contralateral afferents should therefore raise the firing rate of second-order neurons not only by an increase in ipsilateral excitation but also by a decrease in contralateral inhibition (disinhibition). The functional effect of commissural inhibition, therefore, is enhancement of the response of second-order neurons.

Normal function of the vestibular system involves bilateral interaction between the labyrinths and vestibular nuclei. Compensation after hemilabyrinthectomy is due, in part, to the fact that the remaining labyrinth is spontaneously active and can give bidirectional signals. During the development of compensation there is recovery of spontaneous activity in the vestibular nuclei on the damaged side to a level high enough so that activity coming from the intact contralateral labyrinth can be processed and relayed.[100]

Other inputs to vestibular neurons

The vestibular nuclei are an important motor center acting on the musculature. In turn, neurons in the nuclei are influenced by activity arising in many regions of the central nervous system. The vestibular nuclei, therefore, are one locus of interaction between vestibular and other inputs. Somatosensory inputs reach the nuclei by tracts ascending the spinal cord and influencing vestibular neurons directly as well as though the cerebellum or the reticular formation; the latter, by its connections with vestibular neurons, provides a channel for many different inputs. Descending inputs from the cortex, optic tectum, and interstitial nucleus of Cajal modify the activity of vestibular neurons. Activity originating in some or all of these areas may converge on labyrinth-driven neurons, some of which are particularly influenced by activity related to eye movements.[48,71,86] Fuchs and Kimm[48] suggest that the vestibular nuclei may coordinate oculomotor and vestibular inputs to extraocular motoneurons.

Activity originating in a broad spectrum of cutaneous, joint, and muscle receptors reaches the vestibular nuclei by brain stem pathways and via both the fastigial nucleus and the anterior lobe of the cerebellar cortex. Some of this incoming activity may be related to particular motor functions. For example, impulses from neck joint afferents converge on second-order vestibular neurons that project to the abducens nucleus.[65] This provides a pathway for cervico-ocular reflexes and a site at which neck activity can interact with vestibulo-ocular reflexes. Other, more generalized activity provides an excitatory background that is modulated by pathways that reach the lateral nucleus through the cerebellar cortex, principally that of the anterior lobe. As described in Chapter 29, excitation reaches the anterior lobe by both mossy and climbing fibers; the cortical output, consisting of Purkinje cell axons, is inhibitory[67] and distributes mainly dorsally in the lateral vestibular nucleus.[68] Introduction of this cerebellar circuitry into the somatosensory input results in a somatotopic pattern of activity: there is a strong tendency for LVST neurons inhibited via the cerebellum by forelimb nerve stimulation to project no further than the cervical spinal cord and for neurons inhibited by hindlimb nerve stimulation to project as far as the lumbar spinal cord.[27,116]

The inhibition of VST neurons by anterior lobe Purkinje cells not only modifies their response to somatosensory input, but also exerts a controlling action on labyrinth-evoked activity. For example, decerebellate-decerebrate cats have a strong opisthotonus, which is not present in animals that are only decerebrate.[98] The opisthotonus must be due to tonic vestibular reflexes, because it disappears when the labyrinth is destroyed. It is mediated, therefore, by pathways relaying through the vestibular nuclei that must be under cerebellar inhibitory control, since their action is enhanced by decerebellation. As another example, when a decerebrate cat is tilted, those LVST neurons that react have mainly a phasic response during tilt; there is no maintained response. After decerebellation, typical alpha or beta static responses appear.[91]

The vestibulocerebellum (flocculus, nodulus, uvula) receives primary vestibular connections and projects to the vestibular nuclei.[1] It is more closely related to the labyrinth than the anterior lobe, which receives vestibular input mainly by a complex route.[29,41] It might therefore be supposed that the vestibulospinal system would be influenced not only by the anterior lobe but also by the vestibulocerebellum. This is not necessarily so, however. Ito[10,11] has suggested that the vestibulocerebellum, receiving vestibular afferent information directly, acts as a "computer" to control only open-loop vestibular reflexes such as the vestibulo-ocular reflex. As described earlier, when vestibulospinal reflexes result in movement of the head, this movement is detected by the labyrinth; the reflexes are of the closed-loop system type, and vestibulo-cerebellar "computer" control is therefore not needed. In other words, the vestibulocerebellum would be expected to exert an inhibitory action only on vestibulo-ocular reflexes and the anterior lobe only on vestibulospinal reflexes. In agreement with this theoretical formulation, Purkinje cells in the flocculus, at least, have been shown to inhibit vestibular neurons projecting to extraocular motor nuclei, but not VST neurons, which are intermingled with them.[24,49]

Action of the vestibulospinal tracts on spinal neurons

Stimulation of the LVST leads to enhanced activity of the extensor musculature,[3] and stimulation of the lateral nucleus evokes excitatory postsynaptic potentials (EPSPs) in extensor motoneurons innervating neck, back, forelimb, and hind limb muscles.[59,77,130] In neck, some back, and some hind limb motoneurons, particularly those of the knee and ankle extensors quadriceps and gastrocnemius soleus, these EPSPs are monosynaptic (Fig. 30-17, *A* 1).[59,130,132] Connections with other hind limb extensor motoneurons and with forelimb extensors are mainly polysynaptic.[59,130] There is usually disynaptic reciprocal inhibition of flexor motoneurons.[59] Through its endings on interneurons the LVST not only exerts polysynaptic effects, such as disynaptic inhibition, on motoneurons, but it can also modify the activity of various segmental reflexes, such as reciprocal inhibition of antagonists and the crossed extensor reflex.[66,115] The presence of a spinal interneuron in many of these pathways means that their effectiveness can be enhanced or diminished, at the spinal level, by changing excitability of this interneuron.

The MVST contains inhibitory as well as excitatory fibers.[25,131] Its main targets are axial motoneurons: there are monosynaptic inhibitory (Fig. 30-17, *B* 1) and excitatory connections with neck and back motoneurons, but not with limb motoneurons.[25,131,132]

RST neurons, which receive less direct labyrinth input than do VST neurons, also exert very direct actions on spinal motoneurons. RST axons cause monosynaptic excitation of axial motoneurons and of some limb flexor and extensor motoneurons, as well as disynaptic inhibition of other limb motoneurons.[58,60,130,132] There is also monosynaptic inhibition of some neck extensor motoneurons.[97] In addition, there are a variety of less direct effects on either motoneurons or spinal reflex pathways.

As expected from the connections of the VSTs, stimulation of the vestibular nerve evokes disynaptic EPSPs and inhibitory postsynaptic potentials (IPSPs) in axial motoneurons (Fig. 30-17, *A* 2 and *B* 2) but not in limb motoneurons, which receive only polysynaptic input.[130,131] When individual ampullary branches of the ves-

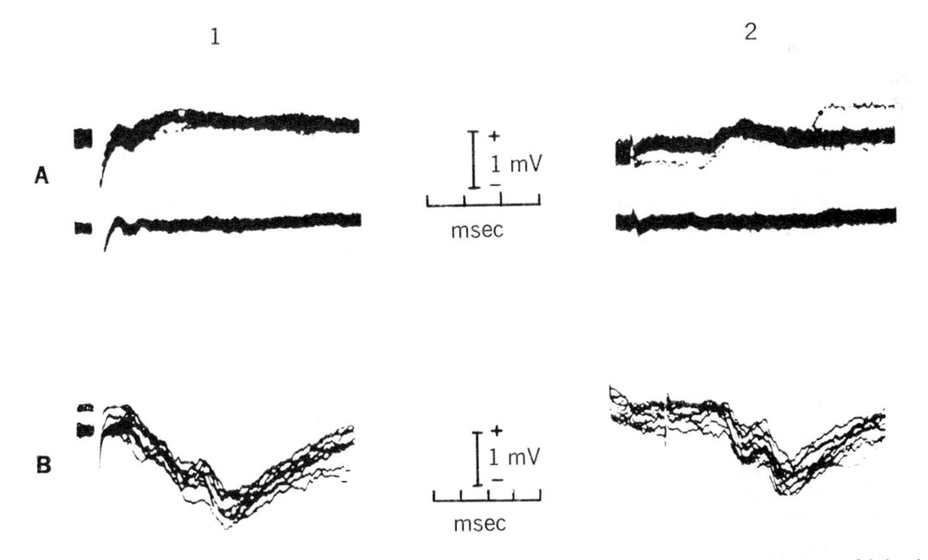

Fig. 30-17. Synaptic potentials evoked in neck extensor motoneurons by stimulation of labyrinth and vestibular nuclei. **A,** EPSPs produced by stimulation of, *1,* lateral vestibular nucleus and, *2,* labyrinth. Upper trace is intracellular record; lower trace is juxtacellular field potential. **B,** IPSPs produced by stimulation of, *1,* medial vestibular nucleus and, *2,* labyrinth. Note in each case the latency difference between labyrinth and nucleus stimulation, corresponding to time required for activation of vestibular nucleus neurons by labyrinth stimulation. (**A** from Wilson and Yoshida[130]; **B** from Wilson and Yoshida.[131])

tibular nerve are stimulated, a very precise pattern of disynaptic potentials is revealed in neck motoneurons.[129] The consistency of this pattern from experiment to experiment means that modifications caused by lesions can be used to detect the contribution made by the two VSTs. This approach shows that the LVST mediates only ipsilateral excitation from the anterior canal. The MVST mediates inhibition bilaterally as well as contralateral excitation. It appears that, for all receptors, the MVST is the main pathway to neck and other axial motoneurons and the LVST the only pathway to limb motoneurons.[25,79,129]

Vestibulospinal tracts and vestibulospinal reflexes

Studies conducted with electrical stimulation emphasize relatively direct connections. More complex pathways may not be activated by single electrical shocks or even by trains of shocks, and they are depressed by anesthetics that are usually employed in experiments. The relative functional importance of direct and complex pathways can therefore not be deduced solely from the effects revealed by electrical stimulation.

The pattern of short-latency connections between canal receptors and neck motoneurons is completely consistent with the head movements expected in response to natural stimulation or produced by ampullary nerve stimulation in behaving cats or monkeys (vestibulocollic reflexes).[113] There is no such consistency between forelimb movements evoked by horizontal angular acceleration and those evoked by stimulation of the horizontal canal nerve with trains of pulses.[79] This suggests that the relatively direct pathways revealed by electrical stimulation form only part of the circuitry involved in production of vestibulospinal reflexes. Indeed, the dynamics of these reflexes, discussed earlier in this chapter, show that direct, short-latency connections do not even fully explain vestibular reflexes of the head.

We have seen (Figs. 30-8 and 30-15) that the phase lag of second-order vestibular neurons is very similar to that of vestibular afferent fibers. The much longer phase lag in the vestibulocollic pathway (Fig. 30-10), similar to that previously seen in vestibulo-ocular reflexes,[106] together with observations on limb muscles,[28,93] shows that complex pathways between vestibular nuclei and motoneurons must be an extremely important part of the neural substrate of the vestibulospinal reflexes described earlier in this chapter. Present thinking is that these complex pathways, which essentially provide one step of integration and therefore transform an input linked to angular velocity to a position response, are located in the reticular formation. Further work is required to determine the relative contribution of the various descending tracts to reflexes evoked over a wide range of conditions, including the frequency of the input acceleration.

CLINICAL SIGNS AND TESTING[13,16,23]

Common signs of vestibular disorders are vertigo, nausea, and nystagmus, as well as disturbances in posture and equilibrium. Vestibular dysfunction may result from peripheral or central lesions of the most varied etiologies. Diseases affecting the labyrinth or vestibular nerve (i.e., Menière's disease, cerebellopontine angle tumors) often also produce tinnitus and hearing loss due to the proximity of the auditory nerve.

The most sensitive tests of labyrinthine and central vestibular function assess vestibulo-ocular rather than vestibulospinal reflexes. One test often used to screen cases of suspected vestibular disturbance is to look for positional nystagmus, that is, nystagmus that occurs when the whole body is moved so as to place the head in various positions in space. Eye movements can be monitored by visual observation or measured by electronystagmography (ENG), with two to six electrodes properly placed around the eyes. Transitory nystagmus is usually indicative of peripheral, persistent nystagmus of central pathology.

The caloric test is the main tool for investigation of semicircular canal function, and the horizontal canal is the one usually tested. The head is elevated approximately 30 degrees from the supine position, placing the horizontal canals in the vertical plane. Nystagmus is produced by irrigating each ear with water above and below body temperature.

This test can be used to examine lateral and anterior, but not posterior, canal function. The stimulus is the temperature gradient across the canal that is in the vertical plane. The resulting change in temperature causes a change in density of endolymph and then vertical movement of fluid under the influence of gravity. This causes deflection of the cupula and therefore eye movement. With water below body temperature, the fast phase of the resulting nystagmus is to the contralateral side; with water above room temperature, it is to the ipsilateral side.

Eye movements evoked by the caloric stimulus can be assessed visually or with ENG. The important parameters of the nystagmus are duration, maximum slow phase speed, and rate of beats at the peak of the response. Unilateral diminution or absence of nystagmus, indicating impairment or loss of canal function in one ear, may be due to peripheral or central disease. There are features of the response that are believed by some to help in distinguishing between these possibilities.

Rotation, in combination with ENG, can also be of value in testing canal function, although it is used rarely. Its main value is for measurement of the threshold of nystagmus and therefore sensitivity of the canal system, for example, when ototoxic drugs have been administered to the patient.

If it is considered necessary to test otolith function, this is done by measuring ocular counterrolling, the reflex conjugate rolling of the eyes observed when the head is tilted. This reflex depends on proper function of the otoliths in at least one labyrinth.

REFERENCES

General reviews

1. Brodal, A.: Anatomy of the vestibular nuclei and their connections. In Kornhuber, H. H., editor: Handbook of sensory physiology: vestibular system, Berlin, 1974, Springer-Verlag, vol. 6, part 1.
2. Brodal, A., and Pompeiano, O., editors: Basic aspects of central vestibular mechanisms, Prog. Brain Res. **37:**entire volume, 1972.
3. Brodal, A., Pompeiano, O., and Walberg, F.: The vestibular nuclei and their connections, Edinburgh, 1962, Oliver & Boyd.
4. Cohen, B.: The vestibulo-ocular reflex arc. In Kornhuber, H. H., editor: Handbook of sensory physiology: vestibular system, Berlin, 1974, Springer-Verlag, vol. 6, part 1.
5. Engstrom, H., Lindeman, H. H., and Ades, H. A.: Anatomical features of the auricular sensory organs. In Second symposium on the role of the vestibular organs in space exploration, NASA SP-115, Washington, D.C., 1966, National Aeronautics and Space Administration.
6. Flock, A.: Sensory transduction in hair cells. In Lowenstein, W. R., editor: Handbook of sensory physiology: principles of receptor physiology, Berlin, 1971, Springer-Verlag, vol. 1.
7. Fregly, A. R.: Vestibular ataxia and its measurement in man. In Kornhuber, H. H., editor: Handbook of sensory physiology: vestibular system, Berlin, 1974, Springer-Verlag, vol. 6, part 2.
8. Gacek, R. R.: Morphological aspects of the efferent vestibular system. In Kornhuber, H. H., editor: Handbook of sensory physiology: vestibular system—basic mechanisms, New York, 1974, Springer-Verlag New York, Inc., vol. 6, part 1.
9. Guedry, F. E.: Psychophysics of vestibular sensation. In Kornhuber, H. H. editor: Handbook of sensory physiology: vestibular system, Berlin, 1974, Springer-Verlag, vol. 6, part 2.
10. Ito, M.: Neural design of the cerebellar motor control system, Brain Res. **40:**81, 1971.
11. Ito, M.: Cerebellar control of the vestibular neurons: physiology and pharmacology. In Brodal, A., and Pompeiano, O., editors: Basic aspects of central vestibular mechanisms, Prog. Brain Res. **37:**377, 1972.
12. Klinke, R., and Galley, N.: Efferent innervation of vestibular and auditory receptors, Physiol. Rev. **54:**316, 1974.
13. Kornhuber, H. H., editor: Handbook of sensory physiology: vestibular system, Berlin, 1974, Springer-Verlag, vol. 6, parts 1 and 2.
14. Lowenstein, O.: The functional significance of the ultrastructure of the vestibular end organs. In Second symposium on the role of the vestibular organs in space exploration, NASA SP-115, Washington, D.C., 1966, National Aeronautics and Space Administration.
15. Melvill Jones, G.: Transfer function of labyrinthine volleys through the vestibular nuclei. In Brodal, A., and Pompeiano, O., editors: Basic aspects of central vestibular mechanisms, Prog. Brain Res. **37:**139, 1972.
16. Naunton, R. F., editor: The vestibular system, New York, 1975, Academic Press, Inc.
17. Roberts, T. D. M.: Neurophysiology of postural mechanisms, New York, 1967, Plenum Publishing Corp.
18. Schaefer, K. P., and Meyer, D. L.: Compensation of vestibular lesions. In Kornhuber, H. H., editor: Handbook of sensory physiology: vestibular system, Berlin, 1974, Springer-Verlag, vol. 6, part 2.
19. Sherrington, C. S.: The integrative action of the nervous system, Cambridge, 1948, Cambridge University Press.
20. Spoendlin, H. H.: Ultrastructure of the vestibular sense organ. In Wolfson, R. J., editor: The vestibular system and its diseases, Philadelphia, 1966, University of Pennsylvania Press.
21. Wilson, V. J.: Physiological pathways through the vestibular nuclei, Int. Rev. Neurobiol. **15:**27, 1972.
22. Wilson, V. J., and Peterson, B. W.: Peripheral and central substrates of vestibulospinal reflexes, Physiol. Rev. **58:**80, 1978.
23. Wolfson, R. J., editor: The vestibular system and its diseases, Philadelphia, 1966, University of Pennsylvania Press.

Original papers

24. Akaike, T., Fanardjian, V. V., Ito, M., and Nakajima, H.: Cerebellar control of vestibulospinal tract cells in rabbit, Exp. Brain Res. **18:**446, 1973.
25. Akaike, T., Fanardjian, V. V., Ito, M., and Ohno, T.: Electrophysiological analysis of the vestibulospinal reflex pathway of rabbit. II. Synaptic actions upon spinal neurones, Exp. Brain Res. **17:**497, 1973.
26. Akaike, T., et al.: Electrophysiological analysis of the vestibulospinal reflex pathway of rabbit. I. Classification of tract cells, Exp. Brain Res. **17:**477, 1973.
27. Allen, G. I., Sabah, N. H., and Toyama, K.: Synaptic actions of peripheral nerve impulses upon Deiters' neurones via the climbing fibre afferents, J. Physiol. (Lond.) **226:**311, 1972.
28. Anderson, J. H., Soechting, J. F., and Terzuolo, C. A.: Dynamic relations between natural vestibular inputs and activity of forelimb extensor muscles in the decerebrate cat. II. Motor output during rotations in the horizontal plane, Brain Res. **120:**17, 1977.
29. Berthoz, A., and Llinás, R.: Afferent neck projections to the cat cerebellar cortex, Exp. Brain Res. **20:**385, 1974.
30. Bizzi, E., Kalil, R. E., and Tagliasco, V.: Eye-head coordination in monkeys: evidence for centrally patterned organization, Science **173:**452, 1971.
31. Byford, G. H.: Eye movements and the optogyral illusion, Aerosp. Med. **34:**119, 1963.
32. Christiansen, J. A.: On hyaluronate molecules in the labyrinth as mechanoelectrical transducers, and as

molecular motors acting as resonators, Acta Otolaryngol. **57:**33, 1964.

33. Corvera, J., Hallpike, C. S., and Schuster, E. H. J.: A new method for the anatomical reconstruction of the human macular planes, Acta Otolaryngol. **49:**4, 1958.
34. Curthoys, I. S., and Markham, C. H.: Convergence of labyrinthine influences on units in the vestibular nuclei of the cat. I. Natural stimulation, Brain Res. **35:**469, 1971.
35. Dichgans, J., Bizzi, E., Morasso, P., and Tagliasco, V.: Mechanisms underlying recovery of eye-head coordination following bilateral labyrinthectomy in monkeys, Exp. Brain Res. **18:**548, 1973.
36. Dohlman, G. F.: Some practical and theoretical points in labyrinthology, Proc. R. Soc. Med. **50:**779, 1935.
37. Duensing, F., and Schaefer, K. P.: Die Aktivatät einzelner Neurone im Bereich der Vestibulariskerne bei Horizontalbeschleunigungen unter besonderer Berücksichtigung des vestibulären Nystagmus, Arch. Psychiatr. Nervenkr. **198:**225, 1958.
38. Duensing, F., and Schaefer, K. P.: Über die Konvergenz verschiedener labyrinthärer Afferenzen auf einzelne Neurone des Vestibulariskerngebietes, Arch. Psychiatr. Nervenkr. **199:**345, 1959.
39. Duensing, F., and Schaefer, K. P.: Die Aktivität einzelner Neurone der Formatio reticularis des necht gefesselten Kaninchens bei Kopfwendungen und vestibulären Reizen, Arch. Psychiatr. Nervenkr. **201:**97, 1960.
40. Ezure, K., Sasaki, S., Uchino, Y., and Wilson, V. J.: A role of upper cervical afferents on vestibular control of neck motor activity. In Homma, S., editor: Understanding the stretch reflex, Prog. Brain Res. **44:**461, 1976.
41. Ferin, M., Grigorian, R. A., and Strata, P.: Mossy and climbing fibre activation in the cat cerebellum by stimulation of the labyrinth, Exp. Brain Res. **12:**1, 1971.
42. Fernandez, C., and Goldberg, J. M.: Physiology of first order afferents innervating the semicircular canals of the squirrel monkey. II. The response to sinusoidal stimulation and the dynamics of the peripheral vestibular system, J. Neurophysiol. **34:**660, 1971.
43. Fernandez, C., and Goldberg, J. M.: Physiology of peripheral neurons innervating otolith organs of the squirrel monkey. I. Response to static tilts and to long duration centrifugal force, J. Neurophysiol. **39:**970, 1976.
44. Fernandez, C., and Goldberg, J. M.: Physiology of peripheral neurons innervating otolith organs of the squirrel monkey. II. Directional selectivity and force-response relations, J. Neurophysiol. **39:**985, 1976.
45. Fernandez, C., and Goldberg, J. M.: Physiology of peripheral neurons innervating otolith organs of the squirrel monkey. III. Response dynamics, J. Neurophysiol. **39:**996, 1976.
46. Fernandez, C., Goldberg, J. M., and Abend, W. K.: Response to static tilts of peripheral neurons innervating otolith organs of the squirrel monkey, J. Neurophysiol. **35:**978, 1972.
47. Flock, A., and Wersäll, J.: A study of the orientation of the sensory hairs of the receptor cells in the lateral line organs of fish with special reference to the function of the receptors, J. Cell Biol. **15:**19, 1962.
48. Fuchs, A., and Kimm, J.: Unit activity in vestibular nucleus of the alert monkey during horizontal angular acceleration and eye movement, J. Neurophysiol. **38:** 1140, 1975.
49. Fukuda, J., Highstein, S. M., and Ito, M.: Cerebellar control of the vestibulo-ocular reflex investigated in rabbit IIId nucleus, Exp. Brain Res. **14:**511, 1972.
50. Gacek, R. R.: Anatomical evidence for an efferent vestibular pathway. In Third symposium on the role of the vestibular organs in space exploration, NASA SP-152, Washington, D.C., 1967, National Aeronautics and Space Administration.
51. Gacek, R. R.: The course and central termination of first order neurons supplying vestibular end organs in the cat, Acta Otolaryngol. (Suppl.) **254:**1, 1969.
52. Gacek, R. R., and Lyon, M.: The localization of vestibular efferent neurons in the kitten with horseradish peroxidase, Acta Otolaryngol. **77:**92, 1974.
53. Goldberg, J. M., and Fernandez, C.: Physiology of first order afferents innervating the semicircular canals of the squirrel monkey. I. Resting discharge and response to constant angular accelerations, J. Neurophysiol. **34:**635, 1971.
54. Goldberg, J. M., and Fernandez, C.: Physiology of first order afferents innervating the semicircular canals of the squirrel monkey, III. Variations among units in their discharge properties, J. Neurophysiol. **34:**676, 1971.
55. Gonshor, A., and Melvill Jones, G.: Extreme vestibulo-ocular adaptation induced by prolonged optical reversal of vision, J. Physiol. (Lond.) **256:**381, 1976.
56. Gray, E. G., and Pumphrey, R. J.: Ultrastructure of the insect ear, Nature **181:**618, 1958.
57. Graybiel, A.: Measurement of otolith function in man. In Kornhuber, H. H., editor: Handbook of sensory physiology: vestibular system, Berlin, 1974, Springer-Verlag, vol. 6, part 2.
58. Grillner, S., and Lund, S.: The origin of a descending pathway with monosynaptic action on flexor motoneurones, Acta Physiol. Scand. **74:**274, 1968.
59. Grillner, S., Hongo, T., and Lund, S.: The vestibulospinal tract. Effects on alpha motoneurones in the lumbosacral spinal cord in the cat, Exp. Brain Res. **10:** 94, 1970.
60. Grillner, S., Hongo, T., and Lund, S.: Convergent effects on alpha motoneurones from the vestibulospinal tract and a pathway descending in the medial longitudinal fasciculus, Exp. Brain Res. **12:**457, 1971.
61. Groen, J. J.: Vestibular stimulation and its effects, from the point of view of theoretical physics, Confin. Neurol. **21:**380, 1961.
62. Groen, J. J., Lowenstein, O., and Vendrik, J. H.: The mechanical analysis of the responses from the end organs of the horizontal semicircular canals in the isolated elasmobranch labyrinth, J. Physiol. (Lond.) **117:** 329, 1952.
63. Hamilton, D. W.: The cilium on mammalian vestibular hair cells, Anat. Rec. **164:**253, 1970.
64. Hardy, M.: Observations on the innervation of the macula saculi in man, Anat. Rec. **59:**403, 1934.
65. Hikosaka, O., and Maeda, M.: Cervical effects on abducens motoneurons and their interaction with vestibulo-ocular reflex, Exp. Brain Res. **18:**512, 1973.
66. Hultborn, H., and Udo, M.: Recurrent depression from motor axons collaterals of supraspinal inhibition in motoneurones, Acta Physiol. Scand. **85:**44, 1972.
67. Ito, M., and Yoshida, M.: The origin of cerebellar-induced inhibition of Deiters' neurones. I. Monosynaptic initiation of the synaptic potentials, Exp. Brain Res. **2:**330, 1966.
68. Ito, M., Kawai, N., and Udo, M.: The origin of cerebellar-induced inhibition of Deiters' neurones. III.

Localization of the inhibitory zone, Exp. Brain Res. **4:**310, 1968.

69. Ito, M., Shiida, T., Yagi, N., and Yamamoto, M.: Visual influence on rabbit horizontal vestibulo-ocular reflex presumably effected via the cerebellar flocculus, Brain Res. **65:**170, 1974.
70. Iurato, S.: Submicroscopic structure of the inner ear, Elmsford, N.Y., 1967, Pergamon Press, Inc.
71. Keller, E., and Daniels, P.: Oculomotor related interaction of vestibular and visual stimulation in vestibular nucleus cells in alert monkey, Exp. Neurol. **46:**187, 1975.
72. Lindsay, K. W., Roberts, T. D. M., and Rosenberg, J.: Asymmetric tonic labyrinth reflexes and their interaction with neck reflexes in the decerebrate cat, J. Physiol. (Lond.) **261:**583, 1976.
73. Loe, P. R., Tomko, D. L., and Werner, G.: The neural signal of angular head position in primary afferent vestibular nerve axons, J. Physiol. (Lond.) **230:**29, 1973.
74. Lowenstein, O., and Roberts, T. D. M.: The equilibrium function of the otolith organs of the thornback ray (Raja clavata), J. Physiol. (Lond.) **110:**392, 1949.
75. Lowenstein, O., and Sand, A.: The mechanisms of the semicircular canal. A study of responses of single fibre preparations to angular accelerations and to rotation of constant speed, Proc. R. Soc. Lond. (Biol.) **129:**256, 1940.
76. Lowenstein, O,. and Wersäll, J.: A functional interpretation of the electron microscopic structure of the sensory hair cells in the cristae of the elasmobranch Raja clavata in terms of directional sensitivity, Nature **184:** 1807, 1959.
77. Lund, S., and Pompeiano, O.: Monosynaptic excitation of alpha motoneurones from supraspinal structures in the cat, Acta Physiol. Scand. **73:**1, 1968.
78. Macadar, O., Wolfe, G. E., O'Leary, D. P., and Segundo, J. P.: Response of the elasmobranch utricle to maintained spatial orientation, transitions and jitter, Exp. Brain Res. **22:**1, 1975.
79. Maeda, M., Maunz, R. A., and Wilson, V. J.: Labyrinthine influence on cat forelimb motoneurons, Exp. Brain Res. **22:**69, 1975.
80. Markham, C. H., and Curthoys, I.: Convergence of labyrinthine influences on units in the vestibular nuclei of the cat. II. Electrical stimulation, Brain Res. **43:** 383, 1972.
81. McCouch, G. P., Deering, I. D., and Ling, T. H.: Location of receptors for tonic neck reflexes, J. Neurophysiol. **14:**191, 1951.
82. Meiry, J. L.: Vestibular and proprioceptive stabilization of eye movements. In Bach-y-Rita, P., Collins, C. C., and Hyde, J. E., editors: The control of eye movements, New York, 1971, Academic Press, Inc.
83. Melvill Jones, G., and Milsum, J. H.: Characteristics of neural transmission from the semicircular canal to the vestibular nuclei of cats, J. Physiol. (Lond.) **209:** 295, 1970.
84. Melvill Jones, G., and Milsum, J. H.: Frequency-response analysis of central vestibular unit activity resulting from rotational stimulation of the semicircular canals, J. Physiol. (Lond.) **219:**191, 1971.
85. Melvill Jones, G., and Watt, D. G. D.: Muscular control of landing from unexpected falls in man, J. Physiol. (Lond.) **219:**729, 1971.
86. Miles, F. A.: Single unit firing patterns in the vestibular nuclei related to voluntary eye movements and passive body rotation in conscious monkeys, Brain Res. **71:** 215, 1971.
87. Money, K. E., et al.: Physical properties of fluids and structures of vestibular apparatus of the pigeon, Am. J. Physiol. **220:**140, 1971.
88. Nashner, L. M.: Sensory feedback in human postural control, Man-vehicle Laboratory, Center for Space Research, Cambridge, Mass., 1970, The M.I.T. Press, MVT-70-30.
89. Nyberg-Hansen, R.: Origin and termination of fibers from the vestibular nuclei descending in the medial longitudinal fasciculus. An experimental study with silver impregnation methods in the cat, J. Comp. Neurol. **122:**355, 1964.
90. Nyberg-Hansen, R., and Mascitti, T. A.: Sites and modes of termination of fibers of the vestibulospinal tract in the cat. An experimental study with silver impregnation methods, J. Comp. Neurol. **122:**369, 1964.
91. Orlovsky, G. N., and Pavlova, G. A.: Vestibular responses of neurons of different descending pathways in cats with intact cerebellum and in decerebellated ones (in Russian), Neirofiziologiia **4:**303, 1972.
92. Outerbridge, J. S., and Melvill Jones, G.: Reflex vestibular control of head movements in man, Aerosp. Med. **42:**935, 1971.
93. Partridge, L. D., and Kim, J. H.: Dynamic characteristics of response in a vestibulomotor reflex, J. Neurophysiol. **32:**485, 1969.
94. Peterson, B. W.: Distribution of neural responses to tilting within vestibular nuclei of the cat, J. Neurophysiol. **33:**750, 1970.
95. Peterson, B. W., and Abzug, C.: Properties of projections from vestibular nuclei to medial reticular formation in the cat, J. Neurophysiol. **38:**1421, 1975.
96. Peterson, B. W., Filion, M., Felpel, L. P., and Abzug, C.: Responses of medial reticular neurons to stimulation of the vestibular nerve, Exp. Brain Res. **22:**335, 1975.
97. Peterson, B. W., Pitts, N. G., Mackel, R., and Fukushima, K.: Reticulospinal excitation and inhibition of neck motoneurons, Exp. Brain Res. **32:**471, 1978.
98. Pollock, L. J., and Davis, L.: The influence of the cerebellum upon the reflex activities of the decerebrate animal, Brain **50:**277, 1927.
99. Precht, W., and Shimazu, H.: Functional connections of tonic and kinetic vestibular neurons with vestibular afferents, J. Neurophysiol. **28:**1014, 1965.
100. Precht, W., Shimazu, H., and Markham, C. H.: A mechanism of central compensation of vestibular function following hemilabyrinthectomy, J. Neurophysiol. **29:**996, 1966.
101. Purdon Martin, J.: Role of the vestibular system in the control of posture and movement in man. In de Reuck, A. V. S., and Knight, J., editors: Myotatic kinesthetic and vestibular mechanisms, Boston, 1967, Little, Brown & Co., pp. 92-95.
102. Ross, D. A.: Electrical studies on the frog's labyrinth, J. Physiol. (Lond.) **86:**117, 1936.
103. Shimazu, H., and Precht, W.: Tonic and kinetic responses of cat's vestibular neurons to horizontal angular acceleration, J. Neurophysiol. **28:**989, 1965.
104. Shimazu, H., and Precht, W.: Inhibition of central vestibular neurons from the contralateral labyrinth and its mediating pathway, J. Neurophysiol. **29:**467, 1966.
105. Shimazu, H., and Smith, C.: Cerebellar and labyrinthine influences on single vestibular neurons identified by natural stimuli, J. Neurophysiol. **34:**493, 1971.
106. Shinoda, Y., and Yoshida, K.: Dynamic characteristics of responses to horizontal head angular acceleration in

vestibulo-ocular pathway in the cat, J. Neurophysiol. **37:**653, 1974.

107. Skavenski, A. A., and Robinson, D. A.: Role of the abducens nucleus in vestibulo-ocular reflex, J. Neurophysiol. **36:**724, 1973.
108. Spoendlin, H.: Organization of the sensory hairs in the gravity receptors in utricle and saccule of the squirrel monkey, Z. Zellforsch **62:**701, 1964.
109. Steer, R. W., Jr., Li, Y. T., Young, L. R., and Meiry, J. L.: Physical properties of the labyrinthine fluids and quantification of the phenomenon of caloric stimulation. In Third symposium on the role of the vestibular organs in space exploration, NASA SP-152, Washington, D.C., 1967, National Aeronautics and Space Administration.
110. Stein, B. M., and Carpenter, M. B.: Central projections of portions of the vestibular ganglia innervating specific parts of the labyrinth in the Rhesus monkey, Am. J. Anat. **120:**281, 1967.
111. Steinhausen, W.: Über die Beobachtung der cupula in den Bogengangsampullen des Labyrinths des lebenden Hechts, Pfluegers Arch. **232:**500, 1933.
112. Suzuki, J.-I.: Vestibular and spinal control of eye movements. In Dichgans, J., and Bizzi, E., editors: Cerebral control of eye movements and motion perception, Basel, 1972, S. Karger.
113. Suzuki, J.-I., and Cohen, B.: Head, eye, body and limb movements from semicircular canal nerves, Exp. Neurol. **10:**393, 1964.
114. Tait, J., and McNally, W. J.: Some features of the action of the utricular maculae (and of the associated action of the semicircular canals) of the frog, Philos. Trans. R. Soc. Lond. (Biol.) **224:**241, 1934.
115. ten Bruggencate, G., and Lundberg, A.: Facilitatory interaction in transmission to motoneurones from vestibulospinal fibres and contralateral primary afferents, Exp. Brain Res. **19:**248, 1974.
116. ten Bruggencate, G., Teichmann, R., and Weller, E.: Neuronal activity in the lateral vestibular nucleus of the cat, Pfluegers Arch. **360:**301, 1975.
117. Trincker, D.: Bestandspotentiale im Bogenganssystem des Meerschweinchens und ihre Anderungen bei experimentellen cupula-Ablenkungen, Pfluegers Arch. **264:**351, 1957.
118. Van Egmond, A. J., Groen, J. J., and Jongkees, L. B. W.: The mechanics of the semicircular canal, J. Physiol. (Lond.) **110:**1, 1949.
119. Vidal, J., et al.: Static and dynamic properties of gravity-sensitive receptors in the cat vestibular system, Kybernetik **9:**205, 1971.
120. Von Békésy, G.: Subjective cupulometry, Arch. Otolaryngol. **61:**16, 1955.
121. Von Békésy, G.: Pressure and shearing forces as stimuli of labyrinthine epithelium, Arch. Otolaryngol. **84:** 122, 1966.
122. Waespe, W., and Henn, V.: Neuronal activity in the vestibular nuclei of the alert monkey during vestibular and optokinetic stimulation, Exp. Brain Res. **27:**523, 1977.
123. Watt, D. G. D.: Responses of cats to sudden falls: an otolith-originating reflex assisting landing, J. Neurophysiol. **39:**257, 1976.
124. Wersäll, J.: Studies on the structure and innervation of the sensory epithelium of the cristae ampullares in the guinea pig, Acta Otolaryngol. (Suppl.)**126:**entire issue, 1956.
125. Wersäll, J., and Bagger-Sjöbäck, D.: Morphology of the vestibular sense organ. In Kornhuber, H. H., editor: Handbook of sensory physiology: vestibular system—basic mechanisms, New York, 1974, Springer-Verlag New York, Inc.
126. Wersäll, J., Flock, Å., and Lundquist, P.-G.: Structural basis for directional sensitivity in cochlear and vestibular sensory receptors, Cold Spring Harbor Symp. Quant. Biol. **30:**115, 1965.
127. Wilson, V. J.: The labyrinth, the brain and posture, Am. Sci. **63:**325, 1975.
128. Wilson, V. J., and Felpel, L. P.: Specificity of semicircular canal input to neurons in the pigeon vestibular nuclei, J. Neurophysiol. **35:**253, 1972.
129. Wilson, V. J., and Maeda, M.: Connections between semicircular canals and neck motoneurons in the cat, J. Neurophysiol. **37:**346, 1974.
130. Wilson, V. J., and Yoshida, M.: Comparison of effects of stimulation of Deiters' nucleus and medial longitudinal fasciculus on neck, forelimb and hindlimb motoneurons, J. Neurophysiol. **32:**743, 1969.
131. Wilson, V. J., and Yoshida, M.: Monosynaptic inhibition of neck motoneurons by the medial vestibular nucleus, Exp. Brain Res. **9:**365, 1969.
132. Wilson, V. J., Yoshida, M., and Schor, R. H.: Supraspinal monosynaptic excitation and inhibition of thoracic back motoneurons, Exp. Brain Res. **11:**282, 1970.
133. Wilson, V. J., Kato, M., Peterson, B. W., and Wylie, R. M.: A single-unit analysis of the organization of Deiters' nucleus, J. Neurophysiol. **30:**603, 1967.

31 The cerebellum

W. T. THACH, Jr.

The cerebellum is a large, discrete part of the central nervous system attached to the brain stem by three pairs of peduncles that contain efferent and afferent nerve fibers. Although much is known about its external connections and intrinsic circuits, its exact functions remain obscure. Historically, attempts to understand its role resemble the approaches directed at other parts of the nervous system: (1) studies of its anatomy and pure conjectures on its function, (2) experiments on the behavioral effect of ablation and electrical stimulation, (3) experiments combining electrical stimulation and recording within the nervous system in a further attempt to define input and output connections, (4) refinement of these last two techniques to examine the behavior of cellular components (and the action at synapses linking them) of the nerve circuits within the cerebellum, and (5) development of techniques to record the electrical discharges of single cerebellar neurons in awake animals during natural behavior. These approaches are used here to outline the knowledge that has accumulated. In brief, anatomic and electroanatomic studies have shown that the cerebellum projects mainly, if not exclusively, to other parts of the nervous system known to control movement and that the cerebellum receives signals from every conceivable source of information, from primary sensory afferents to the most remote parts of cerebral cortex, that could be used to control movement. Ablation studies have shown that cerebellar damage causes defects primarily if not exclusively in movement of the eyes and body and that different parts of the cerebellum control different aspects or kinds of movement. Studies of the intrinsic cerebellar circuitry have revealed that the output is generated by deep nuclear cells that excite distant targets and that nuclear cells are in turn controlled by excitatory afferents that do not come from the cerebellar cortex and by the Purkinje cells in the cerebellar cortex that descend to inhibit the nuclear cells. Recordings from single neurons in the cerebellum have begun to reveal some of the interactions of elements within the basic repeated circuit and at the same time how different parts of the cerebellum may participate in individual motor activities.

ANATOMY AND MORPHOLOGY OF CEREBELLAR CORTEX

Early in the course of its embryologic development[59] the cerebellum is subdivided by *transverse fissures* into a series of *primary lobules,* which are later subdivided into smaller transverse *folia*. The continuity between the medial and lateral parts of each of these lobules is apparent at this time. Later the appearance of deeper, *secondary fissures* and the disproportionate growth of the lateral portions of some lobules obscure this simple arrangement. In the 175 mm human embryo, for example (Fig. 31-1), the continuity of the pyramis and paramedian lobules is still obvious, whereas that of the uvula and paraflocculi is disappearing. Further growth causes great expansion of the lateral lobules of the cerebellum, but vestiges of the original continuity with midline lobules remain in the form of slender cortical bridges or fibrous cords in the depths of the fissures.

Larsell[61] proposed that there are 10 "primary" lobules in mammals. These are shown in Fig. 31-2 as they appear in midline sagittal section and on the surface of the macaque cerebellum. In anteroposterior sequence, they are as follows: I, the *lingula,* a small lobule adjacent to the anterior medullary velum; II and III, a pair, called together the *lobulus centralis;* IV and V, a larger pair called the *culmen;* VI, the *lobulus simplex;* VII, the *folium* and *tuber vermis,* which are continuous laterally with the *ansiform lobules;* VIII, the *pyramis,* which is continuous with the *paramedian lobules;* IX, the *uvula,* which is continuous with the *ventral paraflocculi;* and X, the *nodulus,* adjacent to the posterior medullary velum and continuous with the *flocculi*. It is customary to refer to the unpaired median portion of the cerebellum as the *vermis* and to the lateral masses as the *hemispheres*. In the anterior lobe (lobules I to V) and lobulus simplex the vermis

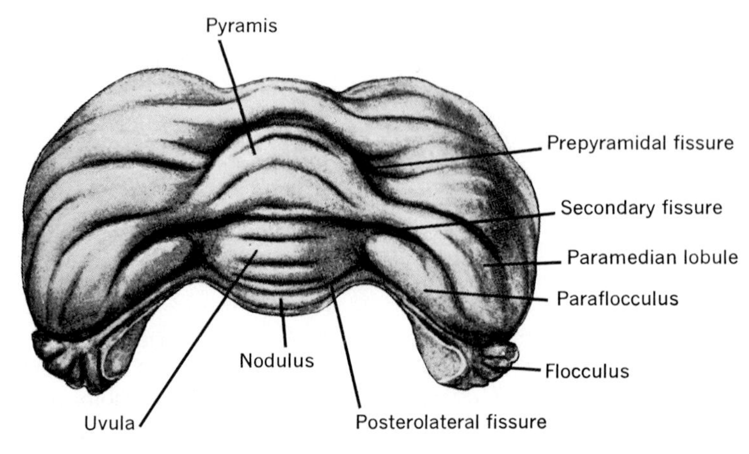

Fig. 31-1. Posterior surface of cerebellum in 175 mm human fetus. (Relabeled from Larsell.[60])

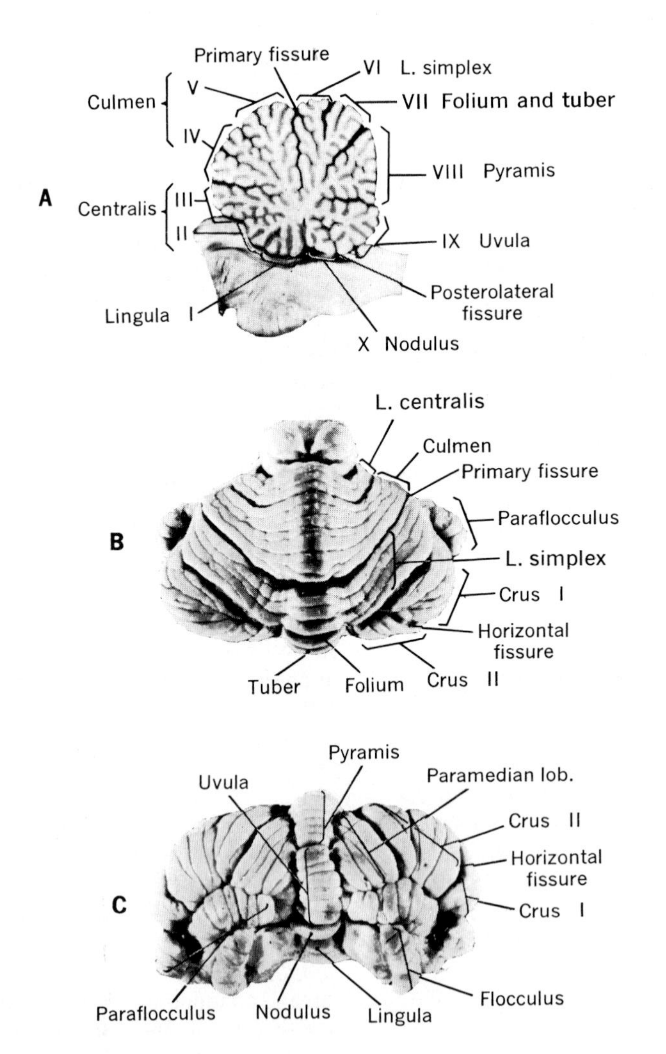

Fig. 31-2. Photographs of three specimens of cerebellum of macaque, prepared to show morphologic subdivisions as described in text. Fissures were opened and some were packed with cotton. **A,** Sagittal section. **B,** Dorsal view. **C,** Ventral view.

is not sharply marked off from the hemispheres as it is elsewhere. The names of the lobules (I to X) apply to both medial and lateral portions.

In the developing embryo,[60] deep horizontal fissures further divide the cerebellum into morphologic subdivisions. The *posterolateral fissure* first separates the *nodulofloccular lobe* (lobule X) from the body of the cerebellum (lobules I to IX). The *primary fissure* (actually the second fissure to appear) next divides the body of the cerebellum into *anterior* (lobules I to V) and *posterior* (lobules VI to IX) *lobes*. The *horizontal* or *intercrural fissure* then divides the ansiform lobules (lateral portion of lobule VII) into *crus I* and *crus II*. These fissures, which sequentially appear in embryologic development, have given rise to ontogenetic and phylogenetic designations[59] for the morphologic cerebellar subdivisions thereby created: the nodulofloccular lobe has been called the *archicerebellum*, the anterior lobe the *paleocerebellum*, and the posterior lobe the *neocerebellum*. Another and in some ways preferable method of subdividing the cerebellum is based on input-output connections.[11,31,55]

Connections and subdivisions based on connections

The cerebellum is joined to the rest of the nervous system by the superior, middle, and inferior *peduncles*, which contain afferent and efferent fibers. The origin and distribution of the fiber systems are summarized, for reference purposes, in Table 31-1 and Fig. 31-3. The wealth of *input fibers* passing to the cerebellum indicates the extent to which the coordination of movement depends on many factors. The impulses carried by these fibers may be thought of as being of two kinds: (1) those arising in sensory receptors and relayed more or less directly to the cerebellum[8,55,80] and (2) those from higher levels of the nervous system.[2,8,29,55] Studies using electrical techniques have revealed in considerable detail how the cerebellar cortex is organized into "receiving areas" for these afferent projections. In turn, the cerebellar cortex sends the axons of its Purkinje cells to the cerebellar nuclei. This corticonuclear system is organized in a sagittal pattern.[54,55] The vermis projects to the medial or *fastigial nucleus* and is represented there in the same rostrocaudal order as the folia. Intermediate areas of cortex send fibers to the *nucleus interpositus* (*n. globosus* and *n. emboliformis* of man). The most lateral portions of the cortex, except the flocculus, project to the lateral or *dentate nucleus*. The *output fibers*[2,29,55] of the cerebellum arise in its nuclei, except those running directly from the nodulofloccular lobe cortex to the vestibular nuclei. The fastigial nucleus sends fibers through the inferior peduncle to the vestibular nuclei and to the reticular formations of the pons and the medulla. The interposed and dentate nuclei project through the superior peduncle to the red nucleus and thalamus. The influence of the cerebellum is therefore exerted on motoneurons through the vestibulospinal, reticulospinal, and rubrospinal pathways and on the precentral motor cortex through the ventrolateral nucleus of the thalamus.

Vestibulocerebellum. The nodulofloccular lobe (archicerebellum) has afferent and efferent connections that are primarily vestibular, justifying another name, vestibulocerebellum. The adjacent uvula also receives some vestibular fibers. It has recently been shown that this part of the cerebellum also receives climbing fiber inputs activated from the retina (visual)[66] and mossy fiber inputs from parts of the brain concerned with eye movements[62,72] or vision[67] while tracking a visual stimulus.

Spinocerebellum. The vermis (projecting to fastigial nuclei) and an intermediate zone (projecting to interposed nuclei) have mossy and climbing fiber inputs from spinocerebellar pathways. These are not the only inputs, but they define the subdivision. There appear to be two receiving areas: *one in the anterior lobe and lobulus simplex*[1,87] and *one in the paramedian lobules of the posterior lobe*[87] (Fig. 31-4). By these input criteria, the anterior lobe, lobulus simplex, and paramedian lobules together constitute a subdivision of the cerebellum that is functionally homogeneous. The fibers conveying somatosensory impulses to the anterior lobe are either of spinal origin or arise in brain stem nuclei that relay spinal impulses. The lobulus simplex receives somatosensory information from the head. The receptors in which this afferent activity arises are located chiefly in muscles, joints, and skin.[1,8,80,92] Impulses from the motor and somesthetic cerebral cortex reach the same areas via the corticopontocerebellar system.* The anterior lobe, lobulus simplex, and paramedian lobules thus receive signals from the central regions in which voluntary movements are thought to originate, as well as from the head and neck, trunk, and limbs that execute them.

The anterior lobe and lobulus simplex have been more intensely studied than the paramedian lobules. "Input mapping" has been done as follows: If physiologic stimuli are applied to the

*See references 1, 2, 13, 29, 55, and 86.

Table 31-1. Afferent systems to cerebellum*

Tract	Origin	Via	Distribution	Impulses transmitted
Dorsal spinocerebellar	Clarke's column (T_1-L_2)	I.C.P.	Chiefly uncrossed to vermis and intermediate part of anterior lobe and pyramis; some fibers to tuber, uvula, and medial part of paramedian lobule	Proprioceptive (muscles and joints) and exteroceptive (skin), from trunk, hind limb, and tail
Ventral spinocerebellar	"Border" cells of ventral horn	S.C.P.	Crossed and uncrossed to vermis of anterior lobe	Proprioceptive (muscles and joints) and exteroceptive (skin), from all parts of body
Cuneocerebellar	External arcuate nucleus	I.C.P.	Uncrossed to vermis and intermediate part of anterior lobe and pyramis; some fibers to uvula and tuber	Proprioceptive, from upper limb and neck
Olivocerebellar	All parts of inferior olive	I.C.P.	Chiefly crossed to all parts of cortex and all intracerebellar nuclei; partly uncrossed to nucleus fastigii	From all levels of spinal cord; from higher nuclei and from cerebral cortex
Pontocerebellar	All parts of pontine gray	M.C.P.	Chiefly crossed to all cortex except nodulofloccular lobe; partly uncrossed to vermis	From cerebral cortex: motor and sensory mostly to intermediate zone cerebellum; "association" mainly to lateral zone cerebellum
Reticulocerebellar	Lateral reticular nucleus	I.C.P.	Uncrossed to entire cerebellar cortex	From all levels of spinal cord and from higher levels
	Paramedian reticular nucleus	I.C.P.	More than half uncrossed to anterior lobe; some to pyramis, uvula, and nucleus fastigii	From higher levels, including cerebral cortex
Vestibulocerebellar	Vestibular nuclei, chiefly medial and descending; some direct vestibular root fibers	I.C.P.	Secondary fibers (crossed and uncrossed) to nodulofloccular lobe, some to uvula and nucleus fastigii; primary fibers to same areas, uncrossed	Vestibular
Perihypoglossocerebellar	Nucleus of Roller Nuclear praepositus Nucleus intercalatus	I.C.P.	More than half uncrossed to anterior lobe; some to pyramis, uvula, and nucleus fastigii	Unknown
Tectocerebellar	Quadrigeminal bodies	S.C.P.	Chiefly crossed, probably to declive, folium, and tuber	Auditory and visual
Rubrocerebellar	Caudal two thirds of red nucleus	S.C.P.	More than half crossed, chiefly to dentate nucleus; some to nucleus fastigii	Cerebral?
Trigeminocerebellar	Direct sensory fibers; secondary fibers from all parts of trigeminal nucleus	I.C.P.	Forming part of commissura cerebelli; to dentate nucleus	Tactile and proprioceptive, from face to jaw
Lateral cervical cerebellar	Lateral cervical nucleus ($C_{1,2}$)	I.C.P.	Unknown	From all levels of spinal cord

*Condensed, with modifications, from a table supplied by Brodal. I am indebted to Professor Brodal for compiling these data on the afferent connections. In addition to older, well-known work, the table contains information derived from more recent investigations, especially those of Professor Brodal and his collaborators, including some of their unpublished results. It should be emphasized that the connections listed are not all equally well-established anatomically. For a detailed description of the afferent systems to the cerebellum and the list of references on which this table is based, the reader may consult Jansen and Brodal.[55]

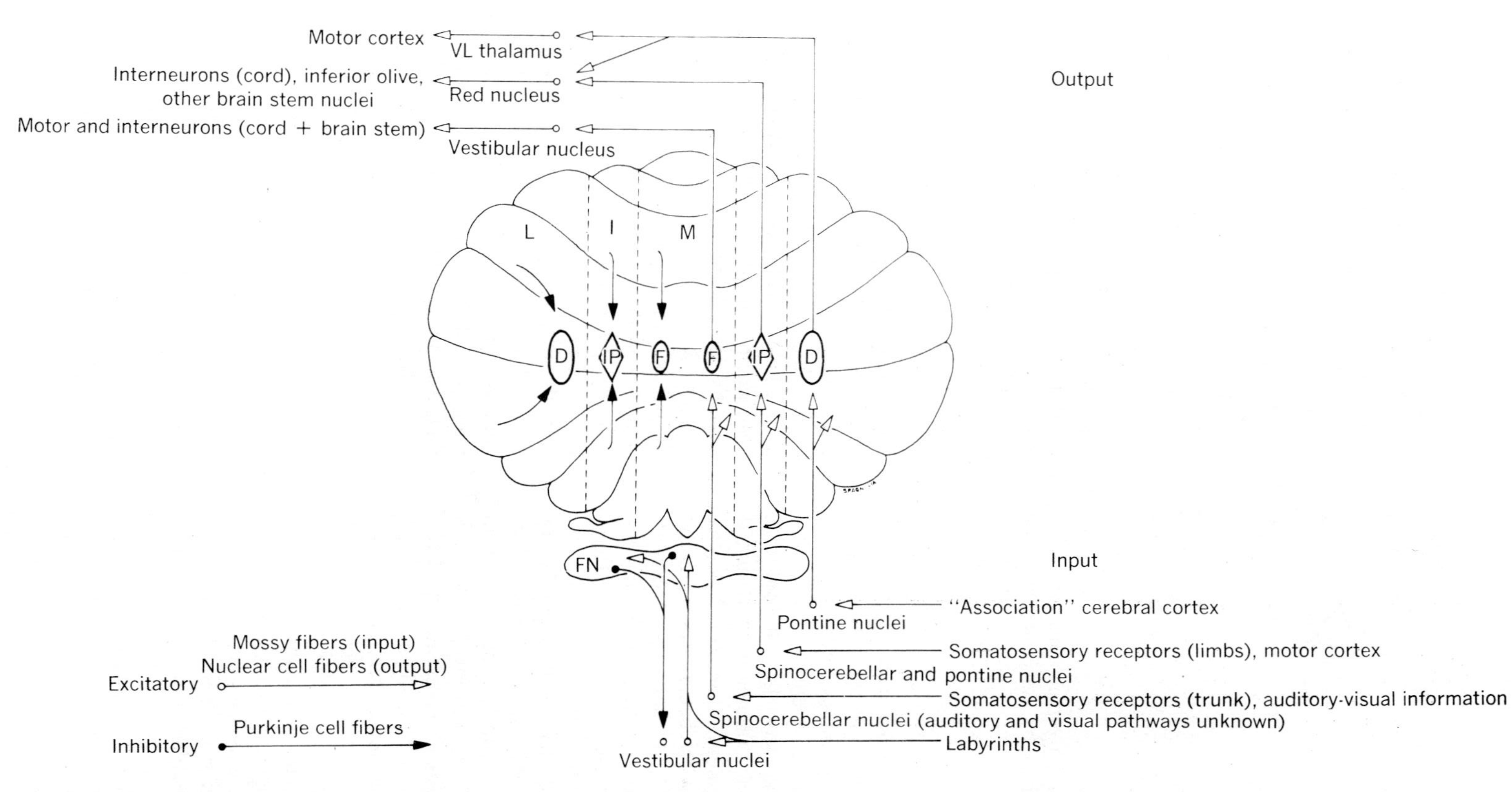

Fig. 31-3. Diagram of gross cerebellar organization. Transverse cortical folds comprise folia, lobules, and lobes. Shown here is longitudinal pattern of projection of cortical Purkinje cells onto deep nuclei and their targets. *Mossy fiber* inputs often branch to reach both nuclei and cortex. Also shown is origin of mossy fibers supplying different subdivisions, and suspected modalities of information they carry. Pontine and medullary tegmental reticular nuclei supply all of cerebellum with *mossy fibers* (not shown). Inferior olivary complex supplies all of cerebellum with *climbing fibers* (not shown). *VL*, Ventrolateral (nucleus of thalamus); *L*, lateral; *I*, intermediate; *M*, medial; *D*, dentate nucleus; *IP*, interposed nucleus; *F*, fastigial nucleus.

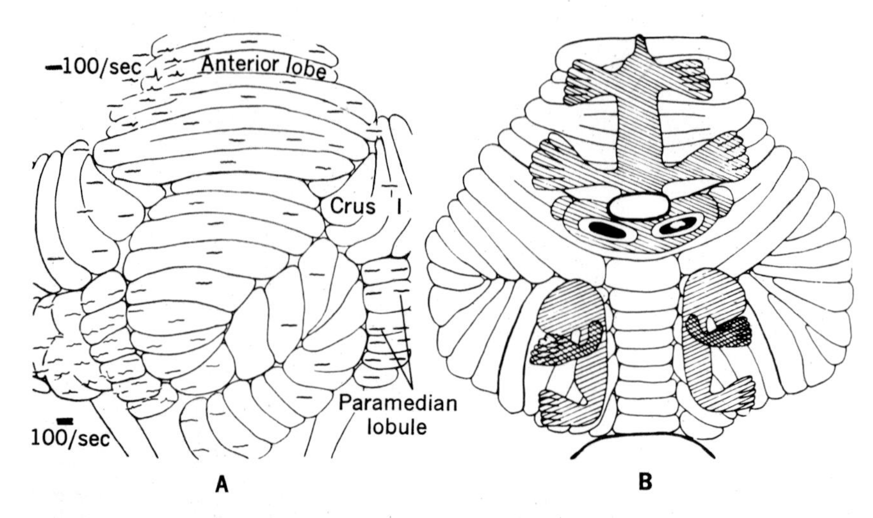

Fig. 31-4. Tactile receiving areas of cerebellum. **A,** Distribution of potentials evoked in cerebellar cortex of cat by movement of hairs around pads of left hind paw (sodium pentobarbital anesthesia). **B,** Pattern of tactile representation in cerebellum of macaque. Representation is ipsilateral in anterior lobe and bilateral in paramedian lobules. (**A** from Snider and Stowell[87]; **B** from Snider.[85])

end-organs of certain sensory systems, it is possible, in animals under a critical level of anesthesia,[21] to record the potentials evoked in small localized areas of the cerebellum. Using this technique, Snider and Stowell[87] showed that movement of a few hairs of a cat's skin elicited responses in discrete areas of anterior lobe and lobulus simplex of the cerebellar cortex. As the example in Fig. 31-4, *A,* indicates, these responses are sufficiently localized to permit one to map the tactile representation of the body in considerable detail on the surface of the cortex. The input "body map" that results is similar for cat and monkey, the tail being represented in lingula, the hind limb in centralis, the forelimb in culmen, and the neck and head in simplex. Each half of the body surface is represented ipsilaterally in the anterior lobe and lobulus simplex, with the trunk medially and the extremities laterally (Fig. 31-4, *B*). The *somatotopic pattern* demonstrated for touch apparently holds also for the projections from muscles and joints.[1,8,80,96] Impulses set up physiologically by pressure on footpads or manipulation of muscles and joints reach the same areas as do those from the overlying skin.[1,96] Electrical stimulation of sensory nerves from these structures also evokes responses in the appropriate areas.[21] In addition, as Adrian showed, the face, arm, and leg subdivisions of the cerebral motor cortex project to the corresponding areas in the cerebellum.[1,42,86] Thus the input via the pons from the motor cortex of the cerebrum is also somatotopically organized, and a second input map overlaps that of the spinocerebellar projection. "Output mapping" has also been performed.[43,44,69,75] In the absence of anesthesia, movements can be elicited by stimulating the cerebellar cortex. Movements of the head, including the face and jaws, are activated from the lobulus simplex, the forelimbs from the culmen, the hind limbs from the centralis, and the tail from the lingula. Trunk muscles are represented medially and extremities laterally. This pattern, discovered in experiments on decerebrate animals,[43] has also been found in intact animals through the use of implanted electrodes.[69] Other studies show that movements elicited from the *cerebral* cortex can be facilitated or inhibited by cerebellar stimulation[75] and thus also reveal a somatotopic arrangement of small effector foci (i.e., an output map) in the cerebellar cortex.

The three kinds of body map demonstrated in anterior lobe and lobulus simplex all overlap. The discrete movements caused by focal stimulation and the localization of cerebral cortical and somatosensory inputs thus suggest that each subdivision of the anterior lobe and lobulus simplex may serve as an effector area for a different part of the body.

In addition to its effects on large motoneurons[41,91] that supply skeletal muscle fibers, the anterior lobe also exerts an influence on the small motoneurons that innervate muscle spindles.* Excitation of its vermian

*See references 33, 34, 39, 41, and 46.

portion inhibits small motor nerve discharge to spindles of extensor muscles, thereby resulting in a decrease in the rate of discharge of these stretch receptors. Stimulation of lateral sites has an opposite effect, producing an increase in the rate of firing from extensor spindles.[39]

The "pyramidoparamedian lobule," as previously stated, has somatosensory input connections resembling those of the anterior lobe and lobulus simplex, and there is evidence that the two areas are interconnected.[6] Spinal inputs come via spinocerebellar fibers and brain stem relays, and there is also a cerebral projection via pons and inferior olive. As in the anterior lobe and lobulus simplex, spinocerebellar inputs are somatotopically organized. Input mapping[87] (Fig. 31-4, *B*) shows that tactile stimulation of the left hind paw elicits potentials in the left paramedian lobule as well as in the anterior lobe. This "second" tactile receiving area is reminiscent of the "second" somatic area in the cerebral cortex (Chapter 11). As in the latter, the cutaneous surface of the body is represented bilaterally and with considerable overlapping. The electrical responses evoked on the side contralateral to stimulation (not shown in Fig. 31-4) are smaller and have a longer latency than those evoked ipsilaterally. The somatotopic pattern, as determined for the monkey, is illustrated in Fig. 31-4, *B*. The face is represented in the superior folia, the arm in the middle folia, and the leg in the inferior folia. The location of the trunk is not well established. It may be represented not as shown but in the pyramis. By stimulating the superior laryngeal nerve,[58] which is the principal sensory supply of the larynx, potentials have been elicited in a small zone located on the medial third of the adjacent lips of folia 3 and 4 of the paramedian lobule bilaterally. This is a logical receiving area for impulses from the larynx, which is derived from the fourth and fifth branchial arches and hence should be represented between the face and upper extremity. A "first" receiving area, located more anteriorly, was also found.

From these observations, and from others not cited, it appears that the same sensory systems that project to the anterior lobe are also represented and are somatotopically arranged in a "second map" in the pyramidoparamedian lobule. Output mapping by electrically stimulating the pyramidoparamedian lobule[43] in cats, dogs, and monkeys elicits ipsilateral and sometimes bilateral movements that are usually less localized than those evoked from the anterior lobe. Movements of facial muscles are evoked from the upper folia, forelimbs from the middle folia, and hindlimbs and tail from the lower folia. Movements of the limbs, generally bilateral, and of the tail are evoked by stimulating the pyramis. To an extent, therefore, the body maps observed by the two techniques would seem to overlap, as seen previously in anterior lobe and lobulus simplex.

One might consider here the remaining vermian structures that give rise in phylogeny to and anatomically connect the ansiform lobes. These are the folium (connecting crus I) and the tuber (connecting crus II). Like the vestibulocerebellar cortex and vermian and intermediate portions of the spinocerebellar cortex, these midline structures also can be excited by stimulation of primary afferents, that is, those of the auditory and visual systems.[87] Using clicks as auditory stimuli, Snider and Stowell[87] recorded evoked potentials in the folium, tuber vermis, and medial part of the lobulus simplex. Within this area there are usually two distinct zones of response: one anteriorly in the lobulus simplex and folium vermis and a second in the tuber vermis. The projection probably reaches the cerebellum via the inferior colliculus and tectocerebellar tract. Destruction of the colliculi abolishes all cerebellar auditory responses, whereas precollicular decerebration does not. By stimulating the retina with brief flashes of light, these investigators[87] also demonstrated a visual receiving area in the same portion of the cerebellum. In some experiments there were two zones of response, one in the lobulus simplex and folium vermis, the other in the tuber vermis and sometimes in the rostral folia of the pyramis. Although the latency of the responses is long, they do not depend on a relay from the cerebrum, for they are obtainable in cats from which all neocortex has been removed. The visual projection probably involves the superior colliculus. Excitation of the folium and tuber vermis results in turning of the head and eyes to the side of stimulation. Thus audio-visual receiving areas appear to overlap those of somatosensory input from the head and neck.

The uvula and paraflocculus have been little studied. The uvula is a relatively constant structure, but the paraflocculus varies greatly in different species. Its dorsal division is usually continuous with the pyramis and its ventral portion with the uvula. In diving mammals such as the whale and porpoise the ventral paraflocculus is enormous; in some whales, it constitutes almost half of the total mass of the cerebellum.[53] The uvula also receives impulses of spinal origin,[23] as well as a contribution from the vestibular system. Its relation to motion sickness is noted later.

Lateral portions of the hemispheres. Unlike the vestibulocerebellum and spinocerebellum, the lateral hemispheric portions of the ansiform lobe do not receive direct or rapidly conducted sensory information.[2,29,55] Rather, as electrophysiologic and anatomic studies have shown, the chief inputs appear to be relayed by the pons from the cerebral cortex.[2,29,55] Unlike the spinocerebellar cortex, which receives its cerebral input chiefly from the motor cortex,[2,29] the lateral hemispheric portions of the ansiform lobe appear to receive inputs chiefly from nonmotor or sensory "association" areas in the frontal, parietal, temporal, and occipital lobes of the cerebral cortex.[2,29] No overt movements have been noted from electrical stimulation of the ansiform lobule.

Cerebellar nuclei

The deep cerebellar nuclei are the major source of output of the cerebellum. The principal exception is the output from the vestibulocerebellum (chiefly the nodulofloccular lobe), which consists of Purkinje cell axons projecting to the vestibular nuclei.[11,55] But this really is no exception, since the vestibular nuclei can be considered as the "deepest" of the cerebellar nuclei; like the cerebellar nuclei proper, they receive a large projection from cerebellar Purkinje cells and also receive inputs that branch to the cerebellar cortex (e.g., primary vestibular afferent fibers). If the vestibular nuclei are included as thus belonging among the cerebellar nuclei, virtually all the cerebellar output is generated by the cerebellar nuclei.

The pattern of projection of cerebellar cortical Purkinje cells onto the cerebellar nuclei has justified a new way of thinking about the regional subdivisions of the cerebellar cortex.[18,54,55] It has been found that in the body of the cerebellum exclusive of the flocculonodular lobe, the cerebellar cortex is arranged in longitudinal zones or strips; midline vermian Purkinje cells project to the fastigial nucleus, lateral hemispheric Purinkje cells (ansiform and to some extent anterior lobes) project to the dentate nucleus, and a longitudinal zone in between the medial and lateral zones projects to the interposed nucleus (globose and emboliform in man). This arrangement—longitudinal cortical zones projecting discretely to the underlying nuclei—is consistent with the way in which somatosensory projections arrive somatotopically organized. The limbs project to an intermediate zone of the hemispheres of not only the anterior lobe, but also the paramedian lobule of the posterior lobe, and it is likely that head, trunk, and tail project to the pyramis as well as to the anterior lobe and lobulus simplex (Fig. 31-4, *B*). The revised view is also consistent with the observation that spinocerebellar projections do *not* go to the lateral hemispheric portions of anterior and ansiform lobes. In the revised view the term "spinocerebellum" would include the medial and intermediate zones of both the anterior lobe and of much of the corpus cerebelli posterior to it. The term "neocerebellum" would include lateral portions of the hemispheres, principally the ansiform lobes. This method of subdivision is often confusing to students, because of Larsell's coronal method of subdivision, in which the neocerebellum included not only the ansiform lobes, but also the simplex and paramedian lobules, and even the midline declive and tuber vermis. The flocculonodular lobe is unique in either system of subdivision, being largely midline but receiving no spinocerebellar afferents and projecting to the vestibular nuclei but not the fastigial.

Stimulation of the cerebellar nuclei results in movements that are similar to, but less localized than, those evoked from the cerebellar cortex.[44] The difference may simply be due to the smaller size of the nuclei and the difficulty of restricting the spread of the stimulating electrical current. Fastigial and globose stimulation usually produces widespread effects, often involving the trunk and all four limbs. A single stimulus appears to excite structures that incorporate the mechanism of reciprocal innervation, so that the flexors in a limb may contract and the extensors relax, with the opposite pattern occurring in the contralateral limb. Emboliform stimulation evokes movements that are generally limited to the ipsilateral forelimb. Dentate stimulation usually evokes no obvious movement.

It is unclear at the present time how the body is "mapped" within the cerebellar nuclei. One view, based on analysis of deficits after focal ablations of the nuclei, is that vestibular nuclei control eye and axial body movements under labyrinthine direction, fastigial nuclei control axial movements, interpositus nuclei control proximal limb fixations, and dentate nuclei control movements of face and distal extremities (hand and foot).[35] A second view, based on areas of cat cerebral cortex activated by focal stimulation of cerebellar nuclei, stipulates just the reverse: that interpositus nuclei control distal musculature and dentate nuclei control more proximal musculature.[68] A third view, based on areas of monkey cerebral cortex activated by stimulation of cerebellar nuclei, proposes that the fastigial nucleus controls tail and hindlimbs, the interpositus controls trunk and proximal parts of the extremities, and the dentate controls distal parts of the forelimb and face.[84] A fourth view is that movements of all parts of the body are represented in each of the deep nuclei (hind limb in the anterior, head and neck in the posterior) and that each nucleus controls different aspects of the movements.[2,19,20,97] More controlled experimentation specifically directed to this issue is needed.

DISORDERS OF MOVEMENT CAUSED BY ABLATION OF CEREBELLAR SUBDIVISIONS

Vestibulocerebellum. Removal of the flocculus or the nodulus causes deficits in the vestibular control of posture and movement. Dow[24,25] has shown that the ablation of the nodulus, with or without the uvula, produces in monkeys a *syndrome of disequilibrium,* characterized by falling, oscillations of the head and trunk, staggering gait on a broad base, and reluctance to move about without artificial support of the body against gravity. Actions in which vestibular control is not involved are apparently unimpaired; for example, voluntary and postural reflex movements of the extremities are well performed if the trunk is supported. Lesions in other portions of the cerebellum do not cause the syndrome. Carrea and Mettler[17] have shown that removals of the flocculus also cause disequilibrium, but that excision of the lingula does not add to the defect. They emphasize that postural abnormalities resembling the attitudes adopted during elicitation of the tonic neck and labyrinthine reflexes also result. Ablation of the nodulus and uvula causes an arms-extended, legs-flexed posture like that observed when the neck is extended, whereas floccular removals result in the opposite pattern. Eye movements are also abnormal[24,25]; there is a nystagmus, usually horizontal, with the slow phase away from and the quick phase toward the side of the lesion. This is opposite the direction of the nystagmus resulting from a lesion of the vestibular nerve on that side, and it is as though the vestibular nuclei on that side, suddenly released from the tonic high-frequency inhibitory discharge coming down from the Purkinje cells, become hyperactive, and the imbalance causes effects similar to those seen when the opposite vestibular nerve or nuclei contain lesions.[4]

Removal of vestibular portions of the cerebellum has another significant effect. According to Bard and collaborators,[5,98] it confers apparent *immunity to motion sickness*. Motion sickness is characterized in dogs by salivation, licking, swallowing, and vomiting. Normal animals who developed motion sickness due to swinging were subjected to various operations and then retested for changes in susceptibility to motion sickness. Ablation of the entire cerebellum rendered dogs outwardly immune to motion sickness. Removals of the nodulus together with the uvula and pyramis had the same effect, whereas excision of other accessible portions of the cerebellum did not alter susceptibility. No significant alterations were detected after bilateral removals of various areas of cerebral cortex. Complete decerebration was equally ineffective; one chronically decerebrate dog continued to vomit promptly in response to swinging throughout a postoperative period of 145 days. The results implicate the nodulus and uvula in the genesis of motion sickness.

A final result of ablation of the vestibulocerebellum is its effect on the plasticity of the vestibulo-ocular reflex. The normal reaction when the head is turned one way is to turn the eyes an equal amount in the opposite direction, which keeps the eyes fixed in space and visually on target. This reaction becomes inappropriate when the visual world is reversed right and left with prisms. Gonshor and Melville Jones[36-38] showed that under such conditions the movement of eyes in a direction opposite that of head movement gradually reduced to zero and then actually reversed so as to be in the same direction as head movement. This reversal of the vestibuloocular reflex was present also in the dark without visual input. The reversal was visually appropriate under the circumstances, but quite altered from the normal vestibulo-ocular reflex, which emphasizes the extreme plasticity of this apparently simple reflex arc. Ito[52] and Robinson[83] have shown that under similar conditions the vestibulo-ocular reflex can be altered in rabbits and cats. But when the vestibulocerebellum is ablated in these animals, the reflex immediately returns to near its initial value and is no longer alterable. These two ablation studies present evidence that the vestibulocerebellum is somehow involved in plasticity of the vestibulo-ocular reflex.

The type of disturbance produced by cerebellar medulloblastomas in children also speaks for localization. These tumors develop from cell rests in the nodulus. In their early stages, before damage is extensive, they cause unsteadiness in walking, loss of balance, and falling. There is otherwise no incoordination of arms and legs. The syndrome is essentially a disorder of equilibrium similar to that which follows removal of the nodulus in monkeys.

Spinocerebellum. In subprimate forms, removal of the anterior lobe in its entirety causes a marked increase in extensor tone throughout the body.[25] Stretch reflexes become hyperactive, with well-defined lengthening and shortening reactions; positive supporting reflexes are accentuated. Lesions restricted to one side of the vermis result in extensor rigidity in the ipsilateral legs and marked flexion in the contralateral limbs. Lateral removals cause opposite changes. Effects are confined to the forelimb if the culmen is removed and to the hind limb if the centralis is damaged.[17] Similar, although less pronounced, alterations in tone and posture occur in monkeys after such removals. Since changes in muscle tone, in its broadest sense, necessarily underlie all movement, reflex or voluntary, lesions that produce such changes should impair any type of muscular performance. Carrea and Mettler[17] found that anterior lobe removals in monkeys caused pronounced ataxia and tremor during all actions. Errors in rate, range, force, and direction similar to cerebellar ataxia in man were seen. Vermian lesions resulted in ataxia of the trunk; lateral removals affected the extremities. It should be stressed that although there was great incapacity due to incoordination, no paralysis was present.

The observations of Carrea and Mettler appear to be in harmony with the experimental facts already noted. Fulton,[31] however, limited the function of the anterior lobe to regulation of posture, stating that ataxia and tremor during voluntary movements result from posterior lobe lesions. The majority of textbook accounts follow his lead in this respect. To the extent that posterior lobe lesions involve the pyramidoparamedian lobule, which appears to resemble the anterior lobe functionally, they may indeed produce ataxia. There is no evidence, however, that voluntary movements are uniquely represented in a separate portion of the cerebellum or that this function is centered exclusively in the posterior lobe. The view that seems most consistent with experimental facts is that the anterior lobe, lobulus simplex, paramedian lobes, and pyramis (spinocerebellum) are prominently, although not exclusively, implicated in the control of all movement. Unilateral ablation of the pyramis causes effects somewhat similar to those that follow unilateral albations of the vermis of the anterior lobe.[88] However, the results of localized albations in the pyramis and paramedian lobule await detailed analysis.

Older clinical evidence regarding *somatotopic localization* in the human spinocerebellum is vague and contradictory. However, one clinicopathologic study by Victor et al.[100] confirms the pattern of representation in the anterior lobe very convincingly. A remarkably uniform cerebellar syndrome was noted in a group of 50 patients with a long history of alcoholism. The most striking abnormalities were severe ataxia of gait and marked incoordination of the legs in voluntary movements. Postmortem examination revealed a degeneration of the cerebellar cortex, confined chiefly to the anterior lobe. The parts most consistently and severely involved were the central lobules, that is, the "leg area." Although the culmen was frequently affected, its lateral extensions, the "arm areas," were spared, a finding that correlates with the relatively mild clinical signs in the arms. Most of the "head and neck area" was intact; correspondingly there were few clinical signs referable to these parts of the body.

Lateral portions of the hemispheres. There is little information on the effects of lesions restricted to the areas under consideration. Several investigators stress the absence of obvious disability after removals that include these areas. Keller et al.[57] found that ablations that included the tuber vermis, paramedian lobules, paraflocculus, and ansiform lobules did not elicit noticeable dysfunction in dogs or monkeys. One of their dogs was able to walk upright on his hindlegs 3 days after operation. The results obtained by Carrea and Mettler[17] in monkeys are in the same tenor: "Removal of the lobulus ansiformis produces no detectable defect." In a group of monkeys studied by Botterell and Fulton[9] after hemispheric excisions there were several animals with a defect perhaps attributable to removal of the ansiform lobe. When allowed to run in a corridor, these animals, which were apparently not ataxic in other ways and had keen vision as judged by other tests, persistently crashed into the end wall as if it were not there. This type of behavior suggested some sort of visuomotor incoordination.

Clinical evidence is in accord with these experimental ablations in animals in that large removals of lateral hemispheric cerebellar cortex (as for tumors) result in little or no deficit detectable by standard neurologic examination. Yet Gordon Holmes, the British neurologist, urged caution in the interpretation of this result. In cryptically describing two patients,[47] both of whom had damage of the left lateral cerebellar hemispheric cortex, he observed that neither had detectable neurologic deficits, yet each was professionally incapacitated: one "with a tumor of the left side of the cerebellum complained that with his left hand he 'could not strike the four notes of a chord in proper sequence or time on the piano' "; another "with a long-standing gunshot wound was no longer able to play the flute, although the movements of his arm were apparently normal to other tests."

Cerebellar nuclei. Ablation of the nuclei causes more severe disability than a lesion of comparable size in cerebellar cortex.[17] Fastigial ablation profoundly disturbs equilibrium and coordination of the trunk muscles and causes a tremor of head and neck (titubation).[17,19] Dentate *and* interpositus ablations cause ipsilateral incoordination and tremor during all voluntary and involuntary movements.[35] Ablation of the interpositus nucleus in monkeys has been claimed to cause failure of fixation of proximal joints[35] and lowered velocity of movement with undershooting of reach.[99] Ablation of the dentate nucleus in monkeys has been claimed to cause incoordination only of distal extremities (hand and foot),[35] increased velocity and overshooting of movements at the elbow,[15] and prolonged duration of movements at the elbow with consequently delayed termination of movement.[22] In other experiments in monkeys a small lesion confined to the dentate nucleus caused a delay in the activation of the motor cortex[70] (to which it projects via the thalamus), and since the motor cortex has long been thought to play a major role in the initiation of volitional movements, the delay in the motor cortex may have accounted for the delay in onset of movement. In man, neurosurgeons required to remove cerebellar cortex are reluctant to encroach on any of the nuclei, since deficits are more severe and less likely to disappear with time.

Nature of cerebellar movement disorders in man

Cerebellar lesions in man cause disorders of movement resembling those produced experimentally in monkeys. In addition, there is usually some degree of *hypotonia,* which is seldom prominent in animals.

The most conspicuous cerebellar signs are the errors in rate, range, force, and direction of movements, known collectively as *ataxia.* They appear only when muscles are in use and are most pronounced in precise actions involving several joints. Neurologists employ a number of simple tests designed to elicit these defects and distinguish between them as follows[25,47,48]: *intention tremor,* an oscillating tremor most marked at the end of fine movements; *asynergia,* a lack of coordination between muscles (e.g., failure of the wrist extensors to contract during flexion of the fingers, allowing the wrist to flex also); *decomposition of movement,* the performance of actions in successive parts rather than as a whole (e.g., in touching the nose, first flexing the forearm and then the arm, and lastly adjusting the wrist and finger); *dysmetria,* errors in the range of movement (e.g., touching a point, arresting the action before reaching it, or shooting past it); *deviation from the line of movement* (e.g., carrying food to the ear instead of the mouth); and *adiadochokinesia,* inability to perform alternating movements (e.g., tapping) rapidly and smoothly. Although they are useful clinically, such distinctions are not fundamental. The form that ataxia takes depends on the particular muscles involved and how their action is tested. In fixing the gaze on some object, for example, the eyes alternately turn quickly toward it and slowly away from it *(nystagmus);* in speech there is a tendency to decompose words into separate slurred syllables pronounced with irregular force and rhythm *(scanning speech);* in walking, the gait is reeling, and associated movements of the arms are lost. Less conspicuous but equally characteristic of cerebellar disease are several related signs that may be defined as *hypotonia.* These include absence of normal resistance to stretch, easy displacement of a limb from a given posture, hyperextensibility of joints, and undamped or pendular reflexes. These abnormalities apparently indicate a lack of normal response to stretch, in which the cerebellum, through its proprioceptive connections, plays an important role. The opposite condition, *hypertonia,* although common in animals after certain lesions, is rarely seen in man. The reason for this difference is not understood.

Methods of diagnosing cerebellar motor deficit logically fall into two categories. Because of the difficulty of distinguishing a cerebellar deficit from other motor deficits, a variety of tests were developed by neurologists around the turn of the century. These were standard postures, perturbations of postures, and volitional movements that revealed stereotyped abnormalities in the presence of cerebellar lesions that were easily recognized and taught. An experienced physician might see subtle signs of a cerebellar defect in the oscillation or just one "bounce" of an extended arm, the irregularity in range, rate, and force of some natural rhythmic movement, or the slowness or clumsiness of manipulation of an object and confirm his suspicions with a series of standard tests. A second diagnostic approach would employ quantifiable tests of *specific cerebellar deficits.* Gordon Holmes[48] made careful measurements of the abnormalities in movement after cerebellar lesions in humans in an attempt to discover which deficits were fundamental and, by inference, the basic characteristics of cerebellar motor control. These deficits included the following:

1. Delay in starting volitional movement
2. Delay in stopping volitional movement
3. Errors in direction of volitional movement
4. Impaired gradation of velocity during movement
5. Errors in the combination and synchronization of joints used in a many-jointed movement
6. Breakdown of repetitive, rhythmic, alternating movements

These were the most outstanding measurable deficits, yet Holmes also demonstrated three more:

7. Reduced stiffness of normal resting muscle to passive displacement
8. Inability to actively hold a posture with steady force, or return to it promptly when passively perturbed
9. Weakness of volitional muscular contraction in maintaining a posture or performing a movement

Many of these deficits had been observed, although less accurately, in animals and man long before Holmes. What they meant about the role of the cerebellum in motor control generated two opposing opinions. One view holds that since deficits 7 through 9 were present after cerebellar damage and all the other deficits could logically be explained as secondary to and deriving from these primary deficits, the main functions of the cerebellum were to control the tone of resting muscle, the stability of actively maintained posture, and the force of muscular contraction during movement and posture. Others were quick to point out that while deficits 7 through 9 could be seen in some circumstances, they were always the least conspicuous and enduring, and some questioned whether 7 and 9 existed at all. This reasoning suggested that since 7 through 9 were slight or nonexistent, the others could not be quantitatively or even qualitatively derived from them and the cerebellum must represent the function of some more elaborate control system concerned with more complicated variables. Carried to its extreme, this argument reduced the role of the cerebellum to one control function—coordination of movement—because incoordination

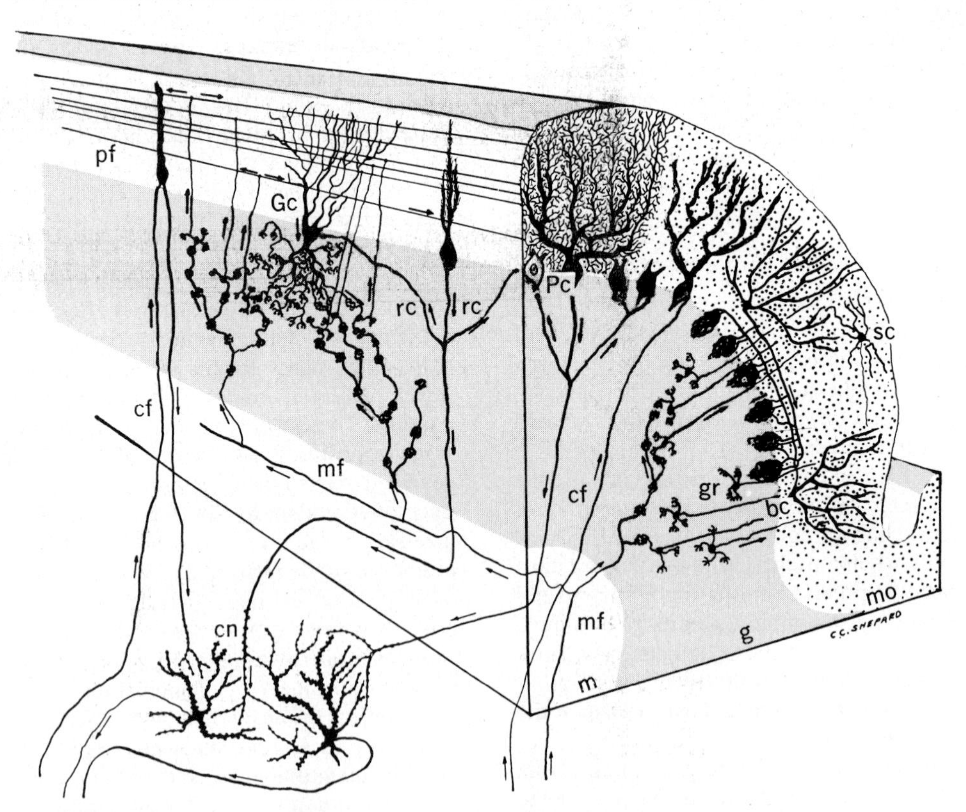

Fig. 31-5. Schematic view of cerebellar folium. *bc,* Basket cells; *cf,* climbing fiber; *cn,* deep cerebellar nuclei; *g,* granular layer; *Gc,* Golgi cell; *gr,* granule cell; *m,* medullary layer (white matter); *mf,* mossy fiber; *mo,* molecular layer; *Pc,* Purkinje cell; *pf,* parallel fiber; *rc,* recurrent collateral; *sc,* stellate cell. (From Fox.[30])

seemed to be the one abstract deficit that best explained all the other deficits. This dispute is clearly semantic. None of the observed deficits provides an unmistakable clue to the nature of cerebellar function. But when and if the fundamental controlled variables and the basic transfer functions of the cerebellar circuitry are discovered, one may hope that simpler, more accurate, and possibly even automated methods of diagnosis may be employed.

In man, as in experimental animals, there is considerable recovery of function (compensation) after cerebellar lesions. A patient may even appear to regain his powers of coordination completely. Careful testing, however, will usually reveal motor deficiencies months or years later. Recovery is less complete after nuclear lesions than after cortical lesions.

Patients with cerebellar lesions report no alterations in sensation. They may reel and fall, but they do not feel dizzy; their movements may deviate from the desired mark, but their sense of position in space is unimpaired. Evidently, the sensory data coming to the cerebellum do not reach consciousness. In comparing equal weights in each hand, that on the affected side is usually judged the heavier. This should not be considered a test of primary sensation, however, because in any weakened limb, weights may feel heavier than they actually are because of the greater effort involved in all performance on the affected side.

CEREBELLAR CIRCUITS

Nuclei. The deep cerebellar nuclei and the vestibular nuclei generate the output of the cerebellum. Signals leaving the cerebellum are apparently all excitatory.[26] Circuits within the deep nuclei have been neglected in most anatomic investigations (Figs. 31-5 and 31-6), although re-

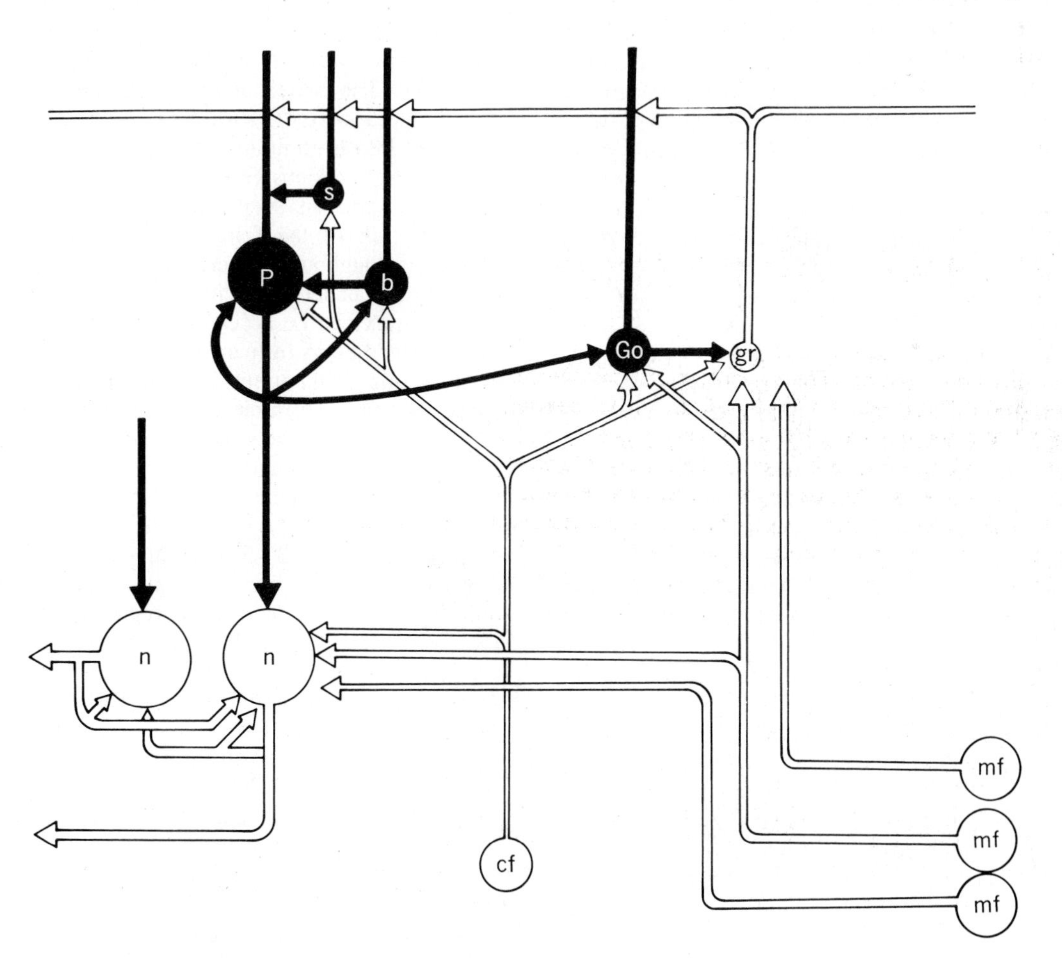

Fig. 31-6. Simplified diagram of cerebellar circuitry. *mf,* Mossy fiber; *cf,* climbing fiber; *gr,* granule cell; *Go,* Golgi cell; *b,* basket cell; *s,* stellate cell; *P,* Purkinje cell; *n,* nuclear cell. White cells are excitatory; black, inhibitory. Diagram shows only what types of cell one type contacts and whether contact is excitatory or inhibitory.

cent studies[20] show a rich pattern of recurrent collateral axons (to the same cell and to neighboring cells—excitatory feedback?) and of smaller nonprojecting interneurons that may be inhibitory[20] (not shown in Figs. 31-5 and 31-6).

Since the time of Ramón y Cajal[82] it had been assumed that the deep nuclei were simply relay stations for the presumably excitatory output of the cerebellar cortex. It therefore came as a surprise when Ito and Yoshida[50,51] found that *Purkinje cells are, apparently without exception,*[26] *inhibitory to the cells on which they synapse.* Since the only axons leaving the cerebellar cortex are those of Purkinje cells, it must be concluded that the net effect of all the elaborate cortical processing is the production of a purely inhibitory output. All areas of the cortex, with the exception of the flocculus, project to the deep nuclei: the vermis to the fastigial nucleus, the intermediate zone to the nucleus interpositus, and the lateral zone to the dentate nucleus. (The vestibular nuclei are, as previously noted, sometimes grouped with the cerebellar nuclei because they receive direct projections from Purkinje cells in the vestibular parts of the cerebellum.) After reaching its target nucleus, each Purkinje axon divides to form a dense, preterminal arborization extending from the upper to the lower surface of the nucleus. These terminals completely envelop the large nuclear cells within their territory, surrounding the cell body and dendritic tree of each with a pericellular nest.

Despite the uniformly inhibitory input that nuclear cells receive from the cortex, they respond actively to a variety of stimuli, perhaps caused by their recently discovered excitatory inputs from outside the cerebellum. The studies of Ito[26,49] indicate that both mossy and climbing fibers give off excitatory collaterals to the cerebellar nuclei before they reach the cortex, and there is other evidence that nuclear cells may receive inputs that do not branch to the cerebellar cortex,[12] just as there are "private inputs" (as opposed to "shared inputs") that go as mossy fibers to the cortex and not to the nuclei[2] (Fig. 31-6).

The corticonuclear projection is very precise. For example, each neuron in the dentate nucleus can be inhibited monosynaptically by stimulation of a certain small area of cortex in the ipsilateral hemisphere.[26] Microelectrode recordings in the cerebellar nuclei[93] reveal that nuclear cells are discharging continuously, even when no movement is occurring. This activity is presumably due in part to the influx of afferent impulses through collateral groups of the same mossy and climbing fibers that carry impulses to the cerebellar cortex. In addition, there are afferent fibers that go exclusively to the nuclei. The available evidence indicates that all these afferent inputs are excitatory to nuclear cells, which fire despite the inhibition they receive from Purkinje cells. It might appear that input to the cerebellum has an initial brief excitatory effect on nuclear cells, followed by a delayed inhibitory effect that has been cortically processed.

To understand the functional significance of this arrangement, one must know where nuclear impulses go and what effects they produce. It may be concluded that all nuclear output is purely excitatory.[26] This is true of the projections from each of the nuclei, which have quite different destinations. The tonic discharges of nuclear cells in general cause steady depolarizations of the brain stem cells on which they impinge. Thus, the waxing and waning flow of inhibitory impulses from Purkinje cells results in less or more excitation of cells in the ventrolateral nucleus of the thalamus, the red nucleus, the vestibular nuclei, and the reticular formation.

Cortex. The cortex of the cerebellum represents nature's supreme effort to pack a maximum of cortex into a minimum of space. This is accomplished by extensive infolding of the cortex to form a series of deep folia (Fig. 30-2). The fine structure of the folia[30,81,82] is remarkably uniform throughout the cerebellum; any folium (Fig. 30-5) can serve as a model and any section of the cortex can be accepted as representative. The cells and fibers form an orderly array, so precisely arranged and oriented in three-dimensional space that they resemble a regular lattice. The repeating pattern of this lattice indicates that although incoming signals may differ from folium to folium, they are subject to the same neural processing throughout the cerebellum.

The cortex (Fig. 31-5) consists of an outer or *molecular* layer separated from an inner or *granular* layer by a single layer of regularly spaced *Purkinje cells*. The Purkinje axons, which descend directly to the cerebellar nuclei, are the only efferent fibers of the cortex. The dendrites of Purkinje cells ascend through the molecular layer, branching repeatedly until they reach the pial surface. The dendritic tree is virtually two dimensional, being flattened in a plane oriented at right angles to the long axis of the folium. The lower, thicker branches of the tree are smooth; the higher, thinner branches (Fig. 31-5) are extremely dense and are covered with spines, providing an enormous surface area for synaptic contacts.

The rest of the cortical mechanism serves to

distribute afferent impulses to the Purkinje cells. Although afferent impulses come from many different parts of the nervous system, there are only two basic types of input to the cortex. The *climbing fibers,* carrying signals mainly from the inferior olive,[3,10,89] emerge from the white matter, cross the granular layer, and climb like ivy up the smooth surface of the primary and secondary branches of the Purkinje dendrites. In electron micrographs, it is apparent that some synaptic contacts are also made with dendritic spines. Each Purkinje cell receives a potent, all-or-none excitatory connection from a single climbing fiber.[26] Whenever the climbing fiber discharges, it fires the Purkinje cell.[28] Although the chief targets of a climbing fiber are the dendrites of a Purkinje cell, recent studies show that its collaterals also end on stellate cells, basket cells, Golgi cells, and even granule cells.[81]

The effect on the Purkinje cell of firing the climbing fiber input is an action potential that is followed by a prolonged depolarization, on which there may be superimposed one or several smaller depolarizing wavelets[27,40,56] (Fig. 31-7), which become more prominent as the Purkinje cell is closely approached by the electrode and possibly injured. This electrical phenomenon has been recorded intra- as well as extracellularly, and has variously been called "complex spike"[92] (because of its complex waveform), "climbing fiber response"[26,27] (because of its presumed cause), and "inactivation response"[40] (because of its presumed effect). There is disagreement[29] as to whether the secondary wavelets are natural, how they are caused, whether they occur in all Purkinje cells, and whether they as well as the initial spike are propagated down the Purkinje cell axon.

Perhaps the greatest peculiarity about the complex spike is that it occurs at such low frequency, 1 to 2/sec,[93] as compared to the ~70/sec spike firing frequency caused by the other excitatory input, the mossy fiber–granule cell system. *The mossy fiber–granule cell system* provides for the cortical distribution of the great majority of impulses that reach the cerebellum. According to Ramón y Cajal,[82] the large mossy fibers divide in the white matter, often sending branches to several folia and, in some instances, to widely separated parts of the cerebellum. On entering a given

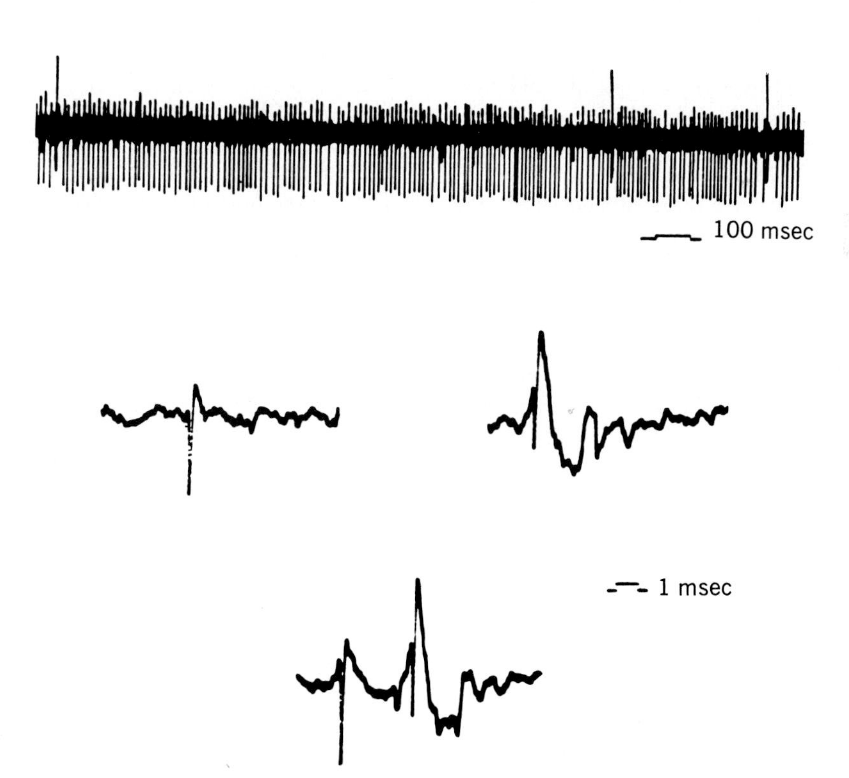

Fig. 31-7. Maintained discharge of Purkinje cell, recorded extracellularly, showing its two different spike potentials, "simple" (left) and "complex" (right). Slow trace (top) shows their different patterns of discharge; fast traces (bottom three), their different shape. Positivity is up. (From Thach.[93])

folium, they give off branches into various parts of the granular layer. Each of these subdivides further and the final branches form bulbous expansions called *rosettes*. Within a complex structure called a *glomerulus,* each rosette makes excitatory synaptic contact with the dendritic claws of approximately five *granule* cells. The terminals of these dendrites are deeply embedded in the surface of the rosette, an arrangement that probably ensures effective synaptic transmission. Through its numerous branches and multiple synapses within the glomeruli, each mossy fiber excites hundreds of granule cells. The latter are extremely small cells with axons less than 1 μ in diameter. According to a recent estimate, there are about 5 million granule cells per cubic millimeter of granular cortex in man. The axons of the granule cells ascend into the molecular layer, where they bifurcate to form a T. The two branches of the T run great distances in the long axis of the folium without crossing or branching and hence are called parallel fibers. They pass through the Purkinje arborization like wires through the crossarms of telegraph poles and make synaptic contacts with the dendritic spines. All these synapses are apparently excitatory.[26,28] Since each mossy fiber excites a large number of scattered granule cells whose axons pass through a long series (300 to 500) of Purkinje arborizations, a single mossy fiber impulse exerts a very small excitatory effect on a great many Purkinje cells spread out over a wide area of cortex.

There are three other types of neurons in the cortex: basket, stellate, and Golgi cells, which are inhibitory interneurons. *Basket cells,* situated in the lower part of the molecular layer just above the Purkinje cells, and similar *stellate cells* in more superficial locations, distribute a powerful inhibitory input to Purkinje cells.[26] Their dendritic trees, which are much less dense than those in Purkinje cells, are oriented in the same transverse plane and receive an excitatory input from the parallel fibers as well as from collaterals of climbing fibers. The axons of basket cells *(transverse fibers)* run for long distances across the folium (i.e., at right angles to the parallel fibers) just above the bodies of the Purkinje cells. As they pass each cell, they give off collaterals that descend to form a dense basketlike network around the Purkinje cell body and axon hillock. One transverse fiber may send collaterals to 15 to 17 Purkinje cells. Right-angle collaterals from transverse fibers extend a distance of five to six Purkinje cell bodies on each side of the transverse fibers. Thus each basket cell impulse may inhibit a rectangular patch of Purkinje cells 10 to 12 cells wide and 10 to 20 cells long. The axons of stellate cells also run across the folium and are believed to form inhibitory synapses with the smooth portions of the Purkinje cell dendrites. *Golgi cells,* usually situated just below the Purkinje cell layer, are the third type of inhibitory interneuron. They are large cells with dendritic trees that extend outward into the molecular layer. Their branches are not as dense as those of Purkinje cells, and they spread out in all directions. Excitatory inputs reach Golgi dendrites via parallel fibers and mossy fibers and the cell body through collaterals of climbing fibers.[81] An inhibitory input also reaches the cell body via recurrent collaterals of Purkinje axons.[81,82] The axon of the Golgi cells forms a dense and often extensive arborization in the granular layer beneath the overlying dendritic field. The terminals of the axons end within the glomeruli, where they make inhibitory synaptic contacts with the outer surface of granule cell dendrites, that is, on the side opposite their contacts with the mossy rosette. Descending dendrites of the Golgi cell may also enter the glomeruli and make contact with the rosette. Thus within a glomerulus the mossy rosette excites the dendrites of both Golgi cells and granule cells, and the latter are inhibited by impulses in Golgi axons.[26,28]

In response to the mossy fiber–granule cell system and its associated intracortical inhibitory neurons, the Purkinje cell is caused to discharge at high, maintained frequencies up to 125/sec.[93] The action potential recorded intra-[26,40,56] and extracellularly is of relatively simple contour as contrasted with the complex spike and has been called the "simple spike"[96] (Fig. 31-7). A Purkinje cell normal has two different kinds of responses to its two different kinds of excitatory input: the complex spike at relatively low frequency and the simple spike at relatively high frequency.

The salient features of cerebellar circuitry can be summarized in four statements: (1) All nuclear output is excitatory, regardless of its destination. (2) The cortical circuits are similar in all parts of the cerebellum, and all cortical output is inhibitory, whether it goes to the deep nuclei or to the vestibular complex. (3) All input to the cerebellum is apparently excitatory, both to the cortex and the nuclei. (4) The Purkinje cell has two types of excitatory input that differ anatomically and physiologically.

For those interested in further study of cerebellar circuits the studies by Brodal[11] and Eccles et al.[26] and the review by Bell and Dow[7] provide further accounts of the subjects discussed in this section and are useful guides to the literature.

ACTIVITY OF CEREBELLAR NEURONS DURING NATURAL BEHAVIOR

The great majority of signals entering the cerebellum are carried by mossy fibers. Hence one way to investigate cerebellar function has been to examine the responses of single mossy fibers to passive peripheral stimulation. Those conveying somatosensory information have received the most attention, chiefly because it is possible to study their receptive fields.* It is apparent in some cases that the information they carry has been transformed before it reaches the cerebellum, so that the receptive fields of individual mossy fibers may no longer resemble those of primary afferent fibers.[4,80] Some mossy fibers are either excited or inhibited by small, ipsilateral, modality-specific fields; others respond to several modalities of sensation and have receptive fields, varying widely in size, that are located in ipsilateral, bilateral, or contralateral portions of the body.[80,92] Recordings from Purkinje cells reveal that, in general, simple spike discharges are rarely changed by somesthetic stimuli that are truly passive and cause no motor response from the animal.[45,96] This suggests that under normal conditions, Purkinje cells respond to a pattern of input more complicated than that caused by passive manipulation of the body and that other overlapping inputs (from motor cortex via pons, from reticular nuclei, etc.) may have to be active as well. The fact that simple spikes may be influenced by electrical stimulation of the peripheral nerves may be useful in studying circuit connections but does not necessarily reveal normal modes of operation.

In contrast, rotation of the head, which naturally stimulates the vestibular afferents, causes marked fluctuations in simple spike activity in the flocculus.[62,64] and in fastigial[32] and vestibular[71] deep nuclear cells. For the vestibular portions of the cerebellum, which is most directly connected to a pure afferent input, the study of cerebellar processing of sensory information would appear particularly appropriate. Nevertheless, neurons in vestibular nuclei have been shown to fire in relation to *volitional saccadic eye movements*[68] as well, raising questions about the purity of input to these nuclei and their function.

Recent advances in recording techniques have made it possible to relate the normal activity of the largest types of cerebellar cells to movement and posture. The cell type whose activity should be most revealing in this respect is the nuclear cell, since it is the final stage of cerebellar output. Recordings[94,95] were therefore made of the activity of individual cells in the dentate and interpositus nuclei of trained monkeys while they moved a lever horizontally by flexion and extension of the wrist in response to light signals. Electromyographic recordings showed that at least 12 muscles in the arm, forearm, shoulder, and trunk participated just in these simple movements. Most cells in the appropriate parts of the interpositus and dentate nuclei discharged continuously even when the wrist was not moving and the animal was at rest. The rate of discharge usually increased or decreased during a movement. The pattern of change was generally different for flexion and extension. Some neurons, predominantly those in the nucleus interpositus, altered their frequency just before, during, or *after* the onset of movement.[94] In these instances the motor cortical input, combined with the classic feedback pathways from active muscles and joints to this part of the cerebellum, may have played a role in *controlling the ongoing movement*. Other neurons, predominantly in the dentate nucleus, altered their frequency as early as 80 msec *before* any muscular activity occurred,[94] eliminating the possibility of feedback. This was consistent with the fact that the spinocerebellar pathways do not project to the dentate nucleus. More recent study[96] has suggested that in a rapid hand movement, triggered by a light signal, dentate neurons may respond before those in the *motor cortex,* to which the dentate nucleus projects via the thalamus. The hypothesis was offered that the *dentate nucleus may play a role in initiating activity in the motor cortex and thereby in muscles*. In support of this suggestion, it may be noted that ablation of the dentate nucleus results in a delay of initiation of activity in motor cortex.[70] The same delay is seen in muscle activity and in movement.[70]

In order to determine the extent to which Purkinje cells control nuclear cells, a similar study was carried out on Purkinje cells.[95] Activity in Purkinje cells was surprisingly similar to that of the interpositus cells to which they project. Changes in spike frequency occurred at similar times before and after movement. Different patterns of discharge related to movement were seen for many cells in both groups, and different frequencies occurred among cells of both groups during maintained flexion and extension of the wrist. For both Purkinje cells and the interpositus cells to which they project, an increase in fre-

*See references 14, 74, 80, 90, and 92.

quency was the earliest detected change related to movement in the majority of them. At first glance, these observations might suggest that Purkinje cells are responsible for the changes in nuclear cells. If this were true, however, the firing rate of nuclear cells should decrease rather than increase in frequency whenever Purkinje cells increase their rate of discharge. It seems more likely that the similarities are due to the fact that nuclear cells receive excitatory inputs from many of the same mossy fibers and climbing fibers that directly or indirectly excite Purkinje cells and that Purkinje cell output acts to restrain through inhibition the output of nuclear cells both at rest and during movements. Whether this is true for all movements is not known.

Complex discharges of the Purkinje cell, unlike simple spikes, were rarely related to movement,[95] continuing to discharge sporadically at the relatively slow rate of 1 to 2/sec. In 20% to 30% of the cells, complex spikes occurred before or after movement in some trials but not in others. These discharges were never as early or as consistent as those of the simple spike.

Mortimer[72] has recorded from nuclear cells in interpositus and dentate and from Purkinje cells projecting mainly to interpositus in awake monkeys subjected to sudden loud noises in the neighborhood of 100 dB. This produces an acoustic startle response, a very rapid reflex movement with electromyographic activity occurring as early as 17 msec after the onset of the stimulus. The earliest changes in cerebellar neuronal discharge occurred in the dentate and interposed nuclei some 7 msec after the onset of the stimulus. These observations show that the earliest nuclear changes could not have been caused by feedback from the later changing muscles and, further, that the dentate must also be involved in activities other than learned or volitional movements, which involve a reaction time of around 200 msec. Mortimer also observed that the earliest Purkinje cell simple spike changes followed those of the nuclear cells by 9 msec, supporting the notion that (at least in this situation) Purkinje cells did not *cause* the initial changes in nuclear cell activity, but rather *modified* the changes after their onset. Complex spikes did not occur until 30 msec after the stimulus.

The question of *time of change* of cerebellar neuronal discharge has also been studied in relation to eye movement. Cells in the dentate nucleus have been found to fire in relation to saccadic eye movements, preceding them.[32] Purkinje cells in the posterior vermis have been found to fire in relation to and prior to the saccades and the quick phases of vestibular and optokinetic nystagmus.[63] These and the previously mentioned single unit recording studies, combined with the fact that lesions may delay onset of movement, argue strongly that at least some parts of the cerebellum may play a role in the initiation of movement. Other parts, in particular those intermediate portions receiving somesthetic input and medial portions receiving vestibular input (which led Sherrington to generalize the cerebellum as the "head ganglion of the proprioceptive system"), appear to change activity later and may serve to control movement after it has been initiated elsewhere.[76-79]

The single unit studies thus far described have been concerned with *whether* there are consistent changes of neural activity related to some aspect of behavior (sensory stimulation, active movement) and *when* the changes occur. Holmes[48] identified a number of aspects of movement that were disturbed by cerebellar damage; we now ask whether cells in the cerebellum uniquely code for and control one or several aspects to the exclusion of others. Purkinje cells in the monkey flocculus appear to discharge in relation to the *target velocity* during smooth pursuit eye movements, and the question has been raised of their possibly unique role in controlling this type of eye movement.[72] In the interposed nucleus, one study found unit discharge frequency closely correlated with *velocity* of movement at the cat's elbow;[16] another study showed a close correlation of unit discharge with *force* exerted by the monkey's wrist.[97] In the dentate nucleus, when a monkey's wrist was held in various positions under various loads, some units discharged in relation to the *pattern of muscles* used regardless of wrist position, whereas others discharged related to *position* of the wrist regardless of the pattern of muscular activity used to hold the position. A third set of neurons appeared to discharge in relation to the *direction of the next anticipated movement*.[97] Although these approaches offer the promise of a better understanding of what the cerebellum does and how it does it, they have as yet produced no unified explanation.

SPECULATIONS ON FUNDAMENTAL CEREBELLAR MECHANISMS

Animals with ablations of the spinocerebellum have difficulty maintaining a stable stance. This, together with consideration of the large size and high conduction velocities of the spinocerebellar pathways, long ago led to the view that the cerebellum plays a critical role in maintaining stable stance through the use of *feedback*. The process is formally referred to as "regulation," where position (posture) is the regulated variable. Explicit in some and implicit in all such formulations is the idea that feedback from peripheral receptors sends a message to the central nervous system as to the *actual* posture and that within the nervous system (possibly in the cerebellum) it is *compared* with a neural replica of the *intended* posture. If a discrepancy (error) exists between the intended and the actual posture, an error signal is sent to the neuromusculoskeletal apparatus to correct the discrepancy. Whether spinocerebellar pathways operate in this fashion is still an open question.

A function that has often been considered similar to stance in the feedback regulation of position is the vestibulo-ocular reflex: fast labyrinthine inputs from the semicircular canals activate extraocular muscles via the vestibular nuclei to cause the eyes to move despite head movements and so maintain eye position in fixed space. Nevertheless, the vestibulo-ocular reflex is not a feedback regulation of posture, but a feedforward process, since the afferent information comes from a different structure (head) than that whose position is being controlled (eye). Another point that the comparison between body stance and the vestibulo-ocular reflex brings up is the absence of a clear-cut dichotomy between posture and movement. In both instances, when a perturbation occurs, the parts in question have to move to maintain the intended position.

Certainly, movement per se seems to be a primary concern of the cerebellum, since cerebellar ablations may cause gross incoordination of movement. What aspects of movement are controlled? Clearly, in the vestibulo-ocular reflex (if, again, the vestibular nuclei are considered the deepest of the cerebellar nuclei), movement of the head *initiates* movement of the eyes. There is also the evidence just presented from ablation and single-unit studies that the lateral cerebellum and dentate nucleus play roles in the initiation of limb and eye movements in volitional tasks (those involving choice, with relatively long reaction times). There are also the very short latency responses in the dentate nucleus preceding movement in the acoustic startle reflex.

Besides *initiation* of movement there have been several ideas presented about how its *duration* might be controlled. One proposes that a dentate–thalamus–motor cortex–pons–dentate reciprocal loop exists that reverberates. Reverberation means that an action potential or train of action potentials, once set in motion, travels around the reciprocal loop until terminated by fatigue or inhibition impinging on the loop. Duration of movement would depend on the timing of one or more passages around the loop, and "stops" could thus be made contingent on "starts." Another idea is that durations of movement are determined by the delays caused by the inherently slow conduction (0.5 m/sec) of the parallel fibers.

Cerebellar control of *stops* might have quite a different mechanism. One might imagine a neural replica of a central intent that the stop position occur at a certain position and that a comparison between intended stop position and sensed position over rapidly conducting spinocerebellar pathways might trigger a neural output causing stop in a follow-up servomechanism much like the regulator proposed for postural control.

It is probably naive to assume that the cerebellum "controls" any one or several of these aspects of posture or movement in toto, since there is no paralysis even after acute total cerebellar removals. One alternative that has been suggested is that the cerebellum is simply a "damping and clamping" system that smooths irregularities out of starts and stops. Some characteristics of the circuitry suggest how this might operate; nuclear cells fire in a nonmoving animal at a high frequency, despite the inhibitory input from Purkinje cells, which also fire at a high frequency. If tonically firing mossy fibers were to suddenly increase their activity in relation to movement onset, the nuclear cells might be expected to suddenly increase their activity, then decrease it as the slower conducting cerebellar cortex Purkinje cell mechanisms built up a descending inhibitory crescendo. This output from the nuclear cells:

might be useful in the initiation of movement by overcoming the inertia of the component to be moved by the larger initial burst. As a corollary, if the high maintained activity in the mossy fibers relating to onset and duration of the movement were suddenly to return to their prior level, signaling for a stop, the lag in the slowly conducting cerebellar Purkinje cell circuit would produce a transient response in nuclear cell output:

which might be useful in smoothly stopping (clamping) the nuclear cells driving the agonists. Such a mechanism would assume that Purkinje cells and the nuclear cells to which they project are ultimately driven by a shared set of mossy fibers, and there is evidence that, to some extent, this is true (Fig. 31-6).

But there is also increasing evidence that cerebellar cortex and nuclear cells may both have "private" inputs. This would give the inhibitory Purkinje cell the options of initiating an output from the tonically firing nuclear cell (a decrease in Purkinje cell firing frequency causing an increase in nuclear cell firing frequency and vice versa) or of "gating" on or off some response mediated through the mossy fiber–nuclear cell "main line" by exerting more or less inhibition.

Finally, there have been various proposals that the role of the cerebellar cortex is to adjust whatever parameters are controlled by the mossy fiber–nuclear cell "main line." They all assume that the firing of the climbing fiber coincident with a certain subset of active granule cells and associated inhibitory neurons makes those granule cells more (or less) effective in driving the Purkinje cell through some presumed heterosynaptic change. The cerebellum would thus be an "adaptive controller" or even a "learner of new movements"; its ability to change input-output relationship is the very essence of its operation and contribution to stable (or changing) motor behavior.

These are but a few of the many speculations that have been made about cerebellar function. The point of presenting them here in such detail is that they try to assimilate the enormous amount of new knowledge about the cerebellum that has accrued. Any one or all of the speculations may be correct. They all can be subjected to experimentation, and they are already generating much enthusiasm and work toward answering the old questions of what the cerebellum does and how it does it.

SUMMARY

A long history of physiologic experiments of different types has provided much useful information about the cerebellum. The cerebellum is concerned with control of movements of the body and the eyes; a sole exception to this is its apparent role in motion sickness. The nature of the behavioral deficit after cerebellar injury is a particular kind of incoordination without paralysis. Damage to different parts of the cerebellum affects motor behavior in different ways, and the deficits are to some extent explained by the input and output connections of each area. This knowledge may serve to help clinically localize cerebellar lesions. There is now greater knowledge of the cerebellum's intrinsic circuit organization—its wiring and the actions at its synapses—than for any other part of the central nervous system, and this has given rise to a number of testable theories about its operation. Some observations on the natural performance of neurons within the intrinsic circuitry have begun to test these theories and to give clues to cerebellar function. These recent advances in anatomy and physiology warrant hope that, within our time, at least a general picture may emerge of the fundamental mechanisms underlying cerebellar function.

REFERENCES

1. Adrian, E. D.: Afferent areas in the cerebellum connected with the limbs, Brain **66:**289, 1943.
2. Allen, G. I., and Tsukahara, N.: Cerebrocerebellar communication systems, Physiol. Rev. **54:**957, 1974.
3. Armstrong, D. M.: Functional significance of connections of the inferior olive, Physiol. Rev. **54:**358, 1974.
4. Arshavsky, Y. I., et al.: Origin of modulation in neurones of the ventral spinocerebellar tract during locomotion, Brain Res. **43:**276, 1972.
5. Bard, P., et al.: Delimitation of central nervous mechanisms involved in motion sickness, Fed. Proc. **6:**72, 1947.
6. Barnard, J. W., and Woolsey, C. N.: Interconnections between the anterior lobe and the paramedian lobule of the cerebellum, Arch. Neurol. Psychiatry **65:**238, 1951.
7. Bell, C. C., and Dow, R. S.: Cerebellar circuitry, Neurosci. Res. Symp. Sum. **2:**515, 1967.
8. Bloedel, J. R.: Cerebellar afferent systems: a review, Prog. Neurobiol. **2:**3, 1973.
9. Botterell, E. H., and Fulton, J. F.: Functional localization in the cerebellum of primates; lesions of hemispheres (neocerebellum), J. Comp. Neurol. **69:**63, 1938.
10. Brodal, A.: Experimentelle Untersuchungen über die olivo-cerebellare Lokalisation, Z. Neurol. Psychiatry **169:**1, 1940.
11. Brodal, A.: Neurological anatomy in relation to clinical medicine, ed. 2, New York, 1969, Oxford University Press, Inc.
12. Brodal, A., and Gogstad, A. C.: Rubro-cerebellar connections: an experimental study in the cat, Anat. Rec. **118:**455, 1954.
13. Brodal, A., and Jansen, J.: The ponto-cerebellar projection in the rabbit and cat; experimental investigations, J. Comp. Neurol. **84:**31, 1946.
14. Brookhart, J. M., Moruzzi, G., and Snider, R. S.: Spike discharges of single units in the cerebellar cortex, J. Neurophysiol. **13:**465, 1950.
15. Brooks, V. B., et al.: Effects of cooling dentate nucleus on tracking-task performance in monkeys, J. Neurophysiol. **36:**974, 1973.
16. Burton, J. E., and Onoda, N.: Interpositus neuron discharge in relation to a voluntary movement, Brain res. **121:**167, 1977.
17. Carrea, R. M. E., and Mettler, F. A.: Physiologic consequences following extensive removals of the cerebellar cortex and deep cerebellar nuclei and effect of secondary cerebral ablations in the primate, J. Comp. Neurol. **87:**169, 1947.

18. Chambers, W. W., and Sprague, J. M.: Functional localization in the cerebellum. I. Organization into longitudinal corticonuclear zones and their contribution to the control of posture, both extrapyramidal and pyramidal, J. Comp. Neurol. **103:**105, 1955.
19. Chambers, W. W., and Sprague, J. M.: Functional localization in the cerebellum. II. Somatotopic organization in cortex and nuclei, Arch. Neurol. Psychiatry **74:**653, 1955.
20. Chan-Palay, V.: Cerebellar dentate nucleus: organization, cytology and transmitters, New York, 1977, Springer-Verlag New York, Inc.
21. Combs, C. M.: Electro-anatomical study of cerebellar localization; stimulation of various afferents, J. Neurophysiol. **17:**123, 1954.
22. Conrad, B., and Brooks, V. B.: Effects of dentate cooling on rapid alternating arm movements, J. Neurophysiol. **37:**792, 1974.
23. Dow, R. S.: The fiber connections of the posterior parts of the cerebellum in the cat and rat, J. Comp. Neurol. **63:**527, 1936.
24. Dow, R. S.: Effect of lesions in the vestibular part of the cerebellum in primates, Arch. Neurol. Psychiatry **40:**500, 1938.
25. Dow, R. S., and Moruzzi, G.: The physiology and pathology of the cerebellum, Minneapolis, 1958, University of Minnesota Press.
26. Eccles, J. C., Ito, M., and Szentagóthai, J.: The cerebellum as a neuronal machine, New York, 1967, Springer-Verlag New York, Inc.
27. Eccles, J. C., Llinas, R., and Sasaki, K.: The excitatory synaptic action of climbing fibers on the Purkinje cells of the cerebellum, J. Physiol. **182:**268, 1966.
28. Eccles, J. C., Llinas, R., and Sasaki, K.: The mossy fiber granule cell relay of the cerebellum and its inhibitory control by Golgi cells, Exp. Brain Res. **1:**82, 1966.
29. Evarts, E. V., and Thach, W. T.: Motor mechanisms of the CNS: cerebrocerebellar inter-relations, Annu. Rev. Physiol. **31:**451, 1969.
30. Fox, C. A.: The cerebellum. In Crosby, E. C., Humphrey, T., and Lauer, E. W., editors: Correlative anatomy of the nervous system, New York, 1962, The Macmillan Co.
31. Fulton, J. F., and Connor, G.: The physiological basis of three major cerebellar syndromes, Trans. Am. Neurol. Assoc. **65:**53, 1939.
32. Gardner, E. P., and Fuchs, A. F.: Single unit responses to natural vestibular stimuli and eye movements in deep nuclei of alert rhesus monkey, J. Neurophysiol. **38:** 627, 1975.
33. Gilman, S., and McDonald, W. I.: Cerebellar facilitation of muscle spindle activity, J. Neurophysiol. **30:** 1494, 1967.
34. Glaser, G. H., and Higgins, D. C.: Motor stability, stretch responses, and the cerebellum. In Granit, R., editor: Nobel Symposium I. Muscular afferents and motor control, Stockholm, 1966, Almquist & Wiksell, p. 121.
35. Goldberger, M. E., and Growden, J. H.: Pattern of recovery following deep nuclear lesions in monkeys, Exp. Neurol. **39:**307, 1973.
36. Gonshor, A., and Melville Jones, G.: Plasticity in the adult vestibuloocular reflex arc, Proc. Can. Fed. Biol. Sci. **14:**11, 1971.
37. Gonshor, A., and Melville Jones, G.: Extreme vestibulo-ocular adaptation induced by prolonged optical reversal of vision, J. Physiol. (Lond.) **256:**381, 1976.
38. Gonshor, A., and Melville Jones, G.: Short term adaptive changes in human vestibulo-ocular reflex, J. Physiol. (Lond.) **256:**361, 1976.
39. Granit, R., and Kaada, B. R.: Influence of stimulation of central nervous structures on muscle spindles in cat, Acta Physiol. Scand. **27:**130, 1952.
40. Granit, R., and Phillips, C. G.: Excitatory and inhibitory processes acting upon individual Purkinje cells of the cerebellum in cats, J. Physiol. (Lond.) **133:**520, 1956.
41. Granit, R., Holmgren, B., and Merton, P. A.: The two routes for excitation of muscle and their subservience to the cerebellum, J. Physiol. **130:**213, 1955.
42. Hampson, J. L.: Relationships between cat cerebral and cerebellar cortices, J. Neurophysiol. **12:**37, 1949.
43. Hampson, J. L., Harrison, C. R., and Woolsey, C. N.: Cerebro-cerebellar projections and the somatotopic localization of motor function in the cerebellum, Res. Publ. Assoc. Res. Nerv. Ment. Dis. **30:**299, 1950.
44. Hare, W. K., Magoun, H. W., and Ransom, S. W.: Localization within the cerebellum of reactions to faradic cerebellar stimulation, J. Comp. Neurol. **67:** 145, 1937.
45. Harvey, R. J., Porter, R., and Rawson, J. A.: The natural discharges of Purkinje cells in paravermal regions of lobules V and VI of the monkey's cerebellum, J. Physiol. **271:**515, 1977.
46. Henatsch, H. D., Manni, E., Wilson, J. H., and Dow, R. S. Linked and independent responses of tonic alpha and gamma hindlimb motoneurons to deep cerebellar stimulation, J. Neurophysiol. **27:**172, 1964.
47. Holmes, G.: The Croonian lectures on the clinical symptoms of cerebellar disease and their interpretation, Lancet **42:**59, 111, 1177, 1231, 1922.
48. Holmes, G.: The cerebellum of man, Brain **62:**1, 1939.
49. Ito, M.: Cited in Bell, C. C., and Dow, R. S.: Cerebellar circuitry, Neurosci. Res. Symp. Sum. **2:**569, 1967.
50. Ito, M., and Yoshida, M.: The cerebellar-evoked monosynaptic inhibition of Deiters neurones, Experientia **20:**515, 1964.
51. Ito, M., and Yoshida, M.: The origin of cerebellar-induced inhibition of Deiters neurones. I. Monosynaptic initiation of the inhibitory postsynaptic potentials, Exp. Brain Res. **2:**330, 1966.
52. Ito, M., Shiida, T., Yagi, N., and Yamamoto, M.: The cerebellar modification of rabbits horizontal vestibulo-ocular reflex induced by sustained head rotation combined with visual stimulation, Proc. Jpn. Acad. **50:**85, 1974.
53. Jansen, J.: Studies on the cetacean brain, hvalradets skrifter (scientific results of marine biological research), no. 37, Nor. Videnskaps-Akad. I.
54. Jansen, J., and Brodal, A.: Experimental studies on the intrinsic fibers of the cerebellum; the cortico-nuclear projection in the rabbit and the monkey, Avhandl. Nor. Videnskaps-Akad. I, Mat. Naturv. Kl. no. 3, p. 1.
55. Jansen, J., and Brodal, A.: Aspects of cerebellar anatomy, Oslo, 1954, Johan Grundt Tanum Forlag.
56. Jansen, J., Jr., and Fangel, C.: Observations on cerebro-cerebellar evoked potentials in the cat, Exp. Neurol. **3:**160, 1961.
57. Keller, A. D., Roy, R. S., and Chase, W. P.: Extirpation of the neocerebellar cortex without eliciting so-called cerebellar signs, Am. J. Physiol. **118:**720, 1937.
58. Lam, R. L., and Ogura, J. H.: An afferent representation of the larynx in the cerebellum, Laryngoscope **62:** 486, 1952.

59. Larsell, O.: The cerebellum: a review and interpretation, Arch. Neurol. Psychiatry **38**:580, 1937.
60. Larsell, O.: The development of the cerebellum in man in relation to its comparative anatomy, J. Comp. Neurol. **87**:85, 1947.
61. Larsell, O.: The cerebellum of the cat and the monkey, J. Comp. Neurol. **99**:135, 1953.
62. Lisberger, S. G., and Fuchs, A. F.: Responses of flocculus Purkinje cells to adequate vestibular stimulation in the alert monkey: fixation vs. compensatory eye movements, Brain Res. **69**:347, 1974.
63. Llinas, R., and Wolf, J. W.: Single cell responses from the cerebellum of rhesus preceding voluntary, vestibular, and optokinetic saccadic eye movements, Proc. Soc. Neurosci. **2**:201, 1972.
64. Llinas, R., Precht, W., and Clarke, M.: Cerebellar Purkinje cell responses to physiological stimulation of the vestibular system in the frog, Exp. Brain Res. **13**: 408, 1971.
65. Luschei, E. S., and Fuchs, A. F.: Activity of brain stem neurons during eye movements in alert monkeys, J. Neurophysiol. **35**:445, 1973.
66. Maekawa, K., and Simpson, J. I.: Climbing fiber responses evoked in vestibulocerebellum of rabbit from visual system, J. Neurophysiol. **36**:649, 1973.
67. Maekawa, K., and Takeda, T.: Mossy fiber responses evoked in the cerebellar flocculus of rabbits by stimulation of the optic pathway, Brain Res. **98**:590, 1975.
68. Massion, J., and Rispal-Padel, L.: Spatial organization of the cerebello-thalamo-cortical pathway, Brain Res. **40**:61, 1972.
69. McDonald, J. V.: Responses following electrical stimulation of anterior lobe of cerebellum in cat, J. Neurophysiol. **16**:69, 1953.
70. Meyer-Lohmann, J., Hore, J., and Brooks, V. B.: Cerebellar participation in generation of prompt arm movements, J. Neurophysiol. **40**:1038, 1977.
71. Miles, F. A.: Single unit firing patterns in the vestibular nuclei related to voluntary eye movements and passive body rotation in conscious monkeys, Brain Res. **71**: 215, 1974.
72. Miles, F. A., and Fuller, J. H.: Visual tracking and the primate flocculus, Science **189**:1000, 1975.
73. Mortimer, J. A.: Temporal sequence of cerebellar Purkinje and nuclear activity in relation to the acoustic startle response, Brain Res. **50**:457, 1973.
74. Murphy, J. T., MacKay, W. A., and Johnson, F.: Responses of cerebellar cortical neurons to dynamic proprioceptive inputs from forelimb muscles, J. Neurophysiol. **36**:711, 1973.
75. Nulsen, F. E., Black, S. P. W., and Drake, C. G.: Inhibition and facilitation of motor activity by the anterior cerebellum, Fed. Proc. **7**:86, 1948.
76. Orlovsky, G. N.: Activity of rubrospinal neurons during locomotion, Brain Res. **46**:99, 1972.
77. Orlovsky, G. N.: Activity of vestibulospinal neurons during locomotion, Brain Res. **46**:85, 1972.
78. Orlovsky, G. N.: Work of the cerebellar nuclei during locomotion, Biophysics **17**:1177, 1972 (translated from Biofizika **17**:1119, 1972).
79. Orlovsky, G. N.: Work of the Purkinje cells during locomotion, Biophysics **17**:935, 1972 (translated from Biofizika **17**:891, 1972).
80. Oscarsson, O.: Functional organization of the spino-cerebellar paths. In Iggo, A., editor: Handbook of sensory physiology, Berlin, 1973, Springer-Verlag, p. 340.
81. Palay, S., and Chan-Palay, V.: Cerebellar cortex cytology and organization, Berlin, 1974, Springer-Verlag.
82. Ramón y Cajal, S.: Histologie du système nerveux de l'homme et des vertebres, Paris, 1909, Librairie Maloine, vol. 1.
83. Robinson, D. A.: Adaptive gain control of the vestibulo-ocular reflex by the cerebellum, J. Neurophysiol. **39**:954, 1976.
84. Sasaki, K., et al.: Electrophysiological studies on the cerebellocerebral projections in monkey, Exp. Brain Res. **24**:495, 1976.
85. Snider, R. S.: Interrelations of cerebellum and brain stem, Res. Publ. Assoc. Res. Nerv. Ment. Dis. **30**: 267, 1950.
86. Snider, R. S., and Eldred, E.: Cerebro-cerebellar relationships in the monkey, J. Neurophysiol. **15**:27, 1952.
87. Snider, R. S., and Stowell, A.: Receiving areas of the tactile, auditory, and visual systems in the cerebellum, J. Neurophysiol. **7**:331, 1944.
88. Sprague, J. M., and Chambers, W. W.: Regulation of posture in intact and decerebrate cat; cerebellum, reticular formation, and vestibular nuclei, J. Neurophysiol. **16**:451, 1953.
89. Szentágothai, J., and Rajkovitz, K.: Über den Ursprung der Kletterfasern des Kleinhirns, Z. Anat. Entwicklungsgesch. **121**:130, 1959.
90. Tarnecki, R., and Konorski, J.: Patterns responses of Purkinje cells in cats to passive displacements of limbs, squeezing and touching, Acta Neurobiol. Exp. **30**:95, 1970.
91. Terzuolo, C., and Terzian, H.: Cerebellar increase of postural forms after deafferentation and labyrinthectomy, J. Neurophysiol. **16**:551, 1953.
92. Thach, W. T.: Somatosensory receptive fields of single units in cat cerebellar cortex, J. Neurophysiol. **30**: 657, 1967.
93. Thach, W. T.: Discharge of Purkinje and cerebellar nuclear neurons during rapidly alternating arm movement in the monkey, J. Neurophysiol. **31**:785, 1968.
94. Thach, W. T.: Discharge of cerebellar neurons related to two maintained postures and two prompt movements. I. Nuclear cell output, J. Neurophysiol. **33**:527, 1970.
95. Thach, W. T.: Discharge of cerebellar neurons related to two maintained postures and two prompt movements. II. Purkinje cell output and input, J. Neurophysiol. **33**: 537, 1970.
96. Thach, W. T.: Timing of activity in cerebellar dentate nucleus and cerebral motor cortex during prompt volitional movement, Brain Res. **88**:237, 1975.
97. Thach, W. T.: Correlation of neural discharge with pattern and force of muscular activity, joint position, and direction of the intended next movement in motor cortex and cerebellum, J. Neurophysiol. **41**:654, 1978.
98. Tyler, D. B., and Bard, P.: Motion sickness, Physiol. Rev. **29**:311, 1949.
99. Uno, M., Kozlovskaya, I. B., and Brooks, V. B.: Effects of cooling interposed nuclei on tracking-task performance in monkeys, J. Neurophysiol. **36**:1973.
100. Victor, M., Adams, R. D., and Mancall, E. L.: A restricted form of cerebellar cortical degeneration occurring in alcoholic patients, Arch. Neurol. **1**:579, 1959.

32

ELWOOD HENNEMAN

Motor functions of the cerebral cortex

The existence of an area in the cerebral cortex specifically concerned with motor functions was suggested as early as 1691 by Robert Boyle, who described the case of a knight who had sustained a depressed fracture of the skull and developed what Boyle called a "dead palsy" of the arm and leg on one side of the body. A barber surgeon elevated the depressed bone and promptly relieved the paralysis. The first notion of an orderly "representation" of motor control came in 1864 from the studies of Hughlings Jackson, an English neurologist, who made careful observations on the development of local seizures in his epileptic patients. As these attacks spread from their original focus, the order in which the parts of the body began to jerk suggested that a disturbance of some kind was spreading by local extension over a region of the brain in which movements of the parts of the body were represented in an orderly plan. Jackson's brilliant clinical deduction was confirmed in 1870 by Fritsch and Hitzig in Germany, who reported that electrical stimulation of the frontal cortex of cat and dog caused movements of the limbs on the opposite side of the body. In a monograph that appeared in 1874, Hitzig accurately defined the limits of the "motor area" from which these responses could be elicited in the dog and monkey. David Ferrier, who was working at about the same time in England as a physician in the West Riding Lunatic Asylum, made a remarkably accurate motor map of the monkey cortex (Fig. 32-1). Ferrier borrowed the term "localization" from Jackson to describe his findings. In further studies, Ferrier first delimited a specific part of the motor area (e.g., the "hand area") by means of electrical stimulation and then removed this area surgically, causing a paralysis of the hand. Ferrier took some of his "hemiplegic" monkeys to the International Medical Congress of 1881 in London. Their resemblance to hemiplegic patients was so striking that the famous French neurologist Charcot, seeing one of the monkeys, declared "It is a patient." The motor cortices of several types of apes were stimulated by Sherrington and his collaborators in the late 19th century, but there was no systematic study of the human motor cortex until the German neurosurgeon Foerster published his report in 1925.[29]

As clinical and experimental studies have progressed, it has become apparent that certain parts of the cerebral cortex such as the precentral region are concerned exclusively with motor functions; other parts such as the somatic sensory areas have both sensory and motor functions; and still other regions, not usually regarded as having any direct role in movement, are doubtless involved in the volitional aspects of motor activity. Closely associated with the cortex in its motor functions are a group of nuclei known as the basal ganglia, which occupy a large volume of brain. They have important connections with the cortex, and their function appears to be chiefly motor. The little that is known about their functions has been described in Chapter 29.

Very little can be said about the origin of the neural commands that lead to movement. No precise "command center" has been identified, and no particular cortical or subcortical region seems to be essential for the formulation of the orders to which the motor cortex responds. Most of the available information concerns the systems that execute these commands. There are two main subdivisions of motor control, the pyramidal and extrapyramidal systems. They work together to produce movement, muscle tone, and posture, but they play very different roles—so much so that the clinical neurologist classifies his patients' motor "signs" as either pyramidal or extrapyramidal.

Pyramidal system

The pyramidal system is much the easier to deal with because it is far simpler to define ana-

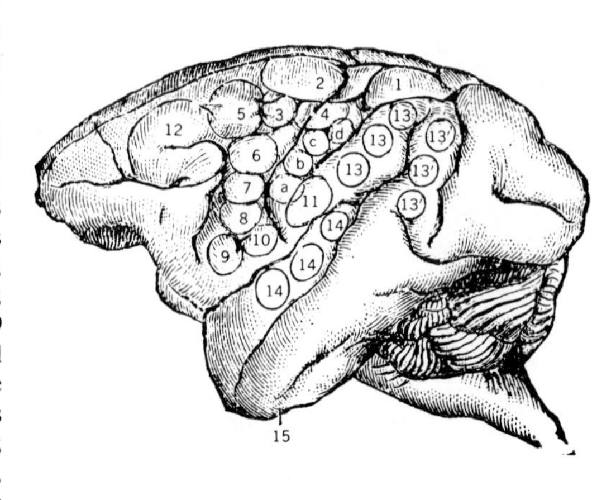

Fig. 32-1. Ferrier's original (1875) motor map of left hemisphere of monkey describing effects of electrical stimulation at series of numbered sites. Ferrier's original legend follows: "The left hemisphere of the monkey. 1, The opposite hind limb is advanced as in walking; 2, flexion with outward rotation of the thigh, rotation inwards of the leg, with flexion of the toes; 3, the tail; 4, the opposite arm is adducted, extended, and retracted, the hand pronated; 5, extension forwards of the opposite arm; *a, b, c, d,* movements of fingers and wrist; 6, flexion and supination of the forearm; 7, retraction and elevation of the angle of the mouth; 8, elevation of the ala of the nose and upper lip; 9 and 10, opening of the mouth, with protrusion (9) and retraction (10) of the tongue; 11, retraction of the angle of the mouth; 12, the eyes open widely, the pupils dilate, and head and eyes turn to the opposite side; 13 and 13′, the eyes move to the opposite side; 14, pricking of the opposite ear, head and eyes turn to the opposite side, pupils dilate widely." (From Ferrier.[24])

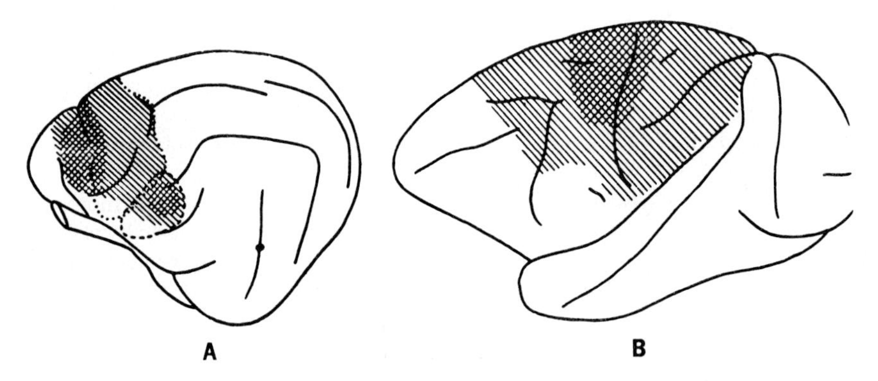

Fig. 32-2. Dorsolateral views of left cerebral cortex of cat, **A,** and monkey, **B,** showing distribution of potentials (hatching and cross-hatching) evoked by antidromic volleys of impulses set up by stimulation of medullary pyramid. (From Woolsey and Chang.[73])

tomically. It consists of neurons whose cell bodies lie in the cerebral cortex and whose axons pass through the *pyramid* of the medulla to form the pyramidal or corticospinal tracts of the spinal cord. In man the longest of these fibers runs at least 1 m. There are about 1 million fibers in each human pyramid. About 60% of them arise in precentral cortex and 40% in postcentral cortex. Only about 2% of the fibers come from the large Betz cells. Since 90% of them are small (1 to 4 μm) and about half of the total are unmyelinated, the pyramidal tract is, on the whole, a slowly conducting pathway.

Some pyramidal axons end in the motor nuclei of the brain stem; the majority pass through the pyramids and form three corticospinal tracts. The largest of the three in primates is the crossed corticospinal tract, which makes up three fourths of the total. Two uncrossed tracts, somewhat variable in size, run in the lateral and ventral parts of the spinal cord and end ipsilaterally. With a special technique for staining terminals of degenerating axons, it has been shown that 80% to 90% of the corticospinal fibers in primates make synaptic contacts with internuncial cells in the spinal cord, whereas 10% to 20% end directly on motoneurons.

Anatomic studies carried out by cutting the pyramidal fibers and studying the retrograde degeneration that follows show that they arise from a very extensive region in front of and behind the central fissure of Rolando. The areas from which they arise include Brodmann's areas 8, 6, and 4 in the precentral region and areas 3, 1, 2, 5, and 7 postcentrally (Fig. 9-3). Since only about 10% of the pyramidal fibers are large enough (>5 μm) to show up well in Weigert and Marchi preparations, neuroanatomists were uncertain about the origin of the smaller fibers. The neatest way of determining the cortical origin of the pyramidal tract, and certainly one of the quickest, is by an electrophysiologic method devised by Woolsey and Chang.[73] They exposed the pyramid in the medulla and stimulated it electrically with single shocks, taking special precautions to avoid exciting other fiber tracts nearby. An antidromic volley that was set up in the pyramidal fibers was conducted back to the cortex wherever the cells of origin were located. The pattern of antidromic responses recorded in this way precisely defined the origins of the pyramidal tract in the cat and monkey. As shown in Fig. 32-2, responses were detected over most of the parietal lobe and a great deal of the precentral cortex. This pattern of origin agrees well with anatomic studies but shows somewhat more extended boundaries in areas 8 and 6. It includes all four of the cortical areas now known to have somatic motor functions. A stimulating electrode placed anywhere within this large area will elicit a movement of some part of the body.

The pyramidal tract is late to develop in evolution. Its size and functional importance increase progressively in moving from mammals to monkeys to the higher apes and to man. This suggests that it developed to serve the needs of recently acquired cortical areas for a more direct and autonomous control over motoneurons than that provided by the extrapyramidal system. Differences in the degree of development of the pyramidal system sometimes make comparisons of experimental results in different species difficult.

Extrapyramidal system

The pyramidal tract is a compact bundle of fibers arising in the cortex that is easily recognized as an anatomic entity. In contrast, the extrapyramidal system consists of many separate components arising at different levels of the nervous system, often without obvious relation to each other, and hardly justifying the term "system." The pyramidal tract transmits impulses directly from cortex to spinal cord; the extrapyramidal system transmits signals indirectly from the cortex through one or more relays to the cord. Cortical extrapyramidal fibers take origin from all the sensory-motor cortex, including the supplementary motor area and the "second" motor areas, and from many other parts of the cerebral cortex as well.[13] It is not known whether they are the axons of separate extrapyramidal cells in the cortex or are collaterals of other corticofugal fibers. Impulses in these cells may take several different routes to the motoneurons. There are two relatively direct routes available: one from the sensorimotor cortex to the reticular formation (RF) in the brain stem and from there to the spinal cord via the reticulospinal tracts; the other from the motor cortex to the red nucleus and from there to the spinal cord via the rubrospinal tract. In addition, a considerable number of cortical fibers descend to the basal ganglia, and from them a series of connections are established to the red nucleus and RF.

MOTOR AREAS OF CEREBRAL CORTEX—RESULTS OF STIMULATION

Motor effects can be elicited by adequate excitation of any site within the crosshatched region shown in Fig. 32-2. Electrical stimulation is the easiest and most readily controlled type of stimulus to use. The excitable region extends into the depths of the central fissure and other fissures within this extensive area and includes a large portion of the medial surface of the brain as far down as the cingulate sulcus. In man and in animals with highly fissured brains, one half to two thirds of the total motor area is buried and rather inaccessible to study unless the buried portions are exposed by removal of one bank of a fissure.

Responses to electrical stimulation

The stimulus most frequently used is a 60 Hz alternating current of a few volts' intensity and 1 to 3 sec duration. If the stimulus is not too strong and the cortex is not too depressed by anesthesia, the typical response generally consists of a brief, twitchlike contracture of one or several muscles. With the electrodes placed over the "thumb" area, for example, a just-threshold stimulus may result in a brief abduction or adduction of the thumb with no other movement. With stronger or longer stimulation the response may spread to include other muscles, perhaps those supplying adjacent fingers or the wrist. These movements occur on the opposite side of the body. In anesthetized monkeys the lowest threshold and the most discrete movements are usually produced by stimulating just in front of the central fissure in area 4. In general, responses elicited from cortex behind the central fissure require a stronger stimulus, a phenomenon that suggests a lower density of the neurons that project to motor centers. Under deep anesthesia, responses are essentially simple contractions of one or more muscles. With anesthesia light enough to permit sustained tonic innervation of muscles, it can be seen that stimulation causes relaxation of some muscles and contraction of others. The pattern of such effects is one of reciprocal innervation (i.e., opposing effects in antagonistic muscles). If the stimulus applied to a cortical site is too intense, the response spreads to involve a whole group of muscles, and there may be alternating jerks due to contractions in opposing muscles. This is really a local seizure, quite comparable to local epilepsy in a patient. If the stimulus is strong enough or the cortex is excitable enough, the spread of activity may continue until the seizure involves the entire body.

Stimulation of human cortex

The movements elicited by stimulating the motor cortex of a conscious patient during surgery are similar to those obtained in animals but they are more discrete, owing to finer control of distal musculature in man. In a conscious, cooperative patient, it can also be shown that stimulation produces paralysis of the voluntary control over muscles. For example, a patient gripping a rod tightly with both hands will relax his grip on one side when the contralateral hand area is excited. A patient who is speaking will stop in the middle of a word if the motor area controlling his laryngeal muscles is stimulated. During stimulation a conscious subject may report merely an awareness of the movement as it occurs. Occasionally a desire to move is noted. Sensations of numbness or tingling are occasionally reported from precentral stimulation but generally only from postcentral sites. Frequently a patient is quite surprised to realize that his hand, for example, has moved as though it were under outside control. From a long experience with stimulation of the human cortex, Penfield and Rasmussen[49] report that the motor area of a young child is like that of a man of 60 years. The type of response is the same, moreover, in an accomplished pianist and a manual laborer, that is, no evidence of any acquisition of motor skills is revealed by this method.

Mechanism of cortically elicited movements

Relatively little is known about the mechanisms set in operation by the cortical stimulus. In most instances the electrode excites both pyramidal and extrapyramidal cortical efferents. The delicacy and discreteness of the responses, however, stamp them as primarily pyramidal. When the pyramids were cut in the medulla, leaving the remainder of the medulla undamaged, Tower[62] reported that somatotopically organized, discrete movements were largely abolished. Stimulation continued to elicit motor responses, but they had lost their fineness and delicacy. Removal of the contralateral motor areas does not abolish these remaining responses; hence they must be mediated by extrapyramidal connections. It is apparent from the electrical studies of Lloyd[40] that a single shock applied to the motor cortex of a cat will evoke descending volleys in at least two systems. There is a widespread activation of the RF in the brain stem 4 msec after the cortical shock. This reticular activity is in turn projected by high-velocity tract fibers to the ventral horn of the spinal cord. A calculation based on the known conduction rates and delays involved shows that volleys relayed through the reticulospinal system could reach the local segmental mechanism at the same time or a little earlier than the more slowly conducted but more direct pyramidal impulses.

Pattern of representation in motor areas

The earliest investigations with electrical stimuli showed clearly that muscles in the different parts of the body are represented in an orderly sequence in area 4, just anterior to the central fissure. This sequence is a reflection of the segmental origin of the nervous system. As illustrated to the left of Fig. 32-3, the toes, ankle, and leg are represented on the medial wall of the brain, with the knee, hip, and trunk following in order on the convexity of the cortex; then the shoulder, elbow, wrist, and fingers are succeeded by the neck, brow, face, lips, jaw, and tongue most laterally. At the oral and anal ends of this sequence the parts of the body that are "turned inward" are apparently represented in the sylvian fissure and the cingulate fissure.

The amounts of cortex devoted to various parts of the body are very unequal. The thumb, for example, is controlled by an area 10 times as great as that which controls the whole thigh. The parts of the body that are capable of fine or delicate movement have a large cortical area devoted to them; those that perform relatively gross movements (e.g., trunk muscles) have comparatively little cortical representation. The area of cortex devoted to a part is related to the density of the innervation of that part in the periphery. From an analysis of their studies on human subjects, Penfield and Boldrey[48] constructed a small figure of a man called a homunculus (Fig. 32-4). The figure was drawn so that the size of each part corresponded to the amount of cortex given over to it. The result was an extremely distorted figure of a man with a very large thumb, a relatively big face, and an enormous tongue. It is not surpris-

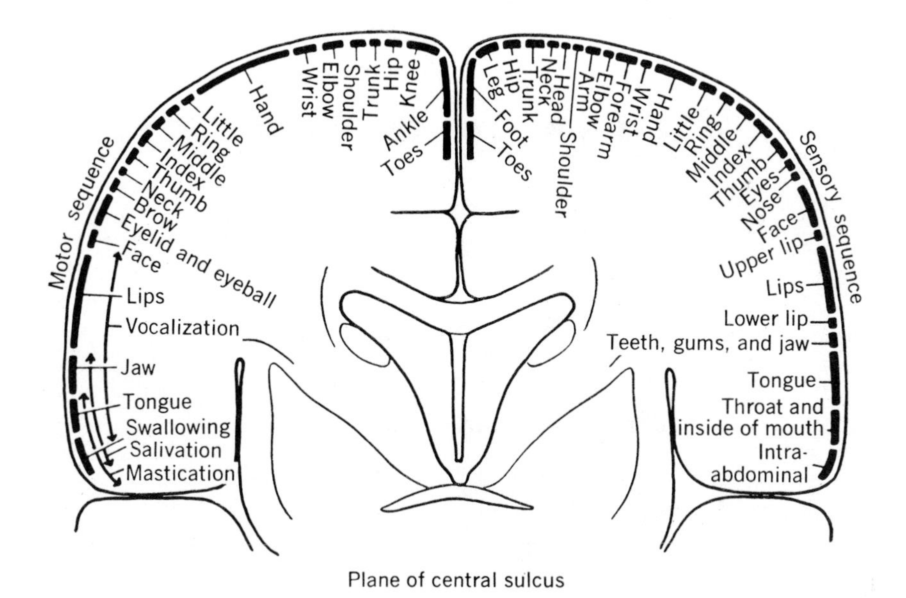

Fig. 32-3. Diagram of sensory and motor sequences in cerebral cortex of man as determined by electrical stimulation. Extent of cortex devoted to each part of body is indicated by length of heavy black lines. (From Rasmussen and Penfield.[52])

ing that focal epilepsy most commonly occurs in these parts of the body, since they comprise a large percentage of the total motor representation.

Despite thousands of cortical stimulations by a great many investigators, the complete and now rather obvious plan of motor representation eluded everyone until 1947. Then, as a result of studies by Mountcastle and Henneman[45] on the localization of tactile responses in the thalamus, it became apparent that within the three dimensions of one nucleus, the entire surface of the contralateral side of the cat was represented in continuity and with no missing parts. This study suggested that a complete representation of skeletal muscles might be demonstrable within the confines of the motor cortex. Woolsey et al.[75] first examined the motor cortex of the rat, which had the advantage of being smooth and unfissured. It soon became evident that all previous maps of the motor cortex were incomplete and partially erroneous. The essence of their study was that the muscles of the opposite side of the body (Fig. 32-5) are represented without separation of face and trunk areas by arm (as previously believed), with the axial muscles arranged in sequence from neck to tail along the rostral and medial portions of the brain and with the limb areas more laterally. As shown in Fig. 32-6, the somatic sensory and motor representations are essentially mirror images.

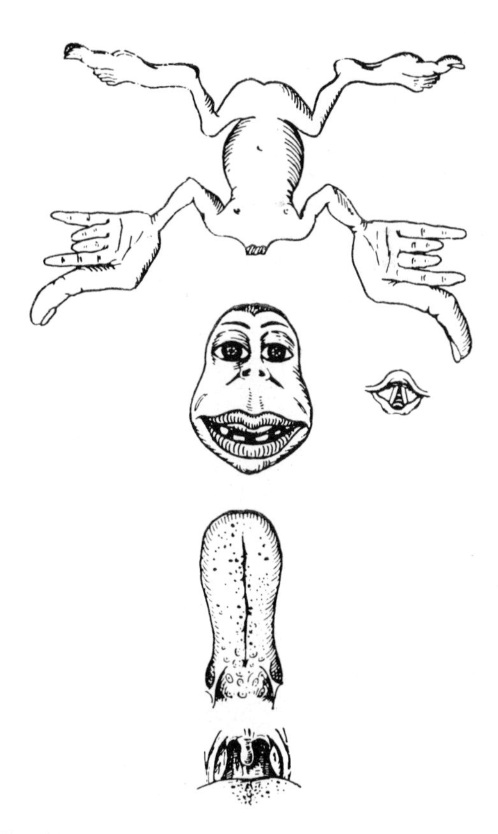

Fig. 32-4. Penfield and Boldrey's homunculus. Parts of body are drawn to illustrate relative amounts of motor cortex devoted to each. (From Penfield and Boldrey.[48])

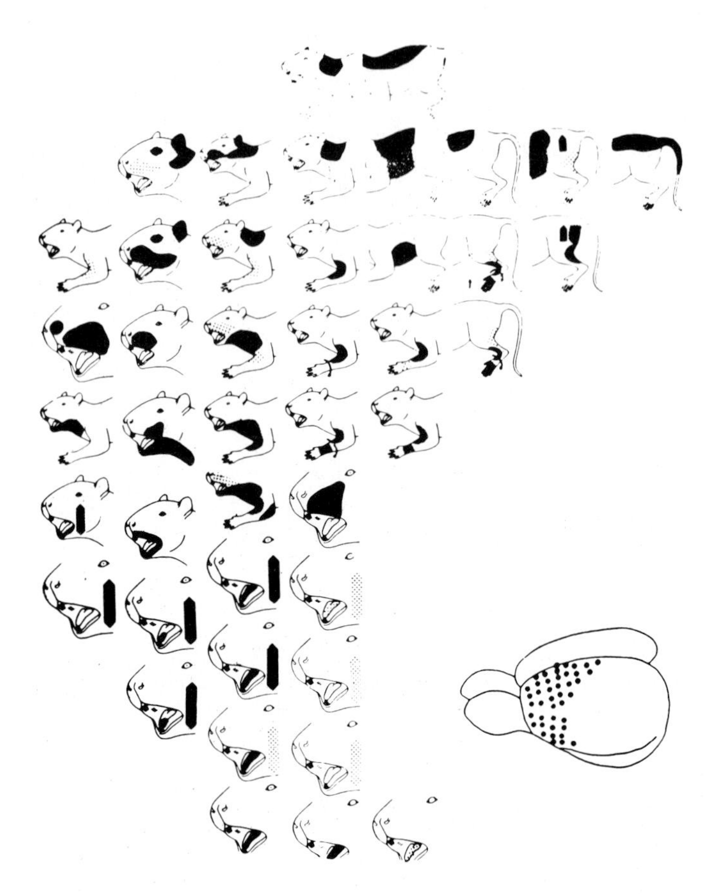

Fig. 32-5. Pattern of representation in left cortical motor area of rat. Figurines correspond to points marked on inset drawing. Intervals between points are approximately 1 mm. Each figurine indicates peripheral location of musculature activated by electrically stimulating cortical point to which figurine corresponds. Regions of maximal responses are indicated in black; those of weaker muscular actions in cross-hatching and stippling. Reacting muscles are in right half of body, contralateral to cortex stimulated, but to maintain desired orientation within pattern, left-sided figurines are used. Motor pattern is essentially a mirror image of cortical pattern of tactile representation in somatic sensory area I of this same species. (After Settlage et al.[54a]; from Woolsey et al.[75])

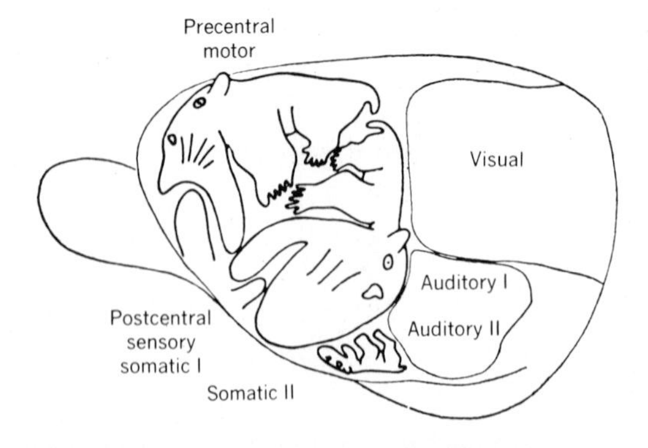

Fig. 32-6. Diagram of rat cortex showing mirror-image representation in precentral motor cortex and somatic sensory area I (SI). (From Woolsey.[72])

The results obtained in the simple brain of the rat led to a reexamination of the precentral motor area of the monkey. Experiments were carried out under barbiturate anesthesia. Cortical sites were stimulated systematically at 0.5 to 1.0 mm intervals, and several investigators made careful observations of the resulting movements, which were recorded by shading or crosshatching the appropriate part of an outline drawing of the monkey. Fig. 32-7 is a composite chart derived from several experiments showing the cortical location of each stimulation and the part of the body that moved in response. The portion of the cerebral cortex shown in Fig. 32-7 is outlined in

Fig. 32-7. Composite figurine charts of precentral and supplementary motor areas of monkey brain, derived from several experiments in which left cortex was mapped by systematic punctate electrical stimulation. Except for responses from points in ipsilateral motor face area (at extreme left of **A**), muscle reactions are in right half of body, but (as in Fig. 32-5) to maintain desired orientation within total pattern, left-sided figurines are used; to appreciate laterality of movement, reader should imagine that he is looking through each figurine to its opposite side. Strongest and earliest movements are indicated in black; cross-hatching signifies intermediate, and stippling, weakest effects. In many cases, these shadings show nature of movement. Symbols with crosses on ankles indicate eversion of foot; those with open centers on hip and ankle signify adduction and inversion; curved lines with arrows designate rotation. Supplementary motor area, situated almost entirely on medial surface of hemisphere and on upper bank of sulcus cinguli, is quite separate from precentral motor area but adjoins medial and rostromedial limits of the latter. Extents and spatial relationships of the two areas are shown diagrammatically in Fig. 32-8; general arrangement of patterns in Fig. 32-9. Note that both representations are partly on medial surface and that important portions of each are located in one or more deep sulci. Boundaries of these sulci are marked in manner indicated in legend of Fig. 32-8. (From Woolsey et al.[75])

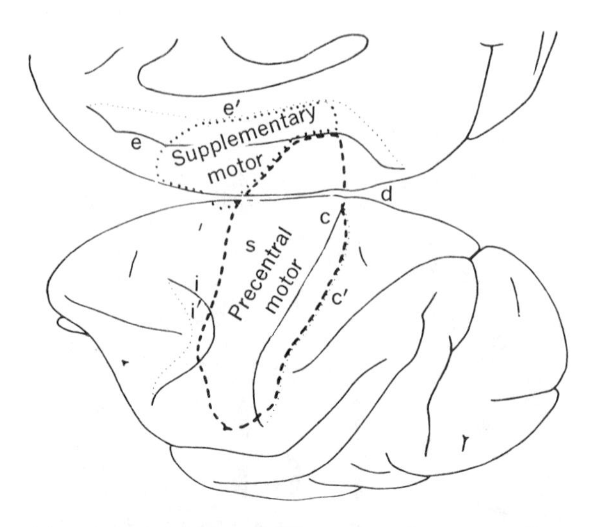

Fig. 32-8. Drawing of dorsolateral (below) and medial (above) views of left hemisphere of macaque showing extents of precentral and supplementary motor areas. (Patterns of localization within each are shown in diagrammatic form in Fig. 32-9.) Caudal bank of central sulcus, anterior bank of inferior precentral sulcus, and lower bank of sulcus cinguli have been cut away to show extension of excitable cortex into opposite banks of these sulci. Different sulci are indicated as follows: *c*, central sulcus; *c'*, bottom of central sulcus; *i*, inferior precentral sulcus; *i'*, bottom of inferior precentral sulcus; *d*, medial edge of hemisphere; *e*, sulcus cinguli (on medial surface of hemisphere between rim and corpus callosum); *e'*, bottom of sulcus cinguli; *s*, superior precentral sulcus. (From Woolsey et al.[75])

Fig. 32-8, in which the fissures and markings are identified. As illustrated in these figures, Woolsey et al.[75] found a complete and continuous pattern of motor representation extending from the bottom (c') of the central fissure posteriorly forward into the inferior precentral sulcus (i) anteriorly and from the bottom of the cingulate fissure (e') on the medial wall to the lateral end of the central fissure. Within this region, two complete motor representations could be distinguished, as shown in Fig. 32-9. The larger and more detailed of the two patterns, occupying the precentral gyrus (area 4), is the "motor cortex" of classic description; the smaller pattern of the medial wall is a more recent addition called the "supplementary motor area," which will be described later in this chapter. Examination of Fig. 32-7 reveals that the individual responses fit together like the pieces of a jigsaw puzzle, with adjacent muscles located contiguously. The muscles of the axial parts of the body, the back, neck, and head, are represented most anteriorly; those controlling the digits of the hind limb and forelimb are represented most posteriorly on the anterior bank of the central fissure. The unnatural proximities of certain parts of the body such as the thumb and lower lip or the hallux and little finger result because all the cortex is used for representation, and there are no unoccupied areas to separate different portions of the body. All the movements elicited by stimulation within this primary motor area occurred contralaterally, except for those evoked from the most lateral sites where movements of the *ipsilateral* face, lips, and tongue were produced (Fig. 32-9). The existence of this ipsilateral face area is probably related to the need for bilateral control and coordination of these particular muscles.

Fig. 32-7 gives some indication of the extent to which the representation of one muscle overlaps with that of others nearby. Due to the depth of anesthesia used in these experiments and the uncertain amount of current spread during stimulation, it was not possible to determine whether the true overlap was greater or less than that revealed in Fig. 32-7. More recent and refined observations on this matter will be described later in this chapter.

Supplementary motor area

As already noted, Woolsey et al.[75] confirmed the existence of a second complete motor representation located almost entirely on the medial surface of the hemisphere. It lies on the upper bank of the sulcus cinguli and the medial surface of the brain dorsal to this fissure, with the tip of

Fig. 32-9. Diagram illustrating general arrangement of patterns of representation of somatic musculature in precentral and supplementary motor areas of monkey. Sulci and medial edge of hemisphere (left) are indicated by lines as in Fig. 32-8. Bodily distortion displayed by each simiusculus (simian equivalent of homunculus) is due to very unequal representation of different parts of musculature in the two areas. Ipsilateral motor face area is indicated at bottom of drawing. (From Woolsey et al.[75])

the thumb area extending up on the convexity. This representation adjoins the hind limb and tail areas of the precentral motor area, but is quite separate from it. By stimulating this region, Penfield and Welch[50] had previously elicited movements of the contralateral extremities in anesthetized monkeys. In patients under local anesthesia, they produced vocalization with associated movements of face and jaw, combinations of movements in the extremities, trunk, and head, inhibition of voluntary activity, and certain autonomic effects. They named this region the "supplementary motor area." Woolsey et al.[75] worked out the complete pattern of representation shown in Fig. 32-9. Instead of the rapid, phasic motor responses observed in conjunction with precentral cortical stimulations, they noted that the contractions lasted much longer and that postures assumed as a result of brief stimulation were often maintained for many seconds. Although movements were less discrete, the whole animal was represented in continuity—much of it in the depths of the sulcus cinguli.

Postcentral motor areas

In 1900 Schaefer[54] published a study of the monkey brain showing that movements could be elicited by stimulating not only the frontal areas now identified as the precentral and supplementary motor areas but also the postcentral gyrus. Similar results were reported by the Vogts[67] in animals and by Foerster[28] and Penfield and Rasmussen[49] in man. However, as Woolsey has pointed out, ". . . since the work of Leyton and Sherrington,[38] the preeminence of the precentral gyrus in motor function has dominated teaching and thinking concerning cortical control of the somatic musculature and the motor effects of postcentral stimulation generally have been explained as the result of the spread of excessive stimulating currents to the precentral area, or on the basis of corticocortical connections with this area."[72]

In 1943 Kennard and McCulloch[33] stimulated the postcentral gyrus of infant monkeys after removal of the precentral motor areas. They found that focal movements like those usually evoked from the precentral gyrus were easily produced. Later Woolsey et al.[74] stimulated the postcentral gyrus of adult monkeys months after removal of both precentral and supplementary motor areas. In these experiments, they demonstrated a "well-organized postcentral motor outflow after complete degeneration of the motor pathways from both frontal lobes."[74] The detailed pattern of motor representation that was found apparently coincided with the somatic sensory pattern in the same area and appeared to be a mirror image of the precentral motor pattern just in front of the central fissure. The chief difference between the pre- and postcentral responses was that the electrical threshold was 2 to 3 times higher for the latter. From these results, it appears that there is a well-organized postcentral motor system that ordinarily functions in cooperation with the precentral systems but can function independently of them.

In addition to the sensorimotor area in the postcentral gyrus, there is another parietal motor area coinciding closely with the *second somatic sensory area* first described by Adrian.[1] This is a receiving area for somatic sensory signals from both sides of the body. In monkeys and man, it is located on the superior bank of the sylvian fissure (Fig. 32-10), with its face area immediately adjacent to the face area of the postcentral somatic sensory area. Sugar et al.[59] found that strong stimulation in this area elicited movements on both sides of the body, and they named it the *second motor area*. Later, it was shown that the sensory and motor patterns in this small area coincide somatotopically.[7,69]

• • •

In summary, there are four functionally distinguishable motor areas, as shown in Fig. 32-10, two in front of the central fissure and two behind it. Three of them definitely contribute fibers to the pyramidal tract (the supplementary motor area has not been examined in this respect). It is probable that all four areas send connections to extrapyramidal centers as well.

Special motor areas

In addition to the four somatotopically organized cortical areas just described, each of which exercises some degree of control over the skeletal muscles of the entire body, there are several cortical areas concerned with control of the muscles associated with the special senses.

Early in the development of our knowledge of cortical motor areas, it was found that stimulation of a localized region in the frontal lobe of the primate produced movements of the eyes, eyelids, and pupils. According to W. K. Smith,[56] the "frontal eye field" of the monkey is extensive, and on the basis of its responses to electrical stimulation, it can be divided into five parts, as shown in Fig. 32-11: (1) an area situated just caudal to and ventral to the angle of the inferior precentral (arcuate) sulcus, from which closure of the eyes is obtained; (2) a field lying just across the sulcus, which yields nystagmus bilaterally (fast component to the opposite side) and opening of the eyes if they are closed; (3) an adjacent area, situated more medially within the upper arm of the arcuate sulcus, which yields conjugate turning of the eyes to the opposite side; (4) a field centered on the upper extremity of the arcuate sulcus, from which an "awakening reaction" (opening of the eyes with slow deviation to the opposite side, blinking, and pupillary dilatation) can be obtained; and (5) an area lying between the medial end of the arcuate sulcus and some point ou the medial surface of the hemisphere, excitation of which results in equal dilatation of both pupils. Somewhat similar results have been reported for both chimpanzee and man. The frontal eye fields may represent an elaboration and outgrowth of the precentral motor area controlling the eyes, or it may be a much more specialized motor area functionally quite distinct from it.

Another motor eye field is located in the occipital lobe within or adjacent to the visual receiving areas. Stimulation of this region generally causes conjugate deviation of the eyes to the opposite side. When the electrodes were applied to the cortex of area 17 superior to the calcarine fissure, the movements were lateral and downward. When the stimuli were applied below the calcarine fissure, the deviation was lateral and upward. The frontal and occipital eye fields are

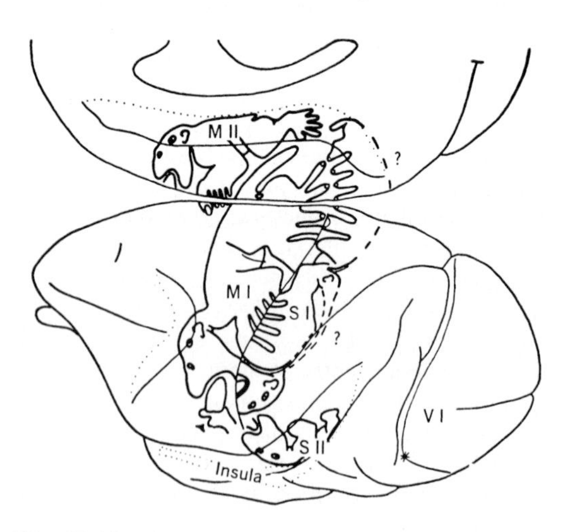

Fig. 32-10. Diagram of monkey cortex showing locations of four principal motor areas: precentral motor, *M I;* supplementary motor, *M II;* somatic sensory I, *S I;* and second sensory, *S II. S II* lies on upper bank of sylvian fissure next to insula and auditory area on lower bank. (From Woolsey.[72])

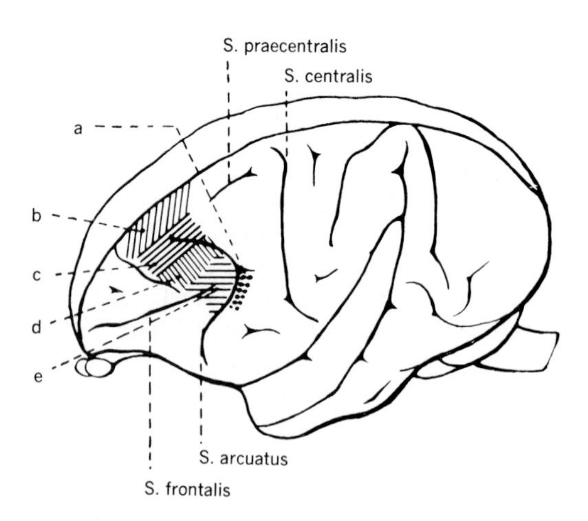

Fig. 32-11. Subdivision of frontal eye field and area yielding closure of eyes in monkey. *a,* Closure of eyes; *b,* pupillary dilatation; *c,* "awakening"; *d,* conjugate deviation to opposite side; *e,* nystagmus to opposite side. (From Smith.[56])

interconnected by long corticocortical fibers, but responses to stimulation in either area do not depend on the integrity of the other.

Adjacent to the auditory receiving area in the cat and monkey is a small region whose stimulation causes movements of the ear in these species. In the cat, electrical responses to clicks can be recorded within this "pinna area," suggesting that it responds reflexly to auditory stimuli by turning the ear toward the source of the sound.

From this brief account, it can be seen that each of the main sensory receiving areas has a motor area ajdacent to it. Whether this rule applies to the receiving areas for taste and olfaction is not yet known.

ORGANIZATION OF MOTOR CORTEX

In order to understand how the motor cortex is organized, it is necessary to know (1) how the efferent cells and fibers are arranged and grouped functionally, (2) what types of signals the cortex receives and how they are distributed to the efferent cells, and (3) how cortical input and output are related. These topics will be considered in order.

EFFERENT ORGANIZATION

Columnar structure of cortex

Sections of the cortex cut perpendicularly to the surface reveal a columnar arrangement of the cells in the precentral motor cortex similar to that found in other regions. Lorente de Nó's studies with Golgi stains showed a wealth of radial interconnections between the cells in the different layers of these columns and led him to suggest that these "vertical chains" of cells might be considered as elementary units of the cortex.[41] Mountcastle's discovery that the cells in somatic sensory columns all receive signals of the same sensory modality from the same part of the body provided the first physiologic demonstration that these cortical columns are functional entities.[44] Since his initial work, several studies have shown that the columns in the motor areas are the "building blocks" around which the cortex is organized, for input as well as output.[2,3,10]

Asanuma and Sakata[2] have developed a technique of intracortical microstimulation that has been very useful in investigating the efferent projections of individual motor columns. Monosynaptic reflexes were used as sensitive indicators of the effects of these stimulations. Electrical thresholds for definite but minimal effects were as low as $1/100$ of the threshold required for surface stimulation and therefore entailed far less spread of current. Facilitation and inhibition were generally evoked from distinct, nonoverlapping regions, with facilitation having its lowest thresholds 1.0 to 1.4 mm from the surface and inhibition 1.0 to 1.5 mm deeper. When the effects of stimulations at various depths in numerous penetrations were correlated with the exact site of each in histologic reconstructions of the electrode tracks, it was found that the lowest thresholds for facilitation of a particular pool of motoneurons were distributed in a radially aligned column estimated to be about 1 mm in diameter. As one would expect with electrical stimuli, some overlapping of the effects obtained from adjacent efferent zones was noted, but not enough to invalidate the clear indication of cylindric arrays of cells with a common spinal projection. Section of the bulbar pyramid abolished these effects, indicating that they were purely pyramidal. Extrapyramidal effects have not yet been identified with this technique.

In further studies with intracortical microstimulation,[3] trains of pulses at 300/sec were used for periods up to several seconds, and the contractions of a pair of antagonistic muscles at the cat's wrist were recorded. With the electrode in layers V and VI, it was possible to produce sustained contractions of one of these muscles lasting as long as the stimulation continued. The regions from which sustained contractions could be elicited were very limited in extent. Stimulations in superficial layers produced only phasic contractions, which died out before the stimulation ended. Sustained contractions elicited from one site could be enhanced or reduced by intracortical stimulation of another site. The cortical zone from which contraction of one muscle was elicited was not coextensive with the zone that inhibited its antagonist. Sometimes stimulation at one site caused two antagonistic muscles to contract simultaneously, indicating that their contractions were not mutually exclusive. With two electrodes, it was possible to produce well-maintained contractions of antagonists. These observations indicate that there must be discrete, spatially separate zones for the excitation and inhibition of each muscle. Furthermore, the corticofugal path that produces contraction of one muscle is different from the path that inhibits contraction of its antagonist. Although the excitatory and inhibitory areas for a given muscle are spatially separate, they are located close to each other, and both of them are also near the excitatory zone of the direct antagonist.

Activity of single cortical neurons during movements

Very little is known about the functional organization of the cells and circuits comprising a motor column, but recordings from individual cortical cells are beginning to reveal how their activity is related to movement. In unanesthetized monkeys, Evarts[18] has examined the discharge patterns of cortical cells during a variety of spontaneous movements of the contralateral arm. Pyramidal cells, identified by backfiring them from the medullary pyramid, were divided into two groups on the basis of their axonal conduction velocities and discharge patterns. The larger cells with rapidly conducting axons were active only during limb movements. Small cells were continuously active even when the arm was at rest, but their rate of firing increased or decreased during a movement. On many occasions, it was possible to record simultaneously from these two types of cells with the same microelectrode. Although these pairs of adjacent cells were obviously in the same motor column, their firing patterns were often completely different. During a particular movement of the arm, one cell might discharge while the other was silent; during a different movement, the two units might discharge in parallel; occasionally their firing patterns were completely uncorrelated. In the great majority of pairs there was no fixed relationship between the two firing patterns, which varied with the type of movement being carried out. This changing or "plastic" relationship is what one would expect if the two cortical cells projected to motoneurons of two different muscles that acted sometimes as synergists, sometimes as antagonists, and at other times independently of each other. The flexors and extensors of the wrist, for example, would discharge in parallel in fixing the wrist during grasping and would discharge reciprocally during simple flexions and extensions of the wrist. It appears, then, that adjacent pyramidal cells usually project to different pools of motoneurons. In view of the fact that there are thousands of cells in a single column, it is surprising that only a few pairs displayed a fixed, positive correlation such as one would expect if both of them projected to the same pool of motoneurons. Although Evarts does not comment on this feature of the results, it suggests that cells projecting to the same motorneurons are systematically separated in the motor columns. It is not surprising that pairs of adjacent cells were never found to have a fixed, reciprocal relationship to each other because such relationships between muscles do not occur, even between direct antagonists.

The precise nature of the relationships between pyramidal cells in the same column did not become apparent until the movements to which the discharges were related were carefully studied. Then it was discovered that when one neuron of a pair was related to movement at a particular joint, the other was almost always related to the same joint. Evarts pointed out that the "common denominator is the joint to whose movements the (pyramidal neurons) are related rather than the way in which they are related to the movement."[19] The same observations were made on the series of cells encountered in a single penetration of the cortex. If the most superficial cell discharged during shoulder movements, the deeper units usually had some relation to the shoulder also.

These observations indicate that the cortical cells controlling a given muscle are not assembled in one compact group, as some have imagined. The cells have different projections even within the narrow limits of a single column. A column, then, is a functional entity responsible for directing a group of muscles acting on a single joint. This conclusion supports the original notion of Hughlings Jackson that movements, not muscles, are represented in the motor cortex. Evarts emphasizes that Jackson "did not hold that muscles were not represented at all, but rather that they were represented again and again and again, in different combinations. . . ."[22]

Recent studies of Phillips and his colleagues give an indication of the extent to which a muscle may be represented in many columns. In one group of experiments,[35] single shocks were applied to the surface of the motor cortex and the excitatory volleys reaching the motoneurons monosynaptically from the pyramidal tract neurons were detected by intracellular recording. With this technique the extent of the cortical field controlling single muscles in a baboon's hand or forearm was determined. The cortical fields were often so extensive that they eliminated the possibility that spread of current to a single "best point" within the area was the cause of a large projection. One motoneuron received projections from a cortical area of 8 × 2.5 mm, and another was found to receive monosynaptic excitation from corticospinal neurons lying 10 mm from the stimulating electrode in all directions! These results indicate that a single pool of motoneurons receives projections from cells in many motor columns and that cortical cells projecting to motoneurons of different muscles are extensively intermingled. Each column is presumably a slightly different combination of these intermingled cells. As Phillips points out, fine control of movements "must require a marvelously subtle routing of activity in the outer cortical layers to pick up, in significant functional grouping, the required corticofugal neurons, which are scattered and intermingled with unwanted ones which may be suppressed."[51] At present there is no evidence to indicate whether entire columns normally contribute as units to the output or whether individual cells within columns are activated selectively. The striking differences in the firing patterns of large and small pyramidal cells have not yet been satisfactorily explained. Evarts suggested that these two types of cells may be the pyramidal counterparts of large and small alpha motoneurons, which tend to discharge phasically and tonically. If this were so, both types of cells should be silent when the arm is completely at rest. Another possibility is that the two types of cortical cells control alpha and gamma motoneurons. The latter are small and have a strong tendency to fire in the absence of any movement or muscle tone just as the small cortical cells do; the former are large and discharge only during active contraction of a muscle, as do the large cortical cells. It has often been suggested that there are independent pathways from the cortex to alpha and gamma motoneurons, but they have not yet been identified.

Inputs to motor areas*

A great many factors control and regulate movement; hence it is not surprising that the primary motor cortex receives inputs from many sources. Somatic sensory impulses from skin, joints, and muscles are relayed to the cortex by way of the ventrobasal nucleus of the thalamus. These signals provide the cortical cells with the sensory cues that are required to guide and direct movements, and they supply the feedback in a long loop that includes the motor cortex and its efferent projections back to the muscles. Cerebellar signals are relayed to the motor cortex via the brachium conjunctivum, red nucleus, and ventrolateral nucleus of the thalamus. This projection may also be regarded as part of a closed loop linking the cerebellum and motor cortex. Input from the "nonspecific" thalamic nuclei reaches the motor cortex, presumably influencing its general level of excitability as it does that of other cortical areas. Signals from the opposite hemisphere, transmitted via the corpus callosum, are important in coordinating motor activity on the two sides of the body. Finally, signals from other, ipsilateral cortical areas reach the motor cortex directly via corticocortical fibers and indirectly via several subcortical centers. The four cortical motor areas are interconnected with each other, the U-shaped fibers linking the pre- and postcentral motor areas being particularly plentiful. Among other inputs are those from the visual cortex, which are of great importance in visually guided movements. The origins of the signals that convey voluntary "commands" to the motor cortex are as yet unidentified.

Afferent signals are distributed to layers I to IV of the motor cortex; efferent projections take origin chiefly from layers V and VI. Fibers from the somatic sensory relay nuclei of the thalamus end as a dense terminal plexus in the fourth layer. These "specific" afferents do not give off collaterals in the deeper layers. "Cerebellar" input also ends in the fourth layer. Thalamocortical fibers from the "nonspecific" thalamic nuclei pass up to layer I, giving off a few collaterals to other layers on the way. "Association" fibers from other cortical areas end chiefly in the upper four layers. The two deepest layers of the cortex receive little direct sensory input. They are influenced chiefly by the discharges of cells in the four superficial layers and by the numerous internuncial cells of the cortex, whose axons are distributed only within the cortex. In attempting to assess the efficacy of various types of input to cortical cells, as investigators are now doing, it is essential to determine the type of neuron that is responding. An input may "drive" a cell in one of the superficial layers or an internuncial cell without having a comparable effect on a pyramidal tract neuron.

Distribution of somatic sensory input

Early studies on the distribution of sensory input to the motor cortex, carried out under chloralose anesthesia and often involving electrical rather than natural stimulation, led to the conclusion that there is "widespread convergence of different modalities onto individual cells . . . with no recognizable pattern of spatial localization based either on modality of input or on the production of excitation versus inhibition."[42] Demonstrations that "one single pyramidal tract cell may react to somatosensory stimuli from wide receptive fields of the body surface, to acoustic stimuli and visual input"[71] made it "difficult to imagine that local sensory inflow could

*Further discussion of the problem of somatic sensory input to motor areas will be found on p. 364. The difference between cortical areas that are "sensory" in function and those that merely receive sensory input is carefully defined, and the reasons for some of the discrepancies and confusion in the current literature are explained.

have a dominant role in providing sufficiently precise information to accurately initiate or guide limb movements.''[10] It now appears that a large proportion of the responses recorded under chloralose, an agent that renders the nervous system abnormally sensitive to sensory stimuli, was mediated by the nonspecific projection system. As its name implies, this system responds to all types of stimuli without regard to modality and evokes cortical activity that may on occasion be confused with that evoked by the specific sensory projections. Responses evoked by the nonspecific projections are not regarded as ''sensory.''

Recent studies carried out under local anesthesia and with natural stimulation are beginning to resolve the confusion that has arisen. In a series of studies on cats, Brooks et al.[11,12] and collaborators[70] have extracellularly recorded the discharges of single cortical cells in the motor cortex in response to natural stimulation of skin, joints, and deeper tissues. With this technique, inputs of sufficient intensity to cause frank discharge of cells or inhibition of existing activity were detected, but subthreshold inputs may have been overlooked. In the somatic sensory cortex the cells in a particular column all respond to inputs of the same sensory modality coming from the same part of the body.[44] In a single motor column, however, different cells respond to different types of input. According to Brooks,[9] only about 6% of cells respond to more than one modality of ''specific'' sensory input. About three fourths of all cells have small (<25 cm^2) contralateral receptive fields that are sensitive to light touch. Examples of the sizes of these fields are shown in Fig. 32-12, *A* to *C*. The remaining cells have wide (>25 cm^2) contralateral or bilateral receptive fields, as shown in Fig. 32-12, *D* to *F*, or they respond to joint movement. By careful reconstruction of the electrode tracks in the cortex, it was shown that all the cells within a particular column having local receptive fields received signals from the same part of the body. The wide fields of the other neurons in the same column consistently overlapped the local fields. About half of all wide field units had *focal areas* within the wide fields, in which stimulation produced more intense effects. The location of these focal areas was usually identical with that of the local fields, or it overlapped them. In general, receptive fields were smallest at the periphery of the limb and increased progressively toward the elbow and then decreased somewhat toward the shoulder. Of 59 cells receiving superficial input exclusively from the forepaw, 52 responded only to stimulation of the ventral surface, 3 only to stimulation of the dorsal surface, and 4 to stimulation of both surfaces. This ventral preponderance existed chiefly for the forepaw area.

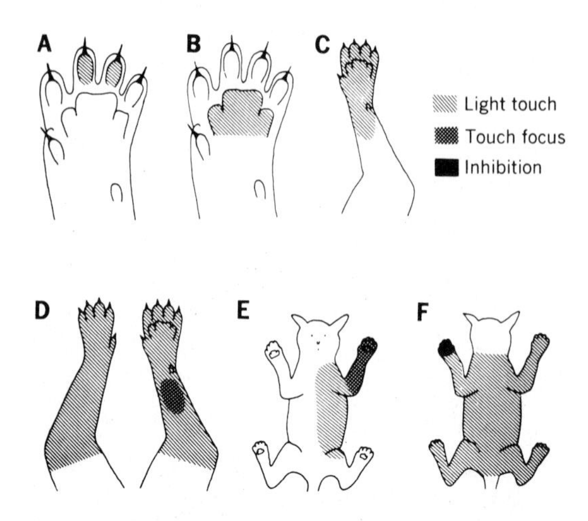

Fig. 32-12. Some typical peripheral ''receptive fields'' of neurons in cat pericruciate (motor-sensory) cortex. **A** to **C,** Fixed local fields of different types. **D** to **F,** Fixed nonlocal fields (areas greater than 25 cm^2). **D,** Stockinglike field with ''focus.'' **E** and **F,** Fixed wide fields. (From Welt et al.: In Neurophysiological basis of normal and abnormal motor activities, edited by M. D. Yahr and D. P. Purpura, © 1967 by Raven Press, New York.)

These findings indicate that the cortical cells in a single motor column receive different kinds of somatic sensory signals that all come from a single local part of the body. Thus far the experimental observations have not revealed significant differences in the receptive fields of the large and small pyramidal cells or in the fields of pyramidal and nonpyramidal cells within a column. The distribution of adequate stimuli is about 60% from superficial receptors (hair bending or light touch), 30% from deep receptors (deep pressure or joint movements), and about 10% from unidentified sources. These data on the cat assign a surprisingly small role to input from joints and muscles. In the monkey, signals from joints and muscles are much more important. Fetz and Baker[26] found that 189 of 233 units in the precentral cortex responded to passive movement of one or more joints of the contralateral leg. Of these, 148 fired only during movement of the joint; 4 fired tonically at rates proportional to the maintained angle of the joint, and 37 discharged both tonically and phasically. Units responding to cutaneous stimulation made up a small percentage of the total in the monkey.

In addition to its excitatory effects, sensory

input produces three types of inhibitory effects according to Brooks et al.[11] "Cross-modality" inhibition, a cessation of response to one type of stimulus when another type is applied, was observed frequently. Pressure or joint movement, for example, frequently inhibited cutaneous responses. In one good example a nonpyramidal cell could be driven by touch of the right forepaw only while the right elbow was partly extended. Flexion inhibited both evoked and spontaneous responses of this unit. "Surround" inhibition, a reduction in spontaneous or evoked firing to sensory stimulation when a stimulus of the same type is applied to a peripheral area surrounding or adjacent to that giving rise to excitation, was observed in 13 of 40 cases in which careful testing was done. "Complementary" inhibition, a form of interaction in which the two receptive areas are located on opposite limbs (right and left forepaws or right and left hind paws), was observed occasionally.

• • •

In **summary,** it appears that motor columns are the basic multicellular units around which sensory input is organized. Some of the cells within a column receive a specific excitatory input from skin or joint muscle in a particular part of the body. The receptive fields of these cells are small and modality specific. Other cells in the column receive convergent inputs from widespread areas of the body and a few may respond to more than one type of sensory stimulation. Inhibitory inputs from the same region, from surrounding areas, and even from contralateral limbs interact with these excitatory inputs. Afferents from the nonspecific thalamic nuclei modulate the excitability of cells in all parts of the column. Intracolumnar circuits permit extensive interaction between cells in different layers. The corticofugal cells, both pyramidal and nonpyramidal, presumably respond to the integrated sum of all these inputs and reach them via the cortical internuncial circuits.

Relation between cortical input and output

It is obvious that there is a close spatial relationship between the peripheral origin of the sensory input to a motor column and the muscles to which the column projects. The important question is how input and output are related.

By using the same microelectrode alternately for recording and stimulating, Asanuma et al.[4] have investigated the relationship between input and output in greater detail. In order to identify minimally effective outputs with maximal precision, they recorded the electromyograms (EMGs) of eight muscles in the cat's forelimb and reduced the stimulus strength until the EMG response was limited to one muscle. To ensure that the stimulating current excited only those pyramidal tract cells whose activity could be recorded by the same electrode (i.e., within a radius of about 90 μm), only sites with thresholds of 10 μA or less were selected for study. In Fig. 32-13, all the cutaneous receptive fields of cells located within the efferent zone of one muscle are shown projected onto a common map. Although the fields varied considerably in size, they shared a common skin area. Density of overlap is indicated by shading: black for maximum, lined for medium, dotted for minimum. As shown in the map at the upper left, the pyramidal cells that control EDCL (a muscle that dorsiflexes the paw and digits) had receptive fields mainly on the dorsal surface of the paw and digits. The area of maximal overlap is restricted to the dorsolateral paw, where 11 of the 17 fields overlapped. In contrast, the cells controlling PAL (a muscle that ventroflexes the paw and digits) had receptive fields with maximal overlap (20 out of 24 fields) on the ventral surface of the paw. Cells controlling ECU (a muscle that causes dorsiflexion of the paw) received maximal input from the ventrolateral paw. TRB (which extends the elbow) is controlled by cells that receive input from a wide area of skin with maximal overlap on the ventrolateral surface of the forearm. *These findings indicate that each efferent zone receives cutaneous inputs predominantly from skin regions that lie in the pathway of limb movement produced by contraction of the muscle to which the zone projects.* Similarly, cortical cells receiving their input from a joint were found to be located in efferent zones controlling muscles that moved that joint.

Using similar techniques, Rosén and Asanuma[53] have recently carried out more detailed studies on the "hand area" of the monkey. In this species, only 45% of the cells studied could be influenced by peripheral stimulation clearly enough to permit description of the receptive field and the adequate stimulus. In this part of the motor cortex the motor columns project to a single distal forelimb muscle or to a few muscles. As in the cat, a single column receives a polymodal input from deep as well as superficial receptors. Fig. 32-14 illustrates the input-output relations of cells that were encountered in a sagittal row of penetrations through the "thumb area." Four adjacent radial columns projecting

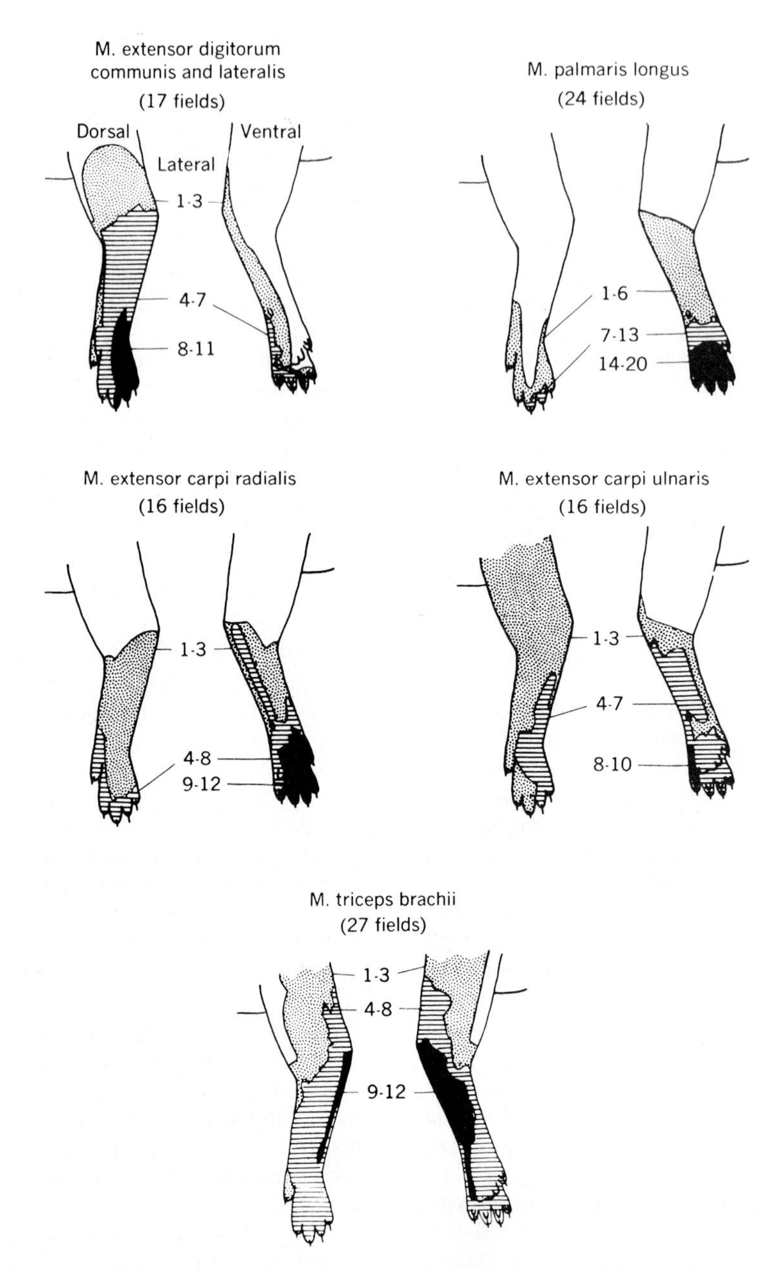

Fig. 32-13. Localization of cutaneous receptive fields of cells in sensorimotor cortex of cat. Receptive fields of all cells responding to light touch or hair bending within efferent zones of each five muscles have been superimposed on individual figurines. Density of overlap is indicated by shading: minimum, dotted; medium, lined; maximum, black. Total number of receptive fields used is shown below each figurine. Numbers associated with dotted, lined, and black areas give number of overlapping fields. (From Asanuma et al.[4])

to different thumb muscles were identified. Cells responding to superficial stimulation were arranged as follows. Two cells in the "thumb flexion" column were both activated from the ventral aspect of the thumb. The cells in the columns projecting to the extensor, adductor, and abductor of the thumb were activated by stimulation of the distal tip, medial aspect, and lateral aspect of the thumb, respectively. The cells in these penetrations responding to stimulation of deeper structures were all activated by passive movements of the thumb or by thenar pressure. The efferent

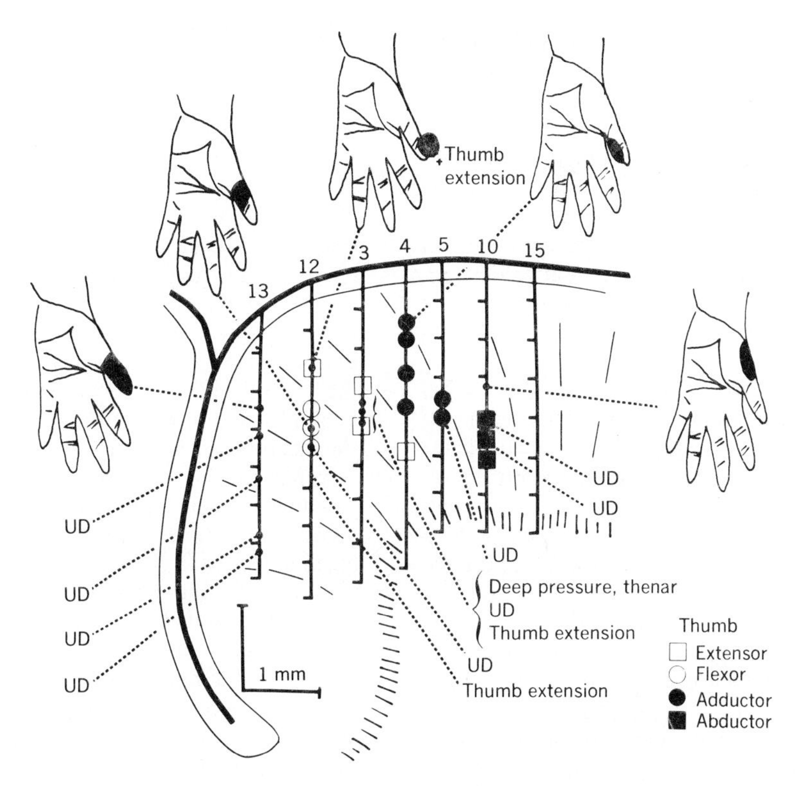

Fig. 32-14. Input-output relations of cells in "thumb area" of monkey. Reconstruction of electrode tracks and cell locations from experiment in which several electrode penetrations (solid lines, identified by numbers) passed through efferent zones projecting to various thumb muscles. Peripheral motor effects produced by intracortical microstimulations are indicated by symbols. Cortical sites stimulated with 5 μA without evoking motor effects are shown by short horizontal lines perpendicular to electrode tracks. Positions of cells encountered are indicated by dots and are connected to descriptions of receptive fields and adequate stimuli. (From Rosén and Asanuma.[53])

zones in this part of the cortex received tactile input chiefly from that side of the finger or hand that was in the direction of the movement produced by intracortical microstimulation. This finding applies to the zones causing thumb flexion, thumb adduction, and finger flexion, all of which would tend to be stimulated during grasping or manipulatory movements of the hand. Zones less likely to be involved in these activities, such as those projecting to extensor muscles of the fingers or to wrist muscles, apparently receive much less input from superficial receptors.

Of 56 cells in this study responding to stimulation of deep structures, 42 were activated by passive movement of joints (38 of them by movement of the joint involved in the motor effect) and the remainder by pressure applied to deep tissues. In almost all cases the direction of joint movement required for activation was the same as that produced by stimulation. Only two cells responded to joint movements in more than one direction. In general, it was found that cells driven by movements in the opposite direction were responding to impulses from muscle receptors. These results suggest that cortical cells receive information from joint receptors activated by the movements they produce and from receptors in the muscles contracting during the movement.

Most cells driven from the periphery were located in the superficial and intermediate layers of the cortex, with the largest number in layer IV. There was apparently no difference in the location of cells responding to superficial and deep receptors. Undriven cells were most frequent in layers V and VI, the sites most responsive to microstimulation.

The results of Rosén and Asanuma add considerably to our understanding of the coupling between sensory input and motor output. The distribution of the cells responding to specific afferent inputs (chiefly in layers I to IV) indicates that most of them are intracortical interneurons. Lorente de Nó's studies of cortical architecture and the results described previously indicate that the activity evoked in these superficial layers is directed to deeper layers, where the corticospinal and other efferent neurons are located. Motor effects evoked by intracortical stimulation in superficial layers are presumably relayed synaptically to the corticospinal cells that lie in deeper layers. In support of this view is the recent observation, already cited, that phasic and tonic contractions of muscles result from stimulation of superficial and deep layers of the cortex, respectively.

The functional role of the tight coupling between input and output is not yet fully apparent. Presumably the sensory input is used as feedback to modulate the output of individual cortical columns during the performance of movement. Under certain circumstances, it elicits cortical reflexes that are important in normal support and locomotion, as described in Chapter 29. According to Bard, "If a cat is held in the air with the legs free and dependent and with the head held up (so that it cannot see its forefeet or any object below and in front), the slightest contact of any aspect of either pair of feet with the edge of a table results in an immediate and accurate placing of the feet, soles down, on the table close to its edge."[5] Similarly, if a cat is held with one paw in a normal supporting position on a table and then is moved in any direction so that the limb is displaced, a rapid "hopping reaction" will be carried out in the appropriate direction to restore the normal supporting position. Both these reactions depend on cortical mechanisms. "Placing reactions" in the monkey are abolished by removal of either the pre- or postcentral gyrus on the side opposite the stimulus but are retained if all other cortex is removed. Hopping reactions require only the integrity of the precentral gyrus, which indicates the importance of muscle and joint input to area 4.

Another cortically mediated reaction in which tight coupling between input and output is evidently essential is the "instinctive tactile grasping reaction." According to Denny-Brown, this is an "orientation of the hand or foot in space such as to bring a light contact stimulus into the palm (or sole) when a very facile grasping then ensues."[16] This is an example of positive feedback, with the limb moving toward the stimulus. It apparently occurs because the skin regions that project most heavily to pyramidal neurons "lie in the pathway of muscle action and therefore, in the course of a manipulatory sequence, they are likely to be excited following muscle contraction."[53]

ROLE OF FEEDBACK IN CONTROL OF MOVEMENT BY HIGHER CENTERS

Whenever a movement occurs, it is accompanied by changes in activity at all levels of the nervous system, in the mechanical state of the muscles and joints that execute the movement, and often in the relation between the organism and its environment. As the movement progresses, there are extensive interactions between these different components. If the control centers are to adapt their output to the changes occurring throughout the control and effector systems, they must be kept informed of them. They must receive a continuous flow of signals from the parts of the body carrying out the movement and from the receptors that provide information about the outside world. This "sensory feedback" has been recognized as essential for control. It has not been as obvious that "internal feedback" from the motor centers themselves is required to monitor their performance. Anatomic evidence suggests that samples of the motor signals are fed back to control centers from many levels of the nervous system and may become an essential part of the information used for control. In this section the evidence concerning these different types of feedback and the role they play will be described. Feedback from sensory receptors will be considered prior to a discussion of feedback from the central nervous system (CNS).

Sensory feedback

One of the most direct ways of assessing the functional significance of a particular type of feedback is to cut or block the input pathway and observe the resulting defect in movement. If the nerves carrying visual, auditory, or vestibular signals into the CNS are interrupted, the resulting loss of motor control is confined to a specific sphere of activity; other types of activity are unaffected. If the sensory fibers from the skin, joints, and muscles of a limb are sectioned, ability to use that limb is impaired for all types of skilled movements. For example, patients with syphilitic lesions of the dorsal root ganglia (tabes dorsalis) supplying the lower limbs have great difficulty in walking and often develop a characteristic gait in which the feet are brought down hard with each step in an effort to increase the number of afferent signals in the remaining nerve fibers. To control their ataxia, they walk with eyes glued to the ground and become so dependent on visual cues that they may fall over if they lift their eyes or close them. When the upper limb is involved, all fine and precise movements suffer. The eyes are no longer able to guide the arm or fingers to their goal, and delicate manipulations are impossible. In 1895 Mott and Sherrington[43] prepared monkeys with unilateral sections of dorsal roots from C4 to T4 or from L2 to S4, thus eliminating all sensory input from one forelimb or hind limb. From the time of the section and for as long as the animals were kept, almost all movements of the hand or foot were abolished. Grasping movements were completely absent. Movements of the elbow or knee were somewhat less impaired, but neither the hind limb nor the forelimb was used thereafter in locomotion or in climbing. If the intact limbs were restrained, a hungry animal would not make the slightest effort to use the operated limb to obtain food. During struggling, some movements occurred at both proximal and distal joints, but no grasping was observed. Associated movements were impaired less than prime actions. Despite the profound loss of motility, there was no change in the response to cortical stimulation. This suggests that the pathways from cortex to motoneurons were able to transmit normally, and that the cause of the paralysis was a failure of input to the motor cortex and other centers. If only C7 and C8 or T1 and T2 dorsal roots were cut, there was obvious clumsiness in the deafferented region but no loss of ability to move. Even section of all except one dorsal root to a limb did not prevent the animal from using that limb. These experiments do not explain the cause of the near-total paralysis (removal of the entire "arm area" does not result in such a complete loss of motility), but they clearly indicate the importance of somatic input for initiation and control of movements. It is possible that complete absence of input results in loss of awareness of a limb, somewhat like the loss of "body image" in lesions of the parietal cortex. Discharge of pyramidal neurons may require convergence of afferent impulses from several sources; elimination of sensory input removes one of the most important of these inputs. However, as will be noted later in this section, monkeys with total sensory denervation can make a variety of simple movements if they have previously been conditioned to do so.

Although many types of motor performance depend on sensory input for control, some movements are too fast to allow time for feedback to influence performance. Stetson and Bouman[58] pointed out that rapid, repetitive movements carried out by expert typists or pianists are essentially ballistic in nature and cannot be modified during their course. According to Brooks and Stoney, "Alternations of opposing agonists and antagonists, which may last about 0.1 sec, are probably beyond control by feedback loops, since minimal feedback times from the periphery to higher centers in man are about 0.1 sec."[10] It is not surprising that errors associated with rapid movements are greater than with slow actions.[47] Although sensory input is continuously available to the CNS, it probably cannot be used continuously by higher centers for control of rapid movements. A number of observations indicate that when movement becomes too rapid for continuous control, it is subject to "intermittent" control. If intput is predictable, as in certain repetitive tracking tasks, intermittent control is sufficient to allow movement to follow the target accurately at high tracking rates. With unpredictable inputs, however, control deteriorates.[57]

Parameters of movement controlled by motor cortex

At the conscious level, voluntary actions are formulated in terms of goals, positions, and postures. The motor cortex, with its direct and indirect connections to the spinal cord, must convert these and other "orders" into command signals suitable for the motoneurons. The nature of these control signals has been investigated by recording the activity of single neurons in the motor cortex of monkeys conditioned to move a lever back and forth by movements of the wrist. The forces exerted on the lever and the displacements occurring during these movements have been measured with transducers, so that the correlations between cortical activity and specific parameters of movement could be examined. Although the conditioned movements are relatively simple, usually involving alternate flexion and extension of the wrist, electromyographic recordings have shown that more than a dozen different muscles participate either as prime movers and antagonists or in fixation of adjacent joints. Unfortunately, it has usually been difficult to determine which of the active muscles the particular cortical unit under study was controlling, so that interpretation of experimental data has been handicapped.

In the initial studies carried out by Evarts,[19] monkeys were trained to respond to a light with an extension of the wrist. Many of the pyramidal tract neurons (PTNs) in the hand area of the precentral motor cortex discharged repetitively just before extension began, an indication that the cortical discharge was not due to feedback from active muscles or joints. Other PTNs showed abrupt cessation of firing prior to extension, suggesting inhibition of cells controlling muscles antagonistic to the prime movers. PTNs in the postcentral hand area generally discharged after movement had commenced, suggesting that they were responding to feedback from the active limb.[23] The minimal latency of PTN discharge in the precentral region following the onset of the light was 100 msec, or 70 msec more than the minimum latency required by the anatomic connections between the retina and motor cortex. If the sequence of events occurring during this 70 msec interval could be identified, many clues regarding the origins of movement would be provided.

In a later study, Evarts[20] attempted to determine whether the discharge of PTNs is related to the *force* exerted by the moving part or to the *displacement* resulting from this force. The basic conditioning experiment was modified so that at certain times a load opposed flexion of the wrist; as a result, both flexor and extensor displacements involved activity of the flexor muscles and exertion of flexor force. At other times the load opposed extension, so that both flexor and extensor displacements were associated with extensor force. With this technique, force was dissociated from direction of displacement and it was possible to determine whether PTN activity was related to direction of force or direction of displacement. For the majority of PTNs, cortical discharge was related primarily to the force (F) and was only secondarily related to the direction of displacement. For example, when flexor or extensor displacements were associated with flexor force of sufficient magnitude, the unit whose activity is reproduced in the upper records of Fig. 32-15 showed intense activity *regardless of the direction of displacement;* when displacements were associated with extensor force (lower records), the unit was virtually silent *regardless of the direction of displacement*. In many instances, it was apparent that the cortical discharge was related both to force and the rate of change of force (dF/dt). A strong relation to dF/dt brings to mind the dynamic response of the Ia afferents of the "nuclear bag" fibers in the muscle spindle. As noted previously, group I muscle afferents have been shown to project to the motor cortex.

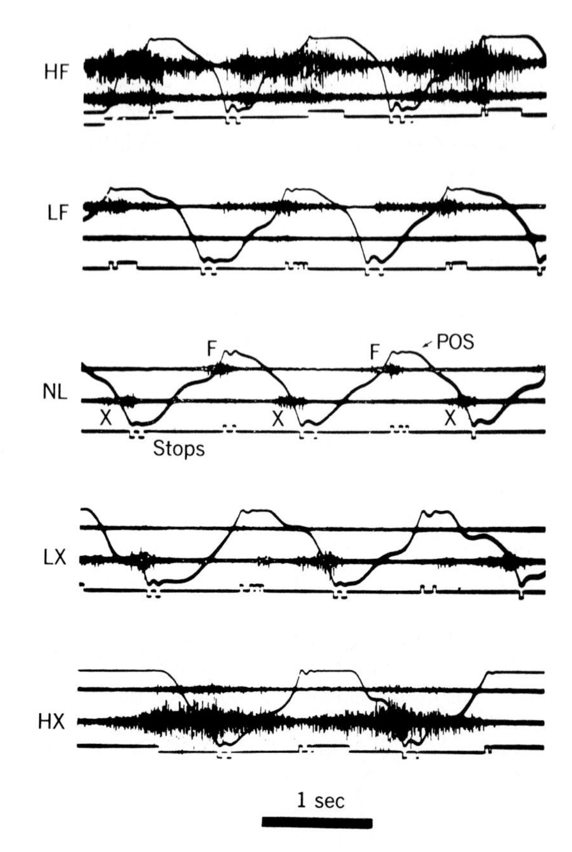

Fig. 32-15. EMG tracings and records of displacement for five of the different loads used in experiment. In middle set of traces, *NL,* the line labeled *POS* is potentiometer output indicating wrist position. Potentiometer output is up for wrist flexion and down for wrist extension. Line labeled *Stops* can assume one of three positions: down for wrist maximally extended, intermediate for wrist in intermediate position (handle not contacting either of stops), and up for wrist maximally flexed. *X* indicates EMG from extensor musculature; *F* indicates flexor musculature. When heavy load (400 gm) opposed flexion, *HF,* flexor muscles had predominant activity. When heavy weight opposed extension, *HX,* predominant activity was in extensor musculature. When no load, *NL,* opposed movement, there was alternate activity of flexor and extensor musculature. With heavy flexor, *HF,* load there was predominant activity in flexor musculature but also considerable activity in extensor musculature. Sets of traces labeled *LF* (100 gm opposing flexion) and *LX* (100 gm opposing extension) show EMG patterns at intermediate loads. (From Evarts.[20])

It is possible that the parameters of movement to which cortical activity is related may change with the task or the type of motor behavior. In the experiments just described, PTN activity was related to force under conditions in which accurate joint displacement was rewarded. In another set of experiments, Evarts[21] trained monkeys to maintain a fixed wrist position while supporting a load that sometimes required predominant activity of flexor muscles and at other times required predominant activity of extensors. In this task the monkey tried to avoid wrist movement regardless of the load that opposed maintenance of this position. It was found that PTNs involved

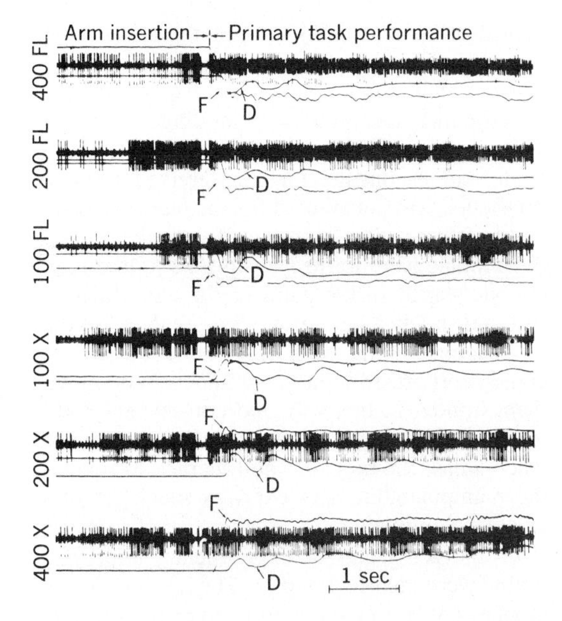

Fig. 32-16. Activity of two pyramidal cells recorded with same microelectrode during insertion of arm through tube leading to lever and during performance of task (maintenance of fixed wrist position). Unit with larger spike was more active during insertion; unit with smaller spike was more active during primary task performance with flexor loads than during insertion. One unit was more active with flexor loads and the other with extensor loads. Numbers at left of records indicate number of grams of flexor load, *FL,* or extensor load, *X,* that wrist was supporting during task. (From Evarts.[21])

in maintaining the required wrist position showed marked variations in discharge frequency for different loads. During a fixed posture, PTN activity was related to the pattern of muscular contraction maintaining the posture rather than to the posture per se. The records in Fig. 32-16 illustrate the variability encountered in experiments of this type that makes interpretation difficult. The microelectrode recorded the activity of two adjacent PTNs. The unit with the smaller spike showed little activity during insertion of the arm through the tube leading to the lever but discharged throughout the performance of the primary task. The unit with the larger spike discharged somewhat more intensely during arm insertion than during the task. The most striking thing about this pair of adjacent units, both of which were related to performance of the task, was that they were related in opposite ways: one discharged maximally against a flexor load and the other discharged maximally against an extensor load.

According to Evarts, all these studies point to one main conclusion, "that the output of precentral motor cortex PTNs is related to the muscular activity of the moving part rather than to the joint displacement or steady-state joint position which the subject may intend to achieve as a result of this muscular activity."[21] This conclusion was not anticipated by Evarts, who expected a correlation between cortical activity and joint position, but it appears almost inevitable when certain basic facts are considered. There is no unique relationship between the muscle forces controlling a joint and the position of the joint. Whenever the load changes, the muscle forces must change accordingly in order to maintain a given position. Since motoneurons control these forces and PTNs control motoneurons, the activity of a single PTN could scarcely correlate with joint position for a wide range of loads. It is not surprising that the activity of PTNs usually correlates with force or rate of change of force. If PTNs do not have the primary responsibility regarding joint position, the centers that do probably lie elsewhere and must use the PTNs to produce the forces required for this task.

Recognizing that thousands of cortical cells must be involved in simple movements, Humphrey et al.[32] attempted to improve the correlation between cortical activity and the parameters of movement by recording from five cells at a time. By suitable weighting and summing of the discharge frequencies of these cells the time course of certain response measures could be accurately predicted. Despite this empirical success in prediction, these investigators concluded that their data did not reveal the response variable that was most closely related to and controlled by the activity of cortical neurons. All these studies were severely handicapped because the investigators did not know whether a cortical cell influences a single muscle, a pair of antagonistic muscles, or a group of muscles. Recently Fetz and Finocchio[27] advised caution in interpreting temporal correlations as evidence for functional relations. They have found that although the activity of a cortical cell may be strongly correlated with activity of one or more muscles in the forelimb, operant conditioning will allow reinforcement of cortical activity along with suppression of all EMG activity in the muscles. Attempts to reinforce muscle activity along with suppression of the cortical unit were only partially successful. They conclude that a "consistent temporal correlation between two events, such as precentral cell activity and some component of the motoneuron response, is necessary but not sufficient evidence for a causal relationship between the correlated events."[27]

Internal feedback

In designing complex control systems, engineers have often found it necessary to monitor the performance of the individual elements and to make use of internal feedback to correct errors before the final output of the system emerges. In the nervous system, where control is exerted through chains of neurons and relatively long delays may occur before sensory feedback can be utilized to correct errors, internal feedback might have great advantages. There are, in fact, many claims that this type of feedback is utilized in various parts of the CNS (for a review of this subject, see Evarts[22]). Although there are a number of studies that suggest the operation of internal feedback, none offers satisfactory evidence regarding the actual neural mechanisms involved. One of the most important series of investigations in this area will be described in detail to illustrate the approaches and problems.

As noted previously, Mott and Sherrington's study[43] on monkeys with deafferented limbs led them to conclude that somatic sensation is essential for the performance of voluntary movement. In later studies by Sherrington, Denny-Brown,[16] Lassek,[36] and Twitchell,[66] similar results were obtained and the same general conclusions were drawn. All these experiments tended to reinforce the belief, widespread since Sherrington's work, that spinal reflexes were the basic elements out of which movements were synthesized and that all movement was reflex in origin. In recent years, however, a series of investigations by Taub and Berman and their colleagues on conditioned movements of deafferented limbs has provided a new set of experimental data with different implications that have led to reevaluation of the original observations of Mott and Sherrington. In the first experiments, monkeys were trained to avoid a shock to their right arm by flexing their left forearm in response to a buzzer.[34] Both forelimbs were hidden from view. After deafferentation of the executive limb (C2 to T3), all animals showed an initial deficit in retention of the conditioned response, but in each case reconditioning back to acquisition criterion was possible. An additional group of monkeys was able to learn the avoidance response without any preoperative training. As previous investigators had reported, none of these animals used their deafferented limbs effectively in the freedom of their cages. If their intact forelimb was immobilized, however, these monkeys learned to push their deafferented limbs through the bars of their cages to obtain food when no other access to it was possible.[34] In the next set of experiments the experimental technique was modified in several ways in order to reduce the amount of feedback from sources outside the deafferented limb. A brief click was substituted for the buzzer without altering the results. Instead of flexing the forearm the animals were trained to squeeze a fluid-filled cylinder taped to the palm of the hand in order to avoid shock. The arm was immobilized so that no response other than flexion of the fingers could exert pressure on the cylinder. With movement limited in this way and consequently with less possibility of feedback from nearby innervated areas, the monkeys still learned to squeeze the manipulandum and exert as much pressure as normal animals.

In another group of experiments, both forelimbs were deafferented.[60] The ability of the monkeys to carry out conditioned responses similar to those already described was not affected, but to the surprise of the investigators, bilateral deafferentation resulted in far less impairment of forelimb movement than the unilateral procedure. During the first week or two after operation the deafferented limbs were virtually useless. During the next 2 months there was considerable recovery of function, so that eventually the animals were able to use the forelimbs rhythmically and in good coordination with the hind limbs during slow and even moderately rapid ambulation. The forelimbs were usually placed palms down on the floor and they bore weight. Grasping was possible, and the animals could climb to the top of an 8 ft bank of cages with reasonable speed. Several monkeys were able to pick up raisins between thumb and forefinger. The investigators emphasized the degree to which the actions of these animals approximated normal patterns of movement. Finally, the animals with bilateral deafferentation of forelimbs were reoperated in two stages to section all remaining dorsal roots in the spinal cord. Total deafferentation was accomplished in three animals. These monkeys showed little or no decrease in the use of their forelimbs. They were not able to use their lower limbs effectively, but none of them survived total spinal deafferentation long enough for recovery such as that observed previously in the forelimbs to take place in the legs.

How can animals with deafferented limbs that they cannot see learn to repeat certain movements until they become conditioned when they do not receive the normal sensory signals that inform them of where the limb is and whether it has moved? Taub and Berman conclude that since the required information was not available from the peripheral nervous system, it must have been provided by purely central mechanisms. They propose, as have others,[22] that signals from motor centers and descending motor pathways provide information about future movements to the CNS before the impulses that will produce these movements have reached the periphery. They suggest that this internal feedback would allow an animal to determine the general position of its limbs in the absence of peripheral sensation.

Anatomic and physiologic evidence of internal feedback

Judging from the connections of the descending motor pathways, there are many examples of internal feedback in the CNS. Some of them may have little to do with the capacities of deafferented limbs that have been described, but others may feed back samples of "command" or "executive" signals and thus provide information regarding intended movements. The pyramidal tract may be taken as a convenient example of a motor pathway that gives off collaterals at many levels of the nervous system. Anatomic and physiologic information regarding these collaterals is summarized in the following paragraphs.

At the cortical level, signals from pyramidal cells are fed back by recurrent collaterals to two different types of stellate cells that exert inhibitory and excitatory effects back on pyramidal cells. These effects may result in alterations in the receptive fields of pyramidal cells. The functional significance of this arrangement has been the subject of considerable speculation.[17] At a subcortical level, fibers from the motor areas of the cerebral cortex send terminals to the caudate nucleus and putamen as well as to other nuclei of the basal ganglia. Some of these fibers may be collaterals of pyramidal neurons. The caudate and putamen project to the pallidum, which sends its efferents to thalamic nuclei that project to the motor cortex. The complexity of this cortico-strio-pallido-thalamo-cortical circuit has defied functional analysis to date. Electrophysiologic experiments indicate that pyramidal fibers send collaterals to the nucleus ventralis lateralis of the thalamus[15] and to the ventrobasal complex.[55] Activity in these collaterals is excitatory to thalamocortical cells and thus might result in rapid feedback to the precentral and postcentral motor areas.

In the brain stem, pyramidal fibers or analogous corticobulbar fibers establish connections with many different cell groups. The most important are the cranial motor nuclei, the trigeminal complex, the mesencephalic and medullary core of the RF, and the lateral reticular nucleus.

Several anatomic studies utilizing different techniques have shown a considerable contribution from the pyramidal tract to the dorsal column nuclei.[14,39] After transsection of the pyramidal tract, degenerated terminals are seen throughout the gracile and cuneate nuclei, particularly in their rostroventral portions. The degeneration is chiefly contralateral to the lesion, and terminals are found on cell bodies and dendrites. An increasing number of physiologic studies indicate that corticofugal activity arising in the pre- and postcentral cortex exerts excitatory and inhibitory effects on a substantial number of cells in these nuclei. Levitt et al.[37] have shown that the excitatory effect is produced by activity in fibers descending from somatic sensory areas I and II. Lesions of the brain stem that spared the pyramidal tract but interrupted other descending fibers revealed that this excitation was pyramidal in origin. Complete severance of the bulbar pyramids without major damage to other structures indicated that the inhibitory effects originated in the precentral motor cortex and were extrapyramidal. Fig. 32-17 illustrates the excitatory effects observed by Levitt and the postexcitatory depression that may follow them. In the experimental situation the inhibitory effects may completely block the relay of sensory signals from lower levels, attenuate the spontaneous activity, or increase the signal-to-noise ratio of sensory activity. Under normal conditions the excitatory projection may have a dual function—modulating the sensory feedback from the periphery and supplying samples of the pyramidal outflow for internal feedback.

Although some pyramidal neurons send terminals directly to motoneurons in primates, the majority of them end on cells at the base of the dorsal horn and in the "intermediate zone" of the spinal gray matter on cells, which, like those in the dorsal column nuclei, respond to passive limb movements and cutaneous stimuli. Nyberg-Hansen and Brodal[46] used silver techniques to show that pyramidal fibers from the motor area differed in their terminal distribution from fibers arising in the somatic sensory areas. Fibers from the motor cortex end chiefly in lamina VI of the spinal gray matter and in the dorsal part of lamina VII, as shown in Fig. 32-18. Fibers from the somatic sensory cortex terminate more dorsally, chiefly in laminae IV and V. A corresponding distribution or pre- and postcentral pyramidal

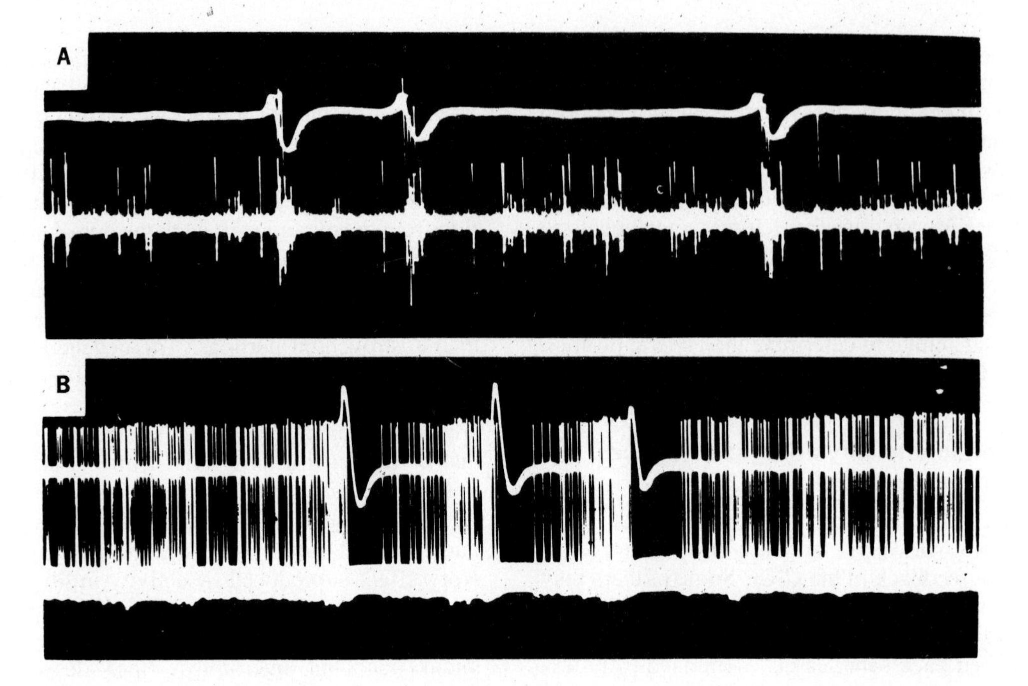

Fig. 32-17. Evocation of unit discharges in gracile nucleus of cat (lower records in **A** and **B**) by synchronous firing of cells in sensorimotor cortex (upper records in **A** and **B**). **A,** Each cortical volley (provoked by application of strychnine) elicits burst of increased firing from units in gracile nucleus that also respond to touch of ipsilateral hind leg. **B,** Similar effect in tactile unit of gracile nucleus that was discharging vigorously, with well-marked postexcitatory depression following it. (From Levitt et al.[37])

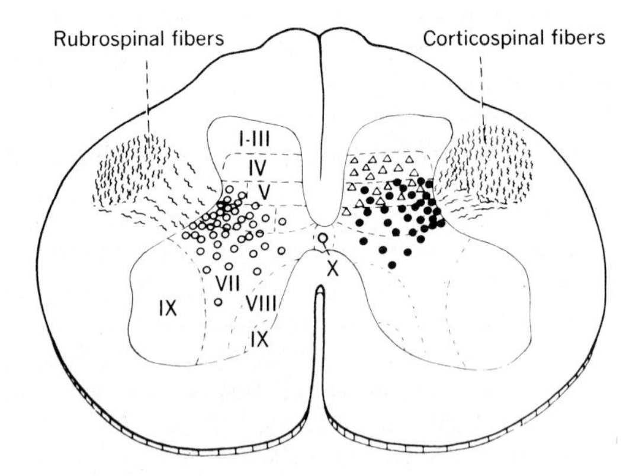

Fig. 32-18. Diagram of cross section of cat spinal cord showing more dorsal terminations of corticospinal fibers from sensory cortex and more ventral terminations of those arising in motor cortex. ○ = Sites of termination of rubrospinal fibers. ● = Sites of termination of corticospinal fibers from "motor" cortex. △ = Sites of termination of corticospinal fibers from "sensory" cortex. (Based on studies by Nyberg-Hansen and Brodal[46]; from Brodal.[8])

fibers has been noted in the monkey. Wall[68] has reported that the receptive fields and thresholds of neurons in this part of the spinal cord are altered during pyramidal stimulation. According to Fetz[25] the more dorsal cells of laminae IV and V are chiefly inhibited by pyramidal stimuli, whereas the more ventral cells in laminae V and VI are predominantly excited. Pyramidal tract fibers also send terminals to neurons of the spinocerebellar tracts. Hongo et al.[31] report that neurons of the dorsal spinocerebellar tract can be excited or inhibited by pyramidal stimuli, depending on their input. Neurons excited monosynaptically by group I afferents are mainly inhibited, whereas neurons responding to volleys in flexor reflex afferents but not to group I are usually excited. In addition to these effects on spinal interneurons, pyramidal fibers are believed to exert a presynaptic inhibitory control over somatosensory input. The details of this circuit are not clear, but pyramidal stimulation results in depolarization of the terminals of primary afferent fibers. This reduces the afferent input to segmental reflex paths, to ascending paths, and to relay nuclei.

As this brief review indicates, the terminals of pyramidal tract neurons are directed chiefly to cells that form different parts of the somatic sensory pathways from the spinal cord to the cerebral cortex or to cells forming parts of other afferent systems projecting to the cerebral cortex and cerebellum. It would appear that the pyramidal tract is organized primarily to control input to motor centers even at spinal levels and perhaps only secondarily to cause discharge of motoneurons. The well-established observations that the pyramidal tract controls gamma motoneurons that regulate sensory input from muscle spindles lends strong support to this view. Pyramidal influences on presynaptic terminals of primary afferent fibers are also consistent with it. Without additional information, it is impossible to say precisely how or why the pyramidal tract influences sensory projections. Modulation of sensory input by impulses arising in the motor cortex suggests that this particular portion of the input is used for "motor" purposes rather than for conscious sensation. The absence of sensory deficits in patients with lesions limited to the pyramidal tract is consistent with this conclusion. The existence of pyramidal collaterals at every level of the nervous system indicates that the same set of signals may be fed back into different portions of the sensory projections with slightly different delays. This arrangement might be of particular advantage when the time relations between ascending sensory signals and descending pyramidal signals are critical.

Although the experimental data regarding the operation of internal feedback circuits are not very revealing, the available evidence indicates that they play important roles. Analysis of these roles will be a fruitful area for future research.

EFFECTS OF LESIONS

It should be obvious from the preceding sections that understanding of the motor systems has not progressed far enough to permit satisfactory explanation of the effects of lesions on movement and posture. Despite extensive clinical and experimental studies there is a surprising amount of disagreement regarding the exact effects of various lesions and their significance. This disagreement is due in part to differences in the species of animals used in experiments, the duration of survival after surgery, the general condition of the animal and his opportunities for maximal recovery of function, and in some cases to failure to distinguish between early and late effects of lesions. Without careful anatomic studies of the extent and severity of lesions, misleading conclusions can easily be drawn. Inability to carry out such studies at all or within a reasonable interval after clinical observations have been made has limited the usefulness of human material. In addition, there is a basic difficulty in the evaluation of lesions that is not always apparent, that is, a cortical area or a pathway may be *necessary* to a particular function but not *sufficient* for it.

Despite these difficulties and limitations, the study of lesions has contributed significantly to the understanding of motor systems. In this section, emphasis will be placed primarily on the value of cortical lesions in revealing important aspects of cortical motor function and only secondarily on their usefulness in localizing clinical disorders in man.

Contribution of cerebral cortex to movement

As Bard and Rioch[6] and others have pointed out, a decorticate cat or dog is capable of a great many complex motor activities. After a short time, he regains the ability to right himself, walk, and run, although defects such as the loss of placing and hopping reactions are permanent. These facts indicate that righting and locomotion can be managed fairly well by subcortical mechanisms in carnivores. A primate is more seriously incapacitated by total decortication, but as Travis and Woolsey[65] have demonstrated, if the surgery is carried out in several steps and sufficient care is taken postoperatively to allow maximal recovery of function after each partial removal of cortex, the decorticate monkey regains the ability to right itself and to move about in an awkward but effective manner. These observations reveal that the cerebral cortex is not essential for many motor performances in primates, and they raise the question of what the cortex does contribute to the control of movement.

In discussing this difficult question, it is useful to conceive of voluntary actions as a sequence of events occurring in several stages. In the first stage the notion of an act is formed in the mind in response to internal stimuli such as thoughts or emotions or in response to an external stimulus of some kind. In the next stage, this "idea" of a movement is presumably translated into patterns of neural signals in a part of the brain in which motor "programs" acquired by learning or practice are stored for use. In the third stage the appropriate "program" is perhaps executed by assemblies of neurons in the principal motor areas of the cortex where the final cortical precision and delicacy are added with the help of various feedback circuits. Pyramidal and extrapyramidal systems working cooperatively transmit these cortical commands to the segmental mechanisms in the spinal cord. This formulation is based on clinical studies of *apraxia* in man, which indicate that the events leading to movement actually occur in this order. A brief description of apraxia is appropriate at this point, because it will help to provide a conceptual framework for a consideration of the types of effects produced by cortical lesions.

Apraxia is the inability to perform an act in the absence of any significant paralysis, sensory loss, or deficit in comprehension. It may take several forms. In one type a patient appears to understand simple commands but has apparently lost his memory of how to perform them, especially if they are called for in unnatural settings. This is called "ideational apraxia." If asked to show "how to wave goodbye" or "how to light his pipe," the patient cannot do so, but if a situation appropriate for these actions arises naturally, he can perform them easily. In this type of disorder the basic defect seems to be in the conception of the act, not in its execution. The patient may make surprising mistakes, such as scratching a cigarette against a match box, without being aware of them. In another type of apraxia a patient may have a clear idea of what he would like to do but be unable to translate this idea into a precise, well-executed act. This "ideomotor apraxia" is characterized by an inability to carry out a simple command, to imitate a gesture, or to perform these actions under natural circumstances. A patient with this defect apparently cannot draw on his store of previously acquired movement patterns. He is aware of the errors he makes, but he cannot correct them and is often irritated with himself. A third and more common form of apraxia is characterized by awkward, inept use of some part of the body for all of its actions. This is "motor" or "kinetic" apraxia. In this form of the disorder the patient knows exactly what he wishes to do, but he cannot carry out the movements properly.

This brief account of apraxia indicates that even relatively simple actions, such as waving goodbye or lighting a pipe, cannot be performed de novo; they utilize acquired patterns of movement or motor memories to a greater extent than we often recognize. There is also considerable evidence that voluntary movements depend on and perhaps even evolve out of basic reflex responses to peripheral stimuli. These reflex elements are most apparent in goal-directed and manipulative activity. Denny-Brown has made an extensive investigation of this aspect of movement in a series of studies on clinical and experimental material. The following account of his findings is drawn largely from his monograph *The Cerebral Control of Movement.*[16]

In patients with lesions of the frontal lobes, automatic prehensile movements of two types could be distinguished, both of which were called "forced grasping."[16] In one type a distally moving tactile stimulus applied to any part of the palm of the hand elicits a simple closing of the hand. After closure begins, traction on the fingers reinforces the reaction, which is termed a "grasp reflex." A similar response occurs in the foot. A second, more complex type of forced grasping can be elicited by contact with a much larger area of skin, including parts of the lateral and dorsal aspects of the hand and wrist. The response consists of extension of the arm followed by flexion, associated with other movements of the forearm that serve to bring further contacts nearer the palm of the hand. When the contact reaches the palm, the hand closes with a rapid grasp reflex. This is an automatic reaction best elicited from patients who are in a deep stupor or have their attention directed elsewhere. This sequence of movements, designed to orient the hand to a stimulus in the environment, is called an "instinctive grasp reaction."

In patients with lesions of the parietal lobes, two essentially opposite reactions to contact, called "tactile avoiding reactions," can be evoked.[16] Contact with any part of the terminal phalanges results in extension of the fingers and wrist (i.e., a simple "avoiding reflex"). A more complex withdrawal of the whole hand from the stimulus, that is, an "instinctive tactile avoiding reaction," can sometimes be evoked from wider areas of skin.

These grasping and avoiding reactions can also be elicited in blindfolded monkeys with frontal or parietal lesions. Each type of response appears to be the converse of the other. If both frontal and parietal cortex are removed, heavy contact on the palm may still evoke a basic grasp reflex, and stimulation of the terminal phalanges may still elicit a coarse avoiding reflex, indications that both responses are, at least in part, subcortical. Denny-Brown believes that there is a natural equilibrium between these two types of cortical reflexes. Damage to areas 8, 6, and 24 destroys tactile avoiding and releases tactile grasping, whereas parietal lesions abolish tactile grasping and release tactile avoiding.

Grasping and avoiding reflexes appear to be the basic elements from which voluntary, goal-directed activity develops. According to Denny-Brown, the ability of the human infant "to reach out with the hand toward a desired object in the field of vision is the outcome of a slow learning process that is preceded by the appearance of automatic reflex grasping and later reflex avoiding and still later by those grasping movements projected into space in response to contact that we have called 'instinctive grasping.' The ability to make a gesture without any visual or tactile stimulus is learned only much later after a long period of random movements."[16] After cerebral lesions, recovery of the ability to make a willed movement is apparently preceded by the appearance of the same movement as a reflex response to a specific stimulus. For example, a patient who was unable for some time to flex his forearm found that he could do so shortly after he recovered a similar type of reflex response, and after a grasp reflex returned, voluntary flexion of the fingers reappeared. From observations such as these, Denny-Brown concluded that "purposive movement is indeed the utilization of reflex function as part of the response to a more elaborate stimulus situation."[16] The disuse of deafferented limbs described by Mott and Sherrington may thus be due to the loss of a reflex basis for movement.

Pyramidal tract lesions

By sectioning the pyramidal tract selectively, it is possible to study motor function in animals whose movements are carried out almost entirely by the extrapyramidal system. The deficits that result indicate what role the pyramidal tract normally plays, whereas the capacities that remain indicate the extent of the extrapyramidal contribution. The only way of producing a pure pyramidal deficit experimentally is to cut the pyramid in the medulla where there is no admixture of extrapyramidal fibers. This has been done by Tower in experiments on cats[61] and monkeys.[62] The results are of great interest. According to Tower, "The most conspicuous result of unilateral pyramidal lesion in the monkey is diminished general usage and loss of initiative in the opposite extremities. The loss of initiative is grave, but not complete. When both sides are free to act, initiative of almost every sort is delegated to the normal side, but if the normal side is restrained the affected side can, with sufficiently strong excitation, be brought to act."[62] This loss of initiative is somewhat like the effect of deafferentation previously described, although not as dramatic. On the affected side there is a striking loss of fine or discrete control of movement. This *paresis,* as it is called, involves movement in proportion to its delicacy and skill. All fine usage is abolished, especially in the distal muscles of the extremities. The usage that survives, such as postural activity, progression, reaching, and grasping, is stripped of its finer qualities of control (e.g., aim and precision). These remaining performances are still very useful but can hardly be called skilled. Some of them may require intense attention and effort. Obviously, voluntary actions are not the exclusive function of the pyramidal tract. The extrapyramidal systems, contrary to some opinion, can be employed just as voluntarily as the pyramidal.

Section of the pyramid also results in hypotonia on the contralateral side. This is demonstrable as diminished resistance to passive movements and is accompanied by slow, full tendon reflexes. It is sometimes apparent on direct palpation of muscles; after a time there may be loss of muscle bulk due to atrophy.

Superficial reflexes such as local reactions to pinprick as well as the abdominal and cremasteric reflexes are raised in threshold and become slow and full because they are unchecked by antagonistic contraction. Contact and visual placing reactions are abolished in the paretic limbs. Proprioceptive placing and hopping reactions are difficult to elicit. Although stereotyped reaching and grasping remain, the ability to hold onto, grasp, and manipulate objects is greatly impaired. The unopposed action of the extrapyramidal system is apparent in the animal's inability to terminate a grasp while there is tension on the flexor muscles.

In the chimpanzee, pyramidal lesions cause a similar set of deficits, but since the pyramidal tract is relatively larger, the deficits are more severe. Discrete control of the digits is more impaired and use of the affected limb is greatly reduced. The grasp reflex, normally more prominent than in the monkey, becomes so hyperactive that there is considerable difficulty in disengaging the hand from the bars of the cage. In the chimpanzee as in man there is a characteristic response to stimulation of the plantar surface of the foot, which alters radically after a lesion of the pyramid. Stimulation with a stick, a key, or a fingernail drawn along the lateral edge of the sole causes plantar flexion of the toes in a normal subject. After a pyramidal lesion the great toe is dorsiflexed and the other toes fan out. This is called the *sign of Babinski*. Its significance is obscure, but it is very useful clinically as an indication of a pyramidal lesion.

Thus far there is no record of an uncomplicated case of a pure pyramidal lesion in man. A few nearly pure pyramidal lesions have been reported, however, and the parallel between them and the studies on the chimpanzee is striking.

Tower's observations have been confirmed and extended by Denny-Brown[16] and his colleagues, whose studies lead them to conclude that "the pyramidal system is concerned not so much with 'discrete' movements of individual muscles or individual joints as with those spatial adjustments that accurately adapt the movement to the spatial attributes of the stimulus. Thus grasping is adapted to the shape of the thing to be grasped, whether a particle of food, a pen or a surface, only in the presence of the pyramidal tract.''[16] This emphasis on the sensory aspects of exploratory or manipulative activity brings to mind the collaterals of pyramidal fibers that establish connections with somatic afferent systems at all levels of the nervous system.

Extrapyramidal lesions

The extrapyramidal system was described in Chapter 29. Lesions of the various portions of this system result in a distinctive group of motor disabilities. These include disorders of muscle tone (spasticity and rigidity), involuntary movements (tremor, chorea, and athetosis), and postures, hyper- and hypokinesis, and paresis. In this chapter, only the effects of lesions of the cortical portion of the extrapyramidal system will be considered.

After cutting the pyramids in the medulla, stimulation of the cerebral cortex still results in movements, indicating the existence of cortical connections to extrapyramidal motor centers. The four principal cortical motor areas are probably the most important sources of these fibers, but recent anatomic studies indicate that *all parts of the cerebral cortex give efferent fibers to the basal ganglia.*[13] From this, it may be inferred that all cortical lesions result in some deficit in motor function, defined in the broadest terms. Quite often, however, no motor deficit can be identified by the neurologist. Lesions in visual or auditory receiving areas, for example, result in no loss of primary motor function, but they remove an area that supplies important inputs to the motor cortex and they impair a particular sphere of motor controls. Deficits resulting from lesions of areas concerned with intellectual function may be extremely subtle and their influence on motor function may be difficult to detect. In the following sections, some of the best-known disabilities that result from cortical lesions will be briefly described.

Precentral motor cortex (area 4) lesions

Since the earliest observations on the effects of lesions in the precentral cortex, there has been a long series of confusing and contradictory reports on the amount of paralysis and paresis that results, on the occurrence of flaccidity or spasticity, and on the relation between these effects and the location of the lesion. Until the studies of Woolsey and his associates established the existence of separate precentral and supplementary motor areas and defined their boundaries, experimental ablations often included parts of both areas. Until the patterns of representation in these areas were fully worked out, investigators did not know exactly which part of the body was likely to be affected by a discrete lesion. Furthermore, until the studies of Hines and Tower,[30] showing that section of the pyramidal tracts caused hypotonia and subsequent ablation of area 4 caused spasticity, it was not clear that spasticity was extrapyramidal in origin and due to the release of subcortical centers from cortical inhibition.

As Woolsey et al.[75] have demonstrated, any given part of the skeletal musculature will show some degree of *paralysis and spasticity* when its representation in the precentral motor cortex is removed. With very small lesions in the caudal part of the precentral area where the digits are represented, the spasticity may be quite minimal and difficult to detect in an active, uncooperative monkey. Slightly larger removals of cortex just in front of the central fissure result in mild spasticity of the toes and ankle with a positive Babinski reaction or in mild spasticity of the fingers and wrist with a positive Hoffman sign, according to Denny-Brown.[16] Removal of the entire precentral hand area in the chimpanzee caused persistent spastic flexion of the fingers, initial paralysis of the hand with a return of clumsy grasping after 25 days, and poor orientation of the hand to new surfaces.[16] No specific movements of the hand were lost, but actions were slow and inept and seemed to require visual guidance in the absence of tactile orientation. If the removal was extended forward to include the anterior part of area 4, spasticity occurred in the more proximal joints as well. If the lesion was extended medially or laterally, the distribution of spasticity began to resemble that seen in hemiplegia, with the development of hypertonia in the flexors of the upper limb and the extensors of the lower limb. After removal of all the cortex of area 4, paralysis and spasticity were maximal. Very localized spasticity and paralysis of an elbow, knee, or hip could not be produced by small lesions, perhaps because muscles are represented in overlapping fashion over wide areas of cortex.[16]

Removal of the anterior half of the precentral cortex also results in the appearance of a reaction wherein stretch of the adductors and retractors of the shoulder leads to increased tone in the flexors of the elbow, wrist, and fingers. This *traction reaction* is enhanced if the flexors of the fingers are stretched at the same time. A similar reflex is often seen in hemiplegic patients, but according to Denny-Brown,[16] it should be distinguished from the previously described "grasp reflex," which is due to lesions of areas 8, 6, and 24.

After a lesion of area 4, ability to flex the limbs in response to a painful stimulus reappears at an early stage of recovery. At first it occurs only in response to pinprick, but later light contact is sufficient. Ability to reach out voluntarily is recovered more slowly. After 3 to 4 weeks a monkey learns to pick up small objects by opposing the thumb to the other fingers, but these digital movements remain clumsy. Denny-Brown concludes that "the precentral gyrus is therefore essential for movement directed into space that accurately orients the hand or foot to the object. It is not essential for retraction from a contact, or a painful or visual stimulus. Exploratory reactions with the hand and fingers, foot or lips to contact stimulus do not recover after area 4 lesions."[16]

Supplementary motor cortex lesions

Relatively little attention has been given to lesions of the supplementary motor cortex compared with those of the precentral motor area. The following brief account will doubtless have to be amended when more extensive studies are carried out. Early reports indicated that lesions confined to this area caused considerable spasticity,[63,64] but later studies by Denny-Brown revealed only a flexed posture of the arms and legs and a soft resistance to extension. The most notable effect of bilateral removal was the appearance of pronounced grasp reflexes.[16] In these preparations, instinctive tactile avoiding was abolished, and all contacts with the hand or foot led to instinctive grasping. The least contact was sufficient to elicit the reaction. The slight withdrawal that is usually observed after an unexpected contact was completely absent, and even noxious stimuli elicited grasping. These observations led Denny-Brown to conclude that "the supplementary area is the focal point of motor projection of avoiding reactions . . . though it is likely that avoidance behavior requires much more widespread cortical areas for its full elaboration."[16]

Postcentral gyrus (areas 3, 1, and 2) lesions

As noted earlier in this chapter, the somatic sensory receiving area in the postcentral gyrus makes a substantial contribution to the pyramidal tract and, on stimulation, yields discrete movements. Removal of the postcentral gyrus alone does not cause spasticity but greatly intensifies the spasticity that follows precentral lesions. This latter effect indicates that extrapyramidal fibers arise in areas 3, 1, and 2. Most of the motor deficits that follow lesions are attributable to somatic sensory defects. The loss of position sense results in complete lack of awareness of movement in the contralateral limbs, in unnatural positions of the limbs, and in reluctance to use them. Under visual control a wide variety of movements can be carried out, but they are clumsy and awkward. Without visual guidance, goal-directed and manipulative activity is essentially abolished. Denny-Brown points out that there is a severe defect in all exploratory reactions directed into space.[16] Tactile placing reactions cannot be elicited, and hopping reactions are depressed. Although movements of the hands and feet are inept, the limbs show well-developed avoiding reactions to contact and visual threats. As part of this picture, monkeys appear to be extremely timid and much of their behavior is dominated by complex visual avoiding reactions. Bilateral ablations of the postcentral gyrus leave animals so dependent on vision that without it they become entirely inactive and unresponsive to any form of tactile stimulation for more than 6 weeks. The instinctive grasp reflex is lost and never returns. Avoiding reactions to pinprick are exaggerated and tend to spread to other parts of the body. It is clear that in the absence of the postcentral gyrus, visually directed movements of the hands and legs and visual avoiding can be managed by the basal ganglia.

Second somatic sensory area (somatic area II) lesions

As noted in Chapter 12, no sensory abnormalities have yet been detected following removal of the second somatic sensory area in man or monkey. Furthermore, such lesions apparently do not accentuate the recognized sensory defects resulting from removal of the postcentral gyrus. To date there is no clear and unequivocal evidence that they impair motor performance or significantly affect motor behavior. Denny-Brown[16] reports some minor effects on instinctive grasping and tactile avoiding, but nothing that points to a distinctive role for this area.

REFERENCES

1. Adrian, E. D.: Double representation of the feet in the sensory cortex of the cat, J. Physiol. **98:**16P, 1940.
2. Asanuma, H., and Sakata, H.: Functional organization of a cortical efferent system examined with focal depth stimulation in cats, J. Neurophysiol. **30:**35, 1967.
3. Asanuma, H., and Ward, J. E.: Patterns of contraction

of distal forelimb muscles produced by intracortical stimulation in cats, Brain Res. **27:**97, 1971.
4. Asanuma, H., Stoney, S. D., and Abzug, C.: Relationship between afferent input and motor outflow in cat motorsensory cortex, J. Neurophysiol. **31:**670, 1968.
5. Bard, P.: Studies on the cerebral cortex. I. Localized control of placing and hopping reactions in the cat and their normal management by small cortical remnants, Arch. Neurol. Psychiatry **30:**40, 1933.
6. Bard, P., and Rioch, D. McK.: A study of four cats deprived of neocortex and additional portions of the forebrain, Bull. Johns Hopkins Hosp. **60:**73, 1937.
7. Benjamin, R. M., and Welker, W. I.: Somatic receiving areas of cerebral cortex of squirrel monkey (Saimiri sciureus), J. Neurophysiol. **20:**286, 1957.
8. Brodal, A.: Neurological anatomy, ed. 2, New York, 1969, Oxford University Press.
9. Brooks, V. B.: Personal communication, 1971.
10. Brooks, V. B., and Stoney, S. D., Jr.: Motor mechanisms: the role of the pyramidal system in motor control, Ann. Rev. Physiol. **33:**337, 1971.
11. Brooks, V. B., Rudomin, P., and Slayman, C. L.: Peripheral receptive fields of neurons in the cat's cerebral cortex, J. Neurophysiol. **24:**302, 1961.
12. Brooks, V. B., Rudomin, P., and Slayman, C. L.: Sensory activation of neurons in the cat's cerebral cortex, J. Neurophysiol. **24:**286, 1961.
13. Carman, J. B., Cowan, W. M., and Powell, T. P. S.: The organization of the cortico-striate connexions in the rabbit, Brain **86:**525, 1963.
14. Chambers, W. W., and Liu, C. N.: Corticospinal tract of the cat. An attempt to correlate the pattern of degeneration with deficits in reflex activity following neocortical lesions, J. Comp. Neurol. **108:**23, 1957.
15. Clare, M. H., Landau, W. M., and Bishop, G. H.: Electrophysiological evidence of a collateral pathway from the pyramidal tract to the thalamus in the cat, Exp. Neurol. **9:**262, 1964.
16. Denny-Brown, D.: The cerebral control of movement, Liverpool, 1966, Liverpool University Press.
17. Eccles, J. C.: Cerebral synaptic mechanisms. In Eccles, J. C., editor: Brain and conscious experience, Berlin, 1966, Springer Verlag.
18. Evarts, E. V.: Relation of discharge frequency to conduction velocity in pyramidal tract neurons, J. Neurophysiol. **28:**216, 1965.
19. Evarts, E. V.: Pyramidal tract activity associated with a conditioned hand movement in the monkey, J. Neurophysiol. **29:**1011, 1966.
20. Evarts, E. V.: Relation of pyramidal tract activity to force exerted during voluntary movement, J. Neurophysiol. **31:**14, 1968.
21. Evarts, E. V.: Activity of pyramidal tract neurons during postural fixation, J. Neurophysiol. **32:**375, 1969.
22. Evarts, E. V.: Feedback and corollary discharge: a merging of concepts, Neurosci. Res. Program Bull. **9:**86, 1971.
23. Evarts, E. V.: Contrasts between activity of pre- and postcentral neurons of cerebral cortex during movement in the monkey. In Buser, P., et al., editors: Neural control of motor performance, Amsterdam, 1971, Elsevier Publishing Co.
24. Ferrier, D.: The functions of the brain, London, 1876, Smith, Elder & Co.
25. Fetz, E. E.: Pyramidal tract effects on interneurons in the cat lumbar dorsal horn, J. Neurophysiol. **31:**69, 1968.
26. Fetz, E. E., and Baker, M. A.: Response properties of precentral neurons in awake monkeys, Physiologist **12:** 223, 1969.
27. Fetz, E. E., and Finocchio, D. V.: Operant conditioning of specific patterns of neural and muscular activity, Science **174:**431, 1971.
28. Foerster, O.: Motorische Felder und Bahnen. In Bumke, O., and Foerster, O., editors: Handbuch der Neurologie, Berlin, 1936, J. Springer, vol. 6
29. Foerster, O.: The motor cortex in man in the light of Hughlings Jackson's doctrines, Brain **59:**135, 1936.
30. Hines, M.: Control of movements by the cerebral cortex in primates, Biol. Rev. **18:**1, 1943.
31. Hongo, T., Okada, Y., and Sato, M.: Corticofugal influences on transmission to the dorsal spinocerebellar tract from hindlimb primary afferents, Exp. Brain Res. **3:**135, 1967.
32. Humphrey, D. R., Schmidt, E. M., and Thompson, W. D.: Predicting measures of motor performance from multiple cortical spike trains, Science **170:**758, 1970.
33. Kennard, M. A., and McCulloch, W. S.: Motor responses to stimulation of cerebral cortex in absence of areas 4 and 6 (Macaca mulatta), J. Neurophysiol. **6:** 181, 1943.
34. Knapp, H. D., Taub, E., and Berman, A. J.: Movements in monkeys with deafferented forelimbs, Exp. Neurol. **7:**305, 1963.
35. Landgren, S., Phillips, C. G., and Porter, R.: Cortical fields of origin of the monosynaptic pyramidal pathways to some alpha motoneurons of the baboon's hand and forearm, J. Physiol. **161:**112, 1962.
36. Lassek, A. M.: Inactivation of voluntary motor function following rhizotomy, J. Neuropathol. Exp. Neurol. **3:** 83, 1953.
37. Levitt, M., Carreras, M., Liu, C. N., and Chambers, W. W.: Pyramidal and extrapyramidal modulation of somatosensory activity in gracile and cuneate nuclei, Arch. Ital. Biol. **102:**197, 1964.
38. Leyton, A. S. F., and Sherrington, C. S.: Observations on the excitable cortex of the chimpanzee, orang-utan and gorilla, Q. J. Exp. Physiol. **11:**135, 1917.
39. Liu, C. N., and Chambers, W. W.: An experimental study of the cortico-spinal system in the monkey (Macaca mulatta). The spinal pathway and preterminal distribution of degenerating fibres following discrete lesions of the pre- and postcentral gyri and bulbar pyramid, J. Comp. Neurol. **123:**257, 1964.
40. Lloyd, D. P. C.: Functional organization of the spinal cord, Physiol. Rev. **24:**1, 1944.
41. Lorente de Nó, R.: Cerebral cortex: architecture, intracortical connections, motor projections. In Fulton, J. F., editor: Physiology of the nervous system, ed. 3, New York, 1949, Oxford University Press.
42. Marchiafava, P. L.: Activities of the central nervous system: motor, Ann. Rev. Physiol. **30:**359, 1968.
43. Mott, F. W., and Sherrington, C. S.: Experiments upon the influence of sensory nerves upon movement and nutrition of the limbs. Preliminary communication, Proc. R. Soc. Lond. (Biol.) **57:**481, 1895.
44. Mountcastle, V. B.: Modality and topographic properties of single neurons of cat's somatic sensory cortex, J. Neurophysiol. **20:**408, 1957.
45. Mountcastle, V. B., and Henneman, E.: The representation of tactile sensibility in the thalamus of the monkey, J. Comp. Neurol. **97:**409, 1952.
46. Nyberg-Hansen, R., and Brodal, A.: Sites of termination of corticospinal fibers in the cat. An experimental study with silver impregnation methods, J. Comp. Neurol. **120:**369, 1963.

47. Partridge, L. D.: Motor control and the myotatic reflex, Am. J. Phys. Med. **40:**96, 1961.
48. Penfield, W. G., and Boldrey, E.: Somatic motor and sensory representation in the cerebral cortex of man as studied by electrical stimulation, Brain **60:**389, 1937.
49. Penfield, W., and Rasmussen, T.: The cerebral cortex of man, ed. 1, New York, 1950, The Macmillan Co.
50. Penfield, W., and Welch, K.: The supplementary motor area of the cerebral cortex, a clinical and experimental study, Arch. Neurol. Psychiatry **66:**289, 1951.
51. Phillips, C. G.: Changing concepts of the precentral motor area. In Eccles, J. C., editor: Brain and conscious experience, New York, 1966, Springer-Verlag New York, Inc.
52. Rasmussen, T., and Penfield, W.: Further studies of sensory and motor cerebral cortex of man, Fed. Proc. **6:**452, 1947.
53. Rosén, I., and Asanuma, H.: Peripheral afferent inputs to the forelimb area of the monkey cortex: input-output relations, Exp. Brain Res. **14:**257, 1972.
54. Schaefer, E. A.: Textbook of physiology, New York, 1900, The Macmillan Co.
54a. Settlage, P. H., et al.: The pattern of localization in the motor cortex of the rat, Fed. Proc. **8:**144, 1949.
55. Shimazu, H. N., Yanagisawa, N., and Garoutte, B.: Corticopyramidal influences on thalamic somatosensory transmission in the cat, Jpn. J. Physiol. **15:**101, 1965.
56. Smith, W. K.: The frontal eye fields. In Bucy, P. C., editor: The precentral motor cortex, ed. 2, Urbana, 1949, University of Illinois Press.
57. Stark, L.: Neurological control systems—studies in bioengineering, New York, 1968, Plenum Press.
58. Stetson, R. H., and Bouman, H. D.: The coordination of simple skilled movements, Arch. Neerl. Physiol. **20:** 179, 1935.
59. Sugar, O., Chusid, J. G., and French, J. D.: A second motor cortex in the monkey (Macaca mulatta), J. Neuropathol. Exp. Neurol. **7:**182, 1948.
60. Taub, E., and Berman, A. J.: Movement and learning in the absence of sensory feedback. In Freedman, S. J., editor: The neuropsychology of spatially oriented behavior, Homewood, Ill., 1968, Dorsey Press.
61. Tower, S. S.: The dissociation of cortical excitation from cortical inhibition by pyramid section, and the syndrome of that lesion in the cat, Brain **58:**238, 1935.
62. Tower, S. S.: Pyramidal lesion in the monkey, Brain **63:**36, 1940.
63. Travis, A. M.: Neurological deficiencies after ablation of the precentral motor area in Macaca mulatta, Brain **78:**155, 1955.
64. Travis, A. M.: Neurological deficiencies following supplementary motor area lesions in Macaca mulatta, Brain **78:**174, 1955.
65. Travis, A. M., and Woolsey, C. N.: Motor performance of monkeys after bilateral partial and total cerebral decortications, Am. J. Phys. Med. **35:**273, 1956.
66. Twitchell, T. E.: Sensory factors in purposive movement, J. Neurophysiol. **17:**239, 1954.
67. Vogt, C., and Vogt, O.: Allegemeinere Ergebnisse unserer Hirnforschung, J. Phychol. Neurol. (Leipzig) **25:**277, 1919.
68. Wall, P. D.: The laminar organization of dorsal horn and effects of descending impulses, J. Physiol. **188:** 403, 1967.
69. Welker, W. I., Benjamin, R. M., Miles, R. C., and Woolsey, C. N.: Motor effects of cortical stimulation in squirrel monkey (Saimiri sciureus), J. Neurophysiol. **20:**347, 1957.
70. Welt, C., Aschoff, J. C., Kameda, K., and Brooks, V. B.: Intracortical organization of cat's motorsensory neurons. In Yahr, M. D., and Purpura, D. P., editors: The neurophysiological basis of normal and abnormal motor activities, New York, 1967, Raven Press.
71. Wiesendanger, M.: The pyramidal tract. Recent investigations on its morphology and function, Ergeb. Physiol. **61:**73, 1969.
72. Woolsey, C. N.: Organization of somatic sensory and motor areas of the cerebral cortex. In Harlow, H. F., and Woolsey, C. N., editors: Biological and biochemical bases of behavior, Madison, 1958, University of Wisconsin Press.
73. Woolsey, C. N., and Chang, H. T.: Activation of the cerebral cortex by antidromic volleys in the pyramidal tract, Res. Publ. Res. Assoc. Nerv. Ment. Dis. **27:** 146, 1947.
74. Woolsey, C. N., Travis, A. M., Barnard, J. W., and Ostenso, R. S.: Motor representation in the postcentral gyrus after chronic ablation of precentral and supplementary motor areas, Fed. Proc. **12:**160, 1953.
75. Woolsey, C. N., et al.: Patterns of localization in precentral and "supplementary" motor areas and their relation to the concept of a premotor area, Res. Publ. Res. Assoc. Nerv. Ment. Dis. **30:**238, 1950.

VIII

THE AUTONOMIC NERVOUS SYSTEM, HYPOTHALAMUS, AND INTEGRATION OF BODY FUNCTIONS

33

KIYOMI KOIZUMI and CHANDLER M. BROOKS

The autonomic system and its role in controlling body functions

The autonomic nervous system is a part of the central nervous system (CNS), not a distinct entity as the term might suggest. It is an efferent outflow, a complex of motoneurons innervating tissues and organs. Although afferent nerves are included in autonomic nerve trunks, these afferents serve the somatic as well as the autonomic system, except in a few cases to be discussed later.

The autonomic outflow is segmental and parallels to some degree the somatic motor outflow to skeletal muscles of the body. Directives from the brain and reflex responses initiated by general and specific stimuli course out these two motor pathways to affect visceral organs and other body parts as reactions appropriate to the occasion are organized and executed. Neither autonomic nor somatic reactions occur in isolation, and it can be said that the autonomic nervous system organizes the visceral support of somatic behavior. The CNS integrates the activities of the body through these two complexes.

The history of our knowledge of the autonomic nervous system is extensive. The earliest accounts of man's physical experiences contain descriptions of autonomically regulated reactions. It was observed that these became most conspicuous under conditions of emotional stress, pain, and injury. Thus the autonomic system has long interested those who have been concerned with psychosomatic and emotional behavior. As a matter of fact, the terminology used today had its origin in these earlier concepts of the control of emotional behavior and visceral reactions. It was observed that activities in and injuries to one part of the body were accompanied by reactions in other organs, as though a sympathetic relationship existed between the two. When the early anatomists located nerves that supplied the heart and other organs, they concluded quite correctly that these nerves mediated the associated reactions. Winslow in 1732 discovered the lateral and collateral ganglia, fibers extending therefrom to visceral organs, and the rami of the thoracolumbar outflow. He called these structures parts of a sympathetic system. Others have spoken of the involuntary nervous system (Gaskell) or the vegetative nervous system (Myer and L. R. Müller), but for the most part we have adopted the terminology of Langley,[80] to whom we owe much of our knowledge of the structure and function of the autonomic nervous system.

During the first quarter of this century the relationships of the adrenal medulla to the autonomic system and its role in the function of this complex were clarified.[24] Shortly thereafter Loewi,[89] Cannon and Bacq,[26] and Dale[32] demonstrated chemical mediation of autonomic nervous system action on peripheral organs. The existence of this humoral transmission of nerve action had been suggested much earlier by Elliot (1904) on the basis of the observation that sympathetic nerve stimulation and injection of epinephrine had much the same effect. Langley, who described many of the functions and effects of autonomic nerves, introduced the concept of receptors. Just what these receptors are and what role they play are still questions of major concern. These same men demonstrated denervation sensitization, the action of blocking agents on transmission, and began many investigations that have resulted in some of the most notable advances of recent times.

A number of the major concepts of physiology were developed by those who studied the autonomic nervous system: the importance of tonic activity, the maintenance of homeostasis, the response of the total organism to conditions of stress, and adaptation phenomena are examples of the work of those who pioneered in this field. There are numerous accounts of the development of our knowledge of the anatomy and physiology of the autonomic nervous system.*

*See references 1, 2, 4, 76, 92, 96, and 114.

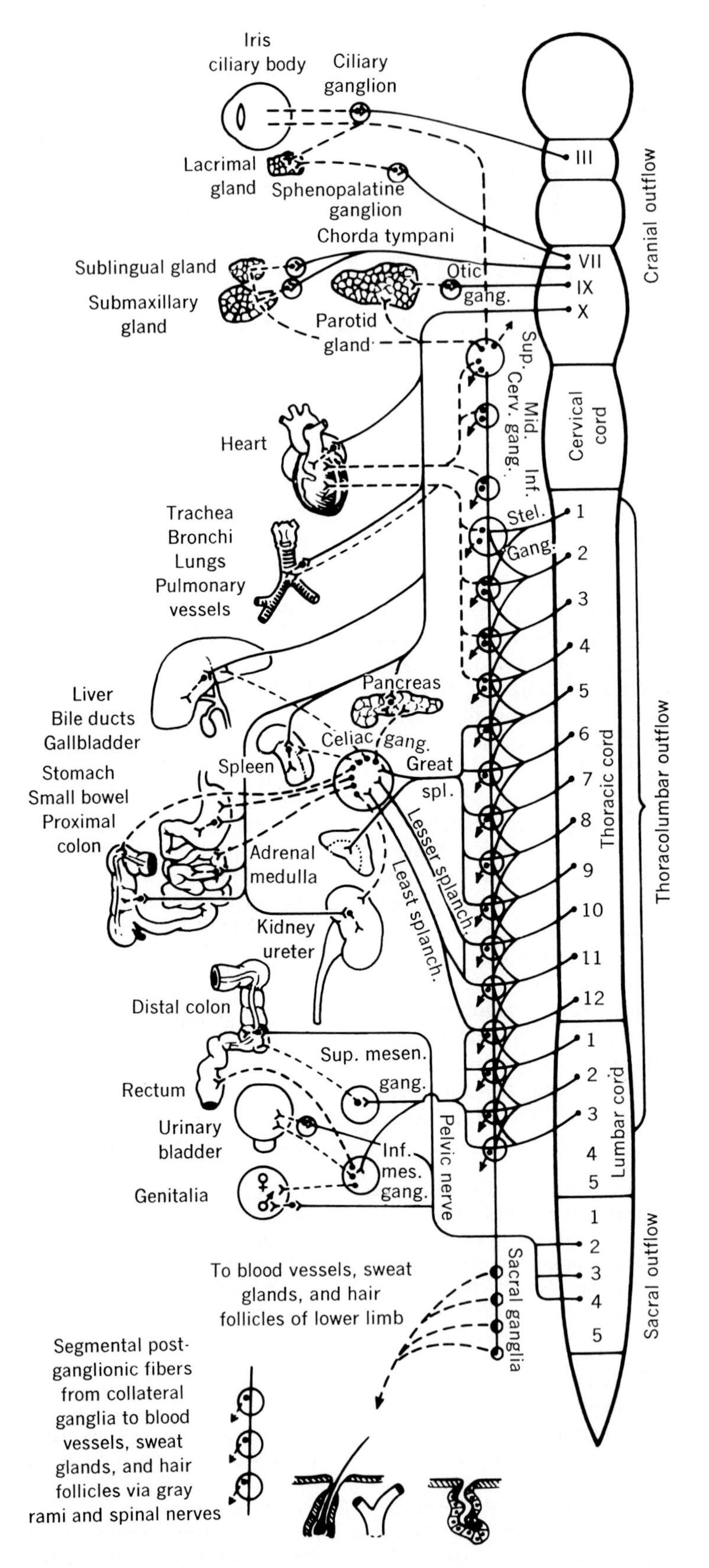

Fig. 33-1. Diagram of general arrangement of autonomic nervous system. Brain and spinal cord are represented at right, but nerves of somatic system are now shown. Preganglionic fibers are indicated by solid lines and postganglionic fibers by broken lines. For further description, see text. (Modified from Goodman and Gilman.[53a])

These are well worth reading because they describe the formation of ideas and present the knowledge that must become familiar to all physiologists.

ANATOMIC ORGANIZATION OF THE AUTONOMIC NERVOUS SYSTEM

Langley spoke of the autonomic system as consisting of two major divisions, the sympathetic, or thoracolumbar outflow, and the parasympathetic, or cranial and sacral outflow. These two divisions have a somewhat reciprocal relationship and perform unique functions.

Fig. 33-1 shows in diagrammatic form the subdivisions of the system, their anatomic characteristics, the relationships and distribution of fibers within the system, and the innervations of visceral organs.

Sympathetic division (thoracolumbar outflow)

There is a preganglionic fiber outflow from each segment of the spinal cord from the first thoracic to the third lumbar level. The white rami that carry the preganglionic fibers from the spinal cord to chains of lateral ganglia are much shorter than indicated in Fig. 33-1. They are more accurately portrayed in Fig. 33-2, in which preganglionic neurons are shown to correspond to interneurons of the somatic reflex pathway. Their cell bodies lie in the lateral horns and not in the ventral horns, which contain somatic motoneuron pools. The preganglionic fibers are included in the ventral roots and accompany the somatic nerve trunks as they leave the spinal column. Shortly thereafter, these thinly myelinated fibers branch off to form the white rami. Most of these fibers make synaptic connections in the lateral ganglia, but some extend to the collateral ganglia before synapsing. The innervation of the adrenal medulla is preganglionic, but this is not exceptional, since cells of the adrenal medulla are modified postganglionic neurons.

One of the features shown diagrammatically in Fig. 33-1 is the considerable divergence of preganglionic fibers from each segment. Some of them extend forward in the chain, making synaptic connections at levels three or four segments above the point of exit. The caudal extension is

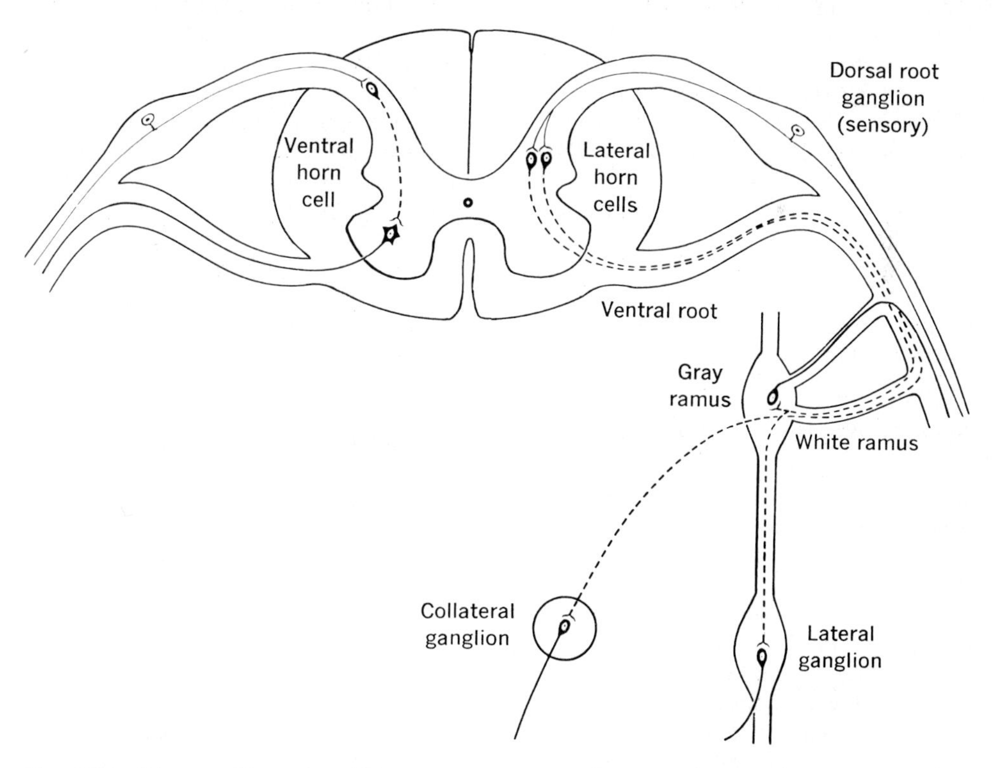

Fig. 33-2. Diagram illustrating different arrangements of neurons in somatic and autonomic nervous systems. Single polysynaptic spinal reflex arc of somatic system is shown at left; that of sympathetic division of autonomic system is shown at right. These segmental arcs are bilateral and superimposed. At right preganglionic neurons are represented by broken lines and postganglionic neurons by solid lines. Note that preganglionic fibers coming out through white ramus make synaptic connection in more than one lateral chain ganglion or a collateral ganglion.

even greater, since the lateral chain extends to the sacral levels despite the fact that the preganglionic outflow terminates in the upper lumbar regions. The phenomenon of convergence is also illustrated here, since a single ganglion receives fibers from several segments. Many preganglionic fibers converge to form with postganglionic neurons the large celiac and inferior mesenteric collateral ganglia.

At the upper end of the lateral chain there is a large stellate ganglion. Some preganglionic fibers pass through the stellate ganglion to form a cervical sympathetic nerve and eventually synapse in an inferior, a medial, and a superior cervical ganglion, which terminates the chain. It appears that during development a number of segmental contributions fuse to produce these larger aggregates of postganglionic neurons. Except for those preganglionic fibers that comprise the splanchnic nerves and terminate in the adrenal glands and collateral and other ganglia near sacral organs, the preganglionic fibers are short. Postganglionic fibers leave the lateral and collateral ganglia and pass by way of discrete nerve trunks to the visceral organs that they innervate. These are unmyelinated fibers of small diameter, and are generally considerably longer than the preganglionic fibers. They innervate practically all organs, smooth muscle, and glandular tissues. The distribution of this postganglionic outflow is shown in Fig. 33-1.

The ratio between post- and preganglionic fibers is large; a single preganglionic fiber innervates many postganglionic neurons. In the superior cervical ganglia of the cat the ratio is 1 preganglionic to between 11 and 17 postganglionic fibers. In other ganglia the ratios are somewhat smaller.

Parasympathetic division (cranial and sacral outflows)

In the parasympathetic system, preganglionic fibers end in ganglia close to or within the organ innervated, and very short postganglionic fibers extend to the tissues. There are no interconnections between the components of the cranial and sacral outflows. Furthermore, the parasympathetic system does not innervate all smooth muscle tissues or visceral organs.

In the cranial division, preganglionic neurons originating from the oculomotor nuclei travel out the third cranial nerve to the ciliary ganglia. The postganglionic fibers comprise the short ciliary nerves that innervate the ciliary muscle and the sphincter muscle of the iris. Preganglionic fibers from the superior and inferior salivary nuclei course through the seventh and ninth cranial nerves to ganglia that supply the innervation of the salivary glands. Submaxillary and sublingual glands are innervated from the superior salivary nuclei through the chorda tympani. The parotid glands receive their innervation from the inferior salivary nuclei by way of the ninth cranial nerve and otic ganglia.

A major component of the cranial parasympathetic outflow is the vagus nerve. This originates in the vagus nuclei of the medulla and comprises the tenth cranial nerve. It consists of preganglionic fibers that synapse in ganglia within the heart, lung, esophagus, stomach, pancreas, liver, intestine, and upper colon. Vagal fibers synapse in ganglia of the pulmonary plexus, and postganglionic neurons therefrom innervate the bronchi and blood vessels of the lung. None of these parasympathetic ganglia are readily visible, and their postganglionic fibers are confined within the tissues of the organs innervated.

The pelvic viscera are innervated from a parasympathetic sacral outflow originating from the second, third, and fourth sacral segments of the cord. Preganglionic axons form the nervi erigentes that end in ganglia adjacent to blood vessels of the erectile tissues. Other components of the outflow connect through ganglia with postganglionic fibers that innervate the uterus, intestines, bladder, lower colon, and rectum as well as adjacent tissues. It should be noted that there is very little overlap, convergence, or divergence in the parasympathetic system except in the ganglia, where one preganglionic fiber makes contact with a number of postganglionic fibers (1:2 is the ratio in the ciliary ganglion).

Another anatomic feature of considerable physiologic significance is the fact that the eye, the salivary glands, the heart, the digestive system, and the pelvic viscera receive dual innervation. Both sympathetic and parasympathetic divisions converge on these organs, although not on the same cells in all instances. In contrast, as shown in Fig. 33-1, the sweat glands, adrenal medulla, piloerectors, and the majority of blood vessels receive only sympathetic innervation.

Autonomic synapses

There are three remaining anatomic considerations that should be reviewed briefly. These are the central, ganglionic, and peripheral synaptic structural relationships. Although these matters have been discussed elsewhere in this text (Chapters 4 and 8), the following review is appropriate. The autonomic neurons located in the lateral horns of the cord form a distinct group. They are considerably smaller than motoneurons of the ventral horns and vary in form, being either spindle shaped or polygonal. Structural details of the synaptic connections between afferent fibers and these neurons are not well known. It is assumed that the same structural arrangements occur as in the case of afferent connections to interneurons and motoneurons of the somatic system.

Much more work has been done on ganglionic synapses,[41,91,96] where preganglionic fibers terminate on postganglionic neurons. A ganglion is composed of many postganglionic cells. These are usually multipolar, possess numerous long dendrites, and are embedded in a fine fibrillary meshwork. The cells range in size from 14 to 55 μm, a majority being 20 to 30 μm in diameter. Studies of fine structure have shown that synapses are found on the cell body as well as on neuronal processes. The presynaptic endings contain numerous clear vesicles and a few dense-core vesicles. A specialized zone of increased density is found in the postsynaptic membrane where synaptic contacts occur.

At the sites where the peripheral terminals of postganglionic fibers make contact with effector cells there appear to be no specialized end-plates, as in the case of somatic neuromuscular junctions; rather, fibers end in close contiguity with the membranes of the cells innervated. In intestinal smooth muscles, in which the fine structures have been carefully studied,[15,64] autonomic fibers become varicose and form bulbous expansions every 1 to 3 μm. The varicosities contain vesicles that are thought to be the storage sites for transmitters. It has been suggested that transmitter is released not only from the varicosity in which the nerve terminates but also from other varicosities that are found along the axons and that make contact with the smooth muscle fibers en passant.

NATURE OF AUTONOMIC FUNCTIONS

Autonomic activity adjusts body states and supports somatic reactions. It is not essential that the system be primarily involved in the initial phase of a response, nor is its action necessarily secondary to somatic activities; it tends to parallel and support these functions. The autonomic system also prepares the body for behavioral reactions, and it effects anticipatory adjustments in emotional states. It can increase cardiac output, adjust blood flow, and make sources of energy more readily available, thus preparing the organism for violent activity. These visceral reactions are basically supportive or anticipatory of body need.

The autonomic nerves are composed of small myelinated and unmyelinated fibers. The preganglionic fibers for the most part are myelinated, have diameters of less than 3 μm, and conduct at a speed of approximately 2 to 14 m/sec. The postganglionic fibers are largely unmyelinated, are approximately 0.3 to 1.3 μm in diameter, and conduct at speeds of less than 2 m/sec. The system is therefore not equipped for the speed of response demonstrated by the somatic motor outflow.

Another characteristic of autonomic fibers that is appropriate to the function of this system is their tonic activity. The origin of this tonic discharge will be discussed later, but it is significant in that because of it the visceral organs are held in a state of intermediate activity and can be controlled by either diminution or augmentation of the rate of fiber firing. This tonic action is not entirely an intrinsic property of these neurons but is in part a contribution of central control and peripheral reflex mechanisms that maintain a state of low-level activity in many of the neurons of this system.

Actions of sympathetic nerves on specific effectors

Early investigators stimulated pre- and postganglionic fibers of the autonomic system innervating various organs of the body in order to determine the actions of these fibers.

Stimulation of sympathetic fibers to the eye causes the eyeball to protrude (exophthalmos), the pupil to dilate (mydriasis), and the nictitating membrane to constrict. Vasoconstriction has also been observed. Claude Bernard was among the first to describe the effects of stimulating and cutting the superior cervical sympathetic trunk. Section of these nerves causes pupillary constriction (miosis), relaxation or protrusion of the nictitating membrane in those species such as cats that possess this structure, drooping of the eyelid (ptosis) associated with retraction of the eyeball, local vasodilatation of the skin, and anhidrosis (less sweating) (Horner-Bernard or Horner's syndrome). The effects of nerve section reveal that the dilator muscles of the pupil are normally held in a state of partial contraction by the tonically active postganglionic sympathetic fibers that innervate them. Sympathetic fibers from the superior cervical ganglion also innervate salivary glands. They cause vasoconstriction in these glands and may reduce salivary secretion from its maximum by a reduction of blood supply, but they also act directly on secretory cells, at least in certain glands, causing secretion of saliva. Other sympathetic fibers coursing to the head and neck tissues in the cervical sympathetic trunk cause sweating, piloerection, and vasoconstriction of skin vessels as well as those of the lacrimal glands.

Postganglionic fibers from the stellate (inferior cervical) ganglia and ganglia of the sympathetic chain innervate the heart. They produce acceleration of pacemaker activity in nodal tissues, speed up conduction of impulses, and increase the strength of contraction. Thus the autonomic fibers produce an augmentation of stroke volume and cardiac output. These cardiac fibers are also tonically active. Postganglionic fibers from the stellate and upper thoracic ganglia pass to the lungs and innervate muscle fibers of the bronchi, causing them to dilate. They produce a degree of constriction of pulmonary blood vessels.

Many preganglionic fibers from the midthoracic region pass through the lateral ganglia to the celiac and other small collateral ganglia. They connect there with postganglionic fibers that innervate abdominal visceral organs. Fibers innervating the liver cause vasoconstriction and inhibit contraction of the gallbladder. These visceral efferents, by a combination of direct and indirect action through the release of epinephrine, produce glycogenolysis and a consequent liberation of glucose. There is also a reduction in blood clotting time consequent to augmented sympathetic activity. The sympathetic nerves and transmitters released therefrom elevate the metabolic rate. Carbohydrate metabolism is much influenced not only by action in the liver, but also by an inhibition of insulin secretion and an augmentation of glucagon output. The neurogenic control of pancreatic endocrines, an old concept once rejected, has recently been confirmed.[53]

Fibers to the spleen cause it to contract, discharging pooled erythrocytes and other components of blood into the vascular system. The mechanical as well as the secretory activities of the stomach and intestines are inhibited, and the cardiac and pyloric sphincters are constricted by sympathetic activity; digestive functions cease. Sympathetic fibers innervate the kidney, causing vasoconstriction of renal vessels.

Another striking effect of visceral sympathetic fiber action is vasoconstriction, which shunts blood away from the viscera, thus making it available to skeletal muscle and the CNS. Sympathetic nerve action also causes the blood vessels supplying the skin of the trunk, abdomen, and limbs to constrict. All sympathetic vasoconstrictor nerves are tonically active; they increase the rigidity of large blood vessel walls and hold many arterioles and other small blood vessels in a state of partial or total occlusion. These nerves are highly important in the control of peripheral resistance and thus blood pressure. Vasoconstrictors also act on the veins and can effect a constriction therein that reduces the capacity of the venous reservoir (Chapter 41).

Sympathetic nerves to blood vessels of the skeletal musculature contain some cholinergic vasodilator fibers along with adrenergic constrictor fibers. These cholinergic fibers are said to have no tonic action and are not involved in regulation of blood pressure or cardiovascular reflexes. They are said to act during exercise and emotional states and to be controlled by pathways in the CNS that are distinct from those that regulate the vasoconstrictors.[60,110] The physiologic significance of these fibers has not been fully demonstrated, but it is considered that they share with the metabolites that accumulate during muscle activity the power to produce vasodilatation.

The sympathetic supply to sweat glands (eccrine glands) is cholinergic. The apocrine sweat glands that are present in some regions (e.g., the axilla) do not participate in thermoregulatory sweating but are thought to secrete in response to mental stress. It appears that they are influenced humorally by epinephrine. It has been claimed that sympathectomy in man does not abolish secretion from the apocrine glands. Cholinergic sweat secretion in heat and adrenergic piloerection on exposure to cold or in anger are produced by a rather generalized discharge of the sympathetic system.

Innervation of the adrenal medulla is of great significance, since sympathetic activity causes liberation of norepinephrine and epinephrine (adrenaline). These catecholamines reinforce the action of all postganglionic sympathetic fibers, except those going to sweat glands and those vasodilator fibers that are cholinergic.

The sympathetic supply to pelvic visceral organs tends to have a mixed action on the bladder. It causes contraction of the internal sphincter and relaxation of the bladder wall. There is, however, an increase in the frequency of micturition and a tendency to void at a lower bladder volume as a consequence of emotional disturbance.

Peristalsis in the lower colon and rectum is inhibited by sympathetic nerve action. Vasoconstriction occurs in most of the pelvic organs, including the uterus, when the sympathetic fibers are active, but a low level of erection can be effected in the male genitalia. Sympathetic fibers can produce an ejection of semen and are also essential. Thus the effects of the sympathetic system on pelvic viscera are not all inhibitory nor vasoconstricting.

In concluding this analysis, it can be said that practically all effects of sympathetic nerve activity are physiologically compatible, and, if evoked simultaneously, they mimic rage and stress reactions. It appears that the thoracolumbar sympathetic complex or the sympathoadrenal system can and often does discharge as a whole. Pupillary dilation, retraction of the nictitating membrane, piloerection, vasoconstriction, sweating, acceleration of the heart, elevation of blood pressure, increase in blood glucose, decrease in blood clotting time, and inhibition of the digestive system are all sympathetically induced phenomena and produce a picture that we have all observed in angry or frightened animals. The anatomic unification of this system would lead one to predict a generalized discharge. This seems to be less of a mystery than the channeling of activity through this system, which permits rather discrete reflex actions and the occurrence of tonic activity in only certain of the sympathetic fibers. The centers that control this system can, to a great degree, determine whether it discharges selectively or totally. Under basal conditions, selective tonic and reflex activities occur, but under stress or in anger the system discharges as a whole.

Actions of parasympathetic nerves on specific effectors

The principal effect of the parasympathetic supply to the eye is pupillary constriction. These fibers have a tonic activity, and section of the third cranial nerve through which they pass produces a degree of pupillary dilation. The lacrimal glands of the eye are innervated by parasympathetic fibers of the seventh cranial nerve, which cause lacrimation. The chorda tympani and other parasympathetic nerves running to the salivary glands produce copious salivation and dilatation of the glands' blood vessels. It is presently thought that vasodilatation is caused by dilator fibers and indirectly by secretory fibers that liberate active dilator substances[42] (Chapter 41).

The vagus or tenth cranial nerve, which is a component of the cranial parasympathetic division, has an inhibitory effect on the heart. Its basic action is on the pacemaker; it retards the depolarization process and decreases heart rate as a consequence. It slows conduction and can actually block transmission of impulses; this block occurs in the atrioventricular nodal region. Thus the vagus tends to produce a reduction in cardiac output and blood pressure. Vagus nerves innervating the bronchi and lungs cause constriction of the bronchioles and possibly an increased secretion from the bronchial glands.

Vagal fibers also innervate the stomach, intestine, and upper colon and have an excitatory action on these organs, increasing peristaltic activities, shortening emptying time, liberating gastrin, and increasing secretion from the gastric and other digestive glands. This parasympathetic fiber action relaxes all sphincters of the digestive tract. It stimulates pancreatic secretion, the secretion of insulin, and the release of bile by causing contraction of the gallbladder.

These vagus fibers to the heart and to the gastrointestinal system are tonically active. However, since smooth muscles of the stomach and gut are intrinsically active, the vagi as well as the sympathetic fibers are modulators of gastrointestinal activity.

The sacral parasympathetic system is responsible for bladder contraction, reflex micturition, and defecation. These fibers increase peristaltic activity in the colon and rectum and cause relaxation of sphincter muscles during micturition and defecation. Other fibers of the system have a vasodilator action, contributing to engorgement of reproductive organs and erection of external genitalia.

It is obvious that there could be no physiologic rationale for discharge of the parasympathetic system as a whole. Simultaneous dilation of the pupil, salivation, slowing of the heart, increased activity of the gut, defecation, urination, and erection of the penis have no single functional objective. These phenomena are associated only in abnormal circumstances that produce a mass reflex. The components of the parasympathetic system behave independently, participating in specific reflexes or well-integrated reactions.

Interaction of the two divisions

The somatic musculature is controlled by an on-off mechanism, that is, impulses go to muscle fibers to produce contractions and, in the absence of such impulses, the muscles relax. Due to reactions through central mechanisms, some somatic motor nerves show tonic activity that can be augmented or minimized, depending on the intensity of afferent stimulation. Activity of the antigravity system provides an example of this phenomenon. Control of those visceral tissues that receive a single autonomic innervation is quite similar to the regulation that occurs in the somatic system. Sympathetic nerves to piloerectors, sweat glands, visceral and skin blood vessels, and the spleen and liver operate by an on-off action. Vasomotor nerves are tonically active; vasoconstriction can be produced by an increase in tonic activity, and vasodilatation can result from a decrease or cessation of tonic activity.

The other visceral organs, which receive a dual innervation, are controlled in a much more complex fashion. There are numerous variants of interactions between sympathetic and parasympathetic fibers that innervate common organs. These are best described individually.

In the eye, parasympathetic nerves act on the sphincter muscle of the pupil to produce pupillary constriction and on the ciliary muscle to evoke accommodation for near vision. The sympathetic innervation causes pupillary dilation and adjustments for distance focusing; this is antagonistic to the parasympathetic action. However, the sympathetics act on radial fibers of the ciliary muscles, whereas parasympathetics act on circular constrictor muscles. Here we have a situation very much like the antagonism seen in flexion and extension of a limb. Also similar is the reciprocal action occurring in reflex responses; when the pupillary constrictor nerve is reflexly excited, the dilator nerve is simultaneously inhibited. Both these fiber outflows are tonically active; thus pupillary constriction and dilation can be affected by increasing or decreasing parasympathetic tonic activity. Similar reactions can be evoked by decreasing or increasing sympathetic tonic activity. Normally, however, both outflows are reciprocally controlled.

The dual innervation observed in certain salivary glands has a slightly different consequence. Parasympathetic and sympathetic fibers end on different types of secretory cells in some salivary glands, but in other glands they innervate the same cells; however, both produce secretory activity. When both nerves are stimulated together in laboratory experiments, no more saliva is produced than when the parasympathetic nerves alone are maximally excited. There thus is a dual secretory but not always a synergistic action. There is, however, an antagonistic action between the two autonomic nerves on the blood vessels of the salivary glands; the sympathetic fibers constrict and the parasympathetic fibers dilate these vessels. Blood flow, of course, affects salivary secretion, and under certain conditions the sympathetic fibers can reduce salivation by decreasing blood volume flow through the glands. The innervation that is dominant in controlling salivary secretion depends on the nature of the stimulation. Thus the composition and quantity of saliva produced depend on the balance of the contributions made by each of these two innervations. Tonic activity of these fibers may be involved in controlling the blood flow to the salivary glands, but vasomotor changes appear not to be significant in the normal regulation of salivary secretion.

The heart provides the most beautiful example of control by an antagonistic innervation. Both postganglionic sympathetic and parasympathetic fibers act on the same tissues. Here we have a true excitatory sympathetic action that can accelerate pacemaker firing and conduction of excitatory impulses and increase the strength of contraction (positive inotropic effect). Vagal parasympathetic fibers, on the other hand, exert a true inhibitory action, which is well described in Chapter 36. Vagal nerve fibers, under most circumstances, slow spontaneous depolarization in the pacemaker cells, and they can also retard and block conduction. There is no definite negative inotropic action, and slowing of the heart may not decrease cardiac output because it may increase stroke volume. The parasympathetic and sympathetic fibers have antagonistic actions on the coronary circulation (sympathetic stimulation increases coronary flow), but the physiologic importance of this is not certain (Chapter 44).

As stated earlier, the lung is also innervated by both the parasympathetic and sympathetic divisions of the autonomic nervous system. Parasympathetic impulses constrict and sympathetic impulses dilate the bronchi. Here we have the two systems playing a role that is somewhat the reverse of that played in the heart.

There is also dual innervation of the digestive system. Fibers from the two divisions are tonically active and antagonistic in their effects on the same musculature. In a first approximation the situation may be summarized as follows: The parasympathetic innervation, through the vagus, augments the rhythm and strength of contraction of the muscles in the walls of the esophagus, stomach, intestine, and upper colon; the sphincters are relaxed by parasympathetic action, and this constitutes the major motor control of the gastrointestinal tract. The sympathetic innervations appear to play a major role only during states of stress or emotional disturbance. At such times, activity in various segments of the gastrointestinal tract is reduced by active inhibition through catecholamine release. The sphincters tend to close more securely. Sympathetic activity also produces vasoconstriction in these tissues.

With respect to secretory activity in component parts of the digestive system, the parasympathetic has a definitely positive action. The opposite effects produced by sympathetic activity probably are secondary to change in blood flow rather than to a direct secretory cell inhibition.

The interrelationship of sympathetic and parasympathetic actions on the pelvic viscera is somewhat less clear. In the colon and rectum the sacral parasympathetic fibers are activators, whereas the sympathetic supply is inhibitory. These effects are similar to those occurring in the small intestine and stomach. Sacral parasympathetic fibers exert a powerful contraction-inducing action on the bladder that initiates micturition, whereas sympathetic nerve effects are mixed; although sympathetic nerves generally cause relaxation of the bladder, it has been reported that under certain circumstances they can cause contraction. The parasympathetic and sympathetic fibers have a more clearly antagonistic action on the sphincters, the parasympathetic relaxing and the sympathetic contracting these muscles. Not much is known about the significance of autonomic innervation of the uterus and associated tissues. With respect to erection in the male, both parasympathetic and sympathetic nerves contribute to some degree, whereas ejaculation is under the control of sympathetic fibers; thus cooperation or supplementation is illustrated here.

Evidence has been produced recently to confirm the idea that peripheral interactions between sympathetic and parasympathetic nerves do occur both in ganglia and in terminal complexes. Fig. 33-3, *A* and *B*, illustrates in diagrammatic form two concepts of such interconnections. One cannot say that the sympathetic division is invariably inhibitory (Fig. 33-3). Like the parasympathetic division, it has inhibitory effects on some tissues and excitatory effects on others. It also must not be forgotten that the CNS plays a major role in adjusting the balance of action of these two outflows.

There are true parasympathetic cholinergic vasodilators running in the nervi erigentes to blood vessels of the external genitalia; similarly, there are cholinergic dilators to salivary gland and cerebral blood vessels. The literature relative to vasodilator actions in various organs in the body is extensive and controversial, but this topic is dealt with in Chapter 41.

Overaction and underaction of the autonomic system

The autonomic system, particularly the sympathetic division, is not essential to survival of an individual, but absence or underaction of autonomic nerves seriously interferes with many body functions. Impairments following denervation are particularly severe until tissues recover from their dependence on these nerves. Overaction of the autonomic system also has major and debilitating physiologic consequences.

Low vasomotor tone produces the condition of orthostatic hypotension. If a person with this deficiency is suddenly brought to an upright position on a tilt table, fainting results. Fainting, which occurs because of enforced immobility

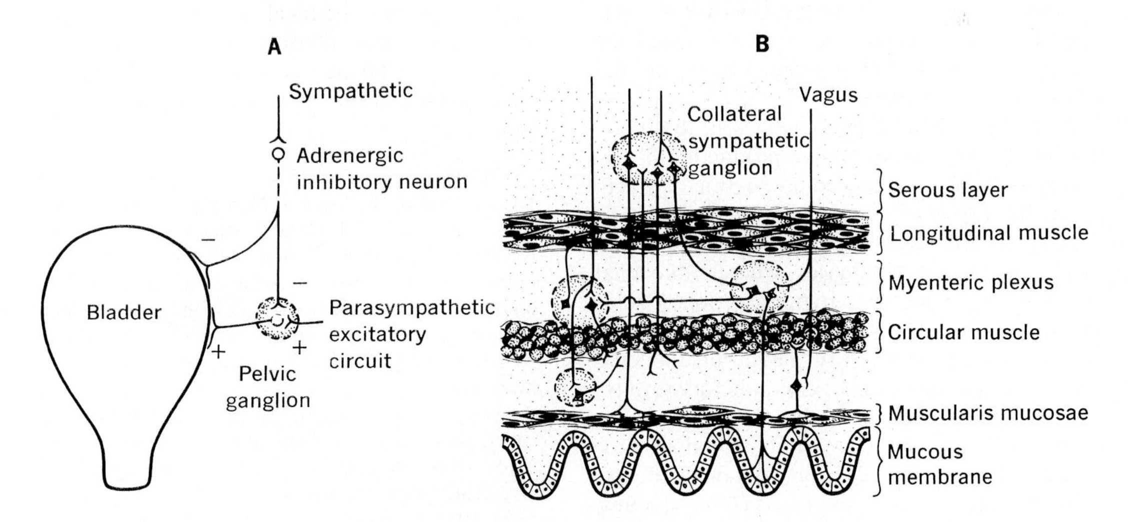

Fig. 33-3. A, Diagram of autonomic system's innervation of bladder, indicating interconnections permitting interactions between sympathetic and parasympathetic fibers in pelvic ganglia. **B,** Autonomic system's innervation of intestinal wall, showing interconnections in myenteric plexuses, which may permit interactions in control of intestinal function. (Modified from de Groat[34] and Schofield.[106a])

while erect, is due to insufficient vasomotor compensation and reduction of blood flow through the brain. Surgical removal of the sympathetic system or blocking the system with drugs produces hypotension. A condition described as familial dysautonomia is occasionally reported in the clinical literature. In such cases there is defective lacrimation, vasomotor instability, and defective temperature control.

Overactivity of vasomotor nerves can result in hypertension. One sees frequent reference to neurogenic hypertension and certainly a hyperresponsiveness of vasomotor reactions can develop. A paroxysmal type of hypertension that occurs in individuals with tumors of the adrenal medulla is occasioned by massive discharges of catecholamine from the gland. The hypertension commonly encountered, however, cannot be explained on the basis of overactivity of the sympathetic division of the autonomic system.

In Raynaud's disease there is a spasmodic contraction chiefly of blood vessels supplying the digits, which consequently receive an insufficient blood supply. The hand, for example, becomes cold and pale in color, and gangrene may actually set in. Since sympathetic denervation of the extremities in such cases is often highly beneficial, an overaction may be involved in causing the disease. Preganglionic denervation is preferable, since it prevents the denervation hypersensitivity that is a consequence of postganglionic fiber removal.

Hyperhidrosis (excessive sweating) is another example of overactivity of the sympathetic system. The sweat glands are overactive, and the condition is quite debilitating. Cases are described in the literature.[8,76,114]

It has been thought that changes in the blood flow in the mucosa, excessive production of hydrochloric acid, and reduced production of mucin might be due to abnormal autonomic neuronal actions. Gastric ulcers are thought to result from reduction of mucosal defenses, whereas duodenal ulcers are associated with hypersecretion of acid. Studies of the possible role of the autonomic system in the production of gastric or duodenal ulcers have produced rather contradictory results. It does appear, however, that vagus resection combined with a partial gastrectomy does produce relief, especially of chronic duodenal ulcers.

In the late 1920s, complete sympathectomy was attempted by bilateral removal of the lateral chains of ganglia,[29] a procedure that also disconnected the collateral ganglia from the CNS. It was found that sympathectomized animals had certain inabilities and vulnerabilities, such as hypersensitivity to heat and cold and to hypoglycemic agents such as insulin. Although these animals could live normal lives, the male was rendered infertile but the female was not; there was also an initial hypotension and some susceptibility to fatigue. The totally denervated heart was found to function effectively, and most organs assumed an autonomy of function that enabled them to play an adequate physiologic role after severance of the external innervation. This work merely provided evidence in support of the concept that although the autonomic system plays a very important physiologic role, neither division is completely essential to the maintenance of basic body functions.

In general, it can be said that the parasympathetic system has a conserving, protective function and tends to promote emptying of the hollow viscera. The sympathoadrenal system functions in emergency situations, coming into strong action during stress, under conditions producing fear and rage, or when an animal suffers pain. Both systems preserve an essential balance of body states; they react strongly to correct imbalances and support behavioral activities. To use the terminology of Cannon, they illustrate the wisdom of the body by reacting to meet emergencies and by maintaining homeostasis.

CHEMICAL TRANSMISSION OF NERVE IMPULSE

One of the most exciting developments of the first third of this century was the discovery and analysis of the chemical transmission of nerve action. It seems advisable to refer students to previously published accounts,[20,27,112] which do more justice to this story than can be attempted here.

The major initial discoveries were those of Otto Loewi, Walter B. Cannon, Henry H. Dale, and their associates. In 1914 Dale,[31] studying acetylcholine (ACh), a derivative of choline that he had found in ergot extracts, showed that this chemical in minute amounts mimicked very accurately the effects of parasympathetic nerve stimulation. In 1921 Loewi[89] demonstrated that the perfusate from an innervated isolated heart would inhibit a second denervated heart, perfused thereby, when the vagus nerve was stimulated. He called this substance *Vagusstoff* and recognized that it had to be protected by anticholinergic agents such as eserine (physostigmine).

In 1931 Cannon and Bacq[26] initiated a series of classic experiments that they and others carried out at Harvard. They showed that an epinephrine-like substance, which they originally called sympathin, is liberated from most postganglionic sympathetic fibers on stimulation. It took some time for others to demonstrate that these postganglionic fibers liberate norepinephrine. They are therefore adrenergic. Parasympathetics that release ACh as their transmitter are cholinergic. We owe this terminology to Dale,[32] who, with his associates, also showed that somatic motor fibers are cholinergic in their action on the skeletal muscle. His work and that of many others made possible the recognition that ganglionic transmission in both the parasympathetic and sympathetic divisions is cholinergic[33] and that some sympathetic postganglionic fibers, such as those innervating sweat glands, are cholinergic.

It was von Euler[111] who first produced convincing evidence that norepinephrine is the normal adrenergic transmitter of the postganglionic sympathetic neuron. The adrenal gland produces both norepinephrine and epinephrine. Usually more epinephrine than norepinephrine is released, but the proportions may change according to the nature of the physiologic stimulus. Also, the proportions of these catecholamines stored and released from the medulla vary from species to species. Tissues differ in their sensitivity to these two adrenergic compounds released by the autonomic system. Norepinephrine is much more effective in producing peripheral vasoconstriction than is epinephrine, but the two compounds have a quantitatively similar effect on the heart. Epinephrine is the more potent agent in producing effects on some tissues (e.g., the cecum of the fowl).[111]

Recent studies of adrenergic transmitters

There has been a tremendous amount of work in recent years on the analysis, synthesis, storage, metabolism, release, and mechanism of action of adrenergic transmitters.[9,10,113]

The individual terminals of adrenergic neurons innervate many smooth muscle fibers. As previously described, these fibers show swellings or varicosities along their course, each of which serves as an ending en passant. Within these swellings are granules or granulated vesicles approximately 300 to 1,000 Å in diameter; each contains an electron-dense core. On isolation by selective centrifugation, fractions containing granules are found to contain dopamine-β-hydroxylase, adenosine triphosphate, and norepinephrine. Unquestionably, norepinephrine is the transmitter agent stored in the en passant or terminal varicosities. There is a rather continual release of norepinephrine from these sites presumably by exocytosis. This discharge is greatly accelerated by the arrival of a nerve impulse. Once released, norepinephrine acts on the receptor of the effector cell and is removed from synaptic sites in a manner that will be discussed later.

Norepinephrine is synthesized in fiber terminals and the cell bodies of the neurons. It is moved down to the axon terminals by axonal transport processes. The process of norepinephrine synthesis within the nerve cell is now known. The precursor is considered to be tyrosine, and the steps of synthesis are shown in Fig. 33-4, *A*. This figure indicates that dopamine is beta-hydroxylated to form the norepinephrine stored in the granules. A feature shown here, which is of considerable interest, is that above a critical concentration, norepinephrine appears to block new production by a negative feedback process that controls the conversion of tyrosine to dopa. It should be pointed out that during activity and as a result of norepinephrine discharge the rate of its new synthesis is increased.

Another process worthy of specific mention is the reabsorption of preformed norepinephrine. This compound may enter the cell from the surrounding tissue fluid. Thus nerve terminals that are depleted of stored norepinephrine either by intense activity or by action of a drug can be charged up again by norepinephrine injected into the circulation or introduced into the surrounding medium. This uptake of norepinephrine into the postganglionic sympathetic neuron is accomplished by an active transport system requiring sodium and potassium ions that can act even against concentration gradients as high as 10,000 to 1.

Norepinephrine undergoes a complex fate after it is discharged from the nerve. A fraction is removed by diffusion into the bloodstream and a small amount is metabolized locally by catechol-*O*-methyltransferase (COMT), but the greatest proportion is taken up again by the nerves and is bound again in granules. Some of the reabsorbed norepinephrine may move directly to storage sites, but much is broken down by the monoamine oxidase (MAO) in the mitochondria to deaminated metabolites, which can be discharged or used in resynthesis. This uptake maintains what appears to be a dynamic equilibrium between granules and extragranular, intra-axonal stores.

Norepinephrine that escapes into the circulation is mainly metabolized in the liver by COMT. The ultimate fate and pathways for metabolic degradation of norepinephrine are shown in Figs. 33-4, *B*, and 33-5.

Studies of the release of norepinephrine during nerve activity have yielded some interesting figures of physiologic significance. It has been found that a stimulus frequency of 30 impulses/sec causes the maximum release per stimulus from splenic nerves. At a faster rate of stimulus,

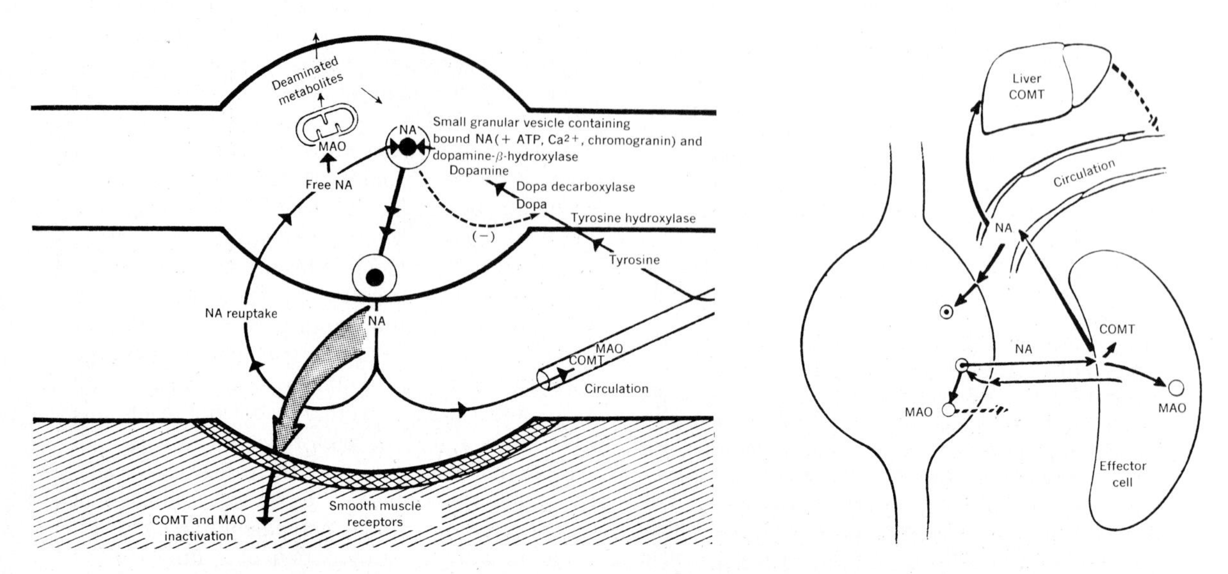

Fig. 33-4. A, Schematic diagram of release, peripheral action, metabolism, reuptake, and resynthesis of norepinephrine at an adrenergic neuromuscular junction. **B,** Fate of norepinephrine released in sympathetic nerves and tissues. *COMT,* Catechol-*O*-methyltransferase; *MAO,* monoamine oxidase; *NA,* norepinephrine. (**A** modified from Burnstock[22]; **B** from Axelrod.[9])

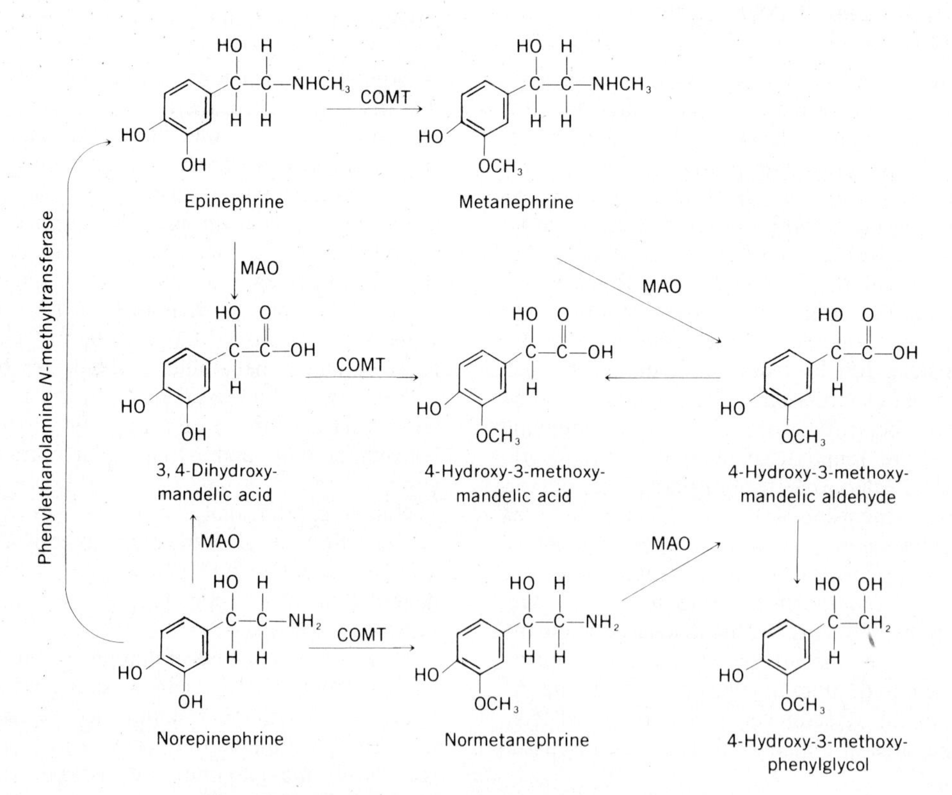

Fig. 33-5. Catabolism of norepinephrine and epinephrine.

less is released. Calculation of amount released per stimulus shows that approximately 8×10^{-10} gm or 0.01% of the total amount present (8 μg in 5 gm of tissue) at the nerve terminals in the spleen is released. This corresponds to the release of 18% of the total tissue content of norepinephrine per minute if a nerve is stimulated at a rate of 30/sec. In a muscle the rate of release is slightly less. It is obvious that if resynthesis were inhibited, the store of norepinephrine would be depleted by 50% in a few minutes' time.

Cholinergic transmitter

ACh is the transmitter for the preganglionic fibers of the total autonomic system, postganglionic fibers of parasympathetic nerves, and some of the sympathetic postganglionic neurons. It is stored in vesicles in the nerve terminals. These vesicles are 400 to 500 Å in size and occur in clumps or heavy concentrations near specialized areas of the terminal membrane. They appear to move progressively in the cytoplasm toward sites of transmitter release.

ACh is synthesized in the terminals and probably also in the cell body and transported down the axons to the nerve terminals where it can be stored in quantity in vesicles accumulated there. Choline acetylase is present wherever ACh is found and is involved in its synthesis from acetate and choline. There is a constant leakage of ACh as vesicles move down to the terminal membrane and release the transmitter. The arrival of an impulse greatly augments this discharge. Estimations of the quantity of ACh released have been made in studies of transmission in toad sympathetic ganglia. A single preganglionic impulse liberates 2.6×10^{-16} gm of ACh per ganglion cell. A single quantum of transmitter release (ACh) is calculated to be 8,000 to 12,000 molecules.[94] The released ACh diffuses across the synaptic cleft to the subsynaptic membrane, where it combines with a receptor substance to produce permeability changes and a depolarization of the postsynaptic membrane. Thus a response is initiated in the effector cell, where this action is excitatory.

When ACh has an inhibitory action, the effect is the opposite. For example, the action of ACh in the heart is inhibitory and there is a hyperpolarization of muscle cell membrane. In the pacemaker cell, this hyperpolarization is associated with an increase in membrane permeability to potassium. The atrial muscle action potential is shortened due to the fact that the rate of membrane repolarization is greatly increased (Chapter 36).

ACh is hydrolyzed and inactivated by acetylcholinesterase (AChE), which is present in high concentrations in synaptic regions. This action is extremely rapid and prevents diffusion of liberated ACh away from the site of release. A diffuse action of this compound would, of course, have debilitating physiologic effects. The action of ACh is greatly augmented by anticholinesterases that prolong and augment the action of ACh by retarding its breakdown.

As stated previously, Dale suggested that neurons of the autonomic system might be classified as adrenergic and cholinergic on the basis of their production and release of one or the other compound. It has been suggested that ACh is involved in the release of catecholamines from postganglionic neuron endings, but the evidence for this is indirect and largely pharmacologic.[47] Of course, preganglionic neurons initiate release of catecholamine from the adrenal medulla by the liberation of ACh, which depolarizes the chromaffin cell membrane, thus increasing its permeability to sodium and calcium ions. These ions initiate an unknown process that moves the granules toward the cell surface and releases catecholamines from storage.[37]

Noncholinergic, nonadrenergic transmitters

Numerous physiologists now agree that noncholinergic, nonadrenergic neurons are involved in the regulation of the alimentary tract. These have been found in many vertebrate species.[22,23] Evidence suggests that the active material they release (the transmitter) is a purine nucleotide. Therefore the term *"purinergic transmission"* is used. Although Langley suggested the existence of a third type of autonomic neuron, not blocked by atropine, the study of purinergic transmission "is in its infancy." It is agreed that these fibers are inhibitory, and it has been suggested that they, rather than adrenergic nerves, comprise the main inhibitory system opposing the action of cholinergic excitatory nerves on the gut.[23] It is certain, however, that there are adrenergic inhibitors of gastrointestinal tract activity, and the question of purinergic fiber existence and function will require further studies of neurons, transmitters, and receptor substances.

Central synaptic and ganglionic transmission

Preganglionic sympathetic neurons have been studied by direct intracellular and extracellular recording methods. The reactions of these lateral horn cells of the cord to antidromic excitation by stimulation of white rami and to orthodromic excitation through the dorsal root or splanchnic nerve afferents have been determined. Intracellular recordings have shown responses to be similar to but slightly longer in duration than those of anterior horn cells.[46] These preganglionic neurons are readily activated from higher centers, but fewer cells of a pool appear to be responsive to peripheral nerve stimulation. This suggests that some of these neurons are available to participate in spinal reflex action, but that many are activated directly by input from higher centers.

Parasympathetic preganglionic neurons of the sacral spinal cord have been studied in a similar manner. They also resemble anterior horn cells in their responses. An interesting observation is that incoming impulses from somatic afferents that excite the anterior horn cells may cause inhibition of parasympathetic neurons or a mixture of hyperpolarization and depolarization in these lateral horn preganglionic cells.[35] Such excitatory and inhibitory influences on spinal preganglionic neurons by afferent impulses may be involved in normal reflex actions. Obviously, some integrative control of function from higher centers must be exerted on these cells to relate their activity to peripheral requirements.

Not much is known about the transmitters operating at central synapses involved in control of the autonomic system. It appears that in the CNS there are some adrenergic, some cholinergic, and still other synapses employing different transmitter substances, for example, 5-hydroxytryptamine (5-HT) and certain amino acids.[55,74] Direct iontophoretic applications of ACh or norepinephrine have thus far failed to detectably affect the preganglionic autonomic lateral horn cells.

The events of ganglionic transmission are much better understood, but although this is presently an active field of study, the situation is quite complicated (Chapter 8). Basically preganglionic fibers act on postganglionic neurons by the release of ACh, which produces postganglionic cell discharges. The preganglionic fiber endings are packed with vesicles; the miniature end-plate potentials that have been recorded are similar to those of the neuromuscular junction. AChE is present in high quantity at these terminals. ACh escapes from the ganglia during preganglionic neuron stimulation and can be detected if anticholinesterase is used. ACh applied to the ganglia will evoke postganglionic responses.

The realization that other events can occur within the ganglia has come from the observation that in addition to the regular excitatory postsynaptic potential there are also slow excitatory (sEPSP) and inhibitory (sIPSP) potentials. They are thought to be elicited by transmitter action on a second set of membrane receptors.[85] Catecholamines also have a depressant action on ganglionic transmission, and it is assumed that the ganglia contain adrenergic terminals as well as cholinergic terminals. Chromaffin-containing cells are found within mammalian ganglia, and there may be dopaminergic interneurons or recurrent adrenergic collaterals that contain these compounds and are responsible for the inhibitory reactions evoked by orthodromic and antidromic stimulation of pre- and postganglionic fibers.[64,85,94]

It has been known for many years that stimulation of some postganglionic fibers can evoke activities in other fibers. This phenomenon, which was identified as an axon reflex, implies a high branching of a neuron, which permits a stimulus to travel antidromically in one branch back to a neuron and out to the periphery orthodromically in the other branch. Axonal branches also synapse on adjacent ganglionic neurons; thus "ganglionic reflexes" occur, and it now appears that there is much more interaction within ganglia due to axon collaterals than was previously realized.

The physiologic significance of interaction within the ganglia and of the possible dopaminergic inhibition occurring there is not fully understood. For general purposes, it suffices to say that the impulses entering the ganglia over preganglionic fibers liberate ACh, which acts on receptor substances and initiates activity in postganglionic neurons. These cholinergic receptors differ somewhat from those found elsewhere because they have a different susceptibility to blocking agents.

The receptor concept

The mode of action of neurotransmitters on target cells has been much discussed, and the concept of the "receptor" has been considerably modified[13,36] since its introduction by Langley[79] in 1905. It is generally assumed that a neurotransmitter has to unite in some way with the target cell before exerting its action. Often the sites of binding between the cell and the active molecule are referred to as receptors. These postulated receptors, in or on the cell membrane, occupy only a very small portion of the cell surface. Although there is frequent mention of receptor sites, morphologic evidence of specific receptor "patches" on the cell surface is still lacking.

Pharmacologic evidence indicates that *cholinergic receptors* of the various tissues differ markedly. Blocking agents are assumed to compete with the transmitter substances for the binding site in the receptor. Curare has a powerful blocking action on the neuromuscular junction and prevents the action of nerve impulses or injected ACh. Curare, however, is relatively ineffective in the autonomic ganglia and does not block inhibitory action of the vagus on the heart.

Another experimentally employed compound, nicotine, has little effect on the neuromuscular junction but excites and then blocks postganglionic neurons, interrupting synaptic transmission through autonomic ganglia. Cells excited by and then blocked by nicotine are said to possess *nicotinic* receptors, and the ACh that excites such cells is said to have a nicotinic action.

ACh is said also to have a muscarinic action in that it acts on some cells that are excited by the alkaloid responsible for the toxicity of certain mushrooms, muscarine. These cells thus possess *muscarinic* receptors. Atropine is a drug, obtained originally from the plant *Atropa belladonna,* which blocks the action of muscarine and some ACh actions. Atropine blocks muscarinic receptors, but has little effect on ganglionic or somatic neuromuscular transmission. It principally blocks the action of postganglionic parasympathetic fibers. There are still other variants in the actions of ACh and cholinergic receptors; some cells, for example, possess both nicotinic and muscarinic receptors.

It has long been recognized that there are *adrenergic receptors,* which may also differ, since the potencies of sympathomimetic agents and adrenergic transmitters vary widely when tested on different tissues and organs. The effectiveness of blocking agents also varies from tissue to tissue. It has been concluded that there are at least two types of receptors, α- and β-adrenergic receptors.[51] Both α- and β-receptors are responsive to various catecholamines, norepinephrine, and epinephrine, but they have different sensitivities. For example, α-receptors are blocked by phenoxybenzamine, phentolamine, and ergot alkaloids. The β-receptors are blocked by propranolol. The α-receptors mediate vasoconstriction, contraction of pupillary dilators, constriction of the gastric, intestinal, and bladder sphincters, and contracture of the spleen. The β-receptors mediate such actions as an increase in cardiac rate and strength of contraction and are involved in the inhibition of gastric motility. The study of receptors is expanding rapidly.[23,74,81]

Denervation hypersensitivity of autonomic effector organs

When autonomic effector organs are denervated, they become increasingly sensitive to transmitters and to chemical agents.[25,28] This sensitivity is much greater when postganglionic fibers are destroyed. In the course of degeneration, transected autonomic fibers release stored transmitters, and this tends to give a brief paradoxical overaction. After removal of the superior cervical ganglia, for example, the nictitating membrane initially relaxes; then after a few hours it may contract strongly, only to relax again. After 7 to 14 days, maximal denervation sensitivity develops, and denervated organs respond to minute quantities of ACh or catecholamine to which they previously would not have been responsive.

The mechanism of denervation hypersensitivity is not fully understood, but at least two contributing factors are known.[43] In the absence of nerve terminals, transmitters are not reabsorbed and inactivated quickly. Furthermore, following denervation, cholinesterase and MAO, which normally inactivate transmitters, disappear. It has been shown, particularly for cholinergic transmission, that in denervated tissue, hypersensitivity is due to spread and a greater diffuseness of receptor sites rather than to an increase in receptor sensitivity at any one locus.[75] This is only a partial explanation, because denervated tissues become sensitive to many compounds, both drugs and ions. A development of hypersensitivity has been observed as a result of attempts to effect surgical relief of Raynaud's disease and other autonomic hyperactivities. Laboratory experiments have indicated that preganglionic denervation should be much more beneficial, since peripheral hypersensitivity is minimal; the sensitization is cholinergic and confined largely to the ganglia. Following removal of postganglionic fibers, sensitivity becomes maximal, and, on exposure to cold or excitement, the catecholamines released from the adrenal glands or adjacent sympathetic fibers cause a strong overaction in the denervated tissues.

Immunosympathectomy and chemical sympathectomy

Those who are interested in autonomic nervous system physiology should be aware of certain relatively recent studies that have yielded very striking results, the physiologic significance of which is not yet fully understood. It has been found that immune reactions to nerve growth factors essential to development of the autonomic system can be used to produce immunosympathectomy.

In the late 1940s and early 1950s it was found that certain tumors release substances that serve as a nerve growth factor. These factors affect autonomic fiber growth much more than they do somatic neurons. Snake venom, which was then being used in attempts to block certain metabolic processes in tumors, was also found to be rich in autonomic growth factor. Venoms come from modified salivary glands, and it was soon learned that the salivary glands of mice and other mammals contain a protein that is an autonomic nerve growth factor. When this agent is injected into rabbits, it elicits antibody formation, and serum collected from such animals can be used to produce immunosympathectomy in newborn animals.[56,82,83] It appears that this antibody neutralizes a nerve growth factor essential to the development of sympathetic fibers. Ganglia of the sympathetic chain are most affected. The cell population is found to be only 2% to 10% of normal. Organs in immunosympathectomized animals contain much lower amounts of norepinephrine than are normally present.

A relatively long-lasting suspension of sympathetic function can be produced by drugs such as reserpine that deplete stores and prevent storage of newly synthesized or administered catecholamines. A truer chemical sympathectomy can be effected by 6-hydroxy-dopamine (6-OHDA), which not only depletes norepinephrine stores but also destroys adrenergic nerve terminals in the adult. Some 3 to 4 months are required for their regeneration in some tissues. Administration of 6-OHDA to neonatal animals results in a profound loss of adrenergic neurons, and an almost complete sympathetomy can be produced in this way. Other drugs such as bretylium and guanethidine can also cause a chemical sympathectomy in neonatal animals. They also deplete and prevent catecholamine storage in adults and even produce axonal retraction in adrenergic neurons. Present use of these agents is largely confined to the laboratory, but they have become very useful experimental tools and have contributed greatly to the advance in our knowledge of the autonomic system.[56,109]

TONIC ACTIVITY WITHIN THE AUTONOMIC NERVOUS SYSTEM

It has already been pointed out that many, but not all, of the components of the sympathetic and parasympathetic divisions of the autonomic nervous system are tonically active. In speaking of nerve function the term "tonic activity" is used to imply a continuing discharge in the nerve fiber. This discharge determines the level of activity in the effector organ or tissue. Tonic activity occurs in fibers that have an excitatory action and in fibers that are inhibitory. In the latter case, activity in an organ can be increased by diminution of tone, for example, a decrease in vagal tone accelerates the heart. In the control of blood vessels, vasoconstriction is maintained by sympathetic fiber tonic discharge. A diminution of tone produces vasodilatation and a decrease in peripheral resistance that, if sufficiently generalized, causes a fall in blood pressure.

Frequency of discharge in tonically active fibers

A number of methods have been employed in an attempt to determine the levels of tonic activity in various nerves and to assay the frequency of firing that produces certain intensities of peripheral effects. Folkow[48,49] has stimulated sympathetic trunks innervating blood vessels of the leg and of visceral organs while assaying peripheral resistance by measuring volume flow under a constant perfusion pressure. By varying the frequency of stimulation of sympathetic trunks, he obtained different degrees of vasoconstriction that reduced the flow accordingly. Thus he was able to estimate the relationship between the frequency of stimulation and the effector organ response. From such data, one can calculate the rates of overall tonic activity that are required to maintain normal peripheral resistance and elevate it to the maximum (see Chapter 41).

Another method involves recording tonic activity directly from single nerve fibers, fine strands of fibers, or preganglionic cell bodies in the cord. It is not possible, of course, to determine accurately the total outflow of impulses to an effector organ in a unit of time by measuring activity in a single fiber. However, one does obtain a very accurate picture of rates and changes in rates in individual neurons.

The rates of tonic discharge in individual pre- and postganglionic sympathetic fibers have been measured in white rami, in fibers of the cervical sympathetic trunk, and in fibers innervating skeletal muscles, skin, and heart.* The rate varies greatly in individual fibers even of the same trunk under basic conditions, and it ranges between 1 impulse in 10 sec to several impulses in 1 sec. The average rate in most postganglionic fibers is 2 to 3/sec. In individual fibers, rates tend to remain relatively constant under the same experimental conditions. There are some sympathetic fibers that do not discharge unless they receive a

*See references 4, 18, 54, 65, and 67.

particular stimulus. Rates are affected by internal as well as external influences. Some sympathetic fibers show a very clear respiratory rhythm, that is, the rate of discharges is augmented during inspiration and reduced during expiration. These same fibers may also show an arterial pulse–related rhythm. Tonic discharges of sympathetic nerve fibers change during reflex action; under conditions of stress such as asphyxia and hemorrhage, the frequency of discharge can increase to 20 or even 30 impulses/sec. Of course, the maximum frequency attained by sympathetic fibers during augmentation varies from one fiber to another.

Parasympathetic tonic activity has been recorded from vagal fibers innervating the heart and stomach and from sacral parasympathetic neurons innervating the bladder.[35,62,63,77] In this system the activity is highly dependent on intrinsic or extrinsic stimuli. Vagal discharge to the heart normally is relatively continuous and quite regular in rhythm, but varies with blood pressure levels. The majority of fibers innervating the stomach and bladder are relatively silent until reflexly activated by distention of these organs or by other more extraneous stimuli. It is therefore difficult to state the exact rates of tonic discharges. In general, they are low, approximately those recorded from sympathetic fibers. As in sympathetic nerves, some parasympathetic fiber discharges may increase to 30 impulses/sec under maximal reflex drive.

Origins of tonic activity in autonomic fibers

In many fibers of the autonomic system, tonic activity is of reflex origin, but in other instances, it may depend on some instrinsic automaticity. The tonic discharge in the vagus nerve supply to the heart is highly dependent on baroreceptor activity, since section of baroreceptor afferents causes a decrease in rate of discharge. Some vagal fibers show arterial pulse–related bursts of activity. Activity in vagus efferents is also related to the respiratory rhythm. Contrary to the respiration-related sympathetic activity, vagal discharges increase during expiration and decrease during inspiration. This fluctuation in vagal tone contributes to the cardiac arrhythmia of respiratory origin; sympathetics are also involved.

In the sympathetic division, pulse-related and respiration-related surges in activity occur in efferents innervating the cardiovascular system. After denervation of baroreceptors and abolition of respiratory cycles the rhythm of discharge is changed, but tonic activity remains, indicating involvement of other factors in the establishment of firing rhythms.[72]

Tonic activity recorded in peripheral autonomic nerves originates from tonic activity in the preganglionic neurons. The rates recorded from pre- and postganglionic fibers, however, may not be the same because of the events occurring within the ganglia. The tonic activity of the spinal preganglionic neurons is normally under control of higher centers, mainly the medulla, at least in the case of those fibers responsible for cardiac and vasomotor tone. Higher centers may modulate activity in the medulla and the spinal cord, as will be discussed later.

It is difficult to ascertain whether tonic activity initiated in the medulla indicates automaticity or whether it is imposed by intrinsic or extrinsic stimuli. Certainly it can be modified by afferent signals and changes in the local cellular environment. There may be a degree of intrinsic rhythmicity due to some internal reverberating circuits or to some type of pacemaker action similar to that occurring in cardiac cells. Although we have no direct evidence that this latter phenomenon occurs in mammalian neurons, such a process is seen in the neurons of primitive animals.

Following transection of the spinal cord there is a marked reduction of sympathetic tonic vasomotor discharge and a consequent fall in blood pressure. There is a gradual return of tone, and in species such as the cat, this recovery becomes evident within a few hours. It appears that the spinal preganglionic fibers develop their own tonic activity by either assuming an intrinsic rhythm of discharge or a rhythm initiated by unknown local processes.[97]

AUTONOMIC REFLEXES

The components of the autonomic system can be reflexly activated.[4] Even in those reflexes normally considered to be somatic there is an autonomic involvement. For example, painful stimuli that cause flexion or withdrawal of a limb also produce a rise in blood pressure and modulation of cardiac rate. In some reflexes, particularly those to the heart, both parasympathetic and sympathetic outflows are involved; a reciprocal action mediated centrally may occur. The origin of reflex responses observed in sympathetic and parasympathetic fibers is not altogether easy to explain, however, and these responses will be discussed in greater detail.

Baroreceptor reflex

Afferents from the carotid sinus that traverse the ninth cranial nerve to the medulla are sensitive to blood pressure changes.[57] The receptors involved are probably stretch receptors, but we speak of pressor receptors, since firing is initiated with each pulse pressure wave. Also, an elevation of arterial pressure increases the basic level of afferent discharge.

These afferent signals act on the vagus nuclei to increase vagal tone, which slows the heart. Simultaneously there is a reciprocal inhibition of the vasomotor and cardiac accelerator "centers" located chiefly in the medulla and thus a reduction of sympathetic vasoconstrictor and cardiac accelerator nerve discharges. It has been shown that carotid sinus afferents make synaptic connections in or near the nucleus of the tractus solitarius and in the reticular formation (RF) of the medulla, but the central organization of this reflex has not been fully analyzed.[93,106]

A rise of pressure in the aorta stimulates receptors in the aortic wall, and impulses that travel upward in a depressor nerve tend to bring blood pressure back to normal again by exerting an inhibitory effect on the heart and causing a diminution of peripheral resistance. This depressor reflex, originally discovered by Cyon and Ludwig in 1866, provided the first known example of a negative feedback mechanism. The carotid and depressor reflexes are mutually supportive. Since they tend to maintain blood pressure within normal limits, we speak of them as barostatic reflexes (see Chapter 42).

The afferents responsible for these reflexes provide the best examples of "autonomic afferents," since they have little effect on the somatic system except for possible interactions in the medulla between barostatic and somatic system reflex pathways. Practically all other afferents evoke both autonomic and somatic responses, with the possible exception of afferents from muscle stretch receptors. Stimulation of large afferents from the muscle spindle has little effect on sympathetic neuron discharge, at least under normal conditions.

Cardiac reflexes

It has been known for many years that a stretch of the right atrium produces acceleration of the heart. It was assumed by Bainbridge that an increase in venous return caused reflex cardiac acceleration by acting through vagus afferents to inhibit vagal discharge and augment accelerator fiber tone. Recent studies have shown that a reflex of this nature is present.[72a] However, stretch of these atrial fibers, the sinoatrial pacemaker in particular, will accelerate the heart in the absence of extrinsic innervation. Stretched isolated nodal tissues accelerate their action, but the mechanism of this is not fully understood.[21]

Stretch of the left atrium has a biphasic action on the heart but no myogenic counterpart. There is an initial inhibitory action followed by an acceleration. The facilitory and inhibitory influences of the right and left atrial reflexes have a generalized influence, affecting tonic and reflex activities in other organs, the kidney most particularly.[70,86] More detailed descriptions of the nature and functions of atrial and ventricular receptors are given in Chapter 42.

Sympathetic reflexes from somatic afferents

Stimulation of the skin, manipulation of the viscera, or any kind of nociceptive and other stimuli resulting from pressure, stretch, temperature changes, etc. applied to various parts of the body will evoke sympathetic reflex actions.[4] Recent studies of sympathetic reflex responses, using action potential recordings directly from sympathetic fibers, have shown that two pathways are involved in the production of sympathetic reflexes.[71,102] Supraspinal reflexes that depend on structures superior to the spinal cord are recognizable because of their long latency (since impulses must travel to the medulla and back) and because they are abolished by disconnection of the brain from the cord. Spinal reflexes are of short latency (Fig. 33-6).

Supraspinal sympathetic reflexes are bilateral and can be evoked from any region of the body. They affect the entire thoracolumbar outflow in a similar manner and have a characteristic "silent period" or "postexcitatory depression" following the initial phase of augmented discharge. It has been found that practically all myelinated fibers, except large group Ia and Ib fibers from muscle and tendon receptors, can evoke both spinal and supraspinal reflexes. Unmyelinated (group IV, C fiber) afferents, when activated, produce sympathetic reflex discharge with a very long latency.[69,104] This reflex has a supraspinal and a spinal component.[59]

It has been known for many years that stimulations of somatic afferents at varying intensities and frequencies produce diverse vasomotor changes. It has now been shown that at low rates of stimulation of myelinated afferents the silent periods tend to fuse in such a fashion that there is a diminution of total discharge during periods of repetitive reflex action. This expresses itself in a depressor response indicating peripheral vasodilatation. More rapid rates of stimulation produce a predominance of active discharge and a pressor response is evoked. Excitation of unmyelinated afferents alone produces a pressor response, even at a low rate of stimulation (less

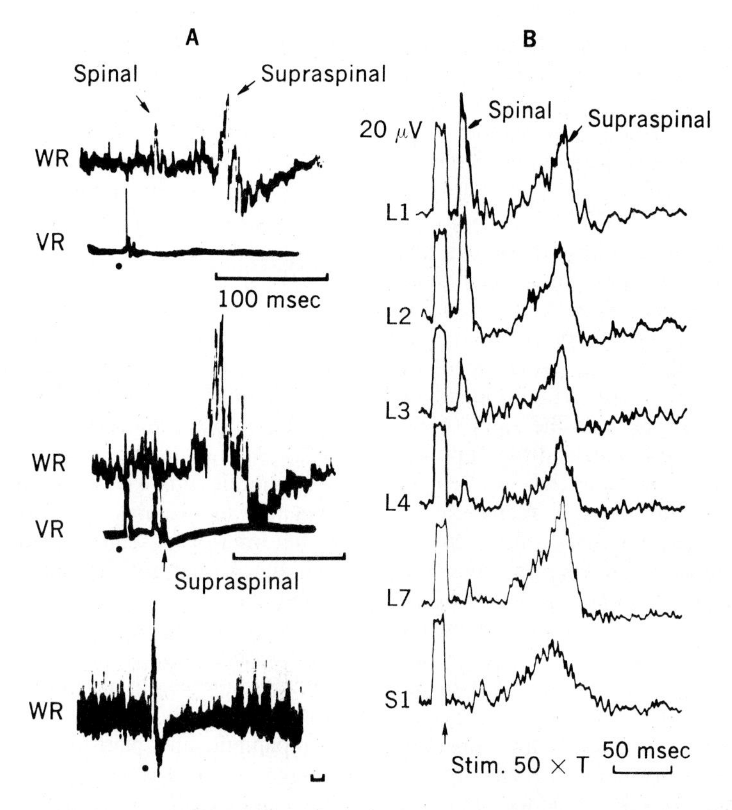

Fig. 33-6. A, Sympathetic reflex responses recorded from lumbar white ramus, *WR*, following single-shock stimulations (marked by dots) of sciatic nerve. Tracings labeled *VR* are somatic reflex responses simultaneously recorded from lumbar seventh ventral root. Sympathetic reflex consists of short-latency spinal and long-latency supraspinal response. Somatic reflex shows mono- and polysynaptic components. In middle tracings, stronger stimulus is shown to evoke greater autonomic reflexes and larger somatic spinal reflex (mono- and polysynaptic reflexes fused in initial response) as well as somatic supraspinal reflex (second large deflection, *VR*). Bottom record was taken at slow-sweep speed from white ramus and shows silent period that follows initial responses.

B, Sympathetic reflex recorded from L1 white ramus (L1WR) but stimuli were applied to spinal nerves at various segmental levels at times indicated by arrows. Each record is average of 10 individual reflexes. Initial rectangular deflection is calibration pulse (20 μV). Early spinal reflex tends to be segmental and is reduced as afferent impulses enter more distant segments, whereas supraspinal reflex remains constant. (**A** from Koizumi et al.[71]; **B** from Sato and Schmidt.[101])

than 1/sec).[69] Stimulation of a mixed population of afferents also tends to produce a pressor response, since the long-latency reflex discharges caused by activation of the unmyelinated afferents fall during the silent period and reverse the inhibitory influence, which results from the excitation of myelinated afferents. Such stimulations of nerve trunks and admixtures of fibers reveal potentialities rather than reactions, which normally are evoked by physiologic stimuli.

Spinal sympathetic reflexes are evoked by stimulation of somatic afferents in intact and decerebrate as well as in spinal preparations. This reflex response has a very short latency, and initial discharges are followed by less conspicuous silent periods. Reflexes are bilateral, but most spinal reflexes are confined to a few segments of the cord, as is the case in spinal somatic reflexes[101] (Fig. 33-6, *B*). These spinal sympathetic reflexes are much more conspicuous in chronic spinal than in normal animals. In conclusion, it should be pointed out that stimuli eliciting sympathetic supraspinal and spinal reflex responses also evoke spinal and supraspinal somatic reflexes.

Chemoreceptors and other visceral receptors

Stimulation of the carotid, aortic, and possibly airway chemoreceptors, mechano- and chemoreceptors in the pulmonary circuit, and enteroreceptors of various types influences autonomic system functions. There are viscerosomatic and viscerovisceral reflexes. All parts of the body are apparently equipped to detect a variety of stimuli that regulate body behavior and the visceral support thereof by autonomic system regulation.

Parasympathetic reflexes

Cranial parasympathetic reflexes do occur, and baroreceptor reflexes can be so classified. Excitation of somatic afferents produces changes in cardiac rate along with pressor or depressor responses; both sympathetic and parasympathetic nerves cooperate in this reflex action. Others are reflex salivation in response to the sight or the presence of food in the mouth and the pupillary constriction induced by shining light in the eye. The light reflex certainly involves reciprocal inhibition of sympathetic pupillary dilator fibers. Vagal parasympathetic effects on secretory glands of the stomach and on the motility of the gastrointestinal wall are reflexly evoked by the smell and taste of appetizing food and by the chemical and mechanical action of food in the tract.

The chief reflexes of the sacral parasympathetic outflow are those controlling the bladder and rectum. Distention of the bladder and the rectum initiate impulses that reach the sacral portion of the spinal cord; these evoke micturition due to contraction of the bladder wall and relaxation of the urethral sphincter muscle and defecation due to contraction of the distal colon and rectum and the relaxation of anal sphincters. Control of the acts of micturition and defecation involves the somatic system as well as the autonomic system; therefore they are described in some detail.

Micturition

Fig. 33-7 is a diagrammatic representation of the motor innervation of the urinary bladder. The sympathetic innervation of the bladder wall (body, neck, and internal sphincter) originates from the upper lumbar segments of the cord. The preganglionic fibers pass down the abdominal sympathetic chains before diverging to form synaptic connections with postganglionic fibers in the inferior mesenteric or inferior hypogastric ganglia. These preganglionic and postganglionic fibers comprise the hypogastric nerves.

The parasympathetic fiber supply comes from the second, third, and fourth sacral segments, and these preganglionic trunks unite on each side to form the pelvic nerves (nervi erigentes, etc.) Sympathetic and parasympathetic nerves intermingle in the hypogastric, pelvic, and vesical plexuses. Short postganglionic parasympathetic fibers that innervate the body and neck of the bladder originate in the pelvic and vesical ganglia.

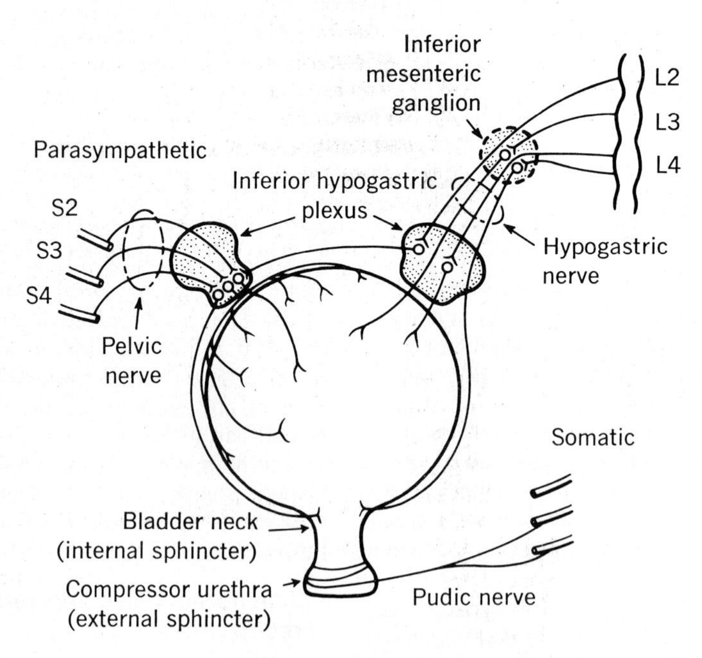

Fig. 33-7. Diagram of somatic and autonomic motor innervation of urinary bladder showing possible interconnections permitting sympathetic-parasympathetic neuron interactions in hypogastric ganglia and plexuses.

The powerful external sphincter, or compressor urethrae, is composed of striated muscle and innervated by the pudendal (pudic) nerves. These sacral somatic nerves originate in the first and second sacral segments of the cord and are under voluntary and reflex control.

The bladder musculature is normally under the influence of low-level tonic discharge from parasympathetic nerves. However, this does not prevent the well-known bladder wall plasticity, which enables this organ to increase greatly in volume without change in intravesicle pressure. Normally the bladder maintains an intravesical pressure below 10 cm H_2O. As the volume increases, the bladder wall accommodates with very little increase in pressure until a volume of 400 to 500 ml (in man) is reached. This is considered to be the micturition threshold under normal circumstances. A desire to void may begin to develop above volumes of 150 ml, but micturition normally does not occur much below volumes of 300 to 400 ml. An increase in volume to 700 ml creates pain and often loss of control. Attainment of a threshold volume initiates a sudden rise in pressure and a micturition reflex. As volume increases and the threshold is approached, small rhythmic or periodic fluctuations in pressure tend to appear. It has been shown that the absolute volume of the bladder is not the stimulus for the micturition reflex but the intravesical tension. The cystometrogram in Fig. 33-8 shows an experimental demonstration of filling without immediate increase in pressure and the ultimate initiation of the micturition reflex in man.

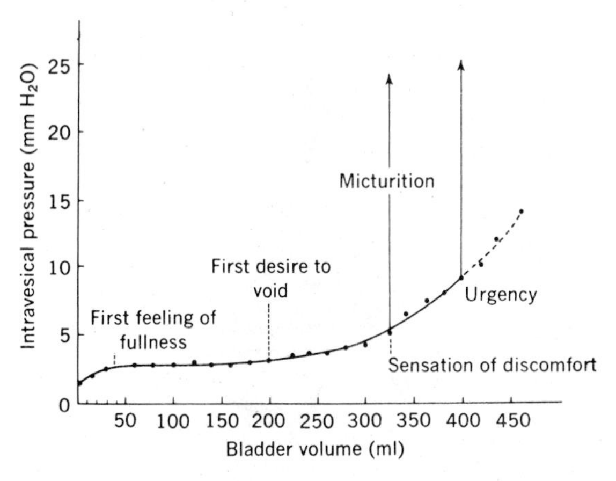

Fig. 33-8. Cystometrogram showing effects of progressive filling of bladder. Note absence of marked change in intravesical pressure until volume of approximately 300 ml is attained. (Modified from Simeone and Lampson.[107])

It has been suggested that the ability of the bladder to accommodate large volumes of fluid without an increase in pressure is not purely myogenic but is due in part to an inhibitory reflex involving its sympathetic innervation.[44,45] Section of the hypogastric nerves as well as the lumbar ganglionic chains in experimental animals and sympathectomy in man produces increased frequency of micturition. The bladder becomes hypertonic and contracts at a lower volume.

In emotional states, bladder tone is frequently raised and the micturition threshold lowered; thus greater frequency of voiding occurs at a lower volume. Sex hormones also have a marked effect on micturition threshold and frequency of urination. Following parturition the bladder may reach an enormous volume before voiding occurs. This may be the result of an estrogen-progesterone "deprivation" effect.

The events of the micturition reflex are as follows: When, through distention, intravesical pressure reaches a critical level, stretch receptors in the bladder wall initiate afferent impulses that travel mostly up the pelvic nerves, ultimately reaching the sacral cord. There they evoke reflex contraction of the detrusor muscle and relaxation of the internal sphincter, chiefly through the pelvic nerves. Fluid flowing through the urethra also causes a bladder contraction through parasympathetic nerves, a phenomenon that is important for continuation of micturition. Higher centers, however, are essential to the normal micturition reflex, which involves relaxation of the external sphincter as well as bladder contraction. Facilitory as well as inhibitory influences on micturition are exerted by structures in the midbrain, pons, and medulla. These "centers" integrate normal association of bladder contraction and internal and external sphincter relaxation. They are also responsible for the very important cooperative action of associated muscles. In voluntary micturition there is normally relaxation of the perineal and levator ani and contraction of abdominal muscles and of the diaphragm.

Sympathetic nerves are not essential to micturition. The force of contraction is largely generated by the parasympathetic supply. Pressures as high as 150 cm H_2O normally develop during evacuation. The powerful abdominal and diaphragmatic muscles that are involved can generate still higher pressure to overcome any resistance to flow. Hypertrophy of bladder musculature can occur in cases of obstruction.

Higher centers have some control over the brain stem–mediated automatic reflex actions.

The cortex, which is assumed to be responsible for voluntary directives, can prevent these reflexes from occurring and maintain a closed external sphincter. This voluntary control actually inhibits the parasympathetic reflex and is not exerted solely on the external sphincter. After transection of the cord at levels above the sacral segments, a degree of bladder paralysis results. Low sacral reflexes may eventually be reestablished, but they do not completely empty the bladder. Residual urine remains, and bladder infections easily develop. In the "mass reflexes" of chronic spinal animals and man, micturition as well as defecation may occur along with somatic and other autonomic discharges. Destruction of the sacral cord or section of the external innervation to the bladder produces an "automatic bladder." The bladder musculature acquires some intrinsic tone and automatic action. Periodic voiding may occur. More elaborate analyses of the bladder reflex and its central control can be found in various reviews.[34,78,99]

Defecation

Fig. 33-9 is a diagram of the innervation of the colon and rectum and shows the pathways involved in the defecation reflex. This external innervation and the intrinsic myenteric plexus of the colon wall are important to the initiation and control of the activity of the colon.

The sympathetic supply delivered through hypogastric nerves and lumbar colonic nerves tends to inhibit activity in the colon but causes the internal sphincter to contract. The parasympathetic supply from the sacral cord through the pelvic nerves has an excitatory action on motility of the colon and an inhibitory action on the internal anal sphincter. It was once thought that megacolon occasioned by relaxation and failures of the viscus to empty might be due to overaction of the sympathetic innervation, but it is now thought to be due to degeneration within Auerbach's myenteric plexus.

Activity within the colon is essential to defecation, since peristaltic activity within it forces material into the rectum. The internal and external anal sphincters prevent escape of the accumulated fecal mass. Tonic contraction of the external sphincter is maintained by the pudendal (pudic), which is a somatic nerve; thus this sphincter is under voluntary control.

The sequence of action in defecation is as follows: Distention of the colon initiates, through afferents in the pelvic nerves, a reflex that opens the internal anal sphincter by inhibiting its tone. Simultaneous contraction of the colon distends the rectum and propels the fecal material into the rectum. The desire to defecate is felt when the rectum is considerably distended and feces begin to be propelled through it by mass peristaltic movement. The sympathetic nerves do not contribute to this reflex, but their activity may be

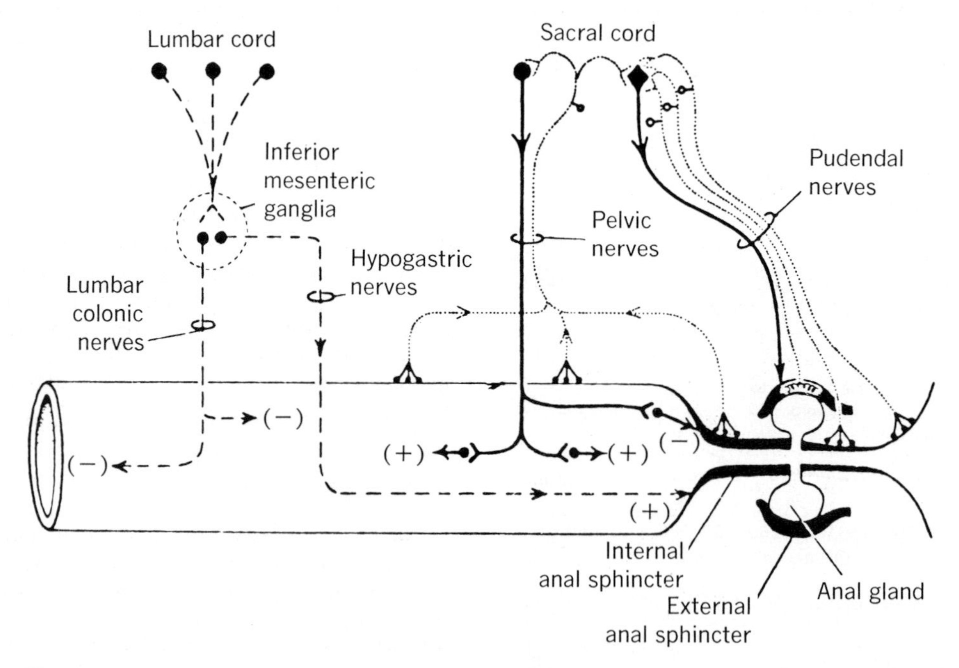

Fig. 33-9. Diagram of innervation of distal colon, rectum, and anal sphincters of cat. (From Garry et al.[52])

inhibited reciprocally as the parasympathetic outflow comes into action. The distention of the rectum and associated stimuli evoke a reflex opening of the anal sphincter and associated contraction of abdominal muscles and diaphragm, much as in the case of the micturition reflex. Although higher centers can inhibit various elements in the defecatory reflex, the act of defecation does not depend on higher centers but rather on lower sacral segmental levels. Destruction of the sacral cord does not prevent defecation, and reflexlike contractions of the bowel seem to be initiated by the peripheral neural plexus. However, the sacral cord segments and parasympathetic fibers are essential to a strong and effective defecatory reflex.[105]

Sexual reflexes

The major remaining sacral parasympathetic reflexes are those resulting in erection and vasodilatation in the external genitalia. Under conditions of sexual excitation the discharge of parasympathetic fibers causing erection is unquestionably of cerebral origin. However, erection can occur in the spinal animal, and it can be evoked reflexly by genital stimulation.

It appears that the action of parasympathetic erector fibers of the pelvic nerves can be supplemented by sympathetic vasodilator nerves to the penis. The sympathetic fibers are still capable of producing erection in the absence of the parasympathetic component, but of the two supplies the parasympathetic is the more potent in its effect. Complete sympathectomy does not significantly alter the erectile response, although it does prevent emission of semen in the sexual act.[8,14]

AUTONOMIC CONTROL OF RECEPTORS

Physiologists have long been aware of the fact that reactions occur that affect sensory perception.[47] Reflex adjustments of the pupil that are autonomically mediated affect visual acuity. There are reactions that modify the acuteness of hearing by actions on the tensor tympani. Only in recent years, however, have we become fully aware of feedback mechanisms that modify the responsiveness of sensory receptors. The best known example of this is the control of muscle spindles by the fusimotor fiber (gamma efferents) system.

Recent evidence that has begun to accumulate shows that there is an autonomic supply to receptors. These components of the autonomic system appear in some cases to affect sensitivity and response characteristics of the receptors. As in any new field, however, there is a good deal of controversy; evidence for and against various hypotheses cannot be fully presented here, but the best examples of this process will be described.

Chemoreceptors. There is a full description of the carotid body and its role as a chemoreceptor in the control of respiration in Chapter 70. The blood flow through the carotid body and oxygen uptake by cells of the glomus are very high. It is also known that sympathetic fibers innervate the carotid body; these are postganglionic fibers originating in the superior cervical ganglia. When the cervical sympathetic trunk is stimulated, blood vessels in the carotid body and surrounding tissues are constricted and there is a decrease in blood flow to this area. This tends to produce augmentation in chemoreceptor activity as estimated from the rate of afferent discharges in the carotid nerve. This mechanism is not a part of normal respiratory control, and section of the cervical sympathetic nerves does not affect respiration nor the rate of chemoreceptor afferent discharge. However, during hemorrhage, asphyxia, or other conditions such as strenuous exercise that greatly augment generalized sympathetic system activity the increase in chemoreceptor discharges is due in part to sympathetic nerve action on the carotid body. Evidence supporting this conclusion is provided by the observation that in normal animals there is a quick rise in chemoreceptor activity and tidal volume within 1 to 2 sec after exercise is begun. If cervical sympathetic nerves are cut, this rise is much delayed and it takes 15 to 20 sec before a significant change is detectable.[98]

This seeming "sensitization" of the chemoreceptors by sympathetic nerve activity may be due entirely to blood flow changes and a decreased oxygen supply, but the possibility of a direct catecholamine action also cannot be excluded, since norepinephrine and epinephrine applied to the carotid boyd cause the rate of discharge to increase.

Mechanoreceptors. A sympathetic influence on the mechanoreceptor of the frog skin has been demonstrated. Stimulation of sympathetic fibers supplying this receptor in vitro lowers the threshold and slows adaptation to stimuli. Epinephrine directly applied has a similar effect, and these changes have been found to be due to an increase in amplitude and rate of rise of generator potential.[87,88] A similar action on the pacinian corpuscles of the mammalian mesentery and mesocolon has been reported.[103]

Baroreceptors. Sympathetic fibers innervate the carotid sinus as well as the carotid body. There is some evidence that this innervation may affect the sensitivity and response of baroreceptors. Stimulation of the sympathetic fibers to the sinus or reflex augmentation of their activity increases the rate of discharge in baroreceptor afferents.[16,66,100] It is an attractive idea that intensified sympathetic activity, causing a rise in blood pressure, might also sensitize those receptors that tend to restore normality.

Muscle spindles. Sympathetic fibers innervate muscle spindles. Repetitive stimulation of these nerves increases the afferent discharge from this receptor.

Epinephrine has effects similar to those of sympathetic stimulation.[40,61] To what degree a vasomotor effect is involved in this change in response has not been settled.

Finally, there is work showing that a sympathetic innervation may influence olfactory and taste receptors.[30]

CENTRAL CONTROL OF AUTONOMIC FUNCTIONS

The autonomic nervous system is composed of efferent fibers supplying practically all tissues of the body. It affects not only those organs generally described as viscera, but also somatic sensory receptors, adipose tissues,[11] blood vessels throughout the body, and even the endocrine glands. One recent development has been the discovery of an optic-hypothalamic-sympathetic-pineal-neuroendocrine loop that governs certain body functions[95] (Chapter 34). Another rapidly developing interest relative to control of tissue and organ functions has been the study of interactions within the autonomic ganglia[64,85] and between sympathetic and parasympathetic nerve terminals within organs and plexuses.[84]

Although these peripheral interactions are important to control, it is the CNS that regulates activity of the autonomic systems, body functions, body states, and the complex of behavioral reactions aroused by environmental impingements.

Earlier, when knowledge of the CNS was less extensive than it is now, it was thought that there were autonomic centers or special regions within the CNS that were responsible for the control of autonomic functions. It is true that there are central regions that are more important to the mediation of autonomic reflex action and tonic activity than are others. But these are parts of total complexes that integrate somatic and autonomic reactions in meeting behavioral requirements, and it is no longer feasible to think in terms of single centers that are solely responsible for specific actions.

Autonomic and somatic reactions cooperate to maintain essential body states and to effect necessary adaptations. Emphasis is now placed on the interrelationships of centers and the organization of the nervous system as a whole. There is, however, a hierarchy in the control mechanism, and by using experimental procedures or observing the consequences of injury and spontaneously occurring abnormalities, one can observe the subdivisions of the patterns of control, which parallel those seen in the somatic system. For the purpose of analysis the role played by various divisions of the CNS in the regulation of autonomic system activity will be discussed in the following sections, although separation of somatic and autonomic control systems is arbitrarily made.[12]

Spinal cord

Transection of the spinal cord causes a state of "shock." Spinal shock is characterized by the absence of reflexes, by low blood pressure, and by other evidence of complete or partial paralysis of all motor systems, both somatic and autonomic (Chapter 28). Eventually, some degree of activity returns, blood pressure rises to normal, and patterns of somatic and autonomic responses can be evoked. Within a few days after transection of the cord at the sixth cervical level in lower forms the isolated cord is able to mediate reflex action in response to afferent nerve stimulation. Increments in blood pressure and heart rate, together with the contraction of nictitating membranes and evidence of adrenomedullary secretion (increases in blood sugar level and decreases in clotting time), can be evoked.[19] These changes, however, are not as great as those obtained in anesthetized normal or unanesthetized decerebrate animals when the same stimuli are applied. As stated previously, more refined techniques have shown that spinal reflexes are segmental and tend to be localized unless massive stimuli are used. Generalized responses are very difficult to evoke in the absence of higher centers, particularly the medulla. Tonic discharges recorded from preganglionic sympathetic fibers are greatly diminished but not completely abolished by cord transection. Tonic activity returns to virtually normal levels within a few days in lower forms; this is responsible at least in part for recovery of blood pressure. Destruction of the cord of the spinal animal again reduces the blood pressure to very low levels, an indication that the peripheral resistance is maintained by spinal sympathetic activity.

Deficiency and release phenomena are seen after cord transection. Failure of the bladder to empty completely is due to an ineffective reflex response or a failure in the coordination of the reactions that are normally organized by higher centers. This is a deficiency. The "mass reflex" best seen in spinal man is an example of a release phenomenon.

In man, spinal shock may last as long as 2 months. Autonomic reflexes are even more completely suppressed during shock than are the somatic reactions. During the first month or two the skin is completely dry and sweating is absent, but skin surfaces may be warm and pink because of vasodilatation. Somatic reactions apparently recover first. Flexor reflexes may be elicited within a few days after injury, and by the end of the third or fourth week the withdrawal reflex tends to become more vigorous. The mass reflex that requires several months to develop is elicited by rather weak stimulation or may even occur spontaneously without obvious stimulation. If the plantar surface is scratched, both legs may withdraw violently. The patient may sweat profusely, and both bladder and rectum may contract. Such flexor contractions and their autonomic concomitants are very disturbing and may even interrupt sleep.

A number of explanations have been offered for the hyperreflexia that eventually develops in chronic spinal preparations. Obviously, reciprocal inhibition from higher centers is not present to provide feedback control. It has also been claimed that new collaterals may sprout from dorsal root fibers to produce new synaptic connections in the absence of trophic influence from higher centers. Finally, these phenomena may be examples of denervation hypersensitivity. It has been reported that in the semitransected cord the ipsilateral reflex has greater responsiveness to afferent impulses, ACh, and other excitant drugs than does the contralateral reflex of the normal side. There is convincing evidence that postganglionic fibers become hypersensitive to transmitters and to drugs when preganglionic neurons are cut. It is reasonable to assume that preganglionic neurons develop a similar sensitivity when they are deprived of their innervation from medullary and other higher centers.

Medulla oblongata

The major reflex actions of the somatic and autonomic system depend on the medulla.

The function of the medulla in the regulation of heart action first became known about 1845. In 1873 Dittmar demonstrated its role by showing that transection of the brain stem above the midpontine region had little effect on blood pressure. Transection further down in the medulla or below the medulla caused a precipitous fall in arterial pressure. In 1916 Ranson and Billingsley found that both pressor and depressor responses could be elicited by stimulating various areas in the medulla. Subsequent work defined pressor and depressor areas in this structure[7] (see Chapter 42). Studies of other reactions involving the autonomic system revealed major dependencies on the medulla such as reflex changes in blood sugar level, reflex excitation of the adrenal medulla, production of salivation, and the initiation of vomiting, swallowing, and micturition.

The baroreceptor reflex illustrates the importance of the medulla to reciprocal control of the cardiovascular system. Respiratory center activity and the respiratory reflexes affect the autonomic system. Control of respiration is mediated largely through the medulla, and the central complex responsible lies close to the central mechanisms regulating autonomic system functions.

In summary, the medulla oblongata is chiefly responsible for tonic activities of the autonomic system and their normal modulations. It also organizes reciprocal actions between the two divisions of the system and maintains the potential for initiating a massive discharge of the sympathoadrenal system. The bulbospinal preparation has very considerable deficiencies in its responses, and the normality of integrated reactions of somatic and autonomic systems requires the presence of higher centers.

Midbrain

It should be pointed out that the brain functions as a whole and that the subdivisions described are defined by purely arbitrary decision and chiefly on the grounds of anatomic configurations and structural features. The midbrain is a narrow isthmus in which ascending and descending fiber tracts are abundant; it does contain extensions of the RF and discrete nuclear structures.

Most of the evidence indicating that the midbrain is involved in the regulation of autonomic function is based on stimulation and ablation experiments. It plays an important role in control of the urinary bladder; inhibition and facilitation of micturition can be effected by stimulation of this region. Injury to the midbrain interferes with the normality of micturition. Changes in skin resistance due to sweating or vasomotor reactions, namely, the galvanic skin reflex, are also affected by midbrain lesions. Stimulation of the midbrain RF causes modifications of the blood pressure and heart rate with or without obvious modification of somatic responses.

The difficulty of estimating the role of this region in regulation of the autonomic nervous system is due to the fact that neurons from the higher centers that regulate the system must traverse this region on their way to the medullary or spinal integrative centers. One of these higher centers that plays a major role in the control of autonomic functions is the hypothalamus.

Hypothalamus

The hypothalamus has been referred to frequently as the center for regulation of the autonomic system. It is certainly of great importance as an integrator of somatic, autonomic, somatoautonomic, and endocrine functions. It confers abilities essential to the adjustment of body states, but these are considerably refined by higher centers. The hypothalamus integrates responses to heat and cold, it mediates those reactions that maintain homeostasis, it plays an essential role in the resistance to stress, and it organizes the visceromotor and other reactions of rage and emotional expression. The hypothalamus is the region of the brain that possesses the means to relate humoral and neural control of body processes. Control of the endocrine system is affected by regulation of the secretory activity of the hypophysis. In recent years there has been a growing appreciation of the major role played by the hypothalamus in regulating functions of the hypophysis and endocrine and exocrine glands. The role of the hypothalamus in integration and control of body functions is discussed in Chapter 34.

Subsequent to removal of those brain structures superior to the hypothalamus, autonomic functions remain relatively adequate. Refinements of control, however, are lacking, and reactions do not attain normal levels of effectiveness. Temperature regulation, for example, is somewhat deficient; thresholds of body temperature–regulating reactions are high, and responses tend to overshoot. Other centers act on the hypothalamus directly or in association with the hypothalamus to influence autonomic system functions.

Limbic system

The limbic system has a close anatomic and functional relationship to the hypothalamus. It is intimately concerned not only with emotional expression but probably also with the genesis of emotions. It does play a role in the control of the autonomic system. Evidence of this has been provided by stimulation as well as ablation experiments.[115]

This system consists of the limbic cortex (orbitofrontal areas, cingulate gyrus, hippocampal gyrus, and pyriform area) and the structures with which it has primary connections, namely, the preoptic area, septum, amygdala, and adjacent areas. The hypothalamus is closely connected with this system and is included in it by some investigators.

Stimulation of the amygdala produces a mixture of parasympathetic and sympathetic reactions as well as hormone secretions (gonadotropins and corticotropins) and respiratory responses. There have been reports of changes in arterial pressure, cardiac rate, gastrointestinal motility and secretion evoked by stimulation of this structure. Such stimuli also can produce defecation and micturition, pupillary dilation and constriction, and piloerection. Sexual activity and ragelike reactions can be evoked from the amygdala, although mediation by the hypothalamus apparently is involved. Stimulation of the hippocampus likewise elicits involuntary movements, rage reactions, sexual phenomena, and hyperexcitability. Application of stimuli to the septum affects blood pressure and apparently produces a "sense of reward" that is basic to the well-known drive for "self-stimulation." Such effects, however, are not elicited exclusively from the septum.

Ablation of the amygdala produces placidity, and thus this structure is thought by some to be involved in the control of emotions. Injury to the septum likewise reduces susceptibility to fear and anxiety. These structures are said to have the power to evoke the autonomic components of behavioral reactions.

Stimulation of the limbic cortex produces visceral effects, such as vasomotor changes and gastrointestinal activity, as well as somatic and autonomic reactions that typify excitement and rage. Removal causes behavioral changes, a loss of emotional expression, and associative powers.

The limbic system plays a role in olfaction and the regulation of feeding behavior; it apparently is involved in sensing of pain and contains high concentrations of the newly discovered opiate receptors. In addition, it is concerned with motivation, the control of sexual behavior, and the expression of rage and fear. It exerts control over the autonomic system, which is superimposed on that organized by the hypothalamus.

Cerebellum

The cerebellum is also involved to some degree in control of the autonomic nervous system.[38,108]

Stimulation of both anterior and posterior lobes produces changes in blood pressure and heart rate. Pupillary constriction and dilation as well as changes in intestinal and colonic motility also result from cerebellar stimulation. Chemical stimulation of the posterior lobe by local application of glutamate augments both sympathetic and parasympathetic tonic discharges.[68] Stimulation of the fastigial nuclei produces reactions of the cardiovascular system, bladder, colon, and intestine, also indicating a cerebellar involvement in control of the autonomic system.[6,90] Injury to or ablation of the cerebellum modifies autonomic reactions. The bladder reflexes are markedly changed, and an increase in frequency of micturition is reported, at least in some species. The refinements of cardiovascular regulation may also be impaired.

The significance of this cerebellar autonomic representation is not fully understood, and all that can be said at the present time is that in addition to regulating somatic behavior, the cerebellum can also modulate functions within the autonomic nervous system. During motion sickness there is salivation, vomiting, sweating, and other evidence of abnormal autonomic action. It has been shown that removal of the cerebellum abolishes experimentally produced motion sickness (Chapter 31), a result that may also be interpreted to indicate some involvement of the cerebellum in control of autonomic system function.

Cerebral cortex

The cerebral cortex, neocortex to be specific, contributes a refinement of control over autonomic as well as somatic system reactions. The significance of the role of the cerebral cortex varies remarkably within the mammalian kingdom. In man and the primates, it has a major responsibility, and its removal abolishes reactions that occur quite adequately in lower species following decortication.

Autonomic reactions are much less affected by cortical lesions in man than are somatic activities. There is evidence, however, of a higher degree of cephalization of control of the autonomic system in man than in lower forms. Certainly the cortex plays a role in elicitation of autonomic involvements in emotional reactions and in the support of voluntarily selected patterns of response. Evidence that the cerebral cortex is involved in the control of autonomic nervous system function has been accumulating for well over 100 years. Much information has been provided by observation of the effects of injury or ablation and by experiments utilizing electrical stimulation techniques. It is claimed also that a considerable degree of voluntary control over cardiovascular activities can be established by operant conditioning.[17]

Ablation experiments have shown that removal of the cerebral cortex produces at least minimal abnormalities in many autonomic reactions. Micturition frequency is changed, and temperature regulation is altered to the extent that compensatory autonomic reactions are delayed in onset and termination. Ablation of the frontal cortex definitely interferes with the normality of blood pressure control, and the occurrence of both hypertension and hypotension has been reported. In the absence of the cerebral cortex there is loss of refined control of autonomic reactions.

A survey of stimulation studies[5,58] reveals that there are discontinuous loci on the dorsolateral surfaces of both hemispheres that, on stimulation, yield a rise or fall of blood pressure and an increase or decrease in heart rate. Stimulation of other areas can evoke dilation of the pupils, retraction of the nictitating membrane, piloerection, salivation, sweat secretion, and secretory and motor functions of the stomach. For the most part, autonomic localization in the motor and premotor regions corresponds closely with related somatic representations.

A significant role of the cortex in regulation of the autonomic system is that of controlling the distribution of blood to various parts of the body. For example, blood can be shunted from the renal circuit to the limbs by cortical stimulation, and it is also known that the cortex is involved in shunting the blood to and from the skin as an aspect of temperature regulation. Unquestionably the cerebral cortex plays an important role in adjustments of autonomic system function.

CONCLUSION

The various parts of the CNS play specific roles in the control of somatic and autonomic reactions. Students of neurophysiology should seek to identify the exact sites and pathways of control, but they also should bear in mind that every part of the CNS relates to all other parts directly or indirectly. Some regions are primarily responsible for specific functions, but that does not mean that other structures are not involved.

The purpose of any textbook in a field, it seems to us, is to review its significant features and present a conceptual approach. The arguments for and against specific conclusions, the evidence provided by experimentation, and the contents of recent papers cannot be included. However, readers must be informed of the problems that exist and the apparent current directions of interest. There are numerous reviews and monographs that may be consulted by those who wish to pursue this subject.[1-5,39,50,73]

REFERENCES

General reviews

1. Brooks, C. M., Koizumi, K., and Pinkston, J. O., editors: The life and contributions of Walter Bradford Cannon 1871-1945. His influence on the development of physiology in the twentieth century, Albany, 1975, State University of New York Press.
2. Cannon, W. B.: The wisdom of the body, ed. 2, New York, 1939, W. W. Norton & Co., Inc.
3. Ingram, W. R.: Central autonomic mechanisms. In Field, J., editor: Handbook of physiology. Section 1, Baltimore, 1960, The Williams & Wilkins Co., vol. 2, chap. 37.
4. Koizumi, K., and Brooks, C. M.: The integration of autonomic system reactions: a discussion of autonomic reflexes, their control and their association with somatic reactions, Ergeb. Physiol. **67**:57, 1972.
5. Monnier, M.: Functions of the nervous system. General physiology: autonomic functions (neurohumoral regulations), Amsterdam, 1968, Elsevier Publishing Co.

Original papers

6. Achari, N. K., Al-Ubaidy, S., and Downman, C. B. B.: Cardiovascular responses elicited by fastigial and hypothalamic stimulation in conscious cats, Brain Res. **60**:439, 1973.
7. Alexander, R. S.: Tonic and reflex functions of medullary sympathetic cardiovascular centers, J. Neurophysiol. **9**:205, 1946.
8. Appenzeller, O.: The autonomic nervous system, an introduction to basic and clinical concepts, ed. 2, Amsterdam, 1976, North Holland Publishing Co.
9. Axelrod, J.: noradrenaline: fate and control of its biosynthesis, Science **173**:598, 1971.
10. Axelsson, J.: Catecholamine functions, Annu. Rev. Physiol. **33**:1, 1971.
11. Ballard, K., Malmfors, T., and Rosell, S.: Adrenergic innervation and vascular patterns in canine adipose tissue, Microvascular Res. **8**:164, 1974.
12. Bard, P.: The central representation of the sympathetic system, Arch. Neurol. Psychiatry **22**:230, 1929.
13. Bealleau, B.: Steric effects in catecholamine interactions with enzymes and receptors, Pharmacol. Rev. **18**: 131, 1966.
14. Bell, C.: Autonomic nervous control of reproduction: circulatory and other factors, Pharmacol. Rev. **24**:657, 1972.
15. Bennett, M. R., and Burnstock, G.: Electrophysiology of the innervation of intestinal smooth muscle. In Code,

C. F., editor: Handbook of physiology. Section 6, Baltimore, 1968, The Williams & Wilkins Co., vol. 4, chap. 84.
16. Bolter, C. P., and Ledsome, J. R.: Effect of cervical sympathetic nerve stimulation on canine carotid sinus reflex, Am. J. Physiol. **230:**1026, 1976.
17. Brener, J.: Factors influencing the specificity of voluntary cardiovascular control. In DiCara, L. V., editor: Limbic and autonomic nervous systems research, New York, 1974, Plenum Publishing Corp., p. 335.
18. Bronk, D. W., Ferguson, L. K., Margaria, R., and Solandt, D. Y.: The activity of the cardiac sympathetic centers, Am. J. Physiol. **117:**237, 1936.
19. Brooks, C. M.: Reflex activation of the sympathetic system in the spinal cat, Am. J. Physiol. **106:**251, 1933.
20. Brooks, C. M.: Chemical mediation of the neural control of peripheral organs and the humoral transmission of mediators. In Brooks, C. M., Gilbert, J. L., Levey, H. A., and Curtis, D. R., editors: Humors, hormones, and neurosecretions, Albany, 1962, State University of New York Press.
21. Brooks, C. M., and Lu, H. H.: Sinoatrial pacemaker of the heart, Springfield, Ill., 1972, Charles C Thomas, Publisher.
22. Burnstock, G.: Purinergic nerves, Pharmacol. Rev. **24:** 509, 1972.
23. Burnstock, G., and Bell, C.: Peripheral autonomic transmission. In Hubbard, J. L., editor: The peripheral nervous system, New York, 1974, Plenum Publishing Corp., p. 277.
24. Cannon, W. B.: Bodily changes in pain, hunger, fear and rage, ed. 2, New York, 1929, Appleton-Century-Crofts.
25. Cannon, W. B.: A law of denervation, Am. J. Med. Sci. **198:**737, 1939.
26. Cannon, W. B., and Bacq, Z. M.: Studies on the conditions of activity in endocrine organs. XXVI. A hormone produced by sympathetic action on smooth muscle, Am. J. Physiol. **96:**392, 1931.
27. Cannon, W. B., and Rosenblueth, A.: Autonomic neuro-effector systems, New York, 1937, The Macmillan Co.
28. Cannon, W. B., and Rosenblueth, A.: The supersensitivity of denervated structures: a law of denervation, New York, 1949, The Macmillan Co.
29. Cannon, W. B., et al.: Some aspects of the physiology of animals surviving complete exclusion of sympathetic impulses, Am. J. Physiol. **89:**84, 1929.
30. Chernetski, K. E.: Sympathetic enhancement of peripheral sensory input in the frog, J. Neurophysiol. **27:**493, 1967.
31. Dale, H. H.: The action of certain esters and ethers of choline, and their relation to muscarine, J. Pharmacol. Exp. Ther. **6:**147, 1914.
32. Dale, H. H.: Chemical transmission of the effects of nerve impulses, Br. Med. J. **1:**834, 1934.
33. Dale, H. H.: Transmission of nervous effects by acetylcholine, Harvey Lect. **32:**229, 1937.
34. de Groat, W. C.: Nervous control of the urinary bladder of the cat, Brain Res. **87:**201, 1975.
35. de Groat, W. C., and Ryall, R. W.: Reflexes to sacral parasympathetic neurons concerned with micturition in the cat, J. Physiol. **200:**87, 1969.
36. de Robertis, E.: Molecular biology of synaptic receptors, Science **171:**963, 1971.
37. Douglas, W. W.: Stimulus-secretion coupling: the concept and clues from chromaffin and other cells, Br. J. Pharmacol. **34:**451, 1968.
38. Dow, R., and Morruzzi, G.: Physiology and pathology of the cerebellum, Minneapolis, 1958, University of Minnesota Press.
39. Eichna, L. W., and McQuarrie, D. G., editors: Central nervous control of circulation, Physiol. Rev. **40**(suppl. 4):1, 1960.
40. Eldred, E., Schnitzlein, H., and Buchwald, J.: Response of muscle spindles to stimulation of sympathetic trunk, Exp. Neurol. **2:**13, 1960.
41. Elfvin, L. G.: The ultrastructures of the superior cervical sympathetic ganglion in the cat, J. Ultrastruct. Res. **8:**403, 1963.
42. Emmelin, N.: Nervous control of salivary glands. In Code, C. F., editor: Alimentary canal section: handbook of physiology, Baltimore, 1967, The Williams & Wilkins Co., vol. 2, chap. 37.
43. Emmelin, N., and Trendelenburg, U.: Degeneration activity after parasympathetic or sympathetic denervation, Ergeb. Physiol. **66:**147, 1972.
44. Evardsen, P.: Nervous control of urinary bladder in cats. I. The collecting phase, Acta Physiol. Scand. **72:** 157, 1968.
45. Evardsen, P.: Nervous control of urinary bladder in cats. II. The expulsion phase, Acta Physiol. Scand. **72:** 172, 1968.
46. Fernandez de Molina, A., Kuno, M., and Perl, E. R.: Antidromically evoked responses from sympathetic preganglionic neurons, J. Physiol. **180:**321, 1965.
47. Ferry, C. B.: Cholinergic link hypothesis in adrenergic neuroeffector transmission, Physiol. Rev. **46:**420, 1966.
48. Folkow, B.: Impulse frequency in sympathetic vasomotor fibres correlated to the release and elimination of the transmitter, Acta Physiol. Scand. **25:**49, 1952.
49. Folkow, B.: Nervous control of blood vessels, Physiol. Rev. **35:**629, 1955.
50. Folkow, B., and Neil, E.: Circulation, New York, 1971, Oxford University Press, Inc.
51. Furchgott, R. I.: The receptors for epinephrine and norepinephrine (adrenergic receptors), Pharmacol. Rev. **11:**429, 1959.
52. Garry, R. C., Bishop, B., Roberts, T. D. M., and Todd, J. K.: Control of external sphincter of the anus in the cat, J. Physiol. **134:**230, 1956.
53. Gerich, J. E., Charles, M. A., and Grodsky, G. M.: Regulation of pancreatic insulin and glucagon secretion, Annu. Rev. Physiol. **38:**353, 1976.
53a. Goodman, L. S., and Gilman, A.: Pharmacological basis of therapeutics, ed. 5, New York, 1975, Macmillan Publishing Co., Inc.
54. Grosse, M., and Jänig, W.: Vasoconstrictor and pilomotor fibres in skin nerves to the cat's tail, Pfluegers Arch. **361:**221, 1976.
55. Hebb, C.: CNS at the cellular level: identity of transmitter agents, Annu. Rev. Physiol. **32:**165, 1970.
56. Hendry, I. A.: Control in the development of the vertebrate sympathetic nervous system. In Ehrenpreis, S., and Kopin, I. J., editors: Reviews of neuroscience, New York, 1976, Raven Press, vol. 2, p. 149.
57. Heymans, C., and Neil, E.: Reflexogenic areas of the cardiovascular system, Boston, 1958, Little, Brown & Co.
58. Hoff, E. C., Kell, J. F., and Carroll, M. N.: Effects of cortical stimulation and lesions on cardiovascular function, Physiol. Rev. **43:**68, 1963.
59. Horeyseck, G., and Jänig, W.: Reflex activity in postganglion fibres within skin and muscle nerves elicited by somatic stimuli in chronic spinal cats, Exp. Brain Res. **21:**155, 1974.

60. Horeyseck, G., Jänig, W., Kirchner, F., and Thämer, V.: Activation and inhibition of muscle and cutaneous postganglionic neurons to hind limb during hypothalamically induced vasoconstriction and atropine-sensitive vasodilation, Pfluegers Arch. **361:**231, 1976.
61. Hunt, C.: Effect of sympathetic stimulation on mammalian muscle spindles, J. Physiol. **151:**332, 1960.
62. Iggo, A., and Leek, B. F.: An electrophysiological study of single vagal efferent units associated with gastric movements in sheep, J. Physiol. **191:**177, 1967.
63. Iriuchijima, J., and Kumada, M.: Activity of single vagal fibers efferent to the heart, Jpn. J. Physiol. **14:** 479, 1964.
64. Jacobowitz, D. M.: The peripheral autonomic system. In Hubbard, J. I., editor: The peripheral nervous system, New York, 1974, Plenum Publishing Corp., p. 87.
65. Jänig, W., and Schmidt, R.: Single unit responses in the cervical sympathetic trunk upon somatic nerve stimulation, Pfluegers Arch. **314:**199, 1970.
66. Koizumi, K., and Sato, A.: Influence of sympathetic innervation on carotid sinus baroreceptor activity, Am. J. Physiol. **216:**321, 1969.
67. Koizumi, K., and Sato, A.: Reflex activity of single sympathetic fibres to skeletal muscle produced by electrical stimulation of somatic and vasodepressor afferent nerves in the cat. Pfluegers Arch. **332:**283, 1972.
68. Koizumi, K., and Suda, I.: Induced modulation in autonomic efferent neuron activity, Am. J. Physiol. **205:** 738, 1963.
69. Koizumi, K., Collin, R., Kaufman, A., and Brooks, C. M.: Contribution of unmyelinated afferent excitation of sympathetic reflexes, Brain Res. **20:**99, 1970.
70. Koizumi, K., Ishikawa, T., Nishino, H., and Brooks, C. M.: Cardiac and autonomic system reactions to stretch of the atria, Brain Res. **87:**247, 1975.
71. Koizumi, K., Sato, A., Kaufman, A., and Brooks, C. M.: Studies of sympathetic neuron discharges modified by central and peripheral stimulation, Brain Res. **11:**212, 1968.
72. Koizumi, K., Seller, H., Kaufman, A., and Brooks, C. M.: Pattern of sympathetic discharges and their relation to baroreceptor and respiratory activities, Brain Res. **27:**281, 1971.

72a. Kollai, M., et al.: Study of cardiac sympathetic and vagal activity during reflex responses produced by stretch of the atria, Brain Res. **150:**519, 1978.

73. Korner, P. I.: Integrative neural cardiovascular control, Physiol. Rev. **51:**312, 1971.
74. Krnjević, K.: Chemical nature of synaptic transmission in vertebrates, Physiol. Rev. **54:**418, 1974.
75. Kuffler, S. W., Dennis, M. J., and Harris, A. J.: The development of chemosensitivity in extrasynaptic areas of the neuronal surface after denervation of parasympathetic ganglion cells in the heart of the frog, Proc. R. Soc. Lond. (Biol.) **177:**555, 1971.
76. Kuntz, A.: The autonomic nervous system, ed. 4, Philadelphia, 1953, Lea & Febiger.
77. Kunze, D. L.: Reflex discharge patterns of cardiac vagal efferent fibers, J. Physiol. **222:**1, 1972.
78. Kuru, M.: Nervous control of micturition, Physiol. Rev. **45:**425, 1965.
79. Langley, J. N.: On the reaction of cells and of nerve endings to certain poisons (nerve endings and receptive substance), J. Physiol. **33:**374, 1905.
80. Langley, J. N.: The autonomic nervous system, Cambridge, 1921, W. Heffer & Sons Ltd.
81. Lefkowitz, R. J.: The β-adrenergic receptor, Life Sci. **18:**461, 1976.
82. Levi-Montalcini, R.: The nerve growth factor: its mode of action on sensory and sympathetic nerve cells, Harvey Lect. **60:**217, 1965.
83. Levi-Montalcini, R., and Angeletti, P. U.: Immunosympathectomy, Pharmacol. Rev. **18:**619, 1966.
84. Levy, M. N.: Sympathetic and parasympathetic interactions in the heart, Circ. Res. **29:**437, 1971.
85. Libet, B.: Generation of slow inhibitory and excitatory postsynaptic potentials, Fed. Proc. **29:**1945, 1970.
86. Linden, R. J.: Reflex from the heart, Prog. Cardiovasc. Dis. **18:**201, 1975.
87. Loewenstein, W.: Modulation of cutaneous mechanoreceptors by sympathetic stimulation, J. Physiol. **132:** 40, 1956.
88. Loewenstein, W., and Altamirano-Orrego, R.: Enhancement of activity in a pacinian corpuscle by sympathomimetic agents, Nature **178:**1292, 1956.
89. Loewi, A.: Über humorale Übertragbarkeit der Herznervenwirkung, Pfluegers Arch. **189:**239, 1921; **193:** 201, 1922.
90. Martner, J.: Influences on colonic and small intestinal motility by the cerebellar fastigial nucleus, Acta Physiol. Scand. **94:**82, 1975.
91. Matthews, M. R.: Ultrastructure of ganglionic junctions. In Hubbard, J. I., editor: The peripheral nervous system, New York, 1974, Plenum Publishing Corp., p. 111.
92. Mitchell, G. A. G.: Anatomy of the autonomic nervous system, Edinburgh, 1953, E. & S. Livingstone, Ltd.
93. Miura, M., and Reis, D. J.: Termination and secondary projections of carotid sinus nerves in the cat brainstem, Am. J. Physiol. **217:**142, 1969.
94. Nishi, S.: Ganglionic transmission. In Hubbard, J. I., editor: The peripheral nervous system, New York, 1974, Plenum Publishing Corp., p. 225.
95. Nishino, H., Koizumi, K., and Brooks, C. M.: The role of suprachiasmatic nuclei of the hypothalamus in the production of circadian rhythm, Brain Res. **112:**45, 1976.
96. Pick, J.: The autonomic nervous system, morphological, comparative, clinical and surgical aspects, Philadelphia, 1970, J. B. Lippincott Co.
97. Polosa, C.: Spontaneous activity of sympathetic preganglionic neurons, Can. J. Physiol. Pharmacol. **46:** 887, 1968.
98. Purves, M. J., and Biscoe, T. J.: Cervical sympathetic activity and the sensitivity of the carotid body chemoreceptors. In Torrance, P. W., editor: Arterial chemoreceptors (proceedings of the Wates Foundation), Oxford, 1968, Blackwell Scientific Publications Ltd., p. 325.
99. Ruch, T. C.: Central control of the bladder. In Field, J., editor: Handbook of physiology. Section 1, Baltimore, 1960, The Williams & Wilkins Co., vol. 2, chap. 48.
100. Sampson, S. R., and Mills, E.: Effects of sympathetic stimulation on discharge of carotid sinus baroreceptors, Am. J. Physiol. **218:**1650, 1970.
101. Sato, A., and Schmidt, R.: Spinal and supraspinal components of the reflex discharges into lumbar and thoracic white rami, J. Physiol. **212:**839, 1971.
102. Sato, A., Tsushima, N., and Fujimori, B.: Further observation of the reflex potential in the lumbar sympathetic trunk, Jpn. J. Physiol. **15:**532, 1965.
103. Schiff, J. D.: Role of the sympathetic innervation of the pacinian corpuscle, J. Gen. Physiol. **63:**601, 1974.
104. Schmidt, R. F., and Weller, E.: Reflex activity in the cervical and lumbar sympathetic trunk induced by unmyelinated somatic afferents, Brain Res. **24:**207, 1970.

105. Schuster, M. M.: Motor action of rectum and anal sphincters in continence and defecation. In Code, C. F., editor: Handbook of physiology. Section 6, Baltimore, 1968, The Williams & Wilkins Co., vol. 4, chap. 103.
106. Seller, H., and Illert, M.: The localization of the first synapse in the carotid sinus baroreceptor reflex pathway and its alteration of the afferent input, Pfluegers Arch. **306:**1, 1969.
106a. Schofield, G. C. In Code, C., editor: Handbook of physiology. Section 6, Washington, D.C., 1968, The Williams & Wilkins Co., vol. 4, chap. 80.
107. Simeone, F. A., and Lampson, R. S.: A cystometric study of the function of the urinary bladder, Ann. Surg. **106:**413, 1937.
108. Snider, R. S.: Some cerebellar influences on autonomic function. In Hockman, C. H., editor: Limbic system mechanism and autonomic function, Springfield, Ill., 1972, Charles C Thomas, Publisher, p. 87.
109. Thonen, H., and Tranzer, J. P.: The pharmacology of 6-hydroxydopamine, Annu. Rev. Pharmacol. **13:**169, 1973.
110. Uvnäs, B.: Central cardiovascular control. In Field, J., editor: Handbook of physiology. Section 1, Baltimore, 1960, The Williams & Wilkins Co., vol. 2, chap. 44.
111. von Euler, U. S.: Noradrenaline: chemistry, physiology, pharmacology and clinical aspects, Springfield, Ill., 1956, Charles C Thomas, Publisher.
112. von Euler, U. S.: Autonomic neuroeffector transmission. In Field, J., editor: Handbook of physiology. Section 1, Baltimore, 1959, The Williams & Wilkins Co., vol. 1, chap. 7.
113. von Euler, U. S.: Adrenergic neuroeffector transmission. In Bourne, G. H., editor: The structure and function of the nervous tissue, New York, 1969, Academic Press, Inc., vol. 2.
114. White, J. C., Smithwick, R. H., and Simeone, F. A.: The autonomic nervous system, ed. 3, New York, 1952, The Macmillan Co.
115. Willis, W. D., and Grossman, R. G.: Medical neurobiology: neuroanatomical and neurophysiological principles basic to clinical neuroscience, ed. 2, St. Louis, 1977, The C. V. Mosby Co.

34

CHANDLER M. BROOKS and KIYOMI KOIZUMI

The hypothalamus and control of integrative processes

The hypothalamus is unique in that it confers on animals the ability to maintain essential body states, to react to conditions of stress, and to carry out certain specific components of functions such as reproduction, locomotion, and postural adjustment. It is involved in the process of sleep and awakening, maintenance of body water and weight balance, and regulation of body temperature. The function of virtually no organ system can be adequately discussed without mention of this complex structure identified as the hypothalamus. It is an entity but also a way station that affects those regions of the nervous system lying caudal to it and those that lie above and laterally. Therefore the hypothalamus should not be considered an independent entity; neither should it be associated exclusively with the autonomic nor the endocrine systems, because it plays an equally important part in the control of somatic reactions.

In lower vertebrates, integrated behavior and most of the responses characteristic of an intact animal are present as long as the hypothalamus retains its connections with lower structures and remains intact. The cerebral cortex and subcortical regions do contribute additional abilities and exert some control over these hypothalamic functions. In man and the primates a higher degree of cephalization has occurred, but the hypothalamus still retains many of the same integrative functions it performs in subprimates.

Prior to this century the hypothalamus was known chiefly as an area lying beneath the thalamus. It was not considered to be an independent functional entity, as it is now, and it is quite probable that in order to emphasize its very important role we have tended to isolate it too much in our thinking from other portions of the brain. In 1842 Rokitanski pointed out that lesions involving the base of the brain were commonly associated with gastric hemorrhage and duodenal ulcers. In 1887 a Dr. Story, of Dublin, described a case in which a tumor of the pituitary fossa caused stoutness, irregularity in menstruation, and unusual drowsiness. A year later Bramwell, in a book called *Intracranial Tumours,* reported the association of obesity, presence of sugar in the urine, simple polyuria (diabetes insipidus), and other symptoms now known to be due to hypothalamic injury caused by the presence of tumors in the pituitary or basal areas of the brain. Mauthner in 1890 associated somnolence with infections involving the base of the brain lying adjacent to the sella turcica. Thus much of our early knowledge of the hypothalamus was obtained by observing the symptoms produced by pituitary tumors or tumors and lesions occurring in this basal region of the brain.

These initial observations have been confirmed both clinically and experimentally. In early attempts to remove hypophyseal tumors or perform hypophysectomies to determine the functions of this gland, hypothalamic lesions were inadvertently produced. Between 1901 and 1912, Paulesco in Rumania and Harvey Cushing at the Johns Hopkins Medical School developed the procedures that finally enabled them to perform hypophysectomies. They observed the effects of pituitary deficiency but also many other symptoms due to injury to the hypothalamus. It was Fröhlich who first described a case of adiposogenital dystrophy in 1901. Cushing observed similar symptoms caused by tumors and brain lesions. Aschner (1909) and Erdheim (1904) must be given credit for recognizing that the adiposogenital syndrome is due not to pituitary but to hypothalamic disturbances.

In the course of studies of these phenomena, Camus and Roussy (1920) found that punctures of the brain that left the pituitary intact produced not only the adiposogenital syndrome but also diabetes insipidus. In 1921 Bailey and Bremer were able to publish a monograph on diabetes insipidus that identified the hypothalamic as well as pituitary malfunctions responsible for this disease.

Studies of the role of the hypothalamus in emotional expression and in control of the autonomic system might be said to have begun in 1892 with the work of Goltz on chronic decorticate dogs in which he observed anger or its characteristic phenomena. Woodworth and Sherrington in 1904 described "pseudoaffective reflexes'' in animals following high decerebrations. These reflexes involved reactions characteristically seen in rage. In 1909 Karplus and Kreidl, by stimulation of the hypothalamus, produced various visceral and autonomically mediated reactions, some of which gave the picture of rage. Cannon and Britton in 1925 produced the phenomenon of sham rage by removing cerebral cortex and a portion of the thalamus, and in 1928 Bard showed that this release phenomenon depended on caudal portions of the diencephalon, the posterior hypothalamus.

One of the greatest aids to the study of hypothalamic function has been the sterotaxic instrument and methodology introduced by Horsley and Clarke in 1908.[55] With this instrument, it has been possible to record from, stimulate, and produce lesions in any discrete area or nucleus of the hypothalamus.

In 1940 a symposium was held and a monograph published that very adequately reviewed all anatomic and physiologic studies to date. Much of the work done since then has been an elaboration of findings reported in or suggested by contributors to that symposium. However, progress has been very great, particularly in the field of neuroendocrinology, autonomic system physiology, and the study of emotions. This latter area involves the relationship of the hypothalamus to the limbic system. Both the early work and that of more contemporary times is described in recently published monographs.[1-6].

ANATOMIC COMPONENTS OF THE HYPOTHALAMUS

The hypothalamus is a relatively small area of the brain. It is rather uniform in size and structure within the mammalian kingdom. Consequently, in lower forms it occupies a much greater proportion of the total brain mass than it does in man. It lies beneath the thalamic nuclei and extends from the mesencephalon forward to the region of the optic chiasm and the preoptic area.

Subdivisions

Very elaborate anatomic studies have been made of the component structures of the hypothalamus and of their afferent and efferent connections with other portions of the brain stem.[29,74,81] For most physiologic purposes, it suffices to speak of the anterior, medial, posterior, and lateral regions of the hypothalamus (Fig. 34-1). Only in a few cases can function be ascribed to specific nuclei; in most instances we merely know that a certain area of the hypothalamus is essential to the performance of certain functions or the

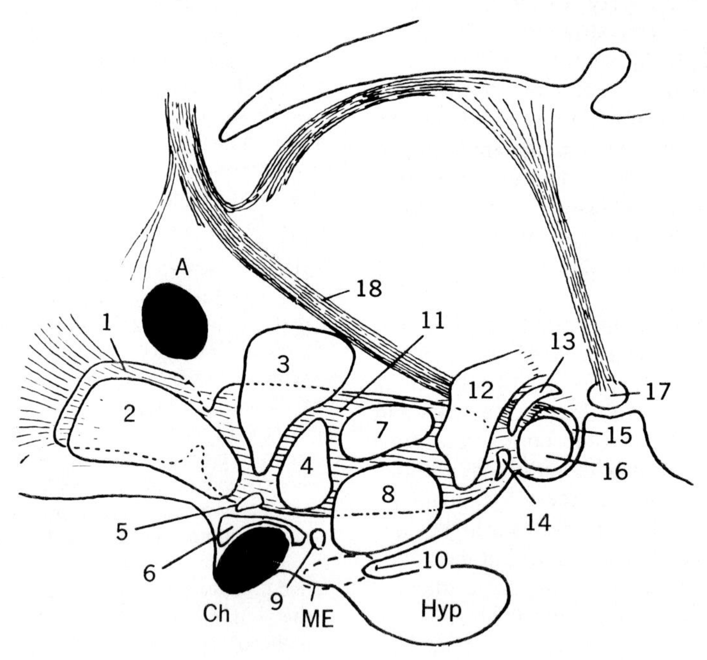

Fig. 34-1. Diagram showing relative positions of hypothalamic nuclei of typical mammalian brain and some interconnecting pathways. *A,* Anterior commissure; *Ch,* optic chiasm; *Hyp,* hypophysis; *ME,* median eminence; *1,* lateral preoptic nucleus (permeated by medial forebrain bundle); *2,* medial preoptic nucleus; *3,* paraventricular nucleus; *4,* anterior hypothalamic nucleus; *5,* suprachiasmatic nucleus; *6,* supraoptic nucleus; *7,* dorsomedial hypothalamic nucleus; *8,* ventromedial hypothalamic nucleus; *9,* arcuate nucleus; *10,* tuber cinereum; *11,* lateral hypothalamic nucleus (permeated by medial forebrain bundle); *12,* posterior hypothalamic area; *13,* supramamillary nucleus; *14,* premamillary nucleus; *15,* lateral mamillary nucleus; *16,* medial mamillary nucleus; *17,* interpeduncular nucleus (a mesencephalic element in which habenular-peduncular tract terminates); *18,* fornix. (Modified from Clark et al.[29])

production of the specific compounds. The anterior region is considered to include the preoptic, supraoptic, suprachiasmatic, and paraventricular nuclei. The medial region, sometimes called the tuberal region, surrounds the third ventricle and is highly cellular; it contains distinct groups such as the ventromedial and dorsomedial nuclei. Ventrally, it is composed of tuberal nuclei and the tuber cinereum, to which is attached the hypophyseal stalk. The lateral hypothalamic region contains cells that are more diffusely arranged and interspersed among fibers of the medial forebrain bundle. This highly complex bundle extends throughout the entire lateral hypothalamus and continues rostrally through the preoptic area to the olfactory regions and caudally to the midbrain. It is composed of many short relays as well as longer fibers, and most of the extrinsic connections to the hypothalamus are made through this medial forebrain bundle. The posterior hypothalamic region includes principally the posterior hypothalamic nuclei and mamillary nuclei.

These hypothalamic nuclei and fiber tracts are bilateral. In physiologic studies, it appears that one nucleus of the pair suffices to maintain a function. To produce maximum diabetes insipidus, for example, the supraoptic and paraventricular nuclei of both sides must be destroyed. The principal nuclei and subdivisions of the hypothalamus are listed here, and their relative positions are diagrammed in Fig. 34-1.

Anterior hypothalamus
- Lateral preoptic nuclei *(1)*
- Medial preoptic nuclei *(2)*
- Supraoptic nuclei *(6)*
- Paraventricular nuclei *(3)*
- Suprachiasmatic nuclei *(5)*
- Anterior hypothalamic nuclei *(4)*

Medial hypothalamus
- Dorsomedial hypothalamic nuclei *(7)*
- Ventromedial hypothalamic nuclei *(8)*
- Arcuate nuclei (infundibular nuclei) *(9)*
- Tuberal nuclei—tuber cinereum *(10)*

Lateral hypothalamus
- Lateral hypothalamic nuclear masses *(11)*
- Medial forebrain bundle

Posterior hypothalamus
- Posterior hypothalamic area *(12)*
- Supramamillary nuclei *(13)*
- Premamillary nuclei *(14)*
- Lateral mamillary nuclei *(15)*
- Medial mamillary nuclei *(16)*

There are other nuclear groups not identified in Fig. 34-1. Their functions are relatively unknown or unimportant to this discussion.

Principal connections

The principal connections between the hypothalamus and other regions of the brain are shown diagrammatically in Fig. 34-2. Details of these afferent and efferent pathways can be obtained elsewhere.[65,74,81]

In summary, it may be said that recent work has shown the hypothalamus to be influenced directly or indirectly by almost all parts of the brain, while it in turn can modify virtually all body functions.

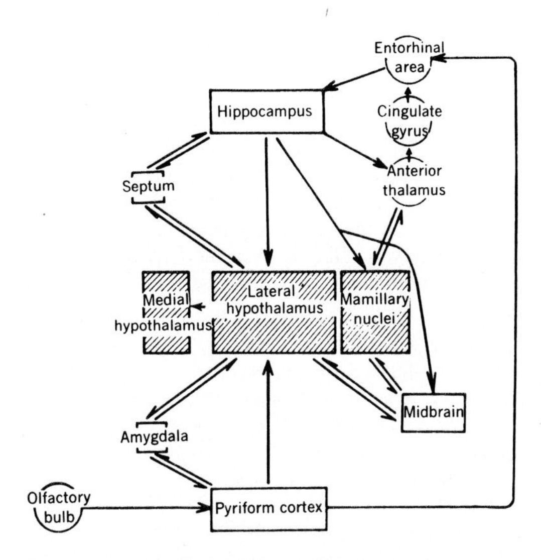

Fig. 34-2. Schematic diagram of principal fiber connections between hypothalamus and other regions of brain. (From Raisman.[81])

Hypothalamicohypophyseal tract

Ramón y Cajal, an early neurohistologist, described nerve fibers passing from the hypothalamus to the hypophysis in the hypophyseal stalk. This neuroconnection brings the pituitary gland, or at least its posterior lobe, under control of the hypothalamus. According to Ranson and his collaborators,[39] this hypothalamicohypophyseal tract consists of two parts. The major portion of the tract is composed of fibers originating in the supraoptic and paraventricular nuclei; it loops over the chiasm and, after coursing medially along the base of the anterior hypothalamus, enters the median eminence, the infundibular stem, and terminates in the posterior lobe. The second component is the tuberohypophyseal tract, which is thought to be derived from the arcuate (infundibular) and tuberal nuclei of the tuber cinereum. Fibers comprising this tract originate caudolaterally, occupying the posterior section of the stalk; they end chiefly on the capillary loops in the median eminence and proximal part of the stalk and serve as connectors from the tuberal region to the portal system serving the anterior lobe.

These stalk fibers are all unmyelinated and relatively numerous; there are about 60,000 in the monkey and 100,000 in man. The supraoptic and paraventricular nuclei contain many more cells than there are fibers in the lower part of the stalk, an indication that the axons of some neurons, chiefly those of the paraventricular nuclei, end in the median eminence or at the base of the stalk and do not course to the neural lobe. Additional evidence that all cells of the nuclei do not send fibers into the stalk and hypophysis is provided by the fact that section of the stalk near the posterior lobe causes fewer cells of the nuclei to degenerate than when a higher section involving the median eminence is made. About 90% of the cells degenerate when the entire stalk and median eminence are destroyed.

The medial anterior portion of the tuber cinereum is referred to as the median eminence. Portal system capillaries are abundant in this region, and nerve terminals containing many granules and vesicles concentrate there. Despite the fact that it has a structural continuity with the tuber, it appears to be part of the neurohypophysis rather than of the hypothalamus.

In the subsequent discussion the following nomenclature will be used:

Adenohypophysis
- Pars infundibularis (pars tuberalis)
- Pars intermedia (intermediate lobe)
- Pars distalis (anterior lobe)

Neurohypophysis
- Infundibulum (median eminence)
- Infundibular stem
- Infundibular process (pars neuralis, neural lobe)

Posterior lobe: Infundibular process (neural lobe) + Pars intermedia

Pituitary stalk: Infundibulum + Pars infundibularis adenohypophysis + Infundibular stem

Blood supply to the hypophysis

A simple diagram of the humoral link between the hypothalamus and hypophysis is shown in Fig. 34-3. In 1930 Popa and Fielding first described the system of vessels that we now call the hypothalamicohypophyseal portal system. They found that the vessels coursing along the stalk originated in a capillary bed in the hypothalamus and ended in a capillary bed in the hypophysis. They assumed that blood flowed from the hypophysis into the hypothalamus. In 1936 Houssay, Biasotti, and Sammartino observed that portal flow in the toad was downward toward the anterior lobe from the brain; this was confirmed in mammals in 1936 by Wislocki and King, who demonstrated conclusively that blood flows from the hypothalamus to the pituitary gland. During the last 30 years, it has become clear that the secretory activities of the cells of the anterior lobe are largely controlled by substances formulated in the hypothalamus and transported humorally to the hypophysis via this portal system.[31,46] The anterior lobe receives virtually all its blood supply via the hypophyseal portal vessels. The superior hypophyseal arteries are branches of the internal carotid arteries; they supply the upper part of the stalk and form the primary capillary beds that drain into the portal vessels. The capillaries comprising these beds are of two types: some are of the simple tubular form seen elsewhere in the nervous system, but others have a most unusual pattern. These capillaries are looped, coiled, and arranged in a highly complicated manner, suggesting that blood flows slowly through them and that they may provide the sites at which neurohumors are transferred from nerve terminals into the bloodstream. The other primary capillary bed is in the lower part of the infundibular stem and derives its blood supply mainly from the inferior hypophyseal arteries. This bed is also composed of coiled capillaries. The short portal vessels originating from this second capillary bed also carry neurohumors from stalk fibers to certain cells in the anterior lobe. They ensure the survival of a well-defined core of anterior lobe cells, even after pituitary stalk section.

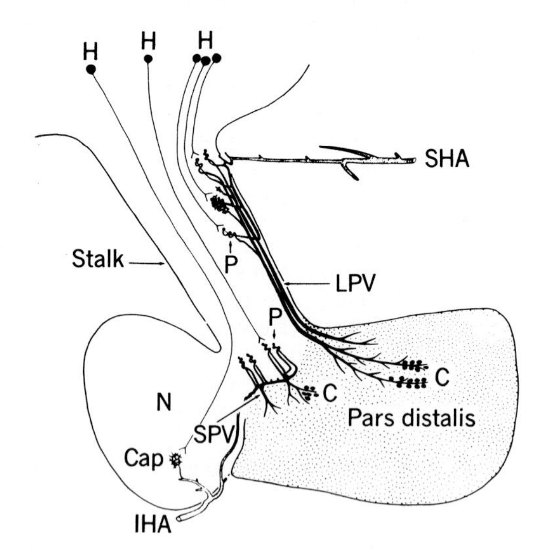

Fig. 34-3. Diagram of midsagittal section of human pituitary gland and stalk showing vessels of portal system. Note complicated coiled capillary loops and nets adjacent to nerve terminals. *H,* Hypothalamic neurons; *N,* posterior lobe; *SHA* and *IHA,* superior and inferior hypophyseal arteries; *P,* upper and lower capillary beds; *LPV* and *SPV,* long and short portal vessels; *C,* representation of secretory epithelial cells adjacent to terminal capillaries; *Cap,* representation of the capillary bed through which hormones enter systemic circulation. (From Adams et al.[8])

The inferior hypophyseal arteries are the sole source of blood supply to the capillary bed serving the posterior lobe. These capillaries have a simple branching pattern in the young, but in adults the capillaries become markedly convoluted and tend to be arranged in lobular units. This pattern is thought to facilitate the transfer of hormones from the fibers of this lobe to the bloodstream.

Another matter of major interest relative to the hypothalamicohypophyseal system is the fact that the supraoptic and paraventricular nuclei have the richest capillary bed density of any nucleus or group of cells in the CNS.[38] Each individual nerve cell is surrounded by a network of capillaries; there are approximately 2,600 mm of capillaries/mm^3 of tissue in the supraoptic nucleus and 1,650 mm/mm^3 in the paraventricular nucleus, as compared with 440 mm/mm^3 in the motor cortex. The blood supply to the white and gray matter of other parts of the hypothalamus is not markedly different from that in other areas of the CNS.

FUNCTIONAL ROLE OF THE HYPOTHALAMUS

General nature of function

As stated previously, the hypothalamus integrates many functions and is involved in the flow of reactions essential to the maintenance of homeostasis and to the initiation and control of many behavioral responses. These numerous roles must be described separately. It should be remembered, however, that the reactions do not occur in isolation. For example, reproductive phenomena are not carried out independently of metabolic and water-balance adjustments. The hypothalamus executes its control over body functions through the endocrine, autonomic, hypothalamicohypophyseal, and somatic neural efferent systems.

Before proceeding with analyses of its role in specific functional processes there should be some discussion of three other matters in which the hypothalamus is involved:

1. Central chemical transmission
2. Cyclic activity
3. Control of pineal gland

Central chemical transmission

In recent years there has been much interest in humoral transmission at central as well as peripheral synapses. It is known that certain cells and nuclei of the brain are sensitive to acetylcholine, whereas others contain and are sensitive to catecholamines. In these studies the hypothalamus has not been neglected.

Attempts to isolate possible transmitters have shown that acetylcholine, catecholamines, dopamine, serotonin, substance P, histamine, gamma aminobutyric acid (GABA), and other endogenous peptides are present in the hypothalamus.[27,89] The use of microiontophoresis has shown that some hypothalamic neurons are cholinergic, whereas others are sensitive to the adrenergic compounds.[63] Recently developed autoradiographic and histochemical fluorescent techniques[41,96] have also been used to demonstrate the presence and distribution of catecholamines within the hypothalamus. Stereotaxic mappings of the serotonin, norepinephrine, and dopamine systems within the brain have shown that the hypothalamus is included in these aminergic pathways.[91] The catecholamines in the hypothalamus are chiefly located in neuron terminals, but their cell bodies are generally located elsewhere. There are some dopaminergic cells, however, that are located near and with axons terminating in the median eminence.[54,96] Cholinergic and adrenergic transmitters certainly are involved in hypothalamic functions, as will be illustrated later.

Cyclic activity

It is well recognized that there are very short (10 to 15 min), intermediate, and diurnal or circadian as well as seasonal rhythms in hormone release and in behavior patterns.[42,49,79] Recording of activity within hypothalamic nuclei has shown that the hypothalamus is involved in these cyclic phenomena. Neurons within the ventromedial nuclei and lateral hypothalamic areas show both fast oscillations and circadian variations in firing rates.[59] These hypothalamic rhythms in activity, eating, sleeping, hormonal secretion, etc. are not readily disturbed by hypothermia, anesthetics, other drugs, or even brain lesions that do not involve the hypothalamus. Injury to the hypothalamus, however, causes regular cycles to disappear and periodicities to become randomized. One of the nuclei that has attracted much attention recently is the suprachiasmatic, which has been shown to be influenced by the optic system and to have an effect on the pineal gland, diurnal rhythm in the release of anterior lobe hormones, and rodent behavior.[76,79]

Control of the pineal gland

Evidence has been obtained recently to show that the pineal body, or epiphysis, does play a role in the control of endocrine functions and that its activity is influenced by brain centers acting through the autonomic system. Immature animals kept in the dark show evidence of a depression of growth and reproductive development as though release of growth and reproductive hormones were inhibited. Removal of the pineal gland or section of the cervical sympathetics abolishes this effect. The current interpretation of these observations is that light and darkness modulate activity of the suprachiasmic nuclei and thus production of catecholamines by the sympathetic supply to the pineal. In the light, pineal peptide levels are low due to inhibition of catecholamine production. In the dark, more catecholamines are present in the pineal gland, and this catalyzes peptide synthesis and release. Melatonin and/or other pineal peptides either suppress liberation of releasing factors from the hypothalamus, and thus growth and reproductive hormone output, or act otherwise to modify some hormone-dependent functions.[83,97] The pineal gland is now thought to be a neurochemical transducer controlled by the hypothalamus and acting on hypothalamic functions.[12]

Specific functional roles

The functions in which the hypothalamus is primarily involved are the following:

1. Control of the endocrine system through the pituitary gland—neurohypophysis and adenohypophysis (anterior and intermediate lobe)

2. Reproduction
3. Thirst and control of water balance
4. Control of body weight (hunger and satiation)
5. Temperature regulation
6. Reactions to stress
7. Control of emotional reactions
8. Sleep and arousal
9. Control of somatic reactions

Control of endocrine system–pituitary gland

The hypothalamus is involved in the regulation of endocrine system functions. It has a close relationship with the hypophysis, and this hypothalamicohypophyseal axis regulates the activity of nearly all endocrine organs. The neural lobe of the hypophysis is, of course, an evagination from the CNS and might be considered a part thereof. Anatomically, this neural lobe is closely related to the adenohypophysis, which originates as an evagination from the buccal mucosa and differentiates to form the intermediate and anterior lobes of the pituitary gland. A direct functional relationship between these parts of the hypophysis has been suggested, but the mechanism is obscure. Unquestionably, however, all parts and functions of the hypophysis are under the control of the hypothalamus.

Discussion of the role of the hypothalamus in the control of the body economy through the endocrine system can be subdivided into at least three parts: its role in the control of reproductive functions, water balance, and metabolic processes. The first thing to consider, however, is how the hypothalamus regulates endocrine activities. It is well recognized that certain of its nuclei act through neural pathways to release hormones from the neural lobe. Stimuli act on other hypothalamic structures to liberate chemical agents that are humorally transported to the adenohypophysis, releasing the trophic hormones that affect the functions of peripheral endocrine glands. These control processes should be considered in greater detail.

Control of release of hormones from the neurohypophysis

Antidiuretic hormone or vasopressin. As early as 1924, Abel demonstrated the presence of vasopressin not only in the posterior lobe of the pituitary but also in the tuber cinereum of the hypothalamus. In the 1930s E. Scharrer, Roussy, and Mosinger postulated on the basis of histologic work that the neurons of the supraoptic and paraventricular nuclei have a secretory or endocrine role. It was Bargmann and his associates, however, who in the late 1940s and early 1950s accumulated morphologic evidence indicating that neurons of the hypothalamus synthesize vasopressin and oxytocin.[15,16] These compounds are transported down the fibers of the hypophyseal tract and stored in the nerve endings of the neurohypophysis, from which they are secreted into the bloodstream. The idea that there are neurosecretions implies that a neuron can synthesize specific compounds and transport, store, and release them much as can cells of an endocrine gland.[28] These neurons also possess behavior typical of nerve cells, in that they are electrically and chemically excitable and can conduct action potentials along axons. These action potentials release transmitters or trigger the release of other synthesized materials, neurosecretions. Practically all nerve cells are thought to release transmitter agents, but the term ''neurosecretion'' refers to the processes by which unique materials are produced by specific groups of neurons and eventually released to act on distant target organs, not merely on those with which the neurons make contact.

Bargmann and his associates, using chrome alum hematoxylin (Gomori stain) or other basic dyes, were able to stain secretory materials within the cell bodies and the axons. The concentration of these granules was found to coincide with concentrations of antidiuretic hormone (ADH) in special hypothalamic nuclei and the neurohypophysis. That these neurosecretory granules are accumulated in or released from the supraoptic and paraventricular nuclei has been demonstrated by hydration and dehydration experiments. If the stalk is sectioned, neurosecretory materials that normally would have been transported down the axon accumulate above the site of section, an indication that they are synthesized in the cell body.

Immunocytochemical techniques have now provided more direct in situ evidence that oxytocin and vasopressin are produced by this magnocellular system of the hypothalamus. These hormones are loosely bound to a carrier protein called neurophysin. This material is identified by histochemical and immunochemical markers. There appear to be two neurophysins, one binding oxytocin and the other vasopressin. These can be secreted independently in response to physiologic stimuli. The newer techniques have shown that neurons containing oxytocin-neurophysin and vasopressin-neurophysin are more widely distributed than previously thought.[101,102]

Lesion experiments have produced diabetes insipidus of intensities proportional to the degree of posterior lobe and/or supraoptic nucleus destruction. This disease is clearly the result of an absence of production or secretion of ADH from the supraoptic nuclei. Some ADH is also liberated from cells of the paraventricular nuclei, but these nuclei are not totally devoted to its production. Fig. 34-4 shows changes in urine output following stalk section.

In recent years, it has been possible to make electrical recordings from cells of the supraoptic nuclei. Since most of these neurosecretory cells send axons to the posterior lobe, electrical stimulation of this lobe evokes antidromic excitation of neurosecretory cells in the hypothalamus. The method has been used for identification of neurosecretory neurons.[60] Many cells of the supraoptic nuclei have a basic rhythm of firing (less than 1 to 5/sec), but this rate varies greatly and at least some cells are periodically silent. Action potentials originating in the soma of these neurosecretory cells are conducted along axons down the stalk to the pituitary gland at a speed of 0.3 to 1.4 m/sec and release hormone from their endings.[35,56] It appears that nerve impulses produce a change in cell membrane permeability and some type of perturbation of hormone-containing granules resulting in exocytosis. Calcium has been found necessary for this process. It is thought that calcium ions are involved in producing an approximation of granules to plasma membranes and a chemical action that releases the stored materials.[32]

Secretion of ADH is influenced by many neural and humoral stimuli that excite or inhibit supraoptic neuron activity. Increase in osmotic pressure of the blood augments ADH secretion. Injections of hypertonic saline solution into a carotid artery produce effective antidiuresis, thus revealing the operation of this osmosensitive system.[93] Such stimuli produce an increase in the firing rate of neurosecretory cells,[60,61] as well as an increase in plasma ADH level as determined by bioassay[34] (Fig. 34-5). Under normal conditions, plasma ADH levels are approximately 1 to 3 μU/ml. Plasma osmolarity is maintained at about 290 mOsm/L, and ADH output is significantly increased when this value is changed by as little as 2%. These cells are also responsive to acetylcholine (ACh) and tend to be inhibited by epinephrine. The excitatory and inhibitory

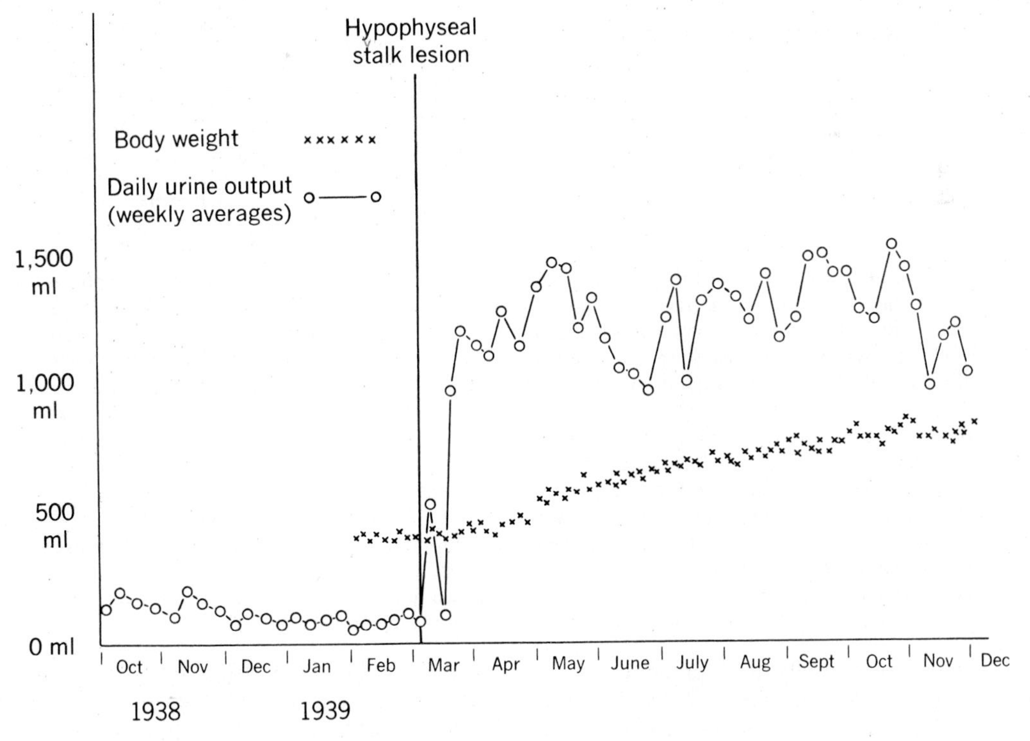

Fig. 34-4. Production of diabetes insipidus and obesity by pituitary stalk section and injury to hypothalamus. Note initial polyuria, thought to be due to transection of "secretory fibers," followed by a brief period of reduced urine output occasioned by release of stored ADH from degenerating cells. Maximum degree of diabetes insipidus developed as cells degenerated and no more ADH was produced. (From Brooks.[22])

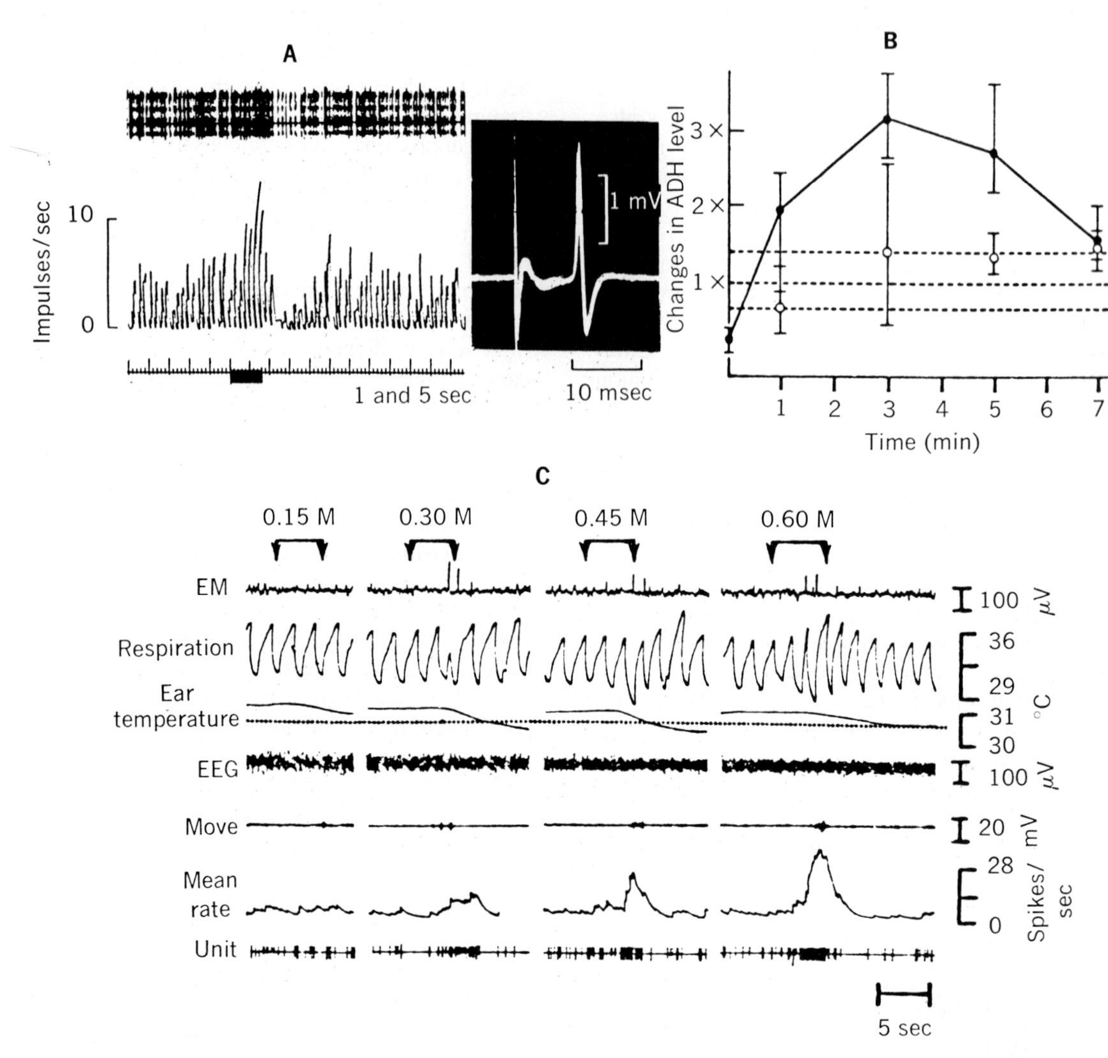

Fig. 34-5. Change in rate of firing of supraoptic nucleus cell and plasma ADH levels in response to osmotic stimulation by intracarotid injection of NaCl (in anesthetized cat). **A,** Upper trace shows unit activity; middle trace is integrated recording (impulses per sec as shown by spike heights); lower trace shows time and interval of injection (0.3 ml of 1M NaCl). Insert shows stimulus artifact and typical antidromic potential induced by stalk stimulation (supraoptic nucleus cell). **B,** Changes in plasma ADH levels following intracarotid injection (at zero time) of hypertonic NaCl (filled circles) and isotonic NaCl (open circles). Bars indicate 95% confidence limits of estimates; middle dashed line represents mean concentration, whereas lower and upper dashed lines indicate 95% confidence limits of ADH before NaCl injection (average value indicated by 1×, whereas double and triple concentrations are shown by 2× and 3×). **C,** Effects of intracarotid injections, in the conscious monkey, of different concentrations of NaCl (0.15 to 0.60M) on eye movements *(EM),* respiration, skin temperature of ear, EEG, body movement *(move)*, mean rate of supraoptic nucleus unit discharge, and record of unit firing *(unit)*. Isotonic NaCl (0.15M) judged to have no effect. (**B** from Dyball[34]; **C** from Hayward and Vincent.[48])

stimuli delivered neurally to supraoptic nuclei come from various other parts of the nervous system and not from within the hypothalamus alone.

Intracellular recordings from these neurosecretory cells have shown that they possess resting potentials of the same order of magnitude found in other neurons. They resemble other neurons in electrical behavior.[60] Unquestionably, these cells of the supraoptic nuclei are neurons that produce a hormone (ADH) essential to the conservation of body fluids through reabsorption, a process that occurs in the kidney tubules (Chapter 50).

In conclusion, the cells of the supraoptic nuclei and some in the paraventricular nuclei are sensitive to changes in the surrounding environment, as in long-term hydration and dehydration. There is a unanimity of opinion that changes in blood volume and blood pressure modify ADH levels in the blood. Decreases in blood pressure, even those produced by rising to a standing position, result in an ADH increase (1.9 to 3.0 μU/ml), whereas increases in blood volume reduce ADH and produce a diuresis. Activation of baroreceptors has been shown to reduce activity of the supraoptic nuclei, decreasing plasma ADH level.[98] Distention of the left atrium by artificial means reduces ADH output through action of the vagus nerve on the supraoptic nuclei.[60a] ADH output increases during exercise and emotional states and as a result of hemorrhage.

Oxytocin. In 1909 Dale reported that posterior pituitary extracts cause the uterus to contract. This was called oxytocic action, and the existence of an oxytocic hormone was suggested. In 1910 Ott and Scott showed that pituitary extracts increased the output of milk and played a role in the milk "let-down" reaction, or milk ejection response. Separation of the oxytocic from the vasopressor principle was first accomplished by Kamm and co-workers in 1928. Du Vigneaud and his colleagues in 1953 separated the two compounds (oxytocin and ADH), determined their exact chemical structures, and synthesized them. They were found to be octopeptides (Chapter 61).[33]

Oxytocin is primarily produced by the paraventricular and to a lesser degree by the supraoptic nuclei. Presumably this hormone, like vasopressin, is transported down axons of the pituitary stalk, stored, and released from the neural lobe. It has been shown that stimulation of the hypophyseal tract liberates oxytocin as determined by milk ejection assay.[30] Injury to the stalk or paraventricular nuclei interferes with normal lactation and the milk ejection response.

Oxytocin normally is liberated by a reflex action following stimulation of the female genital tract, distention of the uterus and vagina being quite effective. However, the most remarkable oxytocin release response occurs during suckling. The afferent pathway of this reflex has not been clearly identified. It is known, however, that stimulation of the nipples produces excitation of afferent nerves, and impulses travel through the spinal cord to reach the hypothalamus.

Recent combinations of electrical recording from single units in the paraventricular nuclei with measurements of oxytocin release during physiologic stimulation have illustrated very clearly the quantitative correspondence between paraventricular cell activity and oxytocin output[26,94] (Fig. 34-6). It has been found that similar stimuli produce an increase in the electrical activity recorded from nerve fibers in the pituitary stalk.[56] Apparently, excitation of the hypothalamic neurons generates impulses that course along the hypophyseal tract to the pituitary gland to release stored oxytocin.

Activity of neurons in the paraventricular nuclei, like that of neurons in the supraoptic nuclei, is much affected by excitatory and inhibitory influences from other regions of the nervous system. It is well known that psychic stimuli greatly modify milk ejection. Emotional disturbances affect lactation in women as well as the milk ejection reflexes of many species of animals. There are certainly very strong central inhibitory mechanisms as well as excitatory drives. The osmosensitivity of these neurons appears to be similar to that of supraoptic nuclear cells. Close arterial injection of hypertonic sodium chloride solution or of small amounts of ACh increases activity of these cells and evokes an accompanying milk ejection response.[26]

Oxytocin has a strong effect on uterine motility, and its secretion may contribute to normal labor and parturition, but it probably is not indispensable thereto. Sensitivity of the uterus to oxytocin is enhanced by estrogen and inhibited to some degree by progesterone. Late in pregnancy the uterus becomes very sensitive to oxytocin. The administration of physiologic doses of oxytocin is effective in producing labor in women and animals. Dilation of the cervix and the descent of the fetus down the birth canal probably initiates, through the paraventricular nuclei, a reflex secretion of oxytocin, which may enhance labor and increase concentration of oxytocin in the bloodstream during labor.

It was once thought that oxytocin might act to facilitate the passage of sperm up the female genital tract to the fallopian tubes, but the evidence for this is not convincing. There is a release of oxytocin following coitus and in some species also during uterine contraction. Secretion of oxytocin, like that of ADH, is strongly inhibited by alcohol.

Control of anterior lobe hormone release

Many phenomena indicate a neural control over the release of various hormones from the anterior pituitary. Ovulation associated with

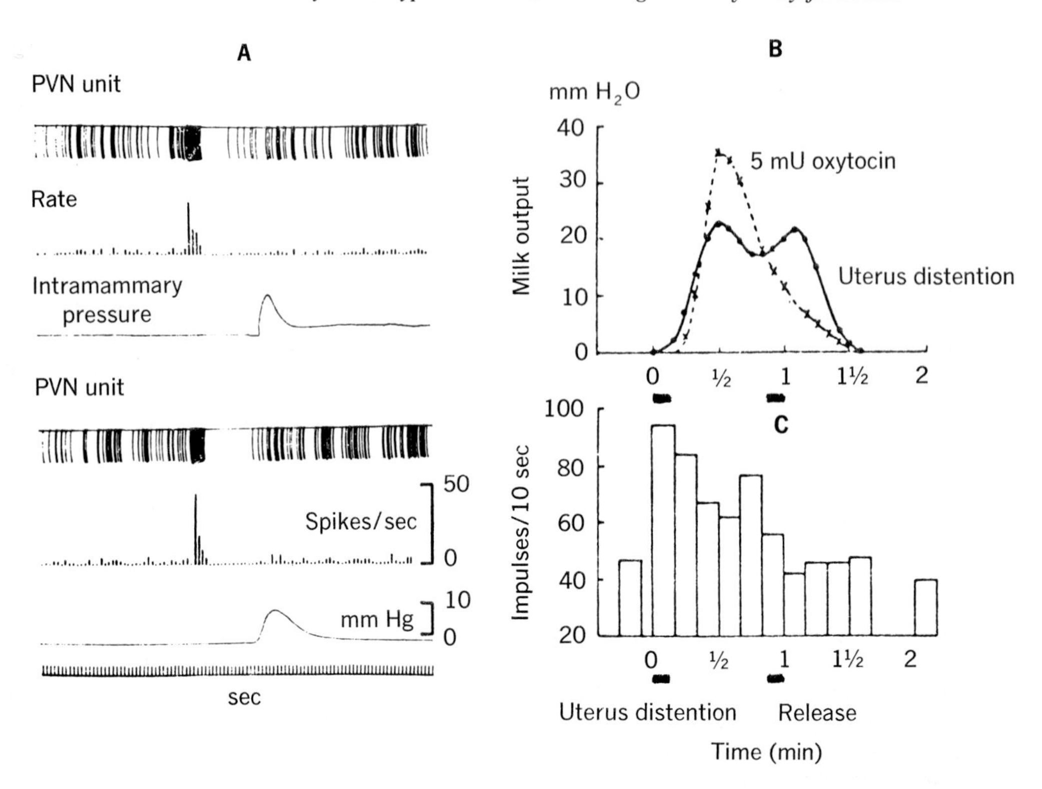

Fig. 34-6. Increase in rate of firing of paraventricular neuron *(PVN unit)* produced by suckling and distention of uterus. **A,** Recordings from two units showing effects of suckling of young in rats. Upper tracings, records of unit firing showing abrupt onset of response with beginning of suckling. Middle tracing, rate histograms of spike firing (number per 0.5 sec). Lower tracing, record of pressure developed in one mammary gland. Note latency between abrupt increment in PVN unit firing and onset of milk ejection response (increment in mammary pressure). **B,** Comparison of changes in mammary duct pressures resulting from intravenous oxytocin injection and from oxytocin release due to distention of uterus in postpartum anesthetized cat. **C,** Discharge rate of PVN unit before, during, and after distention of uterus that gave milk ejection response shown in **B.** (**A** from Wakerley and Lincoln[94]; **B** and **C** from Brooks et al.[26])

coitus, the effect of light on the development of ovaries, reversal of seasonal breeding habits when animals are transferred from one hemisphere to another, and the phenomenon of induced pseudopregnancy all indicate a neural influence over the secretion of gonadotropins. Experimental exposure of animals to cold has given evidence that the hypothalamus is involved in an increased thyroid activity. In the late 1930s hypothalamic stimulation was found to increase the discharge of various anterior pituitary hormones now known as luteinizing hormone (LH), adrenocorticotropin (ACTH), and thyrotropin (TSH). Hypothalamic lesions cause a decrease in the production of certain anterior lobe hormones and a block of their release by usually effective stimuli, but the secretion of certain other hormones, such as prolactin, is increased.

Initially a search was made for a neural connection between the hypothalamus and anterior lobe that might mediate these actions, and it was shown that transection of the pituitary stalk did abolish some of them. However, no very convincing anatomic evidence was found to support the idea that there is a functionally important innervation of the anterior lobe from the hypothalamus. Attention then turned to the hypothalamicohypophyseal portal system as a possible pathway for this relationship.[46] It was soon shown that this portal supply is essential to the maintenance of normal anterior lobe function. It was then not much of a step to conclude that special neurosecretions formulated in the hypothalamus are transported humorally to the anterior lobe, where they act to release hormones produced and stored there. These compounds produced in the hypothalamus become known as "releasing factors." Some of those which have been identified chemically and synthesized are now called releasing hormones.[87a]

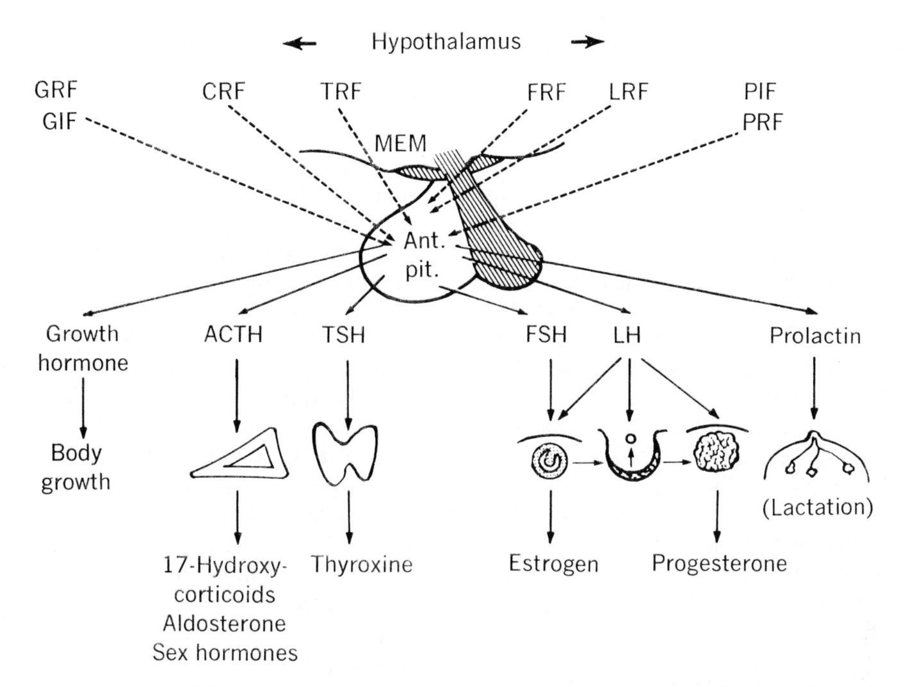

Fig. 34-7. Diagram of releasing factors that control output of hormones from anterior lobe of pituitary gland and ultimate hormone release from peripheral glands. *GRF,* Growth hormone–releasing factor; *GIF,* growth hormone release–inhibitory factor (somatostatin); *CRF,* corticotropin-releasing factor; *TRF,* thyrotropin-releasing factor; *FRF,* follicle-stimulating hormone–releasing factor; *LRF,* luteinizing hormone–releasing factor; *PIF,* prolactin release–inhibitory factor; *PRF,* prolactin-releasing factor; *MEM,* median eminence. Note that combination of FSH and LH produce maturation of follicle, ovulation, and corpus luteum formation.

It has been claimed that there are at least 10 and possibly more releasing factors or releasing hormones produced by neurosecretory cells, hypophysiotrophic neurons of the hypothalamus, but the existence of a number of these is still questioned. Some inhibit release but most of those definitely identified cause release of hormones from the anterior lobe.[18,43,82] They are small polypeptides that are relatively species nonspecific. Unquestionably, active compounds with very special releasing actions have been extracted from the hypothalamus. Ideally, evidence in support of present concepts of the formation and transport of releasing factors requires the demonstration of their presence in portal vessel blood in higher concentration than in the systemic circulation. There must also be a correlation between the concentration in the portal blood and the release of trophic hormone from the anterior lobe. At present, only a few releasing factors have been assayed in portal blood.[37,80]

Presumably, releasing factors are produced by certain neurons of the hypothalamus. These hypophysiotropic neurons lie in or near the median eminence, or the releasing factors are synthesized elsewhere and carried to the median eminence by axonal transport. The median eminence where they appear to be held in the highest concentration and from which they are released has relatively few cells but does contain many nerve fiber endings that lie close to the capillary loops from which the portal vessels originate. The releasing factors are probably secreted by these endings. It is assumed that these specific compounds are produced by different cells and accumulate in separate zones within the median eminence, but evidence for this is not as convincing as one might wish.[27,101,102] Fig. 34-7 shows one concept of the various releasing factors and the hormones they cause to be secreted from the anterior lobe.

Recently, sensitive bioassays and immunoassay techniques have shown that releasing factors are found and presumably produced in other parts of the nervous system.[82] Probably, concepts of the role of these compounds will eventually be modified, but since they are highly concentrated in the hypothalamus and since there is evidence of their participation in numerous reactions, the nature and role of each suggested releasing factor will be discussed. Until the question of their functional roles is clarified, we prefer to speak of these substances as factors rather than hormones.

Corticotropin-releasing factor (CRF). The first of these compounds to be actively investigated in detail was CRF. In 1955 Guillemin and his collaborators obtained an extract that caused release of ACTH from the pituitary gland. This substance has now been extracted from the pituitary stalk and median eminence. It stimulates not only the release of ACTH but also the synthesis of this trophic hormone in the anterior lobe.

CRF is probably a polypeptide, but little progress has been made in determining its chemistry since its discovery.[87,99] It is known to be a distinct entity, even though it has some of the effects of α-melanocyte–stimulating hormone (α-MSH) and though vasopressin in high concentrations can mimic its action in causing ACTH liberation from the anterior lobe.[47] CRF is liberated from the hypothalamus under conditions of stress or by noxious stimuli. There is a diurnal rhythm in its output, which is possibly responsible for the diurnal variation in plasma corticosteroid levels.

Stimulation of the amygdala, septal area, reticular formation, and hypothalamus is known to cause ACTH release, probably by augmenting the output of CRF. The adrenal cortical control system, however, retains its function in the presence of extensive damage to the forebrain and hypothalamus. High levels of corticosteroids in plasma block ACTH release from the anterior lobe by an inhibitory feedback mechanism acting directly on that gland. It is not definitely known, however, whether ACTH or corticosteroids also act on the hypothalamus itself to control CRF release, although this has been suggested frequently.[87,100]

Thyrotropin-releasing factor (TRF or TRH). In 1960 Schreiber and associates found that hypothalamic extracts administered to anterior pituitary grafts in the eyes of hypophysectomized rats prevented the involution of the thyroid glands that otherwise occurred. Guillemin and his co-workers in 1962 obtained conclusive evidence that a hypothalamic extract could cause increased liberation of a thyroid-stimulating hormone (TSH) from the anterior lobe. These investigators obtained highly active extracts of sheep hypothalamus that caused a definite release of TSH in the normal animal. This finding was soon confirmed by others, and it is now accepted that there is a thyrotropin- (TSH) releasing factor (TRF) produced in the hypothalamus of the mammal.

By 1966 TRF was isolated and found to contain histidine, glutamic acid, and proline. A compound that was synthesized in 1969 and 1970 had an action identical with that of TRF.[87,87a,100] This polypeptide, like other releasing factors, has little or no species specificity. It is most highly concentrated in the median eminence and certain nuclei of the hypothalamus, but it has also been found in other regions of the CNS. Despite this extrahypothalamic production, hypothalamic lesions significantly reduce TRF and TSH production.

TRF is liberated under various conditions of stress, but exposure to cold is a specific stimulus for its release and production. Thyroxine increase is known to inhibit TSH release at the pituitary level, but whether TRF production and release are affected at the hypothalamic level has not as yet been clarified. This releasing factor can now be considered a hormone (TRH).

Control of growth hormone release (GRF and GIF). The existence of growth-promoting and growth-inhibiting compounds has been well substantiated. Growth hormone is referred to frequently as somatotropin. The material that promotes growth hormone release from the anterior lobe is called GRF, or somatotropin-releasing factor (SRF). The material that inhibits growth hormone release is GIF, somatotropin release–inhibitory factor (SIF), or somatostatin.

The earliest evidence that the hypothalamus might contain compounds affecting growth hormone release from the anterior pituitary was obtained in 1964 and 1965. It was found that intracarotid injection of an extract of the hypothalamus depleted growth hormone content of the pituitary gland and that such an extract evoked increased secretion of growth hormone from incubated pituitary tissue.

Highly active GRF-containing extracts have been obtained from the hypothalamic tissues of pigs and sheep. GRF is most concentrated in the median eminence, as are other releasing factors. It is also a polypeptide containing a high proportion of glutamic acid and alanine. The concentration of growth hormone in plasma is at present measured by radioimmunoassay methods, a technique that has greatly facilitated the study of factors controlling GRF and growth hormone release.

It is interesting that growth hormone, presumably under the drive of GRF, is liberated in remarkably greater quantity during "slow-wave sleep" than in the waking state. The release of the hormone is also related to the concentration of metabolites such as glucose and arginine in the blood. Insulin-induced hypoglycemia and arginine infusions are standard and reliable clinical tests for the growth hormone–releasing function of the brain and the anterior pituitary. GRF release is also influenced by thyroid hormone, glucocorticoid, and estrogen levels in the blood. Catecholamines affect GRF output; norepinephrine or dopamine injected into the lateral ventricles release the hormone. The importance of such actions to normal growth hormone release has not yet been determined.

It should be said also that the structure of GRF has not been determined, nor has GRF been synthesized, although decapeptides possessing some of its actions have been obtained.[87] Nonetheless, it is considered that growth hormone is released from the pituitary gland under the drive of a GRF. Growth hormone induces formation of somatomedin from peripheral tissues, principally the liver, which is involved in the production of growth. It has been suggested that somatomedin rather than growth hormone itself exerts a feedback control on the hypothalamus or hypophysis to prevent excessive growth hormone production.

The presence of a compound within the hypothalamus that had an inhibiting effect on growth hormone release (GIF) was first reported in 1972.[64] A year later, during attempts to isolate GRF, Guillemin and his associates also obtained from the hypothalamus an inhibitory substance. This group quickly determined its structure and synthesized a compound with identical actions, which they called somatostatin. In a relatively short time, it has been found that somatostatin has a surprising range of effects and an unexpectedly wide distribution in the CNS and other parts of the body.[44a,82]

Control of prolactin release (PIF and PRF). Much evidence indicates that the hypothalamus can both stimulate and inhibit prolactin secretion.[75,87] A prolactin-releasing factor (PRF) has been found in birds and apparently is of primary importance in that species. Such a compound is also produced by the mammalian hypothalamus, but an inhibitory influence predominates in man and other mammals.

In 1965 an extract of the mammalian hypothalamus that inhibited prolactin release was obtained. This compound (PIF) is found in high concentration in the median eminence. The isolated anterior lobe secretes prolactin spontaneously, and this secretion is inhibited by PIF. Prolactin release is increased by estrogens and is altered during the normal estrous cycles; this may be due to the inhibition of PIF by estrogens. PIF concentration in the hypothalamus is reduced by suckling and increased by tranquilizers. Dopamine increases PIF levels in hypophyseal portal blood, thus reducing prolactin levels in man and other mammals. High prolactin blood levels inhibit further release of that hormone, but whether or not an increase in PIF action or release is involved is unknown. Catecholamines inhibit and TRF releases prolactin, but TRF and PRF are not identical. The chemical nature of PIF and PRF remain unknown, but they are probably small polypeptides.[87]

Gonadotropin-releasing factors: follicle-stimulating hormone–releasing factor (FRF or FRH) and luteinizing hormone–releasing factor (LRF or LRH). In 1964, at approximately the same time that other releasing factors were being identified, crude extracts of the hypothalamus were found that depleted the anterior pituitary of follicle-stimulating hormone (FSH) and initiated ovarian development and estrogen release. When added to pituitary tissue cultures, these extracts increased FSH concentration in the medium, a phenomenon that raised the possibility of a releasing factor produced in the hypothalamus that increases FSH output from the anterior pituitary. It has been called FSH-RF or just FRF.

In 1960 minute quantities of another active material were also obtained from the hypothalamus; this substance caused the release of LH from the pituitary glands of animals in estrus.[70] Stepwise purifications of these crude LH-RF or LRF extracts were made, and the compound has now been obtained in pure form, isolated, and identified.[18,82,87] Recently, it has been claimed by some that the LH- and FSH-releasing factors are parts of one molecule, and certainly their actions are usually supplementary.[87,87a] This is supported by the fact that antibodies to LRF block both LH and FSH release. Other observations, however, still support the concept that FRF and LRF are physiologically separate entities. It is not yet known which cells produce these compounds, but they are present in highest concentration in the median eminence. Smaller amounts are found in other parts of the hypothalamus and CNS.[82,87a] By our criteria, these compounds can be called hormones (gonadotropin-releasing hormone, FRH, or LRH).

FRF is thought to initiate cyclic changes in the ovaries and production of estrogens therefrom by causing a release of FSH from the anterior lobe. Subsequent discharge of LRF liberates LH, which produces ovulation and development of the corpus luteum. Ovulation requires both FSH and LH and thus a combined FRF and LRF action. A cyclic change in the LRF content of the median eminence has been demonstrated; it is highest during proestrus and diestrus and low throughout estrus. LRF concentrations are particularly high in the median eminence at the time of puberty. It is common practice to refer to these synergistic releasing factors, FRF and LRF, as gonadotropin-releasing factors.

Ovarian steroids (estrogens and progesterone) are known to have multiple negative feedback effects that block FSH and LH release. This blocking action on release occurs in the basal medial region of the hypothalamus as well as in the pituitary gland. Reviews also mention a stimulatory effect of steroids on FSH and LH release.[69] FSH and LH can also inhibit their own release through inhibitory feedback effects on the brain. These effects, however, seem not to be an essential feature of their normal functions (Chapter 61).

Control of melanocyte-stimulating hormone (MSH). The hypothalamus plays a role in the release of MSH from the pars intermedia. Visual stimuli cause release of MSH and the chromatophore adjustments that occur in fish and amphibia. MSH can affect many body functions, but its importance to mammals is unknown.[57] Releasing factors are involved in regulation of all MSH functions.

MSH-releasing factors (MRF) and MSH release–inhibitory factors (MIF) have been obtained from the hypothalamic structures of amphibia and mammals. The pars intermedia receives nerve fibers through the hypophyseal stalk. Their endings appear to contain neurosecretory granules. It is thought that releasing as well as release-inhibitory factors (MRF and MIF) are transported through the nerve fibers that connect the hypothalamus with the pars intermedia rather than through portal vessels. These hypothalamic hormones have been identified as small peptides.[87]

In considering the control of chromatophores and melanophores, the existence of melatonin, a melanophore-contracting principle found in high concentration in the pineal gland and the basal brain structures, should not be ignored. It has been suggested that melatonin acts on MIF secretion.[85] In lower vertebrates, blanching as well as melanocyte expansions are seen in emotional and visually induced responses. Very complex reciprocal reactions must occur, involving the hypothalamus as well as other structures of the nervous system, in forming the complex patterns of skin coloration that are visually controlled.[36,57]

In summary, it can be said that the secretory responses of cells sensitive to the hypothalamic releasing factors are very prompt in onset and immediately cease when these peptides are removed. The initial step in the action of such compounds probably involves interaction with plasma membrane receptors. Binding sites to TRF and LRF have been found in pituitary gland cell membranes. The concept of action of the hypothalamic releasing factors on receptors is now generally accepted.[92]

A discussion of hypothalamicohypophyseal functions would be incomplete without reference to the recently accomplished separation from whole brain or hyopthalamicohypophyseal portions of peptides that possess opiate activity. In 1964 C. H. Li isolated from pituitary tissue a compound that stimulated lipolytic activity. This he called lipotropin (LPH). Beta lipotropin is now considered to be a prohormone for the endorphins and enkephalins that have a morphinelike action. A number of enkephalins and endorphins are now known that can act on opiate receptors, thus affecting brain function and behavior. The limbic system may be more extensively involved than the hypothalamus,[89] but these compounds are released during stress and evidently play a physiologic role in the reactions of the organism.[44a,62]

Reproduction

Sexual behavior and reproduction involve the hypothalamus and the hypophysis. Stimuli that initiate seasonal development of the gonads and the resulting sexual expressions depend on the hypothalamicohypophyseal complex, as do the automatic "lunar" cycles of the primates and the estrous cycles of other mammalian species.

Estrogens and androgens affect the behavior of animals by action on the hypothalamus. Implants of estrogens in the mamillary region produce the full complex of natural estrus. This estrogen-sensitive region extends anteriorly into the preoptic area. Radioactive-labeled estrogens accumulate in highest concentrations in the anterior hypothalamus, but they also accumulate in the amygdaloid complex. Even electrical recordings from the hypothalamus show changes in cellular activity on administration of estrogens, although these changes are not confined to any specific circumscribed area. The estrogens and presumably the androgens affect structures of the nervous system other than the hypothalamus. Estrogens cause some signs of estrus even in decerebrate animals, but the hypothalamus is essential to the full expression of estrous behavior.

In its simplest description, sexual activity and the reproductive cycle in seasonal breeders begins as follows. With a change in season there is a gradual change in the intensity and duration of daylight. This initiates, through the hypothalamus, a discharge of gonadotropin-releasing factor, which in turn causes a release of gonadotropins. The gonads, as they develop, release ovarian hormones that heighten secondary sexual characteristics and evoke, again by action on the hypothalamus and other parts of the brain, the sexual display and mating behavior of the species. Many other factors are known to be involved, and species differ markedly in their responsiveness to one or another environmental condition. The precise occurrences that initiate the cycle are not fully known, but the hypothalamus is certainly essential.

In the case of periodic reproductive and sex cycles the trigger of the initial gonadotropin-releasing factor output and the consequent gonadotropin release that produces ovarian development of the menstrual and estrous cycles is also not known. The most that can be said is that an intrinsic rhythm or "clock" has been established in these species by some fundamental natural influence. The medial preoptic area is considered to be responsible for the cyclic nature of the female reproductive system. It evidently acts on the anterior part of the medial-basal hypothalamic region, which is capable of sustaining an acyclic, continuous gonadotropin release that would otherwise produce a continuous estrus.

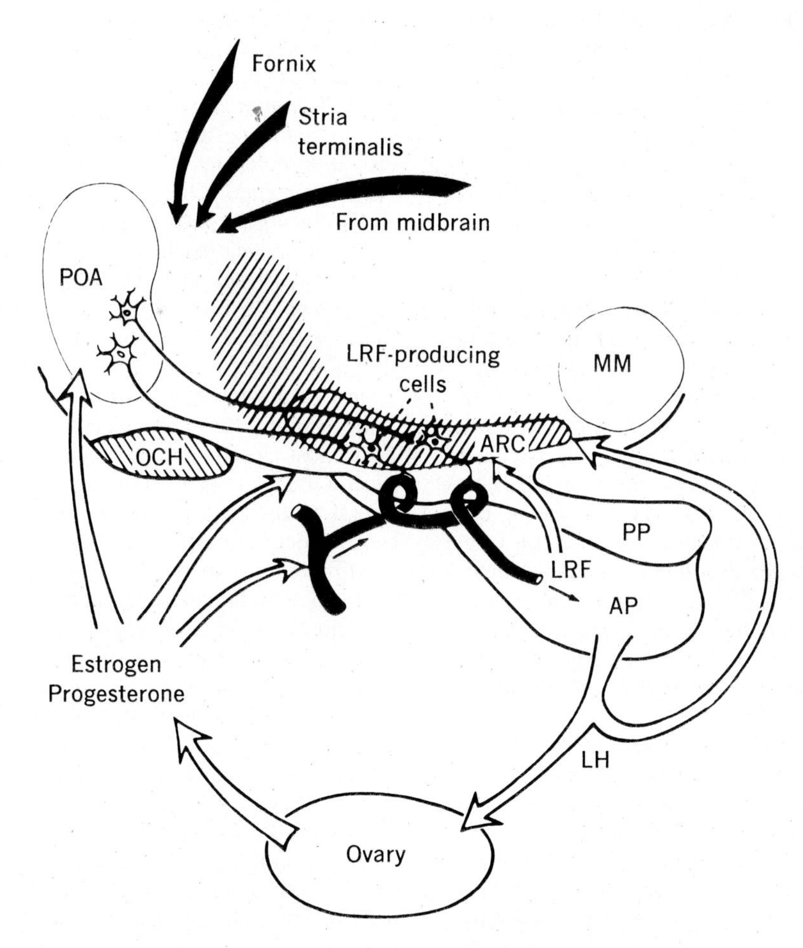

Fig. 34-8. Concepts of interrelationships between brain, pituitary gland, and ovary. *LRF,* produced by neurons and transported to pituitary gland by portal vessels, is shown to release LH from anterior pituitary *(AP).* Both LRF and LH act in hypothalamus. LH is shown to cause release from ovary of steroids, which enter blood vessels to have positive or negative effects on pituitary gland, preoptic area *(POA)* neurons, and hypothalamus. Solid arrows show that other parts of the brain can affect these interactions. *ARC,* Arcuate nucleus; *OCH,* optic chiasm; *MM,* mamillary body; *PP,* posterior pituitary. (From Sawyer.[86])

The inhibitory action of estrogens on gonadotropin-releasing factor liberation is thought to occur in this latter region.[43,86] Lesions in the hypothalamus that destroy these structures essential to gonadotropin release cause genital dystrophy and the abolition of cycles and sexual behavior.

Ovulation and impregnation constitute the next major events of the reproductive cycle. In some species, ovulation is evoked reflexly by special stimuli and/or an emotional reaction. In other species, it occurs spontaneously. Electrical stimuli applied to the cervix or the hypothalamus can, at proper times in the cycle, provoke ovulation and persistence of the corpus luteum (pseudopregnancy in rodents). It appears that release of the proper admixture of FSH and LH triggers escape of the ovum from the follicle and the formation of a corpus luteum. Since the hypothalamus is involved, release of FRF and LRF must occur.

Fig. 34-8 shows recently published concepts of interrelationships between the brain, pituitary gland, and female reproductive glands. It indicates that the estrogen and progesterone secreted by the ovary exert a feedback reaction on the neural centers that terminate releasing factor discharge.[44,86]

The pituitary hormones and the hypothalamic releasing factors are essential to the nurture and growth of the fetus, but the next dynamic action of the hypothalamus relates to the initiation of parturition and the control of lactation. Evidence of the involvement of oxytocin release from paraventricular nuclei cells during parturition has already been cited. The effects of PIF prevent a continuous liberation of prolactin from the anterior hypophysis. Suckling and the emotional stimuli evoked by nursing are thought to block PIF output; prolactin is released and milk secretion is augmented. The anterior lobe is found to be depleted of prolactin by nursing.

Finally, the suckling stimulus causes a reflex release of oxytocin, which has a "let-down reaction," facilitating the ejection of milk. Simultaneous recording of unit activity within the paraventricular nuclei and measurement of mammary gland ejection pressures during suckling have provided confirmatory evidence of the existence of this hypothalamic oxytocin-releasing reaction.[26,94] The information just given, although obtained chiefly from laboratory experimentation with lower mammalian forms, is applicable to man and the treatment of reproduction abnormalities.[44]

Thirst and control of water balance

Unquestionably the hypothalamus is of importance to the perception of thirst and the control of water intake. It is also involved in the regulation of body water loss, particularly via the kidney. Lesions in the hypothalamus have long been known to produce the polydipsia and polyuria of diabetes insipidus.

Analysis of the control of water intake presents many complexities. External and internal sensors undoubtedly contribute to the sensation of thirst; various theories of thirst have been adequately described elsewhere.[40,90] Dryness of the mouth produced by mouth breathing, ingestion of dry foods, and removal of the salivary glands all create thirst. Dehydration produces the desire to ingest liquids; a loss of body water equivalent to 0.5% body weight in the dog and 0.8% body weight in man appears to be the threshold for a strong water-intake drive. A hyperosmotic state also causes the urge to drink, as does an increase in extracellular fluid osmotic pressure of 1% to 2% in both dog and man. High temperature is also a stimulus to drink, even before there has been a detectable change in body water balance.

The receptors involved have not been identified, but unquestionably the hypothalamus plays an essential role. The evidence for this is much as follows: Lesions in the lateral hypothalamus that impair food intake also create adipsia. Furthermore, such lesions abolish the drinking response to an intraperitoneal injection of hypertonic sodium chloride, a phenomenon that has been most clearly demonstrated in the rat. In the goat and dog, lesions in the dorsal hypothalamus lateral to the paraventricular nuclei seem to be most effective in the production of adipsia.

Stimulation experiments have identified areas in the hypothalamus from which excessive drinking can be evoked. Injection of hypertonic solutions into these same areas has a similar effect. There appear to be some species differences with respect to the exact location of the regions most crucially concerned, but even in a given species the areas from which the effects can be obtained are not very clearly delimited.

The most that can be said at the present time is that stimuli applied to the dorsal hypothalamic area, lateral and somewhat caudal to the paraventricular nuclei, initiate drinking. Also, stimuli applied to a more caudal part of the dorsal hypothalamus, a region lying between the mamillothalamic tracts and descending columns of the fornix, cause polydipsia. It has been reported also that application of ACh to these lateral and/ or more dorsal centers causes vigorous and prolonged drinking. Also, renin and angiotensin have been found to produce symptoms of thirst by action on the limbic system, preoptic areas, and anterolateral hypothalamus.

Satiation of thirst can be accomplished by hydration of tissues, gastric distention, and hydration or moistening of the mucous membrane of the mouth and throat. Thus far, however, no area within the hypothalamus that might be called a "thirst satiation center" has been located. Finally, some success has been claimed in attempts to separate centers involved in control of hypovolemia and osmolarity of body fluids, but there has been a growing realization that the neural mechanisms controlling water intake and feeding behavior are widespread anatomically[71]; the midbrain, cerebral cortex, and limbic system as well as the hypothalamus are involved.[40,95]

Body water balance is maintained by relating water intake to water loss. Thirst regulates intake; water loss is controlled to some degree by the kidney. Loss in respiration or by perspiration is not controlled by the mechanisms regulating water balance; control operates primarily on the kidney. The role of the supraoptic nuclei in ADH production and release has been discussed previously. Here it suffices to say that cells of these hypothalamic nuclei, or neighboring osmoreceptors that can control their action, monitor the salt concentration or osmotic state of the blood and tissue fluids. These cells respond appropriately; if maximum water conservation is necessary, they release more ADH and the kidney reabsorbs all the water not required for excretion of materials in the urine (Chapters 49 and 50).

Control of body weight (hunger and satiation)

Energy imbalances, obesity and emaciation, may have numerous causes, but the hypothalamus is certainly involved in the maintenance of normality.[68] Early clinical studies resulted in controversy as to whether obesity was the result of pituitary abnormality or hypothalamic injury. We now know that the latter position is correct. Philip E. Smith in 1927 produced obesity by injecting chromic acid into the hypothalamus, supposedly without injury to the hypophysis. Keller and his associates in 1933 made small surgical lesions in the hypothalamus of dogs that produced an "enhanced appetite." The use of the Horsley-Clarke stereotaxic instrument eventually enabled others to localize those regions of the hypothalamus that are specifically responsible for the control of energy balance.[21,23,24,52]

There is now general agreement that hyperphagia and failure of body weight–regulating mechanisms can be produced by bilateral destruction of the ventromedial hypothalamic nuclei. Obesity apparently will also result from lesions that interrupt fibers passing caudally from these nuclei to the mesencephalic tegmentum. There is still some discussion as to the neural structures that are most crucial to the maintenance of body weight balance, and considerable uncertainty as to the abnormality created by their destruction remains.[53] Behavior of such animals suggests that release or deficiency phenomena, or a combination of both, may be involved (Chapter 57).

Almost immediately after production of an effective lesion, animals begin to eat in an abnormal fashion. The diurnal pattern is lost; meals are larger and more frequent. There is a doubling or tripling of the amount of food eaten at any one time. Certainly the use of the term "hyperphagia" is justified. Some investigators have felt that a loss of the sense of satiety or the mechanism producing that sense, rather than an augmentation of appetite, is responsible for the increase in food intake.

In addition to these seemingly primary abnormalities, there are other factors that must contribute to the accumulation of body fat and weight gain. A reduction in metabolic rate is often, if not invariably, present; after ingestion of food the respiratory quotient is elevated above unity, a phenomenon suggesting a high rate of conversion of carbohydrate to fat; activity is reduced. These changes could explain why a rat rendered potentially obese by a hypothalamic lesion can outgain a litter mate when, by paired feeding, both animals ingest the same amount of food. Such work indicates that obesity is not due simply to hyperphagia and loss of a sense of satiation.

Still other abnormalities are produced by the hypothalamic tumors and lesions that result in obesity. The ventromedial nuclear regions are involved in the control of emotional reactions, and lesions that create obesity may also convert placid animals into rather fierce, intractable individuals. Sexual dystrophy is a frequent accompaniment of obesity, as are disturbances in carbohydrate metabolism and water balance. Large lesions in a small structure interfere with multiple functions; thus it is hard to determine specificities.

Means have been sought to destroy specific cells by using their unique affinities. For example, ventromedial nuclear cells appear to have a specific affinity for glucose. Systemically injected gold thioglucose produces obesity in mice, apparently because this highly toxic compound accumulates in and destroys the ventromedial nuclei. Certainly these nuclei are important to the maintenance of body weight balance, but they are not solely responsible.

It has been demonstrated by Anand and Brobeck[9] that lesions in the lateral hypothalamic regions close to the ventromedial nuclei create hypophagia. The general concept at present is that such lesions destroy bilateral centers essential to hunger and appetite. These are reciprocally related to the ventromedial nuclei, which are responsible for satiation. Those centers of the lateral hypothalamus essential to the food intake drive have not yet been exactly located. It has been demonstrated recently by simultaneous recording from units in the lateral hypothalamic area and the ventromedial nuclei that stimuli which augment activity in one simultaneously diminish firing rates in the other. This reciprocal action has been observed by a number of investigators.[10,25,59,78]

The conditions that activate cells of the ventromedial nuclei and those of the lateral hypothalamus are not fully known at present. One suggestion offered is that these cells are glucose "receptors," which sense the level of glucose within or outside the cell. On the other hand, the rate of glucose utilization may be the determinant of activity of these cells, and this in turn may depend on the availability of glucose and the amount of insulin in the circulating blood. This latter dependency would indicate that these cells, unlike other neural tissues, require insulin for the metabolism of glucose. The special affinity of ventromedial nucleus cells for glucose is indicated by their greater uptake of radioactive glucose and the gold thioglucose experiments mentioned previously. Their dependency on glucose levels is indicated also by the fact that the rate of discharge of ventromedial nucleus neurons increases and that of the lateral hypothalamic cells simultaneously decreases when the level of glucose in the surrounding tissue fluid is raised. Vagus afferent fiber stimulations can affect the firing rate of both groups of neurons, but their role in the control of food intake has not been determined.[25] Finally, iontophoretic application of glucose to cells in the ventromedial nucleus increases their rate of firing, but its effect on cells of the lateral hypothalamus is inconclusive.[77]

It has long been known that obese animals have very low postprandial blood sugar levels. This tendency toward hypoglycemia was thought to be due to the fact that the livers of obese animals are so loaded with fat that the capacity for carbohydrate storage and release is impaired. Modern work has shown, however, that there is an abnormally high blood insulin level in the obese animal; the level remains high and does not follow the normal course of postprandial rise and decline.[58] The level of circulating insulin may determine the level of activity of these hypothalamic neurons, but it should also be remembered that the metabolisms of carbohydrate and lipids are closely interlinked. The level of nonesterified fatty acid (NEFA) may also be a determinant of hunger and satiety. Another idea that has been advanced is that the ventromedial nuclei control the level of fat storage in the body and that injury to these nuclei either abolishes this control or establishes a new level of fat storage.[58] Very fat animals rendered thus by forced feeding do not show as marked a hyperphagia and tendency toward weight gain following hypothalamic lesions as do normal animals with the same lesions. This suggests that there is an upper limit to fat storage and that the body weight limit is established by the fat storage potentiality of an individual.

In a cold environment, animals eat more and eat more continuously. Application of heat or even heating the anterior hypothalamus inhibits food intake. The anterior hypothalamic areas that sense and control body temperature may normally act in conjunction with those centers that control food intake. It is thought by some that the specific dynamic action of food materials produces a sense of body overheating that may contribute to the feeling of satiation. Hypothalamic lesions may impair the sense of satiation by interfering with this temperature-detecting mechanism. Exposure to heat does not depress food intake in animals with obesity-producing lesions to the same degree as in normal animals.

Growth hormone levels in the blood tend to be low after eating; they then rise progressively. This hormone is known to participate in making fat available as an energy source to the body. Hypothalamic lesions that disturb growth hormone output may also contribute to metabolic abnormalities and changes in feeding behavior that result in obesity.

Attempts have recently been made to determine whether cells in the ventromedial nuclei and lateral hypothalamic areas, which are concerned with regulation of food intake and body weight, are under the control of cholinergic or adrenergic transmitters.[53,71,72] Local applications of these chemicals, their blocking agents, and many other drugs normally used in the study of synaptic transmission have been shown to affect food intake and behavior.

The limbic system and other extrahypothalamic areas cannot be ignored. There are those who think that most feeding control is extrahypothalamic,[73] particularly in those species having a greater encephalization of body control.[71] Unquestionably there is a growing conviction that psychologic factors, conditioned behavior, and motivations are important in the control of food intake. Olfaction, taste, and other gustatory sensations are involved in determining the strengths of food intake drives. It may be pointed out again that the functions of the nervous system and the neuroendocrine system are well integrated and that no single isolated part is exclusively involved in the determination of a body state and the maintenance of normality.

Temperature regulation

Chronic spinal and decerebrate animals, although their tissues produce quantities of heat relative to their degree of activity, have no ability to regulate body temperature and are therefore poikilothermic. Removals of the cerebral cortex and structures superior to the hypothalamus somewhat impair the niceties of temperature regulation, but such animals remain essentially homeothermic.

Richet in 1885 and Ott in 1887 held that there were thermosensitive centers in the brain involved in detecting and correcting changes in body temperature. It is now generally agreed that the hypothalamus is the region of the brain chiefly responsible for thermosensitivity, maintenance of body temperature homeostasis, and the required reactions to extremes of heat and cold.

Although individuals vary slightly in body temperature and in their diurnal temperature cycles, there is a rather exact "set point" in all homeotherms. This is considered to be 37° C (98.6° F) in man, and the compensatory reactions organized by the hypothalamus normally prevent fluctuations of more than 1° or 2° above or below this set level. The set point for core temperature and the thresholds for regulatory reactions do undergo adaptive change and are modified in diseased states. Except when at temperatures of thermal neutrality, which individuals rarely experience, they are compensating either to maintain an internal consistency despite the hyperthermic effects of activity or to prevent hypothermia when cooling conditions exist.

Peripheral receptors can initiate generalized reactions to heat and cold. The reflexes evoked by general or localized heating and cooling must act through the hypothalamus to arouse the appropriate body responses. In the absence of the hypothalamus, such stimuli merely provoke localized or ineffective reflex actions. In addition to the peripheral receptors of the skin, there are thermal sensors and/or receptors within the spinal cord, medulla, and hypothalamus that are sensitive to heat and cold. All these can evoke effective responses from the hypothalamus, but even in the absence of this crucial structure, localized reaction to thermal stimuli occurs.

Changes in the temperature of the arterial blood reaching the brain have been shown to evoke temperature-regulating reactions in the absence of any excitation of peripheral afferents. The first experiments providing evidence of central temperature receptors were performed in dogs by Sherrington in 1924 and in man by Gibbon and Landis in 1932. By cooling or heating the denervated lower extremities of spinal dogs and spinal men, it was found that the upper innervated parts of the body showed temperature-regulating changes within a very few minutes after warmed or cooled blood reached the brain. Heating of the denervated extremities evoked sweating and vasodilatation of the innervated parts, whereas cooling produced vasoconstriction, piloerection, and shivering. It was later shown by other techniques that the hypothalamus was the region chiefly responsible for sensing blood temperature variations and for evoking the appropriate somatic and autonomic reactions.

Ablation experiments have shown that various regions of the hypothalamus play specific roles in performing the required adjustments. Destructive lesions in the posterior hypothalamus create deficiencies in responses to cold, whereas lesions in the anterior hypothalamus impair responses to heat. Lateral hypothalamic regions are thought to participate in the adjustment to both heat and cold.

Lesions in the preoptic and anterior hypothalamic areas cause animals to lose their regulatory responses to heat but do not significantly impair reactions to cold. Panting, sweating, vasodilatation of skin vessels, and the normal behavioral reactions to overheating do not occur. In unanesthetized animals, stimulations through electrodes or applications of heat through small thermodes, permanently implanted in the rostal hypothalamus, cause sweating, panting, and cutaneous vasodilatation with a resulting drop in body temperature. Stimulations also tend to stop reactions to cold.

It has been possible in recent years to investigate the localization of thermosensitive elements in the hypothalamus.[19,50] The procedures involved heating or cooling very minute areas and accurately recording the temperature of a neuron as well as its activity. Only one third of the neurons from which recordings were made proved to be sensitive to local temperature changes. Practically all the sensitive cells are located in the preoptic-anterior hypothalamus, even those responding to cold; the ratio of heat to cold sensors is 3:1. Some cells are sensitive only to direct heating or cooling, whereas others were also responsive to peripheral, visceral, and spinal cord thermal stimuli. Local cooling of the preoptic anterior hypothalamus produces the reactions to cold—shivering, constriction of blood vessels of the skin, and piloerection—and the body temperature rises.

Electrical stimulation of the anterior hypothalamic structures evokes those reactions initiated by local heating. Although local cooling of the posterior hypothalamus is ineffectual, electrical stimulation of this region induces muscular tremors that resemble shivering and other reactions that normally occur in response to cold. There is a resultant temperature increase.

Reactions to extremes of ambient temperature involve much more than autonomic system responses. The hypothalamus is also involved in organizing appropriate behavioral activities. It has been shown that a thermode placed in the preoptic-anterior hypothalamic area of a monkey, previously trained to effect changes in ambient temperature by using a mechanical device, could be used to cause the animal to modify the temperature of its surround. Heating caused the monkey to select a lower room temperature; local cooling caused the reverse.[7] Such heating or cooling, if sustained, eventually changes the set point; subsequently, deviations produce reactions quite different from those which normally would occur at specific temperatures. For example, if the set point is changed from 38° to 41° C, a drop of "thermode-created temperature" to 40° C is interpreted as cooling and the animal raises the room temperature. Hypothalamic temperature, not body core temperature, dominates an animal's behavior. Local hypothalamic cooling will actually cause a rise in body temperature to at least a febrile level of 40.5° C when anterior hypothalamic temperature is held at 36° C.

Febrile reactions to pyrogens depend on the integrity of the posterior hypothalamus. The fever results in part from increased heat production and conservation for which the animal does not compensate. Loss of compensatory ability occurs because the pyrogens also affect the set point. Even in normal activity, some heat loss is necessary. Thus a pyrogen-induced insensitivity of those preoptic-anterior hypothalamic sensors that determine the set point results in hyperthermia. It has been suggested that pyrogens liberate prostaglandins and that antipyretics act by blocking their production or action. Certainly, minute quantities of prostaglandin E injected into the anterior hypothalamus or third ventricle do cause a fever.

Various attempts have been made recently to determine the significance of transmitter substances to hypothalamic temperature regulation.[19,45] Direct injections into the anterior hypothalamus of 5-hydroxytryptamine (5-HT), or serotonin, have been found to produce a fever in conscious primates, whereas similarly applied catecholamines produce hypothermia. Responses of various species differ, and this is confusing. Some investigators have believed the actions of injected amines to be nonspecific, since they produce blood flow changes and also directly affect the membranes of many types of neurons. Additional information concerning the role of the hypothalamus in body temperature regulation is given in Chapter 59.

Reactions to stress

Individuals are exposed to a wide variety of stimuli and conditions that tend to drastically alter body homeostasis in a general rather than a specific fashion. Injury, surgical trauma, great extremes of heat and cold, infections, and intense emotional disturbances all create conditions of stress. These injurious influences are termed "stressor" stimuli. A great deal has been written about reaction to stress, and Selye has described them as producing a "general adaptation syndrome." It has also been pointed out that long-continued reactions to stress eventually have harmful effects; it is important to alleviate the causative situation as quickly as possible.

Defense against stress occurs in two phases and the hypothalamus is involved in both. Initially the stressor stimuli evoke, by way of the hypothalamus, a strong sympathicoadrenal discharge. Stored catecholamines are released and produce their characteristic cardiovascular and metabolic reactions. This may suffice to restore or maintain an acceptable degree of normality, particularly if the stress is not severe or of only short duration.

Severe and long-lasting stress must be counteracted by additional reactions, and the most significant of these is a release of corticosteroids from the adrenal cortex. The stressor stimuli act through afferent neurons or directly on the hypothalamus to cause discharge of CRF and a liberation of ACTH from the pituitary gland. It was once thought that the released epinephrine caused ACTH discharge, but now it is generally held that these stimuli evoke reactions in the hypothalamus through the limbic and reticular systems that organize the resistive responses. Animals have a greatly impaired resistance to infection and injury if these resistive reactions to stress are abolished by hypothalamic lesions or removal of the adrenal glands.[47,88]

Situations that create long-lasting tensions or low-level stress tend to produce gastric ulcers. These stimuli presumably act by influencing the hypothalamic control of the digestive tract. It has been repeatedly demonstrated in animals that stimulation of the anterior hypothalamus increases secretory and motor activity in the stomach and intestine. Chronic stimulation produces ulceration.[66] Lesions in the anterior hypothalamus are chiefly responsible for the production of gastric bleeding, but posterior hypothalamic lesions also produce a similar condition. The adrenal gland is thought by some to be involved in the production of a contributory hyperacidity, but the most that can be said at present is that a hypothalamic disorder created by intrinsic or extrinsically imposed disturbances is responsible for this unfortunate consequence of stress.

There is a psychic and emotional contribution to the initiation of reactions to stressor situations and to compensatory adjustments.[20] It should not be assumed that other centers of the brain play no role just because the hypothalamus appears to be essential to the organization of the early and late neuroendocrine reactions to conditions causing stress.

Control of emotional reactions

As early as 1872, Charles Dawson pointed out that a distinction should be made between *emotional feeling* and *emotional expression*. In studies of animals, we can be sure only of the latter. We know, however, that the subjective and behavioral aspects of emotional states are organized within the brain because human individuals with broken necks and high cord transections do smile and show the signs and report the experiencing of pleasure, anger, fear, and sadness. Furthermore, these observations indicate that extensive sensory deprivation does not prevent the arousal of emotional reactions. On the other hand, it is generally recognized that a great many types of sensors can contribute triggering stimuli under certain circumstances. Expressions of the emotions are rather stereotyped and induce patterns of behavior as well as autonomic system–controlled reactions such as blanching or flushing of the skin, cardiac acceleration, pupillary dilatation, piloerection, and sweating. It is obvious that the hypothalamus must be involved. Experimental work has supported this conclusion, but again, it should be pointed out that the hypothalamus is not the sole nervous system structure involved in emotional expression. Ablation experiments have revealed release as well as deficiency phenomena.[67]

Studies of animals such as cats and dogs that have long survived the removal of cerebral cortex have shown conclusively that the capacity to display anger or rage reactions depends on subcortical mechanisms. Goltz found that decorticate dogs, in addition to demonstrating rage, revealed a high degree of emotional instability. As stated previously, it was finally shown in studies of animals with both acute and chronic lesions that the hypothalamus is necessary for the vigorous expression of rage reactions. After transection of the brain behind the mamillary bodies, only fragmentary elements of emotional response can be initiated.[13,14]

Subsequent to these early ablation experiments, it was found by others that small localized lesions also can produce a release of the rage phenomenon. Ingram (1939) and Wheatley (1944) showed that injury to the ventromedial nuclei in cats produced enduring savage behavior; this also occurs in rats. Bilateral lesions in the lateral part of the hypothalamus, however, tend to create placidity in the characteristically wild and excitable rhesus monkey. This work has certainly demonstrated that regions in the medial and caudal hypothalamus integrate emotional expression and suggests that a degree of reciprocal action operates in the control of emotional phenomena.

Electrical stimulation has also produced evidence that the hypothalamus plays some role in the arousal and expression of anger. Hess and Brügger (1943) were among the earliest to perfect a technique for brain stimulation in awake, unrestrained animals.[51] They concluded that a center exists in the perifornical region of the anterior hypothalamus that is responsible for anger and defensive behavior, at least in cats. Hunsperger, using both stimulation and ablation procedures, confirmed and extended the work of Hess. He claimed that the essential neural apparatus for integrating an expression of anger is not located solely in the posterior hypothalamus but involves two central zones, one in the perifornical region described by Hess and the other posterior to the hypothalamus in the central gray of the midbrain.

In an attempt to differentiate between centers and pathways to which current might spread during electrical stimulation, various workers have implanted crystalline ACh or injected other drugs into these areas under discussion. Positive reactions were induced from the lateral hypothalamic area and from the dorsal part of the ventromedial nuclei and the periventricular region of the posterior hypothalamus. This technique, however, did not give a more definite localization of the centers primarily involved in emotional expression than did the stimulation and lesion experiments.

Rage reactions are not the only signs of emotional expression that can be evoked from decorticate animals and those with a further reduction of brain components. Behavior characteristic of fear or terror has been evoked by high-pitched sounds or other stimuli in cats and dogs from which all parts of the forebrain except the preoptic and hypothalamic areas have been removed. Even in chronic mesencephalic animals, high-pitched sounds evoke cringing and other signs indicative of fear. Hess also found points in the hypothalamus from which an escape reaction could be elicited by electrical stimulation. These areas are close to those from which rage responses are evoked.

Reactions of pleasure and those associated with sexual functions are generally considered to be emotional in nature. Unquestionably the hypothalamus is involved, and the present consensus is as follows. There appear to be two hormone-sensitive systems controlling female sexual or reproductive functions. One is located in the median eminence, which regulates the pituitary-gonadal axis. The other, which primarily controls behavior, is located in the preoptic-anterior hypothalamic area. Only in the rabbit is the hormone-sensitive mechanism that organizes the lordosis response located in the posterior hypothalamus.[17]

Many physiologic studies in both animals and man, using stimulation as well as recordings of evoked activity, have demonstrated that limbic system–hypothalamic cooperation is involved in a wide variety of somatic and autonomic phenomena closely related to the broad range of behavioral activities conventionally associated with emotional expression.

Heath and co-workers (1954) observed that stimulation of the septum frequently yielded pleasurable sensations in patients. Shortly thereafter, others observed the phenomenon of self-stimulation in rats. These animals voluntarily pressed levers to cause the delivery of stimuli through electrodes implanted in the septal area, as though they received some intrinsic "reward" therefrom. Positive loci for self-stimulation could be followed along the medial forebrain bundle into the lateral parts of the hypothalamus.

There appears to be an extensive overlap in the control of the diverse responses dependent on this limbic system. The very considerable mass of clinical and experimental observations, however, does not yield a fully satisfactory picture of the integrative role of the limbic system in the control of emotional manifestations, which most certainly involves the hypothalamus.

Sleep and arousal

The older literature contains rather elaborate discussions of evidence that lesions in the posterior hypothalamus produce sleep, whereas stimulations of that area arouse a sleeping animal. Stimulations of the dorsal hypothalamus at critical frequencies and for relatively long periods have been reported to produce sleep and sleep-preparation behavior. Following stimulation, cats walk to an "appropriate place," turn around a few times, curl up, and go to sleep as normal animals do. It is now thought, however, that there are no specific "sleep centers" in the hypothalamus, but that sleep-producing stimuli actually act through the general thalamocortical system. Posterior hypothalamic stimuli act on the reticular activating system to cause arousal, whereas lesions in the posterior hypothalamus interrupt these pathways (Chapter 10).

There is some evidence, however, that the hypothalamus is not completely quiescent during sleep; cardiac and vasomotor surges occur, indicating activity of the autonomic system and its hypothalamic representation. There is also a well-documented release of growth hormone during sleep, and it is thought that GRF secretion is augmented during slow-wave sleep particularly. A diurnal fluctuation in ACTH secretion, indicative of a periodic CRF release, also occurs. In man, ACTH and blood cortisol levels are highest before waking in the morning (6:00 AM) and lowest at night, a fluctuation not caused by stress or external circumstances.

The limbic and other portions of the CNS are probably involved in the control of sleep cycles, but the brain lesions affecting the biologic clock that is responsible for setting sleep-wakefulness rhythms are those that involve the hypothalamus.[11,84] It is clear that this internal clock depends on the hypothalamus for its existence, but the nature of the clock and the mechanisms of its control and actions are unknown.

Control of somatic reactions

The hypothalamus plays a role not only in the control of visceral functions and activity of the endocrine system but also in the regulation of somatic behavior. When the hypothalamus is intact, there is a better integration of reflex activity than that found in a decerebrate preparation. Activities that one would define as behavioral reactions rather than reflex pattern responses are seen. Walking, running, and righting reactions are merely slightly deficient rather than imperfectly displayed and coordinated, as they are in the absence of the hypothalamus.

Stimulations of the hypothalamus are said to produce directional movements and rather complex patterns of response that involve the total body.[51] Also, as pointed out in an earlier section, hypothalamic stimuli elicit a picture of emotional arousal that involves somatic as well as autonomic and endocrine reactions. On exposure to heat and cold, the hypothalamus organizes compensatory somatic as well as autonomic and endocrine reactions. The hypothalamus functions as a part of the brain, not merely as an adjunct to the autonomic or endocrine systems. It is integrative in function, influencing the entire body and all processes therein. Higher centers contribute refinements of control and special sensitivities, but the basic processes of integration require participation of the hypothalamus.

REFERENCES

General reviews

1. Haymaker, W., Anderson, E., and Nauta, W. J. H., editors: The hypothalamus, Springfield, Ill., 1969, Charles C Thomas, Publisher.
2. Lederis, K., and Cooper, K. E., editors: Recent studies of hypothalamic function, Basel, 1974, S. Karger.
3. Martini, L., Motta, M., and Fraschini, F., editors: The hypothalamus, New York, 1970, Academic Press, Inc.
4. Recent studies on the hypothalamus, Br. Med. Bull. **22:**195, 1966.
5. Reichlin, S., Baldessarini, R. J., and Martin, J. B., editors: The hypothalamus, Res. Publ. Assoc. Res. Nerv. Ment. Dis. **56:**entire volume, 1978.
6. Swaab, D. F., and Schade, J. P., editors: Integrative hypothalamic activity, Prog. Brain Res., vol. 41, 1974.

Original papers

7. Adair, E. R., Casby, J. U., and Stolwijk, J. A. J.: Behavioral temperature regulation in the squirrel monkey: changes induced by shifts in hypothalamic temperature, J. Comp. Physiol. Psychol. **72:**17, 1970.
8. Adams, J. H., Daniels, P. M., and Pritchard, M. M.: Distribution of hypophysial portal blood in the anterior lobe of the pituitary gland, Endocrinology **75:**120, 1964.
9. Anand, B. K., and Brobeck, J. R.: Hypothalamic control of food intake in rats and cats, Yale J. Biol. Med. **24:**123, 1951.
10. Anand, B. K., et al.: Activity of single neurons in the hypothalamic feeding centers: effect of glucose, Am. J. Physiol. **207:**1146, 1964.
11. Aschoff, J., and Wever, R.: Human circadian rhythms: a multioscillatory system, Fed. Proc. **35:**2326, 1976.
12. Axelrod, J.: The pineal gland: a neurochemical transducer, Science **184:**1341, 1974.
13. Bard, P.: A diencephalic mechanism for the expression of rage with special reference to the sympathetic nervous system, Am. J. Physiol. **84:**490, 1928.
14. Bard, P.: Central nervous mechanisms for emotional behavior patterns in animals, Res. Publ. Assoc. Res. Nerv. Ment. Dis. **19:**190, 1939.
15. Bargmann, W.: Neurosecretion, Int. Rev. Cytol. **19:** 183, 1966.
16. Bargmann, W., Hild, W., Ortmann, R., and Schiebler, T. H.: Morphologische und experimentelle Untersuchungen über das hypothalamisch-hypophysäre System, Acta Neuroveg. **1:**16, 1950.
17. Beach, F. A.: Some effects of gonadal hormones on sexual behavior. In Martini, L., Motta, M., and Fraschini, F., editors: The hypothalamus, New York, 1970, Academic Press, Inc., p. 617.
18. Blackwell, R. E., and Guillemin, R.: Hypothalamic control of adenohypophysial secretions, Annu. Rev. Physiol. **35:**357, 1973.
19. Bligh, J.: Temperature regulation in mammals and other vertebrates, New York, 1973, Elsevier North-Holland, Inc.
20. Brady, J. V.: Ulcers in "executive" monkeys, Sci. Am. **199:**95, 1958.
21. Brobeck, J. R., Tepperman, J., and Long, C. N. H.: Experimental hypothalamic hyperphagia in the albino rat, Yale J. Biol. Med. **15:**831, 1943.
22. Brooks, C. M.: Relation of the hypothalamus to gonadotrophic function of the hypophysis, Res. Publ. Assoc. Res. Nerv. Ment. Dis. **20:**525, 1940.
23. Brooks, C. M.: The relative importance of changes in activity in the development of experimentally produced obesity in the rat, Am. J. Physiol. **147:**708, 1946.
24. Brooks, C. M., and Lambert, E. F.: A study of the effect of limitation of food intake and the method of feeding on the rate of weight gain during hypothalamic obesity in the albino rat, Am. J. Physiol. **147:**695, 1946.
25. Brooks, C. M., Koizumi, K., and Zeballos, G. A.: A study of factors controlling activity of neurons within the paraventricular, supraoptic and ventromedial nuclei of the hypothalamus, Acta Physiol. Lat. Am. **16:**83, 1966.
26. Brooks, C. M., Ishikawa, T., Koizumi, K., and Lu, H. H.: Activity of neurons in the paraventricular nucleus of the hypothalamus and its control, J. Physiol. **182:**217, 1966.
27. Brownstein, M. J., Palkovits, M., Saavedra, J. M., and Kitzer, J. S.: Distribution of hypothalamic hormones and neurotransmitters within the diencephalon. In Martini, L., and Ganong, W. F., editors: Frontiers in neuroendocrinology, New York, 1976, Raven Press, vol. 4, p. 1.
28. Burn, H. A., and Knowles, F. G. W.: Neurosecretion. In Martini, L., and Ganong, W. F., editors: Neuroendocrinology, New York, 1966, Academic Press, Inc., vol. 1, p. 139.
29. Clark, W. E. L., Beattie, J., Riddoch, G., and Dott, N. M.: The hypothalamus: morphological, functional, clinical and surgical aspects, Edinburgh, 1938, Oliver & Boyd.
30. Cross, B. A., and Harris, G. W.: The role of the neurohypophysis in the milk-ejection reflex, J. Endocrinol. **8:**148, 1952.
31. Daniel, P. M.: The blood supply of the hypothalamus and pituitary gland, Br. Med. Bull. **22:**202, 1966.
32. Douglas, W. W.: Mechanism of release of neurohypophysial hormones: stimulus-secretion coupling. In Greep, R. O., and Astwood, E. B., editors: Handbook of physiology. Section 7, Baltimore, 1974, The Williams & Wilkins Co., vol. 4, part 1, chap. 9, p. 191.
33. Du Vigneaud, V.: Hormones of the posterior pituitary gland: oxytocin and vasopressin, Harvey Lect. **50:**1, 1954.
34. Dyball, R. E. J.: Oxytocin and ADH secretion in relation to electrical activity in antidromically identified supraoptic and paraventricular units, J. Physiol. **214:** 245, 1971.

35. Dyball, R. E. J., and Koizumi, K.: Electrical activity in the supraoptic and paraventricular nuclei associated with neurohypophysial hormone release, J. Physiol. **201:**711, 1969.
36. Fingermann, M.: Comparative physiology: chromatophores, Annu. Rev. Physiol. **32:**345, 1970.
37. Fink, G., Naller, R., and Worthington, W. C., Jr.: The demonstration of luteinizing hormone releasing factor in hypophysectomized rats, J. Physiol. **191:**407, 1967.
38. Finley, K. H.: Angio-architecture of the hypothalamus and its peculiarities, Res. Publ. Assoc. Res. Nerv. Ment. Dis. **20:**286, 1940.
39. Fisher, C., Ingram, W. R., and Ranson, S. W.: Diabetes insipidus and the neurohormonal control of water balance: a contribution to the structure and function of the hypothalamico-hypophysial system, Ann Arbor, 1938, Edwards Brothers, Inc.
40. Fitzsimons, J. T.: Thirst, Physiol. Rev. **52:**468, 1972.
41. Fuxe, K., and Hökfelt, T.: Catecholamines in the hypothalamus and the pituitary gland. In Ganong, W. F., and Martini, L., editors: Frontiers in neuroendocrinology, New York, 1969, Oxford University Press, Inc., p. 47.
42. Goodner, C. J., et al.: Insulin, glucagon, and glucose exhibit synchronous, sustained oscillations in fasting monkeys, Science **195:**177, 1977.
43. Greep, R. O., and Astwood, E. B., editors: Handbook of physiology. Section 7, Baltimore, 1974, The William & Wilkins Co., vol. 4, part 2.
44. Greep, R. O., Koblinsky, M. A. and Jaffe, F. S.: Reproduction and human welfare: a challenge to research, Cambridge, Mass., 1976, The M.I.T. Press, p. 622.
44a. Guillemin, R.: Peptides in the brain: the new endocrinology of the neuron, Science **202:**390, 1978.
45. Hammel, H. T.: Regulation of internal body temperature, Annu. Rev. Physiol. **30:**641, 1968.
46. Harris, G. W.: Neural control of the pituitary gland, London, 1955, Edward Arnold (Publishers) Ltd.
47. Harris, G. W., and George, R.: Neurohumoral control of the adenohypophysis and the regulation of the secretion of TSH, ACTH and growth hormone. In Haymaker, W., Anderson, E., and Nauta, W. J. H., editors: The hypothalamus, Springfield, Ill., 1969, Charles C Thomas, Publisher, p. 326.
48. Hayward, J. N., and Vincent, J. D.: Osmosensitive single neurones in the hypothalamus of unanaesthetized monkeys, J. Physiol. **210:**947, 1970.
49. Hedlund, L. W., Franz, J. M., and Kenny, A. D.: Biological rhythms and endocrine function: advances in experimental medicine and biology, New York, 1974, Plenum Publishing Corp., vol. 54.
50. Hensel, H.: Thermoreceptors, Annu. Rev. Physiol. **36:**233, 1974.
51. Hess, W. R.: The functional organization of the diencephalon, New York, 1957, Grune & Stratton, Inc.
52. Hetherington, A. W., and Ranson, S. W.: Hypothalamic lesions and adiposity in the rat, Anat. Rec. **78:** 149, 1940.
53. Hoebel, B. G.: Feeding: neural control of intake, Annu. Rev. Physiol. **33:**533, 1971.
54. Hökfelt, T.: Morphological contributions to monoamine pharmacology, Fed. Proc. **33:**2177, 1974.
55. Horsley, V., and Clarke, R. H.: The structure and functions of the cerebellum examined by a new method, Brain **31:**45, 1908.
56. Ishikawa, T., Koizumi, K., and Brooks, C. M.: Electrical activity recorded from the pituitary stalk of the cat, Am. J. Physiol. **210:**427, 1966.
57. Kastin, A. J., Viosca, S., and Schally, A. V.: Regulation of melanocyte-stimulating hormone release. In Greep, R. O., and Astwood, E. B., editors: Handbook of physiology. Section 7, Baltimore, 1974, The Williams & Wilkins Co., vol. 4, part 2, chap. 42.
58. Kennedy, G. C.: Food intake, energy balance and growth, Br. Med. Bull. **22:**216, 1966.
59. Koizumi, K., and Nishino, H.: Circadian and other rhythmic activity of neurones in the ventromedial nuclei and lateral hypothalamic area, J. Physiol. **263:** 331, 1976.
60. Koizumi, K., and Yamashita, H.: Studies of antidromically identified neurosecretory cells of the hypothalamus by intracellular and extracellular recordings, J. Physiol. **221:**683, 1972.
60a. Koizumi, K., and Yamashita, H.: Influence of atrial stretch receptors on hypothalamic neurosecretory neurones, J. Physiol. **285:**341, 1978.
61. Koizumi, K., Ishikawa, T., and Brooks, C. M.: Control of activity of neurons in the supraoptic nucleus, J. Neurophysiol. **27:**878, 1964.
62. Kosterlitz, H. W., editor: Opiates and endogenous opioid peptides, New York, 1976, Elsevier North-Holland, Inc.
63. Krnjevec, K.: Chemical nature of synaptic transmission in vertebrates, Physiol. Rev. **54:**418, 1974.
64. Krulich, L., et al.: Dual hypothalamic regulation of growth hormone secretion. In Pecile, A., and Muller, E. E., editors: Growth and growth hormone, Amsterdam, 1972, Excerpta Medica, p. 306.
65. Lammers, H. J., and Lohman, A. H. M.: Structure and fiber connections of the hypothalamus in mammals, Prog. Brain Res. **41:**61, 1974.
66. Long, D. M., Leonard, A. S., Chou, S. N., and French, L. A.: Hypothalamus and gastric ulceration. I. Gastric effects of hypothalamic lesions. II. Production of gastrointestinal ulceration by chronic hypothalamic stimulation, Arch. Neurol. **7:**167, 176, 1962.
67. MacLean, P. D.: The hypothalamus and emotional behavior. In Haymaker, W., Anderson, E., and Nauta, W. J. H., editors: The hypothalamus, Springfield, Ill., 1969, Charles C Thomas, Publisher, p. 659.
68. Mayer, J.: Symposium on experimental models for the study of obesity, Fed. Proc. **36:**137, 1977.
69. McCann, S. M.: Regulation of secretion of follicle-stimulating hormone and luteinizing hormone. In Greep, R. O., and Astwood, E. B., editors: Handbook of physiology. Section 7, Baltimore, 1974, The Williams & Wilkins Co., vol. 4, part 2, chap. 40, p. 48.
70. McCann, S. M., Talesnik, S., and Friedman, H. M.: LH releasing activity in hypothalamic extracts, Proc. Soc. Exp. Biol. Med. **104:**432, 1960.
71. Mogenson, G. J.: Changing views of the role of the hypothalamus in the control of ingestive behaviors. In Lederis, K., editor: Recent studies of hypothalamic function, Basel, 1974, S. Karger, p. 268.
72. Morgane, P. J., editor: Neural regulation of food and water intake, Ann. N. Y. Acad. Sci. **157:**531, 1969.
73. Morrison, S. D.: The hypothalamic syndrome in rats, Fed. Proc. **36:**139, 1977.
74. Nauta, W. J. H., and Haymaker, W.: Hypothalamic nuclei and fiber connections. In Haymaker, W., Anderson, E., and Nauta, W. J. H., editors: The hypothalamus, Springfield, Ill., 1969, Charles C Thomas, Publisher, p. 136.
75. Nicoll, C. S., Fiorindo, R. P., McKennee, C. T., and Parsons, J. A.: Assay of hypothalamic factors which regulate prolactin secretion. In Meites, J., editor: hypophysiotropic hormones of the hypothalamus: assay and

chemistry. Baltimore, 1970, The Williams & Wilkins Co., p. 115.
76. Nishino, H., Koizumi, K., and Brooks, C. M.: The role of the suprachiasmatic nuclei of the hypothalamus in the production of circadian rhythm, Brain Res. **112:** 45, 1976.
77. Oomura, Y., Ono, T., Ooyama, H., and Wayner, M. J.: Glucose and osmosensitive neurons of the rat hypothalamus, Nature **222:**282, 1969.
78. Oomura, Y., et al.: Reciprocal activities of the ventromedial and lateral hypothalamic areas of cats, Science **143:**484, 1964.
79. Pittendrigh, C. S., section editor: Circadian oscillations and organization in nervous system. In Schmitt, F. O., and Worden, F. G., editors: The neurosciences, Third Study Program, Cambridge, Mass., 1974, The M.I.T. Press, p. 435.
80. Porter, J. C., Goldman, B. D., and Wilber, J. F.: Hypophysiotropic hormones in portal vessel blood. In Meites, J., editor: Hypophysiotropic hormones of the hypothalamus: assay and chemistry, Baltimore, 1970, The Williams & Wilkins Co., p. 282.
81. Raisman, G.: Neural connexions of the hypothalamus, Br. Med. Bull. **22:**197, 1966.
82. Reichlin, S., et al.: Hypothalamic hormones, Annu. Rev. Physiol. **38:**389, 1976.
83. Reiter, R. J.: Comparative physiology: pineal gland, Annu. Rev. Physiol. **35:**305, 1973.
84. Richter, C. P.: Sleep and activity: their relation to the 24-hour clock, Res. Publ. Assoc. Res. Nerv. Ment. Dis. **45:**8, 1967.
85. Rust, C. C., and Meyer, R. K.: Hair color, molt and testis size in male, short-tailed weasels treated with melatonin, Science **165:**921, 1969.
86. Sawyer, C. H.: First Geoffrey Harris Memorial Lecture. Some recent developments in brain-pituitary-ovarian physiology, Neuroendocrinology **17:**97, 1975.
87. Schally, A. V., Arimura, A., and Kastin, A. J.: Hypothalamic regulatory hormones, Science **179:**341, 1973.
87a. Schally, A. V.: Aspects of hypothalamic regulation of the pituitary gland, Science **202:**18, 1978.
88. Selye, H.: Homeostasis and the reactions to stress: a discussion of Walter B. Cannon's contributions. In Brooks, C. M., Koizumi, K., and Pinkston, J. O., editors: The life and contributions of Walter Bradford Cannon, Albany, 1975, State University of New York Press, p. 89.
89. Snyder, S. H., Simantov, R., and Pasternak, G. W.: The brain's own morphine, "Enkephalin": a peptide neurotransmitter? In Ferrendelli, J. A., McEwen, B. S., and Snyder, S. H., editors: Neurotransmitters, hormones and receptors: novel approaches, Neuroscience Symposia vol. 1, Bethesda, Md., 1976, The Society for Neuroscience, p. 82.
90. Stevenson, J. A. F.: Neural control of food and water intake. In Haymaker, W., Anderson, E., and Nauta, W. J. H., editors: The hypothalamus, Springfield, Ill., 1969, Charles C Thomas, Publisher, p. 524.
91. Ungerstedt, U.: Stereotaxic mapping of the monoamine pathways in the rat brain, Acta Physiol. Scand. **82** (suppl. 367):1, 1971.
92. Vale, W., Rivier, C., and Brown, M.: Regulatory peptides of the hypothalamus, Annu. Rev. Physiol. **39:** 423, 1977.
93. Verney, E. B.: The antidiuretic hormone and the factors which determine its release, Proc. R. Soc. Lond. (Biol.) **135:**25, 1947.
94. Wakerley, J. B., and Lincoln, D. W.: The milk-ejection reflex of the rat: a 20-40-fold acceleration in the firing of paraventricular neurones during oxytocin release, J. Endocrinol. **57:**477, 1973.
95. Wayner, M. J.: The lateral hypothalamus and adjunctive drinking, Prog. Brain Res. **41:**371, 1974.
96. Wurtman, J., editor: Brain monoamines and endocrine function, Neurosci. Res. Program Bull. **9**(2):172, 1971.
97. Wurtman, R. J., Axelrod, J., and Kelly, D. E.: The pineal, New York, 1968, Academic Press, Inc.
98. Yamashita, H.: Effect of baro- and chemoreceptor activation on supraoptic nuclei neurons in the hypothalamus, Brain Res. **126:**551, 1977.
99. Yates, F. E., and Maran, J. W.: Stimulation and inhibition of adrenocorticotropin release. In Greep, R. O., and Astwood, E. B., editors: Handbook of physiology. Section 7, Baltimore, 1974, The Williams & Wilkins Co., vol. 4, part 2, chap. 36, p. 367.
100. Yates, F. E., Russell, S. M., and Maran, J. W.: Brain-adenohypophysial communication in mammals, Annu. Rev. Physiol. **33:**393, 1971.
101. Zimmerman, E. A.: Localization of hypothalamic hormones by immunocytochemical techniques. In Martini, L., and Ganong, W. F., editors: Frontiers in neuroendocrinology, New York, 1976, Raven Press, vol. 4, p. 25.
102. Zimmerman, E. A.: Localization of neurosecretory peptides in neuroendocrine tissues. In Naftolin, F., Ryan, K. J., and Davis, J., editors: Subcellular mechanisms in reproductive neuroendocrinology, Amsterdam, 1976, Elsevier/North Holland Biomedical Press, p. 81.

Index*

*Page numbers in *italics* indicate illustrations; those followed by *t* indicate tables; those followed by *n* indicate footnotes.

B

D

E

F

G

J

K

L

M

N

P

U

W

X

Y

Z